Contents—1998 REFRIGERATI[illegible]

Contents—1997 FUNDAMENTALS

2000 ASHRAE® HANDBOOK

Heating, Ventilating, and Air-Conditioning SYSTEMS AND EQUIPMENT

Inch-Pound Edition

American Society of Heating, Refrigerating and Air-Conditioning Engineers, Inc.

1791 Tullie Circle, N.E., Atlanta, GA 30329

(404) 636-8400 http://www.ashrae.org

DEDICATED TO THE ADVANCEMENT OF

THE PROFESSION AND ITS ALLIED INDUSTRIES

Volunteer members of ASHRAE Technical Committees and others compiled the information in this handbook, and it is generally reviewed and updated every four years. Comments, criticisms, and suggestions regarding the subject matter are invited. Any errors or omissions in the data should be brought to the attention of the Editor. Additions and corrections to Handbook volumes in print will be published in the Handbook published the year following their verification and as soon as verified on the ASHRAE Internet website.

DISCLAIMER

ISBN 1-883413-80-X

CONTENTS

HEATING EQUIPMENT

GENERAL COMPONENTS

UNITARY EQUIPMENT

ADDITIONS AND CORRECTIONS

INDEX

CONTRIBUTORS

In addition to the Technical Committees, the following individuals contributed significantly to this volume. The appropriate chapter numbers follow each contributor's name.

Howard J. McKew (1)
Sebesta-Blomberg & McKew

David M. Elovitz (2)
Energy Economics, Inc.

Harvey Brickman (3)
Tishman Realty & Construction, Inc.

Gil Avery (4)
Kele & Associates

Gene R. Strehlow (4)
Johnson Controls, Inc.

Mark W. Fly (5)
Governair Corporation

Gary B. Hayden (6, 15)
Burnham Radiant Heating Corporation

Ainul Abedin (7)
Ainul Abedin Consulting Engineers

Ronald L. Davis (7)
AEP

William E. Stewart, Jr. (7)
InterMountain Research

R.L. Douglas Cane (8)
Caneta Research, Inc.

James W. Earley (8)
Georgia Power Company

Joseph A. Pietsch (8)

Frank J. Pucciano (8)
Energy Advantage Corporation

John W. Andrews (9, 28)
Brookhaven National Laboratory

Paul M. Haydock (9, 28)
Carrier Corporation

Frank E. Jakob (9, 28)
Battelle Columbus

Roy C.E. Ahlgren (10, 39)
ITT Fluid Handling

Richard A. Hegberg (10, 31)
Hegberg & Associates

Vernon P. Meyer (11)
Dept. of the Army Corps of Engineers, Missouri River Division

Victor L. Penar (11)
Perma-Pipe, Inc.

Gary E. Phetteplace (11)
U.S. Army Cold Regions Research and Engineering Laboratory

Clinton W. Phillips (11)

Kathleen M. Posteraro (11)
Pittsburg Corning Corporation

Kevin D. Rafferty (11)
Geo-Heat Center, Oregon Institute of Technology

Steven M. Tredinnick (11)
Kattner/FVB

William J. Coad (12, 13)
McClure Engineering Associates

Alexander H. Sleiman (14)
District Energy St. Paul, Inc.

Kirby S. Chapman (15)
Kansas State University

i. Birol Kilkis (15, 32)
Heatway Radiant Floor Heating

John H. Stratton (16)
SMACNA

William J. Waeldner (17)

J. Barrie Graham (18)
Graham Consultants

Gursuran D. Mathur (19)
Zexel USA Corporation

Leon E. Shapiro (19)
ADA Systems

Patricia T. Thomas (19)
Munters Corporation

Roger M. Pasch (20)
Research Products Corporation

Roland A. Ares (21, 23)
Ares Corporation

Alfi Helmy Malek (21)
CETIM

Gary W. Price (21, 23)
Frigid Coil

William C. Griffiths (22)
Kathabar, Inc.

Lewis G. Harriman III (22)
Mason-Grant

Todd Amsey Michael (23)
The Trane Company

Larry C. Holcomb (24)
Holcomb Environmental Services

Eugene L. Valerio (24)
Technical Marketing Services

Derald G. Welles (24)
Healthway Products, Inc.

David J. Murphy (25)
Air Filter Testing Labs

Jeanette M. Murphy (25)
Air Filter Testing Labs

Leonard J. O'Dell (25)

Gregory R. Thiel (25)
Thiel Consulting, Inc.

Douglas W. Vanosdell (25)
Research Triangle Institute

Douglas W. DeWerth (26, 30)

Hall Virgil (26, 30)
Carrier Corporation

Kevin J. Hoey (27)
EASCO Boiler Corporation

Hassan M. Bagheri (31)
P2S Engineering

Michael P. O'Rourke (32)
Mestek

John I. Woodworth (32)

Albert A. Domingorena (34, 38)
Hussmann Corporation

Dean H. Rizzo (34)
Copeland Corporation

John H. Roberts (34, 38)
The Trane Company

Rudy Stegmann (34)
The Enthalpy Exchange

Jaroslav Wurm (34, 47)
Institute of Gas Technology

Zahid H. Ayub (35)
Isotherm, Inc.

John C. Chato (35)
University of Illinois at Urbana-Champaign

CONTRIBUTORS (*Concluded*)

Jon M. Edmonds (35, 36)
Edmonds Engineering Company

Michel LeCompte (35)
RefPlus Co.

William R. Leizear, Sr. (35, 36)
Claude Laval Corporation

Robert P. Miller (35, 36)
Baltimore Aircoil Company

Manmohan Mongia (35)
Heatcraft, Inc.

Michael M. Ohadi (35)
University of Maryland

G. Robert Shriver (35, 36)
Evapco, Inc.

Neel S. Gupte (37)
Carrier Corporation

James W. Larson (37)
The Trane Company

Robert L. Bates (38)
E.I. duPont de Nemours and Company

Richard C. Niess (38)
Gilbert & Associates

Kenneth R. Luther (39)
ITT Fluid Handling

Joseph A. Thuman (39, 41)
Wallace Eannace Associates Inc.

Thomas F. Lowery (40)
Rockwell Automation-Reliance Electric

Sherri L. Peterson (40)
The Trane Company

Edward J. Swan (40)
Rockwell Automation

Richard E. Batherman (41)
VICO, Inc.

Paul A. Bourquin (41)

Randall J. Amerson (42)
Siemens Building Technologies, Inc.

Carl H. Jordan (42)

John C. Glunt (43)
Blackmore and Glunt

Gerald L. Martin (44)
Xetex, Inc.

Bert G. Phillips (44)
UNIES, Ltd.

Ephraim M. Sparrow (44)
University of Minnesota

Van D. Baxter (45, 46)
Lockheed Martin Energy Corporation

Piotr A. Domanski (45, 46)
National Institute of Standards and Technology

Hugh I. Henderson (45)
CDH Energy Corporation

Gregory J. Rosenquist (46)
Lawrence Berkeley National Laboratory

John Brogan (47)

Robert H. Henninger (47)
GARD Analytics, Inc.

ASHRAE HANDBOOK COMMITTEE

Eugene L. Valerio, Chair

2000 Systems and Equipment Volume Subcommittee: **Dennis J. Wessel,** Chair

Edwin I. Griggs **Bradley A. Gustafson** **Gideon Shavit** **Derald G. Welles** **Lynn F. Werman**

ASHRAE HANDBOOK STAFF

Robert A. Parsons, Editor **Christina Tate Nichols,** Associate Editor

Scott A. Zeh, Nancy F. Thysell, and **Jayne E. Jackson,** Publishing Services

W. Stephen Comstock,
Director, Communications and Publications
Publisher

ASHRAE TECHNICAL COMMITTEES AND TASK GROUPS

SECTION 1.0—FUNDAMENTALS AND GENERAL

1.1 Thermodynamics and Psychrometrics
1.2 Instruments and Measurements
1.3 Heat Transfer and Fluid Flow
1.4 Control Theory and Application
1.5 Computer Applications
1.6 Terminology
1.7 Operation and Maintenance Management
1.8 Owning and Operating Costs
1.9 Electrical Systems
1.10 Energy Resources
TG Litigation Risk

SECTION 2.0—ENVIRONMENTAL QUALITY

2.1 Physiology and Human Environment
2.2 Plant and Animal Environment
2.3 Gaseous Air Contaminants and Gas Contaminant Removal Equipment
2.4 Particulate Air Contaminants and Particulate Contaminant Removal Equipment
2.6 Sound and Vibration Control
2.7 Seismic Restraint Design
TG Buildings' Impacts on the Environment
TG Global Climate Change

SECTION 3.0—MATERIALS AND PROCESSES

3.1 Refrigerants and Secondary Coolants
3.2 Refrigerant System Chemistry
3.3 Refrigerant Contaminant Control
3.4 Lubrication
3.5 Desiccant and Sorption Technology
3.6 Water Treatment
3.8 Refrigerant Containment

SECTION 4.0—LOAD CALCULATIONS AND ENERGY REQUIREMENTS

4.1 Load Calculation Data and Procedures
4.2 Weather Information
4.3 Ventilation Requirements and Infiltration
4.4 Building Materials and Building Envelope Performance
4.5 Fenestration
4.6 Building Operation Dynamics
4.7 Energy Calculations
4.10 Indoor Environmental Modeling
4.11 Smart Building Systems
4.12 Integrated Building Design
TG Mechanical Systems Insulation

SECTION 5.0—VENTILATION AND AIR DISTRIBUTION

5.1 Fans
5.2 Duct Design
5.3 Room Air Distribution
5.4 Industrial Process Air Cleaning (Air Pollution Control)
5.5 Air-to-Air Energy Recovery
5.6 Control of Fire and Smoke
5.7 Evaporative Cooling
5.8 Industrial Ventilation
5.9 Enclosed Vehicular Facilities
5.10 Kitchen Ventilation

SECTION 6.0—HEATING EQUIPMENT, HEATING AND COOLING SYSTEMS AND APPLICATIONS

6.1 Hydronic and Steam Equipment and Systems
6.2 District Energy
6.3 Central Forced Air Heating and Cooling Systems
6.4 In Space Convection Heating
6.5 Radiant Space Heating and Cooling
6.6 Service Water Heating
6.7 Solar Energy Utilization
6.8 Geothermal Energy Utilization
6.9 Thermal Storage
6.10 Fuels and Combustion

SECTION 7.0—PACKAGED AIR-CONDITIONING AND REFRIGERATION EQUIPMENT

7.1 Residential Refrigerators and Food Freezers
7.4 Combustion Engine Driven Heating and Cooling Equipment
7.5 Mechanical Dehumidification Equipment and Heat Pipes
7.6 Unitary and Room Air Conditioners and Heat Pumps

SECTION 8.0—AIR-CONDITIONING AND REFRIGERATION SYSTEM COMPONENTS

8.1 Positive Displacement Compressors
8.2 Centrifugal Machines
8.3 Absorption and Heat Operated Machines
8.4 Air-to-Refrigerant Heat Transfer Equipment
8.5 Liquid-to-Refrigerant Heat Exchangers
8.6 Cooling Towers and Evaporative Condensers
8.7 Humidifying Equipment
8.8 Refrigerant System Controls and Accessories
8.10 Pumps and Hydronic Piping
8.11 Electric Motors and Motor Control

SECTION 9.0—AIR-CONDITIONING SYSTEMS AND APPLICATIONS

9.1 Large Building Air-Conditioning Systems
9.2 Industrial Air Conditioning
9.3 Transportation Air Conditioning
9.4 Applied Heat Pump/Heat Recovery Systems
9.5 Cogeneration Systems
9.6 Systems Energy Utilization
9.7 Testing and Balancing
9.8 Large Building Air-Conditioning Applications
9.9 Building Commissioning
9.10 Laboratory Systems
9.11 Clean Spaces
9.12 Tall Buildings
TG Combustion Gas Turbine Inlet Air Cooling Systems

SECTION 10.0—REFRIGERATION SYSTEMS

10.1 Custom Engineered Refrigeration Systems
10.2 Automatic Icemaking Plants and Skating Rinks
10.3 Refrigerant Piping, Controls, and Accessories
10.4 Ultra-Low Temperature Systems and Cryogenics
10.5 Refrigerated Distribution and Storage Facilities
10.6 Transport Refrigeration
10.7 Commercial Food and Beverage Cooling Display and Storage
10.8 Refrigeration Load Calculations
10.9 Refrigeration Application for Foods and Beverages
TG Mineral Oil Circulation

ASHRAE Research: Improving the Quality of Life

The American Society of Heating, Refrigerating and Air-Conditioning Engineers is the world's foremost technical society in the fields of heating, ventilation, air conditioning, and refrigeration. Its members worldwide are individuals who share ideas, identify the need for and support research, and write the industry's standards for testing and practice. The result of these efforts is that engineers are better able to keep indoor environments safe and productive while protecting and preserving the outdoors for generations to come.

One of the ways that ASHRAE supports its members' and industry's need for information is through ASHRAE Research. Thousands of individuals and companies support ASHRAE Research annually, enabling ASHRAE to report new data about material properties and building physics and to promote the application of innovative technologies.

The chapters in ASHRAE Handbooks are updated through the experience of members of ASHRAE technical committees and through results of ASHRAE Research reported at ASHRAE meetings and published in ASHRAE special publications and in *ASHRAE Transactions*.

For information about ASHRAE Research or to become a member contact, ASHRAE, 1791 Tullie Circle, Atlanta, GA 30329; telephone: 404-636-8400; www.ashrae.org.

The 2000 *ASHRAE Handbook*

The 2000 *ASHRAE Handbook—HVAC Systems and Equipment* describes both the combinations of equipment and the components or assemblies that perform a particular function either individually or in combination. The information helps the system designer select and operate equipment, although it does not describe how to design components.

Major topic areas include

- Air-conditioning and heating systems
- Air-handling equipment
- Heating equipment
- General HVAC components
- Unitary equipment

Most of the chapters in this Handbook update material in the 1996 *ASHRAE Handbook* to reflect current technology. The following chapters contain major revisions or new material.

- Chapters 1 through 5 are reorganized and tightened for clarity. Chapter 2, Building Air Distribution, contains expanded coverage on VAV systems, including more information on terminal units. Chapter 4, Central Cooling and Heating, is a new chapter that introduces central plant equipment and options.
- Chapter 7, Cogeneration Systems and Engine and Turbine Drives, has more information on combustion turbine inlet air cooling as well as updates for other sections.
- Chapter 9, Design of Small Forced-Air Heating and Cooling Systems, includes revised duct fitting data and new information on duct efficiency testing. In addition, information on system performance from Chapter 28, Furnaces, was moved to this chapter and clarified and updated to be consistent with emerging duct efficiency technology.
- Chapter 11, District Heating and Cooling, includes new information on the effect of moisture on underground pipe insulation, calculation of undisturbed soil temperatures, cathodic protection of direct buried conduits, and leak detection. Also, the information on consumer interconnections and metering is expanded.
- Chapter 13, Condenser Water Systems, has additional information on water hammer, freeze protection, water treatment, overpressure, and free cooling.
- Chapter 16, Duct Construction, refers to new flexible duct, fiber reinforced plastic duct, acoustical rating of duct liner, and blanket duct insulation standards.
- Chapter 19, Evaporative Air Cooling Equipment, contains more information on humidification and dehumidification.
- Chapter 22, Desiccant Dehumidification and Pressure Drying Equipment, has been expanded, particularly the section on liquid desiccant systems.
- Chapter 23, Air-Heating Coils, provides a broader scope of the types of heating coils and their selection and application.
- Chapter 24, Air Cleaners for Particulate Contaminants, discusses the new ASHRAE *Standard* 52.2 for testing of air-cleaning devices and includes new sections on residential air cleaners and bioaerosols.
- Chapter 25, Industrial Gas Cleaning and Air Pollution Control, received a substantial update and includes revised information on electrostatic precipitators and fabric filters,
- Chapter 26, Automatic Fuel-Burning Equipment, is updated and now discusses low oxides of nitrogen emission from oil-fired equipment.
- Chapter 27, Boilers, includes new information on increasing boiler efficiency and on wall hung boilers for residential use.
- Chapter 36, Cooling Towers, has more information about (1) a combined flow coil/fill evaporative cooling tower, (2) precautions about using variable frequency drives. and (3) the increased use and acceptance of thermal performance certification.
- Chapter 40, Motors, Motor Controls, and Variable-Speed Drives, includes a significant new section on variable-speed drives.
- Chapter 43, Heat Exchangers, is a new chapter that provides an overview of the topic.
- Chapter 44, Air-to-Air Energy Recovery, contains new information on ideal air-to-air energy exchange, rating of performance, comparison of devices, and twin tower enthalpy recovery loops.
- Chapter 46, Room Air Conditioners, Packaged Terminal Air Conditioners, and Dehumidifiers, combines information from two previous chapters.
- Chapter 47, Engine-Driven Heating and Cooling Equipment, is a new chapter that includes information on engine-driven chillers, air conditioners, heat pumps, and refrigeration equipment. The chapter also includes and updates information from an old chapter on unitary heat pumps.

This Handbook is published both as a bound print volume and in electronic format on a CD-ROM. It is available in two editions—one contains inch-pound (I-P) units of measurement, and the other contains the International System of Units (SI).

Look for corrections to the 1997, 1998, and 1999 Handbooks on the Internet at **http://www.ashrae.org**. Any changes in this volume will be reported in the 2001 *ASHRAE Handbook* and on the ASHRAE web site.

If you have suggestions for improving a chapter or you would like more information on how you can help revise a chapter, e-mail ashrae@ashrae.org; write to Handbook Editor, ASHRAE, 1791 Tullie Circle, Atlanta, GA 30329; or fax 404-321-5478.

Robert A. Parsons, ASHRAE Handbook Editor

CHAPTER 1

HVAC SYSTEM ANALYSIS AND SELECTION

AN HVAC SYSTEM maintains desired environmental conditions in a space. In almost every application, a myriad of options are available to the design engineer to satisfy this basic goal. In the selection and combination of these options, the design engineer must consider all criteria defined here to achieve the functional requirements associated with the goal.

HVAC systems are categorized by the method used to control heating, ventilation, and air conditioning in the conditioned area. This chapter addresses the procedures associated with selecting the appropriate system for a given application. It also describes and defines the design concepts and characteristics of basic HVAC systems. Chapters 2 through 5 of this volume describe specific systems and their attributes, based on their heating and cooling medium and commonly used variations.

SELECTING A SYSTEM

The design engineer is responsible for considering various systems and recommending one or two that will satisfy the goal and perform as desired. It is imperative that the design engineer and the owner collaborate on identifying and rating the criteria associated with the design goal. Some criteria that may be considered are

- Temperature, humidity, and space pressure requirements
- Capacity requirements
- Redundancy
- Spatial requirements
- First cost
- Operating cost
- Maintenance cost
- Reliability
- Flexibility
- Life cycle analysis

Because these factors are interrelated, the owner and design engineer must consider how these criteria affect each other. The relative importance of factors, such as these, differs with different owners and often changes from one project to another for the same owner. For example, typical concerns of owners include first cost compared to operating cost, the extent and frequency of maintenance and whether that maintenance requires entering the occupied space, the expected frequency of failure of a system, the impact of a failure, and the time required to correct the failure. Each of these concerns has a different priority, depending on the owner's goals.

Additional Goals

In addition to the primary goal to provide the desired environment, the design engineer must be aware of and account for other goals the owner may require. These goals may include

- Supporting a process, such as the operation of computer equipment
- Promoting a germ-free environment
- Increasing sales
- Increasing net rental income
- Increasing the salability of a property

The owner can only make appropriate value judgments if the design engineer provides complete information on the advantages and disadvantages of each option. Just as the owner does not usually know the relative advantages and disadvantages of different systems, the design engineer rarely knows all the owner's financial and functional goals. Hence, the owner must be involved in the selection of a system.

System Constraints

Once the goal criteria and additional goal options are listed, many constraints must be determined and documented. These constraints may include

- Performance limitations (i.e., temperature, humidity, and space pressure)
- Available capacity
- Available space
- Availability utility source
- Building architecture
- Construction budget

Few projects allow detailed quantitative evaluation of all alternatives. Common sense, historical data, and subjective experience can be used to narrow choices to one or two potential systems.

Heating and air conditioning loads often contribute to the constraints, narrowing the choice to systems that will fit in the available space and be compatible with the building architecture. Chapter 28 of the 1997 *ASHRAE Handbook—Fundamentals* describes methods used to determine the size and characteristics of the heating and air conditioning loads. By establishing the capacity requirement the size of equipment can be determined, and the choice may be narrowed to those systems that work well on projects within a size range.

Loads vary over time due to the time of day/night, changes in the weather, occupancy, activities, and solar exposure. Each space with a different use and/or exposure may require a different control zone to maintain space comfort. Some areas with special requirements may need individual systems. The extent of zoning, the degree of control required in each zone, and the space required for individual zones also narrow the system choices.

No matter how efficiently a particular system operates or how economical it is to install, it can only be considered if it (1) maintains the desired building space environment within an acceptable tolerance under all conditions and occupant activities and (2) physically fits into the building without being objectionable.

The preparation of this chapter is assigned to TC 9.1, Large Building Air-Conditioning Systems.

Cooling and humidity control are often the basis of sizing HVAC components and subsystems, but the system may also be determined based on the ventilation criteria. For example, if large quantities of outside air are required for ventilation or to replace air exhausted from the building, only systems that transport large air volumes need to be considered.

Effective delivery of heat to an area may be an equally important factor in the selection. A distribution system that offers high efficiency and comfort for cooling may be a poor choice for heating. This performance compromise may be small for one application in one climate, but may be unacceptable in another that has more stringent heating requirements.

HVAC systems and the associated distribution systems often **occupy a significant amount of space**. Major components may also require special support from the structure. The size and appearance of terminal devices (i.e., diffusers, fan-coil units, radiant panels, etc.) have an affect on the architectural design because they are visible in the occupied space.

Other architectural factors that limit the selection of some systems include

- Acceptable noise levels in the occupied space
- Space available to house equipment and its location relative to the occupied space
- Space available for horizontal and/or vertical distribution pipes and ducts
- Acceptability of components visible in the occupied space.

Construction budget constraints can also influence the choice of HVAC systems. Based on historical data, some systems may be economically out of reach for an owner's building program.

Narrowing the Choices

Chapters 2 through 5 cover building air distribution, in-room terminal systems, central cooling and heating, and decentralized cooling and heating. Each chapter briefly summarizes the positive and negative features of various systems. One or two systems that best satisfy the project goal can usually be identified by comparing the criteria, other factors and constraints, and their relative importance. In making subjective choices, notes should be kept on all systems considered and the reasons for eliminating those that are unacceptable.

Each selection may require combining a primary system with a secondary system (or distribution system). The **primary system** converts energy from fuel or electricity into a heating and/or cooling media. The **secondary system** delivers heating, ventilation, and/or cooling to the occupied space. The two systems, to a great extent, are independent, so several secondary systems may work with a particular primary system. In some cases, however, only one secondary system may be suitable for a particular primary system.

Once subjective analysis has identified one or two HVAC systems (sometimes only one choice may remain), detailed quantitative evaluations must be made. All systems considered should provide satisfactory performance to meet the owner's essential goals. The design engineer should provide the owner with specific data on each system to make an informed choice. The following chapters in the ASHRAE Handbooks should be consulted to help narrow the choices:

- Chapter 8, 1997 *ASHRAE Handbook—Fundamentals* covers physiological principles, comfort, and health.
- Chapter 30, 1997 *ASHRAE Handbook—Fundamentals* covers methods for estimating annual energy costs.
- Chapter 34, 1999 *ASHRAE Handbook—Applications* covers methods for energy management.
- Chapter 35, 1999 *ASHRAE Handbook—Applications* covers owning and operating cost.
- Chapter 37, 1999 *ASHRAE Handbook—Applications* covers mechanical maintenance.
- Chapter 46, 1999 *ASHRAE Handbook—Applications* covers sound and vibration control.

Selection Report

As the last step of selection, the design engineer should prepare a summary report that addresses the following:

- The goal
- Criteria for selection
- Important factors
- Other goals

A brief outline of each of the final selections should be provided. In addition, those HVAC systems deemed inappropriate should be noted as having been considered but not applicable to meet the owner's primary HVAC goal.

Table 1 Sample HVAC System Selection Matrix (0 to 10 Score)

Goal: Furnish and install an HVAC system that provides moderate space temperature control with minimum humidity control at an operating budget of 70,000 Btu/h per square foot per year				
Categories	**System #1**	**System #2**	**System #3**	**Remarks**
1. Criteria for Selection: • 75°F space temperature with ±3°F control during occupied cycle • 20% relative humidity with ± 5% rh control during heating season. • First cost • Equipment life cycle				
2. Important Factors: • First class office space stature • Individual tenant utility metering				
3. Other Goals: • Engineered smoke control system • ASHRAE *Standard* 62 ventilation rates • Direct digital control building automation				
4. System Constraints: • No equipment on the first floor • No exterior louvers below the perimeter windows				
5. Other Constraints: • No perimeter finned tube radiation				
TOTAL SCORE				

The report should include an HVAC system selection matrix that identifies the one or two suggested HVAC system (primary and secondary when applicable) selections, system constraints, and other constraints. In completing this matrix assessment, the engineer should have the owner's input to the analysis. This input can also be applied as weighted multipliers.

Many grading methods are used to complete an analytical matrix analysis. Probably the simplest grading method is to rate each item Excellent/Very Good/Good/Fair/Poor. A numerical rating system such as 0 to 10, with 10 equal to Excellent and 0 equal to Poor, can provide a quantitative result. The HVAC system with the highest numerical value then becomes the recommended HVAC system to accomplish the goal.

The system selection report should include a summary that provides an overview followed by a more detailed account of the HVAC system analysis and system selection. This summary should highlight the key points and findings that led to the recommendation(s). The analysis should refer to the system selection matrix (such as in Table 1) and the reasons for scoring.

A more detailed analysis, beginning with the owner's goal, should immediately following the summary. With each HVAC system considered, the design engineer should note the criteria associated with each selection. Issues such as close temperature and humidity control may eliminate some HVAC systems from being considered. System constraints and other constraints, noted with each analysis, should continue to eliminate HVAC systems. Advantages and disadvantages of each system should be noted with the scoring from the HVAC system selection matrix. This process should reduce the HVAC selection to one or two optimum choices to present to the owner. Examples of installations for other owners should be included with this report to endorse the design engineer's final recommendation.This third party endorsement allows the owner to inquire about the success of these other HVAC systems.

HVAC SYSTEMS AND EQUIPMENT

HVAC systems may be central or decentralized. In addressing the primary equipment location, the design engineer may locate this equipment in a **central** plant (either inside or outside the building) and distribute the air and/or water for HVAC needs from this plant. The other option is to **decentralize** the equipment, with the primary equipment located throughout the building, on the building, or adjacent to the building.

Central System Features

Some of the criteria associated with this concept are as follows:

Temperature, humidity, and space pressure requirements. A central system may be able to fulfill any or all of these design parameters.

Capacity requirements. A central system usually allows the design engineer to consider HVAC diversity factors that reduce the installed equipment capacity. In turn, this offers some attractive first cost and operating cost benefits.

Redundancy. A central system can accommodate standby equipment of equal size or of a preferred size that decentralized configurations may have trouble accommodating.

Spatial requirements. The equipment room for a central system is normally located outside the conditioned area—in a basement, penthouse, service area, or adjacent to or remote from the building. A disadvantage with this approach may be the additional cost to furnish and install secondary equipment for the air and/or water distribution. A second consideration is the access and physical constraints throughout the building to furnish and install this secondary distribution network of ducts and/or pipes.

First cost. A central system may not be the least costly when compared to decentralized HVAC systems. Historically, central system equipment has a longer equipment service life to compensate for this shortcoming. Thus, a life cycle cost analysis is very important when evaluating central versus decentralized systems.

Operating cost. A central system usually has the advantage of larger, more energy efficient primary equipment when compared to decentralized system equipment.

Maintenance cost. The equipment room for a central system provides the benefit of maintaining its HVAC equipment away from the occupants in an appropriate service work environment. Access to the building occupant workspace is not required, thus eliminating disruption to the space environment, product, or process. Another advantage may be that because of its larger capacity, there is less HVAC equipment to service.

Reliability. Central system equipment can be an attractive benefit when considering its long service life.

Flexibility. Redundancy can be a benefit when selecting standby equipment that provides an alternative source of HVAC or backup.

Among the largest central systems are those HVAC plants serving groups of large buildings. These plants provide improved diversity and generally operate more efficiently with lower maintenance costs than individual central plants. The economics of these larger central systems require extensive analysis. The utility analysis considers multiple fuels and may also include gas and steam turbine-driven equipment. Multiple types of primary equipment using multiple fuels and types of HVAC generating equipment (i.e., centrifugal and absorption chillers) may be installed in combination in one plant. Chapter 12, Chapter 13, and Chapter 14 provide design details for central plants.

Decentralized System Features

Some of the criteria associated with this concept are as follows:

Temperature, humidity, and space pressure requirements. A decentralized system may be able to fulfill any or all of these design parameters.

Capacity requirements. A decentralized system usually requires each piece of equipment to be sized for the maximum capacity. Depending on the type and location of the equipment, decentralized systems cannot take as much benefit of equipment sizing diversity when compared to the central system diversity factor potential.

Redundancy. A decentralized system may not have the benefit of backup or standby equipment. This limitation may need review.

Space requirements. A decentralized system may or may not have in equipment rooms. Due to the space restrictions imposed on the design engineer or architect, equipment may be located on the roof and/or the ground adjacent to the building.

First cost. A decentralized system probably has the best first cost benefit. This feature can be enhanced by phasing in the purchase of decentralized equipment on an as-needed basis (i.e., purchasing equipment as the building is being leased/occupied).

Operating cost. A decentralized system can emphasize this as a benefit when strategically starting and stopping multiple pieces of equipment. When comparing energy consumption based on peak energy draw, decentralized equipment may not be as attractive when compared to larger, more energy efficient central equipment.

Maintenance cost. A decentralized system can emphasize this as a benefit when equipment is conveniently located and the equipment size and associated components (i.e., filters) are standardized. When equipment is located on a roof, maintainability may be difficult because it is difficult to access during bad weather.

Reliability. A decentralized system historically has reliable equipment, although the estimated equipment service life may be less than that of centralized equipment.

Flexibility. A decentralized system may be very flexible because it may be placed in numerous locations.

Primary Equipment

The type of central and decentralized equipment selected for large buildings depends on a well-organized HVAC analysis and selection report. The choice of primary equipment and components depends on factors presented in the selection report with such factors as those presented in the section on Selecting a System. Primary HVAC equipment includes heating equipment, air and water delivery equipment, and refrigeration equipment.

Many HVAC designs recover internal heat from lights, people, and equipment to reduce the size of the heating plant. In large buildings with core areas that require cooling while perimeter areas require heating, one of several heat reclaim systems can heat the perimeter to save energy. Chapter 8 describes some heat recovery arrangements, Chapter 33 describes solar energy equipment, and Chapter 44 introduces air-to-air energy recovery. In the 1999 *ASHRAE Handbook—Applications*, Chapter 34 covers energy management and Chapter 39 covers building energy monitoring.

The search for energy savings has extended to **cogeneration** or total energy systems, in which on-site power generation has been added to the HVAC project. The economics of this function is determined by gas and electric rate differentials and by the ratio of electric to heating demands for the project. In these systems, waste heat from generators can be transferred to the HVAC equipment (i.e., to drive the turbines of centrifugal compressors, to serve an absorption chiller, etc.). Chapter 7 covers cogeneration or total energy systems.

Thermal storage is another energy savings concept, which provides the possibility of off-peak generation of air conditioning with chilled water or ice. Thermal storage of hot water can be used in heating. Many electric utilities impose severe charges for peak summer power use or offer incentives for off-peak use. The storage capacity installed to level the summer load may also be available for use in winter, thus making heat reclaim a viable option. Chapter 33 of the 1999 *ASHRAE Handbook—Applications* has more information on thermal storage.

With ice storage, colder supply air can be provided than that available from a conventional air conditioning. This colder air allows the use of smaller fans and ducts, which reduces first cost and operating cost that can offset the energy cost required to make ice. Similarly, the greater water temperature difference from hot water thermal storage allows smaller pumps and pipe to be used.

Heating Equipment

Steam boilers or hot water boilers are the primary means of heating a space. These boilers are (1) used both for heating and process heating; (2) manufactured to produce high or low pressure; and (3) fired with coal, oil, electricity, gas, and sometimes, waste material. Low-pressure boilers are rated for a working pressure of either 15 or 30 psig for steam and 160 psig for water, with a temperature limit of 250°F. Package boilers, with all components and controls assembled as a unit, are available. Electrode or resistance-type electric boilers that generate either steam or hot water are also available. Chapter 27 has further information.

Where steam or hot water is supplied from a central plant, as on university campuses and in downtown areas of large cities, the utility service entering the building must conform to the utility's standards. The utility provider should be contacted at the system analysis and selection phase of the project to determine availability, cost, and the specific requirements of the service.

When the primary heating equipment is selected, the fuels considered must ensure maximum efficiency. Chapter 26 discusses the design, selection, and operation of the burners for different types of primary heating equipment. Chapter 17 of the 1997 *ASHRAE Handbook—Fundamentals* describes types of fuel, fuel properties, and proper combustion factors.

Air Delivery Equipment

Primary air delivery equipment for HVAC systems are classified as packaged equipment, manufactured and custom manufactured equipment, or field fabricated equipment. Most ventilation equipment for large systems use centrifugal or axial fans; however, plug or plenum fans are becoming more popular. Centrifugal fans are frequently used in packaged and manufactured HVAC equipment. Axial fans are more often part of a custom unit or a field-fabricated unit. Both types of fans can be used as industrial process and high-pressure blowers. Chapter 18 describes fans, and Chapters 16 through 25 provide information about ventilation components.

Refrigeration Equipment

The section on Refrigeration Equipment in Chapter 4 summarizes the primary refrigeration equipment for HVAC systems designed to maintain desired environmental conditions in a space.

SPACE REQUIREMENTS

In the initial phase of a building's design, the engineer seldom has sufficient information to render the HVAC design. As noted in the section on Space Requirements in Chapter 4, the final design is usually a compromise between what the engineer recommends and what the architect can accommodate. At other times, final design and space requirements may be dictated by the building owner who may have a preference for a central or decentralized system. The following paragraphs discuss some of these requirements.

Equipment Rooms

The total mechanical and electrical space requirements range between 4 and 9% of the gross building area with most buildings falling within the 6 to 9% range. These ranges include space for HVAC, electrical, plumbing, and fire protection equipment. These percentages also include vertical shaft space for mechanical and electrical equipment.

Most facilities should be centrally located to (1) minimize long duct, pipe, and conduit runs and sizes; (2) simplify shaft layouts; and (3) centralize maintenance and operation. A central location also reduces pump and fan motor power, which reduces building operating costs. But, for many reasons, not all the mechanical and electrical facilities can be centrally located in the building. In any case, the equipment should be kept together whenever possible to minimize space requirements, centralize maintenance and operation, and simplify the electrical system.

Equipment rooms generally require clear ceiling height ranging from 10 to 18 ft, depending on equipment sizes and the complexity of air and/or water distribution.

The main electrical transformer and switchgear rooms should be located as close to the incoming electrical service as practical. If there is an emergency generator, it should be located considering (1) proximity to emergency electrical loads and sources of combustion and cooling air and fuel, (2) ease of properly venting exhaust gases to the outdoors, and (3) provisions for noise control.

HVAC Facilities

The heating equipment room houses the boiler(s) and may also house a boiler feed unit, chemical treatment equipment, pumps, heat exchangers, pressure-reducing equipment, control air compressors, and miscellaneous equipment. The refrigeration equipment room houses the chiller(s) and may also house chilled water and condenser water pumps, heat exchangers, air-conditioning equipment, control air compressors, and miscellaneous equipment. The design of these rooms needs to consider (1) the size and weight of the equipment; (2) installation and replacement when locating and arranging the room to accept this large equipment; and (3) applicable regulations relative to combustion air and ventilation air criteria.

In addition, ASHRAE *Standard* 15, Safety Code for Mechanical Refrigeration should be consulted for special equipment room requirements.

Most air-conditioned buildings require a cooling tower or condenser unit. If the cooling tower or air-cooled or water-cooled condenser is located on the ground, it should be at least 100 ft away from the building (1) to reduce tower noise in the building, (2) to keep discharge air and moisture carry-over from fogging the building's windows and discoloring the facade of the building, and (3) to keep discharge air and moisture carry-over from contaminating outdoor air being introduced into the building. Cooling towers should be kept the same distance from parking lots to avoid staining car finishes with water treatment chemicals. Chapter 35 and Chapter 36 have further information on this equipment.

It is often economical to locate the heating plant and/or refrigeration plant at an intermediate floor or on the roof. The electrical service and structural costs are higher, but these may be offset by reduced costs for heating piping, condenser and chilled water piping, energy consumption, and a chimney through the building. Also, the initial cost of equipment may be less because the operating pressure is lower.

Applicable regulations relative to both gas and fuel oil systems must be followed. Gas fuel may be more desirable than fuel oil. Fuel oil storage has specific environmental and safety concerns. In addition, the cost of oil leak detection and prevention may be substantial. Oil pumping presents added design and operating problems.

Energy recovery systems can reduce the size of the heating plant and/or refrigeration plant. Well-insulated buildings and electric and gas utility rate structures may encourage the design engineer to consider several energy conservation concepts such as limiting demand, free cooling and thermal storage.

Fan Rooms

The fan rooms house the HVAC fan equipment and may include other miscellaneous equipment. The room must have space for removal of the fan shaft and coil. Installation, replacement, and maintenance of this equipment should be considered when locating and arranging the room.

Fan rooms may be placed in a basement that has an airway for intake of outdoor air. In this situation the placement of air intake louver(s) is a concern because of debris from leaves and snow may fill the area. Also, if parking areas are close to the building, the quality of outdoor air may be compromised.

Fan rooms on the second floor and above, have easier access for outdoor air, exhaust air, and equipment replacement. The number of fan rooms required depends largely on the total floor area and whether the HVAC system is centralized or decentralized. Buildings with large floor areas may have multiple decentralized fan rooms on each floor or a large central fan unit serving the entire area. High-rise buildings may also opt for decentralized fan rooms for each floor; or they may have a more central concept with one fan room serving the lower 10 to 20 floors, one serving the middle floors of the building, and one at the roof serving the top floors.

Life safety is a very important factor in fan room location. Chapter 51 of the 1999 *ASHRAE Handbook—Applications* discusses fire and smoke management. In addition, state and local codes have additional fire and smoke detection and damper criteria.

Vertical Shafts

Vertical shafts provide space for air distribution and water and steam (pipe) distribution. Air distribution includes HVAC supply air, return air, and exhaust air ductwork. If the shaft is used as a return air plenum, close coordination with the architect is necessary to insure that the shaft is airtight. Pipe distribution includes hot water, chilled water, condenser water, and steam supply and condensate return. Other mechanical and electrical distribution found in vertical shafts are electric conduits/closets, telephone cabling/closets, plumbing piping, fire protection piping, pneumatic tubes, and conveyers.

Vertical shafts should be clear of stairs and elevators on at least two sides to permit access to ducts, pipes, and conduit that enter and exit the shaft while allowing maximum headroom at the ceiling. In general, duct shafts having an aspect ratio of 2:1 to 4:1 are easier to develop than large square shafts. The rectangular shape also makes it easier to go from the equipment in the fan rooms to the shafts.

In multistory buildings a vertical distribution system with minimal horizontal branch ductwork is desirable because it is (1) usually less costly; (2) easier to balance; (3) creates less conflict with pipes, beams, and lights; and (4) enables the architect to design lower floor-to-floor heights. These advantages also hold for vertical water and steam pipe distribution systems.

The number of shafts is a function of building size and shape. In larger buildings, it is usually more economical in cost and space to have several small shafts rather than one large shaft. Separate HVAC supply air, return air, and exhaust air duct shafts may be desired to reduce the number of duct crossovers. The same can be said for steam supply and condensate return pipe shafts because the pipe must be pitched in the direction of flow. From 10% to 15% additional shaft space should be allowed for future expansion and modifications. This additional space may also reduce the initial installation cost.

Equipment Access

Properly designed mechanical and electrical equipment rooms must allow for the movement of large, heavy equipment in, out, and through the building. Equipment replacement and maintenance can be very costly if access is not planned properly.

Because systems vary greatly, it is difficult to estimate space requirements for refrigeration and boiler rooms without making block layouts of the system selected. Block layouts allow the engineer to develop the most efficient arrangement of the equipment with adequate access and serviceability. Block layouts can also be used in preliminary discussions with the owner and architect. Only then can the engineer obtain verification of the estimates and provide a workable and economical design.

AIR DISTRIBUTION

Ductwork should deliver conditioned air to an area as directly, quietly, and economically as possible. Structural features of the building generally require some compromise and often limit the depth of the space available for ducts. Chapter 9 discusses air distribution design for small heating and cooling systems. Chapter 32 of the 1997 *ASHRAE Handbook—Fundamentals* discusses space air distribution and duct design.

The designer must coordinate duct design with the structure as well as other mechanical, electrical, and communication systems. In commercially developed projects, a great effort is made to reduce floor-to-floor dimensions. The resultant decrease in the available interstitial space for ductwork is a major design challenge. In institutional buildings, higher floor-to-floor heights are required due to the sophistication and complexity of the mechanical, electrical, and communication distribution systems.

Air Terminal Units

In some instances, such as in low velocity, all-air systems, the air may enter from the supply air ductwork directly into the conditioned space through a grille or diffuser. In medium and high velocity air systems, an intermediate device normally controls air volume, reduces duct pressure, or both. Various devices are available, including (1) a fan-powered terminal unit, which uses an integral fan to accomplish the mixing rather than depending on the induction principle; (2) a variable air volume (VAV) terminal unit, which varies the amount of air delivered to the space (this air may be delivered to low-pressure ductwork and then to the space, or the terminal may contain an integral air diffuser); (3) an all-air induction terminal unit, which controls the volume of primary air, induces return air, and distributes the mixture through low-velocity ductwork to the space; and (4) an air-water induction terminal, which includes a coil in the induced airstream. Chapter 17 has more information about air terminal units.

Insulation

In new construction and renovation upgrade projects, HVAC supply air ductwork should be insulated in accordance with energy code requirements. ASHRAE *Standard* 90.1, Section 9.4, and Chapter 32 of the 1997 *ASHRAE Handbook—Fundamentals* have more information about insulation and the calculation methods.

Ceiling Plenums

Frequently, the space between the suspended ceiling and the floor slab above it is used as a return air plenum to reduce the distribution ductwork. Refer to existing regulations before using this approach in new construction or a renovation because most codes prohibit combustible material in a ceiling return air plenum.

Some ceiling plenum applications with lay-in panels do not work well where high-rise elevators or the stack effect of a high-rise building create a negative pressure. If the plenum leaks to the low-pressure area, the tiles may lift and drop out when the outside door is opened and closed.

Raised floors with a plenum space directly below are another way to provide horizontal air distribution and/or a return air plenum.

The return air temperature in a return air plenum directly below a roof deck is substantially higher during the air conditioning season than in a ducted return. This can be an advantage to the occupied space below because the heat gain to the space is reduced. Conversely, return air plenums directly below a roof deck have substantially lower return air temperatures during the heating season than a ducted return and may require supplemental heat in the plenum.

PIPING

Piping should deliver refrigerant, hot water, chilled water, condenser water, condensate drains, fuel oil, gas, steam, and condensate to and from HVAC equipment as directly, quietly, and economically as possible. Structural features of the building generally require mechanical and electrical coordination to accommodate pipe pitch, draining of low points in the system, and venting of high point in the system. Chapters 33 of the 1997 *ASHRAE Handbook—Fundamentals* covers pipe distribution and pipe design.

Pipe Systems

HVAC piping systems can be divided into two parts; (1) the piping in the central plant equipment room and (2) the piping required to deliver refrigerant, hot water, chilled water, condenser water, condensate drain, fuel oil, gas supply, steam supply, and condensate return to and from HVAC and process equipment throughout the building. Chapters 10 through 14 discuss piping for various heating and cooling systems. Chapters 1 through 4 and 32 of the 1998 *ASHRAE Handbook—Refrigeration* discuss refrigerant piping practices.

The major piping in the central plant equipment room includes refrigerant, hot water, chilled water, condenser water, condensate drains, fuel oil, gas supply, steam supply, and condensate return connections.

Insulation

In new construction and renovation upgrade projects, HVAC piping may or may not be insulated based on existing code criteria. ASHRAE *Standard* 90.1 and Chapters 24 and 33 of the 1997 *ASHRAE Handbook—Fundamentals* have information regarding insulation and the calculation methods.

SYSTEM MANAGEMENT

System management is an important factor in choosing the optimum HVAC system. It can be as simple as a time clock to start and stop the equipment or as sophisticated as a computerized facility management software system serving large centralized HVAC multiple systems, decentralized HVAC systems, a large campus, etc.

Automatic Controls

Basic HVAC system management is available in electric, pneumatic, or electronic temperature control systems. Depending on the application, the design engineer may recommend a simple and basic management strategy as a cost-effective solution to an owner's heating, ventilation, and refrigeration needs. Chapter 45 of the 1999 *ASHRAE Handbook—Applications* and Chapter 37 of the 1997 *ASHRAE Handbook—Fundamentals* discuss automatic control in more detail.

The next level of HVAC system management is direct digital control either with or without pneumatic control damper and valve actuators. This automatic control enhancement may include energy monitoring and energy management software. The configuration may also be accessible by the building manager via telephone modem to a remote computer at an off-site location. Chapter 40 of the 1999 *ASHRAE Handbook—Applications* covers building operating dynamics.

Using computer technology and associated software the design engineer and the building manager can provide complete facility management. This comprehensive building management system may include HVAC system control, energy management, operation and maintenance management, fire alarm system control, and other reporting and trending software. This system may also be integrated and accessible from the owner's information technology computer network and the Internet.

System Management Interface

Today, system management includes the purchasing of automatic controls that come prepackaged and prewired on the HVAC equipment. In the analysis and selection of a system, the design engineer needs to include the merits of purchasing prepackaged automation versus traditional building automation systems. Current HVAC controls and their capabilities need to be compatible with other new and existing automatic controls. Chapter 38 of the 1999 *ASHRAE Handbook—Applications* discusses computer applications and ASHRAE *Standard* 135 discusses interfacing building automation systems.

Other interfaces to be considered include the interface and compatibility of other mechanical and electric control and management systems. Building systems, such as the fire alarm, medical gas systems, and communication systems are just three of the management interfaces that an owner may want to work in unison with the HVAC control system. Predictive and preventive maintenance using computerized maintenance management software (CMMS) also enhances the management and should be considered.

STANDARDS

ASHRAE *Standard* 15-1994. Safety code for mechanical refrigeration.

ASHRAE/IESNA *Standard* 90.1-1999. Energy efficient design of new buildings except low-rise residential buildings.

ASHRAE *Standard* 135-1995. BACnet—A data communication protocol for building automation and control networks.

CHAPTER 2

BUILDING AIR DISTRIBUTION

ALL-AIR SYSTEMS

AN ALL-AIR SYSTEM provides complete sensible and latent cooling, preheating, and humidification capacity in the air supplied by the system. No additional cooling or humidification is required at the zone, except in the case of certain industrial systems. Heating may be accomplished by the same airstream, either in the central system or at a particular zone. In some applications, heating is accomplished by a separate heater. The term zone implies the provision of, or the need for, separate thermostatic control, while the term room implies a partitioned area that may or may not require separate control.

The basic all-air system concept is to supply air to the room at such conditions that the sensible heat gain and latent heat gain in the space, when absorbed by the supply air flowing through the space, will bring the air to the desired room conditions. Since the heat gains in the space will vary with time, a mechanism to vary the energy removed from the space by the supply air is necessary. There are two such basic mechanisms: vary the amount of supply air delivered to the space, either by varying the flow rate or supplying air intermittently, or vary the temperature of the air being delivered to the space, either by modulating the temperature or conditioning the air intermittently.

All-air systems are classified in two categories:

- Single-duct systems, which contain the main heating and cooling coils in a series flow air path; a common duct distribution system at a common air temperature feeds all terminal apparatus. Either capacity varying mechanism (varying temperature or varying volume) can be used with single-duct systems.
- Dual-duct systems, which contain the main heating and cooling coils in parallel flow or series-parallel flow air paths with either (1) a separate cold and warm air duct distribution system that blends the air at the terminal apparatus (dual-duct systems), or (2) a separate supply air duct to each zone with the supply air blended to the required temperature at the main unit mixing dampers (multizone). Dual-duct systems generally vary the supply air temperature by mixing two airstreams of different temperatures, but can also vary the volume of supply air in some applications.

These categories may be further divided as follows:

Single duct
- Constant volume
  - Single zone
  - Multiple-zone reheat
  - Bypass VAV
- Variable air volume (VAV)
  - Throttling
  - Fan-powered
  - Reheat
  - Induction
  - Variable diffusers

Dual duct
- Dual duct
  - Constant volume
  - Variable air volume
  - Dual conduit
- Multizone
  - Constant volume
  - Variable air volume
  - Three-deck or Texas multizone

All-air systems may be adapted to many applications for comfort or process work. They are used in buildings that require individual control of multiple zones, such as office buildings, schools and universities, laboratories, hospitals, stores, hotels, and even ships. All-air systems are also used virtually exclusively in special applications for close control of temperature and humidity, including clean rooms, computer rooms, hospital operating rooms, research and development facilities, as well as many industrial/manufacturing facilities.

All-air systems have the following **advantages**:

- The location of the central mechanical room for major equipment allows operation and maintenance to be performed in unoccupied areas. In addition, it allows the maximum range of choices of filtration equipment, vibration and noise control, and the selection of high quality and durable equipment.
- Keeping piping, electrical equipment, wiring, filters, and vibration and noise-producing equipment away from the conditioned area minimizes service needs and reduces potential harm to occupants, furnishings, and processes.
- These systems offer the greatest potential for use of outside air for economizer cooling instead of mechanical refrigeration for cooling.
- Seasonal changeover is simple and adapts readily to automatic control.
- A wide choice of zoning, flexibility, and humidity control under all operating conditions is possible, with the availability of simultaneous heating and cooling even during off-season periods.
- Air-to-air and other heat recovery may be readily incorporated.
- They permit good design flexibility for optimum air distribution, draft control, and adaptability to varying local requirements.
- The systems are well suited to applications requiring unusual exhaust or makeup air quantities (negative or positive pressurization, etc.).
- All-air systems adapt well to winter humidification.
- By increasing the air change rate and using high-quality controls, it is possible for these systems to maintain the closest operating condition of ±0.25°F dry bulb and ±0.5% rh. Today, some systems can maintain essentially constant space conditions.

The preparation of this chapter is assigned to TC 9.1, Large Building Air-Conditioning Systems.

All-air systems have the following **disadvantages**:

- They require additional duct clearance, which reduces usable floor space and increases the height of the building.
- Depending on layout, larger floor plans are necessary to allow enough space for the vertical shafts required for air distribution.
- Ensuring accessible terminal devices requires close cooperation between architectural, mechanical, and structural designers.
- Air balancing, particularly on large systems, can be more difficult.
- Perimeter heating is not always available to provide temporary heat during construction.

Heating and Cooling Calculations

Basic calculations for airflow, temperatures, relative humidity, loads, and psychrometrics are covered in Chapters 6 and 28 of the 1997 *ASHRAE Handbook—Fundamentals*. The designer should understand the operation of the various components of a system, their relationship to the psychrometric chart, and their interaction under various operating conditions and system configurations. The HVAC designer should work closely with the architect to optimize the building envelope design. Close cooperation of all parties during design can result in reduced building loads, which often allows the use of smaller mechanical systems.

Zoning—Exterior

Exterior zones are affected by varying weather conditions—wind, temperature, and sun—and, depending on the geographic area and season, may require both heating and cooling at different times. While the engineer has many options in choosing a system, the system must respond to these variations. The considerable flexibility to meet such variations enables the greatest advantages from VAV systems to be realized. The need for separate perimeter zone heating is determined by

- Severity of the heating load (i.e., geographic location).
- Nature and orientation of the building envelope.
- Effects of downdraft at windows and the radiant effect of the cold glass surface (i.e., type of glass, area, height, and U-factor).
- Type of occupancy (i.e., sedentary versus transient).
- Operating costs (e.g., in buildings such as offices and schools that are unoccupied for considerable periods, fan operating cost can be reduced by heating with perimeter radiation during unoccupied periods rather than operating the main supply fans or local unit fans.)

Separate perimeter heating can operate with any all-air system. However, its greatest application has been in conjunction with VAV systems for cooling-only service. Care in design minimizes simultaneous heating and cooling. The section on Variable Air Volume has further details.

Zoning—Interior

Interior spaces have relatively constant conditions because they are isolated from external influences. Cooling loads in interior zones may vary with changes in the operation of equipment and appliances in the space and changes in occupancy, but usually interior spaces require cooling throughout the year. A VAV system has limited energy advantages for interior spaces, but it does provide simple temperature control. Interior spaces with a roof exposure, however, may require similar treatment to perimeter spaces requiring heat.

Space Heating

Although steam is an acceptable medium for central system preheat or reheat coils, low-temperature hot water provides a simple and more uniform means of perimeter and general space heating. Individual automatic control of each terminal provides the ideal space comfort. A control system that varies the water temperature inversely with the change in outdoor temperature provides water temperatures that produce acceptable results in most applications. To produce the best results, the most satisfactory ratio can be set after the installation is completed and actual operating conditions are ascertained.

Multiple perimeter spaces on one exposure served by a central system may be heated by supplying warm air from the central system. Areas that have heat gain from lights and occupants and no heat loss require cooling in winter, as well as in summer. In some systems, very little heating of the return and outdoor air is required when the space is occupied. Local codes dictate the amount of outside air required (see ASHRAE *Standard* 62 for recommended optimum outside air ventilation). For example, with return air at 75°F and outside air at 0°F, the temperature of a 25% outdoor/75% return air mixture would be 56°F, which is close to the temperature of the air supplied to cool such a space in summer. In this instance, a preheat coil installed in the minimum outdoor airstream to warm the outdoor air can produce overheating, unless it is sized so that it does not heat the air above 35 to 40°F. Assuming good mixing, a preheat coil located in the mixed airstream, prevents this problem. The outdoor air damper should be kept closed until room temperatures are reached during warm-up. A return air thermostat can terminate the warm-up period.

When a central air-handling unit supplies both perimeter and interior spaces, the supply air must be cool to handle the interior zones. Additional control is needed to heat the perimeter spaces properly. Reheating the air is the simplest solution, but it is not acceptable by most energy codes. An acceptable solution is to vary the volume of air to the perimeter and combine it with a terminal heating coil or a separate perimeter heating system, either baseboard, overhead air heating, or a fan-powered terminal unit with supplemental heat. The perimeter heating should be individually controlled and integrated with the cooling control. Resetting the supply water temperature downward when less heat is required generally improves temperature control. For further information, refer to Chapter 12 in this volume and Chapter 45 of the 1999 *ASHRAE Handbook—Applications*.

Temperature Versus Air Quantity

Designers have considerable flexibility in selecting the supply air temperature and corresponding air quantity within the limitations of the procedures for determining heating and cooling loads. ASHRAE *Standard* 55 also addresses the effect of these variables on comfort. In establishing the supply air temperature, the initial cost of lower airflow and low air temperature (smaller fan and duct systems) must be calculated against the potential problems of distribution, condensation, air movement, and the presence of increased odors and gaseous or particulate contaminants. Terminal devices that use low-temperature air can reduce the air distribution cost. These devices mix room and primary air to maintain reasonable air movement in the occupied space. Because the amount of outside air needed is the same for any system, the percentage in low-temperature systems is high, requiring special care in design to avoid freezing of preheat or cooling coils. Also, the low-temperature air supply reduces humidity in the space. Lower humidity during cooling cycles costs more in energy because the equipment runs longer. Also, if the humidity is too low, it may cause respiratory problems.

Other Considerations

All-air systems operate by maintaining a temperature differential between the supply air and the space. Any load that affects this differential and the associated airflow must be considered. Among these loads are the following:

- All fans (supply, return, and supplemental) add heat. The effect of these gains can be considerable, particularly in process work. If the fan motor is in the airstream, the inefficiencies in the motor must also be counted as heat gain to the air. If the fan is placed after the cooling coil (draw-through,) the total pressure (static plus velocity pressure) must be calculated. If the fan is placed before the coil (blow-through), only the velocity pressure needs to be considered, and the amount of supply air can be reduced. The heat gain in medium-pressure systems is about 0.5°F per inch of water static pressure.
- The supply duct may gain or lose heat from the surroundings. Most energy codes require that the supply duct be insulated, which is usually good practice regardless of code requirements.
- Attempting to control humidity in a space can affect the quantity of air and become the controlling factor in the selection of supply airflow rate. VAV systems provide only limited humidity control, so if humidity is critical, extra care must be taken in design.

First, Operating, and Maintenance Costs

As with all systems, the initial cost, or first cost, of an air-handling system varies widely depending on location, condition of the local economy, and preference of the contractor—even for identical systems. For example, a dual-duct system is more expensive because it requires essentially twice the amount of material for ducts as that of a comparable single-duct system. Systems requiring extensive use of terminal units are also comparatively expensive. The operating cost depends on the system selected, the skill of the designer in selecting and correctly sizing components, the efficiency of the duct design, and the effect of the building design and type on the operation. All-air systems have the greatest potential to minimize operating cost.

Because an all-air system separates the air-handling equipment from the occupied space, maintenance on major components in a central location is more economical. Also, central air-handling equipment requires less maintenance than other comparable equipment. The many terminal units used in an all-air system do, however, require periodic maintenance. Because these units (including reheat coils) are usually installed throughout a facility, maintenance costs for these devices must be considered.

Energy

The engineer's early involvement in the design of any facility can have a considerable effect on the energy consumed by the building. If designed carefully, a system keeps energy costs to a minimum. In practice, however, a system is usually selected based on a low first cost or for the performance of a particular task. In general, single-duct systems consume less energy than dual-duct systems, and VAV systems are more energy efficient than constant air volume systems. Savings from a VAV system come from the savings in fan power and because the system does not overheat or overcool spaces, nor does it cool and heat at the same time like a reheat system.

The air distribution system for an all-air system consists of two major subsystems: (1) air-handling units that generate conditioned air under sufficient positive pressure to circulate it through (2) a distribution system that carries air from the air-handling unit to the space being conditioned. The air distribution subsystem often includes means to control the amount or temperature of the air delivered to each space.

AIR-HANDLING UNITS

The basic secondary system is an all-air, single-zone, air-conditioning system consisting of an air-handling unit and an air distribution system. The air-handling unit may be designed to supply a constant air volume or a variable air volume for low-, medium-, or high-velocity air distribution. Normally, the equipment is located outside the conditioned area in a basement, penthouse, or service area. It can, however, be installed in the area if conditions permit. The equipment can be adjacent to the primary heating and refrigeration equipment or at considerable distance from it by circulating refrigerant, chilled water, hot water, or steam for energy transfer.

Figure 1 shows a typical draw-through central system that supplies conditioned air to a single zone or to multiple zones. A blow-through configuration may also be used if space or other conditions dictate. The quantity and quality of supplied air are fixed by space requirements and determined as indicated in Chapters 27 and 28 of the 1997 *ASHRAE Handbook—Fundamentals*. Air gains and loses heat by contacting heat transfer surfaces and by mixing with air of

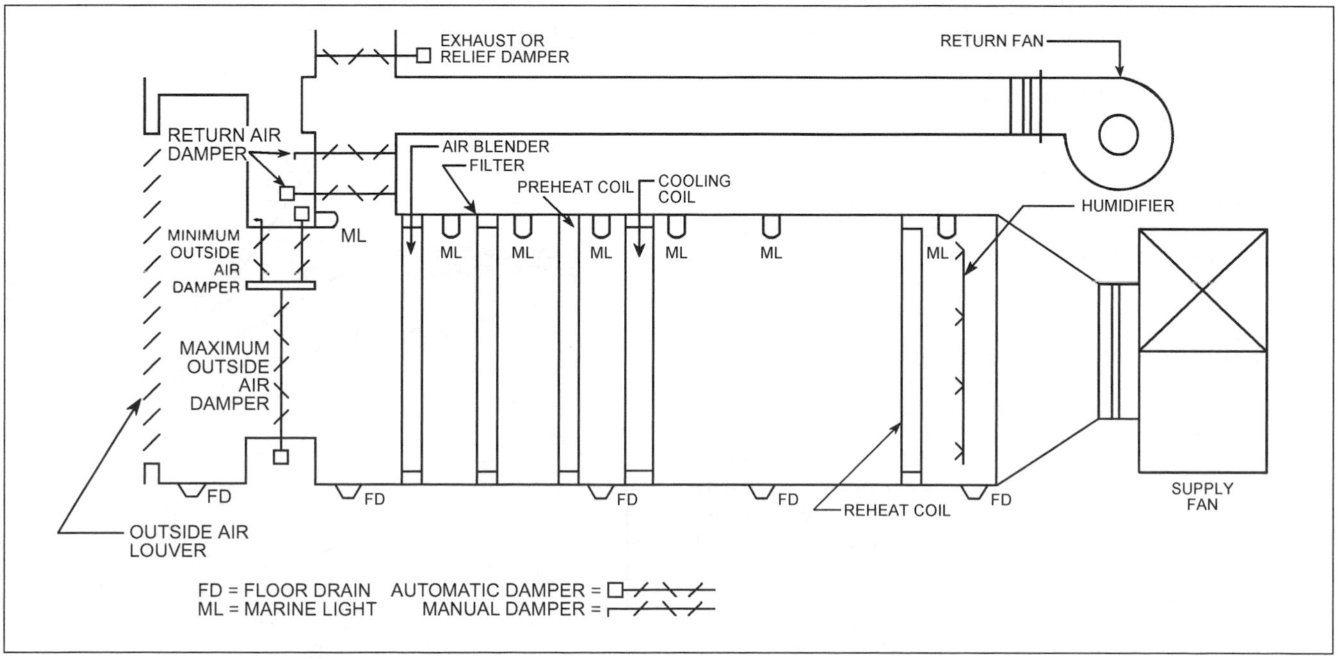

Fig. 1 Equipment Arrangement for Central System Draw-Through Unit

another condition. Some of this mixing is intentional, as at the outdoor air intake; other mixing is the result of the physical characteristics of a particular component, as when untreated air passes through a coil without contacting the fins (bypass factor).

All treated and untreated air must be well mixed for maximum performance of heat transfer surfaces and for uniform temperatures in the airstream. Stratified, parallel paths of treated and untreated air must be avoided, particularly in the vertical plane of systems using double inlet or multiple-wheel fans. Because these fans do not completely mix the air, different temperatures can occur in branches coming from opposite sides of the supply duct.

Primary Equipment

Refrigeration. Either central station or localized equipment, depending on the application, can provide cooling. Most large systems with multiple central air-handling units use a central refrigeration plant. Small, individual air-handling equipment can (1) be supplied from central chilled water generators, (2) use direct expansion cooling with a central condensing (cooling tower) system, or (3) be air cooled and totally self-contained. The decision on whether to provide a central plant or local equipment is based on factors similar to those for air-handling equipment.

Heating. The criteria described for cooling are usually used to determine whether a central heating plant or a local one is desirable. Usually, a central, fuel-fired plant is more desirable for heating. In small facilities, electric heating is a viable option and is often economical, particularly where care has been taken to design energy-efficient systems and buildings.

Air-Handling Equipment

Air-handling equipment is available as packaged equipment in many configurations using any desired method of cooling, heating, humidification, filtration, etc. In large systems (over 50,000 cfm), air-handling equipment is usually custom-designed and fabricated to suit a particular application. Air handlers may be either centrally located or decentralized.

Central Mechanical Equipment Rooms (MER). Usually the type of facility determines where the air-handling equipment is located. Central fan rooms today are more common in laboratory or industrial facilities, where maintenance is kept isolated from the conditioned space.

Decentralized Mechanical Equipment Rooms. Many office buildings locate air-handling equipment at each floor. This not only saves floor space for equipment but minimizes the space required for distribution ductwork and shafts. The reduced size of equipment as a result of duplicated systems allows the use of less expensive packaged equipment and reduces the need for experienced operating and maintenance personnel.

Fans

Both packaged and built-up air-handling units can use any type of fan. Centrifugal fans may be forward curved, backward inclined, or airfoil, and single width-single inlet or double-width double inlet. It is quite common for packaged air handlers to use multiple double-width, double-inlet (DWDI) centrifugal fans on a common shaft with a single drive motor. Plug fans—single-width, single-inlet (SWSI) centrifugal fan wheels without a scroll—are sometimes used on larger packaged air handlers to make them more compact. Vaneaxial fans, both adjustable pitch and variable pitch during operation, are often used on very large air-handling units. Fan selection should be based on efficiency and sound power level throughout the anticipated range of operation as well as the ability of the fan to provide the required flow at the anticipated static pressure. Chapter 18 discusses fans and their selection.

AIR-HANDLING UNIT PROCESSES

Cooling

The basic methods used for cooling include

- **Direct expansion**, which takes advantage of the latent heat of the fluid, as shown in the psychrometric diagram in Figure 2.
- **Fluid-filled coil**, where temperature differences between the fluid and the air cause an exchange of energy by the same process as in Figure 2 (see the section on Dehumidification).
- **Direct spray of water** in the airstream (Figure 3) in which an adiabatic process uses the latent heat of evaporation of water to reduce dry bulb temperature while increasing moisture content. Both sensible and latent cooling is also possible by spraying chilled water. Air can be cooled and greatly humidified by spraying heated water. A conventional evaporative cooler, uses the adiabatic process, by spraying or dripping recirculated water onto a filter pad (see the section on Humidification). The wetted duct or

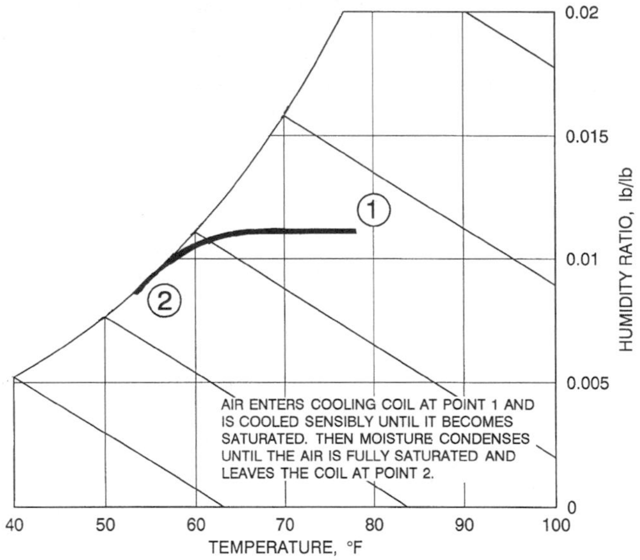

Fig. 2 Direct-Expansion or Chilled Water Cooling and Dehumidification

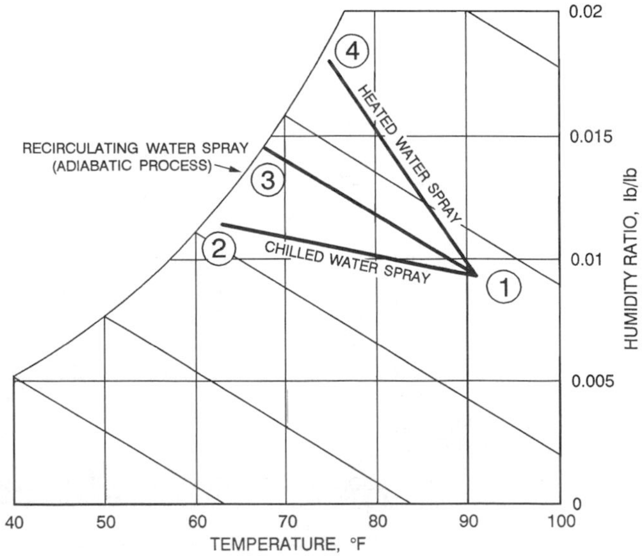

Fig. 3 Water Spray Cooling

supersaturated system is a variation on direct spray. In this system, tiny droplets of free moisture are carried by the air into the conditioned space where they evaporate, providing additional cooling. This reduces the amount of air needed for a given space load (Figure 4).

- **Indirect evaporative cooling** adiabatically cools outdoor air or exhaust air from the conditioned space by spraying water, then passes that cooled air through one side of a heat exchanger, while the air to be supplied to the space is cooled by passing through the other side of the heat exchanger. Chapter 19 has further information on this method of cooling.

Chapter 6 of the 1997 *ASHRAE Handbook—Fundamentals* detail the psychrometric process of these methods.

Heating

The basic methods used for heating include

- **Steam**, which uses the latent heat of the fluid
- **Fluid-filled coils**, which use temperature differences between the warm fluid and the cooler air
- **Electric heating**, which also uses the temperature difference between the heating coil and the air to exchange energy

The effect on the airstream for each of these processes is the same and is shown in Figure 5. For basic equations, refer to Chapter 6 of the 1997 *ASHRAE Handbook—Fundamentals*.

Humidification

The methods used to humidify air include

- **Direct spray of recirculated water** into the airstream (air washer) reduces the dry-bulb temperature while maintaining an almost constant wet bulb in an adiabatic process [see Figure 3, Paths (1) to (3)]. The air may also be cooled and dehumidified, or heated and humidified by changing the temperature of the spray.

  In one variation, the surface area of water exposed to the air is increased by spraying water onto a cooling/heating coil. The coil surface temperature determines the leaving air conditions. Another method is to spray or distribute water over a porous medium, such as those in evaporative coolers and commercial greenhouses. This method requires careful monitoring of the water condition to keep biological contaminants from the airstream (Figure 6).

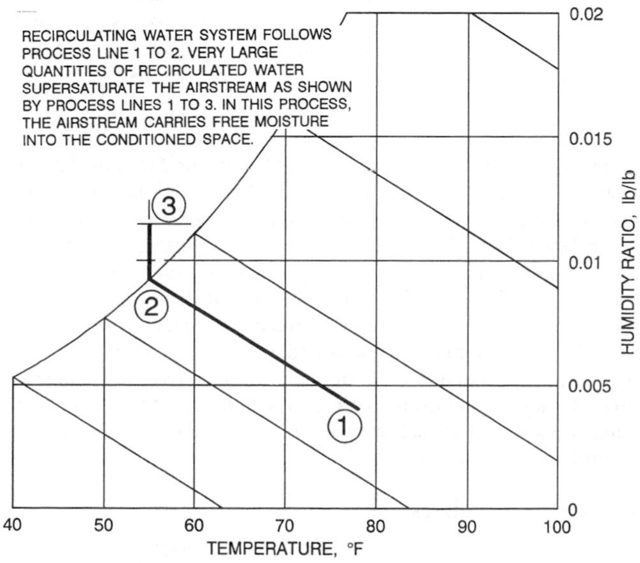

Fig. 4 Water Spray Humidifier

- **Compressed air that forces water** through a nozzle into the airstream is essentially a constant wet-bulb (adiabatic) process. The water must be treated to keep particulates from entering the airstream and contaminating or coating equipment and furnishings. Many types of nozzles are available.
- **Steam injection**, which is a constant dry-bulb process (Figure 7). However, as the steam injected becomes superheated, the leaving dry-bulb temperature increases. If live steam is injected into the airstream, the boiler water treatment chemical must be nontoxic to the occupants and, if the air is supplying a laboratory, to the research under way.

Dehumidification

Moisture condenses on a cooling coil when its surface temperature is below the dew point of the air, thus reducing the humidity of the air. In a similar manner, air will also be dehumidified if a fluid with a temperature below the airstream dew point is sprayed into the

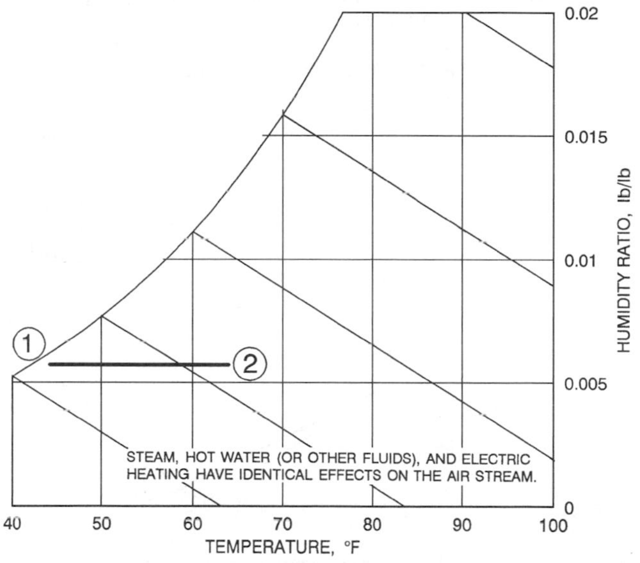

Fig. 5 Steam, Hot Water, and Electric Heating

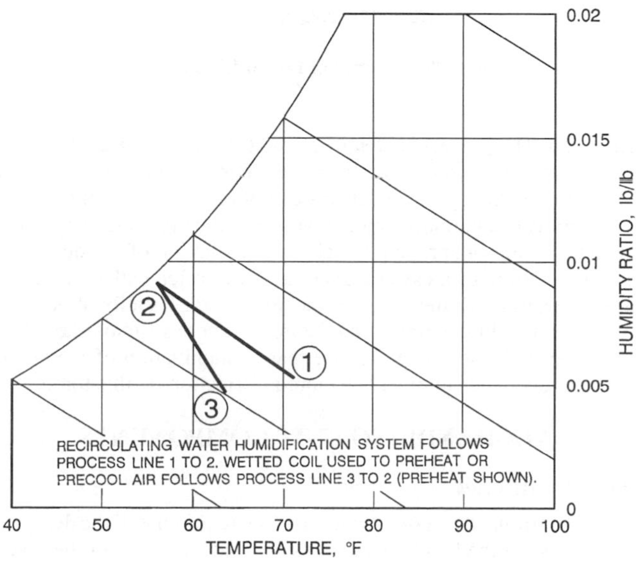

Fig. 6 Humidification

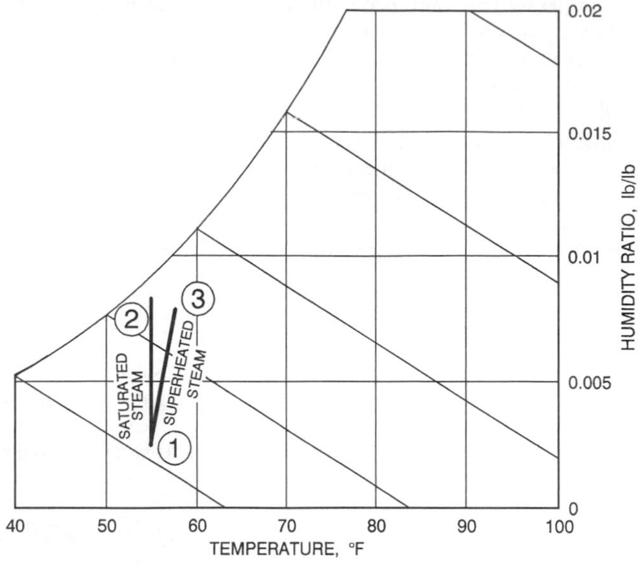

Fig. 7 Steam Humidification

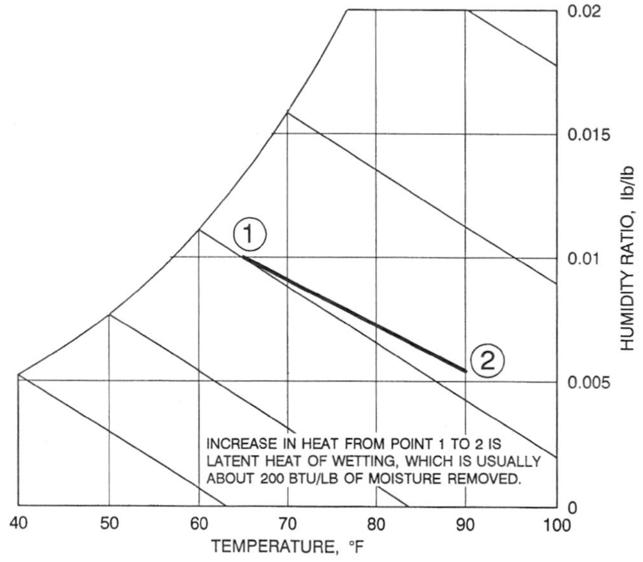

Fig. 8 Chemical Humidification

airstream. The process is identical to that shown in Figure 2, except that the moisture condensed from the airstream condenses on, and dissolves in, the spray droplets instead of on the solid coil surface.

Chemical dehumidification involves either passing air over a solid desiccant or spraying the air with a solution of the desiccant and water. Both of these processes add heat, often called the latent heat of wetting, to the air being dehumidified. Usually about 200 Btu/lb of moisture is removed (Figure 8). These systems should be reviewed with the user to ensure that contamination of the space does not occur. Chapter 22 has more information on this topic.

AIR-HANDLING UNIT COMPONENTS

Air Conditioner

To determine the system's air-handling requirement, the designer must consider the function and physical characteristics of the space to be conditioned and the air volume and thermal exchange capacities required. Then, the various components may be selected and arranged by considering the fundamental requirements of the central system—equipment must be adequate, accessible for easy maintenance, and not too complex in arrangement and control to produce the required conditions. Further, the designer should consider economics in component selection. Both initial cost and operating costs affect design decisions. The designer should not arbitrarily design for a 500 fpm face velocity, which has been common for selection of cooling coils and other components. Filter and coil selection at 300 to 400 fpm, with resultant lower pressure loss, could produce a substantial payback on constant volume systems. Chapter 35 of the 1999 *ASHRAE Handbook—Applications* has further information on energy and life-cycle costs.

Figure 1 shows a general arrangement of the air-handling unit components for a single-zone, all-air central system suitable for year-round air conditioning. With this arrangement, close control of temperature and humidity are possible. All these components are seldom used simultaneously in a comfort application. Although Figure 1 indicates a built-up system, most of the components are available from many manufacturers completely assembled or in subassembled sections that can be bolted together in the field. When selecting central system components, specific design parameters must be evaluated to balance cost, controllability, operating expense, maintenance, noise, and space. The sizing and selection of primary air-handling units substantially affect the results obtained in the conditioned space.

Return Air Fan

A return air fan is optional on small systems but is essential for the proper operation of air economizer systems for free cooling from outside air if the return path has a significant pressure drop (greater than about 0.3 in. of water) It provides a positive return and exhaust from the conditioned area, particularly when mixing dampers permit cooling with outdoor air in intermediate seasons and winter. The return air fan ensures that the proper volume of air returns from the conditioned space. It prevents excess pressure when economizer cycles introduce more than the minimum quantity of outside air. It also reduces the static pressure the supply fan has to work against. (The use of a return fan can increase system energy use if the system is so arranged that the return fan works against more static pressure than the incremental static pressure the supply fan would have to provide to overcome the return path pressure drop) The supply fan(s) must be carefully matched with the return fans, particularly in variable air volume (VAV) systems. The return air fan should handle a slightly smaller amount of air to account for fixed exhaust systems, such as the toilet exhaust, and to ensure a slight positive pressure in the conditioned space. Chapter 45 of the 1999 *ASHRAE Handbook—Applications* has design details.

Relief Air Fan

In many situations, a relief (or exhaust) air fan may be used instead of a return fan. A relief air fan relieves ventilation air introduced during air economizer operation and operates only when this control cycle is in effect. When a relief air fan is used, the supply fan must be designed for the total supply and return static pressure in the system, since the relief air fan does not operate during the non-economizer mode of operation. During the economizer mode of operation, the relief fan must be controlled to exhaust at a rate that tracks the quantity of outside air introduced, to ensure a slight positive pressure in the conditioned space, as with the return air fan system cited previously. The section on Economizers describes the required control for relief air fans.

Automatic Dampers

Opposed blade dampers for the outdoor, return, and relief airstreams provides the highest degree of control. The section on Mixing Plenum covers the conditions that dictate the use of parallel blade dampers. Pressure relationships between various sections must be considered to ensure that automatic dampers are properly sized for wide open and modulating pressure drops.

Relief Openings

Relief openings in large buildings should be constructed similarly to outdoor air intakes, but they may require motorized or self-acting backdraft dampers to prevent high wind pressure or stack action from causing the airflow to reverse when the automatic dampers are open. The pressure loss through relief openings should be 0.10 in. of water or less. Low-leakage dampers, such as those for outdoor intakes, prevent rattling and minimize leakage. Relief dampers sized for the same air velocity as the maximum outdoor air dampers facilitate control when an air economizer cycle is used. The relief air opening should be located so that the exhaust air does not short-circuit to the outdoor air intake.

Return Air Dampers

The negative pressure in the outdoor air intake plenum is a function of the resistance or static pressure loss through the outside air louvers, damper, and duct. The positive pressure in the relief air plenum is, likewise, a function of the static pressure loss through the exhaust or relief damper, the exhaust duct between the plenum and outside, and the relief louver. The pressure drop through the return air damper must accommodate the pressure difference between the positive-pressure relief air plenum and the negative pressure outside air plenum. Proper sizing of this damper facilitates both air balancing and mixing. An additional manual damper may be required for proper air balancing.

Outdoor Air Intakes

Resistance through outdoor intakes varies widely, depending on construction. Frequently, architectural considerations dictate the type and style of louver. The designer should ensure that the louvers selected offer minimum pressure loss, preferably not more than 0.10 in. of water. High-efficiency, low-pressure louvers that effectively limit carryover of rain are available. Flashing installed at the outside wall and weep holes or a floor drain will carry away rain and melted snow entering the intake. Cold regions may require a snow baffle to direct fine snow particles to a low-velocity area below the dampers. Outdoor dampers should be low-leakage types with special gasketed edges and special end treatment. Separate damper sections with separate damper operators are strongly recommended for the minimum outdoor air needed for ventilation. The maximum outdoor air needed for economizer cycles is then drawn through the entire outside air damper.

Economizers

An economizer uses outside air to reduce the refrigeration requirement. A logic circuit maintains a fixed minimum of ventilation outside air. The air side economizer is an attractive option for reducing energy costs when the climate allows. The air-side economizer takes advantage of cool outdoor air to either assist mechanical cooling or, if the outdoor air is cool enough, provide total system cooling. It is necessary to include some method of variable volume relief when air-side economizers are employed, to exhaust the extra outdoor air intake to outdoors. The relief volume may be controlled by several different methods, including fan tracking (operating the supply and return fans to maintain a constant difference in airflow between them,) or relief air discharge dampers which modulate in response to building space pressure. The relief system is off and relief dampers are closed when the air-side economizer is inactive. In systems with large return air static requirements, return fans or exhaust fans may be necessary to properly exhaust building air and take in outside air.

Advantages of Air-Side Economizers

- Substantially reduces compressor, cooling tower, and condenser water pump energy requirements
- Has a lower air-side pressure drop than a water-side economizer
- Has a higher annual energy savings than a water-side economizer
- Reduces tower makeup water and related water treatment

Disadvantages of Air-Side Economizers

- Humidification may be required during winter operation
- Equipment room is generally placed along the building's exterior wall
- Installed cost may be higher than that for a water-side economizer if the cost of providing the exhaust system requirements exceeds the costs of piping, pump, and heat exchanger

Mixing Plenum

If the equipment is closely coupled to outdoor louvers in a wall, the minimum outdoor air damper should be located as close as possible to the return damper connection. An outside air damper sized for 1500 fpm gives good control. Low-leakage outdoor air dampers minimize leakage when closed during shutdown. The pressure difference between the relief plenum and outdoor intake plenum must be measured through the return damper section. A higher velocity through the return air damper—high enough to cause this loss at its full open position—facilitates air balance and creates good mixing. To create maximum turbulence and mixing, return air dampers should be set so that any deflection of air is toward the outside air. Parallel blade dampers may aid mixing. Mixing dampers should be placed across the full width of the unit, even though the location of the return duct makes it more convenient to return air through the side. When return dampers are placed at one side, return air passes through one side of a double-inlet fan, and cold outdoor air passes through the other. If the air return must enter the side, some form of air blender should be used.

Although opposed blade dampers offer better control, properly proportioned parallel blade dampers are more effective for mixing airstreams of different temperatures. If parallel blades are used, each damper should be mounted so that its partially opened blades direct the airstreams toward the other damper for maximum mixing. Baffles that direct the two airstreams to impinge on each other at right angles and in multiple jets create the turbulence required to mix the air thoroughly. In some instances, unit heaters or propeller fans have been used for mixing, regardless of the final type and configuration of dampers. Otherwise, the preheat coil will waste heat, or the cooling coil may freeze.

Filter Section

A system's overall performance depends heavily on the filter. Unless the filter is regularly maintained, system resistance increases and airflow diminishes. Accessibility is the primary consideration in filter selection and location. In a built-up system, there should be at least 3 ft between the upstream face of the filter bank and any obstruction. Other requirements for filters can be found in Chapter 24 and in ASHRAE *Standard* 52.1.

Good mixing of outdoor and return air is also necessary for good filter performance. A poorly placed outdoor air duct or a bad duct connection to the mixing plenum can cause uneven loading of the filter and poor distribution of air through the coil section. Because of the low resistance of the clean areas, the filter gauge may not warn of this condition.

Preheat Coil

The preheat coil should have wide fin spacing, be accessible for easy cleaning, and be protected by filters. If the preheat coil is located in the minimum outdoor airstream rather than in the mixed airstream as shown in Figure 1, it should not heat the air to an exit temperature above 35 to 45°F; preferably, it should become inoperative at outdoor temperatures of 45°F. Inner distributing tube or integral face and bypass coils are preferable with steam. Hot water preheat coils should have a constant flow recirculating pump and should be piped for parallel flow so that the coldest air will contact the warmest part of the coil surface first. Chapter 23 provides more detailed information on heating coils.

Cooling Coil

In this section, sensible and latent heat are removed from the air. In all finned coils, some air passes through without contacting the fins or tubes. The amount of this bypass can vary from 30% for a four-row coil at 700 fpm to less than 2% for an eight-row coil at 300 fpm. The dew point of the air mixture leaving a four-row coil might satisfy a comfort installation with 25% or less outdoor air, a small internal latent load, and sensible temperature control only. For close control of room conditions for precision work, a deeper coil may be required. Chapter 21 provides more information on cooling coils and their selection.

Coil freezing can be a serious problem with chilled water coils. Full flow circulation of chilled water during freezing weather, or even reduced flow with a small recirculating pump, minimizes coil freezing and eliminates stratification. Further, continuous full flow circulation can provide a source of off-season chilled water in air-and-water systems. Antifreeze solutions or complete coil draining also prevent coil freezing. However, because it is difficult, if not impossible, to drain most cooling coils completely, caution should be exercised if this option is considered.

Reheat Coil

Reheat systems are strongly discouraged, unless recovered energy is used (see ASHRAE *Standard* 90.1). Reheating is limited to laboratory, health care, or similar applications where temperature and relative humidity must be controlled accurately. Heating coils located in the reheat position, as shown in Figure 1, are frequently used for warm-up, although a coil in the preheat position is preferable. Hot water heating coils provide the highest degree of control. Oversized coils, particularly steam, can stratify the airflow; thus, where cost-effective, inner distributing coils are preferable for steam applications. Electric coils may also be used. Chapter 23 has more information.

Humidifiers

Humidifiers may be installed as part of the central station air-handling unit, or in terminals at the point of use, or both. Where close humidity control of selected spaces is required, the entire supply airstream may be humidified to a lower humidity level in the air handler, with terminal humidifiers located in the supply ducts serving just those selected spaces bringing them up to their required humidity levels. For comfort installations not requiring close control, moisture can be added to the air by mechanical atomizers or point-of-use electric or ultrasonic humidifiers. Proper location of this equipment prevents stratification of moist air in the system.

In this application, the heat of evaporation should be replaced by heating the recirculated water, rather than by increasing the size of the preheat coil. Steam grid humidifiers with dew-point control usually are used for accurate humidity control. It is not possible to add moisture to saturated air, even with a steam grid humidifier. Air in a laboratory or other application that requires close humidity control must be reheated after leaving a cooling coil before moisture can be added. The capacity of the humidifying equipment should not exceed the expected peak load by more than 10%. If the humidity is controlled from the room or the return air, a limiting humidistat and fan interlock may be needed in the supply duct. This prevents condensation and mold or mildew growth in the ductwork when temperature controls call for cooler air. Humidifiers add some sensible heat that should be accounted for in the psychrometric evaluation. Chapter 20 has additional information.

Dehumidifiers

Many dehumidification systems are available. Dust can be a problem with solid desiccants, and lithium contamination is a concern with spray equipment. Chapter 21 discusses dehumidification by cooling coils and Chapter 22 discusses desiccant dehumidifiers.

Odor Control

Most devices control odors and other contaminants with activated carbon or potassium permanganate as a basic filter medium. Other systems use electronic control methods. Chapters 12 and 13 of the 1997 *ASHRAE Handbook—Fundamentals* have more information on odor control.

Supply Air Fan

Either axial flow, centrifugal, or plug fans may be chosen as supply air fans for straight-through flow applications. In factory-fabricated units, more than one centrifugal fan may be tied to the same shaft. If headroom permits, a single-inlet fan should be chosen when air enters at right angles to the flow of air through the equipment. This permits a direct flow of air from the fan wheel into the supply duct without abrupt change in direction and loss in efficiency. It also permits a more gradual transition from the fan to the duct and increases the static regain in the velocity pressure conversion. To minimize inlet losses, the distance between the casing walls and the fan inlet should be at least the diameter of the fan wheel. With a single-inlet fan, the length of the transition section should be at least half the width or height of the casing, whichever is longer. If fans blow through the equipment, the air distribution through the downstream components needs analyzing, and baffles should be used to ensure uniform air distribution. Chapter 18 has more information.

AIR-HANDLING UNIT CONCERNS

Outside Air Requirements

A common complaint regarding buildings is the lack of outside air. This problem is especially a concern in VAV systems where outside air quantities are established for peak loads and are then reduced in proportion to the air supplied during periods of reduced load. A simple control added to the outside air damper can eliminate this problem and keep the amount of outside air constant, regardless of the operation of the VAV system. However, the need to preheat the outside air must be considered if this control is added.

Another problem is that some codes require as little as 5 cfm per person [about 0.05 cfm/ft^2] of outside air. This amount is far too low for a building in which modern construction materials are used. Higher outside air quantities may be required to reduce odors, VOC's, and potentially dangerous pollutants. ASHRAE *Standard* 62 provides information on ventilation for acceptable indoor air quality. Air quality (i.e., the control or reduction of contaminants such as volatile organic compounds, formaldehyde from furnishings, and dust) must be reviewed by the engineer.

Heat recovery devices are becoming more popular as the requirements for outside air increase. They are used extensively in research and development facilities and in hospitals and laboratories where HVAC systems supply 100% outside air. Many types are available, and the type of facility usually determines which is most suitable. Many countries with extreme climates provide heat exchangers on outside/relief air, even for private homes. This trend is now appearing in larger commercial buildings worldwide. Heat recovery devices such as air-to-air plate heat exchangers can save energy and reduce the required capacity of primary cooling and heating plants by 20% and more under certain circumstances.

The **location of intake and exhaust louvers** should be carefully considered; in some jurisdictions, location is governed by codes. Louvers must be separated enough to avoid short-circuiting of air. Furthermore, intake louvers should not be near a potential source of contaminated air such as a boiler stack or hood exhaust. Relief air should also not interfere with other systems. If heat recovery devices are used, intake and exhaust airstreams may need to run together, such as through air-to-air plate heat exchangers.

Equipment Isolation

Vibration and sound isolation equipment is required for most central fan installations. Standard mountings of fiberglass, ribbed rubber, neoprene mounts, and springs are available for most fans and prefabricated units. The designer must account for seismic restraint requirements for the seismic zone in which the particular project is located. In some applications, the fans may require concrete inertia blocks in addition to non-enclosed spring mountings. Steel springs require sound-absorbing material inserted between the springs and the foundation. Horizontal discharge fans operating at a high static pressure frequently require thrust arrestors. Ductwork connections to fans should be made with fireproof fiber cloth sleeves having considerable slack, but without offset between the fan outlet and rigid duct. Misalignment between the duct and fan outlet can cause turbulence, generate noise, and reduce system efficiency. Electrical and piping connections to vibration-isolated equipment should be made with flexible conduit and flexible connections. Special considerations are required in seismic zones.

Equipment noise transmitted through the ductwork can be reduced by sound-absorbing units, acoustical lining, and other means of attenuation. Sound transmitted through the return and relief ducts should not be overlooked. Acoustical lining sufficient to adequately attenuate any objectionable system noise or locally generated noise should be considered. Chapter 46 of the 1999 *ASHRAE Handbook—Applications*, Chapter 7 of the 1997 *ASHRAE Handbook—Fundamentals*, and ASHRAE *Standard* 68 have further information on sound and vibration control. Noise control, both in the occupied spaces and outside near intake or relief louvers, must be considered. Some local ordinances may limit external noise produced by these devices.

AIR DISTRIBUTION

Ductwork should deliver conditioned air to an area as directly, quietly, and economically as possible. Structural features of the building generally require some compromise and often limit depth. Chapter 32 of the 1997 *ASHRAE Handbook—Fundamentals* describes ductwork design in detail and gives several methods of sizing duct systems. It is imperative that the designer coordinate duct design with the structure. In commercially developed projects, great effort is made to reduce floor-to-floor dimensions. The resultant decrease in the available interstitial space left for ductwork is a major design challenge.

Room Terminals

In some instances, such as in low-velocity, all-air systems, the air may enter from the supply air ductwork directly into the conditioned space through a grille or diffuser. In medium and high-velocity air systems, air terminal units normally control air volume, reduce duct pressure, or both. Various types are available, including

1. A variable air volume (VAV) terminal unit, which varies the amount of air delivered. This air may be delivered to low-pressure ductwork and then to the space, or the terminal may contain an integral air diffuser;
2. A fan-powered VAV terminal unit, which varies the amount of primary air delivered, but also uses a fan to mix ceiling plenum or return air with primary supply air before it is delivered to the space;
3. An all-air induction terminal unit, which controls the volume of primary air, induces ceiling plenum or space air, and distributes the mixture through low-velocity ductwork to the space; and
4. An air-water induction terminal unit, which includes a coil or coils in the induced airstream to condition the return air before it mixes with the primary air and enters the space.

Chapter 17 has more information on room terminals and air terminal units.

Ductwork Design

Chapter 32 of the 1997 *ASHRAE Handbook—Fundamentals* covers calculation methods for duct design. The air distribution ductwork and terminal devices selected must be compatible or the system will either fail to operate effectively or incur high first, operating, and maintenance costs.

Duct Sizing. All-air duct systems can be designed for high or low velocity. A high-velocity system has smaller ducts, which save space but require higher pressures. In some low-velocity systems, medium or high fan pressures may be required for balancing or to overcome high pressure drops from terminal devices. In any variable flow system, changing operating conditions can cause airflow in the ducts to differ from design flow. Thus, varying airflow in the supply duct must be carefully analyzed to ensure that the system performs efficiently at all loads. This precaution is particularly true with high-velocity air. Return air ducts are usually sized by the equal friction method.

In many applications, the space between a hung ceiling and the floor slab above it is used as a return air plenum, so that the return air is collected at a central point. Governing codes should be consulted before using this approach in new design, because most codes prohibit combustible material in a ceiling space used as a return air plenum. For example, the *National Electrical Code Handbook* (NFPA 1996) requires that either conduit or PTFE insulated wire (often called plenum rated cable) be installed in a return air plenum. In addition, regulations often require that return air plenums be divided into smaller areas by fire walls and that fire dampers be installed in ducts, which increases first cost.

In research and some industrial facilities, return ducting must be installed to avoid contamination and growth of biological contaminants in the ceiling space. Lobby ceilings with lay-in panels may not work well as return plenums where negative pressure from high-rise elevators or stack effects of high-rise buildings may occur. If the plenum leaks to the low-pressure area, the tiles may lift and drop out when the outside door is opened and closed. Return plenums directly below a roof deck have substantially higher return air gain than a ducted return, but have the advantage of reducing the heat gain to or loss from the space.

Corridors should not be used for return air because they spread smoke and other contaminants. While most codes ban returning air through corridors, the method is still used in many older facilities.

All ductwork should be sealed. Energy waste due to leaks in the ductwork and terminal devices can be considerable. The ductwork installed in many commercial buildings can have leakage of 20% or more.

SINGLE-DUCT SYSTEMS

Constant Volume

While maintaining constant airflow, single-duct constant volume systems change the supply air temperature in response to the space load (Figure 9).

Single-Zone Systems. The simplest all-air system is a supply unit serving a single-temperature control zone. The unit can be installed either in or remote from the space it serves, and it may

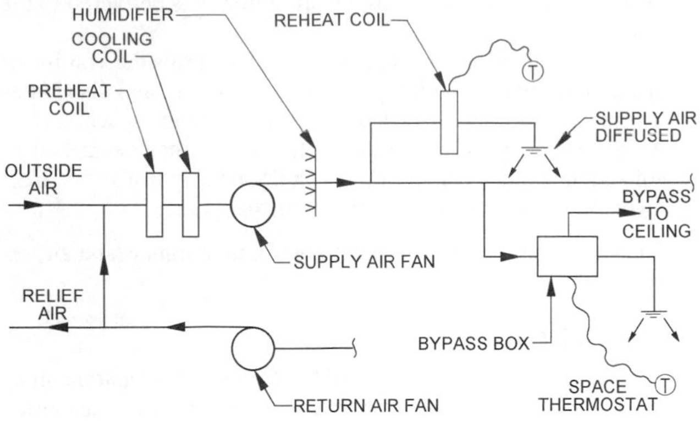

Fig. 9 Constant Volume with Reheat and Bypass Terminal Devices

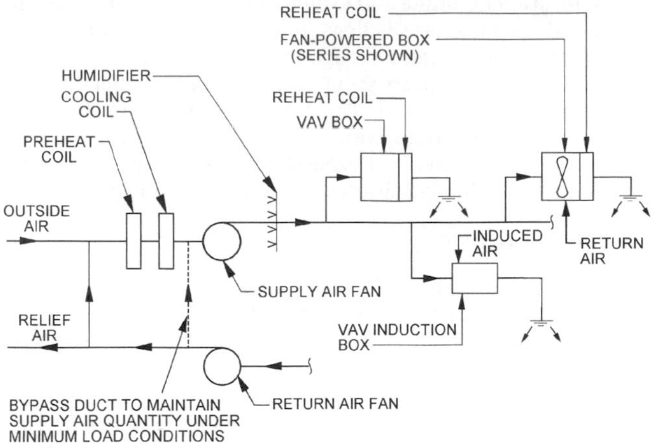

Fig. 10 Variable Air Volume with Reheat, Induction and Fan-Powered Devices

operate with or without distribution ductwork. Ideally, this system responds completely to the space needs, and well-designed control systems maintain temperature and humidity closely and efficiently. Single zone systems often involve short ductwork with low pressure drop and thus low fan energy, and single-zone systems can be shut down when not required without affecting the operation of adjacent areas, offering further energy savings. A return or relief fan may be needed, depending on the capacity of the system and whether 100% outdoor air is used for cooling as part of an economizer cycle. Relief fans can be eliminated if provisions are made to relieve over pressurization by other means, such as gravity dampers.

Multiple-Zone Reheat. The multiple-zone reheat system is a modification of the single-zone system. It provides (1) zone or space control for areas of unequal loading; (2) simultaneous heating or cooling of perimeter areas with different exposures; and (3) close tolerance of control for process or comfort applications. As the word reheat implies, heat is added as a secondary simultaneous process to either preconditioned (cooled, humidified, etc.) primary air or recirculated room air. Relatively small low-pressure systems place reheat coils in the ductwork at each zone. More complex designs include high-pressure primary distribution ducts to reduce their size and cost and pressure reduction devices to maintain a constant volume for each reheat zone.

The system uses conditioned air from a central unit generally at a fixed cold air temperature that is low enough to meet the maximum cooling load. Thus, all supply air is always cooled the maximum amount, regardless of the current load. Heat is added to the airstream in each zone to match the cooling capacity to the current load in that zone. The result is very high energy use. However, the supply air temperature from the unit can be varied, with proper control, to reduce the amount of reheat required and the associated energy consumption. Care must be taken to avoid high internal humidity when the temperature of air leaving the cooling coil is permitted to rise during the spring and fall.

When a reheat system heats a space with an exterior exposure in cold weather, the reheat coil must not only replace the heat lost from the space, but also must offset the cooling of the supply air (enough cooling to meet the peak load for the space), further increasing energy consumption compared to other systems. If a constant volume system is oversized, the reheat cost becomes excessive.

Bypass. A variation of the constant volume reheat system is the use of a bypass terminal unit instead of reheat. This system is essentially a constant volume primary system with a VAV secondary system. The quantity of room supply air is varied to match the space load by dumping excess supply air into the return ceiling plenum or return air duct, i.e., by bypassing the room. When bypass terminal units dump to a return air plenum, the return air plenum temperature is reduced and the plenum must be kept at a lower pressure than the room so that cooler air does not spill into the room through the return air grilles. A return fan is often used to create that negative pressure. While this reduces the air volume supplied to the space, the system air volume and fan energy remains constant. Refrigeration or heating at the air-handling unit is reduced due to the lower return air temperature. A bypass system is generally restricted to small installations where a simple method of temperature control is desired, a modest initial cost is desired, and energy conservation is less important.

Variable Air Volume

A VAV system, as shown in Figure 10, controls temperature in a space by varying the quantity of supply air rather than varying the supply air temperature. A VAV terminal unit at the zone varies the quantity of supply air to the space. The supply air temperature is held relatively constant: while supply air temperature can be moderately reset depending on the season, it must always be low enough to meet the cooling load in the most demanding zone, and to maintain appropriate humidity. Variable air volume systems can be applied to interior or perimeter zones, with common or separate fans, with common or separate air temperature control, and with or without auxiliary heating devices. The greatest energy saving associated with VAV occurs at the perimeter zones, where variations in solar load and outside temperature allow the supply air quantity to be reduced. If the peak room load is not determined accurately, an oversized VAV system will partially throttle at full load and may become excessively noisy throttling at part loads.

Humidity control is a potential problem with VAV systems. If humidity is critical, as in certain laboratories, process work, etc., systems may have to be limited to constant volume airflow. Particular care should be taken in areas where the sensible heat ratio (ratio of sensible heat to sensible plus latent heat to be removed) is low, such as in conference rooms. In these situations, the minimum set point of the VAV terminal unit can be set at about 50% and reheat added as necessary to keep humidity low during reduced load conditions.

Other measures may also be used to maintain enough air circulation through the room to absorb sufficient moisture to achieve acceptable humidity levels. The human body is more sensitive to elevated air temperatures when there is little air movement. Minimum air circulation can be maintained during reduced load by (1) raising the supply air temperature of the entire system, which increases space humidity, or supplying reheat on a zone-by-zone basis; (2) providing auxiliary heat in each room independent of the air system; (3) using individual zone recirculation and blending varying amounts of supply and room air or supply and ceiling plenum air with fan-powered VAV terminal units, or, if the design permits, at the air-handling unit; (4) recirculating air with a VAV induction unit, or (5) providing a dedicated recirculation fan to increase airflow.

Variable Diffuser. The discharge aperture of this diffuser is reduced to keep the discharge velocity relatively constant while reducing the conditioned supply airflow. Under these conditions, the induction effect of the diffuser is kept high, cold air mixes in the space, and the room air distribution pattern is more nearly maintained at reduced loads. These devices are of two basic types—one has a flexible bladder that expands to reduce the aperture, and the other has a diffuser plate that physically moves. Both devices are pressure-dependent, which must be considered in the design of the duct -distribution system. They are either powered by the system or pneumatically or electrically driven.

DUAL-DUCT SYSTEMS

A dual-duct system conditions all the air in a central apparatus and distributes it to the conditioned spaces through two parallel mains or ducts. One duct carries cold air and the other carries warm air. In each conditioned space or zone, a valve mixes the warm and cold air in proper proportion to satisfy the load of the space. A dual-duct system uses more energy than a single-duct VAV system. Dual-duct systems may be designed as constant volume or variable air volume. As with other VAV systems, certain primary air configurations can cause high relative humidity in the space during the spring and fall.

Constant Volume

Dual-duct constant volume systems are of two types:

Single Fan, No Reheat. This is similar to a single-duct system, except that it contains a face-and-bypass damper at the cooling coil, which is arranged to bypass a mixture of outdoor and recirculated air as the internal heat load fluctuates in response to a zone thermostat (Figure 11). A problem occurs during periods of high outdoor humidity, when the internal heat load falls, causing the space humidity to rise rapidly (unless reheat is added). It is identical in concept to face-and-bypass cooling coils. This system has limited use in most modern buildings because most occupants demand more consistent temperature and humidity. In terms of energy use, the single-fan, no reheat dual-duct system does not use any extra energy for reheat, but fan energy is constant regardless of space load since it is a constant volume system.

Single Fan, Reheat. This method is similar in effect to conventional terminal reheat. There are two differences: reheat is applied at a central point instead of at individual zones (Figure 12) and only part of the supply air is cooled by the cooling coil, the other part being heated by the hot deck coil. This uses less heating and cooling energy than the terminal reheat system where all the air is cooled to full cooling capacity, then all of it is reheated as required to match the space load. Fan energy is constant since airflow is constant.

Variable Air Volume

Dual-duct VAV systems blend cold and warm air in various volume combinations. These systems may include single-duct VAV terminal units connected to the cold air chamber of the air handler (cold deck) for cooling only of interior spaces. This saves reheat energy for the air for those cooling only zones because control is by varying volume, not supply air temperature, which saves some fan energy to the extent that the airflow to the interior zones is reduced at lower loads.

Another VAV dual-duct variation uses terminal units that function like a single-duct VAV cooling terminal unit and a single-duct VAV heating terminal unit in one physical package. This arrangement allows the use of significant minimum airflow levels, providing temperature control by means of conventional dual-duct box mixing at minimum airflow. This variation saves both heating and cooling energy and fan energy because less air goes through each coil, and the total air flow is reduced below full load.

Single fan. In this system, a single supply fan is sized for the coincident peak of the hot and cold decks. Control of the fan is from two static pressure controllers, one located in the hot deck and the other in the cold deck. The duct requiring the highest pressure governs the airflow. Usually the cold deck is maintained at a fixed temperature, although some central systems permit the temperature to rise during warmer weather to save refrigeration. The temperature of the hot deck is often adjusted higher during periods of low outside temperature and high humidity to increase the flow over the cold deck for dehumidification. Other systems, particularly in laboratories, use a precooling coil to dehumidify the total airstream or just the outside air to limit humidity in the space. Return air quantity can be controlled either by flow-measuring devices in the supply and return duct or by static pressure controls that maintain space static pressure.

Dual Fan. The volume of each supply fan is controlled independently by the static pressure in its respective duct (Figure 13). The return fan is controlled based on the sum of the hot and cold fan volumes using flow-measuring stations. Each fan is sized for the anticipated maximum coincident hot or cold volume, not the sum of the instantaneous peaks. The cold deck can be maintained at a constant temperature either with mechanical refrigeration when minimum outside air is required or with an economizer when the outside air is below the temperature of the cold deck set point. This operation does not affect the hot deck, which can recover heat from the return air, and the heating coil need operate only when heating requirements cannot be met using return air. Outdoor air can provide ventilation air via the hot duct when the outdoor air is warmer than the return air. However, controls should be used to prohibit the introduction of

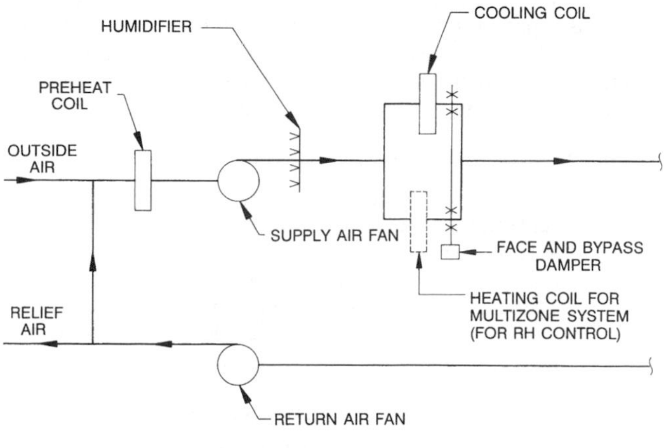

Fig. 11 Single Fan—No Reheat

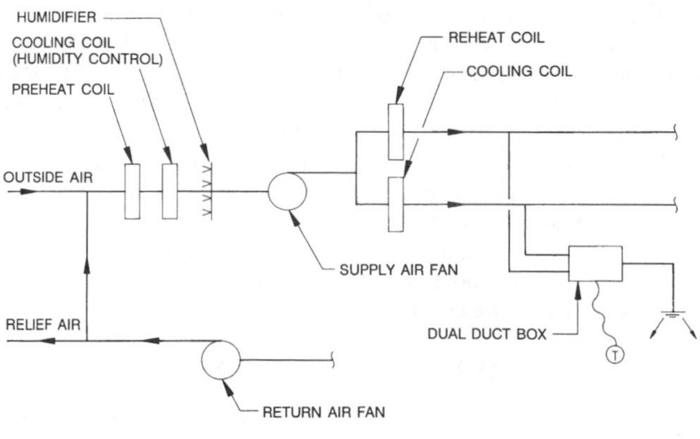

Fig. 12 Dual Duct—Single Fan

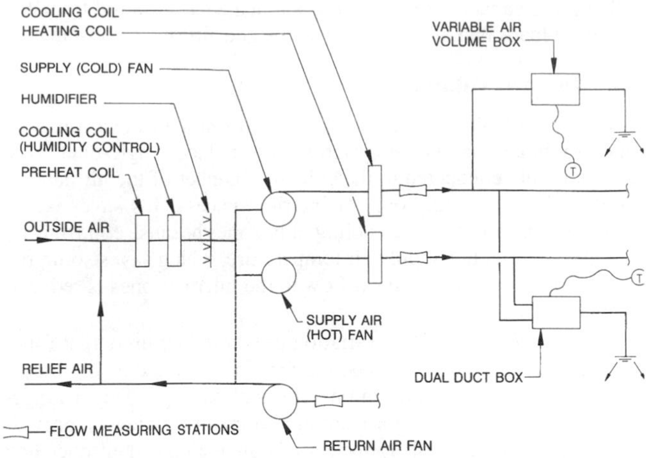

Fig. 13 Variable Air Volume, Dual Duct, Dual Fan

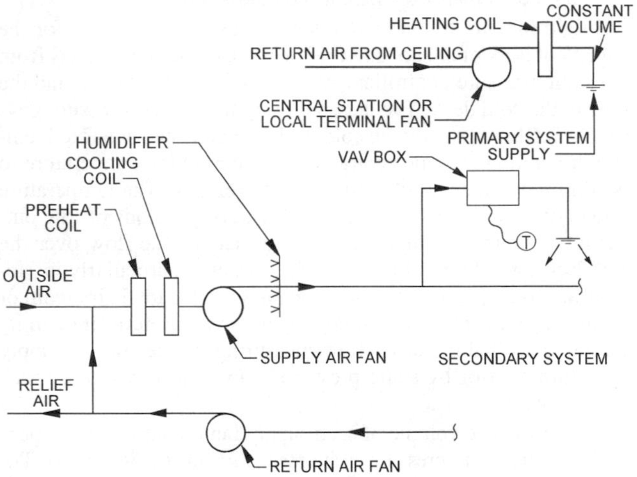

Fig. 14 Dual Conduit System

excessive amounts of outdoor air beyond the required minimum when that air is more humid than the return air.

Dual Conduit. This method has two air supply paths—one offsets exterior transmission cooling or heating loads, and the other provides cooling throughout the year (Figure 14). The first airstream, the primary air, operates as a constant volume system, and the air temperature is varied to offset transmission only (i.e., it is warm in winter and cool in summer). Often, however, the primary air fan is limited to operating only during peak heating and cooling periods to further save energy. When calculating the heating requirements for this system, the cooling effect of the secondary air must be included, even though the secondary system is operating at minimum flow. The other airstream, or secondary air, is cool year-round and varies in volume to match the load due to solar heating, lights, power, and occupants. It serves both perimeter and interior spaces.

While Figure 14 shows the system with separate delivery paths to the space from each conduit, a variation on the dual conduit systems uses dual-duct mixing boxes and a single delivery path to the space. In this variation, the first airstream is always either return air or heated air, never cooled air, and the other airstream has enough capacity for the entire cooling load.

MULTIZONE SYSTEMS

The multizone system (Figure 11) supplies several zones from a single, centrally located air-handling unit. Different zone requirements are met by mixing cold and warm air through zone dampers at the air handler in response to zone thermostats. The mixed, conditioned air is distributed throughout the building by single-zone ducts. The return air is handled in a conventional manner. The multizone system is similar to the dual-duct system. In operation, it has the same potential problem with high humidity levels. This system can provide a smaller building with the advantages of a dual-duct system, and it uses packaged equipment, which is less expensive.

Multizone packaged equipment is usually limited to about 12 zones, while built-up systems can include as many zones as can be physically incorporated in the layout. From an energy standpoint, a multizone system is somewhat more energy efficient than a terminal reheat system because not all the air goes through the cooling coil, which reduces the amount of reheat required. But a multizone system uses essentially the same fan energy as terminal reheat, because the airflow is constant.

Two common variations on the multizone system are called the three-deck multizone and the Texas multizone. A three-deck multizone system is similar to the standard multizone system, but neutral deck (return air) zone dampers are installed in the air-handling unit parallel with the hot deck and cold deck dampers. In the Texas multizone system, the hot deck heating coil is removed from the air handler and replaced with an air resistance plate matching the pressure drop of the cooling coil. Individual heating coils are placed only in each perimeter zone duct. These heating coils are usually located in the equipment room for ease of maintenance. This system is common in humid climates where the cold deck often produces 48 to 52°F air for humidity control. Supply air is then mixed with neutral deck (return) air rather than heated air using the air-handling units' zone dampers to maintain zone conditions. Heat is added only if the zone served cannot be maintained by the delivery of return air alone. These arrangements can save considerable reheat energy.

SPECIAL SYSTEMS

Primary/Secondary Systems

Some processes use two interconnected all-air systems (Figure 15). In these situations, space gains are very high and a large number of air changes are required (e.g., in sterile or clean rooms or where close-tolerance conditions are needed for process work). The primary systems supply the conditioned outside air requirements for the process either to the secondary system or directly to the space. The secondary system provides additional cooling and humidification (and heating, if required) to offset space and fan power gains. Normally, the secondary cooling coil is designed to be dry (i.e., sensible cooling only) to reduce the possibility of bacterial growth, which can create air quality problems.

Up Air System

In this system, also known as the **displacement ventilation** system, air is supplied from the floor rather than from overhead. Special terminal devices installed in a raised floor or as part of the furniture provide individual control for computer workstations, for example. This system is becoming common, particularly in Europe.

Wetted Duct/Supersaturated System

Some industries spray water into the airstream at central air-handling units in sufficient quantities to create a controlled carryover (Figure 16). This supercools the supply air, normally equivalent to an oversaturation of about 10 grains per pound of air and allows less air to be distributed for a given space load. These are used where high humidity is desirable, such as in the textile or tobacco industry, and in climates where adiabatic cooling is sufficient.

Compressed Air and Water Spray System

This system is similar to the wetted duct system, except that the water is atomized with compressed air. The nozzles are sometimes

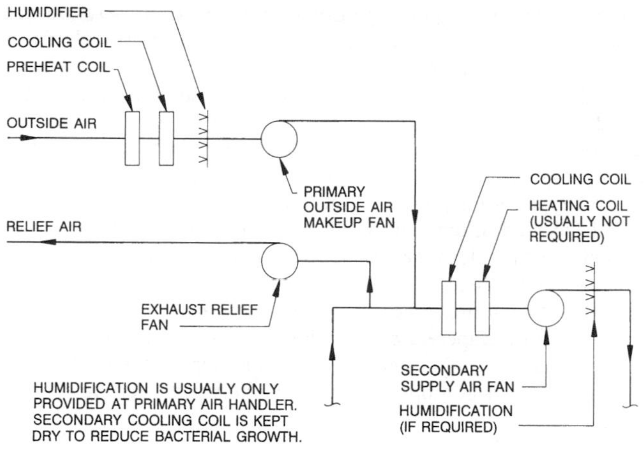

Fig. 15 Primary/Secondary System

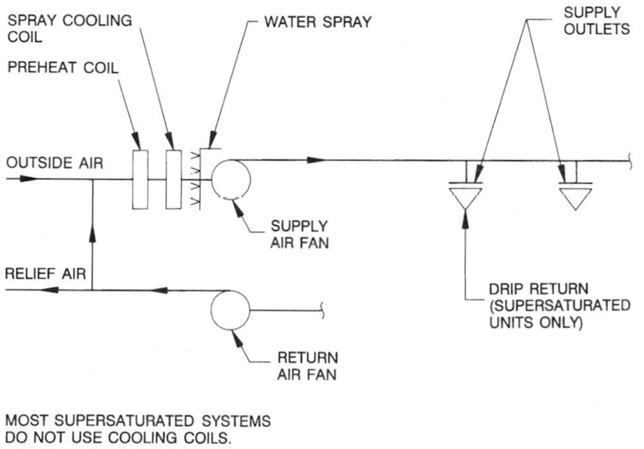

Fig. 16 Supersaturated/Wetted Coil

placed inside the conditioned space and independent of the cooling air supply. Several types of nozzle are available, and the designer should understand their advantages and disadvantages. Depending on the type of nozzle, compressed air and water spray systems can require large and expensive air compressors. The extra first cost may or may not be offset by energy cost savings, depending on the application.

Low-Temperature Systems

Because ice storage is being used to reduce peak electrical demand, low-temperature systems (where air is supplied at as low as 40°F) are sometimes used. Because the quantity of air supplied to a zone is so small, special terminal units are used to maintain minimum airflow for comfort. These devices induce return or room air to be mixed with the cold supply air to increase the air circulation rate within the space.

Smoke Management

Air-conditioning systems are often used for smoke control during fires. Controlled air flow can provide smoke-free areas for occupant evacuation and fire fighter access. Space pressurization creates a low-pressure area at the smoke source, surrounding it with high-pressure spaces. For more information, see Chapter 51 of the 1999 *ASHRAE Handbook—Applications*. *Design of Smoke Management Systems* (Klote and Milke 1992) also has detailed information on this topic.

TERMINAL UNITS

Terminal units are the devices between the primary air distribution systems and the conditioned space. Air distribution systems have two different types of terminal units: (1) supply outlets (registers or diffusers) and return inlets (grilles) and (2) terminal units. The function of the register or diffuser is to deliver the supply air throughout the conditioned space without occupants sensing a draft and without creating excessive noise. The function of the terminal unit is to control the quantity and/or temperature of the supply air to maintain the desired conditions in the space. Both types of terminal units are discussed in Chapter 17 of this volume and Chapter 31 of the 1997 *ASHRAE Handbook—Fundamentals*. Terminal units are discussed here briefly in terms of how they fit into the system concept, but Chapter 17 in this volume should be consulted for more complete information.

Constant Volume Reheat Terminal Units

Constant volume reheat terminal boxes are used mainly in terminal reheat systems with medium- to high-velocity ductwork. The unit serves as a pressure reducing valve and constant volume regulator to maintain a constant supply air quantity to the space, and is generally fitted with an integral reheat coil that controls temperature in the space. The constant supply air quantity is selected to provide cooling to match the peak load in the space, and the reheat coil is controlled to raise the supply air temperature as necessary to maintain the desired space temperature at reduced loads.

VAV Terminal Units

VAV terminal units are available in many configurations, all of which control the space temperature by varying the volume of cool supply air from the air handler to match the actual cooling load. VAV terminal units are fitted with automatic controls that are either pressure dependent or pressure independent. Pressure dependent units control damper position in response to room temperature and the flow may increase and decrease as the pressure at that point in the main duct varies.

Pressure independent units measure actual supply airflow and control that flow in response to room temperature. Pressure dependent units may sometimes be fitted with a velocity limit control that overrides the room temperature signal to limit the measured supply velocity to some selected maximum. Velocity limit control can be used to prevent excess airflow through units nearest the air handler. Excessive airflow at units close to the air handler can draw so much supply air that units far from the air handler do not get enough air.

Throttling Unit Without Reheat. The throttling (or pinch-off) box without reheat is essentially an air valve or damper that reduces the supply airflow to the space in response to falling space temperature. The unit usually includes some means of sound attenuation to reduce the air noise created by the throttling action. It is the simplest and least expensive of the VAV terminal units, but is suitable for use only where no heat is required and the unit can go to the completely closed position at reduced cooling loads. If this type of unit is set up with a minimum position, it will always be providing cooling to the space, whether the space needs it or not, and can overcool the space. This approach offers the lowest fan energy use, because it minimizes airflow to just the amount required by the cooling load.

Throttling Box With Reheat. This simple VAV system integrates heating at the terminal unit with the same type of throttling box. It is applied in interior and exterior zones where full heating and cooling flexibility is required. These terminal units can be set to maintain a predetermined minimum throttling ratio, which is established as the lowest air quantity necessary to (1) offset the heating load, (2) limit the maximum humidity, (3) provide reasonable air movement in the space, and (4) provide required ventilation air. The reheat coil is most commonly hot water or electric, but can be steam.

Variable air volume with reheat permits airflow to be reduced as the first step in control; heat is then initiated as the second step. Compared to constant volume reheat, this procedure reduces operating cost appreciably because the amount of primary air to be cooled and secondary air to be heated is reduced. Many types of controls can provide control sequences with more than one minimum airflow. This type of control allows the box to go to a lower flow rate that just meets ventilation requirements at the lightest cooling loads, and then increase to a higher flow rate when the heating coil is energized, further reducing the use of reheat energy. A feature can be provided to isolate the availability of reheat during the summer, except in situations where even low airflow would overcool the space and should be avoided or where an increase in humidity causes discomfort (e.g., in conference rooms when the lights are turned off).

Because the reheat coil requires some minimum airflow to deliver heat to the space, and because the reheat coil must absorb all of the cooling capacity of that minimum airflow before it starts to deliver heat to the space in cold weather, energy use can be significantly higher than with throttling boxes that go fully closed.

Induction. The VAV induction system uses a terminal unit to reduce cooling capacity by simultaneously reducing primary air and inducing room or ceiling air (replaces the reheat coil) to maintain a relatively constant room supply volume. This operation is the reverse of the bypass box described previously. The primary air quantity decreases with load, retaining the savings of VAV, while the air supplied to the space is kept relatively constant to avoid the effect of stagnant air or low air movement. VAV induction units require a higher inlet static pressure, which requires more fan energy, in order to achieve the velocities necessary for induction.

Fan-Powered. Fan-powered systems are available in either parallel or series airflow. In parallel flow units, the fan is located outside the primary airstream to allow intermittent fan operation. A backdraft damper on the terminal fan prevents conditioned air from escaping into the return air plenum when the terminal fan is off. In series units, the fan is located in the primary airstream and runs continuously when the zone is occupied. These constant airflow fan boxes in a common return plenum can help maintain indoor air quality by recirculating "unvitiated" air from overventilated zones to zones with greater outdoor air ventilation requirements.

Fan-powered systems, both series and parallel, are often selected because they maintain higher air circulation through a room at low loads while still retaining the advantages of VAV systems. As the cold primary air valve modulates from maximum to minimum (or closed), the unit recirculates more plenum air. In a perimeter zone, a hot water heating coil, electric heater, baseboard heater, or remote radiant heater can be sequenced with the primary air valve to offset external heat losses. Between heating and cooling operations, the fan only recirculates ceiling air. This operation permits heat from lights to be used for space heating for maximum energy saving. During unoccupied periods, the main supply air-handling unit remains off and individual fan-powered heating zone terminals are cycled to maintain required space temperature, thereby reducing operating cost during unoccupied hours.

Fans for fan-powered air-handling units operated in series are sized and operated to maintain the minimum static pressures at the unit inlet connections. This reduces the fan energy for the central air handler, but the small fans in the fan-powered units are less efficient than the large air handler fans. As a result, the series type fan powered unit (where those small fans operate continuously) may use more fan energy than a throttling unit system. However, that extra fan energy may be more than offset by the reduction in reheat through the recovery of plenum heat and the ability to operate a small fan to deliver heat during unoccupied hours only where heat is needed.

Because fan-powered boxes involve an operating fan, they may generate higher sound levels than throttling type boxes. Acoustical ceilings generally are not very effective sound barriers, so extra care should be taken in considering the sound level in critical spaces below fan-powered terminal units.

Both parallel and series fan-powered terminal units may be provided with filters. The constant (series) fan VAV terminal can accommodate minimum (down to zero) flow at the primary air inlet while maintaining constant airflow to the space.

Both types of fan powered units and induction terminal units are usually located in the ceiling plenum to recover heat from lights. This allows these terminals to be used without reheat coils in internal spaces. Perimeter zone units are sometimes located above the ceiling of an interior zone where heat from the lights maintains a higher plenum temperature. Provisions must still be made for morning warm-up and night heating. Also, interior spaces with a roof load must have heat supplied either separately in the ceiling or at the terminal.

Terminal Humidifiers

Most projects requiring humidification use steam. This can be centrally generated as part of the heating plant, where potential contamination from the water treatment of the steam is more easily handled and therefore of less concern. Where there is a concern, local generators (e.g., electric or even gas) that use treated water are used. Compressed air and water humidifiers are used to some extent, and supersaturated systems are used exclusively for special circumstances, such as industrial processes. Spray-type washers and wetted coils are also more common in industrial facilities. When using water directly, particularly in recirculating systems, the water must be treated to avoid the accumulation of dust during evaporation and the buildup of bacterial contamination.

Terminal Filters

In addition to the air-handling unit filters, terminal filters may used at the supply outlets to protect particular conditioned spaces where an extra clean environment is desired. Chapter 24 discusses this topic in detail.

AIR DISTRIBUTION SYSTEM CONTROLS

Controls should be automatic and simple for best operating and maintenance efficiency. Operations should follow a natural sequence—depending on the space need, one controlling thermostat closes a normally open heating valve, opens the outdoor air mixing dampers, or opens the cooling valve. In certain applications, an enthalpy controller, which compares the heat content of outdoor air to that of return air, may override the temperature controller. This control opens the outdoor air damper when conditions reduce the refrigeration load. On smaller systems, a dry-bulb control saves the cost of the enthalpy control and approaches these savings when an optimum changeover temperature, above the design dew point, is established. Controls are discussed in more detail in Chapter 45 of the 1999 *ASHRAE Handbook—Applications*.

Air-handling systems, especially variable air volume systems, should include means to measure and control the amount of outside air being brought in to assure adequate ventilation for acceptable indoor air quality. Strategies used to assure adequate outside air ventilation include

- Separate constant volume 100% outside air ventilation systems
- Direct measurement of the flow rate of outside air
- Modulating the return damper to maintain a constant pressure drop across a fixed outside air orifice
- Fan tracking systems that maintain a constant difference in airflow between supply and return fans
- CO_2 and/or VOC based demand controlled ventilation

A minimum outdoor air damper with separate motor, selected for a velocity of 1500 fpm, is preferred to one large outdoor air damper with minimum stops. A separate damper simplifies air balancing. The proper selection of outside, relief, and return air dampers is critical for efficient operation. Most dampers are grossly oversized and are, in effect, unable to control. One method of solving this problem is to provide maximum and minimum dampers. A high enough velocity across a wide open damper is essential to its providing effective control.

A mixed air temperature control can reduce operating costs and also reduce temperature swings from load variations in the conditioned space. Chapter 45 of the 1999 *ASHRAE Handbook—Applications* shows control diagrams for various arrangements of central system equipment. Direct digital control (DDC) is common, and most manufacturers offer either a standard or an optional DDC package for equipment including air-handling units, terminal units, etc. These controls offer a considerable degree of flexibility. DDC controls offer the additional advantage of the ability to record easily actual energy consumption or other operating parameters of the various components of the system, which can be useful for optimizing control strategies.

Constant Volume Reheat. This system typically uses two subsystems for control—one controls the discharge air conditions from the air-handling unit, and the other maintains the space conditions by controlling the reheat coil.

Variable Air Volume. Air volume can be controlled by duct-mounted terminal units serving a number of air outlets in a control zone or by units integral to each supply air outlet.

Pressure-independent volume regulator units control flow in response to the thermostat's call for heating or cooling. The required flow is maintained regardless of fluctuation of the VAV unit inlet or system pressure. These units can be field or factory adjusted for maximum and minimum (or shutoff) air settings. They operate at inlet static pressures as low as 0.2 in. of water for maximum unit design volume.

Pressure-dependent devices control air volume in response to a unit thermostatic (or relative humidity) device, but the flow varies with the inlet pressure variation. Generally, airflow oscillates when pressure varies. These thermostatic units do not regulate the flow but position the volume-regulating device in response to the thermostat. These units are the least expensive and should only be used where there is no need for maximum or minimum limit control and when the pressure is stable.

The type of controls available for VAV units varies with the terminal device. Most use either pneumatic or electric controls and may be either self-powered or system air actuated. Self-powered controls position the regulator by using liquid- or wax-filled power elements. System-powered devices use air from the air supplied to the space to power the operator. Components for both control and regulation are usually contained in the terminal device.

To conserve power and limit noise, especially in larger systems, fan-operating characteristics and system static pressure should be controlled. Many methods are available, including fan speed control, variable inlet vane control, fan bypass, fan discharge damper, and variable pitch fan control. The location of the pressure-sensing devices depends, to some extent, on the type of VAV terminal unit used. Where pressure-dependent units without controllers are used, the system pressure sensor should be near the static pressure midpoint of the duct run to ensure minimum pressure variation in the system. Where pressure-independent units are installed, pressure controllers may be at the end of the duct run with the highest static pressure loss. This sensing point ensures maximum fan power savings while maintaining the minimum required pressure at the last terminal.

As the flow through the various parts of a large system varies, so does the static pressure. Some field adjustment is usually required to find the best location for the pressure sensor. In many systems, the initial location is two-thirds to three-fourths of the distance from the supply fan to the end of the main trunk duct. As the pressure at the system control point increases due to the closing of terminal units, the pressure controller signals the fan controller to position the fan volume control, which reduces flow and maintains constant pressure. Many present-day systems measure flow rather than pressure and, with the development of economical DDC, each terminal unit (if necessary) can be monitored and the supply and return air fans modulated to exactly match the demand.

Dual-Duct. As dual-duct systems are generally more energy-intensive than single-duct systems, their use is less widespread. The use of DDC, with its ability to maintain set points and flow accurately, can make dual-duct systems worthwhile for certain applications. They should be seriously considered as alternatives to single-duct systems.

Personnel. The personnel operating and maintaining the air conditioning and controls should be considered. In large research and development or industrial complexes, experienced personnel are available for maintenance. On small and sometimes even large commercial installations, however, office managers are often responsible, so designs must be in accordance with their capabilities.

Valves. Most valves are sized reasonably well, although the same problems can occur as with dampers. On large hydraulic installations where direct blending is used to maintain the secondary water temperature, the system must be accurately sized for proper control. Many designers use variable flow for hydronic as well as air systems, so the design must be compatible with the air system to avoid operating problems.

Relief Fans. In many applications, relief or exhaust fans can be started in response to a signal from the economizer control or to a space pressure controller. The main supply fan must be able to handle the return air pressure drop when the relief fan is not running.

REFERENCES

ASHRAE. 1992. Gravimetric and dust spot procedures for testing air cleaning devices used in general ventilation for removing particulate matter. ANSI/ASHRAE *Standard* 52.1-1992.

ASHRAE. 1992. Thermal environmental conditions for human occupancy. ANSI/ASHRAE *Standard* 55-1992.

ASHRAE. 1997. Laboratory method to determine the sound power in a duct. ANSI/ASHRAE *Standard* 68-1997.

ASHRAE. 1999. Energy efficient design of new buildings except low-rise residential. *Standard* 90.1-1999.

ASHRAE. 1999. Ventilation for acceptable indoor air quality. *Standard* 62-1999.

Kirkpatrick, A.T. and J.S. Elleson. 1996. *Cold air distribution system design guide*. ASHRAE.

Klote, J.H. and J.A. Milke. 1992. *Design of smoke management systems*. ASHRAE.

Lorsch, H.G., ed. 1993. *Air-conditioning systems design manual*. ASHRAE.

NFPA. 1999. *National electrical code handbook*, 8th ed. National Fire Protection Association, Quincy, MA.

CHAPTER 3

IN-ROOM TERMINAL SYSTEMS

IN-ROOM TERMINAL SYSTEMS condition spaces by distributing air and water sources to terminal units installed in habitable spaces throughout a building. In some systems the air is distributed to the space directly, not through the terminal unit. The air and water are cooled or heated in central equipment rooms. The air supplied is called primary or ventilation air; the water supplied is called secondary water. Sometimes a separate electric heating coil is included in lieu of a hot water coil. This chapter describes induction units and fan-coil units used in in-room terminal unit systems.

In-room terminal unit systems are applied primarily to exterior spaces of buildings with high sensible loads and where close control of humidity is not required. In some cases they may be applied to interior zones. They work well in office buildings, hospitals, hotels, schools, apartment buildings, and research laboratories. In most climates, these systems are installed in exterior building spaces and are designed to provide (1) all required space heating and cooling, (2) outside air for ventilation, and (3) simultaneous heating and cooling in different parts of the building during intermediate seasons.

EXTERIOR ZONE AIR-CONDITIONING LOADS

The variation in air-conditioning load for exterior building spaces causes significant variations in space cooling and heating requirements, even in rooms that have the same exposure. Accordingly, accurate environmental control in exterior building spaces requires individual control. The following basic loads must be considered.

Internal Loads. Heat gain from lights is always a cooling load, and in most buildings (other than residential) it is relatively constant during the day. Lights may be turned off manually or automatically when not required, which makes lighting loads more variable. Heat gain from occupants is also a cooling load and is commonly the only room load that contains a latent component. Heat gain from computers and other equipment can vary greatly and is an important factor in building design.

External Loads. Solar heat gain is always a cooling load. It is often the chief cooling load and is highly variable. For a given space, solar gain always varies during the day. The magnitude and rate of change of this load depend on building orientation, glass area, capacity to store heat, and cloud cover. Constantly changing shade patterns from adjacent buildings, trees, or exterior columns and nonuniform overhangs cause significant variations in solar load between adjacent offices on the same solar exposure.

Transmission load can be either a heat loss or a heat gain, depending on the outdoor temperature.

Moderate pressurization of the building with ventilation air is normally sufficient to offset summer infiltration provided it is uniformly positive. In winter, however, infiltration can cause a significant heat loss, particularly on the lower floors of high-rise buildings. The magnitude of this component varies with the wind and stack effect, as well as with the temperature difference across the outside wall.

To perform successfully, an air-conditioning system must satisfy these load variations on a room-by-room basis and fulfill all other performance criteria, such as humidity control, filtration, air movement, ventilation, and noise.

SYSTEM DESCRIPTION

An in-room terminal unit system includes central air-conditioning equipment, a duct and water distribution system, and a room terminal. The terminal may be an **induction unit** or a **fan-coil unit**. Larger spaces may be handled by several terminals. Generally, the supply air volume is constant. (As noted in the first paragraph, supply air is called primary air or ventilation air to distinguish it from room air or secondary air that has been recirculated.) The quantity of primary air supplied to each space is determined by (1) the amount of outside air required for ventilation and (2) if used for sensible cooling, the required sensible cooling capacity at maximum room cooling load. In some low cost units with a fan coil, air for ventilation can be introduced through building apertures, which eliminates the primary air system.

In the cooling season, the air is dehumidified sufficiently in the central conditioning unit to maintain comfortable humidity conditions and to prevent condensation on the room cooling coil from the normal room latent load. In winter, moisture can be added centrally to limit dryness. As the air is dehumidified, it is also cooled to offset a portion of the room sensible loads. The air may be from outdoors, or may be a mixture of outdoor and return air. A preheater is usually required in areas with freezing weather. High-efficiency filters keep the induction unit nozzles clean. Whether or not reheaters are included in the air-handling equipment depends on the type of system, as explained in the section on Primary Air Systems.

In the ideal in-room terminal unit design, the secondary cooling coil is always dry; this greatly extends terminal unit life and eliminates odors and the possibility of bacterial growth. The primary air normally controls the space humidity. Therefore, the moisture content of the supply air must be low enough to offset the room latent heat gain and to maintain a room dew point low enough to preclude condensation on the secondary cooling surface. But even though some systems operate successfully with little or no condensate, a condensate drain is recommended for in-room terminal units.

In existing induction systems, the energy conservation technique of raising the chilled water temperature on central air-handling cooling coils can damage the terminal cooling coil, causing it to be used constantly as a dehumidifier. Unlike fan-coil units, the induction unit is not designed or constructed to handle condensation. Therefore, it is critical that an induction terminal operates dry.

The water side, in its basic form, consists of a pump and piping to convey water to the heat transfer surface in each conditioned space. The water can be cooled by direct refrigeration, but it is more commonly cooled either by bleeding chilled water from the primary

The preparation of this chapter is assigned to TC 9.1, Large Building Air-Conditioning Systems.

cooling system into the secondary water system or by transferring heat through a water-to-water exchanger. To distinguish it from the primary chilled water circuit, the water side is usually referred to as the secondary water loop or system.

In-room terminals are categorized as two-pipe, three-pipe, or four-pipe systems. They are basically similar in function and include both cooling and heating capabilities for year-round air conditioning. Sections at the end of this chapter discuss these piping arrangements in detail. Chapter 12 discusses the arrangements of the secondary water circuits and their control systems, which differ depending on the type of secondary water.

INDUCTION UNITS

Figure 1 shows a basic arrangement for an induction unit terminal. Centrally conditioned primary air is supplied to the unit plenum at medium to high pressure. The acoustically treated plenum attenuates part of the noise generated in the unit and duct. A balancing damper adjusts the primary air quantity within limits.

Medium- to high-velocity air flows through the induction nozzles and induces secondary air from the room through the secondary coil. Thus the primary air provides the energy required to circulate the secondary air over the coil in the terminal unit. This secondary air is either heated or cooled at the coil, depending on the season, the room requirement, or both. Ordinarily, the room coil does no latent cooling, but a drain pan without a piped drain collects condensed moisture from temporary latent loads such as at startup. This condensed moisture then re-evaporates when the temporary latent loads are no longer present. Primary and secondary air is mixed and discharged to the room.

A lint screen is normally placed across the face of the secondary coil. Induction units are installed in custom enclosures designed for the particular installation or in standard cabinets provided by the manufacturer. These enclosures must permit proper flow of secondary air and discharge of mixed air without imposing excessive pressure loss. They must also allow easy servicing.

Induction units are usually installed under a window at a perimeter wall, although units designed for overhead installation are available. During the heating season, the floor-mounted induction unit can function as a convector during off-hours, with hot water to the coil and without a primary air supply. A number of induction unit configurations are available, including units with low overall height or with larger secondary coil face areas to suit particular space or load needs.

Advantages

- Individual room temperature control allows the adjustment of each thermostat for a different temperature at relatively low cost.

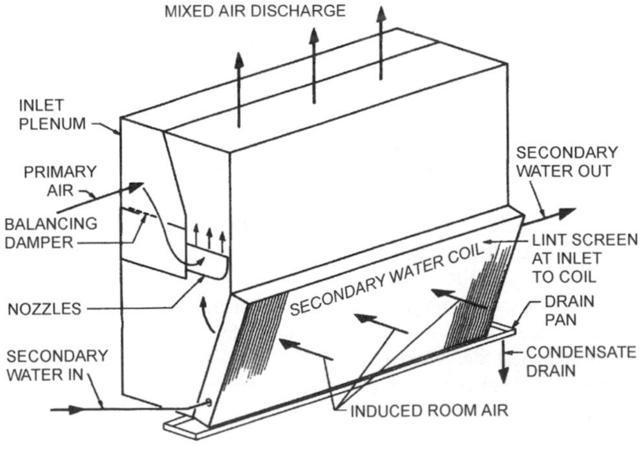

Fig. 1 Induction Unit

- Separate heating and cooling sources in the primary air and secondary water give the occupant a choice of heating or cooling.
- Less space is required for the distribution system when the air supply is reduced by using secondary water for cooling and high-velocity air. The return air duct is smaller and can sometimes be eliminated or combined with the return air system for other areas, such as the interior spaces.
- The size of the central air-handling apparatus is smaller than that of an all air system because little air must be conditioned.
- Dehumidification, filtration, and humidification are performed in a central location remote from conditioned spaces.
- Ventilation air supply is positive and may accommodate recommended outside air quantities.
- Space can be heated without operating the air system via the secondary water system. Nighttime fan operation is avoided in an unoccupied building. Emergency power for heating, if required, is much lower than for most all-air systems.
- Components are long-lasting. Room terminals operated dry have an anticipated life of 15 to 25 years. The piping and ductwork longevity should equal that of the building. Individual induction units do not contain fans, motors, or compressors. Routine service is generally limited to temperature controls, cleaning of lint screens, and infrequent cleaning of the induction nozzles.

Disadvantages

- For most buildings, these systems are limited to perimeter space; separate systems are required for other areas.
- More controls are needed than for many all-air systems.
- Secondary airflow can cause the induction unit coils to become dirty enough to affect performance. Lint screens used to protect these terminals require frequent in-room maintenance and reduce unit thermal performance.
- The primary air supply usually is constant with no provision for shutoff. This is a disadvantage in residential applications, where tenants or hotel room guests may prefer to turn off the air conditioning, or where management may desire to do so to reduce operating expense.
- A low primary chilled water temperature is needed to control space humidity adequately.
- The system is not appropriate for spaces with high exhaust requirements (e.g., research laboratories) unless supplementary ventilation air is provided.
- Central dehumidification eliminates condensation on the secondary water heat transfer surface under maximum design latent load. However, abnormal moisture sources (e.g., from open windows or people congregating) can cause condensation that can have annoying or damaging results.
- Energy consumption for induction systems is higher than for most other systems due to the increased power required by the primary air pressure drop in the terminal units.
- The initial cost for a four-pipe induction system is greater than for most all-air systems.

FAN-COIL UNITS

Fan-coil unit systems include cooling as well as heating, normally move air by forced convection through the conditioned space, filter the circulating air, and introduce outside ventilation air. Fan-coil units with chilled water coils, heating coils, blowers, replaceable air filters, drain pans for condensate, etc., are designed for these purposes. These units are available in various configurations to fit under window sills, above furred ceilings, in vertical pilasters built into walls, etc. These units must be properly controlled by thermostats for heating and cooling temperature control, by humidistats for humidity control, by blower control or other means for regulating air quantity, and they must have a method for adding ventilation air into the building.

Basic elements of fan-coil units are a finned-tube coil, filter, and fan section (Figure 2). The fan recirculates air continuously from the space through the coil, which contains either hot or chilled water. The unit may contain an additional electric resistance, steam, or hot water heating coil. The electric heater is often sized for fall and spring to avoid changeover problems in two-pipe systems.

A cleanable or replaceable 35% efficiency filter, located upstream of the fan, prevents clogging of the coil with dirt or lint entrained in the recirculated air. It also protects the motor and fan, and reduces the level of airborne contaminants in the conditioned space. The fan-coil unit is equipped with an insulated drain pan. The fan and motor assembly is arranged for quick removal for servicing. Most manufacturers furnish units with cooling performance certified as meeting Air-Conditioning and Refrigeration Institute (ARI) standards. The prototypes of the units have been tested and labeled by Underwriters' Laboratories (UL), or Engineering Testing Laboratories (ETL), as required by some codes.

Fan-coil units, with a dampered opening for connection to apertures in the outside wall, are available. These units are not suitable for commercial buildings because wind pressure allows no control over the amount of outside air that is admitted. Also, freeze protection may be required in cold climates. However they are often used in residential construction because of simplicity of operation, low first cost, and the operable windows can unbalance a duct ventilation air system. Fan-coil units for the domestic market are generally available in nominal sizes of 200, 300, 400, 600, 800, and 1200 cfm, often with multispeed, high-efficiency fan motors. Where units do not have individual outside air intakes, means must be provided to introduce retreated outside air through a duct system that engages each room or space.

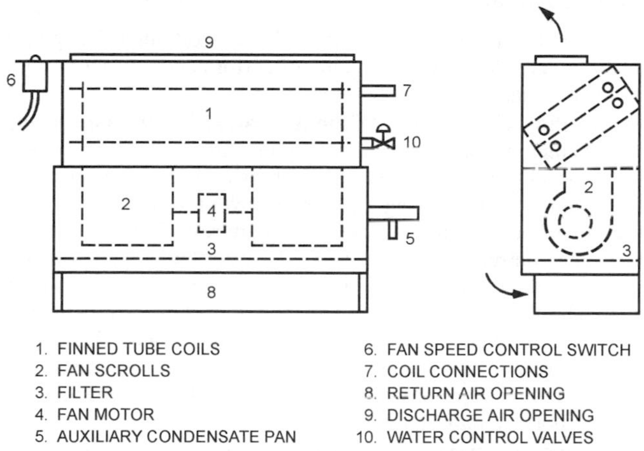

Fig. 2 Typical Fan-Coil Unit

Types and Location

Fan-coil units are available in many configurations. Figure 3 shows several **vertical units**. Low vertical units are available for use under windows with low sills; however, in some cases, the low silhouette is achieved by compromising such features as filter area, motor serviceability, and cabinet style.

Floor-to-ceiling, **chase-enclosed units** are available in which the water and condensate drain risers are part of the factory-furnished unit. These units are used extensively in hotels and other residential buildings. The supply and return air must be isolated from each other to prevent air and sound interchange between rooms.

Vertical models or chase enclosed models located at the perimeter give better results in climates or buildings with high heating requirements. Heating is enhanced by under-window or exterior wall locations. Vertical units can be operated as convectors with the fans turned off during night setback.

Horizontal overhead units may be fitted with ductwork on the discharge to supply several outlets. A single unit may serve several rooms (e.g., in an apartment house where individual room control is not essential and a common air return is feasible). Units must have larger fan motors designed to handle the higher pressure drops of ductwork.

Horizontal models conserve floor space and usually cost less, but when located in furred ceilings, they create problems such as condensate collection and disposal, mixing of return air from other rooms, leakage of pans causing damage to ceilings, difficulty of access for filter and component removal, and air quality concerns.

When outside air is introduced from a central ventilation system it may be connected directly to the inlet plenums of horizontal units or introduced directly into the space. If introduced directly, provisions

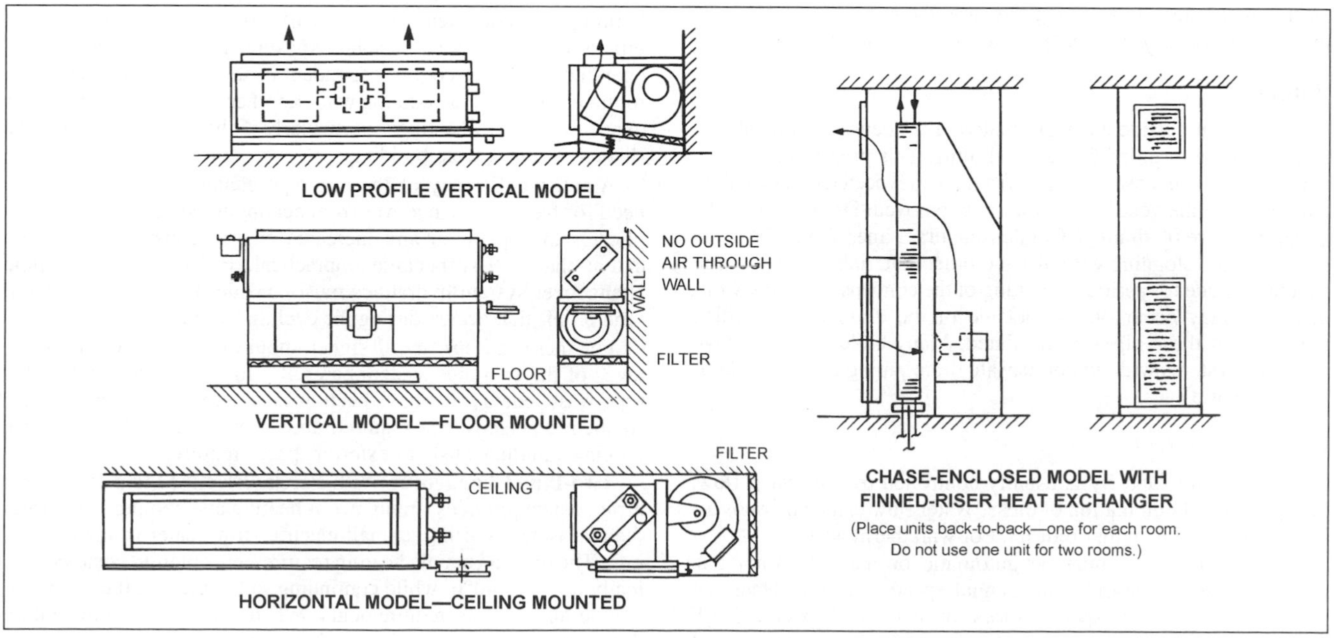

Fig. 3 Typical Fan-Coil Unit Arrangements

should be made to ensure that this air is pretreated and held at a temperature equal to the room temperature so as not to cause occupant discomfort when the unit is off. One way to prevent air leakage is to provide a spring-loaded motorized damper that closes off the ventilation air whenever the unit's fan is off. Coil selection must be based on the temperature of the entering mixture of primary and recirculated air, and the air leaving the coil must satisfy the room sensible and latent cooling and heating requirements.

Selection

Some designers size fan-coil units for nominal cooling at the medium speed setting when a three-speed control switch is provided. This method ensures quieter operation in the space and adds a safety factor, in that capacity can be increased by operating at high speed. Sound power ratings are available from many manufacturers.

Only the internal space heating and cooling loads need to be handled by the terminal fan-coil units when outside air is pretreated by a central system to a neutral air temperature of about 70°F. This pretreatment should reduce the size and cost of the terminal units. All loads must be considered when outside air is introduced directly through building apertures.

Wiring

Fan-coil conditioner fans are driven by small motors, generally shaded pole or capacitor start with inherent overload protection. Operating wattage of even the largest sizes rarely exceeds 300 W at the high speed setting. Running current rarely exceeds 2.5 A. Almost all motors on units in the United States are 120 V, single-phase, 60 Hz current, and they provide multiple (usually three) fan speeds and an off position. Other voltages and power characteristics may be encountered, depending on location, and should be investigated before determining the fan motor characteristics.

In planning the wiring circuit, required codes must be followed. Wiring methods generally provide separate electrical circuits for fan-coil units and do not connect them into the lighting circuit.

Separate electrical circuits connected to a central panel allow the building operator to turn off unit fans from a central point during unoccupied hours. While this panel costs more initially, it can lower operating costs in buildings that do not have 24 hour occupancy. In hot and humid climates care must be taken to avoid excess humidity when units are off to avoid formation of mildew. Use of separate electrical circuits allows a single remote thermostat to be mounted in a well-exposed perimeter space to operate unit fans.

Piping

Even when outside air is pretreated, a condensate removal system should be installed for fan-coil units. This precaution ensures that moisture condensed from air from an unexpected open window that bypasses the ventilation system is removed. Drain pans are an integral feature of all units. Condensate drain lines should be oversized to avoid clogging with dirt and other materials, and provision should be made for periodic cleaning of the condensate drains. Condensation may occur on the outside of the drain piping, which requires that these pipes be insulated. Many building codes have outlawed systems without condensate drain piping due to the damage they could cause.

Capacity Control

Fan-coil unit capacity is usually controlled by coil water flow, fan speed, or a combination of these. Water flow can be thermostatically controlled by either return air or wall thermostats.

Fan speed control may be automatic or manual. Automatic control is usually on-off with manual speed selection. Units are available with variable-speed motors for modulated speed control. Room thermostats are preferred where automatic fan speed control is used. Return air thermostats do not give a reliable index of room temperature when the fan is off. Residential fan-coil units have manual three-speed fan control with the water temperature, both heating and cooling, scheduled by outside temperature. On-off speed control is poor because (1) alternating shifts in fan noise level are more obvious than the sound of a constantly running fan, and (2) air circulation patterns within the room are noticeably affected.

Maintenance

Room fan-coil units are equipped with either filters that should be cleaned or replaced when dirty. Good filter maintenance improves sanitation and provides full airflow, ensuring full capacity. The frequency of cleaning varies with the application. Units in apartments, hotels, and hospitals usually require more frequent filter service because of lint.

Fan-coil unit motors require periodic lubrication. Motor failures are not common, but when they occur, the entire fan can be quickly replaced with minimal interruption in the conditioned space. The condensate drain pan and drain system should be cleaned or flushed periodically to prevent overflow and microbial buildup. Drain pans should be trapped to prevent any gaseous backup.

Water Distribution

Chilled and hot water must run to the fan-coil units. The piping arrangement determines the quality of performance, ease of operation, and initial cost of the system.

Two-Pipe Changeover Without Central Ventilation. In this system either hot or cold water is supplied through the same piping. The fan-coil unit has a single coil. The simplest system with the lowest initial cost is the two-pipe changeover fan coil with (1) outside air introduced through building apertures, (2) manual three-speed fan control, and (3) hot and cold water temperatures scheduled by outdoor temperatures. This system is generally used in residential buildings with operable windows and relies on the occupant to control fan speed and open or close windows. The changeover temperature is set at some predetermined set point. If a thermostat is used to control water flow, it must reverse its action depending on whether hot or cold water is available.

The two-pipe system can not simultaneously heat or cool, which is required for most projects during intermediate seasons when some rooms need cooling and others need heat. This problem can be especially troublesome if a single piping zone supplies the entire building. This deficiency may be partly overcome by dividing the piping into zones based on solar exposure. Then each zone may be operated to heat or cool, independent of the others. However, one room may still require cooling while another room on the same solar exposure requires heating—particularly if the building is partially shaded by an adjacent building.

Another deficiency of the two-pipe changeover system is the need for frequent changeover from heating to cooling, which complicates the operation and increases energy consumption to the extent that it may become impractical. For example, two-pipe changeover system hydraulics must consider the water expansion (and relief) that occurs during the cycling from cooling to heating.

For these reasons, the designer should consider the disadvantages of the two-pipe system carefully; many installations of this type waste energy and have been unsatisfactory in climates where frequent changeover is required and where interior loads require cooling simultaneously as exterior spaces require heat.

Two-Pipe Changeover With Partial Electric Strip Heat. This arrangement provides simultaneous heating and cooling in intermediate seasons by using a small electric strip heater in the fan-coil unit. The unit can handle heating requirements in mild weather, typically down to 40°F, while continuing to circulate chilled water to handle any cooling requirements. When the outdoor temperature drops sufficiently to require heating in excess of the electric strip heater capacity, the water system must be changed over to hot water.

Two-Pipe Nonchangeover with Full Electric Strip Heat. This system is not recommended for energy conservation, but it may be practical in areas with a small heating requirement.

Four-Pipe Distribution

The four-pipe distribution of secondary water has cold water supply, cold water return, warm water supply, and warm water return pipes. The four-pipe system generally has the highest initial cost, but it has the best fan-coil system performance. It provides (1) all-season availability of heating and cooling at each unit, (2) no summer/winter changeover requirement, (3) simpler operation, and (4) use of any heating fuel, heat recovery, or solar heat. In addition, it can be controlled to maintain a "dead band" between heating and cooling so simultaneous heating and cooling cannot occur. For discussion of two, three, and four pipe systems with central ventilation see later sections of this chapter.

Central Plant Equipment

Central equipment size is based on the block load of the entire building at the time of the building peak load, not on the sum of individual in-room terminal unit peak loads. Cooling load should include appropriate diversity factors for lighting and occupant loads. Heating load is based on maintaining the unoccupied building at design temperature, plus an additional allowance for pickup capacity if the building temperature is set back at night.

If water supply temperature or quantities are to be reset at times other than at peak load, the adjusted settings must be adequate for the most heavily loaded space in the building. An analysis of individual room load variations is required.

If the side of the building exposed to the sun or interior zone loads require chilled water in cold weather, the use of condenser water with a water-to-water heat exchanger should be considered. Varying refrigeration loads require the water chiller to operate satisfactorily under all conditions.

Ventilation

The only reason to use central fan equipment for an in-room terminal unit system is to provide the correct amount of ventilation or makeup air to the various spaces being served by terminal units.

Ventilation air is generally the most difficult factor to control and represents a major load component. The designer must select the method that meets local codes, performance requirements, cost constraints, and health requirements.

A central, outside air pretreatment system, which maintains neutral air at about 70°F, best controls ventilation air with the greatest freedom from problems related to the building's stack effect and infiltration. Ventilation air may then be introduced to the room through the fan-coil unit, or directly into the room as shown in Figure 4. Any type of fan-coil unit in any location may be used if the ventilation system has separate air outlets.

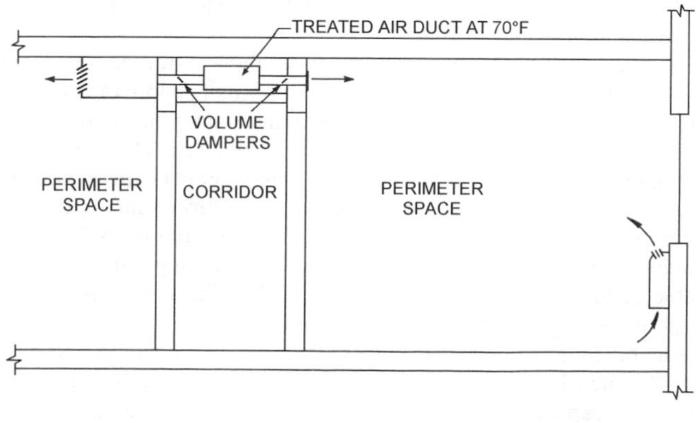

Fig. 4 Ventilation from Separate Duct System

Ventilation air contributes significantly to the room latent cooling load, so a dehumidifying coil should be installed in the central ventilation system to reduce room humidities during periods of high outside moisture content.

An additional advantage of the ventilation unit is that, if it is sized for the internal latent load, the terminal cooling coil remains dry, extending the life of the unit. However a piped condensate drain is recommended. This neutral temperature removes the outside air load from the terminal unit, so it can switch from heating to cooling and vice versa without additional internal or external heat loads.

In buildings where fan-coil units only serve exterior zones and a separate all-air system serves interior zones, it is possible to provide exterior zone ventilation air through the interior zone system. This arrangement can provide desirable room humidity control, as well as temperature control of the ventilation air. In addition, ventilation air held in the neutral zone of 70°F at 50% rh can be introduced into any fan-coil unit without affecting the comfort conditions maintained by the terminal units.

Applications

Fan-coil systems are best applied where individual space temperature control is needed. Fan-coil systems also prevent cross-contamination from one room to another. Suitable applications are hotels, motels, apartment buildings, and office buildings. Fan-coil systems are used in many hospitals, but they are less desirable because of the low-efficiency filtration and difficulty in maintaining adequate cleanliness in the unit and enclosure.

Advantages

A major advantage of the fan-coil unit system is that the delivery system (piping versus duct systems) requires less building space—it requires a smaller or no central fan room and little duct space. The system has all the benefits of a central water chilling and heating plant, while retaining the ability to shut off local terminals in unused areas. It gives individual room control with little cross-contamination of recirculated air from one space to another. Extra capacity for quick pull down response may be provided. Because this system can heat with low-temperature water, it is particularly suitable for solar or heat recovery refrigeration equipment. For existing building retrofit, it is often easier to install the piping and wiring for a fan-coil unit system than the large ductwork required for an all-air system.

Disadvantages

Fan-coil unit systems require much more maintenance than central all-air systems, and this work must be done in occupied areas. Units that operate at low dew points require condensate pans and a drain system that must be cleaned and flushed periodically. Condensate disposal can be difficult and costly. It is also difficult to clean the coil. Filters are small, low in efficiency, and require frequent changing to maintain air volume. In some instances, drain systems can be eliminated if dehumidification is positively controlled by a central ventilation air system.

Rooms are often ventilated by opening windows or by outside wall apertures, if not handled by a central system. Ventilation rates are affected by stack effect and wind direction and speed.

Summer room humidity levels tend to be relatively high, particularly if modulating chilled water control valves are used for room temperature control. Alternatives are two-position control with variable-speed fans and the bypass unit variable chilled water temperature control.

PRIMARY AIR SYSTEMS

Figure 5 illustrates the primary air system for in-room terminal systems. The components are described in Chapter 2. Some primary air systems operate with 100% outdoor air at all times. Systems using return air should have provision for operating with 100% out-

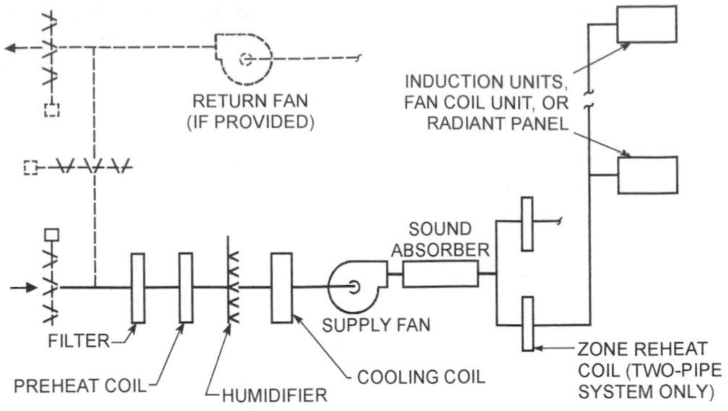

Fig. 5 Primary Air System

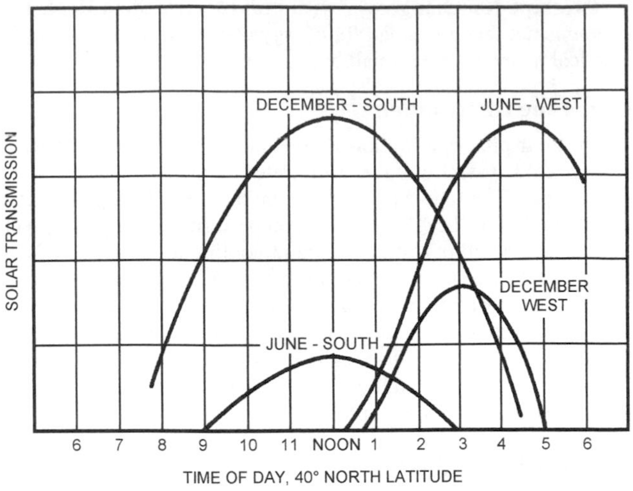

Fig. 6 Solar Radiation Variations with Seasons

door air to reduce operating cost during certain seasons. In some systems, when the quantity of primary air supplied exceeds the ventilation or exhaust required, the excess air is recirculated by a return system common with the interior system. A quality filter is desirable in the central air treatment apparatus. If it is necessary to maintain a given humidity level in cold weather, a humidifier can usually be installed. Steam humidifiers have been used successfully. The water sprays must be operated in conjunction with (1) the preheat coil elevating the temperature of the incoming air or (2) heaters in the spray water circuit.

The cooling coil is usually selected to provide primary air at a dew point low enough to dehumidify the system totally. The air leaves the cooling coil at about 50°F or less, and is almost completely saturated.

The supply fan should be selected at a point near maximum efficiency to reduce power consumption, heating of the supply air, and noise. Sound absorbers may be required at the fan discharge to attenuate fan noise.

Reheat coils are required in a two-pipe system. Reheat is not required for the primary air supply of four-pipe systems. Formerly, many primary air distribution systems for induction units were designed with 8 to 10 in. of water gage static pressure. With energy use restrictions, this is no longer economical. Good duct design and elimination of unnecessary restrictions (for example, sound traps) can result in primary systems that operate at 4.5 to 6.0 in. of water gage. Primary air distribution systems servicing fan-coil systems can operate at pressures 1.0 to 1.5 in. lower. Careful selection of the primary air cooling coil and induction units for reasonably low air pressure drops is necessary to achieve a medium-velocity primary air system. Distribution for fan-coil systems may be low velocity or a combination of low- and medium-velocity systems. See Chapter 32 in the 1997 *ASHRAE Handbook—Fundamentals* for a discussion of duct design. Variations in pressure between the first and last terminals should be minimized to limit the pressure drop across balancing dampers.

Room sound characteristics vary depending on unit selection, air system design, and the manufacturer. Units should be selected by considering the unit manufacturer's sound power ratings, the desired maximum room noise level, and the acoustical characteristics of the room. Limits of sound power level can then be specified to obtain acceptable acoustical performance.

PERFORMANCE UNDER VARYING LOAD

Under peak load conditions, the psychrometrics of induction unit and fan-coil unit systems are essentially identical for two-and four-pipe systems. The primary air mixes with secondary air conditioned by the room coil in an induction unit prior to delivery to a room. Mixing also occurs in a fan-coil unit with a direct-connected air supply. If the primary air is supplied to the space separately, as in fan-coil systems with independent primary air supplies, the same effect would occur in the space. The same room conditions result from two physically independent processes as if the air was directly connected to the unit.

During cooling, the primary air system provides a portion of the sensible capacity and all of the dehumidification. The remainder of the sensible capacity is accomplished by the cooling effect of the secondary water circulating through the unit cooling coils. In winter, primary air is provided at a low temperature, and if humidity control is provided, the air is humidified. All room heating is supplied by the secondary water system. All factors that contribute to the cooling load of perimeter space in the summer, with the exception of the transmission, add heat in the winter. The transmission factor becomes negative when the outdoor temperature falls below the room temperature. Its magnitude is directly proportional to the difference between the room and outdoor temperatures.

For in-room terminal unit systems, it is important to note that in applications where primary air enters at the terminal unit, the primary air is provided at summer design temperature in winter. For systems where primary air does not enter at the terminal unit, the primary air should be reset to room temperature in winter. A limited amount of cooling can be accomplished by the primary air operating without supplementary cooling from the secondary coil. As long as internal heat gains are not high, this amount of cooling is usually adequate to satisfy east and west exposures during the fall, winter, and spring, because the solar heat gain is reduced during these seasons. The north exposure is not a significant factor because the solar gain is very low. For the south, southeast, and southwest exposures, the peak solar heat gain occurs in winter, coincident with a lower outdoor temperature (Figure 6).

In buildings with large areas of glass, the transmitted heat from indoors to the outside, coupled with the normal supply of cool primary air, does not balance internal heat and solar gains until an outdoor temperature well below freezing is reached. Double-glazed windows with clear or heat-absorbing glass aggravate this condition because this type of glass permits constant inflow of solar radiation during the winter. However, the insulating effect of the double glass reduces the reverse transmission; therefore, cooling must be available at lower outdoor temperatures. In buildings with very high internal heat gains from lighting or equipment, the need for cooling from the room coil, as well as from the primary air, can extend well into winter. In any case, the changeover temperature at which the cooling capacity of the secondary water system is no longer required for a given space is an important calculation.

CHANGEOVER TEMPERATURE

For all systems using a primary air system for outdoor air, there is an outdoor temperature, a balance temperature, at which secondary cooling is no longer required. The system can cool by using outdoor air at lower temperatures, and heating rather than cooling is needed. For all-air systems operating with up to 100% outdoor air, mechanical cooling is seldom required at outdoor temperatures below 55°F. An important characteristic of in-room terminal unit systems, however, is that secondary water cooling may continue to be needed, even when the outdoor temperature is considerably less than 50°F. This cooling may be provided by the mechanical refrigeration unit or by a thermal economizer cycle. Full-flow circulation of the primary air-cooling coil below 50°F often provides all necessary cooling while preventing coil freeze-up and reducing the preheat requirement. Alternatively, secondary-water-to-condenser-water heat exchangers function well. Some systems circulate condenser water directly. This system should be used with caution, recognizing that the vast secondary water system is being operated as an open recirculating system with the potential hazards that may accompany improper water treatment.

The outdoor temperature at which the heat gain to every space can be satisfied by the combination of cold primary air and the transmission loss is termed the **changeover temperature**. Below this temperature, cooling is not required.

The following empirical equation approximates the changeover temperature at sea level, and it should be adjusted after installation (Carrier 1965):

$$t_{co} = t_r - \frac{q_{is} + q_{es} - 1.1\,Q_p(t_r - t_p)}{\Delta q_{td}} \quad (1)$$

where

t_{co} = temperature of changeover point, °F
t_r = room temperature at time of changeover, normally 72°F
t_p = primary air temperature at unit after system is changed over, normally 56°F
Q_p = primary air quantity, cfm
q_{is} = internal sensible heat gain, Btu/h
q_{es} = external sensible heat gain, Btu/h
Δq_{td} = heat transmission per degree of temperature difference between room and outdoor air

In two-pipe changeover systems, the entire system is usually changed from winter to summer operation at the same time, so the room with the lowest changeover point should be identified. In northern latitudes, this room usually has a south, southeast, or southwest exposure because the solar heat gains on these exposures reach their maximum during the winter months.

If the calculated changeover temperature is below approximately 48°F, an economizer cycle should operate to allow the refrigeration plant to shut down.

Although the factors controlling the changeover temperature of induction unit systems are understood by the design engineer, the basic principles are often more difficult for system operators to understand. Therefore, it is important that the concept and the calculated changeover temperature are clearly explained in operating instructions given prior to operating the system. Some increase from the calculated changeover temperature is normal in actual operation. Also, a range or band of changeover temperatures, rather than a single value, is a necessary to preclude frequent change in the seasonal cycles and to grant some flexibility in the operating procedure. The difficulties associated with operator understanding and the need to change over several times a day in many areas have severely limited the acceptability of the two-pipe changeover system.

REFRIGERATION LOAD

The design refrigeration load is determined by considering the entire portion or block of the building served by the air-and-water system at the same time. Because the load on the secondary water system depends on the simultaneous demand of all spaces, the sum of the individual room or zone peaks is not considered.

The peak load time is influenced by the outdoor wet-bulb temperature, the period of building occupancy, and the relative amounts of the east, south, and west exposures. Where the magnitude of the solar load is about equal for each of the above exposures, the building peak usually occurs in midsummer afternoon when the west solar load and outdoor wet-bulb temperature are at or near concurrent maximums.

At sea level, the refrigeration load equals the primary air-cooling coil load plus the secondary system heat pickup:

$$q_{re} = q_s + 4.5\,Q_p(h_{ea} - h_{la}) - 1.1\,Q_p(t_r - t_s) \quad (2)$$

where

q_{re} = refrigeration load, Btu/h
q_s = block room sensible heat for all spaces at time of peak, Btu/h
h_{ea} = enthalpy of primary air upstream of cooling coil at time of peak, Btu/lb
h_{la} = enthalpy of primary air leaving cooling coil, Btu/lb
Q_p = primary air quantity, cfm
t_r = average room temperature for all exposures at peak time, °F
t_s = average primary air temperature at point of delivery to rooms, °F

Because the latent load is absorbed by the primary air, the resultant room relative humidity can be determined by calculating the block room latent load of all spaces at the time of the peak load. Then, recalling that there are 7000 grains in a pound, the rise in moisture content of the primary air at sea level is

$$W = \frac{7000 v_a q_L}{60 h_{fg} Q_p} = 1.48\frac{q_L}{Q_p} \quad (3)$$

where

W = moisture content rise per lb dry air, grains
v_a = specific volume of air = 13.3 ft^3/lb at sea level
q_L = block room latent load of all spaces at time of peak load, Btu/h
h_{fg} = latent heat of vaporization = 1050 Btu/lb

Then a psychrometric analysis can be made.

The secondary water cooling load may be determined by subtracting the primary air-cooling coil load from the total refrigeration load.

TWO-PIPE SYSTEMS

With Central Ventilation

Two-pipe systems for induction and fan-coil systems derive their name from the water distribution circuit, which consists of one supply and one return pipe. Each unit or conditioned space is supplied with secondary water from this distribution system and with conditioned primary air from a central apparatus. The system design and the control of the primary air and secondary water temperatures must be such that all rooms on the same system (or zone, if the system is separated into independently controlled air and water zones) can be satisfied during both heating and cooling seasons. The heating or cooling capacity of any unit at a particular time is the sum of the primary air output plus the secondary water output of that unit.

The primary air quantity is fixed, and the primary air temperature is varied in inverse proportion to the outside temperature to provide the necessary amount of heating during summer and intermediate seasons. During winter operation, the primary air is preheated and supplied at approximately 50°F to provide a source of cooling. All

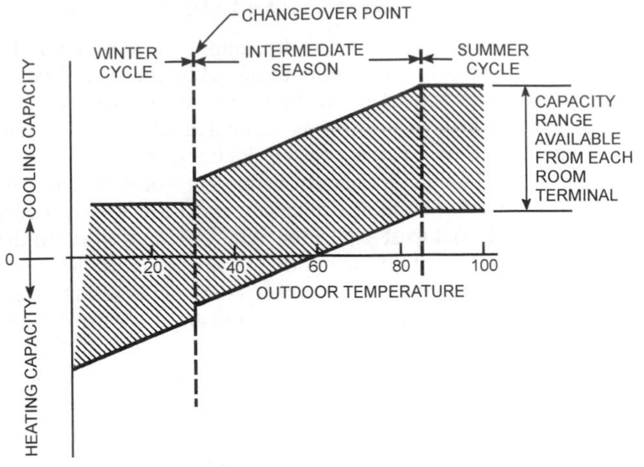

Fig. 7 Capacity Ranges of In-Room Terminal Operating on Two-Pipe System

room terminals in a given primary air reheater zone must be selected to operate satisfactorily with the common primary air temperature.

The secondary water coil (cooling-heating) within each space is controlled by a space thermostat and can vary from 0 to 100% of coil capacity, as required to maintain space temperature. The secondary water is cold in summer and intermediate seasons and warm in winter. All rooms on the same secondary water zone must operate satisfactorily with the same water temperature.

Figure 7 shows the capacity ranges available from a typical two-pipe system. On a hot summer day, loads varying from about 25 to 100% of the design space cooling capacity can be satisfied. On a 50°F intermediate season day, the unit can satisfy a heating requirement by closing off the secondary water coil and using only the output of the warm primary air. A lesser heating or net cooling requirement is satisfied by the cold secondary water coil output, which offsets the warm primary air to obtain cooling. In winter, the unit can provide a small amount of cooling by closing the secondary coil and using only the cold primary air. Smaller cooling loads and all heating requirements are satisfied by using the warm secondary water.

Critical Design Elements

The most critical design elements of a two-pipe system are the calculation of primary air quantities and the final adjustment of the primary air temperature reset schedule. All rooms require a minimum amount of heat from the primary air supply during the intermediate season. The concept of the ratio of primary air to transmission per degree (A/T ratio) to maintain a constant relationship between the primary air quantity and the heating requirements of each space fulfills this need. The A/T ratio determines the primary air temperature and changeover point. This concept is fundamental to proper design and operation of a two-pipe system.

Transmission per Degree. The relative heating requirement of every space is determined by calculating the transmission heat flow per degree temperature difference between the space temperature and the outside temperature (assuming steady-state heat transfer). This is the sum of (1) the glass heat transfer coefficient times the glass areas, (2) the wall heat transfer coefficient times the wall area, and (3) the roof heat transfer coefficient times the roof area.

Air-to-Transmission Ratio. The A/T ratio is the ratio of the primary airflow to a given space divided by the transmission per degree of that space: A/T ratio = Primary air/Transmission per degree.

Spaces on a common primary air zone must have approximately the same A/T ratios. The design base A/T ratio establishes the primary air reheat schedule during intermediate seasons. Spaces with A/T ratios higher than the design base A/T ratio tend to be overcooled during light cooling loads at an outdoor temperature in the 70 to 90°F range, while spaces with an A/T ratio lower than the design ratio lack sufficient heat during the 40 to 60°F outdoor temperature range when the primary air is warm for heating and the secondary water is cold for cooling.

The minimum primary air quantity that satisfies the requirements for ventilation, dehumidification, and both summer and winter cooling (as explained in the section on System Description) is used to calculate the minimum A/T ratio for each space. If the system will be operated with primary air heating during cold weather, the heating capacity can also be the primary air quantity determinant for two-pipe systems.

The design base A/T ratio is the highest A/T ratio obtained, and the primary airflow to each space is increased, as required, to obtain a uniform A/T ratio in all spaces. An alternate approach is to locate the space with the highest A/T ratio requirement by inspection, establish the design base A/T ratio, and obtain the primary airflow for all other spaces by multiplying this A/T ratio by the transmission per degree of all other spaces.

For each A/T ratio, there is a specific relationship between the outdoor air temperature and the temperature of the primary air that maintains the room at 72°F or more during conditions of minimum room cooling load. Figure 8 illustrates this variation based on an assumed minimum room load equivalent to 10°F times the transmission per degree. A primary air temperature over 122°F at the unit is seldom used. The reheat schedule should be adjusted for hospital rooms or other applications where a higher minimum room temperature is desired, or where a space has no minimum cooling load.

Deviation from the A/T ratio is sometimes permissible. A minimum A/T ratio equal to 0.7 of the maximum A/T is suitable, if the building is of massive construction with considerable heat storage effect (Carrier 1965). The heating performance when using warm primary air becomes less satisfactory than that for systems with a uniform A/T ratio. Therefore, systems designed for A/T ratio deviation should be suitable for changeover to warm secondary water for heating whenever the outdoor temperature falls below 40°F. A/T ratios should be more closely maintained on buildings with large glass areas, with curtain wall construction, or on systems with low changeover temperature.

Changeover Temperature Considerations

Transition from summer operations to intermediate season operation is done by gradually raising the primary air temperature as the outdoor temperature falls to keep rooms with small cooling loads from becoming too cold. The secondary water remains cold during both summer and intermediate seasons. Figure 9 illustrates the psychrometrics of summer cycle operation near the changeover temperature. As the outdoor temperature drops further, the changeover temperature will be reached. The secondary water system can then be changed over to provide hot water for heating.

If the primary airflow is increased to some spaces to elevate the changeover temperature, the A/T ratio for the reheat zone will be affected. Adjustments in the primary air quantities to other spaces on that zone will probably be necessary to establish a reasonably uniform ratio.

System changeover can take several hours and usually temporarily upsets room temperatures. Good design, therefore, includes provision for operating the system with either hot or cold secondary water over a range of 15 to 20°F below the changeover point. This range makes it possible to operate with warm air and cold secondary water when the outside temperature rises above the daytime changeover temperature. Changeover to hot water is limited to times of extreme or protracted cold weather.

Optional hot or cold water operation below the changeover point is provided by increasing the primary air reheater capacity to provide adequate heat at a colder outside temperature. Figure 10 shows

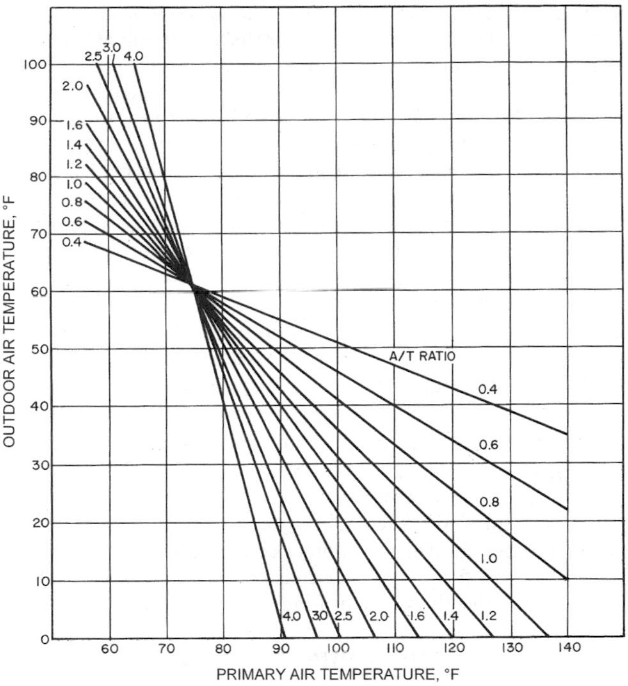

Note: These temperatures are required at the units, and thermostat settings must be adjusted to allow for duct heat gains or losses. Temperatures are based on

1. Minimum average load in the space, equivalent to 10°F multiplied by the transmission per degree.
2. Preventing the room temperature from dropping below 72°F. These values compensate for the radiation and convection effect of the cold outdoor wall.

Fig. 8 Primary Air Temperature Versus Outdoor Air Temperature

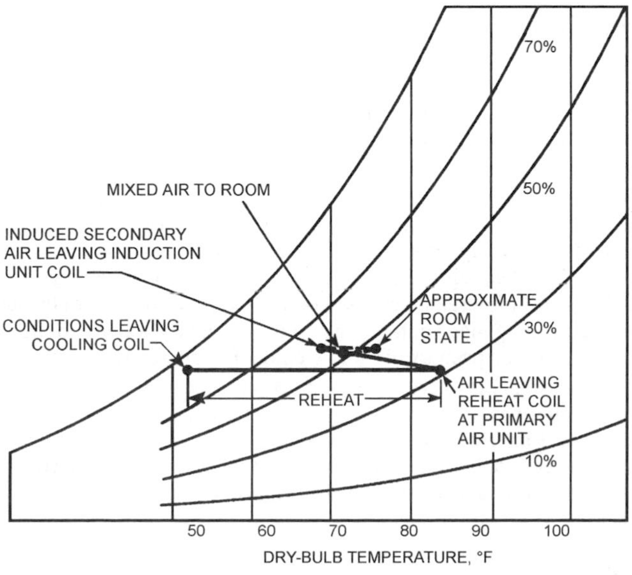

Note: Based on A/T ratio = 1.2, 52°F outdoor, reheat schedule on.

Fig. 9 Psychrometric Chart, Two-Pipe System, Off-Season Cooling

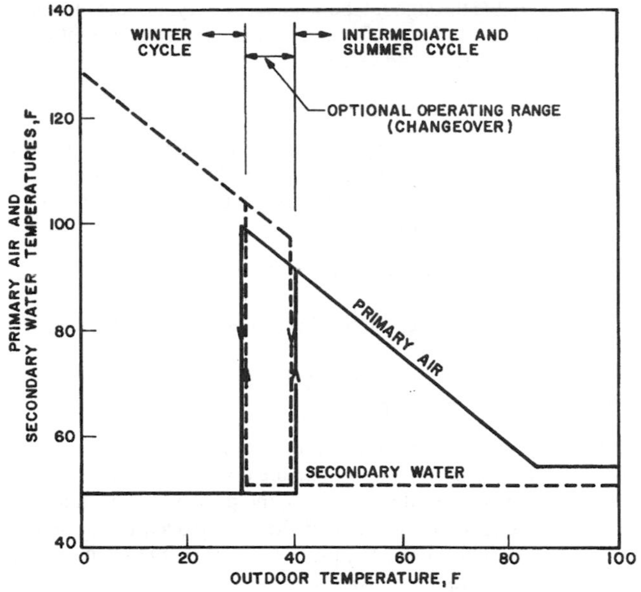

Fig. 10 Typical Changeover, System Variation

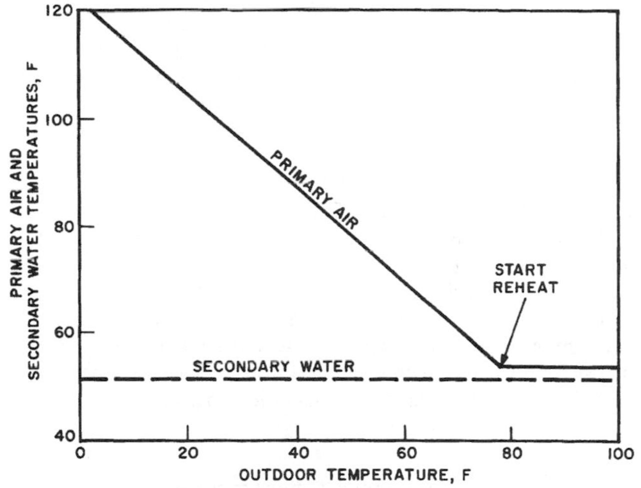

Fig. 11 Typical Nonchangeover System Variations

temperature variation for a system operating with changeover. This indicates the relative temperature of the primary air and secondary water throughout the year and the changeover temperature range. The solid arrows show the temperature variation when changing over from the summer to the winter cycle. The open arrows show the variation when going from the winter to the summer cycle.

Nonchangeover Design

Nonchangeover systems should be considered to simplify operation for buildings with mild winter climates, or for south exposure zones of buildings with a large winter solar load. A nonchangeover system operates on an intermediate season cycle throughout the heating season, with cold secondary water to the terminal unit coils and with warm primary air satisfying all the heating requirements. Typical temperature variation is shown in Figure 11.

Spaces may be heated during unoccupied hours by operating the primary air system with 100% return air. This feature is necessary, because the nonchangeover design does not usually include the ability to heat the secondary water. In addition, cold secondary water must be available throughout the winter. Primary air duct insulation and observance of close A/T ratios for all units are essential for proper heating during cold weather.

Zoning

A properly designed single-zone, two-pipe system can provide good temperature control on all exposures during all seasons of the year. Initial cost or operating cost can be improved by zoning in several ways, such as the following:

- Zoning primary air to permit different A/T ratios on different exposures
- Zoning primary air to permit solar compensation of primary air temperature
- Zoning both air and water to permit a different changeover temperature for different exposures

All spaces on the same primary air zone must have the same A/T ratio. The minimum A/T ratios often are different for spaces on different solar exposures, thus requiring the primary air quantities on some exposures to be increased if they are placed on a common zone with other exposures. The primary air quantity to units serving the north, northeast, or northwest exposures can usually be reduced by using separate primary air zones with different A/ T ratios and reheat schedules. Primary air quantity should never be reduced below minimum ventilation requirements.

The peak cooling load for the south exposure occurs during the fall or winter months when outside temperatures are lower. If present or future shading patterns from adjacent buildings or obstructions will not be present, primary air zoning by solar exposure can reduce air quantities and unit coil sizes on the south. Units can be selected for peak capacity with cold primary air instead of reheated primary air. Primary air zoning and solar compensators save operating cost on all solar exposures by reducing primary air reheat and the secondary water refrigeration penalty.

Separate air and water zoning save operating cost by permitting north, northeast, or northwest exposures to operate on the winter cycle with warm secondary water at outdoor temperatures as high as 60°F during winter months. Systems with a common secondary water zone must operate with cold secondary water to cool south exposures. Primary airflow can be lower because of separate A/T ratios, resulting in reheat and refrigeration cost savings.

Room Control

During summer, the thermostat must increase the output of the cold secondary coil when the room temperature rises. During winter, the thermostat must decrease the output of the warm secondary coil when the room temperature rises. Changeover from cold water to hot water in the unit coils requires changing the action of the room temperature control system. Room control for nonchangeover systems does not require the changeover action, unless it is required to provide gravity heating during shutdown.

Evaluation

The following are characteristics of two-pipe in-room terminal unit systems:

- Usually less expensive to install than four-pipe systems
- Less capable of handling widely varying loads or providing a widely varying choice of room temperatures than four-pipe systems
- Present operational and control changeover problems, increasing the need for competent operating personnel
- More costly to operate than four-pipe systems

Electric Heat for Two-Pipe Systems

Electric heat can be supplied with a two-pipe in-room terminal unit system by a central electric boiler and hot water terminal coils or by individual electric resistance heating coils in the terminal units. One method uses small electric resistance terminal heaters for intermediate season heating and a two-pipe changeover chilled water/hot water system. The electric terminal heater heats when outdoor temperatures are above 40°F, and cooling is available with chilled water in the chilled water/hot water system. System or zone reheating of the primary air is reduced greatly or eliminated entirely. When the outdoor temperature falls below this point, the chilled water/hot water system is switched to hot water, providing greater heating capacity. Changeover is limited to a few times per season, and simultaneous heating/cooling capacity is available, except in extremely cold weather, when little, if any, cooling is needed. If electric resistance terminal heaters are used, they should be prevented from operating whenever the secondary water system is operated with hot water.

Another method is to size electric resistance terminal heaters for the peak winter heating load and to operate the chilled water system as a nonchangeover cooling-only system. This avoids the operating problem of chilled water/hot water system changeover. In fact, this method functions like a four-pipe system, and, in areas where the electric utility establishes a summer demand charge and has a low unit energy cost for high winter consumption, it may have a lower life-cycle cost than hydronic heating with fossil fuel.

THREE-PIPE SYSTEMS

Three-pipe air-and-water systems for induction and fan-coil unit systems have three pipes to each terminal unit. These pipes are a cold water supply, a warm water supply, and a common return. These systems are rarely used today because they consume excess energy. Existing three-pipe systems are ideal candidates for energy conservation retrofit.

FOUR-PIPE SYSTEMS

Four-pipe systems have a cold water supply, cold water return, warm water supply, and warm water return. The terminal unit usually has two independent secondary water coils: one served by hot water, the other by cold water. The primary air is cold and remains at the same temperature year-round. During peak cooling and heating, the four-pipe system performs in a manner similar to the two-pipe system, with essentially the same operating characteristics. Between seasons, any unit can be operated at any level from maximum cooling to maximum heating, if both cold and warm water are being circulated. Any unit can be operated at or between these extremes without regard to the operation of other units.

In-room terminal units are selected on their peak capacity. The A/T ratio design concept for two-pipe systems does not apply to four-pipe systems. There is no need to increase primary air quantities on north or shaded units beyond the amount needed for ventilation and to satisfy cooling loads. The available net cooling is not reduced by heating the primary air. Attention to the changeover point is still important, because cooling of spaces on the south side of the building may continue to require secondary water cooling to supplement the primary air at low outdoor temperatures.

Because the primary air is supplied at a constant cool temperature at all times, it is sometimes feasible for fan-coil unit systems to extend the interior system supply to the perimeter spaces, eliminating the need for a separate primary air system. The type of terminal unit and the characteristics of the interior system are determining factors.

Zoning

Zoning of primary air or secondary water systems is not required. All terminal units can heat or cool at all times, as long as both hot and cold secondary pumps are operated and sources of heating and cooling are available.

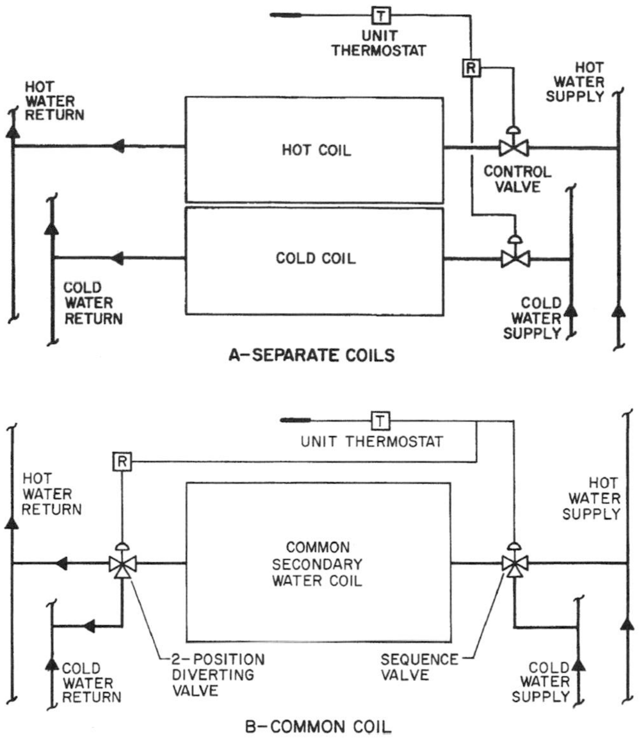

Fig. 12 Four-Pipe System Room Unit Control

Room Control

The four-pipe terminal usually has two completely separated secondary water coils—one receiving hot water and the other receiving cold water. The coils are operated in sequence by the same thermostat; they are never operated simultaneously. The unit receives either hot water or cold water in varying amounts or else no flow is present, as shown in Figure 12A. Adjustable, dead-band thermostats further reduce operating cost.

Figure 12B illustrates another unit and control configuration. A single secondary water coil at the unit and three-way valves located at the inlet and outlet admit water from either the hot or cold water supply, as required, and divert it to the appropriate return pipe. This arrangement requires a special three-way modulating valve, originally developed for one form of the three-pipe system. It controls the hot or cold water selectively and proportionally but does not mix the streams. The valve at the coil outlet is a two-position valve open to either the hot or cold water return, as required.

When all aspects are considered, the two-coil arrangement provides a superior four-pipe system. The operation of the induction and fan-coil unit controls is the same year-round.

Evaluation

Compared to the two-pipe system, the four-pipe air-and-water system has the following characteristics:

- More flexible and adaptable to widely differing loads, responding quickly to load changes
- Simpler to operate
- Operates without the summer-winter changeover and the primary air reheat schedule
- Efficiency is greater and operating cost is lower, though initial cost is generally higher
- Can be designed with no interconnection of the hot and cold water secondary circuits, and the secondary system can be completely independent of the primary water piping

SECONDARY WATER DISTRIBUTION

Secondary water system design applies to induction and fan-coil systems. The secondary water system includes the portion of the water distribution system that circulates water to room terminal units when the cooling (or heating) of such water has been accomplished either by extraction from or heat exchange with another source in the primary circuit. In the primary circuit, water is cooled by flow through a chiller or is heated by a heat input source. Primary water is limited to the cooling cycle and is the source of the secondary water cooling. Water flow through the unit coil performs secondary cooling when the room air (secondary air) gives up heat to the water. The design of the secondary water system differs for two- and four-pipe systems. Secondary water systems are discussed in Chapter 12.

BIBLIOGRAPHY

Carrier Air Conditioning Company. 1965. *Handbook of air conditioning system design.* McGraw-Hill, New York.

Menacker, R. 1977. Electric induction air-conditioning system. *ASHRAE Transactions* 83(1):664.

McFarlan, A.I. 1967. Three-pipe systems: Concepts and controls. *ASHRAE Journal* 9(8):37.

CHAPTER 4

CENTRAL COOLING AND HEATING

THE ENERGY used to heat and cool many buildings often comes from a central location in the facility. The energy input may be any combination of electricity, oil, gas, coal, solar, geothermal, etc. This energy is typically converted into hot or chilled water or steam that is distributed throughout the facility for heating and air conditioning (cooling). Centralizing this function keeps the conversion equipment in one location and distributes the heating and cooling in a more readily usable form. Also, a central cooling and heating plant provides higher diversity and generally operates more efficiently with lower maintenance and labor costs than a decentralized plant. However it does require space at a central location and a potentially large distribution system.

This chapter addresses the design alternatives that should be considered when centralizing the cooling and heating sources in a facility. Distribution system options and equipment are discussed when they relate to the central equipment, but more information on distribution systems can be found in Chapters 10 through 14.

The economics of these systems require extensive analysis. Boilers; gas, steam, and electric turbine-driven centrifugal refrigeration units; and absorption chillers may be installed in combination in one plant. In large buildings with core areas that require cooling and perimeter areas require heating, one of several heat reclaim systems may heat the perimeter to save energy. Chapter 8 gives details of these combinations, and Chapters 12, 13, and 14 give design details for central plants.

The choice of equipment for a central heating and cooling plant depends on the following:

- Required capacity and type of use
- Cost and types of energy available
- Location and space for the equipment room
- A stack for any exhaust gases and its environmental impact
- Safety requirements for the mechanical room and surrounding areas
- Type of air distribution system
- Owning and operating costs

Many electric utilities impose very high charges for peak power use or, alternatively, offer incentives for off-peak use. This policy has renewed interest in both water and ice thermal storage. The storage capacity installed for summer load leveling may also be available for use in the winter, making heat reclaim a more viable option. With ice storage, the low-temperature ice water can provide colder air than that available from a conventional system that supplies air at 50 to 60°F. The use of a higher water temperature rise and a lower supply air temperature, lowers the required pumping and fan energy and, in some instances, offsets the energy penalty due to the lower temperature required to make ice.

The preparation of this chapter is assigned to TC 9.1, Large Building Air-Conditioning Systems.

REFRIGERATION EQUIPMENT

The major types of refrigeration equipment in large systems use reciprocating compressors, helical rotary compressors, centrifugal compressors, and absorption chillers. Chapter 34 has information about compressors and Chapters 41 and 43 of the 1998 *ASHRAE Handbook—Refrigeration* have information about refrigeration equipment, including the size ranges of typical equipment.

These large compressors are driven by electric motors, gas and diesel engines, and gas and steam turbines. The compressors may be purchased as part of a refrigeration chiller that includes the compressor, drive, chiller, condenser, and necessary safety and operating controls. Reciprocating and helical rotary compressor units are frequently field assembled, and include air-cooled or evaporative condensers arranged for remote installation. Centrifugal compressors are usually included in packaged chillers.

Absorption chillers are water-cooled. They use a lithium bromide/water or water/ammonia cycle and are generally available in the following four configurations: (1) direct fired, (2) indirect generated by low-pressure steam or hot water, (3) indirect generated by high-pressure steam or hot water, and (4) indirect generated by hot exhaust gas. Small direct-fired chillers are single-effect machines with capacities of 3.5 to 25 tons. Larger, direct-fired, double-effect chillers in the 100 to 2000 ton range are also available.

Low-pressure steam at 15 psig or low temperature hot water heats the generator of single-effect absorption chillers with capacities from 50 to 1600 tons. Double-effect machines use high-pressure steam up to 150 psig or high-temperature hot water at an equivalent temperature. Absorption chillers of this type are available from 350 to 2000 tons.

In large installations the absorption chiller is sometimes combined with steam turbine-driven centrifugal compressors. Steam from the noncondensing turbine is piped to the generator of the absorption machine. When a centrifugal unit is driven by a gas turbine or an engine, an absorption machine generator may be fed with steam or hot water from the engine jacket. A heat exchanger that transfers the heat of the exhaust gases to a fluid medium may increase the cycle efficiency. Chapter 41 of the 1998 *ASHRAE Handbook—Refrigeration* covers absorption air-conditioning and refrigeration equipment in more detail.

Cooling Towers

Water from condensers is usually cooled by the atmosphere. Either natural draft or mechanical draft cooling towers or spray ponds are used to reject the heat to the atmosphere. Of these, the mechanical draft tower, which may be of the forced draft, induced draft, or ejector type, can be designed for most conditions because it does not depend on the wind. Air-conditioning systems use towers ranging from small package units of 5 to 500 tons to field-erected towers with multiple cells in unlimited sizes.

If the cooling tower is on the ground, it should be at least 100 ft away from the building for two reasons; (1) to reduce tower noise in the building and (2) to keep discharge air from fogging the building's windows. Towers should be kept the same distance from parking lots to avoid staining car finishes with water treatment chemicals. When the tower is on the roof, its vibration and noise must be isolated from the building. Some towers are less noisy than others, and some have attenuation housings to reduce noise levels. These options should be explored before selecting a tower. Adequate room for airflow inside screens should be provided to prevent recirculation.

The bottom of many towers, especially larger ones, must be set on a steel frame 4 to 5 ft above the roof to allow room for piping and proper tower and roof maintenance. Pumps below the tower should be designed for an adequate net positive suction head, but they must be installed to prevent draining the piping on shutdown.

The tower must be winterized if required for cooling at outdoor temperatures below 35°F. Winterizing includes the ability to bypass water directly into the tower basin or return line (either automatically or manually, depending on the installation) and to heat the tower pan water to a temperature above freezing. Heat may be added by steam or hot water coils or by electric resistance heaters in the tower pan. Also, an electric heating cable on the condenser water and makeup water pipes and insulation on these heat-traced sections is needed to keep the pipes from freezing. Special controls are also required when a cooling tower is operated near freezing conditions. Where the cooling tower will not operate in freezing weather, provisions for draining the tower and piping must be made. Draining is the most effective way to prevent the tower and piping from freezing.

Careful attention must also be given to water treatment to minimize the maintenance required by the cooling tower and refrigeration machine absorbers and/or condenser.

Cooling towers may also cool the building in off-seasons by filtering and directly circulating the condenser water through the chilled water circuit, by cooling the chilled water in a separate heat exchanger, or by using the heat exchangers in the refrigeration equipment to produce thermal cooling. Towers are usually selected in multiples so that they may be run at reduced capacity and shut down for maintenance in cool weather. Chapter 36 includes further design and application details.

Air-Cooled Condensers

Air-cooled condensers pass outdoor air over a dry coil to condense the refrigerant. This process results in a higher condensing temperature and, thus, a larger power input at peak condition; however, over 24 hours this peak time may be relatively short. The air-cooled condenser is popular in small reciprocating systems because of its low maintenance requirements. Chapter 35 has further design and application details.

Evaporative Condensers

Evaporative condensers pass outdoor air over coils sprayed with water, thus taking advantage of adiabatic saturation to lower the condensing temperature. As with the cooling tower, freeze prevention and close control of water treatment are required for successful operation. The lower power consumption of the refrigeration system and the much smaller footprint of the evaporative versus the air-cooled condenser are gained at the expense of the cost of the water used and increased maintenance cost. Chapter 35 includes further design and application details.

HEATING EQUIPMENT

A boiler is the most common device used to add heat to the working medium, which is then distributed throughout the facility. Although steam is an acceptable medium for transferring heat between buildings or within a building, low-temperature hot water provides the most common and more uniform means of perimeter and general space heating. The working medium may be either water or steam, which can further be classified by its temperature and pressure range. The term hot-water boiler applies to fuel-fired units that heat water for heating systems. Water heaters differ in that they usually do not have enough space in the top section for use as a steam boiler, but in many respects a water heating boiler is the same as a steam heating boiler of the same type of construction. Many steam heating boilers may serve as water heaters if properly arranged, fitted, and installed. Steam and hot water boilers use gas, oil, coal, electricity, and sometimes, waste material for fuel. Chapter 27 has more information on boiler classification (steam, water, pressure, temperature, fuel, materials, and condensing/noncondensing).

Fuel Section

Functionally a boiler has two major sections. For all fuel burning boilers, that fuel must first be burned in the combustion section of the boiler. Control of both the fuel and air greatly effect the efficiency of the combustion process. This is also the primary point at which the heating capacity of the boiler is regulated. On smaller boilers that only turn on or off, the air/fuel ratio is fixed and is not actively controlled. A next step of regulation may have one or two intermediate steps of fuel input, such a low fire-high fire, with the air-fuel ratio fixed at each level of fuel input. A larger or proportionally regulated boiler may have its fuel flow regulated to any level. Then a complimentary air regulation method is also required to ensure sufficient oxygen for efficient fuel combustion without either heating excess air that is wasted up the exhaust stack, or creating hazardous or polluting compounds.

Heat Transfer Section

The second major section of the boiler is the heat transfer section where heat from the combustion process is transferred to the working medium that will distribute the heat. The effectiveness of these heat transfer surfaces depends on the temperature difference between the fluid, water, or steam on one side and the gases on the other side, and on the circulation rate of both the water and the hot gases. Heating surfaces are classified as direct and indirect. Direct surfaces are those upon which the light of the fire is present; they are very effective in transferring heat to the boiler water because the high furnace temperatures promote heat transfer both by radiation and convection. Indirect heating surfaces are those in contact with flue gases only and are progressively less effective from the standpoint of heat transfer as the flue gases become cooled. Boiler manufactures generally apply a combination of these heat transfer surfaces for boilers. In **fire tube boilers** the products of combustion flow inside tubes and the transfer medium (water) flows outside. In **water tube boilers** the products of combustion flow outside tubes and the transfer medium flows inside the tubes. **Cast iron sectional boilers** transfer the heat from the combustion gases to the water through cast iron sections. More types of construction are described in Chapter 27.

Working Pressure

Pressures in the boiler also affect its classification. The combustion section can be regulated at a slightly negative pressure relative to the atmosphere to ensure that the combustion gases do not flow anywhere except up the stack. In a **natural draft boiler**, the chimney effect of the stack draws the combustion products up the stack. Alternately, the boiler may use a fan to force air through the burner and boiler in a **forced draft** arrangement. In this case the combustion chamber and boiler itself are at a positive pressure relative to the surrounding atmosphere. Chapter 26 has more information on the operation of fuel burning equipment.

The pressure of the working medium (steam or water) can be low, medium, or high as described in Chapter 27. For hot water boilers more so then steam boilers the location in the facility as well as its elevation relative to the rest of the facility impacts the working pressure. Pumping in or out of a water boiler can effect its operating/dynamic pressure; and the boiler's location, either in a basement or on a roof, can effect its standby/static pressure.

DISTRIBUTION

The steam or hot water is the working medium of the heating system that transfers the heat produced by the boiler to the areas where it will be used. Steam typically is distributed by its inherent pressure; but once it is condensed, it must rely on gravity or pumps to return to the boiler. The issue of condensate traps and the return is one of the important design and operational concerns for steam. Chapter 10 covers steam systems in greater detail. If water is the working medium, pumps distribute it from the boiler. The options for water distribution are many and are covered in detail in Chapters 11 and 12.

Pumps

HVAC pumps are usually centrifugal pumps. Pumps for large, heavy-duty systems have a horizontal split case with a double-suction impeller for easier maintenance and higher efficiency. End suction pumps, either close coupled or flexible connected, may be used for smaller tasks. Pumps can be installed in-line or be base mounted.

Pumps are used to distribute or circulate the following fluids:

- Primary and secondary chilled water
- Primary and secondary hot water
- Condenser water
- Condensate
- Boiler feed
- Fuel oil

When pumps handle hot liquids or have a high inlet pressure drop, the required net positive suction head (NPSH) must not exceed the NPSH available at the pump. It is common practice to provide spare pumps to maintain continuity in case of a pump failure. The chilled water and condenser water system characteristics may permit using one spare pump for both systems, if they are properly valved and connected to both manifolds. Chapters 12 and 13 includes design recommendations, and Chapter 38 provides information on the selection of pumps.

Piping

HVAC piping systems can be divided into two parts—the piping in the main equipment room and the piping required to deliver chilled or hot water/steam to the air-handling equipment throughout the building. The air-handling system piping follows procedures detailed in Chapters 10 through 14.

The major piping in the equipment room includes fuel lines, refrigerant piping, and steam and water connections. Chapter 10 includes more information on piping for steam systems, and Chapter 12 has information on chilled and dual-temperature water systems. Chapter 33 of the 1997 *ASHRAE Handbook—Fundamentals* gives details for sizing and installing these piping systems.

Chilled water systems may be designed for constant flow to cooling coils with three-way valves or to coils with balancing valves. Because flow is constant the load on the system is proportional to the temperature difference across the main supply and return lines.

Multiple chillers on constant flow systems are generally piped in series. Piping the chillers in parallel is not recommended because under light load conditions it is difficult to maintain the design supply water temperature when return water through the inactive chiller(s) is mixing with cold water from the active chiller(s).

Very few constant flow systems are designed today because of the increased cost to continuously circulate the design flow and the difficulty in controlling the supply water temperature.

Variable flow chilled water systems, serve cooling coils with two-way modulating control valves so that the flow is proportional to the coil load. In order to maintain a constant flow through the online chiller(s), a primary-secondary pumping arrangement is generally used. Chapter 12 has more information on this technique.

The thermal load is proportional to the flow times the temperature difference across the secondary supply and return mains. For this reason both flow and temperature must be measured accurately to determine the load on the system. An accurate measurement of the load is essential in order to operate the correct number of chillers and/or boilers to optimize the system for peak efficiency. The temperature sensors should be matched to 0.1°F and the flow sensor should be accurate within 1% of the range.

The crossover line on primary-secondary systems may be sized for the flow through the largest chiller rather than the total flow of all chillers.

INSTRUMENTATION

All equipment must have adequate pressure gages, thermometers, flowmeters, balancing devices, and dampers for effective performance, monitoring, and commissioning. In addition, capped thermometer wells, gage cocks, capped duct openings, and volume dampers should be installed at strategic points for system balancing. Chapter 36 of the 1999 *ASHRAE Handbook—Applications* indicates the locations and types of fittings required.

Microprocessors and the related software to establish control sequences are replacing pneumatic and electric relay logic. Output signals are converted to pneumatic or electric commands to actuate HVAC hardware.

A central control console to monitor the many system points and overall system performance should be considered for any large, complex air-conditioning system. A control panel permits a single operator to monitor and perform functions at any point in the building to increase occupant comfort and to free maintenance staff for other duties. Chapter 45 of the 1999 *ASHRAE Handbook—Applications* describes the design and application of controls.

The coefficient of performance for the entire chilled water plant can now be monitored and allows the chilled water plant operator to determine the overall operating efficiency of a plant.

All instrument operations where heat or cooling output are measured should have instrumentation whose calibration is traceable to NIST, the National Institute of Standards and Technology.

SPACE REQUIREMENTS

In the initial phases of building design, the engineer seldom has sufficient information to render the design. Therefore, most experienced engineers have developed rules of thumb to estimate the building space needed. The air-conditioning system selected, the building configuration, and other variables govern the space required for the mechanical system. The final design is usually a compromise between what the engineer recommends and what the architect can accommodate. The design engineer should keep the owner and facility engineer informed, whenever possible, about the HVAC analysis and system selection. Space criteria should satisfy both the architect and the owner or the owner's representative. Where designers cannot negotiate a space large enough to suit the installation, the judgement of the owners should be called upon.

Although few buildings are identical in design and concept, some basic criteria apply to most building and help allocate space that approximates the final requirements. These space requirements are often expressed as a percentage of the total building floor area. For example, the total mechanical and electrical space requirements range from 4 to 9% of the gross building area, with most buildings falling within the 6 to 9% range.

The arrangement and strategic location of the mechanical spaces during planning impact the percentage of space required. The relationship of outdoor air intakes with respect to loading docks, exhaust, and other contaminating sources should be considered during architectural planning.

The main electrical transformer and switchgear rooms should be located as close to the incoming electrical service as practical. If there is an emergency generator, it should be located considering (1) proximity to emergency electrical loads, (2) sources of combustion and cooling air, (3) fuel sources, (4) ease of properly venting exhaust gases to the outdoors, and (5) provisions for noise control.

The main plumbing equipment usually contains gas and domestic water meters, the domestic hot water system, the fire protection system, and such elements as compressed air, special gases, and vacuum, ejector, and sump pumps. Some water and gas utilities require a remote outdoor meter location.

The heating and air-conditioning equipment room houses (1) the boiler, or pressure-reducing station, or both; (2) the refrigeration ma-chines, including the chilled water and condensing water pumps; (3) converters for furnishing hot or cold water for air conditioning; control air compressors; (4) vacuum and condensate pumps; and (5) miscellaneous equipment. For both chillers and boilers, especially in a centralized application, full access is needed on all sides for extended operation, maintenance, annual inspections of tubes, and tube replacement and repair. Local codes and ASHRAE *Standard* 15 should be consulted for special equipment room requirements. Many local jurisdictions require separation of refrigeration and fuel fired equipment.

It is often economical to locate the refrigeration plant at the top of the building, on the roof, or on intermediate floors. These locations closer to the load allow the equipment to operate at a lower pressure. The electrical service and structural costs will be greater, but these may be offset because the energy consumption and the condenser and chilled water piping cost may be less. The boiler plant may also be placed on the roof, which eliminates the need for a chimney through the building.

CENTRAL PLANT LOADS

The design cooling and heating loads are determined by considering the entire portion or block of the building served by the air-and-water system at the same time. Because the load on the secondary water system depends on the simultaneous demand of all spaces, the sum of the individual room or zone peaks is not considered.

The peak cooling load time is influenced by the outdoor wet-bulb temperature, the period of building occupancy, and the relative amounts of the east, south, and west exposures. Where the magnitude of the solar load is about equal for each of the above exposures, the building peak usually occurs on a midsummer afternoon when the west solar load and outdoor wet-bulb temperature are at or near concurrent maximums.

If the side exposed to the sun or interior zone loads require chilled water in cold weather, the use of condenser water with a water-to-water heat exchanger should be considered. Varying refrigeration loads require the water chiller to operate satisfactorily under all conditions.

If water supply temperature or quantities are to be reset at times other than at peak load, the adjusted settings must be adequate for the most heavily loaded space in the building. An analysis of individual room load variations is required.

The peak heating load may occur when the facility must be warmed to occupancy conditions after an unoccupied weekend setback. The peak demand may also occur during unoccupied conditions when the ambient environment is at its harshest and there is little internal heat gain to assist the heating system. Another possible maximum may occur during occupied times if significant outdoor air must be preconditioned or some other process (maybe non-HVAC) requires significant heat. The designer must analyze how the facility will be used.

CHAPTER 5

DECENTRALIZED COOLING AND HEATING

PACKAGED UNIT SYSTEMS are applied to almost all classes of buildings. They are especially suitable for smaller projects with no central plant where low initial cost and simplified installation are important. These units are installed in office buildings, shopping centers, manufacturing plants, hotels, motels, schools, medical facilities, nursing homes, and other multiple-occupancy dwellings. They are also suited to air conditioning existing buildings with limited life or income potential. Applications also include facilities requiring specialized high performance levels, such as computer rooms and research laboratories.

SYSTEM CHARACTERISTICS

These systems are characterized by several separate air-conditioning units, each with an integral refrigeration cycle. The components are factory designed and assembled into a package that includes fans, filters, heating coil, cooling coil, refrigerant compressor(s), refrigerant-side controls, air-side controls, and condenser. The equipment is manufactured in various configurations to meet a wide range of applications. Examples include window air conditioners, through-the-wall room air conditioners, unitary air conditioners for indoor and outdoor locations, air-source heat pumps, and water-source heat pumps. Specialized packages for computer rooms, hospitals, and classrooms are also available.

Commercial grade unitary equipment packages are available only in pre-established increments of capacity with set performance parameters, such as the sensible heat ratio at a given room condition or the cubic feet of air per minute per ton of refrigeration. Components are matched and assembled to achieve specific performance objectives. These limitations make the manufacture of low cost, quality-controlled, factory-tested products practical. For a particular kind and capacity of unit, performance characteristics vary among manufacturers. All characteristics should be carefully assessed to ensure that the equipment performs as needed for the application. Several trade associations have developed standards by which manufacturers may test and rate their equipment. Chapter 45 and Chapter 46 describe the equipment used in multiple-packaged unitary systems and the pertinent industry standards.

Large commercial/industrial grade equipment can be custom designed by the factory to meet specific design conditions and job requirements. This equipment carries a higher first cost and is not readily available in smaller sizes.

Self-contained units can use multiple compressors to control refrigeration capacity. For variable air volume systems, compressors are turned on or off or unloaded to maintain the discharge air temperature. As airflow is decreased, the temperature of the air leaving the unit can often be reset upward so that a minimum ventilation rate can be maintained. Resetting the discharge air temperature is a way of demand limiting the unit, thus saving energy.

Although the equipment can be applied as a single unit, this chapter covers the application of multiple units to form a complete air-conditioning system for a building. Multiple, packaged-unit systems for perimeter spaces are frequently combined with a central all-air or floor-by-floor system. These combinations can provide better humidity control, air purity, and ventilation than packaged units alone. Air-handling systems may also serve interior building spaces that cannot be conditioned by wall or window-mounted units.

Advantages of Packaged Systems

- Heating and cooling capability can be provided at all times, independent of the mode of operation of other spaces in the building.
- Manufacturer-matched components have certified ratings and performance data.
- Assembly by a manufacturer helps ensure better quality control and reliability.
- Manufacturer instructions and multiple-unit arrangements simplify the installation through repetition of tasks.
- Only one unit conditioner and one zone of temperature control are affected if equipment malfunctions.
- System is readily available.
- One manufacturer is responsible for the final equipment package.
- For improved energy control, equipment serving vacant spaces can be turned off locally or from a central point, without affecting occupied spaces.
- System operation is simple. Trained operators are not required.
- Less mechanical and electrical room space is required than with central systems.
- Initial cost is usually low.
- Equipment can be installed to condition one space at a time as a building is completed, remodeled, or as individual areas are occupied, with favorable initial investment.
- Energy can be metered directly to each tenant.

Disadvantages of Packaged Systems

- Limited performance options may be available because airflow, cooling coil size, and condenser size are fixed.
- A larger total building installed cooling capacity is usually required because the diversity factors used for moving cooling needs do not apply to dedicated packages.
- Temperature and humidity control may be less stable especially with mechanical cooling at very low loads.
- Standard commercial units are not generally suited for large percentages of outside air or for close humidity control. Custom equipment or special purpose equipment such as packaged units for computer rooms or large custom units may be required.
- Energy use is usually greater than for central systems, if efficiency of the unitary equipment is less than that of the combined central system components.
- Low cost cooling by outdoor air economizers is not always available.
- Air distribution control may be limited.
- Operating sound levels can be high.
- Ventilation capabilities are fixed by equipment design.

The preparation of this chapter is assigned to TC 9.1, Large Building Air-Conditioning Systems.

- Overall appearance can be unappealing.
- Air filtration options may be limited.
- Discharge temperature varies because control is either on or off or in steps.
- Maintenance may be difficult because of the many pieces of equipment and their location.

ECONOMIZERS

Air-Side Economizers

The air-side economizer takes advantage of cool outdoor air to either assist mechanical cooling or, if the outdoor air is cool enough, provide total cooling. It requires a mixing box designed to allow 100% of the air to be drawn from outside. It can be field-installed accessory that includes an outdoor air damper, relief damper, return air damper, filters, actuator, and linkage. Controls are usually a factory-installed option.

Self-contained units usually do not include return air fans. A variable-volume relief fan must be installed with the air-side economizer. The relief fan is off and discharge dampers are closed when the air-side economizer is inactive.

Advantages of Air-Side Economizers

- Substantially reduces compressor, cooling tower, and condenser water pump energy requirements.
- Has a lower air-side pressure drop than a water-side economizer.
- Reduces tower makeup water and related water treatment.

Disadvantages of Air-Side Economizers

- In systems with larger return air static pressure requirements, return fans or exhaust fans are needed to properly exhaust building air and intake outside air.
- If the unit's leaving air temperature is also reset up during the air-side economizer cycle, humidity control problems may occur and the fan may use more energy.
- Humidification may be required during winter operation.

Water-Side Economizer

The water-side economizer is another option for reducing energy use. ASHRAE *Standard* 90.1 addresses its application. The water-side economizer consists of a water coil located in the self-contained unit upstream of the direct-expansion cooling coil. All economizer control valves, piping between the economizer coil, and the condenser and economizer control wiring can be factory installed (Figure 1).

The water-side economizer takes advantage of the low cooling tower or evaporative condenser water temperature to (1) either precool the entering air, (2) assist mechanical cooling, or, (3) if the cooling water is cold enough, provide total system cooling. If the economizer is unable to maintain the supply air set point for variable-air-volume units or zone set point for constant-volume units, factory-mounted controls integrate economizer and compressor operation to meet cooling requirements.

Cooling water flow is controlled by two valves (Figure 1), one at the economizer coil inlet (A) and one in the bypass loop to the condenser (B). Two control methods are common—constant water flow and variable water flow.

Constant water flow control allows constant condenser water flow during unit operation. The two control valves are factory wired for complementary control, where one valve is driven open while the other is driven closed. This keeps water flow through the unit relatively constant.

Variable modulating control allows variable condenser water flow during unit operation. The valve in the bypass loop (B) is an on-off valve and is closed when the economizer is enabled. Water flow through the economizer coil is modulated by valve A, thus allowing variable cooling water flow. As the cooling load increases,

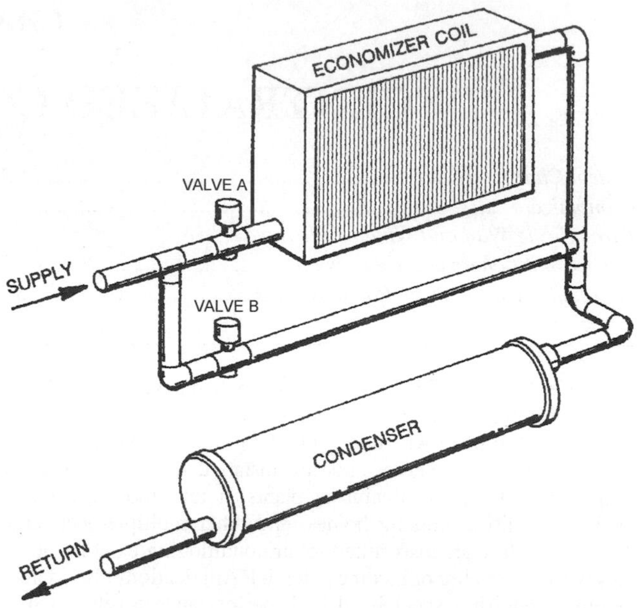

Fig. 1 Water-Side Economizer and Valves

valve A opens, increasing water flow through the economizer coil. If the economizer is unable to satisfy the cooling requirements, factory-mounted controls integrate economizer and compressor operation. In this operating mode, valve A is fully open. When the self-contained unit is not in the cooling mode, both valves are closed. Reducing or eliminating cooling water flow reduces pumping energy.

Advantages of Water-Side Economizers

- Reduces compressor energy by precooling entering air. Often the building load can be completely satisfied with an entering condenser water temperature of less than 55°F. Because the wet-bulb temperature is always less than or equal to the dry-bulb temperature, a lower discharge air temperature is often times available.
- Building humidification may not be required if return air contains sufficient humidity to satisfy the winter requirement.
- No external wall penetration is required for exhaust or outdoor air ducts.
- Mechanical equipment rooms can be centrally located in a building.
- Controls are less complex than for air-side economizers, because they often reside inside the packaged unit.
- Coil can be mechanically cleaned.
- More net usable floor area is available because large fresh air and relief air ducts are unnecessary.

Disadvantages of Water-Side Economizers

- Cooling tower water treatment cost is greater.
- Air-side pressure drop may be increased.
- Condenser water pump may see slightly higher pressure.
- Cooling tower must be designed for winter operation.
- The increased operation (including winter operation) required of the cooling tower may reduce its life.

THROUGH-THE-WALL AND WINDOW-MOUNTED AIR CONDITIONERS AND HEAT PUMPS

A window air conditioner (air-cooled room conditioner), which is further described in Chapter 46, is designed to cool or heat individual room spaces. Window units are used where low initial cost, quick installation, and other operating or performance criteria outweigh the advantages of more sophisticated systems. Room units are also available in through-the-wall sleeve mountings. Sleeve-installed units are popular in low-cost apartments, motels, and homes.

Window units may be used as auxiliaries to a central heating or cooling system or to condition selected spaces when the central system is shut down. In such applications, these units usually serve only part of the spaces conditioned by the basic system. Both the basic system and the window units should be sized to cool the space adequately without the other operating.

A through-the-wall air-cooled room air conditioner is designed to cool or heat individual room spaces. Design and manufacturing parameters vary widely. Specifications range from appliance grade through heavy-duty commercial grade, the latter known as packaged terminal air conditioners (ARI *Standard* 310). With proper maintenance, manufacturers project a life expectancy of 10 to 15 years for these units.

Advantages

- Initial cost is generally less than for a central system adapted to heat or cool each room under the control of the room occupants.
- Because no energy is needed to transfer air or chilled water from mechanical equipment rooms, the energy consumption may be lower than for central systems. However, this advantage may be offset by the lower efficiency of this type of equipment.
- Building space is conserved because ductwork and mechanical rooms are not required.
- Installation is simple. It usually only requires a hole in the wall or displacement of a window to mount the unit and connection to electrical power.
- Generally, the system is well-suited to spaces requiring many zones of individual temperature control.
- Designers can specify electric, hydronic, or steam heat or use an air-to-air heat pump design.
- Service can be quickly restored by replacing a defective chassis with a spare.

Disadvantages

- Equipment life is relatively short, typically 10 years; window units are built to appliance standards, rather than building equipment standards.
- May have relatively high energy use.
- Requires outside air; thus, cannot be used for interior rooms. Packaged terminal air conditioners must be installed on the perimeter of the building.
- The louver and wall box must stop wind-driven rain from collecting in the wall box and leaking into the building. The wall box should drain to the outside.
- Condensate removal can cause dripping on walls, balconies, or sidewalks.
- Temperature control is usually two-position, which causes swings in room temperature.
- Air distribution control is limited.
- Ventilation capabilities are fixed by equipment design.
- Overall appearance can be unappealing.
- Air filtration options are limited.
- Humidification, when required, must be provided by separate equipment.
- Noise levels vary considerably and are not generally suitable for critical applications.
- Routine unit maintenance is required to maintain capacity. Condenser and cooling coils must be cleaned, and filters must be changed regularly.

Design Considerations

Units are usually furnished with individual electric controls. However, when several units are used in a single space, the controls should be interlocked to prevent simultaneous heating and cooling. In commercial applications (e.g., motels), centrally operated switches can de-energize units in unoccupied rooms.

A through-the-wall or window-mounted air-conditioning unit incorporates a complete air-cooled refrigeration and air-handling system in an individual package. Each room is an individual occupant-controlled zone. Cooled or warmed air is discharged in response to thermostatic control to meet room requirements. The section on Controls describes how controls allow the use of individual room systems during off-schedule hours, but automatically return all systems to normal schedule use.

Each packaged terminal air conditioner has a self-contained, air-cooled direct-expansion cooling system; a heating system (electric, hot water, or steam); and controls. Two general configurations are shown—Figure 2 shows a wall box, an outdoor louver, heater section, cooling chassis, and cabinet enclosure; Figure 3 shows a combination wall sleeve cabinet, plus combination heating and cooling chassis with outdoor louver.

A through-the-wall air conditioner or heat pump system is installed in buildings requiring many temperature control zones such as (1) office buildings, (2) motels and hotels, (3) apartments and dormitories, (4) schools and other education buildings, and (5) areas of nursing homes or hospitals where air recirculation is allowed.

These units can be used for renovation of existing buildings, because existing heating systems can still be used. The equipment

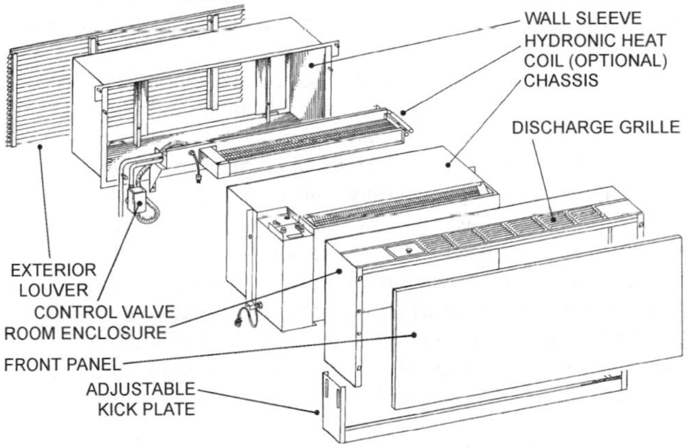

Fig. 2 Packaged Terminal Air Conditioner with Heating Section Separate from Cooling Chassis

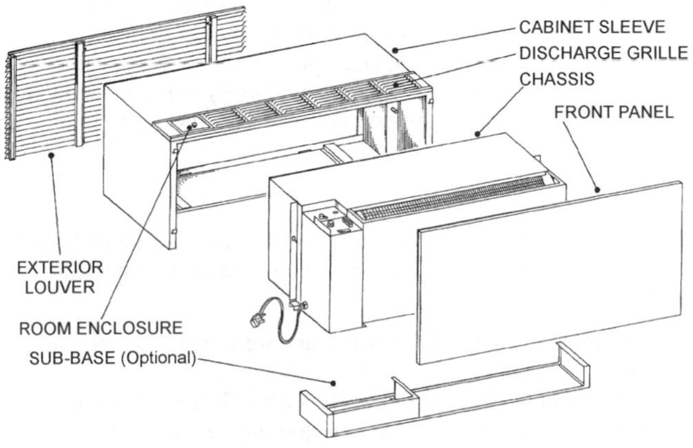

Fig. 3 Packaged Terminal Air Conditioner with Combination Heating and Cooling Chassis

can be used in both low- and high-rise buildings. In buildings where a stack effect is present, use should be limited to those areas that have dependable ventilation and a tight wall of separation between the interior and exterior.

Room air conditioners are often used in parts of buildings primarily conditioned by other systems, especially where spaces to be conditioned are (1) physically isolated from the rest of the building and (2) occupied on a different time schedule (e.g., clergy offices in a church and ticket offices in a theater).

Ventilation air through each terminal may be inadequate in many situation, particularly in high-rise structures because of the stack effect. Chapter 25 of the 1997 *ASHRAE Handbook—Fundamentals* explains combined wind and stack effects. Electrically operated outdoor air dampers, which close automatically when the equipment is stopped, reduce heat losses in winter.

Refrigeration Equipment

Room air conditioners are generally supplied with hermetic reciprocating, or scroll compressors. Capillary tubes are used in place of expansion valves in most units.

Some room air conditioners have only one motor to drive both the evaporator and condenser fans. The unit circulates air through the condenser coil whenever the evaporator fan is running, even during the heating season. The annual energy consumption of a unit with a single motor is generally higher than one with a separate motor, even when energy efficiency ratio (coefficient of performance) is the same for both. The year-round continuous flow of air through the condenser increases dirt accumulation on the coil and other components, which increases maintenance costs and reduces equipment life.

Because through-the-wall conditioners are seldom installed with drains, they require a positive and reliable means of condensate disposal. Conditioners are available that spray the condensate in a fine mist over the condenser coil. These units dispose of more condensate than can be developed without any drip, splash, or spray. In heat pumps, provision must be made for disposal of condensate generated from the outside coil during the defrost cycle.

Many air-cooled room conditioners experience evaporator icing and become ineffective when the outdoor temperature falls below about 65°F. Units that ice at a lower outdoor temperature may be required to handle the added load created by the high lighting levels and high solar radiation found in contemporary buildings.

Heating Equipment

The air-to-air heat pump cycle described in Chapter 45 of this volume is available in through-the-wall room air conditioners. Application considerations are quite similar to conventional units without the heat pump cycle, which is used for space heating when the outdoor temperature is above 35 to 40°F. Electric resistance elements supply heating below this level and during defrost cycles.

The prime advantage of the heat pump cycle is that it reduces the annual energy consumption for heating. Savings in heat energy over conventional electric heating ranges from 10 to 60%, depending on the climate.

Controls

All controls for through-the-wall air conditioners are included as a part of the conditioner. The following control configurations are available:

Thermostat Control. Thermostats are either unit mounted or remote wall mounted.

Guest Room Control for Motels and Hotels. This has provisions for starting and stopping the equipment from a central point.

Office Building and School Controls. These controls (for occupancies of less than 24 h) start and stop the equipment at preset times with a time clock. The conditioners operate normally with the unit thermostat until the preset cutoff time. After this point, each conditioner has its own reset control, which allows the occupant of the conditioned space to reset the conditioner for either cooling or heating, as required.

Master/Slave Control. This type of control is used when multiple conditioners are operated by the same thermostat.

Emergency Standby Control. Standby control allows a conditioner to operate during an emergency, such as a power failure, so that the roomside blowers can operate to provide heating. Units must be specially wired to allow operation on emergency generator circuits.

Acoustics

The noise from these units may be objectionable and should be checked to ensure it meets sound level requirements.

INTERCONNECTED ROOM-BY-ROOM SYSTEMS

Multiple-unit systems generally use single-zone unitary air conditioners with a unit for each zone (Figure 4). Zoning is determined by (1) cooling and heating loads, (2) occupancy considerations, (3) flexibility requirements, (4) appearance considerations, and (5) equipment and duct space availability. Multiple-unit systems are popular for office buildings, manufacturing plants, shopping centers, department stores, and apartment buildings. Unitary self-contained units are excellent for renovation.

A common condensing and heat source loop connects all units together. Heat pumped into the loop by units in the cooling mode can be reclaimed by units in a heat pump heating mode. During moderate weather and in buildings with high diversity in cooling and heating loads, these systems will tend to balance the building load. Heat from warm areas are in effect moved to cooler areas of the building. This minimizes the amount of auxiliary loop heating and cooling required. Figure 5 shows a typical unit with some commonly used components.

Advantages

- Installation is simple. Equipment is readily available in sizes that allow easy handling.
- Relocation of units to other spaces or buildings is practical, if necessary.

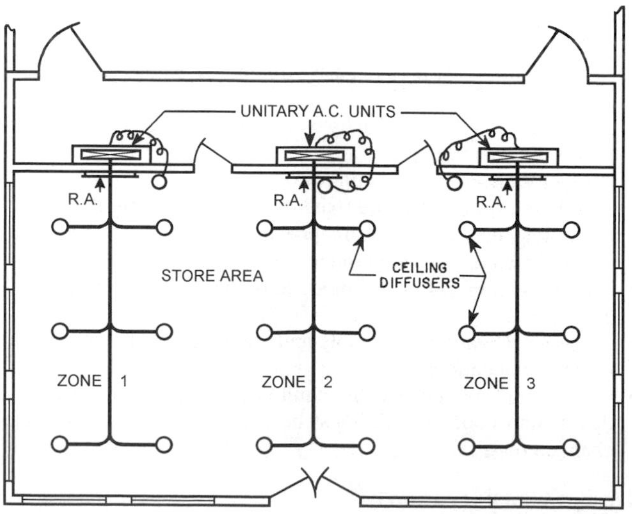

Fig. 4 Multiple Packaged Units

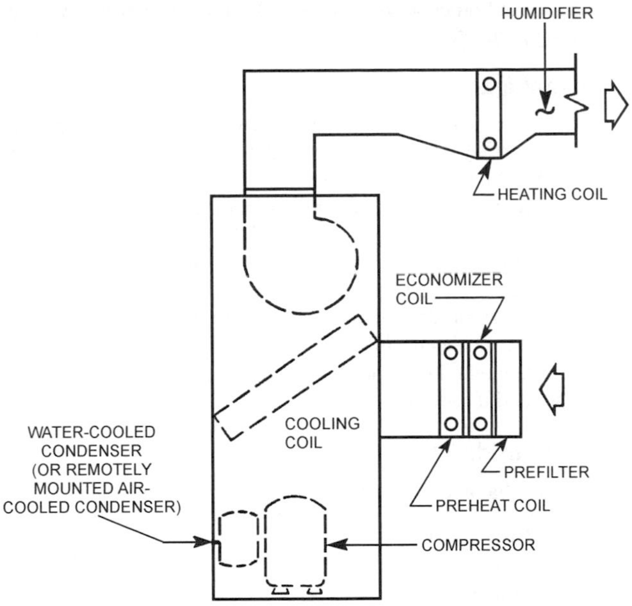

Fig. 5 Unitary Packaged Unit with Accessories

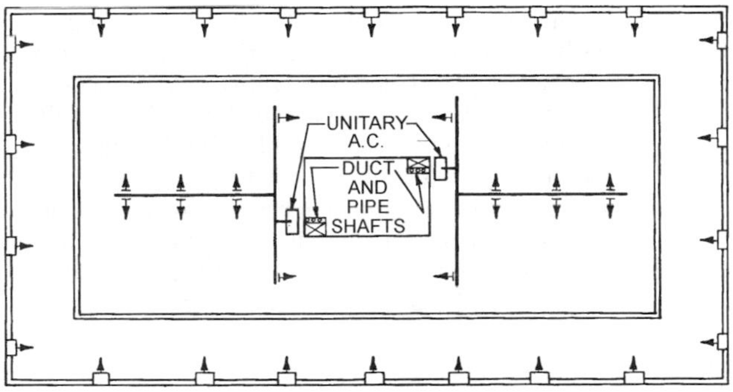

Fig. 6 Multiroom, Multistory Office Building with Unitary Core and Through-the-Wall Perimeter Air Conditioners

- Units are available with complete, self-contained control systems that include variable volume control, economizer cycle, night set-back, and morning warm-up.
- Easy access to equipment facilitates routine maintenance.

Disadvantages

- Fans may have limited static pressure ratings.
- Integral air-cooled units must be located along outside walls.
- Multiple units and equipment closets or rooms may occupy rentable floor space.
- Close proximity to building occupants may create noise problems.
- Discharge temperature may vary too much because of on-off or step control.

Design Considerations

Unitary systems can be used throughout a building or to supplement perimeter area packaged terminal units (Figure 6). Because core areas frequently have little or no heat loss, unitary equipment with water-cooled condensers can be applied with water-source heat pumps serving the perimeter.

In this multiple-unit system, one unit may be used to precondition outside air for a group of units (Figure 7). This all-outdoor air

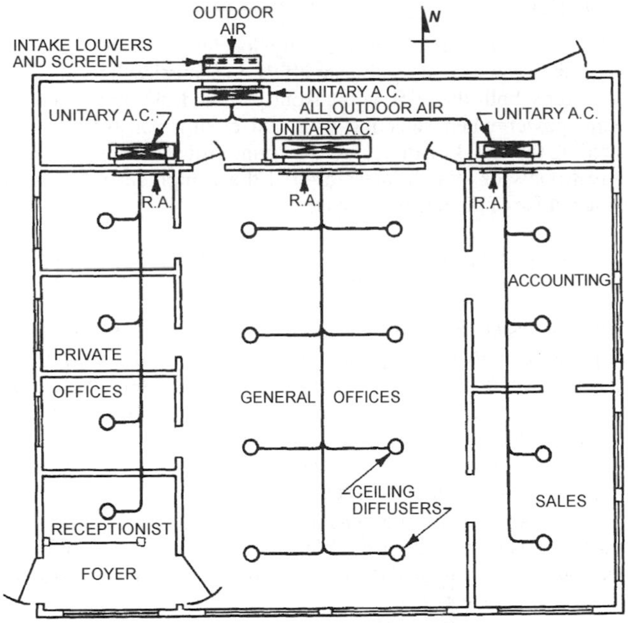

Fig. 7 Multiple-Packaged Units with Separate Outdoor Air Makeup Unit

unit prevents hot, humid air from entering the conditioned space under periods of light load. The outdoor unit should have sufficient capacity to cool the required ventilation air from outdoor design conditions to interior design dew point. Zone units are then sized to handle only the internal load for their particular area.

Special-purpose unitary equipment is frequently used to cool, dehumidify, humidify, and reheat to maintain close control of space temperature and humidity in computer areas. *Chapter 16 of the 1999 ASHRAE Handbook—Applications* has more information.

Refrigeration Equipment

Compressors are usually hermetic, reciprocating, or scroll compressors. Capillary tubes are used for expansion. Condensers are water cooled and connected to a central loop. A cooling tower or fluid cooler supplies supplemental cooling to the central loop as required.

Heating Equipment

The heating cycle on interconnected systems primarily uses an air-to-water heat pump. Supplemental heat to the loop is supplied with a central boiler

Controls

Units under 20 tons of cooling are typically constant-volume units. Variable-air-volume distribution is accomplished with a bypass damper that allows excess supply air to bypass to the return air duct. The bypass damper ensures constant airflow across the direct-expansion cooling coil to avoid coil freeze-up due to low air flow. The damper is usually controlled by the supply duct pressure.

Economizer Cycle. When the outdoor temperature permits, energy use can be reduced in many locales by cooling with outdoor air in lieu of mechanical refrigeration. Units must be located close to an outside wall or outside air duct shafts. Where this is not possible, it may be practical to add an economizer cooling coil adjacent to the preheat coil (Figure 5). Cold water is obtained by cooling the condenser water through a winterized cooling tower. Chapter 36 has further details.

Acoustics

Because these units are typically located near the occupied space, they can have a significant effect on acoustics. The designer must study both the airflow breakout path and the unit's radiated sound power when selecting wall and ceiling construction surrounding the unit. Locating units over non-critical work spaces such as restrooms or storage areas around the equipment room helps reduce noise in the occupied space.

RESIDENTIAL AND LIGHT-COMMERCIAL SPLIT SYSTEMS

A split system consists of an indoor unit with air distribution and temperature control with either a water-cooled condenser, integral air-cooled condenser, or remote air-cooled condenser. These units are commonly used in single-story or low-rise buildings, and residential applications where condenser water is not readily available. Commercial split systems are well-suited to office environments with variable occupancy schedules.

The indoor equipment is generally installed in service areas adjacent to the conditioned space. When a single unit is required, the indoor unit and its related ductwork constitute a central air system, as described in Chapter 4.

Typical components of a split-system air conditioner include an indoor unit with evaporator coils, economizer coils, heating coils, filters, valves, and a condensing unit with the compressors and condenser coils.

Advantages

- Unitary split-system units (Figure 8) allow air-handling equipment to be placed close to the air-conditioning load, which allows ample air distribution to the air-conditioned space with a minimum amount of ductwork and fan power.
- Heat rejection through a remote air-cooled condenser allows the final heat rejector to be located remote from the air-conditioned space.
- A floor-by-floor arrangement can offer reduced fan power consumption because the air handlers are located close to the air-conditioned space.
- Large vertical duct shafts and fire dampers are eliminated.
- Commercial split systems generally reduce mechanical room area. Equipment is generally located in the building interior near elevators and other service areas and does not interfere with the building perimeter.

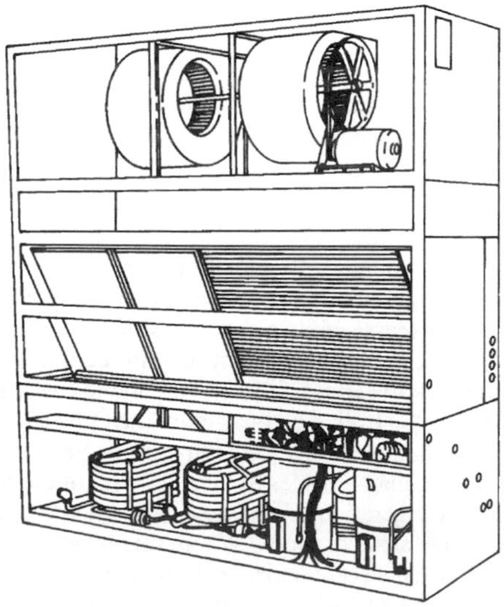

Fig. 8 Vertical Self-Contained Unit

Disadvantages

- The close proximity of the air handler to the air-conditioned space requires special attention to unit inlet and outlet airflow and to building acoustics around the unit.
- Ducting ventilation air to the unit and removing condensate from the cooling coil should be considered. Separate outdoor air systems are commonly applied in conjunction with the split system.
- A unit that uses an air-side economizer must be located near an outside wall or outdoor air shaft. Split-system units do not generally include return air fans.
- A separate method of handling and controlling relief air must be incorporated if required.

Design Considerations

Building characteristics that favor split systems include

- Multiple floors or zones
- Renovation work
- Historic structures
- Multiple tenants per floor with variable schedules
- Common return air paths

The modest space requirements of split-system equipment make it an excellent choice for renovations or for providing air conditioning to previously non-air-conditioned, historic structures. Control is usually one- or two-step cool and one- or two-step or modulating heat. Variable-air-volume operation is possible with a supply air bypass. Some commercial units are capable of airflow modulation with additional cooling modulation with a hot gas bypass.

Commercial split-system equipment can incorporate an integral water-side economizer coil and controls, thus allowing an interior location. In this configuration, the outdoor air shaft is reduced to meet only ventilation and space pressurization requirements.

Refrigeration Equipment

Compressors supplied on these units are usually hermetic or semihermetic reciprocating or scroll compressors. Smaller units have capillary tubes while larger units system have thermostatic expansion valves. Larger systems may also use multiple- or two-speed compressors. Compressors are usually located in the remote condensing unit instead of with the indoor equipment.

Heating Equipment

Heating systems available include electric resistance, indirect fired gas, or air-to-air heat pumps. Larger systems may also have available hot water or steam heating coils.

Controls

Commercial split-system units are available as constant-volume equipment for use in atriums, public areas, and industrial applications. Unit options include fan modulation and variable-air-volume control. When applied with variable-air-volume terminals, commercial split systems provide excellent comfort and individual zone control.

Acoustics

As with any unit with components in the occupied space, acoustics are a concern. Many commercial split-system units include a factory-engineered acoustical discharge plenum, which facilitates a smooth supply air discharge from the equipment room. This allows lower fan power and lower sound power levels. This discharge arrangement also reduces the size of the equipment room.

COMMERCIAL SELF-CONTAINED (FLOOR-BY-FLOOR) SYSTEMS

Commercial, self-contained systems provide central air distribution, refrigeration, and system control on a zone or floor-by-floor basis. Typical components include compressors, water cooled condensers, evaporator coils, economizer coils, heating coils, filters, valves, and controls (Figure 9). To complete the system, a building needs cooling towers and condenser water pumps.

Advantages

- Units are well-suited for office environments with variable occupancy schedules.
- The floor-by-floor arrangement can offer reduced fan power consumption.
- Large vertical duct shafts and fire dampers are eliminated.
- Electrical wiring, condenser water piping, and condensate removal are centrally located.
- Commercial self-contained systems generally require less mechanical room area.
- Equipment is generally located in the building interior near elevators and other service areas and does not interfere with the building perimeter.
- Integral water-side economizer coil and controls allow an interior equipment location and eliminate large outdoor air and exhaust ducts and relief fans.
- This equipment integrates refrigeration, air-handling, and controls into a factory package, thus eliminating many field integration problems.
- An acoustical discharge plenum allows lower fan power and lower sound power levels.
- Cost of operating cooling equipment after normal building hours is reduced by operating only units in occupied areas.

Disadvantages

- Units must be located near an outside wall or outdoor air shaft to incorporate an air-side economizer.
- A separate relief air system and controls must be incorporated if an air-side economizer is used.
- Close proximity to building occupants requires careful analysis of space acoustics.
- Filter options may be limited.
- Discharge temperature varies because of on-off or step control.

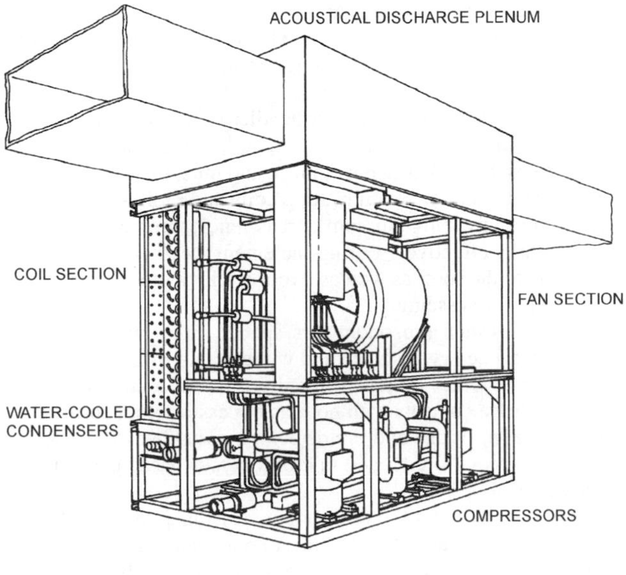

Fig. 9 Commercial Self-Contained Unit with Discharge Plenum

Design Considerations

Commercial self-contained units can serve either variable-air-volume or constant-volume systems. These units contain one or two fans—either forward curved (FC), backward inclined (BI), or airfoil (AF)—inside the cabinet. The fans are commonly configured in a draw-through arrangement.

The size and diversity of the zones served often dictate which system is optimal. For comfort applications, variable-air-volume self-contained units coupled with terminal boxes or fan powered terminal boxes are popular for their energy savings, individual zone control, and acoustic benefits. Constant-volume self-contained units have a low installation cost and are often used in non-comfort or industrial air-conditioning applications or in single-zone comfort applications.

Unit airflow is reduced in response to terminal boxes closing. Several common methods used to modulate the airflow delivered by the fan to match system requirements include inlet guide vanes, fan speed control, inlet/discharge dampers, and multiple speed fan motors.

Appropriate outside air and exhaust fans and dampers work in conjunction with the self-contained unit. Their operation must be coordinated with the unit operation to maintain design air exchange and building pressurization.

Refrigeration Equipment

Self-contained units may control capacity with multiple compressors. For variable-air-volume systems, compressors are turned on or off or unloaded to maintain discharge air temperature. Hot gas bypass is many time incorporated to provide additional capacity control. As system airflow decreases, the temperature of air leaving the unit is often reset upward so that a minimum ventilation rate can be maintained. Resetting the discharge air temperature limits the unit's demand, thus saving energy. However, the increased air temperature and volume will increase fan energy. Thermostatic or electronic expansion valves are used on these systems.

Heating Equipment

In many applications, heating is handled by perimeter radiation, heating installed in the terminal boxes or other such systems when floor by floor units are used. If heating is incorporated in these units it is usually provided by hot water coils or electric resistance heat.

Controls

Self-contained units typically have built-in capacity controls for refrigeration, economizers, and fans. Although units under 15 tons of cooling tend to have basic on-off/automatic controls, many larger systems have sophisticated microprocessor controls that monitor and take action based on local or remote programming. These controls provide for stand-alone operation, or they can be tied to a building automation system.

A building automation system allows for more sophisticated unit control by time-of-day scheduling, optimal start/stop, duty cycling, demand limiting, custom programming, etc. This control can keep the units operating at their peak efficiency by alerting the operator to conditions that could cause substandard performance.

The unit's control panel can sequence the modulating valves and dampers of an economizer. A water-side economizer is located upstream of the evaporator coil, and when condenser water temperature is lower than entering air temperature to the unit, water flow is directed through the economizer coil to either partially or fully meet building load. If the coil alone cannot meet design requirements, but the entering condenser water temperature remains cool enough to provide some useful precooling, the control panel can keep the economizer coil active as stages of compressors are activated. When entering condenser water exceeds entering air temperature to the unit, the coil is valved off, and water is circulated through the unit's condensers only.

Typically, in an air-side economizer an enthalpy switch or dry-bulb temperature energizes the unit to bring in outside air as the first stage of cooling. An outside air damper modulates the flow to meet a design temperature, and when outside air can no longer provide sufficient cooling, compressors are energized.

A temperature input to the control panel, either from a discharge air sensor or a zone sensor, provides information for integrated economizer and compressor control. Supply air temperature reset is commonly applied to variable-air-volume systems.

In addition to capacity controls, units have safety features for the refrigerant-side, air-side, and electrical systems. Refrigeration protection controls typically consist of high and low refrigerant pressure sensors and temperature sensors wired into a common control panel. The controller then cycles compressors on and off or activates hot-gas bypass to meet system requirements.

Constant-volume units typically have high-pressure cut-out controls, which protect the unit and ductwork from high static pressure. Variable-air-volume units typically have some type of static pressure probe inserted in the discharge duct downstream of the unit. As terminal boxes close, the control modulates airflow to meet the set point, which is determined by calculating the static pressure required to deliver design airflow to a zone farthest from the unit.

Acoustics

Because self-contained units are typically located near occupied space, their performance can significantly affect tenant comfort. Units of less than 15 tons of cooling are often placed inside a closet with a discharge grille penetrating the common wall to the occupied space. Larger units have their own equipment room and duct system. Three common sound paths to consider are:

- Fan inlet and compressor sound which radiates through the unit casing to enter the space through the separating wall.
- Fan discharge sound is airborne through the supply duct and enters the space through duct breakout and diffusers.
- Airborne fan inlet sound finds its way to the space through the return air ducts.

Discharge air transition off the self-contained unit is often accomplished with a plenum located on top of the unit. This plenum facilitates multiple duct discharges that reduce the amount of airflow over a single occupied space adjacent to the equipment room (Figure 9). Reducing the airflow in one direction reduces the sound that breaks out from the discharge duct. Several feet of internally lined round duct immediately off the discharge plenum significantly reduces noise levels in adjacent areas.

In addition to the airflow breakout path, the system designer must study unit radiated sound power when determining equipment room wall and door construction. A unit's air-side inlet typically has the highest radiated sound. The inlet space and return air ducts should be located away from the critically occupied area to reduce the effect of this sound path.

Selecting a fan that operates near its peak efficiency point helps in the design of quiet systems. Fans are typically dominant in the first three octave bands, and selections at high static pressures or near the fan's surge region should be avoided.

Units may be isolated from the structure with neoprene pads or spring isolators. Manufacturers often isolate the fan and compressors internally, which generally reduces external isolation requirements.

COMMERCIAL OUTDOOR PACKAGED EQUIPMENT

Outdoor packaged equipment consists of a complete system that includes unitary equipment, ducted air distribution, and temperature control. The equipment is generally mounted on the roof, but it can also be mounted at grade level. Rooftop units are designed as central station equipment for single-zone, multizone, and variable-air-volume applications.

Advantages

- The location of equipment allows for shorter duct runs, reduced duct space requirements, and lower initial cost.
- Installation is simplified.
- Valuable building space for mechanical equipment is conserved.
- Suitable for floor-by-floor control in low-rise office buildings.

Disadvantages

- Maintenance or servicing of outdoor units is sometimes difficult.
- Frequent removal of panels for access may destroy the weatherproofing of the unit, causing electrical component failure, rusting, and water leakage.
- Corrosion of casings is a potential problem. Many manufacturers prevent rusting with galvanized or vinyl coatings and other protective measures.
- Equipment life is reduced by outdoor installation.
- Depending on building construction, sound levels and transmitted vibration may be excessive.

Design Considerations

Centering the rooftop unit over the conditioned space results in reduced fan power, ducting, and wiring. Avoid installation directly above spaces where noise level is critical.

All outdoor ductwork should be insulated. In addition, the ductwork should be (1) sealed to prevent condensation in the insulation during the heating season and (2) weatherproofed to keep it from getting wet.

Use multiple single-zone, not multizone, units where feasible. For large areas such as manufacturing plants, warehouses, gymnasiums, and so forth, single-zone units are less expensive and provide protection against total system failure.

Use units with return air fans whenever return air static pressure loss exceeds 0.5 in. of water or the unit introduces a large percentage of outdoor air via an economizer.

Units are also available with relief fans for use with an economizer in lieu of continuously running a return fan. Relief fans can be initiated by static pressure control.

In a rooftop application, the air handler is outdoors and needs to be weatherproofed against rain, snow, and, in some areas, sand. In cold climates, fuel oil does not atomize and must be warmed to burn properly. Hot water or steam heating coils and piping must be protected against freezing. In some areas, enclosures are needed to maintain units effectively during inclement weather. A permanent safe access to the roof, as well as a roof walkway to protect against roof damage, are essential.

Accessories such as economizers, special filters, and humidifiers are available. Factory-installed and wired economizer packages are also available. Other options offered are return and exhaust fans, variable volume controls with hot-gas bypass or other form of coil frost protection, smoke and fire detectors, portable external service enclosures, special filters, and microprocessor-based controls with various control options.

Rooftop units are generally mounted using integral frames or lightweight steel structures. Integral support frames are designed by the manufacturer to connect to the base of the unit. No duct openings are required for supply and return ducts. The completed installation must adequately drain condensed water. Lightweight steel structures allow the unit to be installed above the roof using separate, flashed duct openings. Any condensed water can be drained through the roof drains.

Refrigeration Equipment

The compressors in large systems are reciprocating, screw or scroll compressors. Chapter 34 has information about compressors and Chapters 41 and Chapter 43 of the 1998 *ASHRAE Handbook—Refrigeration* discusses refrigeration equipment, including the general size ranges of available equipment. Air-cooled or evaporative condensers are built integral to the equipment.

Air-cooled condensers pass outdoor air over a dry coil to condense the refrigerant. This results in a higher condensing temperature and, thus, a larger power input at peak conditions. However, this peak time may be relatively short over a 24-hour period. The air-cooled condenser is popular in small reciprocating systems because of its low maintenance requirements.

Evaporative condensers pass air over coils sprayed with water, thus taking advantage of adiabatic saturation to lower the condensing temperature. As with the cooling tower, freeze prevention and close control of water treatment are required for successful operation. The lower power consumption of the refrigeration system and the much smaller footprint from the use of the evaporative versus the air-cooled condenser are gained at the expense of the cost of the water used and increased maintenance costs.

Heating Equipment

Natural gas, propane, oil, electricity, hot water, steam, and refrigerant gas heating options are available.

Controls

Multiple outdoor units are usually the single-zone, constant-volume, or variable-air-volume units. Zoning for temperature control determines the number of units; each zone has a unit. The zones are determined by the cooling and heating loads for the space served, occupancy, allowable roof loads, flexibility requirements, appearance, duct size limitations, and equipment size availability. Multiple units are installed in manufacturing plants, warehouses, schools, shopping centers, office buildings, and department stores. These units also serve core areas of buildings whose perimeter spaces are served by packaged terminal air conditioners. These systems are usually applied to low-rise buildings of one or two floors, but have been used for conditioning multistory buildings as well.

Most operating and safety controls are provided by the equipment manufacturer. Although remote monitoring panels are optional, they are recommended to allow operating personnel to monitor performance.

Acoustics

Most unitary equipment is available with limited separate vibration isolation of the rotating equipment. Isolation of the entire unit casing is not always required; however care should be taken when mounting on lightweight structures.

Outdoor noise from unitary equipment should be reduced to a minimum. Sound power levels at all property lines must be evaluated. Airborne noise can be attenuated by silencers in the supply and return air ducts or by acoustically lined ductwork. Avoid installation directly above spaces where noise level is critical.

INDUSTRIAL/COMMERCIAL CUSTOM PACKAGED SYSTEMS

An outdoor unitary equipment system can be designed to cool or heat an entire building. On large projects and highly demanding systems, the additional cost of a custom packaged unit can be justified. These systems offer a large degree of flexibility and can be configured to satisfy almost any requirement. Special features such as heat recovery, service vestibules, boiler, chillers and space for other mechanical equipment can be designed into the unit.

The equipment is generally mounted on the roof, but it can also be mounted at grade level. Units usually ship in multiple pieces and require field assembly on the roof.

These units can be designed as central station equipment for single-zone, multizone, and variable-air-volume applications. Multiple, separate air handlers can be incorporated into a single unit.

Advantages

- Equipment location allows for shorter duct runs, reduced duct space requirements, lower installed cost, and ease of service access.
- Construction costs are offset toward the end of the project because the unit can be one of the last items installed.
- Field labor costs are reduced because most components are assembled and tested in a controlled factory environment.
- A single source has responsibility for the design and operation of all major mechanical systems in a building.
- Installation is simplified.
- Valuable building space for mechanical equipment is conserved.

Disadvantages

- Higher first cost than commercial equipment.
- Unit design must be coordinated with structural design because it represents a significant building structural load.
- Overall appearance can be unappealing.
- Maintenance or servicing of outdoor units is sometimes more difficult.
- Corrosion of casings is a potential problem. Many manufacturers prevent rusting with galvanized or vinyl coatings and other protective measures.
- Equipment life is reduced by outdoor installation.
- Depending on building construction, sound levels and transmitted vibration may be excessive.

Design Considerations

Centering the rooftop unit over the conditioned space results in reduced fan power, ducting, and wiring. Avoid installation directly above spaces where noise level is critical.

All outdoor ductwork should be insulated. In addition, the ductwork should be (1) sealed to prevent condensation in the insulation during the heating season and (2) weatherproofed to keep it from getting wet.

Use units with return air fans whenever return air static pressure loss exceeds 0.5 in. of water or the unit introduces a large percentage of outdoor air via an economizer.

Units are also available with relief fans for use with an economizer in lieu of continuously running a return fan. Relief fans can be initiated by static pressure control.

In a rooftop application, the air handler is outdoors and must be of a weatherproof design. In cold climates, fuel oil does not atomize and must be warmed to burn properly. Hot water or steam heating coils and piping must be protected against freezing. A permanent safe access to the roof, as well as a roof walkway to protect against roof damage, is essential.

Accessories such as economizers, special filters, and humidifiers are available. Factory-installed and wired economizer packages are also available. Other options offered are return and exhaust fans, variable volume controls with hot-gas bypass or other form of coil frost protection, smoke and fire detectors, portable external service enclosures, special filters, and microprocessor-based controls with various control options.

These units are generally mounted using integral support frames or steel structures. Integral support frames are designed by the manufacturer to connect to the base of the unit. The installation must allow any condensed water to drain. Lightweight steel structures allow the unit to be installed above the roof using separate, flashed duct openings. Any condensed water can be drained through the roof drains.

Because each unit is custom designed to a the specific application, it may be desirable to require additional witnessed factory testing to insure the performance and quality of the final product.

Refrigeration Equipment

The major types of refrigeration equipment used in large systems are hermetic and semihermetic reciprocating, screw, and scroll compressors. See Chapter 34 for information about compressors and Chapter 41 and Chapter 43 of the 1998 *ASHRAE Handbook—Refrigeration* for further discussion of refrigeration equipment, including the general size ranges of available equipment.

Air-cooled or evaporative condensers are built integral to the equipment. Air-cooled condensers pass outdoor air over a dry coil to condense the refrigerant. This results in a higher condensing temperature and, thus, a larger power input at peak condition; however, over 24 hours this peak time may be relatively short. The major advantage of air-cooled condensers are the low maintenance requirements.

Evaporative condensers pass air over coils sprayed with water, thus taking advantage of adiabatic saturation to lower the condensing temperature. As with the cooling tower, freeze prevention and close control of water treatment are required for successful operation. The lower power consumption of the refrigeration system and the much smaller footprint from the use of the evaporative versus the air-cooled condenser are gained at the expense of the cost of the water used and increased maintenance costs.

Heating Equipment

Natural gas, propane, oil, electricity, hot water, steam, and refrigerant gas heating options are available. These can be incorporated directly into the air-handling sections or a separate pre-piped boiler and circulating systems can be used.

Controls

The equipment manufacturer provides all operating and safety controls. Although remote monitoring panels are optional, they are recommended to permit operating personnel to monitor performance.

Acoustics

This equipment is available with several (optional) degrees of internal vibration isolation of the rotating equipment. Isolation of the entire unit casing is rarely required; however, care should be taken when mounting on lightweight structures. If external isolation is required, it should be coordinated with the unit manufacture to insure proper separation of internal verse external isolation deflection.

Outdoor noise from unitary equipment should be reduced to a minimum. Sound power levels at all property lines must be evaluated. Special attenuated condenser sections are available where required. Airborne noise can be attenuated by silencers in the supply and return air ducts or by acoustically lined ductwork. Avoid installation directly above spaces where noise is critical.

REFERENCES

AHAM. 1992. Room air conditioners. *Standard* RAC-1. Association of Home Appliance Manufacturers, Chicago, IL.

ARI. 1990. Packaged terminal air conditioners. *Standard* 310-90. Air-Conditioning and Refrigeration Institute, Arlington, VA.

CHAPTER 6

PANEL HEATING AND COOLING

PANEL heating and cooling uses controlled-temperature surfaces on the floor, walls, or ceiling; the temperature is maintained by circulating water, air, or electric current through a circuit embedded in the panel. A controlled-temperature surface is called a **radiant panel** if 50% or more of the heat transfer is by radiation to other surfaces seen by the panel. Radiant panel systems may be combined either with a central station air system of one-zone, constant temperature, constant volume design or with dual-duct, reheat, multizone or variable volume systems. These combined systems are called **hybrid HVAC systems**.

This chapter covers controlled-temperature surfaces that are the primary source of heating and cooling within the conditioned space. For snow-melting applications see Chapter 49 of the 1999 *ASHRAE Handbook—Applications*. Chapter 15 covers high-temperature surface radiant panels over 300°F energized by gas, electricity, or high-temperature water.

PRINCIPLES OF THERMAL RADIATION

Thermal radiation (1) is transmitted at the speed of light; (2) travels in straight lines and can be reflected; and (3) elevates the temperature of solid objects by absorption but does not noticeably heat the air through which it travels.

Thermal radiation is exchanged continuously between all bodies in a building environment. The rate at which radiant heat is transferred depends on the following factors:

- Temperature (of the emitting surface and receiver)
- Emittance (of the radiating surface)
- Reflectance, absorptance, and transmittance (of the receiver)
- View factor between the emitting surface and receiver (viewing angle of the occupant to the radiant source)

A critical factor is the structure of the body surface. In general, rough surfaces have low reflectance and high emittance/absorptance characteristics. Conversely, smooth or polished metal surfaces have high reflectance and low absorptance/emittance characteristics.

One example of radiant heating is the feeling of warmth when standing in the sun's rays on a cool, sunny day. Some of the rays come directly from the sun and include the entire electromagnetic spectrum. Other rays from the sun are absorbed by or reflected from surrounding objects. This generates secondary rays that are a combination of the wavelength produced by the temperature of the objects and the wavelength of the reflected rays. If a cloud passes in front of the sun, there is an instant sensation of cold. This sensation is caused by the decrease in radiant heat received from the sun, although there is little, if any, change in the ambient air temperature.

The preparation of this chapter is assigned to TC 6.5, Radiant Space Heating and Cooling.

Thermal comfort, as defined in ASHRAE *Standard* 55, is "that condition of mind which expresses satisfaction with the thermal environment." No system is completely satisfactory unless the three main factors controlling heat transfer from the human body (radiation, convection, and evaporation) result in thermal neutrality. Maintaining correct conditions for human comfort by radiant heat transfer is possible for even the most severe climatic conditions (Buckley 1989). Chapter 8 of the 1997 *ASHRAE Handbook—Fundamentals* has more information on thermal comfort.

Panel heating and cooling systems provide an acceptable thermal environment by controlling surface temperature in an occupied space, thus affecting the radiant heat transfer. With a properly designed system, a person should not be aware that the environment is being heated or cooled. The **mean radiant temperature (MRT)** has a strong influence on human comfort. When the temperature of the surfaces comprising the building (particularly outside walls with large amounts of glass) deviates excessively from the ambient temperature, convective systems sometimes have difficulty in counteracting the discomfort caused by cold or hot surfaces. Heating and cooling panels neutralize these deficiencies and minimize radiation losses or gains by the body.

Most building materials have surfaces with relatively high emittance factors and, therefore, absorb and reradiate radiant heat from active panels. Warm ceiling panels are effective because radiant heat is absorbed and reflected by the irradiated surfaces and not transmitted through the construction. Glass is opaque to the wavelengths emitted by active panels and, therefore, transmits little of the long-wave radiant heat to the outside. This is significant because all surfaces in the room tend to assume temperatures that result in an acceptable thermal comfort condition.

GENERAL EVALUATION

Principal **advantages** of radiant panel systems are the following:

- Comfort levels can be better than those of other conditioning systems because radiant loads are satisfied directly and air motion in the space is at normal ventilation levels.
- Space-conditioning equipment is not needed at the outside walls; this simplifies the wall, floor, and structural systems.
- Almost all mechanical equipment may be centrally located, simplifying maintenance and operation.
- No space within the conditioned room is required for mechanical equipment. This feature is especially valuable in hospital patient rooms and other applications where space is at a premium, where maximum cleanliness is essential, or where it is dictated by legal requirements.
- Draperies and curtains can be installed at the outside wall without interfering with the space-conditioning system.
- When four-pipe systems are used, cooling and heating can be simultaneous, without central zoning or seasonal changeover.
- Supply air quantities usually do not exceed those required for ventilation and dehumidification.

- The modular panel provides flexibility to meet changes in partitioning.
- A 100% outdoor air system may be installed with smaller penalties in terms of refrigeration load because of reduced air quantities.
- A common central air system can serve both the interior and perimeter zones.
- Wet surface cooling coils are eliminated from the occupied space, reducing the potential for septic contamination.
- The panel system can use the automatic sprinkler system piping (see NFPA *Standard* 13, Chapter 3, Section 3.6). The maximum water temperature must not fuse the heads.
- Radiant panel heating and cooling and minimum supply air quantities provide a draft-free environment.
- Noise associated with fan-coil or induction units is eliminated.
- Peak loads are reduced as a result of thermal energy storage in the panel structure, exposed walls, and partitions.
- Panels can be coupled with other conditioning systems for heat loss (gain) compensation for cold or hot floors, windows, etc.

Disadvantages are similar to those listed in Chapter 3 of the 1997 *ASHRAE Handbook—Fundamentals*. In addition:

- Response time can be slow if controls and/or heating elements are not selected or installed correctly
- Improper installation of pipe or element spacing and/or incorrect sizing of heat source can cause nonuniform surface temperatures or insufficient heating capacity

HEAT TRANSFER BY PANEL SURFACES

A heated or cooled panel transfers heat to or from a room by radiation and natural convection.

Radiation Transfer

The basic equation for a multisurface enclosure with gray, diffuse isothermal surfaces is derived by radiosity formulation methods (Chapter 3 of the 1997 *ASHRAE Handbook—Fundamentals*). This equation may be written as

$$q_r = J_p - \sum_{j=1}^{n} F_{pj} J_j \tag{1}$$

where

q_r = net radiation heat transferred by panel surface, Btu/h·ft²
J_p = total radiosity that leaves panel surface, Btu/h·ft²
J_j = radiosity from another surface in room, Btu/h·ft²
F_{pj} = radiation angle factor between panel surface and another surface in room (dimensionless)
n = number of surfaces in room other than panel

Equation (1) can be applied to simple and complex enclosures with varying surface temperatures and emittances. The net radiation transferred by the panels can be found by determining unknown J_j if the number of surfaces is small. More complex enclosures require computer calculations.

Radiation angle factors can be evaluated using Figure 6 in Chapter 3 of the 1997 *ASHRAE Handbook—Fundamentals*. Fanger (1972) shows room-related angle factors; they may also be developed from algorithms in ASHRAE's *Energy Calculations* 1 (1976).

Several methods have been developed to simplify Equation (1) by reducing a multisurface enclosure to a two-surface approximation. In the **MRT method**, the radiant interchange in a room is modeled by assuming that the surfaces radiate to a fictitious surface that has an area emittance and temperature giving about the same heat transfer as the real multisurface case (Walton 1980). In addition, angle factors do not need to be determined in the evaluation of a two-surface enclosure. The MRT equation may be written as

$$q_r = \sigma F_r [T_p^4 - T_r^4] \tag{2}$$

where

σ = Stefan-Boltzmann constant = 0.1713×10^{-8} Btu/h·ft²·°R⁴
F_r = radiation exchange factor (dimensionless)
T_p = effective temperature of heated (cooled) panel surface, °R
T_r = temperature of fictitious surface (unheated or uncooled), °R

The temperature of the fictitious surface is given by an area emittance weighted average of all surfaces other than the panel:

$$T_r = \frac{\sum_{j \neq p}^{n} A_j \varepsilon_j T_j}{\sum_{j \neq p}^{n} A_j \varepsilon_j} \tag{3}$$

where

A_j = area of surfaces other than panel
ε_j = thermal emittance other than panel (dimensionless)

When the emittances of an enclosure are nearly equal, and surfaces exposed to the panel are marginally unheated (uncooled), then Equation (3) becomes the area-weighted average temperature (AUST) of unheated (uncooled) surfaces exposed to the panels.

The radiation interchange factor for two-surface radiation heat exchange is given by the Hottel equation:

$$F_r = \frac{1}{\dfrac{1}{F_{p-r}} + \left(\dfrac{1}{\varepsilon_p} - 1\right) + \dfrac{A_p}{A_r}\left(\dfrac{1}{\varepsilon_r} - 1\right)} \tag{4}$$

where

F_{p-r} = radiation angle factor from panel to fictitious surface (1.0 for flat panel)
A_p, A_r = area of panel surface and fictitious surface, respectively
ε_p, ε_r = thermal emittance of panel surface and fictitious surface, respectively (dimensionless)

In practice, the thermal emittance ε_p of nonmetallic or painted metal nonreflecting surfaces is about 0.9. When this emittance is used in Equation (4), the radiation exchange factor F_r is about 0.87 for most rooms. Substituting this value in Equation (2), σF_r becomes about 0.15×10^{-8}. Min et al. (1956) showed that this constant was 0.152×10^{-8} in their test room. The radiation equation for heating or cooling becomes

$$q_r = 0.15 \times 10^{-8} [(t_p + 460)^4 - (\text{AUST} + 460)^4] \tag{5}$$

where

t_p = effective panel surface temperature, °F
AUST = area-weighted average temperature of uncontrolled surfaces in room, °F

Equation (5) establishes the general sign convention for this chapter, which states that heating by the panel is positive and cooling by the panel is negative.

Radiation exchange calculated from Equation (5) is given in Figure 1. The values apply to ceiling, floor, or wall panel output. Radiation removed by a cooling panel for a range of normally encountered temperatures is given in Figure 2.

In many specific instances where normal multistory commercial construction and fluorescent lighting are used, the room temperature at the 5 ft level closely approaches the average uncooled surface temperature (AUST). In structures where the main heat gain is through the walls or where incandescent lighting is used, the wall

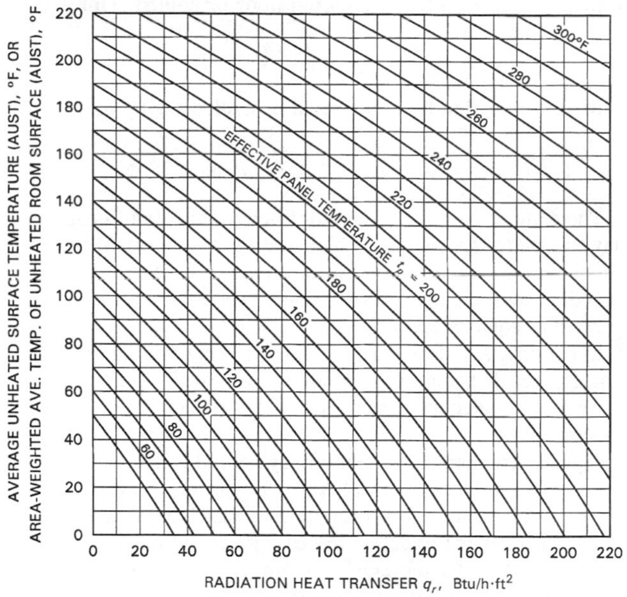

Fig. 1 Radiation Heat Transfer from Heated Ceiling, Floor, or Wall Panel

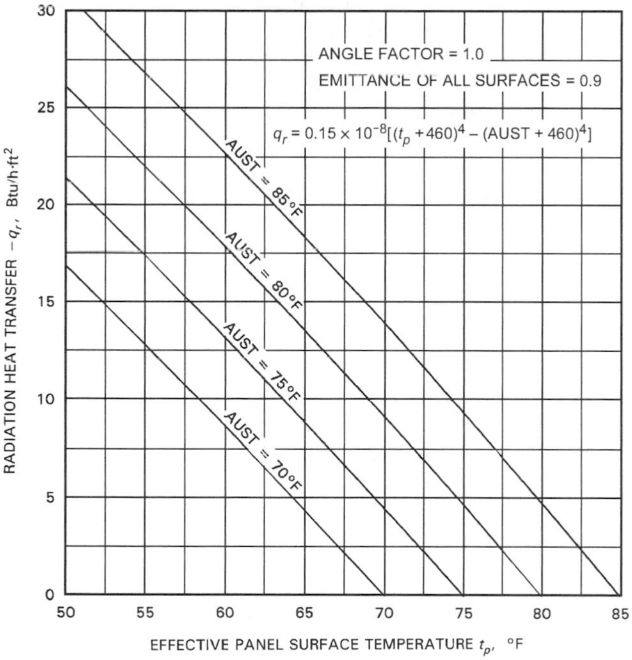

Fig. 2 Heat Removed by Radiation to Cooled Ceiling or Wall Panel

surface temperatures tend to rise considerably above the room air temperature.

Convection Transfer

The convection coefficient q_c is defined as the heat transferred by convection between the air and the panel. Heat transfer convection values are not easily established. Convection in panel systems is usually natural; that is, air motion is generated by the warming or cooling of the boundary layer of air. In practice, however, many factors, such as a room's configuration, interfere with or affect natural convection. Infiltration, the movement of persons, and mechanical ventilating systems can introduce some forced convection that disturbs the natural process.

Parmelee and Huebscher (1947) included the effect of forced convection on heat transfer from panels as an increment to be added to the natural convection coefficient. However, increased heat transfer from forced convection should not be used because the increments are unpredictable in pattern and performance, and forced convection does not significantly increase the total capacity of the panel system.

Convection in a panel system is a function of the panel surface temperature and the temperature of the airstream layer directly below the panel. The most consistent measurements are obtained when the air layer temperature is measured close to the region where the fully developed stream begins, usually 2 to 2.5 in. below the panels.

Min et al. (1956) determined natural convection coefficients 5 ft above the floor in the center of a 12 ft by 24.5 ft room (D_e = 16.1 ft). Equations (6) through (11), derived from this research, can be used to calculate heat transfer from panels by natural convection.

Natural convection from an all-heated ceiling

$$q_c = 0.041\frac{(t_p - t_a)^{1.25}}{D_e^{0.25}} \tag{6}$$

Natural convection from a heated floor or cooled ceiling

$$q_c = 0.39\frac{|t_p - t_a|^{0.31}(t_p - t_a)}{D_e^{0.08}} \tag{7}$$

Natural convection from a heated or cooled wall panel

$$q_c = 0.29\frac{|t_p - t_a|^{0.32}(t_p - t_a)}{H^{0.05}} \tag{8}$$

where

q_c = heat transfer by natural convection, Btu/h·ft²
t_p = effective temperature of panel surface, °F
t_a = temperature of air, °F
D_e = equivalent diameter of panel (4 × area/perimeter), ft
H = height of wall panel, ft

Schutrum and Humphreys (1954) measured panel performance in furnished test rooms that did not have uniform panel surface temperatures and found no variation in performance large enough to be significant in heating practice. Schutrum and Vouris (1954) established that the effect of room size was usually insignificant except for very large spaces like hangars and warehouses. In these cases Equations (6) and (7) should be used. Otherwise, Equations (6), (7), and (8) can be simplified to the following by D_e = 16.1 ft and H = 8.85 ft:

Natural convection from an all-heated ceiling

$$q_c = 0.020(t_p - t_a)^{0.25}(t_p - t_a) \tag{9a}$$

Natural convection from a heated ceiling may be enhanced by leaving cold strips (unheated ceiling sections) between ceiling panels. These strips help initiate the natural convection. In this case, Equation (9a) may be replaced by Equation (9b) (Kollmar and Liese 1957):

$$q_c = 0.13(t_p - t_a)^{0.25}(t_p - t_a) \tag{9b}$$

For large spaces such as aircraft hangars where panels are side by side, Equation (9b) should be adjusted with the multiplier $(16.1/D_e)^{0.25}$.

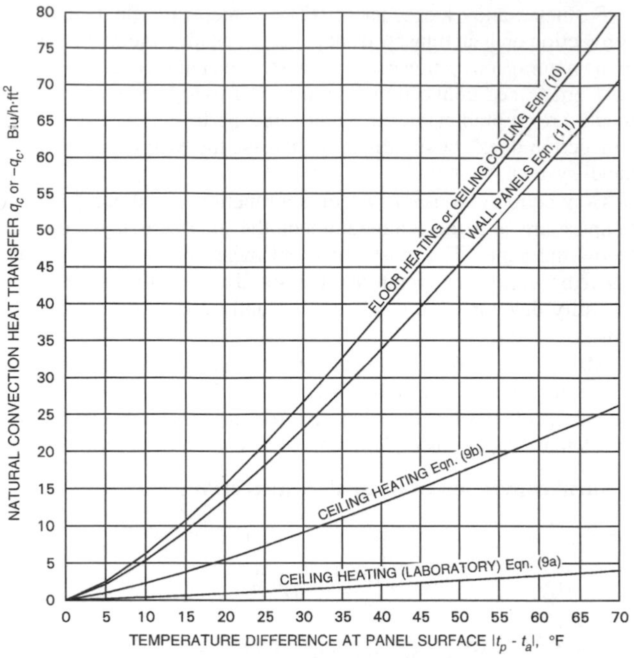

Fig. 3 Natural-Convection Heat Transfer at Floor, Ceiling, and Wall Panel Surfaces [Equations (9a), (9b), (10), and (11)]

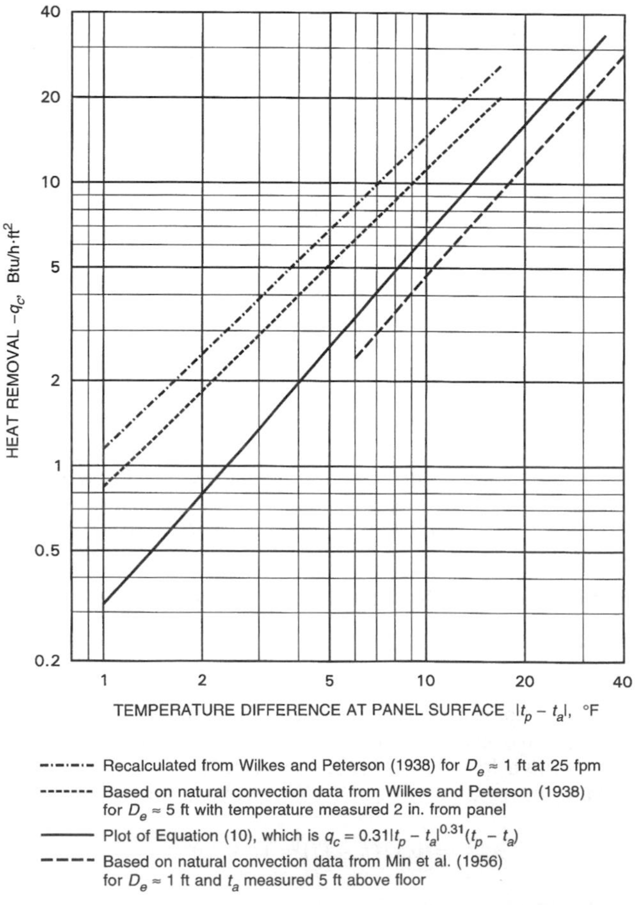

Fig. 4 Heat Removal by Ceiling Cooling Panels with Natural Convection [Equation (10)]

Natural convection from a heated floor or cooled ceiling

$$q_c = 0.31|t_p - t_a|^{0.31}(t_p - t_a) \tag{10}$$

Natural convection from a heated or cooled wall panel

$$q_c = 0.26|t_p - t_a|^{0.32}(t_p - t_a) \tag{11}$$

There are no confirmed data for floor cooling, but Equation (9b) may be used for approximate calculations. Under normal conditions t_a is the indoor air design temperature. In floor-heated or ceiling-cooled spaces with large proportions of exposed fenestration, t_a may be taken as equal to AUST.

In cooling, t_p is less than t_a, so q_c is negative. Figure 3 shows heat transfer by natural convection at floor, wall, and ceiling heating panels as calculated from Equations (10), (11), (9a), and (9b), respectively.

Figure 4 compares heat removal by natural convection at cooled ceiling panel surfaces, as calculated by Equation (10), with data from Wilkes and Peterson (1938) for specific panel sizes. An additional curve illustrates the effect of forced convection on the latter data. Similar adjustment of the data from Min et al. (1956) is inappropriate, but the effects would be much the same.

Combined Heat Transfer (Radiation and Convection)

The combined heat transfer from a panel surface can be determined by adding the radiant heat transfer q_r as calculated by Equation (5) (or from Figure 1 and Figure 2) to the convective heat transfer q_c as calculated from Equations (9a), (9b), (10), or (11) or from Figure 3 or Figure 4, as appropriate.

Equation (5) requires the AUST for the room. In calculating the AUST, the surface temperature of interior wall is assumed to be the same as the room air temperature. The inside surface temperature t_w of outside walls and exposed floors or ceilings can be calculated from the following heat transfer relationship:

$$h(t_a - t_u) = U(t_a - t_o) \tag{12}$$

or

$$t_u = t_a - \frac{U}{h}(t_a - t_o) \tag{13}$$

where

h = inside surface conductance of exposed wall or ceiling
U = overall wall heat transfer coefficient of wall, ceiling, or floor, Btu/h·ft²·°F
t_a = room air temperature, °F
t_u = inside surface temperature of outside wall, °F
t_o = outside air temperature, °F

From Table 1 in Chapter 24 of the 1997 *ASHRAE Handbook—Fundamentals:*

h = 1.63 Btu/h·ft²·°F for a horizontal surface with heat flow up
h = 1.46 Btu/h·ft²·°F for a vertical surface (wall)
h = 1.08 Btu/h·ft²·°F for a horizontal surface with heat flow down

Figure 5 is a plot of Equation (13) for a vertical outdoor wall with 70°F room air temperature and h = 1.46 Btu/h·ft²·°F. For rooms with temperatures above or below 70°F, the values in Figure 5 can be corrected by the factors plotted in Figure 6.

Tests by Schutrum et al. (1953a, 1953b) and simulations by Kalisperis (1985) based on a program developed by Kalisperis and Summers (1985) show that the AUST and room temperature are almost equal, if there is little or no outdoor exposure. Steinman et al. (1989) noted that this argument may not be appropriate for enclosures with large glass areas or a high percentage of outside wall and/or ceiling surface area. These cold surfaces have a lower AUST, which increases the radiant heat transfer.

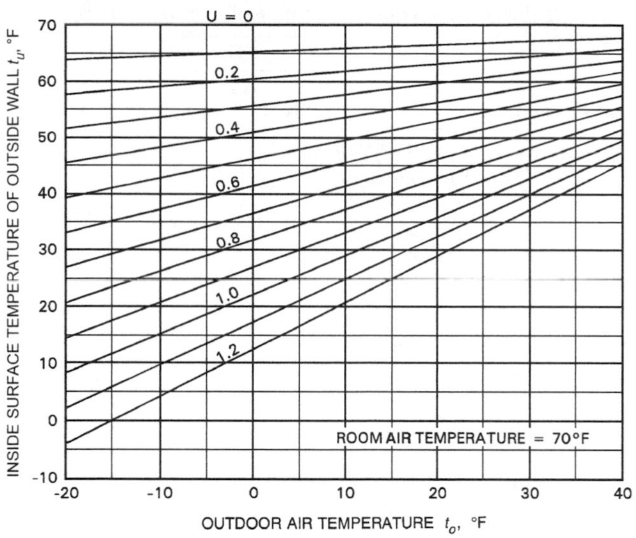

Fig. 5 Relation of Inside Surface Temperature to Overall Coefficient of Heat Transfer

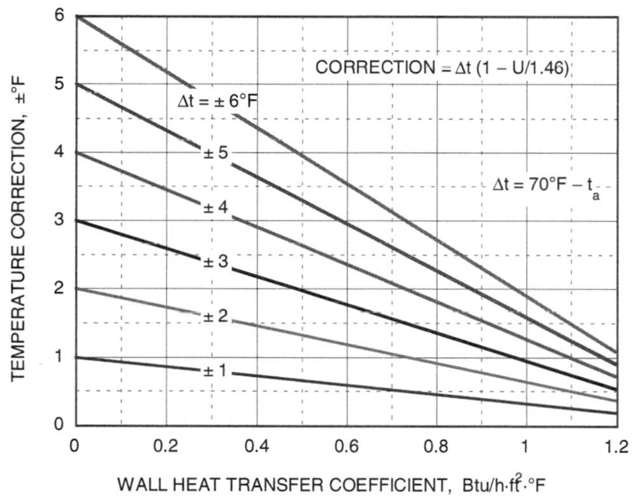

Fig. 6 Inside Surface Temperature Correction for Exposed Wall Air Temperatures Other than 70°F

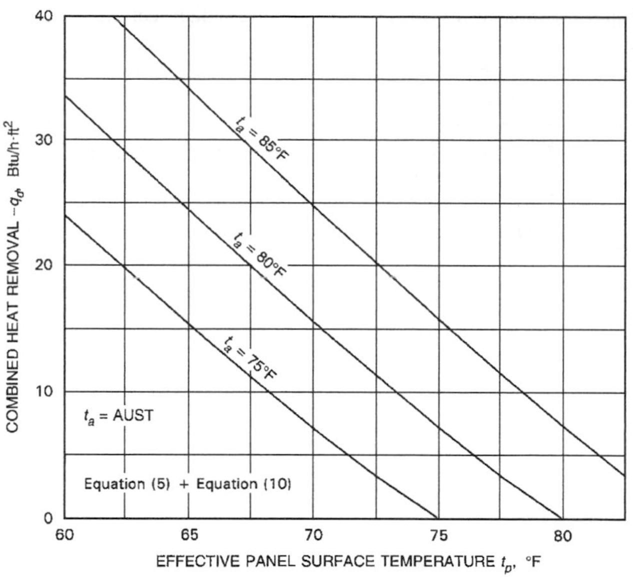

Fig. 7 Cooled Ceiling Panel Performance in Uniform Environment with No Infiltration and No Internal Heat Sources

Figure 7 shows the combined radiation and convection transfer for cooling, as given in Figure 2 and Figure 4. The data in Figure 7 do not include heat gains from sun, lights, people, or equipment; these data are available from manufacturers.

In suspended-ceiling panel systems, heat transfers from the back of the ceiling panel to the floor slab above (heating) and vice versa (cooling). The ceiling panel surface temperature is affected because of heat transfer to or from the panel and the slab by radiation and, to a much smaller extent, by convection. The radiation component can be approximated using Equation (5) or Figure 1. The convection component can be estimated from Equation (9b) or (10) or from Figure 3 or Figure 4. In this case, the temperature difference is that between the top of the ceiling panel and the midspace of the ceiling. The temperature of the ceiling space should be determined by testing, since it varies with different panel systems. However, much of this heat transfer is nullified when insulation is placed over the ceiling panel, which, for perforated metal panels, also provides acoustical control.

If lighting fixtures are recessed into the suspended ceiling space, radiation from the top of the fixtures raises the overhead slab temperature and transfers heat to the space by convection. This energy is absorbed at the top of the cooled ceiling panels by radiation, in accordance with Equation (5) or Figure 2, and by convection, generally in accordance with Equation (9b). The amount the top of the panel absorbs depends on the system. Most manufacturers have information available. Similarly, panels installed under a roof absorb additional heat, depending on configuration and insulation.

GENERAL DESIGN CONSIDERATIONS

Radiant panel systems and hybrid HVAC systems are similar to other air-water systems in the arrangement of the system components. With radiant panel systems, room thermal conditions are maintained primarily by direct transfer of radiant energy, rather than by convection heating and cooling. The room heating and cooling loads are calculated in the conventional manner. Manufacturers' ratings generally are for total performance and can be applied directly to the calculated room load. With hybrid HVAC systems, the latent load is assigned to a convective system, and a large portion of the sensible load is assigned to a radiant panel system. In a hybrid system, the room air temperature and the MRT can be controlled independently (Kilkis et al. 1995).

Because the mean radiant temperature (MRT) in a panel-heated space increases as the heating load increases, the controlled air temperature during this increase may be lowered without affecting comfort. In ordinary structures with normal infiltration loads, the required reduction in air temperature is small, enabling a conventional room thermostat to be used.

In panel heating systems, lowered night temperature can produce less satisfactory results with heavy panels such as concrete floors. These panels cannot respond to a quick increase or decrease in heating demand within the relatively short time required, resulting in a very slow reduction of the space temperature at night and a correspondingly slow pickup in the morning. Light panels, such as plaster or metal ceilings and walls, may respond to changes in demand quickly enough for satisfactory results from lowered night temperatures. Berglund et al. (1982) demonstrated the speed of response on a metal ceiling panel to be comparable to that of convection systems. However, very little fuel savings can be expected even with light panels unless the lowered temperature is maintained for long periods. If temperatures are lowered when the area is unoccupied, a means of providing a higher-than-normal rate of heat input for rapid warm-up (e.g., fast-acting radiant ceiling panels) is necessary.

Metal radiant heating panels, hydronic and electric, are applied to building perimeter spaces for heating in much the same way as finned-tube convectors. Metal panels are usually installed in the ceiling and are integrated into the ceiling design. They provide a fast response system (Watson et al. 1998).

Partitions may be erected to the face of hydronic panels but not to the active heating portion of electric panels because of possible element overheating and burnout. Electric panels are often sized to fit the building module with a small removable filler or dummy panel at the window mullion to accommodate future partitions. Hydronic panels can run continuously. Hydronic panels also may be cut and fitted in the field; however, modification should be kept to a minimum to keep installation costs down.

Panel Thermal Resistance

Any hindrance in the panel to heat transfer from or to its surface will reduce the performance of the system. Thermal resistance to the flow of heat may vary considerably among different panels, depending on the type of bond between the tubing (wiring) and the panel material. Factors such as corrosion or adhesion defects between lightly touching surfaces and the method of maintaining contact may change the bond with time. The actual thermal resistance of any proposed system should be verified by testing. Specific resistance and performance data, when available, should be obtained from the manufacturer. Panel thermal resistances include

r_t = thermal resistance of tube wall per unit tube spacing in a hydronic system, ft²·h·°F/Btu·ft
r_s = thermal resistance between tube (electric cable) and panel per unit spacing, ft²·h·°F/Btu·ft
r_p = thermal resistance of panel, ft²·h·°F/Btu
r_c = thermal resistance of panel covers, ft²·h·°F/Btu
r_u = characteristic panel thermal resistance, ft²·h·°F/Btu

For tube spacing M,

$$r_u = r_t M + r_s M + r_p + r_c \tag{14}$$

When the tubes (electric cables) are embedded in the slab, r_s may be neglected. However, if they are attached to the panel, r_s may be significant, depending on the quality of bonding. Table 1 gives typical r_s values for various ceiling panels.

The value of r_p may be calculated if the characteristic panel thickness x_p and the thermal conductivity k_p of the panel material are known.

If the tubes (electric cables) are embedded in the panel,

$$r_p = \frac{x_p - D_o/2}{k_p} \tag{15a}$$

where D_o = outside diameter of the tube (electric cable). Hydronic floor heating by a heated slab and gypsum-plaster ceiling heating are typical examples.

If the tubes are attached to the panel,

$$r_p = x_p/k_p \tag{15b}$$

Metal ceiling panels (see Table 1) and tubes under subfloor (see Figure 23) are typical examples.

Thermal resistance per unit spacing of a circular tube with an inside diameter D_i and thermal conductivity k_t is

$$r_t = \frac{\ln(D_o/D_i)}{2\pi k_t} \tag{16}$$

Table 1 Thermal Resistance of Ceiling Panels

Type of Panel	Thermal Resistance r_p, ft²·h·°F/Btu	Thermal Resistance r_s, ft²·h·°F/Btu·ft
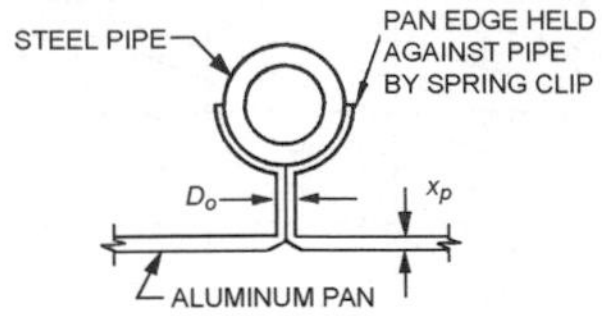STEEL PIPE PAN EDGE HELD AGAINST PIPE BY SPRING CLIP ALUMINUM PAN	$\frac{x_p}{k_p}$	0.55
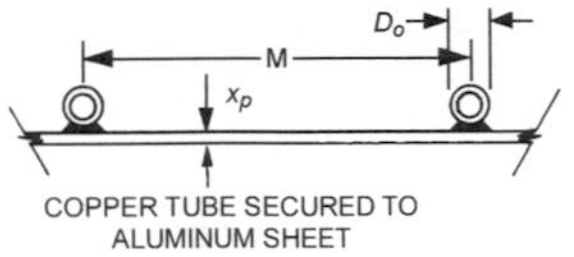COPPER TUBE SECURED TO ALUMINUM SHEET	$\frac{x_p}{k_p}$	0.22
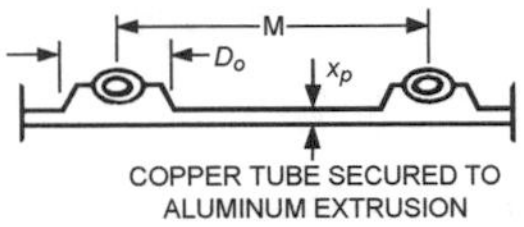COPPER TUBE SECURED TO ALUMINUM EXTRUSION	$\frac{x_p}{k_p}$	0.17
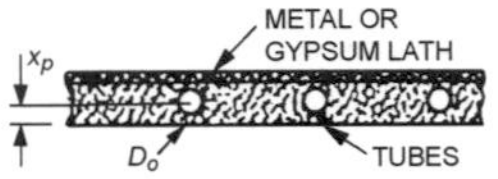METAL OR GYPSUM LATH TUBES	$\frac{x_p - D_o/2}{k_p}$	≈ 0
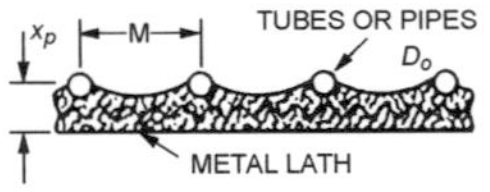TUBES OR PIPES METAL LATH	$\frac{x_p - D_o/2}{k_p}$	≤ 0.20

Table 2 Thermal Conductivity of Typical Tube Material

Material	Thermal Conductivity k_t, Btu/h·ft·°F
Carbon steel (AISI 1020)	30
Copper (drawn)	225
Red brass (85 Cu-15 Zn)	92
Stainless steel (AISI 202)	10
Low-density polyethylene (LDPE)	0.18
High-density polyethylene (HDPE)	0.24
Cross-linked polyethylene (VPE or PEX)	0.22
Textile-reinforced rubber heat transfer hose (HTRH)	0.17
Polypropylene block copolymer (PP-C)	0.13
Polypropylene random copolymer (PP-RC)	0.14
Polybutylene (PB)	0.13

In an electric cable, $r_t = 0$.

In metal pipes, r_t is virtually the fluid-side thermal resistance:

$$r_t = 1/hD_i \tag{16a}$$

Table 6 in Chapter 3 of the 1997 *ASHRAE Handbook—Fundamentals* may be used to calculate the forced-convection heat transfer coefficient h. Table 2 gives values of k_t for different tube and pipe materials.

Effect of Floor Coverings

Panel coverings like carpets and pads on the floor can have a pronounced effect on the performance of a panel system. The added thermal resistance r_c reduces the panel surface heat transfer. In order to reestablish the required performance, the temperature of the

Table 3 Thermal Resistance of Floor Coverings

Description	Thermal Resistance r_c, ft²·h·°F/Btu
Bare concrete, no covering	0
Asphalt tile	0.05
Rubber tile	0.05
Light carpet	0.60
Light carpet with rubber pad	1.00
Light carpet with light pad	1.40
Light carpet with heavy pad	1.70
Heavy carpet	0.80
Heavy carpet with rubber pad	1.20
Heavy carpet with light pad	1.60
Heavy carpet with heavy pad	1.90
3/8 in. hardwood	0.54
5/8 in. wood floor (oak)	0.57
1/2 in. oak parquet and pad	0.68
Linoleum	0.12
Marble floor and mudset	0.18
Rubber pad	0.62
Prime urethane underlayment, 3/8 in.	1.61
48 oz. waffled sponge rubber	0.78
Bonded urethane, 1/2 in.	2.09

Notes:
1. Carpet pad should be no more than 1/4 in. thick.
2. Total resistance of the carpet is more a function of thickness than of fiber type.
3. A general rule for approximating the R-value is 2.6 times the total carpet thickness in inches.
4. Before carpet is installed, it should be established that the backing is resistant to long periods of continuous heat up to 120°F.

water must be increased (decreased in cooling). Thermal resistance of a panel covering is

$$r_c = x_c / k_c \quad (17)$$

where

x_c = thickness of each panel covering, ft
k_c = thermal conductivity of each panel covering, Btu/h·ft·°F

If the panel is covered by more than one cover, individual r_c values should be added. Table 3 gives typical r_c values for floor coverings.

Where covered and bare floor panels exist in the same system, it may be possible to maintain a high enough water temperature to satisfy the covered panels and balance the system by throttling the flow to the bare slabs. In some instances, however, the increased water temperature required when carpeting is applied over floor panels makes it impossible to balance floor panel systems in which only some rooms have carpeting unless the piping is arranged to permit zoning using more than one water temperature.

Panel Heat Losses or Gains

Heat transferred from the upper surface of ceiling panels, the back surface of wall panels, the underside of floor panels, or the exposed perimeter of any panel is considered a panel heat loss. Panel heat losses are part of the building heat loss if the heat is transferred outside of the building. If the heat is transferred to another heated space, the panel loss is a source of heat for that space instead. In either case, the magnitude of panel loss should be determined.

Panel heat loss to space outside the room should be kept to a reasonable amount by insulation. For example, a floor panel may overheat the basement below, and a ceiling panel may cause the temperature of a floor surface above it to be too high for comfort

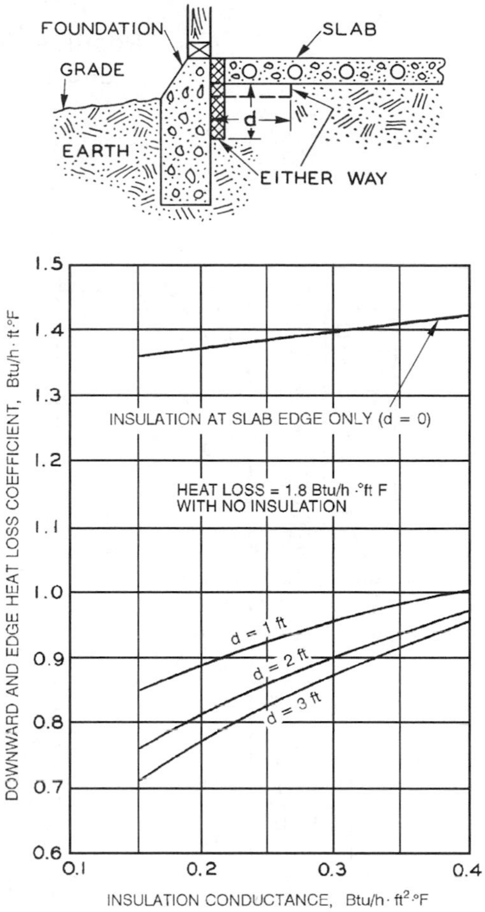

Fig. 8 Downward and Edgewise Heat Loss Coefficient for Concrete Floor Slabs on Grade

The heat loss from most panels can be calculated by using the coefficients given in Table 4 in Chapter 24 of the 1997 *ASHRAE Handbook—Fundamentals*. These coefficients should not be used to determine the downward heat loss from panels built on grade because the heat flow from them is not uniform (Sartain and Harris 1956; ASHAE 1956, 1957). The heat loss from panels built on grade can be estimated from Figure 8 or Equation (6) in Chapter 27 of the 1997 *ASHRAE Handbook—Fundamentals*.

DESIGN OF PANELS

Panel surface temperature is controlled by either hydronic or electric circuits. The required effective surface temperature t_p for a combined heat transfer q (where $q = q_r + q_c$) can be calculated by using applicable heat transfer equations for q_r and q_c depending on the position of the panel. At a given t_a, AUST must be predicted first. Figure 9 and Figure 10 can also be used to find t_p when q and AUST are known. The next step is to determine the required mean water (brine) temperature t_w in a hydronic system. It depends primarily on t_p, tube spacing M, and the characteristic panel thermal resistance r_u. Figure 9 provides design information for ceiling and cooling panels, positioned either at the ceiling or on the floor.

The combined heat transfer for ceiling and floor panels used to heat or cool rooms can be read directly from Figure 9. Here q_u is the combined heat transfer on the floor panel and q_d is the combined heat transfer on the ceiling panel. For an electric resistance heating system, t_w scales correspond to the skin temperature of the cable.

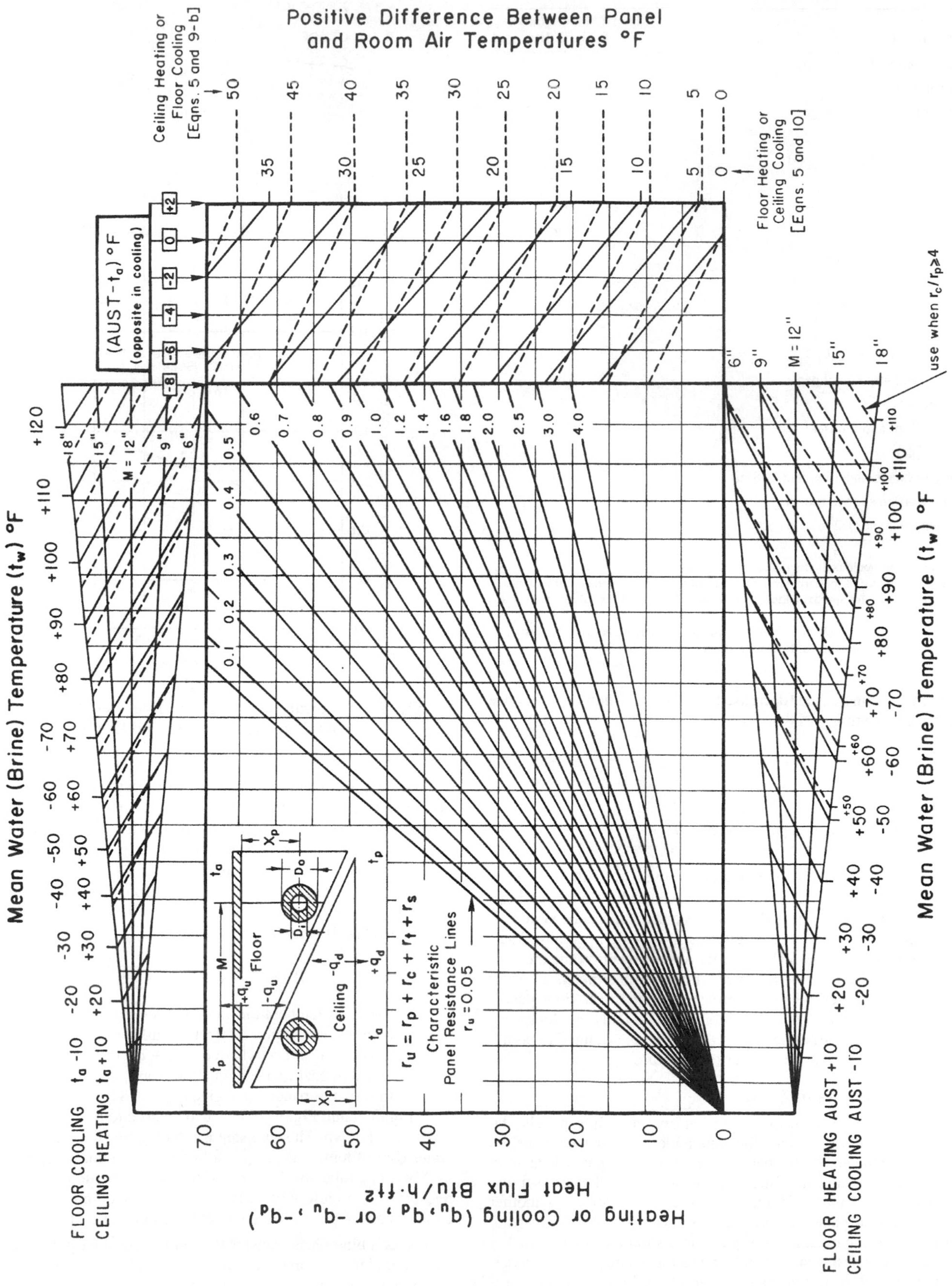

Fig. 9 Design Graph for Heating and Cooling with Floor and Ceiling Panels

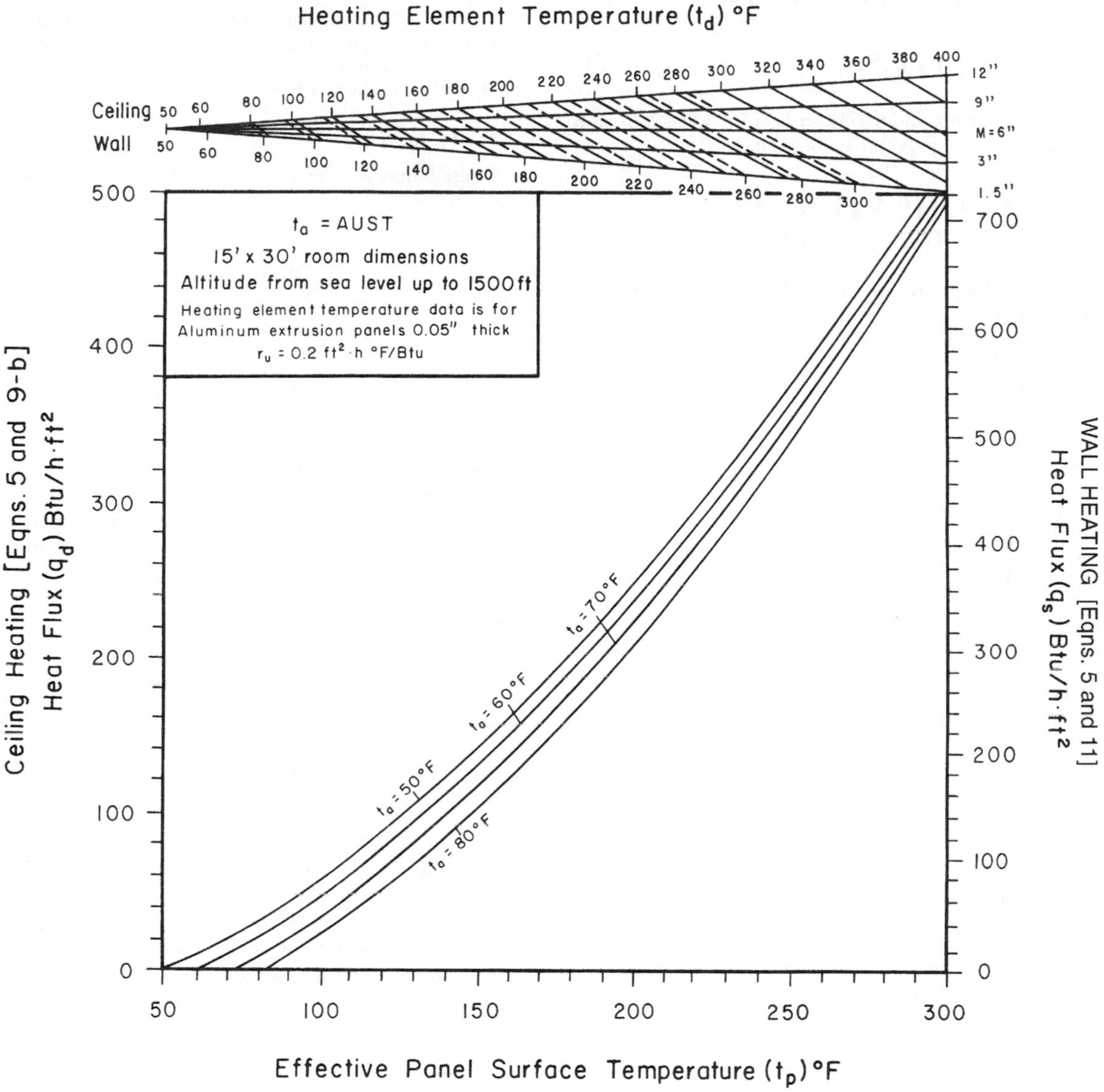

Fig. 10 Design Graph for Heating with Aluminum Ceilings and Wall Panels

The following algorithm (TSI 1994) may also be used to design and analyze panels:

$$t_d \approx t_a + \frac{(t_p - t_a)M}{2W\eta + D_o} + q(r_p + r_c + r_s M) \tag{18}$$

where

t_d = mean skin temperature of tubing (electric cable), °F
q = combined heat transfer flux on panel surface, Btu/h·ft²
t_a = design indoor air temperature, °F
In floor-heated or ceiling-cooled spaces with large exposed fenestration, t_a may be replaced with AUST.
D_o = outside diameter of tube or characteristic contact width of the tube with the panel (see Table 1), ft
M = on-center spacing of circuit, ft
$2W$ = net spacing between tubing (electric cables), $M - D_o$, ft
η = fin efficiency, dimensionless

$$\eta = \frac{\tanh(fW)}{fW} \tag{19a}$$

$$\eta \approx 1/fW \qquad \text{for } fW > 2 \tag{19b}$$

The following equation, which includes transverse heat diffusion in the panel and surface covers, may be used to calculate the fin coefficient:

$$f \approx \left[\frac{q}{m(t_p - t_a)\sum_{i=1}^{n} k_i x_i}\right]^{1/2} \qquad \text{for } t_p \neq t_a \tag{20}$$

where

m = $2 + r_c/2r_p$
n = number of layers with different materials, including panel and surface covers
x_i = characteristic thickness of each layer i, ft
k_i = thermal conductivity of each layer i, Btu/h·ft·°F

For a hydronic system, the required mean water (brine) temperature is

$$t_w = (q + q_b)Mr_t + t_d \quad (21)$$

where q_b is the flux of back and perimeter heat losses (positive) in a heated panel or gains (negative) in a cooled panel.

PANEL HEATING AND COOLING SYSTEMS

The following are the most common forms of panels applied in panel heating systems:

- Metal ceiling panels
- Embedded piping in ceilings, walls, or floors
- Electric ceiling panels
- Electrically heated ceilings or floors
- Air-heated floors

Residential heating applications usually consist of pipe coils or electric elements embedded in masonry floors or plaster ceilings. This construction is suitable where loads are stable and solar effects are minimized by building design. However, in buildings where glass areas are large and load changes occur faster, the slow response, lag, and override effect of masonry panels are unsatisfactory. Light metal panel ceiling systems respond quickly to load changes (Berglund et al. 1982).

Radiant panels are often located in the ceiling because it is exposed to all other surfaces and objects in the room. It is not likely to be covered as are the floors, and higher surface temperatures can be used. Also, its smaller mass enables it to respond more quickly to load changes. Figure 10 gives design data for radiant ceiling and wall panels up to 300°F.

Example 1. An in-slab, on-grade panel (see Figure 20) will be used for both heating and cooling. M = 1 ft (12 in.), r_u = 0.5 ft²·h·°F/Btu, and r_c/r_p is less than 4. t_a is 68°F in winter and 76°F in summer.

AUST is expected to be 2°F less than t_a in winter heating and 1°F higher than t_a in summer cooling.

What is the mean water temperature and effective floor temperature (1) for winter heating when q_u = 40 Btu/h·ft², and (2) for summer cooling when $-q_u$ = 15 Btu/h·ft²?

Solution:

Winter heating

To obtain the mean water temperature using Figure 9, start on the left axis where q_u = 40 Btu/h·ft². Proceed right to the intersect r_u = 0.5 and then down to the M = 12 in. line. The reading is AUST + 56, which is the solid line value because r_c/r_p < 4. As stated in the initial problem statement, AUST = t_a − 2 or AUST = 68 − 2 = 66°F. Therefore the mean water would be t_w = 66 + 56 = 122°F.

To obtain the effective floor temperature, start at q_u = 40 Btu/h·ft² in Figure 9 and proceed right to AUST = t_a − 2°F. The solid line establishes 21°F as the difference between the panel and the room air temperatures. Therefore, the effective floor temperature $t_p = t_a + 21$ or t_p = 68 + 21 = 89°F.

Summer cooling

Using Figure 9, start at the left axis at $-q_u$ = 15 Btu/h·ft². Proceed to r_u = 0.5, and then up (for cooling) to M = 12 in., which reads t_a − 23 or 76 − 23 = 53°F mean water temperature for cooling.

To obtain the effective floor temperature at $-q_u$ = 15 Btu/h·ft², proceed to AUST − t_a = +1°F, which reads −11°F. Therefore, the effective floor temperature is 76 − 11 = 65°F.

Example 2. An aluminum extrusion panel 0.05 in. thick with pipe spacing of M = 0.5 ft (6 in.) is used in the ceiling for winter heating. If a ceiling heat flux q_d of 400 Btu/h·ft² is required to maintain room temperature t_a at 70°F, what is the required heating element temperature t_d and effective panel surface temperature t_p?

Solution:

Using Figure 10, enter the left axis heat flux q_d at 400 Btu/h·ft². Proceed to the t_a = 70°F line and then up to the M = 6 in. line. The ceiling heating element temperature t_d is 320°F. From the bottom axis of Figure 10, the effective panel surface temperature t_p is 265°F.

Special Cases

Figure 9 may also be used for panels with tubing not embedded in the panel:

- x_p is 0 if tubes are externally attached.
- In spring-clipped external tubing, D_i is 0 and D_o is the thickness of the clip.

Warm air and electric heating elements are two design concepts used in systems influenced by local factors. The warm air system has a special cavity construction where air is supplied to a cavity behind or under the panel surface. The air leaves the cavity through a normal diffuser arrangement and is supplied to the room. Generally, these systems are used as floor radiant panels in schools and in floors subject to extreme cold, such as in an overhang. Cold outdoor temperatures and heating medium temperatures must be analyzed with regard to potential damage to the building construction. Electric heating elements embedded in the floor or ceiling construction and unitized electric ceiling panels are used in various applications to provide both full heating and spot heating of the space.

HYDRONIC PANEL SYSTEMS

Design Considerations

Hydronic radiant panels can be used with two- and four-pipe distribution systems. Figure 11 shows the arrangement of a typical system. It is common to design for a 20°F temperature drop for heating across a given grid and a 5°F rise for cooling, but larger temperature differentials may be used, if applicable.

Panel design requires determining panel area, panel type, supply water temperature, water flow rate, and panel arrangement. Panel performance is directly related to room conditions. Air-side design also must be established. Heating and cooling loads may be calculated by

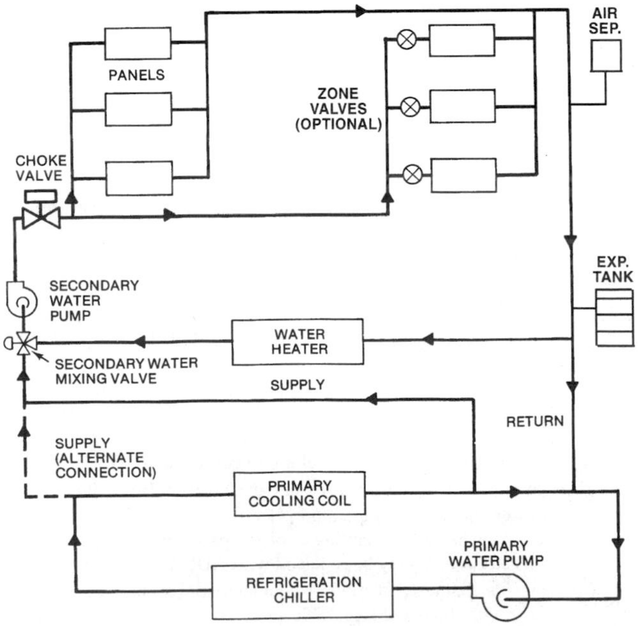

Fig. 11 Primary/Secondary Water Distribution System with Mixing Control

procedures covered in Chapters 24 through 29 of the 1997 *ASHRAE Handbook—Fundamentals*. The procedure is as follows:

A. **Cooling**

1. Determine room design dry-bulb temperature, relative humidity, and dew point.
2. Calculate room sensible and latent heat gains.
3. Select mean water temperature for cooling.
4. Establish minimum supply air quantity.
5. Calculate latent cooling available from air.
6. Calculate sensible cooling available from air.
7. Determine panel cooling load.
8. Determine panel area for cooling.

B. **Heating**

1. Designate room design dry-bulb temperature for heating.
2. Calculate room heat loss.
3. Select mean water temperature for heating.
4. Determine surface temperatures of unheated surfaces. Use Equation (13) to find surface temperatures of exterior walls and exposed floors and ceilings. Interior walls are assumed to have surface temperatures equal to the room air temperature.
5. Determine AUST of surfaces in room.
6. Determine surface temperature of heated radiant surface. Refer to Figure 9 and Figure 10 if AUST does not greatly differ from room air temperature. Use Equations (5), (9a), (9b), (10), and (11) or refer to Figure 1 and Figure 3 otherwise.
7. Determine panel area for heating. Refer to Figure 9 and Figure 10 if AUST does not vary greatly from room air temperature. Refer to manufacturers' data for panel surface temperatures higher than those given in Figure 9 and Figure 10.
8. Design the panel arrangement.

C. **Both Heating and Cooling**

1. Check thermal comfort requirements in the following steps [see Chapter 8 of the 1997 *ASHRAE Handbook—Fundamentals* and NRB (1981)].

   (a) Determine occupant's clothing insulation value and metabolic rate (see Tables 4, 7, and 8 in Chapter 8 of the 1997 *ASHRAE Handbook—Fundamentals*).

   (b) Determine the optimum operative temperature at the coldest point in the room (see Figure 15 in Chapter 8 of the 1997 *ASHRAE Handbook—Fundamentals* for other values).

   (c) Determine the MRT at the coldest point in the room [see Fanger (1972)].

   (d) From the definition of operative temperature, establish the optimum room design temperature at the coldest point in the room. If the optimum room design temperature varies greatly from the designated room design temperature, designate a new temperature.

   (e) Determine the MRT at the hottest point in the room.

   (f) Calculate the operative temperature at the hottest point in the room.

   (g) Compare the operative temperatures at the hottest and coldest points in the room. For light activity and normal clothing, the acceptable operative temperature range is 68 to 75°F [see NRB (1981) for other ranges]. If the range is not acceptable, the heating system must be modified.

   (h) Calculate radiant temperature asymmetry (NRB 1981). Acceptable ranges are less than 18°F for windows and less than 9°F for warm ceilings.

2. Determine water flow rate and pressure drop. Refer to manufacturers' guides for specific products, or use the guidelines in pages 33.1 through 33.6 of the 1997 *ASHRAE Handbook—Fundamentals.* Chapter 12 of this volume also has information on hydronic heating and cooling systems.

The supply and return manifolds need to be carefully designed. If there are circuits of unequal coil lengths, the following equations may be used (Hansen 1985; Kilkis 1998) for a circuit *i* connected to a manifold with *n* circuits:

$$Q_i = (L_{eq}/L_i)^{1/r} Q_{tot} \tag{22}$$

where

$L_{eq} = \left[\sum_{i=1}^{n} L_i^{-1/r} \right]^{-r}$, ft

Q_i = flow rate in circuit *i*, gpm

Q_{tot} = total flow rate in supply manifold, gpm

L_i = coil length of circuit *i*, ft

r = 1.75 for hydronic panels (Siegenthaler 1995)

The application, design, and installation of panel systems have certain requirements and techniques:

1. As with any hydronic system, look closely at the piping system design. Piping should be designed to ensure that water of the proper temperature and in sufficient quantity is available to every grid or coil at all times. Proper piping and system design should minimize the detrimental effects of oxygen on the system. Reverse-return systems should be considered to minimize balancing problems.
2. Individual panels can be connected for parallel flow using headers, or for sinuous or serpentine flow. To avoid flow irregularities within a header-type grid, the water channel or lateral length should be greater than the header length. If the laterals in a header grid are forced to run in a short direction, this problem can be solved by using a combination series-parallel arrangement. Serpentine flow will ensure a more even panel surface temperature throughout the heating or cooling zone.
3. Noise from entrained air, high-velocity or high-pressure-drop devices, or pump and pipe vibrations must be avoided. Water velocities should be high enough to prevent separated air from accumulating and causing air binding. Where possible, avoid automatic air venting devices over ceilings of occupied spaces.
4. Design piping systems to accept thermal expansion adequately. Do not allow forces from piping expansion to be transmitted to panels. Thermal expansion of the ceiling panels must be considered.
5. In circulating water systems, plastic, rubber, steel, and copper pipe or tube are used widely in ceiling, wall, or floor panel construction. Where coils are embedded in concrete or plaster, no threaded joints should be used for either pipe coils or mains. Steel pipe should be the all-welded type. Copper tubing should be soft-drawn coils. Fittings and connections should be minimized. Changes in direction should be made by bending. Solder-joint fittings for copper tube should be used with a medium-temperature solder of 95% tin, 5% antimony, or capillary brazing alloys. All piping should be subjected to a hydrostatic test of at least three times the working pressure. Maintain adequate pressure in embedded piping while pouring concrete.
6. Placing the thermostat on a side wall where it can see the outside wall and the warm panel should be considered. The normal thermostat cover reacts to the warm panel, and the radiant effect of the panel on the cover tends to alter the control point so that the thermostat controls 2 to 3°F lower when the outdoor temperature is a minimum and the panel temperature is a maximum. Experience indicates that radiantly heated rooms are more comfortable under these conditions than when the thermostat is located on a back wall.
7. If throttling valve control is used, either the end of the main should have a fixed bypass, or the last one or two rooms on the mains should have a bypass valve to maintain water flow in the

main. Thus, when a throttling valve modulates, there will be a rapid response.

8. When selecting heating design temperatures for a ceiling panel surface, the design parameters are as follows:

   (a) Excessively high temperatures over the occupied zone cause the occupant to experience a "hot head effect."
   (b) Temperatures that are too low can result in an oversized, uneconomical panel and a feeling of coolness at the outside wall.
   (c) Locate ceiling panels adjacent to perimeter walls and/or areas of maximum load.
   (d) With normal ceiling heights of 8 to 9 ft, panels less than 3 ft wide at the outside wall can be designed for 235°F surface temperature. If panels extend beyond 3 ft into the room, the panel surface temperature should be limited to the values as given in Figure 16. The surface temperature of concrete or plaster panels is limited by construction.

9. Floor panels are limited to surface temperatures of less than 84°F in occupied spaces for comfort reasons. Subfloor temperature may be limited to the maximum exposure temperature specified by the floor covering manufacturer.
10. When the panel chilled water system is started, the circulating water temperature should be maintained at room temperature until the air system is completely balanced, the dehumidification equipment is operating properly, and building humidity is at design value.
11. When the panel area for cooling is greater than the area required for heating, a two-panel arrangement (Figure 12) can be used. Panel HC (heating and cooling) is supplied with hot or chilled water year-round. When chilled water is used, the controls function activate panel CO (cooling only), and both panels are used for cooling.
12. To prevent condensation on the room side of cooling panels, the panel water supply temperature should be maintained at least 1°F above the room design dew-point temperature. This minimum difference is recommended to allow for the normal drift of temperature controls for the water and air systems, and also to provide a factor of safety for temporary increase in space humidity.
13. Selection of summer design room dew point below 50°F generally is not economical.
14. The most frequently applied method of dehumidification uses cooling coils. If the main cooling coil is six rows or more, the dew point of the leaving air will approach the temperature of the leaving water. The cooling water leaving the dehumidifier can then be used for the panel water circuit.
15. Several chemical dehumidification methods are available to control latent and sensible loads separately. In one application, cooling tower water is used to remove heat from the chemical drying process, and additional sensible cooling is necessary to cool the dehumidified air to the required system supply air temperature.
16. When chemical dehumidification is used, hygroscopic chemical-type dew-point controllers are required at the central apparatus and at various zones to monitor dehumidification.
17. When cooled ceiling panels are used with a variable air volume (VAV) system, the air supply rate should be near maximum volume to assure adequate dehumidification before the cooling ceiling panels are activated.

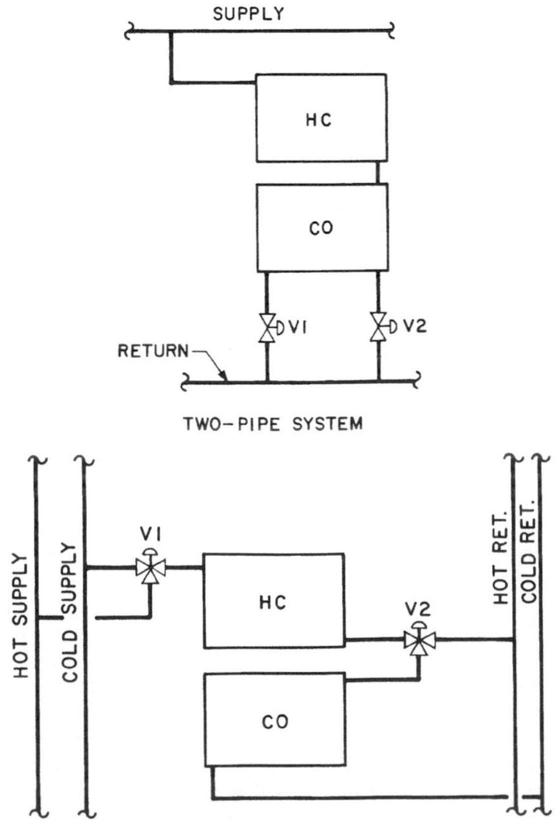

Fig. 12 Split Panel Piping Arrangement for Two-Pipe and Four-Pipe Systems

Other factors to consider when using panel systems are

1. Evaluate the panel system to take full advantage in optimizing the physical building design.
2. Select recessed lighting fixtures, air diffusers, hung ceilings, and other ceiling devices to provide the maximum ceiling area possible for use as radiant panels.
3. The air-side design must be able to maintain humidity levels at or below design conditions at all times to eliminate any possibility of condensation on the panels. This becomes more critical if space dry- and wet-bulb temperatures are allowed to drift as an energy conservation measure, or if duty cycling of the fans is used.
4. Do not place cooling panels in or adjacent to high-humidity areas.
5. Anticipate thermal expansion of the ceiling and other devices in or adjacent to the ceiling.
6. Design operable windows to discourage unauthorized opening.

HYDRONIC METAL CEILING PANELS

Metal ceiling panels can be integrated into a system that heats and cools. In such a system, a source of dehumidified ventilation air is required in summer, so the system is classed as an air-and-water system. Also, various amounts of forced air are supplied year-round. When metal panels are applied for heating only, a ventilation system may be required, depending on local codes.

Ceiling panel systems are an outgrowth of the perforated metal, suspended acoustical ceilings. These radiant ceiling systems are usually designed into buildings where the suspended acoustical ceiling can be combined with panel heating and cooling. The panels can be designed as small units to fit the building module, which provides extensive flexibility for zoning and control; or the panels can be arranged as large continuous areas for maximum economy. Some ceiling installations require active panels to cover only a portion of the room and compatible matching acoustical panels for the remaining ceiling area.

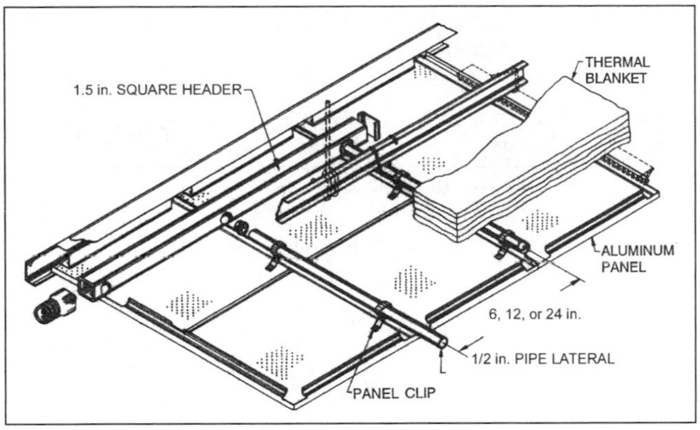

Fig. 13 Metal Ceiling Panels Attached to Pipe Laterals

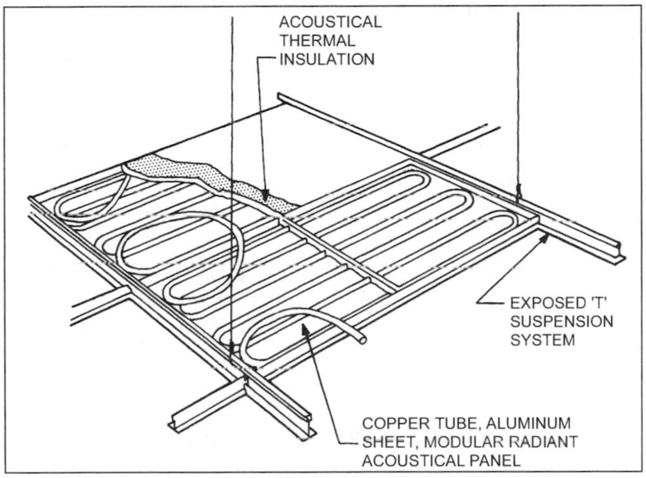

Fig. 14 Metal Ceiling Panels Bonded to Copper Tubing

Three types of metal ceiling systems are available. The first consists of light aluminum panels, usually 12 in. by 24 in., attached in the field to 0.5 in. galvanized pipe coils. Figure 13 illustrates a metal ceiling panel system that uses 0.5 in. pipe laterals on 6, 12, or 24 in. centers, hydraulically connected in a sinuous or parallel flow welded system. Aluminum ceiling panels are clipped to these pipe laterals and act as a heating panel when warm water is flowing or as a cooling panel when chilled water is flowing.

The second type of panel consists of a copper coil secured to the aluminum face sheet to form a modular panel. Modular panels are available in sizes up to about 36 in. by 60 in. and are held in position by various types of ceiling suspension systems, most typically a standard suspended T-bar 24 in. by 48 in. exposed grid system. Figure 14 illustrates metal panels using a copper tube pressed into an aluminum extrusion, although other methods of securing the copper tube have proven equally effective.

Metal ceiling panels can be perforated so that the ceiling becomes sound absorbent when acoustical material is installed on the back of the panels. The acoustical blanket is also required for thermal reasons, so that the reverse loss or upward flow of heat from the metal ceiling panels is minimized.

The third type of panel is an aluminum extrusion face sheet with a copper tube mechanically fastened into a channel housing on the back. Extruded panels can be manufactured in almost any shape and size. Extruded aluminum panels are often used as long, narrow panels at the outside wall and are independent of the ceiling system. Panels 15 or 20 in. wide usually satisfy the heating requirements of a typical office building. Lengths up to 20 ft are available. Figure 15 illustrates metal panels using a copper tube pressed into an aluminum extrusion.

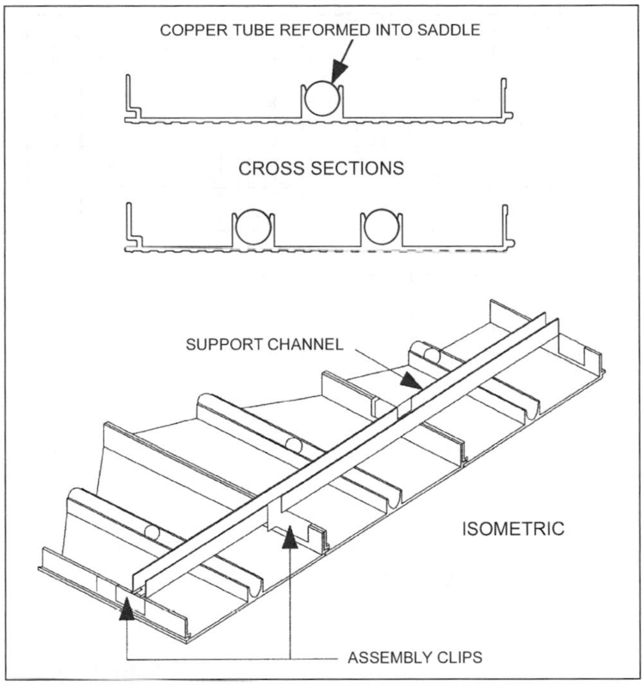

Fig. 15 Extruded Aluminum Panels with Integral Copper Tube

Performance data for extruded aluminum panels vary with the copper tube/aluminum contact and test procedures used. Hydronic ceiling panels have a low thermal resistance and respond quickly to changes in space conditions. Table 1 shows thermal resistance values for various ceiling constructions.

Metal radiant ceiling panels can be used with any of the all-air cooling systems described in Chapter 2. Chapters 25 through 28 of the 1997 *ASHRAE Handbook—Fundamentals* describe how to calculate heating loads. Double glazing and heavy insulation in outside walls have reduced transmission heat losses. As a result, infiltration and reheat have become of greater concern. Additional design considerations are as follows:

1. Perimeter radiant heating panels not extending more than 3 ft into the room may operate at higher temperatures, as described under item 8d in the section on Hydronic Panel Systems.
2. Hydronic panels operate efficiently at low temperature and are suitable for condenser water heat reclaim systems.
3. Locate ceiling panels adjacent to the outside wall and as close as possible to the areas of maximum load. The panel area within 3 ft of the outside wall should have a heating capacity equal to or greater than 50% of the wall transmission load.
4. Ceiling system designs based on passing return air through perforated modular panels into the plenum space above the ceiling are not recommended because much of the panel heat transfer is lost to the return air system.
5. When selecting heating design temperatures for a ceiling panel surface or mean water temperature, the design parameters are as follows:

   (a) Excessively high temperatures over the occupied zone will cause the occupant to experience a "hot head effect."

   (b) Temperatures that are too low can result in an oversized, uneconomical panel and a feeling of coolness at the outside wall.

   (c) Give the technique in item 3 priority.

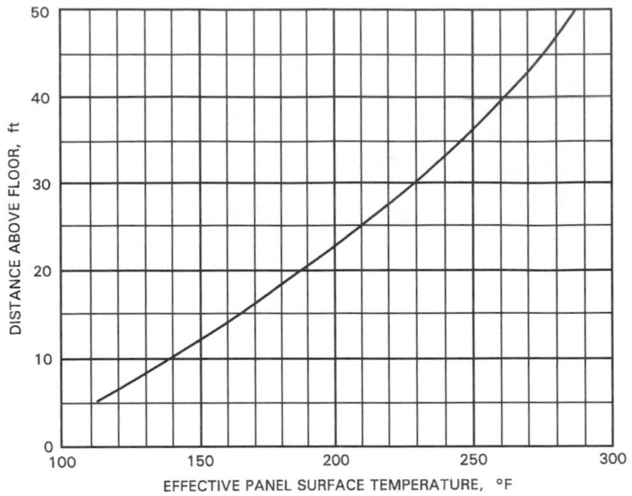

Fig. 16 Permitted Design Ceiling Surface Temperatures at Various Ceiling Heights

(d) With normal ceiling heights of 8 to 9 ft, panels less than 3 ft wide at the outside wall can be designed for 235°F surface temperature. If panels extend beyond 3 ft into the room, the panel surface temperature should be limited to the values as given in Figure 16.

6. Allow sufficient space above the ceiling for installation and connection of the piping that forms the radiant panel ceiling.

Metal radiant acoustic panels provide heating, cooling, sound absorption, insulation, and unrestricted access to the plenum space. They are easily maintained, can be repainted to look new, and have a life expectancy in excess of 30 years. The system is quiet, comfortable, draft-free, and easy to control, and it responds quickly. The system is a basic air-and-water system. First costs are competitive with other systems, and a life-cycle cost analysis often shows that the long life of the equipment makes it the least expensive in the long run. The system has been used in hospitals, schools, office buildings, colleges, airports, and exposition facilities.

Metal radiant panels can also be integrated into the ceiling design to provide a narrow band of radiant heating around the perimeter of the building. The radiant system offers advantages over baseboard or overhead air in appearance, comfort, operating efficiency and cost, maintenance, and product life.

DISTRIBUTION AND LAYOUT

Chapter 3 and Chapter 12 apply to radiant panels. Layout and design of metal radiant ceiling panels for heating and cooling begin early in the job. The type of ceiling chosen influences the radiant design, and conversely, thermal considerations may dictate what ceiling type to use. Heating panels should be located adjacent to the outside wall. Cooling panels may be positioned to suit other elements in the ceiling. In applications with normal ceiling heights, heating panels that exceed 160°F should not be located over the occupied area. In hospital applications, valves should be located in the corridor outside patient rooms.

One of the following types of construction is generally used:

1. Pipe or tube is embedded in the lower portion of a concrete slab, generally within 1 in. of its lower surface. If plaster is to be applied to the concrete, the piping may be placed directly on the wood forms. If the slab is to be used without plaster finish, the piping should be installed not less than 0.75 in. above the undersurface of the slab. Figure 17 shows this method of construction.

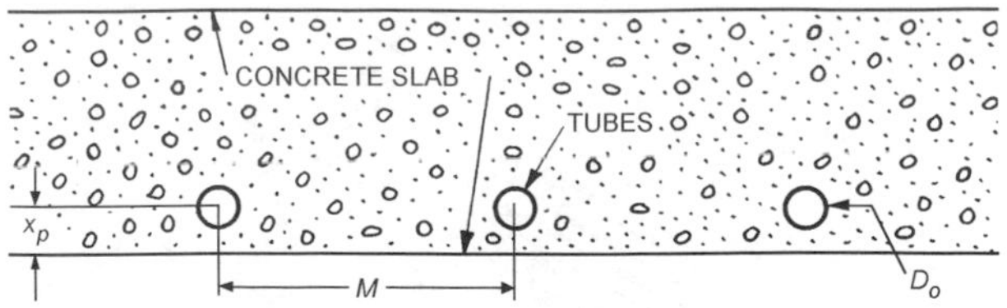

Fig. 17 Coils in Structural Concrete Slab

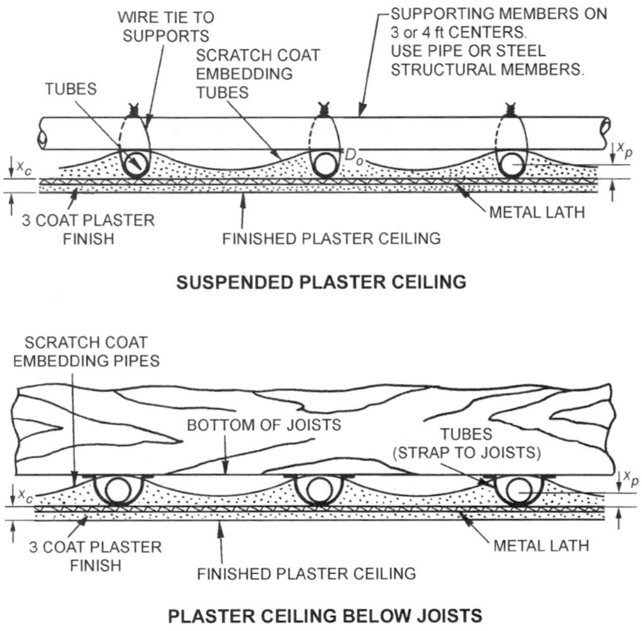

Fig. 18 Coils in Plaster above Lath

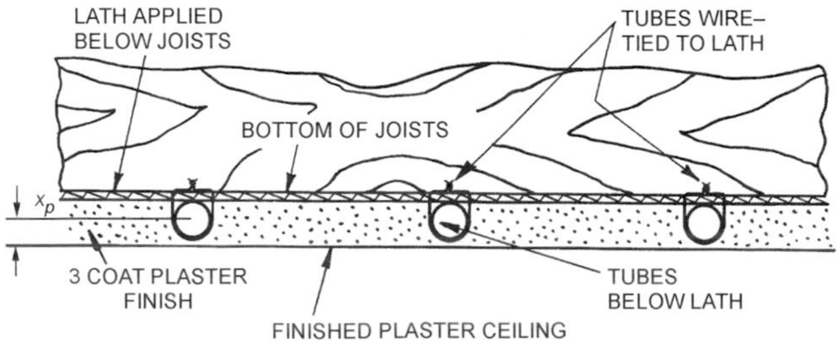

Fig. 19 Coils in Plaster below Lath

The minimum coverage must comply with local building code requirements.

2. Pipe or tube is embedded in a metal lath and plaster ceiling. If the lath is suspended to form a hung ceiling, the lath and heating coils are securely wired to the supporting members so that the lath is below, but in good contact with, the coils. Plaster is then applied to the metal lath, carefully embedding the coil as shown in Figure 18.
3. Smaller diameter copper or plastic tube is attached to the underside of wire lath or gypsum lath. Plaster is then applied to the lath to embed the tube, as shown in Figure 19.
4. Other forms of ceiling construction are composition board, wood paneling, etc., with warm water piping, tube, or channels built into the panel sections.

Coils are usually the sinuous type, although some header or grid-type coils have been used in ceilings. Coils may be plastic, ferrous, or nonferrous pipe or tube, with coil pipes spaced from 4.5 to 9 in. on centers, depending on the required output, pipe or tube size, and other factors.

Where plastering is applied to pipe coils, a standard three-coat gypsum plastering specification is followed, with a minimum of 3/8 in. of cover below the tubes when they are installed below the lath. Generally, the surface temperature of plaster panels should not exceed 120°F. This can be accomplished by limiting the water temperature in the pipes or tubes in contact with the plaster to a maximum temperature of 140°F. Insulation should be placed above the coils to reduce back loss, the difference between heat supplied to the coil and net useful output to the heated room.

To protect the plaster installation and to ensure proper air drying, heat must not be applied to the panels for two weeks after all plastering work has been completed. When the system is started for the first time, the water supplied to the panels should not be higher than 20°F above the prevailing room temperature at that time and not in excess of 90°F. Water should be circulated at this temperature for about two days, then increased at a rate of about 5°F per day to 140°F.

During the air-drying and preliminary warm-up periods, there should be adequate ventilation to carry moisture from the panels. No paint or paper should be applied to the panels before these periods have been completed or while the panels are being operated. After paint and paper have been applied, an additional shorter warm-up period, similar to first-time starting, is also recommended.

Hydronic Wall Panels

Although piping embedded in walls is not as widely used as floor and ceiling panels, it can be constructed by any of the methods outlined for ceilings or floors. Its design is similar to other hydronic panels [see Equations (18) to (21)]. Heat transfer at the surface of wall panels is given by Equations (5) and (11).

Hydronic Floor Panels

Interest has increased in radiant floor heating with the introduction of nonmetallic tubing and new design, application, and control techniques. Whichever method is used for optimum floor output and comfort, it is important that the heat be evenly distributed over the floor. Spacing is generally 4 to 12 in on centers for the coils. Wide spacing under tile or bare floors can cause uneven surface temperatures.

Embedded Piping in Concrete Slab. Plastic, rubber, ferrous, and nonferrous pipe and tube are used in floor slabs that rest on grade. The coils are constructed as sinuous-continuous pipe coils or arranged as header coils with the pipes spaced from 6 to 18 in. on centers. The coils are generally installed with 1.5 to 4 in. of cover above them. Insulation is recommended to reduce the perimeter and back losses. Figure 20 shows the application of pipe coils in slabs resting on grade. Coils should be embedded completely and should not rest on an interface. Any supports used for positioning the heating coils should be nonabsorbent and inorganic. Reinforcing steel, angle iron, pieces of pipe or stone, or concrete mounds can be used. No wood, brick, concrete block, or similar materials should support coils. A waterproofing layer is desirable to protect insulation and piping.

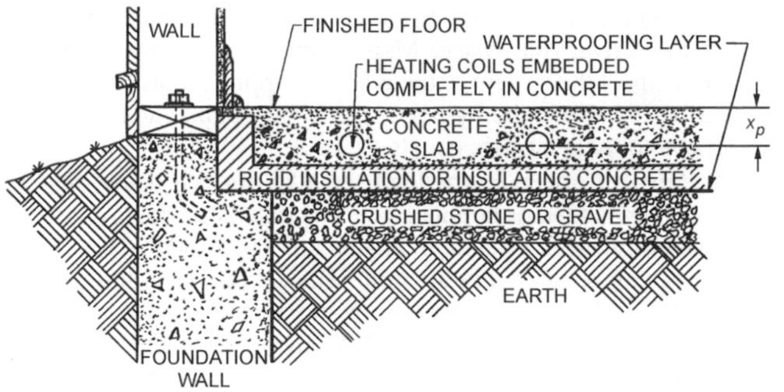

Fig. 20 Coils in Floor Slab on Grade

Where coils are embedded in structural load-supporting slabs above grade, construction codes may affect their position. Otherwise, the coil piping is installed as described for slabs resting on grade.

The warm-up and start-up period for concrete panels are similar to those outlined for plaster panels.

Embedded systems may fail sometime during their life. Adequate valves and properly labeled drawings will help isolate the point of failure.

Suspended Floor Piping. Piping may be applied on or under suspended wood floors using several methods of construction. Piping may be attached to the surface of the floor and embedded in a layer of concrete or gypsum, mounted in or below the subfloor, or attached directly to the underside of the subfloor using metal panels to improve heat transfer from the piping. An alternate method is to install insulation with a reflective surface and leave an air gap of 2 to 4 in. to the subfloor. Whichever method is used for optimum floor output and comfort, it is important that the heat be evenly distributed throughout the floor. Pipes are generally spaced 4 to 12 in. apart. Wide spacing under tile or bare floors can cause uneven surface temperatures.

Figure 21 illustrates construction with piping embedded in concrete or gypsum. The thickness of the embedding material is generally 1 to 2 in. when applied to a wood subfloor. Gypsum products specifically designed for floor heating can generally be installed 1 to 1.5 in. thick because they are more flexible and crack-resistant than concrete. When concrete is used, it should be of structural quality to reduce cracking due to movement of the wood frame or shrinkage. The embedding material must provide a hard, flat, smooth surface that can accommodate a variety of floor coverings.

As illustrated in Figure 22, tubing may also be installed in the subfloor. The tubing is installed on top of the rafters between the subflooring members. Heat diffusion and surface temperature can be improved uniformly by the addition of metal heat transfer plates,

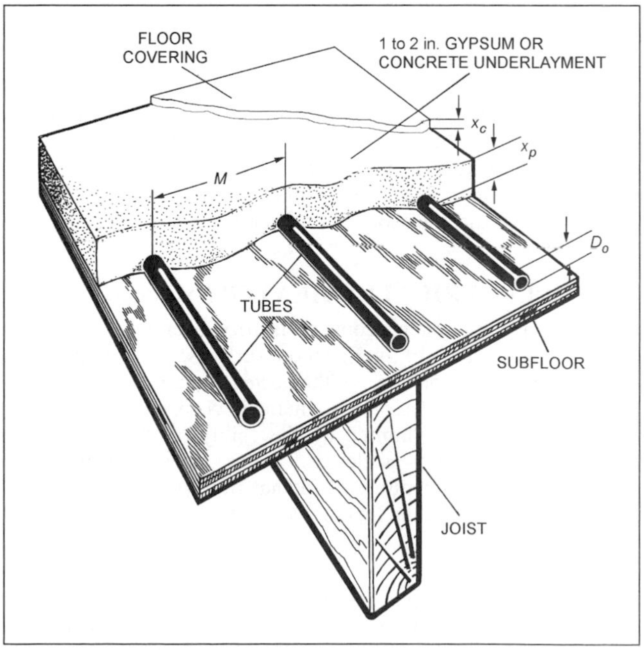

Fig. 21 Embedded Tube in Thin Slab

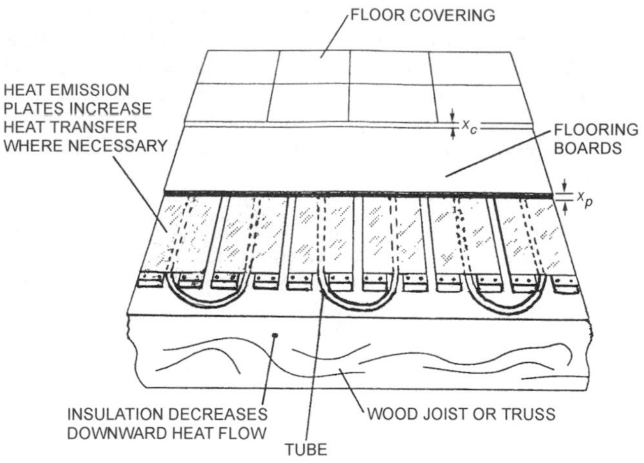

Fig. 22 Tube in Subfloor

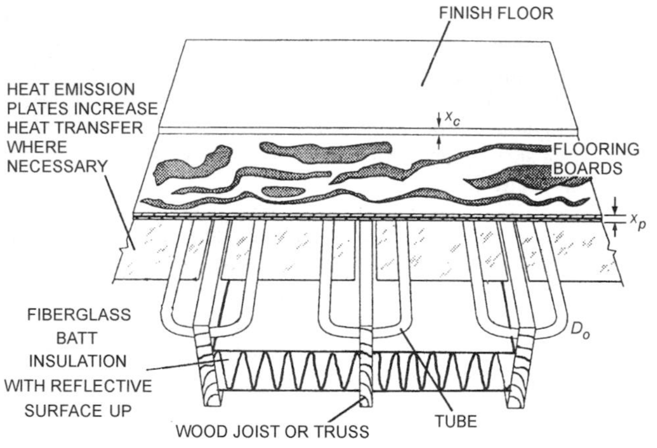

Fig. 23 Tube under Subfloor

which spread the heat beneath the finished flooring. This construction is illustrated in Figure 22.

A third construction option is to attach the tube to the underside of the subfloor with or without metal heat transfer plates. The construction is illustrated in Figure 23.

Transfer from the hot water tube to the surface of the floor is the important consideration in all cases. The floor surface temperature affects the actual heat transfer to the space. Any hindrance between the heated water tube and the floor surface reduces the effectiveness of the system. The method that transfers and spreads heat evenly through the subfloor with the least resistance produces the best results.

ELECTRICALLY HEATED SYSTEMS

Several heating systems convert electrical energy to heat, which raises the temperature of interior room surfaces. These systems are classified by the temperature of the heated system. Higher temperature surfaces require less area to maintain occupant comfort. Surface temperatures are limited by the ability of the materials to maintain their integrity at elevated temperatures. The maximum effective surface temperature of radiant floor panels is limited to what is comfortable to the feet of occupants.

Ceiling Systems

Prefabricated Electric Ceiling Panels. These panels are available in sizes 1 to 6 ft wide by 2 to 12 ft long by 0.5 to 2 in. thick. They are constructed with metal, glass, or semirigid fiberglass board or vinyl. Heated surface temperatures range from 100 to 300°F, with

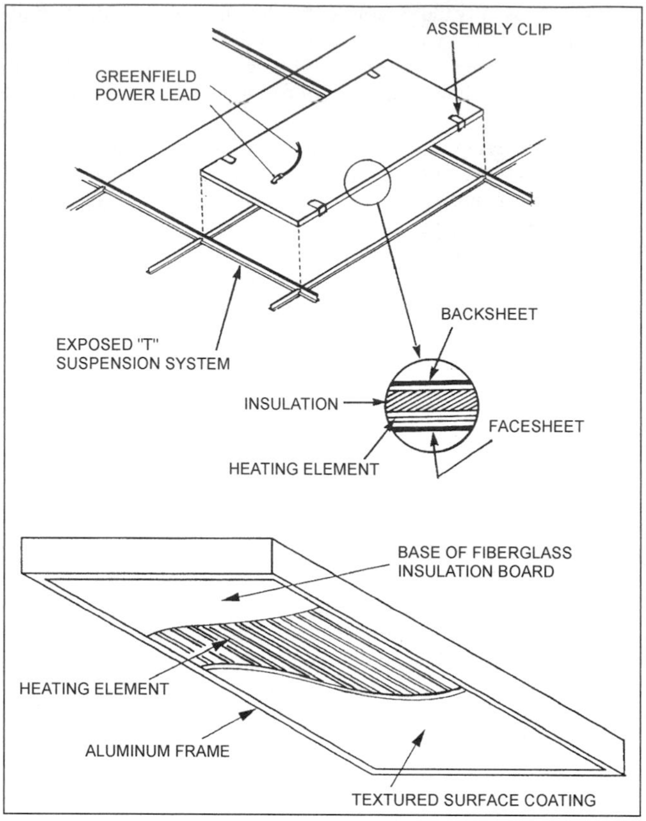

Fig. 24 Electric Heating Panels

corresponding power densities ranging from 25 to 100 W/ft^2 for 120 to 480 V services.

A panel of gypsum board embedded with insulated resistance wire is also available. It is installed as part of the ceiling or between joists in contact with a ceiling. Power density is limited to 22 W/ft^2 to maintain the integrity of the board by keeping the heated surface temperature below 100°F. Nonheating leads are furnished as part of the panel.

Some panels can be cut to fit; others must be installed as received. Panels may be either flush or surface mounted, and in some cases, they are finished as part of the ceiling. Rigid 2 ft by 4 ft panels for lay-in ceilings (Figure 24) are about 1 in. thick and weigh from 6 to 25 lb. Typical characteristics of an electrical radiant panel are listed in Table 4. Panels may also be (1) surface mounted on gypsum board and wood ceilings or (2) recessed between ceiling joists. Panels range in size from 4 ft wide to 8 ft long. Their maximum power output is 95 W/ft^2.

Electric Ceiling Systems. These systems are laminated conductive coatings, printed circuits, or etched elements nailed to the bottom of ceiling joists and covered by 1/2 in. gypsum board. Power density is limited to 18 W/ft^2. In some cases, the heating element can be cut to fit available space. Manufacturers' instructions specify how to connect the system to the electric supply. Appropriate codes should be followed when placing partitions, lights, and air grilles adjacent to or near electric panels.

Electrical Cables Embedded in Ceilings. Electric heating cables for embedded or laminated ceiling panels are factory-assembled units furnished in standard lengths of 75 to 1800 ft. These cable lengths cannot be altered in the field. The cable assemblies are normally rated at 2.75 W per linear foot and are supplied in capacities from 200 to 5000 W in roughly 200 W increments. Standard cable assemblies are available for 120, 208, and 240 V. Each cable unit is supplied with 7 ft nonheating leads for connection at the thermostat or junction box.

Table 4 Characteristics of Typical Electric Panel Heater

Resistor material	Graphite or nichrome wire
Relative heat intensity	Low, 50 to 125 W/ft^2
Resistor temperature	180 to 350°F
Envelope temperature (in use)	160 to 300°F
Radiation-generating ratio[a]	0.7 to 0.8
Response time (heat-up)	240 to 600 s
Luminosity (visible light)	None
Thermal shock resistance	Excellent
Vibration resistance	Excellent
Impact resistance	Excellent
Resistance to drafts or wind[b]	Poor
Mounting position	Any
Envelope material	Steel alloy or aluminum
Color blindness	Very good
Flexibility	Good—wide range of power density, length, and voltage practical
Life expectancy	Over 10,000 h

[a]Ratio of radiant output to power input (elements only).
[b]May be shielded from wind effects by louvers, deep-drawn fixtures, or both.

Electric cables for panel heating have electrically insulated coverings resistant to medium temperature, water absorption, aging effects, and chemical action with plaster, cement, or ceiling lath material. This insulation is normally a polyvinyl chloride (PVC) covering, which may have a nylon jacket. The outside diameter of the insulation covering is usually about 0.12 in.

For plastered ceiling panels, the heating cable may be stapled to gypsum board, plaster lath, or similar fire-resistant materials with rust-resistant staples (Figure 25). With metal lath or other conducting surfaces, a coat of plaster (brown or scratch coat) is applied to completely cover the metal lath or conducting surface before the cable is attached. After the lath is fastened on and the first plaster coat is applied, each cable is tested for continuity of circuit and for insulation resistance of at least 100 kΩ measured to ground.

The entire ceiling surface is finished with a covering of thermally noninsulating sand plaster about 0.50 to 0.75 in. thick or other approved noninsulating material applied according to manufacturer's specifications. The plaster is applied parallel to the heating cable rather than across the runs. While new plaster is drying, the system should not be energized, and the range and rate of temperature change should be kept low by other heat sources or by ventilation until the plaster is thoroughly cured. Vermiculite or other insulating plaster causes cables to overheat and is contrary to code provisions.

For laminated drywall ceiling panels, the heating cable is placed between two layers of gypsum board, plasterboard, or other thermally noninsulating fire-resistant ceiling lath. The cable is stapled directly to the first (or upper) lath, and the two layers are held apart by the thickness of the heating cable. It is essential that the space between the two layers of lath be completely filled with a noninsulating plaster or similar material. This fill holds the cable firmly in place and improves heat transfer between the cable and the finished ceiling. Failure to fill the space between the two layers of plasterboard completely may allow the cable to overheat in the resulting voids and may cause cable failure. The plaster fill should be applied according to manufacturer's specifications.

Electric heating cables are ordinarily installed with a 6 in. nonheating border around the periphery of the ceiling. An 8 in. clearance must be provided between heating cables and the edges of the outlet or junction boxes used for surface-mounted lighting fixtures. A 2 in. clearance must be provided from recessed lighting fixtures, trim, and ventilating or other openings in the ceiling.

Heating cables or panels must be installed only in ceiling areas that are not covered by partitions, cabinets, or other obstructions. However, it is permissible for a single run of isolated embedded cable to pass over a partition.

The *National Electrical Code* (NFPA *Standard* 70) requires that all general power and light wiring be run above the thermal insulation or at least 2 in. above the heated ceiling surface, or that the wiring be derated.

In drywall ceiling construction, the heating cable is always installed with the cable runs parallel to the joist. A 2.5 in. clearance between adjacent cable runs must be left centered under each joist for nailing. Cable runs that cross over the joist must be kept to a minimum. Where possible, these crossings should be in a straight line at one end of the room.

For cable having a power density of 2.75 W/ft, the minimum permissible spacing is 1.5 in. between adjacent runs. Some manufacturers recommend a minimum spacing of 2 in. for drywall construction.

The spacing between adjacent runs of heating cable can be determined using the following equation:

$$M = 12A_n/C \tag{23}$$

where

M = cable spacing, in.
A_n = net panel heated area, ft^2
C = length of cable, ft

Net panel area A_n in Equation (23) is the net ceiling area available after deducting the area covered by the nonheating border, lighting fixtures, cabinets, and other ceiling obstructions. For simplicity, Equation (23) contains a slight safety factor, and small lighting fixtures are usually ignored in determining net ceiling area.

Resistance of the electric cable must be adjusted according to its temperature at design conditions (Ritter and Kilkis 1998):

$$R' = R\frac{[1 + \alpha_e(t_d - 68)]}{[1 + \alpha_o(t_d - 68)]} \tag{24}$$

where

R = electrical resistance of electric cable at standard temperature (68°F), Ω/ft
α_e = thermal coefficient for material resistivity, °F^{-1}
α_o = thermal expansion coefficient, °F^{-1}
t_d = surface temperature of electric cable at operating conditions [see Equation (18)], °F

The 2.5 in. clearance required under each joist for nailing in drywall applications occupies one-fourth of the ceiling area if the joists are 16 in. on centers. Therefore, for drywall construction, the net area A_n must be multiplied by 0.75. Many installations have a spacing of 1.5 in. for the first 2 ft from the cold wall. Remaining cable is then spread over the balance of the ceiling.

Electrically Heated Wall Panels

Cable embedded in walls similar to ceiling construction is used in Europe. Because of possible damage from nails driven for hanging pictures or from building alterations, most codes in the United States prohibit such panels. Some of the prefabricated panels described in the preceding section are also used for wall panel heating.

Electrically Heated Floors

Electric heating cable assemblies such as those used for ceiling panels are sometimes used for concrete floor heating systems. Because the possibility of cable damage during installation is greater for concrete floor slabs than for ceiling panels, these assemblies must be carefully installed. After the cable has been placed, all unnecessary traffic should be eliminated until the concrete covering has been placed and hardened.

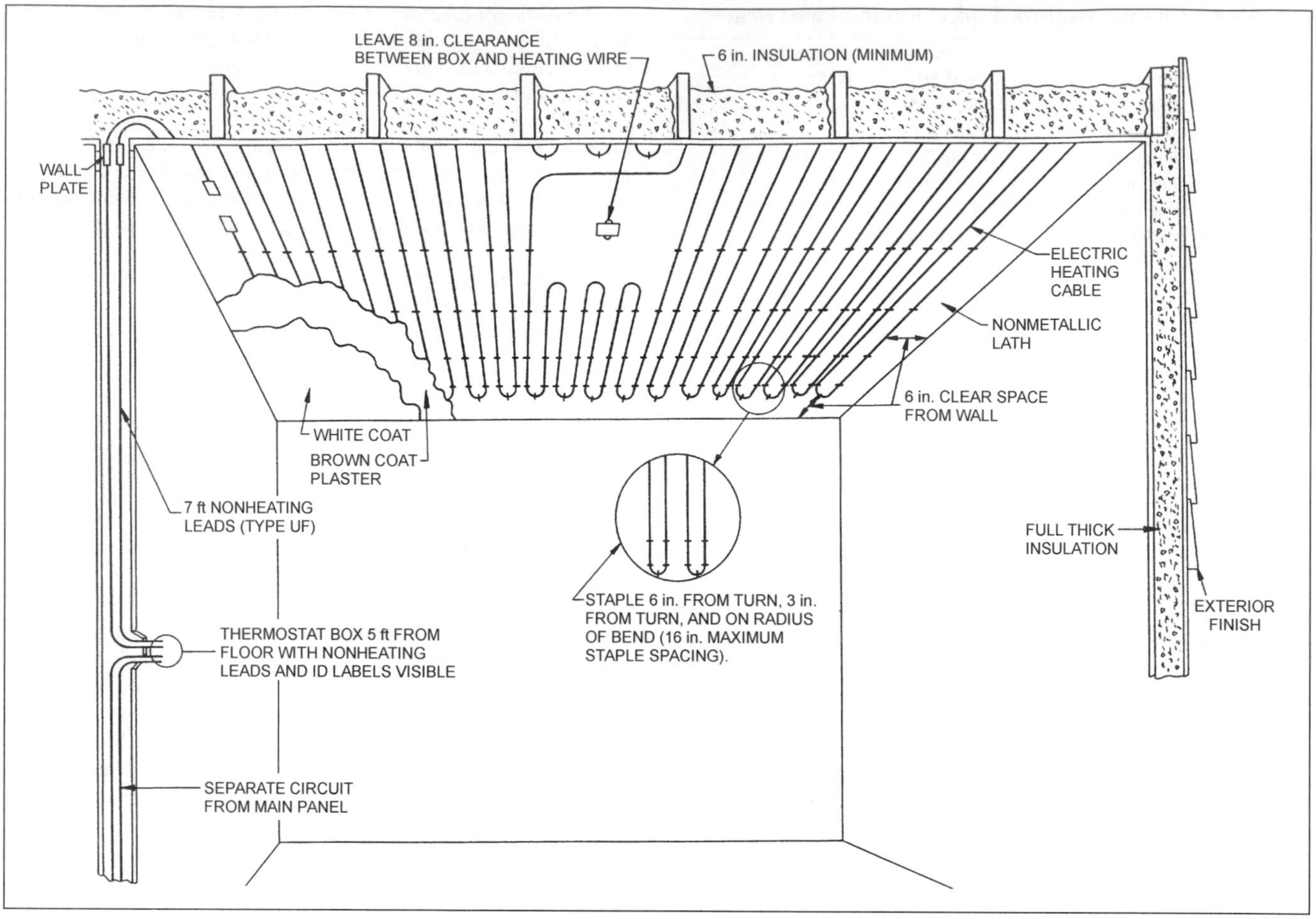

Fig. 25 Electric Heating Panel for Wet Plaster Ceiling

Preformed mats are sometimes used for electric floor slab heating systems. These mats usually consist of PVC-insulated heating cable woven into or attached to metallic or glass fiber mesh. Such mats are available as prefabricated assemblies in many sizes from 2 to 100 ft^2 and with power densities ranging from 15 to 25 W/ft^2. When mats are used with a thermally treated cavity beneath the floor, a heat storage system is provided, which may be controlled for off-peak heating.

Mineral-insulated (MI) heating cable is another effective method of slab heating. MI cable is a small-diameter, highly durable, flexible heating cable composed of solid electric-resistance heating wire or wires surrounded by tightly compressed magnesium oxide electrical insulation and enclosed by a metal sheath. MI cable is available in stock assemblies in a variety of standard voltages, power densities, and lengths. A cable assembly consists of the specified length of heating cable, waterproof hot-cold junctions, 7 ft cold sections, UL-approved end fittings, and connection leads. Several standard MI cable constructions are available, such as single conductor, twin conductor, and double cable. Custom-designed MI heating cable assemblies can be ordered for specific installations.

Other outer-covering materials that are sometimes specified for electric floor heating cable include (1) silicone rubber, (2) lead, and (3) tetrafluoroethylene (Teflon).

For a given floor heating cable assembly, the required cable spacing is determined from Equation (23). In general, cable power density and spacing should be such that floor panel power density is not greater than 15 W/ft^2. Check the latest edition of the *National Electrical Code* (NFPA *Standard* 70) and other applicable codes to obtain information on maximum panel power density and other required criteria and parameters.

Floor Heating Cable Installation. When PVC-jacketed electric heating cable is used for floor heating, the concrete slab is laid in two pourings. The first pour should be at least 3 in. thick and, where practical, should be insulating concrete to reduce downward heat loss. For a proper bond between the layers, the finish slab should be placed within 24 h of the first pour, with a bonding grout applied. The finish layer should be at least 1.5 in. and no more than 2 in. thick. This top layer must not be insulating concrete. At least 1 in. of perimeter insulation should be installed as shown in Figure 26.

The cable is installed on top of the first pour of concrete no closer than 2 in. from adjoining walls and partitions. Methods of fastening the cable to the concrete include the following:

1. Staple the cable to wood nailing strips fixed in the surface of the rough slab. The predetermined cable spacing is maintained by daubs of cement, plaster of paris, or tape.
2. In light or uncured concrete, staple the cable directly to the slab using hand-operated or powered stapling machines.
3. Nail special anchor devices to the first slab to hold the cable in position while the top layer is being poured.

Preformed mats can be embedded in the concrete in a continuous pour. The mats are positioned in the area between expansion and/or construction joints and electrically connected to a junction box. The slab is poured to within 1.5 to 2 in. of the finished level. The surface is rough-screeded and the mats placed in position. The final cap is

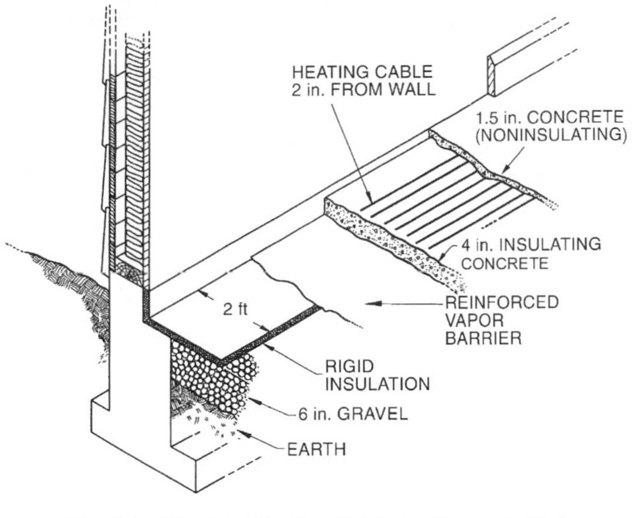

Fig. 26 Electric Heating Cable in Concrete Slab

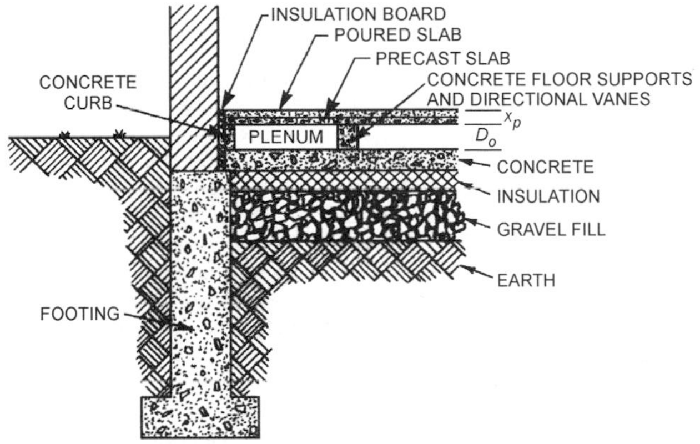

Fig. 27 Warm Air Floor Panel Construction

applied immediately. Because the first pour has not set, there is no adhesion problem between the first and second pours, and a monolithic slab results. A variety of contours can be developed by using heater wire attached to glass fiber mats. Allow for circumvention of obstructions in the slab.

MI electric heating cable can be installed in concrete slab using either one or two pours. For single-pour applications, the cable is fastened to the top of the reinforcing steel before the pour is started. For two-layer applications, the cable is laid on top of the bottom structural slab and embedded in the finish layer. Proper spacing between adjacent cable runs is maintained by using prepunched copper spacer strips nailed to the lower slab.

AIR-HEATED OR AIR-COOLED PANELS

Several methods have been devised to warm interior room surfaces by circulating heated air through passages in the floor. In some cases, the heated air is recirculated in a closed system. In others, all or a part of the air is passed through the room on its way back to the furnace to provide supplementary heating and ventilation. Figure 27 indicates one common type of construction. Compliance with applicable building codes is important.

In principle, the heat transfer equations for the panel surface and the design algorithm explained in the section on Design of Panels apply. In these systems, however, the fluid (air) moving in the duct has a virtually continuous contact with the panel. Therefore, $\eta \approx 1$, $D_o = 0$, and $M = 1$. Equation (18) gives the required surface temperature t_d of the plenum. The design of the air side can be carried out by following the principles given in Chapters 25 and 32 of the 1997 *ASHRAE Handbook—Fundamentals*.

CONTROLS

Automatic controls for panel heating may differ from those for convective heating because of the thermal inertial characteristics of the panel and the increase in the mean radiant temperature within the space under increasing loads. However, low-mass systems using thin metal panels or thin underlay with low thermal heat capacity may be successfully controlled with conventional control technology using indoor sensors. Many of the control principles for hot water heating systems described in Chapter 12 and Chapter 14 also apply to panel heating. Because radiant panels do not depend on air-side equipment to distribute energy, many control methods have been used successfully; however, a control interface between heating and cooling should be installed to prevent simultaneous heating and cooling.

High-mass panels such as concrete radiant slabs require a control approach different from that for low-mass panels. Because of thermal inertia, significant time is required to bring such massive panels from one operating point to another, say from vacation setback to standard operating conditions. This will result in long periods of discomfort from low temperature, then possibly periods of uncomfortable and wasteful overshoot. Careful economic analysis may reveal that a nighttime setback strategy is not warranted.

Once a slab is at operating conditions, the control strategy should be to supply the slab with heat at the rate that heat is being lost from the space (MacCluer et al. 1989). For hydronic slabs with constant circulator flow rate, this means modulating the temperature difference between the outgoing water and the returning water; this is accomplished via mixing valves, via fuel modulation, or, for constant thermal power sources, via pulse-width modulation (on-off control). Slabs with embedded electric resistance cable can be controlled by pulse-width modulators such as the common round thermostat with anticipator or its solid-state equivalent.

A related approach, outdoor reset control, has enjoyed wide acceptance. An outdoor reset control measures the outdoor air temperature, calculates the supply water temperature required for steady operation, and operates a mixing valve or boiler to achieve that supply water temperature. If the heating load of the controlled space is primarily a function of the outdoor air temperature, or indoor temperature measurement of the controlled space is impractical, then outdoor reset control alone is an acceptable control strategy. When other factors such as solar or internal gains are also significant, indoor temperature feedback should be added to the outdoor reset.

In all radiant panel applications, precautions must be taken to prevent excessive temperatures. A manual boiler bypass or other means of reducing the water temperature may be necessary to prevent new panels from drying out too rapidly.

Cooling Controls

The panel water circuit temperature is typically controlled by mixing, by heat exchange, or by using the water leaving the dehumidifier. Other considerations are listed in the section on General Design Considerations. It is imperative to dry out the building space before starting the panel water system, particularly after extended down periods such as weekends. Such delayed starting action can be controlled manually or by a device.

Panel cooling systems require the following basic areas of temperature control: (1) exterior zones, (2) areas under exposed roofs, to compensate for transmission and solar loads, and (3) each typical interior zone, to compensate for internal loads. For optimum results, each exterior corner zone and similarly loaded face zone should be treated as a separate subzone. Panel cooling systems may also be zoned to control temperature in individual exterior offices, particularly in applications where there is a high lighting load or for corner rooms with large glass areas on both walls.

The temperature control of the interior air and panel water supply should not be a function of the outdoor weather. The normal thermostat drift is usually adequate compensation for the slightly lower temperatures desirable during winter weather. This drift should be limited to result in a room temperature change of no more than 1.5°F. Control of the interior zones is best accomplished by devices that reflect the actual presence of the internal load elements. Frequently, time clocks and current-sensing devices are used on lighting feeders.

Because air quantities are generally small, constant volume supply air systems should be used. With the apparatus arranged to supply air at an appropriate dew point at all times, comfortable indoor conditions can be maintained throughout the year with a panel cooling system. As with all systems, to prevent condensation on window surfaces, the supply air dew point should be reduced during extremely cold weather according to the type of glazing installed.

Electric Heating Slab Controls

For comfort heating applications, the surface temperature of a floor slab is held to a maximum of 80 to 84°F. Therefore, when the slab is the primary heating system, thermostatic controls sensing air temperature should not be used to control temperature; instead, the heating system should be wired in series with a slab-sensing thermostat. The remote sensing thermostat in the slab acts as a limit switch to control maximum surface temperatures allowed on the slab. The ambient sensing thermostat controls the comfort level. For supplementary slab heating, as in kindergarten floors, a remote sensing thermostat in the slab is commonly used to tune in the desired comfort level. Indoor-outdoor thermostats are used to vary the floor temperature inversely with the outdoor temperature. If the heat loss of the building is calculated for an outdoor temperature of 70 to 0°F, and the floor temperature range is held from 70 to 84°F with a remote sensing thermostat, the ratio of the change in outdoor temperature to the change in slab temperature is 70:14, or 5:1. This means that a 5°F drop in outdoor temperature requires a 1°F increase in the slab temperature. An ambient sensing thermostat is used to vary the ratio between outdoor and slab temperatures. A time clock is used to control each heating zone if off-peak slab heating is desirable.

REFERENCES

ASHAE. 1956. Thermal design of warm water ceiling panels. *ASHAE Transactions* 62:71.

ASHAE. 1957. Thermal design of warm water concrete floor panels. *ASHAE Transactions* 63:239.

ASHRAE. 1976. *Energy Calculations* 1—Procedures for determining heating and cooling loads for computerized energy calculations.

ASHRAE. 1992. Thermal environmental conditions for human occupancy. ANSI/ASHRAE *Standard* 55-1992.

Berglund, L., R. Rascati, and M.L. Markel. 1982. Radiant heating and control for comfort during transient conditions. *ASHRAE Transactions* (88): 765-75.

Buckley, N.A. 1989. Application of radiant heating saves energy. *ASHRAE Journal* 31(9):17-26.

Fanger, P.O. 1972. Thermal comfort analysis and application in environmental engineering. McGraw-Hill, New York.

Hansen, E.G. 1985. *Hydronic system design and operation*. McGraw-Hill, New York.

Kalisperis, L.N. 1985. Design patterns for mean radiant temperature prediction. Department of Architectural Engineers, Pennsylvania State University, University Park, PA.

Kalisperis, L.N. and L.H. Summers. 1985. MRT33GRAPH—A CAD program for the design evaluation of thermal comfort conditions. Tenth National Passive Solar Conference, Raleigh, NC.

Kilkis, ì.B. 1998. Equipment oversizing issues with hydronic heating systems. *ASHRAE Journal* 40 (1):25-31.

Kilkis, ì.B., A.S.R. Suntur, and M. Sapci. 1995. Hybrid HVAC systems. *ASHRAE Journal* 37(12):23-28.

Kollmar, A. and W. Liese. 1957. *Die Strahlungsheizung*, 4th ed. R. Oldenburg, Munich.

MacCluer, C.R., M. Miklavcic, and Y. Chait. 1989. The temperature stability of a radiant slab-on-grade. *ASHRAE Transactions* 95(1):1001-09.

Min, T.C., L.F. Schutrum, G.V. Parmelee, and J.D. Vouris. 1956. Natural convection and radiation in a panel heated room. *ASHAE Transactions* 62:337.

NFPA. 1999. Installation of sprinkler systems. *Standard* 13-99. National Fire Protection Association, Quincy, MA.

NFPA. 1999. *National Electrical Code*. *Standard* 70-99.

NRB. 1981. Indoor climate. *Technical Report* No. 41. The Nordic Committee on Building Regulations, Stockholm, Sweden.

Parmelee, G.V. and R.G. Huebscher. 1947. Forced convection, heat transfer from flat surfaces. *ASHVE Transactions* 53:245.

Ritter, T.L. and ì.B. Kilkis. 1998. An analytical model for the design of in-slab electric heating panels. *ASHRAE Transactions* 104(1B):1112-15.

Sartain, E.L. and W.S. Harris. 1956. Performance of covered hot water floor panels, Part I—Thermal characteristics. *ASHAE Transactions* 62:55.

Schutrum, L.F. and C.M. Humphreys. 1954. Effects of non-uniformity and furnishings on panel heating performance. *ASHVE Transactions* 60:121.

Schutrum, L.F. and J.D. Vouris. 1954. Effects of room size and non-uniformity of panel temperature on panel performance. *ASHVE Transactions* 60:455.

Schutrum, L.F., G.V. Parmelee, and C.M. Humphreys. 1953a. Heat exchangers in a ceiling panel heated room. *ASHVE Transactions* 59:197.

Schutrum, L.F., G.V. Parmelee, and C.M. Humphreys. 1953b. Heat exchangers in a floor panel heated room. *ASHVE Transactions* 59:495.

Siegenthaler, J. 1995. *Modern hydronic heating*. Delmar Publishers, Boston.

Steinman, M., L.N. Kalisperis, and L.H. Summers. 1989. The MRT-correction method—An improved method for radiant heat exchange. *ASHRAE Transactions* 95(1):1015-27.

TSI. 1994. Fundamentals of design for floor heating systems (in Turkish). Turkish *Standard* 11261. Turkish Standards Institute, Ankara.

Walton, G.N. 1980. A new algorithm for radiant interchange in room loads calculations. *ASHRAE Transactions* 86(2):190-208.

Watson, R.D., K.S. Chapman, and J. DeGreef. 1998. Case study: Seven-system analysis of thermal comfort and energy use for a fast-acting radiant heating system. *ASHRAE Transactions* 104(1B):1106-11.

Wilkes, G.B. and C.M.F. Peterson. 1938. Radiation and convection from surfaces in various positions. *ASHVE Transactions* 44:513.

BIBLIOGRAPHY

Buckley, N.A. and T.P. Seel. 1987. Engineering principles support an adjustment factor when sizing gas-fired low-intensity infrared equipment. *ASHRAE Transactions* 93(1):1179-91

Chapman, K.S. and P. Zhang. 1995. Radiant heat exchange calculations in radiantly heated and cooled enclosures. *ASHRAE Transactions* 101(2):1236-47.

Hogan, R.E., Jr., and B. Blackwell. 1986. Comparison of numerical model with ASHRAE designed procedure for warm-water concrete floor-heating panels. *ASHRAE Transactions* 92(1B):589-601.

Jones, B.W. and K.S. Chapman. 1994. Simplified method to factor mean radiant temperature (MRT) into building and HVAC system design. *Final Report* of ASHRAE Research Project 657.

Kilkis, ì.B. 1993. Computer-aided design and analysis of radiant floor heating systems. *Paper* No. 80. Proceedings of Clima 2000, London (Nov. 1-3).

Kilkis, ì.B. 1993. Radiant ceiling cooling with solar energy: Fundamentals, modeling, and a case design. *ASHRAE Transactions* 99(2):521-33.

Kilkis, ì.B., S.S. Sager, and M. Uludag. 1994. A simplified model for radiant heating and cooling panels. *Simulation Practice and Theory Journal* 2:61-76.

For additional references on radiant heating, see the section on Bibliography in Chapter 15.

CHAPTER 7

COGENERATION SYSTEMS AND ENGINE AND TURBINE DRIVES

COGENERATION is the simultaneous production of electrical or mechanical energy (power) and useful thermal energy from a single energy stream, such as oil, coal, natural or liquefied gas, biomass, or solar. By capturing and applying heat from an effluent energy stream that would otherwise be rejected to the environment, cogeneration systems can operate at efficiencies greater than those achieved when heat and power are produced in separate or distinct processes. Recovering this thermal energy for a useful purpose from reciprocating engines or steam or combustion turbines can take the following forms:

- Direct heating (e.g., a drying process)
- Indirect heating to generate a flow of steam or hot water for remote heating devices or to generate shaft power in a steam turbine
- Extracting the latent heat of condensation from a recovered flow of steam when the load served permits condensation (e.g., a steam-to-water exchanger) instead of rejecting the latent heat to a cooling tower (e.g., a full condensing turbine with a cooling tower)

Cogeneration can be a topping, bottoming, or combined cycle. In a **topping cycle**, the fuel generates power first, and the resulting thermal energy is recovered and productively used. In a **bottoming cycle**, the power is generated last from the thermal energy left over after the higher level thermal energy has been used to satisfy thermal loads. A typical topping cycle recovers heat from the operation of a prime mover and uses this thermal energy for the process (cooling or heating). A bottoming cycle recovers heat from the process to generate power. A **combined cycle** uses the thermal output from a prime mover to generate additional shaft power (e.g., combustion turbine exhaust generates steam for a steam turbine generator).

When a prime mover generates power, and the thermal output is used productively, it is a cogeneration cycle; if the thermal energy is wasted, it is not. The discussion in this chapter about the prime mover and its driven device applies to both cases, the only difference being that one uses the available thermal energy (cogeneration) while the other does not.

Isolated cogeneration systems, the electrical output of which is used on site to satisfy all site power requirements, are referred to as **total energy systems**. A cogeneration system that is actively tied (paralleled) to the utility grid can, on a contractual basis or on a tariff basis, exchange power with the public utility. This lessens the need for redundant on-site generating capacity and allows operation at maximum thermal efficiency when satisfying the facility's thermal load requires more electric power than the facility needs.

System feasibility and design depend on the magnitude, duration, and coincidence of electrical and thermal loads, as well as on the selection of the prime mover and waste heat recovery system. Integrating the design of the project's electrical and thermal requirements with the cogeneration plant is required for optimum economic benefit. The basic components of the cogeneration plant are (1) prime mover and its fuel supply system, (2) generator, (3) waste heat recovery system, (4) control system, (5) electrical and thermal transmission and distribution systems, and (6) connections to building mechanical and electrical services.

The prime mover converts fuel or thermal energy to shaft energy. The conversion devices normally used are reciprocating internal combustion engines, combustion turbines, expansion turbines, and steam boiler-turbine combinations.

This chapter describes prime movers for a variety of uses, including generators and compressors for refrigerants and other gases. Heat recovery, electrical power and control systems, design concepts, plant auxiliaries, installation, feasibility, and utilization systems are also discussed. Thermal distribution systems are covered in Chapter 11 and Chapter 12.

The preparation of this chapter is assigned to TC 9.5, Cogeneration Systems.

RECIPROCATING ENGINES

Reciprocating engines are the most common prime mover used in smaller (i.e., under 15 MW) cogeneration plants. These engines are available in sizes up to 27,000 brake horsepower (bhp) and use all types of liquid and gaseous fuels, including methane from landfills or sewage treatment plant digesters. Internal combustion engines that use the **diesel (compression ignition) cycle** can be fueled by a wide range of petroleum products (up to No. 6 oil), although No. 2 diesel oil is the most commonly used. Diesel cycle engines can also be fired with gaseous fuel in combination with liquid fuel. In such **dual-fuel engines**, the liquid acts as the ignition agent and is called pilot oil.

Spark ignition Otto cycle engines are produced in sizes up to 18,000 bhp and use natural gas, liquefied petroleum gas (LPG), and other gaseous and volatile liquid fuels. Engines usually operate in the range of 360 to 1200 rpm, with some up to 1800 rpm. The specific operating speed selected depends on the size and brand of machine, the generator, and the desired length of time between complete engine overhauls. Engines are usually selected to provide a minimum of 15,000 to 30,000 operating hours between minor overhauls, and up to 50,000 operating hours between major overhauls; the higher operating hours correspond to the larger, lower speed engines.

The engine components include the starter system; fuel input systems; fuel-air mixing cycle; ignition system; combustion chamber; exhaust gas collection and removal system; lubrication systems; and power transmission gear with pistons, connecting rods, crankshafts, and flywheel. Some engines also increase power output using turbochargers (engine exhaust, turbine-driven compressors) that increase the mass flow of air delivered to the combustion chamber. The engines are generally 20 to 40% efficient in converting fuel to shaft energy.

The **four-stroke** diesel cycle engine, also known as a compression ignition (CI) engine, involves the following four piston strokes for each power stroke:

Intake stroke—Piston travels from top dead center to bottom dead center and takes fresh air into the cylinder. The intake valve is open during this stroke.

Compression stroke—Piston travels from the cylinder bottom to the top with all valves closed. As the air is compressed, its temperature increases. Shortly before the end of the stroke, a measured quantity of diesel fuel is injected into the cylinder. Combustion of the fuel begins just before the piston reaches top dead center.

Power stroke—Burning gases exert pressure on the piston, pushing it to bottom dead center. All valves are closed until shortly before the end of the stroke, when the exhaust valves are opened.

Exhaust stroke—Piston returns to top dead center, venting products of combustion from the cylinder through the exhaust valves.

A four-stroke Otto cycle engine, also known as a spark ignition (SI) engine, operates through the same cycle with two variations. First, a fuel-air mixture rather than pure air flows into the cylinder during the intake stroke, and second, an externally supplied spark is used to initiate combustion at the end of the compression stroke.

The **two-stroke cycle** (Figure 1) requires only two piston strokes for each power stroke. The intake and compression functions of the four-stroke cycle are combined into a single stroke, as are the power and exhaust functions. The intake/compression stroke starts as the piston nears bottom dead center, uncovering intake ports located in the cylinder walls. When the intake port is uncovered in a CI engine, a "scavenging" blower forces air into the cylinder. That intake air forces the combustion products out of the cylinder through the exhaust valves, which were opened at the end of the downward stroke. When the upward moving piston passes the intake ports, sealing them, the exhaust valves are closed and compression occurs. When the piston approaches top dead center and the air temperature is greater than the fuel's ignition temperature, the fuel is injected into the cylinder, where it initiates combustion. During the second or power/exhaust stroke, the piston is forced toward the cylinder bottom, and the intake/exhaust ports are opened to allow the intake/exhaust function.

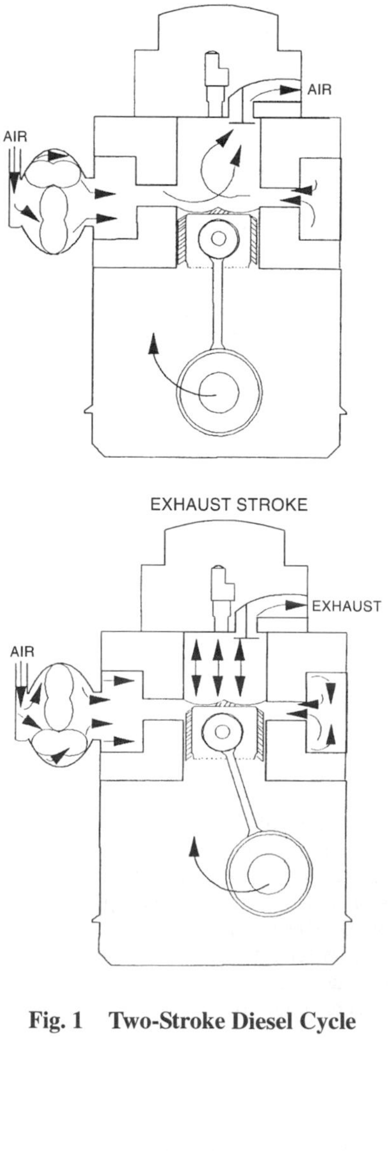

Fig. 1 Two-Stroke Diesel Cycle

Two-stroke engines cost less than four-stroke engines of the same capacity. However, two-stroke engines are less efficient, so they reject more heat. As a result, the heat recovery system tends to be larger and more costly than it would be for the four-stroke engine. Moreover, the effective compression ratio of the two-stroke engine is lower than that of a four-stroke engine of the same size. The lower compression ratio is a result of the placement of the air intake ports in the cylinder walls, which reduces the effective volume of the cylinder. Additionally, the scavenging blower is a parasitic load on the engine, and the opening of the intake port and exhaust valve prior to the end of the power stroke causes a loss of power. Two-stroke engines also require 40% more combustion air than do four-stroke engines. In general, two-stroke engines are used in standby or peak shaving duty where low efficiency is not as critical to project economics.

Pistons are usually arranged in an in-line or a V configuration. However, a particularly interesting two-stroke design is the **opposed-piston engine** (Figure 2). These engines contain two crankshafts, one on top of the engine and the other at the bottom. Each cylinder contains two pistons, and there are no cylinder heads. Opposed-piston engines are especially compact and for this reason are widely used for marine propulsion, especially submarines, where space is limited. For marine and other stationary applications, they are typically operated on diesel oil, but for cogeneration applications, dual-fuel CI or SI operation is usually the most cost-effective.

The cycle begins with the pistons at the outer end of their stroke and a fresh charge of air in the cylinder. As the pistons move inward, the air charge rapidly reaches a high temperature as it is compressed. At approximately 9° before the lower piston reaches inner dead center, injection of fuel oil begins. The high temperature of the air charge causes the fuel oil to ignite, and combustion takes place as the pistons pass through their inner centers. The pressure from combustion forces the pistons apart, thereby delivering power to the crankshafts.

The gases expand until nearly the end of the power stroke as the lower piston begins to uncover the exhaust ports, allowing the burned gases to escape to the atmosphere through the exhaust system. At about the time the pressure in the cylinder has dropped to almost atmospheric, the upper piston starts uncovering the inlet ports. Scavenging air in the air receiver rushes into the cylinder under pressure supplied by the blower.

The cylinder is swept clean of the remaining exhaust gases and refilled with fresh air for the next compression stroke. This cycle is

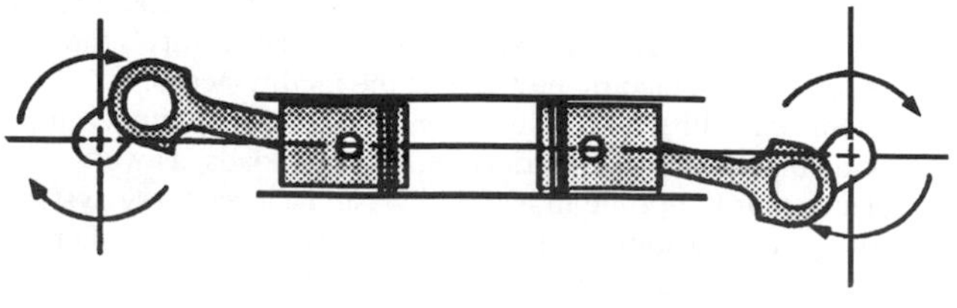

Fig. 2 Opposed-Piston Engine

the same for both diesel and dual-fuel cycles except during fuel admission and combustion. When gas is the primary fuel, shortly after the lower piston covers the exhaust ports, gas is admitted and continues to be admitted for approximately one-quarter revolution of the crankshaft. After the combined air and gas charge is compressed, a small amount of fuel oil (pilot oil) is injected into the cylinder. Ignition of the pilot oil in turn ignites the air-gas charge, and combustion occurs.

PERFORMANCE CHARACTERISTICS

Some of the more important performance characteristics of an engine are its power rating, fuel consumption, and thermal output. Manufacturers base their engine ratings on the engine duty: prime power, standby operations, and peak shaving. Because a cogeneration system is most cost-effective when operating at its baseload, the rating at prime power is usually of greatest interest. Prime power implies that the engine is the primary source of power. This rating is based on providing extended operating life with minimum maintenance. When used for standby, the engine produces continuously for 24 h per day for the length of the primary source outage. Peak power implies an operation level for only a few hours per day to meet peak demand in excess of the prime power capability.

Many manufacturers rate engine capacities according to ISO *Standard* 3046-1. This standard specifies that continuous net brake power under standard reference conditions (total barometric pressure 14.5 psi, corresponding to approximately 330 ft above sea level, air temperature 77°F, and relative humidity 30%) can be exceeded by 10% for 1 h, with or without interruptions, within a period of 12 h of operation. ISO *Standard* 3046-1 defines prime power as power available for continuous operation under varying load factors and 10% overload as previously described. The standard defines standby power as power available for operation under normal varying load factors, not overloadable (for applications normally designed to require a maximum of 300 h of service per year).

However, the basis of the manufacturer's ratings (ambient temperature, altitude, and atmospheric pressure of the test conditions) must be known to determine the engine rating at site conditions. Various derating factors are used. Naturally aspirated engine output typically decreases 3% for each 1000 ft increase in altitude, while turbocharged engines lose 2% per 1000 ft. Output decreases 1% per 10°F increase in ambient temperature, so it is important to avoid the use of heated air for combustion. In addition, an engine must be derated for those fuels with a heating value that is significantly greater than the base specified by the manufacturer. For cogeneration applications, natural gas is usually the fuel of choice.

Power rating is determined by a number of engine design characteristics, the most important of which is displacement; but it also depends on rotational speed, method of ignition, compression ratio, aspiration, cooling system, jacket water temperature, and intercooler temperature. Most engine designs are offered in a range of displacements achieved by different bore and stroke, but with the same number of cylinders in each case. Many larger engine designs retain the same basic configuration, and displacement increases are achieved by simply lengthening the block and adding more cylinders.

Figure 3 illustrates the capacity of some typical SI natural gas engine/generator sets operating at the prime power rating and at 1200 and 900 rpm. The lower value at each speed represents a naturally aspirated design; the upper value represents either an intercooled, turbocharged design or a naturally aspirated design with a higher compression ratio.

Fuel consumption is the greatest contributor to operating cost and should be carefully considered in the planning and design phases of a cogeneration system. It is influenced by combustion cycle, speed, compression ratio, and type of aspiration. It is often expressed in terms of power (Btu/h) for natural gas engines, but for purposes of comparison, it may be expressed as a ratio such as Btu/h per brake horsepower or Btu/kWh. The latter is known as the **heat rate** and equals 3412/efficiency.

The heat rates of several SI engines are shown in Figure 4. Heat rate for an engine of a given size is affected by design and operating factors other than displacement. The most efficient (lowest heat rate) of these engines is naturally aspirated and achieves its increased performance due to the slightly higher compression ratio.

The **thermal-to-electric ratio** is a measure of the useful thermal output for the electrical power being generated. For most reciprocating engines, the recoverable thermal energy is that of the exhaust and jacket. Figure 5 shows the thermal-to-electric ratio of the SI engines shown in Figure 3 and Figure 4. In these example engines, all jacket heat was recovered, and the exhaust gases were cooled to 325°F.

Figure 6 and Figure 7 show the capacities of a large four-stroke and an opposed-piston engine. Capacity varies by the number of cylinders, so performance characteristics relative to output are constant. When the four-stroke engine operates at a heat rate of 10,690 Btu/kWh, the thermal-to-electric ratio is 3880 Btu/kWh. When the two-stroke, opposed-piston engine operates at 720 rpm and a heat rate of 11,030 Btu/kWh, the thermal-to-electric ratio is 2360 Btu/kWh. The large amount of air used to scavenge the exhaust gases from two-stroke engines reduces their exhaust temperatures, thus reducing their usefulness for cogeneration, in which exhaust gas heat recovery is important.

Ideally, a cogeneration plant should operate at full output to achieve maximum cost-effectiveness. In plants that must operate at part load some of the time, part-load fuel consumption and thermal output are important factors that must be considered in the overall economics of the plant. Figure 8 shows the part-load heat rate and

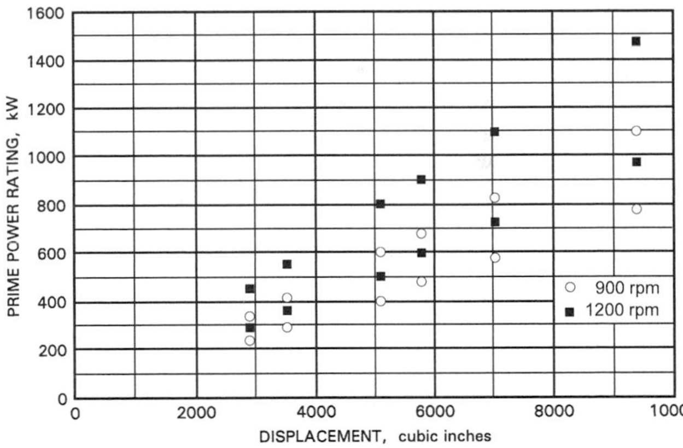

Fig. 3 Capacity of Spark Ignition Engine

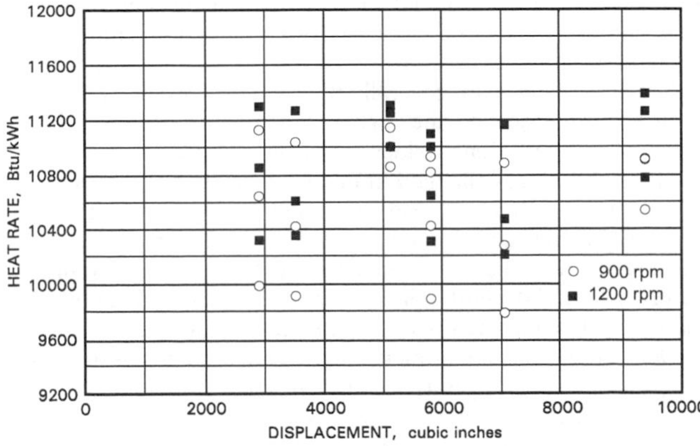

Fig. 4 Heat Rate of Spark Ignition Engines

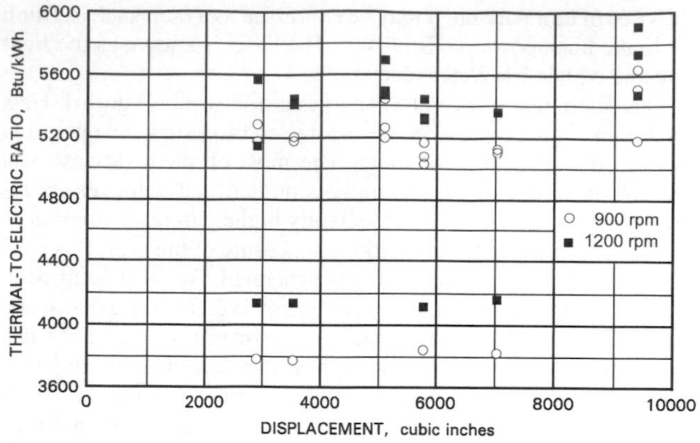

Fig. 5 Thermal-to-Electric Ratio of Spark Ignition Engines

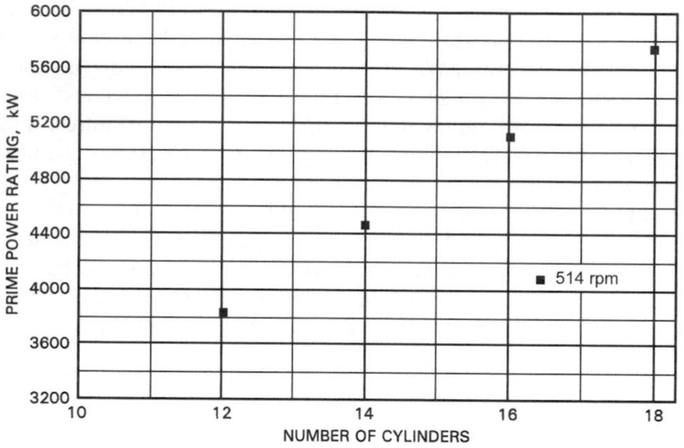

Fig. 6 Capacity of Large Four-Stroke Dual-Fuel Engine

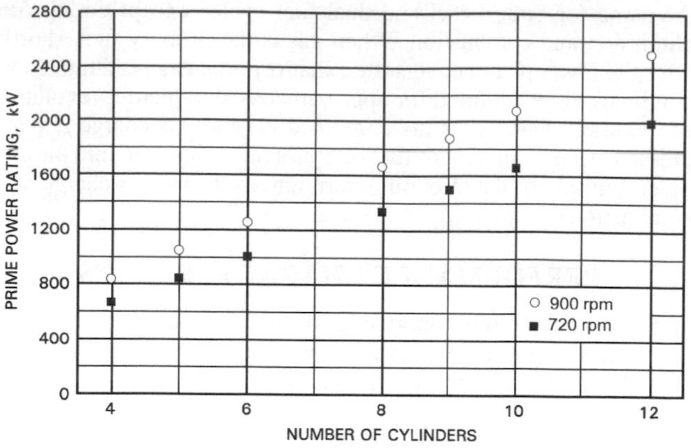

Fig. 7 Capacity of Opposed-Piston, Dual-Fuel Engine

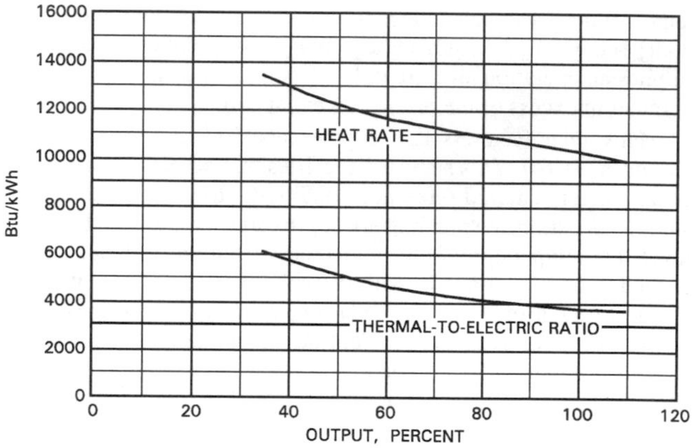

Fig. 8 Part-Load Performance of 725 kW Gas Engine

thermal-to-electric ratio as a function of load for a 725 kW naturally aspirated gas engine, assuming a final exhaust temperature of 375°F.

FUELS

Fuel Selection

Fuel specifications, grade, and characteristics have a marked effect on engine performance. Fuel standards for internal combustion engines are designated by the American Society for Testing and Materials (ASTM) and are substantially different from those for heating use.

Engines may be fueled with gasoline, natural gas, propane, sludge gas, or diesel and heavier oils. Multifuel engines using diesel oil as one of the fuels are available in sizes larger than 220 hp. A small amount of diesel oil is used as the compression ignition agent in gas-fueled engines.

Gasoline engines are generally not used because of fuel storage hazards, fuel cost, and the higher maintenance required due to deposits of combustion products on internal parts.

Methane-rich gas obtained from sewage treatment processes can be used as a fuel for both engines and other heating services. The fuel must be dried and cleaned prior to its injection into the engines. Because of the lower heat content of methane-rich gas (approximately 600 to 700 Btu/ft^3), it is sometimes mixed with natural gas, and the engine must be fitted with a larger carburetor.

Dual carburetors are often installed, with sewage gas as the primary fuel and natural gas as backup. The large amount of hydrogen sulfide in the fuel requires the use of special materials, such as aluminum in the bearings and bushings and low-friction plastic in the O rings and gaskets.

The final choice of fuel should be based on fuel availability, cost, storage requirements, emissions requirements, and fuel rate. Except for gasoline engines, maintenance costs tend to be similar for all engines.

Fuel Heating Value

Fuel consumption data may be reported in terms of either high heating value (HHV) or low heating value (LHV). HHV is used by the gas utility industry and is the basis for evaluating most gaseous fuel. Most natural gases have an LHV/HHV factor of 0.9 to 0.95. This factor ranges from 0.87 for hydrogen to 1.0 for carbon monoxide. For fuel oils, LHV/HHV ranges from 0.96 (heavy oils) to 0.93 (light oils). The HHV is customarily used for oil (including pilot oil in dual-fuel engines).

Many manufacturers base engine power ratings on an LHV of approximately 900 Btu/ft^3 for natural gas. The LHV deducts the energy necessary to keep the water produced during combustion in the vapor state. This latent heat does no work on the piston. Manufacturers sometimes suggest derating engine output about 2% per 10% decrease in fuel heating value below the base specified by the manufacturer. Additional power required to drive auxiliary equipment like compressors, pumps, or generators should not be counted.

FUEL SYSTEMS

Fuel Oil Systems

Storage, handling, and cleaning of liquid fuels are covered in Chapter 26. Most oil-fueled cogeneration is with No. 2 diesel fuel, which is much simpler to handle than the heavier grades. Residual oil is used only for large industrial projects with low-speed engines.

The fuel injection system is the heart of the diesel cycle. Performance functions are as follows:

- Meter a constant quantity to each cylinder at any load during each combustion cycle.
- Inject fuel (1) with a precise and rapid beginning and ending at the correct timing point in each cycle and (2) at a rate needed for controlled combustion and pressure rise.
- Atomize the fuel and distribute it evenly through the air in the combustion chamber.

The two methods for atomization are high-pressure air injection and mechanical injection. Older systems used air injection until satisfactory mechanical injection systems were developed that avoid the high initial cost and parasitic operating cost of the air compressor.

Earlier mechanical injection pressurized a header to provide a common fuel pressure near 5000 psi. A camshaft opened a spray nozzle in each cylinder, with the length of spray time proportioned to the load through a governor or throttle control. With this system, a leaky valve allowed a steady drip into the cylinder throughout the cycle, which caused poor fuel economy and smoking.

Currently, three injection designs are in use:

- Individual plunger pumps for each cylinder, with controlled bypass, controlled suction, variable-suction orifice, variable stroke, or port-and-helix metering
- Common high-pressure metering pump with a separate distribution line to each cylinder that delivers fuel to each cylinder in firing order sequence
- Common low-pressure metering pump and distributor with a mechanically operated, high-pressure pump and nozzle at each cylinder

Spark Ignition Gas Systems

Fuels vary widely in composition and cleanliness, from pipeline natural gas requiring only a meter, a pressure regulating valve, and safety devices to those from sewage or biomass, which may require scrubbers and holding tanks in addition. The following are gas system accessories.

Line-Type Gas Pressure Regulators. Turbocharged (and aftercooled) engines, as well as many naturally aspirated units, are equipped with line regulators designed to control the gas pressure to the engine regulator, as shown in Table 1. The same regulators (both line and engine) used on naturally aspirated gas engines may be used on turbocharged equipment.

Line-type gas pressure regulators are commonly called service[b] regulators (and field regulators). They are usually located just upstream of the engine regulator to ensure that the required pressure range exists at the inlet to the engine regulator. A remote location is sometimes specified; authorities having jurisdiction should be consulted. Although this intermediate regulation does not constitute a safety device, it does permit initial regulation (by the gas utility at the meter inlet) at a higher outlet pressure, thus allowing an extra cushion of gas between the line regulator and the meter for both full gas flow at engine start-ups and delivery to any future branches from the same supply line. The engine manufacturer specifies the size, type, orifice size, and other regulator characteristics based on the anticipated gas-pressure range.

Table 1 Line Regulator Pressures

Line Regulator	Turbocharged Engine[a]	Naturally Aspirated Engine[a]
Inlet	14 to 20 psig	2 to 30 psig
Outlet[b]	12 to 15 psig[c]	7 to 10 in. of water

[a]Overall ranges—not the variation for individual installations.
[b]Also inlet to engine regulator.
[c]Turbocharger boost plus 7 to 10 in. of water.

Engine-Type Gas Pressure Regulators. This engine-mounted pressure regulator, also called a carburetor regulator (and sometimes a secondary regulator or a B regulator), controls the fuel pressure to the carburetor. Regulator construction may vary with the fuel used. The unit is similar to a zero governor.

Air-Fuel Control. The flow of air-fuel mixtures must be controlled in definite ratios under all load and speed conditions required of engines.

Air-Fuel Ratios. High-rated, naturally aspirated, spark ignition engines require closely controlled air-fuel ratios. Excessively lean mixtures cause excessive lubricating oil consumption and engine overheating. Engines using pilot oil ignition can run at rates above 0.18 ppm/hp without misfiring. Air rates may vary with changes in compression ratio, valve timing, and ambient conditions.

Carburetors. In these venturi devices, the airflow mixture is controlled by a governor-actuated butterfly valve. This air-fuel control has no moving parts other than the butterfly valve. The motivating force in naturally aspirated engines is the vacuum created by the intake strokes of the pistons. Turbocharged engines, on the other hand, supply the additional energy as pressurized air and pressurized fuel.

Ignition. An electrical system or pilot oil ignition may be used. Electrical systems are either low-tension (make-and-break) or high-tension (jump spark). Systems with breakerless ignition distribution are also in use.

Dual-Fuel and Multifuel Systems

Engines using gas and pilot oil for diesel ignition are commonly classified as dual-fuel engines, while those using natural gas (NG) with LPG standby are called multifuel engines. Dual-fuel engines operate either on full oil or on gas and pilot oil, with automatic online switchover when appropriate, while NG/LPG multifuel engines operate safely on one fuel or the other. In sewage gas systems, a blend with natural gas may be used to maintain a minimum LHV or to satisfy fuel demand when sewage gas production is short.

JACKET WATER SYSTEM

Various jacket water circuits to control the engine's operating temperature while recovering its conducted thermal waste are illustrated and described in the section on Engine Jacket Heat Recovery on page 7.20. Ultimately, this heat is (1) recovered and productively used, (2) carried away by an air-cooled radiator, (3) dissipated in a cooling tower or raw water stream (e.g., groundwater), or (4) ejected into the engine room.

LUBRICATING SYSTEMS

All engines use the lubricating system to remove some heat from the machine. Some configurations cool only the piston skirt with oil; other designs remove more of the engine heat with the lubricating system. The operating temperature of the engine may be significant in determining the proportion of engine heat removed by the lubricant. Between 5 and 10% of the total fuel input is converted to heat that must be extracted from the lubricating oil; this may warrant using oil coolant at temperatures high enough to permit economic use in a process such as domestic water heating.

Radiator-cooled units generally use the same fluid to cool the engine water jacket and the lubricant; thus, the temperature difference between the oil and the jacket coolant is not significant. If the oil temperature rises in one area (such as around the piston skirts), the heat may be transferred to other engine oil passages and then removed by the jacket coolant. When the engine jacket temperatures are much higher than the lubricant temperatures, the reverse process occurs, and the oil removes heat from the engine oil passages.

Determining the lubricant cooling effect is necessary in the design of heat exchangers and coolant systems. Heat is dissipated to the lubricant in a four-cycle engine with a high-temperature (225 to

250°F) jacket water coolant at a rate of about 7 or 8 Btu/min·bhp; oil heat is rejected in the same engine at 3 to 4 Btu/min·bhp. However, this engine uses more moderate (180°F) coolant temperatures for both lubricating oil and engine jacket.

The characteristics of each lubricant, engine, and application are different, and only periodic laboratory analysis of oil samples can establish optimum lubricant service periods. The following factors should be considered in selecting an engine:

- High-quality lubricating oils are generally required for operation at temperatures between 160 and 200°F, with longer oil life expected at lower temperatures. Moisture may condense in the crankcase if the oil is too cool, which reduces the useful life of the oil.
- Copper piping should be avoided in oil-side surfaces in oil coolers and heat exchangers to reduce the possibility of oil breakdown caused by contact with copper.
- A full-flow filter provides better security against oil contamination than one that filters only a portion of circulated lubricating oil and bypasses the rest.

COMPRESSED AIR SYSTEMS

Larger engines are frequently started with compressed air, either by direct cylinder injection or by air-driven motors. In large plants, one of the smallest of multiple compressors is usually engine driven for a "black start" (dead plant start-up). The same procedure is true for fuel oil systems when the main storage tank cannot gravity feed the day tank. However, the storage tanks must have the capacity for several starting procedures on any one engine, in case of repeated failure to start.

Another start-up concept eliminates all auxiliary engine drives and powers the motor-driven auxiliaries directly through a segregated circuit served by a portable, engine-driven emergency generator. This circuit can be sized for the black start power and control requirements as well as for emergency lighting and receptacles for power tools and welding devices. For a black start, or after any major damage causing a plant failure, this circuit can be used for required repairs at the plant and at other buildings in a given complex.

EXHAUST GAS SYSTEMS

The exhaust stream from the engine can (1) exit directly to the atmosphere through a silencer; (2) pass through a jacket water-cooled exhaust manifold; (3) drive a jacket water-cooled turbocharger; or (4) flow through a heat recovery/silencer device. The section on Reciprocating Engines under Heat Recovery on page 7.20 discusses these various methods of exhaust gas heat recovery.

SUPERCHARGERS

A supercharged engine is generally less costly than a larger, naturally aspirated engine for a given engine output. The turbocharger must match the engine to provide the required pressure ratio and mass flow under all conditions of engine operation, while staying out of the field of instability of any centrifugal blower.

Centrifugal blowers operate at 10,000 to 50,000 rpm to attain pressure ratios up to 3:1 in a single stage. Gear trains for these speeds are much more elaborate than for rotary blowers; therefore, such blowers are normally driven by small exhaust gas turbines called turbochargers.

Rotary blowers are positive displacement blowers using rotary lobes that mesh together with small clearances and have no direct contact with one another. They operate at 2000 to 6000 rpm and can be directly driven by the engine or by a separate motor drive.

Turbochargers may be air or water cooled, and they have an aftercooler on the discharge air side to avoid feeding hot, less dense air to the engine. Larger ones may have their own lubricating systems and oil cooler, while smaller ones operate with the main engine oil.

In two-cycle naturally aspirated engines, the turbocharger increases the output by about 50%, while in four-cycle engines the increase can be from 30% to 150% or more, depending on the pressure ratio. Aftercooling following compression can raise performance approximately another 17%.

The turbocharger extends the optimum fuel consumption curve because the usual limitation on larger gas-fueled engines is the volume of combustion air that can be inserted in the available combustion chamber. Many stationary reciprocating engines were developed for use with diesel fuel. The energy ratings with the liquid fuel are higher than with naturally aspirated gaseous fuel. Turbocharging on diesel service applies more air pressure in the cylinder so that larger quantities of the separately injected fuel can be burned efficiently.

In gaseous fuel systems, the fuel must always have a pressure high enough to enter the carburetor and mix with the combustion air, which is at a boosted pressure. Because the gas and air mixture ignite at a specific temperature-pressure relationship, the lower the inlet air temperature is, the higher the compression ratio can be before spontaneous combustion (preignition) occurs.

COMBUSTION AIR SYSTEMS

All internal combustion engines require clean, cool air for optimum performance. Highly humid air does not hurt performance, and it may even help by slowing combustion and reducing cylinder pressure and temperature. Provisions must be made to silence air noise, provide an adequate amount of air for combustion, and, in the case of two-cycle engines, provide enough air to scavenge the combusted fuel.

Smaller engines generally use engine-mounted impingement filters, often designed for some silencing, while larger engines commonly use a cyclone filter or various oil-bath filters. Selection considerations are (1) efficiency or dirt removal capacity; (2) airflow resistance (high intake pressure drop affects performance); (3) ease, frequency, and cost of cleaning or replacement; and (4) first cost. Many of the filter types and media commonly found in HVAC systems are used, but they are designed specifically for engine use. Air piping is designed for low pressure drop to maintain high engine performance. A low pressure drop is more important for naturally aspirated than for supercharged engines. For engine intakes, conventional velocities range from 3000 to 7200 fpm, governed by the engine manufacturer's recommended pressure drop of approximately 5.5 in. of water.

Evaporative coolers are sometimes used to cool the air before it enters the engine intake, while a recirculated water coolant recovery system is used for aftercooling.

INSTRUMENTS AND CONTROLS

Starting Systems

The start-stop control may include manual or automatic activation of the engine fuel supply, the engine cranking cycle, and establishment of the engine heat removal circuits. Stop circuits always shut off the fuel supply, and, for spark ignition engines, the ignition system is generally grounded as a precaution against incomplete fuel valve closing.

Alarm and Shutdown Controls

The prime mover is protected from malfunction by alarms that warn of unusual conditions and by safety shutdown under unsafe conditions. The control system must protect against failure of (1) speed control (underspeed or overspeed); (2) lubrication (low oil pressure, high oil temperature); (3) heat removal (high coolant temperature or lack of coolant flow); (4) combustion process (fuel, ignition); (5) lubricating oil level; and (6) water level.

Controls for alarms preceding shutdown are provided as needed. Monitored alarms without shutdown include lubricating oil and fuel filter, lubricating oil temperature, manifold temperature, jacket water temperature, etc. Automatic start-up of the standby engine when an alarm/shutdown sequence is triggered is often provided.

Both a low-lubrication pressure switch and a high jacket water temperature cutout are standard for most gas engines. Other safety controls used include (1) an engine speed governor, (2) ignition current failure shutdown (battery-type ignition only), and (3) the safety devices associated with a driven machine. These devices shut down the engine to protect it against mechanical damage. They do not necessarily shut off the gas fuel supply unless they are specifically set to do so.

Governors

A governor senses speed (and sometimes load), either directly or indirectly, and acts by means of linkages to control the flow of gas and air through engine carburetors or other fuel-metering devices to maintain a desired speed. Speed control with electronic, hydraulic, or pneumatic governors extends engine life by minimizing forces on engine parts, permits automatic throttle response without operator attention, and prevents destructive overspeeding. A separate overspeed device, sometimes called an overspeed trip, prevents runaway in the event of a failure that renders the governor inoperative. Both constant and variable engine speed controls are available. For constant speed, the governor is set at a fixed position, which can be reset manually.

Gas Leakage Prevention

The first method of avoiding gas leakage due to engine regulator failure is to install a solenoid shutdown valve with a positive cutoff either upstream or downstream of the engine regulator. The second method is a sealed combustion system that carries any leakage gas directly to the outdoors (i.e., all combustion air is ducted to the engine directly from the outdoors).

NOISE AND VIBRATION

Because engine exhaust must be muffled to reduce ambient noise levels, most recovery units also act as silencers. Figure 9 illustrates a typical noise curve. Figure 10 shows typical attenuation curves for various silencers. Table 34 in Chapter 46 of the 1999 *ASHRAE Handbook—Applications* lists acceptable noise level criteria for various applications. The section on Noise and Vibration Control on page 7.39 has further information.

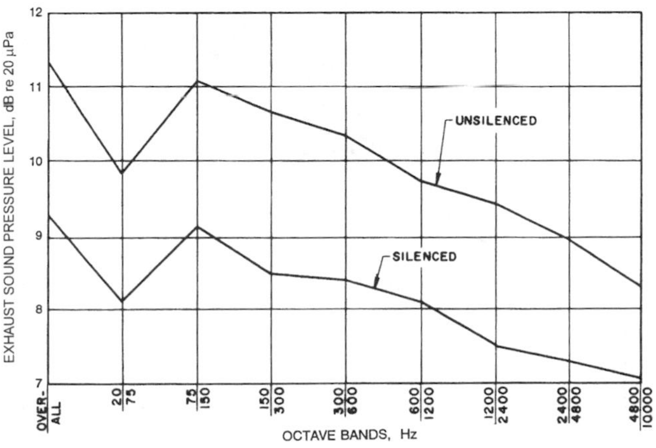

Fig. 9 Typical Reciprocating Engine Exhaust Noise Curves

MAINTENANCE

Engines require periodic servicing and replacement of parts, depending on usage and the type of engine. Transmission drives require periodic gearbox oil changes and the operation and care of external lubricating pumps. Log records should be kept of all servicing; checklists should be used for this purpose.

Table 2 shows ranges of typical maintenance routines for both diesel and gas-fired engines, based on the number of hours run. The actual intervals vary according to the cleanliness of the combustion

Table 2 Recommended Engine Maintenance

	Hours Between Procedures	
Procedure	Diesel Engine	Gas Engine
1. Take lubricating oil sample	Once per month plus once at each oil change	Once per month plus once at each oil change
2. Change lubricating oil filters	350 to 750	500 to 1000
3. Clean air cleaners, fuel	350 to 750	350 to 750
4. Clean fuel filters	500 to 750	n.a.
5. Change lubricating oil	500 to 1000	1000 to 2000
6. Clean crankcase breather	350 to 700	350 to 750
7. Adjust valves	1000 to 2000	1000 to 2000
8. Lubricate tachometer, fuel priming pump, and auxiliary drive bearings	1000 to 2000	1000 to 2000 (fuel pump n.a.)
9. Service ignition system; adjust breaker gap, timing, spark plug gap, and magneto	n.a.	1000 to 2000
10. Check transistorized magneto	n.a.	6000 to 8000
11. Flush lubrication oil piping system	3000 to 5000	3000 to 5000
12. Change air cleaner	2000 to 3000	2000 to 3000
13. Replace turbocharger seals and bearings	4000 to 8000	4000 to 8000
14. Replace piston rings, cylinder liners (if applicable), connecting rod bearings, and cylinder heads; recondition or replace turbochargers; replace gaskets and seals	8000 to 12,000	8000 to 12,000
15. Same as item 14, plus recondition or replace crankshaft; replace all bearings	24,000 to 36,000	24,000 to 36,000

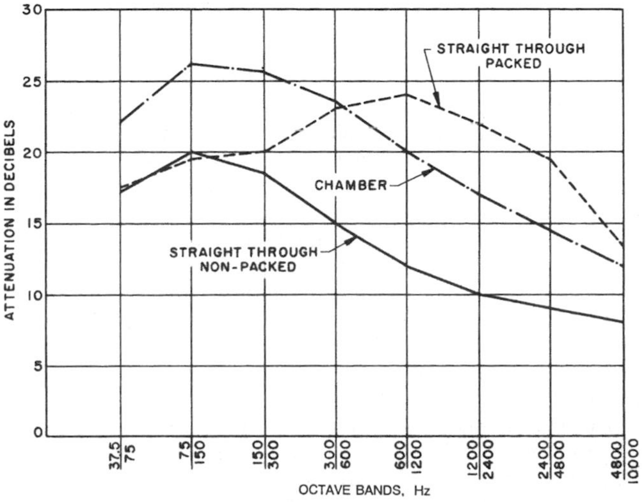

Fig. 10 Typical Attenuation Curves for Engine Silencers

air, the cleanliness of the engine room, the engine manufacturer's recommendations, the number of engine starts and stops, and lubricating conditions indicated by the oil analysis. With some engines and some operating conditions, the intervals between procedures listed in Table 2 may be extended.

A preventive maintenance program should include inspections for

- Leaks (a visual inspection, which is facilitated by a clean engine)
- Abnormal sounds and odors
- Unaccountable speed changes
- Condition of fuel and lubricating oil filters

Daily logs should be kept on all pertinent operating parameters, such as

- Water and lubricating oil temperatures
- Individual cylinder compression pressures, which are useful in indicating blowby
- Changes in valve tappet clearance, which indicate the extent of wear in the valve system

Lubricating oil analysis is a low-cost method of determining the physical condition of the engine and a guide to maintenance procedures. Commercial laboratories providing this service are widely available. The analysis should measure the concentration of various elements found in the lubricating oil, such as bearing metals, silicates, and calcium. It should also measure the dilution of the oil, suspended and nonsuspended solids, water, and oil viscosity. The laboratory can often assist in interpreting the readings and alert the user to impending problems.

Lubricating oil manufacturers' recommendations should be followed. Both the crankcase oil and oil filter elements should be changed at least once every six months.

COMBUSTION TURBINES

TYPES

Combustion gas turbines, although originally used for aircraft propulsion, have been developed for stationary use as prime movers. Turbines are available in sizes from 32 to 173,000 bhp and can burn a wide range of liquid and gaseous fuels. Turbines, and some dual-fuel engines, are capable of shifting from one fuel to another without loss of service.

Combustion turbines consist of an air compressor section to boost combustion air pressure, a combination fuel-air mixing and combustion chamber (combustor), and an expansion power turbine section that extracts energy from the combustion gases.

In addition to these components, some turbines use regenerators and recuperators as heat exchangers to preheat the combustion air entering the combustion chamber with heat from the turbine discharge gas, thereby increasing machine efficiency. Most turbines are the single-shaft type, (i.e., air compressor and turbine on a common shaft). However, split-shaft machines that use one turbine stage on the same shaft as the compressor and a separate power turbine driving the output shaft are available.

Turbines rotate at speeds varying from 3600 to 60,000 rpm and often need speed reduction gearboxes to obtain shaft speeds suitable for generators or other machinery. Turbine motion is completely rotary and relatively vibration-free. This feature, coupled with low mass and high power output, provides an advantage over reciprocating engines with regard to space, foundation requirements, and ease of start-up.

However, in smaller sizes, the combustion gas turbine has a heat rate efficiency of 12 to 30%. Larger turbines approach a heat rate efficiency above 35%, with advanced cycles above 50%.

Combustion turbines have the following advantages over other internal combustion engine drivers:

- Small size, high power-to-weight ratio
- Ability to burn a variety of fuels, though more limited than reciprocating engines
- Ability to meet stringent pollution standards
- High reliability
- Available in self-contained packages
- Instant power—no warm-up required
- No cooling water required
- Vibration-free operation
- Easy maintenance
- Low installation cost
- Clean, dry exhaust
- Lubricating oil not contaminated by combustion oil

Gas Turbine Cycle

The basic gas turbine cycle (Figure 11) is the **Brayton cycle (open cycle)**, which consists of adiabatic compression, constant pressure heating, and adiabatic expansion. Figure 11 shows that the thermal efficiency of a gas turbine falls below the ideal value because of inefficiencies in the compressor and turbine and because of duct losses. Increases in entropy occur during the compression and expansion processes, and the area enclosed by points 1, 2, 3, and 4 is reduced. This loss of area is a direct measure of the loss in efficiency of the cycle.

Nearly all turbine manufacturers present gas turbine engine performance in terms of power and specific fuel consumption. A comparison of fuel consumption in specific terms is the quickest way to compare overall thermal efficiencies of gas turbines.

Components

Figure 12 shows the major components of the gas turbine unit, which include the air compressor, the combustor, and the power turbine. Atmospheric air is compressed by the air compressor. Fuel is then injected into the airstream and ignited in the combustor, with the leaving gases attaining temperatures between 1800 and 2300°F. These high-pressure hot gases are then expanded through a turbine, which provides not only the power required by the air compressor, but also power to drive the load.

Gas turbines are available in two major classifications—single-shaft (Figure 12) and split-shaft (Figure 13). The **single-shaft turbine** has the air compressor, the gas-producer turbine, and the power turbine on the same shaft. The **split-shaft or dual-shaft turbine** has the section required for air compression on one shaft and the section producing output power on a separate shaft. For a split-shaft turbine, the portion that includes the compressor, combustion

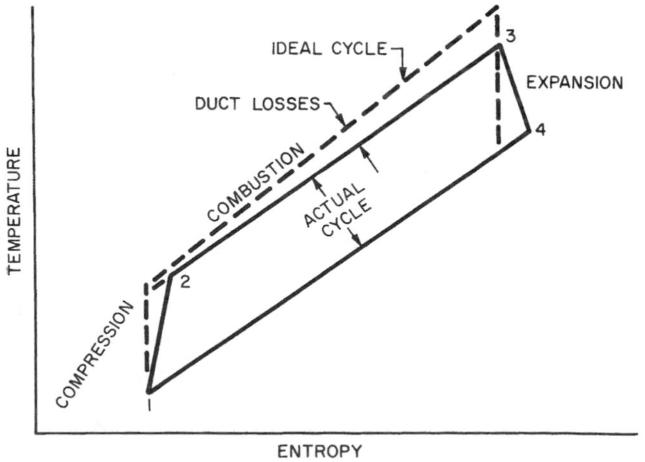

Fig. 11 Temperature-Entropy Diagram for Brayton Cycle

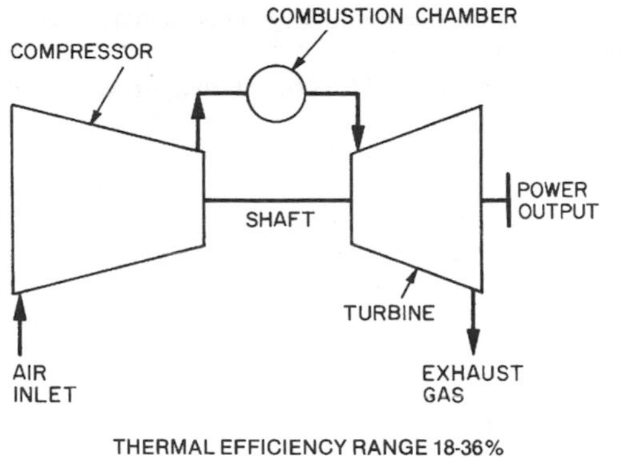

Fig. 12 Single-Shaft Turbine

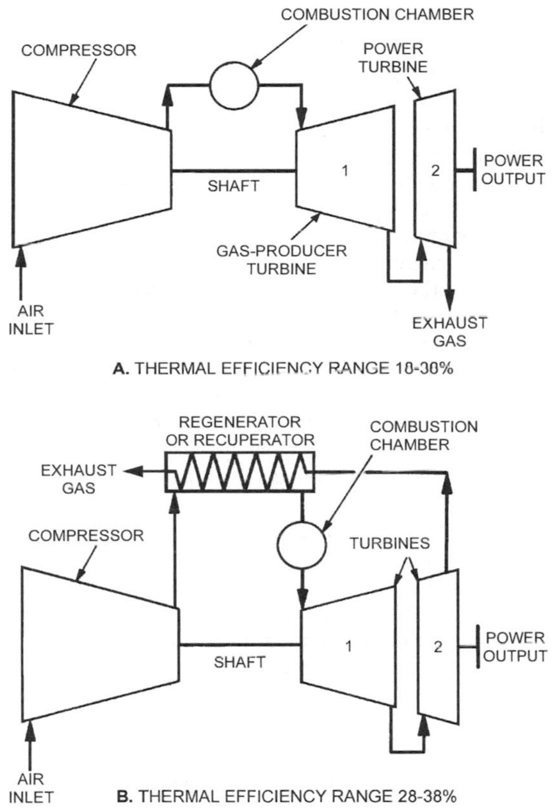

Fig. 13 Split-Shaft Turbines

chamber, and first turbine section is the hot gas generator or gas producer. The second turbine section is the power turbine.

The turbine used depends on job requirements. Single-shaft engines are usually selected when a constant-speed drive is required, as in generator drives, and when starting torque requirements are low. A single-shaft engine can be used to drive centrifugal compressors, but the starting system and the compressor match point must be considered. Split-shaft engines allow for variable speed at full load. Additional advantages of the split-shaft engine are that it can easily be started with a high torque load connected to the power output shaft, and the power turbine can be more optimally configured to match load requirements.

FUELS AND FUEL SYSTEMS

The ability to burn almost any combustible fluid is a key advantage of the gas turbine. Natural gas is the fuel of choice over other gaseous fuels because it is readily available, has good combustion characteristics, and is relatively easy to handle. A typical fuel gas control system is a two-stage system that uses pressure control in combination with flow control to achieve a turndown ratio of about 100:1. Other fuel gases include liquefied petroleum gases, which are considered "wet" gases because they can form condensables at normal gas turbine operating conditions; and a wide range of refinery waste and coal-derived gases, which have a relatively high fraction of hydrogen. Both of these features lead to problems in fuel handling and preparation, as well as in gas turbine operation. Heat tracing to heat the piping and jacketing of valves is required to prevent condensation at start-up. Piping runs should be designed to eliminate pockets where condensate might drop out and collect.

Distillate oil is the most common liquid fuel, and except for a few installations where natural gas is not available, it is primarily used as a backup and alternate start-up fuel. Crude oils are common as primary fuels in many oil-producing countries because of their abundance. Both crude and residual oils require treatment for sodium salts and vanadium contamination. The most common multiple-fuel combination is natural gas and distillate. The combustion turbine may be started on either fuel, and transfers from one fuel to the other may be initiated by the operator at any time after completion of the starting sequence without interrupting operation.

Steam and demineralized water injection are currently used in gas turbines for NO_x abatement in quantities up to 2% of compressor inlet airflow. An additional 3% steam may be injected independently at the compressor discharge for power augmentation. The required steam conditions are 300 to 350 psig and a temperature that is no more than 150°F above compressor discharge temperature, but not less than 50°F of superheat. Steam contaminants should be guarded against, and the steam supply system should be designed to supply dry steam under all operating conditions.

LUBRICATING OIL SYSTEMS

Lubricating systems provide filtered and cooled oil to the gas turbine, the driven equipment, and the gear reducer. Lubricating systems include a motor-driven, start-up/coast-down oil pump, a primary oil pump mounted on and driven by the gear reducer, filters, an oil reservoir, an oil cooler, and automatic controls. Along with lubricating the gas turbine bearings, gear reducer, and driven equipment bearings, the lubrication system provides hydraulic oil to the gas turbine control system.

The start-up/coast-down oil pump circulates oil until the gas turbine reaches a speed at which the primary pump can take over. Where emergency alternating current (ac) power is not available in case of lost outside power, an emergency direct current (dc) motor-driven pump may be required to provide lubricant during start-up and coast-down.

Oil filters serve the full flow of the pumps. Two filters are provided so that one can be changed while the other remains in operation.

The oil reservoir is mounted in the base of the gas turbine's supporting structure. Heater systems in the reservoir maintain the oil temperature above a minimum level. A combination mechanical/coalescer filter in the reservoir's vent removes oil from the vent. The oil cooler can be either water cooled or air cooled, depending on the availability of cooling water.

STARTING SYSTEMS

Starting systems can use pneumatic, hydraulic, or electric motor starters. A pneumatic starting system uses either compressed air or fuel gas to power a pneumatic starter motor or a starting subsystem integrated directly with the rotating components. A hydraulic system uses a hydraulic motor for starting. Hydraulic fluid is provided to the hydraulic motor by either an ac motor-driven pump or a diesel engine-driven pump. An electric motor system couples an electric

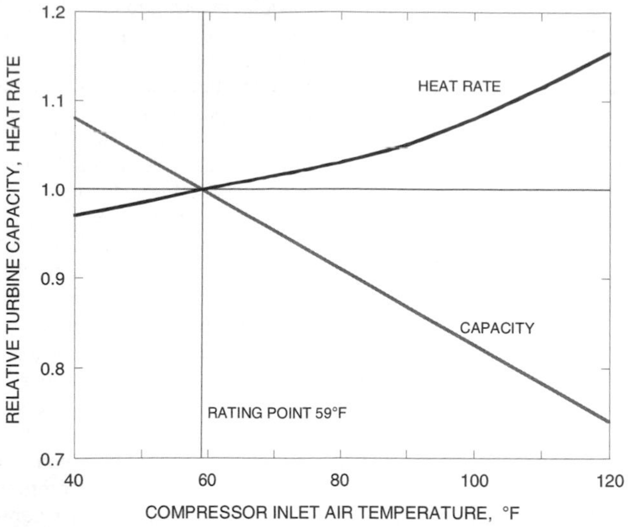

Fig. 14 Relative Turbine Power Output and Heat Rate Versus Inlet Air Temperature

motor directly to the gas engine for starting. All systems use a one-way clutch to couple the starter motor to the gas engine so that as the engine accelerates above the start speed, the starter can shut down. Black starts can be accomplished if a pneumatic or diesel/hydraulic starting system is used.

EXHAUST GAS SYSTEMS

Exhaust systems of gas turbines used in cogeneration systems consist of gas ducts, expansion joints, an exhaust silencer, a dump (or bypass) stack, and a diverter valve (or damper). The exhaust silencer is installed in a dump stack. The diverter valve is used to modulate the flow of exhaust gas into the heat recovery equipment or divert 100% of the exhaust gas to the dump stack when heat recovery is not required.

COMBUSTION TURBINE INLET AIR COOLING

Combustion turbine inlet air cooling (CTIAC) systems increase the capacity of turbine-generators by increasing the density of the combustion air. Since the volumetric flow to most turbines is constant, increasing the air density increases the mass flow rate. As the inlet air temperature increases, as on hot summer days, the capacity decreases (MacCracken 1994). Cooling inlet air increases the power and typically decreases the heat rate, after all parasitic cooling power usage is considered (Figure 14). Factors that affect CTIAC installation and operation include the turbine type, climate, hours of operation, ratio of airflow rate to power generated, ratio of generation increase with increased airflow, and monetary value of power generated.

Cooling Methods

Some CTIAC designs are for turbines that operate only a few hours per year, to demonstrate power reserve or provide peak demand power. **Peaking** combustion turbines are operated when utilities experience the greatest demand. Both inlet evaporative cooling systems and thermal energy storage (TES) systems allow CTIAC during turbine operation with no coincident parasitic power usage except for pumps. TES systems allow the use of small-capacity refrigeration systems, operated only during off-peak hours. Utilities and independent power producers may operate turbines at **baseload** and experience a need for continuous cooling for a significant number of hours per year. For turbines operating continuously or for several hours per day, fuel cost and availability are important factors that favor on-line cooling systems such as direct refrigeration without thermal storage.

The most prevalent CTIAC system is **evaporative cooling** using wetted media, due to low installation and operating costs. Ideal evaporative cooling occurs at a constant wet-bulb temperature, cooling the air to near 100% relative humidity. The typical evaporative cooling system allows the air-water vapor mixture to reach 85 to 95% of the difference between the dry-bulb air temperature and the wet-bulb temperature. Evaporative cooling can be used before or after indirect (secondary fluid) cooling. If a combination of sensible cooling (cooling coils) and evaporative cooling is used, sensible cooling should be used first, and then evaporative cooling, to reach the minimum temperature without latent cooling by the cooling coils.

Chilled water or direct refrigeration can also be used. These processes decrease the enthalpy (and temperature) of the air-water vapor mixture. The water vapor content (humidity ratio) remains constant as the mixture cools to near the dew-point temperature. Continued cooling follows the cooling coil performance curve, lowering the humidity ratio by forcing part of the water vapor to condense out from the mixture, while holding the mixture's relative humidity near 100%.

Chilled water systems can be used in conjunction with either ice or chilled water TES (Ebeling 1994). From a cost standpoint, the TES system is usually justified only for turbines that operate a few hours per week or to increase reserve power. The TES system typically has a higher capital cost and uses more energy than a direct refrigeration cooling system because of the secondary fluid loop, the pumping required, and the increased size of cooling coils. The chilled water system, however, requires less refrigerant piping and inventory and is therefore less susceptible to refrigerant leakage. Thermal storage allows the reduction of refrigeration equipment size and on-peak parasitic energy usage, which can decrease costs. Parasitic loads for a TES system that operates only a few hours per week usually do not severely affect the economic value of a CTIAC system.

A **direct** refrigerant cooling system consists of either a vapor compression system or an absorption system where the liquid refrigerant is used directly in air-cooling coils. The cooling process is identical to that of a chilled water system. A direct system can provide cooling during all hours of turbine operation but must be sized to meet the peak cooling required; therefore, it is larger than a TES system (Stewart 1997).

Advantages of CTIAC

Capacity Enhancement. Use of CTIAC for newer turbines, with lower airflow rates per unit of power generated, is even more economical than for older turbines. The lower flow rates require less cooling capacity to lower inlet air temperatures and therefore smaller evaporative coolers or refrigeration equipment, including TES systems.

Heat Rate Improvement. Fuel mass flow rates increase with inlet airflow and turbine output, but typically at a lower rate. CTIAC systems may be used primarily for decreased heat rate and corresponding fuel cost savings.

Turbine Life Extension. Turbines operating at lower inlet air temperatures have extended life and reduced maintenance. Lower and constant turbine inlet air temperatures reduce the wear on turbines and turbine components.

Increased Combined-Cycle Efficiency. Lower inlet air temperatures result in lower exhaust gas temperatures, potentially decreasing the capacity of the heat recovery steam generator to provide heat to steam turbines and absorption equipment. However, the greater airflow rate of a CTIAC system usually produces an overall increase in capacity because the effect of increased exhaust mass flow rate exceeds the effect of decreased temperature.

Delayed Capacity Addition. With the increased generation capacity provided by a CTIAC system, the addition of actual or reserve generation capacity can be delayed.

Baseload Efficiency Improvements. An ice or chilled water TES system can help level the baseload of a power generation facility by storing energy using electric chiller equipment during off-peak periods; this tends to increase the efficiency of power production. Electric chillers operated at cooler nighttime temperatures are more efficient and operate at reduced condenser temperatures, which can also use less source energy.

When maximum power is desired every hour of the year, a continuous CTIAC system is justified in warm climates to maximize turbine output and minimize heat rate.

Other Benefits. Other advantages include the following:

- Evaporative media filter the inlet air.
- CTIAC systems that reduce the air temperature below saturation can produce a significant amount of condensed water, a potentially valuable resource that can also provide makeup water for cooling towers or evaporative condensers.
- CTIAC systems are simple, energized only when required.
- Emissions can decrease due to increased overall efficiency.
- A CTIAC system can match the inlet air temperature to the required turbine generating capacity, allowing 100% open inlet guide vanes, which eliminate inlet guide vane pressure loss penalties.

Disadvantages

- CTIAC systems require additional space and increase maintenance.
- Evaporative media or cooling coils pose a constant inlet air pressure loss.

INSTRUMENTS AND CONTROLS

Control systems are typically microprocessor based. The control system sequences all systems during starting and running, monitors performance, and protects the equipment. The operator interface is a monitor and keyboard with analog gages provided for redundancy.

Where operating the gas turbine engine at the unit's maximum rating is desirable, the load is controlled based on the temperature of the combustion gases in the turbine section and on the ambient air temperature. When the engine combustion gas temperature reaches a set value, the control system begins to control the engine so that the load (and therefore the temperature) does not increase further. With changes in the ambient air temperature, the control system adjusts the load to maintain the set temperature value in the gas engine's turbine section. Where maintaining a constant load level is desirable, the control system allows the operator to dial in any load, and the system controls the engine accordingly.

PERFORMANCE CHARACTERISTICS

The rating of a gas turbine is greatly affected by altitude, ambient temperature, inlet pressure to the air compressor, and exhaust pressure from the turbine. In most applications, filters and silencers must be installed in the air inlet. Silencers, waste heat boilers, or both are used on the exhaust. The pressure drop of these accessories and piping losses must be considered when determining the power output of the unit.

Gas turbine ratings are usually given at standard conditions defined by the International Organization for Standardization (ISO): 59°F, 60% rh, and sea level pressure at the inlet flange of the air compressor and the exhaust flange of the turbine. Corrections for other conditions must be obtained from the manufacturer, as they vary with each model depending primarily on gas turbine efficiency. Inlet air cooling has been used to increase capacity. The following approximations may be used for design:

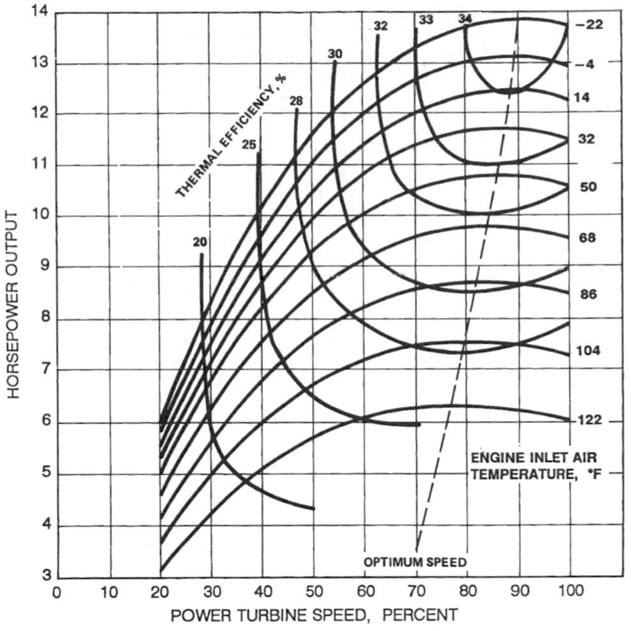

Fig. 15 Turbine Engine Performance Characteristics

- Each 18°F rise in inlet temperature decreases power output by 9%.
- An increase of 1000 ft in altitude decreases power output by approximately 3.5%.
- Inlet pressure loss in filter, silencer, and ducting decreases power output by approximately 0.5% for each inch of water pressure loss.
- Discharge pressure loss in boiler, silencer, and ducting decreases power output by approximately 0.3% for each inch of water pressure loss.

Gas turbines operate with a wide range of fuels. For refrigeration service, a natural gas system is usually provided with an option of a standby No. 1 or 2 grade fuel oil system.

Figure 15 shows a typical performance curve for a 10,000 hp turbine engine. For example, at an air inlet temperature of 86°F, the engine develops its maximum power at about 82% of maximum speed. The shaft thermal efficiency of the prime mover is 18 to 36% with exhaust gases from the turbine ranging from 806 to 986°F. If the exhaust heat can be used, overall thermal efficiency can increase.

Figure 13B shows a regenerator that uses the heat of the exhaust gases to heat the air from the compressor prior to combustion. Overall shaft efficiency can be increased to between 28 and 38% by using a regenerator or recuperator.

If process heat is required, the exhaust can satisfy a portion of that heat, and the combined system is a cogeneration system. The exhaust can be used (1) directly as a source of hot air, (2) in a large boiler or furnace as a source of preheated combustion air (the exhaust contains about 16% oxygen), or (3) to heat a process or working fluid such as the steam system shown in Figure 16. Overall thermal efficiency is [(shaft energy + heat energy) × 100] / (fuel energy). Thermal efficiencies of these systems vary from 50% to greater than 90%. The exhaust of a gas turbine has about 4000 to 8000 Btu of available heat per horsepower.

Additionally, because of the high oxygen content, the exhaust stream can support the combustion of an additional 30,000 Btu of fuel per horsepower. This additional heat can then be used in general manufacturing processes.

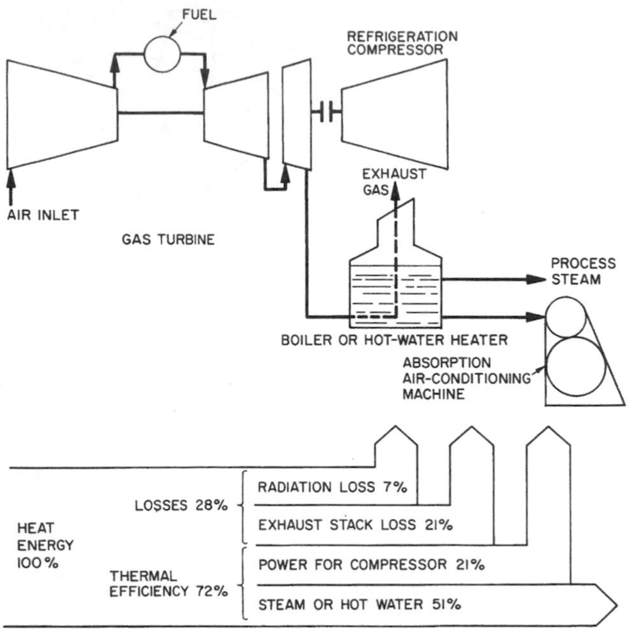

Fig. 16 Gas Turbine Refrigeration System Using Exhaust Heat

EMISSIONS AND NOISE

Emissions. Gas turbine power plants emit relatively low levels of CO_x and NO_x compared to other internal combustion engines; however, for each application, the gas turbine manufacturer should be consulted to ensure that applicable codes are met. Special care should be taken if high-sulfur fuel is being used because gas turbine exhaust stacks are typically not high, and dilution is not possible.

Noise. Gas turbine manufacturers have developed sound-attenuated enclosures that cover the turbine and gear package. Turbine drivers, when properly installed with a sound-attenuated enclosure, an inlet silencer, and an exhaust silencer, meet the strictest noise standards. The turbine manufacturer should be consulted for detailed noise level data and recommendations on the least expensive method of attenuation for a particular installation.

MAINTENANCE

Industrial gas turbines are designed to operate for 12,000 to 30,000 h between overhauls, with normal maintenance. Normal maintenance includes checking filters and oil level, inspecting for leaks, and so forth, all of which can be done by the operator with ordinary mechanics' tools. However, factory-trained service personnel are required to inspect engine components such as combustors and nozzles. These inspections, depending on the manufacturer's recommendations, are required as frequently as every 4000 h of operation.

Most gas turbines are maintained by condition monitoring and inspection rather than by specific overhaul intervals (called predictive maintenance). Gas turbines specifically designed for industrial applications may have an indefinite life for the major housings, shafts, and low-temperature components. Hot-section repair intervals for combustor and turbine components can vary from 10,000 to 100,000 h. The total cost of maintaining a gas turbine includes (1) cost of operator time, (2) normal parts replacement, (3) lubricating oil, (4) filter changes (combustion inlet air, fuel, and lubricating oil), (5) overhauls, and (6) factory service time (to conduct engine inspections). The cost of all these items can be estimated by the manufacturer and must be taken into account to determine the total operating cost.

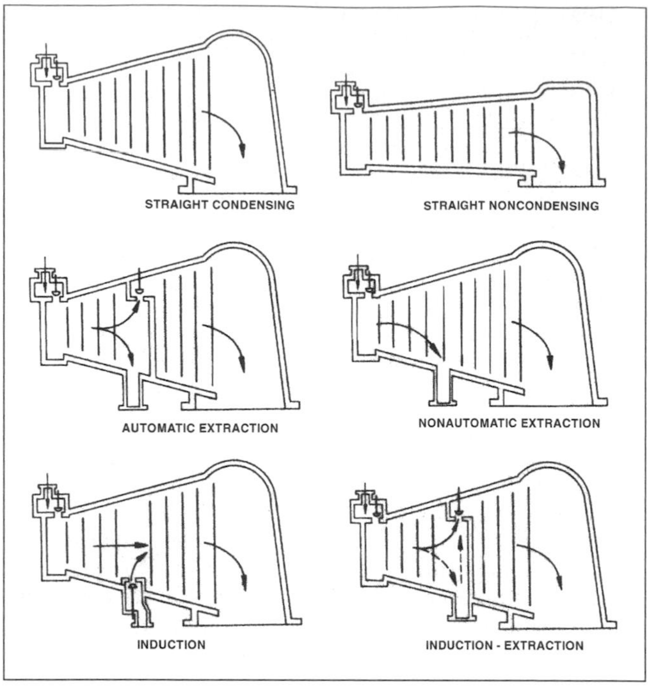

Fig. 17 Basic Types of Axial Flow Turbines

STEAM TURBINES

TYPES

Axial Flow Turbines

Conventional axial flow steam turbines direct steam axially through the peripheral blades of one or more staged turbine wheels (much like a pinwheel) one after another on the same shaft. Figure 17 shows basic types of axial turbines. NEMA *Standard* SM 24 defines these and further subdivisions of their basic families as follows:

Noncondensing (back pressure) turbine. A steam turbine designed to operate with an exhaust steam pressure at any level that may be required by a downstream process, where all condensing takes place.

Condensing turbine. A steam turbine with an exhaust steam pressure below atmospheric pressure, such that steam is directly and completely condensed.

Automatic extraction turbine. A steam turbine that has opening(s) in the turbine casing for the extraction of steam and means for directly regulating the extraction steam pressure.

Nonautomatic extraction turbine. A steam turbine that has opening(s) in the turbine casing for the extraction of steam without a means for controlling its pressure.

Induction (mixed pressure) turbine. A steam turbine with separate inlets for steam at two pressures, with an automatic device for controlling the pressure of the secondary steam induced into the turbine and means for directly regulating the flow of steam to the turbine stages below the induction opening.

Induction-extraction turbine. A steam turbine with the capability of either exhausting or admitting a supplemental flow of steam through an intermediate port in the casing, thereby maintaining a process heat balance. Turbines of the extraction and induction-extraction type may have several casing openings, each passing steam at a different pressure.

The necessary rotative force for shaft power in a turbine may be imposed on the turbine through the velocity of the steam, the pressure energy of the steam, or both. If velocity energy is used, the movable wheels are usually fitted with crescent-shaped blades. A row of fixed nozzles in the steam chest increases the steam velocity into the blades with little or no steam pressure drop across them and causes wheel rotation. Such combinations of nozzles and velocity-powered wheels are characteristic of an **impulse turbine**.

A **reaction turbine** uses alternate rows of fixed and moving blades generally of an airfoil shape. Steam velocity increases in the fixed nozzles and drops in the movable ones, while the steam pressure drops through both.

The power capability of a reaction turbine is maximum when the moving blades travel at about the velocity of the steam passing through them; in the impulse turbine, maximum power is produced with a blade velocity of about 50% of steam velocity. Steam velocity is related directly to pressure drop. To achieve the desired relationship between steam velocity and blade velocity without resorting to large wheel diameters or high rotative speeds, most turbines include a series of impulse or reaction stages or both, thus dividing the total steam pressure drop into manageable increments. A typical commercial turbine may have two initial rows of rotating impulse blading with an intervening stationary row (called a Curtis stage), followed by several alternating rows of fixed and movable blading of either the impulse or the reaction type. Most multistage turbines use some degree of reaction.

Construction. Turbine manufacturers' standards prescribe casing materials for various limits of steam pressure and temperature. The choice between built-up or solid rotors depends on turbine speed or inlet steam temperature. Water must drain from pockets within the turbine casing to prevent damage caused by condensate accumulation. Carbon rings or closely fitted labyrinths prevent leakage of steam between pressure stages of the turbine, outward steam leakage, and inward air leakage at the turbine glands. The erosive and corrosive effect of moisture entering with the supply steam must be considered. Heat loss is controlled by installation (often at the manufacturer's plant) of thermal insulation and protective metal jacketing on the hotter portions of the turbine casing.

Radial Inflow Turbines

Radial inflow turbines have a radically different configuration from axial flow machines. Steam enters through the center or eye of the impeller and exits from the periphery, much like the path of fluid through a compressor or pump; but in this case the steam actuates the wheel, instead of the wheel actuating the air or the water.

Radial, multistage arrangements comprise separate, single-stage wheels connected with integral reduction gearing in a factory-assembled package. Induction, extraction, and moisture elimination are accomplished in the piping between stages, giving the radial turbine a greater tolerance of condensate.

LUBRICATION SYSTEMS

Small turbines often have only simple oil rings to handle bearing lubrication, but most turbines for cogeneration service have a complete pressure lubrication system. Basic components include a shaft-driven oil pump, an oil filter, an oil cooler, a means of regulating oil pressure, a reservoir, and interconnecting piping. Turbines having a hydraulic governor may use oil from the lubrication circuit or, with certain types of governors, use a self-contained governor oil system. To ensure an adequate supply of oil to bearings during acceleration and deceleration periods, many turbines include an auxiliary motor or turbine-driven oil pump. Oil pressure-sensing devices act in two ways: (1) to stop the auxiliary pump once the shaft-driven pump has attained proper flow and pressure or (2) to start the auxiliary pump if the shaft-driven pump fails or loses pressure when decelerating. In some industrial applications, the lubrication systems of the turbine and the driven compressor are integrated. Proper oil pressure, temperature, and compatibility of the lubricant qualities must be maintained.

INSTRUMENTS AND CONTROLS

Starting Systems

Unlike reciprocating engines and combustion turbines, steam turbines do not require auxiliary starting systems. Steam turbines are started through controlled opening of the main steam valve, which is in turn controlled by the turbine governing system. Larger turbines with multiple stages and/or split shafting arrangements are started in a gradual manner to allow for controlled expansion and thermal stressing. Many of these turbines are provided with electrically powered turning gears that slowly rotate the shaft(s) during the initial stages of start-up.

Governing Systems

The wide variety of available governing systems permits the selection of a governor ideally matched to the characteristics of the driven machine and the load profiles. The principal and most common function of a fixed-speed steam turbine governing system is to maintain constant turbine speed despite load fluctuations or minor variations in supply steam pressure. This arrangement assumes that close control of the output of the driven component, such as a generator in a power plant, is primary to plant operation and that the generator can adjust its capacity to varying loads.

Often it is desirable to vary the turbine speed in response to an external signal. In centrifugal water chilling systems, for example, reduced speed generally reduces steam rate at partial load. An electric, electronic, or pneumatic device responds to the system load or the temperature of the fluid leaving the water chilling heat exchanger (evaporator). To avoid compressor surge and to optimize the steam rate, the speed is controlled initially down to some part load, then controlled in conjunction with the compressor's built-in capacity control (e.g., inlet vanes).

Process applications frequently require placing an external signal on the turbine governing system to reset the speed control point. Such external signals may be necessary to maintain a fixed compressor discharge pressure, regardless of load or condenser water temperature variations. Plants relying on a closely maintained heat balance may control turbine speed to maintain an optimum pressure level of steam entering, being extracted from, or exhausting from the turbine. One example is the combination turbine absorption plant, where control of pressure of the steam exhausting from the turbine (and feeding the absorption unit) is an integral part of the plant control system.

Components. The steam turbine governing system consists of (1) a speed governor (mechanical, hydraulic, electrical, or electronic); (2) a speed control mechanism (relays, servomotors, pressure- or power-amplifying devices, levers, and linkages); (3) governor-controlled valve(s); (4) a speed changer; and (5) external control devices, as required.

The **speed governor** responds directly to turbine speed and initiates action of the other parts of the governing system. The simplest speed governor is the direct-acting flyball, which depends on changes in centrifugal force for proper action. Capable of adjusting speeds through an approximate 20% range, it is widely used on single-stage, mechanical-drive steam turbines with speeds of up to 5000 rpm and steam pressure of up to 600 psig.

The speed governor used most frequently on centrifugal water chilling system turbines is the oil pump type. In its direct-acting form, oil pressure, produced by a pump either directly mounted on the turbine shaft or in some form responsive to turbine speed, actuates the inlet steam valve.

The oil relay hydraulic governor, as shown in Figure 18, has greater sensitivity and effective force. Here, the speed-induced oil pressure changes are amplified in a **servomotor** or **pilot-valve relay** to produce the motive effort required to reposition the steam inlet valve or valves.

The least expensive turbine has a single governor-controlled steam admission **throttle valve**, perhaps augmented by one or more small auxiliary valves (usually manually operated), which close off nozzles supplying the turbine steam chest for better part-load efficiency. Figure 19 shows the effect of auxiliary valves on part-load turbine performance.

For more precise speed governing and maximum efficiency without manual valve adjustment, multiple automatic nozzle control is used (Figure 20). Its principal application is in larger turbines where a single governor-controlled steam admission valve would be too large to permit sensitive control. The greater power required to actuate the multiple-valve mechanism dictates the use of hydraulic servomotors. **Speed changers** adjust the setting of the governing systems while the turbine is in operation. Usually, they comprise either a means of changing spring tension or a means of regulating oil flow by a needle valve. The upper limit of a speed changer's capability should not exceed the rated turbine speed. Such speed changers, while usually mounted on the turbine, may sometimes be remotely located at a central control point.

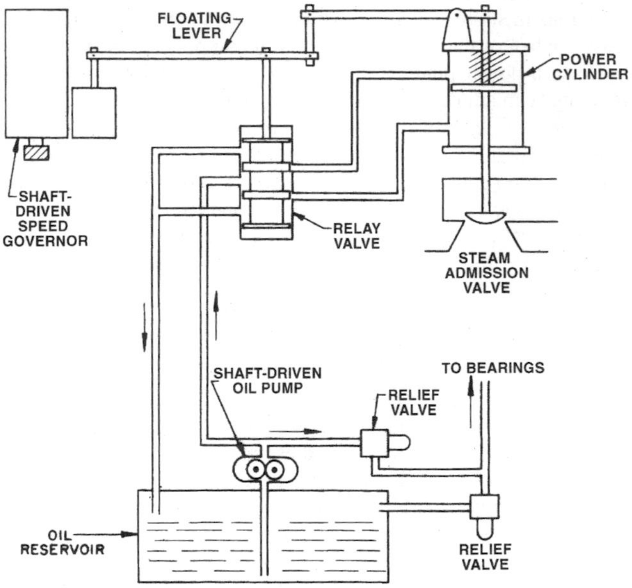

Fig. 18 Oil Relay Governor

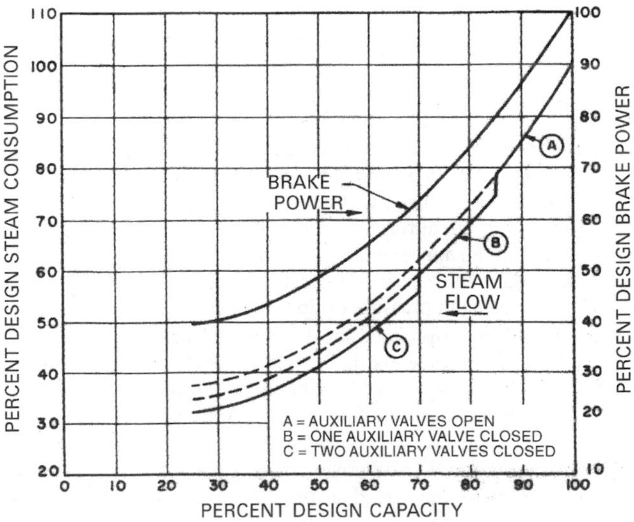

Fig. 19 Part-Load Turbine Performance Showing Effect of Auxiliary Valves

As stated previously, **external control devices** are often used when some function other than turbine speed is controlled. In such cases, a signal overrides the turbine speed governor's action, and the latter assumes a speed-limiting function. The external signal controls steam admission either by direct inlet valve positioning or by adjustment of the speed governor setting. The valve-positioning method either exerts mechanical force on the valve-positioning mechanism or, if power has to be amplified, regulates the pilot valve in a hydraulic servomotor system.

Where more precise control is required, the speed governor adjusting method is preferred. Although the external signal continually resets the governor as required, the speed governor always provides ideal turbine speed control. Thus, it maintains the particular set speed, regardless of load or steam pressure variations.

Classification. The National Electrical Manufacturers Association (NEMA) classifies steam turbine governors as shown in Table 3. **Range of speed changer adjustment**, expressed as a percentage of rated speed, is the range through which the turbine speed may be adjusted downward from rated speed by the speed changer, with the turbine operating under the control of the speed governor and

Table 3 NEMA Classification of Speed Governors

Class of Governor	Range of Speed Changer Adjustment, %	Maximum Steady-State Speed Regulation, %	Maximum Speed Variation, % Plus or Minus	Maximum Speed Rise, %	Trip Speed, % Above Rated Speed
A	10 to 65	10	0.75	13	15
B	10 to 80	6	0.50	7	10
C	10 to 80	4	0.25	7	10
D	10 to 90	0.50	0.25	7	10

Source: NEMA *Standard* SM 24.

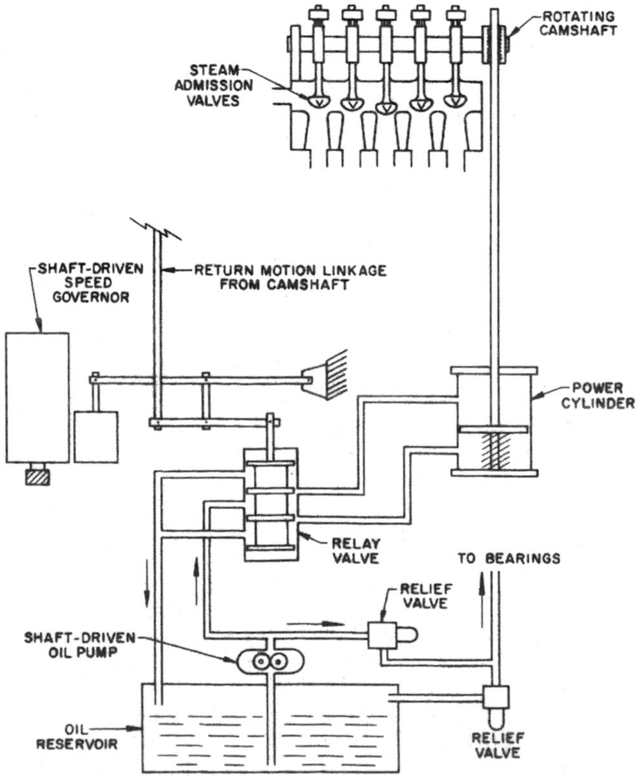

Fig. 20 Multivalve Oil Relay Governor

passing a steam flow equal to the flow at rated power, output, and speed. The range of the speed changer adjustment, expressed as a percentage of rated speed, is derived from the following equation:

$$\text{Range (\%)} = \frac{\left(\begin{matrix}\text{Rated}\\\text{speed}\end{matrix}\right) - \left(\begin{matrix}\text{Minimum}\\\text{speed setting}\end{matrix}\right)}{\text{Rated speed}} \times 100$$

Steady-state speed regulation, expressed as a percentage of rated speed, is the change in sustained speed when the power output of the turbine is gradually changed from rated power output to zero power output under the following conditions:

- Steam conditions (initial pressure, initial temperature, and exhaust pressure) are set at rated values and held constant.
- Speed changer is adjusted to give rated speed with rated power output.
- Any external control device is rendered inoperative and blocked in the open position to allow the free flow of steam to the governor-controlled valve(s).

The steady-state speed regulation, expressed as a percentage of rated speed, is derived from the following equation:

$$\text{Regulation (\%)} = \frac{\left(\begin{matrix}\text{Speed at zero}\\\text{power output}\end{matrix}\right) - \left(\begin{matrix}\text{Speed at rated}\\\text{power output}\end{matrix}\right)}{\text{Speed at rated power output}} \times 100$$

Steady-state speed regulation of automatic extraction or mixed pressure turbines is derived with zero extraction or induction flow and with the pressure-regulating system(s) inoperative and blocked in the position corresponding to rated extraction or induction pressure(s) at rated power output.

Speed variation, expressed as a percentage of rated speed, is the total magnitude of speed change or fluctuations from the speed setting. It is defined as the difference in speed variation between the governing system in operation and the governing system blocked to be inoperative, with all other conditions constant. This characteristic includes dead band and sustained oscillations. Expressed as a percentage of rated speed, the speed variation is derived from the following equation:

$$\begin{matrix}\text{\% Speed}\\\text{Variation}\end{matrix} = \frac{\left(\begin{matrix}\text{Speed change}\\\text{above set speed}\end{matrix}\right) - \left(\begin{matrix}\text{Speed change}\\\text{below set speed}\end{matrix}\right)}{\text{Rated speed}} \times 100$$

Dead band, also called wander, is a characteristic of the speed-governing system. It is the insensitivity of the speed-governing system and the total speed change during which the governing valve(s) do not change position to compensate for the speed change.

Stability is a measure of the ability of the speed-governing system to position the governor-controlled valve(s); thus, sustained oscillations of speed are not produced during a sustained load demand or following a change to a new load demand. Speed oscillations, also called hunt, are characteristics of the speed-governing system. The ability of a governing system to keep sustained oscillations to a minimum is measured by its stability.

Maximum speed rise, expressed as a percentage of rated speed, is the maximum momentary increase in speed obtained when the turbine is developing rated power output at rated speed and the load is suddenly and completely reduced to zero. The maximum speed rise, expressed as a percentage of rated speed, is derived from the following equation:

$$\text{Speed rise (\%)} = \frac{\left(\begin{matrix}\text{Maximum speed at}\\\text{zero power output}\end{matrix}\right) - \left(\begin{matrix}\text{Rated}\\\text{speed}\end{matrix}\right)}{\text{Rated speed}} \times 100$$

Protective Devices

In addition to speed-governing controls, certain safety devices are required on steam turbines. These include an overspeed mechanism, which acts through a quick-tripping valve independent of the main governor valve to shut off the steam supply to the turbine, and a pressure relief valve in the turbine casing. Overspeed trip devices may act directly, through linkages to close the steam valve, or hydraulically, by relieving oil pressure to allow the valve to close. Also, the turbine must shut down if other safety devices, such as oil pressure failure controls or any of the driven system's protective controls, so dictate. These devices usually act through an electrical interconnection to close the turbine trip valve mechanically or hydraulically. To shorten the coast-down time of a tripped condensing turbine, a vacuum breaker in the turbine exhaust opens to admit air on receiving the trip signal.

PERFORMANCE CHARACTERISTICS

The topping cogeneration cycle steam turbine is typically either a back pressure or an extraction condensing type that makes downstream, low-pressure thermal energy available for process use. Bottoming cycles commonly use condensing turbines because these yield more power, having lower grade throttle energy to begin with.

The highest steam plant efficiency is obtainable with a back pressure turbine when 100% of its exhaust steam is used for thermal process. The only inefficiencies are the gear drive, alternator, and inherent steam-generating losses. A large steam system topping cycle using an efficient water-tube boiler, economizer, and preheater can easily achieve an overall efficiency (fuel to end use) of more than 90%.

Full condensing turbine heat rates (Btu/hp·h) are the highest in the various steam cycles because the turbine's exhaust condenses, rejecting the latent heat of condensation (1036 Btu/lb at a condensing pressure of 1 psia and 101°F) to a waste heat sink (e.g., a cooling tower or river).

Conversely, the incremental heat that must be added to a low-pressure [e.g., 15 psig (30 psia)] steam flow to produce high-pressure, superheated steam for a topping cycle is only a small percentage of its latent heat of vaporization. For example, to produce 250°F, 30 psia saturated steam for a single-stage absorption chiller in a low-pressure boiler requires 1164 Btu/lb, of which 945 Btu/lb is the heat of vaporization. To boost this to 600°F, 320 psia requires an additional 146 Btu/lb, which is only 15% of the latent heat at 30 psia, for an enthalpy of 1310 Btu/lb.

A low-pressure boiler generating 30 psia steam to the absorber, directly, has a 75% fuel-to-steam efficiency, which is 15% lower than the 90% efficiency of a high-pressure boiler used in the cogeneration cycle. Therefore, from the standpoint of fuel cost, the power generated by the back pressure turbine is virtually free when its 30 psia exhaust is discharged into the absorption chiller.

The potential power-generating capacity and size of the required turbine are determined by its efficiency and steam rate (or water rate). This capacity is, in turn, the system's maximum steam load, if the turbine is sized to satisfy this demand. Efficiencies range from 55 to 80% and are the ratio of actual to theoretical steam rate, or actual to theoretical enthalpy drop from throttle to exhaust conditions.

NEMA *Standard* SM 24 defines theoretical steam rate as the quantity of steam per unit of power required by an ideal Rankine cycle, which is an isentropic or reversible adiabatic process of expansion. This can best be seen graphically on a enthalpy-entropy (Mollier) chart. Expressed algebraically, the steam rate is

$$w_t = \frac{2546}{h_i - h_e} \quad (1)$$

where

w_t = theoretical steam rate, lb/hp·h
h_e = enthalpy of steam at exhaust pressure and inlet entropy, Btu/lb
h_i = enthalpy of steam at throttle inlet pressure and temperature, Btu/lb
2546 = Btu/hp·h

This isentropic expansion through the turbine represents 100% conversion efficiency of heat energy to power. An example is shown on the Mollier chart in Figure 21 as the vertical line from 320 psia, 600°F, 1310 Btu/lb to 30 psia, 250°F, 93% quality, 1100 Btu/lb.

On the other hand, zero efficiency is a throttling, adiabatic, nonreversible horizontal line terminating at 30 psia, 552°F, 1310 Btu/lb. An actual turbine process would lie between 0 and 100% efficiency, such as the one shown at h_a actual exhaust condition of saturated steam at 30 psia, 250°F, 1164 Btu/lb; the actual turbine efficiency is

$$E_a = \frac{h_i - h_a}{h_i - h_e} \quad (2)$$

and the actual steam rate is

$$w_a = \frac{3412}{h_i - h_a} \quad (3)$$

where

w_a = actual steam rate, lb/kWh
h_a = enthalpy of steam at actual exhaust conditions, Btu/lb
3412 = Btu/kWh

For the case described,

$$E_a = (1310 - 1164)/(1310 - 1100) = 0.69 \text{ or } 69\%$$

As a cogeneration cycle, if the previously described absorption chiller has a capacity of 2500 tons, which requires 45,000 lb/h of 30 psia saturated steam, it can be provided by the 69% efficient turbine at an actual steam rate of 3412/146 = 23.4 lb/kWh; the potential power generation is

$$45{,}000/23.4 = 1923 \text{ kWh}$$

The incremental turbine heat rate to generate this power is only

$$(146 \times 45{,}000)/1923 = 3416 \text{ Btu/kWh}$$

instead of a typical 9000 Btu/kWh (thermal efficiency of 38%) for an efficient steam power plant with full condensing turbines and cooling towers.

Turbine performance tests should be conducted in accordance with the appropriate American Society of Mechanical Engineers (ASME) *Performance Test Code*: PTC 6, PTC 6S Report, or PTC 6A. The steam rate of a turbine is reduced with higher turbine speeds, a greater number of stages, larger turbine size, and a higher difference in heat content between entering and leaving steam

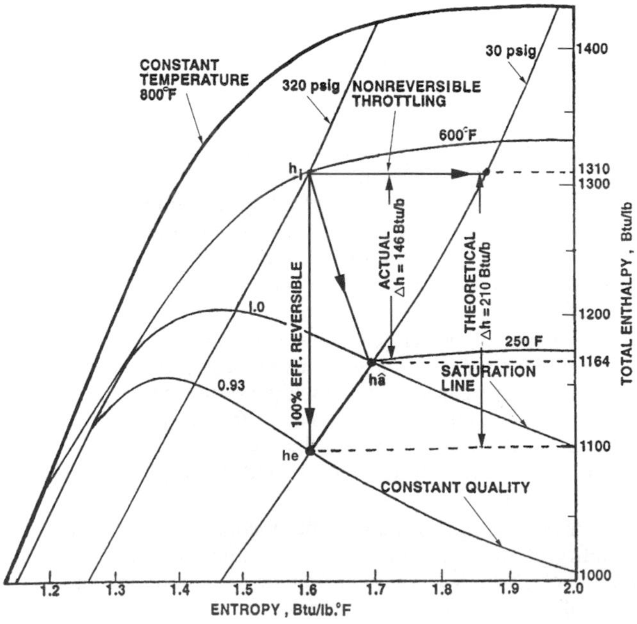

Fig. 21 Isentropic Versus Actual Turbine Process

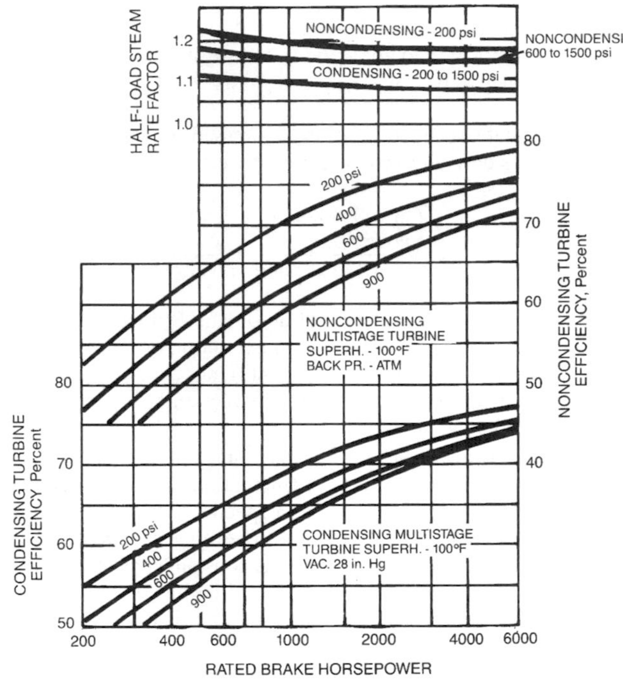

Fig. 22 Efficiency of Typical Multistage Turbines

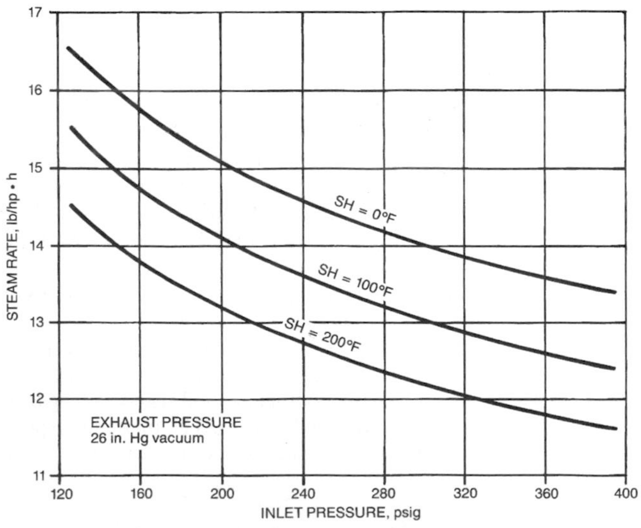

Fig. 23 Effect of Inlet Pressure and Superheat on Condensing Turbine

conditions. Often, one or more of these factors can be improved with only a nominal increase in initial capital cost. Cogeneration applications range, with equal flow turbines, from approximately 100 to 10,000 hp and from 3000 to 10,000 rpm, with the higher speeds generally associated with lower power outputs, and lower speeds with higher power outputs. (Some typical characteristics of turbines driving centrifugal water chillers are shown in Figure 15 and Figure 22 through Figure 26.)

While initial steam pressures for small turbines commonly fall in the 100 to 250 psig range, wide variations are possible. Turbines in the range of 2000 hp and above commonly have throttle pressures of 400 psig or greater.

Back pressure associated with noncondensing turbines generally ranges from 50 psig to atmospheric, depending on the use for the exhaust steam. Raising the initial steam temperature by superheating improves steam rates.

NEMA *Standards* SM 23 and SM 24 govern allowable deviations from design steam pressures and temperatures. Because of possible unpredictable variations in steam conditions and load requirements, turbines are selected for a power capability of 105 to 110% of design shaft output and speed capabilities of 105% of design rpm.

Because no rigid standards prevail for the turbine inlet steam pressure and temperature, fixed design conditions proposed by ASME/IEEE should be used to size the steam system initially. These values are 400 psig at 750°F, 600 psig at 825°F, 850 psig at 900°F, and 1250 psig at 950 or 1000°F.

Table 4 lists theoretical steam rates for steam turbines at common conditions. If project conditions dictate different throttle/exhaust conditions from the steam tables, theoretical steam rate tables or graphical Mollier chart analysis may be used.

Steam rates for multistage turbines depend on many variables and require extensive computation. Manufacturers provide simple tables and graphs to estimate performance, and these data are good guides for preliminary sizing of turbines and associated auxiliaries for the complete system.

If the entire exhaust steam flow from a base-loaded back pressure turbine is fully used, the maximum efficiency of a steam turbine cogeneration cycle is achieved. However, if the facility's thermal/electrical load cannot absorb the fully loaded output of the turbine, whichever profile is lower can be tracked, and the reduced power output or steam flow has to be accepted unless the output remaining is exported. The annual efficiency can still be high if the machine operates at significant combined loads for substantial periods. Straight steam condensing turbines offer no opportunity for topping cycles but are not unusual in bottoming cycles because the waste steam from the process can be most efficiently used by full-condensing turbines when there is no other use for low-pressure steam. Either back pressure or extraction condensing turbines may be used as extraction turbines.

The steam in an extraction turbine expands part of the way through the turbine until the pressure and temperature required by the external thermal load are attained. The remaining steam continues through the low-pressure turbine stages; however, it is easier to adjust for noncoincident electrical and thermal loads.

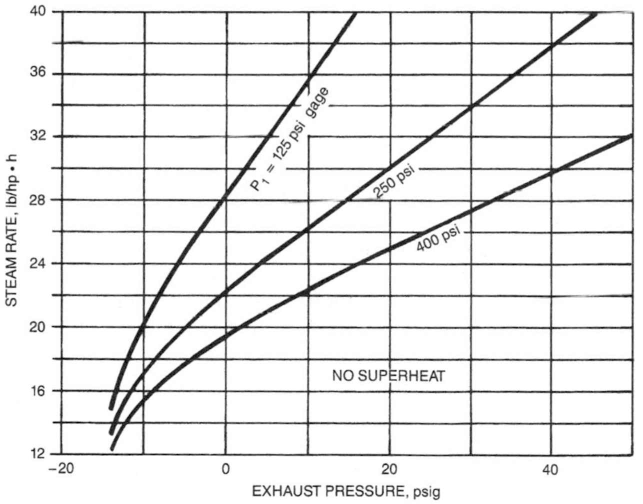

Fig. 24 Effect of Exhaust Pressure on Noncondensing Turbine

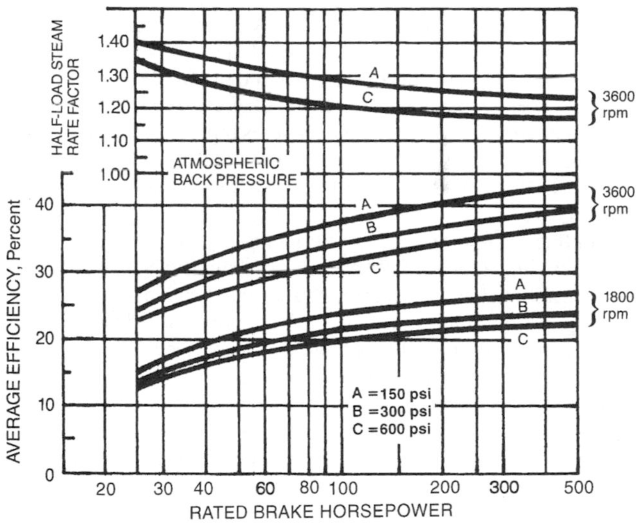

Fig. 25 Single-Stage Noncondensing Turbine Efficiency

Table 4 Theoretical Steam Rates for Steam Turbines at Common Conditions, lb/kWh

	Throttle Steam Conditions							
Exhaust Pressure	150 psig, 366°F, Saturated	200 psig, 388°F, Saturated	250 psig, 500°F, 94°F Superheat	400 psig, 750°F, 302°F Superheat	600 psig, 750°F, 261°F Superheat	600 psig, 825°F, 336°F Superheat	850 psig, 825°F, 298°F Superheat	850 psig, 900°F, 373°F Superheat
2 in. Hg (abs.)	10.52	10.01	9.07	7.37	7.09	6.77	6.58	6.28
4 in. Hg (abs.)	11.76	11.12	10.00	7.99	7.65	7.28	7.06	6.73
0 psig	19.37	17.51	15.16	11.20	10.40	9.82	9.31	8.81
10 psig	23.96	21.09	17.90	12.72	11.64	10.96	10.29	9.71
30 psig	33.6	28.05	22.94	15.23	13.62	12.75	11.80	11.07
50 psig	46.0	36.0	28.20	17.57	15.36	14.31	13.07	12.21
60 psig	53.9	40.4	31.10	18.75	16.19	15.05	13.66	12.74
70 psig	63.5	45.6	34.1	19.96	17.00	15.79	14.22	13.25
75 psig	69.3	48.5	35.8	20.59	17.40	16.17	14.50	13.51

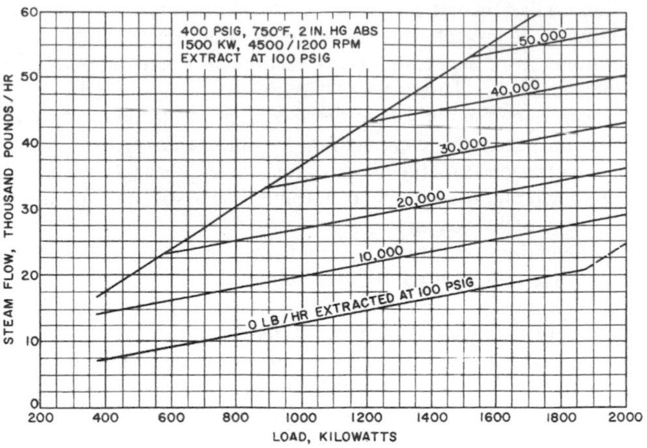

Fig. 26 Effect of Extraction Rate on Condensing Turbine

Because steam cycles operate at pressures exceeding those allowed by ASME and local codes for unattended operation, their use in cogeneration plants is limited to large systems where attendants are required for other reasons or the labor burden of operating personnel does not seriously affect overall economics.

Figure 26 shows the performance of a 1500 kW extraction condensing turbine, indicating the effect of various extraction rates on total steam requirements as follows: at zero extraction and 1500 kW, 17,500 lb/h or a water rate of 11.67 lb/kWh is required. When 45,000 lb/h is extracted at 100 psig, only 4000 lb/h more (49,000 − 45,000) is required at the throttle condition of 400 psig to develop the same 1500 kW, chargeable to the generation of electric power. The portion of input energy chargeable to the power is represented by the sum of the enthalpy of this 4000 lb/h at throttle conditions and the difference in enthalpy between the throttle and extraction conditions of the extracted portion of steam.

In effect, as the extraction rate increases, the overall efficiency increases. However at "full" extraction, a significant flow of "cooling" steam must still pass through the final turbine stages for condensing. At this condition, a simple back pressure turbine would be more efficient, if all of the exhaust steam could be used.

Full-condensing steam turbines have a maximum plant shaft efficiency (power output as a percentage of input fuel to the boiler) ranging from 20 to 36%, but they have no useful thermal output. Therefore, with overall plant efficiencies no better than their shaft efficiencies, they are unsuitable for topping cogeneration cycles.

At maximum extraction, the heat/power ratio of extraction turbines is relatively high. This high ratio makes it difficult to match facility loads, except those with very high base thermal loads, if reasonable annual efficiencies are to be achieved. As extraction rates decrease, plant efficiency approaches that of a condensing turbine, but can never reach it. Thus, the 17,500 lb/h (11.67 lb/kWh) illustrated for Figure 26 at 1500 kW and zero extraction represents a steam-to-electric efficiency of 28.6%, but a fuel-to-electric efficiency of 25%, with a boiler plant efficiency of 85%, developed as follows:

Isentropic Δh from 400 psig steam to 2 in. Hg (absolute) condensing pressure = 1022 Btu/lb output.

$$\text{Actual } \Delta h = \frac{3413 \text{ Btu/kWh}}{\text{Actual steam rate (lb/kWh)}}$$

Actual Δh = 3413/11.67 = 292 Btu/lb

Plant shaft efficiency = $(100 \times 292 \times 0.85)/1022$

= 24.3% (fuel input to electric output)

Radial inflow turbines are more efficient than single-stage axial flow turbines of the same output. They are available up to 15,000 hp from several manufacturers, with throttle steam up to 2100 psig, wheel speeds up to 60,000 rpm, and output shaft speeds as low as 3600 or 1800 rpm. It is these high wheel speeds that yield efficiencies of 70 to 80%, compared with single-stage axial turbines spinning at only 10,000 rpm with efficiencies of up to 40%.

MAINTENANCE

Maintenance requirements for steam turbines vary greatly with complexity of design, throttle pressure rating, duty cycle, and steam quality (both physical and chemical). Typically, several common factors can be attributed to operational problems with steam turbines. First are problems resulting from solid particle erosion. Turbines that undergo frequent cycling may see significant deterioration of initial pressure stages. Solid particles from steam lines, superheaters, or reheats can become dislodged and enter the turbine. These solids, over time, erode the nozzle and blade materials and initiate cracks and weakening of the rotating blades. Turbines subject to cyclical operation should be examined carefully every 18 to 36 months. Usually, nondestructive testing is used to establish material loss trends and predictable maintenance requirements for sustained planned outages.

Cycling duty as well as sustained operation at low loads can create problems in the lower pressure section of the turbine. Chemicals can concentrate in these sections and corrode the blades, especially where steam becomes saturated. Both stress and corrosion fatigue are common in lower pressure stages.

The best way to minimize both corrosion and erosion in steam turbines is to maintain proper feedwater/steam chemistry. The complexity of feedwater chemistry increases with steam pressures. To protect steam turbines from unnecessary damage, both mechanical and vapor carryover from the steam generator must be minimized.

Turbine seals, glands, and bearings are also common areas of deterioration and maintenance. Bearings require frequent examination, especially in systems that experience cyclical duty. Oil samples from the lubrication system should be taken regularly to determine concentration of solid particle contamination and changes in viscous properties. Filters and oil should be recycled according to manufacturers recommendations and the operational history of the turbine system.

Large multistage steam turbines usually contain instrumentation that monitors vibration within the casing. As deposits build on the blades, blade material erodes or corrodes; or as mechanical tolerance of bearing surfaces increases, nonuniform rotation creates increasing turbine vibration. Vibration instrumentation, consequently, is used to determine maintenance intervals for turbines, especially those subject to extensive base-load operations where visual examination is not feasible.

ECONOMICS

The thermal efficiency associated with the steam turbine prime mover (specifically the Rankine cycle) is relatively low. The cycle depends mostly on steam throttle conditions (temperature/pressure), regeneration, and condensing condition. Consequently, the capital cost per unit of output is relatively high when compared to combustion turbines or reciprocating engines. Furthermore, plants that operate with the elevated steam pressures required by turbines require attended operation, which adds to overall operational costs. Consequently, the steam turbine prime movers are not economically appealing in small cogeneration applications (i.e., less than 15 to 20 MW). Exceptions may include the following:

- Applications where the fuel source is inexpensive, such as municipal waste, process gas, or waste streams in which incineration with waste heat recovery can be applied
- District heating/cooling plants that have high process loads, thermal loads, or both

THERMAL OUTPUT AND RECOVERY

THERMAL OUTPUT CHARACTERISTICS

Cogeneration provides an opportunity to use the fuel energy that the prime mover does not convert into shaft energy. If the heat cannot be used effectively, the plant efficiency is limited to that of the prime mover. However, if the site heat energy requirements can be met effectively by the normally wasted heat at the level it is available from the prime mover, this salvaged heat will reduce the normal fuel requirements of the site and increase overall plant efficiency. The prime mover furnishes (1) mechanical energy from the shaft and (2) unused heat energy that remains after the fuel or steam has acted on the shaft. Shaft loads (generators, centrifugal chillers, compressors, and process equipment) require a given amount of rotating mechanical energy. Once the prime mover is selected to provide the required shaft output, it has a fixed relationship to heat availability and system efficiency, depending on the prime mover fuels versus heat balance curves. The ability to use the prime mover waste heat determines overall system efficiency and is one of the critical factors in economic feasibility.

Reciprocating Engines

In all reciprocating internal combustion engines except small air-cooled units, heat can be reclaimed from the jacket cooling system, lubricating system, turbochargers, exhaust, and aftercoolers. These engines require extensive cooling to remove excess heat not conducted into the power train during combustion and the heat resulting from friction. Coolant fluids and lubricating oil are circulated to remove this engine heat. Some engines permit the coolant to reach 250°F at above atmospheric pressure, which allows the coolant to flash into low-pressure steam (15 psig) after it has left the engine jacket (ebullient cooling).

Waste heat in the form of hot water or low-pressure steam is recovered from the engine jacket manifolds and exhaust, and additional heat can be recovered from the lubrication system (see Figure 30 through Figure 34).

Provisions similar to those used with gas turbines are necessary if supplemental heat is required, except an engine exhaust is rarely fired with a booster because it contains insufficient oxygen. If electrical supplemental heat is used, the additional electrical load is reflected back to the prime mover, which reacts accordingly by producing additional waste heat. This action creates a feedback effect, which can stabilize system operation under certain conditions. The approximate distribution of input fuel energy under selective control of the thermal demand for an engine operating at rated load is as follows:

Shaft power	33%
Convection and radiation	7%
Rejected in jacket water	30%
Rejected in exhaust	30%

These amounts vary with engine load and design. Four-cycle engine heat balance for naturally aspirated (Figure 27) and turbocharged gas engines (Figure 28) show typical heat distribution. The exhaust gas temperature for these engines is about 1200°F at full load and 1000°F at 60% load.

Two-cycle lower speed (900 rpm and below) engines operate at lower exhaust gas temperatures, particularly at light loads, because the scavenger air volumes remain high through the entire range of

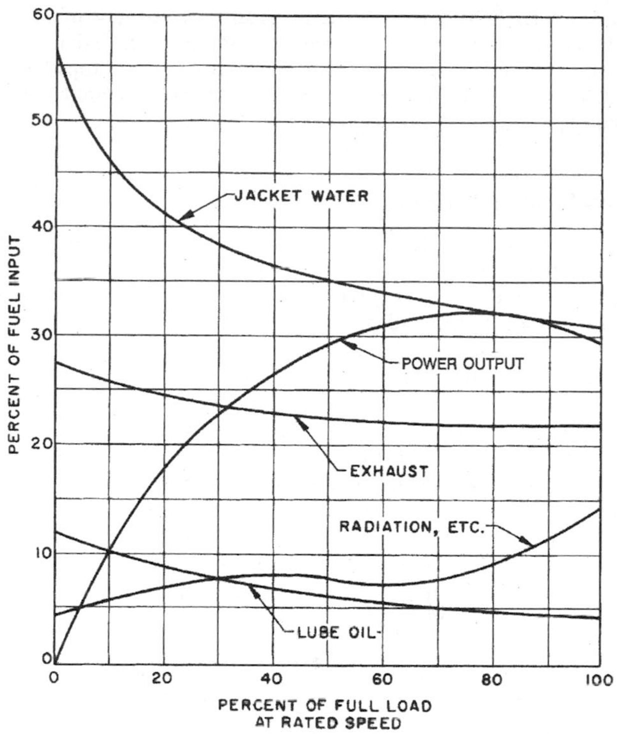

Fig. 27 Heat Balance for Naturally Aspirated Engine

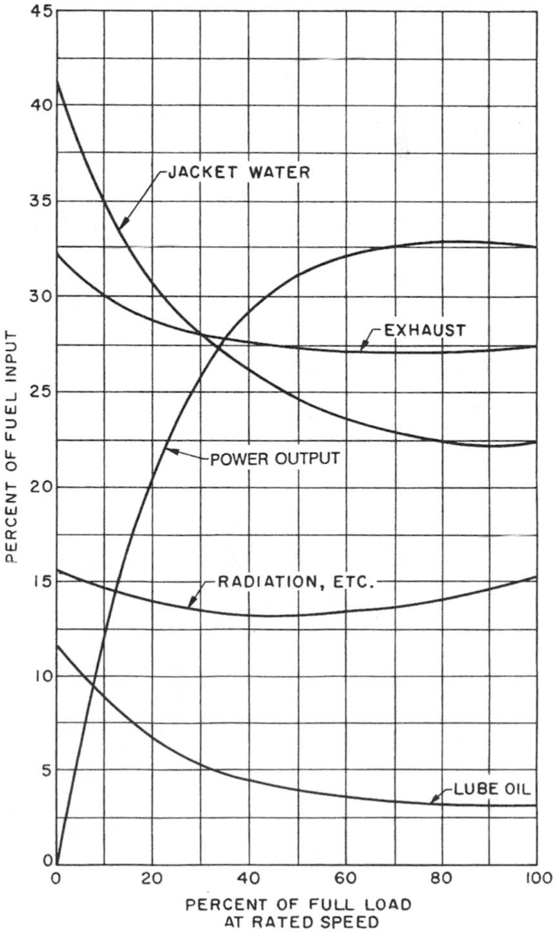

Fig. 28 Heat Balance for Turbocharged Engine

capacity. High-volume, lower temperature exhaust gas offers less efficient exhaust gas heat recovery possibilities. The exhaust gas temperature is approximately 700°F at full load and drops below 500°F at low loads. Low-speed engines generally cannot be ebulliently cooled because their cooling systems often operate at 170°F and below.

Combustion Turbines

In the gas turbine cycle, the average fuel to electrical shaft efficiency ranges from approximately 12% to above 35%, with the remainder of the fuel energy discharged in the exhaust and through radiation or internal coolants in large turbines. A minimum stack exhaust temperature of about 300°F is required to prevent condensation. Because the heat rate efficiency is lower, the quantity of heat recoverable per unit of power is greater for a gas turbine than for a reciprocating engine. This heat is generally available at the higher temperature of the exhaust gas. The net result is an overall thermal efficiency of 60% and higher. Because gas turbine exhaust contains a large percentage of excess air, afterburners or boost burners may be installed in the exhaust to create a supplementary boiler system. This system can provide additional steam or level the steam production during reduced turbine loads. Boost burning can increase cycle efficiency to a maximum of 93%.

Absorption chillers capable of operating directly off turbine exhausts with coefficients of performance of 1.14 or more are also available. The conventional method of controlling steam or hot water production in a heat recovery system at part load is to by-pass a portion of the exhaust gases around the boiler tubes and out the exhaust stack through a gas bypass valve assembly (Figure 29).

Steam Turbines

Steam turbine exhaust or extracted steam, reduced in both pressure and temperature from throttle conditions as it generates the prime mover shaft power, may be fed to heat exchangers, absorption chillers, steam turbine-driven centrifugal chillers, pumps, or other equipment. Its value as recovered energy is the same as that of the fuel otherwise needed to generate the same steam flow at the condition leaving the prime mover in an independent boiler plant cycle.

Heat/Power Ratio and Quality of Heat

Selection of the prime mover depends on the thermal and power profiles required by the end user and on the contemporaneous relationship of these profiles. Ideally, the recoverable heat is fully used as the prime mover follows the power load, but this ideal never occurs over extended periods.

For maximum equipment use and least energy waste, the following methods are used to produce only the power and thermal energy that is required on site:

- Match the heat/power ratio of the prime mover to that of the user's hourly load profile.
- Store excess power as chilled water or ice when the thermal demand exceeds the coincident power demand.
- Store excess thermal production as heat when the power demand exceeds the heat demand. Either cool or heat storage must be able to productively discharge most of its energy before it is dissipated to the environment.
- Sell excess power or heat on a mutually acceptable contract basis to a user outside of the host facility. Usually the buyer is the local utility, but sometimes it is a nearby facility.

The quality of recovered energy is the second major determinant in selecting the prime mover. If the quantity of high-temperature (above 260°F) recoverable heat available from an engine's exhaust is significantly less than that demanded by end users, a combustion or steam turbine may offer greater opportunity.

Low heat/power ratios of 1 to 3 lb/h steam per horsepower output of the prime mover indicate the need for one with a high shaft efficiency of 30 to 45% (shaft energy/fuel input). This efficiency is a good fit for an engine because its heat output is available as 15 psig steam or 250°F water. Higher temperature/pressures are available, but only from a exhaust gas recovery system, separated in that case from the low-temperature jacket water system. However, for a typical case in which 30% of the fuel energy is in the exhaust, approximately 50% of this energy is recoverable (with 300°F final exhaust gas temperature); less than 50% is recoverable if higher steam pressures are required.

Medium heat/power ratios of 4 to 11 lb steam/hp·h can be provided by combustion turbines, which are inherently low in shaft efficiency. Smaller turbines, for example, are only 20 to 25% efficient, with 75 to 80% of their fuel energy released into the exhaust.

High heat/power ratios of 8 to 40 lb steam/hp·h, provided by various steam turbine configurations make this prime mover highly flexible for higher thermal demands. The designer can vary throttle, exhaust and/or extraction conditions, and turbine efficiency to attain the most desirable ratio for varying heat/power loads in many applications; thus furnishing a wide variety of thermal energy quality levels.

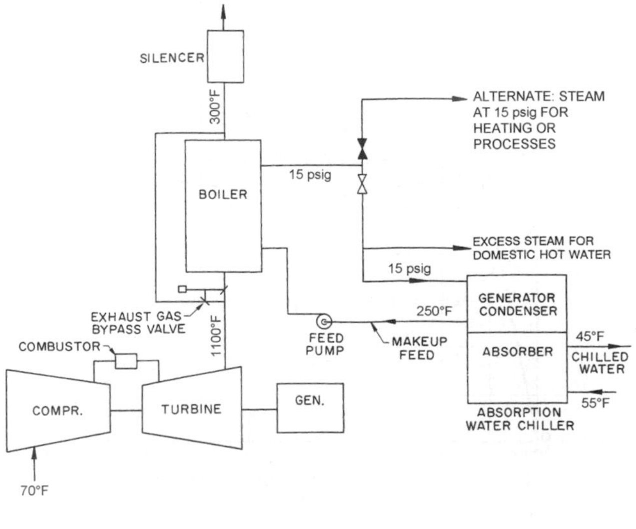

Fig. 29 Typical Heat Recovery Cycle for Gas Turbine

HEAT RECOVERY

Reciprocating Engines

Engine Jacket Heat Recovery. Engine jacket cooling passages for reciprocating engines, including the water cooling circuits in the block, heads, and exhaust manifolds, must remove about 30% of the heat input to the engine. If the machine operates at above 180°F coolant temperature, condensation of combustion products should produce no ill effects. Some engines have modified gaskets and seals to enable satisfactory operation up to 250°F at 30 psia. To avoid thermal stress, the temperature rise through the engine jacket should not exceed 15°F. Flow rates must be kept within the engine manufacturers' design limitations to avoid erosion from excessive flow or inadequate distribution within the engine from low flow rates.

Engine-mounted, water circulating pumps driven from an auxiliary shaft can be modified with proper shaft seals and bearings to give good service life at the elevated temperatures. Configurations that have a circulating pump for each engine increase reliability because the remaining engines can operate if one engine pump assembly fails. An alternative design uses an electric-drive pump battery to circulate water to several engines and has a standby pump

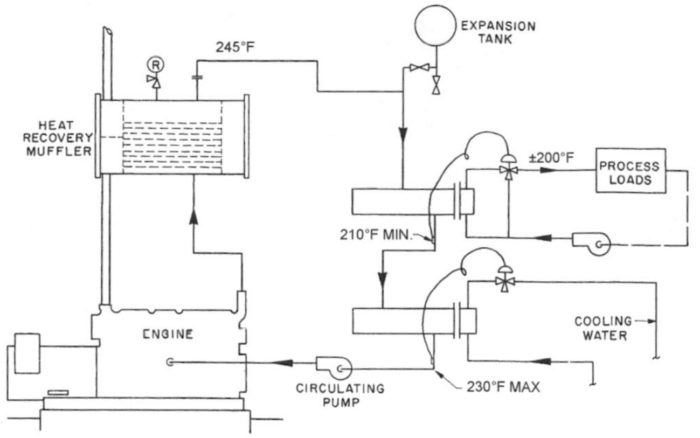

Fig. 30 Hot Water Heat Recovery

assembly in reserve, interconnected so that any engine or pump can be cut off without disabling the jacket water system.

Forced-circulation hot water, at 250°F and 30 psia, can be used for many process loads, including water heating systems for comfort and process loads, absorption refrigeration chillers, and service water heating. Distribution of the engine jacket coolant must be limited to reduce the risk of leaks, contamination, or other failures in downstream equipment that could prevent engine cooling.

One solution is to confine each engine circuit to its individual engine, using a heat exchanger to transfer the salvaged heat to another circuit that serves several engines via an extensive distribution system. An additional heat exchanger is needed in each engine circuit to remove heat whenever the waste heat recovery circuit does not extract all the heat produced (Figure 30). The reliability of this approach is high, but because the salvage circuit must be at a lower temperature than the engine operating level, it requires either larger heat exchangers, piping, and pumps or a sacrifice of system efficiency that might result from these lower temperatures.

Low-temperature limit controls prevent excessive system heat loads from seriously reducing the operating temperature of the engines. Engine castings may crack if they are suddenly cooled because a large demand for heating temporarily overloads the system. A heat storage tank is an excellent buffer for such occasions because it can provide heat at a very high rate for short periods to protect the machinery serving the heat loads. The heat level can be controlled with supplementary heat input such as an auxiliary boiler.

Another method of protecting the engine heat recovery system from instantaneous load is to flash the recovered heat into steam (**ebullient cooling** system). These pressurized, reciprocating engine cooling, heat recovery systems are also called high-temperature systems because they operate with a few degrees of temperature differential in the range of 212 to 250°F.

Engine builders have approved various conventionally cooled models for ebullient cooling. Azeotropic antifreeze solutions should be used to ensure constant coolant composition in boiling. However, the engine power outputs are generally derated by the manufacturer when the jackets are cooled ebulliently.

Ebullient cooling or hot water cooling equipment usually replaces the radiator, the belt-driven fan, and the gear-driven water pump. Many heat recovery systems use the jacket water in water-to-liquid exchangers when the thermal loads are at temperature levels below 180°F, particularly when their load profiles can absorb most or all of the jacket heat. Such systems are easier to design and have historically developed fewer problems than those with ebullient cooling.

Ebullient systems use the steam as the distribution fluid (Figure 31). By using a back pressure regulator, the steam pressure on the engine can be kept uniform, and the auxiliary steam boilers can supplement the distribution steam header. The same engine coolant

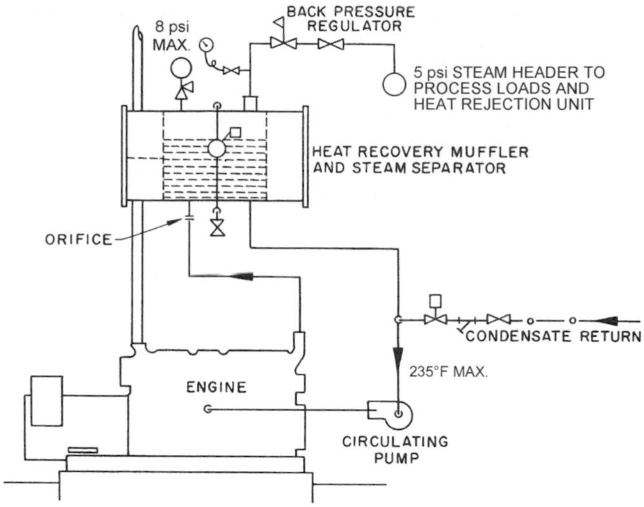

Fig. 31 Hot Water Engine Cooling with Steam Heat Recovery (Forced Recirculation)

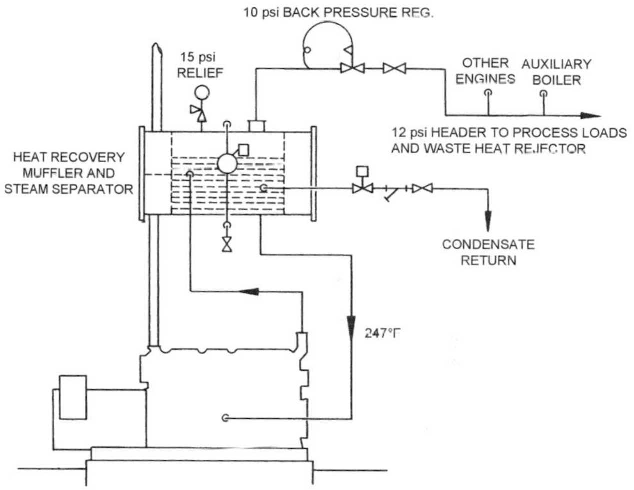

Fig. 32 Engine Cooling with Gravity Circulation and Steam Heat Recovery

system using forced-circulation water at 235°F entering and 250°F leaving the engine can produce steam at 235°F, 8 psig. The engine cooling circuit is the same as in other high-temperature water systems in that an adequate static pressure is maintained above the circulating pump inlet to avoid cavitation at the pump, and flow is restricted at the entrance to the steam flash chamber. The flow restriction prevents flash steam, which can severely damage the engine.

A limited distribution system that distributes steam through nearby heat exchangers can salvage heat without contaminating the main engine cooling system. The salvage heat temperature is kept high enough for most low-pressure steam loads. Returning condensate must be treated to prevent engine oxidation. The flash tank can accumulate the sediment from treatment chemicals.

Some engines can use natural-convection ebullient cooling, in which water circulated by gravity flows to the bottom of the engine. There, it is heated by the engine to form steam bubbles (Figure 32). This heat lowers the density of the fluid, which then rises to a separating chamber, or flash tank, where the steam is released and the water is recirculated to the engine.

Maintaining the flash tank at a static pressure slightly above that of the highest part of the engine causes a rapid flow of coolant through the engine, so only a small temperature difference is maintained between inlet and outlet. A constant back pressure must be maintained at the steam outlet of the flash chamber to prevent sudden lowering of the operating pressure. If the operating pressure changes rapidly, steam bubbles in the engine could quickly expand, interfering with heat transfer and causing overheating at critical points in the engine. This method of engine cooling is suitable for operation at up to 250°F, 15 psig steam, using components built to meet Section VIII of the ASME *Boiler and Pressure Vessel Code.*

The engine coolant passages must be designed for gravity circulation and must eliminate the formation of freely bubbling steam. It is important to consider how coolant flows from the engine heads and exhaust manifolds. Each coolant passage must vent upward to encourage gravity flow and the free release of steam toward the steam separating chamber. Engine heat removal is most satisfactory when these circuits are used because fluid temperatures are uniform throughout the machine. The free convection cooling system depends less on mechanical accessories and is more readily arranged for completely independent assembly of each engine with its own coolant system than other engine cooling systems.

Any system that generates steam from an engine has inherent design hazards. It is preferable, when the downstream facility thermal profiles and system arrangements permit, to use conventional hot water heat recovery such as shown in Figure 30.

Lubricant Heat Recovery. Lubricant heat exchangers should maintain oil temperatures at 190°F, with the highest coolant temperature consistent with the economical use of the salvaged heat. Engine manufacturers usually size their oil cooler heat transfer surface on the basis of 130°F entering coolant water and without provision for additional lubricant heat gains that occur with high engine operating temperatures. The cost of obtaining a reliable supply of lower temperature cooling water must be compared with the cost of increasing the size of the oil cooling heat exchanger and operating at a cooling water temperature of 165°F. In applications where engine jacket coolant temperatures are above 220°F and where there is use for heat at 155 to 165°F, the heat from the lubricant can be salvaged profitably.

The coolant should not foul the oil cooling heat exchanger. Untreated water should not be used unless it is free of silt, calcium carbonates, sand, and other contaminants. A good solution is a closed-circuit, treated water system using an air-cooled heat transfer coil with freeze-protected coolant. A domestic water heater can be installed on the closed circuit to act as a reserve heat exchanger and to salvage some useful heat when needed. Inlet air temperatures are not as critical with diesel engines as with turbocharged natural gas engines, and the aftercooler water on diesel engines can be run in series with the oil cooler (Figure 33). A double-tube heat exchanger can also be used to prevent contamination from a leaking tube.

Turbocharger Heat Recovery. Turbochargers on natural gas engines require medium fuel gas pressure (12 to 20 psi) and rather low aftercooler water temperatures (90°F or less) for high compression ratios and best fuel economy. Aftercooler water at 90°F is a premium coolant in many applications because the usual sources are raw domestic water and evaporative cooling systems such as cooling towers. Aftercooler water at temperatures as high as 135°F can be used, although the engine is somewhat derated. Using a domestic water solution may be expensive because (1) the coolant is continuously needed while the engine is running, and (2) the available heat exchanger designs require a large amount of water even though the load is less than 200 Btu/hp·h. A cooling tower can be used, but unless required for other reasons, it can increase initial costs. Also, it requires winter freeze protection and water quality control. If a cooling tower is used, the lubricant cooling load must be included in the tower design load for periods when there is no use for salvage lubricant heat.

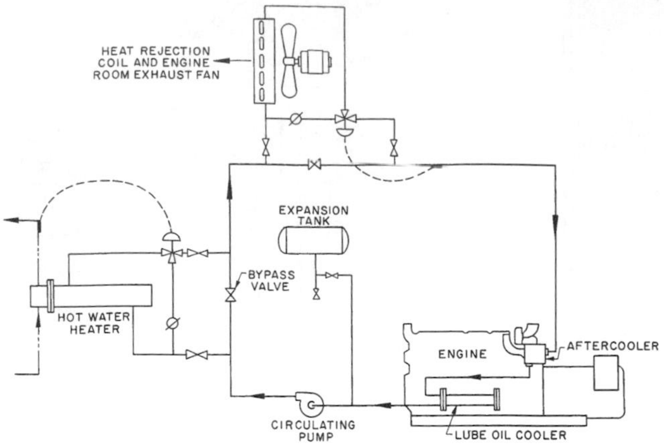

Fig. 33 Lubricant and Aftercooler System

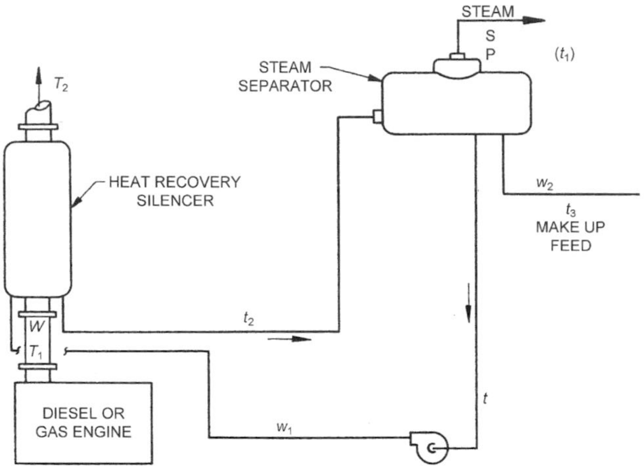

Fig. 34 Exhaust Heat Recovery with Steam Separator

Exhaust Gas Heat Recovery. Almost all heat transferred to the engine jacket cooling system can be reclaimed in a standard jacket cooling process or in combination with exhaust gas heat recovery. However, only part of the exhaust heat can be salvaged due to the limitations of heat transfer equipment and the need to prevent flue gas condensation (Figure 34). Energy balances are often based on standard air at 60°F; however, exhaust temperature cannot easily be reduced to this level.

A minimum exhaust temperature of 250°F was established by the Diesel Engine Manufacturers Association (DEMA). Many heat recovery boiler designs are based on a minimum exhaust temperature of 300°F to avoid condensation and acid formation in the exhaust piping. Final exhaust temperature at part load is important on generator sets that operate at part load most of the time. Depending on the initial exhaust temperature, approximately 50 to 60% of the available exhaust heat can be recovered.

A complete heat recovery system, which includes jacket water, lubricant, turbocharger, and exhaust, can increase the overall thermal efficiency from 30% for the engine generator alone to approximately 75%. Exhaust heat recovery equipment is available in the same categories as standard fire-tube and water-tube boilers.

Other heat recovery silencers and boilers include coil-type water heaters with integral silencers, water-tube boilers with steam separators for gas turbine, and engine exhausts and steam separators for high-temperature cooling of engine jackets. Recovery boiler design should facilitate inspection and cleaning of the exhaust gas and water sides of the heat transfer surface. Diesel engine units should

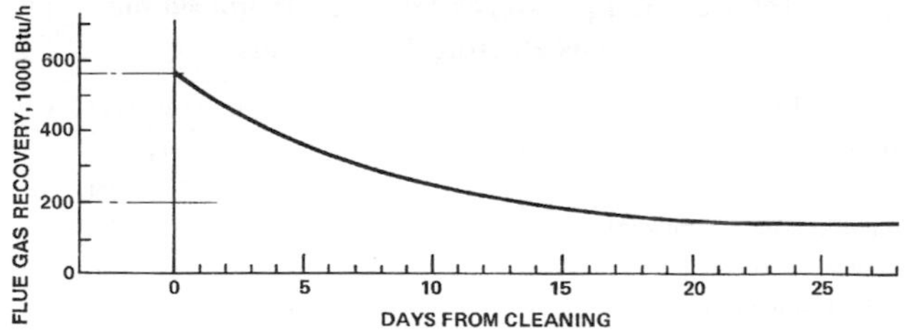

Fig. 35 Effect of Soot on Energy Recovery from Flue Gas Recovery Unit on Diesel Engine

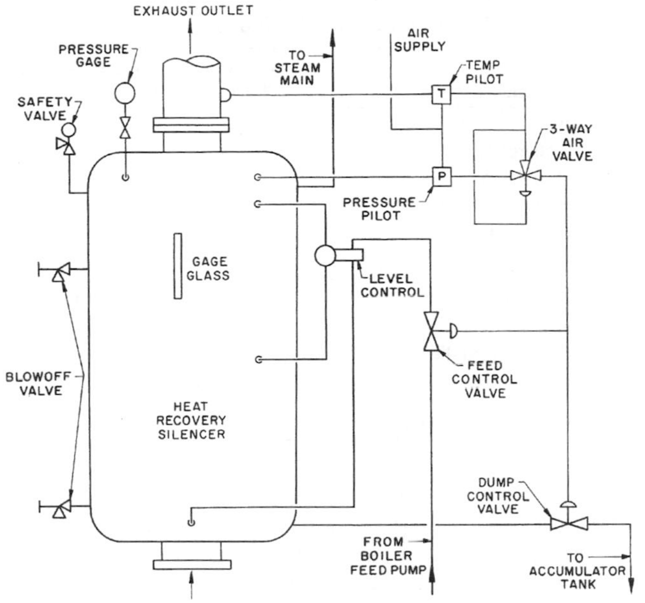

Fig. 36 Automatic Boiler System with Overriding Exhaust Temperature Control

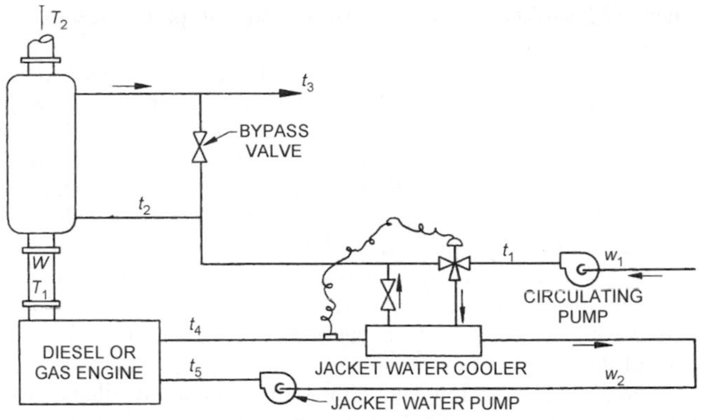

Fig. 37 Combined Exhaust and Jacket Water Heat Recovery System

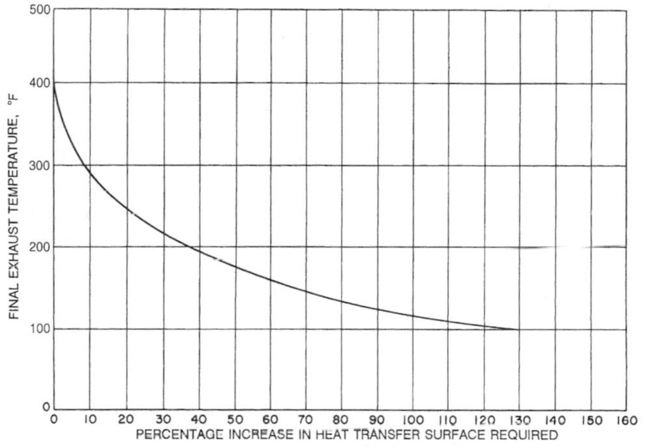

Fig. 38 Effect of Lowering Exhaust Temperature below 300°F

have a means of soot removal because soot deposits can quickly reduce the heat exchanger effectiveness (Figure 35). These recovery boilers can also serve other requirements of the heat recovery system, such as surge tanks, steam separators, and fluid level regulators.

In many applications for heat recovery equipment, the demand for heat requires some method of **automatic control**. In vertical recovery boilers, control can be achieved by varying the water level in the boiler. Figure 36 shows a control system using an air-operated pressure controller with diaphragm or bellows control valves. When steam production begins to exceed demand, the feed control valve begins to close, throttling the feedwater supply. Concurrently, the dump valve begins to open, and the valves reach an equilibrium position that maintains a level in the boiler to match the steam demand.

This system can be fitted with an overriding exhaust temperature controller that regulates the boiler output to maintain a preset minimum exhaust temperature at the outlet. This type of automatic control is limited to vertical boilers because the ASME *Boiler and Pressure Vessel Code* does not permit horizontal boilers to be controlled by varying the water level. Instead, a control condenser, radiator, or thermal storage can be used to absorb excess steam production.

In hot water units, a temperature-controlled bypass valve can divert the water or the exhaust gas to achieve automatic modulation with heat load demand (Figure 37). If the water is diverted, precautions must be taken to limit the temperature rise of the lower flow of water in the recovery device, which could otherwise cause steaming. The heat recovery equipment should not adversely affect the primary function of the engine to produce work. Therefore, the design of waste heat recovery boilers should begin by determining the back pressure imposed on the engine exhaust gas flow. Limiting back pressures vary widely with the make of engine, but the typical value is 6 in. of water gage. The next step is to calculate the heat transfer area that gives the most economical heat recovery without reducing the final exhaust temperature below 300°F.

Heat recovery silencers are designed to adapt to all engines; efficient heat recovery depends on the **initial exhaust temperature**. Most designs can be modified by adding or deleting heat transfer surface to suit the initial exhaust conditions and to maximize heat recovery down to a minimum temperature of 300°F.

Figure 38 illustrates the effect of lowering the exhaust temperature below 300°F. This curve is based on a specific heat recovery silencer design with an initial exhaust temperature of 1000°F. Lowering the final temperature from 300°F to 200°F increases heat recovery 14% but requires a 38% surface increase. Similarly, a reduction from 300°F to 100°F increases heat recovery 29% but requires a 130% surface increase. Therefore, the cost of heat transfer surface is a factor that must be considered when determining the final temperature.

Another factor to consider is the problem of water vapor condensation and acid formation if the exhaust gas temperature falls below the dew point. This point varies with fuel and atmospheric conditions and is usually in the range of 125 to 150°F. This gives an adequate margin of safety for the 250°F minimum temperature recommended by DEMA. Also, it allows for other conditions that could cause condensation, such as an uninsulated boiler shell or other cold surface in the exhaust system, or part loads on an engine.

The quantity of water vapor varies with the type of fuel and the intake air humidity. Methane fuel, under ideal conditions and with only the correct amount of air for complete combustion, produces 2.25 lb of water vapor for every pound of methane burned. Similarly, diesel fuel produces 1.38 lb of water vapor per pound of fuel. In the gas turbine cycle, these relationships would not hold true because of the large quantities of excess air. The condensates formed at low exhaust temperatures can be highly acidic. Sulfuric acid from diesel fuels and carbonic acid from natural gas fuels can cause severe corrosion in the exhaust stack as well as in colder sections of the recovery device.

If engine exhaust flow and temperature data are available, and maximum recovery to 300°F final exhaust temperature is desired, the basic **exhaust recovery equation** is

$$q = \dot{m}_e (c_p)_e (t_e - t_f) \tag{4}$$

where

q = heat recovered, Btu/h
$\dot{m}_e$ = mass flow rate of exhaust, lb/h
t_e = exhaust temperature, °F
t_f = final exhaust temperature, °F
$(c_p)_e$ = specific heat of exhaust gas = 0.25 Btu/lb·°F

Equation (4) applies to both steam and hot water units. To estimate the quantity of steam obtainable, the total heat recovered q is divided by the latent heat of steam at the desired pressure. The latent heat value should include an allowance for the temperature of the feedwater return to the boiler. The basic equation is

$$q = \dot{m}_s (h_s - h_f) \tag{5}$$

where

$\dot{m}_s$ = mass flow rate of steam, lb/h
h_s = enthalpy of steam, Btu/lb
h_f = enthalpy of feedwater, Btu/lb

Similarly, the quantity of hot water can be determined by

$$q = \dot{m}_w (c_p)_w (t_o - t_i) \tag{6}$$

where

$\dot{m}_w$ = mass flow rate of water, lb/h
$(c_p)_w$ = specific heat of water = 1.0 Btu/lb·°F
t_o = temperature of water out, °F
t_i = temperature of input water, °F

If the shaft power is known but engine data are not, the heat available from the exhaust is about 1000 Btu/h per horsepower output or 1 lb/h steam per horsepower output. The exhaust recovery equations also apply to gas turbines, although the rate of flow will be much greater. The estimate for gas turbine boilers is 8 to 10 lb/h of steam per horsepower output. These values are reasonably accurate for steam pressures in the range of 15 to 150 psig.

The normal procedure is to design and fabricate heat recovery boilers to the ASME *Boiler and Pressure Vessel Code* (Section VIII) for the working pressure required. Because the temperatures in most exhaust systems are not excessive, it is common to use flange or firebox quality steels for the pressure parts and low-carbon steels for the nonpressure components. Wrought iron or copper can be used for extended-fin surfaces to improve heat transfer capacities.

In special applications such as sewage gas engines, where exhaust products are highly corrosive, wrought iron or special steels are used to improve corrosion resistance. Exhaust heat may be used to make steam, or it may be used directly for drying or other processes. The steam provides space heating, hot water, and absorption refrigeration, which may supply air conditioning and process refrigeration. Heat recovery systems generally involve equipment specifically tailored for the job, although conventional fire-tube boilers are sometimes used. Exhaust heat may be recovered from reciprocating engines by a muffler-type exhaust heat recovery unit. Table 5 gives the temperatures normally required for various heat recovery applications.

Table 5 Temperatures Normally Required for Various Heating Applications

Application	Temperature, °F
Absorption refrigeration machines	190 to 245
Space heating	120 to 250
Water heating (domestic)	120 to 200
Process heating	150 to 250
Evaporation (water)	190 to 250
Residual fuel heating	212 to 330
Auxiliary power producers, with steam turbines or binary expanders	190 to 350

Table 6 Full-Load Exhaust Mass Flows and Temperatures for Various Engines

Type of Engine	Mass Flow, lb/bhp·h	Temperature, °F
Two-cycle		
Blower-charged gas	16	700
Turbocharged gas	14	800
Blower-charged diesel	18	600
Turbocharged diesel	16	650
Four-cycle		
Naturally aspirated gas	9	1200
Turbocharged gas	10	1200
Naturally aspirated diesel	12	750
Turbocharged diesel	13	850
Gas turbine, nonregenerative	18 to 48[a]	800 to 1050[a]

[a]Lower mass flows correspond to more efficient gas turbines.

In some engines, exhaust heat rejection exceeds jacket water rejection. Generally, gas engine exhaust temperatures run from 700 to 1200°F, as shown in Table 6. About 50 to 75% of the sensible heat in the exhaust may be considered recoverable. The economics of exhaust heat boiler design often limits the temperature differential between exhaust gas and generated steam to a minimum of 100°F. Therefore, in low-pressure steam boilers, the gas temperature can be reduced to 300 to 350°F; the corresponding final exhaust temperature range in high-pressure steam boilers is 400 to 500°F.

Because higher airflows are required to purge the cylinders of two-cycle engines, they have lower temperatures than four-cycle engines. As a result, they are less desirable for heat recovery. Gas turbines have even larger flow rates, but at high enough temperatures to make heat recovery worthwhile when the recovered heat can be efficiently used.

Combustion Turbines

The information in the preceding section on Exhaust Gas Heat Recovery applies to combustion turbines as well.

Steam Turbines

Noncondensing Turbines. The back pressure or noncondensing turbine is the simplest turbine. It consists of a turbine in which the steam is exhausted at atmospheric pressure or higher. These turbines are generally used when there is a process need for high-pressure steam, and all steam condensation takes place downstream of the turbine cycle and in the process. Figure 39 illustrates the steam path for a noncondensing turbine. The back pressure steam turbine operates on the enthalpy difference between steam inlet and exhaust conditions. The noncondensing turbine's Carnot cycle efficiency

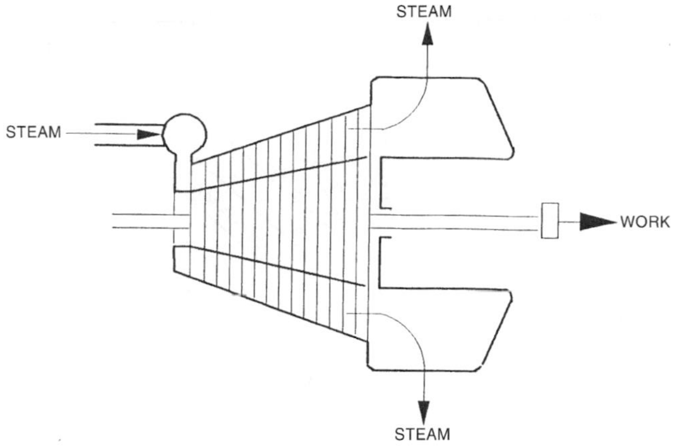

Fig. 39 Back Pressure Turbine

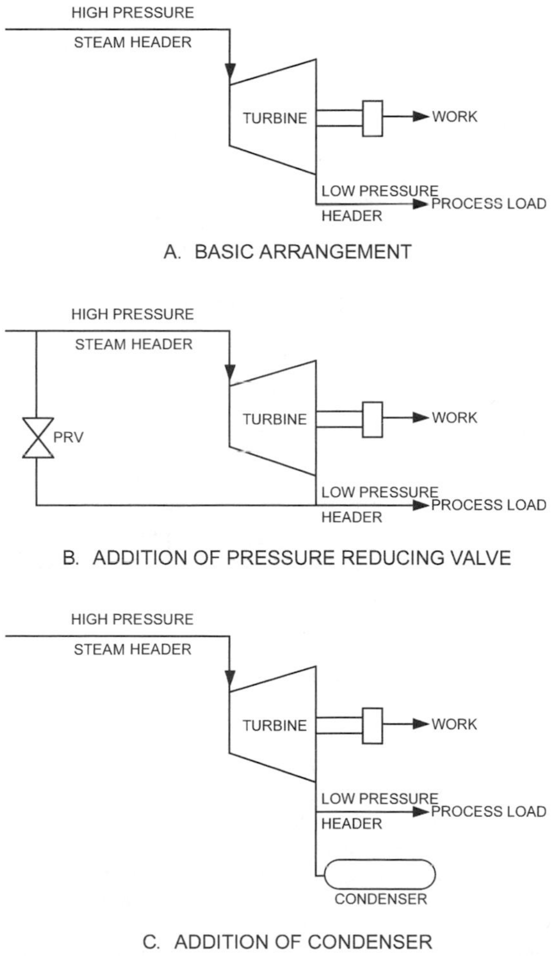

Fig. 40 Integration of Back Pressure Turbine with Facility

tends to be lower than is possible with other turbines because the difference between the turbine inlet and exhaust temperatures tends to be lower. Because much of the steam's heat, including the latent heat of vaporization, is exhausted and then used in a process, the back pressure cogeneration system process efficiency or total energy efficiency can be very high. One potential application for back pressure turbines is as a substitute for pressure reducing valves; they provide the same function (pressure regulation), but also produce a useful product—power.

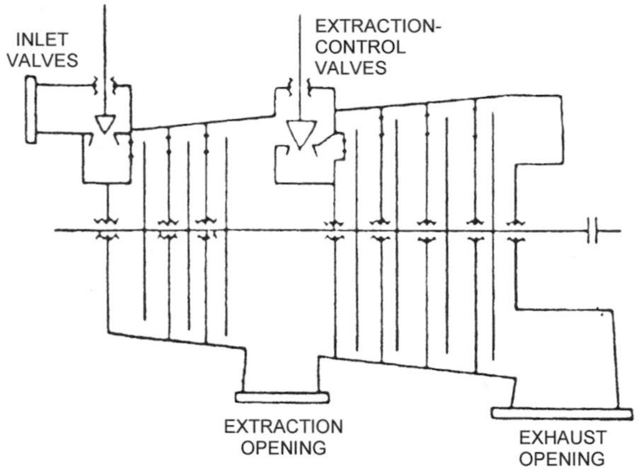

Fig. 41 Condensing Automatic Extraction Turbine

The back pressure turbine has one major disadvantage in cogeneration applications. Because the process load is the heat sink for the steam, the amount of steam passed through the turbine depends on the heat load at the site. Thus, the back pressure turbine provides little flexibility in directly matching electrical output to electrical requirements; electrical output is controlled by the thermal load. The direct linkage of the site steam requirements and electrical output can result in electric utility charges for standby service or increased supplemental service unless some measures are taken to increase system flexibility.

Figure 40 illustrates several ways to achieve some flexibility in performance when the electrical and thermal loads do not match the capability of the back pressure turbine. Figure 40A shows the basic arrangement of a noncondensing steam turbine and its relationship to the facility. Figure 40B illustrates the addition of a pressure reducing valve (PRV) to bypass some or all of the steam around the turbine. Thus, if the process steam demand exceeds the capability of the turbine, the additional steam can be provided through the PRV. Figure 40C illustrates the use of a load condenser to permit the generation of electricity, even when there is no process steam demand. The use of either of these techniques to match thermal and electrical loads is very inefficient, and operating time at these off-design conditions must be minimized by careful analysis of the coincident, time-varying process steam and electrical demands.

Process heat recovered from the noncondensing turbine can be easily estimated using steam tables combined with knowledge of the steam flow, steam inlet conditions, steam exit pressure, and turbine isentropic efficiency.

Extraction Turbines. Figure 41 illustrates the internal arrangement of an automatic extraction turbine where steam is exhausted from the turbine at one or more stations along the steam flow path. Conceptually, the extraction turbine is a hybrid combination of condensing and noncondensing turbines. The advantage of this turbine is that it allows extraction of the quantity of steam required at each temperature or pressure needed by the industrial process. Multiple extraction ports allow great flexibility in matching the cogeneration cycle to the thermal requirements at the site. Extracted steam can also be used in the power cycle for feedwater heating or powerhouse pumps. Depending on cycle constraints and process requirements, the extraction turbine's final exit conditions can be either back pressure or condensing.

A diaphragm in the automatic extraction turbine separates the high-pressure section from the low-pressure section. All the steam passes through the high-pressure section just as it does in a single back pressure turbine. A throttle controls steam flow into the low-pressure section. This throttle is controlled by the pressure in the process steam line. If the pressure drops, the throttle closes, allowing more steam to the process. If the pressure rises, the throttle

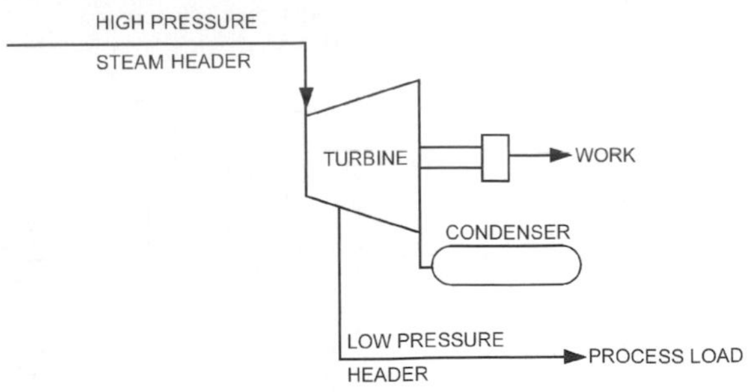

Fig. 42 Automatic Extraction Turbine Cogeneration System

opens to allow steam to flow through the low-pressure section, where additional power is generated.

An automatic extraction turbine (Figure 42) is uniquely designed to meet the specific power and heat capability of a given site; therefore, no simple relationship generally applies. For preliminary design analyses, the procedures presented by Newman (1945) can be used to estimate performance. The product of such an analysis is a performance map similar to that shown in Figure 43 for a 5000 kW generator. The performance map provides the steam flow to the turbine as a function of generator output with extraction flow as a parameter.

HEAT-ACTIVATED CHILLERS

Waste heat may be converted and used to produce chilled water by several methods. The conventional method is to use hot water (>200°F) or low-pressure steam (<15 psig) in single-stage absorption chillers. These single-stage absorbers have a COP of 0.6 or less; 18,000 Btu/h of recovered heat can produce about 1 ton of cooling (12,000 Btu/h).

If a direct exhaust, two-stage absorption chiller is used, the equation to estimate cooling produced from recoverable heat is

$$q = \frac{\dot{m}_e c_p (t_1 - t_2)(1.14 \times 0.97)}{12,000} \tag{7}$$

where

q = cooling produced, tons
$\dot{m}_e$ = mass flow of exhaust gas, lb/h
c_p = specific heat of gas = 0.268 Btu/lb·°F
t_1 = exhaust temperature in, °F
t_2 = exhaust temperature out = 375°F
1.14 = coefficient of performance
0.97 = connecting duct system efficiency
12,000 = Btu/ton·h

If it is available, the manufacturer's COP rating should be used to replace the assumed value.

For internal combustion engines, jacket water heat at 180 to 210°F may be added to the recovered heat of the engine exhaust to produce chilled water in a single-stage absorption machine. The equation to estimate the cooling produced from the heat recovered from the water is

$$q = \frac{0.6 \dot{m}_w (c_p)_w (t_1 - t_2)}{12,000} \tag{8}$$

where

$\dot{m}_w$ = mass flow of water, lb/h
$(c_p)_w$ = specific heat of water, 1.0 Btu/lb·°F
t_1 = water temperature out of engine, °F
t_2 = water temperature returned to engine, °F [Typically $(t_1 - t_2)$ = 15°F.]
0.6 = COP

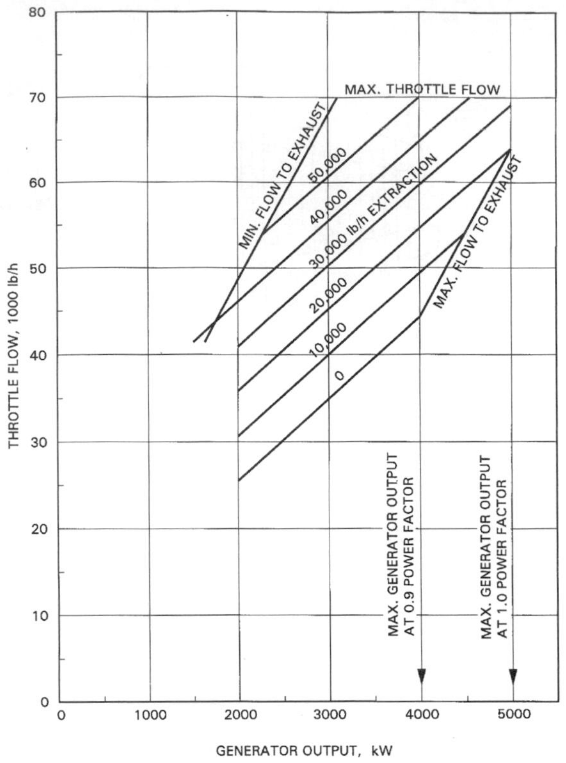

Fig. 43 Performance Map of Automatic Extraction Turbine

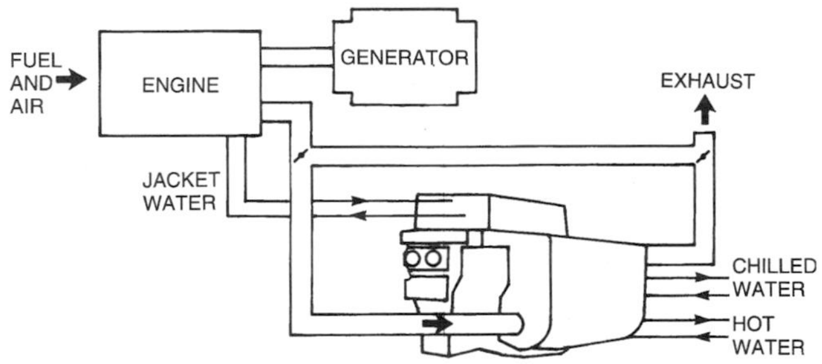

Fig. 44 Exhaust Gas Chiller-Heater

Heat may be recovered from engines and gas turbines as high-pressure steam, depending on exhaust temperature. Steam pressures from 15 to 200 psig are common. When steam is produced at pressures over 43 psig, two-stage steam absorption chillers can also be considered. The COP of a two-stage absorption chiller is 1.14 or greater. The steam input required is 9.3 to 9.7 lb/ton·h. This compares to 18 lb/ton·h for the single-stage absorption machine using 15 psig steam. Two-stage absorption chillers can take advantage of the dual temperatures available from the engine exhaust and jacket water.

The use of the engine exhaust heat to provide cooling is described in the section on Exhaust Gas Heat Recovery, and some absorption machines have been designed specifically for heat recovery in cogeneration applications. These units use both the jacket water and the exhaust gas directly. Ebullient cooling of the engine is not required (Figure 44). These chillers range in capacity from 50 to 200 tons.

Another type of absorption chiller uses gas engine or turbine exhaust directly in a waste heat absorption chiller. These units are available in sizes ranging from 100 to 1500 tons. Using oil/diesel exhaust for this purpose has not been successful due to fouling and corrosion problems in such direct-fired or waste exhaust-fired chillers.

ELECTRICAL SYSTEMS

UTILITY INTERFACING

All electric utility interfacing requires safety on the electric grid and the ability to meet the operating problems of the electric grid and its generating system. Additional control functions depend on the desired operating method during loss of interconnection. For example, the throttle setting on a single generator operating in parallel with the utility is determined by either the heat recovery requirements or the power requirements, whichever governs, and its exciter current, which is set by the reactive power flow through the interconnection. When the interconnection with the utility is lost, the generator control system must detect that loss, assume voltage and frequency control, and immediately disconnect the intertie to prevent an unsynchronized reconnection.

With the throttle control now determined solely by the electrical load, the heat produced may not match the requirements for supplemental or discharge heat from the system. When the utility source is reestablished, the system must be manually or automatically synchronized, and the control functions restored to normal operation.

Loss of the utility source may be sensed through the following factors: overfrequency, underfrequency, overcurrent, overvoltage, undervoltage, or any combination of these. The most severe condition occurs when the generator is delivering all electrical requirements of the system up to the point of disconnection, whether it is on the electric utility system or at the plant switchgear. Under such conditions, the generator tends to operate until the load changes.

At this time, it either speeds up or slows down, allowing the over- or underfrequency device to sense loss of source and to reprogram the generator controls to isolated system operation. The interconnection is normally disconnected during such a change and automatically prevented from reclosing to the electric system until the electric source is reestablished and stabilized and the generator is brought back to synchronous speed. Additional utility interfacing aspects are covered in following sections.

GENERATORS

Criteria influencing the selection of alternating current (ac) generators for cogeneration systems are (1) machine efficiency in converting mechanical input into electrical output at various loads; (2) electrical load requirements, including frequency, power factor, voltage, and harmonic distortion; (3) phase balance capabilities; (4) equipment cost; and (5) motor-starting current requirements.

Generator speed is a direct function of the number of poles and the output frequency. For 60 Hz output, the speed can range from 3600 rpm for a two-pole machine to 900 rpm for an eight-pole machine. A wide latitude exists in matching generator speed to prime mover speed without reducing the efficiency of either unit. This range in speed and resultant frequency suggests that electrical equipment with improved operation at a special frequency might be accommodated.

Induction generators are similar to induction motors in construction and control requirements. The generator draws excitation current from the utility's electrical system and produces power when driven above its synchronous speed. In the typical induction generator, full output occurs at 5% above synchronous speed.

In order to prevent large transient overvoltage in the induction generator circuit, special precautions are required to prevent the generator from being isolated from the electrical system while connected to power factor correction capacitors. Also, an induction generator cannot operate without excitation current from the utility; only the **synchronous generator** has its own excitation.

Generator efficiency is a nonlinear function of the load and is usually at its maximum at or near the rated load (see Figure 45). The rated load estimate should include a safety factor to cover such

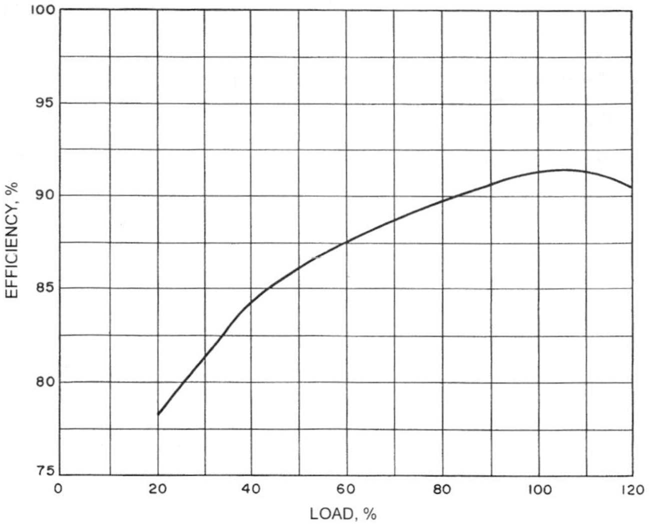

Fig. 45 Typical Generator Efficiency

transient conditions as short-term peaking and equipment start-up. Industrial generators are designed to handle a steady-state overload of 20 to 25% for several hours of continuous operation. If sustained overloads are possible, the generator ventilation system must be able to relieve the temperature rise of the windings, and the prime mover must be able to accommodate the overload.

Proper phase balance is extremely important. Driving three-phase motors from the three-phase generator presents the best phase balance, assuming that power factor requirements are met. Driving single-phase motors and building lighting or distribution systems may cause an unbalanced distribution of the single-phase loads that leads to harmonic distortion, overheating, and electrical imbalance of the generator. In practice, maximum phase imbalance can be held within 5 to 10% by proper distribution system loading.

Voltage is regulated by using static converters or rotating dc generators to excite the alternator. Voltage regulation should be within 0.5% from full load to no load during static conditions. Good electronic three-phase voltage sensing is necessary to control the system response to load changes and the excitation of paralleled alternators to ensure reactive load division.

The system power factor is reflected to the generator and should be no less than 0.8 for reasonable generator efficiency. To fall within this limit, the planned electrical load may have to be adjusted so the combined leading power factor substantially offsets the combined lagging power factor. Although more expensive, individual power factor correction at each load with properly sized capacitors is preferred to total power factor correction on the bus. On-site generators can correct some power factor problems, and for those cogeneration systems interconnected to the grid, they improve the power factor seen by the grid.

Generators operating in **parallel** with the utility system grid have different control requirements than those that operate **isolated** from the utility grid. A system that operates in parallel and provides emergency standby power if a utility system source is lost must also be able to operate in the same control mode as the system that normally operates isolated from the electric utility grid.

Control requirements for systems that provide electricity and heat for equipment and electronic processors differ depending on the number of energy sources and the type of operation relative to the electric utility grid. Isolated systems generally use more than one prime mover during normal operation to allow for load following and redundancy.

Table 7 shows the control functions required for systems operating isolated from the utility grid and systems operating in parallel

Table 7 Generator Control Functions

	Isolated		Parallel	
Control Functions	**One Engine**	**Two or More Engines**	**One Engine**	**Two or More Engines**
Frequency	Yes	Yes	No	No
Voltage	Yes	Yes	No	No
Power	No (Load following)	Yes (Division of load)	No	Yes (Division of load)
Reactive kVAR	No (Load following)	Yes (Division of kVAR)	Yes	Yes
Heat t_1	Supplement only	Supplement only	Load following	Load following
Heat $t_2 - t_x$	Reduce from t_1 or supplement	Reduce from t_1 or supplement	Reduce from t_1 and load following	Reduce from t_1 and load following
Cooling	Remove excess heat (tower, fan, etc.)	Remove excess heat (tower, fan, etc.)	Normally no (Emergency yes)	Normally no (Emergency yes)
Synchronizing	No	Yes	Yes	Yes
Black start	Yes	Yes (one engine)	Emergency use	Emergency use (one engine)

with the grid, both with single and with multiple prime movers. Frequency and voltage are directly controlled in a single-engine isolated system. The power is determined by the load characteristics and is met by automatic adjustment of the throttle. Reactive power is also determined by the load and is automatically met by the exciter in conjunction with voltage control.

In parallel operation, both frequency and voltage are determined by the utility service. The power output is determined by the throttle setting, which responds to the system heat requirements if thermal tracking governs, or to system power load if that governs. Only the reactive power flow is independently controlled by the generator controls.

When additional generators are added to the system, there must be a means for controlling the power division between multiple prime movers and for continuing to divide and control the reactive power flow. All units require synchronizing equipment.

The generator system must be protected from overload, overheating, short-circuit faults, and reverse power. The minimum protection is a properly sized circuit breaker with a shunt-trip coil for immediate automatic disconnect in the event of low voltage, overload, or reverse power. The voltage regulation control must prevent overvoltage. Circuit breakers for low voltage (below 600V) may be air-type, and circuit breakers for medium voltage (up to 12,000 V) should be vacuum-type and should operate within 5 Hz.

Voltage must be held to close tolerances by the voltage regulator from no load to full load. A tolerance of 0.5% is realistic for steady-state conditions from no load to full load. The voltage regulator must allow the system to respond to load changes with minimum transient voltage variations. During parallel operation, the reactive load must be divided through the voltage regulator to maintain equal excitation of the alternators connected to the bus. True reactive load sensing is of prime importance to good reactive load division. An electronic voltage control responds rapidly and, if all three phases are sensed, better voltage regulation is obtained even if the loads are unbalanced on the phases. The alternator construction of a well-designed voltage regulator dictates the transient voltage variation.

Engine sizing can be influenced by the control system's accuracy in dividing real load. If one engine lags another in carrying its share of the load, the capacity that it lags is never used. Therefore, if the load sharing tolerance is small, the engines can be sized more closely to the power requirements. A load-sharing tolerance of less than 5% of unit rating is necessary to use the engine capacity to good advantage.

A load-sharing tolerance of 5% is also true for reactive load sharing and alternator sizing. If reactive load sharing is not close, a circulating current results between the alternators. The circulating current uses up alternator capacity, which is determined by the heat generated by the alternator current. The heat generated, and thus the alternator capacity, is proportional to the square of the current. Therefore, a precise control system should be installed; the added cost is justified by the possible installation of smaller engines and alternators.

POWER QUALITY

Electrical energy can be delivered to the utility grid, directly to the user, or to both. Generators for on-site power plants can deliver electrical energy equal in quality to that provided by the electric utility in terms of voltage regulation, frequency control, harmonic content, reliability, and phase balance. They can be more capable than the utility of satisfying stringent requirements imposed by user computer applications, medical equipment, high-frequency equipment, and emergency power supplies because other end users on the utility grid can create quality problems. The generator's electrical interface should be designed according to user or utility electrical characteristics.

COGENERATION REGULATORY ISSUES

REGULATORY AUTHORITIES

In the interest of energy conservation, a number of laws were developed in the United States to encourage the use of cogeneration and renewable energy sources by removing or limiting barriers to such plants.

United States Laws and Regulations

Public Utilities and Regulatory Policies Act (PURPA) mandates the following benefits for a qualifying cogenerator:

- Electric utilities must purchase power from cogenerators and small power producers (SPPs) at the price that the utility avoids by not producing it (avoided cost); and a utility must allow cogenerators to be paralleled (interconnected) with its grid if the cogenerator or SPP pays for it and complies with the utility's safety and protective requirements.
- The utility must provide supplemental power for facilities that do not cogenerate 100% of their power, standby power for emergency power during a cogenerator's outage, and maintenance power at nondiscriminatory rates during a planned outage for maintenance.
- Cogenerators and SPPs selling power to a utility are exempt from the "rate of return" (allowable return on investment) of utility regulations of the Federal Energy Regulatory Commission (FERC) and the Securities and Exchange Commission (SEC).
- Utilities may wheel (transmit) the cogenerated power over its grid to a location remote from the cogeneration plant and, under certain conditions, can be ordered by FERC to wheel it.

Qualifying Facility. FERC extends these benefits if the cogenerator becomes a qualifying facility (QF) by meeting the following qualifications:

- No more than 50% of a facility may be owned by an electric utility.
- At least 5% of its annual useful output must be useful thermal energy.
- Its efficiency must be at least 45% for topping cycles, and if more than 15% of its output is thermal, the minimum efficiency is 42.5%. There is no efficiency standard for bottoming cycles or for renewable resource plants. FERC efficiency is defined as

(Power Output + Thermal Output/2) / LHV of Fuel Input

- A cogenerator's retail power and steam rates may be regulated by state bodies under a broad authority and subject to local sales or gross receipts taxes.
- An SPP plant must be no greater than 80 MW unless powered by waste products or renewable energy.

State Laws and Regulations

PURPA provides FERC with the obligation to encourage cogeneration through rule making. FERC further charges each state with the responsibility of implementing PURPA rules. While these rules require electric utilities to buy from QFs that are interconnected or that wheel the power to the utility, FERC rules do not require purchase of QF power if the purchase will result in the utility's bearing a cost greater than if it had generated the power itself.

During the design of a facility, the purchasing utility should be consulted regarding its operation. For example, the facility may need to be designed to reduce the generation output at times when the purchasing utility is at reduced load.

While PURPA provides for rules that require all electric utilities to purchase from QFs, there are local cooperatives and municipalities in which the utility may ignore the purchase requirements due to lack of jurisdiction.

PURPA regulations provide for the sale of the power at an avoided cost. The methodologies and options for pricing vary from state to state. The local utility or the state Public Service Commission or Utility Commission can also provide some assistance.

In order to obtain QF certification, an application must be submitted to FERC. This process is typically undertaken at the early stages of the facility design. FERC may reject the certification if it finds that the representations in the application for certification were not obtained. For this reason, it is not uncommon to amend the application with FERC when the design of the facility changes.

Emission Regulations

Number 6 fuel oil, particularly in developing countries, is sometimes used in internal combustion engines and combustion turbines; but it can cause serious environmental pollution hazards, especially because it has a high sulfur content. Air pollution control authorities may require engine exhaust gases to be conditioned or may require special engines or fuels. Turbines may need water injection systems or selective catalyst reduction (SCR) systems. Requirements are based on the best available control technology (BACT). Typically the requirements become more stringent as the technology becomes available. The permitting agency can provide the most current requirements.

Reciprocating engines may be required to have catalytic converters or SCR. Boilers for steam turbine cogeneration normally require low NO_x burner design even if they are operating on high-quality natural gas. In most cases, the cost involved with the monitoring and control of emissions from larger systems is substantial. In addition, many cogenerators sustain substantial costs in permitting related to the emissions of the plant. Emissions trading allows one company that has an excess of emissions to purchase emission credits from another that is below its allocated emission level.

DISTRIBUTION SYSTEMS

Cogeneration system designers have a greater number of systems to choose from when the three major subsystems—production, distribution, and utilization (by the end user)—are all new. Significant limits are imposed on design choices when any one of these three elements already exists, has a substantial remaining life expectancy, and can carry out its assigned task(s) in a fairly efficient manner. Only when replacement is cost-effective, or necessary for a reason such as significant load increase, can options be broadened.

DESIGN CONSIDERATIONS

The geographical density of new or existing loads and the variety and type of existing systems to be connected to the cogeneration system have a profound effect on economic feasibility and choice of distribution.

In a totally new multifunction or multibuilding complex for a single owner or tenant who has total decision control, all energy systems can be designed for best compatibility with basic cogeneration concepts and criteria. However, the design is more difficult for an existing or new complex with multiple owners or tenants and diverse energy systems. In that situation, the cogeneration developer must (1) trail the various needs to cost-effectively pick up all possible services, (2) negotiate agreements with some of the occupants to make their own provisions, or (3) establish criteria to be followed by new facility participants.

In an industrial park or shopping center, for example, the developer might be required to serve one buyer with cooling and power only while another may need heating and cooling but no power. Or the developer may mandate that all buyers take all services and use compatible building energy systems. An example might be that all chilled water air handlers contain coils that meet the users' requirements with a Δt of 18°F (from 45 to 63°F). In any event, the developer's choices of plant design and distribution means will be limited by the extent or the variety of individual production, distribution, and terminal criteria desired by each buyer.

Optimized cogeneration mandates that maximum use be made of all forms of output energy from the production plant and from any existing mechanical/electrical equipment and systems. The prime mover outputs must be either in a form and quality of energy compatible with systems to be serviced by its main output or in some form suitable for conversion, such as from steam to absorption chilling.

OUTPUT ENERGY DISTRIBUTION MEDIA

Interconnects may have to be made to electrical systems of one or more voltages; to low-, medium-, or high-pressure steam systems; to chilled water or secondary coolant systems; to low-, medium-, or high-temperature hot water or thermal fluid systems; or to thermal energy storage systems. Each variation should be addressed in the planning stages.

Electrical

Electrical energy can produce work, heating, or cooling; it is the most transmittable form of energy. As a cogeneration output, it can be used to refrigerate or to supplement the prime mover's thermal output during periods of high thermal, low cooling demand. Mechanical aspects of a cogeneration system must be coordinated with electrical system designers who are familiar with power plant switchgear and utility and building interface requirements. See the section on Electrical Systems on page 7.27 for more information.

Steam

Engines and combustion and steam turbines can provide a range of pressure/temperature characteristics encountered in almost all steam systems. Their selection is basically a matter of choosing the prime mover and heat recovery steam generator combination that best suits the economic and physical goals.

Steam can also provide work, heating, or cooling, but with somewhat less range and flexibility than electric power. Distribution to remote users is more expensive with steam than with electricity and is less adaptable for remote production of work.

Economics limit the pressure and/or temperature (P/T) of steam available from gas turbine exhaust because the incremental cost/benefit ratio of increasing the heat recovery generator surface to yield a higher P/T is limited by a relatively fixed exhaust gas temperature, unless the turbine is refired with an auxiliary duct burner. However, steam turbines are not similarly limited, except by throttle conditions, because extraction can be accomplished from any point in the P/T reduction process of the turbine. Chapter 10 and Chapter 11 have further information on steam systems.

Chilled Water

The entire output of any prime mover can be converted to refrigeration and then chilled water, serving the wide variety of terminal units in conventional systems. In widely spread service distribution systems, choices must be made whether to serve outlying facilities with electric, steam, or hot water. All three can be used directly, for building or process heating and/or cooling, or indirectly, through heat exchangers and mechanical or thermally activated chillers located at remote facilities.

Central chilled water production and distribution to existing individual or multibuilding complexes is most practical if a chilled water network already exists and all that is required is an interconnect at or near the cogeneration plant. If the user building(s) already have one or more types of chillers in good condition, cogeneration and chilled water distribution may have diminished economic prospects unless applied on a small scale to individual buildings.

If chilled water distribution is feasible, the central cogeneration concept is easier to justify, and several techniques can be used to improve the viability of a cogenerated chilled water system by significantly reducing the owning and operating costs of piping and pumping systems and their associated components (e.g., valves, insulation, etc.). Such systems have been widely discussed, successfully developed, modified, and specified by many firms (Avery et al. 1990; Mannion 1988; Becker 1975).

The following are cost reduction concepts that are embodied in variable-flow water systems:

- Let the main pump(s) and the primary distribution system flow rate match the instantaneous sum of the demand flows from all cooling coils served by the primary loop. The chilled water should not be pumped off the primary loop in such a way as to circulate more chilled water through the secondary pump of the outlying buildings than the flow that it draws from the primary loop.
- Use two-way throttling control valves on all coils. Avoid three-way control valves for coil control or for bypassing chilled water supply into the chilled water return (e.g., end-of-line bypass to maintain a constant pump flow or system pressure differential). Valves must have suitable control characteristics for the system and full shutoff capability at the maximum pressure differential encountered.
- Select and circuit cooling coils for a large chilled water temperature difference (Δt as much as 24°F) while maintaining coil tube velocities of 5 to 10 fps and the required supply air conditions off the coil. Such coils may require 8 to 10 rows, but the additional pressure drop and cost are offset by the lower cost of the pump and distribution piping of long distribution systems. A system with Δt of 24°F requires only one-third the flow rate required by one with a Δt of 8°F.
- Care must be exercised by the designer for successful implementation of these concepts. The *Air-Conditioning Systems Design Manual* (ASHRAE 1993) has further design information.

Hot Water

Cogeneration thermal output is well adapted to low-, medium-, or high-temperature water (LTW, MTW, HTW) distribution systems. The major difference between chilled water and hot water systems is that even LTW systems (up to 250°F) can be designed with a Δt as high as 100°F with low flow by using different series-parallel terminal circuiting, as described in Chapter 12. This way, even equipment that is limited to Δt = 20°F (e.g., radiators, convectors) can be adapted to large system temperature differences. For example, besides those circuits given in Chapter 12, unit heaters can be piped in series and parallel on a single hot water building loop without a conventional supply and return line. Parallel runs of five heaters each can drop in 20°F increments. The first group drops the temperature from 250 to 230°F, the last from 170 to 150°F, and all are sized at the 170 to 150°F range. Fan cycling off each local thermostat maintains control despite the different temperatures.

Similarly, larger heaters with conventional small-row coils (not metal cores) can be fitted with three-way modulating bypass valves, sized for a 10 to 40°F drop, with the through-flow and bypass flow from the first flowing to the second, and so forth, using only one primary loop.

Medium- (250 to 350°F) and high- (350°F and higher) temperature water systems are designed with an even higher Δt, but they are not customarily connected directly to the primary loop. These systems can be connected to steam generators in outlying buildings that have steam distribution and steam terminal devices.

When a choice can be made, a prime mover's thermal output should be used according to the following priorities. Apply the output first for useful work, second for an efficient form of thermal conversion, and third for productive thermal use. For example, if a combustion turbine's exhaust can cost-effectively produce shaft power or, if not, heat some process, it should be considered for these functions rather than for a low-pressure absorption function, which is the least efficient application.

For both hot water and chilled water distribution systems, a common approach is to lower the hot water supply temperature as the ambient temperature rises and to raise chilled water supply temperature as loads are reduced. Both techniques reduce pipe transmission and fuel or electrical costs for heating or cooling, and stabilize valve control; but the impact of increased pumping costs is often overlooked.

Below some part-load condition in both hot water and chilled water systems, the cost of reducing flow by varying pump speed and air volume may be more than the saving in energy. This is especially true for chilled water systems, when the cascading effect of VAV fan power reduction from lower supply air temperatures, together with pumping savings, becomes more significant than the low-load chiller savings from raising the chilled water system temperature. Also, humidity control for the space limits how much the chilled water temperature can rise.

These phenomena need to be examined in determining the part-load condition at which hot water and chilled water system scheduling might be advantageously modified.

COGENERATION UTILIZATION SYSTEMS

Good cogeneration planning responds to the requirements of the end user and strives to use to the maximum the equipment and the energy it produces. In order for cogeneration to be economically feasible, the energy recovered must match the site requirements well and avoid as much waste as possible. Depending on the design and operating decisions, users may tie into the electric utility grid for some or all of their electric energy needs.

The selection of prime movers is based partially on user thermal requirements. For maximum heat recovery, the thermal load must remove sufficient energy from the heat recovery medium to lower its temperature to that required to cool the prime mover effectively. A supplementary means for rejecting heat from the prime mover may be required if the thermal load does not provide adequate cooling during all modes of operation or as a backup to thermal load loss.

Internal combustion reciprocating engines have the lowest heat/power ratio, yielding most of the heat at a maximum temperature of 200 to 250°F. This jacket water heat can be used by applications requiring low-temperature heat or to preheat inlet air. Higher temperature levels from the exhaust are limited to approximately one-third the engine's recoverable thermal energy.

Gas turbines can provide a larger quantity and a better quality of heat per unit of power, while extraction steam turbines can provide even greater flexibility in both the quantity and the quality (temperature and pressure) of heat delivered. Externally fired prime mover cycles, such as the Stirling cycle, gas turbines with steam injection (Cheng cycle), and steam-operated absorption chillers for inlet air cooling are also flexible in the quantity and quality of thermal energy they can provide.

If a gas turbine plant is designed to serve a variety of loads (e.g., direct drying, steam generation for thermal heating or cooling, and shaft power), it is even more flexible than one that serves only one or two such loads. Of course, such diverse equipment service must be economically justified.

AIR SYSTEMS

Large, central air handlers with deep and/or suitably circuited coils that operate with a large cooling and heating Δt are available to reduce distribution piping and pumping costs. These units serve a multiplicity of control zones or large single-zone spaces with air distribution ductwork. Smaller units such as perimeter fan-coil units directly condition spaces with small lengths of duct or no ductwork. Thus, they are totally decentralized from the air side of the system. The maximum Δt through the coil is only 12 to 14°F.

Both central and decentralized air handlers can be coupled with cogeneration in mildly cold climates in a two-pipe changeover configuration with a small, intermediate season electric heating coil. This arrangement can heat or cool different zones simultaneously during the intermediate seasons. Chilled water is available to cool any zone. Zones needing heat can cut off the coil and turn on the electric heat. A four-pipe system is unnecessary in these conditions.

When the building's balance point is reached (i.e., when all zones need heat), the pumping system is changed over to hot water. The concept applies best (and mostly) to perimeter zone layouts and to large air handlers with economizer cycles that do not need chilled water below the ambient changeover point (i.e., when economizer cooling can satisfy their loads). However, the application must have a high enough thermal demand for process or other non-space-heating loads to absorb the extra thermal energy produced by the engine generator for this additional electric heating load.

If the predominant thermal load during this period is for space heating and cooling (both being at a low demand level), it makes no sense to exacerbate the already low heat/power ratio by designing for more electrical load with no use for the heat generated. Hospitals and apartment houses with high process heating demands are examples of suitable applications, but single-function office buildings are not.

The significance of this cogeneration design is that the prime mover's electrical output can be swung from a motor-driven refrigeration load, which is less during intermediate seasons, to the electric heating function as long as the additional thermal energy can be absorbed. This can work well with engine-generators and electric refrigeration.

An absorption chiller might be a better match where the site's heat/power ratio is low, such as for an office building. But a mechanical chiller without cogeneration may offer an even better return than an absorption system.

HYDRONIC SYSTEMS

Hydronics, particularly in buildings with no need for process or high-pressure steam, is a much more widely used transport medium than steam. Information on the various types of terminals and systems may be found in Chapter 3. Loosely defined, hydronics covers all liquid transport systems, including (1) chilled water, hot water, and thermal fluids that convey energy to locations where space and process heating or cooling occur; (2) domestic or service hot water; (3) coolant for refrigeration or a process; (4) fresh or raw water for potable or process purposes; and (5) wastewater.

From the standpoint of cogeneration design, all these applications are relevant, but some are not HVAC applications. Each application may offer an opportunity to improve the cogeneration system. For example, a four-pipe, two-coil system and a two-pipe, common coil system offer similar options in plant design, with the four-pipe system offering superior flexibility for individual control of space conditions. All-electric, packaged terminal air conditioners offer little opportunity for a sizable plant, unless a substantial thermal demand (e.g., for process heat) exists in addition to the normal comfort space-conditioning needs. Without the thermal demand, the only option is to install a plant that generates a fraction of the total electrical demand while satisfying service water heating requirements, for example, and to buy the bulk of the electricity required. If the site's heat/power ratio is so low that it cannot support the lowest ratio prime mover for a large portion of the electrical demand, then a smaller plant that can operate close to a base-loaded electric generating condition might be considered.

The temperature required by the site loads also influences the feasibility of cogeneration. If a temperature of 110 to 120°F can satisfy most of the site's heating requirements (with air handlers, fan-coil units, or multitiered finned radiators), a central motor-driven heat pump might offer a more cost-effective alternative to a prime mover in a cogeneration plant that produces more heat than required.

Even if refrigeration from the heat pump is not used, the heat pump takes only 42% of the source fuel from the electric utility's boiler to produce the same heat energy as an 80% efficient on-site boiler. The section on Economic Coefficient of Performance (ECOP) on page 7.33 has further information. (ECOP is a dimensionless performance index representing the energy output of a system per unit energy input of the source fuel measured in the ratio of costs of different energy streams per unit of energy.)

The heat pump is even more effective if there is a simultaneous demand for both refrigeration and heating. To the extent that refrigeration is in excess of demand, it can be used instead of air handler economizer cooling or for fan-coil units not equipped with economizers. If a cogeneration plant incorporates a heat pump, it may produce too much heat for the site to absorb, thereby reducing the heat pump utilization factor. More information on heat pumps may be found in Chapter 8.

UNITARY AND PACKAGED HVAC SYSTEMS

When a building has a predominance of self-contained HVAC units for cooling and/or heat pump applications, as discussed in the previous section on Hydronic Systems, cogeneration is not feasible unless the site has another, substantial demand for heat.

SERVICE HOT WATER SYSTEMS

Service hot water systems can be a major and preferred user of thermal energy from prime movers and can often constitute a fairly level year-round load, when averaged over a 24 to 48 h period. Service hot water use in hospitals, domiciliary facilities, etc., is usually quite variable in a 24 h weekday or 48 h weekend profile; heat storage permits expanded use of the thermal output and justification for larger prime movers.

The service hot water demand often provides a strong case for consuming the entire thermal output with packages sized for the 24 h demand, instead of for space cooling or heating. For the larger prime movers with ebullient cooling, and particularly on sites with wide climate variations, the incremental benefits of using the thermal output for space heating as well as for service hot water may not be great, and the annual utilization factor of the incremental thermal output may be relatively low.

DISTRICT HEATING/COOLING SYSTEMS

The high cost of a central plant and distribution system generally mandates that the economic returns develop soon after the plant and distribution systems are complete. Because of the high risk, the developer must have satisfactory assurance that there are enough buyers for the product. Furthermore, the developer must know what distribution media and quality are best for connection to existing buyers and must install a system that is flexible enough for future buyers.

Generally, the load on a district system tends to level out because of the great diversity factor of the many loads and noncoincident peaking. This variety also makes plant sizing and consumption estimate aspects difficult. Statistical data from case studies and broad assumptions may be the only source of information. Chapter 11 has further information on district systems.

HYBRID, UNCONVENTIONAL PROCESS SYSTEMS

Commercial and Industrial HVAC

Residential and domiciliary facilities without a high ratio of occupants to ground floor area (e.g., single-family residences) are less desirable for large cogeneration projects than those with a high geographical load density or those that can use a substantial portion of the plant output for comfort or process heating and cooling.

Most applications do not have heat/power ratios that match a prime mover sized to meet the electrical peak demand. Without a suitable market for excess electricity or heat, the only rational option is to downsize the generator so that all the heat generated is totally consumed by highly efficient equipment. Any electrical demand not generated by the plant must be purchased, and any shortage in thermal capacity must be purchased or generated separately.

Desiccant Systems

Both solid and liquid desiccants are used for dehumidification and process drying. All of them produce heat in the process, and some or all of this heat must be removed. All reusable desiccants must be reactivated with heat. Some or all of this cooling or heating requires an external energy source, which can be incorporated in the cogeneration process. Chapter 22 has further information on desiccants.

Impact of Other Conservation Measures

A feasibility study for cogeneration should be based on the projected baseload of the site. If actual consumption is substantially lower as a result of other conservation measures, the savings will not be as great as projected unless the excess energy can be sold to others. On the other hand, a partial cogeneration installation is not oversized for reduced demands, but it might not be economically feasible because of the unfavorable economy of scale.

COGENERATION SYSTEM CONCEPTS

PACKAGED SYSTEMS

Packaged cogeneration systems are available from 6 to over 1000 kW. Both direct-drive engine-generators and engine-chillers with heat recovery exchangers are available. Standard designs are available, while some units are designed for specific applications.

Engine-Generators. The major advantage of these packages is the relatively simple hookup to building services; but proper planning for economic matching of the output of the package and the requirements of the facility still requires engineering skills. A standard package tends to eliminate the problems of interactions between components.

Engine-Driven Compressors and Chillers. Coupling a reciprocating engine to a reciprocating or rotating compressor has inherent torsional vibration effects that can be corrected. Such defects can be more easily eliminated in the factory design and assembly than on site. These defects do not occur when gas turbine prime movers are used.

CUSTOM-ENGINEERED SYSTEMS

The high cost of engineering and the risk of technical, economic, and environmental failure, as well as regulatory restraints, mandate extraordinary care and skill to successfully design and build custom-engineered cogeneration systems.

LOAD PROFILING AND PRIME MOVER SELECTION

The selection of a prime mover is determined by the facility's thermal or electrical load profile. The choice depends on (1) the ability of the heat/power ratio of the prime mover to match the facility loads; (2) the decision whether to parallel with the public utility or be totally independent; (3) the decision whether to sell excess power to the utility; or (4) the desire to size to the thermal baseload.

No matter which basis is used to choose the prime mover, the degree of use of the available heat determines the overall system efficiency; this is the critical factor in economic feasibility. Therefore, each prime mover's heat/power characteristics must be analyzed over its range of operating loads to make the best choice. Maximizing efficiency may not be as important as maximizing the total economic value of the cogeneration system output at all loads.

Cogeneration systems paralleled with the utility grid can operate at peak efficiency (1) if the electric generator can be sized to meet the valley of the thermal load profile, operate at a base electrical load (100% full load) at all times, and purchase the balance of the site's electric needs from the utility; or (2) if the electric generators are sized for 100% of the site's electrical demand and the heat recovered can be fully used at that condition, with thermal demands in excess of the recovered heat provided by supplementary means and excess power sold to the utility.

The heat output to the primary process is determined by the engine load. It must be balanced with actual requirements by being supplemented or by having excess heat rejected through peripheral devices such as cooling towers. Similarly, if more than one level of heat is required, controls are needed to (1) reduce heat from a higher level when it is required at the primary level, (2) supplement heat if it is not available, or (3) reject heat when availability exceeds the requirements.

In an isolated plant with more than one prime mover, controls must be added to balance the power output of the prime movers and to balance the reactive power flow between the generators.

Generally, an isolated system requires that the prime movers supply the needed electrical output, with the heat availability controlled by the electrical output requirements. Any imbalance in heat requirements results in burning supplemental fuels or wasting surplus recoverable heat through the heat rejection system.

Supplemental firing and heat loss can be minimized during parallel operation of the generators and the electric utility system grid by adjusting the prime mover throttle for the required amount of heat. This procedure is called thermal load following. The amount of electrical energy generated then depends on heat requirements; imbalances between the thermal load and the electrical load are carried by the electric utility either through absorption of excess generation or through the delivery of supplemental electrical energy to the electrical system.

Similarly, electrical load tracking controls the electric output of the generator(s) to follow the site's electrical load, while using, selling, storing, or discarding (or any combination of these methods) the thermal energy output. To minimize waste of thermal energy, the plant can be sized to track the electrical load profile up to the generator capacity, which is selected for a thermal output that matches the valley of the thermal profile. Supplemental electric power is purchased and/or thermal energy generated by other means when the thermal load exceeds the generator's profile valley.

These tracking scenarios require either a fairly accurate set of coincident electric and thermal profiles that are typical for a variety of repetitious operating modes or a set of typical daily, weekend, and holiday accumulated thermal consumption requirements for the sizing of an appropriate thermal energy storage plant.

PEAK SHAVING

High electrical demand charges in many areas, ratchet rates (minimum demand charge for 11 months = x% of the highest annual peak, leading to an actual payment that in many months is more than the metered charge), and utility capacity shortages have led to demand side management (DSM) by utilities and their consumers.

COGENERATION SYSTEM PERFORMANCE

Power Plant Incremental Heat Rate

Typically, cogeneration power plants are rated against incremental heat rates (and thus incremental thermal efficiency) by comparing the incremental fuel requirements with the base case energy needs of a particular site. For example, if a gas engine generator with a design heat rate of 10,000 Btu/kWh (34% thermal efficiency) provided steam or hot water through waste heat recovery to a particular system that would save 4000 Btu/h energy input, the incremental heat rate of the cogeneration power plant would be only 6000 Btu/kWh, which translates to a thermal efficiency of 57%. If the same system is applied to another site where only 2000 Btu/h of the recovered heat can be used (against the availability of 4000 Btu/h), the incremental heat rate for the same power plant rises to 8000 Btu/kWh, with thermal efficiency dropping to 43%.

Thus, the cogeneration power plant performance really depends on the required heat/power ratio for a particular application, and it is only according to this ratio that the type of cogeneration configuration should be chosen.

A system that requires 1000 kWh electrical energy and 7000 lb/h low-pressure steam at 30 psig (heat/power ratio of about 0.5, or 7 lb/h steam per kilowatt-hour) can be used to further illustrate the above requirement for measuring cogeneration system performance. In this example, a 1000 kW gas engine-generator cogeneration system provides a maximum of 1500 lb/h steam, and the balance of 5500 lb/h would have to be met by a conventional boiler with 75% thermal efficiency. Thus, the total energy requirement for this example cogeneration system is about 10×10^6 Btu/h (34% efficient) for the gas engine and 7.4×10^6 Btu/h for the boiler input for a total of 17.4×10^6 Btu/h.

If instead a gas turbine cogeneration system with the same power and heat requirements is used, the gas turbine-generator with a heat rate of 13,650 Btu/h per kilowatt (thermal efficiency of 25%) would supply both 1000 kW of power and 7000 lb/h steam with only 13.65×10^6 Btu/h fuel input. Thus, the gas engine-generator, although having a very high cogeneration thermal efficiency, is not suitable for the combination system because it would use 3.75×10^6 Btu/h (nearly 28%) more energy than the gas turbine for the same total output.

Economic Coefficient of Performance (ECOP)

The normal procedure for evaluating COP does not provide a workable method for comparing efficiencies of dissimilar energy streams. Instead, each energy stream should be valued on the same energy basis and at prevailing rates, with 1 kWh of electrical energy taken as 3412 Btu and fuel input as the low heating value (LHV) in Btu. Then a direct comparison by means of an economic coefficient of performance (ECOP) can be made. Rates with step charges based on the load factor must be carefully evaluated to be sure the appropriate incremental cost is used.

For example, with energy from the utility at $0.08/kWh and natural gas supply at $2.75 per thousand cubic feet (LHV of 900 Btu/ft³), 1000 Btu of electrical energy costs $0.023 (0.08/3.412), and 1000 Btu of natural gas costs $0.003 (2.75/900). Thus, the ECOP can be defined as all output energy in desired output forms, converted in terms of economic costs, divided by all energy input, again converted in terms of economic costs of each energy stream. For this example, the electrical energy costs (0.023/0.003) = 7.67 times more than the equivalent energy from natural gas. This ratio is used to calculate the ECOP in the following examples.

Example 1. Calculate the ECOP of a low-pressure steam absorption chiller with motor auxiliaries totaling 25 hp per 1000 tons. The on-site boiler generates 19 lb/ton·h steam at 15 psig (1164 Btu/lb enthalpy) at 80% efficiency from feedwater at 0 psig, 212°F (180 Btu/lb enthalpy).

Solution:

On-site fuel input = 19(1164 – 180)/0.8
= 23,400 Btu/ton·h

The electrical input generated off site supplies 25 hp/1000 tons at a motor efficiency of 0.87.

Off-site electrical input = (0.746 kW/hp × 0.025 hp/ton)/0.87
= 0.0214 kW input/ton
= 3412 × 0.0214 = 73 Btu/ton·h

The equivalent total input per unit of output (cooling only) is

23,400 + (73 × 7.67) = 23,960 Btu/ton·h (equivalent energy)

Thus, the ECOP for 12,000 Btu/ton·h output and 23,960 Btu/ton·h equivalent input (at the preceding energy costs) is

12,000/23,960 = 0.501

Example 2. Calculate and compare the ECOP for the same cooling output, using an engine-driven, vapor-compression chiller and piggyback absorption chiller. The engine has an 8600 Btu/hp·h heat rate (30% shaft thermal efficiency) and 3470 Btu/hp·h of saturated steam at 15 psig heat recovery (40% heat recovery rate).

Solution: The total cooling output is

From engine-chiller at 1 hp/ton cooling	12,000 Btu
From the absorption chiller at 19(1164 – 180) = 18,700 Btu/ton·h and for 3470 Btu/ton·h (cooling), 12,000 × 3470/18,700 =	2,227 Btu
Total cooling	14,227 Btu

Off-site electrical input for absorption chiller auxiliaries at 25 hp per 1000 tons, as detailed in Example 1, is 73 Btu/ton·h. The equivalent total input energy is

$$8600 + (73 \times 7.67) = 9160 \text{ Btu}$$

Thus, the ECOP for above is 14,227/9,160 = 1.553, which is more than three times that of the conventional system covered in Example 1.

A similar approach can produce ECOPs for different configurations and with different electrical and fuel (gas or oil) costs. But an ECOP should be considered an indicator only and should be followed with a life-cycle cost analysis to make a final decision.

THERMAL ENERGY STORAGE TECHNOLOGIES

Thermal energy storage (TES) can decouple power generation from the production of process heat, allowing the production of dispatchable power while fully using the thermal energy available from the prime mover. Thermal energy from the prime mover exhaust can be stored as sensible or latent heat and used during peak demand periods to produce electric power or process steam/hot water. However, the additional material and equipment necessary for a TES system add to the capital costs. As a result, the economic benefits of adding TES to a conventional cogeneration system must outweigh the increased cost of the combined system.

The selection of a specific storage system depends on the quality and quantity of recoverable thermal energy and on the nature of the thermal load to be supplied from the storage system. Chapter 33 of the 1999 *ASHRAE Handbook—Applications* has more information on thermal storage. The TES systems and technologies that are being considered for power generation applications can be categorized by storage temperature.

High-temperature storage can be used to store thermal energy from sources like the gas turbine exhaust stream at high temperatures (**heat storage**). High-temperature TES options such as oil/rock storage, molten nitrate salt storage, and combined molten salt and oil/rock storage are well developed and commercially available.

Low-temperature TES technologies store thermal energy at temperatures below ambient temperature (**cool storage**) and can be used for cooling the air entering gas turbines. Examples of cool storage include commercially available options such as diurnal ice storage, as well as more advanced schemes represented by complex, compound chemisorption TES systems.

A number of emerging issues may limit the number of useful applications of cogeneration. One of these is a mismatch between the demand for electricity and the demand for thermal energy on a daily basis. Increasingly, utilities are requiring cogenerators to provide dispatchable power, while most industrial thermal loads are relatively constant during the day. **Diurnal** TES can decouple the generation of electricity from the production of thermal energy, allowing the cogeneration facility to supply dispatchable power. Diurnal TES stores thermal energy recovered from the exhaust of the prime mover (gas turbine) to meet daily variations in the demand for electric power and in thermal loads.

COMBINED CYCLES

Combined-Cycle Power Plants

Combined-cycle power plants are based on combination gas turbine and steam turbine generators to achieve high efficiency in power generation. High-temperature exhaust from the gas turbine produces high-temperature, high-pressure steam that operates the steam turbine generator. Normally, a combined-cycle station consists of the following major components (Figure 46):

- Gas turbine generator set
- Steam turbine generator set
- Waste heat recovery boiler
- Steam and feedwater makeup system
- Condenser
- Electrical equipment
- Control, safety, and instrumentation systems

Gas turbines, with normal efficiencies in the range of 28 to 38%, can provide approximately 25 to 30% more energy in the form of high-quality steam from the exhaust stream. This energy can be recovered in waste heat boilers, and for large sizes, such boilers have separate drums for high-pressure and low-pressure steam. For smaller systems, only a high-pressure steam circuit may be provided to lower costs. The low-pressure steam heat exchanger in the waste heat boiler can be used for feedwater heating at the required temperatures. The superheated steam is normally fed to condensing steam turbines. For higher overall

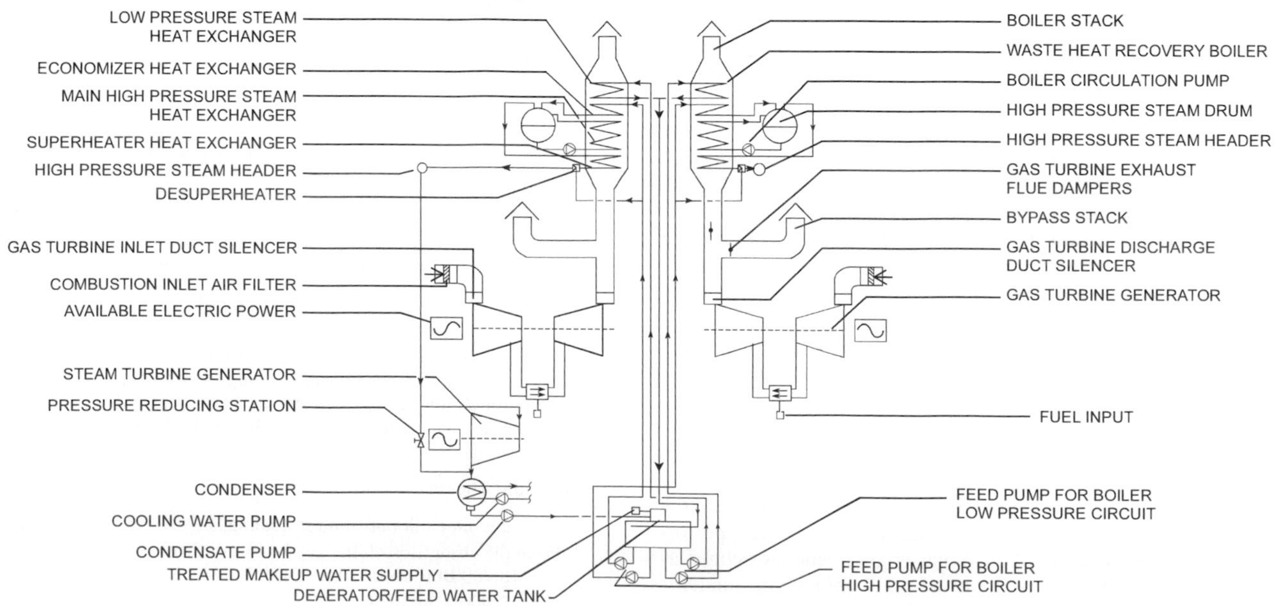

Fig. 46 Typical Combined-Cycle Power Plant Schematic

thermal efficiency, the steam turbine can be of the extraction condensing type when there is a considerable, continuous low-pressure steam demand.

Combined-cycle power plants offer the following advantages compared to conventional steam thermal power plants:

- Higher thermal efficiencies, presently up to 57% on natural gas fuel, compared to peak efficiencies of only 38 to 39% for large steam turbine power stations
- Shorter starting time, especially for gas turbines that can be loaded to nearly 70% of station capacity in only 15 to 20 min
- Better part-load performance with multiple gas turbine installations, allowing flexibility in meeting load demands
- Shorter completion time normally, with 70% power available within 1 year with simple gas turbine cycle operation, and the addition of waste heat recovery boiler and steam turbine cycle taking 1 more year
- Lower capital costs
- Lower water consumption/cooling requirements because the condensing steam cycle is for only one-third of the final power station capacity
- Lower environmental pollution

UNCONVENTIONAL SYSTEMS

Fuel Cells

Fuel cells convert the chemical energy of a hydrogen-based fuel directly into electricity without combustion. In the cell, a hydrogen-rich fuel passes over the anode, while an oxygen-rich gas (air) passes over the cathode. The catalysts help split the hydrogen into hydrogen ions and electrons. The hydrogen ions move through an external circuit, thus providing a direct current at a fixed voltage potential. A typical packaged fuel cell power plant consists of a fuel processor, which chemically combines the supply fuel with steam to form a hydrogen-rich gas; a fuel cell section, which consists of many individual cells; and a power conditioner, which transforms the dc power into ac power.

An advanced phosphoric acid fuel cell, the molten carbonate fuel cell, the solid oxide fuel cell, and the alkaline fuel cell are continuing to be developed. Fuel cells are also being developed small enough to power hybrid city buses. The heat generated by and efficiency of several types of fuel cells are as shown in Table 8. Emissions from fuel cells are very low; NO_x emissions are less than 20 ppm. Large phosphoric acid fuel cells are commercially available. For example, one packaged fuel cell power plant has the following characteristics:

Electrical output	200 kW
Thermal energy available	760,000 Btu/h @ 165°F
	or 350,000 Btu/h @ 250°F
	or 350,000 Btu/h @ 140°F
Electrical efficiency	40%
Overall efficiency	85%
Size	10 ft × 10 ft × 18 ft
Weight	40,000 lb

Table 8 Heating Value and Efficiency of Several Fuel Cells

Fuel Cell Type	Lower Heating Value, Btu/kWh	Efficiency, $kW_{out}/kW_{fuel\ in}$
1st generation phosphoric acid	8500	40%
Advanced phosphoric acid	7600	45%
Molten carbonate	6600	52%

Continuous-Duty Standby

An engine that drives a refrigeration compressor can be switched over automatically to drive an electrical generator in the event of a

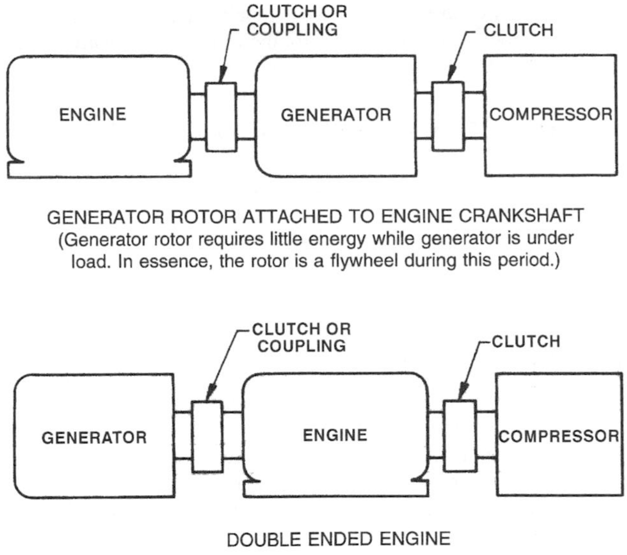

Fig. 47 Dual-Service Applications

power failure (Figure 47). This plan assumes that loss of the compressor service can be tolerated in an emergency.

If the engine capacity of such a dual-service systems equals 150 or 200% of the compressor load, the power available from the generator can be delivered to the utility grid during normal operation. While induction generators may be used for this application, a synchronous generator is required for emergency operation if there is a grid outage.

Dual-service arrangements have the following advantages:

- In comparison to two engines in single service, lower capital investment and reduced space and maintenance are required, even after allowing for the additional controls needed with dual-service installations.
- Because they operate continuously or on a regular basis, dual-service engines are more reliable than emergency (reserved) single-service engines.
- Engines that are in service and running can be switched over to emergency power generation with minimal loss of continuity. Hospital authorities recognize this and prefer dual-service to the cold start of an emergency generator.

Expansion Engines/Turbines

Expansion turbines are turbines that use compressed gases available from gas-producing processes to produce shaft power. One application compresses natural gas from high-pressure pipelines in much the same way that a steam turbine replaces a pressure reducing valve. These engines are used mainly, however, for cryogenic applications to about –320°F (e.g., oxygen for steel mills, low-temperature chemical processes, and the space program). Relatively high-pressure air or gas expanded in an engine drives a piston and is cooled in the process. At the shaft, about 42 Btu/hp is removed. Available units, developing as much as 600 hp, handle flows ranging from 100 to 10,000 cfm. Throughput at a given pressure is controlled by varying the cutoff point, the engine speed, or both. The conversion efficiency of heat energy to shaft work ranges from 65 to 85%. A 5 to 1 pressure ratio and an inlet pressure of 3000 psig are recommended. Outlet temperatures as low as –450°F have been handled satisfactorily.

Such a process consumes no fuel, but it does provide shaft energy, as well as expansion refrigeration, which fits the definition of cogeneration. Similarly, a back pressure steam turbine, used instead of a pressure reducing valve, generates productive shaft power and permits the use of the residual thermal energy.

Natural gas from a high-pressure pipeline can be run through a turbine to produce shaft energy and then burned downstream. See the section on Combustion Turbines for more information.

DESIGN AND INSTALLATION

ENGINES

Additional information on engines may be found at the beginning of this chapter in the section on Reciprocating Engines. Specific information on the application of engines to chillers, heat pumps, and refrigeration equipment may be found in Chapter 47.

Jacket Water Systems

The circulating water and oil systems must be kept clean because the internal coolant passages of the engine are not readily accessible for service. The installation of piping, heat exchangers, valves, and accessories must include provisions for internal cleaning of these circuits before they are placed in service and, when possible, for maintenance access afterwards.

Coolant fluids must be noncorrosive and free from salts, minerals, or chemical additives that can deposit on hot engine surfaces or form sludge in relatively inactive fluid passages. Generally, engines cannot be drained and flushed effectively without major disassembly, making any chemical treatment of the coolant fluid that can produce sediment or sludge undesirable.

An initial step toward maintaining clean coolant surfaces is to limit fresh water makeup. The coolant system should be tight and leak-free. Softened water or mineral-free water is effective for initial fill and makeup. Forced-circulation hot water systems may require only minor corrosion-inhibiting additives to ensure long, trouble-free service. This feature is one of the major assets of hot water heat recovery systems.

Water-Cooled Engines

Heat in the engine coolant should be removed by heat exchange to a separate water system. Recirculated water can then be cooled in open-circuit cooling towers, where water is added to make up for evaporation. Closed-circuit coolant of all types (e.g., for closed-circuit evaporative coolers, radiators, or engine-side circuits of shell and tube heat exchangers) should be treated with a rust inhibitor and/or antifreeze to protect the engine jacket. Because engine coolant is best kept in a protected closed loop, it is usually circulated on the shell side of an exchanger. A minimum fouling factor of 0.002 should be assigned to the tube side.

Jacket water outlet and inlet temperature ranges of 175 to 190°F and 165 to 175°F, respectively, are generally recommended, except when the engines are used with a heat recovery system. These temperatures are maintained by one or more thermostats that bypass water as required. A 10 to 15°F temperature rise is usually accompanied by a circulating water rate of about 0.5 to 0.7 gpm per engine horsepower.

Installation involves (1) sizing water piping according to the engine manufacturer's recommendations, (2) avoiding restrictions in the water pump inlet line, (3) never connecting piping rigidly to the engine, and (4) providing shutoffs to facilitate maintenance.

If the cost of pumping the water at conventional jacket water temperature can be absorbed in the external system, a hot water system is preferred. However, the cooling tower must be sized for the full jacket heat rejection if there are periods when no load is available to absorb it (see Chapter 36 for information on cooling tower design). Most engines are designed for forced circulation of the coolant.

Where several engines are used in one process, independent coolant systems for each machine can be used to avoid complete plant shutdown from a common coolant system component failure. Such independence can be a disadvantage in that the unused engines are not maintained at operating temperature, as they are when all units are in a common circulating system. If the idle machine temperature drops below the dew point of the combustion products, corrosive condensate may form in the exhaust gas passages each time the idle machine is started.

When substantial water volume and machinery mass must be heated up to operating temperature, the condensate volume is quite significant and must be drained. Some contaminants will get into the lubricant and reduce the service life. If the machinery gets very cold, it may be difficult to start. Units that are started and stopped frequently require an off-cycle heating sequence to lessen the exposure to corrosion.

Forced lubrication by an auxiliary, externally powered lubricant pump is beneficial to the engine when the unit is off. Continuously bathing the engine parts in oil and maintaining engine temperatures near their normal operating temperature improves engine life and lowers the engine maintenance costs. This practice should be evaluated against the pumping losses and the radiation heat losses to the engine room environment.

Still another concept for avoiding a total plant shutdown, and even avoiding the probability of any single engine shutdown, requires the following arrangement:

- A common, interconnected jacket water piping system for all the engines.
- An extra standby device for all auxiliaries such as pumps and heat exchangers. Valves must be installed to isolate any auxiliary that fails or is out for preventive maintenance while permitting continued operation.
- Header isolation valves to permit continued plant operation while any section of the common piping is serviced or repaired.

This common piping arrangement permits continuous full-load plant operation if any one auxiliary suffers an outage or needs maintenance; no more than one engine in the battery can have a forced outage if the headers suffer a problem. On the other hand, independent, dedicated auxiliaries for each engine can force an engine outage whenever an auxiliary is down, unless each such auxiliary is provided with a standby, which is not a practical option. Furthermore, the common header concept avoids the possibility of a second engine outage when any of the second engine's support components fails while the first is out for major repair. It also permits a warm start of any engine by the circulation of a moderate flow of the hot jacket water through any idle engine.

Ebullient Cooling

Ebullient cooling systems of any size are not favored by many designers and manufacturers of large industrial engines, but they are commonly used. This method presents some special hazards for water treatment. Evaporation, which should not take place in the engine, sometimes does, and the concentration of dissolved minerals increases at this critical location. This problem can be minimized by using gravity or forced circulation to encourage high flow rates and by reducing minerals added to the system to a negligible amount.

Elevating the flash tank or using pumps to maintain a high static pressure also reduces the tendency for vaporization in the engine, which can cause engine failure or fouling. The water treatment must be adequate to prevent corrosion from free oxygen brought in with returned condensate and makeup water in the flash tank. Additional treatment should control corrosion in the heat recovery muffler, the steam separator, and the condensate system.

The water level of separate steam-producing engine cooling systems must be controlled to prevent backflow through the steam nozzle of idle recovery apparatus. A back pressure regulating valve, a steam check valve, or an equalizing line can prevent this problem.

Exhaust Systems

Engine exhaust must be safely conveyed from the engine through piping and any auxiliary equipment to the atmosphere at an allowable pressure drop and noise level. Allowable back pressures, which vary with engine design, run from 2 to 25 in. of water. For low-speed engines, this limit is typically 6 in. of water; for high-speed engines, it is typically 12 in. Adverse effects of excessive pressure drops include power loss, poor fuel economy, and excessive valve temperatures, all of which result in shortened service life and jacket water overheating.

General installation recommendations include the following:

- Install a high-temperature, flexible connection between the engine and the exhaust piping. Exhaust gas temperature does not normally exceed 1200°F, but 1400°F may be reached for short periods. An appropriate stainless steel connector may be used.
- Adequately support the exhaust system downstream from the connector. At the maximum operating temperature, no weight should be exerted on the engine or its exhaust outlet.
- Minimize the distance between the silencer and the engine.
- Use a 30 to 45° tail pipe angle to reduce turbulence.
- Specify tail pipe length (in the absence of other criteria) in odd multiples of $12.5(T_e^{0.5}/P)$, where T_e is the temperature of the exhaust gas (Rankine), and P is exhaust frequency (pulses per second). The value of P is calculated as follows:

  P = rpm/120 = (rev/s)/2 for four-stroke engines

  P = rpm/60 = rev/s for two-stroke engines

  Note that for V-engines with two exhaust manifolds, rpm or rev/s equals engine speed.
- A second, but less desirable, exhaust arrangement is a Y-connection with branches entering the single pipe at about a 60° angle; never use a T-connection, as the pulses of one branch will interfere with the pulses from the other.
- Use an engine-to-silencer pipe length that is 25% of the tail pipe length.
- Install a separate exhaust for each engine to reduce the possibility of condensation in the engine that is not running.
- Install individual silencers to reduce the condensation that results from an idle engine.
- Limit heat radiation from exhaust piping with a ventilated sleeve around the pipe or with high-temperature insulation.
- Use large enough fittings to minimize pressure drop.
- Allow for thermal expansion in exhaust piping, which is about 0.09 in. per foot of length.
- Specify muffler pressure drops to be within the back pressure limits of the engine.
- Do not connect the engine exhaust pipe to a chimney that serves natural-draft gas appliances.
- Slope the exhaust away from the engine to prevent condensate backflow. Drain plugs in silencers and drip legs in long, vertical exhaust runs may also be required. Raincaps may prevent the entrance of moisture but might add back pressure and prevent adequate upward ejection velocity.

Proper effluent discharge and weather protection can be maintained in continuously operated systems by maintaining sufficient discharge velocity (in excess of 2500 fpm) through a straight stack; in intermittently operated systems, protection can be maintained by installing drain-type stacks. Drain-type stacks effectively eliminate rainfall entry into a vertical stack terminal without destroying the upward ejection velocity, as a raincap does.

Drain-Type Stack. This design places a stack head, rather than a stack cap, over the discharge stack. The height of the upper section is important for adequate rain protection, just as the height of the stack is important for adequate dispersal of effluent. Stack height should be great enough to discharge above the building eddy zone. Bolts for inner stack fastening should be soldered, welded, or brazed, depending on the tack material.

Table 9 Exhaust Pipe Diameter[a]

Output Power, hp	Minimum Pipe Diameter, in. — Equivalent Length of Exhaust Pipe: 25 ft	50 ft	75 ft	100 ft
25	2	2	3	3
50	2.5	3	3	3.5
75	3	3.5	3.5	4
100	3.5	4	4	5
200	4	5	5	6
400	6	6	8	8
600	6	8	8	8
800	8	8	10	10

[a]Minimum exhaust pipe diameter to limit engine exhaust back pressure to 8 in. of water.

Powerhouse Stack. In this design, a fan discharge intersects the stack at a 45° angle. A drain lip and drain are added in the fume discharge version.

Offset Design. This design is recommended for round ductwork and can be used with sheet metal or glass fiber reinforced polyester ductwork.

The exhaust pipe may be routed between an interior engine installation and a roof-mounted muffler through (1) an existing unused flue or one serving power-vented gas appliances only (this should not be used if exhaust gases may be returned to the interior); (2) an exterior fireproof wall with provision for condensate drip to the vertical run; or (3) the roof, provided that a galvanized thimble with flanges having an annular clearance of 4 to 5 in. is used. Sufficient clearance is required between the flue terminal and the rain cap on the pipe to permit the venting of the flue. A clearance of 30 in. between the muffler and roof is common. Vent passages and chimneys should be checked for resonance.

When interior mufflers must be used, minimize the distance between the muffler and the engine, and insulate inside the muffler portion of the flue. Flue runs exceeding 25 ft may require power venting, but vertical flues help to overcome the pressure drop (natural draft).

The following design and installation features should be used for flexible connections:

- Material—Convoluted steel (Grade 321 stainless steel) is favored for interior installation.
- Location—Principal imposed motion (vibration) should be at right angles to the connector axis.
- Assembly—The connector (not an expansion joint) should not be stretched or compressed; it should be secured without bends, offsets, or twisting (the use of float flanges is recommended).
- Anchor—The exhaust pipe should be rigidly secured immediately downstream of the connector in line with the downstream pipe.
- Exhaust piping—Some alloys and standard steel alloy or steel pipe may be joined by fittings of malleable cast iron. Table 9 shows exhaust pipe sizes. The exhaust pipe should be at least as large as the engine exhaust connection. Stainless steel double-wall liners may be used.

Combustion Air Systems

The following factors apply to combustion air requirements:

- Supply 2 to 5 cfm per brake horsepower, depending on type, design, and size. Two-cycle units consume about 40% more air than four-cycle units.

- Avoid heated air because power output varies by $(T_r/T_a)^{0.5}$, where T_r is the temperature at which the engine is rated and T_a is the engine air intake temperature, both in Rankine.
- Locate the intake away from contaminated air sources.
- Install properly sized air cleaners that can be readily inspected and maintained (pressure drop indicators are available). Air cleaners minimize cylinder wear and piston ring fouling. About 90% of valve, piston ring, and cylinder wall wear is the result of dust. Both dry and wet cleaners are used. If wet cleaners are undersized, oil carryover may reduce filter life. Filters may also serve as flame arresters.
- Engine room air-handling systems may include supply and exhaust fans, louvers, shutters, bird screens, and air filters. The total static pressure opposing the fan should be 0.35 in. of water maximum.\

The following sections on Heating and Ventilating Systems, Off-Engine Instruments and Controls, Noise and Vibration Control, Maintenance, and Engine Applications for internal combustion engines also apply, with slight modifications, to combustion and steam turbines.

Heating and Ventilating Systems

Methods of dissipating heat from the jacket water system, exhaust system, lubrication and piston cooling oil, turbocharger, and air intercooler have been covered in previous sections. In addition, radiation and convection losses from the surfaces of the engine components and accessories and piping must be dissipated by ventilation. If the radiated heat is more than 8 to 10% of the fuel input, an air cooler may be required. In some cases, this rejected heat can be productively applied as tempered makeup air in an adjacent space, with consideration given to life/fire safety requirements, but in most instances it is simply vented.

This heat must be removed to maintain acceptable working conditions and to avoid overloading the electrical systems from high ambient conditions. Heat can be removed by outdoor air ventilation systems that include dampers and fans and thermostatic controls regulated to prevent overheating or excessively low temperatures in extreme weather. The manner and amount of heat rejection varies with the type, size, and make of engine and the extent of engine loading.

An **air-cooled engine** installation includes the following:

- An outside air entrance at least as large as the radiator face and 25 to 50% larger if protective louvers impede airflow.
- Auxiliary means (e.g., a hydraulic, pneumatic, or electric actuator) to open the louvers blocking the heated air exit, rather than a gravity-operated actuator.
- Control of jacket water temperature by radiator louvers in lieu of a bypass for freeze protection.
- Thermostatically controlled shutters that regulate airflow to maintain the desired temperature range. In cold climates, louvers should be closed when the engine is shut down to help maintain engine ambient temperature at a safe level. A crankcase heater can be installed on backup systems located in unheated spaces.
- Positioning of the engine so that the face of the radiator is in a direct line with an air exit leeward of the prevailing wind.
- An easily removable shroud so that exhaust air cannot reenter the radiator.
- Separated units in a multiple installation to avoid air short-circuiting among them.
- Low-temperature protection against snow and ice formation. Propeller fans cannot be attached to long ducts because they can only achieve low static pressure. Radiator cooling air directed over the engine promotes good circulation around the engine; thus, the engine runs cooler than for airflow in the opposite direction.
- Adequate sizing to dissipate the other areas of heat emissions to the engine room.

Table 10 Ventilation Air for Engine Equipment Rooms

Room Air Temperature Rise,[a] °F	Airflow, cfm/hp		
	Muffler and Exhaust Pipe[b]	Muffler and Exhaust Pipe[c]	Air- or Radiator-Cooled Engine[d]
10	140	280	550
20	70	140	280
30	50	90	180

[a]Exhaust minus inlet.
[b]Insulated or enclosed in ventilated duct.
[c]Not insulated.
[d]Heat discharged in engine room.

Sufficient ventilation must also be provided to protect against minor fuel supply leaks (not rupture of the supply line). Table 10 is sometimes used for minimum ventilation air requirements. Ventilation may be provided by a fan that induces the draft through a sleeve surrounding the exhaust pipe. A slight positive pressure should be maintained in the engine room.

Ventilation efficiency for operator comfort and equipment reliability is improved by (1) taking advantage of a full wiping effect across sensitive components (e.g., electrical controls and switchgear) with the coolest air; (2) taking cool air in as low as possible and forcing it to travel at occupancy level; (3) letting cooling air pass subsequently over the hottest components; (4) exhausting from the upper, hotter strata; (5) avoiding any short-circuiting of the cool air directly to the exhaust while bypassing equipment; and (6) arranging equipment locations, when possible, to permit such an airflow path.

Larger engines with off-engine filters should accomplish the following:

- Temper cold outside combustion air when its temperature is low enough to delay ignition timing and inhibit good combustion, which leads to a smoky exhaust.
- Permit the silencer and/or recovery device's hot surfaces to warm this cold air to an automatically controlled temperature. As this air enters the machine room it also provides some cooling to a hot machine room or heating to a cold room.
- Manipulate the dampers with a thermostat at the inlet to the machine room, which is reset by room temperature.
- Cool hot combustion air with a cooling device downstream of the air filter to increase engine performance. This practice is particularly helpful with large, slow-speed, naturally aspirated engines. The Diesel Engine Manufacturers Association (DEMA) recommended that engines rated at 90°F and 1500 ft above sea level be derated in accordance with the particular manufacturer's ratings. In a naturally aspirated engine, a rating of 100 hp at 1500 ft drops to 50 hp at 16,000 ft. Also, a rating can drop from 100 hp at 90°F to 88 hp at 138°F.

Off-Engine Instruments and Controls

Controls for cogeneration systems are required for (1) system output, (2) safety, (3) prime mover automation, and (4) waste heat recovery and disposal. Building requirements determine the level of automation. Every system must have controls to regulate the output energy and to protect the equipment. Independent power plants require constantly available control energy to actuate cranking motors, fuel valves, circuit breakers, alarms, and emergency lighting.

Smaller engines and combustion turbines generally use battery systems for these functions because the stored energy is available if the generating system malfunctions. Automatic equalizer battery chargers maintain the energy level with minimum battery maintenance.

Larger plants can use either batteries or an emergency generator technique as described in the section on Compressed Air Systems on page 7.6. Generally, a simple control system is adequate where labor is available to make minor adjustments and to oversee system operation. Fully automatic, completely unattended systems have the same advantages as automatic temperature control systems and are gaining acceptance.

Fully automated generator controls should be considered for office buildings, hotels, motels, apartments, large shopping centers, some schools, and manufacturing plants where close control of frequency and voltage is needed. The control system must be highly reliable, regardless of load change or malfunction, and it must protect the system equipment from electrical transients and malfunctions. To provide this reliability, each generator must be controlled.

Generators that operate in **parallel** require interconnecting controls. The complete system must be integrated to the building use, give properly sequenced operation, and provide overall protection. In addition to those controls required for a single prime mover installation, the following further controls are required for multiple-generator installations:

- Simultaneous regulation of fuel flow to each prime mover to maintain required shaft output
- Load division and frequency regulation of generators by a signal to the fuel controls
- System voltage regulation and reactive load division by maintaining the generator output at the required level
- Automatic starting and stopping of each unit for protection in the event of a malfunction
- Load demand and unit sequencing by determining when a unit should be added or taken off the bus as a function of total load
- Automatic paralleling of the oncoming unit to the bus after it has been started and reaches synchronous speed
- Safety protection for prime movers, generators, and waste heat recovery equipment in the event of overload or abnormal operation, including a means of load dumping (automatic removal of building electrical loads) in case the prime mover overloads or fails

When the system is operated in **isolation** of the utility grid, the engine speed control must maintain frequency within close tolerances, both at steady-state and transient conditions. Generators operating in parallel require a speed control to determine frequency and to balance real load between operating units. To obtain precise control, an electronic governing system is generally used to throttle the engine; the engine responds quickly and is capable of responding to control functions as follows:

- Sensing engine speed and controlling to operating frequency (60 Hz). See the section on Generators on page 7.27 for other generator control information.
- Synchronizing the oncoming generating unit to the bus so it can be paralleled.
- Sensing true real load, rather than current, to divide real load between parallel units. (This must be done through the speed control to the throttle.)
- Proper throttle operation during start-up to ensure engine starting and prevention of overspeed as the engine approaches rated speed.
- Time reference control is desirable to maintain clock accuracy to within 60 s per month.

When starting a unit, the following actions must be performed in sequence: (1) initiate engine crank; (2) open fuel valve and throttle; (3) when engine starts, terminate cranking; (4) when desired speed is reached, eliminate overshoot; (5) if the engine refuses to start in a given period, register a malfunction and start the next unit in sequence; and (6) reaching the desired speed, synchronize the started engine-alternator to the bus to be paralleled without causing severe mechanical and electrical transients. Some time should be allowed for synchronization; if it does not occur during the allotted time, a malfunction should register, and the next unit in sequence should start.

Once a unit is online, real and reactive load division should be effective immediately. The throttle of the unit coming online must be advanced so that it accepts its share of the load. The unit or units that were carrying all the load have their throttle(s) retarded so that they give up load to the oncoming unit until all units share equally or, if they are sized differently, in proportion to unit size. During the load-sharing process, control frequency must be maintained, and once paralleled, the load-sharing control must correct continuously to maintain the load balance. Reactive load sharing is similar but is done by the voltage regulator varying the excitation of the alternator exciter field.

Problems that increase installation cost can arise during installation. These can be minimized by carefully checking out the cogeneration system, particularly the engine-alternator units. Typical problems arising during installation include (1) engine actuators placed too near exhaust pipes, causing excessive wear of actuator bearings and loose parts in the actuator; (2) improper wiring; (3) sloppy linkage between engine actuator and carburetor; and (4) difficulties due to a failure to coordinate electrical design with the utility.

Sensors used to send a malfunction signal to the control system can also cause trouble when not properly installed. It is appropriate to use two sensors in sequence—one for alarm indication of an abnormal condition and the other for shutdown when a malfunction grows worse (e.g., lubricant reaches a shutdown temperature).

Noise and Vibration Control

Engine-driven machines installed indoors, even where the background noise level is high, usually require noise attenuation and isolation from adjoining areas. Air-cooled radiators, noise radiated from surroundings, and exhaust heat recovery boilers may also require silencing. Boilers that operate dry do not require separate silencers. Installations in more sensitive areas may be isolated, receive sound treatment, or both.

Basic attenuation includes (1) turning air intake and exhaust openings away (usually up) from the potential listener; (2) limiting blade-tip speed (if forced-draft air cooling is used) to 12,000 fpm for industrial applications, 10,000 fpm for commercial applications, and 8000 fpm for critical locations; (3) acoustically treating the fan shroud and plenum between blades and coils; (4) isolating (or covering) moving parts, including the unit, from their shelter (where used); (5) properly selecting the gas meter and regulator(s) to prevent singing; and (6) adding sound traps or silencers on ventilation air intake, exhaust, or both.

Further attenuation means include (1) lining the intake and exhaust manifolds with sound-absorbing materials; (2) mounting the unit, particularly a smaller engine, on vibration isolators, thereby reducing foundation vibration; (3) installing a barrier between the prime mover and the listener (often a concrete block enclosure); (4) enclosing the unit with a cover of absorbing material; and (5) locating the unit in a building constructed of massive materials, paying particular attention to the acoustics of the ventilating system and the doors.

Noise levels must meet legal requirements (see Chapter 46 of the 1999 *ASHRAE Handbook—Applications* for details).

Foundations. Multicylinder, medium-speed engines may not require massive concrete foundations, although concrete offers advantages in cost and in maintaining alignment for some driven equipment. Fabricated steel bases are satisfactory for direct-coupled, self-contained units, such as electric sets. Steel bases mounted on steel spring or equal-vibration isolators are adequate and need no special foundation other than a floor designed to accommodate the weight. Concrete bases are also satisfactory for such units, provided the bases are equally well isolated from the supporting floor or subfloor.

Glass fiber blocks are effective as isolation material for concrete bases, which should be thick enough to prevent deflection. Excessively thick bases only increase subfloor or soil loading, and they still should be supported by a concrete subfloor. In addition, some acceptable isolation material should be placed between the base and the floor. To avoid the transmission of vibration, an engine base or foundation should never rest directly on natural rock formations. Under some conditions, such as shifting soil on which an outlying cogeneration plant might be built, a single, very thick concrete pad for all equipment and auxiliaries may be required to avoid a catastrophic shift between one device and another.

Alignment and Couplings. Proper alignment of the prime mover to the driven device is necessary to prevent undue stresses to the shaft, coupling, and seals of the assembly. Installation instructions usually suggest that the alignment of the assembly be performed and measured at maximum load condition and maximum heat input to the turbine or engine.

Torsional vibrations can be a major problem when matching these components, particularly when matching a reciprocating engine to any higher speed centrifugal device. The stiffness of a coupling and its dynamic response to small vibrations from an angular misalignment at critical speed(s) affect the natural frequencies that exist in the assembly. Encotech (1992) describes how changing the mass, stiffness, or damping of the coupling can alter the natural frequency during start-up, synchronization, or load change or when a natural frequency exists within the assembly's operating envelope. Proper grouting is needed to preserve the alignment.

Flexible Connections. Greater care is required in the design of piping connections to turbines and engines than for other HVAC equipment because the larger temperature spread causes greater expansion. (See Chapter 41 for further information on piping.)

Maintenance

Preventive Maintenance. One of the most important provisions for healthy and continuous plant operation is implementation of a comprehensive preventive maintenance program. This should include written schedules of daily wipedown and observation of equipment, weekly and periodic inspection for replacement of degradable components, engine oil analysis, and maintenance of proper water treatment. Immediate access to repair services may be furnished by subcontract or by in-house plant personnel. An inventory of critical parts should be maintained on site. (See Chapters 34 through 41 of the 1999 *ASHRAE Handbook—Applications* for further information on building operation and maintenance.)

Predictive Maintenance. Given the tremendous advancement and availability of both fixed and portable instrumentation for monitoring sound, vibration, temperatures, pressures, flow, and other online characteristics, many key aspects of equipment and system performance can be logged manually or by computer to observe trends. Such factors as fuel rate, heat exchanger approach, and cylinder operating condition can be compared against new and/or optimized baseline conditions to indicate when maintenance may be required. This monitoring permits periods between procedures to be longer, catches incipient problems before they create outages or major repairs, and avoids unnecessary maintenance.

Equipment Rotation. When more than one engine, pump, or other component is serving a given distribution system, it is undesirable to operate the units with the equal life approach mode, which puts each unit in the same state of wear and component deterioration. If only one standby unit (or none) is available to a given battery of equipment, and one unit suffers a major failure or shutdown, all units now needed to carry the full load would be prone to additional failure while the first failed unit is undergoing repair.

The preferred operating procedure is to keep one unit in continuous reserve, with the shortest possible running hours between overhauls or major repair, and to schedule operation of all others for unequal running hours. Thus any two units would have a minimum statistical chance of a simultaneous failure. All units, however, should be exercised for several hours in any week.

Engine Applications

Engine Sizing. In sizing an engine, proper evaluation of the various electrical loads, building heat gains and losses, and their time of occurrence avoids overestimating the actual simultaneous load. For existing buildings, an energy audit should be performed to access correct load.

Consumption rates must be known to determine operating costs. Specific data can be obtained from the manufacturer; however, a range of consumption rates is given in Table 11. Atmospheric corrections included in the following equation may be used to match prime movers to loads at various ambient conditions.

$$P_u = P_{max} \times (\text{Rating or derating factor}) \tag{9}$$

where

P_u = usable power, kW
P_{max} = power rating under ideal conditions, kW
P_{max} is based on manufacturer's performance data (usually a dynamometer test) and corrected in accordance with the manufacturer's specifications. Examples of specifications are

60°F and 29.92 in. Hg
80°F and 1000 ft elevation (NEMA)
85°F and 500 ft elevation (SAE)
90°F and 1500 ft elevation (DEMA)

$$\text{Rating factor} = \frac{100 - C_a - C_t - C_{hv} - C_r}{100} \tag{10}$$

where

C_a = percent altitude correction
= 3% per 1000 ft above a specific level for naturally aspirated engines
= 2% per 1000 ft for turbocharged engines
C_t = percent temperature correction
= 1% per 10°F rise above a specified base
C_{hv} = percent fuel heating value correction to account for the air intake temperature (see the section on Fuel Heating Value on page 7.4)
C_r = percent reserve, which is an allowance (safety factor) to permit design output under unforeseen operating conditions that would reduce output, such as a dusty environment, poor maintenance, higher ambient temperature, or lowered cooling efficiency. Recommended values are in Table 12.

Table 11 Fuel Consumption Rates

Fuel	Heating Value, Btu/gal	Range of Consumption, Btu/hp·h
Fuel oil	137,000 to 156,000	7,000 to 9,000
Gasoline	130,000	10,000 to 14,000
Type of Gas Engine	**Compression Ratio**	**Typical Gas Consumption, Btu/hp·h**
Turbocharged	10.5:1	8,100
Naturally aspirated	10.5:1	9,200
Naturally aspirated	7.5:1	10,250

Table 12 Percent Minimum Engine Reserves for Air Conditioning and Refrigeration

Altitude, ft	Naturally Aspirated		Turbocharged Aftercooled	
	Air Cond.	Refrigeration	Air Cond.	Refrigeration
Sea level	15	20	20	30
1000	12	17	18	28
2000	10	14	16	26
3000	10	11	14	24
4000	10	10	12	22
5000	10	10	10	20
10,000	10	10	10	10

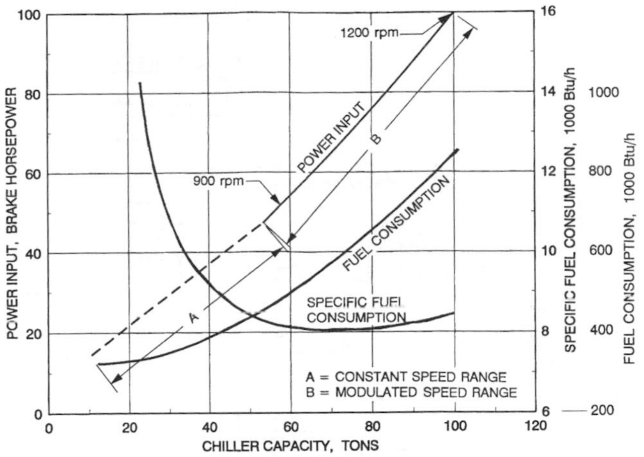

Fig. 48 Performance Curve for Typical 100 Ton, Gas Engine-Driven, Reciprocating Chiller

Engine-Driven Reciprocating Compressors. Engine-driven reciprocating compressor water chiller units may be packaged or field assembled from commercially available equipment for comfort service, low-temperature refrigeration, and heat pump applications. Both direct-expansion and flooded chillers are used. Some models achieve a low operating cost and a high degree of flexibility by combining speed variation with cylinder unloading. These units achieve capacity control by reducing engine speed to about 30 to 50% of rated speed; further capacity modulation may be achieved by unloading the compressor in increments. Engine speed should not be reduced below the minimum specified by the manufacturer for adequate lubrication or good fuel economy.

Most engine-driven reciprocating compressors are equipped with a cylinder loading mechanism for idle (unloaded) starting. This arrangement may be required because the starter may not have sufficient torque to crank both the engine and the loaded compressor. With some compressors, not all the cylinders (e.g., four out of 12) unload; in this case, a bypass valve must be installed for a fully unloaded start. The engine first speeds to one-half or two-thirds of full speed. Then, a gradual cylinder load is added, and the engine speed increases over a period of 2 to 3 min. In some applications, such as an engine-driven heat pump, low-speed starting may cause oil accumulation and sludge. As a result, a high-speed start is required.

These systems operate at specific fuel consumptions (SFCs) of approximately 8 to 13 ft^3/h of pipeline quality natural gas (HHV = 1000 Btu/ft^3) per horsepower in sizes down to 25 tons. Comparable heat rates for diesel engines run from 7000 to 9000 Btu/hp·h. Smaller units are also available. Coolant pumps can also be driven by the engine. These direct-connected pumps never circulate tower water through the engine jacket. Figure 48 illustrates the fuel economy effected by varying prime mover speed with reciprocating compressor load until the machine is operating at about half its capacity. Below this level, the load is reduced at essentially constant engine speed by unloading the compressor cylinders.

Frequent operation at low engine idling speed may require an auxiliary oil pump for the compressor. To reduce wear and assist in starts, a tank-type lubricant heater or a crankcase heater and a motor-driven auxiliary oil pump should be installed to lubricate the engine with warm oil when it is not running. Refrigerant piping practices for engine-driven units are the same as for motor-driven units.

Engine-Driven Centrifugal Compressors. Packaged, engine-driven centrifugal chillers that do not require field assembly are available in capacities up to 2100 tons. Automotive derivative engines modified for use on natural gas are typical of these smaller packages because of their compact size and mass. These units may be equipped with either manual or automatic start-stop systems and engine speed controls.

Larger open-drive centrifugal chillers are usually field assembled and normally include a compressor mounted on an individual base and coupled by means of flanged pipes to an evaporator and a condenser. The centrifugal compressor is driven through a speed increaser. Many of these compressors operate at about six times the speed of the engines; compressor speeds of up to 14,000 rpm have been used.

To effect the best compromise between the initial cost of the equipment (engine, couplings, and transmission) and the maintenance cost, engine speeds between 900 and 1200 rpm are generally used. Engine output can be modulated by reducing engine speed. If the operation at 100% of rated speed produces 100% of rated output, approximately 60% of rated output is available at 75% of rated speed. Capacity control of the centrifugal compressor can be achieved by either variable inlet guide vane control with constant compressor speed or a combination of variable-speed control and inlet guide vane control, the latter providing the greatest operating economy.

Engine-Driven Heat Pumps. An additional economic gain can result from operating an engine-driven refrigeration cycle as a heat pump, provided that the facility has a thermal load profile that can adequately absorb its 100 to 120°F low-quality heat. Using the same equipment for both heating and cooling reduces capital investment. A gas engine drive for heat pump operation also makes it possible to operate in a cogeneration mode, which requires a somewhat larger thermal load. Unless a major portion of this larger thermal recovery can be absorbed, the cycle may not be economical.

For larger projects, the economics of engine or turbine generators combined with motor-driven chillers should be compared with the economics of generators plus combustion engine or turbine-driven chillers. Figure 49 shows the total energy available from a typical engine-driven heat pump. Table 13 lists the economic coefficients of performance (ECOPs) for various configurations. The values in the table can be used to illustrate the difference between the COP, the ECOP, and the thermal efficiency of this cycle. Here, the ECOP is based upon the engine fuel input of 10,000 Btu/ton·h, yielding an ECOP of 14,545/10,000 = 1.45; while the COP, based upon the work input (not the fuel input), is 14,545/2,545 = 5.72.

Table 13 Coefficient of Performance for Engine-Driven Heat Pump

Item	Heat Source		
	Refrigerant Condenser Only, Btu/ton·h	Refrigerant Condenser and Jacket Water, Btu/ton·h	Refrigerant Condenser, Jacket Water, Exhaust Gas, Btu/ton·h
Total heat input to engine	10,000	10,000	10,000
Cooler heat rejection to condenser (from building load)	12,000	12,000	12,000
Heat of compression	2,545	2,545	2,545
Heat from engine jacket water heater	—	2,500	2,500
Heat from exhaust gas heater	—	—	3,000
Total heat to heating circuit	14,545	17,045	20,045
Economic coefficient of performance	1.45	1.70	2.00

The classic definition of reverse-cycle performance is (heat out)/(work in). The definition does not recognize the fuel input to the engine, just as it ignores the fuel input to generate the electricity for the motor of a motor-driven compressor. No COP is really defined for the cycle that captures the jacket and exhaust heat, but

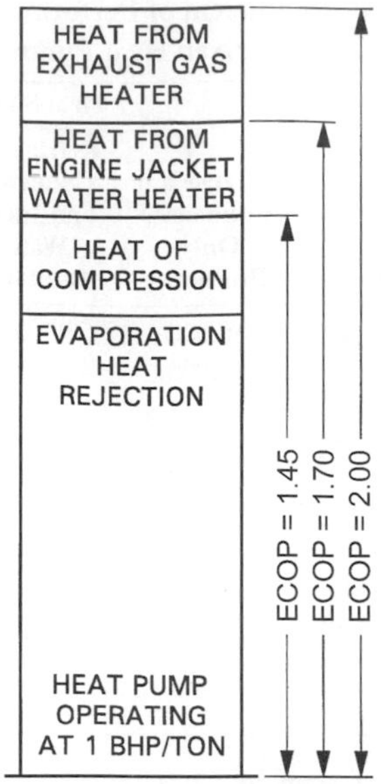

Fig. 49 Heat Balance for Engine-Driven Heat Pump

input) is indeed 17,045/10,000 = 1.70 for jacket recovery and 20,045/10,000 = 2.0 for the jacket and exhaust cycles, as shown in Figure 49.

Engine-Driven Screw Compressors. Chiller packages with these compressors are available for refrigeration applications. Manufacturers offer water chillers that use screw compressors driven directly by natural gas engines. Capacity control is achieved by varying the engine speed and adjusting the slide valve on the compressor. Units have a cooling COP near 5.72 without heat recovery and an ECOP near 1.45 at rated cooling load.

COMBUSTION TURBINE APPLICATIONS

The gas turbine has achieved an increasingly important position as a prime mover for electric power generation up to more than 240 MW and for shaft power drives up to more than 108,000 hp. Figure 50 shows a typical gas turbine refrigeration cycle, with optional combustion air precooling. A gas turbine must be brought up to speed by an auxiliary starter. With a single-shaft turbine, the air compressor, turbine, speed reducer gear, and refrigeration compressor must all be started and accelerated by this starter. The refrigeration compressor must also be unloaded to ease the starting requirement. Sometimes, this may be done by making sure the capacity control vanes close tightly. At other times, it may be necessary to depressurize the refrigeration system to get started.

With a split-shaft design, only the air compressor and the gas producer turbine must be started and accelerated. The rest of the unit starts rotating when enough energy has been supplied to the blades of the power turbine. At this time, the gas producer turbine is up to speed, and the fuel supply is ignited. Electric starters are usually available as standard equipment. Reciprocating engines, steam turbines, and hydraulic or pneumatic motors may also be used. The output shaft of the gas turbine must rotate in the direction required by the refrigeration compressor; in many cases, the manufacturers of split-shaft engines can provide the power turbine with either direction of rotation.

At low loads, both the gas turbine unit and the centrifugal refrigeration machine are affected by surge, a characteristic of all centrifugal and axial flow compressors. At a certain pressure ratio, a minimum flow through the compressor is necessary to maintain stable operation. In the unstable area, a momentary backward flow of gas occurs through the compressor. Stable operation can be maintained, however, by the use of a hot gas bypass valve.

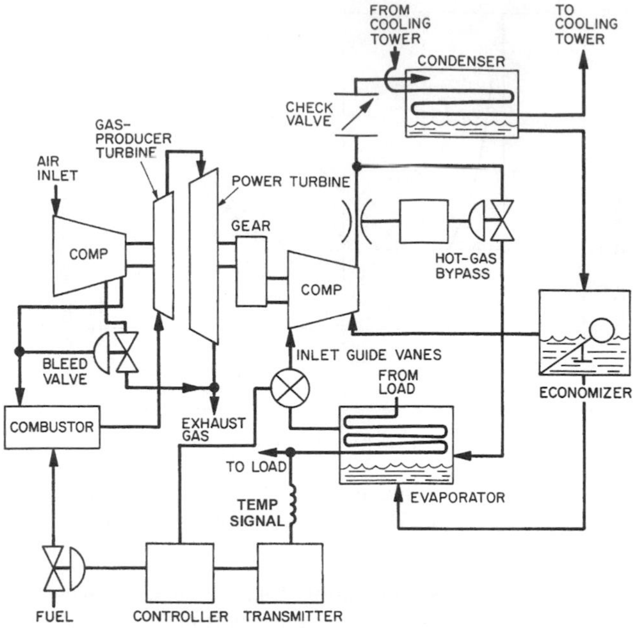

Fig. 50 Typical Gas Turbine Refrigeration Cycle

The turbine manufacturer normally includes automatic surge protection, either as a bleed valve that bypasses a portion of the air directly from the axial compressor into the exhaust duct or by providing for a change in the position of the axial compressor stator vanes. Both methods are used in some cases.

The assembly should be prevented from rotating backward, which may occur if the unit is suddenly stopped by one of the safety controls. The difference in pressure between the refrigeration condenser and cooler can make the compressor suddenly become a turbine and cause it to rotate in the opposite direction. This rotation can force hot turbine gases back through the air compressor, causing considerable damage. Reverse flow through the refrigeration compressor may be prevented in a variety of ways, depending on the system's components.

When there is no refrigerant receiver, quick-closing inlet guide vanes are usually satisfactory because there is very little high-pressure refrigerant to cause reverse rotation. However, when there is a receiver, a substantial amount of energy is available to cause reverse rotation. This can be reduced by opening the hot gas bypass valve on shutdown and installing a discharge check valve on the compressor.

The following safety controls are usually supplied with a gas turbine:

- Overspeed
- Compressor surge
- Overtemperature during operation under load
- Low oil pressure
- Failure to light off during start cycle
- Underspeed during operation under load

A fuel supply regulator can maintain a single-shaft gas turbine at a constant speed. With the split-shaft design, the output shaft of the turbine unit runs at the speed required by the refrigeration compressor. The temperature of the chilled water or brine leaving the cooler of the refrigeration machine controls the fuel. See also the section on Fuels and Fuel Systems on page 7.9.

STEAM TURBINE APPLICATIONS

Steam turbines in the air-conditioning and refrigeration field are principally used to drive centrifugal compressors. Such compressors

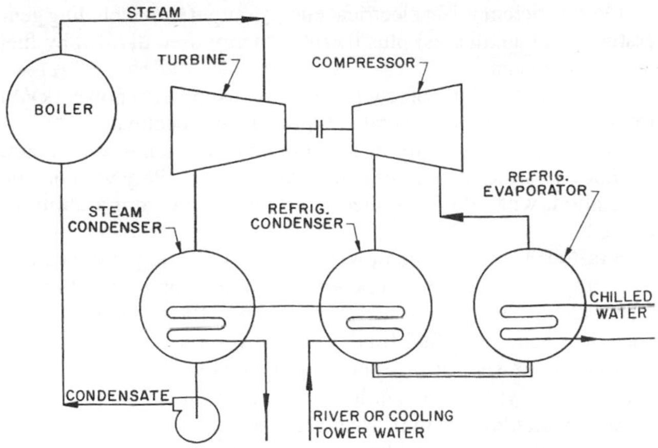

Fig. 51 Condensing Turbine-Driven Centrifugal Compressor

are usually part of a water or secondary coolant chilling system using one of the newer or the halogenated hydrocarbon refrigerants. In addition, many industrial processes employ turbine-driven centrifugal compressors with a variety of other refrigerants such as ammonia, propane, and butane, or other process gases.

Related applications of steam turbines include driving chilled water and condenser water circulating pumps and serving as prime movers for electrical generators in cogeneration systems. In industrial applications, the steam turbine may be advantageous, serving either as a work-producing steam pressure reducer or as a scavenger using otherwise wasted low-pressure steam.

Many steam turbines are used in urban areas where commercial buildings are served with steam from a central public utility or municipal source. Institutions where large central plants serve a multitude of buildings with heating and cooling also use steam turbine-driven equipment.

Most steam turbines driving centrifugal compressors for air conditioning are of the multistage condensing type (Figure 51). Such a turbine provides good steam economy at reasonable initial cost. Usually, steam is available at 50 psig or higher, and there is no demand for exhaust steam. However, turbines may work equally well where an abundance of low-pressure steam is available. The wide range of application of this turbine is shown by at least one industrial firm that drives a sizable capacity of water chilling centrifugal compressors with an initial steam pressure of less than 4 psig, thus balancing summer cooling against winter heating with steam from generator-turbine exhausts.

Aside from wide industrial use, the noncondensing (back pressure) turbine is most often used in water chilling plants to drive a centrifugal compressor that shares the cooling load with one or more absorption units (Figure 52). The exhaust steam from the turbine, commonly at about 15 psig, serves as the heat source for the absorption unit's generator (concentrator). This dual use of the heat energy in the steam generally results in a lower energy input per unit of refrigeration output than is attained by either machine operating alone. An important aspect in the design of such combined systems is the need to balance the turbine exhaust steam flow with the absorption input steam requirements over the full range of load.

Extraction and mixed pressure turbines are used mainly in industry or in large central plants. Extracted steam is often used for boiler feedwater heating or other processes where steam with lower heat content is needed. Most motor-driven centrifugal refrigeration compressors are driven at constant speed (some with variable-frequency drives). However, governors on steam turbines can maintain a constant or variable speed without the need for expensive variable-frequency drives.

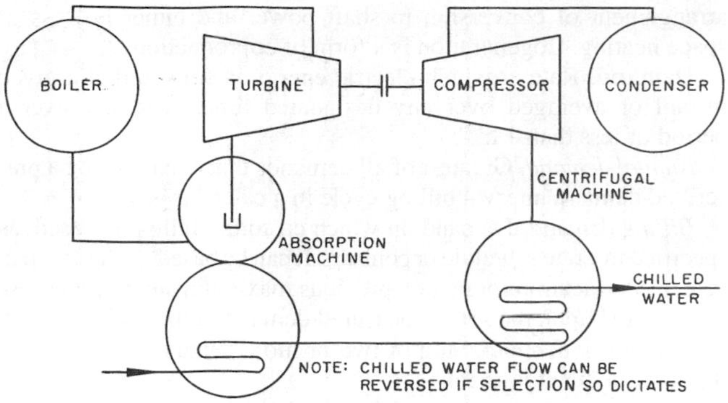

Fig. 52 Combination Centrifugal-Absorption System

ECONOMIC FEASIBILITY

TERMINOLOGY

Avoided cost. Incremental cost for the electric utility to generate or purchase electricity that is avoided through the purchase of power from a cogeneration facility.

Backup power. Electric energy available from or to an electric utility during an unscheduled outage to replace energy ordinarily generated by the facility or the utility. Frequently referred to as standby power.

Baseload. Minimum electric or thermal load generated or supplied over one or more periods.

Bottoming cycle. Cogeneration facility in which the energy input to the system is first applied to another thermal energy process; the rejected heat that emerges from the process is then used for power production.

Capability. Maximum load that a generating unit, generating station, or other electrical apparatus can carry under specified conditions for a given period of time without exceeding approved limits of temperature and stress.

Capacity. Load for which a generating unit, generating station, or other electrical apparatus is rated.

Capacity credits. Value included in the utility's rate for purchasing energy, based on the savings accrued through the reduction or postponement of new generation capacity that results from purchasing power from cogenerators.

Capacity factor. Ratio of the actual annual plant electricity output to the rated output.

Central cooling. Same as central heating except that cooling (heat removal) is supplied instead of heating; usually a chilled water distribution system supply and return for air conditioning.

Central heating. Supply of thermal energy from a central plant to multiple points of end use, usually by steam or hot water, for space and/or service water heating; central heating includes large-scale plants serving university campuses, medical centers, or military installations and central building systems serving multiple zones; also district heating plants.

Coefficient of performance (COP). Refrigeration or refrigeration plus thermal output energy divided by the energy input to the refrigeration compressor or absorption device.

Cogeneration. Sequential production of electrical or mechanical energy and useful thermal energy from a single energy stream. To qualify as a cogeneration facility in the United States, a plant must meet certain energy efficiency standards.

Coproduction. Conversion of energy from a fuel (possibly including solid or other wastes) into shaft power (which may be used to generate electricity) and a second or additional useful form of energy. The process generally entails a series topping and bottoming arrangement of conversion to shaft power and either process or space heating. Cogeneration is a form of coproduction.

Demand. Rate at which electric energy is delivered at a given instant or averaged over any designated time, generally over a period of less than 1 h.

Annual demand. Greatest of all demands that occur during a prescribed demand interval billing cycle in a calendar year.

Billing demand. Demand on which customer billing is based, as specified in a rate schedule or contract. It can be based on the contract year, a contract minimum, or a previous maximum and is not necessarily based upon the actual measured demand of the billing period.

Coincident demand. Sum of two or more demands occurring in the same demand interval.

Instantaneous peak demand. Maximum demand at the instant of greatest load.

Demand charge. Specified charge for electrical capacity on the basis of the billing demand.

Demand factor. Average demand over a specific period divided by the maximum demand over the same period (e.g., monthly demand factor, annual demand factor).

Economic coefficient of performance (ECOP). Energy in desired output units converted in terms of economic costs divided by fuel input in units of energy purchased or produced, where energy is measured in equivalent, consistent units again converted in terms of economic costs of each energy stream (e.g., a ton of cooling output is 12,000 Btu/h, while shaft horsepower output is 2545 Btu/h and electrical kilowatt output is 3412 Btu/h.)

Energy charge. That portion of the billed charge for electric service based on the electric energy (kilowatt-hours) supplied, as contrasted with the demand charge.

FERC efficiency. Electrical output plus one-half the thermal heat utilized divided by the energy input; based on the low heating value of the fuel (defined by the Federal Energy Regulatory Commission).

Generating efficiency. Electrical energy output from the engine or turbine generator divided by the energy input to the prime mover, in consistent units.

Grid. System of interconnected transmission lines, substations, and generating plants of one or more utilities.

Grid interconnection. Intertie of a cogeneration plant to an electric utility's distribution network to allow electricity flow in either direction.

Harmonics. Wave forms whose frequencies are multiples of the fundamental (60 Hz or 50 Hz) wave. The combination of harmonics and the fundamental wave causes a nonsinusoidal, periodic wave. Harmonics in power systems are the result of nonlinear effects. Typically, harmonics are associated with rectifiers and inverters, arc furnaces, arc welders, and transformer magnetizing current. Both voltage and current harmonics occur.

Heat rate. Measure of generating station thermal efficiency, generally expressed in Btu per net kilowatt-hour, or lb steam/kWh.

Heating value. Energy content in a fuel that is available as useful heat. The **high heating value** (HHV) includes the energy transmitted to vapor formed during combustion, whereas the **low heating value** (LHV) deducts this energy because it does no work on the piston.

Interruptible power. Electric energy supplied by an electric utility subject to interruption by the electric utility under specified conditions.

Load factor. Ratio of the average load supplied or required during a designated period to the peak or maximum load occurring in that period.

Maintenance power. Electric energy supplied by an electric utility during scheduled outages of the cogenerator.

Off-peak. Time periods when power demands are below average; for electric utilities, generally nights and weekends; for gas utilities, summer months.

Plant efficiency. Net electrical energy output (not including generating plant auxiliaries) plus thermal energy used divided by fuel input to the plant.

Power factor. Ratio of real power (kW) to apparent power (kVA) for any load and time; generally expressed as a decimal.

Selective energy systems. Form of cogeneration in which part, but not all, of the site's electrical needs are met solely with on-site generation, with additional electricity purchased from a utility as needed.

Shaft efficiency. Prime mover's shaft energy output divided by its energy input in equivalent, consistent units. For a steam turbine, input can be the thermal value of the steam or the fuel value to produce the steam. For a fuel-fired prime mover, it is the fuel input.

Standby power. Electric energy supplied by a utility to a cogenerator or vice-versa during either one's unscheduled outage.

Supplemental thermal. Heat required when recovered engine heat is insufficient to meet thermal demands.

Supplementary firing. Injection of fuel into an exhaust gas stream to raise its energy content (heat).

Supplementary power. Electric energy supplied by an electric utility in addition to the energy the facility generates.

Thermal capacity. Maximum amount of instantaneous heat that a system can produce.

Thermal efficiency (First Law efficiency). Electrical or shaft plus thermal output divided by the energy input in consistent units.

Topping cycle. Cogeneration facility in which the energy input to the facility is first used to produce useful power, and the rejected heat from production is used for other purposes.

Total energy system. Form of cogeneration in which all electrical and thermal energy needs are met by on-site systems. A total energy system can be completely isolated or switched over to a normally disconnected electrical utility system for backup.

Voltage flicker. Significant fluctuation of voltage.

Wheeling. Use of the transmission facilities of one system to transmit gas or power for another system.

SIZING AND OPERATING OPTIONS

Selecting a cogeneration system requires an evaluation of a large number of factors. Aside from the selection of the type of prime mover, the most important step in evaluating or planning a cogeneration system involves determining the cogeneration plant capacity and operating options. The following are possible sizing criteria:

- Peak electrical demand
- Minimum (or base-load) electrical demand
- Peak thermal demand
- Minimum (or base-load) thermal demand
- Maximum economic return with available funds
- Marginal cost/benefit ratio

The following are plant operating options:

- Rated output
- Thermally dispatched (tracks facility thermal load)
- Electrically dispatched (tracks facility electrical load)
- Maximum economic return
- Utility dispatched
- Peak shaving

The significance of each of these options is explained in the following discussion.

PRELIMINARY FEASIBILITY

Planning a cogeneration system is considerably more involved than planning an HVAC system. HVAC systems must be sized to meet peak loads, while cogeneration systems need not. Also, HVAC systems do not have to be coordinated and integrated with other energy systems as extensively as do cogeneration systems.

First Estimates

Becker (1988) suggested a quick way to determine whether a study should be undertaken: if the cost of electricity expressed in \$/kWh is more than 0.013 times the cost of fuel expressed in \$/10^6 Btu, a study should be considered. If it is 0.026 times or more, the chances are excellent for a 3 year or less simple payback.

Load Duration Curve Analysis

A much more comprehensive energy analysis, combined with an economic analysis, must be used to select a cogeneration system that maximizes the efficiency and economic return on investment. For better identification and screening of potential candidates, a simplified but accurate performance analysis must be conducted that considers the dynamics of the electrical and thermal loads of the facility, as well as the size and fuel consumption of the prime mover.

The need for an accurate analysis is especially important for commercial and institutional cogeneration applications because of the large time-dependent changes in magnitude of load and the non-coincident nature of the power and thermal loads. A facility containing a generator sized and operated to meet the thermal demand may occasionally have to purchase supplemental power and sometimes may produce power in excess of facility demand.

Even in the early planning stages, a reasonably accurate estimate of the following must be determined:

- Fuel consumed (if it is a topping cycle)
- Amount of supplemental electricity that must be purchased
- Amount of supplemental boiler fuel (if any) that must be purchased
- Amount of excess power available for sale
- Electrical capacity required from the utility for supplemental and standby power.
- Electrical capacity represented by any excess power if the utility offers capacity credits

Obtaining estimates of the these performance values for multiple time-varying loads is difficult and is further complicated by utility rate structures that may be based on time-of-day or time-of-year purchase and sale of power. Data must be collected at intervals short enough to give the desired levels of accuracy, yet taken over a long period and/or a well-selected group of sampling periods.

A basic method for analyzing the performance of HVAC systems is the bin method. The basic tool for sizing and evaluating performance of power systems is the load duration curve. This curve contains the same information as bins, but the load data are arranged in a slightly different manner. The load duration curve is a plot of hourly averaged instantaneous load data over a period; the plot is rearranged to indicate the frequency, or hours per period, that the load is at or below the stated value. The load duration curve is constructed by sorting the hourly averaged load values of the facility into descending order. Large volumes of load data can be easily sorted with desktop computers and electronic spreadsheets or databases. The load duration curve produces a visually intuitive tool for sizing cogeneration systems and for accurately estimating system performance.

Figure 53 shows a hypothetical steam load profile for a plant operating with two shifts each weekday and one shift each weekend day; no steam is consumed during nonworking hours for this example. The data provide little information for thermally sizing a generator, except to indicate that the peak demand for steam is about 46,000 lb/h and the minimum demand is about 13,000 lb/h

Figure 54 is a load duration curve (a descending order sort) of the steam load data in Figure 53. Mathematically, the load duration curve shows the frequency with which load equals or exceeds a given value; the curve is one minus the integral of the frequency distribution function for a random variable. Because the frequency distribution is a continuous representation of a histogram, the load duration curve is simply another arrangement of bin data.

In the frequency domain, or load duration curve form, the baseload and peak load can be readily identified. Note that the practical baseload at the "knee" of the curve is about 21,000 lb/h rather than the 13,000 lb/h absolute minimum identified on the load profile curve.

The cogenerator sized at the baseload achieves the greatest efficiency and best use of capital. However, it may not offer the shortest payback because of the high value of electrical power. An appropriately sized cogeneration plant might be sized somewhat larger than baseload to achieve minimum payback through increased electrical savings. A combination of load analysis and economic analysis must be performed to determine the most economical plant. For this example, the maximum economical plant size is arbitrarily assumed to be 28,000 lb/h. Cogeneration systems sized this way produce high equipment utilization and depend on the utility to serve the peak loads in excess of the plant's 28,000 lb/h capacity.

The load duration curve allows the designer to estimate the total amount of steam generated within the interval. Because the total amount of steam is the area under the curve, the calculation may be performed by using either formulae for rectangles and triangles or an appropriate curve analysis program. For a thermally tracked

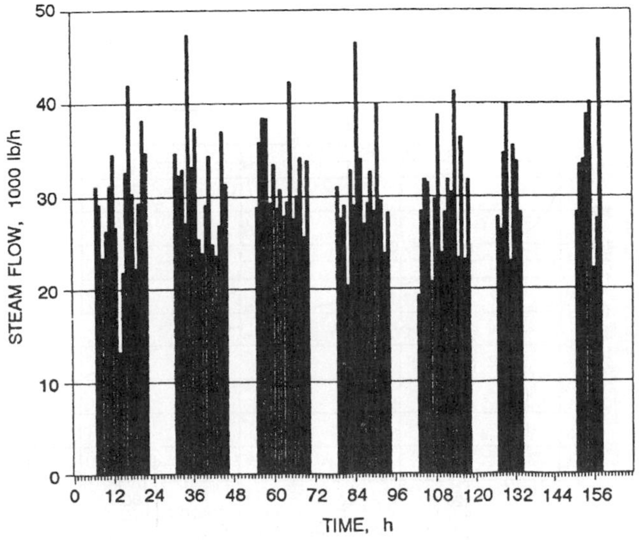

Fig. 53 Hypothetical Steam Load Profile

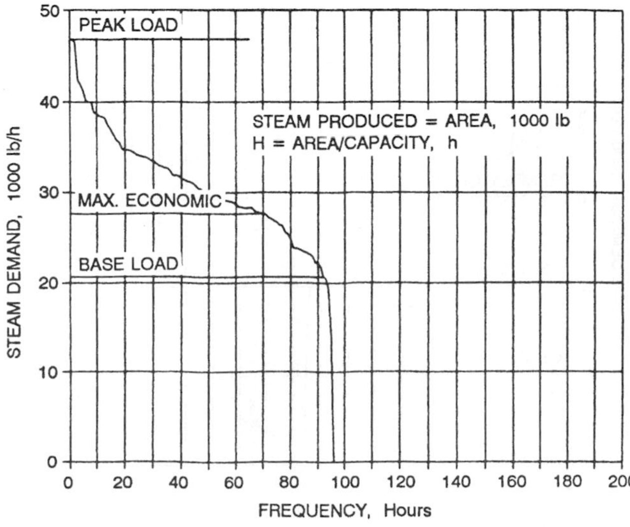

Fig. 54 Load Duration Curve

plant sized at the 21,000 lb/h baseload, the hours of operation are about 93 h/wk. Therefore, based on the area under the rectangle, the total steam produced by the cogeneration plant is

$$\text{Cogenerated steam} = (93 \text{ h/wk})(21 \times 10^3 \text{ lb/h}) = 1.953 \times 10^6 \text{ lb/wk}$$

If the plant is shut down during nonworking hours, no steam is wasted. In addition, if the electrical load profile is always above that needed to produce the 21,000 lb/h steam, the plant can run at full capacity for power and steam, while an electric utility provides peak power and a supplementary boiler provides peak steam. However, if the facility's steam load profile is unable to absorb the steam produced at continuous full power, the electrical output can be reduced accordingly to avoid steam waste.

In cases where it is more cost-effective to generate excess power than to suffer the parasitic cost of condensing the steam, the plant can still be operated at full power. Therefore, it is important to examine whether the electrical load profile matches and/or exceeds the electrical output. If it does, then the load duration curve reveals the quality of boiler-generated steam required. This value, which can be estimated by calculating the size of the triangular area above the baseload, is

$$\text{Boiler steam required} = 93(47{,}000 - 21{,}000)/2 = 1.21 \times 10^6 \text{ lb/wk}$$

$$\text{Cogenerated kW} = (21 \times 10^3 \text{ lb/h})/(\text{Steam-to-Electric Ratio})$$

where the ratio is that of the prime mover, in lb of steam/kWh. The total weekly electrical production is

$$\text{Cogenerated kWh} = 1.953 \times 10^6/\text{Ratio}$$

Fuel consumption equals cogenerated kilowatt-hours times the full-load heat in Btu, to which is added the boiler fuel consumption.

The **equivalent full-load hours** (EFLH) of both steam and electric cogeneration production in this base-load sizing and operating mode is 93 h/wk, so that

$$\text{Electric production} = \text{EFLH} \times \text{Rated capacity (kW) at full load}$$

If the plant had been sized at the maximum economic return, the cogenerated steam would be the area under the 28×10^3 lb/h level, or

$$\text{Cogenerated steam} = 1.953 \times 10^6 + 70(28{,}000 - 21{,}000) + (93 - 70)(28{,}000 - 21{,}000)/2 = 2.52 \times 10^6 \text{ lb/wk}$$

For this larger size, there is a higher electrical and lower boiler production, but a portion of the steam is condensed without any productive use.

Estimating the fuel consumption and electrical energy output of a system sized for the peak load or sized above the baseload is not as simple as for the base-load design because changes in the performance of the prime mover at part load must be considered. In this case, average value estimates of the performance at part load must be used for preliminary studies.

In some cases, an installation has only one prime mover; however, several smaller units operating in parallel provide increased reliability and performance during part-load operation. Figure 55 illustrates the use of three prime movers rated at 10,000 lb/h each. For this example, generators #1 and #2 operate fully loaded, and #3 operates between full-load capacity and 50% capacity while tracking the facility thermal demand. Because operation at less than 50% load is inefficient, further reduction in total output must be achieved by part-load operation of generators #1 and #2.

Offsetting the advantages of multiple units are their higher specific investment and maintenance costs, the control complexity, and the usually lower efficiency of smaller units.

Facilities such as hospitals often seek to reduce their utility costs by using existing standby generators to share the electrical peak (peak shaving). Such an operation is not strictly cogeneration because heat recovery is rarely justified. Figure 56 illustrates a hypothetical electrical load duration curve with frequency as a percent of total hours in the year (8760). A generator rated at 1000 kW, for example, reduces the peak demand by 500 kW if it is operated between 500 and 1000 kW to avoid extended hours of low-efficiency operation.

Some electrical energy saving as well as a reduction in demand are obtained by operating the generator. However, this saving is idealistic because it can only be obtained if the operators or the control system can anticipate in advance when facility demand will exceed 1400 kW in sufficient time to bring the generator up to operating condition. Nor should the peak shaver be started too early because it would waste fuel.

Also, many utilities include ratchet clauses in their rate schedules. As a result, if the peak shaving generator is inoperative for any reason when the facility monthly peak occurs, the ratchet is set for a year hence, and the demand savings potential of the peaking generator

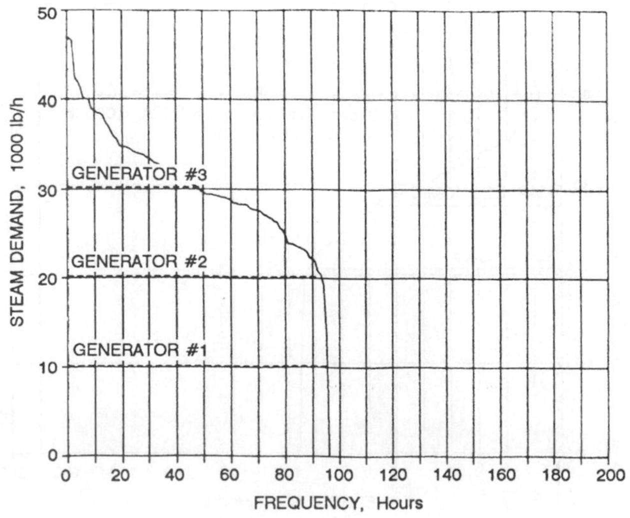

Fig. 55 Load Duration Curve with Multiple Generators

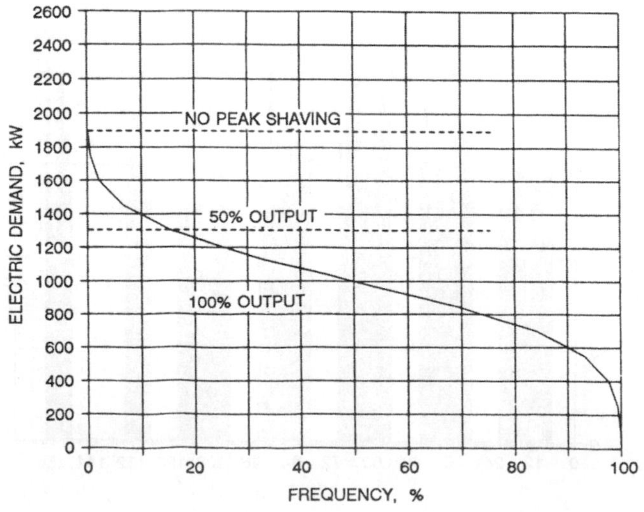

Fig. 56 Hypothetical Peaking Generator

will not be realized until a year later. Even though an existing standby generator may seem to offer "free" peak shaving capacity, careful planning and operation are required to secure its full potential.

Conversely, continuous-duty/standby systems offer the benefits of heat recovery while satisfying the standby requirements of the facility. During emergencies, the generator load is switched from its normal nonessential load to the emergency load.

Two-Dimensional Load Duration Curve

A two-dimensional load duration curve becomes necessary when the designer must consider simultaneous steam and electrical load variations. Such a situation occurs, for example, when the facility electrical demand drops below the output of a steam-tracking generator, and the excess power capacity cannot be exported. During these periods, the generator is throttled to curtail electrical output to that of facility demand. In other words, it is now operating in an electrical tracking mode.

To develop the two-dimensional load duration curve method for simultaneous loads, either the electrical demand or the steam demand must be broken into discrete periods defined by the number of hours per year when the electric or steam demand is within a certain range of values, or bins.

Duration curves for the remaining load are then created for each period as before, using coincident values. Figure 57 illustrates such a representation for three bins. The electrical load values indicated in the figure represent the center value of each bin. In general, a large number of periods gives a more accurate representation of facility loads. Also, the greater the load fluctuation, the greater the number of periods required for accurate representation. Note that the total number of hours for all periods adds up to 8760, the number of hours in a year.

Another situation that requires a two-dimensional load duration curve is when the facility buys or sells power, the price of which depends on the time of day or time of the year. Many electric rate structures contain explicit time periods for the purchase or sale of power. In some cases, there is only a summer-winter distinction. Other rate schedules may have several periods to reflect time-of-use or time-of-sale rates. The two-dimensional analysis is required to consider these rate schedule periods when defining the load bins because the operating schedule with the greatest annual savings is influenced by energy prices as well as energy demands.

Using Figure 57 as an illustration of a summer-peaking utility, the first two bins might coincide with the winter-fall-spring off-peak rate and the third bin with the on-peak rate. Because the analysis becomes burdensome when there are large load swings and several rate periods, the calculations are run by computers.

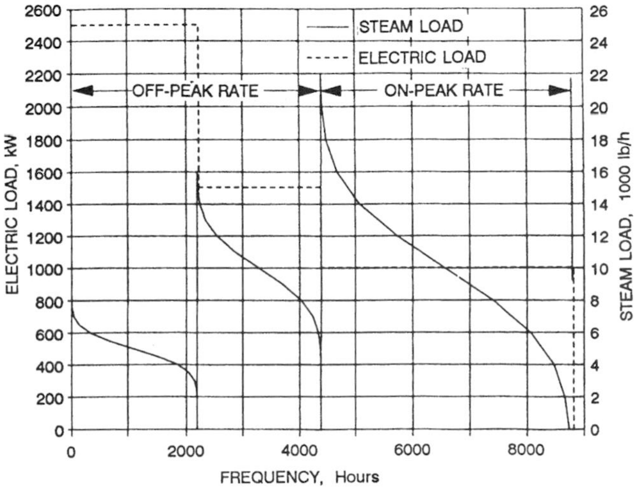

Fig. 57 Example of Two-Dimensional Load Duration Curve

Analysis by Simulations

The load duration curve is a convenient, intuitive graphical tool for preliminary sizing and analysis of a cogeneration system; it lacks the capability, however, for detailed analysis. A commercial or institutional facility, for example, can have as many as four different loads that must be considered simultaneously. These are cooling, noncooling electrical, steam or high-temperature hot water for space heating, and low-temperature service hot water. These loads are never in balance at any instant, which complicates sizing equipment, establishing operating modes, and determining the quantity of heat rejected from the cogenerator that is usefully applied to the facility loads.

Further complicating the evaluation of commercial cogeneration systems is the fact that prime movers rarely operate at full rated load; therefore, part-load operating characteristics such as fuel consumption, exhaust mass flow, exhaust temperature, and heat rejected from the jacket and intercooler of internal combustion engines must be considered. If the prime mover is a combustion gas turbine, then the effects of ambient temperature on full-load capacity and part-load fuel consumption and exhaust characteristics (or on cooling capacity if gas turbine inlet air cooling is used) should be considered.

In addition to the prime movers, a commercial or institutional cogeneration system may include absorption chillers that use jacket heat and exhaust heat to produce chilled water at two different COPs or ECOPs. Some commercial cogeneration systems use thermal energy storage to store hot water, chilled water, or ice to reduce the recovered thermal energy that must be dumped during periods of low demand. Internal combustion engines coupled with steam compressors and steam-injected gas turbines have been produced to allow some variability of the output heat/power ratio from a single package.

Computer programs that analyze cogeneration systems are available. Many of these programs emphasize the financial aspects of cogeneration systems with elaborate rate structures, energy price forecasts, and economic models. However, equipment part-load performance, load schedules, and other technical characteristics that greatly affect system economics are modeled only superficially in many programs. Also, many computer codes contain built-in equipment data and perform system selection and sizing automatically, thereby excluding the designer from important design decisions that could affect system viability. While these programs have their place, they should only be used with care by those who know the program limitations.

The primary consideration in selecting a computer program for analyzing commercial cogeneration technical feasibility is the ability to handle multiple, time-varying loads. The four methods of modeling the thermal and electrical loads are as follows:

- Hourly average values for a complete year
- Monthly average values
- Truncated year consisting of hourly averaged values for one or more typical (usually working and nonworking) days of each month
- Bin methods

Cogeneration simulation using hourly averaged values for load representation provides the greatest accuracy; however, the 8760 values for each type of load can make the management of load data a formidable task. Guinn (1987) describes a public domain simulation that can consider a full year of multiple-load data.

The data needed to perform a monthly averaged load representation can often be obtained from utility billings. This model should only be used as an initial analysis; best accuracy is obtained when thermal and electric profiles are relatively consistent.

The truncated year model is a compromise between the accuracy offered by the full hourly model and the minimal data-handling requirements of the monthly average model. It involves the development of hourly values for each load over a typical average day of each month. Usually, two typical days are considered—a working day and a nonworking day. Thus, instead of 8760 values to represent a load over a year, only 576 values are required. This type of load model is often used in cogeneration computer programs. Pedreyra (1988) and Somasundarum (1986) describe programs that use this method of load modeling.

Bin methods are based on the frequency distribution, or histogram, of load values. The method determines the number of hours per year the load was in different ranges, or bins. This method of representing weather data is widely used to perform building energy analysis. It is a convenient way to condense a large database into a smaller set of values, but it is no more accurate than the time resolution of the original set. Furthermore, bin methods become unacceptably cumbersome for cogeneration analysis if more than two loads must be considered.

CODES AND STANDARDS

In addition to applicable local codes, the following codes and standards should be consulted.

National Electrical Code (NFPA *Standard* 70-98)

Article 440, Air Conditioning and Refrigeration Equipment
Article 445, Generators
Article 700-12(b), Emergency System Set
Articles 701 and 702, Stand-by Power Gas Codes
(These articles cover service at all pressures.)

ASHRAE

Standard 15-1994, Safety Code for Mechanical Refrigeration
(This standard covers the refrigeration section of the system.)

National Fire Protection Association (NFPA)

30-96 Flammable and Combustible Liquids Code
31-97 Installation of Oil-Burning Equipment
37-98 Installation and Use of Stationary Combustion Engines and Gas Turbines
54-99 National Fuel Gas Code
58-98 Liquefied Petroleum Gas Code
59-98 Storage and Handling of Liquefied Petroleum Gases at Utility Gas Plants
59A-96 Production, Storage, and Handling of Liquefied Natural Gas (LNG)
90A-96 Installation of Air Conditioning and Ventilating Systems
211-96 Chimneys, Fireplaces, Vents, and Solid Fuel-Burning Appliances

Other Standards Organizations

National Engine Manufacturers Association (NEMA)
American Society for Testing and Materials (ASTM)
American Refrigeration Institute (ARI)

REFERENCES

ASHRAE. 1993. *Air-conditioning systems design manual.*

ASHRAE. 1996. *Cogeneration design guide.*

ASME. 1982. Appendix A to Test code for steam turbines. ANSI/ASME *Standard* PTC 6A-82 (R 1995). American Society of Mechanical Engineers, New York.

ASME. 1985. Guidance for evaluation of measurement uncertainty in performance tests of steam turbines. ANSI/ASME *Standard* PTC 6 *Report*-85 (R 1997).

ASME. 1988. Procedures for routine performance tests of steam turbines. ANSI/ASME *Standard* PTC 6S *Report*-88 (R 1995).

ASME. 1996. Steam turbines. ANSI/ASME *Standard* PTC 6-96.

ASME. 1998. Rules for the construction of pressure vessels. *Boiler and Pressure Vessel Code*, Section VIII-98.

Avery, G., W.C. Stethem, W.J. Coad, R.A. Hegberg, F.L. Brown, and R. Petitjean. 1990. The pros and cons of balancing a variable flow water system. *ASHRAE Journal* 32(10):32-59.

Becker, H.P. 1975. Energy conservation analysis of pumping systems. *ASHRAE Journal* 17(4):43-51.

Becker, H.P. 1988. Is cogeneration right for you? *Hotel and Resort Industry* (Sept.). New York.

DEMA. 1972. *Standard practices for low and medium stationary diesel and gas engines*, 6th ed. Diesel Engine Manufacturers Association.

Ebeling, J.A. 1994. Combustion turbine inlet air cooling alternatives and case histories. *Proceedings* of the 27th Annual Frontiers of Power Conference, Stillwater, OK (Oct. 24-25).

Encotech. 1992. Coupling effects on the torsional vibration of turbine generator power trains. *Encotech Technical Topics* 7(1). Schenectady, NY.

Guinn, G.R. 1987. Analysis of cogeneration systems using a public domain simulation. *ASHRAE Transactions* 93(2):333-66.

ISO. 1995. Reciprocating internal combustion engines—Performance—Part 1: Standard reference conditions, declarations of power, fuel, and lubricating oil consumptions, and test methods. *Standard* 3046-1-95. International Organization for Standardization, Geneva, Switzerland.

MacCracken, C.D. 1994. Overview of the progress and the potential of thermal storage in off-peak turbine inlet cooling. *ASHRAE Transactions* 100(1):569-71.

Mannion, G.F. 1988. High temperature rise piping design for variable volume pumping systems: Key to chiller energy management. *ASHRAE Transactions* 94(2):1427-43.

National Engine Use Council. 1967. NEUC engine criteria. Chicago.

NEMA. 1991. Land based steam turbine generator sets. *Standard* SM 24-91. National Electrical Manufacturers Association, Rosslyn, VA.

NEMA. 1991. Steam turbines for mechanical drive service. ANSI/NEMA *Standard* SM 23-91.

Newman, L.E. 1945. Modern extraction turbines, Part II—Estimating the performance of a single automatic extraction condensing steam turbine. *Power Plant Engineering* (Feb.).

Pedreyra, D.C. 1988. A microcomputer version of a large mainframe program for use in cogeneration analysis. ASHRAE *Transactions* 94(1): 1617-25.

Somasundarum, S. 1986. An analysis of a cogeneration system at T.W.U. *The Cogeneration Journal* 1(4):16-26.

Stewart, W.E., Jr. 1997. *Design guide for combustion turbine inlet air cooling systems*. ASHRAE.

CHAPTER 8

APPLIED HEAT PUMP AND HEAT RECOVERY SYSTEMS

TERMINOLOGY

Balanced heat recovery. Occurs when internal heat gain equals recovered heat and no external heat is introduced to the conditioned space. Maintaining balance may require raising the temperature of the recovered heat.

Break-even temperature. The outdoor temperature at which the total heat losses from conditioned spaces equal the internally generated heat gains.

Changeover temperature. The outdoor temperature the designer selects as the point of changeover from cooling to heating by the HVAC system.

External heat. Heat generated from sources outside the conditioned area. This heat from gas, oil, steam, electricity, or solar sources supplements internal heat and internal process heat sources. Recovered internal heat can reduce the demand for external heat.

Internal heat. The total passive heat generated within the conditioned space. It includes heat generated by lighting, computers, business machines, occupants, and mechanical and electrical equipment such as fans, pumps, compressors, and transformers.

Internal process heat. Heat from industrial activities and sources such as wastewater, boiler flue gas, coolants, exhaust air, and some waste materials. This heat is normally wasted unless equipment is included to extract it for further use.

Pinch technology. An energy analysis tool that uses vector analysis to evaluate all heating and cooling utilities in a process. Composite curves created by adding the vectors allow identification of a "pinch" point, which is the best thermal location for a heat pump.

Recovered (or reclaimed) heat. Comes from internal heat sources. It is used for space heating, domestic or service water heating, air reheat in air conditioning, process heating in industrial applications, or other similar purposes. Recovered heat may be stored for later use.

Stored heat. Heat from external or recovered heat sources that is held in reserve for later use.

Usable temperature. The temperature or range of temperatures at which heat energy can be absorbed, rejected, or stored for use within the system.

Waste heat. Heat rejected from the building (or process) because its temperature is too low for economical recovery or direct use.

APPLIED HEAT PUMP SYSTEMS

A heat pump extracts heat from a source and transfers it to a sink at a higher temperature. According to this definition, all pieces of refrigeration equipment, including air conditioners and chillers with refrigeration cycles, are heat pumps. In engineering, however, the term **heat pump** is generally reserved for equipment that heats for beneficial purposes, rather than that which removes heat for cooling only. Dual-mode heat pumps alternately provide heating or cooling. Heat reclaim heat pumps provide heating only, or simultaneous heating and cooling. An applied heat pump requires competent field engineering for the specific application, in contrast to the use of a manufacturer-designed unitary product. Applied heat pumps include built-up heat pumps (field- or custom-assembled from components) and industrial process heat pumps. Most modern heat pumps use a vapor compression (modified Rankine) cycle or an absorption cycle. Any of the other refrigeration cycles discussed in Chapter 1 of the 1997 *ASHRAE Handbook—Fundamentals* are also suitable. Although most heat pump compressors are powered by electric motors, limited use is also made of engine and turbine drives. Applied heat pump systems are most commonly used for heating and cooling buildings, but they are gaining popularity for efficient domestic and service water heating, pool heating, and industrial process heating.

Applied heat pumps having capacities ranging from 24,000 to 150,000,000 Btu/h operate in many facilities. Some of these machines are capable of output water temperatures up to 220°F and steam pressures up to 60 psig.

Compressors in large systems vary from one or more reciprocating or screw types to staged centrifugal types. A single or central system is often used, but in some instances, multiple heat pump systems are used to facilitate zoning. Heat sources include the ground, well water, surface water, gray water, solar energy, the air, and internal building heat. Compression can be single-stage or multistage. Frequently, heating and cooling are supplied simultaneously to separate zones.

Decentralized systems with water loop heat pumps are common, using multiple water-source heat pumps connected to a common circulating water loop. They can also include ground coupling, heat rejectors (cooling towers and dry coolers), supplementary heaters (boilers and steam heat exchangers), loop reclaim heat pumps, solar collection devices, and thermal storage. The initial cost is relatively low, and building reconfiguration is easily accommodated.

Community and district heating and cooling systems based on both centralized and distributed heat pump systems are feasible.

HEAT PUMP CYCLES

Several types of applied heat pumps (both open and closed cycle) are available; some reverse their cycles to deliver both heating and cooling in HVAC systems, and others are for heating only in HVAC and industrial process applications. The following are the four basic types of heat pump cycles:

1. **Closed vapor compression cycle** (Figure 1). This is the most common type used in both HVAC and industrial processes. It employs a conventional, separate refrigeration cycle that may be single-stage, compound, multistage, or cascade.
2. **Mechanical vapor recompression (MVR) cycle with heat exchanger** (Figure 2). Process vapor is compressed to a temperature and pressure sufficient for reuse directly in a process. Energy consumption is minimal, because temperatures are optimum for the

---

The preparation of this chapter is assigned to TC 9.4, Applied Heat Pump/Heat Recovery Systems.

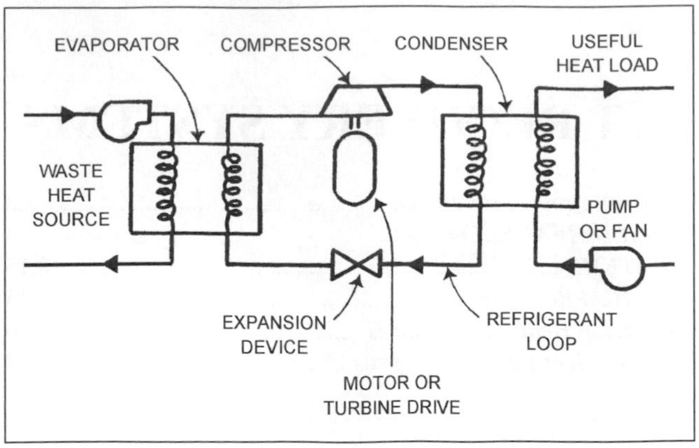

Fig. 1 Closed Vapor Compression Cycle

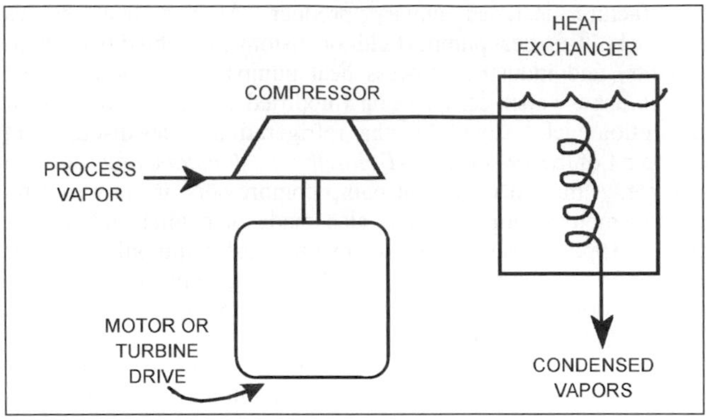

Fig. 2 Mechanical Vapor Recompression Cycle with Heat Exchanger

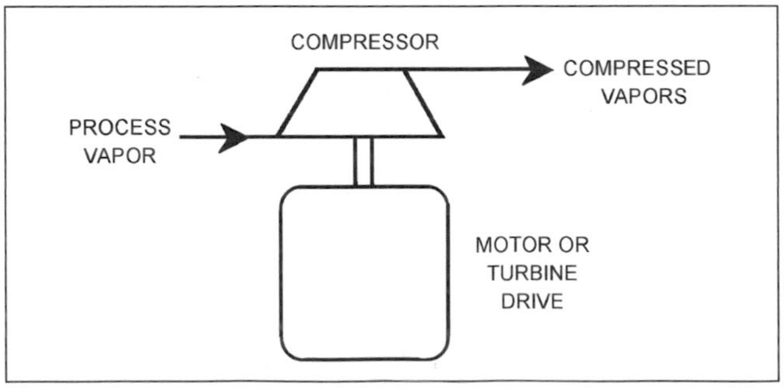

Fig. 3 Open Vapor Recompression Cycle

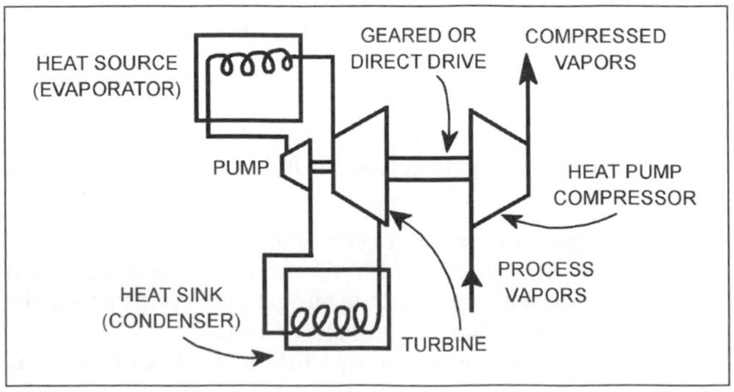

Fig. 4 Heat-Driven Rankine Cycle

process. Typical applications for this cycle include evaporators (concentrators) and distillation columns.

3. **Open vapor recompression cycle** (Figure 3). A typical application for this cycle is in an industrial plant with a series of steam pressure levels and an excess of steam at a lower-than-desired pressure. The heat is pumped to a higher pressure by compressing the lower pressure steam.
4. **Heat-driven Rankine cycle** (Figure 4). This cycle is useful where large quantities of heat are wasted and where energy costs are high. The heat pump portion of the cycle may be either open or closed, but the Rankine cycle is usually closed.

HEAT SOURCES AND SINKS

Table 1 shows the principal media used as heat sources and heat sinks. Selection of a heat source and sink for an application is primarily influenced by geographic location, climate, initial cost, availability, and type of structure. Table 1 presents various factors to be considered for each medium.

Air

Outdoor air is a universal heat-source and heat-sink medium for heat pumps and is widely used in residential and light commercial systems. Extended-surface, forced-convection heat transfer coils transfer heat between the air and the refrigerant. Typically, the surface area of outdoor coils is 50 to 100% larger than that of indoor coils. The volume of outdoor air handled is also greater than the volume of indoor air handled by about the same percentage. During heating, the temperature of the evaporating refrigerant is generally 10 to 20°F less than the outdoor air temperature. The performance of air heating and cooling coils is given in more detail in Chapter 21 and Chapter 23.

When selecting or designing an air-source heat pump, two factors in particular must be considered: (1) the outdoor air temperature in the given locality and (2) frost formation.

As the outdoor temperature decreases, the heating capacity of an air-source heat pump decreases. This makes equipment selection for a given outdoor heating design temperature more critical for an air-source heat pump than for a fuel-fired system. The equipment must be sized for as low a balance point as is practical for heating without having excessive and unnecessary cooling capacity during the summer. A procedure for finding this balance point, which is defined as the outdoor temperature at which heat pump capacity matches heating requirements, is discussed in Chapter 45.

When the surface temperature of an outdoor air coil is 32°F or less, with a corresponding outside air dry-bulb temperature 4 to 10°F higher, frost may form on the coil surface. If allowed to accumulate, the frost inhibits heat transfer; therefore, the outdoor coil must be defrosted periodically. The number of defrosting operations is influenced by the climate, air-coil design, and the hours of operation. Experience shows that, generally, little defrosting is required when outdoor air conditions are below 17°F and 60% rh. This can be confirmed by psychrometric analysis using the principles given in Chapter 21. However, under very humid conditions, when small suspended water droplets are present in the air, the rate of frost deposit may be about three times as great as predicted from psychrometric theory. Then, the heat pump may require defrosting after as little as 20 min of operation. The loss of available heating capacity due to frosting should be taken into account when sizing an air-source heat pump.

Following commercial refrigeration practice, early designs of air-source heat pumps had relatively wide fin spacing of 4 to 5 fins/in., based on the theory that this would minimize the frequency of defrosting. However, experience has proven that effective hot-gas defrosting permits much closer fin spacing and reduced size and bulk of the system. In current practice, fin spacings of 10 to 20 fins/in. are widely used.

Table 1 Heat Pump Sources and Sinks

		Suitability		Availability		Cost		Temperature		Common Practice	
Medium	Examples	Heat Source	Heat Sink	Location Relative to Need	Coincidence with Need	Installed	Operation and Maintenance	Level	Variation	Use	Limitations
AIR											
Outdoor	Ambient air	Good, but efficiency and capacity in heating mode decrease with decreasing outdoor air temperature	Good, but efficiency and capacity in cooling mode decrease with increasing outdoor air temperature	Universal	Continuous	Low	Moderate	Variable	Generally extreme	Most common, many standard products	Defrosting and supplemental heat usually required
Exhaust	Building ventilation	Excellent	Fair	Excellent if planned for in building design	Excellent	Low to moderate	Low unless exhaust is laden with dirt or grease	Excellent	Very low	Excellent as energy-conservation measure	Insufficient for typical loads
WATER											
Well[a]	Ground-water well may also provide a potable water source	Excellent	Excellent	Poor to excellent, practical depth varies by location	Continuous	Low if existing well used or shallow wells suitable; can be high otherwise	Low, but periodic maintenance required	Generally excellent, varies by location	Extremely stable	Common	Water disposal and required permits may limit; may require double-wall exchangers; may foul or scale
Surface	Lakes, rivers, oceans	Excellent for large water bodies or high flow rates	Excellent for large water bodies or high flow rates	Limited, depends on proximity	Usually continuous	Depends on proximity and water quality	Depends on proximity and water quality	Usually satisfactory	Depends on source	Available, particularly for fresh water	Often regulated or prohibited; may clog, foul, or scale
Tap (city)	Municipal water supply	Excellent	Excellent	Excellent	Continuous	Low	Low energy cost, but water use and disposal may be costly	Excellent	Usually very low	Use is decreasing due to regulations	Use or disposal may be regulated or prohibited; may corrode or scale
Condensing	Cooling towers, refrigeration systems	Excellent	Poor to good	Varies	Varies with cooling loads	Usually low	Moderate	Favorable as heat source	Depends on source	Available	Suitable only if heating need is coincident with heat rejection
Closed loops	Building water-loop heat pump systems	Good, loop may need supplemental heat	Favorable, may need loop heat rejection	Excellent if designed as such	As needed	Low	Low to moderate	As designed	As designed	Very common	Most suitable for medium or large buildings
Waste	Raw or treated sewage, gray water	Fair to excellent	Fair, varies with source	Varies	Varies, may be adequate	Depends on proximity, high for raw sewage	Varies, may be high for raw sewage	Excellent	Usually low	Uncommon, practical only in large systems	Usually regulated; may clog, foul, scale, or corrode
GROUND[a]											
Ground-coupled	Buried or submerged fluid loops	Good if ground is moist, otherwise poor	Fair to good if ground is moist, otherwise poor	Depends on soil suitability	Continuous	High to moderate	Low	Usually good	Low, particularly for vertical systems	Rapidly increasing	High initial costs for ground loop
Direct-expansion	Refrigerant circulated in ground coil	Varies with soil conditions	Varies with soil conditions	Varies with soil conditions	Continuous	High	High	Varies by design	Generally low	Extremely limited	Leak repair very expensive; requires large refrigerant quantities
SOLAR ENERGY											
Direct or heated water	Solar collectors and panels	Fair	Poor, usually unacceptable	Universal	Highly intermittent, night use requires storage	Extremely high	Moderate to high	Varies	Extreme	Very limited	Supplemental source or storage required
INDUSTRIAL PROCESS											
Process heat or exhaust	Distillation, molding, refining, washing, drying	Fair to excellent	Varies, often impractical	Varies	Varies	Varies	Generally low	Varies	Varies	Varies	May be costly unless heat need is near rejected source

[a]Groundwater-source heat pumps are also considered ground-source heat pump systems.

In many institutional and commercial buildings, some air must be continuously exhausted year-round. This exhaust air can be used as a heat source, although supplemental heat is generally necessary.

The high humidity due to indoor swimming pools causes condensation on ceiling structural members, walls, windows, and floors and causes discomfort to spectators. Traditionally, outside air and dehumidification coils with reheat from a boiler that also heats the pool water are used. This application is ideal for air-to-air and air-to-water heat pumps because energy costs can be reduced. Suitable materials must be chosen so that the heat pump components are resistant to corrosion from chlorine and high humidity.

Water

Water can be a satisfactory heat source, subject to the considerations listed in Table 1. City water is seldom used because of cost and municipal restrictions. Groundwater (well water) is particularly attractive as a heat source because of its relatively high and nearly constant temperature. The water temperature is a function of source depth and climate, but, in the United States, generally ranges from 40°F in northern areas to 70°F in southern areas. Frequently, sufficient water is available from wells for which the water can be reinjected into the aquifer. The use is nonconsumptive and, with proper design, only the water temperature changes. The water quality should be analyzed, and the possibility of scale formation and corrosion should be considered. In some instances, it may be necessary to separate the well fluid from the equipment with an additional heat exchanger. Special consideration must also be given to filtering and settling ponds for specific fluids. Other considerations are the costs of drilling, piping, pumping, and a means for disposal of used water. Information on well water availability, temperature, and chemical and physical analysis is available from U.S. Geological Survey offices in many major cities.

Heat exchangers may also be submerged in open ponds, lakes, or streams. When surface or stream water is used as a source, the temperature drop across the evaporator in winter may need to be limited to prevent freeze-up.

In industrial applications, waste process water (e.g., spent warm water in laundries, plant effluent, and warm condenser water) may be a heat source for heat pump operation.

Sewage, which often has temperatures higher than that of surface or groundwater, may be an acceptable heat source. Secondary effluent (treated sewage) is usually preferred, but untreated sewage may used successfully with proper heat exchanger design.

Use of water during cooling follows the conventional practice for water-cooled condensers.

Water-to-refrigerant heat exchangers are generally direct-expansion or flooded water coolers, usually of the shell-and-coil or shell-and-tube type. Brazed plate heat exchangers may also be used. In large applied heat pumps, the water is usually reversed instead of the refrigerant.

Ground

The ground is used extensively as a heat source and sink, with heat transfer through buried coils. Soil composition, which varies widely from wet clay to sandy soil, has a predominant effect on thermal properties and expected overall performance. The heat transfer process in soil depends on transient heat flow. Thermal diffusivity is a dominant factor and is difficult to determine without local soil data. Thermal diffusivity is the ratio of thermal conductivity to the product of density and specific heat. The soil moisture content influences its thermal conductivity.

There are three primary types of ground-source heat pumps: (1) groundwater, which is discussed in the previous section; (2) direct-expansion, in which the ground-to-refrigerant heat exchanger is buried underground; and (3) ground-coupled (also called closed-loop ground-source), in which a secondary loop with a brine connects the ground-to-water and water-to-refrigerant heat exchangers (see Figure 5).

Ground loops are can be placed either horizontally or vertically. A horizontal system consists of single, or multiple, serpentine heat exchanger pipes buried 3 to 6 ft apart in a horizontal plane at a depth 3 to 6 ft below grade. Pipes may be buried deeper, but excavation costs and temperature must be considered. Horizontal systems that use coiled loops referred to as **slinky coils** are also used. A vertical system uses a concentric tube or U-tube heat exchanger. The design of ground-coupled heat exchangers is covered in Chapter 31 of the 1999 *ASHRAE Handbook—Applications*.

Solar Energy

Solar energy may be used either as the primary heat source or in combination with other sources. Air, surface water, shallow groundwater, and shallow ground-source systems all use solar energy indirectly. The principal advantage of using solar energy directly as a heat source for heat pumps is that, when available, it provides heat at a higher temperature than the indirect sources, resulting in an increase in the heating coefficient of performance. Compared to solar heating without a heat pump, the collector efficiency and capacity are increased because a lower collector temperature is required.

Research and development of solar-source heat pumps has been concerned with two basic types of systems—direct and indirect. The direct system places refrigerant evaporator tubes in a solar collector, usually a flat-plate type. Research shows that a collector without glass cover plates can also extract heat from the outdoor air. The same surface may then serve as a condenser using outdoor air as a heat sink for cooling.

An indirect system circulates either water or air through the solar collector. When air is used, the collector may be controlled in such a way that (1) the collector can serve as an outdoor air preheater, (2) the outdoor air loop can be closed so that all source heat is derived from the sun, or (3) the collector can be disconnected from the outdoor air serving as the source or sink.

TYPES OF HEAT PUMPS

Heat pumps are classified by (1) heat source and sink, (2) heating and cooling distribution fluid, (3) thermodynamic cycle, (4) building structure, (5) size and configuration, and (6) limitation of the source and the sink. Figure 5 shows the more common types of closed vapor compression cycle heat pumps for heating and cooling service.

Air-to-Air Heat Pumps. This type of heat pump is the most common and is particularly suitable for factory-built unitary heat pumps. It is widely used in residential and commercial applications (see Chapter 45). The first diagram in Figure 5 is typical of the refrigeration circuit used.

In air-to-air heat pump systems, the air circuits can be interchanged by motor-driven or manually operated dampers to obtain either heated or cooled air for the conditioned space. In this system, one heat exchanger coil is always the evaporator, and the other is always the condenser. The conditioned air passes over the evaporator during the cooling cycle, and the outdoor air passes over the condenser. The positioning of the dampers causes the change from cooling to heating.

Water-to-Air Heat Pumps. These heat pumps rely on water as the heat source and sink, and use air to transmit heat to, or from, the conditioned space. See the second diagram in Figure 5. They include the following:

- ***Groundwater heat pumps***, which use groundwater from wells as a heat source and/or sink. These systems can either circulate the source water directly to the heat pump or use an intermediate fluid in a closed loop, similar to the ground-coupled heat pump.

Heat Source and Sink	Distribution Fluid	Thermal Cycle	Diagram Heating Cooling Heating and Cooling
Air	Air	Refrigerant changeover	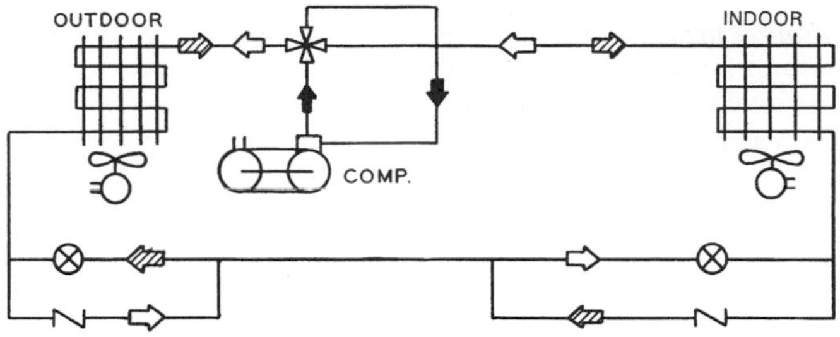
Water	Air	Refrigerant changeover	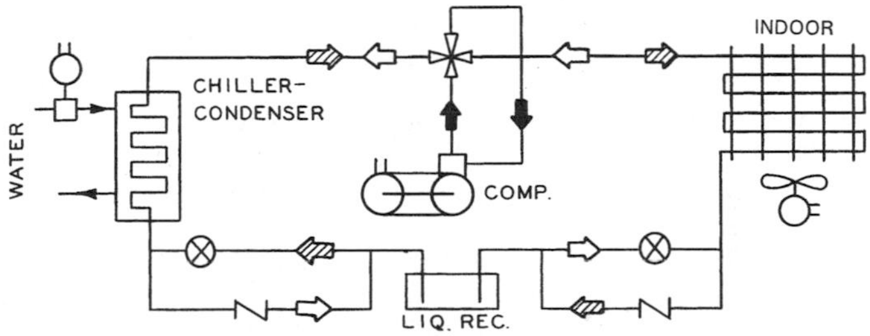
Water	Water	Water change-over	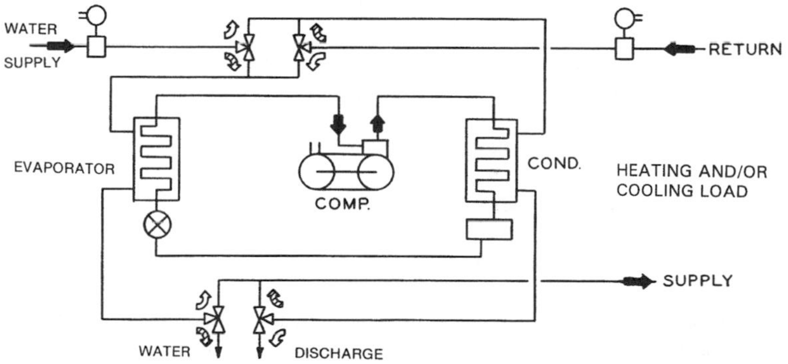
Ground-coupled (or Closed-loop ground-source)	Air	Refrigerant changeover	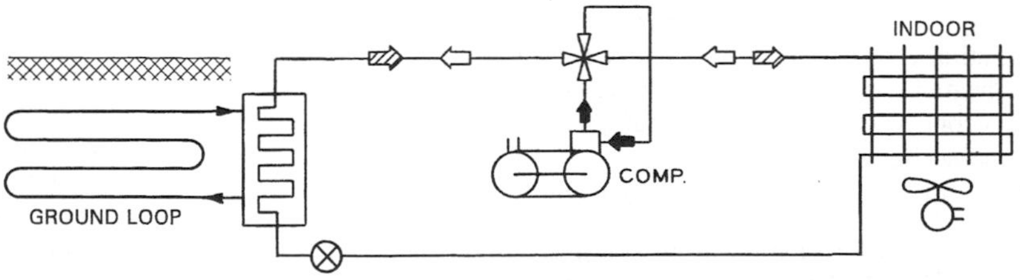
Ground-source, Direct-expansion	Air	Refrigerant changeover	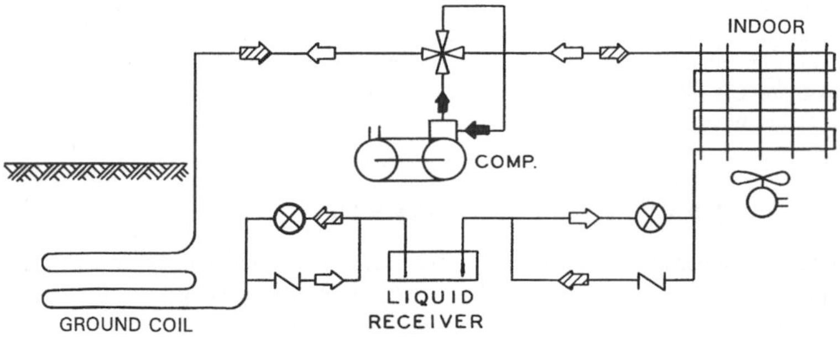

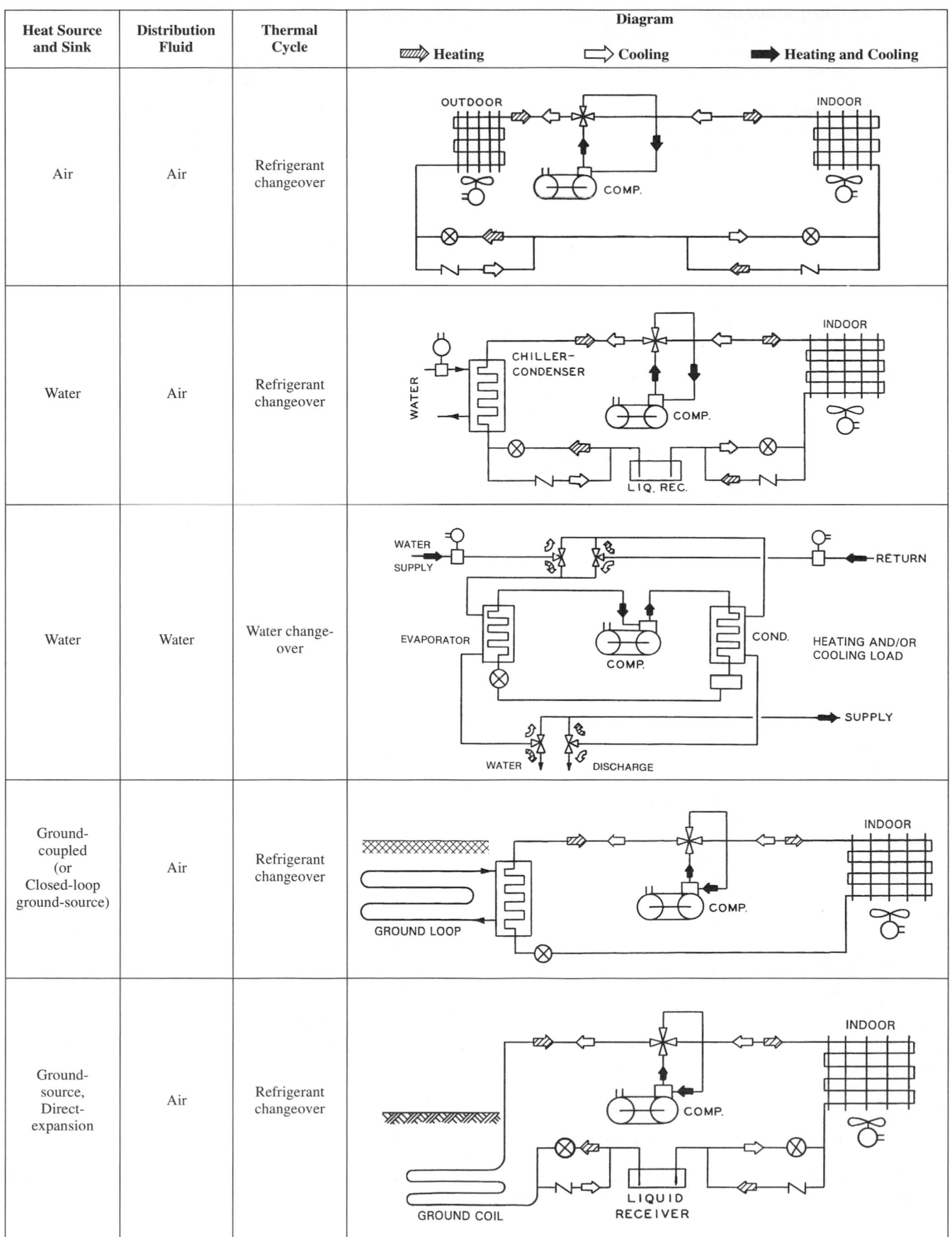

Fig. 5 Heat Pump Types

- ***Surface water heat pumps***, which use surface water from either a lake, pond, or stream as a heat source or sink. Similar to the ground-coupled and groundwater heat pumps, these systems can either circulate the source water directly to the heat pump or use an intermediate fluid in a closed loop.
- ***Internal-source heat pumps***, which use the high internal cooling load generated in modern buildings either directly or with storage. These include water loop heat pumps.
- ***Solar-assisted heat pumps***, which rely on low-temperature solar heat as the heat source. Solar heat pumps may resemble water-to-air, or other types, depending on the form of solar heat collector and the type of heating and cooling distribution system.
- ***Wastewater-source heat pumps***, which use sanitary waste heat or laundry waste heat as a heat source. The waste fluid can be introduced directly into the heat pump evaporator after waste filtration, or it can be taken from a storage tank, depending on the application. An intermediate loop may also be used for heat transfer between the evaporator and the waste heat source.

Water-to-Water Heat Pumps. These heat pumps use water as the heat source and sink for cooling and heating. Heating-cooling changeover can be done in the refrigerant circuit, but it is often more convenient to perform the switching in the water circuits, as shown in the third diagram of Figure 5. Although the diagram shows direct admittance of the water source to the evaporator, in some cases, it may be necessary to apply the water source indirectly through a heat exchanger (or double-wall evaporator) to avoid contaminating the closed chilled water system, which is normally treated. Another method uses a closed-circuit condenser water system.

Ground-Coupled Heat Pumps. These use the ground as a heat source and sink. A heat pump may have a refrigerant-to-water heat exchanger or may be of the direct-expansion (DX) type. Both types are shown in Figure 5. In systems with refrigerant-to-water heat exchangers, a water or antifreeze solution is pumped through horizontal, vertical, or coiled pipes embedded in the ground. Direct-expansion ground-coupled heat pumps use refrigerant in direct-expansion, flooded, or recirculation evaporator circuits for the ground pipe coils.

Soil type, moisture content, composition, density, and uniformity close to the surrounding field areas affect the success of this method of heat exchange. With some piping materials, the material of construction for the pipe and the corrosiveness of the local soil and underground water may affect the heat transfer and service life. In a variation of this cycle, all or part of the heat from the evaporator plus the heat of compression are transferred to a water-cooled condenser. This condenser heat is then available for uses such as heating air or domestic hot water.

Additional heat pump types include the following:

Air-to-Water Heat Pumps Without Changeover. These are commonly called heat pump water heaters.

Refrigerant-to-Water Heat Pumps. These condense a refrigerant by the cascade principle. Cascading pumps the heat to a higher temperature, where it is rejected to water or another liquid. This type of heat pump can also serve as a condensing unit to cool almost any fluid or process. More than one heat source can be used to offset those times when insufficient heat is available from the primary source.

HEAT PUMP COMPONENTS

For the most part, the components and practices associated with heat pumps evolved from work with low-temperature refrigeration. This section outlines the major components and discusses characteristics or special considerations that apply to heat pumps in combined room heating and air-conditioning applications or in higher temperature industrial applications.

Compressors

The principal types of compressors used in applied heat pump systems are briefly described in this section. For more details on these compressor types and others, refer to Chapter 34.

- **Centrifugal compressors.** Most centrifugal applications in heat pumps have been limited to large water-to-water or refrigerant-to-water heat pump systems, heat transfer systems, storage systems, and hydronically cascaded systems. With these applications, the centrifugal compressor enables the use of heat pumps in industrial plants, as well as in large multistory buildings; many installations with double-bundle condensers have been made. The transfer cycles permit low pressure ratios, and many single- and two-stage units with various refrigerants are operational with high COPs. Centrifugal compressor characteristics do not usually meet the needs of air-source heat pumps. High pressure ratios, or high lifts, associated with low gas volume at low load conditions cause the centrifugal compressor to surge.
- **Screw compressors.** Screw compressors offer high pressure ratios at low to high capacities. Capacity control is usually provided by such means as variable porting or sliding vanes. Generally, large oil separators are required, because many compressors use oil injection. Screw compressors are less susceptible to damage from liquid spillover and have fewer parts than do reciprocating compressors. They also simplify capacity modulation.
- **Rotary vane compressors.** These compressors can be used for the low stage of a multistage plant; they have a high capacity but are generally limited to lower pressure ratios. They also have limited means for capacity reduction.
- **Reciprocating compressors.** These compressors are used more than any other type on systems ranging from 0.5 to 100 tons.
- **Scroll compressors.** These are a type of orbital motion positive-displacement compressor. Their use in heat pump systems is increasing. Their capacity can be controlled by varying the drive speed or the compressor displacement. Scroll compressors have low noise and vibration levels.

Compressor Selection. A compressor used for comfort cooling usually has a medium clearance volume ratio (the ratio of gas volume remaining in the cylinder after compression to the total swept volume). For an air-source heat pump, a compressor with a smaller clearance volume ratio is more suitable for low-temperature operation and provides greater refrigerating capacity at lower evaporator temperatures. However, this compressor has somewhat more power demand under maximum cooling load conditions than a compressor with medium clearance.

More total heat capacity can be obtained at low outdoor temperatures by deliberately oversizing the compressor. When this is done, some capacity reduction for operation at higher outdoor temperatures can be provided by multispeed or variable-speed drives, cylinder cutouts, or other methods. The disadvantage of this arrangement is that the greater number of operating hours that occur at the higher suction temperatures must be served with the compressor in the unloaded condition, which generally causes lower efficiency and higher annual operating cost. The additional initial cost of the oversized compressor must be economically justified by the gain in heating capacity.

One method proposed for increasing the heating output at low temperatures uses **staged compression**. For example, one compressor may compress from −30°F saturated suction temperature (SST) to 40°F saturated discharge temperature (SDT), and a second compressor compresses the vapor from 40°F SST to 120°F SDT. In this arrangement, the two compressors may be interconnected in parallel, with both pumping from about 45°F SST to 120°F SDT for cooling. Then at some predetermined outdoor temperature on heating, they are reconnected in series and compress in two successive stages.

Figure 6 shows the performance of a pair of compressors for units of both medium and low clearance volume. At low suction temperatures, the reconnection in series adds some capacity. However, the motor for this case must be selected for the maximum loading conditions for summer operation, even though the low-stage compressor has a greatly reduced power requirement under the heating condition.

Compressor Floodback Protection. A suction line separator, similar to that shown in Figure 7, in combination with a liquid-gas heat exchanger, can be used to minimize migration and harmful liquid floodback to the compressor. The solenoid valve should be controlled to open when the compressor is operating, and the hand valve should then be adjusted to provide an acceptable bleed rate into the suction line.

Heat Transfer Components

Refrigerant-to-air and refrigerant-to-water heat exchangers are similar to heat exchangers used in air-conditioning refrigeration systems.

Defrosting Air-Source Coils. Frost accumulates rather heavily on outdoor air-source coils when the outdoor air temperature is less than approximately 40°F, but lessens somewhat with the simultaneous decrease in outdoor temperature and moisture content. Most systems defrost by reversing the cycle.

Another method of defrosting is spraying heated water over an outdoor coil. The water can be heated by the refrigerant or by auxiliaries.

Draining Heat Source and Heat Sink Coils. Direct-expansion indoor and outdoor heat transfer surfaces serve as condensers and evaporators. Both surfaces, therefore, must have proper refrigerant distribution headers to serve as evaporators and have suitable liquid refrigerant drainage while serving as condensers.

In a flooded system, the normal float control to maintain the required liquid refrigerant level is used.

Liquid Subcooling Coils. A refrigerant subcooling coil can be added to the heat pump cycle, as illustrated in Figure 8, to preheat the ventilation air during the heating cycle and, at the same time, to lower the temperature of the liquid refrigerant. Depending on the refrigerant circulation rate and the quantity and temperature of ventilation air, the heating capacity can be increased as much as 15 to 20%, as indicated in Figure 9.

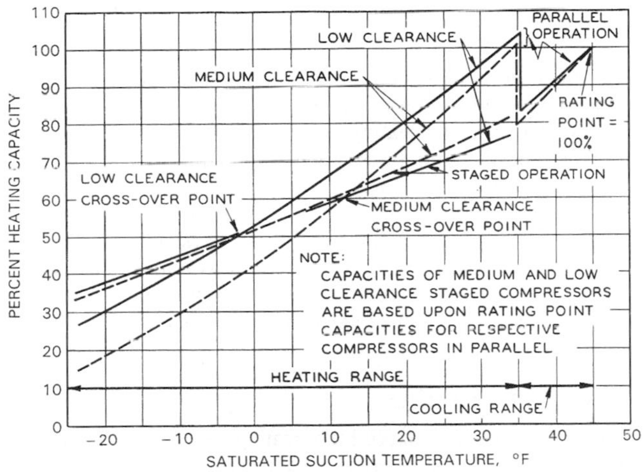

Fig. 6 Comparison of Parallel and Staged Operation for Air-Source Heat Pumps

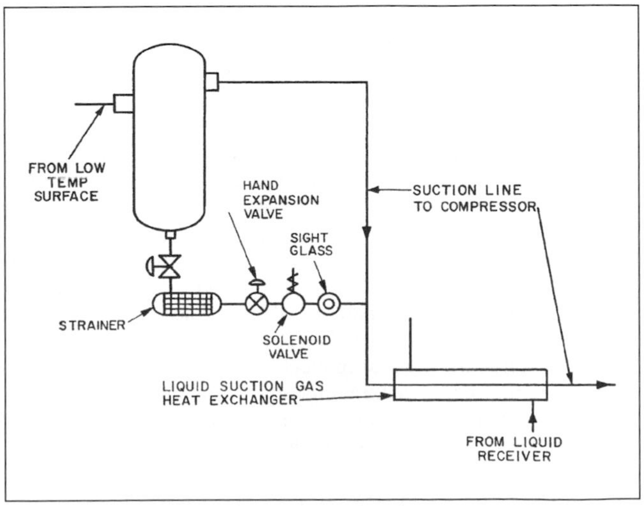

Fig. 7 Suction Line Separator for Protection Against Liquid Floodback

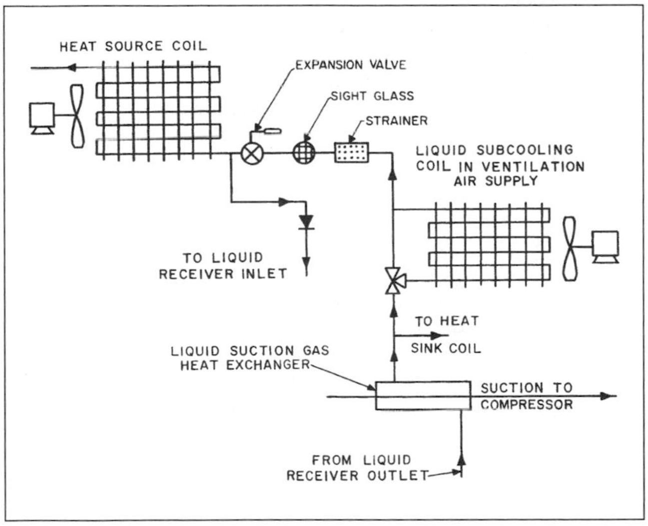

Fig. 8 Liquid Subcooling Coil in Ventilation Air Supply to Increase Heating Effect and Heating COP

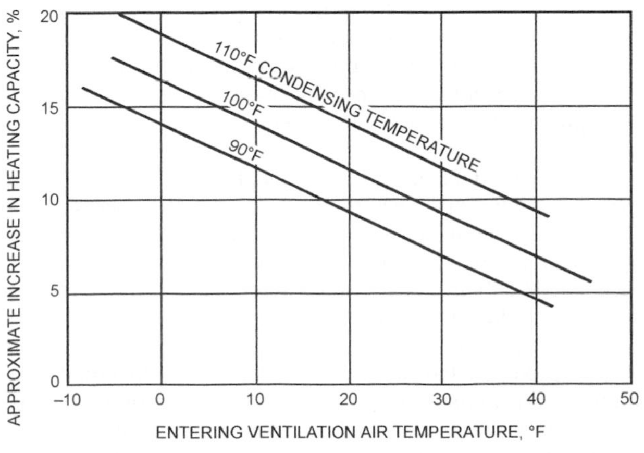

Fig. 9 Typical Increase in Heating Capacity Resulting from Use of Liquid Subcooling Coil

Refrigeration Components

Refrigerant piping, receivers, expansion devices, and refrigeration accessories in heat pumps are usually the same as components found in other types of refrigeration and air-conditioning systems.

A **reversing valve** changes the system from the cooling to the heating mode. This changeover requires the use of a valve(s) in the refrigerant circuit, except where the change occurs in fluid circuits external to the refrigerant circuit. Reversing valves are usually pilot-operated by solenoid valves, which admit compressor head and suction pressures to move the operating elements.

The **expansion device** for controlling the refrigerant flow is normally a **thermostatic expansion valve**, which is described in Chapter 45 of the 1998 *ASHRAE Handbook—Refrigeration.* The control bulb must be located carefully. If the circuit is arranged so that the refrigerant line on which the control bulb is placed can become the compressor discharge line, the resulting pressure developed in the valve power element may be excessive, requiring a special control charge or a pressure-limiting element. When a thermostatic expansion valve is applied to an outdoor air coil, a special cross-charge to limit the superheat at low temperatures allows better use of the coil. In its temperature-sensing element, a cross-charged thermostatic expansion valve uses a fluid, or mixture of fluids, that is different than the refrigerant. An **electronic expansion valve** provides improved control of refrigerant flow.

A **capillary tube** (used as a metering device for an air-source heat pump that operates over a wide range of evaporating temperatures) may pass refrigerant at an excessive rate at low back pressures, causing liquid floodback to the compressor. In some cases, suction line accumulators or charge-control devices are added to minimize this effect. Suction line accumulators may also prevent liquid refrigerant, which has migrated to the evaporator, from entering the compressor on start-up.

When separate metering devices are used for the two heat exchangers, a **check valve** allows the refrigerant to bypass the metering device of the heat exchanger serving as the condenser.

A **refrigerant receiver**, which is commonly used as a storage place for liquid refrigerant, is particularly useful in a heat pump to take care of the unequal refrigerant requirements of heating and cooling. The receiver is usually omitted on heat pumps used for heating only.

Controls

Defrost Control. A variety of defrosting control schemes can sense the need for defrosting air-source heat pumps and initiate (and terminate) the defrost cycle. The cycle is usually initiated by demand rather than a timer. Termination of the cycle may be timer-controlled.

The defrost cycle can also be terminated either by a control sensing the coil pressure or a thermostat that measures the temperature of the liquid refrigerant in the outdoor coil. The completion of defrosting is ensured when the temperature (or corresponding saturation pressure) of the liquid leaving the outdoor coil rises to about 70°F.

A widely used means of starting the defrost cycle on demand is a pressure control that reacts to the air pressure drop across the coil. When frost accumulates, the airflow is reduced and the increased pressure drop across the coil initiates the defrost cycle. This method is applicable only in a clean outdoor environment where fouling of the air side of the outdoor coil is not expected.

Another method for sensing frost formation and initiating defrosting on demand involves a temperature differential control and two temperature-sensing elements. One element is responsive to outdoor-air temperature and the other to the temperature of the refrigerant in the coil. As frost accumulates, the differential between the outdoor temperature and the refrigerant temperature increases, initiating a defrost cycle. The system is restored to operation when the refrigerant temperature in the coil reaches a specified temperature, indicating that defrosting has been completed. When the outdoor air temperature decreases, the differential between outdoor air temperature and refrigerant temperature decreases due to reduced heat pump capacity. Thus, unless compensation is provided, the defrost cycle is not initiated until a greater amount of frost has built up.

The following controls may be used to change from heating to cooling:

- **Conditioned space thermostats** on residences and small commercial applications.
- **Outdoor air thermostats** (with provision for manual overriding for variable solar and internal load conditions) on larger installations, where it may be difficult to find a location in the conditioned space that is representative of the total building.
- **Manual changeover** via a heat-off-cool position on the indoor thermostat. A single thermostat is used for each heat pump unit.
- **Sensing devices**, which respond to the greater load requirement, heating or cooling, are generally applied on simultaneous heating and cooling systems.
- **Dedicated microcomputers** to automate changeover and perform all the other control functions needed, as well as to simultaneously monitor the performance of the system. This may be a stand-alone device, or it may be incorporated as part of a larger building automation system.

On the heat pump system, it is important that space thermostats are interlocked with ventilation dampers so that both operate on the same cycle. During the heating cycle, the fresh air damper should be positioned for the minimum required ventilation air, with the space thermostat calling for increased ventilation air only if the conditioned space becomes too warm. Fan and/or pump interlocks are generally provided to prevent the heat pump system from operating if the use of the accessory equipment is not available. On commercial and industrial installations, some form of head pressure control is required on the condenser when cooling at outdoor air temperatures below 60°F.

Supplemental Heating

Heating needs may exceed the heating capacity available from equipment selected for the cooling load, particularly if outdoor air is used as the heat source. When this occurs, supplemental heating or additional compressor capacity should be considered. The additional compressor capacity or the supplemental heat is generally used only in the severest winter weather and, consequently, has a low usage factor. Both possibilities must be evaluated to determine the most economical selection.

When supplemental heaters are used, the elements should always be located in the air or water circuit downstream from the heat pump condenser. This permits the heat pump system to operate at a lower condensing temperature, increasing the system heating capacity and improving the coefficient of performance. The controls should sequence the heaters so that they are energized after all heat pump compressors are fully loaded. An outdoor thermostat is recommended to limit or prevent energizing heater elements during mild weather when they are not needed. Where 100% supplemental heat is provided for emergency operation, it may be desirable to keep one or more stages of the heaters locked out whenever the compressor is running. In this way, the cost of electrical service to the building is reduced by limiting the maximum coincidental demand.

A flow switch should be used to prevent operation of the heating elements and the heat pump when there is no air or water flow.

INDUSTRIAL PROCESS HEAT PUMPS

Heat recovery in industry offers numerous opportunities for applied heat pumps. The two major classes of industrial heat pump systems are closed-cycle and open-cycle. Factory-packaged, closed-cycle machines have been built to heat fluids to 120°F and as high as 220°F. Skid-mounted, open-cycle and semi-open-cycle machines have been used to produce low-level, saturated and superheated steam.

In industrial applications, heat pumps are generally used for process heating rather than space heating. Each heat pump system must be designed for the particular application. Rather than being dictated by weather or design standards, the selection of size and output temperature for a system is often affected by economic restraints, environmental standards, or desired levels of product quality or output. This gives the designer more flexibility in equipment selection because the systems are frequently applied in conjunction with a more traditional process heating system such as steam.

ASHRAE Research Project 656 (Cane et al. 1994) gathered information on energy performance, economics of operation, operating difficulties, operator and management reactions, and design details for various **heat recovery heat pump** (HRHP) systems.

The most common reason given for installing HRHP systems was reduction of energy cost. Other reasons cited were

- Need to eliminate bacterial growth in storage tanks
- Need to increase ammonia refrigeration system capacity
- Reduction of makeup water use
- Need for flexibility in processing
- Year-round processing (drying) possible
- Superior drying quality compared with conventional forced-air kilns
- Process emissions eliminated without the need for costly pollution control equipment
- Recovery and reuse of product from process
- Reduced effluent into the environment

Economics associated with energy reductions were calculated for most of the test sites. Economic justifications for the other reasons for installation were difficult to estimate. Only half the survey sites reported actual payback periods. Half of these had simple paybacks of less than 5 years.

The most frequently cited problem was widely fluctuating heat source or sink flow rates or temperatures. Considerable differences between design parameters and actual conditions were also mentioned frequently. In some cases, the difference resulted in oversizing, which caused poor response to load variation, nuisance shutdowns, and equipment failure. Other problems were significant process changes after installation and poor placement of the HRHP.

The number of projects that had overstated savings based on overstated run hours demonstrated the need for accurate prediction method. While the low hours of use in some cases resulted from first-year start-up and balancing issues, in most cases it was due to plant capacity reductions, process modifications, or other factors that were not understood during the design phase.

Closed-Cycle Systems

Closed-cycle systems use a suitable working fluid, usually a refrigerant in a sealed system. They can use either the absorption or the vapor compression principle. Traditionally, vapor compression systems have used R-11, R-12, R-113, and R-114 to obtain the desired temperature; however, these refrigerants are being phased out. Refrigerants R-22 and R-134a can be used in vapor compression systems if the output temperature requirements are adjusted to meet the pressure restrictions of the refrigerant and machine selected. Heat is transferred to and from the system through heat exchangers similar to refrigeration system heat exchangers. Closed-cycle heat pump systems are often classed with industrial refrigeration systems except that they operate at higher temperatures.

Heat exchangers selected for the system must comply with federal and local codes. For example, some jurisdictions require a double separation between potable water and refrigerant. Heat exchangers must be resistant to corrosion and fouling conditions of the source and sink fluids. The refrigerant and oil must be (1) compatible with component materials and (2) mutually compatible at the expected operating temperatures. In addition, the viscosity and foaming characteristics of the oil and refrigerant mixtures must be consistent with the lubrication requirements at the specific mechanical load imposed on the equipment. Proper oil return and heat transfer at the evaporators and condensers must also be considered.

The specific application of a closed-cycle heat pump frequently dictates the selection considerations. In this section, the different types of closed-cycle systems are reviewed, and factors important to the selection process are addressed.

Air-to-air heat pumps or **dehumidification heat pumps** (Figure 10) are most frequently used in industrial operations to dry or cure products. For example, dehumidification kilns are used to dry lumber to improve its value. Compared to conventional steam kilns, the heat pump provides two major benefits: improved product quality and reduced percent degrade. With dehumidification, lumber can be dried at a lower temperature, which reduces warping, cracking, checking, and discoloration. The system must be selected according to the type of wood (i.e., hard or soft) and the required dry time. Dehumidification heat pumps can also be used to dry agricultural products; poultry, fish, and meat; textiles; and other products.

Air-to-water heat pumps, which are also called **heat pump water heaters**, are a special application of closed-cycle systems that are usually unitary. See Chapter 45 of this volume and Chapter 48 of the 1999 *ASHRAE Handbook—Applications* for more information.

Water-to-water heat pumps may have the most widespread application in industry. They can use cooling tower water, effluent streams, and even chilled water makeup streams as heat sources. The output hot water can be used for product rinse tanks, equipment cleanup water systems, and product preheaters. The water-to-water heat pump system may be simple, such as that shown in Figure 11, which recovers heat from a process cooling tower to heat water for another process. Figure 11 also shows the integration of a heat exchanger for preheating the process water. Typically, the heat pump COP is in the range of 4 to 6, and the system COP reaches 8 to 15.

Water-to-water heat pump systems may also be complicated, such as the cascaded HRHP system (Figure 12) for a textile dyeing and finishing plant. Heat recovered from various process effluent streams is used to preheat makeup water for the processes. The effluent streams may contain materials, such as lint and yarn, and

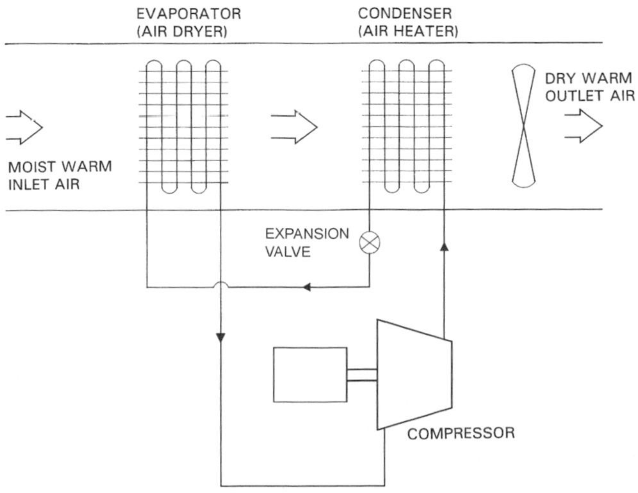

Fig. 10 Dehumidification Heat Pump

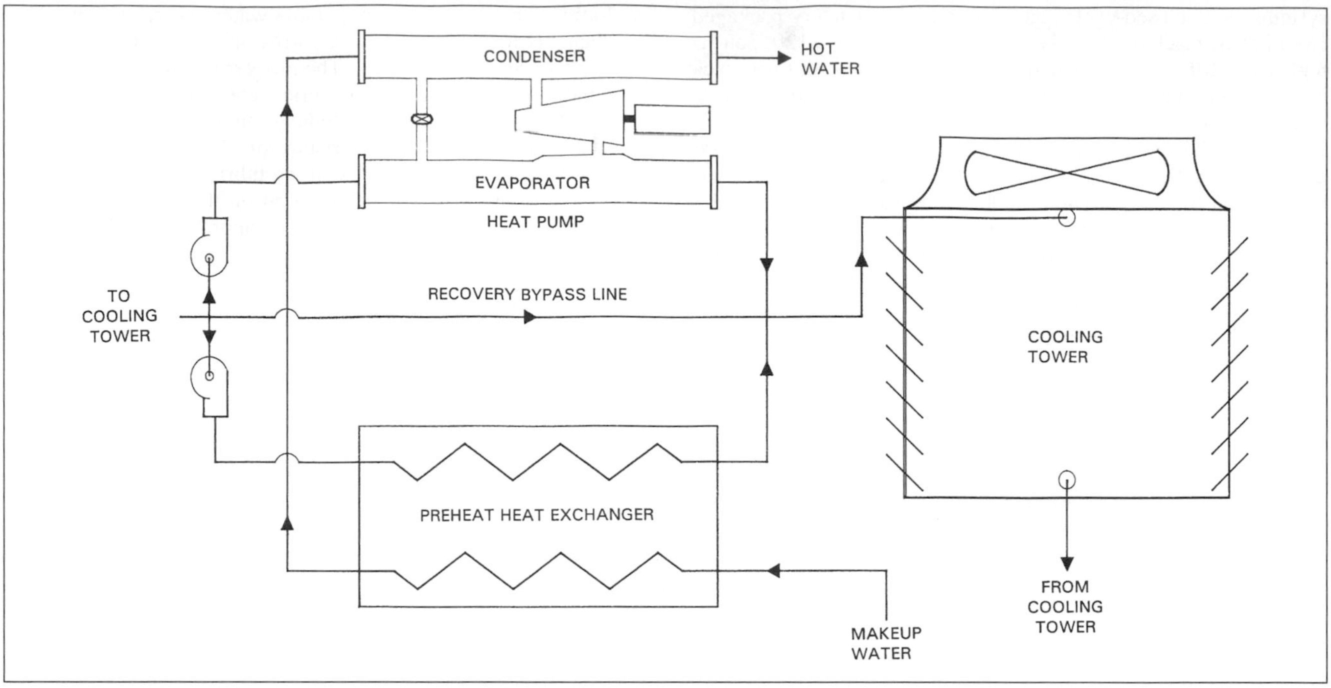

Fig. 11 Cooling Tower Heat Recovery Heat Pump

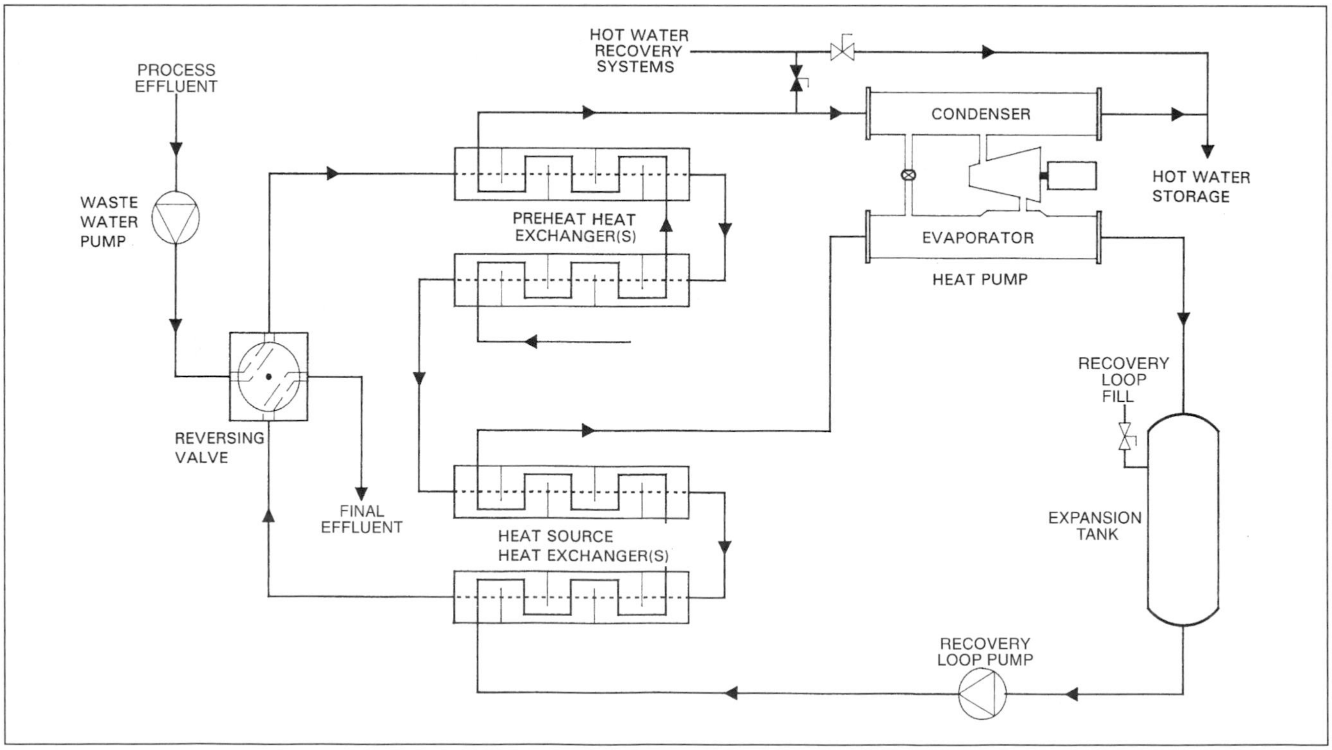

Fig. 12 Effluent Heat Recovery Heat Pump

highly corrosive chemicals that may foul the heat exchanger; therefore, special materials and antifouling devices may be needed for the heat exchanger for a successful design.

Most food-processing plants, which use more water than a desuperheater or a combination desuperheater/condenser can provide, can use a water-to-water heat pump (Figure 13). A water-cooled condenser recovers both sensible and latent heat from the high-pressure refrigerant. The water heated in the condenser is split into two streams—one as a heat source for the water-to-water heat pump, the other to preheat the makeup process and cleanup water. The preheated water is blended with hot water from the storage tank to limit the temperature difference across the heat pump condenser to about 20°F, which is standard for many chiller applications. The heated water is then piped back to the storage tank, which is typically sized for 1.5 to 5 h of holding capacity. Because the refrigeration load may be insufficient to provide all the water heating,

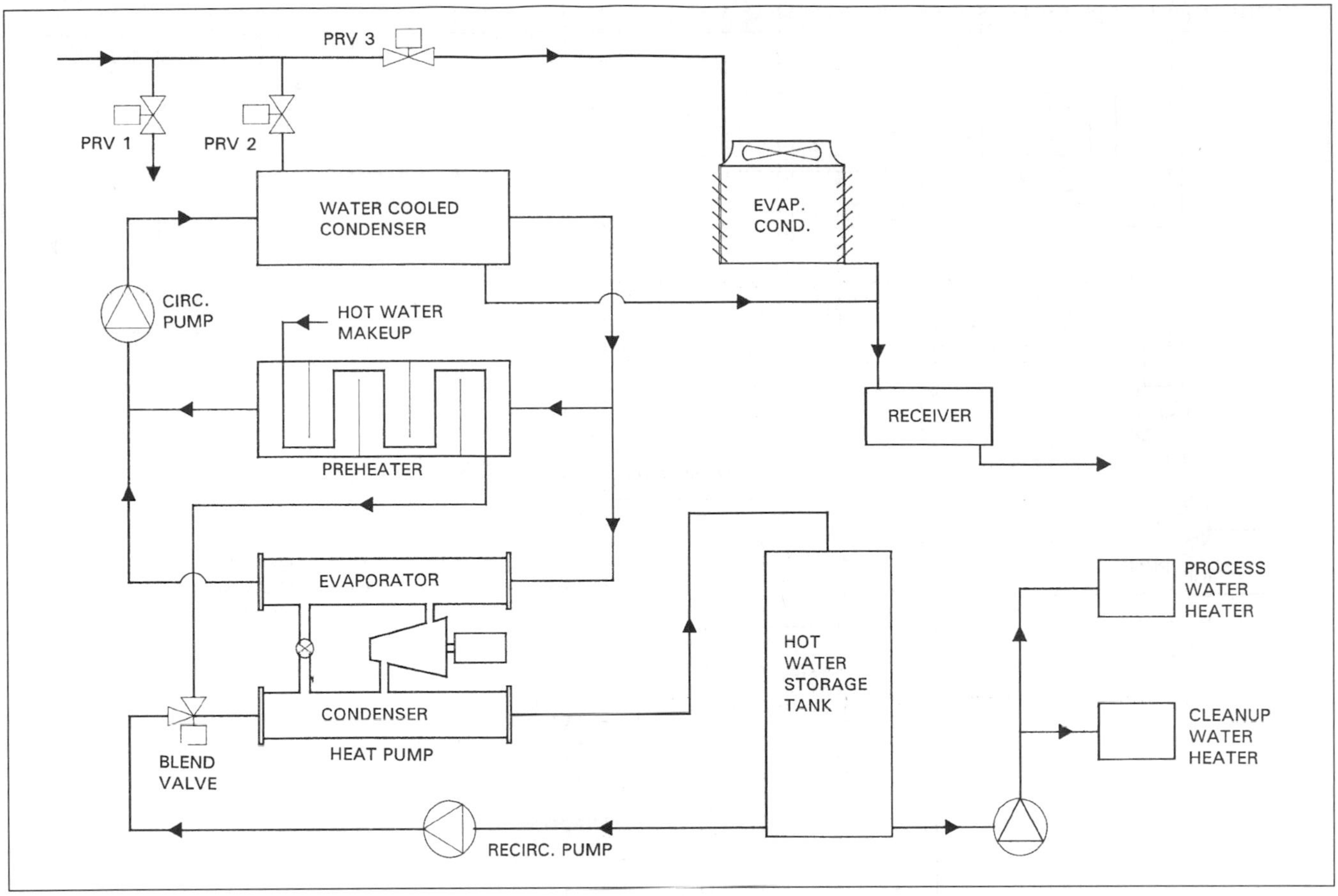

Fig. 13 Refrigeration Heat Recovery Heat Pump

existing water heaters (usually steam) provide additional heat and control for process water at the point of use.

Three major factors tasks must be addressed when adding heat recovery to existing plant refrigeration systems: (1) forcing the hot-gas refrigerant to flow to the desired heat exchanger, (2) scheduling the refrigeration processes to provide an adequate heat source over time while still meeting process requirements, and (3) integrating water-cooled, shell-and-tube condensers with evaporative condensers. The refrigerant direction can be controlled either by series piping of the two condenser systems or by three pressure-regulating valves—one to the hot-gas defrost (the lowest pressure setting), one to the recovery system (the medium-pressure setting), and one to the evaporative condensers (the highest pressure setting). The pressure-regulating valves offer good control but can be mechanically complicated. The series piping is simple but can cost more because of the pipe size required for all the hot gas to pass through the water-cooled condenser.

Each refrigeration load should be reviewed for its required output and production requirements. For example, ice production can frequently be scheduled during cleanup periods, and blast freezing for the end of shifts.

In multisystem integration, as in expanded system integration, equalization lines, liquid lines, receivers, and so forth must be designed according to standard refrigeration practices.

Process-fluid-to-process-fluid heat pumps can be applied to evaporation, concentration, crystallization, and distillation process fluids that contain chemicals that would destroy a steam compressor. A closed-cycle vapor compression system (Figure 14) is used for the separation of a solid and a liquid in an evaporator system and for the separation of two liquids in a distillation system. Both systems frequently have COPs of 8 to 10 and have the added benefit of cooling tower elimination. These systems can frequently be specified and supplied by the column or evaporator manufacturer as a value-added system.

The benefits of water-to-water, refrigerant-to-water, and other fluid-to-fluid systems may include the following:

- Lower energy costs due to the switch from fuel-based water heating to heat recovery water heating.
- Reduced costs for water-treatment chemicals for the boiler due to the switch from a steam-based system.
- Reduced emissions of NO_x, SO_x, CO, CO_2, and other harmful chemicals due to reduced boiler loading.
- Decreased effluent temperature, which improves the effectiveness of the water treatment process that breaks down solids.
- Increased production due to the increased temperature of the water available at the start of process cycles. For processes requiring a cooler start temperature, the preheated water may be blended with ambient water.
- Higher product quality due to rinsing with water at a higher temperature. Blending the preheated water with ambient water may be necessary if a cooler temperature is required.
- Reduced water and chemical consumption at cooling towers and evaporative condensers.
- More efficient process cooling if the heat recovery can be used to reduce refrigeration pressure or cooling tower return temperature.
- Reduced scaling of heat exchangers due to the lower surface temperatures in heat pump systems as compared to steam coils, forced-air gas systems, and resistance heaters.

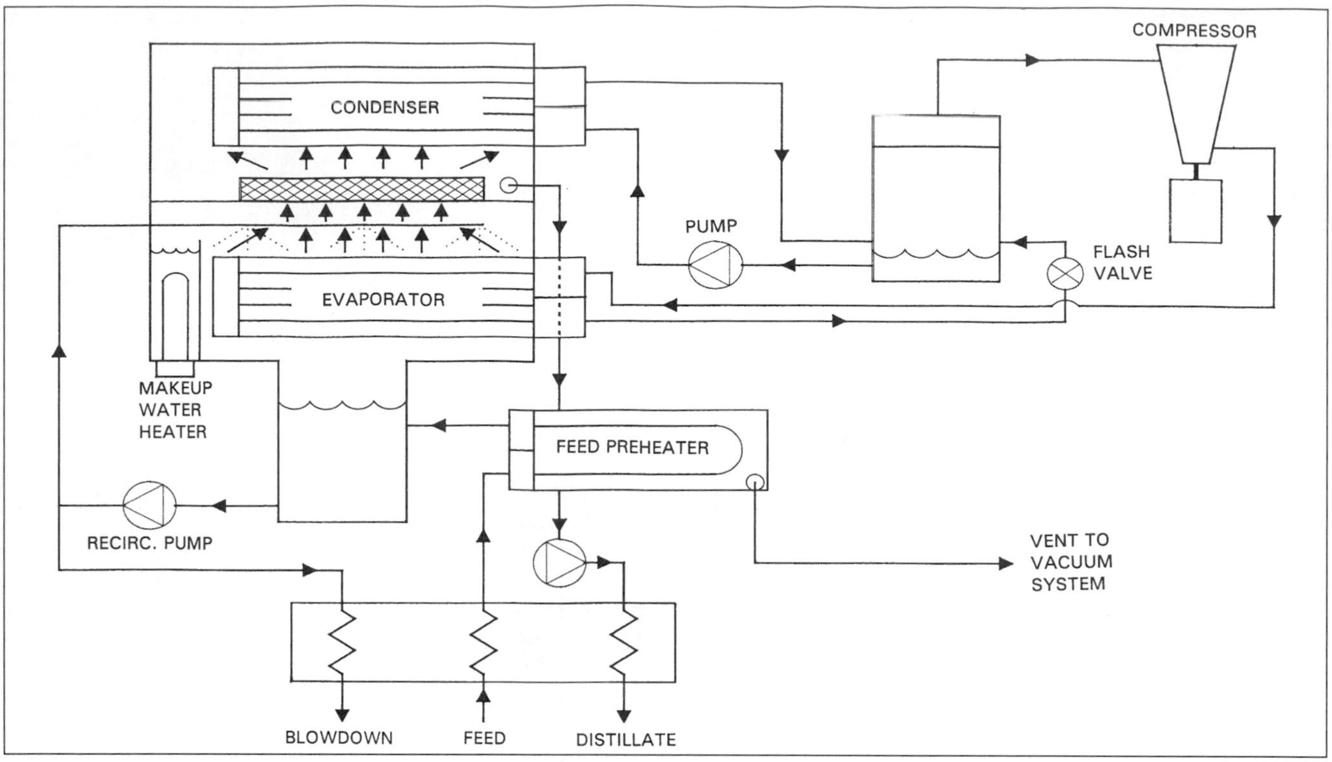

Fig. 14 Closed-Cycle Vapor Compression System

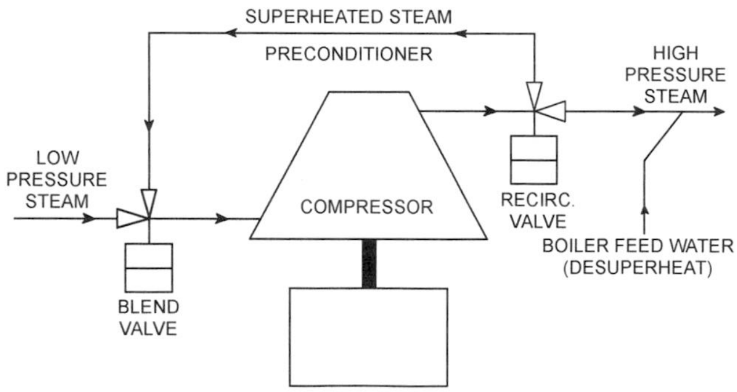

Fig. 15 Recompression of Boiler-Generated Process Steam

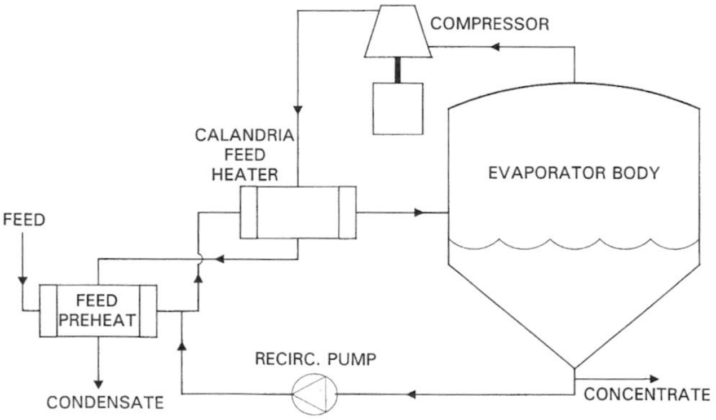

Fig. 16 Single-Effect Heat Pump Evaporator

Open-Cycle and Semi-Open-Cycle Heat Pump Systems

Open-cycle heat pump systems and **semi-open-cycle heat pump systems** use the process fluid to raise the temperature of the available heat energy by vapor compression, thus eliminating the need for chemical refrigerants. The most important class of applications is steam recompression. Compression can be provided with a mechanical compressor or by a thermocompression ejector driven by the required quantity of high-pressure steam. The three main controlling factors for this class of systems are vapor quality, boiling point elevation, and chemical makeup.

Recompression of boiler-generated process steam (Figure 15) has two major applications: (1) large facilities with substantial steam pressure drop due to line losses and (2) facilities with a considerable imbalance between steam requirements at low, medium, and high levels. Boiler-generated process steam usually conforms to cleanliness standards that ensure corrosion-free operation of the compression equipment. Evaluation of energy costs and steam value can be complicated with these applications, dictating the use of analytical tools such as pinch technology.

Application of open-cycle heat pumps to evaporation processes is exceptionally important. Single-effect **evaporators** (Figure 16) are common when relatively small volumes of water and solid need to be separated. The most frequent applications are in the food and dairy industries. The overhead vapors of the evaporator are compressed, and thus heated, and piped to the system heater (i.e., calandria). The heat is transferred to the dilute solution, which is then piped to the evaporator body. Flashing occurs upon entry to the evaporator body, sending concentrate to the bottom and vapors to the top. System COPs reach 10 to 20, and long-term performance of the evaporator improves as less of the product (and thus less buildup) occurs in the calandria. Multiple-effect evaporators (Figure 17) employ the same basic principles, only on a much larger scale. They are usually applied in the paper and chemical industries.

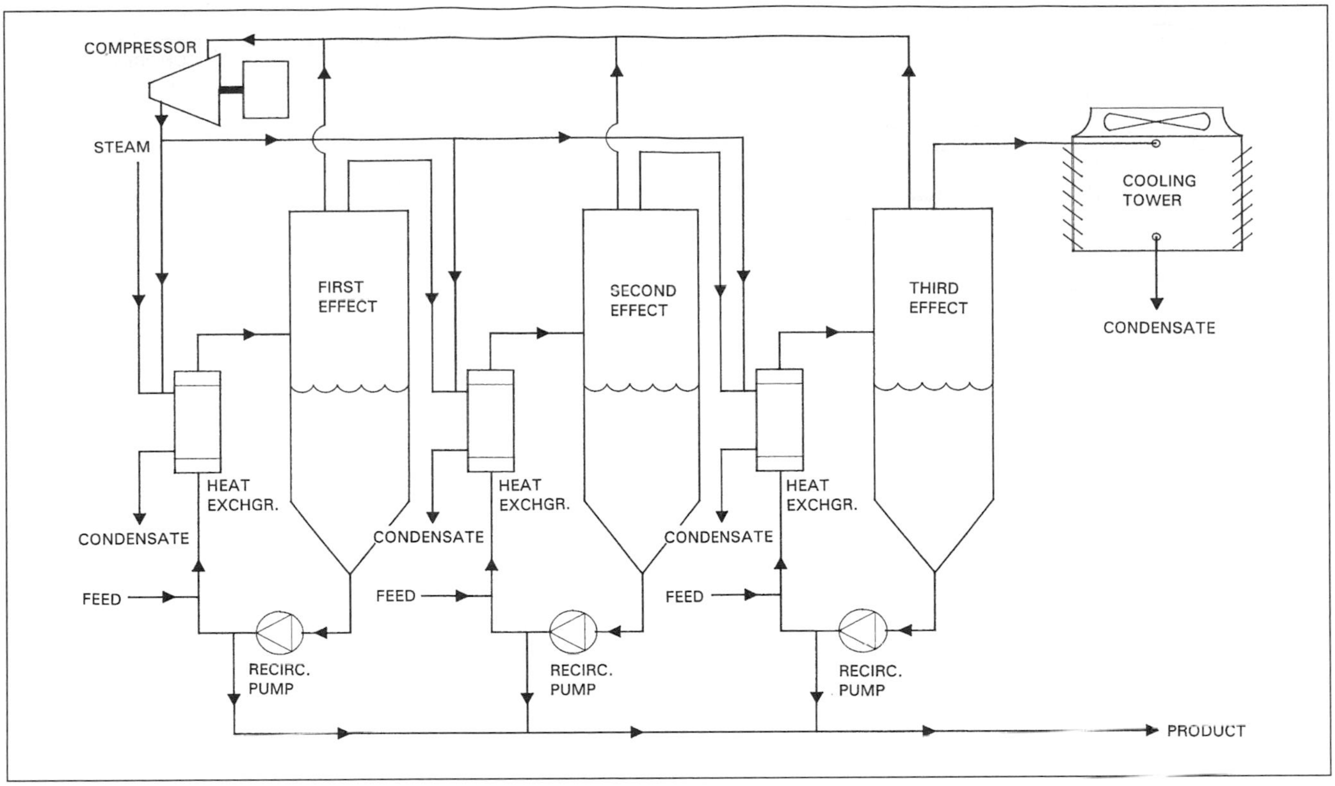

Fig. 17 Multiple-Effect Heat Pump Evaporator

The multiple evaporator bodies can be piped in series or in parallel. System COPs can exceed 30.

Application of open-cycle heat pumps to distillation processes (Figure 18) is similar to evaporation. However, compression of flammable gaseous compounds can be dangerous, so great care must be taken. The overhead vapors are compressed to a higher pressure and temperature, and then condensed in the reboiler. This eliminates the need for boiler steam in the reboiler and reduces overall energy consumption. As the pressure of the condensed vapor is reduced through the expansion valve, some of the liquid at a lower temperature is returned to the column as reflux, and the balance forms the overhead product, both as liquid and as vapor. Vapor is recycled as necessary. System COPs can reach 30 or 40.

Process emission-to-steam heat pump systems can be used for cooking, curing, and drying systems that operate with low-pressure saturated or superheated steam. The vapors from a cooking process, such as at a rendering plant, can be recovered to generate the steam required for cooking (Figure 19). The vapors are compressed with a screw compressor because noncondensable materials removed from the process by the vapor could erode or damage the reciprocating or centrifugal compressors. The compressed steam is supplied as the heat source to the cooker, and the steam condenses. The condensate is then supplied to a heat exchanger to heat process water. The noncondensables are usually scrubbed or incinerated, and the water is treated or discharged to a sewer.

The contaminants in some processes require the use of a semi-open-cycle heat pump (Figure 20). This system uses a heat exchanger, frequently called a reboiler, to recover heat from the stack gases. The reboiler produces low-pressure steam, which is compressed to the desired pressure and temperature. A clean-in-place (CIP) system may be used if the volume of contaminants is substantial.

Variable-speed drives can be specified with all these systems for closer, more efficient capacity control. Additional capacity for emergencies can be made available by temporarily over-speeding the drive, while sizing for nominal operating conditions to retain optimal efficiency. Proper integration of heat pump controls and system controls is essential.

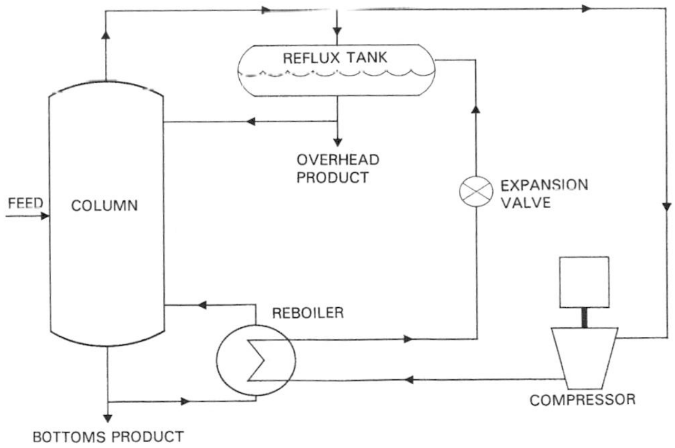

Fig. 18 Distillation Heat Pump System

Heat Recovery Design Principles

The following basic principles should be applied when designing heat recovery systems:

- Use Second Law, pinch technology, or other thermodynamic analysis methods, especially for complex processes, before detailed design to ensure proper thermodynamic placement of the HRHP.
- Design for base-load conditions. Heat recovery systems are designed for reduced operating costs. Process scheduling and thermal storage (usually hot water) can be used for better system

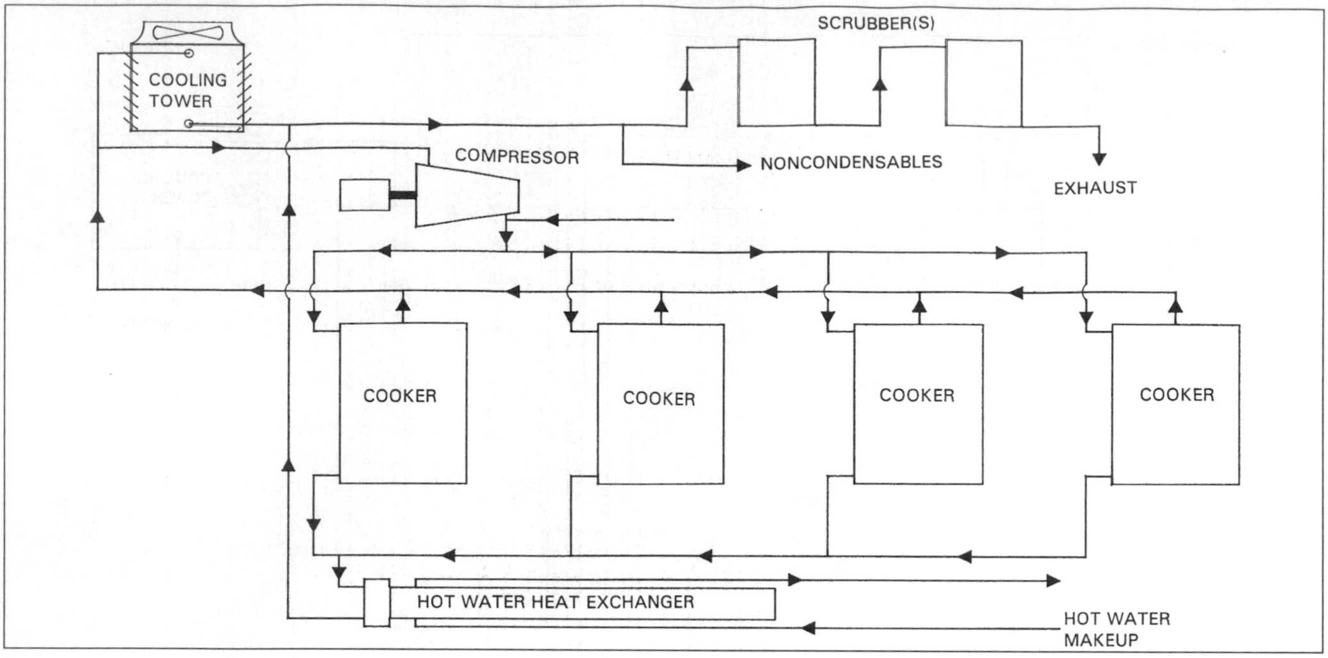

Fig. 19 Heat Recovery Heat Pump System in a Rendering Plant

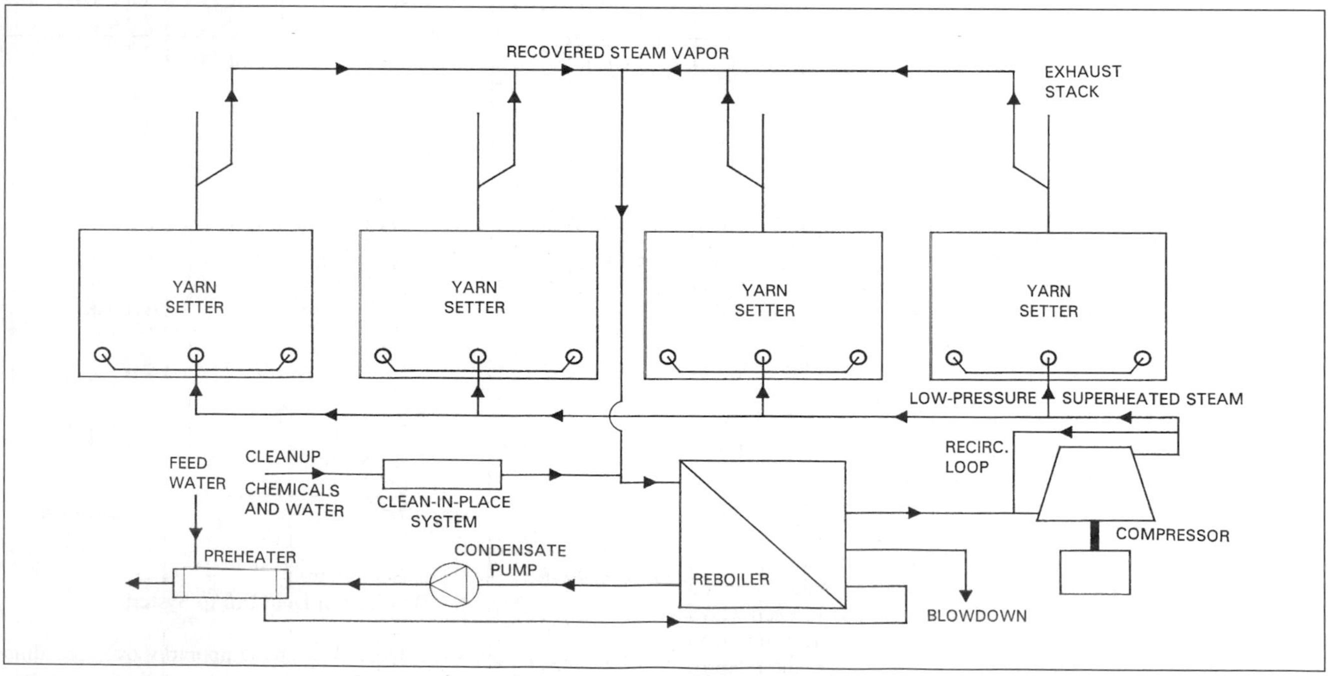

Fig. 20 Semi-Open Cycle Heat Pump in a Textile Plant

balancing. Existing water heating systems can be used for peak load periods and for better temperature control.

- Exchange heat first, then pump the heat. If a heat exchanger is to provide the thermal work in a system, it should be used by itself. If additional cooling of the heat source stream and/or additional heating of the heat sink stream are needed, then a heat pump should be added.
- Do not expect a heat pump to solve a design problem. A design problem such as an unbalanced refrigeration system may be exacerbated by the addition of a heat pump.
- Make a complete comparison of the heat exchanger system and the heat pump system. Compared to heat exchange alone, a heat pump system has additional first and operating costs. If feasible, heat exchange should be added to the front end of the heat pump system.
- Evaluate the cost of the heat displaced by the heat pump system. If the boiler operation is not changed, or if it makes the boiler less efficient, the heat pump may not be economical.
- Investigate standard fuel-handling and heat exchanger systems already in use in an industry before designing the heat pump.

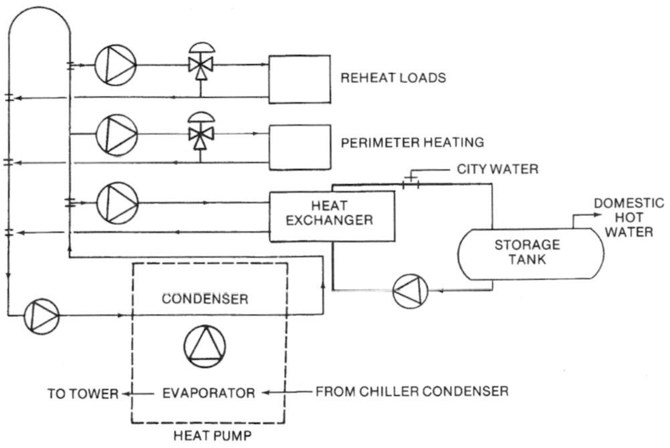

Fig. 21 Heat Recovery Heat Pump

- Measure the flow and temperature profile of both the heat source and heat sink over an extended period of time. The data may help prevent overestimating the requirements and economics of the system.
- Investigate future process changes that may affect the thermal requirements and/or availability of the system. Determine whether the plant is changing to a cold water cleanup system or changing a process to produce less effluent at lower temperatures.
- Design the system to give plant operators the same or better control. Thoroughly review manual versus automatic controls.
- Determine whether special material specifications are needed for handling any process flows. Obtain a chemical analysis for any flow of unknown makeup.
- Inform equipment suppliers of the full range of ambient conditions to which the equipment will be exposed and the expected loading requirements.

APPLIED HEAT RECOVERY SYSTEMS

WASTE HEAT RECOVERY

In many large buildings, internal heat gains require year-round chiller operation. The chiller condenser water heat is often wasted through a cooling tower. Figure 21 illustrates an HRHP installed in the water line from the chiller's condenser before rejection at the cooling tower. This arrangement uses the otherwise wasted heat to provide heat at the higher temperatures required for space heating, reheat, and domestic water heating.

Prudent design may dictate cascade systems with chillers in parallel or series. Custom components are available to meet a wide range of load and temperature requirements. The double-bundle condenser working with a reciprocating or centrifugal compressor is most often used in this application. Figure 22 shows the basic configuration of this system, which makes heat available in the range of 100 to 130°F. The warm water is supplied as a secondary function of the heat pump and represents recovered heat.

Figure 23 shows a similar cycle, except that a storage tank has been added, enabling the system to store heat during occupied hours by raising the temperature of the water in the tank. During unoccupied hours, water from the tank is gradually fed to the evaporator providing load for the compressor and condenser that heats the building during off hours.

Figure 24 shows a heat transfer system capable of generating 130 to 140°F or warmer water whenever there is a cooling load, by cascading two refrigeration systems. In this configuration, Machine No. I can be considered as a chiller only and Machine No. II as a heating-only heat pump.

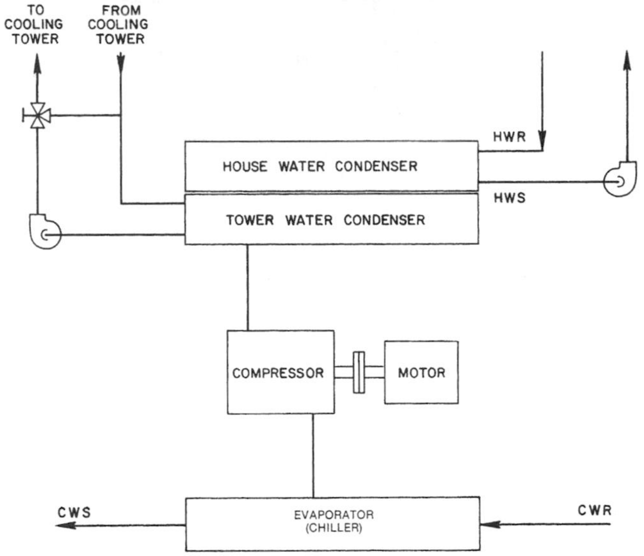

Fig. 22 Heat Recovery Chiller with Double-Bundle Condenser

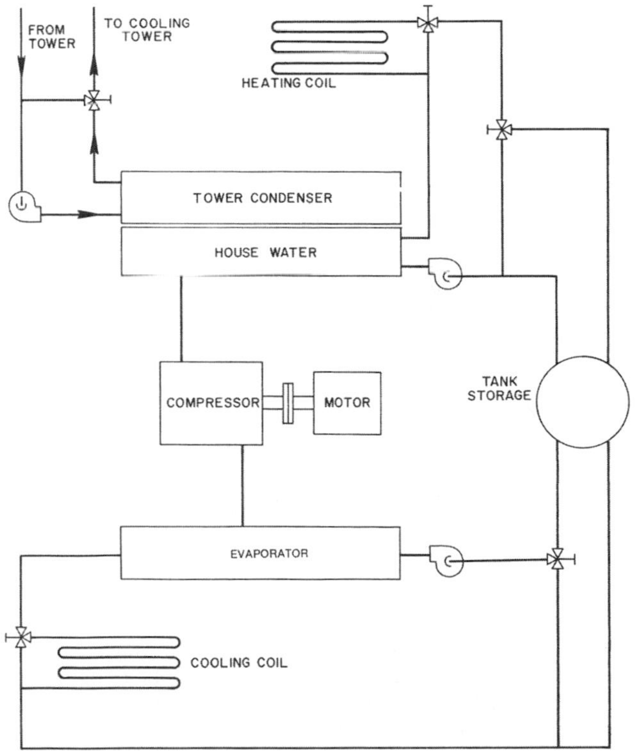

Fig. 23 Heat Recovery Chiller with Storage Tank

An indication of the magnitude of the recoverable heat in a modern multistory office building is shown in Figures 25, 26, and 27. Figures 25 and 26 show the gross heat loss and the internal heat gain of the exterior and interior zones during occupied periods. Figure 27 shows the total amount of heat available for recovery. Heat recovered from internal zones can be used to provide all or a portion of the heating requirements of the external zones. The excess recovered heat can be diverted to thermal storage for later use during unoccupied periods. During the occupied periods, no outside heat source or supplemental heat is needed at outdoor temperatures of 23°F or above for this hypothetical building.

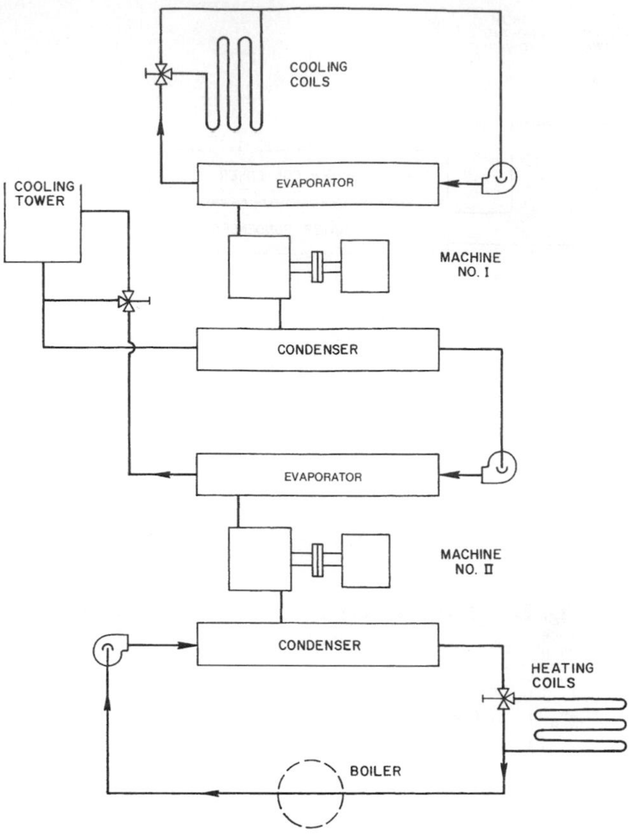

Fig. 24 Multistage (Cascade) Heat Transfer System

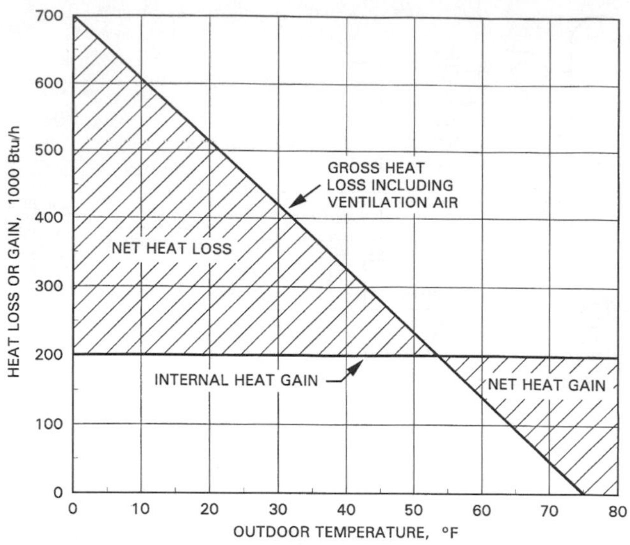

Fig. 25 Heat Loss and Heat Gain for Exterior Zones During Occupied Periods

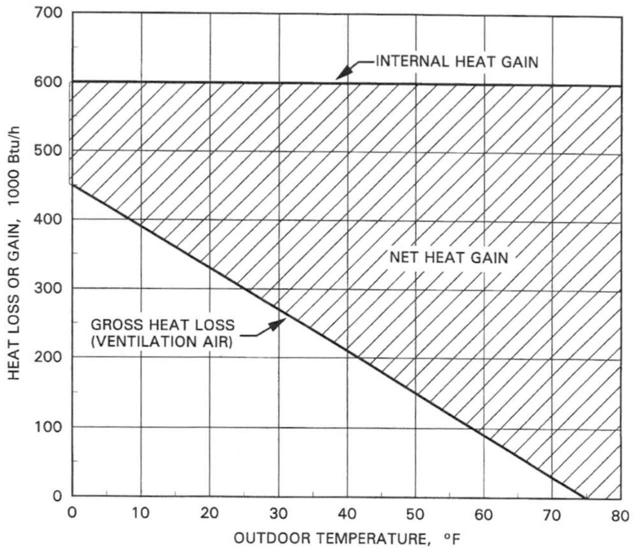

Fig. 26 Heat Loss and Heat Gain for Interior Zones During Occupied Periods

It must be understood that relative performances are indicated in these figures to illustrate general cases. Of course, each building will have parameters that differ from those of the building used for the figures.

WATER LOOP HEAT PUMP SYSTEMS

Description

A water loop heat pump (WLHP) system combines load transfer characteristics with multiple water-to-air heat pump units (Figure 28). Each zone, or space, has one or more water-to-air heat pumps. The units in both the building core and perimeter areas are connected hydronically with a common two-pipe system. Each unit cools conventionally, supplying air to the individual zone and rejecting the heat removed to the two-pipe system through its integral condenser. The excess heat gathered by the two-pipe system is expelled through a common heat rejection device. This device often includes a closed-circuit evaporative cooling tower with an integral spray pump. If and when some of the zones, particularly on the northern side of the building, require heat, the individual units switch (by means of reversing refrigerant valves) into the heating cycle. These units then extract heat from the two-pipe water loop, a relatively high-temperature source, that is totally or partially maintained by heat rejected from the condensers of the units that provide cooling to other zones. When only heating is required, all units are in the heating cycle and, consequently, an external heat input to the loop is needed to maintain the loop temperature. The water loop temperature is usually maintained in the range of 60 to 90°F and, therefore, seldom requires piping insulation.

Any number of water-to-air heat pumps may be installed in such a system. Water circulates through each unit via the closed loop. The water circuit usually includes two circulating pumps (one pump is 100% standby) and a means for adding and rejecting heat to and from the loop. Each heat pump can either heat or cool to maintain the comfort level in each zone.

Units operating in the heating mode extract heat from the circulated water, while units operating in the cooling mode reject heat to the water. Thus, the system recovers and redistributes heat, where needed. Unlike air-source heat pumps, heating output for this system does not depend on the outdoor temperature. The water loop conveys rejected heat, but a secondary heat source, typically a boiler, is usually provided.

Another version of the WLHP system uses a coil buried in the ground as a heat source and heat sink. This ground-coupled system does not normally need the boiler and cooling tower incorporated in conventional WLHP systems to keep the circulating water within acceptable temperature limits. However, ground-coupled heat pumps may operate at lower entering water (or antifreeze solution)

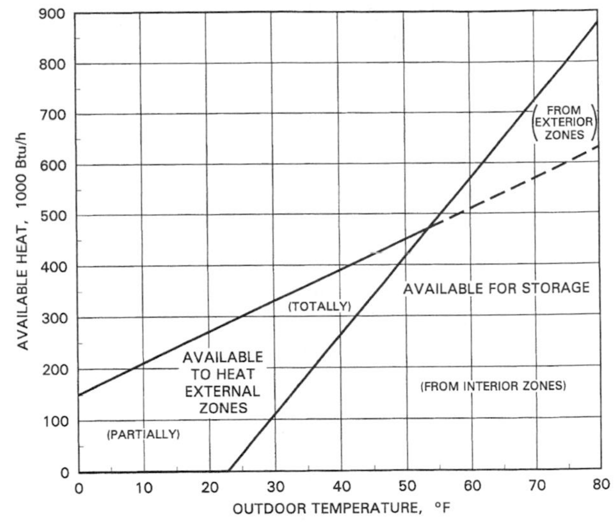

Fig. 27 Internal Heat Available for Recovery During Occupied Periods

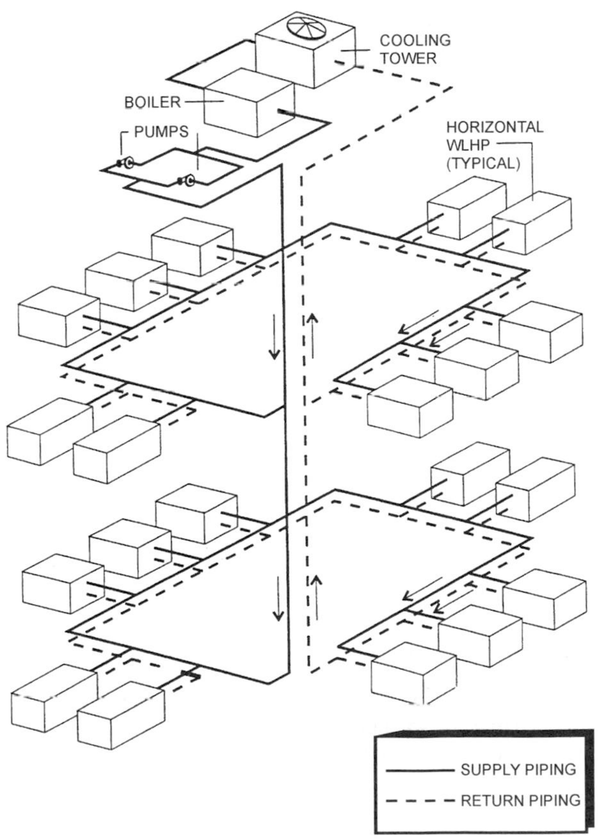

Fig. 28 Heat Recovery System Using Water-to-Air Heat Pumps in a Closed Loop

temperatures. Some applications may require a heat pump tolerant of entering water temperatures ranging from 25 to 110°F. In climates where air conditioning dominates, a cooling tower is sometimes combined with ground coupling to reduce overall installed costs.

Figure 29 illustrates a system with a storage tank and solar collectors. The storage tank in the condenser circuit can store excess heat during occupied hours and provide heat to the loop during unoccupied hours. During this process of storing and draining heat from the tank, solar heat may also be added within the limitations of the temperatures of the condenser system. The solar collectors are more effective and efficient under these circumstances because of the temperature ranges involved.

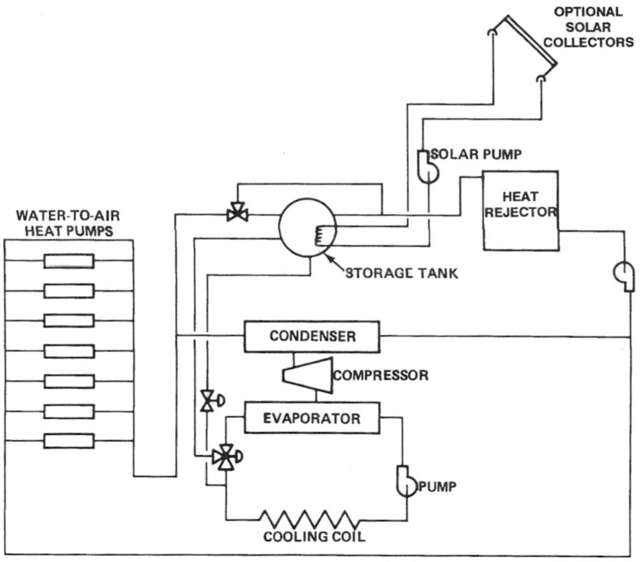

Fig. 29 Closed-Loop Heat Pump System with Thermal Storage and Optional Solar-Assist Collectors

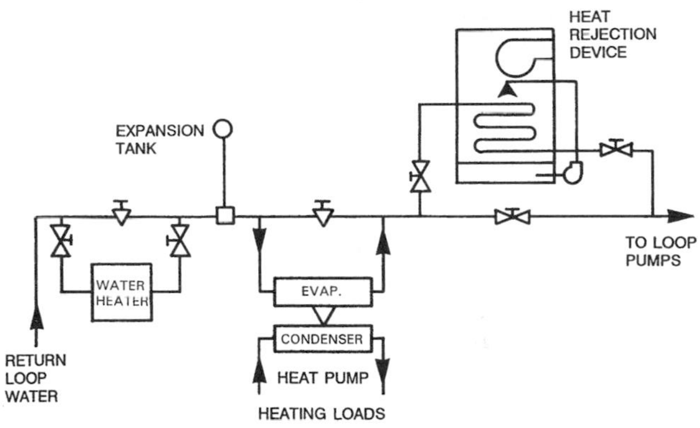

Fig. 30 Secondary Heat Recovery from WLHP System
(Adapted from *Marketing the Industrial Heat Pump*, Edison Electric Institute 1989)

For further heat reclaim, a water-to-water heat pump can be added in the closed water loop before the heat rejection device. This heat pump provides domestic hot water or elevates loop water temperatures in a storage tank so the water can be bled back into the loop when needed during the heating cycle. Figure 30 illustrates such a system.

Many facilities require large cooling loads (e.g., for interior zones, lights, people, business machines, computers, switchgear, and production machinery) that result in net loop heat rejection during all or most of the year, particularly during occupied hours. This rejection of waste heat often occurs while other heating loads in the facility (e.g., ventilation air, reheat, and domestic hot water) are using external purchased energy to supply heat. By including a secondary water-to-water heat pump in the system design (as shown in Figure 30), additional balanced heat recovery can be economically achieved. The secondary heat pump can effectively be used to reclaim this otherwise rejected heat, amplify its temperature, and use it to serve other heating loads, thus minimizing the use of purchased energy.

In another WLHP system variation, the building sprinkler system is used as part of the loop water distribution system.

Some aspects of the systems described in this section may be proprietary and should not be used without appropriate investigation.

Design Considerations

Water loop heat pump (WLHP) systems are used in many types of multiroom buildings. A popular application is in office buildings, where heat gains from interior zones can be redistributed to the perimeter during winter. Other applications include hotels and motels, schools, apartment buildings, nursing homes, manufacturing facilities, and hospitals. Operating costs for these systems are most favorable in applications with simultaneous heating and cooling requirements.

Unit Types. Chapter 45 describes various types and styles of units and the control options available.

Zoning. The WLHP system offers excellent zoning capability. Because equipment can be placed within the zones, future relocation of partitions can be accommodated with minimum duct changes. Some systems use heat pumps for perimeter zones and the top floor, with cooling-only units serving interior zones; all units are connected into the same loop water circuit.

Heat Recovery and Heat Storage. The WLHP system lends itself well to heat storage. Installations, such as schools, that cool most of the day in winter and heat at night make excellent use of heat storage. The water may be stored in a large tank in the closed-loop circuit ahead of the boiler. In this application, the loop temperature is allowed to build up to 90°F during the day. The stored water at 90°F can be used during unoccupied hours to maintain heat in the building, with the loop temperature allowed to drop to 60°F. The water heater (or boiler) would not be used until the loop temperature had dropped the entire 30°F. The storage tank operates as a flywheel to prolong the period of operation where neither heat makeup nor heat rejection is required.

Concealed Units. Equipment installed in ceiling spaces must have access for maintenance and servicing filters, control panels, compressors, and so forth. Adequate condensate drainage must be provided.

Ventilation. Outdoor air for ventilation may be (1) ducted from a ventilation supply system to the units or (2) drawn in directly through a damper into the individual units. To operate satisfactorily, the air entering the water-source heat pumps should be above 60°F. In cold climates, the ventilation air must be preheated. The quantity of ventilation air entering directly through individual units can vary greatly due to the stack effect, wind, and balancing difficulties.

Secondary Heat Source. The secondary heat source for heat makeup may be electric, gas, oil, ground, solar energy, and/or waste heat. Normally, a water heater or boiler is used; however, electric resistance heat in the individual heat pumps, with suitable controls, may also be used. The control may be an aquastat set to switch from heat pump to resistance heaters when the loop water approaches the minimum 60°F.

An electric boiler is readily controllable to provide 60°F outlet water and can be used directly in the loop. With a gas-fired or oil-fired combustion boiler, a heat exchanger may be used to transfer heat to the loop or, depending on the type of boiler used, a modulating valve may blend hot water from the boiler into the loop.

Solar or ground energy can supply part or all of the secondary heat. Water or antifreeze solution circulated through collectors can add heat to the system directly or indirectly via a secondary heat exchanger.

A building with night setback may need a supplementary heater boiler sized for the installed capacity, not the building heat loss, because the morning warm-up cycle may require every heat pump to operate at full heating capacity until the building is up to temperature. In this case, based on a typical heating COP of 4.0 for a water-source heat pump, the boiler would be sized to provide about 75% of the total heating capacity of all water-source heat pumps installed in the building.

Heat Rejector Selection. A closed-loop circuit requires a heat rejector that is either a heat exchanger (loop water to cooling tower water), a closed-circuit evaporative cooler, or a ground coil. The heat rejector is selected in accordance with manufacturer's selection curves, using the following parameters:

- **Water flow rates.** Manufacturers' recommendations on water flow rates vary between 2 and 3 gpm per ton of installed cooling capacity. The lower flow rates are generally preferred in regions having a relatively low summer outdoor design wet-bulb temperature. In more humid climates, a higher flow rate allows a higher water temperature to be supplied from the heat rejector to the heat pumps, without a corresponding increase in temperature leaving the heat pumps. Thus, cooling tower or evaporative cooler size and cost are minimized without penalizing the performance of the heat pumps.
- **Water temperature range.** Range (the difference between the leaving and entering water temperatures at the heat rejector) is affected by heat pump cooling efficiency, water flow rate, and diversity. It is typically between 10 and 15°F.
- **Approach.** Approach is the difference between the water temperature leaving the cooler and the wet-bulb temperature of the outside air. The maximum water temperature expected in the loop supply is a function of the design wet-bulb temperature.
- **Diversity.** Diversity is the maximum instantaneous cooling load of the building divided by the installed cooling capacity:

$$D = Q_m / Q_i \tag{1}$$

where

D = diversity
Q_m = maximum instantaneous cooling load
Q_i = total installed cooling capacity

Diversity times the average range of the heat pumps is the applied range of the total system (the rise through all units in the system and the drop through the heat rejector). For systems with a constant pumping rate regardless of load,

$$R_s = DR_p \tag{2}$$

where

R_s = range of system
R_p = average range of heat pumps

For systems with a variable pumping rate and with each pump equipped with a solenoid valve to start and stop water flow through the unit with compressor operation,

$$R_s \approx R_p \tag{3}$$

The average leaving water temperature of the heat pumps is the entering water temperature of the heat rejector. The leaving water temperature of the heat rejector is the entering water temperature of the heat pumps.

- **Winterization.** For buildings with some potential year-round cooling (e.g., office buildings), loop water may be continuously pumped through the heat rejector. This control procedure reduces the danger of freezing. However, it is important to winterize the heat rejector to minimize the heat loss.

In northern climates, the most important winterization step for evaporative coolers is the installation of a discharge air plenum with positive-closure, motorized, ice-proof dampers. The entire casing that houses the tube bundle and the discharge plenum may be insulated. The sump, if outside the heated space, should be equipped with electric heaters. The heat pump equipment manufacturer's instructions will help in the selection and control of the heat rejector.

If sections of the water circuit will be exposed to freezing temperatures, the addition of an antifreeze solution should be considered. In a serpentine pipe circuit having no automatic valves that could totally isolate individual components, an antifreeze solution prevents bursting pipes, with minimal effect on system performance.

An open cooling tower with a separate heat exchanger is a practical alternative to the closed water cooler (Figure 31). An additional pump is required to circulate the tower water through the heat exchanger. In such an installation, where no tubes are exposed to the atmosphere, it may not be necessary to provide freeze protection on the tower. The sump may be indoors or, if outdoors, may be heated to keep the water from freezing. This arrangement allows the use of small, remotely located towers. Temperature control necessary for tower operation is maintained by a sensor in the water loop system controlling operation of the tower fan(s).

The combination of an open tower, heat exchanger, and tower pump frequently is lower in first cost than an evaporative cooler. In addition, operating costs are lower because no heat is lost from the loop in winter and, frequently, less power is required for the cooling tower fans.

Ductwork Layout. Often, a WLHP system has ceiling-concealed units, and the ceiling area is used as the return plenum. Troffered light fixtures are a popular means of returning air to the ceiling plenum.

The air supply from the heat pumps should be designed for quiet operation. Heat pumps connected to ductwork must be capable of overcoming external static pressure. The heat pump manufacturers' recommendations should be consulted for the maximum and minimum external static pressure allowable with each piece of equipment.

Piping Layout. Reverse-return piping should be used wherever possible with the WLHP system. This is particularly true where all units have essentially the same capacity. Balancing is then minimized except for each of the system branches. If direct-return piping is used, balancing the water flow is required at each individual heat pump. The entire system flow may circulate through the boiler and heat rejector in series. Water makeup should be at the constant pressure point of the entire water loop. Piping system design is similar to the secondary water distribution of air-and-water systems.

Pumping costs for a WLHP system can be significant. Because loads on the system vary considerably, variable-speed pumping should be considered. This requires an automatic valve at each heat pump that allows water flow through the heat pump coil only during compressor operation.

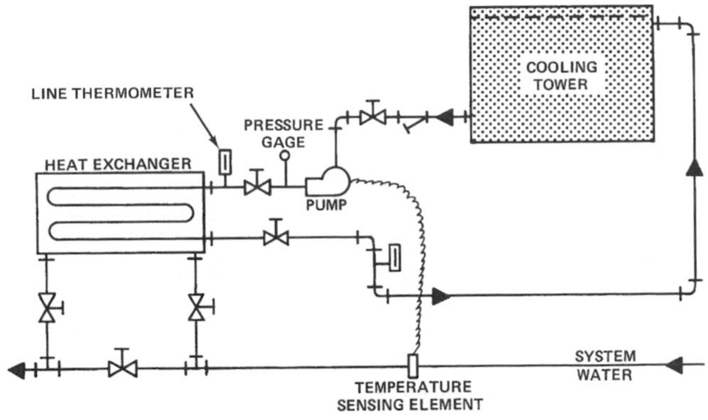

Fig. 31 Cooling Tower with Heat Exchanger

Clean piping is vital to successful performance of the water-source heat pump system. The pipe should be clean when installed, kept clean during construction, and thoroughly cleaned and flushed upon completion of construction. Start-up water filters in the system bypass (pump discharge to suction) should be included on large, extensive systems.

Controls

The WLHP system has simpler controls than totally central systems. Each heat pump is controlled by a thermostat in the zone. There are only two centralized temperature control points: one to add heat when the water temperature approaches a prescribed lower temperature (typically about 60°F), and the other to reject heat when the water temperature approaches a prescribed upper temperature (typically about 90°F).

The boiler controls should be checked to be sure that outlet water is controlled at 60°F, because controls normally supplied with boilers are in a much higher range.

An evaporative cooler should be controlled by increasing or decreasing heat rejection capacity in response to the loop water temperature leaving the cooler. A reset schedule that operates the system at a lower water temperature (to take advantage of lower outdoor wet-bulb temperatures) can save energy when heat from the loop storage is not likely to be used.

Abnormal condition alarms typically operate as follows:

- Upon a fall in loop temperature to 50°F, initiate an alert. Open heat pump control circuits at 45°F.
- Upon a rise in loop temperature to 105°F, initiate an alert. Open heat pump control circuits at 110°F.
- Upon sensing insufficient system water flow, a flow switch initiates an alert and opens the heat pump control circuits.

An outside ambient control should be provided to prevent operation of the cooling tower sump pump at freezing temperatures.

Optional system control arrangements include the following:

- Night setback control
- Automatic unit start/stop, with after-hour restart as a tenant option
- Warm-up cycle
- Pump alternator control

Advantages of a WLHP System

- Affords opportunity for energy conservation by recovering heat from interior zones and/or waste heat and by storing excess heat from daytime cooling for nighttime or other heating uses.
- Allows recovery of solar energy at a relatively low fluid temperature where solar collector efficiency is likely to be greater.
- The building does not require wall penetrations to provide for the rejection of heat from air-cooled condensers.
- Provides environmental control in scattered occupied zones during nights or weekends without the need to start a large central refrigeration machine.
- Units are not exposed to outdoor weather, which allows installation in coastal and other corrosive atmospheres.
- Units have a longer service life than air-cooled heat pumps.
- Noise levels can be lower than those of air-cooled equipment because condenser fans are eliminated and the compression ratio is lower.
- Two-pipe boiler/chiller systems are potentially convertible to this system.
- The entire system is not shut down when a unit fails. However, loss of pumping capability, heat rejection, or secondary heating could affect the entire system.

- Energy usage by the heat pumps can be metered for each tenant. However, this metering would not include energy consumed by the central pump, heat rejector, or boiler.
- Total life-cycle cost of this system frequently compares favorably to that of central systems when considering installed cost, operating costs, and system life.
- Units can be installed as space is leased or occupied.

Limitations of a WLHP System

- Space is required for the boiler, heat exchangers, pumps, and heat rejector.
- Initial cost may be higher than for systems that use multiple unitary HVAC equipment.
- Reduced airflow can cause the heat pump to overheat and cut out. Therefore, periodic filter maintenance is imperative.
- Cleanliness of the piping loop must be maintained.

BALANCED HEAT RECOVERY SYSTEMS

Definition

In an ideal heat recovery system, all components work year-round to recover all the internal heat before adding external heat. Any excess heat is either stored or rejected. Such an idealized goal is identified as a balanced heat recovery system.

When the outdoor temperature drops significantly, or when the building is shut down (e.g., on nights and weekends), internal heat gain may be insufficient to meet the space heating requirements. Then, a balanced system provides heat from storage or an external source. When internal heat is again generated, the external heat is automatically reduced to maintain proper temperature in the space. There is a time delay before equilibrium is reached. The size of the equipment and the external heat source can be reduced in a balanced system that includes storage. Regardless of the system, a heat balance analysis establishes the merits of balanced heat recovery at various outdoor temperatures.

Outdoor air less than 55 to 65°F may be used to cool building spaces with an air economizer cycle. When considering this method of cooling, the space required by ducts, air shafts, and fans, as well as the increased filtering requirements to remove contaminants and the hazard of possible freeze-up of dampers and coils must be weighed against alternatives such as the use of deep row coils with antifreeze fluids and efficient heat exchange. Innovative use of heat pump principles may give considerable energy savings and more satisfactory human comfort than an air economizer. In any case, hot and cold air should not be mixed (if avoidable) to control zone temperatures because it wastes energy.

Heat Redistribution

Many buildings, especially those with computers or large interior areas, generate more heat than can be used for most of the year. Operating cost is kept to a minimum when the system changes over from net heating to net cooling at the break-even outdoor temperature at which the building heat loss equals the internal heat load. If heat is unnecessarily rejected or added to the space, the changeover temperature varies from the natural break-even temperature, and operating costs increase. Heating costs can be reduced or eliminated if excess heat is stored for later distribution.

Heat Balance Concept

The concept of ideal heat balance in an overall building project or a single space requires that one of the following takes place on demand:

- Heat must be removed.
- Heat must be added.
- Heat recovered must exactly balance the heat required, in which case heat should be neither added nor removed.

In small air-conditioning projects serving only one space, either cooling or heating satisfies the thermostat demand. If humidity control is not required, the operation is simple. Assuming both heating and cooling are available, the automatic controls will respond to the thermostat to supply either. A system should not heat and cool the same space simultaneously.

Multiroom buildings commonly require heating in some rooms and cooling in others. Optimum design considers the building as a whole and transfers excess internal heat from one area to another, as required, without introducing external heat that would require waste heat disposal at the same time. The heat balance concept is violated when this occurs.

Humidity control must also be considered. Any system should add or remove only enough heat to maintain the desired temperature and control the humidity. Large percentages of outdoor air with high wet-bulb temperatures, as well as certain types of humidity control, may require reheat, which could upset the desirable balance. Usually, humidity control can be obtained without upsetting the balance. When reheat is unavoidable, internally transferred heat from heat recovery should always be used to the extent it is available, before using an external heat source such as a boiler. However, the effect of the added reheat must be analyzed, because it affects the heat balance and may have to be treated as a variable internal load.

When a building requires heat and the refrigeration plant is not in use, dehumidification is not usually required and the outdoor air is dry enough to compensate for any internal moisture gains. This should be carefully reviewed for each design.

Heat Balance Studies

The following analytical examples illustrate situations that can occur in nonrecovery and unbalanced heat recovery situations. Figure 32 graphically shows the major components that comprise the total air-conditioning load of a building. Values above the zero line are cooling loads, and values below the zero line are heating loads. On an individual basis, the ventilation and conduction loads cross the zero line, which indicates that these loads can be a heating or a cooling load, depending on outdoor temperature. The solar load and internal loads are always a cooling load and are, therefore, above the zero line.

Figure 33 combines all the loads shown in Figure 32. The graph is obtained by plotting the conduction load of a building at various outdoor temperatures, and then adding or subtracting the other loads at each temperature. The project load lines, with and without solar effect, cross the zero line at 16 and 30°F, respectively. These are the outdoor temperatures for the plotted conditions when the naturally created internal load exactly balances the loss.

As plotted, this heat balance diagram includes only the building loads with no allowance for additional external heat from a boiler or other source. If external heat is necessary because of system design, the diagram should include the additional heat.

Figure 34 illustrates what happens when heat recovery is not employed. It is assumed that at a temperature of 70°F, heat from an external source is added to balance conduction through the skin of the building in increasing amounts down to the minimum outdoor temperature winter design condition. Figure 34 also adds the heat required for the outdoor air intake. The outdoor air comprising part or all of the supply air must be heated from outdoor temperature to room temperature. Only the temperature range above the room temperature is effective for heating to balance the perimeter conduction loss.

These loads are plotted at the minimum outdoor winter design temperature, resulting in a new line passing through points A, D, and E. This line crosses the zero line at −35°F, which becomes the artificially created break-even temperature rather than 30°F, when not allowing for solar effect. When the sun shines, the added solar heat at the minimum design temperature would further drop the −35°F break-even temperature. Such a design adds more heat than the overall project requires and does not employ balanced heat recovery to use the available internal heat. This problem is most evident during mild weather on systems not designed to take full advantage of internally generated heat year-round.

The following are two examples of situations that can be shown in a heat balance study:

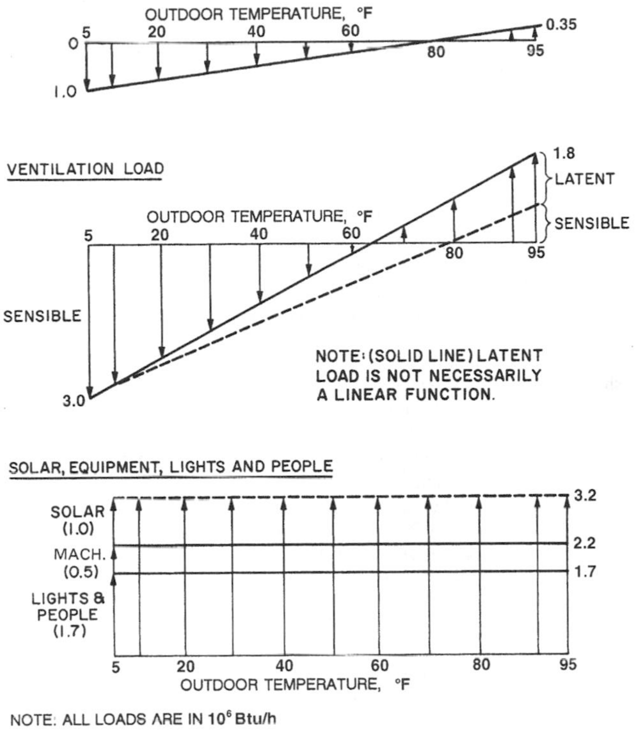

Fig. 32 Major Load Components

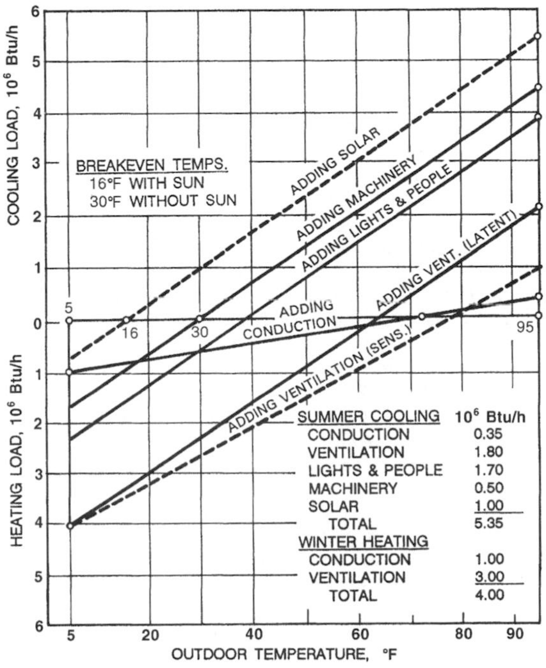

Fig. 33 Composite Plot of Loads in Figure 32 (Adjust for Internal Motor Heat)

1. As the outdoor air wet-bulb temperature drops, the total heat of the air falls. If a mixture of outdoor and recirculated air is cooled to 55°F in summer and the same dry-bulb temperature is supplied by an economizer cycle for interior space cooling in winter, there will be an entirely different result. As the outdoor wet-bulb temperature drops below 55°F, each unit volume of air introduced does more cooling. To make matters more difficult, this increased cooling is latent cooling, which requires the addition of latent heat to prevent too low a relative humidity, yet this air is intended to cool. The extent of this added external heat for free cooling is shown to be very large when plotted on a heat balance analysis at 0°F outside temperature.

   Figure 34 is typical for many current non-heat-recovery systems. There may be a need for cooling, even at the minimum design temperature, but the need to add external heat for humidification can be eliminated by using available internal heat. When this asset is thrown away and external heat is added, operation is inefficient.

   Some systems recover heat from exhaust air to heat the incoming air. When a system operates below its natural break-even temperature t_{be} such as 30 or 16°F (shown in Figure 33), the heat recovered from exhaust air is useful and beneficial. This assumes that only the available internal heat is used and that no supplementary heat is added at or above t_{be}. Above t_{be}, the internal heat is sufficient and any recovered heat would become excessive heat to be removed by more outdoor air or refrigeration.

   If heat is added to a central system to create an artificial t_{be} of −35°F as in Figure 34, any recovered heat above −35°F requires an equivalent amount of heat removal elsewhere. If the project were in an area with a minimum design temperature of 0°F, heat recovery from exhaust air could be a liability at all times for the conditions stipulated in Figure 33. This does not mean that the value of heat recovered from exhaust air should be forgotten. The emphasis should be on recovering heat from exhaust air rather than on adding external heat.

2. A heat balance shows that insulation, double glazing, and so forth can be extremely valuable on some projects. However, these practices may be undesirable in certain regions during the heating season, when excess heat must usually be removed from large buildings. For instance, for minimum winter design temperatures of approximately 35 to 40°F, it is improbable that the interior core of a large office building will ever reach its break-even temperature. The temperature lag for shutdown periods,

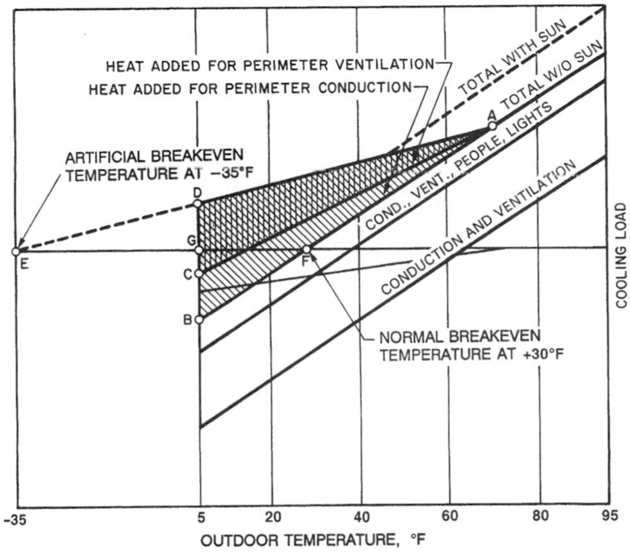

Fig. 34 Non-Heat-Recovery System

such as nights and weekends, at minimum design conditions could never economically justify the added cost of double-pane windows. Therefore, double-pane windows merely require the amount of heat saved to be removed elsewhere.

General Applications

A properly applied heat reclaim system automatically responds to make a balanced heat recovery. An example is a reciprocating water chiller with a hot-gas diverting valve and both a water-cooled and an air-cooled condenser. Hot gas from the compressor is rejected to the water-cooled condenser. This hot water provides internal heat as long as it is needed. At a predetermined temperature, the hot gas is diverted to the air-cooled condenser, rejecting excess heat from the total building system. Larger projects with centrifugal compressors use double-condenser chiller units, which are available from many manufacturers. For typical buildings, chillers normally provide hot water for space heating at 105 to 110°F.

Many buildings that run chillers all or most of the year reclaim some of the condenser heat to provide domestic hot water.

Designers should include a source of external heat for backup. The control system should ensure that backup heat is not injected unless all internal heat has been used. For example, if electric backup coils are in series with hot water coils fed from a hot water storage tank, they may automatically start when the system restarts after the building temperature has dropped to a night low-limit setting. An adjustable time delay in the control circuit gives the stored hot water time to warm the building before energizing the electric heat.

This type of heat reclaim system is readily adaptable to smaller projects using a reciprocating chiller with numerous air terminal units or a common multizone air handler. The multizone air handler should have individual zone duct heating coils and controls arranged to prevent simultaneous heating and cooling in the same zone.

Properly applied heat reclaim systems not only meet all space heating needs, but also provide hot water required for showers, food service facilities, and reheat in conjunction with dehumidification cycles.

Heat reclaim chillers or heat pumps should not be used with air-handling systems that have modulating damper economizer control. This free cooling may result in a higher annual operating cost than a minimum fresh air system with a heat reclaim chiller. Careful study shows whether the economizer cycle violates the heat balance concept.

Heat reclaim chillers or heat pumps are available in many sizes and configurations. Combinations include (1) centrifugal, reciprocating, and screw compressors; (2) single- and double-bundle condensers; (3) cascade design for higher temperatures (up to 220°F); and (4) air- or water-cooled, or both.

The designer can make the best selection after both a heating and cooling load calculation and a preliminary economic analysis, and with an understanding of the building, processes, operating patterns, and available energy sources.

The application of heat reclaim chillers or heat pumps ranges from simple systems with few control modes to complex systems having many control modes and incorporating two- or four-pipe circulating systems. Certain systems using double-bundle condensers and single-bundle condensers coupled with exterior closed-circuit coolers have been patented. Potential patent infringements should be checked early in the planning stage.

A successful heat recovery design depends on the performance of the total system, not just the chiller or heat pump. A careful and thorough analysis is often time-consuming and requires more design time than a nonrecovery system. The balanced heat recovery concept should guide all phases of planning and design, and the effects of economic compromise should be studied. There may be little difference between the initial cost (installed cost) of a heat recovery system and a nonrecovery system, especially in larger projects. Also, in view of energy costs, life-cycle analysis usually shows dramatic savings when using balanced heat recovery.

Multiple Buildings

A multiple-building complex is particularly suited to heat recovery. Variations in occupancy and functions provide an abundance of heat sources and uses. Applying the balanced heat concept to a large multibuilding complex can result in large energy savings. Each building captures its own total heat by interchange. Heat rejected from one building could possibly heat adjacent buildings.

BIBLIOGRAPHY

ASHRAE. 1989. Heat recovery. *Technical Data Bulletin* 5(6).

Ashton, G. J., H.R. Cripps, and H.D. Spriggs. 1987. Application of "pinch" technology to the pulp and paper industry. *Tappi Journal* (August):81-85.

Becker, F.E. and A.I. Zakak. 1985. Recovering energy by mechanical vapor recompression. *Chemical Engineering Progress* (July):45-49.

Bose, J.E., J.D. Parks, and F.C. McQuiston. 1985. Design/data manual for closed-loop ground-coupled heat pump systems. ASHRAE.

Cane, R.L.D., S.B. Clernes, and D.A. Forgas. 1994. Heat recovery heat pump operating experiences. *ASHRAE Transactions* 100(2):165-72.

Edison Electric Institute. 1989. *Marketing the industrial heat pump*. Washington, D.C.

Ekroth, I.A. 1979. Thermodynamic evaluation of heat pumps working with high temperatures. Proceedings of the Inter-Society Energy Conversion Engineering Conference 2:1713-19.

Eley Associates. 1992. *Water-loop heat pump systems, Volume 1: Engineering design guide.* Electric Power Research Institute, Palo Alto, CA.

EPRI. 1984. Heat pumps in distillation processes. *Report* No. EM-3656. Electric Power Research Institute, Palo Alto, CA.

EPRI. 1986. Heat pumps in evaporation processes. *Report* No. EM-4693. Electric Power Research Institute, Palo Alto, CA.

EPRI. 1988. Industrial heat pump manual. *Report* No. EM-6057. Electric Power Research Institute, Palo Alto, CA.

EPRI. 1989. Heat pumps in complex heat and power systems. *Report* No. EM-4694. Electric Power Research Institute, Palo Alto, CA.

IEA. 1992. *Heat Pump Centre Newsletter* 10(1). International Energy Agency Heat Pump Centre, Sittard, The Netherlands.

IEA. 1993. *Heat Pump Centre Newsletter* 11(1).

Karp, A. 1988. Alternatives to industrial cogeneration: A pinch technology perspective. Presented at the International Energy Technology Conference (IETC), Houston, TX.

Linnhoff, B. and G.T. Polley. 1988. General process improvement through pinch technology. *Chemical Engineering Progress* (June):51-58.

Mashimo, K. 1992. Heat pumps for industrial process and district heating. Annual Meeting of International Users Club of Absorption Systems, Thisted, Denmark.

Oil and Gas Journal. 1988. Refiners exchange conservation experiences. *Oil and Gas Journal* (May 23).

Phetteplace, G.E. and H.T. Ueda. 1989. Primary effluent as a heat source for heat pumps. *ASHRAE Transactions* 95(1):141-46.

Pucciano, F.J. 1995. Heat recovery in a textile plant. IEA *Heat Pump Centre Newsletter* 13(2):34-36.

Tjoe, T.N. and B. Linnhoff. 1986. Using pinch technology for process retrofit. *Chemical Engineering* (April 28):47-60.

Zimmerman, K.H., ed. 1987. Heat pumps—Prospects in heat pump technology and marketing. Proceedings of the 1987 International Energy Agency Heat Pump Conference, Orlando, FL.

CHAPTER 9

DESIGN OF SMALL FORCED-AIR HEATING AND COOLING SYSTEMS

THIS CHAPTER describes the basics of design and component selection of small forced-air heating and cooling systems, explains their importance, and describes the parametric effects of the system on energy consumption. It also gives an overview of test methods for duct efficiency. This chapter pertains to residential and certain small commercial systems; large commercial systems are beyond the scope of this chapter.

COMPONENTS

Forced-air systems are heating and/or cooling systems that use motor-driven blowers to distribute heated, cooled, and otherwise treated air for the comfort of individuals within confined spaces. A typical residential or small commercial system includes (1) a heating and/or cooling unit, (2) accessory equipment, (3) supply and return ductwork, (4) supply and return registers and grilles, and (5) controls. These components are described briefly in the following sections and are illustrated in Figure 1.

Heating and Cooling Units

Three types of forced-air heating and cooling devices are (1) furnaces, (2) air conditioners, and (3) heat pumps.

Furnaces are the basic component of most forced-air heating systems. They are augmented with an air-conditioning coil when cooling is included, and they are manufactured to use specific fuels such as oil, natural gas, or liquefied petroleum gas. The fuel used dictates installation requirements and safety considerations (see Chapter 28).

The common air-conditioning system uses a split configuration with an air handling unit, such as a furnace. In this application, the air-conditioning evaporator coil (indoor unit) is installed on the discharge air side of the air handler. The compressor and condensing coil (outdoor unit) are located outside the structure, and refrigerant lines connect the outdoor and indoor units.

Self-contained air conditioners are another type of forced-air system. These units contain all necessary air-conditioning components, including circulating air blowers, and may or may not include fuel-fired heat exchangers or electric heating elements.

The heat pump cools and heats using the refrigeration cycle. It is available in split and packaged (self-contained) configurations. Generally, the air-source heat pump requires supplemental heating; therefore, electric heating elements are usually included with the heat pump as part of the forced-air system. Heat pumps are also combined with fossil fuel furnaces to take advantage of their high efficiency at mild temperatures as a means to minimize heating cost.

Accessory Equipment

Forced-air systems may be equipped to humidify and dehumidify the indoor environment, remove contaminants from recirculated air, and provide for circulation of outside air in an economizer operation. The following accessories affect airflow and pressure requirements. Losses must be taken into account when selecting the heating and cooling equipment and sizing ductwork.

Humidifiers. Several types of humidifiers are available, including self-contained steam, atomizing, evaporative, and heated pan. Chapter 20 describes the units in detail. The Air-Conditioning and Refrigeration Institute (ARI) rates various humidifiers in ARI *Standard* 610.

Humidifiers must match the heating unit. Discharge air temperatures on heating systems vary, and some humidifiers do not provide their own heat source for humidification. These humidifiers should be applied with caution to heat pumps and other heating units with low air temperature rise.

Structures with complete vapor retarders (walls, ceilings, and floors) normally require no supplemental moisture during the heating season because internally generated moisture maintains an acceptable relative humidity of 20 to 60%.

Electronic air cleaners. These units attract oppositely charged particles, fine dust, smoke, and other particulates to collecting plates in the air cleaner. While these plates remove finer particles, larger particles are trapped in an ordinary throwaway or permanent filter. A nearly constant pressure drop and efficiency can be expected unless the cleaner becomes severely loaded with dust.

Custom accessories. Solar, off-peak storage, and other custom systems are not covered in this chapter. However, their components may be classified as duct system accessories.

Heat recovery ventilators. These devices provide ventilation air to the conditioned space and recover heat from the air being exhausted outdoors. They can be operated as stand-alone devices but are more commonly installed with forced-air distribution.

Economizer control. This device monitors outdoor temperature and humidity and automatically shuts down the air-conditioning unit when a preset outdoor condition is met. Damper motors open outdoor return air dampers, letting outside air enter the system to provide comfort cooling. When outdoor air conditions are no longer acceptable, the outdoor air dampers close and the air-conditioning unit comes back on.

Ducts

In small commercial applications and residential work, ductwork design depends on the air-moving characteristics of the blower included with the selected equipment. It is important to recognize this difference between small commercial or residential systems and large commercial and industrial systems. The designer of smaller systems must determine resistances to the movement of air and adjust the duct sizes to limit the static pressure against which the blower operates. Manufacturers publish static pressure versus flow rate information so the designer can determine the maximum static pressure against which the blower will operate while delivering the proper volume of air (see Chapter 18).

The preparation of this chapter is assigned to TC 6.3, Central Forced Air Heating and Cooling Systems.

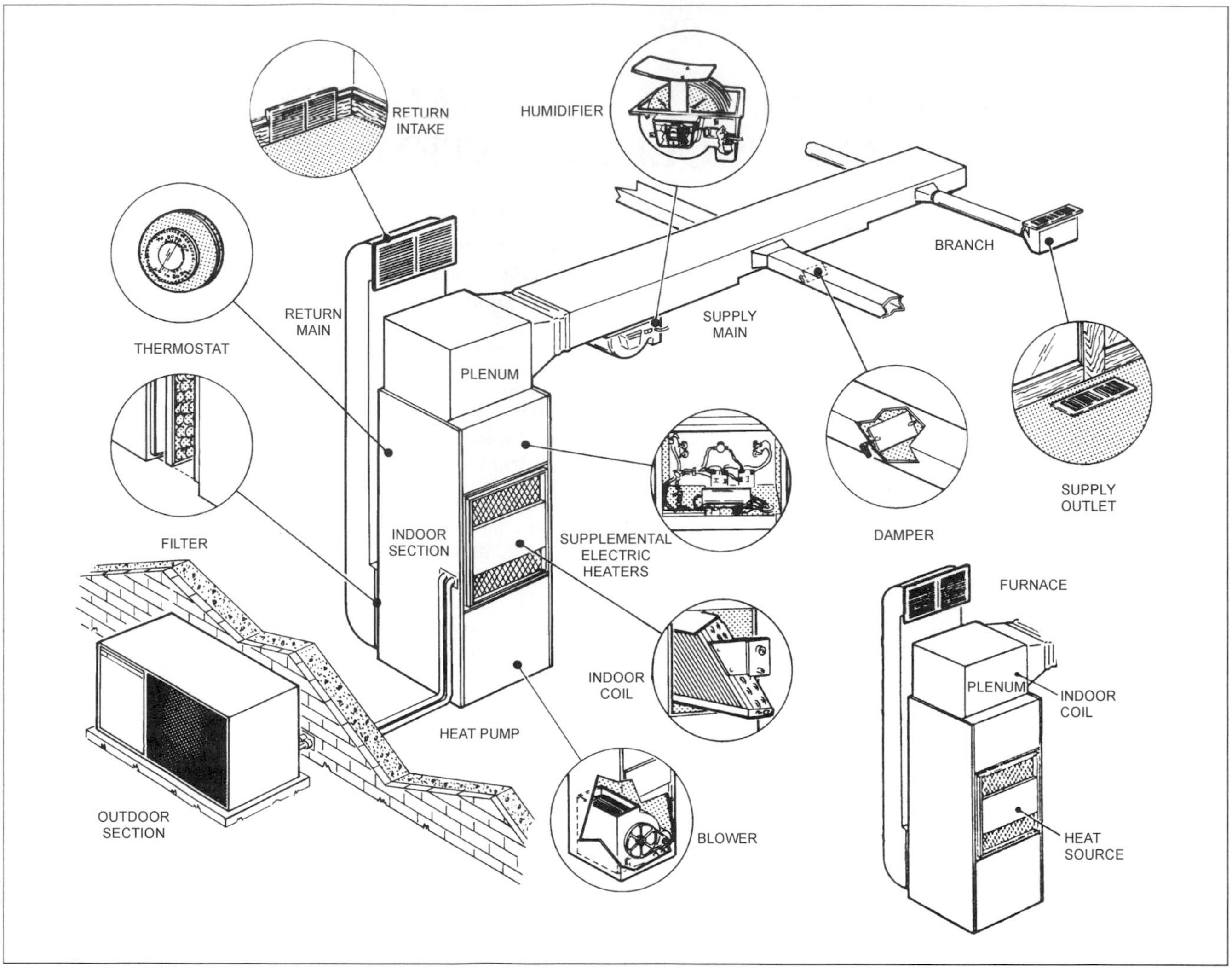

Fig. 1 Heating and Cooling Components

Supply and Return Registers and Grilles

Supply air should be directed to the sources of greatest heat loss and/or heat gain to offset their effects. Registers and grilles for the supply should accommodate all aspects of the air distribution patterns such as throw, spread, and drop. The register and grille velocities must be held within recommended limits, as noise generated is of equal or greater importance than duct noise.

Controls

Forced-air heating and/or cooling systems may be adequately controlled in several ways. Simple on/off cycling of central equipment is frequently adequate to maintain comfort. Spaces with large load variations may require zone control or multiple units with separate ductwork. Systems with minimal load variations in the space may function adequately with one central wall thermostat or a return air thermostat. Residential conditioning systems of 60,000 Btu/h capacity or less are typically operated with one central thermostat.

Forced-air control may require several devices, depending on complexity of the system and the accessories used (see Chapter 45 of the 1999 *ASHRAE Handbook—Applications*). Energy conservation has increased the importance of control, so methods that were once considered too expensive for small systems may now be cost-effective.

Temperature control, the primary consideration in forced-air systems, may be accomplished by a single-stage thermostat. When properly located, and in some cases with correctly adjusted heat anticipators, this device accurately controls temperature. Multistage thermostats are required on many systems (e.g., a heat pump with auxiliary heating) and may improve temperature regulation. Outdoor thermostats, in series with indoor control, can stage heating increments adequately.

Indoor thermostats may incorporate many control capabilities within one device, including continuous or automatic fan control and automatic or manual changeover between heating and cooling. Where more than one system conditions a common space, manual control is preferred to prevent simultaneous heating and cooling.

Thermostats with programmable temperature control allow the occupant to vary the temperature set point for different periods. These devices save substantial energy by applying automatic night setback and/or daytime temperature setback for all systems when appropriately matched.

Two-speed fan control may be desirable for fossil-fueled equipment, but it should not be applied to heat pumps unless recommended by the manufacturer. Humidistats should be specified for humidifier control. When applying an unusual control, manufacturer recommendations should be followed to prevent equipment damage or misuse.

DESIGN

The size and performance characteristics of components are interrelated, and the overall design should proceed in the organized manner described. For example, furnace selection depends on heat gain and loss; however, sizing is also affected by duct location (attic, basement, and so forth), duct materials, night setback, and humidifier use. Here is a recommended procedure:

1. Estimate heating and cooling loads including target values for duct losses.
2. Determine preliminary ductwork location and materials.
3. Determine heating and cooling unit location.
4. Select accessory equipment. Accessory equipment is not generally provided with initial construction; however, the system may be designed for the later addition of these components.
5. Select control components.
6. Determine maximum airflow (cooling or heating) for each supply and return location.
7. Determine airflow at reduced heating and cooling loads (two-speed and variable-speed fan).
8. Select heating/cooling equipment.
9. Select control system.
10. Finalize duct design and size.
11. Select supply and return grilles.
12. When the duct system is in place, measure the duct leakage and compare the results with the target values used in step 1.

This procedure requires certain preliminary information such as location, weather conditions, and architectural considerations. The following sections cover the preliminary considerations and discuss how to follow this recommended procedure.

Locating Outlets, Ducts, and Equipment

The characteristics of a residence determine the appropriate type of forced-air system and where it can be installed. The presence or absence of particular areas in a residence has a direct influence on equipment and duct location. The structure's size, room or area use, and air distribution system determine how many central systems will be needed to maintain comfort temperatures in all areas.

For maximum energy efficiency, ductwork should be installed within the conditioned space. ASHRAE *Standard* 90.2 gives a credit for installation in this location. The next best location for ductwork is in a full basement, which is also a good place to install equipment. If a residence has a crawl space, the ductwork and equipment can be located there, or the equipment can be placed in a closet or utility room. The equipment's enclosure must meet all fire and safety code requirements; adequate service clearance must also be provided. In a home built on a concrete slab, equipment could be located in the conditioned space (for systems that do not require combustion air), in an unconditioned closet, in an attached garage, in the attic space, or outdoors. The ductwork normally is located in a furred-out space, in the slab, or in the attic.

Duct construction must conform to local code requirements, which often reference NFPA *Standard* 90B or the Residential Comfort Systems Installation Standards Manual (SMACNA 1998).

Weather should be considered when locating equipment and ductwork. Packaged outdoor units for houses in severely cold climates must be installed according to manufacturer recommendations. Most houses in cold climates have basements, making them well suited for indoor furnaces and split-system air conditioners or heat pumps. In mild and moderate climates, the ductwork is frequently in the attic or crawl space. Ductwork located outdoors, in attics, in crawl spaces, and in basements must be insulated, as outlined in Chapter 23 of the 1997 *ASHRAE Handbook—Fundamentals*. All duct sections should be properly connected and all corrections and seams properly sealed to minimize duct leakage.

Although the principles of air distribution discussed in Chapter 31 of the 1997 *ASHRAE Handbook—Fundamentals* apply in forced-air system design, simplified methods of selecting outlet size and location are generally used.

Supply outlets fall into four general groups, defined by their air discharge patterns: (1) horizontal high, (2) vertical nonspreading, (3) vertical spreading, and (4) horizontal low. Straub and Chen (1957) and Wright et al. (1963) describe these types and their performance characteristics under controlled laboratory and actual residence conditions. Table 1 lists the general characteristics of supply outlets. It includes the performance of various outlet types for cooling as well as heating, since one of the advantages of forced-air systems is that they may be used for both heating and cooling. However, as indicated in Table 1, no single outlet type is best for both heating and cooling.

The best outlets for heating are located near the floor at outside walls and provide a vertical spreading air jet, preferably under windows, to blanket cold areas and counteract cold drafts. Called perimeter heating, this arrangement mixes the warm supply air with both the cool air from the area of high heat loss and the cold air from infiltration, preventing drafts.

The best outlet types for cooling are located in the ceiling and have a horizontal air discharge pattern. For year-round systems, supply outlets are located to satisfy the more critical load.

Figure 2 illustrates preferred return locations for different supply outlet positions and system functions and typical temperature profiles. These return locations are based on the presence of the stagnant layer in a room, which is beyond the influence of the supply outlet and thus experiences little air motion (e.g., cigarette smoke "hanging" in a spot in a room is evidence of a stagnant region). The stagnant layer degrades room comfort.

The stagnant layer develops near the floor during heating and near the ceiling during cooling. Returns help remove air from this region if the return face is placed within the stagnant zone. Thus, for heating, returns should be placed low; for cooling, returns should be placed high. If a central return is used, the airflow between supply registers and the return should not be impeded even when interior doors are closed.

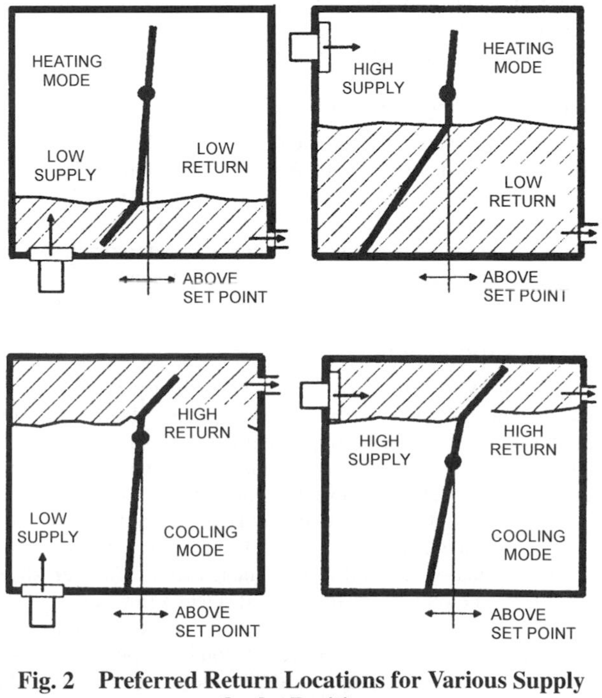

Fig. 2 Preferred Return Locations for Various Supply Outlet Positions

Table 1 General Characteristics of Supply Outlets

Group	Outlet Type	Outlet Flow Pattern	Conditioning Mode	Most Effective Application	Selection Criteria (see Figure 2)
1	Ceiling and high sidewall	Horizontal	Cooling	*Ceiling outlets*	
				Full circle or widespread type	Select for throw equal to distance from outlet to nearest wall at design flow rate and pressure limitations.
				Narrow spread type	Select for throw equal to 0.75 to 1.2 times distance from outlet to nearest wall at design flow rate and pressure limitations.
				Two adjacent ceiling outlets	Select each so that throw is about 0.5 times distance between them at design flow rate and pressure limits.
				High sidewall outlets	Select for throw equal to 0.75 to 1.2 times distance to nearest wall at design flow rate and pressure limits. If pressure drop is excessive, use several smaller outlets rather than one large one to reduce pressure drop.
2	Floor diffusers, baseboard, and low sidewall	Vertical, nonspreading	Cooling and heating		Select for 6 to 8 ft throw at design flow rate and pressure limitations.
3	Floor diffusers, baseboard, and low sidewall	Vertical, spreading	Heating and cooling		Select for 4 to 6 ft throw at design flow rate and pressure limitations.
4	Baseboard, and low sidewall	Horizontal	Heating only		Limit face velocity to 300 fpm.

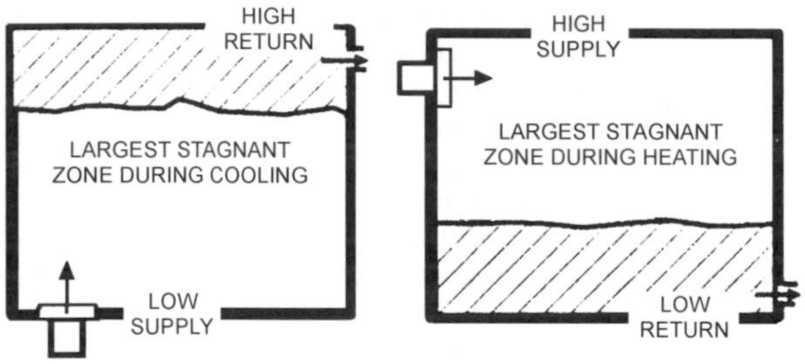

Fig. 3 Best Compromise Return Locations for Year-Round Heating and Cooling

In a year-round heating and cooling application, a compromise must be made by placing returns where the largest stagnant zone develops (Figure 3). With low supply outlets, the largest stagnant zone develops during cooling, so returns should be placed high or opposite the supply locations. Conversely, high supply outlets do not perform as well during heating; therefore, returns should be placed low to be of maximum benefit.

Determining Heating and Cooling Loads

Design heating and cooling loads can be calculated by following the procedures outlined in Chapters 27, 28, and 29 of the 1997 *ASHRAE Handbook—Fundamentals* or by using ACCA *Manual* J. When calculating design loads, heat losses or gains from the air distribution system must be included in the total load for each room. In residential applications, local codes often require outdoor air ventilation, which is added to the building load.

Selecting Equipment

Furnace heating output should match or slightly exceed the estimated design load. A 40% limit on oversizing has been recommended by the Air Conditioning Contractors of America (ACCA) for fossil fuel furnaces. This limit minimizes venting problems associated with oversized equipment and improves part-load performance. Note that the calculated load must include duct loss, humidification load, and night setback recovery load, as well as building conduction and infiltration heat losses. Chapter 28 has detailed information on how to size and select a furnace.

To help conserve energy, manufacturers have added features to improve furnace efficiency. Electric ignition has replaced the standing pilot; vent dampers and more efficient motors are also available. Furnaces with fan-assisted combustion systems (FACSs) and condensing furnaces also improve efficiency. Two-stage heating and cooling, variable-speed heat pumps, and two-speed and variable-speed blowers are also available.

A system that is designed to both heat and cool and that cycles the cooling equipment on and off by sensing dry-bulb temperature alone should be sized to match the design heat gain as closely as possible. Oversizing under this control strategy could lead to higher than desired indoor humidity levels. Chapter 27 of the 1997 *ASHRAE Handbook—Fundamentals* recommends that cooling units not be oversized by more than 25% of the sensible load. Other sources suggest limiting oversizing to 15% of the sensible load if it is not an air-source heat pump application.

Airflow Requirements

After the equipment is selected and prior to duct design, the following decisions must be made:

- Determine the air quantities required for each room or space during heating and cooling based on each room's heat loss or heat gain. The air quantity selected should be the greater of the heating or cooling requirement.
- Determine the number of supply outlets needed for each space to supply the selected air quantity, considering discharge velocity, spread, throw, terminal velocity, occupancy patterns, location of heat gain and heat loss sources, and register or diffuser design.
- Determine the type of return (multiple or central), availability of space for the grilles, filtering, maximum velocity limits for sound, efficient filtration velocity, and space use limitations.

Duct Configuration

The next major decision is to select a generic duct system. In order of decreasing efficiency, the three main types are

- Ducts within the conditioned space
- Minimum-area ductwork
- Traditional designs

In the first option, all ductwork is kept in the conditioned space. Generally this involves the use of trunks and branches located between floors of a two-story house or along the wall-ceiling intersections in a single-story dwelling. It is important that the duct does not contact the outside, as could happen, for example, if the space between floors were used as a plenum.

The second option applies to buildings whose envelopes are well insulated or have sufficient thermal integrity so that the supply registers do not have to be located next to exterior walls. Placing registers in the interior walls can reduce duct surface area by 50% or more, with similar reductions in leakage and conductive losses.

Traditional designs place ductwork in unconditioned spaces and supply registers in or near exterior walls. Because this is the least efficient option overall, particular care should be taken to seal and insulate the ductwork.

Duct Materials

Duct materials affect both thermal and mechanical performance. A typical installation may use combinations of sheet metal, fiberglass duct board, and flexible ducting. Chapter 16 briefly discusses some of the materials used for ducts. Information manuals and design guides are available from the Air Conditioning Contractors of America (ACCA), Air Diffusion Council (ADC), National Association of Home Builders (NAHB), North American Insulation Manufacturers Association (NAIMA), and Sheet Metal and Air Conditioning Contractors National Association (SMACNA). Installation codes should also be reviewed for accepted duct materials and installation practices.

Distribution Design

The required airflow and the static pressure limitation of the blower are the parameters around which the duct system is designed. The heat loss or gain for each space determines the proportion of the total airflow supplied to each space. The static pressure drop in supply registers should be limited to about 0.03 in. of water. The required pressure drop must be deducted from the static pressure available for duct design.

The flow delivered by a single supply outlet should be determined by considering (1) the space limitations on the number of registers that can be installed, (2) the pressure drop for the register at the flow rate selected, (3) the adequacy of the air delivery patterns for offsetting heat loss or gain, and (4) the space use pattern.

Manufacturer's specifications include blower airflow for each blower speed and external static pressure combination. Determining static pressure available for duct design should include the possibility of adding accessories in the future (e.g., electronic air cleaners or humidifiers). Therefore, the highest available fan speed should not be used for design.

For systems that heat only, the blower rate may be determined from the manufacturer's data. The temperature rise of air passing through the heat exchanger of a fossil fuel furnace must be within the manufacturer's recommended range (usually from 40 to 80°F). The possible later addition of cooling should also be considered by selecting a blower that operates in the midrange of the fan speed and settings.

For cooling only, or for heating and cooling, the design flow can be estimated by the following equation:

$$Q = \frac{q_s}{60\rho c_p \Delta t} = \frac{q_s}{1.1\Delta t} \qquad (1)$$

where

Q = flow rate, cfm
ρ = air density assumed to equal 0.075 lb/ft^3
c_p = specific heat of air = 0.24 Btu/lb·°F
q_s = sensible load, Btu/h
Δt = dry-bulb temperature difference between air entering and leaving equipment, °F

For preliminary design, an approximate Δt is as follows:

Sensible Heat Ratio (SHR)	Δt, °F
0.75 to 0.79	21
0.80 to 0.85	19
0.85 to 0.90	17

SHR = Calculated sensible load/Calculated total load

For example, if a calculation indicates the sensible load is 23,000 Btu/h and the latent load is 4900 Btu/h, the SHR is calculated as follows:

$$\text{SHR} = 23{,}000/(23{,}000 + 4900) = 0.82$$

and

$$Q = \frac{23{,}000}{1.1 \times 19} = 1100 \text{ cfm}$$

This value is the estimated design flow. The exact design flow can only be determined after the cooling unit is selected. The unit that is ultimately selected should supply an airflow that is in the range of the estimated flow, and it must also have adequate sensible and latent cooling capacity when operating at design conditions.

Duct Design Recommendations

Because of the competitive nature of the residential construction business and the practical necessity of using less sophisticated design techniques for these systems, the modified equal friction duct loss calculation method is recommended. This method is satisfactory as long as the designer understands its strengths and weaknesses. It should be applied only to those systems requiring a normal air-moving capacity of less than 2250 cfm or a cooling capacity of less than 60,000 Btu/h.

With the equal friction method, the pressure available for supply and return duct losses is found by deducting coil, filter, grille, and accessory losses from the manufacturer's specified blower pressure. The remaining pressure is divided between the supply and return as shown in Table 2.

Using duct calculators and the friction chart (Figure 9 in Chapter 32 of the 1997 *ASHRAE Handbook—Fundamentals*) simplifies the calculations. Branch ducts are sized to balance the available pressure without exceeding recommended maximum velocities. Low-resistance duct runs may require dampers. The velocity reduction method provides similar results, but without the need to check on resulting duct velocities. Hand calculators and computer programs simplify the number of manual calculations required.

The ductwork distributes the air to spaces according to the space heating and/or cooling requirements. The return air system may be single, multiple, or any combination that returns air to the equipment within design static pressure and with satisfactory air movement patterns.

Some general rules in duct design are as follows:

- Keep main ducts as straight as possible.
- Streamline transitions.
- Design elbows with an inside radius of at least one-third the duct width. If this inside radius is not possible, include turning vanes.
- Make ducts tight and sealed to limit air loss.
- Insulate and/or line ducts, where necessary, to conserve energy and limit noise.
- Locate branch duct takeoffs at least 4 ft downstream from a fan or transition, if possible.
- Isolate the air-moving equipment from the duct using flexible connectors to isolate noise.

Table 2 Recommended Division of Duct Pressure Loss

	System Characteristics	Supply, %	Return, %
A	Single return at blower	90	10
B	Single return at or near equipment	80	20
C	Single return with appreciable return duct run	70	30
D	Multiple return with moderate return duct system	60	40
E	Multiple return with extensive return duct system	50	50

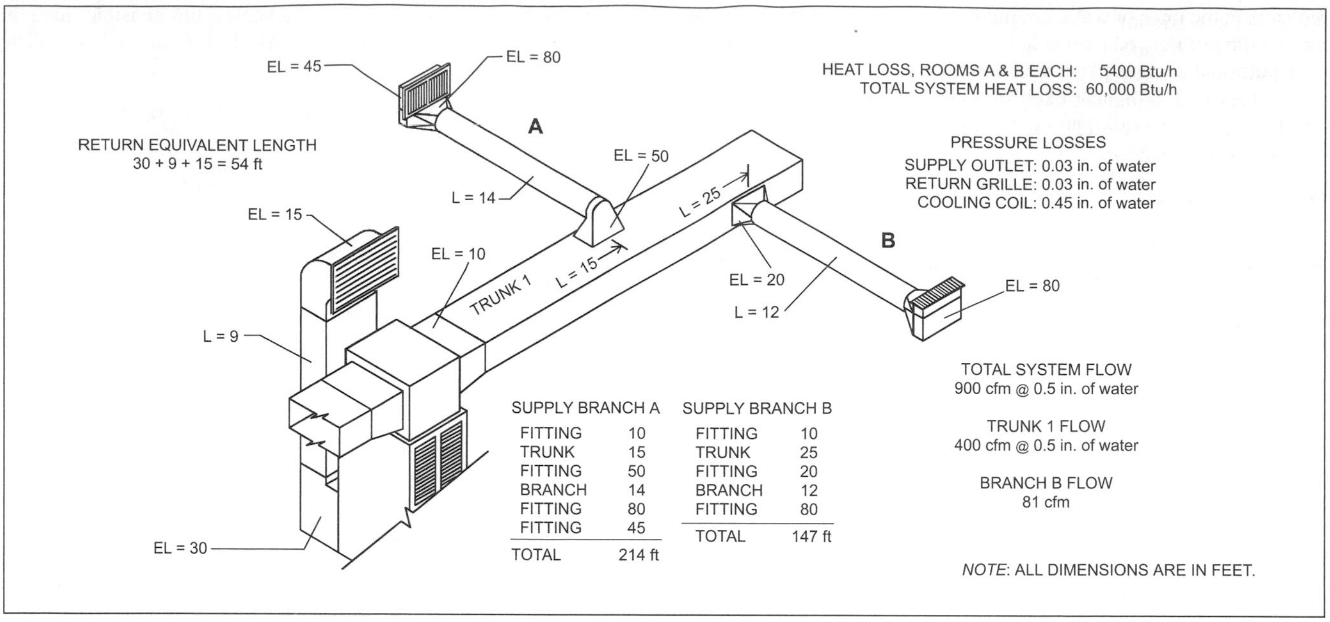

Fig. 4 Equal Friction Duct Design Example

Large air distribution systems are designed to meet specific noise criteria (NC) decibel levels. Small systems are usually kept quiet by limiting airflow velocities in mains and branches to the following:

Main ducts	700 to 900 fpm
Branch ducts	600 fpm
Branch risers	500 fpm

Considerable difference may exist between the cooling and heating flow requirements. Because many systems cannot be rebalanced seasonally, a compromise must be made in the duct design to accommodate the most critical need. For example, a kitchen may require 165 cfm for cooling but only 65 cfm for heating. Because the kitchen may be used heavily during design cooling periods, the cooling flow rate should be used. Normally, the maximum design flow should be used, as register dampers do allow some optional reduction in airflows.

Duct Design Procedures—Equal Friction

Refer to Figure 4 for numerical example.

1. Determine the heating and cooling load to be supplied by each outlet using the procedure outlined in the previous sections or in ACCA *Manual* J. Include duct loss or gain.
2. Make a simple diagram of the supply and return duct systems—at least one outlet in a room or area for each 8000 Btu/h loss or 4000 Btu/h sensible heat gain, whichever requires the greater number of outlets.
3. Label all fittings and transitions to show equivalent lengths on the drawing. (Refer to Figure 7 through Figure 14 for approximate equivalent lengths of various fittings. ACCA *Manual* D includes more fittings.)
4. Show measured lengths of ductwork on the drawing.
5. Determine the total effective length of each branch supply. (Begin at the air handler, and add all equivalent lengths and measured lengths to each outlet: Effective length = Equivalent length + Measured length.)

   Example: Branch B in Figure 4 is 147 ft.

6. Proportion the total airflow to each supply outlet for both heating and cooling.

   Example: Supply outlet flow =

$$\frac{\text{Outlet heat loss (gain)}}{\text{Total heat loss (gain)}} \times \text{Total system flow}$$

   Branch B in Figure 4 is $(5400/60{,}000) \times 900 = 81$ cfm

7. The flow required for each supply outlet equals the heating or cooling flow requirement (normally the larger of the two rates).
8. Label the supply outlet flow requirement on the drawing for each outlet.
9. Determine the total external static pressure available from the unit at the selected airflow.
10. Subtract the supply and return register pressure, external coil pressure, filter pressure, and box plenum (if used) from the available static pressure to determine the static pressure available for the duct design.

   Example: Figure 4

Equipment	0.50 in. water
Cooling coil	–0.24 in. water
Supply outlet	–0.03 in. water
Return grille	–0.03 in. water
Available static pressure	0.20 in. water

11. Proportion the available static pressure between the supply and return.

   Example:

Supply (75%)	0.15 in. water
Return (25%)	0.05 in. water
Total	0.20 in. water

Sizing the Branch Supply Air Duct

12. Use the supply static pressure available to calculate each branch design static pressure for 100 ft of equivalent length.

   Example: Branch B design static pressure =

$$\frac{100\text{ (Supply static pressure available)}}{\text{Effective length of each branch supply}}$$

$$100 \times 0.15/147 = 0.102 \text{ in. of water}/100 \text{ ft (Branch B)}$$

13. Enter the friction chart (Figure 9 in Chapter 32 of the 1997 *ASHRAE Handbook—Fundamentals*) at the branch design static pressure (0.102) opposite the flow rate for each supply, and read the round duct size (6 in.) and velocity (575 fpm).
14. If velocity exceeds the maximum recommended value, increase the size and specify a branch damper.
15. Convert the round duct to rectangular, where needed.

Sizing the Supply Trunk

16. Determine the branch supply with the longest total effective length, and from this determine the static pressure to size the supply trunk duct.

Example: Supply trunk design static pressure =

$$\frac{100 \text{ (Total supply static pressure available)}}{\text{Longest effective length of branch duct supplies}}$$

$$0.15 \times 100/214 \text{ (Branch A)} = 0.070 \text{ in. water}/100 \text{ ft}$$

17. Total the heating airflow and the cooling airflow for each trunk duct section. Select the larger of the two flows for each section of duct between branches or groups of branches.
18. Design each supply trunk duct section by entering the friction chart at the supply trunk static pressure (0.070) and sizing each trunk section for the airflow handled by that section of duct. (Trunk 1—400 cfm; duct size = 10.25 in. at 680 fpm.)

    Trunks should be checked for size after each branch and reduced, as required, to maintain velocity above branch duct design velocity.
19. Convert round duct size to rectangular, where needed.

Sizing the Return Air Duct

20. Select the number of return air openings to be used.
21. Determine flow rate of air returned by each return air opening.
22. From Step 11, select the return air static pressure.
23. Determine the static pressure available per 100 ft effective length for each return run, and design the same as for the supply trunk system.

Example: Supply trunk design static pressure =

$$\frac{100 \text{ (Total return static pressure available)}}{\text{Longest effective length of return duct runs}}$$

$$0.05 \times 100/54 = 0.093 \text{ in. water}/100 \text{ ft}$$

24. The trunk design for the return air is the same as the trunk design for the supply system.

Example: Return duct in Figure 4 is as follows—enter friction chart at 900 cfm and 0.093 in. of water. Duct size is 13.1 in. The velocity, however, is too high at 950 fpm, so use a larger duct.

Sizing Using Velocity Reduction

The velocity reduction method assigns duct velocities throughout the system in decreasing order from furnace to farthest branch. The duct sizes needed to achieve these assigned values are determined using the friction chart, which also provides the pressure drop in each run of duct. This method is fast, and noise control is relatively assured.

Example:

Mains sized for 800 fpm velocity
Branches sized for 400 fpm velocity
Total airflow: 1200 cfm
Step 1: Determine equivalent length of fittings (listed in Figure 5).
Step 2: Measure actual length of duct in mains and branches (listed in Figure 5).
Step 3: Determine duct size and pressure loss using friction chart.
Step 4: Compute duct static loss in longest run (Figure 6).
The loss through the supply outlet and return grille must be added to the duct loss to determine the design static pressure at 1200 cfm.

Zone Control for Small Systems

In residential applications, some complaints about rooms that are too cold or too hot are related to the limitations of the system. No matter how carefully a single-zone system is designed, problems will occur if the control is unable to accommodate the various load conditions that occur simultaneously throughout the house at any time of day and/or during any season.

A single-zone control works as long as the various rooms are open to each other. In this case, room-to-room temperature differences are minimized by the convection currents between the rooms. For small rooms, an open door is adequate. For large rooms, openings in partitions should be large enough to ensure adequate air interchange for a single-zone control.

When rooms are isolated from each other, temperature differences cannot be moderated by convection currents, and the conditions in the room with the thermostat may not be representative of the conditions in the other rooms. In this situation, comfort can be

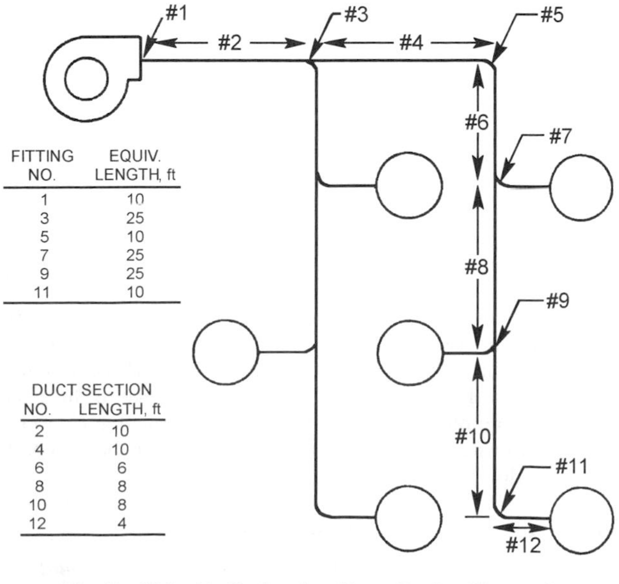

FITTING NO.	EQUIV. LENGTH, ft
1	10
3	25
5	10
7	25
9	25
11	10

DUCT SECTION NO.	LENGTH, ft
2	10
4	10
6	6
8	8
10	8
12	4

Fig. 5 Velocity Reduction Duct Design Example

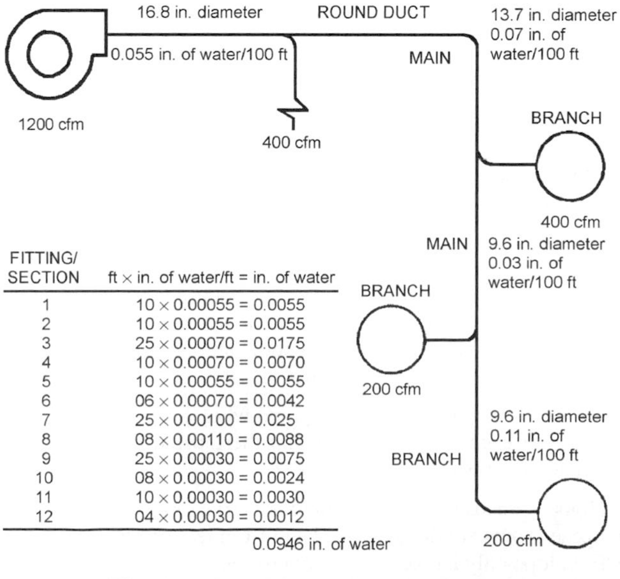

FITTING/ SECTION	ft × in. of water/ft = in. of water
1	10 × 0.00055 = 0.0055
2	10 × 0.00055 = 0.0055
3	25 × 0.00070 = 0.0175
4	10 × 0.00070 = 0.0070
5	10 × 0.00055 = 0.0055
6	06 × 0.00070 = 0.0042
7	25 × 0.00100 = 0.025
8	08 × 0.00110 = 0.0088
9	25 × 0.00030 = 0.0075
10	08 × 0.00030 = 0.0024
11	10 × 0.00030 = 0.0030
12	04 × 0.00030 = 0.0012
	0.0946 in. of water

Fig. 6 Example Duct Pressure Losses

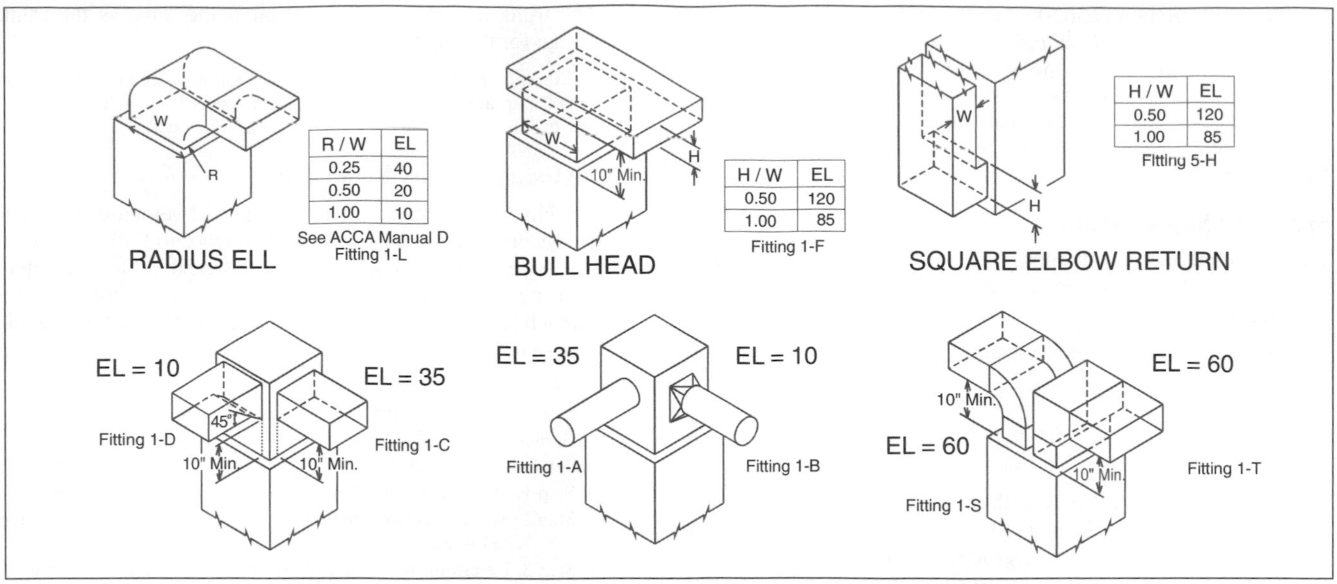

Fig. 7 Equivalent Length (EL in Feet) of Supply and Return Air Plenum Fittings (ACCA 1995)

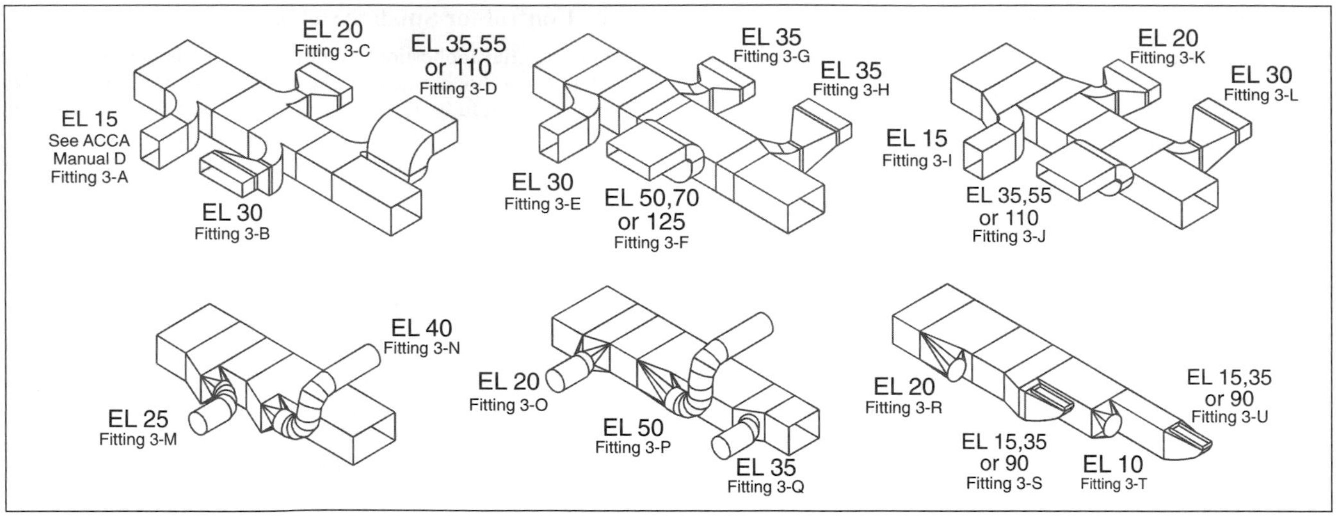

Fig. 8 Equivalent Length (EL in Feet) of Reducing Trunk Duct Fittings (ACCA 1995)
Note: No adjustment is made for downstream location of fittings.

improved by continuous blower operation, but this strategy may not completely solve the problem.

Zone control is required when the conditions at the thermostat are not representative of all the rooms. This situation will almost certainly occur if any of the following conditions exists:

- House has more than one level
- One or more rooms are used for entertaining large groups
- One or more rooms have large glass areas
- House has an indoor swimming pool and/or hot tub
- House has a solarium or atrium

In addition, zoning may be required when several rooms are isolated from each other and from the thermostat. This situation is likely to occur when

- House spreads out in many directions (wings)
- Some rooms are distinctly isolated from rest of house
- Envelope only has one or two exposures
- House has a room or rooms in a finished basement
- House has a room or rooms in a finished attic
- House has one or more rooms with slab or exposed floor

Zone control can be achieved by installing

- Discrete heating/cooling ducts for each zone requiring control
- Automatic zone damper in a single heating/cooling duct system

The rate of airflow delivered to each room must be able to offset the peak room load during cooling. The peak room load can be determined using Chapter 27 of the 1997 *ASHRAE Handbook—Fundamentals*. Then the same supply air temperature difference used to size equipment can be substituted into Equation (1) to find airflow. The design flow rate for any zone is equal to the sum of the peak room flow rates assigned to a zone.

Duct Sizing for Zone Damper Systems

The following guidelines are proposed in ACCA *Manual* D to size the various duct runs.

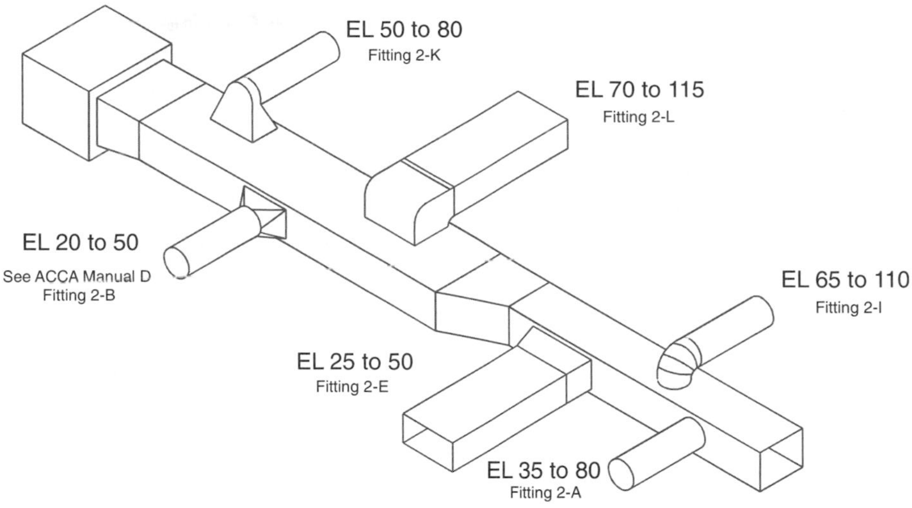

Fig. 9 Equivalent Length (EL in Feet) of Extended Plenum Fittings (ACCA 1995)
Note: No adjustment is made for downstream location of fittings.

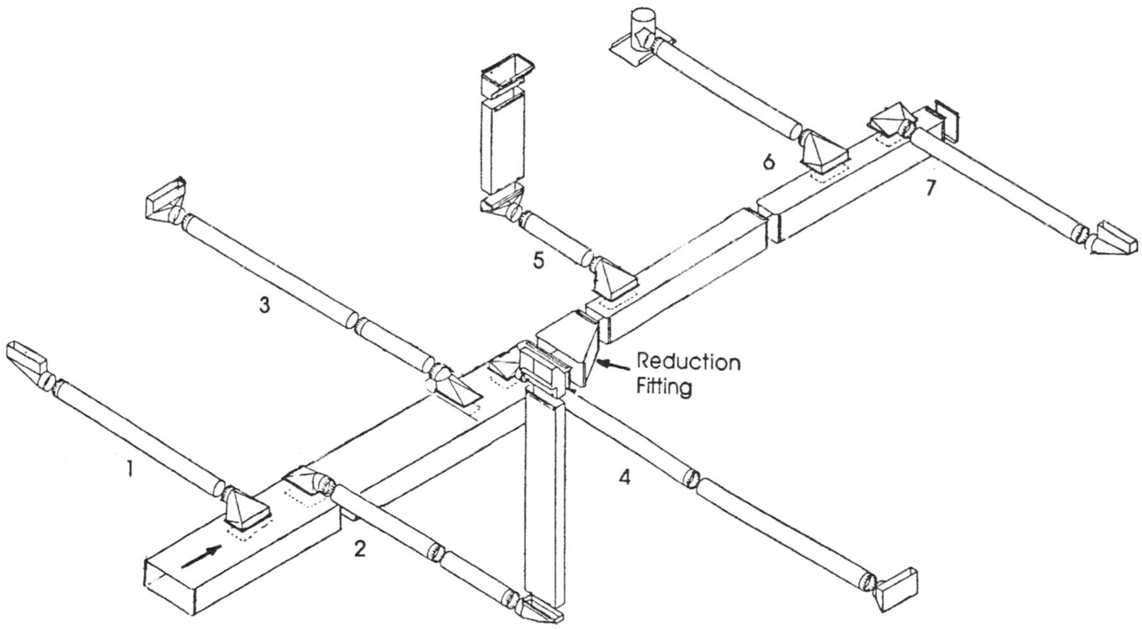

This pressure loss (extended in equivalent length) of a fitting in a branch takeoff of an extended plenum depends on the location of the branch in the plenum. Takeoff branches nearest the air handler impose a higher loss, and those downstream, a lower loss. The lowest equivalent lengths listed in Figure 9 are based on the assumption that the takeoff is the last (or only) branch in the plenum.

To correct these values, add velocity factor increments of 10 ft to each upstream branch, depending on the number of branches remaining downstream. For example, if two branches are downstream from a takeoff, add 2 ×10 = 20 ft of equivalent length to the takeoff loss listed.

In addition, consider a long extended plenum that has a reduction in width as two (or more) separate plenums—one before the reduction and one after. The loss for each top takeoff fitting in this example duct system is shown in the following table.

Branch Number	Takeoff Loss, ft	Downstream Branches	Velocity Factor, ft	Design Equiv. Length, ft
Before plenum reduction				
1	50	3	3 × 10 = 30	80
2	50	2	2 × 10 = 20	70
3	50	1	1 × 10 = 10	60
4	50	0	0	50
After plenum reduction				
5	50	2	2 × 10 = 20	70
6	50	1	1 × 10 = 10	60
7	50	0	0	50

Fig. 10 Example Duct System and Equivalent Length and Extended Plenum Fittings

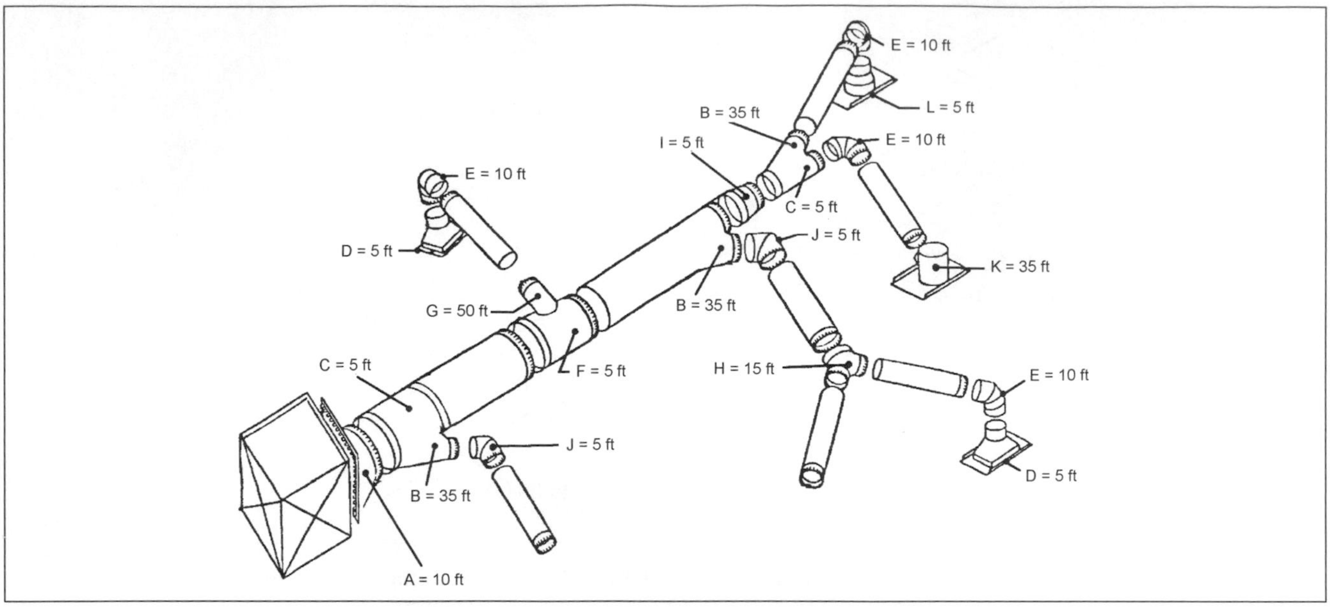

Fig. 11 Equivalent Length of Round Supply System Fittings

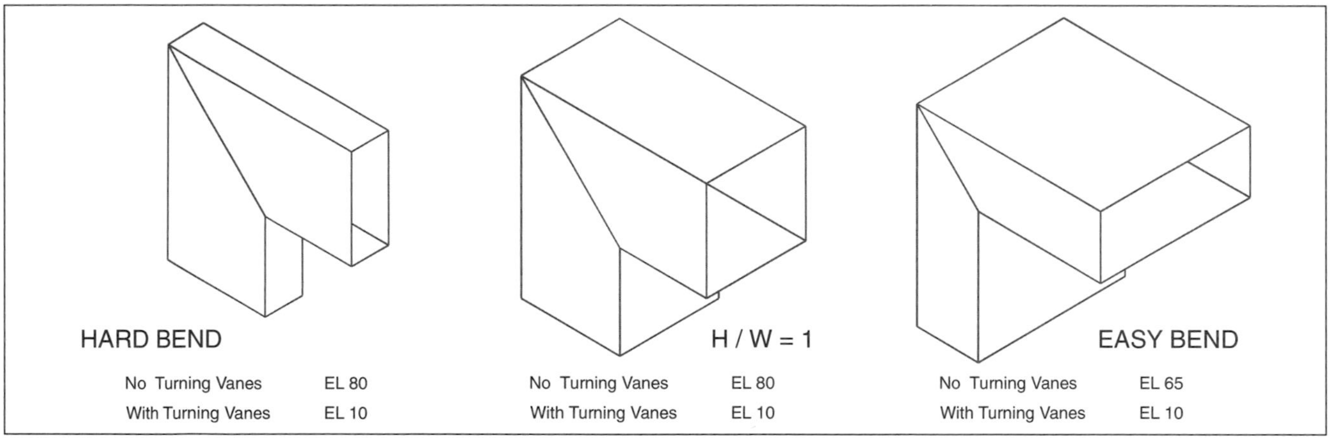

Fig. 12 Equivalent Length (EL in Feet) of Angles and Elbows for Trunk Ducts
(ACCA 1995)

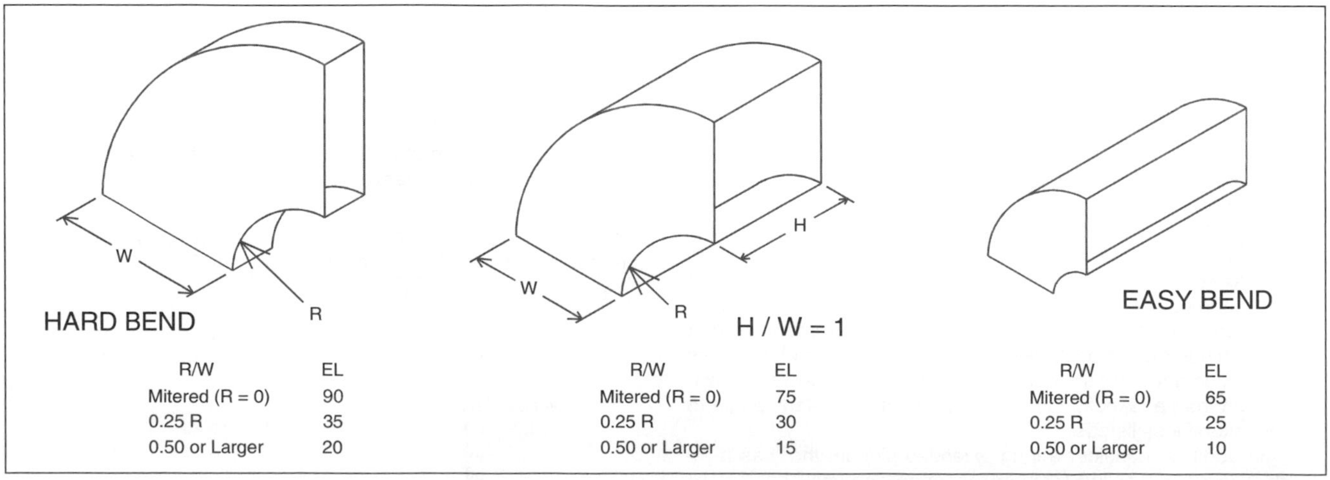

Fig. 13 Equivalent Length (EL in Feet) of Angles and Elbows for Branch Ducts
(ACCA 1995)

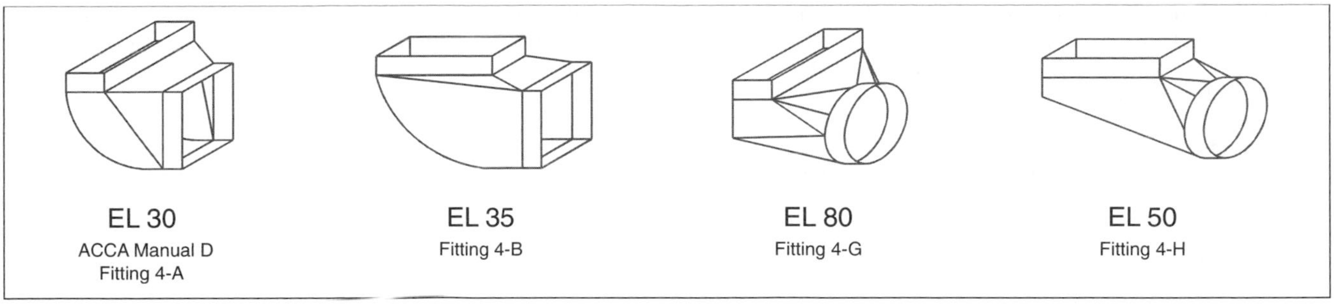

Fig. 14 Equivalent Length (EL in Feet) of Boot Fittings
(ACCA 1995)

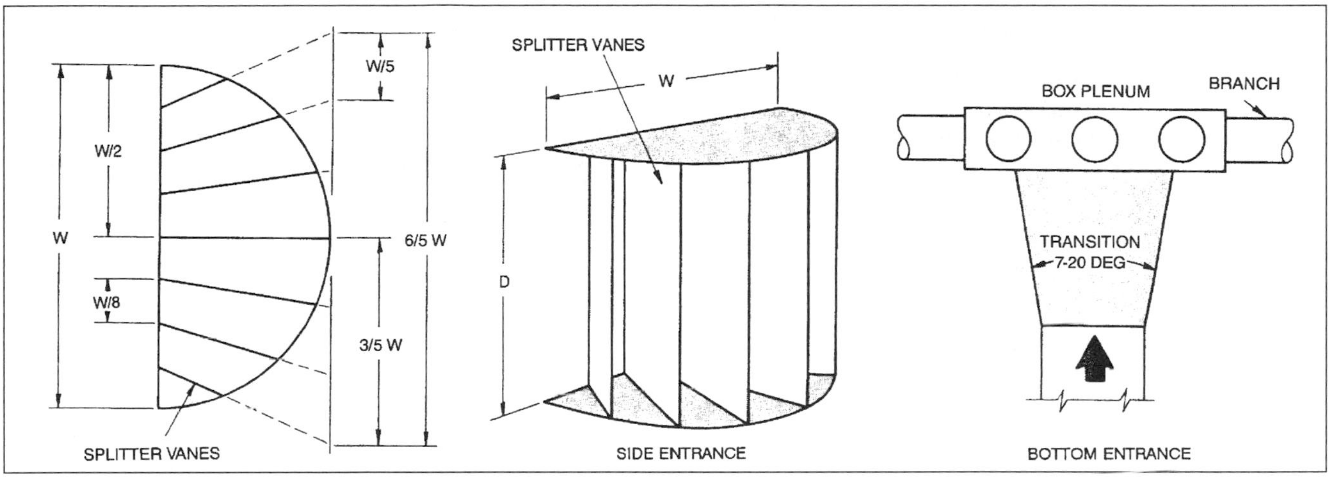

Fig. 15 Entrance Fittings to Eliminate Unstable Airflow in Box Plenum

1. Use the design blower airflow rate to size a plenum or a main trunk that feeds the zone trunks. Size plenum and main trunk ducts at 800 fpm.
2. Use the zone airflow rates (those based on the sum of the peak room loads) to size the zone trunk ducts. Size all zone trunks at 800 fpm.
3. Use the peak room airflow rate (those based on the peak room loads) to size the branch ducts or runouts. Size all branch runouts at a friction rate of 0.10 in. of water per 100 ft.
4. Size return ducts for 600 fpm air velocity.

Box Plenum Systems Using Flexible Duct

In some climates, an overhead duct with a box plenum feeding a series of individual, flexible duct, branch runouts is popular. The pressure drop through a flexible duct is higher than through a rigid sheet metal duct, however. Recognizing this larger loss is important when designing a box plenum/flexible duct system.

The design of the box plenum is critical to avoid excessive pressure loss and to minimize unstable air rotation within the plenum, which can change direction between blower cycles. This in turn may change the air delivery through individual branch takeoffs. Unstable rotation can be avoided by having the air enter the box plenum from the side and by using a special splitter entrance fitting.

Gilman et al. (1951) proposed box plenum dimensions and entrance fitting designs to minimize unstable conditions as summarized in Figure 15 and Figure 16. For residential systems with less than 2250 cfm capacity, the pressure loss through the box plenum is approximately 0.05 in. of water. This loss should be deducted from the available static pressure to determine the static pressure available for duct branches. In terms of equivalent length, add approximately 50 equivalent feet to the measured branch runs.

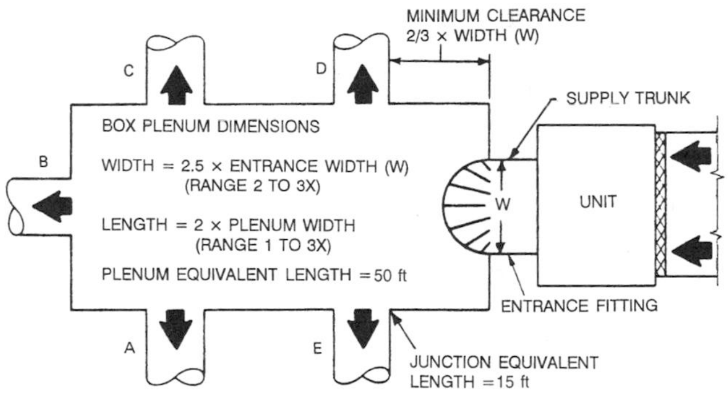

Fig. 16 Dimensions for Efficient Box Plenum

Embedded Loop Ducts

In cold climates, floor slab construction requires that the floor and perimeter of the slab be heated to provide comfort and prevent condensation. The temperature drop (or rise) in the supply air is significant, and special design tables must be used to account for the different supply air temperatures at distant registers. Because duct heat losses may cause a large temperature drop, feed ducts need to be placed at critical points in the loop.

A second aspect of a loop system is installation. The building site must be well drained and the surrounding grade sloped away from the structure. A vapor retarder must be installed under the slab. The bottom of the embedded duct must not be lower than the finished grade. Because a concrete slab loses heat from its edges outward through the foundation walls and downward through the earth, the edge must be properly insulated.

A typical loop duct is buried in the slab 2 to 18 in. from the outside edge and about 2.5 in. beneath the slab surface. If galvanized sheet metal is used for the duct, it must be coated on the outside to comply with federal specifications (SS-A-701). Other special materials used for ducts must be installed according to the manufacturer's instructions. In addition, care must be taken when the slab is poured not to puncture the vapor retarder or to crush or dislodge the ducts.

SELECTING SUPPLY AND RETURN GRILLES AND REGISTERS

Grilles and registers are selected from a manufacturer's catalog with appropriate engineering data after the duct design is completed. Rule-of-thumb selection should be avoided. Carefully determine the suitability of the register or grille selected for each location according to its performance specification for the quantity of air to be delivered and the discharge velocity from the duct.

Generally, in small commercial and residential applications, the selection and application of registers and grilles is particularly important because system size and air-handling capacity are small in energy-efficient structures. Proper selection ensures the satisfactory delivery of heating and/or cooling. Table 1 summarizes selection criteria for common types of supply outlets. Pressure loss is usually limited to 0.03 in. of water or less.

Return grilles are usually sized to provide a face velocity of 400 to 600 fpm, or 2.7 to 4.1 cfm per square inch of free area. Some central return grilles are designed to hold an air filter. This design allows the air to be filtered close to the occupied area and also allows easy access for filter maintenance. Easy access is important when the furnace is located in a remote area such as a crawl space or attic. The velocity of the air through a filter grille should not exceed 300 fpm, which means that the volume of air should not exceed 2.1 cfm per square inch of free filter area.

COMMERCIAL SYSTEMS

The duct design procedure described in this chapter can be applied to small commercial systems using residential equipment, provided that the application does not include moisture sources that create a large latent load.

In commercial applications that do not require low noise, air duct velocities may be increased to reduce duct size. Long throws from supply outlets are also required for large areas, and higher velocities may be required for that reason.

Commercial systems with significant variation in airflow for cooling, heating, and large internal loads (e.g., kitchens and theaters) should be designed in accordance with Chapters 31 and 32 of the 1997 *ASHRAE Handbook—Fundamentals* and Chapters 17, 18, and 19 of this volume.

TESTING FOR DUCT EFFICIENCY

ASHRAE is developing methods for testing the design and seasonal efficiencies of residential duct systems in the heating and cooling modes. Although duct leakage is a major cause of duct inefficiency, other factors such as heat conduction through the duct walls, influence of the fans on pressure in the house, and partial regain of lost heat must also be taken into account. The following summary of information needed to evaluate the efficiency of a duct system also describes the results that a test method should provide.

Data Inputs

The variables that are known or can be measured to provide the basis for calculating duct system efficiency include the following:

Local climate data. Three outdoor design temperatures are needed to describe an area's climate: one dry-bulb and one wet-bulb temperature for cooling, and one dry-bulb for heating.

Dimensions of living space. The volume of the conditioned space must be know to estimate the impact of the duct system on air infiltration. Typical, average values have been developed and if default options are used, the floor area of the conditioned space must be known.

Surface areas of ducts and R-values of insulation. The total surface area of supply and return ducts, and the insulation R-value of each are needed for the calculation of conductive heat losses through the duct walls. Also needed is the fraction of supply and return ducts that are located in each type of buffer zone; for example, an attic, basement or crawl-space.

Fan flow rate. The airflow through the fan must be measured to determine duct leakage (a major factor of efficiency) as a percentage of fan flow. An adjustable, calibrated fan flowmeter is the most accurate device for measuring airflow. Other measurement methods include a pitot tube traverse, calculations based on the temperature rise caused by a known heat input, and measurement of the concentration of a tracer gas.

Duct leakage to the outdoors. Air leakage from the supply ducts to the outdoors and from the outdoors and buffer spaces into the return ducts is another major factor that affects efficiency. Typically, 17% or more of the total airflow is leakage.

Data Output

Distribution efficiency is the main output of a test method. This figure of merit is the ratio of the input energy that would be needed to heat or cool the house if the duct system had no losses to the actual energy input required. The distribution efficiency also accounts for the effect the duct system has on equipment efficiency and the space conditioning load. For this reason the distribution efficiency differs from the simple output to input ratio for a duct system known as delivery effectiveness.

Four types of distribution efficiencies are typically considered. They relate to efficiency during either heating or cooling operation and for either design conditions or seasonal averages.

- Design distribution efficiency, heating
- Seasonal distribution efficiency, heating
- Design distribution efficiency, cooling
- Seasonal distribution efficiency, cooling

The design values of distribution efficiency are peak-load values that should be used when sizing equipment. The seasonal values should be used for determining annual energy use and subsequent costs.

SYSTEM PERFORMANCE

Both the performance of the furnace and the interaction of the furnace and the distribution system in the building determine how much fuel energy input to the furnace beneficially heats the conditioned space. Performance depends on the definition of the space in which it applies. Conditioned space is that space having a temperature actively controlled by a thermostat. A building can contain other space (attic, basement, or crawl space) that may influence the thermal performance of the conditioned space, but it is not defined as part of the conditioned space.

For houses with basements, it is important to decide whether the basement is part of the conditioned space because it typically receives some fraction of the HVAC output. In this analysis the basement is part of the conditioned space only if it is under active thermostat control and warm-air registers are provided to maintain comfort. Otherwise, the basement is not part of the conditioned space even if some heat is provided with fixed open registers, for example. The following performance examples show designs for improving efficiency, along with their effect on temperature in the unconditioned basement.

"HOUSE" Dynamic Simulation Model

The dynamic response and interactions between components of central forced warm-air systems are sufficiently complex that the effects of system options on annual fuel use are not easily evaluated. ASHRAE Special Project 43 (SP43) assessed the effects of system component and control mode options. The resulting simulation model accounts for the dynamic and thermal interactions of equipment and loads in response to varying weather patterns. Both single-zone (HOUSE-I) and multizone (HOUSE-II) versions of the model have been developed.

Fischer et al. (1984) described the HOUSE simulation model. Jakob et al. (1986a), Herold et al. (1987), and Jakob et al. (1987) described the validation model for the heating mode through field experiments in two houses. Herold et al. (1986) summarized both the project and the model.

Jakob et al. (1986b) and Locklin et al. (1987) presented the model's predictions of the overall performance of the forced warm-air heater. These variables include furnace and venting types, furnace installation location and combustion air source, furnace sizing, night setback, thermostat cycling rate, blower operating strategy, basement insulation, duct sealing and insulation, house and foundation type, and climate.

An assumption of the HOUSE models is that duct leakage must be specified as an input. Subsequent research has developed methods for calculating losses from leakage areas and operating pressures. Moreover, in all the SP43 runs, supply and return leakage values were set equal to each other, a situation that often does not hold in practice. Nevertheless, the generic results generally confirm the usefulness of AFUE as a measure of furnace efficiency and point the way toward improvements in efficiency. Also, the parameters used to characterize energy flows and efficiency ratios are the basis for a proposed standard test method for thermal distribution efficiency. ASHRAE Research Project 852 (Gu et al. 1998) provided insights on modeling the performance of residential duct systems.

SYSTEM PERFORMANCE FACTORS

A series of system performance factors, consisting of both efficiency factors and dimensionless energy factors, describe dynamic performance of the individual components and the overall system over any period of interest. Jakob et al. (1986b) and Locklin et al. (1987) described the factors in detail.

Table 3 identifies the performance factors and their mathematical definitions in four categories: (1) equipment-component efficiency factors, (2) equipment-system performance factors, (3) equipment-load interaction factors, and (4) energy cost factors. The detailed results of the analysis may be found in the references mentioned previously. The following discussion focuses on insights gained from SP43.

Equipment-Component Efficiency Factors

Furnace Efficiency E_F. This factor is the ratio of the energy that is delivered to the plenum during cyclic operation of the furnace to the total input energy on an annual basis. E_F includes summertime pilot losses and blower energy.

This factor is similar in concept to the AFUE in ASHRAE *Standard* 103, which provides an estimate of annual energy, taking into account assumed system dynamics. However, EF differs importantly from AFUE in that

- The AFUE for a given furnace is defined by a single predetermined cyclic condition with standard dynamics; E_F is based on the integrated cyclic performance over a year.
- The AFUE does not include auxiliary electric input, and it gives credit for jacket losses, except when the furnace is an outdoor unit; E_F and the other efficiency factors defined here include auxiliary electric input.
- Several other effects regarding combustion induced infiltration and dynamics were investigated, but they had little effect on performance.

Duct Efficiency E_D. This is the ratio of the energy intentionally delivered to the conditioned space through the supply registers to the energy delivered to the furnace plenum, on an annual basis.

Equipment-System Performance Factors

Heat Delivery Efficiency E_{HD}. This factor is the product of the furnace efficiency E_F and duct efficiency E_D. It is the ratio of the energy intentionally delivered to the conditioned space to the total input energy. It is a measure of how effectively the HVAC delivers heat directly to the conditioned space on an annual basis.

Miscellaneous Gain Factor F_{MG}. This factor is the total heat delivered to the conditioned space divided by the energy intentionally delivered to the conditioned space through the duct registers.

System Efficiency E_S. This efficiency factor is the product of E_{HD} and F_{MG}. It is the total energy delivered to the conditioned space divided by the total energy input to the furnace. Thus, E_S includes intentional and unintentional energy gains.

Equipment-Load Interaction Factors

Load Modification Factor F_{LM}. This factor is the ratio of the total heat delivered for a base case to the total heat delivered for a case of interest. It adjusts E_S to account for the effect of operation on the heating load. It accounts for the effect of the combustion-induced infiltration and off-period infiltration due to draft hood flow, as well as for the effects of changes in temperature of unconditioned spaces adjacent to the conditioned space.

System Index I_S. This index is the product of system efficiency E_S and the load modification factor F_{LM}. I_S is an energy-based figure of merit that adjusts E_S for any credits or debits due to system induced loads relative to a base case load. I_S is a powerful tool for comparing alternative systems. However, high values of I_S are sometimes associated with a low basement temperature because a greater portion of the furnace output is delivered directly to the conditioned space. The ratio of the system indexes for two systems being compared is the inverse of the ratio of their annual energy use AEU.

Energy Cost Factors

Table 3 also defines cost factors, which are discussed in the cited literature and not reviewed here.

Implications

The following implications apply to the definitions for the various performance factors.

- The defined conditioned space is important to the comparisons of system index I_S. Because the load modification factor F_{LM} and I_S are based on the same reference equipment and house configuration, the performance of various furnaces installed in basements or furnaces installed in the conditioned space (i.e., closet installations) may be compared. However, the performance of a furnace installed in a basement or crawl space cannot be compared with that of heating systems installed only in the conditioned space.

  Because I_S depends on a reference equipment and house configuration, it may be used only as a ranking index from which the relative benefits of different features can be derived. That is, it can be used to compare the costs and savings of various features in specific applications to those of a base case.
- The miscellaneous gain factor F_{MG} includes only those heating losses that go *directly* to the conditioned space.
- The equipment-system performance factors relate strictly to the subject equipment, whereas the equipment-load interaction factors draw comparisons between the subject equipment and an

Table 3 Definitions of System Performance Factors

	Comments
Equipment-Component efficiency factors	
E_F = Furnace Efficiency = $100\dfrac{\text{Furnace Output}}{\text{Total Energy Input}} = 100\dfrac{\text{Duct Input}}{\text{Total Energy Input}}$	• Integrated energy over all operating cycles
E_D = Duct Efficiency = $100\dfrac{\text{Duct Output}}{\text{Duct Input}}$	
Equipment-System performance factors	
E_{HD} = Heat Delivery Efficiency = $\dfrac{E_F \times E_D}{100} = \dfrac{\text{Duct Output}}{\text{Total Energy Input}}$	• Efficiency of the furnace/duct subsystem
F_{MG} = Miscellaneous Gain Factor = $\dfrac{\text{Total Heat Delivered}^a}{\text{Duct Output}}$	• Accounts for heating by fugitive gains
E_S = System Efficiency = $E_{HD} \times F_{MG} = 100\dfrac{\text{Total Heat Delivered}}{\text{Total Energy Input}}$	• Efficiency of the combined HVAC system
Equipment-Load interaction factors	
F_{IL} = Induced Load Factor[b] = $\dfrac{\text{System Induced Load}^c}{\text{Total Heat Delivered}}$	• Accounts for added loads due to equipment operation
F_{LM} = Load Modification Factor = $1.0 - F_{IL} = \dfrac{\text{Total Heat Delivered} - \text{System Induced Load}}{\text{Total Heat Delivered}}$	
I_S = System Index[d] = $\dfrac{E_S \times F_{LM}}{100} = \dfrac{\text{Total Heat Delivered} - \text{System Induced Load}}{\text{Total Energy Input}}$	• Common index for ranking system. **Not** an efficiency.
Energy cost factors	
R_{AE} = Auxiliary Energy Ratio = $\dfrac{\text{Auxiliary Energy Input}}{\text{Primary Energy Input}} = \dfrac{\text{Electrical Energy Input}}{\text{Fuel Energy Input}}$	• System energy characteristics
R_{CL} = Local Energy Cost Ratio = $\dfrac{\text{Electrical Cost per Energy Unit}}{\text{Reference Fuel Cost per Energy Unit}}$ (in common units)	
F_{CR} = Cost Ratio Factor = $\dfrac{\text{Fuel Energy Input} + \text{Electric Energy Input}}{\text{Fuel Energy Input} + R_{CL}(\text{Electric Energy Input})} = \dfrac{1.0 + R_{AE}}{1.0 + R_{CL} \times R_{AE}}$ Special Case (Fuel = 0): $F_{CR} = 1/R_{CL}$	• Economics
I_{SCM} = Cost-Modified System Index[d] = $I_S \times F_{CR} = \dfrac{\text{Total Heat Delivered} - \text{System Induced Load}}{\text{Primary Energy Input} + R_{CL}(\text{Auxiliary Energy Input})}$	• Common economic index for ranking systems
Annual energy use	
AEU = Annual Energy Use (fuel and electricity) predicted by the HOUSE model, in common energy units	
Annual Fuel Used = AEU / $(1.0 + R_{AE})$	
Annual Electricity Used = AEU / $(1.0 + 1/R_{AE})$	
Percent savings	
% Energy Saving = $100\,[I_S - (I_S)_{BC}]/I_S$, where $(I_S)_{BC} = I_S$ for base case	• Saving relative to base case
% Cost Saving = $100\,[I_{SCM} - (I_{SCM})_{BC}]/I_{SCM}$, where $(I_{SCM})_{BC}$ = Cost-modified I_S for base case	
Other factors for dynamic performance	
AFUE = Annual Fuel Utilization Efficiency by ANSI/ASHRAE *Standard* 103 efficiency rating, applicable to specific furnaces. Values in this chapter are for generic furnaces.	
SSE = Steady-State Efficiency value for a given furnace by ANSI Z21.47/ CSA 2.3 test procedure.	

Note: Energy inputs and outputs are integrated over an annual period. Efficiencies (E) are expressed as percents. Indexes (I), factors (F), and ratios (R) are expressed as fractions.

[a]The *Total Heat Delivered* is the integration over time of all the energy supplied to the conditioned space by the HVAC equipment. By definition, it is exactly equal to the space-heating load.

[b]The *Induced Load Factor* may be positive or negative, depending on the value of the load relative to the selected base case.

[c]The *System Induced Load* is the difference between the space-heating load for a particular case and the space-heating load for the base case. For the base case, the System Induced Load is, by definition, zero.

[d]Indexes are referenced to as "base cases" from which improvements are measured.

Table 4 System Performance Examples

Performance Factor	Base Case: Typical Conventional, Natural-Draft Furnace with IID	Alternative Case: Typical Noncondensing Fan-Assisted Furnace
ASHRAE 103-93 AFUE (indoor) per DOE rules, %	69	81.5
Furnace efficiency E_F, %	75.5	85.5
Duct efficiency E_D, %	60.9	59.3
Heat delivery efficiency E_{HD}, %	46.0	50.7
Miscellaneous gain factor F_{MG}	1.004	0.983
System efficiency E_S, %	46.1	49.8
Load modification factor F_{LM}	1.000 (Base case)	1.099
System Index I_S (Base case = 1.00)	1.000	1.189
Annual energy use AEU, 10^6 Btu	73.0	61.5
Auxiliary energy ratio R_{AE}	0.027	0.028
Energy savings from base case, %	—	15.9
Cost savings from base case, % (with R_{CL} = 4)	—	15.6

Note: The values presented here do not represent only this class of equipment; electric furnaces and heat pumps in a similar installation and under similar conditions would incur similar losses. The system index for any central air system can be improved, in comparison to the above examples, by insulating the ducts, locating the ductwork inside the conditioned space, or both.

explicitly defined base case. This base case is a specific load and equipment configuration to which all alternatives are compared.

Systems with the best total energy economy have the highest I_S. Those with leaky and uninsulated ducts could have a higher efficiency, even though fuel use would be higher, if basement duct losses that become gains to the conditioned space were included in the miscellaneous gain factor F_{MG} The foregoing definitions prevent this possibility.

System Performance Examples

The ASHRAE SP43 study was limited to certain house configurations and climates and to gas-fired equipment. Electric heat pump and zoned baseboard systems were not studied. For this reason, these data should not be used to compare systems or select a heating fuel. Several of the factors addressed are not references to performance; instead, they are figures of merit, which represent the effect of various components on a system. As such, these factors should not be applied outside the scope of these examples.

The following examples of overall thermal performance illustrate how the furnace, vent, duct system, and building can interact. Table 4 summarizes HOUSE-I simulation model predictions of the annual system performance for a base case (a conventional, natural-draft gas furnace with an intermittent ignition device) and an example case (a noncondensing, fan-assisted combustion furnace). Each is installed in a typical three-bedroom, ranch-style house of frame construction, located in Pittsburgh, Pennsylvania constructed according to HUD minimum property standards circa 1980. Table 5 shows the base case operating assumption for the simulation predictions. other assumptions for these predictions.

Base Case. Referring to Table 4, the annual furnace efficiency E_F of the conventional, natural-draft furnace is predicted to be 75.5%. Air leakage and heat loss from the uninsulated duct result in a duct efficiency E_D of 60.9%. The heat delivery efficiency E_{HD}, which is the ratio of the energy intentionally delivered to the conditioned space through the supply registers to the total input energy, is 46.0%. The miscellaneous gain factor F_{MG}, which is 1.004, accounts for the small heat gain to the conditioned space from the heated masonry chimney passing through the conditioned space. The system efficiency E_S (the ratio of the total heat

Table 5 Base Case Assumptions for Simulation Predictions

	Base Case
Furnace, Adjustments, and Controls	
Furnace size	1.4 × DHL
Circulating air temperature rise	60°F
Thermostat set point	68°F
Thermostat cycling rate at 50% on-time	6 cycles/h
Night setback, 8 h	None
Blower control	
On	80 s
Off	90°F
Duct-Related Factors	
Insulation	None
Leakage, relative to duct flow	10%
Location	Basement
Load-Related Factors	
Nominal infiltration[a]	
Conditioned space	0.75 ACH
Basement	0.25 ACH
Occupancy, persons	3 during evening and night, 1 during day
Internal loads	Typical appliances, day and evening only (20 kWh/day)
Shading by adjacent trees or houses	None

[a]Model runs used variable infiltration, as driven by indoor-outdoor temperature differences, wind, and burner operation. Values shown above are nominal.
ACH = Air changes per hour
DHL = Design heat loss

delivered or the space-heating load to the total energy input) is 46.1%. Because this case is designated as the base case, the load modification factor F_{LM} is 1.0. Thus, the system index I_S is also 1.000 (by definition).

The duct loss and jacket loss are accounted for in the energy balance on the basement air and in the energy flow between the basement and the conditioned space. The increase in infiltration caused by the need for combustion air and vent dilution air is also accounted for in the energy balances on the living space air and basement air. In the base case, the temperature in the unconditioned basement is nearly the same (68°F) as in the first floor where the thermostat is located. This condition is caused by heat loss of the exposed ducts in the basement and by the low outdoor infiltration into the basement that was achieved by sealing cracks associated with construction.

Alternative Case. Again referring to Table 4, the furnace efficiency E_F for the noncondensing fan-assisted combustion furnace being compared is 85.5%. The duct efficiency E_D is 59.3%, slightly lower than that for the base case. Therefore, the heat delivery efficiency E_{HD} is 50.7%, which reflects the higher furnace efficiency.

The miscellaneous gain factor F_{MG} is 0.983, reflecting the small heat loss from the conditioned space to the colder masonry chimney (due to reduced off-cycle vent flow). The system efficiency E_S (i.e., $E_{HD} \times F_{MG}$) is 49.8%.

For this furnace system, compared to the base case system of the conventional, natural-draft furnace, the load modification factor F_{LM} is 1.099. Therefore, the space-heating load for the house with the noncondensing fan-assisted combustion furnace is 1/1.099, or 91% of the space-heating load for the house with the conventional, natural-draft furnace. This reduction in heating load is mainly due to the reduction in off-cycle vent flow.

The system index I_S for the example case is 1.189. Note that I_S is also the inverse of the ratio of the annual energy use AEU ($61.5 \times 10^6 / 73 \times 10^6 = 0.842$).

Table 6 Effect of Furnace Type on Annual Heating Performance

	Furnace Characterization Typical Values,[a] AFUE/SSE	Predicted by HOUSE-2 Model: Annual Performance Factors							Auxiliary Energy Ratio	Average Basement, °F
		E_F	E_D	F_{MG}	F_{LM}	I_S	I_{SCM} (R_{CL} = 4)	AEU, 10^6 Btu		
Conventional, natural-draft										
Pilot	64.5/77	72.9	60.9	1.006	1.000	0.970	0.971	75.5	0.026	67.9
Intermittent ignition device (Base case)[b]	69/77	75.5	60.9	1.004	1.000	1.000	1.000	73.0	0.027	67.8
IID + Thermal vent damper	78/77	75.4	61.0	1.002	1.086	1.087	1.085	67.3	0.027	68.2
IID + Electric vent damper	78/77	75.4	61.2	0.988	1.105	1.093	1.091	66.9	0.027	68.3
Fan-assisted types										
Noncondensing	81.5/82.5	85.5	59.3	0.983	1.099	1.189	1.185	61.5	0.028	67.6
Condensing[c]	92.5/93.1	95.5	62.0	1.000	1.050	1.349	1.322	54.2	0.034	66.9
Electric furnace	n.a.	99.5	60.6	1.000	1.079	1.412	0.380	51.8	0.020[d]	67.1

Note: The values in this table are figures of merit to be considered within the confines of the SP43 project, and they should not be applied outside the scope of these examples.

[a]AFUE = Annual Fuel Utilization Efficiency by ANSI/ASHRAE *Standard* 103.
SS = Steady-state efficiency by ANSI Z21.47/CSA 2.3 test procedure.

[b]Ranch-type house with basement in Pittsburgh, PA, climate and base conditions of 60°Fcirculating air temperature rise, 6 cycles/h, no setback, 10% duct air leakage.

[c]Direct vent uses outdoor air for combustion (includes preheat).

[d]Blower energy is treated as auxiliary energy.

Table 7 Effect of Climate and Night Setback on Annual Heating Performance

Furnace Type and Location	Setback,[a] °F	Avg. % On-Time	Avg. Bsmt., °F	Furnace Eff. E_F, %	Duct Eff. E_D, %	System Index I_S	AEU,[b] 10^6 Btu	% Energy Saved by Setback
Conventional, Natural-Draft (Base Case)								
Nashville	0	12.7	64.6	73.9	56.2	1.000	55.6	
	10	10.7	63.1	74.8	58.9	1.192	46.7	16.0
Pittsburgh (base city)	0	18.0	67.8	75.5	60.9	1.000	73.0	
	10	15.6	65.7	75.9	62.4	1.154	63.4	13.2
Minneapolis	0	20.9	68.0	76.8	63.2	1.000	99.1	
	10	18.7	65.7	77.0	62.4	1.121	88.4	10.7
Direct, Condensing								
Nashville	0	11.1	63.9	94.0	57.0	1.000	41.0	
	10	9.5	62.6	93.7	59.5	1.174	35.0	14.8
Pittsburgh	0	16.2	67.8	93.3	61.7	1.000	55.1	
	10	14.2	65.0	93.0	63.2	1.144	48.1	12.6
Minneapolis	0	18.2	66.8	95.1	64.5	1.000	73.7	
	10	16.4	64.7	94.8	65.6	1.115	66.2	10.2

Note: Ranch-type house, basement, and base conditions: 60°F circulating air temperature rise, 6 cycles/h, 10% duct air leakage. Thermal envelope typical of each city; for example, no basement insulation in Nashville, TN.

The values in this table are figures of merit to be considered within the confines of the SP43 project, and they should not be applied outside the scope of these examples.

[a]The base case is 0°F setback in each city.

[b]AEU = Annual energy use.

Effect of Furnace Type

Table 6 summarizes the energy effects of several furnaces. Note that the system indexes I_S for both thermal and electric vent dampers are similar, although thermal vent dampers are slower reacting and less effective at blocking the vent. Also, the ratio of a furnace's AFUE to the base case AFUE closely matches the I_S values for the furnaces. The exceptions are the vent damper cases, where the improvement in I_S suggests a smaller AFUE credit. In general, the study found that the furnace AFUE is a good indication of relative annual performance of furnaces in typical systems.

The results reported in Table 6 are for homes that do not include the basement in the conditioned space; that is, energy lost to the basement contributes only indirectly to the useful heating effect. If the basement is considered to be a part of the conditioned space, the miscellaneous gain factor, load modification factor, and subsequent calculated efficiencies and indexes are adjusted to account for the beneficial effects of equipment (furnace jacket and duct system) heat losses that contribute to heating the basement. The system performance factors increase by 60% when the basement is considered to be part of the conditioned space. The indices however retain their same relative ranking of systems.

Effect of Climate and Night Setback

Table 7 covers the effects of climate (insulation levels change by location) and night setback on performance for two furnaces. The improvements in I_S with higher percent on-time (colder climates) follow improvements in duct efficiency. Furnace efficiency E_F appears to be relatively uniform in houses representative of typical construction practices in each city and where the furnace is sized at 1.4 times the design heat loss. Also, the saving due to night setback increases in magnitude with a warmer climate. The percent energy saved in the three climates varies with the magnitude of energy use (from 10 to 16% for the natural-draft cases).

Effect of Furnace Sizing

Furnace sizing affects I_S depending on how the venting and ducting are designed. As Table 8 indicates, if the ducts and vent are sized according to the furnace size (referred to as the *new* case), I_S drops about 10% as the furnace capacity is varied between 1.0 and 2.5

Table 8 Effect of Sizing, Setback, and Design Parameters on Annual Heating Performance—Conventional, Natural-Draft Furnace

Furnace Multiplier[a]	Duct Design	Setback, °F	Annual Performance Factors E_F	E_D	F_{MG}	F_{LM}	I_S[e]	Temperature Swing, °F	Average Room, °F	Recovery Time, h[b]	Average Basement, °F
1.00	New[c]	0	76.1	59.2	1.016	1.104	1.095	3.4	67.7	n.a.[d]	67.9
1.15	New	0	75.6	59.0	1.009	1.059	1.032	4.0	67.8	n.a.	67.9
	Retrofit	0	75.5	59.0	1.002	1.032	0.998	4.0	67.8	n.a.	67.9
	Retrofit	10	76.1	60.7	0.999	1.155	1.155	4.1	65.8	2.02	65.7
1.40	New	0	75.5	60.8	1.010	1.029	1.034	4.8	68.1	n.a.	67.9
	Retrofit	0	75.5	60.9	1.004	1.000	1.000[e]	4.9	68.0	n.a.	67.8
	Retrofit	10	75.9	62.4	1.001	1.120	1.152	5.2	66.0	1.02	65.7
1.70	New	0	74.8	61.3	1.013	1.017	1.023	5.9	68.2	n.a.	68.1
	Retrofit	0	75.0	61.7	1.006	0.983	0.992	5.9	68.2	n.a.	67.9
	Retrofit	10	75.6	63.5	1.003	1.095	1.143	6.3	66.2	0.54	65.8
2.50	New	0	74.9	63.4	1.008	0.943	0.978	8.2	68.6	n.a.	68.1
	Retrofit	0	75.2	64.9	1.009	0.937	1.000	8.6	68.6	n.a.	67.8
	Retrofit	10	75.7	66.5	1.006	1.042	1.144	9.3	66.6	0.24	65.8

Notes:
1. Ranch house with basement in Pittsburgh, PA, climate with base conditions of 60°F circulating air temperature rise, 6 cycles/h, 10% duct air leakage.
2. The values in this table are figures of merit to be considered within the confines of the SP43 project, and they should not be applied outside the scope of these examples.

[a]Furnace output rating or heating capacity (Furnace multiplier) × (Design heat loss).
[b]Longest recovery time during winter (lowest outdoor temperature = 5°F).
[c]Retrofit case was not run for furnace multiplier 1.00.
[d]n.a. indicates not applicable.
[e]Base case = 1.0.

Table 9 Effect of Furnace Sizing on Annual Heating Performance—Condensing Furnace with Preheat

Furnace Multiplier[a]	Annual Performance Factors E_F	E_D	F_{MG}	F_{LM}	I_S[b]	Temperature Swing, °F	Average Room, °F	Average Basement, °F
1.15	95.1	60.9	1.000	1.076	1.351	3.5	67.9	66.7
1.40	95.5	62.0	1.000	1.050	1.347	4.3	68.1	66.7
2.50	95.4	64.8	1.000	0.990	1.325	7.3	68.7	66.7

Notes:
1. Ranch house with basement in Pittsburgh climate with base conditions of 60°F circulating air temperature rise, 6 cycles/h, 10% duct air leakage.
2. The values in this table are figures of merit to be considered within the confines of the SP43 project, and they should not be applied outside the scope of these examples.

[a]Furnace output rating or heating capacity = (Furnace multiplier) × (Design heat loss).
[b]Base case = 1.0.

times the design heat loss for a given application. In the retrofit case, where the vent and duct are sized at a furnace capacity of 1.4 times the design heat loss, I_S changes little with increased furnace capacity, indicating little energy savings. In a new case, where the duct is designed for cooling and the vent size does not change between furnace capacities, the SP43 study indicates that there is essentially no effect on I_S.

Table 9 shows similar results in condensing furnaces for the new case of ducts and vents sized according to the furnace capacity. In this case, the decrease in I_S is smaller, about 2% over the range of 1.15 to 2.5 times the design heat loss.

Finally, both Table 8 and Table 9 show that duct efficiency E_D increases with furnace capacity because higher capacity furnaces are on less than lower capacity furnaces.

Effects of Furnace Sizing and Night Setback

Table 8 also shows the relationship between furnace sizing and night setback for the retrofit case. The energy saving due to night setback, 8 h per day at 10°F, is nearly constant at 15% and independent of furnace size. Table 7 covers the effect of climate variation on energy saving due to night setback.

Duct Treatment. Table 10 shows the effect of duct treatment on furnace performance. Duct treatment includes sealants to reduce leaks and interior or exterior insulation to reduce heat loss due to conduction. Sealing and insulation improve system performance as indicated by I_S. For cases with no duct insulation, reducing duct leakage from 10% to zero increases I_S by 2.6%. R5 insulation on the exterior of the ducts increases I_S by 4.4%, and R5 insulation on the interior of the duct increases I_S by 8.5%.

Basement Configuration. Table 11 covers the effect of basement configuration and duct treatment on system performance. Insulating and sealing the ducts reduces the basement

Table 10 Effect of Duct Treatment on System Performance

Case	Duct Configuration 1	2[a]	3	4	5	6	7
Condition							
Duct insulation	None	None	None	R5	R5	R5	R5[b]
Duct leakage, %	0	10	20	0	10	20	10
Basement insulation							
Ceiling	None	None	None	None	None	None	None
Wall	R8	R8	R8	R8	R8	R8	R8
Performance							
Burner on-time, %	17.5	18.0	18.6	16.8	17.2	17.8	16.6
Blower on-time, %	23.8	24.3	24.9	23.0	23.4	24.1	22.9
Average basement temperature, °F	66.8	67.8	68.8	65.2	66.3	67.4	65.1
Furnace efficiency E_F, %	75.4	75.5	75.7	75.0	75.2	75.4	75.0
Duct efficiency E_D, %	66.8	60.9	54.8	77.4	70.4	63.2	79.6
Load modification factor	0.94	1.00	1.07	0.85	0.91	0.97	0.84
I_S (base case = 1.0)	1.026	1.000	0.970	1.070	1.044	1.010	1.085

[a]Case 2 is the base case.
[b]Case 7 is interior insulation (liner); Cases 1 through 6 are exterior insulation (wrap).

temperature. More heat is then required in the conditioned space to make up for losses to the colder basement. Where ducts pass through the attic or ventilated crawl space, insulation and sealing improve duct performance, although the total system performance is poorer. On the other hand, installing ducts in the conditioned space significantly improves F_{MG} because the duct losses are added directly to the conditioned space. In this case, I_S would also improve.

Table 11 Effect of Duct Treatment and Basement Configuration on System Performance

Case	Duct Configuration				
	1	2	3	4	5
Condition					
Duct insulation	None	None	None	None	R5
Duct leakage, %	20	10	10	10	0
Basement insulation					
Ceiling	None	None	R11	R11	R11
Wall	None	None	None	R8	R8
Performance					
Burner on-time, %	23.6	22.7	21.8	18.0	16.3
Blower on-time, %	29.8	29.2	28.0	24.2	22.2
Average basement temperature, °F	64.4	63.3	62.4	67.9	64.3
Furnace efficiency E_F, %	75.3	75.2	75.2	75.7	75.0
Duct efficiency E_D, %	50.9	56.9	56.7	61.7	77.0
Load modification factor	0.91	0.85	0.89	0.99	0.88
I_S (base case = 1.0)	0.765	0.794	0.825	1.001	1.103

Note: See Table 10 for base case.

REFERENCES

ACCA. 1999. Residential duct systems. *Manual* D. Air Conditioning Contractors of America, Washington, DC.

ACCA. 1995. Residential load calculations. *Manual J*.

ARI. 1996. Central system humidifiers for residential applications. *Standard* 610-96. Air-Conditioning and Refrigeration Institute, Arlington, VA.

ASHRAE. 1993. Energy-efficient design of new low-rise residential buildings. *Standard* 90.2-1993.

ASHRAE. 1993. Methods of testing for annual fuel utilization efficiency of residential central furnaces and boilers. *Standard* 103-1993.

CSA. 1998. Gas-fired central furnaces. ANSI Z21.47-1998/CSA 2.3-M98. CSA International, Cleveland, OH.

Fischer, R.D., F.E. Jakob, L.J. Flanigan, D.W. Locklin, and R.A. Cudnik. 1984. Dynamic performance of residential warm-air heating systems—Status of ASHRAE Project SP43. *ASHRAE Transactions* 90(2B):573-90.

Gilman, S.F., R.J. Martin, and S. Konzo. 1951. Investigation of the pressure characteristics and air distribution in box-type plenums for air conditioning duct systems. University of Illinois Engineering Experiment Station *Bulletin* No. 393 (July), Urbana-Champaign, IL.

Gu, L., J.E. Cummings, P.W. Fairey, and M.V. Swami. 1998. Comparison of duct computer models that could provide input to the thermal distribution standard method of test (SPC152P). *ASHRAE Transactions* 104(1).

Herold, K.E., F.E. Jakob, and R.D. Fischer. 1986. The SP43 simulation model: Residential energy use. Proceedings of ASME Conference, Anaheim, CA (pp. 81-87).

Herold, K.E., R.A. Cudnik, L.J. Flanigan, and R.D. Fischer. 1987. Update on experimental validation of the SP43 simulation model for forced-air heating systems. *ASHRAE Transactions* 93(2):1919-33.

Jakob, F.E., R.D. Fischer, and L.J. Flanigan. 1987. Experimental validation of the duct submodel for the SP43 simulation model. *ASHRAE Transactions* 93(1):1499-1515.

Jakob, F.E., R.D. Fischer, L.J. Flanigan, D.W. Locklin, K.E. Harold, and R.A. Cudnik. 1986a. Validation of the ASHRAE SP43 dynamic simulation model for residential forced-warm air systems. *ASHRAE Transactions* 92(2B):623-43.

Jakob, F.E., D.W. Locklin, R.D. Fischer, L.J. Flanigan, and R.A. Cudnik. 1986b. SP43 evaluation of system options for residential forced-air heating. *ASHRAE Transactions* 92(2B):644-73.

Locklin, D.W., K.E. Herold, R.D. Fischer, F.E. Jakob, and R.A. Cudnik. 1987. Supplemental information from SP43 evaluation of system options for residential forced-air heating. *ASHRAE Transactions* 93(2):1934-58.

NFPA. 1993. Installation of warm air heating and air conditioning systems. *Standard* 90B-93. National Fire Protection Association, Quincy, MA.

SMACNA. 1998. Residential comfort system installation standards manual, 7th ed. Sheet Metal and Air Conditioning Contractors' National Association, Chantilly, VA.

Straub, H.E. and M. Chen. 1957. Distribution of air within a room for year-round air conditioning—Part II. University of Illinois Engineering Experimental Station *Bulletin* No. 442 (March), Urbana-Champaign, IL.

Wright, J.R., D.R. Bahnfleth, and E.G. Brown. 1963. Comparative performance of year-round systems used in air conditioning research No. 2. University of Illinois Engineering Experimental Station *Bulletin* No. 465 (January), Urbana-Champaign, IL.

BIBLIOGRAPHY

ARI. Updated annually. Directory of certified unitary products; unitary air conditioners; unitary air-source heat pumps; and sound rated outdoor unitary equipment.

Crisafulli, J.C., R.A. Cudnik, and L.R. Brand. 1989. Investigation of regional differences in residential HVAC system performance using the SP43 simulation model. *ASHRAE Transactions* 95(1):915-29.

Fischer, R.D. and R.A. Cudnik. 1993. The HOUSE-II computer model for dynamic and seasonal performance simulation of central forced-air systems in multi-zone residences. *ASHRAE Transactions* 99(1):614-26.

GAMA. Updated annually. Consumer's directory of certified efficiency ratings for residential heating and water heating equipment. Gas Appliance Manufacturers Association, Arlington, VA.

Herold, K.E., R.D. Fischer, and L.J. Flanigan. 1987. Measured cooling performance of central forced-air systems and validation of the SP43 simulation model. *ASHRAE Transactions* 93(1):1443-57.

Hise, E.C. and A.S. Holman. 1977. Heat balance and efficiency measurements of central, forced-air, residential gas furnaces. *ASHRAE Transactions* 83(1):865-80.

Modera, M.P., J. Andrews, and E. Kweller. 1992. *A comprehensive yardstick for residential thermal distribution efficiency.* Proc. 1992 ACEEE Summer Study on Energy Efficiency in Buildings. American Council for and Energy Efficient Economy, Washington, DC. August.

Rutkowski, H. and J.H. Healy. 1990. Selecting residential air-cooled cooling equipment based on sensible and latent performance. ASHRAE *Transactions* 96(2):851-56.

CHAPTER 10

STEAM SYSTEMS

A STEAM system uses the vapor phase of water to supply heat or kinetic energy through a piping system. As a source of heat, steam can heat a conditioned space with suitable terminal heat transfer equipment such as fan coil units, unit heaters, radiators, and convectors (finned tube or cast iron), or steam can heat through a heat exchanger that supplies hot water or some other heat transfer medium to the terminal units. In addition, steam is commonly used in heat exchangers (shell-and-tube, plate, or coil types) to heat domestic hot water and supply heat for industrial and commercial processes such as in laundries and kitchens. Steam is also used as a heat source for certain cooling processes such as single-stage and two-stage absorption refrigeration machines.

ADVANTAGES

Steam offers the following advantages:

- Steam flows through the system unaided by external energy sources such as pumps.
- Because of its low density, steam can be used in tall buildings where water systems create excessive pressure.
- Terminal units can be added or removed without making basic changes to the design.
- Steam components can be repaired or replaced by closing the steam supply, without the difficulties associated with draining and refilling a water system.
- Steam is pressure-temperature dependent; therefore, the system temperature can be controlled by varying either steam pressure or temperature.
- Steam can be distributed throughout a heating system with little change in temperature.

In view of these advantages, steam is applicable to the following facilities:

- Where heat is required for process and comfort heating, such as in industrial plants, hospitals, restaurants, dry-cleaning plants, laundries, and commercial buildings
- Where the heating medium must travel great distances, such as in facilities with scattered building locations, or where the building height would result in excessive pressure in a water system
- Where intermittent changes in heat load occur

FUNDAMENTALS

Steam is the vapor phase of water and is generated by adding more heat than required to maintain its liquid phase at a given pressure, causing the liquid to change to vapor without any further increase in temperature. Table 1 illustrates the pressure-temperature relationship and various other properties of steam.

Temperature is the thermal state of both liquid and vapor at any given pressure. The values shown in Table 1 are for dry saturated steam. The vapor temperature can be raised by adding more heat, resulting in superheated steam, which is used (1) where higher temperatures are required, (2) in large distribution systems to compensate for heat losses and to ensure that steam is delivered at the desired saturated pressure and temperature, and (3) to ensure that the steam is dry and contains no entrained liquid that could damage some turbine-driven equipment.

Enthalpy of the liquid h_f (sensible heat) is the amount of heat in Btu required to raise the temperature of a pound of water from 32°F to the boiling point at the pressure indicated.

Enthalpy of evaporation h_{fg} (latent heat of vaporization) is the amount of heat required to change a pound of boiling water at a given pressure to a pound of steam at the same pressure. This same amount of heat is released when the vapor is condensed back to a liquid.

Enthalpy of the steam h_g (total heat) is the combined enthalpy of liquid and vapor and represents the total heat above 32°F in the steam.

Specific volume, the reciprocal of density, is the volume of unit mass and indicates the volumetric space that 1 lb of steam or water occupies.

An understanding of the above helps explain some of the following unique properties and advantages of steam:

- Most of the heat content of steam is stored as latent heat, which permits large quantities of heat to be transmitted efficiently with little change in temperature. Because the temperature of saturated steam is pressure-dependent, a negligible temperature reduction occurs from the reduction in pressure caused by pipe friction losses as steam flows through the system. This occurs regardless of the insulation efficiency, as long as the boiler maintains the initial pressure and the steam traps remove the condensate. Conversely, in a hydronic system, inadequate insulation can significantly reduce fluid temperature.
- Steam, as all fluids, flows from areas of high pressure to areas of low pressure and is able to move throughout a system without an external energy source. Heat dissipation causes the vapor to condense, which creates a reduction in pressure caused by the dramatic change in specific volume (1600:1 at atmospheric pressure).
- As steam gives up its latent heat at the terminal equipment, the condensate that forms is initially at the same pressure and temperature as the steam. When this condensate is discharged to a lower

The preparation of this chapter is assigned to TC 6.1, Hydronic and Steam Equipment and Systems.

Table 1 Properties of Saturated Steam

Pressure, psi		Saturation Temperature, °F	Specific Volume, ft³/lb		Enthalpy, Btu/lb		
Gage	Absolute		Liquid v_f	Steam v_g	Liquid h_f	Evaporation h_{fg}	Steam h_g
25 in. Hg vac.	2.47	134	0.0163	142.2	101	1018	1119
9.6 in. Hg vac.	10.0	193	0.0166	38.4	161	982	1143
0	14.7	212	0.0167	26.8	180	970	1150
2	16.7	218	0.0168	23.8	187	966	1153
5	19.7	227	0.0168	20.4	195	961	1156
15	29.7	250	0.0170	13.9	218	946	1164
50	64.7	298	0.0174	6.7	267	912	1179
100	114.7	338	0.0179	3.9	309	881	1190
150	164.7	366	0.0182	2.8	339	857	1196
200	214.7	388	0.0185	2.1	362	837	1200

Note: Values are rounded off or approximated to illustrate the various properties discussed in the text. For calculation and design, use values of thermodynamic properties of water shown in Chapter 6 of the 1997 *ASHRAE Handbook—Fundamentals* or a similar table.

pressure (as when a steam trap passes condensate to the return system), the condensate contains more heat than necessary to maintain the liquid phase at the lower pressure; this excess heat causes some of the liquid to vaporize or "flash" to steam at the lower pressure. The amount of liquid that flashes to steam can be calculated as follows:

$$\% \text{ Flash Steam} = \frac{100(h_{f1} - h_{f2})}{h_{fg2}} \quad (1)$$

where

h_{f1} = enthalpy of liquid at pressure p_1
h_{f2} = enthalpy of liquid at pressure p_2
h_{fg2} = latent heat of vaporization at pressure p_2

Flash steam contains significant and useful heat energy that can be recovered and used (see the section on Heat Recovery). This reevaporation of condensate can be controlled (minimized) by subcooling the condensate within the terminal equipment before it discharges into the return piping. The volume of condensate that is subcooling should not be so large as to cause a significant loss of heat transfer (condensing) surface.

EFFECTS OF WATER, AIR, AND GASES

The enthalpies shown in Table 1 are for dry saturated steam. Most systems operate near these theoretically available values, but the presence of water and gases can affect enthalpy, as well as have other adverse operating effects.

Dry saturated steam is pure vapor without entrained water droplets. However, some amount of water usually carries over as condensate forms because of heat losses in the distribution system. **Steam quality** describes the amount of water present and can be determined by calorimeter tests.

While steam quality might not have a significant effect on the heat transfer capabilities of the terminal equipment, the backing up or presence of condensate can be significant because the enthalpy of condensate h_f is negligible compared with the enthalpy of evaporation h_{fg}. If condensate does not drain properly from pipes and coils, the rapidly flowing steam can push a slug of condensate through the system. This can cause water hammer and result in objectionable noise and damage to piping and system components.

The presence of air also reduces steam temperature. Air reduces heat transfer because it migrates to and insulates heat transfer surfaces. Further, oxygen in the system causes pitting of iron and steel surfaces. Carbon dioxide (CO_2) traveling with steam dissolves in condensate, forming **carbonic acid**, which is extremely corrosive to steam heating pipes and heat transfer equipment.

The combined adverse effects of water, air, and CO_2 necessitate their prompt and efficient removal.

HEAT TRANSFER

The quantity of steam that must be supplied to a heat exchanger to transfer a specific amount of heat is a function of (1) the steam temperature and quality, (2) the character and entering and leaving temperatures of the medium to be heated, and (3) the heat exchanger design. For a more detailed discussion of heat transfer, see Chapter 3 of the 1997 *ASHRAE Handbook—Fundamentals*.

BASIC STEAM SYSTEM DESIGN

Because of the various codes and regulations governing the design and operation of boilers, pressure vessels, and systems, steam systems are classified according to operating pressure. **Low-pressure systems** operate up to 15 psig, and **high-pressure systems** operate over 15 psig. There are many subclassifications within these broad classifications, especially for heating systems such as one- and two-pipe, gravity, vacuum, or variable vacuum return systems. However, these subclassifications relate to the distribution system or temperature-control method. Regardless of classification, all steam systems include a source of steam, a distribution system, and terminal equipment, where steam is used as the source of power or heat.

STEAM SOURCE

Steam can be generated directly by boilers using oil, gas, coal, wood, or waste as a fuel source, or by solar, nuclear, or electrical energy as a heat source. Steam can be generated indirectly by recovering heat from processes or equipment such as gas turbines and diesel or gas engines. The cogeneration of electricity and steam should always be considered for facilities that have year-round steam requirements. Where steam is used as a power source (such as in turbine-driven equipment), the exhaust steam may be used in heat transfer equipment for process and space heating.

Steam can be provided by a facility's own boiler or cogeneration plant or can be purchased from a central utility serving a city or specific geographic area. This distinction can be very important. A facility with its own boiler plant usually has a closed-loop system and requires the condensate to be as hot as possible when it returns to the boiler. Conversely, condensate return pumps require a few degrees of subcooling to prevent cavitation or flashing of condensate to vapor at the suction eye of pump impellers. The degree of subcooling varies, depending on the hydraulic design or characteristics of the pump in use. (See the section on Pump Suction Characteristics (NPSH) in Chapter 38.)

Central utilities often do not take back condensate, so it is discharged by the using facility and results in an open-loop system. If a utility does take back condensate, it rarely gives credit for its heat content. If condensate is returned at 180°F, and a heat recovery system reduces this temperature to 80°F, the heat remaining in the condensate represents 10 to 15% of the heat purchased from the utility. Using this heat effectively can reduce steam and heating costs by 10% or more (see the section on Heat Recovery).

Boilers

Fired and waste heat boilers are usually constructed and labeled according to the ASME *Boiler and Pressure Vessel Code* because pressures normally exceed 15 psig. Details on design, construction, and application of boilers can be found in Chapter 27. Boiler selection is based on the combined loads, including heating processes and equipment that use steam, hot water generation, piping losses, and pickup allowance.

The Hydronics Institute standards (HYDI 1989) are used to test and rate most low-pressure heating boilers that have net and gross ratings. In smaller systems, selection is based on a net rating. Larger system selection is made on a gross load basis. The occurrence and nature of the load components, with respect to the total load, determine the number of boilers used in an installation.

Heat Recovery and Waste Heat Boilers

Steam can be generated by waste heat, such as exhaust from fuel-fired engines and turbines. Figure 1 schematically shows a typical exhaust boiler and heat recovery system used for diesel engines. A portion of the water used to cool the engine block is diverted as preheated makeup water to the exhaust heat boiler to obtain maximum heat and energy efficiency. Where the quantity of steam generated by the waste heat boiler is not steady or ample enough to satisfy the facility's steam requirements, a conventional boiler must generate supplemental steam.

Heat Exchangers

Heat exchangers are used in most steam systems. Steam-to-water heat exchangers (sometimes called converters or storage tanks with steam heating elements) are used to heat domestic hot water and to supply the terminal equipment. These heat exchangers are the plate type or the shell-and-tube type, where the steam is admitted to the shell and the water is heated as it circulates through the tubes. Condensate coolers (water-to-water) are sometimes used to subcool the condensate while reclaiming the heat energy.

Water-to-steam heat exchangers (steam generators) are used in high-temperature water (HTW) systems to provide process steam. Such heat exchangers generally consist of a U-tube bundle, through which the HTW circulates, installed in a tank or pressure vessel.

All heat exchangers should be constructed and labeled according to the applicable ASME *Boiler and Pressure Vessel Code*. Many jurisdictions require double-wall construction in shell-and-tube heat exchangers between the steam and potable water. Chapter 43 discusses heat exchangers in detail.

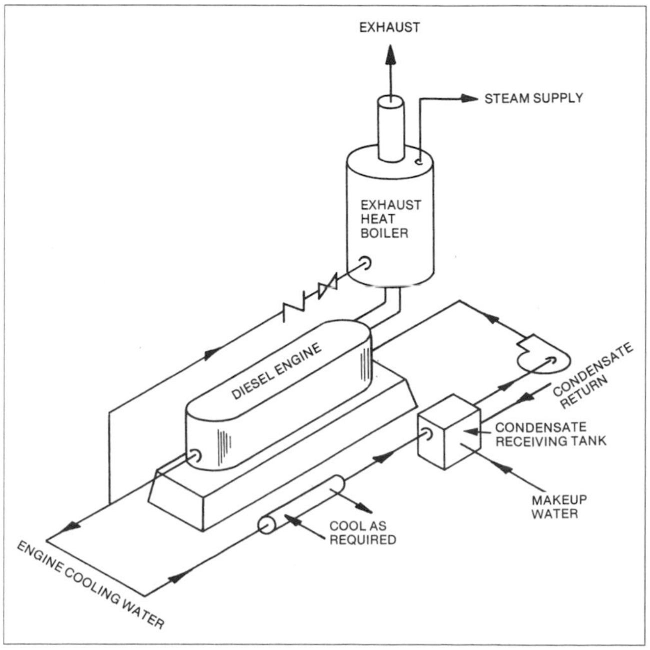

Fig. 1 Exhaust Heat Boiler

BOILER CONNECTIONS

Figure 2 shows recommended boiler connections for pumped and gravity return systems; local codes should be checked for specific legal requirements.

Supply Piping

Small boilers usually have one steam outlet connection sized to reduce steam velocity to minimize carryover of water into supply lines. Large boilers can have several outlets that minimize boiler water entrainment. The boiler manufacturer's recommendations concerning near-boiler piping should be followed because this piping may act as a steam/liquid separator for the boiler.

Figure 2 shows piping connections to the steam header. Although some engineers prefer to use an enlarged steam header for additional storage space, if there is no sudden demand for steam except during the warm-up period, an oversized header may be a disadvantage. The boiler header can be the same size as the boiler connection or the pipe used on the steam main. The horizontal runouts from the boiler(s) to the header should be sized by calculating the heaviest load that will be placed on the boiler(s). The runouts should be sized on the same basis as the building mains. Any change in size after the vertical uptakes should be made by reducing elbows.

Return Piping

Cast-iron boilers have return tappings on both sides, and steel boilers may have one or two return tappings. Where two tappings are provided, both should be used to effect proper circulation through the boiler. Condensate in boilers can be returned by a pump or a gravity return system. Return connections shown in Figure 2 for a multiple-boiler gravity return installation may not always maintain the correct water level in all boilers. Extra controls or accessories may be required.

Recommended return piping connections for systems using gravity return are detailed in Figure 3. Dimension A must be at least 28 in. for each 1 psig maintained at the boiler to provide the pressure required to return the condensate to the boiler. To provide a reasonable safety factor, make dimension A at least 14 in. for small systems with piping sized for a pressure drop of 1/8 psi, and at least 28 in. for larger systems with piping sized for a pressure drop of 0.5 psi. The **Hartford loop** protects against a low water condition, which can occur if a leak develops in the wet return portion of the piping. The Hartford loop takes the place of a check valve on the wet return; however, certain local codes require check valves. Because of hydraulic pressure limitations, gravity return systems are only suitable for systems operating at a boiler pressure between 0.5 and 1 psig. However, since these systems have minimum mechanical equipment and low initial installed cost, they are appropriate for many small systems. Kremers (1982) and Stamper and Koral (1979) provide additional design information on piping for gravity return systems.

Recommended piping connections for steam boilers with pump-returned condensate are shown in Figure 2. Common practice provides an individual condensate or boiler feedwater pump for each

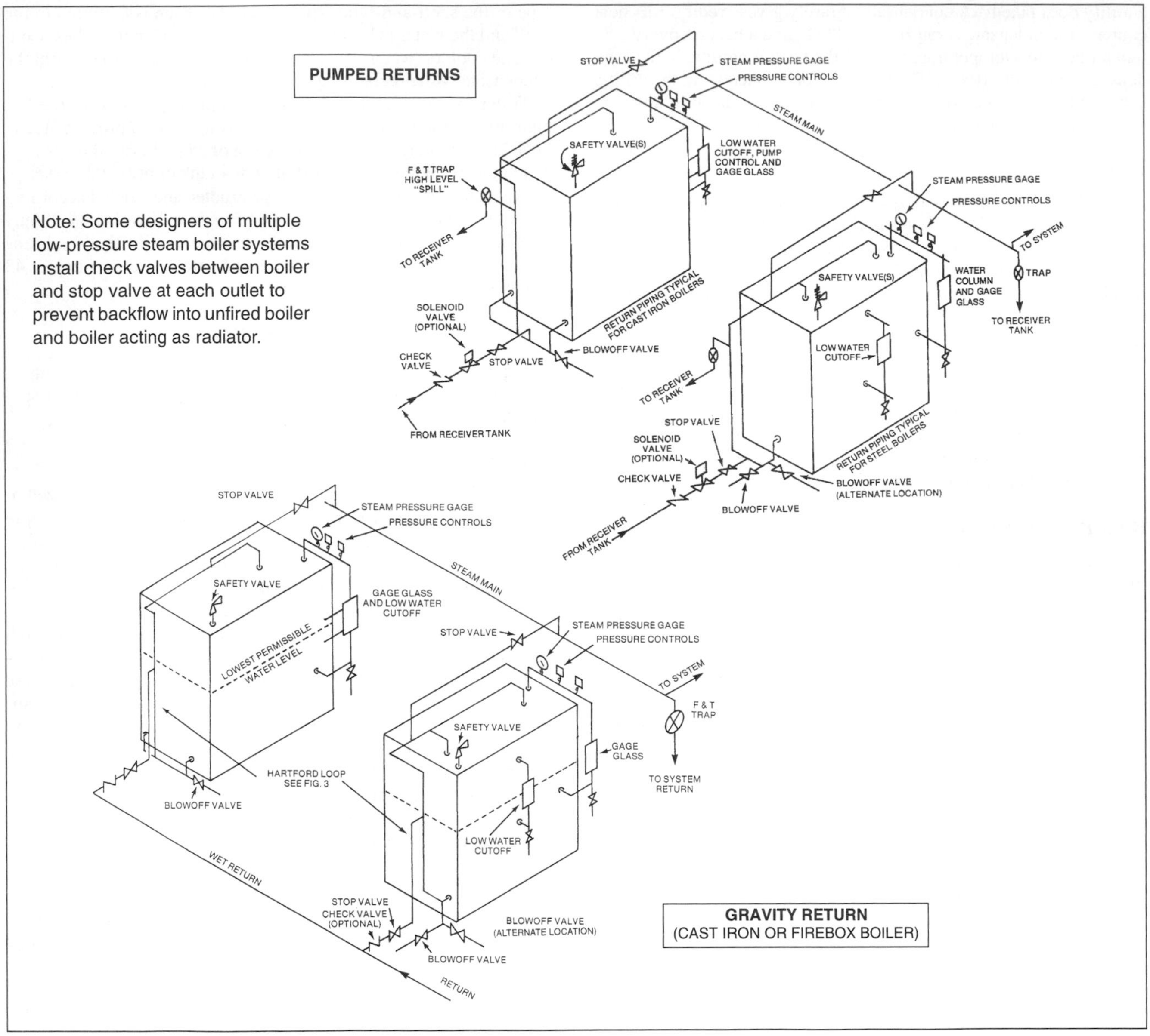

Fig. 2 Typical Boiler Connections

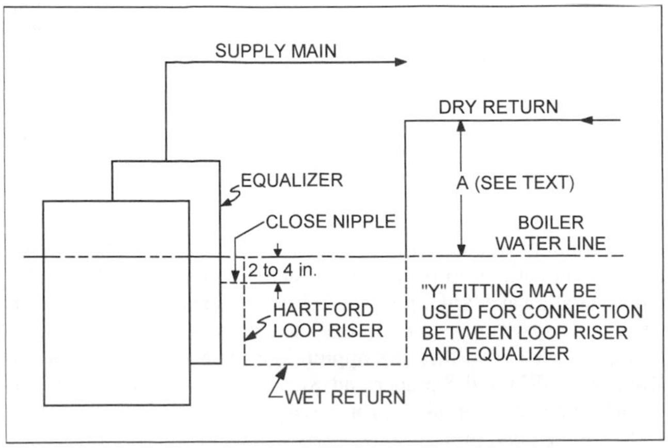

Fig. 3 Boiler with Gravity Return

boiler. Pump operation is controlled by the boiler water level control on each boiler. However, one pump may be connected to supply the water to each boiler from a single manifold by using feedwater control valves regulated by the individual boiler water level controllers. When such systems are used, the condensate return pump runs continuously to pressurize the return header.

Return piping should be sized based on total load. The line between the pump and boiler should be sized for a very small pressure drop and the maximum pump discharge flow rate.

DESIGN STEAM PRESSURE

One of the most important decisions in the design of a steam system is the selection of the generating, distribution, and utilization pressures. Considering investment cost, energy efficiency, and control stability, the pressure should be held to the minimum values above atmospheric pressure that accomplish the required heating task, unless detailed economic analysis indicates advantages in higher pressure generation and distribution.

The first step in selecting pressures is to analyze the load requirements. Space heating and domestic water heating can best be served, directly or indirectly, with low-pressure steam less than 15 psig or 250°F. Other systems that can be served with low-pressure steam include single-stage absorption units (10 psig), cooking, warming, dishwashing, and snow melting heat exchangers. Thus, from the standpoint of load requirements, high-pressure steam (above 15 psig) is required only for loads such as dryers, presses, molding dies, power drives, and other processing, manufacturing, and power requirements. The load requirement establishes the pressure requirement.

When the source is close to the load(s), the generation pressure should be high enough to provide the (1) load design pressure, (2) friction losses between the generator and the load, and (3) control range. Losses are caused by flow through the piping, fittings, control valves, and strainers. If the generator(s) is located remote from the loads, there could be some economic advantage in distributing the steam at a higher pressure to reduce pipe size. When this is considered, the economic analysis should include the additional investment and operating costs associated with a higher pressure generation system. When an increase in the generating pressure requires a change from below to above 15 psig, the generating system equipment changes from low-pressure class to high-pressure class and there are significant increases in both investment and operating cost.

Where steam is provided from a nonfired device or prime mover such as a diesel engine cooling jacket, the source device can have an inherent pressure limitation.

PIPING

The piping system distributes the steam, returns the condensate, and removes air and noncondensable gases. In steam heating systems, it is important that the piping distribute steam, not only at full design load, but at partial loads and excess loads that can occur on system warm-up. The usual average winter steam demand is less than half the demand at the lowest outdoor design temperature. However, when the system is warming up, the load on the steam mains and returns can exceed the maximum operating load for the coldest design day, even in moderate weather. This load comes from raising the temperature of the piping to the steam temperature and that of the building to the indoor design temperature. Supply and return piping should be sized according to Chapter 33 of the 1997 *ASHRAE Handbook—Fundamentals*.

Supply Piping Design Considerations

1. Size pipe according to Chapter 33 of the 1997 *ASHRAE Handbook—Fundamentals*, taking into consideration pressure drop and steam velocity.
2. Pitch piping uniformly down in the direction of steam flow at 0.25 in. per 10 ft. If piping cannot be pitched down in the direction of the steam flow, refer to Chapter 33 of the 1997 *ASHRAE Handbook—Fundamentals* for rules on pipe sizing and pitch.
3. Insulate piping well to avoid unnecessary heat loss (see Chapters 22 and 23 of the 1997 *ASHRAE Handbook—Fundamentals*).
4. Condensate from unavoidable heat loss in the distribution system must be removed promptly to eliminate water hammer and degradation of steam quality and heat transfer capability. Install drip legs at all low points and natural drainage points in the system, such as at the ends of mains and the bottoms of risers, and ahead of pressure regulators, control valves, isolation valves, pipe bends, and expansion joints. On straight horizontal runs with no natural drainage points, space drip legs at intervals not exceeding 300 ft when the pipe is pitched down in the direction of the steam flow and at a maximum of 150 ft when the pipe is pitched up, so that condensate flow is opposite of steam flow. These distances apply to systems where valves are opened manually to remove air and excess condensate that forms during warm-up conditions. Reduce these distances by about half in systems that are warmed up automatically.

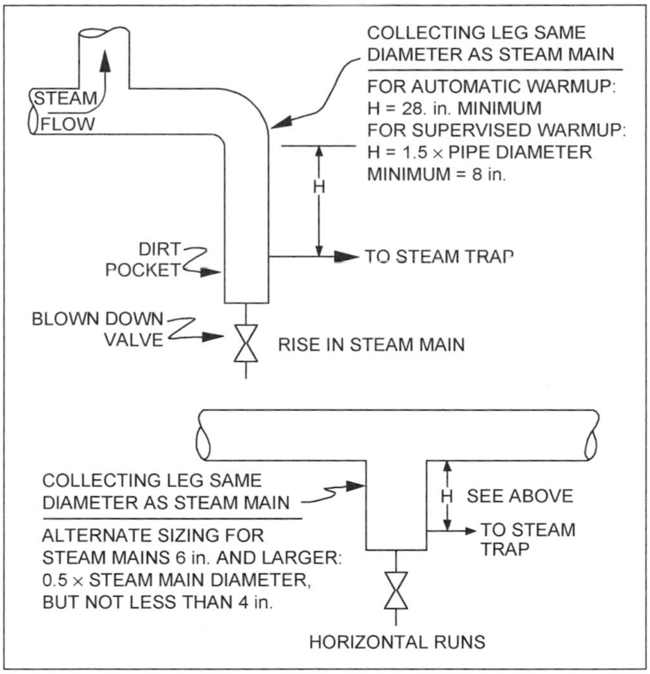

Fig. 4 Method of Dripping Steam Mains

5. Where horizontal piping must be reduced in size, use eccentric reducers that permit the continuance of uniform pitch along the bottom of piping (in downward pitched systems). Avoid concentric reducers on horizontal piping, because they can cause water hammer.
6. Take off all branch lines from the top of the steam mains, preferably at a 45° angle, although vertical 90° connections are acceptable.
7. Where the length of a branch takeoff is less than 10 ft, the branch line can be pitched back 0.5 in. per 10 ft, providing drip legs as described previously in (4).
8. Size drip legs properly to separate and collect the condensate. Drip legs at vertical risers should be full-size and extend beyond the riser, as shown in Figure 4. Drip legs at other locations should be the same diameter as the main. In steam mains 6 in. and over, this can be reduced to half the diameter of the main, but to no less than 4 in. Where warm-up is supervised, the length of the collecting leg is not critical. However, the recommended length is one and a half times the pipe diameter and not less than 8 in. For automatic warm-up, collecting legs should always be the same size as the main and should be at least 28 in. long to provide the hydraulic pressure differential necessary for the trap to discharge before a positive pressure is built up in the steam main.
9. Condensate should flow by gravity from the trap to the return piping system. Where the steam trap is located below the return line, the condensate must be lifted. In systems operating above 40 psig, the trap discharge can usually be piped directly to the return system (Figure 5). However, back pressure at the trap discharge (return line pressure plus hydraulic pressure created by height of lift) must not exceed steam main pressure, and the trap must be sized after considering back pressure. A collecting leg must be used and the trap discharge must flow by gravity to a vented condensate receiver, from which it is pumped to the overhead return in systems (1) operating under

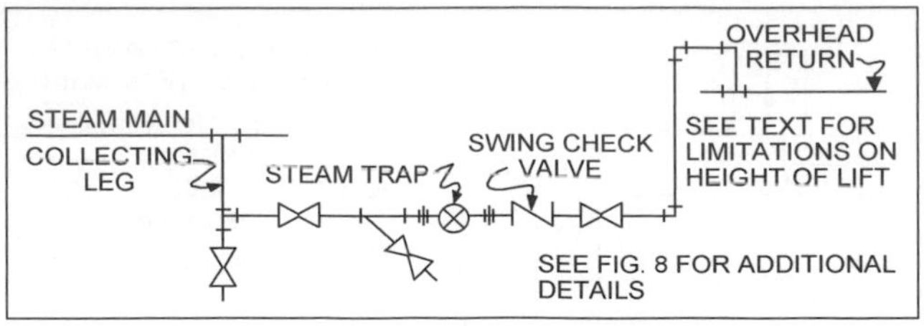

Fig. 5 Trap Discharging to Overhead Return

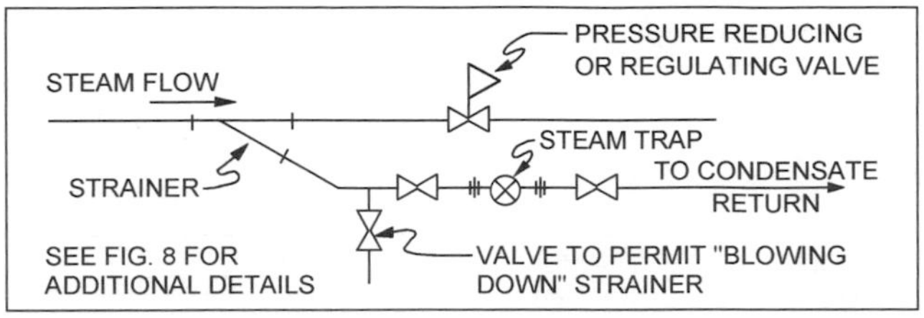

Fig. 6 Trapping Strainers

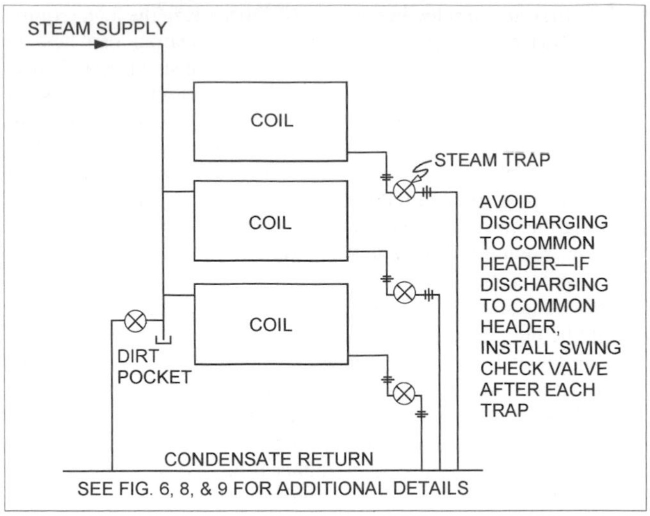

Fig. 7 Trapping Multiple Coils

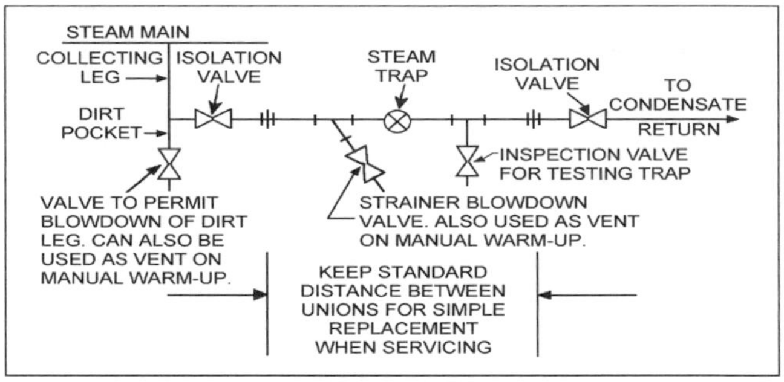

Fig. 8 Recommended Steam Trap Piping

40 psig, (2) where the temperature is regulated by modulating the steam control valves, or (3) where the back pressure at the trap is close to system pressure. The trap discharge should flow by gravity to a vented condensate receiver from which it is pumped to the overhead return.

10. Strainers installed before the pressure-reducing and control valves are a natural water collection point. Since water carry-over can erode the valve seat, install a trap at the strainer blow-down connection (Figure 6).

Terminal Equipment Piping Design Considerations

1. Size piping the same as the supply and return connections of the terminal equipment.
2. Keep equipment and piping accessible for inspection and maintenance of the steam traps and control valves.
3. Minimize strain caused by expansion and contraction with pipe bends, loops, or three elbow swings to take advantage of piping flexibility, or with expansion joints or flexible pipe connectors.
4. In multiple-coil applications, separately trap each coil for proper drainage (Figure 7). Piping two or more coils to a common header served by a single trap can cause condensate backup, improper heat transfer, and inadequate temperature control.
5. Terminal equipment, where temperature is regulated by a modulating steam control valve, requires special consideration. Refer to the section on Condensate Removal from Temperature-Regulated Equipment.

Return Piping Design Considerations

1. Flow in the return line is two-phase, consisting of steam and condensate. See Chapter 33 of the 1997 *ASHRAE Handbook—Fundamentals* for sizing considerations.
2. Pitch return lines downward in the direction of the condensate flow at 0.5 in. per 10 ft to ensure prompt condensate removal.
3. Insulate the return line well, especially where the condensate is returned to the boiler or the condensate enthalpy is recovered. In vacuum systems, the return lines are not insulated since condensate subcooling is required.
4. Where possible and practical, use heat recovery systems to recover the condensate enthalpy. See the section on Heat Recovery. In vacuum systems, the return lines are not insulated since condensate subcooling is required.
5. Equip dirt pockets of the drip legs and strainer blowdowns with valves to remove dirt and scale.
6. Install steam traps close to drip legs and make them accessible for inspection and repair. Servicing is simplified by making the pipe sizes and configuration identical for a given type and size of trap. The piping arrangement shown in Figure 8 facilitates inspection and maintenance of steam traps.
7. When elevating condensate to an overhead return, consider the pressure at the trap inlet and the fact that it requires approximately 1 psi to elevate condensate 2 ft. See (9) in the section on Supply Piping Design Considerations for a complete discussion.

CONDENSATE REMOVAL FROM TEMPERATURE-REGULATED EQUIPMENT

When air, water, or another product is heated, the temperature or heat transfer rate can be regulated by a modulating steam pressure control valve. Because pressure and temperature do not vary at the same rate as load, the steam trap capacity, which is determined by the pressure differential between the trap inlet and outlet, may be adequate at full load, but not at some lesser load.

Analysis shows that steam pressure must be reduced dramatically to achieve a slight lowering of temperature. In most applications, this can result in subatmospheric pressure in the coil, while as much as 75% of the full condensate load has to be handled by the steam trap. This is especially important for coils exposed to outside air, because subatmospheric conditions can occur in the coil at outside temperatures below 32°F, and the coil will freeze if the condensate is not removed.

Armers (1985) provides detailed methods for determining condensate load under various operating conditions. However, in most cases, this load need not be calculated if the coils are piped as shown in Figure 9 and this procedure is followed:

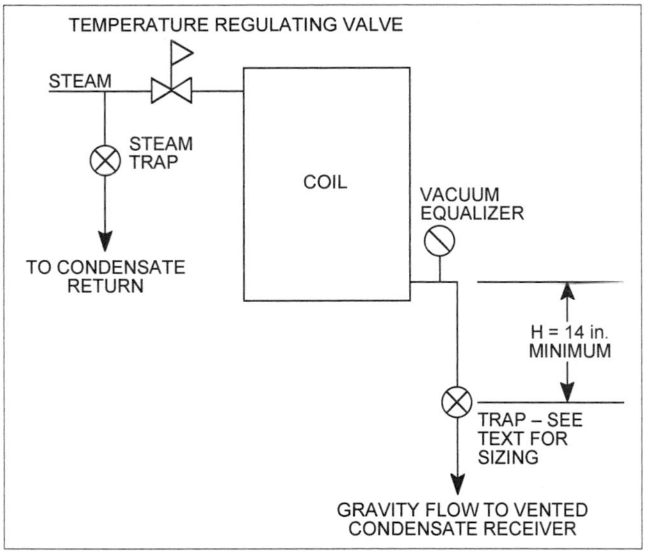

Fig. 9 Trapping Temperature-Regulated Coils

1. Place the steam trap 1 to 3 ft below the bottom of the steam coil to provide a pressure of approximately 0.5 to 1.5 psig. Locating the trap at less than 12 in. usually results in improper drainage and operating difficulties.
2. Install vacuum breakers between the coil and trap inlet to ensure that the pressure can drain the coil when it is atmospheric or subatmospheric. The vacuum breaker should respond to a differential pressure of no greater than 3 in. of water. For atmospheric returns, the vacuum breaker should be opened to the atmosphere, and the return system must be designed to ensure no pressurization of the return line. In vacuum return systems, the vacuum breaker should be piped to the return line.
3. Discharge from the trap must flow by gravity, without any lifts in the piping, to the return system, which must be vented properly to the atmosphere to eliminate any back pressure that could prevent the trap from draining the coil. Where the return main is overhead, the trap discharge should flow by gravity to a vented receiver, from which it is then pumped to the overhead return.
4. Design traps to operate at maximum pressure at the control valve inlet and size them to handle the full condensate load at a pressure differential equal to the hydraulic pressure between the trap and coil. The actual condensate load can vary from the theoretical design load because of the safety factors used in coil selection and the fact that condensate does not always form at a uniform steady rate; therefore, size steam traps according to the following:

   For an actual steam pressure p_1 in the coil at full condensate load w, the proportion X of full load needing atmospheric pressure at the coil is

$$X = \frac{212 - t_c}{t_s - t_c} \quad (2)$$

   where

   t_c = control temperature, °F
   t_s = steam temperature at p_1, °F

   Then, the steam trap must be sized both to pass the full load w at a differential pressure equal to p_1 and to pass $X \cdot w$ (the proportion of full load) at 0.5 psi.
5. To reduce the possibility of a steam coil freezing, the temperature-regulating valve is often left wide open, and the leaving air temperature is controlled by a face-and-bypass damper on the steam coil.
6. For air temperatures below freezing, traps selected for draining steam coils should fail open (e.g., bucket traps) or two traps in parallel should be used.

STEAM TRAPS

Steam traps are an essential part of all steam systems, except one-pipe steam heating systems. Traps discharge condensate, which forms as steam gives up some of its heat, and direct the air and noncondensable gases to a point of removal. Condensate forms in steam mains and distribution piping because of unavoidable heat losses through less-than-perfect insulation, as well as in terminal equipment such as radiators, convectors fan-coil units, and heat exchangers, where steam gives up heat during normal operation. Condensate must always be removed from the system as soon as it accumulates for the following reasons:

- Although condensate contains some valuable heat, using this heat by holding the condensate in the terminal equipment reduces the heat transfer surface. It also causes other operating problems because it retains air, which further reduces heat transfer, and noncondensable gases such as CO_2, which cause corrosion. As discussed in the section on Steam Source, recovering condensate heat is usually only desirable when the condensate is not returned to the boiler. Methods for this are discussed in the section on Heat Recovery.
- Steam moves rapidly in mains and supply piping, so when condensate accumulates to the point where the steam can push a slug of it, serious damage can occur from the resulting water hammer.

Ideally, the steam trap should remove all condensate promptly, along with air and noncondensable gases that might be in the system, with little or no loss of live steam. A steam trap is an automatic valve that can distinguish between steam and condensate or other fluids. Traps are classified as follows:

- **Thermostatic traps** react to the difference in the temperatures of steam and condensate.
- **Mechanical traps** depend on the difference in the densities of steam and condensate.
- **Kinetic traps** rely on the difference in the flow characteristics of steam and condensate.

The following points apply to all steam traps:

- No single type of steam trap is best suited to all applications, and most systems require more than one type of trap.
- Steam traps, regardless of type, should be carefully sized for the application and condensate load to be handled, because both undersizing and oversizing can cause serious problems. Undersizing can result in undesirable condensate backup and excessive cycling, which can lead to premature failure. Oversizing might appear to solve this problem and make selection much easier because fewer different sizes are required, but if the trap fails, excessive steam can be lost.

Steam traps should be installed between two unions to facilitate maintenance and/or replacement.

Thermostatic Traps

In thermostatic traps, a bellows or bimetallic element operates a valve that opens in the presence of subcooled condensate and closes in the presence of steam (Figure 10). Because condensate is initially at the same temperature as the steam from which it was condensed, the thermostatic element must be designed and calibrated to open at a temperature below the steam temperature; otherwise, the trap would blow live steam continuously. Therefore, the condensate is subcooled by allowing it to back up in the trap and in a portion of the upstream drip leg piping, both of which are left uninsulated. Some thermostatic traps operate with a continuous water leg behind the trap so there is no steam loss; however, this prohibits the discharge of air and noncondensable gases, and can cause excessive condensate to back up into the mains or terminal equipment, thereby resulting in operating problems. Devices that operate without significant backup can lose steam before the trap closes.

Although both bellows and bimetallic traps are temperature-sensitive, their operations are significantly different. The **bellows thermostatic trap** has a fluid with a lower boiling point than water. When the trap is cold, the element is contracted and the discharge port is open. As hot condensate enters the trap, it causes the contents of the bellows to boil and vaporize before the condensate temperature rises to steam temperature. Because the contents of the bellows boil at a lower temperature than water, the vapor pressure inside the bellows element is greater than the steam pressure outside, causing the element to expand and close the discharge port.

Assuming the contained liquid has a pressure-temperature relationship similar to that of water, the balance of forces acting on the bellows element remains relatively constant, no matter how the steam pressure varies. Therefore, this is a balanced pressure device that can be used at any pressure within the operating range of the device. However, this device should not be used where superheated steam is present, because the temperature is no longer in step with the pressure and damage or rupture of the bellows element can occur.

Bellows thermostatic traps are best suited for steady light loads on low-pressure service. They are most widely used in radiators and convectors in HVAC applications.

The **bimetallic thermostatic trap** has an element made from metals with different expansion coefficients. Heat causes the element to change shape, permitting the valve port to open or close. Because a bimetallic element responds only to temperature, most traps have the valve on the outlet so that steam pressure is trying to open the valve. Therefore, by properly designing the bimetallic element, the trap can operate on a pressure-temperature curve approaching the steam saturation curve, although not as closely as a balanced pressure bellows element.

Unlike the bellows thermostatic trap, bimetallic thermostatic traps are not adversely affected by superheated steam or subject to damage by water hammer, so they can be readily used for high-pressure applications. They are best suited for steam tracers, jacketed piping, and heat transfer equipment, where some condensate backup is tolerable. If they are used on steam main drip legs, the element should not back up condensate.

Mechanical Traps

Mechanical traps are buoyancy operated, depending on the difference in density between steam and condensate. The **float and thermostatic trap** (Figure 10) is commonly called the F&T trap and is actually a combination of two types of traps in a single trap body: (1) a bellows thermostatic element operating on temperature difference, which provides automatic venting, and (2) a float portion, which is buoyancy operated. Float traps without automatic venting should not be used in steam systems.

On start-up, the float valve is closed and the thermostatic element is open for rapid air venting, permitting the system or equipment to rapidly fill with steam. When steam enters the trap body, the thermostatic element closes and, as condensate enters, the float rises and the condensate discharges. The float regulates the valve opening so it continuously discharges the condensate at the rate at which it reaches the trap.

The F&T trap has large venting capabilities, continuously discharges condensate without backup, handles intermittent loads very well, and can operate at extremely low pressure differentials. Float

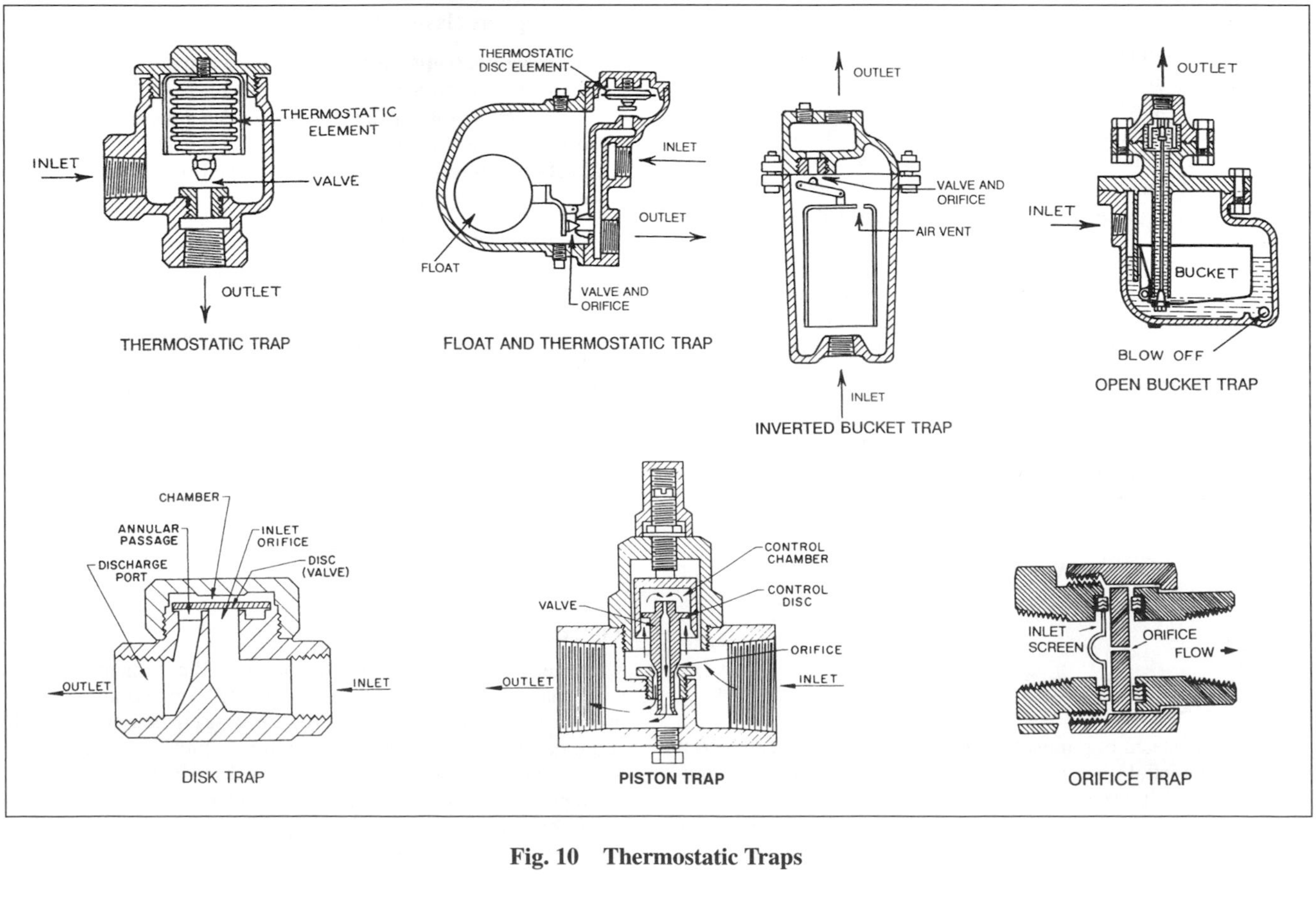

Fig. 10 Thermostatic Traps

and thermostatic traps are suited for use with temperature-regulated steam coils. They also are well suited for steam main and riser drip legs on low-pressure steam heating systems. Although F&T traps are available for pressures to 250 psig or higher, they are susceptible to water hammer, so other traps are usually a better choice for high-pressure applications.

Bucket traps operate on buoyancy, but they use a bucket that is either open at the top or inverted instead of a closed float. Initially, the bucket in an open bucket trap is empty and rests on the bottom of the trap body with the discharge vented, and, as condensate enters the trap, the bucket floats up and closes the discharge port. Additional condensate overflows into the bucket, causing it to sink and open the discharge port, which allows steam pressure to force the condensate out of the bucket. At the same time, it seals the bottom of the discharge tube, prohibiting air passage. Therefore, to prevent air binding, this device has an automatic air vent, as does the F&T trap.

Inverted bucket traps eliminate the size and venting problems associated with open bucket traps. Steam or air entering the submerged inverted bucket causes it to float and close the discharge port. As more condensate enters the trap, it forces air and steam out of the vent on top of the inverted bucket into the trap body where the steam condenses by cooling. When the mass of the bucket exceeds the buoyancy effect, the bucket drops, opening the discharge port, and steam pressure forces the condensate out, and the cycle repeats.

Unlike most cycling-type traps, the inverted bucket trap continuously vents air and noncondensable gases. Although it discharges condensate intermittently, there is no condensate backup in a properly sized trap. Inverted bucket traps are made for all pressure ranges and are well suited for steam main drip legs and most HVAC applications. Although inverted bucket traps can be used for temperature-regulated steam coils, the F&T trap is usually better because it has the high venting capability desirable for such applications.

Kinetic Traps

Numerous devices operate on the difference between the flow characteristics of steam and condensate and on the fact that condensate discharging to a lower pressure contains more heat than necessary to maintain the liquid phase. This excess heat causes some of the condensate to flash to steam at the lower pressure.

Thermodynamic traps or **disk traps** are simple devices with only one moving part. When air or condensate enters the trap on start-up, it lifts the disk off its seat and is discharged. When steam or hot condensate (some of which "flashes" to steam upon exposure to a lower pressure) enters the trap, the increased velocity of this vapor flow decreases the pressure on the underside of the disk and increases the pressure above the disk, causing it to snap shut. Pressure is then equalized above and below the disk, but because the area exposed to pressure is greater above than below it, the disk remains shut until the pressure above is reduced by condensing or bleeding, thus permitting the disk to snap open and repeat the cycle. This device does not cycle open and shut as a function of condensate load, it is a time-cycle device that opens and shuts at fixed intervals as a function of how fast the steam above the disk condenses. Because disk traps require a significant pressure differential to operate properly, they are not well suited for low-pressure systems or for systems with significant back pressure. They are best suited for high-pressure systems and are widely applied to steam main drip legs.

Impulse traps or **piston traps** continuously pass a small amount of steam or condensate through a "bleed" orifice, changing the pressure positions within the piston. When live steam or very hot condensate that flashes to steam enters the control chamber, the increased pressure closes the piston valve port. When cooler condensate enters, the pressure decreases, permitting the valve port to open. Most impulse traps cycle open and shut intermittently, but some modulate to a position that passes condensate continuously.

Impulse traps can be used for the same applications as disk traps; however, because they have a small "bleed" orifice and close piston tolerances, they can stick or clog if dirt is present in the system.

Orifice traps have no moving parts. All other traps have discharge ports or orifices, but in the traps described previously, this opening is oversized, and some type of closing mechanism controls the flow of condensate and prevents the loss of live steam.

The orifice trap has no such closing mechanism, and the flow of steam and condensate is controlled by two-phase flow through an orifice. A simple explanation of this theory is that an orifice of any size has a much greater capacity for condensate than it does for steam because of the significant differences in their densities and because "flashing" condensate tends to choke the orifice. An orifice is selected larger than required for the actual condensate load; therefore, it continuously passes all condensate along with the air and noncondensable gases, plus a small controlled amount of steam. The steam loss is usually comparable to that of most cycling-type traps.

Orifice traps must be sized more carefully than cycling-type traps. On light condensate loads, the orifice size is small and, like impulse traps, tends to clog. Orifice traps are suitable for all system pressures and can operate against any back pressure. They are best suited for steady pressure and load conditions such as steam main drip legs.

PRESSURE-REDUCING VALVES

Where steam is supplied at pressures higher than required, one or more pressure-reducing valves (pressure regulators) are required. The pressure-reducing valve reduces pressure to a safe point and regulates pressure to that required by the equipment. The district heating industry refers to valves according to their functional use. There are two classes of service: (1) the steam must be shut off completely (dead-end valves) to prevent buildup of pressure on the low-pressure side during no load (single-seated valves should be used) and (2) the low-pressure lines condense enough steam to prevent buildup of pressure from valve leakage (double-seated valves can be used). Valves available for either service are direct-operated, spring-loaded, weight-loaded, air-loaded, or pilot-controlled, using either the flowing steam or auxiliary air or water as the operating medium. The direct-operated, double-seated valve is less affected by varying inlet steam pressure than the direct-operated, single-seated valve. Pilot-controlled valves, either single- or double-seated, tend to eliminate the effect of variable inlet pressures.

Installation

Pressure-reducing valves should be readily accessible for inspection and repair. There should be a bypass around each reducing valve equal to the area of the reducing-valve seat ring. The globe valve in a bypass line should have plug disk construction and must have an absolutely tight shutoff. Steam pressure gages, graduated up to the initial pressure, should be installed on the low-pressure side and on the high-pressure side. The low-pressure gage should be ahead of the shutoff valve because the reducing valve can be adjusted with the shutoff valve closed. A similar gage should be installed downstream from the shutoff valve for use during manual operation. Typical service connections are shown in Figure 11 for low-pressure service and Figure 12 for high-pressure service. In the smaller sizes, the standby pressure-regulating valve can be removed, a filler installed, and the inlet stop valves used for manual pressure regulation until repairs are made.

Strainers should be installed on the inlet of the primary pressure-reducing valve and before the second-stage reduction if there is considerable piping between the two stages. If a two-stage reduction is made, it is advisable to install a pressure gage immediately before

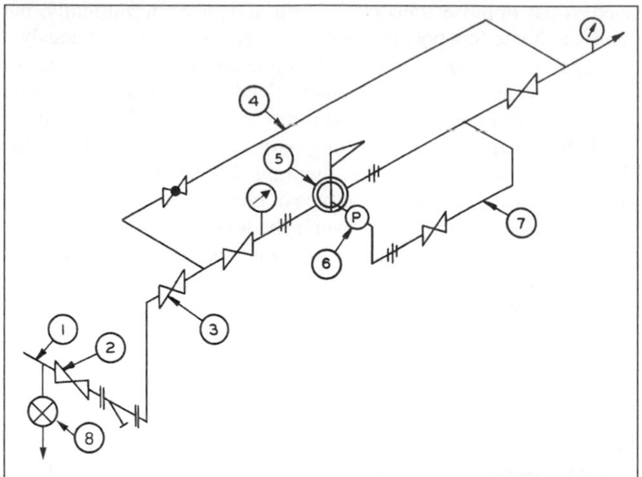

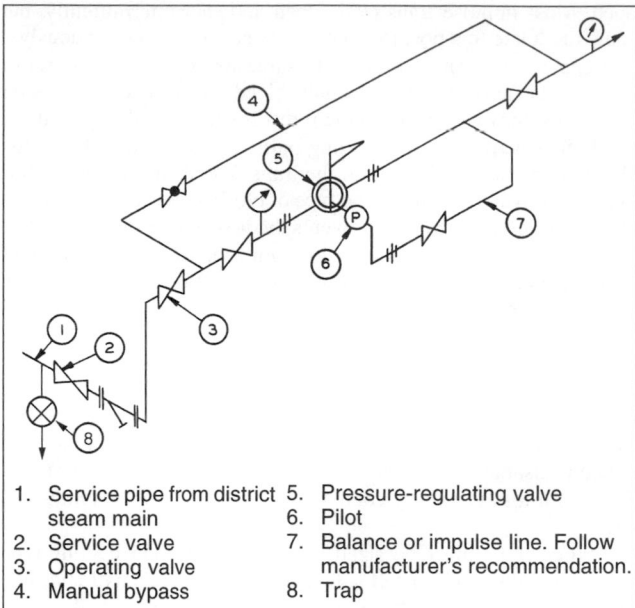

1. Service pipe from district steam main
2. Service valve
3. Operating valve
4. Manual bypass
5. Pressure-regulating valve
6. Pilot
7. Balance or impulse line. Follow manufacturer's recommendation.
8. Trap

Fig. 11 Pressure-Reducing Valve Connections—Low Pressure

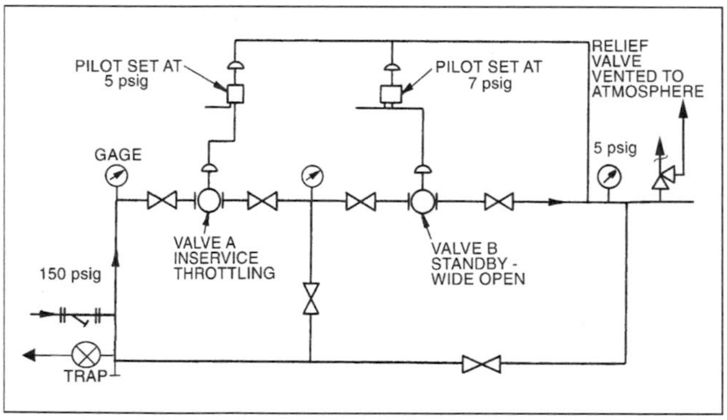

Fig. 12 Pressure-Reducing Valve Connections—High Pressure

the reducing valve of the second-stage reduction to set and check the operation of the first valve. A drip trap should be installed before the two reducing valves.

Where pressure-reducing valves are used, one or more relief devices or safety valves must be provided, and the equipment on the low-pressure side must meet the requirements for the full initial pressure. The relief or safety devices are adjoining or as close as possible to the reducing valve. The combined relieving capacity must be adequate to avoid exceeding the design pressure of the low-pressure system if the reducing valve does not open. In most areas, local codes dictate the safety relief valve installation requirements.

Safety valves should be set at least 5 psi higher than the reduced pressure if the reduced pressure is under 35 psig and at least 10 psi higher than the reduced pressure if the reduced pressure is above 35 psig or the first-stage reduction of a double reduction. The outlet from relief valves should not be piped to a location where the discharge can jeopardize persons or property or is a violation of local codes.

Figure 13 shows a typical service installation, with a separate line to various heating zones and process equipment. If the initial pressure is below 50 psig, the first-stage pressure-reducing valve

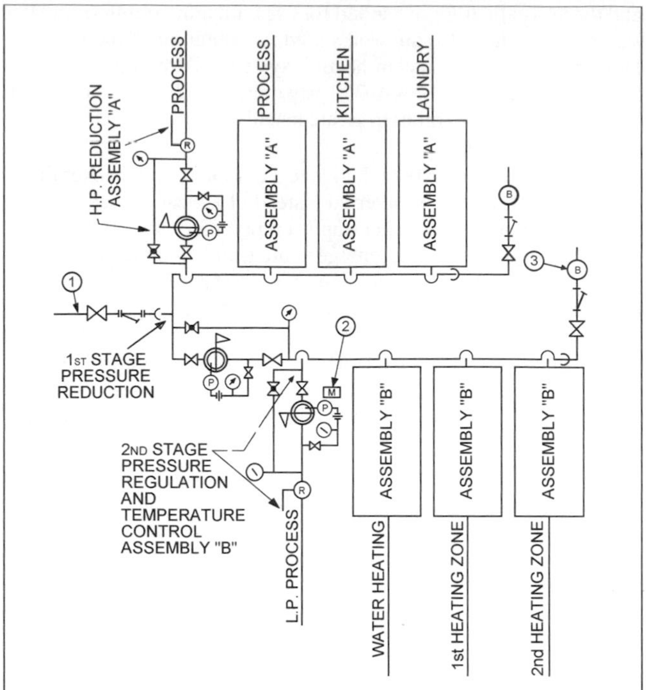

1. Service pipe from district steam main
2. Motorized pilot on pressure-regulating valve
3. Bucket trap on header drip line

Note: Where process equipment is individually controlled, temperature control on header may be omitted. All fittings are American Standard Class 250 cast iron or 300 steel in both reduction and regulation assemblies.

Fig. 13 Steam Supply

can be omitted. In Assembly A (Figure 13), the single-stage pressure-reducing valve is also the pressure-regulating valve.

When making a two-stage reduction (e.g., 150 to 50 psig and then 50 to 2 psig), allow for the expansion of steam on the low-pressure side of each reducing valve by increasing the pipe area to about double the area of a pipe the size of the reducing valve. This also allows steam to flow at a more uniform velocity. It is recommended that the valves be separated by a distance of at least 20 ft to reduce excessive hunting action of the first valve, although this should be checked with the valve manufacturer.

Figure 14 shows a typical double-reduction installation where the pressure in the district steam main is higher than can safely be applied to building heating systems. The first or pressure-reducing valve effects the initial pressure reduction. The pressure-regulating valve regulates the steam to the desired final pressure.

Pilot-controlled or air-loaded direct-operated reducing valves can be used without limitation for all reduced-pressure settings for all heating, process, laundry, and hot water services. Spring-loaded direct-operated valves can be used for reduced pressures up to 50 psig, providing they can pass the required steam flow without excessive deviation in reduced pressure.

Weight-loaded valves may be used for reduced pressures below 15 psig and for moderate steam flows.

Pressure-equalizing or impulse lines must be connected to serve the type of valve selected. With direct-operated diaphragm valves having rubber-like diaphragms, the impulse line should be connected to the reduced-pressure steam line to allow for maximum condensate on the diaphragm and in the impulse or equalizing line. If it is connected to the top of the steam line, a

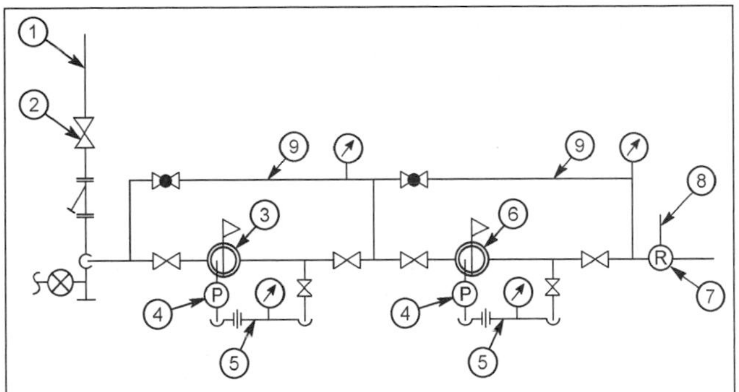

1. Service pipe from district steam main
2. Service valve
3. Pressure-reducing valve (stage 1)
4. Pilot
5. Balance line
6. Pressure-regulating valve (stage 2)
7. Safety relief valve
8. Relief valve discharge to outside
9. Bypass, one size smaller than pressure-reducing valve

Note: All fittings are American Standard Class 250 cast iron and/or 300 steel.

Fig. 14 Two-Stage Pressure-Regulating Valve
(Used where high-pressure steam is supplied for low-pressure requirements)

condensate accumulator should be used to reduce variations in the pressure of condensate on the diaphragm. The impulse line should not be connected to the bottom of the reduced pressure line since it could become clogged. Equalizing or impulse lines for pilot-controlled and direct-operated reducing valves using metal diaphragms should be connected into the expanded outlet piping approximately 2 to 4 ft from the reducing valve and pitched away from the reducing valve to prevent condensate accumulation. Pressure impulse lines for externally pilot-controlled reducing valves using compressed air or fresh water should be installed according to manufacturer's recommendations.

Valve Size Selection

Pressure-regulating valves should be sized to supply the maximum steam requirements of the heating system or equipment. Consideration should be given to rangeability, speed of load changes, and regulation accuracy required to meet system needs, especially with temperature control systems using intermittent steam flow to heat the building.

The reducing valve should be selected carefully. The manufacturer should be consulted. Piping to and from the reducing valve should be adequate to pass the desired amount of steam at the desired maximum velocity. A common error is to make the size of the reducing valve the same as the service or outlet pipe size; this makes the reducing valve oversized and causes wiredrawing or erosion of the valve and seat because of the high-velocity flow caused by the small lift of the valve.

On installations where the steam requirements are large and variable, wiredrawing and cycling control can occur during mild weather or during reduced demand periods. To overcome this condition, two reducing valves are installed in parallel, with the sizes selected on a 70 and 30% proportion of maximum flow. For example, if 10,000 lb of steam per hour are required, the size of one valve is based on 7000 lb of steam per hour, and the other is based on 3000 lb of steam per hour. During mild weather (spring and fall), the larger valve is set for slightly lower reduced pressure than the smaller one and remains closed as long as the smaller one can meet the demand. During the remainder of the heating season, the valve settings are reversed to keep the smaller one closed, except when the larger one is unable to meet the demand.

TERMINAL EQUIPMENT

A variety of terminal units are used in steam heating systems. All are suited for use on low-pressure systems. Terminal units used on high-pressure systems have heavier construction, but are otherwise similar to those on low-pressure systems. Terminal units are usually classified as follows:

1. **Natural convection units** transfer most heat by convection and some heat by radiation. The equipment includes cast-iron radiators, finned-tube convectors, and cabinet and baseboard units with convection-type elements (see Chapter 32).
2. **Forced-convection units** employ a forced air movement to increase heat transfer and distribute the heated air within the space. The devices include unit heaters, unit ventilators, induction units, fan-coil units, the heating coils of central air-conditioning units, and many process heat exchangers. When such units are used for both heating and cooling, there is a steam coil for heating and a separate chilled water or refrigerant coil for cooling. See Chapters 1, 2, 3, 4, 21, 23, and 31.
3. **Radiant panel systems** transfer some heat by convection. Because of the low-temperature and high-vacuum requirements, this type of unit is rarely used on steam systems.

Selection

The primary consideration in selecting terminal equipment is comfortable heat distribution. The following briefly describes suitable applications for specific types of steam terminal units.

Natural Convection Units

Radiators, convectors, convection-type cabinet units, and baseboard convectors are commonly used for (1) facilities that require heating only, rather than both heating and cooling, and (2) in con junction with central air conditioning as a source of perimeter heating or for localized heating in spaces such as corridors, entrances, halls, toilets, kitchens, and storage areas.

Forced-Convection Units

Forced-convection units can be used for the same types of applications as natural convection units but are primarily used for facilities that require both heating and cooling, as well as spaces that require localized heating.

Unit heaters are often used as the primary source of heat in factories, warehouses, and garages and as supplemental freeze protection for loading ramps, entrances, equipment rooms, and fresh air plenums.

Unit ventilators are forced-convection units with dampers that introduce controlled amounts of outside air. They are used in spaces with ventilation requirements not met by other system components.

Cabinet heaters are often used in entranceways and vestibules that can have high intermittent heat loads.

Induction units are similar to fan-coil units, but the air is supplied by a central air system rather than individual fans in each unit. Induction units are most commonly used as perimeter heating for facilities with central systems.

Fan-coil units are designed for heating and cooling. On water systems, a single coil can be supplied with hot water for heating and chilled water for cooling. On steam systems, single- or dual-coil units can be used. Single-coil units require steam-to-water heat exchangers to provide hot water to meet heating requirements. Dual-coil units have a steam coil for heating and a separate coil for cooling. The cooling media for dual-coil units can be chilled water from a central chiller or refrigerant provided by a self-contained compressor. Single-coil units require the entire system to be either in a heating or cooling mode and do not permit simultaneous heating and cooling to satisfy individual space requirements. Dual-coil units can eliminate this problem.

Central air-handling units are used for most larger facilities. These units have a fan, a heating and cooling coil, and a compressor if chilled water is not available for cooling. Multizone units are arranged so that each air outlet has separate controls for individual space heating or cooling requirements. Large central systems distribute either warm or cold conditioned air through duct systems and employ separate terminal equipment such as mixing boxes, reheat coils, or variable volume controls to control the temperature to satisfy each space. Central air-handling systems can be factory assembled, field erected, or built at the job site from individual components.

CONVECTION STEAM HEATING

Any system that uses steam as the heat transfer medium can be considered a steam heating system; however, the term "steam heating system" is most commonly applied to convection systems using radiators or convectors as terminal equipment. Other types of steam heating systems use forced convection, in which a fan or air-handling system is used with a convector or steam coil such as unit heaters, fan-coil units and central air-conditioning and heating systems.

Convection-type steam heating systems are used in facilities that have a heating requirement only, such as factories, warehouses, and apartment buildings. They are often used in conjunction with central air-conditioning systems to heat the perimeter of the building. Also, steam is commonly used with incremental units that are designed for cooling and heating and have a self-contained air-conditioning compressor.

Steam heating systems are classified as one-pipe or two-pipe systems, according to the piping arrangement that supplies steam to and returns condensate from the terminal equipment. These systems can be further subdivided by (1) the method of condensate return (gravity flow or mechanical flow by means of condensate pump or vacuum pump) and (2) by the piping arrangement (upfeed or downfeed and parallel or counterflow for one-pipe systems).

One-Pipe Steam Heating Systems

The one-pipe system has a single pipe through which steam flows to and condensate is returned from the terminal equipment (Figure 15). These systems are designed as gravity return, although a condensate pump can be used where there is insufficient height above the boiler water level to develop enough pressure to return condensate directly to the boiler.

A one-pipe system with gravity return does not have steam traps; instead it has air vents at each terminal unit and at the ends of all supply mains to vent the air so the system can fill with steam. In a system with a condensate pump, there must be an air vent at each terminal unit and steam traps at the ends of each supply main.

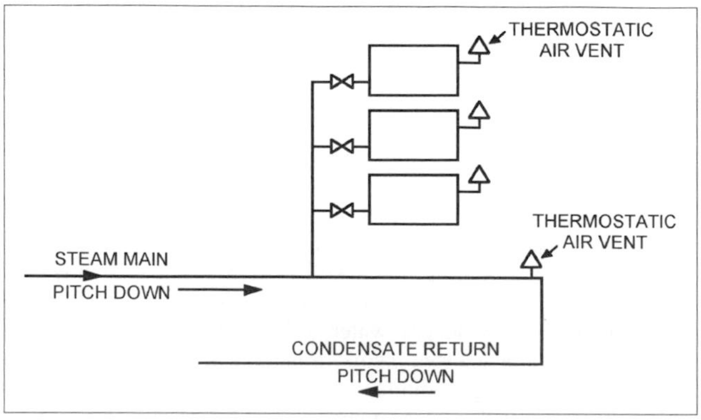

Fig. 15 One-Pipe System

The one-pipe system with gravity return has low initial cost and is simple to install, because it requires a minimum of mechanical equipment and piping. One-pipe systems are most commonly used in small facilities such as small apartment buildings and office buildings. In larger facilities, the larger pipe sizes required for two-phase flow, problems of distributing steam quickly and evenly throughout the system, the inability to zone the system, and difficulty in controlling the temperature make the one-pipe system less desirable than the two-pipe system.

The heat input to the system is controlled by cycling the steam on and off. In the past, temperature control in individual spaces has been a problem. Many systems have adjustable vents at each terminal unit to help balance the system, but these are seldom effective. A practical approach is to use a self-contained thermostatic valve in series with the air vent (as explained in the section on Temperature Control) that provides limited individual thermostatic control for each space.

Many designers do not favor one-pipe systems because of their distribution and control problems. However, when a self-contained thermostatic valve is used to eliminate the problems, one-pipe systems can be considered for small facilities, where initial cost and simple installation and operation are prime factors.

Most one-pipe gravity return systems are in facilities that have their own boiler, and, because returning condensate must overcome boiler pressure, these systems usually operate from a fraction of a psi to a maximum of 5 psig. The boiler hookup is critical and the Hartford loop (described in the section on Boiler Connections) is used to avoid problems that can occur with boiler low-water condition. Stamper and Koral (1979) and Hoffman give piping design information for one-pipe systems.

Two-Pipe Steam Heating Systems

The two-pipe system uses separate pipes to deliver the steam and return the condensate from each terminal unit (Figure 16). Thermostatic traps are installed at the outlet of each terminal unit to keep the steam in the unit until it gives up its latent heat, at which time the trap cycles open to pass the condensate and permits more steam to enter the radiator. If orifices are installed at the inlet to each terminal unit, as discussed in the sections on Steam Distribution and Temperature Control, and if the system pressure is precisely regulated so only the amount of steam is delivered to each unit that it is capable of condensing, the steam traps can be omitted. However, omitting steam traps is generally not recommended for initial design.

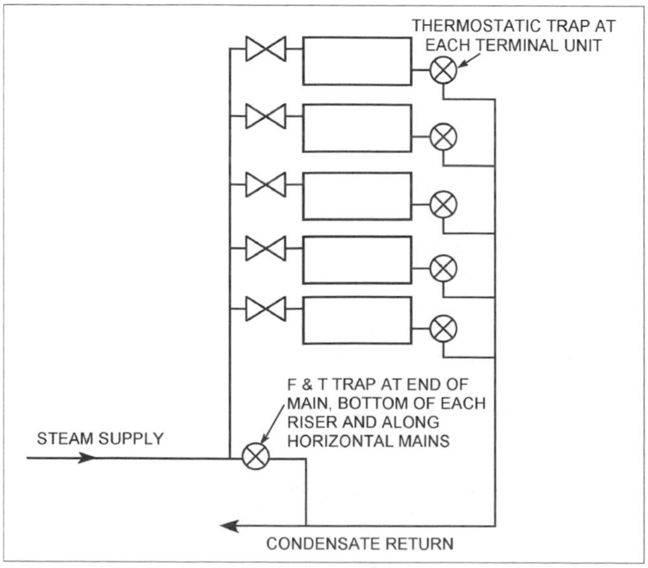

Fig. 16 Two-Pipe System

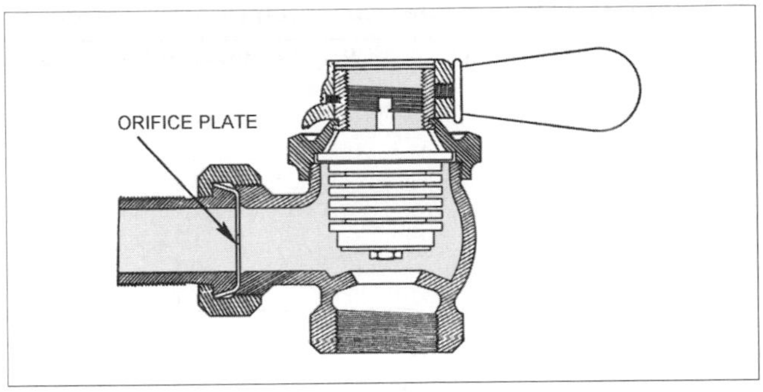

Fig. 17 Inlet Orifice

Two-pipe systems can have either gravity or mechanical returns; however, gravity returns are restricted to use in small systems and are generally outmoded. In larger systems that require higher steam pressures to distribute steam, some mechanical means, such as a condensate pump or vacuum pump, must return condensate to the boiler. A **vacuum return system** is used on larger systems and has the following advantages:

- The system fills quickly with steam. The steam in a gravity return system must push the air out of the system, resulting in delayed heat-up and condensate return that can cause low-water problems. A vacuum return system can eliminate these problems.
- The steam supply pressure can be lower, resulting in more efficient operation.

Variable vacuum or subatmospheric systems are a variation of the vacuum return system in which a controllable vacuum is maintained in both the supply and return sides. This permits using the lowest possible system temperature and prompt steam distribution to the terminal units. The primary purpose of variable vacuum systems is to control temperature as discussed in the section on Temperature Control.

Unlike one-pipe systems, two-pipe systems can be simply zoned where piping is arranged to supply heat to individual sections of the building that have similar heating requirements. Heat is supplied to meet the requirements of each section without overheating other sections. The heat also can be varied according to factors such as the hours of use, type of occupancy, and sun load.

STEAM DISTRIBUTION

Steam supply piping should be sized so that the pressure drops in all branches of the same supply main are nearly uniform. Return piping should be sized for the same pressure drop as supply piping for quick and even steam distribution. Because it is impossible to size piping so that pressure drops are exactly the same, the steam flows first to those units that can be reached with the least resistance, resulting in uneven heating. Units farthest from the source of steam will heat last, while other spaces are overheated. This problem is most evident when the system is filling with steam. It can be severe on systems in which temperature is controlled by cycling the steam on and off. The problem can be alleviated or eliminated by balancing valves or inlet orifices.

Balancing valves are installed at the unit inlet and contain an adjustable valve port to control the amount of steam delivered. The main problem with such devices is that they are seldom calibrated accurately, so that variations in orifice size as small as 0.003 in^2 can make a significant difference.

Inlet orifices are thin brass or copper plates installed in the unit inlet valve or pipe unions (Figure 17). Inlet orifices can solve distribution problems, because they can be drilled for appropriate size and changed easily to compensate for unusual conditions. Properly sized inlet orifices can compensate for oversized heating units, reduce energy waste and system control problems caused by excessive steam loss from defective steam traps, and provide a means of temperature control and balancing within individual zones. Manufacturers of orifice plates provide data for calculating the required sizes for most systems. Also, Sanford and Sprenger (1931) developed data for sizing orifices for low-pressure, gravity return systems See Figure 18 for capacities of orifices at various pressure differentials.

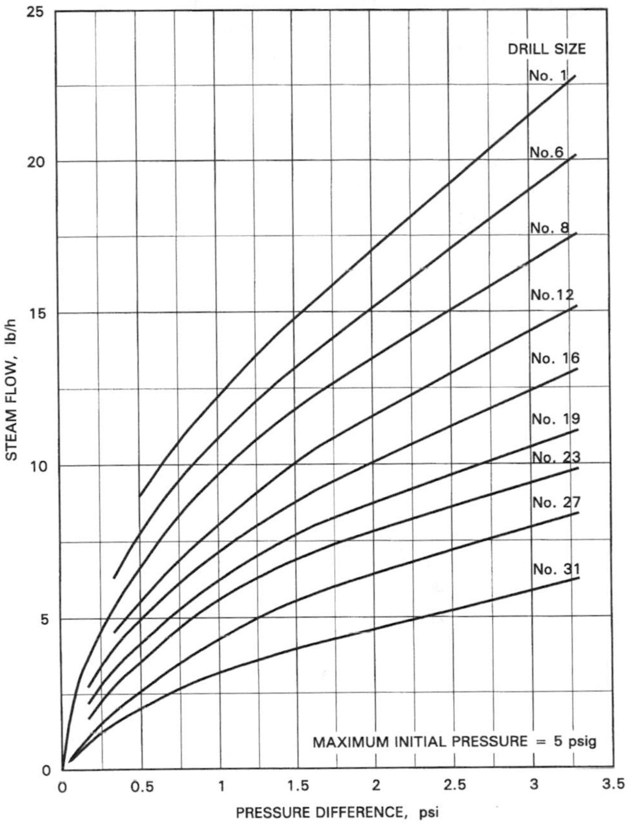

Fig. 18 Orifice Capacities for Different Pressure Differentials
(Schroeder 1950)

If orifices are installed in valves or pipe unions that are conveniently accessible, minor rebalancing among individual zones can be accomplished through replacement of individual orifices.

TEMPERATURE CONTROL

All heating systems require some means of temperature control to achieve the desired comfort conditions and operating efficiencies. In convection-type steam heating systems, the temperature and resulting heat output of the terminal units must be increased or decreased. This can be done by (1) permitting the steam to enter the heating unit intermittently, (2) varying the steam temperature delivered to each unit, and (3) varying the amount of steam delivered to each unit.

There are two types of controls: those that control the temperature or heat input to the entire system and those that control the temperature of individual spaces or zones. Often, both types are used together for maximum control and operating efficiency.

Intermittent flow controls, commonly called **heat timers**, control the temperature of the system by cycling the steam on and off during a certain portion (or fraction) of each hour as a function of the outdoor and/or indoor temperature. Most of these devices provide for night setback, and computerized or electronic models optimize morning start-up as a function of outdoor and indoor temperature and make anticipatory adjustments. Used independently, these devices do not permit varying heat input to different parts of the building or spaces, so they should be used with zone control valves or individual thermostatic valves for maximum energy efficiency.

Zone control valves control the temperature of spaces with similar heating requirements as a function of outdoor and/or indoor temperature, as well as permit duty cycling. These intermittent flow control devices operate full open or full closed. They are controlled by an indoor thermostat and are used in conjunction with a heat timer, variable vacuum, or pressure differential control that controls heat input to the system.

Individual thermostatic valves installed at each terminal unit can provide the proper amount of heat to satisfy the individual requirements of each space and eliminate overheating. Valves can be actuated electrically, pneumatically, or by a self-contained mechanism that has a wax- or liquid-filled sensing element requiring no external power source.

Individual thermostatic valves can be and are often used as the only means of temperature control. However, these systems are always "on," resulting in inefficiency and no central control of heat input to the system or to the individual zones. Electronic, electric, and pneumatic operators allow centralized control to be built into the system. However, it is desirable to have a supplemental system control with self-contained thermostatic valves in the form of zone control valves, or a system that controls the heat input to the entire system, relying on self-contained valves as a local high-temperature shutoff only.

Self-contained thermostatic valves can also be used to control temperature on one-pipe systems that cycle on and off when installed in series with the unit air vent. On initial start-up, the air vent is open, and inherent system distribution problems still exist. However, on subsequent on-cycles where the room temperature satisfies the thermostatic element, the vent valve remains closed, preventing the unit from filling with steam.

Variable vacuum or subatmospheric systems control the temperature of the system by varying the steam temperature through pressure control. These systems differ from the regular vacuum return system in that the vacuum is maintained in both supply and return lines. Such systems usually operate at a pressure range from 2 psig to 25 in. Hg vacuum. Inlet orifices are installed at the terminal equipment for proper steam distribution during mild weather.

Design-day conditions are seldom encountered, so the system usually operates at a substantial vacuum. Because the specific volume of steam increases as the pressure decreases, it takes less steam to fill the system and operating efficiency results. The variable vacuum system can be used in conjunction with zone control valves or individual self-contained valves for increased operating efficiency.

Pressure differential control (two-pipe orifice) systems provide centralized temperature control with any system that has properly sized inlet orifices at each heating unit. This method operates on the principle that flow through an orifice is a function of the pressure differential across the orifice plate. The pressure range is selected to fill each heating unit on the coldest design day; on warmer days, the supply pressure is lowered so that heating units are only partially filled with steam, thereby reducing their heat output. An advantage of this method is that it virtually eliminates all heating unit steam trap losses because the heating units are only partially filled with steam on all but the coldest design days.

The required system pressure differential can be achieved manually with throttling valves or with an automatic pressure differential controller. Table 2 shows the required pressure differentials to maintain 70°F indoors for 0°F outdoor design, with orifices according to Figure 18. Required pressure differential curves can be established for any combination of supply and return line pressures by sizing orifices to deliver the proper amount of steam for the coldest design day, calculating the amount of steam required for warmer days using Equation (3), and then selecting the pressure differential that will provide this flow rate.

Table 2 Pressure Differential Temperature Control

Outdoor Temperature, °F	Required Pressure Differential[a], in. Hg.
0	6.0
10	4.5
20	3.2
30	2.1
40	1.2
50	0.5
60	0.1

[a] To maintain 70°F indoors for 0°F outdoor design.

$$Q_r = Q_d \frac{(t_i)_{design} - t_o}{(t_i)_{design} - (t_o)_{design}} \tag{3}$$

where

Q_r = required flow rate
Q_d = design day flow rate
$(t_i)_{design}$ = design indoor temperature
$(t_o)_{design}$ = design outdoor temperature
t_o = outside temperature

Pressure differential systems can be used with zone control valves or individual self-contained thermostatic valves to increase operating efficiency.

HEAT RECOVERY

Two methods are generally employed to recover heat from condensate: (1) the enthalpy of the liquid condensate (sensible heat) can be used to vaporize or "flash" some of the liquid to steam at a lower pressure, or (2) it can be used directly in a heat exchanger to heat air, fluid, or a process.

The particular methods used vary with the type of system. As explained in the section on Basic Steam System Design, facilities that purchase steam from a utility generally do not have to return condensate and, therefore, can recover heat to the maximum extent possible. On the other hand, facilities with their own boiler generally want the condensate to return to the boiler as hot as possible, limiting heat recovery because any heat removed from condensate has to be returned to the boiler to generate steam again.

Flash Steam

Flash steam is an effective use for the enthalpy of the liquid condensate. It can be used in any facility that has a requirement for steam at different pressures, regardless of whether steam is purchased or generated by a facility's own boiler. Flash steam can be used in any heat exchange device to heat air, water, or other liquids or directly in processes with lower pressure steam requirements. Equation (1) may be used to calculate the amount of flash steam generated, and Figure 19 provides a graph for calculating the amount of flash steam as a function of system pressures.

Although flash steam can be generated directly by discharging high-pressure condensate to a lower pressure system, most designers prefer a **flash tank** to control flashing. Flash tanks can be mounted either vertically or horizontally, but the vertical arrangement shown in Figure 20 is preferred because it provides better separation of steam and water, resulting in the highest possible steam quality.

The most important dimension in the design of vertical flash tanks is the internal diameter, which must be large enough to ensure a low upward velocity of flash to minimize water carryover. If this velocity is low enough, the height of the tank is not important, but

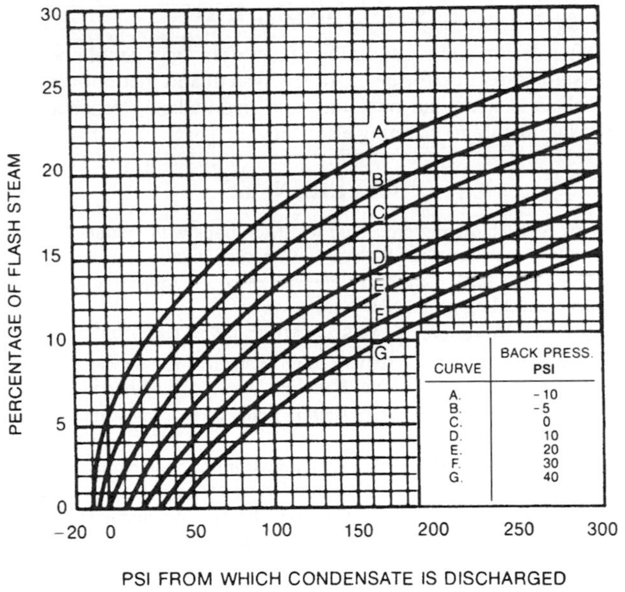

Fig. 19 Flash Steam

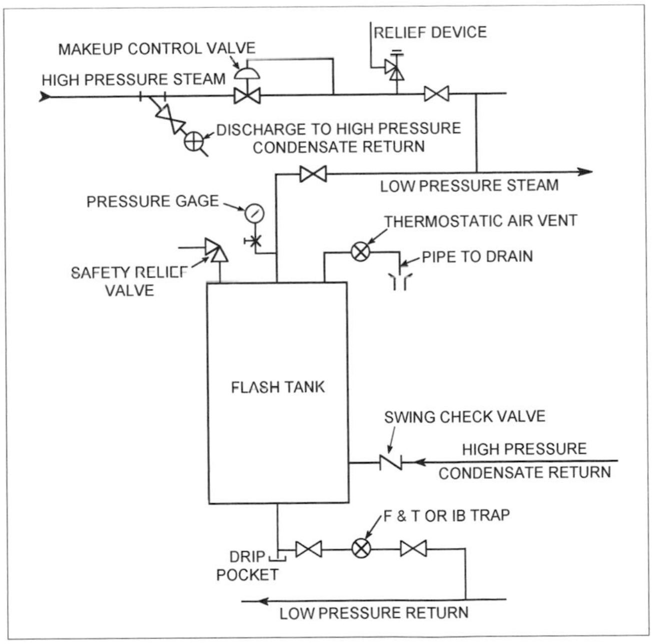

Fig. 20 Vertical Flash Tank

it is good practice to use a height of at least 2 to 3 ft. The graph in Figure 21 can be used to determine the internal diameter and is based on a steam velocity of 10 ft/s, which is the maximum velocity in most systems.

Installation is important for proper flash tank operation. Condensate lines should pitch towards the flash tank. If more than one condensate line discharges to the tank, each line should be equipped with a swing check valve to prevent backflow. Condensate lines and the flash tank should be well insulated to prevent any unnecessary heat loss. A thermostatic air vent should be installed at the top of the tank to vent any air that accumulates. The tank should be trapped at the bottom with an inverted bucket or float and thermostatic trap sized to triple the condensate load.

The demand load must always be greater than the amount of flash steam available to prevent the low-pressure system from

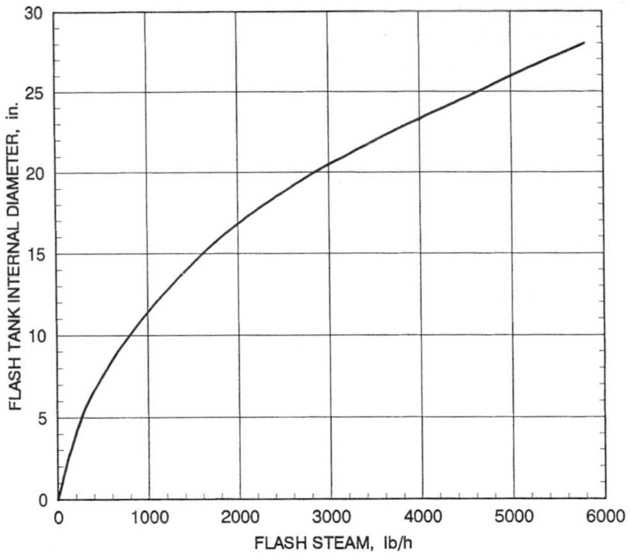

Fig. 21 Flash Tank Diameters

becoming overpressurized. A safety relief valve should always be installed at the top of the flash tank to preclude such a condition.

Because the flash steam available is generally less than the demand for low-pressure steam, a makeup valve ensures that the low-pressure system maintains design pressure.

Flash tanks are considered pressure vessels and must be constructed in accordance with ASME and local codes.

Direct Heat Recovery

Direct heat recovery that uses the enthalpy of the liquid in some type of heat exchange device is appropriate when condensate is not returned to a facility's own boiler; any lowering of condensate temperature below saturation temperature requires reheating at the boiler to regenerate steam.

The enthalpy of the condensate can be used in fan-coil units, unit heaters, or convectors to heat spaces where temperature control is not critical such as garages, ramps, loading docks, and entrance halls or in a shell-and-tube heat exchanger to heat water or other fluids.

In most HVAC applications, the enthalpy of the liquid condensate may be used most effectively and efficiently to heat domestic hot water with a shell-and-tube or plate-type heat exchanger, commonly called an economizer. Many existing economizers do not use the enthalpy of the condensate effectively because they are designed only to preheat makeup water flowing directly through the heat exchanger at the time of water usage. Hot water use seldom coincides with condensate load, and most of the enthalpy is wasted in these preheat economizer systems.

A storage-type water heater can be effectively used for heat recovery. This can be a shell-and-tube heat exchanger with a condensate coil for heat recovery and a steam or electric coil when the condensate enthalpy cannot satisfy the demand. Another option is a storage-type heat exchanger incorporating only a coil for condensate with a supplemental heat exchanger to satisfy peak loads. Note that many areas require a double-wall heat exchanger between steam and the potable water.

Chapter 48 of the 1999 *ASHRAE Handbook—Applications* provides useful information on determining hot water loads for various facilities. In general, however, the following provide optimum heat recovery:

- Install the greatest storage capacity in the available space. Although all systems must have supplemental heaters for peak load conditions, with ample storage capacity these heaters may

seldom function if all the necessary heat is provided by the enthalpy of the condensate.
- For maximum recovery, permit stored water to heat to 180°F or higher, using a mixing valve to temper the water to the proper delivery temperature.

COMBINED STEAM AND WATER SYSTEMS

Combined steam and water systems are often used to take advantage of the unique properties of steam, which are described in the sections on Advantages and Fundamentals.

Combined systems are used where a facility must generate steam to satisfy the heating requirements of certain processes or equipment. They are usually used where steam is available from a utility and economic considerations of local codes preclude the facility from operating its own boiler plant. There are two types of combined steam and water systems: (1) steam is used directly as a heating medium; the terminal equipment must have two separate coils, one for heating with steam and one for cooling with chilled water, and (2) steam is used indirectly and is piped to heat exchangers that generate the hot water for use at terminal equipment; the exchanger for terminal equipment may use either one coil or separate coils for heating and cooling.

Combined steam and water systems may be two-, three-, or four-pipe systems. Chapter 3 and Chapter 4 have further descriptions.

COMMISSIONING

After design and construction of a system, care should be taken to ensure correct performance of the system and that building personnel understand the operating and maintenance procedure required to maintain operating efficiency.

REFERENCES

Armers, A. 1985. Sizing procedure for steam traps with modulating steam pressure control. Spirax Sarco Inc., Concord, Ontario.

ASME. 1998. *Boiler and pressure vessel code*. American Society of Mechanical Engineers, New York.

ASME. 1998. Power piping. ANSI/ASME *Standard* B31.1-98.

Hoffman steam heating systems design manual. *Bulletin* No. TES-181, Hoffman Specialty ITT Fluid Handling Division, Indianapolis, IN.

HYDI. 1989. *Testing and rating heating boilers*. Hydronics Institute, Berkeley Heights, NJ.

Kremers, J.A. 1982. Modulating steam pressure in coils compound steam trap selection procedures. *Armstrong Trap Magazine* 50(1). Available from Armstrong Machine Works, Three Rivers, Michigan 49093.

Sanford, S.S. and C.B. Springer. 1931. Flow of steam through orifices. *ASHVE Transactions* 37:371-94.

Schroeder, D.E. 1950. Balancing a steam heating system by the use of orifices. *ASHVE Transactions* 56:325-40.

Stamper, E. and R.L. Koral. 1979. *Handbook of air conditioning, heating and ventilating*, 3rd ed. Industrial Press, New York.

CHAPTER 11

DISTRICT HEATING AND COOLING

A DISTRICT heating and cooling (DHC) system distributes thermal energy from a central source to residential, commercial, and/or industrial consumers for use in space heating, cooling, water heating, and/or process heating. The energy is distributed by steam or hot or chilled water lines. Thus, thermal energy comes from a distribution medium rather than being generated on site at each facility.

Whether the system is a public utility or user owned, such as a multi-building campus, it has economic and environmental benefits depending somewhat on the particular application. Political feasibility must be considered, particularly if a municipality or governmental body is considering a DHC installation. Historically, successful DHC systems have had the political backing and support of the community.

Applicability

District heating and cooling systems are best used in markets where (1) the thermal load density is high and (2) the annual load factor is high. A high load density is needed to cover the capital investment for the transmission and distribution system, which usually constitutes most of the capital cost for the overall system, often ranging from 50 to 75% of the total cost for district heating systems (normally lower for district cooling applications).

The annual load factor is important because the total system is capital intensive. These factors make district heating and cooling systems most attractive in serving (1) industrial complexes, (2) densely populated urban areas, and (3) high-density building clusters with high thermal loads. Low-density residential areas have usually not been attractive markets for district heating, although there have been some successful applications. District heating is best suited to areas with a high building and population density in relatively cold climates. District cooling applies in most areas that have appreciable concentrations of cooling loads, usually associated with tall buildings.

Components

District heating and cooling systems consist of three primary components: the central plant, the distribution network, and the consumer systems (Figure 1).

The **central source** or **production plant** may be any type of boiler, a refuse incinerator, a geothermal source, solar energy, or thermal energy developed as a by-product of electrical generation. The last approach, called cogeneration, has a high energy utilization efficiency; see Chapter 7 for information on cogeneration.

Chilled water can be produced by

- Absorption refrigeration machines

The preparation of this chapter is assigned to TC 6.2, District Energy.

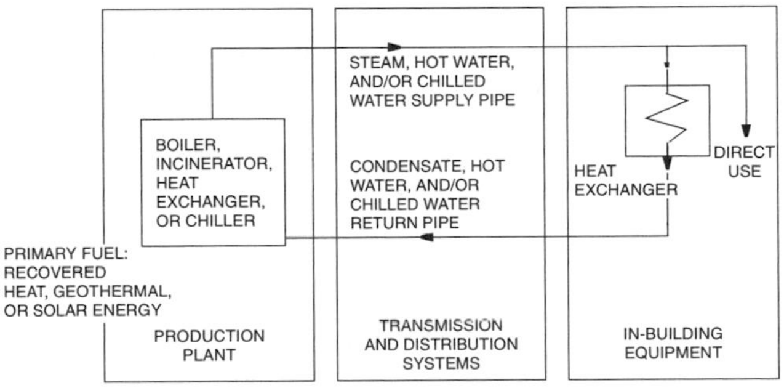

Fig. 1 Major Components of District Heating System

- Electric-driven compression equipment (reciprocating, rotary screw or centrifugal chillers)
- Gas/steam turbine- or engine-driven compression equipment
- Combination of mechanically driven systems and thermal energy driven absorption systems

The second component is the **distribution** or **piping network** that conveys the energy. The piping is often the most expensive portion of a district heating or cooling system. The piping usually consists of a combination of preinsulated and field-insulated pipe in both concrete tunnel and direct burial applications. These networks require substantial permitting and coordinating with nonusers of the system for right-of-way if not on the owner's property. Because the initial cost is high, it is important to optimize use.

The third component is the **consumer system**, which includes in-building equipment. When steam is supplied, (1) it may be used directly for heating; (2) it may be directed through a pressure-reducing station for use in low-pressure (0 to 15 psig) steam space heating, service water heating, and absorption cooling; or (3) it may be passed through a steam-to-water heat exchanger. When hot water or chilled water is supplied, it may be used directly by the building systems or isolated by a heat exchanger (see the section on Consumer Interconnections).

BENEFITS

Environmental Benefits

Emissions from central plants are easier to control than those from individual plants. A central plant that burns high-sulfur coal can economically remove noxious sulfur emissions, where individual combustors could not. Similarly, the thermal energy from municipal wastes can provide an environmentally sound system. Cogeneration of heat and electric power allows for combined efficiencies of energy use that greatly reduce emissions and also allow for fuel flexibility. In addition, refrigerants and other CFCs can be monitored and controlled more readily in a central plant. Where site conditions allow, remote location of the plant reduces many of the concerns with use of ammonia systems for cooling.

Consumer Economic Benefits

A district heating and cooling system offers the following economic benefits. Even though the basic costs are still borne by the central plant owner/operator, because the central plant is large the customer can realize benefits of economies of scale.

Operating Personnel. One of the primary advantages to a building owner is that operating personnel for the HVAC system can be reduced or eliminated. Most municipal codes require operating engineers to be on site when high-pressure boilers are in operation. Some older systems require trained operating personnel to be in the boiler/mechanical room at all times. When thermal energy is brought into the building as a utility, depending on the sophistication of the building HVAC controls, there may be opportunity to reduce or eliminate operating personnel.

Insurance. Both property and liability insurance costs are significantly reduced with the elimination of a boiler in the mechanical room since risk of a fire or accident is reduced.

Usable Space. Usable space in the building increases when a boiler and/or chiller and related equipment are no longer necessary. The noise associated with such in-building equipment is also eliminated. Although this space usually cannot be converted into prime office space, it does provide the opportunity for increased storage or other use.

Equipment Maintenance. With less mechanical equipment, there is proportionately less equipment maintenance, resulting in less expense and a reduced maintenance staff.

Higher Thermal Efficiency. A larger central plant can achieve higher thermal and emission efficiencies than can several smaller units. When strict regulations must be met, additional pollution control equipment is also more economical for larger plants. Cogeneration of heat and electric power results in much higher overall efficiencies than is possible from separate heat and power plants.

Partial load performance of central plants may be more efficient than that of many isolated small systems because the larger plant can operate one or more capacity modules as the combined load requires and can modulate output. Central plants generally have efficient base-load units and less costly peaking equipment for use in extreme loads or emergencies.

PRODUCER ECONOMICS

Available Fuels. Smaller heating plants are usually designed for one type of fuel, which is generally gas or oil. Central DHC plants can operate on less expensive coal or refuse. Larger facilities can often be designed for more than one fuel (e.g., coal and oil).

Energy Source Economics. If an existing facility is the energy source, the available temperature and pressure of the thermal fluid is predetermined. If exhaust steam from an existing electrical generating turbine is used to provide thermal energy, the conditions of the bypass determine the maximum operating pressure and temperature of the DHC system. A trade-off analysis must be conducted to determine what percentage of the energy will be diverted for thermal generation and what percentage will be used for electrical generation. Based on the marginal value of energy, it is critical to determine the operating conditions in the economic analysis.

If a new central plant is being considered, a decision of whether to cogenerate electrical and thermal energy or to generate thermal energy only must be made. An example of cogeneration is a diesel or natural gas engine-driven generator with heat recovery equipment. The engine drives a generator to produce electricity, and heat is recovered from the exhaust, cooling, and lubrication systems. Other systems may use one of several available steam turbine designs for cogeneration. These turbine systems combine the thermal and electrical output to obtain the maximum amount of available energy. Chapter 7 has further information on cogeneration.

The selection of temperature and pressure is crucial because it can dramatically affect the economic feasibility of a DHC system design. If the temperature and/or pressure level chosen is too low, a potential customer base might be eliminated. On the other hand, if there is no demand for absorption chillers or high-temperature industrial processes, a low-temperature system usually provides the lowest delivered energy cost.

The availability and location of fuel sources must also be considered in optimizing the economic design of a DHC system. For example, a natural gas boiler might not be feasible where abundant sources of natural gas are not available.

Initial Capital Investment

The initial capital investment for a DHC system is usually the major economic driving force. Normally, the initial capital investment includes the four components of (1) concept planning, (2) design, (3) construction, and (4) consumer interconnections.

Concept Planning. In concept planning, three areas are generally reviewed. First, the **technical feasibility** of a DHC system must be considered. Conversion of an existing heat source, for example, usually requires the services of an experienced power plant or DHC engineering firm.

Financial feasibility is the second consideration. For example, a municipal or governmental body must consider availability of bond financing. Alternative energy choices for potential customers must be reviewed because consumers are often asked to sign long-term contracts in order to justify a DHC system.

Design. The distribution system accounts for a significant portion of the initial investment. Distribution design depends on the heat transfer medium chosen, its operating temperature and pressure, and the routing. Failure to consider these key variables results in higher-than-planned installation costs. An analysis must be done to optimize insulating properties. The section on Economical Thickness for Pipe Insulation discusses determining insulation values.

Construction. The construction costs of the central plant and distribution system depend on the quality of the concept planning and design. Although the construction cost usually accounts for most of the initial capital investment, neglect in any of the other three areas could mean the difference between economic success and failure. Field changes usually increase the final cost and delay start-up. Even a small delay in start-up can adversely affect both economics and consumer confidence.

Lead time needed to obtain equipment generally determines the time required to build a DHC system. In some cases, lead time on major components in the central plant can be over a year.

Installation time of the distribution system depends in part on the routing interference with existing utilities. A distribution system in a new industrial park is simpler and requires less time to install than a system being installed in an established business district.

Consumer Interconnection. These costs are usually borne by the consumer. High interconnection costs may favor an in-building plant instead of a DHC system. For example, if an existing building is equipped for steam service, interconnection to a hot water DHC system may be too costly, even though the cost of energy is lower.

CENTRAL PLANT

The central plant may include equipment to provide heat only, cooling only, both heat and cooling, or any of these three options in conjunction with electric power generation. In addition to the central plant, small so-called satellite plants are sometimes used in situations where a customer's building is located in an area where distribution piping is not yet installed.

HEATING AND COOLING PRODUCTION

Heating Medium

In plants serving hospitals, industrial customers, or those also generating electricity, steam is the usual choice for production in the plant and, often, for distribution to customers. For systems serving largely commercial buildings, hot water is an attractive medium. From the standpoint of distribution, hot water can accommodate a greater geographical area than steam due to the ease with which booster pump stations can be installed. The common attributes and relative merits of hot water and steam as heat-conveying media are as follows.

Heat Capacity. Steam relies primarily on the latent heat capacity of water rather than on sensible heat. The net heat content for saturated steam at 100 psig (338°F) condensed and cooled to 180°F is approximately 1040 Btu/lb. Hot water cooled from 350 to 250°F has a net heat effect of 103 Btu/lb, or only about 10% as much as that of steam. Thus, a hot water system must circulate about 10 times more mass than a steam system of similar heat capacity.

Pipe Sizes. Despite the fact that less steam is required for a given heat load, and flow velocities are greater, steam usually requires a larger pipe size for the supply line due to its lower density (Aamot and Phetteplace 1978). This is compensated for by a much smaller condensate return pipe. Therefore, piping costs for steam and condensate are often comparable with those for hot water supply and return.

Return System. Condensate return systems require more maintenance than hot water return systems. Corrosion of piping and other components, particularly in areas where feedwater is high in bicarbonates, is a problem. Nonmetallic piping has been used successfully in some applications, such as systems with pumped returns, where it has been possible to isolate the nonmetallic piping from live steam.

Similar concerns are associated with condensate drainage systems (steam traps, condensate pumps, and receiver tanks) for steam supply lines. Condensate collection and return should be carefully considered when designing a steam system. Although similar problems with water treatment occur in hot water systems, they present less of a concern because makeup rates are lower.

Pressure and Temperature Requirements. Flowing steam and hot water both incur pressure losses. Hot water systems may use intermediate booster pumps to increase the pressure at points between the plant and the consumer. Due to the higher density of water, pressure variations caused by elevation differences in a hot water system are much greater than for steam systems. This can adversely affect the economics of a hot water system by requiring the use of a higher pressure class of piping and/or booster pumps.

Regardless of the medium used, the temperature and pressure used for heating should be no higher than needed to satisfy the consumer requirements. Higher temperatures and pressures require additional engineering and planning to avoid higher heat losses. Safety and comfort levels for operators and maintenance personnel also benefit from lower pressure. Higher temperatures may require higher pressure ratings for piping and fittings and may preclude the use of materials such as polyurethane foam insulation and nonmetallic conduits.

Hot water systems are divided into three temperature classes. High-temperature systems supply temperatures over 350°F; medium-temperature systems supply temperatures in the range of 250 to 350°F; and low-temperature systems supply temperatures of 250°F or lower.

The temperature drop at the consumer end should be as high as possible, preferably 40°F or greater. A large temperature drop allows the fluid flow rate through the system, pumping power, return temperatures, return line heat loss, and condensing temperatures in cogeneration power plants to be reduced. A large customer temperature drop can often be achieved through the cascading of loads operating at different temperatures.

In many instances, existing equipment and processes require the use of steam. See the section on Consumer Interconnections for further information.

Heat Production

Fire-tube and water-tube boilers are available for gas/oil firing. If coal is used, either package-type coal-fired boilers in small sizes (less than 20,000 to 25,000 lb/h) or field-erected boilers in larger sizes are available. Coal-firing underfeed stokers are available up to a 30,000 to 35,000 lb/h capacity; traveling grate and spreader stokers are available up to 160,000 lb/h capacity in single-boiler installations. Larger coal-fired boilers are typically multiple installations of the three types of stokers or larger, pulverized fired or fluidized bed boilers. Generally, the complexity of fluidized bed or pulverized firing does not lend itself to small central heating plant operation.

Cooling Supply

Chilled water may be produced by an absorption refrigeration machine or by vapor-compression equipment driven by electric, turbine (steam or combustion), or internal combustion engines. The chilled water supply temperature for a conventional system ranges from 40 to 44°F. A 12°F temperature difference (Δt) results in a flow rate of 2 gpm/ton of refrigeration. Due to the cost of the distribution system piping, large chilled water systems are sometimes operated at lower supply water temperatures to allow a larger Δt to be achieved, thereby reducing chilled water flow per ton of capacity. For systems involving chilled water storage, a practical lower limit is 39°F due to water density considerations. For ice storage systems, temperatures as low as 34°F have been used.

Multiple air-conditioning loads interconnected with a central chilled water system provide some economic advantages and energy conservation opportunities. In addition, central plants afford the opportunity to consider the use of refrigerants such as ammonia that may be impractical for use in individual buildings. The size of air-conditioning loads served, as well as the diversity among the loads and their distance from the chilling plant, are principal factors in determining the feasibility of central plants. The distribution system pipe capacity is directly proportional to the operating temperature difference between the supply and return lines, and it benefits additionally from increased diversity in the connected loads.

An economic evaluation of piping and pumping costs versus chiller power requirements can establish the most suitable supply water temperature. When sizing piping and calculating pumping cost the heat load on the chiller generated by the frictional heating of the flowing fluid should be considered because most of the pumping power adds to the system heat load. For high chiller efficiency it is often more efficient to use isolated auxiliary equipment for special process requirements and to allow the central plant supply water temperature to float up at times of lower load. As with heating plants, optimum chilled water control may require a combination of temperature modulation and flow modulation. However, the designer must investigate the effects of higher chilled water supply temperatures on chilled water secondary system distribution flows and air-side system performance (humidity control) before applying this to individual central water plants.

Thermal Storage

Both hot and chilled water thermal storage can be implemented for district systems. In North America, the current economic situation primarily results in chilled water storage applications. Depending on the plant design and loading, thermal storage can reduce chiller equipment requirements and lower operating costs. By shifting a part of the chilling load, chillers can be sized closer to the average load than the peak load. Shifting the entire refrigeration load to off peak requires the same (or slightly larger) chiller machine capacity, but removes all of the electric load from the peak period. Since many utilities offer lower rates during off peak periods, operating costs for electric-driven chillers can be substantially reduced.

Both ice and chilled water storage have been applied to district-sized chiller plants. In general, the largest systems (>20,000 ton-hour capacity) use chilled water storage and small-to-moderate sized systems use ice storage. Storage capacities in the 10,000 to 30,000 ton-hour range are now common and systems have been installed up to 125,000 ton-hour for district cooling systems.

Selection of the storage configuration (chilled water steel tank above-grade, chilled water concrete tank below grade, ice direct, ice indirect) is often influenced by space limitations. Chilled water storage requires 4 to 6 times the volume of ice storage for the same capacity. For chilled water storage, the footprint of steel tanks (depending on height) can be less than concrete tanks for the same volume (Andrepont 1995). Chapter 33 of the 1999 *ASHRAE Handbook—Applications* has information on thermal storage.

Auxiliaries

Numerous pieces of auxiliary support equipment related to the boiler and chiller operations are not unique to the production plant of a DHC system and are found in similar installations. Some components of a DHC system deserve special consideration due to their critical nature and potential effect on operations.

Although instrumentation can be either electronic or pneumatic, electronic instrumentation systems offer the flexibility of combining control systems with data acquisition systems. This combination brings improved efficiency, better energy management, and reduced operating staff for the central heating and/or cooling plant. For systems involving multiple fuels and/or thermal storage, computer-based controls are indispensable for accurate decisions as to boiler and chiller operation.

Boiler feedwater treatment has a direct bearing on equipment life. Condensate receivers, filters, polishers, and chemical feed equipment must be accessible for proper management, maintenance, and operation. Depending on the temperature, pressure, and quality of the heating medium, water treatment may require softeners, alkalizers, and/or demineralizers for systems operating at high temperatures and pressures.

Equipment and layout of a central heating and cooling plant should reflect what is required for proper plant operation and maintenance. The plant should have an adequate service area for equipment and a sufficient number of electrical power outlets and floor drains. Equipment should be placed on housekeeping pads. Figure 2 presents a layout for a large hot water/chilled water plant.

Expansion Tanks and Water Makeup. The expansion tank is usually located in the central plant building. To control pressure, either air or nitrogen is introduced to the air space in the expansion tank. To function properly, the expansion tank must be the single point of the system where no pressure change occurs. Multiple, air-filled tanks may cause erratic and possibly harmful movement of air though the piping. Although diaphragm expansion tanks eliminate air movement, the possibility of hydraulic surge should be considered. On large chilled water systems, a makeup water pump generally is used to makeup water loss. The pump is typically controlled from level switches on the expansion tank or from a desired pump suction pressure.

A conventional water meter on the makeup line can show water loss in a closed system. This meter also provides necessary data for water treatment. The fill valve should be controlled to open or close and not modulate to a very low flow, so that the water meter can detect all makeup.

Emission Control. Environmental equipment, including electrostatic precipitators, baghouses, and scrubbers, is required to meet emission standards for coal-fired or solid-waste-fired operations.

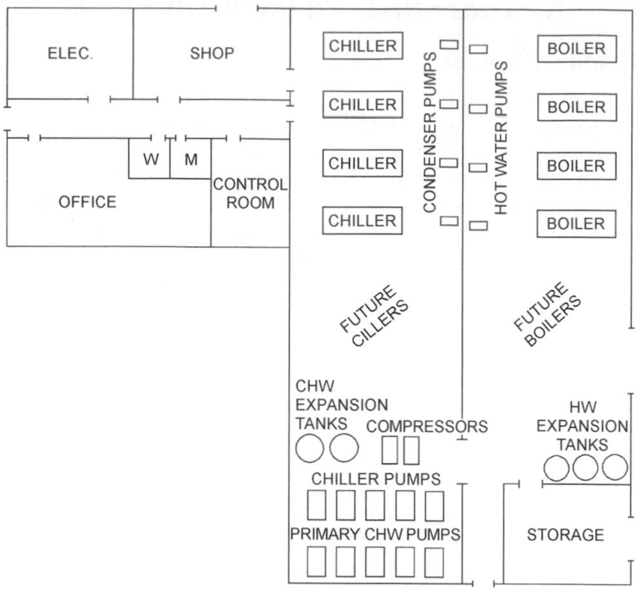

Fig. 2 Layout for Hot Water/Chilled Water Plant

Proper control is critical to equipment operation, and it should be designed and located for easy access by maintenance personnel.

A baghouse gas filter provides good service if gas flow and temperature are properly maintained. Because baghouses are designed for continuous on-line use, they are less suited for cyclic operation. Heating and cooling significantly reduces the useful life of the bags due to acidic condensation. The use of an economizer to preheat boiler feedwater and help control flue gas temperature may enhance baghouse operation. Contaminants generated by plant operation and maintenance, such as wash-down of floors and equipment, may need to be contained.

DISTRIBUTION DESIGN CONSIDERATIONS

Water distribution systems are designed for either constant flow (variable return temperature) or variable flow (constant return temperature). The design decision between constant or variable volume flow affects the (1) selection and arrangement of the chiller(s), (2) design of the distribution system, and (3) design of the customer connection to the distribution system. Unless very unusual circumstances exist, most systems large enough to be considered in the district category are likely to benefit from variable flow design.

Constant Flow

Constant flow is generally applied only to smaller systems where simplicity of design and operation are important and where distribution pumping costs are low. Chillers are usually arranged in series. Flow volume through a full-load distribution system depends on the type of constant flow system used. One technique connects the building and its terminals across the distribution system. The central plant pump circulates chilled water through three-way valve controlled air-side terminal units. Balancing problems may occur in this design when many separate flow circuits are interconnected (Figure 3).

Constant flow distribution is also applied to in-building circuits with separate pumps. This arrangement isolates the flow balance problem between buildings. In this case, the flow through the distribution system can be significantly lower than the sum of the flows needed by the terminal if the in-building system supply temperature is higher than the distribution system supply temperature (Figure 3). The water temperature rise in the distribution system is determined by the connected in-building systems and their controls.

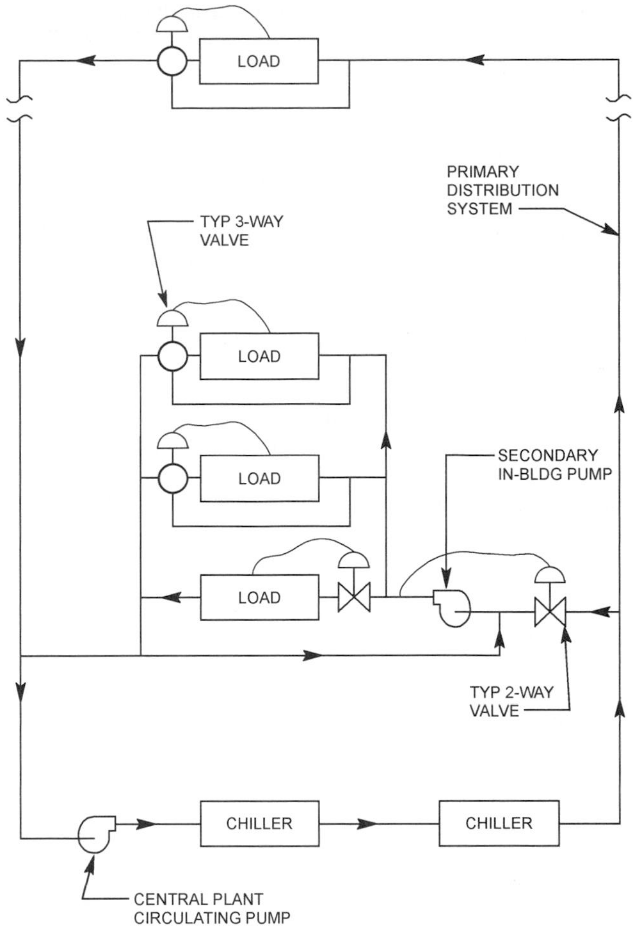

Fig. 3 Constant Flow Primary Distribution with Secondary Pumping

In constant flow design, chillers arranged in parallel have decreased entering water temperatures at part load; thus, several machines may need to run simultaneously, each at a reduced load, to produce the required chilled water flow. In this case, chillers in series are better because constant flow can be maintained though the chilled water plant at all times, with only the chillers required for producing chilled water energized. Constant flow systems should be analyzed thoroughly when considering multiple chillers in a parallel arrangement because the auxiliary electric loads of condenser water pumps, tower fans, and central plant circulating pumps are a significant part of the total energy input.

Variable Flow

Variable flow design can improve energy use and expand the capacity of the distribution system piping by using diversity. To maintain a high temperature differential at part load, the distribution system flow rate must track the load imposed on the central plant. Multiple parallel pumps or more commonly, variable-speed pumps can reduce flow and pressure, and lower pumping energy at part load. Terminal device controls should be selected to ensure that variable flow objectives are met. Flow-throttling (two-way) valves provide the continuous high return temperature needed to correlate the system load change to a system flow change.

Systems in each building are usually two-pipe, with individual in-building pumping. In some cases, the pressure of the distribution system may cause flow through the in-building system without in-building pumping. Distribution system pumps can provide total building system pumping if (1) the distribution system pressure drops are minimal, and (2) the distribution system is relatively short-coupled (3000 ft or less). To implement this pumping method, the total flow must be pumped at a pressure sufficient to meet the requirements of the building with the largest pressure requirement. If the designer has control over the design of each in-building system, this pumping method can be achieved in a reasonable manner. In retrofit situations where existing buildings under different ownership are connected to a new central plant, coordination is difficult and individual building pumps are more practical.

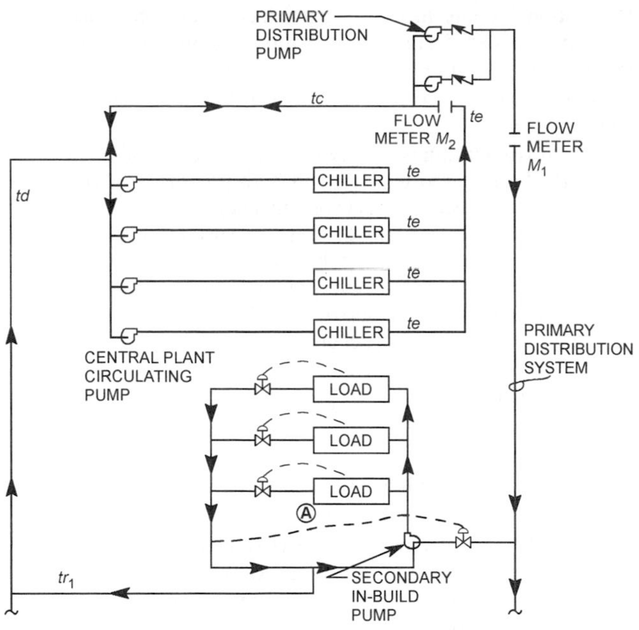

Fig. 4 Variable Flow Primary/Secondary Systems

When buildings have separate circulating pumps, hydraulic isolating piping and pumping design should be used to assure that two-way control valves are subjected only to the differential pressure established by the in-building pump. Figure 4 shows a connection using in-building pumping with hydraulic isolation from the primary loop.

When in-building pumps are used, all series interconnections between the distribution system pump and the in-building pumps must be removed. A series connection can cause the distribution system return to operate at a higher pressure than the distribution system supply and disrupt flow to adjacent buildings. Series operation often occurs during improper use of three-way mixing valves in the distribution to building connection.

In very large systems, a design known as **distributed pumping** may be used. Under this approach, the distribution pumps in the central plant are eliminated. Instead, the distribution system pumping load is borne by the pumps in the user buildings. In cases where the distribution network piping constitutes a significant pressure loss (systems covering a large area), this design allows the distributed pumps in the buildings to be sized for just the pressure loss imposed at that particular location. Ottmer and Rishel (1993) found that this approach reduces total chilled water pump power by 20 to 25% in very large systems. It is best applied in new construction where the central plant and distributed building systems can be coordinated.

Usually, a positive pressure must be maintained at the highest point of the system at all times. This height determines the static pressure at which the expansion tank operates. Excessively tall buildings that do not isolate the in-building systems from the distribution system can impose unacceptable static pressure on the distribution system. To prevent excessive operating pressure in distribution systems, heat exchangers have been used to isolate the in-building system from the distribution system. To ensure reasonable temperature differentials between supply and return temperatures, flow must be controlled on the distribution system side of the heat exchanger.

In high-rise buildings, all piping, valves, coils, and other equipment may be required to withstand high pressure. Where system static pressure exceeds safe or economical operating pressure, either the heat exchanger method or pressure sustaining valves in the return line with check valves in the supply line may be used to minimize pressure. However, the pressure sustaining/check valve arrangement may over pressurize the entire distribution system if a malfunction of either valve occurs.

Design Guidelines

Guidelines for plant design and operation include the following:

- Variable-speed pumping saves energy and should be considered for distribution system pumping.
- Design chilled water systems for a reasonable temperature differential of 12 to 14°F. A 12 to 20°F maximum temperature differential with a 10 to 12°F minimum temperature differential can be achieved with this design.
- Limit the use of constant flow systems to relatively small central chilled water plants. Chillers should be in series.
- Larger central chilled water plants can benefit from primary/secondary or primary/secondary/tertiary pumping with constant flow in central plant and variable flow in the distribution system. Size the distribution system for a low overall total pressure loss. Short-coupled distribution systems (3000 ft or less) can be used for a total pressure loss of 20 to 40 ft of water. With this maximum differential between any points in the system, size the distribution pumps to provide the necessary pressure to circulate chilled water through the in-building systems, eliminating the need for in-building pumping systems. This decreases the complexity of operating central chilled water systems.
- All two-way valves must have proper close-off ratings and a design pressure drop of at least 20% of the maximum design pressure drop for controllability. Commercial quality automatic temperature control valves generally have low shutoff ratings; but industrial valves can achieve higher ratings. See Chapter 42 for more information on valves.
- The lower practical limit for chilled water supply temperatures is 39°F. Temperatures below that should be carefully analyzed; although, systems with thermal energy storage may operate advantageously at lower temperatures.

DISTRIBUTION SYSTEM

HYDRAULIC CONSIDERATIONS

Objectives of Hydraulic Design

Although the distribution of a thermal utility such as hot water encompasses many of the aspects of domestic hot water distribution, many dissimilarities also exist; thus the design should not be approached in the same manner. Thermal utilities must supply sufficient energy at the appropriate temperature and pressure to meet consumer needs. Within the constraints imposed by the consumer's end use and equipment, the required thermal energy can be delivered with various combinations of temperature and pressure. Computer-aided design methods are available for thermal piping networks (Reisman 1985, Rasmussen 1987, COWIconsult 1985, Bloomquist et al. 1999). The use of such methods allows the rapid evaluation of many alternative designs.

General steam system design can be found in Chapter 10, as well as in IDHA (1983). For water systems, consult Chapter 12 and IDHA (1983).

Water Hammer

The term water hammer is used to describe several phenomena that occur in fluid flow. Although these phenomena differ in nature, they all result in stresses in the piping that are higher than normally encountered. Water hammer can have a disastrous effect on a thermal utility by bursting pipes and fittings and threatening life and property.

In steam systems, water hammer is caused primarily by condensate collecting in the bottom of the steam piping. Steam flowing at velocities 10 times greater than normal water flow picks up a slug of condensate and accelerates it to a high velocity. The slug of condensate subsequently collides with the pipe wall at a point where flow changes direction. To prevent this type of water hammer, condensate must be prevented from collecting in steam pipes by the use of proper steam pipe pitch and adequate condensate collection and return facilities.

Water hammer also occurs in steam systems due to rapid condensation of steam during system warm-up. Rapid condensation decreases the specific volume and pressure of steam, which precipitates pressure shock waves. This form of water hammer is prevented by controlled warm-up of the piping. Valves should be opened slowly and in stages during warm-up. Large steam valves should be provided with smaller bypass valves to slow the warm-up.

Water hammer in hot and chilled water distribution systems is caused by sudden changes in flow velocity, which causes pressure shock waves. The two primary causes are pump failure and sudden valve closures. Methods of analysis can be found in Stephenson (1981) and Fox (1977). Preventive measures include operational procedures and special piping fixtures such as surge columns.

Pressure Losses

Friction pressure losses occur at the interface between the inner wall of a pipe and a flowing fluid due to shear stresses. In steam systems, these pressure losses are compensated for with increased steam pressure at the point of steam generation. In water systems, pumps are used to increase pressure at either the plant or intermediate points in the distribution system. The calculation of pressure loss is discussed in Chapter 2 and Chapter 33 of the 1997 *ASHRAE Handbook—Fundamentals*.

Pipe Sizing

Ideally, the appropriate pipe size should be determined from an economic study of the life-cycle cost for construction and operation. In practice, however, this study is seldom done due to the effort involved. Instead, criteria that have evolved from practice are frequently used for design. These criteria normally take the form of constraints on the maximum flow velocity or pressure drop. Chapter 33 of the 1997 *ASHRAE Handbook—Fundamentals* provides velocity and pressure drop constraints. Noise generated by excessive flow velocities is usually not a concern for thermal utility distribution systems outside of buildings. For steam systems, maximum flow velocities of 200 to 250 fps are recommended (IDHA 1983). For water systems, Europeans use the criterion that pressure losses should be limited to 0.44 psi per 100 ft of pipe (Bøhm 1988). Recent studies indicate that higher levels of pressure loss may be acceptable (Stewart and Dona 1987) and warranted from an economic standpoint (Bøhm 1986, Koskelainen 1980, Phetteplace 1989).

When establishing design flows for thermal distribution systems, the diversity of consumer demands should be considered (i.e., the various consumers' maximum demands do not occur at the same time). Thus, the heat supply and main distribution piping may be sized for a maximum load that is somewhat less than the sum of the individual consumers' maximum demands. For steam systems, Geiringer (1963) suggests diversity factors of 0.80 for space heating and 0.65 for domestic hot water heating and process loads. Geiringer also suggests that these factors may be reduced by approximately 10% for high-temperature water systems. Werner (1984) conducted a study of the heat load on six operating low-temperature hot water systems in Sweden and found diversity factors ranging from 0.57 to 0.79, with the average being 0.685.

Network Calculations

Calculating the flow rates and pressures in a piping network with branches, loops, pumps, and heat exchangers can be difficult without the aid of a computer. Methods have been developed primarily for domestic water distribution systems (Jeppson 1977, Stephenson 1981). These may apply to thermal distribution systems with appropriate modifications. Computer-aided design methods usually incorporate methods for hydraulic analysis as well as for calculating heat losses and delivered water temperature at each consumer. Calculations are usually carried out in an iterative fashion, starting with constant supply and return temperatures throughout the network. After initial estimates of the design flow rates and heat losses are determined, refined estimates of the actual supply temperature at each consumer are computed. Flow rates at each consumer are then adjusted to ensure that the load is met with the reduced supply temperature, and the calculations are repeated.

Condensate Drainage and Return

Condensate forms in operating steam lines as a result of heat loss. When a steam system's operating temperature is increased, condensate also forms as steam warms the piping. At system start-up, these loads usually exceed any operating heat loss loads; thus, special provisions should be made.

To drain the condensate, steam piping should slope toward a collection point called a **drip station**. Drip stations are located in access areas or buildings where they are accessible for maintenance. Steam piping should slope toward the drip station at least 1 in. in 40 ft. If possible, the steam pipe should slope in the same direction as steam flow. If it is not possible to slope the steam pipe in the direction of steam flow, increase the pipe size to at least one size greater than would normally be used. This will reduce the flow velocity of the steam and provide better condensate drainage against the steam flow. Drip stations should be spaced no further than 500 ft apart in the absence of other requirements.

Drip stations consist of a short piece of pipe (called a **drip leg**) positioned vertically on the bottom of the steam pipe, as well as a steam trap and appurtenant piping. The drip leg should be the same diameter as the steam pipe. The length of the drip leg should provide a volume equal to 50% of the condensate load from system start-up for steam pipes of 4 in. diameter and larger and 25% of the start-up condensate load for smaller steam pipes (IDHA 1983). Steam traps should be sized to meet the normal load from operational heat losses only. Start-up loads should be accommodated by manual operation of the bypass valve.

Steam traps are used to separate the condensate and noncondensable gases from the steam. For drip stations on steam distribution piping, use inverted bucket or bimetallic thermostatic traps. Some steam traps have integral strainers, while others require separate strainers. Take care to ensure that drip leg capacity is adequate when thermostatic traps are used because they will always accumulate some condensate.

If it is to be returned, condensate leaving the steam trap flows into the condensate return system. If steam pressure is sufficiently high, it may be used to force the condensate through the condensate return system. With low-pressure steam or on systems where a large pressure exists between drip stations and the ultimate destination of the condensate, condensate receivers and pumps must be provided.

Schedule 80 steel piping is recommended for condensate lines because of the extra allowance for corrosion its provides. Steam traps have the potential of failing in an open position, thus nonmetallic piping must be protected from live steam where its temperature/pressure would exceed the limitations of the piping. Nonmetallic piping should not be located so close to steam pipes that heat losses from the steam pipes could overheat it. Information on sizing of condensate return piping may be found in Chapter 33 of the 1997 *ASHRAE Handbook—Fundamentals*.

THERMAL CONSIDERATIONS

Thermal Design Conditions

Three thermal design conditions must be met to ensure satisfactory system performance:

1. The "normal" condition used for the life-cycle cost analysis determines appropriate insulation thickness. Average values for the temperatures, burial depth, and thermal properties of the materials are used for design. If the thermal properties of the insulating material are expected to degrade over the useful life of the system, appropriate allowances should be made in the cost analysis.
2. Maximum heat transfer rate determines the load on the central plant due to the distribution system. It also determines the temperature drop (or increase, in the case of chilled water distribution), which determines the delivered temperature to the consumer. For this calculation, the thermal conductivity of each component must be taken at its maximum value, and the temperatures must be assumed to take on their extreme values, which would result in the greatest temperature difference between the carrier medium and the soil or air. The burial depth will normally be at its lowest value for this calculation.
3. During operation, none of the thermal capabilities of the materials (or any other materials in the area influenced thermally by the system) must exceed design conditions. To satisfy this objective, each component and the surrounding environment must be examined to determine whether thermal damage is possible. A heat transfer analysis may be necessary in some cases.

   The conditions of these analyses must be chosen to represent the worst-case scenario from the perspective of the component being examined. For example, in assessing the suitability of a coating material for a metallic conduit, the thermal insulation is assumed to be saturated, the soil moisture is at its lowest probable level, and the burial depth is maximum. These conditions, combined with the highest anticipated pipe and soil temperatures, give the highest conduit surface temperature to which the coating could be exposed.

Heat transfer in buried systems is influenced by the thermal conductivity of the soil and by the depth of burial, particularly when the insulation has low thermal resistance. Soil thermal conductivity changes significantly with moisture content; for example, Bottorf (1951) indicated that soil thermal conductivity ranges from 0.083 Btu/h·ft·°F during dry soil conditions to 1.25 Btu/h·ft·°F during wet soil conditions.

Thermal Properties of Pipe Insulation and Soil

Uncertainty in heat transfer calculations for thermal distribution systems results from the uncertainty in the thermal properties of the materials involved. Generally, the designer must rely on manufacturers' specifications and handbook data to obtain approximate values. The data in this chapter should only be used as guidance in preliminary calculations until specific products have been identified; then specific data should be obtained from the manufacturer of the product in question.

Table 1 Comparison of Commonly Used Insulations in Underground Piping Systems

	Calcium Silicate Type I/II ASTM C 533	Urethane Foam	Cellular Glass ASTM C 552	Mineral Fiber/ Preformed Glass Fiber Type 1 ASTM C 547
Thermal Conductivity[a] (Values in parenthesis are maximum permissible by ASTM standard listed), Btu/h·ft·°F				
Mean temp. = 100°F	0.028	0.013	0.033 (0.030)	0.022 (0.021)
200°F	0.031 (0.038/0.045)	0.014	0.039 (0.037)	0.025 (0.026)
300°F	0.034 (0.042/0.048)		0.046 (0.045)	0.028 (0.033)
400°F	0.038 (0.046/0.051)		0.053 (0.054)	(0.043)
Density (max.), lb/ft^3	15/22		6.7 to 9.2	3
Maximum temperature, °F	1200	250	800	850
Compressive strength (min)[b], kPa	700 at 5% deformation		450	N/A
Dimensional stability, linear shrinkage at maximum use temperature	2%		N/A	2%
Flame spread	0		5	25
Smoke index	0		0	50
Water absorption	As shipped moisture content, 20% max. (by weight)		0.5	Water vapor sorption, 5% max. (by weight)

[a] Thermal conductivity values included in this table are from previous editions of this chapter and have been retained as they were used in the examples. The thermal conductivity of insulation may vary with temperature, temperature gradient, moisture content, density, thickness, and shape. The ASTM maximum values given are comparative for establishing quality control compliance, and are suggested for preliminary calculations where available. They may not represent the installed performance of insulation under actual conditions that differ substantially from test conditions. The manufacturer should have design values.

[b] The compressive strength for cellular glass shown is for flat material, capped as per ASTM C 240.

Table 2 Effect of Moisture on Underground Piping System Insulations (Chyu et al. 1997a,b; 1998a,b)

Characteristics	Polyurethane[a]	Cellular Glass	Mineral Wool[b]	Fibrous Glass
Heating Test	Pipe temp. 35°F to 260°F Water bath 46°F to 100°F	Pipe temp. 35°F to 420°F Water bath 46°F to 100°F	Pipe temp.35°F to 450°F Water bath 46°F to 100°F	Pipe temp. 35°F to 450°F Water bath 46°F to 100°F
Length of submersion time to reach steady-state *k*-value	70 days	See Note C	10 days	2 hours
Effective *k*-value increase from dry conditions after steady state achieved in submersion	14 to 19 times at steady state. Estimated water content of insulation 70% (by volume).	Avg. 10 times, process unsteady (Note C). Insulation showed evidence of moisture zone on inner diameter	Up to 50 times at steady state. Insulation completely saturated	52 to 185 times. Insulation completely saturated at steady state.
Primary heat transfer mechanism	Conduction	See Note C	Conduction and convection	Conduction and convection
Length of time for specimen to return to dry steady-state *k*-value after submersion	Pipe at 260°F, after 16 days moisture content 10% (by volume) remaining	Pipe at 420°F, 8 hours	Pipe at 450°F, 9 days	Pipe at 380°F, 6 days
Cooling Test	Pipe temp. 37°F Water bath at 52°F	Pipe temp. 36°F Water bath 46°F to 58°F	Pipe temp. 35°F to 45°F Water bath 55°F	Insulation 35°F to 450°F Water bath 46°F to 100°F
Length of submersion time to reach steady-state conditions for *k*-value	16 days	Data recorded at 4 days constant at 12 days	6 days	1/2 hour
Effective *k*-value increase from dry conditions after steady state achieved	2 to 4 times. Water absorption minimal, ceased after 7 days.	None. No water penetration.	14 times. Insulation completely saturated at steady state.	20 times. Insulation completely saturated at steady state.
Primary heat transfer mechanism	Conduction	Conduction	Conduction and convection	Conduction and convection
Length of time for specimen to return to dry steady-state *k*-value after submersion	Pipe at 37°F, data curve extrapolated to 10+ days	Pipe at 33°F, no change	Pipe at 35°F, data curve extrapolated to 25 days	Pipe at 35°F, 15 days

[a] Polyurethane material tested had a density of 2.9 lb/ft^3.

[b] Mineral wool tested was a preformed molded basalt designed for pipe systems operating up to 1200°F. It was specially formulated to withstand the Federal Agency Committee 96 hour boiling water test.

[c] Cracks formed upon heating for all samples of cellular glass insulation tested. Flooded heat loss mechanism involved dynamic two-phase flow of water through cracks; the period of dynamic process was about 20 min. Cracks had negligible effect on the thermal conductivity of dry cellular glass insulation prior to and after submersion. No cracks formed during the cooling test.

Insulation. Insulation provides the primary thermal resistance against heat loss or gain in thermal distribution systems. Thermal properties and other characteristics of insulations normally used in thermal distribution systems are listed in Table 1. Material properties such as thermal conductivity, density, compressive strength, moisture absorption, dimensional stability, and combustibility are typically reported in ASTM standard for the respective material. Some properties have more than one associated standard. For example, thermal conductivity for insulation material in block form may be reported using ASTM C 177, C 518 or C 1114. Thermal conductivity for insulation material fabricated or molded for use on piping is reported using ASTM C 335.

Table 3 Soil Thermal Conductivities

Soil Moisture Content (by mass)	Thermal Conductivity, Btu/h·ft·°F		
	Sand	Silt	Clay
Low, <4%	0.17	0.08	0.08
Medium, 4 to 20%	1.08	0.75	0.58
High, >20%	1.25	1.25	1.25

Chyu et al. (1997a,b; 1998a,b) studied the effect of moisture on the thermal conductivity of insulating materials commonly used in underground district energy systems (ASHRAE Research Project RP721). The results are summarized in Table 2. The insulated pipe was immersed in water maintained at 46 to 100°F to simulate possible conduit water temperatures during a failure. The fluid temperature in the insulated pipe ranged from 35 to 450°F. All insulation materials were tested unfaced and/or unjacketed to simulate installation in a conduit.

Soil. If an analysis of the soil is available or can be done, the thermal conductivity of the soil can be estimated from published data (e.g., Farouki 1981, Lunardini 1981). The thermal conductivity factors in Table 3 may be used as an estimate when detailed information on the soil is not known. Because dry soil is rare in most areas, a low moisture content should be assumed only for system material design, or where it can be validated for calculation of heat losses in the normal operational condition. Values of 0.8 to 1 Btu/h·ft·°F are commonly used where soil moisture content is unknown. Because moisture will migrate toward a chilled pipe, use a thermal conductivity value of 1.25 Btu/h·ft·°F for chilled water systems in the absence of any site-specific soil data. For steady-state analyses, only the thermal conductivity of the soil is required. If a transient analysis is required, the specific heat and density are also required.

METHODS OF HEAT TRANSFER ANALYSIS

Because heat transfer in piping is not related to the load factor, it can be a large part of the total load. The most important factors affecting heat transfer are the difference between earth and fluid temperatures and the thermal insulation. For example, the extremes might be a 6 in., insulated, 400°F water line in 40°F earth with 100 to 200 Btu/h·ft loss; and a 6 in., uninsulated, 55°F chilled water return in 60°F earth with 10 Btu/h·ft gain. The former requires analysis to determine the required insulation and its effect on the total heating system; the latter suggests analysis and insulation needs might be minimal. Other factors that affect heat transfer are (1) depth of burial, related to the earth temperature and soil thermal resistance; (2) soil conductivity, related to soil moisture content and density; and (3) distance between adjacent pipes.

To compute transient heat gains or losses in underground piping systems, numerical methods that approximate any physical problem and include such factors as the effect of temperature on thermal properties must be used. For most designs, numerical analyses may not be warranted, except where the potential exists to thermally damage something adjacent to the distribution system. Also, complex geometries may require numerical analysis. Albert and Phetteplace (1983), Rao (1982), and Minkowycz et al. (1988) have further information on numerical methods.

Steady-state calculations are appropriate for determining the annual heat loss/gain from a buried system if average annual earth temperatures are used. Steady-state calculations may also be appropriate for worst-case analyses of thermal effects on materials. Steady-state calculations for a one-pipe system may be done without a computer, but it becomes increasingly difficult for a two-, three-, or four-pipe system.

The following steady-state methods of analysis use resistance formulations developed by Phetteplace and Meyer (1990) that simplify the calculations needed to determine temperatures within the system. Each type of resistance is given a unique subscript and is defined only when introduced. In each case, the resistances are on a unit length basis so that heat flows per unit length result directly when the temperature difference is divided by the resistance. *Note*: For consistency and simplicity, all thermal conductivities are given in Btu/h·ft·°F with all dimensions in feet (not the more traditional Btu·in/h·ft^2·°F).

Calculation of Undisturbed Soil Temperatures

Before any heat loss/gain calculations may be conducted, the undisturbed soil temperature at the site must be determined. The choice of soil temperature is guided primarily by the type of calculation being conducted, see the previous section on Thermal Design Considerations. For example, if the purpose of the calculation is to determine if a material will exceed its temperature limit, the maximum expected undisturbed ground temperature is used. The appropriate choice of undisturbed soil temperature also depends on the location of the site, time of year, depth of burial, and thermal properties of the soil. Some methods for determining undisturbed soil temperatures and suggestions on appropriate circumstances to use them are as follows:

1. Use the average annual air temperature to approximate the average annual soil temperature. This estimate is appropriate when the objective of the calculation is to yield the average heat loss over the yearly weather cycle. Mean annual air temperatures may be obtained from various sources of climatic data, one source is a table of values from CRREL (1999).
2. Use the maximum/minimum air temperature as an estimate of the maximum/minimum undisturbed soil temperature for pipes buried at a shallow depth. This approximation is an appropriate conservative assumption when checking the temperatures to determine if the temperature limits of any of materials proposed for use will be exceeded. Maximum and minimum expected air temperatures may be calculated from the information found in Chapter 26 of the 1997 *ASHRAE Handbook—Fundamentals*.
3. For systems that are buried at other than shallow depths, maximum/minimum undisturbed soil temperatures may be estimated as a function of depth, soil thermal properties, and prevailing climate. This estimate is appropriate when checking the temperatures in a system to determine if the temperature limits of any of the materials proposed for use will be exceeded. The following equations may be used to estimate the minimum and maximum expected undisturbed soil temperatures:

$$\text{Maximum temperature} = t_{s,z} = t_{ms} + A_s e^{-z\sqrt{\pi/\alpha\tau}} \quad (1)$$

$$\text{Minimum temperature} = t_{s,z} = t_{ms} - A_s e^{-z\sqrt{\pi/\alpha\tau}} \quad (2)$$

where

$t_{s,z}$ = temperature, °F
z = depth, ft
τ = annual period, 365 days
α = thermal diffusivity of the ground, ft^2/day
t_{ms} = mean annual surface temperature, °F
A_s = surface temperature amplitude, °F

CRREL (1999) lists values for the climatic constants t_{ms} and A_s for various regions of the United States.

Thermal diffusivity for soil may be calculated as follows:

$$\alpha = \frac{24k_s}{\rho_s[c_s + c_w(w/100)]} = \frac{24k_s}{\rho_s[c_s + (w/100)]} \quad (3)$$

where

ρ_s = soil density, lb/ft^3
c_s = dry soil specific heat, Btu/lb·°F
c_w = specific heat of water = 1.0 Btu/lb·°F
w = moisture content of soil, % (dry basis)
k_s = soil thermal conductivity, Btu/h·ft·°F

Because the specific heat of dry soil is nearly constant for all types of soil, c_s may be taken as 0.175 Btu/lb·°F.

4. For buried systems the undisturbed soil temperatures may be estimated for any time of the year as a function of depth, soil thermal properties, and prevailing climate. This temperature may be used in lieu of the soil surface temperature normally called for by the steady state heat transfer equations [i.e., Equations (6) and (7)] when estimates of heat loss/gain as a function of time of year are desired. The substitution of the undisturbed soil temperatures at the pipe depth allows the steady state equations to be used as a first approximation to the solution to the actual transient heat transfer problem with its annual temperature variations at the surface. The following equation may be used to estimate the undisturbed soil temperature at any depth at any point during the yearly weather cycle (ASCE 1996). (*Note*: The argument for the sin function is in radians.)

$$t_{s,z} = t_{ms} + A_s e^{-z\sqrt{\pi/\alpha\tau}} \sin\left[\frac{2\pi(\theta - \theta_{lag})}{\tau} - z\sqrt{\frac{\pi}{\alpha\tau}}\right] \quad (4)$$

where

θ = Julian date, days
θ_{lag} = phase lag of soil surface temperature, days

CREL (1999) lists values for the climatic constants t_{ms}, A_s, and θ_{lag} for various regions of the United States. Equation (3) may be used to calculate soil thermal diffusivity.

Equation (4) does not account for latent heat effects due to freezing, thawing, or evaporation. However, for the soil adjacent to a buried heat distribution system the equation provides a good estimate. For simplicity, the ground surface temperature is assumed to equal the air temperature, which is an acceptable assumption for most design calculations. If a more accurate calculation is desired, the following method may be used to compensate for the convective thermal resistance to heat transfer at the ground surface.

Convective Heat Transfer at Ground Surface

Heat transfer between the ground surface and the ambient air occurs by convection. In addition, heat transfer with the soil occurs due to precipitation and radiation. The heat balance at the ground surface is too complex to warrant detailed treatment in the design of buried district heating and cooling systems. However, McCabe et al. (1995) observed significant temperature variations due to the type of surfaces over district heating and cooling systems.

As a first approximation an **effectiveness thickness** of a fictitious soil layer may be added to the burial depth to account for the effect of the convective heat transfer resistance at the ground surface. The effective thickness is calculated as follows:

$$\delta = k_s/h \quad (5)$$

where

δ = effectiveness thickness of fictitious soil layer, ft
k_s = thermal conductivity of soil, Btu/h·ft·°F
h = convective heat transfer coefficient at ground surface, Btu/h·ft^2·°F

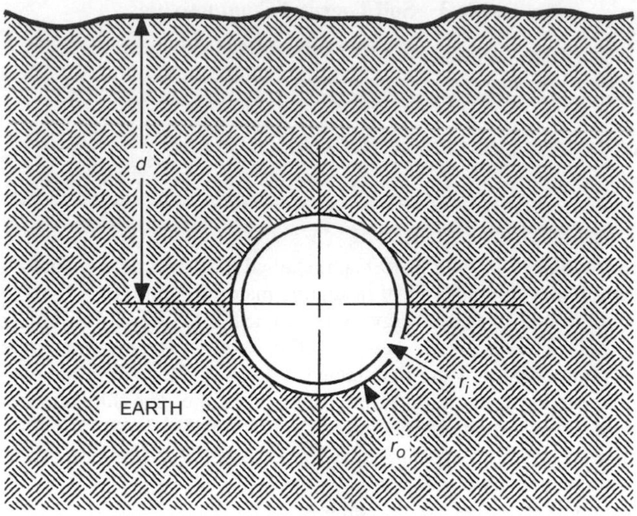

Fig. 5 Single Uninsulated Buried Pipe

The effective thickness calculated with Equation (5) is simply added to the actual burial depth of the pipes in calculating the soil thermal resistance using Equations (6), (7), (20), (21), and (27).

Single Uninsulated Buried Pipe

For this case (Figure 5), an estimate for soil thermal resistance may be used. This estimate is sufficiently accurate (within 1%) for the ratios of burial depth to pipe radius indicated next to Equations (6) and (7). Both the actual resistance and the approximate resistance are presented, along with the depth/radius criteria for each.

$$R_s = \frac{\ln\{(d/r_o) + [(d/r_o)^2 - 1]^{1/2}\}}{2\pi k_s} \qquad \text{for } d/r_o > 1 \quad (6)$$

$$R_s = \frac{\ln(2d/r_o)}{2\pi k_s} \qquad \text{for } d/r_o > 4 \quad (7)$$

where

R_s = thermal resistance of soil, h·ft·°F/Btu
k_s = thermal conductivity of soil, Btu/h·ft·°F
d = burial depth to centerline of pipe, ft
r_o = outer radius of pipe or conduit, ft

The thermal resistance of the pipe is included if it is significant when compared to the soil resistance. The thermal resistance of a pipe or any concentric circular region is given by

$$R_p = \frac{\ln(r_o/r_i)}{2\pi k_p} \quad (8)$$

where

R_p = thermal resistance of pipe wall, h·ft·°F/Btu
k_p = thermal conductivity of pipe, Btu/h·ft·°F
r_i = inner radius of pipe, ft

Example 1. Consider an uninsulated, 3 in. Schedule 40 PVC chilled water supply line carrying 45°F water. Assume the pipe is buried 3 ft deep in soil with a thermal conductivity of 1 Btu/h·ft·°F, and no other pipes or thermal anomalies are within close proximity. Assume the average annual soil temperature is 60°F.

Solution: Calculate the thermal resistance of the pipe using Equation (8):

$$R_p = 0.21 \text{ h} \cdot \text{ft} \cdot \text{°F/Btu}$$

Calculate the thermal resistance of the soil using Equation (7). [*Note*: $d/r_o = 21$ is greater than 4; thus Equation (7) may be used in lieu of Equation (6).]

$$R_s = 0.59 \text{ h} \cdot \text{ft} \cdot \text{°F/Btu}$$

Calculate the rate of heat transfer by dividing the overall temperature difference by the total thermal resistance:

$$q = \frac{t_f - t_s}{R_t} = \frac{(45 - 60)}{0.80 \text{ h} \cdot \text{ft} \cdot \text{°F/Btu}} = -19 \text{ Btu/h} \cdot \text{ft}$$

where

R_t = total thermal resistance, (i.e., $R_s + R_p$ in this case of pure series heat flow), h·ft·°F/Btu
t_f = fluid temperature, °F
t_s = average annual soil temperature, °F
q = heat loss or gain per unit length of system, Btu/h·ft

The negative result indicates a heat gain rather than a loss. Note that the thermal resistance of the fluid/pipe interface has been neglected, which is a reasonable assumption because such resistances tend to be very small for flowing fluids. Also note that in this case, the thermal resistance of the pipe comprises a significant portion of the total thermal resistance. This results from the relatively low thermal conductivity of PVC compared to other piping materials and the fact that no other major thermal resistances exist in the system to overshadow it. If any significant amount of insulation were included in the system, its thermal resistance would dominate, and it might be possible to neglect that of the piping material.

Single Buried Insulated Pipe

Equation (8) can be used to calculate the thermal resistance of any concentric circular region of material, including an insulation layer (Figure 6). When making calculations using insulation thickness, actual thickness rather than nominal thickness should be used to obtain the most accurate results.

Example 2. Consider the effect of adding 1 in. of urethane foam insulation and a 1/8 in. thick PVC jacket to the chilled water line in Example 1. Calculate the thermal resistance of the insulation layer from Equation (8) as follows:

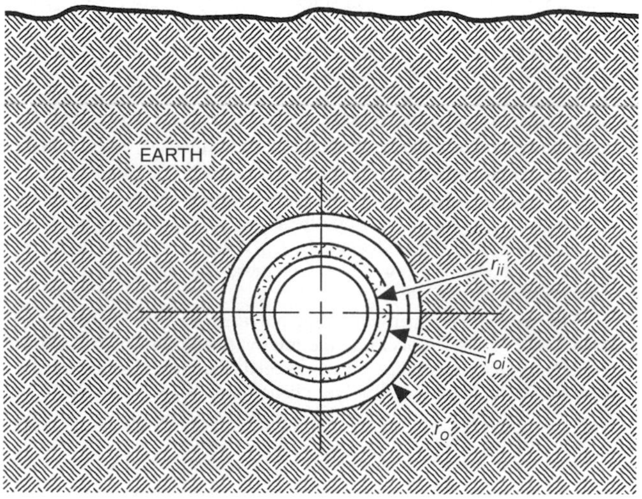

Fig. 6 Single Buried Insulated Pipe

$$R_i = \frac{\ln(0.229/0.146)}{2\pi \times 0.0125} = 5.75 \text{ h} \cdot \text{ft} \cdot \text{°F/Btu}$$

For the PVC jacket material, use Equation (8) again:

$$R_j = \frac{\ln(0.240/0.229)}{2\pi \times 0.10} = 0.07 \text{ h} \cdot \text{ft} \cdot \text{°F/Btu}$$

The thermal resistance of the soil as calculated by Equation (7) decreases slightly to $R_s = 0.51$ h·ft·°F/Btu because of the increase in the outer radius of the piping system. The total thermal resistance is now

$$R_t = R_p + R_i + R_j + R_s = 0.21 + 5.75 + 0.07 + 0.51$$
$$= 6.54 \text{ h} \cdot \text{ft} \cdot \text{°F/Btu}$$

The heat gain by the chilled water pipe is reduced to about 2 Btu/h·ft. In this case, the thermal resistance of the piping material and the jacket material could be neglected with a resultant error of <5%. Considering that the uncertainties in the material properties are likely greater than 5%, it is usually appropriate to neglect minor resistances such as those of piping and jacket materials if insulation is present.

Single Buried Pipe in Conduit with Air Space

Systems with air spaces may be treated by adding an appropriate resistance for the air space. For simplicity, assume a heat transfer coefficient of 3 Btu/h·ft^2·°F (based on the outer surface are a of the insulation), which applies in most cases. The resistance due to this heat transfer coefficient is then

$$R_a = 1/(3 \times 2\pi \times r_{oi}) = 0.053/r_{oi} \tag{9}$$

where

r_{oi} = outer radius of insulation, ft
R_a = resistance of air space, h·ft·°F/Btu

A more precise value for the resistance of an air space can be developed with empirical relations available for convection in enclosures such as those given by Grober et al. (1961). The effect of radiation in the annulus should also be considered when high temperatures are expected within the air space. For the treatment of radiation, refer to Siegel and Howell (1981).

Example 3. Consider a 6 in. nominal diameter (6.625 in. outer diameter) high-temperature water line operating at 375°F. Assume the pipe is insulated with 2.5 in. of mineral wool with a thermal conductivity k_i = 0.026 Btu/h·ft·°F at 200°F and k_i = 0.030 Btu/h·ft·°F at 300°F.

The pipe will be encased in a steel conduit with a concentric air gap of 1 in. The steel conduit will be 0.125 in. thick and will have a corrosion resistant coating approximately 0.125 in. thick. The pipe will be buried 4 ft deep to pipe centerline in soil with an average annual temperature of 60°F. The soil thermal conductivity is assumed to be 1 Btu/h·ft·°F. The thermal resistances of the pipe, conduit, and conduit coating will be neglected.

Solution: Calculate the thermal resistance of the pipe insulation. To do so, assume a mean temperature of the insulation of 250°F to establish its thermal conductivity, which is equivalent to assuming the insulation outer surface temperature is 125°F. Interpolating the data listed previously, the insulation thermal conductivity k_i = 0.028 Btu/h·ft·°F. Then calculate insulation thermal resistance using Equation (8):

$$R_i = \frac{\ln(0.484/0.276)}{2\pi \times 0.028} = 3.19 \text{ h} \cdot \text{ft} \cdot \text{°F/Btu}$$

Calculate the thermal resistance of the air space using Equation (9):

$$R_a = 0.053/0.484 = 0.11 \text{ h} \cdot \text{ft} \cdot \text{°F/Btu}$$

Calculate the thermal resistance of the soil from Equation (7):

$$R_s = \frac{\ln(8.0/0.589)}{2\pi \times 1} = 0.42 \text{ h} \cdot \text{ft} \cdot {}^\circ\text{F/Btu}$$

The total thermal resistance is

$$R_t = R_i + R_a + R_s = 3.19 + 0.11 + 0.42 = 3.72 \text{ h} \cdot \text{ft} \cdot {}^\circ\text{F/Btu}$$

The first estimate of the heat loss is then

$$q = (375 - 60)/3.72 = 84.7 \text{ Btu/h} \cdot \text{ft}$$

Now repeat the calculation with an improved estimate of the mean insulation temperature obtained, and calculate the outer surface temperature as

$$t_{io} = t_{po} - (qR_i) = 375 - (84.7 \times 3.19) = 105^\circ\text{F}$$

where t_{po} = outer surface temperature of pipe, °F.

The new estimate of the mean insulation temperature is 240°F, which is close to the original estimate. Thus, the insulation thermal conductivity changes only slightly, and the resulting thermal resistance is R_i = 3.24 h·ft·°F/Btu. The other thermal resistances in the system remain unchanged, and the heat loss becomes q = 83.8 Btu/h·ft. The insulation surface temperature is now approximately t_{io} = 104°F, and no further calculations are needed.

Single Buried Pipe with Composite Insulation

Many systems are available that use more than one insulating material. The motivation for doing so is usually to use an insulation that has desirable thermal properties or lower cost, but it is unable to withstand the service temperature of the carrier pipe (e.g., polyurethane foam). Another insulation with acceptable service temperature limits (e.g., calcium silicate or mineral wool) is normally placed adjacent to the carrier pipe in sufficient thickness to reduce the temperature at its outer surface to below the limit of the insulation that is desirable to use (e.g., urethane foam). The calculation of the heat loss and or temperature in a composite system is straightforward using the equations presented previously.

Example 4. A high-temperature water line operating at 400°F is to be installed in southern Texas. It consists of a 6 in. nominal diameter (6.625 in. outer diameter) carrier pipe insulated with 1.5 in. of calcium silicate. In addition, 1 in. polyurethane foam insulation will be placed around the calcium silicate insulation. The polyurethane insulation will be encased in a 0.5 in. thick fiberglass reinforced plastic (FRP) jacket. The piping system will be buried 10 ft deep to the pipe centerline. Neglect the thermal resistances of the pipe and FRP jacket.

Calcium silicate thermal conductivity
- 0.038 Btu/h·ft·°F at 200°F
- 0.042 Btu/h·ft·°F at 300°F
- 0.046 Btu/h·ft·°F at 400°F

Polyurethane foam thermal conductivity
- 0.013 Btu/h·ft·°F at 100°F
- 0.014 Btu/h·ft·°F at 200°F
- 0.015 Btu/h·ft·°F at 275°F

Assumed soil properties
- Thermal conductivity = 0.2 Btu/h·ft·°F
- Density (dry soil) = 105 lb/ft^3
- Moisture content = 5% (dry basis)

What is the maximum operating temperature of the polyurethane foam insulation?

Solution: Because the maximum operating temperature of the materials is sought, the maximum expected soil temperature rather than the average annual soil temperature must be found. Also, the lowest anticipated soil thermal conductivity and the deepest burial depth is assumed. These conditions produce the maximum internal temperatures of the components.

To solve the problem, first calculate the maximum soil temperature expected at the installation depth using Equation (1). CRREL (1999) shows the values for climatic constants for this region as t_{ms} = 71.8°F and A_s = 15.0°F. Soil thermal diffusivity may be estimated using Equation (3).

$$\alpha = \frac{24 \times 0.2}{105[0.175 + 1.0(5/100)]} = 0.20 \text{ ft}^2\text{/day}$$

Then Equation (1) is used to calculate the maximum soil temperature at the installation depth as follows:

$$t_{s,z} = 71.8 + 15.0\exp\left[-10\sqrt{\frac{\pi}{0.20 \times 365}}\right] = 73.7^\circ\text{F}$$

Now calculate the first estimates of the thermal resistances of the pipe insulations. For the calcium silicate assume a mean temperature of the insulation of 300°F to establish its thermal conductivity. From the data listed previously, the calcium silicate thermal conductivity k_i = 0.042 Btu/h·ft·°F at this temperature. For the polyurethane foam assume the mean temperature of the insulation is 200°F. From the data listed previously, the polyurethane foam thermal conductivity k_i = 0.014 Btu/h·ft·°F at this temperature. Now the insulation thermal resistances are calculated using Equation (8):

$$\text{Calcium silicate} \quad R_{i,1} = \frac{\ln(0.401/0.276)}{2\pi(0.042)} = 1.42 \text{ h·ft·°F/Btu}$$

$$\text{Polyurethane foam} \quad R_{i,2} = \frac{\ln(0.484/0.401)}{2\pi(0.014)} = 2.14 \text{ h·ft·°F/Btu}$$

Calculate the thermal resistance of the soil from Equation (7):

$$R_s = \frac{\ln(2 \times 10/0.526)}{2\pi(0.2)} = 2.89 \text{ h·ft·°F/Btu}$$

The total thermal resistance is

$$R_t = 1.42 + 2.14 + 2.89 = 6.45 \text{ h·ft·°F/Btu}$$

The first estimate of the heat loss is then

$$q = (400 - 73.7)/6.45 = 50.6 \text{ Btu/h·ft}$$

Next calculate the estimated insulation surface temperature with this first estimate of the heat loss. First find the temperature at the interface between the calcium silicate and polyurethane foam insulations.

$$t_{io,1} = t_{po} - qR_{i,1} = 400 - (50.6 \times 1.42) = 328^\circ\text{C}$$

where t_{po} is the outer surface temperature of the pipe and $t_{io,1}$ is the outer surface temperature of first insulation (calcium silicate).

$$t_{io,2} = t_{po} - q(R_{i,1} + R_{i,2}) = 400 - 50.6(1.42 + 2.14) = 220^\circ\text{C}$$

where $t_{io,2}$ is the outer surface temperature of second insulation (polyurethane foam).

The new estimate of the mean insulation temperature is (400 + 328)/2 = 364°F for the calcium silicate and (328 + 220)/2 = 274°F for the polyurethane foam. Thus, the insulation thermal conductivity for the calcium silicate would be interpolated to be 0.045 Btu/h·ft·°F, and the resulting thermal resistance is $R_{i,1}$ = 1.32 h·ft·°F/Btu.

For the polyurethane foam insulation the thermal conductivity is interpolated to be 0.015 Btu/h·ft·°F, and the resulting thermal resistance is $R_{i,2}$ = 2.00 h·ft·°F/Btu. The soil thermal resistance remains unchanged, and the heat loss is recalculated as q = 52.5 Btu/h·ft.

The calcium silicate insulation outer surface temperature is now approximately $t_{io,1}$ = 331°F while the outer surface temperature of the polyurethane foam is calculated to be $t_{io,2}$ = 226°F. Because these temperatures are within a few degrees of those calculated previously no further calculations are needed.

The maximum temperature of the polyurethane insulation of 331°F occurs at its inner surface, i.e., the interface with the calcium silicate insulation. This temperature clearly exceeds the maximum accepted 30 year service temperature of polyurethane foam of 250°F (EuHP 1991). Thus the amount of calcium silicate insulation needs to be increased significantly in order to achieve a maximum temperature for the polyurethane foam insulation within its long term service temperature limit. Under the conditions of this example it would take about 5 in. of calcium silicate insulation to reduce the insulation interface temperature to less than 250°F.

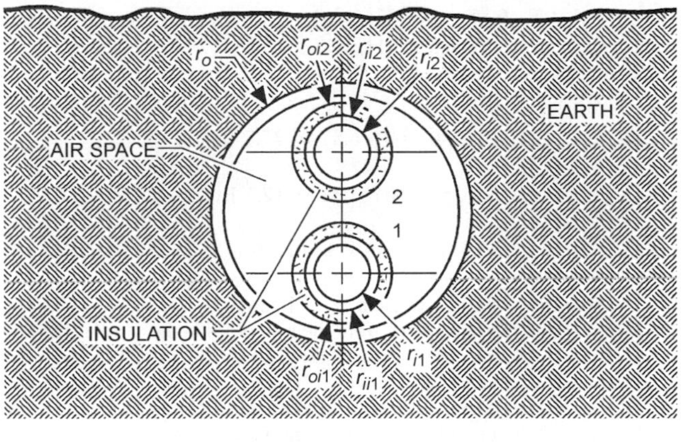

Fig. 7 Two Pipes Buried in Common Conduit with Air Space

Two Pipes Buried in Common Conduit with Air Space

For this case (Figure 7), make the same assumption as made in the previous section, Single Buried Pipe in Conduit with Air Space. For convenience, add some of the thermal resistances as follows:

$$R_1 = R_{p1} + R_{i1} + R_{a1} \tag{10}$$

$$R_2 = R_{p2} + R_{i2} + R_{a2} \tag{11}$$

The subscripts 1 and 2 differentiate between the two pipes within the conduit. The combined heat loss is then given by

$$q = \frac{[(t_{f1} - t_s)/R_1] + [(t_{f2} - t_s)/R_2]}{1 + (R_{cs}/R_1) + (R_{cs}/R_2)} \tag{12}$$

where R_{cs} is the total thermal resistance of conduit shell and soil.

Once the combined heat flow is determined, calculate the bulk temperature in the air space:

$$t_a = t_s + qR_{cs} \tag{13}$$

Then calculate the insulation outer surface temperature:

$$t_{i1} = t_a + (t_{f1} - t_a)(R_{a1}/R_1) \tag{14}$$

$$t_{i2} = t_a + (t_{f2} - t_a)(R_{a2}/R_2) \tag{15}$$

The heat flows from each pipe are given by

$$q_1 = (t_{f1} - t_a)/R_1 \tag{16}$$

$$q_2 = (t_{f2} - t_a)/R_2 \tag{17}$$

When the insulation thermal conductivity is a function of its temperature, as is usually the case, an iterative calculation is required, as illustrated in the following example.

Example 5. A pair of 4 in. NPS medium-temperature hot water supply and return lines run in a common 21 in. outside diameter conduit. Assume that the supply temperature is 325°F and the return temperature is 225°F. The supply pipe is insulated with 2.5 in. of mineral wool insulation, and the return pipe has 2 in. of mineral wool insulation. This insulation has the same thermal properties as those given in Example 3. The pipe is buried 4 ft to centerline in soil with a thermal conductivity of 1 Btu/h·ft·°F. Assume the thermal resistance of the pipe, the conduit, and the conduit coating are negligible. As a first estimate, assume the bulk air temperature within the conduit is 100°F. In addition, use this temperature as a first estimate of the insulation surface temperatures to obtain the mean insulation temperatures and subsequent insulation thermal conductivities.

Solution: By interpolation, estimate the insulation thermal conductivities from the data given in Example 3 at 0.0265 Btu/h·ft·°F for the supply pipe and 0.0245 Btu/h·ft·°F for the return pipe. Calculate the thermal resistances using Equations (10) and (11):

$$R_1 = R_{i1} + R_{a1} = \ln(0.396/0.188)/(2\pi \times 0.0265) + 0.053/0.396$$
$$= 4.48 + 0.13 = 4.61 \text{ h} \cdot \text{ft} \cdot {}^\circ\text{F/Btu}$$

$$R_2 = R_{i2} + R_{a2} = \ln(0.354/0.188)/(2\pi \times 0.0245) + 0.053/0.354$$
$$= 4.11 + 0.15 = 4.26 \text{ h} \cdot \text{ft} \cdot {}^\circ\text{F/Btu}$$

$$R_{cs} = R_s = \ln(8/0.875)/(2\pi \times 1) = 0.352 \text{ h} \cdot \text{ft} \cdot {}^\circ\text{F/Btu}$$

Calculate first estimate of combined heat flow from Equation (12):

$$q = \frac{(325 - 60)/4.61 + (225 - 60)/4.26}{1 + (0.352/4.61) + (0.352/4.26)} = 83.0 \text{ Btu/h} \cdot \text{ft}$$

Estimate bulk air temperature in the conduit with Equation (13):

$$t_a = 60 + (83.0 \times 0.352) = 89.2{}^\circ\text{F}$$

Then revise estimates of the insulation surface temperatures with Equations (14) and (15):

$$t_{i1} = 89.2 + (325 - 89.2)(0.13/4.61) = 95.8{}^\circ\text{F}$$
$$t_{i2} = 89.2 + (225 - 89.2)(0.15/4.26) = 94.0{}^\circ\text{F}$$

These insulation surface temperatures are close enough to the original estimate of 100°F that further iterations are not warranted. If the individual supply and return heat losses are desired, calculate them using Equations (16) and (17).

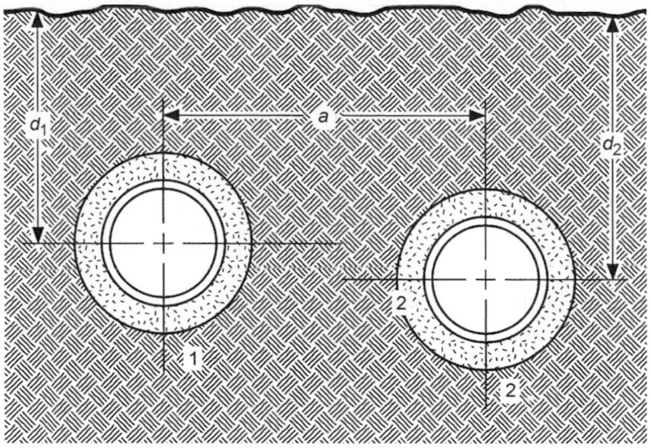

Fig. 8 Two Buried Pipes or Conduits

Two Buried Pipes or Conduits

This case (Figure 8) may be formulated in terms of the thermal resistances used for a single buried pipe or conduit and some correction factors. The correction factors needed are

$$\theta_1 = (t_{p2} - t_s)/(t_{p1} - t_s) \tag{18}$$

$$\theta_2 = 1/\theta_1 = (t_{p1} - t_s)/(t_{p2} - t_s) \tag{19}$$

$$P_1 = \frac{1}{2\pi k_s}\ln\left(\frac{(d_1+d_2)^2+a^2}{(d_1-d_2)^2+a^2}\right)^{0.5} \quad (20)$$

$$P_2 = \frac{1}{2\pi k_s}\ln\left(\frac{(d_2+d_1)^2+a^2}{(d_2-d_1)^2+a^2}\right)^{0.5} \quad (21)$$

where a = horizontal separation distance between centerline of two pipes, ft.

And the thermal resistance for each pipe or conduit is given by

$$R_{e1} = \frac{R_{t1}-(P_1^2/R_{t2})}{1-(P_1\theta_1/R_{t2})} \quad (22)$$

$$R_{e2} = \frac{R_{t2}-(P_2^2/R_{t1})}{1-(P_2\theta_2/R_{t1})} \quad (23)$$

where

θ = temperature correction factor, dimensionless
P = geometric/material correction factor, h·ft·°F/Btu
R_e = effective thermal resistance of one pipe/conduit in two-pipe system, h·ft·°F/Btu
R_t = total thermal resistance of one pipe/conduit if buried separately, h·ft·°F/Btu

Heat flow from each pipe is then calculated from

$$q_1 = (t_{p1}-t_s)/R_{e1} \quad (24)$$

$$q_2 = (t_{p2}-t_s)/R_{e2} \quad (25)$$

Example 6. Consider buried supply and return lines for a low-temperature hot water system. The carrier pipes are 4 in. NPS (4.5 in. outer diameter) with 1.5 in. of urethane foam insulation. The insulation is protected by a 0.25 in. thick PVC jacket. The thermal conductivity of the insulation is 0.013 Btu/h·ft·°F and is assumed constant with respect to temperature. The pipes are buried 4 ft deep to the centerline in soil with a thermal conductivity of 1 Btu/h·ft·°F and a mean annual temperature of 60°F. The horizontal distance between the pipe centerlines is 2 ft. The supply water is at 250°F, and the return water is at 150°F.

Solution: Neglect the thermal resistances of the carrier pipes and the PVC jacket. First, calculate the resistances from Equations (7) and (8) as if the pipes were independent of each other:

$$R_{s1} = R_{s2} = \frac{\ln(8.0/0.333)}{2\pi\times 1.0} = 0.51 \text{ h·ft·°F/Btu}$$

$$R_{i1} = R_{i2} = \frac{\ln(0.313/0.188)}{2\pi\times 0.013} = 6.25 \text{ h·ft·°F/Btu}$$

$$R_{t1} = R_{t2} = 0.51 + 6.25 = 6.76 \text{ h·ft·°F/Btu}$$

From Equations (20) and (21), the correction factors are

$$P_1 = P_2 = \frac{1}{2\pi\times 1}\ln\left(\frac{(4+4)^2+2^2}{(4-4)^2+2^2}\right)^{0.5} = 0.225 \text{ h·ft·°F/Btu}$$

$$\theta_1 = (150-60)/(250-60) = 0.474$$

$$\theta_2 = 1/\theta_1 = 2.11$$

Calculate the effective total thermal resistances as

$$R_{e1} = \frac{6.76-(0.225^2/6.76)}{1-(0.225\times 0.474/6.76)} = 6.87 \text{ h·ft·°F/Btu}$$

$$R_{e2} = \frac{6.76-(0.225^2/6.76)}{1-(0.225\times 2.11/6.76)} = 7.32 \text{ h·ft·°F/Btu}$$

The heat flows are then

$$q_1 = (250-60)/6.87 = 27.7 \text{ Btu/h·ft}$$

$$q_2 = (150-60)/7.32 = 12.3 \text{ Btu/h·ft}$$

$$q_t = 27.7 + 12.3 = 40.0 \text{ Btu/h·ft}$$

Note that when the resistances and geometry for the two pipes are identical, the total heat flow from the two pipes is the same if the temperature corrections are used or if they are set to unity. The individual losses will vary somewhat, however. These equations may also be used with air space systems. When the thermal conductivity of the pipe insulation is a function of temperature, iterative calculations must be done.

Pipes in Buried Trenches or Tunnels

Buried rectangular trenches or tunnels (Figure 9) require several assumptions to obtain approximate solutions for the heat transfer. First, assume that the air within the tunnel or trench is uniform in temperature and that the same is true for the inside surface of the trench/tunnel walls. Field measurements by Phetteplace et al. (1991) on operating shallow trenches indicate maximum spatial air temperature variations of about 10°F. Air temperature variations of this magnitude within a tunnel or trench do not cause significant errors for normal operating temperatures when using the following calculation methods.

Unless numerical methods are used, an approximation must be made for the resistance of a rectangular region such as the walls of a trench or tunnel. One procedure is to assume linear heat flow through the trench or tunnel walls, which yields the following resistance for the walls (Phetteplace et al. 1981):

$$R_w = \frac{x_w}{2k_w(a+b)} \quad (26)$$

where

R_w = thermal resistance of trench/tunnel walls, h·ft·°F/Btu
x_w = thickness of trench/tunnel walls, ft
a = width of trench/tunnel inside, ft
b = height of trench/tunnel inside, ft
k_w = thermal conductivity of trench/tunnel wall material, Btu/h·ft·°F

As an alternative to Equation (26), the thickness of the trench/tunnel walls may be included in the soil burial depth. This approximation is only acceptable when the thermal conductivity of the trench/tunnel wall material is similar to that of the soil.

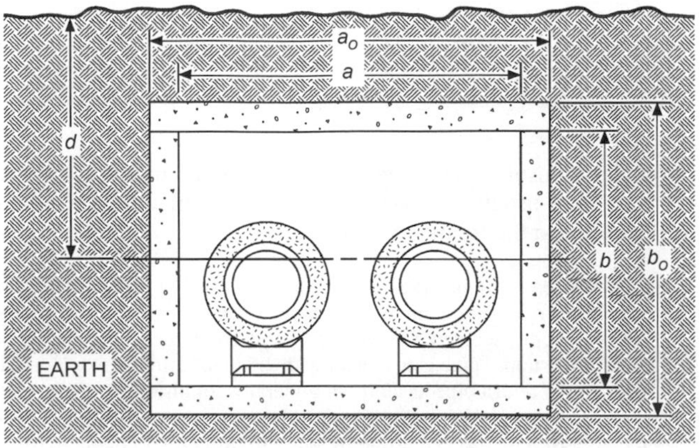

Fig. 9 Pipes in Buried Trenches or Tunnels

The thermal resistance of the soil surrounding the buried trench/tunnel is calculated using the following equation (Rohsenow 1998):

$$R_{ts} = \frac{\ln[3.5d/(b_o^{0.75}a_o^{0.25})]}{k_s[(a_o/2b_o) + 5.7]} \qquad a_o > b_o \tag{27}$$

where

R_{ts} = thermal resistance of soil surrounding trench/tunnel, h·ft·°F/Btu
a_o = width of trench/tunnel outside, ft
b_o = height of trench/tunnel outside, ft
d = burial depth of trench to centerline, ft

Equations (26) and (27) can be combined with the equations already presented to calculate heat flow and temperature for trenches/tunnels. As with the conduits described in earlier examples, the heat transfer processes in the air space inside the trench/tunnel are too complex to warrant a complete treatment for design purposes. The thermal resistance of this air space may be approximated by several methods. For example, Equation (9) may be used to calculate an approximate resistance for the air space.

Alternately, heat transfer coefficients could be calculated using Equations (6) and (7) in Chapter 24 of the 1997 *ASHRAE Handbook—Fundamentals*. Thermal resistances at the pipe insulation/air interface would be calculated from these heat transfer coefficients as done in the section on Pipes in Air. If the thermal resistance of the air/trench wall interface is also included, use Equation (28):

$$R_{aw} = 1/[2h_t(a + b)] \tag{28}$$

where

R_{aw} = thermal resistance of air/trench wall interface, h·ft·°F/Btu
h_t = total heat transfer coefficient at air/trench wall interface, Btu/h·ft²·°F

The total heat loss from the trench/tunnel is calculated from the following relationship:

$$q = \frac{(t_{p1} - t_s)/R_1 + (t_{p2} - t_s)/R_2}{1 + (R_{ss}/R_1) + (R_{ss}/R_2)} \tag{29}$$

where

R_1, R_2 = thermal resistances of two-pipe/insulation systems within trench/tunnel, h·ft·°F/Btu
R_{ss} = total thermal resistance on soil side of air within trench/tunnel, h·ft·°F/Btu

Once the total heat loss has been found, the air temperature within the trench/tunnel may be found as

$$t_{ta} = t_s + qR_{ss} \tag{30}$$

where t_{ta} is the air temperature within trench/tunnel.

The individual heat flows for the two pipes within the trench/tunnel are then

$$q_1 = (t_{p1} - t_{ta})/R_1 \tag{31}$$

$$q_2 = (t_{p2} - t_{ta})/R_2 \tag{32}$$

If the thermal conductivity of the pipe insulation is a function of temperature, assume an air temperature for the air space before starting calculations. Iterate the calculations if the air temperature calculated with Equation (30) differs significantly from the initial assumption.

Example 7. The walls of a buried trench are 6 in. thick, and the trench is 3 ft wide and 2 ft tall. The trench is constructed of concrete, with a thermal conductivity of k_w = 1 Btu/h·ft·°F. The soil surrounding the trench also has a thermal conductivity of k_s = 1 Btu/h·ft·°F. The centerline of the trench is 4 ft below grade, and the soil temperature is assumed be 60°F. The trench contains supply and return lines for a medium-temperature water system with the physical and operating parameters identical to those in Example 5.

Solution: Assuming the air temperature within the trench is 100°F, the thermal resistances for the pipe/insulation systems are identical to those in Example 5, or

$$R_1 = 4.61 \text{ h} \cdot \text{ft} \cdot °\text{F/Btu}$$

$$R_2 = 4.26 \text{ h} \cdot \text{ft} \cdot °\text{F/Btu}$$

The thermal resistance of the soil surrounding the trench is given by Equation (27):

$$R_{ts} = \frac{\ln[(3.5 \times 4)/(3^{0.75} \times 4^{0.25})]}{1[(4/6) + 5.7]} = 0.231 \text{ h} \cdot \text{ft} \cdot °\text{F/Btu}$$

The trench wall thermal resistance is calculated with Equation (26):

$$R_w = 0.5/[2(3 + 2)] = 0.050 \text{ h} \cdot \text{ft} \cdot °\text{F/Btu}$$

If the thermal resistance of the air/trench wall is neglected, the total thermal resistance on the soil side of the air space is

$$R_{ss} = R_w + R_{ts} = 0.050 + 0.231 = 0.281 \text{ h} \cdot \text{ft} \cdot °\text{F/Btu}$$

Find a first estimate of the total heat loss using Equation (29):

$$q = \frac{(325 - 60)/4.61 + (225 - 60)/4.26}{1 + (0.281/4.61) + (0.281/4.26)} = 85.4 \text{ Btu/h} \cdot \text{ft}$$

The first estimate of the air temperature in the trench is given by Equation (30):

$$t_{ta} = 60 + (85.4 \times 0.281) = 84.0°\text{F}$$

Refined estimates of the pipe insulation surface temperatures are then calculated using Equations (14) and (15):

$$t_{i1} = 84.0 + [(325 - 84.0)(0.13/4.61)] = 90.8°\text{F}$$

$$t_{i2} = 84.0 + [(225 - 84.0)(0.15/4.26)] = 89.0°\text{F}$$

From these estimates, calculate the revised mean insulation temperatures to find the resultant resistance values. Repeat the calculation procedure until satisfactory agreement between successive estimates of the trench air temperature is obtained. Calculate the individual heat flows from the pipes with Equations (31) and (32).

If the thermal resistance of the trench walls is added to the soil thermal resistance, the thermal resistance on the soil side of the air space is

$$R_{ss} = \frac{\ln[14/(2^{0.75} \times 3^{0.25})]}{1[(3/4) + 5.7]} = 0.286 \text{ h} \cdot \text{ft} \cdot °\text{F/Btu}$$

The result is less than 2% higher than the resistance previously calculated by treating the trench walls and soil separately. In the event that the soil and trench wall material have significantly different thermal conductivities, this simpler calculation will not yield as favorable results and should not be used.

Pipes in Shallow Trenches

The cover of a shallow trench is exposed to the environment. Thermal calculations for such trenches require the following assumptions: (1) the interior air temperature is uniform as discussed in the section on Pipes in Buried Trenches or Tunnels, and (2) the soil and the trench wall material have the same thermal conductivity. This assumption will yield reasonable results if the thermal conductivity of the trench material is used because most of the heat flows directly through the cover. The thermal resistance of the trench walls and surrounding soil is usually a small portion of the total thermal resistance, and thus the heat losses are not usually highly dependent on this thermal resistance. Using these assumptions, Equations (27) and (29) through (32) may be used for shallow trench systems.

Example 8. Consider a shallow trench having the same physical parameters and operating conditions as the buried trench in Example 7, except that the top of the trench is at grade level. Calculate the thermal resistance of the shallow trench using Equation (27):

$$R_{ts} = R_{ss} = \frac{\ln[(3.5 \times 1.5)/(2^{0.75} \times 3^{0.25})]}{1[(3/4) + 5.7]} = 0.134 \text{ h} \cdot \text{ft} \cdot {}^\circ\text{F/Btu}$$

Use thermal resistances for the pipe/insulation systems from Example 7, and use Equation (24) to calculate q:

$$q = \frac{(325 - 60)/4.61 + (225 - 60)/4.26}{1 + (0.134/4.61) + (0.134/4.26)} = 90.7 \text{ Btu/h} \cdot \text{ft}$$

From this, calculate the first estimate of the air temperature, using Equation (30):

$$t_{ta} = 60 + (90.7 \times 0.134) = 72.2^\circ\text{F}$$

Then refine estimates of the pipe insulation surface temperatures using Equations (14) and (15):

$$t_{i1} = 72.2 + [(325 - 72.2)(0.13/4.61)] = 79.3^\circ\text{F}$$

$$t_{i2} = 72.2 + [(225 - 72.2)(0.15/4.26)] = 77.6^\circ\text{F}$$

From these, calculate the revised mean insulation temperatures to find resultant resistance values. Repeat the calculation procedure until satisfactory agreement between successive estimates of the trench air temperatures are obtained. If the individual heat flows from the pipes are desired, calculate them using Equations (31) and (32).

Another method for calculating the heat losses in a shallow trench assumes an interior air temperature and treats the pipes as pipes in air (see the section on Pipes in Air). Interior air temperatures in the range of 70 to 120°F have been observed in a temperate climate (Phetteplace et al. 1991).

Buried Pipes with Other Geometries

Other geometries that are not specifically addressed by the previous cases have been used for buried thermal utilities. In some instances, the equations presented previously may be used to approximate the system. For instance, the soil thermal resistance for a buried system with a half-round clay tile on a concrete base could be approximated as a circular system using Equation (6) or (7). In this case, the outer radius r_o is taken as that of a cylinder with the same circumference as the outer perimeter of the clay tile system. The remainder of the resistances and subsequent calculations would be similar to those for a buried trench/tunnel. The accuracy of such calculations will vary inversely with the proportion of the total thermal resistance that the thermal resistance in question comprises. In most instances, the thermal resistance of the pipe insulation overshadows other resistances, and the errors induced by approximations in the other resistances are acceptable for design calculations.

Pipes in Air

Pipes surrounded by gases may transfer heat via conduction, convection, and/or thermal radiation. Heat transfer modes depend mainly on the surface temperatures and geometry of the system being considered. For air, conduction is usually dominated by the other modes and thus may be neglected. Equations (6) and (7) in Chapter 24 of the 1997 *ASHRAE Handbook—Fundamentals* give convective and radiative heat transfer coefficients. The thermal resistances for cylindrical systems can be found from these heat transfer coefficients using the equations that follow. Generally the piping systems used for thermal distribution have sufficiently low surface temperatures to preclude any significant heat transfer by thermal radiation.

Example 9. Consider a 6 in. nominal (6.625 in. outside diameter), high-temperature hot water pipe that operates in air at 375°F with 2.5 in. of mineral wool insulation. The surrounding air annually averages 60°F (520°R). The average annual wind speed is 4 mph. The insulation is covered with an aluminum jacket with an emittance $\varepsilon = 0.26$. The thickness and thermal resistance of the jacket material are negligible. As the heat transfer coefficient at the outer surface of the insulation is a function of the temperature there, initial estimates of this temperature must be made. This temperature estimate is also required to estimate mean insulation temperature.

Solution: Assuming that the insulation surface is at 100°F (560°R) as a first estimate, the mean insulation temperature is calculated as 238°F. Using the properties of mineral wool given in Example 3, the insulation thermal conductivity is $k_i = 0.0275$ Btu/h·ft·°F. Using Equation (8), the thermal resistance of the insulation is $R_i = 3.25$ h·ft·°F/Btu. Equation (6) in Chapter 24 of the 1997 *ASHRAE Handbook—Fundamentals* gives the forced convective heat transfer coefficient at the surface of the insulation as

$$h_{cv} = 1.016\left(\frac{1}{d}\right)^{0.2}\left(\frac{2}{t_{oi} + t_a}\right)^{0.181}(t_{oi} - t_a)^{0.266}(1 + 1.277V)^{0.5}$$

$$= 1.016\left(\frac{1}{11.625}\right)^{0.2}\left(\frac{2}{160}\right)^{0.181}(40)^{0.266}(6.11)^{0.5}$$

$$= 1.86 \text{ Btu/h} \cdot \text{ft}^2 \cdot {}^\circ\text{F}$$

where
d = outer diameter of surface, in.
t_a = ambient air temperature, °F
V = wind speed, mph

The radiative heat transfer coefficient must be added to this convective heat transfer coefficient. Determine the radiative heat transfer coefficient from Equation (7) in Chapter 24 of the 1997 *ASHRAE Handbook—Fundamentals* as follows:

$$h_{rad} = \varepsilon\sigma\frac{(T_a^4 - T_s^4)}{T_a - T_s}$$

$$h_{rad} = 0.26 \times 1.713 \times 10^{-9}\frac{(560^4 - 520^4)}{560 - 520} = 0.28 \text{ Btu/h·ft}^2\text{·°F}$$

Add the convective and radiative coefficients to obtain a total surface heat transfer coefficient h_t of 2.14 Btu/h·ft²·°F. The equivalent thermal resistance of this heat transfer coefficient is calculated from the following equation:

$$R_{surf} = \frac{1}{2\pi r_{oi} h_t} = \frac{1}{2\pi \times 0.484 \times 2.14} = 0.15 \text{ h·ft·°F/Btu}$$

With this, the total thermal resistance of the system becomes R_t = 3.40 h·ft·°F/Btu, and the first estimate of the heat loss is q = 92.6 Btu/h·ft.

An improved estimate of the insulation surface temperature is t_{oi} = 375 – (92.6 × 3.25) = 74°F. From this, a new mean insulation temperature, insulation thermal resistance, and surface resistance can be calculated. The heat loss is then 90.0 Btu/h·ft, and the insulation surface temperature is calculated as 77°F. These results are close enough to the previous results that further iterations are not warranted.

Note that the contribution of thermal radiation to the heat transfer could have been omitted with negligible effect on the results. In fact, the entire surface resistance could have been neglected and the resulting heat loss would have increased by only about 4%.

In Example 9, the convective heat transfer was forced. In cases with no wind, where the convection is free rather than forced, the radiative heat transfer is more significant, as is the total thermal resistance of the surface. However, in instances where the piping is well insulated, the thermal resistance of the insulation dominates, and minor resistances can often be neglected with little resultant error. By neglecting resistances, a conservative result is obtained (i.e., the heat transfer is overpredicted).

Economical Thickness for Pipe Insulation

A life-cycle cost analysis may be run to determine the economical thickness of pipe insulation. Because the insulation thickness affects other parameters in some systems, each insulation thickness must be considered as a separate system. For example, a conduit system or one with a jacket around the insulation requires a larger conduit or jacket for greater insulation thicknesses. The cost of the extra conduit or jacket material may exceed that of the additional insulation and is therefore usually included in the analysis. It is usually not necessary to include excavation, installation, and backfill costs in the analysis.

The life-cycle cost of a system is the sum of the initial capital cost and the present worth of the subsequent cost of heat lost or gained over the life of the system. The initial capital cost needs only to include those costs that are affected by insulation thickness. The following equation can be used to calculate the life-cycle cost:

$$\text{LCC} = \text{CC} + (q t_u C_h \text{PWF}) \tag{33}$$

where

LCC = present worth of life-cycle costs associated with pipe insulation thickness, $/ft
CC = capital costs associated with pipe insulation thickness, $/ft
q = annual average rate of heat loss, Btu/h·ft
t_u = utilization time for system each year, h
C_h = cost of heat lost from system, $/Btu
PWF = present worth factor for future annual heat loss costs, dimensionless

The present worth factor is the reciprocal of the capital recovery factor, which is found from the following equation:

$$\text{CRF} = \frac{i(1+i)^n}{(1+i)^n - 1} \tag{34}$$

where

CRF = cost recovery factor, dimensionless
i = interest rate
n = useful lifetime of system, years

If heat costs are expected to escalate, the present worth factor may be multiplied by an appropriate escalation factor and the result substituted in place of PWF in Equation (33).

Example 10. Consider a steel conduit system with an air space. The insulation is mineral wool with thermal conductivity as given in Examples 3 and 4. The carrier pipe is 4 in. NPS and operates at 350°F for the entire year (8760 h). The conduit is buried 4 ft to the centerline in soil with a thermal conductivity of 1 Btu/h·ft·°F and an annual average temperature of 60°F. Neglect the thermal resistance of the conduit and carrier pipe. The useful lifetime of the system is assumed to be 20 years and the interest rate is taken as 10%.

Solution: Find CRF from Equation (34):

$$\text{CRF} = \frac{0.10(1+0.10)^{20}}{(1+0.10)^{20} - 1} = 0.11746$$

The value C_h of heat lost from the system is assumed to be $10 per million Btu. The following table summarizes the heat loss and cost data for several available insulation thicknesses.

Insulation thickness, in.	1.5	2.0	2.5	3.0	3.5	4.0
Insulation outer radius, ft	0.313	0.354	0.396	0.438	0.479	0.521
Insulation k, Btu/h·ft·°F	0.027	0.027	0.027	0.027	0.027	0.027
R_i, Eq. (8), h·ft·°F/Btu	3.00	3.73	4.39	4.99	5.52	6.00
R_a, Eq. (9), h·ft·°F/Btu	0.17	0.15	0.13	0.12	0.11	0.10
Conduit outer radius, ft	0.448	0.448	0.531	0.531	0.583	0.667
R_s, Eq. (7), h·ft·°F/Btu	0.46	0.46	0.43	0.43	0.42	0.40
R_t, h·ft·°F/Btu	3.63	4.34	4.95	5.54	6.04	6.50
q, heat loss rate, Btu/h·ft	79.9	66.8	58.6	52.3	48.0	44.6
Conduit system cost, $/ft	23.00	24.50	28.25	30.00	33.00	40.00
LCC, Eq. (33), $/ft	82.59	74.32	71.95	69.00	68.80	73.27

The table indicates that 3.5 in. of insulation yields the lowest life-cycle cost for the example. Because the results depend highly on the economic parameters used, they must be accurately determined.

EXPANSION PROVISIONS

All piping moves due to temperature changes. The length of the piping increases as the temperature increases. Field conditions and the type of system govern the method used to absorb the movement. Turns where the pipe changes direction must be used to provide flexibility. When the distance between changes in direction becomes too large for the turns to compensate for movement, expansion loops are positioned at appropriate locations. If loops are required, additional right-of-way may be required. If field conditions permit, the flexibility of the piping should be used to allow expansion. Where space constraints do not permit expansion loops and/or changes in direction, mechanical methods, such as expansion joints or ball joints, must be used. However, because ball joints changes the direction of a pipe, a third joint may be required to reduce the length between changes in direction.

Chapter 41 covers the design of the pipe bends, loops, and the use of expansion joints. However, the chapter uses conservative stress values. Computer aided programs that calculate stress due to pressure, thermal expansion, and weight, simultaneously, allow the designer to meet the requirements of ASME *Standard* B31.1. When larger pipe diameters are required, a computer program should be used when the pipe will provide the required flexibility. For example, Table 11 in Chapter 41 indicates that a 12 in. standard weight pipe with 12 in. of movement, requires a 15.5 ft wide by 31 ft high expansion loop. One computer aided program recommends a 13 ft wide by 26 ft high loop; and if an equal height and width is specified, the loop is 23.5 ft in each direction.

Although the inherent flexibility of the piping should be used to handle expansion as much as possible, expansion joints must be used where space is too small to allow a loop to be constructed to handle the required movement. For example, expansion joints are often used in walk-through tunnels because there is seldom space to construct pipe loops. Either pipe loops or expansion joints can be used for above ground, concrete shallow trench, direct-buried, and poured envelope systems. The manufacturer of the conduit or envelope material should design loops and offsets in conduit and poured envelope systems because clearance and design features are critical to the performance of both the loop and the pipe.

All expansion joints require maintenance, therefore they should always be accessible for service. Joints in direct-buried and poured envelope systems and trenches without removable covers should be located in manholes.

Cold springing is normally used when thermal expansion compensation is used. In DHC systems with natural flexibility, cold springing minimizes the clearance required for pipe movement only. The pipes are sprung 50% of the total amount of movement, in the direction toward the anchor. However, ASME *Standard* B 31.1 does not allow cold springing in calculating the stresses in the piping. When expansion joints are used, they are installed in an extended position to achieve maximum movement. However, the manufacturer of the expansion joint should be contacted for the proper amount of extension.

Pipe Supports, Guides, and Anchors

For conduit and poured envelope systems, the system manufacturer usually designs the pipe supports, guides, and anchors in consultation with the expansion joint manufacturer, if such devices are used. For example, the main anchor force of an in-line axial expansion joint is the sum of the pressure thrust (system pressure times the cross-sectional area of the expansion joint and the joint friction or spring force) and the pipe friction forces. The manufacturer of the expansion device should be consulted when determining anchor forces. Anchor forces are normally less when expansion is absorbed through the system instead of with expansion joints.

Pipe guides used with expansion joints should be spaced according to the manufacturer's recommendations. They must permit longitudinal or axial motion and restrict motion perpendicular to the axis of the pipe. Guides with graphite or low-friction fluorocarbon slide surfaces are often desirable for long pipe runs. In addition, these surface finishes do not corrode or increase sliding resistance in above ground installations. Guides should be selected to handle twice the expected movement, so they may be installed in a neutral position without the need for cold springing the pipe.

DISTRIBUTION SYSTEM CONSTRUCTION

The combination of aesthetics, first cost, safety, and life-cycle cost naturally divide distribution systems into two distinct categories—aboveground and underground distribution systems. The materials needed to ensure long life and low heat loss further classify DHC systems into low-temperature, medium-temperature, and high-temperature systems. The temperature range for medium-temperature systems is usually too high for the materials that are used in low-temperature systems; however, the same materials that are used in high-temperature systems are typically used for medium-temperature systems. Because low-temperature systems have a lower temperature differential between the working fluid temperatures and the environment, heat loss is inherently less. In addition, the selection of efficient insulation materials and inexpensive pipe materials that resist corrosion is much greater for low-temperature systems.

The aboveground system has the lowest first cost and the lowest life-cycle cost because it can be maintained easily and constructed with materials that are readily available. Generally, aboveground systems are acceptable where they are hidden from view or can be hidden by landscaping. Poor aesthetics and the risk of vehicle damage to the aboveground system removes it from contention for many projects.

Although the aboveground system is sometimes partially factory prefabricated, more typically it is entirely field fabricated of components such as pipes, insulation, pipe supports, and insulation jackets or protective enclosures that are commercially available. Other common systems that are completely field fabricated include the walk-through tunnel (Figure 10), the concrete surface trench (Figure 11), the deep-burial small tunnel (Figure 12), and underground systems that use poured insulation (Figure 13) or cellular glass (Figure 14) to form an envelope around the carrier pipes.

Field assembled systems must be designed in detail, and all materials must be specified by the project design engineer. Evaluation of the project site conditions indicates which type of system should be considered for the site. For instance, the shallow trench system is best where utilities that are buried deeper than the trench bottom need to be avoided and where the covers can serve as sidewalks. Direct buried conduit, with a thicker steel casing, may be the only system that can be used in flooded sites. The conduit system is used where aesthetics is important. It is often used for short distances between buildings and the main distribution system, and where the owner is willing to accept higher life-cycle costs.

Direct buried conduit, concrete surface trench, and other underground systems must be routed to avoid existing utilities, which requires a detailed site survey and considerable design effort. In the absence of a detailed soil temperature distribution study, direct buried heating systems should be spaced more than 15 ft from other utilities constructed of plastic pipes because the temperature of the soil during dry conditions can be high enough to reduce the strength of plastic pipe to an unacceptable level. Rigid, extruded polystyrene insulation may be used to insulate adjacent utilities from the impact of a buried heat distribution pipe; however, the temperature limit of the extruded polystyrene insulation must not be exceeded. A numerical analysis of the thermal problem may be required to ensure that the desired effect is achieved.

Tunnels that provide walk-through or crawl-through access can be buried in nearly any location without causing future problems because utilities are typically placed in the tunnel. Regardless of the type of construction, it is usually cost effective to route distribution piping through the basements of buildings, but only after liability issues are addressed. In laying out the main supply and return piping, redundancy of supply and return should be considered. If a looped system is used to provide redundancy, flow rates under all possible failure modes must be addressed when sizing and laying out the piping.

Manholes provide access to underground systems at critical points such as where there are

- High or low points on the system profile that vent trapped air or where the system can be drained
- Elevation changes in the distribution system that are needed to maintain the required constant slope
- Major branches with isolation valves
- Steam traps and condensate drainage points on steam lines
- Mechanical expansion devices

To facilitate leak location and repair and to limit damage caused by leaks, access points generally should be spaced no farther than 500 ft apart. Special attention must be given to the safety of personnel who come in contact with distributions systems or who must enter spaces occupied by underground systems. The regulatory authority's definition of a confined space and the possibility of exposure to high-temperature or high-pressure piping can have a significant impact on the access design, which must be addressed by the designer.

Piping Materials and Standards

Supply Pipes for Steam and Hot Water. Adequate temperature and pressure ratings for the intended service should be specified for any piping. All piping, fittings, and accessories should be in accordance with ASME *Standard* B31.1 or with local requirements if more stringent. For steam and hot water all joints should be welded and pipe should conform to either ASTM A 53 seamless or ERW, Grade B; or ASTM A 106 seamless, Grade B. Care should be taken to exclude ASTM A 53, Type F, because of its lower allowable stress and because of the method by which its seams are manufactured. Mechanical joints of any type are not recommended for steam or hot water. Pipe wall thickness is determined by the maximum operating temperature and pressure. In the United States, most piping for steam and hot water is Schedule 40 for 10 in. NPS and below and standard weight for 12 in. NPS and above.

Many European low-temperature water systems that have piping with a wall thickness similar to Schedule 10. Due to reduced piping material in these systems, they are not only less expensive, but they also develop reduced expansion forces and thus require simpler methods of expansion compensation, a point to be considered when making a selection between a high-temperature and a low-temperature system. Welding pipes with thinner walls requires extra care and may require additional inspection. Also, extra care must be taken to avoid internal and external corrosion because the thinner wall provides a much lower corrosion allowance.

Condensate Return Pipes. Condensate pipes require special consideration because condensate is much more corrosive than steam. This corrosive nature is caused by the oxygen that the condensate accumulates. The usual method used to compensate for the corrosiveness of condensate is to select steel pipe that is thicker than the steam pipe. For highly corrosive condensate, stainless steel and/or other corrosion resistant materials should be considered. Materials that are corrosion resistant in air may not be corrosion resistant when exposed to condensate; therefore, a material with good experience handling condensate should be selected. Fiberglass-reinforced plastic (FRP) pipe used for condensate return has not performed well. Failed steam traps, pipe resin solubility, deterioration at elevated temperatures, and thermal expansion are thought to be the cause of premature failure.

Chilled Water Distribution. For chilled water systems, a variety of **pipe materials**, such as steel, ductile iron, PVC, and FRP, have been used successfully. If ductile iron or steel is used, the designer must resolve the internal and external corrosion issue, which may be significant. The soil temperature is usually highest when the chilled water loads are highest; therefore, it is usually life-cycle cost-effective to insulate the chilled water pipe. A life-cycle cost analysis often favors a factory prefabricated product that is insulated with a plastic foam with a waterproof casing or a field fabricated system that is insulated with ASTM C 552 cellular glass insulation. If a plastic product is selected, care must be taken to maintain an adequate distance from any high-temperature underground system that may be near. Damage to the chilled water system would most likely occur from elevated soil temperatures when the chilled water circulation stopped. The heating distribution system should be at least 15 ft from a chilled water system containing plastic unless a detailed study of the soil temperature distribution indicates otherwise. Rigid extruded polystyrene insulation may be used to insulate adjacent chilled water lines from the impacts of a buried heat distribution pipe, however, care must be taken not to excess the temperature limits of the extruded polystyrene insulation; numerical analysis of the thermal problem may be required.

Aboveground System

The aboveground system consists of a distribution pipe, insulation that surrounds the pipe, and a protective jacket that surrounds the insulation. The jacket may have an integral vapor retarder. When the distribution system carries chilled water or other cold media, a vapor retarder is required for all types of insulation except cellular glass. In an ASHRAE test by Chyu et al. (1998a), cellular glass absorbed essentially zero water in a chilled water application. In heating applications the vapor retarder is not needed nor recommended; however, a reasonably watertight jacket is required to keep storm water out of the insulation. The jacket material can be aluminum, stainless steel, galvanized steel, or plastic sheet; a multilayered fabric and organic cement composite; or a combination of these. Plastics and organic cements exposed to sunshine must be ultraviolet light resistant.

Structural columns and supports are typically made of wood, steel, or concrete. A crossbar is often placed across the top of the column when more than one distribution pipe is supported from one column. Sidewalk and road crossings require an elaborate support structure to elevate the distribution piping above traffic. Pipe expansion and contraction is taken up in loops, elbows, and bends. Manufactured expansion joints may be used, but they are usually not recommended because of a shorter life or a higher frequency of required maintenance than the rest of the system. Supports that attach the distribution pipes to the support columns are commercially available as described in MSS *Standard* SP-58 and MSS *Standard* SP-69. The distribution pipes should have welded joints.

The aboveground system has the lowest first cost and is the easiest to inspect and maintain; therefore, it has the lowest life-cycle cost. It is the standard against which all other systems are compared. Its major drawbacks are its poor aesthetics, its safety hazard if struck by vehicles and equipment, and its susceptibility to freezing in cold climates if circulation is stopped or if heat is not added to the working medium. These drawbacks often remove this system from contention as a viable alternative.

Underground Systems

The underground system solves the problems of aesthetics and exposure to vehicles of the aboveground system; however, burying a system causes other problems with materials, design, construction, and maintenance that have historically been difficult to solve. An underground heat distribution system is not a typical utility like gas, domestic water, and sanitary systems. It requires an order of magnitude more design effort and construction inspection accuracy when compared to gas, water, and sanitary distribution projects. The thermal effects and difficulty of keeping the insulation dry make it much more difficult to design and construct when compared to systems operating near the ambient temperature. The underground system costs almost 10 times as much to build, and requires much more to operate and maintain. Segan and Chen (1984) describe the type of premature failures that may occur if this guidance is ignored. Heat distribution systems must be designed for zero leakage and must account for thermal expansion, degradation of material as a function of temperature, high pressure and transient shock waves, heat loss restrictions, and accelerated corrosion. In the past, resolving one problem in underground systems often created a new, more serious problem that was not recognized until premature failure occurred.

Common types of underground systems are the walk-through tunnel, the concrete surface trench, the deep-buried small tunnel, the poured insulation envelope, and the cellular glass, and the conduit system.

Walk-Through Tunnel. This system (Figure 10) consists of a field erected tunnel that is large enough for a someone to walk through after the distribution pipes are in place. It is essentially an aboveground system enclosed with a tunnel. The tunnel is buried deep enough to cover the top with earth. This tunnel is large enough so that routine maintenance and inspection can be done easily without excavation. The preferred construction material for the tunnel walls and top cover is reinforced concrete. Masonry units and metal preformed sections have been used to construct the tunnel and top with less success due to ground water leakage and metal corrosion. The distribution pipes are supported from the tunnel wall or floor with pipe supports that are commonly used on aboveground systems or in buildings. Some ground water will penetrate top and walls of the tunnel; therefore, a water drainage system must be provided. Usually electric lights and electric service outlets are provided for ease of inspection and maintenance. This system has the highest first cost of all underground systems; however, it can have the lowest life-cycle cost due to ease of maintenance, the ability to correct construction errors easily, and an extremely long life.

Shallow Concrete Surface Trench. This system. (Figure 11) is a partially buried system. The floor is usually about 3 ft below surface grade elevation. It is only wide enough for the carrier pipes and the pipe insulation plus some additional width to allow for pipe movement and possibly enough room for a person to stand on the floor. The trench usually is about as wide as it is deep. The top is constructed of reinforced concrete covers that protrude slightly above the surface and may also serve as a sidewalk. The floor and walls are usually cast-in-place reinforced concrete and the top is either precast or cast-in-place concrete. Precast concrete floor and wall sections have not been successful because of the large number of oblique joints and nonstandard sections required to follow the surface topography and to slope the floor to drain. This system is designed to handle the storm water and ground water that enters the

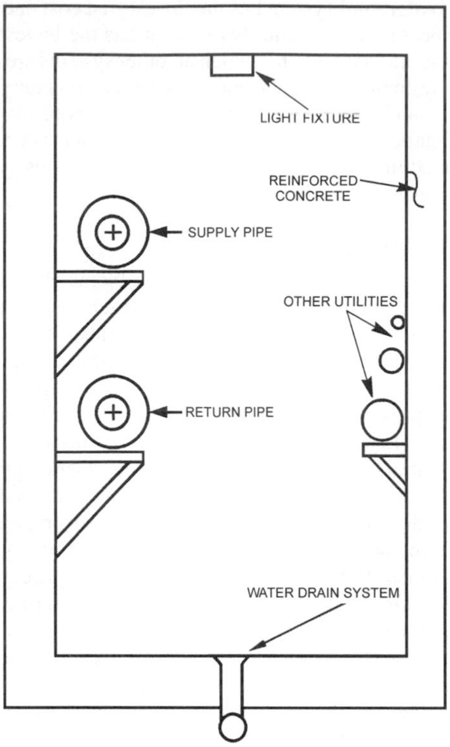

Fig. 10 Walk-Through Tunnel

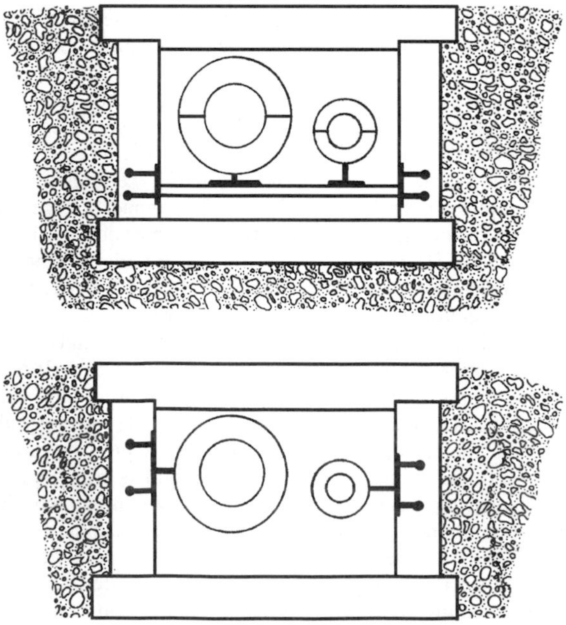

Fig. 11 Concrete Surface Trench

system, so, the floor is always sloped toward a drainage point. A drainage system is required at all floor low points.

Cross beams that attach to the side walls are preferred to support the carrier pipes. This keeps the floor free of obstacles that would interfere with drainage and allows the distribution pipes to be assembled before lowering them on the pipe supports. Also, floor mounted pipe supports tend to corrode.

The carrier pipes, the pipe supports, the expansion loops and bends and the insulation jacket are similar to aboveground systems, with the exception of the pipe insulation. Experience with these

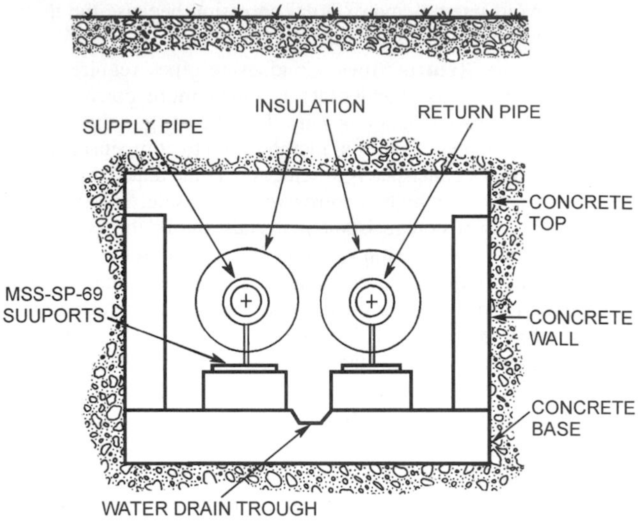

Fig. 12 Deep-Bury Small Tunnel

systems indicates that flooding will occur several times during their design life; therefore, the insulation must be able to survive flooding and boiling and then return to near its original thermal efficiency. The pipe insulation is covered with a metal or plastic jacket to protect the insulation from abuse and from storm water that enters at the top cover butt joints. Small inspection ports of about 12 in. diameter may be cast into the top covers at key locations so that the system can be inspected without removing the top covers. All replaceable elements such as valves, condensate pumps, steam traps, strainers, sump pumps and meters are located in valve vaults. The first cost of this system is among the lowest for underground systems because it uses typical construction techniques and materials. The life-cycle cost is often the lowest because it is easy to maintain, correct construction deficiencies, and repair leaks.

Deep-Bury Tunnel. The tunnel in this system (Figure 12) is only large enough to contain the distribution piping, the pipe insulation, and the pipe supports. One type of deep-bury tunnel is the shallow concrete surface trench covered with earth and sloped independent of the topography. Because the system is covered with earth, it is essentially not maintainable between valve vaults without major excavation. All details of this system must be designed and all materials must be specified by the project design engineer.

Because this system is not maintainable between valve vaults, great care must be taken to select materials that will last for the intended life and to ensure that the ground water drainage system will function reliably. This system is intended to be used on sites where the ground water elevation is typically lower than the bottom of the tunnel. The system can tolerate some ground water saturation depending on the water tightness of the construction and the capacity and reliability of the internal drainage. But even in desert areas, storms occur that expose underground systems to flooding; therefore, as with other types of underground systems, it must be designed to handle ground water or storm water that enters the system. The distribution pipe insulation must be of the type that can withstand flooding and boiling and still retain its thermal efficiency.

Construction of this system is typically started in an excavated trench by pouring a cast-in-place concrete base that is sloped so intruding ground water can drain to the valve vaults. The slope selected must also be compatible with the pipe slope requirements of the distribution system. The concrete base may have provisions for the supports for the distribution pipes, the ground water drainage system, and the mating surface for the side walls. The side walls may have provisions for the pipe supports if the pipes are not bottom supported. If the upper portion is to have cast-in-place concrete walls, the bottom may have reinforcing steel for the walls protruding upward. The pipe supports, the distribution pipes, and the pipe insulation are all installed before the top cover is installed.

The ground water drainage system may be a trough formed into the concrete bottom, a sanitary drainage pipe cast into the concrete bottom, or a sanitary pipe that is located slightly below the concrete base. The cover for the system is typically either of cast-in-place concrete or preformed sections such as precast concrete sections or half-round clay tile sections. The top covers must mate to the bottom and each other as tightly as possible to limit the entry of ground water. After the covers are installed the system is covered with earth to match the existing topography.

Poured Insulation. This system (Figure 13) is buried with the distribution system pipes encased in an envelope of insulating material and the insulation envelope covered with a thick layer of earth as required to match existing topography. This system is used on sites where the ground water is typically below the system. Like other underground systems, experience indicates that it will be flooded because the soil will become saturated with water several times during the design life; therefore, the design must accommodate flooded condition.

The insulation material serves several functions. It may support the distribution pipes and it must support earth loads. The insulation must prevent ground water from entering the interior of the envelope and it must have long term resistance to physical breakdown due to heat and water. The insulation envelope must allow the distribution pipes to expand and contract axially as the pipes change temperature. In elbows, expansion loops, and bends the insulation must allow formed cavities for lateral movement of the pipes; or be able to migrate around the pipe without significant distortion of the insulation envelope while still retaining the required structural load carrying capacity. Special attention must be given to corrosion of metal parts and water penetration at anchors and structural supports that penetrate the insulation envelope.

Hot distribution pipes tend to drive moisture out of the insulation as steam; however, pipes used to distribute a cooling medium tend to condense water in the insulation, which reduces the insulation thermal resistance. A ground water drainage system may be required depending on the insulation material selected and the severity of the ground water; however, if such a drainage is needed, it is a strong indicator that this is not the proper system for the site conditions.

This system is constructed by excavating a trench with a bottom slope that matches the desired slope of the distribution piping. The width of the bottom of the trench is usually the same as the width of the insulation envelope because it serves as a form. The distribution piping is then assembled in the trench and supported at the anchors and by blocks that are removed as the insulation is poured in place. The form for the insulation can be the trench bottom and sides, wooden forms, or sheets of plastic depending on the type of insulation used and the site conditions. The insulation envelope is covered with earth to complete the installation.

The project design engineer is responsible for finding an insulation material that fulfills all of the previously mentioned requirements. At present, no standards have been developed for insulation used in this type of application. **Hydrophobic powders**, which are a special type of pulverized rock that is treated to be water repellent, have been used successfully. The hydrophobic characteristic of this powder prevents water from dampening the powder and has some capability as a barrier for preventing water from entering the insulation envelope. This insulating powder typically has a much higher thermal conductivity than mineral wool or fiberglass pipe insulation; therefore, the thickness of the poured envelope must be significantly greater.

Field-Installed, Direct-Buried, Cellular Glass. In this system (Figure 14) cellular glass insulation is covered with an asphaltic jacket. The insulation supports the pipe. Oversize loops with internal support elements provide for expansion. The project design engineer must make provisions for movement of the pipes in the expansion loops. As shown in Figure 14, a drain should be installed to drain ground water away. A waterproof jacket is recommended for all buried applications.

When used for heating applications, the dry soil condition must be investigated to determine if the temperature of the jacket exceeds the material allowable temperature (see the section on Methods Of Heat Transfer Analysis). This is one of the controlling conditions to determine how high the carrier pipe fluid temperature will be allowed to be without the temperature limits of the jacket material being exceeded. The thickness of insulation depends on the thermal operating parameters of the carrier pipe. For maximum system integrity under extreme operating conditions, such as groundwater flooding, the jacketing may be applied to the insulation segments in the fabrication shop with hot asphalt. As with other underground systems, experience indicates that this system may be flooded several times during its projected life.

Conduits

The term conduit denotes an entire assembly, which consists of a carrier pipe, the pipe insulation, the casing, and the exterior casing coating (Figure 15). The conduit is assembled in a factory and shipped as unit called a **conduit section**. The pipe that carries the

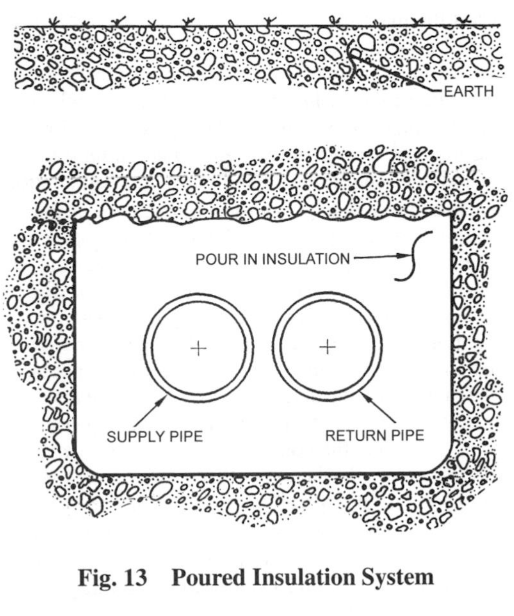

Fig. 13 Poured Insulation System

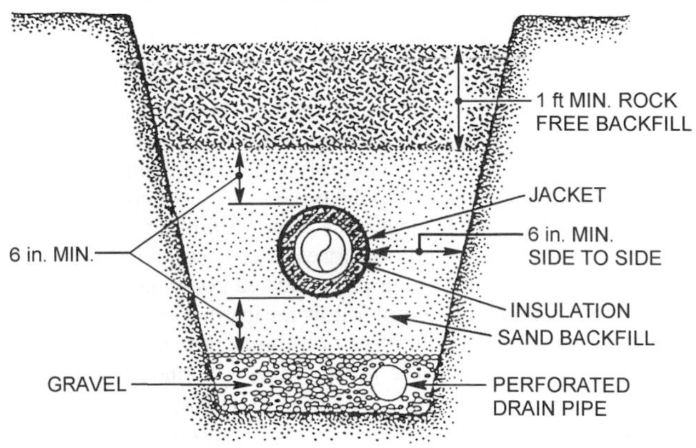

Fig. 14 Field Installed Direct-Buried Cellular Glass Insulated System

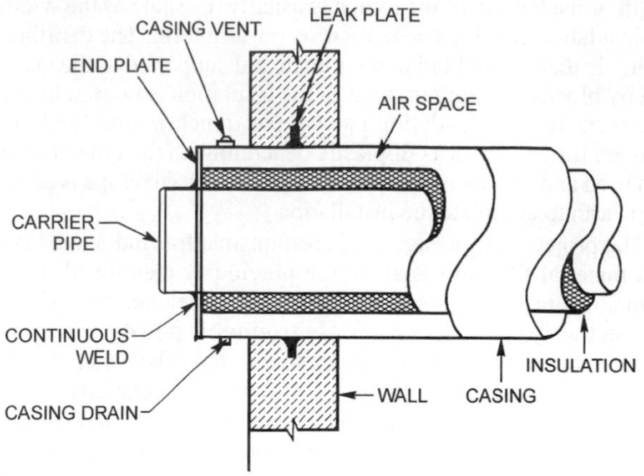

Fig. 15 Conduit System Components

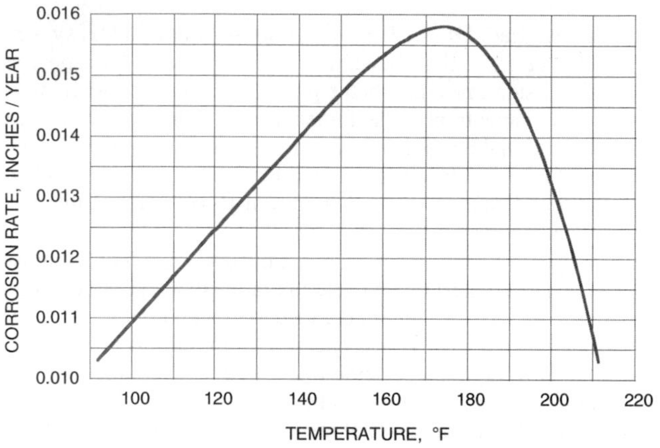

Fig. 16 Corrosion Rate in Aggressive Environment Similar to Mild Steel Casings in Soil

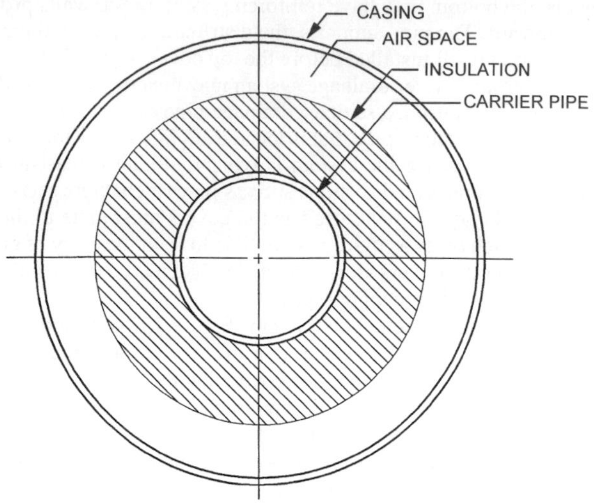

Fig. 17 Conduit System with Annular Air Space and Single Carrier Pipe

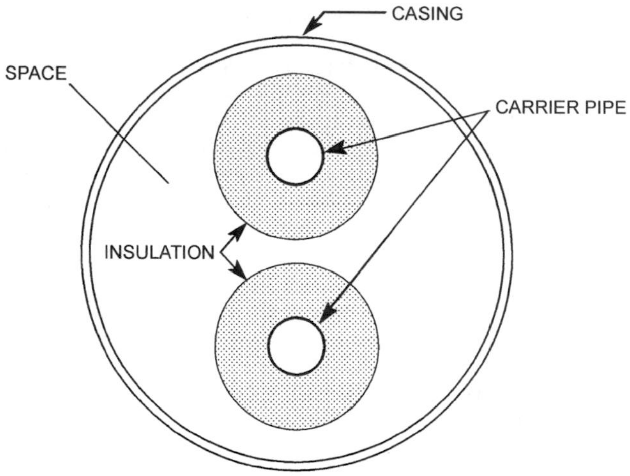

Fig. 18 Conduit System with Two Carrier Pipes and Annular Air Space

working medium is called a **carrier pipe** and the outermost perimeter enclosure is called a **casing**.

Each conduit section is shipped in lengths up to 40 ft. Elbows, tees, loops, and bends are factory prefabricated to match the straight sections. The prefabricated components are assembled at the construction site; therefore, a construction contract is typically required for trenching, backfilling, connecting to buildings, connecting to distribution systems, constructing valve vaults, and performing some electrical work associated with sump pumps, power receptacles, and lights.

Much of the design work is done by the factory that manufactures the prefabricated sections; however, the field work must be designed and specified by the project design engineer or architect. Prefabricated components create a serious problem with accountability. For comparison, when systems are entirely field assembled, the design responsibility clearly belongs to the project design engineer, and the assembly of the system is clearly the responsibility of the construction contractor. When a condition arises where a conduit system cannot be built without modifying prefabricated components, or if the construction contractor does not follow the instructions from the prefabricator, a serious conflict of responsibility arises. For these reasons, it is imperative that the project design engineer or architect clearly delineate the responsibilities of the factory prefabricator.

Crushing loads have been used (erroneously) to size the casing thickness assuming that corrosion was not a factor. But, corrosion rate is usually the controlling factor because the casing temperature can range from less than 100°F to more than 300°F, a range that encompasses the maximum corrosion rate of steel (Figure 16). As shown in the figure, the steel casing of a district heating pipe experiences corrosion rates several times that of domestic water pipes. The temperature of the casing varies with burial depth, soil conditions, the carrier pipe temperature, and the pipe insulation thickness. The casing must be strong enough and thick enough to withstand expansion and contraction forces and corrosion degradation.

All insulation must be kept dry for it to maintain its thermal insulating properties; the exception is cellular glass in cold applications. But because underground systems may be flooded several times during their design life, even on sites that are thought to be dry, a reliable water intrusion removal system is necessary in the valve vaults. Two designs are used to ensure that the insulation performs satisfactorily for the life of the system. In the **air space system**, an annular air space between the pipe insulation and the casing allows the insulation to be dried out if water enters. In the **water spread limiting system** (WSL), which has no air space, the conduit is designed to keep water from entering the insulation. In the event that water enters one section, a WSL system prevents it's spread to adjacent sections of piping.

The air space conduit system (Figure 17 and Figure 18) should have an insulation that can survive short term flooding without damage. The conduit manufacturer usually runs a boiling test with the

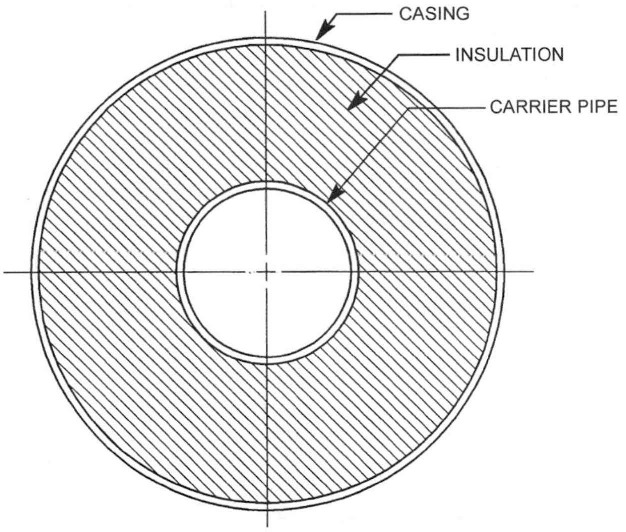

Fig. 19 Conduit System with Single Carrier Pipe and No Air Space

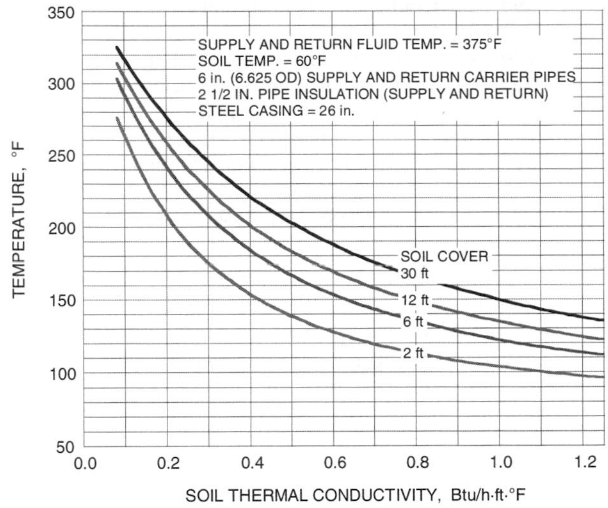

Fig. 20 Conduit Casing Temperature Versus Soil Thermal Conductivity

insulation installed in the typical factory casing. No U.S. standard has been approved for this boiling test; however the U.S. government has been using a Federal Agency Committee 96 hour boiling test for conduit insulation. The insulation must have demonstrated that it can be dried with air flowing through an annular air space, and it must retain nearly new thermal insulating properties when dried.

The insulation fails because the bonding agents, called binders, that hold the principal insulation material in the desired shape degrade. The annular air space around the insulation, typically more than 1 in. wide, allows air to flow outside the insulation to dry it. Unfortunately, the air space has a serious detrimental side effect; it allows unwanted water to flow freely to other parts of the system.

The water spread limiting system (Figure 19) encloses the insulation in an envelope that will not allow water to contact the insulation. The typical insulation is polyurethane foam, which will be ruined if excess water infiltration occurs. Polyurethane foam is limited to a temperature of about 250°F for a service life of 30 years or more. Europe has the most successful of the water spread limiting systems, which are typically used in low-temperature water applications. These systems, which are available in the United States as well, meet European *Standard* EN 253 with regard to all major construction and design details. Standards also have been established for fittings (EN 448), preinsulated valves (EN 488), and the field joint assemblies (EN 489). With this system, the carrier pipe, insulation, and the casing are bonded together to form a single unit. Forces caused by thermal expansion are passed as shear forces to the mating component and ultimately to the soil; thus no additional expansion provisions are required.

This type of WSL system is feasible because of the small temperature differential and because the thinner carrier pipe wall creates smaller expansion forces due to its low cross sectional area. Some systems rely on a watertight field joint between joining casing sections to extend the envelope to a distant envelope termination point where the casing is sealed to the carrier pipe. Other systems form the waterproof insulation envelope in each individual prefabricated conduit section using the casing and the carrier pipe to form part of the envelope and a waterproof bulkhead to seal the casing to the carrier pipe. In another type of construction, a second pipe fits tightly over the carrier pipe and seals the insulation between the second pipe and the casing to achieve a watertight insulation envelope.

Conduit Design Conditions. Three design conditions must be addressed to have reasonable assurance that the system selected will have a satisfactory service life; the wet soil or high heat loss condition, the dry soil or high-temperature condition, and the nominal or average condition.

1. The **maximum heat loss** occurs when the soil is wettest and the conduit is buried the minimum depth allowed, usually with about 2 ft of earth cover. This condition represents the highest gross heat demand and is used to size the distribution piping and equipment in the central heating plant. The relative movement of the carrier pipe with respect to the casing may be maximum during this condition. Also, the casing is coldest during this condition.
2. The **dry soil condition** occurs where the conduit is buried at the maximum allowable burial depth. The soil is quite thick when compared to the pipe insulation thickness; therefore, it plays a significant role when the soil is dry and a better insulator. The highest temperature of the insulation, casing, and casing coating occur during this condition. Paradoxically, the minimum heat loss occurs during this condition because the soil is acting as a good insulator. This condition is used to select temperature sensitive materials and to design for casing expansion. The relative movement of the carrier pipe with respect to the casing may be minimum during this condition; however, if the casing is not restrained, its movement with respect to the soil will be maximum. If restrained, the casing axial stresses and axial forces will be highest and the casing allowable stresses will be lowest because of the high casing temperature.

Analysis of Figure 20 suggests some design solutions that could lower the effects of the dry soil condition. Possible solutions would be to lower the carrier pipe temperature, use thicker carrier pipe insulation, provide a device to keep the soil wet, or minimize the burial depth. However, if these solutions are not feasible or cost effective, a different type of material or an alternate system should be considered.

While it is possible that the soil will never dry out, given the variability of climate in most areas, it is likely that a drought will occur during the life of the system. Only one very dry condition can cause permanent damage to the insulation if the soil thermal conductivity drops below the assumed design value. On-site measurement of the driest soil condition likely to cause insulation damage is not feasible. As a result, the designer is left with the conservative choice of using the lowest thermal conductivity from Table 2 to calculate the highest temperature to be used to select materials.

3. The **nominal or average condition** occurs when the soil is at its average water content and the conduit is buried at the average depth. This condition is used to compute the yearly energy consumption due to heat loss to the soil (see Table 3 for soil thermal conductivity and the previous section on Thermal Properties.

Cathodic Protection of Direct-Buried Conduits

Corrosion is an electrochemical process that occurs when a corrosion cell is formed. A corrosion cell consists of an anode, a cathode, a connecting path between them, and an electrolyte (soil or water). The structure of this cell is the same as a dry cell battery; and, like a battery, it produces a direct electrical current. The anode and cathode in the cell may be dissimilar metals, and due to differences in their natural electrical potentials, a current flows from anode to cathode. When current leaves an anode, it destroys the anode material at that point. The anode and cathode may also be the same material. Differences in composition, environment, temperature, stress, or shape makes one section anodic and an adjacent section cathodic. With a connection path and the presence of an electrolyte, this combination also generates a direct electrical current and causes corrosion at the anodic area.

Cathodic protection is a standard method used by the underground pipeline industry to further protect coated steel against corrosion. Cathodic protection systems are routinely designed for a minimum life of 20 years. Cathodic protection may be achieved by the sacrificial anode method or the impressed current method.

Sacrificial anode systems are normally used with well-coated structures. A direct current is applied to the outer surface of the steel structure with a potential driving force that prevents the current from leaving the steel structure. This potential is created by connecting the steel structure to another metal, such as magnesium, aluminum, or zinc, which becomes the anode and forces the steel structure to be the cathode. The moist soil acts as the electrolyte. These deliberately connected materials become the sacrificial anode and corrode. If they generate sufficient current they adequately protect the coated structure, while their low current output is not apt to corrode other metallic structures in the vicinity.

Impressed current systems use a rectifier to convert an alternating current power source to usable direct current. The current is distributed to the metallic structure to be protected through relatively inert anodes such as graphite or high silicon cast iron. The rectifier allows the current to be adjusted over the life of the system. Impressed current systems, also called rectified systems, are used on long pipelines in existing systems with insufficient coatings, on marine facilities, and on any structure where current requirements are high. They are installed selectively in congested pipe areas to ensure that other buried metallic structures are not damaged.

The design of effective cathodic protection requires information on the diameter of both the carrier pipe and conduit casing, length of run, number of conduits in a common trench, and number of system terminations in access areas, buildings, etc. Soil from the construction area should be analyzed to determine the soil resistivity, or the ease at which current flows through the soil. Areas of low soil resistivity require fewer anodes to generate the required cathodic protection current, but the life of the system depends on the weight of anode material used. The design life expectancy of the cathodic protection must also be defined. All anode material is theoretically used up at the end of the cathodic protection system life. At this point, the corrosion cell reverts to the unprotected system and corrosion occurs at points along the conduit system or buried metallic structure. Anodes may be replaced or added periodically to continue the cathodic protection and increase the conduit life.

A cathodically protected system must be electrically isolated at all points where the pipe is connected to building or access (manhole) piping and where a new system is connected to an existing system. Conduits are generally tied to another building or access piping with flanged connections. Flange isolation kits, including dielectric gaskets, washers, and bolt sleeves, electrically isolate the cathodically protected structure. If an isolation flange is not used, any connecting piping or metallic structure will be in the protection system, *but adequate protection may not be met.*

The effectiveness of cathodic protection can only be determined by an installation survey after the system has been energized. Cathodically protected structures should be tested at regular intervals to determine the continued effectiveness and life expectancy of the system. Sacrificial anode cathodic protection is monitored by measuring the potential (voltage) between the underground metallic structure and the soil versus a stable reference. This potential is measured with a high resistance voltmeter and a reference cell. The most commonly used reference cell material is copper/copper sulfate. One criterion for protection of buried steel structures is a negative voltage of at least 0.85 V as measured between the structure surface and a saturated copper/copper sulfate reference electrode in contact with the electrolyte. Impressed current systems require more frequent and detailed monitoring than sacrificial anode systems. The rectified current and potential output and operation must be verified and recorded at monthly intervals.

NACE *Standard* RPO169 has further information on control of external corrosion on buried, metallic structures.

Leak Detection

The conduit may require excavation to repair construction errors after burial. Various techniques are available for detecting leaks in district heating and cooling piping. They range from performing periodic pressure tests on the piping system to installing a sensor cable along the entire length of the piping to continuously detect and locate leaks. Pressure testing should be performed on all piping to verify integrity during installation and the life of the piping. Chilled water systems should be pressure tested during the winter and hot water and steam systems tested during the summer.

A leak is difficult to locate without the aid of a cable type leak detector. Finding a leak typically involves excavating major sections between valve vaults. Infrared detectors and acoustic detectors can help narrow down the location of a leak, but they do not work equally well for all underground systems. Also, they are not as accurate with underground systems as with an aboveground system.

Chilled Water and Hot Water Systems. Chilled water piping systems are usually insulated with urethane foam with a vapor proof jacket (HDPE, urethane, PVC, CPVC, etc.). Copper wires can be installed during fabrication to aid in detecting and locating liquid leaks. The wires may be insulated or uninsulated depending on the manufacturer. Some systems monitor the entire wire length while others only monitor at the joints of the piping system. The detectors either look for a short in the circuit using Ohm's Law or monitor for impedance change using time domain reflectometry (TDR).

Steam, High-Temperature Hot Water, and Other Conduit Systems. Air-gap designs, which have a gap between the inner wall of the outer casing and the insulation, can have probes installed at the low points of drains or at various points to detect leaks. Leaks can also be detected with a continuous cable that monitors liquid leakage. The cable is installed at the bottom of the conduit with a minimum air gap required, typically 1 in. Pull points or access ports are installed every 400 to 500 ft on straight runs with changes in direction reducing the length between pull points. Systems either monitor by looking for a short on the cable using Ohm's Law or by monitoring the impedance on a coaxial cable using resistance temperature devices (RTD). During installation care must be taken to keep the system clean and dry to keep any contamination from the leak detection system that might cause it to fail. The system must be sealed airtight to prevent condensation from accumulating in the piping at the low points.

Valve Vaults and Entry Pits

Valve vaults allow a user to isolate problems to one area rather than analyze the entire line. This feature is important if the underground distribution system cannot be maintained between valve vaults without excavation. The optimum number of valve vaults is that which affords the lowest life-cycle cost and still meets all design requirements, usually no more than 500 ft apart. Valve vaults provide a space in which to put valves, steam traps, carrier pipe drains, carrier pipe vents, casing vents and drains, condensate pumping units, condensate cooling devices, flash tanks, expansion joints, groundwater drains, electrical leak detection equipment and wiring, electrical isolation couplings, branch line isolation valves, carrier pipe isolation valves, and flowmeters. Valve vaults allow for elevation changes in the distribution system piping while maintaining an acceptable slope on the system; they also allow the designer to better match the topography and avoid unreasonable and expensive burial depths.

Ponding Water. The most significant problem with valve vaults is that water ponds in them. When the hot and chilled water distribution systems share the same valve vault, plastic chilled water lines often fail because ponded water heats the plastic to failure. Water gathers in the valve vault irrespective of climate; therefore, design strives to eliminate the water for the entire life of the underground distribution system. The most successful water removal systems are those that drain to sanitary or storm drainage systems; this technique is successful because the system is affected very little by corrosion and has no moving parts to fail. Backwater valves are recommended in case the drainage system backs up.

Duplex sump pumps with lead-lag controllers and a failure annunciation system are used when storm drains and sanitary drains are not accessible. Because pumps have a history of frequent failures, duplex pumps help eliminate short cycling and provide standby pumping capacity. Steam ejector pumps can be used only if the distribution system is never shut down because the carrier pipe insulation can be severely damaged during even short outages. A labeled, lockable, dedicated electrical service should be used for electric pumps. The circuit label should indicate what the circuit is used for; it should also warn of the damage that will occur if the circuit is de-energized.

Electrical components have experienced accelerated corrosion in the high heat and high humidity of closed, unventilated vaults. A pump that works well at 50°F often performs poorly at 200°F and 100% relative humidity. To resolve this problem, one approach specifies components that have demonstrated high reliability at 200°F and 100% relative humidity with a damp-proof electrical service. The pump should have a corrosion-resistant shaft (when immersed in water) and impeller and have demonstrated 200,000 cycles of successful operation, including the electrical switching components, at the referenced temperature and humidity. The pump must also pass foreign matter; therefore, the requirement to pass a 3/8 in. ball should be specified.

Another method drains the valve vaults into a separate sump adjacent to the vault. Then the pumps are placed in this sump, which is cool and more nearly a sump pump environment. Redundant methods may be necessary if maximum reliability is needed or future maintenance is questionable. The pump can discharge to the sanitary or storm drain or to a splash block near the valve vault. Water pumped to a splash block has a tendency to enter the vault, but this is not a significant problem if the vault construction joints have been sealed properly. Extreme caution must be exercised if the bottom of the valve vault has French drains. These drains allow groundwater to enter the vault and flood the insulation on the distribution system during high groundwater conditions. Adequate ventilation of the valve vault is also important.

Crowding of Components. The valve vault must be laid out in three dimensions, considering standing room for the worker, wrench swings, the size of valve operators, variation between manufacturers in the size of appurtenances, and all other variations that the specifications allow with respect to any item placed in the vault. To achieve desired results, the vault layout must be shown to scale on the contract drawings.

High Humidity. High humidity develops in a valve vault when it has no positive ventilation. Gravity ventilation is often provided in which cool air enters the valve vault and sinks to the bottom. At the bottom of the vault the air warms, becomes lighter, and rises to the top of the vault, where it exits. In the past, some designers used a closed-top valve vault with an exterior ventilation pipe with an elbow that directs the exiting air down. However, the elbowed-down vent hood tends to trap the exiting air and prevent gravity ventilation from working. Open structural grate tops are the most successful covers for ventilation purposes. Open grates allow rain to enter the vault; however, the techniques mentioned in the section on Ponding Water are sufficient to handle the rainwater. Open grates with sump basins have worked well in extremely cold climates and in warm climates. Some vaults have a closed top and screened, elevated sides to allow free ventilation. In this design, the solid vault sides extend slightly above grade; then, a screened window is placed in the wall on at least two sides. The overall above-grade height may be only 18 in.

High Temperatures. The temperature in the valve vault rises when no systematic way is provided to remove heat losses from the distribution system. The gravity ventilation rate is usually not sufficient to transport heat from the closed vaults. Part of this heat transfers to the earth; however, an equilibrium temperature is reached that may be higher than desired. Ventilation techniques discussed in the section on High Humidity can resolve the problem of high temperature if the heat loss from the distribution system is near normal. Typical problems that greatly increase the amount of heat released include

- Leaks from a carrier pipe, gaskets, packings, or appurtenances
- Insulation that has deteriorated due to flooding or abuse
- Standing water in a vault that touches the distribution pipe
- Steam vented to the vault from partial flooding between valve vaults
- Vents from flash tanks
- Insulation removed during routine maintenance and not replaced

To prevent heat release in a new system, a workable ventilation system must be designed. On existing valve vaults, the valve vault must be ventilated properly, all leaks corrected, and all insulation that was damaged or left off replaced. Commercially available insulation jackets that can easily be removed and reinstalled from fittings and valves should be installed. If flooding occurs between valve vaults, portions of the distribution system may have to be excavated and repaired or replaced. Vents from vault appurtenances that exhaust steam into the vault may have to be routed aboveground if the ventilation technique is insufficient to handle the quantity of steam exhausted.

Deep Burial. When a valve vault is buried too deeply, (1) the structure is exposed to groundwater pressures, (2) entry and exit often become a safety problem, (3) construction becomes more difficult, and (4) the cost of the vault is greatly increased. Valve vault spacing should be no more than 500 ft (NAS 1975). If greater spacing is desired, use an accurate life-cycle analysis to determine spacing. The most common way to limit burial depth is to place the valve vaults closer together. Steps in the distribution system slope are made in the valve vault (i.e., the carrier pipes come into the valve vault at one elevation and leave at a different elevation). If the slope of the distribution system is changed to more nearly match the earth topography, the valve vaults will be shallower; however, the allowable range of slope of the carrier pipes restricts this method. In most systems, the slope of the distribution system can be reversed in a valve vault, but not out in the system between valve vaults. The minimum slope for the carrier pipes is 1 in. in 20 ft. Lower slopes are outside the range of normal construction tolerance. If the entire distribution system is buried too deeply, the designer must determine the maximum allowable burial depth of the system and survey the topography of the distribution system to determine where the maximum and minimum depth of burial will occur. All elevations must be adjusted to limit the minimum and maximum allowable burial depths.

Freezing Conditions. Failure of distribution systems due to water freezing in components is common. The designer must consider the coldest temperature that may occur at a site and not the 99% or 99.6% condition used in building design (as discussed in Chapter 26 of the 1997 *ASHRAE Handbook—Fundamentals*). Drain legs or vent legs that allow water to stagnate are usually the cause of failure. Insulation should be on all items that can freeze, and it must be kept in good condition. Electrical heat tape and pipe-type heat tracing can be used under insulation. If part of a chilled water system is in a ventilated valve vault, the chilled water may have to be circulated or be drained if not used in winter.

Safety and Access. Some of the working fluids used in underground distribution systems can cause severe injury and death if accidentally released in a confined space such as a valve vault. The shallow valve vault with large openings is desirable because it allows personnel to escape quickly in an emergency. The layout of the pipes and appurtenances must allow easy access for maintenance without requiring maintenance personnel to crawl underneath or between other pipes. The goal of the designer is to keep clear work spaces for maintenance personnel so that they can work efficiently and, if necessary, exit quickly. Engineering drawings must show pipe insulation thickness; otherwise they will give a false impression of the available space.

The location and type of ladder is important for safety and ease of egress. It is best to lay out the ladder and access openings when laying out the valve vault pipes and appurtenances as a method of exercising control over safety and ease of access. Ladder steps, when cast in the concrete vault walls, may corrode if not constructed of the correct material. Corrosion is most common in steel rungs. Either cast-iron or prefabricated, OSHA-approved, galvanized steel ladders that sit on the valve vault floor and are anchored near the top to hold it into position are best. If the design uses lockable access doors, the locks must be operable from inside or have some keyed-open device that allows workers to keep the key while working in the valve vault.

Vault Construction. The most successful valve vaults are those constructed of cast-in-place reinforced concrete. These vaults conform to the earth excavation profile and show little movement when backfilled properly. Leakproof connections can be made with mating tunnels and conduit casings even though they may enter or leave at oblique angles. In contrast, prefabricated valve vaults may settle and move after construction is complete. Penetrations for prefabricated vaults, as well as the angles of entry and exit, are difficult to locate exactly. As a result, much of the work associated with penetrations is not detailed and must be done by construction workers in the field, which greatly lowers the quality and greatly increases the chances of a groundwater leak.

Construction deficiencies that go unnoticed in the buildings can destroy a heating and cooling distribution system; therefore, the designer must clearly convey to the contractor that a valve vault does not behave like a sanitary manhole. A design that is sufficient for a sanitary manhole will prematurely fail if used for a heating and cooling distribution system.

CONSUMER INTERCONNECTIONS

The thermal energy that is produced at the plant is transported via the distribution network and is finally transferred to the consumer. When thermal energy (hot water, steam, or chilled water) is supplied, it may be used directly by the building HVAC system or process loads, or indirectly via a heat exchanger that transfers energy from one media to another. When energy is used directly, it may be reduced in pressure that is commensurate to the buildings' systems. The design engineer must perform an analysis to determine which connection type is best.

For commercially operated systems, a contract boundary or point of delivery divides responsibilities between the energy provider and the customer. This point can be at a piece of equipment, as in a heat exchanger with an indirect connection, or flanges as in a direct connection. A chemical treatment analysis must be performed (regardless of the type of connection) to determine the compatibility of each side of the system (district and consumer) prior to energizing.

Direct Connection

Because a direct connection offers no barrier between the district water and the building's own system (e.g., air-handling unit cooling and heating coils, fan coils, radiators, unit heaters, and process loads), the water circulated at the district plant has the same quality as the customer's water. Direct connections, therefore, are at a greater risk of incurring damage or contamination based on the poor water quality of either party. Typically district systems have contracts with water treatment vendors and monitor water quality continuously. This may not be the case with all consumers. A direct connection is often more economical than an indirect connection because the consumer is not burdened by the installation of heat exchangers, additional circulation pumps, or water treatment systems; therefore, investment costs are reduced and return temperatures identical to design values are possible.

Figure 21 shows the simplest form of direct connection, which includes a pressure differential regulator, a thermostatic control valve on each terminal unit, a pressure relief valve, and a check valve. Most commercial systems have a flowmeter installed as well as temperature sensors and transmitters to calculate the energy used. The location of each device may vary from system to system, but all of the major components are indicated. The control valve is the capacity regulating device that restricts flow to maintain either a chilled water supply or return temperature on the consumer's side.

Particular attention must be paid to connecting high rise buildings because they induce a static head. Pressure control devices should be investigated carefully. It is not unusual to have a water-based district

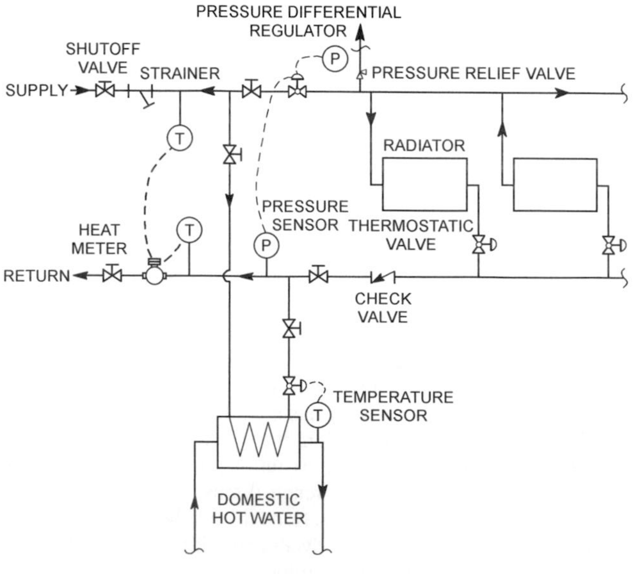

Fig. 21 Direct Connection of Building System to District Hot Water

heating or cooling system with a mixture of direct and indirect connections in which heat exchangers isolate the systems hydraulically.

In a direct system, the pressure in the main distribution system must meet local building codes to protect the customer's installation and the reliability of the district system. To minimize noise, cavitation, and control problems, constant-pressure differential control valves should be installed in the buildings. Special attention should be given to potential noise problems at the control valves. These valves must correspond to the design pressure differential in a system that has constantly varying distribution pressures due to load shifts. Multiple valves may be required in order to serve the load under all flow and pressure ranges. Industrial quality valves and actuators should be used for this application.

If the temperature in the main distribution system is lower than that required in the consumer cooling systems, a larger temperature differential between supply and return occurs, thus reducing the required pipe size. The consumer's desired supply temperature can be attained by mixing the return water with the district cooling supply water. Depending on the size and design of the main system, elevation differences, and types of customers and building systems, additional safety equipment such as automatic shutoff valves on both supply and return lines may be required.

When buildings have separate circulation pumps, primary/secondary piping, and pumping isolating techniques are used (cross-connection between return and supply piping, decouplers, and bypass lines). This ensures that two-way control valves are subjected only to the differential pressure established by the customer's building (tertiary) pump. Figure 4 shows a primary/secondary connection using an in-building pumping scheme.

When tertiary pumps are used, all series connections between the district system pumps must be removed. A series connection can cause the district system return to operate at a higher pressure than the distribution system supply and disrupt normal flow patterns. Series operation usually occurs during improper use of three-way mixing valves in the primary to secondary connection.

Indirect Connection

Many of the components are similar to those used in the direct connection applications with the exception that a heat exchanger performs one or more of the following functions: heat transfer, pressure interception, and buffer between potentially different quality water treatment.

Identical to the direct connection, the rate of energy extraction in the heat exchanger is governed by a control valve that reacts to the building load demand. Once again, the control valve usually modulates to maintain a temperature set point on either side of the heat exchanger depending on the contractual agreement between the consumer and the producer.

The three major advantages of using heat exchangers are (1) the static head influences of a high rise building are eliminated, (2) the two water streams are separated, and (3) consumers must makeup all of their own lost water. The disadvantages of using an indirect connection are the additional cost of the heat exchanger and the loss of temperature and the increased pumping pressure due to the addition of another heat transfer surface.

COMPONENTS

Heat Exchangers

Heat exchangers act as the line of demarcation between ownership responsibility of the different components of an indirect system. They transfer thermal energy and act as pressure interceptors for the water pressure in high-rise buildings. They also keep fluids from each side (which may have different chemical treatments) from mixing. Figure 22 shows a basic building schematic including heat exchangers and secondary systems.

Reliability of the installation is increased if multiple heat exchangers are installed. The number selected depends on the types of loads present and the how they are distributed throughout the year. When selecting all equipment for the building interconnection, but specifically heat exchangers, the designer should:

- Size the unit's capacity to match the given load and estimated load turn-down as close as possible (oversized units may not perform as desired at maximum turn-down; therefore, several smaller units will optimize the installation).
- Assess the critical nature of the load/operation/process to address reliability and redundancy. For example, if a building has 24 h process loads (i.e., computer room cooling, water cooled equipment, etc.), consider adding a separate heat exchanger for this load. Also, consider operation and maintenance of the units. If the customer is a hotel, hospital, casino, or data center, select a minimum of two units at 50% load each to allow one unit to be cleaned without interrupting building service. Separate heat exchangers should be capable of automatic isolation during low load conditions to increase part load performance.
- Determine customer's temperature and pressure design conditions. Some gasket materials for plate heat exchangers have low pressure and temperature limits.
- Evaluate customer's water quality (i.e., use appropriate fouling factor).
- Determine available space and structural factors of the mechanical room.
- Quantify design temperatures. The heat exchanger may require rerating at a higher inlet temperature during off-peak hours.
- Calculate the allowable pressure drop on both sides of heat exchangers. The customer's side is usually the most critical for pressure drop. The higher the pressure drop, the smaller and less expensive the heat exchanger. However, the pressure drop must be kept in reasonable limits (15 psig or below) if the existing pumps are to be reused. Investigate the existing chiller evaporator pressure drop in order to assist in this evaluation.

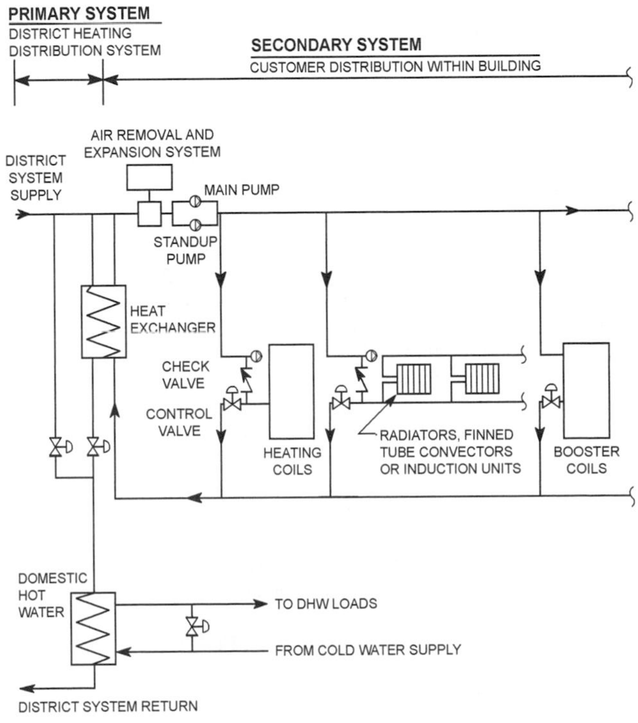

Fig. 22 Basic Heating System Schematic

All heat exchangers should be sized with future expansion in mind. When selecting heat exchangers, be cognizant that closer approach temperatures or low pressure drop require more heat transfer area and hence cost more and take up more space. Strainers should be installed in front of any heat exchanger and control valve to keep debris from fouling surfaces.

Plate, shell-and-coil, and shell-and-tube heat exchangers are all used for indirect connection. Whatever heat transfer device is selected must meet the appropriate temperature and pressure duty, and be stamped/certified accordingly as pressure vessels.

Plate Heat Exchangers (PHEs). These exchangers, which are used for either steam, hot water, or chilled water applications, are available as gasketed units and in two gasket-free designs (brazed and all or semi-welded construction). All PHEs consist of metal plates compressed between two end frames and sealed along the edges. Alternate plates are inverted and the gaps between the plates form the liquid flow channels. Fluids never mix as hot fluid flows on one side of the plate and cool fluid flows countercurrent on the other side. Ports at each corner of the end plates act as headers for the fluid. One fluid travels in the odd numbered plates and the other in the even numbered plates.

Because PHEs require turbulent flow for good heat transfer, pressure drops may be higher than that for a comparable shell and tube model. High efficiency leads to a smaller package. The designer should consider specifying that the frame be sized to hold 20% additional plates. PHEs require very little maintenance because the high velocity of the fluid in the channels tends to keep the surfaces clean. PHEs generally have a cost advantage and require one-third to one-half the surface required by shell-and-tube units for the same operating conditions. PHEs are also capable of closer approach temperatures.

Gasketed PHEs (also called plate-and-frame heat exchangers) consist of a number of gasketed embossed metal plates bolted together between two end frames. Gaskets are placed between the plates to contain the two media in the plates and to act as a boundary. Gasket failure will not cause the two media to mix; instead the media will leak to the atmosphere. Gaskets can be either glued or clip-on. Gasketed PHEs are suitable for steam-to-liquid and liquid-to-liquid applications. Designers should select the appropriate gasket material for the design temperatures and pressures expected. Plates are typically stainless steel; however, plate material can be varied based on the chemical makeup of the heat transfer fluids.

PHEs are typically used for district heating and cooling with water and cooling tower water heat recovery (free cooling). Double wall plates are also available for potable water heating, chemical processes, and oil quenching. PHEs have 3 to 5 times greater heat transfer coefficients than shell and tube units and are capable of achieving 1°F approach temperatures. This type of PHE can be disassembled in the field to clean the plates and replace the gaskets. Typical applications go up to 365°F and 400 psig.

Brazed PHEs are suitable for steam, vapor, or water solutions. They feature a close approach temperature (within 2°F), large temperature drop, compact size, and a high heat transfer coefficient. Construction materials are stainless steel plates and frames brazed together with copper or nickel. Tightening bolts are not required as in the gasketed design. These units cannot be disassembled and cleaned; therefore, adequate strainers must be installed ahead of an exchanger and it must be periodically flushed clean in a normal maintenance program. Brazed PHEs usually peak out at a capacity of under 200,000 Btu/h (about 200 plates and 600 gpm) and suitable for 435 psig and 435°F. Typical applications are district heating using hot water and refrigeration process loads. Double wall plates are also available. Applications where the PHE may be exposed to large, sudden or frequent changes in temperature and load must be avoided due to risk of thermal fatigue.

Welded PHEs can be used in any application that shell-and-tube units are used that are outside the accepted range of gasketed PHE units in liquid-to-liquid, steam-to-liquid, gas-to-liquid, gas-to-gas and refrigerant applications. Construction is very similar to gasketed units except gaskets are replaced with laser welds. Materials are typically stainless steel, but titanium, monel, nickel, and a variety of alloys are available. Models that are offered have design ratings that range from 500°F at 150 psig to 1000°F at 975 psig; however, they are available only in small sizes. Normally these units are used in ammonia refrigeration and aggressive process fluids. They are more suitable to pressure pulsation or thermal cycling because they are thermal fatigue resistant. Semi-welded PHE is a hybrid of the gasketed and the all-welded units in which the plates are alternatively sealed with gaskets and welds.

Shell-and-Coil Heat Exchangers. These European-designed heat exchangers are suitable for steam-to-water and water-to-water applications and feature an all welded and brazed construction. This counter/crossflow heat exchanger consists of a hermetically sealed (no gaskets), carbon-steel pressure vessel with hemispherical heads. Copper or stainless helical tubes within are installed in a vertical configuration. This type of heat exchanger offers a high temperature drop and close approach temperature. It requires less floor space than other designs and has better heat transfer characteristics than shell-and-tube units.

Shell-and-Tube Heat Exchangers. These exchangers are usually a multiple pass design. The shell is usually constructed from steel and the tubes are often of "U" bend construction, usually 3/4 in. (nominal) OD copper, but other materials are available. These units are ASME U-1 stamped for pressure vessels.

Flow Control Devices

In commercial systems, after the flowmeter, control valves are the most important element in the interface with the district energy system because proper valve adjustment and calibration save energy. High-quality, industrial grade control valves provide more precise control, longer service life, and minimum maintenance.

All control valve actuators should take longer than 60 seconds to close from full open to mitigate pressure transients or water hammer, which occurs when valves slam closed. Actuators should also be sized to close against the anticipated system pressure so the valve seats are not forced open, thus forcing water to bypass and degrading temperature differential.

The wide range of flows and pressures expected makes selection of control valves difficult. Typically only one control valve is required; however, for optimal response to load fluctuations and to prevent cavitation, two valves in parallel are often needed. The two valves operate in sequence and for a portion of the load, i.e., one valve is sized for two-thirds of peak flow and the other sized for one third of peak flow. The designer should review the occurrence of these loads to size the proportions correctly. The possibility of overstating customer loads complicates the selection process, so accurate load information is important. It is also important that the valve selected operates under the extreme pressure and flow ranges foreseen. Because most commercial grade valves will not perform well for this installation, industrial quality valves are specified.

Electronic control valves should remain in a fixed position when a power failure occurs and should be manually operable. Pneumatic control valves should close upon loss of air pressure. A manual override on the control valves allows the operator to control flow. All chilled water control valves must fail in the closed position. Then, when any of the secondary in-building systems are de-energized, the valves close and will not bypass chilled water to the return system. All steam pressure reducing valves should close as well.

Oversizing reduces valve life and causes valve hunting. Select control valves having a wide range of control; low leakage; and proportional-plus-integral control for close adjustment, balancing, temperature accuracy, and response time. Control valves should have actuators with enough force to open and close under the maximum pressure differential in the system. The control valve should have a pressure drop through the valve equal to at least 10 to 30% of the static pressure drop of the distribution system. This pressure drop gives the control valve the "authority" it requires to properly control flow. The relationship between valve travel and capacity output should be linear, with an equal percentage characteristic.

In hot water systems, control valves are normally installed in the return line because the lower temperature in the line reduces the risk of cavitation and increases valve life. In chilled water systems, control valves can be installed in either location; typically, however, they are also installed the return line.

Instrumentation

In many systems, where energy to the consumer is measured for billing purposes, temperature sensors assist in calculating the energy consumed as well as in diagnosing performance. Sensors and their transmitters should have an accuracy range commensurate to the accuracy of the flowmeter. In addition, pressure sensors are required for variable speed pump control (water systems) or valve control for pressure reducing stations (steam and water).

Temperature sensors need to be located by the exchangers being controlled rather than in the common pipe. Improperly located sensors will cause one control valve to open and others to close, resulting in unequal loads in the exchangers.

Controller

The controller performs several functions, including recording demand and the amount of energy used for billing purposes, monitoring the differential pressure for plant pump control, energy calculations, alarming for parameters outside normal, and monitoring and control of all components.

Typical control strategies include regulating district flow to maintain the customer's supply temperature (which results in a fluctuating customer return temperature) or maintaining the customer's return temperature (which results in a fluctuating customer supply temperature). When controlling return flow for cooling, the impact on the customer's ability to dehumidify properly with an elevated entering coil temperature should be investigated carefully.

Pressure Control Devices

If the steam or water pressure delivered to the customer is too high for direct use, it must be reduced. Similarly, pressure reducing or sustaining valves may be required if building height creates a high static pressure and influences the district system's return water pressure. Water pressure can also be reduced by control valves or regenerative turbine pumps. The risks of using pressure regulating devices to lower pressure on the return line is that if they fail the entire distribution system is exposed to their pressure and over pressurization will occur.

In high-rise buildings, all piping, valves, coils, and other equipment may be required to withstand higher design pressures. Where system static pressure exceeds safe or economical operating pressure, either the heat exchanger method or pressure sustaining valves in the return line may be used to minimize the impact of the pressure. Vacuum vents should be provided at the top of the building's water risers to introduce air into the piping in case the vertical water column collapses.

HEATING CONNECTIONS

Steam Connections

Although higher pressures and temperatures are sometimes used, most district heating systems supply saturated steam at pressures between 5 and 250 psig to customers' facilities. The steam is pretreated to maintain a neutral pH, and the condensate is both cooled and discharged to the building sewage system or returned back to the central plant for recycling. Many consumers run the condensate

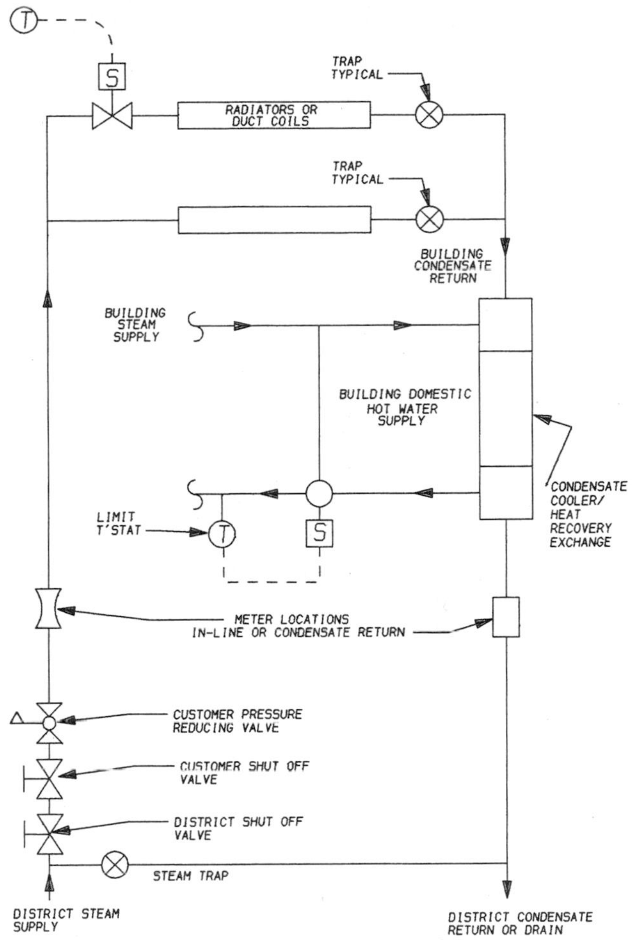

Fig. 23 District/Building Interconnection with Heat Recovery Steam System

through a heat exchanger to heat the domestic hot water supply of the building prior to returning it to the central plant or to the building drains. This process extracts the maximum amount of energy out of the delivered steam.

Interconnection between the district and the building is simple when the building uses the steam directly in heating coils or radiators or for process loads (humidification, kitchen, laundry, laboratory, steam absorption chillers, or turbine driven devices). Other buildings extract the energy from the district steam via a steam to water heat exchanger to generate hot water and circulate it to the air-side terminal units. Typical installations are shown in Figure 23 and Figure 24. Chapter 10 has additional information on building distribution piping, valving, traps, and other system requirements. The type of steam chemical treatment should be considered in applications for the food industry and for humidification.

Other components of the steam connection may include condensate pumps, flowmeters (steam and/or condensate), and condensate conductivity probes, which may dump condensate if they are contaminated by unacceptable debris. Many times, energy meters are installed on both the steam and condensate pipes to allow the district energy supplier to determine how much energy is used directly and how much energy (condensate) is not returned back to the plant. The use of customer energy meters for both steam and condensate is desirable for the following reasons:

- Offers redundant metering (if the condensate meter fails, the steam meter can detect flow or visa versa)

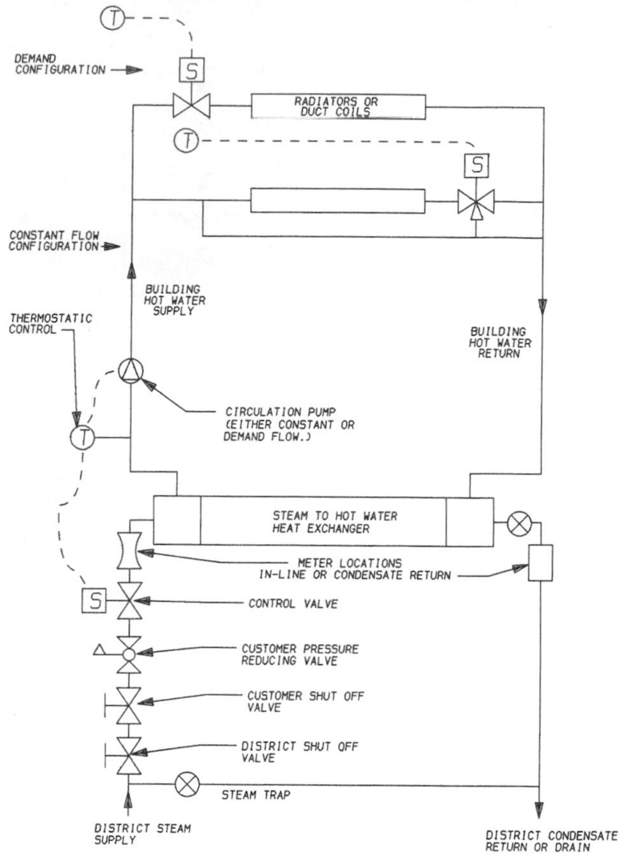

Fig. 24 District/Building Interconnection with Heat Exchange Steam System

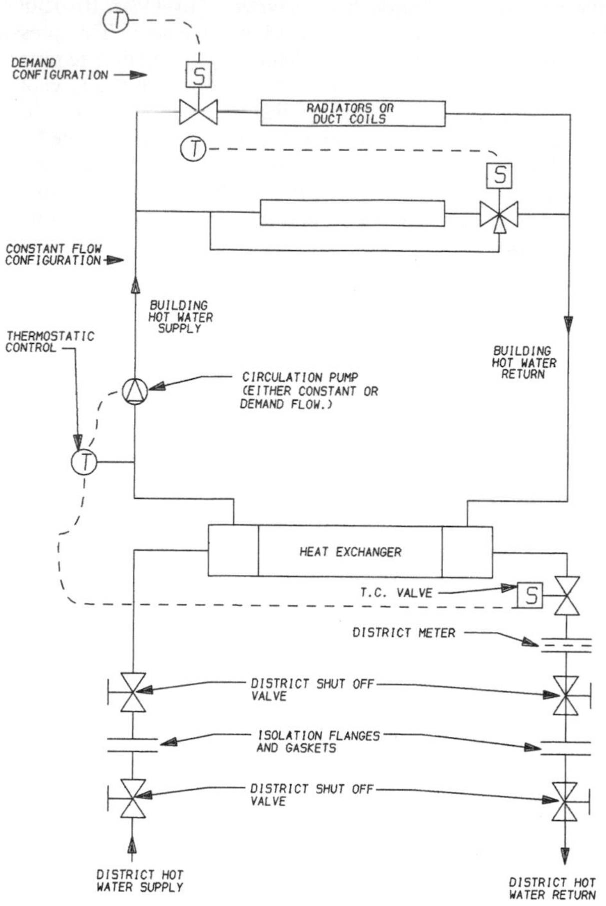

Fig. 25 District/Building Indirect Interconnection Hot Water System

- Bills customer accordingly for makeup water and chemical treatment on all condensate that is not returned or is contaminated
- Meter is in place if customer requires direct use of steam in the future
- Assists in identifying steam and condensate leaks
- Improves customer relations (may ease customer's fears of overbilling due to a faulty meter)
- Provides a more accurate reading for peak demand measurements and charges

Each level of steam pressure reduction should also be monitored as well as the temperature of the condensate. Where conductivity probes are used to monitor the quality of the water returned to the steam plant, adequate drainage and cold water quenching equipment may be required to satisfy local plumbing code requirements (temperature of fluid discharging into a sewer). The probe status should also be monitored at the control panel to communicate high conductivity alarms to the plant and when condensate is being "dumped" to notify the plant a conductivity problem exists at a customer.

Hot Water Connections

Figure 25 illustrates a typical indirect connection using a heat exchanger between the district hot water system and the customer's system. It shows the radiator configurations typically used in both constant flow and variable flow systems. Figure 26 shows a typical direct connection between the district hot water system and the building. It includes the typical configurations for both demand flow and constant flow systems and the additional check valve and piping required for the constant flow system. Figure 27 shows an indirect connection for both space and hot water heating.

Lines on these systems must be sized using the same design used for the main feed lines of an in-building power plant. In general, demand flow systems permit better energy transfer efficiency and smaller line size for a given energy transfer requirement. Line sizing should account for any future loads on the building, etc. To keep the return temperature low, water flow through heating equipment should be controlled according to the heating demand in the space.

The secondary supply temperature must be controlled in a manner similar to the primary supply temperature. To ensure a low primary return temperature and large temperature difference, the secondary system should have a low design return temperature. This design helps lower costs through smaller pipes, pumps, amounts of insulation, and pump motors. A combined return temperature on the secondary side of 130°F or lower is reasonable, although this temperature may be difficult to achieve in smaller buildings with only baseboard or radiator space heating.

A hot water district heating distribution system has many advantages over a steam system. Due to the efficiency of the relatively low temperature, energy can be saved over an equivalent steam system. Low- and medium-temperature hot water district heating allows a greater flexibility in the heat source, lower cost piping materials, and more cost effective ways to compensate for expansion and new customer connections.

Major advantages of a hot water distribution system include the following:

- Changing both temperature and flow can vary the delivered thermal capacity.
- Hot water can be pumped over a greater distance with minimal energy loss. This is a major disadvantage of a steam system.

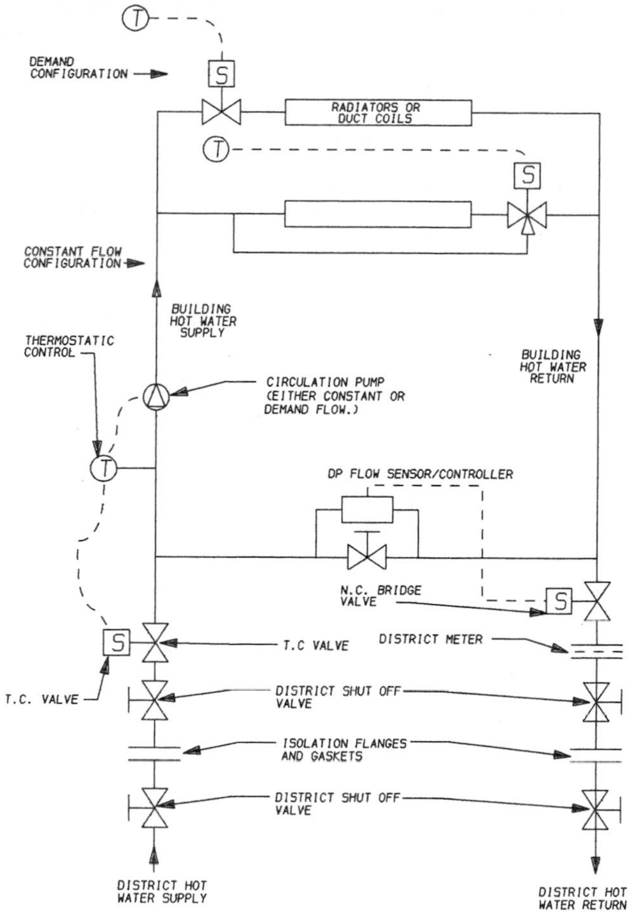

Fig. 26 District/Building Direct Interconnection Hot Water System

- Operates at lower supply temperatures over the year. This results in lower line losses. Steam systems operate at constant pressure and temperature year round, which increases energy production and system losses while decreasing annual system efficiency.
- Has much lower operating and maintenance costs than steam systems (leaks at steam traps, chemical treatment, spares etc.)
- Uses prefabricated insulated piping, which has a lower initial capital cost as compared to steam systems

Building Conversion to District Heating

Table 4 (Sleiman et al. 1990) summarizes the suitability or success rate of converting various heating systems to be served by a district hot water system. As can be seen, the probability is high for water-based systems, lower for steam, and lowest for fuel oil or electric based systems. Systems that are low on suitability usually require the high expense of replacing the entire heating terminal and generating units with suitable water-based equipment including piping, pumps, controls, and heat transfer media.

CHILLED WATER CONNECTIONS

Similar to district hot water systems, chilled water systems can operate with either constant flow or variable flow. Variable-flow systems can interconnect with either building demand or constant-flow systems. Variable flow is best if dehumidification is required with comfort cooling. Figure 28 illustrates a typical configuration for a variable-flow building interconnection. The typical constant-flow system found in older buildings may be converted to a demand system by blocking off the bypass line around the air handler heat

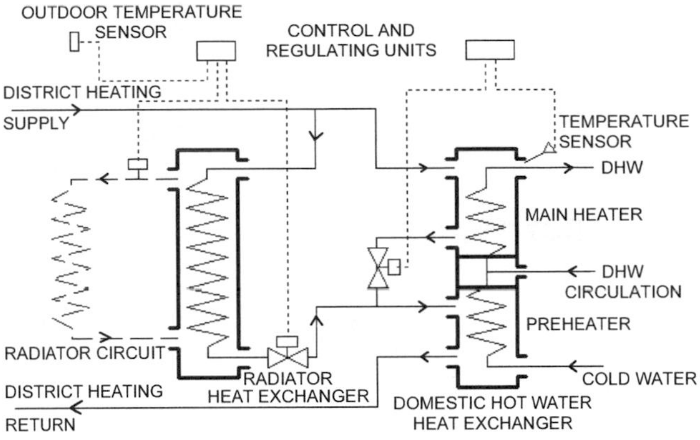

Fig. 27 Building Indirect Connection for Both Heating and Domestic Hot Water

Table 4 Conversion Suitability of Heating System by Type

Type of System	Low	Medium	High
Steam Equipment			
One-pipe cast iron radiation	X		
Two-pipe cast iron radiation		X	
Finned tube radiation		X	
Air-handling unit coils		X	
Terminal unit coils	X		
Hot Water Equipment			
Radiators and convectors			X
Radiant panels			X
Unitary heat pumps			X
Air-handling unit coils			X
Terminal unit coils			X
Gas/Oil Fired Equipment			
Warm air furnaces	X		
Rooftop units	X		
Other systems	X		
Electric Equipment			
Warm air furnaces	X		
Rooftop units	X		
Air-to-air heat pumps	X		
Other systems	X		

exchanger coil. At low operating pressures, this potentially may convert a three-way bypass-type valve to a two-way modulating shutoff valve. Careful analysis of the valve actuator must be undertaken since the shut-off requirements and control characteristics are totally different for a two-way valve compared to a three-way valve.

Peak demand requirements must be determined for the building at maximum design conditions (above the ASHRAE 0.4% design values). These conditions usually include direct bright sunlight on the building, 95 to 100°F dry-bulb temperature, and 73 to 78°F wet-bulb temperature occurring at peak conditions.

The designer must consider the effect of return water temperature control. This is the single most important factor in obtaining a high temperature difference and providing an efficient plant. As the load increases, the return water temperature tends to rise and, with a low-load condition, the supply water temperature rises. Consequently, process or critical humidity control systems may suffer when connected to a system where return water temperature control is used to achieve high temperature differentials. Other techniques, such as separately pumping each chilled water coil, may be used where constant supply water temperatures are necessary year-round.

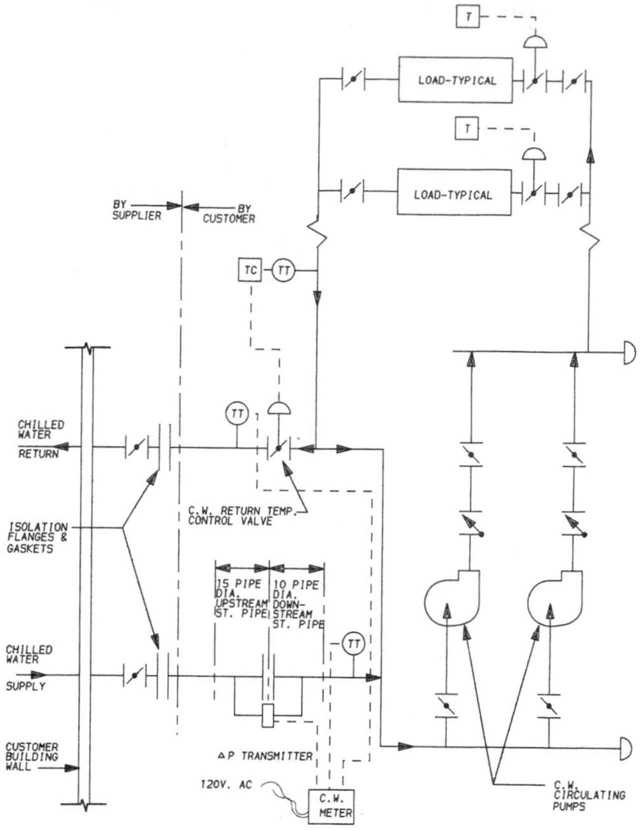

Fig. 28 Typical Chilled Water Piping and Metering Diagram

TEMPERATURE DIFFERENTIAL CONTROL

The success of the district heating and cooling system efficiency is usually measured in terms of the temperature differential for water systems. Proper control of the district heating and cooling temperature differential is not at the plant but at the consumer. If the consumer's system is not compatible with the temperature parameters of the DHC system, operating efficiency will suffer unless components in the consumer's system are modified.

Generally, maintaining a high temperature differential (ΔT) between supply and return lines is most cost effective because it allows smaller pipes to be used in the primary distribution system. These savings must be weighed against any higher building conversion cost that may result from the need for a low primary return temperature. Furthermore, optimization of the ΔT is critical to the successful operation of the district energy system. That is the reason the customer's ΔT must be monitored and controlled.

In order to optimize the ΔT and meet the customer's chilled water demand, the flow from the plant should vary. Varying the flow also saves pump energy. Chilled water flow in the customer's side must be varied as well. Terminal units in the building connected to the chilled water loop (i.e., air-handling units, fan coils, etc.) may require modifications (change 3-way valves to 2-way, etc.) to operate with variable water flow to ensure a maximum return water temperature.

For cooling coils, six-row 12 to 14 fins-per-inch coils are the minimum size coil applied to central station air-handling units to provide adequate performance. With this type of coil, the return water temperature rise should range from 12 to 16°F at full load. Coil performance at reduced loads should be considered as well; therefore, fluid velocity in the tube should remain high to stay in the turbulent flow range. To maintain a reasonable temperature differential at design conditions, fan-coil units are sized for an entering water temperature several degrees above the main chilled water plant supply temperature. This requires that temperature-actuated diversity control valves be applied to the primary distribution cross connection between the supply and return piping.

METERING

All thermal energy or power delivered by a commercially operated district energy system to customers or end users for billing or revenue must be metered. Steam is usually used for direct heating or to power absorption chillers for cooling. It is typically measured by using the differential pressure across calibrated orifices, nozzles or venturi tubes; or by pitot tubes, vortex-shedding meters, or condensate meters. In the United States, the customary commercial unit is pounds of steam per hour with a heat equivalent typically assumed to be 1000 Btu/lb; but for more precise thermal metering, the meters are coupled with steam quality (temperature and pressure) differential heat content measuring devices. Care must be taken to deliver dry steam (superheated or saturated without free water) to the customer. Dry steam is delivered by installing an adequate trap just ahead of the customer's meter (or ahead of the customer's process when condensate meters are used).

When condensate meters are used, care must be taken to ensure that all condensate from the customer's process, but *only* such condensate, goes to the condensate meter. Stultz and Kitto (1992) and IDHA (1969) have more information on steam metering. For steam, as with hot and chilled water system metering, electronic and computer technology provide direct, integrating, and remote input to central control/measurement energy management systems.

Hot and chilled water systems are metered by measuring the temperature differential between the supply and return lines and the flow rate of the energy transfer medium. Thermal (Btu or kWh) meters compensate for the actual volume and heat content characteristics of the energy transfer medium. Thermal transducers, resistance thermometer elements, or liquid expansion capillaries are usually used to measure the differential temperature of the energy transfer medium in supply and return lines.

Water flow can be measured with a variety of meters, usually pressure differential, turbine or propeller, or displacement meters. Chapter 14 of the 1997 *ASHRAE Handbook—Fundamentals*, the *District Heating Handbook* (IDHA 1983), and Pomroy (1994) have more information on measurement. Ultrasonic meters are sometimes used to check performance of installed meters. Various flowmeters are available for district energy billing purposes. Critical characteristics for proper installation include clearances and spatial limitations as well as the attributes presented in Table 5. The data in the table only provide general guidance and the manufacturers of meters should be contacted for data specific to their products.

The meter should be located upstream of the heat exchanger and the control valve(s) should be downstream from the heat exchanger. This orientation minimizes the possible formation of bubbles in the flow stream and provide a more accurate flow indication. The transmitter should be calibrated for zero and span as recommended by the manufacturer.

Wherever possible, the type and size of meters selected should be standardized to reduce the number of stored spare parts, technician training, etc.

Displacement meters are more accurate than propeller meters, but they are also larger. They can handle flow ranges from less than 2% up to 100% of the maximum rated flow with claimed ±1% accuracy. Turbine-type meters require the smallest physical space for a given maximum flow. However, like many meters, they require at least 10 diameters of straight pipe upstream and downstream of the meter to achieve their claimed accuracy.

The United States has no performance standards for thermal meters. ASHRAE *Standard* 125 describes a test method for rating liquid thermal meters. Several European countries have developed

Table 5 Flowmeter Characteristics

Meter Type	Accuracy	Range of Control	Pressure Loss	Straight Piping Requirements (Length in Pipe Diameters)
Orifice plate	±1% to 5% full scale	3:1 to 5:1	High (>5 psi)	10 D to 40 D – Upstream; 2 D to 6 D – Downstream
Electromagnetic	±0.15% to 1% rate	30:1 to 100:1	Low (<3 psi)	5 D to 10 D – Upstream; 3 D – Downstream
Vortex	±0.5% to 1.25% rate	10:1 to 25:1	Medium (3 to 5 psi)	10 D to 40 D – Upstream; 2 D to 6 D – Downstream
Turbine	±0.15% to 0.5% rate	10:1 to 50:1	Medium (3 to 5 psi)	10 D to 40 D – Upstream; 2 D to D – Downstream
Ultrasonic	±1% to 5% rate	>10:1 to 100:1	Low (<3 psi)	10 D to 40 D – Upstream; 2 D to 6 D – Downstream

performance standards and/or test methods for thermal meters and CEN *Standard* EN 1434, developed by the European Community, is a performance and testing standard for heat meters.

District energy plant meters intended for billing or revenue require means for verifying performance periodically. Major meter manufacturers, some laboratories, and some district energy companies maintain facilities for this purpose. In the absence of a single performance standard, meters are typically tested in accordance with their respective manufacturers' recommendations. Primary measurement elements used in these laboratories frequently obtain calibration traceability to the National Institute of Standards and Technology (NIST).

For district energy cogeneration systems that send out and/or accept electric power to or from a utility grid, the kilowatt demand and kilowatt-hour meters must meet the existing utility requirements. For district energy systems that send out electric power directly to customers, the electric kilowatt demand and kilowatt-hour meters must comply with local and state regulations. American National Standards Institute (ANSI) standards are established for all customary electric meters.

REFERENCES

Aamot, H. and G. Phetteplace. 1978. Heat transmission with steam and hot water. *Publication No.* H00128. American Society of Mechanical Engineers, New York.

Albert, M.R. and G.E. Phetteplace. 1983. Computer models for two-dimensional steady-state heat conduction. CRREL *Report* 83-10. U.S. Army Cold Regions Research and Engineering Laboratory, Hanover, NH.

Andrepont, J.S. 1995. Chilled water storage: A suite of benefits for district cooling. *District Energy* 81(1). International District Energy Association, Washington, DC.

ASCE. 1996. Cold regions utilities monograph. American Society of Civil Engineers, Reston, VA.

ASHRAE. 1992. Method of testing thermal energy meters for liquid streams in HVAC systems. ANSI/ASHRAE *Standard* 125-1992.

ASME. 1998. Power piping. *Standard* B31.1. American Society of Mechanical Engineers, New York.

Bloomquist, R.G., R. O'Brien, and M. Spurr, M. 1999. Geothermal district energy at co-located sites. WSU-EEP 99007, Washington State University Energy Office.

Bøhm, B. 1986. On the optimal temperature level in new district heating networks. *Fernwärme International* 15(5):301-306.

Bøhm, B. 1988. Energy-economy of Danish district heating systems: A technical and economic analysis. Laboratory of Heating and Air Conditioning, Technical University of Denmark, Lyngby, Denmark.

Bottorf, J.D. 1951. Summary of thermal conductivity as a function of moisture content. Thesis, Purdue University, West Lafayette, IN.

CEN. 1997. Heat meters. *Standard* EN 1434.

Chyu, M.-C., X. Zeng, and L. Ye. 1997a. Performance of fibrous glass insulation subjected to underground water attack. *ASHRAE Transactions* 103(1):303-308.

Chyu, M.-C., X. Zeng, and L. Ye. 1997b. The effect of moisture content on the performance of polyurethane insulation on a district heating and cooling pipe. *ASHRAE Transactions* 103(1):309-317.

Chyu, M.-C., X. Zeng, and L. Ye. 1998a. Behavior of cellular glass insulation on a DHC pipe subjected to underground water attack. *ASHRAE Transactions* 104(2):161-167.

Chyu, M.-C., X. Zeng, and L. Ye. 1998b. Effect of underground water attack on the performance of mineral wool pipe insulation. *ASHRAE Transactions* 104(2):168-175.

COWIconsult. 1985. Computerized planning and design of district heating networks. COWIconsult Consulting Engineers and Planners AS, Virum, Denmark.

CRREL. 1999. Regional climatic constants for Equation 6 of the Corps of Engineers Guide Spec 02695. (Best fit to mean monthly temperatures averaged for the period 1895-1996). U.S. Army Cold Regions Research and Engineering Laboratory, Hanover, NH. Table may be found at http://www.crrel.usace.army.mil/ard/cegs02695.htm.

EuHP. 1991. Preinsulated bonded pipe systems. EN253, European District Heating Pipe Manufacturers Association, Brussels, Belgium.

Farouki, O.T. 1981. Thermal properties of soils. CRREL *Monograph* 81-1. U.S. Army Cold Regions Research and Engineering Laboratory, Hanover, NH.

Fox, J.A. 1977. Hydraulic analysis of unsteady flow in pipe networks. John Wiley and Sons, New York.

Geiringer, P.L. 1963. High temperature water heating: Its theory and practice for district and space heating applications. John Wiley and Sons, New York.

Grober, H., S. Erk, and U. Grigull. 1961. *Fundamentals of heat transfer.* McGraw-Hill, New York.

IDHA. 1969. Code for steam metering. International District Energy Association, Washington, D.C.

IDHA. 1983. District heating handbook, 4th ed. International District Energy Association, Washington, D.C.

Jeppson, R.W. 1977. *Analysis of flow in pipe networks*. Ann Arbor Science Publishers, Ann Arbor, MI.

Koskelainen, L. 1980. Optimal dimensioning of district heating networks. *Fernwärme International* 9(4):84-90.

Kusuda, T. and P.R. Achenbach. 1965. Earth temperature and thermal diffusivity at selected stations in the United States. *ASHRAE Transactions* 71(1):61-75.

Lunardini, V.J. 1981. *Heat transfer in cold climates*. Van Nostrand Reinhold, New York.

McCabe R.E., J.J. Bender, and K.R. Potter. 1995. Subsurface ground temperature—Implications for a district cooling system. *ASHRAE Journal* 37(12):40-45.

Minkowycz, W.J., E.M. Sparrow, G.E. Schneider, and R.H. Pletcher. 1988. *Handbook of numerical heat transfer.* John Wiley and Sons, New York.

MSS. 1993. Pipe hangers and supports—Materials, design and manufacture. *Standard* MS-58-1993. Manufacturers Standardization Society of the Valve and Fittings Industry, Vienna, VA.

MSS. 1996. Pipe hangers and supports—Selection and application. *Standard* MS-69-1996.

NACE. 1996. Control of external corrosion on underground or submerged metallic piping systems. *Standard* RPO169-96.

NAS. 1975. *Technical Report No.* 66. National Academy of Sciences National Research Council Building Research Advisory Board (BRAB), Federal Construction Council.

Ottmer, J.H. and J.B. Rishel. 1993. Airport pumping system horsepower requirements take a nose dive, *Heating, Piping and Air Conditioning* 65(10).

Phetteplace, G.E. 1989. Simulation of district heating systems for piping design. International Symposium on District Heat Simulation, Reykjavik, Iceland.

Phetteplace, G. and V. Meyer. 1990. Piping for thermal distribution systems. CRREL *Internal Report* 1059. U.S. Army Cold Regions Research and Engineering Laboratory, Hanover, NH.

Phetteplace, G.E., D. Carbee, and M. Kryska. 1991. Field measurement of heat losses from three types of heat distribution systems. CRREL SR 631. U.S. Army Cold Regions Research and Engineering Laboratory.

Phetteplace, G.E., W. Willey, and M.A. Novick. 1981. Losses from the Fort Wainwright heat distribution system. CRREL SR 81-14. U.S. Army Cold Regions Research and Engineering Laboratory.

Pomroy, J. 1994. Selecting flowmeters. *Instrumentation & Control Systems*. Chilton Publications.

Rao, S.S. 1982. *The finite element method in engineering*. Pergamon Press, New York.

Rasmussen, C.H. and J.E. Lund. 1987. Computer aided design of district heating systems. District Heating Research and Technological Development in Denmark, Danish Ministry of Energy, Copenhagen, Denmark.

Reisman, A.W. 1985. *District heating handbook, Volume 2: A handbook of district heating and cooling models*. International District Energy Association, Washington, D.C.

Rohsenow, W.M. 1998. *Handbook of heat transfer.* McGraw-Hill, New York.

Segan, E.G. and C.-P. Chen. 1984. Investigation of tri-service heat distribution systems. CERL *Technical Report* M-347. U.S. Army Construction Engineering Research Laboratory (CERL), Champaign, IL.

Siegel, R. and J.R. Howell. 1981. *Thermal radiation heat transfer.* Hemisphere Publishing Corporation, New York.

Sleimen, A.H. et al. 1990. Guidelines for converting building heating systems for hot water district heating. NOVEM, Sittard, The Netherlands. ISBN 90-72130-12-x.

Stephenson, D. 1981. *Pipeline design for water engineers.* Elsevier Scientific Publishing Company, New York.

Stewart, W.E. and C.L. Dona. 1987. Water flow rate limitations. *ASHRAE Transactions* 93(2):811-25.

Stultz, S.C. and J.B. Kitto, eds. 1992. *Steam: Its generation and use*, 40th ed. Chapter 40-13, Measurement of steam quality and purity. Chapter 40-19, Measurement of steam flow. Babcock & Wilcox, Barberton, OH.

Werner, S.E. 1984. The heat load in district heating systems. Chalmers University of Technology, Göteborg, Sweden.

BIBLIOGRAPHY

The following IEA (International Energy Agency) publications were published by NOVEM, Sittard, The Netherlands.

Bowling, T. 1990. A technology assessment of potential telemetry technologies for district heating. ISBN 90-72130-10-3.

Holtse, C. and P. Randlov, eds. 1989. District heating and cooling R&D project review. ISBN 90-72130-07-3.

Mørck, O. and T. Pedersen, eds. 1989. Advanced district heating production technologies. ISBN 90-90028-76-5.

Ulseth, R., ed. 1990. Heat meters: Report of research activities. ISBN 90-72130-15-4.

U.S. Air Force, Army, and Navy. 1978. Engineering weather data. Dept. of the Air Force *Manual* AFM 88-29, Dept. of the Army *Manual* TM 5-785, and Dept. of the Navy *Manual* NAVFAC P-89.

CHAPTER 12

HYDRONIC HEATING AND COOLING SYSTEM DESIGN

WATER systems that convey heat to or from a conditioned space or process with hot or chilled water are frequently called hydronic systems. The water flows through piping that connects a boiler, water heater, or chiller to suitable terminal heat transfer units located at the space or process.

Water systems can be classified by (1) operating temperature, (2) flow generation, (3) pressurization, (4) piping arrangement, and (5) pumping arrangement.

Classified by flow generation, hydronic heating systems may be (1) gravity systems, which use the difference in density between the supply and return water columns of a circuit or system to circulate water; or (2) forced systems, in which a pump, usually driven by an electric motor, maintains the flow. Gravity systems are seldom used today and are therefore not discussed in this chapter. See the *ASHVE Heating Ventilating Air Conditioning Guide* issued prior to 1957 for information on gravity systems.

Water systems can be either once-through or recirculating systems. This chapter describes forced recirculating systems.

Principles

The design of effective and economical water systems is affected by complex relationships between the various system components. The design water temperature, flow rate, piping layout, pump selection, terminal unit selection, and control method are all interrelated. The size and complexity of the system determine the importance of these relationships to the total system operating success. In the United States, present hydronic heating system design practice originated in residential heating applications, where a temperature drop (Δt) of 20°F was used to determine flow rate. Besides producing satisfactory operation and economy in small systems, this Δt enabled simple calculations because 1 gpm conveys 10,000 Btu/h. However, almost universal use of hydronic systems for both heating and cooling of large buildings and building complexes has rendered this simplified approach obsolete.

TEMPERATURE CLASSIFICATIONS

Water systems can be classified by operating temperature as follows.

Low-temperature water (LTW) system. This hydronic heating system operates within the pressure and temperature limits of the ASME *Boiler and Pressure Vessel Code* for low-pressure boilers. The maximum allowable working pressure for low-pressure boilers is 160 psig, with a maximum temperature limitation of 250°F. The usual maximum working pressure for boilers for LTW systems is 30 psi, although boilers specifically designed, tested, and stamped for higher pressures are frequently used. Steam-to-water or water-to-water heat exchangers are also used for heating low-temperature water. Low-temperature water systems are used in buildings ranging from small, single dwellings to very large and complex structures.

Medium-temperature water (MTW) system. This hydronic heating system operates at temperatures between 250 and 350°F, with pressures not exceeding 160 psi. The usual design supply temperature is approximately 250 to 325°F, with a usual pressure rating of 150 psi for boilers and equipment.

High-temperature water (HTW) system. This hydronic heating system operates at temperatures over 350°F and usual pressures of about 300 psi. The maximum design supply water temperature is usually about 400°F, with a pressure rating for boilers and equipment of about 300 psi. The pressure-temperature rating of each component must be checked against the system's design characteristics.

Chilled water (CW) system. This hydronic cooling system normally operates with a design supply water temperature of 40 to 55°F, usually 44 or 45°F, and at a pressure of up to 120 psi. Antifreeze or brine solutions may be used for applications (usually process applications) that require temperatures below 40°F or for coil freeze protection. Well water systems can use supply temperatures of 60°F or higher.

Dual-temperature water (DTW) system. This hydronic combination heating and cooling system circulates hot and/or chilled water through common piping and terminal heat transfer apparatus. These systems operate within the pressure and temperature limits of LTW systems, with usual winter design supply water temperatures of about 100 to 150°F and summer supply water temperatures of 40 to 45°F.

Terminal heat transfer units include convectors, cast-iron radiators, baseboard and commercial finned-tube units, fan-coil units, unit heaters, unit ventilators, central station air-handling units, radiant panels, and snow-melting panels. A large storage tank may be included in the system to store energy to use when such heat input devices as the boiler or a solar energy collector are not supplying energy.

This chapter covers the principles and procedures for designing and selecting piping and components for low-temperature water, chilled water, and dual-temperature water systems. See Chapter 14 for information on medium- and high-temperature water systems.

CLOSED WATER SYSTEMS

Because most hot and chilled water systems are closed, this chapter addresses only closed systems. The fundamental difference between a closed and an open water system is the interface of the water with a compressible gas (such as air) or an elastic surface

The preparation of this chapter is assigned to TC 6.1, Hydronic and Steam Equipment and Systems.

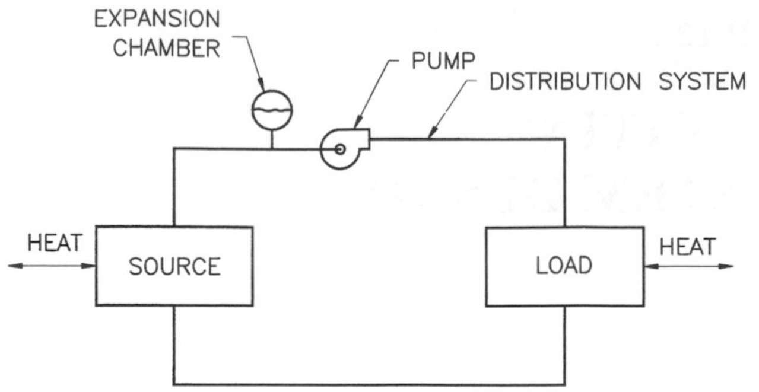

Fig. 1 Hydronic System—Fundamental Components

(such as a diaphragm). A **closed water system** is defined as one with no more than one point of interface with a compressible gas or surface. This definition is fundamental to understanding the hydraulic dynamics of these systems. Earlier literature referred to a system with an open or vented expansion tank as an "open" system, but such a system is actually a closed system; the atmospheric interface of the tank simply establishes the system pressure.

An **open system**, on the other hand, has more than one such interface. For example, a cooling tower system has at least two points of interface: the tower basin and the discharge pipe or nozzles entering the tower. One of the major differences in hydraulics between open and closed systems is that certain hydraulic characteristics of open systems cannot occur in closed systems. For example, in contrast to the hydraulics of an open system, in a closed system (1) flow cannot be motivated by static head differences, (2) pumps do not provide static lift, and (3) the entire piping system is always filled with water.

Basic System

Figure 1 shows the fundamental components of a closed hydronic system. Actual systems generally have additional components such as valves, vents, regulators, etc., but they are not essential to the basic principles underlying the system.

These fundamental components are

- Loads
- Source
- Expansion chamber
- Pump
- Distribution system

Theoretically, a hydronic system could operate with only these five components.

The components are subdivided into two groups—thermal components and hydraulic components. The thermal components consist of the load, the source, and the expansion chamber. The hydraulic components consist of the distribution system, the pump, and the expansion chamber. The expansion chamber is the only component that serves both a thermal and a hydraulic function.

THERMAL COMPONENTS

Loads

The load is the device that causes heat to flow out of or into the system to or from the space or process; it is the independent variable to which the remainder of the system must respond. Outward heat flow characterizes a heating system, and inward heat flow characterizes a cooling system. The quantity of heating or cooling is calculated by one of the following means.

Sensible Heating or Cooling. The rate of heat entering or leaving an airstream is expressed as follows:

$$q = 60 Q_a \rho_a c_p \Delta t \quad (1)$$

where

q = heat transfer rate to or from air, Btu/h
Q_a = airflow rate, cfm
ρ_a = density of air, lb/ft^3
c_p = specific heat of air, Btu/lb·°F
Δt = temperature increase or decrease of air, °F

For standard air with a density of 0.075 lb/ft^3 and a specific heat of 0.24 Btu/lb·°F, Equation (1) becomes

$$q = 1.1 Q_a \Delta t \quad (2)$$

The heat exchanger or coil must then transfer this heat from or to the water. The rate of sensible heat transfer to or from the heated or cooled medium in a specific heat exchanger is a function of the heat transfer surface area; the mean temperature difference between the water and the medium; and the overall heat transfer coefficient, which itself is a function of the fluid velocities, properties of the medium, geometry of the heat transfer surfaces, and other factors. The rate of heat transfer may be expressed by

$$q = UA(\text{LMTD}) \quad (3)$$

where

q = heat transfer rate through heat exchanger, Btu/h
U = overall coefficient of heat transfer, Btu/h·ft^2·°F
A = heat transfer surface area, ft^2
LMTD = logarithmic mean temperature difference, heated or cooled medium to water, °F

Cooling and Dehumidification. The rate of heat removal from the cooled medium when both sensible cooling and dehumidification are present is expressed by

$$q_t = w \Delta h \quad (4)$$

where

q_t = total heat transfer rate from cooled medium, Btu/h
w = mass flow rate of cooled medium, lb/h
Δh = enthalpy difference between entering and leaving conditions of cooled medium, Btu/lb

Expressed for an air-cooling coil, this equation becomes

$$q_t = 60 Q_a \rho_a \Delta h \quad (5)$$

which, for standard air with a density of 0.075 lb/ft^3, reduces to

$$q_t = 4.5 Q_a \Delta h \quad (6)$$

Heat Transferred to or from Water. The rate of heat transfer to or from the water is a function of the flow rate, the specific heat, and the temperature rise or drop of the water as it passes through the heat exchanger. The heat transferred to or from the water is expressed by

$$q_w = \dot{m} c_p \Delta t \quad (7)$$

where

q_w = heat transfer rate to or from water, Btu/h
$\dot{m}$ = mass flow rate of water, lb/h
c_p = specific heat of water, Btu/lb·°F
Δt = water temperature increase or decrease across unit, °F

With water systems, it is common to express the flow rate as volumetric flow, in which case Equation (7) becomes

$$q_w = 8.02\rho_w c_p Q_w \Delta t \quad (8)$$

where

Q_w = water flow rate, gpm
ρ_w = density of water, lb/ft^3

For standard conditions in which the density is 62.4 lb/ft^3 and the specific heat is 1 Btu/lb·°F, Equation (8) becomes

$$q_w = 500 Q_w \Delta t \quad (9)$$

Equation (8) or (9) can be used to express the heat transfer across a single load or source device, or any quantity of such devices connected across a piping system. In the design or diagnosis of a system, the load side may be balanced with the source side using these equations.

Heat Carrying Capacity of Piping. Equations (8) and (9) are also used to express the heat carrying capacity of the piping or distribution system or any portion thereof. The existing temperature differential Δt, sometimes called the temperature range, is identified; for any flow rate Q_w through the piping, q_w is called the **heat carrying capacity**.

Most load devices (in which heat is conveyed to or from the water for heating or cooling the space or process) are a water-to-air finned-coil heat exchanger or a water-to-water exchanger. The specific configuration is usually used to describe the load device. The most common configurations include the following:

Heating load devices
- Preheat coils in central units
- Heating coils in central units
- Zone or central reheat coils
- Finned-tube radiators
- Baseboard radiators
- Convectors
- Unit heaters
- Fan-coil units
- Water-to-water heat exchangers
- Radiant heating panels
- Snow-melting panels

Cooling load devices
- Coils in central units
- Fan-coil units
- Induction unit coils
- Radiant cooling panels
- Water-to-water heat exchangers

Source

The source is the point where heat is added to (heating) or removed from (cooling) the system. Ideally, the amount of energy entering or leaving the source equals the amount entering or leaving through the load. Under steady-state conditions, the load energy and source energy are equal and opposite. Also, when properly measured or calculated, temperature differentials and flow rates across the source and loads are all equal. Equations (8) and (9) are used to express the source capacities as well as the load capacities.

Any device that can be used to heat or cool water under controlled conditions can be used as a source device. The most common source devices for heating and cooling systems are the following:

Heating source devices
- Hot water generator or boiler
- Steam-to-water heat exchanger
- Water-to-water heat exchanger
- Solar heating panels
- Heat recovery or salvage heat device (e.g., water jacket of an internal combustion engine)
- Exhaust gas heat exchanger
- Incinerator heat exchanger
- Heat pump condenser
- Air-to-water heat exchanger

Cooling source devices
- Electric compression chiller
- Thermal absorption chiller
- Heat pump evaporator
- Air-to-water heat exchanger
- Water-to-water heat exchanger

The two primary considerations in selecting a source device are the design capacity and the part-load capability, sometimes called the **turndown ratio**. The turndown ratio, expressed in percent of design capacity, is

$$\text{Turndown ratio} = 100\,\frac{\text{Minimum capacity}}{\text{Design capacity}} \quad (10)$$

The reciprocal of the turndown ratio is sometimes used (for example, a turndown ratio of 25% may also be expressed as a turndown ratio of 4).

The turndown ratio has a significant effect on the performance of a system; lack of consideration of the source system's part-load capability has been responsible for many systems that either do not function properly or do so at the expense of excess energy consumption. The turndown ratio has a significant effect on the ultimate equipment and/or system design selection.

System Temperatures. Design temperatures and temperature ranges are selected by consideration of the performance requirements and the economics of the components. For a cooling system that must maintain 50% rh at 75°F, the dew-point temperature is 55°F, which sets the maximum return water temperature at something near 55°F (60°F maximum); on the other hand, the lowest practical temperature for refrigeration, considering the freezing point and economics, is about 40°F. This temperature spread then sets constraints for a chilled water system. For a heating system, the maximum hot water temperature is normally established by the ASME *Boiler and Pressure Vessel Code* as 250°F, and with space temperature requirements of little over 75°F, the actual operating supply temperatures and the temperature ranges are set by the design of the load devices. Most economic considerations relating to the distribution and pumping systems favor the use of the maximum possible temperature range Δt.

Expansion Chamber

The expansion chamber (also called an expansion or compression tank) serves both a thermal function and a hydraulic function. In its thermal function the tank provides a space into which the noncompressible liquid can expand or from which it can contract as the liquid undergoes volumetric changes with changes in temperature. To allow for this expansion or contraction, the expansion tank provides an interface point between the system fluid and a compressible gas. By definition, a closed system can have only one such interface; thus, a system designed to function as a closed system can have only one expansion chamber.

Expansion tanks are of three basic configurations: (1) a closed tank, which contains a captured volume of compressed air and water, with an air water interface (sometimes called a plain steel tank); (2) an open tank (i.e., a tank open to the atmosphere); and (3) a diaphragm tank, in which a flexible membrane is inserted between the air and the water (another configuration of a diaphragm tank is the bladder tank).

In the plain steel tank and the open tank, gases can enter the system water through the interface and can adversely affect system performance. Thus, current design practice normally employs diaphragm tanks.

Sizing the tank is the primary thermal consideration in incorporating a tank into a system. However, prior to sizing the tank, the control or elimination of air must be considered. The amount of air that will be absorbed and can be held in solution with the water is expressed by Henry's equation (Pompei 1981):

$$x = p/H \tag{11}$$

where

x = solubility of air in water (% by volume)
p = absolute pressure
H = Henry's constant

Henry's constant, however, is constant only for a given temperature (Figure 2). Combining the data of Figure 2 (Himmelblau 1960) with Equation (11) results in the solubility diagram of Figure 3. With that diagram, the solubility can be determined if the temperature and pressure are known.

If the water is not saturated with air, it will absorb air at the air/water interface until the point of saturation has been reached. Once absorbed, the air will move throughout the body of water either by mass migration or by molecular diffusion until the water is uniformly saturated. If the air/water solution changes to a state that reduces solubility, the excess air will be released as a gas. For example, if the air/water interface is at a high pressure point, the water will absorb air to its limit of solubility at that point; if at another point in the system the pressure is reduced, some of the dissolved air will be released.

In the design of systems with open or plain steel expansion tanks, it is common practice to utilize the tank as the major air control or release point in the system.

Equations for sizing the three common configurations of expansion tanks follow (Coad 1980b):

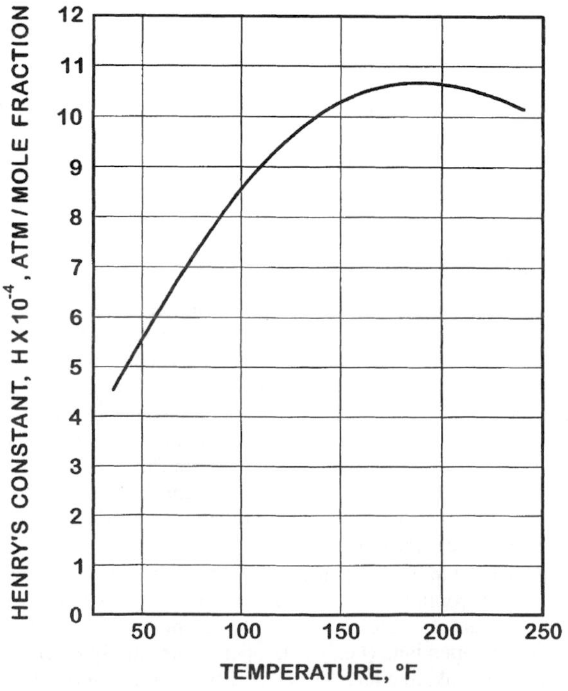

Fig. 2 Henry's Constant Versus Temperature for Air and Water
(Coad 1980a)

For closed tanks with air/water interface,

$$V_t = V_s \frac{[(v_2/v_1) - 1] - 3\alpha\Delta t}{(P_a/P_1) - (P_a/P_2)} \tag{12}$$

For open tanks with air/water interface,

$$V_t = 2V_s\left[\left(\frac{v_2}{v_1} - 1\right) - 3\alpha\Delta t\right] \tag{13}$$

For diaphragm tanks,

$$V_t = V_s \frac{[(v_2/v_1) - 1] - 3\alpha\Delta t}{1 - (P_1/P_2)} \tag{14}$$

where

V_t = volume of expansion tank, gal
V_s = volume of water in system, gal
t_1 = lower temperature, °F
t_2 = higher temperature, °F
P_a = atmospheric pressure, psia
P_1 = pressure at lower temperature, psia
P_2 = pressure at higher temperature, psia
v_1 = specific volume of water at lower temperature, ft^3/lb
v_2 = specific volume of water at higher temperature, ft^3/lb
α = linear coefficient of thermal expansion, in/in · °F
= 6.5×10^{-6} in/in·°F for steel
= 9.5×10^{-6} in/in·°F for copper
Δt = $(t_2 - t_1)$, °F

As an example, the lower temperature for a heating system is usually normal ambient temperature at fill conditions (e.g., 50°F) and the higher temperature is the operating supply water temperature for the system. For a chilled water system, the lower temperature is the design chilled water supply temperature, and the higher temperature is ambient temperature (e.g., 95°F). For a dual-temperature hot/chilled system, the lower temperature is the chilled water

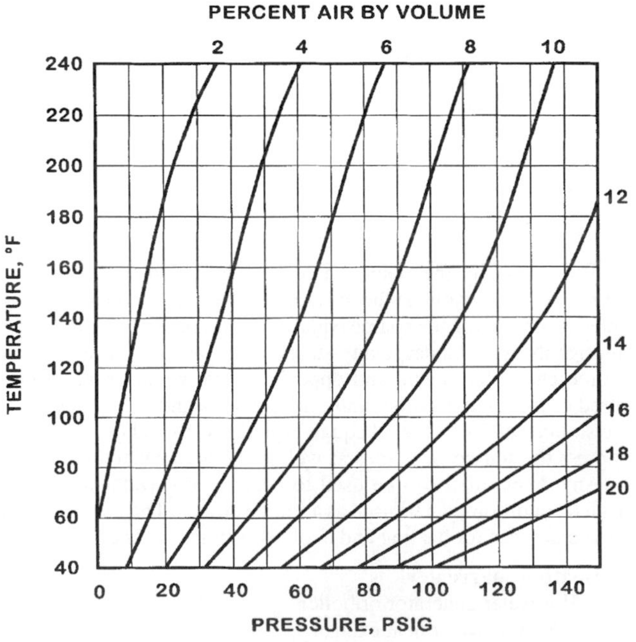

Fig. 3 Solubility Versus Temperature and Pressure for Air/Water Solutions
(Coad 1980a)

design supply temperature, and the higher temperature is the heating water design supply temperature.

For specific volume and saturation pressure of water at various temperatures, see Table 3 in Chapter 6 of the 1997 *ASHRAE Handbook—Fundamentals*.

At the tank connection point, the pressure in closed tank systems increases as the water temperature increases. Pressures at the expansion tank are generally set by the following parameters:

- The lower pressure is usually selected to hold a positive pressure at the highest point in the system (usually about 10 psig).
- The higher pressure is normally set by the maximum pressure allowable at the location of the safety relief valve(s) without opening them.

Other considerations are to ensure that (1) the pressure at no point in the system will ever drop below the saturation pressure at the operating system temperature and (2) all pumps have sufficient net positive suction head (NPSH) available to prevent cavitation.

Example 1. Size an expansion tank for a heating water system that will be operated at a design temperature range of 180 to 220°F. The minimum pressure at the tank is 10 psig (24.7 psia) and the maximum pressure is 25 psig (39.7 psia). (Atmospheric pressure is 14.7 psia.) The volume of water is 3000 gal. The piping is steel.

1. Calculate the required size for a closed tank with an air/water interface.

Solution: For lower temperature t_1, use 40°F.

From Table 3 in Chapter 6 of the 1997 *ASHRAE Handbook—Fundamentals*,

$$v_1(\text{at } 40°\text{F}) = 0.01602 \text{ ft}^3/\text{lb}$$

$$v_2(\text{at } 220°\text{F}) = 0.01677 \text{ ft}^3/\text{lb}$$

Using Equation (12),

$$V_t = 3000 \times \frac{[(0.01677/0.01602) - 1] - 3(6.5 \times 10^{-6})(220 - 40)}{(14.7/24.7) - (14.7/39.7)}$$

$$V_t = 578 \text{ gal}$$

2. If a diaphragm tank were to be used in lieu of the plain steel tank, what tank size would be required?

Solution: Using Equation (14),

$$V_t = 3000 \times \frac{[(0.01677/0.01602) - 1] - 3(6.5 \times 10^{-6})(220 - 40)}{1 - (24.7/39.7)}$$

$$V_t = 344 \text{ gal}$$

HYDRAULIC COMPONENTS

Distribution System

The distribution system is the piping connecting the various other components of the system. The primary considerations in designing this system are (1) sizing the piping to handle the heating or cooling capacity required and (2) arranging the piping to ensure flow in the quantities required at design conditions and at all other loads.

The flow requirement of the pipe is determined by Equation (8) or (9). After Δt is established based on the thermal requirements, either of these equations (as applicable) can be used to determine the flow rate. First-cost economics and energy consumption make it advisable to design for the greatest practical Δt because the flow rate is inversely proportional to Δt; that is, if Δt doubles, the flow rate is reduced by half.

The three related variables in sizing the pipe are flow rate, pipe size, and pressure drop. The primary consideration in selecting a design pressure drop is the relationship between the economics of first cost and energy costs.

Once the distribution system is designed, the pressure loss at design flow is calculated by the methods discussed in Chapter 33 of the 1997 *ASHRAE Handbook—Fundamentals*. The relationship between flow rate and pressure loss can be expressed by

$$Q = C_v\sqrt{\Delta p} \tag{15}$$

where

Q = system flow rate, gpm
Δp = pressure drop in system, psi
C_v = system constant (sometimes called valve coefficient, which is discussed in Chapter 42)

Equation (15) may be modified as follows:

$$Q = C_s\sqrt{\Delta h} \tag{16}$$

where

Δh = system head loss, ft of fluid [$\Delta h = \Delta p/\rho$]
C_s = system constant [$C_s = 0.67C_v$ for water with a density $\rho = 62.4$ lb/ft^3]

Equations (15) and (16) are the system constant form of the Darcy-Weisbach equation. If the flow rate and head loss are known for a system, Equation (16) may be used to calculate the system constant C_v. From this calculation, the pressure loss can be determined at any other flow rate. Equation (16) can be graphed as a system curve (Figure 4).

The system curve changes if anything occurs that changes the flow/pressure drop characteristics. Examples of this are a strainer that starts to block or a control valve closing, either of which increases the head loss at any given flow rate, thus changing the system curve in a direction from curve A to curve B in Figure 4.

Pump or Pumping System

Centrifugal pumps are the type most commonly used in hydronic systems (see Chapter 39). Circulating pumps used in water systems can vary in size from small in-line circulators delivering 5 gpm at 6 or 7 ft head to base-mounted or vertical pumps handling hundreds or thousands of gallons per minute, with pressures limited only by the characteristics of the system. Pump operating characteristics must be carefully matched to system operating requirements.

Pump Curves and Water Temperature. Performance characteristics of centrifugal pumps are described by pump curves, which plot flow versus head or pressure, as well as by efficiency and power information. The point at which a pump operates is the point at which the pump curve intersects the system curve (Figure 5). Chapter 38 discusses system and pump curves.

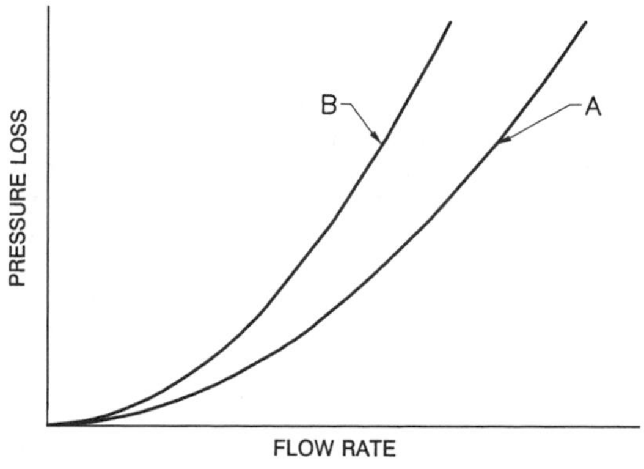

Fig. 4 Typical System Curves for Closed System

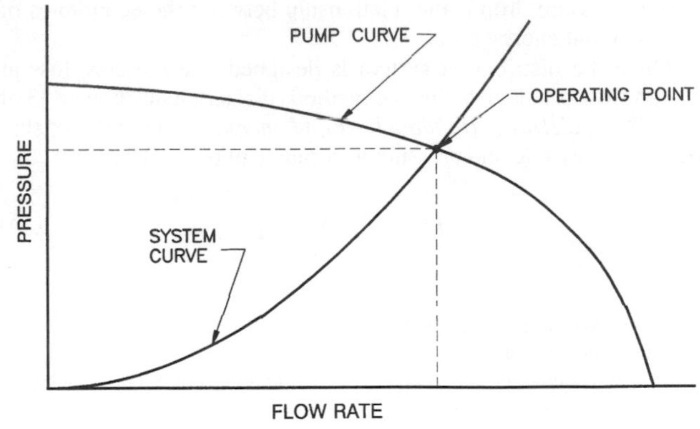

Fig. 5 Pump Curve and System Curve

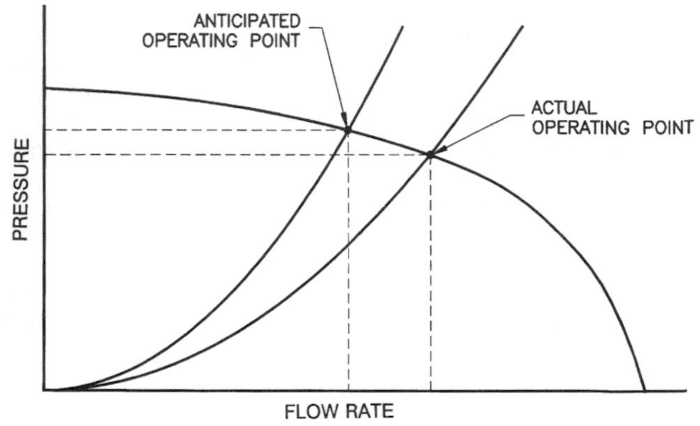

Fig. 6 Shift of System Curve due to Circuit Unbalance

A complete piping system follows the same water flow/pressure drop relationships as any component of the system [see Equation (16)]. Thus, the pressure required for any proposed flow rate through the system may be determined and a system curve constructed. A pump may be selected by using the calculated system pressure at the design flow rate as the base point value.

Figure 6 illustrates how a shift of the system curve to the right affects system flow rate. This shift can be caused by incorrectly calculating the system pressure drop by using arbitrary safety factors or overstated pressure drop charts. Variable system flow caused by control valve operation or improperly balanced systems (subcircuits having substantially lower pressure drops than the longest circuit) can also cause a shift to the right.

As described in Chapter 39, pumps for closed-loop piping systems should have a flat pressure characteristic and should operate slightly to the left of the peak efficiency point on their curves. This characteristic permits the system curve to shift to the right without causing undesirable pump operation, overloading, or reduction in available pressure across circuits with large pressure drops.

Many dual-temperature systems are designed so that the chillers are bypassed during the winter months. The chiller pressure drop, which may be quite high, is thus eliminated from the system pressure drop, and the pump shift to the right may be quite large. For such systems, system curve analysis should be used to check winter operating points.

Operating points may be highly variable, depending on (1) load conditions, (2) the types of control valves used, and (3) the piping circuitry and heat transfer elements. In general, the best selection will be

- For design flow rates calculated using pressure drop charts that illustrate actual closed-loop hydronic system piping pressure drops

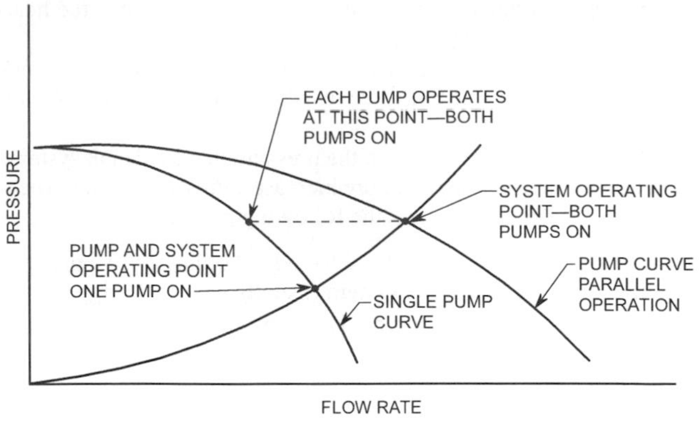

Fig. 7 Operating Conditions for Parallel Pump Installation

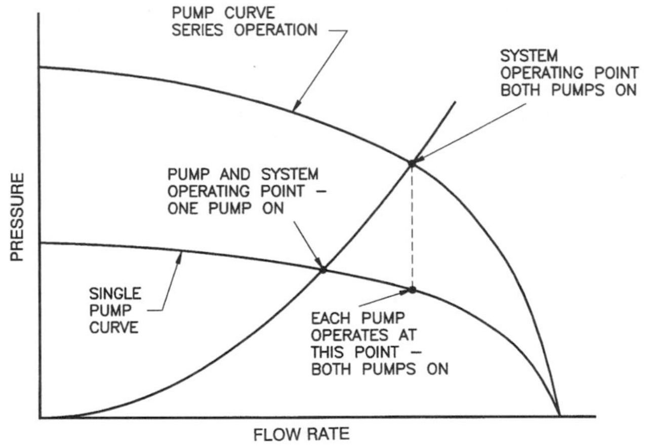

Fig. 8 Operating Conditions for Series Pump Installation

- To the left of the maximum efficiency point of the pump curve to allow shifts to the right caused by system circuit unbalance, direct-return circuitry applications, and modulating three-way valve applications
- A pump with a flat curve to compensate for unbalanced circuitry and to provide a minimum pressure differential increase across two-way control valves

Parallel Pumping. When pumps are applied in parallel, each pump operates at the same head, and provides its share of the system flow at that pressure (Figure 7). Generally, pumps of equal size are used, and the parallel pump curve is established by doubling the flow of the single pump curve (with identical pumps).

Plotting a system curve across the parallel pump curve shows the operating points for both single and parallel pump operation (Figure 7). Note that single pump operation does not yield 50% flow. The system curve crosses the single pump curve considerably to the right of its operating point when both pumps are running. This leads to two important concerns: (1) the pumps must be powered to prevent overloading during single-pump operation, and (2) a single pump can provide standby service of up to 80% of design flow; the actual amount depends on the specific pump curve and system curve.

Series Pumping. When pumps are operated in series, each pump operates at the same flow rate and provides its share of the total pressure at that flow. A system curve plotted across the series pump curve shows the operating points for both single and series pump operation (Figure 8). Note that the single pump can provide up to 80% flow for standby and at a lower power requirement.

Series pump installations are often used in heating and cooling systems so that both pumps operate during the cooling season to provide maximum flow and head, while only a single pump operates during the heating season. Note that both parallel and series pump applications require that the actual pump operating points be used to accurately determine the pumping point. Adding artificial safety factor head, using improper pressure drop charts, or incorrectly calculating pressure drops may lead to an unwise selection.

Multiple-Pump Systems. Care must be taken in designing systems with multiple pumps to ensure that if pumps ever operate in either parallel or series, such operation is fully understood and considered by the designer. Pumps performing unexpectedly in series or parallel have been the cause of performance problems in hydronic systems. Typical problems resulting from pumps functioning in parallel and series when not anticipated by the designer are the following.

Parallel. With pumps of unequal pressures, one pump may create a pressure across the other pump in excess of its cutoff pressure, causing flow through the second pump to diminish significantly or to cease. This phenomenon can cause flow problems or pump damage.

Series. With pumps of different flow capacities, the pump of greater capacity may overflow the pump of lesser capacity, which could cause damaging cavitation in the smaller pump and could actually cause a pressure drop rather than a pressure rise across that pump. In other circumstances, unexpected series operation can cause excessively high or low pressures that can damage system components.

Standby Pump Provision. If total flow standby capacity is required, a properly valved standby pump of equal capacity is installed to operate when the normal pump is inoperable. A single standby may be provided for several similarly sized pumps. Parallel or series pump installation can provide up to 80% standby, which is often sufficient.

Compound Pumping. In larger systems, compound pumping, also known as primary-secondary pumping, is often employed to provide system advantages that would not be available with a single pumping system. The concept of compound pumping is illustrated in Figure 9.

In Figure 9, Pump No. 1 can be referred to as the source or primary pump and Pump No. 2 as the load or secondary pump. The short section of pipe between A and B is called the common pipe because it is common to both the source and load circuits. Other terms used for the common pipe are the decoupling line and the neutral bridge. In the design of compound systems, the common pipe should be kept as short and as large in diameter as practical to minimize the pressure loss between those two points. Care must be taken, however, to ensure adequate length in the common pipe to prevent recirculation from entry or exit turbulence. There should never be a valve or check valve in the common pipe. If these conditions are met and the pressure loss in the common pipe can be assumed to be zero, then neither pump will affect the other. Then, except for the system static pressure at any given point, the circuits can be designed and analyzed and will function dynamically independently of one another.

In Figure 9, if Pump No. 1 has the same flow capacity in its circuit as Pump No. 2 has in its circuit, all of the flow entering Point A from Pump No. 1 will leave in the branch supplying Pump No. 2, and no water will flow in the common pipe. Under this condition, the water entering the load will be at the same temperature as that leaving the source.

If the flow capacity of Pump No. 1 exceeds that of Pump No. 2, some water will flow downward in the common pipe. Under this condition, Tee A is a diverting tee, and Tee B becomes a mixing tee. Again, the temperature of the fluid entering the load is the same as that leaving the source. However, because of the mixing taking place at Point B, the temperature of the water returning to the

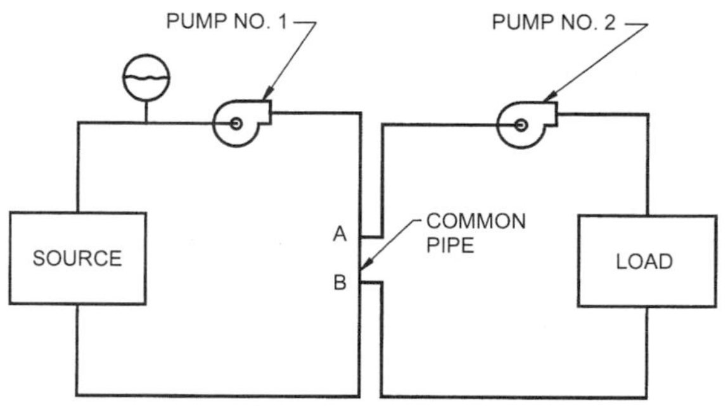

Fig. 9 Compound Pumping (Primary-Secondary Pumping)

source is between the source supply temperature and the load return temperature.

On the other hand, if the flow capacity of Pump No. 1 is less than that of Pump No. 2, then Point A becomes a mixing point because some water must recirculate upward in the common pipe from Point B. Under this condition, the temperature of the water entering the load is between the supply water temperature from the source and the return water temperature from the load.

For example, if Pump No. 1 circulates 25 gpm of water leaving the source at 200°F, and Pump No. 2 circulates 50 gpm of water leaving the load at 100°F, then the water temperature entering the load is

$$t_{load} = 200 - (25/50)(200 - 100) = 150°\text{F}$$

The following are some advantages of compound circuits:

1. They enable the designer to achieve different water temperatures and temperature ranges in different elements of the system.
2. They decouple the circuits hydraulically, thereby making the control, operation, and analysis of large systems much less complex. Hydraulic decoupling also prevents unwanted series or parallel operation.
3. Circuits can be designed for different flow characteristics. For example, a chilled water load system can be designed with two-way valves for better control and energy conservation while the source system operates at constant flow to protect the chiller from freezing.

Expansion Chamber

As a hydraulic device, the expansion tank serves as the reference pressure point in the system, analogous to a ground in an electrical system (Lockhart and Carlson 1953). Where the tank connects to the piping, the pressure equals the pressure of the air in the tank plus or minus any fluid pressure due to the elevation difference between the tank liquid surface and the pipe (Figure 10).

As previously stated, a closed system should have only one expansion chamber. The presence of more than one chamber or of excessive amounts of undissolved air in a piping system can cause the closed system to behave in unexpected (but understandable) ways, causing extensive damage from shock waves or water hammer.

With a single chamber on a system, assuming isothermal conditions for the air, the air pressure can change only as a result of displacement by the water. The only thing that can cause the water to move into or out of the tank (assuming no water is being added to or removed from the system) is expansion or shrinkage of the water in the system. Thus, in sizing the tank, thermal expansion is related to the pressure extremes of the air in the tank [Equations (12), (13), and (14)].

The point of connection of the tank should be based on the pressure requirements of the system, remembering that the pressure at

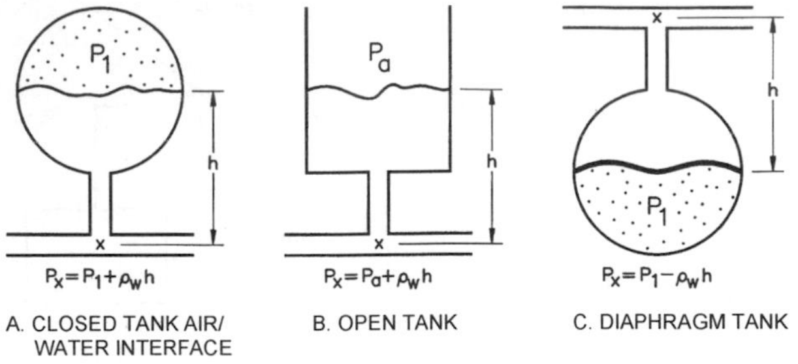

Fig. 10 Tank Pressure Related to "System" Pressure

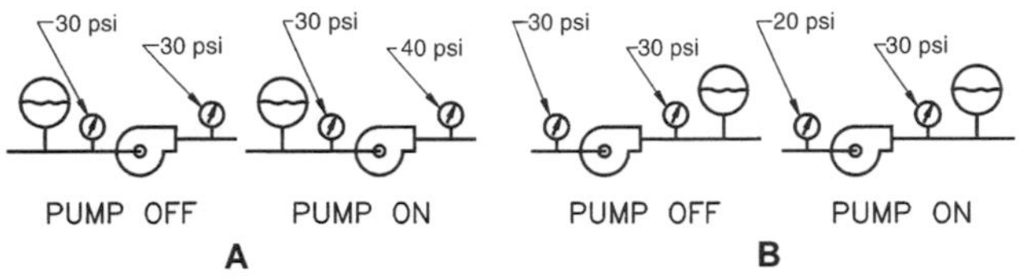

Fig. 11 Effect of Expansion Tank Location with Respect to Pump Pressure

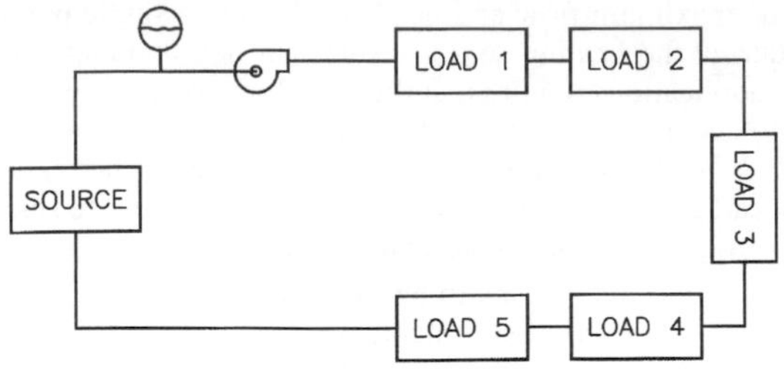

Fig. 12 Flow Diagram of Simple Series Circuit

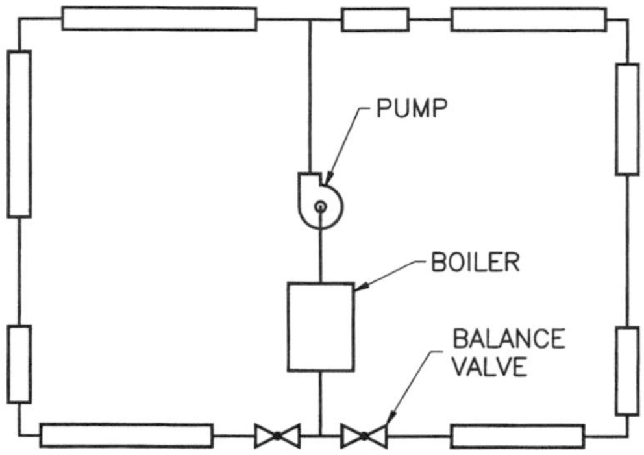

Fig. 13 Series Loop System

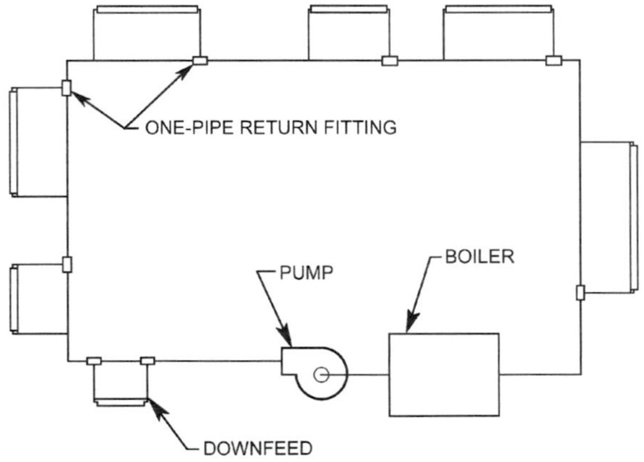

Fig. 14 One-Pipe Diverting Tee System

the tank connection will not change as the pump is turned on or off. For example, consider a system containing an expansion tank at 30 psig and a pump with a pump head of 23.1 ft (10 psig). Figure 11 shows alternative locations for connecting the expansion tank; in either case, with the pump off, the pressure will be 30 psig on both the pump suction and discharge. With the tank on the pump suction side, when the pump is turned on, the pressure increases on the discharge side by an amount equal to the pump pressure (Figure 11A). With the tank connected on the discharge side of the pump, the pressure decreases on the suction side by the same amount (Figure 11B).

Other considerations relating to the tank connection include the following:

- A tank open to the atmosphere must be located above the highest point in the system.
- A tank with an air/water interface is generally used with an air control system that continually revents the air into the tank. For this reason, it should be connected at a point where air can best be released.
- Within reason, the lower the pressure in a tank, the smaller is the tank [see Equations (12) and (14)]. Thus, in a vertical system, the higher the tank is placed, the smaller it can be.

DESIGN CONSIDERATIONS

PIPING CIRCUITS

Hydronic systems are designed with many different configurations of piping circuits. In addition to simple preference by the design engineer, the method of arranging the circuiting can be dictated by such factors as the shape or configuration of the building, the economics of installation, energy economics, the nature of the load, part-load capabilities or requirements, and others.

Each piping system is a network; the more extensive the network, the more complex it is to understand, analyze, or control. Thus, a major design objective is to maintain the highest degree of simplicity.

Load distribution circuits are of four general types:

- Full series
- Diverting series
- Parallel direct return
- Parallel reverse return

Series Circuit. A simple series circuit is shown in Figure 12. Series loads generally have the advantage of both lower piping costs and higher temperature drops that result in smaller pipe size and lower energy consumption. A disadvantage is that the different circuits cannot be controlled separately. Simple series circuits are generally limited to residential and small commercial standing radiation systems. Figure 13 shows a typical layout of such a system with two zones for residential or small commercial heating.

Diverting Series. The simplest diverting series circuit diverts some of the flow from the main piping circuit through a special diverting tee to a load device (usually standing radiation) that has a low pressure drop. This system is generally limited to heating systems in residential or small commercial applications.

Figure 14 illustrates a typical one-pipe diverting tee circuit. For each terminal unit, a supply and a return tee are installed on the main. One of the two tees is a special diverting tee that creates a pressure drop in the main flow to divert part of the flow to the unit. One (return) diverting tee is usually sufficient for upfeed (units above the main) systems. Two special fittings (supply and return

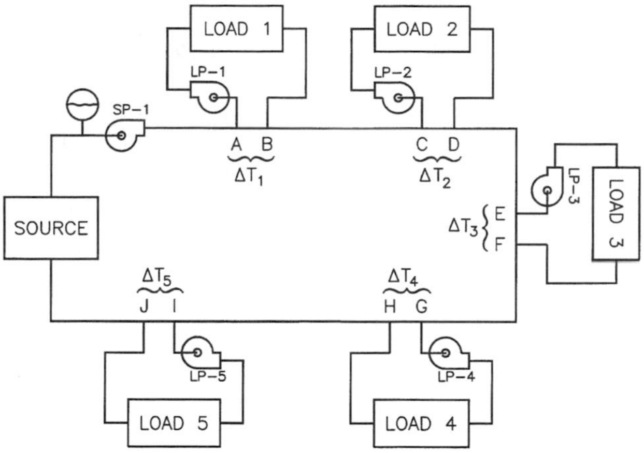

Fig. 15 Series Circuit with Load Pumps

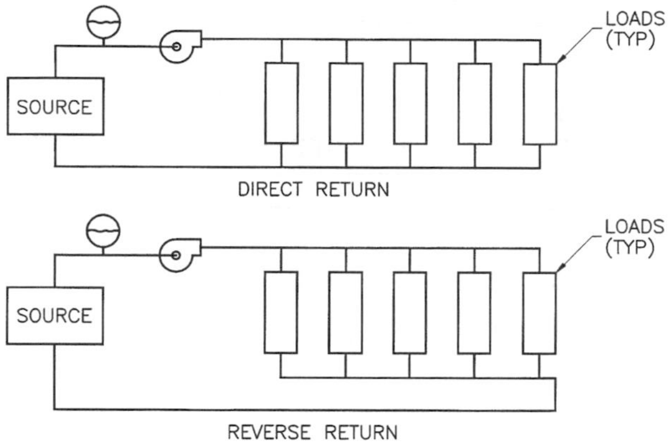

Fig. 16 Direct- and Reverse-Return Two-Pipe Systems

tees) are usually required to overcome thermal pressure in down-feed units. Special tees are proprietary; consult manufacturer's literature for flow rates and pressure drop data on these devices. Unit selection can be only approximate without these data.

One-pipe diverting series circuits allow manual or automatic control of flow to individual heating units. On-off rather than flow modulation control is preferred because of the relatively low pressure drop allowable through the control valve in the diverted flow circuit. This system is likely to cost more than the series loop because extra branch pipe and fittings, including special tees, are required. Each unit usually requires a manual air vent because of the low water velocity through the unit. The length and load imposed on a one-pipe circuit are usually small because of these limitations.

Because only a fraction of the main flow is diverted in a one-pipe circuit, the flow rate and pressure drop are less variable as water flow to the load is controlled than in some other circuits. When two or more one-pipe circuits are connected to the same two-pipe mains, the circuit flow may need to be mechanically balanced. After balancing, sufficient flow must be maintained in each one-pipe circuit to ensure adequate flow diversion to the loads.

When coupled with compound pumping systems, series circuits can be applied to multiple control zones on larger commercial or institutional systems (Figure 15). Note that in the series circuit with compound pumping, the load pumps need not be equal in capacity to the system pump. If, for example, load pump LP-1 circulates less flow (Q_{LP1}) than system pump SP-1 (Q_{SP1}), the temperature difference across Load 1 would be greater than the circuit temperature difference between A and B (i.e., water would flow in the common pipe from A to B). If, on the other hand, the load pump LP-2 is equal in flow capacity to the system pump SP-1, the temperature differentials across Load 2 and across the system from C to D would be equal and no water would flow in the common pipe. If Q_{LP3} exceeds Q_{SP1}, mixing occurs at Point E and, in a heating system, the temperature entering pump LP-3 would be lower than that available from the system leaving load connection D.

Thus, a series circuit using compound or load pumps offers many design options. Each of the loads shown in Figure 15 could also be a complete piping circuit or network.

Parallel Piping. These networks are the most commonly used in hydronic systems because they allow the same temperature water to be available to all loads. The two types of parallel networks are direct return and reverse return (Figure 16).

In the direct-return system, the length of supply and return piping through the subcircuits is unequal, which may cause unbalanced flow rates and require careful balancing to provide each subcircuit with design flow. Ideally, the reverse-return system provides nearly equal total lengths for all terminal circuits.

Direct-return piping has been successfully applied where the designer has guarded against major flow unbalance by

1. Providing for pressure drops in the subcircuits or terminals that are significant percentages of the total, usually establishing pressure drops for close subcircuits at higher values than those for the far subcircuits
2. Minimizing distribution piping pressure drop (In the limit, if the distribution piping loss is zero and the loads are of equal flow resistance, the system is inherently balanced.)
3. Including balancing devices and some means of measuring flow at each terminal or branch circuit
4. Using control valves with a high head loss at the terminals

CAPACITY CONTROL OF LOAD SYSTEM

The two alternatives for controlling the capacity of hydronic systems are on-off control and variable-capacity or modulating control. The on-off option is generally limited to smaller systems (e.g., residential or small commercial) and individual components of larger systems. In smaller systems where the entire building is a single zone, control is accomplished by cycling the source device (the boiler or chiller) on and off. Usually a space thermostat allows the chiller or boiler to run, then a water temperature thermostat (aquastat) controls the capacity of the chiller(s) or boiler(s) as a function of supply or return water temperature. The pump can be either cycled with the load device (usually the case in a residential heating system) or left running (usually done in commercial hot or chilled water systems).

In these single-zone applications, the piping design requires no special consideration for control. Where multiple zones of control are required, the various load devices are controlled first; then the source system capacity is controlled to follow the capacity requirement of the loads.

Control valves are commonly used to control loads. These valves control the capacity of each load by varying the amount of water flow through the load device when load pumps are not used. Control valves for hydronic systems are straight-through (two-way) valves and three-way valves (Figure 17). The effect of either valve is to vary the amount of water flowing through the load device.

With a two-way valve (Figure 17A), as the valve strokes from full-open to full-closed, the quantity of water flowing through the load gradually decreases from design flow to no flow. With a three-way mixing valve (Figure 17B) in one position, the valve is open from Port A to AB, with Port B closed off. In that position, all the flow is through the load. As the valve moves from the A-AB position to the B-AB position, some of the water bypasses the load by

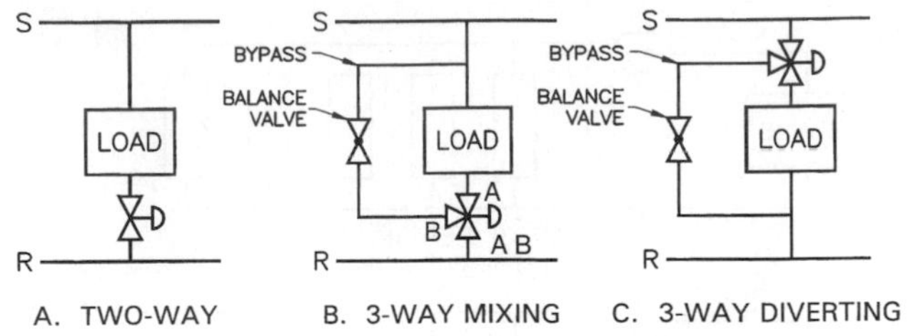

Fig. 17 Load Control Valves

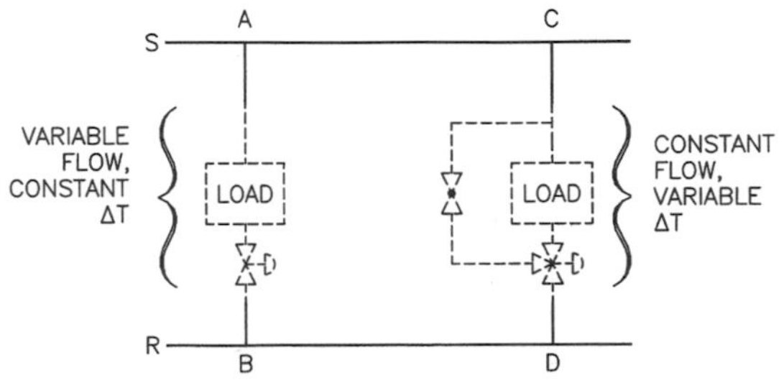

Fig. 18 System Flow with Two-Way and Three-Way Valves

flowing through the bypass line, thus decreasing flow through the load. At the end of the stroke, Port A is closed, and all of the fluid will flow from B to AB with no flow through the load. Thus, the three-way mixing valve has the same effect on the load as the two-way valve—as the load reduces, the quantity of water flowing through the load decreases.

The effect on load control with the three-way diverting valve (Figure 17C) is the same as with the mixing valve in a closed system—the flow is either directed through the load or through the bypass in proportion to the load. Because of the dynamics of valve operation, diverting valves are more complex in design and are thus more expensive than mixing valves; because they accomplish the same function as the simpler mixing valve, they are seldom used in closed hydronic systems.

In terms of load control, a two-way valve and a three-way valve perform identical functions—they both vary the flow through the load as the load changes. The fundamental difference between the two-way valve and the three-way valve is that as the source or distribution system sees the load, the two-way valve provides a variable flow load response and the three-way valve provides a constant flow load response.

According to Equation (9), the load q is proportional to the product of Q and Δt. Ideally, as the load changes, Q changes, while Δt remains fixed. However, as the system sees it, as the load changes with the two-way valve, Q varies and Δt is fixed, whereas with a three-way valve, Δt varies and Q is fixed. This principle is illustrated in Figure 18. An understanding of this concept is fundamental to the design or analysis of hydronic systems.

The flow characteristics of two-way and three-way valve ports are described in Chapter 37 of the 1997 *ASHRAE Handbook—Fundamentals* and must be understood. The equal percentage characteristic is recommended for proportional control of the load flow for two-way and three-way valves; the bypass flow port of three-way valves should have the linear characteristic to maintain a uniform flow during part-load operation.

SIZING CONTROL VALVES

For stable control, the pressure drop in the control valve at the full-open position should be no less than one-half the pressure drop in the branch. For example, in Figure 18, the pressure drop at full-open position for the two-way valve should equal one-half the

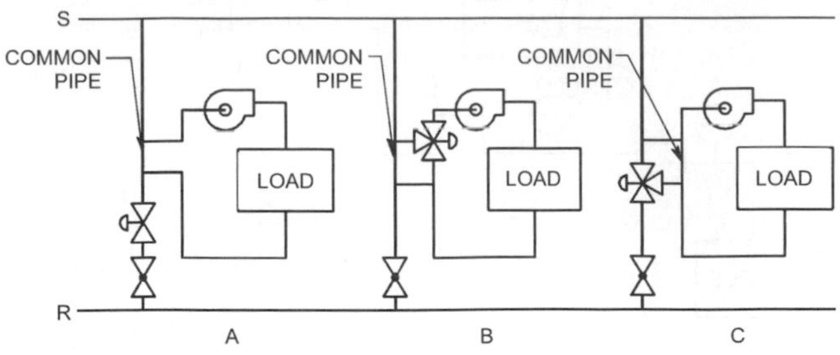

Fig. 19 Load Pumps with Valve Control

pressure drop from A to B, and for the three-way valve, the full-open pressure drop should be half that from C to D. The pressure drop in the bypass balancing valve in the three-way valve circuit should be set to equal that in the coil (load).

Control valves should be sized on the basis of the valve coefficient C_v. For more information, see the section on Control Valve Sizing under Automatic Valves in Chapter 42.

If a system is to be designed with multiple zones of control such that load response is to be by constant flow through the load and variable Δt, control cannot be achieved by valve control alone; a load pump is required.

Several control arrangements of load pump and control valve configurations are shown in Figure 19. Note that in all three configurations the common pipe has no restriction or check valve. In all configurations there is no difference in control as seen by the load. However, the basic differences in control are

1. With the two-way valve configuration (Figure 19A), the distribution system sees a variable flow and a constant Δt, whereas with both three-way configurations, the distribution system sees a constant flow and a variable Δt.
2. Configuration B differs from C in that the pressure required through the three-way valve in Figure 19B is provided by the load pump, while in Figure 19C it is provided by the distribution pump(s).

LOW-TEMPERATURE HEATING SYSTEMS

These systems are used for heating spaces or processes directly, as with standing radiation and process heat exchangers, or indirectly, through air-handling unit coils for preheating, for reheating, or in hot water unit heaters. These systems are generally designed with supply water temperatures from 180 to 240°F and temperature drops from 20 to 100°F.

In the United States, hot water heating systems were historically designed for a 200°F supply water temperature and a 20°F temperature drop. This practice evolved from earlier gravity system designs and provides convenient design relationships for heat transfer coefficients related to coil tubing and finned-tube radiation and for calculations (one gallon per minute conveys 10,000 Btu/h at 20°F Δt). Because many terminal devices still require these flow rates, it is important to recognize this relationship in selecting devices and designing systems.

However, the greater the temperature range (and related lower flow rate) that can be applied, the less costly the system is to install and operate. A lower flow rate requires smaller and less expensive piping, less secondary building space, and smaller pumps. Also, smaller pumps require less energy, so operating costs are lower.

Nonresidential Heating Systems

Possible approaches to enhancing the economics of large heating systems include (1) higher supply temperatures, (2) primary-secondary pumping, and (3) terminal equipment designed for

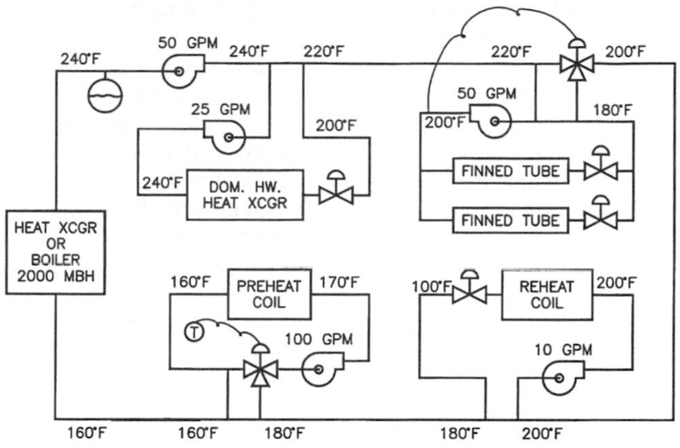

Fig. 20 Example of Series-Connected Loading

smaller flow rates. The three techniques may be used either singly or in combination.

Using higher supply water temperatures achieves higher temperature drops and smaller flow rates. Terminal units with a reduced heating surface can be used. These smaller terminals are not necessarily less expensive, however, because their required operating temperatures and pressures may increase manufacturing costs and the problems of pressurization, corrosion, expansion, and control. System components may not increase in cost uniformly with temperature, but rather in steps conforming to the three major temperature classifications. Within each classification, the most economical design uses the highest temperature in that classification.

Primary-secondary or compound pumping reduces the size and cost of the distribution system and also may use larger flows and lower temperatures in the terminal or secondary circuits. A primary pump circulates water in the primary distribution system while one or more secondary pumps circulate the terminal circuits. The connection between primary and secondary circuits provides complete hydraulic isolation of both circuits and permits a controlled interchange of water between the two. Thus, a high supply water temperature can be used in the primary circuit at a low flow rate and high temperature drop, while a lower temperature and conventional temperature drop can be used in the secondary circuit(s).

For example, a system could be designed with primary-secondary pumping in which the supply temperature from the boiler was 240°F, the supply temperature in the secondary was 200°F, and the return temperature was 180°F. This design results in a conventional 20°F Δt in the secondary zones, but permits the primary circuit to be sized on the basis of a 60°F drop. This primary-secondary pumping arrangement is most advantageous with terminal units such as convectors and finned radiation, which are generally unsuited for small flow rate design.

Many types of terminal heat transfer units are being designed to use smaller flow rates with temperature drops up to 100°F in low-temperature systems and up to 150°F in medium-temperature systems. Fan apparatus, the heat transfer surface used for air heating in fan systems, and water-to-water heat exchangers are most adaptable to such design.

A fourth technique is to put certain loads in series utilizing a combination of control valves and compound pumping (Figure 20). In the system illustrated, the capacity of the boiler or heat exchanger is 2×10^6 Btu/h, and each of the four loads is 0.5×10^6 Btu/h. Under design conditions, the system is designed for an 80°F water temperature drop, and the loads each provide 20°F of the total Δt. The loads in these systems, as well as the smaller or simpler systems in residential or commercial applications, can be connected in a direct-return or a reverse-return piping system. The different features of each load are as follows:

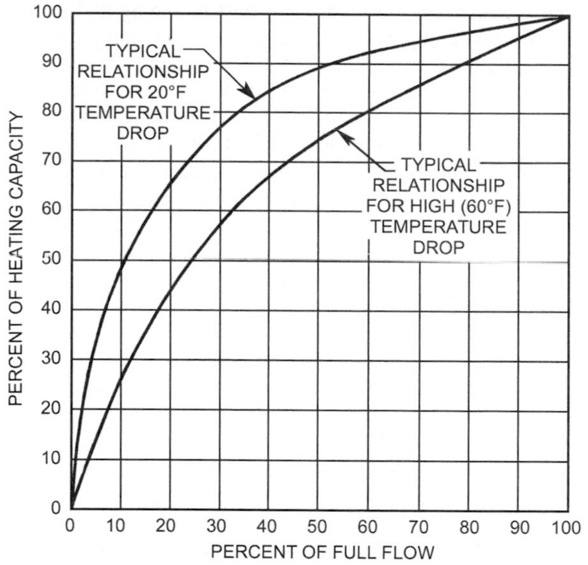

Fig. 21 Heat Emission Versus Flow Characteristic of Typical Hot Water Heating Coil

1. The domestic hot water heat exchanger has a two-way valve and is thus arranged for variable flow (while the main distribution circuit provides constant flow for the boiler circuit).
2. The finned-tube radiation circuit is a 20°F Δt circuit with the design entering water temperature reduced to and controlled at 200°F.
3. The reheat coil circuit takes a 100°F temperature drop for a very low flow rate.
4. The preheat coil circuit provides constant flow through the coil to keep it from freezing.

When loads such as water-to-air heating coils in LTW systems are valve controlled (flow varies), they have a heating characteristic of flow versus capacity as shown in Figure 21 for 20°F and 60°F temperature drops. For a 20°F Δt coil, 50% flow provides approximately 90% capacity; valve control will tend to be unstable. For this reason, proportional temperature control is required, and equal percentage characteristic two-way valves should be selected such that 10% flow is achieved with 50% valve lift. This combination of the valve characteristic and the heat transfer characteristic of the coil makes the control linear with respect to the control signal. This type of control can be obtained only with equal percentage two-way valves and can be further enhanced if piped with a secondary pump arrangement as shown in Figure 19A. See Chapter 45 of the 1999 *ASHRAE Handbook—Applications* for further information on automatic controls.

CHILLED WATER SYSTEMS

Designers have less latitude in selecting supply water temperatures for cooling applications because there is only a narrow range of water temperatures low enough to provide adequate dehumidification and high enough to avoid chiller freeze-up. Circulated water quantities can be reduced by selecting proper air quantities and heat transfer surface at the terminals. Terminals suited for a 12°F rise rather than an 8°F rise reduce circulated water quantity and pump power by one-third and increase chiller efficiency.

A proposed system should be evaluated for the desired balance between installation cost and operating cost. Table 1 shows the effect of coil circuiting and chilled water temperature on water flow and temperature rise. The coil rows, fin spacing, air-side performance, and cost are identical for all selections. Morabito (1960) showed how such changes in coil circuiting affect the overall system. Considering the investment cost of piping and insulation versus the operating cost of refrigeration and pumping motors, higher temperature rises, (i.e., 16 to 24°F temperature rise at about 1.0 to 1.5 gpm per ton of cooling) can be applied on chilled water systems with long distribution piping runs; larger flow rates should be used only where reasonable in close-coupled systems.

For the most economical design, the minimum flow rate to each terminal heat exchanger is calculated. For example, if one terminal can be designed for an 18°F rise, another for 14°F, and others for 12°F, the highest rise to each terminal should be used, rather than designing the system for an overall temperature rise based on the smallest capability.

The control system selected also influences the design water flow. For systems using multiple terminal units, diversity factors can be applied to flow quantities before sizing pump and piping mains if exposure or use prevents the unit design loads from occurring simultaneously and if two-way valves are used for water flow control. If air-side control (e.g, face-and-bypass or fan cycling) or three-way valves on the water side areused, diversity should not be

Table 1 Chilled Water Coil Performance

Coil Circuiting	Chilled Water Inlet Temp., °F	Coil Pressure Drop, psi	Chilled Water Flow, gpm/ton	Chilled Water Temp. Rise, °F
Full[a]	45	1.0	2.2	10.9
Half[b]	45	5.5	1.7	14.9
Full[a]	40	0.5	1.4	17.1
Half[b]	40	2.5	1.1	21.8

Note: Table is based on cooling air from 81°F dry bulb, 67°F wet bulb to 58°F dry bulb, 56°F wet bulb.

[a] Full circuiting (also called single circuit). Water at the inlet temperature flows simultaneously through all tubes in a plane transverse to airflow; it then flows simultaneously through all tubes, in unison, in successive planes (i.e., rows) of the coil.

[b] Half circuiting. Tube connections are arranged so there are half as many circuits as there are tubes in each plane (row) thereby using higher water velocities through the tubes. This circuiting is used with small water quantities.

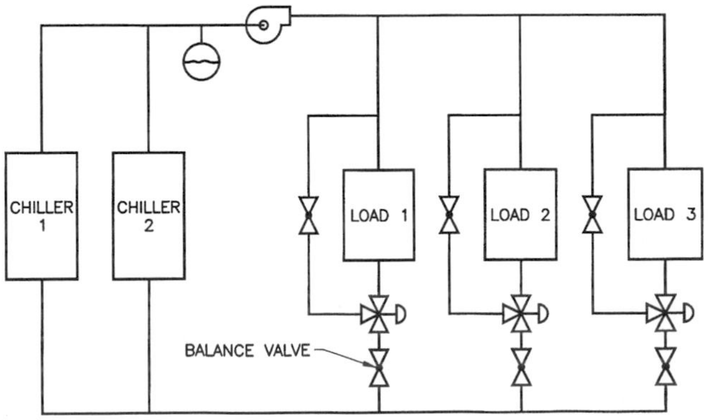

Fig. 22 Constant Flow Chilled Water System

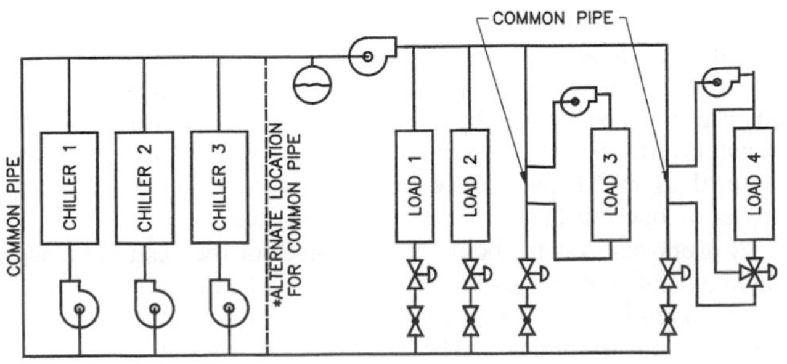

Fig. 23 Variable Flow Chilled Water System

a consideration in pump and piping design, although it should be considered in the chiller selection.

A primary consideration with chilled water system design is the control of the source systems at reduced loads. The constraints on the temperature parameters are (1) a water freezing temperature of 32°F, (2) economics of the refrigeration system in generating chilled water, and (3) the dew-point temperature of the air at nominal indoor comfort conditions (55°F dew point at 75°F and 50% rh). These parameters have led to the common practice of designing for a supply chilled water temperature of 44 to 45°F and a return water temperature between 55 and 64°F.

Historically, most chilled water systems have used three-way control valves to achieve constant water flow through the chillers. However, as systems have become larger, as designers have turned to multiple chillers for reliability and controllability, and as energy economics have become an increasing concern, the use of two-way valves and source pumps for the chillers has greatly increased.

A typical configuration of a small chilled water system using two parallel chillers and loads with three-way valves is illustrated in Figure 22. Note that the flow is essentially constant. A simple energy balance [Equation (9)] dictates that with a constant flow rate, at one-half of design load, the water temperature differential drops to one-half of design. At this load, if one of the chillers is turned off, the return water circulating through the off chiller mixes with the supply water. This mixing raises the temperature of the supply chilled water and can cause a loss of control if the designer does not consider this operating mode.

A typical configuration of a large chilled water system with multiple chillers and loads and compound piping is shown in Figure 23. This system provides variable flow, essentially constant supply temperature chilled water, multiple chillers, more stable two-way control valves, and the advantage of adding chilled water storage with little additional complexity.

One design issue illustrated in Figure 23 is the placement of the common pipe for the chillers. With the common pipe as shown, the chillers will unload from left to right. With the common pipe in the alternate location shown, the chillers will unload equally in proportion to their capacity (i.e., equal percentage).

The **one-pipe chilled water system**, also called the **integrated decentralized chilled water system** is another system that has seen considerable use in campus-type chilled water systems with multiple chillers and multiple buildings (Coad 1976). A single pumped main circulates water in a closed loop through all the connected buildings. Each of the loads and/or chillers is connected to the loop, with the chillers usually downstream from a load connection. The loop capacity is limited only by the fact that the flow capacity for any single load or chiller connection cannot exceed the flow rate of the loop. Because the loads are in series, the cooling coils must be sized for higher entering water temperatures than are normally used.

DUAL-TEMPERATURE SYSTEMS

Dual-temperature systems are used when the same load devices and distribution systems are used for both heating and cooling (e.g., fan-coil units and central station air-handling unit coils). In the design of dual-temperature systems, the cooling cycle design usually dictates the requirements of the load heat exchangers and distribution systems. Dual-temperature systems are basically of three different configurations, each requiring different design techniques:

1. Two-pipe systems
2. Four-pipe common load systems
3. Four-pipe independent load systems

Two-Pipe Systems

In a two-pipe system, the load devices and the distribution system circulate chilled water when cooling is required and hot water when heating is required (Figure 24). Design considerations for these systems include the following:

- Loads must all require cooling or heating coincidentally; that is, if cooling is required for some loads and heating for other loads at a given time, this type of system should not be used.
- When designing the system, the flow and temperature requirements for both the cooling and the heating media must be calculated first. The load and distribution system should be designed for the more stringent, and the water temperatures and temperature differential should be calculated for the other mode.
- The changeover procedure should be designed such that the chiller evaporator is not exposed to damaging high water temperatures and the boiler is not subjected to damaging low water temperatures. To accommodate these limiting requirements, the changeover of a system from one mode to the other requires considerable time. If rapid load swings are anticipated, a two-pipe system should not be selected, although it is the least costly of the three options.

Four-Pipe Common Load Systems

In the four-pipe common load system, load devices are used for both heating and cooling as in the two-pipe system. The four-pipe common load system differs from the two-pipe system in that both heating and cooling are available to each load device, and the changeover from one mode to the other takes place at each individual load device, or grouping of load devices, rather than at the source. Thus, some of the load systems can be in the cooling mode while others are in the heating mode. Figure 25 is a flow diagram of a four-pipe common load system, with multiple loads and a single boiler and chiller.

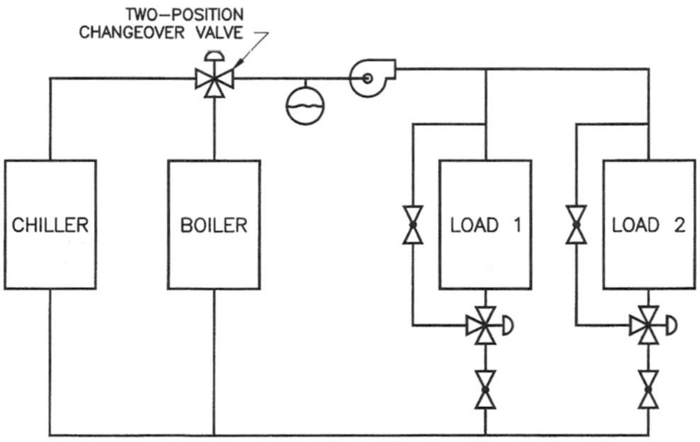

Fig. 24 Simplified Diagram of Two-Pipe System

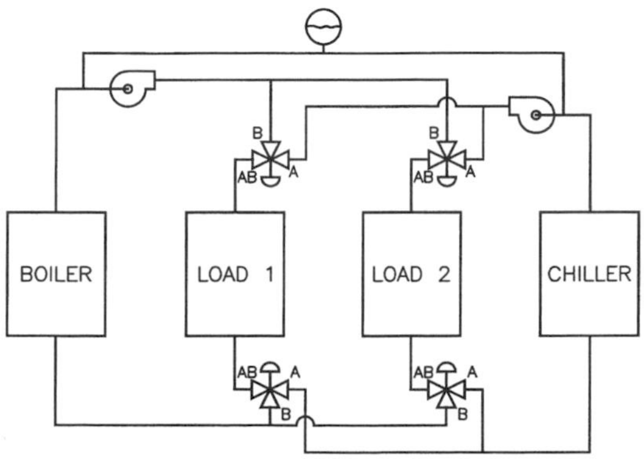

Fig. 25 Four-Pipe Common Load System

Although many of these systems have been installed, many have not performed successfully due to problems in implementing the design concepts.

One problem that must be addressed is the expansion tank connection(s). Many four-pipe systems were designed with two expansion tanks—one for the cooling circuit and one for the heating circuit. However, with multiple loads, these circuits become hydraulically interconnected, thus creating a system with two expansion chambers. The preferred method of handling the expansion tank connection sets the point of reference pressure equal in both circuits (Figure 25).

Another potential problem is the mixing of hot and chilled water. At each load connection, two three-way valves are required—a mixing valve on the inlet and a diverting valve on the outlet. These valves operate in unison in just two positions—opening either Port B to AB or Port A to AB. If, for example, the valve on the outlet does not seat tightly and Load 1 is indexed to cooling and Load 2 is indexed to heating, return heating water from Load 2 will flow into the chilled water circuit, and return chilled water from Load 1 will flow into the heating water circuit. The probability of this occurring increases as the number of loads increases because the number of control valves increases.

Another disadvantage of this system is that the loads have no individual capacity control as far as the water system is concerned. That is, each valve must be positioned to either full heating or full cooling with no control in between.

Because of these disadvantages, four-pipe common load systems should be limited to those applications in which there are no independent load circuits (i.e., radiant ceiling panels or induction unit coils).

Four-Pipe Independent Load Systems

The four-pipe independent load system is preferred for those hydronic applications in which some of the loads are in the heating mode while others are in the cooling mode. Control is simpler and more reliable than for the common load systems, and in many applications, the four-pipe independent load system is less costly to install. Also, the flow through the individual loads can be modulated, providing both the control capability for variable capacity and the opportunity for variable flow in either or both circuits.

A simplified example of a four-pipe independent load system with two loads, one boiler, and two chillers is shown in Figure 26. Note that both hydronic circuits are essentially independent, so that each can be designed with disregard for the other system. Although both circuits in the figure are shown as variable flow distribution systems, they could be constant flow (three-way valves) or one variable flow and one constant flow. Generally, the control modulates the two load valves in sequence with a dead band at the control midpoint.

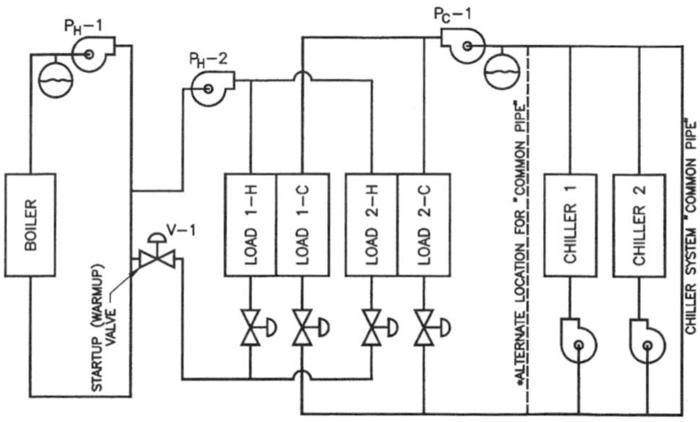

Fig. 26 Four-Pipe Independent Load System

This type of system offers additional flexibility when some selective loads are arranged for heating only or cooling only, such as unit heaters or preheat coils. Then, central station systems can be designed for humidity control with reheat through configuration at the coil locations and with proper control sequences.

OTHER DESIGN CONSIDERATIONS

Makeup and Fill Water Systems

Generally, a hydronic system is filled with water through a valved connection to a domestic water source, with a service valve, a backflow preventer, and a pressure gage. (The domestic water source pressure must exceed the system fill pressure.)

Because the expansion chamber is the reference pressure point in the system, the water makeup point is usually located at or near the expansion chamber.

Many designers prefer to install automatic makeup valves, which consist of a pressure-regulating valve in the makeup line. However, the quantity of water being made up must be monitored to avoid scaling and oxygen corrosion in the system.

Safety Relief Valves

Safety relief valves should be installed at any point at which pressures can be expected to exceed the safe limits of the system components. Causes of excessive pressures include

- Overpressurization from fill system
- Pressure increases due to thermal expansion
- Surges caused by momentum changes (shock or water hammer)

Overpressurization from the fill system could occur due to an accident in filling the system or due to the failure of an automatic fill regulator. To prevent this, a safety relief valve is usually installed at the fill location. Figure 27 shows a typical piping configuration for a system with a plain steel or air/water interface expansion tank. Note that no valves are installed between the hydronic system piping and the safety relief valve. This is a mandatory design requirement if the valve in this location is also to serve as a protection against pressure increases due to thermal expansion.

An expansion chamber is installed in a hydronic system, to allow for the volumetric changes that accompany water temperature changes. However, if any part of the system is configured such that it can be isolated from the expansion tank and its temperature can increase while it is isolated, then overpressure relief should be provided.

The relationship between pressure change due to temperature change and the temperature change in a piping system is expressed by the following equation:

$$\Delta p = \frac{(\beta - 3\alpha)\Delta t}{(5/4)(D/E\Delta r) + \gamma} \tag{17}$$

where

Δp = pressure increase, psi
β = volumetric coefficient of thermal expansion of water, 1/°F
α = linear coefficient of thermal expansion for piping material, 1/°F
Δt = water temperature increase, °F
D = pipe diameter, in.
E = modulus of elasticity of piping material, psi
γ = volumetric compressibility of water, in^2/lb
Δr = thickness of pipe wall, in.

Figure 28 shows a solution to Equation (17) demonstrating the pressure increase caused by any given temperature increase for 1 in. and 10 in. steel piping. If the temperature in a chilled water system with piping spanning sizes between 1 and 10 in. were to increase by 15°F, the pressure would increase between 340 and 420 psi, depending on the average pipe size in the system.

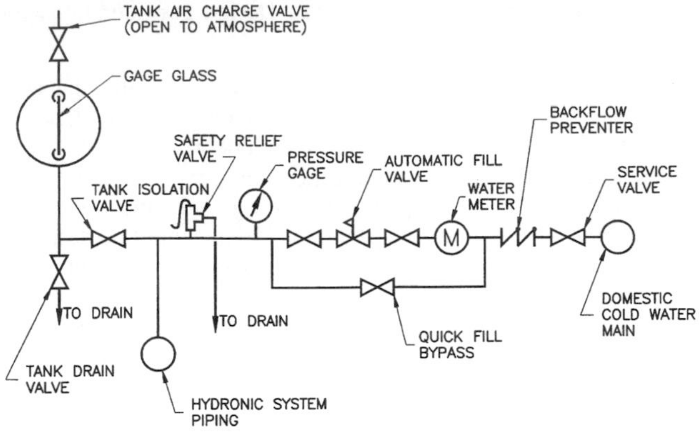

Fig. 27 Typical Makeup Water and Expansion Tank Piping Configuration for Plain Steel Expansion Tank

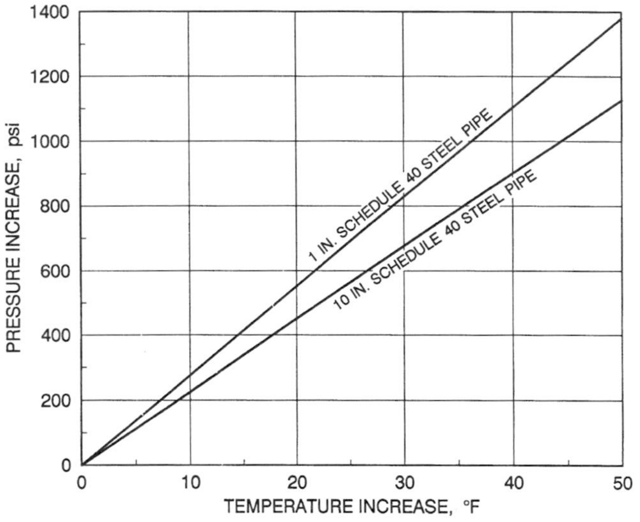

Fig. 28 Pressure Increase Resulting from Thermal Expansion as Function of Temperature Increase

Safety relief should be provided to protect boilers, heat exchangers, cooling coils, chillers, and the entire system when the expansion tank is isolated for air charging or other service. As a minimum, the ASME *Boiler and Pressure Vessel Code* requires that a dedicated safety relief valve be installed on each boiler and that isolating or service valves be provided on the supply and return connections to each boiler.

Potential forces caused by shock waves or water hammer should also be considered in design. Chapter 33 of the 1997 *ASHRAE Handbook—Fundamentals* discusses the causes of shock forces and the methodology for calculating the magnitude of these forces.

Air Elimination

If air and other gases are not eliminated from the flow circuit, they may slow or stop the flow through the terminal heat transfer elements and cause corrosion, noise, reduced pumping capacity, and loss of hydraulic stability (see the section on Principles at the beginning of the chapter). A closed tank without a diaphragm can be installed at the point of the lowest solubility of air in water. When a diaphragm tank is used, air in the system can be removed by an air separator and air elimination valve installed at the point of lowest solubility. Manual vents should be installed at high points to remove all air trapped during initial operation. Shutoff valves should be installed on any automatic air removal device to permit servicing without draining the system.

Drain and Shutoff

All low points should have drains. Separate shutoff and draining of individual equipment and circuits should be possible so that the entire system does not have to be drained to service a particular item. Whenever a device or section of the system is isolated, and the water in that section or device could increase in temperature following isolation, overpressure safety relief protection must be provided.

Balance Fittings

Balance fittings or valves and a means of measuring flow quantity should be applied as needed to permit balancing of individual terminals and subcircuits.

Pitch

Piping need not pitch but can run level, providing that flow velocities exceeding 1.5 fps are maintained or a diaphragm tank is used.

Strainers

Strainers should be used where necessary to protect system elements. Strainers in the pump suction must be checked carefully to avoid cavitation. Large separating chambers can serve as main air venting points and dirt strainers ahead of pumps. Automatic control valves or other devices operating with small clearances require protection from pipe scale, gravel, and welding slag, which may readily pass through the pump and its protective separator. Individual fine mesh strainers may therefore be required ahead of each control valve.

Thermometers

Thermometers or thermometer wells should be installed to assist the system operator in routine operation and troubleshooting. Permanent thermometers, with the correct scale range and separate sockets, should be used at all points where temperature readings are regularly needed. Thermometer wells should be installed where readings will be needed only during start-up and infrequent troubleshooting. If a central monitoring system is provided, a calibration well should be installed adjacent to each sensing point in insulated piping systems.

Flexible Connectors and Expansion Compensation

Flexible connectors are sometimes installed at pumps and machinery to reduce pipe stress. See Chapter 46 of the 1999 *ASHRAE Handbook—Applications* for vibration isolation information. Expansion, flexibility, and hanger and support information is in Chapter 41 of this volume.

Gage Cocks

Gage cocks or quick-disconnect test ports should be installed at points requiring pressure readings. Gages permanently installed in the system will deteriorate because of vibration and pulsation and will, therefore, be unreliable. It is good practice to install gage cocks and provide the operator with several quality gages for diagnostic purposes.

Insulation

Insulation should be applied to minimize pipe thermal loss and to prevent condensation during chilled water operation (see Chapter 22 of the 1997 *ASHRAE Handbook—Fundamentals*). On chilled water systems, special rigid metal sleeves or shields should be installed at all hanger and support points, and all valves should be provided with extended bonnets to allow for the full insulation thickness without interference with the valve operators.

Condensate Drains

Condensate drains from dehumidifying coils should be trapped and piped to an open-sight plumbing drain. Traps should be deep enough to overcome the air pressure differential between drain inlet and room, which ordinarily will not exceed 2 in. of water. Pipe should be noncorrosive and insulated to prevent moisture condensation. Depending on the quantity and temperature of condensate, plumbing drain lines may require insulation to prevent sweating.

Common Pipe

In compound (primary-secondary) pumping systems, the common pipe is used to dynamically decouple the two pumping circuits. Ideally, there is no pressure drop in this section of piping; however, in actual systems, it is recommended that this section of piping be a minimum of 10 diameters in length to reduce the likelihood of unwanted mixing resulting from velocity (kinetic) energy or turbulence.

DESIGN PROCEDURES

Preliminary Equipment Layout

Flows in Mains and Laterals. Regardless of the method used to determine the flow through each item of terminal equipment, the desired result should be listed in terms of mass flow on the preliminary plans or in a schedule of flow rates for the piping system. (In the design of small systems and chilled water systems, the determination may be made in terms of volumetric flow).

In an equipment schedule or on the plans, starting from the most remote terminal and working toward the pump, progressively list the cumulative flow in each of the mains and branch circuits in the distribution system.

Preliminary Pipe Sizing. For each portion of the piping circuit, select a tentative pipe size from the unified flow chart (Figure 1 in Chapter 33 of the 1997 *ASHRAE Handbook—Fundamentals*), using a value of pipe friction loss ranging from 0.75 to 4 ft per 100 ft (approximately 0.1 to 0.5 in/ft).

Residential piping size is often based on pump preselection using pipe sizing tables, which are available from the Hydronics Institute or from manufacturers.

Preliminary Pressure Drop. Using the preliminary pipe sizing indicated above, determine the pressure drop through each portion of the piping. The total pressure drop in the longest circuits determines the maximum pressure drop through the piping, including the terminals and control valves, that must be available in the form of pump pressure.

Preliminary Pump Selection. The preliminary selection should be based on the pump's ability to fulfill the determined capacity requirements. It should be selected at a point left of center on the pump curve and should not overload the motor. Because pressure drop in a flow system varies as the square of the flow rate, the flow variation between the nearest size of stock pump and an exact point selection will be relatively minor.

Final Pipe Sizing and Pressure Drop Determination

Final Piping Layout. Examine the overall piping layout to determine whether pipe sizes in some areas need to be readjusted. Several principal circuits should have approximately equal pressure drops so that excessive pressures are not needed to serve a small portion of the building.

Consider both the initial cost of the pump and piping system and the pump's operating cost when determining final system friction loss. Generally, lower heads and larger piping are more economical when longer amortization periods are considered, especially in larger systems. However, in small systems such as in residences, it may be most economical to select the pump first and design the piping system to meet the available pressure. In all cases, adjust the piping system design and pump selection until the optimum design is found.

Final Pressure Drop. When the final piping layout has been established, determine the friction loss for each section of the piping system from the pressure drop charts (Chapter 33 of the 1997 *ASHRAE Handbook—Fundamentals*) for the mass flow rate in each portion of the piping system.

After calculating the friction loss at design flow for all sections of the piping system and all fittings, terminal units, and control valves, sum them for several of the longest piping circuits to determine the pressure against which the pump must operate at design flow.

Final Pump Selection. After completing the final pressure drop calculations, select the pump by plotting a system curve and pump curve and selecting the pump or pump assembly that operates closest to the calculated design point.

Freeze Prevention

All circulating water systems require precautions to prevent freezing, particularly in makeup air applications in temperate climates (1) where coils are exposed to outdoor air at below-freezing temperatures, (2) where undrained chilled water coils are in the winter airstream, or (3) where piping passes through unheated spaces. Freezing will not occur as long as flow is maintained and the water is at least warm. Unfortunately, during extremely cold weather or in the event of a power failure, water flow and temperature cannot be guaranteed. Additionally, continuous pumping can be energy-intensive and cause system wear. The following are precautions to avoid flow stoppage or damage from freezing:

1. Select all load devices (such as preheat coils) that are subjected to outdoor air temperatures for constant flow, variable Δt control.
2. Position the coil valves of all cooling coils with valve control that are dormant in winter months to the full-open position at those times.
3. If intermittent pump operation is used as an economy measure, use an automatic override to operate both chilled water and heating water pumps in below-freezing weather.
4. Select pump starters that automatically restart after power failure (i.e., maintain-contact control).
5. Select nonoverloading pumps.
6. Instruct operating personnel never to shut down pumps in subfreezing weather.
7. Do not use aquastats, which can stop a pump, in boiler circuits.
8. Avoid sluggish circulation, which may cause air binding or dirt deposit. Properly balance and clean systems. Provide proper air control or means to eliminate air.
9. Install low temperature detection thermostats that have phase change capillaries wound in a serpentine pattern across the leaving face of the upstream coil.

In fan equipment handling outdoor air, take precautions to avoid stratification of air entering the coil. The best methods for proper mixing of indoor and outdoor air are the following:

1. Select dampers for pressure drops adequate to provide stable control of mixing, preferably with dampers installed several equivalent diameters upstream of the air-handling unit.
2. Design intake and approach duct systems to promote natural mixing.
3. Select coils with circuiting to allow parallel flow of air and water.

Freeze-up may still occur with any of these precautions. If an antifreeze solution is not used, water should circulate at all times. Valve-controlled elements should have low-limit thermostats, and sensing elements should be located to ensure accurate air temperature readings. Primary-secondary pumping of coils with three-way valve injection (as in Figures 19B and 19C) is advantageous. Use outdoor reset of water temperature wherever possible.

ANTIFREEZE SOLUTIONS

In systems in danger of freeze-up, water solutions of ethylene glycol and propylene glycol are commonly used. Freeze protection may be needed (1) in snow-melting applications (see Chapter 49 of the 1999 *ASHRAE Handbook—Applications*); (2) in systems subjected to 100% outdoor air, where the methods outlined above may not provide absolute antifreeze protection; (3) in isolated parts or zones of a heating system where intermittent operation or long runs of exposed piping increase the danger of freezing; and (4) in process cooling applications requiring temperatures below 40°F. Although using ethylene glycol or propylene glycol is comparatively expensive and tends to create corrosion problems unless suitable inhibitors are used, it may be the only practical solution in many cases.

Solutions of triethylene glycol, as well as certain other heat transfer fluids, may also be used. However, ethylene glycol and propylene glycol are the most common substances used in hydronic systems because they are less costly and provide the most effective heat transfer.

Effect on Heat Transfer and Flow

Tables 6 through 13 and Figures 9 through 16 in Chapter 20 of the 1997 *ASHRAE Handbook—Fundamentals* show density, specific heat, thermal conductivity, and viscosity of various aqueous solutions of ethylene glycol and propylene glycol. Table 4 and Table 5 of that chapter indicate the freezing points for the two solutions.

System heat transfer rate is affected by relative density and specific heat according to the following equation:

$$q_w = 500Q(\rho/\rho_w)c_p\Delta t \qquad (18)$$

where

q_w = total heat transfer rate, Btu/h
Q = flow rate, gpm
ρ = fluid density, lb/ft^3
ρ_w = density of water at 60°F, lb/ft^3
c_p = specific heat of fluid, Btu/lb·°F
Δt = temperature increase or decrease, °F

Effect on Heat Source or Chiller

Generally, ethylene glycol solutions should not be used directly in a boiler because of the danger of chemical corrosion caused by glycol breakdown on direct heating surfaces. However, properly inhibited glycol solutions can be used in low-temperature water systems directly in the heating boiler if proper operation can be ensured. Automobile antifreeze solutions are not recommended because the silicate inhibitor can cause fouling, pump seal wear, fluid gelation, and reduced heat transfer. The area or zone requiring the antifreeze protection can be isolated with a separate heat exchanger or converter. Glycol solutions are used directly in water chillers in many cases.

Glycol solutions affect the output of a heat exchanger by changing the film coefficient of the surface contacting the solution. This change in film coefficient is caused primarily by viscosity changes. Figure 29 illustrates typical changes in output for two types of heat exchangers, Curve A for a steam-to-liquid converter and Curve B for a refrigerant-to-liquid chiller. The curves are plotted for one set of operating conditions only and reflect the change in ethylene glycol concentration as the only variable. Propylene glycol has a similar effect on heat exchanger output.

Because many other variables, such as liquid velocity, steam or refrigerant loading, temperature difference, and unit construction affect the overall coefficient of a heat exchanger, designers should consult manufacturers' ratings when selecting such equipment. The curves indicate only the magnitude of these output changes.

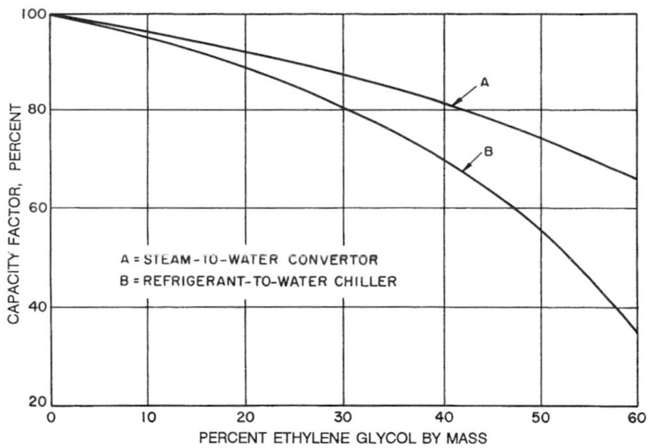

Fig. 29 Example of Effect of Aqueous Ethylene Glycol Solutions on Heat Exchanger Output

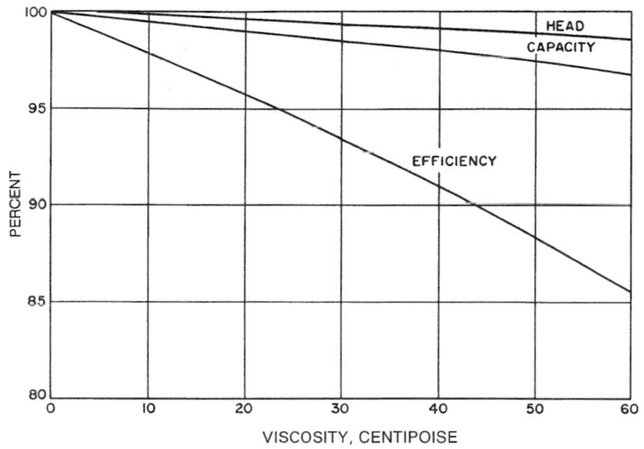

Fig. 30 Effect of Viscosity on Pump Characteristics

Effect on Terminal Units

Because the effect of glycol on the capacity of terminal units may vary widely with temperature, the manufacturer's rating data should be consulted when selecting heating or cooling units in glycol systems.

Effect on Pump Performance

Centrifugal pump characteristics are affected to some degree by glycol solutions because of viscosity changes. Figure 30 shows these effects on pump capacity, head, and efficiency. Figures 12 and 16 in Chapter 20 of the 1997 *ASHRAE Handbook—Fundamentals* plot the viscosity of aqueous ethylene glycol and propylene glycol. Centrifugal pump performance is normally cataloged for water at 60 to 80°F. Hence, absolute viscosity effects below 1.1 centipoise can safely be ignored as far as pump performance is concerned. In intermittently operated systems, such as snow-melting applications, viscosity effects at start-up may decrease flow enough to slow pickup.

Effect on Piping Pressure Loss

The friction loss in piping also varies with viscosity changes. Figure 31 gives correction factors for various ethylene glycol and propylene glycol solutions. These factors are applied to the calculated pressure loss for water [Equation (17)]. No correction is needed for ethylene glycol and propylene glycol solutions above 160°F.

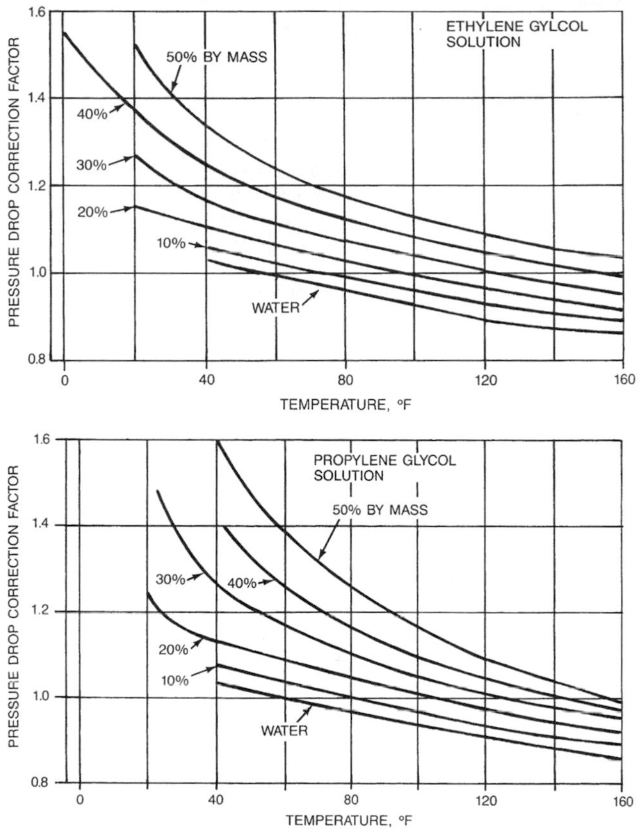

Fig. 31 Pressure Drop Correction for Glycol Solutions

Installation and Maintenance

Because glycol solutions are comparatively expensive, the smallest possible concentrations to produce the desired antifreeze properties should be used. The total water content of the system should be calculated carefully to determine the required amount of glycol (Craig et al. 1993). The solution can be mixed outside the system in drums or barrels and then pumped in. Air vents should be watched during filling to prevent loss of solution. The system and the cold water supply should not be permanently connected, so automatic fill valves are usually not used.

Ethylene glycol and propylene glycol normally include an inhibitor to help prevent corrosion. Solutions should be checked each year using a suitable refractometer to determine glycol concentration. Certain precautions regarding the use of inhibited ethylene glycol solutions should be taken to extend their service life and to preserve equipment:

1. Before injecting the glycol solution, thoroughly clean and flush the system.
2. Use waters that are soft and low in chloride and sulfate ions to prepare the solution whenever possible.
3. Limit the maximum operating temperature to 250°F in a closed hydronic system. In a heat exchanger, limit glycol film temperatures to 300 to 350°F (steam pressures 120 psi or less) to prevent deterioration of the solution.
4. Check the concentration of inhibitor periodically, following procedures recommended by the glycol manufacturer.

REFERENCES

ASME. 1998. *Boiler and Pressure Vessel Codes.* American Society of Mechanical Engineers, New York.

Carlson, G.F. 1981. The design influence of air on hydronic systems. *ASHRAE Transactions* 87(1):1293-1300.
Coad, W.J. 1976. Integrated decentralized chilled water systems. *ASHRAE Transactions* 82(1):566-74.
Coad, W.J. 1980a. Air in hydronic systems. *Heating/Piping/Air Conditioning* (July).
Coad, W.J. 1980b. Expansion tanks. *Heating/Piping/Air Conditioning* (May).
Coad, W.J. 1985. Variable flow in hydronic systems for improved stability, simplicity, and energy economics. *ASHRAE Transactions* 91(1B):224-37.
Craig, N.C., B.W. Jones, and D.L. Fenton. 1993. Glycol concentration requirements for freeze burst protection. *ASHRAE Transactions* 99(2):200-09
Himmelblau, D.M. 1960. Solubilities of inert gases in water. *Journal of Chemical and Engineering Data* 5(1).
Hull, R.F. 1981. Effect of air on hydraulic performance of the HVAC system. *ASHRAE Transactions* 87(1):1301-25.
Lockhart, H.A. and G.F. Carlson. 1953. Compression tank selection for hot water heating systems. *ASHVE Journal* 25(4):132-39. Also in *ASHVE Transactions* 59:55-76.
Morabito, B.P. 1960. How higher cooling coil differentials affect system economics. *ASHRAE Journal* 2(8):60.
Pierce, J.D. 1963. Application of fin tube radiation to modern hot water heating systems. *ASHRAE Journal* 5(2):72.
Pompei, F. 1981. Air in hydronic systems: How Henry's law tells us what happens. *ASHRAE Transactions* 87(1):1326-42.
Stewart, W.E. and C.L. Dona. 1987. Water flow rate limitations. *ASHRAE Transactions* 93(2):811-25.

CHAPTER 13

CONDENSER WATER SYSTEMS

CONDENSER water systems for refrigeration processes are classified as (1) once-through systems, such as city water systems; or (2) recirculating or cooling tower systems.

ONCE-THROUGH CITY WATER SYSTEMS

Figure 1 shows a water-cooled condenser using city water. The return is run higher than the condenser so that the condenser is always full of water. Water flow through the condenser is modulated by a control valve in the supply or discharge line, usually actuated from condenser head pressure to (1) maintain a constant condensing temperature with load variations and (2) close when the refrigeration compressor turns off. City water systems should always include approved backflow prevention devices and open (air gap) drains. When more than one condenser is used on the same circuit, individual control valves are used.

Once-through city water systems are discouraged in most localities because of the waste of city water and the burden on the sewage or wastewater system. Some localities allow their use as a standby or emergency condenser water system for critical refrigeration needs such as for computer rooms, research laboratories, or critical operating room or life support machinery.

Piping materials for these systems are generally nonferrous, usually copper but sometimes high-pressure plastic since corrosion-protective chemicals cannot be used. Scaling can be a problem with higher temperature condensing surfaces in locations in which the water has a relatively high calcium content. In these applications, mechanically cleanable straight tubes should be used.

Piping should be sized according to the principles outlined in Chapter 33 of the 1997 *ASHRAE Handbook—Fundamentals*, with velocities of 5 to 10 fps for design flow rates. A pump is not required where city water is used. **Well water** can be used in lieu of city water, connected on the service side of the pumping/pressure control system. Since most well water has high calcium content, scaling on the condenser surfaces can be a problem.

OPEN COOLING TOWER SYSTEMS

Open systems are those in which there are at least two points of interface between the system water and the atmosphere; they require a different approach to hydraulic design, pump selection, and sizing than do closed hot water and chilled water systems. Some heat conservation systems rely on a split condenser heating system that includes a two-section condenser. One section of the condenser supplies heat for closed-circuit heating or reheat systems; the other section serves as a heat rejection circuit, which is an open system connected to a cooling tower.

In selecting a pump for a cooling tower/condenser water system, consideration must be given to the static head and the system friction loss. The pump inlet must have an adequate net positive suction head (see Chapter 39). In addition, continuous contact with

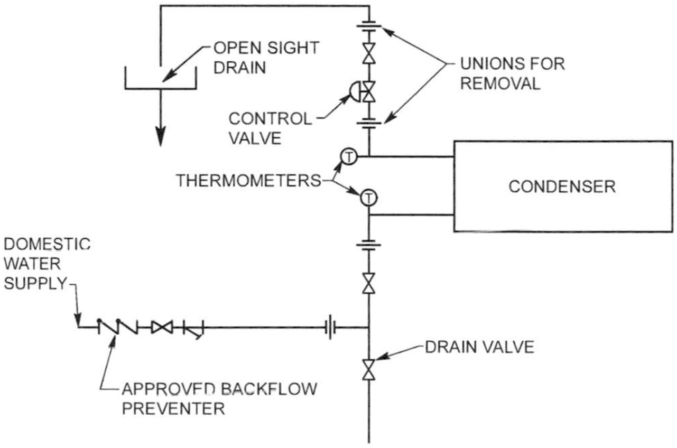

Fig. 1 Condenser Connections for Once-Through City Water System

air introduces oxygen into the water and concentrates minerals that can cause scale and corrosion on a continuing basis. Fouling factors and an increased pressure drop caused by aging of the piping must be taken into account in the condenser piping system design (see Chapter 33 of the 1997 *ASHRAE Handbook—Fundamentals*).

The required water flow rate depends on the refrigeration unit used and on the temperature of the available condenser water. Cooling tower water is available for return to the condenser at a temperature several degrees above the design wet-bulb temperature, depending on tower performance. An approach of 7°F to the design wet-bulb temperature is frequently considered an economically sound design. In city, lake, river, or well water systems, the maximum water temperature that occurs during the operating season must be used for equipment selection and design flow rates and temperature ranges.

The required flow rate through a condenser may be determined with manufacturers' performance data for various condensing temperatures and capacities. With air conditioning refrigeration applications, a return or leaving condenser water temperature of 95°F is considered standard practice. If economic feasibility analyses can justify it, higher leaving water temperatures may be used.

Figure 2 shows a typical cooling tower system for a refrigerant condenser. Water flows to the pump from the tower basin or sump and is discharged under pressure to the condenser and then back to the tower. When it is desirable to control condenser water temperature or maintain it above a predetermined minimum, water is diverted through a control valve directly back to the tower basin.

Piping from the tower sump to the pump requires some precautions. The sump level should be above the top of the pump casing for positive prime, and piping pressure drop should be such that there is always adequate net positive suction head on the pump. All piping must pitch up to the tower basin, if possible, to eliminate air pockets.

The preparation of this chapter is assigned to TC 6.1, Hydronic and Steam Equipment and Systems.

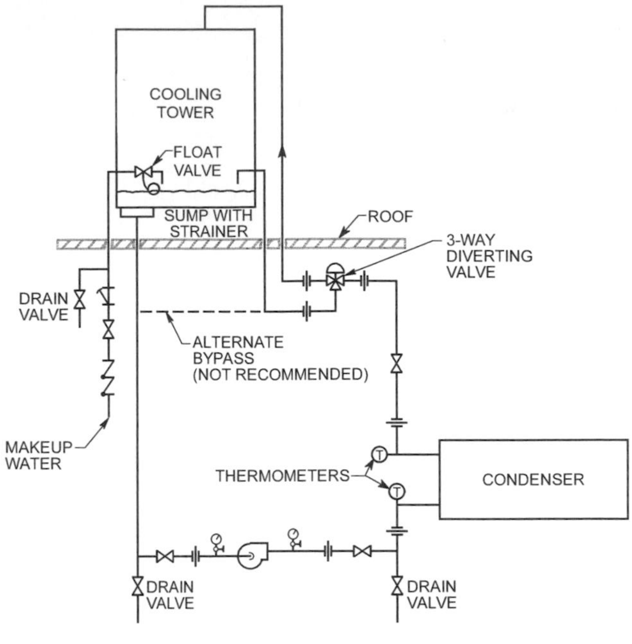

Fig. 2 Cooling Tower Piping System

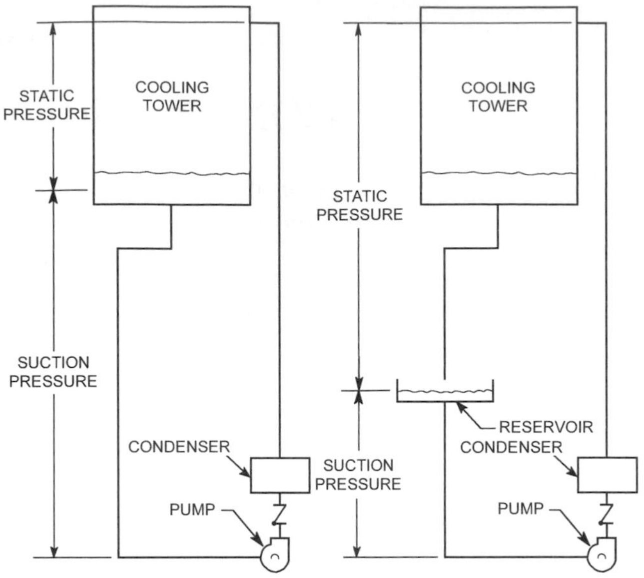

Fig. 3 Schematic Piping Layout Showing Static and Suction Head

If used, suction strainers should be equipped with inlet and outlet gages to indicate when cleaning is required. In-line pipe strainers are not recommended for cooling tower systems because they tend to become blocked and turn into a reliability problem in themselves. Many designers depend on large mesh screens in the tower sump and condenser heads designed with settling volumes to remove particulate matter. If a strainer is deemed necessary, two-large capacity basket strainers, installed in parallel such that they can be alternately put into service and valved out for cleaning, are recommended.

Air and Vapor Precautions

Both vapor and air can create serious problems in open cooling tower systems. Water vaporizes in the pump impeller if adequate net positive suction head is not available. When this occurs, the pump loses capacity, and serious damage to the impeller can result. Equally damaging vaporization can occur in other portions of the system where the pressure in the pipe can drop below the vapor pressure at the temperature of the water. On shutdown, these very low pressures can result from a combination of static pressure and momentum. Vaporization is often followed by an implosion, which causes destructive water hammer. To avoid this problem, all sections of the piping system except the return line to the upper tower basin should be kept below the basin level. When this cannot be achieved, a thorough dynamic analysis of the piping system must be performed for all operating conditions, and a soft start and stop control such as a variable-frequency drive on the pump motor is recommended as an additional precaution.

Air release is another characteristic of open condenser water systems that must be addressed. Since the water/air solution in the tower basin is saturated at atmospheric pressure and cold water basin temperature, the system should be designed to maintain the pressure at all points in the system sufficiently above atmospheric that no air will be released in the condenser or in the piping system (see Figure 2 and Figure 3 in Chapter 12).

Another cause of air in the piping system is "vortexing" at the tower basin outlet. This can be avoided by ensuring that the maximum flow does not exceed that recommended by the tower manufacturer. The release of air in condenser water systems is the major cause of corrosion, and it causes decreased pump flow (similar to cavitation), water flow restrictions in some piping sections, and possible water hammer.

Piping Practice

The elements of required pump head are illustrated in Figure 3. Because there is an equal head of water between the level in the tower sump or interior reservoir and the pump on both the suction and discharge sides, these static heads cancel each other and can be disregarded.

The elements of pump head are (1) static head from tower sump or interior reservoir level to the tower header, (2) friction loss in suction and discharge piping, (3) pressure loss in the condenser, (4) pressure loss in the control valves, (5) pressure loss in the strainer, and (6) pressure loss in the tower nozzles, if used. Added together, these elements determine the required pump total dynamic head.

Normally, piping is sized for water velocities between 5 and 12 fps. Refer to Chapter 33 of the 1997 *ASHRAE Handbook—Fundamentals*, for piping system pressure losses. Friction factors for 15-year-old pipe are commonly used. Manufacturers' data contain pressure drops for the condenser, cooling tower, control valves, and strainers.

If multiple cooling towers are to be connected, the piping should be designed so that the pressure loss from the tower to the pump suction is *exactly* equal for each tower. Additionally, large equalizing lines or a common reservoir can be used to assure the same water level in each tower. However, for reliability and ease of maintenance, multiple basins are often preferred.

Evaporation in a cooling tower concentrates the dissolved solids in the circulating water. This concentration can be limited by discharging a portion of the water as overflow or blowdown.

Makeup water is required to replace the water lost by evaporation, blowdown, and drift. Automatic float valves or level controllers are usually installed to maintain a constant water level.

Water Treatment

Water treatment is necessary to prevent scaling, corrosion, and biological fouling of the condenser and circulating system. The extent and nature of the treatment depends on the chemistry of the available water and on the system design characteristics. On large

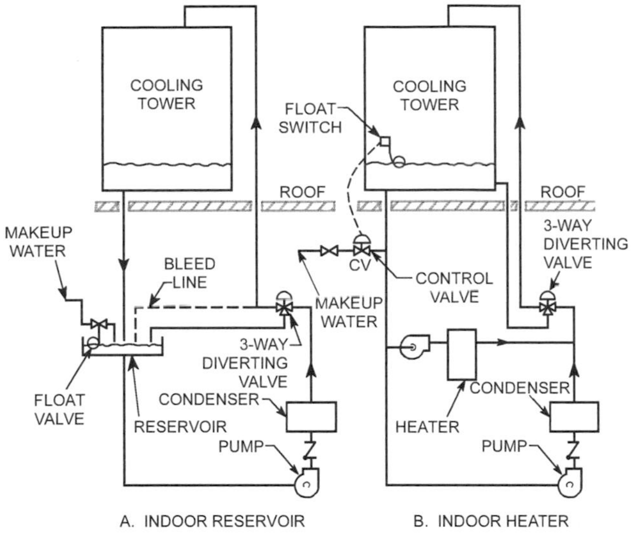

Fig. 4 Cooling Tower Piping to Avoid Freeze-Up

systems, fixed continuous-feeding chemical treatment systems are frequently installed in which chemicals, including acids for pH control, must be diluted and blended and then pumped into the condenser water system. Corrosion-resistant materials may be required for surfaces that come in contact with these chemicals. In the design of the piping system, provisions for feeding the chemicals, blowdowns, drains, and testing must be included. For further information on water treatment, refer to Chapter 36 of this volume and Chapter 47 of the 1999 *ASHRAE Handbook—Applications.*

Freeze Protection

Outdoor piping must be protected or drained when a tower operates intermittently during cold weather. The most satisfactory arrangement is to provide an indoor receiving tank into which the cold water basin drains by gravity as shown in Figure 4A. The makeup, overflow, and pump suction lines are then connected to the indoor reservoir tank rather than to the tower basin.

A control sequence for the piping arrangement of Figure 4A with rising water temperature would be as follows: A temperature sensor measures the temperature of the water leaving the indoor sump. As the temperature starts to rise, the diverting valve begins directing some of the water over the tower. After the water is in full flow over the tower, and the temperature continues to rise above set point, the fan is started on low speed, with the speed increasing until the water temperature reaches set point. With falling water temperature, the opposite sequence occurs.

Tower basin heaters or heat exchangers connected across the supply and return line to the tower can also be selected. Steam or electric basin heaters are most commonly used. An arrangement incorporating an indoor heater is shown in Figure 4B.

LOW-TEMPERATURE (WATER ECONOMIZER) SYSTEMS

When open cooling tower systems are used for generating chilled water directly, through an indirect heat exchanger such as a plate frame heat exchanger, or with a chiller thermocycle circuit, the bulk water temperature is such that there can be icing on or within the cooling tower, with destructive effects. The piping circuit precautions are similar to those described in the section on Open Cooling Tower Systems, but they are much more critical. For precautions in protecting cooling towers from icing damage, see Chapter 36.

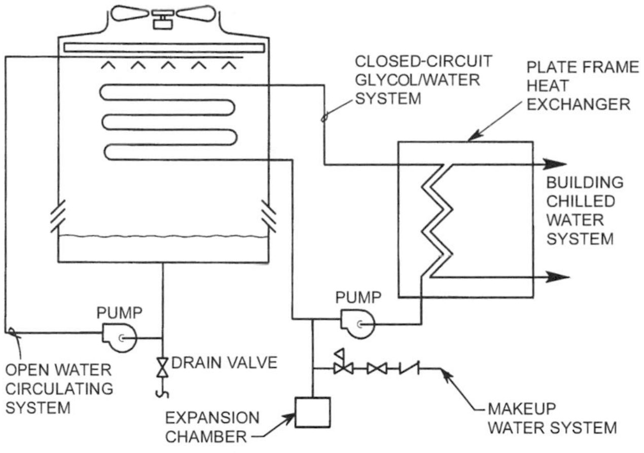

Fig. 5 Closed-Circuit Cooler System

Water from open cooling tower systems should not be piped directly through cooling coils, unitary heat pump condensers, or plate frame heat exchangers unless the water is first filtered through a high-efficiency filtering system and the water treatment system is managed to carefully minimize the dissolved solids. Even with these precautions, cooling coils should have straight tubes arranged for visual inspection and mechanical cleaning; unitary heat pumps should be installed for easy removal and replacement; and plate frame heat exchangers should be installed for ready accessibility for disassembly and cleaning.

CLOSED-CIRCUIT EVAPORATIVE COOLERS

Because of the potential for damaging freezing of cooling towers, some designers prefer to use closed-circuit evaporative/dry coolers for water economizer chilled water systems. There are many different configurations of these systems, one of which is shown in Figure 5. The open water is simply recirculated from the basin to the sprays of the evaporative cooler, and the cooling water system is a closed hydronic system usually using a glycol-water mixture for freeze protection. The open water system is then drained in freezing weather and the cooling heat exchanger unit is operated as a dry heat exchanger. This type of system in some configurations can be used to generate chilled water through the plate frame heat exchanger shown or to remove heat from a closed heat pump circuit. The glycol circuit is generally not used directly either for building cooling or for the heat pump circuit because of the economic penalty of the extensive glycol system. The closed circuit is designed in accordance with the principles and procedures described in Chapter 12.

OVERPRESSURE DUE TO THERMAL FLUID EXPANSION

When open condenser water systems are used at low temperatures for winter cooling, special precautions should be taken to prevent damaging overpressurization due to thermal expansion. This phenomenon has been known to cause severe damage when a section of piping containing water at a lower temperature than the surrounding space is isolated while cold. This isolation could be intentional, such as isolation by two service valves, or it could be as subtle as a section of piping isolated between a check valve and a control valve. If such an isolation occurs when water at 45°F is in a 1 in. pipe passing through a 75°F space, Figure 28 in Chapter 12 reveals that the pressure would increase by 815 psi, which would be destructive to many components of most condenser water systems. Refer to the section on Other Design Considerations in Chapter 12 for a discussion on this phenomenon.

CHAPTER 14

MEDIUM- AND HIGH-TEMPERATURE WATER HEATING SYSTEMS

MEDIUM-TEMPERATURE water systems have operating temperatures below 350°F and permit design to a pressure rating of 125 to 150 psig. High-temperature water systems are classified as those operating with supply water temperatures above 350°F and designed to a pressure rating of 300 psig. The usual practical temperature limit is about 450°F because of pressure limitations on pipe fittings, equipment, and accessories. The rapid pressure rise that occurs as the temperature rises above 450°F increases cost because components rated for higher pressures are required (see Figure 1). The design principles for both medium-temperature and high-temperature systems are basically the same. In this chapter, HTW refers to both systems.

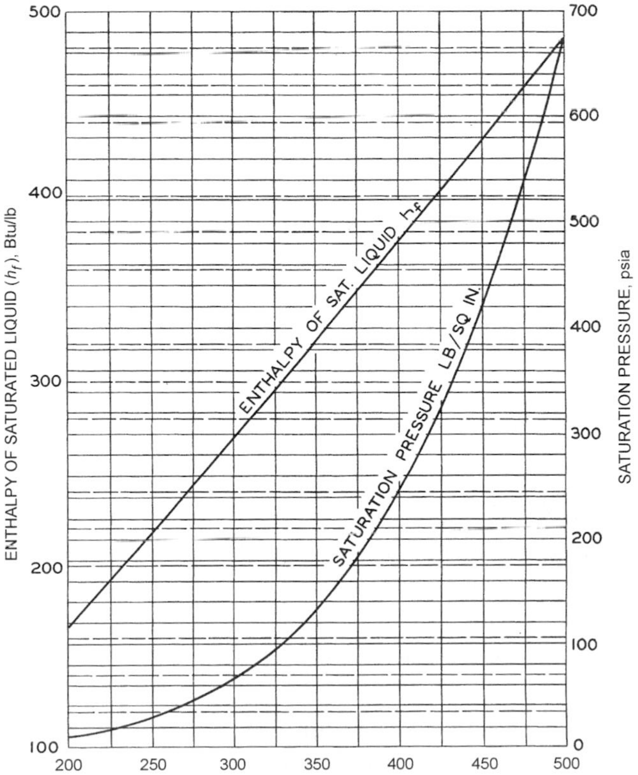

Fig. 1 Relation of Saturation Pressure and Enthalpy to Water Temperature

The preparation of this chapter is assigned to TC 6.1, Hydronic and Steam Equipment and Systems.

This chapter presents the general principles and practices that apply to HTW and distinguishes them from low-temperature water systems operating below 250°F. Refer to Chapter 12 for basic design considerations applicable to all hot water systems.

SYSTEM CHARACTERISTICS

The following characteristics distinguish HTW systems from steam distribution or low-temperature water systems:

- The system is a completely closed circuit with supply and return mains maintained under pressure. There are no losses from flashing, and heat that is not used in the terminal heat transfer equipment is returned to the HTW generator. Tight systems have minimal corrosion.
- Mechanical equipment that does not control the performance of individual terminal units is concentrated at the central station.
- Piping can slope up or down or run at a variety of elevations to suit the terrain and the architectural and structural requirements without provision for trapping at each low point. This may reduce the amount of excavation required and eliminate drip points and return pumps required with steam.
- Greater temperature drops are used and less water is circulated than in low-temperature water systems.
- The pressure in any part of the system must always be well above the pressure corresponding to the temperature at saturation in the system to prevent flashing of the water into steam.
- Terminal units requiring different water temperatures can be served at their required temperatures by regulating the flow of water, modulating the water supply temperature, placing some units in series, and using heat exchangers or other methods.
- The high heat content of the water in the HTW circuit acts as a thermal flywheel, evening out fluctuations in the load. The heat storage capacity can be further increased by adding heat storage tanks or by increasing the temperature in the return mains during periods of light load.
- The high heat content of the heat carrier makes high-temperature water unsuitable for two-pipe dual-temperature (hot and chilled water) applications and for intermittent operation if rapid start-up and shutdown are desired, unless the system is designed for minimum water volume and is operated with rapid response controls.
- Higher engineering skills are required to design a HTW system that is simple, yet safer and more convenient to operate than are required to design a comparable steam or low-temperature water system.
- HTW system design requires careful attention to basic laws of chemistry and physics as these systems are less forgiving than standard hydronic systems.

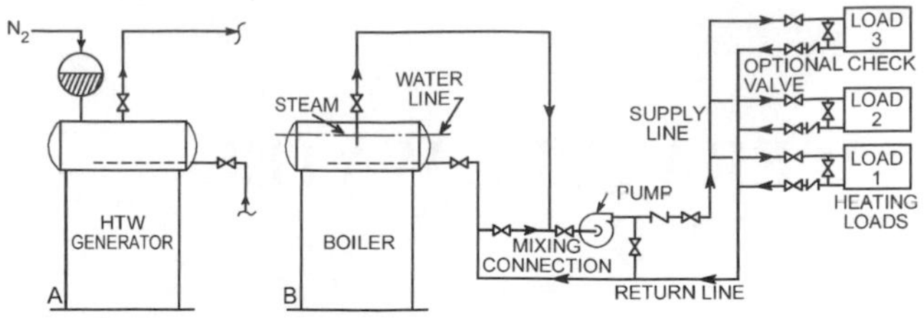

Fig. 2 Elements of High-Temperature Water System

BASIC SYSTEM

High-temperature water systems are similar to conventional forced hot water heating systems. They require a heat source (which can be a direct-fired HTW generator, a steam boiler, or an open or closed heat exchanger) to heat the water. The expansion of the heated water is usually taken up in an expansion vessel, which simultaneously pressurizes the system. Heat transport depends on circulating pumps. The distribution system is closed, comprising supply and return pipes under the same basic pressure. Heat emission at the terminal unit is indirect by heat transfer through heat transfer surfaces. The basic system is shown in Figure 2.

The principal differences of HTW systems from low-temperature water systems are the higher pressure, heavier equipment, generally smaller pipe sizes, and manner in which water pressure is maintained.

Most systems are either (1) a saturated steam cushion system, in which the high-temperature water develops its own pressure, or (2) a gas- or pump-pressurized system, in which the pressure is imposed externally.

HTW generators and all auxiliaries (such as water makeup and feed equipment, pressure tanks, and circulating pumps) are usually located in a central station. Cascade HTW generators sometimes use an existing steam distribution system and are installed remote from the central plant.

DESIGN CONSIDERATIONS

Selection of the system pressure, supply temperature, temperature drop, type of HTW generator, and pressurization method are the most important initial design considerations. The following are some of the determining factors:

- Type of load (space heating and/or process). Load fluctuations during a 24-h period and a 1-year period. Process loads might require water at a given minimum supply temperature continuously, while space heating can permit temperature modulation as a function of outdoor temperature or other climatic influences.
- Terminal unit temperature requirements.
- Distance between heating plant and space or process requiring heat.
- Quantity and pressure of steam used for power equipment in the central plant.
- Elevation variations within the system and the effect of basic pressure distribution.

Usually, distribution piping is the major investment in an HTW system. A distribution system with the widest temperature spread (Δt) between supply and return will have the lowest initial and operating costs. Economical designs have a Δt of 150°F or higher.

The requirements of terminal equipment or user systems determine the system selected. For example, if the users are 10 psig steam generators, the return temperatures would be 250°F. A 300 psig rated system operated at 400°F would be selected to serve the load. In another example, where the primary system serves predominantly 140 to 180°F hot water heating systems, an HTW system that operates at 325°F could be selected. The supply temperature is reduced by blending with 140°F return water to the desired 180°F hot water supply temperature in a direct-connected hot water secondary system. This highly economical design has a 140°F return temperature in the primary water system and a Δt of 185°F.

Because the danger of water hammer is always present when the pressure drops to the point at which pressurized hot water flashes to steam, the primary HTW system should be designed with steel valves and fittings of 150 psi. The secondary water, which operates below 212°F and is not subject to flashing and water hammer, can be designed for 125 psi and standard HVAC equipment.

Theoretically, water temperatures up to about 350°F can be provided using equipment suitable for 125 psi. But in practice, unless push-pull pumping is used, maximum water temperatures are limited by the system design, pump pressures, and elevation characteristics to values between 300 and 325°F.

Many systems designed for self-generated steam pressurization have a steam drum through which the entire flow is taken, and which also serves as an expansion vessel. A circulating pump in the supply line takes water from the tank. The temperature of the water from the steam drum cannot exceed the steam temperature in the drum that corresponds to its pressure at saturation. The point of maximum pressure is at the discharge of the circulating pump. If, for example, this pressure is to be maintained below 125 psig, the pressure in the drum that corresponds to the water temperature cannot exceed 125 psig minus the sum of the pump pressure and the pressure that is caused by the difference in elevation between the drum and the circulating pump.

Most systems are designed for inert gas pressurization. In most of these systems, the pressurizing tank is connected to the system by a single balance line on the suction side of the circulating pump. The circulating pump is located at the inlet side of the HTW generator. There is no flow through the pressurizing tank, and a reduced temperature will normally establish itself inside. A special characteristic of a gas-pressurized system is the apparatus that creates and maintains gas pressure inside the tank.

In designing and operating an HTW system, it is important to maintain a pressure that always exceeds the vapor pressure of the water, even if the system is not operating. This may require limiting the water temperature and thereby the vapor pressure, or increasing the imposed pressure.

Elevation and the pressures required to prevent water from flashing into steam in the supply system can also limit the maximum water temperature that may be used and must therefore be studied in evaluating the temperature-pressure relationships and method of pressurizing the system.

The properties of water that govern design are as follows:

- Temperature versus pressure at saturation (Figure 1)
- Density or specific volume versus temperature
- Enthalpy or sensible heat versus temperature
- Viscosity versus temperature
- Type and amount of pressurization

The relationships among temperature, pressure, specific volume, and enthalpy are all available in steam tables. Some properties of water are summarized in Table 1 and Figure 3.

Direct-Fired High-Temperature Water Generators

In direct-fired HTW generators, the central stations are comparable to steam boiler plants operating within the same pressure range. The generators should be selected for size and type in keeping with the load and design pressures, as well as the circulation requirements peculiar to high-temperature water. In some systems, both steam for power or processing and high-temperature water are supplied from the same boiler; in others, steam is produced in the

Table 1 Properties of Water—212 to 400°F

Temperature, °F	Absolute Pressure, psia[a]	Density, lb/ft³	Specific Heat, Btu/lb·°F	Total Heat above 32°F Btu/lb[a]	Total Heat above 32°F Btu/ft³	Dynamic Viscosity, Centipoise
212	14.70	59.81	1.007	180.07	10,770	0.2838
220	17.19	59.63	1.009	188.13	11,216	0.2712
230	20.78	59.38	1.010	198.23	11,770	0.2567
240	24.97	59.10	1.012	208.34	12,313	0.2436
250	29.83	58.82	1.015	218.48	12,851	0.2317
260	35.43	58.51	1.017	228.64	13,378	0.2207
270	41.86	58.24	1.020	238.84	13,910	0.2107
280	49.20	57.94	1.022	249.06	14,430	0.2015
290	57.56	57.64	1.025	259.31	14,947	0.1930
300	67.01	57.31	1.032	269.59	15,450	0.1852
310	77.68	56.98	1.035	279.92	15,950	0.1779
320	89.66	56.66	1.040	290.28	16,437	0.1712
330	103.06	56.31	1.042	300.68	16,931	0.1649
340	118.01	55.96	1.047	311.13	17,409	0.1591
350	134.63	55.59	1.052	321.63	17,879	0.1536
360	153.04	55.22	1.057	332.18	18,343	0.1484
370	173.37	54.85	1.062	342.79	18,802	0.1436
380	195.77	54.47	1.070	353.45	19,252	0.1391
390	220.37	54.05	1.077	364.17	19,681	0.1349
400	247.31	53.65	1.085	374.97	20,117	0.1308

[a] Reprinted by permission from *Thermodynamic Properties of Steam*, J.H. Keenan and F.G. Keyes, John Wiley and Sons, 1936 edition.

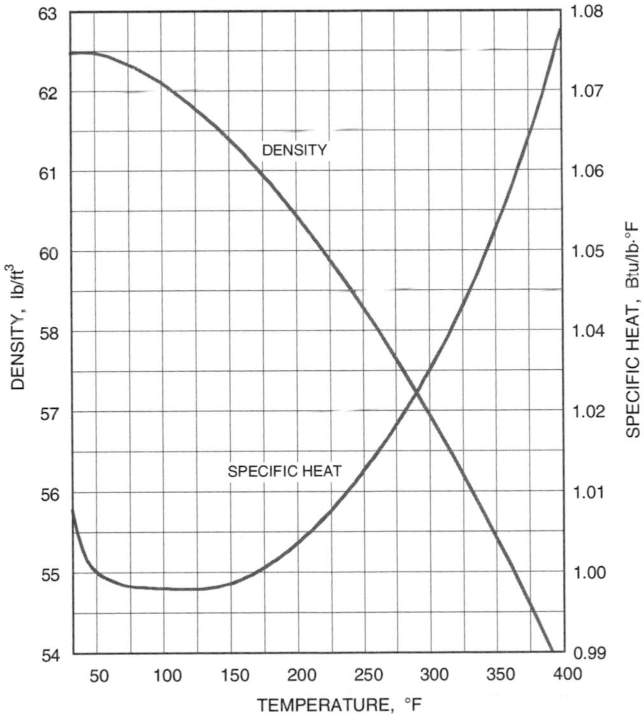

Fig. 3 Density and Specific Heat of Water

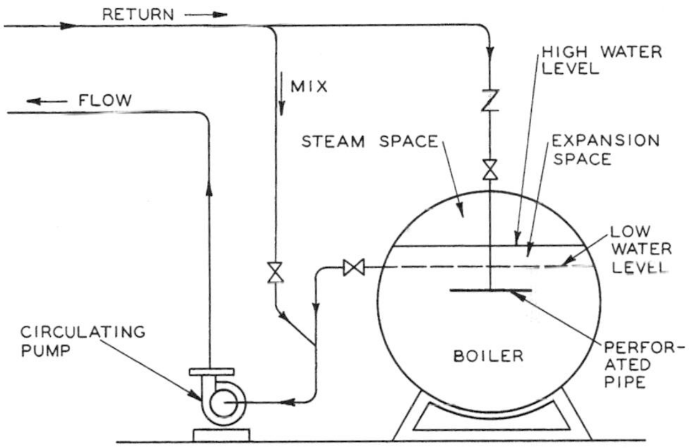

Fig. 4 Arrangement of Boiler Piping

boilers and used for generating high-temperature water; and in many others, the burning fuel directly heats the water.

The HTW generators can be the water-tube or fire-tube type, and can be equipped with any conventional fuel firing apparatus. Water-tube generators can have either forced-circulation, gravity circulation, or a combination of both. The recirculating pumps of forced-circulation generators must operate continuously while the generator is being fired. Steam boilers relying on natural circulation may require internal baffling when used for HTW generation. In scotch marine boilers, thermal shock may occur, caused by a sudden drop in the temperature of the return water or when the Δt exceeds 40°F. Forced-circulation HTW generators are usually the once-through type and rely solely on pumps to achieve circulation. Depending on the design, internal orifices in the various circuits may be required to regulate the water flow rates in proportion to the heat absorption rates. Circulation must be maintained at all times while the generator is being fired, and the flow rate must never drop below the minimum indicated by the manufacturer.

Where gravity circulation steam boilers are used for HTW generation, the steam drum usually serves as an expansion vessel. In steam-pressurized forced-circulation HTW generators, a separate vessel is commonly used for expansion and for maintaining the steam pressure cushion. A separate vessel is always used when the system is cushioned by an inert gas or auxiliary steam. Proper internal circulation is essential in all types of boilers to prevent tube failures due to overheating or unequal expansion.

In early HTW systems, the generator is a steam boiler with an integral steam drum used for pressurization and for expansion of the water level. A dip pipe removes water below the water line (Figure 4) (Applegate 1958). This dip pipe should be installed so that it picks up water at or near the saturation temperature, without too many steam bubbles. If a pipe breaks somewhere in the system, the boiler must not empty to a point where heating surfaces are bared and a boiler explosion occurs. The same precautions must be taken with the return pipe. If the return pipe is connected in the lower part

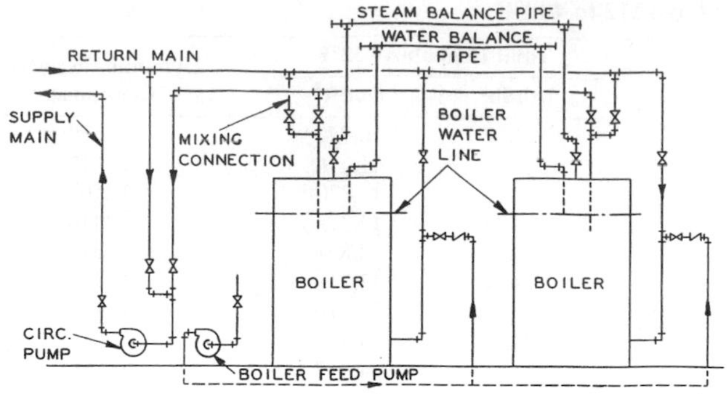

Fig. 5 Piping Connections for Two or More Boilers in HTW System Pressurized by Steam

of the boiler, a check valve should be placed in the connecting line to the boiler to preclude the danger of emptying the boiler.

When two or more such boilers supply a common system, the same steam pressure and water level must be maintained in each. Water and steam balance pipes are usually installed between the drums (Figure 5). These should be liberally sized. The following table shows recommended sizes:.

Boiler Rating, million Btu/h	Balance Pipe Diameter, in.
2.5	3
5	3.5
10	4
15	5
20	6
30	8

A difference of only 0.25 psi in the system pressure between two boilers operated at 100 psig would cause a difference of 9 in. in the water level. The situation is further aggravated because an upset is not self-balancing. Rather, when too high a heat release in one of the boilers has caused the pressure to rise and the water level to fall in this boiler, the decrease in the flow of colder return water into it causes a further pressure rise, while the opposite happens in the other boiler. It is therefore important that the firing rates match the flow through each boiler at all times. Modern practice is to use either flooded HTW heaters with a single external pressurized expansion drum common to all the generators, or the combination of steam boilers and a direct-contact (cascade) heater.

Expansion and Pressurization

In addition to the information in Chapter 12, the following factors should be considered:

- The connection point of the expansion tank used for pressurization greatly affects the pressure distribution throughout the system and the avoidance of HTW flashing.
- Proper safety devices for high and low water levels and excessive pressures should be incorporated in the expansion tank and interlocked with combustion safety and water flow rate controls.

The following four fundamental methods, in which pressure in a given hydraulic system can be kept at a desired level, amplify the discussion in Chapter 12 (Blossom and Ziel 1959, National Academy of Sciences 1959).

1. **Elevating the storage tank** is a simple pressurization method, but because of the great heights required for the pressure encountered, it is generally impractical.

2. **Steam pressurization** requires the use of an expansion vessel separate from the HTW generator. Because firing and flow rates can never be perfectly matched, some steam is always carried. Therefore, the vessel must be above the HTW generators and connected

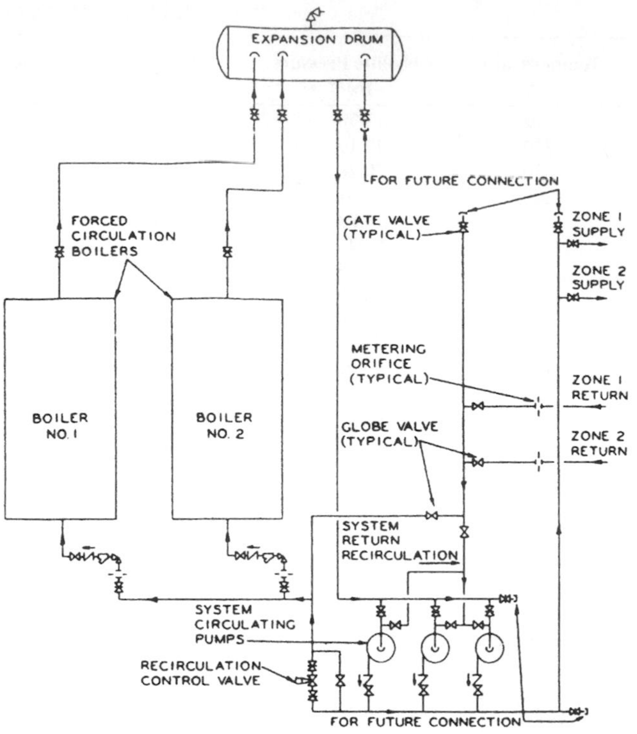

Fig. 6 HTW Piping for Combined (One-Pump) System (Steam Pressurized)

in the supply water line from the generator. This steam, supplemented by flashing of the water in the expansion vessel, provides the steam cushion that pressurizes the system.

The expansion vessel must be equipped with steam safety valves capable of relieving the steam generated by all the generators. The generators themselves are usually designed for a substantially higher working pressure than the expansion drum, and their safety relief valves are set for the higher pressure to minimize their lift requirement.

The basic HTW pumping arrangements can be either single-pump, in which one pump handles both the generator and system loads, or two-pump, in which one pump circulates high-temperature water through the generator and a second pump circulates high-temperature water through the system (see Figure 6 and Figure 7). The circulating pump moves the water from the expansion vessel to the system and back to the generator. The vessel must be elevated to increase the net positive suction pressure to prevent cavitation or flashing in the pump suction. This arrangement is critical. A bypass from the HTW system return line to the pump suction helps prevent flashing. Cooler return water is then mixed with hotter water from the expansion vessel to give a resulting temperature below the corresponding saturation point in the vessel.

In the two-pump system, the boiler recirculation should always exceed the system circulation, because excessive cooling of the water in the drum by the cooler return water entering the drum, in case of over-circulation, can cause pressure loss and flashing in the distribution system. Backflow into the drum can be prevented by installing a check valve in the balance line from the drum to the boiler recirculating pumps. Higher cushion pressure may be maintained by auxiliary steam from a separate generator.

Sizing. Steam-pressurized vessels should be sized for a total volume V_T, which is the sum of the volume V_1 required for the steam space, the volume V_2 required for water expansion, and the volume V_3 required for sludge and reserve. An allowance of 20% of the sum

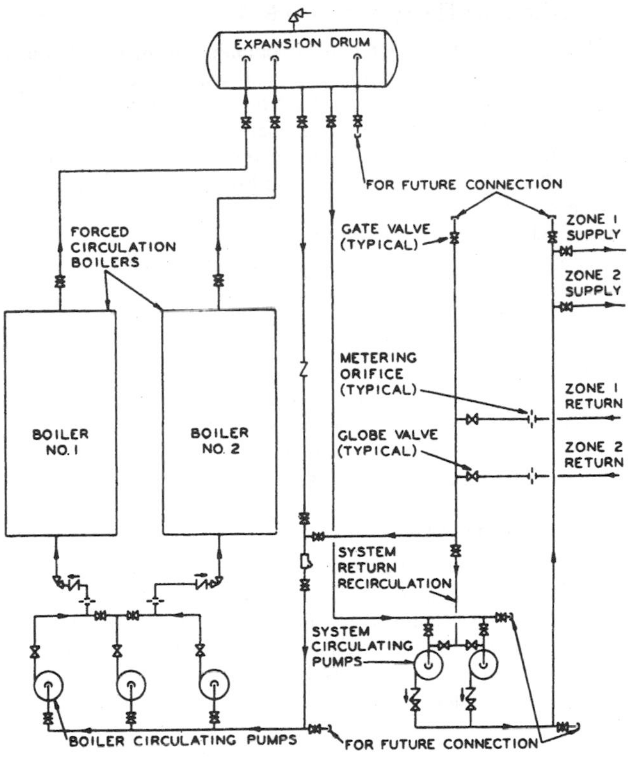

Fig. 7 HTW Piping for Separate (Two-Pump) System (Steam Pressurized)

of V_2 and V_3 is a reasonable estimate of the volume V_1 required for the steam space.

The volume V_2 required for water expansion is determined from the change in water volume from the minimum to the maximum operating temperatures of the complete cycle. It is not necessary to allow for expansion of the total water volume in the system from a cold initial start. It is necessary during a start-up period to bleed off the volume of water caused by expansion from the initial starting temperature to the lowest average operating temperature.

The volume V_3 for sludge and reserve varies greatly depending on the size and design of system and generator capacity. An allowance of 40% of the volume V_2 required for water expansion is a reasonable estimate of the volume required for V_3.

3. **Nitrogen**, the most commonly used inert gas, is used for gas pressurization. Air is not recommended because the oxygen in air contributes to corrosion in the system.

The expansion vessel is connected as close as possible to the suction side of the HTW pump by a balance line. The inert gas used for pressurization is fed into the top of the cylinder, preferably through a manual fill connection using a reducing station connected to an inert gas cylinder. Locating the relief valve below the minimum water line is advantageous, because it is easier to keep it tightly sealed with water on the pressure side. If the valve is located above the water line, it is exposed to the inert gas of the system.

To reduce the area of contact between gas and water and the resulting absorption of gas into the water, the tank should be installed vertically. It should be located in the most suitable place in the central station. Similar to the steam-pressurized system, the pumping arrangements can be either one- or two-pump (see Figure 8 and Figure 9).

The ratings of fittings, valves, piping, and equipment are considered in determining the maximum system pressure. A minimum pressure of about 25 to 50 psi over the maximum saturation pressure can be used. The imposed additional pressure above the vapor

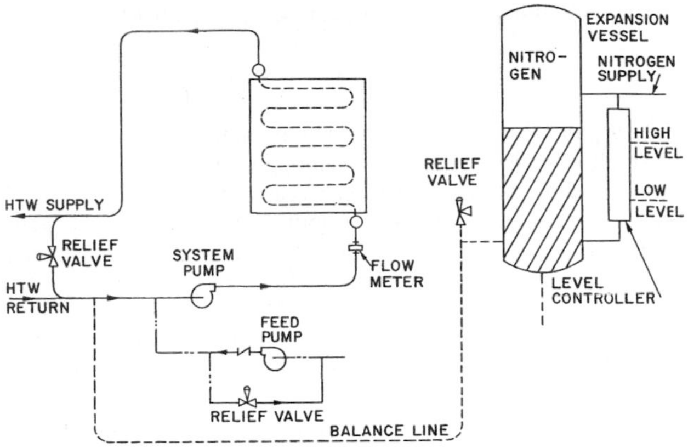

Fig. 8 Inert Gas Pressurization for One-Pump System

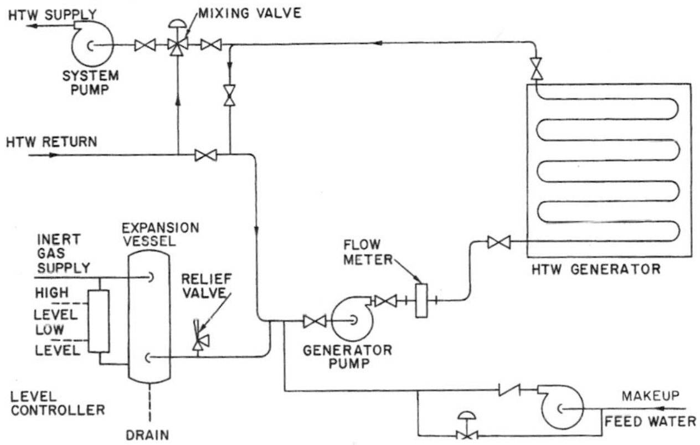

Fig. 9 Inert Gas Pressurization for Two-Pump System

pressure must be large enough to prevent steaming in the HTW generators at all times, even under conditions when flow and firing rates in generators operated in parallel, or flow and heat absorption in parallel circuits within a generator, are not evenly matched. This is critical, because gas-pressurized systems do not have steam separating means and safety valves to evacuate the steam generated.

The simplest type of gas-pressurization system uses a variable gas quantity with or without gas recovery (Figure 10) (National Academy of Sciences 1959). When the water rises, the inert gas is relieved from the expansion vessel and is wasted or recovered in a low-pressure gas receiver from which the gas compressor pumps it into a high-pressure receiver for storage. When the water level drops in the expansion vessel, the control cycle adds inert gas from bottles or from the high-pressure receiver to the expansion vessel to maintain the required pressure.

Gas wastage can significantly affect the operating cost. The gas recovery system should be analyzed based on the economics of each application. Gas recovery is generally more applicable to larger systems.

Sizing. The vessel should be sized for a total volume V_T, which is the sum of the volume V_1 required for pressurization, the volume V_2 required for water expansion, and the volume V_3 required for sludge and reserve.

Calculations made on the basis of pressure-volume variations following Boyle's Law are reasonably accurate, assuming that the tank operates at a relatively constant temperature. The minimum gas volume can be determined from the expansion volume V_2 and from the control range between the minimum tank pressure P_1 and the maximum tank pressure P_2. The gas volume varies from the minimum V_1 to a maximum, which includes the water expansion volume V_2.

The minimum gas volume V_1 can be obtained from

$$V_1 = P_1 V_2 / (P_2 - P_1)$$

where P_1 and P_2 are units of absolute pressure.

An allowance of 10% of the sum of V_1 and V_2 is a reasonable estimate of the sludge and reserve capacity V_3. The volume V_2 required for water expansion should be limited to the actual expansion that occurs during operation through its minimum to maximum operating temperatures. It is necessary to bleed off water during a start-up cycle from a cold start. It is practicable on small systems (e.g., under 1,000,000 Btu/h to 10,000,000 Btu/h) to size the expansion vessel for the total water expansion from the initial fill temperature.

4. **Pump pressurization** in its simplest form consists of a feed pump and a regulator valve. The pump operates continuously, introducing water from the makeup tank into the system. The pressure regulator valve bleeds continuously back into the makeup tank. This method is usually restricted to small process heating systems. However, it can be used to temporarily pressurize a larger system to avoid shutdown during inspection of the expansion tank.

In larger central HTW systems, pump pressurization is combined with a fixed-quantity gas compression tank that acts as a buffer. When the pressure rises above a preset value in the buffer tank, a control valve opens to relieve water from the balance line into the makeup storage tank. When the pressure falls below a preset second value, the feed pump is started automatically to pump water from the makeup tank back into the system. The buffer tank is designed to absorb only the limited expansion volume that is required for the pressure control system to function properly; it is usually small.

To prevent corrosion-causing elements, principally oxygen, from entering the HTW system, the makeup storage tank is usually closed and a low-pressure nitrogen cushion of 1 to 5 psig is maintained. The gas cushion is usually the variable gas quantity type with release to the atmosphere.

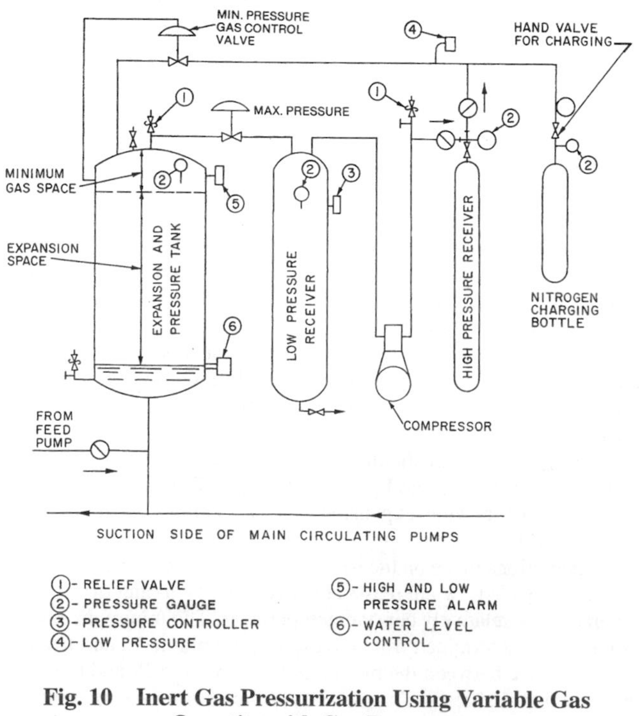

Fig. 10 Inert Gas Pressurization Using Variable Gas Quantity with Gas Recovery

Direct-Contact Heaters (Cascades)

High-temperature water can be obtained from direct-contact heaters in which steam from turbine exhaust, extraction, or steam boilers is mixed with return water from the system. The mixing takes place in the upper part of the heater where the water cascading from horizontal baffles comes in direct contact with steam (Hansen 1966). The basic systems are shown in Figure 11 and Figure 12.

The steam space in the upper part of the heater serves as the steam cushion for pressurizing the system. The lower part of the heater serves as the system's expansion tank. Where the water heater and the boiler operate under the same pressure, the surplus water is usually returned directly into the boiler through a pipe connecting the outlet of the high-temperature water-circulating pump to the boiler.

The cascade system is also applicable where both steam and HTW services are required (Hansen and Liddy 1958). Where heat and power production are combined, the direct-contact heater becomes the mixing condenser (Hansen and Perrsall 1960).

System Circulating Pumps

Forced-circulation boiler systems can be either one-pump or two-pump. These terms do not refer to the number of pumps but to the number of groups of pumps installed. In the one-pump system (see Figure 6), a single group of pumps assures both generator and distribution system circulation. In this system, both the distribution system and the generators are in series (Carter and Sturdevant 1958).

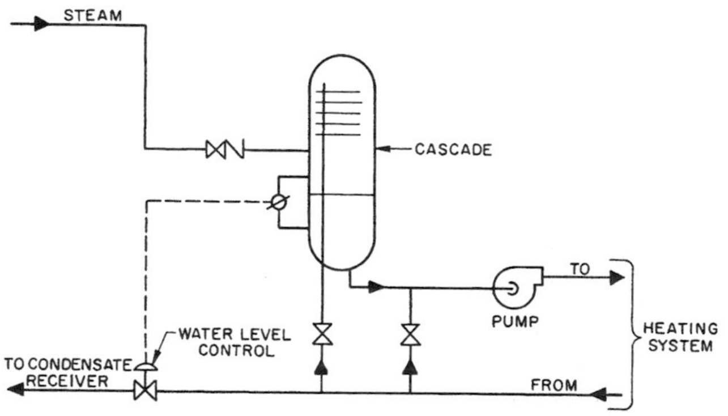

Fig. 11 Cascade HTW System

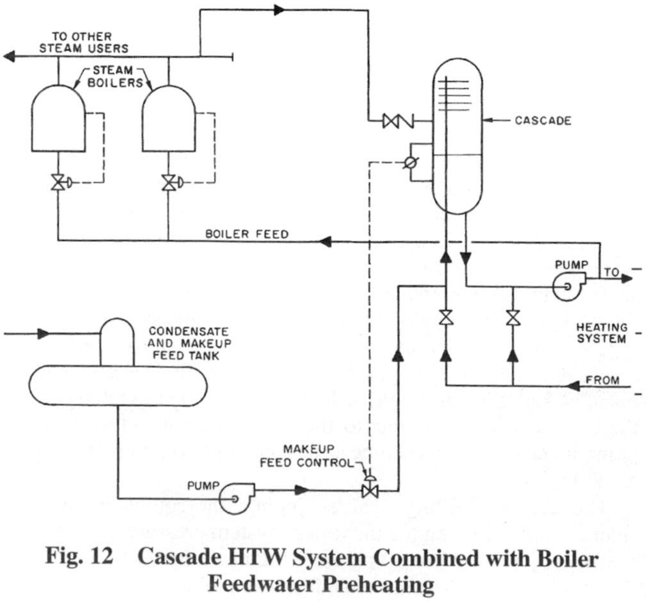

Fig. 12 Cascade HTW System Combined with Boiler Feedwater Preheating

However, to ensure the minimum flow through the boiler at all times, a bypass around the distribution system must be provided. The one-pump method usually applies only to systems in which the total friction pressure is relatively low, because the energy loss of available circulating pressure from throttling in the bypass at times of reduced flow requirements in the district can substantially increase the operating cost.

In the two-pump system (see Figure 7), an additional group of recirculating pumps is installed solely to provide circulation for the generators (Carter and Sturdevant 1958). One pump is often used for each generator to draw water from either the expansion drum or the system return and to pump it through the generator into the expansion drum. The system circulating pumps draw water from the expansion drum and circulate it through the distribution system only. The supply temperature to the distribution system can be varied by mixing water from the return into the supply on the pump suction side. Where zoning is required, several groups of pumps can be used with a different pressure and different temperature in each zone. The flow rate can also be varied without affecting the generator circulation and without using a system bypass.

In steam-pressurized systems, the circulating pump is installed in the supply line to maintain all parts of the distributing system at pressures exceeding boiler pressure. This minimizes the danger of flashing into steam.

It is common practice to install a mixing connection from the return to the pump suction that bypasses the HTW generator. This connection is used for start-up and for modulating the supply temperature; it should not be relied on for increasing the pressure at the pump inlet. Where it is impossible to provide the required submergence by proper design, a separate small-bore premixing line should be provided.

Hansen (1966) describes push-pull pumping, which divides the circulating pressure equally between two pumps in series. One is placed in the supply and is sized to overcome frictional resistance in the supply line of the heat distribution system. The second pump is placed in the return and is sized to overcome frictional resistance in the return. The expansion tank pressure is impressed on the system between the pumps. The HTW generator is either between the pumps or in the supply line from the pumps to the distribution system (see Figure 13).

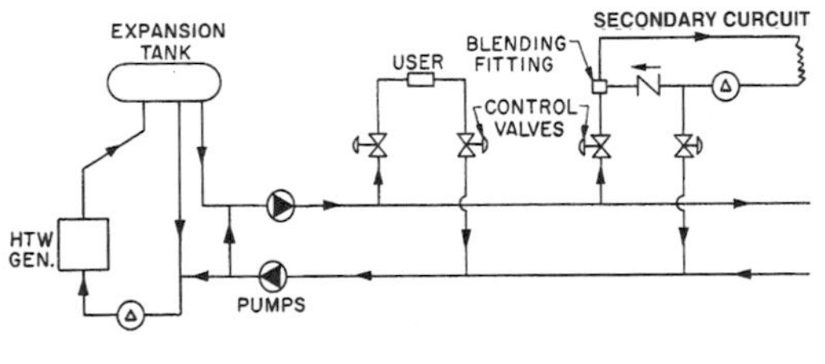

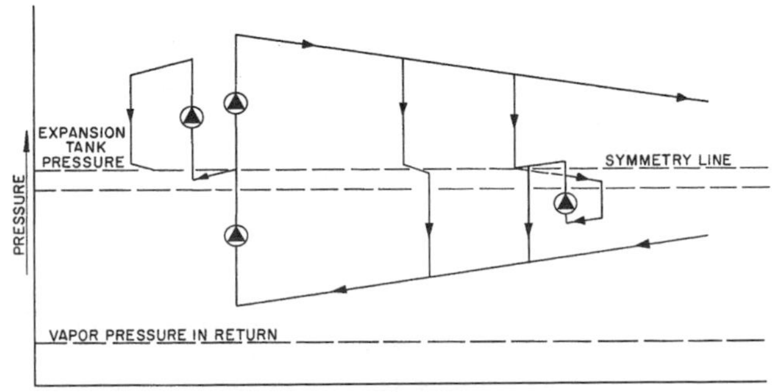

Fig. 13 Typical HTW System with Push-Pull Pumping

In the push-pull system, the pressures in the supply and the return mains are symmetrical in relation to a line representing the pressure imposed on the system by the pressurizing source (expansion tank). This pressure becomes the system pressure when the pumps are stopped. The heat supply to using equipment or secondary circuits is controlled by two equal regulating valves, one on the inlet and the other on the outlet side, instead of the customary single valve on the leaving side. Both valves are operated in unison from a common controller; there are equal frictional resistances on both sides. Therefore, the pressure in the user circuits or equipment is maintained at all times halfway between the pressures in the supply and return mains. Because the halfway point is located on the symmetry line, the pressure in the user equipment or circuits is always equal to that of the pressurizing source (expansion tank) plus or minus static pressure caused by elevation differences. In other words, no system distribution pressure is reflected against the user circuit or equipment.

While the pressure in the supply system is higher than that of the expansion tank, the pressure in the return system, being symmetrical to the former, is lower. Therefore, the push-pull method is applicable only where the temperature in the return is always significantly lower than that in the supply. Otherwise, flashing could occur. This is critical and requires careful investigation of the temperature-pressure relationship at all points. The push-pull method is not applicable in reverse-return systems.

Push-pull pumping permits use of standard 125 psi fittings and equipment in many MTW systems. Such systems, combined with secondary pumping, can be connected directly to low-temperature terminal equipment in the building heating system. Temperature drops normally obtainable only in HTW systems can be achieved with MTW. For example, 330°F water can be generated at 90 psig and distributed at less than 125 psig. Its temperature can be reduced to 200°F by secondary pumping. The pressure in the terminal equipment then is 90 psig and the MTW is returned to the primary system at 180°F. The temperature difference between supply and return in the primary MTW system is 150°F, which is comparable to that of an HTW system. In addition, conventional heat exchangers, expansion tanks, and water makeup equipment are eliminated from the secondary systems.

DISTRIBUTION PIPING DESIGN

Data for pipe friction are presented in Chapter 33 of the 1997 *ASHRAE Handbook—Fundamentals*. These pipe friction and fitting loss tables are for a 60°F water temperature. When applied to HTW systems, the values obtained are excessively high. The data should be used for preliminary pipe sizing only. Final pressure drop calculations should be made using the fundamental Darcy-Weisbach equation [Equation (1) in Chapter 33 of the 1997 *ASHRAE Handbook—Fundamentals*] in conjunction with friction factors, pipe roughness, and fitting loss coefficients presented in the section on Flow Analysis in Chapter 2 of the 1997 *ASHRAE Handbook—Fundamentals*.

The conventional conduit or tunnel distribution systems are used with similar techniques for installation (see Chapter 11). A small valved bypass connection between the supply and the return pipe should be installed at the end of long runs to maintain a slight circulation in the mains during periods of minimum or no demand.

All pipe, valves, and fittings used in HTW systems should comply with the requirements of ASME *Standard* B-31.1, Power Piping, and the *National Fuel Gas Code* (NFPA *Standard* 54 or IAS Z223.1). These codes state that hot water systems should be designed for the highest pressure and temperature actually existing in the piping under normal operation. This pressure equals cushion pressure plus pump pressure plus static pressure. Schedule 40 steel pipe is applicable to most HTW systems with welded steel fittings and steel valves. A minimum number of joints should be used. In many installations, all valves in the piping system are welded or brazed. Flange connections used at major equipment can be serrated, raised flange facing, or ring joint. It is desirable to have back-seating valves with special packing suitable for this service.

The ratings of valves, pipe, and fittings must be checked to determine the specific rating point for the given application. The pressure rating for a standard 300 psi steel valve operating at 400°F is 665 psi. Therefore, it is generally not necessary to use steel valves and fittings over 300 psi ratings in HTW systems.

Since high-temperature water is more penetrating than low-temperature water, leakage caused partly by capillary action should not be ignored because even a small amount of leakage vaporizes immediately. This slight leakage becomes noticeable only on the outside of the gland and stem of the valve where thin deposits of salt are left after evaporation. Avoid screwed joints and fittings in HTW systems. Pipe unions should not be used in place of flange connections, even for small-bore piping and equipment.

Individual heating equipment units should be installed with separate valves for shutoff. These should be readily accessible. If the unit is to be isolated for service, valves are needed in both the supply and return piping to the unit. Valve trim should be stainless steel or a similar alloy. Do not use brass and bronze.

High points in piping should have air vents for collecting and removing air, and low points should have provision for drainage. Loop-type expansion joints, in which the expansion is absorbed by deflection of the pipe loop, are preferable to the mechanical type. Mechanical expansion joints must be properly guided and anchored.

HEAT EXCHANGERS

Heat exchangers or converters commonly use steel shells with stainless steel, admiralty metal, or cupronickel tubes. Copper should not be used in HTW systems above 250°F. Material must be chosen carefully, considering the pressure-temperature characteristics of the particular system. All connections should be flanged or welded. On larger exchangers, water box-type construction is desirable to remove the tube bundle without breaking piping connections. Normally, HTW is circulated through the tubes, and because the heated water contains dissolved air, the baffles in the shell should be constructed of the same material as the tubes to control corrosion.

AIR HEATING COILS

In HTW systems over 400°F, coils should be cupronickel or all-steel construction. Below this point, other materials (e.g., red brass) can be used after determining their suitability for the temperatures involved. Coils in outdoor air connections need freeze protection by damper closure or fan shutdown controlled by a thermostat. It is also possible to set the control valve on the preheat coil to a minimum position. This protects against freezing, as long as there is no unbalance in the tube circuits where parallel paths of HTW flow exist. A better method is to provide constant flow through the coil and to control heat output with face and bypass dampers or by modulating the water temperature with a mixing pump.

SPACE HEATING EQUIPMENT

In industrial areas, space heating equipment can be operated with the available high-temperature water. Convectors and radiators may require water temperatures in the low- and medium-temperature range 120 to 180°F or 200 to 250°F, depending on their design pressure and proximity to the occupants. The water velocity through the heating equipment affects its capacity. This must be considered in selecting the equipment because, if a large water temperature drop is used, the circulation rate is reduced and, consequently, the flow velocity may be reduced enough to appreciably lower the heat transfer rate.

Convectors, specially designed to provide low surface temperatures, are now available to operate with water temperatures from 300 to 400°F.

These high temperatures are suitable for direct use in radiant panel surfaces. Because radiant output is a fourth-power function of the surface temperature, the surface area requirements are reduced over low-temperature water systems. The surfaces can be flat panels consisting of a steel tube, usually 0.38 or 0.5 in., welded to sheet steel turned up at the edges for stiffening. Several variations are available. Steel pipe can also be used with an aluminum or similar reflector to reflect the heat downward and to prevent smudging the surfaces above the pipe.

INSTRUMENTATION AND CONTROLS

Pressure gages should be installed in the pump discharge and suction and at locations where pressure readings will assist operation and maintenance. Thermometers (preferably dial-type) or thermometer wells should be installed in the flow and return pipes, the pump discharge, and at any other points of major temperature change or where temperatures are important in operating the system. It is desirable to have thermometers and gages in the piping at the entrance to each building converter.

On steam-pressurized cycles, the temperature of the water leaving the generator should control the firing rate to the generator. A master pressure control operating from the steam pressure in the expansion vessel should be incorporated as a high-limit override. Inert gas-pressurized systems should be controlled from the generator discharge temperature. Combustion controls are discussed in detail in Chapter 26.

In the water-tube generators most commonly used for HTW applications, the flow of water passes through the generator in seconds. The temperature controller must have a rapid response to maintain a reasonably uniform leaving water temperature. In steam-pressurized units, the temperature variation must not exceed the antiflash pressure margin. At 300°F, a 5°F temperature variation corresponds to a 5 psi variation in the vapor pressure. At 350°F, the same temperature variation results in a vapor pressure variation of 10 psi. At 400°F, the variation increases to 15 psi and at 450°F, to 22 psi. The permissible temperature swing must be reduced as the HTW temperature increases, or the pressure margin must be increased to avoid flashing.

Keep the controls simple. The rapid response through the generator makes it necessary to modulate the combustion rate on all systems with a capacity of over a few million Btu/h. In the smaller size range, this can be done by high-low firing. In large systems, particularly those used for central heating applications, full modulation of the combustion rate is desirable through at least 20% of full capacity. On-off burner control is generally not used in steam-pressurized cycles because the system loses pressurization during the off cycle, which can cause flashing and cavitation at the HTW pumps.

All generators should have separate safety controls to shut down the combustion apparatus when the system pressure or water temperature is high. HTW generators require a minimum water flow at all times to prevent tube failure. Means should be provided to measure the flow and to stop combustion if the flow falls below the minimum value recommended by the generator manufacturer. For inert gas-pressurized cycles, a low-pressure safety control should be included to shut down the combustion system if pressurization is lost. Figure 14 shows the basic schematic control diagram for an HTW generator.

Valve selection and sizing are important because of relatively high temperature drops and smaller flows in HTW systems. The valve must be sized so that it is effective over its full range of travel. The valve and equipment must be sized to absorb, in the control valve at full flow, not less than half the available pressure difference between supply and return mains where the equipment is served. A valve with equal percentage flow characteristics is needed. Some-

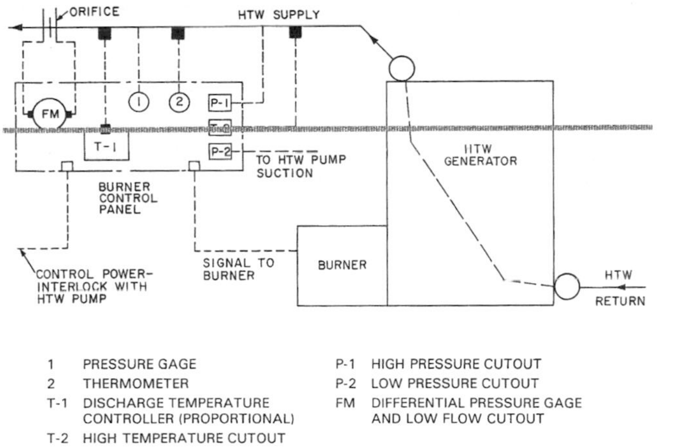

Fig. 14 Control Diagram for HTW Generator

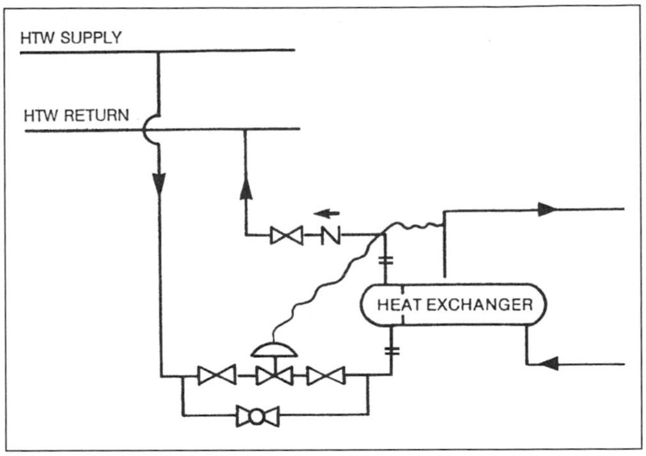

Fig. 15 Heat Exchanger Connections

times two small valves provide better control than one large valve. Stainless steel trim is recommended, and all valve body materials and packing should be suitable for high temperatures and pressures. The valve should have a close-off rating at least equal to the maximum pressure produced by the circulating pump. Generally, two-way valves are more desirable than three-way valves because of equal percentage flow characteristics and the smaller capacities available in two-way valves. Single-seated valves are preferable to double-seated valves because the latter do not close tightly.

Control valves are commonly located in the return lines from heat transfer units to reduce the valve operating temperature and to prevent plug erosion caused by high-temperature water flashing to steam at lower discharge pressure. A typical application is for controlling the temperature of water being heated in a heat exchanger where the heating medium is high-temperature water. The temperature-measuring element of the controller is installed on the secondary side and should be located where it can best detect changes to prevent overheating of outlet water. When the measuring element is located in the outlet pipe, there must be a continuous flow through the exchanger and past the element. The controller regulates the HTW supply to the primary side by means of the control valve in the HTW return. If the water leaving the exchanger is used for space heating, the set point of the thermostat in this water can be readjusted according to outdoor temperature.

Another typical application is to control a low- or medium-pressure steam generator, usually less than 50 psig, using high-temperature water as the heat source. In this application, a proportional pressure controller measures the steam pressure on the secondary side and positions the HTW control valve on the primary side to maintain the desired steam pressure. For general information on automatic controls, refer to Chapter 45 of the 1999 *ASHRAE Handbook—Applications*.

Where submergence is sufficient to prevent flashing in the **vena contracta**, control valves can be in the HTW supply instead of in the return to water heaters and steam generators (see Figure 15). When used in conjunction with a check valve in the return, this arrangement shuts off the high-temperature water supply to the heat exchangers if a tube bundle leaks or ruptures.

WATER TREATMENT

Water treatment for HTW systems should be referred to a specialist. Oxygen introduced in makeup water immediately oxidizes steel at these temperatures, and over a period of time the corrosion can be substantial. Other impurities can also harm boiler tubes. Solids in impure water left by invisible vapor escaping at packings increase maintenance requirements. The condition of the water and the steel surfaces should be checked periodically in systems operating at these temperatures.

For further information, see Chapter 47 of the 1999 *ASHRAE Handbook—Applications,* especially the section on Water Heating Systems under Selection of Water Treatment.

HEAT STORAGE

The high heat storage capacity of water produces a flywheel effect in most HTW systems that evens out load fluctuations. Systems with normal peaks can obtain as much as 15% added capacity through such heat storage. Excessive peak and low loads of a cyclic nature can be eliminated by an HTW accumulator, based on the principle of stratification. Heat storage in an extensive system can sometimes be increased by bypassing water from the supply into the return at the end of the mains, or by raising the temperature of the returns during periods of low load.

SAFETY CONSIDERATIONS

A properly engineered and operated HTW system is safe and dependable. Careful selection and arrangement of components and materials are important. Piping must be designed and installed to prevent undue stress. When high-temperature water is released to atmospheric pressure, flashing takes place, which absorbs a large portion of the energy. Turbulent mixing of the liquid and vapor with room air reduces the temperature well below 212°F. With low mass flow rates, the temperature of the escaping mixture can fall to 125 to 140°F within a short distance, compared with the temperature of the discharge of a low-temperature water system, which remains essentially the same as the temperature of the working fluid (Hansen 1959, Armstrong and Harris 1966).

If large mass flow rates of HTW are released to atmospheric pressure in a confined space (e.g., rupture of a large pipe or vessel) a hazardous condition could exist, similar to that occurring with the rupture of a large steam main. Failures of this nature are rare if good engineering practice is followed in system design.

REFERENCES

Applegate, G. 1958. British and European design and construction methods. *ASHRAE Journal* section, *Heating, Piping & Air Conditioning* 30(3): 169.

Armstrong, C.P. and W.S. Harris. 1966. Temperature distributions in steam and hot water jets simulating leaks. *ASHRAE Transactions* 72(1):147.

ASME. 1998. Power piping. ANSI/ASME *Standard* B31.1-98. American Society of Mechanical Engineers, New York.

Blossom, J.S. and P.H. Ziel. 1959. Pressurizing high-temperature water systems. *ASHRAE Journal* 1(11):47.

Carter, C.A. and B.L. Sturdevant. 1958. Design of high temperature water systems for military installations. *ASHRAE Journal* section, *Heating, Piping & Air Conditioning* 29(2):109.

Hansen, E.G. 1959. Safety of high temperature hot water. *Actual Specifying Engineer* (July).

Hansen, E.G. 1966. Push-pull pumping permits use of MTW in building radiation. *Heating, Piping & Air Conditioning* 38(5):97.

Hansen, E.G. and W. Liddy. 1958. A flexible high pressure hot water and steam boiler plant. *Power* (May):109.

Hansen, E.G. and N.E. Perrsall. 1960. Turbo-generators supply steam for high-temperature water heating. *Air Conditioning, Heating and Ventilating* 32(6):90.

Keenan, J.H. and F.G. Keyes. 1936. *Thermodynamic properties of steam.* John Wiley and Sons, New York.

National Academy of Sciences. 1959. *High temperature water for heating and light process loads.* Federal Construction Council Technical *Report* No. 37. National Research Council Publication No. 753.

NFPA/AGA. 1999. *National fuel gas code.* ANSI/NFPA *Standard* 54-99. National Fire Protection Association, Quincy, MA. ANSI/AGA *Standard* Z223.1-99. American Gas Association, Arlington, VA.

CHAPTER 15

INFRARED RADIANT HEATING

RADIANT principles discussed in this chapter apply to equipment with radiant source temperatures ranging from 300 to 5000°F. Radiant equipment with source temperatures in this range is categorized into three groups as follows:

- Low intensity
- Medium intensity
- High intensity

Radiant equipment with source temperatures from below room temperature to 300°F is classified as panel heating and cooling equipment. See Chapter 6 for further information on panel heating and cooling systems.

Low-intensity source temperatures range from 300 to 1200°F. A typical low-intensity heater is mounted on the ceiling and may be constructed of a 4 in. steel tube 20 to 30 ft long. A gas burner inserted into the end of the tube raises the tube temperature, and because most units are equipped with a reflector, the radiant energy emitted is directed down to the conditioned space.

Medium-intensity source temperatures range from 1200 to 1800°F. Typical sources include porous matrix gas-fired infrared units or metal sheathed electric units.

High-intensity radiant source temperatures range from 1800 to 5000°F. A typical high-intensity unit is an electrical reflector lamp with a resistor temperature of 4050°F.

Low-, medium-, and high-intensity infrared heaters are frequently applied in aircraft hangars, factories, warehouses, foundries, greenhouses, and gymnasiums. They are applied to open areas including loading docks, racetrack stands, under marquees, outdoor restaurants, and around swimming pools. Infrared heaters are also used for snow and ice melting (Chapter 49 of the 1999 *ASHRAE Handbook—Applications*), condensation control, and industrial process heating. Reflectors are frequently used to control the distribution of radiation in specific patterns.

When infrared is used, the environment is characterized by

1. A directional radiant field created by the infrared heaters
2. A radiant field consisting of reradiation and reflection from the walls and/or enclosing surfaces
3. Ambient air temperatures often lower than those found with convective systems

The combined action of these factors determines occupant comfort and the thermal acceptability of the environment.

ENERGY CONSERVATION

Infrared heating units are effective for spot heating. However, due to efficient performance, they are also used for total heating of large areas and entire buildings (Buckley 1989). Radiant heaters transfer energy directly to solid objects. Little energy is lost during transmission because air is a poor absorber of radiant heat. Because an intermediate transfer medium such as air or water is not needed, fans or pumps are not required.

As infrared energy warms floors and objects, they in turn release heat to the air by convection. Reradiation to surrounding objects also contributes to comfort in the area. An energy-saving advantage is that radiant heat can be turned off when it is not needed; when it is turned on again, it is effective in minutes.

Human comfort is determined by the average of mean radiant and dry-bulb temperatures. With radiant heating, the dry-bulb temperature may be kept lower for a given comfort level than with other forms of heating (ASHRAE *Standard* 55). As a result, the heat lost to ventilating air and via conduction through the shell of the structure is proportionally smaller, as is energy consumption. Infiltration loss, which is a function of temperature, is also reduced.

Due to the unique split of radiant and convective components in radiant heating and cooling, air movement and stratification in the conditioned space is minimal. This reduces the infiltration and transmission heat losses.

Buckley and Seel (1987) compared energy savings of infrared heating with those of other types of heating systems. A New York State report (1973) identified annual fuel savings as high as 50%. Recognizing the reduced fuel requirement for these applications, Buckley and Seel (1988) noted that it is common for manufacturers of radiant equipment to recommend installation of equipment with a rated output that is 80 to 85% of the heat loss calculated by methods described in Chapters 27 and 28 of the 1997 *ASHRAE Handbook—Fundamentals*.

Chapman and Zhang (1995) developed a three-dimensional mathematical model to compute radiant heat exchange between surfaces. A building comfort analysis program (BCAP) was developed as part of an ASHRAE-sponsored research project (657-RP) and is available from ASHRAE (Chapman and Jones 1994).

INFRARED ENERGY GENERATORS

Gas Infrared

Modern gas-fired infrared heaters burn gas to heat a specific radiating surface. The surface is heated by direct flame contact or with combustion gases. Studies by the Gas Research Board of London (1944), Plyler (1948), and Haslam et al. (1925) reveal that only 10 to 20% of the energy produced by open combustion of a gaseous fuel is infrared radiant energy. The wavelength span over which radiation from a heated surface is distributed can be controlled by design. The specific radiating surface of a properly designed unit directs radiation toward the load. Heaters are available in the following types (see Table 1 for characteristics).

Indirect infrared radiation units (Figures 1A, 1B, and 1C) are internally fired and have the radiating surface between the hot gases and the load. Combustion takes place within the radiating elements, which operate with surface temperatures up to 1200°F. The elements may be tubes or panels with metal or ceramic components. Indirect infrared radiation units are usually vented and may require eductors.

The preparation of this chapter is assigned to TC 6.5, Radiant Space Heating and Cooling.

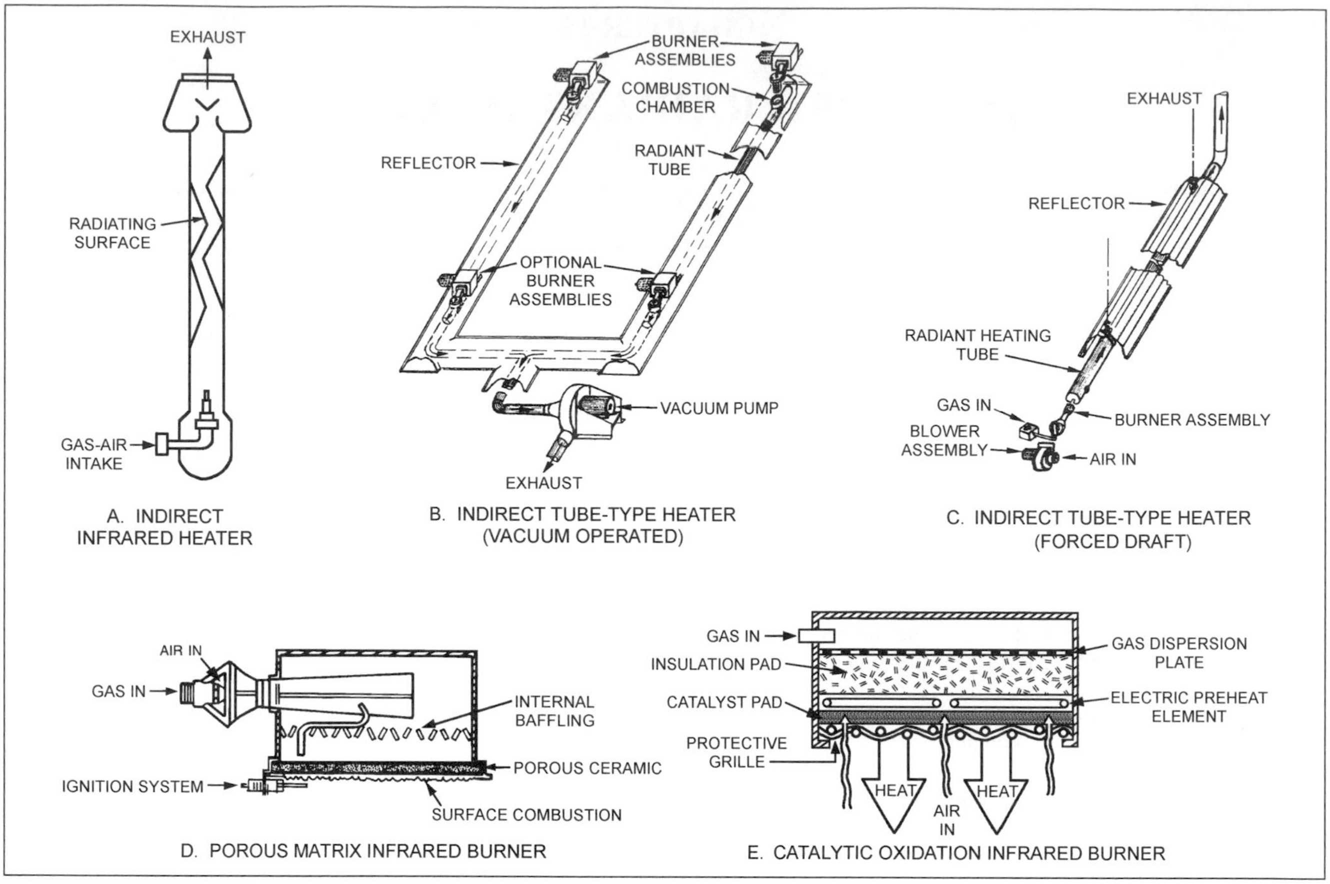

Fig. 1 Types of Gas-Fired Heaters

Table 1 Characteristics of Typical Gas-Fired Infrared Heaters

Characteristics	Indirect	Porous Matrix	Catalytic Oxidation
Operating temperature	To 1200°F	1600 to 1800°F	650 to 700°F
Relative heat intensity,[a] Btu/h·ft^2	Low, to 7500	Medium, 17,000 to 32,000	Low, 800 to 3000
Response time (heat-up)	180 s	60 s	300 s
Radiation-generating ratio[b]	0.35 to 0.55	0.35 to 0.60	No data
Thermal shock resistance	Excellent	Excellent	Excellent
Vibration resistance	Excellent	Excellent	Excellent
Color blindness[c]	Excellent	Very good	Excellent
Luminosity (visible light)	To dull red	Yellow red	None
Mounting height	9 to 50 ft	12 to 50 ft	To 10 ft
Wind or draft resistance	Good	Fair	Very good
Venting	Optional	Nonvented	Nonvented
Flexibility	Good	Excellent—wide range of heat intensities and mounting possibilities available	Limited to low heat intensity applications

[a]Heat intensity emitted at burner surface.
[b]Ratio of radiant output to input.
[c]Color blindness refers to absorptivity by various loads of energy emitted by the different sources.

Porous matrix infrared radiation units (Figure 1D) have a refractory material that may be porous ceramic, drilled port ceramic, stainless steel, or a metallic screen. The units are enclosed, except for the major surface facing the load. A combustible gas-air mixture enters the enclosure, flows through the refractory material to the exposed face, and is distributed evenly by the porous character of the refractory. Combustion occurs evenly on the exposed surface. The flame recedes into the matrix, which adds radiant energy to the flame. If the refractory porosity is suitable, an atmospheric burner can be used, resulting in a surface temperature approaching 1650°F. Power burner operation may be required if refractory density is high. However, the resulting surface temperature may also be higher (1800°F).

Catalytic oxidation infrared radiant units (Figure 1E) are similar to the porous matrix units in construction, appearance, and operation, but the refractory material is usually glass wool, and the radiating surface is a catalyst that causes oxidation to proceed without visible flames.

Electric Infrared

Electric infrared heaters use heat produced by current flowing in a high-resistance wire, graphite ribbon, or film element. The following are the most commonly used types (see Table 2 for characteristics).

Metal sheath infrared radiation elements (Figure 2A) are composed of a nickel-chromium heating wire embedded in an electrical insulating refractory, which is encased by a metal tube.

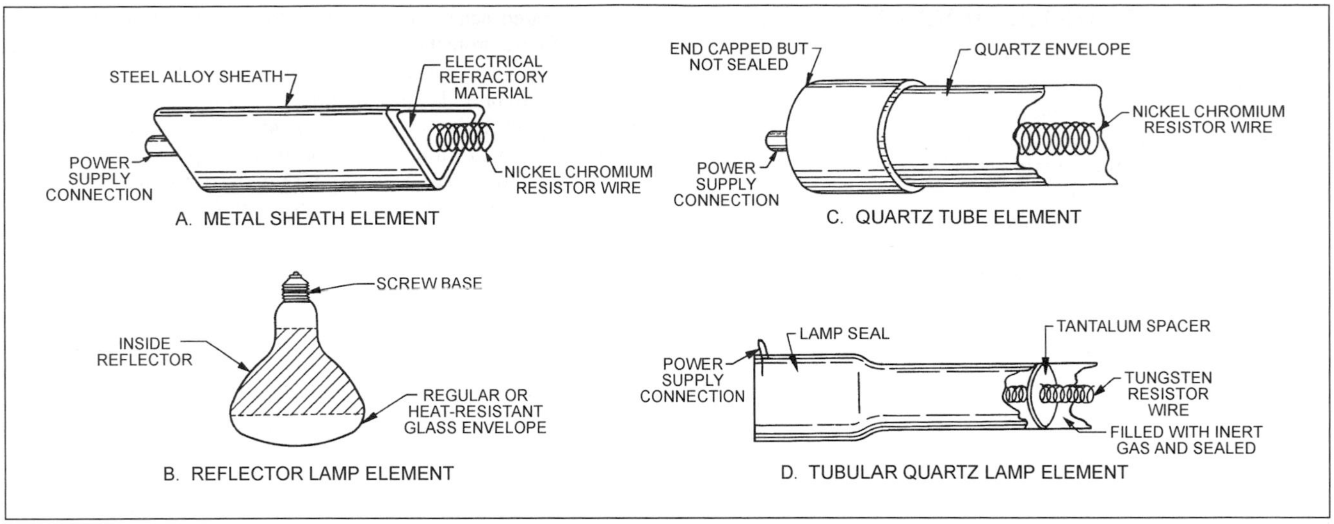

Fig. 2 Common Electric Infrared Heaters

Table 2 Characteristics of Four Electric Infrared Elements

Characteristic	Metal Sheath	Reflector Lamp	Quartz Tube	Quartz Lamp
Resistor material	Nickel-chromium alloy	Tungsten wire	Nickel-chromium alloy	Tungsten wire
Relative heat intensity	Medium, 60 W/in., 0.5 in. dia.	High, 125 to 375 W/spot	Medium to high, 75 W/in., 0.5 in. dia.	High, 100 W/in., 3/8 in. dia.
Resistor temperature	1750°F	4050°F	1700°F	4050°F
Envelope temperature (in use)	1550°F	525 to 575°F	1200°F	1100°F
Radiation-generating ratio[a]	0.58	0.86	0.81	0.86
Response time (heat-up)	180 s	A few seconds	60 s	A few seconds
Luminosity (visible light)	Very low (dull red)	High (8 lm/W)	Low (orange)	High (7.5 lm/W)
Thermal shock resistance	Excellent	Poor to excellent (heat-resistant glass)	Excellent	Excellent
Vibration resistance	Excellent	Medium	Medium	Medium
Impact resistance	Excellent	Medium	Poor	Poor
Wind or draft resistance[b]	Medium	Excellent	Medium	Excellent
Mounting position	Any	Any	Horizontal[c]	Horizontal
Envelope material	Steel alloy	Regular or heat-resistant glass	Translucent quartz	Clear, translucent, or frost quartz and integral red filter glass
Color blindness	Very good	Fair	Very good	Fair
Flexibility	Good—wide range of watt density, length, and voltage practical	Limited to 125-250 and 375 W at 120 V	Excellent—wide range of watt density, diameter, length, and voltage practical	Limited—1 to 3 wattages for each voltage; 1 length for each capacity
Life expectancy	Over 5000 h	5000 h	5000 h	5000 h

[a]Ratio of radiant output to input (elements only).
[b]May be shielded from wind effects by louvers, deep-drawn fixtures, or both.
[c]May be provided with special internal supports for other than horizontal use.

These elements have excellent resistance to thermal shock, vibration, and impact, and they can be mounted in any position. At full voltage, the elements attain a sheath surface temperature of 1200 to 1800°F. Higher temperatures are attained by such configurations as a hairpin shape. These units generally contain a reflector, which directs radiation to the load. Higher radiosity is obtained if the elements are shielded from wind because the surface-cooling effect of the wind is reduced.

Reflector lamp infrared radiation units (Figure 2B) have a coiled tungsten filament, which approximates a point source radiator. The filament is enclosed in a clear, frosted, or red heat-resistant glass envelope, which is partially silvered inside to form an efficient reflector. Units that may be screwed into a light socket are common.

Quartz tube infrared radiant units (Figure 2C) have a coiled nickel-chromium wire lying unsupported within an unevacuated fused quartz tube, which is capped (not sealed) by porcelain or metal terminal blocks. These units are easily damaged by impact and vibration but stand up well to thermal shock and splashing. They must be mounted horizontally to minimize coil sag, and they are usually mounted in a fixture that contains a reflector. Normal operating temperatures are from 1300 to 1800°F for the coil and about 1200°F for the tube.

Tubular quartz lamp units (Figure 2D) consist of a 0.38 in. diameter fused quartz tube containing an inert gas and a coiled tungsten filament held in a straight line and away from the tube by tantalum spacers. Filament ends are embedded in sealing material at the ends of the envelope. Lamps must be mounted horizontally, or nearly so, to minimize filament sag and overheating of the sealed ends. At normal design voltages, quartz lamp filaments operate at about 4050°F, while the envelope operates at about 1100°F.

Oil Infrared

Oil-fired infrared radiation heaters are similar to gas-fired indirect infrared radiation units (Figures 1A, 1B, and 1C). Oil-fired units are vented.

SYSTEM EFFICIENCY

Because many factors contribute to the performance of a specific infrared system, a single criterion should not be used to evaluate comparable systems. Therefore, at least two of the following indicators should be used when evaluating system performance.

Radiation-generating ratio is infrared energy generated divided by total energy input.

Fixture efficiency is an index of a fixture's ability to emit the radiant energy developed by the infrared source; it is usually based on total energy input. The housing, reflector, and other parts of a fixture absorb some infrared energy and convert it to heat, which is lost through convection. A fixture that controls direction and distribution of energy effectively may have a lower fixture efficiency.

Pattern efficiency is an index of a fixture's effectiveness in directing the infrared energy into a specific pattern. This effectiveness, plus effective application of the pattern to the load, influences the total effectiveness of the system (Boyd 1963). Typical radiation-generating ratios of gas infrared generators are indicated in Table 1. Limited test data indicate that the amount of radiant energy emitted from gas infrared units ranges from 35 to 60% of the amount of convective energy. The Stefan-Boltzmann law can be used to estimate the infrared output capability if reasonably accurate values of true surface temperature, emitting area, and surface emittance are available (DeWerth 1960). DeWerth (1962) also addresses the spectral distribution of energy curves for several gas sources.

Table 2 lists typical radiation-generating ratios of electric infrared generators. Fixture efficiencies are typically 80 to 95% of the radiation-generating ratios.

Infrared heaters should be operated at rated input. A small reduction in input causes a larger decrease in radiant output because of the fourth power dependence of radiant output on radiator temperature. As a variety of infrared units with a variety of reflectors and shields are available, the manufacturers' information should be consulted.

REFLECTORS

Radiation from most infrared heating devices is directed by the emitting surface and can be concentrated by reflectors. Mounting height and whether spot heating or total heating is used usually determine which type of reflector will achieve the desired heat-flux pattern at floor level. Four types of reflectors can be used: (1) parabolic, which produce essentially parallel beams of energy; (2) elliptical, which direct all energy that is received or generated at the first focal point through a second focal point; (3) spherical, which are a special class of elliptical reflectors with coincident foci; and (4) flat, which redirect the emitted energy without concentrating or collimating the rays.

Energy data furnished by the manufacturer should be consulted to apply a heater properly.

CONTROLS

Normally, all controls (except the thermostat) are built into gas-fired infrared heaters, whereas electric infrared fixtures usually have no built-in controls. Because of the effects of direct radiation, as well as higher mean radiant temperature (MRT) and decreased ambient temperature compared to warm air systems, infrared heating requires careful selection and location of the thermostat or sensor. Some installers recommend placing the thermostat or sensor in the radiation pattern. The nature of the system, the type of infrared heating units used, and the nature of the thermostat or sensor dictate the appropriate approach. Furthermore, no single location appears to be equally effective during the periods after a cold start and after substantial operation. To reduce high and low temperature swings, a long rather than a short thermostat cycling time is preferred. A properly sized system or modulating or dual-stage operation can improve comfort conditions.

An infrared heater controlled by low-limit thermostats can be used for freeze protection.

On gas-fired infrared units, a thermostat usually controls an automatic valve to provide on-off control of gas flow to all burners. If a unit has a pilot flame, a sensing element prevents the flow of gas to burners (or to both burners and pilot) when the pilot is extinguished. Electrical ignition may be used with provision for manual or automatic reignition of the pilot if it goes out. Electric spark ignition may also be used.

Gas and electric infrared systems for full building heating may have a zone thermostatic control system in which a thermostat representative of one outside exposure operates heaters along that outside wall. Two or more zone thermostats may be required for extremely long wall exposures. Heaters for an internal zone may be grouped around a thermostat representative of that zone. Manual switches or thermostats are usually used for spot or area heating, but input controllers may also be used.

Electric infrared heating units with metal sheath or quartz tube elements are effectively controlled by input controllers. An input controller is a motor-driven cycling device in which *on* time per cycle can be set. A 30 s cycle is normal. When a circuit's capacity exceeds an input controller's rating, the controller can be used to cycle a pilot circuit of contactors adequate for the load.

Input controllers work well with metal sheath heaters because the sheath mass smooths the pulses into even radiation. The control method decreases the efficiency of infrared generation slightly. When controlled with these devices, quartz tube elements, which have a warm-up time of several seconds, have perceptible but not normally disturbing pulses of infrared, with only moderate reduction in generation efficiency.

Input controllers should not be used with quartz lamps because the cycling luminosity would be distracting. Instead, output from a quartz lamp unit can be controlled by changing the voltage to the lamp element. The voltage can be changed by using modulating transformers or by switching the power supply from hot-to-hot to hot-to-ground potential.

Power drawn by the tungsten filament of the quartz lamp varies approximately as the 1.5 power of the voltage, while that of metal sheath or quartz tube elements (using nickel-chromium wire) varies as the square of the voltage. Multiple circuits for electric infrared systems can be manually or automatically switched to provide multiple stages of heat. Three circuits or control stages are usually adequate. For areas with fairly uniform radiation, one circuit should be controlled with input control or voltage variation control on electric units, while the other two are on full on or off control. This arrangement gives flexible, staged control with maximum efficiency of infrared generation. The variable circuit alone provides zero to one-third capacity. Adding another circuit at full *on* provides one-third to two-thirds capacity, and adding the third circuit provides two-thirds to full capacity.

PRECAUTIONS

Precautions for the application of infrared heaters include the following:

- All infrared heaters covered in this chapter have high surface temperatures when they are operating and should, therefore, not be used when the atmosphere contains ignitable dust, gases, or vapors in hazardous concentrations.
- Manufacturers' recommendations for clearance between a fixture and combustible material should be followed. If combustible material is being stored, warning notices defining proper clearances should be posted near the fixture.
- Manufacturers' recommendations for clearance between a fixture and personnel areas should be followed to prevent personnel stress due to local overheating.

- Infrared fixtures should not be used if the atmosphere contains gases, vapors, or dust that decomposes to hazardous or toxic materials in the presence of high temperature and air. For example, infrared units should not be used in an area with a degreasing operation that uses trichloroethylene unless the area has a suitable exhaust system that isolates the contaminant. Trichloroethylene, when heated, forms phosgene (a toxic compound) and hydrogen chloride (a corrosive compound).
- Humidity must be controlled in areas with unvented gas-fired infrared units because water formed by combustion increases humidity. Sufficient ventilation (NFPA/AGA *National Fuel Gas Code*), direct venting, or insulation on cold surfaces helps control moisture problems.
- Lamp holders and grounding for infrared heating lamps should comply with Section 422-15 of the *National Electrical Code* (NFPA *Standard* 70).
- Adequate makeup air (NFPA/AGA *National Fuel Gas Code*) must be provided to replace the air used by combustion-type heaters, regardless of whether units are direct vented.
- If unvented combustion-type infrared heaters are used, the area must have adequate ventilation to ensure that products of combustion in the air are held to an acceptable level (Prince 1962).
- For comfort, personnel should be protected from substantial wind or drafts. Suitable wind shields seem to be more effective than increased radiation density (Boyd 1960).

In the United States, refer to Occupational Safety and Health Administration (OSHA) guidelines for additional information.

MAINTENANCE

Gas- and oil-fired infrared heaters require periodic cleaning to remove dust, dirt, and soot. Reflecting surfaces must be kept clean to remain efficient. An annual cleaning of heat exchangers, radiating surfaces, burners, and reflectors with compressed air is usually sufficient. Chemical cleaners must not leave a film on reflector surfaces.

Both main and pilot air ports of gas-fired units should be kept free of lint and dust. The nozzle, draft tube, and nose cone of oil-fired unit burners are designed to operate in a particular combustion chamber, so they must be replaced carefully when they are removed.

Electric infrared systems require little care beyond the cleaning of reflectors. Quartz and glass elements must be handled carefully because they are fragile, and fingerprints must be removed (preferably with alcohol) to prevent etching at operating temperature, which causes early failure.

DESIGN CONSIDERATIONS FOR BEAM RADIANT HEATERS

Chapter 52 of the 1999 *ASHRAE Handbook—Applications* introduces the principles of design for spot beam radiant heating. The effective radiant flux (ERF) represents the radiant energy absorbed by an occupant from all temperature sources different from the ambient. ERF is defined as

$$\mathrm{ERF} = h_r(\bar{t}_r - t_a) \tag{1}$$

where

ERF = effective radiant flux, Btu/h·ft²
h_r = linear radiation transfer coefficient, Btu/h·ft²·°F
$\bar{t}_r$ = mean radiant temperature affecting occupant, °F
t_a = ambient air temperature near occupant, °F

ERF may be measured as the heat absorbed at the skin-clothing surface from a beam heater treated as a point source:

$$\mathrm{ERF} = \frac{\alpha_K I_K (A_p/d^2)}{A_D} \tag{2}$$

where

α_K = absorptance of skin-clothing surface at emitter temperature (Figure 3), dimensionless
I_K = irradiance from beam heater, Btu/h·sr
A_p = projected area of occupant on plane normal to direction of heater beam, ft²
d = distance from beam heater to center of occupant, ft
A_D = body surface area of occupant, ft²

A_p/d^2 is the solid angle subtended by the projected area of the occupant from the beam heater. See Figure 5 in Chapter 52 of the 1999 *ASHRAE Handbook—Applications* for a representation of these variables. The value of the DuBois area A_D has been defined as follows:

$$A_D = 0.108 W^{0.425} H^{0.725}$$

where

W = mass of occupant, lb
H = height of occupant, in.

Two radiation area factors are defined as

$$f_{eff} = A_{eff}/A_D \tag{3}$$

$$f_p = A_p/A_{eff} \tag{4}$$

where A_{eff} is the effective radiating area of the total body surface. Equation (2) becomes

$$\mathrm{ERF} = \alpha_K f_{eff} f_p I_K/d^2 \tag{5}$$

Fanger (1973) developed precise optical methods to evaluate the angle factors f_{eff} and f_p for both sitting and standing positions and for males and females. An average value for f_{eff} of 0.71 for both sitting and standing is accurate within ±2%. The variations in angle factor f_p over various azimuths and elevations for seated or standing positions are illustrated in Figures 4 and 5, and according to Fanger, they apply equally to both males and females.

Manufacturers of infrared heating equipment usually supply performance specifications for their equipment (Gagge et al. 1967). Design information regarding sizing infrared heating units is also

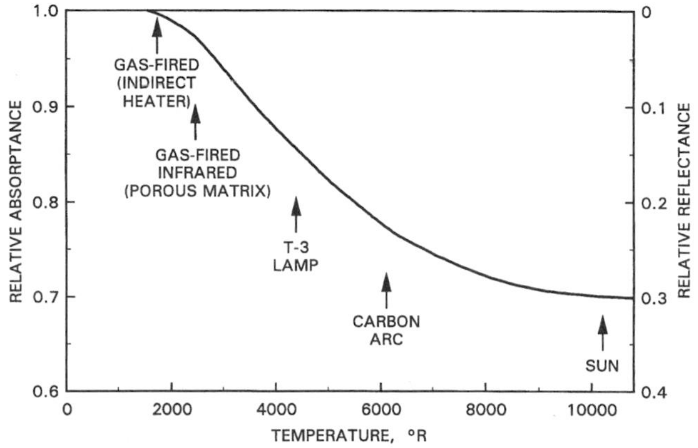

Fig. 3 Relative Absorptance and Reflectance of Skin and Typical Clothing Surfaces

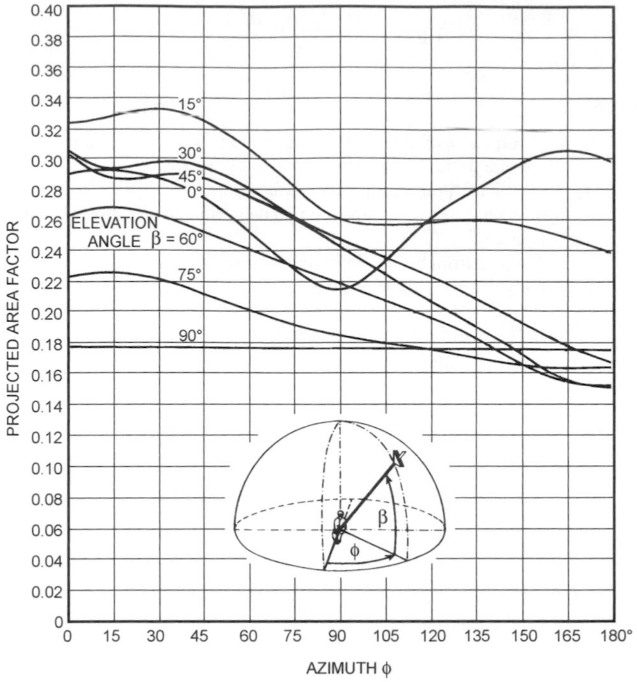

Fig. 4 Projected Area Factor for Seated Persons, Nude and Clothed
(Fanger 1973)

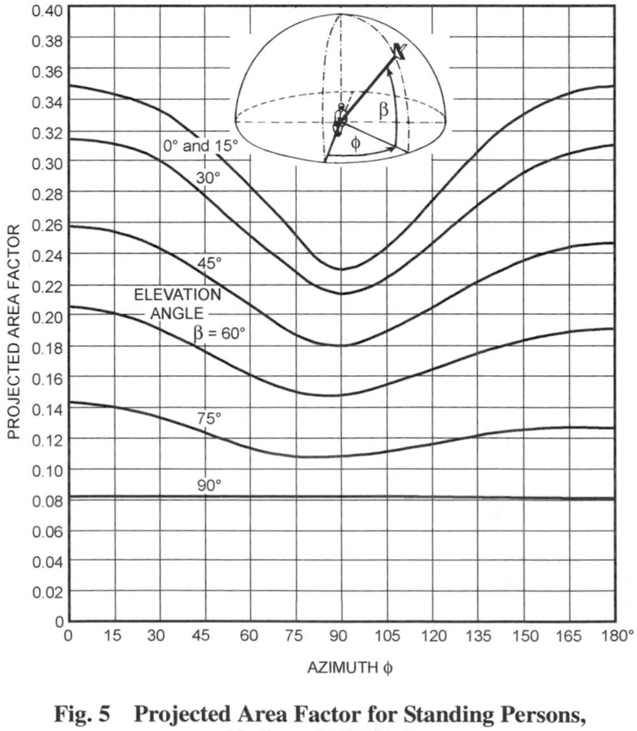

Fig. 5 Projected Area Factor for Standing Persons, Nude and Clothed
(Fanger 1973)

available (Howell and Suryanarayana 1990), as is the relation between color temperature of heaters and the applied voltage or wattage is also available. Gas-fired radiators usually operate at constant emitting temperatures of 1340 to 1700°F (1800 to 2160°R). Figure 3 relates the absorptance α_K to the radiating temperature of the radiant source. Manufacturers also supply the radiant flux

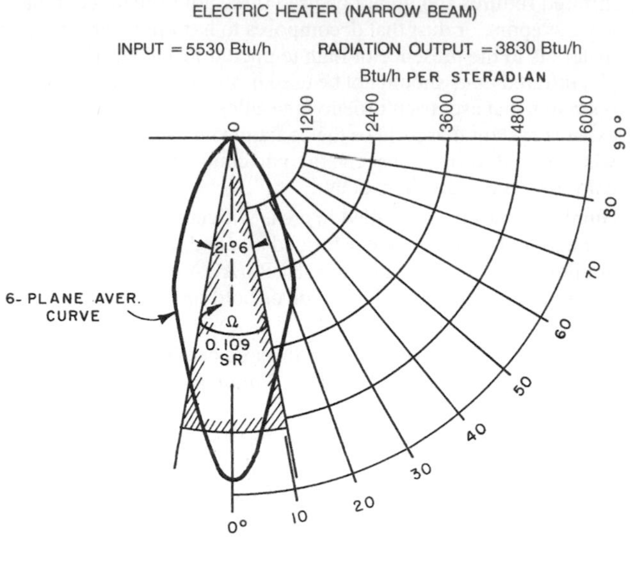

Fig. 6 Radiant Flux Distribution Curve of Typical Narrow-Beam High-Intensity Electric Infrared Heaters

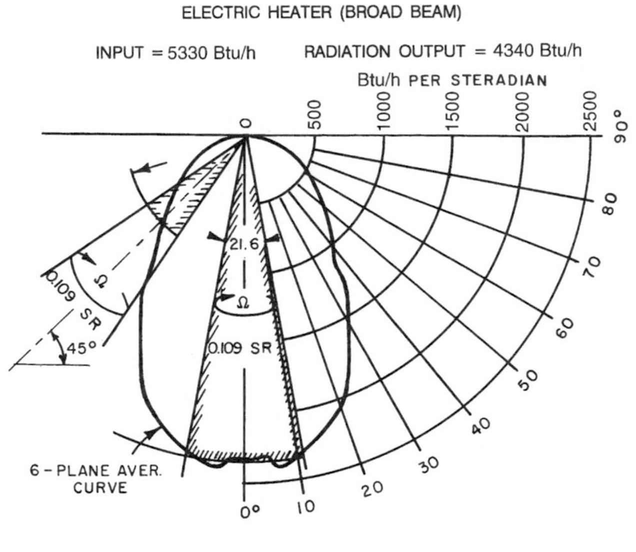

Fig. 7 Radiant Flux Distribution Curve of Typical Broad-Beam High-Intensity Electric Infrared Heaters

distribution of beam heaters with and without reflectors. Figures 6 through 9 illustrate the radiant flux distribution for four typical electric and gas-fired radiant heaters. In general, electrical beam heaters produce as much as 70 to 80% of their total energy output as radiant heat, in contrast to 40% for gas-fired types. In practice, the designer should choose a beam heater that will illuminate the subject with acceptable uniformity. Even with complete illumination by a beam 0.109 sr (21.6°) wide, Figure 6 shows that only 8% (100 × 1225 × 0.109/1620) of the initial input wattage to the heater is usable for specifying the necessary I_K in Equation (5). The corresponding percentages for Figures 7, 8, and 9 are 5%, 4%, and 2%, respectively. The last two are for gas-fired beams.

The input energy to the beam heater not used for directly irradiating the occupant ultimately increases the ambient air temperature and mean radiant temperature of the room. This increase will reduce the original ERF required for comfort and acceptability. The continuing reradiation and convective heating of surrounding walls and

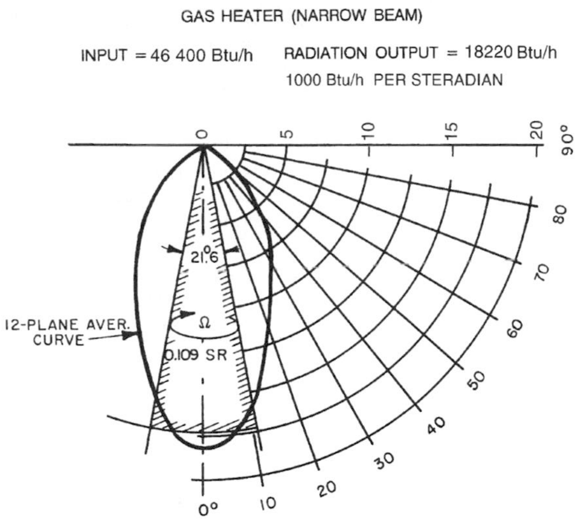

Fig. 8 Radiant Flux Distribution Curve of Typical Narrow-Beam High-Intensity Atmospheric Gas-Fired Infrared Heaters

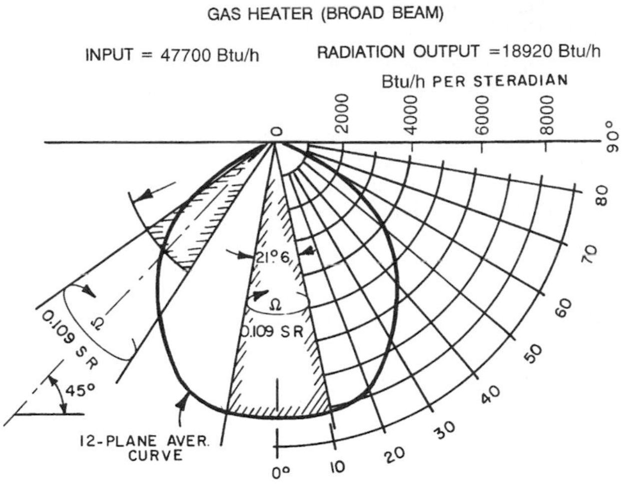

Fig. 9 Radiant Flux Distribution Curve of Typical Broad-Beam High-Intensity Atmospheric Gas-Fired Infrared Heaters

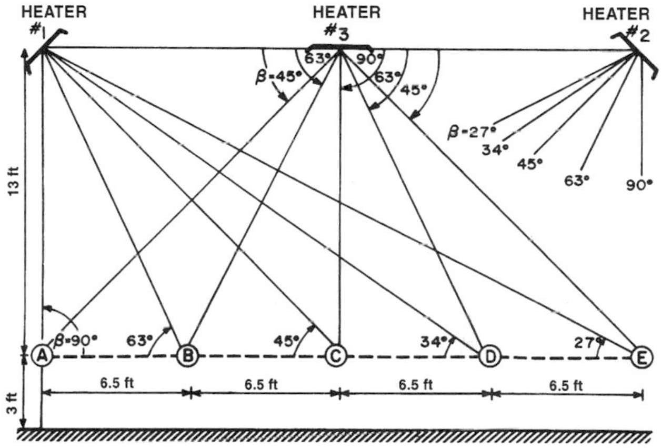

Fig. 10 Calculation of Total ERF from Three Gas-Fired Heaters on Worker Standing at Positions A Through E

the presence of air movement make precise calculations of radiant heat exchange difficult.

The basic principles of beam heating are illustrated by the following examples.

Example 1. Determine the beam radiation intensity required for comfort from a quartz lamp (Figure 6) when the worker is sedentary, lightly clothed (0.5 clo), and seated. The ambient t_a is 59°F, with air movement 30 fpm. The lamp is mounted on the 8 ft high ceiling and is directed at the back of the seated person so that the elevation angle β is 45° and the azimuth angle ϕ is 180°. Assume the ambient and mean radiant temperatures of the unheated room are equal. The lamp operates at 240 V and has an emitter temperature of 4500°R.

Solution: The ERF for comfort can be calculated as 21.5 Btu/h·ft² by procedures outlined in the section on Design Criteria for Acceptable Radiant Heating in Chapter 52 of the 1999 *ASHRAE Handbook—Applications*. α_K = 0.85 at 4500°R (from Figure 3); f_{eff} = 0.71; f_p = 0.17 (Figure 4); d = 8 − 2 = 6 ft, where 2 ft is sitting height of occupant.

From Equation (5), the irradiance I_K from the beam heater necessary for comfort is

$$I_K = \text{ERF} \cdot d^2/\alpha_K f_{eff} f_p$$

$$= 21.5(6)^2/(0.85 \times 0.71 \times 0.17) = 7540 \text{ Btu/h} \cdot \text{sr}$$

Example 2. For the same occupant in Example 1, when two beams located on the ceiling and operating at half of rated voltage are directed downward at the subject at 45° and at azimuth angle 90° on each side, what would be the I_K required from each heater?

Solution: The ERF for comfort from each beam is 21.5/2 or 10.75 Btu/h·ft². The value of f_p is 0.25 (from Figure 4). At half power, V = 170, $R \approx 3600$, and $\alpha_K \approx 0.9$. Hence, the required irradiation from each beam is

$$I_K = 10.75(6)^2/(0.9 \times 0.71 \times 0.25)$$

$$= 2420 \text{ Btu/h} \cdot \text{sr}$$

This estimate indicates that two beams similar to Figure 7, each operating at half of rated power, can produce the necessary ERF for comfort. A comparison between the I_K requirements in Examples 1 and 2 shows that irradiating a sitting person from the back is much less efficient than irradiating from the side.

Example 3. A broad-beam gas-fired infrared heater is mounted 16 ft above the floor. The heater is directed 45° downward toward a standing subject 13 ft away (see Heater #1 in Figure 10, position C).

Question (1): What is the resulting ERF from the beam acting on the subject?

Solution: From Equation (5),

$$\text{ERF} = \alpha_K f_{eff} f_p I_K/d^2 = 4.2 \text{ Btu/h} \cdot \text{ft}^2$$

where

α_K = 0.97 (Figure 3)
f_{eff} = 0.71
f_p = 0.26 (Figure 5 at β = 45° and ϕ = 0°)
I_K = 8000 Btu/h·sr (Figure 9)
d^2 = 13² + 13² = 338 ft² (The center of the standing man is 3 ft above the floor.)

If the heater were 10 ft above the standing subject (13 ft above floor), the ERF would be 5.3 Btu/h·ft².

Question (2): How does the ERF vary along the 0° azimuth, every 6.5 ft beginning at a point directly under Heater #1 (Figure 10) and for elevations β = 90° at A, 63.4° at B, 45° at C, 33.6° at D, and 26.6° at E?

From Figure 5, the values for f_p for the five positions are (A) 0.08, (B) 0.19, (C) 0.26, (D) 0.30, and (E) 0.33.

Solution: Because the beam is directed 45° downward, the respective deviations from the beam center for a person standing at the five positions A through E are 45°, 18.4°, 0°, 11.4°, and 18.4°; the corresponding I_K values from Figure 9 are 5000, 8000, 8000, 8000, and 8000 Btu/h·sr. The respective d^2 are 169, 211, 338, 549, and 845 ft². The ERFs for a person standing in the five positions are (A) 1.6, (B) 4.9, (C) 4.2, (D) 3.0, and (E) 2.1 Btu/h·ft².

Question (3): How will the total ERF at each of the five locations A through E vary if two additional heaters (#2 and #3 in Figure 10) are added 16 ft above the floor over positions C and E? The center heater is directed downward, the outer one directed as above, 45° towards the center of the room.

Solution: At each of five room locations (A, B, C, D, E), add the ERF from each of the three radiators to determine the total ERF affecting the standing person.

A	1.6 + 2.5 + 2.1	or	6.2
B	4.9 + 4.2 + 3.0	or	12.1
C	4.2 + 2.5 + 4.2	or	10.9
D	3.0 + 4.2 + 4.9	or	12.1
E	2.1 + 2.5 + 5	or	6.2

REFERENCES

ASHRAE. 1992. Thermal environmental conditions for human occupancy. ANSI/ASHRAE *Standard* 55-1992.

Boyd, R.L. 1960. What do we know about infrared comfort heating? *Heating, Piping and Air Conditioning* (November):133.

Boyd, R.L. 1963. Control of electric infrared energy distribution. *Electrical Engineering* (February):103.

Buckley, N.A. 1989. Applications of radiant heating saves energy. *ASHRAE Journal* 31(9):17.

Buckley, N.A. and T.P. Seel. 1987. Engineering principles support an adjustment factor when sizing gas-fired low-intensity infrared equipment. *ASHRAE Transactions* 93(1):1179-91.

Buckley, N.A. and T.P. Seel. 1988. Case studies support adjusting heat loss calculations when sizing gas-fired, low-intensity, infrared equipment. *ASHRAE Transactions* 94(1):1848-58.

Chapman, K.S. and B.W. Jones. 1994. Simplified method to factor mean radiant temperature (MRI) into building and HVAC design. 657-RP *Final Report*. ASHRAE.

Chapman, K.S. and P. Zhang. 1995. Radiant heat exchange calculations in radiantly heated and cooled enclosures. *ASHRAE Transactions* 101(1):1236-47.

DeWerth, D.W. 1960. Literature review of infra-red energy produced with gas burners. *Research Bulletin* No. 83. American Gas Association, Cleveland, OH.

DeWerth, D.W. 1962. A study of infra-red energy generated by radiant gas burners. *Research Bulletin* No. 92. American Gas Association, Cleveland, OH.

Fanger, P.O. 1973. *Thermal comfort*. McGraw-Hill, New York.

Gagge, A.P., G.M. Rapp, and J.D. Hardy. 1967. The effective radiant field and operative temperature necessary for comfort with radiant heating. *ASHRAE Transactions* 73(1) and *ASHRAE Journal* 9:63-66.

Gas Research Board of London. 1944. The use of infra-red radiation in industry. *Information Circular* No. 1.

Haslam, W.G. et al. 1925. Radiation from non-luminous flames. *Industrial and Engineering Chemistry* (March).

Howell, R. and S. Suryanarayana. 1990. Sizing of radiant heating systems: Part II—Heated floors and infrared units. *ASHRAE Transactions* 96(1):666-75.

NFPA/AGA. 1996. *National fuel gas code*. ANSI/NFPA *Standard* 54-99. National Fire Protection Association, Quincy, MA. ANSI/IAS *Standard* Z223.1-99. American Gas Association, Cleveland, OH.

NFPA. 1998. *National electrical code*. ANSI/NFPA *Standard* 70-98.

New York State. 1973. *Energy efficiency in large buildings*. Interdepartmental Fuel and Energy Committee Ad Hoc Report.

Plyler, E.K. 1948. Infra-red radiation from bunsen flames. National Bureau of Standards *Journal of Research* 40(February):113.

Prince, F.J. 1962. Selection and application of overhead gas-fired infrared heating devices. *ASHRAE Journal* (October):62.

BIBLIOGRAPHY

Carlson, B.G. and K.D. Lathrop. 1963. *Transport theory—The method of discrete-ordinates in computing methods in reactor physics*, eds. Grenspan, Keller, and Okrent. Gordon and Breach, New York.

Chapman, K.S., J.M. DeGreef, and R.D. Watson. 1997. Thermal comfort analysis using BCAP for retrofitting a radiantly heated residence. *ASHRAE Transactions* 103(1):959-65.

Chapman, K.S., S. Ramadhyani, and R. Viskanta. 1992. Modeling and parametric studies of heat transfer in a direct-fired furnace with impinging jets. Presented at the 1992 ASME Winter Annual Meeting, Anaheim, CA.

Chapman, K.S. and P. Zhang. 1996. Energy transfer simulation for radiantly heated and cooled enclosures. *ASHRAE Transactions* 102(1):76-85.

DeGreef, J.M. and K.S. Chapman. 1998. Simplified thermal comfort evaluation of MRT gradients and power consumption predicted with the BCAP methodology. *ASHRAE Transactions* 104(1B):1090-97.

Fanger, P. 1967. Calculation of thermal comfort: Introduction of a basic comfort equation. *ASHRAE Transactions* 73(2):III.4.1-20.

Fiveland, W.A. 1984. Discrete-ordinates solutions of the radiative transport equation for rectangular enclosures. *Journal of Heat Transfer* 106:699-706.

Fiveland, W.A. 1987. Discrete-ordinates methods for radiative heat transfer in isotropically and anisotropically scattering media. *Journal of Heat Transfer* 109:809-12.

Fiveland, W.A. 1988. Three dimensional radiative heat-transfer solutions by the discrete-ordinates method. *Journal of Thermophysics and Heat Transfer* 2(4):309-16.

Fiveland, W.A. and A.S. Jamaluddin. 1989. Three-dimensional spectral radiative heat transfer solutions by the discrete-ordinates method. ASME Heat Transfer Conference Proceedings, *Heat Transfer Phenomena in Radiation, Combustion, and Fires*. HTD-106:43-48. American Society of Mechanical Engineers, New York.

Gan, G. and D.J. Croome. 1994. Thermal comfort models based on field measurements. *ASHRAE Transactions* 100(1):782-94.

Incropera, F.P. and D.P. DeWitt. 1990. *Fundamentals of heat and mass transfer*. John Wiley and Sons, New York.

Jamaluddin, A.S. and P.J. Smith. 1988. Predicting radiative transfer in rectangular enclosures using the discrete ordinates method. *Combustion Science and Technology* 59:321-40.

Jones, B.W., W.F. Niedringhaus, and M.R. Imel. 1989. Field comparison of radiant and convective heating in vehicle repair buildings. *ASHRAE Transactions* 95(1):1045-51.

Modest, M.F. 1993. *Radiative heat transfer*. McGraw-Hill, New York.

NAHB. 1994. Enerjoy case study: A comparative analysis of thermal comfort conditions and energy consumption for Enerjoy PeopleHeaters™ and a conventional heating system. *Project Report* No. 4159. National Association of Home Builders Research Center, Upper Marlboro, MD.

Özisik, M.N. 1977. *Basic heat transfer*. McGraw-Hill, New York.

Patankar, S.V. 1980. *Numerical heat transfer and fluid flow*. McGraw-Hill, New York.

Sanchez, A. and T.F. Smith. 1992. Surface radiation exchange for two-dimensional rectangular enclosures using the discrete-ordinates method. *Journal of Heat Transfer* 114:465-72.

Siegel, R. and J.R. Howell. 1981. *Thermal radiation heat transfer*. McGraw-Hill, New York.

Truelove, J.S. 1987. Discrete-ordinates solutions of the radiative transport equation. *Journal of Heat Transfer* 109:1048-51.

Truelove, J.S. 1988. Three-dimensional radiation in absorbing-emitting media. *Journal of Quantitative Spectroscopy & Radiative Transfer* 39(1):27-31.

Viskanta, R. and M.P. Mengüc. 1987. Radiation heat transfer in combustion systems. *Progress in Energy and Combustion Science* 13.

Viskanta, R. and S. Ramadhyani. 1988. Radiation heat transfer in directly-fired natural gas furnaces: A review of literature. GRI *Report* GRI-88/0154. Gas Research Institute, Chicago, IL.

Watson, R.D., K.S. Chapman, and J. DeGreef. 1998. Case study: Seven-system analysis of thermal comfort and energy use for a fast-acting radiant heating system. *ASHRAE Transactions* 104(1B):1106-11.

Yücel, A. 1989. Radiative transfer in partitioned enclosures. ASME Heat Transfer Conference Proceedings, *Heat Transfer Phenomena in Radiation, Combustion, and Fires*. HTD-106:35-41. American Society of Mechanical Engineers, New York.

CHAPTER 16

DUCT CONSTRUCTION

THIS CHAPTER covers the construction of heating, ventilating, air-conditioning, and exhaust duct systems for residential, commercial, and industrial applications. Technological advances in duct construction should be judged relative to the construction requirements described here and to appropriate codes and standards. While the construction materials and details shown in this chapter may coincide, in part, with industry standards, they are not in an ASHRAE standard.

BUILDING CODE REQUIREMENTS

In the United States private sector, each new construction or renovation project is normally governed by state laws or local ordinances that require compliance with specific health, safety, property protection, and energy conservation regulations. Figure 1 illustrates relationships between laws, ordinances, codes, and standards that can affect the design and construction of HVAC duct systems; however, Figure 1 may not list all applicable regulations and standards for a specific locality. Specifications for the U.S. federal government construction are promulgated by such agencies as the Federal Construction Council, the General Services Administration, the Department of the Navy, and the Veterans Administration.

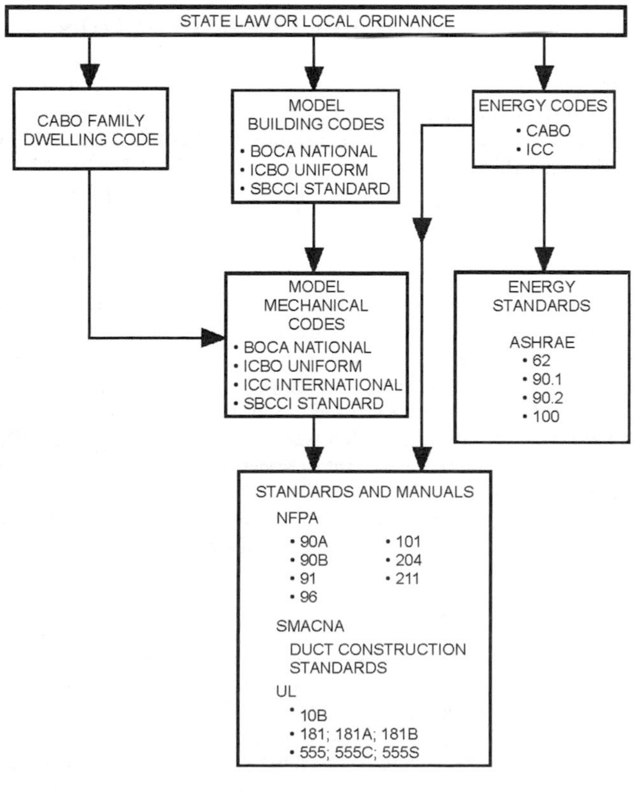

Fig. 1 Hierarchy of Building Codes and Standards

Since the development of safety codes, energy codes and standards proceed independently; the most recent edition of a code or standard may not have been adopted by a local jurisdiction. HVAC designers must know which code compliance obligations affect their designs. If a provision conflicts with the design intent, the designer should resolve the issue with local building officials. New or different construction methods can be accommodated by the provisions for equivalency that are incorporated into codes. Staff engineers from the model code agencies are available to assist in the resolution of conflicts, ambiguities, and equivalencies.

Smoke management is covered in Chapter 51 of the 1999 *ASHRAE Handbook—Applications*. The designer should consider flame spread, smoke development, combustibility, and toxic gas production from duct and duct insulation materials. Code documents for ducts in certain locations within buildings rely on a criterion of limited combustibility (see definitions, NFPA *Standard* 90A), which is independent of the generally accepted criteria of 25 flame spread and 50 smoke development; however, certain duct construction protected by extinguishing systems may be accepted with higher levels of combustibility by code officials.

Combustibility and toxicity ratings are normally based on tests of new materials; little research is reported on ratings of duct materials that are aged or of systems that are poorly maintained for cleanliness.

CLASSIFICATION OF DUCTS

Duct construction is classified by application and pressure. HVAC systems in public assembly, business, educational, general factory, and mercantile buildings are usually designed as commercial systems. Air pollution control systems, industrial exhaust systems, and systems outside the pressure range of commercial system standards are classified as industrial systems.

Classifications are as follows:

Residences	±0.5 in. of water
	±1 in. of water
Commercial Systems	±0.5 in. of water
	±1 in. of water
	±2 in. of water
	±3 in. of water
	±4 in. of water
	±6 in. of water
	±10 in. of water
Industrial Systems	Any pressure

The preparation of this chapter is assigned to TC 5.2, Duct Design.

Air conveyed by a duct imposes both air pressure and velocity pressure loads on the duct's structure. The load resulting from mean static pressure differential across the duct wall normally dominates and is generally used for duct classification. Turbulent airflow introduces relatively low but rapidly pulsating loading on the duct wall.

Static pressure at specific points in an air distribution system is not necessarily the static pressure rating of the fan; the actual static pressure in each duct section must be obtained by computation. Therefore, the designer should specify the pressure classification of the various duct sections in the system. All modes of operation must be taken into account, especially systems used for smoke management and those with fire dampers that must close with the system running.

DUCT CLEANING

Ducts may collect dirt and moisture, which can harbor or transport microbial contaminants. Ducts should be designed, constructed, and maintained to minimize the opportunity for growth and dissemination of microorganisms. Recommended control measures include providing access for cleaning, providing proper filtration, and preventing moisture and dirt accumulation. NADCA (1992) and NAIMA (1993) have specific information and procedures for cleaning ducts. Owners should routinely conduct inspections for cleanliness.

DUCT LEAKAGE

Predicted leakage rates for unsealed and sealed ducts are reviewed in Chapter 32 of the 1997 *ASHRAE Handbook—Fundamentals*. Project specifications should define allowable duct leakage, specify the need for leak testing, and require the ductwork installer to perform a leak test after installing an initial portion of the duct. Ducts should be sealed in compliance with Table 1; duct seal levels are defined in Table 2. Exposed supply ductwork in conditioned spaces should be seal level A to prevent dirt smudges. Leakage classifications for ductwork are given in Table 6 in Chapter 32 of the 1997 *ASHRAE Handbook—Fundamentals*. Procedures in the *HVAC Air Duct Leakage Test Manual* (SMACNA 1985) should be followed for leak testing. If a test indicates excess leakage, corrective measures should be taken to ensure quality.

Table 1 Recommended Duct Seal Levels[a]

	Duct Type			
	Supply			
Duct Location	**≤ 2 in. water**	**> 2 in. water**	**Exhaust**	**Return**
Outdoors	A	A	A	A
Unconditioned spaces	B	A	B	B
Conditioned spaces (concealed ductwork)	C	B	B	C
Conditioned spaces (exposed ductwork)	A	A	B	B

[a]See Table 2 for definition of seal level.

Table 2 Duct Seal Levels[a]

Seal Level	Sealing Requirements
A	All transverse joints, longitudinal seams, and duct wall penetrations
B	All transverse joints and longitudinal seams
C	Transverse joints only

[a]Transverse joints are connections of two ducts oriented perpendicular to flow. Longitudinal seams are joints oriented in the direction of airflow. Duct wall penetrations are openings made by screws, non-self-sealing fasteners, pipe, tubing, rods, and wire. Round and flat oval spiral lock seams need not be sealed prior to assembly, but may be coated after assembly to reduce leakage. All other connections are considered transverse joints, including but not limited to spin-ins, taps and other branch connections, access door frames, and duct connections to equipment.

Table 3 Residential Metal Duct Construction

Shape of Duct and Exposure	Galvanized Steel Minimum Thickness, in.	Aluminum (3003) Nominal Thickness, in.
Enclosed rectangular ducts[a]		
14 in. or less	0.0127	0.016
Over 14 in.	0.0157	0.020
Rectangular and round ducts	Consult SMACNA (1995)	

[a]Data based on nominal thickness, NFPA *Standard* 90B.

Responsibility for proper assembly and sealing belongs to the installing contractor. The most cost-effective way to control leakage is to follow proper installation procedures. However, the incremental cost of achieving 1% or less leakage becomes prohibitively high, particularly for large duct systems. Because access for repairs is usually limited, poorly installed duct systems that must later be resealed can cost more than a proper installation.

RESIDENTIAL DUCT CONSTRUCTION

NFPA *Standard* 90B, CABO's *One- and Two-Family Dwelling Code*, or a local code is used for duct systems in single-family dwellings. Generally, local authorities use NFPA *Standard* 90A for multifamily homes.

Supply ducts may be steel, aluminum, or a material with a UL *Standard* 181 listing. Sheet metal ducts should be constructed of minimum thickness as shown in Table 3 and installed in accordance with *HVAC Duct Construction Standards—Metal and Flexible* (SMACNA 1995). Fibrous glass ducts should be installed in accordance with the *Fibrous Glass Duct Construction Standards* (SMACNA 1992, NAIMA 1997). For return ducts, the use of alternate materials, and other exceptions, consult NFPA *Standard* 90B.

COMMERCIAL DUCT CONSTRUCTION

Materials

NFPA *Standard* 90A is used as a guide standard by many building code agencies. NFPA *Standard* 90A invokes UL *Standard* 181, which classifies ducts as follows:

Class 0—Zero flame spread, zero smoke developed
Class 1—25 flame spread, 50 smoke developed

NFPA *Standard* 90A states that ducts must be iron, steel, aluminum, concrete, masonry, or clay tile. However, ducts may be UL *Standard* 181 Class 1 materials when they are not used as vertical risers of more than two stories or in systems with air temperature higher than 250°F. Many manufactured flexible and fibrous glass ducts are UL approved and listed as Class 1. For galvanized ducts, a G90 coating is recommended (see ASTM *Standard* A 653). The minimum thickness and weight of sheet metal sheets are given in Tables 4A, 4B, and 4C.

External duct-reinforcing members are formed from sheet metal or made from hot-rolled or extruded structural shapes. The size and weights of commonly used members are given in Table 5.

Rectangular and Round Ducts

Rectangular Metal Ducts. *HVAC Duct Construction Standards—Metal and Flexible* (SMACNA 1995) lists construction requirements for rectangular steel ducts and includes combinations of duct thicknesses, reinforcement, and maximum distance between reinforcements. Transverse joints (e.g., standing drive slips, pocket locks, and companion angles) and, when necessary, intermediate structural members are designed to reinforce the duct system. Ducts 85 in. and larger at a pressure of 6 in. of water and greater require

Table 4A Galvanized Sheet Thickness

Galvanized Sheet Gage[a]	Thickness, in. Nominal	Thickness, in. Minimum[b]	Nominal Weight, lb/ft^2
30	0.0157	0.0127	0.656
28	0.0187	0.0157	0.781
26	0.0217	0.0187	0.906
24	0.0276	0.0236	1.156
22	0.0336	0.0296	1.406
20	0.0396	0.0356	1.656
18	0.0516	0.0466	2.156
16	0.0635	0.0575	2.656
14	0.0785	0.0705	3.281
13	0.0934	0.0854	3.906
12	0.1084	0.0994	4.531
11	0.1233	0.1143	5.156
10	0.1382	0.1292	5.781

[a]Galvanized sheet gage is used for convenience. A metric series would conform to ASME *Standard* B32.3M, and the tolerances in ASTM *Standard* A 924 apply.
[b]Minimum thickness is based on thickness tolerances of hot-dip galvanized sheets in cut lengths and coils (per ASTM *Standard* A 924). Tolerance is valid for 48 in. and 60 in. wide sheets.

Table 4B Uncoated Steel Sheet Thickness

Manufacturers' Standard Gage[b]	Thickness, in. Nominal	Minimum[a] Hot-Rolled	Minimum[a] Cold-Rolled	Nominal Weight, lb/ft^2
28	0.0149		0.0129	0.625
26	0.0179		0.0159	0.750
24	0.0239		0.0209	1.000
22	0.0299		0.0269	1.250
20	0.0359		0.0329	1.500
18	0.0478	0.0428	0.0438	2.000
16	0.0598	0.0538	0.0548	2.500
14	0.0747	0.0677	0.0697	3.125
13	0.0897	0.0827	0.0847	3.750
12	0.1046	0.0966	0.0986	4.375
11	0.1196	0.1116	0.1136	5.000
10	0.1345	0.1265	0.1285	5.625

Note: Table is based on 48 in. width coil and sheet stock. 60 in. coil has same tolerance, except that 16 gage is ±0.007 in. in hot-rolled coils and sheets.
[a]Minimum thickness is based on thickness tolerances of hot-rolled and cold-rolled sheets in cut lengths and coils (per ASTM *Standards* A 366, A 568, and A 569).
[b]Manufacturers' standard gage is used for convenience. A metric series would conform to ASME *Standard* B32.3M, and the tolerances in ASTM *Standard* A 568 apply.

Table 4C Stainless Steel Sheet Thickness

Gage[a]	Thickness, in. Nominal	Thickness, in. Minimum[b]	Nominal Weight, lb/ft^2 Stainless Steel 300 Series	Nominal Weight, lb/ft^2 Stainless Steel 400 Series
28	0.0151	0.0131	0.634	0.622
26	0.0178	0.0148	0.748	0.733
24	0.0235	0.0205	0.987	0.968
22	0.0293	0.0253	1.231	1.207
20	0.0355	0.0315	1.491	1.463
18	0.0480	0.0430	2.016	1.978
16	0.0595	0.0535	2.499	2.451
14	0.0751	0.0681	3.154	3.094
13	0.0900	0.0820	3.780	3.708
12	0.1054	0.0964	4.427	4.342
11	0.1200	0.1100	5.040	4.944
10	0.1350	0.1230	5.670	5.562

[a]Stainless sheet gage is not standardized. A metric series would conform to ASME *Standard* B32.3M, and the tolerances in ASTM *Standard* A 480 apply.
[b]Minimum thickness is based on thickness tolerances of hot-rolled sheets in cut lengths and cold-rolled sheets in cut lengths and coils (per ASTM *Standard* A 480).

Table 5 Steel Angle Weight per Unit Length (Approximate)

Angle Size, in.	Weight, lb/ft
3/4 × 3/4 × 1/8	0.59
1 × 1 × 0.0466 (minimum)	0.36
1 × 1 × 0.0575 (minimum)	0.44
1 × 1 × 1/8	0.80
1 1/4 × 1 1/4 × 0.0466 (minimum)	0.45
1 1/4 × 1 1/4 × 0.0575 (minimum)	0.55
1 1/4 × 1 1/4 × 0.0854 (minimum)	0.65
1 1/4 × 1 1/4 × 1/8	1.01
1 1/2 × 1 1/2 × 0.0575 (minimum)	0.66
1 1/2 × 1 1/2 × 1/8	1.23
1 1/2 × 1 1/2 × 3/16	1.80
1 1/2 × 1 1/2 × 1/4	2.34
2 × 2 × 0.0575 (minimum)	0.89
2 × 2 × 1/8	1.65
2 × 2 × 3/16	2.44
2 × 2 × 1/4	3.19
2 1/2 × 2 1/2 × 3/16	3.07
2 1/2 × 2 1/2 × 1/4	4.10

internal tie rods to maintain their structural integrity. For ducts over 36 in. wide, internal midpanel tie rods are an alternative to external reinforcement. *Rectangular Industrial Duct Construction Standards* (SMACNA 1980) gives construction details for ducts up to 168 in. wide at a pressure up to ±30 in. of water and higher.

Fittings must be reinforced similarly to sections of straight duct. On size change fittings, the greater fitting dimension determines material thickness. Where fitting curvature or internal member attachments provide equivalent rigidity, such features may be credited as reinforcement.

Round Metal Ducts. Round ducts are inherently strong and rigid, and are generally the most efficient and economical ducts for air systems. The dominant factor in round duct construction is the ability of the material to withstand the physical abuse of installation and negative pressure requirements. SMACNA (1995) lists construction requirements as a function of static pressure, type of seam (spiral or longitudinal), and diameter.

Nonferrous Ducts. SMACNA (1995) lists construction requirements for rectangular (±3 in. of water) and round (±2 in. of water) aluminum ducts. *Round Industrial Duct Construction Standards* (SMACNA 1999) gives construction requirements for round aluminum duct systems for pressures up to ±30 in. of water.

Flat Oval Ducts

SMACNA (1995) also lists flat oval duct construction requirements. Seams and transverse joints are generally the same as those permitted for round ducts. However, proprietary joint systems are available from several manufacturers. Flat oval duct is for positive pressure applications only, unless special designs are used. Hanger designs and installation details for rectangular ducts generally apply to flat oval ducts.

Fibrous Glass Ducts

Fibrous glass ducts are a composite of rigid fiberglass and a factory-applied facing (typically aluminum or reinforced aluminum), which serves as a finish and vapor barrier. This material is available in molded round sections or in board form for fabrication into rectangular or polygonal shapes. Duct systems of round and rectangular fibrous glass are generally limited to 2400 fpm and ±2 in. of water. Molded round ducts are available in higher pressure ratings. *Fibrous Glass Duct Construction Standards* (SMACNA 1992, NAIMA 1997) and manufacturers' installation instructions give details on fibrous glass duct construction. SMACNA (1992) also covers duct and fitting fabrication, closure, reinforcement, and installation, including installation of duct-mounted HVAC appurtenances (e.g., volume dampers, turning vanes, register and grille connections, diffuser connections, access doors, fire damper connections, and electric heaters). AIA (1996) includes guidelines for the use of fibrous glass duct in hospital and health care facilities.

Flexible Ducts

Flexible ducts connect mixing boxes, light troffers, diffusers, and other terminals to the air distribution system. SMACNA (1995) has an installation standard and a specification for joining, attaching, and supporting flexible duct. ADC (1996) has another installation standard. Because unnecessary length, offsetting, and compression of these ducts significantly increase airflow resistance, they should be kept as short and straight as possible, fully extended, and supported to minimize sagging (see Chapter 32 of the 1997 *ASHRAE Handbook—Fundamentals*).

UL *Standard* 181 covers testing of materials used to fabricate flexible ducts that are separately categorized as air ducts or connectors. NFPA *Standard* 90A defines the acceptable use of these products. The flexible duct connector has less resistance to flame penetration, has lower puncture and impact resistance, and is subject to many restrictions listed in NFPA *Standard* 90A. Only flexible ducts that are air duct rated should be specified. Tested products are listed in the UL *Gas and Oil Directory*, published annually.

Plenums and Apparatus Casings

SMACNA (1995) shows details on field-fabricated plenum and apparatus casings. Sheet metal thickness and reinforcement for plenum and casing pressure outside the range of −3 to +10 in. of water can be based on *Rectangular Industrial Duct Construction Standards* (SMACNA 1980).

Carefully analyze plenums and apparatus casings on the discharge side of a fan for maximum operating pressure in relation to the construction detail being specified. On the suction side of a fan, plenums and apparatus casings are normally constructed to withstand negative air pressure at least equal to the total upstream static pressure loss. The accidental stoppage of intake airflow can apply a negative pressure as great as the fan shutoff pressure. Conditions such as malfunctioning dampers or clogged louvers, filters, or coils can collapse a normally adequate casing. To protect large casing walls or roofs from damage, it is more economical to provide fan safety interlocks, such as damper end switches or pressure limit switches, than to use heavier sheet metal construction.

Apparatus casings can perform two acoustical functions. If the fan is completely enclosed within the casing, the transmission of fan noise through the fan room to adjacent areas is reduced substantially. An acoustically lined casing also reduces airborne noise in connecting ductwork. Acoustical treatment may consist of a single metal wall with a field-applied acoustical liner or thermal insulation, or a double-wall panel with an acoustical liner and a perforated metal inner liner. Double-wall casings are marketed by many manufacturers who publish data on structural, acoustical, and thermal performance and also prepare custom designs.

Acoustical Treatment

Metal ducts are frequently lined with acoustically absorbent materials to reduce aerodynamic noise. Although many materials are acoustically absorbent, duct liners must also be resistant to erosion and fire and have properties compatible with the ductwork fabrication and erection process. For higher-velocity ducts, double-wall construction using a perforated metal inner liner is frequently specified. Chapter 46 of the 1999 *ASHRAE Handbook—Applications* addresses design considerations, including external lagging. ASTM *Standard* C 423 covers laboratory testing of duct liner materials to determine their sound absorption coefficients. ASTM *Standard* E 477 covers laboratory testing of the acoustical insertion loss of duct liner materials. Designers should review all of the tests incorporated in ASTM *Standard* C 1071. A wide range of performance attributes, including erosion resistance, vapor adsorption, temperature resistance, bacteria resistance, and fungi resistance, is covered in the standard. Health and safety precautions are addressed, and manufacturer's certifications of compliance are also covered. AIA (1996) includes guidelines for the use of duct liner in hospital and health care facilities.

Rectangular duct liners should be secured by mechanical fasteners and installed in accordance with *HVAC Duct Construction Standards—Metal and Flexible* (SMACNA 1995). Adhesives should be Type I, in conformance to ASTM *Standard* C 916, and should be applied to the duct, with at least 90% coverage of mating surfaces. Quality workmanship prevents delamination of the liner and possible blockage of coils, dampers, flow sensors, or terminal devices. Avoid uneven edge alignment at butted joints to minimize unnecessary resistance to airflow (Swim 1978).

Rectangular metal ducts are susceptible to rumble from flexure in the duct walls during start-up and shutdown. If a designer wants systems to switch on and off frequently (as an energy conservation measure) during the times in which buildings are occupied, duct construction that reduces objectionable noise should be specified.

Hangers

SMACNA (1995) covers commercial HVAC system hangers for rectangular, round, and flat oval ducts. When special analysis is required for larger ducts or loads or for other hanger configurations than are given, AISC (1989) and AISI (1989) design manuals should be consulted. To hang or support fibrous glass ducts, the methods detailed in *Fibrous Glass Duct Construction Standards* (SMACNA 1992, NAIMA 1997) are recommended. UL *Standard* 181 involves maximum support intervals for UL listed ducts.

INDUSTRIAL DUCT CONSTRUCTION

NFPA *Standard* 91 is widely used for duct systems conveying particulates and removing flammable vapors (including paint-spraying residue), and corrosive fumes. Particulate-conveying duct systems are generally classified as follows:

- Class 1 covers nonparticulate applications, including makeup air, general ventilation, and gaseous emission control.
- Class 2 is imposed on moderately abrasive particulate in light concentration, such as that produced by buffing and polishing, woodworking, and grain handling.
- Class 3 consists of highly abrasive material in low concentration, such as that produced from abrasive cleaning, dryers and kilns, boiler breeching, and sand handling.
- Class 4 is composed of highly abrasive particulates in high concentration, including materials conveying high concentrations of particulates listed under Class 3.
- Class 5 for corrosive applications such as acid fumes has been introduced by SMACNA.

For contaminant abrasiveness ratings, see *Round Industrial Duct Construction Standards* (SMACNA 1999). Consult the 1999 *ASHRAE Handbook—Applications* for specific processes and uses.

Materials

Galvanized steel, uncoated carbon steel, or aluminum are most frequently used for industrial air handling. Aluminum ductwork is not used for systems conveying abrasive materials; when temperatures exceed 400°F, galvanized steel is not recommended. Ductwork material for systems handling corrosive gases, vapors, or mists must be selected carefully. For the application of metals and use of protective coatings in corrosive environments, consult *Accepted Industry Practice for Industrial Duct Construction* (SMACNA 1975), the *Pollution Engineering Practice Handbook*

(Cheremisinoff and Young 1975), and the publications of the National Association of Corrosion Engineers (NACE) and ASM International.

Round Ducts

SMACNA (1999) gives information for the selection of material thickness and reinforcement members for spiral and nonspiral industrial ducts. (Spiral seam ducts are only for Class 1 and 2 applications.) The tables in this manual are presented as follows:

Class. Steel—Classes 1, 2, 3, 4, and 5; Aluminum—Class 1 only. Stainless steel—Classes 1 and 5.

Pressure classes for steels and aluminum. ±2 to ±30 in. of water, in increments of 2 in. of water.

Duct diameter for steels and aluminum. 4 to 96 in., in increments of 2 in. Equations are available for calculating the construction requirements for diameters greater than 96 in.

Software is also available from SMACNA for design with steel, stainless steel, and aluminum. For other spiral duct applications, consult manufacturer's construction schedules, such as those listed in the *Industrial Duct Engineering Data and Recommended Design Standards* (United McGill Corporation 1985).

Rectangular Ducts

Rectangular Industrial Duct Construction Standards (SMACNA 1980) is available for selecting material thickness and reinforcement members for industrial ducts. The data in this manual give the duct construction for any pressure class and panel width. Each side of a rectangular duct is considered a panel. Usually, the four sides of a rectangular duct are built of material with the same thickness. Ducts are sometimes built with the bottom plate thicker than the other three sides (usually in ducts with heavy particulate accumulation) to save material.

The designer selects a combination of panel thickness, reinforcement, and reinforcement member spacing to limit the deflection of the duct panel to a design maximum. Any shape transverse joint or intermediate reinforcement member that meets the minimum requirement of both section modulus and moment of inertia may be selected. The SMACNA data, which may also be used for designing apparatus casings, limit the combined stress in either the panel or structural member to 24,000 psi and the maximum allowable deflection of the reinforcement members to 1/360 of the duct width.

Construction Details

Recommended manuals for other construction details are *Industrial Ventilation: A Manual of Recommended Practice* (ACGIH 1998), NFPA *Standard* 91, and *Accepted Industry Practice for Industrial Ventilation* (SMACNA 1975). For industrial duct Classes 2, 3, and 4, the transverse reinforcing of ducts subject to negative pressure below −3 in. of water should be welded to the duct wall rather than relying on mechanical fasteners to transfer the static load.

Hangers

The *Manual of Steel Construction* (AISC 1989) and the *Cold-Formed Steel Design Manual* (AISI 1989) give design information for industrial duct hangers and supports. The SMACNA standards for rectangular and round industrial ducts (SMACNA 1980, 1999) and manufacturers' schedules include duct design information for supporting ducts at intervals of up to 35 ft.

DUCT CONSTRUCTION FOR GREASE- AND MOISTURE-LADEN VAPORS

The installation and construction of ducts used for the removal of smoke or grease-laden vapors from cooking equipment should be in accordance with NFPA *Standard* 96 and SMACNA's rectangular and round industrial duct construction standards (SMACNA 1980, 1999). Kitchen exhaust ducts that conform to NFPA *Standard* 96 must (1) be constructed from carbon steel with a minimum thickness of 0.054 in. or stainless steel sheet with a minimum thickness of 0.043 in.; (2) have all longitudinal seams and transverse joints continuously welded; and (3) be installed without dips or traps that may collect residues, except where traps with continuous or automatic removal of residue are provided. Since fires may occur in these systems (producing temperatures in excess of 2000°F), provisions are necessary for expansion in accordance with the following table. Ducts that must have a fire resistance rating are usually encased in materials with appropriate thermal and durability ratings.

Kitchen Exhaust Duct Material	Duct Expansion at 2000°F, in/ft
Carbon steel	0.19
Type 304 stainless steel	0.23
Type 430 stainless steel	0.13

Ducts that convey moisture-laden air must have construction specifications that properly account for corrosion resistance, drainage, and waterproofing of joints and seams. No nationally recognized standards exist for applications in areas such as kitchens, swimming pools, shower rooms, and steam cleaning or washdown chambers. Galvanized steel, stainless steel, aluminum, and plastic materials have been used. Wet and dry cycles increase corrosion of metals. Chemical concentrations affect corrosion rate significantly. Chapter 47 of the 1999 *ASHRAE Handbook—Applications* addresses material selection for corrosive environments. Conventional duct construction standards are frequently modified to require welded or soldered joints, which are generally more reliable and durable than sealant-filled, mechanically locked joints. The number of transverse joints should be minimized, and longitudinal seams should not be located on the bottom of the duct. Risers should drain and horizontal ducts should pitch in the direction most favorable for moisture control. ACGIH (1998) covers hood design.

PLASTIC DUCTS

The *Thermoplastic Duct (PVC) Construction Manual* (SMACNA 1994) covers thermoplastic (polyvinyl chloride, polyethylene, polypropylene, acrylonitrile butadiene styrene) and thermosetting (glass-fiber-reinforced polyester) ducts used in commercial and industrial installations. SMACNA's manual provides comprehensive construction details for positive or negative 2, 6, and 10 in. of water polyvinyl chloride ducts. NFPA *Standard* 91 provides construction details and application limitations for plastic ducts. Model code agencies publish evaluation reports indicating terms of acceptance of manufactured ducts and other ducts not otherwise covered by industry standards and codes.

Fiberglass reinforced thermosetting plastic (FRP) ducts are described in the *Thermoset FRP Duct Construction Manual* (SMACNA 1997). The manual covers physical properties, manufacture, construction, installation, and methods of testing. These ducts are intended for air conveyance in corrosive environments as manufactured by hand lay-up, spray-up, and filament winding fabrication techniques. The term FRP also refers to fiber reinforced plastic (fibers other than glass). Other terms for FRP are reinforced thermoset plastic (RTP) and glass reinforced plastic (GRP), which is commonly used in Europe and Australia. SMACNA (1997) has construction standards for pressures up to ±30 in. of water gage and duct sizes from 4 to 72 in. round and 12 to 96 in. rectangular.

UNDERGROUND DUCTS

No comprehensive standards exist for the construction of underground air ducts. Coated steel, asbestos cement, plastic, tile, concrete, reinforced fiberglass, and other materials have been used. Underground duct and fittings should always be round and have a minimum thickness as listed in SMACNA (1995). Thickness above the minimum may be needed for individual applications. Specifications for the construction and installation of underground ducts should account for the following: water tables, ground surface flooding, the need for drainage piping beneath ductwork, temporary or permanent anchorage to resist flotation, frost heave, backfill loading, vehicular traffic load, corrosion, cathodic protection, heat loss or gain, building entry, bacterial organisms, degree of water and air tightness, inspection or testing prior to backfill, and code compliance. Chapter 11 has information on cathodic protection of buried metallic conduits. *Criteria and Test Procedures for Combustible Materials Used for Warm Air Ducts Encased in Concrete Slab Floors* (NRC) provides criteria and test procedures for fire resistance, crushing strength, bending strength, deterioration, and odor. *Installation Techniques for Perimeter Heating and Cooling* (ACCA 1990) covers residential systems and gives five classifications of duct material related to particular performance characteristics. Residential installations may also be subject to the requirements in NFPA *Standard* 90B. Commercial systems also normally require compliance with NFPA *Standard* 90A.

DUCTS OUTSIDE BUILDINGS

The location and construction of ducts exposed to outdoor atmospheric conditions are generally regulated by building codes. Exposed ducts and their sealant/joining systems must be evaluated for the following:

- Waterproofing
- Resistance to external loads (wind, snow, and ice)
- Degradation from corrosion, ultraviolet radiation, or thermal cycles
- Heat transfer
- Susceptibility to physical damage
- Hazards at air inlets and discharges
- Maintenance needs

In addition, support systems must be custom designed for rooftop, wall-mounted, and bridge or ground-based applications. Specific requirements must also be met for insulated and uninsulated ducts.

SEISMIC QUALIFICATION

Seismic analysis of duct systems may be required by building codes or federal regulations. Provisions for seismic analysis are given by the Federal Emergency Management Agency (FEMA 1997) and the Department of Defense (NAVFAC 1982). Ducts, duct hangers, fans, fan supports, and other duct-mounted equipment are generally evaluated independently. Chapter 53 of the 1999 *ASHRAE Handbook—Applications* gives design details. SMACNA (1998) provides guidelines for seismic restraints of mechanical systems. The manual gives bracing details for ducts, pipes, and conduits that apply to the model building codes and ASCE *Standard* 7, Minimum Design Loads for Buildings and Other Structures.

SHEET METAL WELDING

AWS (1990) covers sheet metal arc welding and braze welding procedures. This specification also addresses the qualification of welders and welding operators, workmanship, and the inspection of production welds.

THERMAL INSULATION

Insulation materials for ducts, plenums, and apparatus casings are covered in Chapter 23 of the 1997 *ASHRAE Handbook—Fundamentals*. Codes generally limit factory-insulated ducts to UL *Standard* 181, Class 0 or Class 1. *Commercial and Industrial Insulation Standards* (MICA 1993) gives insulation details. ASTM *Standard* C 1290 gives specifications for fibrous glass blanket external insulation for ducts.

MASTER SPECIFICATIONS

Master specifications for duct construction and most other elements in building construction are produced and regularly updated by several organizations. Two documents are *Masterspec* by the American Institute of Architects (AIA) and *SPECTEXT* by the Construction Sciences Research Foundation (CSRF). These documents are model project specifications that require little editing to customize each application for a project.

Nationally recognized model specifications provide several benefits, including the following:

- They focus industry practice on a uniform set of requirements in a widely known format.
- They reduce the need to prepare new specifications for each project.
- They remain relatively current and automatically incorporate new and revised editions of construction, test, and performance standards published by other organizations.
- They are adaptable to small or large projects.
- They are performance or prescription oriented as the designer desires.
- They provide lists of products and equipment by name and number or descriptions that are deemed equal.
- They are divided into subsections that are coordinated with other sections of related work.
- They are increasingly being used by government agencies to replace separate and often different agency specifications.

REFERENCES

ACCA. 1990. *Installation techniques for perimeter heating and cooling.* Manual No. 4. Air Conditioning Contractors of America, Washington, D.C.

ACGIH. 1998. *Industrial ventilation—A manual of recommended practice*, 23rd ed. American Conference of Governmental Industrial Hygienists, Lansing, MI.

ADC. 1996. Flexible duct performance and installation standards, 3rd ed. Air Diffusion Council, Chicago, IL.

AIA. Updated quarterly. *Masterspec.* American Institute of Architects, Washington, D.C.

AIA. 1996. Guidelines for design and construction of hospital and health care facilities.

AISC. 1989. *Manual of steel construction.* American Institute of Steel Construction, Chicago, IL.

AISI. 1989. *Cold-formed steel design manual.* American Iron and Steel Institute, Washington, D.C.

ASCE. 1995. Minimum design loads for buildings and other structures. ANSI/ASCE *Standard* 7-95. American Society of Civil Engineers, New York.

ASME. 1984. Preferred metric sizes for flat metal products. ANSI/ASME *Standard* B32.3M-84. American Society of Mechanical Engineers, New York.

ASHRAE. 1993. Energy efficient design of new low-rise residential buildings. ANSI/ASHRAE/IESNA *Standard* 90.2-1993.

ASHRAE. 1995. Energy conservation in existing buildings. ANSI/ASHRAE/IESNA *Standard* 100-1995.

ASHRAE. 1999. Energy standard for new buildings except low-rise residential buildings. ASHRAE/IESNA *Standard* 90.1-1999.

ASHRAE. 1999. Ventilation for acceptable indoor air quality. *Standard* 62-1999.

ASTM. 1990. Test method for sound absorption and sound absorption coefficients by the reverberation room method. *Standard* C 423-90A. American Society for Testing and Materials, West Conshohocken, PA.

ASTM. 1995. Standard specification for fibrous glass blanket insulation used to externally insulate HVAC ducts. *Standard* C 1290-95.

ASTM. 1996. Specification for adhesives for duct thermal insulation. *Standard* C 916-96 (R 1996).

ASTM. 1996. Standard test method for measuring acoustical and airflow performance of duct liner materials and prefabricated silencers. *Standard* E 477.

ASTM. 1997. Standard specification for general requirements for flat-rolled stainless and heat-resisting steel plate, sheet, and strip. *Standard* A 480/A 480M-97A.

ASTM. 1997. Standard specification for general requirements for steel, sheet, carbon and high-strength, low-alloy, hot-rolled and cold-rolled. *Standard* A 568/A 568M-97.

ASTM. 1997. Standard specification for general requirements for steel sheet, metallic-coated by the hot-dip process. *Standard* A 924/A 924M-97.

ASTM. 1997. Standard specification for steel, sheet, carbon, cold-rolled, commercial quality. *Standard* A 366/A 366M-97.

ASTM. 1998. Standard specification for steel, carbon (0.15 maximum, percent), hot-rolled sheet and strip, commercial quality. *Standard* A 569/A 569M-98A.

ASTM. 1998. Standard specification for steel sheet, zinc-coated (galvanized) or zinc-iron alloy-coated (galvannealed) by the hot-dip process. *Standard* A 653/A 653M-98C.

ASTM. 1998. Standard specification for thermal and acoustical insulation (glass fiber, duct lining material). *Standard* C 1071-98.

AWS. 1990. Sheet metal welding code. AWS *Standard* D9.1-90. American Welding Society, Miami, FL.

BOCA. 1999. *National building code*. Building Officials and Code Administrators International, Country Club Hills, IL.

BOCA. 1998. *National mechanical code*.

CABO. 1995. *Model energy code*. Council of American Building Officials, Falls Church, VA.

CABO. 1995. *One- and two-family dwelling code*.

Cheremisinoff, P.N. and R.A. Young. 1975. *Pollution engineering practice handbook*. Ann Arbor Science Publishers, Inc., Ann Arbor, MI.

CSRF. Updated quarterly. *SPECTEXT.* Construction Sciences Research Foundation. Available from Construction Specifications Institute, Alexandria, VA.

ASTM. 1996. Standard test method for measuring acoustical and airflow performance of duct liner materials and prefabricated silencers. *Standard* E 477. ASTM, W. Conshohocken, PA.

FEMA. 1997. NEHRP (National Earthquake Hazards Reduction Program) recommended provisions for the development of seismic regulations for new buildings. Building Seismic Safety Council for the Federal Emergency Management Agency, Washington, D.C.

FEMA. 1997. NEHRP (National Earthquake Hazards Reduction Program) guidelines for the seismic rehabilitation of buildings. Building Seismic Safety Council for the Federal Emergency Management Agency, Washington, D.C.

ICBO. 1997. *Uniform building code*. International Conference of Building Officials, Whittier, CA.

ICBO. 1997. *Uniform mechanical code*.

ICC. 1998. *International mechanical code*. International Code Council. Falls Church, VA.

MICA. 1993. *Commercial and industrial insulation standards.* Midwest Insulation Contractors Association, Omaha, NE.

NADCA. 1992. Mechanical cleaning of non-porous air conveyance system components. National Air Duct Cleaners Association, Washington, DC.

NAIMA. 1993. Cleaning fibrous glass insulated air duct systems. North American Insulation Contractors Association, Alexandria, VA.

NAIMA. 1997. Fibrous glass duct construction standards, 2nd ed.

NAVFAC. 1982. *Technical manual—Seismic design for buildings.* Army TM 5-809-10, Navy NAVFAC P-355, Air Force AFM 88-3 (Chapter 13). U.S. Government Printing Office, Washington, D.C.

NFPA. 1996. Chimneys, fireplaces, vents, and solid fuel-burning appliances. ANSI/NFPA *Standard* 211-96. National Fire Protection Association, Quincy, MA.

NFPA. 1997. *Life safety code.* ANSI/NFPA *Standard* 101-97.

NFPA. 1998. Smoke and heat venting. ANSI/NFPA *Standard* 204M-98.

NFPA. 1998. Ventilation control and fire protection of commercial cooking operations. ANSI/NFPA *Standard* 96-98.

NFPA. 1999. Installation of air conditioning and ventilating systems. ANSI/NFPA *Standard* 90A-99.

NFPA. 1999. Installation of warm air heating and air conditioning systems. ANSI/NFPA *Standard* 90B-99.

NFPA. 1999. Exhaust systems for air conveying of vapors, gases, mists, and noncombustible particulate solids. ANSI/NFPA *Standard* 91-99.

NRC. Criteria and test procedures for combustible materials used for warm air ducts encased in concrete slab floors. National Research Council, National Academy of Sciences, Washington, D.C.

SBCCI. 1997. *Standard building code.* Southern Building Code Congress International, Inc., Birmingham, AL.

SBCCI. 1997. *Standard mechanical code.*

SMACNA. 1975. *Accepted industry practice for industrial duct construction.* Sheet Metal and Air Conditioning Contractors' National Association, Chantilly, VA.

SMACNA. 1980. *Rectangular industrial duct construction standards.*

SMACNA. 1985. *HVAC air duct leakage test manual.*

SMACNA. 1992. *Fibrous glass duct construction standards*, 6th ed.

SMACNA. 1994. *Thermoplastic duct (PVC) construction manual.*

SMACNA. 1995. *HVAC duct construction standards—Metal and flexible*, 2nd ed.

SMACNA. 1997. *Thermoset FRP duct construction manual*, 1st ed.

SMACNA. 1998. Seismic restraint manual—Guidelines for mechanical systems, 2nd ed.

SMACNA. 1999. *Round industrial duct construction standards.*

Swim, W.B. 1978. Flow losses in rectangular ducts lined with fiberglass. ASHRAE *Transactions* 84(2).

UL. 1994. Closure systems for use with rigid air ducts and air connectors, 2nd ed. *Standard* 181A. Underwriters Laboratories, Northbrook, IL.

UL. 1995. Closure systems for use with flexible air ducts and air connectors, 1st ed. *Standard* 181B.

UL. 1996. Factory-made air ducts and air connectors, 9th ed. *Standard* 181.

UL. 1996. Ceiling dampers, 2nd ed. *Standard* 555C.

UL. 1997. Fire tests of door assemblies, 9th ed. ANSI/UL *Standard* 10B.

UL. 1999. Fire dampers, 6th ed. *Standard* 555.

UL. 1999. Smoke dampers, 4th ed. *Standard* 555S.

UL. Annual. Gas and oil equipment directory.

United McGill Corporation. 1985. Industrial duct engineering data and recommended design standards. Form No. SMP-IDP (February):24-28. Westerville, OH.

CHAPTER 17

AIR-DIFFUSING EQUIPMENT

SUPPLY air outlets and diffusing equipment introduce air into a conditioned space to obtain a desired indoor atmospheric environment. Return and exhaust air is removed from a space through return and exhaust inlets. Various types of diffusing equipment are available as standard manufactured products. This chapter describes this equipment and details its proper use. Chapter 31 of the 1997 *ASHRAE Handbook—Fundamentals* also covers the subject.

SUPPLY AIR OUTLETS

Supply air outlets, properly sized and located, control the air pattern to obtain proper air motion and equalize temperature in a given space. A supply air outlet may be classified (1) according to its performance, as in Chapter 31 of the 1997 *ASHRAE Handbook—Fundamentals*, Straub and Chen (1957), and Straub et al. (1956); (2) by its physical configuration; or (3) by its mounting or installation location. Because an outlet may fall into more than one category, this chapter classifies outlets according to their general physical configuration, including slight variances of these configurations.

Accessories used with an outlet regulate the volume of supply air and control its flow pattern. For example, an outlet cannot discharge air properly and uniformly unless the air is conveyed to it in a uniform flow. Accessories may also be necessary for proper air distribution in a space, so they must be selected and used in accordance with the manufacturers' recommendations. Outlets should be sized to project air so that its velocity and temperature reach acceptable levels before entering the occupied zone.

Primary airflow from an outlet entrains room air into the jet. This entrained air increases the total air in the jet stream. Because the momentum (MV) of the jet remains constant, as the mass increases the velocity decreases. Also, the temperature in the jet equalizes as the two air masses mix.

The entrainment ratio is shown in Chapter 31 of the 1997 *ASHRAE Handbook—Fundamentals* as

$$\text{Entrainment ratio} = C\frac{V_o}{V_x}$$

where

C = constant = 1.4 for slots and 2.0 for holes
V_o = outlet velocity
V_x = jet terminal velocity at distance x from the outlet

The equation shows that for a given or selected terminal velocity, the ratio depends primarily on the outlet velocity V_o. For a given outlet type and location, the throw (distance the jet travels) is longer with a high V_o or high entrainment ratio.

The preparation of this chapter is assigned to TC 5.3, Room Air Distribution.

Comparisons between outlet locations and patterns also affect the throw, entrainment, and temperature equalization of a jet. Some general characteristics include the following:

- Grilles with straight vanes project the longest throw. When the grille is located close to a ceiling, entrainment is restricted at the ceiling and results in a longer throw. However, when the air pattern is spread horizontally, the throw is reduced and the jet is flatter so its temperature is more uniform.
- Outlets with radial airflow have a short throw and a relatively thin jet, so the temperature in the jet is fairly uniform. Temperature equalizes more readily in a cooling jet, which results in less drop.
- Single jets from ceiling outlets have a longer throw and perform like a grille.

Outlets with high entrainment characteristics can also be used advantageously in air-conditioning systems with low supply air temperatures and consequent high temperature differentials between the room air and the supply air. Radial pattern ceiling diffusers may be used in systems with cooling temperature differentials as high as 30 to 35°F and still provide satisfactory temperature equalization in the space. Linear diffusers may be used with cooling temperature differentials as high as 25°F. Grilles may generally be used in well designed systems with cooling differentials up to 20°F.

Special diffusers are available for use with low temperature air distribution systems (i.e., those with supply air temperatures of 35 to 45°F and cooling differentials as high as 40°F). Those diffusers include special features that mix the cold supply air with room air at the outlet and effectively reduce the temperature differential between the supply and room air.

PROCEDURE FOR OUTLET SELECTION

The following procedure is generally used in selecting and locating an outlet:

1. Determine the amount of air to be supplied to each room. (Refer to Chapters 27 and 28 of the 1997 *ASHRAE Handbook—Fundamentals* to determine air quantities for heating and cooling.)
2. Select the type and quantity of outlets for each room, considering such factors as air quantity required, distance available for throw or radius of diffusion, structural characteristics, and architectural concepts. Table 1, which is based on experience and typical ratings of various outlets, may be used as a guide for using outlets in rooms with various heating and cooling loads. Special conditions, such as ceiling height greater than 8 to 12 ft and/or exposed duct mounting, as well as product modifications and unusual conditions of room occupancy, can modify this table. Manufacturers' rating data should be consulted to determine the suitability of the outlets used.

Table 1 Guide to Use of Various Outlets

Type of Outlet	Air Loading of Floor Space, Max. cfm/ft²	Approx. Max. Air Changes per Hour for 10 ft Ceiling
Grille	0.6 to 1.2	7
Slot	0.8 to 2.0	12
Perforated panel	0.9 to 3.0	18
Ceiling diffuser	0.9 to 6.0	30

3. Locate outlets in the room to distribute air as uniformly as possible. Outlets may be sized and located to distribute air in proportion to the heat gain or loss in various portions of a room.
4. Select the proper size outlet from the manufacturers' rating data according to air quantity, discharge velocity, distribution pattern, and sound level. Note manufacturers' recommendations with regard to use. In an open space, the interaction of airstreams from multiple air outlets may alter a single outlet's throw or air temperature/air velocity. As a result, the data may be insufficient to predict air motion in a particular space. Also, obstructions to the primary air distribution pattern require special study.

Chapter 31 of the 1997 *ASHRAE Handbook—Fundamentals* describes a procedure for selecting the size and location of air outlets.

FACTORS THAT INFLUENCE SELECTION

Surface Effect

An airstream moving adjacent to, or in contact with, a wall or ceiling surface creates a low-pressure area immediately adjacent to that surface, causing the air to remain in contact with the surface substantially throughout the length of throw. This surface effect, commonly referred to as the Coanda effect, counteracts the drop of a horizontally projected cool airstream.

Ceiling diffusers exhibit a surface effect to a high degree because a circular air pattern blankets the entire ceiling area surrounding each outlet. This effect diminishes with a directional discharge that does not blanket the full ceiling surface surrounding the outlet. Linear diffusers, which discharge the airstream in a single direction across the ceiling, exhibit less surface effect than radial pattern discharge; however, the effect is greater with longer diffusers and with diffusers that have spread accessories at the outlet face. Sidewall grilles exhibit varying degrees of surface effect, depending on the spread of the particular air pattern and the proximity and angle of airstream approach to the ceiling.

In many installations, the outlets must be mounted on an exposed duct and discharge the airstream into free space. In this case, the airstream entrains air on both its upper and lower surfaces. As a result, a higher rate of entrainment is obtained and the throw is shortened by 30% of the equivalent throw along a surface. Because there is no surface effect with diffusers installed on exposed ducts, with a cooling differential the supply air drops more rapidly toward the floor unless the outlet surface provides an upward deflection to the discharge stream. Conical and louver face diffusers with radial patterns normally exhibit this upward deflection. Linear and flush perforated face diffusers normally do not possess this characteristic.

Temperature Differential

Heated, horizontally projected air rises and then falls as it cools. Downward projection of cooling air or upward projection of heated air increases with an increase in the temperature difference. Similarly, downward projection of heated air and upward projection of cool air decreases with an increase in the temperature difference.

Low-temperature supply air (i.e., in the range of 38 to 45°F) requires special attention to the environment in which the diffusing and distribution equipment is operating. Cold starts in a saturated environment will cause condensation in all but specially treated equipment. Ramp starts, with a gradual decrease of the relative humidity of the indoor environment, avoid condensation except with an unusually high internal load or high infiltration.

Sound Level

The sound level from an outlet is a function of its discharge velocity and the transmission of system noise. For a given air capacity, a larger outlet has a lower discharge velocity and corresponding lower generated sound. A larger outlet also allows a higher level of sound to pass through the outlet, which may appear as outlet-generated sound. High-frequency sound can be the result of excessive outlet velocity but may also be generated in the duct by the moving airstream. Low-pitched sounds are generally mechanical equipment sound and/or terminal box or balancing damper sound transmitted through the duct and outlet to the room.

The cause of the sound can usually be pinpointed as outlet or system sounds by removing the outlet core during operation. If the sound remains essentially unchanged, the system is at fault. If the sound is significantly reduced, it may be caused by a highly irregular velocity profile at the entrance to the diffuser. The velocity profile should be measured. If the velocity varies less than 10% in the air outlet entrance neck, the outlet is causing the noise. If the velocity profile at the entrance indicates peak velocities significantly higher than average, check the manufacturer's data for the sound at the peak velocity. If this rating approximates the observed sound, the velocity profile in the duct must be corrected to achieve design performance. Note that a high-velocity free stream jet does not cause a high sound level until the jet impinges against an interfering surface or edge.

Smudging

Smudging is the deposition of dirt particles on the air outlet or surface that is contiguous with the outlet. Dirt particles may be either in the room air that is entrained in the discharge or in the air supply to the outlet. Smudging is more prevalent with ceiling diffusers and linear diffusers that discharge the air parallel to the mounting surface than with grilles that discharge air perpendicular to the surface.

Dirt from room air is deposited most frequently at the edge of the stream, where the entrained air comes in contact with the surface, rather than at the center of the stream, which tends to wipe the surface with clean supply air. Edges of the stream occur at interruptions in the discharge stream, such as at a blank section of a linear diffuser or at the corner of a directional rectangular diffuser.

Variable Air Volume

Special diffusers are available for use in variable air volume systems. These outlets include features that maintain a higher discharge air momentum than do conventional outlets as the supply airflow is reduced from design maximum to about 20 to 40% of the maximum. Momentum is maintained by reducing the outlet discharge area as the flow decreases, thus maintaining a constant or increased discharge velocity. Or, the supply air is discharged through two sections of the outlet; one section carries a constant, small fraction of the total at a constant discharge velocity and the second carries the remainder of the airflow, which varies from maximum to minimum turndown. The momentum of the discharge airstream is the sum of the momentum of both streams. At high turndown conditions, the combined momentum of the two streams differs little from a fixed area outlet.

Alternates to these proprietary air outlets, which require either a separate and dedicated damper actuator or a spring biased actuator counterbalanced by duct pressure, are available. Conventional air outlets should function properly over the necessary turndown range, provided that the following factors are recognized and considered in their application:

1. The throw of the jet at a terminal velocity of 50 fpm divided by characteristic length (T_{50}/L) should be equal to or up to 20% higher than listed in Table 2 of Chapter 31 of the 1997 *ASHRAE Handbook—Fundamentals*.
2. As a good estimate, the plan area covered by each outlet should not exceed two times the diffuser height squared; i.e., $A \le (2H)^2$.
3. Concentrated loads should be located under either the air inlet or the supply air outlet, but not at a distance between T_{50} and T_{100} from the supply air outlet. At reduced flow, the thermal plume from the concentrated load causes the supply air to prematurely fall into the occupied zone.

TYPES OF SUPPLY AIR OUTLETS

GRILLE AND REGISTER OUTLETS

A grille consists of a frame enclosing a set of either vertical or horizontal vanes (for a single-deflection grille) or both (for a double-deflection grille). The combination of a grille and a damper is called a register.

Types

Adjustable Bar Grille. This is the most common type of grille used as a supply outlet. The single-deflection grille includes a set of either vertical or horizontal vanes. Vertical vanes deflect the airstream in the horizontal plane; horizontal vanes deflect the airstream in the vertical plane. The double-deflection grille has a second set of vanes installed behind and at right angles to the face vanes. This grille controls the airstream in both the horizontal and vertical planes.

Fixed Bar Grille. This grille is similar to the single-deflection grille, except that the vanes are not adjustable—they may be straight or set at an angle. The angle at which the air is discharged from this grille depends on the type of deflection vanes.

Security Grille. This grille is available for various levels of tamper resistance and access through the grille. Fastening may be concealed or from the rear. When a single-piece stamped plate covers the grille face, the discharge direction is limited to flow normal to the mounting surface as the plate thickness increases and its free area decreases.

Variable Area Grille. This grille is similar to the adjustable double-deflection grille but can vary the discharge area in response to an air volume change. At constant duct static pressure, the variation in throw is minimized for a given change in supply air volume.

Accessories

Various accessories, designed to improve the performance of grille outlets, are available as standard equipment.

- Opposed blade dampers can be attached to the backs of grilles or installed as separate units in the duct (Figure 1A). Adjacent blades of this damper rotate in opposite directions and may receive air from any direction, discharging it in a series of jets without adversely deflecting the airstream to one side of the duct. Refer to Chapter 46 of the 1999 *ASHRAE Handbook—Applications* for the effects of damper location on sound level.
- Multishutter dampers have a series of gang-operated blades that rotate in the same direction (Figure 1B). This uniform rotation deflects the airstream when the damper is partially closed. Most dampers are operated by removable keys or levers.
- Gang-operated turning vanes are installed in collar connections to branch ducts. The device shown in Figure 1C has vanes that pivot and remain parallel to the duct airflow, regardless of the setting. The second device has a set of fixed vanes (Figure 1D). Both devices restrict the area of the duct in which they are installed. They should be used only when the duct is wide enough to allow the device to open to its maximum position without causing undue restriction of airflow in the duct, which would limit downstream airflow and increase the sound level.
- Individually adjusted turning vanes are used in the device shown in Figure 1E. Two sets of vanes are used. The downstream set equalizes flow across the collar, while the upstream set turns the air. The vanes can also be adjusted at various angles to act as a damper, but use as a damper is not practical because balancing requires removing the grille to gain access and then reinstalling it to measure airflow.
- Other miscellaneous accessories available as standard equipment include remote control devices to operate the dampers and maximum to minimum stops to limit damper travel.

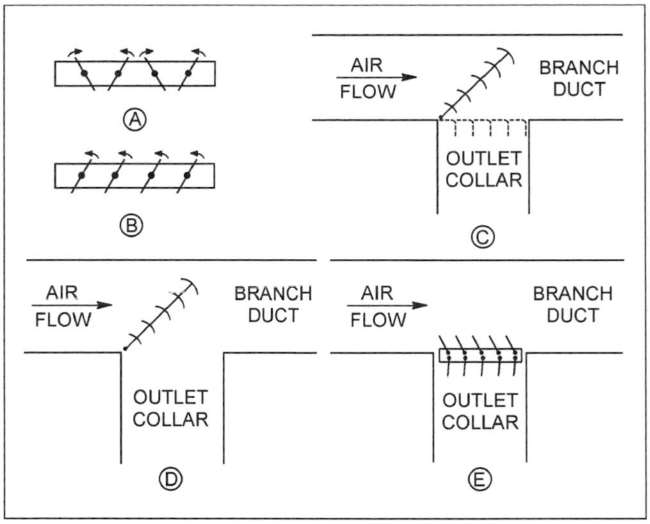

Fig. 1 Accessory Controls for Grille and Register Outlets

Applications

Properly selected grilles operate satisfactorily from high sidewall and perimeter locations in the sill or floor. Grilles installed in 8 to 10 ft high ceilings, which discharge the airstream down, are generally unacceptable in comfort air-conditioning installations. Satisfactory performance can be obtained with higher mounting heights if special allowances are made for the supply air capacities and temperature differentials. Heating and cooling from the same grille must be carefully examined and is normally not recommended.

High Sidewall. A double-deflection grille usually provides the most satisfactory solution. The vertical face louvers of a well-designed grille deflect the air approximately 50° to either side and amply cover the conditioned space. The rear louvers deflect the air at least 15° in the vertical plane, which is ample to control the elevation of the discharge pattern. This upward deflection minimizes the thermal drop in an application with a high cooling differential (15 to 25°F).

Ceiling Installation. Such installation is generally limited to grilles with curved vanes that discharge parallel to the mounting surface. For high mounting locations, generally above 15 to 20 ft, vertical discharge may be the optimum condition.

LINEAR SLOT OUTLETS

A linear slot outlet is a long narrow supply air grille or diffuser outlet with an aspect ratio generally greater than 10 to 1. Linear outlets can be installed in multiple sections to achieve long, continuous lengths or installed as a discrete length in a modular ceiling. It can consist of a single slot or multiple slots. Linear outlets are typically designed for supply applications, but they are also commonly used as a return inlet to provide a consistent architectural appearance.

Types

Linear Bar Grille or Diffuser. This outlet has fixed bars at its face. The bars normally run parallel to the length of the outlet. This device supplies air in a constant direction. Typically, linear bar grilles or diffusers are attached to a separate supply air plenum that has its own inlet.

T-Bar Slot Diffuser. This diffuser is manufactured with an integral plenum. It normally is installed in modular T-bar ceilings and may be concealed behind the ceiling grid suspension. T-bar slot diffusers are available with either fixed deflection or adjustable pattern controllers. These devices are available in configurations that discharge air from fully vertical to fully horizontal.

Linear Slot Diffuser. This diffuser is an elongated air outlet, available with single or multiple linear slot openings. It is commonly used to achieve a long, continuous appearance. Linear slot diffusers are typically available with options such as angle cuts at the ends or mitered corner pieces to allow the device to meet architectural requirements. These devices are available in configurations that provide vertical to horizontal airflow. Typically, a supply air plenum is supplied as a separate device and attached to the linear slot diffuser during installation.

Light Troffer Diffuser. This device is a plenum integrated with an air-handling light fixture. A light troffer diffuser serves as the combined plenum, inlet, and attachment device to the air-handling light fixture, which has a slot opening to receive the diffuser at the face of the lighting device to discharge supply air into the space. Normally, only the air-handling slot opening is visible from the occupied space.

Accessories

Damper. This is an air volume adjuster installed directly on the rear of a linear diffuser outlet, either at the entrance to the outlet plenum or remote near the branch-trunk duct junction. The means to adjust the damper is normally accessible through the face of the linear outlet.

Installation on the outlet, while probably the most convenient, is the least desirable for two reasons: (1) At this location, the damper covers the duct with the largest cross section and requires maximum throttling for flow adjustment; and (2) when the damper is throttled, it causes high-velocity air to impinge unnecessarily on the vanes or elements of the linear diffuser, which increases the sound level. (Chapter 46 of the 1999 *ASHRAE Handbook—Applications* discusses the amount of this increase.) Ideally the damper should be placed at the entrance to the branch duct, with a remote adjustment, so that the air exiting the damper at higher velocity does not impinge on adjacent duct walls or outlet vanes.

Flow Equalizing Vane. This vane is commonly used with linear diffusers to increase the spread or adjust the direction of the air jet from the diffuser. The equalizing vane consists of a series of individually adjustable blades, positioned normal to the longest blades of the linear diffuser, either attached to the rear of the linear or within the throat of the attached supply boot or plenum.

Applications

High sidewall installation with device mounted no more than 6 to 12 in. below ceiling. A slot diffuser with flow perpendicular to the mounting surface is best suited to high sidewall installations. A linear bar diffuser with fixed, 0° deflection may be applied. With the device mounted just below the ceiling, the surface effect of the supply air jet tends to keep the supply air on the ceiling, allowing longer throw distances and less chance of the jet separating from the ceiling and falling into the occupied zone.

High sidewall installation with device mounted 12 in. or more below ceiling. A slot diffuser with some angular deflection is best suited to these high sidewall installations. To keep the supply air jet out of the occupied zone, the device should allow the supply air to be directed toward the ceiling. A linear bar diffuser with a 15 to 30° upward deflection is recommended. A linear slot diffuser with directional adjustability is an alternate choice.

Perimeter ceiling. In this application, the device must handle the exterior surface load as well as the interior zone load generated along the perimeter. The perimeter surface load is especially critical in cold climates during the winter. Diffusers should be installed so they direct air toward both the exterior surface and the interior. Both plenum slot diffusers and linear slot diffusers are able to meet this requirement. These devices can be selected and installed with pattern controllers that allow changes in discharge direction from horizontal to vertical throw.

Interior ceiling. Generally, interior ceiling applications require a device that produces a horizontal pattern along the ceiling. Devices that can meet this criterion include plenum slot diffusers, linear slot diffusers, and light troffer diffusers. These devices should be sized to keep the supply air jet from reaching the occupied zone.

Sill. In sill applications, a linear bar grille works best. The grille should be installed with the supply air jet directed vertically away from the occupied space. When the device is mounted 8 in. or less from the wall, a device with 0° deflection is suitable. The presence of window draperies or blinds and the effect of an impinging airstream must be considered in the selection.

If the device is installed more than 8 in. from the vertical surface, a linear bar grille with a 15 to 30° deflection is recommended. This device should be installed with the jet directed toward the wall. These grilles are typically available with doors or other means of access to mechanical equipment that may be installed within the sill enclosure.

Floor. For floor applications, a linear bar grille works well. The designer must determine the traffic and floor loading on the grille and consult the manufacturer's load limit for the grille.The grille should be placed in low-traffic areas with limited access. A floor-mounted grille is usually selected to bring air up along a wall or exterior surface. A floor grille is appropriate along exterior surfaces for heating in cold climates. Then the airflow performance should be considered with reference to the manufacturer's catalog data. If the device is less than 12 in. from the surface, a 0° deflection grille should be used. If the grille is 12 to 24 in. from the vertical surface, a 15 to 30° deflection toward the surface is recommended.

CEILING DIFFUSER OUTLETS

Ceiling diffuser outlets usually have either a radial or directional discharge parallel to the mounting surface. Diffusers with adjustable deflectors that allow the discharge to be directed perpendicular to the mounting surface are available, as are round, square, and rectangular ceiling diffusers. A ceiling diffuser typically consists of an outer shell, which contains a duct collar, and an internal deflector, which defines the diffuser's performance, including the discharge pattern and direction.

Types

Round Ceiling Diffuser. This diffuser is a series of flaring concentric or expanding truncated conical rings. Typically it is installed either in gypsum-board ceilings or on exposed ducts. Round ceiling diffusers are available in a broad range of sizes and capacities. These devices are available with adjustable deflectors that allow the diffuser to discharge the air either parallel or perpendicular to the ceiling or mounting surface.

Rectangular and Square Ceiling Diffusers. These diffusers are designed to integrate into various ceiling systems. They are most commonly selected for installation in grid suspension ceilings.

Square ceiling diffusers are also available for surface mounting, spline ceiling, and other ceilings. This diffuser is largely selected based on performance, appearance, and cost. Square ceiling diffusers can be categorized as perforated face, louvered face, and plaque face.

Perforated Diffuser. The perforated face of this diffuser typically has a free area of about 50%. This unit is normally selected to meet architectural demands for air outlets that blend into the ceiling. The perforated face tends to create a slightly higher pressure drop and more sound than other square ceiling diffusers. Perforated diffusers also tend to cost less than other devices. Perforated diffusers are available with deflection devices mounted at the neck or on the face plate. The deflectors are adjusted to provide a horizontal air discharge in one, two, three, or four directions. Special faces accommodate adjacent ceiling tile or channel slot grid systems.

Louver Face Diffuser. This diffuser consists of an outer frame, which includes an integral rectangular duct collar, and a series of louvers parallel with the outer frame vanes. Louver face diffusers typically provide a directional discharge pattern with horizontal air discharging perpendicular to the louver length. The louvers may be arranged to provide four-way, three-way, two-way opposite, two-way corner, or one-way discharge to permit the supply air to be directed toward the load source. Louver face diffusers are available with special edges to accommodate adjacent ceiling tile or channel slot grid systems. Some louver face diffusers are available with adjustable louvers that can change the discharge direction from horizontal to vertical.

Rectangular louver face diffusers are available to further match the air discharge direction to the load location. Round-to-square fittings are available to fit the louver face to a round air supply duct.

Plaque Face Diffuser. This diffuser is constructed with a backpan that includes a duct collar and a single plaque that forms the diffuser's face. This air outlet typically has a horizontal, radial discharge pattern. A plaque face diffuser is most often selected for its architectural appearance. The flat plaque face of the diffuser provides a clean, uniform appearance. Typically, the performance of a plaque face diffuser is similar to that of a square face, round neck diffuser.

Square Face, Round Neck Diffuser. This diffuser is constructed of a series of concentric square, drawn louvers that radiate from the center of the diffuser. It is most commonly available with a face that is flush with the ceiling plane. Square face ceiling diffusers are available with various configurations including dropped face and beveled face. These diffusers have a fixed horizontal radial discharge pattern. Some are available with an adjustable discharge pattern that allows the direction to be either horizontal or vertical.

Rectangular Modular Diffuser. This diffuser is available to provide a vertical fan-shaped air discharge pattern. The fan-shaped discharge penetrates the conditioned space perpendicular to the mounting surface. These diffusers are used for high air change, low to moderate cooling differential applications where high airflow is generally required as makeup air for laboratory hood exhaust or a process. Some outlet models are flush to the mounting surface; others intrude into the space below the ceiling. Most function similarly with or without an adjacent ceiling surface. The diffusers normally match 2 ft by 4 ft ceiling modules, with other sizes available. Manufacturers' catalogs list specific characteristics.

Laminar Flow Type Rectangular Modular Perforated Face Diffuser. This diffuser provides a unidirectional or laminar discharge that is perpendicular to the mounting surface. The free area of the perforated face is typically about 10 to 20%. Most outlets include a means to develop a uniform velocity profile over the full face. This minimizes mixing with the surrounding ambient air and reduces the entrainment of any surrounding contaminants. These diffusers are generally used in hospital operating rooms, clean rooms, or laboratories.

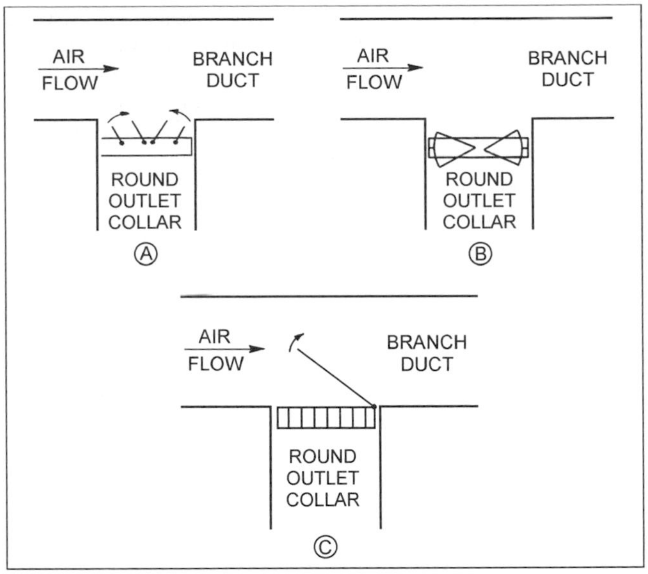

Fig. 2 Accessory Controls for Ceiling Diffuser Outlets

Accessories

Multilouver Damper. Consisting of a series of parallel blades mounted inside a round or square frame, multilouver dampers are installed in the diffuser collar or the takeoff. The blades are usually arranged in two groups rotating in opposite directions and are normally key-operated from the face of the diffuser (Figure 2A). This arrangement equalizes airflow in the diffuser collar.

Opposed Blade Damper. This damper is available for both round and rectangular air diffusers. For a round diffuser, the opposed blade damper consists of a series of vanes mounted inside a round frame and installed on the diffuser collar (Figure 2B). The vanes are typically operated through the face of the diffuser. For diffusers with a rectangular inlet neck or collar, the opposed blade damper consists of a series of parallel blades mounted in a rectangular frame and installed on the neck of the diffuser.

Splitter Damper. A splitter damper is a single-blade device, hinged at one edge and usually located at the branch connection of a duct or outlet (Figure 2C). The device is designed to allow adjustment at the branch connection of a duct or outlet to adjust flow.

Equalizing Device. This device allows adjustment of the airstream to obtain uniform flow to the diffuser.

Other Balancing Devices. Many balancing devices are available, including radial dampers, butterfly dampers, and dampers integral with an air outlet. Manufacturers' literature is a source of information for other air-balancing devices. Table 5 in Chapter 46 of the 1999 *ASHRAE Handbook—Applications* summarizes the effects of damper location on sound level.

Applications

For all the following applications, the manufacturer's rating data should be checked to select the air outlet that meets throw, pressure, and sound requirements.

Perimeter Zone Ceiling. In perimeter ceiling applications, the air outlet must handle the exterior surface load as well as the interior zone load generated along the perimeter. Refer to Chapter 31 of the 1997 *ASHRAE Handbook—Fundamentals* for more details

Interior Zone Ceiling. Generally, interior ceilings require an air outlet that produces a horizontal pattern along the ceiling. Ceiling diffusers are well suited for this application. Selection can be made based on performance, appearance, and cost. The air outlet selected should be sized to keep the supply air jet above the occupied zone.

Vertical Projection. Vertical projection from the outlet is often required in a high-ceiling application or in a high-load area. Vertical projection may be selected to meet an individual's comfort requirements. When selecting an outlet for vertical projection with heating differentials, its performance under both heating and cooling differentials must be taken into consideration. This is especially required on heating systems with an outdoor air supply, or when the air outlet discharge characteristics are not automatically changed as the supply air temperature changes from heating to cooling.

Spot Heating or Cooling. Spot heating or cooling typically requires the use of an outlet that projects the air jet vertically. In this application, the designer can select an air outlet that projects a jet of air to cover the defined area. Most ceiling diffusers are available with a configuration that allows spot heating or cooling. If the diffuser configuration is not changed with a change in supply air from heating to cooling, then the effect of heating and cooling differential must be determined based on manufacturer's data for the respective conditions.

Large-Capacity Area. These areas normally require a ceiling diffuser with a horizontal discharge that handles large air capacities. Applications include open atriums, warehouses, and gymnasiums. Ceiling diffuser outlets typically selected for this application are round and rectangular diffusers.

Exposed Duct. In exposed duct applications, a diffuser outlet is typically selected that discharges the supply air in a horizontal pattern. Most ceiling diffusers can be used, but throw or radius of diffusion distance for exposed duct applications is 70% of the distance listed for a ceiling-mounted rating. The air outlets most commonly selected are diffusers with round discharge patterns.

EQUIPMENT FOR AIR DISTRIBUTION

In high-velocity air distribution systems, special control and acoustical equipment may be required to introduce conditioned air into a space properly. Airflow controls for these systems consist principally of terminal boxes (air terminals), with part of the equipment considered as pressure reducing valves. Terminal boxes may be classified as single or dual duct, with or without reheat, and having either constant or variable primary volume. The constant primary air volume may provide for a variable discharge air volume to the conditioned space. Terminal boxes may also include air induction or fan induction that provides an essentially constant volume with a variable volume primary air capacity.

This section discusses control equipment for single duct and dual duct air conditioning systems. Chapter 16 covers duct construction details. Chapter 46 of the 1999 *ASHRAE Handbook—Applications* includes information on sound control in air-conditioning systems and sound rating for air outlets.

TERMINAL BOXES

The terminal box is a factory-made assembly for air distribution. A terminal box, without altering the composition of the treated air from the distribution system, manually or automatically performs one or more of the following functions: (1) controls velocity, pressure, or temperature of the air; (2) controls rate of airflow; (3) mixes airstreams of different temperatures or humidities; and (4) mixes, within the assembly, air at high velocity and/or high pressure with air from the treated space. To achieve these functions, terminal box assemblies are made from an appropriate selection of the following component parts: casing, mixing section, manual or automatic damper, heat exchanger, induction section (with or without fan), and flow controller.

A terminal box commonly includes a sound chamber to reduce sound generated within it by the damper or flow rate controller. At the same time the terminal box reduces the high-velocity, high-pressure inlet air to low-velocity, low-pressure air. The sound attenuation chamber is typically lined with thermal and sound insulating material and is equipped with baffles.

Additional sound absorption material may be required in the low-pressure distribution ducts connected to the discharge of larger boxes. Smaller boxes may not require additional sound absorption; however, manufacturers' catalogs should be consulted for specific performance information.

Terminal boxes are typically classified according to the function of their flow controllers, which are generally categorized as constant flow or variable flow devices. They are further categorized as either pressure-dependent, where the airflow through the assembly varies in response to changes in system pressure, or pressure-independent (pressure compensating), where the airflow through the device does not vary in response to changes in system pressure.

Constant flow controllers may control the volume mechanically by balancing the static pressure in the primary duct system with an adjustable spring bias for the flow set point. This unit requires no outside source of power, and its static pressure requirement is in the range of 0.3 to 0.8 in. of water. Constant flow controllers may also be pneumatically or electrically actuated, which requires sensing an internal differential pressure. They also require an electric or pneumatic control source to operate the damper actuator.

Variable flow controllers may also be mechanical, pneumatic, electric, or digital. These devices include a means to reset the volume automatically to a different control point in response to an outside signal such as a thermostat. Boxes with this feature are pressure-independent and may be used with reheat components. Variable flow may also be obtained by decreasing the flow through a constant volume regulator with a modulating damper ahead of the regulator. This arrangement typically allows for variations in flow between the high- and low-capacity limits or between a high limit and shutoff. These boxes are pressure-dependent and volume limiting in function.

Terminal boxes can be further categorized as being (1) system powered, in which the assembly derives all the energy necessary for its operation from the supply air within the distribution system; or (2) externally powered, in which the assembly derives part or all of the energy from a pneumatic or electric outside source. In addition, assemblies may be self-contained (when they are furnished with all necessary controls for their operation, including actuators, regulators, motors, and thermostats or space temperature sensors) or non-self-contained assemblies (when part or all of the necessary controls for operation are furnished by someone other than the assembly manufacturer). In the latter case, the controls may be mounted on the assembly by the assembly manufacturer or mounted by others after delivery of the equipment.

The damper or flow controller in the box can be adjusted manually, automatically, or by a pneumatic or electric motor. The unit is actuated by a signal from a thermostat or flow regulator, depending on the desired function of the box.

Air from the box may be discharged through a single opening suitable for connection to a low-pressure rigid branch duct, or through several round outlets suitable for connection to flexible ducts. A single supply air outlet connected directly to the discharge end or bottom of the box is an optional arrangement; however, the acoustic performance of this close-coupled arrangement must be carefully considered.

REHEAT BOXES

Reheat boxes add sensible heat to the supply air. Water or steam coils or electric resistance heaters are placed in or attached directly to the air discharge of the box. These boxes typically are single-duct boxes that can operate as either constant or variable volume units. However, if they are variable volume, they must maintain some minimum airflow for the reheat function. Some are arranged with a dual minimum flow, with one minimum being either zero during the no-occupancy cooling cycle or the airflow required for minimum ventilation during the cooling cycle, and the second minimum being the capacity required for reheat capacity during the reheat cycle. This type of equipment can provide local individual reheat without a central equipment station or zone change.

DUAL DUCT BOXES

Dual duct boxes are typically under control of a room thermostat. They receive warm and cold air from separate air supply ducts in accordance with room requirements to obtain room control without zoning. Pneumatic and electric volume-regulated boxes often have individual modulating dampers and operators to regulate the amount of warm and cool air. When a single modulating damper operator regulates the amount of both warm and cold air, a separate pressure reducing damper or volume controller (either pneumatic or mechanical) is needed in the box to reduce pressure and limit airflow. Specially designed baffles may be required inside the unit or at the box discharge to mix varying amounts of warm and cold air and/or to provide uniform flow and temperature equalization downstream. Dual duct boxes can be equipped with constant or variable flow devices. These are usually pressure-independent in order to provide a number of volume and temperature control functions. Dual duct terminals may also be used as outside air terminals in which the warm air inlet is used to control and maintain the required volumetric flow of ventilation air into the space.

CEILING INDUCTION BOXES

Ceiling induction boxes supply either primary air or a mixture of primary air and relatively warm air to the conditioned space. It achieves this function with a primary air jet and venturi that induces air from the ceiling plenum or from individual rooms via a return duct. A single duct supplies primary air at a temperature cool enough to satisfy all zone cooling loads. The ceiling plenum air induced into the primary air is at a higher temperature than the room because heat from the recessed lighting fixtures enters the plenum directly.

The induction box contains dampers that are actuated in response to a thermostat to control the amount of cool primary air and warm induced air. As less cooling is required, the primary airflow is gradually reduced, and the induced air rate is generally increased.

To meet interior load requirements, water reheat coils can be installed in the primary supply air duct, or electric reheat coils can be installed in the discharge duct.

FAN-POWERED CEILING INDUCTION BOXES

Fan-powered ceiling induction boxes differ from air-to-air induction boxes in that they include a blower. This blower, driven by a small motor, draws air from the space or ceiling plenum (secondary air) to be mixed with the cool primary air. The advantage of fan induction boxes over straight VAV boxes is that for a small energy expenditure and increased terminal maintenance, constant air circulation can be maintained in the space. Fan-powered induction boxes operate at a lower primary air static pressure than air-to-air induction boxes; and perimeter zones can be heated without operating the primary fan during unoccupied periods. The warm air in the plenum can be used for low to medium heating loads (depending on construction of the building envelope). As the load increases, heating coils in the perimeter boxes can be activated to heat the recirculated plenum air to the necessary level.

Fan-powered induction boxes can be divided into two categories: (1) constant volume or series type with all the primary and induced air passing through the continuously operating blower and (2) variable volume or parallel flow type in which the blower operates only on demand when induced air or heat is required. For the most efficient thermal operation, only primary air should be delivered to the conditioned space at peak cooling load because the induced air acts as unnecessary reheat.

In thermal storage and other systems with supply air temperatures in the range of 38 to 48°F, fan induction boxes are used to deliberately mix cold supply air with induced return or plenum air to moderate the supply air temperature. Some boxes are equipped with special insulation to prevent condensation with these low supply temperatures. Manufacturers' catalogs can provide further information on these special features.

Constant volume, or series type, fan-powered induction boxes are used when constant air circulation is desired in the space. The unit has two inlets, one for cool primary air from the central fan system and one for the secondary or plenum air. All air delivered to the space passes through the blower. The blower operates continuously whenever the primary air fan is on and can be cycled to deliver heat, as required, when the primary fan is off.

As the cooling load decreases, a damper throttles the amount of primary air delivered to the blower. The blower makes up for this reduced amount of primary air by drawing air in from the space or ceiling plenum through the return or secondary air opening.

Parallel flow or variable volume fan-powered induction boxes are sometimes called bypass boxes because the cool primary air bypasses the blower portion of the unit and flows directly to the space. The blower section draws in only plenum air and is mounted in parallel with the primary air damper. A backdraft damper keeps primary air from flowing into the blower section when the blower is not energized. The blower in these units is generally energized after the primary air damper is partially or completely throttled. Some electronically controlled units gradually increase the fan speed as the primary air damper is throttled in order to maintain constant airflow while permitting the fan to shut off when it approaches the full cooling mode.

BYPASS OR DUMP BOX

A bypass box handles a constant supply of primary air through its inlet; and, with a diverting damper, it bypasses the primary air to the ceiling plenum so that the amount of cooling delivered to the conditioned space meets the thermal requirement. The bypass air is diverted into the ceiling plenum and returned to the central air handler. The pressure requirement through the supply air path to the conditioned space and through the bypass or dump path is adjusted to be equal so that the fan handles a constant flow. This method provides a low first cost with minimum fan controls; but it is energy-inefficient as compared to a VAV fan system. Its most frequent application is on small systems.

STATIC PRESSURE CONTROL

To prevent static pressure imbalance in dual duct systems and/or to achieve efficient operation with VAV systems, some type of fan discharge capacity and/or pressure control is necessary. Variations in the duct static pressure can be limited by the following:

- Mechanical or electrical variable speed fan controls
- Zoning and changing air supply temperature in response to static pressure changes
- Static pressure controllers operating fan inlet vane dampers

Other control means, such as duct dampers within the system, are not efficient. They merely transfer the location of the pressure loss from the terminal box to another point in the duct system and do not change the overall system pressure requirement.

AIR CURTAINS

In its simplest application, an air curtain is a continuous broad stream of air circulated across a doorway of a conditioned space. It inhibits the penetration of unconditioned air and insects into a conditioned space by forcing an air layer of predetermined thickness and velocity over the entire entrance. The airstream layer moves with a velocity and angle such that any air that tries to penetrate the curtain is entrained. The air layer or jet can be redirected to compensate for pressure changes across the opening. If air is forced inward because of a difference in pressure, the jet can be redirected outward to equalize the pressure differential. Chapter 28 of the 1999 *ASHRAE Handbook—Applications* covers the principles of air curtain design.

Both vertical flow (usually downward) and horizontal flow air curtains are available. The vertical flow air curtain may have either a ducted or a nonducted return.

The air curtain is a high energy user, so this factor should be considered in its application. Air curtain effectiveness in preventing infiltration through an entrance generally ranges from 60 to 80%. The effectiveness is the comparison of infiltration rate or heat flux through an opening when using an air curtain as opposed to the transmission that would take place through a simple opening with no restriction. Lawton and Howell (1995) analyzed the savings of several air curtains versus an open door in a warm, humid climate.

Two important factors influence the pressure differential against which an air curtain must work—the height and the orientation of the structure. In high-rise structures, the possibility of using an air curtain depends mainly on the magnitude of the pressure differential caused by the structure's height or the stack effect. The orientation of the particular building and the location of adjacent buildings should also be studied and considered.

REFERENCES

Elleson, J.S. 1993. Energy use of fan powered mixing boxes with cold air distribution. *ASHRAE Transactions* 99(1):1349-58.

Engel, J.A. 1993. Experimental determination of the airflow performance of a variable area radial diffuser. *ASHRAE Transactions* 99(2):759-69.

Lawton, E.B. and R.H. Howell. 1995. Energy savings using air curtains installed in high-traffic doorways. *ASHRAE Transactions* 101(2): 136-43.

Miller, P. 1991. Diffuser selection for cold air distribution using the air performance index. *ASHRAE Journal* 33(9):32.

Straub, H.E. and M.M. Chen. 1957. Distribution of air within a room for year-round air conditioning—Part II. University of Illinois *Engineering Experiment Bulletin* No. 442.

Straub, H.E., S.F. Gilman, and S. Konzo. 1956. Distribution of air within a room for year-round air conditioning—Part I. University of Illinois *Engineering Experiment Bulletin* No. 435.

CHAPTER 18

FANS

A FAN is an air pump that creates a pressure difference and causes airflow. The impeller does work on the air, imparting to it both static and kinetic energy, which vary in proportion, depending on the fan type.

Fan efficiency ratings are based on ideal conditions; some fans are rated at more then 90% total efficiency. However, actual connections often make it impossible to achieve ideal efficiencies in the field.

TYPES OF FANS

Fans are generally classified as centrifugal or axial flow according to the direction of airflow through the impeller. Figure 1 shows the general configuration of a centrifugal fan. The components of an axial flow fan are shown in Figure 2. Table 1 compares typical characteristics of some of the most common fan types.

Two modified versions of the centrifugal fan are being used but are not listed in Table 1 as separate fan types. Unhoused centrifugal fan impellers are used as circulators in some industrial applications such as heat-treating ovens and are identified as plug fans. In this case, there is no duct connection to the fan since it simply circulates the air within the oven. In some HVAC installations, the unhoused fan impeller is located in a plenum chamber with the fan inlet connected to an inlet duct from the system. Outlet ducts are connected to the plenum chamber. This fan arrangement is identified as a plenum fan.

PRINCIPLES OF OPERATION

All fans produce pressure by altering the velocity vector of the flow. A fan produces pressure and/or flow because the rotating blades of the impeller impart kinetic energy to the air by changing its velocity. Velocity change is in the tangential and radial velocity components for centrifugal fans, and in the axial and tangential velocity components for axial flow fans.

Centrifugal fan impellers produce pressure from (1) the centrifugal force created by rotating the air column contained between the blades and (2) the kinetic energy imparted to the air by virtue of its velocity leaving the impeller. This velocity is a combination of rotative velocity of the impeller and airspeed relative to the impeller. When the blades are inclined forward, these two velocities are cumulative; when backward, oppositional. Backward-curved blade fans are generally more efficient than forward-curved blade fans.

Axial flow fans produce pressure from the change in velocity passing through the impeller, with none being produced by centrifugal force. These fans are divided into three types: propeller, tubeaxial, and vaneaxial. Propeller fans, customarily used at or near free air delivery, usually have a small hub-to-tip ratio impeller mounted in an orifice plate or inlet ring. Tubeaxial fans usually have reduced tip clearance and operate at higher tip speeds, giving them a higher total pressure capability than the propeller fan. Vaneaxial fans are essentially tubeaxial fans with guide vanes and reduced running

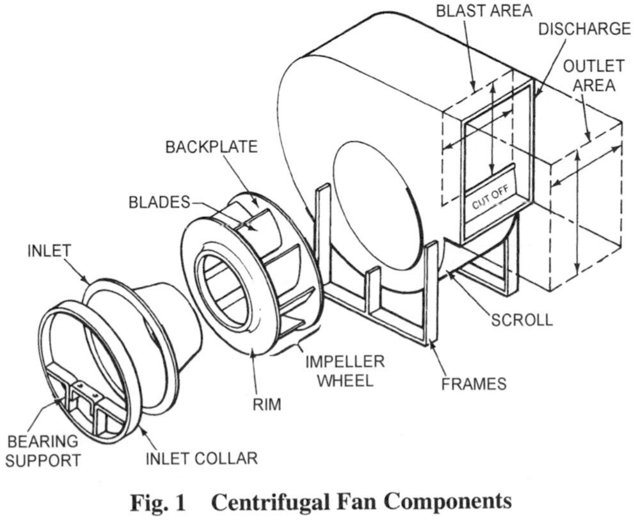

Fig. 1 Centrifugal Fan Components

The preparation of this chapter is assigned to TC 5.1, Fans.

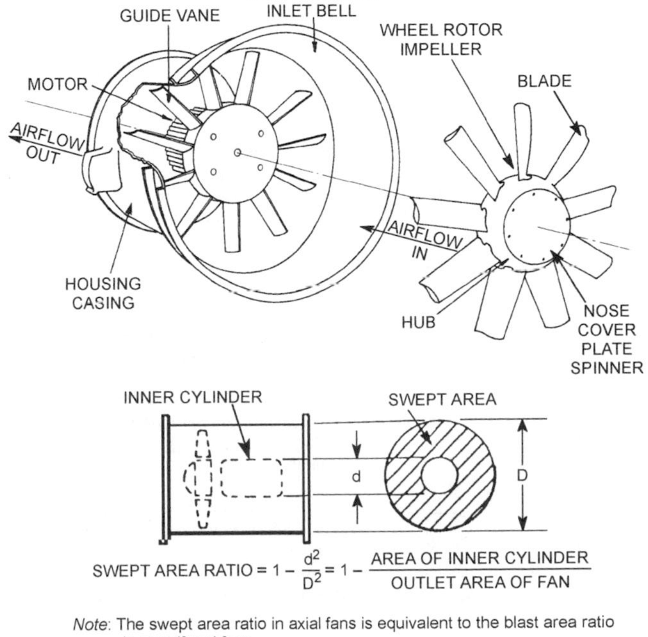

Note: The swept area ratio in axial fans is equivalent to the blast area ratio in centrifugal fans.

Fig. 2 Axial Fan Components

Table 1 Types of Fans

	TYPE	IMPELLER DESIGN		HOUSING DESIGN	
CENTRIFUGAL FANS	AIRFOIL		Highest efficiency of all centrifugal fan designs. Ten to 16 blades of airfoil contour curved away from direction of rotation. Deep blades allow for efficient expansion within blade passages. Air leaves impeller at velocity less than tip speed. For given duty, has highest speed of centrifugal fan designs.	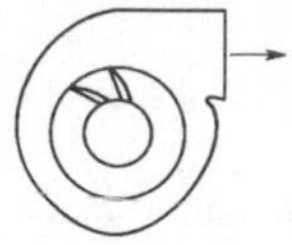	Scroll-type design for efficient conversion of velocity pressure to static pressure. Maximum efficiency requires close clearance and alignment between wheel and inlet.
CENTRIFUGAL FANS	BACKWARD-INCLINED BACKWARD-CURVED	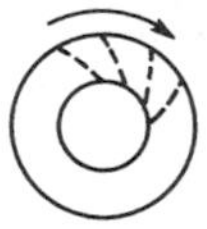	Efficiency only slightly less than airfoil fan. Ten to 16 single-thickness blades curved or inclined away from direction of rotation. Efficient for same reasons as airfoil fan.	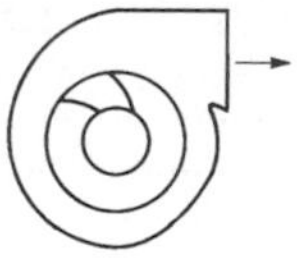	Uses same housing configuration as airfoil design.
CENTRIFUGAL FANS	RADIAL	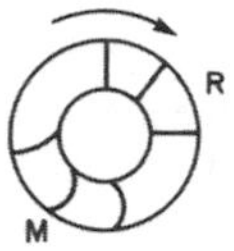	Higher pressure characteristics than airfoil, backward-curved, and backward-inclined fans. Curve may have a break to left of peak pressure and fan should not be operated in this area. Power rises continually to free delivery.	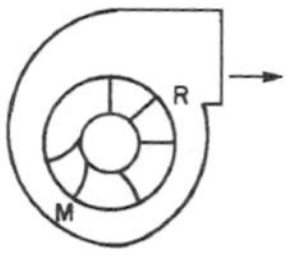	Scroll. Usually narrowest of all centrifugal designs. Because wheel design is less efficient, housing dimensions are not as critical as for airfoil and backward-inclined fans.
CENTRIFUGAL FANS	FORWARD-CURVED		Flatter pressure curve and lower efficiency than the airfoil, backward-curved, and backward-inclined. Do not rate fan in the pressure curve dip to the left of peak pressure. Power rises continually toward free delivery. Motor selection must take this into account.	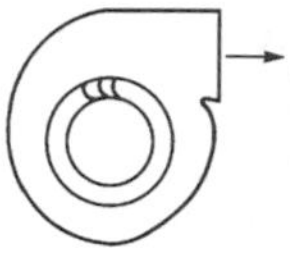	Scroll similar to and often identical to other centrifugal fan designs. Fit between wheel and inlet not as critical as for airfoil and backward-inclined fans.
AXIAL FANS	PROPELLER	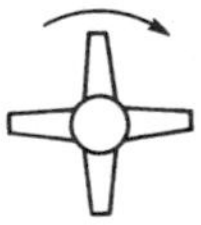	Low efficiency. Limited to low-pressure applications. Usually low cost impellers have two or more blades of single thickness attached to relatively small hub. Primary energy transfer by velocity pressure.	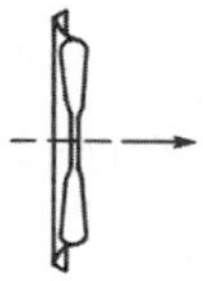	Simple circular ring, orifice plate, or venturi. Optimum design is close to blade tips and forms smooth airfoil into wheel.
AXIAL FANS	TUBEAXIAL	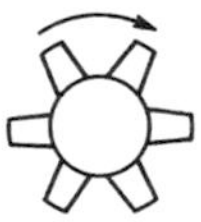	Somewhat more efficient and capable of developing more useful static pressure than propeller fan. Usually has 4 to 8 blades with airfoil or single-thickness cross section. Hub is usually less than half the fan tip diameter.	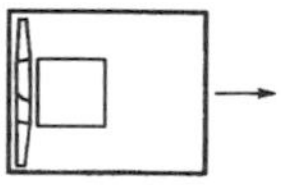	Cylindrical tube with close clearance to blade tips.
AXIAL FANS	VANEAXIAL		Good blade design gives medium- to high-pressure capability at good efficiency. Most efficient of these fans have airfoil blades. Blades may have fixed, adjustable, or controllable pitch. Hub is usually greater than half fan tip diameter.	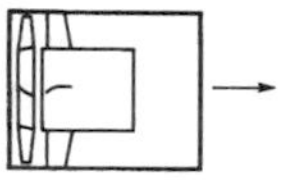	Cylindrical tube with close clearance to blade tips. Guide vanes upstream or downstream from impeller increase pressure capability and efficiency.
SPECIAL DESIGNS	TUBULAR CENTRIFUGAL		Performance similar to backward-curved fan except capacity and pressure are lower. Lower efficiency than backward-curved fan. Performance curve may have a dip to the left of peak pressure.	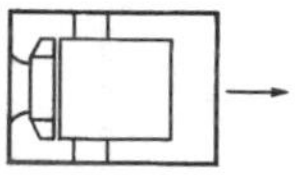	Cylindrical tube similar to vaneaxial fan, except clearance to wheel is not as close. Air discharges radially from wheel and turns 90° to flow through guide vanes.
SPECIAL DESIGNS	POWER ROOF VENTILATORS — CENTRIFUGAL	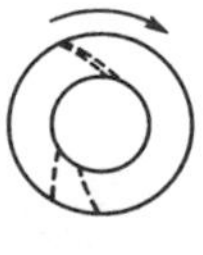	Low-pressure exhaust systems such as general factory, kitchen, warehouse, and some commercial installations. Provides positive exhaust ventilation, which is an advantage over gravity-type exhaust units. Centrifugal units are slightly quieter than axial units.	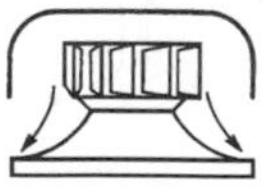	Normal housing not used, since air discharges from impeller in full circle. Usually does not include configuration to recover velocity pressure component.
SPECIAL DESIGNS	POWER ROOF VENTILATORS — AXIAL		Low-pressure exhaust systems such as general factory, kitchen, warehouse, and some commercial installations. Provides positive exhaust ventilation, which is an advantage over gravity-type exhaust units.	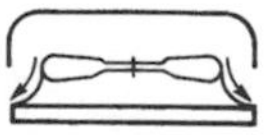	Essentially a propeller fan mounted in a supporting structure. Hood protects fan from weather and acts as safety guard. Air discharges from annular space at bottom of weather hood.

Table 1 Types of Fans (*Concluded*)

PERFORMANCE CURVES[a]	PERFORMANCE CHARACTERISTICS	APPLICATIONS
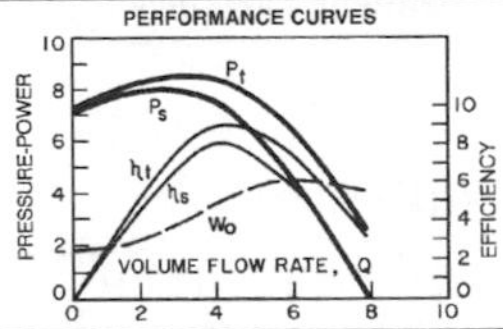	Highest efficiencies occur at 50 to 60% of wide open volume. This volume also has good pressure characteristics. Power reaches maximum near peak efficiency and becomes lower, or self-limiting, toward free delivery.	General heating, ventilating, and air-conditioning applications. Usually only applied to large systems, which may be low-, medium-, or high-pressure applications. Applied to large, clean-air industrial operations for significant energy savings.
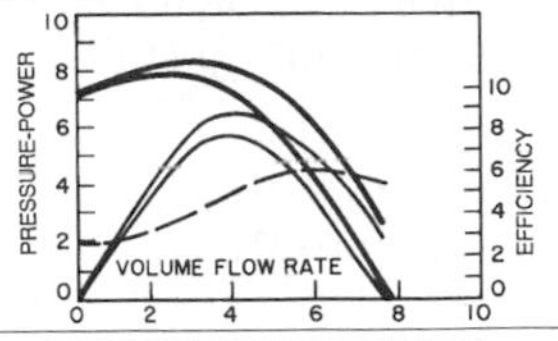	Similar to airfoil fan, except peak efficiency slightly lower.	Same heating, ventilating, and air-conditioning applications as airfoil fan. Used in some industrial applications where airfoil blade may corrode or erode due to environment.
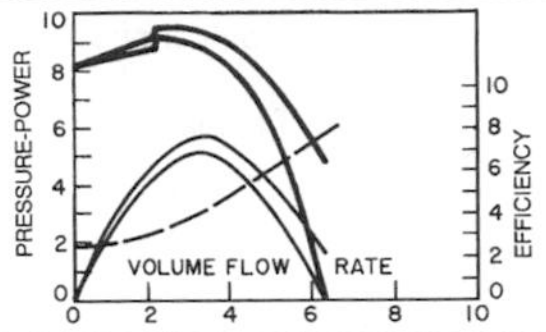	Higher pressure characteristics than airfoil and backward-curved fans. Pressure may drop suddenly at left of peak pressure, but this usually causes no problems. Power rises continually to free delivery.	Primarily for materials handling in industrial plants. Also for some high-pressure industrial requirements. Rugged wheel is simple to repair in the field. Wheel sometimes coated with special material. Not common for HVAC applications.
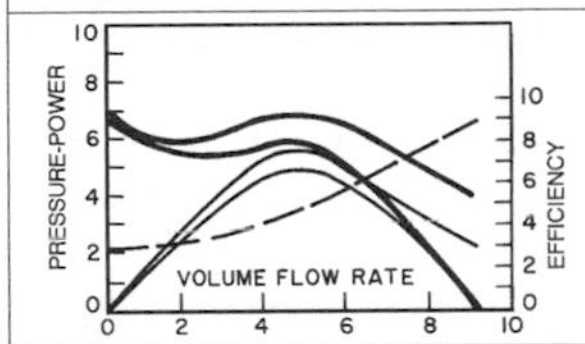	Pressure curve less steep than that of backward-curved fans. Curve dips to left of peak pressure. Highest efficiency to right of peak pressure at 40 to 50% of wide open volume. Rate fan to right of peak pressure. Account for power curve, which rises continually toward free delivery, when selecting motor.	Primarily for low-pressure HVAC applications, such as residential furnaces, central station units, and packaged air conditioners.
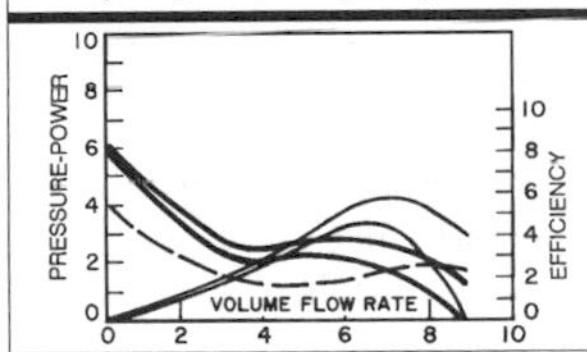	High flow rate, but very low-pressure capabilities. Maximum efficiency reached near free delivery. Discharge pattern circular and airstream swirls.	For low-pressure, high-volume air moving applications, such as air circulation in a space or ventilation through a wall without ductwork. Used for makeup air applications.
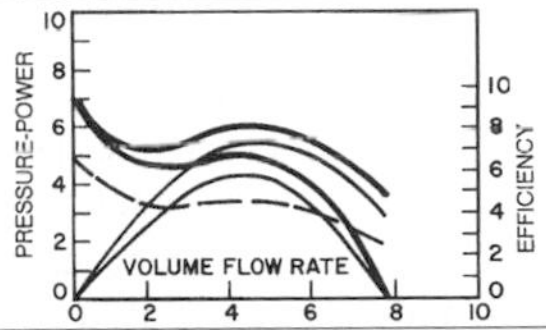	High flow rate, medium-pressure capabilities. Performance curve dips to left of peak pressure. Avoid operating fan in this region. Discharge pattern circular and airstream rotates or swirls.	Low- and medium-pressure ducted HVAC applications where air distribution downstream is not critical. Used in some industrial applications, such as drying ovens, paint spray booths, and fume exhausts.
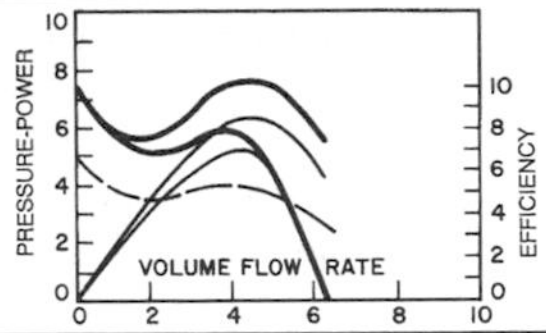	High-pressure characteristics with medium-volume flow capabilities. Performance curve dips to left of peak pressure due to aerodynamic stall. Avoid operating fan in this region. Guide vanes correct circular motion imparted by wheel and improve pressure characteristics and efficiency of fan.	General HVAC systems in low-, medium-, and high-pressure applications where straight-through flow and compact installation are required. Has good downstream air distribution. Used in industrial applications in place of tubeaxial fans. More compact than centrifugal fans for same duty.
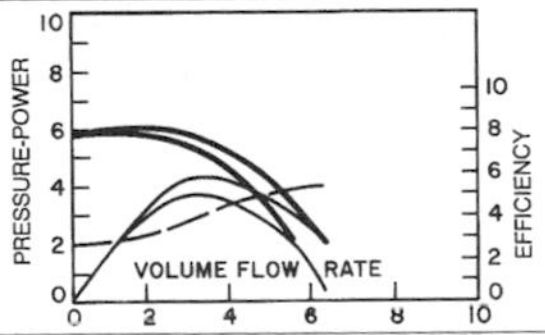	Performance similar to backward-curved fan, except capacity and pressure is lower. Lower efficiency than backward-curved fan because air turns 90°. Performance curve of some designs is similar to axial flow fan and dips to left of peak pressure.	Primarily for low-pressure, return air systems in HVAC applications. Has straight-through flow.
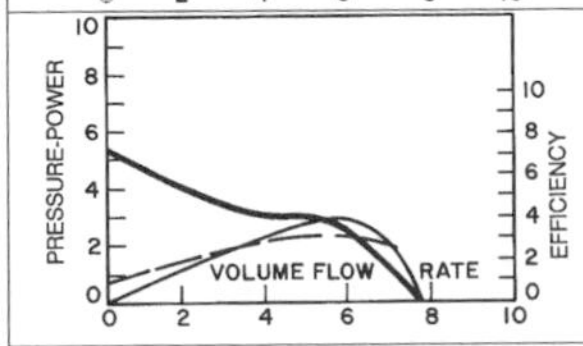	Usually operated without ductwork; therefore, operates at very low pressure and high volume. Only static pressure and static efficiency are shown for this fan.	Low-pressure exhaust systems, such as general factory, kitchen, warehouse, and some commercial installations. Low first cost and low operating cost give an advantage over gravity flow exhaust systems. Centrifugal units are somewhat quieter than axial flow units.
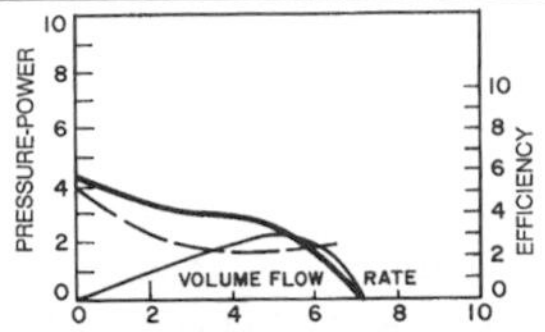	Usually operated without ductwork; therefore, operates at very low pressure and high volume. Only static pressure and static efficiency are shown for this fan.	Low-pressure exhaust systems, such as general factory, kitchen, warehouse, and some commercial installations. Low first cost and low operating cost give an advantage over gravity flow exhaust systems.

[a]These performance curves reflect general characteristics of various fans as commonly applied. They are not intended to provide complete selection criteria, since other parameters, such as diameter and speed, are not defined.

blade tip clearance, which give improved pressure, efficiency, and noise characteristics.

Table 1 includes typical performance curves for various types of fans. These performance curves show the general characteristics of the various fans as they are normally used; they do not reflect the characteristics of these fans reduced to such common denominators as constant speed or constant propeller diameter, since fans are not selected on the basis of these constants. The efficiencies and power characteristics shown are general indications for each type of fan. A specific fan (size, speed) must be selected by evaluating actual characteristics.

TESTING AND RATING

ASHRAE *Standard* 51/AMCA *Standard* 210 specifies the procedures and test setups to be used in testing fans and other air-moving devices. Figure 3 diagrams one of the most common procedures for developing the characteristics of a fan tested from **shutoff** conditions to nearly **free delivery** conditions. At shutoff, the duct is completely blanked off; at free delivery, the outlet resistance is reduced to zero. Between these two conditions, various flow restrictions are placed on the end of the duct to simulate various conditions on the fan. Sufficient points are obtained to define the curve between shutoff and free delivery conditions. Pitot tube traverses of the test duct are performed with the fan operating at constant speed. The point of rating may be any point on the fan performance curve. For each case, the specific point on the curve must be defined by referring to the flow rate and the corresponding total pressure. Other test setups described in ASHRAE *Standard* 51/AMCA *Standard* 210 should produce the same performance curve.

Fans designed for use with duct systems are tested with a length of duct between the fan and the measuring station. This length of duct smooths the flow of the fan and provides stable, uniform flow conditions at the plane of measurement. The measured pressures are corrected back to fan outlet conditions. Fans designed for use without ducts, including almost all propeller fans and power roof ventilators, are tested without ductwork.

Not all sizes are tested for rating. Test information may be used to calculate the performance of larger fans that are geometrically similar, but such information should not be extrapolated to smaller fans. For the performance of one fan to be determined from the known performance of another, the two fans must be dynamically similar. Strict dynamic similarity requires that the important nondimensional parameters vary in only insignificant ways. These nondimensional parameters include those that affect aerodynamic characteristics such as Mach number, Reynolds number, surface roughness, and gap size. (For more specific information, consult the manufacturer's application manual or engineering data.)

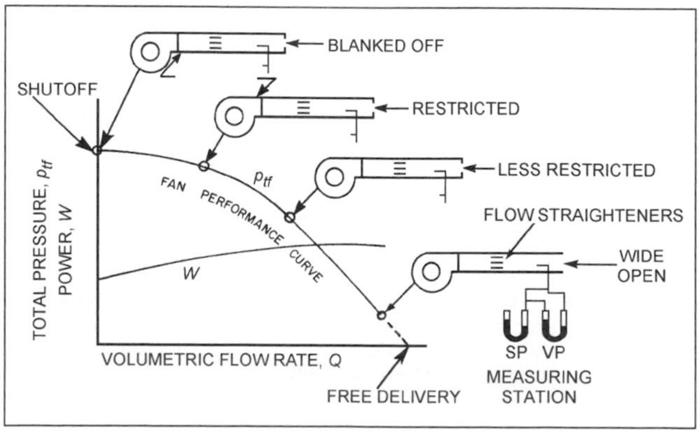

Fig. 3 Method of Obtaining Fan Performance Curves

Table 2 Fan Laws

Law No.	Dependent Variables				Independent Variables
1a	Q_1	=	Q_2	×	$(D_1/D_2)^3\,(N_1/N_2)$
1b	p_1	=	p_2	×	$(D_1/D_2)^2\,(N_1/N_2)^2\,\rho_1/\rho_2$
1c	W_1	=	W_2	×	$(D_1/D_2)^5\,(N_1/N_2)^3\,\rho_1/\rho_2$
2a	Q_1	=	Q_2	×	$(D_1/D_2)^2\,(p_1/p_2)^{1/2}\,(\rho_2/\rho_1)^{1/2}$
2b	N_1	=	N_2	×	$(D_2/D_1)\,(p_1/p_2)^{1/2}\,(\rho_2/\rho_1)^{1/2}$
2c	W_1	=	W_2	×	$(D_1/D_2)^2\,(p_1/p_2)^{3/2}\,(\rho_2/\rho_1)^{1/2}$
3a	N_1	=	N_2	×	$(D_2/D_1)^3\,(Q_1/Q_2)$
3b	p_1	=	p_2	×	$(D_2/D_1)^4\,(Q_1/Q_2)^2\,\rho_1/\rho_2$
3c	W_1	=	W_2	×	$(D_2/D_1)^4\,(Q_1/Q_2)^3\,\rho_1/\rho_2$

Notes:
1. Subscript 1 denotes the variable for the fan under consideration. Subscript 2 denotes the variable for the tested fan.
2. For all fans laws $(\eta_t)_1 = (\eta_t)_2$ and (Point of rating)$_1$ = (Point of rating)$_2$.
3. p equals either p_{tf} or p_{sf}.

FAN LAWS

The fan laws (see Table 2) relate the performance variables for any dynamically similar series of fans. The variables are fan size D; rotational speed N; gas density ρ; volume flow rate Q; pressure p_{tf} or p_{sf}; power W; and mechanical efficiency η_t. **Fan Law 1** shows the effect of changing size, speed, or density on volume flow rate, pressure, and power level. **Fan Law 2** shows the effect of changing size, pressure, or density on volume flow rate, speed, and power. **Fan Law 3** shows the effect of changing size, volume flow rate, or density on speed, pressure, and power.

The fan laws apply only to a series of aerodynamically similar fans at the same point of rating on the performance curve. They can be used to predict the performance of any fan when test data are available for any fan of the same series. Fan laws may also be used with a particular fan to determine the effect of speed change. However, caution should be exercised in these cases, since the laws apply only when all flow conditions are similar. Changing the speed of a given fan changes parameters that may invalidate the fan laws.

Unless otherwise identified, fan performance data are based on dry air at standard conditions—14.696 psi and 70°F (0.075 lb/ft^3). In actual applications, the fan may be required to handle air or gas at some other density. The change in density may be caused by temperature, composition of the gas, or altitude. As indicated by the fan laws, fan performance is affected by gas density. With constant size and speed, the power and pressure vary in accordance with the ratio of gas density to standard air density.

Figure 4 illustrates the application of the fan laws for a change in fan speed N for a specific size fan. The computed p_t curve is derived from the base curve. For example, point E ($N_1 = 650$) is computed from point D ($N_2 = 600$) as follows:

At D,

$$Q_2 = 6000 \text{ cfm and } p_{tf_2} = 1.13 \text{ in. of water}$$

Using Fan Law 1a at point E,

$$Q_1 = 6000 \times 650/600 = 6500 \text{ cfm}$$

Using Fan Law 1b,

$$p_{tf_1} = 1.13(650/600)^2 = 1.33 \text{ in. of water}$$

The total pressure curve p_{tf_1} at $N = 650$ may be generated by computing additional points from data on the base curve, such as point G from point F.

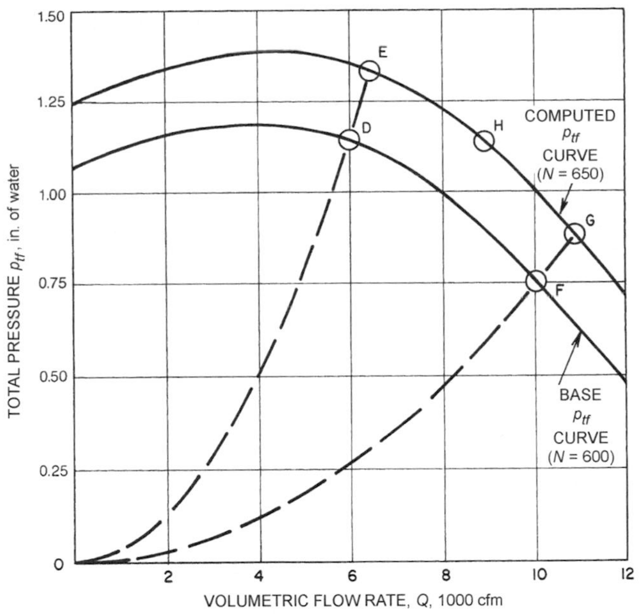

Fig. 4 Example Application of Fan Laws

If equivalent points of rating are joined, as shown by the dotted lines in Figure 4, they form parabolas, which are defined by the relationship expressed in Equation (1).

Each point on the base p_{tf} curve determines only one point on the computed curve. For example, point H cannot be calculated from either point D or point F. Point H is, however, related to some point between these two points on the base curve, and only that point can be used to locate point H. Furthermore, point D cannot be used to calculate point F on the base curve. The entire base curve must be defined by test.

FAN AND SYSTEM PRESSURE RELATIONSHIPS

As previously stated, a fan impeller imparts static and kinetic energy to the air. This energy is represented in the increase in total pressure and can be converted to static or velocity pressure. These two quantities are interdependent; fan performance cannot be evaluated by considering one or the other alone. The conversion of energy, indicated by changes in velocity pressure to static pressure and vice versa, depends on the efficiency of conversion. Energy conversion occurs in the discharge duct connected to a fan being tested in accordance with AMCA *Standard* 210 and ASHRAE *Standard* 51, and the efficiency is reflected in the rating.

Fan total pressure rise p_{tf} is a true indication of the energy imparted to the airstream by the fan. System pressure loss (Δp) is the sum of all individual total pressure losses imposed by the air distribution system duct elements on both the inlet and outlet sides of the fan. An energy loss in a duct system can be defined only as a total pressure loss. The measured static pressure loss in a duct element equals the total pressure loss only in the special case where air velocities are the same at both the entrance and exit of the duct element. By using total pressure for both fan selection and air distribution system design, the design engineer is assured of proper design. These fundamental principles apply to both high- and low-velocity systems. (Chapter 32 of the 1997 *ASHRAE Handbook—Fundamentals* has further information.)

Fan static pressure rise p_{sf} is often used in low-velocity ventilating systems where the fan outlet area essentially equals the fan outlet duct area, and little energy conversion occurs. When fan performance data are given in terms of p_{sf}, the value of p_{tf} may be calculated from the catalog data.

To specify the pressure performance of a fan, the relationship of p_{tf}, p_{sf}, and p_{vf} must be understood, especially when negative pressures are involved. Most importantly, p_{sf} is a term defined in AMCA *Standard* 210 and ASHRAE *Standard* 51 as $p_{sf} = p_{tf} - p_{vf}$. Except in special cases, p_{sf} is not necessarily the measured difference between static pressure on the inlet side and static pressure on the outlet side.

Figure 5 through Figure 8 illustrate the relationships among these various pressures. Note that, as defined, $p_{tf} = p_{t2} - p_{t1}$. Figure 5 illustrates a fan with an outlet system but no connected inlet system. In this particular case, the fan static pressure p_{sf} equals the static pressure rise across the fan. Figure 6 shows a fan with an inlet system but no outlet system. Figure 7 shows a fan with both an inlet system and an outlet system. In both cases, the measured difference in static pressure across the fan ($p_{s2} - p_{s1}$) is not equal to the fan static pressure.

All of the systems illustrated in Figure 5 to Figure 7 have inlet or outlet ducts that match the fan connections in size. Usually the duct size is not identical to the fan outlet or the fan inlet, so that a further complication is introduced. To illustrate the pressure relationships in this case, Figure 8 shows a diverging outlet cone, which is a commonly used type of fan connection. In this case, the pressure relationships at the fan outlet do not match the pressure relationships in the flow section. Furthermore, the static pressure in the cone actually increases in the direction of flow. The static pressure changes throughout the system, depending on velocity. The total pressure, which, as noted in the figure, decreases in the direction of flow, more truly represents the loss introduced by the cone or by flow in the duct. Only the fan changes this trend (i.e., the decrease of total pressure in

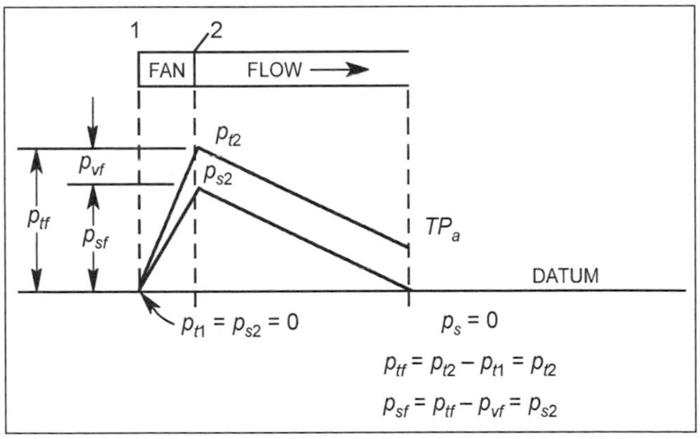

Fig. 5 Pressure Relationships of Fan with Outlet System Only

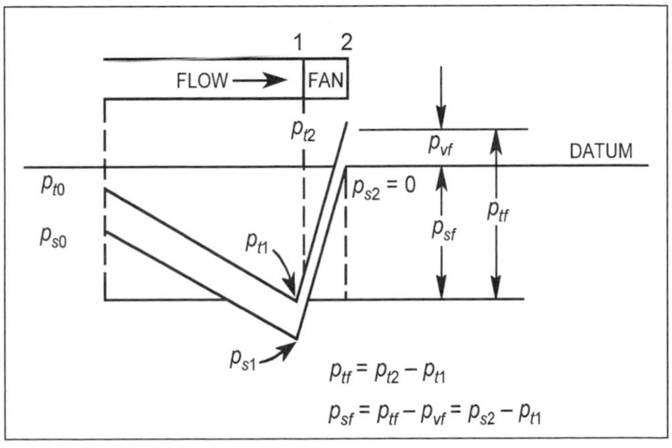

Fig. 6 Pressure Relationships of Fan with Inlet System Only

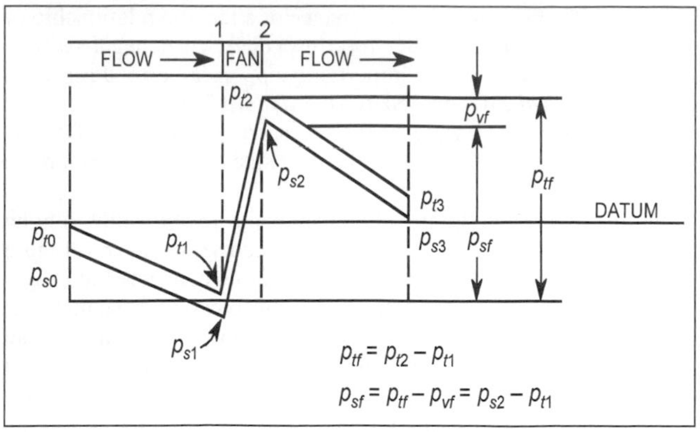

Fig. 7 Pressure Relationships of Fan with Equal-Sized Inlet and Outlet Systems

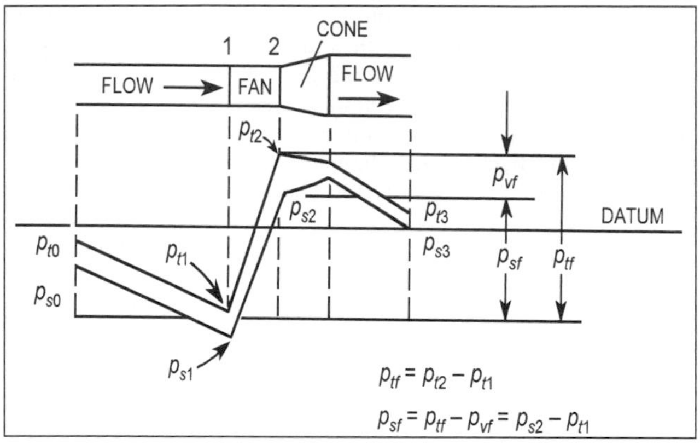

Fig. 8 Pressure Relationships of Fan with Diverging Cone Outlet

the direction of flow). Total pressure, therefore, is a better indication of fan and duct system performance. In this normal fan situation, the static pressure across the fan ($p_{s2} - p_{s1}$) does not equal the fan static pressure p_{sf}.

DUCT SYSTEM CHARACTERISTICS

Figure 9 shows a simplified duct system with three 90° elbows. These elbows represent the resistance offered by the ductwork, heat exchangers, cabinets, dampers, grilles, and other system components. A given rate of airflow through a system requires a definite total pressure in the system. If the rate of flow is changed, the resulting total pressure required will vary, as shown in Equation (1), which is true for turbulent airflow systems. Heating, ventilating, and air-conditioning systems generally follow this law very closely.

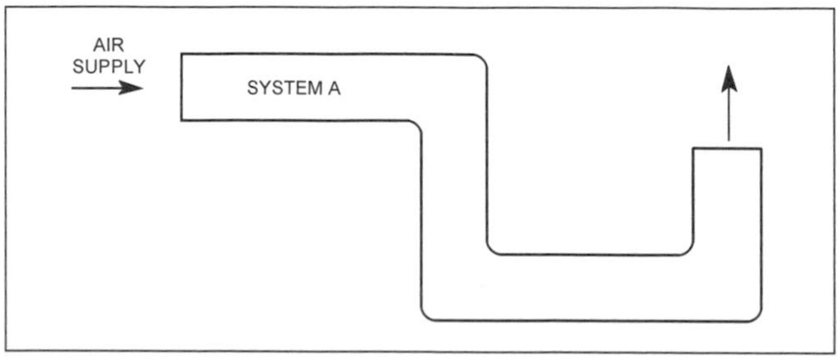

Fig. 9 Simple Duct System with Resistance to Flow Represented by Three 90° Elbows

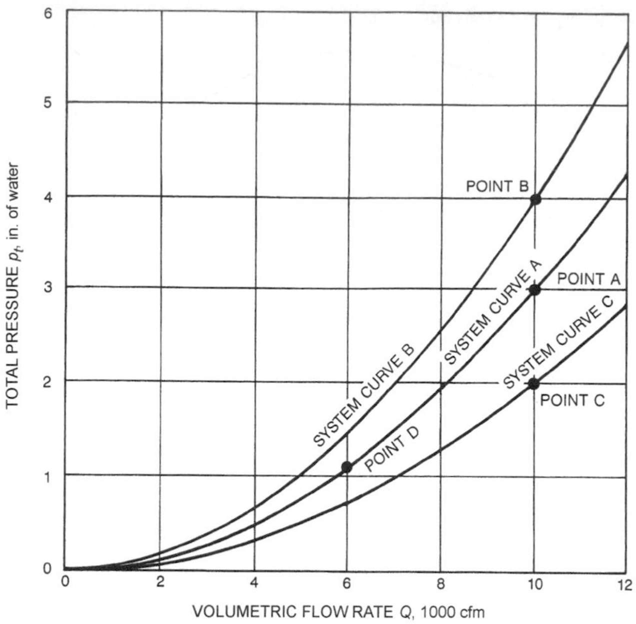

Fig. 10 Example System Total Pressure Loss (Δ*p*) Curves

$$(\Delta p_2/\Delta p_1) = (Q_2/Q_1)^2 \qquad (1)$$

This chapter covers only turbulent flow—the flow regime in which most fans operate. In some systems, particularly constant or variable-volume air conditioning, the air-handling devices and associated controls may produce effective system resistance curves that deviate widely from Equation (1), even though each element of the system may be described by this equation.

Equation (1) permits plotting a turbulent flow system's pressure loss (Δp) curve from one known operating condition (see Figure 4). The fixed system must operate at some point on this system curve as the volume flow rate changes. As an example, in Figure 10, at point A of curve A, when the flow rate through a duct system such as that shown in Figure 9 is 10,000 cfm, the total pressure drop is 3 in. of water. If these values are substituted in Equation (1) for Δp_1 and Q_1, other points of the system's Δp curve (Figure 10) can be determined.

For 6000 cfm (Point D on Figure 10):

$$\Delta p_2 = 3(6000/10{,}000)^2 = 1.08 \text{ in. of water}$$

If a change is made within the system so that the total pressure at design flow rate is increased, the system will no longer operate on the previous Δp curve, and a new curve will be defined.

For example, in Figure 11, an elbow added to the duct system shown in Figure 9 increases the total pressure of the system. If the total pressure at 10,000 cfm is increased by 1.00 in. of water, the system total pressure drop at this point is now 4.00 in. of water, as shown by point B in Figure 10.

If the system in Figure 9 is changed by removing one of the schematic elbows (see Figure 12), the resulting system total pressure drops below the total pressure resistance, and the new Δp curve is curve C of Figure 10. For curve C, a total pressure reduction of 1.00 in. of water has been assumed when 10,000 cfm flows through the system; thus, the point of operation is at 2.00 in. of water, as shown by point C.

These three Δp curves all follow the relationship expressed in Equation (1). These curves result from changes within the system itself and do not change the fan performance. During the design

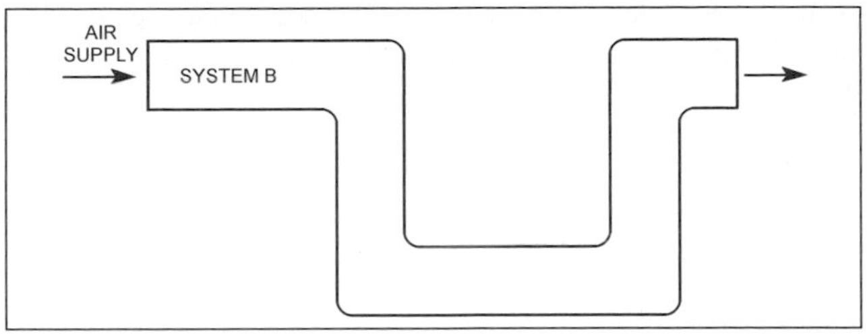

Fig. 11 Resistance Added to Duct System of Figure 9

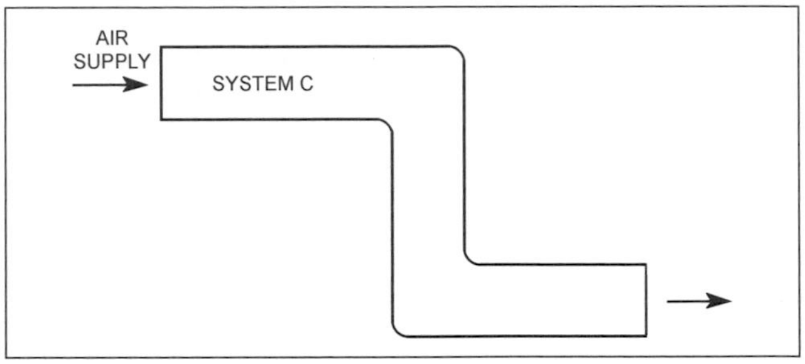

Fig. 12 Resistance Removed from Duct System of Figure 9

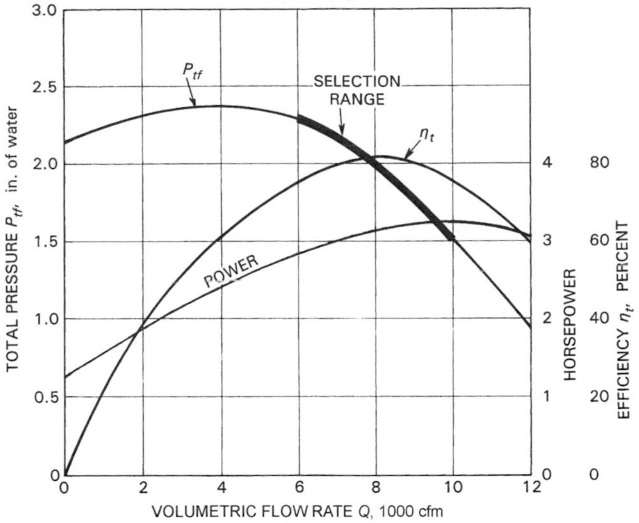

(Curve shows performance of a fixed fan size running at a fixed

Fig. 13 Conventional Fan Performance Curve Used by Most Manufacturers

phase, such system total pressure changes may occur because of studies of alternate duct routing, studies of differences in duct sizes, allowance for future duct extensions, or the effect of the design safety factor being applied to the system.

In an actual operating system, these three Δp curves can represent three system characteristic lines caused by three different positions of a throttling control damper. Curve C is the most open position, and curve B is the most closed position of the three positions illustrated. A control damper forms a continuous series of these Δp curves as it moves from a wide open position to a completely closed position and covers a much wider range of operation than is illustrated here. Such curves can also represent the clogging of turbulent flow filters in a system.

SYSTEM EFFECTS

Normally, a fan is tested with open inlets, and a section of straight duct is attached to the outlet. This setup results in uniform flow into the fan and efficient static pressure recovery on the fan outlet. If good inlet and outlet conditions are not provided in the actual installation, the performance of the fan suffers. To select and apply the fan properly, these effects must be considered and the pressure requirements of the fan, as calculated by standard duct design procedures, must be increased.

These calculated system effect factors are only an approximation, however. Fans of different types and even fans of the same type, but supplied by different manufacturers, do not necessarily react to a system in the same way. Therefore, judgment based on experience must be applied to any design. Chapter 32 of the 1997 *ASHRAE Handbook—Fundamentals* gives information on calculating the system effect factors and lists loss coefficients for a variety of fittings. Clarke et al. (1978) and AMCA *Publication* 201 provide further information.

SELECTION

After the system pressure loss curve of the air distribution system has been defined, a fan can be selected to meet the requirements of the system (Graham 1966, 1972). Fan manufacturers present performance data in either graphic (curve) form (see Figure 13) or tabular form (multirating tables). Multirating tables usually provide only performance data within the recommended operating range. The optimum selection range or peak efficiency point is identified in various ways by different manufacturers.

Performance data as tabulated in the usual fan tables are based on arbitrary increments of flow rate and pressure. In these tables, adjacent data, either horizontally or vertically, represent different points of operation (i.e., different points of rating) on the fan performance curve. These points of rating depend solely on the fan's characteristics; they cannot be obtained one from the other by the fan laws. However, these points of operation listed in the multirating tables are usually close together, so intermediate points may be interpolated arithmetically with adequate accuracy for fan selection.

The selection of a fan for a particular air distribution system requires that the fan pressure characteristics fit the system pressure characteristics. Thus, the total system must be evaluated and the flow requirements, resistances, and system effect factors at the fan inlet and outlet must be known (see Chapter 32 of the 1997 *ASHRAE Handbook—Fundamentals*). Fan speed and power requirements are then calculated, using one of the many methods available from fan manufacturers. These may consist of the multirating tables or of single or multispeed performance curves or graphs.

In using curves, it is necessary that the point of operation selected (see Figure 14) represent a desirable point on the fan curve, so that maximum efficiency and resistance to stall and pulsation can be attained. In systems where more than one point of operation is encountered during operation, it is necessary to look at the range of performance and evaluate how the selected fan reacts within this complete range. This analysis is particularly necessary for variable-volume systems, where not only the fan undergoes a change in performance, but the entire system deviates from the relationships defined in Equation (1). In these cases, it is necessary to look at actual losses in the system at performance extremes.

PARALLEL FAN OPERATION

The combined performance curve for two fans operating in parallel may be plotted by using the appropriate pressure for the ordinates and the sum of the volumes for the abscissas. When two fans having a pressure reduction to the left of the peak pressure point are operated in parallel, a fluctuating load condition may result if one of the fans operates to the left of the peak static point on its performance curve.

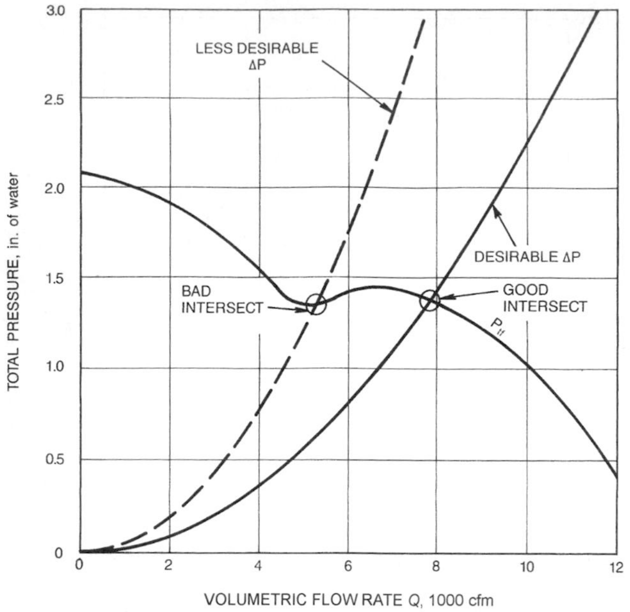

Fig. 14 Desirable Combination of p_{tf} and Δp Curves

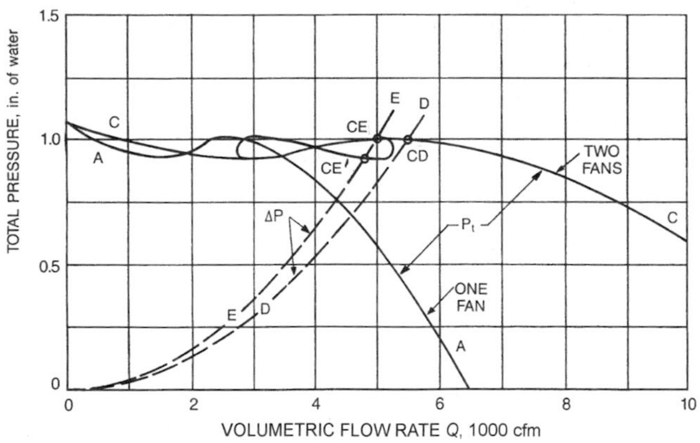

Fig. 15 Two Forward-Curved Centrifugal Fans in Parallel Operation

The p_t curves of a single fan and of two identical fans operating in parallel are shown in Figure 15. Curve A-A shows the pressure characteristics of a single fan. Curve C-C is the combined performance of the two fans. The unique figure-8 shape is a plot of all possible combinations of volume flow at each pressure value for the individual fans. All points to the right of CD are the result of each fan operating at the right of its peak point of rating. Stable performance results for all systems with less obstruction to airflow than is shown on the Δp curve D-D. At points of operation to the left of CD, it is possible to satisfy system requirements with one fan operating at one point of rating while the other fan is at a different point of rating. For example, consider Δp curve E-E, which requires a pressure of 1.00 in. of water and a volume of 5000 cfm. The requirements of this system can be satisfied with each fan delivering 2500 cfm at 1.00 in. of water pressure, Point CE. The system can also be satisfied at Point CE′ with one fan operating at 1400 cfm at 0.9 in. of water, while the second fan delivers 3400 cfm at the same 0.9 in. of water.

Note that system curve E-E passes through the combined performance curve at two points. Under such conditions, unstable operation can result. Under conditions of CE′, one fan is underloaded and operating at poor efficiency. The other fan delivers most of the requirements of the system and uses substantially more power than the underloaded fan. This imbalance may reverse and shift the load from one fan to the other.

NOISE

Fan noise is a function of the fan design, volume flow rate Q, total pressure p_t, and efficiency η_t. After a decision has been made regarding the proper type of fan for a given application (keeping in mind the system effects), the best size selection of that fan must be based on efficiency, because the most efficient operating range for a specific line of fans is normally the quietest. Low outlet velocity does not necessarily ensure quiet operation, so selections made on this basis alone are not appropriate. Also, noise comparisons of different types of fans, or fans offered by different manufacturers, made on the basis of rotational or tip speed are not valid. The only valid basis for comparison are the actual sound power levels generated by the different types of fans when they are all producing the required volume flow rate and total pressure.

The data are reported by fan manufacturers as *sound power levels* in eight octave bands. These levels are determined by using a reverberant room for the test facility and comparing the noise generated by the fan to the noise generated by a noise source of known sound power. The measuring technique is described in AMCA *Standard* 300, Reverberant Room Method for Sound Testing of Fans. ASHRAE *Standard* 68/AMCA *Standard* 330, Laboratory Method of Testing In-Duct Sound Power Measurement Procedure for Fans, describes an alternate test method to determine the sound power a duct fan radiates into a supply and/or return duct terminated by an anechoic chamber. These standards do not fully evaluate the pure tones generated by some fans; these tones can be quite objectionable when they are radiated into occupied spaces. On critical installations, special allowance should be made by providing extra sound attenuation in the octave band containing the tone.

A discussion of the sound generated by fans may be found in Chapter 46 of the 1999 *ASHRAE Handbook—Applications*. Sound power level data should be obtained from the fan manufacturer for the specific fan being considered.

ARRANGEMENT AND INSTALLATION

Direction of rotation is determined from the drive side of the fan. On single-inlet centrifugal fans, the drive side is usually considered the side opposite the fan inlet. AMCA has published standard nomenclature to define positions.

Fan Isolation

In air-conditioning systems, ducts should be connected to fan outlets and inlets with unpainted canvas or other flexible material. Access should be provided in the connections for periodic removal of any accumulations tending to unbalance the rotor. When operating against high resistance or when noise level requirements are low, it is preferable to locate the fan in a room removed from occupied areas or in a room acoustically treated to prevent sound transmission. The lighter building construction that is common today makes it desirable to mount fans and driving motors on resilient bases designed to prevent transmission of vibration through floors to the building structure. Conduits, pipes, and other rigid members should not be attached to fans. Noise that results from obstructions, abrupt turns, grilles, and other items not connected with the fan may be present. Treatments for such problems, as well as the design of sound and vibration absorbers, are discussed in Chapter 46 of the 1999 *ASHRAE Handbook—Applications*.

CONTROL

In many heating and ventilating systems, the volume of air handled by the fan varies. The choice of the proper method for varying

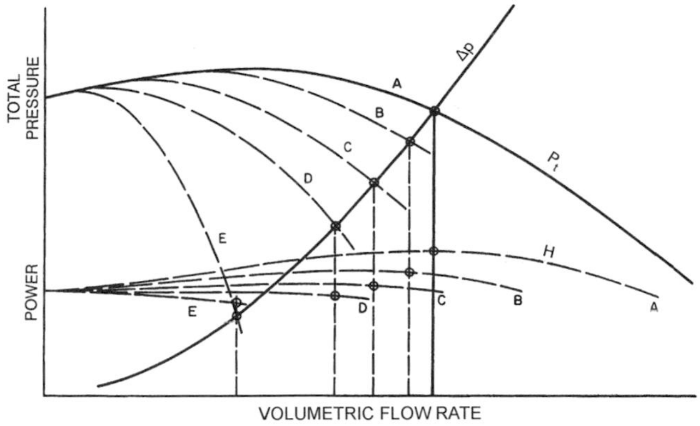

Fig. 16 Effect of Inlet Vane Control on Backward-Curved Centrifugal Fan Performance

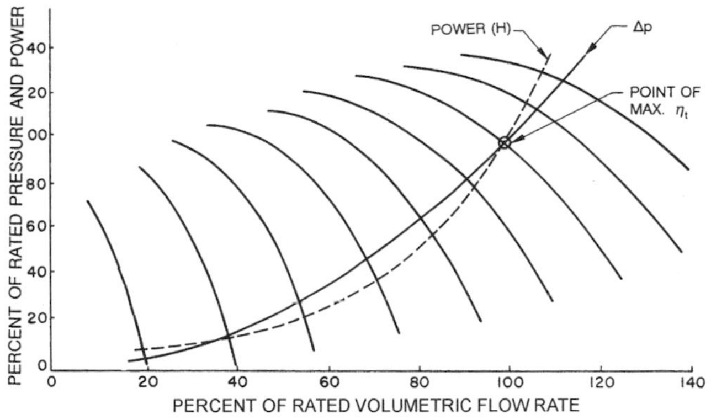

Fig. 17 Effect of Blade Pitch on Controllable Pitch Vaneaxial Fan Performance

flow for any particular case is influenced by two basic considerations: (1) the frequency with which changes must be made and (2) the balancing of reduced power consumption against increases in first cost.

To control flow, the characteristic of either the system or the fan must be changed. The system characteristic curve may be altered by installing dampers or orifice plates. This technique reduces flow by increasing the system pressure required and, therefore, increases power consumption. Figure 10 shows three different system curves, A, B, and C, such as would be obtained by changing the damper setting or orifice diameter. Dampers are usually the lowest first cost method of achieving flow control; they can be used even in cases where essentially continuous control is needed.

Changing the fan characteristic (p_t curve) for control can reduce power consumption. From the standpoint of power consumption, the most desirable method of control is to vary the fan speed to produce the desired performance. If the change is infrequent, belt-driven units may be adjusted by changing the pulley on the drive motor of the fan. Variable-speed motors or variable-speed drives, whether electrical or hydraulic, may be used when frequent or essentially continuous variations are desired. When speed control is used, the revised p_t curve can be calculated with the fan laws.

Inlet vane control is frequently used. Figure 16 illustrates the change in fan performance with inlet vane control. Curves A, B, C, D, and E are the pressure and power curves for various vane settings between wide open (A) and nearly closed (E).

Tubeaxial and vaneaxial fans are made with adjustable pitch blades to permit balancing of the fan against the system or to make infrequent adjustments. Vaneaxial fans are also produced with controllable pitch blades (i.e., pitch that can be varied while the fan is in operation) for frequent or continuous adjustment. Varying pitch angle retains high efficiencies over a wide range of conditions. The performance shown in Figure 17 is from a typical vaneaxial fan with variable pitch blades. From the standpoint of noise, variable speed is somewhat better than variable blade pitch; however, both of these control methods give high operating efficiency control and generate appreciably less noise than inlet vane or damper control.

SYMBOLS

A = fan outlet area, ft^2

D = fan size or impeller diameter

N = rotational speed, revolutions per minute

Q = volume flow rate moved by fan at fan inlet conditions, cfm

p_{tf} = fan total pressure: fan total pressure at outlet minus fan total pressure at inlet, in. of water

p_{vf} = fan velocity pressure: pressure corresponding to average velocity determined from the volume flow rate and fan outlet area, in. of water

p_{sf} = fan static pressure: fan total pressure diminished by fan velocity pressure, in. of water. The fan inlet velocity head is assumed equal to zero for fan rating purposes.

p_{sx} = static pressure at given point, in. of water

p_{vx} = velocity pressure at given point, in. of water

p_{tx} = total pressure at given point, in. of water

V = fan inlet or outlet velocity, fpm

W_o = power output of fan: based on fan volume flow rate and fan total pressure, horsepower

W_i = power input to fan: measured by power delivered to fan shaft, horsepower

η_t = mechanical efficiency of fan (or fan total efficiency): ratio of power output to power input ($\eta_t = W_o/W_i$)

η_s = static efficiency of fan: mechanical efficiency multiplied by ratio of static pressure to fan total pressure, $\eta_s = (p_s/p_t)\eta_t$

ρ = gas (air) density, lb/ft^3

REFERENCES

AMCA. 1986. Drive arrangements for centrifugal fans. *Standards Handbook* 99-2404-86-R1998. Air Movement and Control Association, Arlington Heights, IL.

AMCA. 1986. Laboratory method of testing: In-duct sound power measurement procedure for fans. ANSI/AMCA *Standard* 330-86 (ANSI/ASHRAE *Standard* 68-1986).

AMCA. 1990. Fans and systems. *Application Guide* 201-90.

AMCA. 1996. Reverberant room method for sound testing of fans. *Standard* 300-96.

AMCA. 1999. Laboratory methods of testing fans for aerodynamic performance rating. ANSI/AMCA *Standard* 210-99 (ANSI/ASHRAE *Standard* 51-1999).

ASHRAE. 1986. Laboratory method of testing in-duct sound power measurement procedure forms. ANSI/ASHRAE *Standard* 68-1986 (ANSI/AMCA *Standard* 330-86).

ASHRAE. 1999. Laboratory methods of testing fans for aerodynamic performance rating. ANSI/ASHRAE *Standard* 51-1999 (ANSI/AMCA *Standard* 210-99).

Clark, M.S., J.T. Barnhart, F.J. Bubsey, and E. Neitzel. 1978. The effects of system connections on fan performance. *ASHRAE Transactions* 84(2): 227.

Graham, J.B. 1966. Fan selection by computer techniques. *Heating, Piping and Air Conditioning* (April):168.

Graham, J.B. 1972. Methods of selecting and rating fans. *ASHRAE Symposium Bulletin* SF-70-8, Fan Application—Testing and Selection.

BIBLIOGRAPHY

AMCA. 1986. Motor positions for belt or chain drive centrifugal fans. *Standards Handbook* 99-2407-86-R1998.

AMCA. 1986. Drive arrangements for tubular centrifugal fans. *Standards Handbook* 99-2410-86-R1998.

AMCA. 1986. Designation for rotation and discharge of centrifugal fans. *Standards Handbook* 99-2406-86-R1998.

Buffalo Forge Co. 1983. *Fan engineering*, 8th ed. R. Jorgensen, ed. Buffalo, NY.

CHAPTER 19

EVAPORATIVE AIR COOLING EQUIPMENT

THIS CHAPTER addresses direct and indirect evaporative equipment, air washers, and their associated equipment used for air cooling, humidification, dehumidification, and air cleaning. Residential and industrial humidification equipment are covered in Chapter 20.

Principle advantages of evaporative air conditioning include:

- Substantial energy and cost savings
- Reduced peak power demand
- Improved indoor air quality
- Life cycle cost effectiveness
- Easily integrated into built-up systems
- Wide variety of packages available
- Provide humidification and dehumidification when needed
- Easy to use with direct digital control (DDC)
- Reduced pollution emissions
- No chlorofluorocarbon (CFC) usage

Packaged direct evaporative air coolers, air washers, indirect evaporative air coolers, evaporative condensers, vacuum cooling apparatus, and cooling towers exchange sensible heat for latent heat. This equipment falls into two general categories: (1) apparatus for air cooling and (2) apparatus for heat rejection. This chapter addresses air-cooling equipment.

Adiabatic evaporation of water provides the cooling effect of evaporative air conditioning. In **direct evaporative cooling**, water evaporates directly into the airstream, thus reducing the air's dry-bulb temperature while humidifying the air. Direct evaporative equipment cools air by direct contact with the water, either by an extended wetted-surface material (as in packaged air coolers) or with a series of sprays (as in an air washer).

In **indirect evaporative cooling**, secondary air removes heat from primary air via a heat exchanger. In one indirect method of cooling, water is evaporatively cooled by a cooling tower and circulates through a heat exchanger. Supply air to the space passes over the other side of the heat exchanger. In another common method, one side of an air-to-air heat exchanger is wetted and removes heat from the conditioned supply airstream on the dry side. Even in regions with high wet-bulb temperatures, indirect evaporative cooling can be economically feasible.

Direct and indirect evaporative processes can be **combined (indirect/direct)**. The first stage (indirect) sensibly cools the air, which is then passed through the second stage (direct) and evaporatively cooled further. Combination systems use both direct and indirect evaporative principles as well as secondary heat exchangers and cooling coils. The secondary heat exchangers enhance both cooling and heat recovery (in winter), and the coils provide additional cooling/dehumidification as needed. The secondary heat exchanger has been used in both dual duct systems and in unitary packages. The secondary heat exchanger can also save energy by eliminating the need for terminal reheat in some applications. In such systems, air may exit below the initial wet-bulb temperature.

Direct evaporative coolers for a residences in desert regions typically require 70% less energy than direct expansion air conditioners. For instance, in El Paso, Texas, the typical evaporative cooler consumes 609 kWh per cooling season as compared to 3901 kWh per season for a typical vapor compression air conditioner with a SEER 10. This equates to an average demand of 0.51 kW based on 1200 operating hours, as compared to an average demand of 3.25 kW for a vapor compression air conditioner.

Depending on climatic conditions, many buildings can use indirect/direct evaporative air conditioning to provide comfort cooling. Indirect/direct systems realize a 40 to 50% energy savings in moderate humidity zones (Foster and Dijkstra 1996).

DIRECT EVAPORATIVE AIR COOLERS

In direct evaporative air cooling, air is drawn through porous wetted pads or a spray and its sensible heat energy evaporates some water; the heat and mass transfer between the air and water lowers the air dry-bulb temperature and increases the humidity at a constant wet-bulb temperature. The dry-bulb temperature of the nearly saturated air approaches the ambient air's wet-bulb temperature. The process is adiabatic, so no sensible cooling occurs.

Saturation effectiveness is a key factor in determining the performance of an evaporative cooler. The extent to which the leaving air temperature from a direct evaporative cooler approaches the thermodynamic wet-bulb temperature of the entering air or the extent to which complete saturation is approached is expressed as the **direct saturation effectiveness**, which is defined as:

$$\varepsilon_e = 100\frac{t_1 - t_2}{t_1 - t'} \tag{1}$$

where

ε_e = direct evaporative cooling or saturation effectiveness, %
t_1 = dry-bulb temperature of entering air, °F
t_2 = dry-bulb temperature of leaving air, °F
t' = thermodynamic wet-bulb temperature of entering air, °F

An efficient wetted pad (with a high saturation effectiveness) can reduce the air dry-bulb temperature by as much as 95% of the wet-bulb depression (ambient dry-bulb temperature less wet-bulb temperature), while an inefficient and poorly designed pad may only reduce this by 50%, or less.

Although direct evaporative cooling is simple and inexpensive, it has the disadvantage that if the ambient wet-bulb temperature is higher than about 70°F, the cooling effect is not sufficient for indoor comfort cooling; however, cooling is still sufficient for relief cooling applications (e.g., greenhouses, industrial cooling, etc.). Direct evaporative coolers should not recirculate indoor air.

The preparation of this chapter is assigned to TC 5.7, Evaporative Cooling.

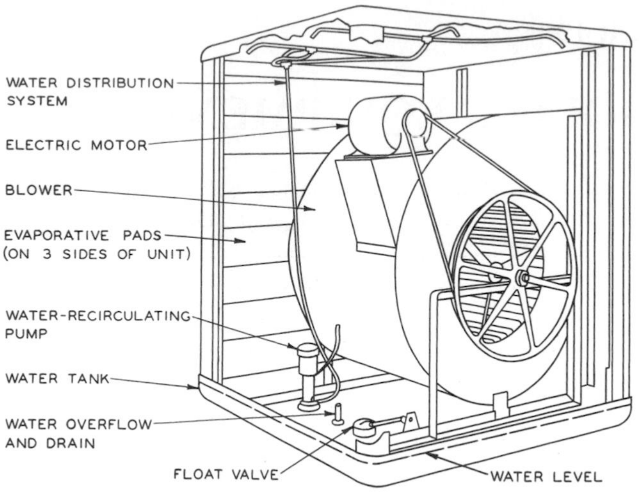

Fig. 1 Typical Random-Media Evaporative Cooler

Random-Media Air Coolers

These coolers contain evaporative pads, usually of aspen wood or absorbent plastic fiber/foam (Figure 1). A water-recirculating pump lifts the sump water to a distributing system, and it flows down through the pads back to the sump.

A fan in the cooler forces air through the evaporative pads and delivers it to the space to be cooled. The fan discharges either through the side of the cooler cabinet or through the sump bottom. Random-media packaged air coolers are made as small tabletop coolers (50 to 200 cfm), window units (100 to 4500 cfm), and standard duct-connected coolers (5000 to 18000 cfm). Cooler selection should be based on a capacity rating from an independent agency.

When clean and well maintained, commercial random-media air coolers operate at an effectiveness of approximately 80% and remove 10 μm and larger particles from the air. In some units, supplementary filters ahead of or following the evaporative pads keep particles from entering the cooler, even when it is operated without water to circulate fresh air. The evaporative pads may be chemically treated to increase wettability. An additive may be included in the fibers to help them resist attack by bacteria, fungi, and other microorganisms.

Random-media coolers are usually designed for an evaporative pad face velocity of 100 to 250 fpm, with a pressure drop of 0.1 in. of water. Aspen fibers are packed to approximately 0.3 to 0.4 lb/ft^2 of face area based on a 2 in. thick pad. Pads are mounted in removable louvered frames, which are usually made of painted galvanized steel or molded plastic. Troughs distribute water to the pads. A centrifugal pump with a submerged inlet pumps the water through tubes that provide an equal flow of water to each trough. It is important that pumps used are thermally protected. The sump or water tank has a water makeup connection, float valve, overflow pipe, and drain. Provisions to bleed water to prevent the buildup of minerals, dirt, and microbial growth are typically incorporated in the design.

The fan is usually a forward-curved, centrifugal fan, complete with motor and drive. The V-belt drive may include an adjustable pitch motor sheave to allow the fan speed to be increased to use the full motor capacity at higher airflow resistance. The motor enclosure may be drip-proof, totally enclosed, or a semi-open type specifically designed for evaporative coolers.

Rigid-Media Air Coolers

Blocks of corrugated material make up the wetted surface of rigid-media direct evaporative air coolers (Figure 2). Materials include cellulose, plastic, and fiberglass that have been treated to absorb water yet resist the weathering effects of water. The medium is cross-corrugated to maximize mixing of air and water. In the direction of airflow, the depth of medium is commonly 12 in., but it may be between 4 and 24 in. The medium has the desirable characteristics of low resistance to airflow, high saturation effectiveness, and self-cleaning by flushing the front face of the pad. The rigid medium is usually designed for a face velocity of 400 to 600 fpm.

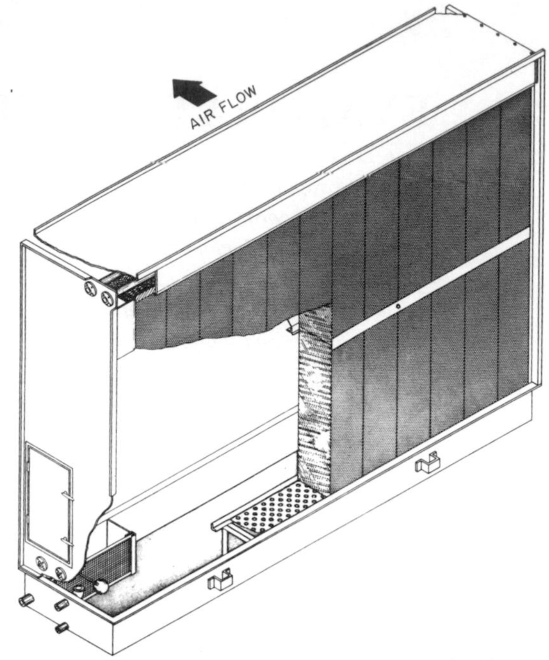

Fig. 2 Typical Rigid-Media Air Cooler

Direct evaporative air coolers using this material are built to handle as much as 600,000 cfm with or without fans. Saturation effectiveness varies from 70 to over 95%, depending on media depth and air velocity. Air flows horizontally while the recirculating water flows vertically over the medium surfaces by gravity feed from a flooding header and water distribution chamber. The header may be connected directly to a pressurized water supply for once through operation (i.e. gas turbines and clean rooms), or a pump may recirculate the water from a lower reservoir, which is constructed of heavy gage corrosion-resistant material. The reservoir is also fitted with overflow and positive flowing drain connections. The upper media enclosure is fabricated of reinforced galvanized steel or other corrosion-resistant sheet metal, or of plastic.

Flanges at the entering and leaving faces allow the unit to be connected to ductwork. In recirculating water systems, a float valve maintains proper water level in the reservoir, makes up water that has evaporated, and supplies fresh water for dilution to prevent an overconcentration of solids and minerals. Because the water recirculation rate is low and because high pressure nozzles are not needed to saturate the medium, pumping power is low when compared to spray-filled air washers with equivalent evaporative cooling effectiveness.

Remote Pad Evaporative Cooling Equipment

Greenhouses, poultry buildings, hog buildings, and similar applications use exhaust fans installed in the wall or roof of the structure. Air is evaporatively cooled as it is drawn through pads located on the other end of the building. The pads are wetted from above by a perforated pipe and excess water is collected for recirculation. In some cases, the pads are wetted with high pressure fogging nozzles. In this case, the fog provides additional cooling. Water from fogging nozzles must never be recirculated. The pad should be sized for an air velocity of approximately 150 fpm for random-media pads, 250 fpm for 4 in. rigid media, and 425 fpm for 6 in. rigid media.

INDIRECT EVAPORATIVE AIR COOLERS

Indirect Packaged Air Coolers

In indirect evaporative air coolers, outdoor air or exhaust air from the conditioned space passes through one side of a heat exchanger. This air (the secondary airstream) is cooled by evaporation by one of several methods: (1) direct wetting of the heat exchanger surface, (2) passing through evaporative cooling media, (3) atomizing spray, (4) disk evaporator, etc. The surfaces of the heat exchanger are cooled by the secondary airstream. On the other side of the heat exchanger surface, the primary airstream (conditioned air to be supplied to the space) is sensibly cooled by the heat exchanger surfaces.

Although the primary air is cooled by secondary air, no moisture is added to the primary air. Hence, the process is known as indirect evaporative air cooling. The supply (primary) air may be recirculated room air, outside air, or a mixture of these. The enthalpy of the primary airstream decreases because no moisture is added to it. This process contrasts with direct evaporative cooling, which is essentially adiabatic (constant wet-bulb temperature). The usefulness of indirect evaporative cooling is related to the depression of the wet-bulb temperature of the secondary air below the dry-bulb temperature of the entering primary air. Because the secondary airstream is evaporatively cooled rather than the primary airstream, indirect evaporative cooling is effective in almost any air conditioning application regardless of geographic location (Mathur 1990, 1991).

Because the enthalpy of the primary air decreases in an indirect evaporative cooler, the leaving dry-bulb temperature of the primary air must always be above the entering wet-bulb temperature of the secondary airstream. Dehumidifying in the primary airstream can occur only when the dew point of the primary airstream is several degrees higher than the wet-bulb temperature of the secondary airstream. This condition exists only when the secondary airstream is drier than the primary airstream, such as when building exhaust air is used for the secondary air.

A packaged indirect evaporative air cooler includes a heat exchanger, a wetting apparatus, a secondary air fan assembly, a secondary air inlet louver, and an enclosure. The heat exchanger may be constructed with folded metal or plastic sheets, with or without a corrosion-resistant or moisture-retaining coating; or it may be constructed with tubes, so that one airstream flows inside the tubes and the other flows over the exterior tube surfaces. Air filters may be placed upstream of the primary and secondary heat exchangers to minimize fouling by dust, insects, or other airborne contaminants.

Continuous bleed-off and fresh water makeup is necessary to keep the concentration of minerals and contaminants in the water from rising. In all evaporative cooling systems water quality should be controlled to avoid scale and other deposits. Water treatment may be necessary to control corrosion of heat exchanger surfaces and other metal parts.

The packaged indirect evaporative air cooler may be either self-contained, with its own primary air supply fan assembly, or part of a built-up or more complete packaged air-handling system. The cooler may include a single stage of indirect evaporative cooling, or it may include indirect evaporative cooling as the first stage, with additional direct evaporative cooling and/or refrigerated (chilled water or direct-expansion) cooling stages.

When the indirect evaporative cooler is placed in series (upstream) with a conventional refrigerated coil, it reduces the sensible load on the coil and refrigeration system (Figure 3). Energy required for the indirect cooling stage includes the pump and secondary air fan motor, as well as some additional fan energy to overcome resistance added in the primary air. The energy consumed by the indirect evaporative cooling stage is less than the energy saved from reducing the load on the refrigeration apparatus. As a result, for existing refrigeration and air conditioning, the overall efficiency may increase because the energy cost and demand are reduced. For new installations the refrigeration unit may be downsized resulting in (1) lower overall cost for the project, and (2) lower operational cost and peak demand. Indirect evaporative cooling may also reduce the total time the refrigeration equipment must be operated during a year, which reduces wear and tear on the refrigeration equipment.

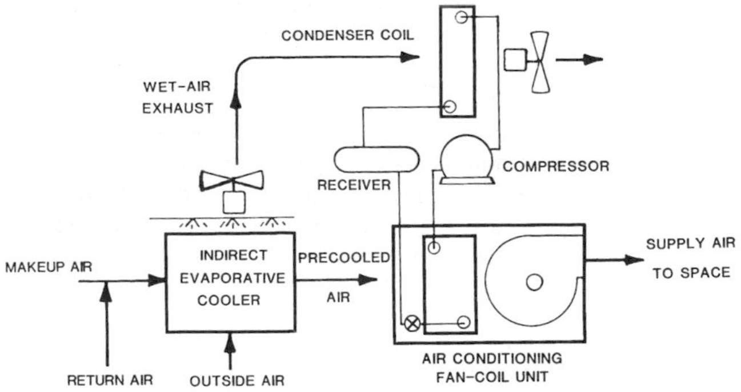

Fig. 3 Indirect Evaporative Cooler Used as Precooler

Evaporatively cooled air can be discharged across air-cooled refrigeration condenser coils to improve the efficiency of the condenser. Chapter 50 of the 1999 *ASHRAE Handbook—Applications* includes sample evaporative cooling calculations. Manufacturers' data should be followed to select equipment for cooling performance, pressure drop, and space requirements.

Indirect evaporative cooling effectiveness ε_{Ie} is defined as

$$\varepsilon_{Ie} = 100\frac{t_1 - t_2}{t_1 - t_s'} \qquad (2)$$

where

ε_e = indirect evaporative cooling effectiveness, %
t_1 = dry-bulb temperature of entering primary air, °F
t_2 = dry-bulb temperature of leaving primary air, °F
t_s' = wet-bulb temperature of entering secondary air, °F

Manufacturers' ratings require careful interpretation. The basis of ratings should be specified because, for the same apparatus, performance is affected by changes in primary and secondary air velocities and mass flow rates, wet-bulb temperature, altitude, and other factors.

Typically, the air resistance on both primary and secondary sections ranges between 0.2 and 2.0 in. of water. The ratio of secondary air to conditioned primary air may range from less than 0.3 to greater than 1.0 and has an effect on performance (Peterson 1993). Based on manufacturers' ratings, available equipment may be selected for indirect evaporative cooling effectiveness ranging from 40 to 80%.

Heat Recovery

Indirect evaporative cooling has been applied to a number of heat recovery systems including plate type heat exchangers (Scofield and DesChamps 1984); heat pipe heat exchangers (Mathur 1991, Scofield 1986); rotary regenerative heat exchangers (Woolridge et al. 1976); two phase thermosiphon loop heat exchangers (Mathur 1990). Indirect evaporative cooling/heat recovery can be used as a retrofit on existing systems, which results in lower operational cost and peak demand. For new installations, the equipment can be downsized, resulting in lower overall cost of the project as well as lower operational cost. Chapter 44 has more information on the use of indirect evaporative cooling with heat recovery.

Cooling Tower/Coil Systems

The combination of a cooling tower or other evaporative water cooler with a water-to-air heat exchanger coil and water circulating pump is another type of indirect evaporative cooling. Water is pumped from the reservoir of the cooling tower to the coil and returns to the upper distribution header of the tower. Both open-water and closed-loop systems are used. Coils in open systems should be cleanable.

The recirculated water is evaporatively cooled to within a few degrees of the wet-bulb temperature as it flows over the wetted surfaces of the cooling tower. As the cooled water flows through the tubes of the coil in the conditioned airstream, it picks up heat from the conditioned air. The temperature of the water increases, and the primary air is cooled without the addition of moisture to the primary air. The water is again cooled as it recirculates through the cooling tower. A float valve controls the fresh-water makeup, which replaces the evaporated water. Bleed-off prevents excessive concentration of minerals in the recirculated water.

One advantage of a cooling tower, especially for retrofit applications, large built-up systems, and dispersed air handlers, is that it may be remotely located from the cooling coil. Also, the tower is more accessible for maintenance. Overall indirect evaporative cooling effectiveness ε_e may range between 55% and 75% or higher. If return air is sent to the cooling tower of an indirect cooling system (before being discharged outside), the cooling tower should be specifically designed for this purpose. These coolers wet a medium that has a high ratio of wetted surface area per unit of medium volume. Performance depends on depth of the medium, air velocity over the medium surface, water flow to airflow ratio, wet-bulb temperature, and water-cooling range. Because of the close approach of the water temperature to the wet-bulb temperature, the overall effectiveness may be higher than that of a conventional cooling tower.

Other Indirect Evaporative Cooling Apparatus

Other combinations of evaporative coolers and heat exchange apparatus can accomplish indirect evaporative cooling. Heat pipes and rotary heat wheels, plate and pleated media, and shell-and-tube heat exchangers have all been applied in this manner. If the conditioned (primary) air and the exhaust or outside (secondary) airstream are side by side, a heat pipe or heat wheel can transfer heat from the warmer air to the cooler air. Evaporative cooling of the secondary airstream by spraying water directly on the surfaces of the heat exchanger or by a direct evaporative cooler upstream of the heat exchanger may cool the primary air indirectly by transferring heat from it to the secondary air.

INDIRECT/DIRECT COMBINATIONS

In a two-stage indirect/direct evaporative cooler, a first-stage indirect evaporative cooler lowers both the dry- and wet-bulb temperature of the incoming air. After leaving the indirect stage, the supply air passes through a second-stage direct evaporative cooler; Figure 4 shows the process on a psychrometric chart. First-stage cooling follows a line of constant humidity ratio because no moisture is added to the primary airstream. The second stage follows the wet-bulb line at the condition of the air leaving the first stage.

The indirect evaporative cooler may be any of the types described previously. Figure 5 shows a cooler using a rotary heat wheel. The secondary air may be exhaust air from the conditioned space or outdoor air. When the secondary air passes through the direct evaporative cooler, the dry-bulb temperature is lowered by evaporative cooling. As this air passes through the heat wheel, the mass of the medium is cooled to a temperature approaching the wet-bulb temperature of the secondary air. The heat wheel rotates so that

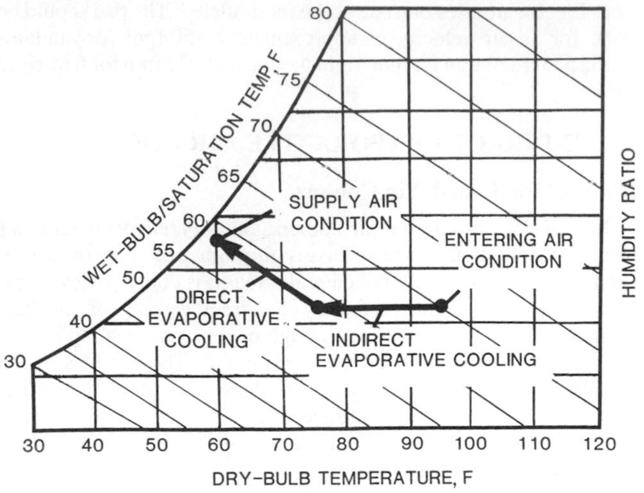

Fig. 4 Combination Indirect/Direct Evaporative Cooling Process

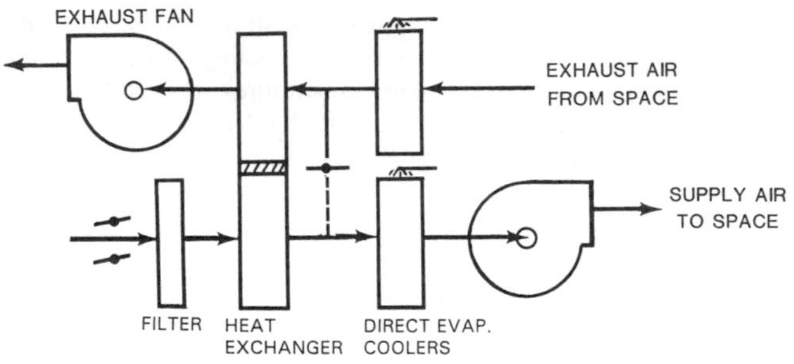

Fig. 5 Indirect/Direct Evaporative Cooler with Heat Exchanger (Rotary Heat Wheel or Heat Pipe)

its cooled mass enters the primary air and, in turn, sensibly cools the primary (supply) air. Following the heat wheel, a direct evaporative cooler further reduces the dry-bulb temperature of the primary air. This method can lower the supply air dry-bulb temperature by 10°F or more below the secondary air wet-bulb temperature.

In areas where the 0.4% mean coincident wet-bulb design temperature is 66°F or lower, average annual cooling power consumption of indirect/direct systems may be as low as 0.22 kW/ton of refrigeration. When the 0.4% mean coincident wet-bulb temperature is as high as 74°F, indirect/direct cooling can have an average annual cooling power consumption as low as 0.81 kW/ton. By comparison, the typical refrigeration system with an air-cooled condenser may have an average annual power consumption greater than 1.0 kW/ton.

In dry environments, indirect/direct evaporative cooling is usually designed to supply 100% outdoor air to the conditioned spaces of a building. In these once-through applications, space latent loads and return air sensible loads are exhausted from the building rather than returned to the conditioning equipment. Consequently, the cooling capacity required from these systems may be less than that required from a conventional refrigerated cooling system.

In areas with a higher wet-bulb design temperature or where the design requires a supply air temperature lower than that attainable using indirect/direct evaporative cooling, a third cooling stage may be required. This stage may be a direct-expansion refrigeration unit or a chilled water coil located either upstream or downstream from the direct evaporative cooling stage, but always downstream from the indirect evaporative stage. Refrigerated cooling is energized only when evaporative stages cannot achieve the required supply air temperature. Figure 6 shows a schematic of a three-stage configuration (indirect/direct, with optional third-stage refrigerated

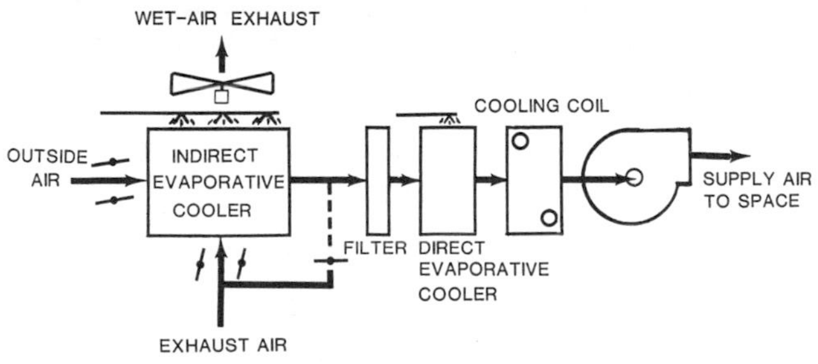

Fig. 6 Three-Stage Indirect/Direct Evaporative Cooler

cooling). The third-stage refrigerated cooling coil is located downstream from the direct evaporative cooler.

A single coil may be used to cool return chilled water with cooling tower water and a plate heat exchanger (a form of indirect evaporative cooling). This hybrid, three-stage configuration permits indirect cooling when the wet-bulb temperature is low and mechanical cooling when the wet bulb is high or when dehumidification is necessary.

The designer should consider options using building exhaust and/or outside air as secondary air for the indirect evaporative cooling stage, whichever has the lower wet-bulb temperature. If possible, the indirect evaporative cooler should be designed to use both outside air and building exhaust as the secondary airstream; whichever source has the lower wet-bulb temperature would be used. Dampers and an enthalpy sensor are used to control this process. If the latent load in the space is significant, the wet-bulb temperature of the building exhaust air in the cooling mode may be higher than that of the outside air. In this case, outside air may be used more effectively as secondary air to the indirect evaporative cooling stage.

Custom indirect/direct and three-stage configurations are available to permit many choices for location of the return, exhaust, and outside air; mixing of airstreams; bypass of components, or variable volume control. The elements that may be controlled include:

- Modulating outside air and return air mixing dampers
- Secondary air fans and recirculating pumps of an indirect evaporative stage
- Recirculating pumps of a direct evaporative cooling stage
- Face and bypass dampers for the direct evaporative stage
- Chilled water or refrigerant flow for a refrigerated stage
- System or individual terminal volume with variable volume terminals, fan variable inlet vanes, adjustable pitch fans, or variable-speed fans.

For sequential control in indirect/direct evaporative cooling, the indirect evaporative cooler is energized for first-stage cooling, the direct evaporative cooler for second-stage cooling, and the refrigeration coil for third-stage cooling. In some applications, reversing the sequence of the direct evaporative cooler and indirect evaporative cooler may reduce the first-stage power requirement.

Precooling and Makeup Air Pretreatment

Evaporative cooling may be used to increase the capacity and reduce the electrical demand of a direct expansion air conditioner or chiller. Both the condenser and the makeup air may be evaporatively cooled by direct and/or indirect means.

The condenser may be cooled by adding a direct evaporative cooler (usually without a fan) to the condenser fan inlet. The direct evaporative cooler must add very little resistance to the airflow to the condenser, and face velocities must be well below velocities that would entrain liquid and carry it to the condenser. Maintenance of condenser coolers should be infrequent and easy to perform. A well-designed direct evaporative cooler can reduce electrical demand and energy consumption of refrigeration units from 10 to 30%.

Makeup air cooling with an indirect/direct evaporative unit can be applied both to standard packaged units and to large built-up systems. Either outside air or building exhaust air can be used as the secondary air source, whichever has the lower wet-bulb temperature. Outside air is generally easier to cool, and in some cases it is the only option because the building exhaust is so remote from the makeup air inlet. If building exhaust air can be used, it has the potential of heat recovery during cold weather. In general, outside air cooling has higher energy savings and lower electrical demand savings than return air cooling. These systems can reduce the outside air load a minimum of 50%, depending on climate and ventilation requirements.

AIR WASHERS

Spray-Type Air Washers

Spray-type air washers consist of a chamber or casing containing spray nozzles, a tank for collecting spray water as it falls, and an eliminator section for removing entrained drops of water from the air. A pump recirculates water at a rate higher than the evaporation rate. Intimate contact between the spray water and the air causes heat and mass transfer between the air and the water (Figure 7). Air washers are commonly available from 2000 to 250,000 cfm capacity, but specially constructed washers can be made in any size. No standards exist; each manufacturer publishes tables giving physical

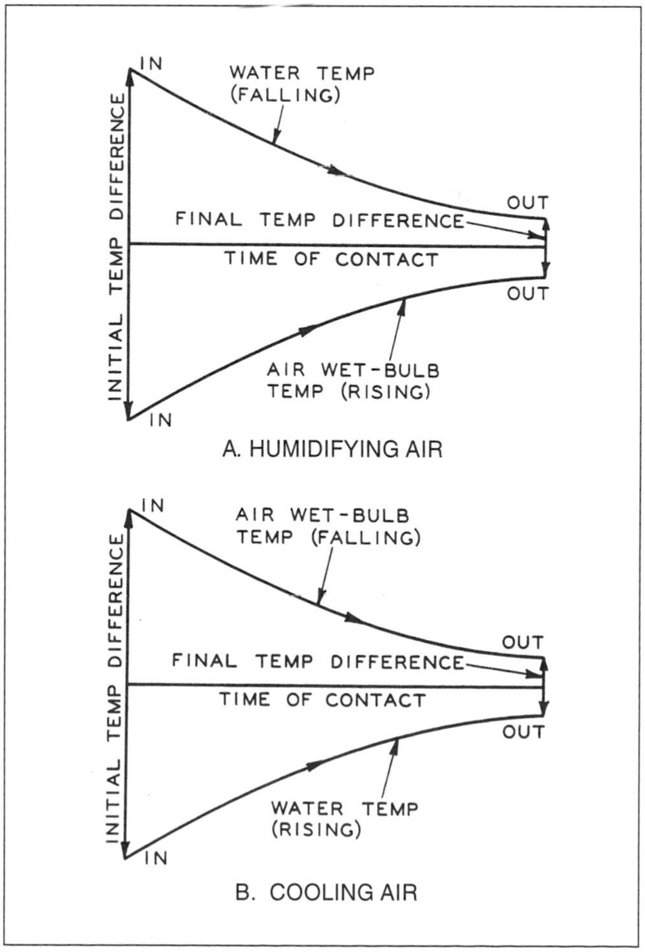

Fig. 7 Interaction of Air and Water in Air Washer Heat Exchanger

data and ratings for specific products. Therefore, air velocity, water-spray density, spray pressure, and other design factors must be considered for each application.

The simplest design has a single bank of spray nozzles with a casing that is usually 4 to 7 ft long. This type of washer is applied primarily as an evaporative cooler or humidifier. It is sometimes used as an air cleaner when the dust is wettable, although the air-cleaning efficiency is relatively low. Two or more spray banks are generally used when a very high degree of saturation is necessary and for cooling and dehumidification applications that require chilled water. Two-stage washers are used for dehumidification when the quantity of chilled water is limited or when the water temperature is above that required for the single-stage design. Arranging the two stages for counter-flow of the water permits a small quantity of water with a greater water temperature rise.

The lengths of washers vary considerably. Spray banks are spaced from 2.5 to 4.5 ft apart; the first and last banks of sprays are located about 1 to 1.5 ft from the entering or leaving end of the washer. In addition, air washers may be furnished with heating or cooling coils in the washer chamber, which may affect the overall length of the washer.

Some water (even very soft water) should always be bled off to prevent mineral buildup and to retard microbial growth. When the unit is shut down, all water should drain from the pipes. Low spots and dead ends must be avoided. Because an air washer is a direct contact heat exchanger, water treatment is critical for proper operation as well as good hygiene. Algae and bacteria can be controlled by a chemical or ozone treatment program and/or regularly scheduled mechanical cleaning.

The resistance to airflow through an air washer varies with the type and number of baffles, eliminators, and wetted surfaces; the number of spray banks and their direction and air velocity; the size and type of other components, such as cooling and heating coils; and other factors, such as air density. Pressure drop may be as low as 0.25 in. of water or as high as 1 in. of water. The manufacturer should be consulted regarding the resistance of any particular washer design combination.

The casing and the tank may be constructed of various materials. One or more doors are commonly provided for inspection and access. An air lock must be provided if the unit is to be entered while it is running. The tank is normally at least 16 in. high with a 14 in. water level; it may extend beyond the casing on the inlet end to make the suction strainer more accessible. The tank may be partitioned by a weir (usually in the entering end) to permit recirculation of spray water for control purposes in dehumidification work. The excess then returns over the weir to the central water chilling machine.

Eliminators consist of a series of vertical plates that are spaced about 0.75 to 2 in. on centers at the exit of the washer. The plates are formed with a number of bends to deflect the air and obtain impingement on the wetted surfaces. Hooks on the edge of the plates improve moisture elimination. Perforated plates may be installed on the inlet end of the washer to obtain more uniform air distribution through the spray chamber. Louvers, which prevent the backlash of spray water, may also be installed for this purpose.

High-Velocity Spray-Type Air Washers

High-velocity air washers generally operate at air velocities in the range of 1200 to 1800 fpm. Some have been applied as high as 2400 fpm, but 1200 to 1600 fpm is the most accepted range. The reduced cross-sectional area of high-velocity air washers allows them to be used in smaller equipment than those operating with lower air velocities. High capacities per unit of space available from high-velocity spray devices permit practical packaging of prefabricated central station units in either completely assembled and transportable form or, for large capacity units, easily handled modules. Manufacturers supply units with capacities of up to 150,000 cfm shipped in one piece, including spray system, eliminators, pump, fan, dampers, filters, and other functional components. Such units are self-housed, pre-wired, pre-piped, and ready for hoisting into place.

The number and arrangement of nozzles vary with different capacities and manufacturers. Adequate values of saturation effectiveness and heat transfer effectiveness are achieved by using higher spray density.

Eliminator blades come in varying shapes, but most are a series of aerodynamically clean, sinusoidal shapes. Collected moisture flows down grooves or hooks designed into their profiles, then drains into the storage tank. Washers may be built with shallow drain pans and connected to a central storage tank. High-velocity washers are rectangular in cross section and, except for the eliminators, are similar in appearance and construction to conventional lower velocity types. Pressure loss is in the range of 0.5 to 1.5 in. of water. These washers are available either as freestanding separate devices for incorporation into field-built central stations or in complete pre-assembled central station packages from the factory.

HUMIDIFICATION/DEHUMIDIFICATION

Humidification with Air Washers and Rigid Media

Air can be humidified with air washers and rigid media in three ways: (1) using recirculated water without prior heating of the air, (2) preheating the air and humidifying it with recirculated water, and (3) preheating recirculated water. Precise humidity control may be achieved by arranging rigid media in one or more banks in depth, height, or width, or by providing a controlled bypass. Each bank is activated independently of the others to achieve the desired humidity. In any evaporative humidification application, air should not be permitted to enter the process with a wet-bulb temperature of less than 39°F, otherwise the water may freeze.

Recirculation Without Preheating. Except for the small amount of energy added by the recirculating pump and the small amount of heat leakage into or from the apparatus (including the pump and its connecting piping), the process is adiabatic. The temperature of the water in the collection basin closely approaches the thermodynamic wet-bulb temperature of the entering air, but it cannot be brought to complete saturation. The psychrometric state point of the leaving air is on the constant thermodynamic wet-bulb temperature line with its end state determined by the saturation effectiveness of the device. Control over leaving humidity conditions may be achieved by controlling the saturation effectiveness of the process by bypassing air around the evaporative process.

Preheating Air. Preheating the air entering an evaporative humidifier increases both the dry-and the wet-bulb temperatures and lowers the relative humidity, but it does not alter the humidity ratio (mass ratio of water vapor to dry air) of the air. As a result, preheating permits more water to be absorbed per unit mass of dry air passing through the process at the same saturation effectiveness. Control is achieved by varying the amount of air preheating at a constant saturation effectiveness. Control precision is a direct function of saturation effectiveness and a high degree of correlation may be achieved between leaving air and leaving dew-point temperatures when high saturation effectiveness devices are used.

Heated Recirculated Water. If heat is added to the water, the process state point of the mixture moves toward the temperature of the heated water (Figure 7A). Elevating the water temperature makes it possible to raise the air temperature, both dry bulb and wet bulb, above the dry-bulb temperature of the entering air with the leaving air becoming fully saturated. Relative humidity of the leaving air can be controlled by (1) bypassing some of the air around the media banks and remixing the two airstreams downstream by using dampers or (2) by automatically reducing the number of operating media banks through pump staging or by operating valves in the different distribution branches.

The following table shows the saturation or humidifying effectiveness of a spray air washer for various spray arrangements. The degree of saturation depends on the extent of contact between air and water. Other conditions being equal, a low-velocity airflow is conducive to higher humidifying effectiveness.

Bank Arrangement	Length, ft	Effectiveness, %
1 downstream	4	50 to 60
1 downstream	6	60 to 75
1 upstream	6	65 to 80
2 downstream	8 to 10	80 to 90
2 opposing	8 to 10	85 to 95
2 upstream	8 to 10	90 to 98

Dehumidification with Air Washers and Rigid Media

Air washers and rigid-media direct evaporative coolers may also be used to cool and dehumidify air (Figure 7B). The heat and moisture removed from the air causes the water temperature to rise. If the entering water temperature is below the entering air dew point, both the dry- and wet-bulb temperatures of the air is reduced, resulting in cooling and dehumidification. The air is typically saturated as it leaves. The vapor pressure difference between the entering air and the water cools the air. Moisture is transferred from the air to the water and condensation occurs. The air leaving an evaporative dehumidifier is typically saturated, usually with less than 1°F difference between leaving dry- and wet-bulb temperatures.

The difference between the leaving air and leaving water temperatures depends on the difference between entering dry- and wet-bulb temperatures and the effectiveness of the process, which may be affected by such factors as length and height of the spray chamber, air velocity, quantity of water flow, and character of the spray pattern. Final water conditions are typically 1 to 2°F below the leaving air temperature, depending on the saturation effectiveness of the device used.

The common design value for the rise in water temperature usually falls between 6 and 12°F for refrigerant chilled water and normal air conditioning applications, although higher rises are possible and have been used successfully. A smaller temperature rise may be considered when water is chilled by mechanical refrigeration. If warmer water is used, less mechanical refrigeration is required; however, a larger quantity of chilled water is needed to do the same amount of sensible cooling. An economic analysis may be required to determine the best alternative. For humidifiers receiving water from a thermal storage or other low-temperature system, a design with a high temperature rise and minimum water flow may be desirable.

Performance Factors. A performance factor of 1.0 is assigned to an evaporative dehumidifier if the device can cool and dehumidify the entering air to a wet-bulb temperature equal to that of the leaving water temperature. This represents a theoretical maximum value which is thermodynamically impossible to achieve. Performance is maximized when both water surface area and air –water contact is maximized. The actual performance factor F_p of any evaporative dehumidifier will be less than one and would be calculated by dividing the actual air enthalpy change by the theoretical maximum air enthalpy change where:

$$F_p = \frac{h_1 - h_2}{h_1 - h_3} \tag{3}$$

where

h_1 = enthalpy at wet bulb of entering air, Btu/lb
h_2 = enthalpy at wet bulb of leaving air at actual condition, Btu/lb
h_3 = enthalpy of air at wet-bulb temperature leaving a dehumidifier with F_p = 1.0, Btu/lb

Air Cleaning

Air washers and rigid-media direct evaporative cooling equipment can remove particulate and gaseous contaminants with varying degrees of effectiveness through wet scrubbing (which is discussed in Chapter 25). The particulate removal efficiency of rigid media and air washers differ due to the differences in equipment construction and principles of operation. Removal also depends largely on the size, density, wettability, and/or solubility of the contaminants to be removed. Large, wettable particles are the easiest to remove. The primary mechanism of separation is by impingement of particles on a wetted surface, which includes eliminator plates in air washers and the corrugations of wetted rigid media. The spray process is relatively ineffective in removing most atmospheric dusts. Because the force of impact increases with the size of the solid, the impact (together with the adhesive quality of the wetted surface) determines the device's usefulness as a dust remover.

In practice, the air-cleaning results of air washers and rigid-media direct evaporative coolers are typical of comparable impingement filters. Air washers are of little use in removing soot particles because of the absence of an adhesive effect from a greasy surface. They are also relatively ineffective in removing smoke because the particles are too small (less than 1 μm) to impact and be retained on the wet surfaces.

Despite its air cleaning performance, rigid media should not be used as a primary filtering device. When a rigid-media cooler is placed in an unfiltered airstream, it can quickly become fouled with airborne dust and fibrous debris. When wet, debris can collect in the recirculation basin and in the media where they can become nutrients for bacteria. Bacteria in the air can propagate in waste materials and debris and cause microbial slimes. Filtering of the entering air is the most effective way to keep debris from accumulating in rigid media. When high efficiency filters are placed upstream from the cells, most microbial agents and nutrients can be removed from the airstream. Rigid media should be replaced if the corrugations are filled with contaminants when they are dry.

MAINTENANCE AND WATER TREATMENT

Regular inspection and maintenance of evaporative coolers, air washers, and ancillary equipment ensures proper service and efficiency. Manufacturers' recommendations for maintenance and operation should be followed to help ensure safe and efficient operation. Water lines, water distribution troughs or sumps, pumps, and pump filters must be clean and free of dirt, scale, and debris. They must be constructed so that they can be easily flushed and cleaned. Inadequate water flow causes dry areas on the evaporative media, which reduces the saturation effectiveness. Motors and bearings should be lubricated and fan drives checked periodically

Water and air filters should be cleaned or replaced as required. The sump water level must be maintained such that the bottom of the pads does not contact the water in the sump, yet high enough to prevent air from short circuiting below the pads. Bleeding off some water is the most practical means to minimize scale accumulation. The bleed rate should be 5 to 100% of the evaporation rate, depending on water hardness and airborne contaminant level. The water circulation pump should be used to bleed off water because suction by a draw-through fan will otherwise prevent the bleed system from operating effectively. A flush-out cycle that runs fresh water through the pad every 24 h when the fan is off may also be used. This water should run for 3 min for every foot of media height.

Regular inspections should be made to ensure that the bleed rate is adequate and is maintained. Some manufacturers provide a purge cycle in which the entire sump is purged of water and accumulated debris. This cycle helps maintain a cleaner system and may actually save water when compared to a standard bleed system. The frequency of the purge cycle depends on the water quality as well as the amount and type of outside contaminants. Sumps should have drain couplings on the bottom rather than on the side in order to drain the sump completely. Additionally, the sump bottom should slope toward the drain (approximately 0.25 in. per foot of sump length) to facilitate complete draining.

Water Treatment. An effective water treatment and biocide program for cooling towers is not necessarily good practice for evaporative coolers. Evaporative coolers and cooling towers differ significantly in that evaporative coolers are directly connected with the supply airstream, whereas cooling towers only indirectly affect the supply air. The affect a biocide may have on the evaporative media (both direct and indirect evaporative systems) as well as the potential for offensive and/or harmful residual off-gassing must be considered.

Pretreatment of a water supply with chemicals intended to hold dissolved material in suspension is best prescribed by a water treatment specialist. Water treated by a zeolite ion exchange softener should not be used because the zeolite exchange of calcium for sodium results in a soft, voluminous scale that may cause dust problems downstream. Any chemical agents used should not promote microbial growth or harm the cabinet, media, or heat exchanger materials. This topic is discussed in more detail in Chapter 47 of the 1999 *ASHRAE Handbook—Applications*. Consider the following factors regarding water treatment

- Use very pure water from reverse osmosis or deionization processing with caution in media-based evaporative coolers. This water does not wet random media well, and it can deteriorate many types of media due to its corrosive nature. The same problem can occur in a once-through water distribution system if the water is very pure.
- Periodically check for algae, slime, and bacterial growth. If required, add a biocide. The biocide must be registered for use in evaporative coolers by an appropriate agency, such as the U.S Environmental Protection Agency (EPA).

Ozone generation systems have been used as an alternative to standard chemical biocide water treatments. Ozone can be produced on-site (eliminating chemical storage) and injected into the water circulation system. It is a fast-acting oxidizer that rapidly breaks down to non-toxic compounds. In low concentrations, ozone is benign to humans and to the materials used in evaporative coolers.

Algae can be minimized by reducing the media and sump exposure to nutrient and light sources (through the use of hoods, louvers and prefilters), by keeping the bottom of the media out of standing water in the sump, and by allowing the media to completely dry out every 24 h.

Scale. Units that have heat exchangers with a totally wetted surface and materials that are not harmed by chemicals can be descaled periodically with a commercial descaling agent and then flushed out. Mineral scale deposits on a wetted indirect evaporative heat exchanger are usually soft and allow wetting through to and evaporation at the surface of the heat exchanger. Excess scale thickness causes a loss in heat transfer and should be removed.

Air Washers. The air washer spray system requires the most attention. Partially clogged nozzles are indicated by a rise in spray pressure, while a fall in pressure is symptomatic of eroded orifices. Strainers can minimize this problem. Continuous operation requires either a bypass around pipeline strainers or duplex strainers. Air washer tanks should be drained and dirt deposits removed regularly. Eliminators and baffles should be periodically inspected and repainted to prevent corrosion damage.

Freeze Protection. In colder climates, evaporative coolers must be protected from freezing. This is usually done seasonally by simply draining the cooler and the water supply line with solenoid valves. Often an outside air temperature sensor initiates this action. It is important that drain solenoid valves be of the zero differential design. If a heat exchanger coil is used, the tubes must be horizontal so that they will drain to the lowest part of their manifold.

Legionnaire's Disease

Legionnaire's disease is contracted by inhaling into the lower respiratory system an aerosol (1 to 5 μm in diameter) laden with sufficient *Legionella pneumophila* bacteria. Evaporative coolers do not provide suitable growth conditions for the bacteria and generally do not release an aerosol. A good maintenance program eliminates potential microbial problems and reduces the concern for disease transmittal (ASHRAE 1998, Puckorius et al. 1995).

The following precautions and maintenance procedures for water systems also improve cooler performance, reduce microbial growth and musty odors, and prolong the life of the equipment:

- Run fans after turning off water until the media completely dries.
- Thoroughly clean and flush the entire cooling water loop regularly (minimum monthly). Include disinfection before and after cleaning.
- Avoid dead end piping, low spots, and other areas in the water distribution system where water may stagnate during shutdown.
- Obtain and maintain the best available mist elimination technology, especially when using misters and air washers.
- Do not locate the inlet of an evaporative cooler near the outlet of a cooling tower
- Maintain system bleedoff and/or purge consistent with makeup water quality.
- Maintain system cleanliness. Deposits from calcium carbonate, minerals, and nutrients may contribute to the growth of molds, slime, and other microbes annoying to building occupants.
- Develop a maintenance checklist, and follow it on a regular basis.
- Consult the equipment or media manufacturer for more detailed assistance in water system maintenance and treatment.

BIBLIOGRAPHY

Anderson, W.M. 1986. Three-stage evaporative air conditioning versus conventional mechanical refrigeration. *ASHRAE Transactions* 92(1B):358-70.

ASHRAE. 1998. Legionellosis: Position statement.

Eskra, N. 1980. Indirect/direct evaporative cooling systems. *ASHRAE Journal* 22(5):21-25.

Foster, R.E. and E. Dijkstra. 1996. Evaporative air-conditioning fundamentals: Environmental and economic benefits worldwide. Refrigeration Science and Technology Proceedings, ISSN 0151 1637. International Institute of Refrigeration, Danish Technological Institute, Danish Refrigeration Association, Aarhus, Denmark, pp. 101-10.

Mathur, G.D. 1991. Indirect evaporative cooling with heat pipe heat exchangers. *ASME* Book No. NE(5):79-85.

Mathur, G.D. 1990. Indirect evaporative cooling with two-phase thermosiphon coil loop heat exchangers. *ASHRAE Transactions* 96(1):1241-49.

Peterson, J.L. 1993. An effectiveness model for indirect evaporative coolers. *ASHRAE Transactions* 99(2):392-99.

Puckorius, P.R., P.T. Thomas, and R.L. Augspurger. 1995. Why evaporative coolers have not caused Legionnaires' disease. *ASHRAE Journal* 37(1):29-33.

Scofield, M. 1986. The heat pipe used for dry evaporative cooling. *ASHRAE Transactions* 92(1B):371-81.

Scofield, M. and N.H. DesChamps. 1980. EBTR compliance and comfort cooling too! *ASHRAE Journal* 22(6):61-63.

Scofield, M. and N.H. DesChamps. 1984. Indirect evaporative cooling using plate type heat exchangers. *ASHRAE Transactions* 90(1):148-53.

Supple, R.G. 1982. Evaporative cooling for comfort. *ASHRAE Journal* 24(8):36.

Watt, J.R. 1986. *Evaporative air conditioning handbook*. Chapman and Hall, London.

Woolridge, M.J., H.L. Chapman, and D. Pescod. 1976. Indirect evaporative cooling systems. *ASHRAE Transactions* 82(1):146-55.

CHAPTER 20

HUMIDIFIERS

IN the selection and application of humidifiers, the designer considers (1) the environmental conditions of the occupancy or process and (2) the characteristics of the building enclosure. Because these may not always be compatible, a compromise solution is sometimes necessary, particularly in the case of existing buildings.

ENVIRONMENTAL CONDITIONS

A particular occupancy or process may dictate a specific relative humidity, a required range of relative humidity, or certain limiting maximum or minimum values. The following classifications explain the effects of relative humidity and provide guidance on the requirements for most applications.

Human Comfort

The complete effect of relative humidity on all aspects of human comfort has not yet been established. For thermal comfort, higher temperature is generally considered necessary to offset decreased relative humidity (see ASHRAE *Standard* 55).

Low relative humidity increases evaporation from the membranes of the nose and throat, drying the mucous membranes in the respiratory system; it also dries the skin and hair. The increased incidence of respiratory complaints during the winter months is often linked to low relative humidity. Epidemiological studies have found lower rates of respiratory illness reported among occupants of buildings with mid-range relative humidity than among occupants of buildings with low humidity.

Extremes of humidity are the most detrimental to human comfort, productivity, and health. Figure 1 shows that the range between 30 and 60% rh (at normal room temperatures) provides the best conditions for human occupancy (Sterling et al. 1985). In this range, both the growth of bacteria and biological organisms and the speed at which chemical interactions occur are minimized.

Prevention and Treatment of Disease

Relative humidity has a significant effect on the control of airborne infection. At 50% rh, the mortality rate of certain organisms is highest, and the influenza virus loses much of its virulence. The mortality rate decreases both above and below this value. High humidity can support the growth of pathogenic or allergenic organisms. Relative humidity in habitable spaces should be maintained between 30 and 60%.

Potential Bacterial Growth

Certain microorganisms are occasionally present in poorly maintained humidifiers. To deter the propagation and spread of these detrimental microorganisms, periodic cleaning of the humidifier and draining of the reservoir (particularly at the end of the humidification season) are required. Cold water reservoir atomizing room humidifiers have been banned in some hospitals because of germ propagation. Research by Unz et al. (1993) on several types of plenum-mounted residential humidifiers showed no evidence of organism transmission originating from the humidifier. Ruud et al. (1993) also determined that humidifiers did not add particles to the heated airstream.

The preparation of this chapter is assigned to TC 8.7, Humidifying Equipment.

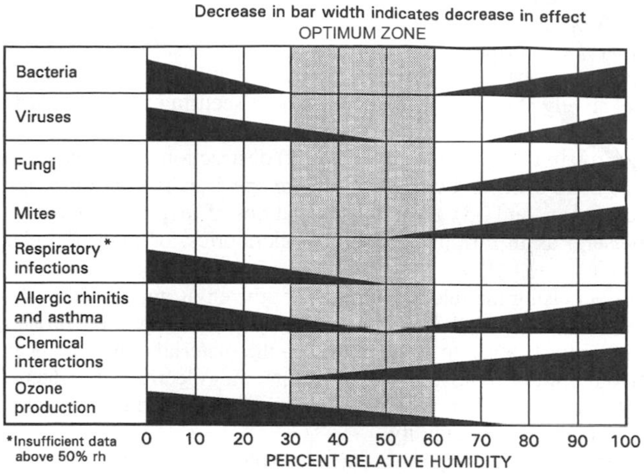

Fig. 1 Optimum Humidity Range for Human Comfort and Health
(Adapted from Sterling et al. 1985)

Electronic Equipment

Electronic data processing equipment requires controlled relative humidity. High relative humidity may cause condensation in the equipment, while low relative humidity may promote static electricity. Also, rapid changes in relative humidity should be avoided due to their effect on bar code readers, magnetic tapes, disks, and data processing equipment. Generally, computer systems have a recommended design and operating range of 35 to 55% rh. However, the manufacturer's recommendations should be adhered to for specific equipment operation.

Process Control and Materials Storage

The relative humidity required by a process is usually specific and related to one or more of several factors:

- Control of moisture content or regain
- Rate of chemical or biochemical reactions
- Rate of crystallization
- Product accuracy or uniformity
- Corrosion
- Static electricity

Typical conditions of temperature and relative humidity for the storage of certain commodities and the manufacturing and processing of others may be found in Chapter 11 of the 1999 *ASHRAE Handbook—Applications.*

Low humidity in winter may cause drying and shrinking of furniture, wood floors, and interior trim. Winter humidification should be considered in order to maintain relative humidity closer to that experienced during manufacture or installation.

For storing hygroscopic materials, maintaining constant humidity is often as important as the humidity level itself. The design of the structure should always be considered. Temperature control is important due to the danger of condensation on products through a transient lowering of temperature.

Static Electricity

Electrostatic charges are generated when materials of high electrical resistance move against each other. The accumulation of such charges may have a variety of results: (1) unpleasant sparks caused by friction between two materials (e.g., stocking feet and carpet fibers); (2) difficulty in handling sheets of paper, fibers, and fabric; (3) objectionable dust clinging to oppositely charged objects (e.g., negatively charged metal nails or screws securing gypsum board to wooden studding in the exterior walls of a building that attract positively charged dust particles); (4) destruction of data stored on magnetic disks and tapes that require specifically controlled environments; and (5) hazardous situations if explosive gases are present, as in hospitals, research laboratories, or industrial clean rooms.

Increasing the relative humidity of the environment reduces the accumulation of electrostatic charges, but the optimum level of humidity depends to some extent on the materials involved. Relative humidity of 45% reduces or eliminates electrostatic effects in many materials, but wool and some synthetic materials may require a higher relative humidity.

Hospital operating rooms, where explosive mixtures of anesthetics are used, constitute a special and critical case. A relative humidity of at least 50% is usually required, with special grounding arrangements and restrictions on the types of clothing worn by occupants. Conditions of 72°F and 55% rh are usually recommended for comfort and safety.

Sound Wave Transmission

The air absorption of sound waves, which results in the loss of sound strength, is worst at 15 to 20% rh, and the loss increases as the frequency rises (Harris 1963). There is a marked reduction in sound absorption at 40% rh; above 50%, the effect of air absorption is negligible. Air absorption of sound does not significantly affect speech but may merit consideration in large halls or auditoriums where optimum acoustic conditions are required for musical performances.

Miscellaneous

Laboratories and test chambers, in which precise control of relative humidity over a wide range is desired, require special attention. Because of the interrelation between temperature and relative humidity, precise humidity control requires equally precise temperature control.

ENCLOSURE CHARACTERISTICS

Vapor Retarders

The maximum relative humidity level to which a building may be humidified in winter depends on the ability of its walls, roof, and other elements to prevent or tolerate condensation. Water vapor migrates from areas of high vapor pressure and lower temperature. Condensed moisture or frost on surfaces exposed to the building interior (visible condensation) can deteriorate the surface finish, cause mold growth and subsequent indirect moisture damage and nuisance, and reduce visibility through windows. If the walls and roof have not been specifically designed and properly protected with vapor retarders on the warm side to prevent the entry of moist air or vapor from the inside, concealed condensation within these constructions is likely to occur, even at fairly low interior humidity, and cause serious deterioration.

Visible Condensation

Condensation forms on an interior surface when its temperature is below the dew-point temperature of the air in contact with it. The maximum relative humidity that may be maintained without condensation is thus influenced by the thermal properties of the enclosure and the interior and exterior environment.

Average surface temperatures may be calculated by the methods outlined in Chapter 22 of the 1997 *ASHRAE Handbook—Fundamentals* for most insulated constructions. However, local cold spots result from high-conductivity paths such as through-the-wall framing, projected floor slabs, and metal window frames that have no thermal breaks. The vertical temperature gradient in the air space and surface convection along windows and sections with a high thermal conductivity result in lower air and surface temperatures at the sill or floor. Drapes and blinds closed over windows lower surface temperature further, while heating units under windows raise the temperature significantly.

In most buildings, windows present the lowest surface temperature and the best guide to permissible humidity levels for no condensation. While calculations based on overall thermal coefficients provide reasonably accurate temperature predictions at mid-height, actual minimum surface temperatures are best determined by test. Wilson and Brown (1964) related the characteristics of windows with a **temperature index**, defined as $(t - t_o)/(t_i - t_o)$, where t is the inside window surface temperature, t_i is the indoor air temperature, and t_o is the outdoor air temperature.

The results of limited tests on actual windows indicate that the temperature index at the bottom of a double, residential-type window with a full thermal break is between 0.55 and 0.57, with natural convection on the warm side. Sealed, double-glazed units exhibit an index from 0.33 to 0.48 at the junction of glass and sash, depending on sash design. The index is likely to rise to 0.53 or greater only 1 in. above the junction.

With continuous under-window heating, the minimum index for a double window with a full break may be as high as 0.60 to 0.70. Under similar conditions, the index of a window with a poor thermal break may be increased by a similar increment.

Figure 2 shows the relationship between temperature index and the relative humidity and temperature conditions at which condensation occurs. The limiting relative humidities for various outdoor temperatures intersect vertical lines representing particular temperature indexes. A temperature index of 0.55 has been selected to represent an average for double-glazed, residential windows; 0.22 represents an average for single-glazed windows. Table 1 shows the limiting relative humidities for both types of windows at various outdoor air temperatures.

Table 1 Maximum Relative Humidity In a Space for No Condensation on Windows

	Limiting Relative Humidity, %	
Outdoor Temperature, °F	Single Glazing	Double Glazing
40	39	59
30	29	50
20	21	43
10	15	36
0	10	30
−10	7	26
−20	5	21
−30	3	17

Note: Natural convection, indoor air at 74°F.

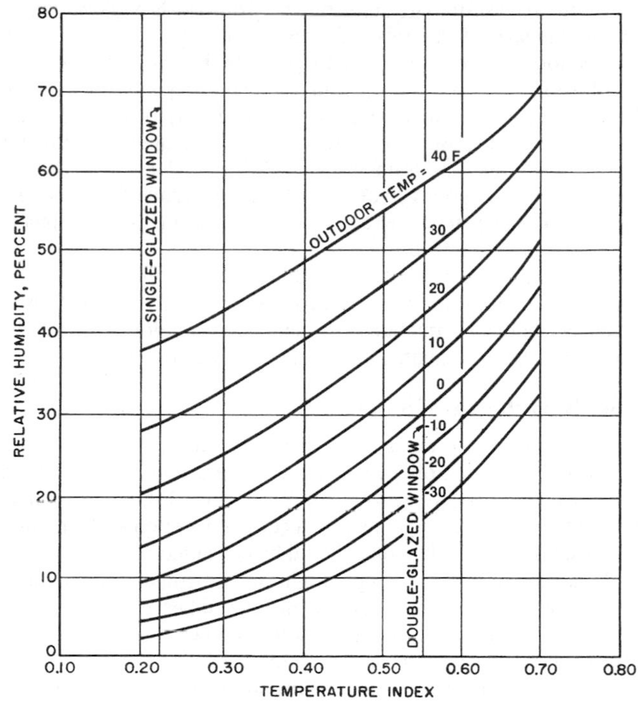

Fig. 2 Limiting Relative Humidity for No Window Condensation

Concealed Condensation

Vapor retarders are imperative in certain applications because the humidity level a building is able to maintain without serious concealed condensation may be much lower than that indicated by visible condensation. The migration of water vapor through the inner envelope by diffusion or air leakage brings the vapor into contact with surfaces at temperatures below its dew point. During the design of a building, the desired interior humidity may be determined by the ability of the building enclosure to handle internal moisture. This is particularly important when planning for building humidification in colder climates.

ENERGY CONSIDERATIONS

When calculating the energy requirement for a humidification system, the effect of the dry air environment on any material supplying it with moisture should be considered. The release of liquid in a hygroscopic material to a vapor state is an evaporative process that requires energy. The source of energy is the heat contained in the air. The heat lost from the air to evaporate moisture equals the heat necessary to produce an equal amount of moisture vapor with an efficient humidifier. If proper humidity levels are not maintained, moisture migration from hygroscopic materials can have destructive effects.

The true energy required for a humidification system must be calculated from the actual humidity level in the building, not from the theoretical level.

A study of residential heating and cooling systems showed a correlation between infiltration and inside relative humidity, indicating a significant energy saving from increasing the inside relative humidity, which reduced infiltration of outside air by up to 50% during the heating season (Luck and Nelson 1977). This reduction is apparently due to the sealing of window cracks by the formation of frost.

To assess accurately the total energy required to provide a desired level of humidity, all elements relating to the generation of humidity and the maintenance of the final air condition must be considered. This is particularly true when comparing different humidifiers. For example, the cost of boiler steam should include generation and distribution losses; costs for an evaporative humidifier include electrical energy for motors or compressors, water conditioning, and the addition of reheat (when the evaporative cooling effect is not required).

Load Calculations

The humidification load depends primarily on the rate of natural infiltration of the space to be humidified or the amount of outside air introduced by mechanical means. Other sources of moisture gain or loss should also be considered. The **humidification load** H can be calculated by the following equations:

For ventilation systems having natural infiltration,

$$H = \rho VR(W_i - W_o) - S + L \qquad (1)$$

For mechanical ventilation systems having a fixed quantity of outside air,

$$H = 60\rho Q_o(W_i - W_o) - S + L \qquad (2)$$

For mechanical systems having a variable quantity of outside air,

$$H = 60\rho Q_t(W_i - W_o)\left(\frac{t_i - t_m}{t_i - t_o}\right) - S + L \qquad (3)$$

where

H = humidification load, lb of water/h
V = volume of space to be humidified, ft^3
R = infiltration rate, air changes per hour
Q_o = volumetric flow rate of outside air, cfm
Q_t = total volumetric flow rate of air (outside air plus return air), cfm
t_i = design indoor air temperature, °F
t_m = design mixed air temperature, °F
t_o = design outside air temperature, °F
W_i = humidity ratio at indoor design conditions, lb of water/lb of dry air
W_o = humidity ratio at outdoor design conditions, lb of water/lb of dry air
S = contribution of internal moisture sources, lb of water/h
L = other moisture losses, lb of water/h
ρ = density of air at sea level, 0.074 lb/ft^3

Design Conditions

Interior design conditions are dictated by the occupancy or the process, as discussed in the preceding sections on Enclosure Characteristics and on Environmental Conditions. Outdoor relative humidity can be assumed to be 70 to 80% at temperatures below 32°F or 50% at temperatures above 32°F for winter conditions in most areas. Additional data on outdoor design data may be obtained from Chapter 26 of the 1997 *ASHRAE Handbook—Fundamentals*. Absolute humidity values can be obtained either from Chapter 6 of the 1997 *ASHRAE Handbook—Fundamentals* or from an ASHRAE Psychrometric Chart.

For systems handling fixed outside air quantities, load calculations are based on outdoor design conditions. Equation (1) should be used for natural infiltration, and Equation (2) for mechanical ventilation.

For economizers that achieve a fixed mixed air temperature by varying outside air, special considerations are needed to determine the maximum humidification load. This load occurs at an outside air temperature other than the lowest design temperature because it is a function of the amount of outside air introduced and the existing moisture content of the air. Equation (3) should be solved for various outside air temperatures to determine the maximum humidification load. It is also important to analyze the energy use of the humidifier (especially for electric humidifiers) when calculating the economizer setting in order to ensure that the energy saved by "free cooling" is greater than the energy consumed by the humidifier.

In residential load calculations, the actual outdoor design conditions of the locale are usually taken as 20°F and 70% rh, while indoor conditions are taken as 70°F and 35% rh. These values yield an absolute humidity difference ($W_i - W_o$) of 0.0040 lb per pound of dry air for use in Equation (1). However, the relative humidity may need to be less than 35% to avoid condensation at low outdoor temperatures (see Table 1).

Ventilation Rate

Ventilation of the humidified space may be due to either natural infiltration alone or natural infiltration in combination with intentional mechanical ventilation. Natural infiltration varies according to the indoor-outdoor temperature difference, wind velocity, and tightness of construction, as discussed in Chapter 25 of the 1997 *ASHRAE Handbook—Fundamentals*. The rate of mechanical ventilation may be determined from building design specifications or estimated from fan performance data (see ASHRAE *Standard* 62).

In load calculations, the water vapor removed from the air during cooling by air-conditioning or refrigeration equipment must be considered. This moisture may have to be replaced by humidification equipment to maintain the desired relative humidity in certain industrial projects where the moisture generated by the process may be greater than that required for ventilation and heating.

Estimates of infiltration rate are made in calculating heating and cooling loads for buildings; these values also apply to humidification load calculations. For residences where such data are not available, it may be assumed that a tight house has an infiltration rate of 0.5 air changes per hour (ACH); an average house, 1 ACH; and a loose house, as many as 1.5 ACH. A tight house is assumed to be well insulated and to have vapor retarders, tight storm doors, windows with weather stripping, and a dampered fireplace. An average house is insulated and has vapor retarders, loose storm doors and windows, and a dampered fireplace. A loose house is generally one constructed before 1930 with little or no insulation, no storm doors, no insulated windows, no weather stripping, no vapor retarders, and often a fireplace without an effective damper. For building construction, refer to local codes and building specifications.

Additional Moisture Losses

Hygroscopic materials, which have a lower moisture content than materials in the humidified space, absorb moisture and place an additional load on the humidification system. An estimate of this load depends on the absorption rate of the particular material selected. Table 2 in Chapter 11 of the 1999 *ASHRAE Handbook—Applications* lists the equilibrium moisture content of hygroscopic materials at various relative humidities.

In cases where a certain humidity must be maintained regardless of condensation on exterior windows and walls, the dehumidifying effect of these surfaces constitutes a load that may need to be considered, if only on a transient basis. The loss of water vapor by diffusion through enclosing walls to the outside or to areas at a lower vapor pressure may also be involved in some applications. The properties of materials and flow equations given in Chapter 22 of the 1997 *ASHRAE Handbook—Fundamentals* can be applied in such cases. Normally, this diffusion constitutes a small load, unless openings exist between the humidified space and adjacent rooms at lower humidities.

Internal Moisture Gains

The introduction of a hygroscopic material can cause moisture gains to the space if the moisture content of the material is above that of the space. Similarly, moisture may diffuse through walls separating the space from areas of higher vapor pressure or move by convection through openings in these walls (Brown et al. 1963).

Moisture contributed by human occupancy depends on the number of occupants and their degree of physical activity. As a guide for residential applications, the average rate of moisture production for a family of four has been taken as 0.7 lb/h. Unvented heating devices produce about 1 lb of vapor for each pound of fuel burned. These values may no longer apply because of changes in equipment as well as in living habits.

Industrial processes constitute additional moisture sources. Single-color offset printing presses, for example, give off 0.45 lb of water per hour. Information on process contributions can best be obtained from the manufacturer of the specific equipment.

Supply Water for Humidifiers

There are three major categories of supply water: potable (untreated) water, softened potable water, and demineralized [deionized (DI) or reverse osmosis (RO)] water. Either the application or the humidifier may require a certain water type; the humidifier manufacturer's literature should be consulted.

In areas with water having a high mineral content, precipitated solids may be a problem. They clog nozzles, tubes, evaporative elements, and controls. In addition, solids allowed to enter the airstream via mist leave a fine layer of white dust over furniture, floors, and carpets. Some wetted media humidifiers bleed off and replace some or all of the water passing through the element to reduce the concentration of salts in the recirculating water.

Dust, scaling, biological organisms, and corrosion are all potential problems associated with water in humidifiers. Stagnant water can provide a fertile breeding ground for algae and bacteria, which have been linked to odor and respiratory ailments. Bacterial slime reacts with sulfates in the water to produce hydrogen sulfide and its characteristic bad odor. Regular maintenance and periodic disinfecting with approved microbicides may be required (Puckorius et al. 1995). This has not been a problem with residential equipment; however, regular maintenance is good practice since biocides are generally used only with atomizing humidifiers.

Scaling

Industrial pan humidifiers, when supplied with water that is naturally low in hardness, require little maintenance, provided a surface skimmer bleedoff is used.

Water softening is an effective means of eliminating mineral precipitation in a pan-type humidifier. However, the concentration of sodium left in a pan as a result of water evaporation must be held below the point of precipitation by flushing and diluting the tank with new softened water. The frequency and duration of dilution depend on the water hardness and the rate of evaporation. Dilution is usually accomplished automatically by a timer-operated drain valve and a water makeup valve.

Demineralized or reverse osmosis (RO) water may also be used. The construction materials of the humidifier and the piping must withstand the corrosive effects of this water. Commercial demineralizers or RO equipment removes hardness and other total dissolved solids completely from the humidifier makeup water. They are more expensive than water softeners, but no humidifier purging is required. Sizing is based on the maximum required water flow to the humidifier and the amount of total dissolved solids in the makeup water.

EQUIPMENT

Humidifiers can generally be classified as either residential or industrial, although residential humidifiers can be used for small industrial applications, and small industrial units can be used in large homes. Equipment designed for use in central air systems also differs from that for space humidification, although some units are adaptable to both.

Air washers and direct evaporative coolers may be used as humidifiers; they are sometimes selected for additional functions such as air cooling or air cleaning, as discussed in Chapter 19.

The capacities of residential humidifiers are generally based on gallons per day of operation; capacities of industrial and commercial humidifiers are based on pounds per hour of operation. Published evaporation rates established by equipment manufacturers through test criteria may be inconsistent. Rates and test methods should be evaluated when selecting equipment. The Air-Conditioning and Refrigeration Institute (ARI) has developed *Standard* 610 for residential central system humidifiers and *Standard* 640 for commercial and industrial humidifiers. Association of Home Appliance Manufacturers (AHAM) *Standard* HU-1 addresses self-contained residential units.

Residential Humidifiers for Central Air Systems

Residential humidifiers designed for central air systems depend on airflow in the heating system for evaporation and distribution. General principles and description of equipment are as follows:

Pan Humidifiers. Capacity varies with temperature, humidity, and airflow.

- *Basic pan.* A shallow pan is installed within the furnace plenum. Household water is supplied to the pan through a control device.
- *Electrically heated pan.* Similar to the basic unit, this type adds an electric heater to increase water temperature and evaporation rate.
- *Pan with wicking plates.* Similar to the basic unit, this type includes fitted water-absorbent plates. The increased area of the plates provides greater surface area for evaporation to take place (Figure 3A).

Wetted Element Humidifiers. Capacity varies with temperature, humidity, and airflow. Air circulates over or through an open-textured, wetted medium. The evaporating surface may be a fixed pad wetted by either sprays or water flowing by gravity, or the surface may be a paddle-wheel, drum, or belt rotating through a water reservoir. The various types are differentiated by the way air flows through them:

- *Fan type.* A small fan or blower draws air from the furnace plenum, through the wetted pad, and back to the plenum. A fixed pad (Figure 3B) or a rotating drum-type pad (Figure 3C) may be used.
- *Bypass type.* These units do not have their own fan, but rather are mounted on the supply or return plenum of the furnace with an air connection to the return plenum (Figure 3D). The difference in static pressure created by the furnace blower circulates air through the unit.
- *Duct-mounted type.* These units are designed for installation within the furnace plenum or ductwork with a drum element rotated by either the air movement in the duct or a small electric motor.

Atomizing Humidifiers. The capacity of an atomizing humidifier does not depend on the air conditions. However, it is important not to oversaturate the air and allow liquid water to form in the duct. The ability of the air to absorb moisture depends on the temperature, flow rate, and moisture content of the air moving through the system. Small particles of water are formed and introduced into the airstream in one of the following ways:

- A spinning disk or cone throws a water stream centrifugally to the rim of the disk and onto deflector plates or a comb, where it is turned into a fine fog (Figure 3E).
- Spray nozzles rely on water pressure to produce a fine spray.
- Spray nozzles use compressed air to create a fine mist.
- Ultrasonic vibrations are used as the atomizing force.

Residential Humidifiers for Nonducted Applications

Many portable or room humidifiers are used in residences heated by nonducted hydronic or electric systems or in residences where the occupant is prevented from making a permanent installation. These humidifiers may be equipped with humidity controllers.

Portable units evaporate water by any of the previously described means, such as heated pan, fixed or moving wetted element, or atomizing spinning disk. They may be tabletop-sized or a larger, furniture-style appliance (Figure 3F). A multispeed motor on the fan or blower may be used to adjust output. Portable humidifiers usually require periodic filling from a bucket or filling hose.

Some portable units are offered with an auxiliary package for semipermanent water supply. This package includes a manual shutoff valve, a float valve, copper or other tubing with fittings, and so forth. Lack of drainage provision for water overflow may result in water damage.

Some units may be recessed into the wall between studs, mounted on wall surfaces, or installed below floor level. These units are permanently installed in the structure and use forced-air circulation. They may have an electric element for reheat when desired. Other types for use with hydronic systems involve a simple pan or pan plate, either installed within a hot water convector or using the steam from a steam radiator.

Industrial and Commercial Humidifiers for Central Air Systems

Humidifiers must be installed where the air can absorb the vapor; the temperature of the air being humidified must exceed the dew point of the space being humidified. When fresh or mixed air is humidified, the air may need to be preheated to allow absorption to take place.

Heated Pan Humidifiers. These units offer a broad range of capacities and may be heated by a heat exchanger supplied with either steam or hot water (see Figure 4A). They may be installed directly under the duct, or they may be installed remotely and feed vapor through a hose. In either case, a distribution manifold should be used.

Steam heat exchangers are commonly used in heated pan humidifiers, with steam pressures ranging from 5 to 15 psig. Hot water heat exchangers are also used in pan humidifiers; a water temperature below 240°F is not practical.

All pan-type humidifiers should have water regulation and some form of drain or flush system. When raw water is used, periodic cleaning is required to remove the buildup of minerals. (Use of softened or demineralized water can greatly extend time between cleanings.) Care should also be taken to ensure that all water is drained off when the system is not in use to avoid the possibility of bacterial growth in the stagnant water.

Direct Steam Injection Humidifiers. These units cover a wide range of designs and capacities. Steam is water vapor under pressure and at high temperature, so the process of humidification can be simplified by adding steam directly into the air. This method is an isothermal process because the temperature of the air remains almost constant as the moisture is added. For this type of humidification system, the steam source is usually a central steam boiler at low pressure. When steam is supplied from a source at a constant supply pressure, humidification responds quickly to system demand. A control valve may be modulating or two-position in response to a humidity sensor/controller. Steam can be introduced into the airstream through one of the following devices:

- *Single or multiple steam-jacketed manifolds* (Figure 4B), depending on the size of the duct or plenum. The steam jacket is designed to reevaporate any condensate droplets before they are discharged from the manifold.

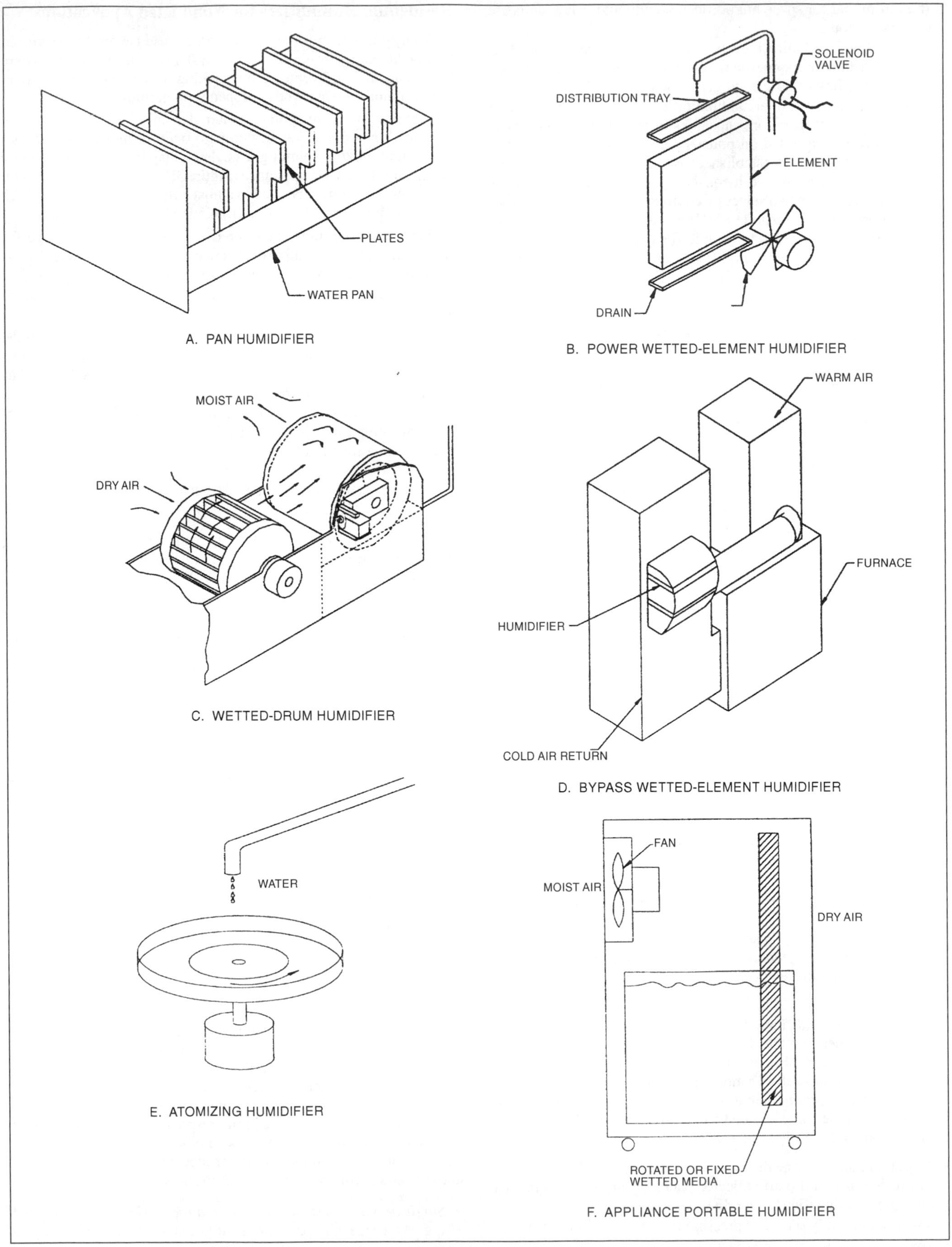

Fig. 3 Residential Humidifiers

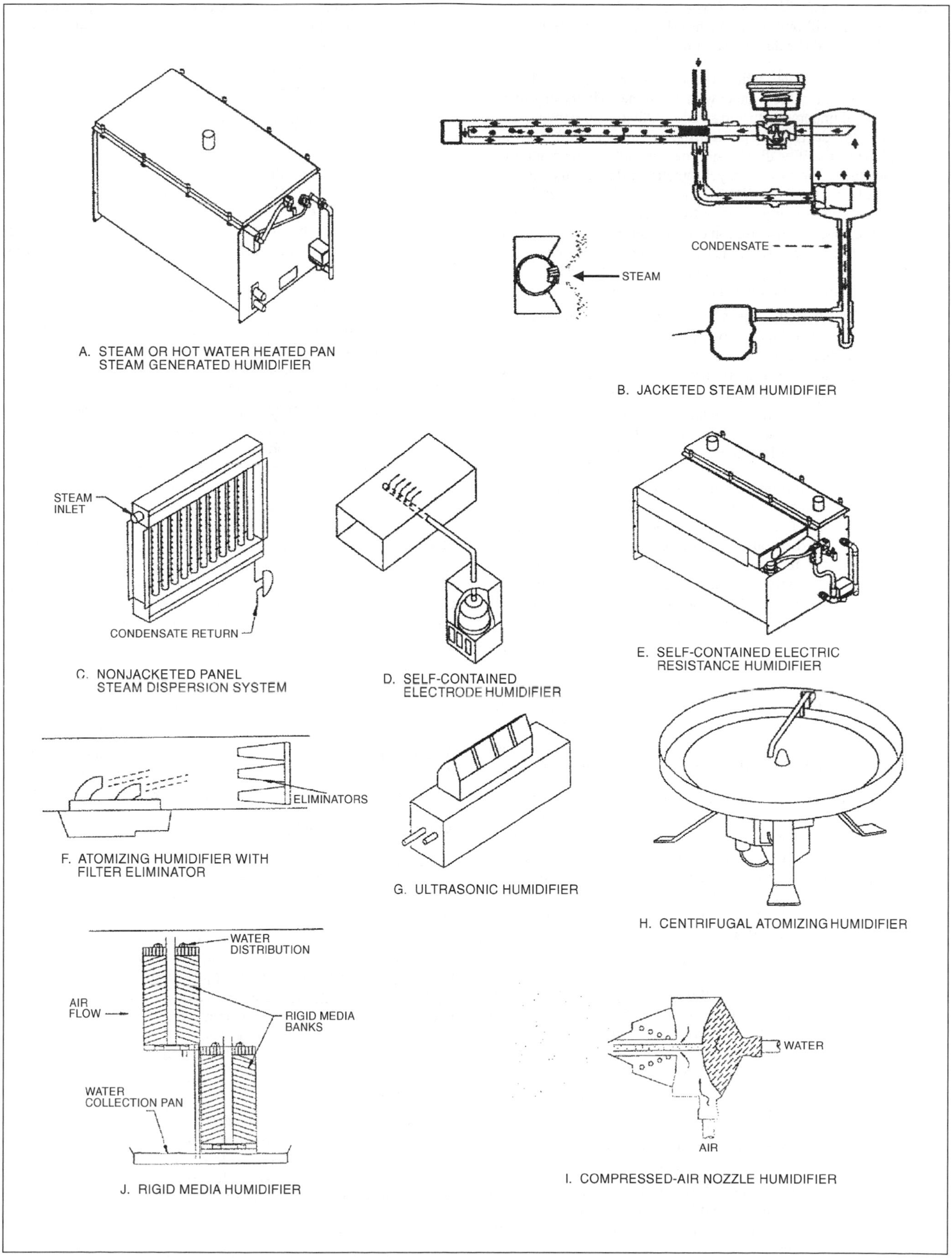
A. STEAM OR HOT WATER HEATED PAN STEAM GENERATED HUMIDIFIER
CONDENSATE
STEAM
B. JACKETED STEAM HUMIDIFIER
STEAM INLET
CONDENSATE RETURN
C. NONJACKETED PANEL STEAM DISPERSION SYSTEM
D. SELF-CONTAINED ELECTRODE HUMIDIFIER
E. SELF-CONTAINED ELECTRIC RESISTANCE HUMIDIFIER
ELIMINATORS
F. ATOMIZING HUMIDIFIER WITH FILTER ELIMINATOR
G. ULTRASONIC HUMIDIFIER
H. CENTRIFUGAL ATOMIZING HUMIDIFIER
WATER DISTRIBUTION
AIR FLOW
RIGID MEDIA BANKS
WATER COLLECTION PAN
J. RIGID MEDIA HUMIDIFIER
WATER
AIR
I. COMPRESSED-AIR NOZZLE HUMIDIFIER

Fig. 4 Industrial Humidifiers

- *Nonjacketed manifold or panel-type distribution systems* (Figure 4C), with or without injection nozzles for distributing steam across the face of the duct or plenum.

Units must be installed where the air can absorb the discharged vapor before it comes into contact with components in the airstream, such as coils, dampers, or turning vanes. Otherwise, condensation can occur in the duct. Absorption distance varies according to the design of the humidifier distribution device and the air conditions within the duct. For proper psychrometric calculations, refer to Chapter 6 of the 1997 *ASHRAE Handbook—Fundamentals*. Because these humidifiers inject steam from a central boiler source directly into the space or distribution duct, boiler treatment chemicals discharged into the air system may compromise indoor air quality. Chemicals should be checked for safety, and care should be taken to avoid contamination from the water or steam supplies.

Electrically Heated, Self-Contained Steam Humidifiers. These units convert ordinary city tap water to steam by electrical energy using either electrodes or resistance heater elements. The steam is generated at atmospheric pressure and discharged into the duct system through dispersion manifolds; if the humidifier is a freestanding unit, the steam is discharged directly into the air space through a fan unit. Some units allow the use of softened or demineralized water, which greatly extends the time between cleanings.

- *Electrode-type humidifiers* (Figure 4D) operate by passing an electric current directly into ordinary tap water, thereby creating heat energy to boil the water. The humidifier usually contains a polypropylene plastic bottle, either throwaway or cleanable, that is supplied with water through a solenoid valve. Water is drained off periodically to maintain a desirable solids concentration and the correct electrical flow. Manufacturers offer humidifiers with several different features, so their data should be consulted.
- *Resistance-type humidifiers* (Figure 4E) utilize one or more electrical elements that heat the water directly to produce steam. The water can be contained in a stainless steel or coated steel shell. The element and shell should be accessible for cleaning out mineral deposits. The high and low water levels should be controlled with either probes or float devices, and a blowdown drain system should be incorporated, particularly for off-operation periods.

Atomizing Humidifiers. Water treatment should be considered if mineral fallout from hard water is a problem. Optional filters may be required to remove the mineral dust from the humidified air (Figure 4F). Depending on the application and the water condition, atomizing humidifiers may require a reverse osmosis (RO) or a deionized (DI) water treatment system to remove the minerals. It is also important to note that wetted parts should be able to resist the corrosive effects of DI and RO water.

There are three main categories of atomizing humidifiers:

- *Ultrasonic humidifiers* (Figure 4G) utilize a piezoelectric transducer submerged in demineralized water. The transducer converts a high-frequency mechanical electric signal into a high-frequency oscillation. A momentary vacuum is created during the negative oscillation, causing the water to cavitate into vapor at low pressure. The positive oscillation produces a high-compression wave that drives the water particle from the surface to be quickly absorbed into the airstream. Because these types use demineralized water, no filter medium is required downstream. The ultrasonic humidifier is also manufactured as a freestanding unit.
- *Centrifugal humidifiers* (Figure 4H) use a high-speed disk, which slings the water to its rim, where it is thrown onto plates or a comb to produce a fine mist. The mist is introduced to the airstream, where it is evaporated.
- *Compressed-air nozzle humidifiers* can operate in two ways:

1. Compressed air and water are combined inside the nozzle and discharged onto a resonator to create a fine fog at the nozzle tip (Figure 4I).
2. Compressed air is passed through an annular orifice at the nozzle tip, and water is passed through a center orifice. The air creates a slight vortex at the tip, where the water breaks up into a fine fog on contact with the high-velocity compressed air.

Wetted Media Humidifiers. *Rigid media humidifiers* (Figure 4J) utilize a porous core. Water is circulated over the media while air is blown through the openings. These humidifiers are adiabatic, cooling the air as it is humidified. Rigid media cores are often used for the dual purpose of winter humidification and summer cooling. They depend on airflow for evaporation: the rate of evaporation varies with air temperature, humidity, and velocity.

The rigid media should be located downstream of any heating or cooling coils. For close humidity control, the element can be broken down into several (usually two to four) banks having separate water supplies. Solenoids controlling water flow to each bank are activated as humidification is required.

Rigid media humidifiers have inherent filtration and scrubbing properties due to the water-washing effect in the filter-like channels. Only pure water is evaporated; therefore, contaminants collected from the air and water must be flushed from the system. A continuous bleed or regular pan flushing is recommended to minimize accumulation of contaminants in the pan and on the media.

Evaporative Cooling. Atomizing and wetted media humidifiers discharge water at ambient temperature. The water absorbs heat from the surrounding air to evaporate the fog, mist, or spray at a rate of 1075 Btu per pound of water. This evaporative cooling effect (see Chapter 19) should be considered in the design of the system and if reheat is required to achieve the final air temperature. The ability of the surrounding air to efficiently absorb the fog, mist, or spray will also depend on its temperature, air velocity, and moisture content.

CONTROLS

Many humidity-sensitive materials are available. Some are organic, such as nylon, human hair, wood, and animal membranes that change length with humidity changes. Other sensors change electrical properties (resistance or capacitance) with humidity.

Mechanical Controls

Mechanical sensors depend on a change in the length or size of the sensor as a function of relative humidity. The most commonly used sensors are synthetic polymers or human hair. They can be attached to a mechanical linkage to control the mechanical, electrical, or pneumatic switching element of a valve or motor. This design is suitable for most human comfort applications, but it may lack the necessary accuracy for industrial applications.

A humidity controller is normally designed to control at a set point selected by the user. Some controllers have a setback feature that lowers the relative humidity set point as outdoor temperature drops to reduce condensation within the structure.

Electronic Controllers

Electrical sensors change electrical resistance as the humidity changes. They typically consist of two conductive materials separated by a humidity-sensitive, hygroscopic insulating material (polyvinyl acetate, polyvinyl alcohol, or a solution of certain salts). Small changes are detected as air passes over the sensing surface. Capacitive sensors use a dielectric material that changes its dielectric constant with relative humidity. The dielectric material is sandwiched between special conducting material that allows a fast response to changes in relative humidity.

Electronic control is common in laboratory or process applications requiring precise humidity control. It is also used to vary fan speed on portable humidifiers in order to regulate humidity in the space more closely and to reduce noise and draft to a minimum.

Electronic controls are now widely used for residential applications because of low-cost, accurate, and stable sensors that can be used with inexpensive microprocessors. They may incorporate methods of determining outside temperature so that relative humidity can be automatically reset to some predetermined algorithm intended to maximize human comfort and minimize any condensation problems (Pasch et al. 1996).

Along with a main humidity controller, the system may require other sensing devices:

- **High-limit sensors** may be required to ensure that duct humidity levels remain below the saturation or dew-point level. Sometimes cooler air is required to offset sensible heat gains. In these cases, the air temperature may drop below the dew point. Operating the humidifier under these conditions causes condensation in the duct or fogging in the room. High-limit sensors may be combined with a temperature sensor in certain designs.
- **Airflow sensors** should be used in place of a fan interlock. They sense airflow and disable the humidifier when insufficient airflow is present in the duct.
- **Steam sensors** are used to keep the control valve on direct-injection humidifiers closed when steam is not present at the humidifier. Either a pneumatic or an electric temperature-sensing switch is fitted between the separator and the steam trap to sense the temperature of the condensate and steam. When the switch senses steam temperature, it allows the control valve to function normally.

Humidity Control in Variable Air Volume (VAV) Systems

Control in VAV systems is much more demanding than in constant volume systems. VAV systems, common in large, central station applications, control space temperature by varying the volume rather than the temperature of the supply air. Continual airflow variations to follow load changes within the building can create wide and rapid swings in space humidity. Due to the fast-changing nature and cooler supply air temperatures (55°F or lower) of most VAV systems, special modulating humidity controls should be applied.

Best results are obtained by using both space and duct modulating-type humidity sensors in conjunction with an integrating device, which in turn modulates the output of the humidifier. This allows the duct sensor to respond quickly to a rapid rise in duct humidity caused by reduced airflow to the space as temperature conditions are satisfied. The duct sensor at times overrides the space humidistat by reducing the humidifier output to correspond to decreasing air volumes. This type of system, commonly referred to as **anticipating control**, allows the humidifier to track the dynamics of the system and provide uniform control. Due to the operating duct static pressures of a VAV system, use of an **airflow proving device** is recommended to detect air movement.

Further information on the evaluation of humidity sensors can be found in ASHRAE (1992a).

Control Location

In centrally humidified structures, the humidity controller is most commonly mounted in a controlled space. Another method is to mount the controller in the return air duct of an air-handling system to sense average relative humidity. Figure 5 shows general recommended locations for the humidistat for a centrally air-conditioned room.

The manufacturer's instructions regarding the use of the controller on counterflow furnaces should be followed because reverse airflow when the fan is off can substantially shift the humidity control point in a home. The sensor should be located where it will not be

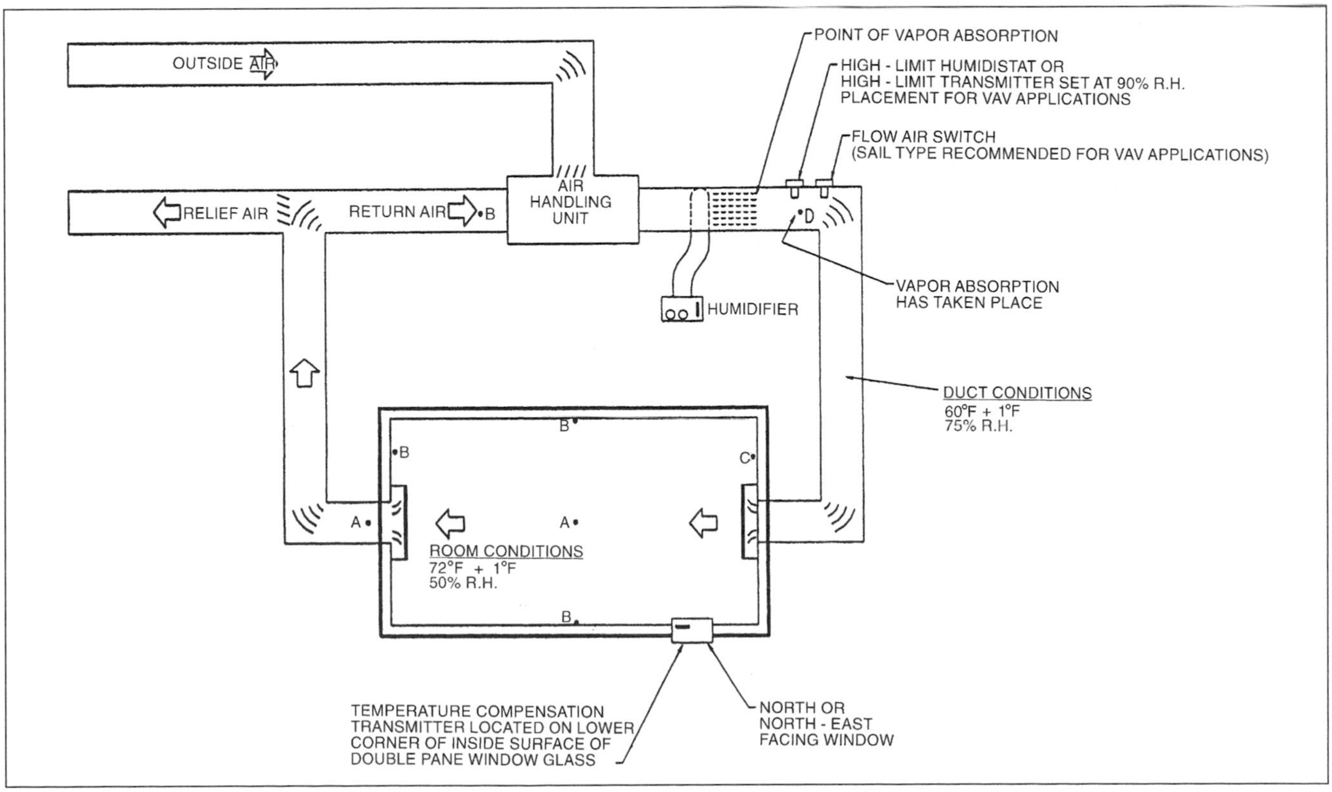

Fig. 5 Recommended Humidity Controller Location

affected by (1) air that exits the bypass duct of a bypass humidifier or (2) drafts or local heat or moisture sources.

REFERENCES

AHAM. 1987. Appliance humidifiers. *Standard* HU-1. Association of Home Appliance Manufacturers, Chicago.

ARI. 1996. Central system humidifiers for residential applications. ANSI/ARI *Standard* 610-96. Air-Conditioning and Refrigeration Institute, Arlington, VA.

ARI. 1996. Commercial and industrial humidifiers. ANSI/ARI *Standard* 640-96.

ASHRAE. 1992a. Control of humidity in buildings. *Technical Data Bulletin* 8(3).

ASHRAE. 1992b. Thermal environmental conditions for human occupancy. ANSI/ASHRAE *Standard* 55-1992.

ASHRAE. 1999. Ventilation for acceptable indoor air quality. *Standard* 62-1999.

Berglund, L.G. 1998. Comfort and humidity. *ASHRAE Journal* 40(8):35-41.

Brown, W.G., K.R. Solvason, and A.G. Wilson. 1963. Heat and moisture flow through openings by convection. *ASHRAE Journal* 5(9):49.

Davis, J.W., C.O. Ruud, and R.F. Unz. 1993. Analysis of furnace-mount humidifier for microbiological and particle emissions—Part I: Test system development. *ASHRAE Transactions* 99(1):1377-86.

Harris, C.M. 1963. Absorption of sound in air in the audio-frequency range. *Journal of the Acoustical Society of America* 35(January).

Luck, J.R. and L.W. Nelson. 1977. The variation of infiltration rate with relative humidity in a frame building. *ASHRAE Transactions* 83(1):718-29.

Pasch, R.M., M. Comins, and J.S. Hobbins. 1996. Field experiences in residential humidification control with temperature-compensated automatic humidistats. *ASHRAE Transactions* 102(2):628-32.

Puckorius, P.R., P.T. Thomas, and R.L. Augspurger. 1995. Why evaporative coolers have not caused Legionnaires' disease. *ASHRAE Journal* 37(1): 29-33.

Ruud, C.O., J.W. Davis, and R.F. Unz. 1993. Analysis of furnace-mount humidifier for microbiological and particle emissions—Part II: Particle sampling and results. *ASHRAE Transactions* 99(1):1387-95.

Sterling, E.M., A. Arundel, and T.D. Sterling. 1985. Criteria for human exposure to humidity in occupied buildings. *ASHRAE Transactions* 91(1B):611-22.

Unz, R.F., J.W. Davis, and C.O. Ruud. 1993. Analysis of furnace-mount humidifiers for microbiological and particle emissions—Part III: Microbiological sampling and results. *ASHRAE Transactions* 99(1):1396-1404.

Wilson, A.G. and W.P. Brown. 1964. Thermal characteristics of double windows. *Canadian Building Digest* No. 58. Division of Building Research, National Research Council, Ottawa, ON.

CHAPTER 21

AIR-COOLING AND DEHUMIDIFYING COILS

THE MAJORITY of the equipment used today for cooling and dehumidifying an airstream under forced convection incorporates a coil section that contains one or more cooling coils assembled in a coil bank arrangement. Such coil sections are used extensively as components in room terminal units; larger factory-assembled, self-contained air conditioners; central station air handlers; and field built-up systems. The applications of each type of coil are limited to the field within which the coil is rated. Other limitations are imposed by code requirements, proper choice of materials for the fluids used, the configuration of the air handler, and economic analysis of the possible alternatives for each installation.

USES FOR COILS

Coils are used for air cooling with or without accompanying dehumidification. Examples of cooling applications without dehumidification are (1) precooling coils that use well water or other relatively high-temperature water to reduce the load on the refrigerating equipment and (2) chilled water coils that remove sensible heat from chemical moisture-absorption apparatus. The heat-pipe coil is also used as a supplementary heat exchanger for preconditioning in air-side sensible cooling (see Chapter 44). Most coil sections provide air sensible cooling and dehumidification simultaneously.

The assembly usually includes a means of cleaning air to protect the coil from accumulation of dirt and to keep dust and foreign matter out of the conditioned space. Although cooling and dehumidification are their principal functions, cooling coils can also be wetted with water or a hygroscopic liquid to aid in air cleaning, odor absorption, or frost prevention. Coils are also evaporatively cooled with a water spray to improve efficiency or capacity. Chapter 19 has more information on indirect evaporative cooling. For general comfort conditioning, cooling, and dehumidifying, the **extended surface (finned) cooling coil** design is the most popular and practical.

COIL CONSTRUCTION AND ARRANGEMENT

In finned coils, the external surface of the tubes is primary, and the fin surface is secondary. The primary surface generally consists of rows of round tubes or pipes that may be staggered or placed in line with respect to the airflow. Flattened tubes or tubes with other nonround internal passageways are sometimes used. The inside surface of the tubes is usually smooth and plain, but some coil designs have various forms of internal fins or turbulence promoters (either fabricated or extruded) to enhance performance. The individual tube passes in a coil are usually interconnected by return bends (or hairpin bend tubes) to form the serpentine arrangement of multipass tube circuits. Coils are usually available with different circuit arrangements and combinations offering varying numbers of parallel water flow passes within the tube core (Figure 1).

Cooling coils for water, aqueous glycol, brine, or halocarbon refrigerants usually have aluminum fins on copper tubes, although

The preparation of this chapter is assigned to TC 8.4, Air-to-Refrigerant Heat Transfer Equipment.

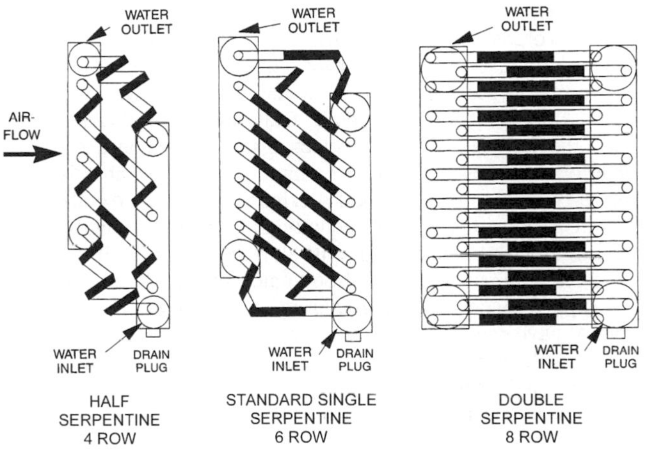

Fig. 1 Typical Water Circuit Arrangement

copper fins on copper tubes and aluminum fins on aluminum tubes (excluding water) are also used. Adhesives are sometimes used to bond header connections, return bends, and fin-tube joints, particularly for aluminum-to-aluminum joints. Certain special-application coils feature an all-aluminum extruded tube-and-fin surface.

Common core tube outside diameters are 5/16, 3/8, 1/2, 5/8, 3/4, and 1 in., with fins spaced 4 to 18 per inch. Tube spacing ranges from 0.6 to 3.0 in. on equilateral (staggered) or rectangular (in-line) centers, depending on the width of individual fins and on other performance considerations. Fins should be spaced according to the job to be performed, with special attention given to air friction; possibility of lint accumulation; and frost accumulation, especially at lower temperatures.

Tube wall thickness and the required use of alloys other than copper are determined mainly by the coil's working pressure and safety factor for hydrostatic burst (pressure). Fin type and header construction also play a large part in this determination. Local job site codes and applicable nationally recognized safety standards should be consulted in the design and application of these coils.

Water and Aqueous Glycol Coils

Good performance of water-type coils requires both the elimination of all air and water traps within the water circuit and the proper distribution of water. Unless properly vented, air may accumulate in the coil tube circuits, reducing thermal performance and possibly causing noise or vibration in the piping system. Air vent and drain connections are usually provided on the coil water headers, but this does not eliminate the need to install, operate, and maintain the coil tube core in a level position. Individual coil vents and drain plugs are often incorporated on the headers (Figure 1). Water traps within the tubing of a properly leveled coil are usually caused by (1) improper nondraining circuit design and/or (2) center-of-coil downward sag.

Such a situation may cause tube failure (e.g., freeze-up in cold climates or tube erosion due to untreated mineralized water).

Depending on performance requirements, the water velocity inside the tubes usually ranges from approximately 1 to 8 fps, and the design water pressure drop across the coils varies from about 5 to 50 ft of water head. For nuclear HVAC applications, ASME *Standard* AG-1, Code on Nuclear Air and Gas Treatment, requires a minimum tube velocity of 2 fps. ARI *Standard* 410 requires a minimum of 1 fps or a Reynolds number of 3100 or greater. This yields more predictable performance.

In certain cases, the water may contain considerable sand and other foreign matter (e.g., in precooling coils using well water or in applications where minerals in the cooling water deposit on and foul the tube surface). It is best to filter out such sediment. Some coil manufacturers offer removable water header plates or a removable plug for each tube that allows the tube to be cleaned, ensuring a continuation of rated performance while the cooling units are in service. Where buildup of scale deposits or fouling of the water-side surface is expected, a scale factor is sometimes included when calculating thermal performance of the coils. Cupronickel, red brass, bronze, and other tube alloys help protect against corrosion and erosion deterioration caused primarily by internal fluid flow abrasive sediment. The core tubes of properly designed and installed coils should feature circuits that (1) have equally developed line length; (2) are self-draining by means of gravity during the coil's off cycle; (3) have the minimum pressure drop to aid in water distribution from the supply header without requiring an excessive pumping head; and (4) have equal feed and return by the supply and return header. Design for the proper in-tube water velocity determines the circuitry style required. Multirow coils are usually circuited to the cross-counterflow arrangement and oriented for top-outlet/bottom-feed connection.

Direct-Expansion Coils

Coils for halocarbon refrigerants present more complex cooling fluid distribution problems than do water or brine coils. The coil should cool effectively and uniformly throughout, with even refrigerant distribution. Halocarbon coils are used on two types of refrigerated systems: flooded and direct-expansion.

A flooded system is used mainly when a small temperature difference between the air and refrigerant is desired. Chapter 3 of the 1998 *ASHRAE Handbook—Refrigeration* describes flooded systems in more detail.

For direct-expansion systems, two of the most commonly used refrigerant liquid metering arrangements are the capillary tube assembly (or restrictor orifice) and the thermostatic expansion valve (TXV) device. The **capillary tube** is applied in factory-assembled, self-contained air conditioners up to approximately 10 ton capacity but is most widely used on smaller capacity models such as window or room units. In this system, the bore and length of a capillary tube are sized so that at full load, under design conditions, just enough liquid refrigerant to be evaporated completely is metered from the condenser to the evaporator coil. While this type of metering arrangement does not operate over a wide range of conditions as efficiently as a thermostatic expansion valve system, its performance is targeted for a specific design condition.

A **thermostatic expansion valve** system is commonly used for all direct-expansion coil applications described in this chapter, particularly field-assembled coil sections and those used in central air-handling units and the larger, factory-assembled hermetic air conditioners. This system depends on the TXV to automatically regulate the rate of refrigerant liquid flow to the coil in direct proportion to the evaporation rate of refrigerant liquid in the coil, thereby maintaining optimum performance over a wide range of conditions. The superheat at the coil suction outlet is continually maintained within the usual predetermined limits of 6 to 10°F. Because the TXV responds to the superheat at the coil outlet, the superheat within the coil is produced with the least possible sacrifice of active evaporating surface.

The length of each coil's refrigerant circuits, from the TXV's distributor feed tubes through the suction header, should be equal. The length of each circuit should be optimized to provide good heat transfer, good oil return, and a complementary pressure drop across the circuit. The coil should be installed level, and coil circuitry should be designed to self-drain by gravity toward the suction header connection.

To ensure reasonably uniform refrigerant distribution in multi-circuit coils, a distributor is placed between the TXV and coil inlets to divide the refrigerant equally among the coil circuits. The refrigerant distributor must be effective in distributing both liquid and vapor because the refrigerant entering the coil is usually a mixture of the two, although mainly liquid by weight. Distributors can be placed in either the vertical or the horizontal position; however, the vertical down position usually distributes refrigerant between coil circuits better than the horizontal position for varying load conditions.

The individual coil circuit connections from the refrigerant distributor to the coil inlet are made of small-diameter tubing; the connections are all the same length and diameter so that the same flow occurs between each refrigerant distributor tube and each coil circuit. To approximate uniform refrigerant distribution, the refrigerant should flow to each refrigerant distributor circuit in proportion to the load on that coil. The heat load must be distributed equally to each of its refrigerant circuits to obtain optimum coil performance. If the coil load cannot be distributed uniformly, the coil should be recircuited and connected with more than one TXV to feed the circuits (individual suction may also help). In this way, the refrigerant distribution is reduced in proportion to the number of distributors that may have less of an effect on overall coil performance when the design must accommodate some unequal circuit loading. Unequal circuit loading may also be caused by such variables as uneven air velocity across the face of the coil, uneven entering air temperature, improper coil circuiting, oversized orifice in distributor, or the TXV's not being directly connected (close-coupled) to the distributor.

Control of Coils

Cooling capacity of water coils is controlled by varying either water flow or airflow. Water flow can be controlled by a three-way mixing, modulating, and/or throttling valve. For airflow control, face and bypass dampers are used. When cooling demand decreases, the coil face damper starts to close, and the bypass damper opens. In some cases, airflow is varied by controlling the fan capacity with speed controls, inlet vanes, or discharge dampers.

Chapter 45 of the 1999 *ASHRAE Handbook—Applications* addresses the control of air-cooling coils to meet system or space requirements and factors to consider when sizing automatic valves for water coils. The selection and application of refrigerant flow control devices (e.g., thermostatic expansion valves, capillary tube types, constant pressure expansion valves, evaporator pressure regulators, suction pressure regulators, and solenoid valves) as used with direct-expansion coils are discussed in Chapter 45 of the 1998 *ASHRAE Handbook—Refrigeration*.

For factory-assembled, self-contained packaged systems or field-assembled systems employing direct-expansion coils equipped with TXVs, a single valve is sometimes used for each coil; in other cases, two or more valves are used. The thermostatic expansion valve controls the refrigerant flow rate through the coil circuits so that the refrigerant vapor at the coil outlet is superheated properly. Superheat is obtained with suitable coil design and proper valve selection. Unlike water flow control valves, standard pressure/temperature-type thermostatic expansion valves alone do not control the refrigeration system's capacity or the temperature of the leaving air, nor do they maintain ambient conditions in specific spaces. However, some electronically controlled TXVs have these attributes.

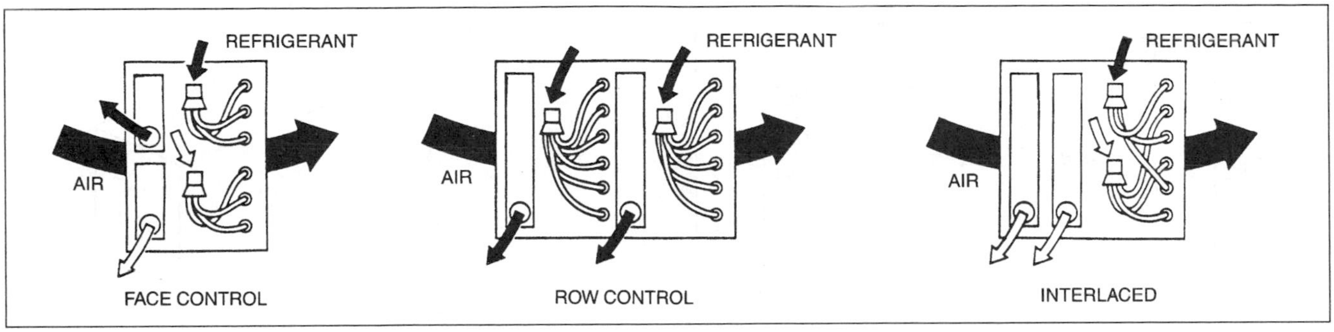

Fig. 2 Arrangements for Coils with Multiple Thermostatic Expansion Valves

In order to match the refrigeration load requirements for the conditioned space to the cooling capacity of the coil(s), a thermostat located in the conditioned space(s) or in the return air temporarily interrupts refrigerant flow to the direct-expansion cooling coils by stopping the compressor(s) and/or closing the solenoid liquid-line valve(s). Other solenoids unload compressors by means of suction control. For jobs with only a single zone of conditioned space, the compressor's on-off control is frequently used to modulate coil capacity. The selection and application of evaporator pressure regulators and similar regulators that are temperature operated and respond to the temperature of the conditioned air are covered in Chapter 45 of the 1998 *ASHRAE Handbook—Refrigeration*.

Applications with multiple zones of conditioned space often use solenoid liquid-line valves to vary coil capacity. These valves should be used where thermostatic expansion valves feed certain types (or sections) of evaporator coils that may, according to load variations, require a temporary but positive interruption of refrigerant flow. This applies particularly to multiples of evaporator coils in a unit where one or more must be shut off temporarily to regulate its zone capacity. In such cases, a solenoid valve should be installed directly upstream of the thermostatic expansion valve(s). If more than one expansion valve feeds a particular zone coil, they may all be controlled by a single solenoid valve.

For a coil controlled by multiple refrigerant expansion valves, there are three arrangements: (1) face control, in which the coil is divided across its face; (2) row control; and (3) interlaced circuitry (Figure 2).

Face control, which is the most widely used because of its simplicity, equally loads all refrigerant circuits within the coil. Face control has the disadvantage of permitting reevaporation of condensate on the coil portion not in operation and bypassing air into the conditioned space during partial load conditions, when some of the TXVs are on an off-cycle. However, while the bottom portion of the coil is cooling, some of the advantages of single-zone humidity control can be achieved with air bypasses through the inactive top portion.

Row control, seldom available as standard equipment, eliminates air bypassing during partial load operation and minimizes condensate reevaporation. Close attention is required for accurate calculation of row-depth capacity, circuit design, and TXV sizing.

Interlaced circuit control uses whole face area and depth of coil when some of the expansion valves are shut off. Without a corresponding drop in airflow, modulating the refrigerant flow to an interlaced coil produces an increased coil surface temperature, thereby necessitating compressor protection (e.g., suction pressure regulators or compressor multiplexing).

Flow Arrangement

In the air-conditioning process, the relation of the fluid flow arrangement within the coil tubes to the coil depth greatly influences the performance of the heat transfer surface. Generally, air-cooling and dehumidifying coils are multirow and circuited for **counterflow** arrangement. The inlet air is applied at right angles to the coil's tube face (coil height), which is also at the coil's outlet header location. The air exits at the opposite face (side) of the coil where the corresponding inlet header is located. Counterflow can produce the highest possible heat exchange within the shortest possible (coil row) depth because it has the closest temperature relationships between tube fluid and air at each (air) side of the coil; the temperature of the entering air more closely approaches the temperature of the leaving fluid than the temperature of the leaving air approaches the temperature of the entry fluid. The potential of realizing the highest possible mean temperature difference is thus arranged for optimum performance.

Most direct-expansion coils also follow this general scheme of thermal counterflow, but the requirements for proper superheat control may necessitate a hybrid combination of parallel flow and counterflow. (Air flows in the same direction as the refrigerant in parallel flow operation.) Quite often, the optimum design for large coils is parallel flow arrangement in the coil's initial (entry) boiling region followed by counterflow in the superheat (exit) region. Such a hybrid arrangement is commonly used for process applications that require a low temperature difference (low TD).

Coil hand refers to either the right hand (RH) or left hand (LH) for counterflow arrangement of a multirow counterflow coil. There is no convention for what constitutes LH or RH, so manufacturers usually establish a convention for their own coils. Most manufacturers designate the location of the inlet water header or refrigerant distributor as the coil hand reference point. Figure 3 illustrates the more widely accepted coil hand designation for multirow water or refrigerant coils.

Applications

Figure 4 shows a typical arrangement of coils in a field built-up central station system. All air should be filtered to prevent dirt, insects, and foreign matter from accumulating on the coils. The cooling coil (and humidifier, when used) should include a drain pan under each coil to catch the condensate formed during the cooling cycle (and the excess water from the humidifier). The drain connection should be on the downstream side of the coils, be of ample size, have accessible cleanouts, and discharge to an indirect waste or storm sewer. The drain also requires a deep-seal trap so that no sewer gas can enter the system. Precautions must be taken if there is a possibility that the drain might freeze. The drain pan, unit casing, and water piping should be insulated to prevent sweating.

Factory-assembled central station air handlers incorporate most of the design features outlined for field built-up systems. These packaged units can generally accommodate various sizes, types, and row depths of cooling and heating coils to meet most job requirements. This usually eliminates the need for field built-up central systems, except on very large jobs.

The design features of the coil (fin spacing, tube spacing, face height, type of fins), together with the amount of moisture on the coil and the degree of surface cleanliness, determine the air velocity at which condensed moisture blows off the coil. Generally, condensate

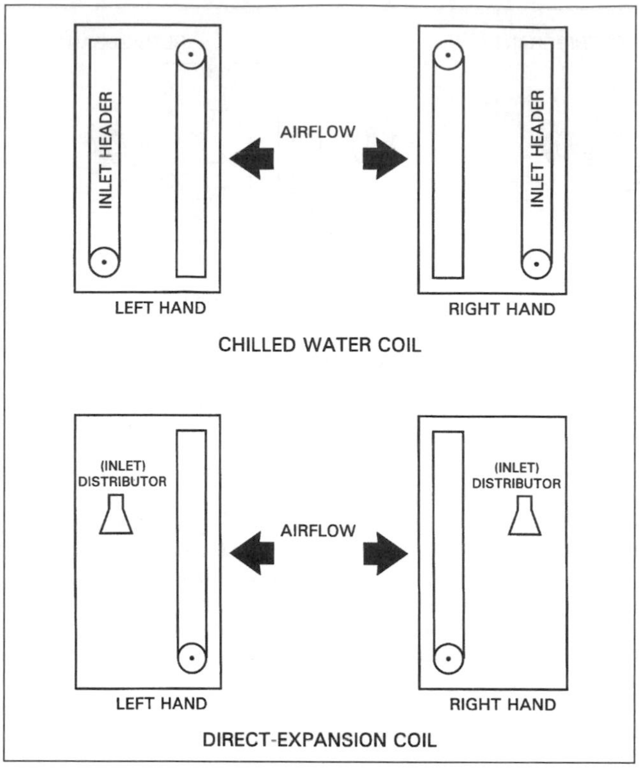

Fig. 3 Typical Coil Hand Designation

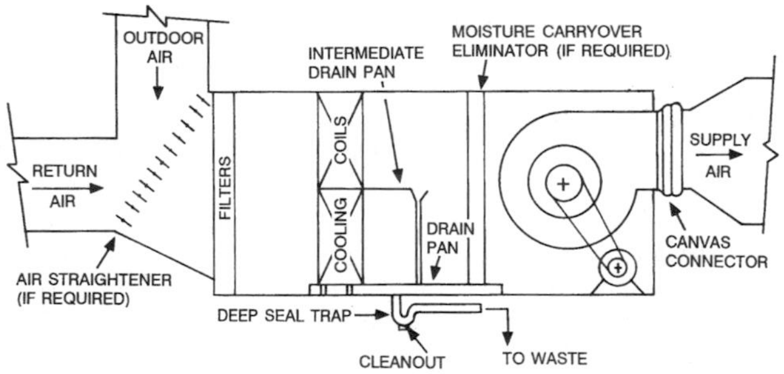

Fig. 4 Typical Arrangement of Cooling Coil Assembly in Built-Up or Packaged Central Station Air Handler

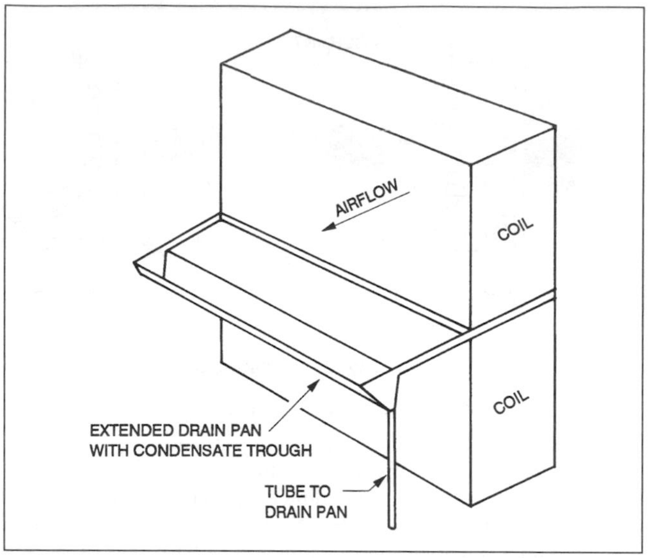

Fig. 5 Coil Bank Arrangement with Intermediate Condensate Pan

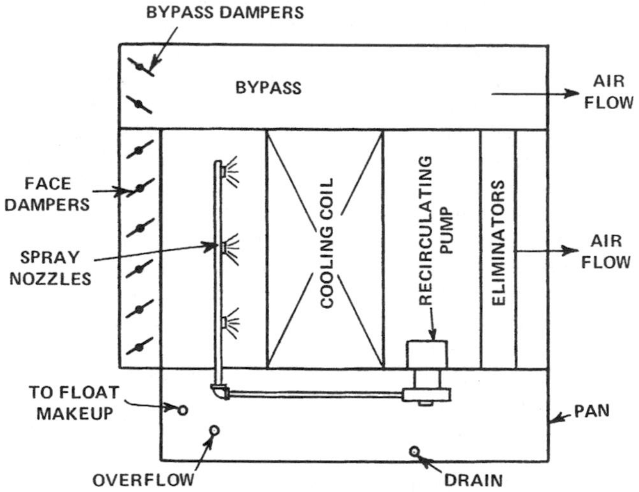

Fig. 6 Sprayed-Coil System with Air Bypass

water begins to be blown off a plate fin coil face at air velocities above 600 fpm. Water blowoff from the coils into air ductwork external to the air-conditioning unit should be prevented. However, water blowoff from the coils is not usually a problem if coil fin heights are limited to 45 in. and the unit is set up to catch and dispose of the condensate. When a number of coils are stacked one above another, the condensate is carried into the airstream as it drips from one coil to the next. A downstream eliminator section could prevent this, but an intermediate drain pan and/or condensate trough (Figure 5) to collect the condensate and conduct it directly to the main drain pan is preferred. Extending downstream of the coil, each drain pan length should be at least one-half the coil height, and somewhat greater when coil airflow face velocities and/or humidity levels are higher.

When water is likely to carry over from the air-conditioning unit into external air ductwork, and no other means of prevention is provided, eliminator plates should be installed on the downstream side of the coils. Usually, eliminator plates are not included in packaged units because other means of preventing carryover, such as space made available within the unit design for longer drain pan(s), are included in the design.

However, on sprayed-coil units, eliminators are usually included in the design. Such cooling and dehumidifying coils are sometimes sprayed with water to increase the rate of heat transfer, provide outlet air approaching saturation, and continually wash the surface of the coil. Coil sprays require a collecting tank, eliminators, and a recirculating pump (see Figure 6). Figure 6 also shows an air bypass, which helps a thermostat control maintain the humidity ratio by diverting a portion of the return air from the coil.

In field-assembled systems or factory-assembled central station air-handling units, the fans are usually positioned downstream from the coil(s) in a draw-through arrangement. This arrangement provides acceptable airflow uniformity across the coil face more often than does the blow-through arrangement. In a blow-through arrangement, fan location upstream from the coils may require air baffles or diffuser plates between the fan discharge and the cooling coil to obtain uniform airflow. This is often the case in packaged multizone unit design. Airflow is considered to be uniform when the measured flow across the entire coil face varies no more than 20%.

Air-cooling and dehumidifying coil frames, as well as all drain pans and troughs, should be of an acceptable corrosion-resistant material suitable for the system and its expected useful service life.

The air handler's coil section enclosure should be corrosion-resistant; be properly double-wall insulated; and have adequate access doors for changing air filters, cleaning coils, adjusting flow control valves, and maintaining motors.

Where suction line risers are used for air-cooling coils in direct-expansion refrigeration systems, the suction line must be sized properly to ensure oil return from coil to compressor at minimum load conditions. Oil return is normally intrinsic with factory-assembled, self-contained air conditioners but must be considered for factory-assembled central station units or field-installed cooling coil banks where suction line risers are required and are assembled at the job site. Sizing, design, and arrangement of suction lines and their risers are described in Chapter 3 of the 1998 *ASHRAE Handbook—Refrigeration.*

COIL SELECTION

When selecting a coil, the following factors should be considered:

- Job requirements—cooling, dehumidifying, and the capacity required to properly balance with other system components (e.g., compressor equipment in the case of direct-expansion coils)
- Temperature conditions of entering air
- Available cooling media and operating temperatures
- Space and dimensional limitations
- Air and cooling fluid quantities, including distribution and limitations
- Allowable frictional resistances in air circuit (including coils)
- Allowable frictional resistances in cooling media piping system (including coils)
- Characteristics of individual coil designs and circuitry possibilities
- Individual installation requirements such as type of automatic control to be used; presence of corrosive atmosphere; design pressures; and durability of tube, fins, and frame material

Chapters 27 and 28 of the 1997 *ASHRAE Handbook—Fundamentals* contains information on load calculation.

Air quantity is affected by such factors as design parameters, codes, space, and equipment. The resistance through the air circuit influences the fan power and speed. This resistance may be limited to allow the use of a given size fan motor, to keep the operating expense low, or because of sound-level requirements. The air friction loss across the cooling coil—in summation with other series air pressure drops for such elements as air filters, water sprays, heating coils, air grilles, and ductwork—determines the static pressure requirement for the complete airway system. The static pressure requirement is used in selecting the fans and drives to obtain the design air quantity under operating conditions. See Chapter 18 for a description of fan selection.

The conditioned air face velocity is determined by economic evaluation of initial and operating costs for the complete installation as influenced by (1) heat transfer performance of the specific coil surface type for various combinations of face areas and row depths as a function of the air velocity; (2) air-side frictional resistance for the complete air circuit (including coils), which affects fan size, power, and sound-level requirements; and (3) condensate water carryover considerations. The allowable friction through the water or brine coil circuitry may be dictated by the head available from a given size pump and pump motor, as well as the same economic factors governing the air side made applicable to the water side. Additionally, the adverse effect of high cooling water velocities on erosion-corrosion of tube walls is a major factor in sizing and circuitry to keep tube velocity below the recommended maximums. On larger coils, water pressure drop limits of 15 to 20 ft usually keep such velocities within acceptable limits of 2 to 4 fps, depending on circuit design.

Coil ratings are based on a uniform velocity. Design interference with uniform airflow through the coil makes predicting coil performance difficult as well as inaccurate. Such airflow interference may be caused by the entrance of air at odd angles or by the inadvertent blocking of a portion of the coil face. To obtain rated performance, the volumetric airflow quantity must be adjusted on the job to correspond to that at which the coil was rated and must be kept at that value. At start-up for air balance, the most common causes of incorrect airflow are the lack of altitude correction to standard air (where applicable) and ductwork problems. At commissioning, the most common causes of an air quantity deficiency are fouling of the filters and collection of dirt or frost on the coils. These difficulties can be avoided through proper design, start-up checkout, and regular servicing.

The required total heat capacity of the cooling coil should be in balance with the capacity of other refrigerant system components such as the compressor, water chiller, condenser, and refrigerant liquid metering device. Chapter 44 of the 1998 *ASHRAE Handbook—Refrigeration* describes methods of estimating balanced system capacity under various operating conditions when using direct-expansion coils for both factory- and field-assembled systems.

In the case of dehumidifying coils, it is important that the proper amount of surface area be installed to obtain the ratio of air-side sensible-to-total heat required for maintaining the air dry-bulb and wet-bulb temperatures in the conditioned space. This is an important consideration when preconditioning is done by reheat arrangement. The method for calculating the sensible and total heat loads and the leaving air conditions at the coil to satisfy the sensible-to-total heat ratio required for the conditioned space is covered in Appendix D of *Cooling and Heating Load Calculation Principles* (Pedersen et al. 1998).

The same room air conditions can be maintained with different air quantities (including outside and return air) through a coil. However, for a given total air quantity with fixed percentages of outside and return air, there is only one set of air conditions leaving the coil that will precisely maintain the room design air conditions. Once the air quantity and leaving air conditions at the coil have been selected, there is usually only one combination of face area, row depth, and air face velocity for a given coil surface that will precisely maintain the required room ambient conditions. Therefore, in making final coil selections it is necessary to recheck the initial selection to ensure that the leaving air conditions, as calculated by a coil selection computer program or other procedure, will match those determined from the cooling load estimate.

Coil ratings and selections can be obtained from manufacturers' catalogs. Most catalogs contain extensive tables giving the performance of coils at various air and water velocities and entering humidity and temperatures. Most manufacturers provide computerized coil selection programs to potential customers. The final choice can then be made based on system performance and economic requirements.

Performance and Ratings

The long-term performance of an extended surface air-cooling and dehumidifying coil depends on its correct design to specified conditions and material specifications, proper matching to other system components, proper installation, and proper maintenance as required.

In accordance with ARI *Standard* 410, Forced-Circulation Air-Cooling and Air-Heating Coils, dry surface (sensible cooling) coils and dehumidifying coils (which both cool and dehumidify), particularly those used for field-assembled coil banks or factory-assembled packaged units using different combinations of coils, are usually rated within the following parameters:

Entering air dry-bulb temperature: 65 to 100°F

Entering air wet-bulb temperature: 60 to 85°F (If air is not dehumidified in the application, select coils based on sensible heat transfer.)

Air face velocity: 200 to 800 fpm

Evaporator refrigerant saturation temperature: 30 to 55°F at coil suction outlet (refrigerant vapor superheat at coil suction outlet is 6°F or higher)

Entering chilled water temperature: 35 to 65°F

Water velocity: 1 to 8 fps

For cold ethylene glycol solution: 1 to 6 fps, 0 to 90°F entering dry-bulb temperature, 60 to 80°F entering wet-bulb temperature, 10 to 60% aqueous glycol concentration by weight

The air-side ratio of sensible to total heat removed by dehumidifying coils varies in practice from about 0.6 to 1.0 (i.e., sensible heat is from 60 to 100% of the total, depending on the application). Sample problems in Chapter 28 of the 1997 *ASHRAE Handbook—Fundamentals* or in Appendix D of *Cooling and Heating Load Calculation Principles* (Pedersen et al. 1998) illustrate the calculation of sensible heat ratio. For a given coil surface design and arrangement, the required sensible heat ratio may be satisfied by wide variations in and combinations of air face velocity, in-tube temperature, flow rate, entering air temperature, coil depth, and so forth, although the variations may be self-limiting. The maximum coil air face velocity should be limited to a value that prevents water carryover into the air ductwork. Dehumidifying coils for comfort application are frequently selected in the range of 400 to 500 fpm air face velocity.

The operating ratings of dehumidifying coils for factory-assembled, self-contained air conditioners are generally determined in conjunction with laboratory testing for the system capacity of the complete unit assembly. For example, a standard rating point has been 33.4 cfm per 1000 Btu/h (or 400 cfm per ton of refrigeration effect), not to exceed 37.5 cfm per 100 Btu/h for unitary equipment. Refrigerant (e.g., R-22) duty would be 6 to 10°F superheat for an appropriate balance at 45°F saturated suction. For water coils, circuitry would operate at 4 fps, 42°F inlet water, 12°F rise (or 2 gpm per ton of refrigeration effect). The standard ratings at 80°F dry bulb and 67°F wet bulb are representative of the entering air conditions encountered in many comfort operations. While the indoor conditions are usually lower than 67°F wet bulb, it is usually assumed that the introduction of outdoor air brings the mixture of air to the cooling coil up to about 80°F dry bulb/67°F wet bulb entering air design conditions.

Dehumidifying coils for field-assembled projects and central station air-handling units were formerly selected according to coil rating tables but are now selected by computerized selection programs. Either way, selecting coils from the load division indicated by the load calculation works satisfactorily for the usual human comfort applications. Additional design precautions and refinements are necessary for more exacting industrial applications and for all types of air conditioning in humid areas. One such refinement, the dual-path air process, uses a separate cooling coil to cool and dehumidify the ventilation air before mixing it with recirculated air. This process dehumidifies what is usually the main source of moisture—makeup outside air. Reheat is another refinement that is required for some industrial applications and is finding greater use in commercial and comfort applications.

Airflow ratings are based on standard air of 0.075 lb/ft^3 at 70°F and a barometric pressure of 29.92 in. Hg. In some developed, mountainous areas with a sufficiently large market, coil ratings and altitude-corrected psychrometrics are available for their particular altitudes.

When checking the operation of dehumidifying coils, climatic conditions must be considered. Most problems are encountered at light-load conditions, when the cooling requirement is considerably less than at design conditions. In hot, dry climates, where the outdoor dew point is consistently low, dehumidifying is not generally a problem, and the light-load design point condition does not pose any special problems. In hot, humid climates, the light-load condition has a higher proportion of moisture and a correspondingly lower proportion of sensible heat. The result is higher dew points in the conditioned spaces during light-load conditions unless a special means for controlling the inside dew points (e.g., reheat or dual path) is used.

Fin surface freezing at light loads should be avoided. Freezing occurs when a dehumidification coil's surface temperature falls below 32°F. Freezing does not occur with standard coils for comfort installations unless the refrigerant evaporating temperature at the coil outlet is below 25 to 28°F saturated; the exact value depends on the design of the coil, its operating dew point, and the amount of loading. With coil and condensing units to balance at low temperatures at peak loads (not a customary design choice), freezing may occur when the load suddenly decreases. The possibility of this type of surface freezing is greater if a bypass is used because it causes less air to be passed through the coil at light loads.

AIRFLOW RESISTANCE

A cooling coil's airflow resistance (air friction) depends on the tube pattern and fin geometry (tube size and spacing, fin configuration, and number of in-line or staggered rows), the coil face velocity, and the amount of moisture on the coil. The coil air friction may also be affected by the degree of aerodynamic cleanliness of the coil core; burrs on fin edges may increase coil friction and increase the tendency to pocket dirt or lint on the faces. A completely dry coil, removing only sensible heat, offers approximately one-third less resistance to airflow than a dehumidifying coil removing both sensible and latent heat.

For a given surface and airflow, an increase in the number of rows or number of fins increases the airflow resistance. Therefore, the final selection involves the economic balancing of the initial cost of the coil against the operating costs of the coil geometry combinations available to adequately meet the performance requirements.

The aluminum fin surfaces of new dehumidifying coils tend to inhibit condensate sheeting action until they have aged for a year. Recently developed hydrophilic aluminum fin surface coatings reduce the water droplet surface tension, producing a more evenly dispersed wetted surface action at initial start-up. Manufacturers have tried different methods of applying such coatings, including dipping the coil into a tank, coating the fin stock material, or subjecting the material to a chemical etching process. Tests have shown as much as a 30% reduction in air pressure drop across a hydrophilic coil as opposed a new untreated coil.

HEAT TRANSFER

The heat transmission rate of air passing over a clean tube (with or without extended surface) to a fluid flowing within it is impeded principally by three thermal resistances. The first, from the air to the surface of the exterior fin and tube assembly, is known as the surface air-side film thermal resistance. The second is the metal thermal resistance to the conductance of heat through the exterior fin and tube assembly. The third is the in-tube fluid-side film thermal resistance, which impedes the flow of heat between the internal surface of the metal and the fluid flowing within the tube. For some applications, an additional thermal resistance is factored in to account for external and/or internal surface fouling. Usually, the combination of the metal and tube-side film resistance is considerably lower than the air-side surface resistance.

For a reduction in thermal resistance, the fin surface is fabricated with die-formed corrugations instead of the traditional flat design. At low airflows or wide fin spacing, the air-side transfer coefficient is virtually the same for flat and corrugated fins. Under normal comfort conditioning operation, the corrugated fin surface is designed to reduce the boundary air film thickness by undulating the passing airstream within the coil; this produces a marked improvement in heat transfer without much airflow penalty. Further fin enhancements, including the louvered and lanced fin designs, have been driven by the desire to duplicate throughout the coil depth the thin boundary air film characteristic of the fin's leading edge. Louvered fin design maximizes the number of fin surface leading edges throughout the entire secondary surface area and increases the external secondary surface area A_s through the multiplicity of edges.

Where an application allows an economical use of coil construction materials, the mass and size of the coil can be reduced when boundary air and water films are lessened. For example, the exterior surface resistance can be reduced to nearly the same as the fluid-side resistance through the use of lanced and/or louvered fins. External as well as internal tube fins (or internal turbulators) can economically decrease overall heat transfer surface resistances. Also, water sprays applied to a particular flat fin coil surface may increase the overall heat transfer slightly, although they may better serve other purposes such as air and coil cleaning.

The transfer of heat between the cooling medium and the airstream across a coil is influenced by the following variables:

- Temperature difference between fluids
- Design and surface arrangement of the coil
- Velocity and character of the airstream
- Velocity and character of the in-tube coolant

With water coils, only the water temperature rises. With coils of volatile refrigerants, an appreciable pressure drop and a corresponding change in evaporating temperature through the refrigerant circuit often occur. Alternative refrigerants to R-22, such as R-407C, which has a temperature glide, will have an evaporation temperature rise of 7 to 12°F through the evaporator. This must be considered in the design and performance calculation of the coil. A compensating pressure drop in the coil may partially, or even totally, compensate for the low-side temperature glide of a zeotropic refrigerant blend. The rating of direct-expansion coils is further complicated by the refrigerant evaporating in part of the circuit and superheating in the remainder. Thus, for halocarbon refrigerants, a cooling coil is tested and rated with a specific distributing and liquid-metering device, and the capacities are stated with the superheat condition of the leaving vapor.

At a given air mass velocity, performance depends on the turbulence of airflow into the coil and the uniformity of air distribution over the coil face. The latter is necessary to obtain reliable test ratings and realize rated performance in actual installations. The air resistance through the coils assists in distributing the air properly, but the effect is frequently inadequate where inlet duct connections are brought in at sharp angles to the coil face. Reverse air currents may pass through a portion of the coils. These currents reduce the capacity but can be avoided with proper inlet air vanes or baffles. Air blades may also be required. Remember that coil performance ratings (ARI *Standard* 410) represent optimum conditions resulting from adequate and reliable laboratory tests (ASHRAE *Standard* 33).

For cases when available data must be extended, for arriving at general design criteria for a single, unique installation, or for understanding the calculation progression, the following material and illustrative examples for calculating cooling coil performance are useful guides.

PERFORMANCE OF SENSIBLE COOLING COILS

The performance of sensible cooling coils depends on the following factors. See the section on Symbols for an explanation of the variables.

1. The overall coefficient U_o of sensible heat transfer between airstream and coolant fluid
2. The mean temperature difference Δt_m between airstream and coolant fluid
3. The physical dimensions of and data for the coil (such as coil face area A_a and total external surface area A_o) with characteristics of the heat transfer surface

The sensible heat cooling capacity q_{td} of a given coil is expressed by the following equation:

$$q_{td} = U_o F_s A_a N_r \Delta t_m \tag{1a}$$

with

$$F_s = A_o / A_a N_r \tag{1b}$$

Assuming no extraneous heat losses, the same amount of sensible heat is lost from the airstream:

$$q_{td} = w_a c_p (t_{a1} - t_{a2}) \tag{2a}$$

with

$$w_a = 60 \rho_a A_a V_a \tag{2b}$$

The same amount of sensible heat is absorbed by the coolant; for a nonvolatile type, it is

$$q_{td} = w_r c_r (t_{r2} - t_{r1}) \tag{3}$$

For a nonvolatile coolant in thermal counterflow with the air, the mean temperature difference in Equation (1a) is expressed as

$$\Delta t_m = \frac{(t_{a1} - t_{r2}) - (t_{a2} - t_{r1})}{\ln[(t_{a1} - t_{r2})/(t_{a2} - t_{r1})]} \tag{4}$$

The proper temperature differences for various crossflow situations are given in many texts, including Mueller (1973). These calculations are based on various assumptions, among them that U for the total external surface is constant. While this assumption is generally not valid for multirow coils, the use of crossflow temperature differences from Mueller (1973) or other texts should be preferable to Equation (4), which applies only to counterflow. However, the use of the log mean temperature difference is widespread.

The overall heat transfer coefficient U_o for a given coil design, whether bare-pipe or finned-type, with clean, nonfouled surfaces, consists of the combined effect of three individual heat transfer coefficients:

1. The **film coefficient** f_a of sensible heat transfer between air and the external surface of the coil
2. The **unit conductance** $1/R_{md}$ of the coil material (i.e., tube wall, fins, tube-to-fin thermal resistance)
3. The **film coefficient** f_r of heat transfer between the internal coil surface and the coolant fluid within the coil

These three individual coefficients acting in series form an overall coefficient of heat transfer in accordance with the material given in Chapters 3 and 22 of the 1997 *ASHRAE Handbook—Fundamentals*.

For a bare-pipe coil, the overall coefficient of heat transfer for sensible cooling (without dehumidification) can be expressed by a simplified basic equation:

$$U_o = \frac{1}{(1/f_a) + (D_o - D_i)/24k + (B/f_r)} \tag{5a}$$

When pipe or tube walls are thin and made of material with high conductivity (as in typical heating and cooling coils), the term $(D_o - D_i)/24k$ in Equation (5a) frequently becomes negligible and is generally disregarded. (This effect in typical bare-pipe cooling coils seldom exceeds 1 to 2% of the overall coefficient.) Thus, the overall coefficient for bare pipe in its simplest form is

$$U_o = \frac{1}{(1/f_a) + (B/f_r)} \quad (5b)$$

For finned coils, the equation for the overall coefficient of heat transfer can be written

$$U_o = \frac{1}{(1/\eta f_a) + (B/f_r)} \quad (5c)$$

where the **fin effectiveness** η allows for the resistance to heat flow encountered in the fins. It is defined as

$$\eta = (EA_s + A_p)/A_o \quad (6)$$

For typical cooling surface designs, the surface ratio B ranges from about 1.03 to 1.15 for bare-pipe coils and from 10 to 30 for finned coils. Chapter 3 of the 1997 *ASHRAE Handbook—Fundamentals* describes how to estimate fin efficiency E and calculate the tube-side heat transfer coefficient f_r for nonvolatile fluids. Table 2 in ARI *Standard* 410 lists thermal conductivity k of standard coil materials.

Estimation of the air-side heat transfer coefficient f_a is more difficult because well-verified general predictive techniques are not available. Hence, direct use of experimental data is usually necessary. For plate fin coils, some correlations that satisfy several data sets are available (McQuiston 1981; Kusuda 1970). Webb (1980) reviewed the air-side heat transfer and pressure drop correlations for various geometries. Mueller (1973) and Chapter 3 of the 1997 *ASHRAE Handbook—Fundamentals* provide guidance on this subject.

For analyzing a given heat exchanger, the concept of **effectiveness** is useful. Expressions for effectiveness have been derived for various flow configurations and can be found in Mueller (1973) and Kusuda (1970). The cooling coils covered in this chapter actually involve various forms of crossflow. However, the case of counterflow is addressed here to illustrate the value of this concept. The air-side effectiveness E_a for counterflow heat exchangers is given by the following equations:

$$q_{td} = w_a c_p (t_{a1} - t_{r1}) E_a \quad (7a)$$

with

$$E_a = \frac{t_{a1} - t_{a2}}{t_{a1} - t_{r1}} \quad (7b)$$

or

$$E_a = \frac{1 - e^{-c_o(1-M)}}{1 - Me^{-c_o(1-M)}} \quad (7c)$$

with

$$c_o = \frac{A_o U_o}{w_a c_p} = \frac{F_s N_r U_o}{60 \rho_a V_a c_p} \quad (7d)$$

and

$$M = \frac{w_a c_p}{w_r c_r} = \frac{60 \rho_a A_a V_a c_p}{w_r c_r} \quad (7e)$$

Note the following two special conditions:

If $M = 0$, then $E_a = 1 - e^{-c_o}$
If $M \geq 1$, then

$$E_a = \frac{1}{(1/c_o) + 1}$$

With a given design and arrangement of heat transfer surface used as cooling coil core material for which basic physical and heat transfer data are available to determine U_o from Equations (5a), (5b), and (5c), the selection, sizing, and performance calculation of sensible cooling coils for a particular application generally fall into either of two categories:

1. The heat transfer surface area A_o or the coil row depth N_r for a specific coil size is required and initially unknown. The sensible cooling capacity q_{td}, flow rates for both air and coolant, entrance and exit temperatures of both fluids, and mean temperature difference between fluids are initially known or can be assumed or determined from Equations (2a), (3), and (4). The required surface area A_o or coil row depth N_r can then be calculated directly from Equation (1a).
2. The sensible cooling capacity q_{td} for a specific coil is required and initially unknown. The face area and heat transfer surface area are known or can be readily determined. The flow rates and entering temperatures of air and coolant are also known. The mean temperature difference Δt_m is unknown, but its determination is unnecessary to calculate the sensible cooling capacity q_{td}, which can be found directly by solving Equation (7a). Equation (7a) also provides a basic means of determining q_{td} for a given coil or related family of coils over the complete rating ranges of air and coolant flow rates and operating temperatures.

The two categories of application problems are illustrated in Examples 1 and 2, respectively:

Example 1. Standard air flowing at a mass rate equivalent to 9000 cfm is to be cooled from 85 to 75°F, using 330 lb/min chilled water supplied at 50°F in thermal counterflow arrangement. Assuming an air face velocity of V_a = 600 fpm and no air dehumidification, calculate coil face area A_a, sensible cooling capacity q_{td}, required heat transfer surface area A_o, coil row depth N_r, and coil air-side pressure drop Δp_{st} for a clean, nonfouled, thin-walled bare copper tube surface design for which the following physical and performance data have been predetermined:

B = surface ratio = 1.07
c_p = 0.24 Btu/lb·°F
c_r = 1.0 Btu/lb·°F
F_s = (external surface area)/(face area)(rows deep) = 1.34
f_a = 15 Btu/h·ft^2·°F
f_r = 800 Btu/h·ft^2·°F
$\Delta p_{st}/N_r$ = 0.027 in. of water/number of coil rows
ρ_a = 0.075 lb/ft^3

Solution: Calculate the coil face area required.

$$A_a = 9000/600 = 15 \text{ ft}^2$$

Neglecting the effect of tube wall, from Equation (5b),

$$U_o = \frac{1}{(1/15) + (1.07/800)} = 14.7 \text{ Btu/h} \cdot \text{ft}^2 \cdot {}^\circ\text{F}$$

From Equations (2a) and (2b), the sensible cooling capacity is

$$q_{td} = 60 \times 0.075 \times 15 \times 600 \times 0.24(85 - 75) = 97{,}200 \text{ Btu/h}$$

From Equation (3),

$$t_{r2} = 50 + 97{,}200/(330 \times 60 \times 1) = 54.9^\circ\text{F}$$

From Equation (4),

$$\Delta t_m = \frac{(85 - 54.9) - (75 - 50)}{\ln[(85 - 54.9)/(75 - 50)]} = 27.5^\circ\text{F}$$

From Equations (1a) and (1b), the surface area required is

$$A_o = 97{,}200/(14.7 \times 27.5) = 240 \text{ ft}^2 \text{ external surface}$$

From Equation (1b), the required row depth is

$$N_r = 240/(1.34 \times 15) = 11.9 \text{ rows deep}$$

The installed 15 ft² coil face, 12 rows deep, slightly exceeds the required capacity. The air-side pressure drop for the installed row depth is then

$$\Delta p_{st} = (\Delta p_{st}/N_r)N_r = 0.027 \times 12 = 0.32 \text{ in. of water at } 70°\text{F}$$

In this example, for some applications where such items as V_a, w_r, t_{r1}, and f_r may be arbitrarily varied with a fixed design and arrangement of heat transfer surface, a trade-off between coil face area A_a and coil row depth N_r is sometimes made to obtain alternate coil selections that produce the same sensible cooling capacity q_{td}. For example, an eight-row coil could be selected, but it would require a larger face area A_a with lower air face velocity V_a and a lower air-side pressure drop Δp_{st}.

Example 2. An air-cooling coil using a finned tube-type heat transfer surface has physical data as follows:

A_a = 10 ft²
A_o = 800 ft² external
B = surface ratio = 20
F_s = (external surface area)/(face area)(rows deep) = 27
N_r = 3 rows deep

Air at a face velocity of V_a = 800 fpm and 95°F entering air temperature is to be cooled by 15 gpm of well water supplied at 55°F. Calculate the sensible cooling capacity q_{td}, leaving air temperature t_{a2}, leaving water temperature t_{r2}, and air-side pressure drop Δp_{st}. Assume clean and nonfouled surfaces, thermal counterflow between air and water, no air dehumidification, standard barometric air pressure, and that the following data are available or can be predetermined:

c_p = 0.24 Btu/lb·°F
c_r = 1.0 Btu/lb·°F
f_a = 17 Btu/h·ft²·°F
f_r = 500 Btu/h·ft²·°F
η = fin effectiveness = 0.9
$\Delta p_{st}/N_r$ = 0.22 in. of water/number of coil rows
ρ_a = 0.075 lb/ft³
ρ_w = 62.4 lb/ft³ (8.34 lb/gal)

Solution: From Equation (5c),

$$U_o = \frac{1}{1/(0.9 \times 17) + (20/500)} = 9.5 \text{ Btu/h} \cdot \text{ft}^2 \cdot °\text{F}$$

From Equations (7d) and (2b),

$$c_o = \frac{800 \times 9.5}{60 \times 0.075 \times 10 + 800 \times 0.24} = 0.88$$

From Equation (7e),

$$M = \frac{60 \times 0.075 \times 10 + 800 \times 0.24}{15 \times 60 \times 8.34 \times 1} = 1.15$$

Substituting in,

$$-c_o(1-M) = -0.88(1-1.15) = 0.132$$

From Equation (7c),

$$E_a = \frac{1-e^{0.132}}{1-1.15e^{0.132}} = 0.452$$

From Equation (7a), the sensible cooling capacity is

$$q_{td} = 60 \times 0.075 \times 10 \times 800 \times 0.24(95-55) \times 0.452 = 156{,}000 \text{ Btu/h}$$

From Equation (2a), the leaving air temperature is

$$t_{a2} = 95 - \frac{156{,}000}{60 \times 0.075 \times 10 \times 800 \times 0.24} = 76.9°\text{F}$$

From Equation (3), the leaving water temperature is

$$t_{r2} = 55 + \frac{156{,}000}{15 \times 60 \times 8.34 \times 1} = 75.8°\text{F}$$

The air-side pressure drop is

$$\Delta p_{st} = 0.22 \times 3 = 0.66 \text{ in. of water}$$

The preceding equations and two illustrative examples demonstrate the method for calculating thermal performance of sensible cooling coils that operate with a dry surface. However, when cooling coils operate wet or act as dehumidifying coils, the performance cannot be predicted without including the effect of air-side moisture (latent heat) removal.

PERFORMANCE OF DEHUMIDIFYING COILS

A dehumidifying coil normally removes both moisture and sensible heat from entering air. In most air-conditioning processes, the air to be cooled is a mixture of water vapor and dry air gases. Both lose sensible heat when in contact with a surface cooler than the air. The removal of latent heat through condensation occurs only on the portions of the coil where the surface temperature is lower than the dew point of the air passing over it. Figure 2 in Chapter 2 shows the assumed psychrometric conditions of this process. As the leaving dry-bulb temperature is lowered below the entering dew-point temperature, the difference between the leaving dry-bulb temperature and the leaving dew point for a given coil, airflow, and entering air condition is lessened.

When the coil starts to remove moisture, the cooling surfaces carry both the sensible and latent heat load. As the air approaches saturation, each degree of sensible cooling is nearly matched by a corresponding degree of dew-point decrease. The latent heat removal per degree of dew-point change is considerably greater. The following table compares the amount of moisture removed from air at standard barometric pressure that is cooled from 60 to 59°F at both wet and dry conditions.

Dew Point	h_s, Btu/lb	Dry Bulb	h_a, Btu/lb
60°F	26.467	60°F	14.415
59°F	25.792	59°F	14.174
Difference	0.675	Difference	0.241

Note: These numerical values conform to Table 2 in Chapter 6 of the 1997 *ASHRAE Handbook—Fundamentals*.

For volatile refrigerant coils, the refrigerant distributor assembly must be tested at the higher and lower capacities of its rated range. Testing at the lower capacities checks whether the refrigerant distributor provides equal distribution and whether the control is able to modulate without hunting. Testing at the higher capacities checks the maximum feeding capacity of the flow control device at the greater pressure drop that occurs in the coil system.

Most manufacturers develop and produce their own performance rating tables using data obtained from suitable tests. *ASHRAE Standard* 33 specifies the acceptable method of lab-testing coils. ARI *Standard* 410 gives a method for rating thermal performance of dehumidifying coils by extending data from laboratory tests on prototypes to other operating conditions, coil sizes, and row depths. To account for simultaneous transfer of both sensible and latent heat from the airstream to the surface, ARI *Standard* 410 uses essentially the same method for arriving at cooling and dehumidifying coil thermal performance as determined by McElgin and Wiley (1940) and described in the context of *Standard* 410 by Anderson (1970). In systems operating at partial flow such as in thermal storage, accurate performance predictions for $2300 < Re < 4000$ flows have been realized by the use of the Gnielinski correlation. This work was presented in detail comparable to *Standard* 410 by Mirth et al. (1993).

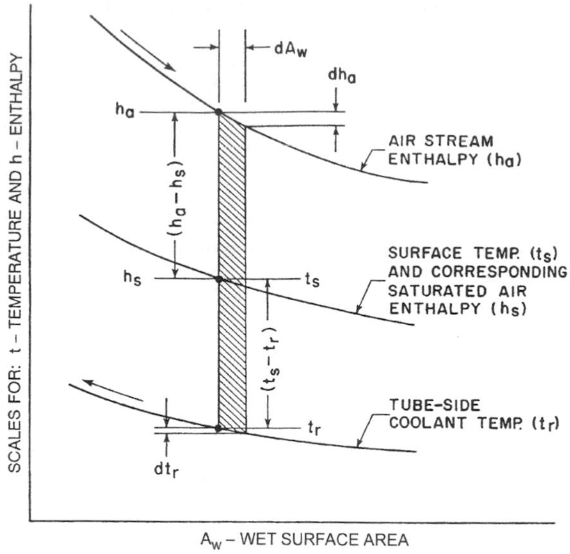

Fig. 7 Two-Component Driving Force Between Dehumidifying Air and Coolant

The potential or driving force for transferring total heat q_t from the airstream to the tube-side coolant is composed of two components in series heat flow: (1) an air-to-surface air enthalpy difference $(h_a - h_s)$ and (2) a surface-to-coolant temperature difference $(t_s - t_r)$.

Figure 7 is a typical thermal diagram for a coil in which the air and a nonvolatile coolant are arranged in counterflow. The top and bottom lines in the diagram indicate, respectively, changes across the coil in the airstream enthalpy h_a and the coolant temperature t_r. To illustrate continuity, the single middle line in Figure 7 represents both surface temperature t_s and the corresponding saturated air enthalpy h_s, although the temperature and air enthalpy scales do not actually coincide as shown. The differential surface area dA_w represents any specific location within the coil thermal diagram where operating conditions are such that the air-surface interface temperature t_s is lower than the local air dew-point temperature. Under these conditions, both sensible and latent heat are removed from the airstream, and the cooler surface actively condenses water vapor.

Neglecting the enthalpy of the condensed water vapor leaving the surface and any radiation and convection losses, the total heat lost from the airstream in flowing over dA_w is

$$dq_t = -w_a(dh_a) \tag{8}$$

This same total heat is transferred from the airstream to the surface interface. According to McElgin and Wiley (1940),

$$dq_t = \frac{(h_a - h_s)dA_w}{c_p R_{aw}} \tag{9}$$

The total heat transferred from the air-surface interface across the surface elements and into the coolant is equal to that given in Equations (8) and (9):

$$dq_t = \frac{(t_s - t_r)dA_w}{R_{mw} + R_r} \tag{10}$$

The same quantity of total heat is also gained by the nonvolatile coolant in passing across dA_w:

$$dq_t = -w_r c_r(dt_r) \tag{11}$$

If Equations (9) and (10) are equated and the terms rearranged, an expression for the coil characteristic C is obtained:

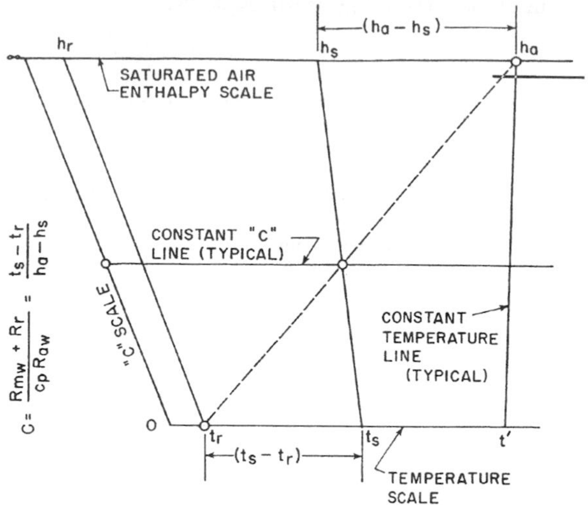

Fig. 8 Surface Temperature Chart

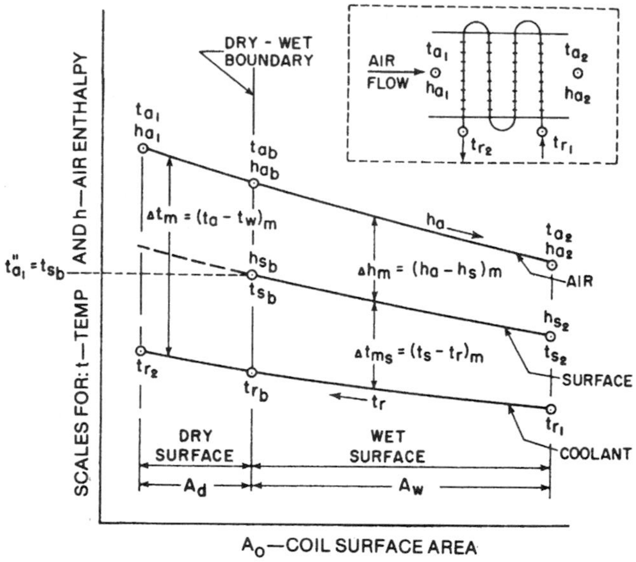

Fig. 9 Thermal Diagram for General Case When Coil Surface Operates Partially Dry

$$C = \frac{R_{mw} + R_r}{c_p R_{aw}} = \frac{t_s - t_r}{h_a - h_s} \tag{12}$$

Equation (12) shows the basic relationship of the two components of the driving force between air and coolant in terms of three principal thermal resistances. For a given coil, these three resistances of air, metal, and in-tube fluid (R_{aw}, R_{mw}, and R_r) are usually known or can be determined for the particular application, which gives a fixed value for C. Equation (12) can then be used to determine point conditions for the interrelated values of airstream enthalpy h_a; coolant temperature t_r; surface temperature t_s; and the enthalpy h_s of saturated air corresponding to the surface temperature. When both t_s and h_s are unknown, a trial-and-error solution is necessary; however, this can be solved graphically by a surface temperature chart such as that shown in Figure 8.

Figure 9 shows a typical thermal diagram for a portion of the coil surface when it is operating dry. The illustration is for counterflow with a halocarbon refrigerant. The diagram at the top of the figure illustrates a typical coil installation in an air duct with tube passes circuited countercurrent to airflow. Locations of the entering and leaving boundary conditions for both air and coolant are shown.

The thermal diagram in Figure 9 is of the same type as that in Figure 7, showing three lines to illustrate local conditions for the air, surface, and coolant throughout a coil. The dry-wet boundary conditions are located where the coil surface temperature t_{sb} equals the entering air dew-point temperature t''_{a1}. Thus, the surface area A_d to the left of this boundary is dry, with the remainder A_w of the coil surface area operating wet.

When using fluids or halocarbon refrigerants in a thermal counterflow arrangement as illustrated in Figure 9, the dry-wet boundary conditions can be determined from the following relationships:

$$y = \frac{t_{r2} - t_{r1}}{h_{a1} - h_{a2}} = \frac{w_a}{w_r c_r} \quad (13)$$

$$h_{ab} = \frac{t''_{a1} - t_{r2} + y h_{a1} + C h''_{a1}}{C + y} \quad (14)$$

The value of h_{ab} from Equation (14) serves as an index of whether the coil surface is operating fully wetted, partially dry, or completely dry, according to the following three limits:

1. If $h_{ab} \geq h_{a1}$, the surface is fully wetted.
2. If $h_{a1} > h_{ab} > h_{a2}$, the surface is partially dry.
3. If $h_{ab} \leq h_{a2}$, the surface is completely dry.

Other dry-wet boundary properties are then determined:

$$t_{sb} = t''_{a1} \quad (15)$$

$$t_{ab} = t_{a1} - (h_{a1} - h_{ab})/c_p \quad (16)$$

$$t_{rb} = t_{r2} - y c_p (t_{a1} - t_{ab}) \quad (17)$$

The dry surface area A_d required and capacity q_{td} are calculated by conventional sensible heat transfer relationships, as follows.

The overall thermal resistance R_o comprises three basic elements:

$$R_o = R_{ad} + R_{md} + R_r \quad (18)$$

with

$$R_r = B/f_r \quad (19)$$

The mean difference between air dry bulb temperature and coolant temperature, using symbols from Figure 9, is

$$\Delta t_m = \frac{(t_{a1} - t_{r2}) - (t_{ab} - t_{rb})}{\ln[(t_{a1} - t_{r2})/(t_{ab} - t_{rb})]} \quad (20)$$

The dry surface area required is

$$A_d = \frac{q_{td} R_o}{\Delta t_m} \quad (21)$$

The air-side total heat capacity is

$$q_{td} = w_a c_p (t_{a1} - t_{ab}) \quad (22a)$$

From the coolant side,

$$q_{td} = w_r c_r (t_{r2} - t_{rb}) \quad (22b)$$

The wet surface area A_w and capacity q_{tw} are determined by the following relationships, using terminology in Figure 9.

For a given coil size, design, and arrangement, the fixed value of the coil characteristic C can be determined from the ratio of the three prime thermal resistances for the job conditions:

$$C = \frac{R_{mw} + R_r}{c_p R_{aw}} \quad (23)$$

Knowing the coil characteristic C for point conditions, the interrelations between the airstream enthalpy h_a, the coolant temperature t_r, and the surface temperature t_s and its corresponding enthalpy of saturated air h_s can be determined by use of a surface temperature chart (Figure 8) or by a trial-and-error procedure using Equation (24):

$$C = \frac{t_{sb} - t_{rb}}{h_{ab} - h_{sb}} = \frac{t_{s2} - t_{r1}}{h_{a2} - h_{s2}} \quad (24)$$

The mean effective difference in air enthalpy between airstream and surface from Figure 9 is

$$\Delta h_m = \frac{(h_{ab} - h_{sb}) - (h_{a2} - h_{s2})}{\ln[(h_{ab} - h_{sb})/(h_{a2} - h_{s2})]} \quad (25)$$

Similarly, the mean temperature difference between surface and coolant is

$$\Delta t_{ms} = \frac{(t_{sb} - t_{rb}) - (t_{s2} - t_{r1})}{\ln[(t_{sb} - t_{rb})/(t_{s2} - t_{r1})]} \quad (26)$$

The wet surface area required, calculated from air-side enthalpy difference, is

$$A_w = \frac{q_{tw} R_{aw} c_p}{\Delta h_m} \quad (27a)$$

Calculated from the coolant-side temperature difference,

$$A_w = \frac{q_{tw}(R_{mw} + R_r)}{\Delta t_{ms}} \quad (27b)$$

The air-side total heat capacity is

$$q_{tw} = w_a[h_{a1} - (h_{a2} + h_{fw})] \quad (28a)$$

The enthalpy h_{fw} of condensate removed is

$$h_{fw} = (W_1 - W_2) c_{pw} (t'_{a2} - 32) \quad (28b)$$

where c_{pw} = specific heat of water = 1.0 Btu/lb$_w$·°F

Note that h_{fw} for normal air-conditioning applications is about 0.5% of the airstream enthalpy difference $(h_{a1} - h_{a2})$ and is usually neglected.

The coolant-side heat capacity is

$$q_{tw} = w_r c_r (t_{rb} - t_{r1}) \quad (28c)$$

The total surface area requirement of the coil is

$$A_o = A_d + A_w \quad (29)$$

The total heat capacity for the coil is

$$q_t = q_{td} + q_{tw} \quad (30)$$

The leaving air dry-bulb temperature is found by the method illustrated in Figure 10, which represents part of a psychrometric chart showing the air saturation curve and lines of constant air enthalpy closely corresponding to constant wet-bulb temperature lines.

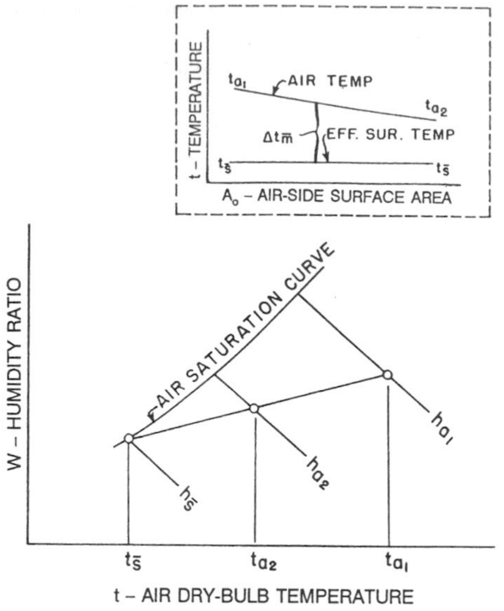

Fig. 10 Leaving Air Dry-Bulb Temperature Determination for Air-Cooling and Dehumidifying Coils

For a given coil and air quantity, a straight line projected through the entering and leaving air conditions intersects the air saturation curve at a point denoted as the effective coil surface temperature $t_{\bar{s}}$. Thus, for fixed entering air conditions t_{a1} and h_{a1} and a given effective surface temperature t_s, the leaving air dry bulb t_{a2} increases but is still located on this straight line if the air quantity is increased or if the coil depth is reduced. Conversely, a decrease in air quantity or an increase in coil depth produces a lower t_{a2} that is still located on the same straight line segment.

An index of the air-side effectiveness is the heat transfer exponent c, which is defined as

$$c = \frac{A_o}{w_a c_p R_{ad}} \tag{31}$$

This exponent c, sometimes called the number of air-side transfer units NTU_a, is also defined as

$$c = \frac{t_{a1} - t_{a2}}{\Delta t_{\bar{m}}} \tag{32}$$

The temperature drop $(t_{a1} - t_{a2})$ of the airstream and the mean temperature difference $\Delta t_{\bar{m}}$ between air and effective surface in Equation (32) are illustrated in the small diagram at the top of Figure 10.

Knowing the exponent c and the entering and leaving enthalpies h_{a1} and h_{a2} for the airstream, the enthalpy of saturated air $h_{\bar{s}}$ corresponding to the effective surface temperature $t_{\bar{s}}$ is calculated as follows:

$$h_{\bar{s}} = h_{a1} - \frac{h_{a1} - h_{a2}}{1 - e^{-c}} \tag{33}$$

After finding the value of $t_{\bar{s}}$ that corresponds to $h_{\bar{s}}$ from the saturated air enthalpy tables, the leaving air dry-bulb temperature can be determined:

$$t_{a2} = t_{\bar{s}} + e^{-c}(t_{a1} - t_{\bar{s}}) \tag{34}$$

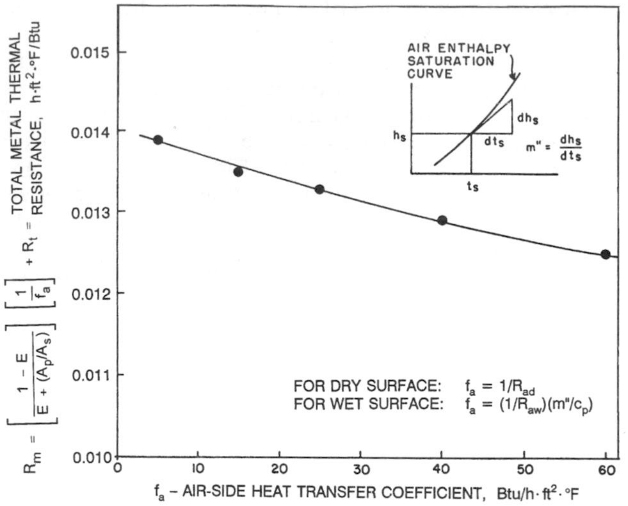

Fig. 11 Typical Total Metal Thermal Resistance of Fin and Tube Assembly

The air-side sensible heat ratio SHR can then be calculated:

$$SHR = \frac{c_p(t_{a1} - t_{a2})}{h_{a1} - h_{a2}} \tag{35}$$

For the thermal performance of a coil to be determined from the foregoing relationships, values of the following three principal resistances to heat flow between air and coolant must be known:

1. The total metal thermal resistances across the fin R_f and tube assembly R_t for both dry R_{md} and wet R_{mw} surface operation
2. The air-film thermal resistances R_{ad} and R_{aw} for dry and wet surfaces, respectively
3. The tube-side coolant film thermal resistance R_r

In ARI *Standard* 410, the metal thermal resistance R_m is calculated based on the physical data, material, and arrangement of the fin and tube elements, together with the fin efficiency E for the specific fin configuration. The metal resistance R_m is variable as a weak function of the effective air-side heat transfer coefficient f_a for a specific coil geometry, as illustrated in Figure 11.

For wetted surface application, Brown (1954), with certain simplifying assumptions, showed that f_a is directly proportional to the rate of change m'' of saturated air enthalpy h_s with the corresponding surface temperature t_s. This slope m'' of the air enthalpy saturation curve is illustrated in the small inset graph at the top of Figure 11.

The abscissa for f_a in the main graph of Figure 11 is an effective value, which, for a dry surface, is the simple thermal resistance reciprocal $1/R_{ad}$. For a wet surface, f_a is the product of the thermal resistance reciprocal $1/R_{aw}$ and the multiplying factor m''/c_p. ARI *Standard* 410 outlines a method for obtaining a mean value of m''/c_p for a given coil and job condition. The total metal resistance R_m in Figure 11 includes the resistance R_t across the tube wall. For most coil designs, R_t is quite small compared to the resistance R_f through the fin metal.

The air-side thermal resistances R_{ad} and R_{aw} for dry and wet surfaces, together with their respective air-side pressure drops $\Delta p_{st}/N_r$ and $\Delta p_{sw}/N_r$, are determined from tests on a representative coil model over the full range in the rated airflow. Typical plots of the experimental data for these four performance variables versus coil air face velocity V_a at 70°F are illustrated in Figure 12.

If water is used as the tube-side coolant, the heat transfer coefficient f_r is calculated from Equation (8) in ARI *Standard* 410. For evaporating refrigerants, many predictive techniques for calculating coefficients of evaporation are listed in Table 2 in Chapter 4 of the

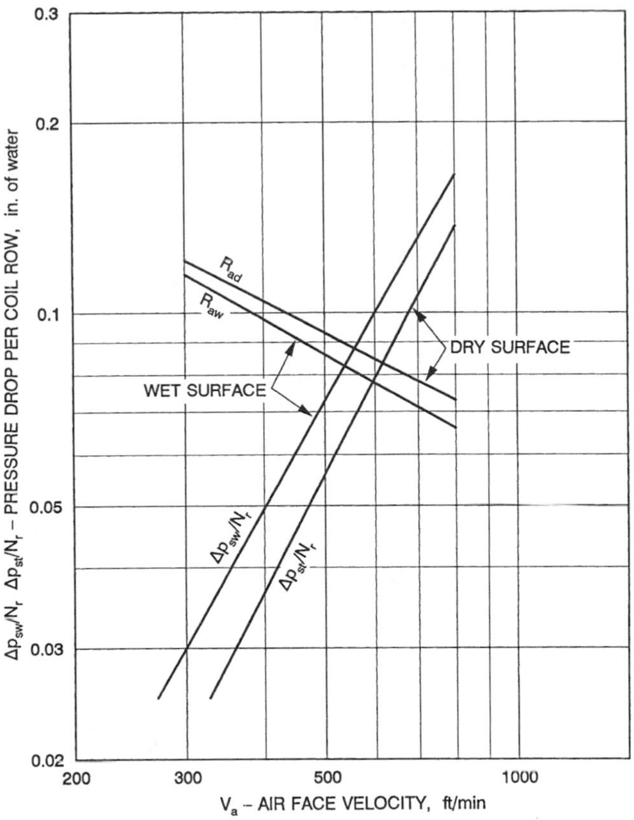

Fig. 12 Typical Air-Side Application Rating Data Determined Experimentally for Cooling and Dehumidifying Water Coils

1997 *ASHRAE Handbook—Fundamentals*. The most verified predictive technique is the Shah correlation (Shah 1976, 1982). A series of tests is specified in ARI *Standard* 410 for obtaining heat transfer data for direct-expansion refrigerants inside tubes of a given diameter.

ASHRAE *Standard* 33 specifies the laboratory apparatus and instrumentation, including the procedure and operating criteria for conducting tests on representative coil prototypes to obtain basic performance data. Procedures are available in ARI *Standard* 410 for reducing these test data to the performance parameters necessary to rate a line or lines of various air coils. This information is published by and/or available on computer disk from various coil manufacturers for use in selecting ARI *Standard* 410 certified coils.

The following example illustrates a method for selecting coil size, row depth, and performance data to satisfy specified job requirements. The application is for a typical cooling and dehumidifying coil selection under conditions in which a part of the coil surface on the entering air side operates dry, with the remaining surface wetted with condensing moisture. Figure 9 shows the thermal diagram, dry-wet boundary conditions, and terminology used in the problem solution.

Example 3. Standard air flowing at a mass rate equivalent of 6700 cfm enters a coil at 80°F dry-bulb temperature t_{a1} and 67°F wet-bulb temperature t'_{a1}. The air is to be cooled to 56°F leaving wet-bulb temperature t'_{a2} using 40 gpm of chilled water supplied to the coil at an entering temperature t_{r1} of 44°F, in thermal counterflow arrangement. Assume a standard coil air face velocity of V_a = 558 fpm and a clean, nonfouled, finned tube heat transfer surface in the coil core, for which the following physical and performance data (such as illustrated in Figures 11 and 12) can be predetermined:

B = surface ratio = 25.9
c_p = 0.243 Btu/lb·°F
F_s = (external surface area)/(face area)(row deep) = 32.4
f_r = 750 Btu/h·ft²·°F
$\Delta p_{st}/N_r$ = 0.165 in. of water/number of coil rows—dry surface
$\Delta p_{sw}/N_r$ = 0.27 in. of water/number of coil rows—wet surface
R_{ad} = 0.073°F·ft²·h/Btu
R_{aw} = 0.066°F·ft²·h/Btu
R_{md} = 0.021°F·ft²·h/Btu
R_{mw} = 0.0195°F·ft²·h/Btu
ρ_a = 0.075 lb/ft³
ρ_w = 8.34 lb/gal

Referring to Figure 9 for the symbols and typical diagram for applications in which only a part of the coil surface operates wet, determine (1) coil face area A_a, (2) total refrigeration load q_t, (3) leaving coolant temperature t_{r2}, (4) dry-wet boundary conditions, (5) heat transfer surface area required for dry A_d and wet A_w sections of the coil core, (6) leaving air dry-bulb temperature t_{a2}, (7) total number N_{ri} of installed coil rows, and (8) dry Δp_{st} and wet Δp_{sw} coil air friction.

Solution: The psychrometric properties and enthalpies for dry and moist air are based on Figure 1 (ASHRAE Psychrometric Chart No. 1) and Tables 2 and 3 in Chapter 6 of the 1997 *ASHRAE Handbook—Fundamentals* as follows:

h_{a1} = 31.52 Btu/lb$_a$	h''_{a1} = 26.67 Btu/lb$_a$
W_1 = 0.0112 lb$_w$/lb$_a$	*h_{a2} = 23.84 Btu/lb$_a$
t''_{a1} = 60.3°F	*W_2 = 0.0095 lb$_w$/lb$_a$

*As an approximation, assume leaving air is saturated (i.e., $t_{a2} = t'_{a2}$).

Calculate coil face area required:

$$A_a = 6700/558 = 12\ \text{ft}^2$$

From Equation (28b), find condensate heat rejection:

$$h_{fw} = (0.0112 - 0.0095)(1)(56 - 32) = 0.04\ \text{Btu/lb}_a$$

Compute the total refrigeration load from the following equation:

$$q_t = 60\rho_a w_a[h_{a1} - (h_{a2} + h_{fw})]$$
$$= 60 \times 0.075 \times 6700[31.52 - (23.84 + 0.04)] = 230{,}000\ \text{Btu/h}$$

From Equation (3), calculate coolant temperature leaving coil:

$$t_{r2} = t_{r1} + q_t/w_r c_r$$
$$= 44 + 230{,}000/(40 \times 60 \times 8.34 \times 1.0) = 55.5\,°\text{F}$$

From Equation (19), determine coolant film thermal resistance:

$$R_r = 25.9/750 = 0.0346\,°\text{F}\cdot\text{ft}^2\cdot\text{h/Btu}$$

Calculate the wet coil characteristic from Equation (23):

$$C = \frac{0.0195 + 0.0346}{0.243 \times 0.066} = 3.37\ \text{lb}_a\cdot°\text{F/Btu}$$

Calculate from Equation (13):

$$y = \frac{55.5 - 44}{31.52 - 23.84} = 1.50\ \text{lb}_a\cdot°\text{F/Btu}$$

The dry-wet boundary conditions are determined as follows: From Equation (14), the boundary airstream enthalpy is

$$h_{ab} = \frac{60.3 - 55.5 + 1.50 \times 31.52 + 3.37 \times 26.67}{3.37 + 1.50}$$
$$= 29.15\ \text{Btu/lb}_a$$

According to limit (2) under Equation (14), a part of the coil surface on the entering air side will be operating dry, because $h_{a1} > 29.15 > h_{a2}$ (see Figure 8).

From Equation (16), the boundary airstream dry-bulb temperature is

$$t_{ab} = 80 - (31.52 - 29.15)/0.243 = 70.25\,°\text{F}$$

The boundary surface conditions are

$$t_{sb} = t''_{a1} = 60.3°\text{F} \quad \text{and} \quad h_{sb} = h''_{a1} = 26.67 \text{ Btu/lb}_a$$

From Equation (17), the boundary coolant temperature is

$$t_{rb} = 55.5 - 1.50 \times 0.243(80 - 70.25) = 51.9°\text{F}$$

The cooling load for the dry surface part of the coil is now calculated from Equation (22b):

$$q_{td} = 40 \times 60 \times 8.34 \times 1.0 \times (55.5 - 51.9) = 72{,}000 \text{ Btu/h}$$

From Equation (18), the overall thermal resistance for the dry surface section is

$$R_o = 0.073 + 0.021 + 0.0346 = 0.129\,°\text{F} \cdot \text{ft}^2 \cdot \text{h/Btu}$$

From Equation (20), the mean temperature difference between air dry bulb and coolant for the dry surface section is

$$\Delta t_m = \frac{(80 - 55.5) - (70.25 - 51.92)}{\ln[(80 - 55.5)/(70.25 - 51.92)]} = 21.3°\text{F}$$

The dry surface area required is calculated from Equation (21):

$$A_d = 72{,}000 \times 0.129/21.3 = 436 \text{ ft}^2$$

From Equation (30), the cooling load for the wet surface section of the coil is

$$q_{tw} = 230{,}000 - 72{,}000 = 158{,}000 \text{ Btu/h}$$

Knowing C, h_{a2}, and t_{r1}, the surface condition at the leaving air side of the coil is calculated by trial and error using Equation (24):

$$C = 3.37 = (t_{s2} - 44)/(23.84 - h_{s2})$$

The numerical values for t_{s2} and h_{s2} are then determined directly by use of a surface temperature chart (as shown in Figure 8 or in Figure 9 of ARI *Standard* 410) and saturated air enthalpies from Table 2 in Chapter 6 of the 1997 *ASHRAE Handbook—Fundamentals*:

$$t_{s2} = 52.03°\text{F} \quad \text{and} \quad h_{s2} = 21.46 \text{ Btu/lb}_a$$

From Equation (25), the mean effective difference in air enthalpy between airstream and surface is

$$\Delta h_m = \frac{(29.15 - 26.67) - (23.84 - 21.46)}{\ln[(29.15 - 26.67)/(23.84 - 21.46)]} = 2.43 \text{ Btu/lb}_a$$

From Equation (27a), the wet surface area required is

$$A_w = 158{,}000 \times 0.066 \times 0.243/2.43 = 1040 \text{ ft}^2$$

From Equation (29), the net total surface area requirement for the coil is then

$$A_o = 436 + 1040 = 1476 \text{ ft}^2 \text{ external}$$

From Equation (31), the net air-side heat transfer exponent is

$$c = 1476/(60 \times 0.075 \times 6700 \times 0.243 \times 0.073) = 2.76$$

From Equation (33), the enthalpy of saturated air corresponding to the effective surface temperature is

$$h_{\bar{s}} = 31.52 - (31.52 - 23.84)/(1 - e^{-2.76}) = 23.32 \text{ Btu/lb}_a$$

The effective surface temperature that corresponds to $h_{\bar{s}}$ is then obtained from Table 2 in Chapter 6 of the 1997 *ASHRAE Handbook—Fundamentals* as $t_{\bar{s}} = 55.15°\text{F}$.

The leaving air dry-bulb temperature is calculated from Equation (34):

$$t_{a2} = 55.15 + e^{-2.76}(80 - 55.15) = 56.7°\text{F}$$

The air-side sensible heat ratio is then determined from Equation (35):

$$\text{SHR} = \frac{0.243(80 - 56.7)}{31.52 - 23.84} = 0.737$$

From Equation (1b), the calculated coil row depth N_{rc} to match job requirements is

$$N_{rc} = A_o/A_a F_s = 1476/(12 \times 32.4) = 3.8 \text{ rows deep}$$

In most coil selection problems of the type illustrated, the initial calculated row depth to satisfy job requirements is usually a noninteger value. In many cases, there is sufficient flexibility in fluid flow rates and operating temperature levels to recalculate the required row depth of a given coil size to match an available integer row depth more closely. For example, if the initial calculated row depth were $N_{rc} = 3.5$ and coils of three or four rows deep were commercially available, operating conditions and/or fluid flow rates or velocities could possibly be changed to recalculate a coil depth close to either three or four rows. Although core tube circuitry has limited possibilities on odd row coils, alternate coil selections for the same job are often made desirable by trading off coil face size and row depth.

Most manufacturers have computer programs to run the iterations needed to predict operating values for the specific row depth chosen.

For this example, assume that the initial coil selection requiring 3.8 rows deep is sufficiently refined that no recalculation is necessary. Then select the next highest integral row depth to provide some safety factor (about 5% in this case) in surface area selection. Thus, the installed row depth N_{ri} is

$$N_{ri} = 4 \text{ rows deep}$$

The completely wetted and completely dry air-side frictions are, respectively,

$$\Delta p_{sw} = (\Delta p_{sw}/N_{ri})N_{ri} = 0.27 \times 4 = 1.08 \text{ in. of water}$$

and

$$\Delta p_{st} = (\Delta p_{st}/N_{ri})N_{ri} = 0.165 \times 4 = 0.66 \text{ in. of water}$$

The amount of heat transfer surface area installed is

$$A_{oi} = A_a F_s N_{ri} = 12 \times 32.4 \times 4 = 1555 \text{ ft}^2 \text{ external}$$

In summary,

A_a = 12 ft² coil face area
N_{ri} = 4 rows installed coil depth
A_{oi} = 1555 ft² installed heat transfer surface area
A_o = 1476 ft² required heat transfer surface area
q_t = 230,000 Btu/h total refrigeration load
t_{r2} = 55.5°F leaving coolant temperature
t_{a2} = 56.7°F leaving air dry-bulb temperature
SHR = air sensible heat ratio = 0.737
Δp_{sw} = 1.08 in. of water wet coil surface air friction
Δp_{st} = 0.66 in. of water dry coil surface air friction

DETERMINING REFRIGERATION LOAD

The following calculation of the refrigeration load shows a division of the true sensible and latent heat loss of the air, which is accurate within the limitations of the data. This division will not correspond to load determination obtained from approximate factors or constants.

The total refrigeration load q_t of a cooling and dehumidifying coil (or air washer) per unit mass of dry air is indicated in Figure 13 and consists of the following components:

1. The sensible heat q_s removed from the dry air and moisture in cooling from entering temperature t_1 to leaving temperature t_2

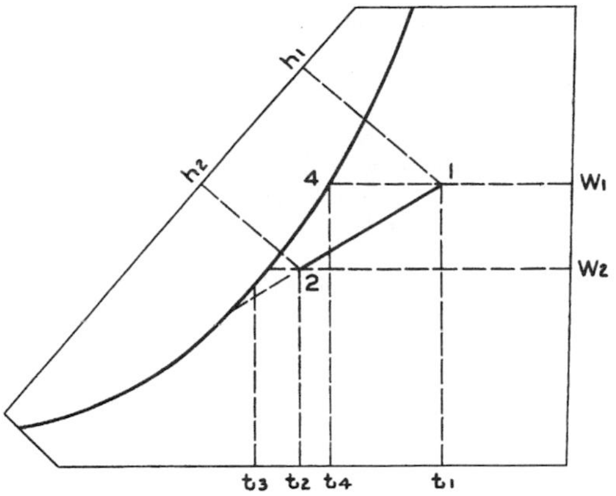

Fig. 13 Psychrometric Performance of Cooling and Dehumidifying Coil

2. The latent heat q_e removed to condense the moisture at the dew-point temperature t_4 of the entering air
3. The heat q_w removed to further cool the condensate from its dew point t_4 to its leaving condensate temperature t_3

Items 1, 2, and 3 are related by the following equation:

$$q_t = q_s + q_e + q_w \tag{36}$$

If only the total heat value is desired, it may be computed by

$$q_t = (h_1 - h_2) - (W_1 - W_2)h_{w3} \tag{37}$$

where

h_1 and h_2 = enthalpy of air at points 1 and 2, respectively
W_1 and W_2 = humidity ratio at points 1 and 2, respectively
h_{w3} = enthalpy of saturated liquid at final temperature t_3

If a breakdown into latent and sensible heat components is desired, the following relations may be used.

The latent heat may be found from

$$q_e = (W_1 - W_2)h_{fg4} \tag{38}$$

where

h_{fg4} = enthalpy representing latent heat of water vapor at the condensing temperature t_4

The sensible heat may be shown to be

$$q_s + q_w = (h_1 - h_2) - (W_1 - W_2)h_{g4} + (W_1 - W_2)(h_{w4} - h_{w3}) \tag{39a}$$

or

$$q_s + q_w = (h_1 - h_2) - (W_1 - W_2)(h_{fg4} + h_{w3}) \tag{39b}$$

where

$h_{g4} = h_{fg4} + h_{w4}$ = enthalpy of saturated water vapor at condensing temperature t_4
h_{w4} = enthalpy of saturated liquid at condensing temperature t_4

The last term in Equation (39a) is the heat of subcooling the condensate from the condensing temperature t_4 to its final temperature t_3. Then,

$$q_w = (W_1 - W_2)(h_{w4} - h_{w3}) \tag{40}$$

The final condensate temperature t_3 leaving the system is subject to substantial variations, depending on the method of coil installation, as affected by coil face orientation, airflow direction, and air duct insulation. In practice, t_3 is frequently the same as the leaving wet-bulb temperature. Within the normal air-conditioning range, precise values of t_3 are not necessary because the heat q_t of the condensate removed from the air usually represents about 0.5 to 1.5% of the total refrigeration cooling load.

Example 4. Air enters a coil at 90°F dry-bulb, 75°F wet-bulb; it leaves at 61°F dry-bulb, 58°F wet-bulb; leaving water temperature is assumed to be 54°F, which is between the leaving air dew point and coil surface temperature. Find the total, latent, and sensible cooling loads on the coil with the air at standard barometric pressure.

Solution: Using Figure 1 or the indicated equations from Chapter 6 of the 1997 *ASHRAE Handbook—Fundamentals*,

h_1 = 38.37 Btu/lb$_a$	(32)
h_2 = 25.06 Btu/lb$_a$	(32)
W_1 = 0.01523 lb$_w$/lb$_a$	(35)
W_2 = 0.00958 lb$_w$/lb$_a$	(35)
t_4 = 69.04°F dew point of entering air	(37)
h_{w4} = 37.04 Btu/lb	(34)
h_{w3} = 22.00 Btu/lb	(34)
h_{g4} = 1091.66 Btu/lb	(31)
h_{fg4} = 1054.61 Btu/lb	(31)

From Equation (37), the total heat is

$$q_t = (38.37 - 25.06) - (0.01523 - 0.00958)22.00 = 13.19 \text{ Btu/lb}_a$$

From Equation (38), the latent heat is

$$q_e = (0.01523 - 0.00958)1054.61 = 5.96 \text{ Btu/lb}_a$$

The sensible heat is therefore

$$q_s + q_w = q_t - q_e = 13.19 - 5.96 = 7.23 \text{ Btu/lb}_a$$

The sensible heat may be computed from Equation (39a) as

$$q_s + q_w = (38.37 - 25.06) - (0.01523 - 0.00958)1091.66 + (0.01523 - 0.00958)(37.04 - 22.00) = 7.22 \text{ Btu/lb}_a$$

The same value is found using Equation (39b). The subcooling of the condensate as a part of the sensible heat is indicated by the last term of the equation, 0.08 Btu/lb$_a$.

MAINTENANCE

If the coil is to deliver its full cooling capacity, both its internal and its external surfaces must be clean. The tubes generally stay clean in pressurized water or brine systems. Should large amounts of scale form when untreated water is used as coolant, chemical or mechanical cleaning of internal surfaces at frequent intervals is necessary. Water coils should be completely drained if freezing conditions are possible. When coils use evaporating refrigerants, oil accumulation is possible, and occasional checking and oil drainage is desirable.

While outer tube surfaces can be cleaned in a number of ways, they are often washed with low-pressure water and mild detergent. The surfaces can also be brushed and cleaned with a vacuum cleaner. In cases of marked neglect—especially in restaurants, where grease and dirt have accumulated—it is sometimes necessary to remove the coils and wash off the accumulation with steam, compressed air and water, or hot water. The best practice, however, is to inspect and service the filters frequently.

SYMBOLS

A_a = coil face or frontal area, ft^2
A_d = dry external surface area, ft^2
A_o = total external surface area, ft^2
A_p = exposed external prime surface area, ft^2

A_s = external secondary surface area, ft^2
A_w = wet external surface area, ft^2
B = ratio of external to internal surface area, dimensionless
C = coil characteristic as defined in Equations (12) and (23), $lb_a \cdot °F/Btu$
c = heat transfer exponent, or NTU_a, as defined in Equations (31) and (32), dimensionless
c_o = heat transfer exponent, as defined in Equation (7d), dimensionless
c_p = specific heat of humid air = 0.243 $Btu/lb_a \cdot °F$ for cooling coils
c_{pw} = specific heat of water = 1.0 $Btu/lb_w \cdot °F$
c_r = specific heat of nonvolatile coolant, $Btu/lb_a \cdot °F$
D_i = tube inside diameter, in.
D_o = tube outside diameter, in.
E = fin efficiency, dimensionless
E_a = air-side effectiveness defined in Equation (7b), dimensionless
F_s = coil core surface area parameter = (external surface area)/(face area) (no. of rows deep)
f = convection heat transfer coefficient, $Btu/h \cdot ft^2 \cdot °F$
h = air enthalpy (actual in airstream or saturation value at surface temperature), Btu/lb_a
Δh_m = mean effective difference of air enthalpy, as defined in Equation (25), Btu/lb_a
k = thermal conductivity of tube material, $Btu/h \cdot ft \cdot °F$
M = ratio of nonvolatile coolant-to-air temperature changes for sensible heat cooling coils, as defined in Equation (7e), dimensionless
m'' = rate of change of air enthalpy at saturation with air temperature, $Btu/lb \cdot °F$
N_r = number of coil rows deep in airflow direction, dimensionless
η = fin effectiveness, as defined in Equation (6), dimensionless
Δp_{st} = isothermal dry surface air-side pressure drop at standard conditions (70°F, 29.92 in. Hg), in. of water
Δp_{sw} = wet surface air-side pressure drop at standard conditions (70°F, 29.92 in. Hg), in. of water
q = heat transfer capacity, Btu/h
q_e = latent heat removed from entering air to condense moisture, Btu/lb_a
q_s = sensible heat removed from entering air, Btu/lb_a
q_t = total refrigeration load of cooling and dehumidifying coil, Btu/lb_a
q_w = sensible heat removed from condensate to cool it to leaving temperature, Btu/lb_a
ρ_a = air density = 0.075 lb/ft^3 at 70°F at sea level
R = thermal resistance, referred to external area A_o, $h \cdot °F \cdot ft^2/Btu$
SHR = ratio of air sensible heat to air total heat, dimensionless
t = temperature, °F
Δt_m = mean effective temperature difference, air dry bulb to coolant temperature, °F
Δt_{ms} = mean effective temperature difference, surface-to-coolant, °F
$\Delta t_{\bar{m}}$ = mean effective temperature difference, air dry bulb to effective surface temperature $t_{\bar{s}}$, °F
U_o = overall sensible heat transfer coefficient, $Btu/h \cdot ft^2 \cdot °F$
V_a = coil air face velocity at 70°F, fpm
W = air humidity ratio, pounds of water per pound of air
w = mass flow rate, lb/h
y = ratio of nonvolatile coolant temperature rise to airstream enthalpy drop, as defined in Equation (13), $lb_a \cdot °F/Btu$

Superscripts

$'$ = wet bulb
$''$ = dew point

Subscripts

1 = condition entering coil
2 = condition leaving coil
a = airstream
b = dry-wet surface boundary
d = dry surface
e = latent
f = fin (with R); saturated liquid water (with h)
g = saturated water vapor
i = installed, selected (with A_o, N_r)
m = metal (with R) and mean (with other symbols)
o = overall (except for A)
r = coolant
s = surface (with t) and saturated (with h)
$\bar{s}$ = effective surface
t = tube (with R) and total (with q)
ab = air, dry-wet boundary
ad = dry air
aw = wet air
md = dry metal
mw = wet metal
rb = coolant dry-wet boundary
sb = surface dry-wet boundary
td = total heat capacity, dry surface
tw = total heat capacity, wet surface
w = water (with ρ), condensate (with h and subscript number), and wet surface (with other symbols)

REFERENCES

Anderson, S.W. 1970. Air-cooling and dehumidifying coil performance based on ARI *Industrial Standard* 410-64. In *Heat and mass transfer to extended surfaces*, ASHRAE Symposium CH-69-3, pp. 22-28.

ARI. 1991. Forced-circulation air-cooling and air-heating coils. ANSI/ARI *Standard* 410-91. Air-Conditioning and Refrigeration Institute, Arlington, VA.

ASHRAE. 1978. Methods of testing forced circulation air-cooling and air heating coils. *Standard* 33-1978.

ASME. 1997. *Code on nuclear air and gas treatment*. ANSI/ASME *Standard* AG-1-97. American Society of Mechanical Engineers, New York.

Brown, G. 1954. Theory of moist air heat exchangers. Royal Institute of Technology *Transactions* No. 77, Stockholm, Sweden, 12.

Kusuda, T. 1970. Effectiveness method for predicting the performance of finned tube coils. In *Heat and mass transfer to extended surfaces*, ASHRAE Symposium CH-69-3, pp. 5-14.

McElgin, J. and D.C. Wiley. 1940. Calculation of coil surface areas for air cooling and dehumidification. *Heating, Piping and Air Conditioning* (March):195.

McQuiston, F.C. 1981. Finned tube heat exchangers: State of the art for air side. *ASHRAE Transactions* 87(1):1077-85.

Mirth, D.R., S. Ramadhyani, and D.C. Hittle. 1993. Thermal performance of chilled water cooling coils operating at low water velocities. *ASHRAE Transactions* 99(1):43-53.

Mueller, A.C. 1973 (Revised 1986). Heat exchangers. *Handbook of heat transfer,* W.M. Rohsenow and J.P. Hartnett, eds. McGraw-Hill Book Company, New York.

Pedersen, C.O., D.E. Fisher, J.D. Spitler, and R.J. Liesen. 1998. *Cooling and heating load calculation principles*. ASHRAE.

Shah, M.M. 1976. A new correlation for heat transfer during boiling flow through pipes. *ASHRAE Transactions* 82(2):66-75.

Shah, M.M. 1978. Heat transfer, pressure drop, visual observation, test data for ammonia evaporating inside pipes. *ASHRAE Transactions* 84(2): 38-59.

Shah, M.M. 1982. CHART correlation for saturated boiling heat transfer: Equations and further study. *ASHRAE Transactions* 88(1):185-96.

Webb, R.L. 1980. Air-side heat transfer in finned tube heat exchangers. *Heat Transfer Engineering* 1(3):33.

CHAPTER 22

DESICCANT DEHUMIDIFICATION AND PRESSURE DRYING EQUIPMENT

DEHUMIDIFICATION is the removal of water vapor from air, gases, or other fluids. There is no limitation of pressure in this definition, and sorption dehumidification equipment has been designed and operated successfully for system pressures ranging from subatmospheric to as high as 6000 psi. The term **dehumidification** is normally limited to equipment that operates at essentially atmospheric pressures and is built to standards similar to other types of air-handling equipment. For drying gases under pressure, or liquids, the term **dryer** or **dehydrator** is normally used.

This chapter mainly covers equipment and systems that dehumidify air rather than those that dry other gases or liquids. Both liquid and solid **desiccant** materials are used in dehumidification equipment. They either adsorb the water on the surface of the desiccant (adsorption) or chemically combine with water (absorption).

Nonregenerative equipment uses hygroscopic salts such as calcium chloride, urea, or sodium chloride. **Regenerative** systems usually use a form of silica or alumina gel, activated alumina, molecular sieves, lithium chloride salt, lithium chloride solution, or glycol solution. In regenerative equipment, the mechanism for water removal is reversible. The choice of desiccant depends on the requirements of the installation, the design of the equipment, and chemical compatibility with the gas to be treated or impurities in the gas. Chapter 21 of the 1997 *ASHRAE Handbook—Fundamentals* has more information on desiccant materials and how they operate.

Some commercial applications of dehumidification include

- Lowering relative humidity to facilitate manufacturing and handling of hygroscopic materials
- Lowering the dew point to prevent condensation on products manufactured in low-temperature processes
- Providing protective atmospheres for the heat treatment of metals
- Controlling humidity in warehouses and caves used for storage
- Preserving ships, aircraft, and industrial equipment that would otherwise deteriorate
- Maintaining a dry atmosphere in a closed space or container, such as the cargo hold of a ship or numerous static applications
- Eliminating condensation and subsequent corrosion
- Drying air to speed the drying of heat-sensitive products, such as candy, seeds, and photographic film
- Drying natural gas
- Drying gases that are to be liquefied
- Drying instrument air and plant air
- Drying process and industrial gases
- Dehydration of liquids
- Frost-free cooling for low-temperature process areas such as brewery fermenting, aging, filtering, and storage cellars; blast freezers; and refrigerated warehouses
- Frost-free dehumidification for processes that require air at a subfreezing dew point humidity

The preparation of this chapter is assigned to TC 3.5, Desiccant and Sorption Technology.

This chapter covers (1) the types of dehumidification equipment for liquid and solid desiccants, including high-pressure equipment; (2) performance curves; (3) variables of operation; and (4) some typical applications. The use of desiccants for the drying of refrigerants is addressed in Chapter 6 of the 1998 *ASHRAE Handbook—Refrigeration*.

METHODS OF DEHUMIDIFICATION

Air may be dehumidified by (1) cooling it or increasing its pressure, which reduces its capacity to hold moisture, or (2) removing moisture by attracting the water vapor with a liquid or solid desiccant. Frequently, systems employ a combination of these methods to maximize operating efficiency and minimize installed cost.

Figure 1 illustrates three methods by which dehumidification with desiccant materials or desiccant equipment may be accomplished. Air in the condition at Point A is dehumidified and cooled to Point B. In a liquid desiccant unit, the air is simultaneously cooled and dehumidified directly from Point A to Point B. In a solid desiccant unit, this process can be completed by precooling and dehumidifying from Point A to Point C, then desiccating from Point C to Point E and, finally, cooling to Point B. It could also be accomplished with solid desiccant equipment by dehumidifying from Point A to Point D and then by cooling from Point D to Point B.

Compression

Compressing air reduces its capacity to hold moisture. The resulting condensation reduces the moisture content of the air in absolute terms, but produces a saturated condition—100% relative humidity at elevated pressure. In applications at atmospheric pressure, this method is too expensive. However, in pressure systems such as instrument air, dehumidification by compression is worthwhile. Other dehumidification equipment, such as coolers or desiccant dehumidifiers, often follows the compressor to avoid

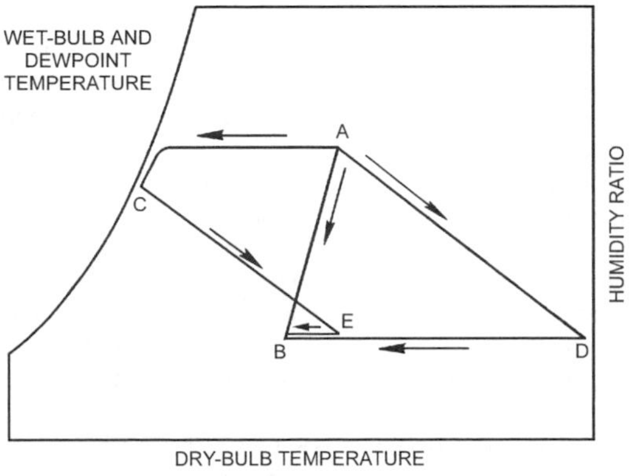

Fig. 1 Methods of Dehumidification

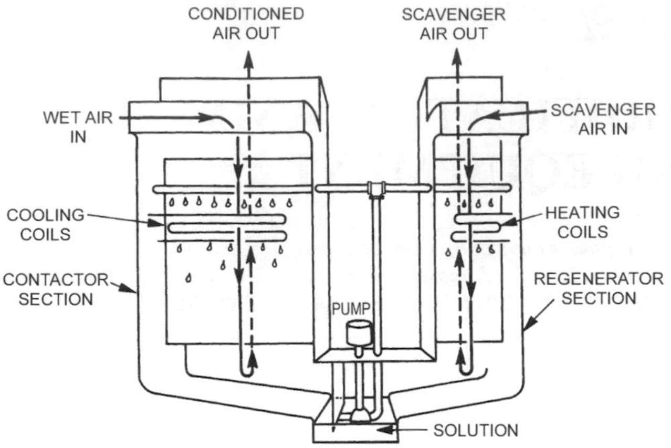

Fig. 2 Flow Diagram for Liquid-Absorbent Dehumidifier

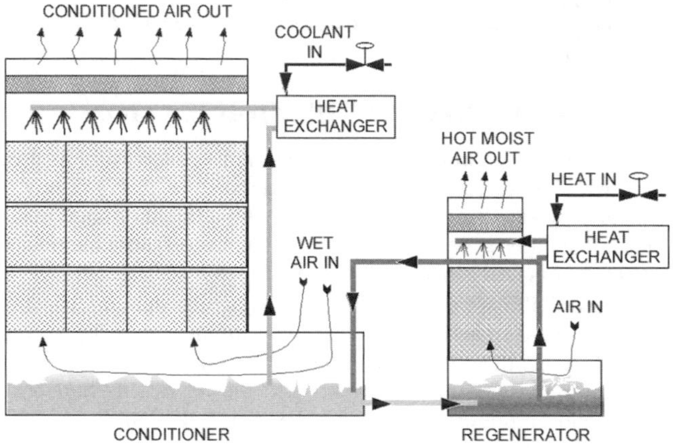

Fig. 3 Flow Diagram for Liquid-Absorbent Unit with Extended Surface Air Contact Media

problems associated with high relative humidity in compressed air lines.

Cooling

Refrigeration of the air below its dew point is the most common method of dehumidification. This method is advantageous when the gas is comparatively warm, has a high moisture content, and when the outlet dew point desired is above 40°F. Frequently, refrigeration is used in combination with desiccant dehumidifiers to obtain an extremely low dew point at minimum cost.

Liquid Desiccant

Liquid desiccant conditioners (absorbers) contact the air with a liquid desiccant, such as lithium chloride or glycol solution (Figures 2 and 3). The water vapor pressure of the solution is a function of its temperature and its concentration. Higher concentrations and lower temperatures result in lower water vapor pressures.

A simple way to show this relationship is to graph the humidity ratio of air in equilibrium with a liquid desiccant as a function of its concentration and temperature. Figure 4 presents this relationship for lithium chloride-water solutions in equilibrium with air at 14.7 psi. The graph has the same general shape as a psychrometric chart, with the relative humidity lines replaced by desiccant concentration lines.

Liquid desiccant conditioners typically have a high contact efficiency, so the air leaves the conditioner at a temperature and humidity ratio very close to the entering temperature and equilibrium

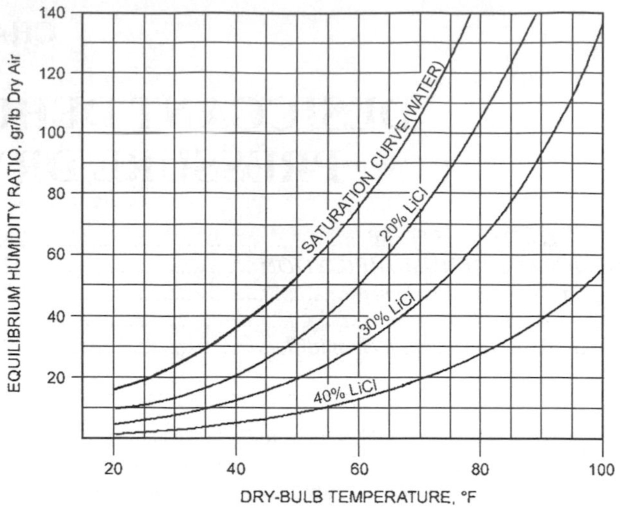

Fig. 4 Lithium Chloride Equilibrium

humidity ratio of the desiccant. When the conditioner is dehumidifying, moisture absorbed from the conditioned airstream dilutes the desiccant solution. The diluted solution is reconcentrated in the regenerator, where it is heated to elevate its water vapor pressure and equilibrium humidity ratio. A second airstream, usually outside air, contacts the heated solution in the regenerator; water evaporates from the desiccant solution into the air, and the solution is reconcentrated. Desiccant solution is continuously recirculated between the conditioner and regenerator to complete the cycle.

Liquid desiccants are typically a very effective antifreeze. As a result, liquid desiccant conditioners can continuously deliver air at subfreezing temperatures without frosting or freezing problems. Lithium chloride-water solution, for example, has a eutectic point of –90°F; liquid desiccant conditioners using this solution can cool air to temperatures as low as –65°F.

Solid Sorption

Solid sorption passes the air through a bed of granular desiccant or through a structured packing impregnated with desiccant. Humid air passes through the desiccant, which in its active state has a vapor pressure below that of the humid air. This vapor pressure differential drives the water vapor from the air onto the desiccant. After becoming loaded with moisture, the desiccant is reactivated (dried out) by heating, which raises the vapor pressure of the material above that of the surrounding air. With the vapor pressure differential reversed, water vapor moves from the desiccant to a second airstream called the reactivation air, which carries the moisture away from the equipment.

DESICCANT DEHUMIDIFICATION

Both types of desiccants, liquid and solid, may be used in equipment designed for the drying of air and gases at atmospheric or elevated pressures. Regardless of pressure levels, basic principles remain the same, and only the desiccant towers or chambers require special design consideration.

Desiccant capacity and actual dew-point performance depend on the specific equipment used, the characteristics of the various desiccants, initial temperature and moisture content of the gas to be dried, reactivation methods, etc. Factory-assembled units are available up to a capacity of about 80,000 cfm. Greater capacities can be obtained with field-erected units.

LIQUID DESICCANT EQUIPMENT

Liquid desiccant dehumidifiers are shown in Figures 2 and 3. In the configuration shown in Figure 2, the liquid desiccant is distributed onto a **cooling coil**, which acts as both a contact surface and a means of removing the heat released when the desiccant absorbs moisture from the air. In the configuration shown in Figure 3, the liquid desiccant is distributed onto an extended heat and mass transfer surface—a **packing** material similar to that used in cooling towers and chemical reactors. The packing provides a great deal of surface for the air to contact the liquid desiccant, and the heat of absorption is removed from the liquid by a heat exchanger located outside the airstream. The air can be passed through the contact surface vertically or horizontally to suit the best arrangement of air system equipment.

Depending on the air and desiccant solution inlet conditions, the air can be simultaneously cooled and dehumidified, heated and dehumidified, heated and humidified, or cooled and humidified. When the enthalpy of the air is to be increased in the conditioner unit, heat must be added either by preheating the air before it enters the conditioner or by heating the desiccant solution with a second heat exchanger. When the air is to be humidified, makeup water is automatically added to the desiccant solution to keep it at the desired concentration.

Moisture is absorbed from the air or desorbed into the air because of the difference in water vapor pressure between the air and the desiccant solution. For a given solution temperature, a higher solution concentration results in a lower water vapor pressure. For a given solution concentration, a lower solution temperature results in a lower water vapor pressure. By controlling the temperature and concentration of the desiccant solution, the conditioner unit can deliver air at a precisely controlled temperature and humidity regardless of inlet air conditions. The unit dehumidifies the air during humid weather and humidifies it during dry weather. Thus, liquid desiccant conditioners can provide accurate humidity control without the need for face-and-bypass dampers or after-humidifiers.

Heat Removal

When a liquid desiccant absorbs moisture, heat is generated. This **heat of absorption** consists of the latent heat of condensation of water vapor at the desiccant temperature and the heat of solution (heat of mixing) of the condensed water and the desiccant. The heat of mixing is a function of the equilibrium relative humidity of the desiccant: a lower equilibrium relative humidity produces a greater heat of mixing.

The total heat that must be absorbed by the desiccant solution consists of (1) the heat of absorption, (2) the sensible heat associated with reducing the dry-bulb temperature of the air, and (3) the residual heat carried to the conditioner by the warm, concentrated desiccant returning from the regenerator unit. This total heat is removed by cooling the desiccant solution in the conditioner heat exchanger (Figure 3). Any coolant can be used, including cooling tower water, ground water, seawater, chilled water or brine, and direct-expansion or flooded refrigerants.

The regenerator residual heat, generally called **regenerator heat dumpback**, can be substantially reduced by using a liquid-to-liquid heat exchanger to precool the warm, concentrated desiccant transferred to the conditioner using the cool, dilute desiccant transferred from the conditioner to the regenerator. This also improves the thermal efficiency of the regenerator, typically reducing the heat input by 10 to 15%.

Regeneration

When the conditioner is dehumidifying, water is automatically removed from the liquid desiccant to maintain the desiccant at the proper concentration. Removal takes place in a separate **regenerator**. A small sidestream of the desiccant solution, typically 8% or less of the flow to the conditioner packing, is transferred to the regenerator unit. In the regenerator, a separate pump continuously circulates the desiccant solution through a heat exchanger and distributes it over the packed bed contactor surface. The heat exchanger heats the desiccant solution so that its water vapor pressure is substantially higher than that of the outside air. This heating is accomplished with low-pressure steam or hot water. Outside air is passed through the packing, and water evaporates into it from the desiccant solution, concentrating the solution. The hot, moist air from the regenerator is discharged to the outdoors. A sidestream of the concentrated solution is transferred to the conditioner to replace the sidestream of weak solution transferred from the conditioner and complete the cycle.

The water removal capacity of the regenerator is controlled to match the moisture load being handled by the conditioner. This load matching is accomplished by regulating the heat flow to the regenerator heat exchanger to maintain a constant desiccant solution concentration. This is most commonly done by maintaining a constant solution level in the system with a level controller, but specific gravity or boiling point controllers are used under some circumstances. The regenerator heat input is regulated to match the instantaneous water removal requirements, so no heat input is required if there is no moisture load on the conditioner. When the conditioner is being used to humidify the air, the regenerator fan and desiccant solution pump are typically stopped to save energy.

Since the conditioner and regenerator are separate units, they can be in different locations and connected by piping. This can substantially lower ductwork cost and required mechanical space. Commonly, a single regenerator services several conditioner units (Figure 5). In the simplest control arrangement, concentrated desiccant solution is metered to each conditioner at a fixed rate. The return flow of weak solution from each conditioner is regulated to maintain a constant operating level in the conditioner. A level controller on the regenerator regulates heat flow to the regenerator solution heater so that a constant volume of desiccant solution, and hence a practically constant solution concentration, is maintained.

The regenerator can be sized to match the dehumidification load of the conditioner unit or units. Regenerator capacity is affected by regenerator heat source temperatures; higher source temperatures produce greater capacity. Regenerator capacity is also affected by desiccant concentration; higher concentrations result in reduced capacity. The relative humidity of the air leaving the conditioner is practically constant for a given desiccant concentration, so the regenerator capacity can be shown as a function of delivered air relative humidity and regenerator heat source temperature. Figure 6 is a normalized graph showing this relationship. For a given moisture load, a variety of regenerator heat sources may be used if the regenerator is sized for the heat source selected. In many cases, the greater capital cost of a larger regenerator is paid back very quickly by reduced operating cost when a lower cost or waste heat source (e.g., process or turbine tailsteam, jacket heat from an engine-driven generator or compressor, or refrigeration condenser heat) is used.

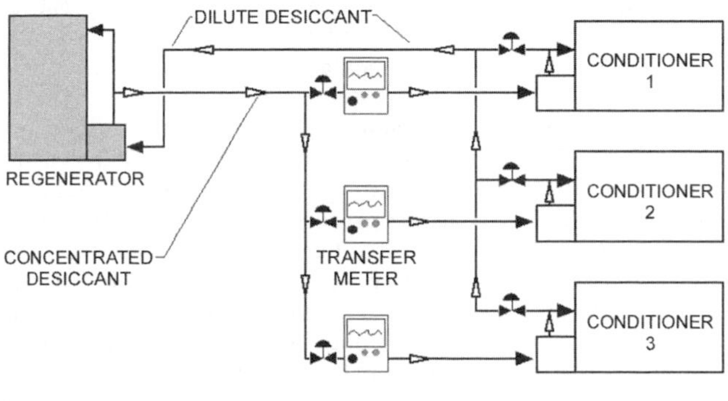

Fig. 5 Liquid Desiccant System with Multiple Conditioners

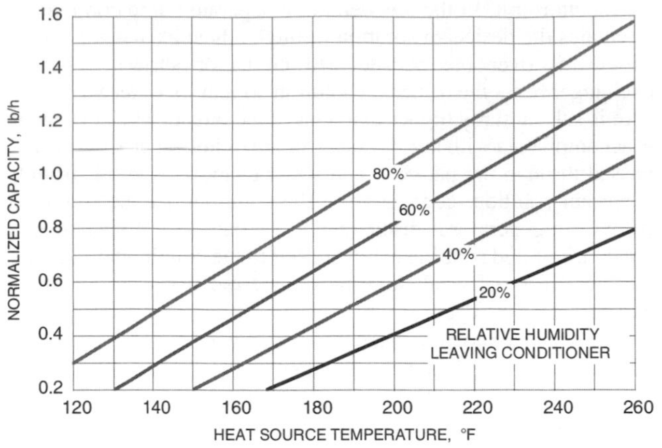

Fig. 6 Liquid Desiccant Regenerator Capacity

SOLID SORPTION EQUIPMENT

Solid desiccants, such as silica gel, zeolites (molecular sieves), activated alumina, or hygroscopic salts, are generally used to dehumidify large volumes of moist air, and in such cases, the desiccant is continuously reactivated. There are also two other methods of dehumidifying air using dry desiccants: (1) nonreactivated, disposable packages and (2) periodically reactivated desiccant cartridges.

Disposable packages of solid desiccant are often sealed into packaging for consumer electronics, pharmaceutical tablets, and military supplies. Disposable desiccant packages rely entirely on vapor diffusion to dehumidify, as air is not forced through the desiccant. This method is used only in applications where there is no anticipated moisture load at all (such as hermetically sealed packages) because the moisture absorption capacity of any nonreactivated desiccant is rapidly exceeded if a continuous moisture load enters the dehumidified space. Disposable packages generally serve as a form of insurance against unexpected, short-term leaks in small, sealed packages.

Periodically reactivated cartridges of solid desiccant are used where the expected moisture load is continuous, but very small. A common example is the **breather**, a tank of desiccant through which air can pass, compensating for changes in liquid volume in petroleum storage tanks or drums of hygroscopic chemicals. The air is dried as it passes through the desiccant, so moisture will not contaminate the stored product. When the desiccant becomes saturated, the cartridge is removed and heated in an oven to restore its moisture sorption capacity. Desiccant cartridges are used where there is no requirement for a constant humidity control level and where the moisture load is likely to exceed the capacity of a small, disposable package of desiccant.

Desiccant dehumidifiers, which are used for drying liquids and gases other than air, often use a variation of this reactivation technique. Two or more pressurized containers of dry desiccant are arranged in parallel, and air is forced through one container for drying, while desiccant in the other container is reactivated. These units are often called **dual-tower** or **twin-tower** dehumidifiers.

Continuous reactivation dehumidifiers are the most common type used in high moisture load applications such as humidity control systems for buildings and industrial processes. In these units, humid air, generally referred to as the **process air**, is dehumidified in one part of the desiccant bed while a different part of the bed is dried for reuse by a second airstream known as the **reactivation air**. The desiccant generally rotates slowly between these two airstreams, so that dry, high-capacity desiccant leaving the reactivation air is always available to remove moisture from the moist process air. This type of solid desiccant equipment is generally called a **rotary desiccant dehumidifier**. It is most commonly used in building air-handling systems, and the section on Rotary Solid Desiccant Dehumidifiers describes its function in greater detail.

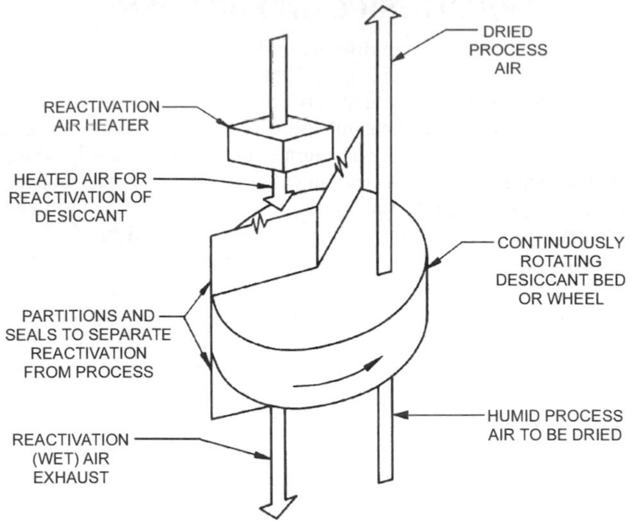

Fig. 7 Typical Rotary Dehumidification Unit

ROTARY SOLID DESICCANT DEHUMIDIFIERS

Operation

Figure 7 illustrates the principle of operation and the arrangement of major components of a typical rotary solid desiccant dehumidifier. The desiccant can be beads of granular material packed into a bed, or it can be finely divided and impregnated throughout a structured media. The structured media resembles corrugated cardboard rolled into a drum, so that air can pass freely through flutes aligned lengthwise through the drum.

In both granular and structured media units, the desiccant itself can be either a single material, such as silica gel, or a combination, such as dry lithium chloride mixed with zeolites. Manufacturers provide options in the choice of desiccants because the wide range of applications for dehumidification systems requires this flexibility to minimize operating and installed costs.

In rotary desiccant dehumidifiers, there are more than 20 variables that affect performance. In general, equipment manufacturers fix most of these in order to provide predictable performance in the more common applications for desiccant systems. Primary variables left to the system designer to define include

Process air

- Inlet air temperature
- Moisture content
- Velocity at the face of the desiccant bed

Reactivation air

- Inlet air temperature
- Moisture content
- Velocity at the face of the desiccant bed

In any system, these variables change because of weather, variations in moisture load, and fluctuations in reactivation energy levels. It is useful for the system designer to understand the effect of these normal variations on the performance of the dehumidifier.

To illustrate these effects, Figures 8 through 12 show changes in the process air temperature and moisture leaving a "generic" rotary desiccant dehumidifier as modeled by a finite difference analysis program (Worek and Zheng 1991). The performance of commercial units differs from this model because such units are generally

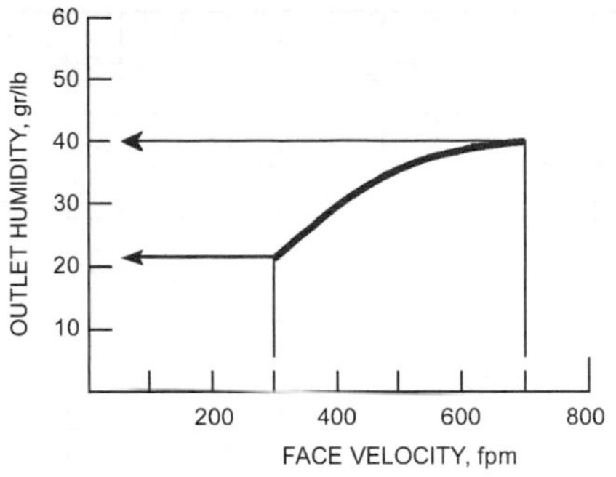

Fig. 8 Effect of Changes in Process Air Velocity on Dehumidifier Outlet Moisture

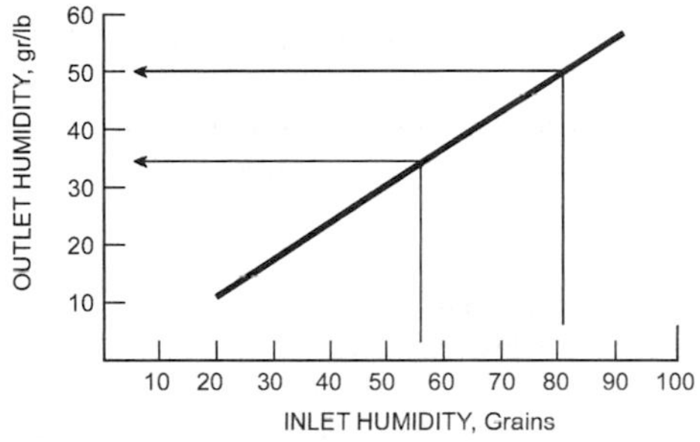

Fig. 9 Effect of Changes in Process Air Inlet Moisture on Dehumidifier Outlet Moisture

optimized for very deep drying. However, for illustration purposes, the model accurately reflects the relationships between the key variables.

The desiccant used for the model is silica gel; the bed is a structured, fluted media; the bed depth is 16 in. in the direction of airflow; and the ratio of process air to reactivation air is approximately 3:1. The process air enters the machine at normal comfort conditions of 70°F, 50% rh (56 grains per pound of dry air).

Process air velocity through the desiccant bed strongly affects the leaving moisture. As shown in Figure 8, if the entering moisture is 56 gr/lb and all other variables are held constant, the outlet moisture varies from 22 gr/lb at 250 fpm to 40 gr/lb at 700 fpm. In other words, air that passes through the bed more slowly is dried more deeply.

For the designer, this relationship means that if air must be dried very deeply, a large unit (slower air velocities) must be used.

Process air inlet moisture content affects the outlet moisture. Thus, if the air is more humid entering the dehumidifier, it will be more humid leaving the unit. For example, Figure 9 indicates that for an inlet humidity of 56 gr/lb, the outlet humidity will be 35 gr/lb. If the inlet moisture content rises to 80 gr/lb, the outlet humidity rises to 50 gr/lb.

For the designer, one consequence is that if a constant outlet humidity is necessary, the dehumidifier will need to have capacity control except in the rare circumstance where the process inlet airstream does not vary in temperature or moisture content throughout the year.

Reactivation air inlet temperature changes the outlet moisture content of the process air. In the range of temperatures from 100 to 250°F, as more heat is added to the reactivation air, the desiccant dries more completely, which means that it can attract more moisture from the process air (Figure 10). If the reactivation air is only heated to 100°F, the process outlet moisture is 50 gr/lb, or only 6 gr/lb lower than the entering humidity. In contrast, if the reactivation air is heated to 200°F, the outlet moisture is 35 gr/lb, so that almost 40% of the original moisture is removed.

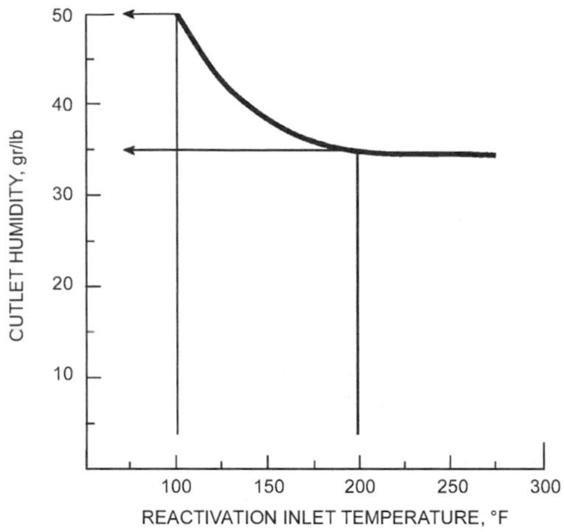

Fig. 10 Effect of Changes in Reactivation Air Inlet Temperature on Dehumidifier Outlet Moisture

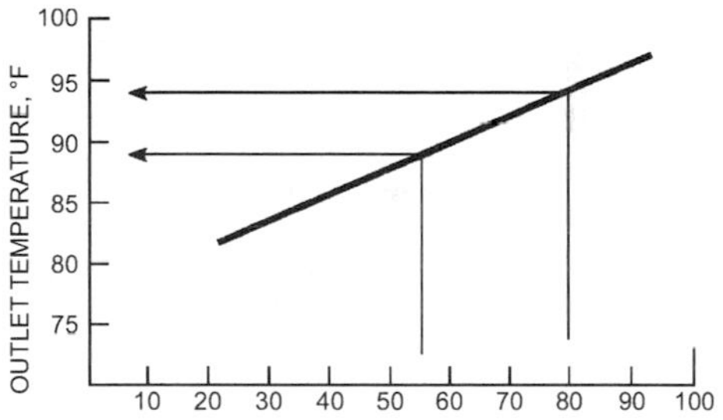

Fig. 11 Effect of Changes in Process Air Inlet Moisture on Dehumidifier Outlet Temperature

Two important consequences follow from this relationship. If the designer needs dry air, it is generally more economical to use high reactivation air temperatures. Conversely, if the leaving humidity from the dehumidifier need not be especially low, inexpensive, low-grade heat sources (e.g., waste heat, cogeneration heat, or rejected heat from refrigeration condensers) can be used to reactivate the desiccant at a low cost.

Process air outlet temperature is higher than the inlet air temperature primarily because the heat of sorption of the moisture removed from the air is converted to sensible heat. The heat of sorption is composed of the latent heat of condensation of the removed moisture, plus additional chemical heat, which varies depending on the type of desiccant and the process air outlet humidity. Additionally, some heat is carried over to the process air from the reactivation sector since the desiccant is warm as it enters the relatively cooler process air. Generally, between 80 and 90% of the temperature rise of the process air is due to the heat of sorption, and the balance is due to heat carried over from reactivation.

Process outlet temperature versus inlet humidity is illustrated in Figure 11. Note that as more moisture is removed (higher inlet humidity), the outlet temperature rises. Air entering at room comfort conditions of 70°F, 56 gr/lb leaves the dehumidifier at a temperature of 89°F. If the dehumidifier removes more moisture, such as when the inlet humidity is 80 gr/lb, the outlet temperature rises to 94°F. The increase in temperature rise is roughly proportional to the increase in moisture removal.

Process outlet temperature versus reactivation air temperature is illustrated in Figure 12, which shows the effect of increasing reactivation temperature when the moisture content of the process inlet air stays constant. If the reactivation sector is heated to elevated temperatures, more moisture is removed on the process side,

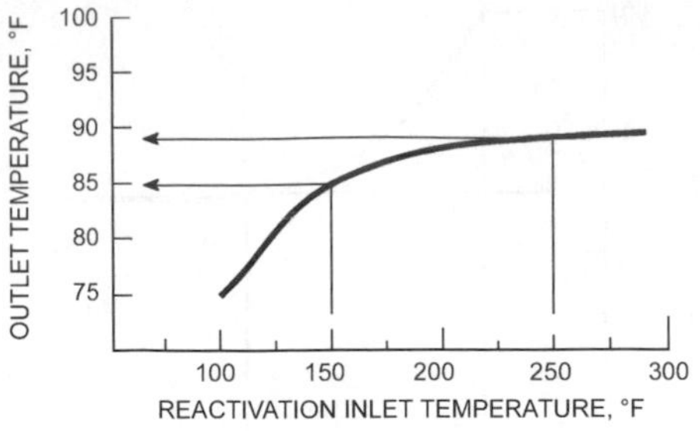

Fig. 12 Effect of Changes in Reactivation Air Inlet Temperature on Dehumidifier Outlet Temperature

so the temperature rise due to latent-to-sensible heat conversion is slightly greater. In this constant-moisture inlet situation, if the reactivation sector is very hot, more heat is carried from reactivation to process as the desiccant mass rotates from reactivation to process. Figure 12 shows that if reactivation air is heated to 150°F, the process air leaves the dehumidifier at 85°F. If reactivation air is heated to 250°F, the process air outlet temperature rises to 89°F. The 4°F increase in process air temperature is due primarily to the increase in heat carried over from reactivation.

One consequence of this relationship is that desiccant equipment manufacturers constantly seek to minimize the "waste mass" in a desiccant dehumidifier, to avoid heating and cooling extra, nonfunctional material such as heavy desiccant support structures or extra desiccant that air cannot reach. Theoretically, the most efficient desiccant dehumidifier has an infinitely large effective desiccant surface combined with an infinitely low mass.

Use of Cooling

In process drying applications, desiccant dehumidifiers are sometimes used without additional cooling because the increase in temperature from the dehumidification process is helpful to the drying process. In semiprocess applications such as controlling frost formation in supermarkets, excess sensible cooling capacity may be present in the system as a whole, so that warm air from a desiccant unit is not a major consideration. However, in most other applications for desiccant dehumidifiers, provision must be made to remove excess sensible heat from the process air following dehumidification.

In a liquid desiccant system, heat is removed by cooling the liquid desiccant itself, so the process air emerges from the desiccant media at the appropriate temperature. In a solid desiccant system, cooling is accomplished downstream of the desiccant bed with cooling coils. The source of this cooling can affect the operating economics of the system.

In some systems, postcooling is accomplished in two stages, with cooling tower water as the primary source followed by compression or absorption cooling. Alternately, various combinations of indirect and direct evaporative cooling equipment are used to cool the dry air leaving the desiccant unit.

In other systems where there is a time difference between the peaks of the latent and sensible loads, the sensible cooling capacity of the basic air-conditioning system is sufficient to handle the process air temperature rise without additional equipment. Systems in moderate climates with high ventilation requirements often combine high latent loads in the morning, evening, and night with high sensible loads at midday, so that desiccant subsystems to handle latent loads are especially economical.

Use of Units in Series

Dry desiccant dehumidifiers are often used to provide air at low dew points. Applications requiring large volumes of air at moisture contents of 5 gr/lb (0°F dew point) are quite common and can be

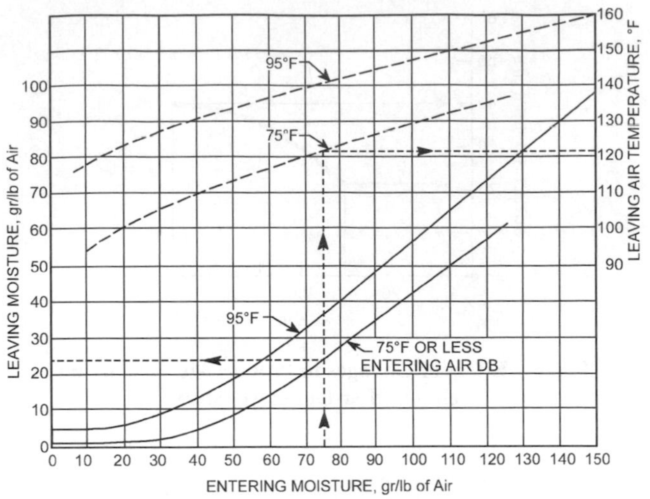

Fig. 13 Typical Performance Data for Rotary Solid Desiccant Dehumidifier

easily achieved by rotary desiccant units in a single pass beginning with inlet moisture contents as high as 45 gr/lb (45°F dew point). Certain types of dry desiccant units commonly deliver air at 2 gr/lb (−18°F dew point) without special design considerations. Where extremely low dew points must be achieved, or where air leakage inside the unit may be a concern, two desiccant dehumidifiers can be placed in series, with the dry air from a first unit feeding both the process and reactivation air to a second unit. The second unit delivers very dry air, since there is reduced risk of moisture being carried over from reactivation to process air when dry air is used to reactivate the second unit.

Industrial Rotary Desiccant Dehumidifier Performance

Figures 8 through 12 are based on the generalized model of a desiccant dehumidifier as described by Worek and Zheng (1991). The model, however, differs somewhat from commercial products. Figure 13 shows the typical performance of an industrial desiccant dehumidifier.

EQUIPMENT OPERATING CONCERNS AND SUGGESTIONS

Desiccant equipment tends to be very durable, often operating at high efficiency 30 years after it was originally installed. But such longevity is not achieved without maintenance. Required maintenance is specific to the type of desiccant equipment, the application, and the installation. Each system requires a somewhat different maintenance and operational routine. The information provided this section does not substitute for or supersede any recommendations of equipment manufacturers, and it is not a substitute for the experience of the owners of a specific application.

Process Air Filters

Owners and manufacturers of desiccant systems agree that clean filters are the most important item in a maintenance routine. If a solid desiccant is clogged with particulates, or if a liquid desiccant's sorption characteristics are changed by entrained particulates, the material may have to be replaced prematurely. Filters are much less expensive and much easier to change than the desiccant. Although each application is different, the desiccant usually must be replaced, replenished, or reconditioned after 5 to 10 years of operation. Without attention to filters, desiccant life can be reduced to 1 or 2 years of operation or less. Filters should be checked at least four times per year, and more frequently when airstreams are heavily laden with particulates.

For designers, the importance of filter maintenance means that the filter racks and doors on desiccant systems must be freely accessible and that enough space must be allowed to inspect, remove, and replace the filter media. The optimal design ensures that filter locations, as well as the current condition of each filter, is clearly visible to maintenance personnel.

Reactivation/Regeneration Filters

Air is filtered before entering the heater of a desiccant unit. If the filters are clogged and airflow is reduced, unit performance may be reduced because there is not enough air to carry all the moisture away from the desiccant. If electrical elements or gas burners are used to heat the air, reducing the airflow may damage the heaters. Consequently, the previous suggestions for the maintenance of process air filters also apply to reactivation/regeneration filters.

Reactivation/Regeneration Ductwork

Air leaving the reactivation/regeneration section is hot and moist. When units first start up in high moisture load applications, the reactivation air may be nearly saturated and may even contain water droplets. Consequently, the ductwork that carries the air away from the unit should be resistant to corrosion, because condensation can occur inside the ducts, particularly if the ducts pass through unheated areas in cool weather. If heavy condensation seems probable, the ductwork should be designed with drains at low points or arranged to let condensation flow out of the duct where the air is vented to the weather. The high temperature and moisture of the leaving air may make it necessary to use dedicated ductwork, rather than combining the air with other exhaust airflows, unless the other flows have similar characteristics.

Leakage

All desiccant units produce dry air in part of the system. If humid air is allowed to leak into either the dry air ductwork or the unit itself, the system efficiency is reduced. Energy is also wasted if dry air is allowed to leak out of the distribution duct connections. Therefore, duct connections for desiccant systems should be sealed tightly. In applications requiring very low dew points (below 10°F), the ductwork and the desiccant system itself are almost always tested for leaks at air pressures above those expected during normal operation. In applications at higher dew points, similar leak testing is considered good practice by experienced end users and is recommended by many equipment manufacturers.

Because desiccant equipment tends to be durably constructed, workers often drill holes in the dehumidifier unit casing to provide support for piping, ductwork, or instruments. Such holes eventually leak air, desiccant, or both. Designers should provide other means of support for external components so contractors do not puncture the system unnecessarily.

Contractors installing desiccant systems should be aware that any holes made in the system must be sealed tightly using both mechanical means and sealant compounds. Sealants must be selected for the working temperatures of the casing walls that have been punctured. For example, reactivation/regeneration sections often operate in a range from a cold winter ambient of −40°F to a heated temperature as high as 300°F. Process sections may operate within a range of −40°F at the inlet to 150°F at the outlet. Sealants should be selected for long life at the working temperatures of the equipment and the application.

Airflow Indication and Control

As explained in the section on Rotary Solid Desiccant Dehumidifiers, the performance of a desiccant unit depends on how quickly the air passes through the desiccant. Changes in air velocity affect the performance. Consequently, it is important to quantify the airflow rate through both the process and the reactivation/regeneration parts of the unit. Unless both airflows are known, it is impossible to determine whether the unit is operating properly. In addition, if the velocity exceeds the maximum design value, the air may carry desiccant particles or droplets out of the unit and into the supply air ductwork. For these reasons, manufacturers often provide airflow gages on larger equipment so the owner can be certain the unit is operating within the intended design parameters.

Smaller equipment is not always provided with airflow indicators because precision in performance may be less critical in applications such as small storage rooms. However, in any system using large equipment, or if performance is critical in smaller systems, the unit airflow should be quantified and clearly indicated. Operating personnel need to be able to compare the current flow rates through the system with the design values.

Many desiccant units are equipped with manual or automatic flow control dampers to control the airflow rate. If these are not provided with the unit, they should be installed elsewhere in the system. Airflows for process and reactivation/regeneration must be correctly set after all ductwork and external components are attached, but before the system is put into use.

Commissioning

Heat and moisture on the dry-air side of desiccant equipment is balanced equally by the heat and moisture on the regeneration/reactivation side. To confirm that a dry desiccant system is operating as designed, the commissioning technician must measure airflow, temperature, and moisture on each side to calculate a mass balance. In liquid systems, these six measurements are taken on the process-air side. On the regenerator side, the liquid temperature is read in the sump and at the spray head to confirm the heat transfer rate of the regenerator at peak load conditions.

If the dehumidification unit does not provide the means, the system should be designed to facilitate taking the readings that are essential to commissioning and troubleshooting. Provisions must be made to measure the flow rates, temperatures, and moisture levels of airstreams as they enter and leave the desiccant. In the case of liquid systems, provisions must be made for measuring the solution temperature and concentration at different points in the system. Four precautions for taking these readings at different points in a desiccant system follow.

Airflow. Airflow instruments measure the actual volumetric flow rate, which must be converted to standard flow rate in order to calculate mass flow. Because the temperatures in a desiccant system are often well above or below standard temperature, these corrections are essential.

Air temperature. Most airstreams in a desiccant system have temperatures between 0 and 300°F. However, there can be wide temperature variation and stratification as the air leaves the desiccant in dry desiccant systems. Air temperature readings must be averaged across the duct for accurate calculations. Readings taken after a fan tend to be more uniform, but corrections must be made for heat added by the fan itself.

Process air moisture leaving dry desiccant. In solid desiccant equipment, the air leaving the desiccant bed or wheel is quite warm as well as rather dry—usually below 20% rh, often below 10%, and occasionally below 2%. Most low-cost instruments have limited accuracy below 15% rh, and all but the most costly instruments have an error of ±2% rh. Consequently, to measure relative humidities near 2%, technicians use very accurate instruments such as manual dew cups or automated optical dew-point hygrometers. ASHRAE *Standard* 41.6 describes these instruments and procedures for their proper use. When circumstances do not allow the use of dew-point instruments, other methods may be necessary. For example, an air sample may need to be cooled to produce a higher, more easily measured relative humidity.

Low humidity readings can be difficult to take with wet-bulb thermometers because the wet wick dries out very quickly, sometimes before the true wet-bulb reading is reached. Therefore, wicks must be monitored for wetness. Additionally, when the wet-bulb temperature is below the freezing point of water, readings take much longer, which may allow the wick to dry out, particularly in dry desiccant systems where there may be considerable heat in the air leaving the desiccant. Many technicians avoid the use of wet-bulb readings in the air leaving a dry desiccant bed, partly for these reasons, and partly because of the difficulty and time required to obtain average readings across the whole bed.

Like the air temperature, the air moisture level leaving a dry desiccant bed varies considerably; if taken close to the bed, readings must be averaged to obtain a true value for the whole air mass.

When very low dew points are expected, the commissioning technician should be especially aware of limitations of the air-sampling system and the sensor. Even the most accurate sensors require more time to come to equilibrium at low dew points than at moderate moisture levels. For example, when taking readings at dew points below −20°F, the sensor and air sample tubing may take many hours rather than a few minutes to come to equilibrium with the air being measured. Time required to come to equilibrium also depends on how much moisture is on the sensor before it is placed into the dry airstream. For example, taking a reading in the reactivation/regeneration outlet essentially saturates the sensor, so it will take much longer than normal to come to equilibrium with the low relative humidity of the process leaving air.

Reactivation/regeneration air moisture leaving the desiccant. Air leaving the reactivation/regeneration side of the desiccant is quite warm and close to saturation. If the humidity measurement sensor is at ambient temperature, moisture may condense on its surface, distorting the reading. It is good practice to warm the sensor (e.g., by taking the moisture reading in the warm, dry air of the process-leaving airstream) before reading moisture in reactivation air. If a wet-bulb instrument is used, the water for the wet bulb must be at or above the dry-bulb temperature of the air, or the instrument will read lower than the true wet-bulb temperature of the air.

Owner's Perspective

Experienced users who were interviewed to gather information for this chapter all agreed on one point: designers and new owners are strongly advised to consult other equipment owners and the manufacturer's service department at an early stage in the design process to gain the useful perspective of direct operating experience (Harriman 1994).

APPLICATIONS FOR ATMOSPHERIC PRESSURE DEHUMIDIFICATION

Preservation of Materials in Storage

Special moisture-sensitive materials are sometimes kept in dehumidified warehouses for long-term storage. Tests by the Bureau of Supplies and Accounts of the U.S. Navy concluded that 40% rh is a safe level to control deterioration of materials. Others have indicated that 60% rh is low enough to control microbiological attack. With storage at 40% rh, no undesirable effects on metals or rubber-type compounds have been noted. Some organic materials such as sisal, hemp, and paper may lose flexibility and strength, but they recover these characteristics when moisture is regained.

Commercial storage relies on similar equipment for applications that include beer fermentation rooms, meat storage, and penicillin processing, as well as storage of machine tools, candy, food products, furs, furniture, seeds, paper stock, and chemicals. For recommended conditions of temperature and humidity, refer to the food refrigeration section of the 1998 *ASHRAE Handbook—Refrigeration*.

Process Dehumidification

The requirements for dehumidification in industrial processes are many and varied. Some of these processes are as follows:

- Metallurgical processes, in conjunction with the controlled atmosphere annealing of metals
- Conveyance of hygroscopic materials
- Film drying
- Manufacture of candy, chocolate, and chewing gum
- Manufacture of drugs and chemicals
- Manufacture of plastic materials
- Manufacture of laminated glass
- Packaging of moisture-sensitive products
- Assembly of motors and transformers
- Solid propellent mixing
- Manufacture of electronic components, such as transistors and microwave components

For information concerning the effect of low dew-point air on drying processes, refer to Chapters 17, 19, 22, and 27 of the 1999 *ASHRAE Handbook—Applications*.

Ventilation Air Preconditioning

Over a full year, ventilation air loads a cooling system with much more moisture than heat. With the exception of desert and high-altitude regions, ventilation moisture loads in the United States exceed sensible loads by at least 3:1, and often by as much as 5:1 (Harriman et al. 1997). Consequently, desiccant systems are used to precondition ventilation air before the air enters the main air-conditioning system.

Preconditioning has gained importance because building codes mandate larger amounts of ventilation air than in the past, in an effort to improve indoor air quality. Large amounts of untreated ventilation air carry enough moisture to upset the operation of "high-efficiency" cooling equipment, which is generally designed to remove more sensible heat than moisture (Kosar et al. 1998). By removing excess moisture from the ventilation air, a desiccant preconditioning system reduces the load on cooling system, which in turn reduces its operating cost. Also, cooling equipment is often oversized to remove ventilation-generated moisture. When that is the case, pretreating with a desiccant system may reduce the building's construction cost, provided that the excess cooling capacity is removed from the design (Spears and Judge 1997).

Ventilation pretreatment is most cost-effective for buildings that have a high ventilation airflow rather than high sensible loads from internal heat or from heat transmitted through the building envelope. As a result, desiccant-based pretreatment is most common in densely occupied buildings such as schools, theaters, elder care facilities, large-scale retail buildings, and restaurants.

Condensation Prevention

Many applications require moisture control to prevent condensation. Moisture in the air will condense on cold cargo in a ship's hold when it reaches a moist climate. Moisture will condense on a ship when the moist air in its cargo hold is cooled by the hull and deck plates as the ship passes from a warm to a cold climate.

A similar problem occurs with military and commercial aircraft when the cold air frame descends from high altitudes into a high dew point at ground level. Desiccant dehumidifiers are used to prevent condensation inside the air frame and avionics that leads to structural corrosion and failure of electronic components.

In pumping stations and sewage lift stations, moisture condenses on piping, especially in the spring when the weather warms and water in the pipes is still cold. Dehumidification is also used to prevent moisture in the air from dripping into oil and gasoline tanks and open fermentation tanks.

Electronic equipment is often cooled by refrigeration, and dehumidifiers are required to prevent internal condensation of moisture. Electronic and instrument compartments in missiles are purged with low dew-point air prior to launching to prevent malfunctioning due to condensation.

Waveguides and radomes are also usually dehumidified, as are telephone exchanges and relay stations. For the proper operation of their components, missile and radar sites depend (to a large extent) on the prevention of condensation on interior surfaces.

Dry Air-Conditioning Systems

Cooling-based air-conditioning systems remove moisture from air by condensing it onto cooling coils, producing saturated air at a lower absolute moisture content. In many circumstances, however, there is a benefit to using a desiccant dehumidifier to remove the latent load from the system, avoiding problems caused by condensation, frost, and high relative humidity in air distribution systems.

For example, low-temperature product display cases in supermarkets operate less efficiently when the humidity in the store is high because condensate freezes on the cooling coils, increasing operating cost. Desiccant dehumidifiers remove moisture from the air, using rejected heat from refrigeration condensers to reduce the cost of desiccant reactivation. Combining desiccants and conventional cooling can lower installation and operating costs (Calton 1985). For information on the effect of humidity on refrigerated display cases, refer to Chapter 47 of the 1998 *ASHRAE Handbook—Refrigeration* and Chapter 2 of the 1999 *ASHRAE Handbook—Applications*.

The air conditioning in hospitals, nursing homes, and other medical facilities is particularly sensitive to biological contamination in condensate drain pans, filters, and porous insulation inside ductwork. These systems often benefit from the use of a desiccant dehumidifier to dry the ventilation air before final cooling. Condensate does not form on cooling coils or drain pans, and filters and duct lining stay dry so that mold and mildew cannot grow inside the system. Refer to ASHRAE *Standard* 62 for guidance concerning maximum relative humidity in air distribution systems. Chapter 7 of the 1999 *ASHRAE Handbook—Applications* has information on ventilation of health care facilities.

Hotels and large condominium buildings historically suffer from severe mold and mildew problems caused by excessive moisture in the building structure. Desiccant dehumidifiers are sometimes used to dry ventilation air so it can act as a sponge to remove moisture from walls, ceilings, and furnishings (AHMA 1991).

Like supermarkets, ice rinks have large exposed cold surfaces that condense and freeze moisture in the air, particularly during spring and summer operations. Desiccant dehumidifiers remove excess humidity from air above the rink surface, preventing fog and improving both the ice surface and operating economics of the refrigeration plant. For recommended conditions of temperature and humidity, refer to Chapter 34 of the 1998 *ASHRAE Handbook—Refrigeration*.

Indoor Air Quality Contaminant Control

Desiccant sorption is not restricted to water vapor. Both liquid and solid desiccants collect both water and large organic molecules at the same time (Hines et al. 1993). As a result, desiccant systems can be used to remove volatile organic compound (VOC) emissions from building air systems.

In addition to preventing the growth of mold, mildew, and bacteria by keeping buildings dry, desiccant systems are used to supplement filters to remove bacteria from the air itself. This is particularly useful for hospitals, medical facilities, and related biomedical manufacturing facilities where airborne microorganisms can cause costly problems. The utility of certain liquid and solid desiccants in such systems stems from their ability to either kill microorganisms or avoid sustaining their growth (Batelle 1971 and SUNY Buffalo School of Medicine).

Testing

Many test procedures require dehumidification with sorption equipment. Frequently, other means of dehumidification may be used in conjunction with sorbent units, but the low moisture content requirements can be obtained only by liquid or solid sorbents. Some of the typical testing applications are as follows:

- Wind tunnels
- Spectroscopy rooms
- Paper and textile testing
- Bacteriological and plant growth rooms
- Dry boxes
- Environmental rooms and chambers

DESICCANT DRYING AT ELEVATED PRESSURE

The same sorption principles that pertain to atmospheric dehumidification apply to drying high-pressure air and process or other gases. The sorbents described previously can be used with equal effectiveness.

EQUIPMENT

Absorption

Solid absorption systems use a calcium chloride desiccant, generally in a single-tower unit that requires periodic replacement of the desiccant that is dissolved by the absorbed moisture. Normally, the inlet air or gas temperature does not exceed 90 to 100°F saturated. The rate of replacement of the desiccant is proportional to the moisture in the inlet process flow. A dew-point depression of 20 to 40°F at pressure can be obtained when the system is operated in the range of 60 to 100°F saturated entering temperature and 100 psig operating pressure. At lower pressures, the ability to remove moisture decreases in proportion to absolute pressure. Such units do not require a power source for operation because the desiccant is not regenerated. However, additional desiccant must be added to the system periodically.

Adsorption

Drying with an adsorptive desiccant such as silica gel, activated alumina, or a molecular sieve usually incorporates regeneration equipment, so the desiccant can be reactivated and reused. These desiccants can be readily reactivated by heat, by purging with dry gas, or by a combination of both. Depending on the desiccant selected, the dew point performance expected is in the range of –40 to –100°F measured at the operating pressure with inlet conditions of 90 to 100°F saturated and 100 psig. Figure 14 shows typical performance using activated alumina or silica gel desiccant.

Equipment design may vary considerably in detail, but most basic adsorption units use twin-tower construction for continuous operation, with an internal or external heat source, with air or process gas as the reactivation purge for liberating moisture adsorbed previously. A single adsorbent bed may be used for intermittent drying requirements. Adsorption units are generally constructed in the same manner as atmospheric pressure units, except that the vessels are suitable for the operating pressure. Units have been operated successfully at pressures as high as 6000 psig.

Prior compression or cooling (by water, brine, or refrigeration) to below the dew point of the gas to be dried reduces the total moisture load to be handled by the sorbent, thus permitting the use of smaller

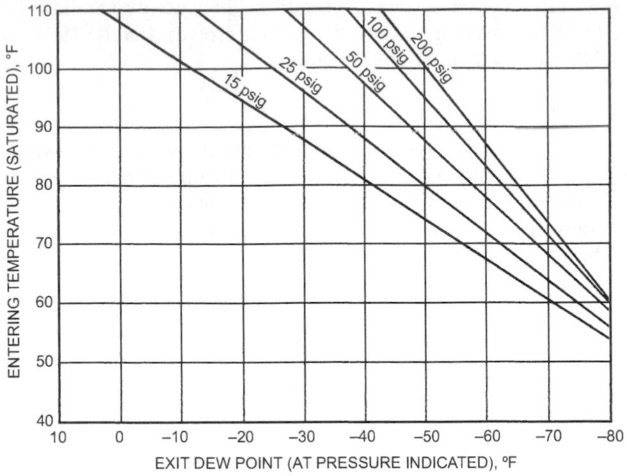

Fig. 14 Typical Performance Data for Solid Desiccant Dryers at Elevated Pressures

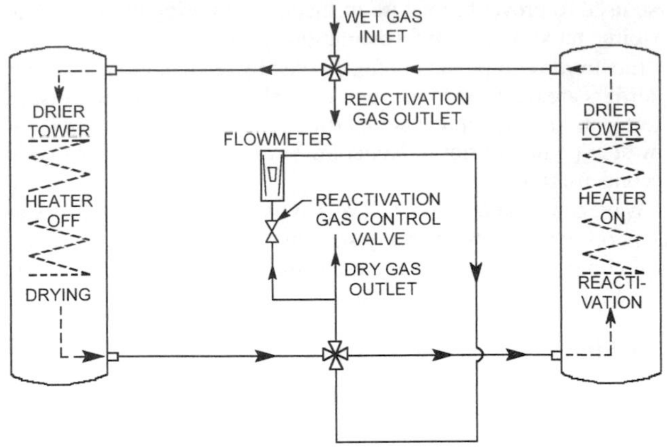

Fig. 15 Typical Adsorption Dryer for Elevated Pressures

drying units. The cost of compression, cooling, or both must be balanced against the cost of a larger adsorption unit.

The many different dryer designs can be grouped into the following basic types:

Heat-reactivated, purge dryers. Normally operating on 4 h (or longer) adsorption periods, these dryers are generally designed with heaters embedded in the desiccant. They use a small portion of the dried process gas as a purge to remove the moisture liberated during reactivation heating. (See Figure 15.)

Heatless dryers. These dryers operate on a short adsorption period (usually 60 to 300 s). Depressurization of the gas in the desiccant tower lowers the vapor pressure, so the adsorbed moisture is liberated from the desiccant and removed by a high purge rate of the dried process gas. The use of an ejector reduces the purge gas requirements.

Convection dryers. These dryers usually operate on 4 h (or longer) adsorption periods and are designed with an external heater and cooler as the reactivation system. Some designs circulate the reactivation process gas through the system by a blower, while other designs divert a portion or all of the process gas flow through the reactivation system prior to adsorption. Both heating and cooling are by convection.

Radiation dryers. Also operating on 4 h (or longer) adsorption periods, radiation-type dryers are designed with an external heater and blower to force heated atmospheric air through the desiccant tower for reactivation. Cooling of the desiccant tower is by radiation to atmosphere.

APPLICATIONS

Preservation of Materials

Generally, materials in storage are preserved at atmospheric pressure, but a few materials are stored at elevated pressures, especially when the dried medium is an inert gas. These materials deteriorate when they are subjected to high relative humidity or oxygen content in the surrounding media. The drying of high-pressure air, which is subsequently reduced to 3.5 to 10 psig, has been used most effectively in pressurizing coaxial cables to eliminate electrical shorts caused by moisture infiltration. This same principle, at somewhat lower pressures, is also used in waveguides and radomes to prevent moisture film on the envelope.

Process Drying of Air and Other Gases

Drying of instrument air to a dew point of –40°F, particularly in areas where the air lines are outdoors or are exposed to temperatures below the dew point of the air leaving the aftercooler, prevents condensation or freeze-up in instrument control lines.

To prevent condensation and freezing, it is necessary to dry the plant air used for pneumatically operated valves, tools, and other equipment in areas where the piping is exposed to low ambient temperatures. Additionally, dry air prevents rusting of the air lines, which produces abrasive impurities, causing excessive wear on tools.

Industrial gases or fuels such as natural gas are dried. For example, fuels (including natural gas) are cleaned and dried before storage underground to ensure that valves and transmission lines do not freeze from condensed moisture during extraordinarily cold weather, when the gas is most needed. Propane must also be clean and dry to prevent ice accumulation. Other gases, such as bottled oxygen, nitrogen, hydrogen, and acetylene, must have a high degree of dryness. In the manufacture of liquid oxygen and ozone, the weather air supplied to the particular process must be clean and dry.

Drying of air or inert gas for conveying hygroscopic materials in a liquid or solid state ensures continuous, trouble-free plant operation. Normally, gases for this purpose are dried to a –40°F dew point. Purging and blanketing operations in the petrochemical industry depend on the use of dry inert gas for reducing problems such as explosive hazards and the reaction of chemicals with moisture or oxygen.

Testing of Equipment

Dry, high-pressure air is used extensively for the testing of refrigeration condensing units to ensure tightness of components and to prevent moisture infiltration. Similarly, dry inert gas is used in the testing of copper tubing and coils to prevent corrosion or oxidation. The manufacture and assembly of solid-state circuits and other electronic components require exclusion of all moisture, and final testing in dry boxes must be carried out in moisture-free atmospheres. The simulation of dry high-altitude atmospheres for testing of aircraft and missile components in wind tunnels requires extremely low dew-point conditions.

REFERENCES

AHMA. 1991. *Mold and mildew in hotels and motels.* Executive Engineers Committee Report. American Hotel and Motel Association, Washington, D.C.

ASHRAE. 1994. Standard method for measurement of moist air properties. ANSI/ASHRAE *Standard* 41.6-1994.

ASHRAE. 1999. Ventilation for acceptable indoor air quality. *Standard* 62-1999.

Batelle Memorial Institute. 1971. Project No. N-0914-5200-1971. Batelle Memorial Institute, Columbus, OH.

Calton, D.S. 1985. Application of a desiccant cooling system to supermarkets. *ASHRAE Transactions* 91(1B):441-46.

Harriman, L.G., III. 1994. Field experience: A look at desiccant systems. *Engineered Systems* (January):63-68. Business News Publishing, Troy, MI.

Harriman, L.G., III., D. Plager, and D. Kosar. 1997. Dehumidification and cooling loads from ventilation air. *ASHRAE Journal* 39(11):37-45.

Hines, A.L., T.K. Ghosh, S.K. Loyalka, and R.C. Warder, Jr. 1993. Investigation of co-sorption of gases and vapors as a means to enhance indoor air quality. Gas Research Institute, Chicago, IL. Available from the National Technical Information Service, Springfield, VA. Order No. PB95-104675.

Kosar, D.R., M.J. Witte, D.B. Shirey, and R.L. Hedrick. 1998. Dehumidification issues of *Standard* 62-1989. *ASHRAE Journal* 40(5):71-75.

Spears, J.W. and J.J. Judge. 1997. Gas-fired desiccant system for retail superstore. *ASHRAE Journal* 39(10):65-69.

SUNY Buffalo School of Medicine. Effects of glycol solutions on microbiological growth. Niagara Blower *Report* No. 03188.

Worek, W. and W. Zheng. 1991. UIC IMPLICIT Rotary desiccant dehumidifier finite difference program. University of Illinois at Chicago, Department of Mechanical Engineering, Chicago, IL.

BIBLIOGRAPHY

ASHRAE. 1975. Symposium on sorption dehumidification. *ASHRAE Transactions* 81(1):606-38.

ASHRAE. 1980. Symposium on energy conservation in air systems through sorption dehumidifier techniques. *ASHRAE Transactions* 86(1):1007-36.

ASHRAE. 1985. Symposium on changes in supermarket heating, ventilating and air-conditioning systems. *ASHRAE Transactions* 91(1B):423-68.

ASHRAE. 1992. *Desiccant cooling and dehumidification.*

Bradley, T.J. 1994. Operating an ice rink year-round by using a desiccant dehumidifier to remove humidity. *ASHRAE Transactions* 100(1):116-31.

Harriman, L.G., III. 1990. *The dehumidification handbook*. Munters Cargocaire, Amesbury, MA.

Harriman, L.G., III. 1996. *Applications engineering manual for desiccant systems*. American Gas Cooling Center, Arlington, VA.

Jones, B.W., B.T. Beck, and J.P. Steele. 1983. Latent loads in low humidity rooms due to moisture. *ASHRAE Transactions* 89(1A):35-55.

Meckler, M. 1994. Desiccant-assisted air conditioner improves IAQ and comfort. *Heating/Piping/Air Conditioning* 66(10):75-84. Penton Publishing, Cleveland.

Mei, V.C. and F.C. Chen. 1991. An assessment of desiccant cooling and dehumidification technology. *Report* No. ORNL/CON-309. Prepared for the Office of Building and Community Services, U.S. Department of Energy, by Energy Division, Oak Ridge National Laboratory, Oak Ridge, TN.

Pesaran, A.A., T.R. Penney, and A.W. Czanderna. 1991. *Desiccant cooling state of the art assessment.* Prepared for the Office of Building and Community Services, U.S. Department of Energy, by Thermal, Fluid and Optical Science Branch, Solar Energy Research Institute, Golden, CO.

CHAPTER 23

AIR-HEATING COILS

AIR-HEATING COILS are used to heat air under forced convection. The total coil surface may consist of a single coil section or several coil sections assembled into a bank. The coils described in this chapter apply primarily to comfort heating and air conditioning using steam, hot water, refrigerant vapor heat reclaim, and electricity.

COIL CONSTRUCTION AND DESIGN

Extended-surface coils consist of a primary and a secondary heat-transfer surface. The primary surface is the external surface of the tubes; generally consisting of rows of round tubes or pipes that may be staggered or parallel (in-line) with respect to the air flow. Flattened tubes or tubes with other nonround internal passageways are sometime used. The inside of the tube is usually smooth and plain, but there are some coil designs that feature various forms of internal fins or turbulence promoters (either fabricated and then inserted, or extruded) to enhance fluid coil performance. The individual tube circuit passes of a coil are usually interconnected by return bends, or hairpin tubes to form the serpentine arrangement of multipass tube circuits. Air heating fluid and steam coils are generally available with different circuit arrangements and combinations that offer varying numbers of parallel water flow passes in the tube core (see Figure 1).

The secondary surface is the external surface of the fins, which consists of thin metal plates or a spiral ribbon uniformly spaced or wound along the length of the primary surface. It is the intimate contact with the primary surface that provides for good heat transfer. This bond must be maintained permanently to ensure continuation of rated performance. The fin designs most frequently used for heating coils are flat plate fins, specially shaped plate fins, and spiral or ribbon fins.

The heat transfer bond between the fin and the tube may be achieved in several ways. Bonding is generally accomplished by expanding the tubes into the tube holes in the fins to obtain a permanent mechanical bond. The tube holes frequently have a formed fin collar, which provides the area of thermal contact and may serve to space the fins uniformly along the tubes. The fins of spiral or ribbon-type fin coils are tension-wound onto the tubes. Tension fit may be enhanced by a thermal conductive adhesive applied during the fin winding process, thus providing a more secure bond of the fin onto the tube. An alloy with a low melting point, such as solder, may be used to provide a metallic bond between copper fins and tube. Some types of spiral fins are knurled into a shallow groove on the exterior of the tube. Sometimes, the fins are formed out of the material of the tube.

Copper and aluminum are the materials most commonly used for extended-surface coils. Tubing made of steel or various copper alloys is used in applications where corrosive forces might attack the coils from either inside or outside. The most common combination for low-pressure applications is aluminum fins on copper tubes. Low-pressure steam coils are usually designed to operate up to 50 psig. Higher strength tube materials such as red brass, admiralty brass, or cupronickel assembled by brazed construction, are usable up to 366°F water or 150 psig saturated steam. Higher operating conditions call for electric welded stainless steel construction, designed to meet Section II and Section VIII requirements of the ASME *Boiler and Pressure Vessel Code*.

Customarily, the coil casing consists of a top and bottom channel (also known as baffles or side sheets), two end supports (also known as end plates or tube sheets), and, on longer coils, intermediate supports (also known as center supports or tube sheets). Designs vary, but most are duct or built-up system mounted. Most often, casing material is spangled zinc-coated (galvanized) steel with a minimum coating designation of G90-U. Some corrosive air conditions may require stainless steel casings or corrosive resistant coating, such as a baked phenolic applied by the manufacturer to the entire coil surfaces. In the case of steam coils, their casings should be designed to accommodate thermal expansion of the tube core during operation. Properly designed steam coils do feature one or more variation of what is referred to as a floating core arrangement.

Common core tube diameters vary from 5/16 up to 1 in. OD and fin spacings from 4 to 18 fins per inch. Fluid heating coils have a tube spacing from 3/4 to 1 3/4 in. and the tube diameter ranges from 5/16 to 5/8 in. OD. Steam coils have a tube spacing from 1 1/4 to 3 in. and the tube diameter ranges from 0.5 up to 1 in. OD. The most common arrangements are one- or two-row steam coils and two- to four-row hot water coils. Fins should be spaced according to the application requirements, with particular attention given to any severe duty conditions, such as inlet temperatures and contaminants in the airstream.

Tube wall thickness and the required use of alloys other than the (standard) copper tube are determined mainly by the coil's specified maximum allowable working pressure (MAWP) requirements. Fin-type, header, and connection construction also play a large part in this determination. All applicable local job site codes and national safety standards should be followed in the design and application of heating coils.

Flow direction can have a large effect on the performance of heat transfer surfaces. In air-heating coils with only one row of tubes, the air flows at right angles relative to the heating medium. Such a crossflow arrangement is common in steam heating coils. The steam temperature in the tubes remains uniform, and the mean temperature difference is the same regardless of the direction of flow relative to the air. The steam supply connection is located either in the center or at the top of the inlet header. The steam condensate outlet (return connection) is always at the lowest point in the return header.

When coils have two or more tube rows in the direction of the airflow, such as hot water coils, the heating medium in the tubes may be circuited in various parallel flow and counterflow arrangements. Figure 1 shows common circuitry of a two-row hot water coil. Counterflow is the arrangement most preferred to obtain the highest possible mean temperature difference. The mean temperature difference determines the heat transfer of the coil. The greater this temperature difference, the greater the heat transfer capacity of the coil.

The preparation of this chapter is assigned to TC 8.4, Air-to-Refrigerant Heat Transfer Equipment.

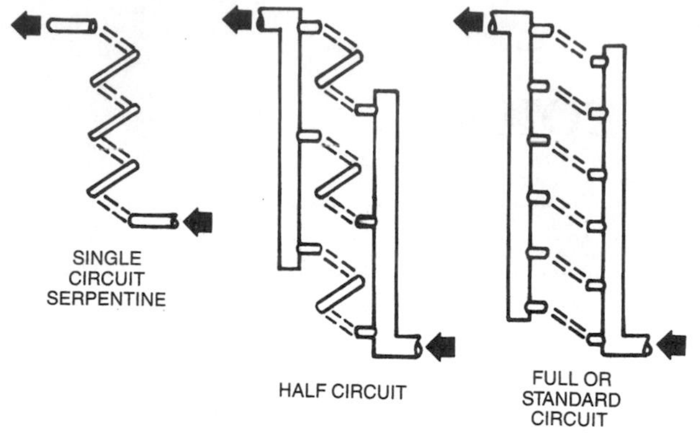

Fig. 1 Water Circuit Arrangements—Two-Row Heating Coils

Multirow coils circuited for counterflow have the water enter the tube row on the leaving air side of the coil.

Steam Coils

Steam coils can be categorized by type as basic steam, steam distributing, or face and bypass.

Basic steam coils generally have smooth tubes with fins on the air side. The steam supply connection is at one end and the tubes are pitched toward the condensate return, which is usually at the opposite end. For horizontal airflow, the tubes can be either vertical or horizontal. In coils with horizontal tubes, the tubes should be pitched within the casing toward the condensate return: This to facilitate condensate removal. Uniform steam distribution to all the tubes is accomplished by careful selection of the header size, its connection locations, and positioning of inlet connection distributor plates. Orifices also may be used at the core tube entrances in the supply header.

Steam distributing coils most often incorporate perforated inner tubes that distribute steam evenly along the entire coil. The perforations perform like small steam ejector jets that, when angle positioned in the inner tube, assist in removing condensate from inside the outer tube. An alternate design for short coils is an inner tube with no distribution holes, but with an open end. On all coils, supply and return connections can be at the same end or at opposite ends of the coil. For long, low-pressure coils, the supply is usually at both ends and the condensate return on one end only.

Face-and- bypass steam coils have short sections of steam coils separated by air bypass openings. Airflow through the coil section or the bypass section is controlled by coil face-and-bypass dampers linked together. As a freeze protection measure, large installations use face-and-bypass steam coils with vertical tubes.

For proper performance of steam heating coils, air or other noncondensables in the steam supply must be eliminated. Equally important, condensate from the steam must easily drain from inside the coil. Air vents are located at a high point of the piping and at the coil's inlet steam header. Whether airflow is horizontal or vertical, the coil's finned section is pitched toward the condensate return connection end of the coil. Installers must give particular care in the selection and installation of piping, controls, and insulation necessary to protect the coil from freeze-up due to incomplete condensate drainage.

When the entering air temperature is at or below 32°F, the steam supply to the coil should not be modulated, but controlled as full on or full off. Coils located in series in the airstream with each coil sized and controlled to be full on or completely off (in a specific sequence, depending on the entering air temperature) are not as likely to freeze. Temperature control with face-and-bypass dampers is also common. During part-load conditions, air is bypassed around the steam coil with full steam flow to the coil. In a face-and-bypass arrangement, high velocity streams of freezing air must not impinge on the coil when the face dampers are partially closed. The section on Overall System Requirements in this chapter and the section on Control of HVAC Elements (Heating Coils) in Chapter 45 of the 1999 *ASHRAE Handbook—Applications* havc more details.

Water/Aqueous Glycol Heating Coils

Normal temperature hot water heating coils can be categorized as booster coils or standard heating coils. Booster (duct-mounted or reheat) coils are commonly found in variable air volume systems. They are one or two rows deep, have minimal water flow, and provide a small air temperature rise. Casings can be either flanged or slip-and-drive construction. Standard heating coils are used in run-around systems, makeup air units, and heating and ventilating systems. All use the standard construction materials of copper tube and aluminum fins.

High-temperature water coils may operate with as high as 400°F water, with pressures comparable or somewhat higher than the saturated vapor temperature of the water supply. The temperature drop across the coil may be as high as 150°F. To safely accommodate these fluid temperatures and thermal stress, the coil requires industrial grade construction that conforms to applicable boiler and safety codes. Such requirements should be listed in detail by the specifying engineer, along with the inspection and certification requirements and a check for compliance prior to coil installation and operation.

Proper performance of a water coil depends on the elimination of air and on good water distribution in the coil and its interconnecting piping. Unless properly vented, air may accumulate in the coil circuits, which will reduce heat transfer and possibly cause noise and vibration in the pipes. For this reason, water coils should be constructed with drainable circuits (Figure 1) that are self-venting. The self-venting design is maintained by field connecting the water supply connection to the bottom and the water return connection to the top of the coil. Ideally, water is supplied at the bottom, flows upward through the coil, and forces any existing air out the return connection. Complete fluid draining at the supply connection is indicative of coils being self-draining and without air or water traps. Such a design ensures that the coil is always filled with water, and it should completely drain when it is required to be empty. Most manufacturers provide vent and drain fittings, located on the supply and return headers of each water coil.

When water does get trapped in the coil core, it is usually caused by a sag in the coil core; or it can be caused by a nondraining circuit design. During freezing periods, even a small accumulation of water in the coil core can rupture a tube. Also, such a static accumulation of either water or glycol can cause the tube to corrode over an extended period. Large multirow, multicircuited coils may not drain rapidly, even with self-draining circuitry; and, if they are not installed level, complete self draining will not take place. This problem can be prevented by including intermediate drain headers and installing the coil so that is pitched toward the connections.

In order to produce desired ratings without an excessive water pressure drop, manufacturers use various circuit arrangements as shown in Figure 1. A single feed serpentine circuit is commonly used on booster coils with low water flows. With this arrangement, a single-feed carrying the entire water flow makes a number of passes across the airstream. The more common circuit arrangement is called a **full row feed** or **standard circuit**. With this design, all the core tubes of a row are fed with an equal amount of water from the supply header. Others, such as quarter, half, and double row feed circuit arrangements may be available, depending on the total number of tubes and rows of the coil. Uniform water flow in each water circuit is obtained by designing each circuit's length as equal to the other as possible.

The selection of the circuit is based on allowable waterside pressure drop and water velocity. Generally, a higher velocity provides a greater capacity and a more even discharge air temperature across the coil face, but with diminishing returns. At higher velocities only modest gains in capacity can be achieved at increasingly higher pumping power penalties. With water velocities above 8 fps, any gain would be negligible. (In order to prevent erosion, a water velocity of 6 fps should not be exceeded for copper coils.)

Water velocities at the lower end of the practical velocity range risk operating in laminar flow, which reduces capacity substantially. Glycol systems reach a practical limit at significantly lower velocities than water only systems (depending on the percent glycol solution and its temperature). Velocities with fluid flow Reynolds numbers between 2000 and 10,000 fall into a transition range where heat transfer capacity predictions are less likely to be accurately computed. Below 2000 Reynolds number the flow is laminar, where heat transfer prediction is again reliable, but the coil capacity is greatly diminished and tube fouling can become a problem. For further insight on the transition flow effect on capacity, refer to Figure 16 of Air-Conditioning and Refrigeration Institute (ARI) *Standard* 410. Methods of controlling water coils to produce a uniform exit air temperature are discussed in Chapter 45 of the 1999 *ASHRAE Handbook—Applications*.

In certain cases, the hot water circulated may contain a considerable amount of sand and other foreign matter such as minerals. Such matter should be filtered from the water circuit. Additionally, some coil manufacturers offer removable water header boxes (some are plates), or a removable plug at the return bends of each tube; allowing the tubes to be rodded clean. In an area where buildup of scale or other types of deposits are expected, the designer should include a fouling factor when computing heating coil performance. Hot water coil ratings to the ARI *Standard* 410 includes a 0.00025 $h \cdot ft^2 \cdot °F/Btu$ fouling factor. Cupronickel, red brass, admiralty, stainless steel, and other tube alloys are usable to protect against corrosion and erosion effects, which are, at times, common in hot water/glycol systems.

Volatile Refrigerant Heat Reclaim Coils

A heat reclaim coil with a volatile refrigerant can function as a condenser either in series or parallel with the primary condenser of a refrigeration system. Heat from condensing vapor or desuperheating vapor warms the airstream. It can be used as a primary source of heat or to assist some other form of heating, such as reheat for humidity control. In the broad sense, a heat reclaim coil functions as half the heat dissipating capacity of its close coupled refrigerant system's condenser. This is why a heat reclaim coil (1) should be piped to be upstream from the condenser and (2) needs to be designed with the assumption that some condensate must be removed from the coil. For these reasons, the outlet of the coil should be located at the lowest point of the coil and trapped if this location is lower than the inlet of the condenser.

Heat reclaim coils are normally circuited for counterflow of air and refrigerant. However, most supermarket heat reclaim coils are two rows deep and use a crossflow design. The section on Air-Cooled Condensers in Chapter 35 has additional information on this topic. Also, because refrigerant heat reclaim involves specialized heating coil design, a refrigerant equipment manufacturer is the best source for information on the topic.

Electric Heating Coils

An electric heating coil consists of a length of resistance wire (commonly nickel/chromium) to which a voltage is applied. The resistance wire may be bare or sheathed. A sheathed coil has the resistance wire encased by an electrically insulating layer such as magnesium oxide, which is compacted inside a finned steel tube. Sheathed coils are more expensive, have a higher air-side pressure drop, and require more space. A usable comparison for sizing is a heat transfer capacity of 41,000 $Btu/h \cdot ft^2$ of face area compared to 100,000 $Btu/h \cdot ft^2$ for bare resistance wire coils. However, the outer surface temperature of sheathed coils is lower, the coils are mechanically stronger, and contact with personnel or a housing is not as dangerous. Coils with sheathed heating elements having an extended finned surface are generally preferred (1) for dust-laden atmospheres, (2) where there is a high probability of maintenance personnel contact, or (3) downstream from a dehumidifying coil that might have moisture carryover. The manufacturers of electric heating coils have further information about them.

COIL SELECTION

The following factors should be considered in coil selection:

- Required duty or capacity considering other components
- Temperature of air entering the coil and air temperature rise
- Available heating media, its operating and maximum pressure(s) and temperature(s)
- Space and dimensional limitations
- Air volume, velocity, distribution, and limitations
- Heating media volume, flow velocity, distribution, and limitations
- Permissible flow resistances for both the air and heating media
- Characteristics of individual designs and circuit possibilities
- Individual installation requirements, such as the type of control and material compatibility
- Specified and applicable codes and standards regulating the design and installation.

Load requirements are discussed in Chapters 27, 28, and 29 of the 1997 *ASHRAE Handbook—Fundamentals*. Much is based on the choice of heating medium, as well as operating temperatures and core tube diameter. Also, proper selection depends on whether the installation is new, one that is being modified, or simply a replacement coil. Dimensional fit is usually the primary concern of modified and replacement coils, while heating capacity is often unknown.

Air quantity is regulated by such factors as design parameters, codes, space, and size of the components. The resistance through the air circuit influences fan power and speed. This resistance may be limited to allow use of a given size fan motor or to keep the operating expense low. It may also be limited because of sound level requirements. All of these will have an effect on the coil selection. The air friction loss across the heating coil (in summation with other series air pressure drops for such system component as air filters, cooling coils, grills, and ductwork) determines the static pressure requirements of the complete airways system. See Chapter 18 for a description on selecting the fan component.

The permissible resistance through the water or glycol coil circuitry may be dictated by the available pressure from a given size pump and motor. This is usually controlled within limits by careful selection of the coil header size and the number of tube circuits. Additionally, the adverse effect of high fluid velocity on erosion/corrosion of the tube wall is a major factor in selecting tube diameter and the circuit. The performance of a heating coil depends on the correct choice of the original equipment and on proper application and installation. For steam coils, proper performance is based first on selecting the correct type of steam coil, and then the proper size and type of steam trap. Critical to heat reclaim coils are properly sized connecting refrigerant lines, risers, and traps.

Heating coil thermal performance is relatively simple to derive. It only involves a dry-bulb temperature and sensible heat—without the complications of latent load and wet-bulb temperature for dehumidifying cooling performance. Even simpler than calculating coil material and fluid thermal resistances is consulting coil manufactures' catalogs for ratings and selection. Most manufacturers provide computerized coil selection and rating programs upon request; some are certified accurate within 5% of an application parameter, representative of a normal application range. Many manufacturers participate in the ARI Coil Certification Program, approving application ratings that conform to all ARI *Standard* 410 requirements, based on qualifying testing to ASHRAE *Standard* 33.

Coil Ratings

Coil ratings are based on uniform face velocity. A nonuniform airflow may be caused by the system, such air entering at odd angles or by inadvertent blocking of part of the coil face. To obtain rated performance, the airflow quantity in the field must correspond to the design requirements and the velocity vary no greater than 20% at any point across the coil face.

The industry accepted method of coil rating is outlined in ARI *Standard* 410. The test requirements for arriving at standard coil ratings is specified in ASHRAE *Standard* 33. ARI application ratings are derived by extending the ASHRAE standard rating test results for other operating conditions, coil sizes, row depths, and fin count for a particular coil design and arrangement. Steam, water, and glycol heating coils are rated within the following limits (listed in Table 2, ARI *Standard* 410), which may be exceeded for special applications.

Air face velocity. 200 to 1500 fpm, based on air density of 0.075 lb/ft^3

Entering air temperature. Steam coils: −20 to 100°F
Water coils: 0 to 100°F

Steam pressure. 2 to 150 psig at the coil steam supply connection (pressure drop through the steam control valve must be considered)

Fluid temperatures. Water: 120 to 250°F
Ethylene glycol: Up to 200°F

Fluid velocities. Water: 0.5 to 8 fps
Ethylene glycol: 0 to 6 fps

Overall Requirements

Individual installations vary widely, but the following values can be used as a guide. The air face velocity is usually between 500 and 1000 fpm. Delivered air temperature varies from about 72°F for ventilation only to about 150°F for complete heating. Steam pressure typically varies from 2 to 15 psig, with 5 psig being the most common. A minimum steam pressure of 5 psig is recommended for systems with entering air temperatures below freezing. Hot water (or glycol) temperature for comfort heating is commonly between 180 and 200°F, with water velocities between 4 and 6 fps. For high-temperature water, water temperatures can range upwards of 400°Fwith operating pressures of 15 to 25 psi over saturated water temperature.

Water quantity is usually based on about 20°F temperature drop through the coil. Air resistance is usually limited to 0.4 to 0.6 in. of water for commercial buildings and to about 1 in. for industrial buildings. High-temperature water systems commonly have a water temperature between 300 and 400°F, with up to 100°F drop through the coil.

Steam coils are selected with dry steam velocities not exceeding 6000 fpm and with acceptable condensate loading per coil core tube depending on the type of steam coil. The following table shows some typical maximum condensate loads.

Tube Outside Diameter	Maximum Allowable Condensate Load, lb/h	
	Basic Coil	Steam Distributing Coil
5/8 in.	68	40
1 in.	168	95

Maximum performance of steam coils is realized when the supply is dry, saturated steam, and condensate is adequately removed from the coil and continually returned to the boiler.

While the steam quality may not have a significant effect on the heat transfer of the coil, the back-up effect of too rapid a condensate rate, augmented by a wet supply stream, can cause a slug of condensate to travel through the coil and condensate return. This situation can result in having a noisy installation and possible damage.

Air resistance is usually limited to 0.4 to 0.6 in. of water gage for commercial building applications and to about 1 in. of water gage for industrial buildings. Normal hot water (or glycol) quantity is usually based on having a 20°F temperature drop through the coil. High temp. water systems commonly have supply water temperatures between 300 and 400°F, and up to 150°F water temperature drop.

Complete mixing of return and outdoor air is essential to the proper operation of a coil. The design of the air mixing damper or ductwork connection section is critical to the proper operation of a system and its air temperature delivery. Systems in which the air passes through a fan before flowing through a coil do not ensure proper air mixing. Dampers located at the inlet air face of a steam coil are preferred to be the opposing blade type. These are better than in-line blades for controlling air volume and reducing individual blade-directed cold airstreams when modulating in low heat mode.

Heat Transfer and Pressure Drops. For air-side heat transfer and pressure drop, the information given in Chapter 21 for sensible cooling coils is applicable. For water (or glycol) coils, the information given in Chapter 21 for water-side heat transfer and pressure drop is also applicable here. For steam coils, the heat transfer coefficient of condensing steam must be calculated. Chapter 3 of the 1997 *ASHRAE Handbook—Fundamentals* lists several equations for this purpose. For estimation of the pressure drop of condensing steam, see Chapter 5 of the 1997 *ASHRAE Handbook—Fundamentals*.

Parametric Effects. The heat transfer performance of a given coil can be changed by varying the air flow rate and/or the temperature of the heating medium; most of which are relatively linear. Understanding the interaction of these parameters is necessary for designing satisfactory coil capacity and control. A review of manufacturers' catalogs and selection programs, many of which are listed in ARI's Directory of Certified Applied Air-Conditioning Products and CHC: Certified Air-Cooling and Air-Heating Coils, shows the effects of varying these parameters.

APPLICATIONS

Steam systems designed to operate at outdoor air temperatures below 32°F should be different from those designed to operate above it. Below 32°F, the steam air-heating system should be designed as a preheat and reheat pair of coils. The preheat coil functions as a nonmodulating basic steam coil, which requires full steam pressure whenever the outdoor temperature is below freezing. The reheat coil, typically a modulated steam distributing coil, provides the heating required to reach the design air temperature. For temperatures above 32°F, the heating coil can be either a basic or steam distributing type as needed for the duty.

When the leaving air temperature is controlled by modulation of the steam supply to the coil, steam distributing tube-type coils provide the most uniform exit air temperature (see the section on Steam Coils). Correctly designed steam distributing tube coils can limit the exit air temperature stratification to a maximum of 6°F over the entire length of the coil, even when the steam supply is modulated to a fraction of the full-load capacity.

Low-pressure steam systems and coils controlled by modulating the steam supply should have a vacuum breaker or be drained through a vacuum-return system to ensure proper condensate drainage. It is good practice to install a closed vacuum breaker (where one is required) connected to the condensate return line through a check valve. This unit breaks the vacuum by equalizing the pressure, yet minimizes the possibility of air bleeding into the system. Steam traps should be located at least 12 in. below the condensate outlet to allow the coil to drain properly. Also, coils supplied with low-pressure steam or controlled by modulating the steam supply should not be trapped directly to an overhead return line. Condensate can be lifted to overhead returns only when sufficient pressure is available to overcome the condensate head and any return line pressure. If overhead returns are necessary, the condensate must be pumped to the higher elevation. (See Chapter 10.)

Water coils for air heating generally have horizontal tubes to avoid air pockets. Where water coils may be exposed to a freezing condition, drainability must be considered. This also applies when glycol fluid is circulated in the coil. If a coil is to be drained and then exposed to below-freezing temperatures, it should first be flushed with a proper nonfreeze solution.

To minimize the danger of freezing in both steam and water heating coils, the outdoor air inlet dampers usually close automatically when the fan is stopped (system shut down). In steam systems with very cold outdoor air conditions (e.g., −20°F or below), it is desirable to have the steam valve go to full open position when the system is shutdown. If outside air is used for proportioning building makeup air, the outside air damper should be an opposing blade design.

Heat reclaim coils depend on a closely located as well as a readily available source of high-side refrigerant vapor. As an example, supermarket rack compressors and the air handler's coil section should be installed close to the store's motor room. Most common is the heat reclaim coil section piped in series with the rack's condenser and sized for 50% heat extraction. Heat reclaim in supermarkets is discussed in Chapter 2 of the 1999 *ASHRAE Handbook—Applications*. Also consult Chapter 46 of the 1998 *ASHRAE Handbook—Refrigeration*.

Heating coils are designed to allow for expansion and contraction resulting from the temperature ranges in which they operate. Care must be taken to prevent imposing strains from the piping to the coil connections. This is particularly important on high-temperature hot water applications. Expansion loops, expansion or three elbow swing joints, or flexible connections usually provide the needed protection. (See Chapter 10.)

Good practice supports banked coils individually in an angle-iron frame or a similar supporting structure. With this arrangement, the lowest coil is not required to support the weight of the coils stacked above. This design also facilitates the removal of individual coils in a multiple-coil bank for repair or replacement.

COIL MAINTENANCE

Both the internal and external surfaces must be clean for coils to deliver their full rated capacity. The tubes generally stay clean in glycol systems, also in water systems if adequately maintained. Should scale be detected in the piping where untreated water is used, then chemical or mechanical cleaning of the internal tube surfaces is required.

The finned surface of heating coils can sometimes be brushed and cleaned with a vacuum cleaner. Coils are commonly surface cleaned annually using pressurized hot water containing a mild detergent. Reheat coils, that contain their refrigerant charge, should never be cleaned with a spray that is above 150°F. In extreme cases of neglect, especially in restaurants where grease and dirt have accumulated, the coil(s) may need to be removed in order to completely clean off the accumulation with steam, compressed air and water, or hot water containing a suitable detergent. Pressurized cleaning is more thorough if first done from the coil's air leaving side, prior to doing a similar spray cleaning from the coil's air entry side. Often, outside makeup air coils have no upstream air filters, so they should be visually checked on a frequent schedule. Overall, coils should be inspected and serviced on a regular preventive maintenance schedule. Visual observation should not be relied on to judge cleaning requirements for coils greater than three rows in depth because airborne dirt has a tendency to pack midway through the depth of the coil.

REFERENCES

ARI 1991. Forced-circulation air-cooling and air-heating coils. *Standard* 410-91. Air-Conditioning and Refrigeration Institute, Arlington, VA.

ARI. 1996. CHC: Certified air-cooling and air-heating coils.

ARI. 1996. Directory of certified applied air-conditioning products.

ASHRAE. 1978. Methods of testing forced circulation air cooling and air heating coils. *Standard* 33-1978.

ASME. 1998. *Boiler and Pressure Vessel Code*. American Society of Mechanical Engineers, New York.

CHAPTER 24

AIR CLEANERS FOR PARTICULATE CONTAMINANTS

THIS chapter discusses the cleaning of particulate contaminants from both ventilation air and recirculated air for the conditioning of building interiors. Complete air cleaning may also require the removal of airborne particles, microorganisms, and gaseous contaminants, but this chapter only covers the removal of airborne particles and briefly discusses bioaerosols.

The total suspended particulate concentration in the applications discussed in the chapter seldom exceeds 2 mg/m^3 and is usually less than 0.2 mg/m^3 of air. This contrasts with flue gas or exhaust gas from processes, etc., where dust concentration typically ranges from 200 to 40,000 mg/m^3. Chapter 44 of the 1999 *ASHRAE Handbook—Applications* covers the removal of gaseous contaminants; Chapter 25 of this volume covers exhaust-gas control.

With certain exceptions, the air cleaners addressed in this chapter are not used in exhaust gas streams, mainly because of the extreme dust concentration and temperature. However, the principles of air cleaning do apply to exhaust streams, and air cleaners discussed in the chapter are used extensively in supplying gases of low particulate concentration to industrial processes.

ATMOSPHERIC DUST

Atmospheric dust is a complex mixture of smokes, mists, fumes, dry granular particles, bioaerosols, and natural and synthetic fibers. (When suspended in a gas, this mixture is called an **aerosol**.) A sample of atmospheric dust usually contains soot and smoke, silica, clay, decayed animal and vegetable matter, organic materials in the form of lint and plant fibers, and metallic fragments. It may also contain living organisms, such as mold spores, bacteria, and plant pollens, which may cause diseases or allergic responses. (Chapter 12 of the 1997 *ASHRAE Handbook—Fundamentals* contains further information on atmospheric contaminants.) A sample of atmospheric dust gathered at any point generally contains materials common to that locality, together with other components that originated at a distance but were transported by air currents or diffusion. These components and their concentrations vary with the geography of the locality (urban or rural), season of the year, weather, direction and strength of the wind, and proximity of dust sources.

Aerosol sizes range from 0.01 µm and smaller for freshly formed combustion particles and radon progeny; to 0.1 µm for aged cooking and cigarette smokes; to 0.1 to 10 µm for airborne dust, microorganisms, and allergens; and up to 100 µm and larger for airborne soil, pollens, and allergens.

Concentrations of atmospheric aerosols are generally at a maximum at submicrometre sizes and decrease rapidly as the particulate size increases above 1 µm. For a given size, the concentration can vary by several orders of magnitude over time and space, particularly in the vicinity of an aerosol source. Such sources include human activities, equipment, furnishings, and pets (McCrone et al. 1967). This wide range of particulate size and concentration makes it impossible to design one cleaner to serve all applications.

The preparation of this chapter is assigned to TC 2.4, Particulate Air Contaminants and Particulate Contaminant Removal Equipment.

VENTILATION AIR CLEANING

Different fields of application require different degrees of air cleaning effectiveness. In industrial ventilation, removing only the larger dust particles from the airstream may be necessary for cleanliness of the structure, protection of mechanical equipment, and employee health. In other applications, surface discoloration must be prevented. Unfortunately, the smaller components of atmospheric dust are the worst offenders in smudging and discoloring building interiors. Electronic air cleaners or medium- to high-efficiency dry filters are required to remove smaller particles, especially the respirable fraction, which often must be controlled for health reasons. In cleanroom applications or when radioactive or other dangerous particles are present, high- or ultrahigh-efficiency filters should be selected. For more information on cleanrooms, see Chapter 15 of the 1999 *ASHRAE Handbook—Applications*.

The characteristics of aerosols that most affect the performance of an air filter include particle size and shape, mass, concentration, and electrical properties. The most important of these is particle size. Figure 1 in Chapter 12 of the 1997 *ASHRAE Handbook—Fundamentals* gives data on the sizes and characteristics of a wide range of airborne particles that may be encountered.

Particle size may be defined in numerous ways. Particles smaller than 2.5 µm in diameter are generally referred to as the **fine mode** and those larger than 2.5 µm as the **coarse mode**. The fine and coarse mode particles typically originate by separate mechanisms, are transformed separately, have different chemical compositions, and require different control strategies. Fine mode particles generally originate from condensation processes or are directly emitted as combustion products. Many microorganisms (bacteria and fungi) either are in this size range or produce parts or components this size. These particles are less likely to be removed by gravitational settling and are just as likely to deposit on vertical surfaces as on horizontal surfaces. Coarse mode particles are typically produced by mechanical actions such as erosion and frictional wear. Coarse particles are more easily removed by gravitational settling, and thus have a shorter lifetime in the airborne state.

For industrial hygiene, particles 5 µm in diameter or smaller are considered **respirable particles** (RSPs) because a large percentage of them may reach the alveolar region of the lungs. Willeke and Baron (1993) describe a detailed aerosol sampling technique. They discuss the need for an impactor with a 2.5 µm cutoff for RSPs that can be deposited in the alveolar region.

The cutoff for particles affecting on respiratory function is considered to be 2.0 or 2.5 µm. See the discussion in the section on Nature of Airborne Contaminants in Chapter 12 of the 1997 *ASHRAE Handbook—Fundamentals.* A cutoff of 5.0 µm includes 80 to 90% of the particles that can reach the functional pulmonary region of the lungs (James et al. 1991, Phalen et al. 1991).

Particle size in this discussion refers to aerodynamic particle size. Therefore, larger particles with lower densities may be found in the alveolar region. Also note that fibers are different than particles in that the fiber's shape, diameter, and density affect where a fiber settles in the body (NIOSH 1973).

Cleaning efficiency with certain media is affected to a negligible extent by the velocity of the airstream. Major factors influencing filter design and selection include (1) degree of air cleanliness required, (2) specific particle size range or aerosols that require filtration, (3) aerosol concentration, and (4) resistance to airflow through the filter.

EVALUATING AIR CLEANERS

In addition to criteria affecting the degree of air cleanliness, factors such as cost (initial investment and maintenance), space requirements, and airflow resistance have encouraged the development of a wide variety of air cleaners. Accurate comparisons of different air cleaners can be made only from data obtained by standardized test methods.

The three operating characteristics that distinguish the various types of air cleaners are efficiency, resistance to airflow, and dust-holding capacity. **Efficiency** measures the ability of the air cleaner to remove particles from an airstream. Average efficiency during the life of the filter is the most meaningful for most filters and applications. However, because the efficiency of many dry-type filters increases with dust load, the initial (clean filter) efficiency should be considered for design in applications with low dust concentrations. **Resistance to airflow** (or simply resistance) is the static pressure drop across the filter at a given airflow rate. The term pressure drop is used interchangeably with resistance. **Dust-holding capacity** defines the amount of a particular type of dust that an air cleaner can hold when it is operated at a specified airflow rate to some maximum resistance value (ASHRAE *Standard* 52.1).

Complete evaluation of air cleaners therefore requires data on efficiency, resistance, dust-holding capacity, and the effect of dust loading on efficiency and resistance. When applied to automatic renewable media devices (roll filters, for example), the evaluation must include the rate at which the media is supplied to maintain constant resistance when standardized dust is fed at a specified rate. When applied to electronic air cleaners, the effect of dust buildup on efficiency must be evaluated.

Air filter testing is complex and no individual test adequately describes all filters. Ideally, performance testing of equipment should simulate the operation of the device under actual conditions and furnish an evaluation of the characteristics important to the equipment user. Wide variations in the amount and type of particles in the air being cleaned make evaluation difficult. Another complication is the difficulty of closely relating measurable performance to the specific requirements of users. Recirculated air tends to have a larger proportion of lint than does outside air. However, these difficulties should not obscure the principle that tests should simulate actual use as closely as possible.

In general, four types of tests, together with certain variations, determine air cleaner performance: (1) arrestance, (2) dust-spot efficiency, (3) fractional efficiency, and (4) efficiency by particle size.

1. **Arrestance.** A standardized synthetic dust consisting of various particle sizes and types is fed into the test air stream to the air cleaner and the weight fraction of the dust removed is determined. In the ASHRAE *Standard* 52.1 test, which is summarized in the section on Air Cleaner Test Methods in this chapter, this measurement is called **synthetic dust weight arrestance** to distinguish it from other efficiency values.

   The indicated weight arrestance of air filters, as determined in the arrestance test, depends greatly on the particle size distribution of the test dust, which, in turn, is affected by its state of agglomeration. Therefore, this filter test requires a high degree of standardization of the test dust, the dust dispersion apparatus, and other elements of test equipment and procedures. This test is particularly suited to low- and medium-efficiency air filters that are most commonly used on recirculating systems. It does not distinguish between filters of higher efficiency.
2. **Dust-spot efficiency.** Atmospheric conditioned air is passed into the air cleaner, and the discoloration effect of the cleaned air on filter paper targets is compared with that of the incoming air. This type of measurement is called atmospheric dust-spot efficiency.

   The dust-spot test measures the ability of a filter to reduce the soiling of fabrics and building interior surfaces. Because these effects depend mostly on fine particles, this test is most useful for high-efficiency filters. The variety and variability of atmospheric dust (McCrone et al. 1967, Whitby et al. 1958, Horvath 1967) may cause the same filter to test at different efficiencies at different locations (or even at the same location at different times). This disadvantage of the test is more strongly apparent in low-efficiency filters.
3. **Fractional efficiency** or **penetration.** Uniform-sized particles are fed into the air cleaner and the percentage removed by the cleaner is determined, typically by a photometer, particle counter, or condensation nuclei counter.

   In fractional efficiency tests, the use of uniform particle size aerosols has resulted in accurate measure of the particle size versus efficiency characteristic of filters over a wide atmospheric size spectrum. The method is time-consuming and has been used primarily in research. However, the dioctyl phthalate (DOP) or Emory 3000 test for HEPA filters is widely used for production testing at a narrow particle size range. For more information on the DOP test, see the section on DOP Penetration Test.
4. **Efficiency by particle size.** A polydispersed challenge aerosol is metered into the test airstream to the air cleaner. Air samples taken upstream and downstream are drawn through an optical particle counter or similar measurement device to obtain removal efficiency versus particle size at a specific airflow rate.

Dust Holding Capacity. The exact measurement of true dust-holding capacity is complicated by the variability of atmospheric dust; therefore, standardized artificial loading dust is normally used. Such a dust also shortens the dust-loading cycle to hours instead of weeks or years.

Artificial dusts are not the same as atmospheric dusts, so dust-holding capacity as measured by these accelerated tests are different from that achieved by tests using atmospheric dust. The exact life of a filter in field use is impossible to determine by laboratory testing. However, tests of filters under standard conditions do provide a rough guide to the relative effect of dust on the performance of various units and can be used to compare them.

Reputable laboratories perform accurate and reproducible filter tests. Differences in reported values generally lie within the variability of the test aerosols, measurement devices, and dusts. Because most media are made of random air- or water-laid fibrous materials, the inherent media variations affect filter performance. Awareness of these variations prevents misunderstanding and specification of impossibly close performance tolerances. Caution must be exercised in interpreting published efficiency data, because the performance of two cleaners tested by different procedures generally cannot be compared. A value of air cleaner efficiency is only a guide to the rate of soiling of a space or of mechanical equipment.

AIR CLEANER TEST METHODS

Air cleaner test methods have been developed in several areas: the heating and air-conditioning industry, the automotive industry, the atomic energy industry, and government and military agencies. Several tests have become standard in general ventilation applications in the United States. In 1968, the test techniques developed by the U.S. National Bureau of Standards (now the National Institute of Standards and Technology, NIST) and the Air Filter Institute (AFI) were unified (with minor changes) into a single test procedure, ASHRAE *Standard* 52-1968. Dill (1938), Whitby et al. (1956), and Nutting and Logsdon (1953) give details of the original codes.

ASHRAE *Standard* 52-1968 was revised and is currently ASHRAE *Standard* 52.1-1992.

In general, the ASHRAE *Weight Arrestance Test* parallels that of the AFI, making use of a similar test dust. The ASHRAE *Atmospheric Dust-Spot Efficiency Test* parallels the AFI and NBS *Atmospheric Dust-Spot Efficiency Tests* and specifies a dust-loading technique.

ASHRAE *Standard* 52.1 includes both a weight test and a dust-spot test and requires that values for both be reported. The results may be used for comparisons among air cleaners. ASHRAE *Standard* 52.2 develops **Minimum Efficiency Reporting Values** (MERVs) for air cleaner particle size efficiency. Table 3 (from ASHRAE *Standard* 52.2) provides an approximate cross-reference for air cleaners tested under ASHRAE *Standards* 52.1 and 52.2.

Arrestance Test

ASHRAE *Standard* 52.1 defines synthetic test dust as a compounded test dust consisting of (by weight) 72% ISO 12 103-A2 fine test dust, 23% powdered carbon, and 5% No. 7 cotton linters.

A known amount of the prepared test dust is fed into the test unit at a known and controlled rate. The concentration of dust in the air leaving the filter is determined by passing the entire airflow through a high-efficiency after-filter and measuring the gain in filter weight. The **arrestance** is calculated using the weights of the dust passing the tested device and the total dust fed.

Atmospheric dust particles range in size from a small fraction of a micrometre to tens of micrometres in diameter. The artificially generated dust cloud used in the ASHRAE weight arrestance method is considerably coarser than typical atmospheric dust. It tests the ability of a filter to remove the largest atmospheric dust particles and gives little indication of the filter performance in removing the smallest particles. But, where the mass of dust in the air is the primary concern, this is a valid test because most of the mass is contained in the larger particles. Where extremely small particles are involved, the weight arrestance method of rating does not differentiate between filters.

Atmospheric Dust-Spot Efficiency Test

One objectionable characteristic of finer airborne dust particles is their capacity to soil walls and other interior surfaces. The discoloring rate of white, filter-paper targets (microfine glass fiber HEPA filter media) filtering samples of air constitutes an accelerated simulation of this effect. By measuring the change in light transmitted by these targets, the efficiency of the filter in reducing the soiling of surfaces may be computed.

ASHRAE *Standard* 52.1 specifies two equivalent atmospheric dust-spot test procedures, each taking a different approach to correct for the nonlinearity of the relation between the discoloration of target papers and their dust load. In the first procedure, called the **intermittent-flow method**, samples of conditioned atmospheric air are drawn upstream and downstream of the tested filter. These samples are drawn at equal flow rates through identical targets of glass fiber filter paper. The downstream sample is drawn continuously; the upstream sample is interrupted in a timed cycle so that the average rate of discoloration of the upstream and downstream targets is approximately equal. The percentage of off-time approximates the efficiency of the filter. (See ASHRAE *Standard* 52.1 for details.)

In the alternate procedure, called the **constant-flow method**, conditioned atmospheric air samples are also drawn at equal flow rates through equal-area glass fiber filter paper targets upstream and downstream, but without interrupting either sample. Discoloration of the upstream target is therefore greater than for the downstream target. Sampling is halted when the upstream target light transmission has dropped by at least 10% but no more than 40%. The **opacities** (the percent change in light transmission) are then calculated for the targets as defined for the intermittent flow method. These opacities are next converted into **opacity indices** to correct for nonlinearity.

The advantage of the constant-flow method is that it takes the same length of time to run regardless of the efficiency of the filter, whereas the intermittent-flow method takes longer for higher efficiency filters. For example, an efficiency test run on a 90% efficient filter using the intermittent-flow method requires ten times as long as a test run using the constant-flow method.

The standard allows dust-spot efficiencies to be taken at intervals during an artificial dust-loading procedure. This characterizes the change of dust-spot efficiency as dust builds up on the filter in service.

Dust-Holding Capacity Test

The synthetic test dust is fed to the filter in accordance with ASHRAE *Standard* 52.1 procedures. The pressure drop across the filter (its resistance) rises as dust is fed. The test is normally terminated when the resistance reaches the maximum operating resistance set by the manufacturer. However, not all filters of the same type retain collected dust equally well. The test, therefore, requires that arrestance be measured at least four times during the dust-loading process and that the test be terminated when two consecutive arrestance values of less than 85%, or one value equal to or less than 75% of the maximum arrestance, have been measured. The ASHRAE **dust-holding capacity** is, then, the integrated amount of dust held by the filter up to the time the dust-loading test is terminated. (See ASHRAE *Standard* 52.1 for more detail.)

A typical set of curves for an ASHRAE air filter test report on a fixed cartridge-type filter is shown in Figure 1. Both synthetic dust weight arrestance and atmospheric dust-spot efficiencies are shown. The standard also specifies how self-renewable devices are to be loaded with dust to establish their performance under standard

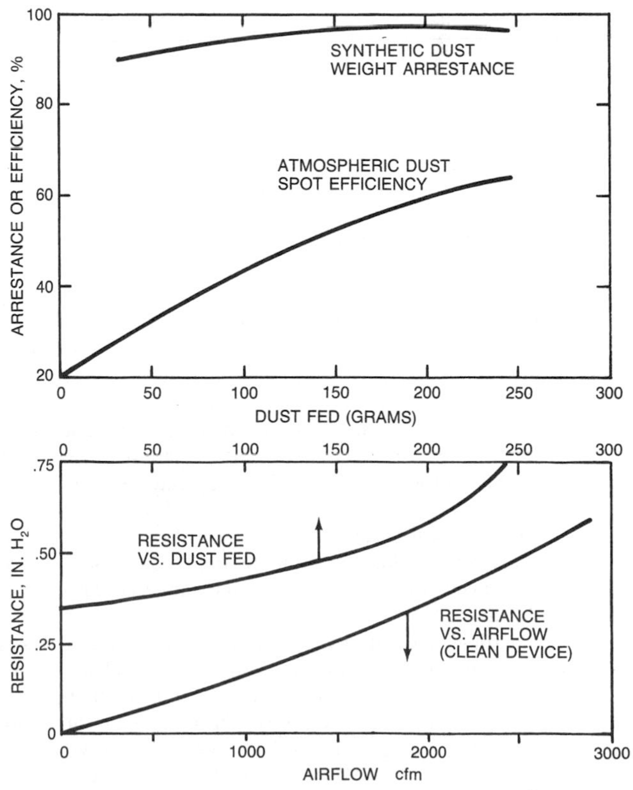

Fig. 1 Typical Performance Curves for Fixed Cartridge-Type Filter According to ASHRAE *Standard* 52.1

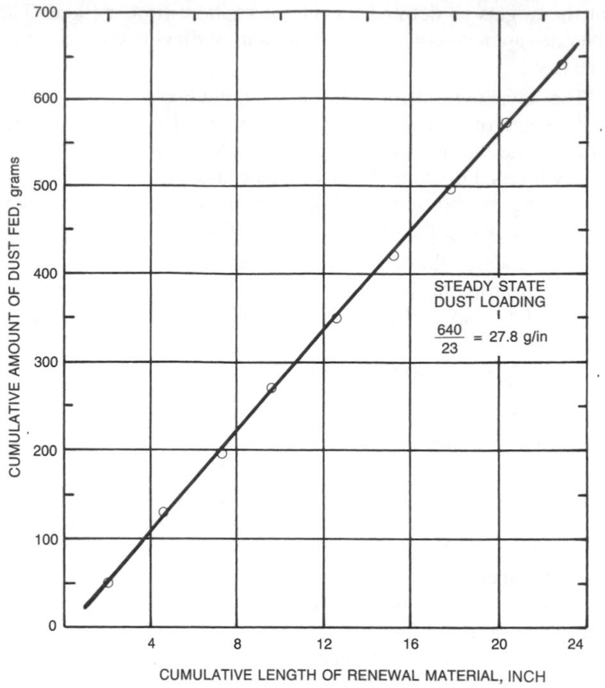

Note: Calculate loading per unit by multiplying steady-state dust loading by 144 and dividing by the media width, which is normally 24 in. For example, 27.8 × 144/24 = 167 g/ft^2.

Fig. 2 Typical Dust-Loading Graph for Self-Renewable Air Filter

conditions. Figure 2 shows the results of such a test on an automatic roll media filter.

Particle Size Removal Efficiency Test

ASHRAE *Standard* 52.2 prescribes a method of testing air-cleaning devices for removal efficiency by particle size while addressing two air cleaner performance characteristics of importance to users. These characteristics are the ability of the device to remove particles from the airstream and its resistance to airflow. In this standard method, air cleaner testing is conducted at a specific airflow based on one of seven face velocities selected by the party paying for the test. Airflow may not be less than 470 cfm nor greater than 2990 cfm in the 24 in. by 24 in. test section (face velocity between 120 and 750 fpm). The test aerosol consists of laboratory-generated potassium chloride particles dispersed in the airstream. An optical particle counter(s) measures and counts the particles in 12 geometric equally distributed particle size ranges both upstream and downstream for efficiency determinations. The size range encompassed by the test is 0.3 to 10 μm polystyrene latex equivalent optical particle size. A method of loading the air cleaner with synthetic dust to simulate field conditions is also specified. The synthetic loading dust is the same as that used in ASHRAE *Standard* 52.1.

A set of particle size removal efficiency performance curves is developed from the test, and together with an initial clean performance curve, is the basis of a composite curve representing performance in the range of sizes. Points on the composite curve are averaged and these averages are used to determine the Minimum Efficiency Reporting Value (MERV) of the air cleaner. A complete test report includes (1) a summary section, (2) removal efficiency curves of the clean devices at each of the loading steps, and (3) a composite minimum removal efficiency curve.

DOP Penetration Test

For high-efficiency filters of the type used in cleanrooms and nuclear applications (HEPA filters), the normal test in the United States is the Thermal DOP method, as outlined in U.S. Military Standard MIL-STD-282 (1956) and U.S. Army document 136-300-175A (1965). DOP is dioctyl phthalate or bis-[2-ethylhexyl] phthalate, which is an oily liquid with a high boiling point. In this method, a smoke cloud of DOP droplets condenses from DOP vapor.

The count median diameter for DOP aerosols is about 0.18 μm, while the mass median diameter is about 0.27 μm with a cloud concentration of approximately 80 mg/m^3 under properly controlled conditions. The procedure is sensitive to the mass median diameter, and DOP test results are commonly referred to as efficiency on 0.3 μm particles.

The DOP smoke cloud is fed to the filter, which is held in a special test chuck. Any smoke that penetrates the body of the filter or leaks through gasket cracks passes into the region downstream from the filter where it is thoroughly mixed. The air leaving the chuck thus contains the average concentration of penetrating smoke. This concentration, as well as the upstream concentration, is measured by a light-scattering photometer. The **filter penetration** *P* in percent is

$$P = 100\left(\frac{\text{Downstream concentration}}{\text{Upstream concentration}}\right) \quad (1)$$

Penetration, not efficiency, is usually specified in the test procedure because HEPA filters have efficiencies so near 100%, for example, 99.97% or 99.99% on 0.3 μm particles. The two terms are related by the equation $E = 100 - P$.

U.S. specifications frequently call for the testing of HEPA filters at both rated flow and 20% of rated flow. This procedure helps detect gasket leaks and pinholes that would otherwise escape notice. Such defects, however, are not located by the DOP penetration test.

The Institute of Environmental Sciences and Technology has published two recommend practices: IEST RP-CC 001.3, HEPA and ULPA Filters, and IEST RP-CC 007.1, Testing ULPA Filters.

Leakage (Scan) Tests

In the case of HEPA filters, leakage tests are sometimes desirable to show that no small "pinhole" leaks exist or to locate any that may exist so they may be patched. Essentially the same technique as used in the DOP Penetration Test is employed, except that the downstream concentration is measured by scanning the face of the filter and its gasketed perimeter with a moving probe. The exact point of smoke penetration can then be located and repaired. This same test (described in IEST RP-CC 001.3) can be performed after the filter is installed; in this case, a portable aspirator-type DOP generator is used instead of the bulky thermal generator. The smoke produced by a portable generator is not uniform in size, but its average diameter can be approximated as 0.6 μm. Particle diameter is less critical for leak location than for penetration measurement.

Specialized Performance Test

AHAM *Standard* AC-1 describes a method for measuring the ability of portable household air cleaners to reduce generated particles suspended in the air in a room-size test chamber. The procedure compares the natural decay of three contaminants (dust, smoke, and pollen) with the reduction in particles by the air cleaner.

Miscellaneous Performance Tests

The European Standardization Institute (Comité Européen de Normalisation or CEN) developed EN 779, a standard for particulate air filters (CEN 1994) and Eurovent/Cecomaf 4/9. In addition, special test standards have been developed for respirator air filters (NIOSH/MSHA 1977) and ULPA filters (IEST RP-CC 007.1).

Environmental Tests

Air cleaners may be subjected to fire, high humidity, a wide range of temperatures, mechanical shock, vibration, and other environmental stress. Several standardized tests exist for evaluating these environmental effects on air cleaners. U.S. Military Standard MIL-STD-282 includes shock tests (shipment rough handling) and filter media water-resistance tests. Several U.S. Atomic Energy Commission agencies (now part of the U.S. Department of Energy) specify humidity and temperature-resistance tests (Peters 1962, 1965).

Underwriters Laboratories has two major standards for air cleaner flammability. The first, for commercial applications, determines flammability and smoke production. UL *Standard* 900 Class 1 filters are those that, when clean, do not contribute fuel when attacked by flame and emit only negligible amounts of smoke. UL *Standard* 900 Class 2 filters are those that, when clean, burn moderately when attacked by flame or emit moderate amounts of smoke, or both. In addition, UL *Standard* 586 for flammability of HEPA filters has been established. The UL tests do not evaluate the effect of collected dust on filter flammability; depending on the dust, this effect may be severe. UL *Standard* 867 applies to electronic air cleaners.

ARI Standards

The Air-Conditioning and Refrigeration Institute has published ARI *Standard* 680 and ARI *Standard* 850 for air filter equipment. These standards establish (1) definitions and classification; (2) requirements for testing and rating (performance test methods are per ASHRAE *Standard* 52.1); (3) specification of standard equipment; (4) performance and safety requirements; (5) proper marking; (6) conformance conditions; and (7) literature and advertising requirements.

MECHANISMS OF PARTICLE COLLECTION

In the collection of particles, air cleaners rely on the following five main principles or mechanisms:

Straining. The coarsest kind of filtration strains particles through a membrane opening smaller than the particle being removed. It is most often observed as the collection of large particles and lint on the filter surface. The mechanism is not adequate to explain the filtration of submicrometre aerosols through fibrous matrices, which occurs through other theoretical mechanisms, as follows.

Direct Interception. The particles follow a fluid streamline close enough to a fiber that the particle contacts the fiber and remains there. The process is nearly independent of velocity.

Inertial Deposition. Particles in the airstream are large enough or of sufficient density that they cannot follow the fluid streamlines around a fiber; thus, they cross over streamlines, contact the fiber, and remain there. At high velocities (where the effect of inertia is most pronounced), the particle may not adhere to the fiber because drag and bounce forces are so high. In this case, a viscous coating is applied to the fiber to improve adhesion of the particles and is critical to the performance of an adhesive-coated, wire screen impingement filter.

Diffusion. Very small particles have random motion about their basic streamlines (Brownian motion), which contributes to deposition on the fiber. This deposition creates a concentration gradient in the region of the fiber, further enhancing filtration by diffusion. The effects increase with decreasing particle size and velocity.

Electrostatic Effects. Particle or media charging can produce changes in the collection of dust.

Some progress has been made in calculating theoretical filter media efficiency from the physical constants of the media by considering the effects of the collection mechanisms (Lee and Liu 1982a, 1982b; Liu and Rubow 1986).

TYPES OF AIR CLEANERS

Common air cleaners are broadly grouped as follows:

Fibrous media unit filters, in which the accumulating dust load causes pressure drop to increase up to some maximum recommended value. During this period, efficiency normally increases. However, at high dust loads, dust may adhere poorly to filter fibers and efficiency drops due to off-loading. Filters in such condition should be replaced or reconditioned, as should filters that have reached their final (maximum recommended) pressure drop. This category includes viscous impingement and dry-type air filters, available in low-efficiency to ultrahigh-efficiency construction.

Renewable media filters, in which fresh media is introduced into the airstream as needed to maintain essentially constant resistance and, consequently, constant average efficiency.

Electronic air cleaners, which, if maintained properly by regular cleaning, have relatively constant pressure drop and efficiency.

Combination air cleaners, which combine the above types. For example, an electronic air cleaner may be used as an agglomerator with a fibrous media downstream to catch the agglomerated particles blown off the plates. Electrode assemblies have been installed in air-handling systems, making the filtration system more effective (Frey 1985, 1986). Also, a renewable media filter may be used upstream of a high-efficiency unit filter to extend its life. Charged media filters are also available that increase particle deposition on media fibers by an induced electrostatic field. In this case, pressure loss increases as it does on a fibrous media filter. The benefits of combining different air cleaning processes vary. ASHRAE *Standard* 52.1 test methods may be used to compare the performance of combination air cleaners.

FILTER TYPES AND THEIR PERFORMANCE

Panel Filters

Viscous Impingement Filters. These are panel filters made up of coarse fibers with a high porosity. The filter media are coated with a viscous substance, such as oil (also known as adhesive), which causes particles that impinge on the fibers to stick to them. Design air velocity through the media is usually in the range of 200 to 800 fpm. These filters are characterized by low pressure drop, low cost, and good efficiency on lint but low efficiency on normal atmospheric dust. They are commonly made 1/2 to 4 in. thick. Unit panels are available in standard and special sizes up to about 24 in. by 24 in. This type of filter is commonly used in residential furnaces and air conditioning and is often used as a prefilter for higher-efficiency filters.

A number of different materials are used as the filtering medium, including coarse (15 to 60 μm diameter) glass fibers, coated animal hair, vegetable fibers, synthetic fibers, metallic wools, expanded metals and foils, crimped screens, random-matted wire, and synthetic open-cell foams.

Although viscous impingement filters usually operate in the range of 300 to 600 fpm they may be operated at higher velocities. The limiting factor, other than increased flow resistance, is the danger of blowing off agglomerates of collected dust and the viscous coating on the filter.

The loading rate of a filter depends on the type and concentration of the dirt in the air being handled and the operating cycle of the system. Manometers, static pressure gages, or pressure transducers are often installed to measure the pressure drop across the filter bank. From the pressure drop, it can be determined when the filter requires servicing. The final allowable pressure drop may vary from one installation to another; but, in general, unit filters are serviced when their operating resistance reaches 0.5 in. of water. The decline in filter efficiency (which is caused by the absorption of the viscous coating by dust, rather than by the increased resistance because of dust load) may be the limiting factor in operating life.

The manner of servicing unit filters depends on their construction and use. Disposable viscous impingement, panel-type filters are constructed of inexpensive materials and are discarded after one period of use. The cell sides of this design are usually a combination of cardboard and metal stiffeners. Permanent unit filters are generally constructed of metal to withstand repeated handling. Various cleaning methods have been recommended for permanent filters; the most widely used involves washing the filter with steam or water (frequently with detergent) and then recoating it with its recommended adhesive by dipping or spraying. Unit viscous filters are also sometimes arranged for in-place washing and recoating.

The adhesive used on a viscous impingement filter requires careful engineering. Filter efficiency and dust-holding capacity depend on the specific type and quantity of adhesive used; this information is an essential part of test data and filter specifications. Desirable adhesive characteristics, in addition to efficiency and dust-holding capacity, are (1) a low percentage of volatiles to prevent excessive evaporation; (2) a viscosity that varies only slightly within the service temperature range; (3) the ability to inhibit growth of bacteria and mold spores; (4) a high capillarity or the ability to wet and retain the dust particles; (5) a high flash point and fire point; and (6) freedom from odorants or irritants.

Typical performance of viscous impingement unit filters operating within typical resistance limits is shown as Group I in Figure 3.

Dry-Type Extended-Surface Filters. The media in dry-type air filters are random fiber mats or blankets of varying thicknesses, fiber sizes, and densities. Bonded glass fiber, cellulose fibers, wool felt, synthetics, and other materials have been used commercially. The media in filters of this class are frequently supported by a wire frame in the form of pockets, or V-shaped or radial pleats. In other designs, the media may be self-supporting because of inherent rigidity or because airflow inflates it into extended form such as with bag filters. Pleating of the media provides a high ratio of media area to face area, thus allowing reasonable pressure drop and low media velocities.

In some designs, the filter media is replaceable and is held in position in permanent wire baskets. In most designs, the entire cell is discarded after it has accumulated its maximum dust load.

The efficiency of dry-type air filters is usually higher than that of panel filters, and the variety of media available makes it possible to furnish almost any degree of cleaning efficiency desired. The dust-holding capacities of modern dry-type filter media and filter configurations are generally higher than those of panel filters.

The placement of coarse prefilters ahead of extended-surface filters is sometimes justified economically by the longer life of the main filters. Economic considerations should include the prefilter material cost, changeout labor, and increased fan power. Generally, prefilters should be considered only if they can substantially reduce the part of the dust that may plug the protected filter. A prefilter usually has an arrestance efficiency of 70% or more. Temporary prefilters are worthwhile during building construction to capture heavy loads of coarse dust. HEPA-type filters of 95% DOP efficiency and greater should always be protected by prefilters of 80% or greater ASHRAE atmospheric dust-spot efficiency. A single filter gage may be installed when a panel prefilter is placed adjacent to a final filter. Because the prefilter is normally changed on a schedule, the final filter pressure drop can be read without the prefilter in place every time the prefilter is changed.

Typical performance of some types of filters in this group, when they are operated within typical rated resistance limits and over the life of the filters, is shown as Groups II and III in Figure 3.

The initial resistance of an extended-surface filter varies with the choice of media and the filter geometry. Commercial designs typically have an initial resistance from 0.1 to 1.0 in. of water. It is customary to replace the media when the final resistance of 0.5 in. of water is reached for low resistance units and 2.0 in. of water for the highest resistance units. Dry media providing higher orders of cleaning efficiency have a higher average resistance to airflow. The operating resistance of the fully dust-loaded filter must be considered in the design, because that is the maximum resistance against which the fan operates. Variable-air-volume and constant air volume system controls prevent abnormally high airflows or possible fan motor overloading from occurring when filters are clean.

Flat panel filters with media velocity equal to duct velocity are possible only in the lowest efficiency units of the dry type (open cell foams and textile denier nonwoven media). Initial resistance of this group, at rated airflow, is mainly between 0.05 and 0.25 in. of water. They are usually operated to a final resistance of 0.50 to 0.70 in. of water.

In extended-surface filters of the intermediate efficiency ranges, the filter media area is much greater than the face area of the filter; hence, velocity through the filter media is substantially lower than the velocity approaching the filter face. Media velocities range from 6 to 90 fpm, although the approach velocities run to 750 fpm. Depth in direction of airflow varies from 2 to 36 in.

Filter media used in the intermediate efficiency range include those of (1) fine glass or synthetic fibers, 0.7 to 10 μm in diameter, in mat form up to 1/2 in. thick; (2) thin nonwoven mats of fine glass fibers, cellulose, or cotton wadding; and (3) nonwoven mats of comparatively large diameter fibers (more than 30 μm) in greater thicknesses (up to 2 in.).

Electret filters are composed of electrostatically charged fibers. The charges on the fibers augment the collection of smaller particles by Brownian diffusion with Coulomb forces caused by the charges on the fibers. There are three types of these filters: resin wool, electret, and an electrostatically sprayed polymer. The charge on the resin wool fibers is produced by friction during the carding process. During production of the electret, a corona discharge injects positive charges on one side of a thin polypropylene film and negative

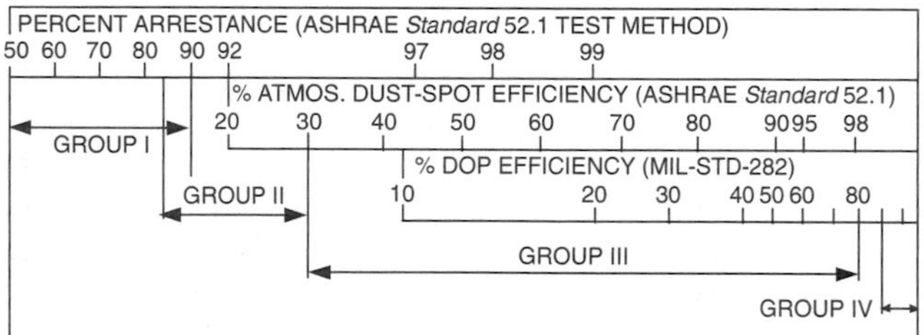

Group I
Panel-type filters of spun glass, open cell foams, expanded metal and screens, synthetics, textile denier woven and nonwoven, or animal hair.

Group II
Pleated panel-type filters of fine denier nonwoven synthetic and synthetic-natural fibers blends, or all natural fiber.

Group III
Extended-surface supported and nonsupported filters of fine glass fibers, fine electret synthetic fibers, or wet-laid paper of cellulose-glass, synthetic, or all-glass fibers.

Group IV
Extended-area pleated HEPA-type filters of wet-laid ultrafine glass fiber paper. Biological grade air filters are generally 95% DOP efficiency; HEPA filters are 99.97% and 99.99%; and ULPA filters are 99.999%.

Notes:
1. Group numbers have no significance other than their use in this figure.
2. Correlation between the test methods shown are approximations for general guidance only.

Fig. 3 Comparative Performance of Viscous Impingement and Dry-Media Filters

Table 1 Performance of Renewable Media Filters (Steady-State Values)

Description	Type of Media	ASHRAE Weight Arrestance, %	ASHRAE Atmospheric Dust-Spot Efficiency, %	ASHRAE Dust-Holding Capacity, g/ft^2	Approach Velocity, fpm
20 to 40 μm glass and synthetic fibers, 2 to 2 1/2 in. thick	Viscous impingement	70 to 82	<20	60 to 180	500
Permanent metal media cells or overlapping elements	Viscous impingement	70 to 80	<20	NA (permanent media)	500
Coarse textile denier nonwoven mat, 1/2 to 1 in. thick	Dry	60 to 80	<20	15 to 70	500
Fine textile denier nonwoven mat, 1/2 to 1 in. thick	Dry	80 to 90	<20	10 to 50	200

charges on the other side. These thin sheets are then shredded into fibers of rectangular cross-section. The third process spins a liquid polymer into fibers in the presence of a strong electric field, which produces the charge separation. The efficiency of the charged-fiber filters is due to both the normal collection mechanisms of a media filter and the strong local electrostatic effects. The effects induce efficient preliminary loading of the filter to enhance the caking process. However, dust collected on the media can reduce the efficiency of electret filters.

Very high-efficiency dry filters, HEPA (high-efficiency particulate air) filters, and ULPA (ultralow penetration air) filters, are made in an extended-surface configuration of deep space folds of submicrometre glass fiber paper. Such filters operate at duct velocities near 250 fpm, with resistance rising from 0.5 to more than 2.0 in. of water over their service life. These filters are the standard for clean-room, nuclear, and toxic-particulate applications.

Membrane filters are used predominately for air sampling and specialized small-scale applications where their particular characteristics compensate for their fragility, high resistance, and high cost. They are available in many pore diameters and resistances and in flat sheet and pleated forms.

Renewable Media Filters

Moving-Curtain Viscous Impingement Filters. Automatic moving-curtain viscous filters are available in two main types. In one type, random-fiber (nonwoven) media is furnished in roll form. Fresh media is fed manually or automatically across the face of the filter, while the dirty media is rewound onto a roll at the bottom. When the roll is exhausted, the tail of the media is wound onto the take-up roll, and the entire roll is thrown away. A new roll is then installed, and the cycle is repeated.

Moving-curtain filters may have the media automatically advanced by motor drives on command from a pressure switch, timer, or media light-transmission control. A pressure switch control measures the pressure drop across the media and switches *on* and *off* at chosen upper and lower set points. This control saves media, but only if the static pressure probes are located properly and unaffected by modulating outside air and return air dampers. Most pressure drop controls do not work well in practice. Timers and media light-transmission controls help to avoid these problems; their duty cycles can usually be adjusted to provide satisfactory operation with acceptable media consumption.

Filters of this replaceable roll design generally have a signal indicating when the roll of media is nearly exhausted. At the same time, the drive motor is deenergized so that the filter cannot run out of media. The normal service requirements involve insertion of a clean roll of media at the top of the filter and disposal of the loaded dirty roll. Automatic filters of this design are not, however, limited in application to the vertical position. Horizontal arrangements are available for use with makeup air units and air-conditioning units. Adhesives must have qualities similar to those for panel-type viscous impingement filters, and they must withstand media compression and endure long storage.

The second type of automatic viscous impingement filter consists of linked metal mesh media panels installed on a traveling curtain that intermittently passes through an adhesive reservoir. In the reservoir, the panels give up their dust load and, at the same time, take on a new coating of adhesive. The panels thus form a continuous curtain that moves up one face and down the other face. The media curtain, continually cleaned and renewed with fresh adhesive, lasts the life of the filter mechanism. The precipitated dirt must be removed periodically from the adhesive reservoir.

The resistance of both types of viscous impingement automatically renewable filters remains approximately constant as long as proper operation is maintained. A resistance of 0.40 to 0.50 in. of water at a face velocity of 500 fpm is typical of this class.

Moving-Curtain Dry-Media Filters. Random-fiber (nonwoven) dry media of relatively high porosity are also used in moving-curtain (roll) filters for general ventilation service. Operating duct velocities near 200 fpm are generally lower than those of viscous impingement filters.

Special automatic dry filters are also available, which are designed for the removal of lint in textile mills and dry-cleaning establishments and the collection of lint and ink mist in pressrooms. The medium used is extremely thin and serves only as a base for the buildup of lint, which then acts as a filter medium. The dirt-laden media is discarded when the supply roll is used up.

Another form of filter designed specifically for dry lint removal consists of a moving curtain of wire screen, which is vacuum cleaned automatically at a position out of the airstream. Recovery of the collected lint is sometimes possible with such a device.

Performance of Renewable Media Filters. ASHRAE arrestance, efficiency, and dust-holding capacities for typical viscous impingement and dry renewable media filters are listed in Table 1.

Electronic Air Cleaners

Electronic air cleaners can be highly efficient filters using electrostatic precipitation to remove and collect particulate contaminants such as dust, smoke, and pollen. The designation electronic air cleaner denotes a precipitator for HVAC air filtration. The filter consists of an ionization section and a collecting plate section.

In the ionization section, small-diameter wires with a positive direct current potential between 6 and 25 kV are suspended equidistant between grounded plates. The high voltage on the wires creates an ionizing field for charging particles. The positive ions created in the field flow across the airstream and strike and adhere to the particles, thus imparting a charge to them. The charged particles then pass into the collecting plate section.

The collecting plate section consists of a series of parallel plates equally spaced with a positive direct current voltage of 4 to 10 kV applied to alternate plates. Plates that are not charged are at ground potential. As the particles pass into this section, they are forced to the plates by the electric field on the charges they carry; thus, they are removed from the airstream and collected by the plates. Particle retention is a combination of electrical and intermolecular adhesion forces and may be augmented by special oils or adhesives on the plates. Figure 4 shows a typical electronic air cleaner cell.

In lieu of positive direct current, a negative potential also functions on the same principle, but more ozone is generated.

With voltages of 4 to 25 kV (dc), safety measures are required. A typical arrangement makes the air cleaner inoperative when the doors are removed for cleaning the cells or servicing the power pack.

Electronic air cleaners typically operate from a 120- or 240-V (ac) single-phase electrical service. The high voltage supplied to the air cleaner cells is normally created with solid-state power supplies. The electrical power consumption ranges from 20 to 40 watts per 1000 cfm of air cleaner capacity.

This type of air filter can remove and collect airborne contaminants with an initial efficiency of up to 98% at low airflow velocities (150 to 350 fpm) when tested according to ASHRAE *Standard* 52.1. Efficiency decreases (1) as the collecting plates become loaded with particulates, (2) with higher velocities, or (3) with nonuniform velocity.

As with most air filtration devices, the duct approaches to and from the air cleaner housing should be arranged so that the airflow is distributed uniformly over the face area. Panel prefilters should also be used to help distribute the airflow and to trap large particles that might short out or cause excessive arcing within the high-voltage section of the air cleaner cell. Electronic air cleaner design parameters of air velocity, ionizer field strength, cell plate spacing, depth, and plate voltage must match the application requirements. These include contaminant type, particle size, volume of air, and required efficiency. Many units are designed for installation into central heating and cooling systems for total air filtration. Other self-contained units are furnished complete with air movers for source control of contaminants in specific applications that need an independent air cleaner.

Electronic air cleaner cells must be cleaned periodically with detergent and hot water. Some designs incorporate automatic wash systems that clean the cells in place; in other designs, the cells are removed for cleaning. The frequency of cleaning (washing) the cell depends on the contaminant and the concentration. Industrial applications may require cleaning every 8 hours, but a residential unit may only require cleaning at one to three month intervals. The timing of the cleaning schedule is important to keep the unit performing at peak efficiency. For some contaminants, special attention must be given to cleaning the ionizing wires.

Optional features are often available for electronic air cleaners. After-filters such as roll filters collect particulates that agglomerate and blow off the cell plates. These are used mainly where heavy contaminant loading occurs and extension of the cleaning cycle is desired. Cell collector plates may be coated with special oils, adhesives, or detergents to improve both particle retention

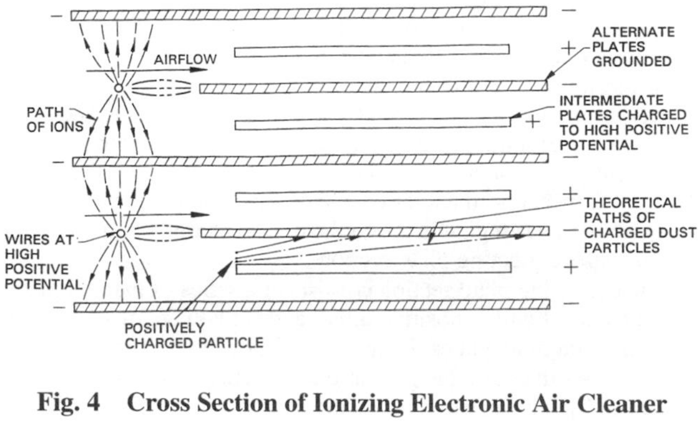

Fig. 4 Cross Section of Ionizing Electronic Air Cleaner

and particle removal during cleaning. High-efficiency dry-type extended media area filters are also used as after-filters in special designs. The electronic air cleaner used in this system improves the service life of the dry filter and collects small particles such as smoke.

Another device, a negative ionizer, uses the principle of particle charging but does not use a collecting section. Particles enter the ionizer of the unit, receive an electrical charge, and then migrate to a grounded surface closest to the travel path.

Space Charge. Particulates that pass through an ionizer and are charged, but not removed, carry the electrical charge into the space. If continued on a large scale, a space charge builds up, which tends to drive these charged particles to walls and interior surfaces. Thus, a low-efficiency electronic air cleaner used in areas of high ambient dirt concentrations, or a malfunctioning unit, can blacken walls faster than if no cleaning device were used (Penney and Hewitt 1949, Sutton et al. 1964).

Ozone. All high-voltage devices are capable of producing ozone, which is toxic and damaging to paper, rubber, and other materials. When properly designed and maintained, an electronic air cleaner produces an ozone concentration that only reach a fraction of the level acceptable for continuous human exposure and is less than that prevalent in many American cities (EPA 1996). Continuous arcing and brush discharge in an electronic air cleaner may yield an ozone level that is annoying or mildly toxic; this is indicated by a strong ozone odor. Although the nose is sensitive to ozone, only actual measurement of the concentration can determine that a hazardous condition exists.

ASHRAE *Standard* 62 defines acceptable concentrations of oxidants, of which ozone is a major contributor. The United States Environmental Protection Agency (EPA) specifies a 1-hour average maximum allowable exposure to ozone of 0.12 ppm for outside ambient air. The United States Department of Health and Human Services specifies a maximum allowable continuous exposure to ozone of 0.05 ppm for contaminants of indoor origin. Sutton et al. (1976) showed that indoor ozone levels are only 30% of the outdoor level with ionizing air cleaners operating, although Weschler et al. (1989) found that this level would increase when outdoor airflow is increased during an outdoor event that creates ozone.

SELECTION AND MAINTENANCE

To evaluate filters and air cleaners properly for a particular application, the following factors should be considered:

- Degree and type of air cleanliness required
- Disposal of dust after it is removed from the air
- Amount and type of dust in the air to be filtered
- Operating resistance to airflow (pressure drop)
- Space available for filtration equipment
- Cost of maintaining or replacing filters
- Initial cost of the system

Savings—from reduction in housekeeping expenses, protection of valuable property and equipment, dust-free environments for manufacturing processes, improved working conditions, and even health benefits—should be credited against the cost of installing and operating an adequate system. The capacity and physical size of the required unit may emphasize the need for low maintenance cost. Operating cost, predicted life, and efficiency are as important as initial cost because air cleaning is a continuing process.

Panel filters do not have efficiencies as high as can be expected from extended-surface filters, but their initial cost and upkeep are generally low. Compared to moving-curtain filters, panel filters require more attention to maintain the resistance within reasonable limits.

If higher efficiency is required, extended-surface filters or electronic air cleaners should be considered. The use of very fine glass

Table 2 Typical Filter Applications Classified by Filter Efficiency and Type[a]

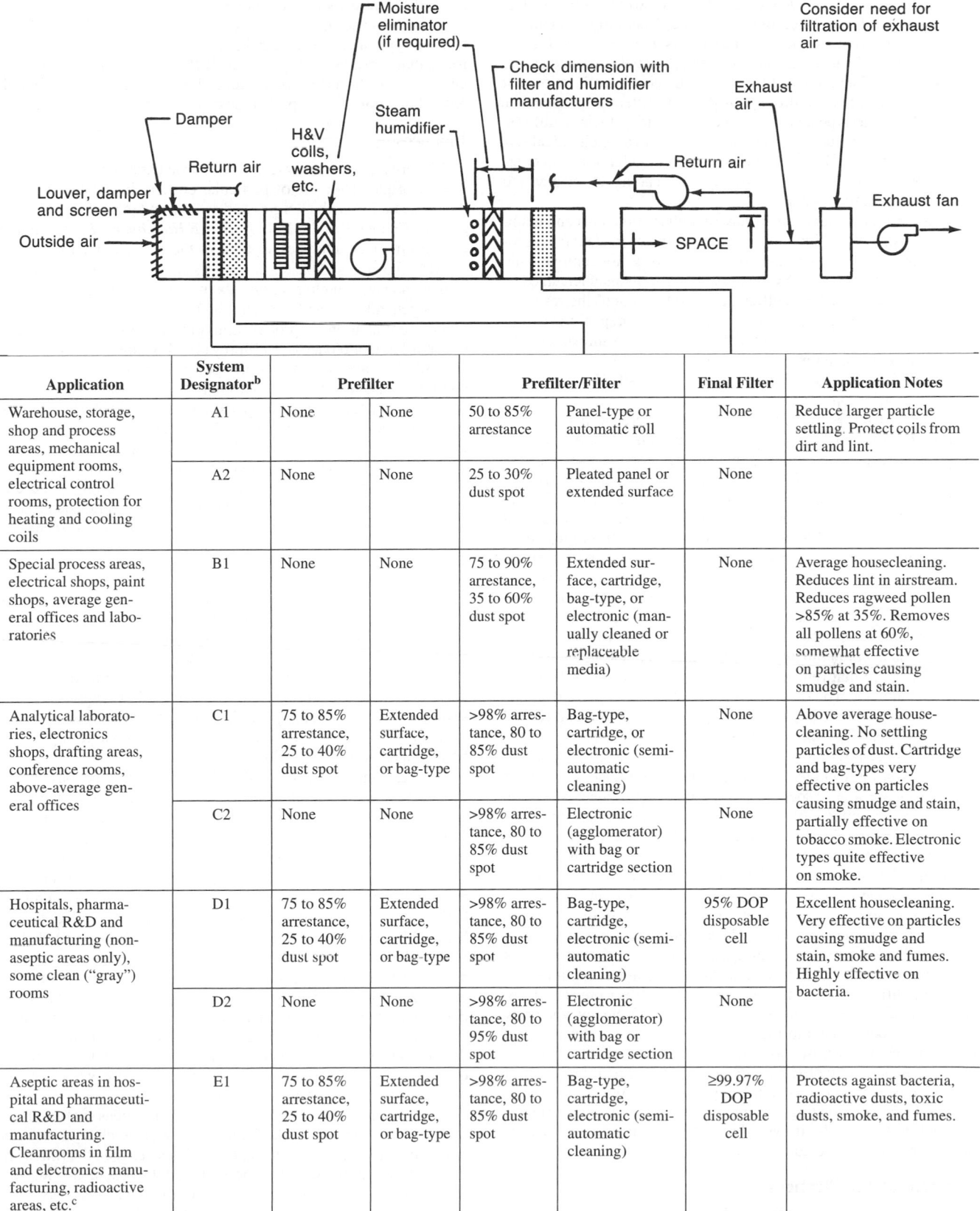

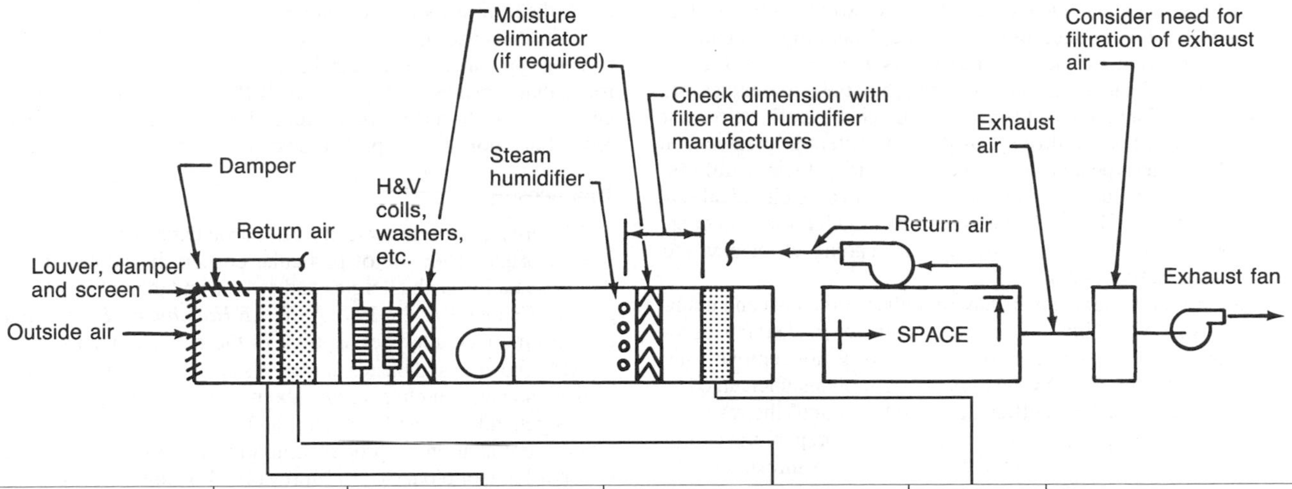

Application	System Designator[b]	Prefilter		Prefilter/Filter		Final Filter	Application Notes
Warehouse, storage, shop and process areas, mechanical equipment rooms, electrical control rooms, protection for heating and cooling coils	A1	None	None	50 to 85% arrestance	Panel-type or automatic roll	None	Reduce larger particle settling. Protect coils from dirt and lint.
	A2	None	None	25 to 30% dust spot	Pleated panel or extended surface	None	
Special process areas, electrical shops, paint shops, average general offices and laboratories	B1	None	None	75 to 90% arrestance, 35 to 60% dust spot	Extended surface, cartridge, bag-type, or electronic (manually cleaned or replaceable media)	None	Average housecleaning. Reduces lint in airstream. Reduces ragweed pollen >85% at 35%. Removes all pollens at 60%, somewhat effective on particles causing smudge and stain.
Analytical laboratories, electronics shops, drafting areas, conference rooms, above-average general offices	C1	75 to 85% arrestance, 25 to 40% dust spot	Extended surface, cartridge, or bag-type	>98% arrestance, 80 to 85% dust spot	Bag-type, cartridge, or electronic (semi-automatic cleaning)	None	Above average housecleaning. No settling particles of dust. Cartridge and bag-types very effective on particles causing smudge and stain, partially effective on tobacco smoke. Electronic types quite effective on smoke.
	C2	None	None	>98% arrestance, 80 to 85% dust spot	Electronic (agglomerator) with bag or cartridge section	None	
Hospitals, pharmaceutical R&D and manufacturing (non-aseptic areas only), some clean ("gray") rooms	D1	75 to 85% arrestance, 25 to 40% dust spot	Extended surface, cartridge, or bag-type	>98% arrestance, 80 to 85% dust spot	Bag-type, cartridge, electronic (semi-automatic cleaning)	95% DOP disposable cell	Excellent housecleaning. Very effective on particles causing smudge and stain, smoke and fumes. Highly effective on bacteria.
	D2	None	None	>98% arrestance, 80 to 95% dust spot	Electronic (agglomerator) with bag or cartridge section	None	
Aseptic areas in hospital and pharmaceutical R&D and manufacturing. Cleanrooms in film and electronics manufacturing, radioactive areas, etc.[c]	E1	75 to 85% arrestance, 25 to 40% dust spot	Extended surface, cartridge, or bag-type	>98% arrestance, 80 to 85% dust spot	Bag-type, cartridge, electronic (semi-automatic cleaning)	≥99.97% DOP disposable cell	Protects against bacteria, radioactive dusts, toxic dusts, smoke, and fumes.

[a] Adapted from a similar table courtesy of E.I. du Pont de Nemours & Company.
[b] System designators have no significance other than their use in this table.
[c] Electronic agglomerators and air cleaners are not usually recommended for cleanroom applications.

fiber mats or other materials in extended-surface filters has made these available at the highest efficiency.

Initial cost of an extended-surface filter is lower than for an electronic unit, but higher than for a panel type. Operating and maintenance costs of some extended-surface filters may be higher than for panel types and electronic air cleaners, but the efficiencies are always higher than for panel types, and the cost/benefit ratio must be considered. Pressure drop of media-type filters is greater than that of electronic-type and slowly increases during their useful life. The advantages include the fact that no mechanical or electrical services are required. Choice should be based on both initial and operating costs, as well as on the degree of cleaning efficiency and maintenance requirements.

While electronic air cleaners have a higher initial cost and maintenance cost, they exhibit high initial efficiencies in cleaning atmospheric air—largely because of their ability to remove fine particulate contaminants. System resistance remains unchanged as particles are collected, and efficiency is reduced until the resulting residue is removed from the collection plates to prepare the equipment for further duty. The manufacturer must supply information on maintenance or cleaning.

Table 2 lists some applications of filters classified according to their efficiencies and type.

Table 3 provides an approximate cross-reference between the ASHRAE *Standard* 52.1 and 52.2 reporting methods. A corollary purpose is to provide application guidance for the user and the HVAC designer. HEPA filters are tested by other than ASHRAE standards, but they have been included in the table by arbitrarily assigning Minimum Efficiency Reporting Values (MERVs) to Institute of Environmental Sciences and Technology (IEST) test standard ratings. Table 3 combines all the parameters into a single reference covering most types of air cleaners and applications. However, a single performance measurement cannot be applied precisely to all types and styles of air cleaners. Each air cleaner has unique characteristics that change during its useful life.

The typical contaminants listed in Table 3 appear in the general reporting group that removes the smallest known size of that specific contaminant. The order in which they are listed has no significance nor is the list complete. The typical applications and typical air cleaners listed are intended to show where and what type of air cleaner has been traditionally used. Traditional usage may not represent the optimum choice, so using the table as a selection guide is not appropriate when a specific performance requirement is needed. An air cleaner application specialist should then be consulted and manufacturers' performance curves should be reviewed.

Some knowledge of how air cleaners work along with common sense also helps the user achieve satisfactory results. Air cleaner performance varies from the time it is first installed until it reaches the end of its service life. Generally, the longer a media-type filter is in service, the better it performs. The accumulation of contaminants begins to close the porous openings, and, therefore, the filter is able to intercept smaller particles. However, there are exceptions that vary with different styles of media-type filters. Electronic air cleaners and charged-fiber media filters start at high efficiency when new (or after proper service in the case of electronic air cleaners) but their efficiency decreases as contaminants accumulate. Some air cleaners, in particular efficiency devices, may begin to shed some of the collected contaminants after being in service. Testing with standardized synthetic loading dust attempts to predict this occurrence, but such testing rarely, if ever, duplicates the air cleaner's performance on atmospheric dust.

Residential Air Cleaners

Filters used for residential applications are often of the spun glass variety that only filter out the largest of particles. These filters may prevent damage to downstream equipment, but they do little to improve air quality in the residence. Offermann (1992) describes a series of tests used to rate residential air cleaners. The tests were run in a test house with environmental tobacco smoke (ETS) as the test particulate (mass mean diameter = 0.5 μm).

In the absence of air cleaning systems in line with the HVAC, console type air cleaners can be used. Steiber and Sparks (1993) found that consoles equipped with HEPA filters or electrostatic precipitators can be effective in controlling particulates in a single room. For whole house applications, in-line units are recommended.

Bioaerosols

Bioaerosols are a diverse class of particulates that are biological in their origin. They are of particular concern in indoor air due to their association with allergies and asthma and their ability to cause disease. Chapter 9 of the 1997 *ASHRAE Handbook—Fundamentals* contains more detailed descriptions of these contaminants.

Airborne viral and bacterial aerosols are generally transmitted by droplet nuclei, which average about 3 μm in diameter. Fungal spores generally range between 2 and 5 μm in diameter (Wheeler 1994). Combinations of proper ventilation and filtration can be used to control indoor bioaerosols. Morey (1994) recommends providing a ventilation rate of 15 to 35 cfm per person, or its equivalent, to control human shed bacteria. ACGIH (1998) recommends dilution with a minimum of 15 cfm per person. It also reports 50 to 70% dust spot efficiency filters as being capable of removing most microbial agents 1 to 2 μm in diameter. Wheeler (1994) states that 60% dust spot efficiency filters will remove 85% or more of particles 2.5 μm in diameter, while 80 to 85% efficiency filters will remove 96% of 2.5 μm particles. Wheeler also worked through several equations to help determine ventilation and filtration needs for various scenarios.

FILTER INSTALLATION

Many air cleaners are available in units of convenient size for manual installation, cleaning, and replacement. A typical unit filter may be 20 to 24 in. square, from 1 to 40 in. thick, and of either the dry or viscous impingement type. In large systems, the frames in which these units are installed are bolted or riveted together to form a filter bank. Automatic filters are constructed in sections offering several choices of width up to 70 ft and generally range in height from 40 to 200 in., in 4 to 6 in. increments. Several sections may be bolted together to form a filter bank.

Several manufacturers provide side-loading filter sections for various types of filters. Filters are changed from outside the duct, making service areas in the duct unnecessary, thus saving cost and space.

Of course, in-service efficiency of an air filter is sharply reduced if air leaks through the bypass dampers or poorly designed frames. The higher the efficiency of a filter, the more attention must be paid to the rigidity and sealing effectiveness of the frame. In addition, high-efficiency filters must be handled and installed with care. Gilbert and Palmer (1965) suggest some of the precautions needed for HEPA filters.

Air cleaners may be installed in the outdoor-air intake ducts of buildings and residences and in the recirculation and bypass air ducts, but the prefilters should be placed ahead of heating or cooling coils and other air-conditioning equipment in the system to protect that equipment from dust. The dust captured in an outdoor-air intake duct is likely to be mostly particles of a greasy nature, while lint may predominate in dust from within the building.

Where high-efficiency filters protect critical areas such as cleanrooms, it is important that the filters be installed as close to the room as possible to prevent the pickup of particles between the filters and the outlet. The ultimate is the unidirectional flow room, in which the entire ceiling or one entire wall becomes the final filter bank.

Published performance data for all air filters are based on straight-through unrestricted airflow. Filters should be installed so that the face area is at right angles to the airflow whenever possible. Eddy currents and dead air spaces should be avoided; air should be

Table 3 Cross-Reference and Application Guidelines (Table E-1, ASHRAE *Standard* 52.2)

Std. 52.2 Minimum Efficiency Reporting Value (MERV)	Approx. *Std.* 52.1 Results		Application Guidelines		
	Duct Spot Efficiency	Arrestance	Typical Controlled Contaminant	Typical Applications and Limitations	Typical Air Filter/Cleaner Type
20 19 18 17	n/a n/a n/a n/a	n/a n/a n/a n/a	**≤0.30 μm Particle Size** Virus (unattached) Carbon dust Sea salt All combustion smoke Radon progeny	Cleanrooms Radioactive materials Pharmaceutical manufacturing Carcinogenic materials Orthopedic surgery	**HEPA/ULPA Filters** ≥99.999% efficiency on 0.1–0.2 μm particles, IEST Type F ≥99.999% efficiency on 0.3 μm particles, IEST Type D >99.99% efficiency on 0.3 μm particles, IEST Type C ≥99.97% efficiency on 0.3 μm particles, IEST Type A
16 15 14 13	n/a >95% 90–95% 80–90%	n/a n/a >98% >98%	**0.3–1.0 μm Particle Size** All bacteria Most tobacco smoke Droplet nuclei (sneeze) Cooking oil Most smoke Insecticide dust Copier toner Most face powder Most paint pigments	Hospital inpatient care General surgery Smoking lounges Superior commercial buildings	**Bag Filters** Nonsupported (flexible) microfine fiberglass or synthetic media. 12 to 36 in. deep, 6 to 12 pockets. **Box Filters** Rigid style cartridge filters 6 to 12 in. deep may use lofted (air laid) or paper (wet laid) media.
12 11 10 9	70–75% 60–65% 50–55% 40–45%	>95% >95% >95% >90%	**1.0–3.0 μm Particle Size** Legionella Humidifier dust Lead dust Milled flour Coal dust Auto emissions Nebulizer drops Welding fumes	Superior residential Better commercial buildings Hospital laboratories	**Bag Filters** Nonsupported (flexible) microfine fiberglass or synthetic media. 12 to 36 in. deep, 6 to 12 pockets. **Box Filters** Rigid style cartridge filters 6 to 12 in. deep may use lofted (air laid) or paper (wet laid) media.
8 7 6 5	30–35% 25–30% <20% <20%	>90% >90% 85–90% 80–85%	**3.0–10.0 μm Particle Size** Mold Spores Hair spray Fabric protector Dusting aids Cement dust Pudding mix Snuff Powdered milk	Commercial buildings Better residential Industrial workplaces Paint booth inlet air	**Pleated Filters** Disposable, extended surface, 1 to 5 in. thick with cotton-polyester blend media, cardboard frame. **Cartridge Filters** Graded density viscous coated cube or pocket filters, synthetic media **Throwaway** Disposable synthetic media panel filters
4 3 2 1	<20% <20% <20% <20%	75–80% 70–75% 65–70% <65%	**>10.0 μm Particle Size** Pollen Spanish moss Dust mites Sanding dust Spray paint dust Textile fibers Carpet fibers	Minimum filtration Residential Window air conditioners	**Throwaway** Disposable fiberglass or synthetic panel filters **Washable** Aluminum mesh, latex coated animal hair, or foam rubber panel filters **Electrostatic** Self charging (passive) woven polycarbonate panel filter

Note: A MERV for other than HEPA/ULPA filters also includes a test airflow rate, but it is not shown here because it is of no significance for the purposes of this table.

distributed uniformly over the entire filter surface using baffles, diffusers, or air blenders, if necessary. Filters are sometimes damaged if higher-than-normal air velocities impinge directly on the face of the filter.

Failure of air filter installations to give satisfactory results can, in most cases, be traced to faulty installation, improper maintenance, or both. The most important requirements of a satisfactory and efficiently operating air filter installation are as follows:

- The filter must be of ample capacity for the amount of air and dust load it is expected to handle. An overload of 10 to 15% is regarded as the maximum allowable. When air volume is subject to future increase, a larger filter bank should be installed initially.
- The filter must be suited to the operating conditions, such as degree of air cleanliness required, amount of dust in the entering air, type of duty, allowable pressure drop, operating temperature, and maintenance facilities.
- The filter type should be the most economical for the specific application. The initial cost of the installation should be balanced against efficiency and depreciation, as well as expense and convenience of maintenance.

The following recommendations apply to filters installed with central fan systems:

- Duct connections to and from the filter should change size or shape gradually to ensure even air distribution over the entire filter area.
- Sufficient space should be provided in front of or behind the filter, or both, depending on its type, to make it accessible for inspection and service. A distance of 20 to 40 in. is required, depending on the filter chosen.
- Access doors of convenient size should be provided to the filter service areas.

- All doors on the clean-air side should be gasketed to prevent infiltration of unclean air. All connections and seams of the sheet-metal ducts on the clean-air side should be as airtight as possible. The filter bank must be caulked to prevent bypass of unfiltered air, especially when high-efficiency filters are used.
- Electric lights should be installed in the plenum in front of and behind the air filter.
- Filters installed close to an air inlet should be protected from the weather by suitable louvers or inlet hood. A large mesh wire bird screen should be placed in front of the louvers or in the hood.
- Filters, other than electronic air cleaners, should have permanent indicators to give a warning when the filter resistance reaches too high a value or is exhausted, as with automatic roll media filters.
- Electronic air cleaners should have an indicator or alarm to indicate when high voltage is off or shorted out.

SAFETY REQUIREMENTS

Safety ordinances should be investigated when the installation of an air cleaner is contemplated. Combustible filtering media may not be permitted in accordance with some existing local regulations. Combustion of dust and lint on filtering media is possible, although the media itself may not burn. This may cause a substantial increase in filter combustibility. Smoke detectors and fire sprinkler systems may be considered for filter bank locations. In some cases, depending on the contaminant, hazardous material procedures must be followed during removal and disposal of the spent filter.

Many air filters are efficient collectors of bioaerosols. When provided moisture and nutrients, the microorganisms can multiply and may become a health hazard for maintenance personnel. Moisture in filters can be minimized by preventing (1) entrance of rain, snow, and fog; (2) carryover of water droplets from coils, drain pans, and humidifiers; and (3) prolonged exposure to elevated humidity. Changing or cleaning filters on a regular basis is important for controlling microbial growth. Good health-safety practices for personnel handling dirty filters include the use of face masks, thorough washing upon completion of the work, and placement of used filters in plastic bags or other containers for disposal.

REFERENCES

AHAM. 1988. Performance standard for room air cleaners. *Standard* AC-1-1988. Association of Home Appliance Manufacturers, Chicago, IL.

ARI. 1993. Commercial and industrial air filter equipment. *Standard* 850-93. Air-Conditioning and Refrigeration Institute, Arlington, VA.

ARI. 1993. Residential air filter equipment. *Standard* 680-93. Air-Conditioning and Refrigeration Institute, Arlington, VA.

ASHRAE. 1992. Gravimetric and dust-spot procedures for testing air-cleaning devices used in general ventilation for removing particulate matter. *Standard* 52.1-1992.

ASHRAE. 1999. Method of testing general ventilation air-cleaning devices for removal efficiency by particle size. *Standard* 52.2-1999.

ASHRAE. 1999. Ventilation for acceptable indoor air quality. *Standard* 62-1999.

CEN. 1994. Particulate air filters for general ventilation—Requirements, testing, marking. Comité Européen de Normalisation, Central Secretariat, rue de Stassart 36, B-1050 Brussels, Belgium.

Dill, R.S. 1938. A test method for air filters. *ASHRAE Transactions* 44:379.

EPA. Air quality criteria for photochemical oxidants. Environmental Protection Agency, Office of Air Noise and Radiation, Office of Air Quality Planning and Standards, Research Triangle Park, NC.

EPA. 1996. National air quality and emissions trends reports. U.S. Environmental Protection Agency, Office of Air Quality Planning and Standards, Research Triangle Park, NC 454/R-96.005.

Eurovent/Cecomaf. 1996. Method of testing air filters used in general ventilation for determination of fractional efficiency. *Document* 4/9. European Committee of Air Handling and Refrigeration Equipment Manufacturers, Brussels (www.eurovent-cecomaf.org).

Frey, A.H. 1985. Modification of aerosol size distribution by complex electric fields. *Bulletin of Environmental Contamination and Toxicology* 34:850-57.

Frey, A.H. 1986. The influence of electrostatics on aerosol deposition. *ASHRAE Transactions* 92(1B):55-64.

Gilbert, H. and J. Palmer. 1965. High efficiency particulate air filter units. USAEC, TID-7023. Available from NTIS, Springfield, VA.

Horvath, H. 1967. A comparison of natural and urban aerosol distribution measured with the aerosol spectrometer. *Environmental Science and Technology* (August):651.

IEST. 1992. Testing ULPA filters. IES RP-CC 007.1. Institute of Environmental Sciences and Technology, Mount Prospect, IL.

IEST. 1993. HEPA and ULPA filters. IES RP-CC 001.3. Institute of Environmental Sciences and Technology, Mount Prospect, IL.

Lee, K.W. and B.T. Liu. 1982a. Experimental study of aerosol filtration by fibrous filters. *Aerosol Science and Technology* (1):35-46.

Lee. K.W. and B.T. Liu. 1982b. Theoretical study of aerosol filtration by fibrous filters. *Aerosol Science and Technology* (1):147-61.

Liu, B.Y. and K.L. Rubow. 1986. Air filtration by fibrous filter media. In *Fluid-Filtration: Gas*. ASTM STF973. American Society for Testing and Materials, Philadelphia, PA.

McCrone, W.C., R.G. Draftz, and J.G. Delley. 1967. *The particle atlas*. Ann Arbor Science Publishers, Ann Arbor, MI.

National Air Pollution Control Administration. 1969. Air quality criteria for particulate matter. Available from NTIS, Springfield, VA.

NIOSH. 1973. *The industrial environment—Its evaluation and control*. S/N 017-001-00396-4, U.S. Government Printing Office, Washington, D.C.

NIOSH/MSHA. 1977. U.S. Federal mine safety and health act of 1977, Title 30 CFR Parts 11, 70. National Institute for Occupational Safety and Health, Columbus, OH, and Mine Safety and Health Administration, Department of Labor, Washington, D.C.

Nutting, A. and R.F. Logsdon. 1953. New air filter code. *Heating, Piping and Air Conditioning* (June):77.

Penney, G.W. and G.W. Hewitt. 1949. Electrically charged dust in rooms. *AIEE Transactions* (68):276-82.

Peters, A.H. 1962. Application of moisture separators and particulate filters in reactor containment. USAEC-DP812, U.S. Department of Energy, Washington, D.C.

Peters, A.H. 1965. Minimal specification for the fire-resistant high-efficiency filter unit. USAEC *Health and Safety Information* (212). U.S. Department of Energy, Washington, D.C.

Sutton, D.J., H.A. Cloud, P.E. NcNall, Jr., K.M. Nodolf, and S.H. McIver. 1964. Performance and application of electronic air cleaners in occupied spaces. *ASHRAE Journal* 6(6):55.

Sutton, D.J., K.M. Nodolf, and H.K. Makino. 1976. Predicting ozone concentrations in residential structures. *ASHRAE Journal* 18(9):21.

UL. 1989. Electrostatic air cleaners. *Standard* 867-89. Underwriters Laboratories, Inc., Northbrook, IL.

UL. 1990. High efficiency, particulate, air filter units. *Standard* 586-90. Underwriters Laboratories, Inc., Northbrook, IL.

UL. 1994. Air filter units. *Standard* 900-94. Underwriters Laboratories, Inc., Northbrook, IL.

U.S. Army. 1965. Instruction manual for the installation, operation, and maintenance of penetrometer, filter testing, DOP, Q107. Document 136-300-175A. Edgewood Arsenal, MD.

U.S. Military *Standard* MIL-STD-282 (1956). DOP-Smoke penetration and air resistance of filters. Department of Navy-Defense Printing Service, Philadelphia, PA.

Weschler, C.J., H.C. Shields, and D.V. Noik. 1989. Indoor ozone exposures. *Journal of the Air Pollution Control Association* 39(12):1562.

Whitby, K.T., A.B. Algren, and R.C. Jordan. 1956. The dust spot method of evaluating air cleaners. *Heating, Piping and Air Conditioning* (December):151.

Whitby, K.T., A.B. Algren, and R.C. Jordan. 1958. Size distribution and concentration of airborne dust. *ASHRAE Transactions* 64:129.

Willeke, K. and P.A. Baron, eds. 1993. *Aerosol measurement: Principles, techniques and applications*. Van Nostrand Reinhold, New York.

BIBLIOGRAPHY

Bauer, E.J., B.T. Reagor, and C.A. Russell. 1973. Use of particle counts for filter evaluation. *ASHRAE Journal* 15(10):53.

Davies, C.N. 1973. *Aerosol filtration*. Academic Press, London.

Dennis, R. 1976. *Handbook on aerosols*. Technical Information Center, Energy Research and Development Administration. Oak Ridge, TN.

Dorman, R.G. 1974. *Dust control and air cleaning*. Pergamon Press, New York.

Duncan, S.F. 1964. Effect of filter media microstructure on dust collection. *ASHRAE Journal* 6(4):37.

Engle, P.M., Jr. and C.J. Bauder. 1964. Characteristics and applications of high performance dry filters. *ASHRAE Journal* 6(5):72.

Gieseke, J.A., E.R. Blosser, and R.B. Rief. 1975. Collection and characterization of airborne particulate matter in buildings. *ASHRAE Transactions* 84(1):572.

Hinds, W.C. 1982. *Aerosol technology: Properties, behavior, and measurement of airborne particles*. John Wiley and Sons, New York.

Hunt, C.M. 1972. An analysis of roll filter operation based on panel filter measurements. *ASHRAE Transactions* 78(2):227.

James, A.C., W. Stahlhofen, G. Rudolf, M.J. Egan, W. Nixon, P. Gehr, and J.K. Briant. 1991. The respirator tract deposition model proposed by the ICRP task group. *Radiation Protection Dosimetry* 38:159-65.

Jorgensen, R., ed. 1961. *Fan engineering*, 6th ed. Buffalo Forge Company, Buffalo, NY.

Kemp, S.J., T.H. Kuehn, D.Y.H. Pui, D. Vesley, and A.J. Streifel. 1995. Filter collection efficiency and growth of the microorganisms on filters loaded with outdoor air. *ASHRAE Transactions* 101(1):228-38.

Kemp, S.J., T.H. Kuehn, D.Y.H. Pui, D. Vesley, and A.J. Streifel. 1995. Growth of microorganisms on HVAC filters under controlled temperature and humidity conditions. *ASHRAE Transactions* 101(1):305-16.

Kuehn, T.H., D.Y.H. Pui, and D.Vesley. 1991. Matching filtration to health requirements. *ASHRAE Transactions* 97(2):164-69.

Licht, W. 1980. *Air pollution control engineering: Basic calculations for particulate collection*. Marcel Dekker, New York.

Lioy, P.J. and M.J.Y. Lioy, eds. 1983. *Air sampling instruments for evaluation of atmospheric contaminants*, 6th ed. American Conference of Governmental Industrial Hygienists, Cincinnati, OH.

Liu, B.Y.H. 1976. *Fine particles*. Academic Press, New York.

Lundgren, D.A., F.S. Harris, Jr., W.H. Marlow, M. Lippmann, W.E. Clark, and M.D. Durham, eds. 1979. *Aerosol measurement*. University Presses of Florida, Gainesville, FL.

Matthews, R.A. 1963. Selection of glass fiber filter media. *Air engineering* 5(October):30.

McNall, P.E., Jr. 1986. Indoor air quality—A status report. *ASHRAE Journal* 28(6):39.

NAFA.1993. NAFA Guide to air filtration. National Air Filtration Association, Washington, DC.

NAFA. 1997. Installation, operation and maintenance of air filtration systems. National Research Council, Washington, DC.

NRC. 1981. *Indoor pollutants*. National Research Council, National Academy Press, Washington, D.C.

Nazaroff, W.W. and G.R. Cass. 1989. Mathematical modeling of indoor aerosol dynamics. *Environmental Science and Technology* 23(2):157-66.

Ogawa, A. 1984. *Separation of particles from air and gases*, Volumes I and II. CRC Press, Boca Raton, FL.

Penney, G.W. and N.G. Ziesse. 1968. Soiling of surfaces by fine particles. *ASHRAE Transactions* 74(1).

Phalen, R.F., R.G. Cuddihy, G.I. Fisher, O.R. Moss, R.B. Schlesinger, D.L. Swift, and H.-C. Yeh. 1991. Main features of the proposed NCRP respiratory tract model. *Radiation Protection Dosimetry* 38:179-84.

Rivers, R.D. 1988. Interpretation and use of air filter particle-size-efficiency data for general-ventilation applications. *ASHRAE Transactions* 88(1).

Rose, H.E. and A.J. Wood. 1966. *An introduction to electrostatic precipitation*. Dover Publishing, New York.

Stern, A.C., ed. 1977. *Air pollution*, 3rd ed. Vol. 4, *Engineering control of air pollution*. Academic Press, New York.

Swanton, J.R., Jr. 1971. Field study of air quality in air-conditioned spaces. *ASHRAE Transactions* 77(1):124.

U.S. Government. 1988. *Clean room* and work station requirements, controlled environments. Federal *Standard* 209D, amended 1991. Available from GSA, Washington, D.C.

USAEC/ERDA/DOE. Air cleaning conference proceedings. Available from NTIS, Springfield, VA.

Walsh, P.J., C.S. Dudney, and E.D. Copenhaver, eds. 1984. *Indoor air quality*. CRC Press, Boca Raton, FL.

Whitby, K.T. 1965. Calculation of the clean fractional efficiency of low media density filters. *ASHRAE Journal* 7(9): 56.

White, P.A.F. and S.E. Smith, eds. 1964. *High efficiency air filtration*. Butterworth, London, England.

Yocom, J.E. and W.A. Cote. 1971. Indoor/Outdoor air pollutant relationships for air-conditioned buildings. *ASHRAE Transactions* 77(1):61.

Ziesse, N.G. and G.W. Penney. 1968. The effects of cigarette smoke on space charge soiling of walls when air is cleaned by a charging type electrostatic precipitator. *ASHRAE Transactions* 74(2).

CHAPTER 25

INDUSTRIAL GAS CLEANING AND AIR POLLUTION CONTROL

AN INDUSTRIAL gas-cleaning installation performs one or more of the following functions:

- Maintains compliance of an industrial process with the laws or regulations for air pollution
- Reduces nuisance or physical damage from contaminants to individuals, equipment, products, or adjacent properties
- Prepares cleaned gases for processes
- Reclaims usable materials, heat, or energy
- Reduces fire, explosion, or other hazards

Equipment that removes particulate matter from a gas stream may also remove or create some gaseous contaminants; on the other hand, equipment that is primarily intended for removal of gaseous pollutants might also remove or create objectionable particulate matter to some degree. In all cases, gas-cleaning equipment changes the process stream, and it is therefore essential that the engineer evaluate the consequences of those changes to the plant's overall operation.

Equipment Selection

In selecting industrial gas-cleaning equipment, plant operations and the use or disposal of materials captured by the gas-cleaning equipment must be considered. Because the cost of gas-cleaning equipment affects manufacturing costs, alternative processes should be evaluated early to minimize the impact the equipment may have on the total cost of a product. An alternate manufacturing process may reduce the cost of or eliminate the need for gas-cleaning equipment. However, even when gas-cleaning equipment is required, process and system control should minimize the load on the collection device.

An industrial process may be changed from dirty to clean by substituting a process material (e.g., switching to a cleaner burning fuel or pretreating the existing fuel). Equipment redesign, such as enclosing pneumatic conveyors or recycling noncondensable gases, may also clean the process. Occasionally, additives (e.g., chemical dust suppressants used in quarrying or liquid animal fat applied to dehydrated alfalfa prior to grinding) reduce the potential for air pollution or concentrate the pollutants so that a smaller, more concentrated process stream may be treated.

Gas streams containing contaminants should not usually be diluted with extraneous air unless the extra air is required for cooling or to condense contaminants to make them collectible. The volume of gas to be cleaned is a major factor in the owning and operating costs of control equipment. Therefore, source capture ventilation, where contaminants are kept concentrated in relatively small volumes of air, is generally preferable to general ventilation, where pollutants are allowed to mix into and be diluted by much of the air in a plant space. Chapters 28 and 29 of the 1999 *ASHRAE Handbook—Applications* address local and general ventilation of industrial environments. Regulatory authorities generally require the levels of emissions to be corrected to standard conditions taking into account temperature, pressure, moisture content, and factors related to combustion or production rate. However, the air-cleaning equipment must be designed using the actual conditions of the process stream as it will enter the equipment.

In this chapter, each generic type of equipment is discussed on the basis of its primary method for gas or particulate abatement. The development of systems that incorporate several of the devices discussed here for specific industrial processes is left to the engineer.

The preparation of this chapter is assigned to TC 5.4, Industrial Process Air Cleaning (Air Pollution Control).

REGULATIONS AND MONITORING

Gas-Cleaning Regulations

In the United States, industrial gas-cleaning installations that exhaust to the outdoor environment are regulated by the U.S. Environmental Protection Agency (EPA); those that exhaust to the workplace are regulated by the Occupational Safety and Health Administration (OSHA) of the U.S. Department of Labor.

The EPA has established Standards of Performance for New Stationary Sources [New Source Performance Standards (NSPSs), GPO] and more restrictive State Implementation Plans (SIP 1991) and local codes as a regulatory basis to achieve air quality standards. Information on the current status of the NSPSs can be obtained through the Semi-Annual Regulatory Agenda, as published in the Federal Register, and through the regional offices of the EPA. Buonicore and Davis (1992) and Sink (1991) provide additional design information for gas-cleaning equipment.

Where air is not affected by combustion, solvent vapors, and toxic materials, it may be desirable to recirculate the air to the workplace to reduce energy costs or to balance static pressure in a building. High-efficiency fabric or cartridge filters, precipitators, or special-purpose wet scrubbers are typically used in general ventilation systems to reduce particle concentrations to levels acceptable for recirculated air.

The Industrial Ventilation Committee of the American Conference of Governmental Industrial Hygienists (ACGIH) and the National Institute of Occupational Safety and Health (NIOSH) have established criteria for recirculation of cleaned process air to the work area (ACGIH 1998; NIOSH 1978). Fine particle control by various dust collectors under recirculating airflows has been investigated by Bergin et al. (1989).

Public complaints may occur even when the effluent concentrations discharged to the atmosphere are below the maximum permissible emission rates and opacity limits. Thus, in addition to codes or regulations, the plant location, the contaminants involved, and the meteorological conditions of the area must be evaluated.

In most cases, emission standards require a higher degree of gas cleaning than necessary for economical recovery of process products (if this recovery is desirable). Gas cleanliness is a priority, especially where toxic materials are involved and cleaned gases might be recirculated to the work area.

Measuring Gas Streams and Contaminants

Stack sampling is often required to fulfill requirements of operating and installation permits for gas-cleaning devices, to establish conformance with regulations, and to commission new equipment. In addition, it can be used to establish specifications for gas-cleaning equipment and to certify that the equipment is functioning properly. The tests determine the composition and quantity of gases and particulate matter at selected locations along the process stream. The following general principles apply to a stack sampling program:

- The sampling location(s) must be acceptable to all parties who will use the results.
- The sampling location(s) must meet acceptable criteria with respect to temperature, flow distribution and turbulence, and distance from disturbances to the process stream. Exceptions based on physical constraints must be identified and reported.
- Samples that are withdrawn from a duct or stack must represent typical conditions in the process stream. Proper stack traverses must be made and particulate samples withdrawn isokinetically.
- Stack sampling should be performed in accordance with approved methods and established protocols whenever possible.
- Variations in the volumetric flow, temperature, and particulate or gaseous pollutant emissions, along with upset conditions, should be identified.
- A report from stack sampling should include a summary of the process, which should identify any deviations from normal process operations that occurred during testing. The summary should be prepared during the testing phase and certified by the process owner at completion of the tests.
- Disposal methods for waste generated during testing must be identified before testing begins. This is especially critical where pilot plant equipment is being tested because new forms of waste are often produced.
- The regulatory basis for the tests should be established so that the results can be presented in terms of process mass rate, consumption of raw materials, energy use, and so forth.

Analyses of the samples can provide the following types of information about the emissions:

- Physical characteristics of the contaminant—solid dust, liquid mist or "smoke," waxy solids, or a sticky mixture of liquid and solid.
- Distribution of particle sizes—optical, physical, aerodynamic, etc.
- Concentration of particulate matter in the gases, including average and extreme values, and a profile of concentrations in the duct or stack.
- Volumetric flow of gases, including average and extreme values, and a profile of this flow in the duct or stack. The volumetric flow is commonly expressed at actual conditions and at various standard conditions of temperature, pressure, moisture, and process state.
- Chemical composition of gases and particulate matter, including recovery value, toxicity, solubility, acid dew point, etc.
- Particle and bulk densities of particulate matter.
- Handling characteristics of particulate matter, including erosive, corrosive, abrasive, flocculative, or adhesive/cohesive qualities.
- Flammable or explosive limits.
- Electrical resistivity of deposits of particulate matter under stack and laboratory conditions. These data are useful for assessments of electrostatic precipitators and other electrostatically augmented technology.

EPA Reference Methods. The EPA has developed methods to measure the particulate and gaseous components of emissions from many industrial processes and has incorporated these in the NSPS by reference. Appendix A of the New Source Performance Standards lists the reference methods (GPO 1996). These are updated regularly in the Federal Register. Guidance for using these reference methods can be found in the *Quality Assurance Handbook for Air Pollution Measurement Systems* (EPA 1994).

The EPA (1984) has promulgated fine particle standards for ambient air quality. Known as PM-10 and PM-2.5 Standards, these revised standards focus particulate abatement efforts toward the collection and control of airborne particles smaller than 10 μm and 2.5 μm, respectively. Fine particles (with aerodynamic particle size smaller than 2.5 μm) are of concern because they penetrate deeply into the lungs. With the development of these standards, concern has arisen over the efficiency of industrial gas cleaners at various particle sizes.

ASTM Methods. The EPA has also approved ASTM test methods when cited in the NSPSs or the applicable EPA reference methods.

Other Methods. Sometimes, the reference methods must be modified, with the consent of regulatory authorities, to achieve representative sampling under the less-than-ideal conditions of industrial operations. Modifications to test methods should be clearly identified and explained in test reports.

Gas Flow Distribution

The control of gas flow through industrial gas-cleaning equipment is important for good system performance. Because of the large gas flows commonly encountered and the frequent need to retrofit equipment to existing processes, space allocations often preclude gradual expansion and long-radius turns. Instead, elbow splitters, baffles, etc., are used. These components must be designed to limit dust buildup and corrosion to acceptable levels.

Monitors and Controls

Current regulatory trends anticipate or demand continuous monitoring and control of equipment to maintain optimum performance against standards. Under regulatory data requirements, operating logs provide the owner with process control information and others with a baseline for the development or service of equipment.

Larger systems include programmable controllers and computers for control, energy management, data logging, and diagnostics. Increasing numbers of systems have modem connections to support monitoring and service needs from remote locations.

PARTICULATE CONTAMINANT CONTROL

A large range of equipment for the separation of particulate matter from gaseous streams is available. Typical concentration ranges for this equipment are summarized in Table 1.

High-efficiency particulate air (HEPA) and ultralow penetration air (ULPA) filters are often used in clean rooms to maintain nearly particulate-free environments. In commercial and residential buildings, air cleaners are used to remove nuisance dust. In other instances, air cleaners are selected to control particulate matter that constitutes a health hazard in the workplace (e.g., radioactive particles, beryllium-containing particles, or biological airborne wastes). Recirculation of air in industrial plants could use air cleaners or may require more heavy-duty equipment. Secondary filtration systems

Table 1 Intended Duty of Gas-Cleaning Equipment

	Maximum Concentration
Air cleaners	<0.002 gr/ft³
Clean rooms	
Commercial/residential buildings	
Plant air recirculation	
Industrial gas cleaners	<35 gr/ft³
Product capture in pneumatic conveying	<3500 gr/ft³

Table 2 Principal Types of Particulate Control Equipment

Gravity and momentum collectors	• Settling chambers • Louvers and baffle chambers
Centrifugal collectors	• Cyclones and multicyclones • Rotating "centrifugal" mist collectors
Electrostatic precipitators	• Tubular or plate-type, wet or dry, high-voltage (single-stage) precipitators • Plate-type, wet or dry, low-voltage (two-stage) mist and smoke precipitators
Fabric filters	• Baghouses; fabric collectors; cartridge filters • Disposable media filters (for dust and/or mist)
Granular bed filters	• Fixed bed • Moving bed
Particulate scrubbers	• Spray scrubbers • Impingement scrubbers • Centrifugal-type scrubbers • Orifice-type scrubbers • Venturi scrubbers • Packed towers • Mobile bed scrubbers • Electrostatically augmented scrubbers

(typically HEPA filter systems) are sometimes required with recirculation systems.

Air cleaners are discussed in Chapter 24. This chapter is concerned with heavy-duty equipment for the control of emissions from industrial processes. The particulate emissions from these processes generally have a concentration in the range of 0.01 to 35 gr/ft³. The gas-cleaning equipment is usually installed for air pollution control.

Particulate control technology is selected to satisfy the requirements for specific processes. Available technology differs in basic design, removal efficiency, first cost, energy requirements, maintenance, land use, operating costs, and ability of the collectors to handle various types and sizes of contaminant particles without requiring excessive maintenance. Some of the principal types of particulate control equipment are listed in Table 2 and discussed in the following sections.

Collector Performance

Particulate collectors may be evaluated for their ability to remove particulate matter from a gas stream and for their ability to reduce the emissions of selected particle sizes. The degree to which particulate matter is separated from a gas stream is known as the efficiency of a collector; the fraction of material escaping collection is the penetration. The reduction of particulate matter in a selected particle size range is known as the fractional efficiency of a particulate collector.

The **efficiency** η of a collector is generally expressed as a percent of the mass flow rate of material entering and exiting the collector.

$$\eta = 100(w_i - w_o)/w_i = 100w_c/w_i \qquad (1)$$

where

η = efficiency of collector, %
w_i = mass flow rate of contaminant in gases entering collector
w_o = mass flow rate of contaminant in gases exiting collector
w_c = mass flow rate of contaminant captured by collector

Alternately, the efficiency of a collector can be expressed in terms of the concentrations of particulate matter entering and exiting the equipment. This approach can be unsatisfactory because of changes that occur in the gas stream due to air leakage in the system, condensation, and temperature and pressure changes.

Penetration *P* is usually measured, and the efficiency for a collector calculated, using the following equations:

$$P = 100w_o/w_i \qquad (2)$$

with

$$\eta = 100 - P \qquad (3)$$

The **fractional efficiency** of a particulate collector is determined by measuring the mass rate of contaminants entering and exiting the collector in selected particle size ranges. Methods for measuring the fractional efficiency of particulate collectors in industrial applications are only beginning to emerge, largely because of the need to compare the fine particle performance of collectors used under the PM-10 and PM-2.5 regulations for ambient air quality.

Measures of performance other than efficiency should also be considered in designing industrial gas-cleaning systems. Table 3 compares some of these factors for typical equipment. Note that this table contains only nominal values and is no substitute for experience, trade studies, and an engineering assessment of requirements for specific installations.

Table 4 summarizes the types of collectors that have been used in industrial applications.

MECHANICAL COLLECTORS

Settling Chambers

Particulate matter will fall from suspension in a reasonable time if the particles are larger than about 40 μm. Plenums, dropout boxes, or gravitational settling chambers are thus used for the separation of coarse or abrasive particulate matter from gas streams.

Settling chambers are occasionally used in conjunction with fabric filters or electrostatic precipitators to reduce overall system cost. The settling chamber serves as a precollector to remove coarse particles from the gas stream.

Settling chambers sometimes contain baffles to distribute gas flow and to serve as surfaces for the impingement of coarse particles. Other designs use baffles to change the direction of the gas flow, thereby allowing coarse particles to be thrown from the gas stream by inertial forces.

The fractional efficiency for a settling chamber with uniform gas flow may be estimated by

$$\eta = 100u_t L/HV \qquad (4)$$

where

η = efficiency of collector for particles with settling velocity u_t, %
u_t = settling velocity for selected particles, fps
L = length of chamber, ft
H = height of chamber, ft
V = superficial velocity of gases through chamber, fps

The **superficial velocity** of gases through the chamber is determined from measurements of the volumetric flow of gases entering and exiting the chamber and the cross-sectional area of the chamber. This average velocity must be low enough to prevent reentrainment

Table 3 Measures of Performance for Gas-Cleaning Equipment

Type of Particle Collector	Particle Diameter,[a] µm	Max. Loading, gr/ft³	Collection Efficiency, % by mass	Pressure Loss: Gas, in. of water	Pressure Loss: Liquid, psi	Utilities per 1000 cfm (gas)	Comparative Energy Requirement	Superficial Velocity,[b] fpm	Capacity Limits, 1000 cfm	Space Required (Relative)
Dry inertial collectors										
Settling chamber	>40	>5	50	0.1 to 0.5	—	—	1	300 to 600	None	Large
Baffle chamber	>20	>5	50	0.5 to 1.5	—	—	1.5	1000 to 2000	None	Medium
Skimming chamber	>20	>1	70	<1.0	—	—	3.0	2000 to 4000	50	Small
Louver	>10	>1	80	0.3 to 2.0	—	—	1.5 to 6.0	2000 to 4000	30	Medium
Cyclone	>15	>1	85	0.5 to 3.0	—	—	1.5 to 9.0	2000 to 4000	50	Medium
Multicyclone	>5	>1	95	2.0 to 10.0	—	—	6.0 to 20	2000 to 4000	200	Small
Impingement	>10	>1	90	1.0 to 2.0	—	—	3.0 to 6.0	2000 to 4000	None	Small
Dynamic	>10	>1	90	Provides pressure	—	1.0 to 2.0 hp	10 to 20	—	50	—
Electrostatic precipitators										
High-voltage	>0.01	>0.1	99	0.2 to 1.0	—	0.1 to 0.6 kW	0.8 to 20	60 to 400	10 to 2000	Large
Low-voltage	>0.001	0.5	90 to 99	0.2 to 0.5	—	0.03 to 0.06 kW	0.5 to 1.0	200 to 700	0.1 to 100	Medium
Fabric filters										
Baghouses	>0.08	>0.5	99	2.0 to 6.0	—	—	6.0 to 20	1.0 to 20	200	Large
Cartridge filters	>0.05	>0.1	99+	2.0 to 8.0	—			0.5 to 5	40 to 50	Medium
Wet scrubbers										
Gravity spray	>10	>1	70	0.1 to 1.0	20 to 100	0.5 to 2.0 gpm	5.0	100 to 200	100	Medium
Centrifugal	>5	>1	90	2.0 to 8.0	20 to 100	1 to 10 gpm	12 to 26	2000 to 4000	100	Medium
Impingement	>5	>1	95	2.0 to 8.0	20 to 100	1 to 5 gpm	9.0 to 31	3000 to 6000	100	Medium
Packed bed	>5	>0.1	90	0.5 to 10	5 to 30	5 to 15 gpm	4.0 to 34	100 to 300	50	Medium
Dynamic	>2	>1	95	Provides pressure	5 to 30	1 to 5 gpm 3 to 20 hp	30 to 200	3000 to 4000	50	Small
Submerged orifice	>2	>0.1	90	2.0 to 6.0	None	No pumping	9.0 to 21	3000	50	Medium
Jet	>2	>0.1	90	Provides pressure	50 to 100	50 to 100 gpm	15 to 30	2000 to 20,000	100	Small
Venturi	>0.1	>0.1	95 to 99	10 to 60	10 to 30	3 to 10 gpm	30 to 300	12,000 to 42,000	100	Small

Source: IGCI (1964). Information updated by ASHRAE Technical Committee 5.4.
[a]The minimum particle diameter for which the device is effective.
[b]The average speed of gases flowing through the equipment's collection region.

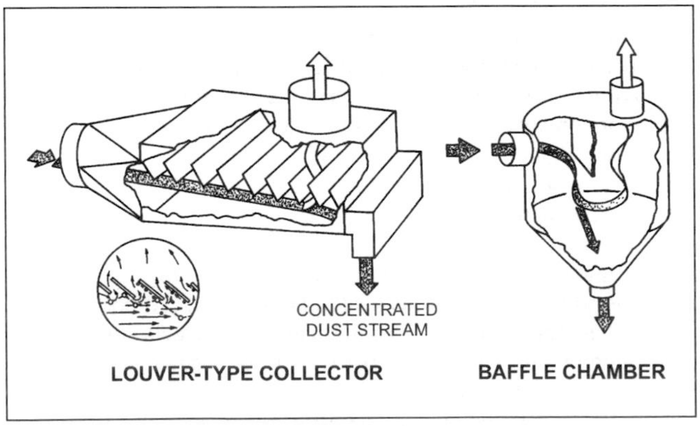

Fig. 1 Typical Louver and Baffle Collectors

of the deposited dust; a superficial velocity below 60 fpm is satisfactory for many materials.

Typical data on settling can be found in Table 5. Because of air inclusions in the particle, the density of a dust particle can be substantially lower than the true density of the material from which it is made.

Inertial Collectors

Louver and Baffle Collectors. Louvers are widely used to control particles larger than about 15 µm in diameter. The louvers cause a sudden change in direction of gas flow. By virtue of their inertia, particles move away from high-velocity gases and are either collected in a hopper or trap or withdrawn in a concentrated sidestream. The sidestream is cleaned using a cyclone or high-efficiency collector, or it is simply discharged to the atmosphere. In general, the pressure drop across inertial collectors with louvers or baffles is greater than that for settling chambers, but this loss is balanced by higher collection efficiency and more compact equipment.

Inertial collectors are occasionally used to control mist. In some applications, the interior of the collector may be irrigated to prevent reentrainment of dry dust and to remove soluble deposits.

Typical louver and baffle collectors are shown in Figure 1.

Cyclones and Multicyclones. A cyclone collector transforms a gas stream into a confined vortex, from which inertia moves suspended particles to the wall of the cyclone's body. The inertial effect of turning the gas stream, as employed in the baffle collector, is used continuously in a cyclone to gain improved collection efficiency. Cyclone collectors are often used as precleaners to reduce the loading of more efficient pollution control devices. Figure 2 shows some typical cyclone collectors.

A low-efficiency cyclone operates with a static pressure drop from 1 to 1.5 in. of water between its inlet and outlet and can remove 50% of the particles from 5 to 10 µm. High-efficiency cyclones operate with static pressure drops from 3 to 8 in. of water between their inlet and outlet and can remove 70% of the particulates of approximately 5 µm.

The efficiency of a cyclone depends on particle density, shape, and size (aerodynamic size D_p, which is the average of the size range). Cyclone efficiency may be estimated from Figure 3.

Table 4 Collectors Used in Industry

Operation	Concentration	Particle Size	Cyclone	High-Efficiency Centrifugal	Rotating Centrifugal Mist	Wet Collectors: Medium-Pressure	Wet Collectors: High-Energy	Self-Cleaning Fabric Filter	Disposable Media Filter	Electrostatic Precipitators: High-Voltage	Electrostatic Precipitators: Low-Voltage	Notes
Ceramics												
a. Raw product handling	Light	Fine	Rare	Seldom	N/A	Frequent	N/U	Frequent	N/A	N/U	N/A	1
b. Fettling	Light	Fine to medium	Rare	Occasional	N/A	Frequent	N/U	Frequent	N/A	N/U	N/A	2
c. Refractory sizing	Heavy	Coarse	Seldom	Occasional	N/A	Frequent	Rare	Frequent	N/A	N/U	N/A	3
d. Glaze and vitreous enamel spray	Moderate	Medium	N/U	N/U	N/A	Usual	N/U	Occasional	N/A	N/A	N/A	—
e. Glass melting	Light	Fine	N/A	N/A	N/A	Occasional	N/U	Occasional	N/A	Usual	N/A	—
f. Frit smelting	Light	Fine	N/A	N/A	N/A	N/U	Often	Often	N/A	Often	N/A	—
g. Fiberglass forming and curing	Light	Fine	N/A	N/A	N/A	Occasional	N/U	N/U	Rare	Usual	N/A	—
Chemicals												
a. Material handling	Light to moderate	Fine to medium	Occasional	Frequent	N/A	Frequent	Frequent	Frequent	N/A	N/U	N/A	4
b. Crushing, grinding	Moderate to heavy	Fine to coarse	Often	Frequent	N/A	Frequent	Occasional	Frequent	N/A	N/U	N/A	5
c. Pneumatic conveying	Very heavy	Fine to coarse	Usual	Occasional	N/A	Rare	Rare	Usual	N/A	N/U	N/A	6
d. Roasters, kilns, coolers	Heavy	Medium to coarse	Occasional	Usual	N/A	Usual	Frequent	Rare	N/A	Often	N/A	7
e. Incineration	Light to medium	Fine	N/U	N/U	N/A	N/U	Frequent	Rare	N/A	Frequent	N/A	8
Coal Mining and Handling												
a. Material handling	Moderate	Medium	Rare	Occasional	N/A	Occasional	N/U	Usual	N/A	N/A	N/A	9
b. Bunker ventilation	Moderate	Fine	Occasional	Frequent	N/A	Occasional	N/U	Usual	N/A	N/A	N/A	10
c. Dedusting, air cleaning	Heavy	Medium to coarse	Occasional	Frequent	N/A	Occasional	N/U	Usual	N/A	N/A	N/A	11
d. Drying	Moderate	Fine	Rare	Occasional	N/A	Frequent	Occasional	N/U	N/A	N/A	N/A	12
Combustion Fly Ash												
a. Coal burning:												
Chain grate	Light	Fine	N/A	Rare	N/A	N/U	N/U	Frequent	N/A	N/U	N/A	13
Spreader stoker	Moderate	Fine to coarse	Rare	Rare	N/A	N/U	N/U	Frequent	N/A	Rare	N/A	14
Pulverized coal	Heavy	Fine	N/A	Frequent	N/A	N/U	N/U	Frequent	N/A	Usual	N/A	14
Fluidized bed	Moderate	Fine	Usual	—	N/A	—	—	Frequent	N/A	Frequent	N/A	—
Coal slurry	Light	—	—	—	N/A	—	—	Often	N/A	Often	N/A	—
b. Wood waste	Varied	Coarse	Usual	Usual	N/A	N/U	N/U	Occasional	N/A	Often	N/A	15
c. Municipal refuse	Light	Fine	N/U	N/U	N/A	Occasional	N/U	Usual	N/A	Frequent	N/A	—
d. Oil	Light	Fine	N/U	N/U	N/A	N/U	N/U	Usual	N/A	Often	N/A	—
e. Biomass	Moderate	Fine to coarse	N/U	N/U	N/A	Occasional	N/U	Usual	N/A	Frequent	N/A	—
Foundry												
a. Shakeout	Light to moderate	Fine	Rare	Rare	N/A	Rare	Seldom	Usual	N/A	N/U	N/A	16
b. Sand handling	Moderate	Fine to medium	Rare	Rare	N/A	Usual	N/U	Rare	N/A	N/U	N/A	17
c. Tumbling mills	Moderate	Medium to coarse	N/A	N/A	N/A	Frequent	N/U	Usual	N/A	N/U	N/A	18
d. Abrasive cleaning	Moderate to heavy	Fine to medium	N/A	Occasional	N/A	Frequent	N/U	Usual	N/A	N/U	N/A	19
Grain Elevator, Flour and Feed Mills												
a. Grain handling	Light	Medium	Usual	Occasional	N/A	Rare	N/U	Frequent	N/A	N/A	N/A	20
b. Grain drying	Light	Coarse	N/A	N/A	N/A	N/U	N/U	See Note 20	N/A	N/A	N/A	21
c. Flour dust	Moderate	Medium	Rare	Often	N/A	Occasional	N/U	Usual	N/A	N/A	N/A	22
d. Feed mill	Moderate	Medium	Often	Often	N/A	Occasional	N/U	Frequent	N/A	N/A	N/A	23
Metal Melting												
a. Steel blast furnace	Heavy	Varied	Frequent	Rare	N/A	Frequent	Frequent	N/U	N/A	Frequent	N/A	24
b. Steel open hearth, basic oxygen furnace	Moderate	Fine to coarse	N/A	N/A	N/A	N/A	Often	Rare	N/A	Frequent	N/A	25
c. Steel electric furnace	Light	Fine	N/A	N/A	N/A	N/A	Occasional	Usual	N/A	Rare	N/A	26
d. Ferrous cupola	Moderate	Varied	N/A	N/A	N/A	Frequent	Often	Frequent	N/A	Occasional	N/A	27
e. Nonferrous reverberatory furnace	Varied	Fine	N/A	N/A	N/A	Rare	Occasional	Usual	N/A	N/U	N/A	28
f. Nonferrous crucible	Light	Fine	N/A	N/A	N/A	Rare	Rare	Occasional	N/A	N/U	N/A	29

Metal Mining and Rock Products												
a. Material handling	Moderate	Fine to medium	Rare	Occasional	N/A	Usual	N/U	Considerable	N/A	N/U	N/A	30
b. Dryers, kilns	Moderate	Medium to coarse	Frequent	Occasional	N/A	Frequent	Occasional	N/U	N/A	Occasional	N/A	31
c. Cement rock dryer	Moderate	Fine to medium	N/A	Frequent	N/A	Occasional	Rare	N/U	N/A	Occasional	N/A	30
d. Cement kiln	Heavy	Fine to medium	N/A	Frequent	N/A	Rare	N/U	Usual	N/A	Usual	N/A	32
e. Cement grinding	Moderate	Fine	N/A	Rare	N/A	N/U	N/U	Usual	N/A	Rare	N/A	33
f. Cement clinker cooler	Moderate	Coarse	N/A	Occasional	N/A	N/U	N/U	Occasional	N/A	N/U	N/A	34
Metal Working												
a. Production grinding, scratch brushing, abrasive cutoff	Light	Coarse	Occasional	Frequent	N/A	Considerable	N/U	Considerable	N/A	N/U	N/A	35
b. Portable and swing frame	Light	Medium	Rare	Frequent	N/A	Frequent	N/U	Considerable	N/A	N/U	N/A	—
c. Buffing	Light	Varied	Frequent	Rare	N/A	Frequent	N/U	Rare	N/A	N/U	N/A	36
d. Tool room	Light	Fine	Frequent	Frequent	N/A	Frequent	N/U	Frequent	N/A	N/U	N/A	37
e. Cast-iron machining	Moderate	Varied	Rare	Frequent	N/A	Considerable	N/U	Considerable	N/A	N/U	N/A	38
f. Steel, brass, aluminum machining	Light to moderate	Submicron smoke, med. mist to solids	N/A	N/A	Frequent	Occasional	N/U	Occasional	Frequent	N/U	Frequent	39
g. Welding	Light to moderate	Submicron fume to medium	N/A	N/A	N/A	Occasional	N/U	Frequent	Frequent	Rare	Occasional	40
h. Plasma and laser cutting	Moderate		N/A	N/A	N/A	Occasional	N/U	Frequent	Rare	N/A	N/U	41
i. Laser welding	Moderate		N/A	N/A	N/A	Occasional	N/U	Frequent	Rare	N/A	N/U	41
j. Abrasive machining	Moderate to heavy	Fine to Submicron	N/A	N/U	Occasional	Occasional	N/U	Rare	Frequent	N/A	Rare	39
k. Milling, turning, cutting tools	Light to moderate	Fine to Submicron	N/A	N/U	Frequent	Occasional	N/U	N/A	Frequent	N/A	Frequent	—
l. Annealing, heat treating, induction heating, quenching	Moderate to heavy	Submicron	N/A	N/U	N/A	Rare	Rare	N/A	Rare	N/A	Frequent	—
Pharmaceutical and Food Products												
a. Mixers, grinders, weighting, blending, bagging, packaging	Light	Medium	Rare	Frequent	N/A	Frequent	N/U	Frequent	Occasional	N/U	N/U	42
b. Coating pans	Varied	Fine to medium	Rare	Rare	N/A	Frequent	N/U	Frequent	Rare	N/U	N/U	43
PLASTICS												
a. Raw material processing	(See comments under CHEMICALS)											44
b. Plastic finishing	Light to moderate	Varied	Frequent	Frequent	N/A	Frequent	N/U	Frequent	Rare	N/U	N/U	45
c. Molding, extruding, curing	Light to moderate	Submicron smoke	N/A	N/A	N/A	Rare	N/U	N/A	Occasional	N/U	Considerable	46
Pulp and Paper												
a. Recovery boilers:												
Direct contact	Heavy	Medium	N/U	N/U	N/A	N/U	N/U	Occasional	N/A	Usual	N/A	—
Low odor	Heavy	Medium	N/U	N/U	N/A	N/U	N/U	Occasional	N/A	Usual	N/A	—
b. Lime kilns	Heavy	Coarse	N/U	N/U	N/A	N/U	N/U	Often	N/A	Often	N/A	—
c. Wood-chip dryers	Varied	Fine to coarse	N/U	N/U	N/A	N/U	N/U	Occasional	N/A	Often	N/A	—
Rubber Products												
a. Mixers	Moderate	Fine	N/A	N/A	N/A	Frequent	N/U	Usual	Rare	N/U	N/U	47
b. Batch-out rolls	Light	Fine	N/A	N/A	N/A	Usual	N/U	Frequent	N/A	N/U	Rare	48
c. Talc dusting and dedusting	Moderate	Medium	N/A	N/A	N/A	Frequent	N/U	Usual	Rare	N/U	N/U	49
d. Grinding	Moderate	Coarse	Often	Often	N/A	Frequent	N/U	Often	Rare	N/U	N/U	50
e. Molding, extruding, curing	Light to moderate	Submicron smoke	N/A	N/A	N/A	Rare	N/U	N/A	Occasional	N/A	Considerable	46
Wood Particle Board and Hard Board												
a. Particle dryers	Moderate	Fine to coarse	Usual	Occasional	N/A	Frequent	Occasional	Rare	N/A	Occasional	Rare	51
Woodworking												
a. Woodworking machines	Moderate	Varied	Usual	Occasional	N/A	Rare	N/U	Frequent	N/A	N/U	N/A	52
b. Sanding	Moderate	Fine	Frequent	Occasional	N/A	Occasional	N/U	Frequent	Rare	N/U	N/A	53
c. Waste conveying, hogs	Heavy	Varied	Usual	Rare	N/A	Occasional	N/U	Occasional	N/A	N/U	N/A	54

Source: Kane and Alden (1982). Information updated by ASHRAE Technical Committee 5.4.

Notes for Table 4

Definitions
N/A—Not applicable because of inefficiency or process incompatibility.
N/U—Not widely used.
Particle size
Fine—50% in 0.5 to 7 μm diameter range
Medium—50% in 7 to 15 μm diameter range
Coarse—50% over 15 μm diameter range
Concentration of particulate matter entering collector (loading)
Light—<2 gr/ft^3
Moderate—2 to 5 gr/ft^3
Heavy—>5 gr/ft^3

1 Dust released from bin filling, conveying, weighing, mixing, pressing, forming. Refractory products, dry pan, and screening operations more severe.
2 Operations found in vitreous enameling, wall and floor tile, pottery.
3 Grinding wheel or abrasive cutoff operation. Dust abrasive.
4 Operations include conveying, elevating, mixing, screening, weighing, packaging. Category covers so many different materials that recommendation will vary widely.
5 Cyclone and high-efficiency centrifugal collectors often act as primary collectors, followed by fabric filters or wet collectors.
6 Usual setup uses cyclone as product collector followed by fabric filter for high overall collection efficiency.
7 Dust concentration determines need for dry centrifugal collector; plant location, product value determines need for final collectors. High temperatures are usual, and corrosive gases not unusual. Liquid smoke emissions may be controlled by condensing precipitator systems using low-voltage, two-stage electrostatic precipitators.
8 Ionizing wet scrubbers are widely used.
9 Conveying, screening, crushing. unloading.
10 Remote from other dust-producing points. Separate collector generally used.
11 Heavy loading suggests final high efficiency collector for all except very remote locations.
12 Loadings and particle sizes vary with different drying methods.
13 Boiler blow-down discharge is regulated, generally for temperature and, in some places, for pH limits; check local environmental codes on sanitary discharge.
14 Collection for particulate or sulfur control usually requires a scrubber (dry or wet) and a fabric filter or electrostatic precipitator.
15 Public nuisance from settled wood char indicates collectors are needed.
16 Hot gases and steam usually involved.
17 Steam from hot sand, adhesive clay bond involved.
18 Concentration very heavy at start of cycle.
19 Heaviest load from airless blasting because of high cleaning speed. Abrasive shattering greater with sand than with grit or shot. Amounts removed greater with sand castings, less with forging scale removal, least when welding scale is removed.
20 Operations such as car unloading, conveying, weighing, storing.
21 Special filters are successful.
22 In addition to grain handling, cleaning rolls, sifters, purifiers, conveyors, as well as storing, packaging operations are involved.
23 In addition to grain handling, bins, hammer mills, mixers, feeders, conveyors, bagging operations need control.
24 Primary dry trap and wet scrubbing usual. Electrostatic precipitators are added where maximum cleaning is required.
25 Air pollution control is expensive for open hearth, accelerating the use of substitute melting equipment, such as basic oxygen process and electric-arc furnace.
26 Fabric filters have found extensive application for this air pollution control problem.
27 Cupola control varies with plant size, location, melt rate, and air pollution emission regulations.
28 Corrosive gases can be a problem, especially in secondary aluminum.
29 Zinc oxide plume can be troublesome in certain plant locations.
30 Crushing screening, conveying, storing involved. Wet ores often introduce water vapor in exhaust airstream.
31 Dry centrifugal collectors are used as primary collectors, followed by a final cleaner.
32 Collectors usually permit salvage of material and also reduce nuisance from settled dust in plant area.
33 Salvage value of collected material is high. Same equipment used on raw grinding before calcining.
34 Coarse abrasive particles readily removed in primary collector types.
35 Roof discoloration, deposition on autos can occur with cyclones and, less frequently, with dry centrifugal. Heavy-duty air filter sometimes used as final cleaner.
36 Linty particles and sticky buffing compounds can cause trouble in high-efficiency centrifugals and fabric filters. Fire hazard is also often present.
37 Unit collectors extensively used, especially for isolated machine tools.
38 Dust ranges from chips to fine floats, including graphitic carbon.
39 Coolant mist and thermal smoke, often with solid swarf particulate entrained.
40 Submicron smoke. Arc welding creates mostly dry metal oxide particulate, sometimes with liquid oil smoke. Resistance welding usually creates only liquid oil smoke, unless done at extremely high currents that vaporize some of the metal being welded.
41 Plasma and laser cutting and welding of clean metals usually creates dry submicron smoke, but oily work pieces frequently generate a sticky mix of liquid and solid submicron smoke or fume.
42 Materials involved vary widely. Collector selection may depend on salvage value, toxicity, sanitation yardsticks.
43 Controlled temperature and humidity of supply air to coating pans makes recirculation from coating pans desirable.
44 Manufacture of plastic compounds involves operations allied to many in chemical field and vanes with the basic process employed.
45 Operations are similar to woodworking, and collector selection involves similar considerations.
46 Submicron liquid smoke is frequently emitted when plastic and rubber products are heated.
47 Concentration is heavy during feed operation. Carbon black and other fine additions make collection and dust-free disposal difficult.
48 Often, no collection equipment is used where dispersion from exhaust stack is good and stack location is favorable.
49 Salvage of collected material often dictates type of high-efficiency collector.
50 Fire hazard from some operations must be considered.
51 Granular bed filters, at times electrostatically augmented, have occasionally been used in this application.
52 Bulky material. Storage for collected material is considerable; bridging from splinters and chips can be a problem.
53 Production sanding produces heavy concentrations of particles too fine to be effectively captured by cyclones or dry centrifugal collectors.
54 Primary collector invariably indicated with concentration and partial size range involved; when used, wet or fabric collectors are employed as final collectors.

Table 5 Terminal Settling Velocities of Particles, fps

Particle Density Relative to Water	Particle or Aggregate Diameter, μm									
	1	2	5	10	20	50	100	200	500	1000
0.05	5.9 E–6	2.3 E–5	1.3 E–4	5.2 E–4	2.3 E–3	1.3 E–3	5.2 E–2	0.18	0.75	1.7
0.1	1.2 E–5	4.6 E–5	2.6 E–4	1.0 E–3	4.6 E–3	2.6 E–3	9.8 E–2	0.36	1.3	2.7
0.2	2.4 E–5	9.2 E–5	5.2 E–4	2.1 E–3	9.2 E–3	5.2 E–2	0.19	0.62	2.1	4.3
0.5	5.9 E–5	2.3 E–4	1.3 E–3	5.2 E–3	2.3 E–2	0.13	0.46	1.4	4.0	8.2
1.0	1.2 E–4	4.6 E–4	2.6 E–3	1.0 E–2	3.9 E–2	0.25	0.82	2.3	6.4	12.8
2.0	2.4 E–4	9.2 E–4	5.2 E–3	2.1 E–2	8.2 E–2	0.46	1.5	3.7	10.2	20.5
5.0	5.9 E–4	2.3 E–3	1.3 E–2	4.9 E–2	0.21	1.1	3.2	7.3	18.9	36.1
10.0	1.2 E–3	4.6 E–3	2.6 E–2	0.10	0.39	2.0	5.4	11.5	29.2	56.8

Note: E–6 = ×10^{-6}, etc.

Source: Billings and Wilder (1970).

The parameter D_{pc}, known as the **cut size**, is defined as the diameter of particles collected with 50% efficiency. The cut size may be estimated using the following equation:

$$D_{pc}\sqrt{\frac{9\mu b}{2N_e V_i(\rho_p-\rho_g)\pi}} \qquad (5)$$

where

D_{pc} = cut size, ft
μ = absolute gas viscosity, centipoise
b = cyclone inlet width, ft
N_e = effective number of turns within cyclone. The quantity N_e is approximately 5 for a high-efficiency cyclone and may be in the range from 0.5 to 10 for other cyclones.
V_i = inlet gas velocity, fpm
ρ_p = density of particle material, lb/ft^3
ρ_g = density of gas, lb/ft^3

At inlet gas velocity above 4800 fpm, internal turbulence limits improvements in the efficiency of a given cyclone. The pressure

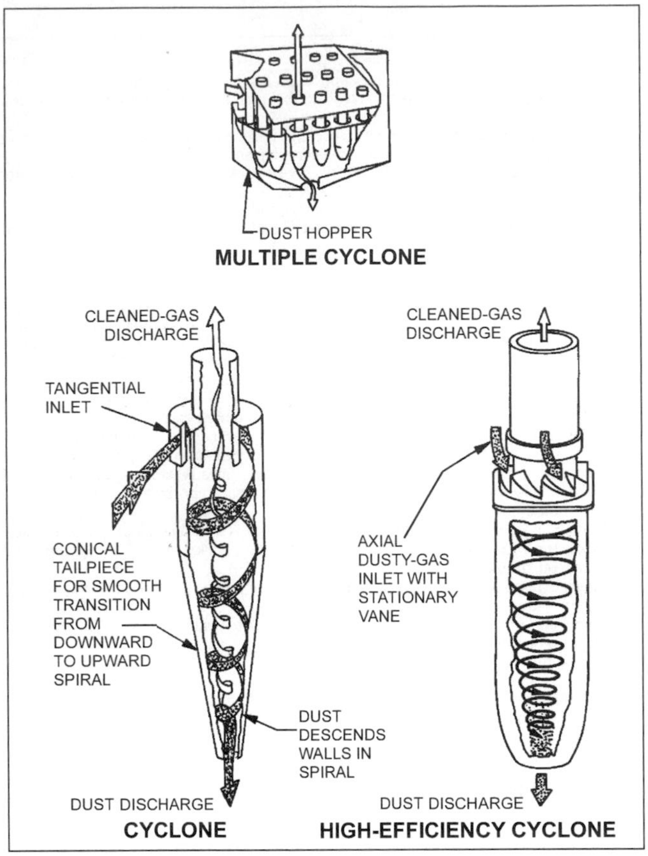

Fig. 2 Typical Cyclone Collectors

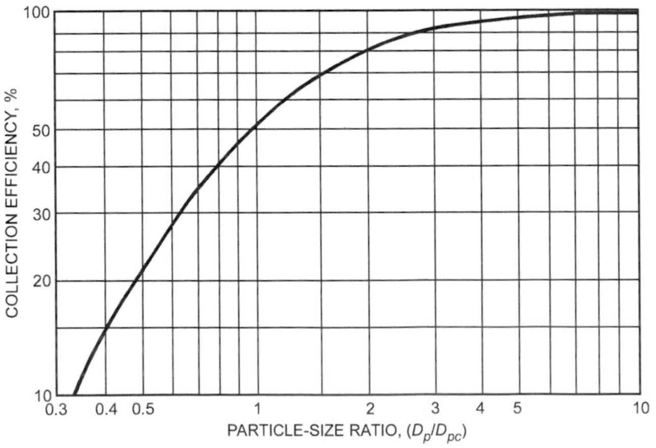

Fig. 3 Cyclone Efficiency
(Lapple 1951; EPA 1973)

drop through a cyclone is proportional to the inlet velocity pressure and hence the square of the volumetric flow.

ELECTROSTATIC PRECIPITATORS

Electrostatic precipitators use the forces acting on charged particles passing through an electric field to separate those particles from the airstream in which they were suspended. In every precipitator, three distinct functions must be accomplished:

1. **Ionization**—Charging the contaminant particles
2. **Collection**—Subjecting the particles to a precipitating force that moves them toward collecting electrodes

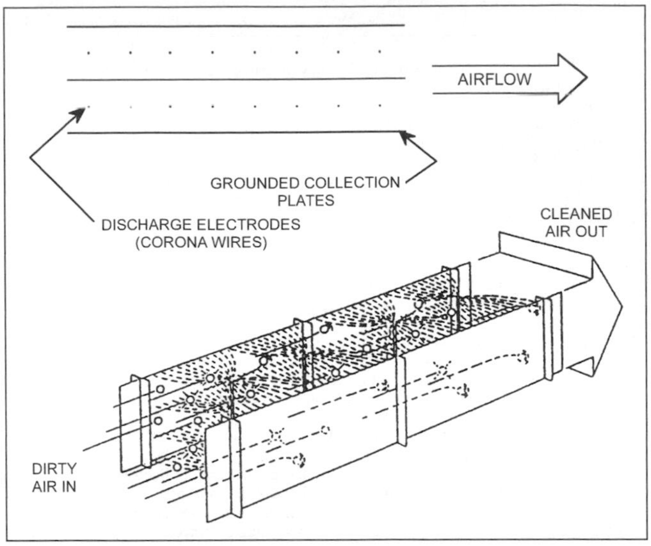

Fig. 4 Typical Single-Stage Electrostatic Precipitator

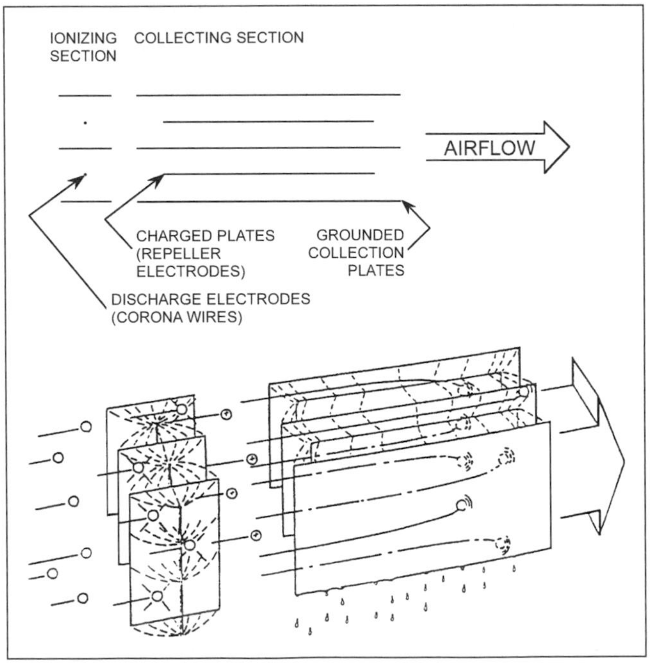

Fig. 5 Typical Two-Stage Electrostatic Precipitators

3. **Collector cleaning**—Removal of collected contaminant from the precipitator

Units in which ionization and collection are accomplished simultaneously in a single structure are called **single-stage precipitators** (Figure 4). They have widely spaced electrodes (3.15 to 6 in.) and typically operate with high voltages (20 to 60 kV) but relatively low (rarely as high as 11 kV/in) field gradients.

In **two-stage precipitators**, ionization and collection are performed independently in discrete charging and precipitating structures (Figure 5). Because their ionizing and collecting electrodes are closely spaced (0.7 to 1.5 in.), two-stage precipitators normally operate with high field gradients (usually more than 10 kV/in) but low voltages (usually 10 kV or less, and never more than 14 kV) (White 1963).

Due to fundamental differences in their ionization processes and practical differences in the way they are usually constructed, high-voltage (single-stage) and low-voltage (two-stage) precipitators are suited for entirely different air-cleaning requirements.

Single-Stage Designs

Figure 6 shows several types of single-stage precipitators. the charging electrodes are located between parallel collecting plates. the gas flows horizontally through the precipitators. High-voltage precipitators collect larger particles better than small particles (they are less efficient at collecting contaminants smaller than 1 to 2 µm). Their precipitation efficiency depends in part on the relative electrical resistivity of the pollutant being collected; most are less efficient when collecting either conductive or highly dielectric contaminants.

Single-stage high-voltage precipitators can easily handle heavy loadings of dry dust. Most are configured to operate continuously (using on-line vibratory or shaker cleaning). They can continuously collect large quantities (hundreds of pounds per hour) of such airborne materials as foundry shakeout, cement, ceramics, chemical dusts, fly ash, blast furnace dust and fumes, and paper mill recovery boiler emissions.

Although they are rarely used to clean exhaust airflows much smaller than 50,000 cfm, single-stage precipitators can be constructed to handle airflows as large as 2,000,000 cfm. Gas velocity through the electrostatic field is ordinarily 60 to 400 fpm, with treatment time in the range of 2 to 10 s. Only special-purpose single-stage "wet" precipitators are configured for collection of liquid contaminants.

Single-stage industrial electrostatic precipitators are distinguished from low-voltage two-stage precipitator designs by several attributes:

1. Separation between the electrodes is larger to secure acceptable collection and electrode cleaning under the high dust loading and hostile conditions of industrial gas streams.
2. Construction is heavy-duty for operation to 850°F and +30 in. of water.
3. They are generally used for exhaust applications where ozone generation is not of concern. Consequently, they may operate with negative ionization, the polarity that gives the maximum electric field strength between the electrodes.
4. They are normally custom-engineered and assembled on location for a particular application.

Single-stage precipitators can be designed to operate at collection efficiencies above 99.9% for closely specified conditions. Properties of the dust, such as particle size distribution and a deposit's electrical resistivity, can affect performance significantly, as can variations in gas composition and flow rate.

Two-Stage Designs

Two-stage precipitators are manufactured in two general forms: (1) electronic air cleaners, which are commonly used for light dust loadings in commercial/residential ventilation and air-conditioning service (see Chapter 24); and (2) heavy-duty industrial electrostatic precipitators designed primarily to handle the heavy loadings of submicron liquid particulates that are emitted from hot industrial processes and machining operations. Two-stage industrial precipitators often include sumps and drainage provisions that encourage continuous gravity runoff of collected liquids (Beck 1975; Shabsin 1985).

Two-stage precipitators are frequently used in industry to collect submicron pollutants that would be difficult (or impossible) to collect in other types of equipment. Two-stage precipitators are frequently sized to handle less than 1000 cfm of air and rarely built to handle more than 50,000 to 75,000 cfm of contaminated air in one unit. Gas velocity through the collecting fields usually ranges from 200 to 700 fpm, with treatment times per pass of 0.015 to 0.05 s in the ionizing fields and 0.06 to 0.25 s in the collecting fields.

Because gravity drainage of precipitated liquid "smoke" or "fog" is a dependable collector-cleaning mechanism, low-voltage

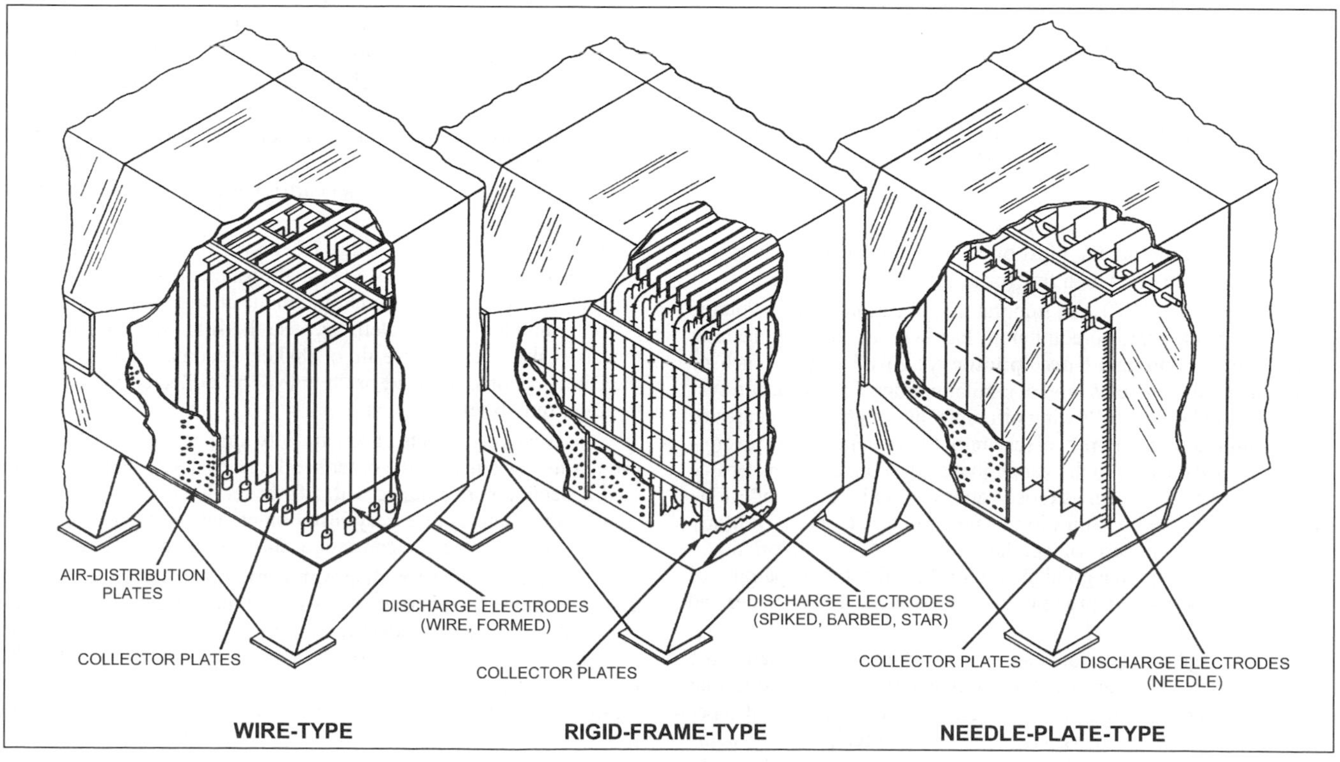

Fig. 6 Typical Single-Stage Precipitators

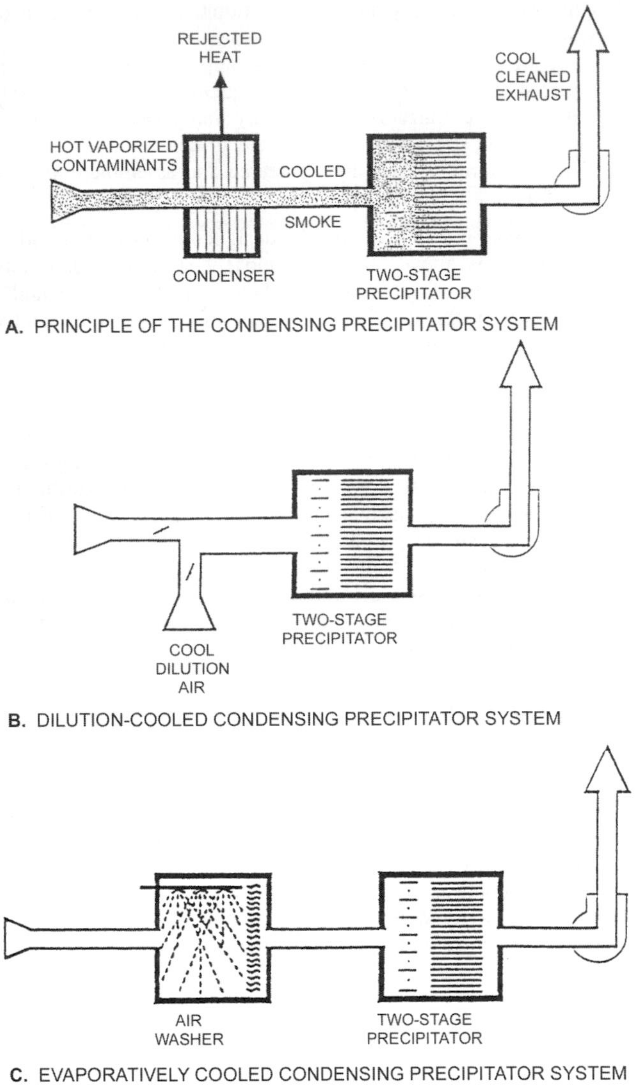

Fig. 7 Condensing Precipitator Systems for Control of Hot Organic Smokes

two-stage precipitators are most often recommended to collect liquefiable pollutants in which few (or no) solids are entrained.

Since the early 1970s, an important application for low-voltage two-stage electrostatic precipitators has been as the gas-cleaning component of **condensing precipitator systems**. Design of such air pollution control systems (and of similar **condensing filtration systems**) is based on the following principle: although the hot gases or fumes emitted by many processes are not easily filtered or precipitated (because they are in the vapor phase), the condensation aerosol fogs or smokes that form as those vaporized pollutants cool can be efficiently collected by filtration or precipitation (Figure 7). Many such condensation aerosol smokes consist of submicron liquid droplets, making them a good match for the collecting capability of low-voltage two-stage precipitators (Rossnagel 1973; Sauerland 1976; Thiel 1977).

Low-voltage precipitators can be very effective at collecting the aerosol particles smaller than 1 μm (down to 0.001 to 0.01 μm) that are responsible for most plume opacity and for virtually all blue-tinted "blue smoke" emissions. Condensing precipitator systems may therefore be a good choice for eliminating the residual opacity of blue smoke formed by condensation aerosol plumes from hot processes, dryers, ovens, furnaces, or other exhaust air cleaning devices (Beltran 1972; Beltran and Surati 1976; Thiel 1977).

When condensation aerosol pollutants are odorous in character, precipitation of the submicron droplets of odorant can prevent the long distance drift of odorous materials, possibly eliminating neighborhood complaints that are associated with submicron particulate smokes. Odors that have been successfully controlled by precipitation include asphalt fumes, food frying smokes, meat smokehouse smoke, plasticizer smokes, rubber curing smoke, tar, and textile smokes (Chopyk and Larkin 1982; Thiel 1983; EPA 1973, p. 163).

FABRIC FILTERS

Fabric filters are dry dust collectors that use a stationary medium to capture the suspended dust and remove it from the gas stream. This medium, called fabric, can be composed of a wide variety of materials, including natural and synthetic cloth, glass fibers, and even paper.

Three types of industrial fabric filters are in common use: pulse jet (and, rarely, reverse air or shaker) cleaned baghouses, pulse jet cleaned cartridge filters, and disposable media filters. Although most commercially available fabric filter systems are currently one of these three types, there are a number of design variations among competing products that can greatly influence the suitability of a particular collector for a specific air-cleaning application.

Most **baghouses** (see Figures 9, 10, 12, and 13) use flexible cloth or other fabric-like filter media in the shape of long cylindrical bags. Bags in pulse jet cleaned systems are rarely more than 6.25 in. in diameter but can be up to 20 ft long. Timed pulses of compressed air flex the bags while blowing collected dust off the media surface. Continuous operation, with no downtime for filter cleaning or dust removal, is common (Figure 12).

Cartridge filters, illustrated in Figure 14, use a nearly identical compressed air media cleaning system. However, their fabric is relatively rigid material, packaged into pleated cylindrical cartridges. Cartridges are self-supported and much easier to handle and replace than are long tubular baghouse bags. Most cartridges are 12.75 in. or less in diameter and rarely longer than 30 in.

With the pleated media construction, a large filter surface area can be packaged in a relatively small housing to reduce both cost and space required. Most cartridge filter units are not nearly as tall as baghouses having comparable capacity. Pleat depth and spacing are critically important variables that determine the suitability and useful lifetime of particular filter cartridges under the conditions of each specific application.

Both baghouse and cartridge filter systems are practical only when airborne contaminants consist almost exclusively of dry dust. The presence of any entrained liquid in the airstream usually creates a severe maintenance problem because the filter self-cleaning systems (i.e., pulse jet, reverse air, or shaker) become less effective, so the filters become plugged or "blinded" by collected material and fail after only brief operation.

Conservatively selected and carefully applied baghouse and cartridge filter systems, on the other hand, can easily provide excellent dust collection performance with a filter service life of more than 1 year. They often require very little maintenance, even when handling heavy dust loads in continuous 24 h, 7 day operation.

Disposable fabric (or **disposable media**) filter collectors are usually simple and economical units that hold enough fabric or similar media to collect modest quantities of almost any particulate pollutant, including liquid, solid, and sticky or waxy materials, regardless of particle size (at least for particles larger than 0.5 to 1 μm). When each filter has accumulated as much material as it can practically hold, it is discarded and replaced by a new element. Both envelope bag arrays and pleated rectangular cartridge elements are popular media forms for use in disposable filter collectors.

When considering using disposable media filter collectors, serious attention must be given to safe, legal, and ethical disposal of spent and/or contaminated filter elements.

Principle of Operation

When contaminated gases pass through a fabric, particles may attach to the fibers and/or other particles and separate from the gas stream. The particles are normally captured near the inlet side of a fabric to form a deposit known as a **dust cake**. In self-cleaning designs, the dust cake is periodically removed from the fabric to prevent excessive resistance to the gas flow. Finer particles may penetrate more deeply into a fabric and, if not removed during cleaning, may blind it.

The primary mechanisms for particle collection are direct interception, inertial impaction, electrostatic attraction, and diffusion (Billings and Wilder 1970).

Direct interception occurs when the fluid streamline carrying the particle passes within one-half of a particle diameter of a fiber. Regardless of the particle's size, mass, or inertia, it will be collected if the streamline passes sufficiently close.

Inertial impaction occurs when the particle would miss the fiber if it followed the streamline, but its inertia overcomes the resistance offered by the flowing gas, and the particle continues in a direct enough course to strike the fiber.

Electrostatic attraction occurs when the particle, the filter, or both possess sufficient electrical charge to cause the particles to precipitate on the fiber.

Diffusion makes particles more likely to pass near fibers and thus be collected. Once a particle resides on a fiber, it effectively increases the size of the fiber.

Self-cleaning fabric filters have several **advantages** over other high-efficiency dust collectors such as electrostatic precipitators and wet scrubbers:

- They provide high efficiency with lower installed cost.
- The particulate matter is collected in the same state in which it was suspended in the gas stream—a significant factor if product recovery is desired.
- Process upsets seldom result in the violation of emission standards. The mass rate of particulate matter escaping collection remains low over the life of the filter media and is insensitive to large changes in the mass-loading of dust entering the collector.

Self-cleaning fabric filter dust collectors also have some **limitations**:

- Liquid aerosols, moist and sticky materials, and condensation blind fabrics and reduce or prevent gas flow. These actions can curtail plant operation.
- Fabric life may be shortened in the presence of acidic or alkaline components in the gas stream.
- Use of fabric filters is generally limited to temperatures below 500°F.
- Should a spark or flame accidently reach the collector, fabrics can contribute to the fire/explosion hazard.
- When a large volume of gas is to be cleaned, the large number of fabric elements (bags, cartridges, or envelopes) required and the maintenance problem of detecting, locating, and replacing a damaged element should be considered. Newer monitoring equipment can detect leakage in an individual row of filters or an individual bag.

Pressure-Volume Relationships

The size of a fabric filter is media based on empirical data on the amount of fabric media required to clean the desired volumetric flow of gas with an acceptable static pressure drop across the media. The appropriate media and conditions for its use are selected by pilot test or from experience with media in similar full-scale installations. These tests provide a recommended range of approach velocities for a specific application.

The **approach velocity** is stated in practical applications as a gas-to-media ratio or filtration velocity (i.e., the velocity at which gas flows through the filter media). The **gas-to-media ratio** is the ratio of the volumetric flow of gases through the fabric filter to the area of fabric that participates in the filtration. It is the average approach velocity of the gas to the surface of the filter media.

It is difficult to estimate the correlation between pressure drop and gas-to-media ratio for a new installation. However, when the flow entering a filter with a dust cake is laminar and uniform, the pressure drop is proportional to the approach velocity:

$$\Delta p = KV_iW = KQW/A \qquad (6)$$

where

Δp = pressure drop, in. of water
K = specific resistance coefficient, pressure drop per unit air velocity and mass of dust per unit area
V_i = approach velocity or gas-to-media ratio, fpm
W = area density of dust cake, oz/ft^2
Q = volumetric flow of gases, cfm
A = area of cloth that intercepts gases, ft^2

Equation (6) suggests that increasing the area of the fabric during initial installation has some advantage. A larger fabric area reduces both the gas-to-media ratio and the thickness of the dust cake, resulting in a decreased pressure drop across the collector and reduced cleaning requirements. A lower gas-to-media ratio generally lowers operational cost for the fabric filter system, extends the useful life of the filter elements, and reduces maintenance frequency and expense. In addition, the lower gas-to-media ratio allows for some expansion of the system and, more importantly, additional surge capacity when upset conditions such as unusually high moisture content occur.

The specific resistance coefficient K is usually higher for fine dusts. The use of a primary collector to remove the coarse fraction seldom causes a significant change in the pressure drop across the collector. In fact, the coarse dust fraction helps reduce operating pressure because it results in a more porous dust cake, which provides better dust cake release.

Figure 8 illustrates the dependence of the pressure drop on time for a single-compartment fabric filter, operating through its cleaning cycle. The volumetric flow, particulate concentration, and distribution of particle sizes are assumed to remain constant. In practice, these assumptions are not usually valid. However, the interval

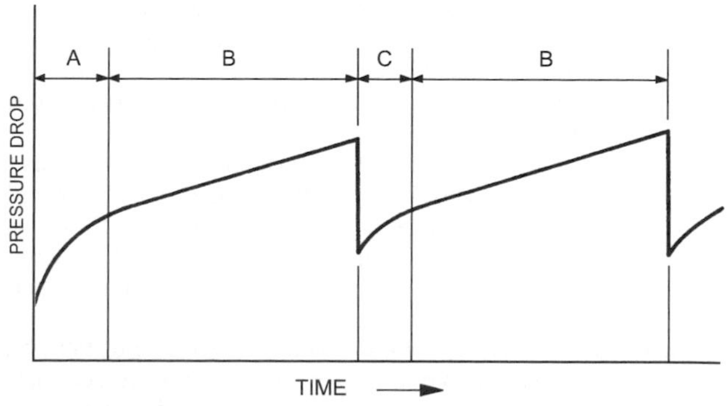

A = Initial cake formation (new fabric)
B = Deposition during homogeneous cake formation
C = Cake removal and initial cake formation
B + C = Interval between cleaning events

Fig. 8 Time Dependence of Pressure Drop Across Fabric Filter

Table 6 Temperature Limits and Characteristics of Fabric Filter Media

	Maximum Continuous Operating Temp., °F	Acid Resistance	Alkali Resistance	Flex Abrasion	Cost Relative to Cotton
Cotton	180	Poor	Very good	Very good	1.00
Wool	200	Good	Poor	Fair to good	2.75
Nylon[a,b]	200	Poor	Excellent	Excellent	2.50
Nomex[a,b]	400	Fair	Very good	Very good	8.00
Acrylic	260	Good	Fair	Good	3.00
Polypropylene	180	Excellent	Excellent	Very good	1.75
Polyethylene	145	Excellent	Excellent	Very good	2.00
Teflon[b]	425	Excellent	Excellent	Good	30.00
Glass fiber	500	Fair to good	Fair to good	Poor	5.00
Polyester[a]	275	Good	Good	Very good	2.50
Cellulose	180	Poor	Good	Good	—

[a] These fibers are subject to hydrolysis when they are exposed to hot, wet atmospheres.
[b] DuPont trademark.

between cleaning events is usually long enough that these variations are insignificant for most systems; between cleaning events, the dependence of pressure drop on time is approximately linear.

The volumetric flow of gases from a process often varies in response to changes in pressure drop across the fabric filter. The degree to which this variation is significant depends on the operating point for the fan and the requirements for the process.

Electrostatic Augmentation

Electrostatic augmentation involves establishing an electric field between the fabric and another electrode, precharging the dust particles, or both. The effect of electrostatic augmentation is that the interstitial openings in the fabric material function as if they were smaller, and hence smaller particles are retained. Its principle advantage has been in the more rapid buildup of the dust layer and somewhat higher efficiency for a given pressure drop. Although tested by many, this technique has not been broadly applied.

Fabrics

Commercially available fabrics, when applied appropriately, will separate 1 μm or larger particles from a gas with an efficiency of 99.9% or better. Particle size is not the major factor influencing efficiency attained from an industrial fabric filter. Most manufacturers of fabric filters will guarantee such efficiencies on applications in which they have prior experience. Lower efficiencies are generally attributed to poor maintenance (torn fabric seams, loose connections, etc.) or the inappropriate selection of lighter/higher permeability fabrics in an effort to reduce the cost of the collector.

Fabric specifications summarize information on such factors as cost, fiber diameter, type of weave, fabric density, tensile strength, dimensional stability, chemical resistance, finish, permeability, and abrasion resistance. Usually, comparisons are difficult, and the supplier must be relied on to select the appropriate material for the service conditions.

Table 6 summarizes experience with the exposure of fabrics to industrial atmospheres. While higher temperatures are acceptable for short periods, reduced fabric life can be expected with continued use above the maximum temperature. The filter media is often protected from high temperatures by thermostatically controlled air bleed-in or collector bypass dampers.

When the gases are moist or the fabric must collect hygroscopic or sticky materials, **synthetic media** are recommended. They are also recommended for high-temperature gases. Polypropylene has become a frequent selection. One limitation of synthetic media is greater penetration of the media during the cleaning cycle.

Woven fabrics are generally porous, and effective filtration depends on prior formation of a dust cake. New cloth collects poorly until particles bridge the openings in the cloth. Once the cake is formed, the initial layers become part of the fabric; they are not destroyed when the bulk of the collected material is dislodged during the cleaning cycle.

Cotton and wool fibers in woven media as well as most felted fabrics accumulate an initial dust cake in a few minutes. Synthetic woven fabrics may require a few hours in the same application because of the smoothness of the monofilament threads. Spun threads in the fill direction, when used, reduce the time required to build up the initial dust cake. Felted fabrics contain no straight-through openings and have a reasonably good efficiency for most particulates, even when clean. After the dust cake builds internally, as well as on the fabric's surface, shaking does not make it porous.

Cartridges manufactured from **pleated paper** and various **synthetic microfibers** (usually spun-bonded, not resin-glued, to form a stiff, nonwoven, microporous media) are also fabric filters because they operate in the same general manner as high-efficiency fabrics. As with cloth filters, the efficiency of the filter is increased by the formation of a dust cake on the medium.

Media used in cartridge type filters is usually manufactured to have many more pores (through which gas flows while being filtered) than do any of the common baghouse fabrics. Initial filtration efficiency of clean new cartridge filters is usually much greater, particularly for submicron-sized dust particles, than that of bare baghouse media (before a significant cake of filtered dust forms) because the pores in cartridge media are much smaller than those in baghouse fabrics. Despite having pores approximately one-tenth the diameter of those in the best baghouse felts, both cellulose (paper) and synthetic (most often polyester) cartridge media have so many pores that their permeability to gas flow is considerably greater than that of commonly used polyester felt baghouse media.

Because cartridge filters pack much more filter media per unit volume into their cartridges than can conventional bag filters, pulse jet cartridges filter collectors usually have less resistance to gas flow and operate with lower pressure drop from inlet to outlet than any of the three baghouse variations.

Pulse jet cleaned cartridge filter dust collectors are usually designed to operate at much lower filtration velocity (typically 1 to 3 fpm) than pulse jet cleaned tubular media baghouses. Submicron dust collection efficiency of cartridge filter media is so high that cartridge-type collectors are often used in applications where cleaned air will be directly recirculated back into the factory to reduce the expense of heating or cooling replacement (makeup) air.

Types of Self-Cleaning Mechanisms for Fabric Dust Collectors

The most common filter cleaning methods are (1) shaking the bags, (2) reversing the flow of gas through the bags, and (3) using an air pulse (pulse jet) to shock the dust cake and break it from the bags. Pulse jet fabric filters are usually cleaned on-line, whereas fabric filters using shakers or reverse air cleaners are usually cleaned off-line. Generally, large installations are compartmented and use off-line cleaning. Other cleaning methods include shake-deflate, a combination of shaker and reverse air cleaning, and acoustically augmented cleaning.

Shaker Collectors. When a single compartment is needed that can be cleaned off-line during shift change or breaks, shaker-type fabric filters are usually the least expensive choice. The fabric medium in a shaker-type fabric filter, whether formed into cylindrical tubes or rectangular envelopes, is mechanically agitated to remove the dust cake. Figures 9 and 10 show typical shaker-type fabric filters employing bag and envelope media, respectively.

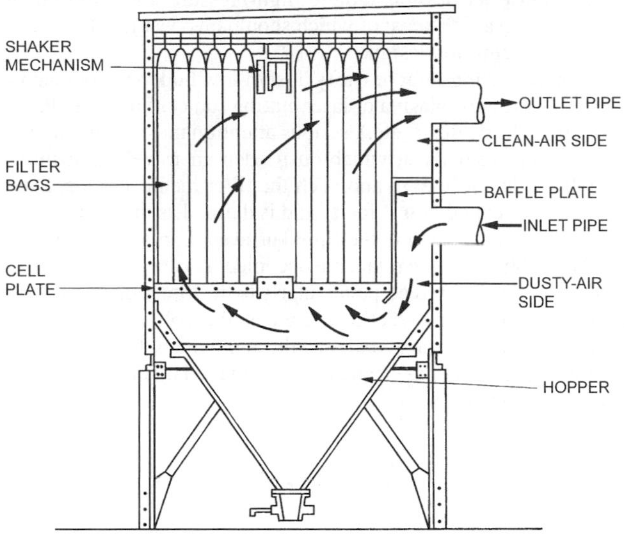

Fig. 9 Bag-Type Shaker Collector

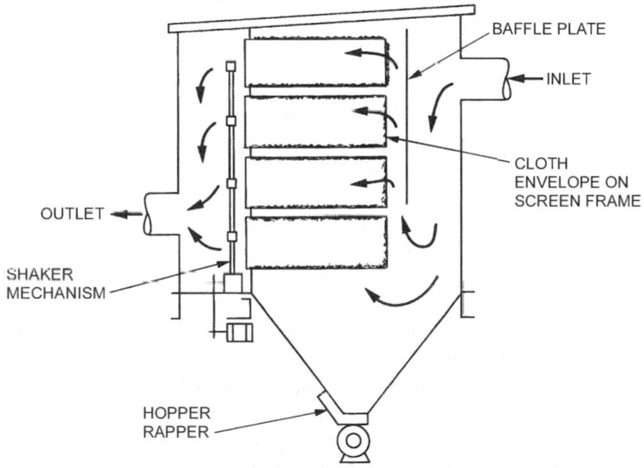

Fig. 10 Envelope-Type Shaker Collector

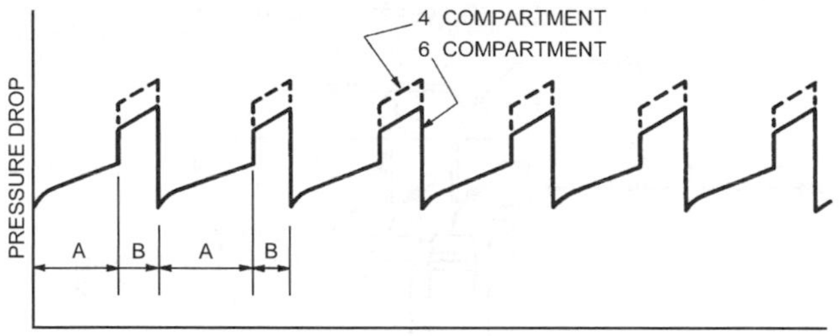

Fig. 11 Pressure Drop Across Shaker Collector Versus Time

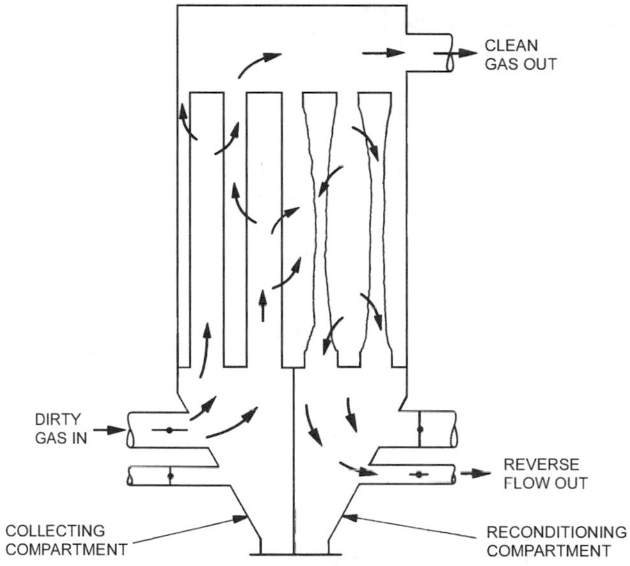

Fig. 12 Draw-Through Reverse Flow Cleaning of Fabric Filter

When the fabric filter cannot be stopped for cleaning, the collector is divided into a number of independent sections that are sequentially taken off-line for cleaning. Because it is usually difficult to maintain a good seal with dampers, relief dampers are often included. The relief dampers introduce a small volume of reverse gas to keep the gas flow at the fabric suitable for cake removal. Use of compartments, with their frequent cleaning cycles, does not permit a substantial increase in flow rates over those of a single-compartment unit cleaned periodically. The best situation for fabric reconditioning is when the system is stopped because even small particles will then fall into the hopper.

Figure 11 shows typical pressure diagrams for four- and six-compartment fabric filters that are continuously cleaned with mechanical shakers. Continuously cleaned units have compartment valves that close for the shaking cycle. This diagram is typical for a multiple-compartment fabric filter where individual compartments are cleaned off-line.

The gas-to-media ratio for a shaker-type dust collector is usually in the range of 2 to 4 fpm; it might be lower where the collector filters particles that are predominantly smaller than 2 μm. The abatement of metallurgical fumes is one example where a shaker collector is used to control particles that are less than 1 μm in size.

The bags are usually 4 to 8 in. in diameter and 10 to 20 ft in length, whereas envelope assemblies may be almost any size or shape. For ambient air applications, a woven cotton or polypropylene fabric is usually selected for shaker-type fabric filters. Synthetics are chosen for their resistance to elevated temperature and to chemicals.

Reverse Flow Collectors. Reverse flow cleaning is generally chosen when the volumetric flow of gases is very large. This method of cleaning inherently requires a compartmented design because the reverse flow needed to collapse the bags entrains dust that must be returned to on-line compartments of the fabric filter. Each compartment is equipped with one main shut-off valve and one reverse gas valve (whether the system is blown-through or drawn-through). A secondary blower and duct system is required to reverse the gas flow in the compartment to be cleaned. When a compartment is isolated for cleaning, the reverse gas circuit increases the volumetric flow and dust loading through the collector's active compartments.

The fabric medium is reconditioned by reversing the direction of flow through the bags, which partially collapse. The cleaning action is illustrated in Figure 12. After cleaning, the reintroduction of gas is delayed to allow dislodged dust to fall into the hopper.

Reverse flow cleaning reduces the number of moving parts in the fabric filter system—a maintenance advantage, especially when

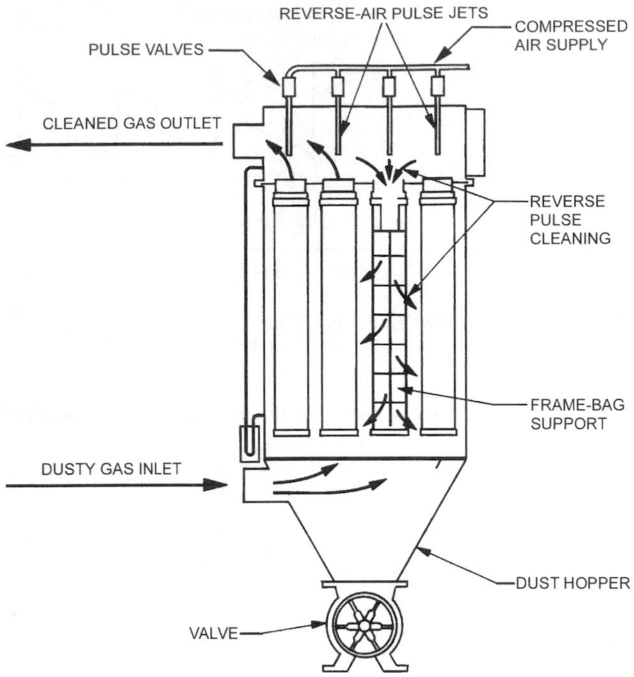

Fig. 13 Typical Pulse Jet Fabric Filter

large volumetric flows are cleaned. However, the cleaning or reconditioning is less vigorous than other methods, and the residual drag of the reconditioned fabric is higher. Reverse flow cleaning is particularly suited for fabrics, like glass cloth, that require gentle cleaning.

Reverse flow bags are usually 8 to 12 in. in diameter and 22 to 33 ft long and are generally operated at flow velocities in the 2 to 4 fpm range. As a consequence, reverse air dust collectors tend to be substantially larger than pulse jet cleaned designs of similar capacity.

For ambient air applications, a woven cotton or polypropylene fabric is the usual selection for reverse flow cleaning. For higher temperatures, woven polyester, glass fiber, or trademarked fabrics are often selected.

Pulse Jet Collectors. Efforts to decrease fabric filter sizes by increasing the flow rates through the fabric have concentrated on methods of implementing frequent or continuous cleaning cycles without taking major portions of the filter surface out of service. In the pulse jet design shown in Figure 13, a compressed air jet operating for a fraction of a second causes a rapid vibration or ripple in the fabric, which dislodges the accumulated dust cake. Simultaneously, outflow of both compressed cleaning air and entrained air from the top clean air plenum helps to sweep pulsed-off dust away from the filter surface. The pulse jet design is predominantly used because (1) it is easier to maintain than the reverse jet mechanism and (2) collectors can be smaller and less costly because the greater useful cleaning energy makes operation with higher filtration velocities practical.

The tubular bags are supported by wire cages during normal operation. The reverse flow pulse breaks up the dust layer on the outside of the bag, and the dislodged material eventually falls to a hopper. Filtration velocities for reverse jet or pulse jet designs range from 5 to 15 fpm for favorable dusts but require greater fabric area for many materials that produce a low-permeability dust cake. Felted fabrics are generally used for these designs because the jet cleaning opens the pores of woven cloth and produces excessive leakage through the filter.

Most pulse jet designs require high-pressure, dry, compressed (up to 100 psi) air, the cost of which should be considered when air-cleaning systems are designed.

When collecting light or fluffy dust of low bulk density (such as arc welding fumes, plasma or laser cutting fumes, finish sanding of wood, fine plastic dusts, etc.), serious attention must be given to the direction and velocity at which dust-laden air travels as it moves from the collector inlet to approach the filter media surface. Dusty air velocity is called **can velocity** and is defined as the actual velocity of airflow approaching the filter surfaces. Can velocity is computed by subtracting the total area occupied by filter elements (bags or cartridges, measured perpendicularly to the direction of gas flow) from the overall cross-sectional area of the collector's dusty air housing to compute the actual area through which dusty gas flows. The total gas flow being cleaned is then divided by that flow area to yield can velocity:

$$V_c = \frac{Q}{A_h - NA_f} \tag{7}$$

where

V_c = can velocity, fpm
Q = gas flow being cleaned, cfm
A_h = cross-sectional area of collector dusty gas housing, ft^2
N = number of filter bags or cartridges in collector
A_f = cross-sectional area, perpendicular to gas flow, of each filter bag or cartridge, ft^2

The maximum can velocity in **upflow** collectors (i.e., those in which dusty gas enters through a plenum or hopper beneath the filter elements) exists at the bottom end of the filter elements, where the entire gas flow must pass between and around the filters. Unless the maximum can velocity is low enough that pulsed-off dust can fall through the upwardly flowing gas, dust will simply redeposit on filter surfaces. The result is that on-line pulse cleaning cannot function, and the collector must be operated in the downtime pulse mode, with filter cleaning done only when there is no airflow through the collector.

Can velocity is sometimes overlooked when attempting to increase upflow collector capacity with improved fabrics or cartridges. Regardless of the theoretical gas-to-media ratio at which a filter operates, if released dust cannot fall through the rising airflow into the hopper, the collector will not be able to clean itself.

Collector designs in which dusty gas flows **downward** around the filters are much less susceptible to problems caused by high can velocity because the downward gas flow sweeps pulsed-off dust down toward (and into) the bottom dust discharge hopper, from which it can easily be removed.

Perhaps the most significant design difference among the many commercially available cartridge filter units is orientation of the filter cartridges. In collectors having **vertical filter surfaces** (Figure 14), all the pulsed-off dust can fall (by gravity) toward the bottom discharge flange, where it can be removed from the system (presuming that can velocity is low enough in upflow collectors). However, in designs having **horizontal or sloped filter surfaces**, nearly half of all pulsed-off dust (i.e., that from the top surfaces of the cartridges) must fall back onto that top surface. As a result, up to 50% of the total media area (i.e., nearly all media above the horizontal centerline of each filter cartridge) rapidly becomes blinded by a dust cake that is so thick that little or no gas can pass through to be filtered. This means that collectors having horizontal filter surfaces require twice as much installed filter area to achieve the same level of performance and useful filter life as collectors having vertical filter surfaces.

This chapter can not adequately cover all the collector design variables and experience-related factors that must be considered when deciding which baghouse or cartridge self-cleaning dust

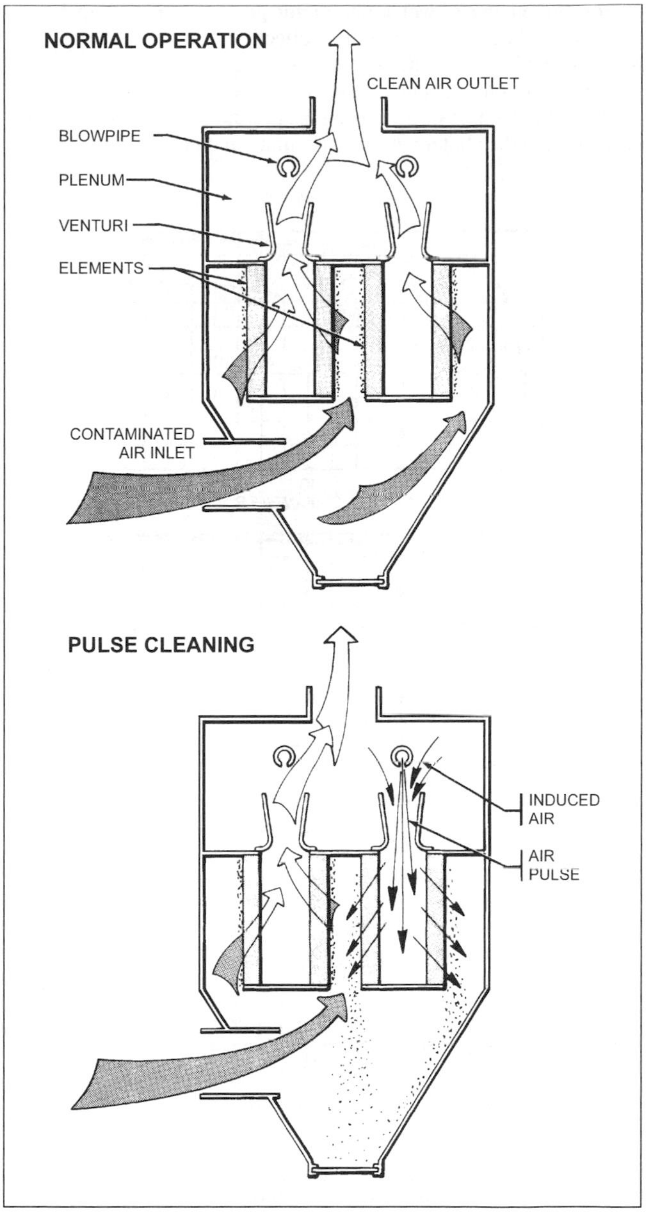

Fig. 14 Pulse Jet Cartridge Filters (Upflow Design with Vertical Filters)

collector design is best suited for each particular application. Engineers making dust collector selections are encouraged to discuss all aspects of each application in detail with all vendors being considered. It is necessary to judge

- The relative expertise of each prospective vendor
- Which dust collector design is most desirable
- How much media surface is needed in each design for each specified gas flow rate
- Which filter media is best suited to the particular application
- In the case of pulse jet cleaned cartridge collectors, what pleat spacing and pleat depth will give optimum or acceptable dust cake removal performance under the particular application conditions

GRANULAR BED FILTERS

Usually, granular bed filters employ a fixed bed of granular material that is periodically cleaned off-line. Continuously moving

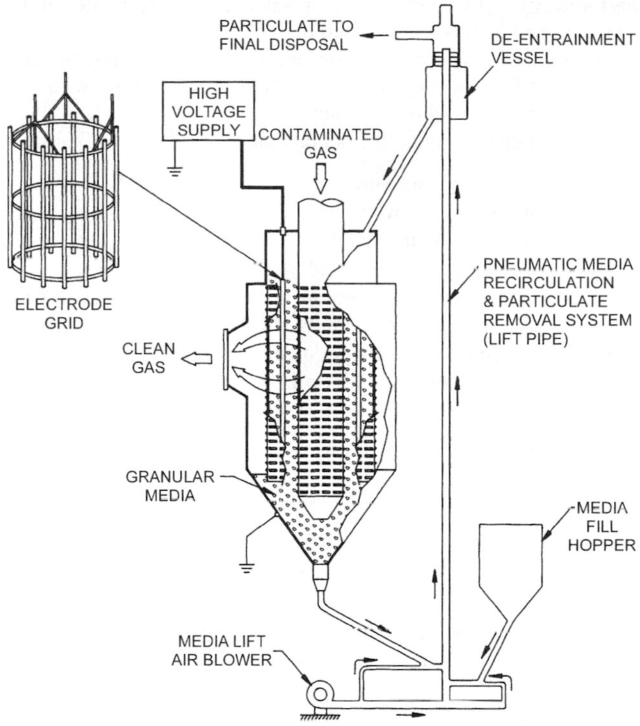

Fig. 15 Typical Granular Bed Filter

beds have been developed. Most commercial systems incorporate electrostatic augmentation to enhance fine particle control and to achieve good performance with a moving bed. Reentrainment in moving granular bed filters still significantly influences overall bed efficiency (Wade et al. 1978).

Principle of Operation

A typical granular bed filter is shown in Figure 15. Particulate-laden gas travels horizontally through the louvers and a granular medium, while the bed material flows downward. The gases typically travel with a superficial velocity near 100 to 150 fpm.

The filter medium moves continuously downward by gravity to prevent a filter cake from forming on the face of the filter and to prevent a high pressure drop. To provide complete cleaning of the louver's face, the louvers are designed so that some of the medium falls through each louver opening, thus preventing any bridging or buildup of particulate material.

Electrostatic augmentation gives the granular bed filter many of the characteristics of a two-stage electrostatic precipitator. The obvious disadvantage of a granular bed filter is in removal of the collected dust, which requires liquid backwash or circulation and cleaning of the filter material.

PARTICULATE SCRUBBERS (WET COLLECTORS)

Wet-type dust collectors use liquid (usually, but not necessarily, water) to capture and separate particulate matter (dust, mist, and fumes) from a gas stream. Some scrubbers operate by spraying the **scrubbing liquid** into the contaminated air. Others bubble air through the scrubbing liquid. In addition, many hybrid designs exist.

Particle sizes, which can be controlled by a wet scrubber, range from 0.3 to 50 μm or larger. Wet collectors can be classified into three categories: (1) **low-energy** (up to 1 W/cfm, 1 to 6 in. of water); (2) **medium-energy** (1 to 3 W/cfm, 6 to 18 in. of water); and (3)

high-energy (>3 W/cfm, >18 in. of water). The performance of typical wet scrubbers is summarized in Table 2.

Wet collectors may be used for the collection of most particulates from industrial process gas streams where economics allow for collection of the material in a wet state.

Advantages of wet collectors include

- Constant operating pressure
- No secondary dust sources
- Small spare parts requirement
- Ability to collect both gases and particulates
- Ability to handle high-temperature and high-humidity gas streams, as well as to reduce the possibility of fire or explosion
- Reasonably small space requirements for scrubbers
- Ability to continuously collect sticky and hygroscopic solids without becoming fouled

Disadvantages include

- High susceptibility to corrosion (corrosion-resistant construction is expensive)
- High humidity in the discharge gas stream, which may give rise to visible and sometimes objectionable fog plumes, particularly during winter
- Large pressure drops and high power requirement for most designs that can efficiently collect fine (particularly submicron) particles
- Possible difficulty or high cost of disposal of waste water or clarification waste
- Rapidly decreasing fractional efficiency for most scrubbers for particles less than 1 μm in size
- Freeze protection required in many applications in northern environments

Principle of Operation

The more important mechanisms involved in the capture and removal of particulate matter in scrubbers are inertial impaction, Brownian diffusion, and condensation.

Inertial impaction occurs when a dust particle and a liquid droplet collide, resulting in the capture of the particle. The resulting liquid/dust particle is relatively large and may be easily removed from the carrier gas stream by gravitation or impingement on separators.

Brownian diffusion occurs when the dust particles are extremely small and have motion independent of the carrier gas stream. These small particles collide with one another, making larger particles, or collide with a liquid droplet and are captured.

Condensation occurs when the gas or air is cooled below its dew point. When moisture is condensed from the gas stream, fogging occurs, and the dust particles serve as condensation nuclei. The dust particles become larger as a result of the condensed liquid, and the probability of their removal by impaction is increased.

Wet collectors perform two individual operations. The first occurs in the **contact zone**, where the dirty gas comes in contact with the liquid; the second is in the **separation zone**, where the liquid that has captured the particulate matter is removed from the gas stream. All well-designed wet collectors use one or more of the following principles:

- High liquid-to-gas ratio
- Intimate contact between the liquid and dust particles, which may be accomplished by formation of large numbers of small liquid droplets or by breaking up the gas flow into many small bubbles that are driven through a bath of scrubbing liquid, to increase the chances that contaminants will be wetted and collected
- Abrupt transition from dry to wet zones to avoid particle buildup where the dry gas enters the collectors

For a given type of wet collector, the greater the power applied to the system, the higher will be the collection efficiency (Lapple and Kamack 1955). This is the contacting power theory. Figure 16 compares the fractional efficiencies of several wet collectors, and Figure 17 shows the relationship between the pressure drop across a venturi scrubber and the abatement of particulate matter.

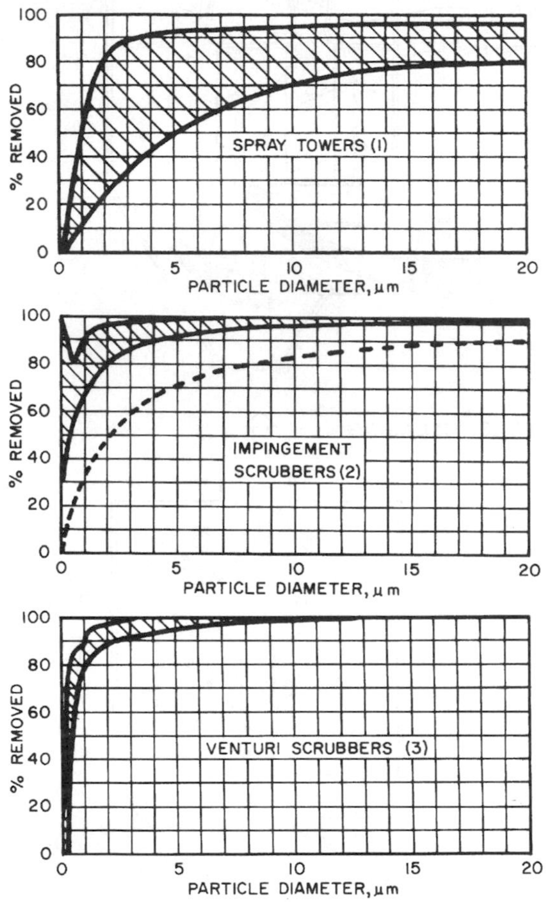

Notes:
1. Efficiency depends on liquid distribution.
2. Upper curve is for packed tower; lower curve is for orifice-type wet collector. Dashed lines indicate performance of low-efficiency, irrigated baffles or rods.
3. Efficiency is directly related to fluid rate and pressure drop.

Fig. 16 Fractional Efficiency of Several Wet Collectors

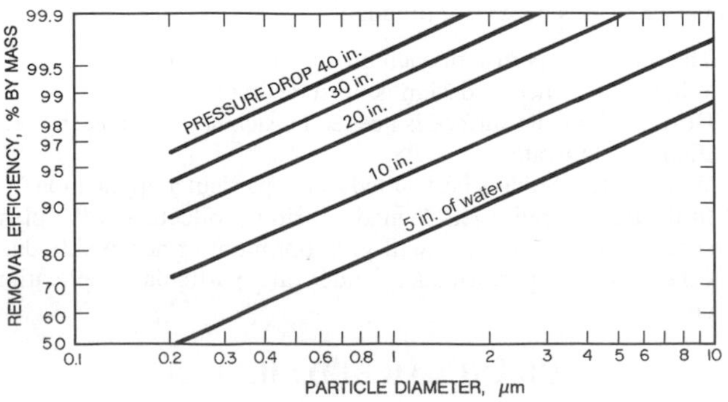

Fig. 17 Efficiency of Venturi Scrubber

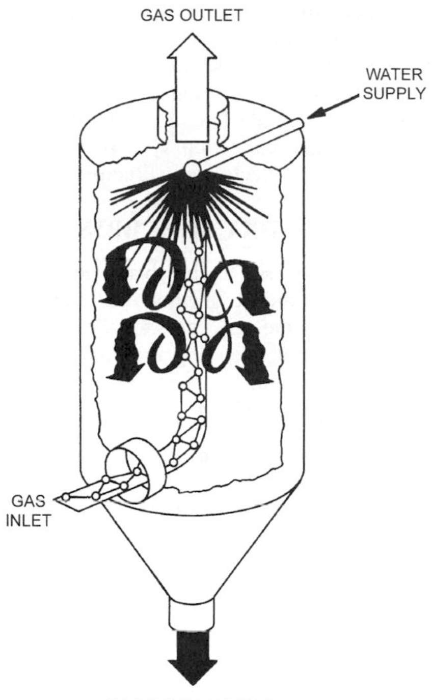

Fig. 18 Typical Spray Tower

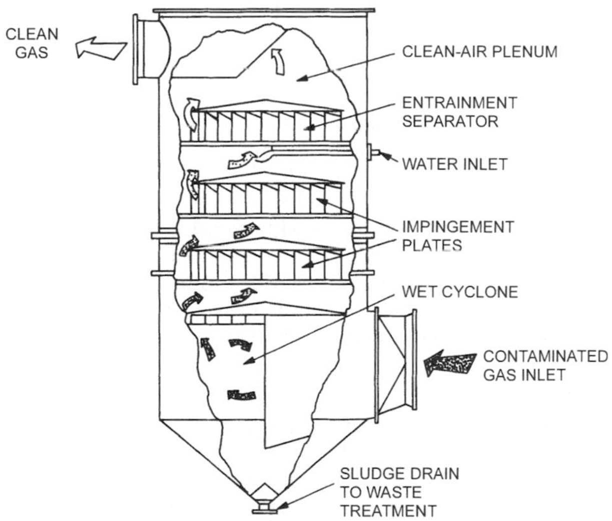

Fig. 19 Typical Impingement Scrubber

Spray Towers and Impingement Scrubbers

Spray towers and impingement scrubbers are available in many different arrangements. The gas stream may be subjected to a single spray or a series of sprays, or the gas may be forced to impinge on a series of irrigated baffles. Except for packed towers, these types of scrubbers are in the low-energy category; thus, they have relatively low particulate removal efficiency. A typical spray tower and an impingement scrubber are illustrated in Figures 18 and 19, respectively.

The efficiency of a spray tower can be improved by adding high-pressure sprays. A spray tower efficiency of 50 to 75% can be improved to 95 to 99% (for dust particles with size near 2 μm) by pressures in the range of 30 to 100 psig.

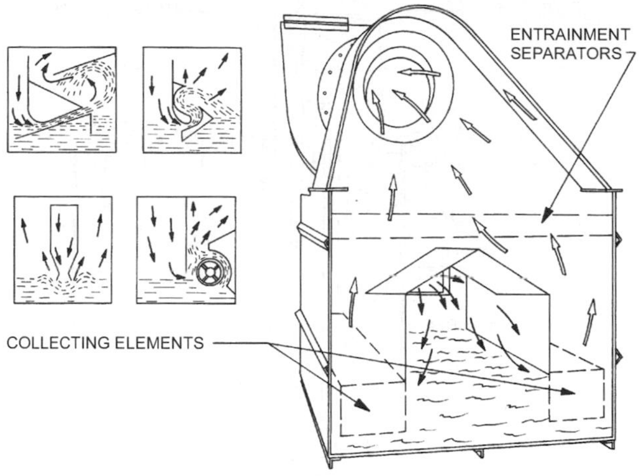

Fig. 20 Typical Orifice-Type Wet Collector

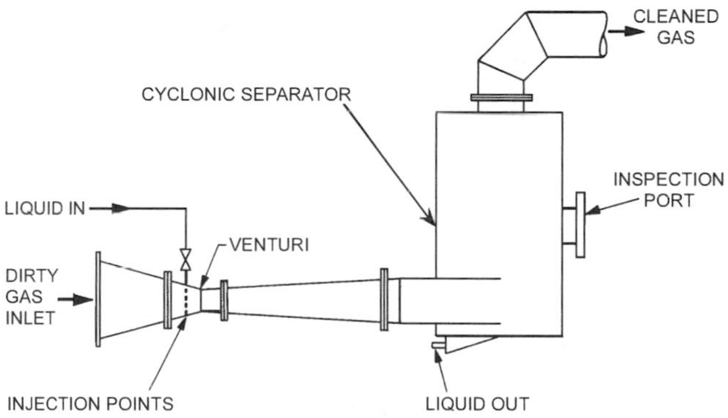

Fig. 21 Typical High-Energy Venturi Scrubber

Centrifugal-Type Collectors

These collectors are characterized by a tangential entry of the gas stream into the collector. They are classed with medium-energy scrubbers. The impingement scrubber shown in Figure 19 is an example of a centrifugal-type wet collector.

Orifice-Type Collectors

Orifice-type collectors are also classified in the medium-energy category. Usually, the gas stream is made to impinge on the surface of the scrubbing liquid and is forced through constrictions where the gas velocity is increased and where the liquid-gaseous-particulate interaction occurs. Water usage for orifice collectors is limited to evaporation loss and removal of collected pollutants. A typical orifice-type wet collector is illustrated in Figure 20.

Venturi Scrubber

A high-energy venturi scrubber passes the gas through a venturi-shaped orifice where the gas is accelerated to 12,000 fpm or more. Depending on the design, the scrubbing liquid is added at, or ahead of, the throat. The rapid acceleration of the gas shears the liquid into a fine mist, increasing the chance of liquid-particle impaction. Yung has developed a mathematical model for the performance and design of venturi scrubbers (Semrau 1977). Subsequent validation experiments (Rudnick et al. 1986) have demonstrated that this model yields a more representative prediction of venturi scrubber performance than other performance models do.

In typical applications, the pressure drop for gases across a venturi is higher than for other types of scrubber. Water circulation is

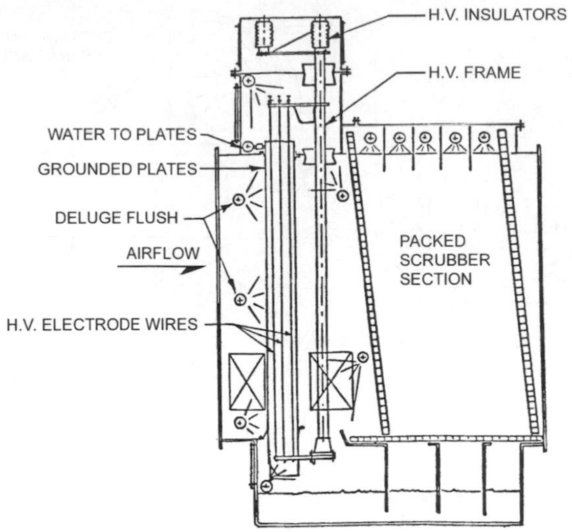

Fig. 22 Typical Electrostatically Augmented Scrubber

also higher; thus, venturi systems use water reclamation systems. One example of a venturi scrubber is illustrated in Figure 21.

Electrostatically Augmented Scrubbers

Several gas-cleaning devices combine electrical charging of particulate matter with wet scrubbing. Electrostatic augmentation enhances fine particle control by causing an electrical attraction between the particles and the liquid droplets. Compared to venturi scrubbers, electrostatically augmented scrubbers remove particles smaller than 1 μm at a much lower pressure drop.

There are three generic designs for electrostatic augmentation:

1. Unipolar charged aerosols pass through a contact chamber containing randomly oriented packing elements of dielectric material. A typical electrostatically augmented scrubber of this design is shown in Figure 22.
2. Unipolar charged aerosols pass through a low-energy venturi scrubber.
3. Unipolar charged aerosols pass into a spray chamber where they are attracted to oppositely charged liquid droplets.

Collection efficiencies of 50 to 90% can be achieved in a single particle charging and collection stage, depending on the mass loading of fine particles and the superficial velocity of gases in the collector. Higher collection efficiencies can be obtained by using two or more stages. Removal efficiency of gaseous pollutants depends on the mass transfer and absorption design of the scrubber section.

In most applications of electrostatically augmented scrubbers, the dirty gas stream is quenched by adiabatic cooling with liquid sprays; thus, it contains a large amount of water vapor that wets the particulate contaminants. This moisture provides a dominant influence on particle adhesion and the electrical resistivity of deposits within the collector.

Electrical equipment for particle charging is similar to that for electrostatic precipitators. The scrubber section is usually equipped with a liquid recycle pump, recycle piping, and a liquid distribution system.

GASEOUS CONTAMINANT CONTROL

Many industrial processes produce large quantities of gaseous or vaporized contaminants that must be separated from gas streams. Removal of these contaminants is usually achieved through absorption into a liquid or adsorption onto a solid medium. Incineration of the exhaust gas (see the section on Incineration of Gases and Vapors) has also been successfully used for the removal of organic gases and vapors. Low vapor pressure odorous materials that condense to form submicron condensation aerosols after being emitted from hot industrial processes can sometimes be successfully controlled by well-designed condensing filter or condensing precipitator submicron particulate collection systems (see the section on Two-Stage Designs under Electrostatic Precipitators).

SPRAY DRY SCRUBBING

Spray dry scrubbing is used to absorb and neutralize acidic gaseous contaminants in hot industrial gas streams. The system uses an alkali spray to react with the acid gases to form a salt. The process heat evaporates the liquid, resulting in a dry particulate that is removed from the gas stream.

Typical industrial applications of spray dry scrubbing are

- Control of hydrochloric acid (HCl) emissions from biological hazardous-waste incinerators
- Control of sulfuric acid and sulfur trioxide emissions from burning high-sulfur coal
- Control of sulfur oxides (SO_x), boric acid, and hydrogen fluoride (HF) gases from glass-melting furnaces.

Principle of Operation

Spray drying involves four operations: (1) atomization, (2) gas droplet mixing, (3) drying of liquid droplets, and (4) removal and collection of a dry product. These operations are carried out in a tower or a specially designed vessel.

In any spray dryer design, good mixing and efficient gas droplet contact are desirable. Dryer height is largely determined by the time required to dry the largest droplets produced by the atomizer. Towers used for acid gas control typically have gas residence times of about 10 s, compared to about 3 s for towers designed for evaporative cooling. The longer residence time is needed because drying by itself is not the primary goal for the equipment. Many of the acid-alkali reactions are accelerated in the liquid state. It is, therefore, desirable to cool the gases to as close as possible to the adiabatic saturation temperature (dew point) without risking condensation in downstream particulate collectors. At these low temperatures, droplets survive against evaporation much longer, thus obtaining a better chance of contacting all acidic contaminants while the alkali compounds are in their most reactive state.

Equipment

The **atomizer** must disperse a liquid containing an alkali compound that will react with acidic components of the gas stream. The liquid must be distributed uniformly within the dryer and mixed thoroughly with the hot gases in droplets of a size that will evaporate before striking a dryer surface.

In typical spray dryers used for acid gas control, the droplets have diameters ranging from 50 to 200 μm. The larger droplets are of most concern because these might survive long enough to impinge on equipment surfaces. In general, a trade-off must be made between the largest amount of liquid that can be sprayed and the largest droplets that can be tolerated by the equipment. The angular distribution or **fan-out** of the spray is also important. In spray drying, the angle is often 60 to 80°, although both lower and higher angles are sometimes required. The fan-out may change with distance from the nozzle, especially at high pressures.

An important aspect of spray dryer design and operation is the production and control of the gas flow patterns within the drying chamber. Because of the importance of the flow patterns, spray dryers are usually classified on the basis of gas flow direction in the chamber relative to the spray. There are three basic designs: (1) **cocurrent**, in which the liquid feed is sprayed with the flow of the hot gas; (2) **countercurrent**, in which the feed is sprayed against the flow of the gas; and (3) **mixed flow** in which there is a combined cocurrent and countercurrent flow.

There are several types of atomizers. High-speed rotating **disks** achieve atomization through centrifugal motion. Although disks are bulky and relatively expensive, they are also more flexible than nozzles in compensating for changes in particle size caused by variations in feed characteristics. Disks are also used when high-pressure feed systems are not available. They are frequently used when high volumes of liquid must be spray dried. Disks are not well suited to counterflow or horizontal flow dryers.

Nozzles are also commonly used. These may be subdivided into two distinct types—centrifugal pressure nozzles and two-fluid (or pneumatic) nozzles. In the **centrifugal pressure nozzle**, energy for atomization is supplied solely by the pressure of the feed liquid. Most pressure nozzles are of the swirl type, in which tangential inlets or slots spin the liquid in the nozzle. The pressure nozzle satisfactorily atomizes liquids with viscosities of 300 centipoise or higher. It is well suited to counterflow spray dryers and to installations requiring multiple atomizers. Capacities up to 10,000 lb/h through a single nozzle are possible. Pressure nozzles have some disadvantages. For example, pressure, capacity, and orifice size are independent, resulting in a certain degree of inflexibility. Moreover, pressure nozzles (particularly those with small passages) are susceptible to erosion in applications involving abrasive materials. In such instances, tungsten carbide or a similarly tough material is mandatory.

In **two-fluid nozzles**, air (or steam) supplies most of the energy required to atomize the liquid. Liquid, admitted under low pressure, may be mixed with the air either internally or externally. Although energy requirements for this atomizer are generally greater than for spinning disks or pressure nozzles, the two-fluid nozzle can produce very fine atomization, particularly with viscous materials.

The density and viscosity of the feed materials and how these might change at elevated temperature should be considered. Some alkali compounds do not form a solution at the concentrations needed for acid gas control. It is not uncommon that pumps and nozzles must be chosen to handle and meter slurries.

Spray dryer systems include metering valves, pumps and compressors, and controls to assure optimal chemical feed and temperature within the gas-cleaning system.

WET-PACKED SCRUBBERS

Packed scrubbers are used to remove gaseous and particulate contaminants from gas streams. Scrubbing is accomplished by impingement of particulate matter and/or by absorption of soluble gas or vapor molecules on the liquid-wetted surface of the packing. There is no limit to the amount of particulate capture, as long as the properties of the liquid film are unchanged.

Gas or vapor removal is more complex than particulate capture. The contaminant becomes a solute and has a vapor pressure above that of the scrubbing liquid. This vapor pressure typically increases with increasing concentration of the solute in the liquid and/or with increasing liquid temperature. Scrubbing of the contaminant continues as long as the partial pressure of contaminant in the gas is above its vapor pressure with respect to the liquid. The rate of contaminant removal is a function of the difference between the partial pressure and vapor pressure, as well as the rate of diffusion of the contaminant.

In most common gas-scrubbing operations, small increases in the superficial velocity of gases through the collector decrease the removal efficiency; therefore, these devices are operated at the highest possible superficial velocities consistent with acceptable contaminant control. Usually, increasing the liquid rate has little effect on efficiency; therefore, liquid flow is kept near the minimum required for satisfactory operation and removal of particulate matter. As the superficial velocity of gases in the collector increases above a certain value, there is a tendency to strip liquid from the surface of the packing and entrain the liquid from the scrubber. If the pressure drop is not the limiting factor for operation of the equipment, maximum scrubbing capacity will occur at a gas rate just below the rate that causes excessive liquid reentrainment.

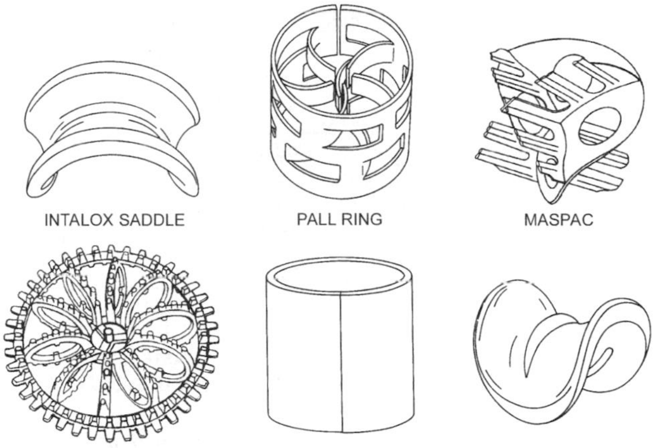

Fig. 23 Typical Packings for Scrubbers

Scrubber Packings

Packings are designed to present a large surface area that will wet evenly with liquid. They should also have high void ratio so that pressure drop will be low. High-efficiency packings promote turbulent mixing of the gas and liquid. Figure 23 illustrates six types of packings that are randomly dumped into scrubbers. Packings are available in ceramic, metal, and thermoplastic materials. Plastic packings are extensively used in scrubbers because of their low mass and resistance to mechanical damage. They offer a wide range of chemical resistance to acids, alkalies, and many organic compounds; however, plastic packing can be deformed by excessive temperatures or by solvent attack.

The relative capacity of tower packings at constant pressure drop can be obtained by calculation from the **packing factor** F. The gas-handling capacity G of a packing is inversely proportional to the square root of F:

$$G = K/\sqrt{F} \tag{8}$$

where

G = mass flow rate of gases through scrubber
F = packing factor (surface area of packing per unit volume of gently poured material)

The smaller the packing factor of a given packing, the greater will be its gas-handling capacity. Typical packing factors for scrubber packings are summarized in Table 7.

Arrangements of Packed Scrubbers

The four generic arrangements for wet-packed scrubbers are illustrated in Figure 24:

1. Horizontal cocurrent scrubber
2. Vertical cocurrent scrubber
3. Cross-flow scrubber
4. Countercurrent scrubber

Cocurrent flow scrubbers can be operated with either horizontal or vertical gas and liquid flows. A **horizontal cocurrent scrubber** depends on the gas velocity to carry the liquid into the packed bed. It operates as a wetted entrainment separator with limited gas and

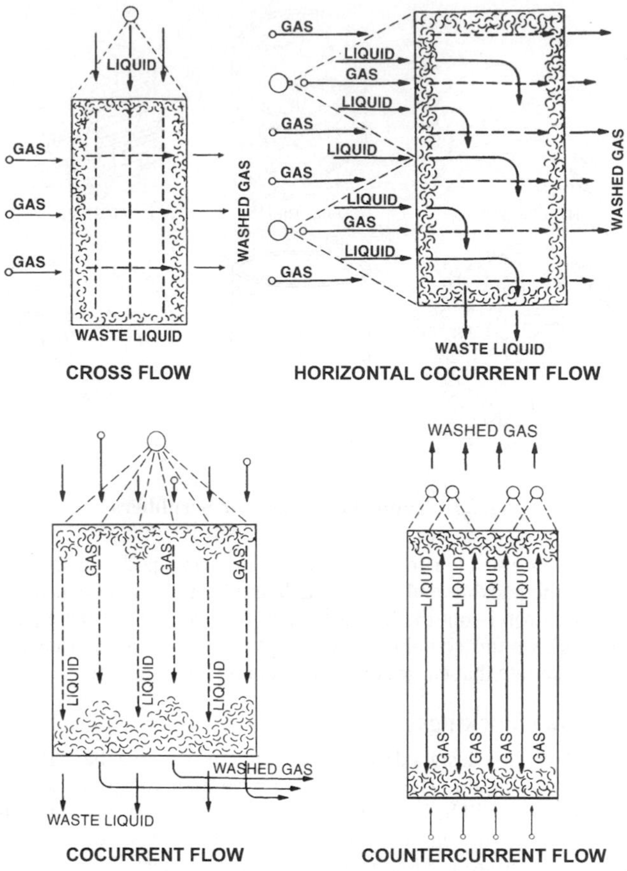

Fig. 24 Flow Arrangements Through Packed Beds

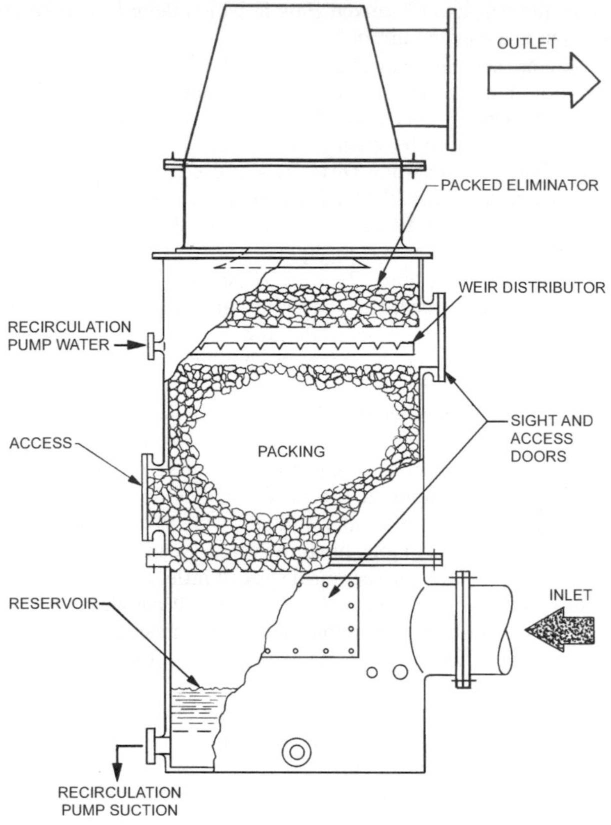

Fig. 25 Typical Countercurrent Packed Scrubber

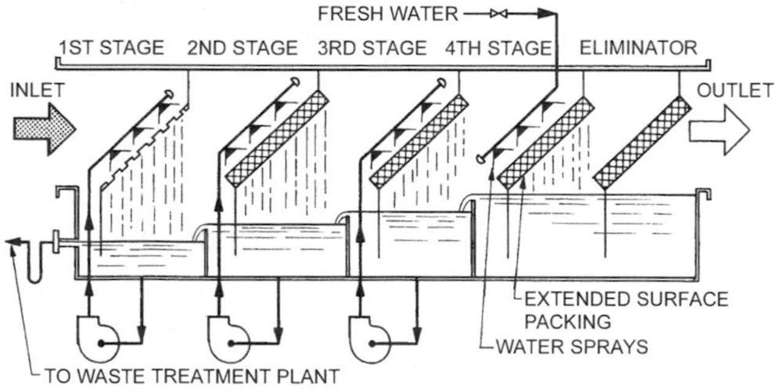

Fig. 26 Horizontal Flow Scrubber with Extended Surface

Table 7 Packing Factor *F* for Various Scrubber Packing Materials

		Nominal Size, in.				
Type of Packing	**Material**	**3/4**	**1**	**1.5**	**2**	**3 or 3.5**
Super Intalox	Plastic		40		21	16
Super Intalox	Ceramic		60		30	
Intalox saddles	Ceramic	145	92	52	40	22
Berl saddles	Ceramic	170	110	65	45	
Raschig rings	Ceramic	255	155	95	65	37
Hy-Pak	Metal		43	26	18	15
Pall rings	Metal		48	33	20	16
Pall rings	Plastic		52	40	24	16
Tellerettes	Plastic		36		18	16
Maspac	Plastic				32	21

liquid contact time. The superficial velocity of gases in the collector is limited by liquid reentrainment to about 650 fpm. A **vertical cocurrent scrubber** can be operated at very high pressure drop (1 to 3 in. of water per foot of packing depth) because there is no flooding limit for the superficial velocity. The contact time in a cocurrent scrubber is a function of bed depth. The effectiveness of absorption processes is lower in cocurrent scrubbers than in the other arrangements because the liquid containing contaminant is in contact with the exit gas stream.

Cross-flow scrubbers use downward-flowing liquid and a horizontally moving gas stream. The effectiveness of absorption processes in cross-flow scrubbers lies between those for cocurrent and countercurrent flow scrubbers.

Countercurrent scrubbers use a downward-flowing liquid and an upward-flowing gas. The gas-handling capacity of countercurrent scrubbers is limited by pressure drop or by liquid entrainment. The contact time can be controlled by the depth of packing used. The effectiveness of absorption processes is maximized because the exiting gas is in contact with fresh scrubbing liquid.

The most broadly used arrangement is the countercurrent packed scrubber. This type of scrubber, illustrated in Figure 25, gives the best removal of gaseous contaminants while keeping liquid consumption to a minimum. The effluent liquid has the highest contaminant concentration.

Extended surface packings have been used successfully for the absorption of highly soluble gases such as HCl because the required contact time is minimal. This type of packing consists of a woven mat of fine fibers of a plastic material that is not affected by chemical exposure. Figure 26 shows an example of a scrubber consisting of three wetted stages of extended surface packing in series with the

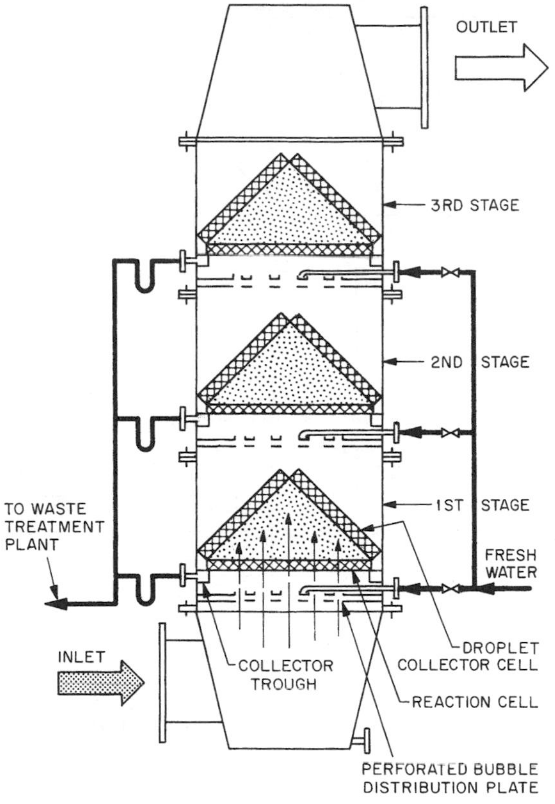

Fig. 27 Vertical Flow Scrubber with Extended Surface

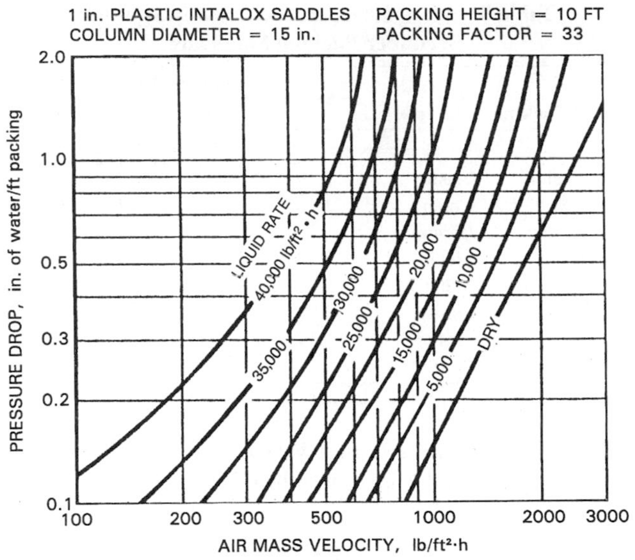

Fig. 28 Pressure Drop Versus Gas Rate for Typical Packing

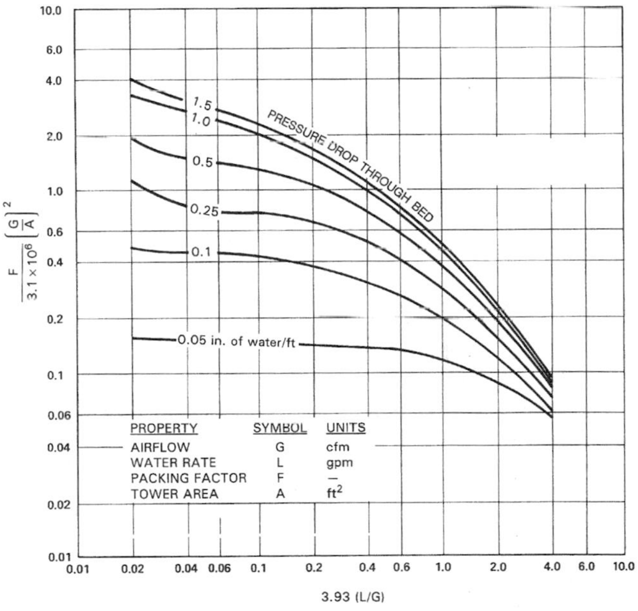

Fig. 29 Generalized Pressure Drop Curves for Packed Beds

gas flow. A final dry mat is used as an entrainment eliminator. If solids are present in the inlet gas stream, a wetted impingement stage precedes the wetted mats to prevent plugging of the woven mats.

Figure 27 shows a scrubber with a vertical arrangement of extended surface packing. This design uses three complete stages in series with the gas flow. The horizontal mat at the bottom of each stage operates as a flooded bed scrubber. The flooded bed is used to minimize water consumption. The two inclined upper mats operate as entrainment eliminators.

Pressure Drop

The pressure drop through a particular packing in countercurrent scrubbers can be calculated from the airflow and water flow per unit area. Charts, such as the one shown in Figure 28, are available from manufacturers of each type and size of packing.

The pressure drop for any packing can also be estimated by using the data on packing factors in Table 7 and the modified, generalized, pressure drop correlation shown in Figure 29. This correlation was developed for a gas stream substantially of air, with water as the scrubbing liquid. It should not be used if the properties of the gas or liquid vary significantly from air or water, respectively. Countercurrent scrubbers are generally designed to operate at pressure drops between 0.25 and 0.65 in. of water per foot of packing depth. Liquid irrigation rates typically vary between 5 and 20 gpm per square foot of bed area.

Absorption Efficiency

The prediction of the absorption efficiency of a packed bed scrubber is much more complex than estimating its capacity because performance estimates involve the mechanics of absorption. Some of the factors affecting efficiency are superficial velocity of gases in the scrubber, liquid injection rate, packing size, type of packing, amount of contaminant to be removed, distribution and amount of absorbent available for reaction, temperature, and reaction rate for absorption.

Practically all commercial packings have been tested for absorption rate (mass transfer coefficient) using standard absorber conditions—carbon dioxide (CO_2) in air and a solution of caustic soda (NaOH) in water. This system was selected because the interaction of the variables is well understood. Further, the mass transfer coefficients for this system are low; thus, they can be determined accurately by experiment. The values of mass transfer coefficients (K_Ga) for various packings under these standard test conditions are given in Table 8.

The vast majority of wet absorbers are used to control low concentrations (less than 0.005 mole fraction) of contaminants in air. Dilute aqueous solutions of NaOH are usually chosen as the scrubbing fluid. These conditions simplify the design of scrubbers somewhat. Mass transfer from the gas to the liquid is then explained by

Table 8 Mass Transfer Coefficients (K_Ga) for Scrubber Packing Materials

Type of Packing	Material	Nominal Size, in. 1	1.5	2	3 or 3.5
		K_Ga, lb·mol/h·ft³·atm			
Super Intalox	Plastic	2.19		1.44	0.887
Intalox saddles	Ceramic	1.96	1.71	1.44	0.820
Raschig rings	Ceramic	1.73	1.50	1.21	
Hy-Pak	Metal	2.20	1.87	1.69	1.09
Pall rings	Metal	2.32	1.87	1.62	0.91
Pall rings	Plastic	1.98	1.73	1.46	0.89
Tellerettes	Plastic	2.19		1.98	
Maspac	Plastic			1.44	0.89

System: CO_2 and NaOH; gas rate: 110 cfm/ft²; liquid rate: 4 gpm/ft².

Table 9 Relative K_Ga for Various Contaminants in Liquid-Film-Controlled Scrubbers

Gas Contaminant	Scrubbing Liquid	K_Ga, lb·mol / h·ft³·atm
CO_2	4% (by mass) NaOH	2.0
H_2S	4% (by mass) NaOH	5.92
SO_2	Water	2.96
HCN	Water	5.92
HCHO	Water	5.92
Cl_2	Water	4.55

Note: Data for 2 in. plastic Super Intalox. Temperatures: from 60 to 75°F; liquid rate: 10 gpm/ft²; gas rate: 215 cfm/ft².

the **two-film theory**: the gaseous contaminant travels by diffusion from the main gas stream through the gas film, then through the liquid film, and finally into the main liquid stream. The relative influences of the gas and liquid films on the absorption rate depend on the solubility of the contaminant in the liquid. Sparingly soluble gases like hydrogen sulfide (H_2S) and CO_2 are said to be liquid film controlled; highly soluble gases such as HCl and ammonia (NH_3) are said to be gas film controlled. In liquid-film-controlled systems, the mass transfer coefficient varies with the liquid injection rate but is only slightly affected by the superficial velocity of the gases. In gas-film-controlled systems, the mass transfer coefficient is a function of both the superficial velocity of the gases and the liquid injection rate.

In the absence of leakage, the percentage by volume of the contaminant removed from the air can be found from the inlet and outlet concentrations of contaminant in the airstream:

$$\%\ \text{Removed} = 100(1 - Y_o/Y_i) \tag{9}$$

where

Y_i = mole fraction of contaminants entering scrubber (dry gas basis)
Y_o = mole fraction of contaminants exiting scrubber (dry gas basis)

The driving pressure for absorption (assuming negligible vapor pressure above the liquid) is controlled by the logarithmic mean of inlet and outlet concentrations of the contaminant:

$$\Delta p_{\ln} = p\frac{Y_i - Y_o}{\ln(Y_i/Y_o)} \tag{10}$$

where

$\Delta p_{\ln}$ = driving or diffusion pressure acting to absorb contaminants on packing
p = inlet pressure

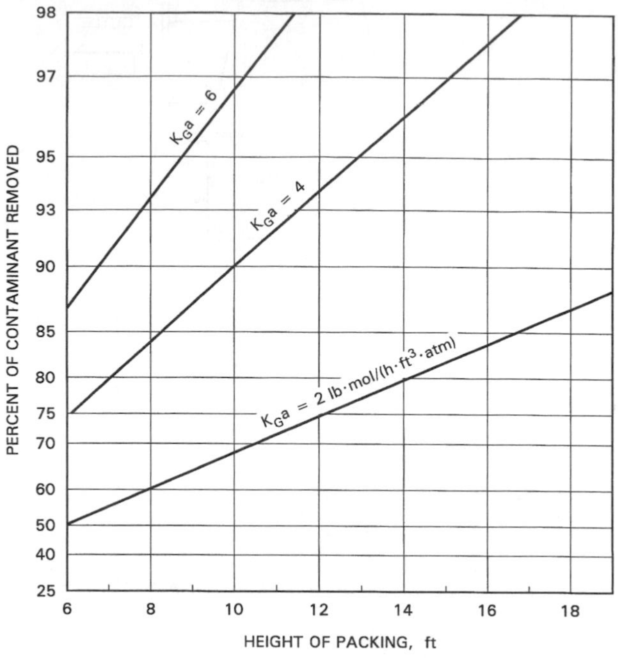

Fig. 30 Contaminant Control at Superficial Velocity = 120 fpm (Liquid Film Controlled)

The rate of absorption of contaminant (mass transfer coefficient) is related to the depth of packing as follows:

$$K_Ga = N/HA\Delta p_{\ln} \tag{11}$$

where

N = solute absorbed, lb·mol/h
H = depth of packing, ft
A = cross-sectional area of scrubber, ft²

The value of N can be determined from

$$N = G(Y_i - Y_o) \tag{12}$$

where G = mass flow rate of gases through scrubber, lb·mol/h.

The superficial velocity of gases is a function of the unit gas flow rate and the gas density:

$$V = 60TGM_v/C_1A \tag{13}$$

where

V = superficial gas velocity, fpm
M_v = molar volume, ft³/lb·mol
T = exit gas temperature, °R = °F + 460
C_1 = 460°R

By combining these equations and assuming ambient pressure, a graphical solution can be derived for both liquid-film- and gas-film-controlled systems. Figures 30, 31, and 32 show the height of packing required versus percent removal for various mass transfer coefficients at superficial velocities of 120, 240, and 360 fpm, respectively, with liquid-film-controlled systems. Figures 33, 34, and 35 show the height of packing versus percent removal for various mass transfer coefficients at the same three superficial velocities with gas film-controlled systems.

These graphs can be used to determine the height of 2 in. plastic Intalox saddles (Figure 23) required to give the desired percentage

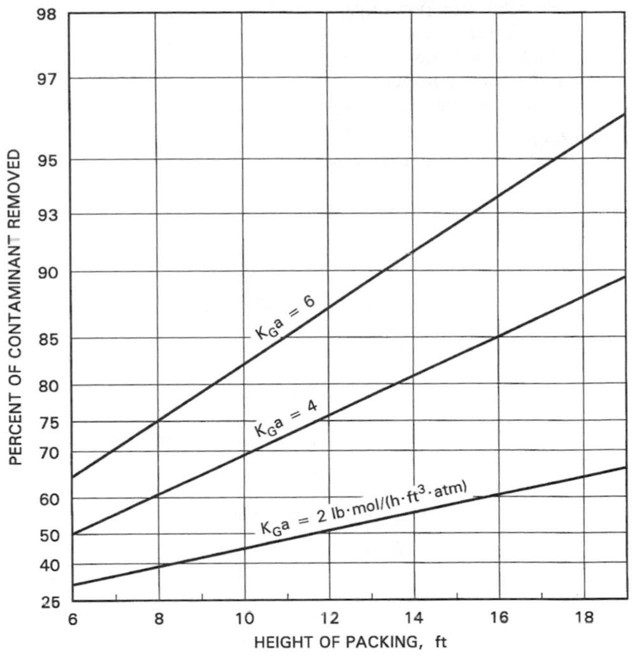

Fig. 31 Contaminant Control at Superficial Velocity = 240 fpm (Liquid Film Controlled)

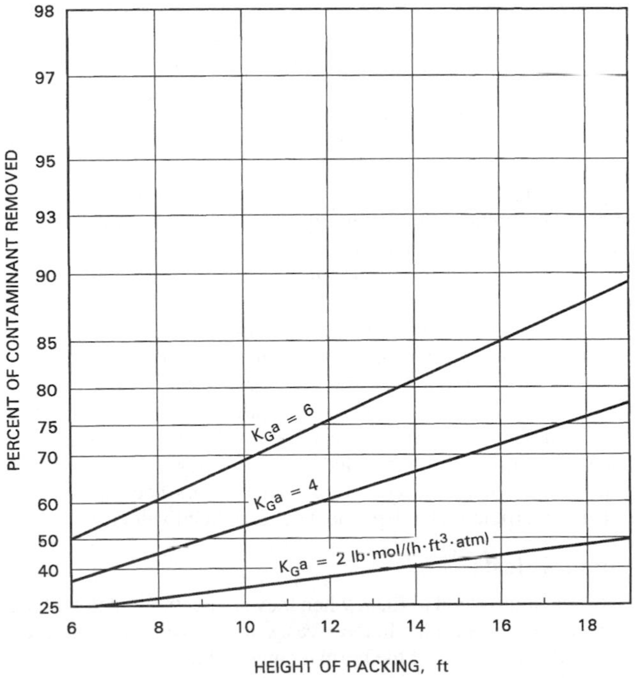

Fig. 32 Contaminant Control at Superficial Velocity = 360 fpm (Liquid Film Controlled)

of contaminant removal. The height for any other type or size of packing is inversely proportional to the ratio of standard K_Ga taken from Table 8. Thus, if 13 ft packing depth were required for 95% removal of contaminants, the same efficiency could be obtained with a 9.5 ft depth of 1 in. plastic pall rings (Figure 23), at the same superficial velocity and liquid injection rate. However, the pressure drop would be higher for the smaller diameter packing.

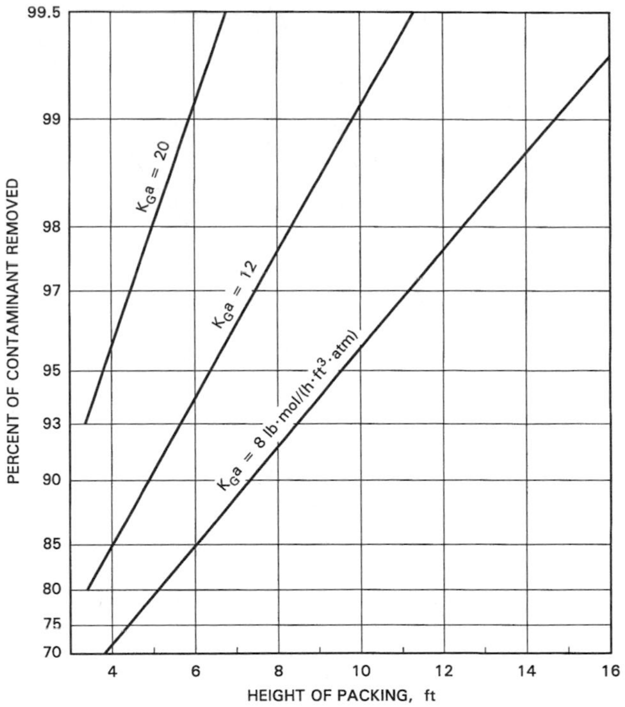

Fig. 33 Contaminant Control at Superficial Velocity = 120 fpm (Gas Film Controlled)

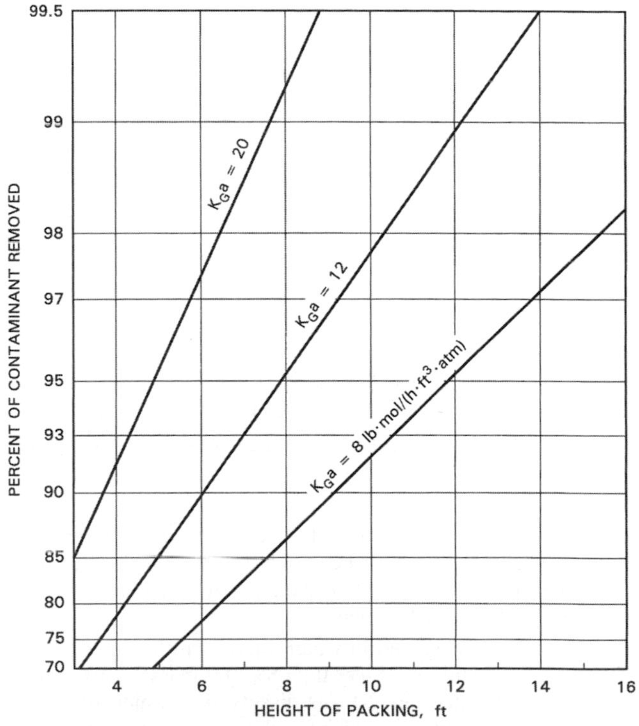

Fig. 34 Contaminant Control at Superficial Velocity = 240 fpm (Gas Film Controlled)

Figures 30 through 35 are useful when the value of the mass transfer coefficient for the particular contaminant to be removed is known. Table 9 contains mass transfer coefficients for 2 in. plastic Intalox saddles in typical liquid-film-controlled scrubbers. These values can be compared with the mass transfer coefficients in Table 10 for the same packing used in gas-film-controlled scrubbers.

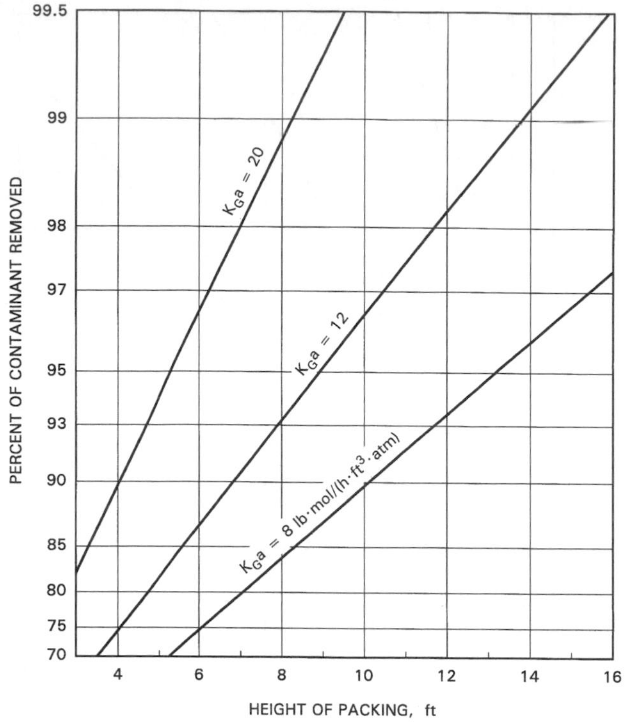

Fig. 35 Contaminant Control at Superficial Velocity = 360 fpm (Gas Film Controlled)

Table 10 Relative K_Ga for Various Contaminants in Gas-Film-Controlled Scrubbers

Gas Contaminant	Scrubbing Liquid	K_Ga, $\frac{\text{lb·mol}}{\text{h·ft}^3\text{·atm}}$
HCl	Water	18.66
HBr	Water	5.92
HF	Water	7.96
NH_3	Water	17.30
Cl_2	8% (by mass) NaOH	14.33
SO_2	11% (by mass) Na_2CO_3	11.83
Br_2	5% (by mass) NaOH	5.01

Note: Data for 2 in. plastic Super Intalox. Temperatures: from 60 to 75°F; liquid rate: 10 gpm/ft^2; gas rate: 215 cfm/ft^2.

When the scrubbing liquid is not water, the mass transfer coefficients in these tables can only be used if the amount of reagent in the solution exceeds by at least 33% the amount needed to completely absorb the gaseous contaminant.

When HCl is dissolved in water, there is little vapor pressure of HCl above solutions of less than 8% (by mass) concentration. On the other hand, when NH_3 is dissolved in water, there is an appreciable vapor pressure of NH_3 above solutions, even at low concentrations. The height of packing needed for NH_3 removal, obtained from Figures 33 through 35, is based on the use of dilute acid to maintain the pH of the solution below 7.

The following is an example of a typical scrubbing problem.

Example 1. Remove 95% of the HF from air at 90°F. The concentration of HF is 600 ppm on a dry gas basis. The concentration of HF in the exhaust gas should not exceed 30 ppm.

The following design conditions apply:

Total volumetric flow of gas	G	=	4600 cfm
Liquid injection rate		=	3.75 gpm/ft^2
Liquid temperature		=	68°F
Packed tower diameter		=	4 ft

Packing material is 2 in. polypropylene Super Intalox

Solution:

Cross-sectional area of absorber

$$A = \pi(4/2)^2 = 12.6 \text{ ft}^2$$

Total liquid flow rate

$$L = 3.75 \times 12.6 = 47.1 \text{ gpm}$$

Packing factor (from Table 7)

$$F = 21$$

Figure 29 may be used to find the pressure drop through the packed tower:

$$\text{x-axis} = 3.93(L/G) = 0.040$$

$$\text{y-axis} = (F/3.1 \times 10^6)(G/A)^2 = 0.90$$

From Figure 29, the pressure drop is about 0.28 in. of water per foot of packing depth.

From Table 10, $K_Ga = 0.35$ for HF. From Figure 33, the depth of packing required for 95% removal is 13 ft. Thus, the total pressure drop is $13 \times 0.28 = 3.6$ in. of water.

General Efficiency Comparisons

Figure 35 indicates that, with $K_Ga = 0.35$, 90% removal of HF could be achieved with 10 ft of packing; this is 23% less packing than needed for 96% removal. Furthermore, with the same superficial velocity, both liquid-film- and gas-film-controlled systems would require a 43% increase in absorbent depth to raise the removal efficiency from 80 to 90%.

A comparison of Figures 34 and 35 shows that increasing the superficial velocity by 50% in a gas-film-controlled scrubber requires only a 12% increase in bed depth to maintain equal removal efficiencies. In the liquid-film-controlled system (Figures 31 and 32), increasing the superficial velocity by 50% requires an approximately 50% increase in bed depth to maintain equal removal efficiencies.

Thus, in a gas-film-controlled system, the superficial velocity can be increased significantly with only a small increase in bed depth required to maintain the efficiency. In practical terms, gas-film-controlled scrubbers of fixed depth can handle an overload condition with only a minor loss of removal efficiency. The performance of liquid-film-controlled scrubbers degrades significantly under similar overload conditions. This occurs because the mass transfer coefficient is independent of superficial velocity.

Liquid Effects

Some liquids tend to foam when they are contaminated with particulates or soluble salts. In these cases, the pressure drop should be kept in the lower half of the normal range—0.25 to 0.40 in. of water per foot of packing depth.

In the control of gaseous pollution, most systems do not destroy the pollutant but merely remove it from the air. When water is used as the scrubbing liquid, the effluent from the scrubber will contain suspended particulate or dissolved solute. Water treatment is often required to alter the pH and/or remove toxic substances before the solutions can be discharged.

ADSORPTION OF GASEOUS CONTAMINANTS

The surface of freshly broken or heated solids often contains van der Waals (London dispersion) forces that are able to physically or chemically adsorb nearby molecules in a gas or liquid. The captured

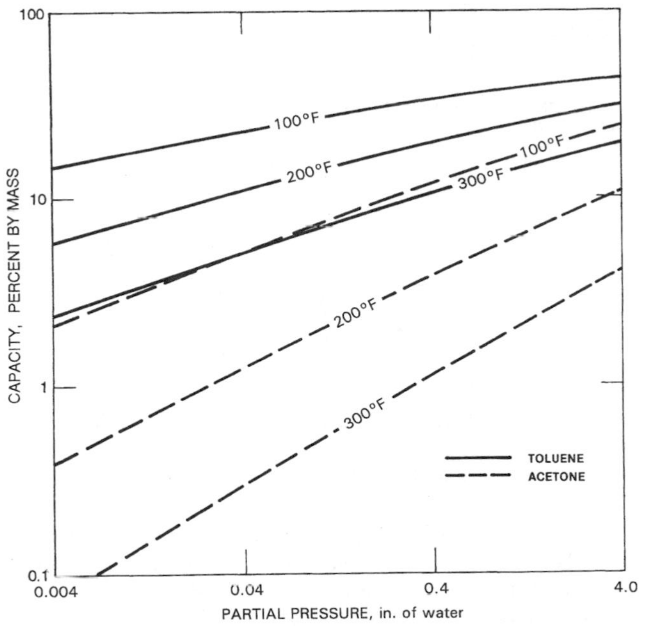

Fig. 36 Adsorption Isotherms on Activated Carbon

molecules form a thin layer on the surface of the solid that is typically one to three molecules thick. Commercial adsorbents are solids with an enormous internal surface area. This large surface area enables them to capture and hold large numbers of molecules. For example, each gram of a typical activated carbon adsorbent contains over 10,000 ft^2 of internal surface area. Adsorbents are used for removing organic vapors, water vapor, odors, and hazardous pollutants from gas streams.

The most common adsorbents used in industrial processes include activated carbons, activated alumina, silica gel, and molecular sieves. Activated carbons are derived from coal, wood, or coconut shells. They are primarily selected to remove organic compounds in preference to water. The other three common gas-phase adsorbents have a great affinity for water and will adsorb it to the exclusion of any organic molecules also present in a gas stream. They are used primarily as gas-drying agents. Molecular sieves also find use in several specialized pollution control applications, including removal of mercury vapor, sulfur dioxide (SO_2), or nitrogen oxides (NO_x) from gas streams.

The capacity of a particular activated carbon to adsorb any organic vapor from an exhaust gas stream is related to the concentration and molecular weight of the organic compound and the temperature of the gas stream. Compounds with a higher molecular weight are usually more strongly adsorbed than those with lower molecular weight. The capacity of activated carbon to adsorb any given organic compound increases with the concentration of that compound. Reducing the temperature also favors adsorption. Typical adsorption capacities of an activated carbon for toluene (molecular weight 92) and acetone (molecular weight 58) are illustrated in Figure 36 for various temperatures and concentrations.

Regeneration. Adsorption is reversible. An increase in temperature causes some or all of an adsorbed vapor to desorb. The temperature of low-pressure steam is sufficient to drive off most of a low boiling point organic compound previously adsorbed at ambient temperature. Higher boiling point organic compounds may require high-pressure steam or hot inert gas to secure good desorption. Compounds with a very high molecular weight can require reactivation of the carbon adsorber in a furnace at 1350°F to drive off all the adsorbed material. Regeneration of the carbon adsorbers can also be accomplished, in some instances, by washing with an aqueous solution of a chemical that will react with the adsorbed organic material, making it water soluble. An example is washing carbon containing adsorbed sulfur compounds with NaOH.

The difference between an adsorbent's capacity under adsorbing and desorbing conditions in any application is its **working capacity**. Activated carbon for air pollution control is found in canisters under the hoods of most automobiles. The adsorber in these canisters captures gasoline vapors escaping from the carburetor (when the engine is stopped) and from the fuel tank's breather vent. Desorption of gasoline vapors is accomplished by pulling fresh air through the carbon canister and into the carburetor when the engine is running. Although there is no temperature difference between adsorbing and desorbing conditions in this case, the outside airflow desorbs enough gasoline vapor to give the carbon a substantial working capacity.

For applications where only traces of a pollutant must be removed from exhaust air, the life of a carbon bed is very long. In these cases, it is often more economical to replace the carbon than to invest in regeneration equipment. Larger quantities can be returned to the carbon manufacturer for high-temperature thermal reactivation. Regeneration in place by steam, hot inert gas, or washing with a solution of alkali is sometimes practiced.

Impregnated (chemically reactive) adsorbents are used when physical adsorption alone is too weak to remove a particular gaseous contaminant from an industrial gas stream. Through impregnation, the reactive chemical is spread over the immense internal surface area of an adsorbent.

Typical applications of impregnated adsorbent include the following:

- Sulfur- or iodine-impregnated carbon removes mercury vapor from air, hydrogen, or other gases by forming mercuric sulfide or iodide.
- Metal oxide-impregnated carbons remove hydrogen sulfide.
- Amine- or iodine-impregnated carbons and silver exchanged zeolites remove radioactive methyl iodide from nuclear power plant work areas and exhaust gases.
- Alkali-impregnated carbons remove acid gases.
- Activated alumina impregnated with potassium permanganate removes acrolein and formaldehyde.

Equipment for Adsorption

Three types of adsorbers are usually found in industrial applications: (1) fixed beds, (2) moving beds, and (3) fluidized beds (Figure 37).

Fixed beds of regenerable or disposable media are most common. Carbon filter elements are a typical example.

Moving beds use granular adsorbers placed on inclined trays or on vertical frames similar to those used in granular bed particulate collectors. Moving beds offer continuous contaminant control and regeneration. Often, moving bed adsorbers and regeneration equipment are integrated components in a process.

Fluidized beds contain a fine granular adsorber, which is continuously mixed with the contaminated gas by suspension in the process gas stream. The bed may be either "fixed" at lower superficial velocities or highly turbulent and conveying (circulating). Illustrations of these types of fluidized beds are shown in Figure 37.

Solvent Recovery

The most common use of adsorption in stationary sources is in recovering solvent vapors from manufacturing and cleaning processes. Typical applications include solvent degreasing, rotogravure printing, dry cleaning, and the manufacture of such products as synthetic fibers, adhesive labels, tapes, coated copying paper, rubber goods, and coated fabrics.

Figure 38 illustrates the components of a typical solvent recovery system using two carbon beds. One bed is used as an adsorber

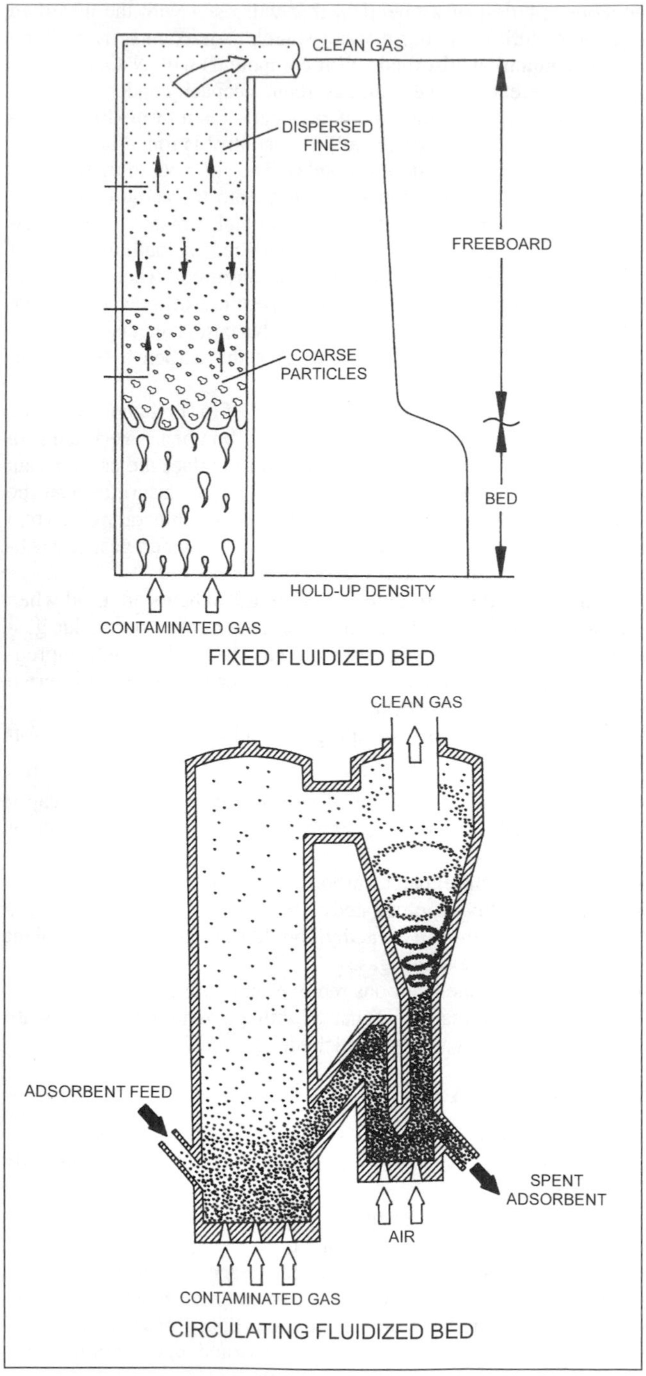

Fig. 37 Fluidized Bed Adsorption Equipment

while the other is regenerated with low-pressure steam. Desorbed solvent vapor and steam are recovered in a water-cooled condenser. If the solvent is immiscible with water, an automatic decanter separates the solvent for reuse. A distillation column is used for water-miscible solvents.

Adsorption time per cycle typically runs from 30 min to several hours. The adsorbing carbon bed is switched to regeneration by (1) an automatic timer shortly before the solvent vapor breaks through from the bed or (2) an organic vapor-sensing control device in the exhaust gas stream immediately after the solvent breaks through from the bed.

Low-pressure steam consumption for regeneration is generally about 3.5 lb/h per pound of solvent recovered (Boll 1976), but it can

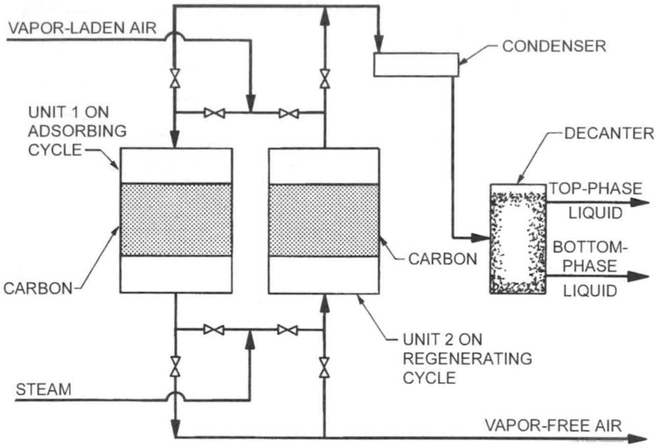

Fig. 38 Schematic of Two-Unit Fixed Bed Adsorber

range from 2 to over 5 lb/h per pound of solvent recovered, depending on the specific solvent and its concentration in the exhaust gas stream being stripped. Steam with only a slight superheat is normally used, so that it will condense quickly and give rapid heat transfer.

After steaming, the hot, moist carbon bed is usually cooled and partially dried before being placed back on stream. Heat for drying is supplied by the cooling of the carbon and adsorber, and sometimes by an external air heater. In most cases, it is desirable to leave some moisture in the bed. When solvent vapors are adsorbed, heat is generated. For most common solvents, the heat of adsorption is 40 to 60 Btu/lb·mol. When high-concentration vapors are adsorbed in a dry carbon bed, this heat can cause a substantial temperature rise and can even ignite the bed, unless it is controlled. If the bed contains moisture, the water absorbs energy and helps to prevent an undue rise in bed temperature. Certain applications may require heat sensors and automatic sprinklers.

Because the adsorptive capacity of activated carbon depends on temperature, it is important that solvent-laden air going to a recovery unit be as cool as is practicable. The exhaust gases from many solvent-emitting processes (such as drying ovens) are at elevated temperatures. Water- or air-cooled heat exchangers must be installed to reduce the temperature of the gas that enters the adsorber.

Solvent at very low vapor concentrations can be recovered in an activated carbon system. The size and cost of the recovery unit, however, depend on the volume of air to be handled; it is thus advantageous to minimize the volume of an exhaust stream and keep the solvent vapor concentration as high as possible, consistent with safety requirements. Insurance carriers specify that solvent vapor concentrations must not exceed 25% of the lower explosive limit (LEL) when intermittent monitoring is used. With continuous monitoring, concentrations as high as 50% of the LEL are permissible.

Solvent recovery systems, with gas-handling capacities up to about 11,000 cfm, are available as skid-mounted packages. Several of these packaged units can be used for larger gas flows. Custom systems can be built to handle 200,000 cfm or more of gas. Materials of construction may be painted carbon steel, stainless steel, Monel, or even titanium, depending on the nature of the gas mixture. The activated carbon is usually placed in horizontal flat beds or vertical cylindrical beds. The latter design minimizes ground space required for the system. Other alternatives are possible; one manufacturer uses a segmented horizontal rotating cylinder of carbon in which one segment is adsorbing while others are being steamed and cooled.

Commercial-scale solvent recovery systems typically recover over 99% of the solvent contained in a gas stream. The efficiency of the collecting hoods at the source of the solvent emission is usually the determining factor in solvent recovery.

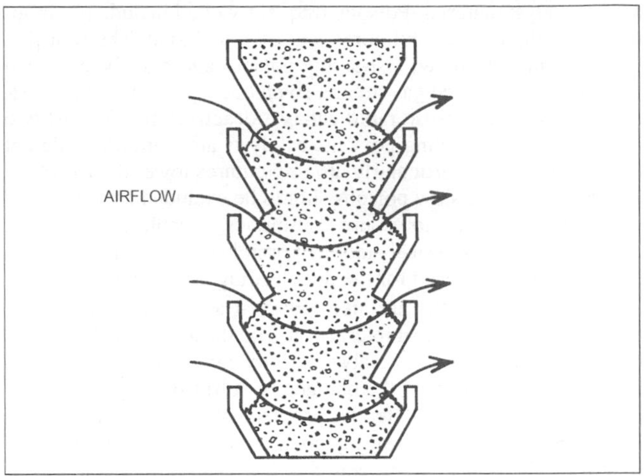

Fig. 39 Moving Bed Adsorber

Dust filters are generally placed ahead of carbon beds to prevent blinding of the adsorber by dust. Occasionally, the carbon is removed for screening to eliminate accumulated dust and fine particles of carbon.

If the solvent mixture contains high-boiling-point components, the working capacity of activated carbon can decrease with time. This occurs when high-boiling-point organic compounds are only partially removed by low-pressure steam. In this situation, two alternatives should be considered:

1. Periodic removal of the carbon and return to its manufacturer for high-temperature furnace reactivation to virgin carbon activity.
2. Use of more rigorous solvent desorbing conditions in the solvent recovery system. High-temperature steam, hot inert gas, or a combination of electrical heating and application of a vacuum may be used. The last method is selected, for example, to recover lithography ink solvents with high boiling points. Note that this method may not remove all of the high-boiling-point compounds.

Odor Control

Incineration and scrubbing are usually the most economical methods of controlling high concentrations of odorous compounds from equipment such as cookers in rendering plants. However, many odors that arise from harmlessly low concentrations of vapors are still offensive. The odor threshold (for 100% response) of acrolein in air, for example, is only 0.21 ppm, while that for ethyl mercaptan is 0.001 ppm and that for hydrogen sulfide is 0.0005 ppm (MCA 1968; AIHA 1989). Activated carbon beds effectively overcome many odor emission problems. Activated carbon is used to control odors from chemical and pharmaceutical manufacturing operations, foundries, sewage treating plants, oil and chemical storage tanks, lacquer drying ovens, food processing plants, and rendering plants. In some of these applications, activated carbon is the sole odor control method; in others, the carbon adsorber is applied to the exhaust from a scrubber.

Odor control systems using activated carbon can be as simple as a steel drum fitted with appropriate gas inlet and outlet ducts, or as complex as a large, vertically moving bed, in which carbon is contained between louvered side panels. A typical moving bed adsorber is shown in Figure 39. In this arrangement, fresh carbon can be added at the top, while spent or dust-laden carbon is periodically removed from the bottom.

Figure 40 shows a fixed bed odor adsorber. Adsorbers of this general configuration are available as packaged systems, complete with motor and blower. Air-handling capacities range from 500 to 12,000 cfm.

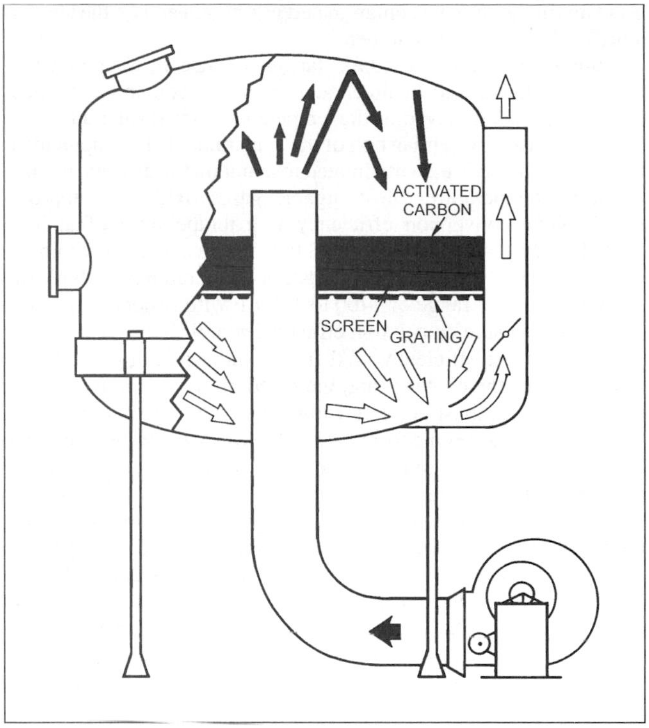

Fig. 40 Typical Odor Adsorber

The life of activated carbon in odor control systems ranges from a few weeks to a year or more, depending on the concentration of the odorous emission.

Applications of Fluidized Bed Adsorbers

The injection of alkali compounds into fluidized bed combustors for control of sulfur-containing compounds is one example of the use of fluidized bed adsorbers. Another example is the control of HF emissions from Søderberg aluminum reduction processes by a fixed or circulating fluidized bed of alumina.

INCINERATION OF GASES AND VAPORS

Incineration is the process by which volatile organic compounds (VOCs), organic aerosols, and most odorous materials in a contaminated gas stream are converted to innocuous carbon dioxide and water vapor using heat energy. Incineration is an effective means for totally eliminating VOCs. The types of incineration commonly used for air pollution control are thermal and catalytic, sometimes with recuperative or regenerative heat recovery.

To differentiate such air-cleaning systems from liquid and solid waste incinerators, the preferred term to describe such gas and aerosol phase air pollution control systems is now **oxidizers**.

Thermal Oxidizers

Thermal oxidizers, also known as **afterburners** or **direct flame incinerators**, consist of an insulated oxidation chamber in which gas and/or oil burners are typically located. The contaminated gas stream enters the chamber and comes into direct contact with the flame, which provides the heat energy necessary to promote oxidation. Under the proper conditions of time, temperature, and turbulence, the gas stream contaminants are oxidized effectively. The contaminated gas stream enters the combustion chamber near the burner, where turbulence-inducing devices are usually installed. The final contaminant conversion efficiency largely depends on good mixing within the contaminated gas stream and on the temperature of the oxidation chamber.

Supplemental fuel is used for start-up, to raise the temperature of the contaminated gas stream enough to initiate contaminant oxidation. Once oxidation begins, the temperature rises further due to the energy released by combustion of the contaminant. The supplementary fuel feed rate is then modulated to maintain the desired oxidizer operating temperature. Most organic gases oxidize to approximately 90% conversion efficiency if a temperature of at least 1200°F and a residence time of 0.3 to 0.5 s are achieved within the oxidation chamber. However, oxidation temperatures are typically maintained in the range of 1400 to 1500°F with residence times of 0.5 to 1 s to ensure conversion efficiencies of 95% or greater.

Although the efficiency of thermal oxidizers can exceed 95% destruction of the contaminant, the reaction may form undesirable products of combustion. For example, oxidation of chlorinated hydrocarbons causes the formation of hydrogen chloride, which can have an adverse effect on equipment. These new contaminants then require additional controls.

Oxidation systems incorporate primary heat recovery to preheat the incoming contaminated gas stream and, in some cases, provide secondary heat recovery for process or building heating. Primary heat recovery is almost always achieved using air-to-air heat exchangers. Use of a regenerable, ceramic medium for heat recovery has increased due to superior heat recovery efficiency. Secondary heat recovery may incorporate an air-to-air heat exchanger or a waste heat boiler (DOE 1979).

Oxidation systems using conventional air-to-air heat exchangers can achieve up to 80% heat recovery efficiency. Regenerative heat exchanger units have claimed as high as 95% heat recovery efficiency and are routinely operated at 85 to 90%. When operated at these high heat recovery efficiencies and with inlet VOC concentrations of 15 to 25% of the LEL, the oxidation process approaches a self-sustaining condition, requiring very little supplementary fuel.

Catalytic Oxidizers

Catalytic oxidizers operate under the same principles as thermal oxidizers, except that they use a catalyst to promote oxidation. The catalyst allows oxidation to occur at lower temperatures than in a thermal oxidizer for the same VOC concentration. Therefore, catalytic oxidizers require less supplemental fuel to preheat the contaminated gas stream and have lower overall operating temperatures.

A catalytic oxidizer generally consists of a preheat chamber followed by the catalyst bed. Residence time and turbulence are not as important as with thermal oxidizers, but it is essential that the contaminated gas stream be heated uniformly to the required catalytic reaction temperature. The required temperature varies, depending on the catalyst material and configuration.

The temperature of the contaminated gas stream is raised in the preheat chamber by a conventional burner. Although the contaminated gas stream contacts the burner flame, the heat input is significantly less than that for a thermal oxidizer, and only a small degree of direct contaminant oxidation occurs. Natural gas is preferred to prevent catalyst contamination, which could occur with sulfur-bearing fuel oils. However, No. 2 fuel oil units have been operated successfully. The most effective catalysts contain noble metals such as platinum or palladium.

Catalysis occurs at the molecular level. Therefore, an available, active catalyst surface area is important for maintaining high conversion efficiencies. If particulate materials contact the catalyst as either discrete or partially oxidized aerosols, they can ash on the catalyst surface and blind it. This problem is usually accompanied by a secondary pollution problem—odorous emissions caused by the partially oxidized organic compounds.

The greatest concern to users of catalytic oxidizers is **catalyst poisoning or deactivation**. Poisoning is caused by specific gas stream contaminants that chemically combine or alloy with the active catalyst material. Poisons frequently cited include phosphorus, bismuth, arsenic, antimony, lead, tin, and zinc. The first five materials are considered fast-acting poisons and must be excluded from the contaminated gas stream. Even trace quantities of the fast-acting poisons can cause rapid catalyst deactivation. The last two materials are slow-acting poisons; catalysts are somewhat tolerant of these materials, particularly at temperatures lower than 1000°F. However, even the slow poisons should be excluded from the contaminated gas stream to ensure continuous, reliable performance. Therefore, galvanized steel, another possible source of the slow poisons, should not be used for the duct leading to the oxidizer.

Sulfur and halogens are also regarded as catalyst poisons. In most cases, their chemical interaction with the active catalyst material is reversible. That is, catalyst activity can be restored by operating the catalyst without the halogen or sulfur-bearing compound in the gas stream. The potential problem of greater concern with respect to the halogen-bearing compounds is the formation of hydrogen chloride or hydrogen fluoride gas, or hydrochloric or hydrofluoric acid emissions.

Some organic compounds, such as polyester amides and imides, are also poisonous.

Catalytic oxidizers generally cost less to operate than thermal oxidizers because of their lower fuel consumption. With the exception of regenerative heat recovery techniques, primary and secondary heat recovery can be incorporated into a catalytic oxidation system to further reduce operating costs. Maintenance costs are usually higher for catalytic units, particularly if frequent catalyst cleaning or replacement is necessary. The concern regarding catalyst life has been the major factor limiting more widespread application of catalytic oxidizers.

Applications of Oxidizers

Odor Control. All highly odorous pollutant gases are combustible or chemically changed to less odorous pollutants when they are sufficiently heated. Often, the concentration of odorous materials in the waste gas is extremely low, and the only feasible method of control is oxidation. Odors from rendering plants, mercaptans, and organic sulfides from kraft pulping operations are examples of effluents that can be controlled by incineration. Other forms of oxidation can achieve the same ends (see Chapter 44 of the 1999 *ASHRAE Handbook—Applications*).

Reduction in Emissions of Reactive Hydrocarbons. Some air pollution control agencies regulate the emission of organic gases and vapors because of their involvement in photochemical smog reactions. Flame afterburning is an effective way of destroying these materials.

Reduction in Explosion Hazard. Refineries and chemical plants are among the factories that must dispose of highly combustible or otherwise dangerous organic materials. The safest method of disposal is usually by burning in flares or in specially designed furnaces. However, special precautions and equipment design must be used in the handling of potentially explosive mixtures.

Adsorption and Oxidation

Alternate cycles of adsorption and desorption in an activated carbon bed are used to concentrate solvent or odor vapors prior to oxidation. This technique greatly reduces the fuel required for burning organic vapor emissions. Fuel savings of 98% compared to direct oxidation are possible. The process is particularly useful in cases where emission levels vary from hour to hour. This technique is common in metal finishing for automotive and office furniture manufacturing.

Contaminated gas is passed through a carbon bed until saturation occurs. The gas stream is then switched to another carbon bed, and the exhausted bed is shut down for desorption. A hot inert gas, usually burner flue gas, is introduced to the adsorber to drive off concentrated organic vapors and to convey them to an oxidizer. The volume of this desorbing gas stream is much smaller than the original contaminated gas volume, so that only a small oxidizer, operating intermittently, is required (Grandjacques 1977).

AUXILIARY EQUIPMENT

DUCTS

Basic duct design is covered in Chapter 32 of the 1997 *ASHRAE Handbook—Fundamentals*. This chapter covers only those duct components or problems that warrant special concern when handling gases that contain particulate or gaseous contaminants.

Duct systems should be designed to allow thermal expansion or contraction as gases move from the process, through gas-cleaning equipment, and on to the ambient environment. Appropriately designed duct expansion joints must be located in proper relation to sliding and fixed duct supports. Besides withstanding maximum possible temperature, duct must also withstand maximum positive or negative pressure, a partial load of accumulated dust, and reasonable amounts of corrosion. Duct supports should be designed to accommodate these overload conditions as well. Bottom entries into duct junctions create a low-velocity area that can allow contaminants to settle out, thus creating a potential fire hazard. Bottom duct entries should be avoided.

Where gases might condense and cause corrosion or sticky deposits, duct should be insulated or fabricated from materials that will survive this environment. A psychrometric analysis of the exhaust gases is useful to determine the dew point. Surface temperatures should then be held above the dew-point temperature by preheating on start-up or by insulating. Slag traps with clean-out doors, inspection doors where direction changes, and dead-end full-sized caps are required for systems having heavy particulate loading. Special attention should be given to high-temperature duct, where the duct might corrode if insulated or become encrusted when molten particulate impacts on cool, uninsulated surfaces. Water-cooled duct or refractory lining is often used where the high operating temperature exceeds the safe limits of low-cost materials.

Gas flow through ducts should be considered as a part of overall system design. Good gas flow distribution is essential for measurements of process conditions and can lead to energy savings and increased system life. The minimum speed of gases in a duct should be sufficiently high to convey the heaviest particulate fraction with a degree of safety.

Slide gates, balance gates, equipment bypass ducts, and clean-out doors should be incorporated in the duct system to allow for maintenance of key gas-cleaning systems. In some cases, emergency bypass circuitry should be included to vent emissions and protect gas-cleaning equipment from process upsets.

Temperature Controls

Control of gas temperature in a gas-cleaning system is often vital to a system's performance and life. In some cases, gases are cooled to concentrate contaminants, condense gases, and recover energy. In other cases, gas-cleaning equipment, such as fabric filters and scrubbers, can only operate at well-controlled temperatures. Cooling exhaust gases through air-to-air heat transfer has been highly successful in many applications. Controlled evaporative cooling is also employed, but it increases the dew point and the danger of acid gas condensation and/or the formation of sticky deposits. However, controlled evaporative cooling to within 50°F of dew point has been used with success. Dilution by the injection of ambient air into the duct is expensive because it increases the volumetric flow of gas and, consequently, the size of collector needed to meet gas-cleaning objectives. Water-cooled duct is often used where the gas temperature exceeds the safe limits of the low-cost materials.

For dilution cooling, louver-type dampers are often used to inject ambient air and provide fine temperature control. Controls can be used to provide full modulation of the damper or to provide open or closed operation. Emergency bypass damper systems and bypass duct/stacks are used where limiting excessive temperature is critical.

Fans

Because the static pressure across a gas-cleaning device varies depending on conditions, the fan should operate on the steep portion of the fan pressure-volumetric flow curve. This tends to provide less variation in the volumetric flow. An undersized fan has a steeper characteristic than an oversized fan for the same duty; however, it will be noisier.

In the preferred arrangement, the fan is located on the clean gas side of gas-cleaning equipment. Advantages of placing the fan at this point include the following:

- A fan on the clean gas side handles clean gases and minimizes abrasive exposure from the collected product.
- High-efficiency backward blade and air foil designs can be selected because accumulation on the fan wheel is not as great a factor.
- Escape of hazardous materials due to leaks is minimized.
- The collector can be installed inside the plant, even near the process, because any leakage in the duct or collector will be into the system and will not increase the potential for exposure. However, the fan itself should be mounted outside, so that the positive pressure duct is exterior to the work environment.

For economic reasons, a fan may be located on the contaminant-laden side of the gas-cleaning equipment if the contaminants are relatively nonabrasive and especially if the equipment can be located outdoors. This arrangement should be avoided because of the potential for leakage of concentrated contaminants to the environment. In some instances, however, the collector housing design, duct design, and energy savings of this arrangement reduce costs.

Most scrubbers are operated on the suction side of the fan. This not only eliminates the leakage of contaminants into the work area, but also allows for servicing the unit while it is in operation. Additionally, such an arrangement minimizes corrosion of the fan. Stacks on the exhaust streams of scrubbers should be arranged to drain condensate rather than allow it to accumulate and reenter the fan.

Fabric filters require special consideration. When new, clean fabric is installed in a collector, the resistance is low, and the fan motor may be overloaded. This overloading may be prevented during start-up by use of a temporary throttling damper in the main duct, for example, on the clean side of the filter in a pull-through system. Overloading may also be prevented by using a backward-curved blade (nonoverloading fan) on the clean side of the collector.

DUST- AND SLURRY-HANDLING EQUIPMENT

Once the particulate matter is collected, new control problems arise from the need to remove, transport, and dispose of the material from the collector. A study of all potential methods for handling the collected material, which might be a hazardous waste, must be an integral part of initial system design.

Hoppers

Dust collector hoppers are intended only to channel collected material to the hopper's outlet, where it is continuously discharged. When hoppers are used to store dust, the plates and charging electrodes of electrostatic precipitators can be shorted electrically, resulting in failure of entire electrical sections. Fabric collectors, particularly those that have the gas inlet in the hopper, are usually designed on the assumption that the hopper is not used for storage and that collected waste will be continuously removed. High dust levels in the hoppers of fabric collectors often result in high dust reentrainment. This reentrainment causes increased operating pressure and the potential for fabric damage. The excessive dust levels not only expose the system to potentially corrosive conditions and fire/explosion hazards, but also place increased structural demands on the system.

Aside from misuse of hoppers for storage, common problems with dust-handling equipment include (1) plugging of hoppers, (2) blockage of dust valves with solid objects, and (3) improper or insufficient maintenance.

Hopper auxiliaries to be considered include (1) insulation, (2) dust level indicators, (3) rapper plates, (4) vibrators, (5) heaters, and (6) "poke" holes.

Hopper Discharge. Dust is often removed continuously from hoppers by means of rotary valves. Alternate equipment includes the double-flap valve, or vacuum system valves. Wet electrostatic precipitators and scrubbers often use sluice valves and drains to ensure that insoluble particulate remains in suspension during discharge.

Dust Conveyors

Larger dust collectors are fitted with one or more conveyors to feed dust to a central discharge location or to return it to a process. Drag, screw, and pneumatic conveyors are commonly used with dust collectors. Sequential start-up of conveyor systems is essential. Motion switches to monitor operation of the conveyor are useful.

Dust Disposal

Several methods are available for disposal of collected dust. It can be emptied into dumpsters in its as-collected dry form or be pelletized and hauled to a landfill. It can also be converted to a slurry and pumped to a settling pond or to clarification equipment. While the advantages and disadvantages of each method are beyond the scope of this chapter, they should be evaluated for each application.

Slurry Treatment

When slurry from wet collectors cannot be returned directly to the process or tailing pond, liquid clarification and treatment systems can be used for recycling the water to prevent stream pollution. Stringent stream pollution regulations make even a small discharge of bleed water a problem. Clarification equipment may include settling tanks, sludge-handling facilities, and, possibly, centrifuges or vacuum filters. Provisions must be provided for handling and disposal of dewatered sludge, so that secondary pollution problems do not develop.

OPERATION AND MAINTENANCE

A planned program for operation and maintenance of equipment is a necessity. Such programs are becoming mandatory because of the need for operators to prove continuous compliance with emission regulations. Good housekeeping and record-keeping will also help prolong the life of the equipment, support a program for positive relations with regulatory and community groups, and aid problem-solving efforts, should nonroutine maintenance or service be necessary.

A typical program includes the following minimum requirements (Stern et al. 1984):

- Central location for filing equipment records, warranties, instruction manuals, etc.
- Lubrication and cleaning schedules
- Planning and scheduling of preventive maintenance (including inspection and major repair)
- Storeroom and inventory system for spare parts and supplies
- Listing of maintenance personnel (including supplier contacts and consultants)
- Costs and budgets for activities associated with operation and maintenance of the equipment
- Storage for special tools and equipment

Guidance on conducting inspections can be found in EPA (1983).

Corrosion

Because high-temperature gas cleaning often involves corrosive materials, chemical attack on system components must be anticipated. This is especially true if the temperature in a gas-cleaning system falls below the moisture or acid dew point. Housing insulation should be such that the internal metal surface temperature is 20 to 30°F greater than the moisture and/or acid dew point at all times. In applications with fabric filters where alkali materials are injected into the gas stream to react with acid gases, care must be taken to protect the clean side of the housing downstream of the fabric from corrosion.

Fires and Explosions

Industrial gas-cleaning systems often concentrate combustible materials and expose them to environments that are hostile and difficult to control. These environments also make fires difficult to detect and stop. Industrial gas-cleaning systems are, therefore, potential fire or explosion hazards (Billings and Wilder 1970; Frank 1981; EEI 1980).

Fires and explosions in industrial process exhaust streams are not generally limited to gas-cleaning equipment. Ignition may take place in the process itself, in the duct, or in exhaust system components other than the gas-cleaning equipment. Once uncontrolled combustion begins, it may propagate throughout the system. Workers around pollution control equipment should never open access doors to gas-cleaning equipment when a fire is believed to be in process; the fire could easily transform into an explosion.

The following devices help maintain a safe particulate control system.

Explosion Doors. An explosion door or explosion relief valve permits instantaneous pressure relief for equipment when the pressure reaches a predetermined level. Explosion doors are mandatory for certain applications to meet OSHA, insurance, or National Fire Protection Association (NFPA) regulations.

Detectors. Temperature-actuated switches or infrared sensors can be used to detect changes in the inlet-to-outlet temperature difference or a localized, elevated temperature that might signal a fire within the gas-cleaning system or a process upset. These detectors can be used to activate bypass dampers, trigger fire alarm/control systems, and/or shut down fans.

Fire Control Systems. Inert gas and water spray systems can be used to control fires in dust collectors. They are of little value in controlling explosions.

REFERENCES

ACGIH. 1998. *Industrial ventilation: A manual of recommended practice*, 23rd ed. Committee on Industrial Ventilation, American Conference of Governmental Industrial Hygienists, Cincinnati, OH.

AIHA. 1989. Odor thresholds for chemicals with established occupational health standards. American Industrial Hygiene Association, Akron, OH.

Beck, A.J. 1975. Heat treat engineering. *Heat Treating* (January).

Beltran, M.R. 1972. Smoke abatement for textile finishers. *American Dyestuff Reporter* (August).

Beltran, M.R. and H. Surati. 1976. Heat recovery vs. evaporative cooling on organic electrostatic precipitators. *Proceedings of the American Institute of Plant Engineers/Rossnagel & Associates Sixth Annual Industrial Air Pollution Control Seminar* (April 6, 1976), Cherry Hill, NJ.

Bergin, M.H., D.Y.H. Pui, T.H. Kuehn, and W.T. Fay. 1989. Laboratory and field measurements of fractional efficiency of industrial dust collectors. *ASHRAE Transactions* 95(2):102-12.

Billings, C.E. and J. Wilder. 1970. *Handbook of fabric filter technology,* Vol. 1: Fabric filter systems study. NTIS *Publication* PB 200 648, 2-201. National Technical Information Service, Springfield, VA.

Boll, C.H. 1976. Recovering solvents by adsorption. *Plant Engineering* (January).

Buonicore, A.J. and W.T. Davis, eds. 1992. *Air pollution engineering manual.* Van Nostrand Reinhold, New York.

Chopyk, J. and M.C. Larkin. 1982. Smoke and odor subdued with two-stage precipitator. *Plant Services* (January).

DOE. 1979. The coating industry: Energy savings with volatile organic compound emission control. *Report* No. TID-28706, U.S. Department of Energy, Washington, D.C.

EEI. 1980. Air preheaters and electrostatic precipitators fire prevention and protection (coal fired boilers). Report of the Fire Protection Committee, No. 06-80-07 (September). Edison Electric Institute, Washington, D.C.

EPA. 1973 *Air pollution engineering manual,* 2nd ed. *Publication* AP-40. Environmental Protection Agency, Washington, D.C.

EPA. 1983. Coal-fired industrial boiler inspection guide. *Report* No. 340/1-83-025 (December).

EPA. 1984. Proposed fine particle standards for ambient air quality. Federal Register 48, 10408, March 30.

EPA. 1994. *Quality assurance handbook for air pollution measurement systems. Report* No. EPA-600R94038A.

Frank, T.E. 1981. Fire and explosion control in bag filter dust collection systems. Proceedings of the Conference on the Hazards of Industrial Explosions from Dusts, New Orleans, LA (October).

Grandjacques, B. 1977. Carbon adsorption can provide air pollution control with savings. *Pollution Engineering* (August).

GPO. Annual. *Code of federal regulations* 40(60). U.S. Government Printing Office, Washington, D.C. This document is revised annually and published each year in July.

IGCI. 1964. Determination of particulate collection efficiency of gas scrubbers. *Publication* No. 1. Industrial Gas Cleaning Institute, Washington, D.C.

Kane, J.M. and J.L. Alden. 1982. *Design of industrial ventilation systems.* Industrial Press, New York.

Lapple, C.E. 1951. Processes use many collection types. *Chemical Engineering* (May):145-51

Lapple, C.E. and H.J. Kamack. 1955. Performance of wet scrubbers. *Chemical Engineering Progress* (March).

MCA. 1968. Odor thresholds for 53 commercial chemicals. Manufacturing Chemists Association, Washington, D.C. (October).

NIOSH. 1978. A recommended approach to recirculation of exhaust air. *Publication* No. 78-124, National Institute of Occupational Safety and Health, Washington, D.C.

Rossnagel, W.B. 1973. Condensing/Precipitator systems on organic emissions. *Proceedings of the Third Annual Industrial Air Pollution Control Seminar* (May 8, 1973), Paramus, NJ.

Rudnick, S.N., J.L.M. Koehler, K.P. Martin, D. Leith, and D.W. Cooper. 1986. Particle collection efficiency in a venturi scrubber: Comparison of experiments with theory. *Environmental Science & Technology* 20(3):237-42.

Sauerland, W.A. 1976. Successful application of electrostatic precipitators on asphalt saturator emissions. *Proceedings of the American Institute of Plant Engineers/Rossnagel & Associates Sixth Annual Industrial Air Pollution Control Seminar* (April 6, 1976), Cherry Hill, NJ.

Semrau, K.T. 1977. Practical process design of particulate scrubbers. *Chemical Engineering* (September):87-91.

Shabsin, J. 1985. Clean plant air PLUS energy conservation. *Fastener Technology* (April).

Sink, M.K. 1991. *Handbook: Control technologies for hazardous air pollutants. Report* No. EPA/625/6-91/014. Environmental Protection Agency, Washington, D.C.

SIP. 1991. State implementation plans and guidance are available from the Regional Offices of the U.S. EPA and the state environmental authorities.

Stern, A.C., R.W. Boubel, D.B. Turner, and D.L. Fox. 1984. *Fundamentals of air pollution control,* 2nd ed. Chapter 25, Control Devices and Systems.

Thiel, G.R. 1977. Advances in electrostatic control techniques for organic emissions. *Proceedings of the Seventh Annual Industrial Air Pollution/Contamination Control Seminar* (March 29, 1977), Paramus, NJ.

Thiel, G.R. 1983. Cleaning and recycling plant air... Improvement of air cleaner performance and recirculation procedures. *Plant Engineering* (January 6).

Wade, G., J. Wigton, J. Guillory, G. Goldback, and K. Phillips. 1978. Granular bed filter development program. U.S. DOE *Report* No. FE-2579-19 (April).

White, H.J. 1963. *Industrial electrostatic precipitation.* Addison-Wesley, Reading, MA.

BIBLIOGRAPHY

Crynack, R.B. and J.D. Sherow. 1984. Use of a mobile electrostatic precipitator for pilot studies. *Proceedings of the Fifth Symposium on the Transfer and Utilization of Particulate Control Technology* 2:3-1.

Deutsch, W. 1922. *Ann. der Physik* 68:335.

DuBard, J.L. and R.F. Altman. 1984. Analysis of error in precipitator performance estimates. *Proceedings of the Fifth Symposium on the Transfer and Utilization of Particulate Control Technology* 2:2-1.

Faulkner, M.G. and J.L. DuBard. 1984. A mathematical model of electrostatic precipitation, 3rd ed. *Publication* No. EPA-600/7-84-069a. Environmental Protection Agency, Washington, D.C.

Hall, H.J. 1975. Design and application of high voltage power supplies in electrostatic precipitation. *Journal of the Air Pollution Control Association* 25(2).

HEW. 1967. Air pollution engineering manual. HEW *Publication* No. 999-AP-40. Department of Health and Human Services (formerly Department of Health, Education, and Welfare), Washington, D.C.

Noll, C.G. 1984a. Electrostatic precipitation of particulate emissions from the melting of borosilicate and lead glasses. *Glass Technology* (April).

Noll, C.G. 1984b. Demonstration of a two-stage electrostatic precipitator for application to industrial processes. *Proceedings of the Second International Conference on Electrostatic Precipitation,* Kyoto, Japan (November):428-34.

Oglesby, S. and G.B. Nichols. 1970. A manual of electrostatic precipitator technology. NTIS PB-196-380. National Technical Information Service, Springfield, VA.

White, H.J. 1974. Resistivity problems in electrostatic precipitation. *Journal of the Air Pollution Control Association* 24(4).

CHAPTER 26

AUTOMATIC FUEL-BURNING EQUIPMENT

GAS-BURNING EQUIPMENT

A GAS burner conveys gas (or a mixture of gas and air) to the combustion zone. Burners are of the atmospheric injection or power burner type.

RESIDENTIAL EQUIPMENT

Residential gas burners are designed for central heating plants or for unit application. Both gas-designed units and conversion burners are available for the different types of central systems and for applications where the units are installed in the heated space.

Central heating appliances include warm-air furnaces and steam or hot water boilers.

Warm-air furnaces are available in different designs. Design choice depends on (1) the force needed to move combustion products, (2) the force needed to move heated supply and return air, (3) the location within a building, and (4) the efficiency required.

The force to move supply and return air is supplied by the natural buoyancy of heated air in a gravity furnace (if the space to be heated is close to and/or above the furnace) or by a blower in a forced-air furnace. Furnaces are available in upflow, downflow, horizontal, and other heated air directions to meet the application requirements.

The force to move the combustion products is supplied by the natural buoyancy of hot combustion products in a natural-draft furnace, by a blower in a forced-draft or induced-draft furnace, or by the thermal expansion forces in a pulse combustion furnace.

The efficiency required determines to a great extent not only some of the characteristics above but also other characteristics, such as the source of combustion air, the use of vent dampers and draft hoods, the need to recover latent heat from the combustion products, and the design of the heat transfer components.

Conversion burners are not normally used in modern furnaces because the burner is integrated with the furnace design for safety and efficiency. Older gravity furnaces can usually use conversion burners more readily than can modern gravity and forced-air furnaces.

Steam or hot water boilers are available in cast iron, steel, and nonferrous metals. In addition to supplying space heating, many boilers provide domestic hot water by means of tankless integral or external heat exchangers.

Some gas furnaces and boilers are available with sealed combustion chambers. These units have no draft hood and are called direct vent appliances. Combustion air is piped from outdoors directly to the combustion chamber.

In some instances, the combustion air intake is in the same location as the flue gas outlet. The air and flue pipes may be constructed as concentric pipes, with a terminal that exposes the air intake and flue gas outlet to a common pressure condition. No chimney or vertical vent is needed with such units. Some induced-draft combustion systems are designed to operate safely when common-vented with other appliances that have natural-draft combustion systems.

For appliances covered by the U.S. National Appliance Energy Conservation Act, the required minimum annual fuel utilization efficiencies (AFUEs) determined by U.S. Department of Energy tests are as follows:

1. Boilers
   - Excluding gas steam: 80% AFUE as of January 1, 1992
   - Gas steam: 75% AFUE as of January 1, 1992
2. Furnaces
   - Excluding mobile home furnaces: 78% AFUE as of January 1, 1992
   - Mobile home furnaces: 75% as of January 1, 1992

Conversion burners are complete burner and control units designed for installation in existing boilers and furnaces. Atmospheric conversion burners may have drilled-port, slotted-port, or single-port burner heads. These burners are either upshot or inshot types. Figure 1 shows a typical upshot gas conversion burner.

Several power burners are available in residential sizes. These are of gun-burner design and are desirable for furnaces or boilers with restricted flue passages or with downdraft passages.

Conversion burners for domestic application are available in sizes ranging from 40,000 to 400,000 Btu/h input, the maximum rate being set by ANSI *Standard* Z21.17/CSA 2.7. However, many such gas conversion burners installed in apartment buildings have input rates up to 900,000 Btu/h or more.

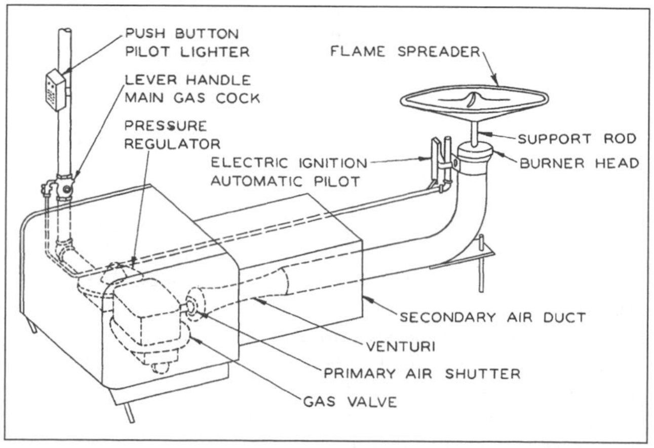

Fig. 1 Typical Single-Port Upshot Gas Conversion Burner

The preparation of this chapter is assigned to TC 6.10, Fuels and Combustion.

As the successful and safe performance of a gas conversion burner depends on numerous factors other than those incorporated in the equipment, installations must be made in strict accordance with current ANSI *Standard* Z21.8. Draft hoods conforming to current ANSI *Standard* Z21.12 should also be installed (in place of the dampers used with solid fuel) on all boilers and furnaces converted to burn gas. Due to space limitations, a converted appliance with a breeching over 12 in. in diameter is often fitted with a double-swing barometric regulator instead of a draft hood.

Pulse combustion systems may be used in residential heating equipment. Pulse systems do not burn gas in a steady flame; instead, the gas-air mixture is ignited at a rate of 60 to 80 times per second. Reported advantages of pulse combustion include reduced oxides of nitrogen (NO_x) emissions and improved heat transfer due to the oscillating flow. Additional discussion on pulse combustion technology can be found in Chapter 17 of the 1997 *ASHRAE Handbook—Fundamentals*.

COMMERCIAL-INDUSTRIAL EQUIPMENT

Many basic types of industrial gas burners are available, but only those commonly used for heating service are covered here. Atmospheric or power burners are available for use in central heating systems, and various gas-designed units such as unit heaters, duct furnaces, infrared heaters, and makeup heaters are available for space heating.

Types of Burners

Atmospheric burners include an air shutter, a venturi tube, a gas orifice, and outlet ports. These burners are of inshot or upshot design. Inshot burners are placed horizontally, making them suitable for firing Scotch-type boilers. Upshot burners are arranged vertically, making them more adapted to firebox-type boilers.

Power burners use a fan to supply and control combustion air. These burners can be natural-draft or forced-draft. In natural-draft installations, a chimney must draw the products of combustion through the boiler or furnace; the burner fan supplies only enough power to move the air through the burner. Many natural-draft power burners have a configuration similar to that of an inshot atmospheric burner to which a fan and windbox have been added. This configuration allows more complex gas-air mixing patterns, and combustion capabilities are thereby improved.

In forced-draft power burners, the fan size and speed are increased, and the combustion process has been modified so that the fan not only moves air through the burner but also forces it through the boiler. Combustion occurs under pressure in controlled airflow. While a vent of only limited height may be needed to convey the combustion products to the outdoors, higher chimneys and vents are usually required to elevate the effluent further. In a forced-draft power burner, air and gas flows can be modulated by suitable burner controls provided by the manufacturer. Gas input is usually controlled by an appliance pressure regulator and a firing rate valve, both piped in series in the gas train. The utility may also install a regulator at the gas service entry at the gas meter location.

The gas is introduced into a controlled airstream designed to produce thorough gas-air mixing but still capable of maintaining a stable flame front. In a ring burner, the gas is introduced into the combustion airstream through a gas-filled ring just ahead of the combustion zone. In a premix burner, gas and primary air are mixed together, and the mixture is then introduced into secondary air in the combustion zone.

The power burner has superior combustion control, particularly in restricted furnaces and under forced draft. Commercial and industrial gas burners frequently operate with higher gas pressures than those intended for residential equipment. It is often necessary to determine both maximum and minimum gas pressures to be applied to the gas regulator and control trains. If the maximum gas pressure exceeds the 4 to 14 in. of water standard for domestic equipment, it is necessary to select appropriate gas controls. All gas-control trains must be rated for the maximum expected gas pressure. The fact that gas pressures in densely populated areas may be significantly lower than in rural areas must be taken into account.

The installation of conversion burners larger than 400,000 Btu/h for use in large boilers is usually engineered by the burner manufacturer, the manufacturer's representative, or a local utility company. Conversion burners are available in several sizes and types. In some cases, the burner may be an assembly of multiple burner heads filling the entire firebox. For conversion burner installation in boilers requiring more than 400,000 Btu/h input, reference should be made to current ANSI *Standard* Z83.3. Conversion power burners above 400,000 Btu/h are available and should conform to UL *Standard* 795.

Pulse combustion technology is available for commercial heating equipment. Boilers using this technology are available with inputs up to 750,000 Btu/h.

Gas-fired systems for packaged fire-tube or water-tube boilers in heating plants consist of specially engineered and integrated combustion and control systems. The burners have forced- and/or induced-draft fans. The burner fan power requirements of these systems differ from those of the forced-draft equipment mentioned previously. The system pressure, up to and including the stack outlet, may be positive. If the boiler flue breeching has positive flue gas pressure, it must be gastight.

The integrated design of burner and boiler makes it possible, through close control of air-fuel ratios and by matching of flame patterns to boiler furnace configurations, to maintain high combustion efficiency over a wide range of loads. Combustion space and heat transfer areas are designed for maximum heat transfer with the specific fuel for which the unit is offered. Most packaged units are fire-tested as complete packages prior to shipment, and all components are inspected to ensure that burner equipment, automatic controls, and so forth, function properly. These units generally bear the Underwriters Laboratories label.

Unitary Heaters

Gas-fired air heaters are generally designed for use in large spaces such as airplane hangars and public garages. They are self-contained; automatically controlled with integral means for air circulation; and equipped with automatic electric ignition of pilots, induced or forced draft, prepurge, and fast-acting combustion safeguards. They also are used with ducts, discharge nozzles, grilles or louvers, and filters.

Unit heaters are used extensively for heating spaces such as stores, garages, and factories. These heaters consist of a burner, a heat exchanger, a fan for distributing the air, a draft hood, an automatic pilot, and controls for burners and fan. They are usually mounted in an elevated position from which the heated air is directed downward by louvers. Some unit heaters are suspended from the ceiling, while others are freestanding floor units of the heat tower type. Unit heaters are classified for use with or without ducts, depending on the applicable ANSI standard under which their design is certified. When connected to ducts, they must have sufficient blower capacity to deliver an adequate air quantity against duct resistance.

Duct furnaces are usually like unit heaters without the fan and are used for heating air in a duct system with blowers provided to move the air through the system. Duct furnaces are tested for operation at much higher static pressures than are obtained in unit heaters (ANSI *Standard* Z83.8/CGA 2.6).

Infrared heaters, vented or unvented, are used extensively for heating factories, foundries, sports arenas, loading docks, garages, and other installations where convection heating is difficult to apply. Following are two general types of gas-fired infrared heaters.

1. Surface combustion radiant heaters, which have a ported refractory or metallic screen burner face through which a self-sufficient mixture of gas and air flows. The gas-air mixture burns on the surface of the burner face, heating it to incandescence and releasing heat by radiation. These units operate at surface temperatures of about 1600°F and generally are unvented. Buildings containing unvented heaters discharging the combustion gases into the space should be adequately ventilated.
2. Internally fired heaters, which consist of a heat exchanger radiating at a surface temperature of approximately 180°F. The exposed surface can be equipped with reflecting louvers or a single large reflector to direct the radiated energy. These units are usually vented.

These infrared heaters are usually mounted in elevated positions and radiate heat downward (see Chapter 15).

Direct-fired makeup air heaters (see Chapter 31) are used to temper the outside air supply, which replaces contaminated exhaust air. The combustion gases of the heater are mixed directly with large volumes of outside air. Such mixing is considered safe because of the high dilution ratio.

ENGINEERING CONSIDERATIONS

With gas-burning equipment, the principal engineering considerations are sizing of gas piping and venting, building pressurization or depressurization, adjustment of the primary air supply, and in some instances (e.g., power burners and conversion burners) adjustment of secondary aeration. Consideration of input rating may be necessary to compensate for high altitude conditions. Chapters 30 and 33 of the 1997 *ASHRAE Handbook—Fundamentals* have details on venting and sizing of gas piping.

Combustion Process and Adjustments

Gas burner adjustment is mainly an adjustment of air supply. Most residential gas burners are of the atmospheric injection (Bunsen) type, in which primary air is introduced and mixed with the gas in the throat of the mixing tube. For normal operation of most atmospheric burners, 40 to 60% of the theoretical air as primary air will give best operation. Slotted-port and ribbon burners may require from 50 to 80% primary air for proper operation. The amount of excess air required depends on several factors, including uniformity of air distribution and mixing, direction of gas travel from the burner, and height and temperature of the combustion chamber. For power burners using motor-driven blowers to provide both primary and secondary air, the excess air can be closely controlled while proper combustion is secured.

Secondary air is drawn into gas appliances by natural draft. Yellow flame burners depend on secondary air for combustion. Air shutter adjustments should be made by closing the air shutter until yellow flame tips appear and then opening the air shutter to a final position at which the yellow tips just disappear. This type of flame obtains ready ignition from port to port and also favors quiet flame extinction.

Gas-designed equipment usually does not incorporate any means of varying the secondary air supply (and hence the CO_2 level). The amount of effective opening and baffling is determined by compliance with ANSI standards. Gas conversion burners, however, do incorporate means for controlling secondary air to permit adjustment over a wide range of inputs. It is desirable, through the use of suitable indicators, to determine whether or not carbon monoxide is present in flue gases. For safe operation, carbon monoxide should not exceed 0.04% (air-free basis).

Compensation for Altitude

Compensation must be made for altitudes higher than 2000 ft above sea level. The typical gas-fired central heating appliance using atmospheric burners, multiple flue ways, and an effective draft diverter must be derated at high altitudes.

All air for combustion is supplied by the chimney effect of the flue way. The geometry of these flue ways allows a given volume of air to pass through the appliance, regardless of its mass. Each burner has a gas orifice and a venturi tube designed to permit a given volume of gas to be introduced and burned during a given period. The entire system is essentially a constant volume device and must, therefore, be derated in accordance with the decrease in mass of these constant volumes at higher altitudes. The current derating factor recommended is 4% per 1000 ft of altitude (Segeler 1965). New options are being developed for atmospheric and power burners.

Many commercial and industrial applications use a forced-draft gas-fired packaged system, including a burner with a forced-draft fan and a fuel-handling system. The burner head, the heat exchanger, and the vent act as a series of orifices downstream of the forced-draft fan. To compensate for the increased volume of air and flue products that must be forced through these fixed restrictions at higher altitudes, it is necessary to increase burner fan capacity until the required volume of air is delivered at a pressure high enough to overcome them. These oversized fans and fan motors are usually offered as options.

One problem that frequently occurs in the selection of commercial gas-fired equipment, particularly for forced-draft gas combustion equipment, relates to the heat content of the gas as delivered at elevated locations. Natural gas usually has a heating value of just over 1000 Btu/ft^3 at standard conditions. At lower elevations, the heating value of the gas, as delivered, is close to the heating value at the standard conditions, and selection of controls is relatively simple. However at 5000 ft elevation, due to the reduced atmospheric pressure, natural gas having a heat content of 1000 Btu/ft^3 at sea level has a heat content of 850 Btu/ft^3 as delivered to the burner-control train. Some gas supplies in the mountains are enriched so that their energy content at standard conditions is higher than 1000 Btu/ft^3. Problems frequently occur in the selection of controls sized to permit the required volume of gas to be delivered for combustion. Gas specifications should indicate whether the energy value and density shown are for standard conditions or for the gas as furnished at a higher elevation.

The problem of gas supply heat content is more significant for the proper application of commercial forced-draft combustion equipment than it is for smaller atmospheric burner units because of the larger gas volumes handled and the higher design pressure drops common to this type of equipment.

For example, a gas-control train at sea level has a pressure of 14 in. of water delivered to the meter inlet. Allowing a 6 in. of water pressure drop through the meter and piping to the burner-control train at full gas flow, 8 in. of water remains for delivery of gas for combustion. To force the gas through the gas pressure regulator, shutoff valve, and associated manifold piping, 4 in. of water is typically applied. The last 4 in. of water overcomes the resistance of the gas flow control valve, the gas ring, and the furnace pressure, which may be about 2 in. of water.

Consider this same system at 5000 ft altitude, starting from the furnace back. Because of the larger combustion gas volume handled, the furnace pressure will have increased to about 2.5 in. of water. Some changes can be made to the gas ring to allow for a larger gas flow, but there is usually some increase in the pressure drop, perhaps to 2.5 in. of water, so that a pressure of 5 in. of water is now required at the gas volume control valve. Because gas pressure drop varies with the square of the volume flow rate, the manifold pressure drop will increase to about 5.5 in. of water, and the piping pressure drop will increase to 8.5 in. of water. A meter inlet pressure of 19 in. of water would now be required. Because higher gas pressures frequently are not available, it may be necessary to increase piping diameters and the size of control manifolds.

OIL-BURNING EQUIPMENT

An oil burner is a mechanical device for preparing fuel oil to combine with air under controlled conditions for combustion. Fuel oil is atomized at a controlled flow rate. Air for combustion is generally supplied with a forced-draft fan, although natural draft or mechanically induced draft can be used. Ignition is typically provided by an electric spark, although gas pilot flames, oil pilot flames, and hot-surface igniters may also be used. Oil burners operate from automatic temperature- or pressure-sensing controls.

Oil burners may be classified by application, type of atomizer, or firing rate. They can be divided into two major groups: residential and commercial-industrial. Further distinction is made based on design and operation; different types include pressure atomizing, air or steam atomizing, rotary, vaporizing, and mechanical atomizing. Unvented, portable kerosene heaters are not classified as residential oil burners or as oil heat appliances.

RESIDENTIAL OIL BURNERS

Residential oil burners ordinarily consume fuel at a rate of 0.5 to 3.5 gph. However, burners up to 7 gph sometimes fall in the residential classification because of basic similarities in controls and standards. (Burners having a capacity of 3.5 gph and above are classified as commercial-industrial.) No. 2 fuel oil is generally used, although burners in the residential size range can also operate on No. 1 fuel oil. Burners in the 0.5 to 2.5 gph range are used not only for boilers and furnaces for space heating, but also for separate tank-type residential water heaters, infrared heaters, space heaters, and other commercial equipment.

Central heating appliances include warm-air furnaces and steam or hot water boilers. Oil-burning furnaces and boilers operate essentially the same way as their gas counterparts. NFPA *Standard* 31, Installation of Oil Burning Equipment, prescribes correct installation practices for oil-burning appliances.

Steam or hot water boilers are available in cast iron and steel. In addition to supplying space heating, many boilers are designed to provide hot water using tankless integral or external heat exchangers. Residential units designed to operate as direct vent appliances are available.

Over 95% of residential burners manufactured today are high-pressure atomizing gun burners with retention-type heads (Figure 2). This type of burner supplies oil to the atomizing nozzle at pressures that range from 100 to 300 psi. A fan supplies air for combustion, and generally a fan inlet damper regulates the air supply at the burner. A high-voltage electric spark ignites the fuel by either **constant ignition** (on when the burner motor is on) or **interrupted ignition** (on only to start combustion). Typically, these burners fire into a combustion chamber in which draft is maintained.

Use of retention heads and residential burner motors operating at 3500 rpm instead of 1750 rpm is widespread. The retention head is essentially a bluff-body flame stabilizer that assists combustion by providing better air-oil mixing, turbulence, and shear (Figure 3). Using 3500 rpm motors (with the retention head) produces a more compact burner with equal capacity and more tolerance for varying draft conditions. In addition, most burner designs now include fans with increased static pressure (to 3 in. of water). Such burners are less sensitive to draft variations and, in most cases, can be used without a flue draft regulator. These higher static burners are particularly useful in systems that include sealed combustion and sidewall venting (induced or forced) rather than being vented with conventional chimneys.

In terms of fuel handling, some new burner designs have incorporated high-pressure pumps to improve atomization, fuel preheaters for cold fuel, and solenoid fuel shutoff or nozzle check valves to inhibit after-drip.

The following designs are still in operation but are not a significant part of the residential market:

- The low-pressure atomizing gun burner differs from the high-pressure type in that it uses air at a low pressure to atomize the oil.
- The pressure atomizing induced-draft burner uses the same type of oil pump, nozzle, and ignition system as the high-pressure atomizing gun burner.
- Vaporizing burners are designed for use with No. 1 fuel oil. Fuel is ignited electrically or by manual pilot.
- Rotary burners are usually of the vertical wall flame type.

Beyond the pressure atomizing burner, considerable developmental effort has been put into advanced technologies such as air atomization, fuel prevaporization, and pulsed fuel flow. Some of these are now used commercially in Europe. Reported benefits include reduced emissions, increased efficiency, and the ability to quickly vary the firing rate. See Locklin and Hazard (1980) for a historical review of technology for residential oil burners. More recent developments are described in the *Proceedings of the Oil Heat Technology Conferences and Workshops* (McDonald 1989-1998).

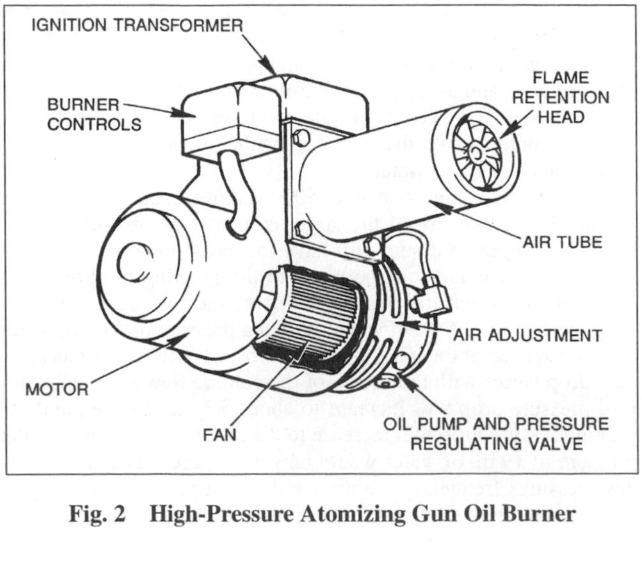

Fig. 2 High-Pressure Atomizing Gun Oil Burner

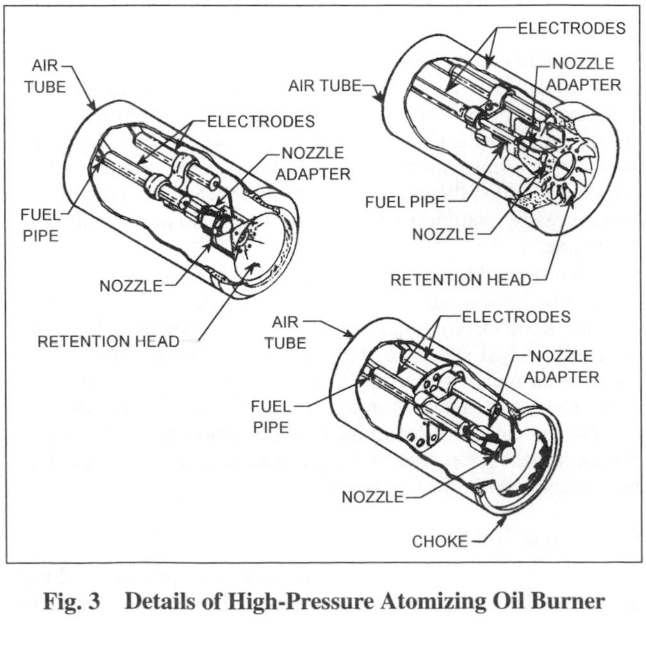

Fig. 3 Details of High-Pressure Atomizing Oil Burner

Table 1 Classification of Atomizing Oil Burners

Type of Oil Burner	Heat Range, 1000 Btu/h	Flow Volume, gph	Fuel Grade	Usual Application
Pressure atomizing	70 to 7000	0.5 to 50	No. 2 (less than 25 gph)	Boilers Warm-air furnaces
			No. 4 (greater than 25 gph)	Appliances
Return-flow pressure atomizing or modulating pressure atomizing	3500 and above	25 and above	No. 2 and heavier	Boilers Warm-air furnaces
Air atomizing	70 to 1000	0.5 to 70	No. 2 and heavier	Boilers Warm-air furnaces
Horizontal rotary cup	750 to 37,000	5 to 300	No. 2 for small sizes	Boilers
			No. 4, 5, or 6 for larger sizes	Large warm-air furnaces
Steam atomizing (register-type)	12,000 and above	80 and above	No. 2 and heavier	Boilers
Mechanical atomizing (register-type)	12,000 and above	80 and above	No. 2 and heavier	Boilers Industrial furnaces
Return-flow mechanical atomizing	45,000 to 180,000	300 to 1200	No. 2 and heavier	Boilers

In Europe, there is increasing use of low-NO_x, **blue flame burners** for residential and commercial applications. Peak flame temperatures (and thermal NO_x emissions) are reduced through the use of high flame zone recirculation rates. In a typical boiler, a conventional yellow flame retention head oil burner has a NO_x emission of 90 to 110 ppm. A blue flame burner emits about 60 ppm. The blue flame burners require flame sensors for the safety control, which responds to the fluctuating light emitted. Blue flame burners are somewhat more expensive than conventional yellow flame burners. See Butcher et al. (1994) for additional discussion of NO_x and small oil burners.

COMMERCIAL-INDUSTRIAL OIL BURNERS

Commercial and industrial oil burners are designed for use with distillate or residual grades of fuel oil. With slight modifications, burners designed for residual grades can use distillate fuel oils.

The commercial-industrial burners covered here have atomizers, which inject the fuel oil into the combustion space in a fine, conical spray with the apex at the atomizer. The burner also forces combustion air into the oil spray, causing an intimate and turbulent mixing of air and oil. Applied for a predetermined time, an electrical spark, a spark-ignited gas, or an oil igniter ignites the mixture, and sustained combustion takes place. Safety controls are used to shut down the burner upon failure to ignite.

All of these burners are capable of almost complete burning of the fuel oil without visible smoke when they are operated with excess air as low as 20% (approximately 12% CO_2 in the flue gases). Atomizing oil burners are generally classified according to the method used for atomizing the oil, such as pressure atomizing, return-flow pressure atomizing, air atomizing, rotary cup atomizing, steam atomizing, mechanical atomizing, or return-flow mechanical atomizing. Descriptions of these burners are given in the following sections, together with usual capacities and applications. Table 1 lists approximate size range, fuel grade, and usual applications. All burners described are available as gas-oil (dual-fuel) burners.

Pressure Atomizing Oil Burners

This type of burner is used in most installations where No. 2 fuel oil is burned. The oil is pumped at pressures of 100 to 300 psi through a suitable burner nozzle orifice that breaks it into a fine mist and swirls it into the combustion space as a cone-shaped spray. Combustion air from a fan is forced through the burner air-handling parts surrounding the oil nozzle and is directed into the oil spray.

For smaller capacity burners, ignition is usually started by an electric spark applied near the discharge of the burner nozzle. For burner capacities above 20 gph, a spark-ignited gas or an oil igniter is used.

Pressure atomizing burners are designated commercially as forced-draft, natural-draft, or induced-draft burners. The forced-draft burner has a fan and motor with capacity to supply all the air for combustion to the combustion chamber or furnace at a pressure high enough to force the gases through the heat-exchange equipment without the assistance of an induced-draft fan or a chimney draft. Mixing of the fuel and air is such that a minimum of refractory material is required in the combustion space or furnace to support combustion. The natural-draft burner requires a draft in the combustion space.

Burner range, or variation in burning rate, is changed by simultaneously varying the oil pressure to the burner nozzle and regulating the airflow by a damper. This range is limited to about 1.6 to 1 for any given nozzle orifice. Burner firing mode controls for various capacity burners differ among manufacturers. Usually, larger burners are equipped with controls that provide variable heat inputs. If the burner capacity is up to 15 gph, an **on-off** control is used; if it is up to 25 gph, a **modulation control** is used. In both cases, the low burning rate is about 60% of the full-load capacity of the burner.

For pressure atomizing burners, no preheating is required for burning No. 2 oil. No. 4 oil must be preheated to about 100°F for proper burning. When properly adjusted, these burners operate well with less than 20% excess air (approximately 12% CO_2); no visible smoke (approximately No. 2 smoke spot number, as determined by ASTM *Standard* D 2156); and only a trace of carbon monoxide in the flue gas in commercial applications. In these applications, the regulation of combustion airflow is typically based on smoke level or flue gas monitoring using analyzers for CO, O_2, or CO_2.

Burners with lower firing rates used to power appliances, residential heating units, or warm-air furnaces are usually set up to operate to about 50% excess air (approximately 10% CO_2).

Good operation of these burners calls for (1) a relatively constant draft (either in the furnace or at the breeching connection, depending on the burner selected), (2) clean burner components, and (3) good quality fuel oil that complies with the appropriate specifications.

Return-Flow Pressure Atomizing Oil Burners

This burner is a modification of the pressure atomizing burner; it is also called a modulating pressure atomizer. It has the advantage of wide load range for any given atomizer, about 3 to 1 turndown (or variation in load) as compared to 1.6 to 1 for the straight pressure atomizing burner.

This wide range is accomplished by means of a return-flow nozzle, which has an atomizing swirl chamber just ahead of the orifice. Good atomization throughout the load range is attained by maintaining a high rate of oil flow and high pressure drop through the swirl chamber. The excess oil above the load demand is returned from the swirl chamber to the oil storage tank or to the suction of the oil pump.

Control of the burning rate is effected by varying the oil pressure in both the oil inlet and oil return lines. Except for the atomizer, load range, and method of control, the information given for the straight pressure atomizing burner applies to this burner as well.

Air Atomizing Oil Burners

Except for the nozzle, this burner is similar in construction to the pressure atomizing burner. Atomizing air and oil are supplied to individual parts within the nozzle. The nozzle design allows the oil to break up into small droplet form as a result of the shear forces created by the atomizing air. The atomized oil is carried from the nozzle through the outlet orifice by the airflow into the furnace.

The main combustion air from a draft fan is forced through the burner throat and mixes intimately with the oil spray inside the combustion space. The burner igniter is similar to that used on pressure atomizing burners.

This burner is well suited for heavy fuel oils, including No. 6, and has a wide load range, or turndown, without changing nozzles. Turndown of 3 to 1 for the smaller sizes and about 6 or 8 to 1 for the larger sizes may be expected. Load range variation is accomplished by simultaneously varying the oil pressure, the atomizing air pressure, and the combustion air entering the burner. Some designs use relatively low atomizing air pressure (5 psi and lower); other designs use air pressures up to 75 psi. The burner uses from 2.2 to 7.7 ft^3 of compressed air per gallon of fuel oil (on an air-free basis).

Because of its wide load range, this burner operates well on modulating control.

No preheating is required for No. 2 fuel oil. The heavier grades of oil must be preheated to maintain proper viscosity for atomization. When properly adjusted, these burners operate well with less than 15 to 25% excess air (approximately 14 to 12% CO_2, respectively, at full load); no visible smoke (approximately No. 2 smoke spot number); and only a trace of carbon monoxide in the flue gas.

Horizontal Rotary Cup Oil Burners

This burner atomizes the oil by spinning it in a thin film from a horizontal rotating cup and injecting high-velocity primary air into the oil film through an annular nozzle that surrounds the rim of the atomizing cup.

The atomizing cup and frequently the primary air fan are mounted on a horizontal main shaft that is motor driven and rotates at constant speed—3500 to 6000 rpm—depending on the size and make of the burner. The oil is fed to the atomizing cup at controlled rates from an oil pump that is usually driven from the main shaft through a worm and gear.

A separately mounted fan forces secondary air through the burner windbox. Secondary air should not be introduced by natural draft. The oil is ignited by a spark-ignited gas or an oil-burning igniter (pilot). The load range or turndown for this burner is about 4 to 1, making it well suited for operation with modulating control. Automatic combustion controls are electrically operated.

When properly adjusted, these burners operate well with 20 to 25% excess air (approximately 12.5 to 12% CO_2, respectively, at full load); no visible smoke (approximately No. 2 smoke spot number); and only a trace of carbon monoxide in the flue gas.

This burner is available from several manufacturers as a package comprising burner, primary air fan, secondary air fan with separate motor, fuel oil pump, motor, motor starter, ignition system (including transformer), automatic combustion controls, flame safety equipment, and control panel.

Good operation of these burners requires relatively constant draft in the combustion space. The main assembly of the burner, with motor, main shaft, primary air fan, and oil pump, is arranged for mounting on the boiler front and is hinged so that the assembly can be swung away from the firing position for easy access.

Rotary burners require some refractory in the combustion space to help support combustion. This refractory may be in the form of throat cones or combustion chamber liners.

Steam Atomizing Oil Burners (Register Type)

Atomization is accomplished in this burner by the impact and expansion of steam. Oil and steam flow in separate channels through the burner gun to the burner nozzle. There, they mix before discharging through an orifice, or series of orifices, into the combustion chamber.

Combustion air, supplied by a forced-draft fan, passes through the directing vanes of the burner register, through the burner throat, and into the combustion space. The vanes give the air a spinning motion, and the burner throat directs it into the cone-shaped oil spray, where intimate mixing of air and oil takes place.

Full-load oil pressure at the burner inlet is generally some 100 to 150 psi, and the steam pressure is usually kept higher than the oil pressure by about 25 psi. Load range is accomplished by varying these pressures. Some designs operate with oil pressure ranging from 150 psi at full load to 10 psi at minimum load, resulting in a turndown of about 8 to 1. This wide load range makes the steam atomizing burner suited to modulating control. Some manufacturers provide dual atomizers within a single register so that one can be cleaned without dropping load.

Depending on the burner design, steam atomizing burners use from 1 to 5 lb of steam to atomize a gallon of oil. This corresponds to 0.5 to 3.0% of the steam generated by the boiler. Where no steam is available for start-up, compressed air from the plant air supply may be used for atomizing. Some designs permit the use of a pressure atomizing nozzle tip for start-up when neither steam nor compressed air is available.

This burner is used mainly on water-tube boilers, which generate steam at 150 psi or higher and at capacities above 12,000,000 Btu/h input.

Oils heavier than grade No. 2 must be preheated to the proper viscosity for good atomization. When properly adjusted, these burners operate well with 15% excess air (14% CO_2) at full load, without visible smoke (approximately No. 2 smoke spot number), and with only a trace of carbon monoxide in the flue gas.

Mechanical Atomizing Oil Burners (Register Type)

Mechanical atomizing, as generally used, describes a technique synonymous with pressure atomizing. Both terms designate atomization of the oil by forcing it at high pressure through a suitable stationary atomizer.

The mechanical atomizing burner has a windbox, which is a chamber into which a fan delivers combustion and excess air for distribution to the burner. The windbox has an assembly of adjustable internal air vanes called an air register. Usually the fan is mounted separately and connected to the windbox by a duct.

Oil pressure of some 90 to 900 psi is used, and load range is obtained by varying the pressure between these limits. The operating range or turndown for any given atomizer can be as high as 3 to 1. Because of its limited load range, this type of burner is seldom selected for new installations.

Return-Flow Mechanical Atomizing Oil Burners

This burner is a modification of a mechanical atomizing burner; atomization is accomplished by oil pressure alone. Load ranges up to 6 or 8 to 1 are obtained on a single-burner nozzle by varying the oil pressure between 100 and 1000 psi.

The burner was developed for use in large installations such as on shipboard and in electric generating stations where wide load range is required and water loss from the system makes the use of atomizing steam undesirable. It is also used for firing large hot water boilers. Compressed air is too expensive for atomizing oil in large burners.

This is a register burner similar to the mechanical atomizing burner. Wide range is accomplished by use of a return-flow nozzle, which has a swirl chamber just ahead of the orifice or sprayer plate. Good atomization is attained by maintaining a high rate of oil flow and a high pressure drop through the swirl chamber. The excess oil above the load demand is returned from the swirl chamber to the oil storage tank or to the oil pump suction. Control of burning rate is accomplished by varying the oil pressure in both the oil inlet and the oil return lines.

DUAL-FUEL GAS/OIL BURNERS

Dual-fuel, combination gas/oil burners are forced-draft burners that incorporate, in a single assembly, the features of the commercial-industrial grade gas and oil burners described in the preceding sections. These burners have a three-position switch that permits the manual selection of gas, oil, or a **center-off** position. This switch contains a positive center stop or delay to ensure that the burner flame relay or programmer cycles the burner through a postpurge and prepurge cycle before starting again on the other fuel. The burner manufacturers of larger boilers design the special mechanical linkages needed to deliver the correct air-fuel ratios at full-fire, low-fire, or any intermediate rate. Smaller burners may have straight **on-off** firing. Larger burners may have low-fire starts on both fuels and use a common flame scanner. Smaller dual-fuel burners usually include pressure atomization of the oil. Air atomization systems are included in large oil burners.

Automatic changeover dual-fuel burners are available for use with gas and No. 2 oil. A special temperature control that is mounted on an outside wall and electrically interlocked with the dual-fuel burner control system senses outdoor temperature. When the outdoor temperature drops to the outdoor control set point, the control changes fuels automatically after putting the burner through a postpurge and a prepurge cycle. The **minimum** additional controls and wiring needed to operate automatically with an outdoor temperature control are (1) burner fuel changeover relay(s) and (2) time delay devices to ensure interruption of the burner control circuit at the moment of fuel changeover. The fuel changeover relays replace the three-position manual fuel selection switch. A manual fuel selection switch can be retained as a manual override on the automatic feature. These control systems require special design and are generally provided by the burner manufacturer.

The dual-fuel burner is fitted with a gas train and oil piping that is connected to a two-pipe oil system following the principles of the preceding sections. A reserve of oil must be maintained at all times for automatic fuel changeover.

Boiler flue chimney connectors are equipped with a double-swing barometric draft regulator or, if required, sequential furnace draft control to operate an automatic flue damper.

Dual-fuel burners and their accessories should be installed by experienced contractors to ensure satisfactory operation.

EQUIPMENT SELECTION

Economic and practical factors (e.g., the degree of operating supervision required by the installation) generally dictate the selection of fuel oil based on the maximum heat input of the oil-burning unit. For heating loads and where only one oil-burning unit is operated at any given time, the relationship is as shown in Table 2 (this table is only a guide). In many cases, a detailed analysis of operating parameters results in the burning of lighter grades of fuel oil at capacities far above those indicated.

Table 2 Guide for Fuel Oil Grades Versus Firing Rate

Maximum Heat Input of Unit, 1000 Btu/h	Volume Flow Rate, gph	Fuel Grade
Up to 3500	Up to 25	2
3500 to 7000	25 to 50	2, 4, 5
7000 to 15,000	50 to 100	5, 6
Over 15,000	Over 100	6

Fuel Oil Storage Systems

All fuel oil storage tanks should be constructed and installed in accordance with NFPA *Standard* 31 and with local ordinances.

Storage Capacity. Dependable and economical operation of oil-burning equipment requires ample and safe storage of fuel oil at the site. Design responsibility should include analysis of specific storage requirements as follows:

1. Rate of oil consumption.
2. Dependability of oil deliveries.
3. Economical delivery lots. The cost of installing larger storage capacity should be balanced against the savings indicated by accommodating larger delivery lots. Truck lots and railcar lots vary with various suppliers, but the quantities are approximated as follows:

Small truck lots in metropolitan area	500 to 2000 gal
Normal truck lots	3000 to 5000 gal
Transport truck lots	5000 to 9000 gal
Rail tanker lots	8000 to 12,000 gal

Tank Size and Location. Standard oil storage tanks range in size from 55 to 50,000 gal and larger. Tanks are usually built of steel; concrete construction may be used only for heavy oil. Unenclosed tanks located in the lowest story, cellar, or basement should not exceed 660 gal capacity each, and the aggregate capacity of such tanks should not exceed 1320 gal unless each 660 gal tank is insulated in an approved fireproof room having a fire resistance rating of at least 2 h.

If the storage capacity at a given location exceeds about 1000 gal, storage tanks should be underground and accessible for truck or rail delivery with gravity flow from the delivering carrier into storage. If the oil is to be burned in a central plant such as a boiler house, the storage tanks should be located, if possible, so that the oil burner pump (or pumps) can pump directly from storage to the burners. In case of year-round operation, except for storage or supply capacities below 2000 gal, at least two tanks should be installed to facilitate tank inspection, cleaning, repairs, and clearing of plugged suction lines.

When the main oil storage tank is not close enough to the oil-burning units for the burner pumps to take suction from storage, a supply tank must be installed near the oil-burning units and oil must be pumped periodically from storage to the supply tank by a transport pump at the storage location. Supply tanks should be treated the same as storage tanks regarding location within buildings, tank design, etc. On large installations, it is recommended that standby pumps be installed as a protection against heat loss in case of pump failure.

Since all piping connections to underground tanks must be at the top, such tanks should not be more than 10.5 ft from top to bottom to avoid pump suction difficulties. (This dimension may have to be less for installations at high altitudes.) At sea level, the total suction head for the oil pump must not exceed 14 ft.

Connections to Storage Tank. All piping connections for tanks over 275 gal capacity should be through the top of the tank. Figure 4 shows a typical arrangement for a cylindrical storage tank with a heating coil as required for No. 5 or No. 6 fuel oils. The heating coil and oil suction lines should be located near one end of the tank. The maximum allowable steam pressure in such a heating coil is 15 psi.

The heating coil is unnecessary for oils lighter than No. 5 unless a combination of high pour point and low outdoor temperature makes heating necessary.

A watertight manhole with internal ladder provides access to the inside of the tank. If the tank is equipped with an internal heating coil, a second manhole is required in order to permit withdrawal of the coil.

The fill line should be vertical and should discharge near the end of the tank away from the oil suction line. The inlet of the fill line must be outside the building and accessible to the oil delivery vehicle unless an oil transfer pump is used to fill the tank. When possible, the inlet of the fill line should be at or near grade level where filling may be accomplished by gravity. For gravity filling, the fill line should be at least 2 in. in diameter for No. 2 oil and 6 in. in diameter for No. 4, 5, or 6 oils. Where filling is done by pump, the fill line for No. 4, 5, or 6 oils may be 4 in. in diameter.

An oil return line bringing recirculated oil from the burner line to the tank should discharge near the oil suction line inlet. Each storage tank should be equipped with a vent line sized and arranged in accordance with NFPA *Standard* 31.

Each storage tank must have a device for determining oil level. For tanks inside buildings, the gaging device should be designed and installed so that oil or vapor will not discharge into a building from the fuel supply system. No storage tank should be equipped with a glass gage or any gage that, if broken, would permit the escape of oil from the tank. Gaging by a measuring stick is permissible for outside tanks or for underground tanks.

Fuel-Handling Systems

The fuel-handling system consists of the pumps, valves, and fittings for moving fuel oil from the delivery truck or car into the storage tanks and from the storage tanks to the oil burners. Depending on the type and arrangement of the oil-burning equipment and the grade of fuel oil burned, fuel-handling systems vary from simple to quite complicated arrangements.

The simplest handling system would apply to a single burner and small storage tank for No. 2 fuel oil similar to a residential heating installation. The storage tank is filled through a hose from the oil delivery truck, and the fuel-handling system consists of a supply pipe between the storage tank and the burner pump. Equipment should be installed on light oil tanks to indicate visibly or audibly when the tank is full; on heavy oil tanks, a remote-reading liquid-level gage should be installed.

Figure 5 shows a complex oil supply arrangement for two burners on one oil-burning unit. For a unit with a single burner, the

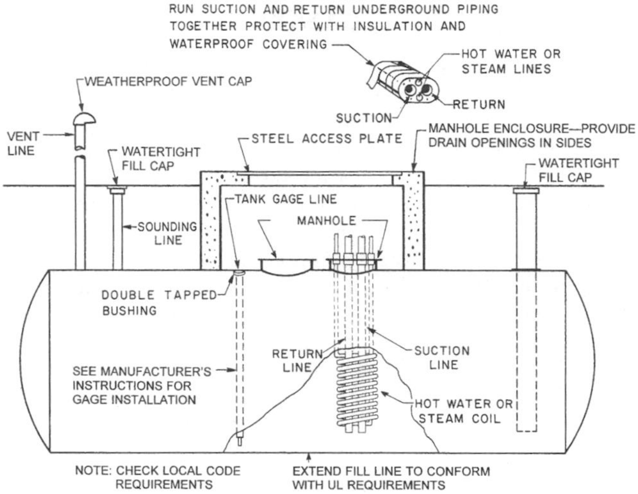

Fig. 4 Typical Oil Storage Tank (No. 6 Oil)

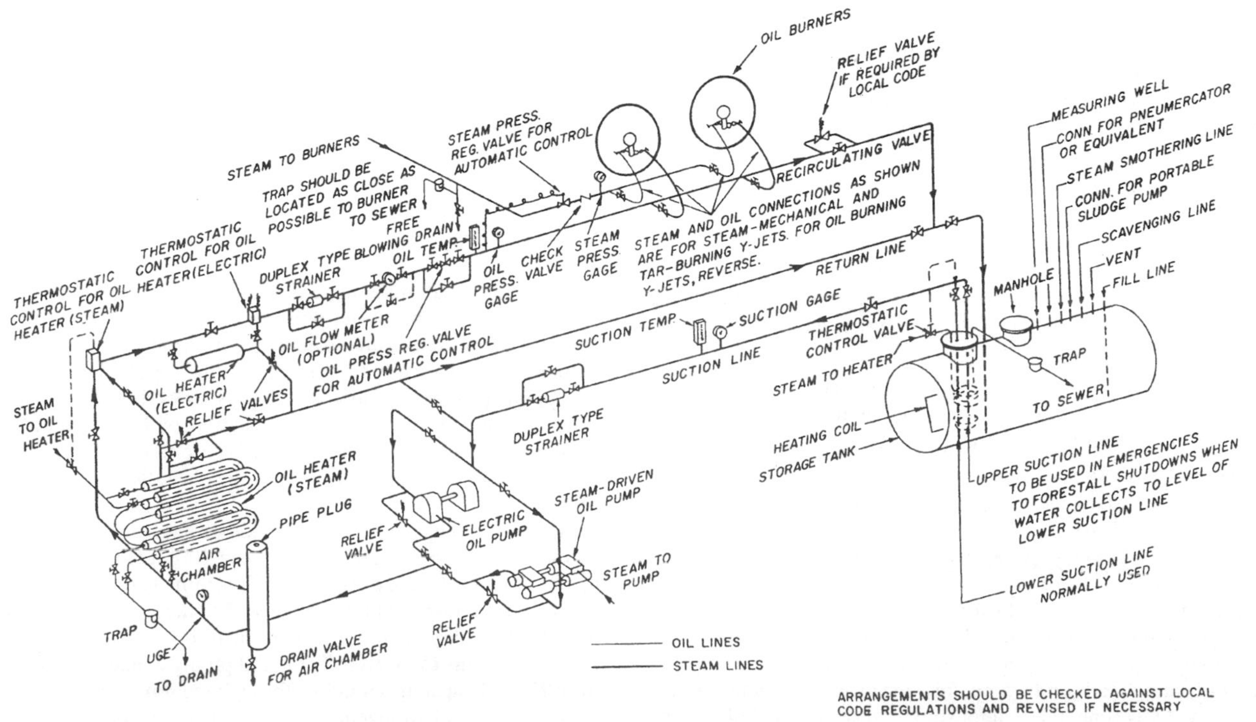

Fig. 5 Industrial Burner Auxiliary Equipment

change in piping is obvious. For a system having two or more units, the oil line downstream of the oil discharge strainer would become a main supply header, and the branch supply line to each unit would include a flowmeter, an automatic control valve, etc. For light oils requiring no heating, all oil heating equipment shown in Figure 5 would be omitted. Both a suction and a return line should be used, except for gravity flow in residential installations.

Oil pumps (steam or electrically driven) should deliver oil at the maximum rate required by the burners (this includes the maximum firing rate, the oil required for recirculating, plus a 10% margin).

The calculated suction head at the entrance of any burner pump should not exceed 10 in. Hg for installations at sea level. At higher elevations, the suction head should be reduced in direct proportion to the reduction in barometric pressure.

The oil temperature at the pump inlet should not exceed 120°F. Where oil burners with integral oil pumps (and oil heaters) are used and the suction lift from the storage tank is within the capacity of the burner pump, each burner may take oil directly from the storage tank through an individual suction line unless No. 6 oil is used.

Where two or more tanks are used, the piping arrangement into the top of each tank should be the same as for a single tank so that any tank may be used at any time; any tank can be inspected, cleaned, or repaired while the system is in operation.

The length of suction line between storage tank and burner pumps should not exceed 100 ft. If the main storage tank(s) are located more than 100 ft from the pumps, a supply tank should be installed near the pumps, and a transfer pump should be installed at the storage tanks for delivery of oil to the supply tank.

Central oil distribution systems comprising a central storage facility, distribution pumps or provision for gravity delivery, distribution piping, and individual fuel meters are used for residential communities—notably mobile home parks. The provisions of NFPA *Standard* 31, Installation of Oil Burning Equipment, should be followed in installing a central oil distribution system.

Fuel Oil Preparation System

Fuel-oil preparation systems consist of oil heater, oil temperature controls, strainers, and associated valves and piping required to maintain fuel oil at the temperatures necessary to control the oil viscosity, facilitate oil flow and burning, and remove suspended matter.

Preparation of fuel oil for handling and burning requires heating the oil if it is No. 5 or 6. This decreases its viscosity so it flows properly through the oil system piping and can be atomized by the oil burner. No. 4 oil occasionally requires heating to facilitate burning. No. 2 oil requires heating only under unusual conditions.

For handling residual oil from the delivering carrier into storage tanks, the viscosity should be about 156×10^6 centistokes (cSt). For satisfactory pumping, the viscosity of the oil surrounding the inlet of the suction pipe must be 444×10^6 cSt or lower; in the case of oil having a high pour point, the temperature of the entire oil content of the tank must be above the pour point.

Storage tank heaters are usually made of pipe coils or grids using steam or hot water at not over 15 psi as the heating medium. Electric heaters are sometimes used. For control of viscosity for pumping, the heated oil surrounds the oil suction line inlet. For heating oils with high pour points, the heater should extend the entire length of the tank. All heaters have suitable thermostatic controls. In some cases, storage tank heating may be accomplished satisfactorily by returning or recirculating some of the oil to the tank after it has passed through heaters located between the oil pump and oil burner.

Heaters to regulate viscosity at the burners are installed between the oil pumps and the burners. When required for small packaged burners, the heaters are either assembled integrally with the individual burners or mounted separately. The source of heat may be electricity, steam, or hot water. For larger installations, the heater is mounted separately and is often arranged in combination with central oil pumps, forming a central oil pumping and heating set. The separate or central oil pumping and heating set is recommended for those installations that burn heavy oils, have a periodical load demand, and require continuous circulation of hot oil during down periods.

Another system of oil heating to maintain pumping viscosity that is occasionally used for small- or medium-sized installations consists of an electrically heated section of oil piping. Low-voltage current is passed through the pipe section, which is isolated by nonconducting flanges.

The oil heating capacity for any given installation should be approximately 10% greater than the maximum oil flow. Maximum oil flow is the maximum oil-burning rate plus the rate of oil recirculation.

Controls for oil heaters must be dependable to ensure proper oil atomization and avoid overheating of oil, which results in coke deposits inside the heaters. In steam or electric heating, an interlock should be included with a solenoid valve or switch to shut off the steam or electricity. During periods when the oil pump is not operating, the oil in the heater can become overheated and deposit carbon. Overheating also can be a problem with oil heaters using high-temperature hot water. Provisions must be made to avoid overheating the oil when the oil pump is not operating.

Oil heaters with low- or medium-temperature hot water are not generally subject to coke deposits. Where steam or hot water is used in oil heaters located after the oil pumps, the pressure of the steam or water in the heaters is usually lower than the oil pressure. Consequently, heater leakage between oil and steam causes oil to flow into the water or condensing steam. To prevent oil from entering the boilers, the condensed steam or the water from such heaters should be discarded from the system, or special equipment should be provided for oil removal.

Hot water oil heaters of double-tube-and-shell construction with inert heat transfer oil and a sight glass between the tubes are available. With this type of heater, oil leaks through an oil-side tube appear in the sight glass, and repairs can be made to the oil-side tube before a water-side tube leaks.

This discussion of oil-burning equipment also applies to oil-fired boilers and furnaces. For more details, see Chapter 27 and Chapter 28.

SOLID-FUEL-BURNING EQUIPMENT

A mechanical stoker is a device that feeds a solid fuel into a combustion chamber. It supplies air for burning the fuel under automatic control and, in some cases, incorporates automatic ash and refuse removal.

CAPACITY CLASSIFICATION OF STOKERS

Stokers are classified according to their coal-feeding rates. Although some residential applications still use stokers, the primary thrust of stoker application is in commercial and industrial areas. The following classification has been made by the U.S. Department of Commerce, in cooperation with the Stoker Manufacturers Association:

Class 1: Capacity under 60 lb of coal per hour
Class 2: Capacity 60 to 100 lb of coal per hour
Class 3: Capacity 100 to 300 lb of coal per hour
Class 4: Capacity 300 to 1200 lb of coal per hour
Class 5: Capacity 1200 lb of coal per hour and over

Class 1 stokers are used primarily for residential heating and are designed for quiet, automatic operation. These stokers are usually underfeed types and are similar to those shown in Figure 6, except that they are usually screw-feed. Class 1 stokers feed coal to the furnace intermittently, in accordance with temperature or pressure

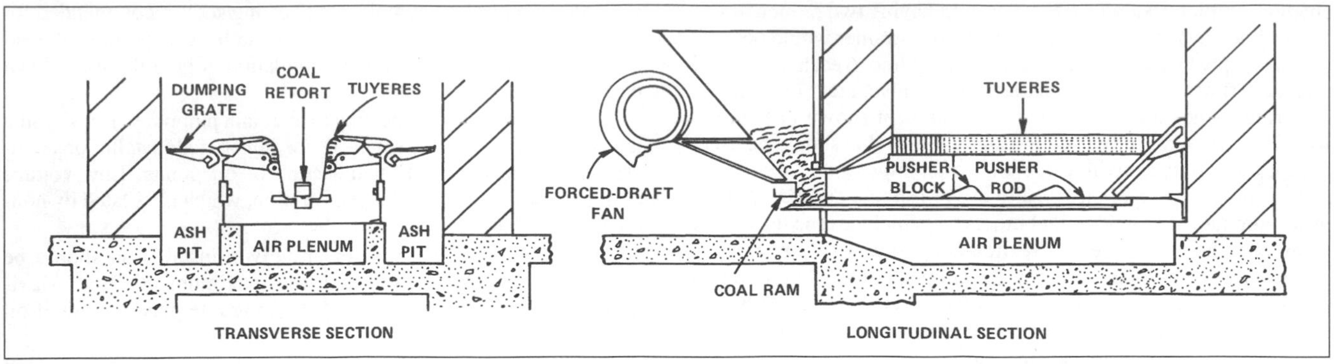

Fig. 6 Horizontal Underfeed Stoker with Single Retort

Table 3 Characteristics of Various Types of Stokers (Class 5)

Stoker Type and Subclass	Typical Capacity Range, lb/h	Maximum Burning Rate, Btu/h·ft^2	Characteristics
Spreader			Capable of burning a wide range of coals; best to follow fluctuating loads; high fly ash carryover; low-load smoke
Stationary and dumping grate	20,000 to 80,000	450,000	
Traveling grate	100,000 to 400,000	750,000	
Vibrating grate	20,000 to 100,000	400,000	
Underfeed	20,000 to 30,000	400,000	Capable of burning caking coals and a wide range of coals (including anthracite); high maintenance, low fly ash carryover; suitable for continuous-load operation
Single- or double-retort			
Multiple-retort	30,000 to 500,000	600,000	
Chain grate and *traveling grate*	20,000 to 100,000	500,000	Low maintenance, low fly ash carryover; capable of burning a wide variety of weakly caking coals; smokeless operation over entire range
Vibrating grate	1,400 to 150,000	400,000	Characteristics similar to chain grate and traveling grate stokers, except that these stokers have no difficulty in burning strongly caking coals

demands. A special control is needed to ensure stoker operation in order to maintain a fire during periods when no heat is required.

Class 2 and 3 stokers are usually of the screw-feed type, without auxiliary plungers or other means of distributing the coal. They are used extensively for heating plants in apartment buildings, hotels, and industrial plants. They are of the underfeed type and are available in both the hopper and bin-feed type. These units are also built in a plunger-feed type with an electric motor, steam, or hydraulic cylinder coal-feed drive.

Class 2 and 3 stokers are available for burning all types of anthracite, bituminous, and lignite coals. The tuyere and retort design varies according to the fuel and load conditions. Stationary grates are used on bituminous models, and the clinkers formed from the ash accumulate on the grates surrounding the retort.

Class 2 and 3 anthracite stokers are equipped with moving grates that discharge the ash into a pit below the grate. This ash pit may be located on one side or both sides of the grate and, in some installations, is of sufficient capacity to hold the ash for several weeks of operation.

Class 4 stokers vary in details of design, and several methods of feeding coal are practiced. Underfeed stokers are widely used, although overfeed types are used in the larger sizes. Bin-feed and hopper models are available in underfeed and overfeed types.

Class 5 stokers are spreader, underfeed, chain grate or traveling grate, and vibrating grate. Various subcategories reflect the type of grate and method of ash discharge.

STOKER TYPES BY FUEL-FEED METHODS

Class 5 stokers are classified according to the method of feeding fuel to the furnace: (1) spreader, (2) underfeed, (3) chain grate or traveling grate, and (4) vibrating grate. The type of stoker used in a given installation depends on the general system design, capacity required, and type of fuel burned. In general, the spreader stoker is the most widely used in the capacity range of 75,000 to 400,000 lb/h because it responds quickly to load changes and can burn a wide variety of coals. Underfeed stokers are principally used with small industrial boilers of less than 30,000 lb/h. In the intermediate range, the large underfeed units, as well as the chain and traveling grate stokers, are being displaced by spreader and vibrating grate stokers. Table 3 summarizes the major features of the different stokers.

Spreader Stokers

Spreader stokers use a combination of suspension burning and grate burning. As illustrated in Figure 7, coal is continually projected into the furnace above an ignited fuel bed. The coal fines are partially burned in suspension. Large particles fall to the grate and are burned in a thin, fast-burning fuel bed. Because this firing method provides extreme responsiveness to load fluctuations and because ignition is almost instantaneous on increased firing rate, the spreader stoker is favored over other stokers in many industrial applications.

The spreader stoker is designed to burn about 50% of the fuel in suspension. Thus, it generates much higher particulate loadings than other types of stokers and requires dust collectors to trap particulate material in the flue gas before discharge to the stack. To minimize carbon loss, fly carbon reinjection systems are sometimes used to return the particles into the furnace for complete burnout. Because this process increases furnace dust emissions, it can be used only with highly efficient dust collectors.

Grates for spreader stokers may be of several types. All grates are designed with high airflow resistance to avoid formation of blowholes through the thin fuel bed. Early designs were simple stationary grates from which ash was removed manually. Later designs allowed intermittent dumping of the grate either manually or by a power cylinder. Both types of dumping grates are frequently used

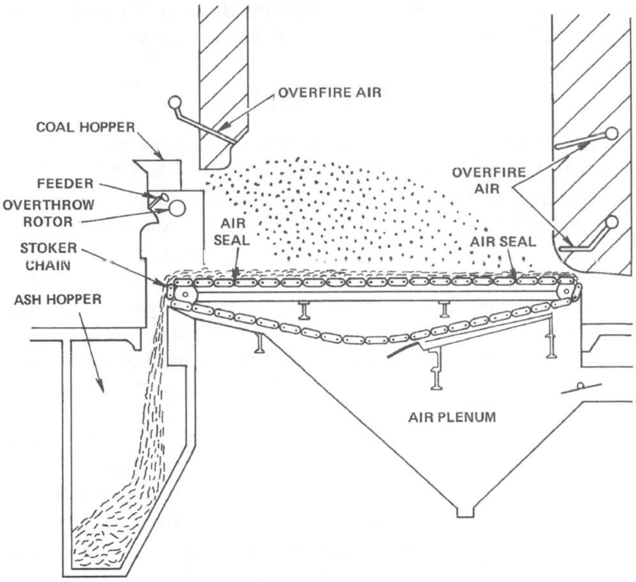

Fig. 7 Spreader Stoker, Traveling Grate Type

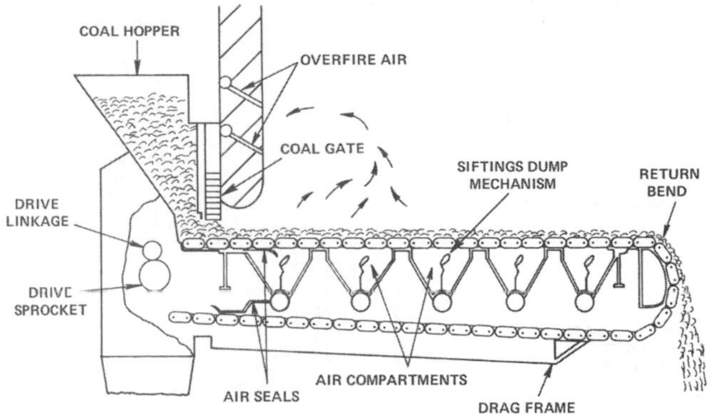

Fig. 8 Chain Grate Stoker

for small and medium-sized boilers (see Table 3). Also, both types are sectionalized, and there is a separate undergrate air chamber for each grate section and a grate section for each spreader unit. Consequently, both the air supply and the fuel supply to one section can be temporarily discontinued for cleaning and maintenance without affecting the operation of other sections of the stoker.

For high-efficiency operation, a continuous ash-discharging grate, such as the traveling grate, is necessary. The introduction of the spreader stoker with the traveling grate increased burning rates by about 70% over the stationary grate and dumping grate types. Although both reciprocating and vibrating continuous ash discharge grates have been developed, the traveling grate stoker is preferred because of its higher burning rates.

Fuels and Fuel Bed. All spreader stokers (particularly those with traveling grates) are able to use fuels with a wide range of burning characteristics, including caking tendencies, because the rapid surface heating of the coal in suspension destroys the caking tendency. High-moisture, free-burning bituminous and lignite coals are commonly burned; coke breeze can be burned in a mixture with a high-volatile coal. However, anthracite, because of its low volatile content, is not a suitable fuel for spreader stoker firing. Ideally, the fuel bed of a coal-fired spreader stoker is 2 to 4 in. thick.

Burning Rates. The maximum heat release rates range from 400,000 Btu/h·ft^2 (a coal consumption of approximately 40 lb/h) on stationary, dumping, and vibrating grate designs to 750,000 Btu/h·ft^2 on traveling grate spreader stokers. Higher heat release rates are practical with certain waste fuels in which a greater portion of fuel can be burned in suspension than is possible with coal.

Underfeed Stokers

Underfeed stokers introduce raw coal into a retort beneath the burning fuel bed. They are classified as horizontal feed and gravity feed. In the horizontal type, coal travels within the furnace in a retort parallel with the floor; in the gravity-feed type, the retort is inclined by 25°. Most horizontal feed stokers are designed with single or double retorts (and rarely, with triple retorts), while gravity-feed stokers are designed with multiple retorts.

In the horizontal stoker (Figure 6), coal is fed to the retort by a screw (for the smaller stokers) or a ram (for the larger units). Once the retort is filled, the coal is forced upward and spills over the retort to form and feed the fuel bed. Air is supplied through tuyeres at each side of the retort and through air ports in the side grates. Over-fire air provides additional combustion air to the flame zone directly above the bed to prevent smoking, especially at low loads.

Gravity-feed units are similar in operating principle. These stokers consist of sloping multiple retorts and have rear ash discharge. Coal is fed into each retort, where it is moved slowly to the rear while simultaneously being forced upward over the retorts.

Fuels and Fuel Bed. Either type of underfeed stoker can burn a wide range of coal, although the horizontal type is better suited for free-burning bituminous coal. These units can burn caking coal, provided there is not an excess amount of fines. The ash-softening temperature is an important factor in selecting coals because the possibility of excessive clinkering increases at lower ash-softening temperatures. Because combustion occurs in the fuel bed, underfeed stokers respond slowly to load change. Fuel-bed thickness is extremely nonuniform, ranging from 8 to 24 in. The fuel bed often contains large fissures separating masses of coke.

Burning Rates. Single-retort or double-retort horizontal stokers are generally used to service boilers with capacities up to 30,000 lb/h. These units are designed for heat release rates of 400,000 Btu/h·ft^2.

Chain Grate and Traveling Grate Stokers

Figure 8 shows a typical chain grate or traveling grate stoker. These stokers are often used interchangeably because they are fundamentally the same, except for grate construction. The essential difference is that the links of chain grate stokers are assembled so that they move with a scissors-like action at the return bend of the stoker, while in most traveling grates there is no relative movement between adjacent grate sections. Accordingly, the chain grate is more suitable for handling coal with clinkering ash characteristics than is the traveling grate unit.

The operation of the two types is similar. Coal, fed from a hopper onto the moving grate, enters the furnace after passing under an adjustable gate that regulates the thickness of the fuel bed. The layer of coal on the grate entering the furnace is heated by radiation from the furnace gases or from a hot refractory arch. As volatile matter is driven off by this rapid radiative heating, ignition occurs. The fuel continues to burn as it moves along the fuel bed, and the layer becomes progressively thinner. At the far end of the grate, where the combustion of the coal is completed, the ash is discharged into the pit as the grates pass downward over a return bend.

Often furnace arches (front and/or rear) are included with these stokers to improve combustion by reflecting heat to the fuel bed. The front arch also serves as a bluff body, mixing rich streams of volatile gases with air to reduce unburned hydrocarbons. A chain grate stoker with overfire air jets eliminates the need for a front arch for burning volatiles. As shown in Figure 9, the stoker was zoned, or sectionalized, and equipped with individual zone dampers to control the pressure and quantity of air delivered to the various sections.

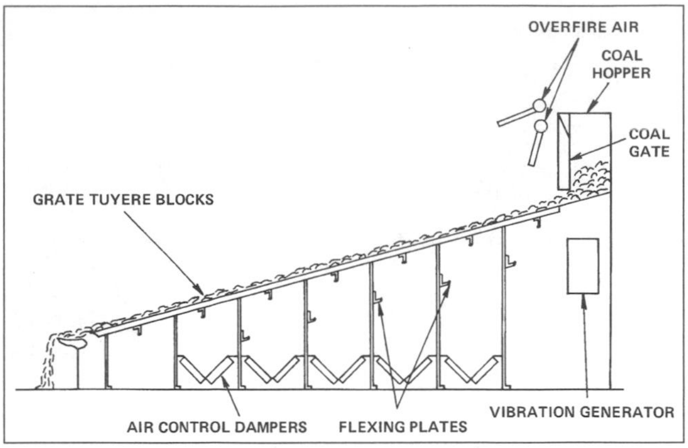

Fig. 9 Vibrating Grate Stoker

Fuels and Fuel Bed. The chain grate and traveling grate stokers can burn a variety of fuels (e.g., peat, lignite, subbituminous coal, free-burning bituminous coal, anthracite coal, and coke), provided that the fuel is sized properly. However, strongly caking bituminous coals have a tendency to mat and prevent proper air distribution to the fuel bed. Also, a bed of strongly caking coal may not be responsive to rapidly changing loads. Fuel-bed thickness varies with the type and size of the coal burned. For bituminous coal, a 5 to 7 in. bed is common; for small-sized anthracite, the fuel bed is reduced to 3 to 5 in.

Burning Rates. Chain grate and traveling grate stokers are offered for a maximum continuous burning rate of 350,000 to 500,000 Btu/ft·ft^2, depending on the type of fuel and its ash and moisture content.

Vibrating Grate Stokers

The vibrating grate stoker, as shown in Figure 9, is similar to the chain grate stoker in that both are overfeed, mass-burning, continuous ash discharge units. However, in the vibrating stoker, the sloping grate is supported on equally spaced vertical plates that oscillate back and forth in a rectilinear direction, causing the fuel to move from the hopper through an adjustable gate into the active combustion zone. Air is supplied to the stoker through laterally exposed areas beneath the stoker formed by the individual flexing of the grate support plates. Ash is automatically discharged into a shallow or basement ash pit. The grates are water cooled and are connected to the boiler circulating system.

The rates of coal feed and fuel-bed movement are determined by the frequency and duration of the vibrating cycles and regulated by automatic combustion controls that proportion the air supply to optimize heat release rates. Typically, the grate is vibrated about every 90 s for durations of 2 to 3 s, but this depends on the type of coal and boiler operation. The vibrating grate stoker has found increasing acceptance because of its simplicity, inherently low fly ash carryover, low maintenance, wide turndown (10 to 1), and adaptability to multiple-fuel firing.

Fuels and Fuel Bed. The water-cooled vibrating grate stoker is suitable for burning a wide range of bituminous and lignite coals. The gentle agitation and compaction of the vibratory actions allow coal having a high free-swelling index to be burned and a uniform fuel bed without blowholes and thin spots to be maintained. The uniformity of air distribution and fuel-bed conditions produce both good response to load swings and smokeless operation over the entire load range. Fly ash emission is probably greater than from the traveling grate because of the slight intermittent agitation of the fuel bed. The fuel bed is similar to that of a traveling grate stoker.

Burning Rates. Burning rates of vibrating grate stokers vary with the type of fuel used. In general, however, the maximum heat release rate should not exceed 400,000 Btu/h·ft^2 (a coal use of approximately 40 lb/h) to minimize fly ash carryover.

CONTROLS

This section covers only the automatic controls necessary for automatic fuel-burning equipment. Chapter 37 of the 1997 *ASHRAE Handbook—Fundamentals* addresses basic automatic control.

Automatic fuel-burning equipment requires a control system that provides a prescribed sequence of operating events and takes proper corrective action in the event of any failure in the equipment or its operation. The basic requirements for oil burners, gas burners, and coal burners (stokers) are the same. The term **burner**, unless otherwise specified, refers to all three types of fuel-burning equipment. However, the details of operation and the components used differ for the three types of burners.

The controls can be classified as operating controls, limit controls, and interlocks. Operating controls include a primary sensor, secondary sensors, actuators, an ignition system, firing-rate controls, draft controls, and programmers. Limit controls include flame safeguard, temperature limit, pressure limit, and water level limit controls. Several control functions are frequently included in a single component. Control systems for domestic burners and the smaller commercial burners are generally operated on low voltage (frequently 24 V). Some domestic gas burner control systems obtain their electrical power from the direct conversion of heat to low-voltage electrical energy by a thermopile. These are called self-generating systems. Line voltage and pneumatic controls are common in the larger commercial and industrial areas.

OPERATING CONTROLS

Operating controls initiate the normal starting and stopping of the burner by acting through appropriate actuators in response to the primary sensor. Secondary sensors, secondary actuators, and the ignition system are all part of the operating controls systems.

A **primary sensor** monitors the effect of the fuel-burning equipment on the controlled variable. Examples of primary sensors are a room thermostat for a residential furnace; a pressure-actuated switch for a steam boiler; or a thermostat for a water heater. Several sensors may be connected together to average the measurement (temperature or pressure) at several points. Control on the basis of a temperature or pressure difference between two points can also be achieved. An example is an outdoor temperature-sensing element used in conjunction with a room thermostat to reset room temperature as a function of outdoor temperature.

Secondary sensors are generally required as well. An example is the fan control used on warm-air furnaces to control the fan as a function of plenum temperature. In some cases, the primary sensor operates a heat control device such as a damper motor or a circulating pump. The burner is then operated from a secondary control, such as an immersion thermostat in a hot water system or a thermostat in the bonnet of the warm-air furnace. For example, the primary sensor in a steam system with individual room control actuates a steam valve supplying that room or zone; the burner is then operated by a secondary control consisting of a pressure sensor located on a boiler.

An **actuator** is a device that converts the control system signal into a useful function. Actuators generally consist of valves, dampers, or relays.

Valves are required for final shutoff of the gas pilot and main gas (or oil) valves in the supply line. The type of valve used depends on the service, fuel, and characteristics of the burner. Valves are classified as follows:

- Solenoid gas or oil valves provide quick opening and closing.
- Motor-operated gas and oil valves may be of the gear train, pneumatic, or hydraulically operated type. They provide relatively slow opening and quick closing. Opening time is not adjustable.

They may have provision for direct- or reverse-acting operating levers and, in some cases, have position indicators.

- Diaphragm valves for gas are usually operated by gas pressure, although some types are steam or air operated. These valves may have provision for adjustable opening time to deliver the desired burner operating characteristics. They may also have provision for direct- or reverse-acting lever arms.
- Manually opened safety shutoff valves for gas or oil can be opened only when power is available. When power is interrupted or any associated interlock is activated, they trip free for fast closure. They are normally used on semiautomatic and manually fired installations.
- Burner input control valves for gas or oil are required in addition to safety shutoff valves by some regulatory agencies. These control valves may be slow opening or may provide for high-low or modulating operation. These valves may or may not have provision for final shutoff. Requirements of local codes and approval bodies should be followed to ensure that the valves selected are approved for the type of burner used and that maximum overall efficiency of the burner is obtained.

Ignition

An automatic burner igniter is necessary for safe operation and is, in most applications, an essential part of the automatic control system. The burner igniter is either an electric spark that directly ignites the main fuel supply or a relatively small fuel burner that ignites the main fuel supply. Most igniters for oil burners are continuous (function continuously while burner is in operation) or intermittent (function only long enough to establish main flame). On small-input gas-fired appliances, ignition of the main fuel is generally achieved by an intermittent ignition device (e.g., electric spark), a hot surface igniter, or a small standing gas pilot that burns continuously.

Electric spark igniters are generally used only for ignition of distillate grades of fuel oil. The spark is generated by a transformer, which supplies up to 10,000 V to the sparking electrodes. This ignition system is used frequently on oil burners having capacities of up to 20 gph.

When pilot burners are used on larger equipment, the fuel burned may be gas, liquid petroleum (LP), or distillate fuel oil. Ignition of the pilot burner is by electric spark.

Safety Controls and Interlocks

Safety controls and interlocks are provided to protect against furnace explosions and other hazards, (e.g., overheating of boilers resulting from low water). These controls shut off the fuel or prevent firing in case of the following:

- Flame failure—main flame or pilot flame
- Combustion air fan failure
- Overheating of warm-air furnace parts
- Low water level in boilers
- Low fuel-oil pressure
- Low fuel-oil temperature when burning heavy oils
- High fuel-oil temperature when burning heavy oils
- Low atomizing air pressure or atomizing steam pressure
- High water temperature in water heaters and hot water boilers
- High steam pressure
- Burner out of firing position (burner position interlock)
- Failure of rotary burner motor (rotary cup interlock)

These interlocks do not affect the normal operation of the unit. The various types of safety devices in normal use and their application and operation are described in the following sections.

Limit Controls. The safety features, which must be incorporated into the control system, function only when the system exceeds prescribed safe operating conditions. Such safety features are embodied in the limit controls. They actuate electrical switches that are normally closed and close the fuel valve in the event of an unsafe condition such as (1) excessive temperature in the combustion chamber or heat exchanger; (2) excessive pressure in a boiler or water heater; (3) low water level in a boiler or in larger commercial and industrial burners; (4) high or low gas pressure; (5) low oil pressure; (6) low atomizing media pressure; and (7) low oil temperature when firing residual fuel oil. Separate limit and operating controls are always recommended. The operating control is adjusted to function within the operating range, while the limit is set at a point above the operating range to stop the burner in the event of failure of the operating control.

Flame Safeguard Controls. Loss of ignition or flame failure while fuel is being injected into a furnace can cause a furnace explosion. Therefore, a flame sensor is used to shut off the fuel supply in case of flame failure. The flame detector of this device senses the presence or absence of flame through temperature, flame conductance of an electric current, or flame radiation. Through suitable electrical devices, the flame detector shuts down the burner when there is loss of flame.

Modern flame sensors of the flame conductance type and flame scanner type supervise the igniter (pilot burner) and the main burner. Thus, at light-off, a safe igniter flame must be established before fuel is admitted to the main burner.

Other Safety Devices. Safety devices to shut off the burner in case of hazardous conditions other than flame failure consist of simple sensors such as temperature elements, water level detectors, and pressure devices. These are all designed, installed, and interlocked to shut off the fuel oil and combustion air to the burner if any of the conditions listed occurs. All safety interlocks are arranged to prevent lighting the burner until satisfactory conditions have been established.

Flame Quality Controls

Butcher (1990) has applied the measured intensity of light being emitted from the flames of fixed-input oil burners to judge basic flame quality. Flame steady-state intensity is compared to a predetermined set-point value. Deviation of flame intensity from of a certain range indicates poor flame quality.

PROGRAMMING

Flame safety systems are interlocked and equipped with timing devices so that automatic lighting occurs in proper sequence with proper time intervals. In the larger units (size dependent on local authorities), these controls comprise a programmer and a flame-sensing device or flame detector.

The programmer maintains the proper safety sequence for lighting the burner, subject to supervision by the flame detector. When light-off is called for, by push button or by the combustion control, the programmer controls the sequence of operation as follows:

1. Starts the burner fan so that the furnace or combustion space, the gas passages of the heat exchanger, and the chimney connector are purged of any unburned combustible gases. This flow of air is maintained until the entire volume of combustion space, gas passages, and chimney connector has been changed at least several times. It is commonly called **prepurge**.
2. After the prepurge, the controller initiates operation of the burner igniter and allows a short time for the flame detector to prove satisfactory operation of the igniter. If the flame detector is not satisfied, the program controller switches back to its initial start position, and an integral switch opens to prevent further operation until the electrical circuit is manually reset.
3. If the igniter operation satisfies the flame detector, the program controller starts fuel flow to the burner. The flame detector is in control from this point on until the burner is shut off. If the flame detector is not satisfied with established stable burning during the start-up, it shuts down the burner and the program controller switches back to its initial position.

4. If the burner is shut off for any reason, the program controller causes the forced-draft fan to continue delivering air for a specific time to postpurge the unit of any atomized oil and combustible gases. This postpurge should amount to at least two air changes in the combustion area, gas passages, and chimney connector. (Some authorities disagree about the necessity of this function.)

Electronic programmers are available that, in addition to programming the entire boiler operating sequence, have data acquisition and storage capability to record operating conditions and flag the cause of burner shutdowns.

COMBUSTION CONTROL

Some modern gas- and oil-burning units are equipped with integrated devices that automatically regulate the flow of fuel and combustion air to satisfy the demand from the units. These combustion control systems consist of a master controller, which senses the heat demands through room temperature or steam pressure, etc., and controlled devices, which receive impulses from the master controller. The signals change the rate of heat input by starting or stopping motors, driving oil pumps and fans, or adjusting air dampers and oil valves. The signal may be electrical or, in the case of some larger boilers, compressed air. Descriptions of various types of combustion control systems follow.

An **on-off control** simply starts and stops the fuel burner to satisfy heat demand and does not control a combustion system's rate of heat input. The burner remains on until the controlled temperature of a room, the outlet temperature of a water heater, or the steam boiler pressure reaches a predetermined point. At a set maximum condition of temperature or steam pressure, the burner shuts off.

The burner ignites again at a predetermined condition of temperature, pressure, etc. On-off control is normal for small units with burners using No. 2 fuel oil at capacities up to about 2,120,000 Btu/h or 15 gph. These small burners have gas controls or a single motor to drive the burner fan or the fuel oil pump. Proportioning of combustion air to fuel is accomplished by the fan design and/or fixed position inlet air damper.

For gas burners, pressure atomizing oil burners, and other types of burners burning No. 2 or No. 4 oil at capacities of 2,120,000 to 4,240,000 Btu/h or 15 to 30 gph, the on-off control system usually has a low-start feature. This device starts the burner at less than full load (perhaps 60%) and then increases to full burning rate after several seconds. This type of control is called **low-high-off**.

A **low-high-low-off** control (often referred to as two-position firing) also provides a low-fire start. In addition, it cycles the burner back to low fire when the maximum control set point is reached.

In operation, the two-position control cycle has considerable advantage with a rapidly fluctuating load. However, for oil firing, combustion conditions are poor at the low-load rate, so this cycle is seldom recommended for cases of sustained low loads, such as heating loads. The two-position control cycle provides somewhat steadier conditions. A positioning motor, through a mechanical linkage, properly positions the oil flow control valve and the burner damper. This permits damper settings for proportioning air to fuel.

A **modulating control** regulates the firing rate to follow the load demands more closely than the on-off or low-high-low-off controls. The fuel oil control valve and the burner damper respond over a range of positions within the operating range capability of the burner. The simplest and least expensive modulating control systems, as frequently used for burner capacities of 4,240,000 to 70,000,000 Btu/h or 30 to 500 gph, use a positioning motor and mechanical linkage, which permit a large number of positions of the fuel control valve and the forced-draft damper. Most dampers larger than 70,000,000 Btu/h or 500 gph require more sophisticated controls that permit a continuous adjustment of valve and damper positions within the range of burner operation.

Modulating control systems that position the fuel oil control valve and the forced-draft damper in a predetermined relationship are called **parallel controls or positioning controls**. These valves maintain the fuel and air in proper proportion throughout the load range. Modulating controls for single-burner boilers usually include an on-off control for loads below the operating range of the burner.

A **full-metering and proportioning control** is a modulating control that maintains the proper ratio of air to fuel throughout the burner range with an air-fuel ratio controller. This system meters the airflow, usually by sensing the air pressure drop through the burner throat, and positions the air controller to suit momentary load demands. The fuel control valve immediately adjusts to match fuel flow to airflow. For some requirements, the ratio controller is reversed to match airflow to fuel flow. In unusual cases, the ratio controller is set to operate one way on load increase and the opposite way during load decrease. This control maintains optimum efficiency throughout the entire burner range, usually holding close to 15% excess air, as compared to 25% or greater with on-off controls.

Draft Controls

Combustion controls function properly with a reasonably constant building depressurization and furnace draft or pressure. The main concern is the varying draft effect of high chimneys. To control this draft, various regulators are used in the flue connections at the gas outlets of the units. Draft hoods or balanced draft dampers are used for smaller units, and draft damper controllers for large boilers. These regulators should be supplied as part of the combustion control equipment. See Chapter 30 for design considerations relating to vent and chimney draft.

REFERENCES

ANSI. 1971. Gas utilization equipment in large boilers. ANSI *Standard* Z83.3-71. Secretariat: CSA International, Cleveland, OH.

ANSI. 1990. Draft hoods. ANSI *Standard* Z21.12-90 (R 1998).

ANSI. 1991. Gas-fired low pressure steam and hot water boilers. ANSI *Standard* Z21.13-91 (R 1998).

ANSI. 1994. Installation of domestic gas conversion burners. ANSI *Standard* Z21.8-94.

ASTM. 1994. Test method for smoke density in flue gases from burning distillate fuels. ANSI/ASTM *Standard* D 2156-94. American Society for Testing and Materials, West Conshohocken, PA.

Butcher, T. 1990. Performance control strategies for oil-fired residential heating systems. BNL *Report* 52250. Brookhaven National Laboratory, Upton, NY.

Butcher, T., L.A. Fisher, B. Kamath, T. Kirchstetter, and J. Batey. 1994. Nitrogen oxides (NO_x) and oil burners. BNL *Report* 52430. Brookhaven National Laboratory, Upton, NY.

CSA International. 1996. Gas-fired duct furnaces and unit heaters. ANSI *Standard* Z83.8-96/CSA 2.6-M96. CSA International, Cleveland, OH.

CSA International. 1998. Domestic gas conversion burners. ANSI *Standard* Z21.17-98/CSA 2.7-M98.

CSA International. 1998. Gas-fired central furnaces. ANSI *Standard* Z21.47-98/CSA 2.3-M98.

Locklin, D.W. and H.R. Hazard. 1980. Technology for the development of high-efficiency oil-fired residential heating equipment. BNL *Report* 51325. Brookhaven National Laboratory, Upton, NY.

McDonald, R.J., ed. 1989-1998. Proceedings of the Oil Heat Technology Conferences and Workshops. BNL *Reports* 52217, 52284, 52340, 52392, 52430, 52475, 52506, 52537, and 52544. Brookhaven National Laboratory, Upton, NY.

NFPA. 1997. Installation of oil-burning equipment. ANSI/NFPA *Standard* 31-97. National Fire Protection Association, Quincy, MA.

Segeler, C.G., ed. 1965. *Gas engineers handbook*, Section 12, Chapter 2. Industrial Press, New York.

UL. 1994. Commercial-industrial gas-heating equipment. *Standard* 795-94. Underwriters Laboratories, Northbrook, IL.

BIBLIOGRAPHY

UL. 1994. Oil burners. ANSI/UL *Standard* 296-94. Underwriters Laboratories, Northbrook, IL.

UL. 1994. Oil-fired central furnaces. *Standard* 727-94.

UL. 1995. Oil-fired boiler assemblies. *Standard* 726-95.

CHAPTER 27

BOILERS

A BOILER is a pressure vessel designed to transfer heat (produced by combustion) to a fluid. The definition has been expanded to include transfer of heat from electrical resistance elements to the fluid or by direct action of electrodes on the fluid. In most boilers, the fluid is usually water in the form of liquid or steam. If the fluid being heated is air, the heat exchange device is called a furnace, not a boiler. The firebox, or combustion chamber, of some boilers is also called a furnace.

Excluding special and unusual fluids, materials, and methods, a boiler is a cast-iron, steel, aluminum, or copper pressure vessel heat exchanger designed to (1) burn fossil fuels (or use electric current) and (2) transfer the released heat to water (in water boilers) or to water and steam (in steam boilers). Boiler heating surface is the area of fluid-backed surface exposed to the products of combustion, or the fire-side surface. Various codes and standards define allowable heat transfer rates in terms of heating surface. Boiler designs provide for connections to a piping system, which delivers heated fluid to the point of use and returns the cooled fluid to the boiler.

Chapters 6, 10, 11, 12, and 14 cover applications of heating boilers. Chapter 7 discusses cogeneration, which may require boilers.

CLASSIFICATIONS

Boilers may be grouped into classes based on working pressure and temperature, fuel used, material of construction, type of draft (natural or mechanical), and whether they are condensing or noncondensing. They may also be classified according to shape and size, application (such as heating or process), and the state of the output medium (steam or water). Boiler classifications are important to the specifying engineer because they affect performance, first cost, and space requirements. Excluding designed-to-order boilers, significant class descriptions are given in boiler catalogs or are available from the boiler manufacturer. The following basic classifications may be helpful.

Working Pressure and Temperature

With few exceptions, boilers are constructed to meet ASME *Boiler and Pressure Vessel Code*, Section IV (SCIV), Rules for Construction of Heating Boilers (low-pressure boilers), or Section I (SCI), Rules for Construction of Power Boilers (high-pressure boilers).

Low-pressure boilers are constructed for maximum working pressures of 15 psig steam and up to 160 psig hot water. Hot water boilers are limited to 250°F operating temperature. Operating and safety controls and relief valves, which limit temperature and pressure, are ancillary devices required to protect the boiler and prevent operation beyond design limits.

High-pressure boilers are designed to operate above 15 psig steam, or above 160 psig and/or 250°F for water boilers. Similarly, operating and safety controls and relief valves are required.

The preparation of this chapter is assigned to TC 6.1, Hydronic and Steam Heating Equipment and Systems.

Steam boilers are generally available in standard sizes up to and above 100,000 lb steam/h (60,000 to over 100,000,000 Btu/h), many of which are used for space heating applications in both new and existing systems. On larger installations, they may also provide steam for auxiliary uses, such as hot water heat exchangers, absorption cooling, laundry, and sterilizers. In addition, many steam boilers provide steam at various temperatures and pressures for a wide variety of industrial processes.

Water boilers are generally available in standard sizes from 35,000 to over 100,000,000 Btu/h, many of which are in the low-pressure class and are used primarily for space heating applications in both new and existing systems. Some water boilers may be equipped with either internal or external heat exchangers for domestic water service.

Every steam or water boiler is rated for a maximum working pressure that is determined by the applicable boiler code under which it is constructed and tested. When installed, it also must be equipped at a minimum with operation and safety controls and pressure/temperature-relief devices mandated by such codes.

Fuel Used

Boilers may be designed to burn coal, wood, various grades of fuel oil, waste oil, various types of fuel gas, or to operate as electric boilers. A boiler designed for one specific fuel type may not be convertible to another type of fuel. Some boilers can be adapted to burn coal, oil, or gas. Several designs accommodate firing oil or gas, and other designs permit firing dual-fuel burning equipment. Accommodating various fuel burning equipment is a fundamental concern of boiler manufacturers, who can furnish details to a specifying engineer. The manufacturer is responsible for performance and rating according to the code or standard for the fuel used (see section on Performance Codes and Standards).

Construction Materials

Most noncondensing boilers are made with cast iron sections or steel. Some small boilers are made of copper or copper-clad steel. Condensing boilers are typically made of stainless steel or aluminum.

Cast-iron sectional boilers generally are designed according to ASME SCIV requirements and range in size from 35,000 to 13,975,000 Btu/h gross output. They are constructed of individually cast sections, assembled into blocks (assemblies) of sections. Push or screw nipples, gaskets, and/or an external header join the sections pressure-tight and provide passages for the water, steam, and products of combustion. The number of sections assembled determines the boiler size and energy rating. Sections may be vertical or horizontal, the vertical design being more common (Figure 1A and Figure 1C).

The boiler may be **dry-base** (the combustion chamber is beneath the fluid backed sections), as in Figure 1B; **wet-base** (the combustion chamber is surrounded by fluid-backed sections, except for necessary openings), as in Figure 2A; or **wet-leg** (the combustion

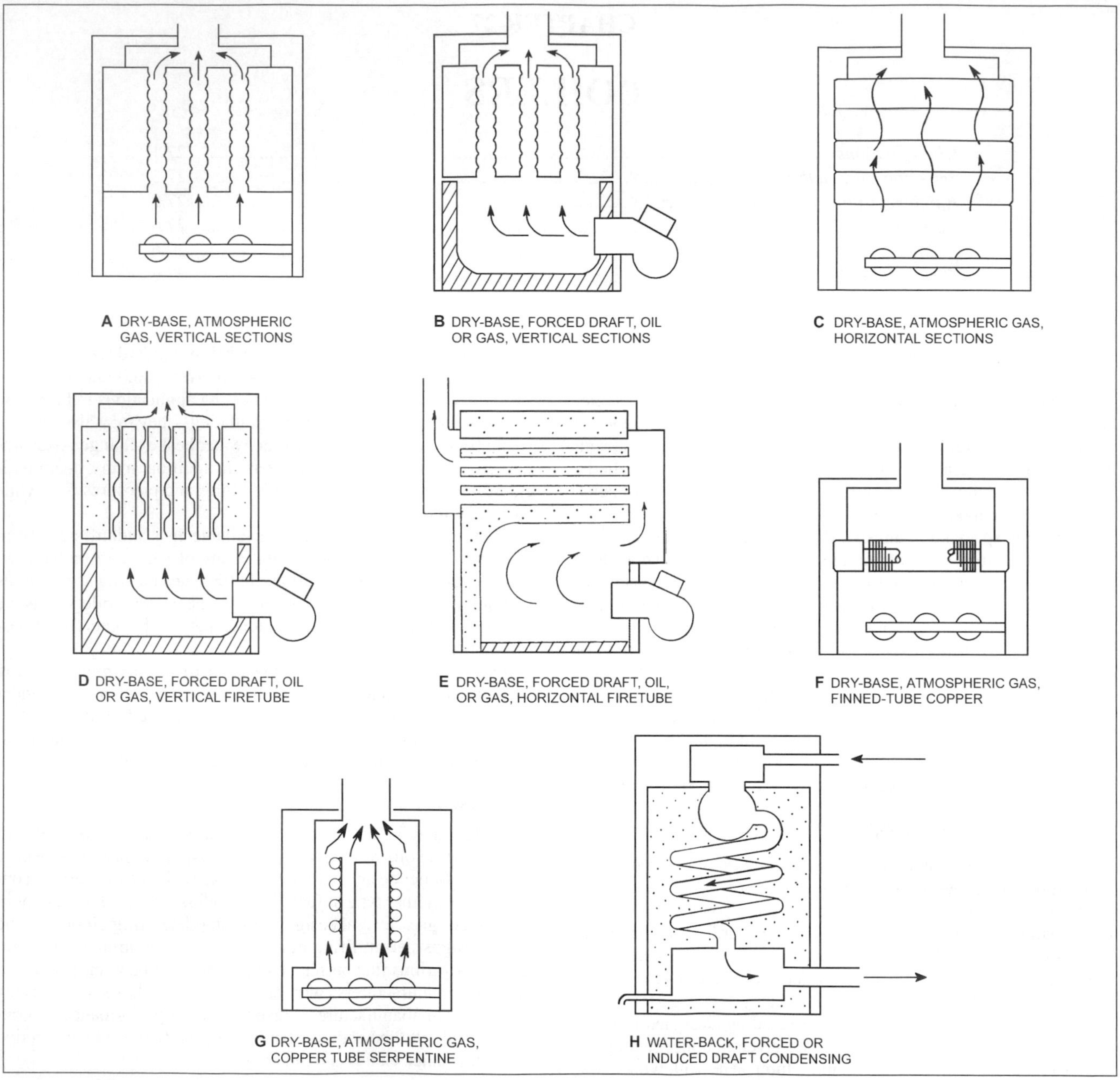

Fig. 1 Residential Boilers

chamber top and sides are enclosed by fluid-backed sections), as in Figure 2B.

The three types of boilers can be designed to be equally efficient. Testing and rating standards apply equally to all three types. The wet-base design is easiest to adapt for combustible floor installations. Applicable codes usually demand a floor temperature under the boiler no higher than 90°F above room temperature. A steam boiler at 215°F or a water boiler at 240°F may not meet this requirement without appropriate floor insulation. Large cast-iron boilers are also made as water-tube units with external headers (Figure 2C).

Steel boilers generally range in size from 50,000 Btu/h to the largest boilers made. Designs are constructed to either ASME SCI or SCIV (or other applicable code) requirements. They are fabricated into one assembly of a given size and rating, usually by welding. The heat exchange surface past the combustion chamber is usually an assembly of vertical, horizontal, or slanted tubes. Boilers of the fire-tube design contain flue gases in tubes completely submerged in fluid (Figure 1D and Figure 1E show residential units, and Figure 3A through Figure 3D and Figure 4A show commercial units). Water-tube boilers contain fluid inside tubes with tube pattern arrangement providing for the combustion chamber (Figure 4C and Figure 4D). The internal configuration may accommodate one or more flue gas passes. As with cast-iron sectional boilers, dry-base, wet-leg, or wet-base designs may be used. Most small steel boilers are of the dry-base, vertical fire-tube type (Figure 1D).

Larger boilers usually incorporate horizontal or slanted tubes; both fire-tube and water-tube designs are used. A popular horizontal fire-tube design for medium and large steel boilers is the **scotch marine**, which is characterized by a central fluid-backed cylindrical combustion chamber, surrounded by fire-tubes accommodating two or more flue gas passes, all within an outer shell (Figure 3A through

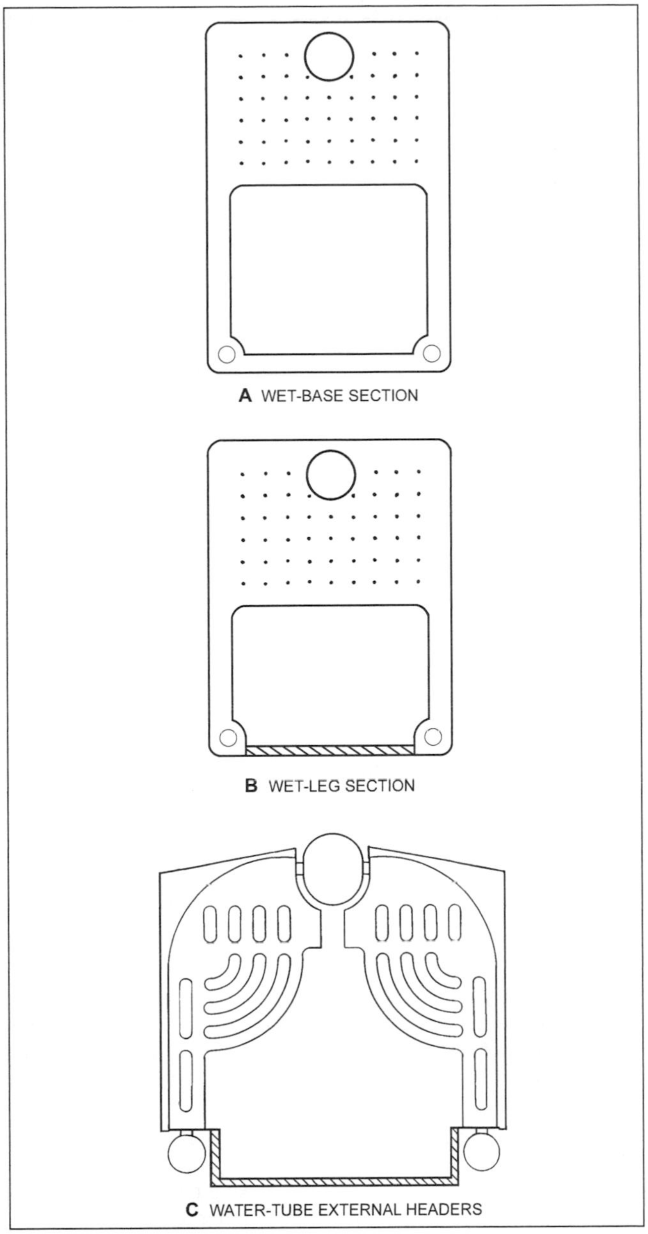

Fig. 2 Cast-Iron Commercial Boilers

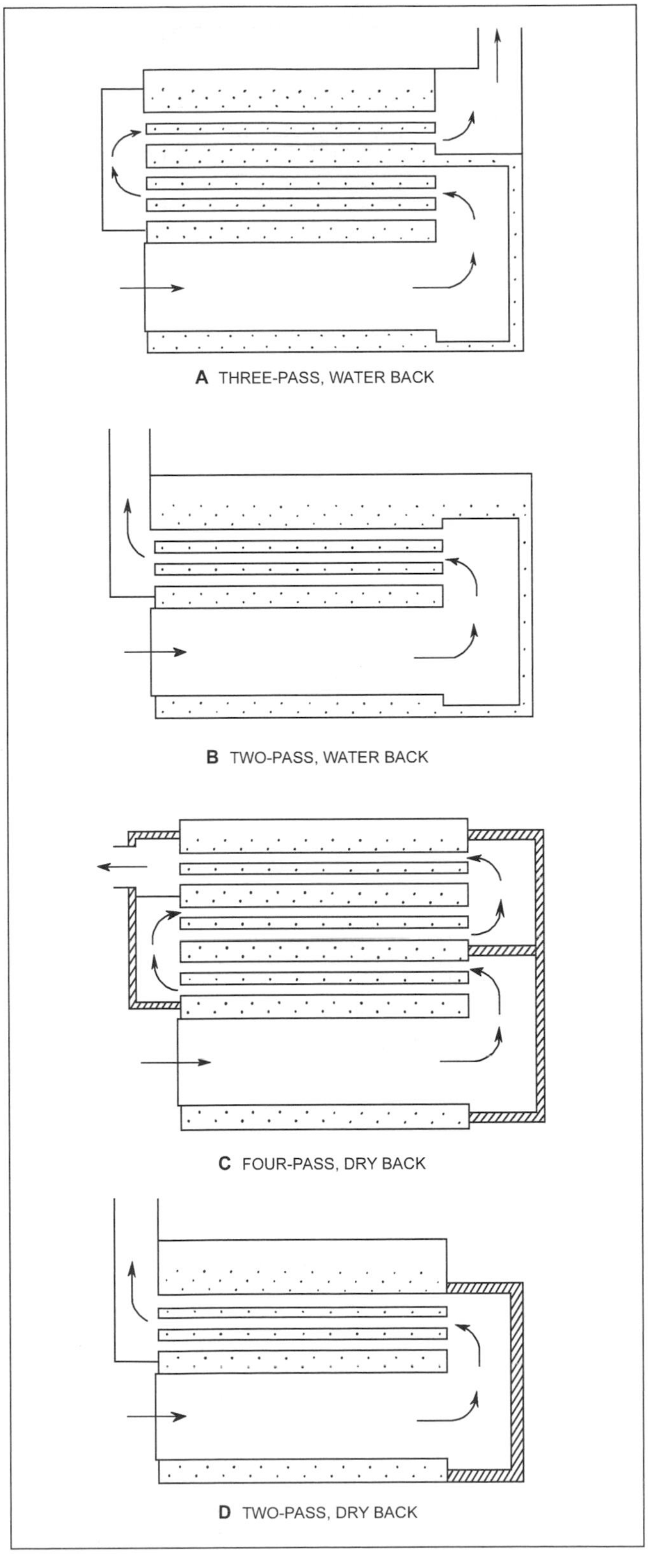

Fig. 3 Scotch Marine Commercial Boilers

Figure 3D). In another horizontal fire-tube design, the combustion chamber has a similar central fluid-backed combustion chamber surrounded by fire tubes accommodating two or more flue gas passes, all within an outer shell. However, this design uses a dry base and wet-leg (or mud leg) (Figure 4A).

Copper boilers are usually some variation of the water-tube boiler. Parallel finned copper tube coils with headers, and serpentine copper tube units are most common (Figure 1F and Figure 1G). Some are offered as wall-hung residential boilers. The commercial bent water-tube design is shown in Figure 4B. Natural gas is the usual fuel for copper boilers.

Stainless steel boilers usually are designed to operate with condensing flue gases. They are often limited to operating temperatures of 210°F or less to avoid problems that might result from stress cracking.

Aluminum boilers are also usually designed to operate with condensing flue gases. Typical designs incorporate either cast aluminum boiler sections or integrally finned aluminum tubing.

Type of Draft

Draft is the pressure difference that causes air and/or fuel to flow through a boiler or chimney. A **natural draft boiler** is designed to operate with a negative pressure in the combustion chamber and in the flue connection. The pressure difference is created by the

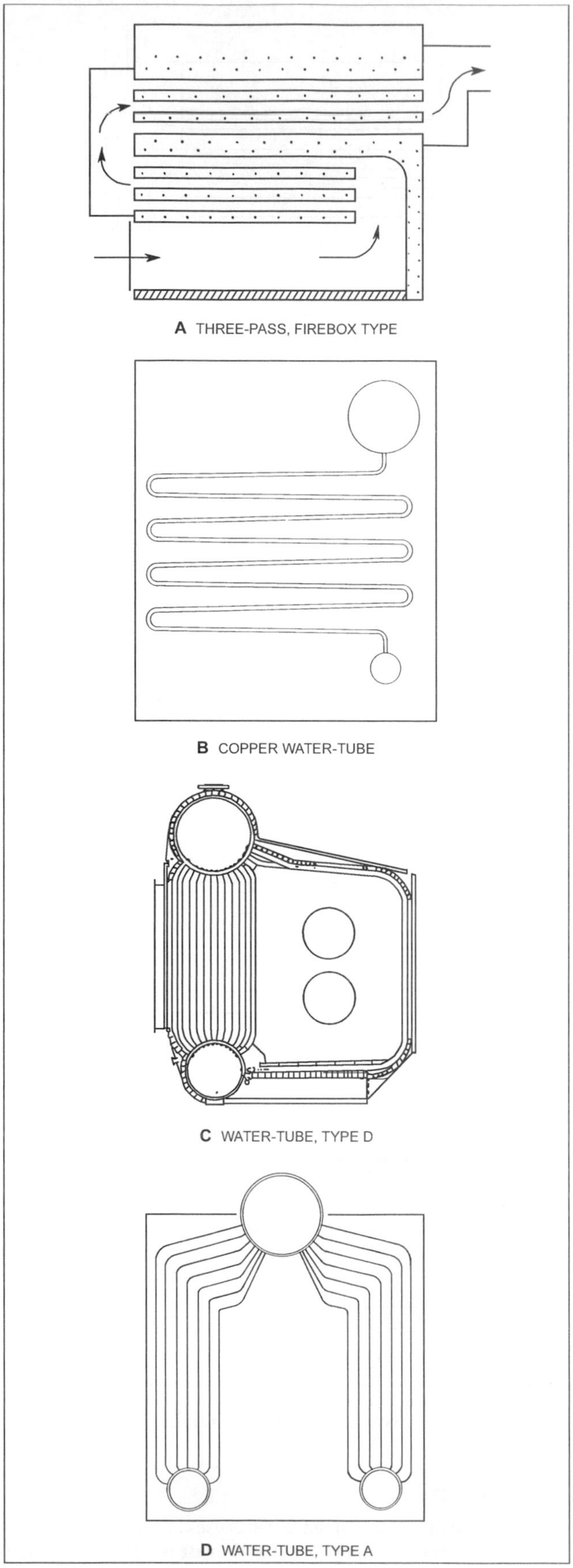

Fig. 4 Commercial Fire-Tube and Water-Tube Boilers

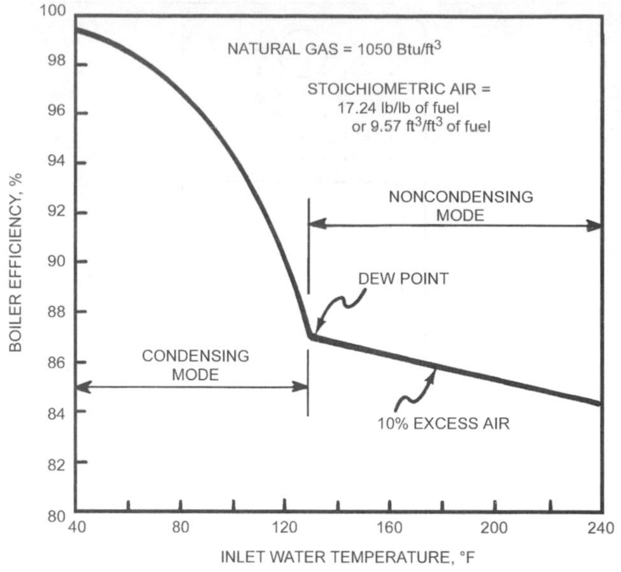

Fig. 5 Effect of Inlet Water Temperature on Efficiency of Condensing Boilers

tendency of hot gases to rise up a chimney or by the height of the boiler up to the draft control device. In a **mechanical draft boiler**, a fan or blower or other machinery creates the required pressure difference. These boilers may be either forced draft or induced draft. In a **forced draft boiler**, air is forced into the combustion chamber to maintain a positive pressure in the combustion chamber and/or the space between the tubing and the jacket (breaching). In an **induced draft boiler**, air is drawn into the combustion chamber to maintain a negative pressure in the combustion chamber.

Condensing or Noncondensing

Until recently, boilers were designed to operate without condensing the flue gas in the boiler. This precaution was necessary to prevent corrosion of cast-iron or steel parts. Hot water units were often operated at 140°F minimum return water temperature to prevent rusting when natural gas was used.

Because a higher boiler efficiency can be achieved with a lower water temperature, the condensing boiler allows the flue gas water vapor to condense and drain. Full condensing boilers are unique in design and may require a very low inlet water temperature, corrosion-resistant materials, chemical treatment of the condensate, and an elevated fresh water makeup for proper operation. Figure 5 shows a typical relationship of overall condensing boiler efficiency to return water temperature. The dew point of 130°F shown in the figure varies with the percentage of hydrogen in the fuel and oxygen-carbon dioxide ratio, or excess air, in the flue gases. A condensing boiler is shown in Figure 1H. Condensing boilers can be of the fire-tube, water-tube, or cast aluminum sectional design.

Condensing boilers with low return water temperatures are very efficient at part-load operation when a high water temperature is not required. For example, a natural gas water heater operating with 80°F return water has a potential overall boiler efficiency of 97% at the conditions shown in Figure 5.

Figure 6 shows how dew point varies with a change in the percentages of oxygen/carbon dioxide for natural gas. Boilers that operate with a combustion efficiency and oxygen and carbon dioxide concentrations in the flue gas such that the flue gas temperature falls between the dew point and the dew point plus 140°F should be avoided, unless the venting is designed for condensation. This temperature typically occurs with boilers operating between 83 and 87% efficiency and the flue gas has an oxygen concentration of 7 to

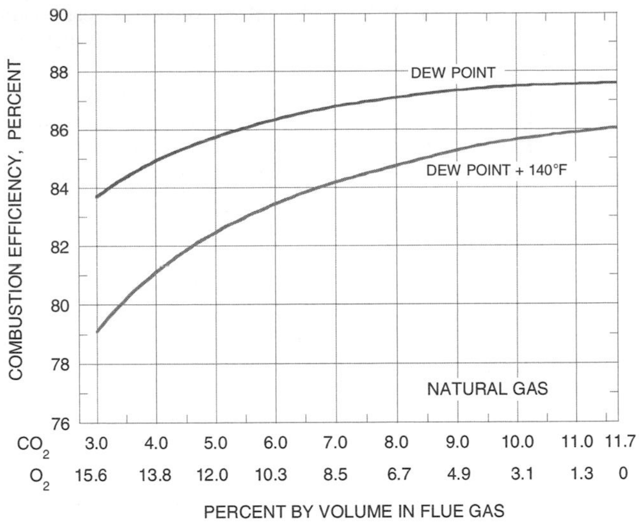

Fig. 6 Relationship of Dew Point, Carbon Dioxide, and Combustion Efficiency for Natural Gas

10% and the carbon dioxide is 6 to 8%. Chapter 30 gives further details on chimneys.

The condensing portion of these boilers requires special material to resist the corrosive effects of the condensing flue gases. Cast iron and carbon steel are not suitable materials for the condensing section of a boiler. Certain stainless steels and aluminum alloys, however, are suitable. Commercial boiler installations can be adapted to condensing operation by adding a condensing heat exchanger in the flue gas vent.

Heat exchangers in the flue gas venting require a condensing medium such as (1) low pressure steam condensate or hot water return, (2) domestic water service, (3) fresh water makeup, or (4) other fluid sources in the 70 to 130°F range. The medium can also be a source of heat recovery in HVAC systems.

Wall Hung Boilers

Wall hung boilers are a type of small residential gas fired boiler developed to conserve space in buildings such as apartments and condominiums. These boilers are popular in Europe. The most common designs are mounted on outside walls. Combustion air enters through a pipe from the outdoors, and flue products are vented directly through another pipe to the outdoors. In some cases, the air intake pipe and vent pipe are concentric. Other designs mount adjacent to a chimney for venting and use indoor air for combustion. These units may be condensing or noncondensing. As these boilers are typically installed in the living space, provisions for proper venting and combustion air supply are very important.

Integrated (Combination) Boilers

Integrated boilers are relatively small, residential boilers that combine space heating and water heating in one appliance. They are usually wall mounted but may also be floor standing. They operate primarily on natural gas and are practical to install and operate. The most common designs have an additional heat exchanger and a storage tank to provide domestic hot water. Some designs (particularly European) do not have a storage tank. Instead they use a larger heat exchanger and the appropriate burner input to provide instantaneous domestic hot water.

Electric Boilers

Electric boilers are a separate class of boiler. Because no combustion occurs, a boiler heating surface and flue gas venting are unnecessary. The heating surface is the surface of the electric elements or electrodes immersed in the boiler water. The design of electric boilers is largely determined by the shape and heat release rate of the electric elements used. Electric boiler manufacturers' literature describes available size, shapes, voltages, ratings, and methods of control.

SELECTION PARAMETERS

Boiler selection should be based on a competent review of the following parameters:

All Boilers

- Applicable code under which the boiler is constructed and tested
- Gross boiler heat output
- Total heat transfer surface area
- Water content weight or volume
- Auxiliary power requirement
- Cleaning and service access provisions for fireside and waterside heat transfer surfaces
- Part-load and full-load efficiency
- Space requirement and piping arrangement
- Water treatment requirement
- Operating personnel capabilities and maintenance/operation requirements
- Regulatory requirements for emissions, fuel usage/storage

Fuel-Fired Boilers

- Combustion chamber (furnace volume)
- Internal flow pattern of combustion products
- Combustion air and venting requirements
- Fuel availability/capability

Steam Boilers

- Steam quality

The codes and standards outlined in the section on Performance Codes and Standards include requirements for minimum efficiency, maximum temperature, burner operating characteristics, and safety control. Test agency certification and labeling, which are published in boiler manufacturers' catalogs and shown on boiler rating plates, are generally sufficient for determining boiler steady-state operating characteristics. However, for commercial and industrial boilers, these ratings typically do not consider part-load or seasonal efficiency, which is less than steady-state efficiency. Some boilers are not tested and rated by a recognized agency, and, therefore, do not bear the label of an agency. Nonrated boilers (rated and warranted only by the manufacturer) are used when jurisdictional codes or standards do not require a rating agency label. As previously indicated, almost without exception, both rated and nonrated boilers are of ASME Code construction and are marked accordingly.

EFFICIENCY: INPUT AND OUTPUT RATINGS

The efficiency of fuel-burning boilers is defined in three ways: combustion efficiency, overall efficiency, and seasonal efficiency.

Combustion efficiency is input minus stack (flue gas outlet) loss, divided by input, and generally ranges from 75 to 86% for most noncondensing boilers. Condensing boilers generally operate in the range of 88 to 95% combustion efficiency.

Overall efficiency is gross energy output divided by energy input. Gross output is measured in the steam or water leaving the boiler and depends on the characteristics of the individual installation. Overall efficiency of electric boilers is generally 92 to 96%. Overall efficiency is lower than combustion efficiency by the percentage of heat lost from the outside surface of the boiler (radiation loss or jacket loss) and by off-cycle energy losses (for applications where the boiler cycles on and off). Overall efficiency can be precisely determined only under controlled laboratory test conditions, directly measuring the fuel input and the heat absorbed by the water or steam of the boiler. Precise efficiency measurements are

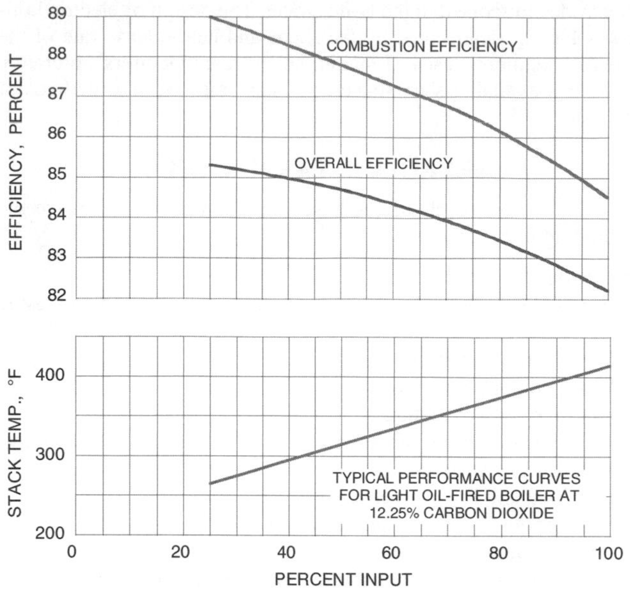

Fig. 7 Boiler Efficiency as Function of Fuel and Air Input

generally not performed under field conditions because of the inability to control the required parameters and the high cost involved in performing such an analysis. An approximate combustion efficiency for noncondensing boilers can be determined under any operating condition by measuring flue-gas temperature and percentage of CO_2 or O_2 in the flue gas and by consulting a chart or table for the fuel being used. The approximate combustion efficiency of a condensing boiler must include the energy transferred by condensation in the flue gas.

Seasonal efficiency is the actual operating efficiency that the boiler will achieve during the heating season at various loads. Because most heating boilers operate at part load, the part-load efficiency, including heat losses when the boiler is off, has a great effect on the seasonal efficiency. The difference in seasonal efficiency between a boiler with an on-off firing rate and one with modulating firing rate can be appreciable if the airflow through the boiler is modulated along with the fuel input. Figure 7 shows how efficiency increases at part load for a typical boiler equipped with a burner that can fire at reduced inputs while modulating both fuel and air. This increase in efficiency is due to the increase in the ratio of heat exchanger surface area to heat input as the firing rate is reduced.

PERFORMANCE CODES AND STANDARDS

Commercial heating boilers (i.e., boilers with inputs of 300,000 Btu/h and larger) at present are only tested for full-load steady-state efficiency according to standards developed by either (1) the Hydronics Institute Division of the Gas Appliance Manufacturers Association (GAMA) [formerly the Institute of Boiler and Radiator Manufacturers (I-B-R) and the Steel Boiler Institute (SBI)], (2) the American Gas Association (AGA), or (3) Underwriters Laboratories (UL).

The Hydronics Institute standard (1990) for rating cast-iron sectional, steel, and copper boilers bases performance on controlled test conditions for fuel inputs of 300,000 Btu/h and larger. The gross output obtained by the test is limited by such factors as flue gas temperature, draft, CO_2 in the flue gas, and minimum overall efficiency. This standard applies primarily to oil-fired equipment; however, it is also applied to forced draft gas fired or dual-fueled units.

Gas boilers are generally design-certified by an accredited testing laboratory based on tests conducted in accordance with ANSI *Standard* Z21.13 or UL *Standard* 795.

Instead of the HI-GAMA, AGA, and UL standards, test procedures for commercial-industrial and packaged fire-tube boilers are often performed based on ASME *Performance Test Code* 4.1 (1991). Units are tested for performance under controlled test conditions with minimum required levels of efficiency. Further, the American Boiler Manufacturers Association (ABMA) publishes several guidelines for the care and operation of commercial and industrial boilers and for control parameters.

Residential heating boilers (i.e., all gas- and oil-fired boilers with inputs less than 300,000 Btu/h in the United States) are rated according to standards developed by the U.S. Department of Energy (DOE). The procedure determines both on-cycle and off-cycle losses based on a laboratory test. The test results are applied to a computer program, which simulates an installation and predicts an annual fuel utilization efficiency (AFUE). The steady-state efficiency developed during the test is similar to combustion efficiency and is the basis for determining DOE **heating capacity**, a term similar to gross output. The AFUE represents the part-load efficiency at the average outdoor temperature and load for a typical boiler installed in the United States. Although this value is useful for comparing different boiler models, it is not meant to represent actual efficiency for a specific installation.

SIZING

Boiler sizing is the selection of boiler output capacity to meet connected load. The boiler gross output is the rate of heat delivered by the boiler to the system under continuous firing at rated input. Net rating (I-B-R rating) is gross output minus a fixed percentage (called the piping and pickup factor) to allow for an estimated average piping heat loss, plus an added load for initially heating up the water in a system (sometimes called **pickup**). This I-B-R piping and pickup factor is 1.15 for water boilers and ranges from 1.27 to 1.33 for steam boilers, with the smaller number applying as the boilers get larger. The net rating is calculated by dividing the gross output by the appropriate piping and pickup factor.

Piping loss is variable. If all piping is in the space defined as load, loss is zero. If piping runs through unheated spaces, heat loss from the piping may be much higher than accounted for by the fixed net rating factor. Pickup is also variable. When the actual connected load is less than design load, the pickup factor may be unnecessary.

On the coldest day, extra output (boiler and radiation) is needed to pickup the load from a shutdown or low night setback. If night setback is not used, or if no extended shutdown occurs, no pickup load exists. Standby capacity for pickup, if needed, can be in the form of excess capacity in baseload boilers or in a standby boiler.

If piping and pickup losses are negligible, the boiler gross output can be considered the design load. If piping loss and pickup load are large or variable, those loads should be calculated and equivalent gross boiler capacity added. Boiler capacity must be matched to the terminal unit and system delivery capacity. That is, if the boiler output is greater than the terminal output, the water temperature will rise and the boiler will cycle on the high-limit control, delivering an average input that is much lower than the boiler gross output. Significant oversizing of the boiler may result in a much lower overall boiler efficiency.

CONTROL OF INPUT AND OUTPUT

Boiler controls regulate the rate of fuel input in response to a signal representing load change, or demand, so that the average boiler output equals the load within some accepted tolerance. Boiler controls include operating and safety controls that shut off fuel flow when operating parameters are met and when unsafe high-limit conditions develop.

Operating Controls

Steam boilers are operated by boiler-mounted, pressure-actuated controls, which vary the input of fuel to the boiler. Common examples of burner controls are on-off, high-low-off, and modulating. Modulating controls infinitely vary the fuel input from 100% down to a selected minimum set point. The ratio of maximum to minimum is the **turndown ratio**. The minimum input is usually between 5 and 33% (i.e., 20 to 1 down to 3 to 1 ratios), and input depends on the size and type of fuel-burning equipment and system. High turndown ratios in noncondensing boilers must be considered carefully in order to prevent condensation at the lower firing rates.

Hot-water boilers are operated by temperature-actuated controls that are usually mounted on the boiler. Burner controls are the same as for steam boilers, i.e. on-off, high-low-off and modulating. Modulating controls typically offer more precise water temperature control and higher efficiency than on-off, or high-low controls, if the airflow through the boiler is modulated along with the input of fuel.

Boiler reset controls can enhance the efficiency of hot water boilers. These controls may operate with any of the burner controls mentioned previously. They automatically change the high-limit set point of the boiler to match the variable building load demands caused by changing outdoor temperatures. By keeping boiler water temperature as low as possible, efficiency is enhanced and standby losses are reduced.

The boiler manufacturer's instructions should be followed if a low operating temperature or frequent cold starts are anticipated. Thermal shock, which is caused by cold water returning to the boiler and impinging on the hot surfaces inside the boiler, should be avoided with steel fire-tube boilers. Many steel boiler manufacturers suggest special piping arrangements, use of boiler water blend pumps, and/or primary/secondary loop controls to prevent thermal shock. Commercial bent water-tube boilers generally can withstand relatively cold water. However, as with any steel boiler, possible corrosion of boiler tubes, breaching, and chimney caused by condensation should be considered.

Basic control designs and diagrams are furnished by boiler manufacturers. Controls diagrams and/or specifications to meet special safety requirements are available from insurance agencies, governmental bodies, and ASME. Such special requirements usually apply to boilers with inputs of 400,000 Btu/h or greater, but they may also apply to smaller sizes. These special requirements may include (but are not limited to) low water cutoff, manual reset high limits, mechanical draft fans, draft control devices, air proving switches, air dampers, and redundant safety and operating controls. The boiler manufacturer may furnish installed and prewired boiler controls with diagrams when the requirements are known.

When a boiler installation code applies, the details of control requirements may be specified by the inspector after boiler installation. Special controls must then be supplied and the control design completed per the inspector's instructions. It is essential that the heating system designer or specifying engineer determine the applicable codes, establish sources of necessary controls, and provide the skills needed to complete the control system. Local organizations are often available to provide these services.

REFERENCES

ABMA. 1999. Packaged firetube ratings. American Boiler Manufacturers Association, Arlington, VA.

ANSI. 1991. Gas-fired low-pressure steam and hot water boilers. *Standard* Z21.13-1991. Secretariat: CSA International (formerly American Gas Association Laboratories), Cleveland, OH.

ASME. 1991. Steam generating units. PTC 4.1. American Society of Mechanical Engineers, New York.

ASME. 1998. Controls and safety devices for automatically fired boilers, *Standard* CSD-1-1998.

ASME. 1998. Rules for construction of power boilers. *Boiler and Pressure Vessel Code*, Section I-1998.

ASME. 1998. Rules for construction of heating boilers. *Boiler and Pressure Vessel Code*, Section IV-1998.

ASME. 1998. Recommended rules for the care and operation of heating boilers. *Boiler and Pressure Vessel Code*, Section VI-1998.

ASME. 1998. Recommended rules for the care of power of boilers. *Boiler and Pressure Vessel Code*, Section VII-1998.

Hydronics Institute. 1990. Testing and rating standard for heating boilers, 6th ed. Hydronics Institute, Berkeley Heights, NJ.

UL. 1994. Commercial-industrial gas heating equipment. *Standard* 795-94. Underwriters Laboratories, Northbrook, IL.

BIBLIOGRAPHY

Strehlow, R.A. 1984. *Combustion fundamentals*. McGraw-Hill, New York.

Woodruff, E.B., H.B. Lammers, and T.F. Lammers. 1984. *Steam-plant operation*, 5th ed. McGraw-Hill, New York.

Operating Controls

CHAPTER 28

FURNACES

RESIDENTIAL FURNACES

RESIDENTIAL furnaces are available in a variety of self-enclosed appliances that provide heated air through ductwork to the space being heated. There are two types of furnaces: (1) fuel-burning furnaces and (2) electric furnaces.

Fuel-Burning Furnaces. Combustion takes place within a combustion chamber. Circulating air passes over the outside surfaces of a heat exchanger such that it does not contact the fuel or the products of combustion, which are passed to the outside atmosphere through a vent.

Electric Furnaces. A resistance-type heating element heats the circulating air either directly or through a metal sheath enclosing the resistance element.

Residential furnaces may be further categorized by (1) type of fuel, (2) mounting arrangement, (3) airflow direction, (4) combustion system, and (5) installation location.

NATURAL GAS FURNACES

Natural gas is the most common fuel supplied for residential heating, and the central system forced-air furnace, such as that shown in Figure 1, is the most common way of heating with natural gas. This type of furnace is equipped with a blower to circulate air

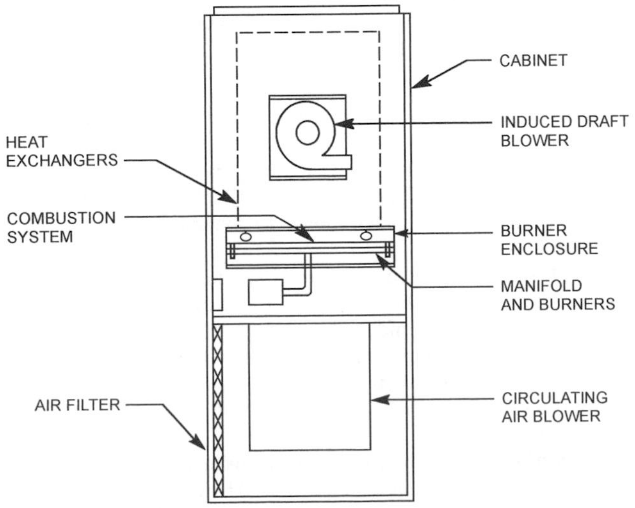

Fig. 1 Induced-Draft Gas Furnace

The preparation of this chapter is assigned to TC 6.3, Central Forced Air Heating and Cooling Systems.

through the furnace enclosure, over the heat exchanger, and through the ductwork distribution system. A typical furnace consists of the following basic components: (1) a cabinet or casing; (2) heat exchangers; (3) a combustion system including burners and controls; (4) a forced-draft blower, induced-draft blower, or draft hood; (5) a circulating air blower and motor; and (6) an air filter and other accessories such as a humidifier, an electronic air cleaner, an air-conditioning coil, or a combination of these elements.

Casing or Cabinet

The furnace casing is most commonly formed from painted cold-rolled steel. Access panels on the front of the furnace allow access to those sections requiring service. The inside of the casing adjacent to the heat exchanger is lined with a foil-faced blanket insulation and/or a metal radiation shield to reduce heat losses through the casing and to limit the outside surface temperature of the furnace. On some furnaces, the inside of the blower compartment is lined with insulation to acoustically dampen the blower noise.

Heat Exchangers

Heat exchangers are normally made of mirror-image formed parts that are joined together to form a clam shell. Heat exchangers made of finless tubes bent into a compact form are also found in some furnaces. Standard indoor furnaces are generally made of cold-rolled steel. If the furnace is exposed to clean air and the heat exchanger remains dry, this material has a long life and does not easily corrode. Some problems of heat exchanger corrosion and failure have been encountered because of exposure to halogen ions in the flue gas. These problems were caused by combustion air contaminated by substances such as laundry bleach, cleaning solvents, and halogenated hydrocarbon refrigerants.

Coated or alloy steel is used in top-of-the-line models and in furnaces for special applications. Common corrosion-resistant materials include aluminized steel, ceramic-coated cold-rolled steel, and stainless steel. Furnaces certified for use downstream of a cooling coil must have corrosion-resistant heat exchangers.

Research has been done on corrosion-resistant materials for use in condensing (secondary) heat exchangers (Stickford et al. 1985). The presence of chloride compounds in the condensate can cause a condensing heat exchanger to fail, unless a corrosion-resistant material is used.

Several manufacturers produce liquid-to-air heat exchangers in which a liquid is heated and is either evaporated or pumped to a condenser section or fan-coil, which heats circulating air.

Burners and Internal Controls

Burners are most frequently made of stamped sheet metal, although cast iron is also used. Fabricated sheet metal burners may be made from cold-rolled steel coated with high-temperature paint or from a corrosion-resistant material such as stainless or aluminized steel. Burner material must meet the corrosion protection requirements of the specific application. Gas furnace burners may be of either the monoport or multiport type; the type used with a particular furnace depends on compatibility with the heat exchanger.

Furnace controls include an ignition device, a gas valve, a fan control, a limit switch, and other components specified by the manufacturer. These controls allow gas to flow to the burners when heat is required. The four most common ignition systems are (1) standing pilot, (2) intermittent pilot, (3) direct spark, and (4) hot surface ignition (ignites either a pilot or the main burners directly). The section on Technical Data has further details on the function and performance of individual control components.

Venting Components

Natural-draft indoor furnaces are equipped with a **draft hood** connecting the heat exchanger flue gas exit to the vent pipe or chimney. The draft hood has a relief air opening large enough to ensure that the exit of the heat exchanger is always at atmospheric pressure. One purpose of the draft hood is to make certain that the natural-draft furnace continues to operate safely without generating carbon monoxide if the chimney is blocked, if there is a downdraft, or if there is excessive updraft. Another purpose is to maintain constant pressure on the combustion system. Residential furnaces built since 1987 are equipped with a blocked vent shutoff switch to shut down the furnace in case the vent becomes blocked.

Fan-assisted combustion furnaces use a small blower to force or induce the flue products through the furnace. Induced-draft furnaces may or may not have a relief air opening, but they meet the same safety requirements regardless.

Research into common venting of natural-draft appliances (water heaters) and fan-assisted combustion furnaces shows that nonpositive vent pressure systems may operate on a common vent. Refer to manufacturers' instructions for specific information.

Direct vent furnaces use outdoor air for combustion. Outdoor air is supplied to the furnace combustion chamber by direct connections between the furnace and the outdoor air. If the vent or the combustion air supply becomes blocked, the furnace control system will shut down the furnace.

ANSI *Standard* Z21.47/CSA 2.3 classifies venting systems. Central furnaces are categorized by temperature and pressure attained in the vent and by the steady-state efficiency attained by the furnace. While ANSI *Standard* Z21.47/CSA 2.3 uses 83% as the steady-state efficiency dividing furnace categories, a general rule of thumb is as follows:

Category I—A central furnace that operates with a nonpositive vent pressure and a flue loss no less than 17%.

Category II—A central furnace that operates with a nonpositive vent pressure and a flue loss less than 17%.

Category III—A central furnace that operates with a positive vent pressure and a flue loss no less than 17%.

Category IV—A central furnace that operates with a positive vent pressure and a flue loss less than 17%.

Furnaces rated in accordance with ANSI *Standard* Z21.47/CSA 2.3 that are not direct vent are marked to show that they are in one of the four venting categories listed here.

Blowers and Motors

Centrifugal blowers with forward-curved blades of the double-inlet type are used in most forced-air furnaces. These blowers overcome the resistance of the furnace air passageways, filters, and ductwork. They are usually sized to provide the additional air requirement for cooling and the static pressure required for the cooling coil. The blower may be a direct-drive type, with the blower wheel attached directly to the motor shaft, or it may be a belt-drive type, with a pulley and V-belt used to drive the blower wheel.

Electric motors used to drive furnace blowers are usually custom designed for each furnace model or model series. Direct-drive motors may be of the shaded pole or permanent split-capacitor type. Speed variation may be obtained by taps connected to extra windings in the motor. Belt-drive blower motors are normally split-phase or capacitor-start. The speed of belt-drive blowers is controlled by adjusting a variable-pitch drive pulley.

Electronically controlled, variable-speed motors are also available. This type of motor reduces electrical consumption when operated at low speeds.

Air Filters

An air filter in a forced-air furnace removes dust from the air that could reduce the effectiveness of the blower and heat exchanger(s). Filters installed in a forced-air furnace are often disposable. Permanent filters that may be washed or vacuum cleaned and reinstalled are also used. The filter is always located in the circulating airstream ahead of the blower and heat exchanger.

Accessories

Humidifiers. These are not included as a standard part of the furnace package. However, one advantage of a forced-air heating system is that it offers the opportunity to control the relative humidity of the heated space at a comfortable level. Chapter 20 addresses various types of humidifiers used with forced-air furnaces.

Electronic Air Cleaners. These air cleaners are much more effective than the air filter provided with the furnace, and they filter out much finer particles, including smoke and pollen. Electronic air cleaners create an electric field of high-voltage direct current in which dust particles are given a charge and collected on a plate having the opposite charge. The collected material is then cleaned periodically from the collector plate by the homeowner. Electronic air cleaners are mounted in the airstream entering the furnace. Chapter 24 has detailed information on filters.

Automatic Vent Dampers. This device closes the vent opening on a draft hood-equipped natural-draft furnace when the furnace is not in use, thus reducing off-cycle losses. More information about the energy-saving potential of this accessory is included in the section on Technical Data.

Airflow Variations

The components of a gas-fired, forced-air furnace can be arranged in a variety of configurations to suit a residential heating system. The relative positions of the components in the different types of furnaces are as follows:

- **Upflow or "highboy" furnace.** In an upflow furnace (Figure 2), the blower is located beneath the heat exchanger and discharges vertically upward. Air enters through the bottom or the side of the blower compartment and leaves at the top. This furnace may be used in closets and utility rooms on the first floor or in basements, with the return air ducted down to the blower compartment entrance.
- **Downflow furnace.** In a downflow furnace (Figure 3), the blower is located above the heat exchanger and discharges downward. Air enters at the top and is discharged vertically at the bottom. This furnace is normally used with a perimeter heating system in a house without a basement. It is also used in upstairs furnace closets and utility rooms supplying conditioned air to both levels of a two-story house.
- **Horizontal furnace.** In a horizontal furnace, the blower is located beside the heat exchanger (Figure 4). The air enters at one end, travels horizontally through the blower and over the heat exchanger, and is discharged at the opposite end. This furnace is used for locations with limited head room such as attics and crawl spaces, or is suspended under a roof or placed above a suspended ceiling. These units are often designed so that the components

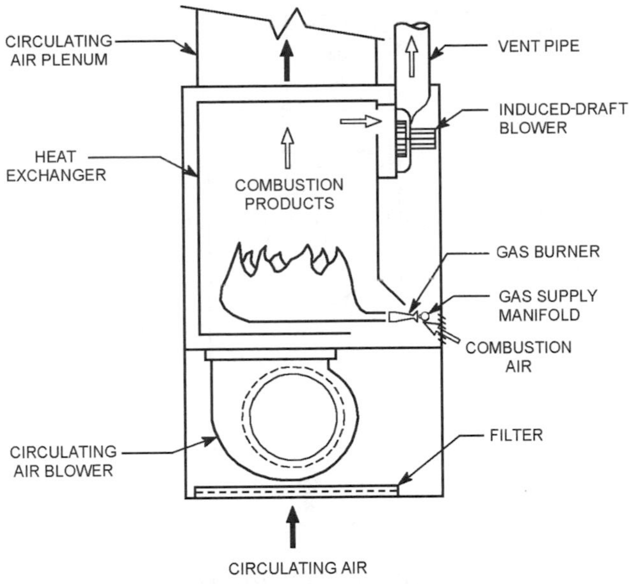

Fig. 2 Upflow Category I Furnace with Induced-Draft Blower

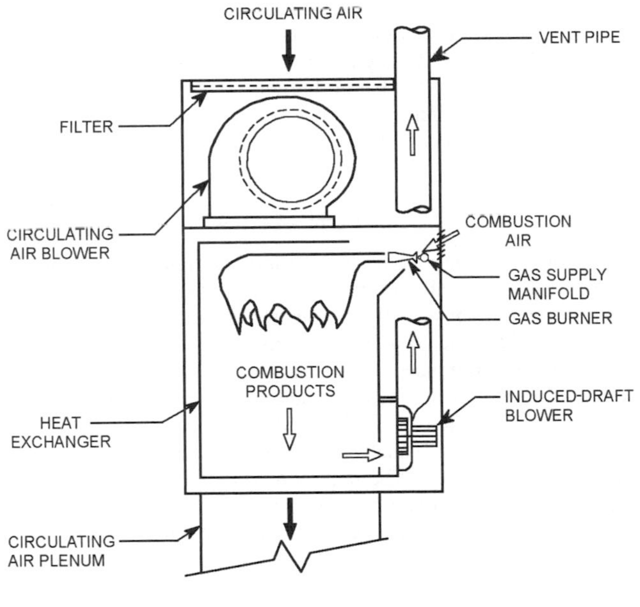

Fig. 3 Downflow (Counterflow) Category I Furnace with Induced-Draft Blower

may be rearranged to allow installation with airflow from left to right or from right to left.

- **Multiposition furnace.** A furnace that can be installed in more than one airflow configuration (e.g., upflow or horizontal; downflow or horizontal; or upflow, downflow, or horizontal) is a multiposition furnace. In some models, a field conversion is necessary to accommodate an alternate installation.
- **Basement or "lowboy" furnace.** The basement furnace (Figure 5) is a variation of the upflow furnace and requires less head room. The blower is located beside the heat exchanger at the bottom. Air enters the top of the cabinet, is drawn down through the blower, is discharged over the heat exchanger, and leaves vertically at the top. In recent years, this type of furnace has become less popular due to the advent of short upflow furnaces.

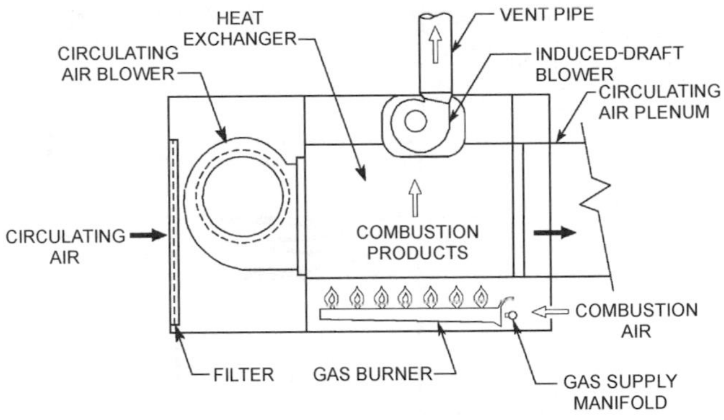

Fig. 4 Horizontal Category I Furnace with Induced-Draft Blower

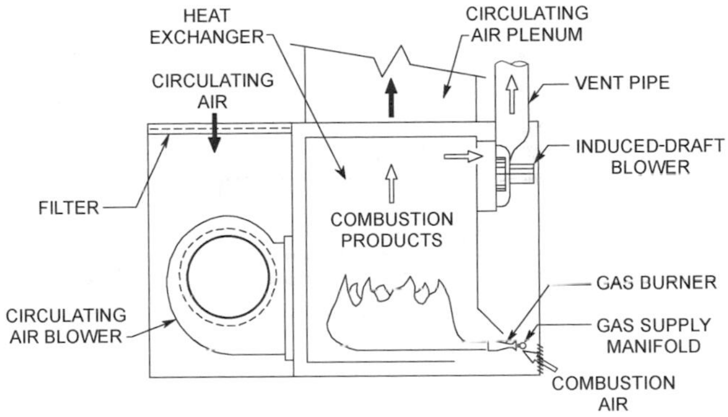

Fig. 5 Basement (Lowboy) Category I Furnace with Induced-Draft Blower

- **Gravity furnace.** These furnaces are no longer available, and they are not common. This furnace has larger air passages through the casing and over the heat exchanger so that the buoyancy force created by the air being warmed circulates the air through the ducts. Wall furnaces that rely on natural convection (gravity) are discussed in Chapter 29.

Combustion System Variations

Gas-fired furnaces use a natural-draft or a fan-assisted combustion system. With a natural-draft furnace, the buoyancy of the hot combustion products carries these products through the heat exchanger, into the draft hood, and up the chimney.

Fan-assisted combustion furnaces have a combustion blower, which may be located either upstream or downstream from the heat exchangers (Figure 6). If the blower is located upstream, blowing the combustion air into the heat exchangers, the system is known as a forced-draft system. If the blower is downstream, the arrangement is known as an induced-draft system. Fan-assisted combustion systems have generally been used with outdoor furnaces; however, with the passage of the 1987 U.S. National Appliance Energy Conservation Act, fan-assisted combustion has become more common for indoor furnaces as well. Fan-assisted combustion furnaces do not require a draft hood, resulting in reduced off-cycle losses and improved efficiency.

Direct vent furnaces may have either natural-draft or fan-assisted combustion. They do not have a draft hood, and they obtain combustion air from outside the structure. Mobile home furnaces must be of the direct vent type.

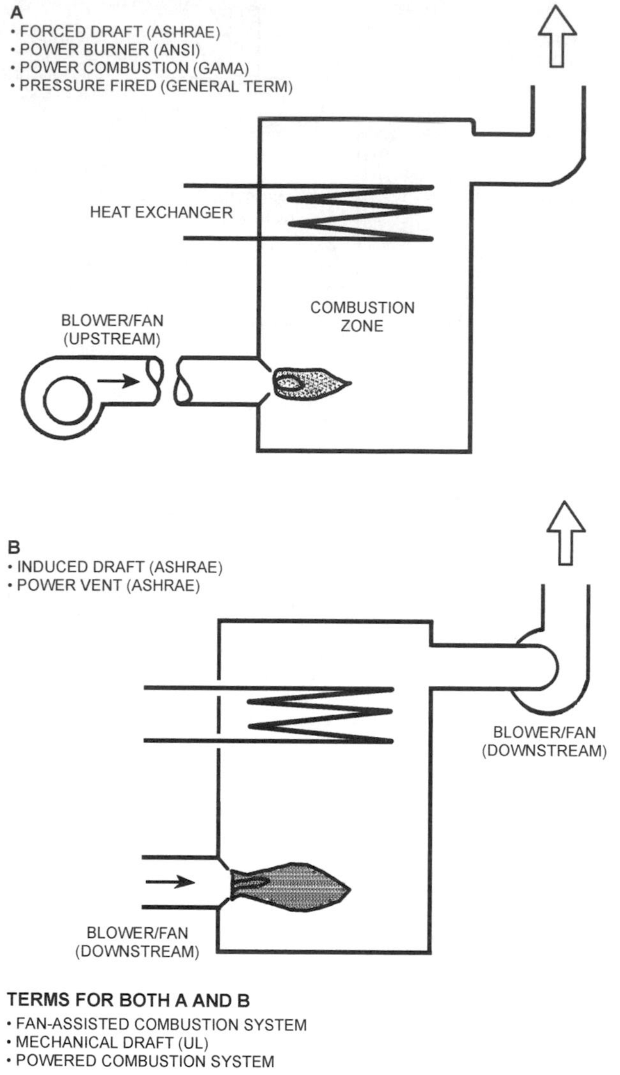

Fig. 6 Terminology Used to Describe Fan-Assisted Combustion

Indoor-Outdoor Furnace Variations

Central system residential furnaces are designed and certified for either indoor or outdoor use. Outdoor furnaces are normally horizontal flow.

The heating-only outdoor furnace is similar to the more common indoor horizontal furnace. The primary difference is that the outdoor furnace is weatherized; the motors and controls are sealed, and the exposed components are made of corrosion-resistant materials such as galvanized or aluminized steel.

A common style of outdoor furnace is the combination package unit. This unit is a combination of an air conditioner and a gas or electric furnace built into a single casing. The design varies, but the most common combination consists of an electric air conditioner coupled with a horizontal gas or electric furnace. The advantage is that much of the interconnecting piping and wiring is included in the unit.

PROPANE FURNACES

Most manufacturers have their furnaces certified for both natural gas and propane. The major difference between natural gas and propane furnaces is the pressure at which the gas is injected from the manifold into the burners. For natural gas, the manifold pressure is

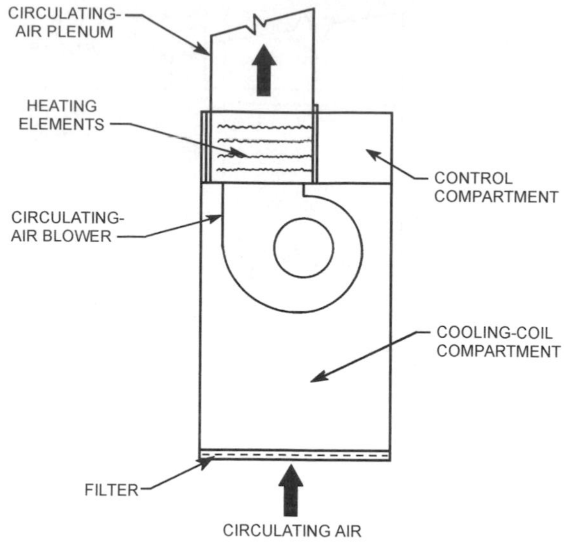

Fig. 7 Electric Forced-Air Furnace

usually controlled at 3 to 4 in. of water; for propane, the pressure is usually 10 to 11 in. of water.

Because of the higher injection pressure and the greater heat content per volume of propane, there are certain physical differences between a natural gas furnace and an propane furnace. One difference is that the pilot and burner orifices must be smaller for propane furnaces. The gas valve regulator spring is also different. Sometimes it is necessary to change burners, but this is not normally required. Manufacturers sell conversion kits containing both the required parts and instructions to convert furnace operation from one gas to the other.

OIL FURNACES

Indoor oil furnaces come in the same configuration as gas furnaces. They are available in upflow, downflow, horizontal, and multiposition lowboy configurations for ducted systems. Oil-fired outdoor furnaces and combination units are not common.

The major differences between oil and gas furnaces are in the combustion system, the heat exchanger, and the barometric draft regulator used in lieu of a draft hood. Ducted system, oil-fired, forced-air furnaces are usually forced-draft and equipped with pressure atomizing burners. The pump pressure and the orifice size of the injection nozzle regulate the firing rate of the furnace. Electric ignition lights the burners. Other furnace controls, such as the blower switch and the limit switch, are similar to those used on gas furnaces.

The heat exchangers of oil-fired furnaces are normally heavy-gage steel formed into a welded assembly. The hot flue products flow through the inside of the heat exchanger into the chimney. The conditioned air flows over the outside of the heat exchanger and into the air supply plenum.

ELECTRIC FURNACES

Electric-powered furnaces come in a variety of configurations and have some similarities to gas- and oil-fired furnaces. However, when a furnace is used with an air conditioner, the cooling coil may be upstream from the blower and heaters. On gas- and oil-fired furnaces, the cooling coil is normally mounted downstream from the blower and heat exchangers.

Figure 7 shows a typical arrangement for an electric forced-air furnace. Air enters the bottom of the furnace and passes through the filter, then flows up through the cooling coil section into the blower. The electric heating elements are immediately above the blower so that the high-velocity air discharging from the blower passes directly through the heating elements.

The furnace casing, air filter, and blower are similar to equivalent gas furnace components. The heating elements are made in modular form, with 5 kW capacity being typical for each module. Electric furnace controls include electric overload protection, contactor, limit switches, and a fan control switch. The overload protection may be either fuses or circuit breakers. The contactor brings the electric heat modules on. The fan control switch and limit switch functions are similar to those of the gas furnace, but one limit switch is usually used for each heating element.

Frequently, electric furnaces are made from modular sections; for example, the coil box, blower section, and electric heat section are made separately and then assembled in the field. Regardless of whether the furnace is made from a single-piece casing or a modular casing, it is generally a multiposition unit. Thus, the same unit may be used for upflow, downflow, or horizontal installations.

When an electric heating appliance is sold without a cooling coil, it is known as an electric furnace. The same appliance is called a fan-coil air handler when it has an air-conditioning coil already installed. When the unit is used as the indoor section of a split heat pump, it is called a heat pump fan-coil air handler. For detailed information on heat pumps, see Chapter 45.

Electric forced-air furnaces are also used with packaged heat pumps and packaged air conditioners.

SYSTEM DESIGN AND EQUIPMENT SELECTION

Warm-Air Furnaces

Two steps are required in selecting a warm-air furnace: (1) determining the required heating capacity of the furnace and (2) selecting a specific furnace to satisfy this requirement.

Heating Capacity. Thc hcating capacity of a warm-air furnacc depends on several variables that operate alone or in combination. The first variable is the design heating requirement of the residence. The heat loss of the structure can be calculated using the procedures outlined in Chapter 27 of the 1997 *ASHRAE Handbook—Fundamentals*.

The additional heating required if the furnace is operating on a night setback cycle should also be considered. During the morning recovery period, additional capacity is required to bring the conditioned space temperature up to the desired level. The magnitude of this recovery capacity depends on weather conditions, the magnitude of the night setback, and the time allowed for the furnace to return room air temperature to the desired level. Another consideration similar to night setback concerns structures that require only intermittent heating, such as houses of worship and auditoriums. Chapter 4 of the 1999 *ASHRAE Handbook—Applications* has further information.

A third variable is the influence of internal loads. Normally, the heat gain from internal loads is neglected when selecting a furnace, but if the internal loads are constant, they should be used to reduce the required capacity of the furnace, especially in nonresidential applications.

The energy required for humidification is a fourth variable. The humidification energy depends on the desired level of relative humidity and the rate at which the moisture must be supplied to maintain the specified level. Net moisture requirements must take into account internal gains due to people, equipment, and appliances; losses through migration in exterior surfaces; and air infiltration. Chapter 20 gives details on how to determine humidification requirements.

A fifth variable is the influence of off-peak storage devices. When used in conjunction with a furnace, a storage device decreases the required capacity of the furnace. The storage device can supply the additional capacity required during the morning recovery of a night setback cycle or reduce the daily peak loads to assist in load shedding. Detailed calculations can determine the contribution of storage devices.

The sixth variable is the furnace's capacity to accommodate air conditioning, even if air conditioning is not planned initially. The cabinet should be large enough to accept a cooling coil that satisfies the cooling load. The blower and motor should have sufficient capacity to provide the increased airflow rates typically required in air-conditioning applications. Chapter 9 includes specific design considerations.

Specific Furnace Selection. The second step in the selection of a warm-air furnace is to choose a specific furnace that satisfies the required design capacity. The final decision depends on numerous parameters, the most significant of which is the fuel type. The second step of the furnace selection process is subdivided by fuel types.

Natural Gas Furnaces

Size Selection. Historically, furnaces have been oversized because (1) the calculation procedure was not exact, especially the estimate of air infiltration; (2) a safety factor was added to account for weather conditions that are more severe than the design conditions used to calculate the required furnace capacity; (3) the additional first cost of a slightly larger furnace was considered a good value in view of possible undersizing, which would be expensive to correct; and (4) adequate airflow for cooling was another consideration. Natural gas was relatively inexpensive, so possible inefficiencies due to oversizing were not considered detrimental. The net result was significant oversizing.

Oversizing can increase overall energy use for new houses where vents and ducts are sized to furnace capacity. However, in retrofits (where fixed vent and duct sizes are assumed), oversizing has little effect on overall energy use. In either situation, oversizing may reduce the comfort level due to wide temperature variations in the conditioned space. In a retrofit, if a higher efficiency furnace is selected, the output capacity must match the original equipment's output. Otherwise, additional furnace oversizing results.

Chapter 27 of the 1997 *ASHRAE Handbook—Fundamentals* recommends oversizing new installations by 40%, if a 10°F night setback is prescribed, to obtain a 1 hour recovery.

Performance Criteria. Performance criteria or a consistent definition of efficiency must be used throughout. Some typical efficiencies encountered are (1) steady-state efficiency, (2) utilization efficiency, (3) annual fuel utilization efficiency.

These efficiencies are generally used by the furnace industry in the following manner:

- **Steady-state efficiency (SSE).** This is the efficiency of a furnace when it is operated under equilibrium conditions based on ASHRAE *Standard* 103. It is calculated by measuring the energy input, subtracting the losses for exhaust gases and flue gas condensate (for condensing furnaces only), and then dividing by the fuel input (cabinet loss not included):

$$\text{SSE}(\%) = \frac{\text{Fuel Input} - \text{Flue Loss} - \text{Condensate Loss}}{\text{Fuel Input}} \times 100$$

For furnaces tested under the isolated combustion system (ICS) method and for outdoor furnaces, cabinet heat loss (jacket loss) must also be deducted from the energy input:

$$\text{SSE(ICS)}(\%) = \frac{\text{Fuel Input} - \text{Flue Loss} - \text{Condensate Loss} - \text{Jacket Loss}}{\text{Fuel Input}} \times 100$$

An ICS is a system installed in the building structure but removed from the space it is heating. Locations include garages, attics, and crawl spaces.

A decreased flue temperature corresponds to increased SSE.

- **Utilization efficiency.** This efficiency is obtained from an empirical equation developed by Kelly et al. (1978) with 100% efficiency and deducting losses for exhausted latent and sensible heat, cyclic effects, infiltration, and pilot burner effect.
- **Annual fuel utilization efficiency (AFUE).** This value is the same as utilization efficiency, except that losses from a standing pilot during the nonheating season are deducted. This equation can also be found in Kelly et al. (1978) or ASHRAE *Standard* 103. AFUE is displayed on each furnace produced in accordance with U.S. Federal Trade Commission requirements for appliance labeling found in *Code of Federal Regulations* 16 Part 305.

  The AFUE is determined for residential fan-type furnaces by using ASHRAE *Standard* 103. The test procedure is also presented in *Code of Federal Regulations* Title II, 10 Part 430, Appendix N, in conjunction with the amendments issued by the U.S. Department of Energy in the *Federal Register.* This version of the test method allows the rating of nonweatherized furnaces as indoor combustion systems, ICSs, or both. Weatherized furnaces are rated as outdoor.

Federal law requires manufacturers of furnaces to use AFUE as determined using the isolated combustion system method to rate efficiency. Effective January 1, 1992, all furnaces produced have a minimum AFUE (ICS) level of 78%. Table 1 gives efficiency values for different furnaces.

Annual fuel energy savings may be compared using the following formula:

$$\text{Annual energy reduction (AER)} = \frac{\text{AFUE}_2 - \text{AFUE}_1}{\text{AFUE}_2}$$

where AFUE_2 is greater than AFUE_1. For example, compare items 3 and 2 of Table 1:

$$\text{AER} = \frac{78 - 69}{78} = \frac{9}{78} = 11.5\%$$

The ASHRAE SP43 work (see the section on System Performance in Chapter 9) confirms that this is a reasonable estimate.

Table 1 Typical Values of Efficiency

	AFUE, %	
Type of Gas Furnace	**Indoor**	**ICS[a]**
1. Natural-draft with standing pilot	64.5	63.9[b]
2. Natural-draft with intermittent ignition	69.0	68.5[b]
3. Natural-draft with intermittent ignition and auto vent damper	78.0	68.5[b]
4. Fan-assisted combustion with standing pilot or intermittent ignition	80.0	78.0
5. Same as 4, except with improved heat transfer	82.0	80.0
6. Direct vent, natural-draft with standing pilot, preheat	66.0	64.5[b]
7. Direct vent, fan-assisted combustion, and intermittent ignition	80.0	78.0
8. Fan-assisted combustion (induced-draft)	80.0	78.0
9. Condensing	90.0	88.0
Type of Oil Furnace	**Indoor**	**ICS[a]**
1. Standard—pre-1992	71.0	69.0[b]
2. Standard—post-1992	80.0	78.0
3. Same as 2, with improved heat transfer	81.0	79.0
4. Same as 3, with automatic vent damper	82.0	80.0
5. Condensing	91.0	89.0

[a]Isolated combustion system (estimate).
[b]Pre-1992 design (see text).

Construction Features and Limitations. Many indoor furnaces have cold-rolled steel heat exchangers. If the furnace is exposed to clean air, and the heat exchanger remains dry, this material has a long life and does not corrode easily. Many deluxe, noncondensing furnaces have a coated heat exchanger to provide extra protection against corrosion. Research by Stickford et al. (1985) indicates that chloride compounds in the condensate of condensing furnaces can cause the heat exchanger to fail unless it is made of specialty steel. A corrosion-resistant heat exchanger must also be used in a furnace certified for use downstream of a cooling coil.

Design Life. Typically, the heat exchangers made of cold-rolled steel have a design life of approximately 15 years. Special coated or alloy heat exchangers, when used for standard applications, have a design life of as much as 20 years. Coated or alloy heat exchangers are recommended for furnace applications in corrosive atmospheres.

Sound Level. This variable must be considered in most applications. Chapter 46 of the 1999 *ASHRAE Handbook—Applications* outlines the procedures to follow in determining acceptable noise levels.

Safety. Because of open-flame combustion, the following safety items need to be considered: (1) the surrounding atmosphere should be free of dust or chemical concentrations; (2) a path for combustion air must be provided for both sealed and open combustion chambers; and (3) the gas piping and vent pipes must be installed according to the NFPA/AGA *National Fuel Gas Code*, local codes, and the manufacturer's instructions.

Applications. Gas furnaces are primarily applied to residential heating. The majority are used in single-family dwellings but are also applicable to apartments, condominiums, and mobile homes.

Performance Versus Cost. These factors must be considered in selection. Included in life-cycle cost determination are initial cost, maintenance, energy consumption, design life, and the price escalation of the fuel. Procedures for establishing operating costs for use in product labeling and audits are available in the United States from the Department of Energy. For residential furnaces, fact sheets provided by manufacturers are available at the point of sale. In addition, AFUE (ICS), fuel, and electrical energy consumption data are listed semiannually in the Gas Appliance Manufacturers Association (GAMA) Directory.

Other Fuels

The design and selection criteria for propane furnaces are identical to those for natural gas furnaces.

The design criteria for oil furnaces are similar to those for natural gas furnaces, except that oil-fired furnaces should be tested in accordance with UL *Standard* 727 and oil burners in accordance with UL *Standard* 296.

Electric Furnaces

The design criteria for electric furnaces are similar to those for natural gas furnaces. The selection criteria are similar, except that an electric furnace does not have the flue loss and combustion air loss of a gas furnace. For this reason, and since the calculations do not account for electrical generation and transmission losses, seasonal efficiency approaches 100% for electric furnaces. Their AFUE ratings are typically 96% to 99% for ICSs.

The design life of electric furnaces is related to the durability of the contactors and the heating elements. The typical design life is approximately 15 years.

Safety primarily concerns proper wiring techniques. Wiring should comply with the *National Electrical Code* (NEC) (NFPA *Standard* 70) and applicable local codes.

TECHNICAL DATA

Detailed technical data on furnaces are available from manufacturers, wholesalers, and dealers. The data are generally tabulated in product specification bulletins printed by the manufacturer for each furnace line. These bulletins usually include performance information, electrical data, blower and air delivery data, control system information, optional equipment information, and dimensions.

Natural Gas Furnaces

Capacity Ratings. ANSI *Standard* Z21.47/CSA 2.3 requires that the heating capacity be marked on the rating plates of commercial furnaces in the United States. The heating capacity of residential furnaces, less than 225,000 Btu/h input, is required by the Federal Trade Commission and can be found in furnace directories published semiannually by GAMA. Capacity is calculated by multiplying the input by the steady-state efficiency.

Residential gas furnaces with heating capacities ranging from 35,000 to 175,000 Btu/h are readily available. Some smaller furnaces are manufactured for special-purpose installations such as mobile homes. Smaller capacity furnaces are becoming common because new homes are better insulated and have lower heat loads than older homes. Larger furnaces are also available, but these are generally considered for commercial use.

Because of the overwhelming popularity of the upflow furnace, or multiposition including upflow, it is available in the greatest number of models and sizes. Downflow furnaces, horizontal furnaces, and various combinations are also available but are generally limited in model type and size.

Residential gas furnaces are available in heating-only and heat-cool models. The difference is that the heat-cool model is designed to operate as the air-handling section of a split-system air conditioner. The heating-only models typically operate with enough airflow to allow a 60 to 100°F air temperature rise through the furnace. This rise provides good comfort conditions for the heating system with a low-noise blower and low electrical energy consumption. Condensing furnaces may be designed for a lower temperature rise (as low as 35°F).

Heat-cool model furnaces have multispeed blowers with a more powerful motor capable of delivering about 400 cfm per ton of air conditioning. Models are generally available in 2, 3, 4, and 5 ton sizes, but all cooling sizes are not available for every furnace size. For example, a 60,000 Btu/h furnace would be available in models with blowers capable of handling 2 or 3 tons of air conditioning; 120,000 Btu/h models would be matched to 4 or 5 tons of air conditioning. Controls of the heat-cool furnace models are generally designed to operate the multispeed blower motor at the most appropriate speed for either heating or cooling operation when airflow requirements vary for each mode. This feature provides optimum comfort for year-round operation.

Efficiency Ratings. Currently, gas furnaces have steady-state efficiencies that vary from about 78 to 96%. Natural-draft and fan-assisted combustion furnaces typically range from 78 to 80% efficiency, while condensing furnaces have over 90% steady-state efficiency. Koenig (1978), Gable and Koenig (1977), Hise and Holman (1977), and Bonne et al. (1977) found that oversizing residential gas furnaces with standing pilots reduced the seasonal efficiency of heating systems in new installations with vents and ducts sized according to furnace capacity.

The AFUE of a furnace may be improved by ways other than changing the steady-state efficiency. These improvements generally add more components to the furnace. One method replaces the standing pilot with an intermittent ignition device. Gable and Koenig (1977) and Bonne et al. (1976) indicated that this feature can save as much as 5.6×10^6 Btu/year per furnace. For this reason, some jurisdictions require the use of intermittent ignition devices.

Another method of improving AFUE is to take all combustion air from outside the heated space (direct vent) and preheat it. A combustion air preheater incorporated into the vent system draws combustion air through an outer pipe that surrounds the flue pipe. Such systems have been used on mobile home and outdoor furnaces. Annual energy consumption of a direct vent furnace with combustion air preheat may be as much as 9% less than that of a standard furnace of the same design (Bonne et al. 1976). Direct vent without combustion air preheat is not inherently more efficient because the reduction in combustion-induced infiltration is offset by the use of colder combustion air.

An automatic vent damper (thermal or electromechanical) is another device that saves energy on indoor furnaces. This device, which is placed after the draft hood outlet, closes the vent when the furnace is not in operation. It saves energy during the off cycle of the furnace by (1) reducing exfiltration from the house and (2) trapping residual heat from the heat exchanger within the house rather than allowing it to flow up the chimney. These savings approach 11% under ideal conditions, where combustion air is taken from the heated space, which is under thermostat control. However, these savings are much less (estimates vary from 0 to 4%) if combustion air is taken from outside the heated space. The ICS method of determining AFUE gives no credit to vent dampers installed on indoor furnaces because it assumes the use of outdoor combustion air with the furnace installed in an unconditioned space.

The AFUE of fan-assisted combustion furnaces is higher than for standard natural-draft furnaces. Fan-assisted combustion furnaces normally have such a high internal flow resistance that combustion airflow stops when the combustion blower is off. This characteristic results in greater energy savings than those from a vent damper. Computer studies by Gable and Koenig (1977), Bonne et al. (1976), and Chi (1977) have estimated annual energy savings up to 16% for fan-assisted combustion furnaces with electric ignition as compared to natural-draft furnaces with standing pilot.

Control. Externally, the furnace is controlled by a low-voltage room thermostat. Control can be heating-only, combination heat-cool, multistage, or night setback. Chapter 37 of the 1997 *ASHRAE Handbook—Fundamentals* addresses thermostats in more detail. A night setback thermostat can reduce the annual energy consumption of a gas furnace. Dual setback (setting the temperature back during the night and during unoccupied periods in the day) can save even more energy. Gable and Koenig (1977) and Nelson and MacArthur (1978) estimated that energy savings of up to 30% are possible, depending on the degree and length of setback and the geographical location. The percentage of energy savings is greater in regions with mild climates; however, the total energy savings is greatest in cold regions.

Several types of gas valves perform various operating functions within the furnace. The type of valve available relates closely to the type of ignition device used. Two-stage valves, available on some furnaces, operate at full gas input or at a reduced rate and are controlled by either a two-stage thermostat or a software algorithm programmed in the furnace control system. They provide less heat at the reduced input and, therefore, less temperature variation and greater comfort during mild weather conditions when full heat output is not required. Two-stage control is used frequently for zoning applications. Fuel savings with two-stage firing rate systems may not be realized unless both the gas and the combustion air are controlled.

The fan control switch controls the circulating air blower. This switch may be temperature-sensitive and exposed to the circulating airstream within the furnace cabinet, or it may be an electronically operated relay. Blower start-up is typically delayed about 1 min after the start-up of the burners. This delay gives the heat exchangers time to warm up and eliminates the excessive flow of cold air when the blower comes on. Blower shutdown is also delayed several minutes after burner shutdown to remove residual heat from the heat exchangers and to improve the annual efficiency of the furnace. Constant blower operation throughout the heating season has been encouraged to improve air circulation and provide even temperature distribution throughout the house. However, constant blower operation increases electrical energy consumption and overall operating cost in many instances. Electronic motors that provide continuous but variable airflow use less energy. Both strategies may be considered when air filtering performance is important.

The limit switch prevents overheating in the event of severe reduction in circulating airflow. This temperature-sensitive switch is exposed to the circulating airstream and shuts off the gas valve if the temperature of the air leaving the furnace is excessive. The fan control and limit switches are sometimes incorporated in the same housing and are sometimes operated by the same thermostatic element. In the United States, the blocked vent shutoff switch and flame rollout switch have been required on residential furnaces produced since November 1989; they shut off the gas valve if the vent is blocked or when insufficient combustion air is present.

Furnaces using fan-assisted combustion feature a pressure switch to verify the flow of combustion air prior to the opening of the gas valve. The ignition system has a required pilot gas shutoff feature in case the pilot ignition fails.

Propane Furnaces

Most residential natural gas furnaces are also available in a propane version with identical ratings. The technical data for these two furnaces are identical, except for the gas control and the burner and pilot orifice sizes. Orifice sizes on propane furnaces are much smaller because propane has a higher density and may be supplied at a higher manifold pressure. The heating value and specific gravity of typical gases are listed as follows:

Gas Type	Heating Value, Btu/ft^3	Specific Gravity (Air = 1.0)
Natural	1030	0.60
Propane	2500	1.53
Butane	3175	2.00

As in natural gas furnaces, the ignition systems have a required pilot gas shutoff feature in case the pilot ignition fails. Pilot gas leakage is more critical with propane or butane gas because both are heavier than air and can accumulate to create an explosive mixture within the furnace or furnace enclosure.

Since 1978, ANSI *Standard* Z21.47/CSA 2.3 has required a gas pressure regulator as part of the propane furnace. Prior to that, the pressure regulator was provided only with the propane supply system.

Besides natural and propane, a furnace may be certified for manufactured gas, mixed gas, or propane-air mixtures; however, furnaces with these certifications are not commonly available. Mobile home furnaces are certified as convertible from natural gas to propane.

Oil Furnaces

Oil furnaces are similar to gas furnaces in size, shape, and function, but the heat exchanger, burner, and combustion control are significantly different.

Input ratings are based on the oil flow rate (gph), and the heating capacity is calculated by the same method as that for gas furnaces. The heating value of oil is 140,000 Btu/gal. Fewer models and sizes are available for oil than are available for gas, but residential furnaces in the range of 64,000 to 150,000 Btu/h heating capacity are common. Air delivery ratings are similar to gas furnaces, and both heating-only and heat-cool models are available.

The efficiency of an oil furnace can drop during normal operation if the burner is not maintained and kept clean. In this case, the oil does not atomize sufficiently to allow complete combustion, and energy is lost up the chimney in the form of unburned hydrocarbons. Because most oil furnaces use power burners and electric ignition, the annual efficiency is relatively high.

Oil furnaces are available in upflow, downflow, and horizontal models. The thermostat, fan control switch, and limit switch are similar to those of a gas furnace. Oil flow is controlled by a pump and burner nozzle, which sprays the oil-air mixture into a single-chamber drum-type heat exchanger. The heat exchangers are normally heavy-gage cold-rolled steel. Humidifiers, electronic air cleaners, and night setback thermostats are available as accessories.

Electric Furnaces

Residential electric resistance furnaces are available in heating capacities of 5 to 35 kW. Air-handling capabilities are selected to provide sufficient air to meet the requirement of an air conditioner of a reasonable size to match the furnace. Small furnaces supply about 800 cfm, and large furnaces about 2000 cfm.

The only loss associated with an electric resistance furnace is in the cabinet—about 2% of input. Both the steady-state efficiency and the annual fuel utilization efficiency of an electric furnace are greater than 98%, and if the furnace is located within the heated space, the seasonal efficiency is 100%.

Although the efficiency of an electric furnace is high, electricity is a relatively expensive form of energy. The operating cost may be reduced substantially by using an electric heat pump in place of a straight electric resistance furnace. Heat pump systems are discussed in Chapter 8.

Humidifiers and electronic air cleaners are available as accessories for both electric resistance furnaces and heat pumps.

Conventional setback thermostats are recommended to save energy. The electric demand used to recover from the setback, however, may be quite significant. Bullock (1977) and Schade (1978) addressed this problem by using (1) a conventional two-stage setback thermostat with staged supplemental electric heat or (2) solid-state thermostats with programmed logic to inhibit supplemental electric heat from operating during morning recovery. Benton (1983) reported energy savings of up to 30% for these controls, although the recovery time may be extended up to several hours.

Electric furnaces are available in upflow, downflow, or horizontal models. Internal controls include overload fuses or circuit breakers, overheat limit switches, a fan control switch, and a contactor to bring on the heating elements at timed intervals.

COMMERCIAL FURNACES

The basic difference between residential and commercial furnaces is the size and heating capacity of the equipment. The heating capacity of a commercial furnace may range from 150,000 to over 2,000,000 Btu/h. Generally, furnaces with output capacities less than 320,000 Btu/h are classified as light commercial, and those above 320,000 Btu/h are considered large commercial equipment. In addition to the difference in capacity, commercial equipment is constructed from material with increased structural strength and has more sophisticated control systems.

EQUIPMENT VARIATIONS

Light commercial heating equipment comes in almost as many flow arrangements and design variations as residential equipment. Some are identical to residential equipment, while others are unique to commercial applications. Some commercial units function as a part of a ducted system, and others operate as unducted space heaters.

Ducted Equipment

Upflow Gas-Fired Commercial Furnaces. These furnaces are available up to 300,000 Btu/h and supply enough airflow to handle up to 10 tons of air conditioning. They may have high static pressure and belt-driven blowers, and frequently they consist of two standard upflow furnaces tied together in a side-by-side arrangement. They are normally incorporated into a system in conjunction with a commercial split-system air-conditioning unit and are available in either propane or natural gas. Oil-fired units may be available on a limited basis.

Horizontal Gas-Fired Duct Furnaces. Available for built-up light commercial systems, this type of furnace is not equipped with its own blower but is designed for uniform airflow across the entire furnace. Duct furnaces are normally certified for operation either upstream or downstream of an air conditioner cooling coil. If a combination blower and duct furnace is desired, a package called a blower unit heater is available. Duct furnaces and blower unit heaters are available in natural gas, propane, oil, and electric models.

Electric Duct Furnaces. These furnaces are available in a large range of sizes and are suitable for operation in upflow, downflow, or horizontal positions. These units are also used to supply auxiliary heat with the indoor section of a split-heat pump.

Combination Package Units. The most common commercial furnace is the combination package unit, sometimes known as a combination rooftop unit. These are available as air-conditioning units with propane and natural gas furnaces, electric resistance heaters, or heat pumps. Combination oil heat/electric cool units are not commonly available. Combination units come in a full range of sizes covering air-conditioning ratings from 5 to 50 tons with matched furnaces supplying heat-to-cool ratios of approximately 1.5 to 1.

Combination units of 15 tons and under are available as single-zone units. The entire unit must be in either heating mode or cooling mode. All air delivered by the unit is at the same temperature. Frequently, the heating function is staged so that the system operates at reduced heat output when the load is small.

Large combination units in the 15 to 50 ton range are available as single-zone units, as are small units; however, they are also available as multizone units. A multizone unit supplies conditioned air to several different zones of a building in response to individual thermostats controlling those zones. These units are capable of supplying heating to one or more zones at the same time that cooling is supplied to other zones.

Large combination units are normally available only in a curbed configuration; that is, the units are mounted on a rooftop over a curbed opening in the roof. The supply and return air enters through the bottom of the unit. Smaller units may be available for either curbed or uncurbed mounting. In either case, the unit is usually connected to ductwork within the building to distribute the conditioned air.

Unducted Heaters

Three types of commercial heating equipment are used as unducted space heaters. One is the **unit heater**, which is available from about 25,000 to 320,000 Btu/h. These heaters are normally mounted from ceiling hangers and blow air across the heat exchanger into the heated space. Natural gas, propane, and electric unit heaters are available. The second unducted heater used in commercial heating is the **infrared heater**. These units are mounted from ceiling hangers and transmit heat downward by radiation. Both gas and electric infrared heaters are available.

Finally, **floor (standing) furnaces** (Figure 8) are used as large area heaters and are available in capacities ranging from 200,000 to 2,000,000 Btu/h. Floor furnaces direct heated air through nozzles for task heating or use air circulators to heat large industrial spaces. Residential floor-suspended furnaces are described in Chapter 29.

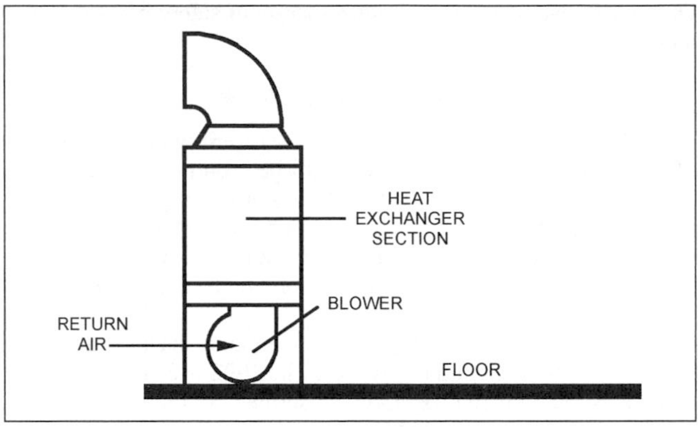

Fig. 8 Standing Floor Furnace

SYSTEM DESIGN AND EQUIPMENT SELECTION

The procedure for design and selection of a commercial furnace is similar to that for a residential furnace. First, the design capacity of the heating system must be determined, considering heat loss from the structure, recovery load, internal load, humidification, off-peak storage, waste heat recovery, and backup capacity. Because most commercial buildings use setback during weekends, evenings, or other long periods of inactivity, the recovery load is important, as are internal loads and waste heat recovery.

Selection criteria differ from those for a residential furnace in some respects and are identical in others. Sizing criteria are essentially the same, and it is recommended that the furnace be oversized 30% above total load. Because combination units must be sized accurately for the cooling load, it is possible that the smallest gas-fired capacity available will be larger than the 30% value. This is especially true for the warmer climates of the United States.

Efficiency of commercial units is about the same as for residential units. Two-stage gas valves are frequently used with commercial furnaces, but the efficiency of a two-stage system may be lower than for a single-stage system. At a reduced firing rate, the excess combustion airflow through the burners increases, decreasing the steady-state operating efficiency of the furnace. Multistage furnaces with multistage thermostats and controls are commonly used to provide more uniform distribution of heat within the building.

The design life of commercial heating and cooling equipment is about 20 years. Most gas furnace heat exchangers are either coated steel or stainless steel. Because most commercial furnaces are made for outdoor application, the cabinets are made from corrosion-resistant coated steel (e.g., galvanized or aluminized). Blowers are usually belt-driven and capable of delivering air at high static pressure.

The noise level of commercial heating equipment is important with some products and less important with others. Unit heaters, for example, are used primarily in industrial applications where noise is less important. Most other commercial equipment is used in schools, office buildings, and other commercial buildings where noise level is important. In general, the larger the furnace, the more air it handles, and the more noise it generates. However, commercial systems with longer and larger ductwork result in more sound attenuation. The net result is that quality commercial heating systems produce about the same noise level in the heated space as do residential systems.

Safety requirements are the same for light commercial systems as they are for residential systems. Above 400,000 Btu/h gas input, the ANSI *Standard* Z21.47/CSA 2.3 requirements for gas controls are more stringent. A large percentage of commercial heating systems are located on rooftops or some other location outside a building.

Outdoor furnaces always provide a margin of safety beyond that of an indoor furnace.

TECHNICAL DATA

Technical data for commercial furnaces are supplied by the manufacturer. Furnaces are available with heat outputs ranging from 150,000 to more than 2,000,000 Btu/h. For heating-only commercial heaters, the airflow is set to supply air with an 85°F temperature rise. Heat-cool combination units supply air equal to about 400 cfm per ton of cooling capacity. Heat-to-cool ratios are generally held at about 1.5 to 1.

The steady-state efficiency for commercial furnaces is about the same as that for residential furnaces. The 1992 U.S. Energy Policy and Conservation Act (EPCA) prescribes minimum efficiency requirements for commercial furnaces based on ASHRAE *Standard* 90.1. Some efficiency improvement components, such as intermittent ignition devices, are common in commercial furnaces.

GENERAL CONSIDERATIONS

INSTALLATION PRACTICES

Installation requirements call for a forced-air heating system to meet two basic criteria: (1) the system must be safe, and (2) it must provide comfort for the occupants of the conditioned space.

Indoor furnaces are sometimes installed as isolated combustion systems (ICSs): a furnace is installed indoors, and all combustion and ventilation air is admitted through grilles or is ducted from outdoors and does not interact with air in the conditioned space. Examples of ICS installations include interior enclosures with air from an attic or ducted from outdoors, exterior enclosures with air from outdoors through grilles, or enclosures in garages or carports attached to the building (NFPA *Standard* 54). This type of installation presents special considerations in determining efficiency.

Generally, the following three categories of installation guidelines must be followed to ensure the safe operation of a heating system: (1) the equipment manufacturer's installation instructions, (2) local installation code requirements, and (3) national installation code requirements. While local code requirements may or may not be available, the other two are always available. Depending on the type of fuel being used, one of the following U.S. national code requirements will apply:

- NFPA 54-99 (also AGA Z223.1-99) — *National Fuel Gas Code*
- NFPA 70-99 — *National Electrical Code*
- NFPA 31-97 — Standard for the Installation of Oil-Burning Equipment

Comparable Canadian standards are

- CAN/CGA-B149.1-M95 — *Natural Gas Installation Code*
- CAN/CGA-B149.2-M95 — *Propane Installation Code*
- CSA C22.1-98 — *Canadian Electrical Code*
- CAN/CSA B139-M91 — *Installation Code for Oil Burning Equipment*

An additional source is the *International Fuel Gas Code* (IFGC) (ICC 1997). These regulations provide complete information about construction materials, gas line sizes, flue pipe sizes, wiring sizes, and so forth.

Proper design of the air distribution system is necessary for both comfort and safety. Chapter 32 of the 1997 *ASHRAE Handbook—Fundamentals*, Chapter 1 of the 1999 *ASHRAE Handbook—Applications*, and Chapter 9 of this volume provide information on the design of ductwork for forced-air heating systems. Forced-air furnaces provide design airflow at a static pressure as low as 0.12 in. of water for a residential unit to above 1.0 in. of water for a commercial unit. The air distribution system must handle the required volumetric flow rate within the pressure limits of the equipment. If the system is a combined heating-cooling installation, the air distribution system must meet the cooling requirement because more air is required for cooling than for heating. It is also important to include the pressure drop of the cooling coil. The Air-Conditioning and Refrigeration Institute (ARI) maximum allowable pressure drop for residential cooling coils is 0.3 in. of water.

AGENCY LISTINGS

The construction and performance of furnaces is regulated by several agencies.

The Gas Appliance Manufacturers Association (GAMA), in cooperation with its industry members, sponsors a certification program relating to gas- and oil-fired residential furnaces and boilers. This program uses an independent laboratory to verify the furnace and boiler manufacturers' certified AFUEs and heating capacities, as determined by testing in accordance with the U.S. Department of Energy's *Uniform Test Method for Measuring the Energy Consumption of Furnaces and Boilers* (CFR Title II, 10 Part 30, Subpart B, Appendix N). Gas and oil furnaces with input ratings less than 225,000 Btu/h and gas and oil boilers with input ratings less than 300,000 Btu/h are currently included in the program.

Also included in the program is the semiannual publication of the GAMA consumers directory, which identifies certified products and lists the input rating, certified heating capacity, and AFUE for each furnace. Participating manufacturers are entitled to use the GAMA Certification Symbol (seal). These directories are published semiannually and distributed to the reference departments of public libraries in the United States.

ANSI *Standard* Z21.47/CSA 2.3, Gas-Fired Central Furnaces (CSA America is secretariat), gives minimum construction, safety, and performance requirements for gas furnaces. The CSA maintains laboratories to certify furnaces and operates a factory inspection service. Furnaces tested and found to be in compliance are listed in the CSA Directory and carry the Seals of Certification. Underwriters Laboratories (UL) and other approved laboratories can also test and certify equipment in accordance with ANSI *Standard* Z21.47/CSA 2.3.

Gas furnaces may be certified for standard, alcove, closet, or outdoor installation. Standard installation requires clearance between the furnace and combustible material of at least 6 in. Furnaces certified for alcove or closet installation can be installed with reduced clearance, as listed. Furnaces certified for either sidewall venting or outdoor installation must operate properly in a 31 mph wind. Construction materials must be able to withstand natural elements without degradation of performance and structure. Horizontal furnaces are normally certified for installation on combustible floors and for attic installation and are so marked, in which case they may be installed with point or line contact between the jacket and combustible constructions. Upflow and downflow furnaces are normally certified for alcove or closet installation. Gas furnaces may be listed to burn natural gas, mixed gas, manufactured gas, propane, or propane-air mixtures. A furnace must be equipped and certified for the specific gas to be used because different burners and controls, as well as orifice changes, may be required.

Sometimes oil burners and control packages are sold separately; however, they are normally sold as part of the furnace package. Pressure-type or rotary burners should bear the Underwriters Laboratory label showing compliance with UL *Standard* 296. In addition, the complete furnace should bear markings indicating compliance with UL *Standard* 727. Vaporizing burner furnaces should also be listed under UL *Standard* 727.

Underwriters Laboratories *Standard* 1995 gives requirements for the listing and labeling of electric furnaces and heat pumps.

The following list summarizes important standards issued by the International Approval Service, Underwriters Laboratories, the Canadian Gas Association, and the Canadian Standards Association that apply to space-heating equipment:

ANSI/ASHRAE 103-1993	Method of Testing for Annual Fuel Utilization Efficiency of Residential Central Furnaces and Boilers
ANSI Z21.66-96/CGA 6.14-M96	Automatic Vent Damper Devices for Use with Gas-Fired Appliances
ANSI Z83.4-91 (R 1998)	Direct Gas-Fired Makeup Air Heaters
ANSI Z83.6-90 (R 1998)	Gas-Fired Infrared Heaters
ANSI Z83.8-96/CGA 2.6-M96	Gas-Fired Duct Furnaces and Unit Heaters
ANSI Z21.47-98/CSA 2.3-M98	Gas-Fired Central Furnaces
ANSI/UL 296-94	Oil Burners
ANSI/UL 307A-95	Liquid Fuel-Burning Heating Appliances for Manufactured Homes and Recreational Vehicles
UL 307B-95	Gas-Burning Heating Appliances for Manufactured Homes and Recreational Vehicles
UL 727-94	Oil-Fired Central Furnaces
UL 1995-95 • CAN/CSA C22.2 No. 236	Heating and Cooling Equipment
CGA 3.2-1976	Industrial and Commercial Gas-Fired Package Furnaces
CAN1-3.7-77 (R 1986)	Direct Gas-Fired Non-Recirculating Makeup Air Heaters
CAN/CGA-2.5-M86 (R 1996)	Gas-Fired Gravity and Fan Type Vented Wall Furnaces
CAN/CGA-2.16-M81 (R 1996)	Gas-Fired Infra-Red Radiant Heaters
CAN/CGA-2.19-M81 (R 1996)	Gas-Fired Gravity and Fan Type Direct Vent Wall Furnaces
CSA-B140.4-1974 (R 1998)	Oil-Fired Warm Air Furnaces

REFERENCES

ASHRAE. 1993. Method of testing for annual fuel utilization efficiency of residential central furnaces and boilers. ANSI/ASHRAE *Standard* 103-1993.

ASHRAE. 1999. Energy standard for buildings except low-rise residential buildings. *Standard* 90.1-1999.

Benton, R. 1983. Heat pump setback: Computer prediction and field test verification of energy savings with improved control. *ASHRAE Transactions* 89(1B):716-34.

Bonne, J., J.E. Janssen, A.E. Johnson, and W.T. Wood. 1976. Residential heating equipment HFLAME evaluation of target improvements. National Bureau of Standards. *Final Report*, Contract No. T62709. Center of Building Technology, Honeywell, Inc. Available from NIST, Gaithersburg, MD.

Bonne, U., J.E. Janssen, and R.H. Torborg. 1977. Efficiency and relative operating cost of central combustion heating systems: IV—Oil-fired residential systems. *ASHRAE Transactions* 83(1):893-904.

Bullock, E.C. 1977. Energy saving through thermostat setback with residential heat pumps. Workshop on Thermostat Setback. National Bureau of Standards. Available from NIST, Gaithersburg, MD.

Code of Federal Regulations. FTC appliance labeling. CFR 16 Part 305.

Code of Federal Regulations. Uniform test method for measuring the energy consumption of furnaces and boilers. CFR Title II, 10 Part 430, Subpart B, Appendix N.

Chi, J. 1977. DEPAF—A computer model for design and performance analysis of furnaces. AICHE-ASME Heat Transfer Conference, Salt Lake City, UT (August).

CSA International. 1998. Gas-fired central furnaces. ANSI *Standard* Z21.47/CSA 2.3-M98. Cleveland, OH.

Gable, G.K. and K. Koenig. 1977. Seasonal operating performance of gas heating systems with certain energy-saving features. *ASHRAE Transactions* 83(1):850-64.

GAMA. Published semiannually. Consumers directory of certified efficiency rating for residential heating and water heating equipment. Gas Appliance Manufacturers Association, Arlington, VA.

Hise, E.C. and A.S. Holman. 1977. Heat balance and efficiency measurements of central, forced-air, residential gas furnaces. *ASHRAE Transactions* 83(1):865-80.

ICC. 1997. *International fuel gas code*. International Code Council, Falls Church, VA.

Kelly, G.E., J.G. Chi, and M. Kuklewicz. 1978. Recommended testing and calculation procedures for determining the seasonal performance of residential central furnaces and boilers. Available from National Technical Information Service, Springfield, VA (Order No. PB289484).

Koenig, K. 1978. Gas furnace size requirements for residential heating using thermostat night setback. *ASHRAE Transactions* 84(2):335-51.

Nelson, L.W. and W. MacArthur. 1978. Energy saving through thermostat setback. *ASHRAE Journal* (September).

NFPA. 1999. *National electrical code*. *Standard* 70-99. National Fire Protection Association, Quincy, MA.

NFPA/AGA. 1999. *National fuel gas code*. ANSI/NFPA *Standard* 54-99. National Fire Protection Association, Quincy, MA. ANSI/AGA *Standard* Z223.1-99. American Gas Association, Arlington, VA.

Schade, G.R. 1978. Saving energy by night setback of a residential heat pump system. *ASHRAE Transactions* 84(1):786-98.

Stickford, G.H., B. Hindin, S.G. Talbert, A.K. Agrawal, and M.J. Murphy. 1985. Technology development for corrosion-resistant condensing heat exchangers. GRI-85-0282. Battelle Columbus Laboratories to Gas Research Institute. Available from National Technical Information Service, Springfield, VA.

BIBLIOGRAPHY

Deppish, J.R. and D.W. DeWerth. 1986. GATC studies common venting. *Gas Appliance and Space Conditioning Newsletter* 10 (September).

McQuiston, F.C. and Spitler, J.D. 1992. *Cooling and heating load calculation manual*. ASHRAE.

CHAPTER 29

RESIDENTIAL IN-SPACE HEATING EQUIPMENT

IN-SPACE heating equipment differs from central heating in that fuel is converted to heat in the space to be heated. In-space heaters may be either permanently installed or portable and may transfer heat by a combination of radiation, natural convection, and forced convection. The energy source may be liquid, solid, gaseous, or electric.

GAS IN-SPACE HEATERS

Room Heaters

A **vented circulator room heater** is a self-contained, freestanding, nonrecessed gas-burning appliance that furnishes warm air directly to the space in which it is installed, without ducting (Figure 1). It converts the energy in the fuel gas to convected and radiant heat without mixing flue gases and circulating heated air by transferring heat from flue gases to a heat exchanger surface.

A **vented radiant circulator** is equipped with high-temperature glass panels and radiating surfaces to increase radiant heat transfer. Separation of flue gases from circulating air must be maintained. Vented radiant circulators range from 10,000 to 75,000 Btu/h.

Gravity-vented radiant circulators may also have a circulating air fan, but they perform satisfactorily with or without the fan. Fan-type vented radiant circulators are equipped with an integral circulating air fan, which is necessary for satisfactory performance.

Vented room heaters are connected to a vent, chimney, or single-wall metal pipe venting system engineered and constructed to develop a positive flow to the outside atmosphere. Room heaters should not be used in a room that has limited air exchange with adjacent spaces because combustion air is drawn from the space.

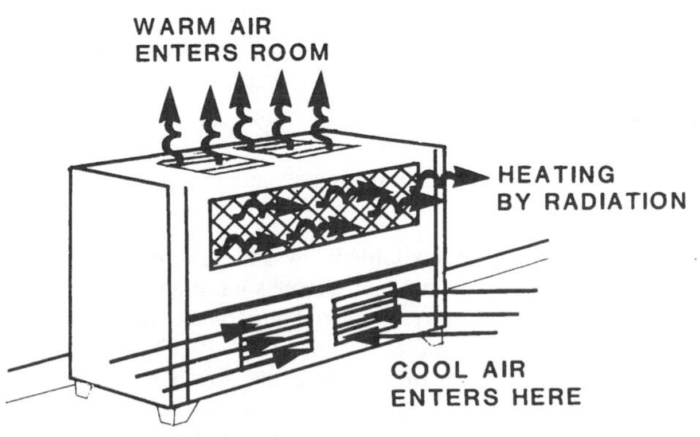

Fig. 1 Room Heater

The preparation of this chapter is assigned to TC 6.4, In-Space Convection Heating.

Unvented radiant or convection heaters range in size from 10,000 to 40,000 Btu/h and can be freestanding units or wall-mounted, nonrecessed units of either the radiant or closed-front type. Unvented room heaters require an outside air intake. The size of the fresh air opening required is marked on the heater. To ensure adequate fresh air supply, unvented gas-heating equipment must, according to voluntary standards, include a device that shuts the heater off if the oxygen in the room becomes inadequate. Unvented room heaters may not be installed in hotels, motels, or rooms of institutions such as hospitals or nursing homes.

Catalytic room heaters are fitted with fibrous material impregnated with a catalytic substance that accelerates the oxidation of a gaseous fuel to produce heat without flames. The design distributes the fuel throughout the fibrous material so that oxidation occurs on the surface area in the presence of a catalyst and room air. Catalytic heaters transfer heat by low-temperature radiation and by convection. The surface temperature is below a red heat and is generally below 1200°F at the maximum fuel input rate. The flameless combustion of catalytic heaters is an inherent safety feature not offered by conventional flame-type gas-fueled burners. Catalytic heaters have also been used in agriculture and for industrial applications in combustible atmospheres.

Unvented household catalytic heaters are used in Europe. Most of these are portable and mounted on casters in a casing that includes a cylinder of liquefied petroleum gas (LPG) so that they may be rolled from one room to another. LPG cylinders holding more than 2 lb of fuel are not permitted for indoor use in the United States. As a result, catalytic room heaters sold in the United States are generally permanently installed and fixed as wall-mounted units. Local codes and the *National Fuel Gas Code* (NFPA *Standard* 54/ANSI Z223.1) should be reviewed for accepted combustion air requirements.

Wall Furnaces

A wall furnace is a self-contained vented appliance with grilles that are designed to be a permanent part of the structure of a building (Figure 2). It furnishes heated air that is circulated by natural or forced convection. A wall furnace can have boots, which may not extend more than 10 in. beyond the horizontal limits of the casing through walls of normal thickness, to provide heat to adjacent rooms. Wall furnaces range from 10,000 to 90,000 Btu/h. Wall furnaces are classified as conventional or direct vent.

Conventional vent units require approved B-1 vent pipes and are installed to comply with the *National Fuel Gas Code*. Some wall furnaces are counterflow units that use fans to reverse the natural flow of air across the heat exchanger. Air enters at the top of the furnace and discharges at or near the floor. Counterflow systems reduce heat stratification in a room. As with any vented unit, a minimum of inlet air for proper combustion must be supplied.

Vented-recessed wall furnaces are recessed into the wall, with only the decorative grillwork extending into the room. This leaves more usable area in the room being heated. Dual-wall furnaces are

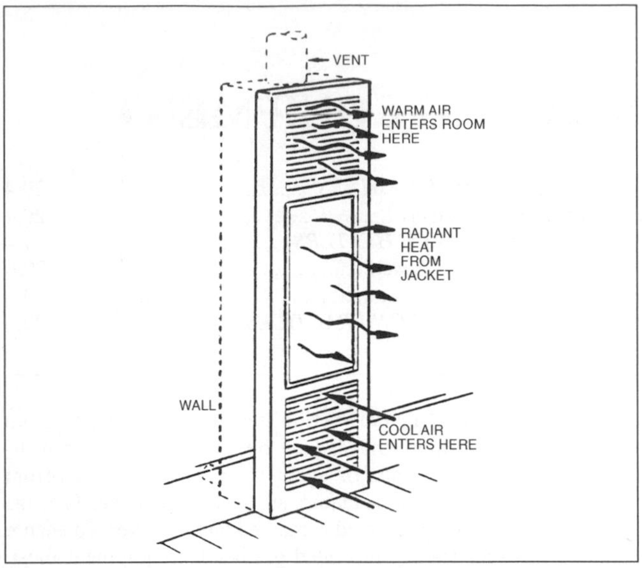

Fig. 2 Wall Furnace

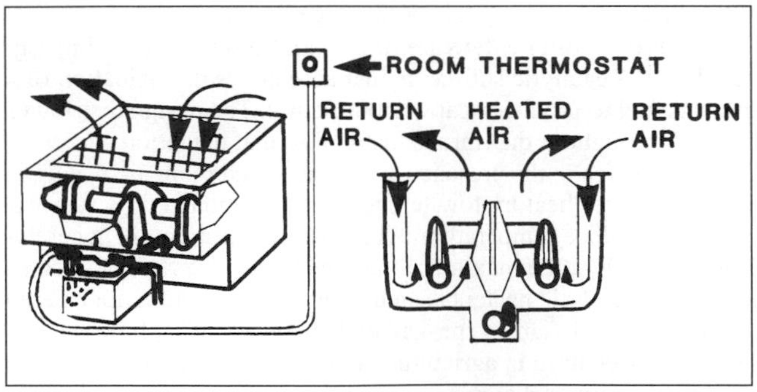

Fig. 3 Floor Furnace

two units that fit between the studs of adjacent rooms, thereby using a common vent.

Both vented-recessed and dual-wall furnaces are usually natural convection units. Cool room air enters at the bottom and is warmed as it passes over the heat exchanger, entering the room through the grillwork at the top of the heater. This process continues as long as the thermostat calls for the burners to be on. Accessory fans assist in the movement of air across the heat exchanger and help minimize air stratification.

Direct vent wall furnaces are constructed so that combustion air comes from outside, and all flue gases discharge into the outside atmosphere. These appliances include grilles or the equivalent and are designed to be attached to the structure permanently. Direct vent wall heaters are normally mounted on walls with outdoor exposure.

Direct vent wall furnaces can be used in extremely tight (well-insulated) rooms because combustion air is drawn from outside the room. There are no infiltration losses for dilution or combustion air. Most direct vent heaters are designed for natural convection, although some may be equipped with fans. Direct vent furnaces are available with inputs of 6000 to 65,000 Btu/h.

Floor Furnaces

Floor furnaces are self-contained units suspended from the floor of the heated space (Figure 3). Combustion air is taken from outside, and flue gases are also vented outside. Cold air returns at the periphery of the floor register, and warm air comes up to the room through the center of the register.

Table 1 Efficiency Requirements in the United States for Gas-Fired Direct Heating Equipment

Input, 1000 Btu/h	Minimum AFUE, %	Input, 1000 Btu/h	Minimum AFUE, %
Wall Furnace (with fan)		*Floor Furnace*	
<42	73	<37	56
>42	74	>37	57
Wall Furnace (gravity type)			
<10	59	*Room Heaters*	
10 to 12	60	<18	57
12 to 15	61	18 to 20	58
15 to 19	62	20 to 27	63
19 to 27	63	27 to 46	64
27 to 46	64	>46	65
>46	65		

United States Minimum Efficiency Requirements

The National Appliance Energy Conservation Act (NAECA) of 1987 mandates minimum annual fuel utilization efficiency (AFUE) requirements for gas-fired direct heating equipment (Table 1). The minimums (effective as of January 1, 1990) are measured using the U.S. Department of Energy test method (DOE 1984) and must be met by manufacturers of direct heating equipment (i.e., gas-fired room heaters, wall furnaces, and floor furnaces).

CONTROLS

Valves

Gas in-space heaters are controlled by four types of valves:

The full on-off, **single-stage valve** is controlled by a wall thermostat. Models are available that are powered by a 24 V supply or from energy supplied by the heat of the pilot light on the thermocouple (self-generating).

The **two-stage control valve** (with hydraulic thermostat) fires either at full input (100% of rating) or at some reduced step, which can be as low as 20% of the heating rate. The amount of time at the reduced firing rate depends on the heating load and the relative oversizing of the heater.

The **step-modulating control valve** (with a hydraulic thermostat) steps on to a low fire and then either cycles off and on at the low fire (if the heating load is light) or gradually increases its heat output to meet any higher heating load that cannot be met with the low firing rate. This control allows an infinite number of fuel firing rates between low and high fire.

The **manual control valve** is controlled by the user rather than by a thermostat. The user adjusts the fuel flow and thus the level of fire to suit heating requirements.

Thermostats

Temperature controls for gas in-space heaters are of the following two types.

- **Wall thermostats** are available in 24 and millivolt systems. The 24 V unit requires an external power source and a 24 V transformer. Wall thermostats respond to temperature changes and turn the automatic valve to either full-on or full-off. The millivolt unit requires no external power because the power is generated by multiple thermocouples and may be either 250 or 750 mV, depending on the distance to the thermostat. This thermostat also turns the automatic valve to either full-on or full-off.
- **Built-in hydraulic thermostats** are available in two types: (1) a snap-action unit with a liquid-filled capillary tube that responds to changes in temperature and turns the valve to either full-on or full-off; and (2) a modulating thermostat, which is similar to the

Table 2 Gas Input Required for In-Space Supplemental Heaters

			Gas Consumption per Unit House Volume, Btu/h per ft^3					
		Steady State	Outside Air Temperature, °F					
	Average	Effi-	Older Bungalow[a]			Energy-Efficient House[b]		
Heater Type	AFUE, %	ciency, %	5	30	50	5	30	50
Vented	54.6	73.1	6.5	3.8	1.6	2.8	1.6	0.7
Unvented	90.5	90.5	6.0	3.5	1.5	2.6	1.5	0.6
Direct vent	76.0	78.2	5.9	3.4	1.5	2.1	1.2	0.5

[a]Tested bungalow total heated volume = 6825 ft^3 and $U \approx 0.3$ to 0.5 Btu/h·°F.
[b]Tested energy-efficient house total heated volume = 11,785 ft^3 and $U \approx 0.2$ to 0.3 Btu/h·°F.

snap-action unit, except that the valve comes on and shuts off at a preset minimum input. Temperature alters the input anywhere from full-on to the minimum input. When the heating requirements are satisfied, the unit shuts off.

VENT CONNECTORS

Any vented gas-fired appliance must be installed correctly to vent combustion products. A detailed description of proper venting techniques is found in the *National Fuel Gas Code* and Chapter 30.

SIZING UNITS

The size of the unit selected depends on the size of the room, the number and direction of exposures, the amount of insulation in the ceilings and walls, and the geographical location. Heat loss requirements can be calculated from procedures described in Chapter 27 of the 1997 *ASHRAE Handbook—Fundamentals*.

DeWerth and Loria (1989) studied the use of gas-fired, in-space supplemental heaters in two test houses. They proposed a heater sizing guide, which is summarized in Table 2. The energy consumption in Table 2 is for unvented, vented, and direct vent heaters installed in (1) a bungalow built in the 1950s with average insulation, and (2) a townhouse built in 1984 with above-average insulation and tightness.

OIL AND KEROSENE IN-SPACE HEATERS

Vaporizing Oil Pot Heaters

These heaters have an oil-vaporizing bowl (or other receptacle) that admits liquid fuel and air in controllable quantities; the fuel is vaporized by the heat of combustion and mixed with the air in appropriate proportions. Combustion air may be induced by natural draft or forced into the vaporizing bowl by a fan. Indoor air is generally used for combustion and draft dilution. Window-installed units have the burner section outdoors. Both natural- and forced-convection heating units are available. A small blower is sold as an option on some models. The heat exchanger, usually cylindrical, is made of steel (Figure 4). These heaters are available as room units (both radiant and circulation), floor furnaces, and recessed wall heaters. They may also be installed in a window, depending on the cabinet construction. The heater is always vented to the outside. A 3 to 5 gal fuel tank may be attached to the heater, or a larger outside tank can be used.

Vaporizing pot burners are equipped with a single constant-level and metering valve. Fuel flows by gravity to the burner through the adjustable metering valve. Control can be manual, with an off pilot and variable settings up to maximum, or it can be thermostatically

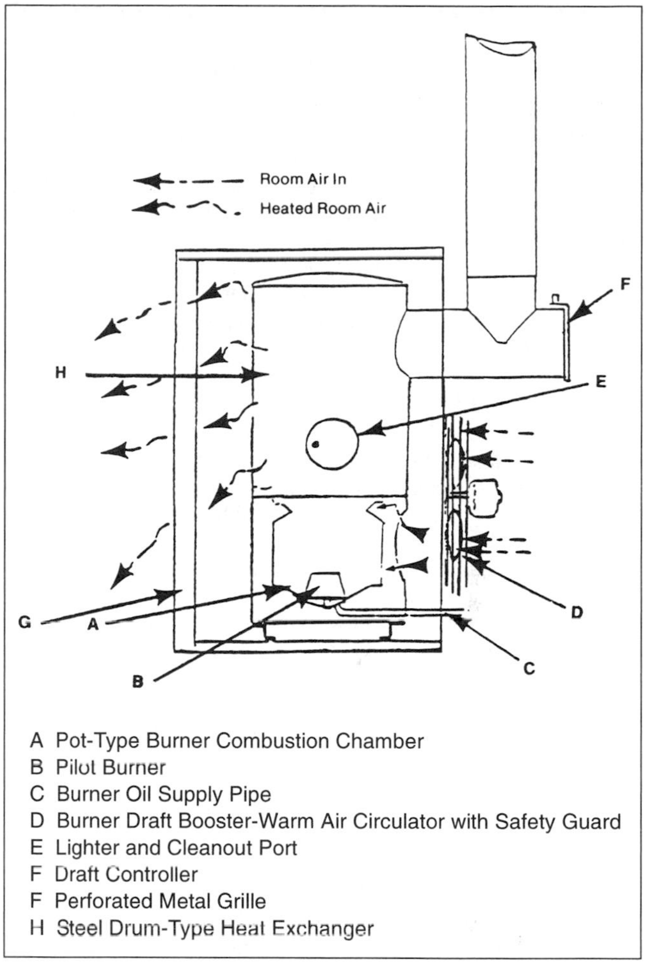

Fig. 4 Oil-Fueled Heater with Vaporizing Pot-Type Burner

controlled, with the burner operating at a selected firing rate between pilot and high.

Powered Atomizing Heaters

Wall furnaces, floor furnaces, and freestanding room heaters are also available with a powered gun-type burner using No. 1 or No. 2 fuel oil. For more information, refer to Chapter 26.

Portable Kerosene Heaters

Because kerosene heaters are not normally vented, precautions must be taken to provide sufficient ventilation. Kerosene heaters are of four basic types: radiant, natural-convection, direct-fired, forced-convection, and catalytic.

The radiant kerosene heater has a reflector, while the natural convection heater is cylindrical in shape. Fuel vaporizes from the surface of a wick, which is immersed in an integral fuel tank of up to 2 gal capacity similar to that of a kerosene lamp. Fuel-burning rates range from about 5000 to 22,500 Btu/h. Radiant heaters usually have a removable fuel tank to facilitate refueling.

The direct-fired, forced-convection portable kerosene heater has a vaporizing burner and a heat-circulating fan. These heaters are available with thermostatic control and variable heat output.

The catalytic type uses a metal catalyst to oxidize the fuel. It is started by lighting kerosene at the surface; however, after a few moments, the catalyst surface heats to the point that flameless oxidation of the fuel begins.

ELECTRIC IN-SPACE HEATERS

Wall, Floor, Toe Space, and Ceiling Heaters

Heaters for recessed or surface wall mounting are made with open wire or enclosed, metal-sheathed elements. An inner liner or reflector is usually placed between the elements and the casing to promote circulation and minimize the rear casing temperature. Heat is distributed by both convection and radiation; the proportion of each depends on unit construction.

Ratings are usually 1000 to 5000 W at 120, 208, 240, or 277 V. Models with air circulation fans are available. Other types can be recessed into the floor. Electric convectors should be placed so that air moves freely across the elements.

Baseboard Heaters

These heaters consist of a metal cabinet containing one or more horizontal, enclosed, metal-sheathed elements. The cabinet is less than 6 in. in overall depth and can be installed 18 in. above the floor; the ratio of the overall length to the overall height is more than two to one.

Units are available from 2 to 12 ft in length, with ratings from 100 to 400 W/ft, and they fit together to make up any desired continuous length or rating. Electric hydronic baseboard heaters containing immersion heating elements and an antifreeze solution are made with ratings of 300 to 2000 W. The placement of any type of electric baseboard heater follows the same principles that apply to baseboard installations (see Chapter 32) because baseboard heating is primarily perimeter heating.

RADIANT HEATING SYSTEMS

Heating Panels and Heating Panel Sets

These systems have electric resistance wire or etched or graphite elements embedded between two layers of insulation. High-density thermal insulation behind the element minimizes heat loss, and the outer shell is formed steel with baked enamel finish. Heating panels provide supplementary heating by convection and radiation. They can be recessed into or surface mounted on hard surfaces or fit in standard T-bar suspended ceilings.

Units are usually rated between 250 and 1000 W in sizes varying from 24 in. by 24 in. to 24 in. by 96 in. in standard voltages of 120, 208, 240, and 277 V.

Embedded Cable and Storage Heating Systems

Ceiling and floor electric radiant heating systems that incorporate embedded cables are covered in Chapter 6. Electric storage systems, including room storage heaters and floor slab systems, are covered in Chapter 33 of the 1999 *ASHRAE Handbook—Applications*.

Cord-Connected Portable Heaters

Portable electric heaters are often used in areas that are not accessible to central heat. They are also used to maintain an occupied room at a comfortable level independent of the rest of the residence.

Portable electric heaters for connection to 120 V, 15 A outlets are available with outputs of 2050 to 5100 Btu/h (600 to 1500 W), the most common being 1320 and 1500 W. Many heaters are available with a selector switch for three wattages (e.g., 1100-1250-1500 W). Heavy-duty heaters are usually connected to 240 V, 20 A outlets with outputs up to 13,700 Btu/h (4000 W), while those for connection to 240 V, 30 A outlets have outputs up to 19,100 Btu/h (5600 W). All electric heaters of the same wattage produce the same amount of heat.

Portable electric heaters transfer heat by one of two predominant methods: radiation and convection. Radiant heaters provide heat for people or objects. An element in front of a reflector radiates heat outward in a direct line. Conventional radiant heaters have ribbon or wire elements. Quartz radiant heaters have coil wire elements encased in quartz tubes. The temperature of a radiant wire element usually ranges between 1200 and 1600°F.

Convection heaters warm the air in rooms or zones. Air flows directly over the hot elements and mixes with room air. Convection heaters are available with or without fans. The temperature of a convection element is usually less than 930°F.

An adjustable, built-in bimetal thermostat usually controls the power to portable electric heaters. Fan-forced heaters usually provide better temperature control because the fan, in addition to cooling the case, forces room air past the thermostat. One built-in control uses a thermistor to signal a solid logic circuit that adjusts wattage and fan speed. Most quartz heaters use an adjustable control that operates the heater for a percentage of total cycle time from 0 (off) to 100% (full-on).

Controls

Low-voltage and line-voltage thermostats with on-off operation are used to control in-space electric heaters. Low-voltage thermostats, operating at 30 V or less, control relays or contactors that carry the rated voltage and current load of the heaters. Because the control current load is small (usually less than 1 A), the small switch can be controlled by a highly responsive sensing element.

Line-voltage thermostats carry the full load of the heaters at rated voltage directly through their switch contacts. Most switches carry a listing by Underwriters Laboratories (UL) at 22 A (resistive), 277 V rating. While most electric in-space heating systems are controlled by remote wall-mounted thermostats, many are available with integral or built-in line-voltage thermostats.

Most low-voltage and line-voltage thermostats use small internal heaters, either fixed or adjustable in heat output, that provide heat anticipation by energizing when the thermostat contacts close. The cycling rate of the thermostat is increased by the use of anticipation heaters, resulting in more accurate control of the space temperature.

Droop is an apparent shift or lowering of the control point and is associated with line-voltage thermostats. In these thermostats, switch heating caused by large currents can add materially to the amount of droop. Most line-voltage thermostats in residential use control room heaters of 3 kW (12.5 A at 240 V) or less. At this moderate load and with properly sized anticipation heaters, the droop experienced is acceptable. Cycling rates and droop characteristics have a significant effect on thermostat performance.

SOLID-FUEL IN-SPACE HEATERS

Most wood-burning and coal-burning devices, except central wood-burning furnaces and boilers, are classified as solid-fuel in-space heaters (see Table 3). An in-space heater can be either a fireplace or a stove.

FIREPLACES

Simple Fireplaces

Simple fireplaces, especially all-masonry and noncirculating metal built-in fireplaces, produce little useful heat. They lend atmosphere and a sense of coziness to a room. Freestanding fireplaces are slightly better heat producers. Simple fireplaces have an average efficiency of about 10%. In extreme cases, the chimney draws more heated air than the fire produces.

The addition of glass doors to the front of a fireplace has both a positive and a negative effect. The glass doors restrict the free flow of indoor heated air up the chimney, but at the same time they restrict the radiation of the heat from the fire into the room.

Table 3 Solid-Fuel In-Space Heaters

Type[a]	Approximate Efficiency[a], %	Features	Advantages	Disadvantages
Simple fireplaces, masonry or prefabricated	−10 to +10	Open front. Radiates heat in one direction only.	Visual beauty.	Low efficiency. Heats only small areas.
High-efficiency fireplaces	25 to 45	Freestanding or built-in with glass doors, grates, ducts, and blowers.	Visual beauty. More efficient. Heats larger areas. Long service life. Maximum safety.	Medium efficiency.
Box stoves	20 to 40	Radiates heat in all directions.	Low initial cost. Heats large areas.	Fire hard to control. Short life. Wastes fuel.
Airtight stoves	40 to 55	Radiates heat in all directions. Sealed seams, effective draft control.	Good efficiency. Long burn times, high heat output. Longer service life.	Can create creosote problems.
High-efficiency catalytic wood heaters	65 to 75	Radiates heat in all directions. Sealed seams, effective draft control.	Highest efficiency. Long burn times, high heat output. Long life.	Creosote problems. High purchase price.

[a]Product categories are general; product efficiencies are approximate.

Factory-Built Fireplaces

A factory-built fireplace consists of a fire chamber, chimney, roof assembly, and other parts that are factory made and intended to be installed as a unit in the field. These fireplaces have fireboxes of refractory-lined metal rather than masonry. Factory-built fireplaces come in both radiant and heat-circulating designs. Typical configurations are open front designs, but corner-opening, three-sided units with openings either on the front or side; four-sided units; and see-through fireplaces are also available.

Radiant Design. The radiant system transmits heat energy from the firebox opening by direct radiation to the space in front of it. These fireplaces may also incorporate such features as an outside air supply and glass doors. Radiant design factory-built fireplaces are primarily used for aesthetic wood burning and typically have efficiencies similar to those of masonry fireplaces (0 to 10%).

Heat-Circulating Design. This unit transfers heat by circulating air around the fire chamber and releasing to the space to be heated. The air intake is generally below the firebox or low on the sides adjacent to the opening, and the heated air exits through grilles or louvers located above the firebox or high on the sides adjacent to it. In some designs, ducts direct heated air to spaces other than the area near the front of the fireplace. Some circulating units rely on natural convection, while others have electric fans or blowers to move air. These energy-saving features typically boosts efficiency 25 to 60%.

Freestanding Fireplaces

Freestanding fireplaces are open-combustion wood-burning appliances that are not built into a wall or chase. One type of freestanding fireplace is a fire pit in which the fire is open all around; smoke rises into a hood and then into a chimney. Another type is a prefabricated metal unit that has an opening on one side. Because they radiate heat to all sides, freestanding fireplaces are typically more efficient than radiant fireplaces.

STOVES

Conventional Wood Stoves

Wood stoves are chimney-connected, solid-fuel-burning room heaters designed to be operated with the fire chamber closed. They deliver heat directly to the space in which they are located. They are not designed to accept ducts and/or pipes for heat distribution to other spaces. Wood stoves are controlled-combustion appliances. Combustion air enters the firebox through a controllable air inlet; the air supply and thus the combustion rate are controlled by the user. Conventional controlled-combustion wood stoves manufactured prior to the mid-1980s typically have overall efficiencies ranging from 40 to 55%.

Most **controlled-combustion** appliances are constructed of steel, cast iron, or a combination of the two metals; others are constructed of soapstone or masonry. Soapstone and masonry have lower thermal conductivities but greater specific heats (the amount of heat that can be stored in a given mass). Other materials such as special refractories and ceramics are used in low-emission appliances. Wood stoves are classified as either radiant or convection (sometimes called circulating) heaters, depending on the way they heat interior spaces.

Radiant wood stoves are generally constructed with single exterior walls, which absorb radiant heat from the fire. This appliance heats primarily by infrared radiation; it heats room air only to the extent that air passes over the hot surface of the appliance.

Convection wood stoves have double vertical walls with an air space between the walls. The double walls are open at the top and bottom of the appliance to permit room air to circulate through the air space. The more buoyant hot air rises and draws in cooler room air at the bottom of the appliance. This air is then heated as it passes over the surface of the inner radiant wall. Some radiant heat from the inner wall is absorbed by the outer wall, but the constant introduction of room temperature air at the bottom of the appliance keeps the outer wall moderately cool. This characteristic generally allows convection wood stoves to be placed closer to combustible materials than radiant wood stoves. Fans in some wood stoves augment the movement of heated air. Convection wood stoves generally provide more even heat distribution than do radiant types.

Advanced Design Wood Stoves

Strict air pollution standards have prompted the development of new stove designs. These clean-burning wood stoves use either catalytic or noncatalytic technology to achieve very high combustion efficiency and to reduce creosote and particulate and carbon monoxide emission levels.

Catalytic combustors are currently available as an integral part of many new wood-burning appliances and are also available as add-on or retrofit units for most existing appliances. The catalyst may be platinum, palladium, rhodium, or a combination of these elements. It is bonded to a ceramic or stainless steel substrate. A catalytic combustor's function in a wood-burning appliance is to substantially lower the ignition temperatures of unburned gases, solids, and/or liquid droplets (from approximately 1000°F to 500°F). As these unburned combustibles leave the main combustion chamber and pass through the catalytic combustor, they ignite and burn rather than enter the atmosphere.

For the combustor to efficiently burn the gases, the proper amount of oxygen and a sufficient temperature to maintain ignition are required; further, the gases must have sufficient residence time in the combustor. A properly operating catalytic combustor has a temperature in the range of 1000 to 1700°F. Catalyst-equipped wood stoves have a default efficiency, as determined by the U.S. Environmental Protection Agency (EPA), of 72%, although many stoves are considerably more efficient. This EPA default efficiency is the value one standard deviation below the mean of the efficiencies from a database of stoves.

Another approach to increasing combustion efficiency and meeting emissions requirements is the use of technologically advanced internal appliance designs and materials. Generally, noncatalytic, low-emission wood-burning appliances incorporate high-temperature refractory materials and have smaller fireboxes than conventional appliances. The fire chamber is designed to increase temperature, turbulence, and residence time in the primary combustion zone. Secondary air is introduced to promote continued burning of the gases, solids, and liquid vapors in a secondary combustion zone. Many stoves add a third and fourth burn area within the firebox. The location and design of the air inlets is critical because proper air circulation patterns are the key to approaching complete combustion. Noncatalytic wood stoves have an EPA default efficiency of 63%; however, many models approach 80%.

Fireplace Inserts

Fireplace inserts are closed-combustion wood-burning room heaters that are designed to be installed in an existing masonry fireplace. They combine elements of both radiant and convection wood stove designs. They have large radiant surfaces that face the room and circulating jackets on the sides that capture heat that would otherwise go up the chimney. Inserts may use either catalytic or noncatalytic technology to achieve clean burning.

Pellet-Burning Stoves

Pellet-burning stoves burn small pellets made from wood byproducts rather than burning logs. An electric auger feeds the pellets from a hopper into the fire chamber, where air is blown through, creating very high temperatures in the firebox. The fire burns at such a high temperature that the smoke is literally burned up, resulting in a very clean burn, and no chimney is needed. Instead, the waste gases are exhausted to the outside through a vent. An air intake is operated by an electric motor; another small electric fan blows the heated air from the area around the fire chamber into the room. A microprocessor controls the operation, allowing the pellet-burning stove to be controlled by a thermostat. Pellet-burning stoves typically have the lowest emissions of all wood-burning appliances and have an EPA default efficiency of 78%. Because of the high air-fuel ratios used by pellet-burning stoves, these stoves are excluded from EPA wood stove emissions regulations.

GENERAL INSTALLATION PRACTICES

The criteria to ensure safe operation are normally covered by local codes and ordinances or, in rare instances, by state and federal requirements. Most codes, ordinances, or regulations refer to the following building codes and standards for in-space heating:

Building Codes	
BOCA/National Building Code	BOCA
CABO One- and Two-Family Dwelling Code	CABO
International Building Code	ICC
National Building Code of Canada	NBCC
Standard Building Code	SBCCI
Uniform Building Code	ICBO
Mechanical Codes	
National Mechanical Code	BOCA
Uniform Mechanical Code	ICBO/IAPMO
International Mechanical Code	ICC
Standard Mechanical Code	SBCCI
Electrical Codes	
National Electrical Code	NFPA 70
Canadian Electrical Code	CSA C22.1
Chimneys	
Chimneys, Fireplaces, Vents and Solid Fuel-Burning Appliances	NFPA 211
Chimneys, Factory-Built Residential Type and Building Heating Appliance	UL 103
Solid-Fuel Appliances	
Factory-Built Fireplaces	UL 127
Room Heaters, Solid-Fuel Type	UL 1482

The chapter on Codes and Standards has further information, including the names and addresses of these agencies. Safety and performance criteria are furnished by the manufacturer.

Safety with Solid Fuels

The evacuation of combustion gases is a prime concern in the installation of solid-fuel-burning equipment. NFPA *Standard* 211, Chimneys, Fireplaces, Vents and Solid Fuel-Burning Appliances, lists requirements that should be followed. Because safety requirements for connector pipes (stovepipes) are not always readily available, these requirements are summarized as follows:

- Connector pipe is usually black (or blue) steel single-wall pipe; thicknesses are shown in Table 4. Stainless steel is a corrosion-resistant alternative that does not have to meet the thicknesses listed in Table 4.
- Connectors should be installed with the crimped (male) end of the pipe toward the stove, so that creosote and water drip back into the stove.
- The pipe should be as short as is practical, with a minimum of turns and horizontal runs. Horizontal runs should be pitched 1/4 in. per foot up toward the chimney.
- Chimney connectors should not pass through ceilings, closets, alcoves, or concealed spaces.
- When passing through a combustible interior or exterior wall, connectors must be routed through a listed wall pass-through that has been installed in accordance with the conditions of the listing, or they must follow one of the home-constructed systems recognized in NFPA *Standard* 211 or local building codes. Adequate clearance and protection of combustible materials is extremely important. In general, listed devices are easier to install and less expensive than home-constructed systems.

Creosote forms in all wood-burning systems. The rate of formation is a function of the quantity and type of fuel burned, the appliance in which it is burned, and the manner in which the appliance is operated. Thin deposits in the connector pipe and chimney do not interfere with operation, but thick deposits (greater than 1/4 in.) may ignite. Inspection and cleaning of chimneys connected to wood-burning appliances should be performed on a regular basis (at least annually).

Only the solid fuel that is listed for the appliance should be burned. Coal should be burned only in fireplaces or stoves designed specifically for coal burning. The chimney used in coal-fired applications must also be designed and approved for coal and wood.

Solid-fuel appliances should be installed in strict conformance with the clearance requirements established as part of their safety listing. When clearance reduction systems are used, stoves must remain at least 12 in. and connector pipe at least 6 in. from combustibles, unless smaller clearances are established as part of the listing.

Table 4 Chimney Connector Wall Thickness[a]

Diameter	Gage	Minimum Thickness, in.
Less than 6 in.	26	0.019
6 to 10 in.	24	0.023
10 to 16 in.	22	0.029
16 in. or greater	16	0.056

[a]Do not use thinner connector pipe. Replace connectors as necessary. Leave at least 18 in. clearance between the connector and a wall or ceiling, unless the connector is listed for a smaller clearance or an approved clearance reduction system is used.

Utility-Furnished Energy

Those systems that rely on energy furnished by a utility are usually required to comply with local utility service rules and regulations. The utility usually provides information on the installation and operation of the equipment using their energy. Bottled gas (LPG) equipment is generally listed and tested under the same standards as natural gas. LPG equipment may be identical to natural gas equipment, but it always has a different orifice and sometimes has a different burner and controls. The listings and examinations are usually the same for natural, mixed, manufactured, and liquid petroleum gas.

Products of Combustion

The combustion chamber of equipment that generates products of combustion must be connected by closed piping to the outdoors. Gas-fired equipment may be vented through masonry stacks, chimneys, specifically designed venting, or, in some cases, venting incorporating forced- or induced-draft fans. Chapter 30 covers chimneys, gas vents, and fireplace systems in more detail.

Agency Testing

The standards of several agencies contain guidelines for the construction and performance of in-space heaters. The following list summarizes the standards that apply to residential in-space heating; they are coordinated or sponsored by ASHRAE, the American National Standards Institute (ANSI), Underwriters Laboratories (UL), the American Gas Association (AGA), and the Canadian Gas Association (CGA). Some CGA standards have a CAN1 prefix.

ANSI Z21.11.1	Gas-Fired Room Heaters, Vented
ANSI Z21.11.2	Gas-Fired Room Heaters, Unvented
ANSI Z21.44	Gas-Fired Gravity and Fan-Type Direct-Vent Wall Furnaces
ANSI Z21.48	Gas-Fired Gravity and Fan-Type Floor Furnaces
ANSI Z21.49	Gas-Fired Gravity and Fan-Type Vented Wall Furnaces
ANSI Z21.60/ CSA 2.26-M96	Decorative Gas Appliances for Installation in Solid-Fuel Burning Fireplaces
ANSI Z21.76	Gas-Fired Unvented Catalytic Room Heaters for use with Liquefied Petroleum (LP) Gases
ANSI Z21.50/ CSA 2.22-M98	Vented Gas Fireplaces
ANSI Z21.86/ CSA 2.32-M98	Vented Gas-Fired Space Heating Appliances
ANSI Z21.88/ CSA 2.33-M98	Vented Gas Fireplace Heaters
CAN1-2.1-M86	Gas-Fired Vented Room Heaters
CAN/CGA-2.5-M86	Gas-Fired Gravity and Fan Type Vented Wall Furnaces
CAN1/CGA-2.19-M81	Gas-Fired Gravity and Fan Type Direct Vent Wall Furnaces
ANSI/UL 127	Factory-Built Fireplaces
UL 574	Electric Oil Heaters
UL 647	Unvented Kerosene-Fired Heaters and Portable Heaters
ANSI/UL 729	Oil-Fired Floor Furnaces
ANSI/UL 730	Oil-Fired Wall Furnaces
UL 737	Fireplace Stoves
ANSI/UL 896	Oil-Burning Stoves
ANSI/UL 1042	Electric Baseboard Heating Equipment
ANSI/UL 1482	Heaters, Room Solid-Fuel Type
ASHRAE 62	Ventilation for Acceptable Indoor Air Quality

REFERENCES

DeWerth, D.W. and R.L. Loria. 1989. In-space heater energy use for supplemental and whole house heating. *ASHRAE Transactions* 95(1).

DOE. 1984. Uniform test method for measuring the energy consumption of vented home heating equipment. *Federal Register* 49:12, 169 (March).

GAMA. 1995. Directory of gas room heaters, floor furnaces and wall furnaces. Gas Appliance Manufacturers Association, Arlington, VA.

MacKay, S., L.D. Baker, J.W. Bartok, and J.P. Lassoie. 1985. *Burning wood and coal.* Natural Resources, Agriculture, and Engineering Service, Cornell University, Ithaca, NY.

NFPA. 1996. Standard for chimneys, fireplaces, vents and solid fuel-burning appliances. *Standard* 211-1996. National Fire Protection Association, Quincy, MA.

NFPA/AGA. 1999. *National fuel gas code*. NFPA *Standard* 54-99. National Fire Protection Association, Quincy, MA. ANSI *Standard* Z223.1-99. American Gas Association, Arlington, VA.

Wood Heating Education and Research Foundation. 1984. Solid fuel safety study manual for Level I solid fuel safety technicians. Washington, D.C.

CHAPTER 30

CHIMNEY, GAS VENT, AND FIREPLACE SYSTEMS

A PROPERLY designed chimney controls draft and removes flue gas. This chapter describes the design of chimneys that discharge flue gas from appliance-chimney and fireplace-chimney systems.

In this chapter, **appliance** refers to any furnace, boiler, or incinerator (including the burner). Unless the context indicates otherwise, the term **chimney** includes specialized vent products such as gas vents. **Draft** is negative static pressure, measured relative to atmospheric pressure; thus positive draft is negative static pressure. **Flue gas** is the mixture of gases discharged from the appliance and conveyed by the chimney or vent system.

Appliances can be grouped by draft configuration as follows (Stone 1971):

1. Those that require draft applied at the appliance flue gas outlet to induce air into the appliance
2. Those that operate without draft applied at the appliance flue gas outlet (e.g., a gas appliance with a draft hood in which the combustion process is isolated from chimney draft variations)
3. Those that produce positive pressure at the appliance outlet collar so that no chimney draft is needed; appliances that produce some positive outlet pressure but also need some chimney draft

In the first two configurations, hot flue gas buoyancy, induced-draft chimney fans, or a combination of both produces draft. The third configuration may not require chimney draft, but it should be considered in the design if a chimney is used. If the chimney system is undersized, draft inducers in the connector or chimney may supply draft needs. If the connector or chimney pressure requires control, draft control devices must be used.

The draft Δp needed to overcome chimney flow resistance is as follows:

$$\Delta p = \text{Theoretical draft} - \text{Available draft} = D_t - D_a$$

This equation is based on a neutral (zero) pressure difference between the space surrounding the appliance or fireplace and the atmosphere. If the space surrounding the appliance or fireplace is at a lower pressure than the atmosphere (space depressurized), the pressure difference D_p should also be subtracted from D_t when calculating draft Δp and vice versa. This equation applies to the three above appliance draft configurations; for example, in the second configuration with zero draft requirement at the appliance outlet, available draft is zero, so theoretical draft of the chimney equals the chimney flow resistance.

The preparation of this chapter is assigned to TC 6.10, Fuels and Combustion.

Available draft D_a is the draft needed at the appliance outlet. If increased chimney height and flue gas temperatures provide surplus available draft, draft control is required.

Theoretical draft D_t is the natural draft produced by the buoyancy of hot gases in the chimney relative to cooler gases in the atmosphere. It depends on chimney height, local barometric pressure, and the **mean chimney flue gas temperature difference Δt_m**, which is the difference in temperature between the flue and atmospheric gases. Therefore, cooling by heat transfer through the chimney wall is a key variable in chimney design. Precise evaluation of theoretical draft is not necessary for most design calculations due to the availability of design charts, computer programs, capacity tables in the references, building codes, and manufacturers' data sheets.

Chimney temperatures and acceptable combustible material temperatures must be known in order to determine safe clearances between the chimney and combustible materials. Safe clearances for some chimney systems, such as Type B gas vents, are determined by standard tests and/or specified in building codes.

The following sections cover the basis of chimney design for average operating conditions. For gas-fired appliances, a rigorous evaluation of the flue gas and material surface temperatures in the chimney vent system can be obtained using the VENT-II computer program (Rutz and Paul 1991). For oil-fired appliances, chimney flue gas and material surface temperature evaluations can be obtained using the OHVAP computer program (Strasser et al. 1997).

CHIMNEY FUNCTIONS

The proper chimney can be selected by evaluating such factors as draft, configuration, size, and operating conditions of the appliance; construction of surroundings; appliance usage classification; residential, low, medium, or high heat (NFPA *Standard* 211); and height of building. The chimney designer should know the applicable codes and standards to ensure acceptable construction.

In addition to chimney draft, the following factors must be considered for safe and reliable operation: adequate air supply for combustion; building depressurization effects; draft control devices; chimney materials (corrosion and temperature resistance); flue gas temperatures, composition, and dew point; wind eddy zones; and particulate dispersion. Chimney materials must resist oxidation and condensation at both high and low fire levels.

Start-Up

The equations and design chart (Figure 1) may be used to determine vent or chimney size based on steady-state operating conditions. The equations and chart, however, do not consider modulation, cycling, or time to achieve equilibrium flow conditions from a cold start. While mechanical draft systems can start gas flow,

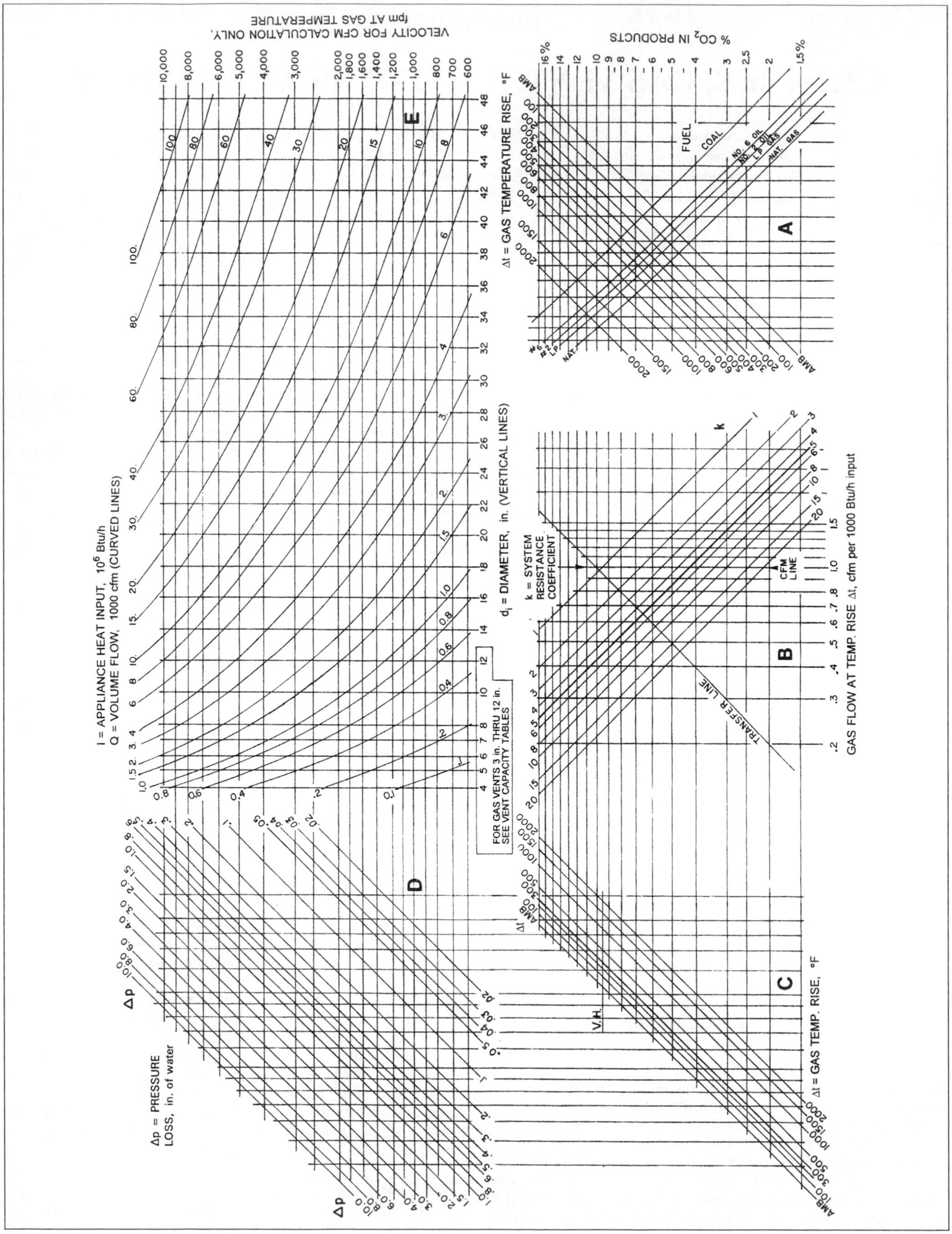

Fig. 1 Design Chart for Vents, Chimneys, and Ducts
(This solution of Equations (11) and (13) applies to combustion products and air.)

gravity systems rely on the buoyancy of hot gases as the sole force to displace the cold air in the chimney. Priming follows Newton's laws of motion. The time to fill a system with hot gases, displace the cold air, and start flow is reasonably predictable and is usually a minute or less. But unfavorable thermal differentials, building-chimney interaction, mechanical equipment (e.g., exhaust fans), or wind forces that oppose the normal flow of vent gases can overwhelm the buoyancy force. Then, rapid priming cannot be obtained solely from correct system design. The VENT-II computer program contains detailed analysis of gas vent and chimney priming and other cold-start considerations and allows for appliance cycling and pressure differentials that affect performance (Rutz and Paul 1991). A copy of the solution methodology (Rutz 1991), including equations, may be ordered from ASHRAE Customer Service.

Air Intakes

All rooms or spaces containing fuel-burning equipment must have a constant supply of combustion air at adequate static pressure to ensure proper combustion. In addition, outside air is required to replace the air entering chimney systems through draft hoods and barometric draft regulators and to ventilate closely confined boiler and furnace rooms.

Because of the variable air requirements, the presence of other air-moving equipment in the building, and the variations in building construction and arrangement, there is no universally accepted rule specifying outdoor air openings. The minimum combustion air opening size depends on burner input, with the practical minimum being the area of the vent connector or chimney. Any design must consider flow resistance of the combustion air supply, including register-louver resistance. Air supply openings that meet most building code requirements have a negligible flow resistance.

Vent Size

Small residential and commercial natural-draft gas appliances need vent diameters of 3 to 12 in. The NFPA/AGA *National Fuel Gas Code* recommends sizes or input capacities for most acceptable gas appliance venting materials. These sizes also apply to gas appliances with integral automatic vent dampers, as well as to appliances with field-installed automatic vent dampers. Field-installed automatic vent dampers should be certified for use with a specific appliance by a recognized testing agency and installed by qualified installers.

Draft Control

Pressure, temperature, and other draft controls have replaced draft hoods in many residential furnaces and boilers to attain higher steady-state and seasonal efficiencies. Appliances that use pulse combustion or forced- or induced-draft fans, as well as those designed for sealed or direct venting, do not have draft hoods but may require special venting and special vent terminals. If fan-assisted burners deliver fuel and air to the combustion chamber and also overcome the appliance flow resistance, draft hoods or other control devices may be installed, depending on the design of the appliance. Category III and IV and some Category I and II appliances, as defined in ANSI *Standard* Z21.47/CSA 2.3, do not use draft hoods; in such cases, the appliance manufacturer's vent system design requirements should be followed. The section on Vent and Chimney Accessories has information on draft hoods, barometric regulators, draft fans, and other draft control devices.

Frequently, a chimney must produce excess flow or draft. For example, dangerously high flue gas outlet temperatures from an incinerator may be reduced by diluting the air in the chimney with excess draft. The section on Mass Flow Based on Fuel and Combustion Products briefly addresses draft control conditions.

Pollution Control

Where control of pollutant emissions is impossible, the chimney should be tall enough to ensure dispersion over a wide area to prevent objectionable ground level concentrations. The chimney can also serve as a passageway to carry flue gas to pollution control equipment. This passageway must meet the building code requirements for a chimney, even at the exit of pollution control equipment, because of possible exposure to heat and corrosion. A bypass chimney should also be provided to allow continued exhaust in the event of pollution control equipment failure, repair, or maintenance.

Equipment Location

Chimney materials may permit the installation of appliances at intermediate or all levels of a high-rise building without imposing penalties due to weight. Some gas vent systems permit individual apartment-by-apartment heating systems.

Wind Effects

Wind and eddy currents affect the discharge of gases from vents and chimneys. A vent or chimney must expel flue gas beyond the cavity or eddy zone surrounding a building to prevent reentry through openings and fresh air intakes. A chimney and its termination can stabilize the effects of wind on appliances and their equipment rooms. In many locations, the equipment room air supply is not at neutral pressure under all wind conditions. Locating the chimney outlet well into the undisturbed wind stream and away from the cavity and wake zones around a building both counteracts wind effects on the air supply pressure and prevents reentry through openings and contamination of fresh air intakes.

Chimney outlets below parapet or eaves level, nearly flush with the wall or roof surface, or in known regions of stagnant air may be subjected to downdrafts and are undesirable. Caps for downdraft and rain protection must be installed according to either their listings and the cap manufacturer's instructions or the applicable building code.

Wind effects can be minimized by locating the chimney terminal and the combustion air inlet terminal close together in the same pressure zone.

Safety Factors

Safety factors allow for uncertainties of vent and chimney operation. For example, flue gas must not spill from a draft hood or barometric regulator, even when the chimney has very low available draft. The Table 2 design condition for gas vents (i.e., 300°F rise in the system at 5.3% CO_2) allows gas vents to operate with reasonable safety above or below the suggested temperature and CO_2 limits.

Safety factors may also be added to the system friction coefficient to account for a possible extra fitting, soot accumulation, and air supply resistance. The specific gravity of flue gas can vary depending on the fuel burned. Natural gas flue gas, for example, has a density as much as 5% less than air, while coke flue gas has a density as much as 8% greater. However, these density changes are insignificant relative to other uncertainties, so no compensation is needed.

STEADY-STATE CHIMNEY DESIGN EQUATIONS

Chimney design balances the forces that produce flow against those that retard flow (friction). **Theoretical draft** is the pressure that produces flow in gravity or natural-draft chimneys. It is defined as the static pressure resulting from the difference in densities between a stagnant column of hot flue gas and an equal column of ambient air. In the design or balancing process, theoretical draft may not equal friction loss because the appliance is frequently built to operate with some specific pressure (positive or negative) at the appliance flue gas exit. This exit pressure, or **available draft**, depends on appliance operating characteristics, fuel, and type of draft control.

Flow losses caused by friction may be estimated by several formulas for flow in pipes or ducts, such as the equivalent length method or the loss coefficient (velocity head) method. Chapter 32 of the 1997 *ASHRAE Handbook—Fundamentals* covers computation of flow losses. This chapter emphasizes the loss coefficient method because fittings usually cause the greater portion of system pressure drop in chimney systems, and conservative loss coefficients (which are almost independent of piping size) provide an adequate basis for design.

Rutz and Paul (1991) developed a computer program entitled VENT-II: An Interactive Personal Computer Program for Design and Analysis of Venting Systems for One or Two Gas Appliances, which predicts flows, temperatures, and pressures in venting systems. Similarly, Strasser et al. (1997) developed a computer program entitled OHVAP: Oil-Heat Vent Analysis Program.

For large gravity chimneys, available draft D_a may be calculated from Equation (1) or (2). Both equations use the equivalent length approach, as indicated by the symbol L_e in the flow-loss term (ASHVE 1941). These equations permit consideration of the density difference between chimney gases and ambient air and compensation for shape factors. Mean flue gas temperature T_m must be estimated separately.

For a cylindrical chimney,

$$D_a = 2.96HB\left(\frac{\rho_o}{T_o} - \frac{\rho_c}{T_m}\right) - \frac{0.000315w^2 T_m f L_e}{1.3 \times 10^7 d_f^5 B \rho_c} \quad (1)$$

For a rectangular chimney,

$$D_a = 2.96HB\left(\frac{\rho_o}{T_o} - \frac{\rho_c}{T_m}\right) - \frac{0.000097w^2 T_m f L_e (x+y)}{1.3 \times 10^7 (xy)^3 B \rho_c} \quad (2)$$

where

H = height of vent or chimney system above grade or system inlet, ft
B = existing or local barometric pressure, in. Hg
ρ_o = density of air at 0°F and 29.92 in. Hg, lb/ft^3
ρ_c = density of chimney gas at average temperature and local barometric pressure, lb/ft^3
T_o = ambient temperature, °R
T_m = mean flue gas temperature at average conditions in system, °R
w = mass flow of gas, lb/h
f = Darcy friction factor
L_e = total piping length L plus equivalent length of elbows, ft
d_f = inside diameter, ft
x = length of one internal side of rectangular chimney cross section, ft
y = length of other internal side of rectangular chimney cross section, ft

In these equations, the first term determines theoretical draft, based on applicable gas and ambient density. The second term defines draft loss based on the factors for flow in a circular or rectangular duct system. To use these equations, the mass flow w for a variety of fuels and situations and the available draft needs of various types of appliances must be determined.

Equations (1) and (2) may be derived and expressed in a form that is more readily applied to the problems of chimney design, size, and capacity by considering the following factors, which are the steps used to solve the problems in the section on Chimney Capacity Calculation Examples.

1. Mass flow of combustion products
2. Chimney gas temperature and density
3. Theoretical and available draft
4. System pressure loss due to flow
5. Chimney gas velocity
6. System resistance coefficient
7. Final input-volume relationships

For applications to system design, the chimney gas velocity step is eliminated; however, actual velocity can be found readily, if needed.

1. Mass Flow of Combustion Products

Mass flow in a chimney or venting system may differ from that in the appliance, depending on the type of draft control or number of appliances operating in a multiple-appliance system. Mass flow is preferred to volumetric flow because it remains constant in any continuous portion of the system, regardless of changes in temperature or pressure. For the chimney gases resulting from any combustion process, mass flow can be expressed as

$$w = IM/1000 \quad (3)$$

where

w = mass flow rate, lb/h
I = appliance heat input, Btu/h
M = ratio of mass flow to heat input, lb of chimney products per 1000 Btu of fuel burned. M depends on fuel composition and percentage excess air (or CO_2) in the chimney.

2. Mean Chimney Gas Temperature and Density

Chimney gas temperature, which is covered in the section on Chimney Gas Temperature and Heat Transfer, depends on the fuel, appliance, draft control, chimney size, and configuration.

Density of gas within the chimney and theoretical draft both depend on gas temperature. Although the gases flowing in a chimney system lose heat continuously from entrance to exit, a single mean gas temperature must be used in either the design equation or the chart. Mean chimney gas density is essentially the same as air density at the same temperature. Thus, density may be found as

$$\rho_m = \rho_a \frac{T_s B}{T_m B_o} = 1.325 \frac{B}{T_m} \quad (4)$$

where

ρ_m = gas density, lb/ft^3
T_s = standard temperature = 518.67°R
ρ_a = air density at T_s and B_o = 0.0765 lb/ft^3
B = local barometric pressure, in. Hg
T_m = mean flue gas temperature at average system conditions, °R
B_o = standard pressure = 29.92 in Hg

The density ρ_a in Equation (4) is a compromise value for typical humidity. The subscript m for density and temperature requires that these properties be calculated at mean gas temperature or vertical midpoint of a system (inlet conditions can be used where temperature drop is not significant).

3. Theoretical Draft

The theoretical draft of a gravity chimney or vent is the difference in weight between a given column of warm (light) chimney gas and an equal column of cold (heavy) ambient air. Chimney gas density or temperature, chimney height, and barometric pressure determine theoretical draft; flow is not a factor. The equation for theoretical draft assumes chimney gas density is the same as that of air at the same temperature and pressure; thus,

$$D_t = 0.2554BH\left(\frac{1}{T_o} - \frac{1}{T_m}\right) \quad (5)$$

where

D_t = theoretical draft, in. of water
H = height of chimney above grade or inlet, ft
T_o = ambient temperature, °R

Theoretical draft thus increases directly with height and with the difference in density between the hot and cold columns.

4. System Pressure Loss due to Flow

In any chimney system, flow losses, expressed as pressure drop Δp in inches of water, absorb the difference between theoretical and available draft:

$$\Delta p = D_t - D_a \quad (6)$$

Available draft D_a is the static pressure defined by the appliance operating requirements as follows:

- Positive draft (negative chimney pressure) appliances: D_a is positive in Equation (6), thus $\Delta p < D_t$.
- Draft hood (neutral draft) appliances: $D_a = 0$, and $\Delta p = D_t$.
- Negative draft (positive pressure, above atmospheric, forced draft) appliances: D_a is negative, so $\Delta p = D_t - (-D_a) = D_t + D_a$, and $\Delta p > D_t$.

Regardless of the sign of D_a, Δp is always positive.

In any duct system, flow losses resulting from velocity and resistance can be determined from the Bernoulli equation:

$$\Delta p = \frac{k\rho_m V^2}{5.2(2g)} \quad (7)$$

where

k = dimensionless system resistance coefficient of piping and fittings
V = system gas velocity at mean conditions, fps
g = gravitational constant = 32.1740 ft/s^2

Pressure losses are thus directly proportional to the resistance factor and to the square of the velocity.

5. Chimney Gas Velocity

Velocity in a chimney or vent varies inversely with gas density ρ_m and directly with mass flow rate. The equation for gas velocity at mean gas temperature in the chimney is

$$V = \frac{144 \times 4w}{3600\pi\rho_m d_i^2} \quad (8)$$

where

V = gas velocity, fps
d_i = inside diameter, in.
ρ_m = gas density, lb/ft^3

To express velocity as a function of input and chimney gas composition, w in Equation (8) is replaced by using Equation (3):

$$V = \frac{144 \times 4}{3600\pi\rho_m d_i^2} \times \frac{IM}{1000} \quad (9)$$

Thus, chimney velocity depends on the product of heat input I and the ratio M of mass flow to input.

6. System Resistance Coefficient

The velocity head method of determining resistance losses assigns a fixed numerical coefficient (independent of velocity) or k factor to every fitting or turn in the flow circuit, as well as to piping.

7. Input, Diameter, and Temperature Relationships

To obtain a design equation in which all terms are readily defined, measured, or predetermined, the gas velocity and density terms must be eliminated. Using Equation (4) to replace ρ_m and Equation (9) to replace V in Equation (7) gives

$$\Delta p = \frac{k\rho_m V^2}{5.2(2g)} = \frac{k}{5.2(2g)}\left(\frac{T_m}{1.325B}\right)\left(\frac{144 \times 4IM}{3.6\times10^6\pi d_i^2}\right)^2 \quad (10)$$

Rearranging to solve for I and including the values of π and $2g$ gives

$$I = 4.13\times10^5\,\frac{d_i^2}{M}\left(\frac{\Delta pB}{kT_m}\right)^{0.5} \quad (11)$$

Solving for input using Equation (11) is a one-step process, given the diameter and configuration of the chimney. More frequently, however, input, available draft, and height are given and the diameter d_i must be found. Because system resistance is a function of the chimney diameter, a trial resistance value must be assumed to calculate a trial diameter. This method allows for a second (and usually accurate) solution for the final required diameter.

8. Volumetric Flow in Chimney or System

Volumetric flow Q may be calculated in a chimney system for which Equation (11) can be solved by solving Equation (7) for velocity at mean density (or temperature) conditions:

$$V = 18.3\sqrt{\Delta p/k\rho_m} \quad (12)$$

This equation can be expressed terms of the same variables as Equation (11) by using the density value ρ_m of Equation (4) in Equation (12) and then substituting Equation (12) for velocity V. Area is expressed in terms of d_i. Multiplying area and velocity and adjusting for units,

$$Q = 5.2d_i^2(\Delta pT_m/kB)^{0.5} \quad (13)$$

where Q = volumetric flow rate, cfm. The volumetric flow obtained from Equation (13) is at mean gas temperature T_m and at local barometric pressure B.

Equation (13) is useful in the design of forced-draft and induced-draft systems because draft fans are usually specified in terms of volumetric flow rate at some standard ambient or selected gas temperature. An induced-draft fan is necessary for chimneys that are undersized, that are too low, or that must be operated with draft in the manifold under all conditions.

Figure 1 is a graphic solution for Equations (11) and (13) that is accurate enough for most problems. However, to use either the equations or the design chart, the details in the sections to follow should be understood so that proper choices can be made for mass flow, pressure loss, and heat transfer effects. Neither Figure 1 nor the equations contain the same number or order of steps as the derivation; for example, a step disappears when theoretical and available draft are combined into Δp. Similarly, the examples selected vary in their sequence of solution, depending on which parameters are known and on the need for differing answers, such as diameter

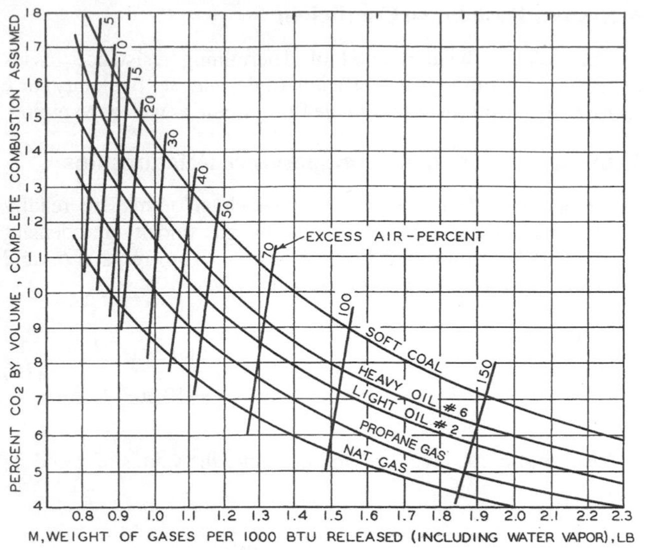

Fig. 2 Graphical Evaluation of Rate of Vent Gas Flow from Percent CO_2 and Fuel Rate

for a given input, diameter versus height, or the amount of pressure boost D_b from a forced-draft fan.

MASS FLOW BASED ON FUEL AND COMBUSTION PRODUCTS

In chimney system design, the composition and flow rate of the flue gas must be assumed to determine the ratio M of mass flow to input. The mass flow equations in Table 1 illustrate the influence of fuel composition; however, additional guidance is needed for system design. The information provided with many heat-producing appliances is limited to whether they have been tested, certified, listed, or approved to comply with applicable standards. From this information and from the type of fuel and draft control, certain inferences can be drawn regarding the flue gas. Table 2 suggests typical values for the vent or chimney systems for gaseous and liquid fuels when specific outlet conditions for the appliance are not known. When combustion conditions are given in terms of excess air, Figure 2 can be used to estimate CO_2.

Figure 3 can be used with Figure 1 to estimate mass and volumetric flow. Flow conditions within the chimney connector, manifold, vent, and chimney vary with configuration and appliance design and are not necessarily the same as boiler or appliance outlet conditions. The equations and design chart (Figure 1) are based on the fuel combustion products and temperatures within the chimney system. If a gas appliance with draft hood is used, Table 2 recommends that dilution air through the draft hood reduce the CO_2 percentage to 5.3%. For equipment using draft regulators, the dilution and temperature reduction is a function of the draft regulator gate opening, which depends on excess draft. If the chimney system produces the exact draft necessary for the appliance, little dilution takes place.

For manifolded gas appliances that have draft hoods, the dilution through draft hoods of inoperative appliances must be considered in precise system design. However, with forced-draft appliances having wind box or inlet air controls, dilution through inoperative appliances may be unimportant, especially if pressure at the outlet of inoperative appliances is neutral (atmospheric level).

Mass flow within incinerator chimneys must account for the probable heat value of the waste, its moisture content, and the use of additional fuel to initiate or sustain combustion. Classifications of waste and corresponding values of M in Table 3 are based on recommendations from the Incinerator Institute of America. Combustion data given for Types 0, 1, and 2 waste do not include any additional fuel. Where constant burner operation accompanies the combustion of waste, the additional quantity of products should be considered in the chimney design.

The system designer should obtain exact outlet conditions for the maximum rate operation of the specific appliance. This information can reduce chimney construction costs. For appliances with higher seasonal or steady-state efficiencies, however, special attention should be given to the manufacturer's venting recommendations because the flue products may differ in composition and temperature from conventional values.

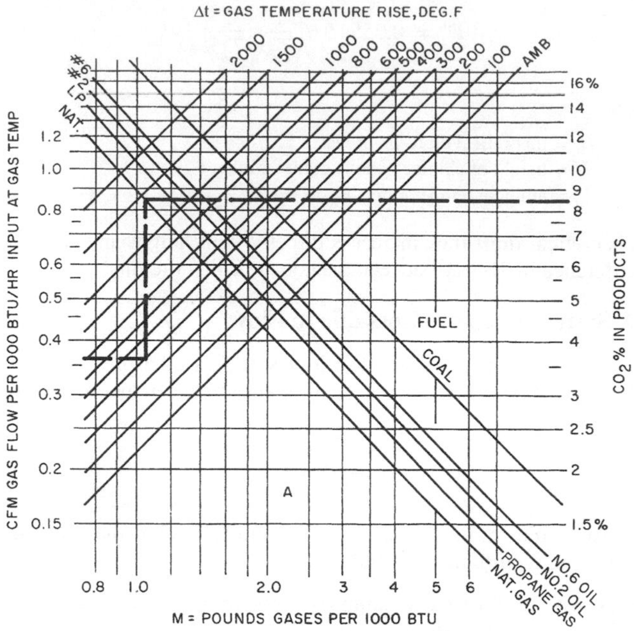

EXAMPLE: NAT. GAS AT 8.5% CO_2 AND 300° RISE
READ M = 1.05 POUNDS GASES PER 1000 BTU
CFM = .36 PER 1000 BTUH INPUT

Fig. 3 Flue Gas Mass and Volumetric Flow

Table 1 Mass Flow Equations for Common Fuels

Fuel	Ratio M of Mass Flow to Input[a]: $M = \frac{\text{lb Total Products}^b}{\text{1000 Btu Fuel Input}}$
Natural gas	$0.705\left(0.159 + \frac{10.72}{\%CO_2}\right)$
LPG (propane, butane, or mixture)	$0.706\left(0.144 + \frac{12.61}{\%CO_2}\right)$
No. 2 oil (light)	$0.72\left(0.12 + \frac{14.4}{\%CO_2}\right)$
No. 6 oil (heavy)	$0.72\left(0.12 + \frac{15.8}{\%CO_2}\right)$
Bituminous coal (soft)	$0.76\left(0.11 + \frac{18.2}{\%CO_2}\right)$
Type 0 waste or wood	$0.69\left(0.16 + \frac{19.7}{\%CO_2}\right)$

[a]Percent CO_2 is determined in products with water condensed (dry basis).
[b]Total products include combustion products and excess air.

Table 2 Typical Chimney and Vent Design Conditions[a]

Fuel	Appliance	% CO_2	Temperature Rise, °F	*M* (Mass Flow/Input), lb Total Products[b] per 1000 Btu Fuel Input	Vent Gas Density,[c] lb/ft³	Flow Rate/Unit Heat Input,[c] cfm per 1000 Btu/h at Gas Temperature
Natural gas	Draft hood	5.3	300	1.54	0.0483	0.531
Propane gas	Draft hood	6.0	300	1.59	0.0483	0.549
Natural gas						
Low-efficiency	No draft hood	8.0	400	1.06	0.0431	0.410
High-efficiency	No draft hood	7.0	240	1.19	0.0522	0.381
No. 2 oil	Residential	9.0	500	1.24	0.0389	0.532
Oil	Forced-draft, over 400,000 Btu/h	13.5	300	0.86	0.0483	0.293
Waste, Type 0	Incinerator	9.0	1340	1.62	0.0213	1.268

[a]The values tabulated are for appliances with flue losses of 17% or more. For appliances with lower flue losses (high-efficiency types), see appliance installation instructions or ask manufacturer for operating data.
[b] Total products include combustion products and excess air.
[c] At sea level and 60°F ambient temperature.

Table 3 Mass Flow for Incinerator Chimneys

Type of Waste	Heat Value of Waste,[a] Btu/lb	Auxiliary Fuel[a] per Unit Waste, Btu/lb	Combustion Products: cfm/lb Waste at 1400°F[b]	Combustion Products: lb per lb Waste[c]	Combustion Products: *M*, lb Products per 1000 Btu
Type 0	8500	0	10.74	13.76	1.62
Type 1	6500	0	8.40	10.80	1.66
Type 2	4300	0	5.94	7.68	1.79
Type 3	2500	1500	4.92	6.25	2.50
Type 4	1000	3000	4.14	5.33	5.33

[a]Auxiliary fuel may be used with any type of waste, depending on incinerator design.
[b]Specialized units may produce higher or lower outlet gas temperatures, which must be considered in sizing the chimney, using Equation (11), Equation (13), or Figure 1.
[c]Multiply these values by pounds of waste burned per hour to establish mass flow.

Table 4 Mean Chimney Gas Temperature for Various Appliances

Appliance Type	Mean Temperature t_m[a] in Chimney, °F
Natural gas-fired heating appliance with draft hood (low-efficiency)	360
LP gas-fired heating appliance with draft hood (low-efficiency)	360
Gas-fired heating appliance, no draft hood	
Low-efficiency	460
High-efficiency	300
Oil-fired heating appliance (low-efficiency)	560
Conventional incinerator	1400
Controlled air incinerator	1800 to 2400
Pathological incinerator	1800 to 2800
Turbine exhaust	900 to 1400
Diesel exhaust	900 to 1400
Ceramic kiln	1800 to 2400

[a]Subtract 60°F ambient to obtain temperature rise for use with Figure 1.

CHIMNEY GAS TEMPERATURE AND HEAT TRANSFER

Figure 1 is based on a design ambient temperature of 60°F (520°R), and all temperatures given are in terms of rise above this ambient. Thus, the 300°F line indicates a 360°F observed vent gas temperature. Using a reasonably high ambient (such as 60°F) for design ensures improved operation of the chimney when ambient temperatures drop because temperature differentials and draft increase.

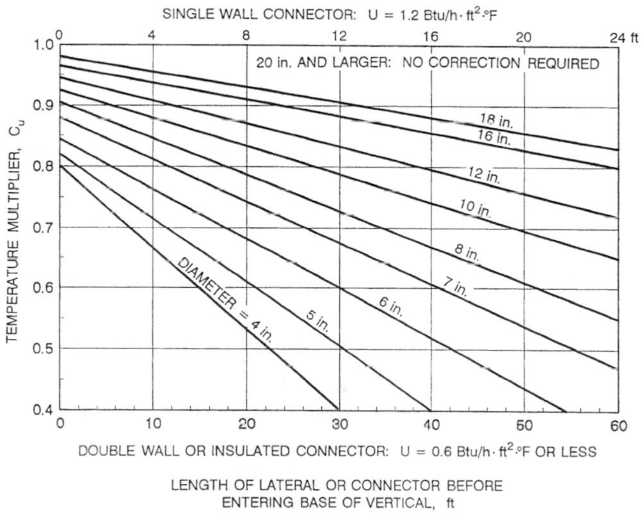

Fig. 4 Temperature Multiplier C_u for Compensation of Heat Losses in Connector

A design requires assuming an initial or inlet chimney gas temperature. In the absence of specific data, Table 4 provides a conservative temperature. For appliances capable of operating over a range of temperatures, size should be calculated at both extremes to ensure an adequate chimney.

The drop in vent gas temperature from appliance to exit reduces capacity, particularly in sizes of 12 in. or less. In gravity Type B gas vents, which may be as small as 3 in. in diameter, and in other systems used for venting gas appliances, capacity is best determined from the *National Fuel Gas Code*. In this code, the tables compensate for the particular characteristics of the chimney material involved, except for very high single-wall metal pipe. Between 12 and 18 in. diameters, the effect of heat loss diminishes greatly because there is greater gas flow relative to system surface area. For 20 in. and greater diameters, cooling has little effect on final size or capacity.

A straight vertical vent or chimney directly off the appliance requires little compensation for cooling effects, even with smaller sizes. However, a horizontal connector running from the appliance to the base of the vent or chimney has enough heat loss to diminish draft and capacity. Figure 4 is a plot of temperature correction C_u, which is a function of connector size, length, and material for either conventional single-wall metal connectors or double-wall metal connectors.

To use Figure 4, estimate connector size and length and read the temperature multiplier. For example, 16 ft of 7 in. diameter single-

Table 5 Overall Heat Transfer Coefficients of Various Chimneys and Vents

Material	U, Btu/h · ft² · °F[a] Observed	Design	Remarks
Industrial steel stacks	—	1.3	Under wet wind
Clay or iron sewer pipe	1.3 to 1.4	1.3	Used as single-wall material
Asbestos-cement gas vent	0.72 to 1.42	1.2	Tested per UL *Standard* 441
Black or painted steel stove pipe	—	1.2	Comparable to weathered galvanized steel
Single-wall galvanized steel	0.31 to 1.38	1.0	Depends on surface condition and exposure
Single-wall unpainted pure aluminum	—	1.0	No. 1100 or other bright-surface aluminum alloy
Brick chimney, tile-lined	0.5 to 1.0	1.0	For gas appliances in residential construction per NFPA *Standard* 211
Double-wall gas vent, 1/4 in. air space	0.37 to 1.04	0.6	Galvanized steel outer pipe, pure aluminum inner pipe; tested per UL *Standard* 441
Double-wall gas vent, 1/2 in. air space	0.34 to 0.7	0.4	
Insulated prefabricated chimney	0.34 to 0.7	0.3	Solid insulation meets UL *Standard* 103 when chimney is fully insulated

[a]U-factors based on inside area of chimney.

wall connector has a multiplier of 0.61. If inlet temperature rise Δt_e above ambient is 300°F, operating mean temperature rise Δt_m will be 0.61 × 300 = 183°F. This factor adequately corrects the temperature at the midpoint of the vertical vent for heights up to 100 ft.

The temperature multiplier must be applied to Grids A and C of Figure 1 as follows:

1. In Grid A, the entering temperature rise Δt_e must be multiplied by 0.61. For an appliance with an outlet temperature rise above ambient of 300°F, flow in the vent is based on $\Delta t_m = 300 \times 0.61 = 183$°F rise.
2. This same 183°F rise must be used in Grid C.
3. Determine Δp using a 183°F rise for theoretical draft to be consistent with the other two temperatures. (It is incorrect to multiply theoretical draft pressure by the temperature multiplier.)

The first trial solution for diameter, using Figure 1 or Equations (11) and (13), need not consider the cooling temperature multiplier, even for small sizes. A first approximate size can be used for the temperature multiplier for all subsequent trials because capacity is insensitive to small changes in temperature.

The correction procedure includes the assumption that the overall heat transfer coefficient of a vertical chimney is approximately 0.6 Btu/h · ft² · °F or less—the value for double-wall metal. This procedure does not correct for cooling in very high stacks constructed entirely of single-wall metal, especially those exposed to cold ambient temperatures. For severe exposures or excessive heat loss, a trial calculation assuming a conservative operating temperature shows whether capacity problems will be encountered.

For more precise heat loss calculations, Table 5 suggests overall heat transfer coefficients for various constructions installed in typical environments at usual flue gas flow velocities (Segeler 1965). For masonry, any additional thickness beyond the single course of brick plus tile liner used in residential chimneys decreases the coefficient.

Equations (11) and (13) do not account for the effects of heat transfer or cooling on flow, draft, or capacity. Equations (14) and (15) describe flow and heat transfer, respectively, within a venting system with $D_a = 0$.

$$q_m = 3600 T_s \sqrt{2gc_p} \frac{B}{B_o} \frac{A}{T_m} \left(\frac{H}{kT_o}\right)^{0.5} \Delta t_m^{\;1.5} \rho_m \qquad (14)$$

$$\frac{q}{q_m} = \frac{\Delta t_e}{\Delta t_m} = \exp\left(\pi \bar{U} d_f \frac{L_m \Delta t_m}{q_m}\right) \qquad (15)$$

where

A = area of passage cross section, ft²
d_f = inside diameter, ft
$\exp x = e^x$
H = height of chimney above grade or inlet, ft
L_m = length from inlet to location (in the vertical) of mean gas temperature, ft
q = heat flow rate at vent inlet, Btu/h
q_m = heat flow rate at midpoint of vent, Btu/h
$\Delta t_e = (T - T_o)$ = temperature difference entering system, °F
$\Delta t_m = (T_m - T_o)$ = chimney gas mean temperature rise, °F
T_o = 520°R
$\bar{U}$ = heat transfer coefficient, Btu/h · ft² · °F
ρ_m = mean gas density from Equation (4)

Assuming reasonable constancy of $\bar{U}$, the overall heat transfer coefficient of the venting system material, Equations (14) and (15) provide a solution for maximum vent gas capacity. They can also be used to develop cooling curves or calculate the length of pipe where internal moisture condenses. Kinkead (1962) details methods of solution and application to both individual and combined gas vents.

THEORETICAL DRAFT, AVAILABLE DRAFT, AND ALTITUDE CORRECTION

Equation (5) for theoretical draft is the basis for Figure 5, which can be used up to 1000°F and 7000 ft elevation. Theoretical draft should be estimated and included in system calculations, even for appliances producing considerable positive outlet static pressure, to achieve the economy of minimum chimney size. Equation (5) may be used directly to calculate exact values for theoretical draft at any altitude. For ease of application and consistency with Figure 1, Table 6 lists approximate theoretical draft for typical gas temperature rises above 60°F ambient.

Appliances with fixed fuel beds, such as hand-fired coal stoves and furnaces, require positive available draft (negative gage pressure). Small oil heaters with pot-type burners, as well as residential furnaces with pressure atomizing oil burners, need positive available draft, which can usually be set by following the manufacturer's instructions for setting the draft regulator. Available draft requirements for larger packaged boilers or equipment assembled from components may be negative, zero (neutral), or positive.

Compensation of theoretical draft for altitude or barometric pressure is usually necessary for appliances and chimneys functioning at elevations greater than 2000 ft. Depending on the design, one of the following approaches to pressure or altitude compensation is necessary for chimney sizing.

1. Figure 1: Use sea level theoretical draft.
2. Equation (11): Use local theoretical draft with actual energy input, or use sea level theoretical draft with energy input multiplied by ratio of sea level to local barometric pressure (Table 7 factor).
3. Equation (13): Use local theoretical draft and barometric pressure with volumetric flow at the local density.

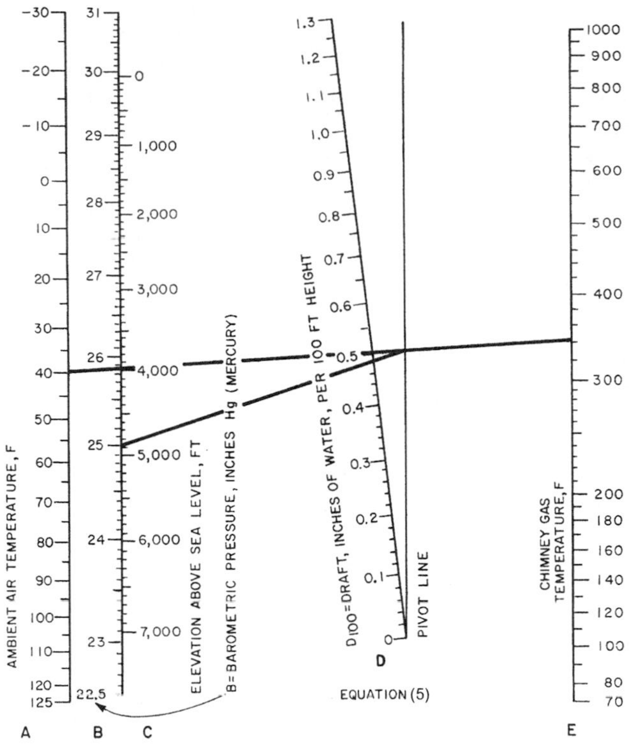

Fig. 5 Theoretical Draft Nomograph

Table 6 Approximate Theoretical Draft of Chimneys

Vent Gas Temperature Rise, °F	D_t per 100 ft, in. of water
100	0.2
150	0.3
200	0.4
300	0.5
400	0.6
500	0.7
600	0.8
800	0.9
1100	1.0
1600	1.1
2400	1.2

Notes: Ambient temperature = 60°F = 520°R
Chimney gas density = air density
Sea level barometric pressure = 29.92 in. Hg
Equation (5) may be used to calculate exact values for D_t at any altitude.

Theoretical draft must be estimated at sea level to calculate for use with Figure 1. The altitude correction multiplier for input (Table 7) is the only method of correcting to other elevations. Reducing theoretical draft imposes an incorrect compensation on the chart.

Figure 1 may be corrected for altitude or reduced air density by multiplying the operating input by the factor in Table 7. Gas appliances with draft hoods, for example, are usually derated 4% per 1000 ft of elevation above sea level when they are operated at 2000 ft altitude or above. The altitude correction factor derates the design input so that the vent size at altitude for derated gas equipment is effectively the same as at sea level. For other appliances where burner adjustments or internal changes might be used to adjust for reduced density at altitude, the same factors produce an adequately compensated chimney size. For example, an appliance operating at 6000 ft elevation at 10×10^6 Btu/h input, but requiring the same draft as at sea level, should have a chimney selected on the basis of 1.25 times the operating input, or 12.5×10^6 Btu/h.

Table 7 Altitude Correction

Altitude, ft	Barometric Pressure *B*, in. Hg	Factor[a]
Sea level	29.92	1.00
2,000	27.82	1.08
4,000	25.82	1.16
6,000	23.98	1.25
8,000	22.22	1.34
10,000	20.58	1.45

[a]Multiply operating input by the factor to obtain design input.

Table 8 Pressure Equations for Δp

Required Appliance Outlet Pressure or Available Draft D_a	Δp Equation[a]	
	Gravity Only	Gravity plus Inducer[b]
1. Negative, needs positive draft	$\Delta p = D_t - D_a$	$\Delta p = D_t - D_a + D_b$
2. Zero, vent with draft hood or balanced forced draft	$\Delta p = D_t$	$\Delta p = D_t + D_b$
3. Positive, causes negative draft	$\Delta p = D_t + D_a$	$\Delta p = D_t + D_a + D_b$

[a]Equations use absolute pressure for D_a.
[b]D_b = static pressure boost of inducer at flue gas temperature and rated flow.

SYSTEM FLOW LOSSES

Theoretical draft D_t is always positive (unless chimney gases are colder than ambient air); however, available draft D_a can be negative, zero, or positive. The pressure difference Δp, or theoretical minus available draft, overcomes the flow losses. Table 8 lists the pressure components for three draft configurations.

The table applies to still air (no wind) conditions and a neutral (zero) pressure difference between the space surrounding the appliance or fireplace and the atmosphere.

The effect of a nonneutral pressure difference on capacity or draft may be included by imposing a static pressure (either positive or negative). One way to circumvent a space-to-atmosphere pressure difference is to use a sealed-combustion system or direct vent system (i.e., all combustion air is taken directly from the outdoor atmosphere, and all flue gas is discharged to the outdoor atmosphere with no system openings such as draft hoods). The effect of wind on capacity or draft may be included by imposing a static pressure (either positive or negative) or by changing the vent terminal resistance loss. However, a properly designed and located vent terminal should cause little change in Δp at typical wind velocities.

Although small static draft pressures can be measured at the entrance and exit of gas appliance draft hoods, D_a at the appliance is effectively zero. Therefore, all theoretical draft energy produces chimney flow velocity and overcomes chimney flow resistance losses.

CHIMNEY GAS VELOCITY

The input capacity or diameter of a chimney may usually be found without determining flow velocity. Internal or exit velocity must occasionally be known, however, in order to ensure effluent dispersal or avoid flow noise. Also, the flow velocity of incinerator chimneys, turbine exhaust systems, and other appliances with high outlet pressures or velocities is needed to estimate piping loss coefficients.

Equations (7), (8), (9), and (12) can be applied to find velocity. Figure 1 may also be used to determine velocity; the right-hand scale of Grid E reads directly in velocity for any combination of flow and diameter. For example, at 10,000 cfm and 34 in. diameter (6.305 ft^2 area), the indicated velocity is about 1600 fpm. The velocity may also be calculated by dividing volumetric flow rate Q by chimney area A.

A similar calculation may be performed when the energy input is known. For example, a 34 in. chimney serving a 10×10^6 Btu/h natural gas appliance, at 8.5% CO_2 and 300°F above ambient in the chimney, produces (from Figure 3) 0.36 cfm per 1000 Btu/h. Chimney gas flow rate is

$$Q = 10 \times 10^6 (0.36/1000) = 3600 \text{ cfm}$$

Dividing by area to obtain velocity, $V = 3600/6.305 = 571$ fpm. The right-hand scale of Grid E, Figure 1, may also be multiplied by the cfm per 1000 Btu/h to find velocity. For the same chimney design conditions, the scale velocity value of 1600 fpm is multiplied by 0.36 to yield a velocity of 576 fpm in the chimney.

Chimney gas velocity affects the piping friction factor k_L and also the roughness correction factor. The section on Resistance Coefficients has further information, and Example 2 illustrates how these factors are used in the velocity equations.

Chimney systems can operate over a wide range of velocities, depending on modulation characteristics of the burner equipment or the number of appliances in operation. The typical velocity in vents and chimneys ranges from 300 to 3000 fpm. A chimney design developed for maximum input and maximum velocity should be satisfactory at reduced input because theoretical draft is roughly proportional to flue gas temperature rise, while flow losses are proportional to the square of the velocity. Thus, as input is reduced, flow losses decrease more rapidly than system motive pressures.

Effluent dispersal may occasionally require a minimum upward chimney outlet velocity, such as 3000 fpm. A tapered exit cone can best meet this requirement. For example, to increase the outlet velocity from the 34 in. chimney (A = 6.305 ft^2 area) from 1600 to 3000 fpm, the cone must have a discharge area of $6.305 \times 1600/3000 = 3.36$ ft^2 and a 24.8 in. diameter.

An exit cone avoids excessive flow losses because the entire system operates at the lower velocity, and a resistance factor is only added for the cone. In this case, the added resistance for a gradual taper approximates the following (see Table 9):

$$k = (d_{i1}/d_{i2})^4 - 1 = (34/24.8)^4 - 1 = 2.53$$

Noise in chimneys may be caused by turbulent flow at high velocity or by combustion-induced oscillations or resonance. Noise is seldom encountered in gas vent systems or in systems producing positive available draft, but it may be a problem with forced-draft appliances. Turbulent flow noise can be avoided by designing for lower velocity, which may entail increasing the chimney size above the minimum recommended by the appliance manufacturer. Chapter 46 of the 1999 *ASHRAE Handbook—Applications* has more information on noise control.

RESISTANCE COEFFICIENTS

The resistance coefficient k that appears in Equations (11) and (13) and Figure 1 summarizes the friction loss of the entire chimney system, including piping, fittings, and configuration or interconnection factors. Capacity of the chimney varies inversely with the square root of k, while diameter varies as the fourth root of k. The insensitivity of diameter and input to small variations in k simplifies design. Analyzing such details as pressure regain, increasers and reducers, and gas cooling junction effects is unnecessary if slightly high resistance coefficients are assigned to any draft diverters, elbows, tees, terminations, and, particularly, piping.

Table 9 Resistance Loss Coefficients

Component	Suggested Design Value, Dimensionless[a]	Estimated Span and Notes
Inlet acceleration (k_1)		
Gas vent with draft hood	1.5	1.0 to 3.0
Barometric regulator	0.5	0.0 to 0.5
Direct connection	0.0	Also dependent on blocking damper position
Round elbow (k_2)		
90°	0.75	0.5 to 1.5
45°	0.3	—
Tee or 90° connector (k_3)	1.25	1.0 to 4.0
Y connector	0.75	0.5 to 1.5
Cap, top (k_4)		
Open straight	0.0	—
Low-resistance (UL)	0.5	0.0 to 1.5
Other	—	1.5 to 4.5
Spark screen	0.5	—
Converging exit cone	$(d_{i1}/d_{i2})^4 - 1$	System designed using d_{i1}
Tapered reducer (d_{i1} to d_{i2})	$1 - (d_{i2}/d_{i1})^4$	System designed using d_{i2}
Increaser		See Chapter 2, 1997 *ASHRAE Handbook—Fundamentals.*
Piping (k_L)	$0.4\frac{L, \text{ft}}{d_i, \text{in.}}$	Numerical coefficient (friction factor F) varies from 0.2 to 0.5; see Figure 13, Chapter 2, 1997 *ASHRAE Handbook—Fundamentals* for size, roughness, and velocity effects.

[a]Initial assumption when size is unknown:
k = 5.0 for entire system, for first trial
k = 7.5 for combined gas vents only

Note: For combined gravity gas vents serving two or more appliances (draft hoods), multiply total k [components + piping—see Equations (16), (17), and (18)] by 1.5 to obtain gravity system design coefficient. (This rule does not apply to forced- or induced-draft vents or chimneys.)

The flow resistance of a fitting such as a tee with gases entering the side and making a 90° turn is assumed to be constant at $k = 1.25$, independent of size, velocity, orientation, inlet or outlet conditions, or whether the tee is located in an individual vent or in a manifold. Conversely, if the gases pass straight through a tee, as in a manifold, assumed resistance is zero, regardless of any area changes or flow entry from the side branch. For any chimney with fittings, the total flow resistance is a constant plus variable piping resistance—the latter being a function of centerline length divided by diameter. Table 9 suggests moderately conservative resistance coefficients for common fittings. Elbow resistance may be lowered by long-radius turns; however, corrugated 90° elbows may have resistance values at the high end of the scale. Table 9 shows resistance as a function of inlet diameter d_{i1} and outlet diameter d_{i2}.

System resistance k may be expressed as follows:

$$k = k_f + k_L \tag{16}$$

with

$$k_f = k_1 + n_2 k_2 + n_3 k_3 + k_4 + \text{ etc.} \tag{17}$$

and

$$k_L = FL/d_i \tag{18}$$

where

k_f = fixed fitting loss coefficient
k_L = piping resistance loss function (Figure 13, Chapter 2 of the 1997 *ASHRAE Handbook—Fundamentals*, adjusted for units)
k_1 = inlet acceleration coefficient
k_2 = elbow loss coefficient, n_2 = number of elbows

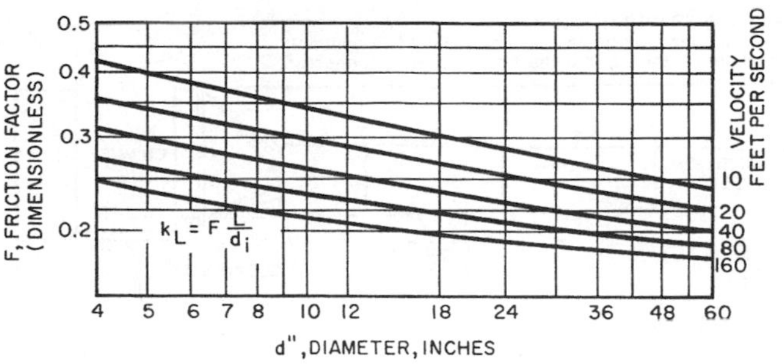

Fig. 6 Friction Factor for Commercial Iron and Steel Pipe
(Lapple 1949)

k_3 = tee loss coefficient, n_3 = number of tees
k_4 = cap, top, or exit cone loss coefficient
F = friction factor
L = length of all piping in chimney system, ft

For combined gas vents using appliances with draft hoods, the summation k must be multiplied by a diversity factor of 1.5 (see Table 9 note and Example 5). This multiplier does not apply to forced- or induced-draft vents or chimneys.

When size is unknown, the following k values may be used to run a first trial estimate:

k = 5.0 for the entire system
k = 7.5 for combined gas vents only

The resistance coefficient method adapts well to systems in which the fittings cause significant losses. Even for extensive systems, an initial assumption of $k = 5.0$ gives a tolerably accurate vent or chimney diameter in the first trial solution. Using this diameter with the piping resistance function [Equation (18)] in a second trial normally yields the final answer.

The minimum system resistance coefficient in a gas vent with a draft hood is always 1.0 because all gases must accelerate through the draft hood from almost zero velocity to vent velocity.

For a system connected directly to the outlet of a boiler or other appliance where the capacity is stated as full-rated heat input against a positive static pressure at the chimney connection, minimum system resistance is zero, and no value is added for existing velocity head in the system.

For simplified design, a value of 0.4 for F in Equation (18) applies for all sizes of vents or chimneys and for all velocities and temperatures. As diameter increases, this function becomes increasingly conservative, which is desirable because larger chimneys are more likely to be made of rough masonry construction or other materials with higher pressure losses. The 0.4 constant also introduces an increasing factor of safety for flow losses at greater lengths and heights.

Figure 6 is a plot of friction factor F versus velocity and diameter for commercial iron and steel pipe at a gas temperature of 300°F above ambient (Lapple 1949). The figure shows, for example, that a 48 in. diameter chimney with a gas velocity of 80 fps may have a friction factor as low as 0.2. In most cases, $k_L = 0.3L/d_i$ gives reasonable design results for chimney sizes 18 in. and larger because systems of this size usually operate at gas velocities greater than 10 fps.

At 1000°F or over, the factors in Figure 6 should be multiplied by 1.2. Because Figure 6 is for commercial iron and steel pipe, an additional correction for greater or less surface roughness may be imposed. For example, the factor for a very rough 12 in. diameter pipe may be doubled at a velocity as low as 2000 fpm.

For most chimney designs, a friction factor F of 0.4 gives a conservative solution for diameter or input for all sizes, types, and operating conditions of prefabricated and metal chimneys; alternately, F = 0.3 is reasonable if the diameter is 18 in. or more. Because neither input nor diameter is particularly sensitive to the total friction factor, the overall value of k requires little correction.

Masonry chimneys, including those lined with clay flue tile, may have rough surfaces, tile shape variations that cause misalignment, and joints at frequent intervals with possible mortar protrusions. In addition, the inside cross-sectional area of liner shapes may be less than expected because of local manufacturing variations, as well as differences between claimed and actual size. To account for these characteristics, the estimate for k_L should be on the high side, regardless of chimney size or velocity.

Computations should be made by assuming smooth surfaces and then adding a final size increase to compensate for shape factor and friction loss. Performance or capacity of metal and prefabricated chimneys is generally superior to that of site-constructed masonry.

Configuration and Manifolding Effects

The most common configuration is the individual vent, stack, or chimney, in which one continuous system carries the products from appliance to terminus. Other configurations include the combined vent serving a pair of appliances, the manifold serving several, and branched systems with two or more lateral manifolds connected to a common vertical system. As the number of appliances served by a common vertical vent or chimney increases, the precision of design decreases because of diversity factors (variation in the number of units in operation) and the need to allow for maximum and minimum input operation (Stone 1957). For example, the vertical common vent for interconnected gas appliances must be larger than for a single appliance of the same input to allow for operating diversity and draft hood dilution effects. Connector rise, headroom, and configuration in the equipment room must be designed carefully to avoid draft hood spillage and related oxygen depletion problems.

For typical combined vents, the diversity effect must be introduced into Figure 1 and the equations by multiplying system resistance loss coefficient k by 1.5 (see Table 9 note and Example 5). This multiplier compensates for junction effect and part-load operation.

Manifolds for appliances with barometric draft regulators can be designed without allowing for dilution by inoperative appliances. In this case, because draft regulators remain closed until regulation is needed, dilution under part load is negligible. In addition, flow through any inoperative appliance is negligible because the combustion air inlet dampers are closed and the multiple-pass heat exchanger has a high internal resistance.

Manifold systems of oil-burning appliances, for example, have a lower flow velocity and, hence, lower losses. As a result, they produce reasonable draft at part load or with only one of several appliances in operation. Therefore, diversity of operation has little effect on chimney design. Some installers set each draft regulator at a slightly different setting to avoid oscillations or hunting possibly caused by burner or flow pulsations.

Calculation of the resistance coefficient of any portion of a manifold begins with the appliance most distant from the vertical portion. All coefficients are then summed from its outlet to the vent terminus. The resistance of a series of tee joints to flow passing horizontally straight through them (not making a turn) is the same as that of an equal length of piping (as if all other appliances were off). This assumption holds whether the manifold is tapered (to accommodate increasing input) or of a constant size large enough for the accumulated input.

Coefficients are assigned only to inlet and exit conditions, to fittings causing turns, and to the piping running from the affected appliance to the chimney exit. Initially, piping shape (round, square, or rectangular) and function (for connectors, vertical piping, or both) are irrelevant.

Certain high-pressure, high-velocity packaged boilers require special manifold design to avoid turbulent flow noise. In such cases, manufacturers' instructions usually recommend increaser Y fittings, as shown in Figure 7. The loss coefficients listed in Table 9 for

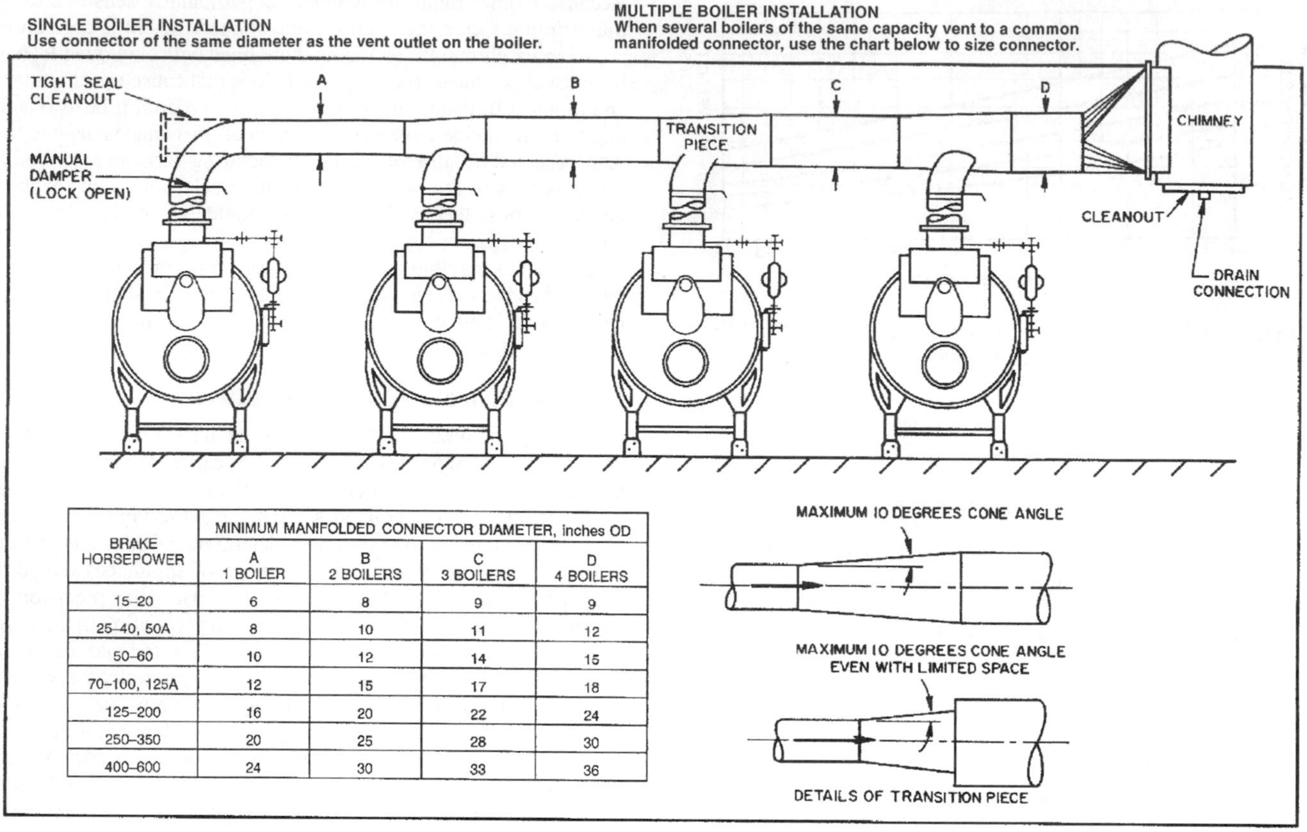

BRAKE HORSEPOWER	MINIMUM MANIFOLDED CONNECTOR DIAMETER, inches OD			
	A 1 BOILER	B 2 BOILERS	C 3 BOILERS	D 4 BOILERS
15–20	6	8	9	9
25–40, 50A	8	10	11	12
50–60	10	12	14	15
70–100, 125A	12	15	17	18
125–200	16	20	22	24
250–350	20	25	28	30
400–600	24	30	33	36

Fig. 7 Connector Design

standard tees and elbows are higher than necessary for long- radius elbows or Y entries. Occasionally, on equipment with high chimneys augmenting boiler outlet pressure, it may appear feasible to reduce the diameter of the vertical portion to below that recommended by the manufacturer. However, any reduction may cause turbulent noise, even though all normal design parameters have been considered.

The manufacturers' sizing recommendations shown in Figure 7 apply to the specific appliance and piping arrangement shown. The values are conservative for long-radius elbows or Y entries. Frequently, the boiler room layout forces the use of additional elbows. In such cases, the size must be increased to avoid excessive flow losses.

With the simplifying assumption that the maximum velocity of the flue gas (which exists in the smaller of the two portions) exists throughout the entire system, the design chart (Figure 1) can be used to calculate the size of a vertical portion smaller in area than the manifold or of a chimney connector smaller than the vertical. This assumption leads to a conservative design, as true losses in the larger area are lower than assumed. Further, if the size change is small, either as a contraction or as an enlargement, the added loss coefficient for this transition fitting (see Table 9) is compensated for by reduced losses in the enlarged part of the system.

These comments on size changes apply more to individual than to combined systems because it is undesirable to reduce the vertical area of the combined type, and, more frequently, it is desirable to enlarge it. If an existing vertical chimney is slightly undersized for the connected load, the complete chart method must be applied to determine whether a pressure boost is needed, as size is no longer a variable.

Sectional gas appliances with two or more draft hoods do not pose any special problems if all sections fire simultaneously. In this case, the designer can treat them as a single appliance. The appliance installation instructions either specify the size of manifold for interconnecting all draft hoods or require a combined area equal to the sum of all attached draft hood outlet areas. Once the manifold has been designed and constructed, it can be connected to a properly sized chimney connector, vent, or chimney. If the connector and chimney size is computed as less than manifold size (as may be the case with a tall chimney), the operating resistance of the manifold will be lower than the sum of the assigned component coefficients because of reduced velocity.

The general rule for conservative system design in which manifold, chimney connector, vent, or chimney are different sizes, can be stated as follows: *Always assign full resistance coefficient values to all portions carrying combined flow, and determine system capacity from the smallest diameter carrying the combined flow.* In addition, horizontal chimney connectors or vent connectors should pitch upward toward the stack at 1/4 in/ft minimum.

The following sample calculation using Equation (11) varies the original order of steps; it illustrates the direct solution for input, velocity, and volume. A calculation for input that is derived from Figure 1 differs because of the chart arrangement.

Example 1. Find the input capacity (Btu/h) of a vertical, double-wall Type B gas vent, 24 in. in diameter, 100 ft high at sea level. This vent is used with draft hood natural-gas-burning appliances.

Solution:

1. Mass flow from Table 2. M = 1.54 lb/1000 Btu for natural gas, if no other data are given.
2. Temperature from Table 4. Temperature rise = 300°F and $T_m = 360 + 460 = 820$°R for natural gas.
3. Theoretical draft from Table 6 or Equation (5). For 100 ft height at 300°F rise, D_t = 0.5 in. of water.
   Available draft for draft hood appliances: $D_a = 0$.
4. Flow losses from Table 8. $\Delta p = D_t = 0.5$; flow losses for a gravity gas vent equal theoretical draft at mean gas temperature.

5. Resistance coefficients from Table 9. For a vertical vent,

Draft hood	$k_1 = 1.5$
Vent cap	$k_4 = 1.0$
100 ft piping	$k_L = 0.4(100/24) = 1.67$
System total	$k = 4.17$

6. Solution for input.
Altitude: Sea level, $B = 29.92$ from Table 7. $d_i = 24$ in. These values are substituted into Equation (11) as follows:

$$I = 4.13\times10^5 \frac{(24)^2}{1.54}\left(\frac{(0.5)(29.92)}{(4.17)(820)}\right)^{0.5}$$

$$I = 10.2\times10^6 \text{ Btu/h input capacity.}$$

7. A solution for velocity requires a prior solution for input to apply to Equation (9). First, using Equation (4),

$$\rho_m = 1.325\frac{29.92}{820} = 0.0483 \text{lb/ft}^3$$

From Equation (9),

$$V = \frac{0.0509}{(0.0483)(24)^2} \times \frac{(10.2\times10^6)(1.54)}{1000} = 28.7 \text{ ft/s}$$

8. Volume flow can now be found because velocity is known. The flow area of 24 in. diameter is 3.14 ft^2, so

$$Q = (60 \text{ s/min})(3.14 \text{ ft}^2)(28.7 \text{ ft/s}) = 5410 \text{ cfm}$$

No heat loss correction is needed to find the new gas temperature because the size is greater than 20 in., and this vent is vertical with no horizontal connector.

For the same problem, Figure 1 requires a different sequence of solution. The ratio of mass flow to input for a given fuel (with parameter *M*) is not used directly; the chart requires selecting a percentage in the chimney, either from Table 2 or from operating data on the appliances. Then, the temperatures are entered only as rise above ambient. The solution path is as follows:

1. Enter Grid A at 5.3% CO_2, and construct line horizontally to left intersecting Nat. gas.
2. From Nat. gas intersection, construct vertical line to $t = 300$.
3. From 300 intersection in Grid *A*, go horizontally left to transfer line to Grid *B*.
4. From transfer line go vertically to $k = 4.17$.
5. From $k = 4.17$ run horizontally left to $\Delta t = 300$ in Grid *C*.
6. From 300 go up to $\Delta p = 0.54$ in Grid D.
7. From $\Delta p = 0.54$ in Grid *D*, go horizontally right to intersection $d_i = 24$ in. in Grid E.
8. Read capacity or input at 24 in. intersection in Grid *E* as 10.2×10^6 Btu/h between curved lines.

If input is known and diameter must be found, the procedure is the same as with Equation (11). A preliminary *k*, usually 5.0 for an individual vent or chimney, must be estimated to find a trial diameter. This diameter is used to find a corrected *k*, and the chart is solved again for diameter.

CHIMNEY CAPACITY CALCULATION EXAMPLES

Figure 8 through Figure 11 show chimney capacity for individually vented appliances computed by the methods presented. These capacity curves may be used to estimate input or diameter for the design chart (Figure 1), Equation (11), or Equation (13). These capacity curves apply primarily to individually vented appliances with a lateral chimney connector; systems with two or more appliances or additional fittings require a more detailed analysis. Figure 8 through Figure 11 assume the length of the horizontal connector is (1) at least 10 ft and (2) no longer than 50% of the height or 50 ft, whichever is less. For chimney heights of 10 to 20 ft, a fixed 10 ft long connector is assumed. Between 20 and 100 ft, the connector is 50% of the height. If the chimney height exceeds 100 ft, the connector is fixed at 50 ft long.

For a chimney of similar configuration but with a shorter connector, the size indicated in the figures is slightly larger than necessary. In deriving the data for Figure 8 through Figure 11, additional conservative assumptions were used, including the temperature correction C_u for double-wall laterals (see Figure 4) and a constant friction factor (0.4) for all sizes.

The loss coefficient k_4 for a low resistance cap is included in Figures 8, 9, and 10. If no cap is installed, these figures indicate a larger size than needed.

Figure 8 applies to a gas vent with draft hood and a lateral that runs to the vertical section. Maximum static draft is developed at the base of the vertical, but friction reduces the observed value to less than the theoretical draft. Areas of positive pressure may exist at the elbow above the draft hood and at the inlet to the cap. The height of the system is the vertical distance from the draft hood outlet to the vent cap.

Figure 9 applies to a typical boiler system requiring both negative combustion chamber pressure and negative static outlet pressure. The chimney static pressure is below atmospheric pressure, except for the minor outlet reversal caused by cap resistance. Height of this system is the difference in elevation between the point of draft measurement (or control) and the exit. (Chimney draft should not be based on the height above the boiler room floor.)

Figure 10 illustrates the use of a negative static pressure connector serving a forced-draft boiler. This system minimizes flue gas leakage in the equipment room. The draft is balanced or neutral, which is similar to a gas vent, with zero draft at the appliance outlet and pressure loss Δp equal to theoretical draft.

Figure 11 applies to a forced-draft boiler capable of operating against a positive static outlet pressure of up to 0.50 in. The chimney system has no negative pressure, so outlet pressure may be combined with theoretical draft to get minimum chimney size. For chimney heights or system lengths less than 100 ft, the effect of adding 0.50 in. positive pressure to theoretical draft causes all curves to fall into a compressed zone. An appliance that can produce 0.50 in. positive forced draft is adequate for venting any simple arrangement with up to 100 ft of flow path and no wind back pressure, for which additional forced draft is required.

The following examples illustrate the use of the design chart (Figure 1) and the corresponding equations.

Example 2. Individual gas appliance with draft hood (see Figure 12). The natural gas appliance is located at sea level and has an input of 980,000 Btu/h. The double-wall vent is 80 ft high with 40 ft lateral. Find the vent diameter.

Solution: Assume $k = 5.0$. The following factors are used successively in Grids A through D of Figure 1: $CO_2 = 5.3\%$ for Nat. Gas; Gas temp. rise = 300°F; Transfer line: $k = 5.0$; $\Delta p = 0.537\ (80/100) = 0.43$. See Example 1 for the solution path.

Preliminary solution: From Grid E at $I = 0.98 \times 10^6$, read diameter $d_i = 8.5$ in. Use next largest diameter, 9 in., to correct temperature rise and theoretical draft; compute new system *k* and Δp. From Figure 4, $C_u = 0.7$, $\Delta t = (0.7)(300) = 210$°F. This temperature rise determines the new value of theoretical draft, found by interpolation between 200 and 300°F in Table 6: $D_t = 0.41$ in. of water/100 ft (80 ft) = 0.33 in. of water. The four fittings have a total fixed $k_f = 4.0$. Adding k_L, the piping component for 120 ft of 9 in. diameter, to k_f gives $k = 4.0 + 0.4(120/9) = 9.3$. System losses $\Delta p = D_t = 0.33$ in. of water.

Final solution: Returning to Figure 1, the factors are $CO_2 = 5.3\%$ for Nat. gas; Gas temperature rise = 210°F; Transfer line: $k = 9.3$; $\Delta p = 0.33$. At $I = 0.98 \times 10^6$, read $d_i = 10$ in., which is the correct answer. Had system resistance been found using 10 in. rather than 9 in., the final size would be less than 10 in. based on a system *k* of less than 9.3.

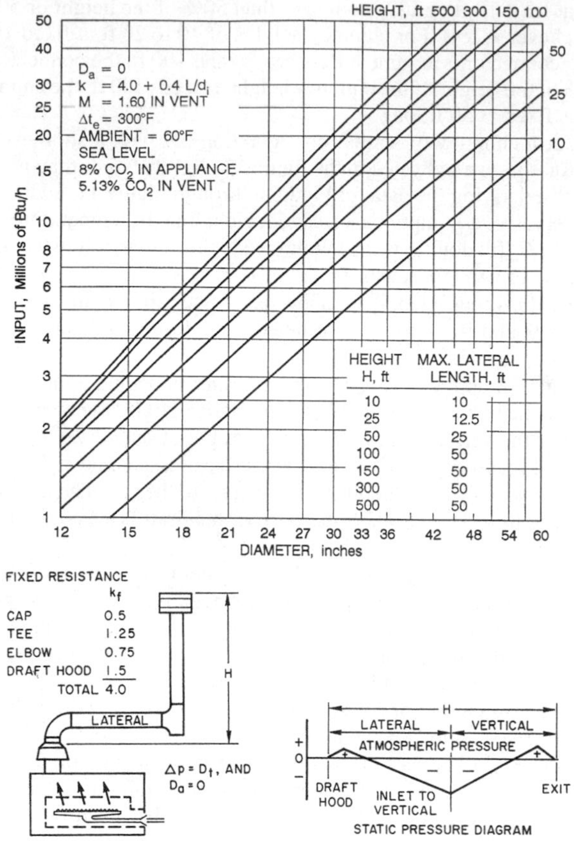

Fig. 8 Gas Vent with Lateral

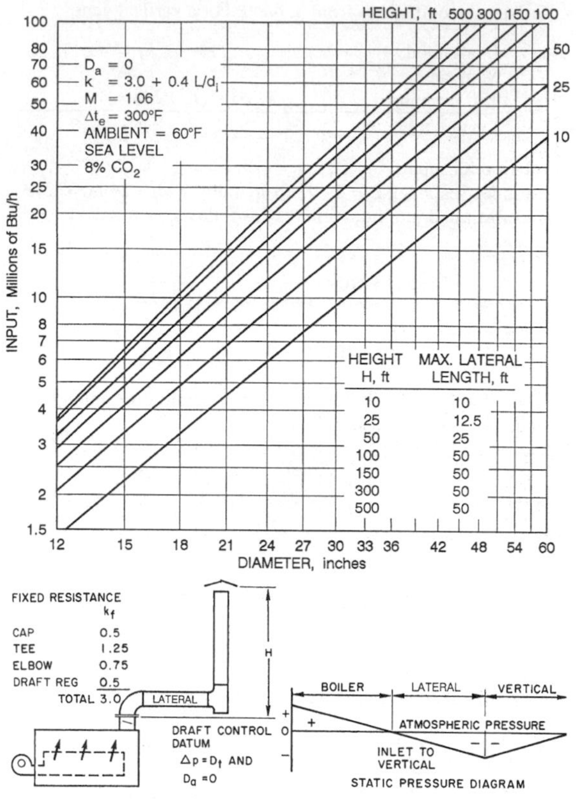

Fig. 10 Forced-Draft Appliance with Neutral (Zero) Draft (Negative Pressure Lateral)

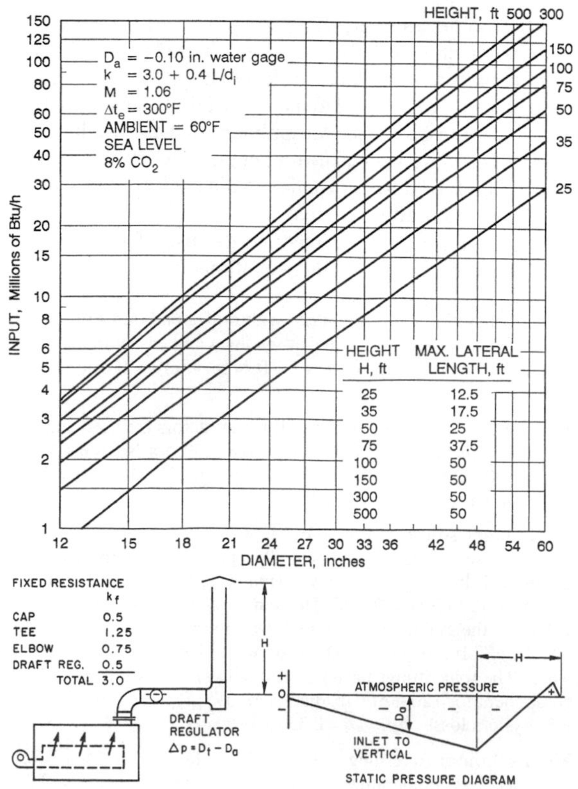

Fig. 9 Draft-Regulated Appliance with 0.10 in. of water gage Available Draft Required

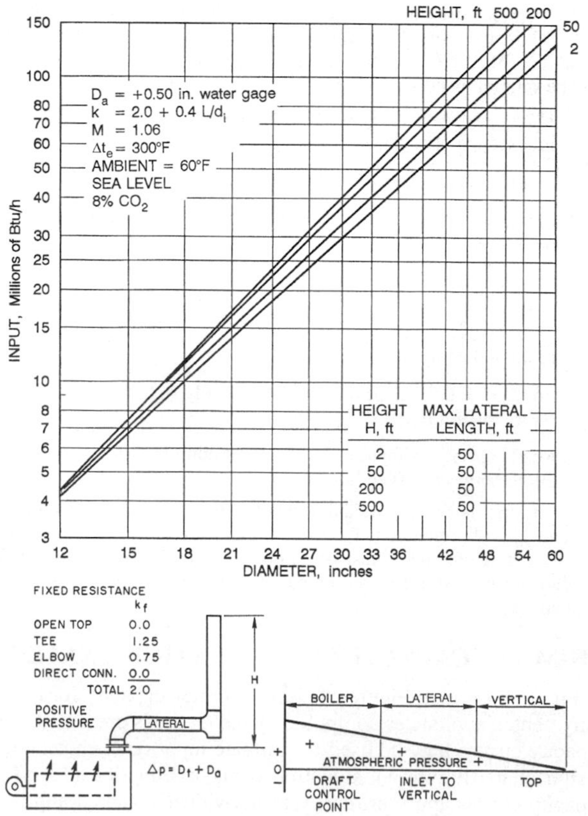

Fig. 11 Forced-Draft Appliance with Positive Outlet Pressure

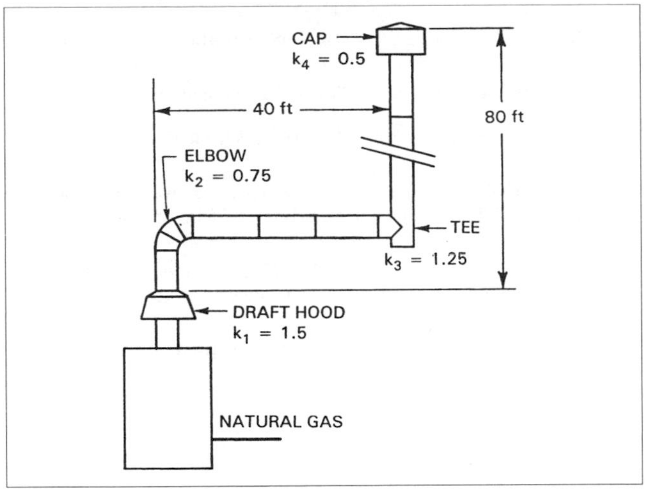

Fig. 12 Illustration for Example 2

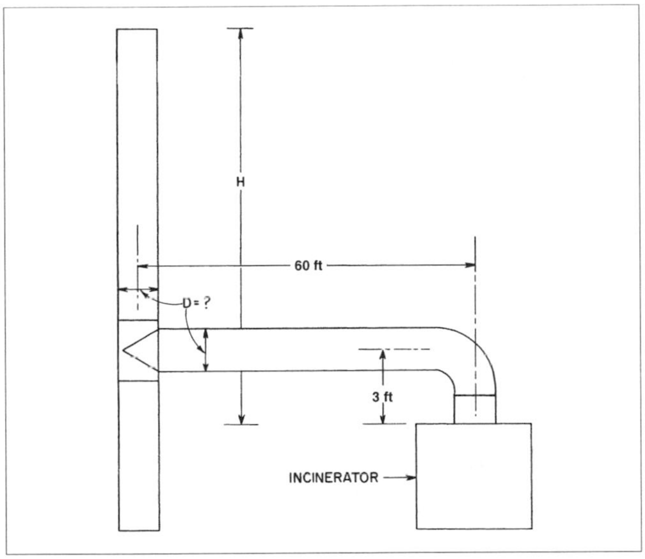

Fig. 13 Illustration for Example 3

Example 3. Gravity incinerator chimney (see Figure 13).

Located at 8000 ft elevation, the appliance burns 600 lb/h of Type 0 waste with 100% excess air at 1400°F outlet temperature. Ambient temperature T_o is 60°F. Outlet pressure is zero at low fire, +0.10 in. of water at high fire. The chimney will be a prefabricated medium-heat type with a 60 ft connector and a roughness factor of 1.2. The incinerator outlet is 18 in. in diameter, and it normally uses a 20 ft vertical chimney. Find the diameter of the chimney and the connector and the height required to overcome flow and fitting losses.

Solution:

1. Find mass flow from Table 3 as 13.76 lb gases per pound of waste, or $w = 600(13.76) = 8256$ total lb/h.
2. Find mean chimney gas temperature. Based on 60 ft length of 18 in. diameter double-wall chimney, $C_u = 0.83$ (see Figure 4). Temperature rise $\Delta t_e = 1400 - 60 = 1340$°F; thus, $\Delta t_m = 1340(0.83) = 1112$°F rise above 60°F ambient. $T_m = 1112 + 60 + 460 = 1632$°R. Use this temperature in Equation (4) to find gas density at 8000 ft elevation (from Table 7, $B = 22.22$ in. Hg):

$$\rho_m = 1.325(22.22/1632) = 0.0180 \text{ lb/ft}^3$$

3. Find the required height by finding theoretical draft per foot from Equation (5) or Figure 5 (Table 6 applies only to sea level).

$$D_t/H = (0.2554)(22.22)\left(\frac{1}{520} - \frac{1}{1632}\right)$$

$$= 0.0074 \text{ in. of water per ft of height}$$

4. Find allowable pressure loss Δp in the incinerator chimney for a positive pressure appliance having an outlet pressure of + 0.1 in. of water. From Table 8, $\Delta p = D_t + D_a$, where $D_t = 0.0074H$, $D_a = 0.10$, and $\Delta p = 0.0074H + 0.1$ in. of water.
5. Calculate flow velocity at mean temperature from Equation (8) to balance flow losses against diameter/height combinations:

$$V = \frac{(0.0509)(8256)}{(0.0180)(18^2)} = 72 \text{ fps}$$

This velocity exceeds the capability of a gravity chimney of moderate height and may require a draft inducer if an 18 in. chimney must be used. Verify velocity by calculating resistance and flow losses by the following steps.

6. From Table 9, resistance coefficients for fittings are

1 Tee	$k_3 = 1.25$
1 Elbow	$k_2 = 0.75$
Spark screen	$k_4 = 0.50$
Fitting total	$k_f = 2.50$

The piping resistance, adjusted for length, diameter, and a roughness factor of 1.2, must be added to the total fitting resistance. From Figure 6, find the friction factor F at 18 in. diameter and 72 fps as 0.22. Assuming 20 ft of height with a 60 ft lateral, piping friction loss is

$$k_L = \frac{1.2FL}{d_i} = \frac{(1.2)(0.22)(80)}{18} = 1.17$$

and total $k = 2.50 + 1.17 = 3.67$

Use Equation (7) to find Δp, which will determine whether this chimney height and diameter are suitable.

$$\Delta p = \frac{(3.67)(0.0180)(72)^2}{(5.2)(64.4)} = 1.02 \text{ in. of water flow losses}$$

For these operating conditions, theoretical draft plus available draft yields

$$\Delta p = 0.0074(20) + 0.1 = 0.248 \text{ in. of water driving force}$$

Flow losses of 1.02 in. exceed the 0.248 in. driving force; thus, the selected diameter, height, or both are incorrect, and this chimney will not work. This can also be shown by comparing draft per foot with flow losses per foot for the 18 in. diameter configuration:

Flow losses per foot of 18 in. chimney = 1.02/80 = 0.0128 in. of water

Draft per foot of height = 0.0074 in. of water

Regardless of how high the chimney is made, losses caused by a 72 fps velocity build up faster than draft.

7. A draft inducer could be selected to make up the difference between losses of 1.03 in. and the 0.25 in. driving force. Operating requirements are

$$Q = w/(60\rho_m) = 8256/(60 \times 0.0180) = 7644 \text{ cfm}$$

$$\Delta p = 1.02 - 0.248 = 0.772 \text{ in. of water at 7644 cfm and 1112°F}$$

If the inducer selected (see Figure 22C) injects single or multiple air jets into the gas stream, it should be placed only at the chimney top or outlet. This location requires no compensation for additional air introduced by an enlargement downstream from the inducer.

Because 18 in. is too small, assume that a 24 in. diameter may work at a 20 ft height and recalculate with the new diameter.

1. As before, $w = 8256$ lb/h.

2. At 24 in. diameter, no temperature correction is needed for the 60 ft connector. Thus, T_m = 1400 + 460 = 1860°R (see Table 4), and density is

$$\rho_m = 1.325(22.22/1860) = 0.0158 \text{ lb/ft}^3$$

3. Theoretical draft per foot of chimney height is

$$\frac{D_t}{H} = 0.2554(22.22)\left(\frac{1}{520} - \frac{1}{1860}\right) = 0.00786 \text{ in. of water per ft}$$

4. Velocity is

$$V = \frac{(0.0509)(8256)}{(0.0158)(24^2)} = 46.2 \text{ fps}$$

5. From Figure 6, the friction factor is 0.225, which, when multiplied by a roughness factor of 1.2 for the piping used, becomes 0.27. For the entire system with k_f = 2.50 and 80 ft of piping, find k = 2.5 + 0.27(80/24) = 3.4. From Equation (7),

$$\Delta p = \frac{(3.4)(0.0158)(46.2)^2}{(5.2)(64.4)} = 0.342 \text{ in. of water}$$

$D_t + D_a$ = (20)(0.00786) + 0.1 = 0.257 in. of water, so driving force is less than losses. The small difference indicates that while 20 ft of height is insufficient, additional height may solve the problem. The added height must make up for 0.085 in. additional draft. As a first approximation,

Added height	=	Additional draft/draft per foot
	=	0.085/0.00786 = 10.8 ft
Total height	=	20 + 10.8 = 30.8 ft

This is less than the actual height needed because resistance changes have not been included. For an exact solution for height, the driving force can be equated to flow losses as a function of H. Substituting H + 60 for L in Equation (16) to find k for Equation (7), the complete equation is

$$0.10 + 0.00786H = \frac{(3.175 + 0.01125H)(0.0158)(46.2)^2}{(5.2)(64.4)}$$

Solving, H = 32.66 ft at 24 in. diameter.

Checking by substitution, total driving force = 0.357 in. of water and total losses, based on a system with 92.66 ft of piping, equal 0.357 in. of water.

The value of H = 32.66 ft is the minimum necessary for proper system operation. Because of the great variation in fuels and firing rate with incinerators, greater height should be used for assurance of adequate draft and combustion control. An acceptable height would be from 40 to 50 ft.

Example 4. Two forced-draft boilers (see Figure 14).

This example shows how multiple-appliance chimneys can be separated into subsystems. Each boiler is rated 100 boiler horsepower on No. 2 oil. The manufacturer states operation at 13.5% CO_2 and 300°F rise against 0.50 in. positive static pressure at the outlet. The 50 ft high chimney has a 20 ft single-wall manifold and is at sea level. Find the size of connectors, manifold, and vertical.

Solution: First, find the capacity or size of the piping and fittings from boiler A to the tee over boiler B. Then, size the boiler B tee and all subsequent portions to carry the combined flow of A and B. Also, check the subsystem for boiler B; however, because its shorter length compensates for greater fitting resistance, its connector may be the same size as for boiler A.

Find the size for combined flow of A and B either by assuming k = 5.0 or by estimating that the size will be twice that found for boiler A operating by itself. Estimate system resistance for the combined portion by including those fittings in the B connector with those in the combined portion.

Data needed for the solution of Equation (11) for boiler A for No. 2 oil at 13.5% CO_2 include the following:

$$M = 0.72\left(0.12 + \frac{14.4}{13.5}\right) = 0.854 \text{ lb/1000 Btu} \qquad \text{(Table 1)}$$

For a temperature rise of 300°F and an ambient of 60°F,

$$T_m = 300 + 60 + 460 = 820°\text{R}$$

From Table 6, theoretical draft = 0.5 in. of water for 100 ft of height; for 50 ft height, D_t = (0.5)50/100 = 0.25 in. of water.

Using D_a = 0.5 in. of water, Δp = 0.25 + 0.5 = 0.75 in. of water.

Assume k = 5.0. Assuming 80% efficiency, input is 41,800 times boiler horsepower (see the section on Conversion Factors):

$$I = 100(41{,}800) = 4.18\times10^6 \text{ Btu/h}$$

Substitute in Equation (11):

$$4.18\times10^6 = 4.13\times10^5 \frac{(d_i)^2}{0.854}\left(\frac{(0.75)(29.92)}{(5.0)(820)}\right)^{0.5}$$

Solving, d_i = 10.81 in. as a first approximation. From Table 9, find correct k using next largest diameter, or 12 in.:

Inlet acceleration (direct connection)	k_1	= 0.0
90° Elbow	k_2	= 0.75
Tee	k_3	= 1.25
70 ft piping	k_L = 0.4(70/12)	= 2.33
System total	k	= 4.33

Note: Assume tee over boiler B has k = 0 in subsystem A.

Corrected temperature rise (see Figure 4 for 20 ft single-wall connector) = 0.75(300) = 225°F, and T_m = 225 + 60 + 460 = 745°R. This corrected temperature rise changes D_t to 0.425 per 100 ft (Table 6), or 0.21 for 50 ft. Δp becomes 0.21 + 0.50 = 0.71 in. of water.

$$4.18\times10^6 = 4.13\times10^5 \frac{(d_i)^2}{0.854}\left(\frac{(0.71)(29.92)}{(4.33)(745)}\right)^{0.5}$$

d_i = 10.35 in., or a 12 in. diameter is adequate.

For size of manifold and vertical, starting with the tee over boiler B, assume 16 in. diameter (see also Figure 11).

System k = 3.9 for the 55 ft of piping from B to outlet:

Inlet acceleration	k_1	= 0.0
Two tees (boiler B subsystem)	k_3	= 2.5
55 ft piping	k_L = 0.4(55/16)	= 1.4
System total	k	= 3.9

Temperature and Δp will be as corrected (a conservative assumption) in the second step for boiler A. Having assumed a size, find input.

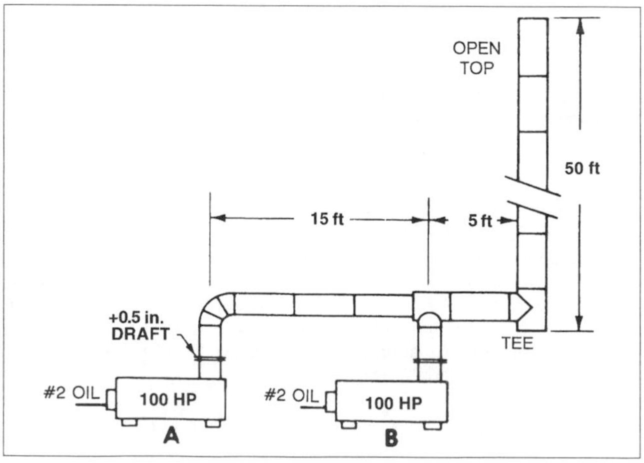

Fig. 14 Illustration for Example 4

$$I = 4.13\times10^5 \frac{16^2}{0.854}\left(\frac{(0.71)(29.92)}{(3.9)(745)}\right)^{0.5} = 10.59\times10^6 \text{ Btu/h}$$

A 16 in. diameter manifold and vertical is more than adequate. Solving for the diameter at the combined input, d_i = 14.2 in.; thus, a 15 or 16 in. chimney must be used.

Note: Regardless of calculations, do not use connectors smaller than the appliance outlet size in any combined system. Applying the temperature correction for a single-wall connector has little effect on the result because positive forced draft is the predominant motive force for this system.

Example 5. Six gas boilers manifolded at 6000 ft elevation (see Figure 15). Each boiler is fired at 1.6×10^6 Btu/h, with draft hoods and an 80 ft long manifold connecting into a 400 ft high vertical. Each boiler is controlled individually. Find the size of the constant-diameter manifold, vertical, and connectors with a 2 ft rise. All are double-wall.

Solution: Simultaneous operation determines both the vertical and manifold sizes. Assume the same appliance operating conditions as in Example 1: CO_2 = 5.3%, natural gas, temperature rise = 300°F. Initially assume k = 5.0 is multiplied by 1.5 for combined vent (see note at bottom of Table 9); thus, design k = 7.5. For a gas vent at 400 ft height, $\Delta p = D_t = 4 \times 0.5 = 2.0$ in. of water (Table 6). At 6000 ft elevation, operating input must be multiplied by an altitude correction (Table 7) of 1.25. Total design input is $1.6 \times 10^6(6)(1.25) = 12 \times 10^6$ Btu/h. From Table 2, M = 1.54 at operating conditions. T_m = 300 + 60 + 460 = 820°R, and B = 29.92 because the 1.25 input multiplier corrects back to sea level. From Equation (11),

$$12\times10^6 = 4.13\times10^5 \frac{(d_i)^2}{1.54}\left(\frac{2.0 \times 29.92}{7.5 \times 820}\right)^{0.5}$$

$$d_i = 21.3 \text{ in.}$$

Because the diameter is greater than 20 in., no temperature correction is needed.

Recompute k for 6 × 80 = 480 ft of 22 in. diameter (Table 9):

Draft hood inlet acceleration	k_1 = 1.5
Two tees (connector and base of chimney)	$2k_3$ = 2.5
Low-resistance top	k_4 = 0.5
480 ft piping	k_L = 0.4(480/22) = 8.7
System total	k = 13.2

Combined gas vent design k = multiple vent factor 1.5 × 13.2 = 19.8. Substitute again in Equation (11):

$$12\times10^6 = 4.13\times10^5 \frac{(d_i)^2}{1.54}\left(\frac{2.0 \times 29.92}{19.8 \times 820}\right)^{0.5}$$

d_i = 27.1 in.; thus use 28 in.

k_f	= 4.5
k_L = 0.4(480/28)	= 6.9
	11.4

Design k = 11.4(1.5)= 17.1

Substitute in Equation (11) to obtain the third trial:

$$12\times10^6 = 4.13\times10^5 \frac{(d_i)^2}{1.54}\left(\frac{2.0 \times 29.92}{17.1 \times 820}\right)^{0.5}$$

$$d_i = 26.2 \text{ in.}$$

The third trial is less than the second and again shows the manifold and vertical chimney diameter to be between 26 and 28 in.

For connector size, see the *National Fuel Gas Code* for double-wall connectors of combined vents. The height limit of the table is 100 ft—do not extrapolate and read the capacity of 18 in. connector as 1,740,000 Btu/h at 2 ft rise. Use 18 in. connector or draft hood outlet, whichever is larger. No altitude correction is needed for connector size; the draft hood outlet size considers this effect.

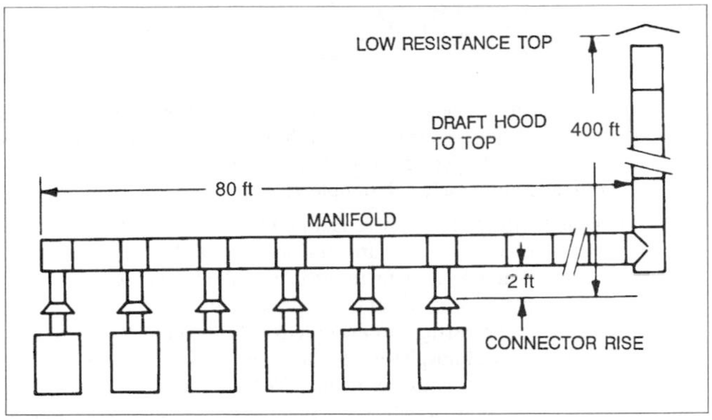

Fig. 15 Illustration for Example 5

Note: Equation (11) can also be solved at local elevation for exact operating conditions. At 6000 ft, local barometric pressure is 23.98 in. Hg (Table 7), and assumed theoretical draft must be corrected in proportion to the reduction in pressure:

D_t = 2.0(23.98/29.92) = 1.60 in. of water. Operating input of $6(1.6 \times 10^6) = 9.6 \times 10^6$ Btu/h is used to find d_i, again taking final k = 17.1:

$$9.6\times10^6 = 4.13\times10^5 \frac{(d_i)^2}{1.54}\left(\frac{1.60 \times 23.98}{17.1 \times 820}\right)^{0.5}$$

$$d_i = 26.2 \text{ (same as above)}$$

This example illustrates the equivalence of the chart method of solution with solution by Equation (11). Equation (11) gives the correct solution using either Method 1, with only the input corrected back to sea level condition, or Method 2, correcting Δp for local barometric pressure and using operating input at altitude. Method 1, correcting input only, is the only choice with Figure 1 because the design chart cannot correct to local barometric pressure.

Example 6. Pressure boost for undersized chimney (not illustrated).

A natural gas boiler at sea level (no draft hood) is connected to an existing 12 in. diameter chimney. Input is 4×10^6 Btu/h operating at 10% CO_2 and 300°F rise above ambient. System resistance loss coefficient k = 5.0 with 20 ft vertical. The appliance operates with neutral outlet static draft, so, D_a = 0.

a. How much draft boost is needed at operating temperature?

b. What fan rating is required at 60°F ambient temperature?

c. Where in the system should the fan be located?

Solution: Combustion data—from Figure 3, 10% CO_2 at 300°F rise indicates 0.31 cfm per 1000 Btu/h.

Total flow rate $Q = (0.31/1000)(4 \times 10^6) = 1240$ cfm at chimney gas temperature. Then, Equation (13) can be solved for the only unknown, Δp:

$$1240 = 5.2(12)^2\left(\frac{\Delta p(300 + 60 + 460)}{(5.0)(29.92)}\right)^{0.5}$$

Δp = 0.50 in. of water needed at 300°F rise. For 20 ft of height at 300°F rise (Table 6),

$$D_t = 0.5(20/100) = 0.10 \text{ in. of water}$$

a. Pressure boost D_b supplied by the fan must equal Δp minus theoretical draft (Table 8) when available draft is zero. $D_b = \Delta p - D_t = 0.5 - 0.10 = 0.40$ in. of water at operating temperature.

b. Draft fans are usually rated for standard ambient (60°F) air. Pressure is inversely proportional to absolute gas temperature. Thus, for ambient air,

$$D_b = 0.40\frac{T_m}{T_o} = 0.40\left(\frac{300+60+460}{60+460}\right) = 0.63 \text{ in. of water}$$

This pressure is needed to produce 0.40 in. at operating temperature. In specifying power ratings for draft fan motors, a safe policy is to select one that operates at the required flow rate at ambient temperature and pressure (see Example 7).

c. A fan can be located anywhere from boiler outlet to chimney outlet. Regardless of location, the amount of boost is the same; however, chimney pressure relative to atmosphere will change. At boiler outlet, the fan pressurizes the entire connector and chimney. Thus, the system should be gastight to avoid leaks. At the chimney outlet, the system is below atmospheric pressure; any leaks will flow into the system and seldom cause problems. With an ordinary sheet metal connector attached to a tight vertical chimney, the fan may be placed close to the vertical chimney inlet. Thus, the connector leaks safely inward, while the vertical chimney is under pressure.

Example 7. Draft inducer selection (see Figure 16).

A third gas boiler must be added to a two-boiler system at sea level with an 18 in. diameter, 15 ft horizontal, and 75 ft of total height. Outlet conditions for natural gas draft hood appliances are 5.3% CO_2 at 300°F rise. Boilers are controlled individually, each with 1.6×10^6 Btu/h, for 4.8×10^6 Btu/h total input. The system is currently undersized for gravity full-load operation. Find capacity, pressure, size, and power rating of a draft inducer fan installed at the outlet.

Solution: Using Equation (13) requires evaluating two operating conditions: (1) full input at 300°F rise and (2) no input with ambient air. Because the boilers are controlled individually, the system may operate at nearly ambient temperature (100°F rise or less) when only one boiler operates at part load. Use the system resistance k for boiler 3 as the system value for simultaneous operation. It needs no compensating increase as with gravity multiple venting because a fan induces flow at all temperatures. From Table 9, the resistance summation is

Inlet acceleration (draft hoods)	$k_1 = 1.50$
Tee above boiler	$k_3 = 1.25$
Tee at base of vertical	$k_3 = 1.25$
90 ft of 18 in. pipe	$k_L = 0.4(90/18) = 2.00$
System total	$k = 6.00$

At full load, $T_m = 300 + 60 + 460 = 820$°R; $B = 29.92$ in. Hg. Flow rate Q must be found for operating conditions of 1.54 lb per 1000 Btu (Table 2) at density ρ_m and full input $(4.8 \times 10^6)/60$ Btu/min.

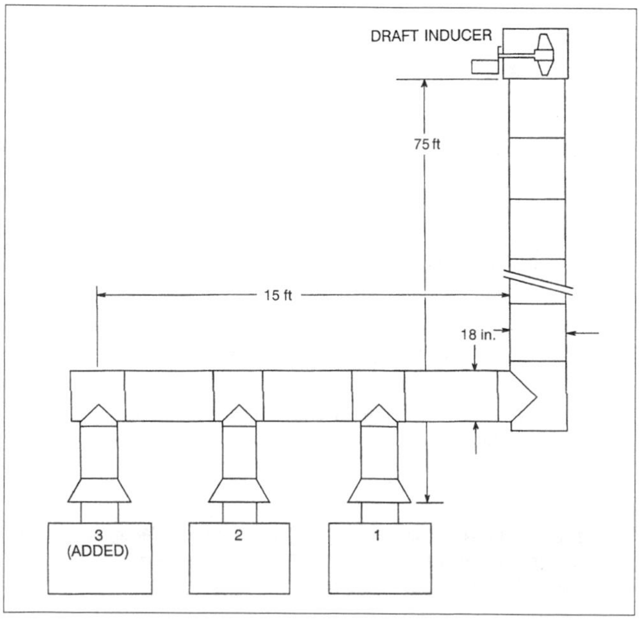

Fig. 16 Illustration for Example 7

From Equation (4),

$$\rho_m = 1.325B/T_m = 1.325(29.92/820) = 0.0483 \text{ lb/ft}^3$$

Volumetric flow rate is

$$Q = \left(\frac{1.54}{1000}\right)\left(\frac{1}{0.0483}\right)\left(\frac{4.8\times10^6}{60}\right) = 2551 \text{ cfm}$$

For 300°F rise, $D_t = 0.5(75/100) = 0.375$ in. of water theoretical draft in the system (Table 6). Solving Equation (13) for Δp,

$$\Delta p = \frac{(6.00)(29.92)(2551)^2}{(5.2)^2(18)^4(820)} = 0.502 \text{ in. of water}$$

Thus, a fan is needed because Δp exceeds D_t. Required static pressure boost (Table 8) is

$D_b = 0.502 - 0.375 = 0.127$ in. of water at 300°F rise
or a density of 0.0483 lb/ft^3

Fans are rated for ambient or standard air (60 to 70°F) conditions. Pressure is directly proportional to density or inversely proportional to absolute temperature. Moving 2551 cfm at 300°F rise against 0.127 in. of water pressure requires the static pressure boost with standard air to be

$$D_b = 0.127(820/520) = 0.20 \text{ in. of water}$$

Thus, a fan that delivers 2551 cfm at 0.20 in. of water at 60°F is required. Figure 17 shows the operating curves of a typical fan that meets this requirement. The exact volume and pressure developed against a system k of 6.0 can be found for this fan by plotting airflow rate versus Δp from Equation (13) on the fan curve. The solution, at Point C, occurs at 1950 cfm, where both the system Δp and fan static pressure equal 0.46 in. of water.

While some fan manufacturers' ratings are given at standard air conditions, the motors selected will be overloaded at temperatures below 300°F rise. Figure 17 shows that power required for two conditions with ambient air is as follows:

1. 1950 cfm at 0.46 in. static pressure requires 0.51 hp
2. 2650 cfm at 0.25 in. static pressure requires 0.50 hp

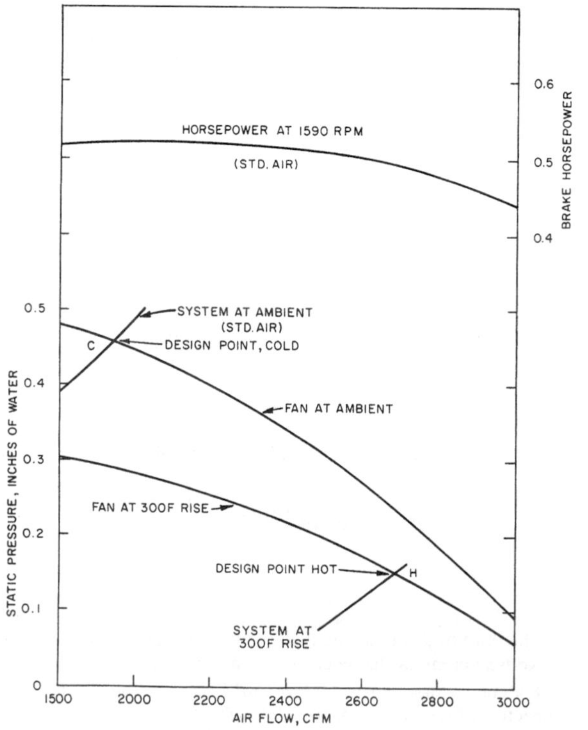

Fig. 17 Typical Fan Operating Data and System Curves

Thus, the minimum size motor will be 0.5 hp and run at 1590 rpm.

Manufacturers' literature must be analyzed carefully to discover whether the sizing and selection method is consistent with appliance and chimney operating conditions. Final selection requires both a thorough analysis of fan and system interrelationships and consultation with the fan manufacturer to verify the capacity and power ratings.

GAS APPLIANCE VENTING

In the United States, gas-burning equipment requiring venting of combustion products must be installed and vented in accordance with the *National Fuel Gas Code*. This standard includes capacity data and definitions for commonly used gas vent systems.

Traditionally, gas appliances were designed with a draft hood or draft diverter and depended on natural buoyancy to vent their products of combustion. The operating characteristics of many different types of appliances were similar, allowing generic venting guidelines to be applied to any gas appliance. These guidelines consisted of tables of maximum capacities, representing the largest appliance input rating that could safely be vented using a certain vent diameter, height, lateral, and material.

Small gas-fired appliance design changed significantly, however, with the increase of furnace minimum efficiency requirements to 78% annual fuel utilization efficiency (AFUE) isolated combustion system (ICS) imposed by the U.S. National Appliance Energy Conservation Act (NAECA) of 1987. Many manufacturers developed appliances with fan-assisted combustion systems (FACSs) to meet efficiency standards, which greatly changed the venting characteristics of gas appliances.

Because FACS appliances do not entrain dilution air, a FACS appliance connected to a given vent system can have a larger maximum capacity than a draft hood-equipped appliance connected to the same vent system. The dew point of the vent gases from a FACS appliance, however, remains high in the absence of dilution air, and the potential for condensation to form and remain in the vent system is much greater than with draft hood-equipped appliances. Excessive condensate dwell time (wet time) in the vent system can cause corrosion failure of the vent system or problems with condensate runoff, either into the appliance or into the structure surrounding the vent.

New venting guidelines (*National Fuel Gas Code*) were developed to address the differences between draft hood-equipped and FACS Category I appliances. In addition to new maximum capacity values for FACS appliances, the guidelines include minimum capacity values for FACS appliances to ensure that vent system wet times do not reach a level that would lead to corrosion or drainage problems. Because corrosion is the principal effect of condensation in the vent connector, whereas drainage is the principal concern in the vertical portion of the vent, different wet time limits were established for the vent connector and the vertical portion of the vent.

Wet time values in the vent system were determined using the VENT-II computer program (Rutz and Paul 1991) to perform the transient analysis with the appliance cycling. The cycle times of 3.87 min on, 13.3 min off, 17.17 min total, were determined based on a design outdoor ambient of 42°F with an oversize factor of 1.7. The vent connector is required to dry before the end of the appliance on-cycle, while the vertical portion of the vent is required to dry out before the end of the total cycle.

The new venting guidelines for FACS appliances severely restrict the use of single-wall metal vent connectors. Also, due to excessive condensation, tile-lined masonry chimneys are not recommended in most typical installations when a FACS appliance is the only appliance connected to the venting system. FACS appliances can be used with a typical masonry chimney, however, if (1) the FACS appliance is common-vented with a draft hood-equipped appliance or (2) a liner listed for use with gas appliances is installed in the masonry chimney.

The *National Fuel Gas Code* lists both the new minimum and maximum vent capacities for FACS gas appliances and the previous maximum vent capacities for gas appliances equipped with draft hoods. Sections 11.2.9 and 11.3.18 of the code provide new criteria for permitting the use of masonry chimneys (either exposed or not exposed to the outdoors below the roof line) and permit "alternative venting designs" and the installation of vents serving listed appliances in accordance with the appliance manufacturer's instructions and the terms of listing.

Vent Connectors

Vent connectors connect gas appliances to the gas vent, chimney, or single-wall metal pipe, except when the appliance flue outlet or draft hood outlet is connected directly to one of these. Materials for vent connectors for conversion burners or other equipment without draft hoods must have resistance to corrosion and heat not less than that of galvanized 0.0276 in. (24 gage) thick sheet steel. Where a draft hood is used, the connector must have resistance to corrosion and heat not less than that of galvanized 0.0187 in. (28 gage) thick sheet steel or Type B vent material.

Masonry Chimneys for Gas Appliances

A masonry chimney serving a gas-burning appliance should have a tile liner and should comply with applicable building codes such as NFPA *Standard* 211. Sections 11.2.9 and 11.3.18 of the *National Fuel Gas Code* have other provisions pertaining to masonry chimneys. An additional chimney liner may be needed to avoid slow priming and/or condensation, particularly for an exposed masonry chimney with high mass and low flue gas temperature. A low-temperature chimney liner may be a single-wall passage of pure aluminum or stainless steel or a double-wall Type B vent.

Type B and Type L Factory-Built Venting Systems

Factory-prefabricated vents are listed by Underwriters Laboratories for use with various types of fuel-burning equipment. These should be installed according to the manufacturer's instructions and their listing.

Type B gas vents are listed for vented gas-burning appliances. They should not be used for incinerators, appliances readily converted to the use of solid or liquid fuel, or combination gas-oil burning appliances. They may be used in multiple gas appliance venting systems.

Type BW gas vents are listed for vented wall furnaces certified as complying with the pertinent ANSI standard.

Type L venting systems are listed by Underwriters Laboratories in 3 through 6 in. sizes and may be used for those oil- and gas-burning appliances (primarily residential) certified or listed as suitable for this type of venting. Under the terms of the listing, a single-wall connector may be used in open accessible areas between the appliance's outlet and the Type L material in a manner analogous to Type B. Type L piping material is recognized in the *National Fuel Gas Code* and NFPA *Standard* 211 for certain connector uses between appliances such as domestic incinerators and chimneys.

Gas Equipment Without Draft Hoods

Figure 1 or the equations may be used to calculate chimney size for nonresidential gas appliances with the draft configurations listed as (1) and (3) at the beginning of this chapter. Draft configurations (1) and (3) for residential gas appliances, such as boilers and furnaces, may require special vent systems. The appliance test and certification standards include evaluation of manufacturers' appliance installation instructions (including the vent system) and of operating and application conditions that affect venting. The instructions must be followed strictly.

Conversion to Gas

The installation of conversion burner equipment requires evaluating for proper chimney draft and capacity by the methods indicated in this chapter or by conformance to local regulations. The physical condition and suitability of an existing chimney must be checked before it is converted from a solid or liquid fuel to gas. For masonry chimneys, local experience may indicate how well the construction will withstand the lower temperature and higher moisture content of natural or liquefied petroleum gas combustion products. The section on Masonry Chimneys for Gas Appliances has more details.

The chimney should be relined, if required, with corrosion-resistant masonry or metal to prevent deterioration. The liner must extend beyond the top of the chimney. The chimney drop-leg (bottom of the chimney) must be at least 4 in. below the bottom of the connection to the chimney. The chimney should be inspected and, if needed, cleaned. The chimney should also have a cleanout at the base.

OIL-FIRED APPLIANCE VENTING

Oil-fired appliances requiring venting of combustion products must be installed and vented in accordance with ANSI/NFPA *Standard* 31. The standard offers recommendations for metal relining of masonry chimneys.

Current recommendations for the minimum chimney areas for oil-fired natural-draft appliances are offered in Tables 3 and 4 in the Testing and Rating Standard for Heating Boilers (HYDI 1989).

The implementation of the U.S. National Appliance Energy Conservation Act (NAECA) of 1987 has brought attention to heating appliance efficiency and the effect of NAECA on existing chimney systems. Oil-fired appliances have maintained a steady growth in efficiency since the advent of the retention-head oil burner and its broad application in both new appliances and the replacement of older burners in existing appliances. Higher appliance efficiencies have brought about lower flue gas temperatures. Reduced firing rates have become more common as heating appliances are more closely matched to the building heating load. Burner excess air levels have also been reduced, which has resulted in lower mass flows through the chimney and additional reductions in the flue gas temperature. However, the improvements in overall appliance efficiency have not been accompanied by upgrades in existing chimney systems, nor are upgraded systems commonly applied in new construction. An upgrade in a vent system probably involves the application of corrosion-resistant materials and/or the reduction in heat loss from the vent system to maintain draft and reduce condensation on interior surfaces of the vent system.

Condensation and Corrosion

Condensation and corrosion within the vent system are of growing concern as manufacturers of oil-fired appliances strive to improve equipment efficiencies. The conditions for condensation of the corrosive components of the flue gas produced in oil-fired appliances involve a complex interaction of the water formed in the combustion process and the sulfur trioxide formed from small quantities of sulfur in the fuel oil. The sulfur is typically less than 0.5% by mass for No. 2 fuel oil. The dew point of the two-component system (sulfuric acid and water) in the flue gas resulting from the combustion of this fuel is about 225 to 240°F. This is similar to the effect on dew point of fuel gas sulfur (see Figure 4 in Chapter 17 of the 1997 *ASHRAE Handbook—Fundamentals*). In determining the proper curve for fuel oil in Figure 4, a value of 18 may be used. This value applies to No. 2 fuel oil with 0.5% sulfur by mass.

The effect of post-combustion air dilution on the dew point characteristics of flue gas from oil-fired appliances is highly dependent on the presence of sulfur in the fuel. Work conducted by Verhoff and Banchero (1974) yielded an equation relating the flue gas dew point to the partial pressures of both water and sulfuric acid present in the flue gas. According to the researchers, the predictions obtained using this equation are in good agreement with experimental data. Applying this equation to the flue gas from the combustion of No. 2 fuel oil without sulfur and with varying sulfur contents reveals that a broad range of dew point temperatures is possible.

For a fuel oil with zero sulfur and 20% excess combustion air, the dew point of the combustion products is calculated as approximately 114°F, typical for the presence of water formed in combustion. With the addition of post-combustion dilution air to a level equivalent to increasing the excess air in the flue gas to 100%, the apparent water dew point is reduced to 99°F. For fuel oil with 0.25% sulfur and 20% excess combustion air, the dew point is elevated to 229°F due to the formation of dilute sulfuric acid in the flue gas. With the addition of post-combustion dilution air, the dew point is 237°F at 100% excess air and 234°F at 200% excess air. The implication of these calculations is that for the combustion of fuel oil, the flue gas dew-point temperature can range from a low at the apparent water dew point (calculated with no fuel sulfur) up to some elevated dew point due to the presence of sulfur in the fuel. With no sulfur in the fuel, the addition of post-combustion dilution air to the flue gas has some effect on depressing the apparent water dew-point temperature. The addition of post-combustion dilution air has no significant effect on the elevated flue gas dew point when sulfur is present in the fuel.

It is difficult to meet the flue gas side material surface temperatures required to exceed the elevated dew point at all points in the venting system of an oil-fired appliance. Even if the dew-point temperature is not reached at these surfaces, however, a surface temperature approaching 200°F will permit the condensation of sulfuric acid at higher concentrations and will result in lower rates of corrosion.

The issue of the condensed acid concentration is critical to the applicability of plain carbon and stainless steels in vent connectors and chimney liners. The flue gas side surface temperatures of conventional connectors and masonry chimney tile liners are often at or below the dew point for some portion of the burner "on" period. During cool down (burner "off" period) these surface temperatures can drop to below the apparent water dew point and, in some cases, to ambient conditions. This is of great concern because for the period of time that the system surfaces are below the apparent water dew point during burner operation, the condensed sulfuric acid is formed in concentrations well below limits acceptable for steel connectors. This can be seen from an interpretation of a sulfuric acid-water phase diagram presented in work by Land (1977). An estimate of the condensed liquid sulfuric acid concentration at a condensing surface temperature of 120°F, for example, shows a concentration for the condensed liquid acid of about 10 to 20%.

According to Fontana and Greene (1967), the relative corrosion of plain carbon steel rises rapidly at sulfuric acid concentrations below 65%. According to Land (1977), for the condensed liquid acid concentration to rise above 65%, the condensing surface temperature must be above 200°F. However, according to Fontana and Greene, at acid concentrations above 65%, corrosion rates increase at metal surface temperatures above 175°F. This presents the designer with a restrictive operating range for steel surfaces (i.e., between 175 and 200°F). This is a compromise that does not completely satisfy either the acid concentration or the metal temperature criterion but should minimize the corrosion rates induced by each.

Another important phenomenon should be mentioned. When the vent system surfaces are at or below the apparent water dew point, a large amount of water condensation occurs on these surfaces. This condensate contains, in addition to the sulfuric acid mentioned above, quantities of sulfurous acid, nitrous acid, carbonic acid and hydrochloric acid. Under these conditions, the corrosion rate of commonly used vent materials is severe. Koebel and Elsener (1989) found that corrosion rates increase by a factor of 10 when the material temperatures on the flue gas side are permitted to fall below the apparent water dew point.

The applicability of ordinary stainless steels is very limited and generally follows that of plain carbon steel. Nickel-molybdenum and nickel-molybdenum-chromium alloys show good corrosion resistance over a wide range of sulfuric acid concentrations for surface temperatures up to 220°F. This is also true for high-silicon cast iron, which has found application in heat exchangers for oil-fired appliances. This latter material might find application as a liner system for masonry and metal chimneys.

Connector and Chimney Corrosion

Water and acid condensation can each result in corrosion of the connector wall and deterioration of the chimney material. While there is little documentation of specific failures, concern within the industry is growing. The volume of anecdotal information regarding corrosive failures is significant and seems well supported by findings from within the heating industry at large.

For oil-fired appliances, the rate of acid corrosion in the connector and chimney is a function of two groups of contributing factors: combustion factors and operational factors. The sulfur content of the fuel and the percent excess combustion air are the major combustion-related factors. The frequency and duration of equipment "on" and "off" periods, draft control dilution air, and the rate of heat loss from the vent system are the major operational factors. In terms of combustion, studies by Butcher et al. (1993) show a direct correlation of acid deposition (condensation) rate and subsequent corrosion to fuel sulfur content and excess combustion air. In general, reductions in fuel sulfur and excess combustion air cause reductions in the amount sulfuric acid produced in combustion and subsequently delivered to condensing surfaces in the appliance heat exchanger and carried over into the vent system. From an operational standpoint, long equipment "on" and short "off" periods and low vent system heat loss result in shorter warm-up transients and higher end-point temperatures for surfaces exposed to the flue gas. Within the limits of frictional loss, reduced vent sizes increase flue gas velocities and vent surface temperatures.

Vent Connectors

An oil-fired appliance is commonly connected to a chimney through a connector pipe. Generally, a draft control device in the form of a barometric damper is included as a component part of the connector assembly. With the advent of new power burners having high static pressure capability, draft control devices in the vent system have become less important, although many local codes still require their use. The portion of the connector assembly between the appliance flue collar and the draft control is called the flue connector; the portion between the draft control and the chimney is called the stack or chimney connector.

Chimney connectors are usually of single-wall galvanized steel. The required wall thickness for these connectors varies as a function of pipe diameter. For example, in accordance with Table 5-2.2.3 in NFPA *Standard* 211, the material thickness for galvanized steel pipe connectors between 6 and 10 in. in diameter is set at 0.023 in. (24 gage).

In accordance with the *National Fuel Gas Code*, Paragraph 7.10.2(b)3, vent, chimney, stack, and flue connectors should not be covered with insulation unless permitted in terms of their listing. Listed insulated connectors should also be installed in accordance with the terms of their listing. Because single-wall connectors must remain uninsulated for inspection, substantial cooling of the connector wall and the flue gas can occur, especially with long connector runs through spaces with low ambient temperatures. Close examination of the connector joints, seams, and surfaces is essential whenever the heating appliance is serviced. If the connector is left unrepaired, corrosion damage can cause a complete separation failure of the connector and leakage of flue gas into the occupied building space.

Where corrosion in the connector has proven to be a chronic problem, consideration should be given to replacement of the connector with a Type L vent pipe or its listed equivalent. One current product configuration consists of connector pipe with a double wall (stainless steel inner and galvanized steel outer with 0.25 in. gap). The insulated gap of this type of double-wall connector elevates inner wall surface temperature and reduces the overall connector heat loss.

Masonry Chimneys for Oil-Fired Appliances

A masonry chimney serving an oil-fired appliance should have a tile liner and should comply with applicable building codes such as NFPA *Standard* 211. An additional listed chimney liner may be needed to improve thermal response (warm-up) of the inner chimney surface, thereby reducing transient low draft during start-up and acid/water condensation during cyclic operation. This is particularly true for exposed exterior high-mass chimneys but does not exclude cold interior chimneys serving oil-fired appliances that produce relatively low flue gas temperatures. The application of insulation around tile liners within masonry chimneys is common in Europe and may be worthy of consideration in chimney replacement or new construction.

A computational analysis by Strasser et al. (1997) using OHVAP (Version 3.1) to analyze a series of masonry chimney systems with various firing rates and exit temperatures revealed that current applications of modern oil-fired heating appliances are in some difficulty in terms of acid/water condensation during winter operation. For residential oil-fired heating appliance firing rates below 1.25 to 1.5 gph with flue-loss steady-state efficiencies of 82% or higher, exterior masonry chimneys may need special treatment to reduce condensation. For conservative design, listed chimney liners and listed Type L connectors may be required for some exterior chimneys serving equipment operating under these conditions.

Replacement of Equipment

The physical condition and suitability of an existing chimney must be checked before the installation of a new oil-fired appliance. The chimney should be inspected and, if necessary, cleaned. In accordance with NFPA *Standard* 211, Section 3-2.6, the chimney drop-leg (bottom of the chimney flue) must be at least 8 in. below the bottom of the appliance connection to the chimney. The liner must be continuous, properly aligned, and intact and must extend beyond the top of the chimney. The chimney should also have a properly installed, reasonably airtight clean-out at the base.

For masonry chimneys, local experience may indicate how well the construction has withstood the lower temperatures produced by a modern oil-fired appliance. Evidence of potential or existing chimney damage can be procured by visual examination of a chimney site. Exterior indicators such as missing or loose mortar/bricks, white deposits (efflorescence) on brickwork, a leaning chimney, or water stains on interior building walls should be investigated further. Interior chimney examination with a mirror or video camera can reveal damaged or missing liner material. Any debris collected in the chimney base, drop-leg, or connector should be removed and examined. If any doubt exists regarding the condition of the chimney, examination by an experienced professional is recommended. Kam et al. (1993) offer specific guidance on the examination and evaluation of existing masonry chimneys in the field.

FIREPLACE CHIMNEYS

Fireplaces with natural-draft chimneys follow the same gravity fluid flow law as gas vents and thermal flow ventilation systems (Stone 1969). All thermal or buoyant energy is converted into flow,

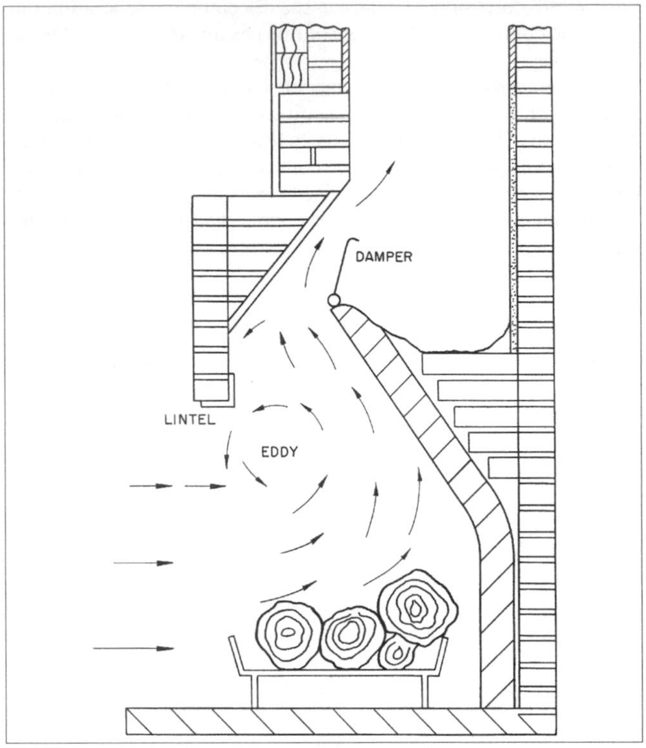

Fig. 18 Eddy Formation

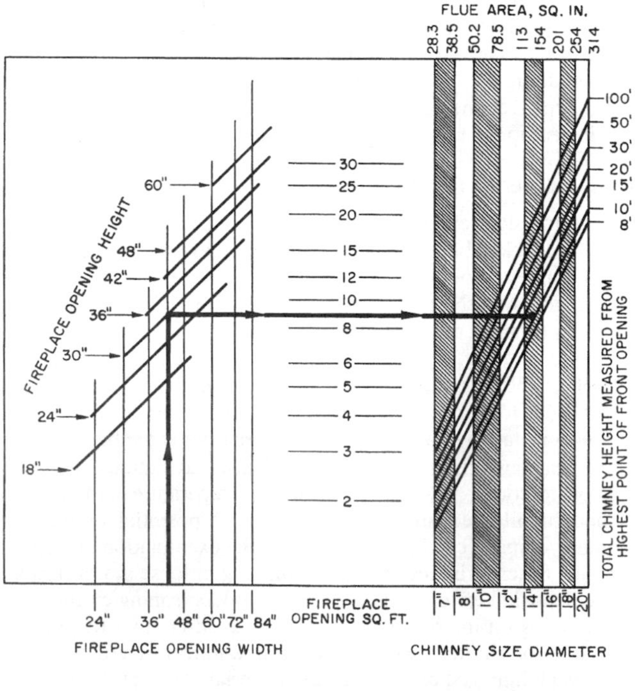

Fig. 19 Fireplace Sizing Chart for Circular Chimneys
Mean Face Velocity = 0.8 fps

and no draft exists over the fire or at the fireplace inlet. Formulas have been developed to study a wide range of fireplace applications, but the material in this section covers general cases only.

Up to some limiting value, mass flow of hot flue gases through a vertical pipe is a function of rate of heat release and the chimney area, height, and system pressure loss coefficient k. A fireplace may be considered as a gravity duct inlet fitting with a characteristic entrance-loss coefficient and an internal heat source. A fireplace functions properly (does not smoke) when adequate intake or face velocity across those critical portions of the frontal opening nullifies external drafts and internal convection effects.

The mean flow velocity into a fireplace frontal opening is nearly constant from 300°F gas temperature rise up to any higher temperature. Local velocities vary within the opening, depending on its design, because the air enters horizontally along the hearth and then flows into the fire and upward, clinging to the back wall (Figure 18). A recirculating eddy forms just inside the upper front half of the opening, induced by the high velocity of flow along the back. Restrictions or poor construction in the throat area between the lintel and damper also increase the eddy. Because the eddy moves smoke out of the zone of maximum velocity, the tendency of this smoke to escape must be counteracted by some minimum inward air movement over the entire front of the fireplace, particularly under the lintel.

A minimum mean frontal inlet velocity of 0.8 fps, in conjunction with a chimney gas temperature of at least 300 to 500°F above ambient, should control smoking in a well-constructed conventional masonry fireplace. The chart in Figure 19 shows fireplace and chimney dimensions for the specific conditions of circular flues at 0.8 fps frontal velocity. This chart solves readily for maximum frontal opening for a given chimney, as well as for chimney size and height with a predetermined opening. Figure 19 assumes no wind or air supply difficulties.

In standard masonry construction, a damper with a free area equal to or less than the required flue area is too restrictive. Most damper information does not list damper free area or effective throat opening dimensions. Further, interference with lintels or other parts may cause dimensions to vary. For best results, a damper should have a free area at least twice the area of the chimney.

Indoor-outdoor pressure differences caused by winds, kitchen or bath exhaust fans, building stack effects, and operation of forced-air heating systems or mechanical ventilation affect the operation of a fireplace. Thus, smoking during start-up can be caused by factors unrelated to the chimney. Frequently, in new homes (especially in high-rise multiple-family construction), fireplaces of normal design cannot cope with mechanically induced reverse flow or shortages of combustion air. In such circumstances, a fireplace should include an induced-draft blower of sufficient capability to overpower other mechanized air-consuming systems. An inducer for this purpose is best located at the chimney outlet and should produce 0.8 to 1.0 fps fireplace face velocity of ambient air in any individual flue or 10 to 12 fps chimney velocity.

Remedies to increase frontal velocity include the following: (1) increase chimney height (using the same flue area) and extend the last tile 6 in. or more upward; (2) decrease frontal opening by lowering the lintel or raising the hearth (glass doors may help); and (3) increase free area through the damper (ensure that it opens fully and without interference).

AIR SUPPLY TO FUEL-BURNING EQUIPMENT

A failure to supply outdoor air may result in erratic or even dangerous operating conditions. A correctly designed gas appliance with a draft hood can function with short vents (5 ft high) using an air supply opening as small in area as the vent outlet collar. Such an orifice, when equal to vent area, has a resistance coefficient in the range of 2 to 3. If the air supply opening is as much as twice the vent area, however, the coefficient drops to 0.5 or less.

The following rules may be used as a guide:

1. Residential heating equipment installed in unconfined spaces in buildings of conventional construction does not ordinarily require ventilation other than normal air infiltration. In any residence or building that has been built or altered to conserve energy or minimize infiltration, the heating appliance area should be considered a confined space. The air supply should be

installed in accordance with the *National Fuel Gas Code* or the following recommendations.

2. Residential heating equipment installed in a confined space having unusually tight construction requires two permanent openings to an unconfined space or to the outdoors. An unconfined space has a volume of at least 50 ft^3 per 1000 Btu/h of the total input rating of all appliances installed in that space. Free opening areas must be greater than 1 in^2 per 4000 Btu/h input with vertical ducts or 1 in^2 per 2000 Btu/h with horizontal ducts to the outdoors. The two openings communicating directly with sufficient unconfined space must be greater than 1 in^2 per 1000 Btu/h. Upper openings should be within 12 in. of the ceiling; lower openings should be within 12 in. of the floor.
3. Complete combustion of gas or oil requires approximately 1 ft^3 of air, at standard conditions, for each 100 Btu of fuel burned, but excess air is usually required for proper burner operation.
4. The size of these air openings may be modified if special engineering ensures an adequate supply of air for combustion, dilution, and ventilation or if local ordinances apply to boiler and machinery rooms.
5. In calculating free area of air inlets, the blocking effect of louvers, grilles, or screens protecting openings should be considered. Screens should not be smaller than 1/4 in. mesh. If the free area through a particular louver or grille is known, it should be used in calculating the size opening required to provide the free area specified. If the free area is not known, it may be assumed that wood louvers have 20 to 25% free area and metal louvers and grilles have 60 to 75% free area.
6. Mechanical ventilation systems serving the fuel-burning equipment room or adjacent spaces should not be permitted to create negative equipment room air pressure. The equipment room may require tight self-closing doors and provisions to supply air to the spaces under negative pressure so the fuel-burning equipment and venting operate properly.
7. Fireplaces may require special consideration. For example, a residential attic fan can be hazardous if it is inadvertently turned on while a fireplace is in use.
8. In buildings where large quantities of combustion and ventilation or process air are exhausted, a sufficient supply of fresh uncontaminated makeup air, warmed if necessary to the proper temperature, should be provided. It is good practice to provide about 5 to 10% more makeup air than the amount exhausted.

VENT AND CHIMNEY MATERIALS

Factors to be considered when selecting chimney materials include (1) the temperature of gases; (2) their composition and propensity to condense (dew point); (3) the presence of sulfur, halogens, and other fuel and air contaminants that lead to corrosion; and (4) the operating cycle of the appliance.

Figure 1 covers materials for vents and chimneys in the 4 to 48 in. size range; these include single-wall metal, various multiwall air- and mass-insulated types, and precast and site-constructed masonry. While each has different characteristics, such as frequency of joints, roughness, and heat loss, the type of materials used for systems 14 in. and larger is relatively unimportant in determining draft or capacity. This does not preclude selecting a safe product or method of construction that minimizes heat loss and fire hazard in the building.

National codes and standards classify heat-producing appliances as low-, medium-, and high-heat, with appropriate reference to chimney and vent constructions permitted with each. These classifications are primarily based on size, process use, or combustion temperature. In many cases, the appliance classification gives little information about the outlet gas temperature or venting need. The designer should, wherever possible, obtain gas outlet temperature conditions and properties that apply to the specific appliance, rather than going by code classification only.

Where building codes permit engineered chimney systems, selection based on gas outlet temperature can save space as well as reduce structural and material costs. For example, in some jurisdictions, approved gas-burning appliances with draft hoods operating at inputs over 400,000 Btu/h may be placed in a heat-producing classification that prohibits use of Type B gas vents. An increase in input may not cause an increase in outlet temperature or in venting hazards, and most building codes recommend correct matching of appliance and vent.

Single-wall uninsulated steel stacks can be protected from condensation and corrosion internally with refractory firebrick liners or by spraying calcium aluminate cement over a suitable interior expanded metal mesh or other reinforcement. Another form of protection applies proprietary silica or other prepared refractory coatings to pins or a support mesh on the steel. The material must then be suitably cured for moisture and heat resistance.

Moisture condensation on interior surfaces of connectors, vents, stacks, and chimneys is a more serious cause of deterioration than heat. Chimney wall temperature and flue gas velocity, temperature, and dew point affect condensation. Contaminants such as sulfur, chlorides, and fluorides in the fuel and combustion air raise the flue gas dew point. Studies by Yeaw and Schnidman (1943), Pray et al. (1942-53), Mueller (1968), and Beaumont et al. (1970) indicate the variety of analytical methods as well as difficulties in predicting the causes and probability of actual condensation.

Combustion products from any fuel containing hydrogen condense onto cold surfaces or condense in bulk if the main flow of flue gas is cooled sufficiently. Because flue gas loses heat through walls, condensation, which first occurs on interior wall surfaces cooled to the flue gas dew point, forms successively a dew and then a liquid film and, with further cooling, causes liquid to flow down into zones where condensation would not normally occur.

Start-up of cold interior chimney surfaces is accompanied by transient dew formation, which evaporates on heating above the dew point. This phenomenon causes little corrosion when very low sulfur fuels are used. Proper selection of chimney dimension and materials minimizes condensation and thus corrosion.

Experience shows a correlation between the sulfur content of the fuel and the deterioration of interior chimney surfaces. Figure 4 in Chapter 17 of the 1997 *ASHRAE Handbook—Fundamentals* illustrates one case, which applies to any fuel gas. The figure shows that the flue gas dew point increases at 40% excess air from 127°F with zero sulfur to 160°F with 15 grains of sulfur per 100 ft^3 of fuel having a heat value of 550 Btu/ft^3.

Figure 4 can also be used to approximate the effect of fuel oil sulfur content on flue gas dew point. For example, fuel oil with a sulfur content of 0.5% (by mass) contains about 252 grains of sulfur per gallon or 25,200 grains of sulfur per 100 gallons. If the fuel heat value is 140,000 Btu/gal, the ratio that defines the curves in Figure 4 gives a curve value of 18. Estimates for lower percentages of sulfur (0.25, 0.05) can be formed as factors of the value 18.

Because the corrosion mechanism is not completely understood, judicious use of resistant materials, suitably insulated or jacketed to reduce heat loss, is preferable to low-cost single-wall construction. Refractory materials and mortars should be acid- resistant, while steels should be resistant to sulfuric, hydrochloric, and hydrofluoric acids; pitting; and oxidation. Where low flue gas temperatures are expected together with low ambients, an air space jacket or mineral fiber lagging, suitably protected against water entry, helps maintain surface and flue gas temperatures above the dew point. The use of low-sulfur fuel, which is required in many localities, reduces both corrosion and air pollution.

Type 1100 aluminum alloy or any other non-copper-bearing aluminum alloy of 99% purity or better provides satisfactory performance in prefabricated metal gas vent products. For chimney service, gas temperatures from appliances burning oil or solid fuels may exceed the melting point of aluminum; therefore, steel is required. Stainless steels such as Type 430 or Type 304 give good service in residential construction and are referenced in UL-listed prefabricated chimneys. Where more corrosive substances (e.g., high-sulfur fuel or chlorides from solid fuel, contaminated air, or refuse) are anticipated, Type 29-4C or equivalent stainless steel offers a good match of corrosion resistance and mechanical properties.

As an alternative to stainless steel, porcelain enamel offers good resistance to corrosion if two coats of acid-resistant enamel are used on all surfaces. A single coat, which always has imperfections, will allow base metal corrosion, spalling, and early failure.

Prefabricated chimneys and venting products are available that use light corrosion-resistant materials, both in metal and masonry. The standardized, prefabricated, double-wall metal Type B gas vent has an aluminum inner pipe and a coated steel outer casing, either galvanized or aluminized. Standard air space from 1/4 to 1/2 in. is adequate for applicable tests and a wide variety of exposures.

Air-insulated all-metal chimneys are available for low-heat use in residential construction. Thermosiphon air circulation or multiple reflective shielding with three or more walls keeps these units cool. Insulated, double-wall residential chimneys are also available. The annulus between metal inner and outer walls is filled with insulation and retained by coupler end structures for rapid assembly.

Prefabricated, air-insulated, double-wall metal chimneys for multifamily residential and larger buildings, classed as building heating appliance chimneys, are available (Figure 20). Refractory-lined prefabricated chimneys (medium-heat type) are also available for this use.

Commercial and industrial incinerators, as well as heating appliances, may be vented by prefabricated metal-jacketed cast refractory chimneys, which are listed in the medium-heat category and are suitable for intermittent gas temperatures to 2000°F. All prefabricated chimneys and vents carrying a listing by a recognized testing laboratory have been evaluated for class of service regarding temperature, strength, clearance to adjacent combustible materials, and suitability of construction in accordance with applicable national standards.

Underwriters Laboratories (UL) standards, listed in Table 10, describe the construction and temperature testing of various classes of prefabricated vent and chimney materials. Standards for some related parts and appliances are also included in Table 10 because a

Table 10 Underwriters Laboratories Test Standards

No.	Subject	Steady-State Appliance Flue Gas Temperature Rise, °F	Fuel
103	Chimneys, factory-built, residential type (includes building heating appliances)	930	All
127	Fireplaces, factory-built	930	Solid or gas
311	Roof jacks for manufactured homes and recreational vehicles	930	Oil, gas
378	Draft equipment (such as regulators and inducers)	—	All
441	Gas vents (Type B, BW)	480	Gas only
641	Low-temperature venting systems (Type L)	500	Oil, gas
959	Chimneys, factory-built, medium-heat	1730	All
1738	Venting systems for gas-burning appliances, Categories II, III, and IV	140 to 480	Gas only

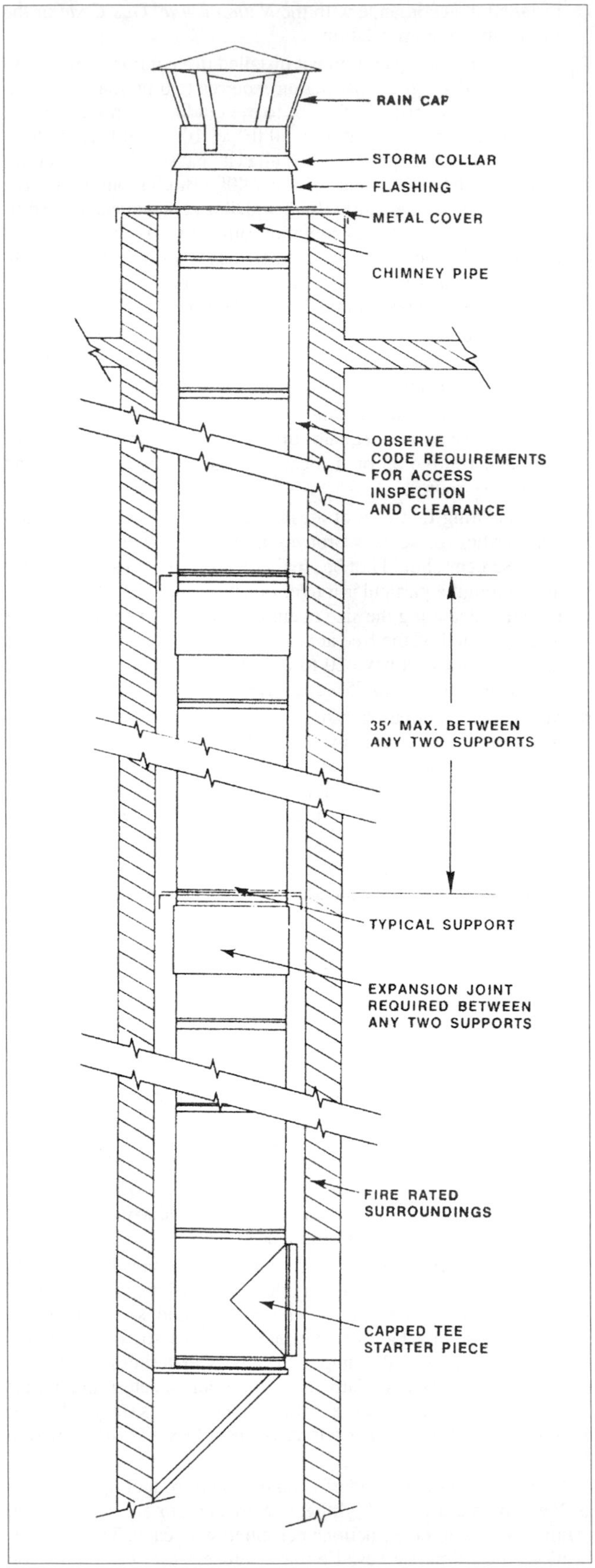

Fig. 20 Building Heating Equipment or Medium-Heat Chimney

listed factory-built fireplace, for example, must be used with a specified type of factory-built chimney. The temperature given for the steady-state operation of chimneys is the lowest in the test sequence. Factory-built chimneys under UL *Standard* 103 are also required to demonstrate adequate safety during a 1 h test at 1330°F rise and to withstand a 10 min. simulated soot burnout at either a 1630°F or a 2030°F rise.

These product tests determine minimum clearance to combustible surfaces or enclosures, based on allowable temperature rise on combustibles. They also ensure that the supports, spacers, and parts of the product that contact combustible materials remain at safe temperatures during operation. Product markings and installation instructions of listed materials are required to be consistent with test results, refer to types of appliances that may be used, and explain structural and other limitations.

VENT AND CHIMNEY ACCESSORIES

The design of a vent or chimney system must consider the existence of or need for such accessories as draft diverters, draft regulators, induced-draft fans, blocking dampers, expansion joints, and vent or chimney terminals. Draft regulators include barometric draft regulators and furnace sequence draft controls, which monitor automatic flue dampers during operation. The design, materials, and flow losses of chimney and vent connectors are covered in previous sections.

Draft Hoods

The draft hood isolates the appliance from venting disturbances (updrafts, downdrafts, or blocked vent) and allows combustion to start without venting action. Suggested general dimensions of draft hoods are given in ANSI *Standard* Z21.12, which describes certification test methods for draft hoods. In general, the pipe size of the inlet and outlet flues of the draft hood should be the same as that of the appliance outlet connection. The vent connection at the draft hood outlet should have a cross-sectional area at least as large as that of the draft hood inlet.

Draft hood selection comes under the following two categories:

1. Draft hood supplied with a design-certified gas appliance—certification of a gas appliance design under pertinent national standards includes its draft hood (or draft diverter). Consequently, the draft hood should not be altered, nor should it be replaced without consulting the manufacturer and local code authorities.
2. Draft hoods supplied separately for gas appliances—installation of listed draft hoods on existing vent or chimney connectors should be made by experienced installers in accordance with accepted practice standards.

Every design-certified gas appliance requiring a draft hood must be accompanied by a draft hood or provided with a draft diverter as an integral part of the appliance. The draft hood is a vent inlet fitting as well as a safety device for the appliance, and certain assumptions can be made regarding its interaction with a vent. First, when the hood is operating without spillage, the heat content of gases (enthalpy relative to dilution air temperature) leaving the draft hood is almost the same as that entering. Second, safe operation is obtained with 40 to 50% dilution air. It is unnecessary to assume 100% dilution air for gas venting conditions. Third, during certification tests, the draft hood must function without spillage, using a vent with not over 5 ft of effective height and one or two elbows. Therefore, if vent heights appreciably greater than 5 ft are used, an individual vent of the same size as the draft hood outlet may be much larger than necessary.

When vent size is reduced, as with tall vents, draft hood resistance is less than design value relative to the vent; the vent tables in the *National Fuel Gas Code* give adequate guidance for such size reductions.

Despite its importance as a vent inlet fitting, the draft hood designed for a typical gas appliance primarily represents a compromise of the many design criteria and tests solely applicable to that appliance. This permits considerable variation in resistance loss; thus, catalog data on draft hood resistance loss coefficients do not exist. The span of draft hood loss coefficients, including inlet acceleration, varies from the theoretical minimum of 1.0 for certain low-loss bell or conical shapes to 3 or 4, where the draft hood relief opening is located within a hot-air discharge (as with wall furnaces), and high resistance is needed to limit sensible heat loss into the vent.

Draft hoods must not be used on equipment having draft configuration (1) or (3) that is operated with either power burners or forced venting, unless the equipment has fan-assisted burners that overcome some or most of the appliance flow resistance and create a pressure inversion ahead of the draft (or barometric) hood.

Gas appliances with draft hoods must have excess chimney draft capacity to draw in adequate draft hood dilution air. Failure to provide adequate combustion air can cause oxygen depletion and spillage of flue gases from the draft hood.

Draft Regulators

Appliances requiring draft at the appliance flue gas outlet generally make use of barometric regulators for combustion stability. A balanced hinged gate in these devices bleeds air into the chimney automatically when pressure decreases. This action simultaneously increases gas flow and reduces temperature. Well-designed barometric regulators provide constant static pressure over a span of impressed draft of about 0.2 in. of water, where impressed draft is that which would exist without regulation. A regulator can maintain a 0.06 in. draft for impressed drafts from 0.06 to 0.26 in. of water. If the chimney system is very high or otherwise capable of generating available draft in excess of the pressure span capability of a single regulator, additional or oversize regulators may be used. Figure 21 shows proper locations for regulators in a chimney manifold.

Barometric regulators are available with double-acting dampers, which also swing out to relieve momentary internal pressures or divert continuing downdrafts. In the case of downdrafts, temperature safety switches actuated by hot gases escaping at the regulator sense and limit malfunctions.

Vent Dampers

Electrically, mechanically, and thermally actuated automatic vent dampers can reduce energy consumption and improve the seasonal efficiency of gas- and oil-burning appliances. Vent dampers reduce the loss of heated air through gas appliance draft hoods and the loss of specific heat from the appliance after the burner has ceased firing. These dampers may be retrofit devices or integral components of some appliances.

Electrically and mechanically actuated dampers must open prior to main burner gas ignition and must not close during operation. These safety interlocks, which electrically interconnect with existing control circuitry, include an additional main control valve, if called for, or special gas pressure-actuated controls.

Vent dampers that are thermally actuated with bimetallic elements and have spillage-sensing interlocks with burner controls are available for draft hood-type gas appliances. These dampers open in response to gas temperature after burner ignition. Because thermally actuated dampers may exhibit some flow resistance, even at equilibrium operating conditions, instructions regarding allowable heat input and minimum required vent or chimney height should be observed.

Special care must be taken to ensure that safety interlocks with appliance controls are installed according to instructions. Spillage-free gas venting after the damper has been installed must be verified with all types.

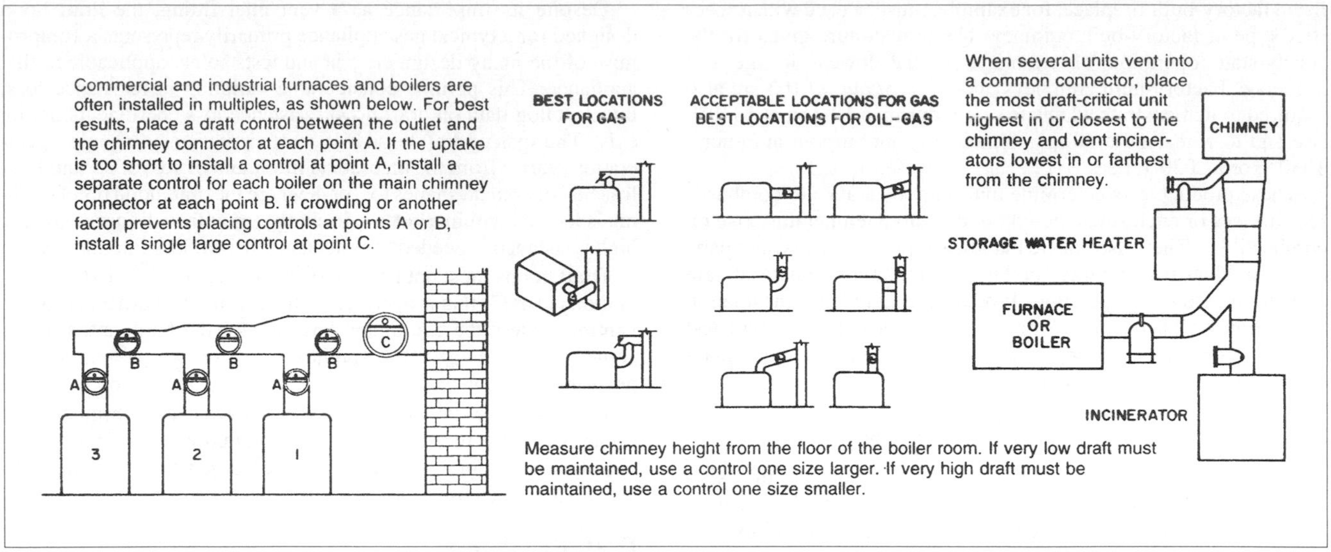

Commercial and industrial furnaces and boilers are often installed in multiples, as shown below. For best results, place a draft control between the outlet and the chimney connector at each point A. If the uptake is too short to install a control at point A, install a separate control for each boiler on the main chimney connector at each point B. If crowding or another factor prevents placing controls at points A or B, install a single large control at point C.
BEST LOCATIONS FOR GAS
ACCEPTABLE LOCATIONS FOR GAS
BEST LOCATIONS FOR OIL-GAS
When several units vent into a common connector, place the most draft-critical unit highest in or closest to the chimney and vent incinerators lowest in or farthest from the chimney.
CHIMNEY
STORAGE WATER HEATER
FURNACE OR BOILER
INCINERATOR
Measure chimney height from the floor of the boiler room. If very low draft must be maintained, use a control one size larger. If very high draft must be maintained, use a control one size smaller.

Fig. 21 Use of Barometric Draft Regulators

The energy savings of a vent damper can vary widely. Dampers reduce energy consumption under one or a combination of the following conditions:

- Heating appliance is oversized.
- Chimney is too high or oversized.
- Appliance is located in heated space.
- Two or more appliances are on the same chimney (but a damper must be installed on each appliance connected to that chimney).
- Appliance is located in building zone at higher pressure than outdoors. This positive pressure can cause steady flow losses through the chimney.

Energy savings may not justify the cost of installing a vent damper if one or more of the following conditions exist:

- All combustion and ventilation air is supplied from outdoors to direct vent appliances or to appliances located in an isolated, unheated room.
- Appliance is in an unheated basement that is isolated from the heated space.
- A one-story flat-roof house has a short vent, which is unlikely to carry away a significant amount of heated air.

For vents or chimneys serving two or more appliances, dampers should be installed on all attached appliances for maximum effectiveness. If only one damper is installed in this instance, loss of heated air through an open draft hood may negate a large portion of the potential energy savings.

Heat Exchangers or Flue Gas Heat Extractors

The sensible heat available in the flue products of properly adjusted furnaces burning oil or gas is about 10 to 15% of the rated input. Small accessory heat exchangers that fit in the connector between the appliance outlet and the chimney can recover a portion of this heat for localized use; however, they may cause some adverse effects.

All gas vent and chimney size or capacity tables assume the gas temperature or heat available to create theoretical draft is not reduced by a heat transfer device. In addition, the tables assume flow resistance for connectors, vents, and chimneys, comprising typical values for draft hoods, elbows, tees, caps, and piping with no allowance for added devices placed directly in the stream. Thus, heat exchangers or flue gas extractors should offer no flow resistance or negligible resistance coefficients when they are installed.

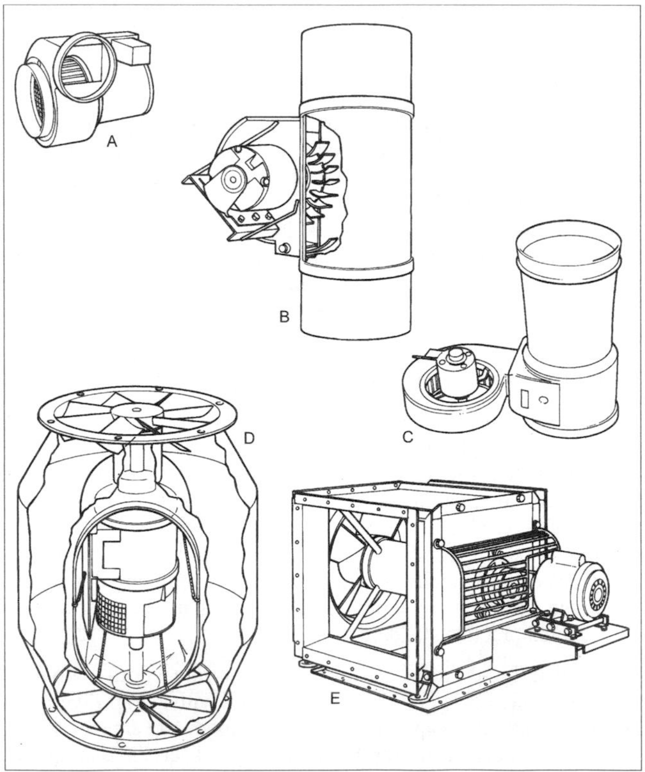

Fig. 22 Draft Inducers

A heat exchanger that is reasonably efficient and offers some flow resistance may adversely affect the system by reducing both flow and gas temperature. This may cause moisture condensation in the chimney, draft hood spillage, or both. Increasing heat transfer efficiency increases the probability of the simultaneous occurrence of both effects. An accessory heat exchanger in a solid-fuel system, especially a wood stove or heater, may collect creosote or cause its formation downstream.

The retrofit of heat exchangers in gas appliance venting systems requires careful evaluation of heat recovered versus both installed cost and the potential for chimney safety and operating problems.

Every heat exchanger installation should undergo the same spillage tests given a damper installation. In addition, the gas temperature should be checked to ensure it is high enough to avoid condensation between the exchanger outlet and the chimney outlet.

DRAFT FANS

The selection of draft fans, blowers, or inducers must consider (1) types and combinations of appliances, (2) types of venting material, (3) building and safety codes, (4) control circuits, (5) gas temperature, (6) permissible location, (7) noise, and (8) power cost. Besides specially designed fans and blowers, some conventional fans can be used if the wheel and housing materials are heat- and corrosion-resistant and if blower and motor bearings are protected from adverse effects of the gas stream.

Small draft inducers for residential gas appliance and unit heater use are available with direct-drive blower wheels and an integral sail switch to sense flow (Figure 22A). The control circuit for such applications must provide adequate vent flow both before and while fuel flows to the main burner. Other types of small inducers are either saddle-mounted blower wheels (Figure 22B) or venturi ejectors that induce flow by jet action (Figure 22C). An essential safety requirement for inducers serving draft hood gas appliances does not permit appliance interconnections on the discharge or outlet side of the inducer. This requirement prevents backflow through an inoperative appliance

With prefabricated sheet metal venting products such as Type B gas vents, the vent draft inducer should be located at or downstream from the point the vent exits the building. This placement keeps the indoor system below atmospheric pressure and prevents gases from escaping through seams and joints. If the inducer cannot be placed on the roof, metal joints must be reliably sealed in all pressurized parts of the system.

Pressure capability of residential draft inducers is usually less than 1 in. of water at rated flow. Larger inducers of the fan, blower, or ejector type have greater pressure capability and may be used to reduce system size as well as supplement available draft. Figure 22D shows a specialized axial flow fan capable of higher pressures. This unit is structurally self-supporting and can be mounted in any position in the connector or stack because the motor is in a well, separated from the gas stream. A right-angle fan, as shown in Figure 22E, is supported by an external bracket and adapts to several inlet and exit combinations. The unit uses the developed draft and an insulated tube to cool the extended shaft and bearings.

Pressure, volume, and power curves should be obtained to match an inducer to the application. For example, in an individual chimney system (without draft hood) in which a directly connected inducer only handles combustion products, calculation of the power required for continuous operation need only consider volume at operating gas temperature. An inducer serving multiple, separately controlled draft hood gas appliances must be powered for ambient temperature operation at full flow volume in the system. At any input, the inducer for a draft hood gas appliance must handle about 50% more standard chimney gas volume than a directly connected inducer. At constant volume with a given size inducer or fan, these demands follow the Fan Laws (Chapter 18) applicable to power venting as follows:

- Pressure difference developed is directly proportional to gas density.
- Pressure difference developed is inversely proportional to absolute gas temperature.
- Pressure developed diminishes in direct proportion to drop in absolute atmospheric pressure, as with altitude.
- Required power is directly proportional to gas density.
- Required power is inversely proportional to absolute gas temperature.

Centrifugal and propeller draft inducers in vents and chimneys are applied and installed the same as any heat-carrying duct system. Venturi ejector draft boosters involve some added consideration. An advantage of the ejector is that motor, bearings, and blower blades are outside the contaminated gas stream, thus eliminating a major source of deterioration. This advantage causes some loss of efficiency and can lead to reduced capacity because undersized systems, having considerable resistance downstream, may be unable to handle the added volume of the injected airstream without loss of performance. Ejectors are best suited for use at the exit or where there is an adequately sized chimney or stack to carry the combined discharge.

If total pressure defines outlet conditions or is used for fan selection, the relative amounts of static pressure and velocity pressure must be factored out; otherwise, the velocity head method of calculation does not apply. To factor total pressure into its two components, either the discharge velocity in an outlet of known area or the flow rate must be known. For example, if an appliance or blower produces a total pressure of 0.25 in. of water at 1500 fpm discharge velocity, the velocity pressure component can be found from Figure 1. Enter at the horizontal line marked V.H. in Grid C, and move horizontally to operating temperature. Read up from the appropriate temperature rise line to the 1500 fpm horizontal velocity line in Grid D. Here the velocity pressure reads 0.14 in. (calculates to 0.143 at ambient standard conditions), so static pressure is $0.25 - 0.14 = 0.11$ in. As this static pressure is part of the system driving force, it combines with theoretical draft to overcome losses in the system.

Draft inducer fans can be operated either continuously or on demand. In either case, a safety switch that senses flow or pressure is needed to interrupt burner controls if adequate draft fails. Demand operation links the thermostat with the draft control motor. Once flow starts, as sensed with a flow or pressure switch, the burner can be started. A single draft inducer, operating continuously, can be installed in the common vent of a system serving several separately controlled appliances. This simplifies the circuitry because only one control is needed to sense loss of draft. However, the single fan increases boiler standby loss and heat losses via ambient air drawn through inoperative appliances.

TERMINATIONS—CAPS AND WIND EFFECTS

The vent or chimney height and method of termination is governed by a variety of considerations, including fire hazard; wind effects; entry of rain, debris, and birds; and operating considerations such as draft and capacity. For example, the 3 ft height required for residential chimneys above a roof is required so that small sparks will burn out before they fall on the roof shingles.

Many vent and chimney malfunctions are attributed to interactions of the chimney termination or its cap with winds acting on the roof or with adjoining buildings, trees, or mountains. Because winds fluctuate, no simple method of analysis or reduction to practice exists for this complex situation. Figure 23 through Figure 25 show some of the complexities of wind flow contours around simple structural shapes.

Figure 23 shows three zones with differing degrees of flue gas dispersion around a rectangular building: the cavity or eddy zone, the wake zone, and the undisturbed flow zone (Clarke 1967). In addition, a fourth flow zone of intense turbulence is located downwind of the cavity. Chimney gases discharged into the wind at a point close to the roof surface in the cavity zone may be recirculated locally. Higher in the cavity zone, wind eddies can carry more dilute flue products to the lee side of the building. Gases discharged into the wind in the wake zone do not recirculate into the immediate vicinity, but may soon descend to ground level. Above the wake zone, dispersal into the undisturbed wind flow carries and dilutes the flue gases over a wider area. The boundaries of these zones vary with building configuration and wind direction and turbulence; they are strongly influenced by surroundings.

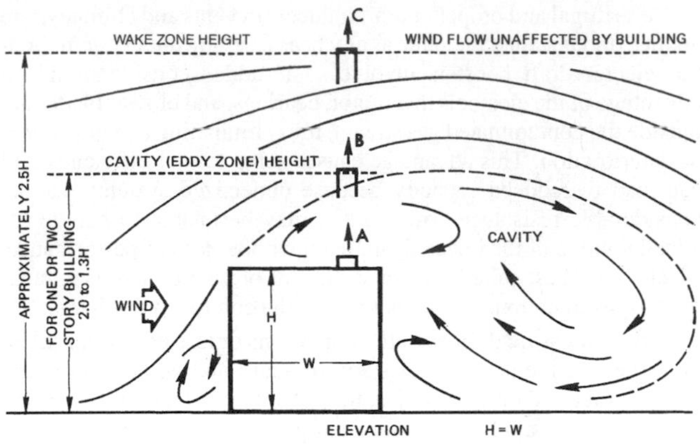

Chimney heights:

A—Discharge into cavity should be avoided because reentry will occur. Dispersion equations do not apply.

B—Discharge above cavity is good. Reentry is avoided, but dispersion may be marginal or poor from standpoint of air pollution. Dispersion equations do not apply.

C—Discharge above wake zone is best; no reentry; maximum dispersion.

Fig. 23 Wind Eddy and Wake Zones for One- or Two-Story Buildings and Their Effect on Chimney Gas Discharge

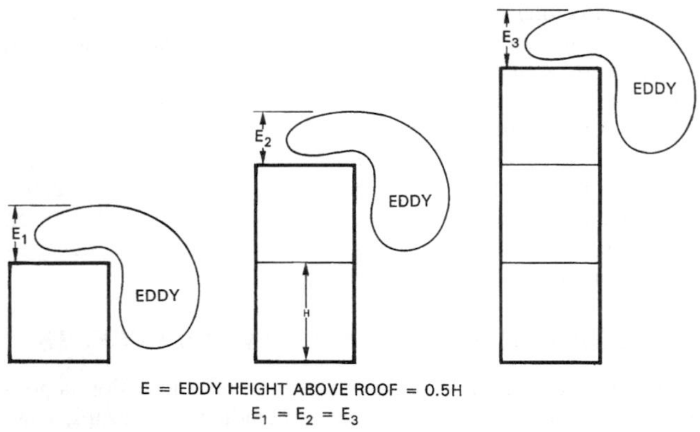

Studies found for a single cube-shaped building (length equals height) that (1) the height of the eddy above grade is 1.5 times the building height and (2) the height of unaffected air is 2.5 times height above grade. The eddy height above the roof equals 0.5H, and it does not change as building height increases in relation to building width.

Fig. 24 Height of Eddy Currents Around Single High-Rise Buildings

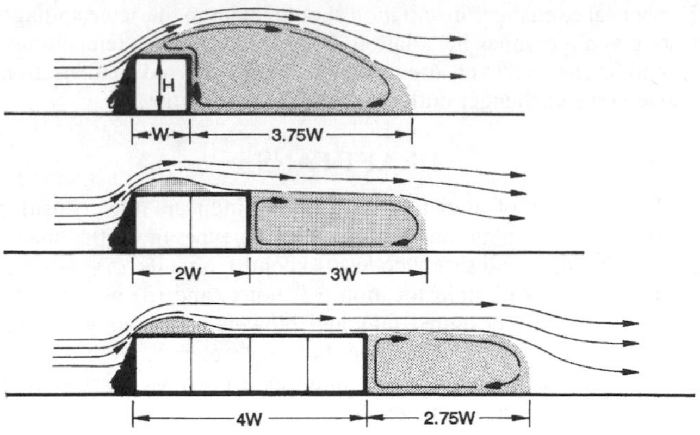

Fig. 25 Eddy and Wake Zones for Low, Wide Buildings

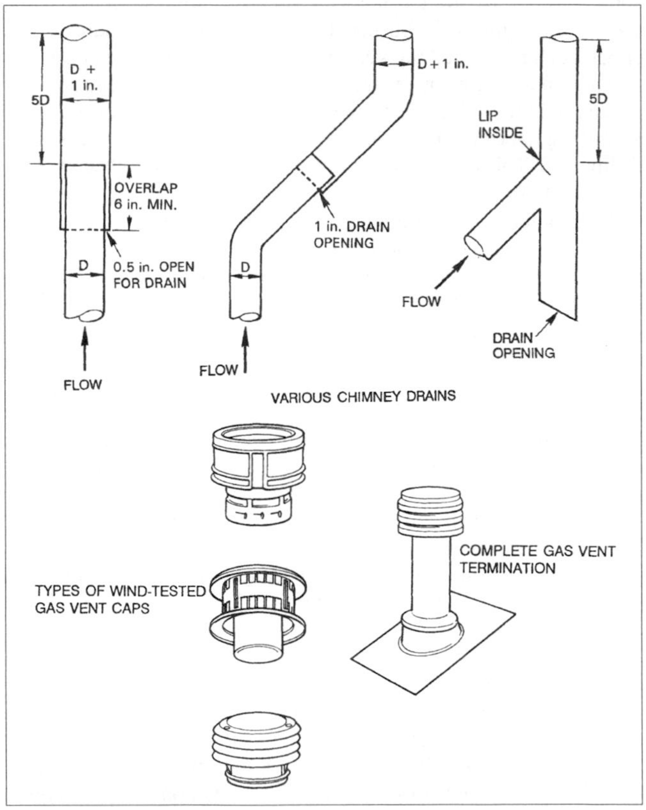

Fig. 26 Vent and Chimney Rain Protection

The possibility of air pollutants reentering the cavity zone due to plume spread or of air pollution intercepting downwind cavities associated with adjacent structures or downwind buildings should be considered. Thus, the design criterion of elevating the stack discharge above the cavity is not valid for all cases. A meteorologist experienced with dispersion processes near buildings should be consulted for complex cases and for cases involving air contaminants.

As chimney height increases through the three zones, draft performance improves dispersion, while additional problems of gas cooling, condensation, and structural wind load are created. As building height increases (Figure 24), the eddy forming the cavity zone no longer descends to ground level. For a low, wide building (Figure 25), wind blowing parallel to the long roof dimension can reattach to the surface; thus, the eddy zone becomes flush with the roof surface (Evans 1957). For satisfactory dispersion with low, wide buildings, chimney height must still be determined as if $H = W$ (Figure 23).

Evans (1957) and Chien et al. (1951) studied pitched roofs in relation to wind flow and surface pressures. Because the typical residence has a pitched roof and probably uses natural gas or a low-sulfur fossil fuel, dispersion is not important because flue products are relatively free of pollutants. For example, the *National Fuel Gas Code* requires a minimum distance between the gas vent termination and any air intake, but it does not require penetration above the cavity zone.

Flow of wind over a chimney termination can impede or assist draft. In regions of stagnation on the windward side of a wall or a steep roof, winds create positive static pressures that impede established flow or cause backdrafts within vents and chimneys. Location of a chimney termination near the surface of a low flat roof can aid draft because the entire roof surface is under negative static pressure. Velocity is low, however, due to the cavity formed as wind sweeps up over the building. With greater chimney height, termination above the low-velocity cavity or negative pressure zone subjects the chimney exit to greater wind velocity, thereby increasing draft from two causes: (1) height and (2) wind aspiration over an open top. As the termination is moved from the center of the building to the sides, its exposure to winds and pressure also varies.

Terminations on pitched roofs may be exposed to either negative or positive static pressure, as well as to variation in wind velocity and direction. On the windward side, pitched roofs vary from complete to partial negative pressure as pitch increases from approximately flat to 30° (Chien et al. 1951). At a 45° pitch, the windward pitched roof surface is strongly positive, and, beyond this slope, pressures approach those observed on a vertical wall facing the wind. Wind pressure varies with its horizontal direction on a pitched roof, and on the lee or sheltered side, wind velocity is very low, and static pressures are usually negative. Wind velocities and pressures vary not only with pitch, but with position between ridge and eaves. Reduction of these observed external wind effects to simple rules of termination for a wide variety of chimney and venting systems requires many compromises.

In the wake zone or any higher location exposed to full wind velocity, an open top can create strong venting updrafts. The updraft effect relative to wind dynamic pressure is related to the Reynolds number. Open tops, however, are sensitive to the wind angle as well as to rain (Clarke 1967), and many proprietary tops have been designed to stabilize wind effects and improve the performance. Because of the many compromises made in the design of a vent termination, this stability is usually achieved by sacrificing some of the updraft created by the wind. Further, the location of a vent cap in a cavity region frequently removes it from the zone where wind velocity could have a significant effect.

Studies of vent cap design undertaken to optimize performance of residential types indicate that the following performance features are important: (1) still-air resistance, (2) updraft ability with no flow, and (3) discharge resistance when vent gases are carried at low velocity in a typical wind (10 fps vent velocity in a 20 mph wind). Tests described in UL *Standard* 441 for proprietary gas vent caps consider these three aspects of performance to ensure adequate vent capacity.

Frequently, the air supply to an appliance room is difficult to orient to eliminate wind effects. Therefore, the vent outlet must have a certain updraft capability, which can help balance a possible adverse wind. When wind flows across an inoperative vent termination, a strong updraft develops. Appliance start-up reduces this updraft, and in typical winds, the vent cap may develop greater resistance than it would have in still air. Certain vent caps can be made with very low still-air resistance, yet exhibit excessive wind resistance, which reduces capacity. Finally, because the appliance operates whether or not there is a wind, still-air resistance must be low.

Some proprietary air ventilators have excessive still-air resistance and should be avoided on vent and chimney systems unless a considerably oversized vent is specified. Vertical-slot ventilators, for example, have still-air resistance coefficients of about 4.5. To achieve low still-air resistance on vents and chimneys, the vertical-slot ventilator must be 50% larger than the diameter of the chimney or vent unless it has been specifically listed for such use.

Freestanding chimneys high enough to project above the cavity zone require structurally adequate materials or guying and bracing for prefabricated products. The prefabricated metal building heating equipment chimney places little load on the roof structure, but guying is required at 8 to 12 ft intervals to resist both overturning and oscillating wind forces. Various other expedients, such as spiral baffles on heavy-gage freestanding chimneys, have been used to reduce oscillation.

The chimney height needed to carry the effluent into the undisturbed flow stream above the wake zone can be reduced by increasing the effluent discharge velocity. A 3000 fpm discharge velocity avoids downward eddying along the chimney and expels the effluent free of the wake zone. Velocity this large can be achieved only with forced or induced draft.

The entry of rain is a problem for open, low-velocity, or inoperative systems. Good results have been obtained with drains that divert the water onto a roof or into a collection system leading to a sump. Figure 26 shows several configurations (Clarke 1967 and Hama and Downing 1963). The runoff from stack drains contains acids, soot, and metallic corrosion products, which can cause roof staining. Therefore, these methods are not recommended for residential use. An alternate procedure is to allow all water to drain to the base of the chimney, where it is piped from a capped tee to a sump.

Rain caps prevent the vertical discharge of high-velocity gases. However, caps are preferred for residential gas-burning equipment because it is easier to exclude rain than to risk rainwater leakage at horizontal joints or to drain it. Also, caps keep out debris and bird nests, which can block the chimney. Satisfactory gas vent cap performance can be achieved in the wind by using one of a variety of standard configurations, including the *A* cap and the wind band ventilator, or one of the proprietary designs shown in Figure 26. Where partial rain protection without excessive flow resistance is desired, and either wind characteristics are unimportant or the wind flow will be horizontal, a flat disk or cone cap 1.7 to 2.0 diameters across located 0.5 diameter above the end of the pipe has a still-air resistance loss coefficient of about 0.5.

Table 11 List of U.S. National Standards Relating to Installation[a]

Subject	Materials Covered	NFPA[b]	ANSI[c]
Oil-burning equipment	Type L listed chimneys, single-wall, masonry	31	—
Gas appliances and gas piping	Type B, L listed chimneys, single-wall, masonry	54	Z223.1[d]
Chimneys, fireplaces, vents, and solid-fuel-burning appliances	All types	211	—
Recreational vehicles	Roof jacks and vents	501C	—
Gas piping and gas equipment on industrial premises	All types	54	Z223.1[d]
Gas conversion burners	Chimneys		Z21.8[e]
	Safe design		Z21.17[e]
Draft hoods	Part dimensions		Z21.12[e]
Automatic vent dampers for use with gas-fired appliances			Z21.66[e]

[a]These standards are subject to periodic review and revision to reflect advances in the industry, as well as for consistency with legal requirements and other codes.
[b]National Fire Protection Association, Quincy, MA.
[c]American National Standards Institute, New York.
[d]Available from American Gas Association, Arlington, VA.
[e]Available from CSA America, Cleveland, OH.

CODES AND STANDARDS

Building and installation codes and standards prescribe the installation and safety requirements of heat-producing appliances and their vents and chimneys. Chapter 48 lists the major national building codes, one of which may be in effect in a given area. Some jurisdictions either adopt a national building code with varying degrees of revisions to suit local custom or, as in many major metropolitan areas, develop a local code that agrees in principle but shares little common text with the national codes. Familiarity with applicable building codes is essential because of the great variation in local codes and adoption of modern chimney design practice.

The national standards listed in Table 11 give greater detail on the mechanical aspects of fuel systems and chimney or vent construction. While these standards emphasize safety aspects, especially clearances to combustibles for various venting materials, they also recognize the importance of proper flow, draft, and capacity.

CONVERSION FACTORS

The following conversion factors have been simplified for chimney design.

$$\text{Btu/h input} = 3{,}347{,}500\,\frac{\text{Boiler Horsepower}}{\text{Percent Efficiency}}$$

$= 44{,}600 \times$ Boiler horsepower (approx. 75% eff.)

$= 41{,}800 \times$ Boiler horsepower (approx. 80% eff.)

$= 37{,}200 \times$ Boiler horsepower (approx. 90% eff.)

$= 140{,}000 \times$ No. 1 and 2 oil (gph)

$= 150{,}000 \times$ No. 4, 5, and 6 oil (gph)

$= 13{,}000 \times$ Coal (lb/h)

$= 1000 \times$ Natural gas (ft^3/h)

$= 3.412 \times$ Watt rating

MBh $=$ 1000 Btu/h

kW input $=$ kW output/efficiency

kW $= 9.81 \times$ Boiler horsepower

SYMBOLS

A = area of passage cross section, ft^2
B = existing or local barometric pressure, in. Hg
B_o = standard pressure, in. Hg (29.92 in. Hg)
C_u = temperature multiplier for heat loss, dimensionless
c_p = specific heat of gas at constant pressure, Btu/lb · °F
d_f = inside diameter, ft
d_i = inside diameter, in.
D_a = available draft, in. of water
D_t = theoretical draft, in. of water
D_b = increase in static pressure by fan, in. of water
F = friction factor for L/d_i
f = Darcy friction factor from Moody diagram (Figure 13, Chapter 2, 1997 *ASHRAE Handbook—Fundamentals*)
g = gravitational constant, 32.174 ft/s^2
H = height of vent or chimney system above grade or system inlet, ft
I = operating heat input, Btu/h
k = system resistance loss coefficient, dimensionless
k_f = fitting friction loss coefficient, dimensionless
k_L = piping friction loss coefficient, dimensionless
L = length of all piping in chimney system from inlet to exit, linear ft
L_e = total equivalent length, L plus equivalent length of elbows, etc. [for use in Equations (1) and (2) only], ft
L_m = length of system from inlet to midpoint of vertical or to location of mean gas temperature, ft
M = ratio of mass flow to heat input, lb of products per 1000 Btu of fuel burned
Δp = system flow losses or pressure drop, in. of water
q = sensible heat at a particular point in vent, Btu/h
q_m = sensible heat at average temperature in vent, Btu/h
Q = volumetric flow rate, cfm
Δt = temperature difference, °F
Δt_e = temperature difference entering system, °F
Δt_m = temperature difference at average temperature location in system, °F [$\Delta t_m = T_m - T_o$]
T = absolute temperature, °R
T_m = mean flue gas temperature at average conditions in system, °R
T_o = ambient temperature, °R
T_s = standard temperature, °R (518.67°R)
U = overall heat transfer coefficient of vent or chimney wall material, referred to inside surface area, Btu/h · ft^2 · °F
$\bar{U}$ = heat transfer coefficient for Equation (15), Btu/h · ft^2 · °F
V = velocity of gas flow in passage, fps
w = mass flow of gas, lb/h
x = length of one internal side of rectangular chimney cross section, ft
y = length of other internal side of rectangular chimney cross section, ft
ρ_a = density of air at 59°F and 29.92 in. Hg, lb/ft^3 (0.0765 lb/ft^3)]
ρ_c = density of chimney gas at 0°F and 29.92 in. Hg, lb/ft^3 [0.09 lb/ft^3 in Equations (1) and (2)]
ρ_m = density of chimney gas at average temperature and local barometric pressure, lb/ft^3
ρ_o = density of air at 0°F and 29.92 in. Hg, lb/ft^3 (0.0863 lb/ft^3)

REFERENCES

ANSI. 1990. Draft hoods. *Standard* Z21.12-1990 (R 1998). Secretariat: CSA International, Cleveland, OH.

ANSI. 1999. Gas-fired central furnaces. *Standard* Z21.47-99/CSA 2.3-M99. Secretariat: CSA International, Cleveland, OH.

ASHVE. 1941. *Heating ventilating air conditioning guide*. American Society of Heating and Ventilating Engineers. Reference available from ASHRAE.

Beaumont, M., D. Fitzgerald, and D. Sewell. 1970. Comparative observations on the performance of three steel chimneys. *Institution of Heating and Ventilating Engineers* (July):85.

Butcher, T. et al. 1993. Fouling of oil-fired boilers and furnaces. Proceedings of the 1993 Oil Heat Technology Conference and Workshop (March):140.

Chien, N., Y. Feng, H. Wong, and T. Sino. 1951. *Wind-tunnel studies of pressure on elementary building forms*. Iowa Institute of Hydraulic Research.

Clarke, J.H. 1967. Air flow around buildings. *Heating, Piping and Air Conditioning* (May):145.

Evans, B.H. 1957. Natural air flow around buildings. Texas Engineering Experiment Station *Research Report* 59.

Fontana, M.G. and N.D. Greene. 1967. *Corrosion Engineering*, p. 223. McGraw-Hill, New York.

Hama, G.M. and D.A. Downing. 1963. The characteristics of weather caps. *Air Engineering* (December):34.

HYDI. 1989. Testing and rating standard for heating boilers. Hydronics Institute, Berkeley Heights, NJ.

Kam, V.P., R.A. Borgeson, and D.W. DeWerth. 1993. Masonry chimney for Category I gas appliances: Inspection and relining. *ASHRAE Transactions* 99(1):1196-1201.

Kinkead, A. 1962. Gravity flow capacity equations for designing vent and chimney systems. Proceedings of the Pacific Coast Gas Association 53.

Koebel, M. and M. Elsener. 1989. Corrosion of oil-fired central heating boilers. *Werkstoffe und Korrosion* 40:285-94.

Land, T. 1977. The theory of acid deposition and its application to the dew-point meter. *Journal of the Institute of Fuel* (June):68.

Lapple, C.E. 1949. Velocity head simplifies flow computation. *Chemical Engineering* (May):96.

Mueller G.R. 1968. Charts determine gas temperature drops in metal flue stacks. *Heating, Piping and Air Conditioning* (January):138.

NFPA. 1996. Chimneys, fireplaces, vents, and solid fuel-burning appliances. ANSI/NFPA *Standard* 211-96. National Fire Protection Association, Quincy, MA.

NFPA. 1997. Installation of oil-burning equipment. ANSI/NFPA *Standard* 31-97.

NFPA/AGA. 1999. *National fuel gas code*. ANSI/NFPA *Standard* 54-99. National Fire Protection Association, Quincy, MA. ANSI/AGA *Standard* Z223.1-99. American Gas Association, Arlington, VA.

Pray, H.R. et al. 1942-1953. The corrosion of metals and materials by the products of combustion of gaseous fuels. *Battelle Memorial Institute Reports* 1, 2, 3, and 4 to the American Gas Association.

Rutz, A.L. 1991. VENT-II *Solution methodology*. Available from ASHRAE Customer Service.

Rutz, A.L. and D.D. Paul. 1991. User's manual for VENT-II (Version 4.1) with diskettes: An interactive personal computer program for design and analysis of venting systems of one or two gas appliances. GRI-90/0178. Gas Research Institute, Chicago, IL. Available as No. PB91-509950 from National Technical Information Services, U.S. Department of Commerce, 5285 Port Royal Road, Springfield, VA 22161.

Segeler, C.G., ed. 1965. *Gas engineers' handbook*, Section 12. Industrial Press, New York, 49.

Stone, R.L. 1957. Design of multiple gas vents. *Air Conditioning, Heating and Ventilating* (July).

Stone, R.L. 1969. Fireplace operation depends upon good chimney design. *ASHRAE Journal* (February):63.

Stone, R.L. 1971. A practical general chimney design method. *ASHRAE Transactions* 77(1):91-100.

Strasser, J. et al. 1997. Development of the oil heat vent analysis program (OHVAP), Version 3.1. An interactive personal computer program for the design and analysis of venting systems for oil-fired appliances. Brookhaven National Laboratory, Upton, NY.

UL. 1995. Factory-built chimneys for residential type and building heating appliances. ANSI/UL *Standard* 103-95. Underwriters Laboratories, Northbrook, IL.

UL. 1996. Gas vents. *Standard* 441-96.

Verhoff, F.H. and J.T. Banchero. 1974. Predicting dew points of flue gases. *Chemical Engineering Progress* (August):71.

Yeaw, J.S. and L. Schnidman. 1943. Dew point of flue gases containing sulfur. *Power Plant Engineering* 47 (I and II).

BIBLIOGRAPHY

Briner, C.F. 1984. Heat transfer and fluid flow analysis of sheet metal chimney systems. ASME *Paper* No. 84-WA/SOL-36. American Society of Mechanical Engineers, New York.

Briner, C.F. 1986. Heat transfer and fluid flow analysis of three-walled metal factory-built chimneys. *ASHRAE Transactions* 92(1B):727-38.

Hampel, T.E. 1956. Venting system priming time. *Research Bulletin* 74, Appendix E, 146. American Gas Association Laboratories, Cleveland, OH.

Jakob, M. 1955. *Heat transfer*, Volume I. John Wiley and Sons, New York.

Paul, D.D. 1992. Venting guidelines for Category I gas appliances. GRI-89/0016. Gas Research Institute, Chicago, IL.

Ramsey, C.G. and H.R. Sleeper. 1970. *Architectural graphic standards*, 6th ed., p. 528. John Wiley and Sons, New York.

Reynolds, H.A. 1960. Selection of induced draft fans for heating boilers. *Air Conditioning, Heating and Ventilating* (December):51.

Sepsy, C.F. and D.B. Pies. 1972. An experimental study of the pressure losses in converging flow fittings used in exhaust systems. Ohio State University College of Engineering, Columbus, OH (December).

CHAPTER 31

UNIT VENTILATORS, UNIT HEATERS, AND MAKEUP AIR UNITS

UNIT VENTILATORS

A **HEATING unit ventilator** is an assembly whose principal functions are to heat, ventilate, and cool a space by introducing outdoor air in quantities up to 100% of its rated capacity. The heating medium may be steam, hot water, gas, or electricity. The essential components of a heating unit ventilator are the fan, motor, heating element, damper, filter, automatic controls, and outlet grille, all of which are encased in a housing.

An **air-conditioning unit ventilator** is similar to a heating unit ventilator; however, in addition to the normal winter function of heating, ventilating, and cooling with outdoor air, it is also equipped to cool and dehumidify during the summer. It is usually arranged and controlled to introduce a fixed quantity of outdoor air for ventilation during cooling in mild weather. The air-conditioning unit ventilator may be provided with a variety of combinations of heating and air-conditioning elements. Some of the more common arrangements include

- Combination hot and chilled water coil (two-pipe)
- Separate hot and chilled water coils (four-pipe)
- Hot water or steam coil and direct-expansion coil
- Electric heating coil and chilled water or direct-expansion coil
- Gas-fired furnace with direct-expansion coil

Unit ventilators are used primarily in schools, meeting rooms, offices, and other areas where the density of occupancy requires controlled ventilation to meet local codes. The typical unit is equipped with controls that permit heating, ventilating, and cooling to be varied while the fans operate continuously. In normal operation, the discharge air temperature from a unit is varied in accordance with the room requirements. The heating unit ventilator can provide **ventilation cooling** by bringing in outdoor air whenever the room temperature is above the room set point. Air-conditioning unit ventilators can provide refrigeration when the outdoor air temperature is too high to be used effectively for ventilation cooling.

DESCRIPTION

Unit ventilators are available for floor mounting, ceiling mounting, and recessed applications. They are available with various airflow and capacity ratings, and the fan can be arranged so that air is either blown through or drawn through the unit. With direct-expansion refrigerant cooling, the condensing unit can either be furnished as an integral part of the unit ventilator assembly or be remotely located.

Figure 1A shows a typical heating unit ventilator. The heating coil can be hot water, steam, or electric. Hot water coils can be provided with face-and-bypass dampers for capacity control, if desired. Valve control of capacity is also available.

Figure 1B shows a typical air-conditioning unit ventilator with a combination hot and chilled water coil for use in a two-pipe system. This type of unit is usually provided with face-and-bypass dampers for capacity control.

Figure 1C illustrates a typical air-conditioning unit ventilator with two separate coils, one used for heating and the other used for cooling with a four-pipe system. The heating coil may be hot water, steam, or electric. The cooling coil can be either a chilled water coil or a direct-expansion refrigerant coil. Heating and cooling coils are sometimes combined in a single coil by providing separate tube circuits for each function. In such cases, the effect is the same as having two separate coils.

Figure 1D illustrates a typical air-conditioning unit ventilator with a fan section, a gas-fired heating furnace section, and a direct-expansion refrigerant coil section.

CAPACITY

Manufacturers publish the heating and cooling capacities of unit ventilators. Table 1 lists typical nominal capacities.

Heating Capacity Requirements

Because a unit ventilator has a dual function of introducing outdoor air for ventilation and maintaining a specified room condition, the required heating capacity is the sum of the heat required to bring outdoor ventilation air to room temperature and the heat required to offset room losses. The ventilation cooling capacity of a unit ventilator is determined by the air volume delivered by the unit and the

Table 1 Typical Unit Ventilator Capacities

Airflow, cfm	Heating Unit Ventilator Total Heating Capacity, Btu/h	A/C Unit Ventilator Total Cooling Capacity, Btu/h
500	38,000	19,000
750	50,000	28,000
1000	72,000	38,000
1250	85,000	47,000
1500	100,000	56,000

The preparation of the sections on Unit Ventilators and Unit Heaters is assigned to TC 6.1, Hydronic and Steam Equipment and Systems. The preparation of the section on Makeup Air Units is assigned to TC 5.8, Industrial Ventilation.

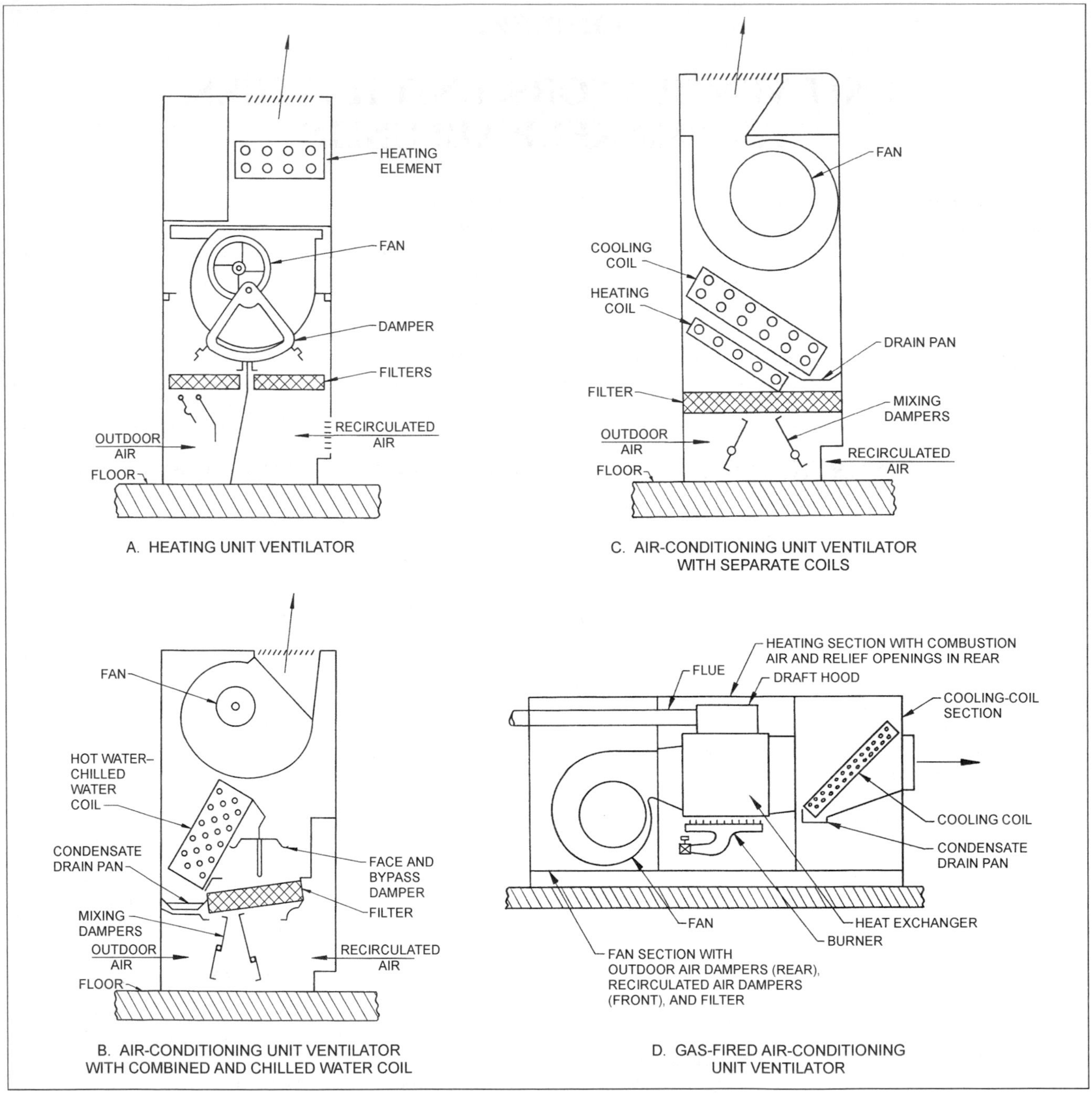

Fig. 1 Typical Unit Ventilators

temperature difference between the unit discharge and the room temperature.

Example. A room has a heat loss of 24,000 Btu/h at a winter outdoor design condition of 0°F and an indoor design of 70°F, with 20% outdoor air. Minimum air discharge temperature from the unit is 60°F. To obtain the specified number of air changes, a 1250 cfm unit ventilator is required. Determine the ventilation heat requirement, the total heating requirement, and the ventilation cooling capacity of this unit with outdoor air temperature below 60°F.

Solution:

Ventilation heat requirement:

$$q_v = 60\rho c_p Q(t_i - t_o)$$

where

q_v = heat required to heat ventilating air, Btu/h
ρ = density of air at standard conditions = 0.075 lb/ft^3
c_p = air specific heat = 0.24 Btu/lb·°F
Q = ventilating airflow, cfm
t_i = required room air temperature, °F
t_o = outdoor air temperature, °F

$$q_v = 60 \times 0.075 \times 0.24 \times 1250(20/100)(70 - 0) = 18{,}900 \text{ Btu/h}$$

Total heating requirement:

$$q_t = q_v + q_s$$

where

q_t = total heat requirement, Btu/h
q_s = heat required to make up heat losses, Btu/h

$$q_t = 18{,}900 + 24{,}000 = 42{,}900 \text{ Btu/h}$$

Ventilation cooling capacity:

$$q_c = 60\rho c_p Q(t_i - t_f)$$

where

q_c = ventilation cooling capacity of unit, Btu/h
t_f = unit discharge air temperature, °F

$$q_c = 60 \times 0.075 \times 0.24 \times 1250(70 - 60) = 13{,}500 \text{ Btu/h}$$

SELECTION

Items to be considered in the application of unit ventilators are

- Unit air capacity
- Percent minimum outdoor air
- Heating and cooling capacity
- Cycle of control
- Location of unit

Mild-weather cooling capacity and the number of occupants in the space are the primary considerations in selecting the unit's air capacity. Other factors to consider are state and local requirements, volume of the room, density of occupancy, and the use of the room. The number of air changes required for a specific application also depends on window area, orientation, and the maximum outdoor temperature at which the unit is expected to prevent overheating.

Rooms oriented to the north (in the northern hemisphere) with small window areas require about six air changes per hour. About nine air changes are required in rooms oriented to the south that have large window areas. As many as 12 air changes may be required for very large window areas and southern exposures. These airflows are based on preventing overheating at outdoor temperatures up to about 55°F. For satisfactory cooling at outdoor air temperatures up to 60°F, the airflow should be increased accordingly.

These airflows apply principally to classrooms. Factories and kitchens may require 30 to 60 air changes per hour (or more). Office areas may need from 10 to 15 air changes per hour.

The minimum amount of outdoor air for ventilation is determined after the total air capacity has been established. It may be governed by local building codes or it may be calculated to meet the ventilating needs of the particular application. ASHRAE *Standard* 62 requires 15 cfm of outdoor air per occupant in a classroom and 20 cfm per occupant in laboratories, cafeterias, and conference rooms.

The heating and cooling capacity of a unit to meet the heating requirement can be determined from the manufacturer's data. Heating capacity should always be determined after selecting the unit air capacity for mild-weather cooling.

CONTROL

Many cycles of control are available. The principal difference in the various cycles is the amount of outdoor air delivered to the room. Usually, a room thermostat simultaneously controls both a valve, damper, or step controller to regulate the heat supply and an outdoor and return air damper. A thermostat in the airstream prevents the discharge of air below the desired minimum temperature. Unit ventilator controls provide the proper sequence for the following stages:

Warm-Up Stage. All control cycles allow rapid warm-up by having the units generate full heat with the outdoor damper closed. Thus 100% of the room air is recirculated and heated until the room temperature approaches the desired level.

Heating and Ventilating Stage. As the room temperature rises into the operating range of the thermostat, the outdoor air damper partially or completely opens to provide ventilation, depending on the cycle used. Auxiliary heating equipment is shut off. As the room temperature continues to rise, the unit ventilator heat supply is throttled.

Cooling and Ventilating Stage. When the room temperature rises above the desired level, the room thermostat throttles the heat supply so that cool air flows into the room. The thermostat gradually shuts off the heat and then opens the outdoor air damper. The airstream thermostat frequently takes control during this stage to keep the discharge temperature from falling below a set level.

The section on Control of Systems in Chapter 45 of the 1999 *ASHRAE Handbook—Applications* describes the three cycles of control commonly used for unit ventilators. These control cycles are as follows:

Cycle I. 100% outdoor air is admitted at all times, except during the warm-up stage.

Cycle II. A minimum amount of outdoor air (normally 20% to 50%) is admitted during the heating and ventilating stage. This percentage is gradually increased to 100%, if needed, during the ventilation cooling stage.

Cycle III. Except during the warm-up stage, a variable amount of outdoor air is admitted, as needed, to maintain a fixed temperature of the air entering the heating element. The amount of air admitted is controlled by the airstream thermostat, which is set low enough—often at 55°F—to provide cooling when needed.

Air-conditioning unit ventilators can include any of the three cycles in addition to the mechanical cooling stage in which a fixed amount of outdoor air is introduced. The cooling capacity is controlled by the room thermostat.

For maximum heating economy, the building temperature is reduced at night and during weekends and vacations. Several arrangements are used to accomplish this. One arrangement takes advantage of the natural convective capacity of the unit when the fans are off. This capacity is supplemented by cycling the fan with the outdoor damper closed as required to maintain the desired room temperature.

LOCATION

Floor-model unit ventilators are normally installed on an outside wall near the centerline of the room. Ceiling models are mounted against either the outside wall or one of the inside walls. Ceiling models discharge air horizontally. Best results are obtained if the unit can be placed so that the airflow is not interrupted by ceiling beams or surface-mounted lighting fixtures.

Downdraft can be a problem in classrooms with large window areas in cold climates. Air in contact with the cold glass is cooled and flows down into the occupied space. Floor-standing units often include one of the following provisions to prevent downdraft along the windows (Figure 2):

- **Window sill heating** uses finned radiators of moderate capacity installed along the wall under the window area. Heated air rises upward by convection and counteracts the downdraft by tempering it and diverting it upward.
- **Window sill recirculation** is obtained by installing the return air intake along the window sill. Room or return air to the unit includes the cold downdrafts, takes them from the occupied area, and eliminates the problem.
- **Window sill discharge** directs a portion of the unit ventilator discharge air into a delivery duct along the sill of the window. The discharge air, delivered vertically at the window sill, is distributed

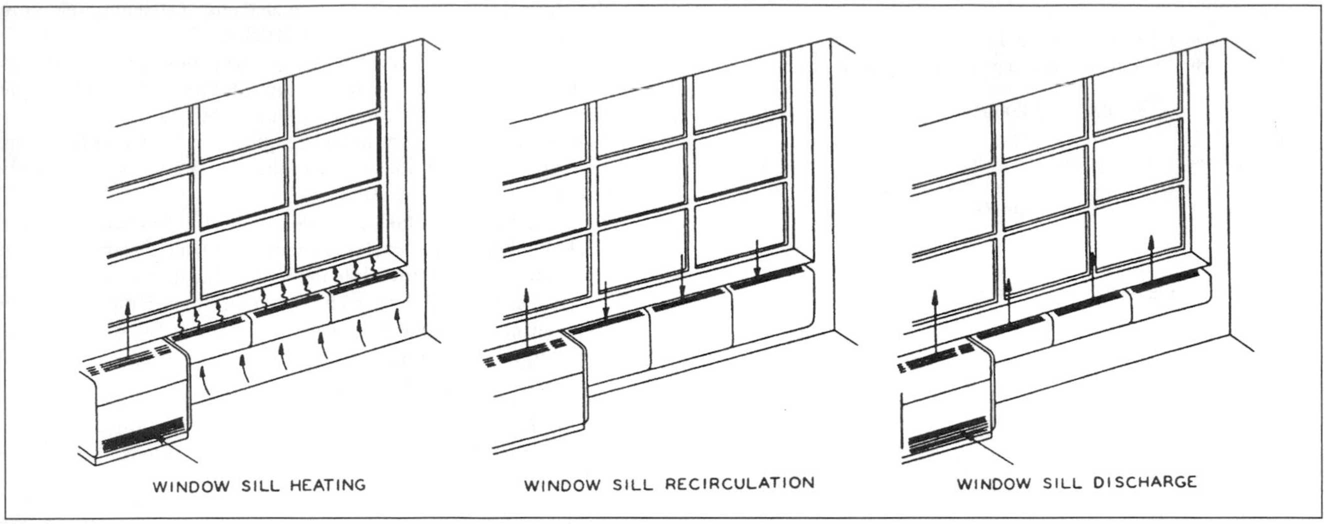

Fig. 2 Methods of Preventing Downdraft along Windows

throughout the room, and the upwardly directed air combats downdraft.

UNIT HEATERS

A unit heater is an assembly of elements, the principal function of which is to heat a space. The essential elements are a fan and motor, a heating element, and an enclosure. Filters, dampers, directional outlets, duct collars, combustion chambers, and flues may also be included. Some types of unit heaters are shown in Figure 3.

Unit heaters can usually be classified in one or more of the following categories.

1. **Heating Medium.** The media include (1) steam, (2) hot water, (3) gas indirect-fired, (4) oil indirect-fired, and (5) electric heating.
2. **Type of Fan.** Three types of fans can be considered: (1) propeller, (2) centrifugal, and (3) remote air mover. Propeller fan units may be arranged to blow air horizontally (horizontal blow) or vertically (downblow). Units with centrifugal fans may be small cabinet units or large industrial units. Units with remote air movers are known as duct unit heaters.
3. **Arrangement of Elements.** Two types of units can be considered: (1) draw-through, in which the fan draws air through the unit; and (2) blow-through, in which the fan blows air through the heating element. Indirect-fired unit heaters are always blow-through units.

APPLICATION

Unit heaters have the following principal characteristics:

- Relatively large heating capacities in compact casings
- Ability to project heated air in a controlled manner over a considerable distance
- Relatively low installed cost per unit of heat output
- Application where sound level is permissible

They are, therefore, usually placed in applications where the heating capacity requirements, the physical volume of the heated space, or both, are too large to be handled adequately or economically by other means. Through the elimination of extensive duct installations, the space is freed for other use.

Unit heaters are principally used for heating commercial and industrial structures such as garages, factories, warehouses, showrooms, stores, and laboratories, as well as corridors, lobbies, vestibules, and similar auxiliary spaces in all types of buildings. Unit heaters may often be used to advantage in specialized applications requiring spot or intermittent heating, such as at outside doors in industrial plants or in corridors and vestibules. Cabinet unit heaters may be used where heated air must be filtered.

Unit heaters may be applied to a number of industrial processes, such as drying and curing, in which the use of heated air in rapid circulation with uniform distribution is of particular advantage. They may be used for moisture absorption applications, such as removing fog in dye houses, or to prevent condensation on ceilings or other cold surfaces of buildings in which process moisture is released. When such conditions are severe, unit ventilators or makeup air units may be required.

SELECTION

The following factors should be considered when selecting a unit heater:

- Heating medium to be employed
- Type of unit
- Location of the unit for proper heat distribution
- Permissible sound level
- Rating of the unit
- Need for filtration

Heating Medium

The proper heating medium is usually determined by economics and involves an examination of initial cost, operating cost, and conditions of use.

Steam or hot water unit heaters are relatively inexpensive but require a boiler and piping system. The unit cost of such a system generally decreases as the number of units increases. Therefore, steam or hot water heating is most frequently used (1) in new installations involving a relatively large number of units, and (2) in existing systems that have sufficient capacity to handle the additional load. High-pressure steam or high-temperature hot water units are normally used only in very large installations or when a high-temperature medium is required for process work. Low-pressure steam and conventional hot water units are usually selected for smaller installations and for those concerned primarily with comfort heating.

Gas and oil indirect-fired unit heaters are frequently preferred in small installations where the number of units does not justify the expense and space requirements of a new boiler system or where

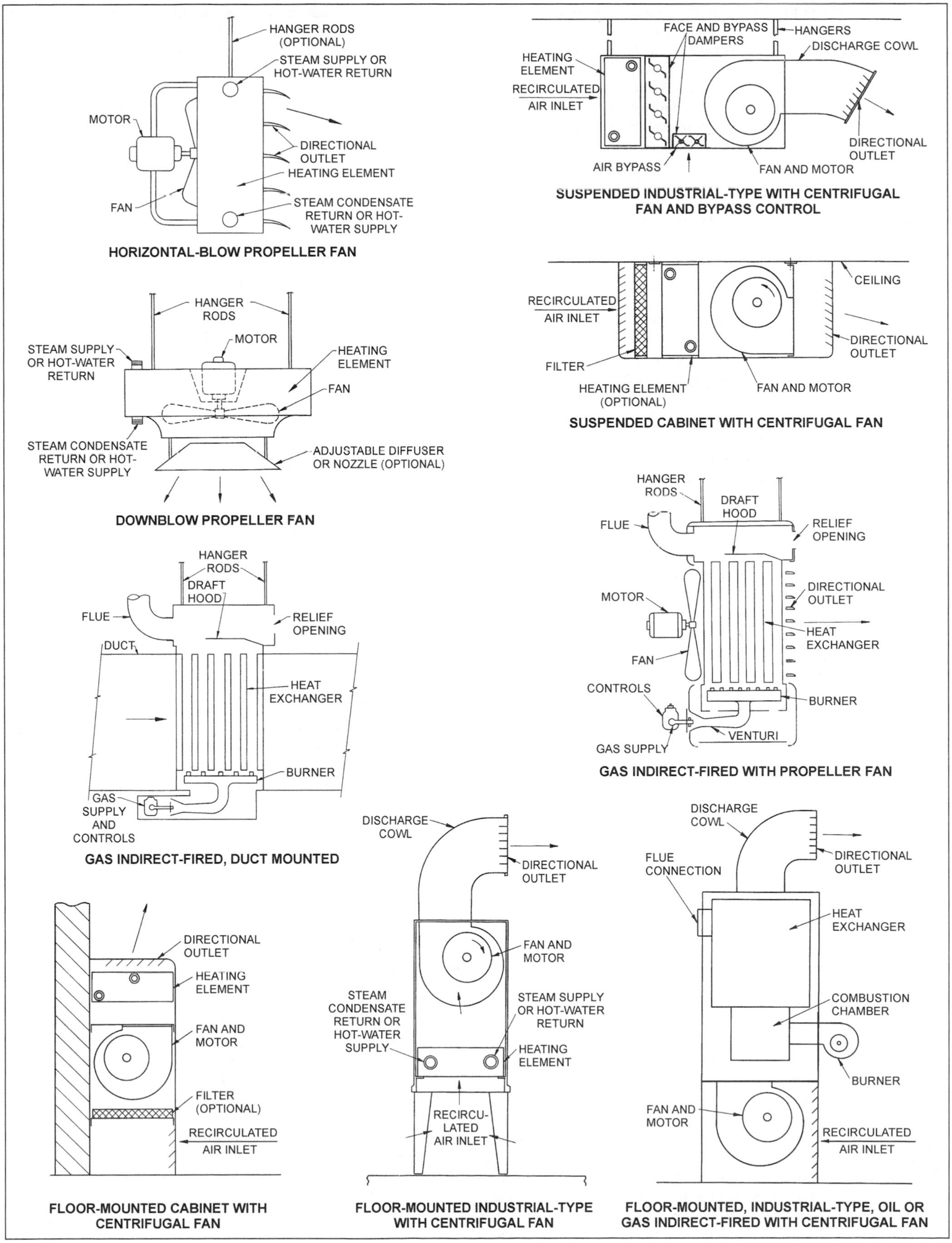

Fig. 3 Typical Unit Heaters

individual metering of the fuel supply is required, as in a shopping center. Gas indirect-fired units usually have either horizontal propeller fans or industrial centrifugal fans. Oil indirect- fired units largely have industrial centrifugal fans. Some codes limit the use of indirect-fired unit heaters in some applications.

Electric unit heaters are used when low-cost electric power is available and for isolated locations, intermittent use, supplementary heating, or temporary service. Typical applications are ticket booths, security offices, factory offices, locker rooms, and other isolated rooms scattered over large areas. Electric units are particularly useful in isolated and untended pumping stations or pits, where they may be thermostatically controlled to prevent freezing.

Type of Unit

Propeller fan units are generally used in nonducted applications where the heating capacity and distribution requirements can best be met by units of moderate output and where heated air does not need to be filtered. Horizontal-blow units are usually installed in buildings with low to moderate ceiling heights. Downblow units are used in spaces with high ceilings and where floor and wall space limitations dictate that the heating equipment be kept out of the way. Downblow units may have an adjustable diffuser to vary the discharge pattern from a high-velocity vertical jet (to achieve the maximum distance of downward throw) to a horizontal discharge of lower velocity (to prevent excessive air motion in the zone of occupancy). Revolving diffusers are also available.

Cabinet unit heaters are used for applications in which a more attractive appearance is desired. They are suitable for free-air delivery or low static pressure duct applications. They may be equipped with filters, and they can be arranged to discharge either horizontally or vertically up or down.

Industrial centrifugal fan units are applied where heating capacities and space volumes are large or where filtration of the heated air or operation against static resistance is required. Downblow or horizontal-blow units may be used, depending on the requirements.

Duct unit heaters are used where the air handler is remote from the heater. These heaters sometimes provide an economical means of adding heating to existing cooling or ventilating systems with ductwork. They require flow and temperature limit controls.

Location for Proper Heat Distribution

Units must be selected, located, and arranged to provide complete heat coverage and, at the same time, maintain acceptable air motion and temperature at an acceptable sound level in the working or occupied zone. Proper application depends on size, number, and type of units; direction of airflow and type of directional outlet used; mounting height; outlet velocity and temperature; and air volumetric flow. Many of these factors are interrelated.

The mounting height may be governed by space limitations or by the presence of equipment such as display cases or machinery. The higher a downblow heater is mounted, the lower the temperature of the air leaving the heater must be to force the heated air into the occupied zone. Also, the distance that air leaving the heater travels depends largely on the air temperature and initial velocity. A high temperature reduces the area of heat coverage.

Unit heaters for high-pressure steam or high-temperature hot water should be designed to produce approximately the same leaving air temperature as would be obtained from a lower temperature heating medium.

To obtain the desired air distribution and heat diffusion, unit heaters are commonly equipped with directional outlets, adjustable louvers, or fixed or revolving diffusers. For a given unit with a given discharge temperature and outlet velocity, the mounting height and heat coverage can vary widely with the type of directional outlet, adjustable louver, or diffuser.

Other factors that may substantially reduce heat coverage include obstructions (such as columns, beams, or partitions) or machinery in either the discharge airstream or the approach area to the unit. The presence of strong drafts or other air currents also reduces coverage. Exposures such as large glass areas or outside doors, especially on the windward side of the building, require special attention; units should be arranged so that they blanket the exposures with a curtain of heated air and intercept the cold drafts.

For area heating, horizontal-blow unit heaters in exterior zones should be placed so that they blow either along the exposure or toward it at a slight angle. When possible, multiple units should be arranged so that the discharge airstreams support each other and create a general circulatory motion in the space. Interior zones under exposed roofs or skylights should be completely blanketed. Downblow units should be arranged so that the heated areas from adjacent units overlap slightly to provide complete coverage.

For spot heating of individual spaces in larger unheated areas, single unit heaters may be used, but allowance must be made for the inflow of unheated air from adjacent spaces and the consequent reduction in heat coverage. Such spaces should be isolated by partitions or enclosures, if possible.

Horizontal unit heaters should have discharge outlets located well above head level. Both horizontal and vertical units should be placed so that the heated airstream is delivered to the occupied zone at acceptable temperature and velocity. The outlet air temperature of free-air delivery unit heaters used for comfort heating should be 50 to 60°F higher than the design room temperature. When possible, units should be located so that they discharge into open spaces, such as aisles, and not directly on the occupants. For further information on air distribution, see Chapter 31 of the 1997 *ASHRAE Handbook—Fundamentals*.

Manufacturers' catalogs usually include suggestions for the best arrangements of various unit heaters, recommended mounting heights, heat coverage for various outlet velocities, final temperatures, directional outlets, and sound level ratings.

Sound Level in Occupied Spaces

Sound pressure levels in workplaces should be limited to values listed in Table 34 in Chapter 46 of the 1999 *ASHRAE Handbook—Applications*. Although the noise level is generated by all equipment within hearing distance, unit heaters may contribute a significant portion of noise level. Both noise and air velocity in the occupied zone generally increase with increased outlet velocities. An analysis of both the diverse sound sources and the locations of personnel stations establishes the limit to which the unit heaters must be held.

Ratings of Unit Heaters

Steam or Hot Water. Heating capacity must be determined at a standard condition. Variations in entering steam or water temperature, entering air temperature, and steam or water flow affect capacity. The following are typical standard conditions for rating of steam unit heaters: dry saturated steam at 2 psig pressure at the heater coil, air at 60°F (29.92 in. Hg barometric pressure) entering the heater, and the heater operating free of external resistance to airflow. The following are standard conditions for rating of hot water unit heaters: entering water at 200°F, water temperature drop of 20°F, entering air at 60°F (29.92 in. Hg barometric pressure), and the heater operating free of external resistance to airflow.

Gas-Fired. Gas-fired unit heaters are rated in terms of both input and output, in accordance with the approval requirements of the American Gas Association.

Oil-Fired. Ratings of oil-fired unit heaters are based on heat delivered at the heater outlet.

Electric. Electric unit heaters are rated based on the energy input to the heating element.

Effect of Airflow Resistance on Capacity. Unit heaters are customarily rated at free-air delivery. Airflow and heating capacity will decrease if outdoor air intakes, air filters, or ducts on the inlet or discharge are used. The reduction in capacity caused by this added resistance depends on the characteristics of the heater and on the type, design, and speed of the fans. As a result, no specific capacity reduction can be assigned for all heaters at a given added resistance. The manufacturer should have information on the heat output to be expected at other than free-air delivery.

Effect of Inlet Temperature. Changes in entering air temperature influence the total heating capacity in most unit heaters and the final temperature in all units. Because many unit heaters are located some distance from the occupied zone, possible differences between the temperature of the air actually entering the unit and that of air being maintained in the heated area should be considered—particularly with downblow unit heaters.

Higher velocity units and units with lower vertical discharge air temperature maintain lower temperature gradients than units with higher discharge temperatures. Valve-controlled or bypass-controlled units with continuous fan operation maintain lower temperature gradients than units with intermittent fan operation. Directional control of the discharged air from a unit heater can also be important in distributing heat satisfactorily and in reducing floor-to-ceiling temperature gradients.

Filters

Air from propeller unit heaters cannot be filtered because the heaters are designed to operate with heater friction loss only. If dust in the building must be filtered, centrifugal fan units or cabinet units should be used. Chapter 24 has further information on air cleaners for particulate contaminants.

AUTOMATIC CONTROL

The controls for a steam or hot water unit heater can provide either (1) on-off operation of the unit fan, or (2) continuous fan operation with modulation of heat output. For on-off operation, a room thermostat is used to start and stop the fan motor or group of fan motors. A limit thermostat, often strapped to the supply or return pipe, prevents fan operation in the event that heat is not being supplied to the unit. An auxiliary switch that energizes the fan only when power is applied to open the motorized supply valve may also be used to prevent undesirable cool air from being discharged by the unit.

Continuous fan operation eliminates both the intermittent blasts of hot air resulting from on-off operation and the stratification of temperature from floor to ceiling that often occurs during off periods. In this arrangement, a proportional room thermostat controls a valve modulating the heat supply to the coil or a bypass around the heating element. A limit thermostat or auxiliary switch stops the fan when heat is no longer available.

One type of control used with downblow unit heaters is designed to automatically return the warm air, which would normally stratify at the higher level, down to the zone of occupancy. Two thermostats and an auxiliary switch are required. The lower thermostat is placed in the zone of occupancy and is used to control a two-position supply valve to the heater. An auxiliary switch is used to stop the fan when the supply valve is closed. The higher thermostat is placed near the unit heater at the ceiling or roof level where the warm air tends to stratify. The lower thermostat automatically closes the steam valve when its setting is satisfied, but the higher thermostat overrides the auxiliary switch so that the fan continues to run until the temperature at the higher level falls below a point sufficiently high to produce a heating effect.

Indirect-fired and electric units are usually controlled by intermittent operation of the heat source under control of the room thermostat, with a separate fan switch to run the fan when heat is being supplied. For more information on automatic control, refer to Chapter 45 of the 1999 *ASHRAE Handbook—Applications*.

Unit heaters can be used to circulate air in summer. In such cases, the heat is shut off and the thermostat has a bypass switch, which allows the fan to run independently of the controls.

PIPING CONNECTIONS

Piping connections for steam unit heaters are similar to those for other types of fan blast heaters. The piping of unit heaters must conform strictly to the system requirements, while at the same time allowing the heaters to function as intended. Basic piping principles for steam systems are discussed in Chapter 10.

Steam unit heaters condense steam rapidly, especially during warm-up periods. The return piping must be planned to keep the heating coil free of condensate during periods of maximum heat output, and the steam piping must be able to carry a full supply of steam to the unit to take the place of condensed steam. Adequate pipe size is especially important when a unit heater fan is operated under on-off control because the condensate rate fluctuates rapidly.

Recommended piping connections for unit heaters are shown in Figure 4. In steam systems, the branch from the supply main to the heater must pitch toward the main and be connected to its top in order to prevent condensate in the main from draining through the heater, where it might reduce capacity and cause noise.

The return piping from steam unit heaters should provide a minimum drop of 10 in. below the heater, so that the pressure of water required to overcome resistances of check valves, traps, and strainers will not cause condensate to remain in the heater.

Dirt pockets at the outlet of unit heaters and strainers with 0.063 in. perforations to prevent rapid plugging are essential to trap dirt and scale that might affect the operation of check valves and traps. Strainers should always be installed in the steam supply line if the heater has steam-distributing coils or is valve controlled.

An adequate air vent is required for low-pressure closed gravity systems. The vertical pipe connection to the air vent should be at least 3/4 in. NPT to allow water to separate from the air passing to the vent. If thermostatic instead of float-and-thermostatic traps are used in vacuum systems, a cooling leg must be installed ahead of the trap.

In high-pressure systems, it is customary to continuously vent the air through a petcock (as indicated in Figure 4C), unless the steam trap has a provision for venting air. Most high-pressure return mains terminate in flash tanks that are vented to the atmosphere. When possible, pressure reducing valves should be installed to permit operation of the heaters at low pressure. Traps must be suitable for the operating pressure encountered.

When piping is connected to hot water unit heaters, it must be pitched to permit air to vent to the atmosphere at the high point in the piping. An air vent at the heater is used to facilitate air removal or to vent the top of the heater. The system must be designed for complete drainage, including placing nipple and cap drains on drain cocks when units are located below mains.

MAINTENANCE

Regular inspection, based on a schedule determined by the amount of dirt in the atmosphere, assures maximum operating economy and heating capacity. Heating elements should be cleaned when necessary by brushing or blowing with high-pressure air or by using a steam spray. A portable sheet metal enclosure may be used to partially enclose smaller heaters for cleaning in place with air or steam jets. In certain installations, however, it may be necessary to remove the heating element and wash it with a mild alkaline solution, followed by a thorough rinsing with water. Propeller units do not have filters and are, therefore, more susceptible to dust buildup on the coils.

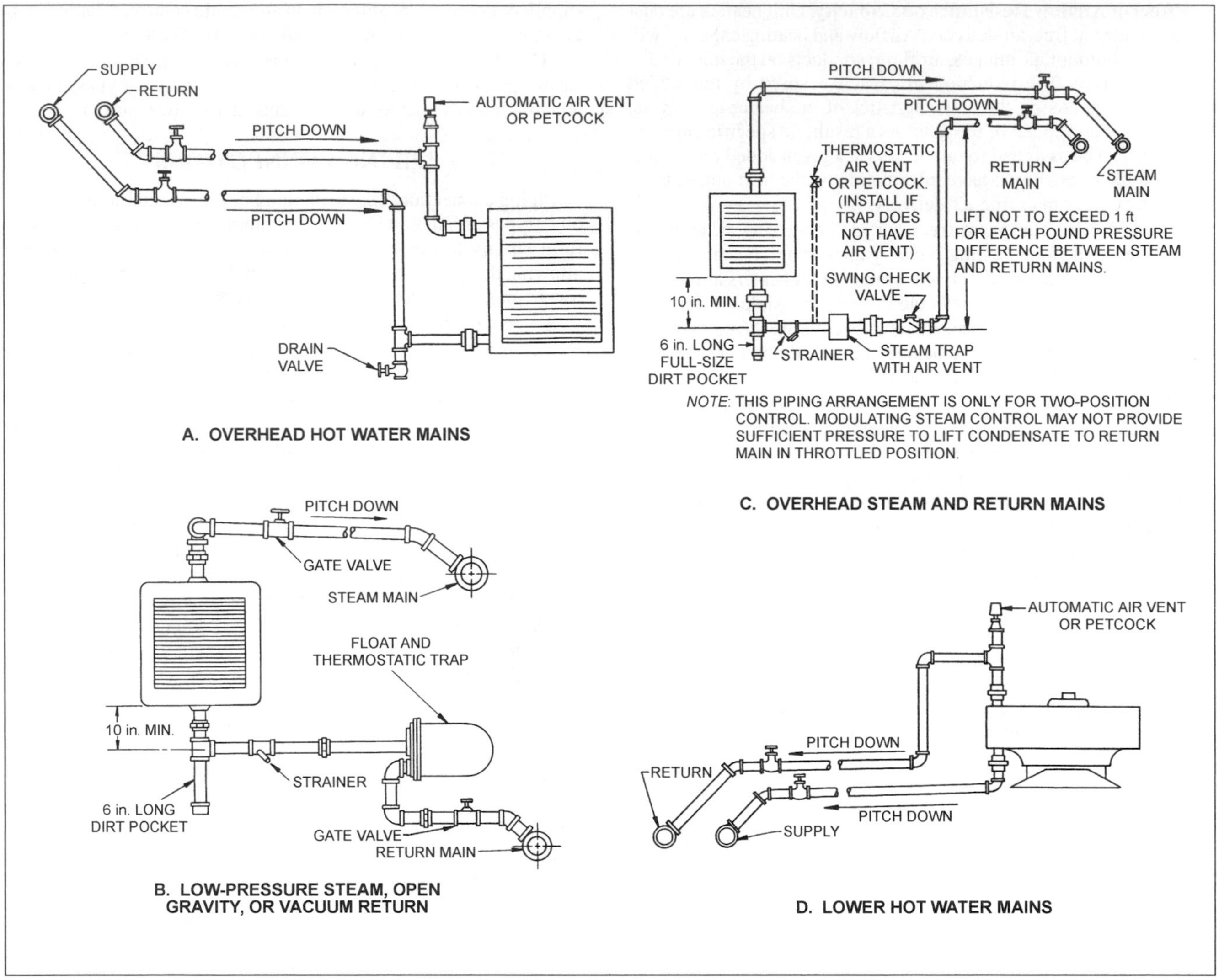

Fig. 4 Hot Water and Steam Connections for Unit Heaters

Dirt on fan blades reduces capacity and may unbalance the blades, which causes noise and bearing damage. Fan blades should be inspected and cleaned when necessary. Vibration and noise may also be caused by improper fan position or loose set screws. A fan guard should be placed on downblow unit heaters that have no diffuser or other device to catch the fan blade if it comes loose and falls from the unit.

The amount of attention required by the various motors used with unit heaters varies greatly. The instructions for lubrication, in particular, must be followed carefully for trouble-free operation. Excess lubrication, for example, may damage the motor. An improper lubricant may cause the bearings to fail. Instructions for care of the motor on any unit heater should be obtained from the manufacturer and kept at the unit.

Fan bearings and drives must be lubricated and maintained according to the instructions specified by the manufacturer. If the unit is direct-connected, the couplings should be inspected periodically for wear and alignment. V-belt drives should have all belts replaced with a matched set if one belt shows wear.

Periodic inspections of traps, inspections of check and air valves, and the replacement of worn fans are other important maintenance functions. Strainers should be cleaned regularly. Filters, if included, must be cleaned or replaced when dirty.

MAKEUP AIR UNITS

Makeup air units are designed to condition ventilation air introduced into a space or to replace air exhausted from a building. The air exhausted may be from a process or general area exhaust, through either powered exhaust fans or gravity ventilators. The units may be used to prevent negative pressure within buildings or to control the contaminant level in the air.

If the temperature or humidity (or both) within the structure are controlled, the makeup system must have the capacity to condition the replacement air. Makeup air units may, therefore, be required to heat, ventilate, cool, humidify, dehumidify, and filter. This equipment may be required to replace air at the space conditions, or it may be used for part or all of the heating, ventilating, or cooling load.

Makeup air can enter at a fixed flow rate or as variable volume outside air, under building pressurization or contamination control, to respond directly to exhaust flow. Makeup air units may also be connected to process exhaust with air-to-air heat recovery units.

Buildings under negative pressure because of inadequate makeup air may have the following symptoms:

- Gravity stacks from unit heaters and processes back-vent.
- Exhaust systems do not perform at rated volume.

- The perimeter of the building is cold in winter due to the high infiltration.
- Severe drafts occur at exterior doors.
- Exterior doors are hard to open.
- Heating systems cannot maintain comfortable conditions throughout the building because the central core area becomes overheated.

SELECTION

Makeup air systems used for ventilation may be (1) sized to balance air exhaust volumes or (2) sized in excess of the exhaust volume in order to dilute contaminants. In applications where contaminant levels vary, variable flow units should be considered so that the supply air varies for contaminant control and the exhaust volume varies to track supply volume. In critical areas, the exhaust volume may be based on requirements to control pressure in the area.

Location

Makeup air units are defined by their location or the use of a key component. Examples are rooftop makeup air units, truss- or floor-mounted units, and sidewall units. Some manufacturers differentiate their units by heating mode, such as steam or direct gas-fired makeup air units.

Rooftop units are commonly used for large single-story industrial buildings because air distribution is simpler. Also, access (via roof walks) is more convenient than access to equipment mounted in the truss; truss units can be easily reached only by installing a catwalk adjacent to the air units. Disadvantages of rooftop units are (1) they increase foot traffic on the roof, which reduces its life and increases the likelihood of leaks; (2) inclement weather reduces the accessibility of equipment; and (3) units are exposed to weather.

Makeup air units can also be placed around the perimeter of a building with air ducted through the sidewall. This approach limits future building expansion, and the effectiveness of ventilating internal areas decreases as the building gets larger. But access to the units is good, and minimum support is required because the units are mounted on the ground.

Heating and Cooling Media

Heating. Makeup air units are often identified by the heating or cooling medium they use. Heaters in makeup air systems may be direct gas-fired burners, electric resistance heating coils, indirect gas-fired heaters, or steam or hot water heating coils. (Chapter 23 covers the design and application of heating coils.) Air distribution systems are often required to direct heat to areas of heat loss.

Natural gas can be used in an indirect-fired burner, which is the same as a large furnace. (Chapter 28 has more information on types of heaters.) In a direct-fired gas burner, natural gas is completely burned so that no harmful products of combustion enter the airstream. This type of burner can only be placed in an airstream with 100% outside air—no room air may be recirculated over the burner.

Hydronic heating sections in areas requiring 100% outside air must be protected from freezing in cold climates. Low-temperature protection includes two-position control of steam coils; careful selection of the water coil heating surface, control valves, and water supply temperature; and use of antifreeze.

Cooling. Mechanical refrigeration with direct-expansion evaporator coils or chilled water coils, direct or indirect evaporative cooling sections, or well water coils may be used. Air distribution systems are often required to direct cooling to specific areas with heat gain.

Because industrial facilities often have high sensible heat loads, evaporative cooling can be particularly effective. An evaporative cooler helps clean the air as well. A portion of the spray water must be bled off to keep the water acceptably clean and to maintain a low solids concentration. Chapter 19 of this volume and Chapter 50 of the 1999 *ASHRAE Handbook—Applications* cover evaporative cooling in more detail.

Chapter 21 provides information on air-cooling coils. If direct-expansion coils are used in conjunction with direct-fired gas coils, the cooling coils must be downstream from the gas burner to assure that flame cannot contact refrigerant in the event of a leak. Also refrigerant coils must not be overheated because overheating can cause refrigerant to discharge through pressure relief valves.

Filters

High-efficiency filters are not normally used in a makeup air unit because the large volume of outside air handled by the unit is relatively clean. It is important that the filters in the unit be easy to change or clean. Cleanable filters should have a space nearby with appropriate washing equipment. Large throwaway filters should not be so large that they are difficult to remove. Chapter 24 and Chapter 25 have more information on air filters and cleaners.

Controls

Controls for a makeup air unit fall into the following categories: (1) local temperature controls, (2) airflow controls, (3) plant-wide controls for proper equipment operation and efficient performance, (4) safety controls for burning gas, and (5) building smoke control systems. For control system information, refer to Chapter 39 and Chapter 45 of the 1999 *ASHRAE Handbook—Applications*.

Safety controls for gas-fired units include components to properly light the burner and to provide a safeguard against flame failure. To light the burner, the safety controls ensure that the pilot for the burner is burning and the unit is adequately purged with outside air before the main burner is lighted. The flame safeguard monitors the operation of the burner and shuts off gas to the burner on sensing a malfunction. Malfunctions include flame failure, supply fan failure, combustion air failure, power failure, too high or low gas pressure, excess air temperature, and gas leaks in the motorized valves.

Applicable Codes and Standards

A gas-fired makeup air unit must be designed and built in accordance with NFPA *Standard* 54, the *National Fuel Gas Code*, and the requirements of the owner's insurance underwriter. Local codes must also be checked when using direct-fired gas makeup air units because some jurisdictions prohibit their use or require exhaust fans to be operated during heating.

The following standards may also apply, depending on the application:

ACCA. 1992. Direct-fired makeup air equipment. *Technical Bulletin* 109. Air-Conditioning Contractors of America, Washington, DC.

ACGIH. 1998. *Industrial ventilation: A manual of recommended practice*, 23rd ed. American Conference of Governmental Industrial Hygienists, Cincinnati, OH.

ARI. 1989. Central-station air-handling units. ANSI/ARI *Standard* 430-89. Air-Conditioning and Refrigeration Institute, Arlington, VA.

ASHRAE. 1999. Ventilation for acceptable indoor air quality. ASHRAE *Standard* 62-1999.

ASHRAE. 1995. Energy conservation in existing buildings. ASHRAE/IESNA *Standard* 100-1995.

CSA International. 1990. Direct gas-fired industrial air heaters, 2nd ed. ANSI *Standard* Z83.18-90 (R 1998). Cleveland, OH.

CSA International. 1991. Direct gas-fired make-up air heaters, 5th ed. ANSI *Standard* Z83.4-91 (R 1998).

ICBO. 1998. *Uniform mechanical code*. International Conference of Building Officials, Whittier, CA. Also, International Association of Plumbing and Mechanical Officials, Los Angeles.

Other Applications

Spot Cooling. High-velocity air jets in the unit may be directed to working positions. During cold weather, the supply air must be tempered or reduced in velocity to avoid overcooling of workers.

Door Heating. Localized air supply at swinging doors or overhead doors such as for loading docks can be provided by makeup air units. Heaters may blanket the door openings with tempered air. The temperature may be reset from the outdoor temperature or with dual temperature air—low when the door is closed and high when the door is open during cold weather. Heating may be arranged to serve a single door or multiple doors by an air distribution system. Door heating systems may also be arranged to minimize entry of insects during warm weather.

COMMISSIONING

Commissioning of makeup air systems is similar to that of other air-handling systems, requiring attention to

- Equipment identification
- Piping system identification
- Belt drive adjustment
- Control system checkout
- Documentation of system installation
- Lubrication
- Electrical system checkout for overload heater size and function
- Cleaning and degreasing of hydronic piping systems
- Pretreatment of hydronic fluids
- Setup of chemical treatment program for hydronic systems and evaporative apparatus
- Start-up of major equipment items by factory-trained technician
- Testing and balancing
- Planning of preventive maintenance program
- Instruction of owner's operating and maintenance personnel

MAINTENANCE

Basic operating and maintenance data required for makeup air systems may be obtained from the *ASHRAE Handbook* chapters covering the components. Specific operating instructions are required for makeup air heaters that require changeover from winter to summer conditions, including manual fan speed changes, air distribution pattern adjustment, or heating cycle lockout.

Operations handling 100% outside air may require more frequent maintenance, such as changing filters, lubricating bearings, and checking the water supply to evaporative coolers/humidifiers. Filters on systems in locations with dirty air require more frequent changing, so a review may determine whether upgrading filter media would be cost-effective. More frequent cleaning of fans blades and heat transfer surfaces may be required in such locations to maintain airflow and heat transfer performance.

BIBLIOGRAPHY

Brown, W.K. 1990. Makeup air systems—Energy saving opportunities. *ASHRAE Transactions* 96(2):609-15.

Bridgers, F.H. 1980. Efficiency study—Preheating outdoor air for industrial and institutional applications. *ASHRAE Journal* 22(2):29-31.

Gadsby, K.J. and T.T. Harrje. 1985. Fan pressurization of buildings—Standards, calibration and field experience. *ASHRAE Transactions* 91(2B): 95-104.

Holness, G.V.R. 1989. Building pressurization control: Facts and fallacies. *Heating/Piping/Air Conditioning* (February).

Persily, A. 1982. Repeatability and accuracy of pressurization testing. In Thermal Performance of the Exterior Envelopes of Buildings II, *Procedures of ASHRAE/DOE Conference*. ASHRAE SP 38:380-90.

CHAPTER 32

HYDRONIC HEAT-DISTRIBUTING UNITS AND RADIATORS

RADIATORS, convectors, and baseboard and finned-tube units are heat-distributing devices used in steam and low-temperature water heating systems. They supply heat through a **combination of radiation and convection** and maintain the desired air temperature and/or mean radiant temperature in a space without fans. Figures 1 and 2 show sections of typical heat-distributing units. In low-temperature systems, radiant panels are also used. Units are inherently self- adjusting in the sense that heat output is based on temperature differentials; cold spaces receive more heat and warmer spaces receive less heat.

DESCRIPTION

Radiators

The term radiator, while generally confined to sectional cast-iron column, large-tube, or small-tube units, also includes flat panel types and fabricated steel sectional types. Small-tube radiators, with a length of only 1.75 in. per section, occupy less space than column and large-tube units and are particularly suited to installation in recesses (see Table 1). Column, wall-type, and large-tube radiators are no longer manufactured, although many of these units are still in use. Refer to Tables 2, 3, and 4 in Chapter 28 of the 1988 *ASHRAE Handbook—Equipment*, Hydronics Institute (1989), or Byrley (1978) for principal dimensions and average ratings of these units.

The following are the most common types of radiators:

Sectional radiators are fabricated from welded sheet metal sections (generally 2, 3, or 4 tubes wide), and resemble freestanding cast-iron radiators.

Panel radiators consist of fabricated flat panels (generally 1, 2, or 3 deep), with or without an exposed extended fin surface attached to the rear for increased output. These radiators are most common in Europe.

Tubular steel radiators consist of supply and return headers with interconnecting parallel steel tubes in a wide variety of lengths and heights. They may be specially shaped to coincide with the building structure. Some are used to heat bathroom towel racks.

Specialty radiators are fabricated of welded steel or extruded aluminum and are designed for installation in ceiling grids or floor-mounting. An array of unconventional shapes is available.

Pipe Coils

Pipe coils have largely been replaced by finned tubes. See Table 5 in Chapter 28 of the 1988 *ASHRAE Handbook—Equipment* for the heat emission of such pipe coils.

Convectors

A convector is a heat-distributing unit that operates with gravity-circulated air (natural convection). It has a heating element with a large amount of secondary surface and contains two or more tubes with headers at both ends. The heating element is surrounded by an enclosure with an air inlet opening below and an air outlet opening above the heating element.

Convectors are made in a variety of depths, sizes, and lengths and in enclosure or cabinet types. The heating elements are available in fabricated ferrous and nonferrous metals. The air enters the enclosure below the heating element, is heated in passing through the element, and leaves the enclosure through the outlet grille located above the heating element. Factory-assembled units comprising a heating element and an enclosure have been widely used. These may be freestanding, wall-hung, or recessed and may have outlet grilles or louvers and arched inlets or inlet grilles or louvers, as desired.

Baseboard Units

Baseboard (or baseboard radiation) units are designed for installation along the bottom of walls in place of the conventional baseboard. They may be made of cast iron, with a substantial portion of the front face directly exposed to the room, or with a finned-tube element in a sheet metal enclosure. They operate with gravity-circulated room air.

Baseboard heat-distributing units are divided into three types: radiant, radiant convector, and finned tube. The **radiant** unit, which is made of aluminum, has no openings for air to pass over the wall side of the unit. Most of this unit's heat output is by radiation.

The **radiant-convector** baseboard is made of cast iron or steel. The units have air openings at the top and bottom to permit circulation of room air over the wall side of the unit, which has extended surface to provide increased heat output. A large portion of the heat emitted is transferred by convection.

The **finned-tube** baseboard has a finned-tube heating element concealed by a long, low sheet metal enclosure or cover. A major portion of the heat is transferred to the room by convection. The output varies over a wide range, depending on the physical dimensions and the materials used. A unit with a high relative output per unit length compared to overall heat loss (which would result in a concentration of the heating element over a relatively small area) should be avoided. Optimum comfort for room occupants is obtained when units are installed along as much of the exposed wall as possible.

Finned-Tube Units

Finned-tube (or fin-tube) units are fabricated from metallic tubing, with metallic fins bonded to the tube. They operate with gravity-circulated room air. Finned-tube elements are available in several tube sizes, in either steel or copper—1 to 2 in. nominal steel or 3/4 to 1 1/4 in. nominal copper—with various fin sizes, spacings, and materials. The resistance to the flow of steam or water is the same as that through standard distribution piping of equal size and type.

The preparation of this chapter is assigned to TC 6.1, Hydronic and Steam Equipment and Systems.

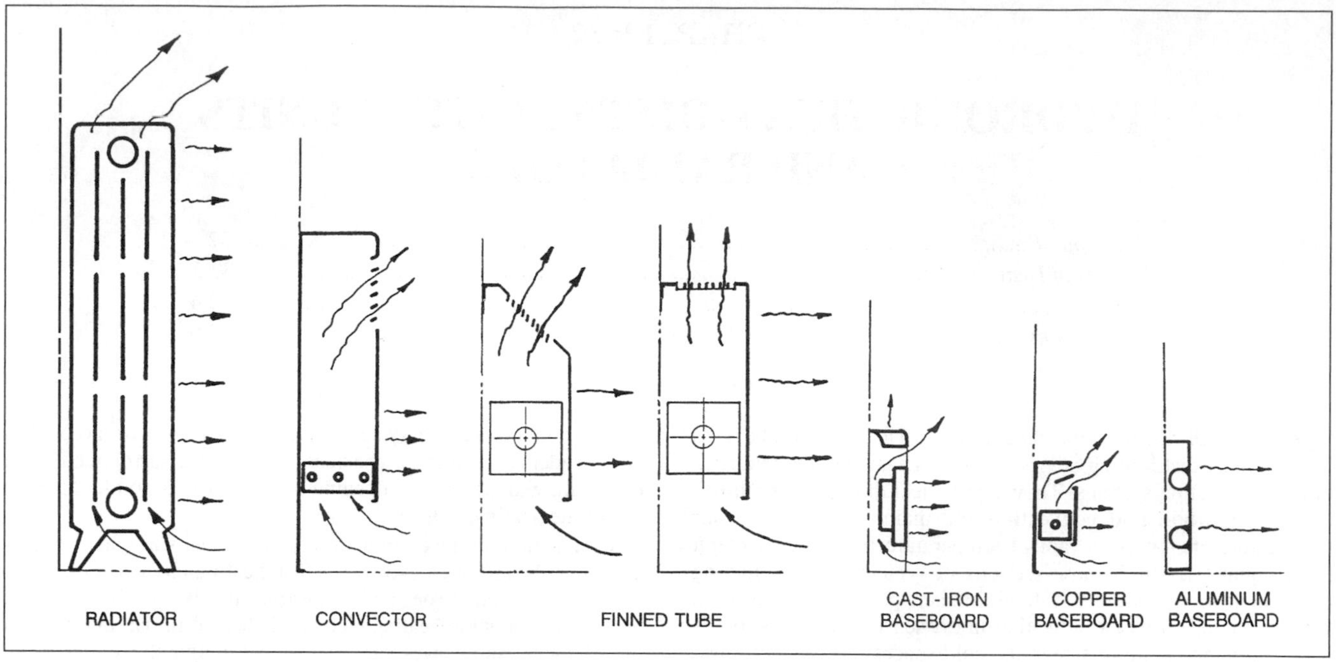

Fig. 1 Terminal Units

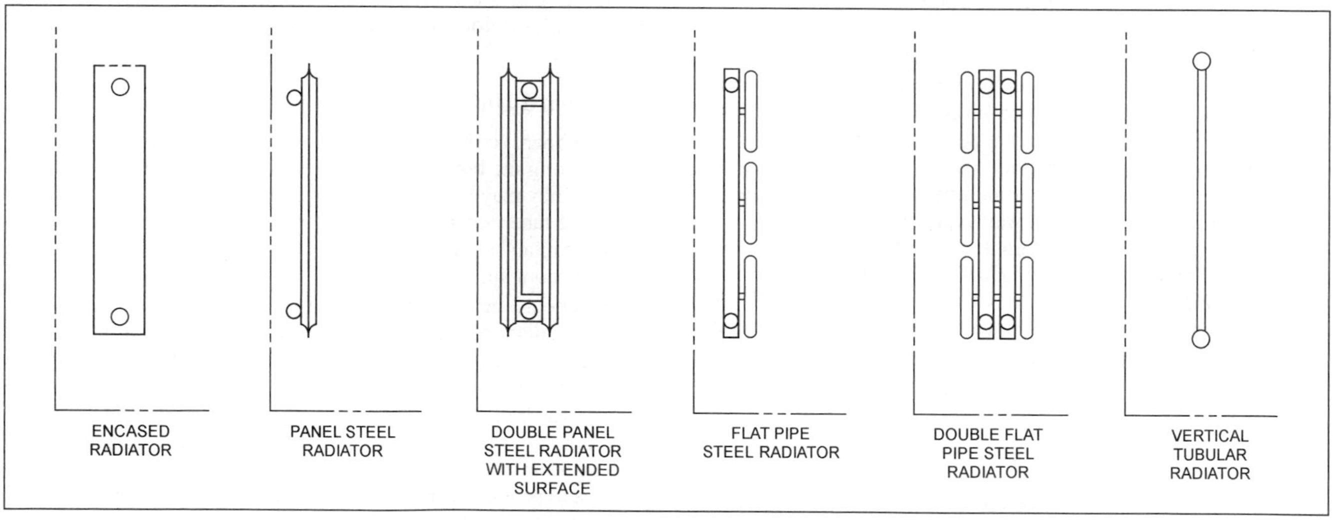

Fig. 2 Typical Radiators

Finned-tube elements installed in occupied spaces generally have covers or enclosures in a variety of designs. When human contact is unlikely, they are sometimes installed bare or provided with an expanded metal grille for minimum protection.

A cover has a portion of the front skirt made of solid material. The cover can be mounted with clearance between the wall and the cover, and without completely enclosing the rear of the finned-tube element. A cover may have a top, front, or inclined outlet. An enclosure is a shield of solid material that completely encloses both the front and rear of the finned-tube element. An enclosure may have an integral back or may be installed tightly against the wall so that the wall forms the back, and it may have a top, front, or inclined outlet.

Heat Emission

These heat-distributing units emit heat by a combination of radiation to the surfaces and occupants in the space and convection to the air in the space.

Chapter 3 of the 1997 *ASHRAE Handbook—Fundamentals* covers the heat transfer processes and the factors that influence them. Those units with a large portion of their heated surface exposed to the space (i.e., radiator and cast-iron baseboard) emit more heat by radiation than do units with completely or partially concealed heating surfaces (i.e., convector, finned-tube, and finned-tube type baseboard). Also, finned-tube elements constructed of steel emit a larger portion of heat by radiation than do finned-tube elements constructed of nonferrous materials.

The heat output ratings of heat-distributing units are expressed in Btu/h, MBh (1000 Btu/h), or in square feet equivalent direct radiation (EDR). By definition, 240 Btu/h = 1 ft^2 EDR with 1 psig steam.

RATINGS OF HEAT-DISTRIBUTING UNITS

For convectors, baseboard units, and finned-tube units, an allowance for heating effect may be added to the **test capacity** (the heat extracted from the steam or water under standard test conditions).

Table 1 Small-Tube Cast-Iron Radiators

Number of Tubes per Section	Catalog Rating per Section,[a]		Section Dimensions				
			A Height, in.[b]	B Width, in.		C Spacing, in.[c]	D Leg Height, in.[b]
	ft²	Btu/h		Min.	Max.		
3	1.6	384	25	3.25	3.50	1.75	2.50
4	1.6	384	19	4.44	4.81	1.75	2.50
	1.8	432	22	4.44	4.81	1.75	2.50
	2.0	480	25	4.44	4.81	1.75	2.50
5	2.1	504	22	5.63	6.31	1.75	2.50
	2.4	576	25	5.63	6.31	1.75	2.50
6	2.3	552	19	6.81	8	1.75	2.50
	3.0	720	25	6.81	8	1.75	2.50
	3.7	888	32	6.81	8	1.75	2.50

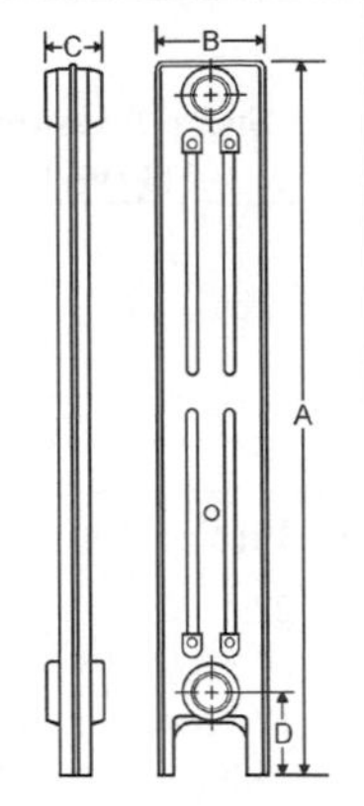

[a] These ratings are based on steam at 215°F and air at 70°F. They apply only to installed radiators exposed in a normal manner, not to radiators installed behind enclosures, behind grilles, or under shelves. For Btu/h ratings at other temperatures, multiply table values by factors found in Table 2.

[b] Overall height and leg height, as produced by some manufacturers, are 1 in. greater than shown in Columns A and D. Radiators may be furnished without legs. Where greater than standard leg heights are required, leg height shall be 4.5 in.

[c] Length equals number of sections multiplied by 1.75 in.

This **heating effect** reflects the ability of the unit to direct its heat output to the occupied zone of a room. The application of a heating effect factor implies that some units use less steam or hot water than others to produce an equal comfort effect in a room.

Radiators

Current methods for rating radiators were established by the U.S. National Bureau of Standards publication, *Simplified Practices Recommendation* R174-65, Cast-Iron Radiators, which has been withdrawn (see Table 1).

Convectors

The generally accepted method of testing and rating ferrous and nonferrous convectors in the United States was given in *Commercial Standard* CS 140-47, Testing and Rating Convectors (Dept. of Commerce 1947), but it has been withdrawn. This standard contained details covering construction and instrumentation of the test booth or room and procedures for determining steam and water ratings.

Under the provisions of *Commercial Standard* CS 140-47, the rating of a top outlet convector was established at a value not in excess of the test capacity. For convectors with other types of enclosures or cabinets, a percentage that varies up to a maximum of 15% (depending on the height and type of enclosure or cabinet) was added for heating effect (Brabbee 1927; Willard et al. 1929). The addition made for heating effect must be shown in the manufacturer's literature.

The testing and rating procedure set forth by *Commercial Standard* CS 140-47 does not apply to finned-tube or baseboard radiation.

Baseboard Units

The generally accepted method of testing and rating baseboards in the United States is covered in the Testing and Rating Standard for Baseboard Radiation (Hydronics Institute 1990a). This standard contains details covering construction and instrumentation of the test booth or room, procedures for determining steam and hot water ratings, and licensing provisions for obtaining approval of these ratings. Baseboard ratings include an allowance for heating effect of 15% in addition to the test capacity. The addition made for heating effect must be shown in the manufacturer's literature.

Finned-Tube Units

The generally accepted method of testing and rating finned-tube units in the United States is covered in the Testing and Rating Standard for Finned-Tube (Commercial) Radiation (Hydronics Institute 1990b). This standard contains details covering construction and instrumentation of the test booth or room, procedures for determining steam and water ratings, and licensing provisions for obtaining approval of these ratings.

The rating of a finned-tube unit in an enclosure that has a top outlet is established at a value not in excess of the test capacity. For finned-tube units with other types of enclosures or covers, a percentage is added for heating effect that varies up to a maximum of 15%, depending on the height and type of enclosure or cover. The addition made for heating effect must be shown in the manufacturer's literature (Pierce 1963).

Other Heat-Distributing Units

Unique radiators and radiators from other countries generally are tested and rated for heat emission in accordance with prevailing-standards. These other testing and rating methods have basically the same procedures as the Hydronics Institute standards, which are used in the United States. See Chapter 6 for information on the design and sizing of radiant panels.

Corrections for Nonstandard Conditions

The heating capacity of a radiator, convector, baseboard, finned-tube heat-distributing unit, or radiant panel is a power function of the temperature difference between the air in the room and the heating medium in the unit, shown as

$$q = c(t_s - t_a)^n \tag{1}$$

where

q = heating capacity, Btu/h
c = constant determined by test
t_s = average temperature of heating medium, °F. For hot water, the arithmetic average of the entering and leaving water temperatures is used.
t_a = room air temperature, °F. Air temperature 60 in. above the floor is generally used for radiators, while the entering air temperature is used for convectors, baseboard units, and finned-tube units.
n = exponent that equals 1.3 for cast-iron radiators, 1.4 for baseboard radiation, 1.5 for convectors, 1.0 for ceiling radiant panels, and 1.1 for floor radiant panels. For finned-tube units, n varies with air and heating medium temperatures. Correction factors to convert heating capacities at standard rating conditions to heating capacities at other conditions are given in Table 2.

Table 2 Correction Factors *c* for Various Types of Heating Units

Steam Pressure (Approx.)		Steam or Water Temp., °F	Cast-Iron Radiator Room Temp., °F					Convector Inlet Air Temp., °F					Finned-Tube Inlet Air Temp., °F					Baseboard Inlet Air Temp., °F				
			80	75	70	65	60	75	70	65	60	55	75	70	65	60	55	75	70	65	60	55
		100											0.10	0.12	0.15	0.17	0.20	0.08	0.10	0.13	0.15	0.18
		110											0.15	0.17	0.20	0.23	0.26	0.13	0.15	0.18	0.21	0.25
		120											0.20	0.23	0.26	0.29	0.33	0.18	0.21	0.25	0.28	0.31
		130											0.26	0.29	0.33	0.36	0.40	0.25	0.28	0.31	0.34	0.38
		140											0.33	0.36	0.40	0.42	0.45	0.31	0.34	0.38	0.42	0.45
in. Hg Vac.	**psia**																					
22.4	3.7	150	0.39	0.42	0.46	0.50	0.54	0.35	0.39	0.43	0.46	0.50	0.40	0.42	0.45	0.49	0.53	0.38	0.42	0.45	0.49	0.53
20.3	4.7	160	0.46	0.50	0.54	0.58	0.62	0.43	0.47	0.51	0.54	0.58	0.45	0.49	0.53	0.57	0.61	0.45	0.49	0.53	0.57	0.61
17.7	6.0	170	0.54	0.58	0.62	0.66	0.69	0.51	0.54	0.58	0.63	0.67	0.53	0.57	0.61	0.65	0.69	0.53	0.57	0.61	0.65	0.69
14.6	7.5	180	0.62	0.66	0.69	0.74	0.78	0.58	0.63	0.67	0.71	0.76	0.61	0.65	0.69	0.73	0.78	0.61	0.65	0.69	0.72	0.78
10.9	9.3	190	0.69	0.74	0.78	0.83	0.87	0.67	0.71	0.76	0.81	0.85	0.69	0.73	0.78	0.81	0.86	0.69	0.73	0.78	0.82	0.86
6.5	11.5	200	0.78	0.83	0.87	0.91	0.95	0.76	0.81	0.85	0.90	0.95	0.77	0.81	0.86	0.90	0.95	0.81	0.86	0.92	0.95	1.00
psig	**psia**																					
1	15.6	215	0.91	0.95	1.00	1.04	1.09	0.90	0.95	1.00	1.05	1.10	0.91	0.94	1.00	1.06	1.11	0.91	0.95	1.00	1.05	1.09
6	21	230	1.04	1.09	1.14	1.18	1.23	1.05	1.10	1.15	1.20	1.26	1.03	1.08	1.14	1.19	1.24	1.04	1.09	1.14	1.19	1.25
15	30	250	1.23	1.28	1.32	1.37	1.43	1.27	1.32	1.37	1.43	1.47	1.20	1.26	1.31	1.37	1.43	1.22	1.27	1.32	1.37	1.43
27	42	270	1.43	1.47	1.52	1.56	1.61	1.47	1.54	1.59	1.67	1.72	1.38	1.44	1.50	1.56	1.62	1.43	1.47	1.52	1.59	1.64
52	67	300	1.72	1.75	1.82	1.89	1.92	1.85	1.89	1.96	2.04	2.08	1.67	1.73	1.79	1.86	1.92	1.75	1.82	1.89	1.92	1.96

Note: Use these correction factors to determine output ratings for radiators, convectors, and finned-tube and baseboard units at operating conditions other than standard. Standard conditions in the United States for a radiator are 215°F heating medium temperature and 70°F room temperature (at the center of the space and at the 5 ft level).

Standard conditions for convectors and finned-tube and baseboard units are 215°F heating medium temperature and 65°F inlet air temperature at 29.92 in. Hg atmospheric pressure. Water flow is 3 fps for finned-tube units. Inlet air at 65°F for convectors and finned-tube or baseboard units represents the same room comfort conditions as 70°F room air temperature for a radiator. Standard conditions for radiant panels are 122°F heating medium temperature and 68°F for room air temperature; *c* depends on panel construction.

To determine the output of a heating unit under conditions other than standard, multiply the standard heating capacity by the appropriate factor for the actual operating heating medium and room or inlet air temperatures.

Equation (1) may also be used to calculate heating capacity at nonstandard conditions.

DESIGN

Effect of Water Velocity

Designing for high temperature drops through the system—drops of as much as 60 to 80°F in low-temperature systems and as much as 200°F in high-temperature systems—can result in low water velocities in the finned-tube or baseboard element. Application of very short runs designed for conventional temperature drops (i.e., 20°F) can also result in low velocities.

Figure 3 shows the effect of water velocity on the heat output of typical sizes of finned-tube elements. The figure is based on work done by Harris (1957) and Pierce (1963) and tests at the Hydronics Institute. The velocity correction factor F_v is

$$F_v = (V/3.0)^{0.04} \tag{2}$$

where V = water velocity, fps.

Heat output varies little over the range from 0.5 to 3 fps, where F_v ranges from 0.93 to 1.00. The factor drops rapidly below 0.5 fps because the flow changes from turbulent to laminar at around 0.1 fps. Such a low velocity should be avoided because the output is difficult to predict accurately when designing a system. In addition, the curve is so steep in this region that small changes in actual flow have a significant effect on output. Not only does the heat transfer rate change, but the temperature drop and, therefore, the average water temperature change (assuming a constant inlet temperature).

The designer should check water velocity throughout the system and select finned-tube or baseboard elements on the basis of velocity as well as average temperature. Manufacturers of finned-tube and baseboard elements offer a variety of tube sizes—ranging from 0.5 in. copper tubes for small baseboard elements to 2 in. for large finned-tube units—to aid in maintenance of turbulent flow conditions over a wide range of flow.

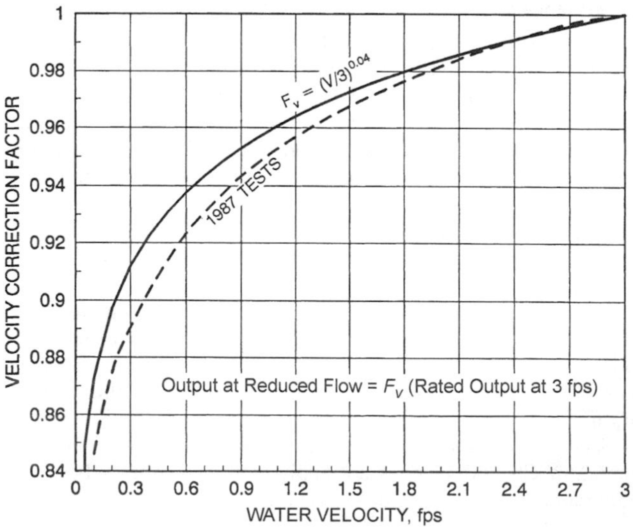

Fig. 3 Water Velocity Correction Factor for Baseboard and Finned-Tube Radiators

Effect of Altitude

The effect of altitude on heat output varies depending on the material used and the portion of the unit's output that is radiant rather than convective. The reduced air density affects the convective portion. Figure 4 shows the reduction in heat output with air density (Sward and Decker 1965). The approximate correction factor F_A for determining the reduced output of typical units is

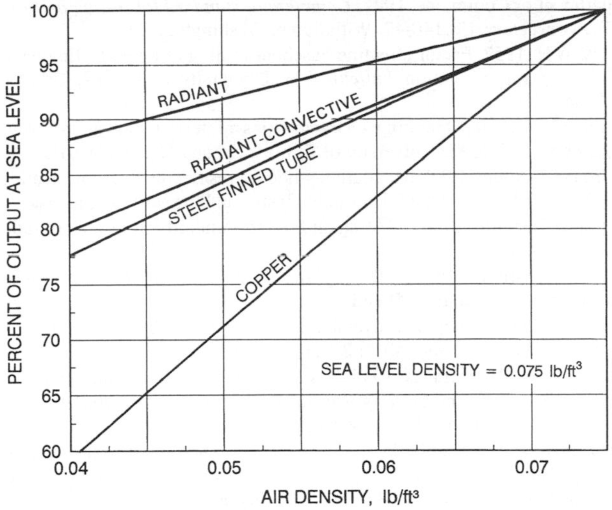

Fig. 4 Effect of Air Density on Radiator Output

$$F_A = (p/p_o)^n \tag{3}$$

where

p = local station atmospheric pressure
p_o = standard atmospheric pressure
n = 0.9 for copper baseboard or finned tube
= 0.5 for steel finned-tube or cast-iron baseboard
= 0.2 for radiant baseboard and radiant panels

The value of p/p_o at various altitudes may be calculated as follows:

$$p/p_o = e^{-3.73 \times 10^{-5} h} \tag{4}$$

where h = altitude, ft. The following are typical values of p/p_o:

Altitude h, ft	p/p_o
2000	0.93
4000	0.86
5000	0.83
6000	0.80

Sward and Harris (1970) showed that some components of heat loss are affected in the same manner.

Effect of Mass

Mass of the terminal unit (typically cast-iron versus copper-aluminum finned element) affects the heat-up and cool-down rates of the equipment. It is important that high- and low-mass radiation not be mixed in the same zone. High-mass systems have historically been favored for best comfort and economy, but tests by Harris (1970) have shown no measurable difference. The thermostat or control can compensate by changing the cycle rate of the burner. The effect of mass is further reduced by constantly circulating modulated temperature water to the unit. The only time that mass can have a significant effect is in response to a massive shift in load. In that situation, the low-mass unit will respond faster.

Performance at Low Water Temperatures

Table 2 summarizes the performance of baseboard and finned-tube units with an average water temperature down to 100°F. Solar-heated water, industrial waste heat in a low- to medium-temperature district heating system, heat pump system cooling water, and ground-source heat pumps are typical applications in this range. In order to compensate for the heating capacity loss, either heat-distributing equipment should be oversized, or additional heat-distributing units should be installed. Capital and operating costs should be minimized (Kilkis 1998).

Effect of Enclosure and Paint

An enclosure placed around a direct radiator restricts the airflow and diminishes the proportion of output resulting from radiation. However, enclosures of proper design may improve the heat distribution within the room compared to the heat distribution obtained with an unenclosed radiator (Willard et al. 1929; Allcut 1933).

For a radiator or cast-iron baseboard, the finish coat of paint affects the heat output. Oil paints of any color give about the same results as unpainted black or rusty surfaces, but an aluminum or a bronze paint reduces the heat emitted by radiation. The net effect may reduce the total heat output of the radiator by 10% or more (Rubert 1937; Severns 1927; Allen 1920).

APPLICATIONS

Radiators

Radiators can be used with steam or hot water. They are installed in areas of greatest heat loss—under windows, along cold walls, or at doorways. They can be installed freestanding, semirecessed, or with decorative enclosures or shields, although the enclosures or shields affect the output (Willard et al. 1929).

Unique and imported radiators are generally not suitable for steam applications, although they have been used extensively in low-temperature water systems with valves and connecting piping left exposed. Various combinations of supply and return locations are possible, which may alter the heat output. Although long lengths may be ordered for linear applications, lengths may not be reduced or increased by field modification. The small cross-sectional areas often inherent in unique radiators require careful evaluation of flow requirements, water temperature drop, and pressure drops.

Convectors

Convectors can be used with steam or hot water. Like radiators, they should be installed in areas of greatest heat loss. They are particularly applicable where wall space is limited, such as in entryways and kitchens.

Baseboard Radiation

Baseboard units are used almost exclusively with hot water. When used with one-pipe steam systems, tube sizes of 1.25 in. NPS must be used to allow drainage of condensate counterflow to the steam flow.

The basic advantage of the baseboard unit is that its normal placement is along the cold walls and under areas where the greatest heat loss occurs. Other advantages are (1) it is inconspicuous, (2) it offers minimal interference with furniture placement, and (3) it distributes the heat near the floor. This last characteristic reduces the floor-to-ceiling temperature gradient to about 2 to 4°F and tends to produce uniform temperature throughout the room. It also makes baseboard heat-distributing units adaptable to homes without basements, where cold floors are common (Kratz and Harris 1945).

Heat loss calculations for baseboard heating are the same as those used for other types of heat-distributing units. The Hydronics Institute (1989) describes a procedure for designing baseboard heating systems.

Finned-Tube Radiation

The finned-tube unit can be used with either steam or hot water. It is advantageous for heat distribution along the entire outside wall, thereby preventing downdrafts along the walls in buildings such as schools, churches, hospitals, offices, airports, and factories. It may be the principal source of heat in a building or a supplementary heater to combat downdrafts along the exposed walls in conjunction with a central conditioned air system. Its placement under or next to windows or glass panels helps to prevent fogging or condensation on the glass.

Normal placement of a finned tube is along the walls where the heat loss is greatest. If necessary, the units can be installed in two or three tiers along the wall. Hot water installations requiring two or three tiers may run a serpentine water flow if the energy loss is not excessive. A header connection with parallel flow may be used, but the design must not (1) permit water to short circuit along the path of least resistance, (2) reduce capacity because of low water velocity in each tier, or (3) cause one or more tiers to become air-bound.

Many enclosures have been developed to meet building design requirements. The wide variety of finned-tube elements (tube size and material, fin size, spacing, fin material, and multiple tier installation), along with the various heights and designs of enclosures, give great flexibility of selection for finned-tube units that meet the needs of load, space, and appearance.

In areas where zone control rather than individual room control can be applied, all finned-tube units in the zone should be in series. In such a series loop installation, however, temperature drop must be considered in selecting the element for each separate room in the loop.

Radiant Panels

Hydronic radiant heating panels are controlled-temperature surfaces on the floor, walls, or ceiling; a heated fluid circulates through a circuit embedded in or attached to the panel. More than 50% of the total heating capacity is transmitted by radiant heat transfer. Usually, 120°F mean fluid temperature delivers enough heat to indoor surfaces. With such low temperature ratings, hydronic radiant panels are suitable for low-temperature heating. See Chapter 6 for more information.

REFERENCES

Allcut, E.A. 1933. Heat output of concealed radiators. School of Engineering *Research Bulletin* 140. University of Toronto, Canada.

Allen, J.R. 1920. Heat losses from direct radiation. *ASHVE Transactions* 26:11.

Brabbee, C.W. 1926. The heating effect of radiators. *ASHVE Transactions* 32:11.

Byrley, R.R. 1978. *Hydronic rating handbook*. Color Art Inc., St. Louis, MO.

Department of Commerce. 1947. *Commercial standard for testing and rating convectors*. CS 140-47. Withdrawn. Washington, D.C.

Harris, W.H. 1957. Factor affecting baseboard rating test results. Engineering Experiment Station *Bulletin* 444. University of Illinois, Urbana-Champaign, IL.

Harris, W.H. 1970. Operating characteristics of ferrous and non-ferrous baseboard. IBR #8. University of Illinois, Urbana-Champaign, IL.

Hydronics Institute. 1989. Installation guide for residential hydronic heating systems. IBR 200, 1st ed. Hydronics Institute, Berkeley Heights, NJ.

Hydronics Institute. 1990a. Testing and rating standard for baseboard radiation, 11th ed.

Hydronics Institute. 1990b. Testing and rating standard for finned-tube (commercial) radiation, 5th ed.

Kilkis, ì.B. 1998. Equipment oversizing issues with hydronic heating systems. *ASHRAE Journal* 40(1):25-31.

Kratz, A.P. and W.S. Harris. 1945. A study of radiant baseboard heating in the IBR research home. Engineering Experiment Station *Bulletin* 358. University of Illinois, Urbana-Champaign, IL.

National Bureau of Standards (currently NIST). 1965. Cast-iron radiators. *Simplified Practices Recommendation* R 174-65. Withdrawn. Available from Hydronics Institute, Berkeley Heights, NJ.

Pierce, J.S. 1963. Application of fin tube radiation to modern hot water systems. *ASHRAE Journal* 5(2):72.

Rubert, E.A. 1937. Heat emission from radiators. Engineering Experiment Station *Bulletin* 24. Cornell University, Ithaca, NY.

Severns, W.H. 1927. Comparative tests of radiator finishes. *ASHVE Transactions* 33:41.

Sward, G.R. and A.S. Decker. 1965. Symposium on high altitude effects on performance of equipment. ASHRAE.

Sward, G.R. and W.S. Harris. 1970. Effect of air density on the heat transmission coefficients of air films and building materials. *ASHRAE Transactions* 76:227-39.

Willard, A.C., A.P. Kratz, M.K. Fahnestock, and S. Konzo. 1929. Investigation of heating rooms with direct steam radiators equipped with enclosures and shields. *ASHVE Transactions* 35:77 or Engineering Experiment Station *Bulletin* 192. University of Illinois, Urbana-Champaign, IL.

BIBLIOGRAPHY

Kratz, A.P. 1931. Humidification for residences. Engineering Experiment Station *Bulletin* 230:20, University of Illinois, Urbana-Champaign, IL.

Laschober, R.R. and G.R. Sward. 1967. Correlation of the heat output of unenclosed single- and multiple-tier finned-tube units. *ASHRAE Transactions* 73(I):V.3.1-15.

Willard, A.C, A.P. Kratz, M.K. Fahnestock, and S. Konzo. 1931. Investigation of various factors affecting the heating of rooms with direct steam radiators. Engineering Experiment Station *Bulletin* 223. University of Illinois, Urbana-Champaign, IL.

CHAPTER 33

SOLAR ENERGY EQUIPMENT

COMMERCIAL and industrial solar energy systems are generally classified according to the heat transfer medium used in the collector loop—namely, air or liquid. While both systems share the basic fundamentals associated with conversion of solar radiant energy, the equipment used in each is entirely different. Air systems are primarily limited to forced-air space heating and industrial and agricultural drying processes. Liquid systems are suitable for a broader range of applications, such as hydronic space heating, service water heating, industrial process water heating, energizing absorption air conditioning, pool heating, and as a heat source for series-coupled heat pumps. Because of this wide range in capability, liquid systems are more common than air systems in commercial industrial applications.

An entirely different class of solar energy equipment called photovoltaic systems converts light from the sun directly into electricity for a wide variety of applications.

SOLAR HEATING SYSTEMS

Solar energy system design requires careful attention to detail. Solar radiation is a low-intensity form of energy, and the equipment to collect and use it is expensive. Imperfections in design and installation can lead to poor cost-effectiveness or to complete system failure. Chapter 32 of the 1999 *ASHRAE Handbook—Applications* covers solar energy use. Books on design, installation, operation, and maintenance are also available (ASHRAE 1988, 1990, 1991a).

Solar energy and HVAC systems often use the same components and equipment. This chapter covers only the following elements, which are exclusive to solar energy applications:

- Collectors and collector arrays
- Thermal energy storage
- Heat exchangers
- Controls

Thermal energy storage is also covered in Chapter 33 of the 1999 *ASHRAE Handbook—Applications*.

AIR HEATING SYSTEMS

Air heating systems circulate air through ducts to and from an air heating collector (Figure 1). Air systems are effective for space-heating applications because a heat exchanger is not required and the collector inlet temperature is low throughout the day (approximately room temperature). Air systems do not need protection from freezing, overheat, or corrosion. Furthermore, air costs nothing and does not cause disposal problems. However, air ducts and air-handling equipment require more space than pipes and pumps, ductwork is hard to seal, and leaks are difficult to detect. Fans consume more power than the pumps of a liquid system, but if the unit is installed in a facility that uses air distribution, only a slight power cost is chargeable against the solar space-heating system.

The preparation of this chapter is assigned to TC 6.7, Solar Energy Utilization.

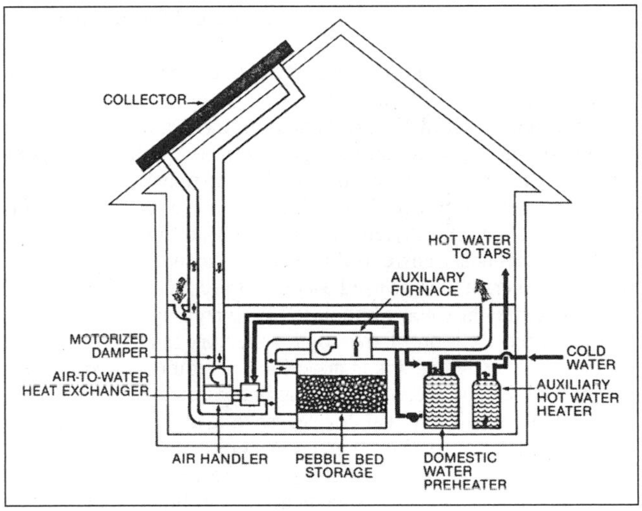

Fig. 1 Air Heating Space and Domestic Water Heater System

Most air space-heating systems also preheat domestic hot water through an air-to-liquid heat exchanger. In this case, tightly fitting dampers are required to prevent reverse thermosiphoning at night, which could freeze water in the heat exchanger coil. If this system heats only water in the summer, the parasitic power consumption must be charged against the solar energy system because no space heating is involved and there are no comparable energy costs associated with conventional water heating. In some situations, solar water-heating systems could be more expensive than conventional water heaters, particularly if electrical energy costs are high. To reduce the parasitic power consumption, some systems use the low speed of a two-speed fan.

LIQUID HEATING SYSTEMS

Liquid heating systems circulate a liquid, often a water-based fluid, through a solar collector (Figure 2 and Figure 3). The liquid in solar collectors must be protected against freezing, which could damage the system.

Freezing is the principal cause of liquid system failure. For this reason, freeze tolerance is an important factor in selecting the heat transfer fluid and equipment in the collector loop. A solar collector radiates heat to the cold sky and freezes at air temperatures well above 32°F. Where freezing conditions are rare, small solar heating systems are often equipped with low-cost, low-reliability protection devices that depend on simple manual, electrical, and/or mechanical components, such as electronic controllers and automatic valves, for freeze protection. Because of the large investment associated with most commercial and industrial installations, solar designers and installers must consider designs providing reliable freeze protection, even in the warmest climates.

Direct and Indirect Systems

In a direct liquid system, city water circulates through the collector. In an indirect system, the collector loop is separated from the high-pressure city water supply by a heat exchanger. In areas of poor water quality, isolation protects the collectors from fouling due to minerals in the water. Indirect systems also offer greater freeze protection, so they are used in commercial and industrial applications almost exclusively.

Freeze Protection

Direct systems, used where freezing is infrequent and not severe, avoid freeze damage by (1) recirculating warm storage water through the collectors, (2) continually flushing the collectors with cold water, or (3) isolating the collectors from the water and draining them. Systems that can be drained to avoid freeze damage are called drain-down systems.

Indirect systems use two methods of freeze protection: (1) nonfreezing fluids and (2) drain-back.

Nonfreezing Fluid Freeze Protection. The most popular solar energy system for commercial application is the indirect system containing a nonfreezing heat transfer fluid to transmit heat from the solar collectors to storage (Figure 2). The most common heat transfer fluids are water/ethylene glycol and water/propylene glycol, although other heat transfer fluids such as silicone oils, hydrocarbon oils, or refrigerants can be used. Because the collector loop is closed and sealed, the only contribution to pump pressure is friction loss; therefore, the location of the solar collectors relative to the heat exchanger and storage tank is not critical. Traditional hydronic sizing methods can be used for selecting pumps, expansion tanks, heat exchangers, and air removal devices, as long as the thermal properties of the heat transfer liquid are considered.

When the control system senses an increase in solar panel temperature, the pump circulates the heat transfer liquid, and energy is collected. The same control also activates a pump on the domestic water side that circulates water through the heat exchanger, where it is heated by the heat transfer fluid. This mode continues until the temperature differential between the collector and the tank is too slight for meaningful energy to be collected. At this point, the control system shuts the pumps off. At low temperatures, the nonfreezing fluid protects the solar collectors and related piping from bursting. Because the heat transfer fluid can affect system performance, reliability, and maintenance requirements, fluid selection should be carefully considered.

Because the collector loop of the nonfreezing system remains filled with fluid, it allows flexibility in routing pipes and locating components. However, a double-separation (double-wall) heat exchanger is generally required (by local building codes) to prevent contamination of the domestic water in the event of a leak. The double-wall heat exchanger also protects the collectors from freeze damage if water leaks into the collector loop. However, the double-wall heat exchanger reduces efficiency by forcing the collector to operate at a higher temperature. The heat exchanger can be placed inside the tank, or an external heat exchanger can be used, as shown in Figure 2. The collector loop is closed and, therefore, requires an expansion tank and pressure relief valve. Air purge is also necessary to expel air during filling and to remove air that has been absorbed into the heat transfer fluid.

Over-temperature protection is necessary to ensure that the system operates within safe limits and to prevent the collector fluid from corroding or decomposing. For maximum reliability, the glycol should be replaced every few years. In some cases, systems have failed because the collector fluid in the loop thermosiphoned and froze the water in the heat exchanger. Such a situation is disastrous and must be avoided by design if the water side is exposed to the city water system because the collector loop eventually fills with water, and all freeze protection is lost.

Drain-Back Freeze Protection. A drain-back solar water-heating system (Figure 3) uses ordinary water as the heat transport medium between the collectors and thermal energy storage. Reverse draining (or back siphoning) the water into a drain-back tank located in a nonfreezing environment protects the system from freezing whenever the controls turn off the circulator pump or a power outage occurs.

For drain-back systems with a large amount of working fluid in the primary loop, a significant decrease in heat loss and an increase in overall efficiency can be obtained by including a tank for storing the heat transfer fluid at night. Using a night storage tank in large systems is an appropriate strategy even for regions with favorable meteorological conditions.

The drain-back tank can be a sump with a volume slightly greater than the collector loop, or it can be the thermal energy storage tank. The collector loop may or may not be vented to the atmosphere. Many designers prefer the nonvented drain-back loop because makeup water is not required and the corrosive effects of the air that would otherwise be ingested into the collector loop are eliminated.

The drain-back system is virtually fail-safe because it automatically reverts to a safe condition whenever the circulator pump stops. Furthermore, a 20 to 30% glycol solution can be added to drain-back loops for added freeze protection in case of controller or sensor failure. Because the glycol is not exposed to stagnation temperatures, it does not decompose.

A drain-back system requires space for the pitching of collectors and pipes necessary for proper drainage. Also, a nearby heated area must have a room for the pumps and the drain-back tank. Plumbing exposed to freezing conditions drains to the drain-back tank, making the drain-back design unsuitable for sites where the collector cannot be elevated above the storage tank.

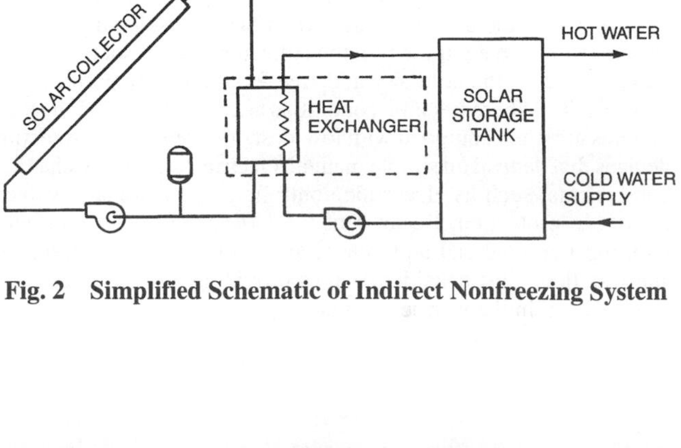

Fig. 2 Simplified Schematic of Indirect Nonfreezing System

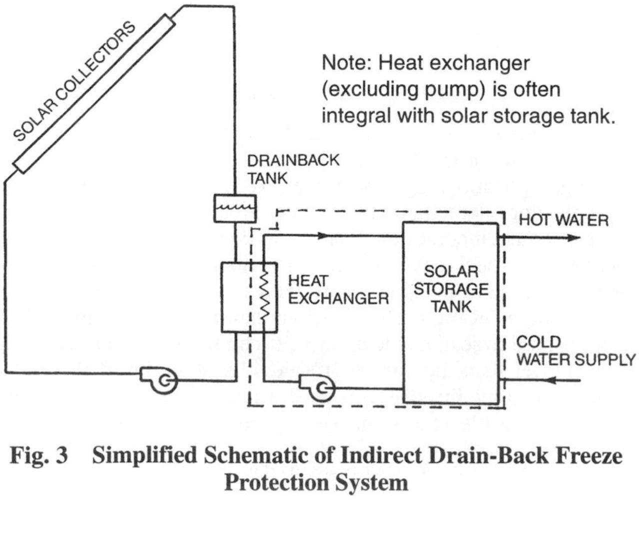

Fig. 3 Simplified Schematic of Indirect Drain-Back Freeze Protection System

Both dynamic and static pressure losses must be considered in the design of a drain-back system. The dynamic pressure loss is due to friction in the pipes, and the static pressure loss is associated with the distance the water must be lifted above the level of the drain-back tank to the top of the collector. There are two distinct designs of drain-back systems: the oversized downcomer (or open drop) and the siphon return. The static head requirement remains constant in the open drop system and decreases in the siphon return system.

Drain-back performs better than other systems in areas that are marginal for solar applications. Drain-back has the advantage that time and energy are not lost in reheating a fluid mass left in the collector and associated piping (as in the case of antifreeze systems). Also, water has a higher heat transfer capacity and is less viscous than other heat transfer fluids, resulting in smaller parasitic energy use and higher overall system efficiency. In closed return (indirect) designs there is also less parasitic energy consumption for pumping because water is the heat transfer fluid. Drain-back systems can be worked on safely under stagnation conditions, but they should not be restarted during peak solar conditions to avoid unnecessary thermal stress on the collector.

SOLAR ENERGY COLLECTORS

Collector Types

Solar collectors depend on liquid heating, air heating, or liquid-vapor phase change to transfer heat. The most common type for commercial, residential, and low-temperature industrial applications (<200°F) is the flat-plate collector.

Liquid Heating Collectors. Figure 4A shows a cross section of a flat-plate liquid collector. A flat-plate collector contains an absorber plate covered with a black surface coating and one or more transparent covers. The covers are transparent to incoming solar radiation and relatively opaque to outgoing (long-wave) radiation, but their principal purpose is to reduce convection heat loss. The collector box is insulated to prevent conduction heat loss from the back and edge of the absorber plate. This type of collector can supply hot water or air at temperatures up to 200°F, although relative efficiency diminishes rapidly above 160°F. The advantages of flat-plate collectors are simple construction, low relative cost, no moving parts, relative ease of repair, and durability. They also absorb diffuse radiation, which is a distinct advantage in cloudy climates.

Another type of collector is the integral collector storage (ICS) system. These collectors incorporate thermal storage within the collector itself. The storage tank surface serves as the absorber surface. Most ICS systems use only one tank, but some use a number of tanks in series. As with flat-plate collectors, insulated boxes enclose the tanks with transparent coverings on the side facing the sun. While the simplicity of ICS systems is attractive, they are generally suitable only for applications in mild climates with small thermal storage requirements. Freeze protection is necessary in colder climates.

Collectors also take the form of concentric glass cylinders, with the space between cylinders evacuated (Figure 5). This vacuum envelope reduces convection and conduction losses, so the cylinders can operate at higher temperatures than flat-plate collectors. Like flat-plate collectors, they collect both direct and diffuse radiation. However, their efficiency is higher at low incidence angles than at 90°. This effect tends to give evacuated tube collectors an advantage over flat-plate collectors in day-long performance. Because of its high-temperature capability, the evacuated tube collector is favored for energizing absorption air-conditioning equipment.

Flat-plate and evacuated-tube collectors are usually mounted in a fixed position. Concentrating collectors are available that must be arranged to track the movement of the sun. These are mainly used for high-temperature industrial applications above 240°F. For more information on concentrating collectors, see Chapter 32 of the 1999 *ASHRAE Handbook—Applications*.

Air Heating Collectors. Air heating collectors, similar to their liquid heating counterparts, are contained in a box, covered with one or more glazings, and insulated on the sides and back. Figure 4B shows a cross section of a flat-plate air collector. The primary differences are in the design of the absorber plate and flow passages. Because the working fluid (air) has poor heat transfer characteristics, it flows over the entire absorber plate, and sometimes on both the front and back of the plate, in order to make use of a larger heat transfer surface. In spite of the larger surface area, air collectors generally have poorer overall heat transfer than liquid collectors. However, they are usually operated at a lower temperature for space heating applications because they require no intervening heat exchangers.

As with liquid collectors, there is a trend toward the use of spectrally selective surface treatments in combination with a single glazing of low-iron glass. Also, air collectors are being designed for flow in the range of 4 scfm per square foot, whereas early models had suggested flows of 2 scfm per square foot. While these modifications increase air collector efficiency and lower manufacturing costs, they also increase fan power consumption and lower output

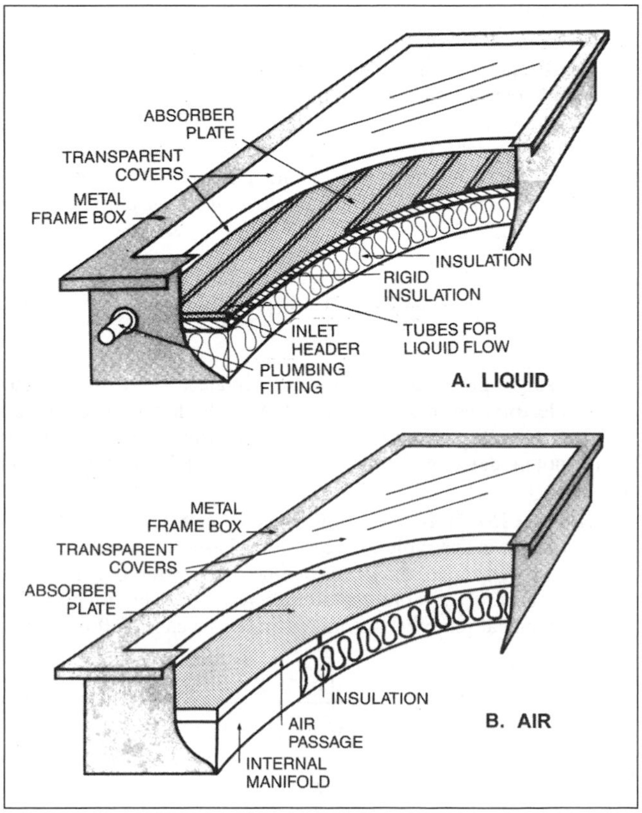

Fig. 4 Solar Flat-Plate Collectors

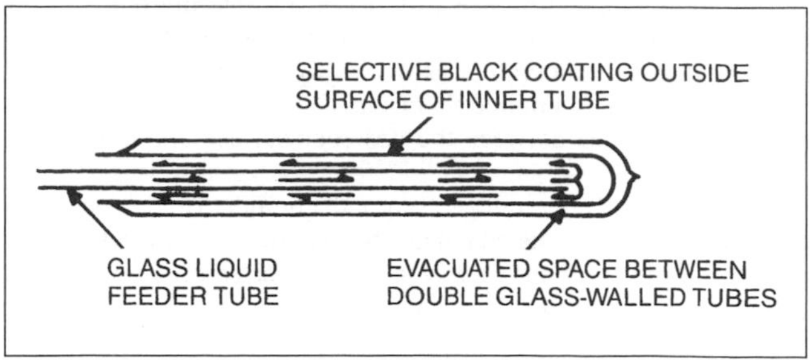

Fig. 5 Evacuated Tube Collector

air temperature. In some applications, natural-convection air collectors are cost-effective. These collectors are self-regulating, can be designed not to reverse at night, and use no fan power.

Liquid-Vapor Collectors. Besides air and liquid collectors, there is a third class of collectors that use liquid-vapor phase change to transfer heat at high efficiency. These collectors feature a heat pipe (a highly efficient thermal conductor) placed inside a vacuum sealed tube. The heat pipe contains a small amount of fluid (e.g., methanol) that undergoes an evaporating-condensing cycle. In this cycle, solar heat evaporates the liquid, and the vapor travels to a heat sink region where it condenses and releases its latent heat. This process is repeated by a return feed of the condensed fluid back to the solar absorber.

Because no evaporation or condensation above the phase-change temperature is possible, the heat pipe offers inherent protection from freezing and overheating. This self-limiting temperature control is a unique feature of the evacuated heat pipe collector.

Collector Performance

Under steady conditions, the useful heat delivered by a solar collector is equal to the energy absorbed in the heat transfer fluid minus the direct and indirect heat losses from the surface to the surroundings. This principle can be stated in the following relationship:

$$q_u = A_c[I_t\tau\alpha - U_L(\bar{t}_p - t_a)] \qquad (1)$$

where

q_u = useful energy delivered by collector, Btu/h
A_c = total aperture collector area, ft^2
I_t = total (direct plus diffuse) solar energy incident on upper surface of sloping collector structure, Btu/h·ft^2
τ = fraction of incoming solar radiation that reaches absorbing surface, transmissivity (dimensionless)
α = fraction of solar energy reaching surface that is absorbed, absorptivity (dimensionless)
U_L = overall heat loss coefficient, Btu/h·ft^2·°F
$\bar{t}_p$ = average temperature of absorbing surface of absorber plate, °F
t_a = atmospheric temperature, °F

With the exception of average plate temperature $\bar{t}_p$, these terms can be readily determined. For convenience, Equation (1) can be modified by substituting inlet fluid temperature for the average plate temperature, if a suitable correction factor is included. The resulting equation is

$$q_u = F_R A_c[I_t\tau\alpha - U_L(t_i - t_a)] \qquad (2)$$

where

F_R = correction factor, or collector heat removal efficiency factor, having a value less than 1.0
t_i = temperature of fluid entering collector, °F

The heat removal factor F_R can be considered the ratio of the heat actually delivered to that delivered if the collector plate were at a uniform temperature equal to that of the entering fluid. An F_R of 1.0 is theoretically possible if (1) the fluid is circulated at such a high rate that its temperature rises a negligible amount, and (2) the heat transfer coefficient and fin efficiency are so high that the temperature difference between the absorber surface and the fluid is negligible.

In Equation (2), the temperature t_i of the inlet fluid depends on the characteristics of the complete solar heating system and the heat demand of the building. However, F_R is affected only by the solar collector characteristics, the fluid type, and the fluid flow rate through the collector.

Solar air heaters remove substantially less heat than liquid collectors. However, their lower collector inlet temperature makes their system efficiency comparable to that of liquid systems for space-heating applications.

Equation (2) may be rewritten in terms of the efficiency of total solar radiation collection by dividing both sides of the equation by $I_t A_c$. The result is

$$\eta = F_R\tau\alpha - F_R U_L \frac{(t_i - t_a)}{I_t} \qquad (3)$$

where η = collector efficiency, dimensionless.

Equation (3) plots as a straight line on a graph of efficiency versus the heat loss parameter $(t_i - t_a)/I_t$. Plots of Equation (3) for various liquid collectors are shown in Figure 11. The intercept (intersection of the line with the vertical efficiency axis) equals $F_R\tau\alpha$. The slope of the line, that is, any efficiency difference divided by the corresponding horizontal scale difference, equals $-F_R U_L$. If experimental data on collector heat delivery at various temperatures and solar conditions are plotted, with efficiency as the vertical axis and $(t_i - t_a)/I_t$ as the horizontal axis, the best straight line through the data points correlates collector performance with solar and temperature conditions. The intersection of the line with the vertical axis is where the temperature of the fluid entering the collector equals the ambient temperature, and collector efficiency is at its maximum. At the intersection of the line with the horizontal axis, collection efficiency is zero. This condition corresponds to such a low radiation level, or to such a high temperature of the fluid into the collector, that heat losses equal solar absorption, and the collector delivers no useful heat. This condition, normally called **stagnation**, usually occurs when no coolant flows to a collector.

Equation (3) includes all important design and operational factors affecting steady-state performance except collector flow rate and solar incidence angle. Flow rate indirectly affects performance through the average plate temperature. If the heat removal rate is reduced, the average plate temperature increases, and more heat is lost. If the flow is increased, collector plate temperature and heat loss decrease.

Solar Incidence Angle. These relationships assume that the sun is perpendicular to the plane of the collector, which rarely occurs. For glass cover plates, specular reflection of radiation occurs, thereby reducing the $\tau\alpha$ product. The incident angle modifier $K_{\tau\alpha}$, defined as the ratio of $\tau\alpha$ at some incidence angle θ to $\tau\alpha$ at normal radiation $(\tau\alpha)_n$, is described by the following simple expression for specular reflection:

$$K_{\tau\alpha} = \frac{\tau\alpha}{(\tau\alpha)_n} = 1 + b_o\left[\frac{1}{\cos\theta} - 1\right] \qquad (4)$$

For a single glass cover, b_o is approximately –0.10. Many flat-plate collectors, particularly evacuated tubes, have some limited focusing capability. The incident angle modifiers for these collectors are not modeled well by Equation (4), which is a linear function of $(1/\cos\theta) - 1$.

Cellular Flow Rate. Equation (3) is not convenient for air collectors when it is desirable to present data based on collector outlet temperature t_{out} rather than inlet temperature, which is the commonly measured variable for liquid systems. The relationship between the heat removal factors for these two cases follows:

$$F_R\tau\alpha = \frac{(F_R\tau\alpha)'}{1 + (F_R U_L)'/\dot{m}c_p} \qquad (5a)$$

$$F_R U_L = \frac{(F_R U_L)'}{1 + (F_R U_L)'/\dot{m}c_p} \qquad (5b)$$

where $\dot{m}c_p$ is mass flow times specific heat of air, $F_R U_L$ and $F_R \tau\alpha$ apply to $t_i - t_a$ in Equation (3), and $(F_R U_L)'$ and $(F_R \tau\alpha)'$ apply to $t_{out} - t_a$.

Testing Methods

ASHRAE *Standard* 93 gives information on testing solar energy collectors using single-phase fluids and no significant internal storage. The data can be used to predict performance in any location and under any conditions where load, weather, and insolation are known.

The standard presents efficiency in a modified form of Equation (3). It specifies that the efficiency be reported in terms of gross collector area A_g rather than aperture collector area A_c. The reported efficiency is lower than the efficiency based on net area, but the total energy collected does not change by this simplification. Therefore, gross collector area must be used when analyzing performance of collectors based on experiments that determine $F_R \tau\alpha$ and $F_R U_L$ according to ASHRAE *Standard* 93.

Standard 93 suggests that testing be done at 0.03 gpm per square foot of gross collector area for liquid systems and that the test fluid be water. While it is acceptable to use lower flow rates or a heat transfer fluid other than water, the designer must adjust the F_R for a different heat removal rate based on the product $\dot{m}c_p$ of mass flow and specific heat. The following approximate approach may be used to estimate small changes in $\dot{m}c_p$.

$$\frac{(F_R U_L)_2}{(F_R U_L)_1} = \frac{1 - \exp[-A_c (F_R U_L)'/(\dot{m}c_p)_2]}{1 - \exp[-A_c (F_R U_L)'/(\dot{m}c_p)_1]} \tag{6}$$

Air collectors are tested at a flow rate of 2 scfm per square foot, and the same relationship applies for adjusting to other flow rates. Annual compilations of collector test data that meet the criteria of ASHRAE *Standard* 93 may be obtained from Solar Rating and Certification Corporation (www.solar-rating.org).

Another source for this information is the collector manufacturer. However, a manufacturer sometimes publishes efficiency data at a much higher flow rate than the recommended design value, so collector data should be obtained from an independent laboratory qualified to conduct testing prescribed by ASHRAE *Standard* 93.

Operational Results

Logee and Kendall (1984) analyzed data from actual sites with regard to collector and collector control performance. The systems were classified with respect to load (service hot water, building space heating, air conditioning); collector design (glazing, absorber surface, form, mounting); and heat transfer medium (air, water, oil, or glycol). Collector efficiency curves generated from the field data represented one month's operation during each of the four seasons. Most of the collectors (18 in the 33 systems studied) performed below test results described in ASHRAE *Standard* 93-1977. Eleven systems had collector efficiencies similar to the test panel results, and four were slightly more efficient.

Collector Construction

Absorber Plates. The key component of a flat-plate collector is the absorber plate. It contains the heat transfer fluid and serves as a heat exchanger by converting radiant energy into thermal energy. It must maintain structural integrity at temperatures ranging from below freezing to well above 300°F. Chapter 32 of the 1999 *ASHRAE Handbook—Applications* illustrates typical liquid collectors and shows the wide variety of absorber plate designs in use.

Materials for absorber plates and tubes are usually highly conductive metals such as copper, aluminum, and steel, although low-temperature collectors for swimming pools are usually made from extruded elastomeric material such as ethylene-propylene terpolymer (EPDM) and PVC. Flow passages and fins are usually copper, but aluminum fins are sometimes inductively welded to copper tubing. Occasionally, fins are mechanically attached, but the potential for corrosion exists with this design. A few manufacturers produce all-aluminum collectors, but they must be checked carefully to determine whether they incorporate corrosion protection in the collector loop.

Figure 6 shows a plan view of typical absorber plates. The serpentine design is used less frequently because it is difficult to drain and imposes a high pressure drop. Most manufacturers use absorber plates similar to those shown in Figure 6D and Figure 6E.

In liquid collectors, manifold selection is important because the design can restrict the array piping configuration. The manifold must be drainable and free-floating, with generous allowance for thermal expansion. Some manufacturers provide a choice of manifold connections to give designers flexibility in designing arrays. One such collector can be obtained with side, back, or end connectors or a combination of these.

In air collectors, most manufacturers increase the heat transfer area via fins, matrices, or corrugated surfaces. Many of these designs increase air turbulence, which improves the collector efficiency (at the expense of increased fan power).

Figure 7 shows cross sections of typical air collectors. Fins on the back of the absorber (Figure 7A) increase the convection heat transfer surface. Air flowing across a corrugated absorber plate (Figure 7B) creates turbulence along the plate, which increases the convective heat transfer coefficient. A box frame (Figure 7C) creates airflow passages between the vanes. The vanes conduct from the absorbing surface plate to the back plate. Heat is transferred to the air by all of the surfaces of each boxed airflow channel. A matrix absorber plate (Figure 7D) is formed by stacking several sheets of metal mesh such as expanded metal plastering lath. Placing the mesh diagonally in the collector forces the air through the matrix so it does not contact the glazing after being heated.

Figure 7 also shows cross sections of typical solar water collectors. Fluid passages can be integrated with the absorber plate to ensure good thermal contact (Figure 7E), or they can be soldered,

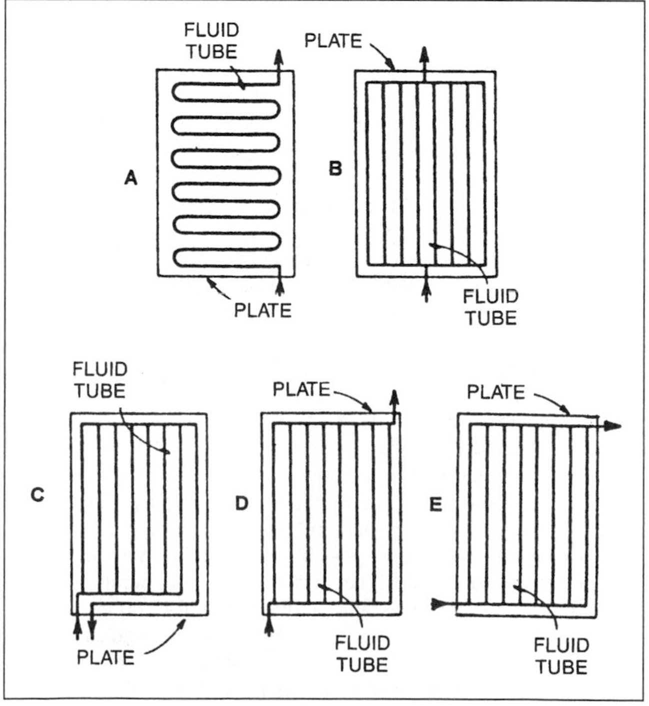

Fig. 6 Plan View of Liquid Collector Absorber Plates

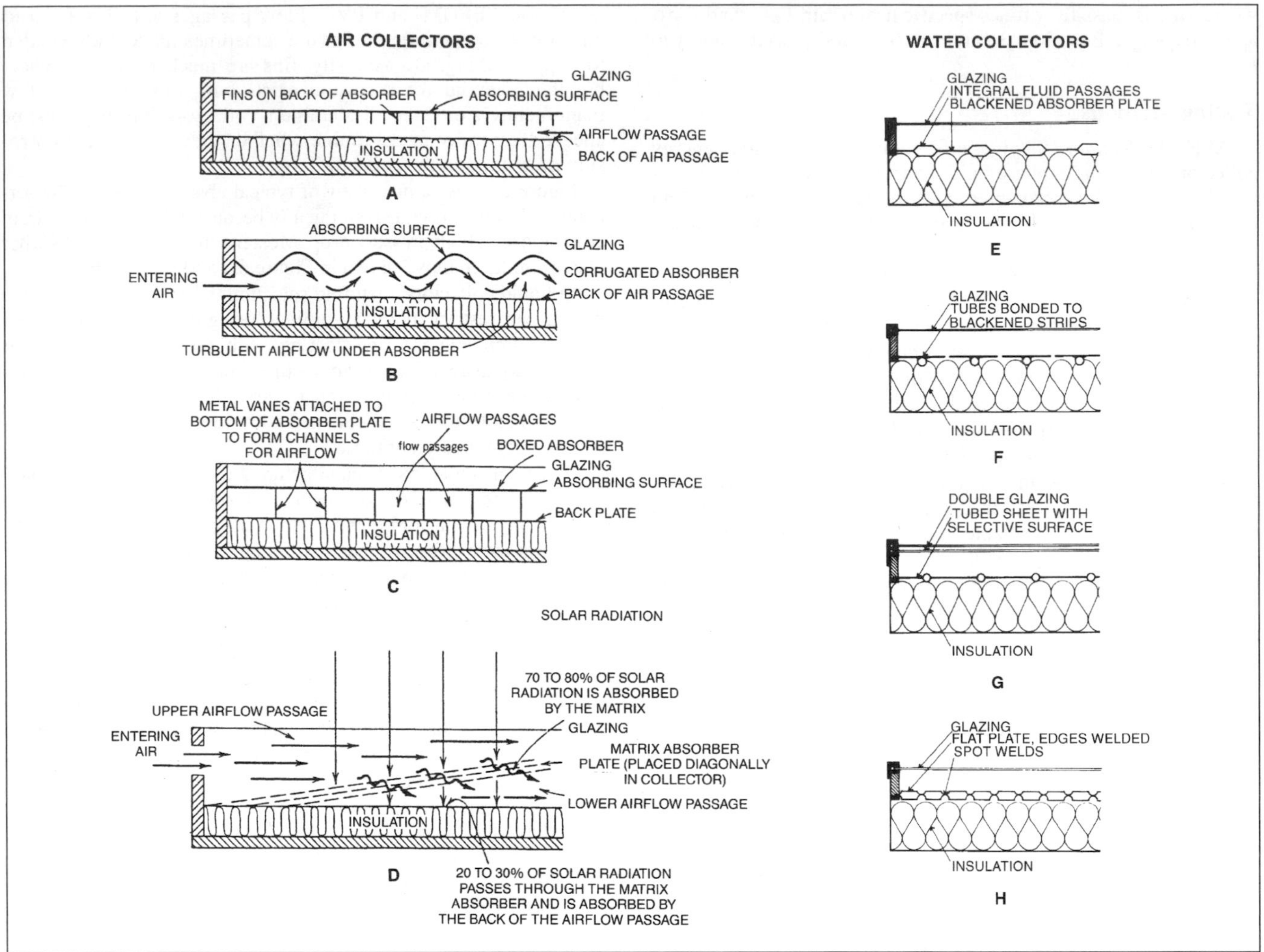

Fig. 7 Cross Sections of Various Solar Air and Water Heaters

brazed, or otherwise fastened to the absorber plate (Figure 7F). Figure 7G shows a double-glazed collector with a tubed copper sheet. In nontubular design, thin parallel sheets of malleable metal (copper, aluminum, or stainless steel) can be seam-welded along their edges and spot-welded at intervals to provide fluid passages that are developed by expansion (Figure 7H). Nontubular collectors, however, are limited to the internal pressure they can sustain and therefore are more adapted to space heating than to domestic hot water heating, which usually requires higher water pressure.

A well-designed collector manifold with a series-parallel connection minimizes leaks and reduces the operating cost of the fans. Manufactured collectors are made in modular sizes, typically 3 ft by 7 ft, and these are connected to form an array. Because most commercial-scale systems involve upwards of 1000 ft^2 of collector, there can be numerous ducting connections, depending on design.

Figure 8 shows the simplest collector manifolding. Each collector has its own inlet and outlet, and each requires two branch connections per collector (and possibly a balancing damper). Some models are specifically designed for either series or parallel connections, and some collectors have built-in manifold connections to simplify their connection as an array. Figure 9 shows an example of internally manifolded collectors in a combination series-parallel arrangement. The number of ducts connecting the collectors to the trunk ducts is reduced to minimize leakage and reduce ducting and installation costs.

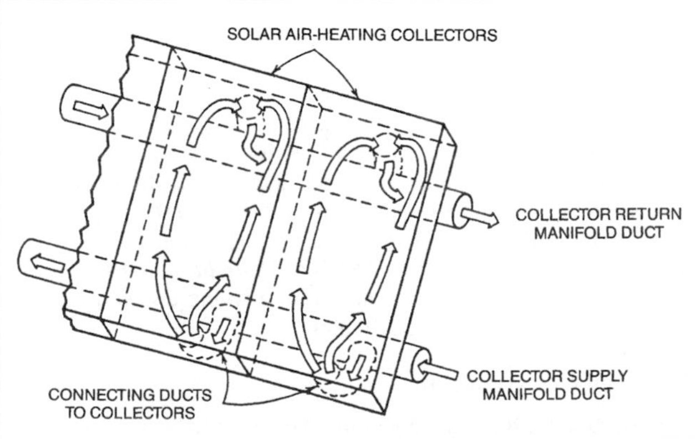

Fig. 8 Air Collectors with External Manifolds

Absorber plates may be coated with spectrally selective or spectrally nonselective materials. Selective coatings are more efficient, but they also cost more than nonselective, or flat black, coatings. The Argonne National Laboratory has published a detailed discussion of the various coating materials used in solar applications (ANL 1979b).

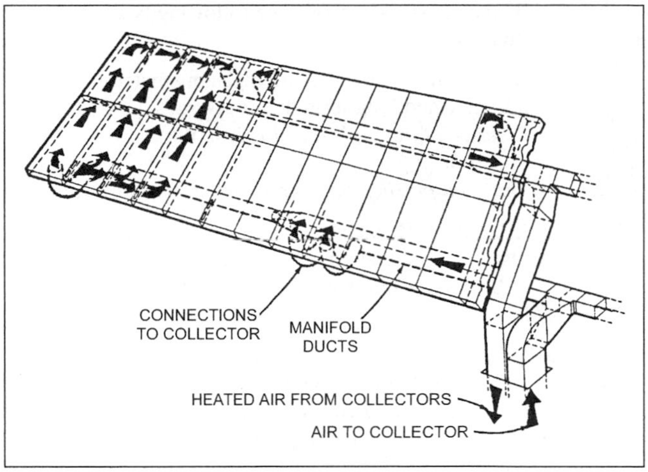

Fig. 9 Air Collectors with Internal Manifolds

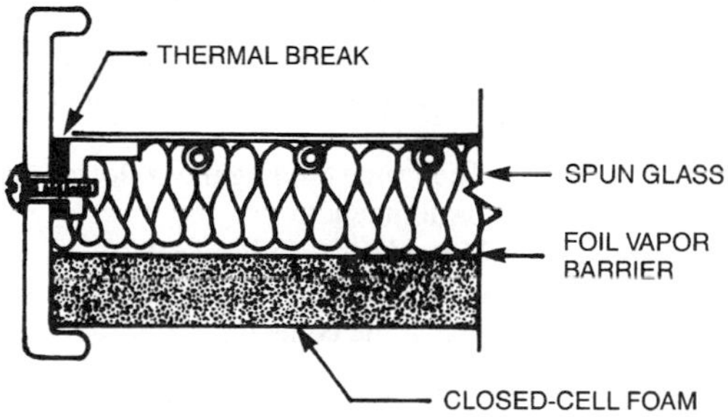

Fig. 10 Cross Section of Suggested Insulation to Reduce Heat Loss from Back Surface of Absorber

Housing. The collector housing is the container that provides structural integrity for the collector assembly. The housing must be structurally sound, weathertight, fire-resistant, and capable of being connected mechanically to a substructure to form an array. Collector housing materials include the following:

- Galvanized or painted steel
- Aluminum folded sheet stock or extruded wall materials
- Various plastics, either molded or extruded
- Composite wood products
- Standard elements of the building

Extruded anodized aluminum, including extruded channels, offers durability and ease of fabrication. Grooves are sometimes included in the extruded channels to accommodate proprietary mounting fixtures. The high temperatures of a solar collector deteriorate wood housings; consequently, wood is often forbidden by fire codes.

Glazing. Solar collectors for domestic hot water are usually single-glazed to reduce absorber plate convective and radiative losses. Some collectors have double glazing to further reduce these losses; however, double glazing should be restricted to applications whose value of $(t_i - t_a)/I_t$ exceeds that of domestic hot water applications (e.g., space heating or activating absorption refrigeration). Glazing materials are plastic, plastic film, or glass. Glass can absorb the long-wave thermal radiation emitted by the absorber coating, but it is not affected by ultraviolet (UV) radiation. Because of their impact tolerance, only tempered, low-iron glass covers should be considered. These covers have a solar transmission rating of 86%. With acid etching, this transmission rating can be increased to 90%.

If the probability of vandalism is high, polycarbonate, which has high impact resistance, should be considered. Unfortunately, its

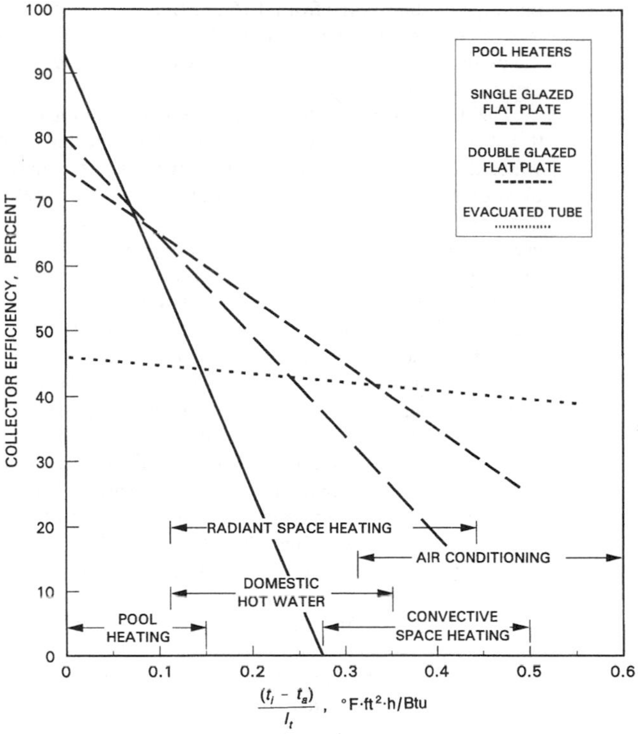

Fig. 11 Collector Efficiencies of Various Liquid Collectors

transmittance is not as high as that of low-iron glass, and it is susceptible to long-term UV degradation.

Insulation. Collector enclosures must be well insulated to minimize heat losses. The insulation must withstand temperatures up to 400°F and, most important, must not produce volatiles within this range. Many insulation materials designed for construction applications are not suitable for solar collectors because the binders outgas volatiles at normal collector operating temperatures.

Solar collector insulation is typically made of mineral fiber, ceramic fiber, glass foam, plastic foam, or fiberglass. Fiberglass is the least expensive insulation and is widely used in solar collectors. For high-temperature applications, rigid fiberglass board with a minimum of binder is recommended. Also, a layer of polyisocyanurate foam in collectors is often used because of its superior R-value. Because it can outgas at high temperatures, the foam must not be allowed to contact the collector plate.

Figure 10 illustrates the preferred method of combining fiberglass and foam insulations to obtain both high efficiency and durability. Note that the absorber plate should be free-floating to avoid thermal stresses. Regardless of attempts to make collectors watertight, moisture is always present in the interior. This moisture can physically degrade mineral wool and reduce the R-value of fiberglass, so drainage and venting are crucial.

Collector Test Results and Initial Screening Methods

Final selection of a collector should be made only after energy analyses of the complete system, including realistic weather conditions and loads, have been conducted for one year. Also, a preliminary screening of collectors with various performance parameters should be conducted in order to identify those that best match the load. The best way to accomplish this is to identify the expected range of heat loss parameter $(t_i - t_a)/I_t$ for the load and climate on a plot of efficiency η as a function of heat loss parameter, as indicated in Figure 11.

Ambient temperature during the swimming season may vary from 18°F below pool temperature to 18°F above. The corresponding parameter values range from 0.15 on cool overcast days (low I_t) to as low as –0.15 on hot overcast days. For most swimming pool heating, the unglazed collector offers the highest performance and is the least expensive collector available.

The heat loss parameter for service water heating can range from 0.05 to 0.35, depending on the climate at the site and the desired hot water delivery temperature. Convective space heating requires an even greater collector inlet temperature than water heating, and the primary load coincides with lower ambient temperature. In many areas of the United States, space-heating operation coincides with low radiation values, which further increases the heat loss parameter. However, many space-heating systems are accompanied by water heating, and some space-heating systems, such as radiant panels, are available at a lower value of $(t_i - t_a)/I_t$.

Air conditioning with solar activated absorption equipment is economical only when the cooling season is long and solar radiation levels are high. These devices require at least 180°F water. Thus, on an 80°F day with radiation at 300 Btu/h·ft^2, the heat loss parameter is 0.3. Higher operating temperatures are desirable to prevent excessive derating of the air conditioner. Only the most efficient (low $F_R U_L$) collector is suitable. Derating may also be minimized by employing convective-radiant hybrid air-conditioning systems.

Collector efficiency curves may be used as an initial screening device. However, efficiency curves illustrate only the instantaneous performance of a collector. They do not include incidence angle effects (which vary throughout the year), heat exchanger effects, probabilities of occurrence of t_i, t_a, and I_t, system heat loss, or control strategies. Final selection requires determining the long-term energy output of a collector as well as performing cost-effectiveness studies. Estimating the annual performance of a particular collector and system requires the aid of appropriate analysis tools such as *f*-Chart (Beckman et al. 1977) or TRNSYS (SEL 1983).

Generic Test Results

While details of construction may vary, it is possible to group collectors into generic classifications. Using the performance characteristics of each classification, the designer can select the category best suited to a particular application.

Huggins and Block (1983) identified eight such classifications for liquid collectors from a sample of 270 collectors tested by various laboratories. Following is a list of the parameters that differentiate the categories:

Number of Cover Plates. The majority of collectors used in the southern and far western regions of the United States have one cover plate. Double-glazed (i.e., two-cover plates), are more common in colder climates. The trend in cold climates has been toward single glazing combined with a selective surface. However, with the advent of reflectivity coatings on plastic films, triple and quadruple glazings may become common.

Cover Plate Material. The following is a list of common cover plate materials:

- *Glass* is the most widely used glazing material. It has a high transmittance and long-term durability.
- *Fiber-reinforced plastic* (FRP) is the second most widely used material.
- *Thin film plastics* serve as a glazing on some collectors.

Absorber Materials. Copper, aluminum, and stainless steel may be used in combinations of tubes and fins or integral tubes in plates.

Absorber Plate Coating. The most common coatings are the following:

- *Selective surface coatings* such as black chrome, black nickel, and copper oxide have high absorptance and low emittance properties. The infrared emissivity of these surfaces is below 0.2.
- *Moderately selective surface coatings* are special paints that have moderately selective surface properties. The emissivity of the surfaces ranges from 0.2 to 0.7.
- *Flat black paints* are nonselective and have high heat resistance. The emissivity of these surfaces ranges from 0.7 to 0.98.

Absorber Configuration. The absorber may be configured with parallel pipes, series or serpentine pipes, a series-parallel combination, or plate flow.

Enclosure. The frame holding the collector components may be either metallic or nonmetallic.

Insulation Materials. Fiberglass insulation, foam insulation, or a combination of both may be used to keep heat from escaping from the back and sides of the collector.

Table 1 lists the eight generic collector classifications in which five or more collectors were tested. The first three columns specify the generic type; the fourth column shows the number of collectors in each category; and the final columns list the mean, the standard deviation, and the maximum and minimum values of the intercept and slope. Logee and Kendall (1984) showed that collectors in actual applications can perform differently than they do on test stands. This difference emphasizes the need for good system design and installation practices.

MODULE DESIGN

Piping Configuration

Most commercial and industrial systems require a large number of collectors. Connecting the collectors with one set of manifolds makes it difficult to ensure drainability, balanced flow, and low pressure drop. An array usually includes many individual groups of collectors, called modules, to provide the necessary flow characteristics. Modules can be grouped into (1) parallel flow or (2) combined

Table 1 Collector Intercept and Slope by Generic Type

Generic Type				Intercept $F_R\tau\alpha$		Slope F_RU_L, Btu/h·ft^2·°F	
Glazing and Cover Material	Absorber Material and Type	Absorber Coating	Number of Collectors	Mean	Standard Deviation	Mean	Standard Deviation
Single glass	Copper tubes and fins	Flat black paint	47	67.2	5.0	–115	14
Single glass	Copper tubes and fins	Moderately selective	9	73.0	3.6	–112	11
Single glass	Copper tubes and fins	Selective surface	58	71.7	3.3	–83	11
Single glass	Copper tubes and aluminum fins	Flat black paint	22	69.1	6.0	–116	12
Single glass	Copper sheet integral tubes	Selective surface	6	70.5	5.1	–89	17
Single FRP	Copper tubes and fins	Flat black paint	26	61.9	5.5	–117	15
Single FRP	Copper tubes and aluminum fins	Flat black paint	11	57.1	6.2	–114	10
Double glass	Copper tubes and fins	Flat black paint	9	59.7	6.7	–84	

Source: Huggins and Block (1983).

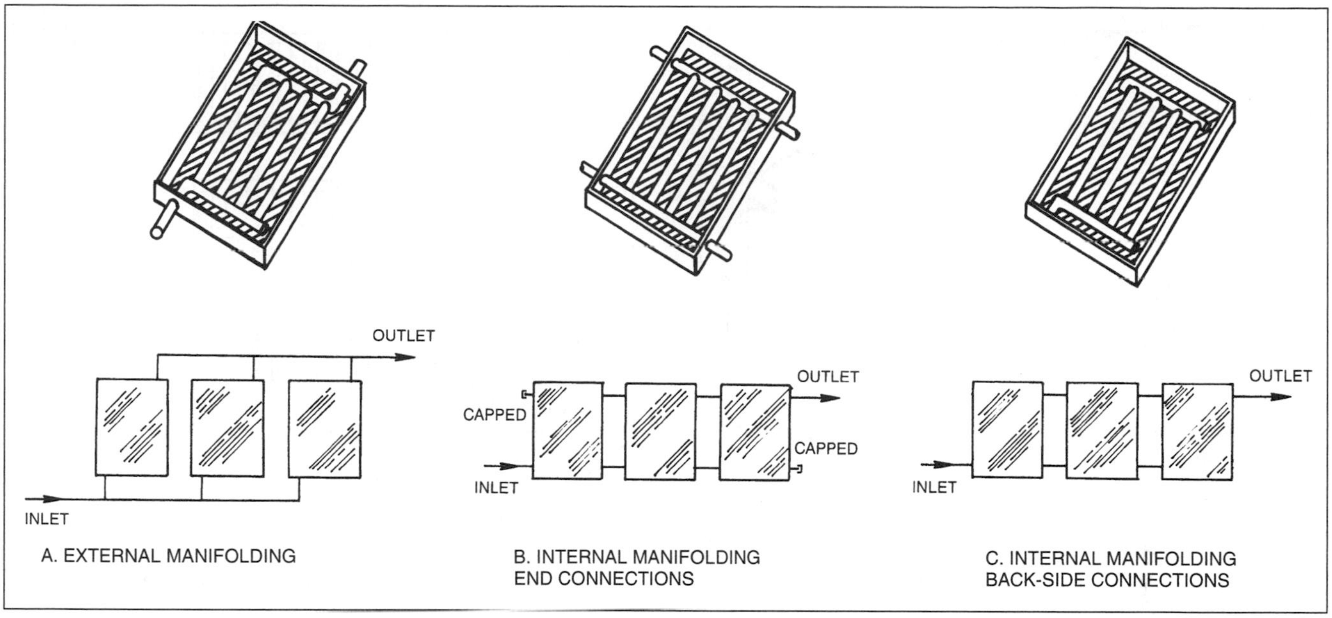

Fig. 12 Collector Manifolding Arrangements for Parallel Flow Module

series-parallel flow. Parallel flow is the most frequently used because it is inherently balanced, has low pressure drop, and is drainable. Figure 12 illustrates various collector header designs and the reverse-return method of forming a parallel flow module (the flow is parallel but the collectors are connected in series). When arrays must be greater than one panel high, a combination of series and parallel flow may be used. Figure 13 illustrates one method of connecting collectors in a two-high module.

Generally, flat-plate collectors are made to connect to the main piping in one of two methods shown in Figure 12. The **external manifold** collector has a small-diameter connection meant to carry only the flow for one collector. It must be connected individually to the manifold piping, which is not part of the collector panel, as depicted in Figure 12A and in Figure 13.

The **internal manifold** collector incorporates the manifold piping integral with each collector (Figure 12B and Figure 12C). Headers at either end of the collector distribute flow to the risers. Several collectors with large headers can be placed side by side to form a continuous supply and return manifold. With 1 in. headers, four to six 40 ft^2 collectors can be placed side by side. Collectors with 1 in. headers can be mounted in a module without producing unbalanced flow. Most collectors have four plumbing connections, some of which may be capped if the collector is located on the end of the array. Internally manifolded collectors have the following advantages:

- Piping costs are lower because fewer fittings and less insulation are required.
- Heat loss is less because less piping is exposed.
- Installation is more attractive.

Some of the disadvantages of the internally manifolded collectors are as follows:

- For drain-back systems, the entire module must be pitched, thus complicating mounting.
- Flow may be imbalanced if too many collectors are connected in parallel.
- Removing the collector for servicing may be difficult.
- Stringent thermal expansion requirements must be met if too many collectors are combined in a module.

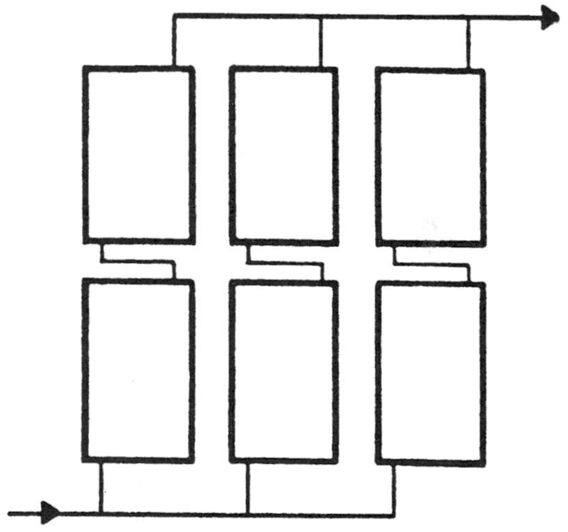

Fig. 13 Example of Collector Manifolding Arrangement for Combined Series-Parallel Flow Modules

Velocity Limitations

Fluid velocity limits the number of internally manifolded collectors that can be contained in a module. For 1 in. headers, up to eight 20 ft^2 collectors can usually be connected for satisfactory performance. If too many are connected in parallel, the middle collectors will not receive enough flow, and performance will decrease. Connecting too many collectors also increases pressure drop. Figure 14 illustrates the effect of collector number on performance and pressure drop for one particular design. Newton and Gilman (1983) describe a general method to determine the number of internally manifolded collectors that can be connected.

Flow restrictors can be used to accommodate a large number of collectors in a row. The flow distribution in the 12 collectors of Figure 15 would not be satisfactory without the flow restrictors shown at the interconnections. The flow restrictors are barriers having a drilled hole of the diameter indicated. Some manufacturers calculate the required hole diameters and provide the predrilled restrictors.

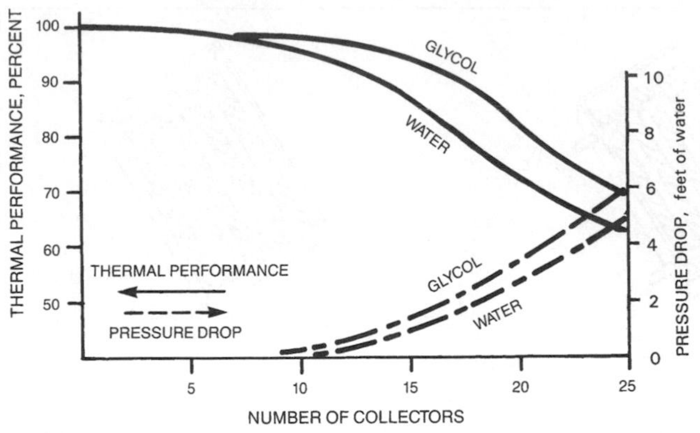

Fig. 14 Pressure Drop and Thermal Performance of Collectors with Internal Manifolds Numbers

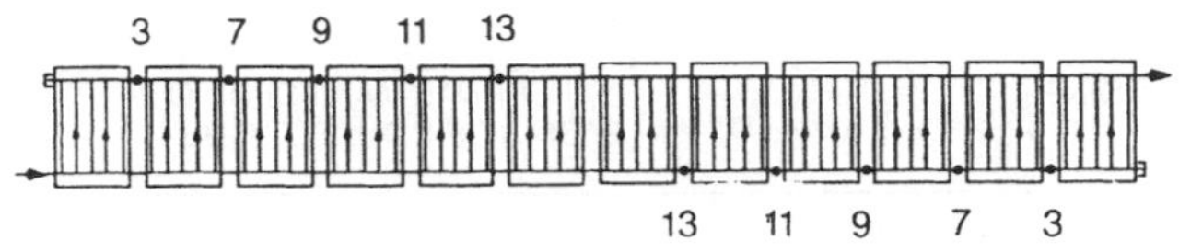

Numbers indicate restrictor hole diameter in sixteenths of an inch.

Fig. 15 Flow Pattern in Long Collector Row with Restrictions

Chapter 33 of the 1997 *ASHRAE Handbook—Fundamentals* gives information on sizing piping for self-balancing flow in externally manifolded collectors. Knowles (1980) provides the following expression for the minimum acceptable header diameter:

$$D = 0.24(Q/\Delta p)^{0.45}N^{0.64} \quad (7)$$

where

D = header diameter, in.
N = number of collectors in module
Q = recommended flow rate for collector, gpm
Δp = pressure drop across collector at recommended flow rate, psi

Because piping is available in a limited number of diameters, selection of the next larger size ensures balanced flow. Usually, the sizes of supply and return piping are graduated to maintain the same pressure drop while minimizing piping cost. Complicated configurations may require a hydraulic static regain calculation.

Thermal Expansion

Thermal expansion will affect the module shown in Figure 16. Thermal expansion (or contraction) of a module of collectors in parallel may be estimated by the following equation:

$$\Delta = 0.000335n(t_c - t_i) \quad (8)$$

where

Δ = expansion or contraction of collector array, in.
n = number of collectors in array
t_c = collector temperature, °F
t_i = installation temperature of collector array, °F

Because absorbers are rigidly connected, the absorber must have sufficient clearance from the side frame to allow the expansion indicated in Equation (8).

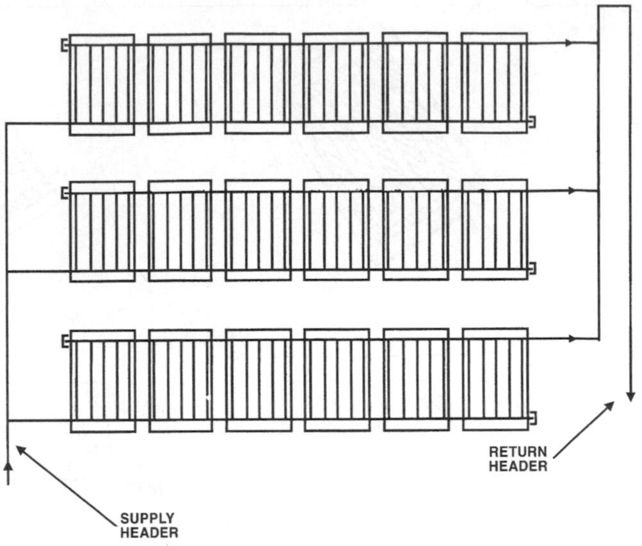

Fig. 16 Reverse-Return Array Piping

ARRAY DESIGN

Piping Configuration

Liquid Systems. To maintain balanced flow, an array or field of collectors should be built from identical modules configured as described in previous sections. Whenever possible, modules must be connected in reverse-return fashion (Figure 16). The reverse return ensures that the array is self-balanced. With proper care, an array can drain, which is an essential requirement for drain-back freeze protection.

Piping to and from the collectors must be sloped properly in a drain-down system. Typically, piping and collectors must slope to drain at 1/4 in. per linear foot. Elevations throughout the array, especially the highest and lowest point of the piping, should be noted on the drawings.

The external manifold collector has different mounting and plumbing considerations from the internal manifold collector (Figure 17). A module of externally manifolded collectors can be mounted horizontally, as shown in Figure 17A. The lower header *must* be pitched as shown. The pitch of the upper header can be either horizontal or pitched toward the collectors so it can drain back through the collectors.

Arrays with internal manifolds pose a greater challenge in designing and installing the collector mounting system. For these collectors to drain, the entire bank must be tilted, as shown in Figure 17B.

Reverse return always implies an extra pipe run. Sometimes, it is more convenient to use direct return (Figure 18). In this case, balancing valves are needed to ensure uniform flow through the modules. The balancing valves *must* be connected at the module outlet to provide the flow resistance necessary to ensure filling of all modules on pump start-up.

It is often impossible to configure parallel arrays because of the presence of rooftop equipment, roof penetrations, or other building-imposed constraints. Although the list is not complete, the following requirements should be considered when developing the array configuration:

- Strive for a self-balancing configuration.
- For drain-back, design the modules, subarrays, and arrays to be individually and collectively drainable.
- Always locate collectors or modules with high flow resistance at the outlet to improve flow balance and ensure filling of the drain-down system.
- Minimize flow and heat transfer losses.

In general, it is easier to configure complex array designs for nonfreezing fluid systems. Newton and Gilman (1983) provide some typical examples. However, with careful attention to the criteria mentioned above, it is also possible to design successful large drain-back arrays.

Air Systems. Air distribution is the most critical feature of an air system because pressure drop has a critical effect on fan power. Each collector and the other components must have proper air distribution for effective operation. Balancing dampers, automatic dampers, backdraft dampers, and fire dampers are usually needed. Air leaks (both into and out of the ducting and from component to component) must be kept to a minimum.

For example, some air collector systems contain a water coil for preheating water. Despite the inclusion of automatic dampers and backdraft dampers, leakage within the system can freeze the coils. One possible solution is to position the coil near the warm end of the storage bin. Another is to circulate an antifreeze solution in the coil. Whatever the solution, the designer must remember that even the lowest leakage dampers will leak if installed or adjusted improperly.

As with liquid systems, the main supply and return ducts should be connected in a reverse-return configuration, with balancing dampers on each supply branch to the collector modules. If reverse return is not feasible, the main ducts should include balancing dampers at strategic locations. Here too, having fewer branch ducts reduces balancing needs and costs.

Unlike liquid systems, air collectors can be built on site. While material and cost savings can be substantial with site-built collectors, extreme care must be taken to ensure long life, low leakage, and proper air distribution. Quality control in the field can be a problem, so well-trained designers and installers are critical to the success of these systems.

Performance. The impact of the array design and air distribution system designs on the overall performance of the system must be considered. Beckman et al. (1977) give standard procedures for estimating the impact of series connection and duct thermal losses. The effect on fan operation and fan power is more difficult to determine. For unique system designs and more detailed performance estimates, including fan power, an hourly simulation like TRNSYS (SEL 1983) can be used.

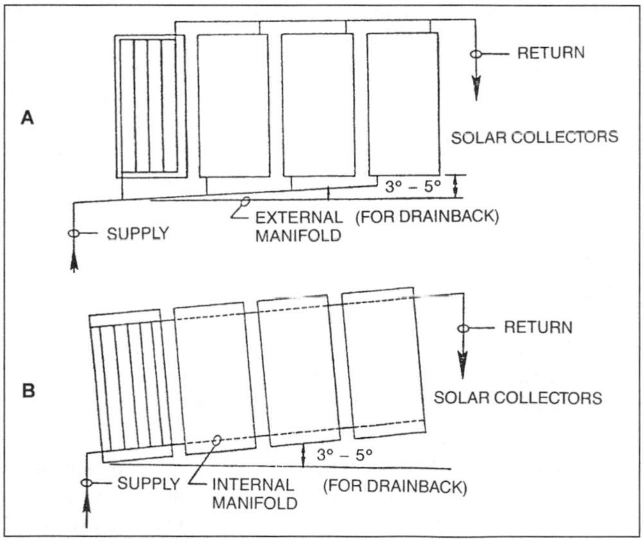

Fig. 17 Mounting for Drain-Back Collector Modules

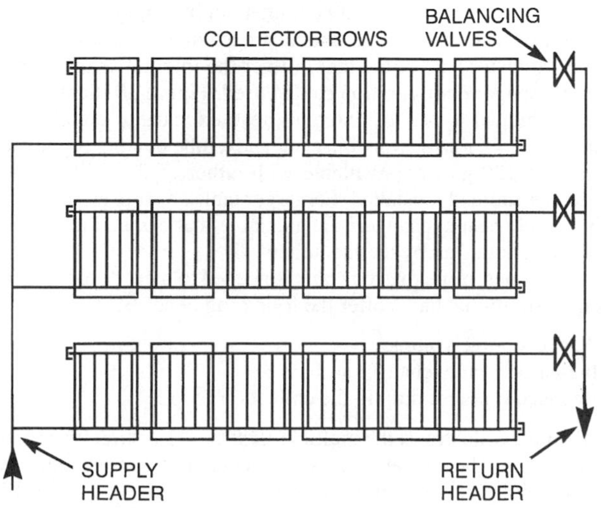

Fig. 18 Direct-Return Array Piping

Shading

When large collector arrays are mounted on flat roofs or level ground, multiple rows of collectors are usually installed in a saw-tooth fashion. These multiple rows should be spaced so that they do not shade each other at low sun angles. However, it is usually not desirable to avoid mutual shading altogether. It is sometimes possible to add additional rows to a roof or other constrained area; this increases the solar fraction but sacrifices efficiency. Kutscher (1982) presents a method of estimating the energy delivered annually when there is some row-to-row shading within the array. Figure 19 provides a factor F_{shad}, which corrects the annual performance (on an unshaded basis) of the field of shaded collectors. Because the first row is unshaded, the following equation is used to compute the average shading factor $F_{shad,field}$ for the entire field.

$$F_{shad,field} = \frac{1 + (n-1)F_{shad}}{n} \tag{9}$$

where n = number of rows in field.

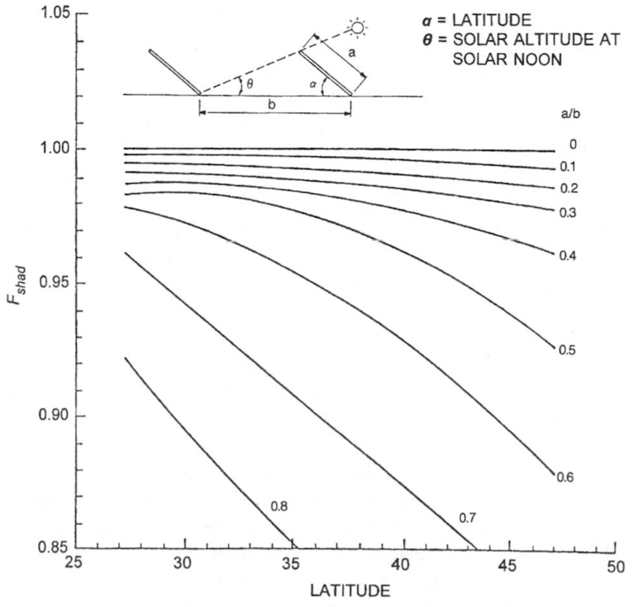

Fig. 19 Annual Row-to-Row Shading Loss Modifiers for Flat-Plate Collectors Tilted at Latitude
(Kutscher 1982)

THERMAL ENERGY STORAGE

Design and selection of the thermal storage equipment is one of the most neglected elements of solar energy systems. In fact, the energy storage system has an enormous influence on overall system cost, performance, and reliability. Furthermore, the storage system design is highly interactive with other system elements such as the collector loop and the thermal distribution system. Thus, it should be considered within the context of the total system.

Energy can be stored in liquids, solids, or phase-change materials (PCMs). Water is the most frequently used liquid storage medium for liquid systems, although the collector loop may contain water, oils, or aqueous glycol as a collection fluid. For service water heating applications and most building space heating, water is normally contained in some type of tank. Air systems typically store heat in rocks or pebbles, but sometimes the structural mass of the building is used. Chapter 32 of the 1999 *ASHRAE Handbook—Applications* and ASHRAE (1991) cover this topic in more detail.

Air System Thermal Storage

The most common storage media for air collectors are rocks or a regenerator matrix made from concrete masonry units (CMUs). Other possible media include PCMs, water, and the inherent building mass. Gravel is widely used as a storage medium because it is plentiful and relatively inexpensive.

In places where large interior temperature swings are tolerable, the inherent mass of the building may be sufficient for thermal storage. Designated storage may also be eliminated where the array output seldom exceeds the concurrent demand. Loads requiring no storage are usually the most cost-effective applications of air collectors, and heated air from the collectors can be distributed directly to the space.

The three main requirements for gravel storage (or any storage system for air collectors) are good insulation, low air leakage, and low pressure drop. Many different designs can fulfill these needs. The container is usually constructed from masonry, frame, or a combination of the two. Airflow can be vertical or horizontal, whichever is most convenient.

A vertical flow bed that has solar-heated air enter at the bottom and exit from the top can work as effectively as a horizontal flow bed. However, it is important to heat the bed with airflow in one direction and to retrieve the heat with flow in the opposite direction. In this manner, pebble beds perform as effective counterflow heat exchangers. Conversely, properly designed and applied wash-through (one-way flow) beds can be effective, as can underflow beds that charge through airflow and discharge by flow radiation. These are less expensive because they do not require elaborate ductwork or complicated controls.

Rocks for pebble beds range from 1 3/8 in. to 4 in. in size, depending on airflow, bed geometry, and desired pressure drop. The volume of rock needed depends on the fraction of the collector output that must be stored. For residential systems, storage volume is typically in the range of 0.5 to 1 ft^3 per square foot of collector area. For commercial systems, these values can be used as guidelines, but a more detailed analysis should be performed. Pebble beds can be quite large for large arrays, and their large mass and size can create location problems.

Although other storage options exist for air systems, practical knowledge of these techniques is limited. Active airflow through the cores of unit-masonry and precast concrete floors has been successfully demonstrated in residential and commercial buildings (Howard 1986, Johnston 1982). Hollow-core concrete decking appears attractive because of the potential to reduce storage and distribution costs. PCMs are also functionally attractive because of their high volumetric heat storage capabilities; they typically require one-tenth of the volume of a pebble bed.

Water can also be used as a storage medium for air collectors through the use of a conventional heating coil to transfer heat from the air to the water in the storage tank. Advantages of water storage include compatibility with hydronic heating systems and relative compactness (roughly one-third the volume requirement of pebble beds).

Liquid System Thermal Storage

For units large enough for commercial liquid systems, the following factors should be considered:

- Pressurized versus unpressurized storage
- External heat exchanger versus internal heat exchanger
- Single versus multiple tanks
- Steel tank(s) versus nonmetallic tank(s)
- Type of service [i.e., service hot water (SHW), building space heating (BSH), or a combination of the two]
- Location, space, and accessibility constraints imposed by architectural limitations
- Interconnect constraints imposed by existing mechanical systems
- Limitations imposed by equipment availability

In the following sections, examples of the more common configurations are presented.

Pressurized Storage. Defined here as storage that is open to the city water supply, pressurized storage is preferred for small service water heating systems because it is convenient and provides an economical way of meeting ASME *Boiler and Pressure Vessel Code* requirements with off-the-shelf equipment. Typical storage size is about 1 to 2 gal per square foot of collector area. The largest size off-the-shelf water heater is 120 gal; however, no more than three of these should be connected in parallel. Hence, the largest storage that can be considered with off-the-shelf water heater tanks is about 360 gal. For larger solar hot water and combined systems, the following concerns are important when selecting storage:

- Higher cost per unit volume of ASME rated tanks in sizes greater than 120 gal
- Handling difficulties due to large mass
- Accessibility to locations suitable for storage
- Interfacing with existing SHW and BSH systems
- Corrosion protection of steel tanks

The choice of pressurized storage for medium-sized systems is based on the availability of suitable low-cost tanks near the site. Identification of a suitable supplier of low-cost tanks can extend the advantages of pressurized storage to larger SHW installations.

Storage pressurized at city water supply pressure is not practical for building space heating, except for small applications such as residences, apartments, and small commercial buildings.

With pressurized storage, the heat exchanger is always located on the collector side of the tank. Either the internal or the external heat exchanger configuration can be used. Figure 20 illustrates the three principal types of internal heat exchanger concepts: an immersed coil, a wraparound jacket, and a tube bundle. Small tanks (less than 120 gal) are available with either of the first two heat exchangers already installed. For larger tanks, a large assortment of tube bundle heat exchangers are available that can be incorporated into the tank design by the manufacturer.

Sometimes, more than one tank is needed to meet design requirements. Additional tanks offer the following benefits:

- Added storage volume
- Increased heat exchanger surface
- Reduced pressure drop in the collection loop

Figure 21 illustrates the multiple-tank configuration for pressurized storage. The exchangers are connected in reverse-return fashion to minimize flow imbalance. A third tank may be added. Additional tanks have the following disadvantages compared to a single tank of the same volume:

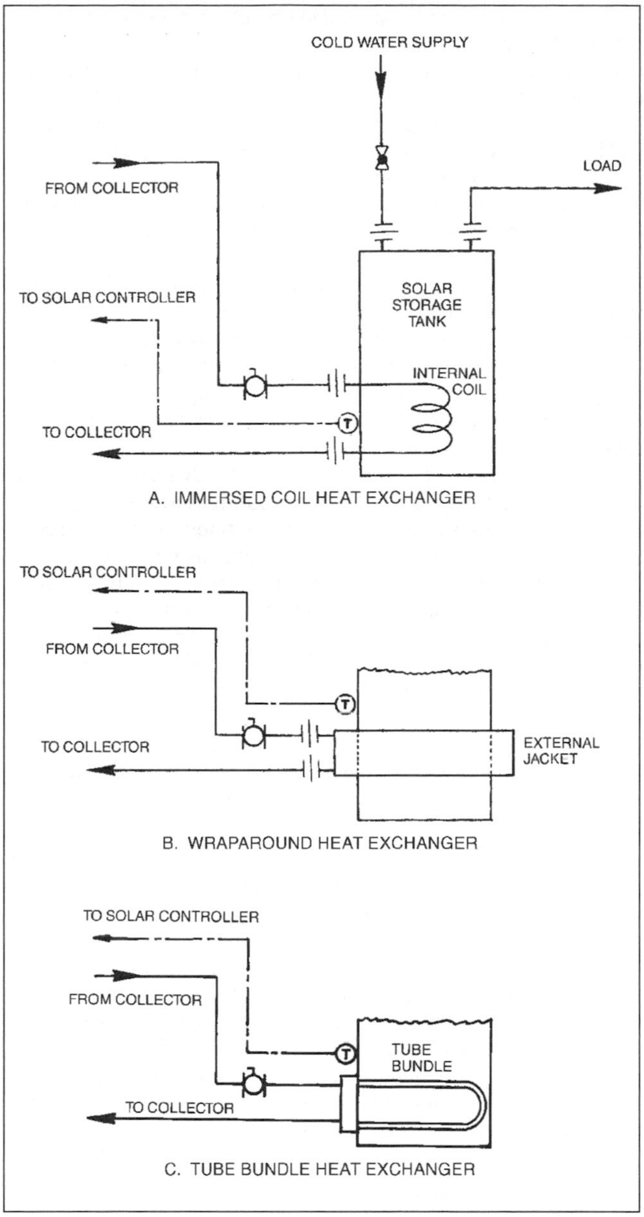

Fig. 20 Pressurized Storage with Internal Heat Exchanger

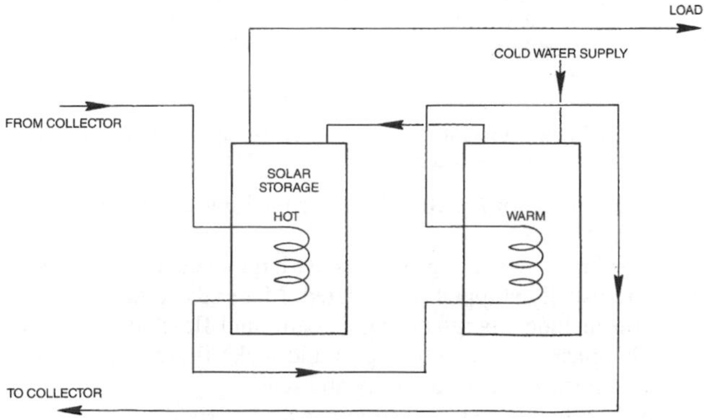

Fig. 21 Multiple Storage Tank Arrangement with Internal Heat Exchangers

- Higher installation costs
- Greater space requirements
- Higher heat losses (reduced performance)

An external heat exchanger provides greater flexibility because the tank and the heat exchanger can be selected independently of each other (Figure 22). Flexibility is not achieved without cost, however, because an additional pump, with its parasitic energy consumption, is required.

When selecting an external heat exchanger for a system protected by a nonfreezing liquid, the following factors related to start-up after at least one night in extremely cold conditions should be considered:

- Freeze-up of the water side of the heat exchanger
- Performance loss due to extraction of heat from storage

For small systems, an internal heat exchanger/tank arrangement prevents the water side of the heat exchanger from freezing. However, the energy required to maintain the water above freezing must be extracted from storage, thereby decreasing overall performance. With the external heat exchanger/tank combination, a bypass can be arranged to divert cold fluid around the heat exchanger until it has been heated to an acceptable level, such as 80°F. When the heat transfer fluid has warmed to this level, it can enter the heat exchanger without causing freezing or extraction of heat from storage. If necessary, this arrangement can also be used with internal heat exchangers to improve performance.

Unpressurized Storage. For systems sized greater than about 1000 ft^3 (1500 gal storage volume minimum), unpressurized storage is usually more cost-effective than pressurized. As used in this chapter, the term **unpressurized** means tanks at or below the pressure expected in an unvented drain-back loop.

Unpressurized storage for water and space heating implies a heat exchanger on the load side of the tank to isolate the high-pressure (potable water) loop from the low-pressure collector loop. Figure 23 illustrates unpressurized storage with an external heat exchanger. In this configuration, heat is extracted from the top of the solar storage

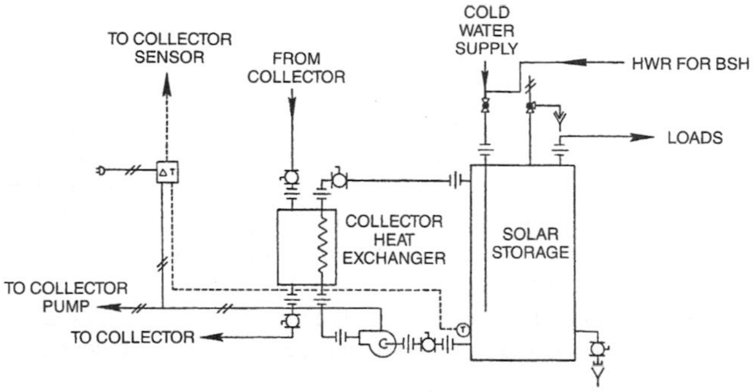

Fig. 22 Pressurized Storage System with External Heat Exchanger

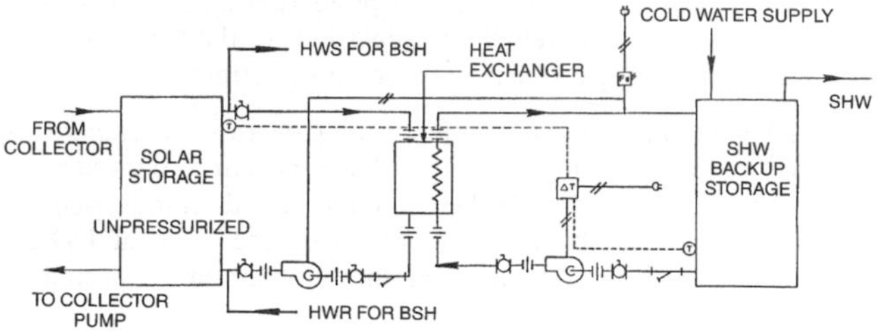

Fig. 23 Unpressurized Storage System with External Heat Exchanger

tank, and the cooled water is returned to the bottom of the tank. On the load side of the heat exchanger, the water to be heated flows from the bottom of the backup storage tank, and heated water returns to the top. The heat exchanger may have a double wall to protect a potable water supply. A differential temperature controller controls the two pumps on either side of the heat exchanger. When small pumps are used, both may be controlled by the same controller without overloading it.

The external heat exchanger shown in Figure 23 provides good system flexibility and freedom in component selection. In some cases, system cost and parasitic power consumption may be reduced by an internal heat exchanger. Some heat exchangers fabricated in the field use coiled soft copper tube. For larger systems, where custom fabrication is more feasible, a specified heat exchanger can be installed at the top of the tanks.

Storage Tank Construction

Steel. For most liquid solar energy systems, steel is the preferred material for storage tank construction. Steel tanks are relatively easy to fabricate to ASME Pressure Vessel Code requirements, readily available, and easily attached by pipes and fittings.

Steel tanks used for pressures of 30 psi and above must be ASME rated. Because water main pressure is usually above this level, open systems must use ASME code tanks. These pressure-rated storage tanks are more expensive than nonpressurized types. Significant cost reduction can usually be realized if a nonpressurized tank can be used.

Steel tanks are subject to corrosion, however. Because corrosion rates increase with temperature, the designer must be particularly aware of corrosion protection methods for solar energy applications. A steel tank must be protected against (1) electrochemical corrosion, (2) oxidation (rusting), and (3) galvanic corrosion (ANL 1979a).

The pH of the liquid and the electric potential of the metal are the primary governing factors of electrochemical corrosion. A sacrificial anode fabricated from a metal more reactive than steel can provide protection. Magnesium is recommended for solar applications. Because protection ends when the anode has dissolved, the anode must be inspected annually.

Oxygen can enter the tank through the air dissolved in the water entering the tank or through an air vent. In pressurized storage, oxygen is continually replenished by the incoming water. Besides causing rust, the oxygen catalyzes other types of corrosion. Unpressurized storage systems are less susceptible to corrosion caused by oxygen because they can be designed as unvented systems, so corrosion is limited to the small amount of oxygen contained in the initial fill water.

Coatings applied to the inside of the tank protect it from oxidation. The following are some of the most commonly used coatings:

- **Phenolic epoxy** should be applied in four coats.
- **Baked-on epoxy** is preferred over painted-on epoxy.
- **Glass lining** offers more protection than the epoxies and can be used under severe water conditions.
- **Hydraulic stone** provides the best protection against corrosion and increases the heat retention capabilities of the tank. Its mass may cause handling problems in some installations.

Whichever lining is used should either be flexible enough to withstand extreme thermal cycling or have the same coefficient of expansion as the steel tank. In the United States, all linings used for potable water tanks should be approved by the Food and Drug Administration (FDA) for the maximum temperature expected in the tank.

Dissimilar materials in an electrolyte (water) are in electrical contact with each other and corrode by galvanic action. Copper fittings screwed into a steel tank corrode the steel, for example. Galvanic corrosion can be minimized by using dielectric bushings to connect pipes to tanks.

Plastic. Fiberglass-reinforced plastic (FRP) tanks offer the advantages of low mass, high corrosion resistance, and low cost. Premium quality resins permit operating temperatures as high as 210°F, well above the temperature imposed by flat-plate solar collectors. Before delivery of an FRP tank is accepted, the tank should be inspected for damage that may have occurred during shipment. The gel coat on the inside must be intact, and no glass fibers should be exposed.

Concrete. Concrete vessels lined with a waterproofing membrane may also be used (in vented systems only) to contain a liquid thermal storage medium. Concrete storage tanks are inexpensive, and they can be shaped to fit almost any retrofit application. Also, they possess excellent resistance to the loading that occurs when they are placed in a below-grade location. Concrete tanks must have smooth corners and edges.

Concrete storage vessels have some disadvantages. Seepage often occurs unless a proper waterproofing surface is applied. Waterproofing paints are generally unsatisfactory because the concrete often cracks after settling. Other problems may occur due to poor workmanship, poor location, or poor design. Usually, the tanks should stand alone and not be integrated with a building or other structure. Careful attention should be given to the expansion joints and seams because they are particularly difficult to seal. Sealing at pipe taps can be a problem. Penetrations should be above the liquid level, if possible.

Finally, concrete tanks are heavy and may be more difficult to support in a proper location. Their weight may make insulation of the tank bottom more expensive and difficult.

Storage Tank Insulation

Heat loss from storage tanks and appurtenances is one of the major causes of poor system performance. The average R-value of storage tanks in solar applications is about half the insulation design value because of poorly insulated supports. Different standards recommend various design criteria for tank insulation. The Sheet Metal and Air Conditioning Contractors National Association (SMACNA) recommendation of a 2% loss in 12 h is generally accepted because it maintains a more stringent requirement than other standards. The following equation can be used to calculate the insulation R-value for this requirement:

$$\frac{1}{R} = \frac{fQ}{A\theta} \frac{1}{(t_{avg} - t_a)} \quad (10)$$

where

R = thermal resistivity of insulation, ft²·h·°F/Btu
f = specified fraction of stored energy that can be lost in time θ
Q = stored energy, Btu
A = exposed surface area of storage unit, ft²
θ = given time period, h
t_{avg} = average temperature in storage unit, °F
t_a = ambient temperature surrounding storage unit during season when it will be heated, °F

The insulation factor $fQ/A\theta$ is found from Table 2 for various tank shapes (ANL 1980).

Most solar water heating systems use large steel pressure vessels, which are usually shipped uninsulated. Materials suitable for field insulation include fiberglass, rigid foam, and flexible foam blankets. Fiberglass is easy to transport and make fire-retardant, but it requires significant labor to apply and seal.

Another widely used insulation consists of rigid sheets of polyisocyanurate foam cut and taped around the tank. Material that is 3 to 4 in. thick can provide R-20 to R-30 insulation value. Rigid foam insulation is sprayed directly onto the tank from a foaming

Table 2 Insulation Factor $fQ/A\theta$ for Cylindrical Water Tanks

	Horizontal Tank Insulation Factor, Btu/h·ft²			
Size, gal	D × D	D × 2D	D × 4D	D × 6D
250	3.63	3.46	3.05	2.77
500	4.57	4.36	4.84	3.49
750	5.24	4.99	4.40	3.99
1000	5.76	5.49	4.84	4.39
1500	6.60	6.28	5.54	5.03
2000	7.26	6.92	6.10	5.53
3000	8.31	7.92	6.98	6.33
4000	9.15	8.71	7.68	6.97
5000	9.86	9.39	8.28	7.51

	Vertical Tank Insulation Factor, Btu/h·ft²			
Size, gal	D to 3D × D	D/2 × D	D/3 × D	D/4 × D
80	2.10	1.88	2.15	1.97
120	2.39	2.15	2.46	2.26
250	3.07	2.74	2.46	2.88
500	3.87	3.46	3.96	3.63
750	4.43	3.96	4.53	4.16
1000	4.87	4.36	4.99	4.57
1500	5.58	4.99	5.71	5.24
2000	6.13	5.49	6.28	5.76
3000	7.03	6.28	7.19	6.60
4000	7.73	6.92	7.92	7.26
5000	8.33	7.45	8.53	7.82

truck or in a shop. It bonds well to the tank surface (no air space between tank and insulation) and insulates better than an equivalent thickness of fiberglass. When most foams are exposed to flames, they ignite and/or produce toxic gases. When located in, or adjacent to, a living space, they often must be protected by a fire barrier and/or a sprinkler system.

Some tank manufacturers and suppliers offer custom tank jackets of flexible foam with zipper-like connections that provide quick installation and neat appearance. Some of the fire considerations for rigid foam apply to these foam blankets.

The supports of a tank are a major source of heat loss. To provide suitable load-bearing capability, the supports must be in direct contact with the tank wall or attached to it. Thermal breaks must be provided between the supports and the tank (Figure 24). If insulating the tank from the support is impractical, the external surface of the supports must be insulated, and the supports must be placed on insulative material capable of supporting the compressive load. Wood, foam glass, and closed-cell foam can be used, depending on the compressive load.

Stratification and Short Circuiting

Because hot water rises and cold water sinks in a vessel, the pipe to the collector inlet should always be connected to the bottom of the tank. The collector then operates at its best efficiency because it is always at the lowest possible temperature. The return from the collector should always run near the top of the tank, the location of the hottest fluid. Similarly, the load should be extracted from the top of the tank, and cold water makeup should be introduced at the bottom. Because of the increased static pressure, vertical storage tanks enhance stratification better than horizontal tanks.

If a system has a rapid tank turnover rate or if the buffer tank of the system is closely matched to the load, thermal stratification does not offer any advantages. However, in most solar energy systems, thermal stratification increases performance because temperature

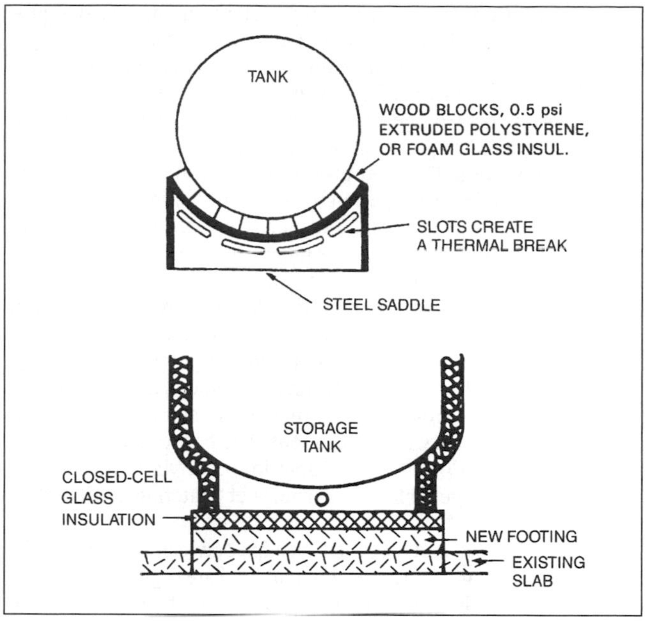

Fig. 24 Typical Tank Support Detail

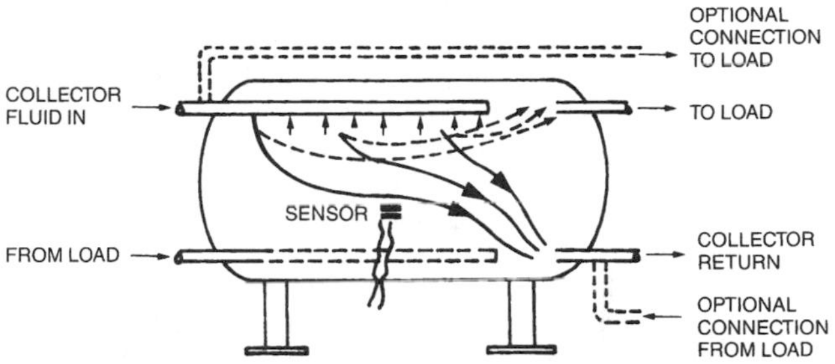

Fig. 25 Tank Plumbing Arrangements to Minimize Short Circuiting and Mixing
(Kreider 1982)

differences are relatively small. Thus, any enhancement of temperature differentials improves heat transfer efficiency.

Thermal stratification can be enhanced by the following:

- Using a tall vertical tank
- Situating the inlet and outlet piping of a horizontal tank to minimize vertical fluid mixing
- Sizing the inlets and outlets such that exhaust flow velocity is less than 2 fps
- Using flow diffusers (Figure 25)
- Plumbing multiple tanks in series

Multiple tanks generally yield the greatest temperature difference between cold water inlet and hot water outlet. With the piping/tank configuration shown in Figure 21, the difference can approach 30°F. In addition to the cost of the second tank, the cost of extra footings, insulation, and sensors must be considered. However, multiple tanks may save labor at the site due to their small diameter and shorter length. The smaller tanks require less demolition to install in retrofit (e.g., access through doorways, hallways, and windows).

Sizing

The following are specific site factors related to system performance or cost.

Solar Fraction. Most solar energy systems are designed to produce about 50% of the energy required to satisfy the load. If a system is sized to produce less than this amount, there must be greater coincidence between load and available solar energy. Therefore, smaller, lower cost storage can be used with low solar fraction systems without impairing system performance.

Load Matching. The load profile of a commercial application may be such that less storage can be used without incurring performance penalties. For example, a company cafeteria or a restaurant serving large luncheon crowds has its greatest hot water usage during, and just after, midday. Because the load coincides with the availability of solar radiation, less storage is required for carryover. For loads that peak during the morning, collectors can be rotated toward the southeast, sometimes improving output for a given array.

Storage Cost. If a solar storage tank is insulated adequately, performance generally improves with increase in storage volume, but the savings must justify the investment. The cost of storage can be kept at a minimum by using multiple low-cost water heaters and unpressurized storage tanks made from such materials as fiberglass or concrete.

The performance and cost relationships between solar availability, collector design, storage design, and load are highly interactive. Furthermore, the designer should maximize the benefits from the investment for at least one year, rather than for one or two months. For this reason, solar energy system performance models combined with economic analyses should be used to optimize storage size for a specific site. Models that can be used to determine system performance include *f*-Chart, SOLCOST, and TRNSYS. *f*-Chart (Beckman et al. 1977) is readily available either in the work-sheet form for hand calculations or as a reasonably priced computer program; however, it contains a built-in daily hot water load profile specifically for residential applications. This limitation is not very serious; studies have shown that the daily hot water profile has little effect on system performance.

For daily profiles with extreme deviations from the residential profile, such as the company cafeteria example used previously, *f*-Chart may not provide satisfactory results. SOLCOST (DOE 1978) can accommodate user-selected daily profiles, but it is not widely available. Neither program can accommodate special days (e.g., holidays and weekends) when no service hot water loads are present. TRNSYS (SEL 1983), which has been modified for operation on personal computers, is readily available and well supported. It is tedious to set up and run for the first time, but after it is set up, various design parameters, such as collector area and storage size, can be easily evaluated for a given system configuration.

Many hydronic solar energy systems are used with a series-coupled heat pump or an absorption air-conditioning system. In the case of a heat pump, lower storage temperatures, which create higher collector efficiencies, are desirable. In this case, the collector size may be greater than recommended for conventional SHW or hydronic building space heating (BSH) systems. In contrast, absorption air conditioners require much higher temperatures (>175°F) than BSH or hydronic BSH systems to activate the generator. Therefore, the storage collector ratio can be much less than recommended for BSH. For absorption air conditioning, thermal energy storage may act as a buffer between the collector and generator, which prevents cycling due to frequent changes in insolation level.

HEAT EXCHANGERS

Requirements

The heat exchanger transfers heat from one fluid to another. In closed solar energy systems, it also isolates circuits operating at different pressures and separates fluids that must not be mixed. Heat exchangers are used in solar applications to separate the heat transfer fluid in the collector loop from the domestic water supply in the storage tank (pressurized storage) or the domestic water supply from the storage (unpressurized storage). Heat exchangers for solar applications may be placed either inside or outside the storage or drain-back tank.

The selection of a heat exchanger involves the following considerations:

Performance. Heat exchangers always degrade the performance of the solar collector; therefore, the selection of an adequate size is important. When in doubt, an oversized heat exchanger should be selected.

Guaranteed separation of fluids. Many code authorities require a vented, double-wall heat exchanger to ensure fluid isolation. System protection requirements may also dictate the need for guaranteed fluid separation.

Thermal expansion. The temperature in a heat exchanger may vary from below freezing to the boiling temperature of water. The design must withstand these thermal cycles without failing.

Materials. Galvanic corrosion is always a concern in liquid solar energy systems. Consequently, the piping, collectors, and other hydronic component materials must be compatible.

Space constraints. Often, limited space is available for mounting and servicing the heat exchanger. Physical size and configuration must be considered when selection is made.

Serviceability. The water side of a heat exchanger is exposed to the scaling effects of dissolved minerals, so the design must provide access for cleaning and scale removal.

Pressure loss. Energy consumed in pumping fluids reduces system performance. The pressure drop through the heat exchanger should be limited to 1 to 2 psi to minimize energy consumption.

Pressure capability. Because the heat exchanger is exposed to cold water supply pressure, it should be rated for pressures above 75 psig.

Internal Heat Exchanger

The internal heat exchanger can be a coil inside the tank or a jacket wrapped around the pressure vessel (Figure 20). Several manufacturers supply tanks with either type of internal heat exchanger. However, the maximum size of pressurized tanks with internal heat exchangers is usually about 120 gal. Heat exchangers may be installed with relative ease inside nonpressurized tanks that open from the top. Figure 26 and Figure 27 illustrate methods of achieving double-wall protection with either type of internal heat exchanger.

For installations having larger tanks or heat exchangers, a tube bundle is required. However, it is not always possible to find a heat exchanger of the desired area that will fit within the tank. Consequently, a horizontal tank with the tube bundle inserted from the tank end can be used. A second option is to place a shroud around the tube bundle and to pump fluid around it (Figure 28). Such an approach combines the performance of an external heat exchanger

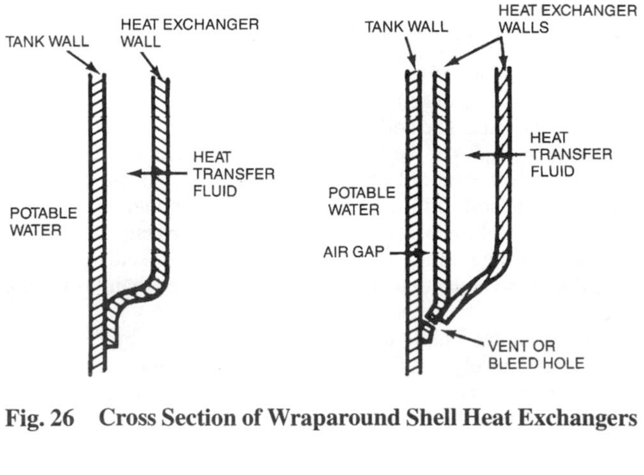

Fig. 26 Cross Section of Wraparound Shell Heat Exchangers

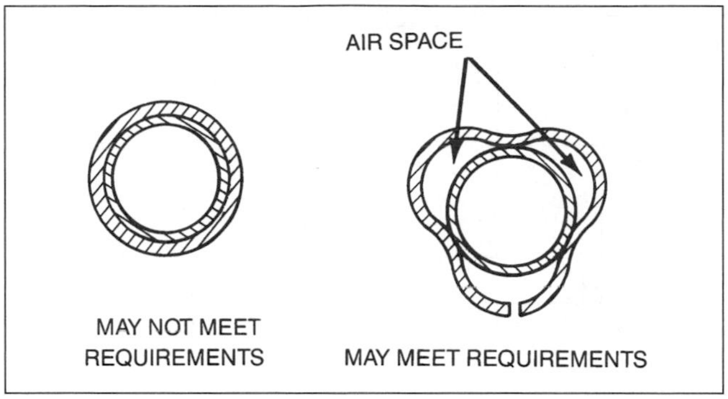

Fig. 27 Double-Wall Tubing

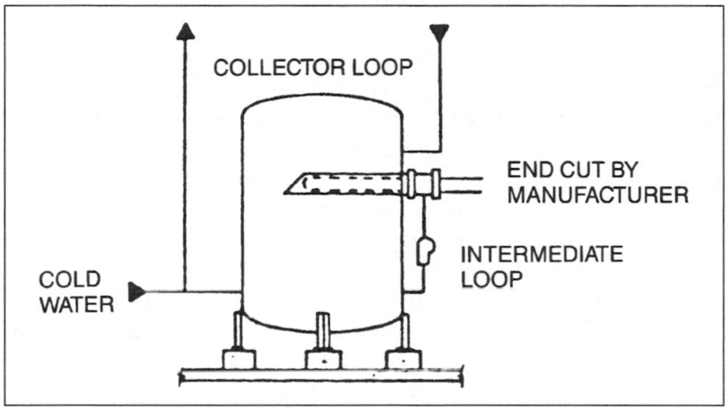

Fig. 28 Tube Bundle Heat Exchanger with Intermediate Loop

with the compactness of an internal heat exchanger. Unfortunately, this approach causes tank mixing and loss of stratification.

External Heat Exchanger

The external heat exchanger offers a greater degree of design flexibility than the internal heat exchanger because it is detached from the tank. For this reason, it is preferred for most commercial applications. Shell-and-tube, tube-and-tube, and plate-and-frame heat exchangers are used in solar applications. Shell-and-tube heat exchangers are found in many solar designs because they are economical, easy to obtain, and constructed with suitable material. One limitation of the shell-and-tube heat exchanger is that it normally does not have double-wall protection, which is often required for potable water heating applications. A number of manufacturers produce tube-and-tube heat exchangers that offer high performance and the double-wall safety required by many code authorities, but they are usually limited in size. The plate-and-frame heat exchanger is more cost-effective for large potable water applications where positive separation of heat transfer fluids is required. These heat exchangers are compact and offer excellent heat transfer performance.

Shell-and-Tube. Shell-and-tube heat exchangers accommodate large heat exchanger areas in a compact volume. The number of shell-side passes and tube-side passes (i.e., the number of times the fluid changes direction from one end of the heat exchanger to the other) is a major variable in shell-and-tube heat exchanger selection. Because the exchanger must compensate for thermal expansion, flow in and flow out of the tube side are generally at the same end of the exchanger. Therefore, the number of tube-side passes is even. By appropriate baffling, two, four, or more tube passes may be created. However, as the number of passes increases, the path length grows, resulting in greater pressure drop of the tube-side fluid. Unfortunately, shell-and-tube heat exchangers are hard to find in the double-wall configuration.

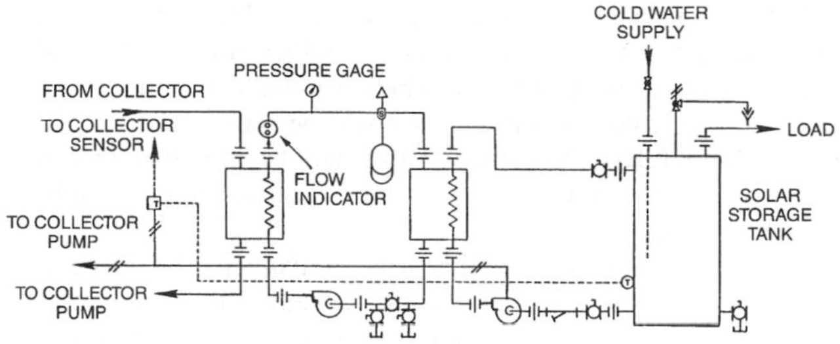

Fig. 29 Double-Wall Protection Using Two Heat Exchangers in Series

Tube-and-Tube. Fluids in the tube-and-tube heat exchanger run counterflow, which gives closer approach temperatures. The exchanger is also compact and only limited by system size. Several may be piped in parallel for higher flow, or in series to provide approach temperatures as close as 15°F. Many manufacturers offer the tube-and-tube with double-wall protection.

The counterflow configuration operates at high efficiency. For two heat exchangers in series, the effectiveness may reach 0.80. For single heat exchangers or multiple heat exchangers in parallel, the effectiveness may reach 0.67. Tube-and-tube heat exchangers are cost-effective for smaller residential systems.

Plate-and-Frame. These heat exchangers are suitable for pressures up to 300 psig and temperatures to 400°F. They are economically attractive when a quality, heavy-duty construction material is required or if a double-wall, shell-and-tube heat exchanger is needed for leak protection. Typical applications include food industries or domestic water heating when any possibility of product contamination must be eliminated. This exchanger also gives added protection to the collector loop.

Contamination is not possible when an intermediate loop is used (Figure 29) and the integrity of the plates is maintained. A colored fluid in the intermediate loop gives a visual means of detecting a leak through changes in color. The sealing mechanism of the plate-and-frame heat exchanger prevents cross-contamination of heat transfer fluids. Plate-and-frame heat exchangers cost the same as or less than equivalent shell-and-tube heat exchangers constructed of stainless steel.

Heat Exchanger Performance

Solar collectors perform less efficiently at high fluid inlet temperatures. Heat exchangers require a temperature difference between the two streams in order to transfer heat. The smaller the heat transfer surface area, the greater the temperature difference must be to transfer the same amount of heat and the higher the collector inlet temperature must be for a given tank temperature. As the solar collector is forced to operate at the progressively higher temperature associated with smaller heat exchanger, its efficiency is reduced.

In addition to size and surface area, the configuration of the heat exchanger is important for achieving maximum performance. The performance of a heat exchanger is characterized by its effectiveness (a type of efficiency), which is defined as follows:

$$E = \frac{q}{(\dot{m}c_p)_{min}(t_{hi} - t_{ci})} \tag{11}$$

where

E = effectiveness
q = heat transfer rate, Btu/h
$\dot{m}c_p$ = minimum mass flow rate times fluid specific heat, Btu/h·°F
t_{hi} = hot (collector loop) stream inlet temperature, °F
t_{ci} = cold (storage) stream inlet temperature, °F

For heat exchangers located in the collector loop, the minimum flow usually occurs on the collector side rather than the tank side.

The effectiveness E is the ratio between the heat actually transferred and the maximum heat that could be transferred for given flow and fluid inlet temperature conditions. The effectiveness is relatively insensitive to temperature, but it is a strong function of heat exchanger design.

A designer must decide what heat exchanger effectiveness is required for the specific application. A method that incorporates heat exchanger performance into a collector efficiency equation [Equation (3)] uses storage tank temperature t_s as the collector inlet temperature with an adjusted heat removal factor F_R. Equation (12) relates F_R and heat exchanger effectiveness:

$$\frac{F'_R}{F_R} = \frac{1}{1 + (F_R U_L / \dot{m}c_p)[A\dot{m}c_p / E(\dot{m}c_p)_{min} - 1]} \quad (12)$$

The heat exchanger effectiveness must be converted into heat transfer surface area. SERI (1981) provides details of shell-and-tube heat exchangers and heat transfer fluids.

CONTROLS

The heart of an active solar energy system is the automatic temperature control. Numerous studies and reports of operational systems show that faulty controls are usually the cause of poor solar system performance. Reliable controllers are available, and with proper understanding of each system function, proper control systems can be designed. In general, control systems should be simple; additional controls are not a good solution to a problem that can be solved by better mechanical design. The following key considerations pertain to control system design:

- Collector sensor location/selection
- Storage sensor location
- Over-temperature sensor location
- On-off controller characteristics
- Selection of reliable solid-state devices, sensors, controllers, etc.
- Control panel location in heated space
- Connection of controller according to manufacturer's instructions
- Design of control system for all possible system operating modes, including heat collection, heat rejection, power outage, freeze protection, auxiliary heating, etc.
- Selection of alarm indicators for pump failure, low temperatures, high temperatures, loss of pressure, controller failure, nighttime operation, etc.

The following control categories should be considered when designing automatic controls for solar energy systems:

- Collection to storage
- Storage to load
- Auxiliary energy to load
- Alarms
- Miscellaneous (e.g., for heat rejection, freeze protection, draining, and over-temperature protection)

Differential Temperature Controller

Most controls used in solar energy systems are similar to those for HVAC systems. The major exception is the differential temperature controller (DTC), which is the basis of solar energy system control. The DTC is a comparing controller with at least two temperature sensors that controls one or several devices. Typically, the sensors are located at the solar collectors and storage tank (Figure 30). On unpressurized systems, other DTCs may control the extraction of heat from the storage tank.

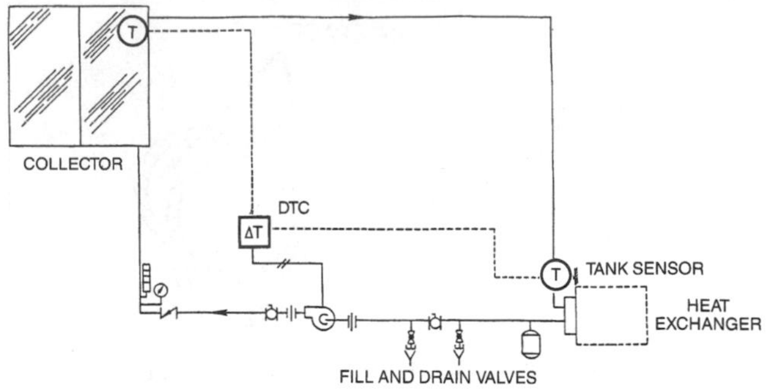

Fig. 30 Basic Nonfreezing Collector Loop for Building Service Hot Water Heating—Nonglycol Heat Transfer Fluid

The DTC monitors the temperature difference, and when the temperature of the panel exceeds that of the storage by the predetermined amount (generally 8 to 20°F), the DTC switches on the actuating devices. When the temperature of the panel drops to 3 to 10°F above the storage temperature, the DTC, either directly or indirectly, stops the pump. Indirect control through a control relay may operate one or more pumps and possibly perform other control functions, such as the actuation of control valves.

The manufacturer's predetermined set point of the DTC may be adjustable or fixed. If the controller set point is a fixed temperature differential, the controller selected should correspond to the requirements of the system. An adjustable differential set point makes the controller more flexible and allows it to be adjusted to the specific system. The optimum *off* temperature differential should be the minimum possible; the minimum depends on whether there is a heat exchanger between the collectors and storage.

If the system requires a heat exchanger, the energy transferred between two fluids raises the differential temperature set point. The minimum, or *off*, temperature differential is the point at which pumping the energy costs as much as the value of the energy being pumped. For systems with heat exchangers, the *off* set point is generally between 5 and 10°F. If the system does not have a heat exchanger, a range of 3 to 6°F is acceptable for the *off* set point. The heat lost in the piping and the power required to operate the pump should also be considered.

The optimum differential *on* set point is difficult to calculate because of the changing variables and conditions. Typically, the *on* set point is 10 to 15°F above the *off* set point. The optimum *on* set point is a balance between optimum energy collection and avoiding short cycling of the pump. ASHRAE's *Active Solar Thermal Design Manual* (Newton and Gilman 1983) describes techniques for minimizing short cycling.

Over-Temperature Protection

Overheating may occur during periods of high insolation and low load; thus, all portions of the solar energy system require protection against overheating. Liquid expansion or excessive pressure may burst piping or storage tanks, and steam or other gases within a system may restrict the liquid flow, making the system inoperable. Glycols break down and become corrosive if subjected to temperatures greater than 240°F. The system can be protected from overheating by (1) stopping circulation in the collection loop until the storage temperature decreases, (2) discharging the overheated water from the system and replacing it with cold makeup water, or (3) using a heat exchanger coil as a means of heat rejection to ambient air.

The following questions should be answered to determine whether over-temperature protection is necessary.

1. Is the load ever expected to be off, such that the solar input will be much higher than the load? The designer must determine

possibilities based on the owner's needs and a computer analysis of system performance.

2. Do individual components, pumps, valves, circulating fluids, piping, tanks, and liners need protection? The designer must examine all components and base the over-temperature protection set point on the component that has the lowest specified maximum operating temperature. This may be a valve or pump with a 180 to 300°F maximum operating temperature. Sometimes, this criterion may be met by selecting components capable of operating at higher temperatures.
3. Is the formation of steam or discharging boiling water at the tap possible? If the system has no mixing valve that mixes cold water with the solar-heated water before it enters the tap, the water must be maintained below boiling temperature. Otherwise, the water will flash to steam as it exits the tap and, most likely, scald the user. Some city codes require a mixing valve to be placed in the system for safety.

Differential temperature controllers are available that sense over-temperature. Depending on the controller used, the sensor may be mounted at the bottom or the top of the storage tank. If it is mounted at the bottom of the tank, the collector-to-storage differential temperature sensor can be used to sense over-temperature. Input to a DTC mounted at the top of the tank is independent of the bottom-mounted sensor, and the sensor monitors the true high temperature.

The normal action taken when the DTC senses an over-temperature is to turn off the pump to stop heat collection. After the panels in a drain-back system are drained, they attain stagnation temperatures. While drain-back is not desirable, the panels used for these systems should be designed and tested to withstand over-temperature. In addition, drain-back panels should withstand the thermal shock of start-up when relatively cool water enters the panels while they are at stagnation temperatures. The temperature difference can range from 75 to 300°F. Such a difference could warp panels made with two or more materials of different thermal expansion coefficients. If the solar panels cannot withstand the thermal shock, an interlock should be incorporated into the control logic to prevent this situation. One method uses a high-temperature sensor mounted on the collector absorber that prevents the pump from operating until the collector temperature drops below the sensor set point.

If circulation stops in a closed-loop antifreeze system that has a heat exchanger, high stagnation temperatures will occur. These temperatures could break down the glycol heat transfer fluid. To prevent damage or injury due to excessive pressure, a pressure relief valve must be installed in the loop, and a means of rejecting heat from the collector loop must be provided. The section on Hot Water Dump describes a common way to relieve pressure. Pressure increases due to the thermal expansion of any fluid; when water-based absorber fluids are used, pressure builds up from boiling.

The pressure relief valve should be set to relieve at or below the maximum operating pressure of any component in the closed-loop system. Typical settings are around 50 psig, corresponding to a temperature of approximately 300°F. However, these settings should be checked. When the pressure relief valve does open, it discharges expensive antifreeze solution. Glycol antifreeze solutions damage many types of roof membranes. The discharge can be piped to large containers to save the antifreeze, but this design can create dangerous conditions because of the high pressures and temperatures involved.

If a collector loop containing glycol stagnates, chemical decomposition raises the fusion point of the liquid, and freezing becomes possible. An alternate method continues fluid circulation but diverts the flow from storage to a heat exchanger that dumps heat to the ambient air or other sink (Figure 31). This wastes energy, but it protects the system. A sensor on the solar collector absorber plate that turns on the heat rejection equipment can provide control. The temperature sensor set point is usually 200 to 250°F and depends on the system components. When the sensor reaches the high-temperature set point, it turns on pumps, fans, alarms, or whatever is necessary to reject the heat and warn of the over-temperature. The dump continues to operate until the over-temperature control in the collector loop DTC senses an acceptable drop in tank temperature and is reset to its normal state.

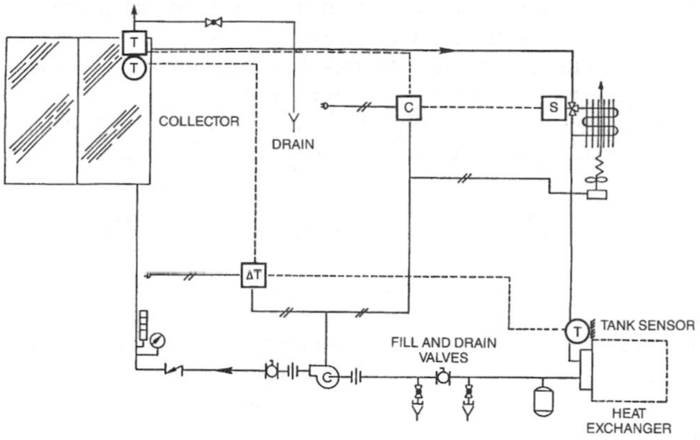

Fig. 31 Heat Rejection from Nonfreezing System Using Liquid-to-Air Heat Exchanger

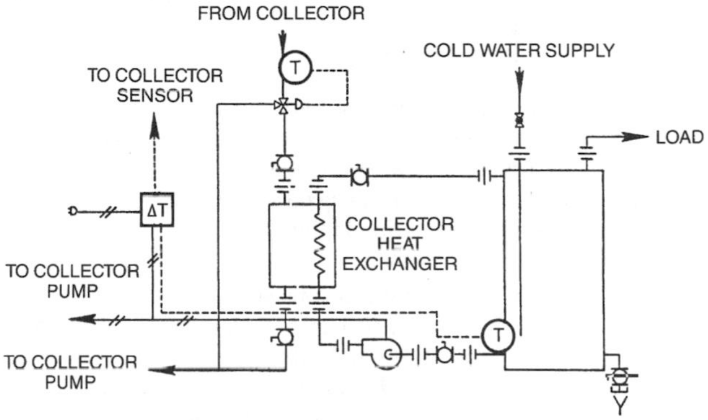

Fig. 32 Nonfreezing System with Heat Exchanger Bypass

Hot Water Dump

If water temperatures above 200°F are allowed, the standard temperature-pressure (210°F, 125 psig) safety relief valve may operate occasionally. If these temperatures are reached, the valve opens, and some of the hot water vents out. However, these valves are designed for safety purposes, and after a few openings, they leak hot water. Thus, they should not be relied on as the only control device. An aquastat that controls a solenoid, pneumatic, or electrically actuated valve should be used instead.

Heat Exchanger Freeze Protection

The following factors should be considered when selecting an external heat exchanger for a system protected by a nonfreezing fluid that is started after an overnight, or longer, exposure to extreme cold.

- Freeze-up of the water side of the heat exchanger
- Performance loss due to extraction of heat from storage

An internal heat exchanger/tank has been placed on the water side of the heat exchanger of small systems to prevent freezing. However, the energy required to maintain the water above freezing must be extracted from storage, which decreases overall performance. With the external heat exchanger/tank combination, a bypass can be installed, as illustrated in Figure 32. The controller positions the valve to bypass the heat exchanger until the fluid in the collector loop attains a reasonable level (e.g., 80°F). When the heat transfer fluid has warmed to this level, it can enter the heat exchanger without freezing or extracting heat from storage. The arrangement illustrated in Figure 32 can also be used with an internal heat exchanger, if necessary, to improve performance.

PHOTOVOLTAIC SYSTEMS

Photovoltaic (PV) devices, or cells, convert light directly into electricity. These cells are packaged into convenient modules to produce a specific voltage and current when illuminated. PV modules can be connected in series or in parallel to produce larger voltages or currents. PV systems rely on sunlight, have few or no moving parts, are modular to match power requirements on any scale, are reliable and long-lived, and are easily produced. Photovoltaic systems can be used independently or in conjunction with other electrical power sources. Applications powered by PV systems include communications, remote power, remote monitoring, lighting, water pumping, battery charging, and cathodic protection.

Fundamentals of Photovoltaics

Photovoltaic Effect. When a photon enters a photovoltaic material, the photon can be reflected, absorbed, or transmitted through. If the photon is absorbed, an electron is knocked lose from an atom. The electron can be removed by an electric field across the front and back of the photovoltaic material. In the absence of a field, the electron recombines with the atom.

Cell Design. A photovoltaic cell consists of the active photovoltaic material, metal grids, anti-reflection coatings, and supporting material. The complete cell is optimized to maximize both the amount of light entering the cell and the power out of the cell. The photovoltaic material can be one of a number of compounds. The metal grids enhance the current collection from the front and back of the solar cell. Grid design varies among manufacturers. The anti-reflective coating is applied to the top of the cell to maximize the light going into the cell; typically, this coating is a single layer optimized for sunlight. As a result, PV cells range in color from black to blue. In some types of photovoltaic cells, the top of the cell is covered with a semitransparent conductor that functions as both the current collector and the antireflection coating. A completed PV cell is a two-terminal device with positive and negative leads.

Current-Voltage Curves. The current from a PV cell depends on the external voltage applied and the amount of light on the cell. When the cell is short-circuited, the current is at maximum (short-circuit current I_{sc}), and the voltage across the cell is zero. When the PV cell circuit is open, with the leads not making a circuit, the voltage is at a maximum (open-circuit voltage V_{oc}), and the current is zero. In either case, at open circuit or short circuit, the power (current times voltage) is zero. Between open circuit and short circuit, the power output is greater than zero. A current-voltage curve (Figure 33) represents the range of combinations of current and voltage.

Power versus voltage can be plotted on the same graph. There exists an operating point P_{max}, I_{max}, V_{max} at which the output power is maximized. Given P_{max}, an additional parameter, fill factor FF, can be calculated such that

$$P_{max} = I_{sc}V_{oc}\mathrm{FF} \quad (13)$$

By illuminating and loading a PV cell so that the voltage equals the PV cell's V_{max}, the output power is maximized. The cell can be loaded using resistive loads, electronic loads, or batteries. The typical parameters of a single-crystal solar cell are current density J_{sc} = 206 mA/in^2; V_{oc} = 0.58 V; V_{max} = 0.47 V; FF = 0.72; and P_{max} = 2273 mW.

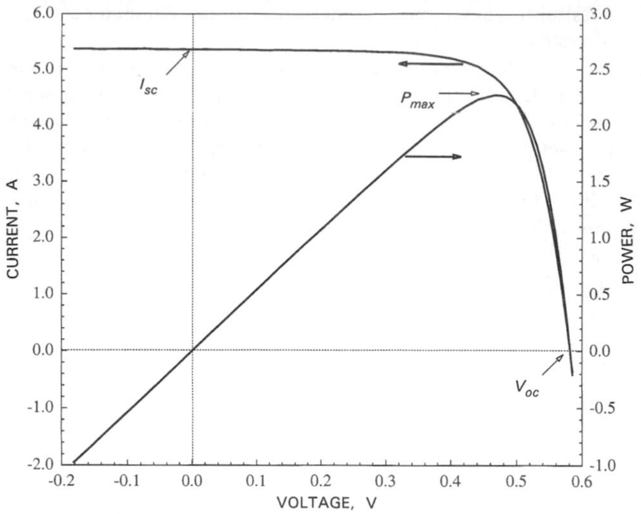

Fig. 33 Representative Current-Voltage and Power-Voltage Curves for Photovoltaic Device

Efficiency and Power. Efficiency is another measure of PV cells that is sometimes reported. Efficiency is defined as the maximum electrical power output divided by the incident light power. Efficiency is commonly reported for a PV cell temperature of 77°F and incident light at an irradiance of 93 W/ft^2 with a spectrum close to that of sunlight at solar noon. For cells designed for concentrator systems, the standard irradiance is 79 W/ft^2.

Photovoltaic Cells and Modules

Photovoltaic Cells. In general, all PV cells perform similarly. However, the choice of the photovoltaically active material can have important effects on system design and performance. Both the composition of the material and its atomic structure are influential. Photovoltaic materials include silicon, gallium arsenide, copper indium diselenide, cadmium telluride, indium phosphide, and many others. The atomic structure of a PV cell can be single-crystal, polycrystalline, or amorphous. The most commonly produced PV material is crystalline silicon, either single-crystal or polycrystalline.

Module Construction. A module is a collection of PV cells that protects the cells and provides a usable operating voltage. PV cells can be fragile and susceptible to corrosion by humidity or fingerprints and can have delicate wire leads. Also, the operating voltage of a single PV cell is less than 1 V, making it unusable for many applications. Depending on the manufacturer and the type of PV material, modules have different appearances and performance characteristics. Also, modules may be designed for specific conditions, such as hot climates.

Photovoltaic cells such as single-crystal and polycrystalline silicon are produced individually. Some amorphous silicon cells, while manufactured in large rolls, are later processed as individual cells. These cells are series-connected to other cells to produce an operating voltage around 14 to 16 V. These strings of cells are then encapsulated with a polymer, a front glass piece, and a back material. Also, a robust junction box is attached to the back of the module for convenient wiring to other modules or other electrical equipment.

Cells made of amorphous silicon, cadmium telluride, or copper indium diselenide are manufactured on large pieces of material that become either the front or the back of the module. A large area of PV material is divided into smaller cells by scribing or cutting the material into electrically isolated cells. Later processing steps automatically series-interconnect the cells. Depending on the manufacturer, a front glass or a back material is glued on using polymers. Again, a robust junction box is attached to the back of the module for convenient wiring to other modules or other electrical equipment.

Module Reliability and Durability. The PV module is reliable and durable when properly made. As with any electrical equipment, the designer must verify that the product is tested or rated for the intended application. A PV module is a large-area semiconductor with no internal moving parts. Even a concentrator module has no internal moving parts, although there are external, easily maintainable parts that keep the module aimed at the sun.

PV modules are designed for outdoor use in such harsh surroundings as marine, tropic, arctic, and desert environments. There are many international standards that test the suitability of PV modules for outdoor use in different climates. A buyer or user of PV modules should check the suitability of the module for a given climate.

Modules are also tested for electrical safety. All modules have a maximum voltage rating that is supplied by the manufacturer or stamped on the module. This becomes important when interconnecting several modules or connecting modules to other electrical equipment. Most modules are tested for ground isolation to prevent personal injury or property damage. Several manufacturers have listed their modules through Underwriters Laboratories (UL).

Related Equipment

Batteries. Batteries are required in many PV systems to supply power at night or when the PV system cannot meet the demand. The selection of battery type and size is primarily dependent on the load and availability requirements. In all cases, the batteries must be located in an area without extreme temperatures and should have some ventilation.

Battery types include lead-acid, nickel cadmium, nickel hydride, lithium, and many others. For a PV system, the main requirement is that the batteries be capable of repeated deep discharges without damage. Starting-lighting-ignition batteries (car batteries) are not designed for repeated deep discharges and should not be used.

Deep cycle lead-acid batteries are commonly used. These batteries can be flooded or valve-regulated (sealed) batteries and are commercially available in a variety of sizes. Flooded, or wet, batteries require greater maintenance but can last longer with proper care. Valve-regulated batteries require less maintenance. For more capacity, batteries can be arranged in parallel.

Battery Charge Controllers. Controllers regulate the power from the PV modules to prevent the batteries from overcharging. The controller can be a shunt type or series type and can also function as a low battery voltage disconnect to prevent the battery from overdischarge. The controller is chosen for the correct capacity and desired features.

Normally, controllers allow the battery voltage to determine the operating voltage of the PV system. However, the battery voltage may not be the optimum PV operating voltage. Some controllers can optimize the operating voltage of the PV modules independently of the battery voltage so that the PV operates at its maximum power point.

Inverters. An inverter is used to convert direct current (dc) into alternating current (ac) electricity. The output of the inverter can be single-phase or multiphase, with a voltage of 120 or 220 V and a frequency of 50 or 60 Hz. Inverters are rated by total power capacity, which ranges from hundreds of watts to megawatts. Some inverters have good surge capacity for starting motors; others have limited surge capacity. The designer should specify both the type and the size of the load the inverter is intended to service.

The output waveform of the inverter, while still ac, can be a square wave, a modified square wave (also called modified sine wave), or a sine wave. For some applications, the modified sine wave inverter has better surge capabilities. Electric utilities supply sine wave output; however, many electrical systems can operate on square or modified square wave outputs. Equipment that contains silicon controlled rectifiers or variable-speed motors, such as some vacuum cleaners and laser printers, can be damaged by square and modified square wave inverter outputs. More manufacturers are producing inverters with sine wave outputs that avoid unexpected equipment damage. Inverters can supply ac power independently or can synchronize the waveform frequency to another ac power supply such as the electric utility or a portable electrical generator.

Peak Power Trackers. Peak power trackers optimize the operating voltage of the PV system to maximize the current. Typically, the PV system voltage is changed automatically. Rudimentary peak power trackers may have fixed, operator-selected set points. Peak power trackers can be purchased separately or specified as an option with battery charge controllers or inverters. The cost and complexity of adding a peak power tracker should be balanced against the expected power gain and the impact on system reliability.

Balance-of-System Components. While the major components of a PV system have been outlined, the balance-of-system components are just as important. These components include the mounting structures and wiring. PV systems should be designed and installed with the intent of only minimal maintenance.

Photovoltaic modules can be mounted on the ground or on a building or can be included as part of a building. A structural engineer can design the support structures in accordance with local building codes. Wind and snow loading are a major design consideration. The PV modules can last 20 or more years; the support structure and building should be designed for at least as long a lifetime.

In the United States, wire sizing, insulation, and use of fuses, circuit breakers, and other protective devices are covered by the *National Electrical Code* (NFPA *Standard* 70).

REFERENCES

ANL. 1979a. *Design and installation manual for thermal energy storage.* ANL-79-15. Argonne National Laboratory, Argonne, IL.

ANL. 1979b. Reliability and materials design guidelines for solar domestic hot water systems. ANL/SDP-9. Argonne National Laboratory, Argonne, IL.

ANL. 1980. *Design and installation manual for thermal storage,* 2nd ed. ANL-79-15. Argonne National Laboratory, Argonne, IL.

ASHRAE. 1988. *Active solar heating systems design manual.*

ASHRAE. 1990. *Guide for preparing active solar heating systems operation and maintenance manuals.*

ASHRAE. 1991a. *Active solar heating systems installation manual.*

ASHRAE. 1991b. Methods of testing to determine the thermal performance of solar collectors. *Standard* 93-1991.

ASME. 1995. *Boiler and pressure vessel code.* American Society of Mechanical Engineers, New York.

Beckman, W. A., S. Klein, and J.A. Duffie. 1977. *Solar heating design by the f-Chart method.* Wiley-Interscience, New York.

DOE. 1978. SOLCOST—Solar hot water handbook: A simplified design method for sizing and costing residential and commercial solar service hot water systems, 3rd ed. DOE/CS-0042/2. U.S. Department of Energy.

Howard, B.D. 1986. Air core systems for passive and hybrid energy-conserving buildings. *ASHRAE Transactions* 92(2B):815-31.

Huggins, J.C. and D.L. Block. 1983. Thermal performance of flat plate solar collectors by generic classification. Proceedings of the ASME Solar Energy Division Fifth Annual Conference, Orlando, FL.

Johnston, S.A. 1982. Passive solar and hybrid manufactured building systems: Precast concrete applications. American Solar Energy Society, Boulder, CO.

Knowles, A.S. 1980. *A simple balancing technique for liquid cooled flat plate solar collector arrays.* International Solar Energy Society, Phoenix, AZ.

Kreider, J. 1982. *The solar heating design process—Active and passive systems.* McGraw-Hill, New York.

Kutscher, C.F. 1982. *Design approaches for industrial process heat systems.* SERI/TR-253-1356. Solar Energy Research Institute, Golden, CO.

Logee, T.L. and P.W. Kendall. 1984. Component report performance of solar collector arrays and collector controllers in the National Solar Data Network. SOLAR/0015-84/32. Vitro Corporation, Silver Spring, MD.

Newton, A.B. and S.H. Gilman. 1983. Solar collector performance manual. ASHRAE SP32.

NFPA. 1998. *National Electric Code. Standard* 70-98. National Fire Protection Association, Quincy, MA.

SEL. 1983. TRNSYS: A transient simulation program. Engineering Experiment Station. *Report* 38-12. Solar Energy Laboratory. University of Wisconsin-Madison, Madison, WI.

SERI. 1981. *Solar design workbook*. Solar Energy Research Institute, Golden, CO.

BIBLIOGRAPHY

Architectural Energy Corporation. 1991. Maintenance and operation of stand-alone photovoltaic systems, Vol. 5. Naval Facilities Engineering Command, Charleston, SC.

Davidson, J. 1990. *The new solar electric home—The photovoltaics how-to handbook*. Aatech Publications, P.O. Box 7119, Ann Arbor, MI 48107.

ICBO. 1991. *Uniform building code*. International Conference of Building Officials, Whittier, CA.

Marion, W. and S. Wilcox. 1994. Solar radiation data manual for flat-plate and concentrating collectors. NREL/TP-463-5607. 252 pp. National Renewable Energy Laboratory, Golden, CO.

Maxwell, E.L. 1990. 1961-1990 Solar radiation data base. SERI/TP-220-3632. 4 pp. Solar Energy Research Institute, Golden, CO.

NREL. 1992. User's manual: National solar radiation data base (1961-1990). 130 pp. National Renewable Energy Laboratory, Golden, CO.

Risser, V. and H. Post, eds. 1995. *Stand-alone photovoltaic systems: A handbook of recommended design practices*. SAND87-7023. Photovoltaic Design Assistance Center, Department 6218, Sandia National Laboratories, Albuquerque, NM.

Strong, S. 1991. The solar electric house: A design manual for home-scale photovoltaic poser systems. Rodale Press, Emmaus, PA.

Wiles, J. 1995. Photovoltaic systems and the *National Electric Code*—Suggested practices. Photovoltaic Design Assistance Center, Department 6218, Sandia National Laboratories, Albuquerque, NM.

CHAPTER 34

COMPRESSORS

THE COMPRESSOR is one of the four essential components of the compression refrigeration system; the others include the condenser, evaporator, and expansion device. The compressor circulates refrigerant through the system in a continuous cycle.

There are two basic types of compressors: positive displacement and dynamic. **Positive-displacement compressors** increase the pressure of refrigerant vapor by reducing the volume of the compression chamber through work applied to the compressor's mechanism: reciprocating, rotary (rolling piston, rotary vane, single-screw, and twin-screw), scroll, and trochoidal.

Dynamic compressors increase the pressure of refrigerant vapor by a continuous transfer of angular momentum from the rotating member to the vapor followed by the conversion of this momentum into a pressure rise. Centrifugal compressors function based on these principles.

This chapter describes the design features of several categories of commercially available refrigerant compressors.

POSITIVE-DISPLACEMENT COMPRESSORS

PERFORMANCE

Compressor performance is the result of design compromises involving physical limitations of the refrigerant, compressor, and motor, while attempting to provide the following:

- Greatest trouble-free life expectancy
- Most refrigeration effect for the least power input
- Lowest applied cost
- Wide range of operating conditions
- Acceptable vibration and sound level

Two useful measures of compressor performance are the coefficient of performance (COP) and the measure of power required per unit of refrigerating capacity (brake horsepower per ton of refrigeration). The COP is a dimensionless number that is the ratio of the compressor refrigerating capacity to the heat rate equivalent of the input power. The COP for a hermetic compressor includes the combined operating efficiencies of the motor and the compressor:

$$\text{COP (hermetic)} = \frac{\text{Capacity, Btu/h}}{\text{Input power to motor, Btu/h}}$$

The COP for an open compressor does not include motor efficiency:

$$\text{COP (open)} = \frac{\text{Capacity, Btu/h}}{\text{Input power to shaft, Btu/h}}$$

Power input per unit of refrigerating capacity (bhp/ton) is a measure of performance that is used to compare different compressors at the same operating conditions. It is primarily used with open-drive industrial equipment.

$$\frac{\text{bhp}}{\text{ton}} = \frac{\text{Power input to shaft, bhp}}{\text{Compressor capacity, ton}}$$

Ideal Compressor

The capacity of a compressor at a given operating condition is a function of the mass of gas compressed per unit time. Ideally, the mass flow is equal to the product of the compressor displacement per unit time and the gas density, as shown in Equation (1):

$$\omega = \rho V_d \tag{1}$$

where

ω = ideal mass flow of gas compressed, lb/h
ρ = density of gas entering compressor, lb/ft^3
V_d = geometric displacement of compressor, ft^3/h

The ideal refrigeration cycle is addressed in Chapter 1 of the 1997 *ASHRAE Handbook—Fundamentals*; the following quantities can be determined from the pressure-enthalpy diagram in Figure 8 of that chapter:

$$Q_{refrigeration\ effect} = (h_1 - h_4)$$
$$Q_{work\ of\ compression} = (h_2 - h_1)$$

Using ω, the mass flow of gas as determined by Equation (1),

$$\text{Ideal capacity} = \omega Q_{refrigeration\ effect}$$
$$\text{Ideal power input} = \omega Q_{work}$$

Actual Compressor Performance

Ideal conditions never occur, however. Actual compressor performance deviates from ideal performance because of various losses, with a resulting decrease in capacity and an increase in power input. Depending on the type of compressor, some or all of the following factors can have a major effect on compressor performance.

The preparation of this chapter is assigned to TC 8.1, Positive Displacement Compressors, and TC 8.2, Centrifugal Machines.

1. **Pressure drops within the compressor unit**
   - Through shutoff valves (suction, discharge, or both)
   - Across suction strainer/filter
   - Across motor (hermetic compressor)
   - In manifolds (suction and discharge)
   - Through valves and valve ports (suction and discharge)
   - In internal muffler
   - Through internal lubricant separator
   - Across check valves
2. **Heat gain to refrigerant** from
   - Hermetic motor
   - Lubricant pump
   - Friction
   - Heat of compression; heat exchange within compressor
3. **Valve inefficiencies** due to imperfect mechanical action
4. **Internal gas leakage**
5. **Oil circulation**
6. **Reexpansion.** The volume of gas remaining in the compression chamber after discharge, which reexpands into the compression chamber during the suction cycle and limits the mass of fresh gas that can be brought into the compression chamber.
7. **Deviation from isentropic compression.** When considering the ideal compressor, an isentropic compression cycle is assumed. In the actual compressor, the compression cycle deviates from isentropic compression primarily because of fluid and mechanical friction and heat transfer within the compression chamber. The actual compression process and the work of compression must be determined from measurements.
8. **Over- and undercompression.** In fixed volume ratio rotary, screw, and orbital compressors, overcompression occurs when pressure in the compression chamber reaches discharge pressure before reaching the discharge port. Undercompression occurs when the compression chamber reaches the discharge port prior to achieving discharge pressure.

Compressor Efficiency, Subcooling, and Heat Rejection

These deviations from ideal performance are difficult to evaluate individually. They can, however, be grouped together and considered by category. Their effect on ideal compressor performance is measured by the following efficiencies:

Volumetric efficiency (e_0) is the ratio of actual volume of gas entering the compressor to the geometric displacement of the compressor.

Compression efficiency (e_2) considers only what occurs within the compression volume and is a measure of the deviation of actual compression from isentropic compression. It is defined as the ratio of the work required for isentropic compression of the gas to the work delivered to the gas within the compression volume (as obtained by measurement).

Mechanical efficiency (e_3) is the ratio of the work delivered to the gas (as obtained by measurement) to the work input to the compressor shaft.

Isentropic (reversible adiabatic) efficiency (e_4) is the ratio of the work required for isentropic compression of the gas to work input to the compressor shaft.

Actual capacity is a function of the ideal capacity and the overall volumetric efficiency e_0 of the actual compressor:

$$\text{Actual capacity} = e_0 \omega Q_{refrigeration\ effect} \quad (2)$$

Actual shaft power is a function of the power input to the ideal compressor and the compression, mechanical, and volumetric efficiencies of the compressor, as shown in the following equation:

$$P_{bp} = \frac{P_{tp} e_0}{e_2 e_3} = \frac{P_{tp} e_0}{e_4} \quad (3)$$

where

P_{bp} = actual shaft power
P_{tp} = ideal power input = ωQ_{work}
e_0 = volumetric efficiency
e_2 = compression efficiency
e_3 = mechanical efficiency
e_4 = isentropic (reversible adiabatic) efficiency

Liquid subcooling is not accomplished by the compressor. However, the effect of liquid subcooling is included in the ratings for compressors by most manufacturers.

Total heat rejection is the sum of the refrigeration effect and the heat equivalent of the power input to the compressor. The quantity of heat rejection must be known in order to size condensers.

PROTECTIVE DEVICES

Compressors are provided with one or more of the following devices for protection against abnormal conditions and to comply with various codes.

1. **High-pressure protection** as required by Underwriters Laboratories and per ARI standards and ASHRAE *Standard* 15. This may include the following:
   (a) A high-pressure cutout.
   (b) A high- to low-side internal relief valve, external relief valve, or rupture member to comply with ASHRAE *Standard* 15. The differential pressure setting depends on the refrigerant used and the operating conditions. Care must be taken to ensure that the relief valve will not accidentally blow on a fast pull-down. Many welded hermetic compressors have an internal high- to low-pressure relief valve to limit maximum pressure in units not equipped with other high-pressure control devices.
   (c) A relief valve assembly on the oil separator of a screw compressor unit.
2. **High-temperature control** devices to protect against overheating and oil breakdown.
   (a) Motor overtemperature protective devices are addressed in the section on Integral Thermal Protection in Chapter 40.
   (b) To protect against lubricant and refrigerant breakdown, a temperature sensor is sometimes used to stop the compressor when discharge temperature exceeds safe values. The switch may be placed internally (near the compression chamber) or externally (on the discharge line).
   (c) On larger compressors, lubricant temperature is controlled by cooling with a heat exchanger or direct liquid injection, or the compressor may shut down on high lubricant temperature.
   (d) Where lubricant sump heaters are used to maintain a minimum lubricant sump temperature, a thermostat may be used to limit the maximum lubricant temperature.
3. **Low-pressure protection** may be provided for
   (a) *Suction gas.* Many compressors or systems are limited to a minimum suction pressure by a protective switch. Motor cooling, freeze-up, or pressure ratio usually determine the pressure setting.
   (b) *Compressor.* Lubricant pressure protectors are used with forced feed lubrication systems to prevent the compressor from operating with insufficient lubricant pressure.

4. **Time delay or lockouts with manual resets** prevent damage to both compressor motor and contactors from repetitive rapid-starting cycles.
5. **Low voltage and phase loss or reversal protection** is used on some systems. Phase reversal protection is used with multi-phase devices to ensure the proper direction of rotation.
6. **Suction line strainer.** Most compressors are provided with a strainer at the suction inlet to remove any dirt that might exist in the suction line piping. Factory-assembled units with all parts cleaned at the time of assembly may not require the suction line strainer. A suction line strainer is normally required in all field-assembled systems.

Liquid Hazard

A gas compressor is not designed to handle liquid. The damage that may occur depends on the quantity of liquid, the frequency with which it occurs, and the type of compressor. Slugging, floodback, and flooded starts are three ways in which liquid can damage a compressor. Generally, reciprocating compressors, because they are equipped with discharge valves, are more susceptible to damage from slugging than various rotary and orbital compressors.

Slugging is the short-term pumping of a large quantity of liquid refrigerant and/or lubricant. It can occur just after start-up if refrigerant accumulated in the evaporator during shutdown returns to the compressor. It can also occur when system operating conditions change radically, such as during a defrost cycle. Slugging can also occur with quick changes on compressor loading.

Floodback is the continuous return of liquid refrigerant mixed with the suction gas. It is a hazard to compressors that depend on the maintenance of a certain amount of lubricant in the compression chamber. A properly sized suction accumulator can be used for protection.

Flooded start occurs when refrigerant is allowed to migrate to the compressor during shutdown. Compressors can be protected with crankcase heaters and automatic pumpdown cycles, where applicable.

Sound Level

An acceptable sound level is a basic requirement of good design and application. Chapter 46 of the 1999 *ASHRAE Handbook—Applications* covers design criteria in more detail.

Vibration

Compressor vibration results from gas-pressure pulses and inertia forces associated with the moving parts. The problems of vibration can be handled in the following ways.

Isolation. With this common method, the compressor is resiliently mounted in the unit by springs, synthetic rubber mounts, etc. In hermetic compressors, the internal compressor assembly is usually spring-mounted within the welded shell, and the entire unit is externally isolated.

Reduction of Amplitude. The amount of movement can be reduced by adding mass to the compressor. Mass is added either by rigidly attaching the compressor to a base, condenser, or chiller, or by providing a solid foundation. When structural transmission is a problem, particularly with large machines, the entire assembly is then resiliently mounted.

Chapter 46 of the 1999 *ASHRAE Handbook—Applications* has further information.

Shock

In designing for shock, three types of dynamic loads are recognized:

- Suddenly applied loads of short duration
- Suddenly applied loads of long duration
- Sustained periodic varying loads

Since the forces are primarily inertial, the basic approach is to maintain low equipment mass and make the strength of the carrying structure as great as possible. The degree to which this practice is followed is a function of the amount of shock loading.

Commercial Units. The major shock loading to these units occurs during shipment or when they operate on commercial carriers. Train service provides a severe test because of low forcing frequencies and high shock load. Shock loads as high as 10*g* have been recorded; 5*g* can be expected.

Trucking service results in higher forcing frequencies, but shock loads can be equal to, or greater than, those for rail transportation. Aircraft service forcing frequencies generally fall in the range of 20 to 60 Hz with shocks to 3*g*.

Military Units. The requirements are given in detail in specifications that exceed anything expected of commercial units. In severe applications, deformation of the supporting members and shock isolators may be tolerated, provided that the unit performs its function.

Basically, the compressor must be made of components rigid enough to avoid misalignment or deformation during shock loading. Therefore, structures with low natural frequencies should be avoided.

Testing

Testing for ratings must be in accordance with ASHRAE *Standard* 23. Compressor tests are of two types: the first determines capacity, efficiency, sound level, motor temperatures, etc.; the second predicts the reliability of the machine.

Standard rating conditions, which are usually specified by compressor manufacturers, include the maximum possible compression ratio, the maximum operating pressure differential, the maximum permissible discharge pressure, and the maximum inlet and discharge temperature.

Lubrication requirements, which are prescribed by the compressor manufacturers, include the type, viscosity, and other characteristics of the lubricants suitable for use with the many different operating levels and the specific refrigerant being used.

Power requirements for compressor starting, pull-down, and operation vary because unloading means differ in the many styles of compressors available. Manufacturers supply full information covering the various methods employed.

MOTORS

Motors for positive-displacement compressors range from fractional horsepower to several thousand horsepower. When selecting a motor for driving a compressor, the following factors should be considered:

- Horsepower and rotational speed
- Voltage and phase
- Starting and pull-up torques
- Ambient and maximum rise temperatures
- Cost and availability
- Insulation
- Efficiency and performance
- Starting currents
- Type of protection required
- Multispeed or variable speed
- Location (high-pressure side versus low-pressure side)

Large, industrial, open compressors can be driven from 800 to 3600 rpm with three-phase, 200 to 4160 V motors. Although all motors can be started across-the-line, local utilities, local codes, or specification by the end user may require that motors be started at reduced power levels. These typically include part-winding, wye-delta, double-delta, autotransformer, and solid-state starting methods all designed to limit inrush starting current. Care must be taken that the starting method will supply enough torque to accelerate the motor and overcome the torque required for compression.

Hermetic motors can be more highly loaded than comparable open motors because of the refrigerant lubricant cooling used.

For effective hermetic motor application, the maximum design load should be as close as possible to the breakdown torque at the lowest voltage used. This approach yields a motor design that operates better at lighter loads and higher voltage. The limiting factor at high loads is normally the motor temperature, while at light loads the limiting factor is the discharge gas temperature. Overdesign of the motor may increase discharge gas temperature at light loads. Consideration must be given to low-side versus high-side motor location.

The single-phase motor presents more design problems than the polyphase, because the relationship between main and auxiliary windings becomes critical together with the necessary starting equipment.

The locked-rotor rate of temperature rise must be kept low enough to prevent excessive motor temperature with the motor protection available. The maximum temperature under these conditions should be held within the limits of the materials used. With better protection (and improved materials), a higher rate of rise can be tolerated, and a less expensive motor can be used.

The materials selected for these motors must have high dielectric strength, resist fluid and mechanical abrasion, and be compatible with an atmosphere of refrigerant and lubricant.

The types of hermetic motors commonly selected for various applications are as follows:

Refrigeration compressors—Single-phase
- Low to medium torque—Split-phase or PSC (permanent split-capacitor)
- High torque—CSCR (capacitor-start/capacitor-run) and CSIR (capacitor-start/induction-run)

Room air conditioner compressors—Single-phase
- PSC or CSCR
- 2 speed, pole switching variable speed

Central air conditioning and commercial refrigeration
- Single-phase, PSC and CSCR to 6 hp
- 3 phase, 2 hp and above, across-the-line start
- 10 hp and above; part winding; wye-delta, double-delta, and across-the-line start
- 2 speed, pole switching variable speed
- Electronically commutated dc motors (ECMs)

For further information on motors and motor protection, see Chapter 40.

RECIPROCATING COMPRESSORS

Most reciprocating compressors are single-acting, using pistons that are driven directly through a pin and connecting rod from the crankshaft. Double-acting compressors that use piston rods, crossheads, stuffing boxes, and oil injection are not used extensively and, therefore, are not covered here.

Single-stage compressors are primarily used for medium temperatures (0 to 30°F) and in air-conditioning applications but can achieve temperatures below –30°F at 95°F condensing temperatures with suitable refrigerants. Chapters 2 and 3 of the 1998 *ASHRAE Handbook—Refrigeration* have information on other halocarbon and ammonia systems.

Booster compressors are typically used for low-temperature applications with R-22 or ammonia. Minus 85°F saturated suction can be achieved by using R-22, and –65°F saturated suction is possible using ammonia.

The booster raises the refrigerant pressure to a level where further compression can be achieved with a high-stage compressor, without exceeding the compression-ratio limits of the respective machines.

Since superheat is generated as a result of compression in the booster, intercooling is normally required to reduce the refrigerant stream temperature to the practical level required at the inlet to the high-stage unit. Intercooling methods include controlled liquid injection into the intermediate stream, gas bubbling through a liquid reservoir, and use of a liquid-to-gas heat exchanger where no fluid mixing occurs.

Integral two-stage compressors achieve low temperatures (–20 to –80°F), using R-22 or ammonia within the frame of a single compressor. The cylinders within the compressor are divided into respective groups so that the combination of volumetric flow and pressure ratios are balanced to achieve booster and high-stage performance effectively. Refrigerant connections between the high-pressure suction and low-pressure discharge stages allow an inter-stage gas cooling system to be connected to remove superheat between stages. This interconnection is similar to the methods used for individual high-stage and booster compressors.

Capacity reduction is typically achieved by cylinder unloading, as in the case of single-stage compressors. Special consideration must be given to maintaining the correct relationship between high- and low-pressure stages.

The most widely used compressor is the halocarbon compressor, which is manufactured in three types of design: (1) open, (2) semi-hermetic or bolted hermetic, and (3) welded-shell hermetic.

Ammonia compressors are manufactured only in the open design because of the incompatibility of the refrigerant and hermetic motor materials.

Open-type compressors are those in which the shaft extends through a seal in the crankcase for an external drive.

Hermetic compressors are those in which the motor and compressor are contained within the same housing, with the motor shaft integral with the compressor crankshaft and the motor in contact with the refrigerant.

A **semihermetic compressor** (bolted, accessible, or serviceable) is a hermetic compressor of bolted construction amenable to field repair.

In **welded-shell hermetic compressors** (sealed) the motor-compressor is mounted inside a steel shell, which, in turn is sealed by welding.

Combinations of design features used are shown in Table 1. Typical performance values for halocarbon compressors are given in Table 2.

Performance Data

Figure 1 presents a typical set of capacity and power curves for a four-cylinder semihermetic compressor, 2.38 in. bore, 1.75 in. stroke, 1740 rpm, operating with Refrigerant 22. Figure 2 shows the heat rejection curves for the same compressor. Compressor curves should contain the following information:

- Compressor identification
- Degrees of subcooling and correction factors for zero or other subcool temperatures
- Degrees of superheat
- Compressor speed
- Refrigerant
- Suction gas superheat and correction factors
- Compressor ambient
- External cooling requirements (if any)
- Maximum power or maximum operating conditions
- Minimum operating conditions at fully loaded and fully unloaded operation

Table 1 Typical Design Features of Reciprocating Compressors

Item	Halocarbon Compressor			Ammonia Compressor
	Open	Semi-hermetic	Welded Hermetic	Open
1. Number of cylinders—one to:	16	12	6	16
2. Power range	0.17 hp and up	0.5 to 150 hp	0.17 to 25 hp	10 hp and up
3. Cylinder arrangement				
a. Vertical, V or W, radial	X	X		
b. Radial, horizontal opposed			X	
c. Horizontal, vertical V or W		X		X
4. Drive				
a. Hermetic compressors, electric motor		X	X	
b. Open compressors—direct drive, V belt chain, gear, by electric motor or engine	X			X
5. Lubrication—splash or force feed, flooded	X	X	X	X
6. Suction and discharge valves—ring plate or ring or reed flexing	X	X	X	X
7. Suction and discharge valve arrangement				
a. Suction and discharge valves in head	X	X	X	X
b. Uniflow—suction valves in top of piston, suction gas entering through cylinder walls; discharge valves in head	X			X
8. Cylinder cooling				
a. Suction gas cooled	X	X	X	X
b. Water jacket cylinder wall, head, or cylinder wall and head	X			X
c. Air cooled	X	X	X	X
d. Refrigerant cooled heads	X			X
9. Cylinder head				
a. Spring loaded	X	X	X	X
b. Bolted head	X	X	X	X

Item	Halocarbon Compressor			Ammonia Compressor
	Open	Semi-hermetic	Welded Hermetic	Open
10. Bearings				
a. Sleeve, antifriction	X	X	X	X
b. Tapered roller	X			X
11. Capacity control, if provided—manual or automatic				
a. Suction valve lifting	X	X	X	X
b. Bypass-cylinder heads to suction	X	X	X	X
c. Closing inlet	X	X		X
d. Adjustable clearance	X	X		X
e. Variable speed	X	X	X	X
12. Materials				
Motor insulations and rubber materials must be compatible with refrigerant and lubricant mixtures; otherwise, no restrictions		X	X	
No copper or brass				X
13. Lubricant return				
a. Crankcase separated from suction manifolds, oil return check valves, equalizers, spinners, foam breakers	X	X		X
b. Crankcase common with suction manifold			X	
14. Synchronous speeds (50 to 60 Hz)	250 to 3600	1500 to 3600	1500 to 3600	250 to 1500
15. Pistons				
a. Aluminum or cast iron	X	X	X	X
b. Ringless	X	X	X	X
c. Compression and oil control rings	X	X	X	X
16. Connecting rod				
Split rod with removable cap or solid eccentric strap	X	X	X	X
17. Mounting				
Internal spring mount		X	X	
External spring mount		X	X	
Rigidly mounted on base	X	X		X

Table 2 Typical Performance Values for Reciprocating Compressors

Compressor Size and Type	Operating Conditions and Refrigerants			
	R-404a	R-134a	R-22	R-22
	Evap. Temp. = –40°F Cond. Temp. = 105°F Suction Gas = 65°F Subcooling = 0°F	Evap. Temp. = 0°F Cond. Temp. = 110°F Suction Gas = 65°F Subcooling = 0°F	Evap. Temp. = 40°F Cond. Temp. = 105°F Suction Gas = 55°F Subcooling = 0°F	Evap. Temp. = 45°F Cond. Temp. = 130°F Suction Gas = 65°F Subcooling = 0°F
Large, over 25 hp				
Open	0.21 tons/hp	0.40 tons/hp	1.05 tons/hp	1.07 tons/hp
Hermetic	3.15 Btu/h per W	6.00 Btu/h per W	14.2 Btu/h per W	10.4 Btu/h per W
Medium, 5 to 25 hp				
Open	0.19 tons/hp	0.37 tons/hp	1.00 ton/hp	1.00 tons/hp
Hermetic	2.89 Btu/h per W	5.60 Btu/h per W	14.0 Btu/h per W	10.2 Btu/h per W
Small, under 5 hp				
Open	—	—	—	—
Hermetic	—	3.80 Btu/h per W	13.8 Btu/h per W	10.0 Btu/h per W

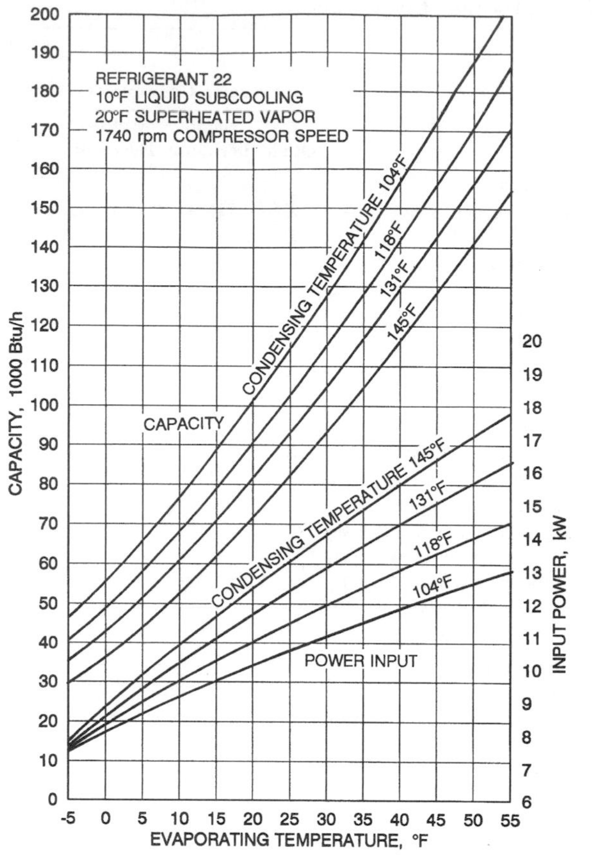

Fig. 1 Capacity and Power-Input Curves for Typical Hermetic Reciprocating Compressor

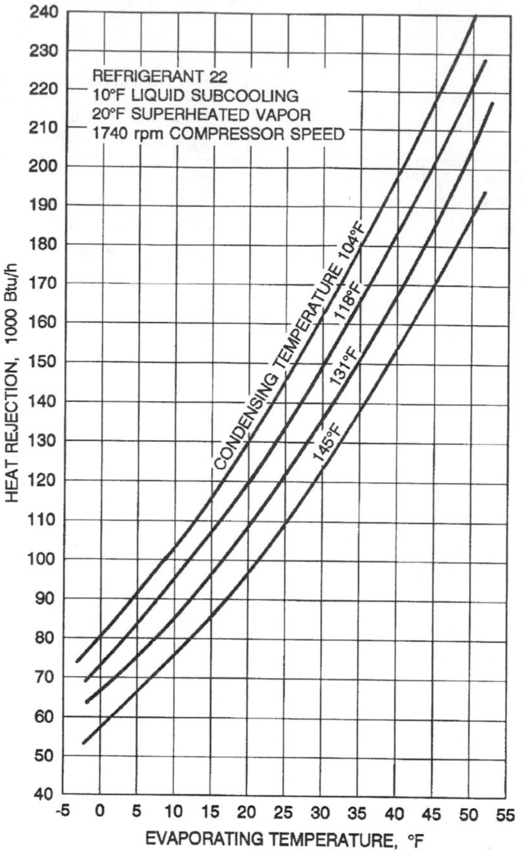

Fig. 2 Heat Rejection Curves for Typical Hermetic Reciprocating Compressor

Motor Performance

The motor efficiency is usually the result of a compromise between cost and size. Generally, the physically larger a motor is for a given rating, the more efficient it can be made. Accepted efficiencies range from approximately 88% for a 3 hp motor to 95% for a 100 hp motor. Uneven loading has a marked effect on motor efficiency. It is important that cylinders be spaced evenly. Also, the more cylinders there are, the smaller the impulses become. Greater moments of inertia of moving parts and higher speeds reduce the impulse effect. Small and evenly spaced impulses also help reduce noise and vibration.

Since many compressors start against load, it is desirable to estimate starting torque. The following equation is for a single cylinder compressor. It neglects friction, the additional torque required to force discharge gas out of the cylinder, and the fact that the tangential force at the crankpin is not always equal to the normal force at the piston. This equation also assumes considerable gas leakage at the discharge valves but little or no leakage past the piston rings or suction valves. It yields a conservative estimate.

$$T_s = \frac{(p_2 - p_1)As}{2N_2/N_1} \qquad (4)$$

where

T_s = starting torque, $lb_f \cdot in.$
p_2 = discharge pressure, psi
p_1 = suction pressure, psi
A = area of cylinder, in^2
s = stroke of compressor, in.
N_2 = motor speed, rpm
N_1 = compressor speed, rpm

Equation (4) shows that when pressures are balanced or almost equal ($p_2 = p_1$), torque requirements are considerably reduced. Thus, a pressure-balancing device on an expansion valve or a capillary tube that equalizes pressures at shutdown allows the compressor to be started without excessive effort. For multicylinder compressors, an analysis must be made of both the number of cylinders that might be on a compression stroke and the position of the rods at start. Since the force needed to push the piston to the top dead center is a function of how far the rod is from the cylinder centerline, the worst possible angles these might assume can be graphically determined by torque diagrams. The torques for some arrangements are shown in the following table.

No. Cylinders	Arrangement of Crank Throws	Angle Between Cylinders	Approx. Torque from Equation (4)
1	Single		T_s
2	Single	90°	$1.025T_s$
2	180° apart	0° or 180°	T_s
3	Single	60°	$1.225T_s$
3	120° apart	120°	T_s
4	180°, 2 rods/crank	90°	$1.025T_s$
6	180°, 3 rods/crank	60°	$1.23T_s$

Pull-up torque is an important characteristic of motor starting strength because it represents the lowest torque capability of the motor and occurs between 25 and 75% of the operating speed. The pull-up torque of the motor must exceed the torque requirement of the compressor or the motor will cease to accelerate and it will trip the safety overload protection device.

Features

Crankcases. The crankcase, or in a welded hermetic compressor, the cylinder block, is usually of cast iron. Aluminum is also used, particularly in small open and welded hermetic compressors. Open and semihermetic crankcases enclose the running gear, oil sump, and, in the latter case, the hermetic motor. Access openings with removable covers are provided for assembly and service purposes. Welded hermetic cylinder blocks are often just skeletons, consisting of the cylinders, the main bearings, and either a barrel into which the hermetic motor stator is inserted or a surface to which the stator can be bolted.

The cylinders can be integral with the crankcase or cylinder block, in which case a material that provides a good sealing surface and resists wear must be provided. In aluminum crankcases, cast-in liners of iron or steel are usual. In large compressors, premachined cylinder sleeves inserted in the crankcase are common. With halocarbon refrigerants, excessive cylinder wear or scoring is not much of a problem and the choice of integral cylinders or inserted sleeves is often based on manufacturing considerations.

Crankshafts. Crankshafts are made of either forged steel with hardened bearing surfaces finished to 8 microinches or iron castings. Grade 25 to 40 (25,000 to 40,000 psi) tensile gray iron can be used where the lower modulus of elasticity can be tolerated. Nodular iron shafts approach the stiffness, strength, and ductility of steel and should be polished in both directions of rotation to 16 microinches maximum for best results. Crankshafts often include counterweights and should be dynamically and/or statically balanced.

While a safe maximum stress is important in shaft design, it is equally important to prevent excessive deflection that can edge-load bearings to failure. In hermetics, deflection can permit motor air gap to become eccentric, which affects starting, reduces efficiency, produces noise, and further increases bearing edge-loading.

Generally, the harder the bearing material used, the harder the shaft. With bronze bearings, a journal hardness of 350 Brinell is usual, while unhardened shafts at 200 Brinell in babbitt bearings are typical. Many combinations of materials and hardnesses have been used successfully.

Main Bearings. Both the crank and drive means may be overhung with the bearings between; however, usual practice places the cylinders between the main bearings and, in a hermetic, overhangs the motor. Main bearings are made of steel-backed babbitt, steel-backed or solid bronze, or aluminum. In an aluminum crankcase, the bearings are usually integral. By automotive standards, unit loadings are low. The oil-refrigerant mixture frequently provides only marginal lubrication, but 8000 h/year operation in commercial refrigeration service is quite possible. For conventional shaft diameters and speeds, 600 psi main bearing loading based on projected area is not unusual. Running clearances average 0.001 in/in of diameter with steel-backed babbitt bearings and a steel or iron shaft. Bearing oil grooves placed in the unloaded area are usual. Feeding oil to the bearing is only one requirement; another is the venting of evolved refrigerant gas and lubricant escape from the bearing to carry away heat.

In most compressors, crankshaft thrust surfaces (with or without thrust washers) must be provided in addition to main bearings. Thrust washers may be steel-backed babbitt, bronze, aluminum, hardened steel, or polymer and are usually stationary. Oil grooves are often included in the thrust face.

Connecting Rods and Eccentric Straps. Connecting rods have the large end split and a bolted cap for assembly. Unsplit eccentric straps require the crankshaft to be passed through the big bore at assembly. Rods or straps are of steel, aluminum, bronze, nodular iron, or gray iron. Steel or iron rods often require inserts of such bearing material as steel-backed babbitt or bronze, while aluminum and bronze rods can bear directly on the crankpin and piston pin. Refrigerant compressor service limits unit loading to 3000 psi based on projected area with a bronze bushing in the rod small bore and a hardened steel piston pin. An aluminum rod load at the piston pin of 2000 psi has been used. Large end unit loads are usually under 1000 psi.

The Scotch yoke type of piston-rod assembly has also been used. In small compressors, it has been fabricated by hydrogen brazing steel components. Machined aluminum components have been used in large hermetic designs.

Piston, Piston Ring, and Piston Pin. Pistons are usually made of cast iron or aluminum. Cast-iron pistons with a running clearance of 0.0004 in/in of diameter in the cylinder will seal adequately without piston rings. With aluminum pistons, rings are required because a running clearance in the cylinder of 0.002 in/in or more of diameter may be necessary, as determined by tests at extreme conditions. A second or third compression ring may add to power consumption with little increase in capacity; however, it may help oil control, particularly if drained. Oil scraping rings with vented grooves may also be used. Cylinder finishes are usually obtained by honing, and a 12 to 40 microinch range will give good ring seating. An effective oil scraper can often be obtained with a sharp corner on the piston skirt.

The minimum piston length is determined by the side thrust and is also a function of running clearance. Where clearance is large, pistons should be longer to prevent slap. An aluminum piston (with ring) having a length equal to 0.75 times the diameter, with a running clearance of 0.002 in/in of diameter, and a rod length to crank arm ratio of 4.5, has been used successfully.

Piston pins are steel, case-hardened to Rockwell C 50 to 60 and ground to an 8 microinch finish or better. Pins can be restrained against rotation in either the piston bosses or the rod small end, be free in both, or be full-floating, which is usually the case with aluminum pistons and rods. Retaining rings prevent the pin from moving endwise and abrading the cylinder wall.

There is no well-defined limit to piston speed; an average velocity of 1200 fpm, determined by multiplying twice the stroke in feet by the rpm, has been used successfully.

Suction and Discharge Valves. The most important components in the reciprocating compressor are the suction and discharge valves. Successful designs provide long life and low pressure loss. The life of a properly made and correctly applied valve is determined by the motion and stress it undergoes in performing its function. Excessive pressure loss across the valve results from high gas velocities, poor mechanical action, or both.

For design purposes, gas velocity is defined as being equal to the bore area multiplied by the average piston speed and divided by the valve area. Permissible gas velocity through the restricted areas of the valve is left to the discretion of the designer and depends on the level of volumetric efficiency and performance desired. In general, designs with velocities up to 12,000 fpm with ammonia and up to 9000 fpm with R-22 have been successful.

A valve should meet the following requirements:

- Large flow areas with shortest possible path
- Straight gas flow path, minimum directional changes
- Low valve mass combined with low lift for quick action
- Symmetry of design with minimum pressure imbalance
- Minimum clearance volume
- Durability
- Low cost
- Tight sealing at ports
- Minimum valve flutter

Most valves in use today fall in one of the following groups:

1. **Free-floating reed valve**, with backing to limit movement, seats against a flat surface with circular or elongated ports. It is simple, and stresses can be readily determined, but it is limited to relatively small ports; therefore, multiples are often used. Totally

backed with a curved stop, it is a valve that can stand considerable abuse.

2. **Reed, clamped at one end**, with full backstop support or a stop at the tip to limit movement, has a more complex motion than a free-floating reed; the resulting stresses are far greater than those calculated from the curvature of the stop. Considerable care must be taken in the design to ensure reliability.
3. **Ring valve** usually has a spring return. A free-floating ring is seldom used because of its high leakage loss. Improved performance is obtained by using spring return, in the form of coil springs or flexing backup springs, with each valve. Ring-type valves are particularly adaptable to compressors using cylinder sleeves.
4. **Valve formed as a ring** has part of the valve structure clamped. Generally, full rings are used with one or more sets of slots arranged in circles. By clamping the center, alignment is ensured and a force is obtained that closes the valve. To limit stresses, the valve proportions, valve stops, and supports are designed to control and limit valve motion.

Lubrication. Lubrication systems range from a simple splash system to the elaborate forced-feed systems with filters, vents, and equalizers. The type of lubrication required depends largely on bearing load and application.

For low to medium bearing loads and factory-assembled systems where cleanliness can be controlled, the splash system gives excellent service. Bearing clearances must be larger, however; otherwise, oil does not enter the bearing readily. Thus, the splashing effect of the dippers in the oil and the freer bearings cause the compressor to operate somewhat noisily. Furthermore, the splash at high speed encourages frothing and oil pumping; this is not a problem in package equipment but may be in remote systems where gas lines are long.

A **flooded system** includes disks, screws, grooves, oil-ring gears, or other devices that lift the oil to the shaft or bearing level. These devices flood the bearing and are not much better than splash systems, except that the oil is not agitated as violently, so that quieter operation results. Since little or no pressure is developed by this method, it is not considered forced feed.

In **forced-feed lubrication**, a pump gear, vane, or plunger develops pressure, which forces oil into the bearing. Smaller bearing clearances can be used because adequate pressure feeds oil in sufficient quantity for proper bearing cooling. As a result, the compressor may be quieter in operation.

Gear pumps are used to a large extent. Spur gears are simple but tend to promote flashing of the refrigerant dissolved in the oil because of the sudden opening of the tooth volume as two teeth disengage. This disadvantage is not apparent in internal-type eccentric gear or vane pumps where a gradual opening of the suction volume takes place. The eccentric gear pump, the vane pump, or the piston pump therefore give better performance than simple gear pumps when the pump is not submerged in the oil.

Oil pumps must be made with minimum clearances to pump a mixture of gas and oil. The discharge of the pump should have provision to bleed a small quantity of oil into the crankcase. A bleed vents the pumps, prevents excess pressure, and ensures faster priming.

A strainer should be inserted in the suction line to keep foreign substances from the pump and bearings. If large quantities of fine particles are present and bearing load is high, it may be necessary to add an oil filter to the discharge side of the pump.

Oil must return from the suction gas into the compressor crankcase. A flow of gas from piston leakage opposes this oil flow, so the velocity of the leakage gas must be low to permit oil to separate from the gas. A separating chamber may be built as part of the compressor to help separate oil from the gas.

In many designs, a check valve is inserted at the bottom of the oil return port to prevent a surge of crankcase oil from entering the suction. This check valve must have a bypass, which is always open, to permit the check valve to open wide after the oil surge has passed. When a separating chamber is used, the oil surge is trapped before it can enter the suction port, thus making a check valve less essential.

Seals. Stationary and rotary seals have been used extensively on open-type reciprocating compressors. Older stationary seals usually used metallic bellows and a hardened shaft for a wearing surface. Their use has diminished because of high cost.

The rotary seal costs less and is more reliable. A synthetic seal tightly fitted to the shaft prevents leakage and seals against the back face of the stationary member of the seal. The front face of this carbon nose seals against a stationary cover plate. This design has been used on shafts up to 4 in. in diameter. The rotary seal should be designed so that the carbon nose is never subjected to the full thrust of the shaft; the spring should be designed for minimum cocking force; and materials should be such that a minimum of swelling and shrinking is encountered.

Special Devices

Capacity Control. An ideal capacity control system would have the following operating characteristics (not all of these benefits can occur simultaneously):

- Continuous adjustment to load
- Full-load efficiency unaffected by the control
- No loss in efficiency at part load
- Reduction of starting torque
- No reduction in compressor reliability
- No reduction of compressor operating range
- No increase in compressor vibration and sound level at part load

Capacity control may be obtained by (1) controlling suction pressure by throttling; (2) controlling discharge pressure; (3) returning discharge gas to suction; (4) adding reexpansion volume; (5) changing the stroke; (6) opening a cylinder discharge port to suction while closing the port to discharge manifold; (7) changing compressor speed; (8) closing off cylinder inlet, and (9) holding the suction valve open.

The most commonly used methods are opening the suction valves by some external force, gas bypassing within the compressor, and gas bypassing outside the compressor.

When capacity control compressors are used, system design becomes more important and the following must be considered:

- Possible increase in compressor vibration and sound level at unloaded conditions
- Minimum operating conditions as limited by discharge or motor temperatures (or both) at part-load conditions
- Good oil return at minimum operating conditions when fully unloaded
- Rapid cycling of unloaders
- Refrigerant feed device capable of controlling at minimum capacity

Crankcase Heaters. During shutdown, refrigerants tend to migrate to the coldest part of the refrigeration system. In cold weather, the compressor oil sump could be the coldest area. When the refrigerant charge is large enough to dilute the oil excessively and cause flooded starts, a crankcase heater should be used. The heater should maintain the oil at least 20°F above the rest of the system at shutdown and well below the breakdown temperature of the oil at any time.

Internal Centrifugal Separators. Some compressors are equipped with antislug devices in the gas path to the cylinders. This device centrifugally separates oil and liquid refrigerant from the flow of foam during a flooded start and thus protects the cylinders. It does not eliminate the other hazards caused by liquid refrigerant in gas compressors.

Application

To operate through the entire range of conditions for which the compressor was designed and to obtain the desired service life, it is important that the mating components in the system be correctly designed and selected. Suction superheat must be controlled, lubricant must return to the compressor, and adequate protection must be provided against abnormal conditions.

Chapters 1 through 4 of the 1998 *ASHRAE Handbook—Refrigeration* cover design. Chapter 6 of that volume gives details of cleanup in the event of a hermetic motor burnout.

Suction Superheat. No liquid refrigerant should be present in the suction gas entering the compressor because it causes oil dilution and gas formation in the lubrication system. If the liquid carryover is severe enough to reach the cylinders, excessive wear of valves, stops, pistons, and rings can occur; liquid slugging can break valves, pistons, and connecting rods.

Suction gas without superheat is not harmful to the compressor as long as no liquid refrigerant is entrained; some systems are even designed to operate this way, although suction superheat is intended in the design of most systems. Measuring suction superheat can be difficult, and the indication of a small amount does not necessarily mean that liquid is not present. An effective suction separator may be necessary to remove all liquid.

High suction superheat may result in dangerously high discharge temperatures and, in hermetics, high motor temperatures.

Automatic Oil Separators. Oil separators are used most often to reduce the amount of oil discharged into the system by the compressor and to return oil to the crankcase. They are recommended for all field-erected systems and on packaged equipment where lubricant contamination will have a negative effect on evaporator capacity and/or where lubricant return at reduced capacity is marginal.

Parallel Operation. Where multiple compressors are used, the trend is toward completely independent refrigerant circuits. This has an obvious advantage in the case of a hermetic motor burnout and with lubricant equalization.

Parallel operation of compressors in a single system has some operational advantage at part load. Careful attention must be given to apportioning returned oil to the multiple compressors so that each always has an adequate quantity. Figure 3 shows the method most widely used. Line A connects the tops of the crankcases and tends to equalize the pressure above the oil, while line B permits oil equalization at the normal level. Lines of generous size must be used. Generally, line A is a large diameter, while line B is a small diameter, which limits the possible blowing of oil from one crankcase to the other.

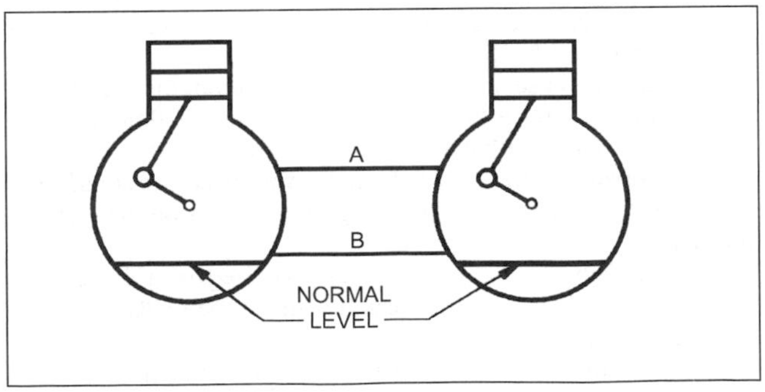

Fig. 3 Modified Oil-Equalizing System

A central reservoir for returned oil may also be used with means (such as crankcase float valves) for maintaining the proper levels in the various compressors. With staged systems, the low-stage compressor oil pump can sometimes deliver a measured amount of oil to the high-stage crankcase. The high-stage oil return is then sized and located to return a slightly greater quantity of oil to the low-stage crankcase. Where compressors are at different elevations and/or staged, the use of pumps in each oil line is necessary to maintain adequate crankcase oil level. In both cases, proper gas equalization must be provided.

ROTARY COMPRESSORS

ROLLING PISTON COMPRESSORS

Rolling piston, or fixed vane, rotary compressors are used in household refrigerators and air-conditioning units in sizes up to about 3 hp (Figure 4). This type of compressor uses a roller mounted on the eccentric of a shaft with a single vane or blade suitably positioned in the nonrotating cylindrical housing, generally called the cylinder block. The blade reciprocates in a slot machined in the cylinder block. This reciprocating motion is caused by the eccentrically moving roller.

Displacement for this compressor can be calculated from

$$V_d = \pi H(A^2 - B^2)/4 \qquad (5)$$

where

V_d = displacement
H = cylinder block height
A = cylinder diameter
B = roller diameter

The drive motor stator and compressor are solidly mounted in the compressor housing. This design feature is possible due to low vibration associated with the rotary compression process, as opposed to reciprocating designs which employ spring isolation between the compressor parts and the housing.

Suction gas is directly piped into the suction port of the compressor, and the compressed gas is discharged into the compressor housing shell. This high-side shell design is used because of the simplicity of its lubrication system and the absence of oiling and compressor cooling problems. Compressor performance is also improved because this arrangement minimizes heat transfer to the suction gas and reduces gas leakage area.

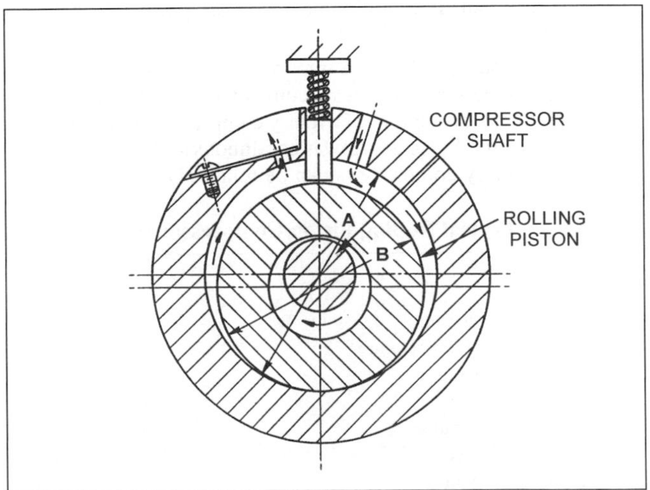

Fig. 4 Fixed Vane, Rolling Piston Rotary Compressor

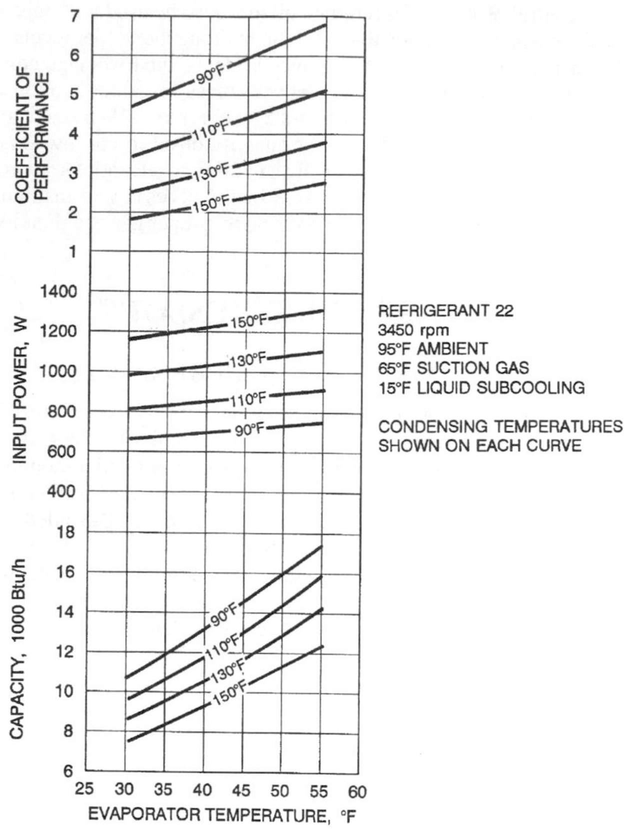

Fig. 5 Performance Curves for Typical Rolling Piston Compressor

Internal leakage is controlled through hydrodynamic sealing and selection of mating parts for optimum clearance. Hydrodynamic sealing depends on clearance, surface speed, surface finish, and oil viscosity. Close tolerance and low surface finish machining is necessary to support hydrodynamic sealing and to reduce gas leakage.

Performance

Rotary compressors have a high volumetric efficiency because of the small clearance volume and correspondingly low reexpansion losses inherent in their design. Figure 5 and Table 3 show performance of a typical rolling piston rotary compressor, commercially available for room air-conditioning and small, packaged heat pump applications.

An acceptable sound level is important in the design of any small compressor. Figure 6 illustrates a convenient method of analyzing and evaluating sound output of a rotary compressor designed for the home refrigerator. Since gas flow is continuous and no suction valve is required, rotary compressors can be relatively quiet. The sound level for a rotary compressor of the same design is directly related to its power input.

Table 3 Typical Rolling Piston Compressor Performance

Compressor speed	3450 rpm
Refrigerant	R-22
Condensing temperature	130°F
Liquid refrigerant temperature	115°F
Evaporator temperature	45°F
Suction pressure	90.7 psia
Suction gas temperature	65°F
Evaporator capacity	12,000 Btu/h
Energy efficiency ratio	11.0 Btu/W·h
Coefficient of performance	3.22
Input power	1090 W

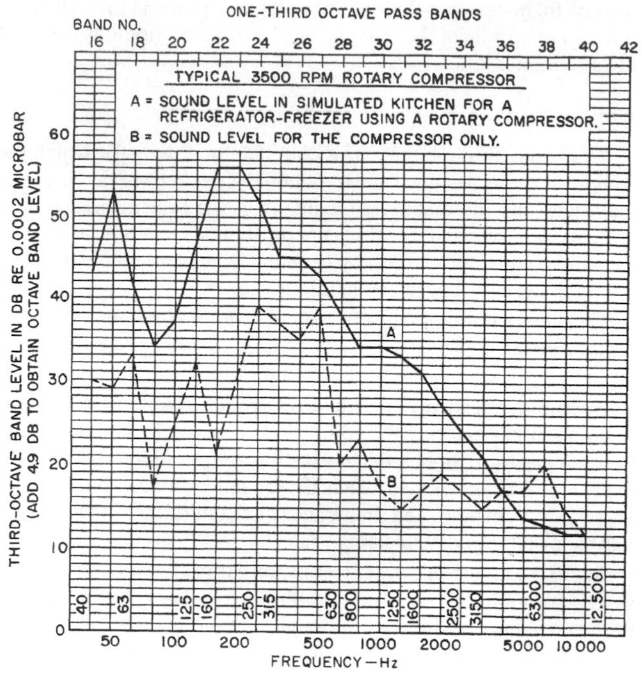

Fig. 6 Sound Level of Combination Refrigerator-Freezer with Typical Rotary Compressor

Features

Shafts and Journals. Shaft deflection, under load, is caused by compression gas loading of the roller and the torsional and side pull loading of the motor rotor. Design criteria must allow minimum oil film under the maximum run and starting loads. The motor rotor should have minimal deflection to eliminate starting problems under extreme conditions of torque.

The shafts are generally made from steel forging and nodular cast iron. Depending on the materials chosen, a relative hardness should be maintained between the bearing and the journal. Journals are ground round to high precision and polished to a finish of 10 microinch or better.

Bearings. The bearing must support the rotating member under all conditions. Powdered metal has been extensively used for these components, due to its porous properties, which help in lubrication. This material can also be formed into complex bearing shapes with little machining required.

Vanes. Vanes are designed for reliability by the choice of materials and lubrication. The vanes are hardened, ground, and polished to the best finish obtainable. Steel, powdered metal, and aluminum alloys have been used. Powder metal vanes have a particular hardness typically in the 60 to 80 Rockwell *C* (R_c) range, and apparent hardness in the 35 to 55 R_c range.

To obtain good sealing and proper lubrication, the shape of the vane tip must conform to the surface of generation along the roller wall in accordance with hydrodynamic theory.

Vane Springs. Vane springs force the vane to stay in contact with the roller during start-up. The spring rate depends on the inertia of the vane.

Valves. Only discharge valves are required by rolling piston rotary compressors. They are usually simple reed valves made of high-grade steel.

Lubrication. A good lubrication system circulates an ample supply of clean oil to all working surfaces, bearings, blades, blade slots, and seal faces. High-side pressure in the housing shell ensures a sufficient pressure differential across the bearings; passageways distribute oil to the bearing surfaces.

Mechanical Efficiency. High mechanical efficiency depends on minimizing friction losses. Friction losses occur in the bearings and between the vane and slot wall, vane tip, and roller wall, and roller and bearing faces. The amount and distribution of these losses vary based on the geometry of the compressor.

Motor Selection. Breakdown torque requirements depend on the displacement of the compressor, the refrigerant, and the operating range. Domestic refrigerator compressors typically require a breakdown torque of about 36 to 38 oz·ft per cubic inch of compressor displacement per revolution. Similarly, larger compressors using R-22 for window air conditioners require about 67 to 69 oz·ft breakdown torque per cubic inch of compressor displacement per revolution.

Rotary machines do not usually require complete unloading for successful starting. The starting torque of standard split-phase motors is ample for small compressors. Permanent split capacitor motors for air conditioners of various sizes provide sufficient starting and improve the power factor to the required range.

ROTARY VANE COMPRESSORS

Rotary vane compressors have a low weight-to-displacement ratio, which, in combination with compact size, makes them suitable for transport application. Small compressors in the 3 to 50 hp range are single-staged for a saturated suction temperature range of –40 to 45°F at saturated condensing temperature up to 140°F. By employing a second stage, low-temperature applications down to –60°F are possible. Currently, R-22, R-404a, and R-717 refrigerants are used.

Figure 7 is a cross-sectional view of an eight-bladed compressor. The eight discrete volumes are referred to as cells. A single shaft rotation produces eight distinct compression strokes. While conventional valves are not required for this compressor, suction and discharge check valves are recommended to prevent reverse rotation and oil logging during shutdown.

Design of the compressor results in a fixed built-in volume ratio. Compressor volume ratio is determined by the relationship between the volume of the cell as it is closed off from the suction port to its volume before it opens to the discharge port. The internal compression ratio can be calculated from the following relationship:

$$p_i = V_i^k \tag{6}$$

where

p_i = internal compression ratio
V_i = compression volume ratio
k = c_p/c_v = isentropic specific heat ratio for refrigerant being compressed

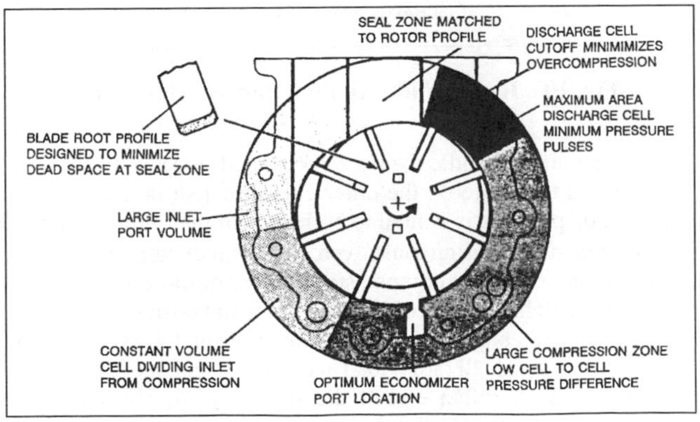

Fig. 7 Rotary Vane Compressor

The compressors currently available are of an oil-flooded, open-drive design, which requires use of an oil separator. Single-stage separators are used in close-coupled, high-temperature, direct-expansion systems, where oil return is not a problem. Two-stage separators with a coalescing-type second stage are used in low-temperature systems, in ammonia systems, and in flooded evaporators likely to trap oil.

SINGLE-SCREW COMPRESSORS

Screw compressors currently in production for refrigeration and air-conditioning applications comprise two distinct types—single-screw and twin-screw. Both are conventionally used with fluid injection where sufficient fluid cools and seals the compressor. Screw compressors have the capability to operate at pressure ratios above 20:1 single stage. The capacity range currently available is from 20 to 1300 tons.

Description

The single-screw compressor consists of a single cylindrical main rotor that works with a pair of gate rotors. Both the main rotor and gate rotors can vary widely in terms of form and mutual geometry. Figure 8 shows the design normally encountered in refrigeration.

The main rotor has helical grooves, with a cylindrical periphery and a globoid (or hourglass shape) root profile. The two identical gate rotors are located on opposite sides of the main rotor. The casing enclosing the main rotor has two slots, which allow the teeth of the gate rotors to pass through them. Two diametrically opposed discharge ports use a common discharge manifold located in the casing.

The compressor is driven through the main rotor shaft, and the gate rotors follow by direct meshing action with the main rotor. The geometry of the single-screw compressor is such that the gas compression power is transferred directly from the main rotor to the gas. No power (other than small frictional losses) is transferred across the meshing points to the gate rotors.

Compression Process

The operation of the single-screw compressor can be divided into three distinct phases: suction, compression, and discharge. With reference to Figure 9, the process is as follows:

Suction. During rotation of the main rotor, a typical groove in open communication with the suction chamber gradually fills with

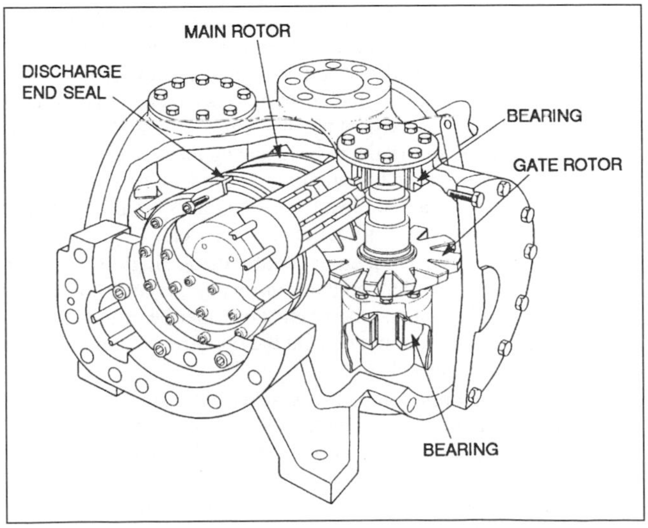

Fig. 8 Section of Single-Screw Refrigeration Compressor

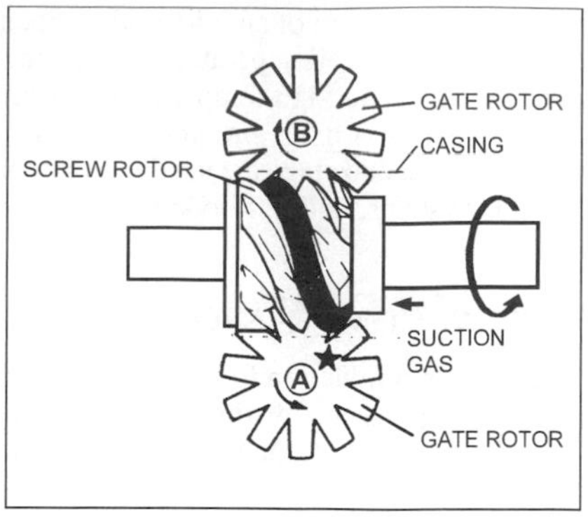

Suction. During rotation of the main rotor, a typical groove in open communication with the suction chamber gradually fills with suction gas. The tooth of the gate rotor in mesh with the groove acts as an aspirating piston.

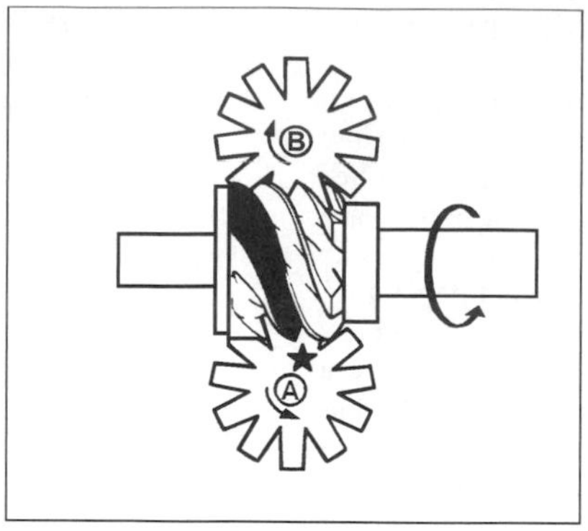

Compression. As the main rotor turns, the groove engages a tooth on the gate rotor and is covered simultaneously by the cylindrical main rotor casing. The gas is trapped in the space formed by the three sides of the groove, the casing, and the gate rotor tooth. As rotation continues, the groove volume decreases and compression occurs.

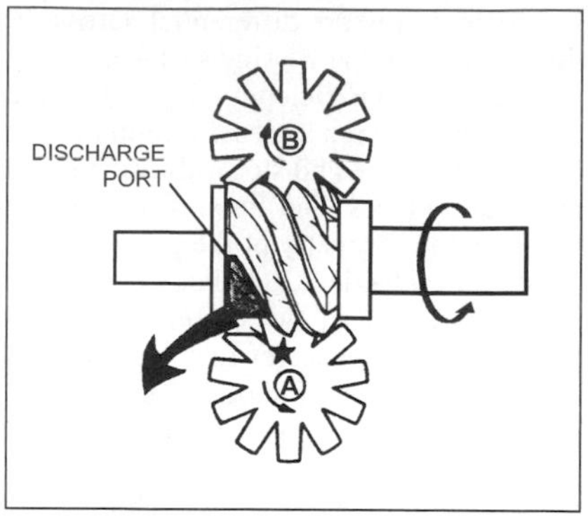

Discharge. At the geometrically fixed point where the leading edge of the groove and the edge of the discharge port coincide, compression ceases, and the gas discharges into the delivery line until the groove volume has been reduced to zero.

Fig. 9 Sequence of Compression Process in Single-Screw Compressor

suction gas. The tooth of the gate rotor in mesh with the groove acts as an aspirating piston.

Compression. As the main rotor turns, the groove engages a tooth on the gate rotor and is covered simultaneously by the cylindrical main rotor casing. The gas is trapped in the space formed by the three sides of the groove, the casing, and the gate rotor tooth. As rotation continues, the groove volume decreases and compression occurs.

Discharge. At the geometrically fixed point where the leading edge of the groove and the edge of the discharge port coincide, compression ceases, and the gas discharges into the delivery line until the groove volume has been reduced to zero.

Mechanical Features

Rotors. The screw rotor is normally made of cast iron, and the mating gate rotors are made from an engineered plastic. The inherent lubricating quality of the plastic, as well as its compliant nature, allow the single-screw compressor to achieve close clearances with conventional manufacturing practice.

The gate rotors are mounted on a metal support designed to carry the differential pressure between discharge pressure and suction pressure. The gate rotor function is equivalent to that of a piston in that it sweeps the groove and causes compression to occur. Furthermore, the gate rotor is in direct contact with the screw groove flanks and thus also acts as a seal. Each gate rotor is attached to its support by a simple spring and dashpot mechanism, allowing the gate rotor, with a low moment of inertia, to have an angular degree of freedom from the larger mass of the support. This method of attachment allows the gate rotor assemblies to be true idlers by allowing them to dampen out transients without damage or wear.

Bearings. In a typical open or semihermetic single-screw compressor, the main rotor shaft contains one pair of angular contact ball bearings (an additional angular contact or roller bearing is used for some heat pump semihermetics). On the opposite side of the screw, one roller bearing is used.

Note that the compression process takes place simultaneously on each side of the main rotor of the single-screw compressor. This balanced gas pressure results in virtually no load on the rotor bearing during full load and while symmetrically unloaded as shown in Figure 10. Should the compressor be unloaded asymmetrically (see economizer operation below 50% capacity), the designer is not restricted by the rotor geometry and can easily add bearings with a

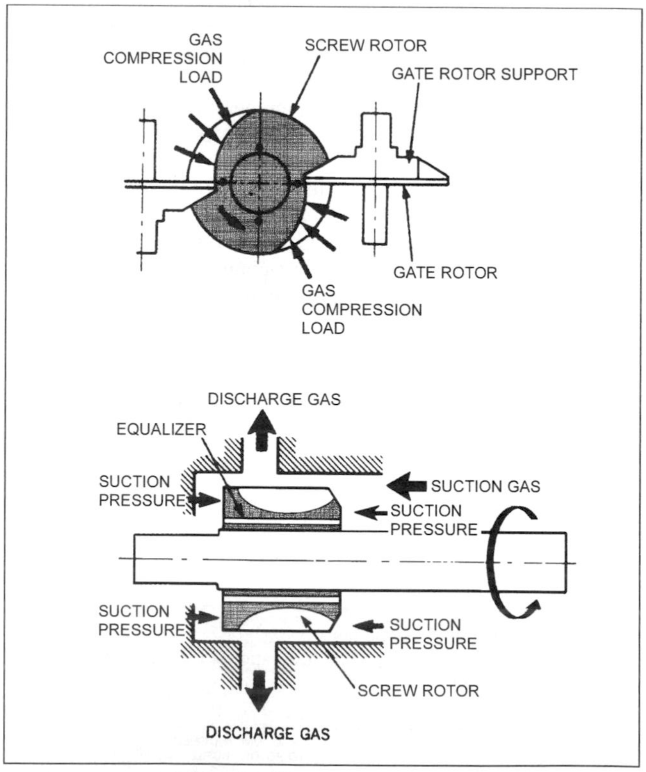

Fig. 10 Radial and Axially Balanced Main Rotor

long design life to handle the load. Axial loads are also low because the grooves terminate on the outer cylindrical surface of the rotor and suction pressure is vented to both ends of the rotor (Figure 10).

The gate rotor bearing must overcome a small moment force due to the gas acting on the compression surface of the gate rotor. Each gate rotor shaft has at least one bearing for axial positioning (usually a single angular contact ball bearing can perform the axial positioning and carry the small radial load at one end), and one roller or needle bearing at the other end of the support shaft also carries the radial load. Since the single-screw compressor's physical geometry places no constraints on bearing size, lives of 200,000 h are typical.

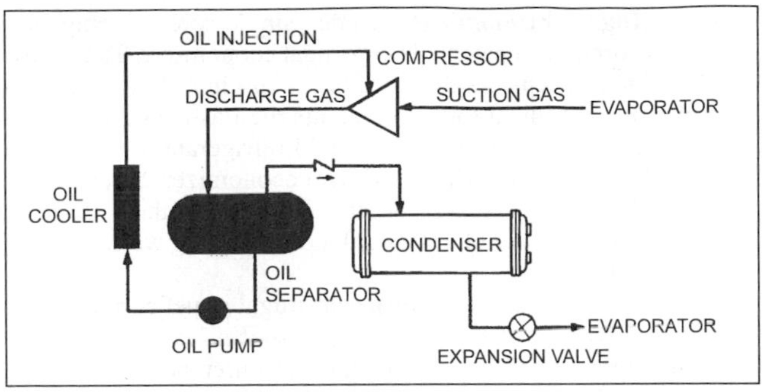

Fig. 11 Oil and Refrigerant Schematic of Oil Injection System

Cooling, Sealing, and Bearing Lubrication. A major function of injecting a fluid into the compression area is the removal of the heat of compression. Also, since the single-screw compressor has fixed leakage areas, the fluid helps seal leakage paths. Fluid is normally injected into a closed groove through ports in the casing or in the moving capacity control slide. Most single-screw compressors allow the use of many different injection fluids, oil being the most common, to suit the nature of the gas being compressed.

Oil-Injected Compressors. Oil is used in single-screw compressors to seal, cool, lubricate, and actuate capacity control. It gives a flat efficiency curve over a wide compression ratio and speed range, thus decreasing discharge temperature and reducing noise. In addition, the compressor can handle some liquid floodback because it tolerates oil.

Oil-injected single-screw compressors operate at high head pressures using common high-pressure refrigerants such as R-22, R-134a, and R-717. They also operate effectively at high pressure ratios because the injected oil cools the compression process. Currently, compressors with capacities in the 20 to 1500 hp range are manufactured.

Oil injection requires an oil separator to remove the oil from the high-pressure refrigerant (see Figure 11). For those applications with exacting demands for low oil carryover, separation equipment is available to leave less than 5 ppm oil in the circulated refrigerant.

With most compressors, oil can be injected automatically without a pump because of the pressure difference between the oil reservoir (discharge pressure) and the reduced pressure in a flute or bearing assembly during compression. A continuously running oil pump is used in some compressors to generate oil pressure 30 to 45 psi over compressor discharge pressure. This pump requires 0.3 to 1.0% of the compressor's motor power.

Methods of oil cooling include the following:

- **Direct injection of liquid refrigerant** into the compression process. Injection is controlled directly from the compressor discharge temperature, and loss of compressor capacity is minimized as injection takes place in a closed flute just before discharge occurs. This method requires very little power (typically less than 5% of compressor power).
- A **small refrigerant pump** draws liquid from the receiver and injects it directly into the compressor discharge line. The injection rate is controlled by sensing discharge temperature and modulating the pump motor speed. The power penalty in this method is the pump power (about 1 hp for compressors up to 1000 hp), which can result in energy savings over refrigerant injected into the compression chamber.
- **External oil cooling** between the oil reservoir and the point of injection is possible. Various heat exchangers are available to cool the oil: (1) separate water supply, (2) chiller water on a package unit, (3) condenser water on a package unit, (4) water from an evaporative condenser sump, (5) forced air-cooled oil cooler, and (6) high-pressure liquid recirculation (thermosiphon).

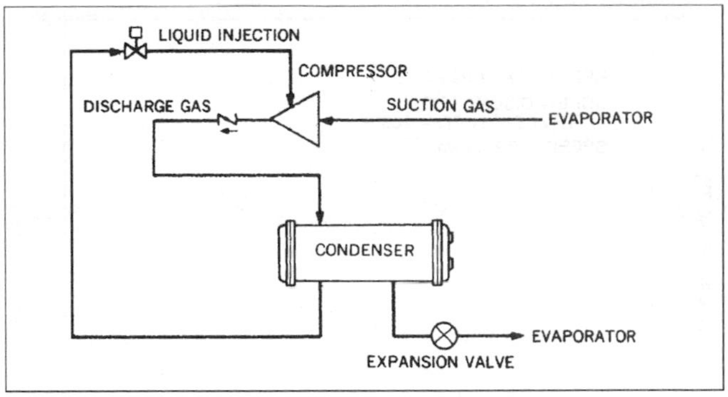

Fig. 12 Schematic of Oil-Injection-Free Circuit

The heat added to the oil during compression is the amount usually removed in the oil cooler.

Oil-Injection-Free Compressors. While the single-screw compressor operates well with oil injection, it also operates with equal or better efficiency in an oil-injection-free (OIF) mode with many common halocarbon refrigerants. This means that the fluid injected into the compression chamber is the condensate of the fluid being compressed. For air conditioning and refrigeration, where pressure ratios are in the range of 2 to 8, the oil normally injected into the casing is replaced by liquid refrigerant. No lubrication is required because the only power transmitted from the screw to the gate rotors is that needed to overcome small frictional losses. Thus, the refrigerant need only cool and seal the compressor. The liquid refrigerant may still contain a small amount of oil to lubricate the bearings (0.1 to 1%, depending on compressor design). A typical OIF circuit is depicted in Figure 12.

Oil-injection-free operation has the following advantages:

- It requires no discharge oil separator.
- Semihermetic compressors require no oil or refrigerant pumps.
- External coolers are not required.
- The compressor tolerates liquid refrigerant entering the suction and significantly reduces the size requirements of direct-expansion evaporators.

Economizers. Screw compressors are available with a secondary suction port between the primary compressor suction and discharge port. This port, when used with an economizer, improves compressor-useful refrigeration and compressor efficiency (Figure 13).

In operation, gas is drawn into the rotor grooves in the normal way from the suction line. The grooves are then sealed off in sequence, and compression begins. An additional charge is added to the closed flute through a suitably placed port in the casing by an intermediate gas source at a slightly higher pressure than that reached in the compression process at that time. The original and the additional charge are then compressed together to discharge conditions. The pumping capacity of the compressor at suction conditions is not affected by this additional flow through the economizer port.

When the port is used with an economizer, the effective refrigerating capacity of the economized compressor is increased over the noneconomized compressor by the increased heat absorption capability H of the liquid entering the evaporator. Furthermore, the only additional mass flow the compressor must handle is the flash gas entering a closed flute, which is above suction pressure. Thus, under most conditions, the capacity improvement is accompanied by an efficiency improvement (Figure 13). Economizers become effective when the pressure ratio is 3.5 and above.

Figure 14 shows a pressure-enthalpy diagram for a flash tank economizer. In it, high-pressure liquid passes through an expansion device and enters a tank at an intermediate pressure between suction and discharge. This pressure is maintained by the pressure in the

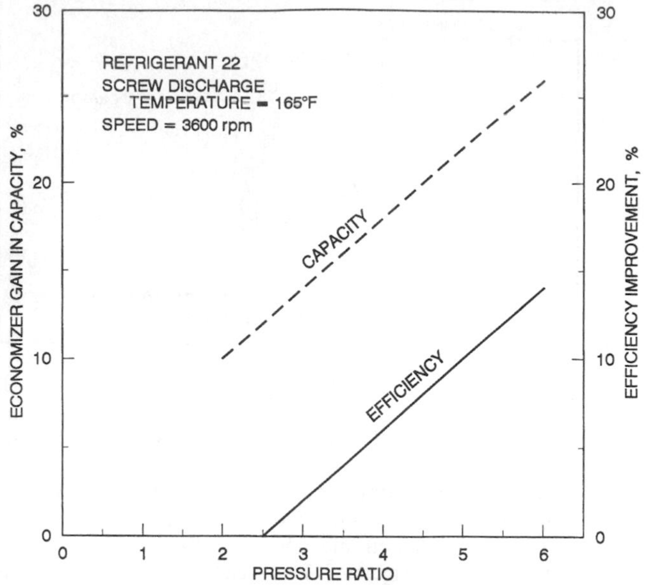

Fig. 13 Typical Improvement in Efficiency and Capacity with Economizer

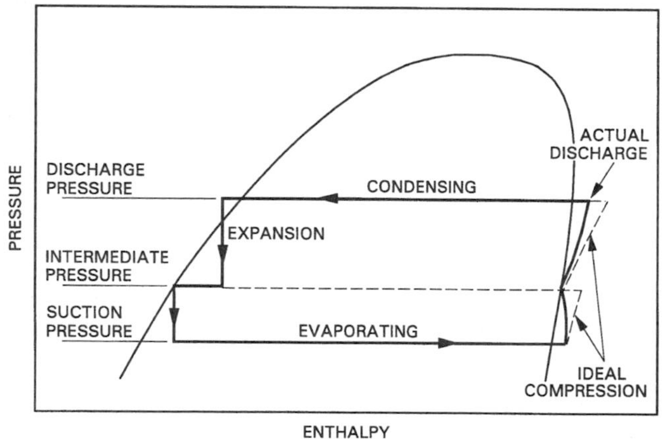

Fig. 14 Theoretical Economizer Cycle

compressor's closed flute (closed from suction). The gas generated from the expansion enters the compressor through the economizer port. When passed to the evaporator, the liquid (which is now saturated at the intermediate pressure) gives a larger refrigeration capacity per pound. In addition, the percentage increase in power input is lower than the percentage capacity increase.

As a screw compressor is unloaded, the economizer pressure falls toward suction pressure. As a result, the additional capacity and improved efficiency of the economizer fall to zero at 70 to 80% of full-load capacity.

The single-screw compressor has two compression chambers, each with its own slide valve. Each slide valve can be operated independently, which allows economizer gas to be introduced into one side of the compressor. By operating the slide independently, the chamber without the economizer gas can be unloaded to 0% capacity (50% capacity of the compressor). The other chamber remains at full capacity and retains the full economizer effect, making the economizer effective below 50% compressor capacity.

The secondary suction port may also be used for (1) a system side load or (2) a second evaporator that operates at a temperature above that of the primary evaporator.

Centrifugal Economizer. Some single-screw compressor designs incorporate a patented centrifugal economizer. The centrifugal economizer replaces the force of gravity in a flash-type economizer with centrifugal force to separate the flash gas generated at an intermediate pressure from the liquid refrigerant prior to liquid entering the evaporator. The centrifugal economizer thereby uses a much smaller pressure vessel and, in some designs, the economizer fits within the envelope of a standard motor housing without having to increase its size.

The separation is achieved by a centrifugal impeller mounted on the compressor shaft (see Figure 22); a special valve maintains a uniformly thick liquid ring around the circumference of the impeller, assuring that no gas leaves with the liquid going to the evaporator. The flash gas is then ducted to a closed groove in the compression cycle. Some designs use the flash gas with a similar liquid refrigerant to cool the motor prior to introduction into the closed compression groove.

Volume Ratio. The degree of compression within the rotor grooves is predetermined for a particular port configuration on screw compressors having fixed suction and discharge ports. A characteristic of the compressor is the volume ratio V_i, which is defined as the ratio of the volume of the groove at the start of compression to the volume of the same groove when it first begins to open to the discharge port. Hence, the volume ratio is determined by the size and shape of the discharge port.

For maximum efficiency, the pressure generated within the grooves during compression should exactly equal the pressure in the discharge line at the moment when the groove opens to it. If this is not the case, either over- or undercompression occurs, both resulting in internal losses. Although such losses cause no harm to the compressor, they increase power consumption and noise and reduce efficiency.

Volume ratio selection should be made according to operating conditions. The built-in pressure ratio of a screw compressor is a function of the volume ratio:

$$p_i = V_i^k \tag{7}$$

where k is the isentropic exponent (specific heat ratio) for the refrigerant being used [see Equation (6)].

Compressors equipped with slide valves (for capacity modulation) usually locate the discharge port at the discharge end of the slide valve. Alternative port configurations yielding the required volume ratios are then designed into the capacity control components, thus providing ease of interchangeability both during construction and after installation (although partial disassembly is required).

Single-screw compressors in refrigeration and process applications are being equipped with a simple slide valve to vary the volume ratio of the compressor while the compressor is running. The slide valve advances or delays the discharge port opening. Note that a separate slide has been designed to modulate the capacity independently of the volume ratio slide (see Figures 16, 17, and 18). Having the independent modulation of volume ratio (through discharge port control) and capacity modulation (through a completely independent slide that only varies the position where compression begins) allows the single-screw compressor to achieve efficient volume ratio control when capacity is less than full load.

Capacity Control. As with all positive-displacement compressors, both speed modulation and suction throttling can be used. Ideal capacity modulation for any compressor includes (1) continuous modulation from 100% to less than 10%, (2) good part-load efficiency, (3) unloaded starting, and (4) unchanged reliability.

Variable compressor displacement, the most common means for meeting these criteria, usually takes the form of two movable slide valves in the compressor casing (the single-screw compressor has two gate rotors forming two compression areas). At part load, each slide valve produces a slot that delays the point at which compression begins. This causes a reduction in groove volume, and hence in compressor throughput. As the suction volume is displaced before compression takes place, little or no thermodynamic loss occurs. However, if no other steps were taken, this mechanism would result in an undesirable drop in the effective volume ratio in undercompression and inefficient part-load operation.

This problem is avoided either by arranging that the capacity modulation valve reduces the discharge port area at the same time as the bypass slot is created (Figure 15) or having one valve control capacity only and a second valve independently modulate volume ratio (Figures 16, 17, and 18). A full modulating mechanism is provided in most large single-screw compressors, while two-position slide valves are used where requirements allow. The specific part-load performance will be affected by a compressor's built-in volume ratio V_i, evaporator temperature, and condenser temperature, and whether the slide valves are symmetrically or asymmetrically controlled.

Detailed design of the valve mechanism differs between makes of compressors but usually consists of an axial sliding valve along each side of the rotor casing (Figure 15). This mechanism is usually operated by a hydraulic or gas piston and cylinder assembly located within the compressor itself or by a positioning motor. The piston is actuated either by oil, discharge gas, or high-pressure liquid refrigerant at discharge pressure driven in either direction according to the operation of a four-way solenoid valve.

Figure 17 shows a capacity slide valve (top) and a variable volume ratio slide (bottom). The capacity slide is in the full-load position, and the volume ratio slide is at a moderate volume ratio. Figure 18 depicts the same system as shown in Figure 17, except that the capacity slide is in a partially loaded position, and the volume ratio slide has moved to a position to match the new conditions.

The single-screw compressor's two compression chambers, each having its own capacity slide valve that can be operated independently, permits one slide valve to be unloaded to 0% capacity (50% compressor capacity) while the other slide valve remains at full capacity. Operation in this manner (asymmetrical) realizes an improvement in part-load efficiency below the 50% capacity point and further part-load efficiency gains are realized when the economizer gas is only entered into a closed groove on the side that is unloaded second (see explanation in the section on Economizers).

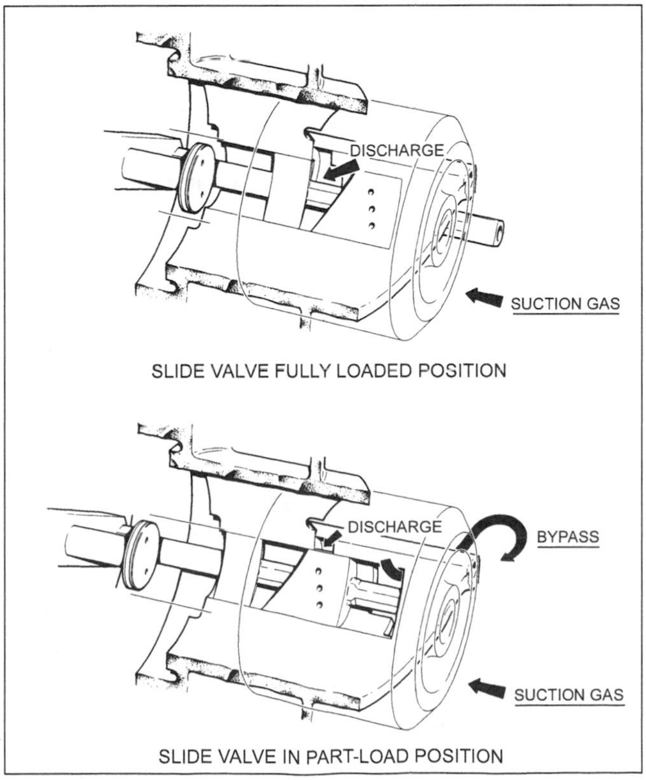

Fig. 15 Capacity Control Slide Valve Operation

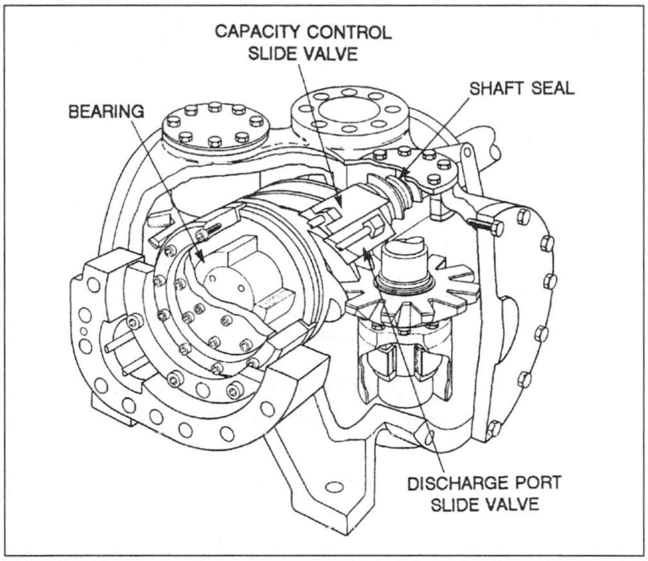

Fig. 16 Refrigeration Compressor Equipped with Variable Capacity Slide Valve and Variable Volume Ratio Slide Valve

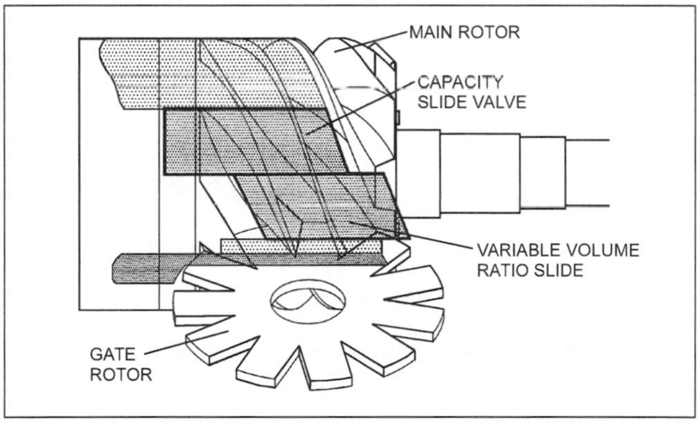

Fig. 17 Capacity Slide in Full-Load Position and Volume Ratio Slide in Intermediate Position

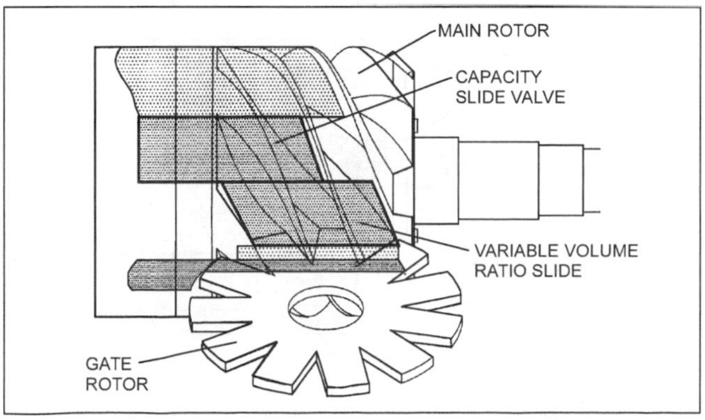

Fig. 18 Capacity Slide in Part-Load Position and Volume Ratio Slide Positioned to Maintain System Volume Ratio

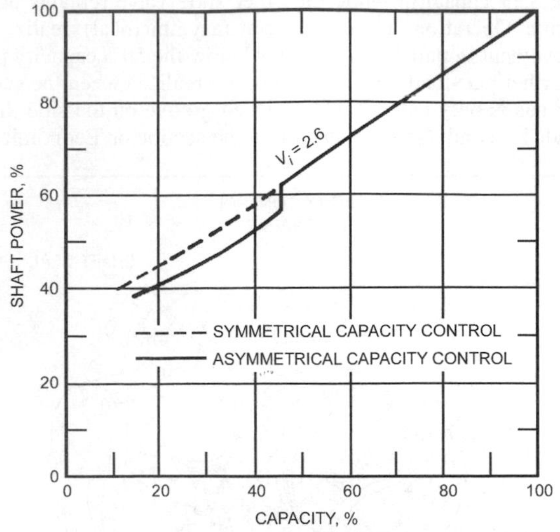

Fig. 19 Part-Load Effect of Symmetrical and Asymmetrical Capacity Control

Figure 19 demonstrates the effect of asymmetrical capacity control of a single-screw compressor.

Performance. Figures 20 and 21 show typical efficiencies of all single-screw compressor designs. High isentropic and volumetric efficiencies are the result of internal compression, the absence of suction or discharge valves and their losses, and extremely small clearance volumes. The curves show the importance of selecting the correct volume ratio in fixed volume ratio compressors.

Manufacturer's data for operating conditions or speed should not be extrapolated. Screw compressor performance at reduced speed is usually significantly different from that specified at the normally rated point. Performance data normally include information about the degree of liquid subcooling and suction superheating assumed in data.

Applications. Single-screw compressors have been widely applied as refrigeration compressors, using halocarbon refrigerants, ammonia, and hydrocarbon refrigerants. A single gate rotor semihermetic version is increasingly being applied in large supermarkets.

Oil-injected and oil-injection-free (OIF) semihermetic compressors are widely used for air-conditioning and heat pump service, with compressor sizes ranging from 40 to 500 tons.

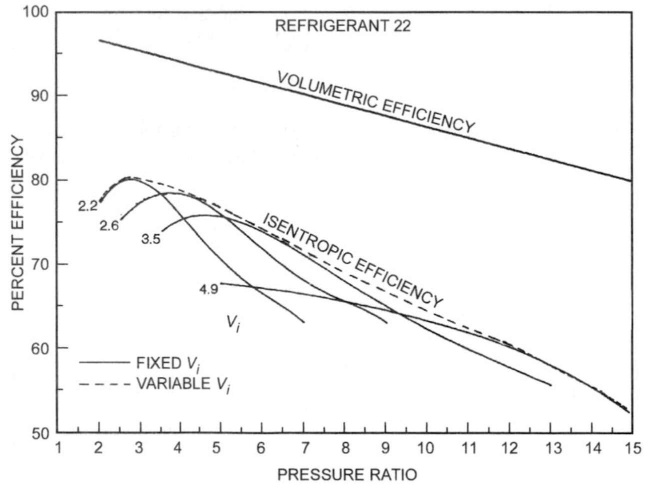

Fig. 20 Typical Compressor Performance on R-22

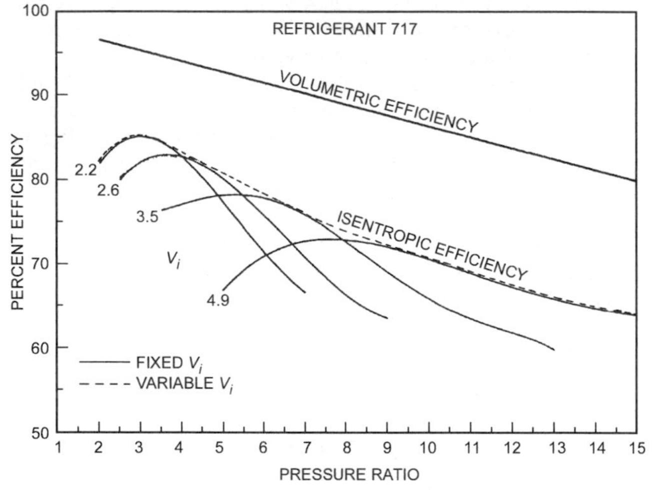

Fig. 21 Typical Compressor Performance on R-717 (Ammonia)

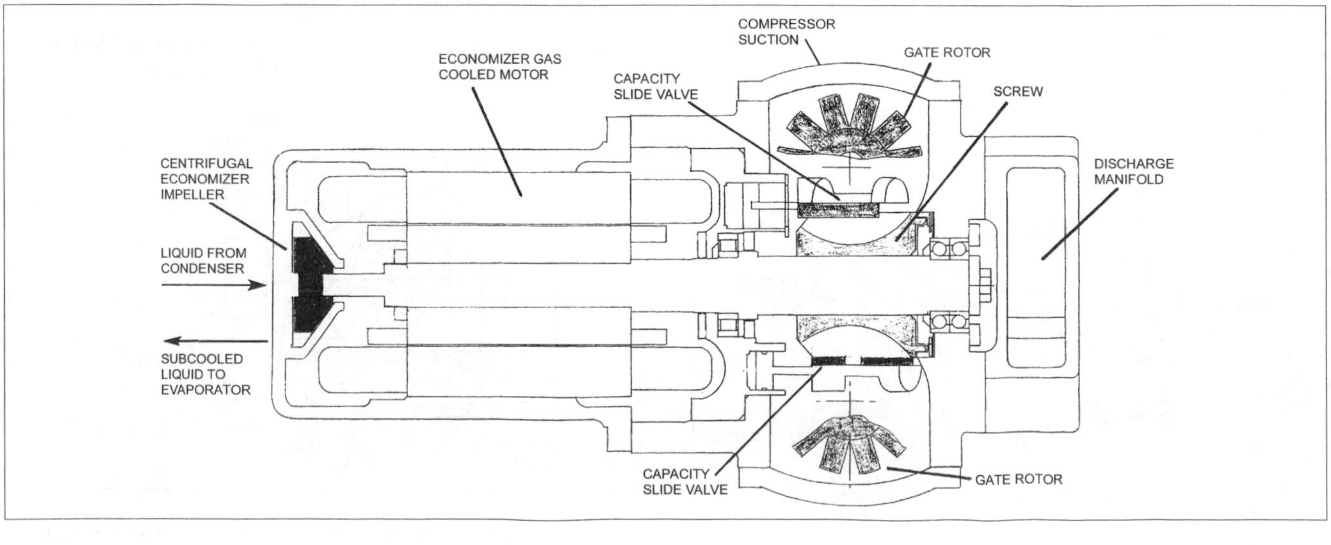

Fig. 22 Typical Semihermetic Single-Screw Compressor

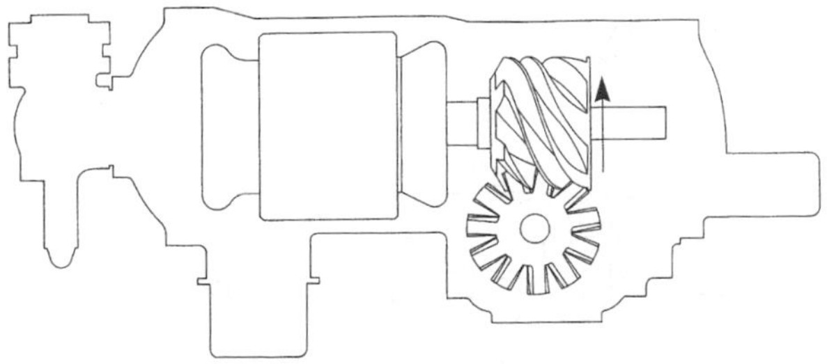

Fig. 23 Single Gate Rotor Semihermetic Single-Screw Compressor

Semihermetic Design. Figure 22 shows a semihermetic single-screw compressor. Figure 23 exhibits a semihermetic single-screw compressor using only one gate rotor. This design has found application in large supermarket rack systems. The single gate rotor compressor exhibits high efficiency and has been designed for long bearing life, which compensates for the unbalanced load on the screw rotor shaft with increasing bearing size.

Noise and Vibration

The inherently low noise and vibration characteristics of single-screw compressors are due to small torque fluctuation and no valving required in the compression chamber. In particular, the advent of OIF technology eliminates the need for oil separators that have traditionally created noise.

TWIN-SCREW COMPRESSORS

Twin screw is the common designation for double helical rotary screw compressors. A twin-screw compressor consists of two mating helically grooved rotors—male (lobes) and female (flutes or gullies) in a stationary housing with inlet and outlet gas ports (Figure 24). The flow of gas in the rotors is mainly in an axial direction. Frequently used lobe combinations are 4 + 6, 5 + 6, and 5 + 7 (male + female). For instance, with a four-lobe male rotor, the driver rotates at 3600 rpm; the six-lobe female rotor follows at 2400 rpm. The female rotor can be driven through synchronized timing gears or directly driven by the male rotor on a light oil film. In some applications, it is practical to drive the female rotor, which results in a 50% speed and displacement increase over the male-driven compressor, assuming a 4 + 6 lobe combination. Geared speed increasers are also used on some applications to increase the capacity delivered by a particular compressor size.

Twin helical screws find application in many air-conditioning, refrigeration, and heat pump applications, typically in the industrial and commercial market. Machines can be designed to operate at high or low pressure and are often applied below 2:1 and above 20:1 compression ratios single-stage. Commercially available compressors are suitable for application on all normally used high-pressure refrigerants.

Compression Process

Compression is obtained by direct volume reduction with pure rotary motion. For clarity, the following description of the three basic compression phases is limited to one male rotor lobe and one female rotor interlobe space (Figure 25).

Suction. As the rotors begin to unmesh, a void is created on both the male side (male thread) and the female side (female thread), and gas is drawn in through the inlet port. As the rotors continue to turn, the interlobe space increases in size, and gas flows continuously into the compressor. Just prior to the point at which the interlobe space leaves the inlet port, the entire length of the interlobe space is completely filled with gas.

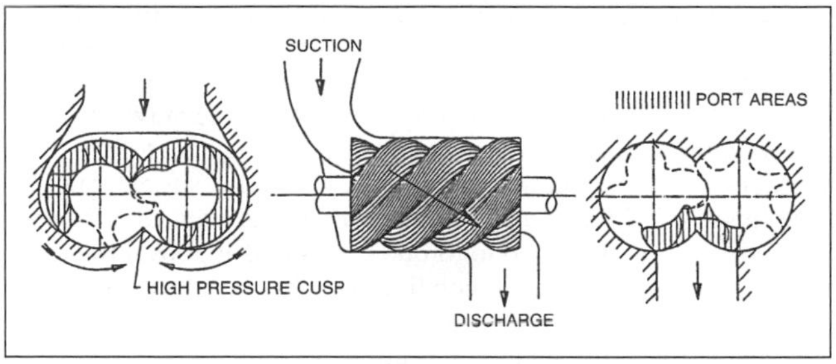

Fig. 24 Twin-Screw Compressor

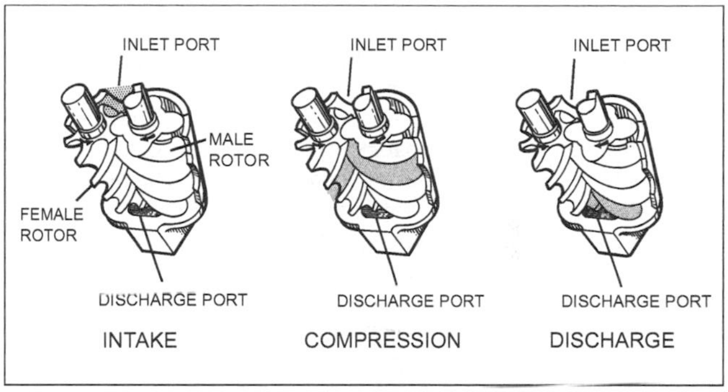

Fig. 25 Compression Process

Compression. Further rotation starts the meshing of another male lobe with another female interlobe space on the suction end and progressively compresses the gas in the direction of the discharge port. Thus, the occupied volume of the trapped gas within the interlobe space is decreased and the gas pressure consequently increased.

Discharge. At a point determined by the designed built-in volume ratio, the discharge port is uncovered and the compressed gas is discharged by further meshing of the lobe and interlobe space.

During the remeshing period of compression and discharge, a fresh charge is drawn through the inlet on the opposite side of the meshing point. With four male lobes rotating at 3600 rpm, four interlobe volumes are filled and give 14,400 discharges per minute. Since the intake and discharge cycles overlap effectively, a smooth continuous flow of gas results.

Mechanical Features

Rotor Profiles. Helical rotor design started with an asymmetrical point-generated rotor profile. This profile was only used in compressors with timing gears (dry compression). The symmetrical, circular rotor profile was introduced because it was easier to manufacture than the preceding profile, and it could be used without timing gears for wet or oil-flooded compression.

Current rotor profiles are normally asymmetrical and line-generated profiles, giving higher performance due to better rotor dynamics and decreased leakage area. This design allowed the possibility for female rotor drive, as well as the conventional male drive. Rotor profile, blowhole, length of sealing line, quality of sealing line, torque transmission between rotors, rotor-housing clearances, interlobe clearances, and lobe combinations are optimized for specific pressure, temperature, speed, and wet or dry operation. Optimal rotor tip speed is 3000 to 8000 fpm for wet operation (oil-flooded) and 12,000 to 24,000 fpm for dry operation.

Rotor Contact and Loading. Contact between the male and female rotors is mainly rolling, primarily at a contact band on each rotor's pitch circle. Rolling at this contact band means that virtually no rotor wear occurs.

Gas Forces. On the driven rotor, the internal gas force always creates a torque in a direction opposite to the direction of rotation. This is known as positive or braking torque. On the undriven rotor, the design can be such that the torque is positive, negative, or zero, except on female drive designs, where zero or negative torque does not occur. Negative torque occurs when internal gas force tends to drive the rotor. If the average torque on the undriven rotor is near zero, this rotor is subjected to torque reversal as it goes through its phase angles. Under certain conditions, this can cause instability. Torque transmitted between the rotors does not create problems because the rotors are mainly in rolling contact.

Male drive. The transmitted torque from male rotor to female rotor is normally 5 to 25% of input torque.

Female drive. The transmitted torque from female rotor to male rotor is normally 50 to 60% of input torque.

Rotor loads. The rotors in an operating compressor are subjected to radial, axial, and tilting loads. Tilting loads are radial loads caused by axial loads outside of the rotor center line. The axial load is normally balanced with a balancing piston for larger high-pressure machines (rotor diameter above 4 in. and discharge pressure above 160 psi). Balancing pistons are typically close-tolerance, labyrinth-type devices with high-pressure oil or gas on one side and low pressure on the other. They are used to produce a thrust load to offset some of the primary gas loading on the rotors, thus reducing the amount of thrust load the bearings support.

Bearings. Twin-screw compressors normally have either four or six bearings, depending on whether one or two bearings are used for the radial and axial loads. Some designs incorporate multiple rows of smaller bearings per shaft to share the loads. Sleeve bearings have been used historically to support radial loads in machines with male rotor diameters larger than 6 in., while antifriction bearings were generally applied to smaller machines. However, improvements in antifriction designs and materials have led to compressors with up to 14 in. rotor diameter with full antifriction bearing designs. Cylindrical and tapered roller bearings and various types of ball bearings are used in screw compressors for carrying radial loads. The most common thrust or axial load-carrying bearings are angular contact ball bearings, although tapered rollers or tilting pad bearings are used in some machines.

General Design. Screw compressors are often designed for particular pressure ranges. Low-pressure compressors have long, high displacement rotors and adequate space to accommodate bearings to handle the relatively light loads. They are frequently designed without thrust balance pistons, since the bearings alone can handle the low thrust loads and still maintain good life.

High-pressure compressors have short and strong rotors (shallow grooves) and, therefore, have space for large bearings. They are normally designed with balancing pistons for high thrust bearing life.

Rotor Materials. Rotors are normally made of steel, but aluminum, cast iron, and nodular iron are used in some applications.

Capacity Control

As with all positive-displacement compressors, both speed modulation and suction throttling can reduce the volume of gas drawn into a screw compressor. Ideal capacity modulation for any compressor would be (1) continuous modulation from 100% to less than 10%, (2) good part-load efficiency, (3) unloaded starting, and (4) unchanged reliability. However, not all applications need ideal capacity modulation. Variable compressor displacement and variable speed are the best means for meeting these criteria. Variable compressor displacement is the most common capacity control method used. Various mechanisms achieve variable displacement, depending on the requirements of a particular application.

Capacity Slide Valve. A slide valve for capacity control is a valve with sliding action parallel to the rotor bores. They are placed within or close to the high-pressure cusp region, face one or both rotor bores, and bypass a variable portion of the trapped gas charge

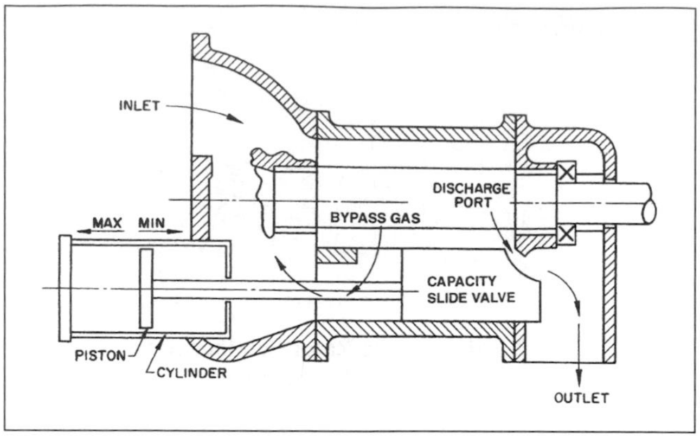

Fig. 26 Slide Valve Unloading Mechanism

back to suction, depending on their position. Within this definition, there are two types of capacity slide valves.

1. *Capacity slide valve regulating discharge port.* This type of slide valve is located within the high-pressure cusp region. It controls capacity as well as the location of the radial discharge port at part load. The axial discharge port is designed for a volume ratio giving good part-load performance without losing full-load performance. Figure 26 shows a schematic view of the most common arrangement.
2. *Capacity slide valve not regulating discharge port.* A slide valve outside the high-pressure cusp region controls only capacity and not the radial discharge port.

The first type is the most common arrangement. It is generally the most efficient of the available capacity reduction methods, due to its indirect correction of built-in volume ratio at part load and its ability to give large volume reductions without large movement of the slide valve.

Capacity Slot Valve. A capacity slot valve consists of a number of slots that follow the rotor helix and face one or both rotor bores. The slots are gradually opened or closed with a plunger or turn valve. These recesses in the casing wall increase the volume of the compression space and also create leakage paths over the lobe tips. The result is somewhat lower full-load performance when compared to a design without slots.

Capacity Lift Valve. Capacity lift valves or plug valves are movable plugs in one or both rotor bores (with radial or axial lifting action) that regulate the actual start of compression. These valves control capacity in a finite number of steps, rather than by the infinite control of a conventional slide valve (Figure 27).

Neither slot valves nor lift valves offer quite as good efficiency at part load as a slide valve, because they do not relocate the radial discharge port. Thus, undercompression losses at part load can be expected if the machines have the correct volume ratio for full-load operation and the compression ratio at part load does not reduce.

Volume Ratio

In all positive-displacement rotary compressors with fixed port location, the degree of compression within the rotor thread is determined by the location of the suction and discharge ports. The built-in volume ratio of screw compressors is defined as the ratio of volume of the thread at the start of the compression process to the volume of the same thread when it first begins to open to the discharge port. The suction port must be located to trap the maximum suction charge; hence, the volume ratio is determined by the location of the discharge port.

Only the suction pressure and volume ratio of the compressor determine the internal pressure achieved before opening to discharge. However, the condensing and evaporating temperatures

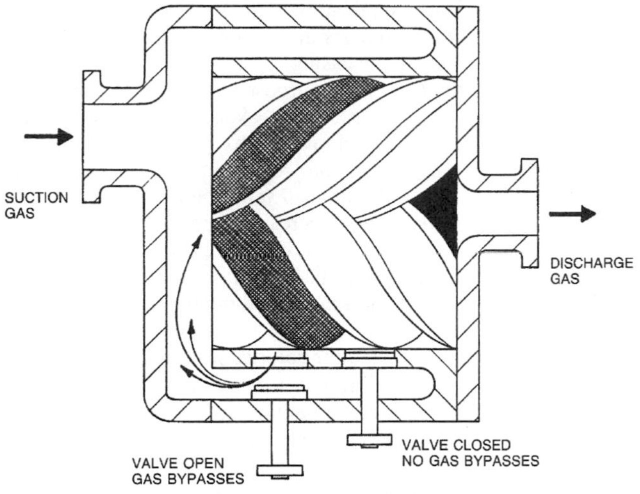

Fig. 27 Lift Valve Unloading Mechanism

determine the discharge pressure and the compression ratio in the piping that leads to the compressor. Any mismatch between the internal and system discharge pressures results in under- or over-compression loss and in lower efficiency.

If the operating conditions of the system seldom change, it is possible to specify a fixed volume ratio compressor that will give good efficiency. Compressor manufacturers normally make compressors with three or four possible discharge port sizes that correspond to system conditions encountered frequently. Generally, the designer is responsible for specifying a compressor that most closely matches expected pressure conditions.

The required volume ratio for a particular application can be determined as follows.

First, determine the compression ratio of a given refrigerating system:

$$CR = p_d/p_s \tag{8}$$

where

CR = compression ratio
p_s = expected suction pressure, absolute
p_d = expected discharge pressure, absolute

Then, determine the internal pressure ratio of the available compressors by approximating compression as an isentropic process as follows:

$$p_i = V_i^k \tag{9}$$

where

p_i = internal pressure ratio
V_i = compressor volume ratio
k = ratio of specific heats (isentropic exponent) for refrigerant used [see Equation (6)]

And, finally, the compressor should be selected to match as nearly as possible the internal pressure ratio of the compressor to the system compression ratio:

$$p_i = CR \tag{10}$$

Usually, in slide valve-equipped compressors, the radial discharge port is located in the discharge end of the slide valve. For a given ratio *L/D* of rotor length to rotor diameter and a given stop position, a short slide valve gives a low volume ratio, and a long slide valve gives a higher volume ratio. The difference in length basically locates the discharge port earlier or later in the compression process. Different length slide valves allow changing the volume ratio of a given compressor, although disassembly is required.

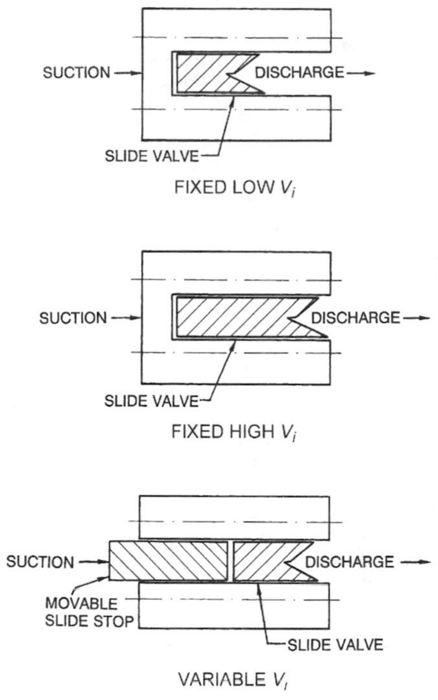

Fig. 28 View of Fixed and Variable Volume Ratio (V_i) Slide Valves from Above

Variable Volume Ratio. While operating, some twin-screw compressors adjust the volume ratio of the compressor to the most efficient ratio for whatever pressures are encountered.

In fixed volume ratio compressors, the motion of the slide valve toward the inlet end of the machine is stopped when it comes in contact with the rotor housing in that area. In the most common of the variable volume ratio machines, this portion of the rotor housing has been replaced with a second slide, the movable slide stop, which can be actuated to different locations in the slide valve bore (Figure 28).

By moving the slides back and forth, the radial discharge port can be relocated during operation to match the compressor volume ratio to the optimum. This added flexibility allows operation at different suction and discharge pressure while still maintaining maximum efficiency. The comparative efficiencies of fixed and variable volume ratio screw compressors are shown in Figure 29 for full-load operation on ammonia and R-22 refrigerants. The figure shows that a variable volume ratio compressor efficiency curve encompasses the peak efficiencies of compressors with fixed volume ratio over a wide range of pressure ratio. Following are other secondary effects of a variable volume ratio:

- Less oil foam in oil separator (no overcompression)
- Less oil carried over into the refrigeration system (because of less oil foam in oil separator)
- Extended bearing life; minimized load on bearings
- Extended efficient operating range with economizer discharge port corrected for flash gas from economizer, as well as gas from suction
- Less noise
- Lower discharge temperatures and oil cooler heat rejection

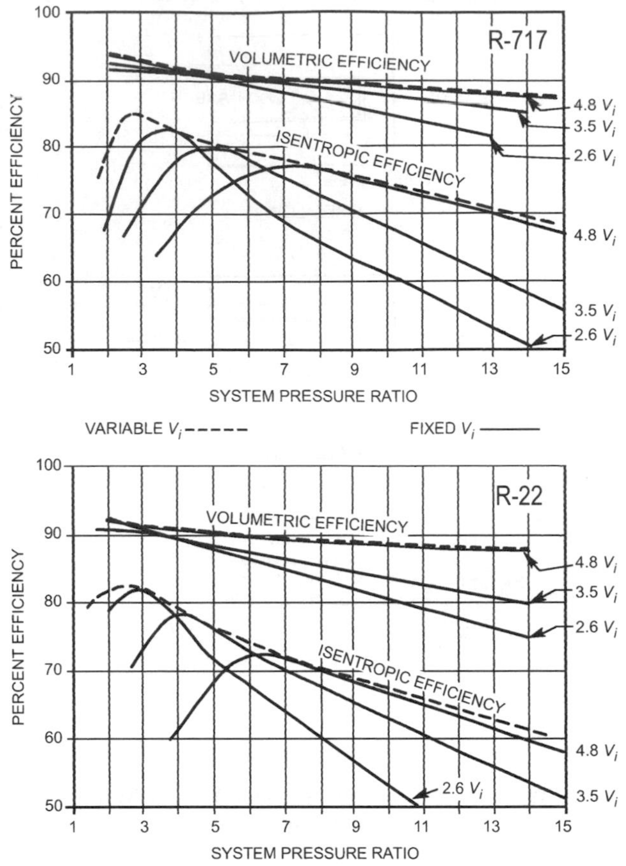

Fig. 29 Twin-Screw Compressor Efficiency Curves

The greater the change in either suction or condensing pressure, the more benefits are possible with a variable volume ratio. Efficiency improvements as high as 30% are possible, depending on the application, refrigerant, and operating range.

Oil Injection

Two primary types of compressor lubrication systems are employed in twin-screw compressors—dry and oil flooded.

Dry Operation (No Rotor Contact). Since the two rotors in twin-screw compressors are parallel, timing gears are a practical means of synchronizing the rotors so that they do not touch each other. Eliminating rotor contact eliminates the need for lubrication in the compression area. Initial screw compressor designs were based on this approach, and dry screws still find application in the gas process industry.

Synchronized twin-screw compressors once required high rotor tip speed and were, therefore, noisy. However, with current profile technology, the synchronized compressor can run at a lower tip speed and higher pressure ratio, giving quieter operation. The added cost of timing gears and internal seals generally make the dry screw more expensive than an oil-flooded screw for normal refrigeration or air conditioning.

Oil-Flooded Operation. The oil-flooded twin-screw compressor is the most common type of screw used in refrigeration and air conditioning. Compressor capacities range from 20 to 6000 cfm. Oil-flooded compressors typically have oil supplied to the compression area at a volume rate of about 0.5% of the displacement volume. Part of this oil is used for lubrication of the bearings and shaft seal prior to injection. Typically, paraffinic or naphthenic-based mineral oils are used, although synthetics are also used as required by the refrigerant or gas application. The oil is normally injected into a closed thread through ports in the moving slide valve and/or through stationary ports in the casing.

The oil fulfills three primary purposes—sealing, cooling, and lubrication. It also tends to fill any leakage paths between and around the rotors. This keeps volumetric efficiency (VE) high, even at high compression ratios. Normal compressor VE exceeds 85% even at 25:1 single stage (ammonia, 7.6 in. rotor diameter). It also gives flat efficiency curves with decreasing speeds and quiet operation. Oil transfers much of the heat of compression from the gas to the oil, keeping the typical discharge temperature below 190°F, which allows high compression ratios without the danger of breaking down the refrigerant or the oil. The lubrication function of the oil protects bearings, seals, and the rotor contact areas.

The ability of a screw compressor to tolerate oil also permits the compressor to handle a certain amount of liquid floodback, as long as the liquid quantity is not large enough to lock the rotors hydraulically.

Oil Separation and Cooling. Oil injection requires an oil separator to remove oil from the high-pressure refrigerant. Coalescing separation equipment routinely gives less than 5 ppm oil in the circulated refrigerant. Some smaller compressors, primarily used on packaged units, have less efficient or no separation capability and may circulate up to 7% lubricant with the discharge gas from the compressor.

Oil injection is normally achieved by one of two methods: (1) with a continuously running oil pump capable of generating an oil pressure of 30 to 45 psi over compressor discharge pressure, representing 0.3 to 1.0% of compressor motor power; or (2) with some compressors, oil can be injected automatically, without a pump because of the pressure difference between the oil reservoir (discharge pressure) and the reduced pressure in a thread during the compression process.

Since the oil absorbs a significant amount of the heat of compression in an oil-flooded operation, it must be cooled to maintain low discharge temperature. One cooling method is by direct injection of liquid refrigerant into the compression process. The amount of injected liquid refrigerant corresponds to about 0.02% of displacement volume. The amount of liquid injected is normally controlled by sensing the discharge temperature and injecting enough liquid to maintain a constant temperature. Some of the injected liquid mixes with the oil and leaks to lower pressure threads, where it tends to raise pressure and reduce the amount of gas the compressor can draw in. Also, any of the liquid that has time to absorb heat and expand to vapor must be recompressed, which tends to raise absorbed power levels. Compressors are designed with the liquid injection ports as late as possible in the compression to minimize capacity and power penalties. Typical penalties for liquid injection are in the 1 to 10% range, depending on the compression ratio.

Another method of oil cooling draws liquid from the receiver with a small refrigerant pump and injects it directly into the compressor discharge line. The power penalty in this method is the pump power (about 1 hp for compressors up to 1000 hp).

In the third method, the oil can be cooled outside the compressor between the oil reservoir and the point of injection. Various configurations of heat exchangers are available for this purpose, and the oil cooler heat rejection can be accomplished by (1) separate water supply, (2) chiller water on a packaged unit, (3) condenser water on a packaged unit, (4) water from an evaporative condenser sump, (5) forced air-cooled oil cooler, (6) liquid refrigerant, and (7) high-pressure liquid recirculation (thermosiphon).

External oil coolers using water or other means from a source independent of the condenser allow for the condenser to be reduced in size by an amount corresponding to the oil cooler capacity. Where oil cooling is carried out from within the refrigerant system by means such as (1) direct injection of liquid refrigerant into the compression process or the discharge line, (2) direct expansion of liquid in an external heat exchanger, (3) using chiller water on a packaged unit, (4) recirculation of high-pressure liquid from the condenser, or (5) water from an evaporative condenser sump, the condenser must be sized for the total heat rejection (i.e., evaporator load plus shaft power for open compressors and input power for hermetic compressors).

With an external oil cooler, the mass flow rate of oil injected into the compressor is usually determined by the desired discharge temperature rather than by the compressor sealing requirements, since the oil acts predominantly as a heat transfer medium. Conversely, with direct liquid injection cooling, the oil requirement is dictated by the compressor lubrication and sealing needs.

Economizers

Twin-screw compressors are available with a secondary suction port between the primary compressor suction and discharge ports. This port can accept a second suction load at a pressure above the primary evaporator, or flash gas from a liquid subcooler vessel, known as an economizer.

In operation, gas is drawn into the rotor thread from the suction line. The thread is then sealed in sequence and compression begins. An additional charge may be added to the closed thread through a suitably placed port in the casing. The port is connected to an intermediate gas source at a pressure slightly higher than that reached in the compression process at that time. Both original and additional charges are then compressed to discharge conditions.

When the port is used as an economizer, a portion of the high-pressure liquid is vaporized at the side port pressure and subcools the remaining high-pressure liquid nearly to the saturation temperature at the operating side port pressure. Since this has little effect on the suction capacity of the compressor, the effective refrigerating capacity of the compressor is increased by the increased heat absorption capacity of the liquid entering the evaporator. Furthermore, the only additional mass flow the compressor must handle is the flash gas entering a closed thread, which is above suction pressure. Thus, under most conditions, the capacity improvement is accompanied by an efficiency improvement.

Economizers become effective when the pressure ratio is equal to about two and above (depending on volume ratio). The subcooling can be made with a direct-expansion shell-and-tube or plate heat exchanger, flash tank, or shell-and-coil intercooler.

As twin-screw compressors are unloaded, the economizer pressure falls toward suction pressure. The additional capacity and improved efficiency of the economizer system is no longer available below a certain percentage of capacity, depending on design.

Hermetic Compressors

Hermetic screw compressors are commercially available through 200 tons of refrigeration effect using R-22. The hermetic motors can operate under discharge, suction, or intermediate pressure. Motor cooling can be with gas, oil, and/or liquid refrigerant.

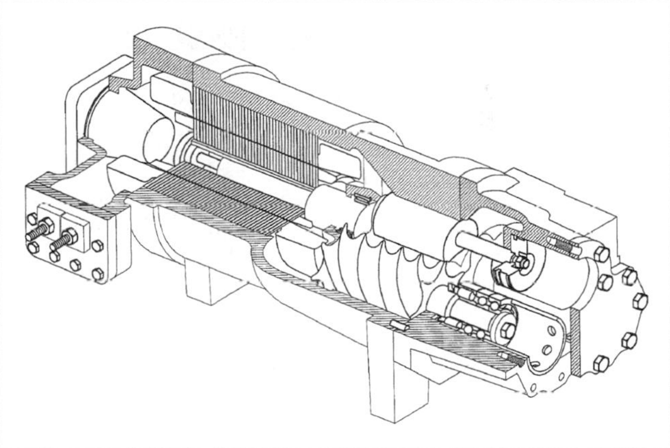

Fig. 30 Semihermetic Twin-Screw Compressor with Suction Gas-Cooled Motor

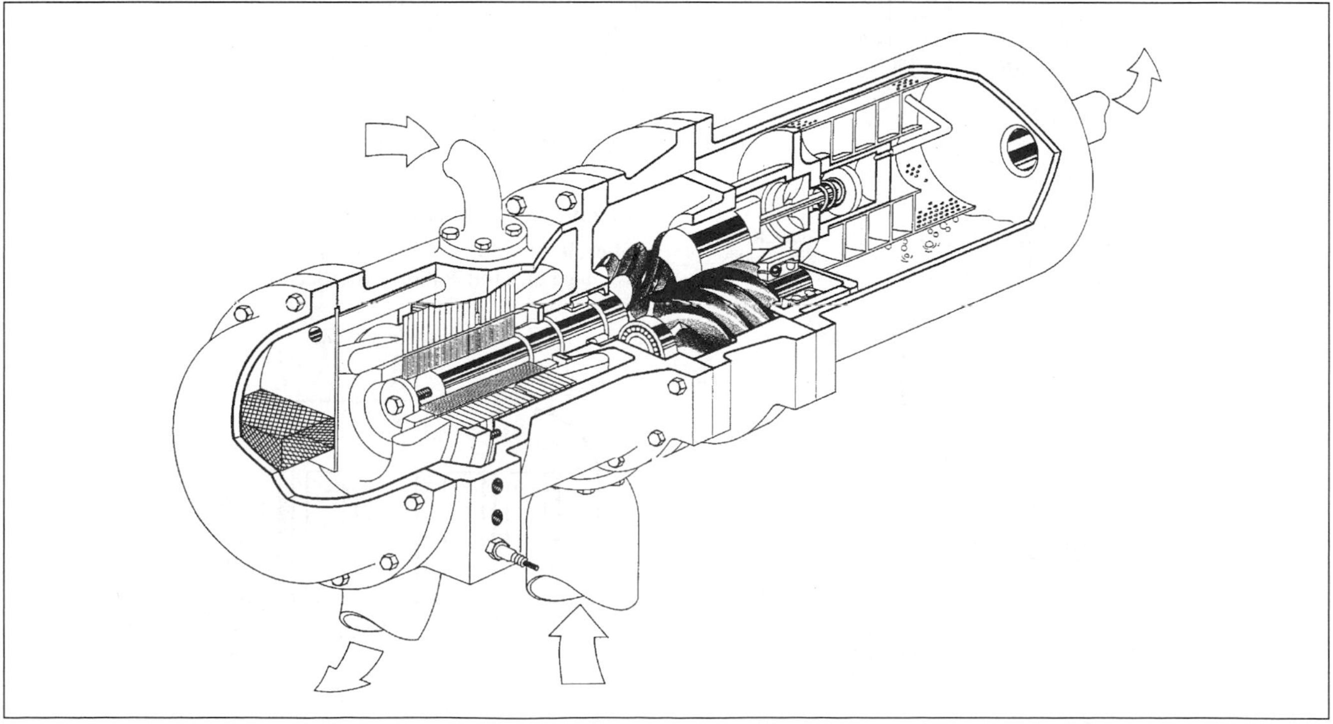

Fig. 31 Semihermetic Twin-Screw Compressor with Motor Housing Used as Economizer; Built-In Oil Separator

Oil separation for these types of compressors may be accomplished either with an integrated oil separator or with a separately mounted oil separator in the system. Figures 30, 31, and 32 show three types of hermetic twin-screw compressors.

Performance Characteristics

Figure 29 shows the full-load efficiency of a modern twin-screw compressor. Both fixed and variable volume ratio compressors without economizers are indicated. High isentropic and volumetric efficiencies are the result of internal compression, the absence of suction or discharge valves, and small clearance volume. The curves show that while volumetric efficiency depends little on the choice of volume ratio, isentropic efficiency depends strongly on it.

Performance data usually note the degree of liquid subcooling and suction superheating assumed. If an economizer is used, the liquid temperature approach and pressure drop to the economizer should be specified.

ORBITAL COMPRESSORS

SCROLL COMPRESSORS

Description

Scroll compressors are orbital motion, positive-displacement machines that compress with two interfitting, spiral-shaped scroll members (Figure 33). They are currently used in residential and commercial air-conditioning, refrigeration, and heat pump applications as well as in automotive air conditioning. Capacities range from 10,000 to 170,000 Btu/h. To function effectively, the scroll compressor requires close tolerance machining of the scroll members, which is possible due to the recent advances in manufacturing technology. This positive-displacement, rotary motion compressor includes performance features, such as high efficiency and low noise.

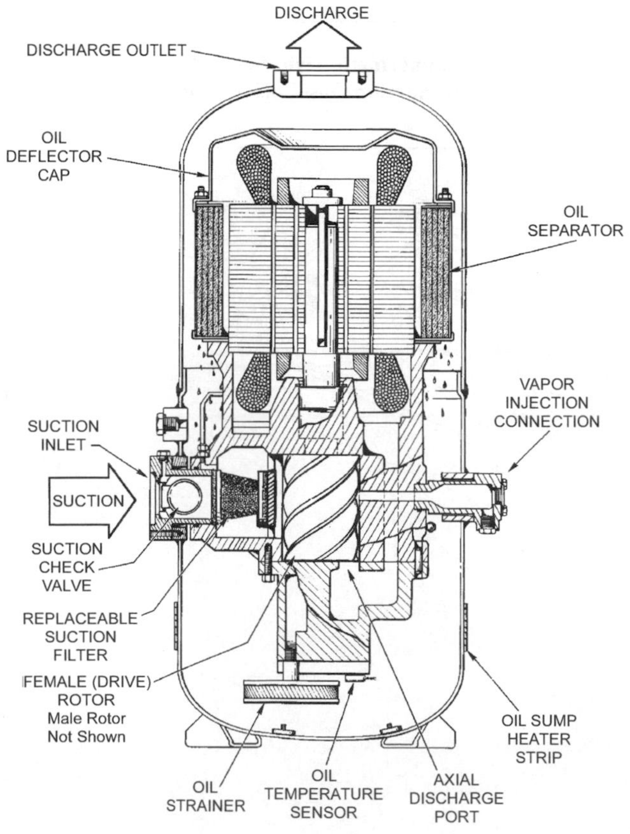

Fig. 32 Vertical, Discharge-Cooled, Semihermetic Twin-Screw Compressor

Scroll members are typically a geometrically matched pair, assembled 180° out of phase. Each scroll member is open on one end of the vane and bound by a base plate on the other. The two scrolls are fitted to form pockets between their respective base plate and various lines of contact between their vane walls. One scroll is held fixed, while the other moves in an orbital path with respect to the first. The flanks of the scrolls remain in contact, although the contact locations move progressively inward. Relative rotation between the pair is prevented by an interconnecting coupling. An alternate approach creates relative orbital motion via two scrolls synchronously rotating about noncoincident axes. As in the former case, an interconnecting coupling maintains a relative angle between the pair of scrolls (Morishita et al. 1988).

Compression is accomplished by sealing suction gas in pockets of a given volume at the outer periphery of the scrolls and progressively reducing the size of those pockets as the scroll relative motion moves them inwards toward the discharge port. Figure 34 shows the sequence of suction, compression, and discharge phases. As the outermost pockets are sealed off (Figure 34A), the trapped gas is at suction pressure and has just entered the compression process. At stages B through F, orbiting motion moves the gas toward the center of the scroll pair, and pressure rises as pocket volumes are reduced. At stage G, the gas reaches the central discharge port and begins to exit from the scrolls. Stages A through H in Figure 34 show that two distinct compression paths operate simultaneously in a scroll set. The discharge process is nearly continuous, since new pockets reach the discharge stage very shortly after the previous discharge pockets have been evacuated.

Scroll compression embodies a fixed, built-in volume ratio that is defined by the geometry of the scrolls and by discharge port location. This feature provides the scroll compressor with different performance characteristics than those of reciprocating or conventional rotary compressors.

Both high-side and low-side shells are available. In the former, the entire compressor is at discharge pressure, except for the outer areas of the scroll set. Suction gas is introduced into the suction port of the scrolls through piping, which keeps it discrete from the rest of the compressor. Discharge gas is directed into the compressor shell, which acts as a plenum. In the low-side type, most of the shell is at suction pressure, and the discharge gas exiting from the scrolls is routed outside the shell, sometimes through a discrete or integral plenum.

Mechanical Features

Scroll Members. Gas sealing is critical to the performance advantage of scroll compressors. Sealing within the scroll set must

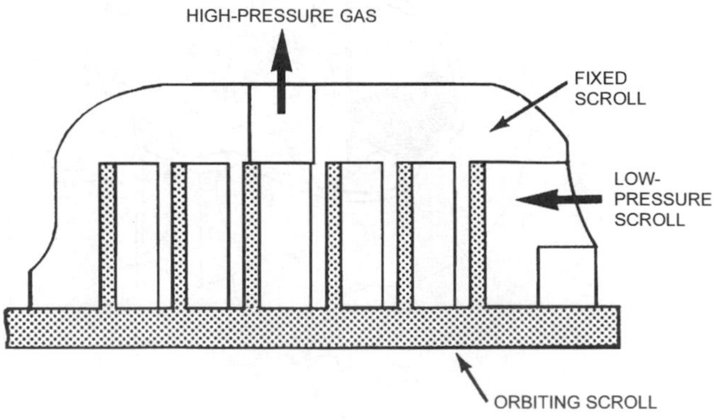

Fig. 33 Interfitted Scroll Members
(Purvis 1987)

be accomplished at the flank contact locations and between the vane tips and bases of the intermeshed scroll pair. Tip/base sealing is generally considered more critical than flank sealing. The method used to seal the scroll members tends to separate scroll compressors into compliant and noncompliant designs.

Noncompliant Designs. In designs lacking compliance, the orbiting scroll takes a fixed orbital path. In the radial direction, sealing small irregularities between the vane flanks (due to flank machining variation) can be accomplished with oil flooding. In the axial direction, the position of both scrolls remains fixed, and flexible seals fitted into machined grooves on the tips of both scrolls accomplish tip sealing. The seals are pressure loaded to enhance uniform contact (McCullough and Shaffer 1976, Sauls 1983).

Radial Compliance. This feature enhances flank sealing and allows the orbiting scroll to follow a flexible path defined by its own contact with the fixed scroll. In one type of radial compliance, a sliding "unloader" bushing is fitted onto the crankshaft eccentric pin in such a way that it directs the radial motion of the orbiting scroll. The orbiting scroll is mounted over this bushing through a drive bearing, and the scroll may now move radially in and out to accommodate variations in orbit radius caused by machining and assembly discrepancies. This feature tends to keep the flanks constantly in contact, and reduces impact on the flanks that can result from intermittent contact. Sufficient clearance in the pin/unloader assembly allows the scroll flanks to separate fully when desired.

In some designs, the mass of the orbiting scroll is selected so that centrifugal force overcomes radial gas compression forces that would otherwise keep the flanks separated.

In some other designs, the drive is designed so that the influence of centrifugal force is reduced, and the drive force overcomes the radial gas compression force (McCullough 1975). Radial compliance has the added benefit of increasing resistance to slugging and contaminants, since the orbiting scroll can "unload" to some extent as it encounters obstacles or nonuniform hydraulic pressures (Bush and Elson 1988).

Axial Compliance. With this feature, an adjustable axial pressure maintains sealing contact between the scroll tips and bases while running. This pressure is released when the unit is shut down, allowing the compressor to start unloaded and to approach full operational speed before a significant load is encountered. This scheme obviates the use of tip seals, eliminating them as a potential source of wear and leakage. With the scroll tips bearing directly on the opposite base plates and with suitable lubrication, sealing tends to improve over time. Axial compliance can either be implemented on the orbiting scroll or the fixed scroll (Tojo et al. 1982, Caillat et al. 1988). The use of axial compliance requires auxiliary scaling of the discharge side with respect to the suction side of the compressor.

Antirotation Coupling. To ensure relative orbital motion, the orbiting scroll must not rotate in response to gas loading. This rotation is most commonly accomplished by an Oldham coupling mechanism, which physically connects the scrolls and permits all planar motion, except relative rotation, between them.

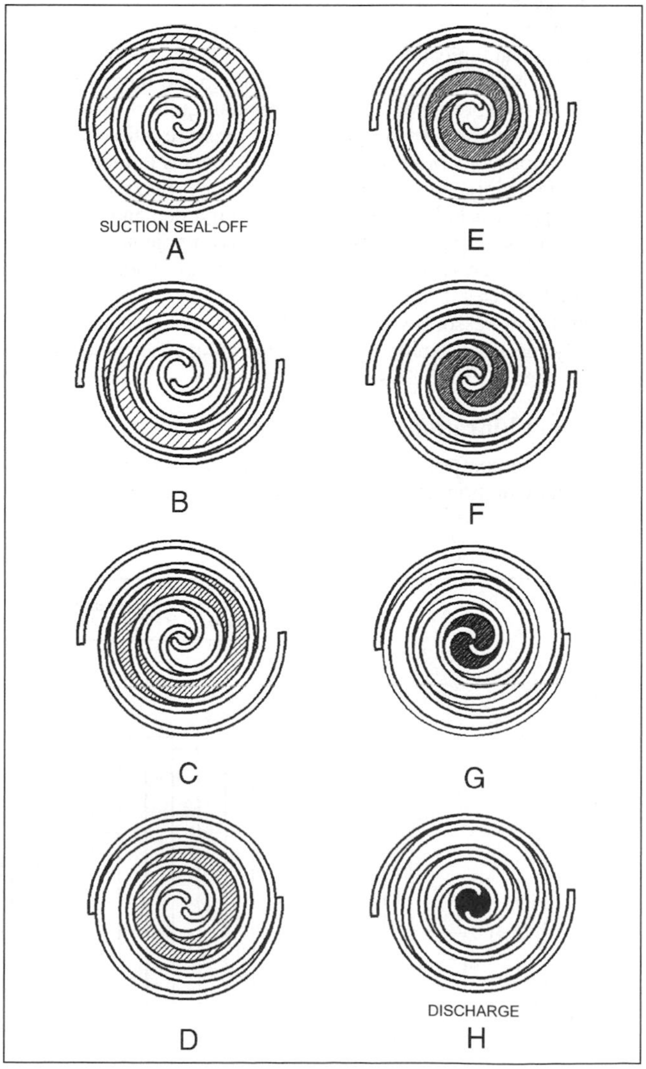

Fig. 34 Scroll Compression Process
(Purvis 1987)

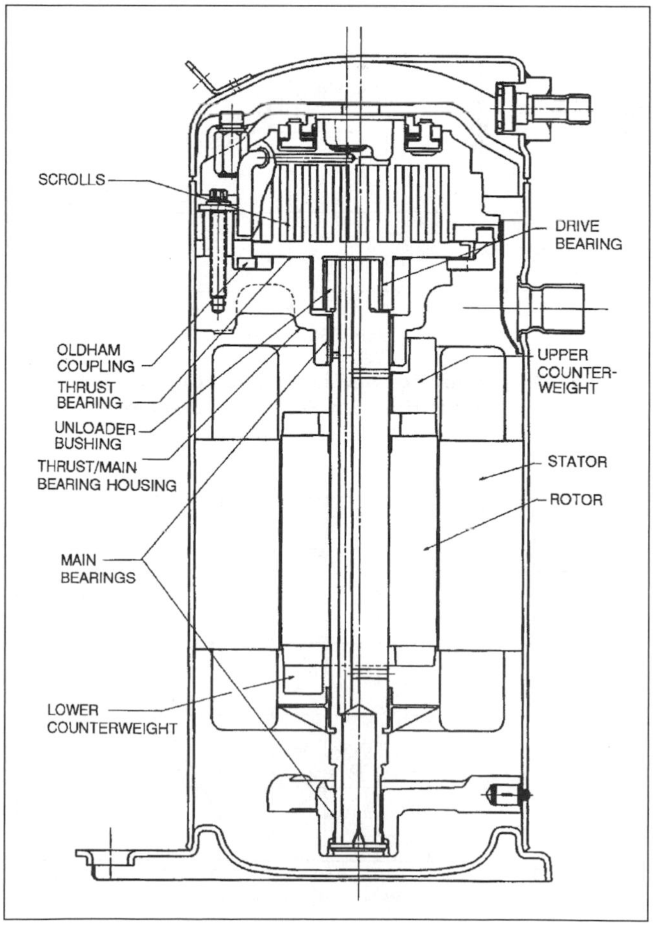

Fig. 35 Bearings and Other Components of Scroll Compressor
(Elson et al. 1990)

Bearing System. The bearing system consists of a drive bearing mounted in the orbiting scroll and generally one of two main bearings. The main bearings are either of the cantilevered type (main bearings on same side of the motor as the scrolls) or consists of a main bearing on either side of the motor (Figure 35). All bearing load vectors rotate through a full 360° due to the nature of the drive load.

The orbiting scroll is supported axially by a thrust bearing on a housing which is part of the internal frame or is mounted directly to the compressor shell.

Capacity Control

Two different capacity control mechanisms are currently being used by the scroll compressor industry.

Variable-Speed Scroll Compressor. Conventional air conditioning uses a constant speed motor to drive the compressor. The variable-speed scroll compressor uses an inverter drive to convert a fixed frequency alternating current into one with adjustable voltage and frequency, which allows the variation of the rotating speed of the compressor motor. The compressor uses either an induction or a permanent magnet motor. Typical operating frequency varies between 15 and 150 Hz. The capacity provided by the machine is nearly directly proportional to its running frequency. Thus, virtually infinite capacity steps are possible for the system with a variable-speed compressor. The variable-speed scroll compressor is now widely used in Japan.

Variable-Displacement Scroll Compressor. This capacity control mechanism incorporates porting holes in the fixed scroll member. The control mechanism disconnects or connects compression chambers to the suction side by respectively closing or opening the porting holes. When all porting holes are closed, the compressor runs at full capacity; opening of all porting holes to the suction side yields the smallest capacity. Thus, by opening or closing a different number of porting holes, variable cooling or heating capability is provided to the system. The number of different capacities and the extent of the capacity reduction available is governed by the locations of the ports in reference to full capacity suction seal-off.

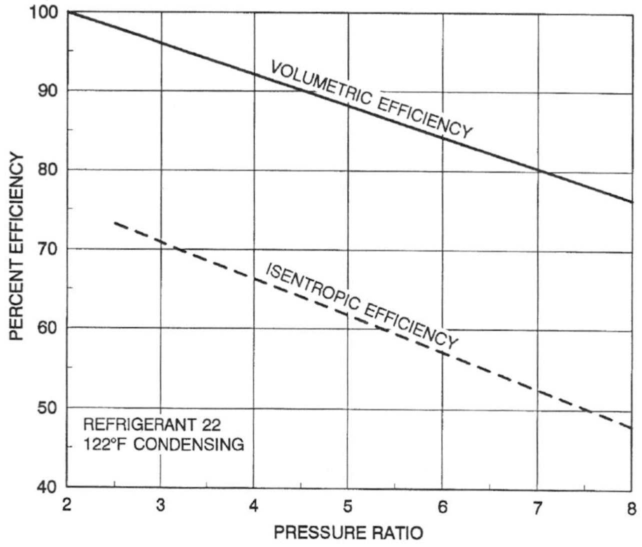

Fig. 36 Volumetric and Isentropic Efficiency Versus Pressure Ratio for Scroll Compressors
(Elson et al. 1990)

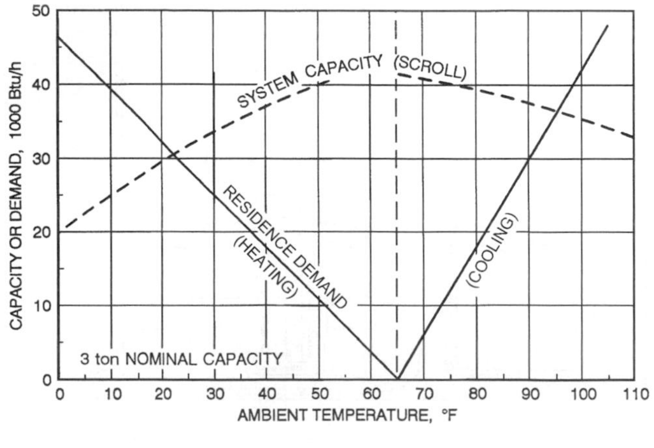

Fig. 37 Scroll Capacity Versus Residence Demand
(Purvis 1987)

Performance

Scroll technology offers an advantage in performance for a number of reasons. Large suction and discharge ports reduce pressure losses incurred in the suction and discharge processes. Also, physical separation of these processes reduces heat transfer to the suction gas. The absence of valves and reexpansion volumes and the continuous flow process results in high volumetric efficiency over a wide range of operating conditions. Figure 36 illustrates this effect. The built-in volume ratio can be designed for lowest over- or under-compression at typical demand conditions (2.5 to 3.5 pressure ratio for air conditioning). Isentropic efficiency in the range of 70% is possible at such pressure ratios, and it remains quite close to the efficiency of other compressor types at high pressure ratio (Figure 36). Scroll compressors offer a flatter capacity versus outdoor ambient curve than reciprocating products, which means that they can more closely approach indoor requirements at high demand conditions. As a result, the heat pump mode requires less supplemental heating; the cooling mode is more comfortable, because cycling is less as demand decreases (Figure 37).

Scroll compressors available for the North American market are typically specified as producing ARI operating efficiencies (COPs) in the range of 3.10 to 3.34.

Noise and Vibration

The scroll compressor inherently possesses a potential for low sound and vibration. It includes a minimal number of moving parts

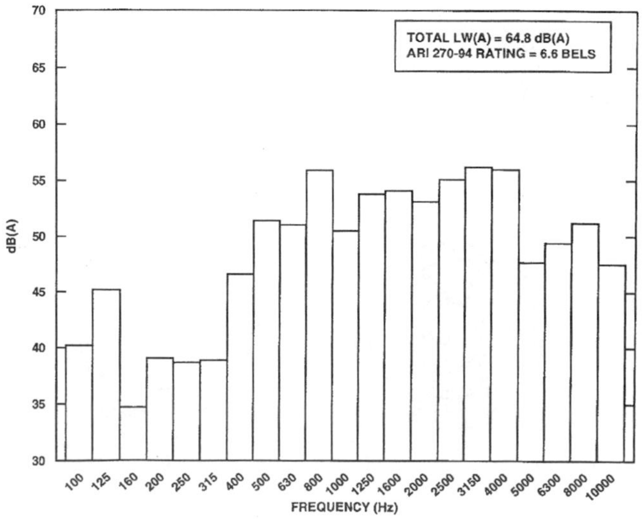

Fig. 38 Typical Scroll Sound Spectrum

compared to other compressor technologies. Since scroll compression requires no valves, impact noise and vibration are completely eliminated. The presence of a continuous suction-compression-discharge process and low gas pressure pulsation help to keep vibration low. A virtually perfect dynamic balancing of the orbiting scroll inertia with counterweights eliminates possible vibration due to the rotating parts.

Also, smooth surface finish and accurate machining of the vane profiles and base plates of both scroll members (requirement for small leakage) aids in minimal impact of the vanes.

A typical sound spectrum of the scroll compressor is shown in Figure 38.

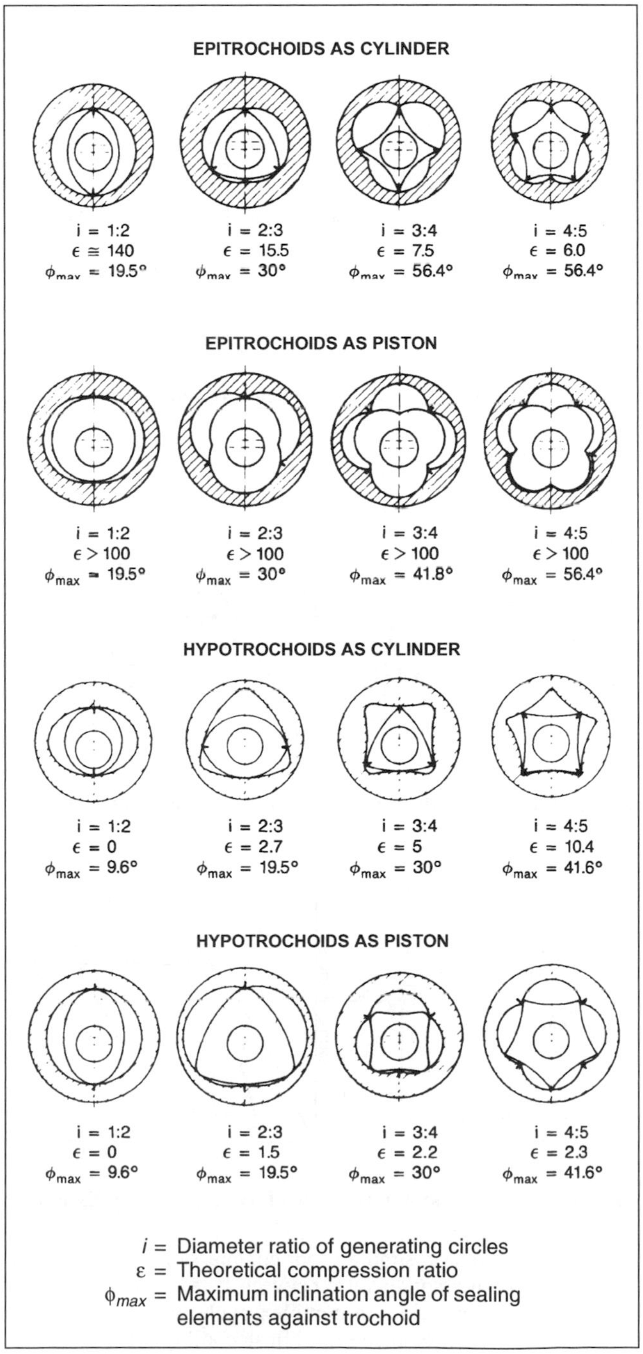

Fig. 39 Possible Versions of Epitrochoidal and Hypotrochoidal Machines

Operation and Maintenance

Most scroll compressors used today are of the hermetic type, which require virtually no maintenance. However, the compressor manufacturer's operation and application manual should be followed.

TROCHOIDAL COMPRESSORS

The trochoidal compressor is a small, rotary, positive-displacement compressor which can run at high speed up to 9000 rpm. They are manufactured in various configurations. Trochoidal curvatures can be produced by the rolling motion of one circle outside or inside the circumference of a basic circle, producing either epitrochoids or hypotrochoids, respectively. Both types of trochoids can be used either as a cylinder or piston form, so that four types of trochoidal machines can be designed (Figure 39).

In each case, the counterpart of the trochoid member always has one apex more than the trochoid itself. In the case of a trochoidal cylinder, the apexes of the piston show a slipping motion along the inner cylinder surface; for trochoidal piston design, the piston shows a gear-like motion. As seen in Figure 40, a built-in theoretical pressure ratio disqualifies many configurations as valid concepts for refrigeration compressor design. Because of additional valve ports, clearances, etc., and the resulting decrease in the built-in maximum theoretical pressure ratio, only the first two types with epitrochoidal cylinders, and all candidates with epitrochoidal pistons, can be used for compressor technology. The latter, however, require sealing elements on the cylinder as well as on the side plates, which does not allow the design of a closed sealing borderline.

In the past, trochoidal machines were designed much like those of today. However, like other positive-displacement rotary concepts that could not tolerate oil injection, the early trochoidal equipment failed because of sealing problems. The invention of a closed sealing border by Wankel changed this (Figure 40). Today, the Wankel trochoidal compressor with a three-sided epitrochoidal piston (motor) and two-envelope cylinder (casing) is built in capacities of up to 2 tons.

Description and Performance

Compared to other compressors of similar capacity, trochoidal compressors have many advantages typical of reciprocating compressors. Because of the closed sealing border of the compression space, these compressors do not require extremely small and expensive manufacturing tolerances; neither do they need oil for sealing,

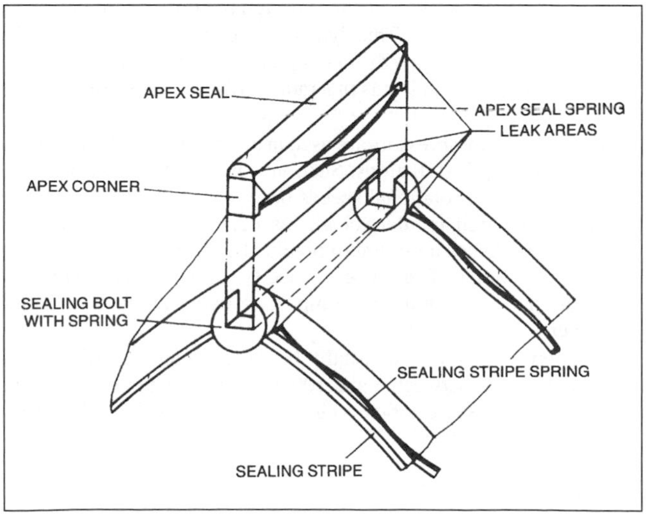

Fig. 40 Wankel Sealing System for Trochoidal Compressors

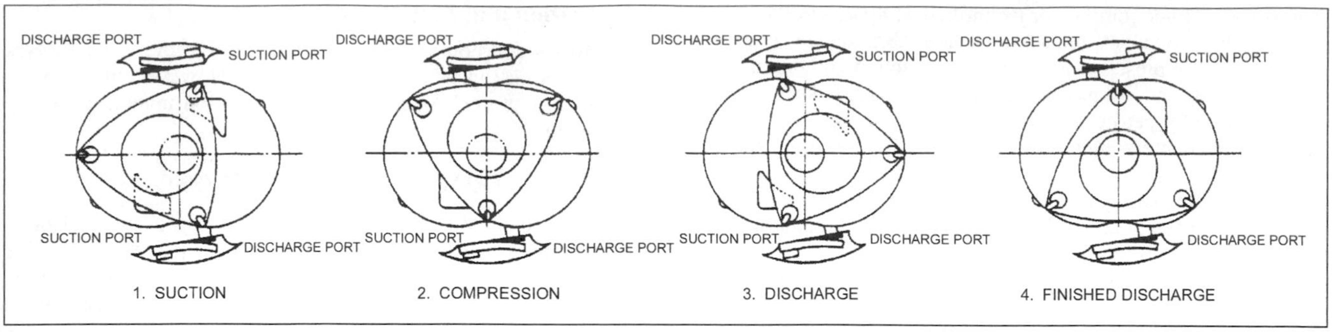

Fig. 41 Sequence of Operation of Wankel Rotary Compressor

keeping them at low pressure side with the advantage of low solubility and high viscosity of the oil-refrigerant mixture. Valves are usually used on a high-pressure side while suction is ported. A valveless version of the trochoidal compressor can also be built. Figure 41 shows the operation of the Wankel rotary compressor (2:3 epitrochoid) with discharge reed valves.

The Wankel compressor performance compares favorably with the reciprocating piston compressors at a higher speed and moderate pressure ratio range. A smaller number of moving parts, less friction, and the resulting higher mechanical efficiency improves the overall isentropic efficiency. This can be observed at higher speed when sealing is better, and in the moderate pressure ratio range when the influence of the clearance volume is limited.

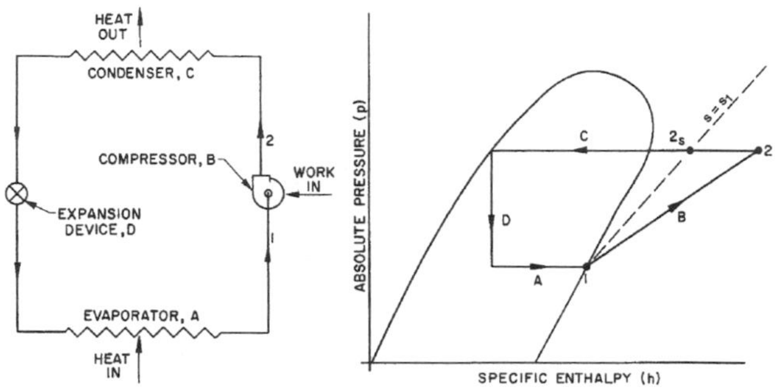

Fig. 42 Simple Vapor Compression Cycle

CENTRIFUGAL COMPRESSORS

Centrifugal compressors, sometimes called turbocompressors, belong to a family of turbomachines that includes fans, propellers, and turbines. These machines continuously exchange angular momentum between a rotating mechanical element and a steadily flowing fluid. Because their flows are continuous, turbomachines have greater volumetric capacities, size for size, than do positive-displacement devices. For effective momentum exchange, their rotating speeds must be higher, but little vibration or wear results because of the steadiness of the motion and the absence of contacting parts.

Centrifugal compressors are well suited for air-conditioning and refrigeration applications because of their ability to produce a high pressure ratio. The suction flow enters the rotating element, or impeller, in the axial direction and is discharged radially at a higher velocity. The change in diameter through the impeller increases the velocity of the gas flow. This dynamic pressure is then converted to static pressure, through a diffusion process, which generally begins within the impeller and ends in a radial diffuser and scroll outboard of the impeller.

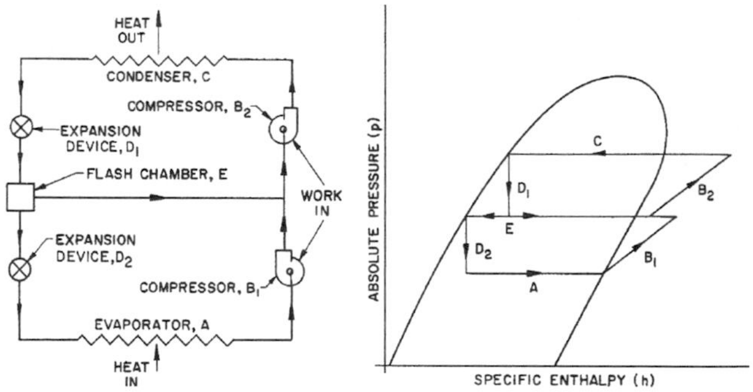

Fig. 43 Compression Cycle with Flash Cooling

Centrifugal compressors are used in a variety of refrigeration and air-conditioning installations. Suction flow ranges between 60 and 30,000 cfm, with rotational speeds between 1800 and 90,000 rpm. However, the high angular velocity associated with a low volumetric flow establishes a minimum practical capacity for most centrifugal applications. The upper capacity limit is determined by physical size, a 30,000 cfm compressor being about 6 or 7 ft in diameter.

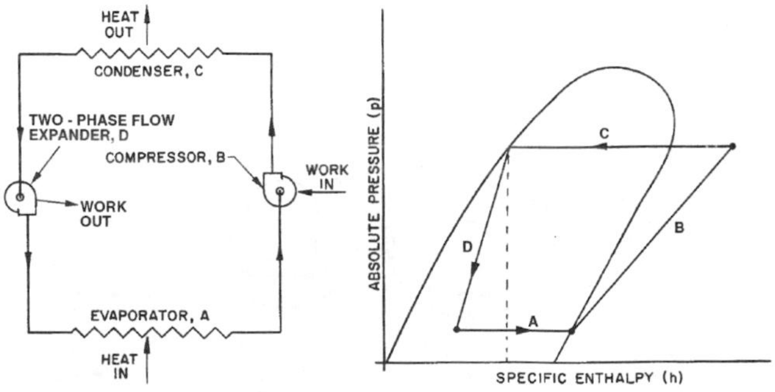

Fig. 44 Compression Cycle with Power Recovery Expander

Suction temperature is usually between 50 and −150°F, with a suction pressure between 2 and 100 psia and discharge pressure up to 300 psia. Pressure ratios range between 2 and 30. Almost any refrigerant can be used.

Refrigeration Cycle

Typical applications might involve a single-, two-, or three-stage halocarbon compressor or a seven-stage ammonia compressor. Figure 42 illustrates a simple vapor compression cycle in which a centrifugal compressor operates between states 1 and 2.

Figure 43 shows a more complex cycle, with two stages of compression and interstage liquid flash cooling. This cycle has a higher coefficient of performance than the simple cycle and is frequently used with two- through four-stage halocarbon and hydrocarbon compressors.

Figure 44 shows a vapor compression cycle in which the expansion device is replaced by a power-recovering two-phase-flow turbine. The power recovered by the turbine is used to reduce the required compressor input work (Brasz 1995). Power recovery during the expansion process reduces the enthalpy of the two-phase flow mixture, thus increasing the refrigeration effect of this cycle.

More than one stage of flash cooling can be applied to compressors with more than two impellers. Liquid subcooling and interstage desuperheating can also be advantageously used. For more information on refrigeration cycles, see Chapter 1 of the 1997 *ASHRAE Handbook—Fundamentals*.

Angular Momentum

The momentum exchange, or energy transfer, between a centrifugal impeller and a flowing refrigerant is expressed by

$$W_i = u_i c_u / g_c \tag{11}$$

where

W_i = impeller work input per unit mass of refrigerant, $ft \cdot lb_f/lb_m$
u_i = impeller blade tip speed, fps
c_u = tangential component of refrigerant velocity leaving impeller blades, fps
g_c = gravitational constant = 32.17 $lb_m \cdot ft/lb_f \cdot s^2$

These velocities are shown in Figure 45, where refrigerant flows out from between the impeller blades with relative velocity b and absolute velocity c. The relative velocity angle β is a few degrees less than the blade angle because of a phenomenon known as slip.

Equation (11) assumes that the refrigerant enters the impeller without any tangential velocity component or swirl. This is generally the case at design flow conditions. If the incoming refrigerant was already swirling in the direction of rotation, the impeller's ability to impart angular momentum to the flow would be reduced. A subtractive term would then be required in the equation.

Some of the work done by the impeller increases the refrigerant pressure, while the remainder only increases its kinetic energy. The ratio of pressure-producing work to total work is known as the impeller reaction. Since this varies from about 0.4 to about 0.7, an appreciable amount of kinetic energy leaves the impeller with magnitude $c^2/2g_c$.

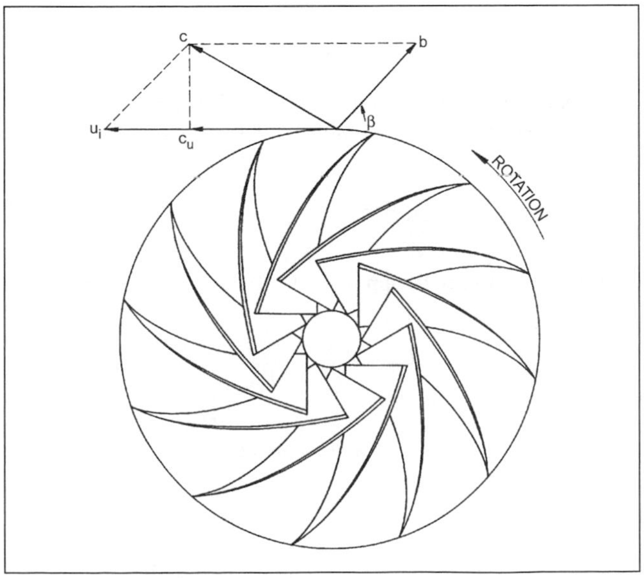

Fig. 45 Impeller Exit Velocity Diagram

To convert this kinetic energy into additional pressure, a diffuser is located after the impeller. Radial vaneless diffusers are most common, but vaned diffusers, scroll diffusers, and conical diffusers are also used.

In a multistage compressor, the flow leaving the first diffuser is guided to the inlet of the second impeller and so on through the machine, as can be seen in Figure 46. The total compression work input per unit mass of refrigerant is the sum of the individual stage inputs:

$$W = \Sigma W_i \tag{12}$$

provided that the mass flow rate is constant throughout the compressor.

Description

A centrifugal compressor can be single-stage, having only one impeller, or it can be multistage, having two or more impellers mounted in the same casing as shown in Figure 46. For process refrigeration applications, a compressor can have as many as ten stages.

The suction gas generally passes through a set of adjustable inlet guide vanes or an external suction damper before entering the impeller. The vanes (or suction damper) are used for capacity control as will be described later.

The high-velocity gas discharging from the impeller enters the radial diffuser which can be vaned or vaneless. Vaned diffusers are typically used in compressors designed to produce high pressure. These vanes are generally fixed, but they can be adjustable. Adjustable diffuser vanes can be used for capacity modulation either in lieu of or in conjunction with the inlet guide vanes.

For multistage compressors, the gas discharged from the first stage is directed to the inlet of the second stage through a return channel. The return channel can contain a set of fixed flow straightening vanes or an additional set of adjustable inlet guide vanes. Once the gas reaches the last stage, it is discharged in a volute or collector chamber. From there, the high-pressure gas passes through the compressor discharge connection.

When multistage compressors are used, side loads can be introduced between stages so that one compressor performs several functions at several temperatures. Multiple casings can be connected in tandem to a single driver. These can be operated in series, in parallel, or even with different refrigerants.

ISENTROPIC ANALYSIS

The static pressure resulting from a compressor's work input or, conversely, the amount of work required to produce a given pressure rise, depends on the efficiency of the compressor and the thermodynamic properties of the refrigerant. For an adiabatic process, the work input required is minimal if the compression is isentropic. Therefore, actual compression is often compared to an isentropic process, and the performance thus evaluated is based on an isentropic analysis.

The reversible work required by an isentropic compression between states 1 and 2_s in Figure 42 is known as the **adiabatic work**, as measured by the enthalpy difference between the two points:

$$W_s = J(h_{2s} - h_1) \tag{13}$$

where J = 778.16 $ft \cdot lb_f/Btu$. Assuming negligible cooling occurs, the irreversible work done by the actual compressor is

$$W_s = J(h_2 - h_1) \tag{14}$$

Flash-cooled compressors cannot be analyzed by this procedure unless they are subdivided into uncooled segments with the cooling

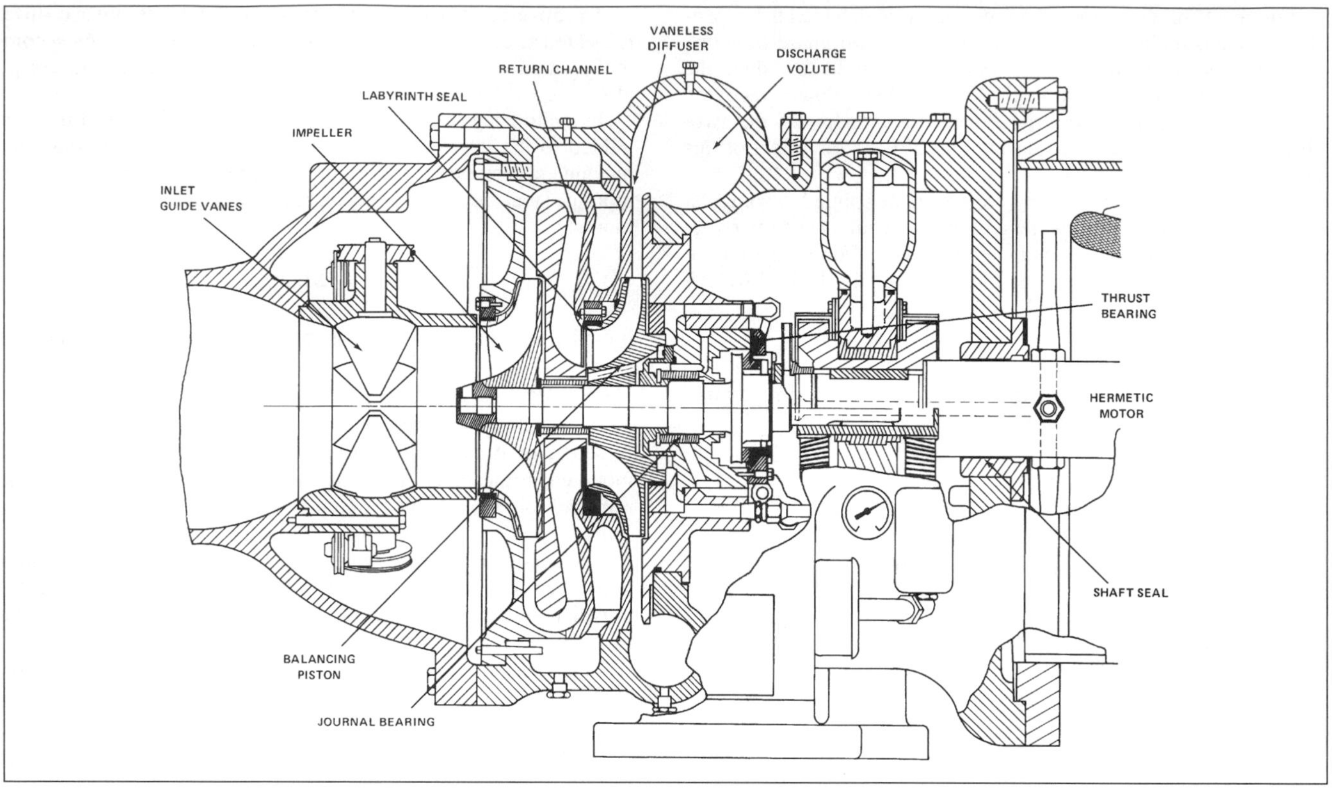

Fig. 46 Centrifugal Refrigeration Unit Cross Section

effects evaluated by other means. Compressors with side flows must also be subdivided. In Figure 43, the two compression processes must be analyzed individually.

Equation (14) also assumes a negligible difference in the kinetic energies of the refrigerant at states 1 and 2. If this is not the case, a kinetic energy term must be added to the equation. All the thermodynamic properties throughout the section on Centrifugal Compressors are static properties as opposed to stagnation properties; the latter includes kinetic energy.

The ratio of isentropic work to actual work is the **adiabatic efficiency**:

$$\eta_s = \frac{h_{2s} - h_1}{h_2 - h_1} \tag{15}$$

This varies from about 0.62 to about 0.83, depending on the application. Because of the thermodynamic properties of gases, a compressor's overall adiabatic efficiency does not completely indicate its individual stage performance. The same compressor produces different adiabatic results with different refrigerants and also with the same refrigerant at different suction conditions.

In spite of its shortcomings, isentropic analysis has a definite advantage in that adiabatic work can be read directly from thermodynamic tables and charts similar to those presented in Chapter 19 of the 1997 *ASHRAE Handbook—Fundamentals*. Where these are unavailable for the particular gas or gas mixture, they can be accurately calculated and plotted using thermodynamic relationships and a computer.

POLYTROPIC ANALYSIS

Polytropic work and efficiency are more consistent from one application to another, because a reversible polytropic process duplicates the actual compression between states 1 and 2 in Figure 42. Therefore, values calculated by the polytropic analysis have greater versatility than those of the isentropic analysis.

The path equation for this reversible process is

$$\eta = v(dp/dh) \tag{16}$$

where η is the **polytropic efficiency** and v is the specific volume of the refrigerant. The reversible work done along the polytropic path is known as the **polytropic work** and is given by

$$W_p = \int_{p_1}^{p_2} v\, dp \tag{17}$$

It follows from Equations (14), (16), and (17) that the polytropic efficiency is the ratio of reversible work to actual work:

$$\eta = \frac{W_p}{h_2 - h_1} \tag{18}$$

Equation (16) can be approximated by

$$\frac{p^m}{T} = \frac{p_1^m}{T_1} = \frac{p_2^m}{T_2} \tag{19}$$

$$pv^n = p_1 v_1^n = p_2 v_2^n \tag{20}$$

where

$$m = \frac{ZR}{c_p}\left(\frac{1}{\eta} + X\right) = \frac{(k - 1/k)(1/\eta + X)Y}{(1 + X)^2} \tag{21}$$

$$n = \frac{1}{Y - (ZR/c_p)(1/\eta + X)(1 + X)}$$

$$= \frac{1 + X}{Y[(1/k)(1/\eta + X) - (1/\eta - 1)]} \quad (22)$$

and

$$X = \frac{T}{v}\left(\frac{\partial v}{\partial p}\right)_p - 1 \quad (23)$$

$$Y = -\frac{p}{v}\left(\frac{\partial v}{\partial p}\right)_T \quad (24)$$

$$Z = \frac{pv}{RT} \quad (25)$$

Also, R is the gas constant and k is the ratio of specific heats; all properties are at temperature T. These equations can be used to permit integration so that Equation (17) can be written as follows:

$$W_p = \frac{n}{(n-1)}p_1 v_1\left[\left(\frac{p_2}{p_1}\right)^{(n-1)/n} - 1\right] \quad (26)$$

Further manipulation eliminates the exponent:

$$W_p = \left[\frac{p_2 v_2 - p_1 v_1}{\ln(p_2 v_2/p_1 v_1)}\right]\ln\left(\frac{p_2}{p_1}\right) \quad (27)$$

For greater accuracy in handling gases with properties known to deviate substantially from those of a perfect gas, a more complex procedure is required. The accuracy with which Equations (19) and (20) represent Equation (16) depends on the constancy of m and n along the polytropic path. Because these exponents usually vary, mean values between states 1 and 2 should be used.

Compressibility functions X and Y have been generalized for gases in corresponding states by Schultz (1962) and their equivalents are listed by Edminster (1961). For usual conditions of refrigeration interest (i.e., for $p < 0.9p_c$, $T < 1.5T_c$, and $0.6 < Z$), these functions can be approximated by

$$X = 0.1846(8.36)^{1/Z} - 1.539 \quad (28)$$

$$Y = 0.074(6.65)^{1/Z} + 0.509 \quad (29)$$

The compressibility factor Z has been generalized by Edminster (1961) and Hougen et al. (1959), among others. Generalized corrections for the specific heat at constant pressure c_p can also be found in these works.

Equations (19) and (20) make possible the integration of Equation (17):

$$W_p = f\left(\frac{n}{n-1}\right)p_1 v_1\left[\left(\frac{p_2}{p_1}\right)^{(n-1)/n} - 1\right] \quad (30)$$

In Equation (30), the polytropic work factor f corrects for whatever error may result from the approximate nature of Equations (19) and (20). Since the value of f is between 1.00 and 1.02 in most refrigeration applications, it is generally neglected.

Once the polytropic work has been found, the efficiency follows from Equation (18). Polytropic efficiencies range from about 0.70 to about 0.84, a typical value being 0.76.

The highest efficiencies are obtained with the largest compressors and the densest refrigerants because of a Reynolds number effect discussed by Davis et al. (1951). A small number of stages is also advantageous because of parasitic loss associated with each stage.

Overall, polytropic work and efficiency are more consistent from one application to another because they represent an average stage aerodynamic performance.

Instead of using Equations (16) through (30), it is easier and often more desirable to determine the adiabatic work by an isentropic analysis and then to convert to polytropic work by

$$\frac{W_p}{W_s} \approx \eta\left[\frac{(p_2/p_1)^{(k-1)/k\eta} - 1}{(p_2/p_1)^{(k-1)/k} - 1}\right] \quad (31)$$

Equation (31) is strictly correct only for an ideal gas, but because it is a ratio involving comparable errors in both numerator and denominator, it is of more general utility. Equation (31) is plotted in Figure 47 for $\eta = 0.76$. To obtain maximum accuracy, the ratio of specific heats k must be a mean value for states 1, 2, and 2_s. If c_p is known, k can be determined by

$$k = \frac{1}{1 - (ZR/c_p)(1 + X)^2/Y} \quad (32)$$

The gas compression power is

$$P = wW \quad (33)$$

where w is the mass flow. To obtain total shaft power, add the mechanical friction loss. Friction loss varies from less than 1% of the gas power to more than 10%. A typical estimate is 3% for compressor friction losses.

Nondimensional Coefficients

Some nondimensional performance parameters used to describe centrifugal compressor performance are flow coefficient, polytropic work coefficient, Mach number, and specific speed.

Flow Coefficient. Desirable impeller diameters and rotational speeds are determined from the blade tip velocity by a dimensionless flow coefficient Q/ND^3 in which Q is the volumetric flow rate.

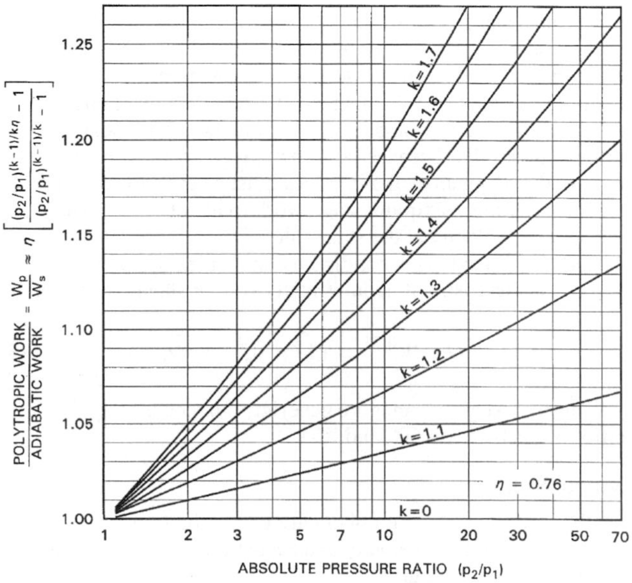

Fig. 47 Ratio of Polytropic to Adiabatic Work

Practical values for this coefficient range from 0.02 to 0.35, with good performance falling between 0.11 and 0.21. Optimum results occur between 0.15 and 0.18. Impeller diameter D_i and rotational speed N follow from

$$Q/ND^3 = \pi Q_i/u_iD_i^2 = \pi^3 Q_iN^2/u_i^3 \quad (34)$$

where u_i is the tip speed.

The maximum flow coefficient in multistage compressors is found in the first stage and the minimum in the last stage (unless large side loads are involved). For high-pressure ratios, special measures may be necessary to increase the last stage (Q/ND^3) to a practical level as stated previously. Side loads are beneficial in this respect, but interstage flash cooling is not.

Polytropic Work Coefficient. Polytropic work and polytropic head can be used interchangeably. Since this chapter is concerned with polytropic work, that term will be used exclusively. Polytropic head is related to it by $H_p = W_p/g$.

Besides the power requirement, polytropic work also determines impeller blade tip speed and number of stages. For an individual stage, the stage work is related to speed by

$$W_{pi} = \mu_i u_i^2/g_c \quad (35)$$

where μ_i is the stage work coefficient.

The overall polytropic work is the sum of the stage works:

$$W_p = \sum W_{pi} \quad (36)$$

and the overall work coefficient is

$$\mu = g_c W_p \sum u_i^2 \quad (37)$$

Values for μ (and μ_i) range from about 0.42 to about 0.74, with 0.55 representative for estimating purposes. Compressors designed for modest work coefficients have backward-curved impeller blades. These impellers tend to have greater part-load ranges and higher efficiencies than do radial-bladed designs.

Maximum tip speeds are limited by strength considerations to about 1400 fps. For cost and reliability, 980 fps is a more common limitation. On this basis, the maximum polytropic work capability of a typical stage is about 15,000 ft·lb$_f$/lb.

A greater restriction on stage work capability is often imposed by the impeller Mach number M_i. For adequate performance, M_i must be limited to about 1.8 for stages with impellers overhung from the ends of shafts and to about 1.5 for impellers with shafts passing through their inlets because flow passage geometries are moved out. For good performance, these values must be even lower. Such considerations limit maximum stage work to about 1.5 a_i^2, where a_i is the acoustic velocity at the stage inlet.

Specific Speed. This nondimensional index of optimum performance characteristic of geometrically similar stages is defined by

$$N_s = N\sqrt{Q_i}/W_{pi}^{0.75} = (1/\pi^3\mu_i^{0.75})\sqrt{Q_i/ND_i^3} \quad (38)$$

The highest efficiencies are generally attained in stages with specific speeds between 600 and 850.

Mach Number

Two different Mach numbers are used. The flow Mach number M is the ratio of flow velocity c to acoustic velocity a at a particular point in the fluid stream:

$$M = c/a \quad (39)$$

Table 4 Acoustic Velocity of Saturated Vapor, fps

	Evaporator Temperature, °F						
Refrigerant	–200	–150	–100	–50	–0	50	100
11			386	410	430	446	456
12		388	413	433	445	447	437
13	390	417	433	434	419	384	
13B1	324	349	369	381	382	370	345
22		471	500	523	535	535	522
23	490	526	550	558	547	516	
113				343	362	377	388
114		315	338	357	372	381	380
123			361	383	402	417	425
124		358	382	403	417	423	417
125	353	381	405	419	421	407	373
134a		418	446	469	482	484	470
142b		421	450	475	494	503	501
152a		530	565	594	614	621	612
500		429	458	479	493	494	483

Source: Gallagher et al. (1993).

where

$$a = v\sqrt{-g(\partial p/\partial v)_s} = \sqrt{n_s gpv} \quad (40)$$

Values of acoustic velocity for a number of saturated vapors at various temperatures are presented in Table 4.

The flow Mach number in a typical compressor varies from about 0.3 at the stage inlet and outlet to about 1.0 at the impeller exit. With increasing flow Mach number, the losses increase because of separation, secondary flow, and shock waves.

The impeller Mach number M_i, which is a pseudo Mach number, is the ratio of impeller tip speed to acoustic velocity a_i at the stage inlet:

$$M_i = u_i/a_i \quad (41)$$

Performance

From an applications standpoint, more useful parameters than μ and (Q/ND^3) are Ω and Θ (Sheets 1952):

$$\Omega = gW_p/a_i^2 = \mu(\Sigma u_i^2/a_i^2) \quad (42)$$

$$\Theta = Q_1/a_1D_1^2 = (M_1/\pi)(Q_1/ND_1{}^3) \quad (43)$$

They are as general as the customary test coefficients and produce performance maps like the one in Figure 48, with speed expressed in terms of first-stage impeller Mach number M_1.

A compressor user with a particular installation in mind may prefer more explicit curves, such as pressure ratio and power versus volumetric flow at constant rotational speed. Plots of this sort may require fixed suction conditions to be entirely accurate, especially if discharge pressure and power are plotted against mass flow or refrigeration effect.

A typical compressor performance map is shown in Figure 48 where percent of rated work is plotted with efficiency contours against percent of rated volumetric flow at various speeds. Point A is the design point at which the compressor operates with maximum efficiency. Point B is the selection or rating point at which the compressor is being applied to a particular system. From the application or user's point of view, Ω and Θ have their 100% values at Point B.

To reduce first cost, refrigeration compressors are selected for pressure and capacity beyond their peak efficiency, as shown in Figure 48. The opposite selection would require a larger impeller and additional stages. Refrigerant acoustic velocity and the ability

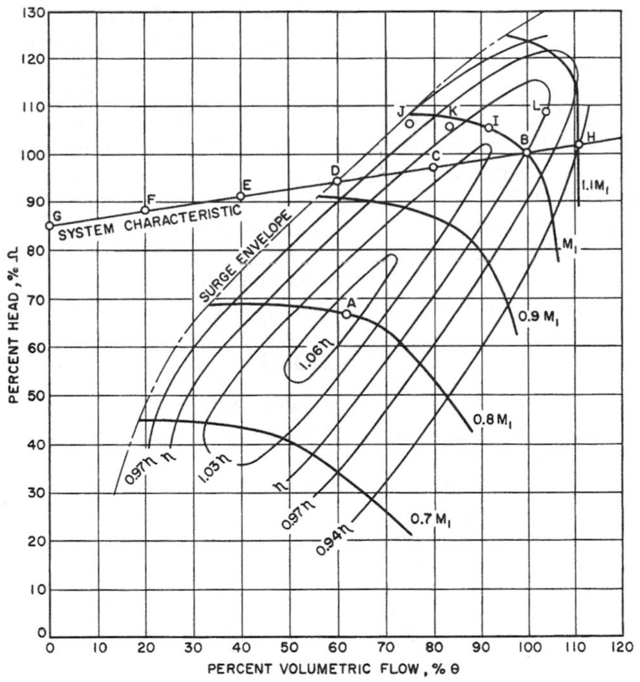

Fig. 48 Typical Compressor Performance Curves

to operate at a high enough Mach number are also of concern. If the compressor shown in Figure 48 were of a multistage design, M_1 would be about 1.2; for a single-stage compressor, it would be about 1.5.

Another acoustical effect is seen on the right of the performance map, where increasing speed does not produce a corresponding increase in capacity. The maximum flow at M_1 and $1.1M_1$ approach a limit determined by the relative velocity of the refrigerant entering the first impeller. As this velocity approaches a sonic value, the flow becomes choked and further increases become impossible. Another commonly used term for this phenomenon is stonewalling; it represents the maximum capacity of an impeller.

Testing

When a centrifugal compressor is tested, overall μ and η versus Q_1/ND_1^3 at constant M_1 are plotted. They are useful because test results with one gas are sometimes converted to field performance with another. When side flows and cooling are involved, the overall work coefficient is found from Equations (36) and (37) by evaluating the mixing and cooling effects between stages separately. The **overall efficiency** in such cases is

$$\eta = \frac{\sum w_i W_{pi}}{\sum w_i W_i} \tag{44}$$

Testing with a fluid other than the design refrigerant is a common practice known as **equivalent performance testing**. Its need arises from the impracticability of providing test facilities for the complete range of refrigerants and input power for which centrifugal compressors are designed. Equivalent testing is possible because a given compressor produces the same μ and η at the same (Q/ND^3) and M_i with any fluids whose volume ratios (v_1/v_2) and Reynolds numbers are the same.

The thermodynamic performance of a compressor can be evaluated according to either the stagnation or the static properties of the refrigerant, and it is important to distinguish between these concepts. The **stagnation efficiency**, for example, may be higher than the **static efficiency**. The safest procedure is to use static properties and evaluate kinetic energy changes separately.

Surging

Part-load range is limited (on the left side of the performance map) by a **surge envelope**. Satisfactory compressor operation to the left of this line is prevented by unstable **surging** or **hunting**, in which the refrigerant alternately flows backward and forward through the compressor, accompanied by increased noise, vibration, and heat. Prolonged operation under these conditions can damage the compressor.

The flow reverses during surging about once every 2 s. Small systems surge at higher frequencies and large systems at lower. Surging can be distinguished from other kinds of noise and vibration by the fact that its flow reversals alternately unload and load the driver. Motor current varies markedly during surging, and turbines alternately speed up and slow down.

Another kind of instability, **rotating stall**, may occur slightly to the right of the true surge envelope. This phenomenon involves the formation of rotating stall pockets or cells in the diffuser. It produces a roaring noise at a frequency determined by the number of cells formed and the impeller running speed. The driver load is steady during rotating stall, which is harmless to the compressor, but it may, however, vibrate components excessively.

System Balance

If a refrigeration system characteristic is superimposed on a compressor performance map, it shows the speed and efficiency at which the compressor operates in that particular application. A typical brine cooling system curve is plotted in Figure 48, passing through Points B, C, D, E, F, G, and H. With increased speed, the compressor at Point H produces more than its rated capacity; with decreased speeds at Points C and D, it produces less. Because of surging, the compressor cannot be operated satisfactorily at Points E, F, or G.

The system can be operated at these capacities, however, by using a hot gas bypass. The volume flow at the compressor suction must be at least that for Point D in Figure 48; this volume flow is reached by adding hot gas from the compressor discharge to the evaporator, or compressor suction piping. When hot gas bypass is used, no further power reduction is realized as the load decreases. The compressor is being artificially loaded to stay out of the surge envelope. The increased volume due to hot gas recirculation performs no useful refrigeration.

Capacity Control

When the driver speed is constant, a common method of altering capacity is to swirl the refrigerant entering one or more impellers. Adjustable inlet guide vanes, or **prerotation vanes** (Figure 46), produce the swirl. Setting these vanes to swirl the flow in the direction of rotation produces a new compressor performance curve without any change in speed. Controlled positioning of the vanes can be accomplished by pneumatic, electrical, or hydraulic means.

Typical curves for five different vane positions are shown in Figure 49 for the compressor in Figure 48 at the constant speed M_1. With the prerotation vanes wide open, the performance curve is identical to the M_1 curve in Figure 48. The other curves are different, as are the efficiency contours and the surge envelope.

The same system characteristic has been superimposed on this performance map, as in Figure 48, to provide a comparison of these two modes of operation. In Figure 49, Point E can be reached with prerotation vanes, Point H cannot. Theoretically, turning the vanes against rotation would produce a performance line passing through Point H, but sonic relative inlet velocities prevent this, except at low Mach numbers. Hot gas bypass is still necessary at Points F and G

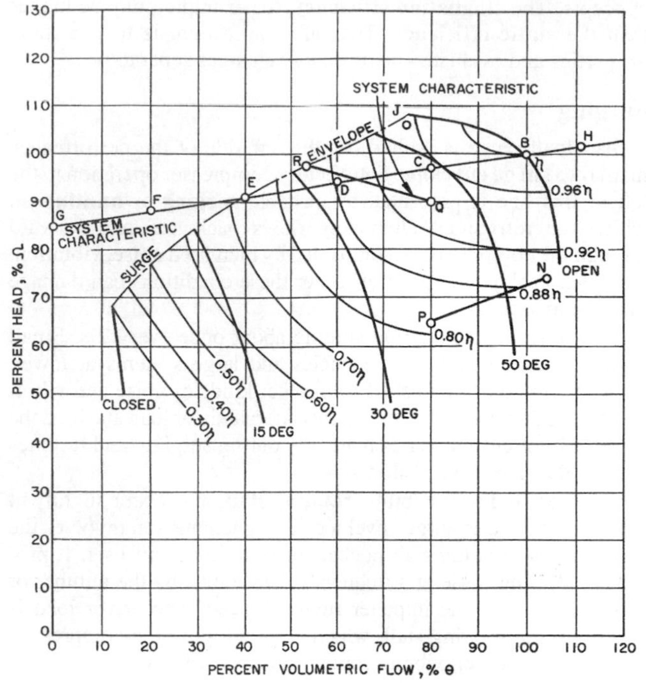

Fig. 49 Typical Compressor Performance with Various Prerotation Vane Settings

Table 5 Typical Part-Load Gas Compression Power Input for Speed and Vane Controls

System Volumetric Flow, %	Power Input, %	
	Speed Control	Vane Control
111	120	—
100	100	100
80	76	81
60	59	64
40	55	50
20	51	46
0	47	43

with prerotation vane control, but to a lesser extent than with variable speed.

The gas compression powers for both control methods are listed in Table 5. For the compressor and system assumed in this example, Table 5 shows that speed control requires less gas compression power down to about 55% of rated capacity. Prerotation vane control requires less power below 55%. For a complete analysis, friction loss and driver efficiency must also be considered.

Variable-speed control requires decreasing polytropic work (pressure) with decreasing flow (capacity) to perform more efficiently. Decreasing polytropic work is generally accomplished by reducing condensing pressure. Impeller tip speed must remain constant if the lift requirements do not change.

Refrigeration capacity is directly related to compressor speed, but the compressor's ability to produce pressure is a function of the square of a change in compressor speed. For example, a 50% reduction in compressor speed will result in a 50% reduction in refrigeration capacity; however, the available pressure for the compressor will be 25% of its value at 100% speed. If this is not consistent with actual operating conditions, hot gas bypass will be required to false load the compressor and prevent it from entering into a surge condition.

In Figure 49, for example, the compressor could operate at 15% of its rated capacity with 28% of its rated power if the polytropic work requirement could be reduced by 29%.

Since fixed-speed motors are the most common drivers of centrifugal compressors, prerotation vane control is more prevalent than speed variation. This is generally the best control mode for applications where pressure requirements do not vary significantly at part load. Less common control methods are (1) suction throttling, (2) adjustable diffuser vanes, (3) movable diffuser walls, (4) impeller throttling sleeve, and (5) combinations of these with prerotation vanes and variable speed. Each method has advantages and disadvantages in terms of performance, complexity, and cost.

SYMBOLS

a = acoustic velocity at a particular point
a_i = acoustic velocity at impeller inlet
c = flow velocity
C_p = specific heat at constant pressure
D = impeller diameter
f = polytropic work factor
g_c = gravitation constant
h = enthalpy at a specific state point
J = mechanical equivalent of heat, 778.1 $\text{ft}\cdot\text{lb}_f/\text{Btu}$
k = ratio of specific heats, Equation (32)
m = exponent, Equation (21)
M = flow Mach number, Equation (39)
M_i = flow Mach number impeller, Equation (41)
n = exponent, Equation (22)
N = rotational speed
N_s = specific speed
P = gas compression power
p = pressure at a specific state point
p^m = pressure raised to the power of m
Q = volumetric flow rate
Q/ND^3 = dimensionless flow coefficient
Q_i = volumetric flow rate in impeller
R = gas constant
T = absolute temperature at a specific state point
u_i = impeller tip speed
v = specific volume
V^n = volume raised to the power of n
W = total work input
w = mass flow
W_i = impeller work input
W_p = polytropic work input
W_{pi} = polytropic work by impeller
W_s = adiabatic work input
X = compressibility function, Equation (23)
Y = compressibility function, Equation (24)
Z = compressibility function, Equation (25)
Ω = head parameter, Equation (42)
η_s = adiabatic efficiency
η = polytropic efficiency
μ = overall work coefficient
Θ = flow parameter, Equation (43)

APPLICATION

Critical Speed

Centrifugal compressors are designed so that the first lateral critical speed is either well above or well below the operating speed. Operation at a speed between 0.8 and 1.1 times the first lateral speed is generally unacceptable from a reliability standpoint. The second lateral critical speed should be at least 25% above the operating speed of the machine.

The operating speeds of hermetic compressors are fixed, and each manufacturer has full responsibility for making sure the critical speeds are not too close to the operating speeds. For open-drive compressors, however, operating speed depends on the required flow of the application. Thus, the designer must make sure that the critical speed is sufficiently far away from the operating speed.

In applying open-drive machines, it is also necessary to consider torsional critical speed, which is a function of the designs of the compressor, the drive turbine or motor, and the coupling(s). In geared systems, the gearbox design is also involved. Manufacturers of centrifugal compressors use computer programs to calculate the torsional natural frequencies of the entire system, including the driver, the coupling(s), and the gears, if any. Responsibility for performing this calculation and ascertaining that the torsional natural frequencies are sufficiently far away from torsional exciting frequencies should be shared between the compressor manufacturer and the designer.

For engine drives, it may be desirable to use a fluid coupling to isolate the compressor (and gear set) from engine torque pulsations. Depending on compressor bearing design, there may be other speed ranges that should be avoided to prevent the nonsynchronous shaft vibration commonly called oil whip or oil whirl.

Vibration

Excessive vibration of a centrifugal compressor is an indication of malfunction, which may lead to failure. Periodic checks of the vibration spectrum at suitable locations or continuous monitoring of vibration at such locations are, therefore, useful in ascertaining the operational health of the machine. The relationship between internal displacements and stresses and external vibration is different for each compressor design. In a given design, this relationship also differs for the various causes of internal displacements and stresses, such as imbalance of the rotating parts (either inherent or caused by deposits, erosion, corrosion, looseness, or thermal distortion), bearing instability, misalignment, distortion because of piping loads, broken motor rotor bars, or cracked impeller blades. It is, therefore, impossible to establish universal rules for the level of vibration considered excessive.

To establish meaningful criteria for a given machine or design, it is necessary to have baseline data indicative of proper operation. Significant increases of any of the frequency components of the vibration spectrum above the baseline will then indicate a deterioration in the machine's operation; the frequency component for which this increase occurs is a good indication of the part of the machine deteriorating. Increases in the component at the fundamental running frequency, for instance, are usually because of deterioration of balance. Increases at approximately one-half the fundamental running frequency are due to bearing instability, and increases at twice the running frequency are usually the result of deterioration of alignment, particularly coupling alignment.

As a general guide to establishing satisfactory vibration, a constant velocity criterion is sometimes used. In many cases, a velocity amplitude of 0.2 in/s constitutes a reasonable criterion for vibration measured on the bearing housing.

Although measurement of the vibration amplitude on the bearing housing is convenient, the value of such measurements is limited because the stiffness of the bearing housing in typical centrifugal compressors is generally considerably larger than that of the oil film. Thus, vibration monitoring systems often use noncontacting sensors, which measure the displacement of the shaft relative to the bearing housing, either instead of, or in addition to, monitoring the vibration of the bearing housing (Mitchell 1977). Such sensors are also useful for monitoring the axial displacement of the shaft relative to the thrust bearing.

In some applications, compressor vibration, which is perfectly acceptable from a reliability standpoint, can cause noise problems if the machine is not isolated properly from the building. Conducting vibration tests of the installed machine under operating conditions gives a base comparison for future reference.

Noise

The satisfactory application of centrifugal compressors requires careful consideration of noise control, especially if compressors are to be located near a noise-sensitive area of a building. The noise of centrifugal compressors is primarily of aerodynamic origin, constituted principally of gas pulsations associated with the impeller frequency and gas flow noise. Most of the predominant noise sources are of a sufficiently high frequency (above 1000 Hz) so that significant noise reduction can be achieved by carefully designed acoustical and structural isolation of the machine. While the noise originates within the compressor proper, most is usually radiated from the discharge line and the condenser shell. Reductions of equipment room noise by up to 10 dB can be obtained by covering the discharge line and the condenser shell with acoustical insulation. In geared compressors, gear-mesh noise may also contribute to the high-frequency noise; however, these frequencies are often above the audible range. This noise can be reduced by the application of sound insulation material to the gear housing.

There are two aspects of importance in noise considerations for applications of centrifugal compressors. In the equipment room, OSHA regulations specify employer responsibilities with regard to exposure to high sound levels. Increasing liability concerns in this regard are making designers more aware of compressor sound level considerations. Another important consideration is noise travel beyond the immediate equipment room.

Noise problems with centrifugal refrigeration equipment can occur in noise-sensitive parts of the building, such as a nearby office or conference room. The cost of controlling the transmission of compressor noise to such areas should be considered in the building layout and weighed against cost factors for alternative locations of the equipment in the building.

If the equipment room is to be located close to noise-sensitive building areas, it is usually cost-effective to have the noise and vibration isolation designed by an experienced acoustical consultant, since small errors in design or execution can make the results unsatisfactory (Hoover 1960).

Blazier (1972) covers general information on typical noise levels near centrifugal refrigeration machines. Data on the noise output of a specific machine should be obtained from the manufacturer; the request should specify that the measurements are to be in accordance with the current edition of ARI *Standard* 575.

Drivers

Centrifugal compressors are driven by almost any prime mover—a motor, turbine, or engine. Power requirements range from 33 to 12,000 hp. Sometimes the driver is coupled directly to the compressor; often, however, there is a gear set between them, usually because of low driver speed. Flexible couplings are required to accommodate the angular, axial, and lateral misalignments that may arise within a drive train. Additional information on prime movers may be found in Chapter 7 and Chapter 40.

Centrifugal refrigeration compressors are used in many special applications. An outstanding example is their use in factory-packaged water chilling systems of 80 to 2100 ton capacity. These units use single-, two-, and three-stage compressors driven by open and hermetic motors. These designs have internal gear and direct drives, both of which are quieter, less costly, and more compact than external gearboxes. Internal gears are used when compressors operate at rotative speeds higher than two-pole motor synchronous speed. Chapter 43 of the 1998 *ASHRAE Handbook—Refrigeration* discusses centrifugal water chilling systems in greater detail.

A hermetic compressor absorbs the heat of the motor because the motor is cooled by the refrigerant. An open motor is cooled by air in the equipment room while the heat rejected by a hermetic motor must be considered in the design of the refrigeration system. However, heat must still be removed from the equipment room, generally by mechanical ventilation. Because they operate at a lower temperature, hermetic motors are generally smaller than open motors for a given power rating. But, if a motor burns out, a hermetic system will require thorough cleaning, while an open motor will not. When serviced or replaced, an open motor must be carefully aligned to ensure reliable performance.

Starting torque must be considered in selecting a driver, particularly a motor or single-shaft gas turbine. Compressor torque is roughly proportional to both speed squared and to the refrigerant density. The latter is often much higher at start-up than at rated operating conditions. If prerotation vanes or suction throttling cannot provide sufficient torque reduction for starting, the standby pressure must be lowered by some auxiliary means.

In certain applications, a centrifugal compressor drives its prime mover backward at shutdown. The compressor is driven backward by refrigerant equalizing through the machine. The extent to which reverse rotation occurs depends on the kinetic energy of the drive train relative to the expansive energy in the system. Large installations with dense refrigerants are most susceptible to running backward, a modest amount of which is harmless if suitable provisions have been made. Reverse rotation can be minimized or eliminated by closing discharge valves, side-load valves, and prerotation vanes at shutdown and opening hot gas bypass valves and liquid refrigerant drains.

Paralleling

The problems associated with paralleling turbine-driven centrifugal compressors at reduced load are illustrated by Points I and J in Figure 48. These represent two identical compressors connected to common suction and discharge headers and driven by identical turbines. A single controller sends a common signal to both turbine governors so that both compressors should be operating at part-load Point K (full load is at Point L). The I machine runs 1% faster than its twin because of their respective governor adjustments, while the J compressor works against 1% more pressure difference because of the piping arrangement. The result is a 20% discrepancy between the two compressor loads.

One remedy is to readjust the turbine governors so that the J compressor runs 0.5% faster than the other unit. A more permanent solution, however, is to eliminate one of the common headers and to provide either separate evaporators or separate condensers. This increases the compression ratio of whichever machine has the greater capacity, decreases the compression ratio of the other, and shifts both toward Point K.

The best solution of all is to install a flow meter in the discharge line of each compressor and to use a master-slave control in which the original controller signals only one turbine, the master, while a second controller causes the slave unit to match the master's discharge flow.

The problem of imbalance, associated with turbine-driven centrifugal compressors, is minimal in fixed-speed compressors with vane controls. A loading discrepancy comparable to the above mentioned example would require a 25% difference in vane positions.

The paralleling of centrifugal compressors offers advantages in redundancy and improved part-load operation. This arrangement provides the capability of efficiently unloading to a lower percentage of total load. When the unit requirement reduces to 50%, one compressor can carry the complete load and will be operating at a higher percent volumetric flow and efficiency than a single large compressor.

Means must be provided to prevent refrigeration flow through the idle compressor to prevent an inadvertent flow of hot gas bypass through the compressor. In addition, isolation valves should be provided on each compressor to allow removal or repair of either compressor.

Other Special Applications

Other specialized applications of centrifugal compressors are found in petroleum refineries, marine refrigeration, and in the chemical industry, as discussed in Chapters 30 and 36 of the 1998 *ASHRAE Handbook—Refrigeration*. Marine requirements are also detailed in ASHRAE *Standard* 26.

MECHANICAL DESIGN

Impellers

Impellers without covers, such as the one shown in Figure 45, are known as open or unshrouded designs. Those with covered blades (Figure 46), are known as shrouded impellers. Open models must operate in close proximity to contoured stationary surfaces to avoid excessive leakage around their vanes. Shrouded designs must be fitted with labyrinth seals around their inlets for a similar purpose. Labyrinth seals behind each stage are required in multistage compressors.

Impellers must be shrunk, clamped, keyed, or bolted to their shafts to prevent loosening from thermal and centrifugal expansions. Generally, they are made of cast or brazed aluminum or of cast, brazed, riveted, or welded steel. Aluminum has a higher strength-weight ratio than steel, up to about 300°F, which permits higher rotating speeds with lighter rotors. Steel impellers retain their strength at higher temperatures and are more resistant to erosion. Lead-coated and stainless steels can be selected in corrosive applications.

Casings

Centrifugal compressor casings are about twice as large as their largest impellers, with suction and discharge connections sized for flow Mach numbers between 0.1 and 0.3. They are designed for the pressure requirements of ASHRAE *Standard* 15. A hydrostatic test pressure 50% greater than the maximum design working pressure is customary. If the casing is listed by a nationally recognized testing laboratory, a hydrostatic test pressure three times the working pressure is required.

Cast iron is the most common casing material, having been used for temperatures as low as −150°F and pressures as high as 300 psia. Nodular iron and cast or fabricated steel are also used for low temperatures, high pressures, high shock, and hazardous applications. Multistage casings are usually split horizontally, although unsplit barrel designs can also be used.

Lubrication

Like motors and gears, the bearings and lubrication systems of centrifugal compressors can be internal or external, depending on whether or not they operate in refrigerant atmospheres. For reasons of simplicity, size, and cost, most air-conditioning and refrigeration compressors have internal bearings, as shown in Figure 46. In addition, they often have internal oil pumps, driven either by an internal motor or the compressor shaft; the latter arrangement is typically used with an auxiliary oil pump for starting and/or backup service.

Most refrigerants are soluble in lubricating oils, the extent increasing with refrigerant pressure and decreasing with oil temperature. A compressor's oil may typically contain 20% refrigerant (by mass) during idle periods of high pressure and 5% during normal operation. Thus, refrigerant will come out of solution and foam the oil when such a compressor is started.

To prevent excessive foaming from cavitating the oil pump and starving the bearings, oil heaters minimize refrigerant solubility during idle periods. Standby oil temperatures between 130 and 150°F are required, depending on pressure. Once a compressor has started, its oil should be cooled to increase oil viscosity and maximize refrigerant retention during the pull-down period.

A sharp reduction in pressure before starting tends to supersaturate the oil. This produces more foaming at start-up than would the same pressure reduction after the compressor has started. Machines designed for a pressure ratio of 20 or more may reduce pressure so rapidly that excessive oil foaming cannot be avoided, except by maintaining a low standby pressure. Additional information on the solubility of refrigerants in oil can be found in Chapters 1, 2, and 7 of the 1998 *ASHRAE Handbook—Refrigeration.*

External bearings avoid the complications of refrigerant-oil solubility at the expense of some oil-recovery problems. Any nonhermetic compressor must have at least one shaft seal. Mechanical seals are commonly used in refrigeration machines because they are leak-tight during idle periods. These seals require some lubricating oil leakage when operating, however. Shaft seals leak oil out of compressors with internal bearings and into compressors with external bearings. Means for recovering seal oil leakage with a minimal loss of refrigerant must be provided in external bearing systems.

Bearings

Centrifugal compressor bearings are generally of a hydrodynamic design, with sleeve bearings (one-piece or split) being the most common for radial loads; tilting pad, tapered land, and pocket bearings are customary for thrust. The usual materials are aluminum, babbit-lined, and bronze.

Thrust bearings tend to be the most important in turbomachines, and centrifugal compressors are no exception. Thrust comes from the pressure behind an impeller exceeding the pressure at its inlet. In multistage designs, each impeller adds to the total, unless some are mounted backward to achieve the opposite effect. In the absence of this opposing balance, it is customary to provide a balancing piston behind the last impeller, with pressure on the piston thrusting opposite to the stages. To avoid axial rotor vibration, some net thrust must be retained in either balancing arrangement.

Accessories

The minimum accessories required by a centrifugal compressor are an oil filter, an oil cooler, and three safety controls. Oil filters are usually rated for 15 to 20 μm or less. They may be built into the compressor but are more often externally mounted. Dual filters can be provided for industrial applications so that one can be serviced while the other is operating.

Single or dual oil coolers usually use condenser water, chilled water, refrigerant, or air as their cooling medium. Water- and refrigerant-cooled models may be built into the compressor, and refrigerant-cooled oil coolers may be built into a system heat exchanger. Many oil coolers are mounted externally for maximum serviceability.

Safety controls, with or without anticipatory alarms, must include a low oil pressure cutout, a high oil temperature switch, and high discharge and low suction pressure (or temperature) cutouts. A high motor temperature device is necessary in a hermetic compressor. Other common safety controls and alarms sense discharge temperature, bearing temperature, oil filter pressure differential, oil level, low oil temperature, shaft seal pressure, balancing piston pressure, surging, vibration, and thrust bearing wear.

Pressure gages and thermometers are useful indicators of the critical items monitored by the controls. Suction, discharge, and oil pressure gages are the most important, followed by suction, discharge, and oil thermometers. Suction and discharge instruments are often attached to components rather than to the compressor itself, but they should be provided. Interstage pressures and temperatures can also be helpful, either on the compressor or on the system. Electronic components may be used for all safety and operating controls. Electronic sensors and displays may be used for pressure and temperature monitoring.

OPERATION AND MAINTENANCE

Reference should be made to the compressor manufacturer's operating and maintenance instructions for recommended procedures. A planned maintenance program, as described in Chapter 37 of the 1999 *ASHRAE Handbook—Applications*, should be established. As part of this program, operating documentation should be kept, tabulating pertinent unit temperatures, pressures, flows, fluid levels, electrical data, and refrigerant added. ASHRAE *Guideline* 4 has further information on documentation. These can be compared periodically with values recorded for the new unit. Gradual changes in data can be used to signify the need for routine maintenance; abrupt changes should indicate system or component difficulty. A successful maintenance program requires the operating engineer to be able to recognize and identify the reason for these data trends. In addition, by having a knowledge of the component parts and their operational interaction, the designer will be able to use these symptoms to prescribe the proper maintenance procedures.

The following items deserve attention in establishing a planned compressor maintenance program:

1. A tight system is important. Leaks on compressors operating at subatmospheric pressures allow noncondensables and moisture to enter the system, adversely affecting operation and component life. Leakage in higher pressure systems allows oil and refrigerant loss. ASHRAE *Guideline* 3 can be used as a guide to ensure system tightness. The existence of vacuum leaks can be detected by a change in operational pressures not supported by corresponding refrigerant temperature data or the frequency of purge unit operation. Pressure leaks are characterized by symptoms related to refrigerant charge loss such as low suction pressures and high suction superheat. Such leaks should be located and fixed to prevent component deterioration.
2. Compliance with the manufacturer's recommended oil filter inspection and replacement schedule allows visual indication of the condition of the compressor lubrication system. Repetitive clogging of filters can mean system contamination. Periodic oil sample analysis can monitor acid, moisture, and particulate levels to assist in problem detection.
3. Operating and safety controls should be checked periodically and calibrated to ensure reliability.
4. The electrical resistance of hermetic motor windings between phases and to ground should be checked (megged) regularly, following the manufacturer's outlined procedure. This will help detect any internal electrical insulation deterioration or the formation of electrical leakage paths before a failure occurs.
5. Water-cooled oil coolers should be systematically cleaned on the water side (depending on water conditions), and the operation of any automatic water control valves should be checked.
6. For some compressors, periodic maintenance, such as manual lubrication of couplings and other external components, and shaft seal replacement, is required. Prime movers and their associated auxiliaries all require routine maintenance. Such items should be made part of the planned compressor maintenance schedule.
7. Vibration analysis, when performed periodically, can locate and identify trouble (e.g., unbalance, misalignment, bent shaft, worn or defective bearings, bad gears, mechanical looseness, and electrical unbalance). Without disassembly of the machine, such trouble can be found in its early stages before machinery failure or damage can occur. Dynamic balancing can restore rotating equipment to its original, efficient and quiet operating mode. Such testing can help avoid costly emergency repairs, pinpoint irregularities before they become major problems, and increase the useful life of components.

8. The necessary steps for preparing the unit for prolonged shutdown (i.e., winter) and specified instruction for starting after this standby period, should both be part of the program. With compressors that have internal lubrication systems, provisions should be made to have their oil heaters energized continuously throughout this period or to have their oil charges replaced prior to putting them back into operation.

REFERENCES

ARI. 1994. Method of measuring machinery sound within an equipment space. *Standard* 575-94. Air-Conditioning and Refrigeration Institute, Arlington, VA.

ASHRAE. 1993. Methods of testing for rating positive displacement refrigerant compressors and condensing units. ANSI/ASHRAE *Standard* 23-1993.

ASHRAE. 1993. Preparation of operating and maintenance documentation for building systems. *Guideline* 4-1993.

ASHRAE. 1994. Safety code for mechanical refrigeration. ANSI/ASHRAE *Standard* 15-1994.

ASHRAE. 1996. Mechanical refrigeration and air conditioning installations aboard ship. *Standard* 26-1996.

ASHRAE. 1996. Reducing emission of fully halogenated chlorofluorocarbon (CFC) refrigerants in refrigeration and air-conditioning equipment and applications. *Guideline* 3-1996.

Blazier, W.E., Jr. 1972. Chiller noise: Its impact on building design. *ASHRAE Transactions* 78(1):268.

Brasz, J.J. 1995. Improving the refrigeration cycle with turbo-expanders. Proceedings of the 19th International Congress of Refrigeration.

Bush, J. and J. Elson. 1988. Scroll compressor design criteria for residential air conditioning and heat pump applications. Proceedings of the 1988 International Compressor Engineering Conference (1):83-97 (July). Office of Publications, Purdue University, West Lafayette, IN.

Caillat, J., R. Weatherston, and J. Bush. 1988. Scroll-type machine with axially compliant mounting. U.S. Patent 4,767,292.

Davis, H., H. Kottas, and A.M.G. Moody. 1951. The influence of Reynolds number on the performance of turbomachinery. *ASME Transactions* (July):499.

Edminster, W.C. 1961. *Applied hydrocarbon thermodynamics*, 22 and 52. Gulf Publishing Company, Houston, TX.

Elson, J., G. Hundy, and K. Monnier. 1990. Scroll compressor design and application characteristics for air conditioning, heat pump, and refrigeration applications. Proceedings of the Institute of Refrigeration (London), 2.1-2.10 (November).

Gallagher, J., M. McLinden, G. Morrison, and M. Huber. 1993. NIST thermodynamic properties of refrigerants and refrigerant mixtures database (REFPROP), Version 4.0. *NIST Standard Reference Database* 23. National Institute of Standards and Technology, Gaithersburg, MD. Available from ASHRAE.

Hoover, R.M. 1960. Noise levels due to a centrifugal compressor installed in an office building penthouse. *Noise Control* (6):136.

Hougen, O.A., K.M. Watson, and R.A. Ragatz. 1959. *Chemical process principles, Part II—Thermodynamics*. John Wiley and Sons, New York, 579 and 611.

McCullough, J. 1975. Positive fluid displacement apparatus. U.S. Patent 3,924,977.

McCullough, J. and R. Shaffer. 1976. Axial compliance means with radial sealing for scroll type apparatus. U.S. Patent 3,994,636.

Mitchell, J.S. 1977. Monitoring machinery health. *Power* 121(I):46; (II):87; (III):38.

Morishita, E., Y. Kitora, T. Suganami, S. Yamamoto, and M. Nishida. 1988. Rotating scroll vacuum pump. Proceedings of the 1988 International Compressor Engineering Conference (July). Office of Publications, Purdue University, West Lafayette, IN.

Purvis, E. 1987. Scroll compressor technology. Heat Pump Conference, New Orleans.

Sauls, J. 1983. Involute and laminated tip seal of labyrinth type for use in a scroll machine. U.S. Patent 4,411,605.

Schultz, J.M. 1962. The polytropic analysis of centrifugal compressors. *ASME Transactions* (January):69 and (April):222.

Sheets, H.E. 1952. Nondimensional compressor performance for a range of Mach numbers and molecular weights. *ASME Transactions* (January):93.

Tojo, K., T. Hosoda, M. Ikegawa, and M. Shiibayashi. 1982. Scroll compressor provided with means for pressing an orbiting scroll member against a stationary scroll member and self-cooling means. U.S. Patent 4,365,941.

CHAPTER 35

CONDENSERS

THE CONDENSER in a refrigeration system is a heat exchanger that generally rejects all the heat from the system. This heat consists of heat absorbed by the evaporator plus the heat from the energy input to the compressor. The compressor discharges hot, high-pressure refrigerant gas into the condenser, which rejects heat from the gas to some cooler medium. Thus, the cool refrigerant condenses back to the liquid state and drains from the condenser to continue in the refrigeration cycle.

The common forms of condensers may be classified on the basis of the cooling medium as (1) water-cooled, (2) air-cooled, and (3) evaporative (air- and water-cooled).

WATER-COOLED CONDENSERS

HEAT REMOVAL

The heat rejection rate in a condenser for each unit of heat removed by the evaporator may be estimated from the graph in Figure 1. The theoretical values shown are based on Refrigerant 22 with 10°F suction superheat, 10°F liquid subcooling, and 80% compressor efficiency. Actually, the heat removed is slightly higher or lower than these values, depending on compressor efficiency. Usually, the heat rejection requirement can be accurately determined by adding the known evaporator load and the heat equivalent of the actual power required for compression (obtained from the compressor manufacturer's catalog). (Note that the heat from the compressor is reduced by any independent heat rejection processes such as oil cooling, motor cooling, etc.)

The volumetric flow rate of condensing water required may be calculated as follows:

$$Q = \frac{q_o}{\rho c_p (t_2 - t_1)} \quad (1)$$

where

Q = volumetric flow rate of water, ft³/h(multiply ft³/h by 0.125 to obtain gpm)
q_o = heat rejection rate, Btu/h
ρ = density of water, lb/ft³
t_1 = temperature of water entering condenser, °F
t_2 = temperature of water leaving condenser, °F
c_p = specific heat of water, Btu/lb·°F

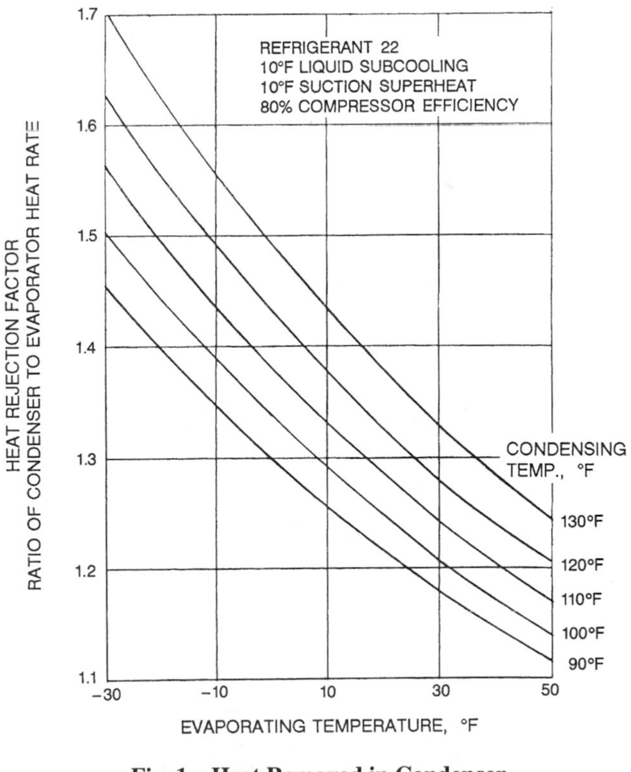

Fig. 1 Heat Removed in Condenser

The preparation of this chapter is assigned to TC 8.5, Liquid-to-Refrigerant Heat Exchangers; TC 8.4, Air-to-Refrigerant Heat Transfer Equipment; and TC 8.6, Cooling Towers and Evaporative Condensers.

Example 1. Estimate volumetric flow rate of condensing water required for the condenser of a Refrigerant 22 water-cooled unit operating at a condensing temperature of 105°F, an evaporating temperature of 40°F,

10°F liquid subcooling, and 10°F suction superheat. Water enters the condenser at 86°F and leaves at 95°F. The refrigeration load is 100 tons.

Solution: From Figure 1, the heat rejection factor for these conditions is about 1.19.

q_o = 100 × 1.19 = 119 tons
ρ = 62.1 lb/ft^3 at 90.5°F
c_p = 1.0 Btu/(lb·°F)

From Equation (1):

$$Q = \frac{1496 \times 119}{62.1 \times 1.0(95-86)} = 319 \text{ gpm}$$

Note: The value 1496 is a unit conversion factor.

HEAT TRANSFER

A water-cooled condenser transfers heat by sensible cooling in the gas desuperheating and condensate subcooling stages and by transfer of latent heat in the condensing stage. Condensing is by far the dominant process in normal refrigeration applications, accounting for 83% of the heat rejection in Example 1. Because the tube wall temperature is normally lower than the condensing temperature at all locations in the condenser, condensation takes place throughout the condenser.

The effect of changes in the entering gas superheat is typically insignificant due to an inverse proportional relationship between temperature difference and heat transfer coefficient. As a result, an average overall heat transfer coefficient and the mean temperature difference (calculated from the condensing temperature corresponding to the saturated condensing pressure and the entering and leaving water temperatures) give reasonably accurate predictions of performance.

Subcooling has an effect on the average overall heat transfer coefficient when tubes are submerged in liquid. The heat rejection rate is then determined as

$$q = UA\Delta t_m \tag{2}$$

where

q = total heat transfer rate, Btu/h
U = overall heat transfer coefficient, Btu/h·ft^2·°F
A = heat transfer surface area associated with U, ft^2
Δt_m = mean temperature difference, °F

Chapter 3 of the 1997 *ASHRAE Handbook—Fundamentals* describes how to calculate Δt_m.

Overall Heat Transfer Coefficient

The overall heat transfer coefficient U_o in a **water-cooled condenser with water inside the tubes** may be computed from calculated or test-derived heat transfer coefficients of the water and refrigerant sides, from physical measurements of the condenser tubes, and from a fouling factor on the water side, by using the following equation:

$$U_o = \frac{1}{\left(\frac{A_o}{A_i}\frac{1}{h_w}\right) + \left(\frac{A_o}{A_i}r_{fw}\right) + \left(\frac{A_o}{A_m}\frac{t}{k}\right) + \left(\frac{1}{h_r \phi_s}\right)} \tag{3}$$

where

U_o = overall heat transfer coefficient, based on external surface and mean temperature difference, between external and internal fluids, Btu/h·ft^2·°F
A_o/A_i = ratio of external to internal surface area
h_w = internal or water-side film coefficient, Btu/h·ft^2·°F
r_{fw} = fouling resistance on water side, ft^2·h·°F/Btu
t = thickness of tube wall, ft
k = thermal conductivity of tube material, Btu/h·ft·°F
A_o/A_m = ratio of external to mean heat transfer surface areas of metal wall
h_r = external, or refrigerant-side coefficient, Btu/h·ft^2·°F
ϕ_s = surface fin efficiency (100% for bare tubes)

For **tube-in-tube condensers or other condensers where refrigerant flows inside the tubes**, the equation for U_o, in terms of water-side surface, becomes

$$U_o = \frac{1}{\left(\frac{A_o}{A_i}\frac{1}{h_r}\right) + r_{fw} + \left(\frac{t}{k}\right) + \left(\frac{1}{h_w}\right)} \tag{4}$$

where

h_r = internal or refrigerant-side coefficient, Btu/h·ft^2·°F
h_w = external or water-side coefficient, Btu/h·ft^2·°F

For **brazed or plate and frame condensers** $A_0 = A_i$; therefore the equation for U_o is

$$U_o = \frac{1}{(1/h_r) + r_{fw} + (t/k) + (1/h_w)} \tag{5}$$

where t is the plate thickness.

Water-Side Film Coefficient

Values of the water-side film coefficient h_w may be calculated from equations in Chapter 3 of the 1997 *ASHRAE Handbook—Fundamentals*. For turbulent flow, at Reynolds numbers exceeding 10,000 in horizontal tubes and using average water temperatures, the general equation (McAdams 1954) is

$$\frac{h_w D}{k} = 0.023\left(\frac{DG}{\mu}\right)^{0.8}\left(\frac{c_p \mu}{k}\right)^{0.4} \tag{6}$$

where

D = inside tube diameter, ft
k = thermal conductivity of water, Btu/h·ft·°F
G = mass velocity of water, lb/h·ft^2
μ = viscosity of water, lb/(ft·h)
c_p = specific heat of water at constant pressure, Btu/lb·°F

The constant (0.023) in Equation (6) reflects plain ID tubes. Pate et al. (1991) and Bergles (1995) discuss numerous water-side enhancement methods that increase the value of this constant.

Because of its strong influence on the value of h_w, a high water velocity should generally be maintained without initiating erosion or excessive pressure drop. Typical maximum velocities from 6 to 10 fps are common with clean water. Experiments by Sturley (1975) at velocities up to approximately 26 fps showed no damage to copper tubes after long operation. Regarding erosion potential, water quality is the key factor (Ayub and Jones 1987). A minimum velocity of 3 fps is good practice when the water quality is such that noticeable fouling or corrosion could result. With clean water, the velocity may be lower if it must be conserved or if it has a low temperature. In some cases, the minimum flow may be determined by a lower Reynolds number limit.

For brazed or plate and frame condensers, the equation is similar to Equation (6). However, the diameter D is replaced by H.

Refrigerant-Side Film Coefficient

Factors influencing the value of the refrigerant-side film coefficient h_r are

- Type of refrigerant being condensed
- Geometry of condensing surface (plain tube OD; finned tube fin spacing, height, and cross-sectional profile; and plate geometry)

- Condensing temperature
- Condensing rate in terms of mass velocity or rate of heat transfer
- Arrangement of tubes in bundle and location of inlet and outlet connections
- Vapor distribution and rate of flow
- Condensate drainage
- Liquid subcooling

Values of the refrigerant-side coefficients may be estimated from correlations in Chapter 4 of the 1997 *ASHRAE Handbook—Fundamentals*. Information on the effects of the type of refrigerant, condensing temperature, and loading (temperature drop across the condensate film) on the condensing film coefficient are in the section on Condensing in the same chapter. Actual values of h_r for a given physical condenser design can be determined from test data by use of a Wilson plot (McAdams 1954, Briggs and Young 1969).

The type of condensing surface has a considerable effect on the condensing coefficient. Most halocarbon refrigerant condensers use finned tubes where the fins are integral with the tube. Water velocities normally used are large enough for the resulting high water-side film coefficient to justify using an extended external surface to balance the heat transfer resistances of the two surfaces. Pearson and Withers (1969) compared the refrigerant condensing performance of some integral finned tubes with different fin spacing. Some other refrigerant-side enhancements are described by Pate et al. (1991) and Webb (1984a). The effect of fin shape on the condensing coefficient is addressed by Kedzierski and Webb (1990).

In the case of brazed or plate and frame condensers inlet nozzle size, chevron angle, pitch, and depth of the nozzles are important design parameters. For a trouble-free operation, refrigerant should flow counter to the water flow. Little specific design information is available; however, film thickness is certainly a factor in plate condensers design due to the falling film nature along the vertical surface. Kedzierski (1997) showed that placing a brazed condenser in a horizontal position improved U_o by 17 to 30% due to the shorter film distance.

Huber et al. (1994a) determined condensing coefficients for R-134a, R-12, and R-11 condensing on conventional finned tubes with a fin spacing of 26 fins per inch and a commercially available tube specifically developed for the condensation of halocarbon refrigerants (Huber et al. 1994b). This tube has a sawtooth-shaped outer enhancement. The data indicated that the condensing coefficients for the sawtooth tube were approximately three times higher than for the conventional finned tube exchanger and two times higher for R-123.

Further, Huber et al. (1994c) found that for tubes with 26 fin/inch the R-134a condensing coefficients are 20% larger than those for R-12 at a given heat flux. However, on the sawtooth tube, the R-134a condensing coefficients are nearly two times larger than those for R-12 at the same heat flux. The R-123 condensing coefficients were 10 to 30% larger than the R-11 coefficients at a given heat flux, with the largest differences occurring at the lowest heat fluxes tested. The differences in magnitude between the R-123 and R-11 condensing coefficients were the same for both the 26 fin/inch tube and the sawtooth tube.

Physical aspects of a given condenser design—such as tube spacing and orientation, shell-side baffle arrangement, orientation of multiple water-pass arrangements, refrigerant connection locations, and the number of tubes high in the bundle—affect the refrigerant-side coefficient by influencing vapor distribution and flow through the tube bundle and condensate drainage from the bundle. Butterworth (1977) reviewed correlations accounting for these variables in predicting the heat transfer coefficient for shell-side condensation. These effects are also surveyed by Webb (1984b). Kistler et al. (1976) developed analytical procedures for design within these parameters. Ghaderi et al. (1995) reviewed in-tube condensation heat transfer correlations for smooth and augmented tubes.

As refrigerant condenses on the tubes, it falls on the tubes in lower rows. Due to the added resistance of this liquid film, the effective film coefficient for lower rows should be lower than that for upper rows. Therefore, the average overall refrigerant film coefficient should decrease as the number of tube rows increase. Webb and Murawski (1990) present row effect data for five tube geometries. However, the additional compensating effects of added film turbulence and direct contact condensation on the subcooled liquid film make actual row effect uncertain.

Huber et al. (1994c, 1994d) determined that the row effect on finned tubes is nearly negligible when condensing low surface tension refrigerants such as R-134a. However, the finned-tube film coefficient for higher surface tension refrigerants such as R-123 can drop by as much as 20% in lower bundle rows. The row effect for the sawtooth condensing tube is quite large for both R-134a and R-123, as the film coefficient drops by nearly 80% from top to bottom in a 30-row bundle.

Honda et al. (1994, 1995) demonstrated that row effects due to condensate drainage and inundation are less for staggered tube bundles than for in-line tubes. In addition, performance improvements as high as 85% were reported for optimized two-dimensional fin profiles relative to conventional fin profiles. The optimized fin profiles differed from the sawtooth profile tested by Huber et al. (1994b) primarily in that external fins were not notched, unlike the sawtooth case. This observation coupled with the differences in row effects between sawtooth and conventional fin profiles reported by Huber et al. (1994c, 1994d) suggested that there is opportunity for the further development and commercialization of two-dimensional fin profiles for large shell-and-tube heat exchanger applications.

Liquid refrigerant may be subcooled by raising the condensate level to submerge a desired number of tubes. The refrigerant film coefficient associated with the submerged tubes is less than the condensing coefficient. If the refrigerant film coefficient in Equation (3) is an average based on all tubes in the condenser, its value decreases as a greater portion of the tubes is submerged.

Tube-Wall Resistance

Most refrigeration condensers, with the exception of those using ammonia, have relatively thin-walled copper tubes. Where these are used, the temperature drop or gradient across the tube wall is not significant. If the tube metal has a high thermal resistance, as does 70/30 cupronickel, a considerable temperature drop occurs or, conversely, an increase in the mean temperature difference Δt_m or the surface area is required to transfer the same amount of heat as copper. Although the tube-wall resistance t/k in Equations (3) and (4) is an approximation, as long as the wall thickness is not more than 14% of the tube diameter, the error will be less than 1%. To improve the accuracy of the wall resistance calculation for heavy tube walls or low-conductivity material, see Chapter 3 of the 1997 *ASHRAE Handbook—Fundamentals*.

Surface Efficiency

For a finned tube, a temperature gradient exists from the root of a fin to its tip because of the thermal resistance of the fin material. The surface efficiency ϕ_s can be calculated from the fin efficiency, which accounts for this effect. For tubes with low-conductivity material, high fins, or high values of fin pitch, the fin efficiency becomes increasingly significant. Young and Ward (1957) and Wolverine Tube (1984) describe methods of evaluating these effects.

Fouling Factor

Manufacturers' ratings are based on clean equipment with an allowance for possible water-side fouling. The fouling factor r_{fw} is a thermal resistance referenced to the water-side area of the heat transfer surface. Thus, the temperature penalty imposed on the condenser equals the heat flux at the water-side area multiplied by the fouling factor. Increased fouling increases overall heat transfer resistance due to the parameter $(A_o/A_i)r_{fw}$ in Equation (3). Fouling increases the Δt_m required to obtain the same capacity—with a corresponding increase in condenser pressure and system power—or lowers capacity.

Allowance for a given fouling factor has a greater influence on equipment selection than simply increasing the overall resistance (Starner 1976). Increasing the surface area results in lowering the water velocity. Consequently, the increase in heat transfer surface required for the same performance is due to both the fouling resistance and the additional resistance that results from a lower water velocity.

For a given tube surface, load, and water temperature range, the tube length can be optimized to give a desired condensing temperature and water-side pressure drop. The solid curves in Figure 2 show the effect on water velocity, tube length, and overall surface required due to increased fouling. A fouling factor of 0.00072 $ft^2 \cdot h \cdot °F/Btu$ doubles the required surface area compared to that with no fouling allowance.

A worse case occurs when an oversized condenser must be selected to meet increased fouling requirements but without the flexibility of increasing tube length. As shown by the dashed lines in Figure 2, the water velocity decreases more rapidly as the total surface increases to meet the required performance. Here, the required surface area doubles with a fouling factor of only 0.00049 $ft^2 \cdot h \cdot °F/Btu$. If the application can afford more pumping power, the water flow may be increased to obtain a higher velocity, which increases the water film heat transfer coefficient. This factor, plus the lower leaving water temperature, reduces the condensing temperature.

Fouling is a major unresolved problem in heat exchanger design (Taborek et al. 1972). The major uncertainty is which fouling factor to choose for a given application or water condition to obtain expected performance from the condenser; use of too low a fouling factor wastes compressor power, while too high a factor wastes heat exchanger material.

Fouling may result from sediment, biological growth, or corrosive products. Scale results from the deposition of chemicals from the cooling water on the warmer surface of the condenser tube. Chapter 47 of the 1999 *ASHRAE Handbook—Applications* discusses water chemistry and water treatment factors that are important in controlling corrosion and scale in condenser cooling water.

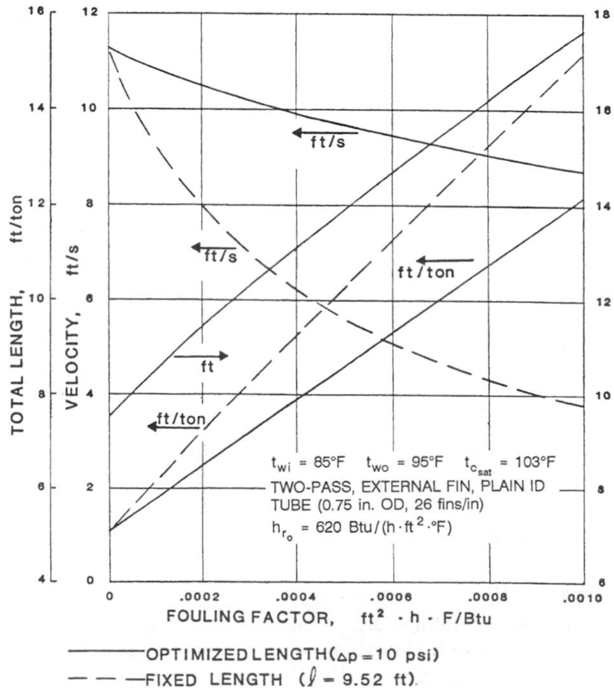

Fig. 2 Effect of Fouling on Condenser

Tables of fouling factors are available; however, in many cases, the values are greater than necessary (TEMA 1988). Extensive research generally found that fouling resistance reaches an asymptotic value with time (Suitor et al. 1976). Much of the fouling research is based on surface temperatures that are considerably higher than those found in air-conditioning and refrigeration condensers. Lee and Knudsen (1979) and Coates and Knudsen (1980) found that, in the absence of suspended solids or biological fouling, long-term fouling of condenser tubes does not exceed 0.0002 $ft^2 \cdot h \cdot °F/Btu$, and short-term fouling does not exceed 0.0001 $ft^2 \cdot h \cdot °F/Btu$ (ASHRAE 1982). These studies have resulted in a standard industry fouling value of 0.00025 $ft^2 \cdot h \cdot °F/Btu$ for condenser ratings (ARI 1997). Periodic cleaning of condenser tubes (mechanically or chemically) usually maintains satisfactory performance, except in severe environments. ARI *Standard* 450 for water-cooled condensers should be referred to when reviewing manufacturers' ratings. This standard describes methods for correcting ratings for different values of fouling.

WATER PRESSURE DROP

Water (or other fluid) pressure drop is important for designing or selecting condensers. Where a cooling tower cools the condensing water, the water pressure drop through the condenser is generally limited to about 10 psi. If the condenser water comes from another source, the pressure drop through the condenser should be lower than the available pressure to allow for pressure fluctuations and additional flow resistance caused by fouling.

Pressure drop through horizontal condensers includes loss through the tubes, tube entrance and exit losses, and losses through the heads or return bends (or both). The effect of the coiling of the tubes must be considered in shell-and-coil condensers. Expected pressure drop through tubes can be calculated from a modified Darcy-Weisbach equation:

$$\Delta p = N_p\left(K_H + f\frac{L}{D}\right)\frac{\rho V^2}{2g} \quad (7)$$

where

Δp = pressure drop, psi
N_p = number of tube passes
K_H = entrance and exit flow resistance and flow reversal coefficient, number of velocity heads ($V^2/2g$)
f = friction factor
L = length of tube, ft
D = inside tube diameter, ft
ρ = fluid density, lb/ft^3
V = fluid velocity, fps
g = gravitational constant =32.17 $lb_m \cdot ft/(lb_f \cdot s^2)$

For tubes with smooth inside diameters, the friction factor may be determined from a Moody chart or various relations, depending on the flow regime and wall roughness (see Chapter 2 of the 1997 *ASHRAE Handbook—Fundamentals*). For tubes with internal enhancement, the friction factor should be obtained from the tube manufacturer.

The value of K_H depends on tube entry and exit conditions and the flow path between passes. A minimum recommended value is 1.5. This factor is more critical with short tubes.

Predicting pressure drop for shell-and-coil condensers is more difficult than it is for shell-and-tube condensers because of the curvature of the coil and the flattening or kinking of the tubes as they are bent. Seban and McLaughlin (1963) discuss the effect of curvature or bending of pipe and tubes on the pressure drop.

For brazed and plate and frame condensers, the total pressure drop is the sum of the pressure drop in the plate region and the pressure drops associated with the entry/exit ports. The general form of equation is similar to Equation (7) except $N_p = 1$.

LIQUID SUBCOOLING

The amount of condensate subcooling provided by the condensing surface in a shell-and-tube condenser is small, generally less than 2°F. When a specific amount of subcooling is required, it may be obtained by submerging tubes in the condensate. Tubes in the lower portion of the bundle are used for this purpose. If the condenser is multipass, then the subcooling tubes should be included in the first pass to gain exposure to the coolest water.

When means are provided to submerge the subcooler tubes to a desired level in the condensate, heat is transferred principally by natural convection. Subcooling performance can be improved by enclosing tubes in a separate compartment within the condenser to obtain the benefits of forced convection over the enclosed tubes.

Segmental baffles may be provided to produce flow across the tube bundle. Kern and Kraus (1972) describe how heat transfer performance can be estimated analytically by use of longitudinal or crossflow correlations, but it is more easily determined by test because of the large number of variables. The refrigerant pressure drop along the flow path should not exceed the pressure difference permitted by the saturation pressure of the subcooled liquid.

CIRCUITING

Varying the number of water-side passes in a condenser can affect the saturated condensing temperature significantly, which affects performance. Figure 3 shows the change in condensing temperature for one, two, or three passes in a particular condenser. As an example, at a loading of 22,000 Btu/h per tube, a two-pass condenser with a 10°F range would have a condensing temperature of 102.3°F. At the same load with a one-pass, 5°F range, this unit would have a condensing temperature of 99.3°F. The one-pass option does, however, require twice the water flow rate with an associated increase in pumping power. A three-pass design may be favorable when the costs associated with water flow outweigh the gains from a lower condensing temperature. Hence, different numbers of passes (if an option) and ranges should be considered against other parameters (water source, pumping power, cooling tower design, etc.) to optimize overall performance and cost.

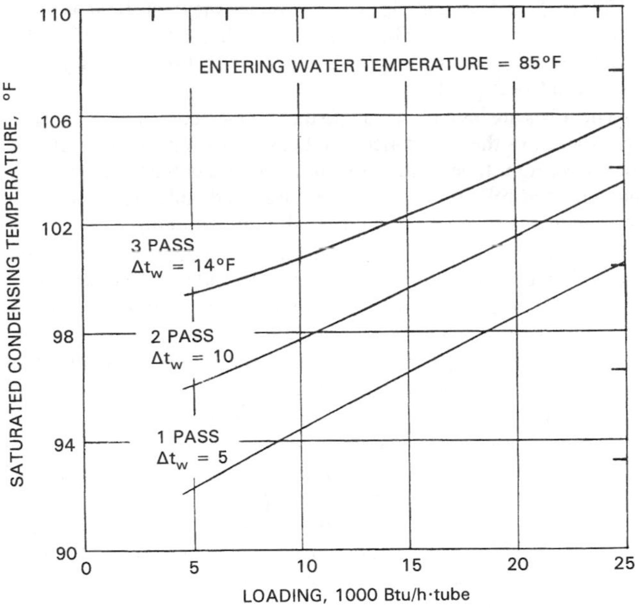

Fig. 3 Effect of Condenser Circuiting

CONDENSER TYPES

The most common types of water-cooled refrigerant condensers are (1) shell-and-tube, (2) shell-and-coil, (3) tube-in-tube, and (4) brazed plate. The type selected depends on the size of the cooling load, the refrigerant used, the quality and temperature of the available cooling water, the amount of water that can be circulated, the location and space allotment, the required operating pressures (water and refrigerant sides), and cost and maintenance concerns.

Shell-and-Tube Condensers

Built in sizes from 1 to 10,000 tons, the refrigerant in these condensers condenses outside the tubes and the cooling water circulates through the tubes in a single or multipass circuit. Fixed tube sheet, straight tube construction is usually used, although U-tubes that terminate in a single tube sheet are sometimes used. Typically, shell-and-tube condenser tubes run horizontally. Where floor installation area is limited, the condenser tubes may be oriented vertically. However, vertical tubes have poor condensate draining, which reduces the refrigerant film coefficient. Vertical condensers with open water systems have been used with ammonia.

Gas inlet and liquid outlet nozzles should be located with care. The proximity of these nozzles may adversely affect condenser performance by requiring excessive amounts of liquid refrigerant to seal the outlet nozzle from inlet gas flow. This effect can be diminished by adding baffles at the inlet and/or outlet connection.

Halocarbon refrigerant condensers have been made with many materials, including all prime surface or finned, ferrous, or nonferrous tubes. Common tubes are nominal 0.75 and 1.0 in. OD copper tubes with integral fins on the outside. These tubes are often available with fin heights from 0.035 to 0.061 in. and fin spacings of 19, 26, and 40 fins/in. For ammonia condensers, prime surface steel tubes, 1.25 in. OD and 0.095 in. average wall thickness, are common.

An increased number of tubes designed for enhanced heat transfer are available (Bergles 1995). On the inside of the tube, common enhancements include longitudinal or spiral grooves and ridges, internal fins, and other devices to promote turbulence and augment heat transfer. On the refrigerant side, condensate surface tension and drainage are important in design of the tube outer surface. Tubes are available with the outsides machined or formed specifically to enhance the condensation and promote drainage. Heat transfer design equations should be obtained from the manufacturer.

The electrohydrodynamics (EHD) technique couples a high-voltage low-current electric field with the flow field in a fluid with low electrical conductivity to achieve higher heat transfer coefficients. Ohadi et al. (1995) demonstrated experimentally enhancement factors in excess of tenfold for both boiling and condensation of such refrigerants as R-134a. For condensation processes, the technique responds equally well to augmentation of condensation over (or inside) both vertical and horizontal tubes. Most of the passive augmentation techniques perform poorly for condensation enhancement in vertical orientation. Additional details of the EHD technique can be found in Chapter 3 of the 1997 *ASHRAE Handbook—Fundamentals*.

Because the water and refrigerant film resistances with enhanced tubes are reduced, the effect of fouling becomes relatively great. Where high levels of fouling occur, the fouling resistance may easily account for over 50% of the total. In such cases, the advantages of enhancement may diminish. On the other hand, water-side augmentation, which creates turbulence, may reduce fouling. The actual value of fouling resistances depends on the particular type of enhancement and the service conditions (Starner 1976, Watkinson et al. 1974).

Similarly, refrigerant-side enhancements may not show as much benefit in very large tube bundles as in smaller bundles. This is due to the row effect addressed previously in the section on Refrigerant-Side Film Coefficient.

The tubes are either brazed into thin copper, copper alloy, or steel tube sheets, or they are rolled into heavier nonferrous or steel tube sheets. Straight tubes with a maximum OD less than the tube hole diameter and rolled into tube sheets are removable. This construction facilitates field repair in the event of tube failure.

The required heat transfer area for a shell-and-tube condenser can be found by solving Equations (1), (2), and (3). The mean temperature difference is the logarithmic mean temperature difference, with the entering and leaving refrigerant temperatures taken as the saturated condensing temperature. Depending on the parameters fixed, an iterative solution may be required.

Shell-and-U-tube condenser design principles are the same as those outlined for horizontal shell-and-tube units, with one exception: the water pressure drop through the U-bend portion of the U-tube is generally less than that through the compartments in the water header where the direction of water flow is reversed. The pressure loss is a function of the inside tube diameter and the ratio of the inside tube diameter to bending centers. Pressure loss should be determined by test.

Shell-and-Coil Condensers

Shell-and-coil condensers circulate the cooling water through one or more continuous or assembled coils contained within the shell. The refrigerant condenses outside the tubes. Capacities range from 0.5 to 15 tons. Due to the type of construction, the tubes are neither replaceable nor mechanically cleanable.

Again, Equations (1), (2), and (3) may be used for performance calculations, with the saturated condensing temperature used for the entering and leaving refrigerant temperatures in the logarithmic mean temperature difference. The values of h_w (the waterside film coefficient) and, especially, the pressure loss on the water side require close attention; laminar flow can exist at considerably higher Reynolds numbers in coils than in straight tubes. Because the film coefficient for turbulent flow is greater than that for laminar flow, values of h_w, as calculated from Equation (6), will be too high if the flow is not turbulent. Once the flow has become turbulent, the film coefficient will be greater than that for a straight tube (Eckert 1963). Pressure drop through helical coils can be much greater than that through smooth straight tubes for the same length of travel. The section on Water Pressure Drop outlines the variables that make an accurate determination of the pressure loss difficult. The pressure loss and heat transfer rate should be determined by test because of the many variables inherent in this condenser.

Tube-in-Tube Condensers

These condensers consist of one or more assemblies of two tubes, one within the other, in which the refrigerant vapor is condensed in either the annular space or the inner tube. These units are built in sizes ranging from 0.3 to 50 tons. Both straight tube and axial tube (coaxial) condensers are available.

Equations (1) and (2) can be used to size a tube-in-tube condenser. Because the refrigerant may undergo a significant pressure loss through its flow path, the refrigerant temperatures used in calculating the mean temperature difference should be selected carefully. The refrigerant temperatures should be consistent with the model used for the refrigerant film coefficient. The logarithmic mean temperature difference for either counterflow or parallel flow should be used, depending on the piping connections. Equation (3) can be used to find the overall heat transfer coefficient when the water flows in the tubes, and Equation (4) may be used when the water flows in the annulus.

Tube-in-tube condenser design differs from those outlined previously, depending on whether the water flows through the inner tube or through the annulus. Condensing coefficients are more difficult to predict when condensation occurs within a tube or annulus, because the process differs considerably from condensation on the outside of a horizontal tube. Where the water flows through the annulus, disagreement exists regarding the appropriate method that should be used to calculate the waterside film coefficient and the water pressure drop. The problem is further complicated if the tubes are formed in a spiral.

The water side is mechanically cleanable only when the water flows inside straight tubes and cleanout access is provided. Tubes are not replaceable.

Brazed Plate and Plate-and-Frame Condensers

Brazed plate condensers are constructed of plates brazed together to form an assembly of separate channels. Capacities range from 0.5 to 100 tons. The plate and frame condenser is a standard design in which the plate pairs are laser welded to form a single cassette. The refrigerant is confined to the space between the welded plates and is exposed to gaskets only at the ports. Such condensers have a higher range of capacity.

The plates, typically stainless steel, are usually configured with a wave pattern, which results in high turbulence and low susceptibility to fouling. The design has some ability to withstand freezing and, because of the compact design, requires a low refrigerant charge. The construction of brazed units does not allow mechanical cleaning, and internal leaks usually cannot be repaired. Plate and frame units can be cleaned on the water side.

Performance calculations are similar to those for other condensers; however, very few correlations are available for the heat transfer coefficients.

NONCONDENSABLE GASES

When first assembled, most refrigeration systems contain gases, usually air and water vapor. As addressed later in this chapter, these gases are detrimental to condenser performance, so it is important to evacuate the entire refrigeration system before operation.

For low-pressure refrigerants, where the operating pressure of the evaporator is less than ambient pressure, even slight leaks can be a continuing source of noncondensables. In such cases, a purge system, which automatically expels noncondensable gases, is recommended. Figure 4 shows some examples of the refrigerant loss associated with the use of purging devices at various operating conditions. As a general rule, purging devices should emit less than one part refrigerant per part of air as rated in accordance with ARI *Standard* 580 (see ASHRAE *Guideline* 3).

When present, noncondensable gases collect on the high-pressure side of the system and raise the condensing pressure above that corresponding to the temperature at which the refrigerant is actually condensing. The increased condensing pressure increases power consumption and reduces capacity. Also, if oxygen is present at a point of high discharge temperature, the oil may oxidize.

The excess pressure is caused by the partial pressure of the noncondensable gas. These gases form a resistance film over some of the condensing surface, thus lowering the heat transfer coefficient. Webb et al. (1980) showed how a small percentage of noncondensables can cause major decreases in the refrigerant film coefficient in shell-and-tube condensers. [See Chapter 4 of the 1997 *ASHRAE Handbook—Fundamentals*.] The noncondensable situation of a given condenser is difficult to characterize because such gases tend to accumulate in the coldest and least agitated part of the condenser or in the receiver. Thus, a fairly high percentage of noncondensables can be tolerated if the gases are confined to areas far from the heat transfer surface. One way to account for noncondensables is to treat them as a refrigerant

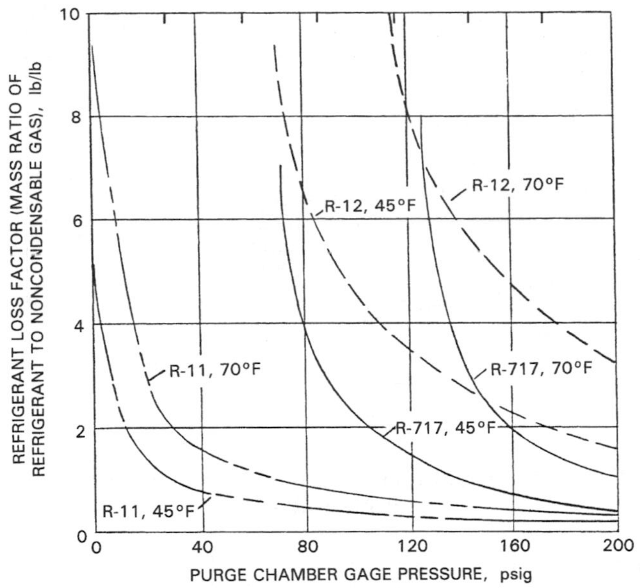

Fig. 4 Loss of Refrigerant During Purging at Various Gas Temperatures and Pressures

or gas-side fouling resistance in Equation (3). Some predictions are presented by Wanniarachchi and Webb (1982).

As an example of the effect on system performance, experiments performed on a 250 ton R-11 chiller condenser reveal that 2% noncondensables by volume caused a 15% reduction in the condensing coefficient. Also, 3% and 8% noncondensables by volume caused power increases of 2.6% and 5%, respectively.

Huber et al. (1994b) determined that noncondensable gases have a more severe effect on the condensing coefficient of tubes with sawtooth enhancements than on conventional finned tubes. For a noncondensable gas concentration of 0.5%, the coefficient for the tube with a fin spacing of 26 fins/in. dropped by 15%, while the coefficient for the sawtooth tube dropped by nearly 40%. At a noncondensable concentration of 5%, the degradation is similar for both tubes.

The presence of noncondensable gases can be tested by shutting down the refrigeration system while allowing the condenser water to flow long enough for the refrigerant to reach the same temperature as the water. If the condenser pressure is higher than the pressure corresponding to the refrigerant temperature, noncondensable gases are present. This test may not be sensitive enough to detect the presence of small amounts of noncondensables, which can, nevertheless, decrease shell-side condensing coefficients.

CODES AND STANDARDS

Pressure vessels must be constructed and tested under the rules of appropriate codes and standards. The introduction of the current ASME *Boiler and Pressure Vessel Code*, Section VIII, gives guidance on rules and exemptions.

The more common applicable codes and standards include the following:

- ARI *Standard* 450 covers industry criteria for standard equipment, standard safety provisions, marking, fouling factors, and recommended rating points for water-cooled condensers.
- ASHRAE *Standard* 22 covers recommended testing methods.
- ARI *Standard* 580 covers methods of testing, evaluating, and rating the efficiency of non-condensate gas purge equipment.
- ASHRAE *Standard* 15 specifies design criteria, use of materials, and testing. It refers to the ASME *Boiler and Pressure Vessel Code*, Section VIII, for refrigerant-containing sides of pressure vessels, where applicable. Factory test pressures are specified, and minimum design working pressures are given by this code. This code requires pressure-limiting and pressure-relief devices on refrigerant-containing systems, as applicable, and defines the setting and capacity requirements for these devices.
- ASME *Boiler and Pressure Vessel Code,* Section VIII covers the safety aspects of design and construction. Most states require condensers to meet the requirements of the ASME code if they fall within the scope of the ASME code. Some of the exceptions from meeting the requirements listed in the ASME code are as follows:
  - Condenser shell ID is 6 in. or less.
  - Working pressure is 15 psig or less.
  - The fluid (water) portion of the condenser need not be built to the requirements of the ASME code if the fluid is water, the design pressure does not exceed 300 psig, and the design temperature does not exceed 210°F.

  Condensers meeting the requirements of the ASME code will have an ASME stamp. The ASME stamp is a U or UM inside a three-leaf clover. The U can be used for all condensers; the UM can be used (considering local codes) for those with net refrigerant-side volume less than 1.5 ft^3 if less than 600 psig, or less than 5 ft^3 if less than 250 psig.
- UL *Standard* 207 covers specific design criteria, use of materials, testing, and initial approval by Underwriters Laboratories. A condenser with the ASME U stamp does not require UL approval.

Design Pressure

Refrigerant-side pressure, as a minimum, should be saturated pressure for the refrigerant used at 105°F or 15 psig, whichever is greater. Standby temperature and temperatures encountered during shipping of units with a refrigerant charge should also be considered.

Required fluid- (water-) side pressure varies, depending largely on the following conditions: static head, pump head, transients due to pump start-up, and valve closing. A common water-side design pressure is 150 psig, although with taller building construction, requirements for 300 psig are not uncommon.

OPERATION AND MAINTENANCE

When a water-cooled condenser is selected, anticipated operating conditions, including water and refrigerant temperatures, have usually been determined. Standard practice allows for a fouling factor in the selection procedure. A new condenser, therefore, operates at a condensing temperature lower than the design point because it has not yet fouled. Once a condenser starts to foul or scale, economic considerations determine how frequently the condenser should be cleaned. As the scale builds up in a condenser, the condensing temperature and subsequent power increase while the unit capacity decreases. This effect can be seen in Figure 5 for a condenser with a design fouling factor of 0.00025 $ft^2 \cdot h \cdot °F/Btu$. At some point, the increased cost of power can be offset by the labor cost of cleaning.

Local water conditions, as well as the effectiveness of chemical water treatment, if used, make the use of any specific maintenance schedule difficult. Cleaning can be done either mechanically with a brush or chemically with an acid solution. In applications where water-side fouling may be severe, online cleaning can be accomplished by brushes installed in cages in the water heads. By using valves, flow is reversed at set intervals, propelling the brushes through the tubes (Kragh 1975). The most effective method depends on the type of scale formed. Competent advice in selecting the particular method of tube cleaning is advisable.

Occasionally, one or more tubes may develop leaks because of corrosive impurities in the water or through improper cleaning. These leaks must be found and repaired as soon as possible; this can normally be done by replacing the leaky tubes, a procedure

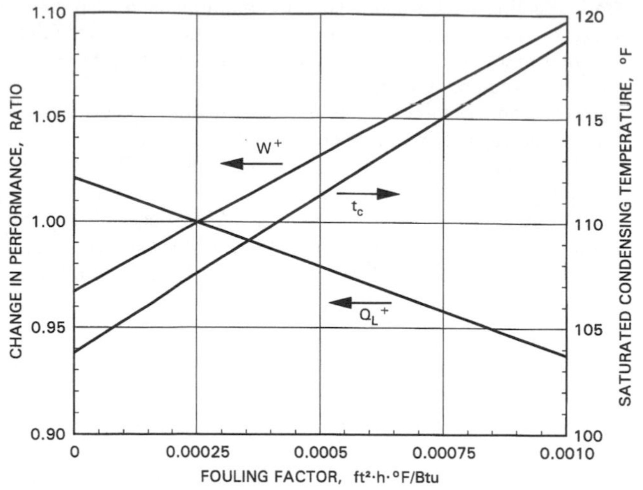

Q_L^+ = chiller actual capacity/chiller design capacity
W^+ = compressor actual kW/compressor design kW
T_c = saturated condensing temperature
Design condenser fouling factor = 0.00025 ft²·h·°F/Btu
Cooler leaving water temperature = 44°F
Condenser entering water temperature = 85°F

Fig. 5 Effect of Fouling on Chiller Performance

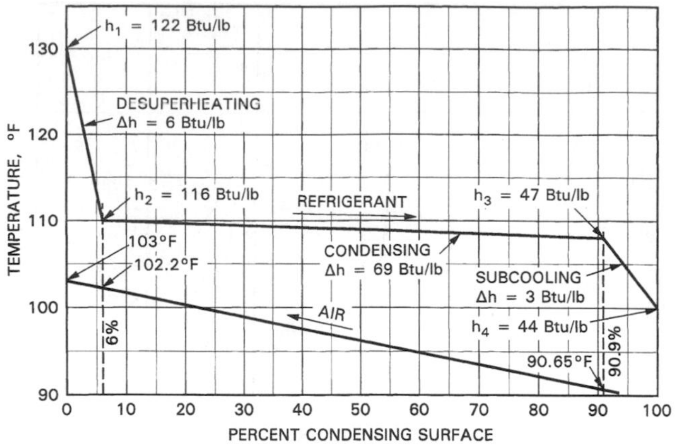

Fig. 6 Temperature and Enthalpy Changes in Air-Cooled Condenser with R-134a

requiring tools and skills that are best found through the original condenser manufacturer. In large condensers, where the contribution of a single tube is relatively insignificant, a simpler approach may be to seal the ends of the leaking tube.

If the condenser is located where the water can freeze during the winter, special precautions should be taken when it is idle. Opening all vents and drains may be sufficient, but water heads should he removed and tubes blown free of water.

If refrigerant vapor is to be released from the condenser and there is water in the tubes, the pumps should be on and the water flowing. Otherwise, freezing can easily occur.

In any case, the condenser manufacturer's installation recommendations on orientation, piping connections, space requirements for tube cleaning or removal, and other important factors should be followed.

AIR-COOLED CONDENSERS

An air-cooled condenser may be located either adjacent to or remote from the compressor. This type of condenser may be designed for indoor or outdoor operation and the air discharge may be either vertical (top) or horizontal (side). Interconnecting piping connects to the compressor unit. Electrical wiring connections are most often evident at the unit.

In an air-cooled condenser, heat is transferred by (1) desuperheating, (2) condensing, and (3) subcooling. Figure 6 shows the changes of state of R-134a passing through the condenser coil and the corresponding temperature change of the cooling air as it passes through the coil. Desuperheating and subcooling zones vary from 5 to 10% of the total heat transfer, depending on the entering gas and the leaving liquid temperatures, whereas Figure 6 is typical of a common application. Condensing takes place in approximately 85% of the condenser area generally at a constant temperature. The drop in saturated condensing temperature through the condenser coil is relative to the coil's refrigerant flow friction loss.

COIL CONSTRUCTION

A condenser coil with optimum circuiting requires the smallest heat transfer surface and has the fewest operational problems (such as trapping oil and refrigerant and a high discharge pressure). Coils are commonly constructed of copper, aluminum, or steel tubes, ranging from 0.25 to 0.75 in. in diameter. Copper is easy to use in manufacturing and requires no protection against corrosion. Aluminum has similar manufacturing methods, but special protection is necessary in the case of aluminum-to-copper joints. Steel tubing requires weather protection.

Tube diameter is chosen as a compromise between such factors as manufacturing tooling, cost, header arrangement, air resistance, and refrigerant flow resistance. Where a choice exists, the smaller diameter gives more flexibility in coil circuit design and results in a lower refrigerant charge.

Occasionally, a bare tube condenser (i.e., without fins) is used where airborne dirt loading is expected to be excessive; and, more so, in situations where access for coil cleaning is severely limited. When conditions are normal, and even in an atmosphere with a high salt content, a finned surface coil is used. The most common is the plate fin with extruded tube collars, each of which is fastened to the coil core tubes by mechanical expansion. This manufacturing process results in a rigid coil assembly that has maximum tube-to-fin thermal conductance. Spiral-fin surface coils as well as the spline fin, each of which are tightly wound onto the individual tubes that make up the coil core, are also used. Common fin spacings for each type range from 8 to 20 fins/in.

Stock fin coatings, as well as entire (completed) coil dip process coatings, are readily available from specialty chemical processors. These coatings are for use in specific corrosive atmospheres.

During the manufacturing process, tubesheets or endplates are fastened to the finned tube core at each end to complete the coil assembly. These provide a means of mounting the condenser enclosure or, in themselves, can form a part of the condenser unit enclosure. Center supports of much the same endplate design also provide condenser cabinetry fastening area as well as fan section compartmentalizing possibilities within the unit.

When a condenser coil fails (develops leaks), it is usually at the point where the tube is in contact with one of the supporting tubesheets. If not neutralized in some way, such wear points tend to leak after a number of years of full-time operation. Air-cooled condenser tube fractures are mainly caused by thermal stresses in the coil core. Excessive vibration also causes many tube core leaks.

Refrigerant containment laws now require that larger air-cooled condensers, most particularly horizontal coil units constructed of lightwall non-ferrous tubing, include certain refrigerant leak prevention construction. These designs either minimize or totally eliminate the tube-to-metal contact wear point evident at tube traverse points of endplates and center supports. Substantially enlarging the endplate and center support tube holes eliminates the tubing's expanded fit in these areas. Depending on the manufacturer's design, either all or a select number of these tube holes are oversized. This design substantially limits refrigerant tube fractures in the coil.

Another failure point is the inlet connecting header. Failure occurs when the connecting pipe is not adequately suspended, and/or discharge pulsing is not properly controlled.

Condenser units may be constructed with some of the coil or an additional smaller coil in the air entry side that is dedicated to refrigerant subcooling. In the latter arrangement, the refrigerant usually exits the main condenser coil and flows to the receiver. From the receiver, the refrigerant liquid travels back to the condenser for its flow through the dedicated subcooling section and then on to the expansion device(s).

Integrating a subcooling section in the condenser's main coil assembly achieves the same subcooling benefit and does not lose any of the main coil's liquid subcooling in a receiver. There the liquid area of the coil is circuited to provide more passes through fewer tube circuits. This design is sometimes called tripod circuitry, where two parallel circuits are joined to form one continuing single circuit. Condenser coils that integrate subcooling passes have several converging circuits located at the coil's air entry. If the system requires a receiver, a surge-type design best fits this method of subcooling.

The amount of subcooling depends on the heat transfer characteristics of the condenser and the temperature difference between the condenser and the entering air. Overall capacity increases about 0.5% per 1°F subcooling at the same suction and discharge pressure. Increasing condensing pressure increases the amount of subcooling, but the resultant liquid temperature is warmer, which usually has a negative effect on net capacity and the energy efficiency ratio (EER). Overall, to ensure proper operation, liquid quality needs to be 100% at the entrance to the evaporator's inlet flow control device; and, preferably, the liquid line should be cool to the touch.

Proper circuit design can increase subcooling with minimal condensing pressure drop. When correctly optimized, in-tube vapor velocities fall well within the higher heat transfer region of turbulent flow. Capacity, which is increased by added subcooling, is reduced to a lesser degree by the high-side pressure. Increasing subcooling within required limits of pressure drop through the condenser keeps the liquid refrigerant temperature at the expansion valve below saturation temperature. For this condition to be realized, the optimum circuitry loading needs to be determined for the specific application's mass flow rate.

FANS AND AIR REQUIREMENTS

Condenser coils can be cooled by natural convection or wind or by propeller, centrifugal, and vaneaxial fans. Because efficiency increases sharply by increasing air speed across the coil, forced convection with fans predominates.

For a given condensing temperature, the lower limit of air quantity occurs when the leaving air temperature approaches the entering saturated refrigerant condensing temperature. The upper limit is usually determined by air resistance in the coil and casing, noise, fan power, or fan size. As with the coil, an optimum must be reached by balancing operating cost and first cost with size and sound requirements. Commonly used values are 600 to 1200 cfm/ton at 400 to 800 fpm. Temperature differences (TD) at the normal cooling condition between condensing refrigerant and entering air range from 10 to 30°F. Fan power requirements generally ranges from 0.1 to 0.2 hp/ton.

The type of fan depends primarily on static pressure and unit shape requirements. Propeller fans are well suited to units with a low static pressure drop and free air discharge. Fan blade speeds are selected in the range of 515 to 1750 rpm, depending on size and sound requirements. Propeller fans are used more as direct-drive motor-to-blade assemblies (and in multiple assemblies) in virtually all sizes of condenser units. Such assemblies have propeller fan blades up to 36 in. in diameter and motors down to 12-pole synchronous speeds. Because of the low starting torque requirements of a direct-drive propeller fan assembly, the most commonly used is a permanent split-capacitor (PSC) fractional horsepower motor, inherently protected and in multiple arrangements. When the fan blade is mounted directly on the motor shaft, proper motor bearings and mounting must be used. Propeller fans at free air discharge are more efficient than centrifugal fans at free air discharge. A belt drive arrangement on propeller and centrifugal fan(s) offers more flexibility but usually requires greater maintenance. Belt drives are used more on centrifugal fans than on propeller fans.

Centrifugal fans perform best at higher static pressures. They are used almost exclusively where any but the simplest duct or damper assembly is required. Vaneaxial fans are often more efficient than centrifugal fans, but they generally do not adapt well to mounting on the condenser unit. A partially obstructed fan inlet or outlet can drastically increase noise level. Support bracket(s) close to the fan inlet or any partial obstruction of the fan inlet or outlet can noticeably effect fan noise.

The condenser manufacturer designs and tests motors and fan combinations as a complete assembly for the specific coil and air flow arrangement offered. Test measurements include motor winding and bearing temperature rise during the highest ambient operating conditions, duty cycling at the lowest ambient, and mounting positions and noise amplitude.

HEAT TRANSFER AND PRESSURE DROP

The overall heat transfer coefficient of an air-cooled condenser can be expressed by Equation (5c) in Chapter 21. The information in Chapter 21 (regarding sensible cooling coils) also applies to condensers for estimating air-side heat transfer coefficients and fin effectiveness. Computing heat transfer on the refrigerant side is somewhat complex. For vapor desuperheating and liquid subcooling (i.e., inlet and outlet regions of the coil) a cross-counterflow arrangement of refrigerant and air produces the best performance. In the condensing portion (i.e., at the center of the coil) the latent heat dissipates as a result of the refrigerant pressure drop. That is, the refrigerant enters the condenser as superheated vapor and cools by singe-phase convection to saturation temperature, after which condensation starts. If the coil's surface temperature is much lower than the saturation temperature, condensation occurs close to the gas entry area of the condenser coil. Eventually, in the last part of the condenser, all the refrigerant vapor condenses to liquid. Because of the change of state, the velocity and volume of the refrigerant, as a gas, drops substantially as it becomes a liquid. In general, most designs provide an adequate heat transfer surface area and complementary pressure drop to cause the refrigerant to subcool an average of 2 to 6°F before it leaves the condenser coil.

Chapter 3 of the 1997 *ASHRAE Handbook—Fundamentals* lists equations for heat transfer coefficients in the single-phase flow sections. For condensation from superheated vapors, no verified general predictive technique is available. For condensation from saturated vapors, one of the best calculation methods is the Shah correlation (Shah 1979, 1981). Also, Ghaderi et al. (1995) reviewed available correlations for condensation heat transfer in smooth and enhanced tubes.

Proper condenser design also involves calculating the pressure drops for the gas and liquid flow areas in the coil. The sum of these pressure drops, when converted to an equivalent saturated condensing temperature drop, approximates the amount of liquid subcooling. Chapter 4 of the 1997 ASHRAE *Handbook—Fundamentals* describes an estimation of two-phase pressure drop.

CONDENSERS REMOTE FROM COMPRESSOR

Remote air-cooled condensers are used for refrigeration from 0.5 ton to over 500 tons. Vertical flow condensers are the most common. The main components consist of (1) a finned condensing coil, (2) one or more fans and motors, (3) an enclosure or frame, and (4) an electrical panel on most large units.

The coil-fan arrangement can take almost any form, ranging from the horizontal coil with upflow air (top discharge) to the vertical coil with horizontal airflow (front discharge). The choice of design depends primarily on the intended application and the surroundings.

Refrigerant piping and electrical wiring are the interconnections between the compressor and the remote condenser unit. Receivers are most often installed close to the compressor and not the condenser. When selecting installation points for any of the high-side equipment, major concerns include direct sunlight (high solar load), warm surroundings, too cold a location (i.e., winter conditions), elevation differences, and length of piping runs.

CONDENSERS AS PART OF CONDENSING UNIT

When the condenser coil is included with the compressor as a packaged assembly, it is called a condensing unit. The condenser coil is then further categorized as an indoor or outdoor unit, depending on its rain-protective cabinet, electrical controls, and code approvals. Factory-assembled condensing units consist of one or more compressors, a condenser coil, electrical controls in a panel, a receiver, shutoff valves, switches, and sometimes other related units. Precharged interconnecting refrigerant lines to and from the evaporator may be offered as a kit with smaller units. Open, semi-hermetic or full-hermetic compressors are used for single or parallel (multiplexing) piping connections, in which case an accumulator, oil separator, and its crankcase oil return is likely to be included in the package. Indoor units lack weather enclosures, thus allowing easy access for service to all components, while outdoor unit access is obtained by removing cabinet panels or opening hinged covers.

Noise is a major concern in both the design and installation of condensing units, mainly because of the compressor. The following can be done to reduce unwanted sound:

- Avoid a straight-line path from the compressor to the listener.
- Acoustical material may be used inside the cabinet and to envelop the compressor. Air passages should be streamlined as much as possible; a top-discharge unit is generally quieter.
- Cushion or suspend the compressor on a suitable base.
- A unit's light gage base can often be superior to a heavier base. A stiff base readily transmits vibration to other panels.
- Natural frequencies of panels and refrigerant lines must be different from basic compressor and fan frequencies.
- Refrigerant lines with many bends produce numerous pulsation forces and natural frequencies that cause audible sounds.
- At times, a condenser with a very low refrigerant pressure drop has one or two passes resonant with the compressor discharge pulsations.
- Wire basket fan assembly supports help isolate noise and vibration.
- Fan selection is a major factor; steep pitch propeller blades and unstable centrifugal fans can be serious noise producers.
- At night, fan speed is often reduced because this is usually the time when air-cooled condenser noise is more objectionable.

Comparisons between water- and air-cooled equipment are often made. Where small units (less than 3 hp) are used and abundant low-cost water is available, both first cost and operating cost may be lower for water-cooled equipment. Up to a 20% larger compressor and/or longer run time is generally required for air-cooled operation; thus operating cost is higher. With water shortages common in many areas, the selection factors change. When cooling towers are used to produce condenser cooling water, the initial cost of water cooling may be higher. Also, the lower operating cost of water cooling may be offset by the cooling tower pump, fan use, and maintenance costs. When making a comparison, operation and maintenance of the complete system as well as local conditions need to be considered.

APPLICATION AND RATING OF CONDENSERS

Condensers are rated in terms of total heat rejection (THR), which is the total heat removed in desuperheating, condensing, and subcooling the refrigerant. This value is the product of the mass flow of the refrigerant and the difference in enthalpies of the refrigerant vapor entering and the liquid leaving the condenser coil.

A condenser may also be rated in terms of net refrigeration effect (NRE) or net heat rejection (NHR), which is the total heat rejection less the heat of compression added to the refrigerant in the compressor. This is the typical expression of the capacity of a refrigeration system.

For open compressors, the THR is the sum of the actual power input to the compressor and the NRE. For hermetic compressors, the THR is obtained by adding the NRE to the total motor power input and subtracting any heat losses from the surface of the compressor and discharge line. The surface heat losses are generally 0 to 10% of the power consumed by the motor. All quantities must be expressed in consistent units. Table 1 recommends factors for converting condenser THR ratings to NRE for both open and hermetic reciprocating compressors.

Ratings of air-cooled condensers are based on the temperature difference (TD) between the dry-bulb temperature of the air entering the coil and the saturated condensing temperature (which corresponds to the refrigerant pressure at the condenser outlet). Typical TD values are 10 to 15°F for low-temperature systems at a –20 to –40°F evaporator temperature, 15 to 20°F for medium-temperature systems at a 20°F evaporator temperature, and 25 to 30°F for air-conditioning systems at a 45°F evaporator temperature. The THR capacity of the condenser is considered proportional to the TD. That is, the capacity at 30°F TD is considered to be double the capacity for the same condenser selected for 15°F TD.

When selecting a condenser, determine the maximum air temperature entering the coil by referring to the weather data in Chapter 26 of the 1997 *ASHRAE Handbook—Fundamentals*. The 1% occurrence value is suggested for summer operation.

The specific design dry-bulb temperature must be selected carefully, especially for refrigeration serving process cooling. An entering air temperature that is higher than expected quickly causes the compressor discharge pressure and power to exceed design. Such conditions can cause an unexpected shutdown, usually at a time when it can least be tolerated. Also, congested or unusual locations may create entering air temperatures higher than general ambient conditions. Recirculated condenser air is usually the result of a poor choice in locating the installation.

The capacity of an air-cooled condenser equipped with an integral subcooling circuit varies depending on the refrigerant charge. The charge is greater when the subcooling circuit is full of liquid, which increases subcooling. When the subcooling circuit is used for condensing, the refrigerant charge is lower, the condensing capacity is greater, and the liquid subcooling is reduced. Laboratory testing should be in accordance with ASHRAE *Standard* 20, which requires liquid leaving the condenser coil (under test). Currently no industry certification program exists for air-cooled condensers.

Condenser ratings are based on THR capacity for a given temperature difference and the specified refrigerant. When application ratings are given in terms of NRE, they must always be defined with respect to the suction temperature, temperature difference, type of compressor (either open or hermetic), and the refrigerant. Condensers should be selected by matching the THR ratings of a particular compressor to the THR effect of the condenser.

Table 1 Net Refrigeration Effect Factors for Air-Cooled and Evaporative Condensers

Saturated Suction Temperature, °F	Condensing Temperature, °F										
	85	90	95	100	105	110	115	120	125	130	135
Open Compressors[a,b]											
–40	0.71	0.70	0.69	0.68	0.67	0.65	0.64	0.63	0.62	0.60	–
–20	0.77	0.76	0.74	0.73	0.72	0.71	0.70	0.69	0.67	0.66	–
0	0.82	0.80	0.79	0.78	0.77	0.76	0.75	0.74	0.73	0.71	–
+20	0.86	0.85	0.84	0.83	0.82	0.81	0.79	0.78	0.77	0.76	0.75
+40	0.91	0.90	0.89	0.87	0.86	0.85	0.84	0.83	0.82	0.81	0.80
Sealed Compressors[a]											
–40	0.55	0.54	0.53	0.52	0.51	0.50	0.49	0.47	0.46	0.44	–
–20	0.65	0.64	0.62	0.61	0.60	0.59	0.58	0.55	0.53	0.51	–
0	0.72	0.71	0.70	0.69	0.67	0.66	0.64	0.62	0.60	0.58	–
+20	0.77	0.76	0.75	0.74	0.72	0.71	0.69	0.68	0.66	0.64	0.62
+40	0.81	0.80	0.79	0.78	0.77	0.75	0.74	0.72	0.71	0.70	0.68
+50	0.83	0.82	0.81	0.80	0.79	0.78	0.76	0.75	0.74	0.73	0.72

[a] For Refrigerant 22, factors are based on 15°F of superheat entering the compressor.
[b] For ammonia (R-717), factors are based on 10°F superheat.

Notes:

1. These factors should be used only for air-cooled and evaporative condensers connected to reciprocating compressors.
2. Condensing temperature is that temperature corresponding to the saturation pressure as measured at the discharge of the compressor.
3. Net refrigeration effect factors are an approximation only. Represent the net refrigeration capacity as *Approximate Only* in published ratings.
4. For more accurate condenser selection and for condensers connected to centrifugal compressors, use the total heat rejection effect of the condenser and the compressor manufacturer's total rating, which includes the heat of compression.

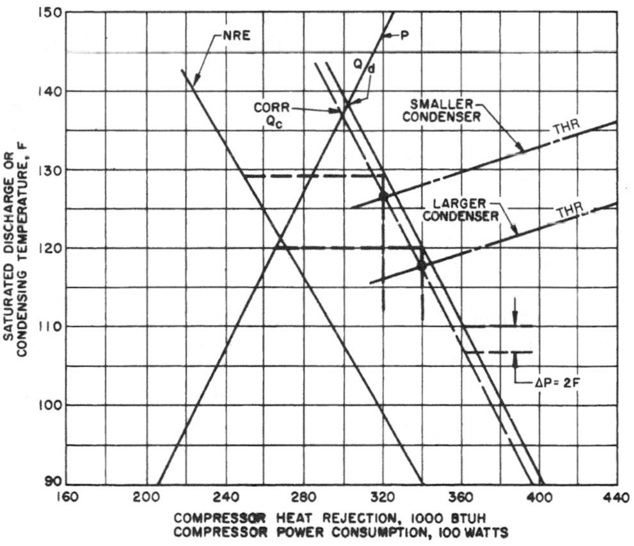

Fig. 7 Performance of Condenser-Compressor Combination

The capacities of available compressors and condensers are seldom the same at a given set of tabulated conditions. The capacity balance point of these two components can be determined either graphically by plotting condenser THR against compressor THR, as shown in Figure 7, or by computer simulation. When graphing the compressor THR, Q_d is plotted for a given saturated suction temperature at various saturated discharge temperatures. The saturated discharge temperature at the compressor must be corrected for the pressure drop in the discharge gas line as well as the condenser's internal pressure drop. This correction is made to estimate the working saturated condensing temperature more accurately. For an R-22 system, the pressure drop through a properly designed discharge line less than 100 ft long equals a saturated temperature drop of about 1 to 2°F, and the condenser internal pressure drop should correspond to about a 3°F saturated temperature change (for the refrigerant used). The corrected Q_c, compressor NRE, and power input are also plotted.

Then the condenser THR is plotted for various saturated condensing temperatures for a given entering air temperature. The balanced compressor and condenser THR and the saturated condensing temperature may be read at the intersection of the corrected compressor curve Q_c and the condenser THR curve. The compressor saturated discharge is read from the compressor curve Q_d for the same THR. The net refrigeration capacity and power input are read at the saturated discharge temperature.

Figure 7 is a plot of a compressor operating at a 90°F ambient with two sizes of an air-cooled condenser. The compressor operates at a saturated discharge temperature of 129°F, with a saturated condensing temperature of 126°F when using the smaller condenser. The THR is 320,000 Btu/h. The NRE is 250,000 Btu/h and the power consumption is 28 kW. With the larger condenser, the compressor operates at a saturated discharge temperature of 120°F and a saturated condensing temperature of 118°F. The THR is 340,000 Btu/h, and the power consumption is 26.5 kW.

The condenser selected must satisfy the anticipated cooling requirements at design ambient conditions at the lowest possible total compressor-condenser system operating cost, including an evaluation of the cost of control equipment.

CONTROL OF AIR-COOLED CONDENSERS

For a refrigeration system to function properly, the condensing pressure and temperature must be maintained within certain limits. An increase in condensing temperature causes a loss in capacity, requires extra power, and may overload the compressor motor. A condensing pressure that is too low hinders flow through conventional liquid feed devices. This hindrance starves the evaporator and causes a loss of capacity, unbalances the distribution of refrigerant in the evaporator, possibly causes strips of ice to form across the face of a freezer coil, and trips off the unit on low pressure.

Some systems use low-pressure-drop thermostatic expansion valves (balance port TXVs). These low head (usually surge-receiving) systems require an additional means of control to ensure that liquid line subcooled refrigerant enters the expansion valve. Supplemental electronic controls, valves, or, at times, liquid pumps are included. This flow control equipment is often found on units that operate year-round to provide medium- and low-temperature food processing refrigeration, where a precise temperature must be held.

To prevent excessively low head pressure during winter operation, two basic control methods are used: (1) refrigerant-side control and (2) air-side control. The most common methods of accomplishing pressure control in each of these two categories are as follows:

Refrigerant-Side Control. Control on the refrigerant side may be accomplished in the following ways:

1. By modulating the amount of active condensing surface available for condensing by flooding the coil with liquid refrigerant. This method requires a receiver and a larger charge of refrigerant. Several valving arrangements give the required amount of flooding to meet the variable needs. Both temperature and pressure actuation are used.
2. By going to one-half condenser operation. The condenser is initially designed with two equal parallel sections, each accommodating 50% of the load during normal summer operation. During the winter, solenoid or 3-way valves block off one of the condenser sections (as well as its pump-down to suction). This saves the flooding overcharge and also allows the shutdown of the fans on the inactive condenser side (Figure 8). Similar splits, such as one-third or two-thirds, are also possible. Anticipated load variations, along with climate conditions, usually dictate the preferred split arrangement for each specific installation.

   This split-condenser design has gained popularity, not only for its cold weather control, but also for its ability to reduce the refrigerant charge. Part-load operating conditions also benefit when using this split condenser option.

Air-Side Control. Control on the air side may be accomplished by one of three methods, or a combination of two of them: (1) fan cycling, (2) modulating dampers, and (3) fan speed control. Any control of airflow must be oriented so that the prevailing wind does not cause adverse operating conditions.

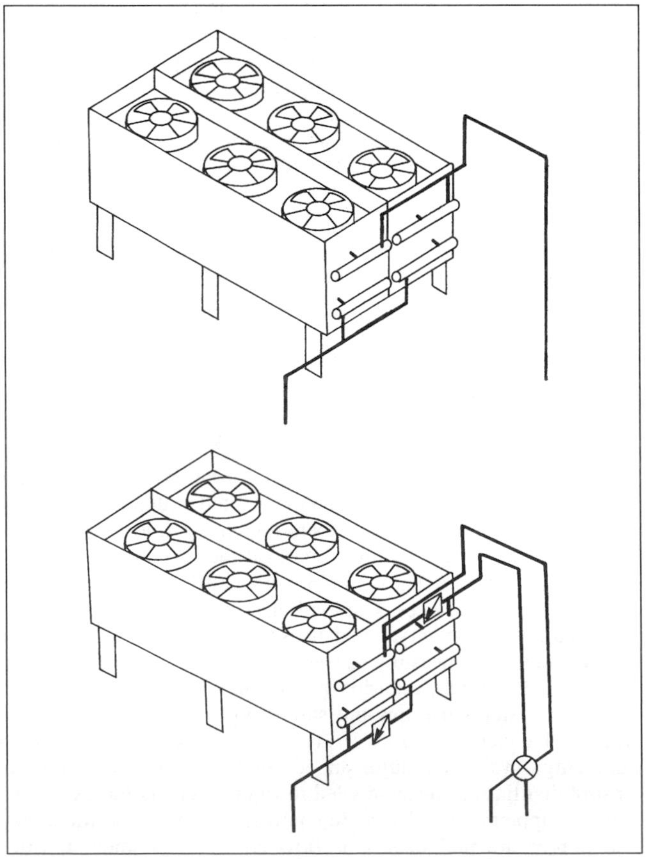

Fig. 8 Equal-Sized Condenser Sections Connected in Parallel and for Half-Condenser Operation During Winter

Fan cycling in response to outdoor ambient temperature eliminates rapid cycling, but is limited to use with multiple fan units or is supplemental to other control methods. A common method for control of a two-fan unit is to cycle only one fan. A three-fan unit may cycle two fans. During average winter conditions, large multi-fan outdoor condensers have electronic controllers programmed to run the first fan (or pair of fans) continuously, at least at low fan speed. The cycle starts with the fan(s) closest to the refrigerant connecting piping as the first fan(s) on, last fan(s) off. The remaining fan(s) may cycle as required on ambient or high-side pressure. Airflow through the condenser may be further reduced by modulating the airflow through the uncontrolled fan section, either by controlling the speed of some or all motors, or by dampers on either the air intake or discharge side (Figure 9).

In multiple, direct-drive motor-propeller arrangements, idle (off-cycle) fans should not be allowed to rotate backward; otherwise air will short circuit through them. Additionally, motor starting torque may be insufficient to overcome this reverse rotation. To eliminate this problem, designs for large units incorporate baffles to separate each fan assembly into individual compartments.

Air dampers controlled in response to either receiver pressure, ambient conditions, or liquid temperatures are also used to control the compressor head pressure. These devices throttle the airflow through the condenser coil from 100% to zero. Motors that drive propeller fans for such an application should have a flat power characteristic or should be fan cooled so that they do not overheat when the damper is nearly closed.

Fan speed can also be used to control compressor pressure. Because fan power increases in proportion to the cube of fan speed, energy consumption can be reduced substantially by slowing the fan at low ambient temperatures and during part-load operation. Solid state controls can modulate frequency along with voltage to vary the speed of synchronous motors. Two-speed fan motors also produce energy savings. Chapter 40 has more information on speed control.

Parallel operation of condensers, especially with capacity control devices, requires careful design. Connecting only identical condensers in parallel reduces operational problems. (The Multiple

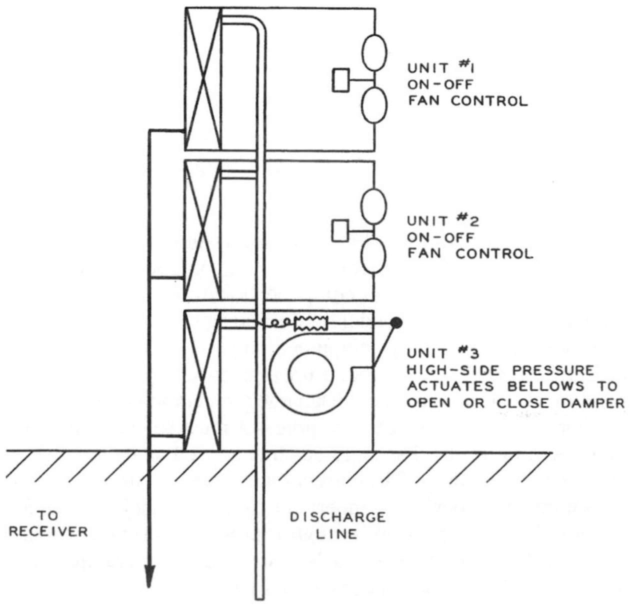

Fig. 9 Unit Condensers Installed in Parallel with Combined Fan Cycling and Damper Control

Condenser Installation section has further information. Figures 14 and 15 also apply to air-cooled condensers.)

Parallel operation of condensers, especially with capacity control devices, requires careful design. Connecting only identical condensers in parallel reduces operational problems. (The section on Multiple-Condenser Installations has further information. Figure 15 and Figure 16 also apply to air-cooled condensers.)

These control methods described maintain sufficient condenser and liquid line pressure for proper expansion valve operation. During off cycles, when the outdoor temperature is lower than the temperature of the indoor space to be conditioned or refrigerated, the refrigerant migrates from the evaporator and receiver and condenses in the cold condenser. The system pressure drops to what may correspond to the outdoor temperature. On start-up, insufficient feeding of liquid refrigerant to the evaporator, because of low compressor head pressure, causes low-pressure cut-out cycling of the compressor until the suction pressure reaches an operating level above the setting of the low pressurestat. At extremely low outdoor temperatures, the pressure in the system may be below the cut-in point of the pressurestat and the compressor will not start. This difficulty is solved by either of two methods (or combination thereof): (1) bypassing the low-pressure switch on start-up, or (2) using the condenser isolation method of control.

- **Low-pressure switch bypass.** On system start-up, a time-delay relay may bypass the low-pressure switch for three to five minutes to permit the head pressure to build up and allow uninterrupted compressor start-up operation.
- **Fan cycle switch.** On system start-up, a high-side pressure sensing switch can delay condenser fan(s) operation until a minimum pre-set head pressure is reached.
- **Condenser isolation.** This method uses a check valve arrangement to isolate the condenser. This prevents the refrigerant from migrating to the condenser coil during the off cycle.

When the liquid temperature coming from the outdoor condenser is lower than the evaporator operating temperature, use of a non-condensable charged thermostatic expansion valve bulb should be considered. This type of bulb prevents erratic operation of the expansion valve by eliminating the possibility of condensation of the thermal bulb charge on the head of the expansion valve.

INSTALLATION AND MAINTENANCE

The installation and maintenance of remote condensers require little labor because of their relatively simple design. Remote condensers are located as close as possible to the compressor, either indoors or outdoors. They may be located above or below the level of the compressor, but always above the level of the receiver. The following are some installation and maintenance concerns:

Indoor condenser. When condensers are located indoors, conduct the warm discharge air to the outdoors. An outdoor air intake opening near the condenser is provided and may be equipped with shutters. Indoor condensers can be used for space heating during the winter and for ventilation during the summer (Figure 10).

Outdoor installations. In outdoor installations of vertical face condensers, ensure that prevailing winds blow toward the air intake, or discharge shields should be installed to deflect opposing winds.

Piping. Piping practice with condensers is identical to that established by experience with other remote condensers. Size discharge piping going to the condenser for a total pressure drop equivalent not to exceed much over 2°F saturated temperature drop. Liquid line drop leg to the receiver should incorporate an adequately sized check valve. Follow standard piping procedures and good piping practices. For pipe sizing, refer to Chapter 33 of the 1997 *ASHRAE Handbook—Fundamentals*.

A pressure tap valve should be installed at the highest point in the discharge piping run so as to facilitate the removal of inadvertently trapped non-condensable gases from the system. The purging process should only be done with the compressor system off and pressures equalized. Note: This is only to be done by qualified personnel having the proper reclaim/recovery equipment as mandated by the appropriate agency, such as the Environmental Protection Agency in the United States.

Receiver. A condenser may have its receiver installed close by, but most often it will be separate and remotely installed. (Most water-cooled systems function without a receiver, because it is an integral part of the water-cooled condenser.) If a receiver is used and located in a comparatively warm ambient temperature, the liquid drain line from the condenser should be sized for a liquid velocity of under 100 fpm and designed for gravity drainage as well as for venting to the condenser. The receiver should be equipped with a pressure relief valve assembly that meets applicable codes and discharge requirements, sight glass(es), and positive action in-and-out shut-off valves. Requirements for such valving are outlined in ASHRAE *Standard* 15.

The manufacturer for the receiver can recommend mechanical subcooling and/or reduced refrigerant charge concepts incorporating improved ambient subcooling for condenser coils.

Maintenance. Schedule periodic inspection and lubrication of fan motor and fan bearings. Lube if required, following the instructions found either on the motor nameplate and/or the operation manual for this equipment. Adjust belt tension as necessary after installation and at yearly intervals.

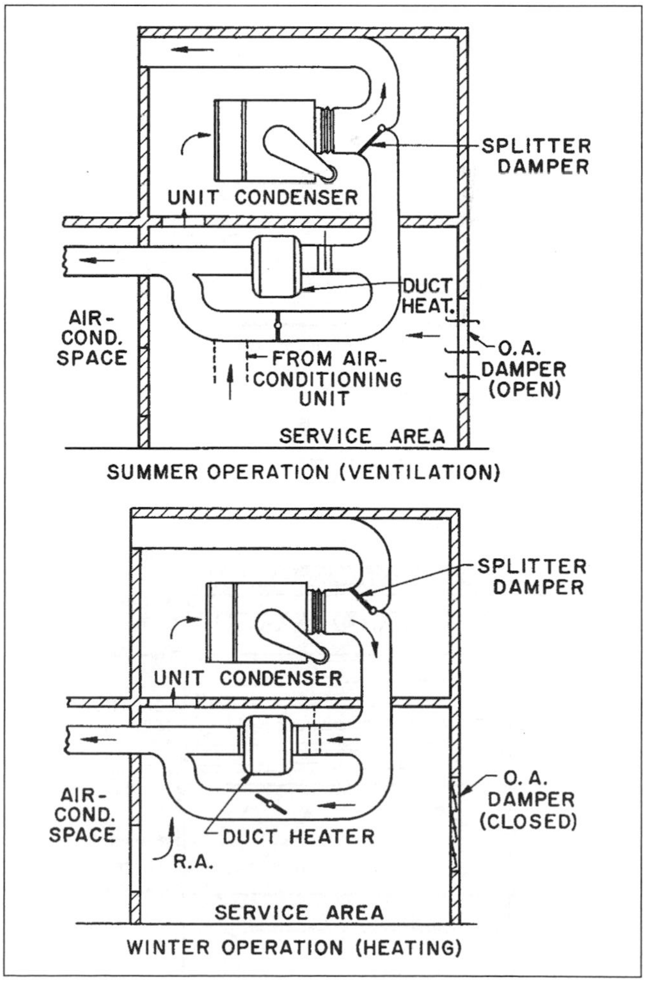

Fig. 10 Air-Cooled Unit Condenser for Winter Heating and Summer Ventilation

Condensers also require periodic removal of lint, leaves, dirt, and other airborne materials from their inlet coil surface by brushing the air entry side of the coil with a long, bristle brush. Washing with a low-pressure water hose or blowing with compressed air is more effectively done from the side opposite the air entry face. A mild cleaning solution is required to remove restaurant grease and other oil films from the fin surface. Indoor condensers in dusty locations, such as processing plants and supermarkets, should have adequate access for unrestricted coil cleaning and fin surface inspection. In all cases, extraneous material on fins reduces equipment performance, shortens operating life, increases running time, and causes more energy use.

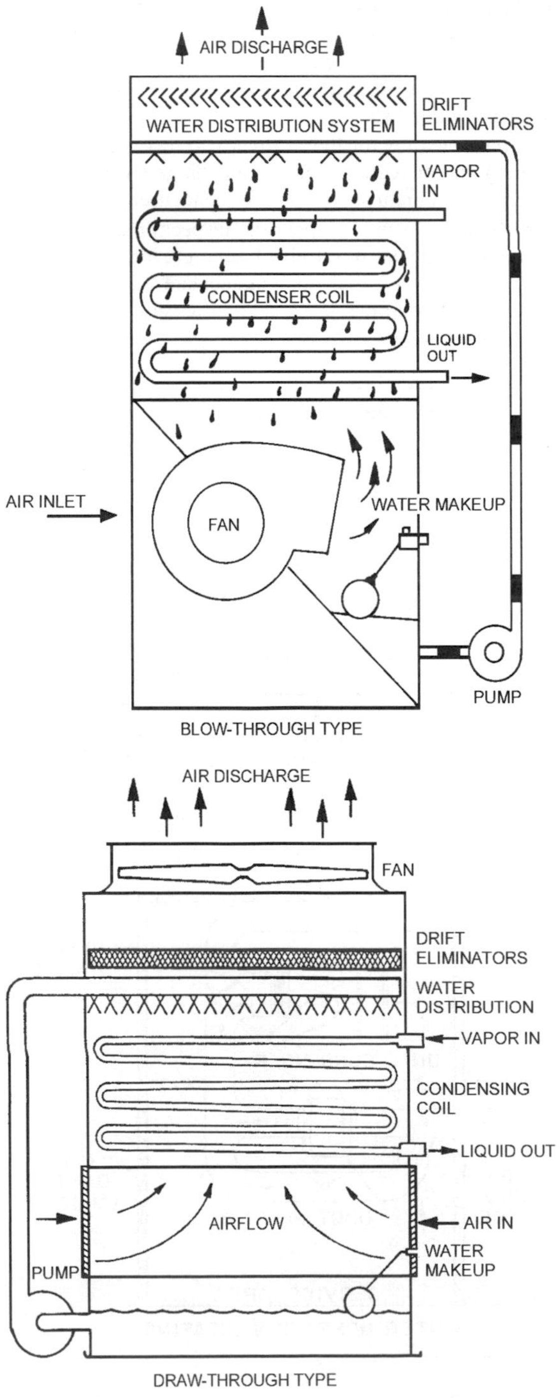

Fig. 11 Functional View of Evaporative Condenser

EVAPORATIVE CONDENSERS

As with water-cooled and air-cooled condensers, evaporative condensers reject heat from a condensing vapor into the environment. In an evaporative condenser, hot, high-pressure vapor from the compressor discharge circulates through a condensing coil that is continually wetted on the outside by a recirculating water system. As seen in Figure 11, air is simultaneously directed over the coil, causing a small portion of the recirculated water to evaporate. This evaporation removes heat from the coil, thus cooling and condensing the vapor.

Evaporative condensers reduce the water pumping and chemical treatment requirements associated with cooling tower/refrigerant condenser systems. In comparison with an air-cooled condenser, an evaporative condenser requires less coil surface and airflow to reject the same heat, or alternatively, greater operating efficiencies can be achieved by operating at a lower condensing temperature.

The evaporative condenser can operate at a lower condensing temperature than an air-cooled condenser because the air-cooled condenser is limited by the ambient dry-bulb temperature. In the evaporative condenser, heat rejection is limited by the ambient wet-bulb temperature, which is normally 14 to 25°F lower than the ambient dry bulb. Also, evaporative condensers typically provide lower condensing temperatures than the cooling tower/water-cooled condenser because the heat and mass transfer steps (between the refrigerant and the cooling water and between the water and ambient air) are more efficiently combined in a single piece of equipment, allowing minimum sensible heating of the cooling water. Evaporative condensers are, therefore, the most compact for a given capacity.

HEAT TRANSFER

In an evaporative condenser, heat flows from the condensing refrigerant vapor inside the tubes, through the tube wall, to the water film outside the tubes, and then from the water film to the air. Figure 12 shows temperature trends in a counterflow evaporative condenser. The driving potential in the first step of heat transfer is the temperature difference between the condensing refrigerant and the surface of the water film, whereas the driving potential in the second step is a combination of temperature and water vapor enthalpy difference between the water surface and the air. Sensible heat transfer between the water stream and the airstream at the water-air interface occurs because of the temperature gradient, while mass transfer (evaporation) of water vapor from the water-air interface to the airstream occurs because of the enthalpy gradient. Commonly, a single enthalpy driving force between the air saturated at the temperature of the water-film surface and the enthalpy of air in contact with that surface is applied to simplify an analytical approach. The exact formulation of the heat and mass transfer process requires consideration of the two driving forces simultaneously.

Because the performance of an evaporative condenser can not be represented solely by a temperature difference or an enthalpy difference, simplified predictive methods can only be used for interpolation of data between test points or between tests of different size units, provided that the air velocity, water flow, refrigerant velocity, and tube bundle configuration are comparable.

The rate of heat flow from the refrigerant through the tube wall and to the water film can be expressed as

$$q = U_s A(t_c - t_s) \tag{8}$$

where

q = rate of heat flow, Btu/h
U_s = overall heat transfer coefficient, Btu/h·ft^2·°F
t_c = saturation temperature at pressure of refrigerant entering condenser, °F
t_s = temperature of water film surface, °F
A = outside surface area of condenser tubes, ft^2

The rate of heat flow from the water-air interface to the airstream can be expressed as

$$q = U_c A(h_s - h_e) \tag{9}$$

where

q = heat input to condenser, Btu/h
U_c = overall transfer coefficient from water-air interface to airstream, Btu/h·ft^2·°F divided by enthalpy difference Δh in Btu/lb
h_s = enthalpy of air saturated at t_c, Btu/lb
h_e = enthalpy of air entering condenser, Btu/lb

Equations (8) and (9) have three unknowns: U_s, U_c, and t_s (h_s is a function of t_s). Consequently, the solution requires an iterative procedure that estimates one of the three unknowns and then solves for the remaining. This process is further complicated because of some of the more recently developed refrigerant mixtures for which t_c changes throughout the condensing process.

Leidenfrost and Korenic (1979, 1982), Korenic (1980), and Leidenfrost et al. (1980) evaluated heat transfer performance of evaporative condensers by analyzing the internal conditions in a coil.

CONDENSER CONFIGURATION

Principal components of an evaporative condenser include the condensing coil, the fan(s), the spray water pump, the water distribution system, cold water sump, drift eliminators, and water makeup assembly.

Coils

Generally, evaporative condensers use condensing coils fabricated from bare pipe or tubing without fins. The high rate of energy transfer from the wetted external surface to the air eliminates the need for extended surface. Furthermore, bare coils sustain performance better because they are less susceptible to fouling and are easier to clean. The high rate of energy transfer from the wetted external surface to the air makes finned coils uneconomical when used exclusively for wet operation. However, partially or wholly finned coils are sometimes used to reduce or eliminate plumes and/or to reduce water consumption by operating the condenser dry during off-peak conditions.

Coils are usually fabricated from steel tubing, copper tubing, iron pipe, or stainless steel tubing. Ferrous materials are generally hot-dip galvanized for exterior protection.

Method of Coil Wetting

The spray water pump circulates water from the cold water sump to the distribution system located above the coil. The water descends through the air circulated by the fan(s), over the coil surface, and eventually returns to the pan sump. The distribution system is designed to completely and continuously wet the full coil surface. Complete wetting ensures the high rate of heat transfer achieved with wet tubes and prevents excessive scaling, which is more likely to occur on intermittently or partially wetted surfaces. Such scaling is undesirable because it decreases the heat transfer efficiency (which tends to raise the condensing temperature) of the unit. Water lost through evaporation, drift, and blowdown from the cold water sump is replaced through a makeup assembly that typically consists of a mechanical float valve or solenoid valve and float switch combination.

Airflow

All commercially available evaporative condensers use mechanical draft; that is, fans move a controlled flow of air through the unit. As illustrated in Figure 11, these fans may be on the air inlet side (forced draft) or the air discharge side (induced draft). The type of fan selected, centrifugal or propeller, depends on the external pressure needs, energy requirements, and permissible sound levels. In many units the recirculating water is distributed over the condensing coil and flows downward to the collecting basin, counterflow to the air flowing up through the unit.

In an alternative configuration (Figure 13) a cooling tower fill augments the thermal performance of the condenser. In this unit, one airstream flows down over the condensing coil, parallel to the recirculating water, and exits horizontally into the fan plenum. The recirculating water then flows down over cooling tower fill where it is further cooled by a second airstream before it is reintroduced over the condenser coil.

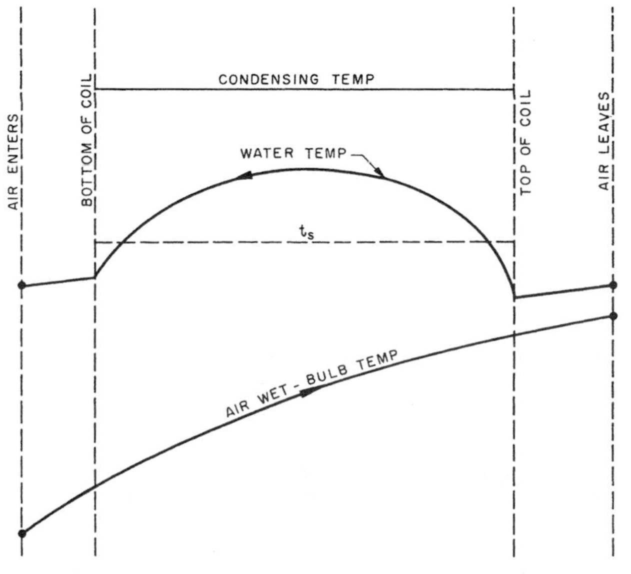

Fig. 12 Heat Transfer Diagram for Evaporative Condenser

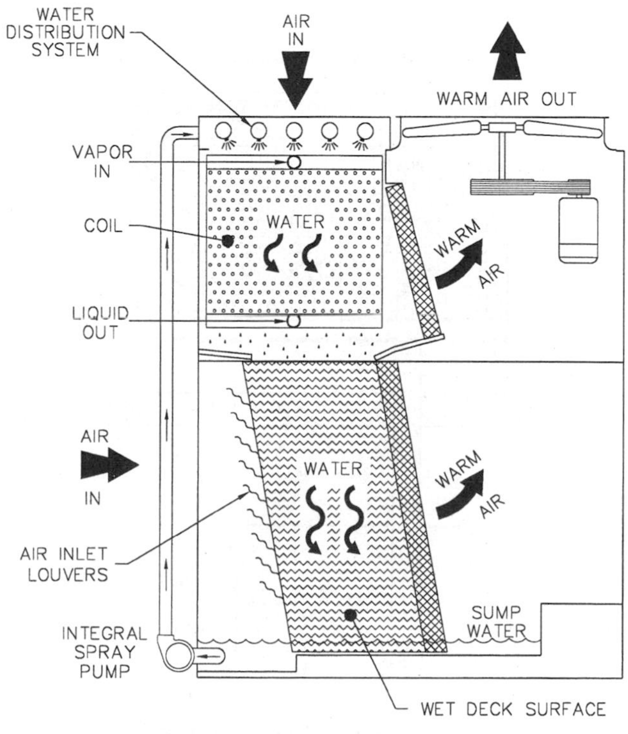

Fig. 13 Combined Coil/Fill Evaporative Condenser

Drift eliminators strip most of the water droplets from the discharge airstream, but some escape as drift. The rate of drift loss from an evaporative condenser is a function of the unit configuration, eliminator design, the airflow rate through the evaporative condenser, and the water flow rate. Generally, an efficient eliminator design can reduce drift loss to a range of 0.001 to 0.2% of the water circulation rate. If the air inlet is near the sump, louvers or deflectors may be installed to prevent water from splashing from the unit.

CONDENSER LOCATION

Most evaporative condensers are located outdoors, frequently on the roofs of machine rooms. They may also be located indoors and ducted to the outdoors. Generally, centrifugal fan models must be used for indoor applications to overcome the static resistance of the duct system.

Evaporative condensers installed outdoors can be protected from freezing in cold weather by a remote sump arrangement in which the water and pump are located in a heated space that is remote to the condensers (Figure 14). Piping is arranged so that whenever the pump stops, all of the water drains back into the sump to prevent freezing. Where remote sumps are not practical, reasonable protection can be provided by sump heaters, such as electric immersion heaters, steam coils, or hot water coils. Water pumps and lines must also be protected, for example with electric heat tracing tape and insulation.

Where the evaporative condenser is ducted to the outdoors, moisture from the warm saturated air can condense in condenser discharge ducts, especially if the ducts pass through a cool space. Some condensation may be unavoidable even with short, insulated ducts. In such cases, the condensate must be drained. Also, in these ducted applications, the drift eliminators must be highly effective.

MULTIPLE-CONDENSER INSTALLATIONS

Large refrigeration plants may have several evaporative condensers connected in parallel or evaporative condensers in parallel with shell-and-tube condensers. In such systems, unless all condensers have the same refrigerant-side pressure drop, condensed liquid refrigerant will flood back into the condensing coils of those condensers with the highest pressure loss. Also, in periods of light load when some condenser fans are off, liquid refrigerant will flood the lower coil circuits of the active condensers.

Trapped drop legs, as illustrated in Figure 15 and Figure 16, provide proper control of such multiple installations. The effective height *H* of drop legs must equal the pressure loss through the condenser at maximum loading. Particularly in cold weather, when condensing pressure is controlled by shutting down some condenser fans, active condensers may be loaded considerably above nominal rating, with higher pressure losses. The height of the drop leg should be great enough to accommodate these greater design loads.

RATINGS

Heat rejected from an evaporative condenser is generally expressed as a function of the saturated condensing temperature and entering air wet bulb. The type of refrigerant has considerable effect on ratings; this effect is handled by separate tables (or curves) for each refrigerant or by a correction factor when the difference is small.

Evaporative condensers are commonly rated in terms of total heat rejection when condensing a particular refrigerant at a specific condensing temperature and entering air wet-bulb temperature. Many manufacturers also provide alternative ratings in terms of evaporator load for a given refrigerant at a specific combination of saturated suction temperature and condensing temperature. Such ratings include an assumed value for the heat of compression, typically based on an open, reciprocating compressor, and should only

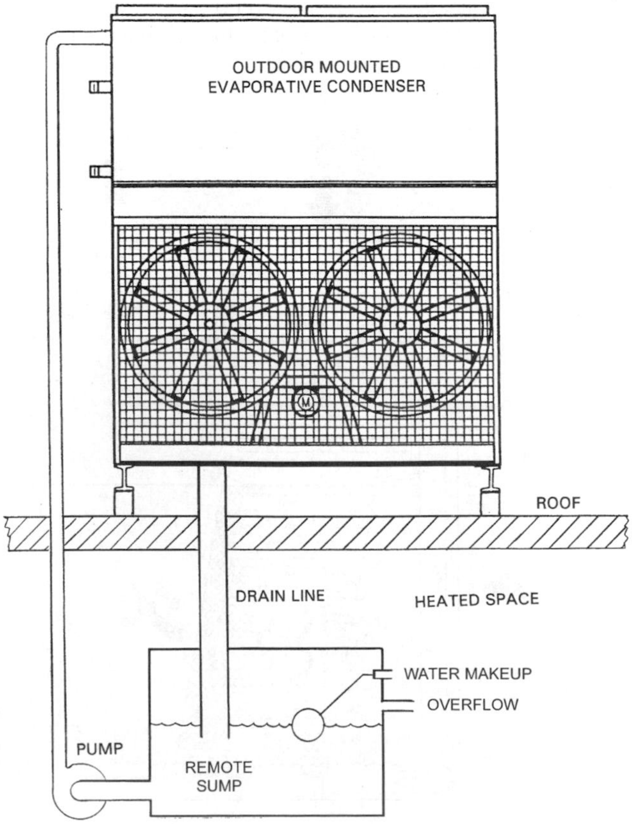

Fig. 14 Evaporative Condenser Arranged for Year-Round Operation

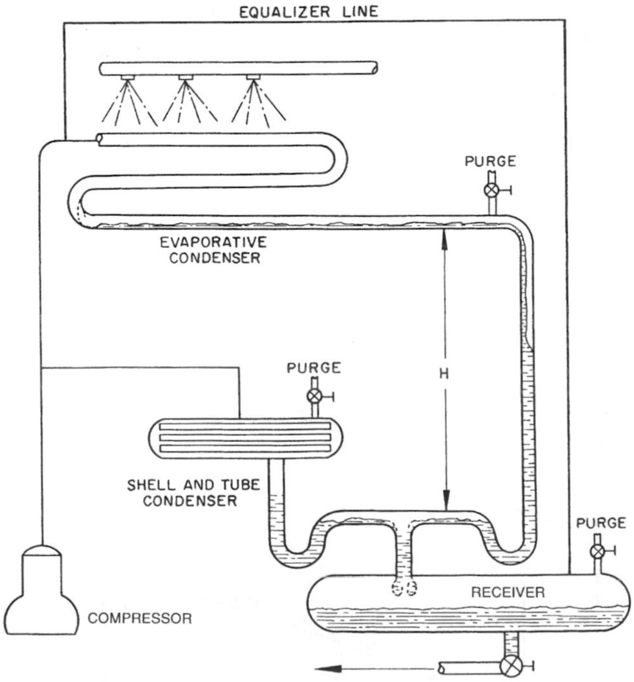

The pressure created by fluid height (H) above the trap must be greater than the internal resistance of the condenser. Note purge connections at both condensers and receiver.

Fig. 15 Parallel Operation of Evaporative and Shell-and-Tube Condenser

be used for that type of unit. Where another type of compression equipment is used, such as a gas-cooled hermetic and semihermetic compressor, the total power input to the compressor(s) should be added to the evaporator load and this value should be used to select the condenser.

Rotary screw compressors use oil to lubricate the moving parts and provide a seal between the rotors and the compressor housing. This oil is subsequently cooled either in a separate heat exchanger or by refrigerant injection. In the former case, the amount of heat removed from the oil in the heat exchanger is subtracted from the sum of the refrigeration load and the compressor brake power to obtain the total heat rejected by the evaporative condenser. When liquid injection is used, the total heat rejection is the sum of the refrigeration load and the brake horsepower. Heat rejection rating data together with any ratings based on refrigeration capacity should be included.

DESUPERHEATING COILS

A desuperheater is an air-cooled finned coil usually installed in the discharge airstream of an evaporative condenser (Figure 17).

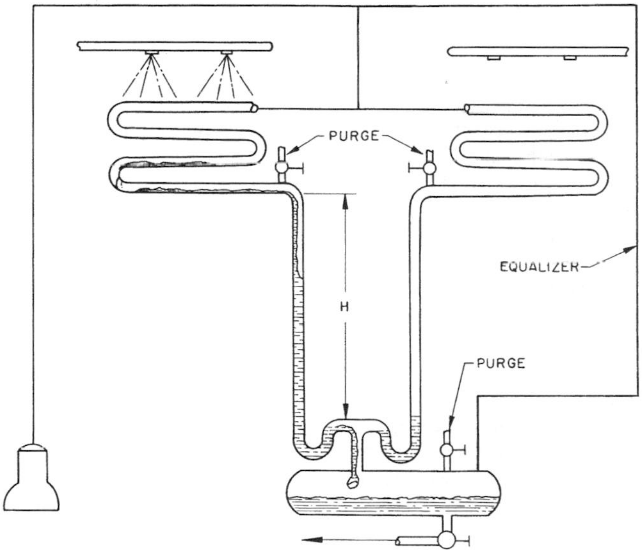

Either condenser can operate independent of the other. Height (H) above the trap, for either condenser, must be not less than the internal resistance of the condenser. Note purge connections at both condensers and the receiver.

Fig. 16 Parallel Operation of Two Evaporative Condensers

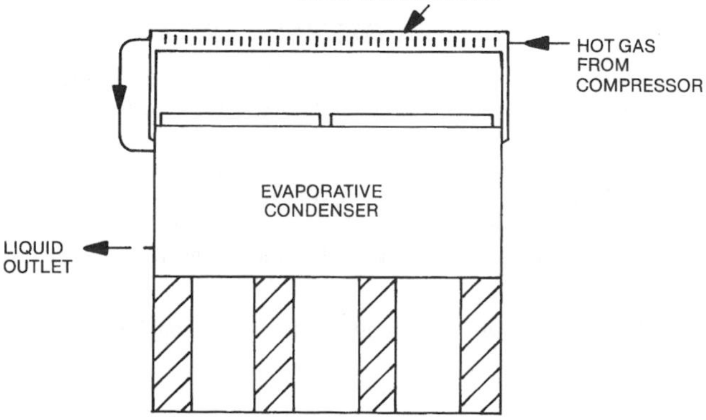

Fig. 17 Evaporative Condenser with Desuperheater Coil

The primary function of the desuperheater is to increase the condenser capacity by removing some of the superheat from the discharge vapor before the vapor enters the wetted condensing coil. The amount of superheat removed is a function of the desuperheater surface, condenser airflow, and the temperature difference between the refrigerant and the air. In practice, a desuperheater is limited to reciprocating compressor ammonia installations where discharge temperatures are relatively high (250 to 300°F).

REFRIGERANT LIQUID SUBCOOLERS

The pressure of the refrigerant at the expansion device feeding the evaporator(s) is lower than that in the receiver because of the pressure drop in the liquid line. If the liquid line is long or if the evaporator is above the receiver (which further decreases the refrigerant pressure at the expansion device), significant flashing can occur in the liquid line. To avoid flashing where these conditions exist, the liquid refrigerant must be subcooled after it leaves the receiver. The minimum amount of subcooling required is the temperature difference between the condensing temperature and the saturation temperature corresponding to the saturation pressure at the expansion device. Subcoolers are often used with halocarbon systems but are seldom used with ammonia systems for the following reasons:

- Because ammonia has a relatively low liquid density, liquid line static pressure loss is small.
- Ammonia has a very high latent heat; thus, the amount of flash gas resulting from typical pressure loss in the liquid line is extremely small.
- Ammonia is seldom used in a direct-expansion feed system where subcooling is critical to proper expansion valve performance.

One method commonly used to supply subcooled liquid for halocarbon systems places a subcooling coil section in the evaporative condenser below the condensing coil (Figure 18). Depending on the design wet-bulb temperature, condensing temperature, and subcooling coil surface area, the subcooling coil section normally furnishes 10 to 15°F of liquid subcooling. As shown in Figure 18, a receiver must be installed between the condensing coil and subcooling coil to provide a liquid seal for the subcooling circuit.

MULTICIRCUIT CONDENSERS AND COOLERS

Evaporative condensers and evaporative fluid coolers, which are essentially the same, may be multicircuited to condense different refrigerants or cool different fluids simultaneously, but only if the difference between the condensing temperature and the leaving fluid temperature is small. Typical multicircuits (1) condense different refrigerants found in food market units, (2) condense and cool different fluids in separate circuits such as condensing the

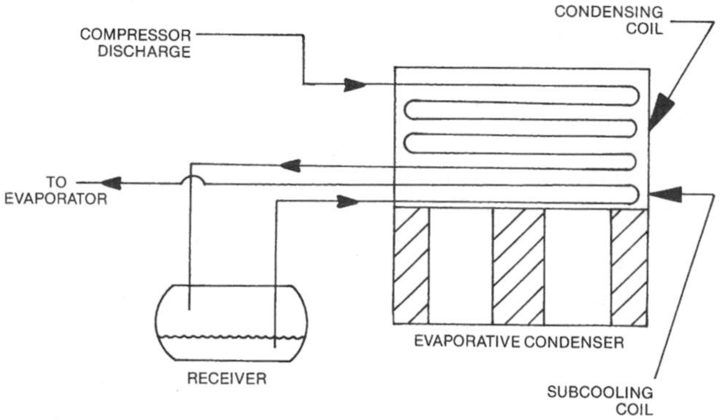

Fig. 18 Evaporative Condenser with Liquid Subcooling Coil

refrigerant from a screw compressor in one circuit and cooling the water or glycol for its oil cooler in another circuit, and (3) cool different fluids in separate circuits such as in many industrial applications where it is necessary to keep the fluids from different heat exchangers in separate circuits.

A multicircuit unit is usually controlled by sensing sump water temperature and using capacity control dampers, fan cycling, or a variable speed drive to modulate the airflow and match the unit capacity to the load. Two-speed motors or separate high- and low-speed motors are also used. Sump water temperature has an averaging effect on control. Because all water in the sump is recirculated once every minute or two, its temperature is a good indicator of changes in load.

WATER TREATMENT

As the recirculated water evaporates in an evaporative condenser, the dissolved solids in the makeup water continually increase as more water is added. Continued concentration of these dissolved solids can lead to scaling and/or corrosion problems. In addition, airborne impurities and biological contaminants are often introduced into the recirculated water. If these impurities are not controlled, they can cause sludge or biological fouling. Simple blowdown (the wasting of a small portion of the recirculating water) may be adequate to control scale and corrosion on sites with good quality makeup water, but it does not control biological contaminants including *Legionellae*. All evaporative condensers should be treated to restrict biological growth and many benefit from chemical treatment to control scale and corrosion. Chapter 47 of the 1999 *ASHRAE Handbook—Applications* covers water treatment in more detail. Specific recommendations on water treatment can be obtained from any competent water treatment supplier.

WATER CONSUMPTION

For the purpose of sizing the makeup water piping, all of the heat rejected by an evaporative condenser is assumed to be latent heat (approximately 1050 Btu/lb of water evaporated). The heat rejected depends on operating conditions, but it can range from 14,000 Btu/h per ton of air conditioning to 17,000 Btu/h per ton of freezer storage. The evaporated water ranges from about 1.6 to 2 gph per ton of refrigeration. In addition, a small amount of water can be lost in the form of drift through the eliminators. With good, quality makeup water, the bleed rates may be as low as one-half the evaporation rate, and the total water consumption would range from 2.4 gph/ton for air conditioning to 3 gph/ton for refrigeration.

CAPACITY MODULATION

To ensure operation of expansion valves and other refrigeration components, extremely low condensing pressures must be avoided. Capacity can be controlled in response to condensing pressure on single-circuit condensers and in response to the temperature of the spray water on multicircuited condensers. Means of controlling capacity include: (1) intermittent operation of the fan(s), (2) a modulating damper in the airstream to reduce the airflow (centrifugal fan models only), and (3) fan speed control using variable-speed motors and/or drives, two-speed motors, or additional lower power pony motors (small modular motors on the same shaft). Two-speed fan motors usually operate at 100 and 50% fan speed, which provides 100% and approximately 60% condenser capacity, respectively. With the fans off and the water pump operating, condenser capacity is approximately 10%.

Often, a two-speed fan motor or supplementary (pony) motor that operates the fan at reduced speed can provide sufficient capacity control, because condensing pressure seldom needs to be held to a very tight tolerance other than to maintain a minimum condensing pressure to ensure refrigerant liquid feed pressure for the low-pressure side and/or sufficient pressure for hot gas defrost requirements. For those applications requiring close control of condensing pressure, virtually infinite capacity control can be provided by using frequency modulating controls to control fan motor speed. These controls may also be justified economically because they save energy and extend the life of the fan and drive assembly as compared to fan cycling with two-speed control. However, special concerns and limitations are associated with their use, which are discussed in Chapter 40 and in the section on Capacity Control in Chapter 36.

Modulating air dampers also offer closer control on condensing pressure, but they do not offer as much fan power reduction at part load as fan speed control. Water pump cycling for capacity control is not recommended because the periodic drying of the tube surface promotes a build-up of scale.

The designer should research applicable codes and standards when applying evaporative condensers. Most manufacturers provide sound level information for their equipment. Centrifugal fans are inherently quieter than axial fans, but axial fans generally have a lower power requirement for a given application. Low-speed operation and multistaging usually lowers sound levels of condensers with axial fans. In some cases, factory-supplied sound attenuators may need to be installed.

PURGING

Refrigeration systems operating below atmospheric pressure and systems that are opened for service may require purging to remove air that causes a high condensing pressure. With the system operating, purging should be done from the top of the condensing coil outlet connection. On multiple-coil condensers or multiple-condenser installations, one coil at a time should be purged. Purging two or more coils at one time equalizes coil outlet pressures and can cause liquid refrigerant to back up in one or more of the coils, thus reducing the operating efficiency.

Purging may also be done from the high point of the evaporative condenser refrigerant feed, but it is only effective when the system is not operating. During normal operation, noncondensables are dispersed throughout the high-velocity vapor, and excessive refrigerant would be lost if purging were done from this location (see Figure 15 and Figure 16). Finally, all codes and ordinances governing the discharge, recovery, and recycling of refrigerants must be followed.

MAINTENANCE

Evaporative condensers are often installed in remote locations and may not receive the routine attention of operating and maintenance personnel. Therefore, programmed maintenance is essential: where manufacturers' recommendations are not available, Table 2 may be used as a guide. Chapter 36 also has information on maintenance that applies to evaporative condensers.

CODES AND STANDARDS

If state or local codes do not take precedence, design pressures, materials, welding, tests, and relief devices should be in accordance with the ASME *Boiler and Pressure Vessel Code*, Section VIII, Division 1. Evaporative condensers are exempt, however, from the ASME Code on the basis of Item (c) of the Scope of the Code, which states that if the inside diameter of the condenser shell is 6 in. or less, it is not governed by the Code. Other rating and testing standards include

- ASHRAE *Standard* 15, Safety Code for Mechanical Refrigeration.
- ASHRAE *Standard* 64, Methods of Testing Remote Mechanical-Draft Evaporative Refrigerant Condensers.
- ARI *Standard* 490, Remote Mechanical-Draft Evaporative Refrigerant Condensers.

Table 2 Maintenance Checklist

Maintenance Item	Frequency
1. Check fan and motor bearings, and lubricate if necessary. Check tightness and adjustment of thrust collars on sleeve-bearing units and locking collars on ball-bearing units.	Q
2. Check belt tension.	M
3. Clean strainer. If air is extremely dirty, strainer may need frequent cleaning.	W
4. Check, clean, and flush sump, as required.	M
5. Check operating water level in sump, and adjust makeup valve, if required.	W
6. Check water distribution, and clean as necessary.	W
7. Check bleed water line to ensure it is operative and adequate as recommended by manufacturer.	M
8. Check fans and air inlet screens and remove any dirt or debris.	D
9. Inspect unit carefully for general preservation and cleanliness, and make any needed repairs immediately.	R
10. Check operation of controls such as modulating capacity control dampers.	M
11. Check operation of freeze control items such as pan heaters and their controls.	Y
12. Check the water treatment system for proper operation.	W
13. Inspect entire evaporative condenser for spot corrosion. Treat and refinish any corroded spot.	Y

D = Daily; W = Weekly; M = Monthly; Q = Quarterly; S = Semiannually; Y = Yearly; R = As required.

REFERENCES

ARI. 1997. Fouling factors: A survey of their application in today's air conditioning and refrigeration industry. *Guideline* E-1997. Air-Conditioning and Refrigeration Institute, Arlington, VA.

ARI. 1994. Remote mechanical-draft air-cooled refrigerant condensers. *Standard* 460-94. Air-Conditioning and Refrigeration Institute.

ASHRAE. 1997. Methods of testing for rating remote mechanical-draft air-cooled refrigerant condensers. *Standard* 20-1997.

ASHRAE. 1996. Reducing emission of halogenated refrigerants in refrigeration and air conditioning equipment and systems. *Guideline* 3-1996.

ASHRAE. 1982. Waterside fouling resistance inside condenser tubes. *Research Note* 31 (RP 106). *ASHRAE Journal* 24(6):61.

Ayub, Z.H. and S.A. Jones. 1987. Tubeside erosion/corrosion in heat exchangers. *Heating/Piping/Air Conditioning* (December):81.

Bergles, A.E. 1995. Heat transfer enhancement and energy efficiency—Recent progress and future trends. In *Advances in Enhanced Heat/Mass Transfer and Energy Efficiency*, M.M. Ohadi and J.C. Conklin, eds. HTD-Vol. 320/PID-Vol. 1, ASME Publications, New York.

Briggs, D.E. and E.H. Young. 1969. Modified Wilson plot techniques for obtaining heat transfer correlations for shell and tube heat exchangers. *Chemical Engineering Progress Symposium Series* 65(92):35.

Butterworth, D. 1977. Developments in the design of shell-and-tube condensers. *ASME Paper No.* 77-WA/HT-24. American Society of Mechanical Engineers, New York.

Coates, K.E. and J.G. Knudsen. 1980. Calcium carbonate scaling characteristics of cooling tower water. *ASHRAE Transactions* 86(2).

Eckels, S.J. and M.B. Pate. 1991. Evaporation and condensation of HFC-134a and CFC-12 in a smooth tube and a micro-fin tube. *ASHRAE Transactions* 97(2):71-81.

Eckert, E.G. 1963. *Heat and mass transfer.* McGraw-Hill, New York, 143.

Ghaderi, M., M. Salehi, and M.H. Saeedi. 1995. Review of in-tube condensation heat transfer correlations for smooth and enhanced tubes. In *Advances in Enhanced Heat/Mass Transfer and Energy Efficiency*, M.M. Ohadi and J.C. Conklin, eds. HTD-Vol. 320/PID-Vol. 1, ASME Publications, New York.

Honda, H., H. Takamatsu, and K. Kim. 1994. Condensation of CFC-11 and HCFC-123 in in-line bundles of horizontal finned tubes: Effect of fin geometry. *Enhanced Heat Transfer* 1(2):197-209.

Honda, H., H. Takamatsu, N. Takada, O. Makishi, and H. Sejimo. 1995. Film condensation of HCFC-123 in staggered bundles of horizontal finned tubes. *Proceedings* ASME/JSME Thermal Engineering Conference 2:415-420.

Huber, J.B., L.E. Rewerts, and M.B. Pate. 1994a. Shell-side condensation heat transfer of R-134a. Part I: Finned tube performance. *ASHRAE Transactions* 100(2):239-47.

Huber, J.B., L.E. Rewerts, and M.B. Pate. 1994b. Shell-side condensation heat transfer of R-134a. Part II: Enhanced tube performance. *ASHRAE Transactions* 100(2):248-56.

Huber, J.B., L.E. Rewerts, and M.B. Pate. 1994c. Shell-side condensation heat transfer of R-134a. Part III: Comparison of R-134a with R-12. *ASHRAE Transactions* 100(2):257- 64.

Huber, J.B., L.E. Rewerts, and M.B. Pate. 1994d. Experimental determination of shell side condenser bundle design factors for refrigerants R-123 and R-134a. *Final Report* for ASHRAE Research Project RP-676.

Kedzierski, M.A. 1997. Effect of inclination on the performance of a compact brazed plate condenser and evaporator. *Heat Transfer Engineering* 18(3):25-38.

Kedzierski, M.A. and R.L. Webb. 1990. Practical fin shapes for surface drained condensation. *Journal of Heat Transfer* 112:479-85.

Kern, D.Q. and A.D. Kraus. 1972. *Extended surface heat transfer*, Chapter 10. McGraw-Hill, New York.

Kistler, R.S., A.E. Kassem, and J.M. Chenoweth. 1976. Rating shell-and-tube condensers by stepwise calculations. *ASME Paper No.* 76-WA/HT-5. American Society of Mechanical Engineers, New York.

Korenic, B. 1980. Augmentation of heat transfer by evaporative coolings to reduce condensing temperatures. PhD *Thesis*, Purdue University, West Lafayette, IN.

Kragh, R.W. 1975. Brush cleaning of condenser tubes saves power costs. *Heating/Piping/Air Conditioning* (September).

Lee, S.H. and J.G. Knudsen. 1979. Scaling characteristics of cooling tower-water. *ASHRAE Transactions* 85(1).

Leidenfrost, W. and B. Korenic. 1979. Analysis of evaporative cooling and enhancement of condenser efficiency and of coefficient of performance. *Wärme und Stoffübertragung* 12.

Leidenfrost, W. and B. Korenic. 1982. Experimental verification of a calculation method for the performance of evaporatively cooled condensers. *Brennstoff-Wärme-Kraft* 34(1):9. VDI Association of German Engineers, Dusseldorf.

Leidenfrost, W., K.H. Lee, and B. Korenic. 1980. Conservation of energy estimated by second law analysis of a power-consuming process. *Energy*.

McAdams, W.H. 1954. *Heat transmission*, 3rd ed. McGraw-Hill, New York, 219 and 343.

Ohadi, M.M., S. Dessiatoun, A. Singh, K. Cheung, and M. Salehi. 1995. EHD enhancement of boiling/condensation heat transfer of alternate refrigerants/refrigerant mixtures, *Progress Report* No. 9, submitted to U.S. DOE, Grant No. DOE/DE-FG02-93CE23803.A000.

Pate, M.B., Z.H. Ayub, and J. Kohler. 1991. Heat exchangers for the air-conditioning and refrigeration industry: State-of-the-art design and technology. *Heat Transfer Engineering* 12(3):56-70.

Pearson, J.F. and J.G. Withers. 1969. New finned tube configuration improves refrigerant condensing. *ASHRAE Journal* 11(6):77.

Seban, R.A. and E.F. McLaughlin. 1963. Heat transfer in tube coils with laminar and turbulent flow. *International Journal of Heat and Mass Transfer* 6:387.

Shah, M.M. 1979. A general correlation for heat transfer during film condensation in tubes. *International Journal of Heat and Mass Transfer* 22(4):547.

Shah, M.M. 1981. Heat transfer during film condensation in tubes and annuli: A review of the literature. *ASHRAE Transactions* 87(1).

Starner, K.E. 1976. Effect of fouling factors on heat exchanger design. *ASHRAE Journal* 18(May):39.

Sturley, R.A. 1975. Increasing the design velocity of water and its effect in copper tube heat exchangers. *Paper* No. 58. The International Corrosion Forum, Toronto.

Suitor, J.W., W.J. Marner, and R.B. Ritter. 1976. The history and status of research in fouling of heat exchangers in cooling water service. *Paper* No. 76-CSME/CS Ch E-19. National Heat Transfer Conference, St. Louis, MO.

Taborek, J., F. Voki, R. Ritter, J. Pallen. and J. Knudsen. 1972. Fouling—The major unresolved problem in heat transfer. *Chemical Engineering Progress Symposium Series* 68, Parts I and 11, Nos. 2 and 7.

TEMA. 1988. *Standards of the Tubular Exchanger Manufacturers Association*, 7th ed. Tubular Exchanger Manufacturers Association, New York.

Wanniarachchi, A.S. and R.L. Webb. 1982. Noncondensible gases in shell-side refrigerant condensers. *ASHRAE Transactions* 88(2):170-84.

Watkinson, A.P., L. Louis, and R. Brent. 1974. Scaling of enhanced heat exchanger tubes. *The Canadian Journal of Chemical Engineering* 52:558.

Webb, R.L. 1984a. Shell-side condensation in refrigerant condensers. *ASHRAE Transactions* 90(1B):5-25.

Webb, R.L. 1984b. The effects of vapor velocity and tube bundle geometry on condensation in shell-side refrigeration condensers. *ASHRAE Transactions* 90(1B):39-59.

Webb, R.L. and C.G. Murawski. 1990. Row effect for R-11 condensation on enhanced tubes. *Journal of Heat Transfer* 112(3).

Webb, R.L., A.S. Wanniarachchi, and T.M. Rudy. 1980. The effect of noncondensible gases on the performance of an R-11 centrifugal water chiller condenser. *ASHRAE Transactions* 86(2):57.

Wolverine Tube, Inc. 1984. Engineering data book 11. Section I, 30.

Young, E.H. and D.J. Ward. 1957. Fundamentals of finned tube heat transfer. *Refining Engineer* I (November).

CHAPTER 36

COOLING TOWERS

MOST air-conditioning systems and industrial processes generate heat that must be removed and dissipated. Water is commonly used as a heat transfer medium to remove heat from refrigerant condensers or industrial process heat exchangers.

In the past, this was accomplished by drawing a continuous stream of water from a utility water supply or a natural body of water, heating it as it passed through the process, and then discharging the water directly to a sewer or returning it to the body of water. Water purchased from utilities for this purpose has now become prohibitively expensive because of increased water supply and disposal costs. Similarly, cooling water drawn from natural sources is relatively unavailable because the disturbance to the ecology of the water source caused by the increased temperature of the discharge water has become unacceptable.

Air-cooled heat exchangers may be used to cool the water by rejecting heat directly to the atmosphere, but the first cost and fan energy consumption of these devices are high. They are capable of economically cooling the water to within approximately 20°F of the ambient dry-bulb temperature. Such temperatures are too high for the cooling water requirements of most refrigeration systems and many industrial processes.

Cooling towers overcome most of these problems and therefore are commonly used to dissipate heat from water-cooled refrigeration, air-conditioning, and industrial process systems. The water consumption rate of a cooling tower system is only about 5% of that of a once-through system, making it the least expensive system to operate with purchased water supplies. Additionally, the amount of heated water discharged (**blowdown**) is very small, so the ecological effect is greatly reduced. Lastly, cooling towers can cool water to within 4 to 5°F of the ambient wet-bulb temperature or about 35°F lower than can air-cooled systems of reasonable size.

PRINCIPLE OF OPERATION

A cooling tower cools water by a combination of heat and mass transfer. The water to be cooled is distributed in the tower by spray nozzles, splash bars, or film-type fill, which exposes a very large water surface area to atmospheric air. Atmospheric air is circulated by (1) fans, (2) convective currents, (3) natural wind currents, or (4) induction effect from sprays. A portion of the water absorbs heat to change from a liquid to a vapor at constant pressure. This heat of vaporization at atmospheric pressure is transferred from the water remaining in the liquid state into the airstream.

Figure 1 shows the temperature relationship between water and air as they pass through a counterflow cooling tower. The curves indicate the drop in water temperature (Point A to Point B) and the rise in the air wet-bulb temperature (Point C to Point D) in their respective passages through the tower. The temperature difference between the water entering and leaving the cooling tower (A minus B) is the **range**. For a system operating in a steady state, the range is the same as the water temperature rise through the load heat exchanger, provided the flow rate through the cooling tower and heat exchanger are the same. Accordingly, the range is determined by the heat load and water flow rate, not by the size or thermal capability of the cooling tower.

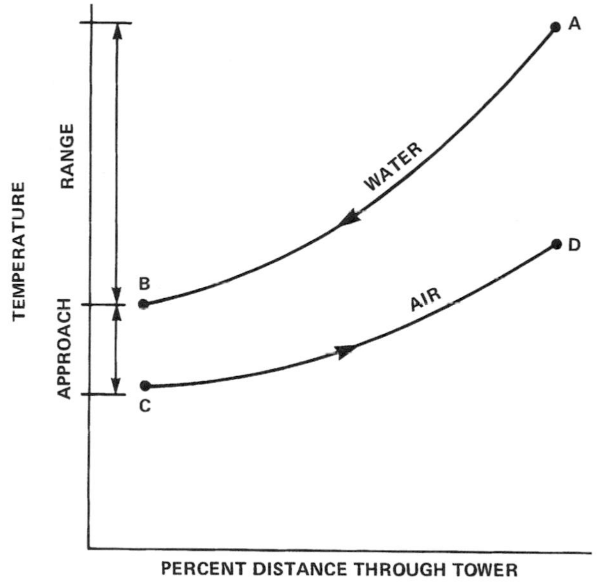

Fig. 1 Temperature Relationship Between Water and Air in Counterflow Cooling Tower

The difference between the leaving water temperature and the entering air wet-bulb temperature (B minus C) in Figure 1 is the **approach to the wet bulb** or simply the **approach** of the cooling tower. The approach is a function of cooling tower capability, and a larger cooling tower produces a closer approach (colder leaving water) for a given heat load, flow rate, and entering air condition. Thus, the amount of heat transferred to the atmosphere by the cooling tower is always equal to the heat load imposed on the tower, while the temperature level at which the heat is transferred is determined by the thermal capability of the cooling tower and the entering air wet-bulb temperature.

The thermal performance of a cooling tower depends principally on the entering air wet-bulb temperature. The entering air dry-bulb temperature and relative humidity, taken independently, have an insignificant effect on thermal performance of mechanical-draft cooling towers, but they do affect the rate of water evaporation within the cooling tower. A psychrometric analysis of the air passing through a cooling tower illustrates this effect (Figure 2). Air enters at the ambient condition Point A, absorbs heat and mass (moisture) from the water, and exits at Point B in a saturated condition (at very light loads, the discharge air may not be saturated). The amount of heat transferred from the water to the air is proportional to the difference in enthalpy of the air between the entering and

The preparation of this chapter is assigned to TC 8.6, Cooling Towers and Evaporative Condensers.

leaving conditions ($h_B - h_A$). Because lines of constant enthalpy correspond almost exactly to lines of constant wet-bulb temperature, the change in enthalpy of the air may be determined by the change in wet-bulb temperature of the air.

The heating of the air, represented by Vector AB in Figure 2, may be separated into component AC, which represents the sensible portion of the heat absorbed by the air as the water is cooled, and component CB, which represents the latent portion. If the entering air condition is changed to Point D at the same wet-bulb temperature but at a higher dry-bulb temperature, the total heat transfer, represented by Vector DB, remains the same, but the sensible and latent components change dramatically. DE represents sensible **cooling** of air, while EB represents latent heating as water gives up heat and mass to the air. Thus, for the same water-cooling load, the ratio of latent heat transfer to sensible heat transfer can vary significantly.

The ratio of latent to sensible heat is important in analyzing the water usage of a cooling tower. Mass transfer (evaporation) occurs only in the latent portion of the heat transfer process and is proportional to the change in specific humidity. Because the entering air dry-bulb temperature or relative humidity affects the latent to sensible heat transfer ratio, it also affects the rate of evaporation. In Figure 2, the rate of evaporation in Case AB ($W_B - W_A$) is less than in Case DB ($W_B - W_D$) because the latent heat transfer (mass transfer) represents a smaller portion of the total.

The evaporation rate at typical design conditions is approximately 1% of the water flow rate for each 12.5°F of water temperature range; however, the average evaporation rate over the operating season is less than the design rate because the sensible component of total heat transfer increases as the entering air temperature decreases.

In addition to water loss from evaporation, losses also occur because of liquid carryover into the discharge airstream and blowdown to maintain acceptable water quality. Both of these factors are addressed later in this chapter.

DESIGN CONDITIONS

The thermal capability of any cooling tower may be defined by the following parameters:

1. Entering and leaving water temperatures
2. Entering air wet-bulb or entering air wet-bulb and dry-bulb temperatures
3. Water flow rate

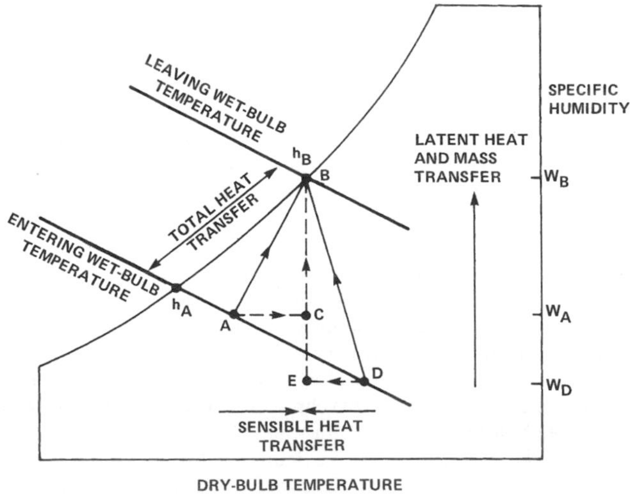

Fig. 2 Psychrometric Analysis of Air Passing Through Cooling Tower

The entering air dry-bulb temperature affects the amount of water evaporated from any evaporative-type cooling tower. It also affects airflow through hyperbolic towers and directly establishes thermal capability within any indirect-contact cooling tower component operating in a dry mode. Variations in tower performance associated with changes in the remaining parameters are covered in the section on Performance Curves.

The thermal capability of a cooling tower used for air-conditioning applications is often expressed in nominal cooling tower tons. A nominal cooling tower ton is defined as cooling 3 gpm of water from 95°F to 85°F at a 78°F entering air wet-bulb temperature. At these conditions, the cooling tower rejects 15,000 Btu/h per nominal cooling tower ton. The historical derivation of this 15,000 Btu/h cooling tower ton, as compared to the 12,000 Btu/h evaporator ton, is based on the assumption that at typical air-conditioning conditions, for every 12,000 Btu/h of heat picked up in the evaporator, the cooling tower must dissipate an additional 3000 Btu/h of compressor heat. For specific applications, however, nominal tonnage ratings are not used, and the thermal performance capability of the tower is usually expressed as a water flow rate at specific operating temperature conditions (entering water temperature, leaving water temperature, entering air wet-bulb temperature).

TYPES OF COOLING TOWERS

Two basic types of evaporative cooling devices are used. The first of these, the **direct-contact** or **open cooling tower** (see Figure 3), exposes water directly to the cooling atmosphere, thereby transferring the source heat load directly to the air. The second type, often called a **closed-circuit cooling tower**, involves **indirect contact** between heated fluid and atmosphere (Figure 4).

Of the direct-contact devices, the most rudimentary is a **spray-filled tower** that exposes water to the air without any heat transfer medium or fill. In this device, the amount of water surface exposed

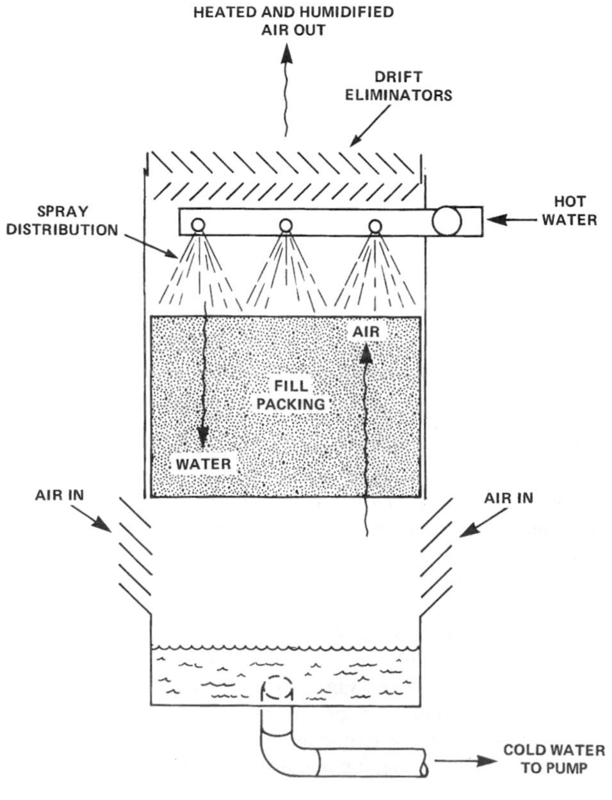

Fig. 3 Direct-Contact or Open Evaporative Cooling Tower

to the air depends on the efficiency of the sprays, and the time of contact depends on the elevation and pressure of the water distribution system.

To increase contact surfaces, as well as time of exposure, a heat transfer medium, or **fill**, is installed below the water distribution system, in the path of the air. The two types of fill in use are splash-type and film-type (Figure 5). **Splash-type fill** maximizes contact area and time by forcing the water to cascade through successive elevations of splash bars arranged in staggered rows. **Film-type fill** achieves the same effect by causing the water to flow in a thin layer over closely spaced sheets, principally polyvinyl chloride (PVC), that are arranged vertically.

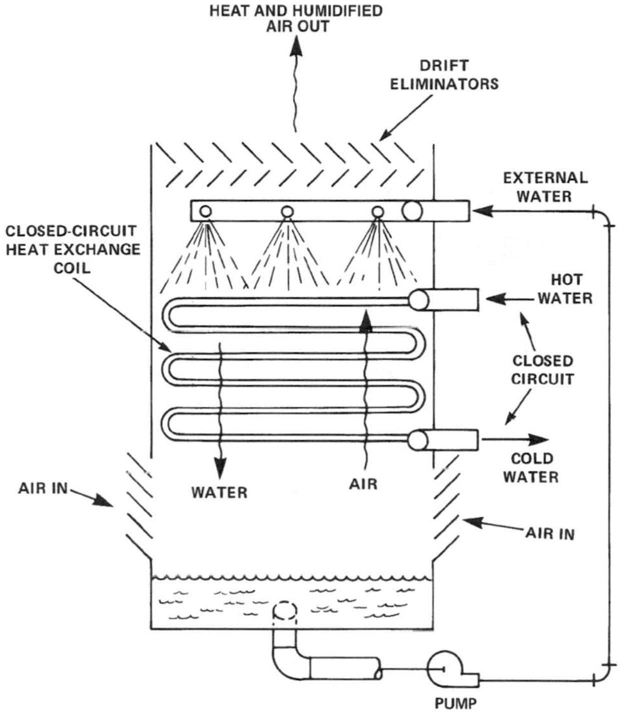

Fig. 4 Indirect-Contact or Closed-Circuit Evaporative Cooling Tower

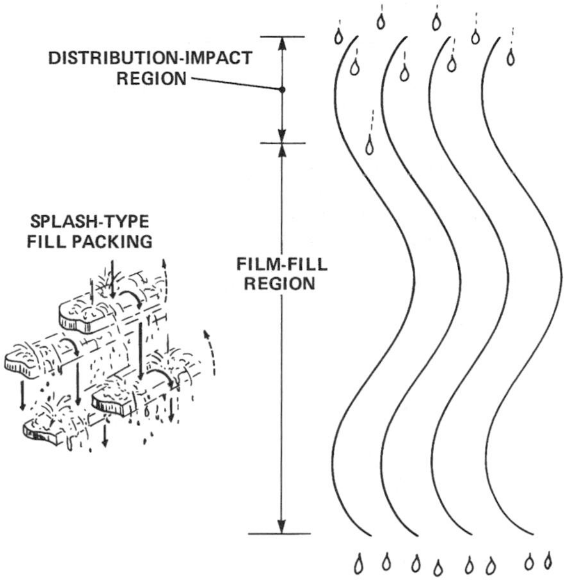

Fig. 5 Types of Fill

Either type of fill is applicable to both counterflow and crossflow towers. For thermal performance levels typically encountered in air conditioning and refrigeration, the tower with film-type fill is usually more compact. However, splash-type fill is less sensitive to the initial air and water distribution and, along with specially configured, more widely spaced film-type fills, is preferred for those applications that may be subjected to blockage by scale, silt, or biological fouling.

Indirect-contact (closed-circuit) cooling towers contain two separate fluid circuits: (1) an external circuit, in which water is exposed to the atmosphere as it cascades over the tubes of a coil bundle, and (2) an internal circuit, in which the fluid to be cooled circulates inside the tubes of the coil bundle. In operation, heat flows from the internal fluid circuit, through the tube walls of the coil, to the external water circuit and then, by heat and mass transfer, to the atmospheric air. As the internal fluid circuit never contacts the atmosphere, this unit can be used to cool fluids other than water and/or to prevent contamination of the primary cooling circuit with airborne dirt and impurities.

An alternative, closed-circuit cooling tower (Figure 6) includes cooling tower fill to augment heat exchange in the coil. In this unit, one airstream flows down over the coil, parallel to the recirculating water, and exits horizontally into the fan plenum. The recirculating water then flows over cooling tower fill where it is further cooled by a second airstream before it is reintroduced over the coil.

Types of Direct-Contact Cooling Towers

Non-Mechanical-Draft Towers. Aspirated by sprays or a differential in air density, these towers do not contain fill and do not use a mechanical air-moving device. The aspirating effect of the water spray, either vertical (Figure 7) or horizontal (Figure 8), induces airflow through the tower in a parallel flow pattern.

Because air velocities for the **vertical spray tower** (both entering and leaving) are relatively low, such towers are susceptible to adverse wind effects and, therefore, are normally used to satisfy a low cost requirement when operating temperatures are not critical to the system. Some **horizontal spray towers** (Figure 8) use high-pressure sprays to induce large air quantities and improve air/water contact. Multispeed or staged pumping systems are normally recommended to reduce energy use in periods of reduced load and ambient conditions.

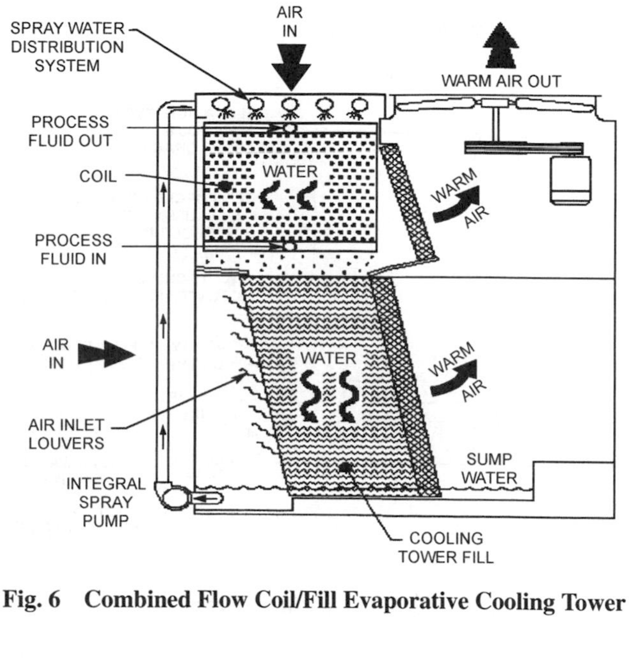

Fig. 6 Combined Flow Coil/Fill Evaporative Cooling Tower

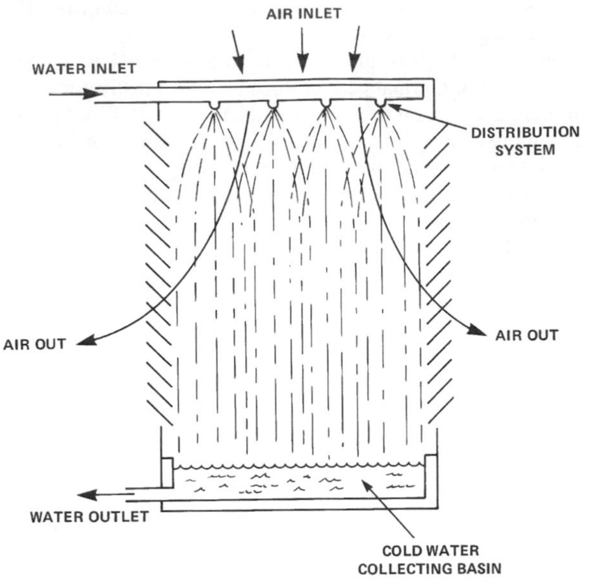

Fig. 7 Vertical Spray Tower

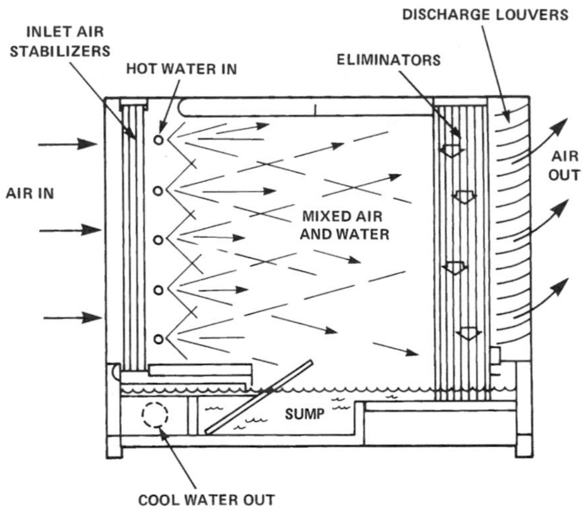

Fig. 8 Horizontal Spray Tower

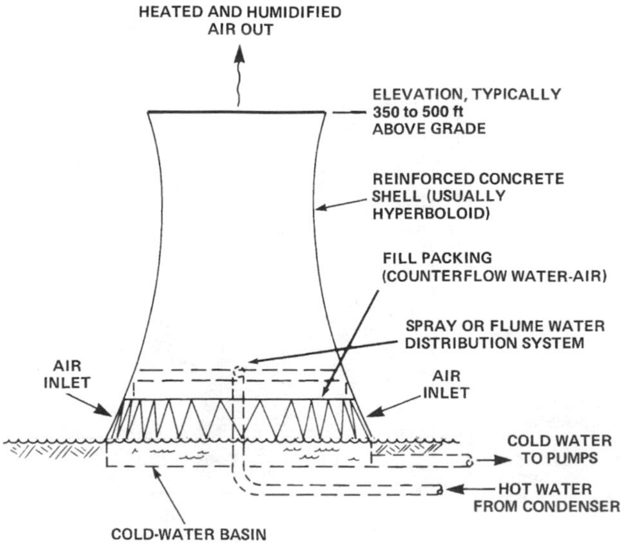

Fig. 9 Hyperbolic Tower

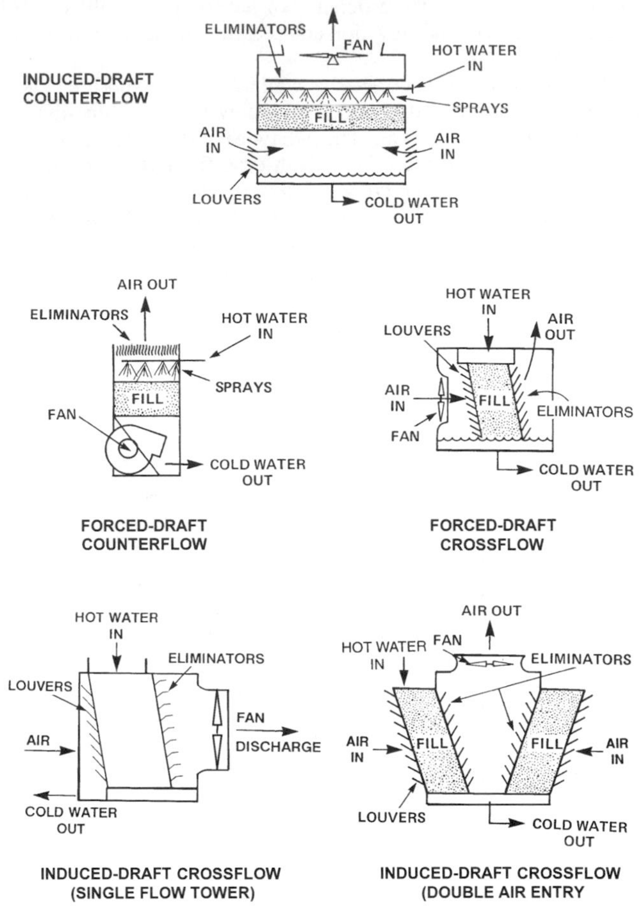

Fig. 10 Conventional Mechanical-Draft Cooling Towers

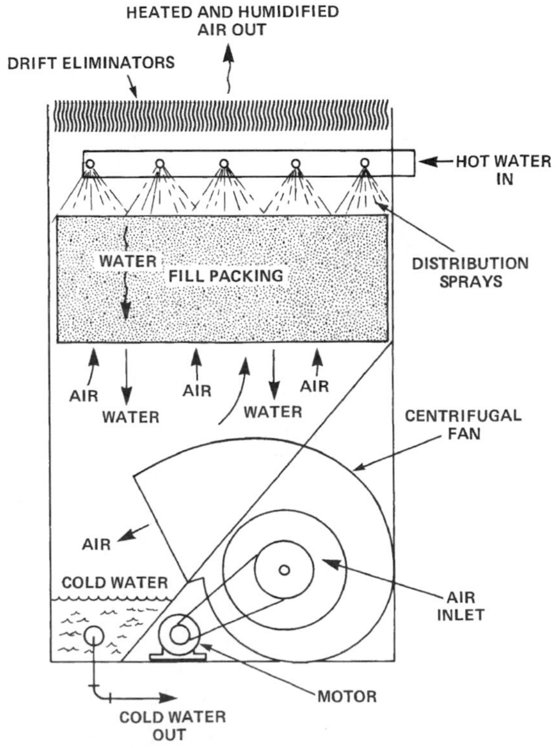

Fig. 11 Factory-Assembled Counterflow Forced-Draft Tower

Chimney (hyperbolic) towers have been used primarily for large power installations, but may be of generic interest (Figure 9). The heat transfer mode may be counterflow, cross-flow, or parallel flow. Air is induced through the tower by the air density differentials that exist between the lighter, heat-humidified chimney air and the outside atmosphere. Fill can be splash or film type.

Primary justification of these high first-cost products comes through reduction in auxiliary power requirements (elimination of fan energy), reduced property acreage, and elimination of recirculation and/or vapor plume interference. Materials used in chimney construction have been primarily steel-reinforced concrete; early-day timber structures had limitations of size.

Mechanical-Draft Towers. Figure 10 shows five different designs for mechanical-draft towers (conventional towers). The fans may be on the inlet air side (forced-draft) or the exit air side (induced-draft). The type of fan selected, either centrifugal or propeller, depends on external pressure needs, permissible sound levels, and energy usage requirements. Water is downflow, while the air may be upflow (counterflow heat transfer) or horizontal flow (cross-flow heat transfer). Air entry may be through one, two, three, or all four sides of the tower. All four combinations (i.e., forced-draft counterflow, induced-draft counterflow, forced-draft cross-flow, and induced-draft cross-flow) have been produced in various sizes.

Towers are typically classified as either factory-assembled (Figure 11), where the entire tower or a few large components are factory-assembled and shipped to the site for installation, or field-erected (Figure 12), where the tower is constructed completely on-site.

Most factory-assembled towers are of metal construction, usually galvanized steel. Other constructions include stainless steel and fiberglass-reinforced plastic (FRP) towers and components. Field-erected towers are predominantly framed of preservative-treated Douglas fir or redwood, with FRP used for special components and the casing materials. Environmental concerns about cutting timber and the wood preservatives leaching to the cooling tower water have given rise to an increased number of cooling towers having FRP structural framing. Field-erected towers may also be constructed of galvanized steel or stainless steel. Coated metals, primarily steel, are used for complete towers or components. Concrete and ceramic materials are usually restricted to the largest towers (see the section on Materials of Construction).

Special-purpose towers containing a conventional mechanical-draft unit in combination with an air-cooled (finned-tube) heat exchanger are **wet/dry towers** (Figure 13). They are used for either vapor plume reduction or water conservation. The hot, moist plumes discharged from cooling towers are especially dense in cooler weather. On some installations, limited abatement of these plumes is required to avoid restricted visibility on roadways, on bridges, and around buildings.

A vapor plume abatement tower usually has a relatively small air-cooled component that tempers the leaving airstream to reduce the relative humidity and thereby minimize the fog-generating potential of the tower. Conversely, a water conservation tower usually requires a large air-cooled component to provide significant savings in water consumption. Some designs can handle heat loads entirely by the nonevaporative air-cooled heat exchangers during reduced ambient temperature conditions.

A variant of the wet/dry tower is an evaporatively precooled/air-cooled heat exchanger. It uses an adiabatic saturator (air precooler/humidifier) to enhance the summer performance of an air-cooled exchanger, thus conserving water in comparison to conventional cooling towers (annualized) (Figure 14). Evaporative fill sections usually operate only during specified summer periods,

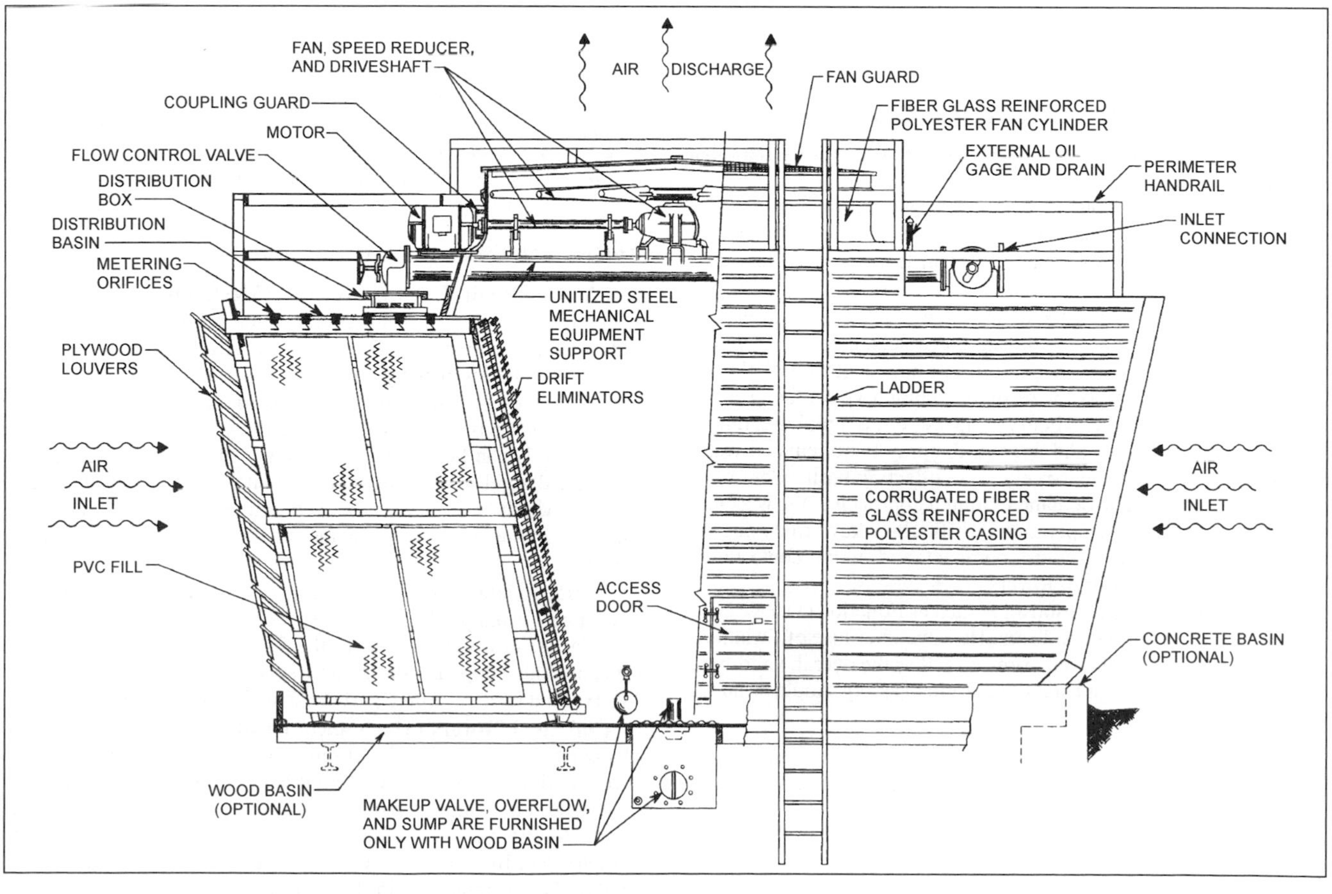

Fig. 12 Field-Erected Cross-Flow Mechanical-Draft Tower

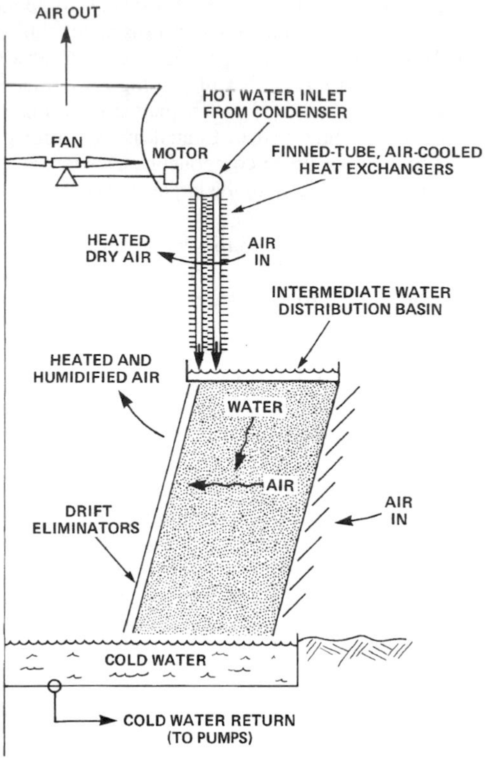

Fig. 13 Combination Wet-Dry Tower

while full dry operation is expected below 50 to 70°F ambient conditions. Integral water pumps return the lower basin water to upper distribution systems of the adiabatic saturators in a manner similar to closed-circuit fluid cooler and evaporative condenser products.

Other Methods of Direct Heat Rejection

Ponds, Spray Ponds, Spray Module Ponds, and Channels. Heat dissipates from the surface of a body of water by evaporation, radiation, and convection. Captive lakes or ponds (man-made or natural) are sometimes used to dissipate heat by natural air currents and wind. This system is usually used in large plants where real estate is not limited.

A pump-spray system above the pond surface improves heat transfer by spraying the water in small droplets, thereby extending the water surface and bringing it into intimate contact with the air. The heat transfer is largely the result of evaporative cooling (see the section on Cooling Tower Theory). The system is a piping arrangement using branch arms and nozzles to spray the circulated water into the air. The pond acts largely as a collecting basin. Control of temperatures, real estate demands, limited approach to the wet-bulb temperature, and winter operational difficulties have ruled out the spray pond in favor of more compact and more controllable mechanical- or natural-draft towers.

Empirically derived relationships such as Equation (1) have been used to estimate cooling pond area. However, because of variations in wind velocity and solar radiation as well as the overall validity of the relationship itself, a substantial margin of safety should be added to the result.

$$w_p = \frac{A(95 + 0.425v)}{h_{fg}}[p_w - p_a] \qquad (1)$$

where

w_p = evaporation rate of water, lb/h
A = area of pool surface, ft^2

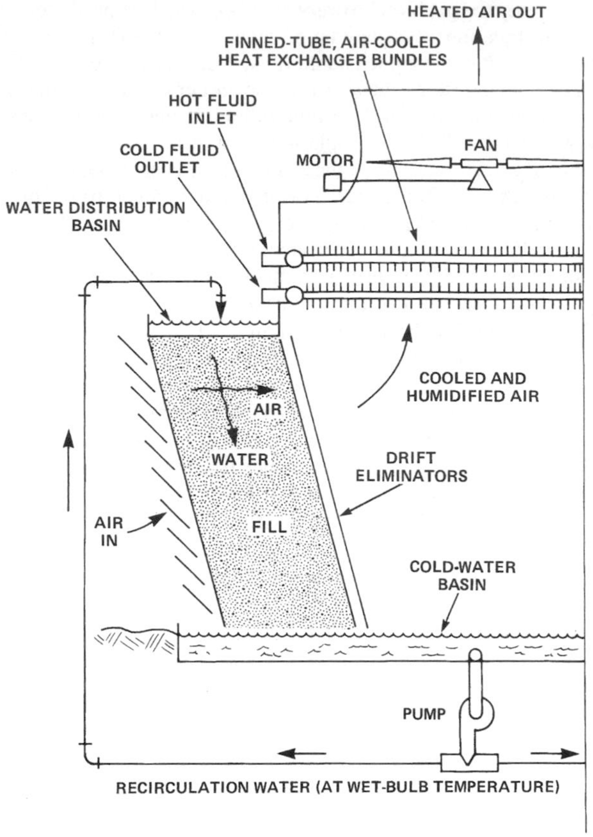

Fig. 14 Adiabatically Saturated Air-Cooled Heat Exchanger

v = air velocity over water surface, fpm
h_{fg} = latent heat required to change water to vapor at temperature of surface water, Btu/lb
p_a = saturation vapor pressure at dew-point temperature of ambient air, in. Hg
p_w = saturation vapor pressure at temperature of surface water, in. Hg

Types of Indirect-Contact Towers

Closed-Circuit Cooling Towers (Mechanical Draft). Both counterflow and cross-flow arrangements are used in forced and induced fan arrangements. The tubular heat exchangers are typically serpentine bundles, usually arranged for free gravity internal drainage. Pumps are integrated in the product to transport water from the lower collection basin to the upper distribution basins or sprays. The internal coils can be fabricated from any of several materials, but galvanized steel and copper predominate. Closed-circuit cooling towers, which are similar to evaporative condensers, are used increasingly on heat pump systems and screw compressor oil pump systems.

Indirect-contact towers (Figure 4) require a closed-circuit heat exchanger (usually tubular serpentine coil bundles) that is exposed to air/water cascades similar to the fill of a cooling tower. Some types include supplemental film or splash fill sections to augment the external heat exchange surface area (Figure 5).

Coil Shed Towers (Mechanical Draft). Coil shed towers usually consist of isolated coil sections (nonventilated) located beneath a conventional cooling tower (Figure 15). Counterflow and cross-flow types are available with either forced or induced fan arrangements. Redistribution water pans, located at the tower's base, feed cooled water by gravity flow to the tubular heat exchange bundles (coils). These units are similar in function to closed-circuit fluid coolers, except that supplemental fill is always required, and the

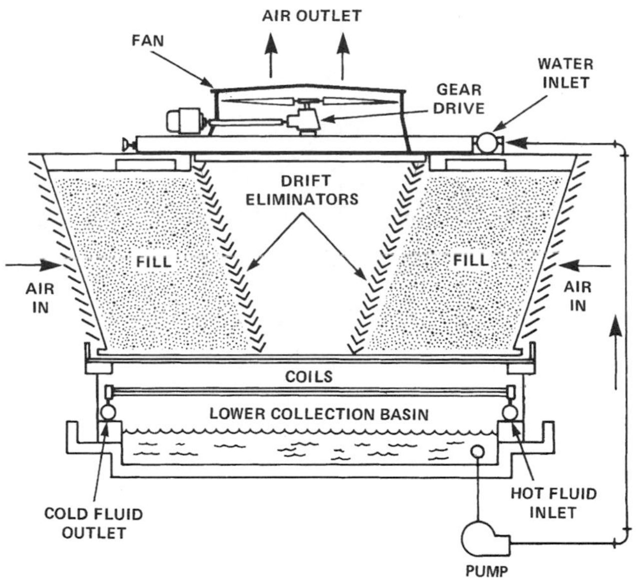

Fig. 15 Coil Shed Cooling Tower

airstream is directed only through the fill regions of the tower. Typically, these units are arranged as field-erected, multifan cell towers and are used primarily in industrial process cooling.

MATERIALS OF CONSTRUCTION

Materials for cooling tower construction are usually selected to resist the corrosive water and atmospheric conditions.

Wood. Wood has been used extensively for all static components except hardware. Redwood and fir predominate, usually with post-fabrication pressure treatment of waterborne preservative chemicals, typically chromated copper arsenate (CCA) or acid copper chromate (ACC). These microbicidal chemicals prevent the attack of wood-destructive organisms such as termites or fungi.

Metals. Steel with galvanized zinc is used for small and medium-sized installations. Hot dip galvanizing after fabrication is used for larger weldments. Hot dip galvanizing and cadmium and zinc plating are used for hardware. Brasses and bronzes are selected for special hardware, fittings, and tubing material. Stainless steels (principally 302, 304, and 316) are often used for sheet metal, drive shafts, and hardware in exceptionally corrosive atmospheres. Cast iron is a common choice for base castings, fan hubs, motor or gear reduction housings, and piping valve components. Metals coated with polyurethane and PVC are used selectively for special components. Two-part epoxy compounds and epoxy-powdered coatings are also used for key components or entire cooling towers.

Plastics. Fiberglass-reinforced plastic (FRP) materials are used for components such as structure, piping, fan cylinders, fan blades, casing, louvers, and structural connecting components. Polypropylene and acrylonitrile butadiene styrene (ABS) are specified for injection-molded components, such as fill bars and flow orifices. PVC is increasingly used as fill, eliminator, and louver materials. Reinforced plastic mortar is used in larger piping systems, coupled by neoprene O-ring-gasketed bell and spigot joints.

Graphite Composites. Graphite composite drive shafts have recently become available for use on cooling tower installations. These shafts offer a strong, corrosion-resistant alternative to steel/stainless steel shafts.

Concrete, Masonry, and Tile. Concrete is typically specified for cold water basins of field-erected cooling towers and is used in piping, casing, and structural systems of the largest towers, primarily in the power industry. Special tiles and masonry are used when aesthetic considerations are important.

SELECTION CONSIDERATIONS

Selecting the proper water-cooling equipment for a specific application requires consideration of cooling duty, economics, required services, environmental conditions, and aesthetics. Many of these factors are interrelated, but they should be evaluated individually.

Because a wide variety of water-cooling equipment may meet the required cooling duty, such items as height, length, width, volume of airflow, fan and pump energy consumption, materials of construction, water quality, and availability influence the final equipment selection.

The optimum choice is generally made after an economic evaluation. Chapter 35 of the 1999 *ASHRAE Handbook—Applications* describes two common methods of economic evaluation—life-cycle costing and payback analysis. Each of these procedures compares equipment on the basis of total owning, operating, and maintenance costs.

Initial cost comparisons consider the following factors:

1. Erected cost of equipment
2. Costs of interface with other subsystems, which include items such as
   - Basin grillage and value of the space occupied
   - Pumps and prime movers
   - Electrical wiring to pump and fan motors
   - Electrical controls and switchgear
   - Piping to and from the tower (Some designs require more inlet and discharge connections than others, thus affecting the cost of piping.)
   - Tower basin, sump screens, overflow piping, and makeup lines, if they are not furnished by the manufacturer
   - Shutoff and control valves, if they are not furnished by the manufacturer
   - Walkways, ladders, etc., providing access to the tower
   - Fire protection sprinkler system

In evaluating owning and maintenance costs, the following are major items to consider:

- System energy costs (fans, pumps, etc.) on the basis of operating hours per year
- Demand charges
- Expected equipment life
- Maintenance and repair costs
- Money costs

Other factors are (1) safety features and safety codes, (2) conformity to building codes, (3) general design and rigidity of structures, (4) relative effects of corrosion, scale, or deterioration on service life, (5) availability of spare parts, (6) experience and reliability of manufacturers, and (7) operating flexibility for economical operation at varying loads or during seasonal changes. In addition, equipment vibration, sound levels, acoustical attenuation, and compatibility with the architectural design are important. The following section details many of these more important considerations.

APPLICATION

This section describes some of the major design considerations, but the manufacturer of the cooling tower should be consulted for more detailed recommendations.

Siting

When a cooling tower can be located in an open space with free air motion and unimpeded air supply, siting is normally not an obstacle to obtaining a satisfactory installation. However, towers are often situated indoors, against walls, or in enclosures. In such cases, the following factors must be considered:

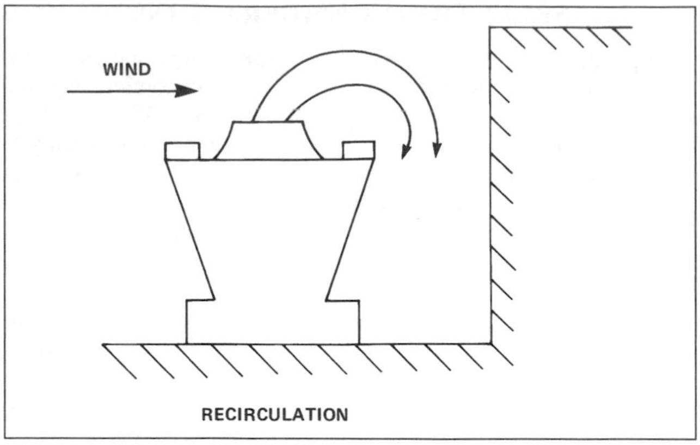

Fig. 16 Discharge Air Reentering Tower

1. Sufficient free and unobstructed space should be provided around the unit to ensure an adequate air supply to the fans and to allow proper servicing.
2. Tower discharge air should not be deflected in any way that might promote recirculation [a portion of the warm, moist discharge air reentering the tower (Figure 16)]. Recirculation raises the entering wet-bulb temperature, causing increased hot water and cold water temperatures, and, during cold weather operation, can promote the icing of air intake areas. The possibility of air recirculation should be considered, particularly on multiple-tower installations.

Additionally, cooling towers should be located to prevent the introduction of the warm discharge air and any associated drift, which may contain chemical and/or biological contaminants, into the fresh air intake of the building that the tower is serving or into those of adjacent buildings.

Location of the cooling tower is usually determined by one or more of the following: (1) structural support requirements, (2) rigging limitations, (3) local codes and ordinances, (4) cost of bringing auxiliary services to the cooling tower, and (5) architectural compatibility. Sound, fog, and drift considerations are also best handled by proper site selection during the planning stage.

Piping

Piping should be adequately sized according to standard commercial practice. All piping should be designed to allow expansion and contraction. If the tower has more than one inlet connection, balancing valves should be installed to balance the flow to each cell properly. Positive shutoff valves should be used, if necessary, to isolate individual cells for servicing.

When two or more towers are operated in parallel, an equalizer line between the tower sumps handles imbalances in the piping to and from the units and changing flow rates that arise from such obstructions as clogged orifices and strainers. All heat exchangers, and as much tower piping as possible, should be installed below the operating water level in the cooling tower to prevent overflowing of the cooling tower at shutdown and to ensure satisfactory pump operation during start-up. Tower basins must carry the proper amount of water during operation to prevent air entrainment into the water suction line. Tower basins should also have enough reserve volume between the operating and overflow levels to fill riser and water distribution lines on start-up and to fulfill the water-in-suspension requirement of the tower.

Capacity Control

Most cooling towers encounter substantial changes in ambient wet-bulb temperature and load during the normal operating season.

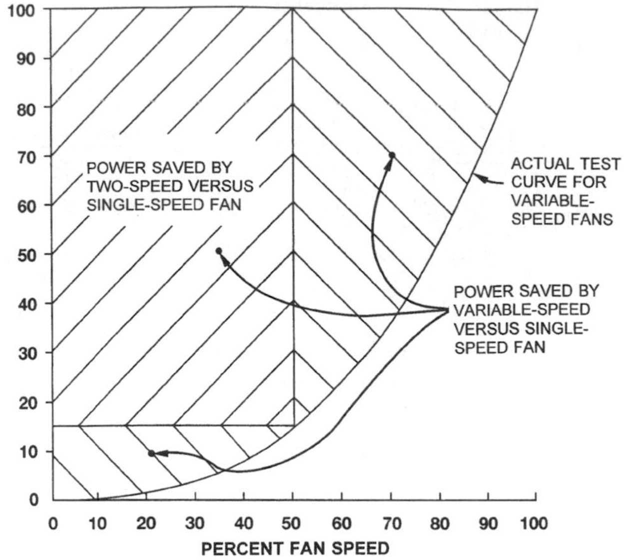

Fig. 17 Cooling Tower Fan Power Versus Speed
(White 1994)

Accordingly, some form of capacity control may be required to maintain prescribed condensing temperatures or process conditions.

Fan cycling is the simplest method of capacity control on cooling towers and is often used on multiple-unit or multiple-cell installations. In nonfreezing climates, where close control of the exit water temperature is not essential, fan cycling is an adequate and inexpensive method of capacity control. However, motor burnout from too-frequent cycling is a concern.

Two-speed fan motors or additional lower horsepower pony motors, in conjunction with fan cycling, can double the number of steps of capacity control compared to fan cycling alone. This is particularly useful on single-fan motor units, which would have only one step of capacity control by fan cycling. Two-speed fan motors are commonly used on cooling towers as the primary method of capacity control, and they provide the added advantage of reduced energy consumption at reduced load.

It is more economical to operate all fans at the same speed than to operate one fan at full speed before starting the next. Figure 17 compares cooling tower fan power versus speed for single-speed, two-speed, and variable-speed fan motors.

Modulating dampers in the discharge of centrifugal blower fans are used for cooling tower capacity control, as well as for energy management. In many cases, modulating dampers are used in conjunction with two-speed motors. Frequency-modulating controls for fan motor speed can provide virtually infinite capacity control and energy management, as can automatic, variable-pitch propeller fans. Variable-frequency fan drives are economical and can save considerable energy as well as extend the life of the fan and drive (gearbox or V-belt) assembly compared to fan cycling with two-speed control. However, there are special considerations that should be discussed with the tower manufacturer and the supplier of the variable-frequency drive (VFD):

- Care must be taken to avoid operating the fan system at a critical speed or a multiple thereof. The tower manufacturer can advise what speeds (if any) must be avoided.
- Some drives, particularly pulse-width modulating (PWM) drives, create overvoltages at the motor that can cause motor and bearing failures. The magnitude of these overvoltages increases significantly with the length of cable between the controller and the motor, so the lead lengths should be kept as short as possible. Special motors, filters, or other corrective measures may be necessary

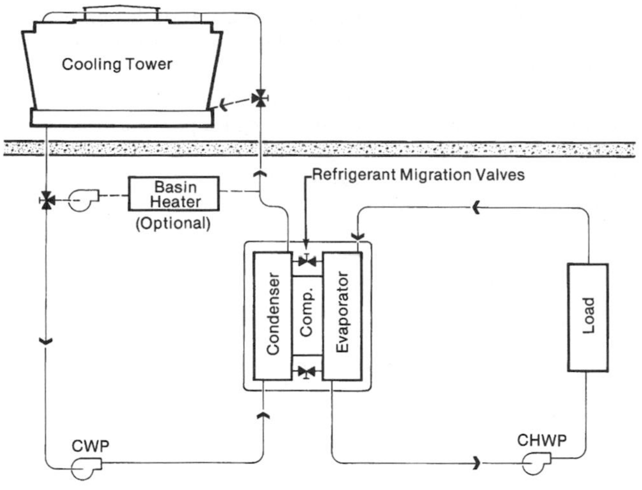

Fig. 18 Free Cooling by Use of Refrigerant Vapor Migration

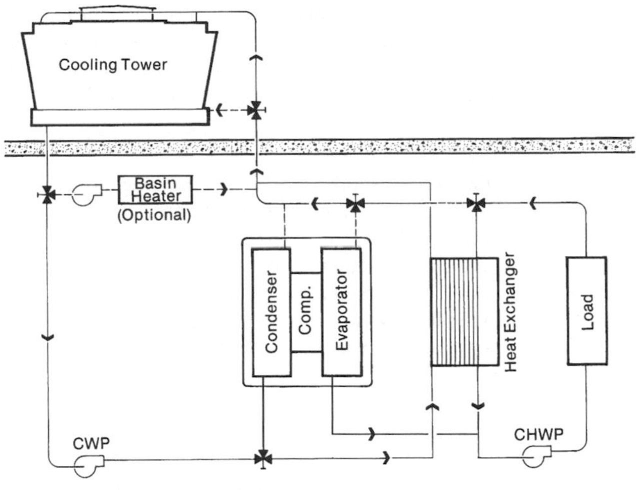

Fig. 19 Free Cooling by Use of Auxiliary Heat Exchanger

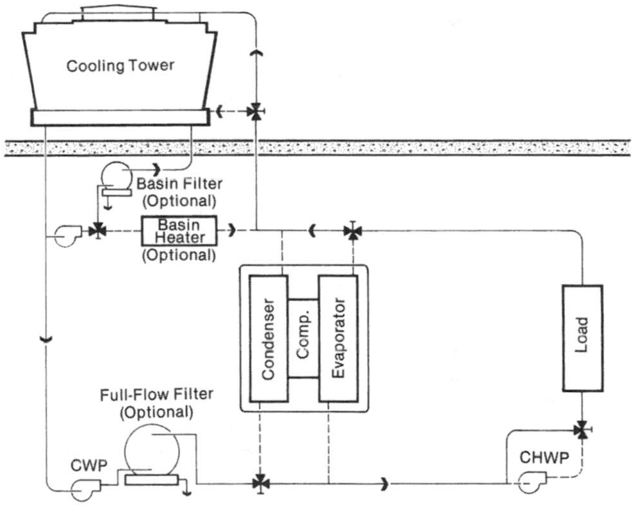

Fig. 20 Free Cooling by Interconnection of Water Circuits

to ensure dependable operation. Chapter 40 has more information on variable-frequency drives.

- Variable-frequency drives should not be operated at speeds below 25% of the full design fan speed. This precaution is particularly true for units with gear speed reducers because low speeds can cause vibration, noise or "gear chatter," and pose a lubrication problem if the gear is not equipped with an electric oil pump. Consult the manufacturer for proper selection and adjustments of the VFD.

Cooling towers that inject water to induce the airflow through the cooling tower have various pumping arrangements for capacity control. Multiple pumps in series or two-speed pumping provide capacity control and also reduce energy consumption.

Modulating water bypasses for capacity control should be used only after consultation with the cooling tower manufacturer. This is particularly important at low ambient conditions in which the reduced water flows can promote freezing within the tower.

Free Cooling

With an appropriately equipped and piped system, using the tower for free cooling during reduced load and/or reduced ambient conditions can significantly reduce energy consumption. Because the tower's cold water temperature drops as the load and ambient temperature drop, the water temperature will eventually be low enough to serve the load directly, allowing the energy-intensive chiller to be shut off. Figure 18, Figure 19, and Figure 20 outline three methods of free cooling but do not show all of the piping, valving, and controls that may be necessary for the functioning of a specific system.

Maximum use of the free-cooling mode of system operation occurs when a drop in the ambient temperature reduces the need for dehumidification. Therefore, higher temperatures in the chilled water circuit can normally be tolerated during the free-cooling season and are beneficial to the system's heating-cooling balance. In many cases, typical 45°F chilled water temperatures are allowed to rise to 55°F or higher in the free-cooling mode of operation. This maximizes tower usage and minimizes system energy consumption. Some applications require a constant chilled water supply temperature, which greatly reduces the hours of free cooling operation.

Indirect Free Cooling. This type of cooling keeps the condenser water and chilled water circuits separate and may be accomplished in the following ways:

1. In the vapor migration system (Figure 18), bypasses between the evaporator and the condenser permit the migratory flow of refrigerant vapor to the condenser; they also permit gravity flow of liquid refrigerant back to the evaporator without operation of the compressor. Not all chiller systems are adaptable to this arrangement, and those that are may offer limited load capability under this mode of operation. In some cases, auxiliary pumps enhance refrigerant flow and, therefore, load capability.
2. A separate heat exchanger in the system (usually of the plate-and-frame type) allows heat to transfer from the chilled water circuit to the condenser water circuit by total bypass of the chiller system (Figure 19).
3. An indirect-contact, closed-circuit evaporative cooling tower (Figure 4 and Figure 6) also permits indirect free cooling. Its use is covered in the following section on Direct Free Cooling.

Direct Free Cooling. This type of cooling involves interconnecting the condenser water and chilled water circuits so the cooling tower water serves the load directly (Figure 20). In this case, the chilled water pump is normally bypassed so design water flow can be maintained to the cooling tower. The primary disadvantage of the direct free-cooling system is that it allows the relatively dirty condenser water to contaminate the clean chilled water system. Although filtration systems (either side-stream or full-flow) minimize this contamination, many specifiers consider it to be an overriding concern. The use of a closed-circuit (indirect-contact) cooling tower eliminates this contamination concern. During summer, the water from the tower is circulated in a closed loop through the condenser. During winter, the water from the tower is circulated in a closed loop directly through the chilled water circuit.

Winter Operation

When a cooling tower is to be used in subfreezing climates, the following design and operating considerations are necessary.

Open Circulating Water. Direct-contact cooling towers that operate in freezing climates can be winterized by a suitable method of capacity control. This capacity control maintains the temperature of the water leaving the tower well above freezing. In addition, during cold weather, regular visual inspections of the cooling tower should be made to ensure all controls are operating properly.

On induced-draft propeller fan towers, fans may be periodically operated in reverse to deice the air intake areas. Forced-draft centrifugal fan towers should be equipped with capacity control dampers or variable-frequency drives to minimize the possibility of icing.

Recirculation of moist discharge air on forced-draft equipment can cause ice formation on the inlet air screens and fans. Installation of vibration cut-out switches can minimize the risk of damage due to ice formation on rotating equipment.

Closed Circulating Water. Precautions in addition to those mentioned for open circulating water must be taken to protect the fluid inside the heat exchanger of a closed-circuit fluid cooler. When system design permits, the best protection is to use an antifreeze solution. When this is not possible, supplemental heat must be provided to the heat exchanger, and the manufacturer should be consulted concerning the amount of heat input required.

All exposed piping to and from the cooler should be insulated and heat traced. In case of a power failure during subfreezing weather, the heat exchanger should include an emergency draining system.

Sump Water. Freeze protection for the sump water in an idle tower or closed-circuit fluid cooler can be obtained by various means. A good method for protecting the sump water is to use an auxiliary sump tank located within a heated space. When a remote sump is impractical, auxiliary heat must be supplied to the tower sump to prevent freezing. Common sources are electric immersion heaters and steam and hot water coils. The tower manufacturer should be consulted for the exact heat requirements to prevent freezing at design winter temperatures.

All exposed water lines susceptible to freezing should be protected by electric tape or cable and insulation. This precaution applies to all lines or portions of lines that have water in them when the tower is shut down.

Sound

Sound has become an important consideration in the selection and siting of outdoor equipment such as cooling towers and other evaporative cooling devices. Many communities have enacted legislation that limits allowable sound levels of outdoor equipment. Even if legislation does not exist, people who live and work near a tower installation may object if the sound intrudes on their environment. Because the cost of correcting a sound problem may exceed the original cost of the cooling tower, sound should be considered in the early stages of system design.

To determine the acceptability of tower sound in a given environment, the first step is to establish a noise criterion for the area of concern. This may be an existing or pending code or an estimate of sound levels that will be acceptable to those living or working in the area. The second step is to estimate the sound levels generated by the tower at the critical area, taking into account the effects of geometry of the tower installation and the distance from the tower to the critical area. Often, the tower manufacturer can supply sound rating data on a specific unit that serve as the basis for this estimate. Lastly, the noise criterion is compared to the estimated tower sound levels to determine the acceptability of the installation.

In cases where the installation may present a sound problem, several potential solutions are available. It is good practice to situate the tower as far as possible from any sound-sensitive areas. Two-speed fan motors should be considered to reduce tower sound levels (by a nominal 12 dB) during light load periods, such as at night, if these correspond to critical sound-sensitive periods. However, fan motor cycling should be held to a minimum because a fluctuating sound is usually more objectionable than a constant sound.

In critical situations, effective solutions may include barrier walls between the tower and the sound-sensitive area or acoustical treatment of the tower. Attenuators specifically designed for the tower are available from most manufacturers. It may be practical to install a tower larger than would normally be required and lower the sound levels by operating the unit at reduced fan speed. For additional information on sound control, see Chapter 46 of the 1999 *ASHRAE Handbook—Applications.*

Drift

Water droplets become entrained in the airstream as it passes through the tower. While eliminators strip most of this water from the discharge airstream, a certain amount discharges from the tower as drift. The rate of drift loss from a tower is a function of tower configuration, eliminator design, airflow rate through the tower, and water loading. Generally, an efficient eliminator design reduces drift loss to between 0.002 and 0.2% of the water circulation rate.

Because drift contains the minerals of the makeup water (which may be concentrated three to five times) and often contains water treatment chemicals, cooling towers should not be placed near parking areas, large windowed areas, or architectural surfaces sensitive to staining or scale deposits.

Fogging (Cooling Tower Plume)

The warm air discharged from a cooling tower is essentially saturated. Under certain operating conditions, the ambient air surrounding the tower cannot absorb all of the moisture in the tower discharge airstream, and the excess condenses as fog.

Fogging may be predicted by projecting a straight line on a psychrometric chart from the tower entering air conditions to a point representing the discharge conditions (Figure 21). A line crossing

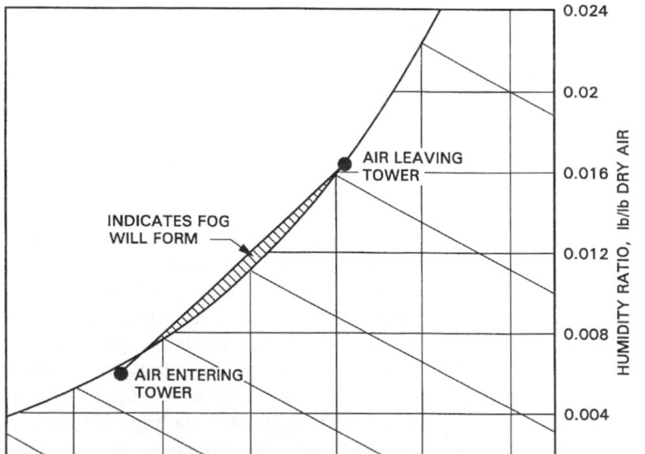

Fig. 21 Fog Prediction Using Psychrometric Chart

the saturation curve indicates fog generation; the greater the area of intersection to the left of the saturation curve, the more intense the plume. Fog persistence depends on its original intensity and on the degree of mechanical and convective mixing with ambient air that dissipates the fog.

Methods of reducing or preventing fogging have taken many forms, including heating the tower exhaust with natural gas burners or steam coils, installing precipitators, and spraying chemicals at the tower exhaust. However, such solutions are generally costly to operate and are not always effective.

On larger, field-erected installations, combination wet-dry cooling towers, which combine the normal evaporative portion of a tower with a finned-tube dry surface heat exchanger section (in series or in parallel), afford a more practical means of plume control. In such units, the saturated discharge air leaving the evaporative section is mixed within the tower with the warm, relatively dry air off the finned-coil section to produce a subsaturated air mixture leaving the tower.

Often, however, the most practical solution to tower fogging is to locate the tower where visible plumes, should they form, will not be objectionable. Accordingly, when selecting cooling tower sites, the potential for fogging and its effect on tower surroundings, such as large windowed areas or traffic arteries, should be considered.

Maintenance

Usually, the tower manufacturer furnishes operating and maintenance manuals that include recommendations for procedures and intervals as well as parts lists for the specific unit. These recommendations should be followed when formulating the maintenance program for the cooling tower. When such instructions are unavailable, the schedule of services in Table 1 and the following discussion can guide the operator in establishing a reasonable inspection and maintenance program.

The efficient operation and thermal performance of a cooling tower depend not only on mechanical maintenance, but also on cleanliness. Accordingly, cooling tower owners should incorporate the following as a basic part of their maintenance program.

1. Periodic inspection of mechanical equipment, fill, and both hot water and cold water basins to ensure that they are maintained in a good state of repair.
2. Periodic draining and cleaning of wetted surfaces and areas of alternate wetting and drying to prevent the accumulation of dirt, scale, or biological organisms, such as algae and slime, in which bacteria may develop.
3. Proper treatment of the circulating water for biological control and corrosion, in accordance with accepted industry practice.
4. Systematic documentation of operating and maintenance functions. This is extremely important because without it, no policing can be done to determine whether an individual has actually adhered to a maintenance policy.

Table 1 Typical Inspection and Maintenance Schedule[a]

	Fan	Motor	Gear Reducer	Drive Shaft	V-Belt Drives	Fan Shaft Bearings	Drift Eliminators	Fill	Cold Water Basin	Distribution System	Structural Members	Casing	Float Value	Bleed Rate	Flow Control Valves	Suction Screen
1. Inspect for clogging							W	W		W						W
2. Check for unusual noise or vibration	D	D	D	D							Y					
3. Inspect keys and set screws		S	S	S	S	S										
4. Lubricate		Q				Q									S	
5. Check oil seals			S													
6. Check oil level			W													
7. Check oil for water and dirt			M													
8. Change oil (at least)			S													
9. Adjust tension					M											
10. Check water level									W	W						
11. Check flow rate														M		
12. Check for leakage									S	S			S			
13. Inspect general condition				S	M		Y	Y	Y		S	Y	Y		S	
14. Tighten loose bolts	S	S	S	S		S					Y	R				
15. Clean	R	S	R	R			R	R	S	R			R		R	W
16. Repaint	R	R	R	R					R		R	R				
17. Completely open and close															S	
18. Make sure vents are open			M													

D—daily; W—weekly; M—monthly; Q—quarterly; S—semiannually; Y—yearly; R—as required.

[a] More frequent inspections and maintenance may be desirable.

Inspections

The following should be checked daily (no less than weekly) in an informal walk-through inspection. Areas requiring attention have been loosely grouped for clarity, although category distinctions are often hazy because the areas are interdependent.

Performance. Optimum performance and safety depend on the operation of each individual component at its designed capability. A single blocked strainer, for instance, can adversely affect the capacity and efficiency of the entire system. Operators should always be alert to any degradation in performance, as this usually is the first sign of a problem and is invaluable in pinpointing minor problems before they become major. Consult the equipment manufacturers to obtain specific information on each piece of equipment (both for maintenance and technical characteristics), and keep manuals handy for quick reference.

Check and record all water and refrigerant temperatures, pump pressures, outdoor conditions, and pressure drops (differential pressure) across condensers, heat exchangers and filtration devices. This record helps operators become familiar with the equipment as it operates under various load conditions and provides a permanent record that can be used to calculate flow rates, assess equipment efficiency, expedite diagnostic procedures, and adjust maintenance and water treatment regimens to obtain maximum performance from the system.

For those units with water-side economizers using plate heat exchangers, check temperature and pressure differentials daily.

Major Mechanical Components. During tower inspections, be alert for any unusual noise or vibration from pumps, motors, fans, and other mechanical equipment. This is often the first sign of mechanical trouble. Operators thoroughly familiar with their equipment generally have little trouble recognizing unusual conditions. Also listen for cavitation noises from pumps, as this can indicate blocked strainers.

Check the tower fan and drive system assembly for loose mounting hardware, condition of fasteners, grease and oil leaks, and noticeable vibration or wobble when the fan is running. Excessive vibration can rapidly deteriorate the tower.

Observe at least one fan start and stop each week. If a fan has a serious problem, lock it out of operation and call for expert assistance. To be safe, do not take chances by running defective fans.

Fan and drive systems should be professionally checked for dynamic balance, alignment, proper fan pitch (if adjustable), and vibration whenever major repair work is performed on the fan or if unusual noises or vibrations are present. It is good practice to have these items checked at least once every third year on all but the smallest towers. Any vibration switches should be checked for proper operation at least annually.

Verify calibration of the fan thermostat periodically to prevent excessive cycling and to ensure that the most economical temperature to the chiller is maintained.

Tower Structure. Check the tower structure and casing for water and air leaks as well as deterioration. Inspect louvers, fill, and drift eliminators for clogging, excessive scale or algal growth. Clean as necessary, using 1000 to 1500 psi water and taking care not to damage fragile fill and eliminator components.

Watch for excessive drift (water carryover), and take corrective action as required. Drift is the primary means of *Legionella* transmittal by cooling towers and evaporative condensers. Deteriorated drift eliminators should be replaced. Many older towers have drift eliminators that contain asbestos. In the United States, deteriorated asbestos-type eliminators should as a rule be designated friable material and be handled and disposed of in a manner approved by the Environment Protection Agency (EPA) and the Occupational Safety and Health Administration (OSHA).

Check the tower basin, structural members and supports, fasteners, safety rails, and ladders for corrosion or other deterioration and repair as necessary. Replace deteriorated tower components as required.

Water Distribution and Quality. Check the hot water distribution system frequently, and clear clogged nozzles as required. Water distribution should be evenly balanced when the system is at rated flow and should be rechecked periodically. Towers with open distribution pans benefit from covers because this retards algal growth.

The sump water level should be within the manufacturer's range for normal operating level; the level should be high enough to allow most solids to settle out, thereby improving water quality to the equipment served by the tower.

Tower water should be clear, and the surface should not have an oily film, excessive foaming, or scum. Oil inhibits heat transfer in cooling towers, condensers, and other heat exchangers and should not be present in tower water. Foam and scum can be an indication of excess organic material that can provide nutrients to bacteria (Rosa 1992). If such conditions are encountered, contact the water treatment specialist, who will take steps to correct the problem.

Check the cold water basin in several places for corrosion, accumulated deposits, and excessive algae, as sediments and corrosion may not be uniformly distributed. Corrosion and microbiological activity often occur under sediments. Tower outlet strainers should be in place and free of clogging.

Do not neglect the strainers in the system. In-line strainers may be the single most neglected component in the average installation. They should be inspected and, if necessary, cleaned each time the tower is cleaned. Pay particular attention to the small, fine strainers used on auxiliary equipment such as computer cooling units and blowdown lines.

Blow down chilled water risers frequently, particularly on systems using direct free cooling. Exercise all valves in the system periodically by opening and closing them fully.

For systems with water-side economizers, maintaining good water quality is paramount to prevent fouling of the heat exchanger or chilled water system, depending on the type of economizer used.

Check, operate, and enable winterization systems well before subfreezing temperatures are anticipated in order to allow time to obtain parts and make repairs as necessary. Take care that tower sediments are not allowed to build up around immersion heater elements because this will cause rapid failure of the elements.

Maintain sand filters in good order, and inspect the media bed for channeling at least quarterly. If channeling is found, either replace or clean the media as soon as possible. Do not forget to carefully clean the underdrain assembly while the media is removed. If replacing the media, use only that which is specified for cooling towers. Do not use swimming pool filter sand in filters designed for cooling towers and evaporative condensers.

Centrifugal separators rarely require service, although they must not be allowed to overfill with contaminants. Verify proper flow rate, pressure drop, and purge operation.

Check bag and cartridge filters as necessary. Clean the tower if it is dirty.

Cleanliness. Towers are excellent air washers, and the water quality in a given location quickly reflects that of the ambient air (Hensley 1985). A typical 200 ton cooling tower operating 1000 hours may assimilate upwards of 600 lb of particulate matter from airborne dust and the makeup water supply (Broadbent et al. 1992). Proximity to highways and construction sites, air pollution, and operating hours are all factors in tower soil loading.

Design improvements in cooling towers that increase thermal performance also increase air scrubbing capability (Hensley 1985). Recommendations by manufacturers regarding cleaning schedules are, therefore, to be recognized as merely guidelines. The actual frequency of cleanings should be determined at each location by careful observation and system history. Do not expect sand filters, bag filters, centrifugal separators, water treatment programs, and so forth, to take the place of a physical cleaning. They are designed to improve water quality and as such should increase the time interval between cleanings. Conversely, regular cleanings should not be expected to replace water treatment.

Towers should not be allowed to become obviously fouled. Instead, they should be cleaned often enough that sedimentation and visible biological activity (algae and slime) are easily controlled by water treatment between cleanings. The tower is the only component in the condenser loop that can be viewed easily without system shutdown, so it should be considered an indicator of total system condition and cleanliness.

Water treatment should not be expected to protect surfaces it cannot reach, such as the metal or wood components under accumulated sediments. Biocides are not likely to be effective unless administered in conjunction with a regular cleaning program. Poorly maintained systems create a greater demand upon the biocide because organic sediments neutralize the biocide and tend to shield bacterial cells from the chemical, thus requiring higher and more frequent doses to keep microbial populations under control (Broadbent et al. 1992, McCann 1988). High concentrations of an oxidizing biocide can contribute to corrosion. Keeping the tower clean reduces the breeding grounds and nutrients available to the microbial organisms (ASHRAE 1989; Broadbent 1989; Meitz 1986, 1988).

Proper cleaning procedures address the entire tower, including not only the cold water basin, but also the distribution system, strainers, eliminators, casing, fan and fan cylinders, and louvers. The water treatment specialist should be advised and consulted prior to and following the cleaning.

In the United States, it is recommended that personnel involved wear high-efficiency particulate air (HEPA) type respirators, gloves, goggles, and other body coverings approved by OSHA/ National Institute for Occupational Safety and Health (NIOSH). This is especially true if the cleaning procedures involve the use of high-pressure water, air, and steam (ASHRAE 1989) or the use of wet-dry vacuum equipment. If any chemicals are used, they must be handled according to their material safety data sheets (MSDSs), which are obtainable from the chemical supplier.

Operation in Subfreezing Weather. During operation in subfreezing weather, the tower should be inspected more frequently, preferably daily, for ice formation on fill, louvers, fans, etc. This is especially true when the system is being operated outside the tower design parameters, such as when the main system is shut down and only supplementary units (e.g., computer cooling equipment) are operating. Ice on fan and drive systems is dangerous and can destroy the fan. Moderate icing on fill is generally not dangerous but can cause damage if allowed to build up.

Follow the manufacturer's specific recommendations both for operation in freezing temperatures and for deicing methods such as reversal of fan direction for short periods of time. Monitor the operation of winterization equipment, such as immersion heaters and heat-tracing tape on makeup lines, to ensure that they are working properly. Check for conditions that could render the freeze protection inoperable, such as tripped breakers, closed valves, and erroneous temperature settings.

Help from Manufacturers. Equipment manufacturers will provide assistance and technical publications on the efficient operation of their equipment; some even provide training. Also, manufacturers can often provide names of reputable local service companies that are experienced with their equipment. Most of these services are free or their cost is nominal.

Water Treatment

The quality of water circulating through an evaporative cooling system has a significant effect on the overall system efficiency, the degree of maintenance required, and the useful life of system components. Because the water is cooled primarily by evaporation of a portion of the circulating water, the concentration of dissolved solids and other impurities in the water can increase rapidly. Also, appreciable quantities of airborne impurities, such as dust and gases, may enter during operation. Depending on the nature of the impurities, they can cause scaling, corrosion, and/or silt deposits.

Simple blowdown (the wasting of a small portion of the recirculating water) may be adequate to control scale and corrosion on sites with good-quality makeup water, but it will not control biological contaminants, including *Legionella pneumophila.* All cooling tower systems should be treated to restrict biological growth, and many benefit from treatment to control scale and corrosion. For a complete and detailed description of water treatment, see Chapter 44 of the 1999 *ASHRAE Handbook—Applications.* Specific recommendations on water treatment can be obtained from any qualified water treatment supplier.

PERFORMANCE CURVES

The combination of flow rate and heat load dictates the range a cooling tower must accommodate. The entering air wet-bulb temperature and required system temperature level combine with cooling tower size to balance the heat rejected at a specified approach. The performance curves in this section are typical and may vary from project to project.

Cooling towers can accommodate a wide diversity of temperature levels, ranging as high as 150 to 160°F hot water temperature in the hydrocarbon processing industry. In the air-conditioning and refrigeration industry, towers are generally applied in the range of 90 to 115°F hot water temperature. A typical standard design condition for such cooling towers is 95°F hot water to 85°F cold water temperature, and 78°F wet-bulb temperature.

A means of evaluating the typical performance of a cooling tower used for a typical air-conditioning system is shown in Figure 22 through Figure 25. The example tower was selected for a flow rate of 3 gpm per nominal ton when cooling water from 95 to 85°F at 78°F entering wet-bulb temperature (Figure 22).

When operating at other wet bulbs or ranges, the curves may be interpolated to find the resulting temperature level (hot and cold water) of the system. When operating at other flow rates (2, 4, and 5 gpm per nominal ton), this same tower performs at the levels described by the titles of Figure 23, Figure 24, and Figure 25, respectively. Intermediate flow rates may be interpolated between charts to find resulting operating temperature levels.

The format of these curves is similar to the predicted performance curves supplied by manufacturers of cooling towers; the difference is that only three specific ranges—80%, 100%, and 120% of design range—and only three charts are provided, covering 90%, 100%, and 110% of design flow. The curves in Figure 22 through Figure 25, therefore, bracket the acceptable tolerance range of test conditions and may be interpolated for any specific test condition within the scope of the curve families and chart flow rates.

The curves may also be used to identify the feasibility of varying the parameters to meet specific applications. For example, the subject tower can handle a greater heat load (flow rate) when operating

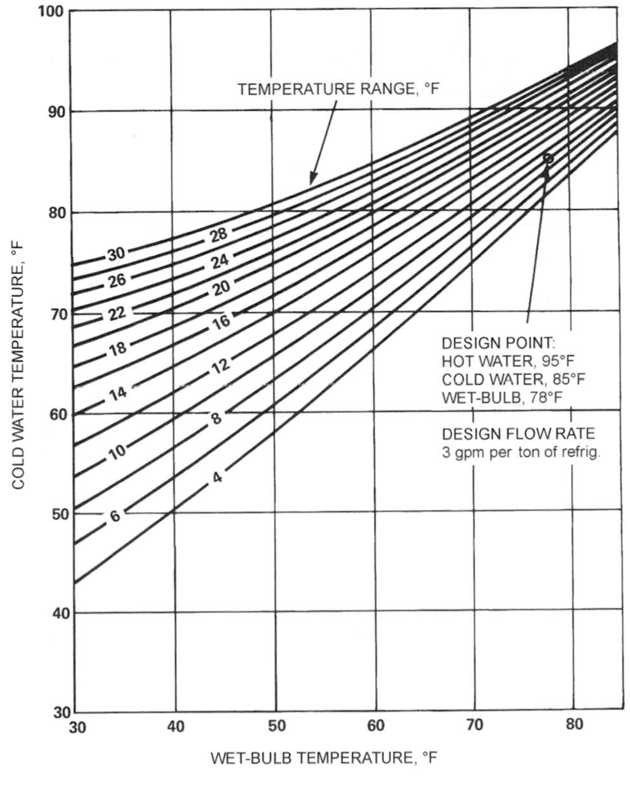

Fig. 22 Cooling Tower Performance—100% Design Flow

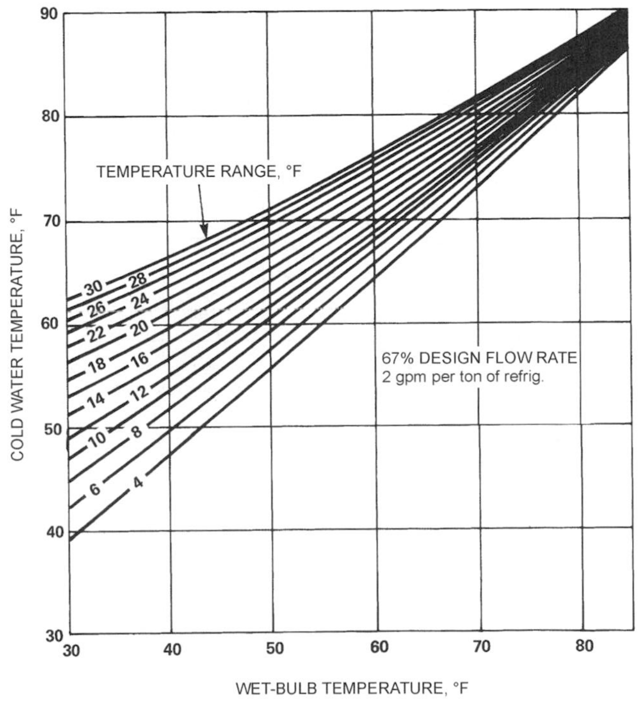

Fig. 23 Cooling Tower Performance—67% Design Flow

in a lower ambient wet-bulb region. This may be seen by comparing the intersection of the 10°F range curve with 73°F wet bulb at 85°F cold water to show the tower is capable of rejecting 33% more heat load at this lower ambient temperature (Figure 24).

Similar comparisons and cross-plots identify relative tower capacity for a wide range of variables. The curves produce accurate comparisons within the scope of the information presented but should not be extrapolated outside the field of data given. Also, the curves are based on a typical mechanical-draft, film-filled, cross-flow, medium-sized, air-conditioning cooling tower. Other types and sizes of towers produce somewhat different balance points of temperature level. However, the curves may be used to evaluate a tower for year-round or seasonal use if they are restricted to the general operating characteristics described. (See specific manufacturer's data for maximum accuracy when planning for test or critical temperature needs.)

As stated, the cooling tower, when selected for a specified design condition, operates at other temperature levels when the ambient temperature is off-design or when heat load or flow rate varies from the design condition. When flow rate is held constant, range falls as heat load falls, causing temperature levels to fall to a closer approach. Hot and cold water temperature levels fall when the ambient wet bulb falls at constant heat load, range, and flow rate. As water loading to a particular tower falls at constant ambient wet bulb and range, the tower cools the water to a lower temperature level or closer approach to the wet bulb.

COOLING TOWER THERMAL PERFORMANCE

Three basic alternatives are available to a purchaser/designer seeking assurance that a cooling tower will perform as specified: (1) certification of performance by an independent third party, (2) an acceptance test performed at the site after the unit is installed, or (3) a performance bond. Codes and standards that pertain to performance certification and field testing of cooling towers are listed in Chapter 48.

Certification. The thermal performance of many commercially available cooling tower lines is certified by the Cooling Tower Institute (CTI) in accordance with their *Standard* STD-201. This standard applies to mechanical-draft, open- and closed-circuit water cooling towers. It is based on entering wet-bulb temperature and certifies the performance of the tower when operating in an open, unrestricted environment.

Field Acceptance Test. As an alternative to certification, the performance of the tower can be verified after installation by conducting a field acceptance test in accordance with one of the two available test standards. Of the two standards, CTI *Standard* ATC-105 is more commonly used, although American Society of Mechanical Engineers (ASME) *Standard* PTC-23 is also used. These standards are similar in their requirements, and both base the performance evaluation on entering wet-bulb temperature. ASME *Standard* PTC-23, however, provides an alternative for evaluation based on ambient wet-bulb temperature as well.

With either procedure, the test consists of measuring the temperature of the hot water in the inlet piping to the tower or in the hot water distribution basin. Preferably, the cold water temperature is measured at the discharge of the circulating pump, where there is much less chance for temperature stratification. The wet-bulb temperature is measured by an array of mechanically aspirated psychrometers. The flow rate of the recirculating water is measured by any of several approved methods, most often a pitot-tube traverse of the piping leading to the tower. Recently calibrated instruments should be used for all measurements and electronic data acquisition is recommended for all but the smallest installations.

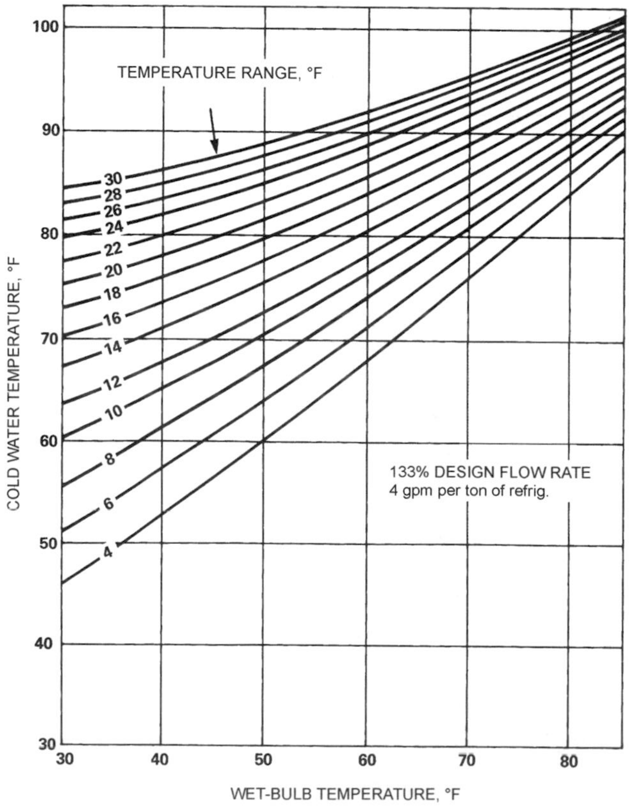

Fig. 24 Cooling Tower Performance—133% Design Flow

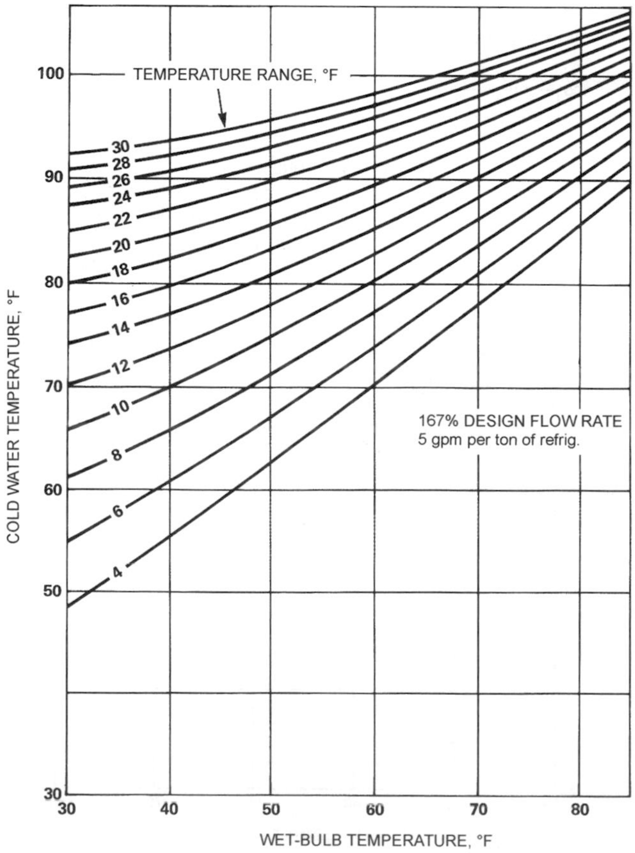

Fig. 25 Cooling Tower Performance—167% Design Flow

For an accurate test, the tower should be running under a steady heat load combined with a steady flow of recirculating water, both as near design as possible. Weather conditions should be reasonably stable, with prevailing winds of 10 mph or less. The tower should be clean and adjusted for proper water distribution, with all fans operating at design speed. Both the CTI and ASME standards specify maximum recommended deviations from design operating conditions of range, flow, wet-bulb temperature, heat load, and fan power.

Performance Bond. A few manufacturers offer a performance bond, which provides for seeking redress from a surety in the event the cooling tower does not meet the manufacturer's rated thermal performance.

COOLING TOWER THEORY

Baker and Shryock (1961) developed the following theory. Consider a cooling tower having one square foot of plan area; cooling volume V, containing extended water surface per unit volume a; and water mass flow rate L and air mass flow rate G. Figure 26 schematically shows the processes of mass and energy transfer. The bulk water at temperature t is surrounded by the bulk air at dry-bulb temperature t_a, having enthalpy h_a and humidity ratio W_a. The interface is assumed to be a film of saturated air with an intermediate temperature t'', enthalpy h'', and humidity ratio W''. Assuming a constant value of 1 Btu/lb·°F for the specific heat of water c_p, the total energy transfer from the water to the interface is

$$dq_w = Lc_p\,dt = K_L a(t - t'')dV \tag{1}$$

where

q_w = rate of total heat transfer, bulk water to interface, Btu/h
L = inlet water mass flow rate, lb/h
K_L = unit conductance, heat transfer, bulk water to interface, Btu/h·ft²·°F
V = cooling volume, ft³
a = area of interface, ft²/ft³

The heat transfer from interface to air is

$$dq_s = K_G a(t'' - t_a)dV \tag{2}$$

where

q_s = rate of sensible heat transfer, interface to airstream, Btu/h
K_G = overall unit conductance, sensible heat transfer, interface to main airstream, Btu/h·ft²·°F

The diffusion of water vapor from film to air is

$$dm = K'a(W'' - W_a)dV \tag{3}$$

where

m = mass transfer rate, interface to airstream, lb/h
K' = unit conductance, mass transfer, interface to main airstream, lb/h·ft²·(lb/lb)
W'' = humidity ratio of interface (film), lb/lb
W_a = humidity ratio of air, lb/lb

The heat transfer due to evaporation from film to air is

$$dq_L = r\,dm = rK'a(W'' - W_a)dV \tag{4}$$

where

q_L = rate of latent heat transfer, interface to airstream, Btu/h
r = latent heat of evaporation (constant), Btu/lb

The process reaches equilibrium when $t_a = t$, and the air becomes saturated with moisture at that temperature. Under adiabatic conditions, equilibrium is reached at the temperature of adiabatic saturation or at the thermodynamic wet-bulb temperature of the air. This is the lowest attainable temperature in a cooling tower. The circulating water rapidly approaches this temperature when a tower operates without heat load. The process is the same when a heat load is applied, but the air enthalpy increases as it moves through the tower so the equilibrium temperature increases progressively. The approach of the cooled water to the entering wet-bulb temperature is a function of the capability of the tower.

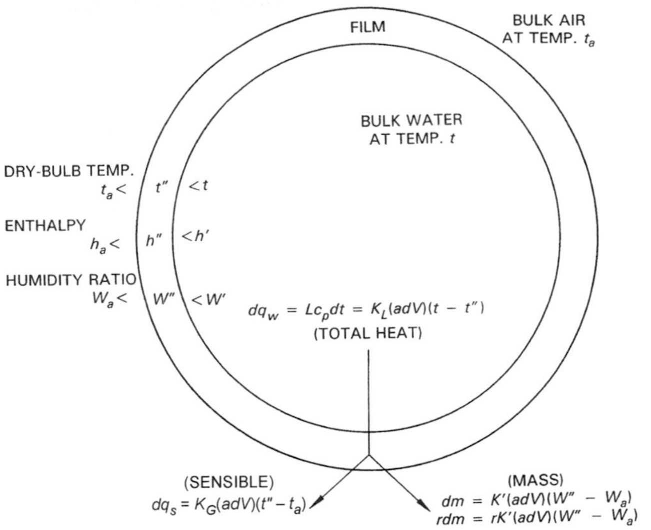

Fig. 26 Heat and Mass Transfer Relationships Between Water, Interfacial Film, and Air
(Baker and Shryock 1961)

Merkel (1925) assumed the Lewis relationship to be equal to one in combining the transfer of mass and sensible heat into an overall coefficient based on enthalpy difference as the driving force:

$$K_G/(K'c_{pm}) = 1 \tag{5}$$

where c_{pm} = humid specific heat of moist air in Btu/lb·°F (dry air basis).

Equation (5) also explains why the wet-bulb thermometer closely approximates the temperature of adiabatic saturation in an air-water vapor mixture. Setting water heat loss equal to air heat gain yields

$$Lc_p\,dt = G\,dh = K'a(h'' - h_a)dV \tag{6}$$

where G = air mass flow rate in lb/h.

The equation considers the transfer from the interface to the airstream, but the interfacial conditions are indeterminate. If the film resistance is neglected and an overall coefficient K' is postulated, based on the driving force of enthalpy h' at the bulk water temperature t, the equation becomes

$$Lc_p\,dt = G\,dh = K'a(h' - h_a)dV \tag{7}$$

or

$$K'aV/L = \int_{t_1}^{t_2} \frac{c_p}{h' - h_a}dt \tag{8}$$

and

$$K'aV/G = \int_{h_1}^{h_2} \frac{dh}{h' - h_a} \tag{9}$$

In cooling tower practice, the integrated value of Equation (8) is commonly referred to as the number of transfer units (NTU). This

Table 2 Counterflow Integration Calculations for Example 1

1 Water Temperature t, °F	2 Enthalpy of Film h', Btu/lb	3 Enthalpy of Air h_a, Btu/lb	4 Enthalpy Difference $h' - h_a$, Btu/lb	5 $\frac{1}{(h' - h_a)}$	6 Δt, °F	7 NTU = $\frac{c_p \Delta t}{(h' - h_a)_{avg}}$	8 ΣNTU	9 Cumulative Cooling Range, °F
85	49.4	38.6	10.8	0.0926				
86	50.7	39.8	10.9	0.0917	1	0.0921	0.0921	1
87	51.9	41.0	10.9	0.0917	1	0.0917	0.1838	2
88	53.2	42.2	11.0	0.0909	1	0.0913	0.2751	3
89	54.6	43.4	11.2	0.0893	1	0.0901	0.3652	4
90	55.9	44.6	11.3	0.0885	1	0.0889	0.4541	5
92	58.8	47.0	11.8	0.0847	2	0.1732	0.6273	7
94	61.8	49.9	12.4	0.0806	2	0.1653	0.7925	9
96	64.9	51.8	13.1	0.0763	2	0.1569	0.9493	11
98	68.2	54.2	14.0	0.0714	2	0.1477	1.097	13
100	71.7	56.6	15.1	0.0662	2	0.1376	1.2346	15

value gives the number of times the average enthalpy potential $(h' - h_a)$ goes into the temperature change of the water (Δt) and is a measure of the difficulty of the task. Thus, one transfer unit has the definition of $c_p\Delta t/(h' - h_a)_{avg} = 1$.

The equations are not self-sufficient and are not subject to direct mathematical solution. They reflect mass and energy balance at any point in a tower and are independent of relative motion of the two fluid streams. Mechanical integration is required to apply the equations, and the procedure must account for relative motion. The integration of Equation (8) gives the NTU for a given set of conditions.

Counterflow Integration

The counterflow cooling diagram is based on the saturation curve for air-water vapor (Figure 27). As water is cooled from t_{w1} to t_{w2}, the air film enthalpy follows the saturation curve from A to B. Air entering at wet-bulb temperature t_{aw} has an enthalpy h_a corresponding to C′. The initial driving force is the vertical distance BC. Heat removed from the water is added to the air, so the enthalpy increase is proportional to water temperature. The slope of the air operating line CD equals *L/G.*

Counterflow calculations start at the bottom of a tower, the only point where the air and water conditions are known. The NTU is calculated for a series of incremental steps, and the summation is the integral of the process.

Example 1. Air enters the base of a counterflow cooling tower at 75°F wet-bulb temperature, water leaves at 85°F, and *L/G* (water-to-air ratio) is 1.2, so $dh = 1.2 \times 1 \times dt$, where 1 Btu/lb·°F is the specific heat c_p of water. Calculate the NTU for various cooling ranges.

Solution: The calculation is shown in Table 2. Water temperatures are shown in Column 1 for 1°F increments from 85 to 90°F and 2°F increments from 90 to 100°F. The corresponding film enthalpies, obtained from psychrometric tables, are shown in Column 2.

The upward air path is in Column 3. The initial air enthalpy is 38.6 Btu/lb, corresponding to a 75°F wet bulb, and increases by the relationship $\Delta h = 1.2 \times 1 \times \Delta t$.

The driving force $h' - h_a$ at each increment is listed in Column 4. The reciprocals $1/(h' - h_a)$ are calculated (Column 5), Δt is noted (Column 6), and the average for each increment is multiplied by $c_p\Delta t$ to obtain the NTU for each increment (Column 7). The summation of the incremental values (Column 8) represents the NTU for the summation of the incremental temperature changes, which is the cooling range given in Column 9.

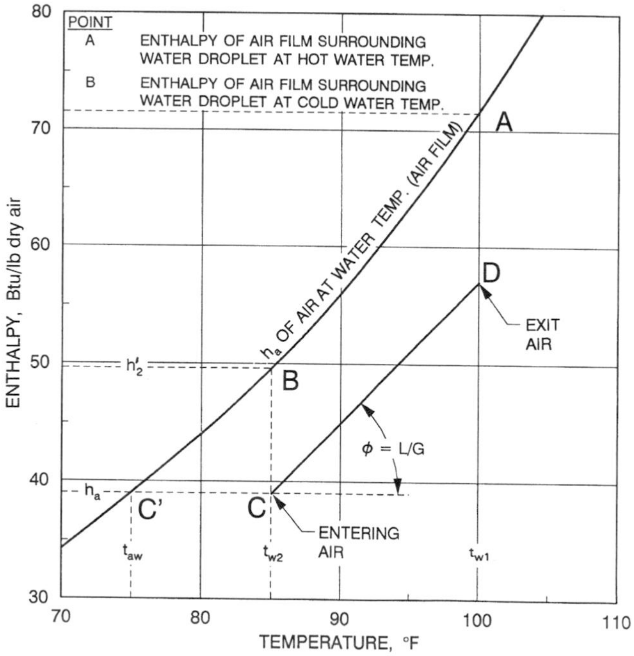

Fig. 27 Counterflow Cooling Diagram

Because of the slope and position of CD relative to the saturation curve, the potential difference increases progressively from the bottom to the top of the tower in this example. The degree of difficulty decreases as this driving force increases, which is reflected as a reduction in the incremental NTU proportional to a variation in incremental height. This procedure determines the temperature gradient with respect to tower height.

The procedure of Example 1 considers increments of temperature change and calculates the coincident values of NTU, which correspond to increments of height. Baker and Mart (1952) developed a unit-volume procedure that considers increments of NTU (representing increments of height) with corresponding temperature changes calculated by iteration. The unit-volume procedure is more cumbersome but is necessary in cross-flow integration because it accounts for temperature and enthalpy change, both horizontally and vertically.

Cross-Flow Integration

In a cross-flow tower, water enters at the top; the solid lines of constant water temperature in Figure 28 show its temperature distribution. Air enters from the left, and the dotted lines show constant enthalpies. The cross section is divided into unit volumes in which dV becomes $dxdy$ and Equation (7) becomes

$$c_pL\ dt\ dx = G\ dh\ dy = Ka(h' - h_a)dx\ dy \tag{10}$$

The overall L/G ratio applies to each unit-volume by considering $dx/dy = w/z$. The cross-sectional shape is automatically considered when an equal number of horizontal and vertical increments are used. Calculations start at the top of the air inlet and proceed down and across. Typical calculations are shown in Figure 29 for water entering at 100°F, air entering at 75°F wet-bulb temperature, and L/G = 1.0. Each unit-volume represents 0.1 NTU. Temperature change vertically in each unit is determined by iteration from

$$\Delta t = 0.1(h' - h_a)_{avg} \tag{11}$$

$c_p(L/G)dt = dh$ determines the horizontal change in air enthalpy. With each step representing 0.1 NTU, two steps down and across equal 0.2 NTU, etc., for conditions corresponding to the average leaving water temperature.

Figure 28 shows that air flowing across any horizontal plane is moving toward progressively hotter water, with entering hot water temperature as a limit. Water falling through any vertical section is moving toward progressively colder air that has the entering wet-bulb temperature as a limit. This is shown in Figure 30, which is a plot of the data in Figure 29. Air enthalpy follows the family of curves radiating from Point A. Air moving across the top of the tower tends to coincide with OA. Air flowing across the bottom of a tower of infinite height follows a curve that coincides with the saturation curve AB.

Water temperatures follow the family of curves radiating from Point B, between the limits of BO at the air inlet and BA at the outlet of a tower of infinite width. The single operating line CD of the

100°F HOT WATER TEMPERATURE

75°F WET-BULB TEMPERATURE	71.73 h′	71.73 h′	71.73 h′	71.73 h′	71.73 h′
38.59 h_a	1,1 2.92 Δt 97.08°F 66.68 h′ 2.92 Δh 41.51 h	1,2 2.66 Δt 97.37°F 67.11 h′ 2.66 Δh 44.17 h	1,3 2.42 Δt 97.58°F 67.52 h′ 2.42 Δh 46.59 h	1,4 2.21 Δt 97.79°F 67.87 h′ 2.21 Δh 48.80 h	1,5 2.02 Δt 97.98°F 68.20 h′ 2.02 Δh 50.82 h
38.59 h_a	2,1 2.49 Δt 94.59°F 62.67 h′ 2.49 Δh 41.08 h	2,2 2.30 Δt 95.07°F 63.42 h′ 2.30 Δh 43.38 h	2,3 2.13 Δt 95.45°F 64.03 h′ 2.13 Δh 45.51 h	2,4 1.97 Δt 95.82°F 64.62 h′ 1.97 Δh 47.48 h	2,5 1.83 Δt 96.15°F 65.15 h′ 1.83 Δh 49.32 h
38.59 h_a	3,1 2.14 Δt 92.45°F 59.43 h′ 2.14 Δh 40.72 h	3,2 2.01 Δt 93.06°F 60.33 h′ 2.01 Δh 42.73 h	3,3 1.89 Δt 93.56°F 61.09 h′ 1.89 Δh 44.62 h	3,4 1.77 Δt 94.05°F 61.84 h′ 1.77 Δh 46.39 h	3,5 1.67 Δt 94.48°F 62.50 h′ 1.67 Δh 48.06 h
38.59 h_a	4,1 1.85 Δt 90.55°F 56.69 h′ 1.85 Δh 40.44 h	4,2 1.77 Δt 91.29°F 57.74 h′ 1.77 Δh 42.21 h	4,3 1.68 Δt 91.88°F 58.59 h′ 1.68 Δh 43.89 h	4,4 1.60 Δt 92.45°F 59.43 h′ 1.60 Δh 45.59 h	4,5 1.51 Δt 92.97°F 60.20 h′ 1.51 Δh 47.00 h
33.59 h_a	5,1 1.62 Δt 88.93°F 54.46 h′ 1.62 Δh 40.21 h	5,2 1.56 Δt 89.73°F 55.55 h′ 1.56 Δh 41.77 h	5,3 1.50 Δt 90.38°F 56.45 h′ 1.50 Δh 43.27 h	5,4 1.45 Δt 91.00°F 57.33 h′ 1.45 Δh 44.72 h	5,5 1.38 Δt 91.59°F 58.17 h′ 1.38 Δh 46.10 h

90.53°F AVERAGE WATER TEMPERATURE

L/G = 1.0 Δt (Vertically) = $0.1(h' - h_a)_{avg}$ Δh (Horizontally) = Δt

Fig. 29 Cross-Flow Calculations
(Baker and Shryock 1961)

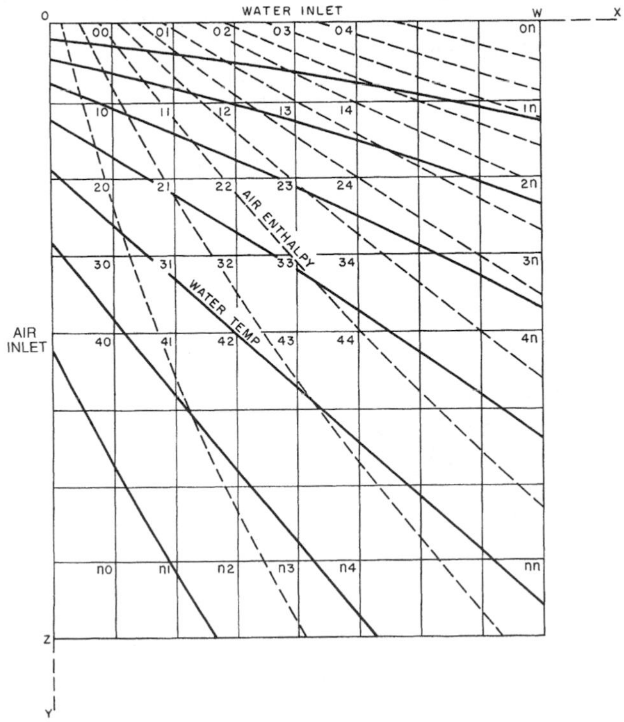

Fig. 28 Water Temperature and Air Enthalpy Variation Through Cross-Flow Cooling Tower
(Baker and Shryock 1961)

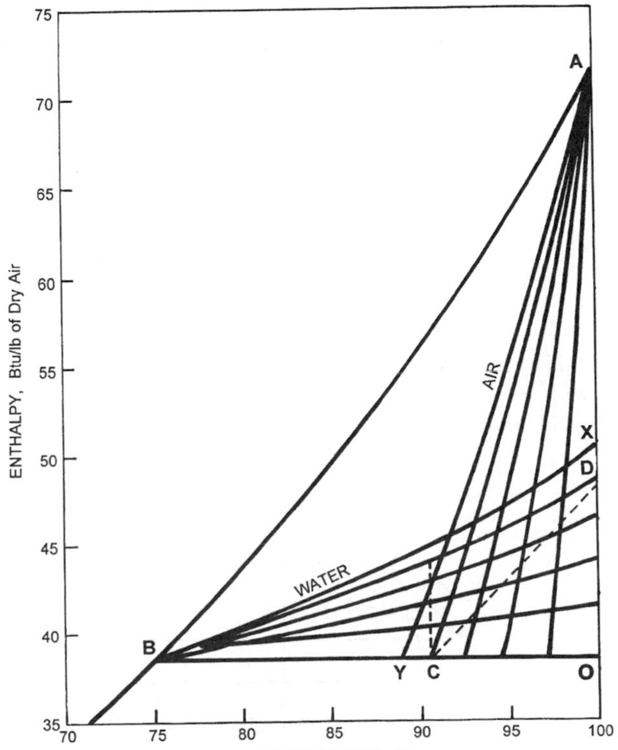

Fig. 30 Cross-Flow Cooling Diagram

counterflow diagram in Figure 27 is replaced in the cross-flow diagram (Figure 30) by a zone represented by the area intersected by the two families of curves.

TOWER COEFFICIENTS

Calculations can reduce a set of conditions to a numerical value representing degree of difficulty. The NTU corresponding to a set of hypothetical conditions is called the required coefficient and evaluates degree of difficulty. When test results are being considered, the NTU represents the available coefficient and becomes an evaluation of the equipment tested.

The calculations consider temperatures and the *L/G* ratio. The minimum required coefficient for a given set of temperatures occurs at *L/G* = 0, corresponding to an infinite air rate. No increase in air enthalpy occurs, so the driving force is maximum and the degree of difficulty is minimum. Decreased air rate (increase in *L/G*) decreases the driving force, and the greater degree of difficulty shows as an increase in NTU. This situation is shown for counterflow in Figure 31. Maximum *L/G* (minimum air rate) occurs when CD intersects the saturation curve. Driving force becomes zero, and NTU is infinite. The point of zero driving force may occur at the air outlet or at an intermediate point because of the curvature of the saturation curve.

Similar variations occur in cross-flow cooling. Variations in *L/G* vary the shape of the operating area. At *L/G* = 0, the operating area becomes a horizontal line, which is identical to the counterflow diagram (Figure 31), and both coefficients are the same. An increase in *L/G* causes an increase in the height of the operating area and a decrease in the width. This continues as the areas extend to Point A as a limit. This maximum *L/G* always occurs when the wet-bulb temperature of the air equals the hot water temperature and not at an intermediate point, as may occur in counterflow.

Both types of flow have the same minimum coefficient at *L/G* = 0, and both increase to infinity at a maximum *L/G*. The maximums are the same if the counterflow potential reaches zero at the air outlet, but the counterflow tower will have a lower maximum *L/G* when the potential reaches zero at an intermediate point, as in Figure 31. A cooling tower can be designed to operate at any point within the two limits, but most applications restrict the design to much narrower limits determined by air velocity.

A low air rate requires a large tower, while a high air rate in a smaller tower requires greater fan power. Typical limits in air velocity are about 300 to 700 fpm in counterflow and 350 to 800+ fpm in cross-flow.

Available Coefficients

A cooling tower can operate over a wide range of water rates, air rates, and heat loads, with variation in the approach of the cold water to the wet-bulb temperature. Analysis of a series of test points shows that the available coefficient is not a constant but varies with the operating conditions, as shown in Figure 32.

Figure 32 is a typical correlation of a tower characteristic showing the variation of available *KaV/L* with *L/G* for parameters of constant air velocity. Recent fill developments and more accurate test methods have shown that some of the characteristic lines are curves rather than a series of straight, parallel lines on logarithmic coordinates.

Ignoring the minor effect of air velocity, a single average curve may be considered:

$$KaV/L \sim (L/G)^n \tag{12}$$

The exponent n varies over a range of about −0.35 to −1.1 but averages between −0.55 and −0.65. Within the range of testing, 0.6 has been considered sufficiently accurate.

The family of curves corresponds to the following relation:

$$KaV/L \sim (L)^n (G)^m \tag{13}$$

where n is as above and m varies slightly from n numerically and is a positive exponent.

The triangular points in Figure 32 show the effect of varying temperature at nominal air rate. The deviations result from simplifying assumptions and may be overcome by modifying the integration procedure. Usual practice, as shown in Equation (9), ignores evaporation and assumes that

$$G\,dh = c_p L\,dt \tag{14}$$

The exact enthalpy rise is greater than this because a portion of the heat in the water stream leaves as vapor in the airstream.

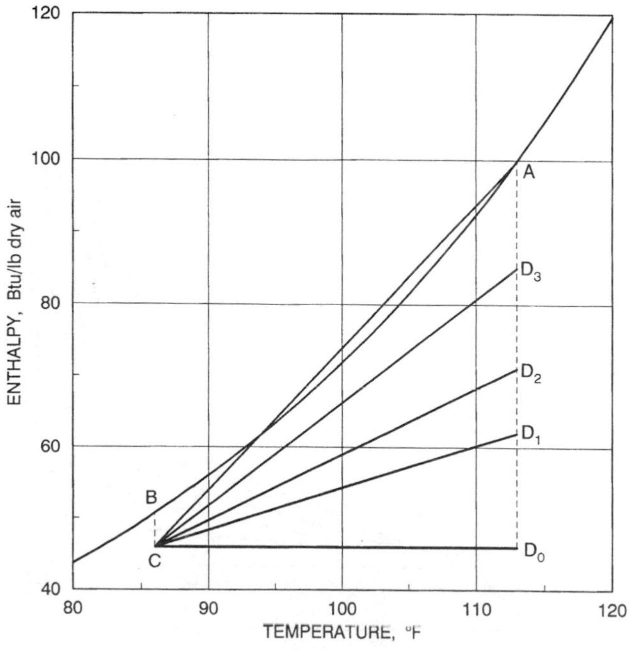

Fig. 31 Counterflow Cooling Diagram for Constant Conditions, Variable *L/G* Ratios

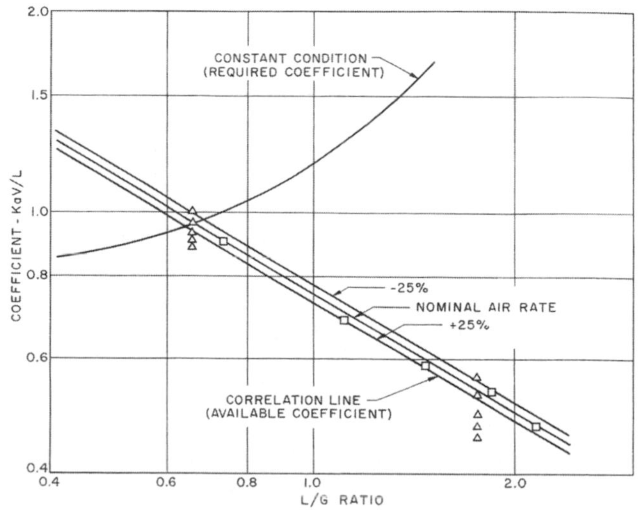

Fig. 32 Tower Characteristic, *KaV/L* Versus *L/G*
(Baker and Shryock 1961)

The correct heat balance is as follows (Baker and Shryock 1961):

$$G\,dh = c_p L\,dt + c_p L_E (t_{w2} - 32) \qquad (15)$$

where L_E = mass flow rate of water that evaporates in lb/h. This reduces the driving force and increases the NTU.

Evaporation causes the water rate to decrease from L at the inlet to $L - L_E$ at the outlet. The water-to-air ratio varies from L/G at the water inlet to $(L - L_E)/G$ at the outlet. This results in an increase in NTU.

Basic theory considers the transfer from the interface to the airstream. As the film conditions are indeterminate, the film resistance is neglected as assumed in Equation (7). The resulting coefficients show deviations closely associated with hot water temperature and may be modified by an empirical hot water correction factor (Baker and Mart 1952).

The effect of film resistance (Mickley 1949) is shown in Figure 33. Water at temperature t is assumed to be surrounded by a film of saturated air at the same temperature at enthalpy h' (Point B on the saturation curve). The film is actually at a lower temperature t'' at enthalpy h'' (Point B′). The surrounding air at enthalpy h_a corresponds to Point C. The apparent potential difference is commonly considered to be $h' - h_a$, but the true potential difference is $h'' - h_a$ (Mickley 1949). From Equations (1) and (6),

$$\frac{h'' - h_a}{t - t''} = \frac{K_L}{K'} \qquad (16)$$

The slope of CB′ is the ratio of the two coefficients. No means to evaluate the coefficients has been proposed, but a slope of −11.1 for cross-flow towers has been reported (Baker and Shryock 1961).

Establishing Tower Characteristics

The performance characteristic of a fill pattern can vary widely due to several external factors. For a given volume of fill, the optimal thermal performance is obtained with uniform air and water distribution throughout the fill pack. Irregularities in either alter the local L/G ratios within the pack and adversely affect the overall thermal performance of the cooling tower. Accordingly, the design of the tower water distribution system, the air inlets, the fan plenum, and so forth is very important in assuring that the tower will perform to its potential.

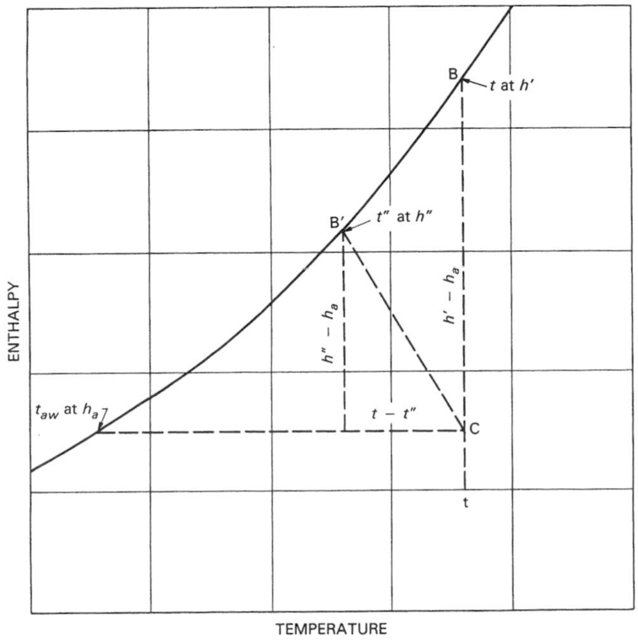

Fig. 33 True Versus Apparent Potential Difference
(Baker and Shryock 1961)

In counterflow towers, some cooling occurs in the spray chamber above the fill and in the open space below the fill. This additional performance is erroneously attributed to the fill itself, which can lead to inaccurate predictions of a tower's performance in other applications. A true performance characteristic for the total cooling tower can only be developed from full-scale tests of the actual cooling tower assembly, typically by the tower manufacturer, and not by combining performance data of individual components.

REFERENCES

ASHRAE. 1989. Legionellosis position statement and Legionellosis position paper.

ASME. 1986. Atmospheric water cooling equipment. ANSI/ASME *Standard* PTC 23-86 (R 97). American Society of Mechanical Engineers, New York.

Baker, D.R. and H.A. Shryock. 1961. A comprehensive approach to the analysis of cooling tower performance. ASME *Transactions, Journal of Heat Transfer* (August):339.

Baker, D.R. and L.T. Mart. 1952. Analyzing cooling tower performance by the unit-volume coefficient. *Chemical Engineering* (December):196.

Broadbent, C.R. 1989. Practical measures to control Legionnaire's disease hazards. *Australian Refrigeration, Air Conditioning and Heating* (July): 22-28.

Broadbent, C.R. et al. 1992. *Legionella* ecology in cooling towers. *Australian Refrigeration, Air Conditioning and Heating* (October):20-34.

CTI. 1996. Acceptance test code for closed circuit cooling. *Standard* ATC-105S-96. Cooling Tower Institute, Houston, TX.

CTI. 1996. Standard for the certification of water-cooling tower thermal performance. *Standard* STD-201-96.

CTI. 1997. Acceptance test code for water-cooling towers. *Standard* ATC-105-97, Vol. 1.

Hensley, J.C., ed. 1985. *Cooling tower fundamentals*, 2nd ed. Marley Cooling Tower Company, Kansas City.

McCann, M. 1988. Cooling towers take the heat. *Engineered Systems* 5(October):58-61.

Meitz, A. 1986. Clean cooling systems minimize *Legionella* exposure. *Heating, Piping and Air Conditioning* 58(August):99-102.

Merkel, F. 1925. Verduftungskühlung. *Forschungarbeiten*, No. 275.

Mickley, H.S. 1949. Design of forced-draft air conditioning equipment. *Chemical Engineering Progress* 45:739.

Rosa, F. 1992. Some contributing factors in indoor air quality problems. *National Engineer* (May):14

White, T.L. 1994. Winter cooling tower operation for a central chilled water system. *ASHRAE Transactions* 100(1):811-16.

BIBLIOGRAPHY

Baker, D.R. 1962. Use charts to evaluate cooling towers. *Petroleum Refiner* (November).

Braun, J.E. and Diderrich, G.T. 1990. Near-optimal control of cooling towers for chilled water systems. *ASHRAE Transactions* 96(2):806-13.

CIBSE. 1991. Minimizing the risk of Legionnaire's disease. *Technical Memorandum* TM13. The Chartered Institute of Building Services Engineers, London.

Fliermans, C.B. et al. 1981. Measure of *Legionella pneumophila* activity in situ. *Current Microbiology* 6:89-94.

Fluor Products Company. 1958. Evaluated weather data for cooling equipment design.

Kohloss, F.H. 1970. Cooling tower application. *ASHRAE Journal* (August).

Landon, R.D. and J.R. Houx, Jr. 1973. Plume abatement and water conservation with the wet-dry cooling tower. Marley Cooling Tower Company, Mission, KS.

Mallison, G.F. 1980. Legionellosis: Environmental aspects. *Ann. New York Academy of Science* 353:67-70.

McBurney, K. 1990. Maintenance suggestions for cooling towers and accessories. *ASHRAE Journal* 32(6):16-26.

Meitz, A. 1988. Microbial life in cooling water systems. *ASHRAE Journal* 30(August):25-30.

CHAPTER 37

LIQUID COOLERS

A LIQUID cooler (hereafter called a cooler) is a component of a refrigeration system in which the refrigerant is evaporated, thereby producing a cooling effect on a fluid (usually water or brine). This chapter addresses the performance, design, and application of coolers. It briefly describes various types of liquid coolers and the refrigerants commonly used, as listed in Table 1.

TYPES OF LIQUID COOLERS

Direct-Expansion

Refrigerant evaporates inside tubes of a direct-expansion cooler. These coolers are usually used with positive-displacement compressors, such as reciprocating, rotary, or rotary screw compressors, to cool water or brine. A shell-and-tube heat exchanger is the most common; tube-in-tube and brazed plate coolers are also available.

Figure 1 shows a typical shell-and-tube cooler. A series of baffles channels the fluid throughout the shell side. The baffles increase the velocity of the fluid, thereby increasing its heat transfer coefficient. The velocity of the fluid flowing perpendicular to the tubes should be at least 2 fps to clean the tubes and less than 10 fps to prevent erosion.

Distribution is critical in direct-expansion coolers. If some tubes are fed more refrigerant than others, they tend to bleed liquid refrigerant into the suction line. Because most direct-expansion coolers are controlled to a given suction superheat, the remaining tubes must produce a higher superheat to evaporate the liquid bleeding through. This unbalance causes poor overall heat transfer. Uniform distribution is usually achieved by a distributor or by keeping the volume of the refrigerant inlet head to a minimum. Both methods create sufficient turbulence to keep a homogeneous mixture so that each tube gets the same mixture of liquid and vapor.

The preparation of this chapter is assigned to TC 8.5, Liquid-to-Refrigerant Heat Exchangers.

The number of refrigerant passes is another important item in the performance of a direct-expansion cooler. A single-pass cooler must evaporate all the refrigerant before it reaches the end of the tubes; this requires long tubes or enhanced inside tube surfaces. A multiple-pass cooler can have less or no surface enhancement, but after the first pass, good distribution is difficult to obtain.

A tube-in-tube cooler is similar to a shell-and-tube design, except that it consists of one or more pairs of coaxial tubes. The fluid usually flows inside the inner tube while the refrigerant flows in the annular space between the tubes. In this way, the fluid side can be mechanically cleaned if access to the header is provided.

Brazed or semiwelded plate coolers are constructed of plates brazed or laser welded together to make up an assembly of separate channels. Space requirements are minimal due to the compactness of these coolers; however, their construction does not allow mechanical cleaning. Internal leaks in brazed plates typically cannot be repaired.

Most direct-expansion coolers are designed for horizontal mounting. If they are mounted vertically, performance may vary

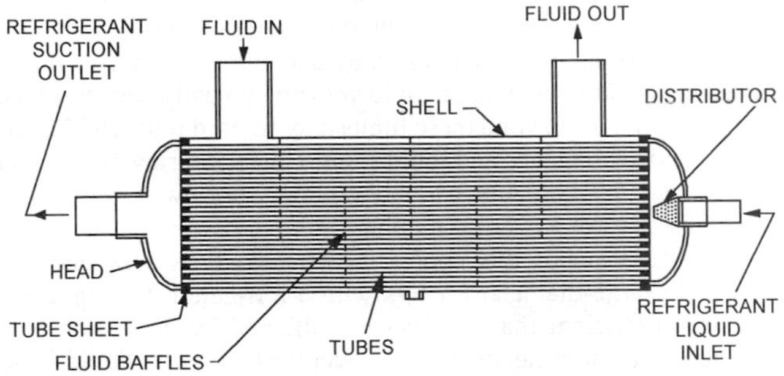

Fig. 1 Direct-Expansion Shell-and-Tube Cooler

Table 1 Types of Coolers

Type of Cooler	Usual Refrigerant Feed Device	Usual Capacity Range, tons	Commonly Used Refrigerants
Flooded shell-and-tube	Low-pressure float High-pressure float Fixed orifice(s) Weir	25 to 2000	11, 12, 22, 113, 114, 123, 134a, 500, 502, 717
Spray-type shell-and-tube[a]	Low-pressure float High-pressure float	50 to 10,000	11, 12, 13B1, 22, 113, 114, 123, 134a
Direct-expansion shell-and-tube	Thermal expansion valve Electronic modulation valve	2 to 1000	12, 22, 134a, 500, 502, 717
Baudelot (flooded)	Low-pressure float	10 to 100	22, 717
Baudelot (direct-expansion)	Thermal expansion valve	5 to 25	12, 22, 134a, 717
Tube-in-tube	Thermal expansion valve	5 to 25	12, 22, 134a, 717
Shell-and-coil	Thermal expansion valve	2 to 10	12, 22, 134a, 717
Brazed or semiwelded plate[b]	Thermal expansion valve Fixed orifice	0.5 to 2000	12, 22, 134a, 500, 502, 717

[a]See the section on Flooded Shell-and-Tube.
[b]See the section on Direct-Expansion.

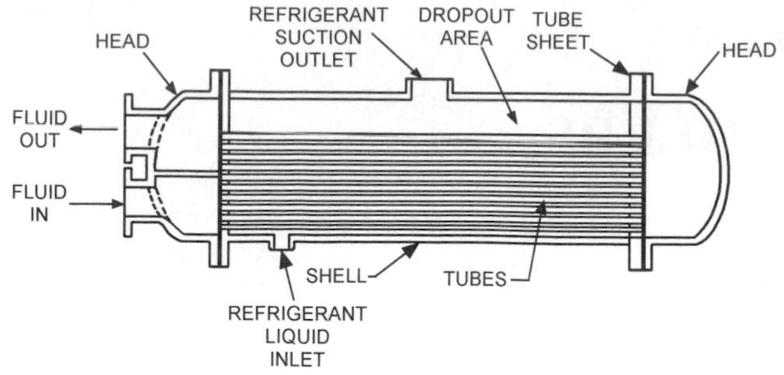

Fig. 2 Flooded Shell-and-Tube Cooler

considerably from that predicted because two-phase flow heat transfer is a direction-sensitive phenomenon.

Flooded Shell-and-Tube

In a flooded cooler, the refrigerant vaporizes on the outside of the tubes, which are submerged in liquid refrigerant within a closed shell. The fluid flows through the tubes as shown in Figure 2. Flooded coolers are usually used with rotary screw or centrifugal compressors to cool water or brine.

Refrigerant liquid/vapor mixture usually feeds into the bottom of the shell through a distributor that distributes the refrigerant vapor equally under the tubes. The relatively warm fluid in the tubes heats the refrigerant liquid surrounding the tubes, causing it to boil. As bubbles rise up through the space between tubes, the liquid surrounding the tubes becomes increasingly bubbly (or foamy, if much oil is present).

The refrigerant vapor must be separated from the mist generated by the boiling refrigerant. The simplest separation method is provided by a dropout area between the top row of tubes and the suction connectors. If this dropout area is insufficient, a coalescing filter may be required between the tubes and connectors. Perry and Green (1984) give additional information on mist elimination.

The size of tubes, number of tubes, and number of passes should be determined to maintain the fluid velocity typically between 3 and 10 fps. Velocities beyond these limits may be used if the fluid is free of suspended abrasives and fouling substances (Sturley 1975; Ayub and Jones 1987). In some cases, the minimum flow may be determined by a lower Reynolds number limit.

One variation of this cooler is the spray-type shell-and-tube cooler. In large-diameter coolers with a refrigerant having a heat transfer coefficient that is adversely affected by the head of the refrigerant, a spray can be used to cover the tubes with liquid rather than flooding them. A mechanical pump circulates liquid from the bottom of the cooler to the spray heads.

Flooded shell-and-tube coolers are generally unsuitable for other than horizontal orientation.

Baudelot

Baudelot coolers (Figure 3) are used to cool a fluid to near its freezing point in industrial, food, and dairy applications. In this cooler, the fluid is circulated over the outside of vertical plates, which are easy to clean. The inside surface of the plates is cooled by evaporating the refrigerant. The fluid to be cooled is distributed uniformly along the top of the heat exchanger and then flows by gravity to a collection pan below. The cooler may be enclosed by insulated walls to avoid unnecessary loss of refrigeration effect.

Refrigerant 717 (ammonia) is commonly used with the Baudelot cooler and arranged for flooded operation, using conventional gravity feed with a surge drum. A low-pressure float valve maintains a suitable refrigerant liquid level in the surge drum. Baudelot coolers using other common refrigerants are generally of the direct-expansion type, with thermostatic expansion valves.

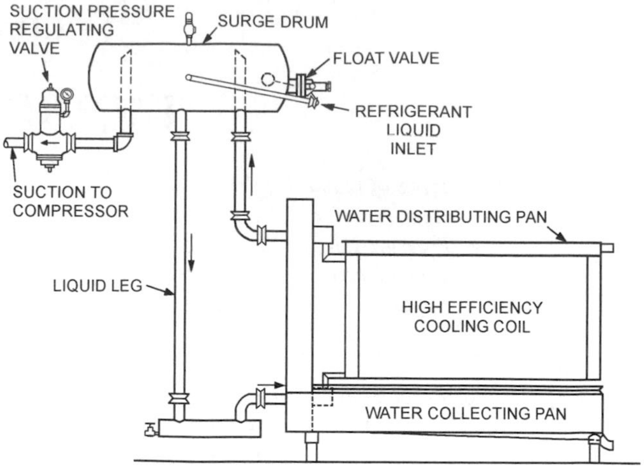

Fig. 3 Baudelot Cooler

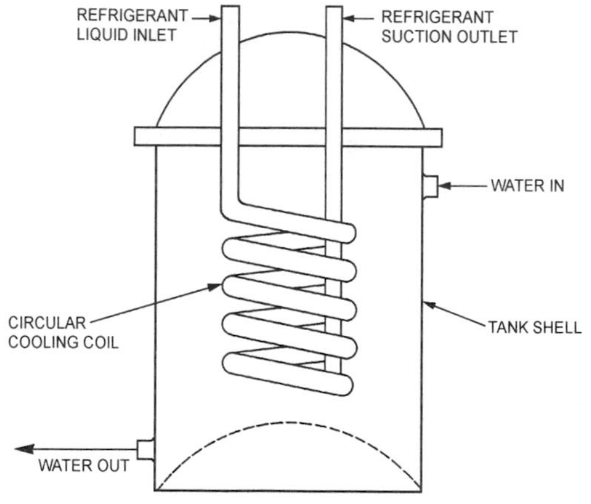

Fig. 4 Shell-and-Coil Cooler

Shell-and-Coil

A shell-and-coil cooler is a tank containing the fluid to be cooled with a simple coiled tube used to cool the fluid. This type of cooler has the advantage of cold water storage to offset peak loads. In some models, the tank can be opened for cleaning. Most applications are at low capacities (e.g., for bakeries, for photographic laboratories, and to cool drinking water).

The coiled tube containing the refrigerant can be either inside the tank (Figure 4) or attached to the outside of the tank in such a manner as to permit heat transfer.

HEAT TRANSFER

Heat transfer for liquid coolers can be expressed by the following steady-state heat transfer equation:

$$q = UA\Delta t_m \tag{1}$$

where

q = total heat transfer rate, Btu/h
Δt_m = mean temperature difference, °F
A = heat transfer surface area associated with U, ft²
U = overall heat transfer coefficient, Btu/h·ft²·°F

The area A can be calculated if the geometry of the cooler is known. Chapter 3 of the 1997 *ASHRAE Handbook—Fundamentals* describes the calculation of the mean temperature difference.

This chapter discusses the components of U, but not in depth. U may be calculated by one of the following equations.

Based on inside surface area

$$U = \frac{1}{1/h_i + [A_i/(A_o h_o)] + (t/k)(A_i/A_m) + r_{fi}} \quad (2)$$

Based on outside surface area

$$U = \frac{1}{[A_o/(A_i h_i)] + 1/h_o + (t/k)(A_o/A_m) + r_{fo}} \quad (3)$$

where

h_i = inside heat transfer coefficient based on inside surface area, Btu/h·ft^2·°F
h_o = outside heat transfer coefficient based on outside surface area, Btu/h·ft^2·°F
A_o = outside heat transfer surface area, ft^2
A_i = inside heat transfer surface area, ft^2
A_m = mean heat transfer area of metal wall, ft^2
k = thermal conductivity of heat transfer material, Btu/h·ft·°F
t = thickness of heat transfer surface (tube wall thickness), ft
r_{fi} = fouling factor of fluid side based on inside surface area, ft^2·h·°F/Btu
r_{fo} = fouling factor of fluid side based on outside surface area, ft^2·h·°F/Btu

Note: If fluid is on inside, multiply r_{fi} by A_o/A_i to find r_{fo}.
If fluid is on outside, multiply r_{fo} by A_i/A_o to find r_{fi}.

Heat Transfer Coefficients

The refrigerant-side coefficient usually increases under the following conditions: (1) increase in cooler load, (2) decrease in suction superheat, (3) decrease in oil concentration, and (4) increase in saturated suction temperature. The amount of increase or decrease varies, depending on the type of cooler. Schlager et al. (1989) discuss the effects of oil in direct-expansion coolers. Flooded coolers have a relatively small change in heat transfer coefficient as a result of a change in load, whereas a direct-expansion cooler shows a significant increase in heat transfer coefficient with an increase in load. A Wilson Plot of test data (McAdams 1954; Briggs and Young 1969) can show actual values for the refrigerant-side coefficient of a given cooler design. Webb (1994) developed additional information on refrigerant-side heat transfer coefficients.

The fluid-side coefficient is determined by cooler geometry, fluid flow rate, and fluid properties (viscosity, specific heat, thermal conductivity, and density) (Palen and Taborek 1969; Wolverine 1984). For a given fluid, the fluid-side coefficient increases with an increase in fluid flow rate due to increased turbulence and an increase in fluid temperature due to improvement of fluid properties as temperature increases.

The heat transfer coefficient in direct-expansion and flooded coolers increases significantly with fluid flow. The effect of flow is smaller for Baudelot and shell-and-coil coolers. Many of the listed references give additional information on fluid-side heat transfer coefficients.

An enhanced heat transfer surface can help increase the heat transfer coefficient of coolers in the following ways:

- It increases heat transfer area, thereby increasing overall heat transfer rate, even if refrigerant-side heat transfer coefficient is unchanged.
- Where the flow of fluid or refrigerant is low, it improves the heat transfer coefficient by increasing turbulence at the surface and mixing the fluid at the surface with fluid away from the surface.

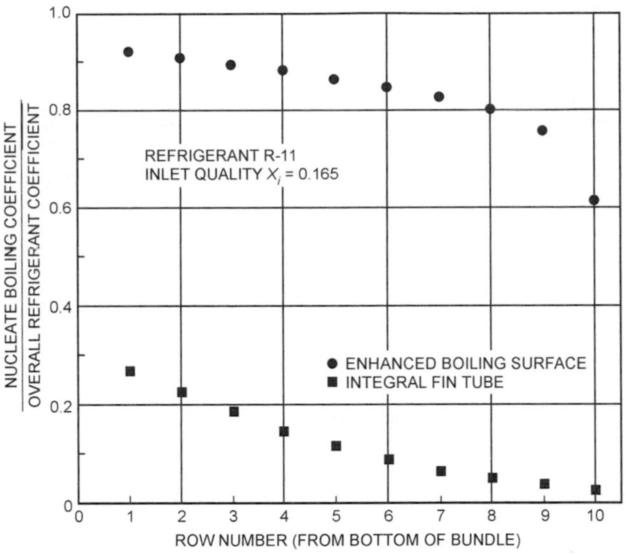

Fig. 5 Nucleate Boiling Contribution to Total Refrigerant Heat Transfer

- In flooded coolers, an enhanced refrigerant-side surface may provide more and better nucleation points to promote boiling of refrigerant.

Pais and Webb (1991) describe many enhanced surfaces used in flooded coolers. The enhanced surface geometries provide substantially higher boiling coefficients than do integral finned tubes. Nucleate pool boiling data are provided by Webb and Pais (1991). The boiling process that occurs in the tube bundle of a flooded cooler may be enhanced by forced-convection effects. This is basically an additive effect, in which the local boiling coefficient is the sum of the nucleate boiling coefficient and the forced-convection effect. Webb and Apparao (1989) describe a theoretical model to predict the performance of flooded coolers recommending row-by-row calculations.

Based on the model, Webb et al. (1990) present the results of calculations using a computer program. These results show some performance differences of various internal and external surface geometries. As an example, Figure 5 shows the contribution of nucleate pool boiling to the overall refrigerant heat transfer coefficient for an integral finned tube and an enhanced tube as a function of the tube row. The forced-convection effect predominates with the integral finned tube.

In ASHRAE research projects 725-RP and 668-RP, Chyu (1995) and Moeykens et al. (1995) studied performance of spray evaporation in ammonia and halocarbon refrigerant systems. Moeykens investigated shell-side heat transfer performance for commercially available enhanced surface tubes in a spray evaporation environment. The study determined that the spray evaporation heat transfer mode can yield shell-side heat transfer coefficients equal to or greater than those found with enhanced nucleate boiling surface tubes in the flooded boiling environment. Moeykens and Pate (1996) describe an enhancement to shell-side heat transfer performance generated with small concentrations of oil (<2.5%) in the spray evaporation mode. They attribute the improvement to foaming, which enhances heat transfer performance in the upper rows of large tube bundles operating in the flooded boiling mode.

Gupte and Webb (1995a, 1995b) investigated convective vaporization in triangular enhanced tube bundles. They proposed a modified Chen superposition model that predicts the overall convective/vaporization coefficient as the sum of the single tube nucleate pool boiling coefficient and a weighted contribution of a single-phase convective coefficient.

Fouling Factors

Most fluids over time foul the fluid-side heat transfer surface, thus reducing the overall heat transfer coefficient of the cooler. If fouling is expected to be a problem, a mechanically cleanable cooler should be used, such as a flooded, Baudelot, or cleanable direct-expansion tube-in-tube cooler. Direct-expansion shell-and-tube, shell-and-coil, and brazed plate coolers can be cleaned chemically. Flooded coolers and direct-expansion tube-in-tube coolers with enhanced fluid-side heat transfer surfaces have a tendency to be self-cleaning due to high fluid turbulence, so a smaller fouling factor can probably be used for these coolers. Water quality in closed chilled water loops has been studied as part of ASHRAE-sponsored research (560-RP). Haider et al. (1991) found little potential for fouling in such systems in a field survey. Experimental work with various tube geometries by Haider et al. (1992) confirmed that negligible fouling occurs in closed loop evaporator tubes at 3 to 5 fps and 7 fps water velocities. ARI *Standard* 480 discusses fouling calculations.

The refrigerant side of the cooler is not subject to fouling, and a fouling factor need not be included for that side.

Tube Wall Resistance

Typically, the t/k term in Equations (2) and (3) is negligible. However, with low thermal conductivity material or thick wall tubing, it may become significant. Refer to Chapter 3 of the 1997 *ASHRAE Handbook—Fundamentals* and to Chapter 35 of this volume for further details.

PRESSURE DROP

Fluid Side

Pressure drop is usually minimal in Baudelot and shell-and-coil coolers but must be considered in direct-expansion and flooded coolers. Both the direct-expansion and flooded coolers rely on turbulent fluid flow to improve heat transfer. This turbulence is obtained at the expense of pressure drop.

For air-conditioning service, the pressure drop is commonly limited to 10 psi to keep pump size and energy cost reasonable. For flooded coolers, see Chapter 35 for a discussion of pressure drop for flow in tubes. Pressure drop for fluid flow in shell-and-tube direct-expansion coolers depends greatly on tube and baffle geometry. The following equation projects the change in pressure drop due to a change in flow.

$$\text{New Pressure Drop} = \text{Original Pressure Drop}\left[\frac{\text{New rate}}{\text{Original rate}}\right]^{1.8} \quad (4)$$

Refrigerant Side

The refrigerant-side pressure drop must be considered for the following coolers: direct-expansion, shell-and-coil, brazed plate, and, sometimes, Baudelot. When there is a pressure drop on the refrigerant side, the refrigerant inlet and outlet pressures and corresponding saturated temperature are different. This difference causes a change in the mean temperature difference, which affects the total heat transfer rate. If the pressure drop is high, operation of the expansion valve may be affected due to reduced pressure drop across the valve. This pressure drop varies, depending on the refrigerant used, operating temperature, and type of tubing (Wallis 1969; Martinelli and Nelson 1948).

VESSEL DESIGN

Mechanical Requirements

Pressure vessels must be constructed and tested under the rules of national, state, and local codes. The introduction of the current ASME *Boiler and Pressure Vessel Code*, Section VIII, gives guidance on rules and exemptions.

The more common applicable codes and standards are as follows:

1. ARI *Standard* 480, Refrigerant-Cooled Liquid Coolers, Remote Type, covers industry criteria for standard equipment, standard safety provisions, marking, and recommended rating requirements.
2. ASHRAE *Standard* 24, Methods of Testing for Rating Liquid Coolers, covers the recommended testing methods for measuring the capacity of liquid coolers.
3. ASHRAE *Standard* 15, Safety Code for Mechanical Refrigeration, involves specific design criteria, use of materials, and testing. It refers to the ASME *Boiler and Pressure Vessel Code*, Section VIII, for refrigerant-containing sides of pressure vessels, where applicable. Factory test pressures are specified, and minimum design working pressures are given. This code requires pressure limiting and pressure relief devices on refrigerant-containing systems, as applicable, and defines the setting and capacity requirements for these devices.
4. ASME *Boiler and Pressure Vessel Code*, Section VIII, Unfired Pressure Vessels, covers the safety aspects of design and construction. Most states require coolers to meet ASME requirements if they fall within scope of the ASME code. Some of the exceptions from meeting the ASME requirements are as follows:
   - Pressure is 15 psig or less.
   - The fluid (water) portion of the cooler need not be built to the requirements of the ASME code if the fluid is water, the design pressure does not exceed 300 psig, and the design temperature does not exceed 210°F.

   Coolers meeting the requirements of the ASME code will have an ASME stamp, which is a *U* or *UM* inside a three-leaf clover. The *U* can be used for all coolers and the *UM* can be used for small coolers.
5. Underwriters Laboratories (UL) *Standard* 207, Refrigerant-Containing Components and Accessories, involves specific design criteria, use of materials, testing, and initial approval by Underwriters Laboratories. A cooler with the ASME *U* stamp does not require UL approval.

Design Pressure. On the refrigerant side, design pressure as a minimum should be the saturated pressure for the refrigerant used at 80°F as per ASHRAE *Standard* 15. Standby temperature and the temperature encountered during shipping of chillers with a refrigerant charge should also be considered.

Required fluid-side (water-side) pressure varies depending largely on the following conditions: (1) static pressure, (2) pump pressure, (3) transients due to pump start-up, and (4) valve closing.

Chemical Requirements

The following chemical requirements are given by Perry and Green (1984) and NACE (1974).

Refrigerant 717 (Ammonia). Carbon steel and cast iron are the most widely used materials for ammonia systems. Stainless steel alloys are satisfactory but more costly. Copper and high-copper alloys are avoided because they are attacked by ammonia when moisture is present. Aluminum and aluminum alloys may be used with caution with ammonia.

Halocarbon Refrigerants. Almost all the common metals and alloys are used satisfactorily with these refrigerants. Exceptions include magnesium and aluminum alloys containing more than 2% magnesium, where water may be present. Zinc is not recommended for use with Refrigerant 113; it is more chemically reactive than other common construction metals and, therefore, it is usually avoided when other halogenated hydrocarbons are used as a refrigerant. Under some conditions with moisture present, halocarbon refrigerants form acids that attack steel and even nonferrous metals. This problem does not commonly occur in properly cleaned and dehydrated systems. ASHRAE *Standard* 15, paragraph 7.1.2, states that aluminum, zinc, or magnesium shall not be used in contact with methyl chloride nor magnesium alloys with any halogenated refrigerant.

Water. Relatively pure water can be satisfactory with both ferrous and nonferrous metals. Brackish water, seawater, and some river water are quite corrosive to iron and steel and also to copper, aluminum, and many alloys of these metals. A reputable water consultant who knows the local water condition should be contacted. Chemical treatment by pH control, inhibitor applications, or both may be required. Where this is not feasible, more noble construction material or special coatings must be used. Aluminum should not be used in the presence of other metals in water circuits.

Brines. Ferrous metal and a few nonferrous alloys are almost universally used with sodium chloride and calcium chloride brines. Copper base alloys can be used if adequate quantities of sodium dichromate are added and caustic soda is used to neutralize the solution. Even with ferrous metal, these brines should be treated periodically to hold the pH value near the neutral point.

Ethylene glycol and propylene glycol are stable compounds that are less corrosive than chloride brines.

Electrical Requirements

When the fluid being cooled is electrically conductive, the system must be grounded to prevent electrochemical corrosion.

APPLICATION CONSIDERATIONS

Refrigerant Flow Control

Direct-Expansion Coolers. The constant superheat thermal expansion valve is the most common control used. It is located directly upstream of the cooler. A thermal bulb strapped to the suction line leaving the cooler senses refrigerant temperature. The valve can be adjusted to produce a constant suction superheat during steady operation.

The thermal expansion valve adjustment is commonly set at a suction superheat of 10°F, which is sufficient to ensure that liquid is not carried into the compressor. Direct-expansion cooler performance is affected greatly by superheat setting. Reduced superheat improves cooler performance; thus, suction superheat should be set as low as possible while avoiding liquid carryover to the compressor.

Flooded Coolers. As the name implies, flooded coolers must have good liquid refrigerant coverage of the tubes to achieve good performance. Liquid level control in a flooded cooler becomes the principal issue in flow control. Some systems are designed critically charged, so that when all the liquid refrigerant is delivered to the cooler, it is just enough for good tube coverage. In these systems, an orifice is often used as the throttling device between condenser and cooler.

A float valve is another method of flooded cooler control. A high-side float valve can be used, with the float sensing condenser liquid level, to drain the condenser. For more exact control of liquid level in the cooler, a low-side float valve is used, where the valve senses cooler liquid level and controls the flow of entering refrigerant.

Freeze Prevention

Freeze prevention must be considered for coolers operating near the freezing point of the fluid. Freezing of the fluid in some coolers causes extensive damage. Two methods can be used for freeze protection: (1) hold the saturated suction pressure above the fluid freezing point or (2) shut the system off if the temperature of the fluid approaches its freezing point.

A suction pressure regulator can hold the saturated suction pressure above the freezing point of the fluid. A low-pressure cutout can shut the system off before the saturated suction pressure drops to below the freezing point of the fluid. The leaving fluid temperature can be monitored to cut the system off before a danger of freezing, usually about 10°F above the fluid freezing temperature. It is recommended that both methods be used.

Baudelot, shell-and-coil, brazed plate, and direct-expansion shell-and-tube coolers are all somewhat resistant to damage caused by freezing and ideal for applications where freezing may be a problem.

If a cooler is installed in an unconditioned area, possible freezing due to low ambient temperature must be considered. If the cooler is used only when the ambient temperature is above freezing, the fluid should be drained from the cooler for cold weather. As an alternate to draining, if the cooler is in use year-round, the following methods can be used:

- A heat tape or other heating device can be used to keep the cooler above freezing.
- For water, adding an appropriate amount of ethylene glycol prevents freezing.
- Continuous pump operation may also prevent freezing.

Oil Return

Most compressors discharge a small percentage of oil in the discharge gas. This oil mixes with the condensed refrigerant in the condenser and flows to the cooler. Because the oil is nonvolatile, it does not evaporate and may collect in the cooler.

In direct-expansion coolers, the gas velocity in the tubes and the suction gas header is usually sufficient to carry the oil from the cooler into the suction line. From there, with proper piping design, it can be carried back to the compressor. At light load and low temperature, oil may gather in the superheat section of the cooler, detracting from performance. For this reason, operation of refrigerant circuits at light load for long periods should be avoided, especially under low-temperature conditions.

In flooded coolers, vapor velocity above the tube bundle is usually insufficient to return oil up the suction line, and oil tends to accumulate in the cooler. With time, depending on compressor oil loss rate, the oil concentration in the cooler may become large. When concentration exceeds about 1%, heat transfer performance may be adversely affected if enhanced tubing is used.

It is common in flooded coolers to take some oil-rich liquid and return it to the compressor on a continuing basis, in order to establish a rate of return equal to compressor oil loss rate.

Maintenance

Maintenance of coolers centers around two areas: (1) safety and (2) cleaning of the fluid side. The cooler should be inspected periodically for any weakening of its pressure boundaries. The inspection should include visual inspection for corrosion, erosion, and any deformities. Any pressure relief device should also be inspected. The insurer of the cooler may require regular inspection. If the fluid side is subjected to fouling, it may require periodic cleaning. Cleaning may be by either mechanical or chemical means. The manufacturer or service organization experienced in cooler maintenance should have details for cleaning.

Insulation

In addition, a cooler operating at a saturated suction temperature lower than the dew point of the surrounding air should be insulated to prevent condensation. Chapter 22 of the 1997 *ASHRAE Handbook—Fundamentals* describes insulation in more detail.

REFERENCES

ARI. 1995. Refrigerant-cooled liquid coolers, remote type. *Standard* 480-95. Air-Conditioning and Refrigeration Institute, Arlington, VA.

ASHRAE. 1989. Methods of testing for rating liquid coolers. ANSI/ASHRAE *Standard* 24-1989.

ASHRAE. 1994. Safety code for mechanical refrigeration. ANSI/ASHRAE *Standard* 15-1994.

ASME. 1998. Rules for construction of pressure vessels. ANSI/ASME *Boiler and Pressure Vessel Code*, Section VIII-98. American Society of Mechanical Engineers, New York.

Ayub, Z.H. and S.A. Jones. 1987. Tubeside erosion/corrosion in heat exchangers. *Heating/Piping/Air Conditioning* (December):81.

Briggs, D.E. and E.H. Young. 1969. Modified Wilson Plot techniques for obtaining heat transfer correlations for shell and tube heat exchangers. *Chemical Engineering Symposium Series* 65(92):35-45.

Chyu, M. 1995. Nozzle-sprayed flow rate distribution on a horizontal tube bundle. *ASHRAE Transactions* 101(2):443-53.

Gupte, N.S. and R.L. Webb. 1995a. Shell-side boiling in flooded refrigerant evaporators—Part I: Integral finned tubes. *International Journal of HVAC&R Research* 1(1):35-47.

Gupte, N.S. and R.L. Webb. 1995b. Shell-side boiling in flooded refrigerant evaporators—Part II: Enhanced tubes. *International Journal of HVAC&R Research* 1(1):48-60.

Haider, S.I., R.L. Webb, and A.K. Meitz. 1991. A survey of water quality and its effect on fouling in flooded water chiller evaporators. *ASHRAE Transactions* 97(1):55-67.

Haider, S.I., R.L. Webb, and A.K. Meitz. 1992. An experimental study of tube-side fouling resistance in water-chiller-flooded evaporators. *ASHRAE Transactions* 98(2):86-103.

Martinelli, R.C. and D.B. Nelson. 1948. Prediction of pressure drop during forced circulation boiling of water. ASME *Transactions* (August):695.

McAdams, W.H. 1954. *Heat transmission*, 3rd ed. McGraw-Hill, New York.

Moeykens, S.A., B.J. Newton, and M.B. Pate. 1995. Effects of surface enhancement, film-feed supply rate, and bundle geometry on spray evaporation heat transfer performance. *ASHRAE Transactions* 101(2): 408-19.

Moeykens, S.A. and M.B. Pate. 1996. Effects of lubricant on spray evaporation heat transfer performance of R-134a and R-22 in tube bundles. *ASHRAE Transactions* 102(1):410-26.

NACE. 1974. *Corrosion data survey*, 5th ed. Compiled by N.E. Hamner for the National Association of Corrosion Engineers, Houston, TX.

Pais, C. and R.L. Webb. 1991. Literature survey of pool boiling on enhanced surfaces. *ASHRAE Transactions* 97(1):79-89.

Palen, J.W. and J. Taborek. 1969. Solution of shell side pressure drop and heat transfer by stream analysis method. *Chemical Engineering Progress Symposium Series* 65(92).

Perry, J.H. and R.H. Green. 1984. *Chemical engineers handbook*, 6th ed. McGraw-Hill, New York.

Schlager, L.M., M.B. Pate, and A.E. Bergles. 1989. A comparison of 150 and 300 SUS oil effects on refrigerant evaporation and condensation in a smooth tube and a micro-fin tube. *ASHRAE Transactions* 95(1).

Sturley, R.A. 1975. Increasing the design velocity of water and its effect on copper tube heat exchangers. Paper No. 58, The International Corrosion Forum, Toronto, Canada.

UL. 1993. Refrigerant-containing components and accessories, nonelectrical. ANSI/UL *Standard* 207-93. Underwriters Laboratories, Northbrook, IL.

Wallis, G.B. 1969. *One dimensional two phase flow*. McGraw-Hill, New York.

Webb, R.L. 1994. *Principles of enhanced heat transfer.* John Wiley and Sons, New York.

Webb, R.L. and T. Apparao. 1990. Performance of flooded refrigerant evaporators with enhanced tubes. *Heat Transfer Engineering* 11(2):29-43.

Webb, R.L. and C. Pais. 1991. Pool boiling data for five refrigerants on three tube geometries. *ASHRAE Transactions* 97(1):72-78.

Webb, R.L., K.-D. Choi, and T. Apparao. 1989. A theoretical model for prediction of the heat load in flooded refrigerant evaporators. *ASHRAE Transactions* 95(1):326-38.

Wolverine Tube, Inc. 1984. *Engineering data book II*.

CHAPTER 38

LIQUID CHILLING SYSTEMS

A LIQUID chilling system cools water, brine, or other secondary coolant for air conditioning or refrigeration. The system may be either factory assembled and wired or shipped in sections for erection in the field. The most frequent application is water chilling for air conditioning, although both brine cooling for low-temperature refrigeration and chilling of fluids in industrial processes are also common.

The basic components of a vapor-compression, liquid chilling system include a compressor, a liquid cooler (evaporator), a condenser, a compressor drive, a liquid refrigerant expansion or flow-control device, and a control center; the system may also include a receiver, an economizer, an expansion turbine, and/or a subcooler. In addition, certain auxiliary components may be used, such a lubricant cooler, lubricant separator, lubricant-return device, purge unit, lubricant pump, a refrigerant transfer unit, refrigerant vents, and/or additional control valves.

GENERAL CHARACTERISTICS

PRINCIPLES OF OPERATION

Liquid (usually water) enters the cooler, where it is chilled by liquid refrigerant evaporating at a lower temperature. The refrigerant vaporizes and is drawn into the compressor, which increases the pressure and temperature of the gas so that it may be condensed at the higher temperature in the condenser. The condenser cooling medium is warmed in the process. The condensed liquid refrigerant then flows back to the evaporator through an expansion device. A fraction of the liquid refrigerant changes to vapor (flashes) as the pressure drops between the condenser and the evaporator. Flashing cools the liquid to the saturated temperature at the evaporator pressure. It produces no refrigeration effect in the cooler. The following modifications (sometimes combined for maximum effect) reduce flash gas and increase the net refrigeration effect per unit of power consumption.

Subcooling. Condensed refrigerant may be subcooled to a temperature below its saturated condensing temperature in either the subcooler section of a water-cooled condenser or a separate heat exchanger. Subcooling reduces the amount of flashing and increases the refrigeration effect in the chiller.

Economizing. This process can occur either in a direct-expansion (DX), an expansion turbine, or a flash-type system. In a **DX system**, the main liquid refrigerant is usually cooled in the shell of a shell-and-tube heat exchanger, at condensing pressure, from the saturated condensing temperature to within several degrees of the intermediate saturated temperature. Before cooling, a small portion of the liquid flashes and evaporates in the tube side of the heat exchanger to cool the main liquid flow. Although subcooled, the liquid will still be at the condensing pressure.

An **expansion turbine** extracts rotating energy as a portion of the refrigerant vaporizes. As in the DX system, the remaining liquid is supplied to the cooler at the intermediate pressure.

In a **flash-type system**, the entire liquid flow is expanded to the intermediate pressure in a vessel that supplies liquid to the cooler at the saturated intermediate pressure; however, the liquid is at the intermediate pressure.

In any case, the flash gas enters the compressor at either an intermediate stage of a multistage centrifugal compressor, at the intermediate stage of an integral two-stage reciprocating compressor, at an intermediate pressure port of a screw compressor, or at the inlet of a high-pressure stage on a multistage reciprocating or screw compressor.

Liquid Injection. Condensed liquid is throttled to the intermediate pressure and injected into the second-stage suction of the compressor to prevent excessively high discharge temperatures and, in the case of centrifugal machines, to reduce noise. In the case of screw compressors, condensed liquid is injected into a port fixed at slightly below discharge pressure to provide lubricant cooling.

COMMON LIQUID CHILLING SYSTEMS

Basic System

The refrigeration cycle of a basic system is shown in Figure 1. Chilled water enters the cooler at 54°F, for example, and leaves at 44°F. Condenser water leaves a cooling tower at 85°F, enters the condenser, and returns to the cooling tower near 95°F. Condensers may also be cooled by air or through evaporation of water. This system, with a single compressor and one refrigerant circuit with a water-cooled condenser, is used extensively to chill water for air conditioning because it is relatively simple and compact.

The preparation of this chapter is assigned to TC 8.1, Positive Displacement Compressors, and TC 8.2, Centrifugal Machines.

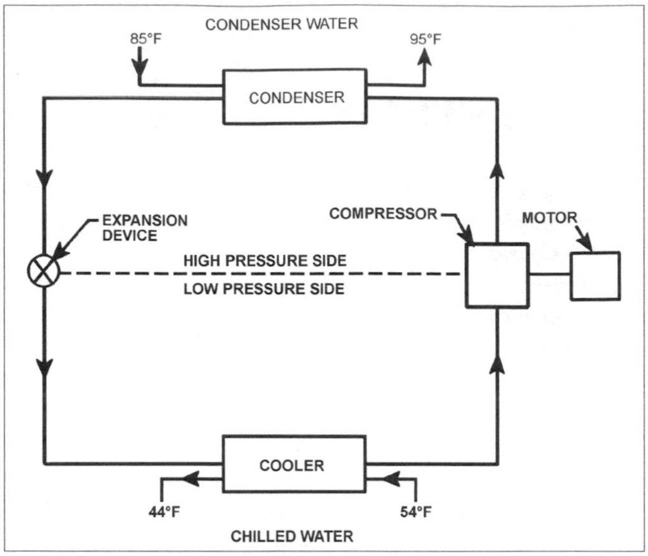

Fig. 1 Equipment Diagram for Basic Liquid Chiller

Multiple Chiller System

A multiple chiller system has two or more chillers connected by parallel or series piping to a common distribution system. Multiple chillers offer operational flexibility, standby capacity, and less disruptive maintenance. The chillers can be sized to handle a base load and increments of a variable load to allow each chiller to operate at its most efficient point.

Multiple chiller systems offer some standby capacity if repair work must be done on one chiller. Starting in-rush current is reduced, as well as power costs at partial-load conditions. Maintenance can be scheduled for one chilling machine during part-load times, and sufficient cooling can still be provided by the remaining unit(s). These advantages require an increase in installed cost and space, however.

Water should flow constantly through the chillers for stable control. Load variation is temperature-related and is easily detected by temperature controls. In contrast, when water flow varies the load becomes flow related. Because a temperature control system cannot sense a variation in flow, it is unable to maintain stable control. However, some applications do have variable water flow through the cooling coils. In this case, a decoupled system is typically used to separate the distribution pumping from the production pumping. It allows variable flow through the cooling coils but maintains constant water flow through the chillers, allowing good control of the multiple chillers.

A typical decoupled system is shown in Figure 2. The multiple pumps are connected by a bypass pipe that connects the return and supply headers. Each chiller-pump combination operates independently from the remaining chillers. Capacity control is simplified and as if each chiller operated alone. Instead of using temperature as an indicator of demand, relative flow is the indicator. If greater flow is demanded than that supplied by the chiller-pumps, return water is forced through the bypass into the supply header. This flow indicates a need for additional chiller capacity and another chiller-pump starts. Bypass flow in the opposite direction indicates overcapacity and the chiller-pumps are turned off.

Two basic multiple chiller systems are used: **parallel** and **series chilled water flow**. In the **parallel arrangement**, liquid to be chilled is divided among the liquid chillers; the multiple chilled streams are combined again in a common line after chilling. As the cooling load decreases, one unit may be shut down. Unless water flow is stopped through the inoperative chiller, the remaining unit(s) provide colder-than-design chilled liquid. The combined streams

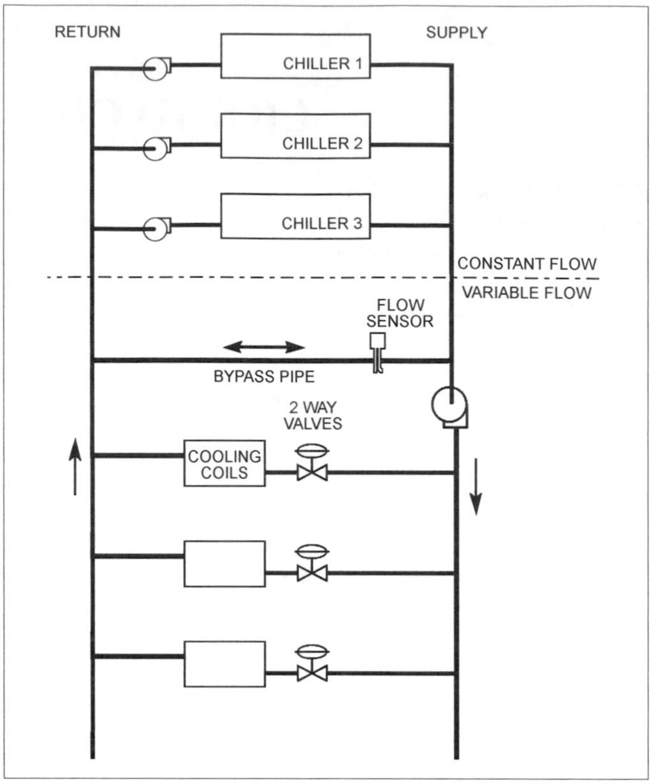

Fig. 2 Decoupled System

(including one from the idle chiller) then supply the chilled water at the design temperature in the common line.

When the design chilled water temperature is above about 45°F, all units should be controlled by the combined exit water temperature or by the return water temperature (RWT), since overchilling will not cause dangerously low water temperature in the operating machine(s). Chilled water temperature can be used to cycle one unit off when it drops below a capacity that can be matched by the remaining units.

When the design chilled water temperature is below about 45°F, each machine should be controlled by its own chilled water temperature, both to prevent dangerously low evaporator temperatures and to avoid frequent shutdowns by the low-temperature cutout. In this case, the temperature differential setting of the RWT must be adjusted carefully to prevent short cycling caused by the step increase in chilled water temperature when one chiller is cycled off. These control arrangements are shown in Figures 3 and 4.

In the **series arrangement**, the chilled liquid pressure drop may be higher if shells with fewer liquid-side passes or baffles are not available. No overchilling by either unit is required, and compressor power consumption is lower than it is for the parallel arrangement at partial loads. Because the evaporator temperature never drops below the design value (because no overchilling is necessary), the chances of evaporator freeze-up are minimized. However, the chiller should still be protected by a low-temperature safety control.

Water cooled condensers in series are best piped in a counterflow arrangement so that the lead machine is provided with warmer condenser and chilled water and the lag machine is provided with colder entering condenser and chilled water. Refrigerant compression for each unit is nearly the same. If about 55% of design cooling capacity is assigned to the lead machine and about 45% to the lag machine, identical units can be used. In this way, either machine can provide the same standby capacity if the other is down, and lead and lag machines may be interchanged to equalize the number of operating hours on each.

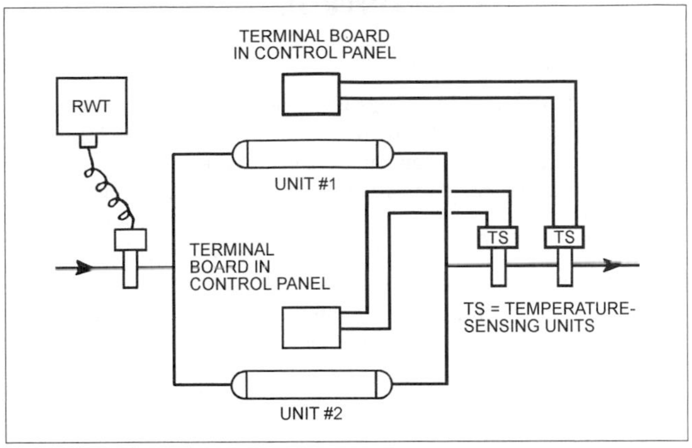

Fig. 3 Parallel Operation High Design Water Leaving Coolers (Approximately 45°F and Above)

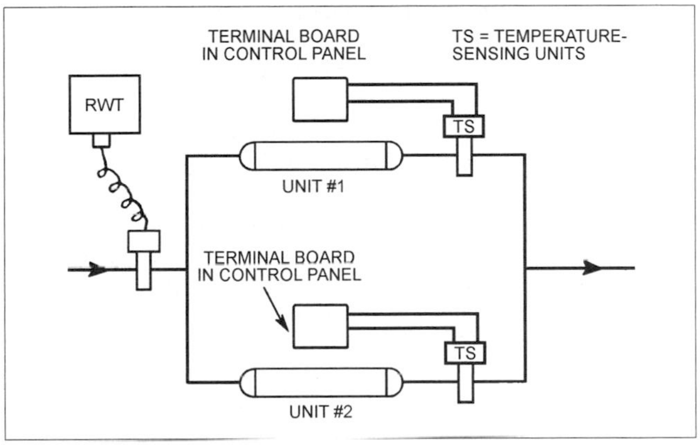

Fig. 4 Parallel Operation Low Design Water Leaving Coolers (Below Approximately 45°F)

A control system for two machines in series is shown in Figure 5. (On reciprocating chillers, RWT sensing is usually used instead of leaving water sensing because it allows closer temperature control.) Both units are modulated to a certain capacity; then, one unit shuts down, leaving less than 100% load on the operating machine.

One machine should be shut down as soon as possible, with the remaining unit carrying the full load. This not only reduces the number of operating hours on a unit, but also leads to less total power consumption because the COP tends to decrease below the full load value when unit load drops much below 50%.

Heat-Recovery Systems

Any building or plant requiring the simultaneous operation of heat-producing and cooling equipment has the potential for a heat-recovery installation. Heat-recovery equipment should be considered for all new or retrofit installations. In some cases, the installed cost may be less because of the elimination or reduction of both heating equipment and the space required for it.

Heat-recovery systems extract heat from chilled liquid and reject some of that heat, plus the energy of compression, to a warm-water circuit for reheat or heating. Air-conditioned spaces thus furnish heating for other spaces in the same building. During the full-cooling season, all heat must be rejected outdoors, usually by a cooling tower. During spring or fall, some heat is required inside, while a portion of the heat extracted from the air-conditioned spaces must be rejected outside simultaneously.

Heat recovery offers a low heating cost and reduces space requirements for equipment. The control system must be designed carefully, however, to take the greatest advantage of recovered heat and to maintain proper temperature and humidity in all parts of the building. Chapter 8 covers balanced heat-recovery systems.

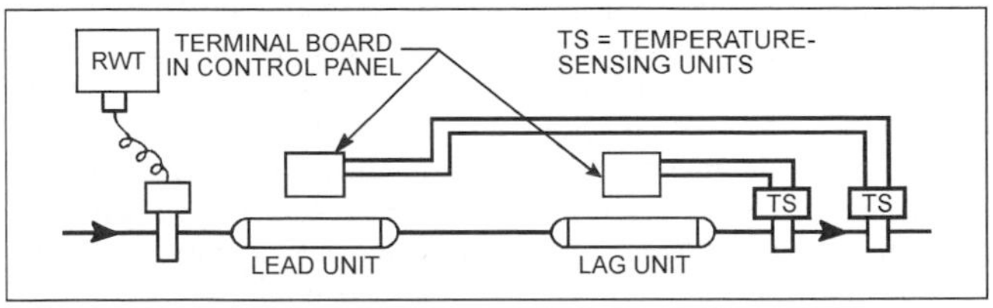

Fig. 5 Series Operation

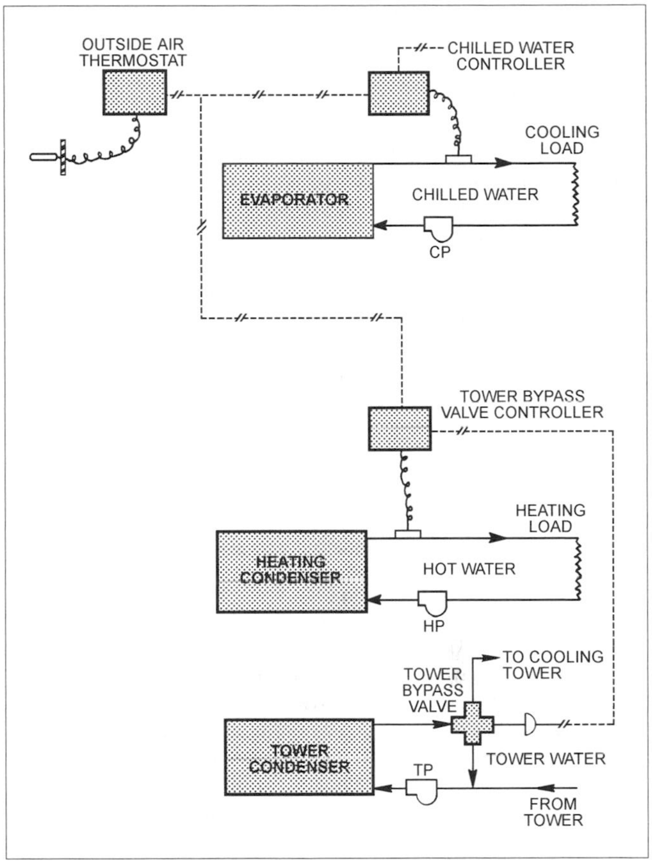

Fig. 6 Heat-Recovery Control System

Since cooling tower water is not satisfactory for heating coils, a separate, closed warm-water circuit with another condenser bundle or auxiliary condenser, in addition to the main water chiller condenser, must be provided. In some cases, it is economically feasible to use a standard condenser and a closed-circuit water cooler.

A suggested control scheme is shown in Figure 6. The heating water temperature is controlled by a cooling tower bypass valve, which modulates the flow of condenser cooling water to the tower. An outside air thermostat resets the hot water control point upward as the outdoor temperature drops and resets the chilled water temperature control point upward on colder days. In this way, extra power is not consumed unnecessarily by the compressor in attempting to maintain summer design coil temperatures during dry, cold outdoor conditions.

EQUIPMENT SELECTION

The largest factor that determines total liquid chiller owning cost is the cooling load size; therefore, the total required chiller capacity should be calculated accurately. The practice of adding 10 to 20% to load estimates is unnecessary because of the availability of accurate load estimating methods, and it proportionately increases costs related to equipment purchase, installation, and the poor efficiency resulting from wasted power. Oversized equipment can also cause operational difficulties such as frequent on-off cycling or surging of centrifugal machines at low loads. The penalty for a small underestimation of cooling load, however, is not serious. On the few design load days of the year, an increase in chilled liquid temperature is often acceptable. However, for some industrial or commercial loads, a safety factor can be added to the load estimate.

The life-cycle cost as discussed in Chapter 35 of the 1999 *ASHRAE Handbook—Applications* should be used to minimize the overall purchase and operating costs. Total owning cost is composed of the following:

- **Equipment price.** Each machine type and/or manufacturer's model should include all the necessary auxiliaries such as starters and vibration mounts. If these are not included, their price should be added to the base price. Associated equipment, such as condenser water pump, tower, and piping, should be included.
- **Installation cost.** Factory-packaged machines are both less expensive to install and usually considerably more compact, resulting in space savings. The cost of field assembly of field-erected chillers must also be evaluated.
- **Energy cost.** Using an estimated load schedule and part-load power consumption curves furnished by the manufacturer, a year's energy cost should be calculated.
- **Water cost.** With water cooled towers, the cost of acquisition, water treatment, tower blowdown, and overflow water should be included.
- **Maintenance cost.** Each bidder may be asked to quote on a maintenance contract on a competitive basis.
- **Insurance and taxes.**

For package chillers that include heat recovery, system cost and performance should be compared in addition to equipment costs. For example, the heat-recovery chiller installed cost should be compared with the installed cost of a chiller plus a separate heating system. The following factors should also be considered: (1) energy costs, (2) maintenance requirements, (3) life expectancy of equipment, (4) standby arrangement, (5) relationship of heating to cooling loads, (6) effect of package selection on sizing, and (7) type of peripheral equipment.

Condensers and coolers are often available with either **liquid heads**, which require the water pipes to be disconnected for tube access and maintenance, or **marine-type water boxes**, which permit tube access with water piping intact. The liquid head is considerably lower in price. The cost of disconnecting piping must be greater than the additional cost of marine-type water boxes to justify their use. Typically, an elbow and union or flange connection is installed only to facilitate the removal of heads.

The following types of liquid chillers are generally used for air conditioning:

Up to 25 tons	—	Reciprocating or scroll
25 to 80 tons	—	Screw, reciprocating, or scroll
80 to 450 tons	—	Screw, reciprocating, or centrifugal
200 to 1000 tons	—	Screw or centrifugal
Above 1000 tons	—	Centrifugal

For air-cooled condenser duty, brine chilling, or other high pressure applications from 80 to about 200 tons, reciprocating and screw liquid chillers are more frequently installed than centrifugals. Centrifugal liquid chillers (particularly multistage machines), however, may be applied quite satisfactorily at high pressure conditions.

Factory packages are available to about 2400 tons and field-assembled machines to about 10,000 tons.

CONTROL

Liquid Chiller Controls

The **chilled liquid temperature sensor** sends an air pressure (pneumatic control) or electrical signal (electronic control) to the control circuit, which then modulates compressor capacity in response to leaving or return chilled liquid temperature change from its set point.

Compressor capacity adjustment is accomplished differently on the following liquid chillers:

Reciprocating chillers use combinations of cylinder unloading and on-off compressor cycling of single or multiple compressors.

Centrifugal liquid chillers, driven by electric motors, commonly use adjustable prerotation vanes, which are sometimes combined with movable diffuser walls. Turbine and engine drives and inverter-driven, variable-speed electric motors allow the use of speed control in addition to prerotation vane modulation, reducing power consumption at partial loads.

Screw compressor liquid chillers include a slide valve that adjusts the length of the compression path. Inverter-driven, variable-speed electric motors and turbine and engine drives can also modulate screw compressor speed to control capacity.

In air-conditioning applications, most centrifugal and screw compressor chillers modulate from 100% to approximately 10% load. Although relatively inefficient, hot-gas bypass can be used to reduce capacity to nearly 0% with the unit in operation.

Reciprocating chillers are available with simple on-off cycling control in small capacities and with multiple steps of unloading down to 12.5% in the largest multiple compressor units. Most intermediate sizes provide unloading to 50, 33, or 25% capacity. Hot-gas bypass can reduce capacity to nearly 0%.

The **water temperature controller** is a thermostatic device that unloads or cycles the compressor(s) when the cooling load drops below minimum unit capacity. An *antirecycle timer* is sometimes used to limit starting frequency.

On centrifugal or screw compressor chillers, a **current limiter** or **demand limiter** limits compressor capacity during periods of possible high power consumption (such as pulldown) to prevent current draw from exceeding the design value; such a limiter can be set to limit demand, as described in the section on Centrifugal Liquid Chillers.

Controls That Influence the Liquid Chiller

Condenser cooling water may need to be controlled to regulate condenser pressure. Normally, the temperature of the water leaving a cooling tower can be controlled by fans, dampers, or a water bypass around the tower. Bypass around the tower allows the water velocity through the condenser tubes to be maintained, which prevents low-velocity fouling.

A flow-regulating valve is another common means of control. The orifice of this valve modulates in response to condenser pressure. For example, a reduction in pressure decreases the water flow, which, in turn, raises the condenser pressure to the desired minimum level.

For air-cooled or evaporative condensers, compressor discharge pressure can be controlled by cycling fans, shutting off circuits, or flooding coils with liquid refrigerant to reduce the heat transfer.

A reciprocating chiller usually has a thermal expansion valve, which requires a restricted range of pressure to avoid starving the evaporator (at low pressure).

An expansion valve(s) usually controls a screw compressor chiller. Cooling tower water temperature can be allowed to fall with decreasing load from the design condition to the chiller manufacturer's recommended minimum limit.

Screw compressor chillers above 150 tons may use flooded-type evaporators and evaporator liquid refrigerant controls similar to those used on centrifugal chillers.

A thermal expansion valve may control a centrifugal chiller at low capacities, while higher capacity machines employ a high-pressure float, orifice(s), or even a low-side float valve to control refrigerant liquid flow to the cooler. These latter types of controls allow relatively low condenser pressures, particularly at partial loads. Also, a centrifugal machine may surge if pressure is not reduced when cooling load decreases. In addition, low pressure reduces compressor power consumption and operating noise. For these reasons, in a centrifugal installation, cooling tower water temperature should be allowed to fall naturally with decreasing load and wet-bulb temperature, except that the liquid chiller manufacturer's recommended minimum limit must be observed.

Safety Controls

Some or all of the cutouts listed below may be provided in a liquid chilling package to stop the compressor(s) automatically. Cutouts may be manual or automatic reset.

- **High condenser pressure.** This pressure switch opens if the compressor discharge pressure exceeds the value prescribed in ASHRAE *Standard* 15.
- **Low refrigerant pressure (or temperature).** This device opens when evaporator pressure (or temperature) reaches a minimum safe limit.
- **High lubricant temperature.** This device protects the compressor if loss of lubricant cooling occurs or if a bearing failure causes excessive heat generation.
- **High motor temperature.** If loss of motor cooling or overloading because of a failure of a control occurs, this device shuts down the machine. It may consist of direct-operating bimetallic thermostats, thermistors, or other sensors embedded in the stator windings; it may be located in the discharge gas stream of the compressor.
- **Motor overload.** Some small, reciprocating compressor hermetic motors may use a directly operated overload in the power wiring to the motor. Some larger motors use pilot-operated overloads. Centrifugal and screw compressor motors generally use starter overloads or current-limiting devices to protect against overcurrent.
- **Low lubricant sump temperature.** This switch is used either to protect against a lubricant heater failure or to prevent starting after a prolonged shutdown before the lubricant heaters have had time to drive off refrigerant dissolved in the lubricant.
- **Low lubricant pressure.** To protect against clogged lubricant filters, blocked lubricant passageways, loss of lubricant, or a lubricant pump failure, a switch shuts down the compressor when lubricant pressure drops below a minimum safe value or if sufficient lubricant pressure is not developed shortly after the compressor starts.
- **Chilled liquid flow interlock.** This device may not be furnished with the liquid chilling package, but it is needed in the external piping to protect against a cooler freeze-up in case the liquid stops flowing. An electrical interlock is typically installed.
- **Condenser water flow interlock.** This device, which is similar to the chilled liquid flow interlock, is sometimes used in the external piping.
- **Low chilled liquid temperature.** Sometimes called **freeze protection**, this cutout operates at a minimum safe value of leaving chilled liquid temperature to prevent cooler freeze-up in the case of an operating control malfunction.
- **Relief valves.** In accordance with ASHRAE *Standard* 15, relief valves, rupture disks, or both, set to relieve at the shell design working pressure, must be provided on most pressure vessels or on piping connected to the vessels. Fusible plugs may also be used in some locations. Pressure relief devices should be vented outdoors or to the low-pressure side, in accordance with regulations or the standard.

STANDARDS

ARI *Standards* 550 and 590 provide guidelines for the rating of centrifugal and reciprocating liquid chilling machines, respectively. The design and construction of refrigerant pressure vessels are governed by the ASME *Boiler and Pressure Vessel Code*, Section VIII, except when design working pressure is 15 psig or less (as is usually the case for R-123 liquid chilling machines). The water-side design and construction of a condenser or cooler is not within the scope of the ASME code unless the design pressure is greater than 300 psi or the design temperature is greater than 210°F.

ASHRAE *Standard* 15 applies to all liquid chillers and new refrigerants on the market. New standards for equipment rooms are included. Methods for the measurement of unit sound levels are described in ARI *Standard* 575.

All tests of reciprocating liquid chillers for rating or verification of rating should be conducted in accordance with ASHRAE *Standard* 30. Centrifugal or screw liquid chiller ratings should be derived and verified by test in accordance with ARI *Standard* 550.

GENERAL MAINTENANCE

The following maintenance specifications apply to reciprocating, centrifugal, and screw chillers. The equipment should be neither overmaintained nor neglected. A preventive maintenance schedule should be established; the items covered can vary with the nature of the application. The list is intended as a guide; in all cases, the manufacturer's specific recommendation should be followed.

Continual Monitoring

- Condenser water treatment—treatment is determined specifically for the condenser water used.
- Operating conditions—daily log sheets should be kept to indicate trends and provide an advanced notice of deteriorating chillers.
- Brine quality for concentration and corrosion inhibitor levels.

Periodic Checks

- Leak check
- Purge operation
- System dryness
- Lubricant level
- Lubricant filter pressure drop
- Refrigerant quantity or level
- System pressures and temperatures
- Water flows
- Expansion valves operation

Regularly Scheduled Maintenance

- Condenser and lubricant cooler cleaning
- Evaporator cleaning on open systems
- Calibrating pressure, temperature, and flow controls
- Tightening wires and power connections
- Inspection of starter contacts and action
- Safety interlocks
- Dielectric checking of hermetic and open motors
- Tightness of hot gas valve
- Lubricant filter and drier change
- Analysis of lubricant and refrigerant
- Seal inspection
- Partial or complete valve or bearing inspection, as per manufacturer's recommendations
- Vibration levels

Extended Maintenance Checks

- Compressor guide vanes and linkage operation and wear
- Eddy current inspection of heat exchanger tubes
- Compressor teardown and inspection of rotating components
- Other components as recommended by manufacturer

RECIPROCATING LIQUID CHILLERS

EQUIPMENT DESCRIPTION

Components and Their Function

The reciprocating compressor described in Chapter 34 is a positive-displacement machine that maintains fairly constant volume flow rate over a wide range of pressure ratios. The following types of compressors are commonly used in liquid chilling machines:

- Welded hermetic, to about 25 tons chiller capacity
- Semihermetic, to about 200 tons chiller capacity
- Direct-drive open, to about 450 tons chiller capacity

Open motor-driven liquid chillers are usually more expensive than hermetically sealed units, but they can be more efficient. Hermetic motors are generally suction gas cooled; the rotor is mounted on the compressor crankshaft.

Condensers may be evaporative, air- or water-cooled. Water-cooled versions may be either tube-in-tube, shell-and-coil, shell-and-tube, or plate type heat exchangers. Most shell-and-tube condensers can be repaired, while others must be replaced if a refrigerant-side leak occurs.

Air-cooled condensers are much more common than evaporative condensers. Less maintenance is needed for air-cooled heat exchangers than for the evaporative type. Remote condensers can be applied with condenserless packages. (Information on condensers can be found in Chapter 35)

Coolers are usually direct-expansion, in which refrigerant evaporates while it is flowing inside tubes and chilled liquid is cooled as it is guided several times over the outside of the tubes by shell-side baffles. Flooded coolers are sometimes used on industrial chillers. Flooded coolers maintain a level of refrigerant liquid on the shell side of the cooler, while the liquid to be cooled flows through tubes inside the cooler. Tube-in-tube coolers are sometimes used with small machines; they offer low cost when repairability and installation space are not important criteria. (Chapter 37 describes coolers in more detail)

The **thermal expansion** valve, capillary, or other expansion device modulates refrigerant flow from the condenser to the cooler to maintain enough suction superheat to prevent any unevaporated refrigerant liquid from reaching the compressor. Excessively high values of superheat are avoided so that unit capacity is not reduced. (For additional information, see Chapter 45 in the 1998 *ASHRAE Handbook—Refrigeration.*)

Lubricant cooling is not usually required for air conditioning. However, if lubricant cooling is necessary, a refrigerant-cooled coil in the crankcase or a water-cooled cooler may be used. Lubricant coolers are often used in conjunction with applications that have a low suction temperature or high-pressure ratio when extra lubricant cooling is needed.

Capacities and Types Available

Available capacities range from about 2 to 450 tons. Multiple reciprocating compressor units have become popular for the following reasons:

- The number of capacity increments is greater, resulting in closer liquid temperature control, lower power consumption, less current in-rush during starting, and extra standby capacity.
- Multiple refrigerant circuits are used, resulting in the potential for limited servicing or maintenance of some components while maintaining cooling.

Selection of Refrigerant

R-12 and R-22 have been the primary refrigerants used in chiller applications. CFC-12 has been replaced with HFC-134a, which has similar properties. However, R-134a requires synthetic lubricants because it is not miscible with mineral oils. R-134a is suitable for both open and hermetic compressors.

R-22 provides greater capacity than R-134a for a given compressor displacement. R-22 is used for most open and hermetic compressors, but as an HCFC, it is scheduled for phaseout in the future. R-717 (ammonia) has similar capacity characteristics to R-22, but, because of odor and toxicity, R-717 is subject to restrictions for use in public or populated areas. However, R-717 chillers are becoming more popular because of bans on CFC and HCFC refrigerants. R-717 units are open-drive compressors and are piped with steel because copper cannot be used in ammonia systems.

PERFORMANCE CHARACTERISTICS AND OPERATING PROBLEMS

A distinguishing characteristic of the reciprocating compressor is its pressure rise versus capacity. Pressure rise has only a slight influence on the volume flow rate of the compressor, and, therefore, a reciprocating liquid chiller retains nearly full cooling capacity, even on above design wet-bulb days. It is well suited for air-cooled condenser operation and low-temperature refrigeration. A typical performance characteristic is shown in Figure 7 and compared with the centrifugal and screw compressors. Methods of capacity control are furnished by the following:

- Unloading of compressor cylinders (one at a time or in pairs)
- On-off cycling of compressors
- Hot-gas bypass
- Compressor speed control
- Combination of the previous methods

Figure 8 illustrates the relationship between system demand and performance of a compressor with three steps of unloading. As

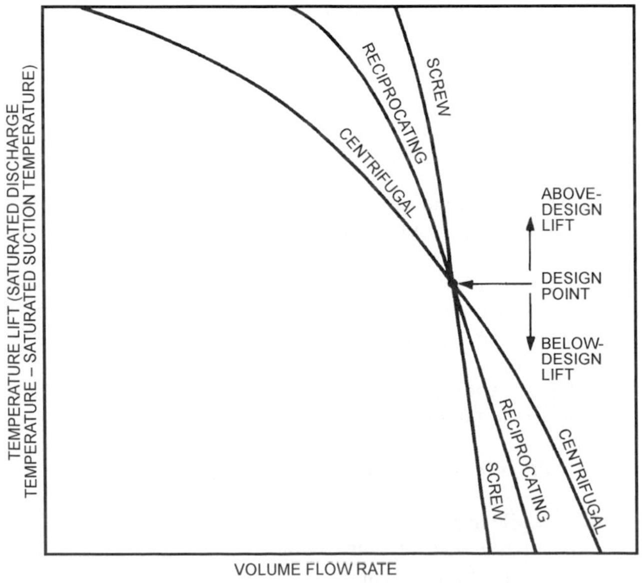

Fig. 7 Comparison of Single-Stage Centrifugal, Reciprocating, and Screw Compressor Performance

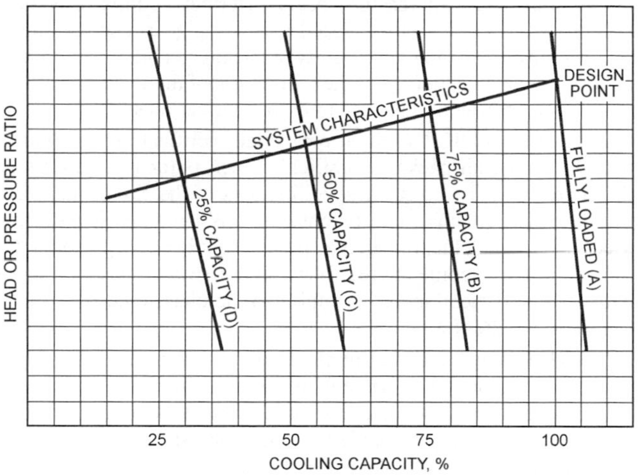

Fig. 8 Reciprocating Liquid Chiller Performance with Three Equal Steps of Unloading

cooling load drops to the left of the fully loaded compressor line (A), compressor capacity is reduced to that represented by line (B), which produces the required refrigerant flow. Since cooling load varies continuously while machine capacity is available in fixed increments, some compressor on-off cycling or successive loading and unloading of cylinders is required to maintain fairly constant liquid temperature. In practice, a good control system minimizes the load-unload or on-off cycling frequency while maintaining satisfactory temperature control.

METHOD OF SELECTION

Ratings

Two types of ratings are published. The first, for a packaged liquid chiller, lists values of capacity and power consumption for many combinations of leaving condenser water and chilled water temperatures (ambient dry-bulb temperatures for air-cooled models). The second type of rating shows capacity and power consumption for different condensing temperatures and chilled water temperatures. This type of rating permits selection with a remote condenser that can be evaporative, water, or air cooled. Sometimes the required rate of heat rejection is also listed to aid in selection of a separate condenser.

Power Consumption

With all liquid chilling systems, power consumption increases as condensing temperature rises. Therefore, the smallest package, with the lowest ratio of input to cooling capacity, can be used when condenser water temperature is low, the remote air-cooled condenser is relatively large, or when leaving chilled water temperature is high. The cost of the total system, however, may not be low when liquid chiller cost is minimized. Increases in cooling tower or fan coil cost will reduce or offset the benefits of reduced compression ratio. Life-cycle costs (initial cost plus operating expenses) should be evaluated.

Fouling

A fouling allowance of 0.00025 $ft^2 \cdot °F \cdot h/Btu$ is included in manufacturers' ratings in accordance with ARI *Standard* 590. However, fouling factors greater than 0.00025 should be considered in the selection if water conditions are other than ideal.

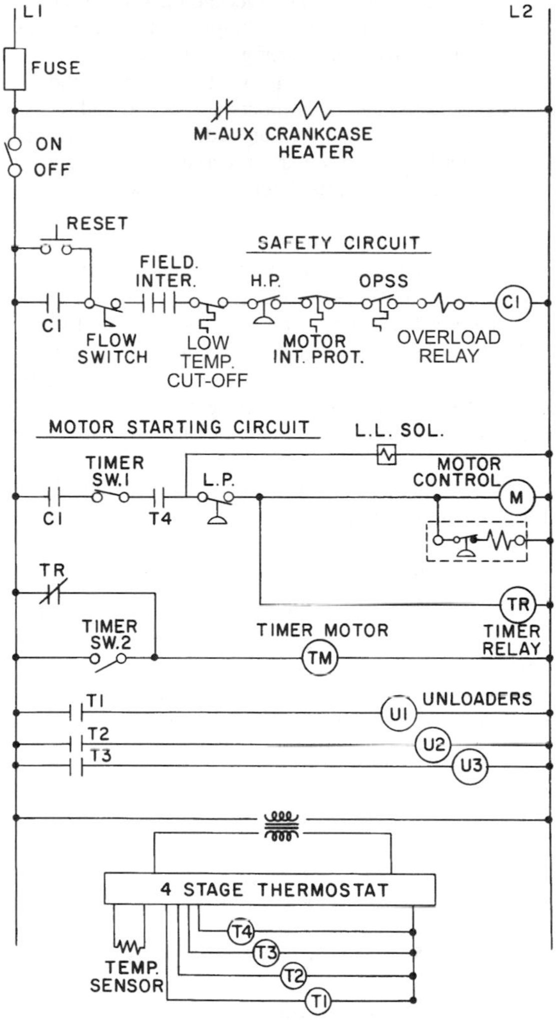

Fig. 9 Reciprocating Liquid Chiller Control System

CONTROL CONSIDERATIONS

A reciprocating chiller is distinguished from centrifugal and screw compressor-operated chillers by its use of increments of capacity reduction rather than continuous modulation. Therefore, special arrangements must be used to establish precise chilled liquid temperature control while maintaining stable operation free from excessive on-off cycling of compressors or unnecessary loading and unloading of cylinders.

To help provide good temperature control, return chilled liquid temperature sensing is normally used by units with steps of capacity control. The resulting flywheel effect in the chilled liquid circuit damps out excessive cycling. Leaving chilled liquid temperature sensing has the advantage of preventing excessively low leaving chilled liquid temperatures if chilled liquid flow falls significantly below the design value. It may not provide stable operation, however, if rapid changes in load are encountered.

An example of a basic control circuit for a single compressor-packaged reciprocating chiller with three steps of unloading is shown in Figure 9. The on-off switch controls start-up and starts the programmed timer. Assuming that the flow switch, field interlocks, and chiller safety devices are closed, pressing the momentarily closed reset button energizes control relay C1, locking in the safety circuit and the motor starting circuit. When the timer completes its program, timer switch 1 closes and timer switch 2 opens. Timer relay TR energizes, stopping the timer motor. When timer switch 1 closes, the motor starting circuit is completed and the motor contactor holding coil is energized, starting the compressor.

The four-stage thermostat controls the capacity of the compressor in response to demand. Cylinders are loaded and unloaded by de-energizing and energizing the unloader solenoids. If the load is reduced so that the return water temperature drops to a predetermined setting, the unit shuts down until the demand for cooling increases.

Opening a device in the safety circuit de-energizes control relay C1 and shuts down the compressor. The liquid line solenoid is also de-energized. Manual reset is required to restart. The crankcase heater is energized whenever the compressor is shut down.

If the automatic reset, low-pressure cutout opens, the compressor shuts down, but the liquid line solenoid remains energized. The timer relay (TR) is de-energized, causing the timer to start and complete its program before the compressor can be restarted. This prevents rapid cycling of the compressor under low-pressure conditions. A time delay low-pressure switch can also be used for this purpose with the proper circuitry.

SPECIAL APPLICATIONS

For multiple chiller applications and a 10°F chilled liquid temperature range, the use of a parallel chilled liquid arrangement is common because of the high cooler pressure drop resulting from the series arrangement. For a large (18°F) range, however, the series arrangement eliminates the need for overcooling during operation of one unit only. Special coolers with low water pressure drop may also be used to reduce total chilled water pressure drop in the series arrangement.

CENTRIFUGAL LIQUID CHILLERS

EQUIPMENT DESCRIPTION

Components and Their Function

Chapter 34 describes the centrifugal compressor. Because it is not a constant displacement machine, it offers a wide range of capacities continuously modulated over a limited range of pressure ratios. By altering built-in design items (including number of stages, compressor speed, impeller diameters, and choice of refrigerant), it can be used in liquid chillers having a wide range of design chilled liquid temperatures and design cooling fluid temperatures. Its ability to vary capacity continuously to match a wide range of load conditions with nearly proportional changes in power consumption makes it desirable for both close temperature control and energy conservation. Its ability to operate at greatly reduced capacity allows it to run most of the time with infrequent starting.

The hour of the day for starting an electric-drive centrifugal liquid chiller can often be chosen by the building manager to minimize peak power demands. It has a minimum of bearing and other contacting surfaces that can wear; this wear is minimized by providing forced lubrication to those surfaces prior to startup and during shutdown. Bearing wear usually depends more on the number of startups than the actual hours of operation. Thus, by reducing the number of startups, the life of the system is extended, and maintenance costs are reduced.

Both open and hermetic compressors are manufactured. Open compressors may be driven by steam turbines, gas turbines or engines, or electric motors, with or without speed-changing gears. (Engine and turbine drives are covered in Chapter 7 and electric motor drives in Chapter 40.)

Packaged electric-drive chillers may be of the open- or hermetic-type and use two-pole, 50- or 60-Hz polyphase electric motors, with or without speed-increasing gears. Hermetic units use only polyphase induction motors. Speed-increasing gears and their bearings, in most open- and hermetic-type packaged chillers, operate in a refrigerant atmosphere, and the lubrication of their contacting surfaces is incorporated in the compressor lubrication system.

Magnetic and SCR (silicon controlled rectifier) motor controllers are used with packaged chillers. When purchased separately, the controller must meet the specifications of the chiller manufacturer to ensure adequate equipment safety. When timed step starting methods are used, the time between steps should be long enough for the motor to overcome the relatively high inertia of the compressor and attain sufficient speed to minimize the electric current drawn immediately after transition.

Flooded coolers are commonly used, although direct-expansion coolers are employed by some manufacturers in the lower capacity ranges. The typical flooded cooler uses copper tubes or copper alloy that are mechanically expanded into the tube sheets, and, in some cases, into intermediate tube supports, as well.

Because liquid refrigerant that flows into the compressor increases power consumption and may cause internal damage, mist eliminators or baffles are often used in flooded coolers to minimize refrigerant liquid entrainment in the suction gas. (Additional information on coolers for liquid chillers can be found in Chapter 37.)

The condenser is generally water cooled, with refrigerant condensing on the outside of copper tubes. Large condensers may have refrigerant drain baffles, which direct the condensate from within the tube bundle directly to the liquid drains, reducing the thickness of the liquid film on the lower tubes.

Air-cooled condensers can be used with units that use higher pressure refrigerants, but with considerable increase in unit energy consumption at design conditions. Operating costs should be compared with systems using cooling towers and condenser water circulating pumps.

System modifications, including subcooling and economizing (described under Principles of Operation) are often used to conserve energy. Some units combine the condenser, cooler, and refrigerant flow control in one vessel; a subcooler may also be incorporated. (Additional information about thermodynamic cycles is in Chapter 1 of the 1997 *ASHRAE Handbook—Fundamentals*. Chapter 35 has information on condensers and subcoolers.)

Capacities and Types Available

Centrifugal packages are currently available from about 80 to 2400 tons at nominal conditions of 44°F leaving chilled water temperature and 95°F leaving condenser water temperature. This upper limit is continually increasing. Field-assembled machines extend to about 10,000 tons. Single-stage and two-stage internally geared machines and two- and three-stage direct-drive machines are commonly used in packaged units. Electric motor-driven machines constitute the majority of units sold.

Units with hermetic motors, cooled by refrigerant gas or liquid, are offered from about 80 to 2000 tons. Open-drive units are not offered by all hermetic manufacturers in the same size increments but are generally available from 80 to 10,000 tons.

Selection of Refrigerant

The centrifugal compressor is particularly suitable for handling relatively large flows of suction vapor. As the volumetric flow of suction vapor increases with higher capacities and lower suction temperatures, the higher pressure refrigerants, for example, R-134a and R-22, are used. The physical size and weight of the refrigerant piping and, often, other components of the refrigeration system are reduced by the use of higher pressure refrigerants. In order of decreasing volumetric flow and increasing pressures are refrigerants R-113, R-123, R-11, R-114, R-134a, R-12 or R-500, and R-22.

The CFC refrigerants R-11, R-12, R-113, R-114, and R-500 have been phased out.

Pressure vessels for use with R-123 usually have a design working pressure of 15 psi on the refrigerant side. The vessel shells are usually stronger than necessary for this requirement to ensure sufficient rigidity and prevent collapse under vacuum.

The thermal stability of the refrigerant and its compatibility with materials it contacts are also important. Selection of elastomers and electrical insulating materials require special attention because many of these materials are affected by the refrigerants. (Additional information concerning refrigerants can be found in Chapters 18 and 19 of the 1997 *ASHRAE Handbook—Fundamentals*.)

PERFORMANCE AND OPERATING CHARACTERISTICS

Figure 10 illustrates a compressor's performance at constant speed with various inlet guide vane settings. Figure 11 illustrates that compressor's performance at various speeds with open inlet guide vanes. Capacity is modulated at constant speed by automatic adjustment of prerotation vanes that whirl the refrigerant gas at the impeller eye. This effect matches demand by shifting the compressor performance curve downward and to the left (as shown in Figure 10). Compressor efficiency, when unloaded in this manner, is superior to suction throttling. Some manufacturers automatically reduce diffuser width or throttle the impeller outlet with decreasing load.

Speed control for a centrifugal compressor offers even lower power consumption. Down to about 50% capacity or at off design conditions, the speed may be reduced gradually without surging. Control is transferred to the prerotation vanes for operation at lower loads. While capacity is related directly to a change in speed, the pressure produced is proportional to the square of the change in speed. Therefore, the pressure produced by reducing the speed may be less than that required by the load. Combined use of gas bypass, prerotation vanes, etc., would then be necessary.

A **gas bypass** allows the compressor to operate down to zero load. This feature is a particular advantage for such intermittent industrial applications as the cooling of quenching tanks. Bypass vapor obtained by either method maintains the power consumption at the same level attained just prior to starting bypass, regardless of load reductions. At light loads, some bypass vapor, if introduced into the cooler below the tube bundle, may increase the evaporating temperature by agitating the liquid refrigerant and thereby more thoroughly wetting the tube surfaces.

Figure 12 shows how **temperature lift** varies with load. A typical reduction in entering condenser water temperature of 10°F helps to reduce temperature lift at low load. Other factors producing lower lift at reduced loads are

- Reduction in condenser cooling water range (the difference between entering and leaving temperatures, resulting from decreasing heat rejection)
- Decrease in temperature difference between condensing refrigerant and leaving condenser water
- Similar decrease between evaporating refrigerant and leaving chilled liquid temperature

In many cases, the actual reduction in temperature lift is even greater because the wet-bulb temperature usually drops with the cooling load, producing a greater decrease in entering condenser water temperature.

As stated earlier, speed control is usually used from 100% down to about 50% load or at off-design conditions; below 50%, inlet vane control is used. Power consumption is reduced when the coldest possible condenser water is used, consistent with the chiller manufacturer's recommended minimum condenser water temperature. In cooling tower applications, minimum water temperatures should be controlled by a cooling tower bypass and/or by cooling tower fan control, not by reducing the water flow through the condenser. Maintaining a high flow rate at lower temperatures minimizes fouling and power requirements.

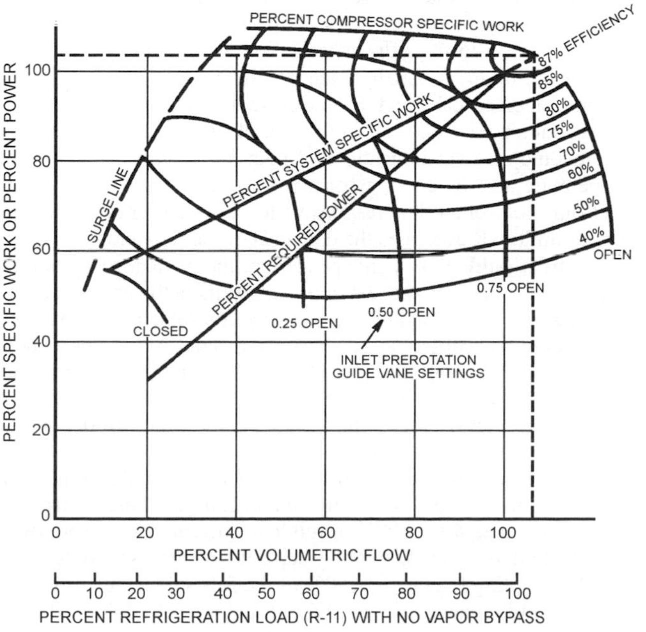

Fig. 10 Typical Centrifugal Compressor Performance at Constant Speed

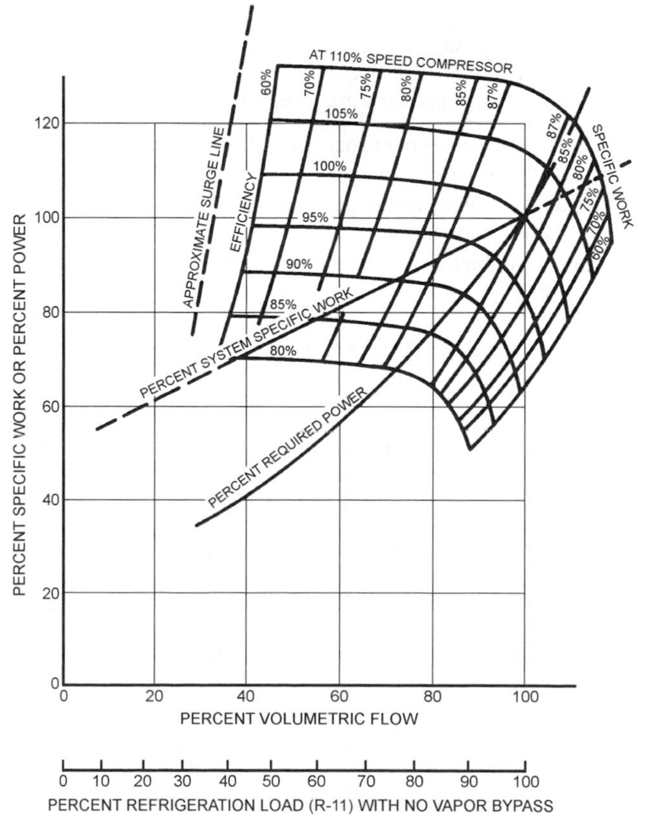

Fig. 11 Typical Centrifugal Compressor Performance at Various Speeds

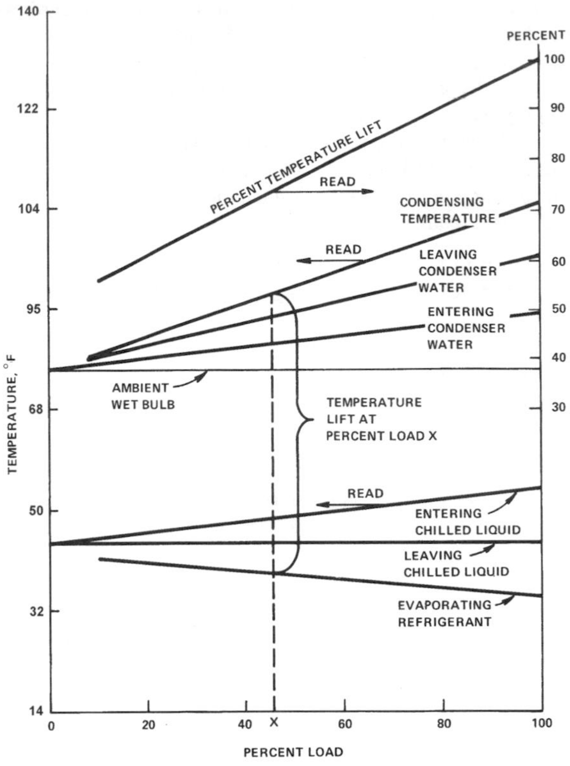

Fig. 12 Temperature Relations in a Typical Centrifugal Liquid Chiller

Surging occurs when the system specific work becomes greater than the compressor developed specific work or above the surge line indicated in Figures 10 and 11. Excessively high temperature lift and corresponding specific work commonly originate from

- Excessive condenser or evaporator water-side fouling beyond the specified allowance
- Inadequate cooling tower performance and higher-than-design condenser water temperature
- Noncondensables in the condenser, which increase condenser pressure.

METHOD OF SELECTION

Ratings

A refrigeration machine with specified details is chosen from selection tables for given capacities and operating conditions, or through computer-generated selection or performance programs. Rating tables differ from selection tables in that they list the capacities and operating data for each refrigeration machine under various operating conditions, often with specific details for the listed conditions. The details specified for centrifugal systems include the number of passes and the water-side pressure drop in each of the heat exchangers, the required power input, electrical characteristics, and part-load performance.

The maximum number of condenser and cooler water passes should be used, without producing excessive water pressure drop. The greater the number of water-side passes, the less the power consumption. Sometimes a slight reduction in condenser water flow (and slightly higher leaving water temperature) allows a better selection (lower power consumption or smaller model) than will the choice of fewer water passes when a rigid pressure-drop limit exists.

Fouling

In accordance with ARI *Standard* 590, a fouling allowance of 0.00025 $ft^2 \cdot °F \cdot h/Btu$ is included in manufacturers' ratings. (Chapter 35 has further information about fouling factors.) To reduce fouling, a minimum water velocity of about 3.3 ft/s is recommended in coolers or condensers. Maximum water velocities exceeding 11 ft/s are not recommended because of potential erosion problems.

Proper water treatment and regular tube cleaning are recommended for all liquid chillers to reduce power consumption and operating problems. Chapter 47 of the 1999 *ASHRAE Handbook—Applications* has water treatment information.

Continuous or daily monitoring of the quality of the condenser water is desirable. Checking the quality of the chilled liquid is also desirable. The intervals between checks become greater as the possibilities for fouling contamination become less—for example, an annual check should be sufficient for closed-loop water-circulating systems for air conditioning. Corrective treatment is required, and periodic, usually annual, cleaning of the condenser tubes usually keeps fouling within the specified allowance. In applications where more frequent cleaning is desirable, an on-line cleaning system may be economical.

Noise and Vibration

The chiller manufacturer's recommendations for mounting should be followed to prevent transmission or amplification of vibration to adjacent equipment or structures. Auxiliary pumps, if not connected with flexible fittings, can induce vibration of the centrifugal unit, especially if the rotational speed of the pump is nearly the same as either the compressor prime mover or the compressor. Flexible tubing becomes less flexible when it is filled with liquid under pressure and some vibration can still be transmitted. General information on noise, measurement, and control may be found in Chapter 7 of the 1997 *ASHRAE Handbook—Fundamentals*, Chapter 46 of the 1999 *ASHRAE Handbook—Applications*, and ARI *Standard* 575.

CONTROL CONSIDERATIONS

In centrifugal systems, the **chilled liquid temperature sensor** is usually placed in thermal contact with the leaving chilled water. In electrical control systems, the electrical signal is transmitted to an electronic control module, which controls the operation of an electric motor(s) positioning the capacity controlling inlet guide vanes. A current limiter is usually included on machines with electric motors. An electrical signal from a current transformer in the compressor motor controller is sent to the electronic control module. The module receives indications of both the leaving chilled water temperature and the compressor motor current. The portion of the electronic control module responsive to motor current is called the current limiter. It overrides the demands of the temperature sensor.

The **inlet guide vanes**, independent of the demands for cooling, do not open more than the position that results in the present setting of the current limiter. Pneumatic capacity controls operate in a similar manner. The chilled liquid temperature sensor provides a pneumatic signal. The controlling module receives both that signal and the motor current electrical signal and controls the operation of a pneumatic motor(s) positioning the inlet guide vanes. Both controlling systems have sensitivity adjustments.

The **current limiter** on most machines can limit current draw during periods of high electrical demand charges. This control can be set from about 40 to 100% of full load current. Whenever power consumption is limited, cooling capacity is correspondingly reduced. If cooling load is only 50% of the full value, the current (or demand) limiter can be set at 50% without loss of cooling. By setting the limiter at 50% of full current draw, any subsequent high demand charges are prevented during pulldown after startup. Even during periods of high cooling load, it may be desirable to limit electrical demand if a small increase in chiller liquid temperature is acceptable. If the temperature continues to decrease after the capacity control has reached its minimum position, a low-temperature control stops the compressor and restarts it when a rise in temperature indicates the need for cooling. Manual controls may also be provided to bypass the temperature control. Provision is included to ensure that the capacity control is at its minimum position when the compressor starts to provide an unloaded starting condition.

Additional operating controls are needed for appropriate operation of lubricant pumps, lubricant heaters, purge units, and refrigerant transfer units. An **antirecycle timer** should also be included to prevent frequent motor starts. Multiple unit applications require additional controls for capacity modulation and proper sequencing of units. (See the section on Multiple Chiller System.)

Safety controls protect the unit under abnormal conditions. Safety cutouts that may be required are for high condenser pressure, low evaporator refrigerant temperature or pressure, low lubricant pressure, high lubricant temperature, high motor temperature, and high discharge temperature. Auxiliary safety circuits are usually provided on packaged chillers. At installation, the circuits are field wired to field-installed safety devices, including auxiliary contacts on the pump motor controllers and flow switches in the chilled water and condenser water circuits. Safety controls are usually provided in a lockout circuit, which will trip out the compressor motor controller and prevent automatic restart. The controls reset automatically, but the circuit cannot be completed until a manual reset switch is operated and the safety controls return to their safe positions.

AUXILIARIES

Purge units are required for centrifugal liquid chilling machines using R-123, R-11, R-113, or R-114 because evaporator pressure is below atmospheric pressure. If a purge unit were not used, air and moisture would accumulate in the refrigerant side. Noncondensables collect in the condenser during operation, reducing the heat-transfer coefficient and increasing condenser pressure as a result of both their insulating effect and the partial pressure of the noncondensables. Compressor power consumption increases, capacity is reduced, and surging may occur.

Moisture may build up as free moisture once the refrigerant becomes saturated. Acids produced by a reaction between the free moisture and the refrigerant will then cause internal corrosion. A purge unit prevents the accumulation of noncondensables and ensures internal cleanliness of the chiller. However, a purge unit does not reduce the need to check for leaks and the need to repair them, which is required maintenance for any liquid chiller. Purge units may be manual or automatic, compressor-operated, or compressorless. To reduce the potential for air leaks when chillers are off, the chillers may be heated externally to pressurize them to atmospheric pressure.

ASHRAE *Standard* 15 requires purge units and rupture disks to be vented outdoors. Because of environmental concerns and the increasing cost of refrigerants, high efficiency (air to refrigerant) purges are available that reduce refrigerant losses during normal purging operations.

Lubricant coolers may be water cooled, using condenser water when the quality is satisfactory, or chilled water when a small loss in net cooling capacity is acceptable. These coolers may also be refrigerant or air cooled, eliminating the need for water piping to the cooler.

A **refrigerant transfer unit** may be provided for centrifugal liquid chillers. The unit consists of a small reciprocating compressor with electric motor drive, a condenser (air- or water-cooled), a lubricant reservoir and separator, valves, and interconnecting piping. Refrigerant transfers in three steps:

1. **Gravity drain.** When the receiver is at the same level as or below the cooler, some liquid refrigerant may be transferred to the receiver by opening valves in the interconnecting piping.
2. **Pressure transfer.** By resetting valves and operating the compressor, refrigerant gas is pulled from the receiver to pressurize the cooler, forcing refrigerant liquid from the cooler to the storage receiver. If the chilled liquid and condenser water pumps can be operated to establish a temperature difference, the migration of refrigerant from the warmer vessel to the colder vessel can also be used to assist in the transfer of refrigerant.
3. **Pump-out.** After the liquid refrigerant has been transferred, valve positions are changed and the compressor is operated to pump refrigerant gas from the cooler to the transfer unit condenser, which sends condensed liquid to the storage receiver. If any chilled liquid (water, brine, etc.) remains in the cooler tubes, pump-out must be stopped before cooler pressure drops below the saturation condition corresponding to the freezing point of the chilled liquid.

If the saturation temperature corresponding to cooler pressure is below the chilled liquid freezing point when recharging, refrigerant gas from the storage receiver must be introduced until the cooler pressure is above this condition. The compressor can then be operated to pressurize the receiver and move refrigerant liquid into the cooler without danger of freezing.

Water-cooled transfer unit condensers provide fast refrigerant transfer. Air-cooled condensers eliminate the need for water, but they are slower and more expensive.

SPECIAL APPLICATIONS

Heat Recovery

Instead of rejecting all heat extracted from the chilled liquid to a cooling tower, a separate, closed condenser cooling water circuit is heated by the condensing refrigerant for such purposes as comfort heating, preheating, or reheating. Some factory packages include an extra condenser water circuit, either in the form of a double-bundle condenser or an auxiliary condenser.

A centrifugal heat-recovery package is controlled as follows:

- **Chilled liquid temperature** is controlled by a sensor in the leaving chilled liquid line signaling the capacity control device.
- **Hot water temperature** is controlled by a sensor in the hot water line that modulates a cooling tower bypass valve. As the heating requirement increases, hot water temperature drops, opening the tower bypass slightly. Less heat is rejected to the tower, condensing temperature increases, and hot water temperature is restored as more heat is rejected to the hot water circuit.

The hot water temperature selected has a bearing on the installed cost of the centrifugal package, as well as on the power consumption while heating. Lower hot water temperatures of 95 to 105°F result in a less expensive machine that uses less power. Higher temperatures require a greater compressor motor output, perhaps higher pressure condenser shells, sometimes extra compression stages, or a cascade arrangement. Installed cost of the centrifugal heat-recovery machine is increased as a result.

Another concern in the design of a central chilled water plant with heat-recovery centrifugal compressors is the relative size of the cooling and heating loads. These loads should be equalized on each machine so that the compressor may operate at optimum efficiency during both the full cooling and full heating seasons. When the heating requirement is considerably smaller than the cooling requirement, multiple packages will lower operating costs and allow standard air-conditioning centrifugal packages of lower cost to be used for the remainder of the cooling requirement. In multiple packages, only one unit is designed for heat recovery and carries the full heating load.

Free Cooling

Cooling without operating the compressor of a centrifugal liquid chiller is called free cooling. When a supply of condenser water is available at a temperature below the needed chilled water temperature, the chiller can operate as a thermal siphon. Low-temperature condenser water condenses refrigerant, which is either drained by gravity or pumped into the evaporator. Higher-temperature chilled water causes the refrigerant to evaporate, and vapor flows back to the condenser because of the pressure difference between the evaporator and the condenser. Free cooling is limited to about 10 to 30% of the chiller design capacity. Free cooling capacity depends on chiller design and the temperature difference between the desired chilled water temperature and the condenser water temperature. Free cooling is also available using either direct or indirect methods as described in Chapter 36.

Air-Cooled System

Two types of air-cooled centrifugal systems are prevalent. One consists of a water-cooled centrifugal package with a closed-loop condenser water circuit. The condenser water is cooled in a water/air heat exchanger. This arrangement results in higher condensing temperature and increased power consumption. In addition, winter operation requires the use of glycol in the condenser water circuit, which reduces the heat-transfer coefficient of the unit.

The other type of unit is directly air-cooled, which eliminates the intermediate heat exchanger and condenser water pumps, resulting in lower power requirements. However, the condenser and refrigerant piping must be kept leak free.

Because a centrifugal machine will surge if it is subjected to a pressure appreciably higher than design, the air-cooled condenser must be designed to reject the required heat. In common practice, the selection of a reciprocating air-cooled machine is based on an outside dry-bulb temperature that will be exceeded 5% of the time. A centrifugal may be unable to operate during such times because of surging, unless the chilled water temperature is raised proportionately. Thus, the compressor impeller(s) and/or speed should be selected for the maximum dry-bulb temperature to ensure that the desired chilled water temperature will be maintained at all times. In addition, the condenser coil must be kept clean.

An air-cooled centrifugal chiller should allow the condensing temperature to fall naturally to about 70°F during colder weather. The resulting decrease in compressor power consumption is greater than that for reciprocating systems controlled by thermal expansion valves.

During winter shutdown, precautions must be taken to prevent freezing of the cooler liquid caused by a free cooling effect from the air-cooled condenser. A thermostatically controlled heater in the cooler, in conjunction with a low refrigerant pressure switch to start the chilled liquid pumps, will protect the system.

Other Coolants

Centrifugal liquid chilling units are most frequently used for water chilling applications. But centrifugals are also used with such coolants as calcium chloride, methylene chloride, ethylene glycol, and propylene glycol. (Chapter 20 of the 1997 *ASHRAE Handbook—Fundamentals* describes the properties of secondary coolants.) Coolant properties must be considered in calculating heat-transfer performance and pressure drop. Because of the greater temperature rise, higher compressor speeds and possibly more stages may be required for cooling these coolants. Compound and/or cascade systems are required for low-temperature applications.

Vapor Condensing

Many process applications condense vapors such as ammonia, chlorine, or hydrogen fluoride. Centrifugal liquid chilling units are used for these applications.

OPERATION AND MAINTENANCE

Proper operation and maintenance are essential for reliability, longevity, and safety. Chapter 37 of the 1999 *ASHRAE Handbook—Applications* includes general information on principles, procedures, and programs for effective maintenance. The manufacturer's operation and maintenance instructions should also be consulted for specific procedures. In the United States, Environmental Protection Agency regulations require (1) certification of service technicians, (2) a statement of minimum pressures necessary during evacuation of the system, and (3) definition of when a refrigerant charge must be removed prior to opening a system for service. All service technicians or operators maintaining systems must be familiar with these regulations.

Normal operation conditions should be established and recorded at initial startup. Changes from these conditions can be used to signal the need for maintenance. One of the most important items is to maintain a leak free unit.

Leaks on units operating at subatmospheric pressures allow air and moisture to enter the unit, which increases the condenser pressure. While the purge unit can remove noncondensables sufficiently to prevent an increase in condenser pressure, continuous entry of air and attendant moisture into the system promotes refrigerant and lubricant breakdown and corrosion. Leaks from units that operate above atmospheric pressure may release environmentally harmful refrigerants. Regulations require that annual leakage not exceed a percentage of the refrigerant charge. It is good practice, however, to find and repair all leaks.

Periodic analysis of the lubricant and refrigerant charge can also be used to identify system contamination problems. High condenser pressure or frequent purge unit operation indicate leaks that should be corrected as soon as possible. With positive operating pressures, leaks result in loss of refrigerant and such operating problems as low evaporator pressure. A leak check should also be included in preparation for a long-term shutdown. (Chapter 6 in the 1998 *ASHRAE Handbook—Refrigeration* discusses the harmful effects of air and moisture.)

Normal maintenance should include periodic lubricant and refrigerant filter changes as recommended by the manufacturer. All safety controls should be checked periodically to ensure that the unit is protected properly.

Cleaning of inside tube surfaces may be required at various intervals, depending on the water condition. Condenser tubes may only need annual cleaning if proper water treatment is maintained. Cooler tubes need less frequent cleaning if the chilled water circuit is a closed loop.

If the refrigerant charge must be removed and the unit opened for service, the unit should be leak-checked, dehydrated, and evacuated properly before recharging. Chapter 46 of the 1998 *ASHRAE Handbook—Refrigeration* has information on dehydrating, charging, and testing.

SCREW LIQUID CHILLERS

EQUIPMENT DESCRIPTION

Components and Their Function

Single- and twin-screw compressors are both positive-displacement machines with nearly constant flow performance. Compressors for liquid chillers can be both lubricant-injected and lubricant-injection-free. (Chapter 35 describes screw compressors in detail.)

The cooler may be flooded or direct-expansion. No particular design has a cost advantage over the other. The flooded cooler is more sensitive to freezing, requires more refrigerant, and requires closer evaporator pressure control, but its performance is easier to predict and it can be cleaned. The direct-expansion cooler requires closer mass flow control, is less likely to freeze, and returns lubricant to the lubricant system rapidly. The decision to use one or the other is based on the relative importance of these factors on a given application.

Screw coolers have the following characteristics: (1) high maximum working pressure, (2) continuous lubricant scavenging, (3) no mist eliminators (flooded coolers), and (4) distributors designed for high turndown ratios (direct-expansion coolers). A suction gas, high-pressure liquid heat exchanger is sometimes incorporated into the system to provide subcooling for increased thermal expansion valve flow and reduced power consumption. (For further information on coolers, see Chapter 37.)

Flooded coolers were once used in units with a capacity larger than about 400 tons. Direct-expansion coolers are also used in larger units up to 800 tons with a servo-operated expansion valve having an electronic controller that measures evaporating pressure, leaving secondary coolant temperature, and suction gas superheat.

The condenser may be included as part of the liquid chilling package when water-cooled, or it may be remote. Air-cooled liquid chilling packages are also available. When remote air-cooled or evaporative condensers are applied to liquid chilling packages, a liquid receiver generally replaces the water-cooled condenser on the package structure. Water-cooled condensers are the cleanable shell-and-tube type (see Chapter 35).

Lubricant cooler loads vary widely, depending on the refrigerant and application, but they are substantial because lubricant injected into the compressor absorbs a portion of the heat of compression. Lubricant is cooled by one of the following methods:

- Water-cooled using condenser water, evaporative condenser sump water, chilled water, or a separate water- or glycol-to-air cooling loop
- Air-cooled using a lubricant-to-air heat exchanger
- Refrigerant-cooled (where lubricant cooling load is low)
- Liquid injection into the compressor
- Condensed refrigerant liquid thermal recirculation (thermosiphon), where appropriate, compressor head pressure is available.

The latter two methods are the most economical both in first cost and overall operating cost because cooler maintenance and special water treatment are eliminated.

Efficient lubricant separators are required. The types and efficiencies of these separators vary according to refrigerant and application. Field built systems require better separation than complete factory-built systems. Ammonia applications are most stringent because no appreciable lubricant returns with the suction gas from the flooded coolers normally used in ammonia applications. However, separators are available for ammonia packages, which do not require the periodic addition of lubricant that is customary on other ammonia systems. The types of separators in use are centrifugal, demister, gravity, coalescer, and combinations of these.

Hermetic compressor units may use a centrifugal separator as an integral part of the hermetic motor while cooling the motor with discharge gas and lubricant simultaneously. A schematic of a typical refrigeration system is shown in Figure 13.

Capacities and Types Available

Screw compressor liquid chillers are available as factory-packaged units from about 30 to 1250 tons. Both open and hermetic styles are manufactured. Packages without water-cooled condensers, with receivers, are made for use with air-cooled or evaporative condensers. Most factory-assembled liquid chilling packages use R-22 and some use R-134a.

Additionally, compressor units, comprised of a compressor, hermetic or open motor, and lubricant separator and system, are available from 20 to 2000 tons. These are used with remote evaporators and condensers for low, medium, and high evaporating temperature applications. Condensing units, similar to compressor units in range and capacity but with water-cooled condensers, are also built. Similar open motor-drive units are available for ammonia, as are booster units.

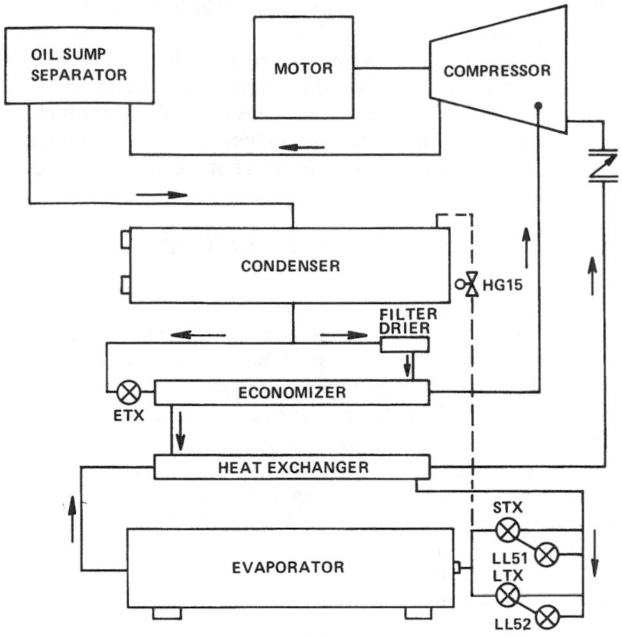

Fig. 13 Refrigeration System Schematic

Selection of Refrigerant

The refrigerants most commonly used with screw compressors on liquid chiller applications are R-22, R-134a, and R-717. The active use of R-12 and R-500 has been discontinued for new equipment.

PERFORMANCE AND OPERATING CHARACTERISTICS

The screw compressor operating characteristic shown in Figure 7 is compared with reciprocating and centrifugal performance. Additionally, since the screw compressor is a positive-displacement compressor, it does not surge. Because it has no clearance volume in the compression chamber, it pumps high volumetric flows at high pressure. Because of this, screw compressor chillers suffer the least capacity reduction at high condensing temperatures.

The screw compressor provides stable operation over the whole working range because it is a positive-displacement machine. The working range is wide because the discharge temperature is kept low and is not a limiting factor because of lubricant injection into the compression chamber. Consequently, the compressor is able to operate single-stage at high pressure ratios.

An economizer can be installed to improve the capacity and lower the power consumption at full-load operation. An example of such an economizer arrangement is shown in Figure 13, where the main refrigerant liquid flow is subcooled in a heat exchanger connected to the intermediate pressure port in the compressor. The evaporating pressure in this heat exchanger is higher than the suction pressure of the compressor.

Lubricant separators must be sized for the size of compressor, type of system (factory-assembled or field-connected), refrigerant, and type of cooler. Direct-expansion coolers have less stringent separation requirements than do flooded coolers. In a direct-expansion system, the refrigerant evaporates in the tubes, which means that the velocity is kept so high that the lubricant rapidly returns to the compressor. In a flooded evaporator, the refrigerant is outside the tubes, and some type of external lubricant-return device must be used to minimize the concentration of lubricant in the cooler. Suction or discharge check valves are used to minimize backflow and lubricant loss during shutdown.

Because the lubricant system is on the high-pressure side of the unit, precautions must be taken to prevent lubricant dilution. Dilution can also be caused by excessive floodback through the suction or intermediate ports; and unless properly monitored, it may go unnoticed until serious operating or mechanical problems are experienced.

METHOD OF SELECTION

Ratings

Screw liquid-chiller ratings are generally presented similarly to those for centrifugal-chiller ratings. Tabular values include capacity and power consumption at various chilled water and condenser water temperatures. In addition, ratings are given for packages without the condenser that list capacity and power versus chilled water temperature and condensing temperature. Ratings for compressors alone are also common, showing capacity and power consumption versus suction temperature and condensing temperature for a given refrigerant.

Power Consumption

Typical part-load power consumption is shown in Figure 14. Power consumption of screw chillers benefits from reduction of condensing water temperature as the load decreases, as well as operating at the lowest practical pressure at full load. However, because direct-expansion systems require a pressure differential, the power consumption saving is not as great at part load as shown.

Fouling

A fouling allowance of 0.00025 $ft^2 \cdot °F \cdot h/Btu$ is incorporated in screw compressor chiller ratings. Excessive fouling (above the design value) increases power consumption and reduces capacity. Fouling of water-cooled lubricant coolers results in higher than desirable lubricant temperatures.

CONTROL CONSIDERATIONS

Screw chillers provide continuous capacity modulation, from 100% capacity down to 10% or less. The leaving chilled liquid temperature is sensed for capacity control. Safety controls commonly required are (1) lubricant failure switch, (2) high discharge pressure cutout, (3) low suction pressure switch, (4) cooler flow switch, (5) high lubricant and discharge temperature cutout, (6) hermetic motor inherent protection, (7) lubricant pump and compressor motor overloads, and (8) low lubricant temperature (flood back/dilution protection). The compressor is unloaded automatically (slide valve driven to minimum position) before starting. Once it starts operating, the slide valve is controlled hydraulically by a temperature-load controller that energizes the load and unload solenoid valves.

The current limit relay protects against motor overload from higher than normal condensing temperatures or low-voltage and also allows a demand limit to be set, if desired. An antirecycle timer is used to prevent overly frequent recycling. Lubricant sump heaters are energized during the off cycle. A hot gas capacity control is optionally available and prevents automatic recycling at no-load conditions such as is often required in process liquid chilling. A suction to discharge starting bypass sometimes aids starting and allows the use of standard starting torque motors.

Some units are equipped with electronic regulators specially developed for the screw compressor characteristics. These regulators include PI control (Proportional-Integrating) of the leaving brine temperature and such functions as automatic/manual control, capacity indication, time circuits to prevent frequent recycling and to bypass the lubricant pressure cutout during startup, switch for unloaded starting, etc. (Typical external connections are shown in Figure 15.)

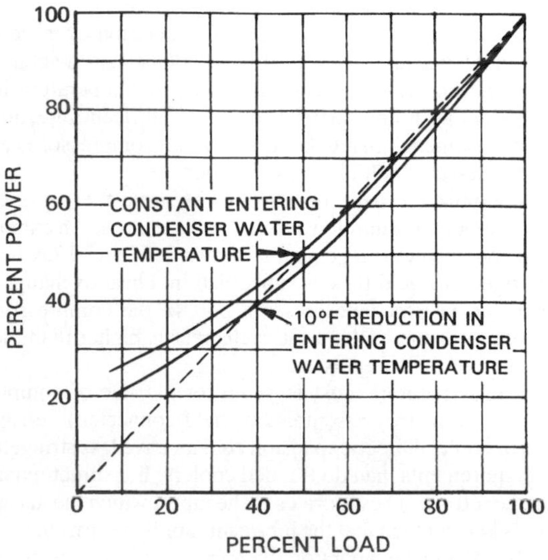

Fig. 14 Typical Screw Compressor Chiller Part-Load Power Consumption

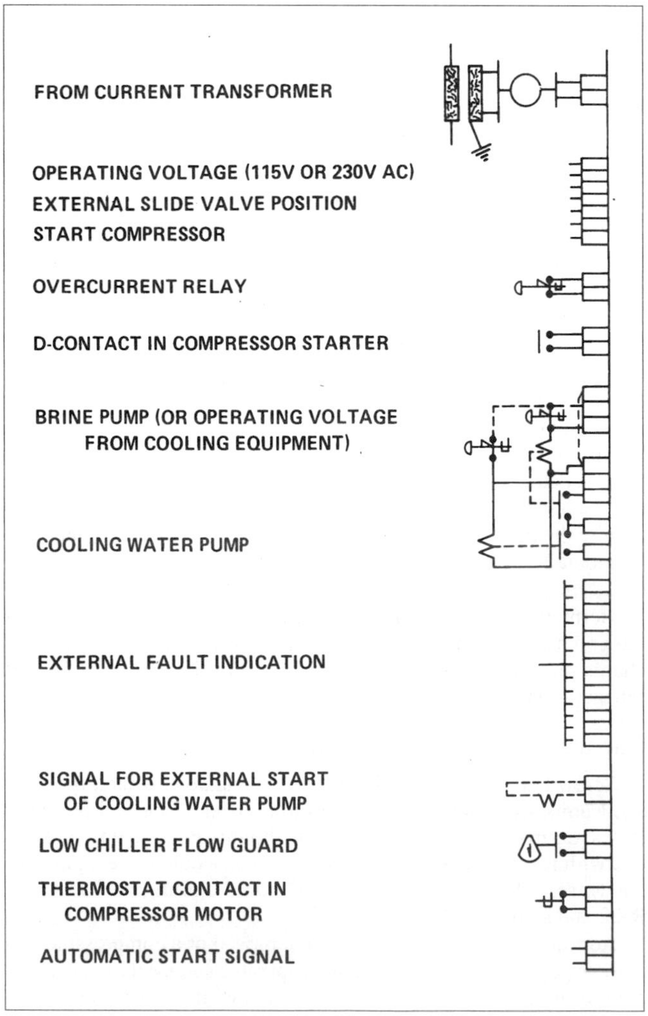

Fig. 15 Typical External Connections for Screw Compressor Chiller

AUXILIARIES

A **refrigerant transfer unit** is similar to the unit described in the section on Auxiliaries under Centrifugal Liquid Chillers. It is designed for R-22 operating pressure. Its flexibility is increased by including a reversible liquid pump on the unit. It is available as a portable unit or mounted on a storage receiver.

A **lubricant-charging pump** is useful for adding lubricant to the pressurized lubricant sump. Two types are used: a manual pump and an electric motor-driven positive-displacement pump.

Acoustical enclosures are available for installations that require low noise levels.

SPECIAL APPLICATIONS

Because of the screw compressor's positive-displacement characteristic and lubricant-injected cooling, its use for high differential applications is limited only by power considerations and maximum design working pressures. Therefore, it is being used for many special applications because of reasonable compressor cost and no surge characteristic. Some of the fastest growing areas are:

- Heat-recovery installations
- Air-cooled
  - Split packages with field-installed interconnecting piping
  - Factory-built rooftop packages
- Low-temperature brine chillers for process cooling
- Ice rink chillers
- Power transmission line lubricant cooling

High temperature compressor and condensing units are being used increasingly for air conditioning because of the higher efficiency of direct air-to-refrigerant heat exchange resulting in higher evaporating temperatures. Many of these installations have air-cooled condensers.

MAINTENANCE

Manufacturer's maintenance instructions should be followed, especially because some items differ substantially from reciprocating or centrifugal units. Water-cooled condensers must be cleaned of scale periodically (see the section on General Maintenance). If the condenser water is also used for the lubricant cooler, this should be considered in the treatment program. Lubricant coolers operate at higher temperatures and lower flows than condensers, so it is possible that the lubricant cooler may have to be serviced more often than the condenser.

Because large lubricant flows are a part of the screw compressor system, the lubricant filter pressure drop should be monitored carefully and the elements changed periodically. This is particularly important in the first month or so after startup of any factory-built package and is essential on field-erected systems. Since the lubricant and refrigeration systems merge at the compressor, much of the loose dirt and fine contaminants in the system eventually find their way to the lubricant sump, where they are removed by the lubricant filter. Similarly, the filter-drier cartridges should be monitored for pressure drop and moisture during initial start and regularly thereafter. Generally, if a system reaches an acceptable dryness level, it stays that way unless it is opened.

It is good practice to check the lubricant for acidity periodically, using commercially available acid test kits. The lubricant does not need to be changed unless it is contaminated by water, acid, or metallic particles. Also, a refrigerant sample should be analyzed yearly to determine its condition.

Certain procedures that should be followed on a yearly basis or during a regularly scheduled shutdown. These include checking and calibrating all operation and safety controls, tightening all electrical connections, inspecting power contacts in starters, dielectric checking of hermetic and open motors, and checking the alignment of open motors.

Leak testing of the unit should be performed regularly. A water-cooled package used for summer cooling should be leak tested annually. A flooded unit with proportionately more refrigerant in it, used for year-round cooling, should be tested every four to six months. A process air-cooled chiller designed for year-round operation 24 hours per day should be checked every one to three months.

Based on 6000 operating hours per year and depending on the above considerations, a typical inspection or replacement timetable is as follows:

Shaft seals	1.5 to 4 yr	Inspect
Hydraulic cylinder seals	1.5 to 4 yr	Replace
Thrust bearings	4 to 6 yr	Check preload via shaft end play every 6 months and replace as required
Shaft bearings	7 to 10 yr	Inspect

CHAPTER 39

CENTRIFUGAL PUMPS

CENTRIFUGAL pumps provide the primary force to distribute and recirculate hot and chilled water in a variety of space-conditioning systems. The pump provides a predetermined flow of water to the space load terminal units or to a thermal storage chamber for release at peak loads. The effect of centrifugal pump performance on the application, control, and operation of various terminal units is discussed in Chapter 12. Other hydronic systems that use pumps include (1) condensing water circuits to cooling towers (Chapter 13 and Chapter 36), (2) water-source heat pumps (Chapter 8), (3) boiler feeds, and (4) condensate returns (Chapter 10). Boiler feed and condensate return pumps for steam boilers should be selected based on boiler manufacturer's requirements.

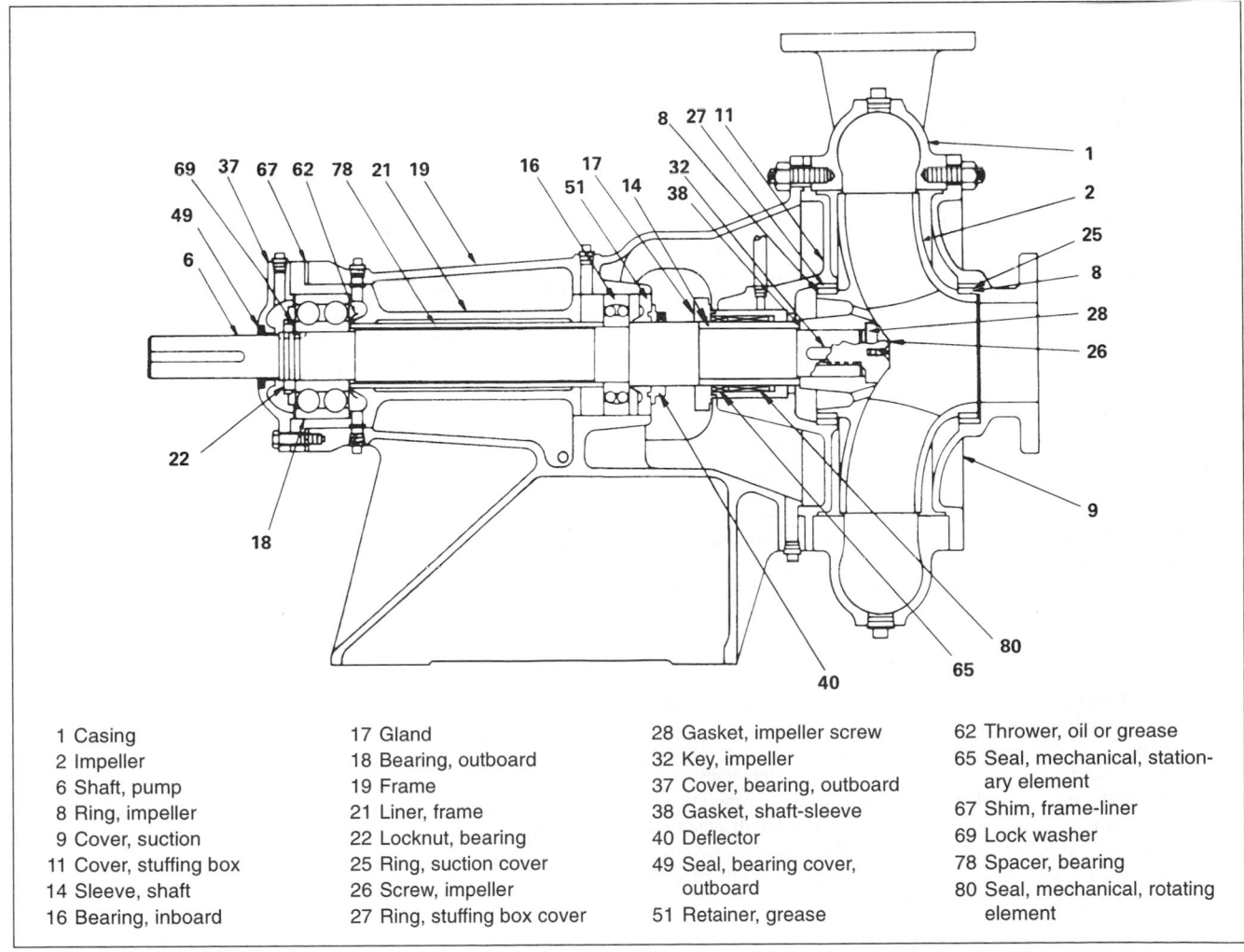

The preparation of this chapter is assigned to TC 8.10, Pumps and Hydronic Piping.

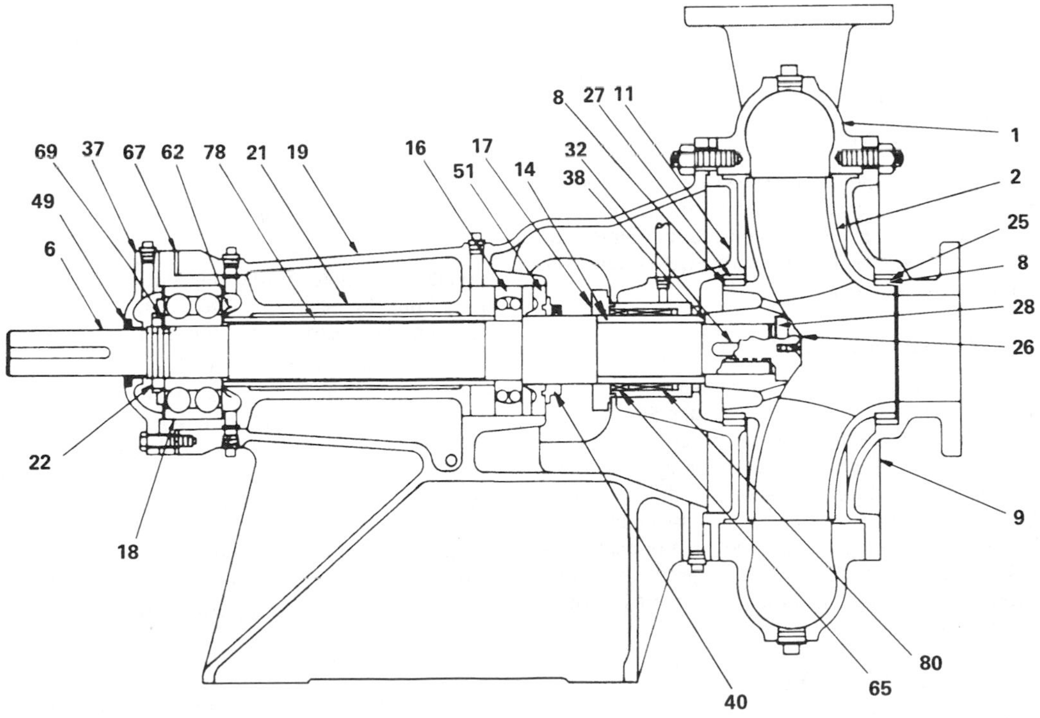

1 Casing
2 Impeller
6 Shaft, pump
8 Ring, impeller
9 Cover, suction
11 Cover, stuffing box
14 Sleeve, shaft
16 Bearing, inboard
17 Gland
18 Bearing, outboard
19 Frame
21 Liner, frame
22 Locknut, bearing
25 Ring, suction cover
26 Screw, impeller
27 Ring, stuffing box cover
28 Gasket, impeller screw
32 Key, impeller
37 Cover, bearing, outboard
38 Gasket, shaft-sleeve
40 Deflector
49 Seal, bearing cover, outboard
51 Retainer, grease
62 Thrower, oil or grease
65 Seal, mechanical, stationary element
67 Shim, frame-liner
69 Lock washer
78 Spacer, bearing
80 Seal, mechanical, rotating element

Fig. 1 Cross Section of Typical Overhung-Impeller End-Suction Pump

CONSTRUCTION FEATURES

The construction features of a typical centrifugal pump are shown in Figure 1. These features vary according to the manufacturer and the type of pump.

Materials. Centrifugal pumps are generally available in bronze-fitted or iron-fitted construction. The choice of material depends on those parts in contact with the liquid being pumped. In bronze-fitted pumps, the impeller and wear rings (if used) are bronze, the shaft sleeve is stainless steel or bronze, and the casing is cast iron. All-bronze construction is often used in domestic water applications.

Seals. The **stuffing box** is that part of the pump where the rotating shaft enters the pump casing. To seal leaks at this point, a mechanical seal or packing is used. **Mechanical seals** are used predominantly in clean hydronic applications, either as unbalanced or balanced (for higher pressures) seals. Balanced seals are used for high-pressure applications, particulate-laden liquids, or for extended seal life at lower pressures. Inside seals operate inside the stuffing box, while outside seals have the rotating element outside the box. Pressure and temperature limitations vary with the liquid pumped and the style of seal. **Packing** is used where abrasive substances (that are not detrimental to operation) could damage mechanical seals. Some leakage at the packing gland is needed to lubricate and cool the area between the packing material and shaft. Some designs use a large seal cavity instead of a stuffing box.

Shaft sleeves protect the motor or pump shaft.

Wear rings prevent wear to the impeller and/or casing and are easily replaced when worn.

Ball bearings are most frequently used, except in low-pressure circulators, where motor and pump bearings are the sleeve type.

A **balance ring** placed on the back of a single-inlet enclosed impeller reduces the axial load, thereby decreasing the size of the thrust bearing and shaft. Double-inlet impellers are inherently axially balanced.

Normal operating speeds of motors may be selected from 600 to 3600 rpm. The pump manufacturer can help determine the optimum pump speed for a specific application by considering pump efficiency, the available pressure at the inlet to prevent cavitation, maintenance requirements, and operating cost.

PUMP OPERATION

In a centrifugal pump, an electric motor or other power source rotates the impeller at the motor's rated speed. Impeller rotation adds energy to the fluid after it is directed into the center or eye of the rotating impeller. The fluid is then acted upon by centrifugal force and rotational or tip speed force, as shown in the vector diagram in Figure 2. These two forces result in an increase in the velocity of the fluid. The pump casing is designed for the maximum conversion of velocity energy of the fluid into pressure energy, either by the uniformly increasing area of the volute or by diffuser guide vanes (when provided).

PUMP TYPES

Most centrifugal pumps used in hydronic systems are single-stage pumps with a single- or double-inlet impeller. Double-inlet pumps are generally used for high-flow applications, but either type is available with similar performance characteristics and efficiencies.

A centrifugal pump has either a volute or diffuser casing. Pumps with volute casings collect water from the impeller and discharge it perpendicular to the pump shaft. Casings with diffusers discharge water parallel to the pump shaft. All pumps described in this chapter have volute casings except the vertical turbine pump, which has a diffuser casing.

Pumps may be classified as close-coupled or flexible-coupled to the electric motor. The close-coupled pump has the impeller mounted on a motor shaft extension, and the flexible-coupled pump has an impeller shaft supported by a frame or bracket that is connected to the electric motor through a flexible coupling.

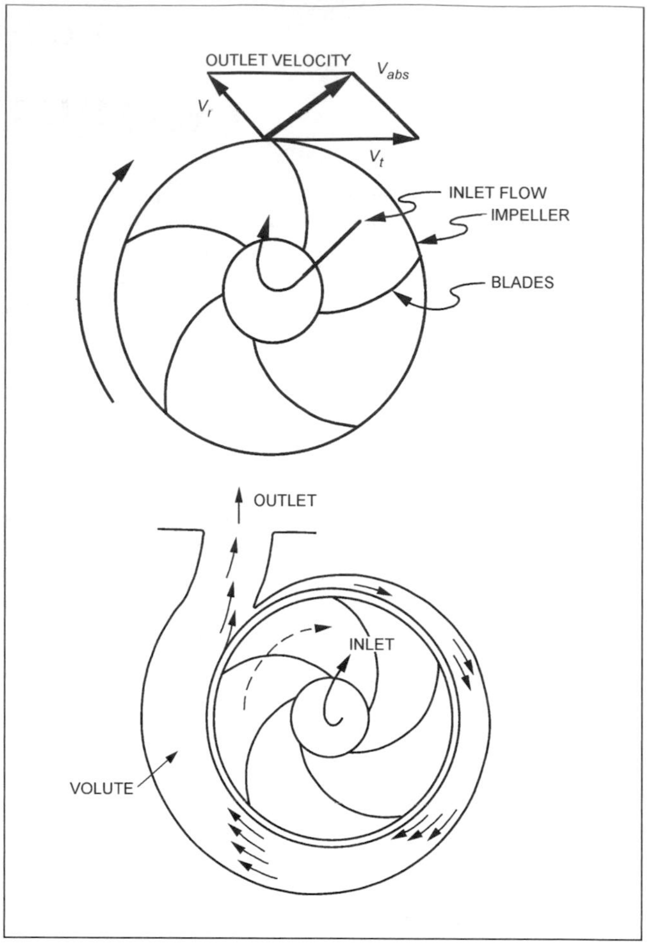

Fig. 2 Impeller and Volute Interaction

Pumps may also be classified by their mechanical features and installation arrangement. One-horsepower and larger pumps are available as close-coupled or base-mounted. Close-coupled pumps have an end-suction inlet for horizontal mounting or a vertical in-line inlet for direct installation in the piping. Base-mounted pumps are (1) end-suction, frame-mounted or (2) double-suction, horizontal or vertical split-case units. Double-suction pumps can also be arranged in a vertical position on a support frame with the motor vertically mounted on a bracket above the pump. Pumps are usually labeled by their mounting position as either horizontal or vertical.

Circulator Pump

Circulator is a generic term for a pipe-mounted, low-pressure, low-capacity pump (Figure 3). This pump may have a wet rotor or may be driven by a close-coupled or flexible-coupled motor. Circulator pumps are commonly used in residential and small commercial buildings to circulate source water and to recirculate the flow of terminal coils to enhance heat transfer and improve the control of large systems.

Close-Coupled, Single-Stage, End-Suction Pump

The close-coupled pump is mounted on a horizontal motor supported by the motor foot mountings (Figure 4). Mounting usually requires a solid concrete pad. The motor is close-coupled to the

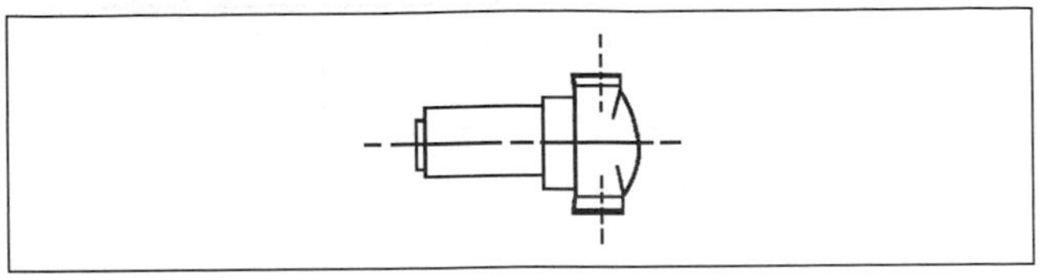

Fig. 3 Circulator Pump (Pipe-Mounted)

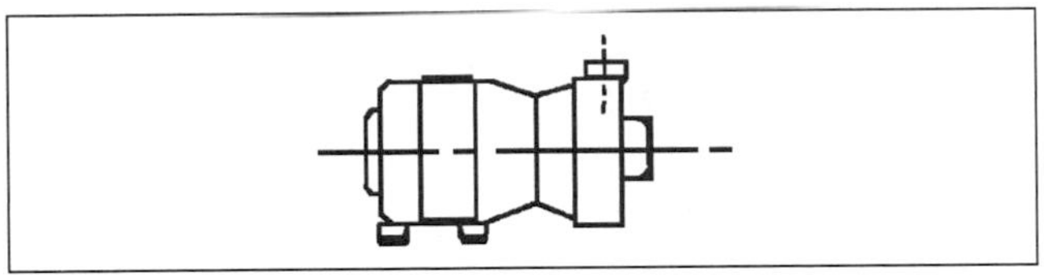

Fig. 4 Close-Coupled End-Suction Pump

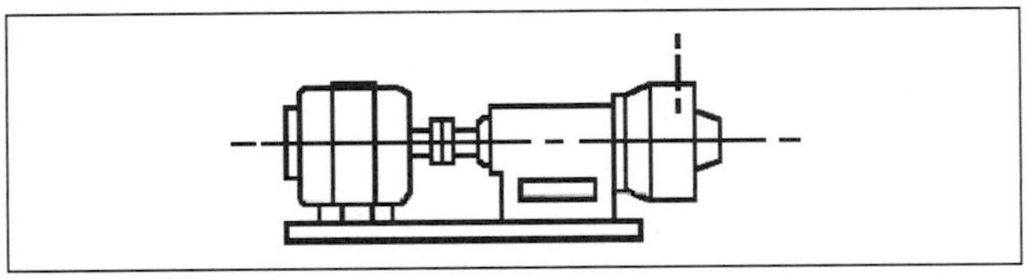

Fig. 5 Frame-Mounted End-Suction Pump on Base Plate

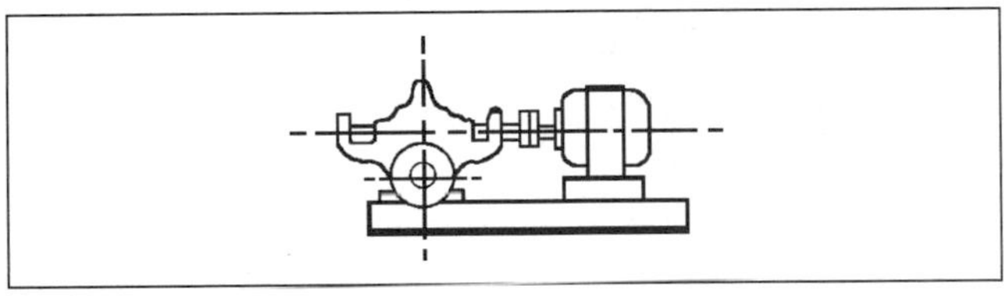

Fig. 6 Base-Mounted, Horizontal (Axial), Split-Case, Single-Stage, Double-Suction Pump

pump shaft. This compact pump has a single horizontal inlet and vertical discharge. It may have one or two impellers.

Frame-Mounted, End-Suction Pump on Base Plate

Typically, the motor and pump are mounted on a common, rigid base plate for horizontal mounting (Figure 5). Mounting requires a solid concrete pad. The motor is flexible-coupled to the pump shaft and should have an OSHA-approved guard. For horizontal mounting, the piping is horizontal on the suction side and vertical on the discharge side. This pump has a single suction.

Base-Mounted, Horizontal (Axial) or Vertical, Split-Case, Single-Stage, Double-Suction Pump

The motor and pump are mounted on a common, rigid base plate for horizontal mounting (Figures 6 and 7). Sometimes axial pumps are vertically mounted with a vertical pump casing and motor mounting bracket. Mounting requires a solid concrete pad. The motor is flexible-coupled to the pump shaft, and the coupling should have an OSHA-approved guard. A split case permits complete access to the impeller for maintenance. This pump may have one or two double suction impellers.

Base-Mounted, Horizontal, Split-Case, Multistage Pump

The motor and pump are mounted on a common, rigid base plate for horizontal mounting (Figure 8). Mounting requires a solid concrete pad. The motor is flexible-coupled to the pump shaft, and the coupling should have an OSHA-approved guard. Piping is horizontal on both the suction and discharge sides. The split case permits

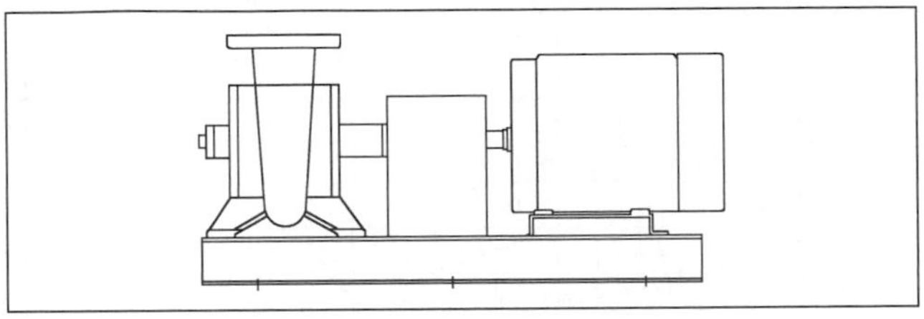

Fig. 7 Base-Mounted, Vertical, Split-Case, Single-Stage, Double-Suction Pump

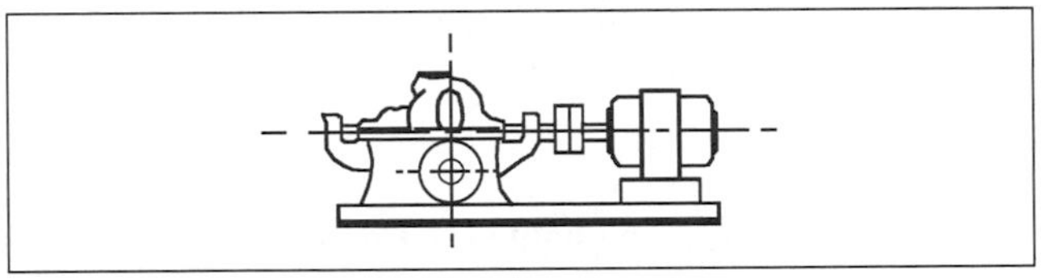

Fig. 8 Base-Mounted, Horizontal, Split-Case, Multistage Pump

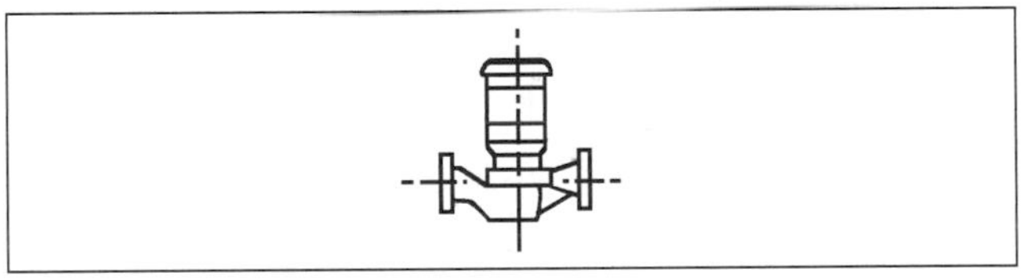

Fig. 9 Vertical In-Line Pump

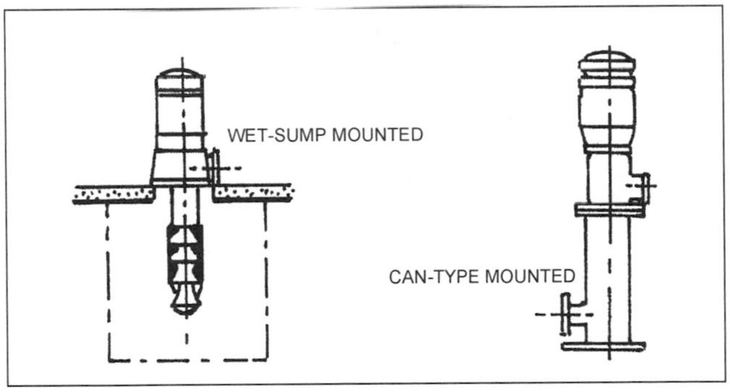

Fig. 10 Vertical Turbine Pumps

complete access to the impellers for maintenance. This pump has a single suction and may have one or more impellers for multistage operation.

Vertical In-Line Pump

This close-coupled pump and motor are mounted on the pump casing (Figure 9). The unit is compact and depends on the connected piping for support. Mounting requires adequately spaced pipe hangers and, sometimes, a vertical casing support. The suction and discharge piping is horizontal. The pump has a single or double suction impeller.

Vertical Turbine, Single- or Multistage, Sump-Mounted Pump

Vertical turbine pumps have a motor mounted vertically on the pump discharge head for either wet-sump mounting or can-type mounting (Figure 10). This single-suction pump may have one or multiple impellers for multistage operation. Mounting requires a solid concrete pad or steel sole plate above the wet pit with accessibility to the screens or trash rack on the suction side for

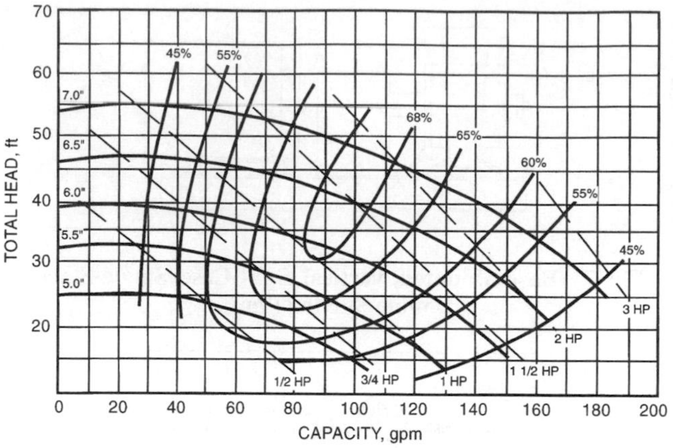

Fig. 11 Typical Pump Performance Curve

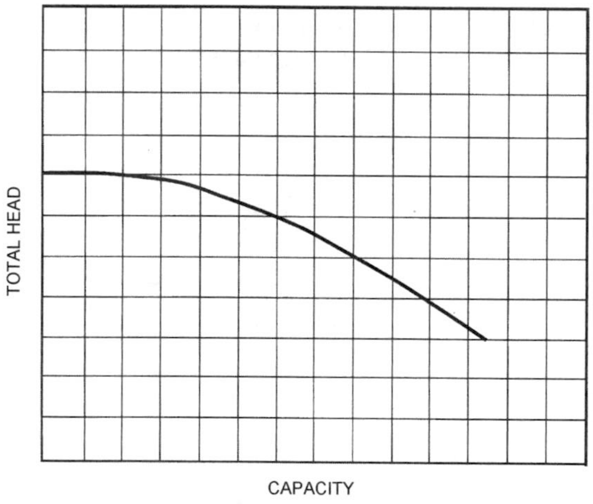

Fig. 12 Typical Pump Curve

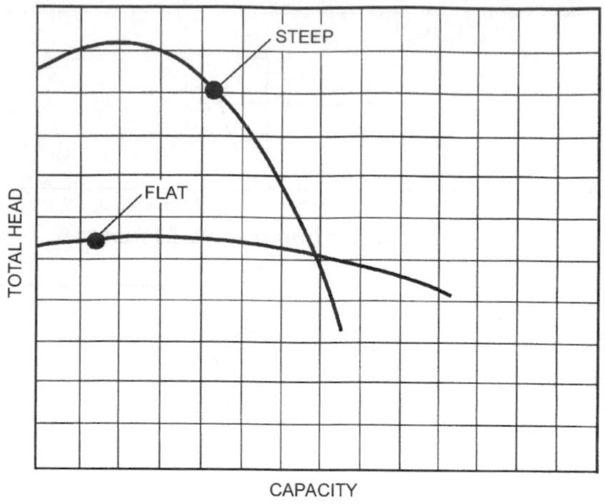

Fig. 13 Flat Versus Steep Performance Curves

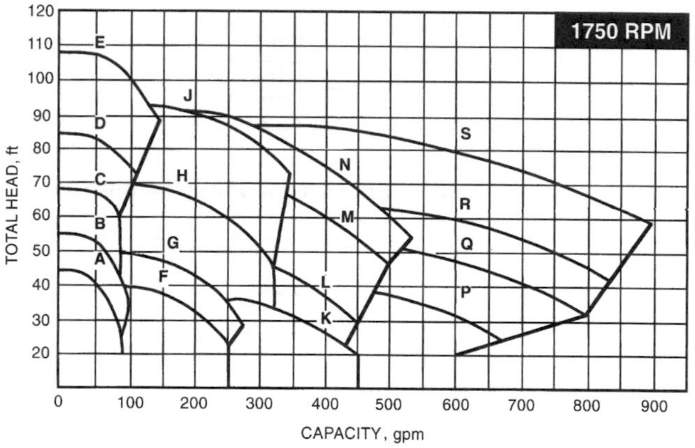

Fig. 14 Typical Pump Manufacturer's Performance Curve Series

maintenance. Can-type mounting requires a suction strainer. Piping is horizontal on the discharge side and on the suction side. The sump should be designed according to Hydraulic Institute (1994) recommendations.

PUMP PERFORMANCE CURVES

Performance of a centrifugal pump is commonly shown by a manufacturer's performance curve (Figure 11). The figure displays the pump power required for a liquid with a specific gravity of 1.0 (water) over a particular range of impeller diameters and flows. The curves are generated from a set of standard tests developed by the Hydraulic Institute (1994). The tests are performed by the manufacturer for a given pump volute or casing and several impeller diameters, normally from the maximum to the minimum allowable in that volute. The tests are conducted at a constant impeller speed for various flows.

Pump curves represent the average results from testing several pumps of identical design under the same conditions. The curve is sometimes called the head-capacity curve (H-Q) for the pump. Typically, the discharge head of the centrifugal pump, sometimes called the total dynamic head (TDH), is measured in feet of water flowing at a standard temperature and pressure. TDH represents the difference in total head between the suction side and the discharge side of the pump. This discharge head decreases as the flow increases (Figure 12). Motors are often selected to be non-overloading at a specified impeller size and maximum flow to ensure safe motor operation at all flow requirements.

The pump characteristic curve may be further described as flat or steep (Figure 13). Sometimes these curves are described as a normal rising curve (flat), a drooping curve (steep), or a steeply rising curve. The pump curve is considered flat if the pressure at shutoff is about 1.10 to 1.20 times the pressure at the best efficiency point. Flat characteristic pumps are usually installed in closed systems with modulating two-way control valves. Steep characteristic pumps are usually installed in open systems, such as cooling towers (see Chapter 13), where higher head and constant flow are usually desired.

Pump manufacturers may compile performance curves for a particular set of pump volutes in a series (Figure 14). The individual curves are shown in the form of an envelope consisting of the maximum and minimum impeller diameters and the ends of their curves. This set of curves is known as a family of curves. A family of curves is useful in determining the approximate size and model required, but the particular pump curve (Figure 11) must then be used to confirm an accurate selection.

Many pump manufacturers and HVAC software suppliers offer electronic versions for pump selection (Figure 14A). Pump selection software typically allows the investigation of different types of pumps and operating parameters. Corrections for fluid specific gravity, temperature, and motor speeds are easily performed.

HYDRONIC SYSTEM CURVES

Pressure drop caused by the friction of a fluid flowing in a pipe may be described by the Darcy-Weisbach equation:

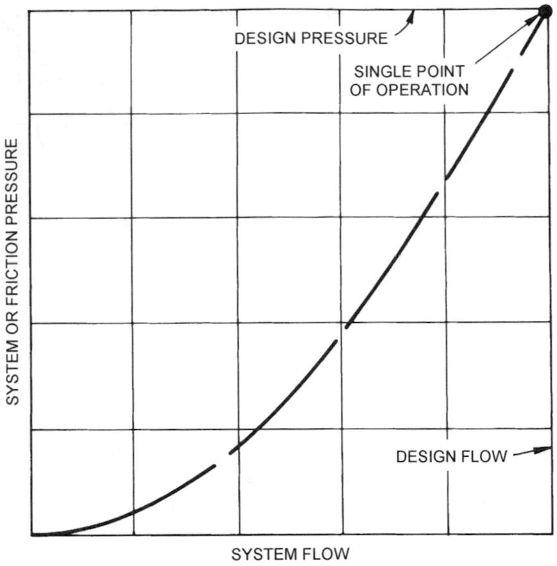

Fig. 15 Typical System Curve

$$\Delta p = f \frac{L}{D} \frac{\rho}{g} \frac{V^2}{2} \qquad (1)$$

Equation (1) shows that pressure drop in a hydronic system (pipe, fittings, and equipment) is proportional to the square of the flow (V^2 or Q^2 where Q is the flow). Experiments show that pressure drop is more nearly proportional to between $V^{1.85}$ and $V^{1.9}$, or a nearly parabolic curve as shown in Figures 15 and 18. The design of the system (including the number of terminals and flows, the fittings and valves, and the length of pipe mains and branches) affects the shape of this curve.

Equation (1) may also be expressed in head or specific energy form:

$$\Delta h = \frac{\Delta p}{\rho} = f \frac{L}{D} \frac{V^2}{2g} \qquad (2)$$

where

Δh = head loss through friction, ft (of fluid flowing)
Δp = pressure drop, lb/ft^2
ρ = fluid density, lb/ft^3
f = friction factor, dimensionless
L = pipe length, ft
D = inside diameter of pipe, ft
V = fluid average velocity, ft/s
g = gravitational acceleration, 32.2 ft/s^2

The system curve (Figure 15) defines the system head required to produce a given flow rate for a liquid and its characteristics in a piping system design. To produce a given flow, the system head must overcome pipe friction, inside pipe surface roughness, actual fitting losses, actual valve losses, resistance to flow due to fluid viscosity, and possible system effect losses. The general shape of this curve is parabolic since, according to the Darcy-Weisbach equation [Equation (2)], the head loss is proportional to the square of the flow.

If static pressure is present due to the height of the liquid in the system or the pressure in a compression tank, this head is sometimes referred to as **independent head** and is added to the system curve (Figure 16).

PUMP AND HYDRONIC SYSTEM CURVES

The pump curve and the system curve can be plotted on the same graph. The intersection of the two curves (Figure 17) is the system **operating point**, where the pump's developed head matches the system's head loss.

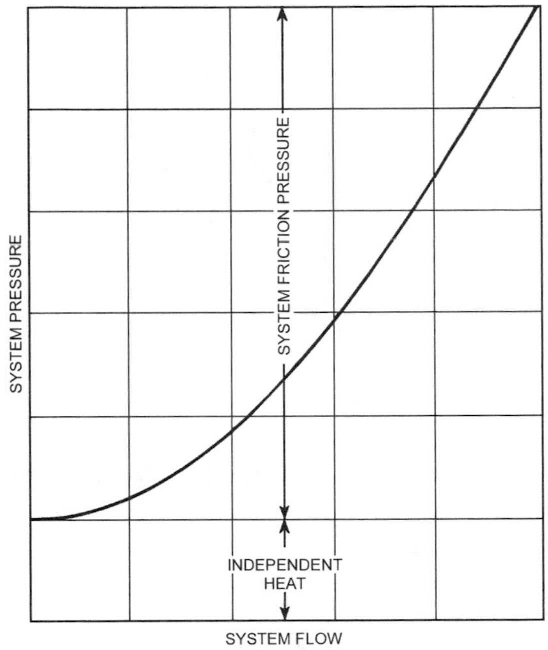

Fig. 16 Typical System Curve with Independent Head

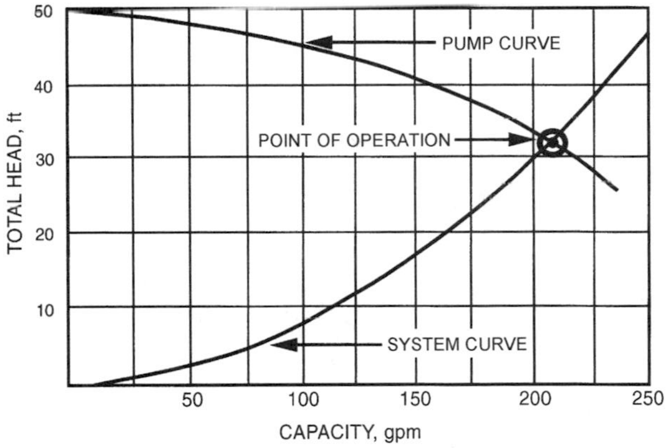

Fig. 17 System and Pump Curves

In a typical hydronic system, a thermostat or controller varies the flow in a load terminal by positioning a two-way control valve to match the load. At full load the two-way valves are wide open, and the system follows curve A in Figure 18. As the load drops, the terminal valves begin closing to match the load (part load). This increases the friction and reduces the flow in the terminals. The system curve gradually changes to curve B.

The operating point of a pump should be considered when the system includes two-way control valves. Point 1 in Figure 19 shows the pump operating at the design flow at the calculated design head loss of the system. But typically, the actual system curve is slightly different than the design curve. As a result, the pump operates at point 2 and produces a flow rate higher than design.

To reduce the actual flow to the design flow at point 1, a balancing valve downstream from the pump can be adjusted while all the terminal valves are in a wide-open position. This pump discharge balancing valve imposes a pressure drop equal to the pressure difference between point 1 and point 3. The manufacturer's pump curve shows that the capacity may be reduced by substituting a new

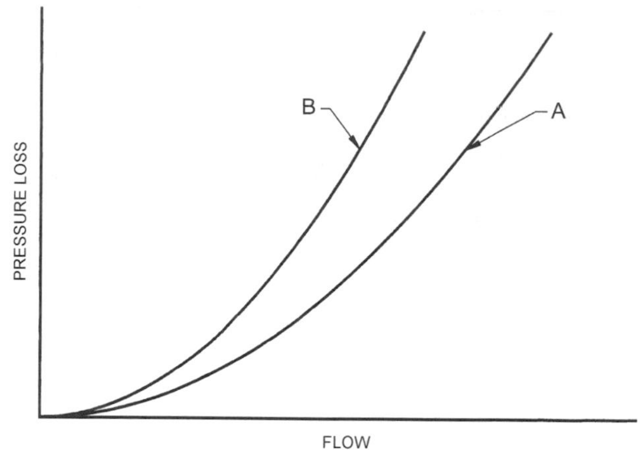

Fig. 18 System Curve Change due to Part-Load Flow

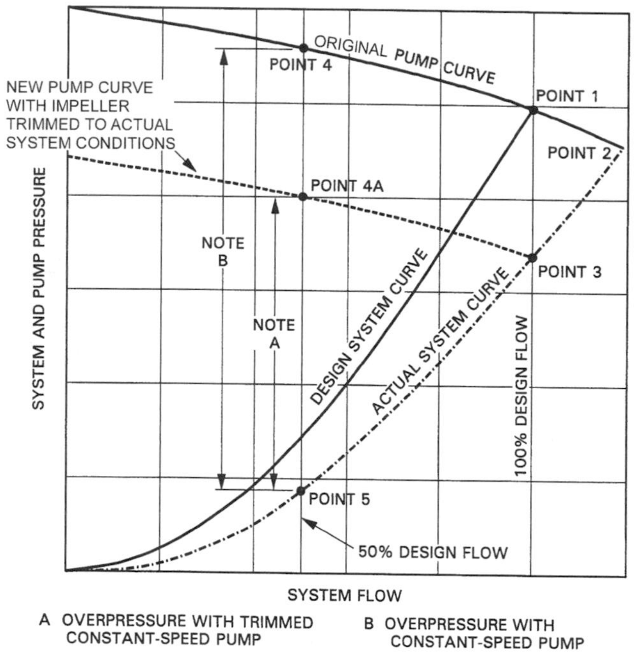

Fig. 19 Pump Operating Points

impeller with a smaller diameter or by trimming the existing pump impeller. After trimming, reopening the balancing valve in the pump discharge then eliminates the artificial drop and the pump operates at point 3. Points 3 and 4A demonstrate the effect a trimmed impeller has on reducing flow.

Figure 20 is an example of a system curve with both fixed head loss and variable head loss. Such a system might be an open piping circuit between a refrigerating plant condenser and its cooling tower. The elevation difference between the water level in the tower pan and the spray distribution pipe creates the fixed head loss. The fixed loss occurs at all flow rates and is, therefore, an independent head as shown.

Most variable-flow hydronic systems have individual two-way control valves on each terminal unit to permit full diversity (random loading from zero to full load). Regardless of the load required in most variable-flow systems, the designer establishes a minimum pressure difference to ensure that any terminal and its control valve receive the design flow at full demand. When graphing a system curve for a nonsymmetrically loaded variable-flow system (Figure 20), the Δh (minimum maintained pressure difference) is treated like a fixed pressure loss (independent head), and becomes the starting datum for the system curve.

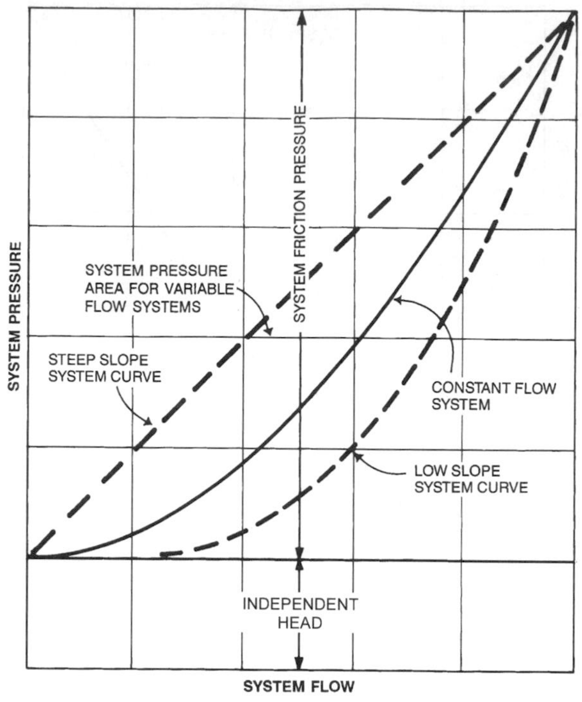

Fig. 20 System Curve with System Static Pressure

The low slope and steep slope curves in Figure 20 represent the boundaries for operation of the system. The net vertical difference between the curves is the difference in friction loss developed by the distribution mains for the two extremes of possible loads. The area in which the system operates depends on the diverse loading or unloading imposed by the terminal units. This area represents the pumping energy that can be conserved with one-speed, two-speed, or variable-speed pumps after a review of the pump power, efficiency, and affinity relationships.

PUMP POWER

The theoretical power to circulate water in a hydronic system is the **water horsepower** (whp) and is calculated as follows:

$$\text{whp} = \frac{\dot{m}\Delta h}{33{,}000} \tag{3}$$

where

$\dot{m}$ = mass flow of fluid, lb/min
Δh = total head, ft of fluid
33,000 = units conversion, ft·lb/min per hp

At 68°F, water has a density of 62.3 lb/ft^3, and Equation (3) becomes

$$\text{whp} = \frac{Q\Delta h}{3960} \tag{4}$$

where

Q = fluid flow rate, gpm
3960 = units conversion, ft·gpm per hp

Figure 21 shows how water power increases with flow. At other water temperatures or for other fluids, Equation (4) is corrected by multiplying by the specific gravity of the fluid.

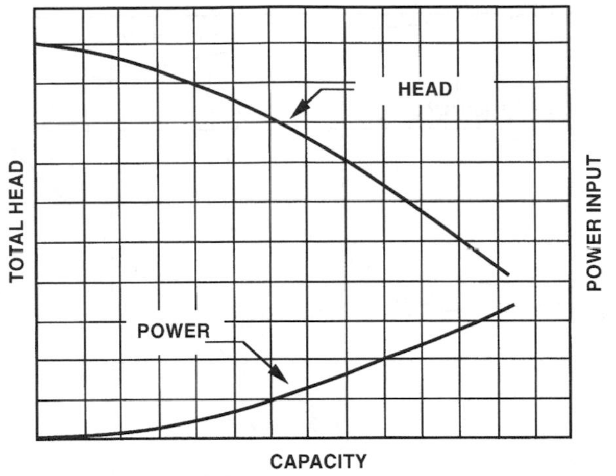

Fig. 21 Typical Pump Water Power Increase with Flow

The **brake horsepower** (bhp) required to operate the pump is determined by the manufacturer's test of an actual pump running under standard conditions to produce the required flow and head as shown in Figure 11.

PUMP EFFICIENCY

Pump efficiency is determined by comparing the output power to the input power:

$$\text{Efficiency} = \frac{\text{Output}}{\text{Input}} = \frac{\text{whp}}{\text{bhp}} \times 100\% \quad (5)$$

Figure 22 shows a typical efficiency versus flow curve.

The pump manufacturer usually plots the efficiencies for a given volute and impeller size on the pump curve to help the designer select the proper pump (Figure 23). The best efficiency point (BEP) is the optimum efficiency for this pump—operation above and below this point is less efficient. The locus of all the BEPs for each impeller size lies on a system curve that passes through the origin (Figure 24).

AFFINITY LAWS

The centrifugal pump, which imparts a velocity to a fluid and converts the velocity energy to pressure energy, can be categorized by a set of relationships called **affinity laws** (Table 1). The laws can be described as similarity processes that follow these rules:

1. Flow (capacity) varies with rotating speed N (i.e., the peripheral velocity of the impeller).
2. Head varies as the square of the rotating speed.
3. Brake horsepower varies as the cube of the rotating speed.

The affinity laws are useful for estimating pump performance at different rotating speeds or impeller diameters D based on a pump with known characteristics. The following two variations can be analyzed by these relationships:

1. By changing speed and maintaining constant impeller diameter, pump efficiency remains the same, but head, capacity, and brake horsepower vary according to the affinity laws.
2. By changing impeller diameter and maintaining constant speed, the pump efficiency for a diffuser pump is not affected if the impeller diameter is changed by less than 5%. However, efficiency changes if the impeller size is reduced enough to affect the clearance between the casing and the periphery of the impeller.

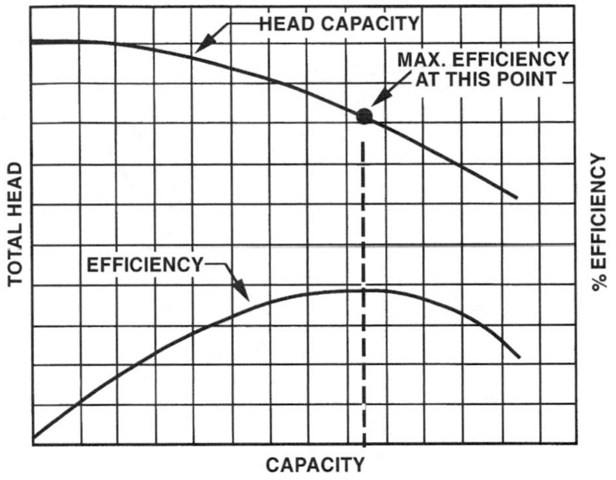

Fig. 22 Pump Efficiency Versus Flow

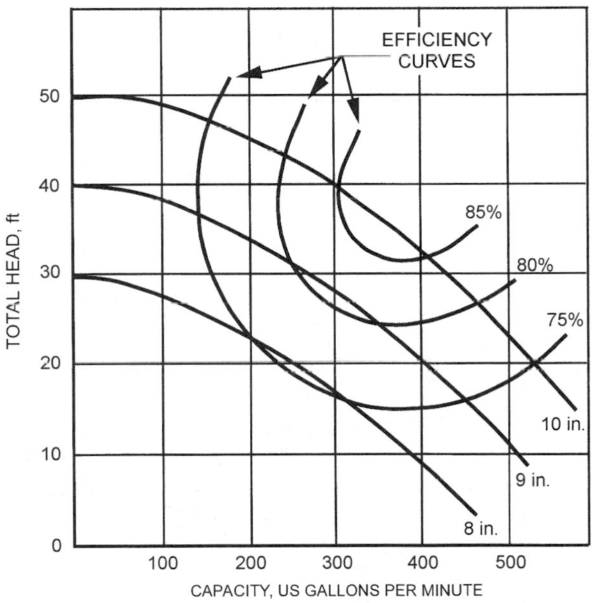

Fig. 23 Pump Efficiency Curves

Table 1 Pump Affinity Laws

Function	Speed Change	Impeller Diameter Change
Flow	$Q_2 = Q_1\left(\frac{N_2}{N_1}\right)$	$Q_2 = Q_1\left(\frac{D_2}{D_1}\right)$
Head	$h_2 = h_1\left(\frac{N_2}{N_1}\right)^2$	$h_2 = h_1\left(\frac{D_2}{D_1}\right)^2$
Horsepower	$\text{bhp}_2 = \text{bhp}_1\left(\frac{N_2}{N_1}\right)^3$	$\text{bhp}_2 = \text{bhp}_1\left(\frac{D_2}{D_1}\right)^3$

The affinity laws assume that the system curve is known and that head varies as the square of flow. The operating point is the intersection of the total system curve and the pump curve (Figure 17). Because the affinity law is used to calculate a new condition due to a flow or head change (e.g., reduced pump speed or impeller diameter), this new condition also follows the same system curve. Figure 25 shows the relationship of flow, head, and power as expressed by the affinity laws.

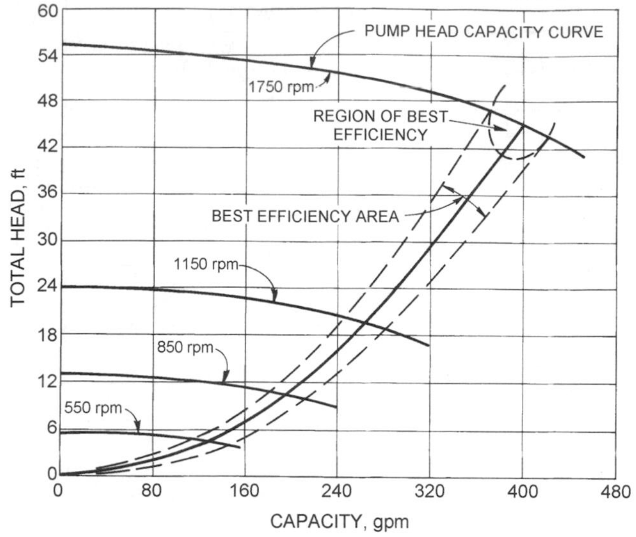

Fig. 24 Pump Best Efficiency Curves

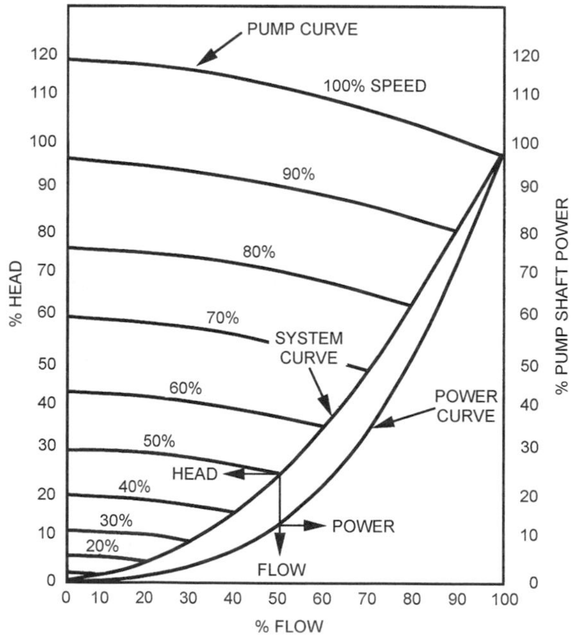

Fig. 25 Pumping Power, Head, and Flow Versus Pump Speed

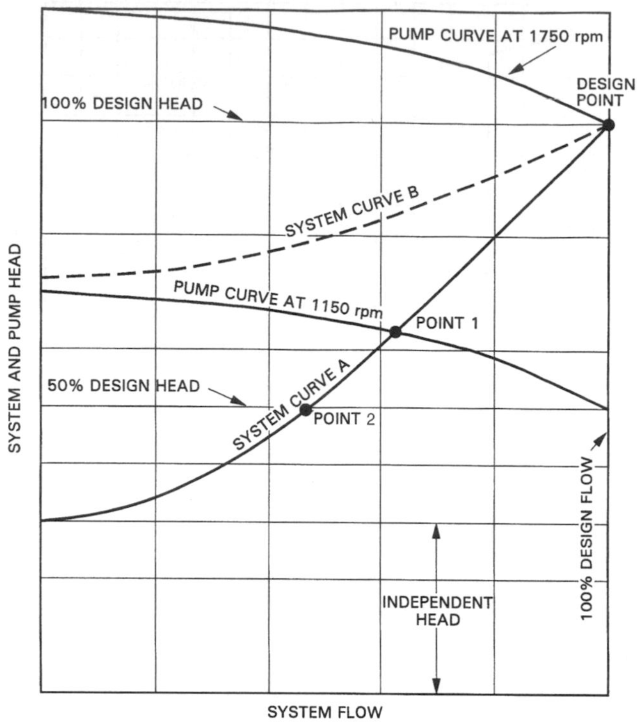

Fig. 26 Example Application of Affinity Law

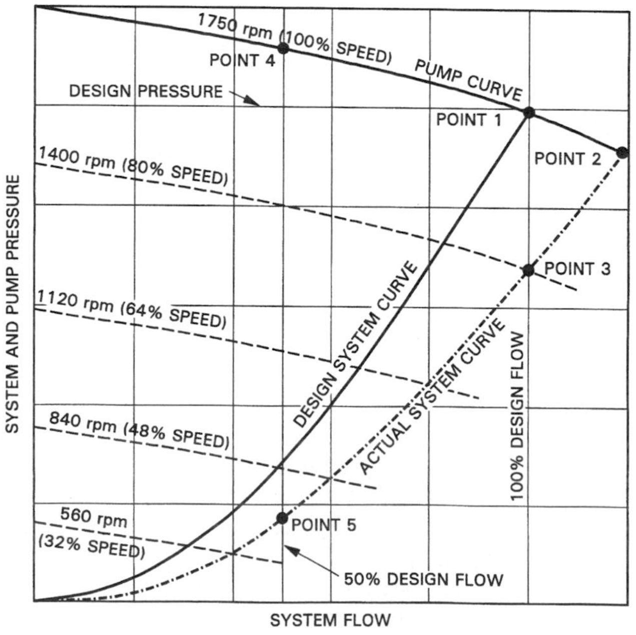

Fig. 27 Variable-Speed Pump Operating Points

The affinity laws can also be used to predict the BEP at other pump speeds. As discussed in the section on Pump Efficiency, the BEP follows a parabolic curve to zero as the pump speed is reduced (Figure 24).

Multiple-speed motors can be used to reduce system overpressure at reduced flow. Standard two-speed motors are available in models with speeds of 1750/1150 rpm, 1750/850 rpm, 1150/850 rpm, and 3500/1750 rpm. Figure 26 shows the performance of a system with a 1750/1150 multiple-speed pump. In the figure, curve A shows a system's response when the pump runs at 1750 rpm. When the pump runs at 1150 rpm, operation is at point 1 and not at point 2 as the affinity laws predict. If the system were designed to operate as shown in curve B, the pump would operate at shutoff and be damaged if run at 1150 rpm. This example demonstrates that the designer must analyze the system carefully to determine the pump limitation and the effect of lower speed on performance.

Variable-speed drives have a similar effect on pump curves. These drives normally have an infinitely variable speed range, so that the pump, with proper controls, follows the system curve without any overpressure. Figure 27 shows operation of the pump in Figure 19 at 100%, 80%, 64%, 48% and 32% of the speed.

Although the variable-speed motor in this example can correct overpressure conditions, about 20% of the operating range of the variable-speed motor is not used. The maximum speed required to provide design flow and pressure is 80% of full speed (1400 rpm) and the practical lower limit is 30% (525 rpm) due to the characteristics of the system curve as well as pump and motor limitations.

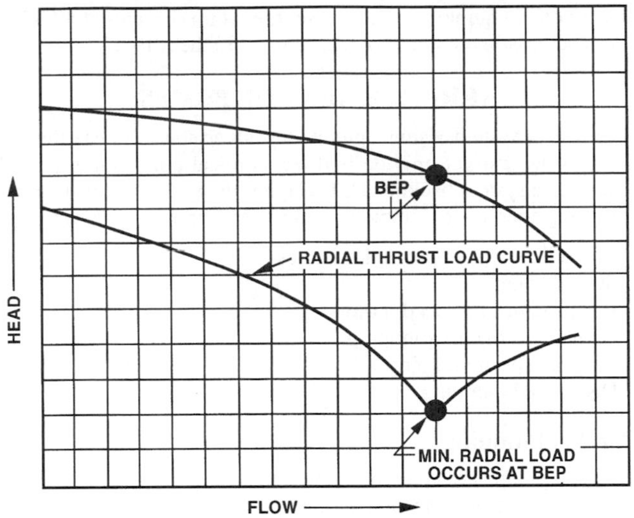

Fig. 28 Radial Thrust Versus Pumping Rate

The variable-speed drive and motor should be sized for actual balanced hydronic conditions.

The affinity laws can also be used to predict the effect of trimming the impeller to reduce overpressure. If, for example, the system shown in Figure 27 were allowed to operate at point 2, an excess flow of 15% to 25% would occur, depending on the shapes of the system and pump curves. The pump affinity laws (Table 1) show that pump capacity varies directly with impeller diameter. In Figure 27, the correct size impeller operates at point 3. A diameter ratio of 0.8 would reduce the overcapacity from 125% to 100%.

The second affinity law shows that head varies as the square of the impeller diameter. For the pump in Figure 27 and an impeller diameter ratio of 0.8, the delivered head of the pump is $(0.8)^2 = 0.64$ or 64% of the design pump head.

The third affinity law states that pump power varies as the cube of the impeller diameter. For the pump in Figure 27 and an impeller diameter ratio of 0.8, the power necessary to provide the design flow is $(0.8)^3 = 0.512$ or 51.2% of the original pump's power.

RADIAL THRUST

In a single-volute centrifugal pump, uniform or near-uniform pressures act on the impeller at design capacity, which coincides with the BEP. However, at other capacities, the pressures around the impeller are not uniform and there is a resultant radial reaction.

Figure 28 shows the typical change in radial thrust with changes in the pumping rate. Specifically, radial thrust decreases from shutoff to the design capacity (if chosen at the BEP) and then increases as flow increases. The reaction at overcapacity is roughly opposite that at partial capacity and is greatest at shutoff. The radial forces at extremely low flow can cause severe impeller shaft deflection and, ultimately, shaft breakage. This danger is even greater with high-pressure pumps.

NET POSITIVE SUCTION CHARACTERISTICS

Particular attention must be given to the pressure and temperature of the water as it enters the pump, especially in condenser towers, steam condensate returns, and steam boiler feeds. If the absolute pressure at the suction nozzle approaches the vapor pressure of the liquid, vapor pockets form in the impeller passages. The collapse of the vapor pockets (**cavitation**) is noisy and can be destructive to the pump impeller.

The amount of pressure in excess of the vapor pressure required to prevent vapor pockets from forming is known as the

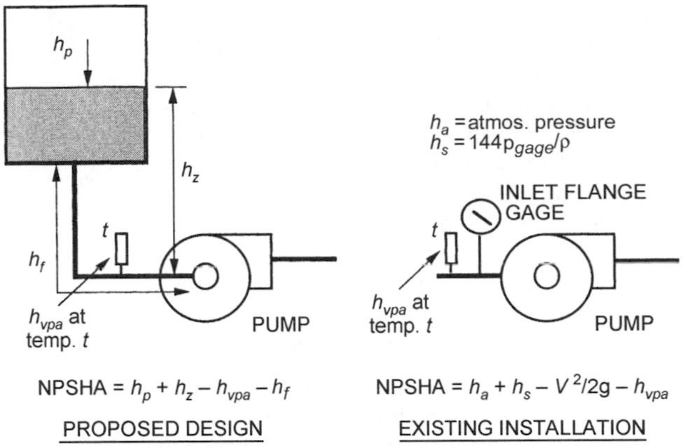

Fig. 29 Net Positive Suction Head Available

net positive suction head required (NPSHR). NPSHR is a characteristic of a given pump and varies with pump speed and flow. It is determined by the manufacturer and is included on the pump performance curve.

NPSHR is particularly important when a pump is operating with hot liquids or is applied to a circuit having a suction lift. The vapor pressure increases with water temperature and reduces the net positive suction head available (NPSHA). Each pump has its NPSHR, and the installation has its NPSHA, which is the total useful energy above the vapor pressure at the pump inlet.

The following equation may be used to determine the NPSHA in a proposed design (see Figure 29):

$$\text{NPSHA} = h_p + h_z - h_{vpa} - h_f \tag{6}$$

where

h_p = absolute pressure on surface of liquid that enters pump, ft of head
h_z = static elevation of liquid above center line of pump (h_z is negative if liquid level is below pump center line), ft
h_{vpa} = absolute vapor pressure at pumping temperature, ft
h_f = friction and head losses in suction piping, ft

To determine the NPSHA in an existing installation, the following equation may be used (see Figure 29):

$$\text{NPSHA} = h_a + h_s + \frac{V^2}{2g} - h_{vpa} \tag{7}$$

where

h_a = atmospheric head for elevation of installation, ft
h_s = head at inlet flange corrected to center line of pump (h_s is negative if below atmospheric pressure), ft
$V^2/2g$ = velocity head at point of measurement of h_s, ft

If the NPSHA is less than the pump's NPSHR, cavitation, noise, inadequate pumping, and mechanical problems will result. **For trouble-free design, the NPSHA must always be greater than the pump's NPSHR.** In closed hot and chilled water systems where sufficient system fill pressure is exerted on the pump suction, NPSHR is normally not a factor. Figure 30 shows pump curves and NPSHR curves. Cooling towers and other open systems require calculations of NPSHA.

SELECTION OF PUMPS

A substantial amount of data is required to ensure that an adequate, efficient, and reliable pump is selected for a particular system. The designer should review the following criteria:

- Design flow
- Pressure drop required for the most resistant loop
- Minimum system flow
- System pressure at maximum and minimum flows
- Type of control valve—two-way or three-way
- Continuous or variable flow
- Pump environment
- Number of pumps and standby
- Electric voltage and current
- Electric service and starting limitations
- Motor quality versus service life
- Water treatment, water conditions, and material selection

When a centrifugal pump is applied to a piping system, the operating point satisfies both the pump and system curves (Figure 17). As the load changes, control valves change the system curve and the operating point moves to a new point on the pump curve. Figure 31 shows the optimum regions to use when selecting a centrifugal pump. The areas bounded by lines AB and AC represent operating points that lie in the preferred pump selection range. But, because pumps are only manufactured in certain sizes, selection limits of 66% to 115% of flow at the BEP are suggested. The satisfactory range is that portion of a pump's performance curve where the combined effect of circulatory flow, turbulence, and friction losses are minimized. Where possible, pumps should be chosen to operate to the left of the BEP because the pressure in the actual system may be less than design due to overstated data for pipe friction and for other equipment. Otherwise, the pump operates at a higher flow and possibly in the turbulent region (Stethem 1988).

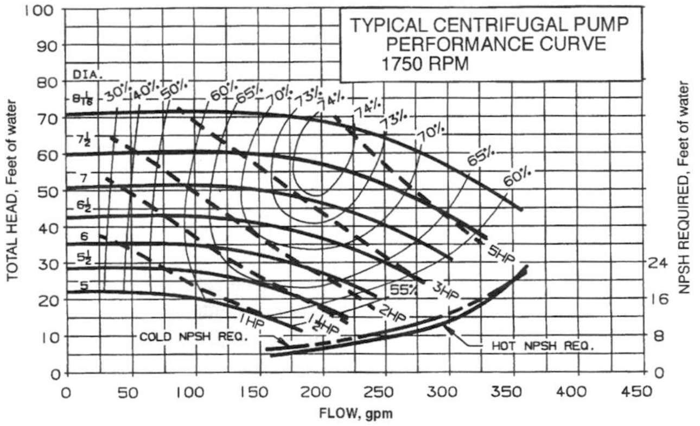

Fig. 30 Pump Performance and NPSHR Curves

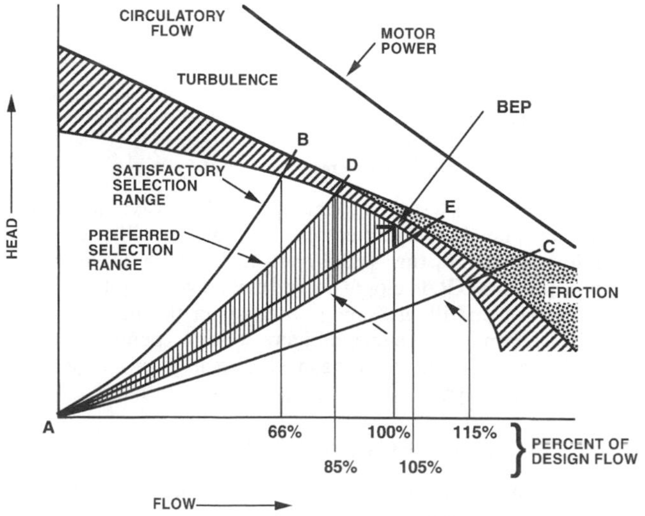

Fig. 31 Pump Selection Regions

ARRANGEMENT OF PUMPS

In a large system, a single pump may not be able to satisfy the full design flow and yet provide both economical operation at partial loads and a system backup. The designer may need to consider the following alternative pumping arrangements and control scenarios:

- Multiple pumps in parallel or series
- Standby pump
- Pumps with two-speed motors
- Primary-secondary pumping
- Variable-speed pumping
- Distributed pumping

Parallel Pumping

When pumps are applied in parallel, each pump operates at the same head and provides its share of the system flow at that head (Figure 32). Generally, pumps of equal size are recommended, and the parallel pump curve is established by doubling the flow of the single pump curve.

Plotting a system curve across the parallel pump curve shows the operating points for both single and parallel pump operation (Figure 33). Note that single pump operation does not yield 50% flow. The system curve crosses the single pump curve considerably to the right of its operating point when both pumps are running. This leads to two important

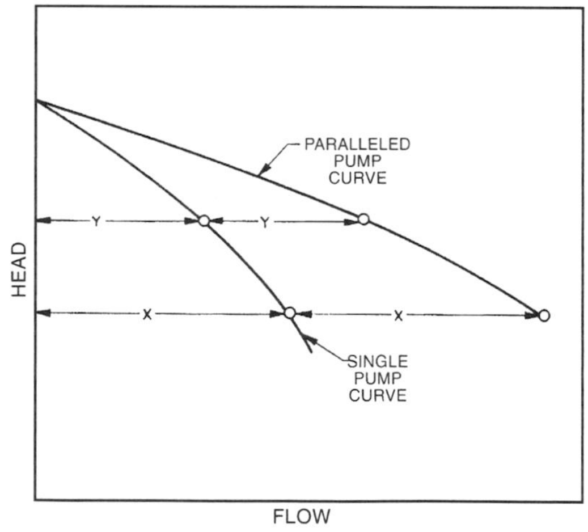

Fig. 32 Pump Curve Construction for Parallel Operation

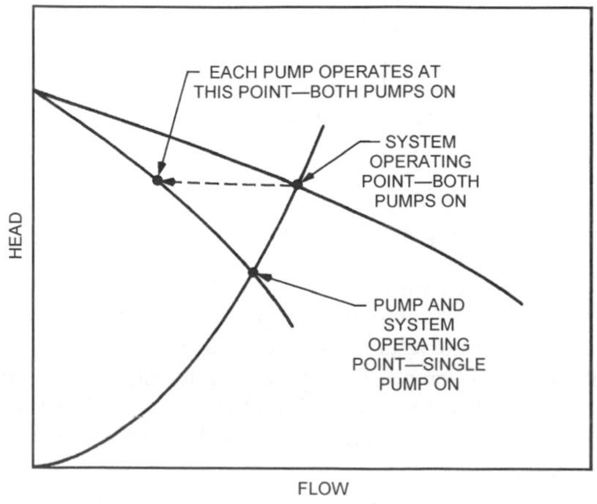

Fig. 33 Operating Conditions for Parallel Operation

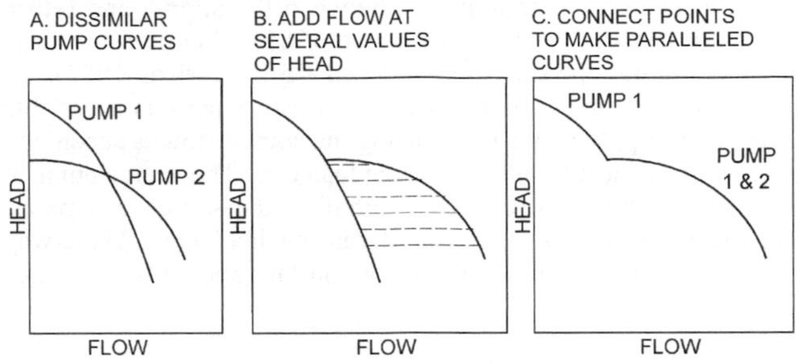

Fig. 34 Construction of Curve for Dissimilar Parallel Pumps

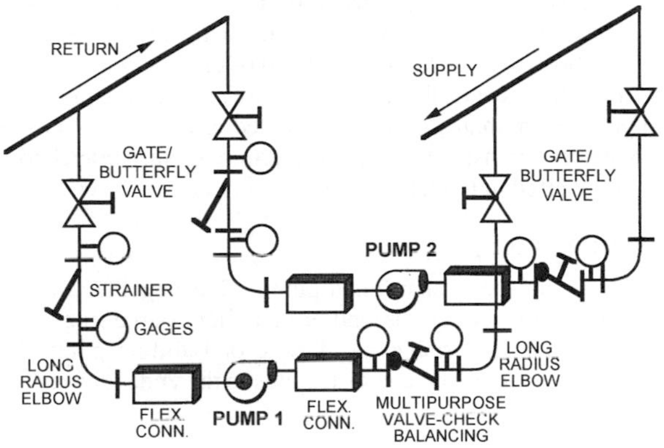

Fig. 35 Typical Piping for Parallel Pumps

concerns: (1) the motor must be selected to prevent overloading during operation of a single pump and (2) a single pump can provide standby service for up to 80% of the design flow, the actual amount depending on the specific pump curve and system curve.

Construction of the composite curve for two dissimilar parallel pumps requires special care; for example, note the shoulder in the composite pump curve in Figure 34.

Operation. The piping of parallel pumps (Figure 35) should permit running either pump. A check valve is required in each pump's discharge to prevent backflow when one pump is shut down. Hand valves and a strainer allow one pump to be serviced while the other is operating. A strainer protects a pump by preventing foreign material from entering the pump. Gages or a common gage with a trumpet valve, which includes several valves as one unit, or pressure taps permits checking pump operation.

Flow can be determined (1) by measuring the pressure increase across the pump and using a factory pump curve to convert the pressure to flow, or (2) by use of a flow-measuring station or multipurpose valve. Parallel pumps are often used for hydronic heating and cooling. In this application, both pumps operate during the cooling season to provide maximum flow and pressure, but only one pump operates during the heating season.

Series Pumping

When pumps are applied in series, each pump operates at the same flow rate and provides its share of the total pressure at that flow (Figure 36). A system curve plot shows the operating points for both single and series pump operation (Figure 37). Note that the single pump can provide up to 80% flow for standby and at a lower power requirement.

As with parallel pumps, piping for series pumps should permit running either pump (Figure 38). A bypass with a hand valve permits servicing one pump while the other is in operation. Operation and flow can be checked the same way as for parallel pumps. A strainer prevents foreign material from entering the pumps.

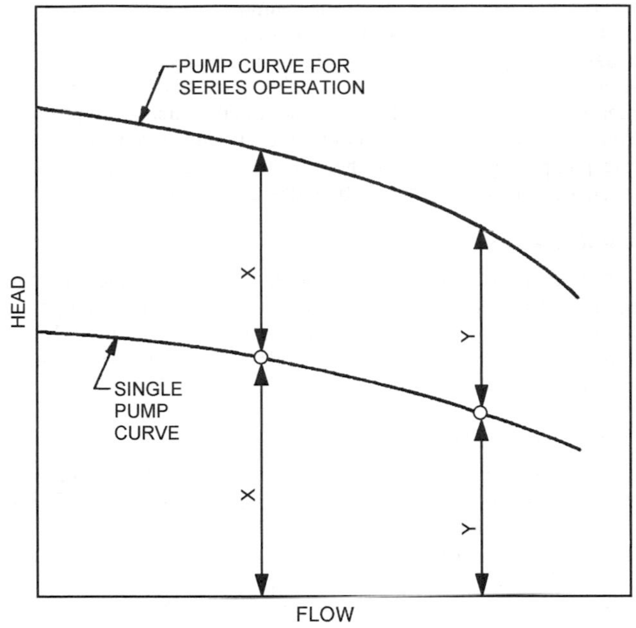

Fig. 36 Pump Curve Construction for Series Operation

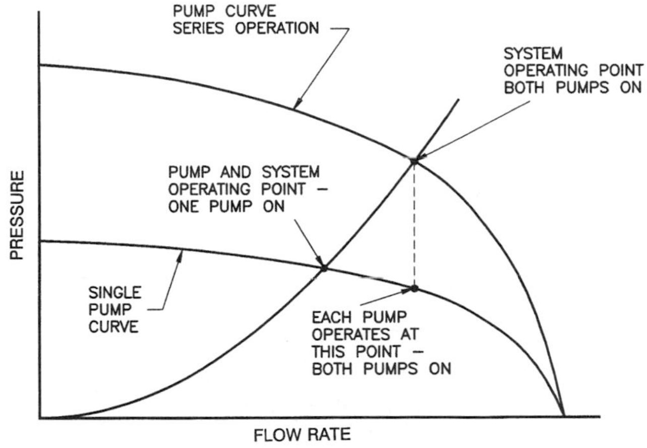

Fig. 37 Operating Conditions for Series Operation

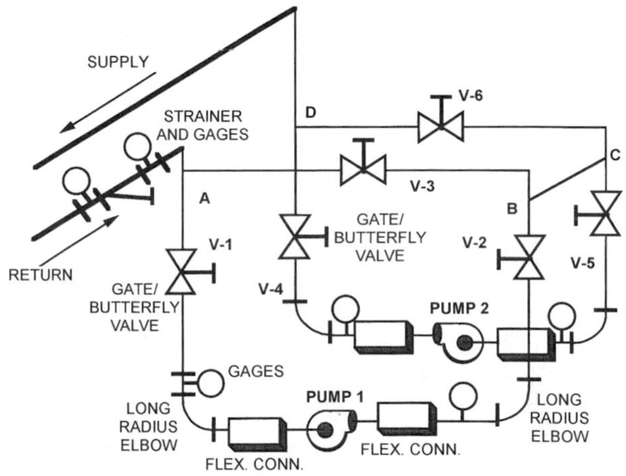

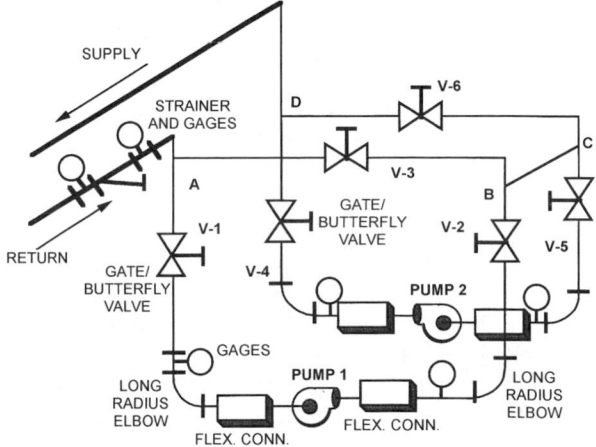

Fig. 38 Typical Piping for Series Pumps

Note that both parallel and series pump applications require that the pump operating points be used to accurately determine the actual pumping points. The manufacturer's pump test curve should be consulted. Adding too great a safety factor for pressure, using improper pressure drop charts, or incorrectly calculating pressure drops may lead to a poor selection. In designing systems with multiple pumps, operation in either parallel or series must be fully understood and considered by both designer and operator.

Standby Pump

A backup or standby pump of equal capacity and pressure installed in parallel to the main pump is recommended to operate during an emergency or to ensure continuous operation when a pump is taken out of operation for routine service. A standby pump installed in parallel with the main pump is shown in Figure 35.

Pumps with Two-Speed Motors

A pump with a two-speed motor provides a simple means of reducing capacity. As discussed in the section on Affinity Laws, pump capacity varies directly with impeller speed. At 1150 rpm, the capacity of a pump with a 1750/1150 rpm motor is 1150/1750 = 0.657 or 66% of the capacity at 1750 rpm.

Figure 39 shows an example (Stethem 1988) with two parallel two-speed pumps providing flows of 2130 gpm at 75 ft of head, 1670 gpm at 50.5 ft, 1250 gpm at 33 ft, and 985 gpm at 26.2 ft of head. Points A, B, C, and D will move left along the pump curve as loading changes.

Primary-Secondary Pumping

In a primary-secondary or compound pumping arrangement, a secondary pump is selected to provide the design flow in the load coil from the common pipe between the supply and return distribution mains (Figure 40) (Coad 1985). The pressure drop in the common pipe should not exceed 1.5 ft (Carlson 1972).

In circuit A of the figure, a two-way valve permits a variable flow in the supply mains by reducing the source flow; a secondary pump provides a constant flow in the load coil. The source pump at the chiller or boiler is selected to circulate the source and mains, and the secondary pump is sized for the load coil. Three-way valves in circuits B and C provide a constant source flow regardless of the load.

Variable-Speed Pumping

In a variable-speed pumping arrangement, constant flow pump(s) recirculate the chiller or boiler source in a primary source loop, and a variable-speed distribution pump located at the source plant draws flow from the source loop and distributes to the load terminals as shown in Figure 41. The speed of the distribution pump is determined by a controller measuring differential pressure across the supply-return mains or across selected critical zones.Two-way control valves are installed in the load terminal return branch to vary the flow required in the load.

Distributed Pumping

In a variable-speed distributed pumping arrangement, constant flow pump(s) recirculate the chiller or boiler source in a primary source loop, and a variable-speed zone or building pump draws flow from the source loop and distributes to the zone load terminals

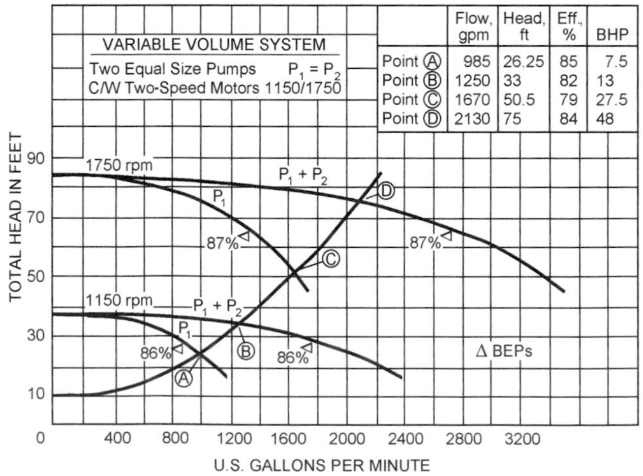

Fig. 39 Example of Two Parallel Pumps with Two-Speed Motors

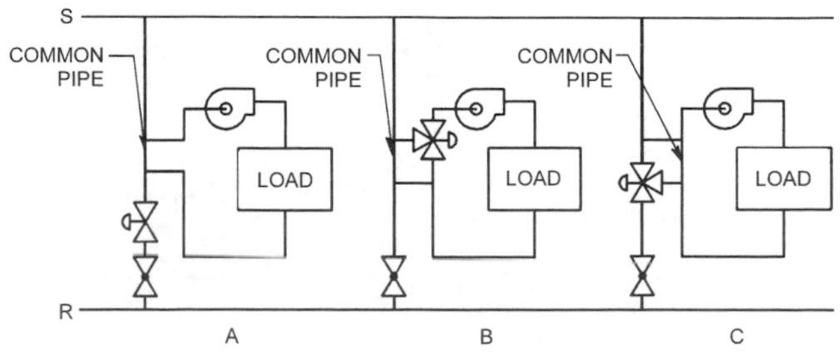

Fig. 40 Primary-Secondary Pumping

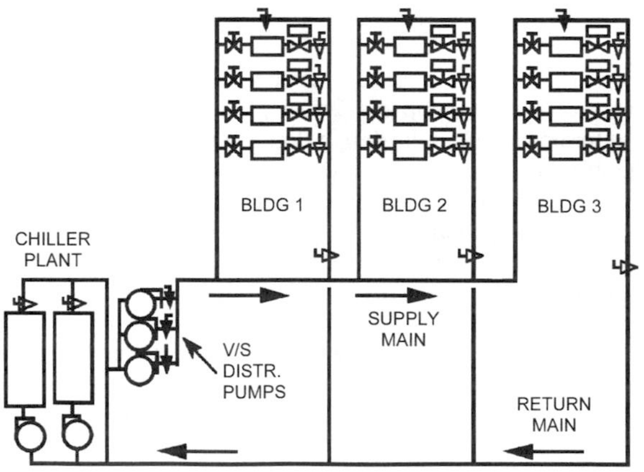

Fig. 41 Variable-Speed Source-Distributed Pumping

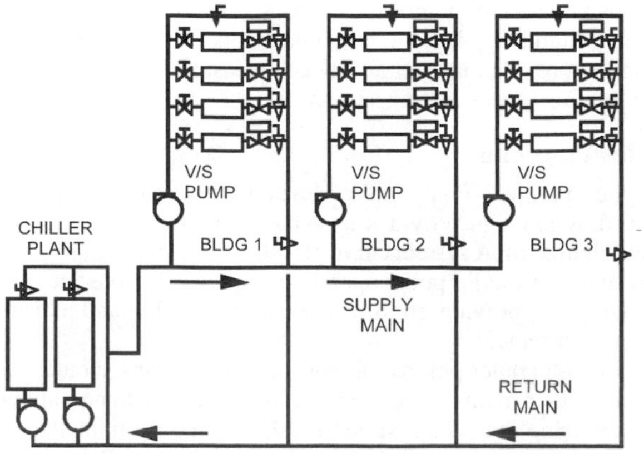

Fig. 42 Variable-Speed Distributed Pumping

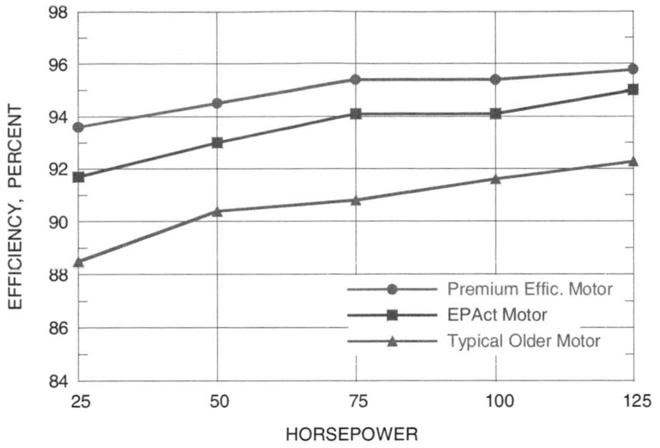

Fig. 43 Efficiency Comparison of Four-Pole Motors

as shown in Figure 42. The speed of the zone or building distribution pump is determined by a controller measuring zone differential pressure across supply-return mains or across selected critical zones. Two-way control valves in the load terminal return branch vary the flow required in the load.

MOTIVE POWER

Figure 43 demonstrates the improvement in efficiency of four-pole (1800 rpm) 25 to 125 hp motors from old, standard efficiency models to current standards and available premium efficiency models.

Electric motors drive most centrifugal pumps for hydronic systems. Internal combustion engines or steam turbines power some pumps, especially in central power plants for large installations. Electric motors for centrifugal pumps can be any of the horizontal or vertical electric motors described in Chapter 40. The sizing of electric motors is critical because of the cost of electric power and the desire for improved efficiency. Non-overloading motors should be used; that is, the motor nameplate rating must exceed the pump brake horsepower (kW) at any point on the pump curve.

Many pumps for hydronic systems are close-coupled, with the pump impeller mounted on the motor shaft extension. Other pumps are flexible-coupled to the electric motor through a pump mounting bracket or frame. A pump on a hydronic system with variable flow has a broad range of power requirements, which results in reduced motor loading at low flow.

Many variable speed drives (VSDs) are available for operating centrifugal pumps. Primarily, these include variable-frequency drives, and, occasionally, direct-current, wound rotor, and eddy current drives. Each drive has specific design features that should be evaluated for use with hydronic pumping systems. The efficiency range from minimum to maximum speed (as shown in Figure 44) should be investigated.

ENERGY CONSERVATION IN PUMPING

Pumps for heating and air conditioning consume appreciable amounts of energy. Economical use of energy depends on the efficiency of pumping equipment and drivers, as well as the use of the pumping energy required. Equipment efficiency (sometimes called the wire to water efficiency) shows how much energy applied to the pumping system results in useful energy distributing the water. For an electric-driven, constant speed pump, the equipment efficiency is

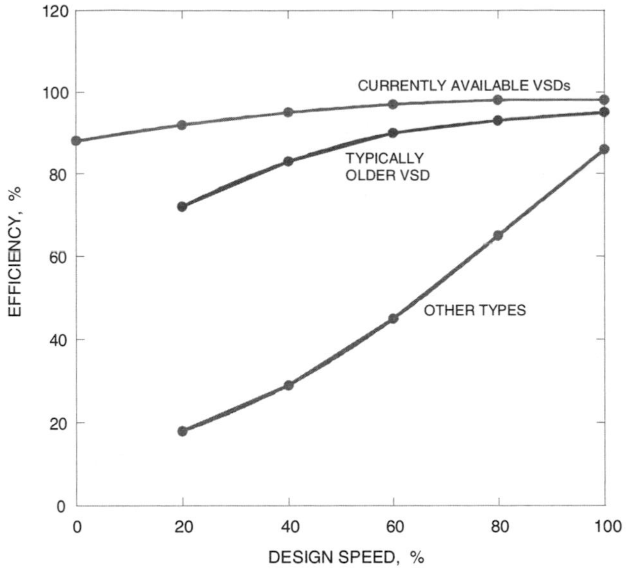

Fig. 44 Typical Efficiency Range of Variable-Speed Drives

$$\eta_e = \eta_p \eta_m \tag{8}$$

where

η_e = equipment efficiency, 0 to 1
η_p = pump efficiency, 0 to 1
η_m = motor efficiency, 0 to 1

For a variable-speed pump, the variable-speed drive efficiency η_v (0 to 1) must be included in the equipment efficiency equation:

$$\eta_e = \eta_p \eta_m \eta_v \tag{9}$$

INSTALLATION, OPERATION, AND COMMISSIONING

1. Pumps may be base plate-mounted (Figure 45), either singly or in packaged sets, or installed in-line directly in the piping system (Figure 46). Packaged sets include multiple pumps, accessories, and electrical controls shipped to the job site on one frame. Packaged pump sets may reduce the requirements for multiple piping and field electrical connections and can be factory tested to ensure specified performance.
2. A concrete pad provides a secure mounting surface for anchoring the pump base plate and raises the pump off the floor to permit housekeeping. The minimum weight of concrete that should be used is 2.5 times the weight of the pump assembly. The pad should be at least 4 in. thick and 6 in. wider than the pump base plate on each side.
3. In applications where the pump bolts rigidly to the pad base, level the pad base, anchor it, and fill the space between pump base and the concrete with a non-shrink grout. Grout prevents the base from shifting and fills in irregularities. Pumps mounted on vibration isolation bases require special installation (see the section on Vibration Isolation and Control in Chapter 46 of the 1999 *ASHRAE Handbook—Applications*).
4. Support in-line pumps independently from the piping so that pump flanges are not overstressed.
5. Once the pump has been mounted to the base, check the alignment of the motor to the pump. Align the pump shaft couplings properly and shim the motor base as required. Incorrect alignment may cause rapid coupling and bearing failure.

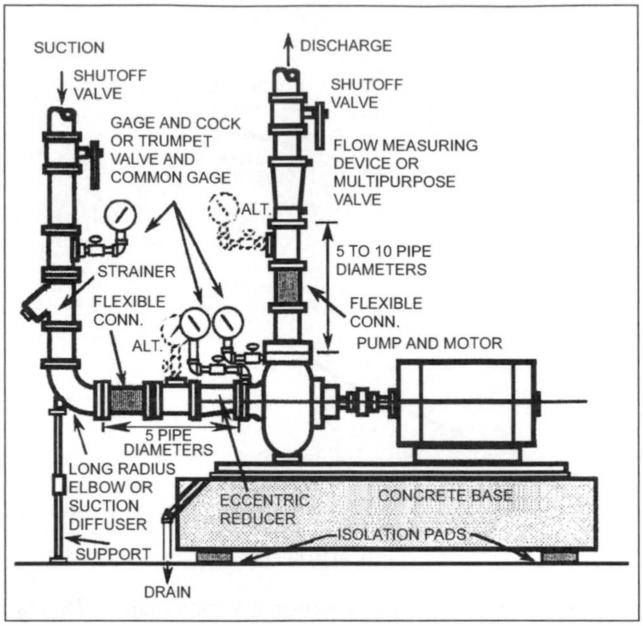

Fig. 45 Base Plate-Mounted Centrifugal Pump Installation

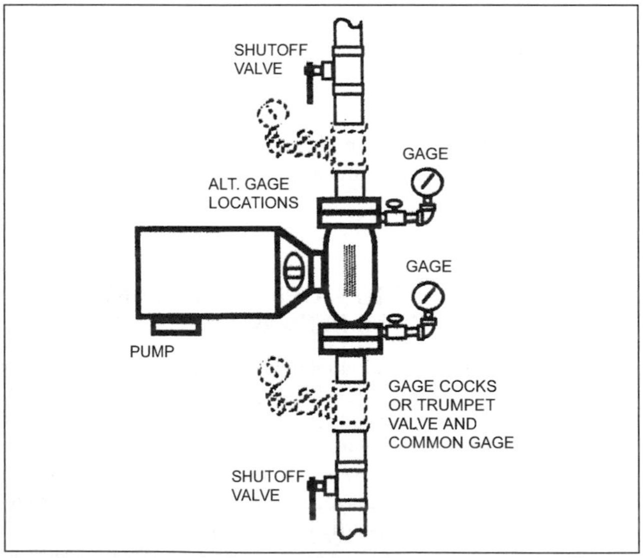

Fig. 46 In-Line Pump Installation

6. Pump suction piping should be direct and as smooth as possible. Install a strainer (coarse mesh) in the suction to remove foreign particles that can damage the pump. Use a straight section of piping at least 5 to 10 diameters long at the pump inlet and long radius elbows to ensure uniform flow distribution. Suction diffusers may be installed in lieu of the straight pipe requirement where spacing is a constraint. Eccentric reducers at the pump flange reduce the potential of air pockets forming in the suction line.
7. If a flow-measuring station (venturi, orifice plate, or balancing valve) is located in the pump discharge, allow 10 diameters of straight pipe between the pump discharge and the flow station for measurement accuracy.
8. Pipe flanges should match the size of pump flanges. Mate flat-face pump flanges with flat-face piping flanges and full-face

Table 2 Pumping System Noise Analysis Guide

Complaint	Possible Cause	Recommended Action
Pump or system noise	Shaft misalignment	• Check and realign
	Worn coupling	• Replace and realign
	Worn pump/motor bearings	• Replace, check manufacturer's lubrication recommendations • Check and realign shafts
	Improper foundation or installations	• Check foundation bolting or proper grouting • Check possible shifting caused by piping expansion/contraction. • Realign shafts
	Pipe vibration and/or strain caused by pipe expansion/contraction	• Inspect, alter, or add hangers and expansion provision to eliminate strain on pump(s)
	Water velocity	• Check actual pump performance against specified, and reduce impeller diameter as required • Check for excessive throttling by balance valves or control valves
	Pump operating close to or beyond end point of performance curve	• Check actual pump performance against specified, and reduce impeller diameter as required
	Entrained air or low suction pressure	• Check expansion tank connection to system relative to pump suction • If pumping from cooling tower sump or reservoir, check line size • Check actual ability of pump against installation requirements • Check for vortex entraining air into suction line

Table 3 Pumping System Flow Analysis Guide

Complaint	Possible Cause	Recommended Action
Inadequate or no circulation	Pump running backward (3-phase)	• Reverse any two motor leads
	Broken pump coupling	• Replace and realign
	Improper motor speed	• Check motor nameplate wiring and voltage
	Pump (or impeller diameter) too small	• Check pump selection (impeller diameter) against specified requirements
	Clogged strainer(s)	• Inspect and clean screen
	Clogged impeller	• Inspect and clean
	System not completely filled	• Check setting of PRV fill valve • Vent terminal units and piping high points
	Balance valves or isolating valves improperly set	• Check setting and adjust as required
	Air-bound system	• Vent piping and terminal units • Check location of expansion tank connection line relative to pump suction • Review provisions to eliminate air
	Air entrainment	• Check pump suction inlet conditions to determine if air is being entrained from suction tanks or sumps
	Insufficient NPSHR	• Check NPSHR of pump • Inspect strainers and check pipe sizing and water temperature

gaskets. Install tapered reducers and increasers on suction and discharge lines to match the pipe size and pump flanges.

9. If fine mesh screen is used in the strainer at initial start-up to remove residual debris, replace it with normal size screen after commissioning to protect the pump and minimize the suction pressure drop.
10. Install shutoff valves in the suction and discharge piping near the pump to permit removing and servicing the pump and strainer without draining the system. Install a check valve in the pump discharge to prevent reverse flow in a non-running pump when multiple pumps are installed.
11. Install vibration isolators in the pump suction and discharge lines to reduce the transmission of vibration noise to building spaces (Figure 45). Properly located pipe hangers and supports can reduce the transmission of piping strains to the pump.
12. Various accessories need to be studied as alternates to conventional fittings. A suction diffuser in the pump inlet is an alternate to an eccentric reducer and it contains a strainer. Separate strainers can be specified with screen size. A multipurpose valve in the pump discharge is an alternate way to combine the functions of shutoff, check, and balancing valves.
13. Each pump installation should include pressure gages and a gage cock to verify system pressures and pressure drop. As a minimum, pressure taps with an isolation valve and common gage should be available at the suction and discharge of the pump. An additional pressure tap upstream of the strainer permits checking for pressure drop.

TROUBLESHOOTING

Table 2 lists possible causes and recommended solutions for pump or system noise. Table 3 lists possible causes and recommended solutions for inadequate circulation.

REFERENCES

Carlson, G.F. 1972. Central plant chilled water systems—Pumping & flow balance, Part I. *ASHRAE Journal* (February):27-34.

Coad, W.J. 1985. Variable flow in hydronic systems for improved stability, simplicity, and energy economics. *ASHRAE Transactions* 91(1B):224-37.

Hydraulic Institute. 1994. Centrifugal pumps. *Standards* HI 1.1 to 1.5. Parsippany, NJ.

Stethem, W.C. 1988. Application of constant speed pumps to variable volume systems. *ASHRAE Transactions* 94(2):1458-66.

BIBLIOGRAPHY

ASHRAE. 1985. Hydronic systems: Variable-speed pumping and chiller optimization. *ASHRAE Technical Data Bulletin* 1(7).

ASHRAE. 1991. Variable-flow pumping systems. *ASHRAE Technical Data Bulletin* 7(2).

ASHRAE. 1998. *Fundamentals of water system design*. Self-Directed Learning Course.

Beaty, F.E., Jr. 1987. *Sourcebook of HVAC details*. ISBN 0-07-004193-8. McGraw-Hill, New York.

Carrier. 1965. *Carrier handbook of air conditioning system design*. McGraw-Hill, New York.

Clifford, G. 1990. *Modern heating, ventilating and air conditioning*. ISBN 0-13-594755-3. Prentice Hall, New York.

Crane Co. 1988. Flow of fluids through valves, fittings, and pipe. Technical Paper 410. Joliet, IL.

Dufour, J.W. and W.E. Nelson. 1993. *Centrifugal pump sourcebook*. ISBN 0-07-018033-4. McGraw-Hill, New York.

Garay, P.N. 1990. *Pump application desk book*. ISBN 0-88173-043-2. Fairmont Press, Lilburn, GA.

Haines, R.W. and C.L. Wilson. 1994. *HVAC systems design handbook*, 2nd ed. McGraw-Hill, New York.

Hegberg, R.A. 1991. Converting constant-speed hydronic pumping systems to variable-speed pumping. *ASHRAE Transactions* 97(1):739-45.

Karassik, I.J. 1989. Centrifugal pump clinic. Marcel Dekker Inc., New York.

Karassik, I.J., W.C. Krutzsch, W.H. Fraser, and J.P. Messina, eds. 1986. *Pump handbook*. McGraw-Hill, New York.

Levenhagen, J. and D. Spethmann. 1993. *HVAC controls and systems*. McGraw-Hill, New York.

Lobanoff, V.S. and R.R. Ross. 1986. *Centrifugal pumps design and application*. Gulf Publishing, Houston, TX

Matley, J., ed. 1989. *Progress in pumps*. Chemical Engineering, McGraw-Hill, New York.

McQuiston, F.C. and J.D. Parker. 1988. *Heating, ventilating and air conditioning*. ISBN 0-471-63757-2. John Wiley and Sons, New York.

Monger, S. 1990. Testing and balancing HVAC air and water systems. ISBN 088173-075. Fairmont Press, Lilburn, GA.

Rishel, J.B. 1994. Distributed pumping for pumping for chilled- and hot-water systems. *ASHRAE Transactions* 100(1):1521-27.

Trane Co. 1965. *Trane manual of air conditioning*. La Crosse, WI.

CHAPTER 40

MOTORS, MOTOR CONTROLS, AND VARIABLE-SPEED DRIVES

MOTORS

MANY TYPES of alternating-current (ac) motors are available, and while the direct-current (dc) motor is also used, it is only to a limited degree. NEMA *Standard* MG 1 provides technical information on all types of ac and dc motors.

ALTERNATING CURRENT POWER SUPPLY

Important characteristics of an ac power supply include (1) voltage, (2) number of phases, (3) frequency, (4) voltage regulation, and (5) continuity of power.

The **nominal voltage** (or **service voltage**) is the value assigned to the circuit or system to designate its voltage class. It is the voltage at the connection between the supplier and the user. **Utilization voltage** is the voltage at the line terminals of the equipment.

Single-phase and three-phase motor and control voltage ratings shown in Table 1 are adapted to the nominal voltages indicated. Motors with these ratings are considered suitable for ordinary use on their corresponding systems; for example, a 230 V motor should generally be used on a nominal 240 V system. A 230 V motor should not be installed on a nominal 208 V system because the utilization voltage is below the tolerance on the voltage rating for which the motor is designed. Such operation generally results in overheating and a serious reduction in torque.

Motors are usually guaranteed to operate satisfactorily and to deliver their full power at the rated frequency and at a voltage 10% above or below rating, or at the rated voltage and plus or minus 5% frequency variation. Table 2 shows the effect of voltage and frequency variation on induction motor characteristics.

The phase voltages of three-phase motors should be balanced. If not, a small voltage imbalance can product a greater current imbalance and a much greater temperature rise, which can result in nuisance overload trips or motor failures. Motors should not be operated where the voltage imbalance is greater than 1% without checking with the manufacturer. Voltage imbalance is defined in NEMA *Standard* MG 1, Paragraph 14.34, as

$$\%\text{ Voltage imbalance} = 100 \times \frac{\text{Maximum voltage deviation from average voltage}}{\text{Average voltage}}$$

In addition to voltage imbalance, current imbalance can be present in a system where Y-Y transformers without tertiary windings are used, even if the voltage is in balance. As stated previously, this current imbalance is not desirable. If this current imbalance exceeds either 10% or the maximum imbalance recommended by the manufacturer, corrective action should be taken (see NFPA *Standard* 70).

$$\%\text{ Current imbalance} = 100 \times \frac{\text{Maximum current deviation from average current}}{\text{Average current}}$$

Another cause of current imbalance is normal winding impedance imbalance, which adds or subtracts from the current imbalance caused by voltage imbalance.

Table 1 Motor and Motor Control Equipment Voltages

System Nominal Voltage	Equipment Nameplate Voltage Ratings: Integral Horsepower, Three-Phase	Integral Horsepower, Single-Phase	Fractional Horsepower, Three-Phase	Fractional Horsepower, Single-Phase
120	—	115	—	115
208	200	—	200	—
240	230	230	230	230
277	—	265	—	265
480	460	—	460	—
600[a]	575	—	575	—
2,400	2,300	—	—	—
4,160	4,000	—	—	—
4,800	4,600	—	—	—
6,900	6,600	—	—	—
13,800	13,200	—	—	—

[a]Certain control and protective equipment have a maximum voltage limit of 600 V. Consult the manufacturer, power supplier, or both to ensure proper application.

CODES AND STANDARDS

The *National Electrical Code* (NEC) (NFPA *Standard* 70) and the *Canadian Electrical Code*, Part I (CSA *Standard* C22.1) are important in the United States and Canada. The *National Electrical Code* contains minimum recommendations considered necessary to ensure safety of electrical installations and equipment. It is referred to in Subpart S (Electrical) of the Occupational Safety and Health Acts (OSHA) of 1970 and, therefore, is part of the OSHA requirements. In addition, practically all communities in the United States have adopted the NEC as a minimum electrical code.

Underwriters Laboratories (UL) promulgates standards for various types of equipment. UL standards for electrical equipment cover construction and performance for the safety of such equipment and interpret requirements to ensure compliance with the intent of the NEC. A complete list of available standards may be obtained from UL, which also publishes lists of equipment that comply with their standards. These listed products bear the UL label and are recognized by local authorities.

The preparation of this chapter is assigned to TC 8.11, Electric Motors and Motor Control.

Table 2 Effect of Voltage and Frequency Variation on Induction Motor Characteristics

Voltage and Frequency Variation		Starting and Maximum Running Torque	Synchronous Speed	% Slip	Full-Load Speed	Efficiency		
						Full Load	0.75 Load	0.5 Load
Voltage variation	120% Voltage	Increase 44%	No change	Decrease 30%	Increase 1.5%	Small increase	Decrease 0.5 to 2%	Decrease 7 to 20%
	110% Voltage	Increase 21%	No change	Decrease 17%	Increase 1%	Increase 0.5 to 1%	Practically no change	Decrease 1 to 2%
	Function of voltage	$Voltage^2$	Constant	$1/Voltage^2$	Synchronous speed slip	—	—	—
	90% Voltage	Decrease 19%	No change	Increase 23%	Decrease 1.5%	Decrease 2%	Practically no change	Increase 1 to 2%
Frequency variation	105% Voltage	Decrease 10%	Increase 5%	Practically no change	Increase 5%	Slight increase	Slight increase	Slight increase
	Function of frequency	$1/Frequency^2$	Frequency	—	Synchronous speed slip	—	—	—
	95% Frequency	Increase 11%	Decrease 5%	Practically no change	Decrease 5%	Slight decrease	Slight decrease	Slight decrease

Voltage and Frequency Variation		Power Factor			Full-Load Current	Starting Current	Temperature Rise, Full Load	Maximum Overload Capacity	Magnetic Noises, No Load in Particular
		Full Load	0.75 Load	0.5 Load					
Voltage variation	120% Voltage	Decrease 5 to 15%	Decrease 10 to 30%	Decrease 15 to 40%	Decrease 11%	Increase 25%	Decrease 9 to 11°F	Increase 44%	Noticeable increase
	110% Voltage	Decrease 3%	Decrease 4%	Decrease 5 to 6%	Decrease 7%	Increase 10 to 12%	Decrease 5 to 7°F	Increase 21%	Increase slightly
	Function of voltage	—	—	—	—	Voltage	—	$Voltage^2$	—
	90% Voltage	Increase 3%	Increase 2 to 3%	Increase 4 to 5%	Increase 11%	Decrease 10 to 12%	Increase 11 to 13°F	Decrease 19%	Decrease slightly
Frequency variation	105% Frequency	Slight increase	Slight increase	Slight increase	Decrease slightly	Decrease 5 to 6%	Decrease slightly	Decrease slightly	Decrease slightly
	Function of frequency	—	—	—	—	$\frac{1}{Frequency}$	—	—	—
	95% Frequency	Slight decrease	Slight decrease	Slight decrease	Increase slightly	Increase 5 to 6%	Increase slightly	Increase slightly	Increase slightly

Note: These variations are general and differ for specific ratings.

The *Canadian Electrical Code*, Part I, is a standard of the Canadian Standards Association (CSA). It is a voluntary code with minimum requirements for electrical installations in buildings of every kind. The *Canadian Electrical Code*, Part II, contains specifications for the construction and performance of electrical equipment, in compliance with Part I. UL and CSA standards for electrical equipment are similar, so equipment designed to meet the requirements of one code may also meet the requirements of the other. However, agreement between the codes is not complete, so individual standards must be checked when designing equipment for use in both countries. The CSA examines and tests material and equipment for compliance with the *Canadian Electrical Code*.

MOTOR EFFICIENCY

The many factors affecting motor efficiency include (1) sizing of the motor to the load, (2) type of motor specified, (3) motor design speed, and (4) type of bearing specified. Oversizing a motor may reduce efficiency. As shown in the performance characteristic curves for single-phase motors in Figures 1, 2, and 3, the efficiency falls off rapidly at loads lighter than the rated full load. Polyphase motors usually reach peak efficiency at loads slightly lighter than full load (Figure 4). Motor performance curves are available from the motor manufacturer and can help in applying the optimum motor for an application. Larger output motors are more efficient at rated load than smaller motors. Also, higher speed induction motors are more efficient.

The type of motor specified is significant because, for example, a permanent split-capacitor motor is more efficient than a shaded-pole fan motor. A capacitor-start/capacitor-run motor is more efficient than either a capacitor-start or a split-phase motor. In polyphase motors, the lower the locked rotor torque specified, the higher the efficiency obtained in the design.

The motor industry offers high-efficiency motors that include more material than a standard-efficiency motor of the same type. NEMA *Standards* MG 10 and MG 11 have more information on motor efficiency.

GENERAL-PURPOSE INDUCTION MOTORS

The electrical industry classifies motors as **small kilowatt (fractional horsepower)** or **integral kilowatt (integral horsepower)**. In this context, kilowatt refers to power output of the motor. Small kilowatt motors have ratings of less than 1 hp at 1700 to 1800 rpm for four-pole and 3500 to 3600 rpm for two-pole machines. Single-phase motors are available through 5 hp and are most common through 0.75 hp because motors larger than 0.75 hp are usually polyphase.

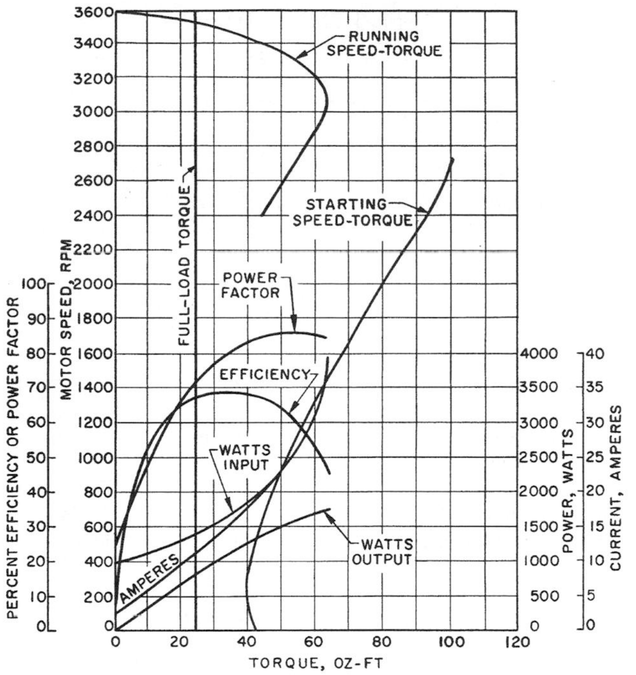

Fig. 1 Typical Performance Characteristics of Capacitor-Start/Induction-Run Two-Pole General-Purpose Motor, 1 hp

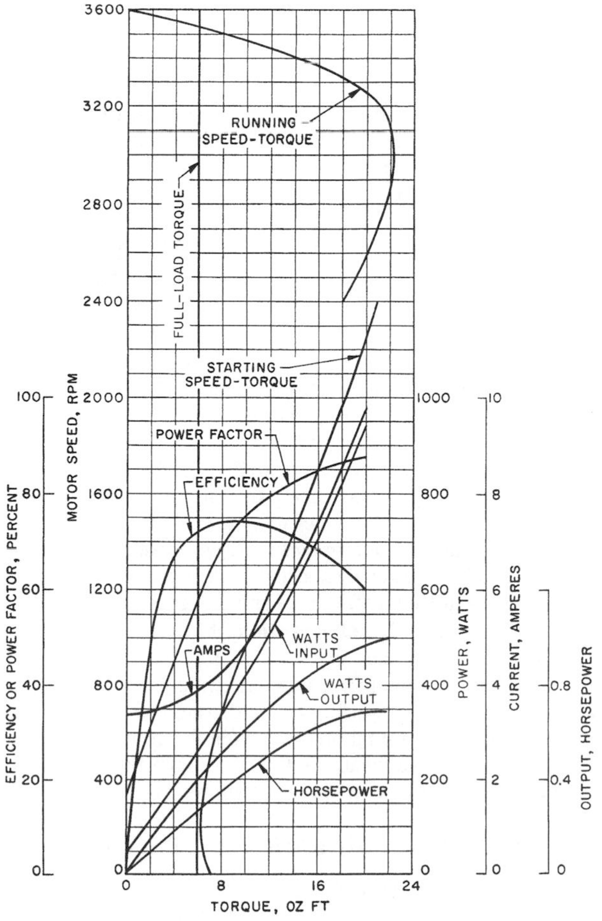

Fig. 2 Typical Performance Characteristics of Resistance-Start Split-Phase Two-Pole Hermetic Motor, 0.25 hp

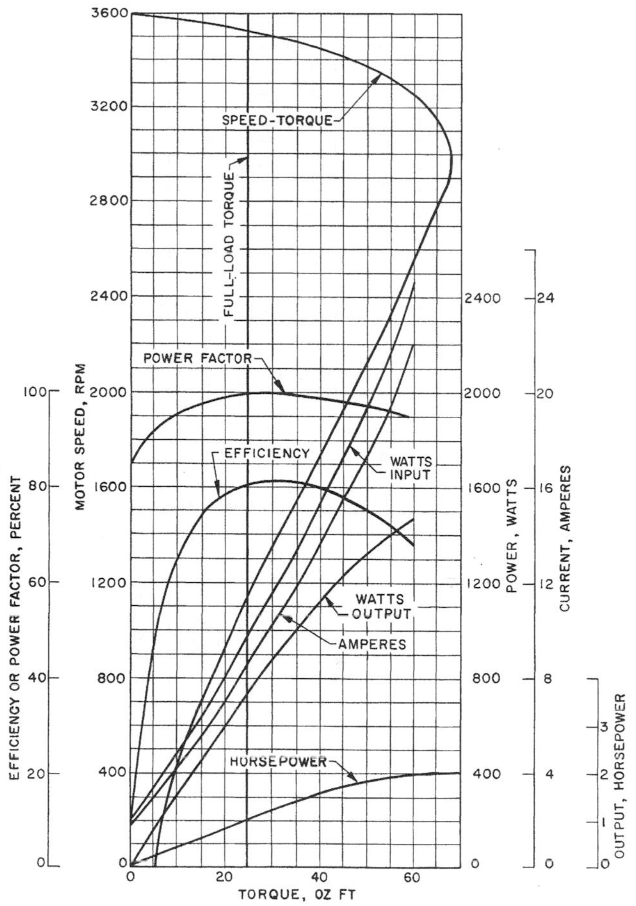

Fig. 3 Typical Performance Characteristics of Permanent Split-Capacitor Two-Pole Motor, 1 hp

Table 3 lists motors by types indicating the normal kilowatt range and the type of power supply used. All motors listed are suitable for either direct or belt drive, except shaded-pole motors (limited by low starting torque).

Application

When applying an electric motor, the following characteristics are important: (1) mechanical arrangement, including position of the motor and shaft, type of bearing, portability desired, drive connection, mounting, and space limitations; (2) speed range desired; (3) power requirement; (4) torque; (5) inertia; (6) frequency of starting; and (7) ventilation requirements. Motor characteristics that are frequently applied are generally presented in curves (see Figures 1 through 4).

Torque. The torque required to operate the driven machine at all times between initial breakaway and final shutdown is important in determining the type of motor. The torque available at zero speed or standstill (**starting torque**) may be less than 100% or as high as 400% of full-load torque, depending on motor design. The **starting current**, or **locked-rotor current**, is usually 400 to 600% of the current at rated full load.

Full-load torque is the torque developed to produce the rated power at the rated speed. **Full-load speed** also depends on the design of the motor. For induction motors, a speed of 1725 rpm is typical for four-pole motors, and a speed of 3450 rpm is typical for two-pole motors at 60 Hz.

Motors have a **maximum or breakdown torque**, which cannot be exceeded. The relation between breakdown torque and full-load torque varies widely, depending on motor design.

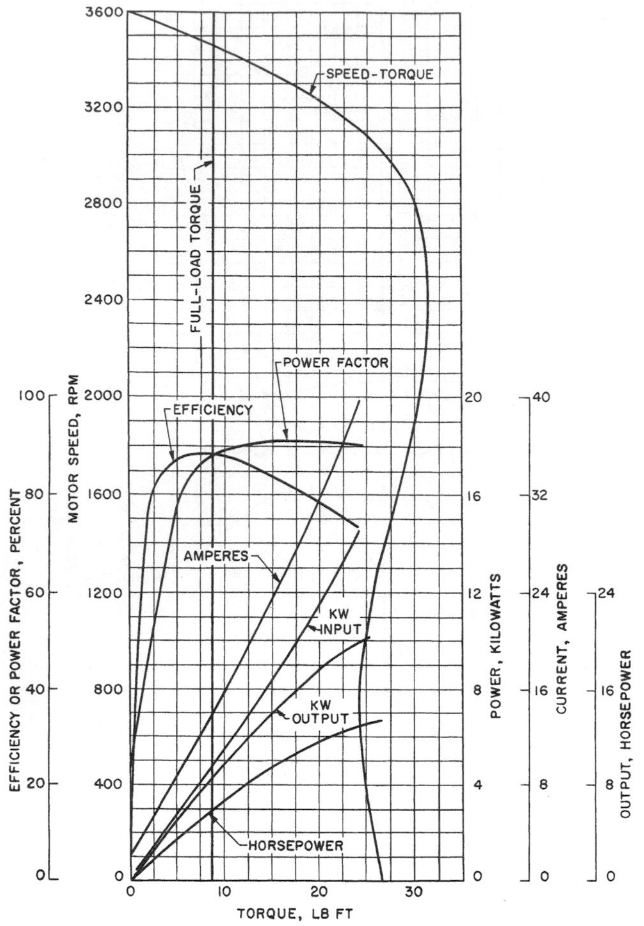

Fig. 4 Typical Performance Characteristics of Polyphase Two-Pole Motor, 5 hp

Power. The power delivered by a motor is a product of its torque and speed. Because a given motor delivers increasing power up to maximum torque, a basis for power rating is needed. The National Electrical Manufacturers Association (NEMA) bases **power rating** on breakdown torque limits for single-phase motors, 10 hp and less. All others are rated at their power capacity within voltage and temperature limits as listed by NEMA.

Full-load rating is based on the maximum winding temperature. If the nameplate marking includes the maximum ambient temperature for which the motor is designed and the insulation designation, the maximum temperature rise of the winding may be determined from the appropriate section of NEMA *Standard* MG 1.

Service Factor. This factor is the maximum overload that can be applied to general-purpose motors and certain definite-purpose motors without exceeding the temperature limitation of the insulation. When the voltage and frequency are maintained at the values specified on the nameplate and the ambient temperature does not exceed 104°F, the motor may be loaded to the power obtained by multiplying the rated power by the service factor shown on the nameplate.

The power rating is normally established on the basis of a test-run in still air. However, most direct-drive, air-moving applications are checked with air flowing over the motor. If the motor nameplate marking does not specify a service factor, refer to the appropriate section of NEMA *Standard* MG 1. Characteristics of alternating current motors are given in Table 4.

Table 3 Motor Types

Type	Range, hp	Type of Power Supply
Fractional Sizes		
Split phase	0.05 to 0.5	Single phase
Capacitor-start	0.05 to 1.5	Single phase
Repulsion-start	0.13 to 1.5	Single phase
Permanent split-capacitor	0.05 to 1.5	Single phase
Shaded-pole	0.01 to 0.25	Single phase
Squirrel cage induction	0.17 to 1.5	Polyphase
Direct current	0.5 to 1.5	DC
Integral Sizes		
Capacitor-start/capacitor-run	1 to 5	Single phase
Capacitor-start	1 to 5	Single phase
Squirrel cage induction (normal torque)	1 and up	Polyphase
Slip-ring	1 and up	Polyphase
Direct current	1 and up	DC
Permanent split-capacitor	1 to 5	Single phase

HERMETIC MOTORS

A hermetic motor is a partial motor usually consisting of a stator and a rotor without shaft, end shields, or bearings. It is for installation in hermetically-sealed refrigeration compressor units. With the motor and compressor sealed in a common chamber, the winding insulation system must be impervious to the action of the refrigerant and lubricating oil. Hermetic motors are used in both welded and accessible hermetic (semihermetic) compressors.

Application

Domestic Refrigeration. Hermetic motors up to 0.33 hp are used. They are split-phase, permanent split-capacitor, or capacitor-start motors for medium or low starting torque compressors and capacitor-start and special split-phase motors for high starting torque compressors.

Room Air Conditioners. Motors from 0.33 to 3 hp are in use. They are permanent split-capacitor or capacitor-start/capacitor-run types. These designs have high power factor and efficiency and meet the need for low current draw, particularly on 115 V circuits.

Central Air Conditioning (Including Heat Pumps). Both single-phase (6 hp and below) and polyphase (1.5 hp and above) motors are used. The single-phase motors are permanent split-capacitor or capacitor-start/capacitor-run types.

Small Commercial Refrigeration. Practically all these units are below 5 hp, with single-phase being the most common. Capacitor-start/induction-run motors are normally used up to 0.75 hp because of starting torque requirements. Capacitor-start/capacitor-run motors are used for larger sizes because they provide high starting torque and high full-load efficiency and power factor.

Large Commercial Refrigeration. Most motors are three-phase and larger than 5 hp.

Power ratings of motors for hermetic compressors do not necessarily have a direct relationship to the thermodynamic output of a compressor. Designs are tailored to match the compressor characteristics and specific applications. Chapter 35 briefly discussed hermetic motor applications for various compressors.

INTEGRAL THERMAL PROTECTION

The *National Electrical Code* and UL standards cover motor protection requirements. Separate, external protection devices include the following:

Thermal Protectors. These protective devices are an integral part of a motor or hermetic motor refrigerant compressor. They

Table 4 Characteristics of AC Motors (Nonhermetic)

	Split-Phase	Permanent Split-Capacitor	Capacitor-Start/ Induction-Run	Capacitor-Start/ Capacitor-Run	Shaded-Pole	Polyphase, 60-Hz
Connection Diagram	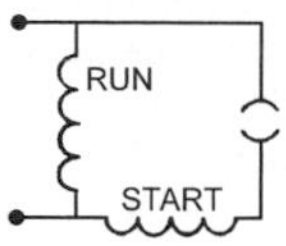	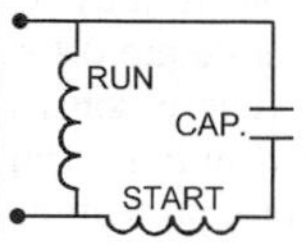	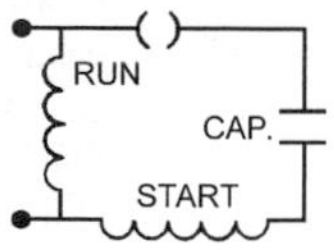	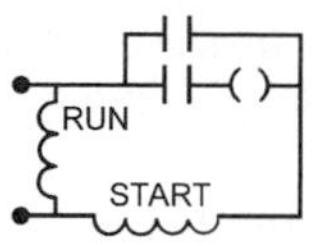		
Speed Torque Curves	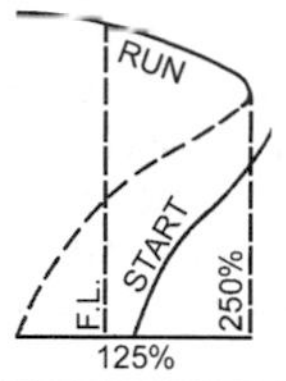	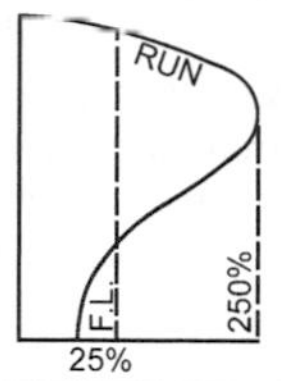	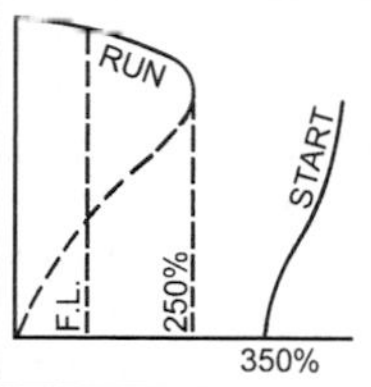	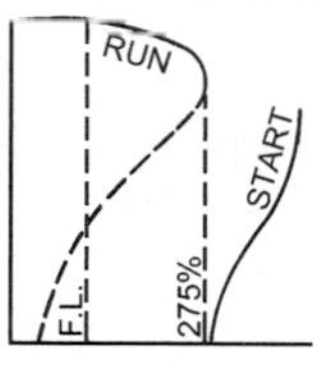	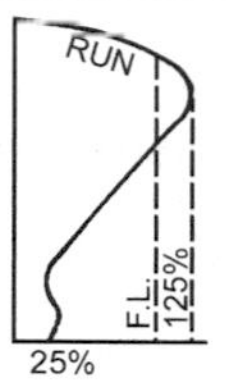	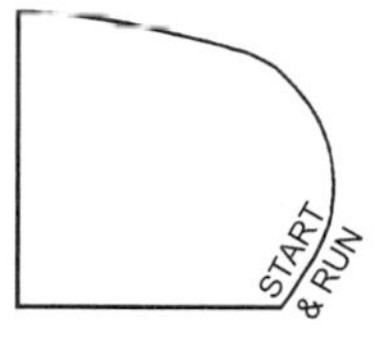
Starting Method	Centrifugal switch	None	Centrifugal switch	Centrifugal switch	None	Motor controller
Ratings, hp	0.05 to 0.5	0.05 to 5	0.05 to 5	0.05 to 5	0.01 to 0.25	0.5 and up
Full-Load Speeds at 60-Hz (Two-Pole, Four-Pole)	3450 to 1725	3450 to 1725	3450 to 1725	3500 to 1750	3100 to 1550	3500 to 1750
Torque[a]						
Locked Rotor	125 to 150%	25%	250 to 350%	250%	25%	150 to 350%
Breakdown	250 to 300%	250 to 300%	250 to 300%	250%	125%	250 to 350%
Speed Classification	Constant	Constant	Constant	Constant	Constant or adjustable	Constant
Full-Load Power Factor	60%	95%	65%	95%	60%	80%
Efficiency	Medium	High	Medium	High	Low	High-Medium

[a] Expressed as percent of rated horsepower torque.

protect the motor against overheating caused by overload, failure to start, or excessive operating current. Thermal protectors are required to protect polyphase motors from overheating because of an open phase in the primary circuit of the supply transformer. Thermal protection is accomplished by either a line break device or a thermal sensing control circuit.

The protector of a hermetic motor-compressor has some unique capabilities compared to nonhermetic motor protectors. The refrigerant cools the motor and compressor, so the thermal protector may be required to prevent overheating from loss of refrigerant charge, low suction pressure and high superheat at the compressor, obstructed suction line, or malfunction of the condensing means.

Article 440 of the NEC limits the maximum continuous current on a motor-compressor to 156% of rated load current if an integral thermal protector is used. NEC Article 430 limits the maximum continuous current on a nonhermetic motor to different percentages of full-load current as a function of size. If separate overload relays or fuses are used for protection, Article 430 limits maximum continuous current to 140% and 125%, respectively, of rated load.

UL *Standard* 984 specifies that the compressor enclosure must not exceed 300°F under any conditions. The motor winding temperature limit is set by the compressor manufacturer based on individual compressor design requirements. UL *Standard* 547 sets the limit for the motor winding temperature for open motors as a function of the class of the motor insulation used.

Line-Break Thermal Protectors. Integral with a motor or motor-compressor, line-break thermal protectors that sense both current and temperature are connected electrically in series with the motor; their contacts interrupt the total motor line-current. These protectors are used in small, single-phase and polyphase motors up through 15 hp.

Protectors installed inside a motor-compressor are hermetically sealed because exposed arcing in the presence of refrigerant cannot be tolerated. They provide better protection than the external type for loss of charge, obstructed suction line, or low voltage on the stalled rotor. This is due to low current associated with these fault conditions, hence the need to sense the motor temperature increase by thermal contact. Protection inside the compressor housing must withstand pressure requirements established by UL.

Protectors mounted externally on motor-compressor shells, sensing only shell temperature and line current, are typically used on smaller compressors, such as those in household refrigerators and small room air conditioners. One benefit occurs during high head pressure starting conditions, which can occur if voltage is lost momentarily or if the user inadvertently turns off the compressor with the temperature control and then turns it back on immediately. Usually, these units do not start under these conditions. When this happens, the protector takes the unit off the line and resets automatically when the compressor cools and pressures have equalized to a level that allows the compressor to start.

Protectors installed in nonhermetic motors may be attached to the windings or may be mounted off the windings but in the motor housing. Those protectors placed on the winding are generally installed prior to varnish dip and bake, and their construction must prevent varnish from entering the contact chamber.

Since the protector carries full motor line current, its size is based on adequate contact capability to interrupt the stalled current of the motor on continuous cycling for periods specified in UL *Standards* 984 and 547.

The compressor or motor manufacturer applies and selects appropriate motor protection in cooperation with the protector manufacturer. Any change in protector rating, by other than the specifying manufacturer after the proper application has been made, may result in either overprotection and frequent nuisance tripouts or underprotection and burnout of the motor windings. Connections to protector terminals, including lead wire sizes, should not be changed, and no additional connections should be made to the terminals. Any change in connection changes the terminal conditions and affects protector performance.

Control circuit thermal protectors approved for use with a motor or motor-compressor, either sensing both current and temperature or sensing temperature only, are used with integral horsepower single-phase and three-phase motors.

The current and temperature protector uses a bimetallic temperature sensor installed in the motor winding in conjunction with thermal overload relays. The sensors are connected in series with the control circuit of a magnetic contactor that interrupts the motor current. Thermostat sensors of this type, which depend on their size and mass, are capable of tracking motor winding temperature for running overloads. When a rotor is locked (when the rate of change in winding temperature is rapid), the temperature lag is usually too great for such sensors to provide protection when they are used alone. However, when the bimetallic sensor is used with separate thermal overload or magnetic time-delay relays that sense motor current, the combination provides protection; on a locked rotor condition, the thermal or magnetic relays protect for the initial heating cycle, and the combined functioning of relay and thermostat protects for subsequent cycles.

The resistance change of a thermistor may be used to provide a switching signal to the electronic circuit, whose output is in series with the control circuit of a magnetic contactor used to interrupt the motor current. The output of the electronic protection circuitry (module) may be an electromechanical relay or a power triac. The sensors may be installed directly on the stator winding end turns or buried inside the windings. Their small size and good thermal transfer allow them to track the temperature of the winding for locked rotor, as well as running overload.

Three types of sensors are available. One type uses a ceramic material with a positive temperature coefficient of resistance; the material exhibits a large, abrupt change in resistance at a particular design temperature. This change occurs at the **anomaly point**, which is inherent in the sensor. The anomaly point remains constant once the sensor is manufactured; sensors are produced with anomaly points at different temperatures to meet different requirements. However, a single module calibration can be supplied for all anomaly temperatures of a given sensor type.

Another type of sensor uses a metal wire, which has a linear increase in resistance with temperature. The sensor assumes a specified value of resistance corresponding to each desired value of response or operating temperature. It is used with an electronic protection module calibrated to a specific resistance. Modules supplied with different calibrations are used to achieve various values of operating temperatures.

A third type is a negative temperature coefficient of resistance sensor, which is integrated with electronic circuitry similar to that used with the metal wire sensor.

More than one sensor may be connected to a single electronic module in parallel or series, depending on design. However, the sensors and modules must be of the same design and intended for use with the particular number of sensors installed and the wiring method used. Electronic protection modules must be paired only with sensors specified by the manufacturer, unless specific equivalency is established and identified by the motor or compressor manufacturer.

MOTOR CONTROL

In general, motor control equipment may (1) disconnect the motor and controller from the power supply, (2) start and stop the motor, (3) protect against short circuits, (4) protect from overheating, (5) protect the operator, (6) control motor speed, and (7) protect motor branch circuit conductors and control apparatus.

Separate Motor Protection

Most air-conditioning and refrigeration motors or motor-compressors, whether open or hermetic, are equipped with integral motor protection by the equipment manufacturer. If this is not the case, separate motor-protection devices, sensing current only, must be used. These consist of thermal or magnetic relays, similar to those used in industrial control, that provide running overload and stalled-rotor protection. Because hermetic motor windings heat rapidly due to the loss of the cooling effect of refrigerant gas flow when the rotor is stalled, **quick-trip devices** must be used.

Thermostats or **thermal devices** are sometimes used to supplement current-sensing devices. Such supplements are necessary (1) when automatic restarting is required after trip or (2) to protect from abnormal running conditions that do not increase motor current. These devices are covered in the section on Integral Thermal Protection.

Protection of Control Apparatus and Branch Circuit Conductors

In addition to protection of the motor itself, Articles 430 and 440 of the *National Electrical Code* require the control apparatus and branch circuit conductors to be protected from overcurrent resulting from motor overload or failure to start. This protection can be given by some thermal protective systems that do not permit a continuous current in excess of required limits. In other cases, a current-sensing device, such as an overload relay, a fuse, or a circuit breaker, is used.

Circuit Breakers. These devices are used for disconnecting as well as circuit protection and are available in ratings for use with small household refrigerators as well as in large commercial installations. Manual switches for disconnecting and fuses for short-circuit protection are also used. For single-phase motors up to 3 hp, 230 V, an attachment plug is an acceptable disconnecting device.

Controllers. The motor control used is determined by the size and type of motor, the power supply, and the degree of automation. Control may be manual, semiautomatic, or fully automatic.

Central air conditioners are generally located some distance from the controlled space environment control, such as room thermostats and other control devices. Therefore, **magnetic controllers** must be used in these installations. Also, all dc and all large ac installations must be equipped with in-rush **current-limiting controllers**, which are discussed later. **Synchronous motors** are sometimes used to improve the power factor. **Multi-speed motors** provide flexibility for many applications.

Manual Control. For an ac or dc motor, manual control is usually located near the motor. If so, an operator must be present to start and stop or change the speed of the motor by adjusting the control mechanism.

Manual control is the simplest and least expensive control method for small ac motors, both single-phase and polyphase, but it is seldom used with hermetic motors. The manual controller usually consists of a set of main line contacts, which are provided with thermal overload relays for motor protection.

Manual speed controllers can be used for large air conditioners using **slip-ring motors**; they may also provide reduced-current starting. Different speed points are used to vary the amount of cooling provided by the compressor.

Across-the-Line Magnetic Controllers. These controllers are widely used for central air conditioning. They may be applied to motors of all sizes, provided power supply and motor are suitable to this type of control. Across-the-line magnetic starters may be used with automatic control devices for starting and stopping. Where push buttons are used, they may be wired for either low-voltage release or low-voltage protection.

Full-Voltage Starting. For motors, full-voltage starting is preferable because of its lower initial cost and simplicity of control. Except for dc machines, most motors are mechanically and electrically designed for full-voltage starting. The starting in-rush current, however, is limited in many cases by power company requirements made because of voltage fluctuations, which may be caused by heavy current surges. Therefore, the starting current must often be reduced below that obtained by across-the-line starting in order to meet the limitations of power supply. One of the simplest ways to make this reduction is to place resistors in the primary circuit. As the motor accelerates, the resistance is cut out by the use of timing or current relays.

Another method of reducing the starting current for an ac motor uses an **autotransformer** motor controller. Starting voltage is reduced, and, when the motor accelerates, it is disconnected from the transformer and connected across-the-line by timing or current relays. Primary resistor starters are generally smaller and less expensive than autotransformer starters for moderate size motors. However, primary resistor starters require more line current for a given starting torque than do autotransformer starters.

Star-Delta (Wye Delta) Motor Controllers. These controllers limit current efficiently, but they require motors designed for this type of starting. They are particularly suited for centrifugal, rotary screw, and reciprocating compressor drives starting without load.

Part-Winding Motor Controllers (or Incremental Start Controllers). These controllers limit line disturbances by connecting only part of the motor winding to the line and connecting the second motor winding to the line after a time interval of 1 to 3 s. If the motor is not heavily loaded, it accelerates when the first part of the winding is connected to the line; if it is too heavily loaded, the motor may not start until the second winding is connected to the line. In either case, the voltage dip will be less than the dip that would result if a standard squirrel cage motor with an across-the-line starter were used. Part-winding motors may be controlled either manually or magnetically. The magnetic controller consists of two contactors and a timing device for the second contactor.

Multispeed Motor Controllers. Multispeed motors provide flexibility in many types of drives in which variation in capacity is needed. Two types of multispeed motors are used: (1) motors with one reconnectable winding and (2) motors with two separate windings. Motors with separate windings need a contactor for each winding, and only one contactor can be closed at any time. Motors with a reconnectable winding are similar to motors with two windings, but the contactors and motor circuits are different.

Slip-Ring Motor Controllers. Slip-ring ac motors provide variable speed. The wound rotor of these motors functions in the same manner as in the squirrel cage motor, except that the rotor windings are connected through slip rings and brushes to external circuits with resistance to vary the motor speed. Increasing the resistance in the rotor circuit reduces motor speed, and decreasing the resistance increases motor speed. When the resistance is shorted out, the motor operates with maximum speed, efficiency, and power factor. On some large installations, manual drum controllers are used as speed-setting devices. Complete automatic control can be provided with special control devices for selecting motor speeds.

Controllers for Direct-Current Motors. These motors have favorable speed-torque characteristics, and their speed is easily controlled. Controllers for dc motors are more expensive than those for ac motors, except for very small motors. Large dc motors are started with resistance in the armature circuit, which is reduced step by step until the motor reaches its base speed. Higher speeds are provided by weakening the motor field.

Single-Phase Motor-Starting Methods

Motor-starting switches and relays for single-phase motors must provide a means for disconnecting the starting winding of split-phase or capacitor-start/induction-run motors or the start capacitor of capacitor-start/capacitor-run motors. Open machines usually have a centrifugal switch mounted on the motor shaft, which disconnects the starting winding at about 70% of full-load speed.

The starting methods by use of relays are as follows:

Thermally Operated Relay. When the motor is started, a contact that is normally closed applies power to the starting winding. A thermal element that controls these contacts is in series with the motor and carries line current. Current flowing through this element heats it until, after a definite time, it is warmed sufficiently to open the contacts and remove power from the starting winding. The running current then heats the element enough to keep the contacts open. The setting of the time for the starting contacts to open is determined by tests on the components (i.e., the relay, the motor, and the compressor) and is based on a prediction of the time delay required to bring the motor up to speed.

An alternate form of a thermally operated relay is a positive temperature coefficient of resistance (PTC) starting device. This device has a ceramic element with low resistance at room temperature that increases about 1000 times when it is heated to a predetermined temperature. It is placed in series with the start winding of split-phase motors and allows current flow when power is applied. After a definite period, the self-heating of the PTC resistive element causes it to reach its high-resistance state, which reduces current flow in the start winding. The small residual current maintains the PTC element in the high-resistance state while the motor is running. A PTC starting device may also be connected in parallel with a run capacitor, and the combination may be connected in series with the starting winding. It allows the motor to start like a split-phase motor and then, when the PTC element reaches the high-resistance state, operate as a capacitor-run motor. When power is removed, the PTC element must be allowed to cool to its low resistance state before restarting the motor.

Current-Operated Relay. In this type of connection, a relay coil carries the line current going to the motor. When the motor is started, the in-rush current to the running winding passes through the relay coil, causes the normally open contacts to close, and applies power to the starting winding. As the motor comes up to speed, the current decreases until, at a definite calibrated value of current corresponding to a preselected speed, the magnetic force of the coil diminishes to a point that allows the contacts to open to remove power from the starting winding. This relay takes advantage of the **main winding current** versus **speed** characteristics of the motor. The current-speed curve varies with line voltage, so that the starting relay must be selected for the voltage range likely to be encountered in service. Ratings established by the manufacturer should not be changed because this may result in undesirable starting characteristics. They are selected to disconnect the starting winding or start capacitor at approximately 70 to 90% of synchronous speed for four-pole motors.

Voltage-Operated Relay. Capacitor-start and capacitor-start/capacitor-run hermetically sealed motors above 0.5 hp are usually started with a normally closed contact voltage relay. In this method of starting, the relay coil is connected in parallel with the starting winding. When power is applied to the line, the relay does not operate because it is calibrated to operate at a higher voltage. As the motor comes up to speed, the voltage across the starting winding and relay coil increases in proportion to the motor speed. At a definite voltage corresponding to a preselected speed, the relay opens, thereby opening the starting winding circuit or disconnecting the starting capacitor. The relay keeps these contacts open because sufficient voltage is induced in the starting winding when the motor is running to hold the relay in the open position.

AIR VOLUME CONTROL

A review of the fan laws (Chapter 18) shows that volume delivered by a fan is directly proportional to its speed, pressure is proportional to the square of the speed, and power is proportional to the cube of the speed. According to these laws, a fan operating at 50% volume requires only 12.5% of the power required at 100% volume.

While the fan in a typical VAV system is sized to handle peak volume, the system operates at reduced volume most of the time. For example, Figure 5 shows the volume levels of a typical VAV system operating below 70% volume over 87% of the time. Thus, adjustable speed operation of the fan for this duty cycle could provide a significant energy saving.

Historically, centrifugal fans have been driven by fixed-speed ac motors, and volume has been varied by outlet dampers, variable inlet guide vanes, or eddy current couplings.

Outlet dampers are mounted in the airstream on the outlet side of the fan. Closing the damper reduces the volume, but at the expense of increased pressure. Points B and C on the fan performance curve in Figure 6 show the modified system curves for two closed damper positions. The natural operating point corresponds to a wide open damper position shown as point A. The input power profile is also shown for the referenced points.

Variable inlet vanes are mounted on the fan inlet to control air volume. Altering the pitch of the vane imparts a spin to the air entering the fan wheel, which results in a family of fan performance curves as shown in Figure 7. With reference to the required power at reduced flows, the inlet vane is more efficient than an outlet damper.

An **eddy current coupling** connects an ac motor driven fixed-speed input shaft to a variable-speed output shaft through a magnetic flux coupling. Reducing the level of flux density in the coupling increases slip between the couplings input and output shafts and reduces speed. **Slip** is wasted energy in the form of heat that must be dissipated by fan cooling or by water cooling for large motors.

Figure 8 shows that reducing fan speed also generates a family of performance curves, but the required input power still remains relatively high because the speed of the induction motor remains relatively constant.

VARIABLE-SPEED DRIVES

An alternative to these VAV speed control methods is the variable-speed drive. [In this section, the term variable-speed drive (VSD) is considered synonymous with variable frequency drive (VFD), pulse width modulated drive (PWM drive), adjustable speed drive (ASD), and adjustable frequency drive (AFD).] An alternating-current adjustable-speed drive consists of a pulse width modulation (PWM) controller with inverter gate bipolar transistors

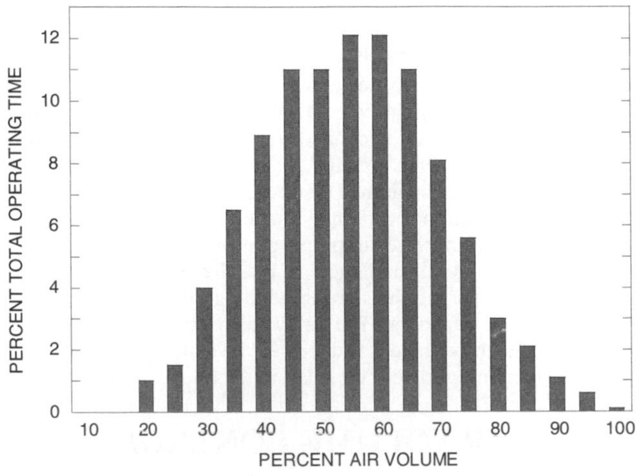

Fig. 5 Typical Fan Duty Cycle for VAV System

Table 5 Comparison of VAV Energy Consumption with Various Volume Control Techniques

	Outlet Damper	Inlet Guide Vane	Eddy Current Coupling	AC PWM Drive
Input Power, hp	91	62	62	28
Annual kWh	340,000	230,000	230,000	105,000

Typical 100 hp application at average of 60% design airflow rate for 5000 h operation.

(IGBTs) and an induction motor. These very fast switching power transistors generate a variable-voltage, variable-frequency waveform that changes the speed of the prime mover. As shown in Figure 9, as speed is reduced, input power is reduced substantially because the power required varies as the cube of the speed (plus losses).

A comparison of Figures 6, 7, 8, and 9 shows that significant energy savings can be achieved by using an ac adjustable-speed drive to achieve variable air volume control. In addition, very high efficiencies can be achieved by using the solid-state controller, which is over 96% efficient, with a state-of-the-art, energy-efficient ac motor with an efficiency of 93 to 95%. Table 5 shows typical annual energy use for the four VAV control techniques.

Power Transistor Characteristics

The key technology used to generate the output waveform is the IGBT. This transistor changes the characteristics of waveforms applied to a motor due to the speed at which the transistor cycles on and off. Pulse width modulation has been used for many years for variable-speed drives; however, as switching speeds increased from 1 or 2 kHz to 8, 15, or as high as 20 kHz to allow higher carrier frequencies, concerns arose over phenomena previously seen only in wave transmission devices like antennae and broadcast signal equipment. The faster switching time began to change the application variables such as drive to motor lead length. These factors must be considered when applying newer IGBT-based variable-speed drives.

Switching Times and d*v*/d*t*. Figure 10 shows the switching of a bipolar junction transistor (BJT) versus an IGBT as an example of how the increased switching speeds affect the turn-on and turn-off times as a ratio of the overall cycle. Note that the BJT switches at 2 kHz and the IGBT switches at 8 kHz, or 4 times faster.

The high rate of change in voltage over a relatively short time is know as the dv/dt of the voltage pulse. The amount of dv/dt is usually 10 to 90%. As the number of pulses increases so must the dv/dt. Note that this voltage waveform is a function of the drive design and is not user-settable. The maximum design carrier frequency sets the limits on how fast a transistor must cycle on and off.

Motor and Conductor Impedance

If the waveform shown at the output of the drive were identical at the motor terminals, a high dv/dt would not be a concern. But impedance or electrical resistance in ac circuits has an impact on the voltage pulse as it travels from the drive to the motor. When the cable impedance closely matches the motor impedance, the voltage pulse is evenly distributed. However, when the motor impedance is much larger than the cable impedance, the pulse reflects at the motor terminals, causing standing waves. Figure 11 shows the surge impedance of both the motor and the cable for different size drives and motors. Note that a relatively small motor (less than 2 hp) has a very high impedance with respect to the typical cable. Larger motors (greater than 100 hp) closely match cable impedance values.

Potential for Damaging Reflected Waves. Reflected waves damage motors because transmitted and reflected pulses can add together, causing a very high voltage. Because these voltage pulses are transmitted through the conductor over specific distances, cable length is a key variable when examining the potential for damaging voltages. Figure 12 shows the relationship between cable distance,

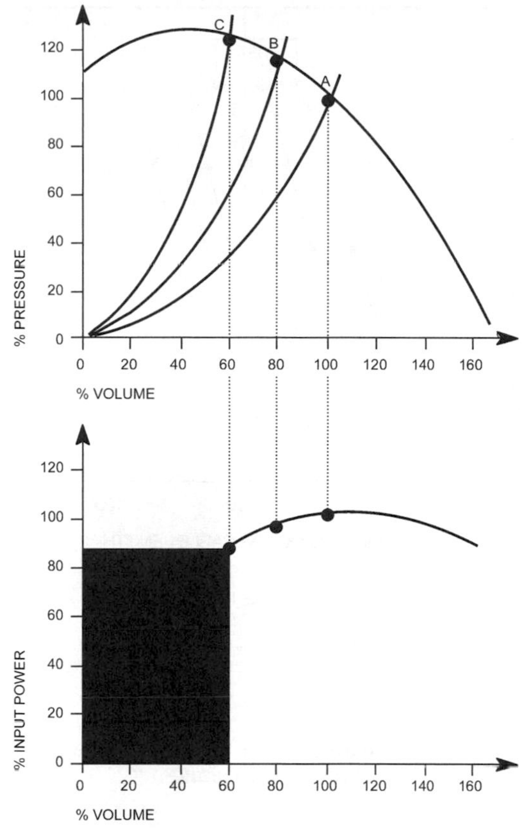

Fig. 6 Outlet Damper Control

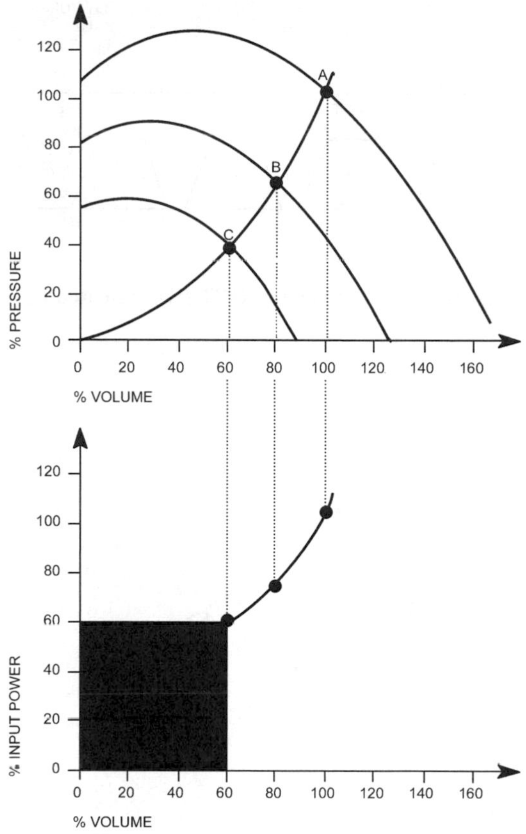

Fig. 8 Eddy Current Coupling Control

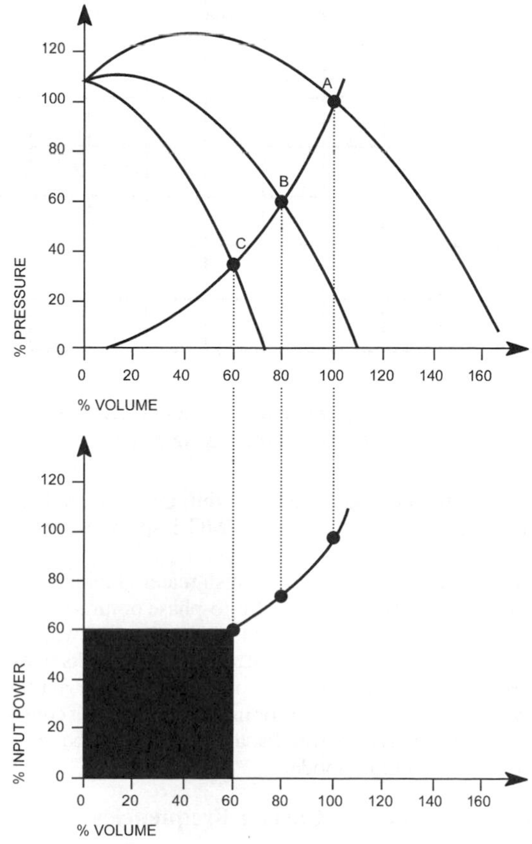

Fig. 7 Variable Inlet Vane Control

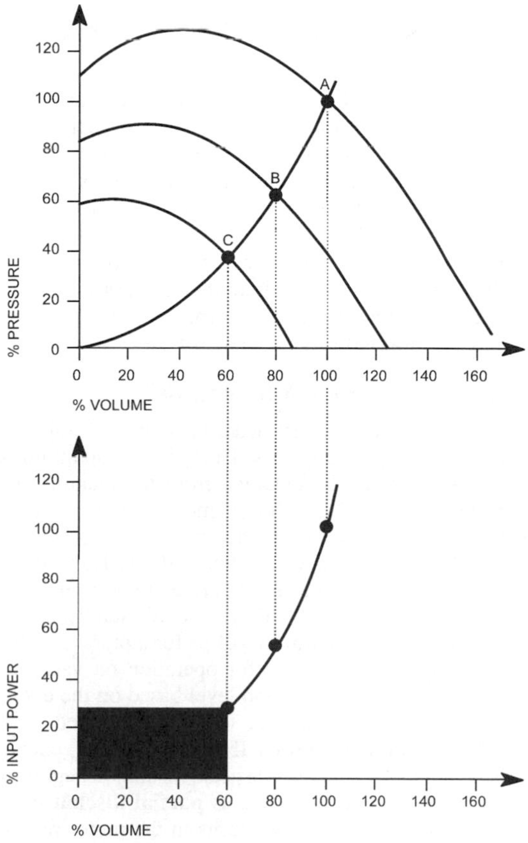

Fig. 9 AC Drive Control

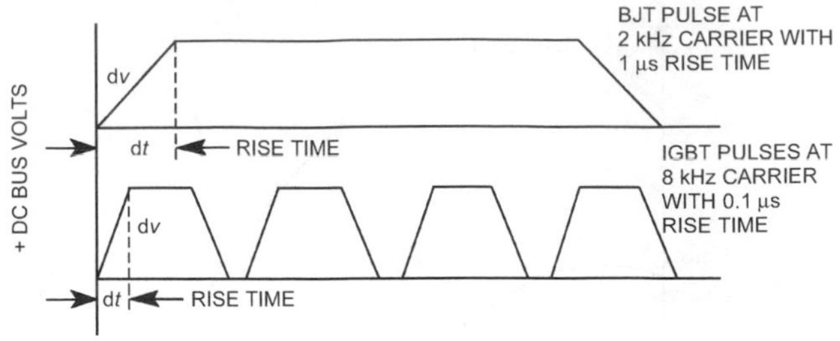

Fig. 10 Bipolar Versus IGBT PWM Switching

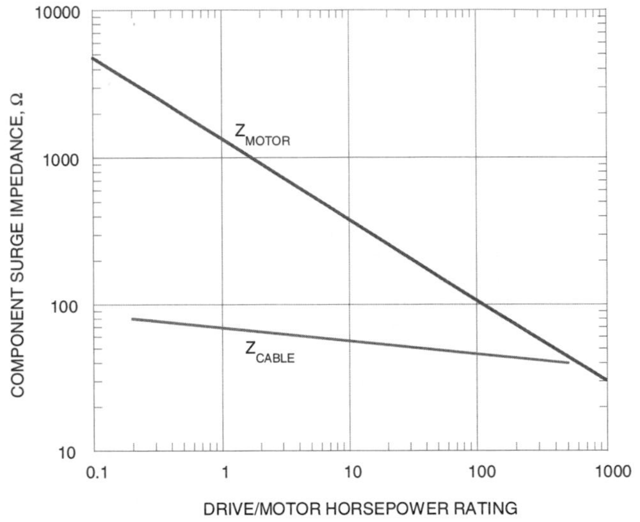

Fig. 11 Motor and Drive Relative Impedance

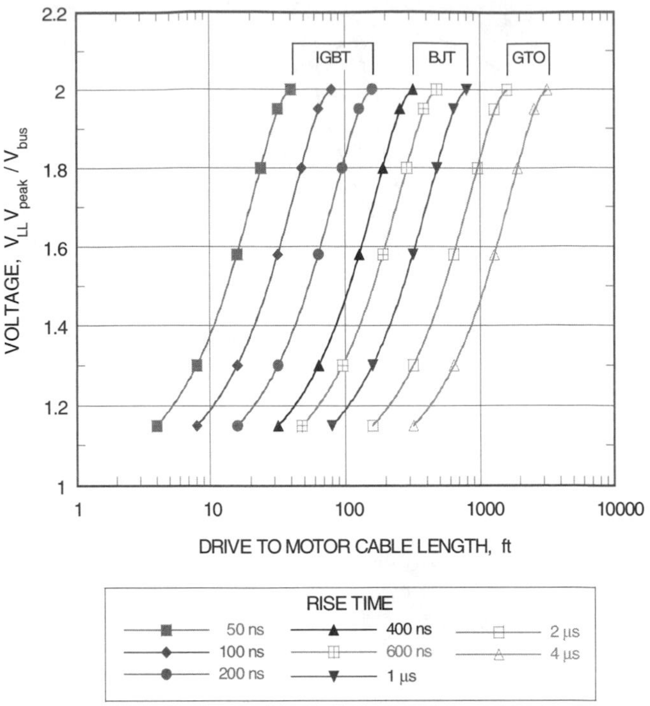

Fig. 12 Switching Times, Cable Distance, and Pulse Peak Voltage

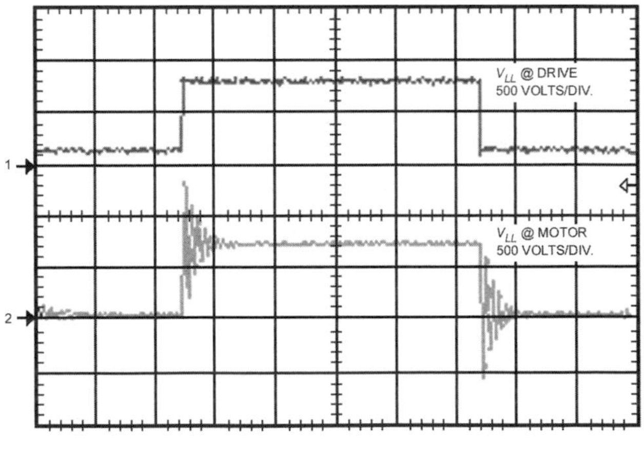

Fig. 13 Reflected Wave Voltage Levels at Drive and Motor Insulation

switching times, and voltage levels of pulses at motor terminals. Damaging reflected waves are more likely to occur in smaller motors because of the mismatch in surge impedance value. Special design techniques are required if multiple small motors are run from a single drive because the potential for reflected waves is high.

Figure 13 shows oscilloscope measurements taken at each end of the drive-to-motor conductor to describe the reflected wave phenomena. The two traces demonstrate the effect of transmitted and reflected pulses adding together to form damaging voltages. The induction motor must be designed to withstand these voltage levels.

Motor Ratings and NEMA Standards

An induction motor is constructed to withstand voltage levels higher than the nameplate suggests. The specific maximum **voltage withstand value** should be obtained from the manufacturer, but typical values for 208 V and 460 V motors range from 1000 to 1800 V. Higher voltage motors, such as 575 V motors, may be rated up to 2000 V peak. NEMA *Standard* MG 1, Revision 1, Part 31.40.4.2 states the established PWM drive motor limits, which are shown graphically in Figure 14. This standard establishes a peak of 1600 V and a minimum rise time of 0.1 μs for motors rated less than 600 V. When specifying motors for operation on variable-speed PWM drives, the voltage withstand level based on the dv/dt of the drive and the known cable distance should be specified.

Motor Insulation Breakdown. If reflected waves generate voltage levels higher than the allowable peak, insulation begins to break down. This phenomenon is known as **partial discharge** (PD) or **corona**. When two phases or two turns in the motor pass next to each other, high voltage peaks can cause a spark plug effect and damage the insulation. The voltage at which this effect begins is referred to as the **corona inception voltage** (CIV) rating of the motor (Figure 15). NEMA *Standard* MG 1 specifies this level at 1600 V.

Eventually air gaps inside the varnish material ionize due to the high voltage gradients, causing phase-to-phase or turn-to-turn short circuits. This causes microscopic insulation breakdown, which is usually detected by the drive current sensors and results in overcurrent trips. Under this short-circuit condition, a motor may operate properly when run across the line or in bypass mode but consistently trip when run from drive power. Factory testing may be required to confirm this PMV failure mode.

Motor Noise and Drive Carrier Frequencies

The first implementation of PWM drive technology caused extreme motor noise at objectionable frequencies. IGBT technology

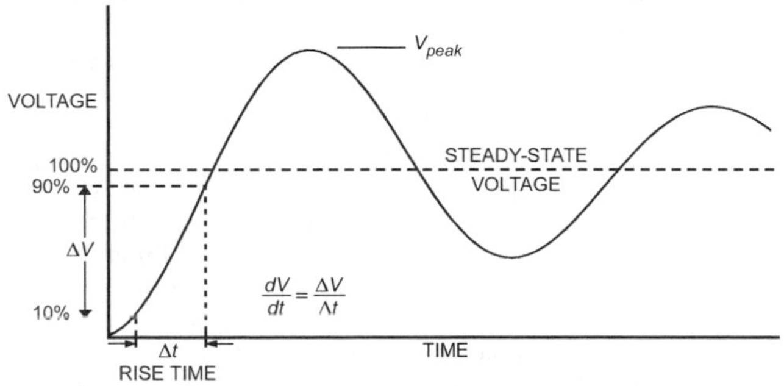

Fig. 14 Motor Voltage Peak and dv/dt Limits
(NEMA *Standard* MG 1, Part 30, Figure 30-5)

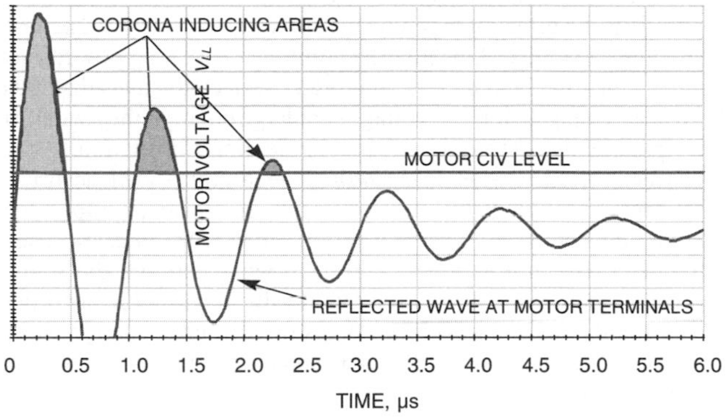

Fig. 15 Damaging Reflected Waves above Motor CIV Levels

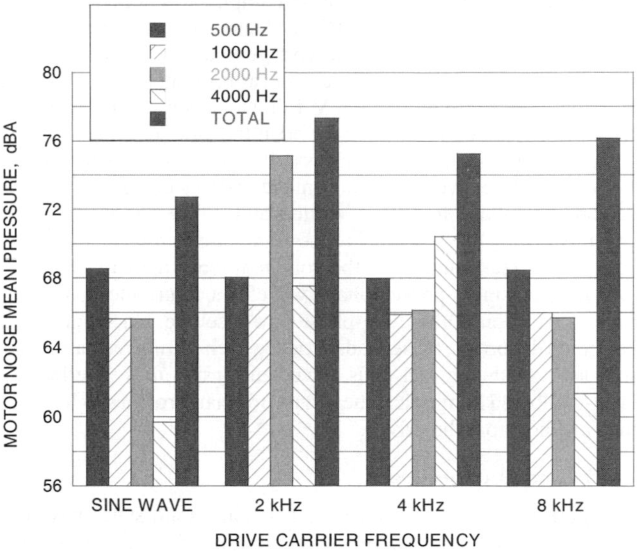

Fig. 16 Motor Audible Noise

allows drive designers to increase the carrier frequency to levels that minimize objectionable noise in the human hearing spectrum. Drive designs can switch up to 20 kHz, if required; however, some engineering compromises must be made to optimize the design. During the transition between turning off and on, the transistor generates heat that must be dissipated. This heat loss rises proportionally with the carrier frequency. While higher carrier frequencies do eliminate objectionable audible noise, they also require larger heat sinks and, consequently, yield lower efficiency.

Audible noise measured in the dBA-weighted scale does not increase proportionally with drive carrier frequency. Additionally, concern with noise may not be over the measured total mean pressure level but a particular frequency band that is objectionable to humans.

Figure 16 shows audible noise test results measured on a 100 hp energy-efficient motor. Note that the dominant octave band is at the drive carrier frequency setting. Sine wave power is used as a reference point on the left side of the table. When running at 2 kHz, the total sound pressure is almost 6 dBA over the sine wave power recordings. This represents 4 times the sound pressure from the motor because the scale is logarithmic and an increase of 3 dBA doubles the mean pressure level. By comparison, running the drive at 4 kHz increases the mean pressure by only 3 dBA, or half the mean pressure of the 2 kHz setting. (For reference, a 10 dB rise in sound pressure is perceived by the human ear as being twice as loud.)

High Carrier Frequencies and Subharmonics. At high (above 5 kHz) carrier frequencies, harmonics can create vibration forces that match the natural mechanical resonant frequency of the stator and cause sound pressure to exceed 85 dB. The likelihood of subharmonics increases as carrier frequency approaches 20 kHz. If subharmonic vibrations appear, the carrier frequency setting should be decreased to lower the sound pressure generated from the motor.

Carrier Frequencies and Drive Ratings

As carrier frequency increases, drive output ampere ratings decrease largely due to the heat that must be dissipated from the IGBT switching. If the rated carrier frequency of a drive is 2 kHz, setting the carrier frequency up to 8 kHz decreases the ampere output. Generally, for every 1 kHz increase in carrier frequency, the drive output current must be derated by 2%. As an example, a 10 hp, 460 V drive rated at 2 kHz may have an output of 14 A. If this drive is run at 10 kHz, or an increase of 8 kHz, it must be derated to 11.76 A, or a 16% decrease in current. If the motor nameplate full load were 14 A, this drive would not generate enough output current to obtain the full 10 hp. In effect, the drive and motor would only generate 8.4 hp continuously. This may not be enough power to drive a fan or pump at the performance specified for the application. For this reason, the specifying engineer should always state the desired running carrier frequency of the drive to ensure proper operation.

POWER DISTRIBUTION SYSTEM EFFECTS

Some concern has been expressed that adjustable frequency drives may cause harmonic disturbances to the basic powerline waveform. Line harmonics are particularly critical to ac drive users for the following reasons:

- Current harmonics cause additional heating in transformers, conductors, and switchgear.
- Voltage harmonics upset the smooth, predictable voltage waveform in a normal sine wave. A line wave severely distorted by voltage harmonics may damage components connected to the line or cause erratic operation of some equipment.
- High-frequency components of voltage distortion can interfere with signals transmitted on the ac line for some control systems.

However, PWM ac drives with built-in bus reactors cause little, if any, disturbance to the input power. A **linear load**, such as a three-phase induction motor operated across the line, may cause a phase displacement between the voltage and current waveforms (phase lag or lead), but the shapes of these waveforms are nearly identical sine waves.

In contrast, a **nonlinear load** draws current only from the peaks of the ac sine wave, flattening the top of the voltage waveform. Many nonlinear loads connected to a power system can inject harmonics. Single-phase equipment (e.g., TVs, VCRs, computers, and electronic lighting) and three-phase equipment (e.g., AFDs, uninterruptible power supplies (UPSs), electric arc furnaces, electric heaters, and welders) convert ac voltage to dc voltage and contain circuitry that draws current in a nonlinear fashion. Figure 17 shows how the current drawn by a PWM full wave rectification AFD distorts the voltage waveform measured at the input terminals.

A single-phase load is not necessarily too small to be of concern. With ac-to-dc converters, the demand current occurs around the peak of the voltage sine wave. A thousand 100 W fluorescent lights consume 100 kW of power. If the lights are nonlinear loads, the peaks add directly and cause the voltage waveform to dip. This distortion in the single-phase voltage waveform contributes to the harmonic distortion of the three-phase power source. On single-phase harmonic distortion, these loads produce even-numbered harmonics such as 2nd, 4th, 6th, etc. Thus, if a balanced system is experiencing even-numbered harmonics, they must originate from a single-phase load and not from the drives.

AFDs and Harmonics

Figure 18 shows the basic elements of any solid-state drive. The converter section (for conversion of ac line power to dc) and the inverter section (for conversion of dc to variable frequency ac) both contain nonlinear devices that cause harmonics on both the input and output lines. Input line harmonics are caused solely by the converter section and are usually referred to as **line-side harmonics**. Output line harmonics are caused solely by the inverter section and are known as **load-side** or **motor harmonics**.

These effects are isolated from each other by a dc bus capacitor and in some designs by a dc choke so that load-side harmonics only affect equipment driven by the AFD and line-side harmonics affect the power system as a whole.

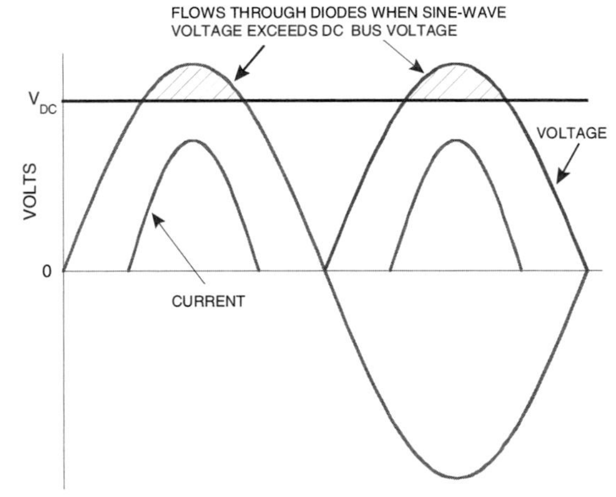

Fig. 17 Voltage Waveform Distortion by Pulse Width Modulated AFD

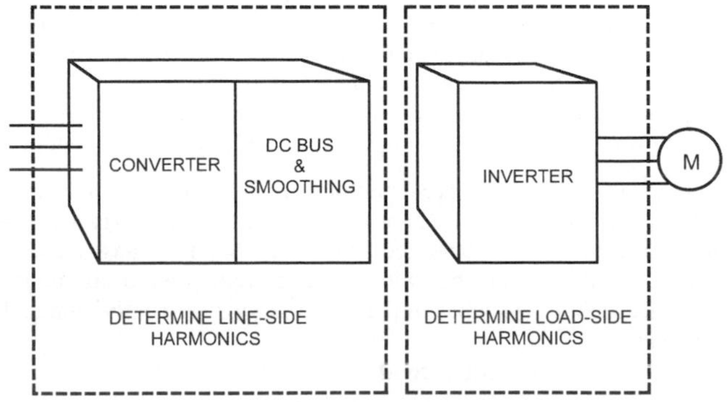

Fig. 18 Basic Elements of Solid-State Drive

Effects of Load-Side Harmonics. Load-side harmonics generated by the inverter section of an AFD are of concern only for the motor. The load-side harmonics can slightly decrease motor life due to the **additional heating** created. However, the use of high-efficiency motors or motors designed specifically for AFDs compensates for any damaging effects. Additionally, hermetic refrigerant-cooled motors, as used in some variable-speed chiller designs, experience insignificant, if any, increases in motor heat. Selection and matching of both the motor and drive should account for these effects and ensure that motor performance and equipment life are not compromised when applying variable speed. Retrofit applications should be engineered to make sure that the motor and drive are capable of providing enough power to the connected load.

As discussed previously in the section on Motor and Conductor Impedance, a second phenomenon associated with inverters on the load side is the effect of high voltage spikes on motor life. The fast-switching capability of the inverter combined with long power lines between the drive and motor can produce reflected waves that have high peak voltages. If these voltages are large enough, they will produce potentially destructive stresses in the motor insulation.

Effects of Line-Side Harmonics. Generally, PWM ac drives that contain internal bus reactors or three-phase ac input line reactors do not create electrical interference with other electrical equipment. But any harmonic current flowing through the source impedance causes a voltage drop that results in harmonic distortion of the supply voltage waveform. A distorted supply voltage waveform can have undesirable effects on some equipment connected to the power line:

- Communications equipment, computers, and diagnostic equipment are "sensitive" equipment having a low tolerance to harmonics. Typical effects include receipt of false commands and data corruption.
- Transformers may experience trouble due to possible additional heating in the core and windings. Many transformer manufacturers rate special transformers by K-factor, which indicates the transformer's ability to withstand degradation due to harmonics. Special cores to reduce eddy currents, specially designed windings that reduce heating, and an oversized neutral bus are some of the special design features found in some K-factor transformers. Other manufacturers simply derate their standard transformers to compensate for harmonic effects.
- Standby generators operate at frequencies that change with load. When an AFD is switched onto generator power, the frequency fluctuation could affect the AFD converter. Standby generators also have voltage regulators that are susceptible to harmonics. In addition, generators have a very high impedance in comparison to the normal power. The harmonic currents flowing in this higher impedance can give rise to harmonic voltages three to four times the normal levels. Compounding this problem is the fact that standby generators are usually installed where sensitive equipment is prevalent (e.g., in hospitals and computer centers). However, because generator power is typically used only during emergencies for short periods, higher harmonic distortion levels may be tolerable.

Any AFD application with standby generators requires careful design, and the following information should be gathered:

- Power output (kW, MW or kVA) of the generator
- Subtransient reactance
- How the generator is being applied in reference to the AFD; what is the worst-case running condition of the drives (number of drives running at one time and the load on these drives)

Additional problems can be caused by resonance that can occur when **power factor correction capacitors** (PFCCs) are installed.

Resonance can severely distort the voltage waveform. PFCCs may fail prematurely, or capacitor fuses may blow. Additionally, because AFDs have an inherent high power factor (typically 0.96 or greater), PFCCs should never be required or used with a drive because they can cause the drive to fail if installed on the load side of the AFD. If an older motor is retrofitted with capacitors, PFCCs should be removed since they are no longer required.

Only the fundamental current transmits power to the load. Harmonic currents increase the equipment input kVA without contributing to input power. Operating with a high harmonic content is much like operating at a low input power factor. High harmonic content means that higher total current is required to deliver a given amount of power due to equipment heat losses. This means that all components of the power distribution system must be oversized to handle the additional current. If the utility meters are capable of measuring the harmonic content and/or power factor, they may assess a distortion (demand) charge or power factor penalty.

Effect of Harmonics on a System. In most applications, no harmonic problems will occur with PWM AFDs that use a series reactor in the dc bus or in the input ac line. With other converter loads (e.g., arc furnaces, dc drives, current source drives) and other high reactive current loads, harmonic problems may exist. The following problems, typically more common on single-phase systems, may indicate a harmonic condition, but they may also indicate line voltage unbalance or overloaded conditions:

- Nuisance input fuse blowing or circuit breaker tripping
- Power factor capacitor overheating, or fuse failure
- Overheating of supply transformers

Problems that are *not* harmonic problems include

- Overcurrent tripping of AFDs
- Interference with AM radio reception
- Wire failure in conduits

CODES AND STANDARDS

CSA. 1998. *Canadian electrical code, Part I. Standard* C22.1-98, p. 8. Canadian Standards Association, Etobicoke, Ont.

NEMA. 1977. Energy management guide for selection and use of single-phase motors. *Standard* MG 11-1977. National Electrical Manufacturers Association, Rosslyn, VA.

NEMA. 1994. Energy management guide for selection and use of polyphase motors. *Standard* MG 10-1994.

NEMA. 1995. Electrical power systems and equipment—Voltage ratings (60 Hz). ANSI/NEMA *Standard* C84.1-1995.

NEMA. 1998. Motors and generators. *Standard* MG 1-1998.

NFPA. 1998. *National electrical code*. ANSI/FFPA *Standard* 70-1998. National Fire Protection Association, Quincy, MA.

BIBLIOGRAPHY

Evon, S., D. Kempke, L. Saunders, and G. Skibinski. 1996. IGBT drive technology demands new motor and cable considerations. IEEE Petroleum & Chemical Industry Conf. IEEE, New York.

Kerkman, R., D. Leggate, and G. Skibinski. 1997. Cable characteristics and their influence on motor over-voltages. IEEE Applied Electronic Conference (APEC). IEEE, New York.

Kerkman, R., D. Leggate, and G. Skibinski. 1996. Interaction of drive modulation & cable parameters on ac motor transients. IEEE Industry Application Society Conf. IEEE, New York.

Malfait A., R. Reekmans, and R. Belmans. 1994. Audible noise and losses in variable speed induction motor drives with IGBT inverters-influence of the squirrel cage design and the switching frequency. IEEE 1194. Proceedings Industry Application. IEEE, New York.

Saunders, L., G. Skibinski, R. Kerkman, D. Schlegel, and D. Anderson. 1996. Modern drive application issues and solutions. IEEE PCIC Conf. Tutorial on Reflected Wave, Motor Failure, CM Electrical Noise, Motor Bearing Current. IEEE, New York.

Sung, J. and S. Bell, S. 1996. Will your motor insulation survive a new adjustable frequency drive? IEEE Petroleum & Chemical Industry Conf. IEEE, New York.

Takahashi, T., G. Wagoner, H. Tsai, and T. Lowery. 1995. Motor lead length issues for IGBT PWM drives. IEEE Pulp and Paper Conf. IEEE, New York.

CHAPTER 41

PIPES, TUBES, AND FITTINGS

THIS CHAPTER covers the selection, application, and installation of pipe, tubes, and fittings commonly used for heating, air-conditioning, and refrigeration. Pipe hangers and pipe expansion are also addressed. When selecting and applying these components, applicable local codes, state or provincial codes, and voluntary industry standards (some of which have been adopted by code jurisdictions) must be followed.

The following organizations in the United States issue codes and standards for piping systems and components:

ASME	—	American Society of Mechanical Engineers
ASTM	—	American Society for Testing and Materials
NFPA	—	National Fire Protection Association
BOCA	—	Building Officials and Code Administrators, International
MSS	—	Manufacturers Standardization Society of the Valve and Fittings Industry, Inc.
AWWA	—	American Water Works Association

Parallel federal specifications also have been developed by government agencies and are used for many public works projects. Chapter IV of ASME *Standard* B31.9 lists applicable U.S. codes and standards for HVAC piping. In addition, it gives the requirements for the safe design and construction of piping systems for building heating and air conditioning. ASME *Standard* B31.5 gives similar requirements for refrigerant piping.

PIPE

Steel Pipe

Steel pipe is manufactured by several processes. Seamless pipe, made by piercing or extruding, has no longitudinal seam. Other manufacturing methods roll a strip or sheet of steel (skelp) into a cylinder and weld a longitudinal seam. A continuous-weld (CW) furnace butt-welding process forces and joins the edges together at high temperature. An electric current welds the seam in electric resistance welded (ERW) pipe. ASTM *Standards* A106 and A53 specify steel pipe. Both standards specify A and B grades. The A grade has a lower tensile strength and is not widely used.

The ASME pressure piping codes require that a longitudinal joint efficiency factor E (Table 1) be applied to each type of seam when calculating the allowable stress. ASME *Standard* B36.10M specifies the dimensional standard for steel pipe. Through 12 in. diameter, nominal pipe sizes (NPS) are used, which do not match the internal or external diameters. For pipe 14 in. and larger, the size corresponds to the outside diameter.

Steel pipe is manufactured with wall thicknesses identified by schedule or weight class. Although schedule numbers and weight class designations are related, they are not constant for all pipe sizes. Standard weight (STD) and Schedule 40 pipe have the same wall thickness through NPS 10. For 12 in. and larger standard weight pipe, the wall thickness remains constant at 0.375 in., while Schedule 40 wall thickness increases with each size. A similar equality exists between Extra Strong (XS) and Schedule 80 pipe through 8 in.; above 8 in., XS pipe has a 0.500 in. wall, while Schedule 80 increases in wall thickness. Table 2 lists properties of representative steel pipe.

Joints in steel pipe are made by welding or by using threaded, flanged, or grooved fittings. Unreinforced welded-in branch connections weaken a main pipeline, and added reinforcement is necessary, unless the excess wall thickness of both mains and branches is sufficient to sustain the pressure.

ASME *Standard* B31.1 gives formulas for determining whether reinforcement is required. Such calculations are seldom needed in HVAC applications because (1) standard weight pipe through NPS 20 at 300 psig requires no reinforcement; full-size branch connections are not recommended; and (2) fittings such as tees and reinforced outlet fittings provide inherent reinforcement.

Type F steel pipe is not permitted for ASME *Standard* B 31.5 refrigerant piping.

Copper Tube

Because of their inherent resistance to corrosion and ease of installation, copper and copper alloys are often used in heating, air-conditioning, refrigeration, and water supply installations. There are two principal classes of copper tube. ASTM *Standard* B88 includes Types K, L, M, and DWV for water and drain service. ASTM *Standard* B280 specifies air-conditioning and refrigeration (ACR) tube for refrigeration service.

Types K, L, M, and DWV designate descending wall thicknesses for copper tube. All types have the same outside diameter for corresponding sizes. Table 3 lists properties of ASTM B88 copper tube. In the plumbing industry, tube of nominal size approximates the inside diameter. The heating and refrigeration trades specify copper tube by the outside diameter (OD). ACR tubing has a different set of wall thicknesses. Types K, L, and M tube may be hard drawn or annealed (soft) temper.

Copper tubing is joined with soldered or brazed, wrought or cast copper capillary socket-end fittings. Table 4 lists pressure-temperature ratings of soldered and brazed joints. Small copper tube is also joined by flare or compression fittings.

Hard-drawn tubing has a higher allowable stress than annealed tubing, but if hard tubing is joined by soldering or brazing, the annealed allowable stress should be used.

Brass pipe and copper pipe are also made in steel pipe thicknesses for threading. High cost has eliminated these materials from the market, except for special applications.

The heating and air-conditioning industry generally uses Types L and M tubing, which have higher internal working pressure ratings than the solder joints used at fittings. Type K may be used with brazed joints for higher pressure-temperature requirements or for

---

The preparation of this chapter is assigned to TC 8.10, Pumps and Hydronic Piping.

Table 1 Allowable Stresses[a] for Pipe and Tube

ASTM Specification	Grade	Type	Manufacturing Process	Available Sizes, in.	Minimum Tensile Strength, psi	Basic Allowable Stress S, psi	Joint Efficiency Factor E	Allowable Stress[b] S_E, psi	Allowable Stress Range[c] S_A, psi
A53 Steel	—	F	Cont. Weld	1/2 to 4	45,000	11,250	0.6	6,800	16,900
A53 Steel	B	S	Seamless	1/2 to 26	60,000	15,000	1.0	15,000	22,500
A53 Steel	B	E	ERW	2 to 20	60,000	15,000	0.85	12,800	22,500
A106 Steel	B	S	Seamless	1/2 to 26	60,000	15,000	1.0	15,000	22,500
B88 Copper	—	—	Hard Drawn	1/4 to 12	36,000	9,000	1.0	9,000	13,500

[a] Listed stresses are for temperatures to 650°F for steel pipe (to 400°F for Type F) and to 250°F for copper tubing.
[b] To be used for internal pressure stress calculations in Equations (1) and (2).
[c] To be used only for piping flexibility calculations; see Equations (3) and (4).

direct burial. Type M should be used with care where exposed to potential external damage.

Copper and brass should not be used in ammonia refrigerating systems. The section on Special Systems covers other limitations on refrigerant piping.

Ductile Iron and Cast Iron

Cast-iron soil pipe comes in XH or service weight. It is not used under pressure because the pipe is not suitable and the joints are not restrained. Cast-iron pipe and fittings typically have bell and spigot ends for lead and oakum joints or elastomer push-on joints. Cast-iron pipe and fittings are also furnished with *no-hub* ends for joining with *no-hub* clamps. Local plumbing codes specify permitted materials and joints.

Ductile iron has now replaced cast iron for pressure pipe. Ductile iron is stronger, less brittle, and similar to cast iron in corrosion resistance. It is commonly used for buried pressure water mains or in other locations where internal or external corrosion is a problem. Joints are made with flanged fittings, mechanical joint (MJ) fittings, or elastomer gaskets for bell and spigot ends. Bell and spigot and MJ joints are not self-restrained. Restrained MJ systems are available. Ductile-iron pipe is made in seven thickness classes for different service conditions. AWWA *Standard* C150/A21.50, Thickness Design of Ductile-Iron Pipe, covers the proper selection of pipe classes.

FITTINGS

The following standards give dimensions and pressure ratings for fittings, flanges, and flanged fittings. This data is also available from manufacturers' catalogs.

Applicable Standards for Fittings

Steel[a]	**ASME *Std.***
Pipe Flanges and Flanged Fittings	B16.5
Factory-Made Wrought Steel Buttwelding Fittings	B16.9
Forged Fittings, Socket-Welding and Threaded	B16.11
Wrought Steel Buttwelding Short Radius Elbows and Returns	B16.28
Cast Iron, Malleable Iron, Ductile Iron[b]	
Cast Iron Pipe Flanges and Flanged Fittings	B16.1
Malleable Iron Threaded Fittings	B16.3
Gray Iron Threaded Fittings	B16.4
Cast Iron Threaded Drainage Fittings	B16.12
Ductile Iron Pipe Flanges and Flanged Fittings, Classes 150 and 300	B16.42
Copper and Bronze[c]	
Cast Bronze Threaded Fittings, Classes 125 and 25	B16.15
Cast Copper Alloy Solder Joint Pressure Fittings	B16.18
Wrought Copper and Copper Alloy Solder Joint Pressure Fittings	B16.22
Cast Copper Alloy Solder Joint Drainage Fittings, DWV	B16.23
Cast Copper Alloy Pipe Flanges and Flanged Fittings, Classes 150, 300, 400, 600, 900, 1500, and 2500	B16.24
Cast Copper Alloy Fittings for Flared Copper Tubes	B16.26
Wrought Copper and Wrought Copper Alloy Solder Joint Drainage Fittings	B16.29
Nonmetallic[d]	**ASTM *Std.***
Threaded PVC Plastic Pipe Fittings, Schedule 80	D2464
Threaded PVC Plastic Pipe Fittings, Schedule 40	D2466
Socket-Type PVC Plastic Pipe Fittings, Schedule 80	D2467
Reinforced Epoxy Resin Gas Pressure Pipe and Fittings	D2517
Threaded CPVC Plastic Pipe Fittings, Schedule 80	F437
Socket-Type CPVC Plastic Pipe Fittings, Schedule 40	F438
Socket-Type CPVC Plastic Pipe Fittings, Schedule 80	F439
Polybutylene (PB) Plastic Hot- and Cold-Water Distribution Systems	D3309
Plastic Insert Fittings for Polybutylene Tubing	F845
Solvent Cements for PVC Plastic Piping Systems	D2564
Solvent Cements for CPVC Plastic Pipe and Fittings	F493

[a] Wrought steel butt-welding fittings are made to match steel pipe wall thicknesses and are rated at the same working pressure as seamless pipe. Flanges and flanged fittings are rated by working steam pressure classes. Forged steel fittings are rated from 2000 to 6000 psi in classes and are used for high-temperature and high-pressure service for small pipe sizes.
[b] The class numbers refer to the maximum working saturated steam gage pressure (in psi). For liquids at lower temperatures, higher pressures are allowed. Groove-end fittings of these materials are made by various manufacturers who publish their own ratings.
[c] The classes refer to maximum working steam gage pressure (in psi). At ambient temperatures, higher liquid pressures are allowed. Solder joint fittings are limited by the strength of the soldered or brazed joint (see Table 4).
[d] Ratings of plastic fittings match the pipe of corresponding schedule number.

JOINING METHODS

Threading

Threading as per ASME *Standard* B1.20.1 is the most common method for joining small-diameter steel or brass pipe. Pipe with a wall thickness less than standard weight should not be threaded. ASME *Standard* B31.5 limits the threading for various refrigerants and pipe sizes.

Soldering and Brazing

Copper tube is usually joined by soldering or brazing socket end fittings. Brazing materials melt above 1000°F and produce a stronger joint than solder. Table 4 lists soldered and brazed joint strengths. ASME *Standard* B16.22 specified wrought copper solder joint fittings and ASME *Standard* B16.18 specified cast copper solder joint fittings are pressure rated the same way as annealed Type L copper tube of the same size. Health concerns have caused many jurisdictions to ban solder containing lead or antimony for joining pipe in potable water systems. Lead-based solder, in particular, must not be used for potable water.

Table 2 Steel Pipe Data

Nominal Size, in.	Pipe OD, in.	Schedule Number or Weight[a]	Wall Thickness t, in.	Inside Diameter d, in.	Surface Area		Cross Section		Weight		Working Pressure[c] ASTM A53 B to 400°F		
					Outside, ft²/ft	Inside, ft²/ft	Metal Area, in²	Flow Area, in²	Pipe, lb/ft	Water, lb/ft	Mfr. Process	Joint Type[b]	psig
1/4	0.540	40 ST	0.088	0.364	0.141	0.095	0.125	0.104	0.424	0.045	CW	T	188
		80 XS	0.119	0.302	0.141	0.079	0.157	0.072	0.535	0.031	CW	T	871
3/8	0.675	40 ST	0.091	0.493	0.177	0.129	0.167	0.191	0.567	0.083	CW	T	203
		80 XS	0.126	0.423	0.177	0.111	0.217	0.141	0.738	0.061	CW	T	820
1/2	0.840	40 ST	0.109	0.622	0.220	0.163	0.250	0.304	0.850	0.131	CW	T	214
		80 XS	0.147	0.546	0.220	0.143	0.320	0.234	1.087	0.101	CW	T	753
3/4	1.050	40 ST	0.113	0.824	0.275	0.216	0.333	0.533	1.13	0.231	CW	T	217
		80 XS	0.154	0.742	0.275	0.194	0.433	0.432	1.47	0.187	CW	T	681
1	1.315	40 ST	0.133	1.049	0.344	0.275	0.494	0.864	1.68	0.374	CW	T	226
		80 XS	0.179	0.957	0.344	0.251	0.639	0.719	2.17	0.311	CW	T	642
1-1/4	1.660	40 ST	0.140	1.380	0.435	0.361	0.669	1.50	2.27	0.647	CW	T	229
		80 XS	0.191	1.278	0.435	0.335	0.881	1.28	2.99	0.555	CW	T	594
1-1/2	1.900	40 ST	0.145	1.610	0.497	0.421	0.799	2.04	2.72	0.881	CW	T	231
		80 XS	0.200	1.500	0.497	0.393	1.068	1.77	3.63	0.765	CW	T	576
2	2.375	40 ST	0.154	2.067	0.622	0.541	1.07	3.36	3.65	1.45	CW	T	230
		80 XS	0.218	1.939	0.622	0.508	1.48	2.95	5.02	1.28	CW	T	551
2-1/2	2.875	40 ST	0.203	2.469	0.753	0.646	1.70	4.79	5.79	2.07	CW	W	533
		80 XS	0.276	2.323	0.753	0.608	2.25	4.24	7.66	1.83	CW	W	835
3	3.500	40 ST	0.216	3.068	0.916	0.803	2.23	7.39	7.57	3.20	CW	W	482
		80 XS	0.300	2.900	0.916	0.759	3.02	6.60	10.25	2.86	CW	W	767
4	4.500	40 ST	0.237	4.026	1.178	1.054	3.17	12.73	10.78	5.51	CW	W	430
		80 XS	0.337	3.826	1.178	1.002	4.41	11.50	14.97	4.98	CW	W	695
6	6.625	40 ST	0.280	6.065	1.734	1.588	5.58	28.89	18.96	12.50	ERW	W	696
		80 XS	0.432	5.761	1.734	1.508	8.40	26.07	28.55	11.28	ERW	W	1209
8	8.625	30	0.277	8.071	2.258	2.113	7.26	51.16	24.68	22.14	ERW	W	526
		40 ST	0.322	7.981	2.258	2.089	8.40	50.03	28.53	21.65	ERW	W	643
		80 XS	0.500	7.625	2.258	1.996	12.76	45.66	43.35	19.76	ERW	W	1106
10	10.75	30	0.307	10.136	2.814	2.654	10.07	80.69	34.21	34.92	ERW	W	485
		40 ST	0.365	10.020	2.814	2.623	11.91	78.85	40.45	34.12	ERW	W	606
		XS	0.500	9.750	2.814	2.552	16.10	74.66	54.69	32.31	ERW	W	887
		80	0.593	9.564	2.814	2.504	18.92	71.84	64.28	31.09	ERW	W	1081
12	12.75	30	0.330	12.090	3.338	3.165	12.88	114.8	43.74	49.68	ERW	W	449
		ST	0.375	12.000	3.338	3.141	14.58	113.1	49.52	48.94	ERW	W	528
		40	0.406	11.938	3.338	3.125	15.74	111.9	53.48	48.44	ERW	W	583
		XS	0.500	11.750	3.338	3.076	19.24	108.4	65.37	46.92	ERW	W	748
		80	0.687	11.376	3.338	2.978	26.03	101.6	88.44	43.98	ERW	W	1076
14	14.00	30 ST	0.375	13.250	3.665	3.469	16.05	137.9	54.53	59.67	ERW	W	481
		40	0.437	13.126	3.665	3.436	18.62	135.3	63.25	58.56	ERW	W	580
		XS	0.500	13.000	3.665	3.403	21.21	132.7	72.04	57.44	ERW	W	681
		80	0.750	12.500	3.665	3.272	31.22	122.7	106.05	53.11	ERW	W	1081
16	16.00	30 ST	0.375	15.250	4.189	3.992	18.41	182.6	62.53	79.04	ERW	W	421
		40 XS	0.500	15.000	4.189	3.927	24.35	176.7	82.71	76.47	ERW	W	596
18	18.00	ST	0.375	17.250	4.712	4.516	20.76	233.7	70.54	101.13	ERW	W	374
		30	0.437	17.126	4.712	4.483	24.11	230.3	81.91	99.68	ERW	W	451
		XS	0.500	17.000	4.712	4.450	27.49	227.0	93.38	98.22	ERW	W	530
		40	0.562	16.876	4.712	4.418	30.79	223.7	104.59	96.80	ERW	W	607
20	20.00	20 ST	0.375	19.250	5.236	5.039	23.12	291.0	78.54	125.94	ERW	W	337
		30 XS	0.500	19.000	5.236	4.974	30.63	283.5	104.05	122.69	ERW	W	477
		40	0.593	18.814	5.236	4.925	36.15	278.0	122.82	120.30	ERW	W	581

[a] Numbers are schedule numbers per ASME *Standard* B36.10M; ST = Standard Weight; XS = Extra Strong.

[b] T = Thread; W = Weld

[c] Working pressures were calculated per ASME B31.9 using furnace butt-weld (continuous weld, CW) pipe through 4 in. and electric resistance weld (ERW) thereafter. The allowance A has been taken as

(1) 12.5% of t for mill tolerance on pipe wall thickness, *plus*

(2) An arbitrary corrosion allowance of 0.025 in. for pipe sizes through NPS 2 and 0.065 in. from NPS 2½ through 20, *plus*

(3) A thread cutting allowance for sizes through NPS 2.

Because the pipe wall thickness of threaded standard pipe is so small after deducting the allowance A, the mechanical strength of the pipe is impaired. It is good practice to limit standard weight threaded pipe pressure to 90 psig for steam and 125 psig for water.

Table 3 Copper Tube Data

Nominal Diameter, in.	Type	Wall Thickness t, in.	Diameter: Outside D, in.	Diameter: Inside d, in.	Surface Area: Outside, ft²/ft	Surface Area: Inside, ft²/ft	Cross Section: Metal Area, in²	Cross Section: Flow Area, in²	Weight: Tube, lb/ft	Weight: Water, lb/ft	Working Pressure[a,b,c] ASTM B88 to 250°F: Annealed, psig	Working Pressure[a,b,c] ASTM B88 to 250°F: Drawn, psig
1/4	K	0.035	0.375	0.305	0.098	0.080	0.037	0.073	0.145	0.032	851	1596
	L	0.030	0.375	0.315	0.098	0.082	0.033	0.078	0.126	0.034	730	1368
3/8	K	0.049	0.500	0.402	0.131	0.105	0.069	0.127	0.269	0.055	894	1676
	L	0.035	0.500	0.430	0.131	0.113	0.051	0.145	0.198	0.063	638	1197
	M	0.025	0.500	0.450	0.131	0.118	0.037	0.159	0.145	0.069	456	855
1/2	K	0.049	0.625	0.527	0.164	0.138	0.089	0.218	0.344	0.094	715	1341
	L	0.040	0.625	0.545	0.164	0.143	0.074	0.233	0.285	0.101	584	1094
	M	0.028	0.625	0.569	0.164	0.149	0.053	0.254	0.203	0.110	409	766
5/8	K	0.049	0.750	0.652	0.196	0.171	0.108	0.334	0.418	0.144	596	1117
	L	0.042	0.750	0.666	0.196	0.174	0.093	0.348	0.362	0.151	511	958
3/4	K	0.065	0.875	0.745	0.229	0.195	0.165	0.436	0.641	0.189	677	1270
	L	0.045	0.875	0.785	0.229	0.206	0.117	0.484	0.455	0.209	469	879
	M	0.032	0.875	0.811	0.229	0.212	0.085	0.517	0.328	0.224	334	625
1	K	0.065	1.125	0.995	0.295	0.260	0.216	0.778	0.839	0.336	527	988
	L	0.050	1.125	1.025	0.295	0.268	0.169	0.825	0.654	0.357	405	760
	M	0.035	1.125	1.055	0.295	0.276	0.120	0.874	0.464	0.378	284	532
1-1/4	K	0.065	1.375	1.245	0.360	0.326	0.268	1.217	1.037	0.527	431	808
	L	0.055	1.375	1.265	0.360	0.331	0.228	1.257	0.884	0.544	365	684
	M	0.042	1.375	1.291	0.360	0.338	0.176	1.309	0.682	0.566	279	522
	DWV	0.040	1.375	1.295	0.360	0.339	0.168	1.317	0.650	0.570	265	497
1-1/2	K	0.072	1.625	1.481	0.425	0.388	0.351	1.723	1.361	0.745	404	758
	L	0.060	1.625	1.505	0.425	0.394	0.295	1.779	1.143	0.770	337	631
	M	0.049	1.625	1.527	0.425	0.400	0.243	1.831	0.940	0.792	275	516
	DWV	0.042	1.625	1.541	0.425	0.403	0.209	1.865	0.809	0.807	236	442
2	K	0.083	2.125	1.959	0.556	0.513	0.532	3.014	2.063	1.304	356	668
	L	0.070	2.125	1.985	0.556	0.520	0.452	3.095	1.751	1.339	300	573
	M	0.058	2.125	2.009	0.556	0.526	0.377	3.170	1.459	1.372	249	467
	DWV	0.042	2.125	2.041	0.556	0.534	0.275	3.272	1.065	1.416	180	338
2-1/2	K	0.095	2.625	2.435	0.687	0.637	0.755	4.657	2.926	2.015	330	619
	L	0.080	2.625	2.465	0.687	0.645	0.640	4.772	2.479	2.065	278	521
	M	0.065	2.625	2.495	0.687	0.653	0.523	4.889	2.026	2.116	226	423
3	K	0.109	3.125	2.907	0.818	0.761	1.033	6.637	4.002	2.872	318	596
	L	0.090	3.125	2.945	0.818	0.771	0.858	6.812	3.325	2.947	263	492
	M	0.072	3.125	2.981	0.818	0.780	0.691	6.979	2.676	3.020	210	394
	DWV	0.045	3.125	3.035	0.818	0.795	0.435	7.234	1.687	3.130	131	246
3-1/2	K	0.120	3.625	3.385	0.949	0.886	1.321	8.999	5.120	3.894	302	566
	L	0.100	3.625	3.425	0.949	0.897	1.107	9.213	4.291	3.987	252	472
	M	0.083	3.625	3.459	0.949	0.906	0.924	9.397	3.579	4.066	209	392
4	K	0.134	4.125	3.857	1.080	1.010	1.680	11.684	6.510	5.056	296	555
	L	0.110	4.125	3.905	1.080	1.022	1.387	11.977	5.377	5.182	243	456
	M	0.095	4.125	3.935	1.080	1.030	1.203	12.161	4.661	5.262	210	394
	DWV	0.058	4.125	4.009	1.080	1.050	0.741	12.623	2.872	5.462	128	240
5	K	0.160	5.125	4.805	1.342	1.258	2.496	18.133	9.671	7.846	285	534
	L	0.125	5.125	4.875	1.342	1.276	1.963	18.665	7.609	8.077	222	417
	M	0.109	5.125	4.907	1.342	1.285	1.718	18.911	6.656	8.183	194	364
	DWV	0.072	5.125	4.981	1.342	1.304	1.143	19.486	4.429	8.432	128	240
6	K	0.192	6.125	5.741	1.603	1.503	3.579	25.886	13.867	11.201	286	536
	L	0.140	6.125	5.845	1.603	1.530	2.632	26.832	10.200	11.610	208	391
	M	0.122	6.125	5.881	1.603	1.540	2.301	27.164	8.916	11.754	182	341
	DWV	0.083	6.125	5.959	1.603	1.560	1.575	27.889	6.105	12.068	124	232
8	K	0.271	8.125	7.583	2.127	1.985	6.687	45.162	25.911	19.542	304	570
	L	0.200	8.125	7.725	2.127	2.022	4.979	46.869	19.295	20.280	224	421
	M	0.170	8.125	7.785	2.127	2.038	4.249	47.600	16.463	20.597	191	358
	DWV	0.109	8.125	7.907	2.127	2.070	2.745	49.104	10.637	21.247	122	229
10	K	0.338	10.125	9.449	2.651	2.474	10.392	70.123	40.271	30.342	304	571
	L	0.250	10.125	9.625	2.651	2.520	7.756	72.760	30.054	31.483	225	422
	M	0.212	10.125	9.701	2.651	2.540	6.602	73.913	25.584	31.982	191	358
12	K	0.405	12.125	11.315	3.174	2.962	14.912	100.554	57.784	43.510	305	571
	L	0.280	12.125	11.565	3.174	3.028	10.419	105.046	40.375	45.454	211	395
	M	0.254	12.125	11.617	3.174	3.041	9.473	105.993	36.706	45.863	191	358

[a] When using soldered or brazed fittings, the joint determines the limiting pressure.

[b] Working pressures were calculated using ASME *Standard* B31.9 allowable stresses. A 5% mill tolerance has been used on the wall thickness. Higher tube ratings can be calculated using the allowable stress for lower temperatures.

[c] If soldered or brazed fittings are used on hard drawn tubing, use the annealed ratings. Full-tube allowable pressures can be used with suitably rated flare or compression-type fittings.

Table 4 Internal Working Pressure for Copper Tube Joints

Alloy Used for Joints	Service Temperature, °F	Internal Working Pressure, psi: Water and Noncorrosive Liquids and Gases[a], Nominal Tube Size (Types K, L, M), in.: 1/4 to 1	1-1/4 to 2	2-1/2 to 4	5 to 8[a]	10 to 12[a]	Sat. Steam and Condensate: 1/4 to 8
50-50 Tin-lead[b] solder (ASTM B32 Gr 50A)	100	200	175	150	130	100	—
	150	150	125	100	90	70	—
	200	100	90	75	70	50	—
	250	85	75	50	45	40	15
95-5 Tin-antimony[c] solder (ASTM B32 Gr 50TA)	100	500	400	300	270	150	—
	150	400	350	275	250	150	—
	200	300	250	200	180	140	—
	250	200	175	150	135	110	15
Brazing alloys melting at or above 1000°F	100 to 200	d	d	d	d	d	—
	250	300	210	170	150	150	—
	350	270	190	150	150	150	120

Source: Based on ASME *Standard* B31.9, Building Services Piping

[a] Solder joints are not to be used for
(1) Flammable or toxic gases or liquids
(2) Gas, vapor, or compressed air in tubing over 4 in., unless max. pressure is limited to 20 psig.

[b] Lead solders must not be used in potable water systems.

[c] Tin-antimony solder is allowed for potable water supplies in some jurisdictions.

[d] Rated pressure for up to 200°F applies to the tube being joined.

Flared and Compression Joints

Flared and compression fittings can be used to join copper, steel, stainless steel, and aluminum tubing. Properly rated fittings can keep the joints as strong as the tube.

Flanges

Flanges can be used for large pipe and all piping materials. They are commonly used to connect to equipment and valves, and wherever the joint must be opened to permit service or replacement of components. For steel pipe, flanges are available in pressure ratings to 2500 psig. High tensile strength bolts must be used for high pressure flanged joints.

For welded pipe, weld neck, slip-on, or socket weld flanges are available. Thread-on flanges are available for threaded pipe.

Flanges are generally flat faced or raised face. Flat-faced flanges with full-faced gaskets are most often used with cast iron and materials that cannot take high bending loads. Raised-face flanges with ring gaskets are preferred with steel pipe because they facilitate increasing the sealing pressure on the gasket to help prevent leaks. Other facings, such as O ring and ring joint, are available for special applications.

All flat-faced, raised-face, and lap-joint flanges require a gasket between the mating flange surfaces. Gaskets are made from rubber, synthetic elastomers, cork, fiber, plastic, teflon, metal, and combinations of these materials. The gasket must be compatible with the flowing media and the temperatures at which the system operates.

Welding

Welded-steel pipe joints offer the following advantages:

- Do not age, dry out, or deteriorate as do gasketed joints
- Can accommodate greater vibration and water hammer and higher temperatures and pressures than other joints
- For critical service, can be tested by any of several nondestructive examination (NDE) methods, such as radiography or ultrasound
- Provide maximum long-term reliability

The applicable sections of the ASME *Standard* B31series and the ASME *Boiler and Pressure Vessel Code* give rules for welding. ASTM *Standard* B31 requires that all welders and welding procedure specifications (WPS) be qualified. Separate WPS are needed for different welding methods and materials. The qualifying tests and the variables requiring separate procedure specifications are set forth in the ASME *Boiler and Pressure Vessel Code*, Section IX. The manufacturer, fabricator, or contractor is responsible for the welding procedure and welders. ASME *Standard* B31.9 requires visual examination of welds and outlines limits of acceptability.

The following welding processes are often used in the HVAC industry:

SMAW—Shielded metal arc welding (stick welding). The molten weld metal is shielded by the vaporization of the electrode coating.

GMAW—Gas metal arc welding, also called MIG. The electrode is a continuously fed wire, which is shielded by argon or carbon dioxide gas from the welding gun nozzle.

GTAW—Gas tungsten arc welding, also called TIG or Heliarc. This process uses a nonconsumable tungsten electrode surrounded by a shielding gas. The weld material may be provided from a separate noncoated rod.

Reinforced Outlet Fittings

Reinforced outlet fittings are used to make branch and take-off connections and are designed to permit welding directly to pipe without supplemental reinforcing. Fittings are available with threaded, socket, or butt-weld outlets.

Other Joints

Grooved joints require special grooved fittings and a shallow groove cut or rolled into the pipe end. These joints can be used with steel, cast iron, ductile-iron, copper, and plastic pipes. A segmented clamp engages the grooves and a special gasket designed so that internal pressure tightens the seal. Some clamps are designed with clearance between tongue and groove to accommodate misalignment and thermal movements, and others are designed to limit movement and provide a rigid system. Manufacturers' data gives temperature and pressure limitations.

Another form of mechanical joint consists of a **sleeve** slightly larger than the outside diameter of the pipe. The pipe ends are inserted into the sleeve, and gaskets are packed into the annular space between the pipe and coupling and held in place by retainer rings. This type of joint can accept some axial misalignment, but it must be anchored or otherwise restrained to prevent axial pullout or lateral movement. Manufacturers provide pressure-temperature data.

Ductile-iron pipe is furnished with a spigot end adapted for a gasket and retainer ring. This joint is also not restrained.

Table 5 Application of Pipe, Fittings, and Valves for Heating and Air Conditioning

Application	Pipe Material	Weight	Joint Type	Fitting Class	Fitting Material	System Temperature, °F	System Maximum Pressure at Temperature[a], psig
Recirculating Water 2 in. and smaller	Steel (CW)	Standard	Thread	125	Cast iron	250	125
	Copper, hard	Type L	Braze or silver solder[b]		Wrought copper	250	200
	PVC	Sch 80	Solvent	Sch 80	PVC	75	
	CPVC	Sch 80	Solvent	Sch 80	CPVC	150	
	PB	SDR-11	Heat fusion		PB	160	
			Insert crimp		Metal	160	
2.5 to 12 in.	A53 B ERW Steel	Standard	Weld	Standard	Wrought steel	250	400
			Flange	150	Wrought steel	250	250
			Flange	125	Cast iron	250	175
			Flange	250	Cast iron	250	400
			Groove		MI or ductile iron	230	300
	PB	SDR-11	Heat fusion		PB	160	
Steam and Condensate 2 in. and smaller	Steel (CW)	Standard[c]	Thread	125	Cast iron		90
			Thread	150	Malleable iron		90
	A53 B ERW Steel	Standard[c]	Thread	125	Cast iron		100
			Thread	150	Malleable iron		125
	A53 B ERW Steel	XS	Thread	250	Cast iron		200
			Thread	300	Malleable iron		250
2.5 to 12 in.	Steel	Standard	Weld	Standard	Wrought steel		250
			Flange	150	Wrought steel		200
			Flange	125	Cast iron		100
	A53 B ERW Steel	XS	Weld	XS	Wrought steel		700
			Flange	300	Wrought steel		500
			Flange	250	Cast iron		200
Refrigerant	Copper, hard	Type L or K	Braze		Wrought copper		
	A53 B SML Steel	Standard	Weld		Wrought steel		
Underground Water							
Through 12 in.	Copper, hard	Type K	Braze or silver solder[b]		Wrought copper	75	350
Through 6 in.	Ductile iron	Class 50	MJ	MJ	Cast iron	75	250
	PB	SDR 9, 11	Heat fusion		PB	75	
		SDR 7, 11.5	Insert crimp		Metal	75	
Potable Water, Inside Building	Copper, hard	Type L	Braze or silver solder[b]		Wrought copper	75	350
	Steel, galvanized	Standard	Thread	125	Galv. cast iron	75	125
				150	Galv. mall. iron	75	125
	PB	SDR-11	Heat fusion		PB	75	
			Insert crimp		Metal	75	

[a] Maximum allowable working pressures have been derated in this table. Higher system pressures can be used for lower temperatures and smaller pipe sizes. Pipe, fittings, joints, and valves must all be considered.

[b] Lead- and antimony-based solders should not be used for potable water systems. Brazing and silver solders should be employed.

[c] Extra strong pipe is recommended for all threaded condensate piping to allow for corrosion.

Unions

Unions allow disassembly of threaded pipe systems. Unions are three-part fittings with a mating machined seat on the two parts that thread onto the pipe ends. A threaded locking ring holds the two ends tightly together. A union also allows threaded pipe to be turned at the last joint connecting two pieces of equipment. Companion flanges (a pair) for small pipe serve the same purpose.

SPECIAL SYSTEMS

Certain piping systems are governed by separate codes or standards, which are summarized below. Generally, any failure of the piping in these systems is dangerous to the public, so some local areas have adopted laws enforcing the codes.

- **Boiler piping.** ASME *Standard* B31.1 and the ASME *Boiler and Pressure Vessel Code* (Section I) specify the piping inside the code-required stop valves on boilers that operate above 15 psig with steam, or above 160 psig or 250°F with water. These codes require fabricators and contractors to be certified for such work. The field or shop work must also be inspected while it is in progress by inspectors commissioned by the National Board of Boiler and Pressure Vessel Inspectors.
- **Refrigeration piping.** ASME *Standard* B31.5, Refrigerant Piping, and ASHRAE *Standard* 15, Standard Safety Code for Mechanical Refrigeration, cover the requirements for refrigerant piping.
- **Plumbing systems.** Local codes cover piping for plumbing.
- **Sprinkler systems.** NFPA *Standard* 13, Installation of Sprinkler Systems, covers this field.
- **Fuel gas.** NFPA *Standard* 70/ANSI Z223.1, *National Fuel Gas Code*, prescribes fuel gas piping in buildings.

SELECTION OF MATERIALS

Each HVAC system and, under some conditions, portions of a system require a study of the conditions of operation to determine suitable materials. For example, because the static pressure of water in a high-rise building is higher in the lower levels than in the upper levels, different materials may be required along vertical zones.

The following factors should be considered when selecting material for piping:

- Code requirements
- Working fluid in the pipe
- Pressure and temperature of the fluid
- External environment of the pipe
- Installation cost

Table 5 lists materials used for heating and air-conditioning piping. The pressure and temperature rating of each component selected must be considered; the lowest rating establishes the operating limits of the system.

PIPE WALL THICKNESS

The primary factors determining pipe wall thickness are hoop stress due to internal pressure and longitudinal stresses due to pressure, weight, and other sustained loads. Detailed stress calculations are seldom required for HVAC applications because standard pipe has ample thickness to sustain the pressure and longitudinal stress due to weight (assuming hangers are spaced in accordance with Table 6).

STRESS CALCULATIONS

Although stress calculations are seldom required, the factors involved should be understood. The main areas of concern are (1) internal pressure stress, (2) longitudinal stress due to pressure and weight, and (3) stress due to expansion and contraction.

ASME B31 standards establish a basic allowable stress S equal to one-fourth of the minimum tensile strength of the material. This value is adjusted, as discussed in this section, because of the nature of certain stresses and manufacturing processes.

Hoop stress caused by internal pressure is the major stress on pipes. As certain forming methods form a seam that may be weaker than the base material, ASME *Standard* B31.9 specifies a joint efficiency factor E which, multiplied by the basic allowable stress, establishes a maximum allowable stress value in tension S_E. (Table A-1 in ASME B31.9 lists values of S_E for commonly used pipe materials.) The joint efficiency factor can be significant; for example, seamless pipe has a joint efficiency factor of 1, so it can be used to the full allowable stress (one-quarter of the tensile strength). In contrast, butt-welded pipe has a joint efficiency factor of 0.60, so its maximum allowable stress must be derated ($S_E = 0.6S$).

Equation (1) determines the minimum wall thickness for a given pressure. Equation (2) determines the maximum pressure allowed for a given wall thickness.

$$t_m = \frac{pD}{2S_E} + A \qquad (1)$$

$$p = \frac{2S_E(t_m - A)}{D} \qquad (2)$$

where

t_m = minimum required wall thickness, in.
S_E = maximum allowable stress, psi
D = outside pipe diameter, in.
A = allowance for manufacturing tolerance, threading, grooving, and corrosion, in.
p = internal pressure, psi

Both equations incorporate an allowance factor A to compensate for manufacturing tolerances, material removed in threading or grooving, and corrosion. For the seamless, butt-welded, and electric resistance welded (ERW) pipe most commonly used in HVAC work, the standards apply a manufacturing tolerance of 12.5%. Working pressure for steel pipe, as listed in Table 2, has been calculated using a manufacturing tolerance of 12.5%, standard allowance for depth of thread (where applicable), and a corrosion allowance of 0.065 in. for pipes 2-1/2 in. and larger and 0.025 in. for pipes 2 in. and smaller. Where corrosion is known to be greater or smaller, pressure rating can be recalculated using Equation (2). Higher pressure ratings than shown in Table 2 can be obtained (1) by using ERW or seamless pipe in lieu of continuous-weld (CW) pipe 4 in. and less and seamless pipe in lieu of ERW pipe 5 in. and greater (due to higher joint efficiency factors), or (2) by using heavier wall pipe.

Longitudinal stresses due to pressure, weight, and other sustained forces are additive, and the sum of all such stresses must not exceed the basic allowable stress S at the highest temperature at which the system will operate. Longitudinal stress due to pressure equals approximately one-half the hoop stress caused by internal pressure, which means that at least one-half the basic allowable stress is available for weight and other sustained forces. This factor is taken into account in Table 6.

Stresses due to expansion and contraction are cyclical, and, because creep allows some stress relaxation, the ASME B31 standards permit designing to an allowable stress range S_A as calculated by Equation (3). Table 1 lists allowable stress ranges for commonly used piping materials.

$$S_A = 1.25S_c + 0.25S_h \qquad (3)$$

where

S_A = allowable stress range, psi
S_c = allowable cold stress at coolest temperature the system will experience, psi
S_h = allowable hot stress at hottest temperature the system will experience, psi

Table 6 Suggested Hanger Spacing and Rod Size for Straight Horizontal Runs

NPS, in.	Hanger Spacing, ft: Standard Steel Pipe[a] Water	Standard Steel Pipe[a] Steam	Copper Tube Water	Rod Size, in.
1/2	7	8	5	1/4
3/4	7	9	5	1/4
1	7	9	6	1/4
1-1/2	9	12	8	3/8
2	10	13	8	3/8
2-1/2	11	14	9	3/8
3	12	15	10	3/8
4	14	17	12	1/2
6	17	21	14	1/2
8	19	24	16	5/8
10	20	26	18	3/4
12	23	30	19	7/8
14	25	32		1
16	27	35		1
18	28	37		1-1/4
20	30	39		1-1/4

Source: Adapted from MSS *Standard* SP-69
[a] Spacing does not apply where span calculations are made or where concentrated loads are placed between supports such as flanges, valves, specialties, etc.

PLASTIC PIPING

Nonmetallic pipe is widely used in HVAC and plumbing. Plastic is light in weight, generally inexpensive, and corrosion-resistant. Plastic also has a low "C" factor (i.e., its surface is very smooth), which results in lower pumping power and smaller pipe

Table 7 Properties of Plastic Pipe Materials[a]

Material: Designation	Material: Type and Grade	Material: Cell No.	Tensile Strength, psi (at 73°F)	Hydrostatic[b] Design Stress, psi (at 73°F): Mfr.	Hydrostatic[b] Design Stress, psi (at 73°F): ASME B31	Upper Temperature Limit, °F: Mfr.	Upper Temperature Limit, °F: ASME B31	HDS[b] Upper Limit, psi	Specific Gravity[c]	Impact Strength, ft·lb/in (at 73°F)	Modulus of Elasticity, psi (at 73°F)	Coefficient of Expansion, $\frac{in}{10^6\ in \cdot °F}$	Thermal Conductivity, $\frac{Btu \cdot in}{h \cdot ft^2 \cdot °F}$	Relative Pipe Cost[d]
Thermoplastics														
PVC 1120	T I,G1	12454-B	7,500	2,000	2,000	140	150	440	1.40	0.8	420,000	30.0	1.1	1.0
PVC 1200	T I,G2	12454-C			2,000		150				410,000	35.0		
PVC 2120	T II,G1	14333-D			2,000		150					30.0		
CPVC 4120	T IV,G1	23447-B	8,000	2,000	2,000	210	210	320	1.55	1.5	423,000	35.0	0.95	2.9
PB 2110	T II,G1		4,800	1,000	1,000	180	210	<500	0.93		38,000	72.0	1.5	2.9
PE 2306	Gr. P23				630		140				90,000	80.0		
PE 3306	Gr. P34				630		160				130,000	70.0		
PE 3406	Gr. P33				630		180				150,000	60.0		
HDPE 3408	Gr. P34	355434-C	5,000	1,600	800	140	180	800	0.96	12	110,000	120.0	2.7	1.1
PP			5,000	705		212	210		0.91	1.3	120,000	60.0	1.3	2.9
ABS	Duraplus	6-3-3	5,500			176			1.06	8.5	240,000	56.0	1.7	3.4
ABS 1210	T I,G2	5-2-2			1,000		180	640			250,000	55.0		
ABS 1316	T I,G3	3-5-5			1,600		180	1,000			340,000	40.0		
ABS 2112	T II,G1	4-4-5			1,250		180	800				40.0		
PVDF			7,000	1,275		280	275	306	1.78	3.8	125,000	79.0	0.8	28.0
Thermosetting														
Epoxy-Glass	RTRP-11AF		44,000	8,000			300	7,000			1,000,000	9 to 13	2.9	
Polyester-Glass	RTRP-12EF		44,000	9,000		200	200	5,000			1,000,000	9 to 11	1.3	
For Comparison														
Steel	A 53 B	ERW	60,000		12,800		800	9,200	7.80	30.0	27,500,000	6.31	344	1.3
Copper	Type L	Drawn	36,000		9,000		400	8,200	8.90		17,000,000	9.5		3.5

[a] The properties listed are for the specific materials listed as each plastic has other formulations. Consult the manufacturer of the system chosen. These values are for comparative purposes.

[b] The hydrostatic design stress (HDS) is equivalent to the allowable design stress.

[c] Relative to water at 62.4 lb/ft^3.

[d] Based on the cost of pipe only, without factoring in fittings, joints, hangers, and labor.

sizes. The disadvantages of plastic pipe include the rapid loss of strength at temperatures above ambient and the high coefficient of linear expansion. The modulus of elasticity of plastics is low, resulting in a short support span. Some jurisdictions do not allow certain plastics in buildings because of toxic products emitted under fire conditions.

Plastic piping materials fall into two main categories—thermoplastic and thermosetting. Thermoplastics melt and are formed by extruding or molding. They are usually used without reinforcing filaments. Thermosets are cured and cannot be reformed. They are normally used with glass fiber reinforcing filaments.

For the purposes of this chapter, thermoplastic piping is made of the following materials:

PVC	polyvinyl chloride
CPVC	chlorinated polyvinyl chloride
PB	polybutylene
PE	polyethylene
PP	polypropylene
ABS	acrylonitrile butadiene styrene
PVDF	polyvinylidene fluoride

Thermosetting piping used in HVAC is called (1) reinforced thermosetting resin (RTR) and (2) fiberglass-reinforced plastic (FRP). RTR and FRP are interchangeable and refer to pipe and fittings commonly made of (1) fiberglass-reinforced epoxy resin, (2) fiberglass-reinforced vinyl ester, and (3) fiberglass-reinforced polyester.

Pipe and fittings made from epoxy resin are generally stronger and operate at a higher temperature than those made from polyester or vinyl ester resins, so they are more likely to be used in HVAC.

Table 7 lists properties of various plastic piping materials. Values for steel and copper are given for comparison.

Allowable Stress

Both thermoplastics and thermosets have an allowable stress derived from a hydrostatic design basis stress (HDBS). The HDBS is determined by a statistical analysis of both static and cyclic stress rupture test data as set forth in ASTM *Standard* D2837 for thermoplastics and ASTM *Standard* D2992 for glass fiber-reinforced thermosetting resins.

The allowable stress, which is called the hydrostatic design stress (HDS), is obtained by multiplying the HDBS by a service factor. The HDS values recommended by some manufacturers and those allowed by the ASME B31, *Code for Pressure Piping*, are listed in Table 7.

The pressure design thickness for plastic pipe can be calculated using the Code stress values and the following formula:

$$t = pD/(2S + p) \qquad (4)$$

where

t = pressure design thickness, in.
p = internal design pressure, psig
D = pipe outside diameter, in.
S = design stress (HDS), psi

The minimum required wall thickness can be found by adding an allowance for mechanical strength, threading, grooving, erosion, and corrosion to the calculated pressure design thickness.

As there are many formulations of the polymers used for piping materials and different joining methods for each, manufacturers' recommendations should be observed. Most catalogs give the pressure ratings for pipe and fittings at various temperatures up to the maximum the material will withstand.

Plastic Material Selection

The selection of a plastic for a specific purpose requires attention. All are suitable for cold water. Plastic pipe should not be used for compressed gases or compressed air if the pipe is made of a material subject to brittle failure. For other liquids and chemicals, refer to charts provided by plastic pipe manufacturers and distributors. Table 7 gives properties of the various plastics discussed in this section. The last column gives the relative cost of small pipe in each category. Table 8 lists some applications pertinent to HVAC. The following are brief descriptions of common uses for the various materials.

Table 8 Manufacturers' Recommendations[a,b] for Plastic Materials

	PVC	CPVC	PB	HDPE	PP	ABS	PVDF	RTRP
Cold water service	R	R	R	R	R	R	R	R
Hot (140°F) water	N	R	R	R	R	R	R	R
Potable water service	R	R	R	R	R	R	R	R
Drain, waste, and vent	R	R	N	—	R	R	—	—
Demineralized water	R	R	—	—	R	R	R	—
Deionized water	R	R	—	—	R	R	R	R
Salt water	R	R	R	R	R	R	—	R
Heating (200°F) hot water	N	N	N	N	N	N	—	R
Natural gas	N	N	N	R	N	N	—	—
Compressed air	N	N	—	R	N	R	—	—
Sunlight and weather resistance	N	N	N	R	—	R	R	R
Underground service	R	R	R	R	R	R	—	R
Food handling	R	R	—	—	R	R	R	R

R = Recommended N = Not recommended — = Insufficient information

[a] Before selecting a material, check the availability of a suitable range of sizes and fittings and of a satisfactory joining method. Also have the manufacturer verify the best material for the purpose intended.

[b] Local building codes should be consulted for compliance of the materials listed.

PVC. Because polyvinyl chloride has the best overall range of properties at the lowest cost, it is the most widely used plastic. It is joined by solvent cementing, threading, or flanging. Gasketed push-on joints are also used for larger sizes.

CPVC. Chlorinated polyvinyl chloride has the same properties as PVC but can withstand a higher temperature before losing strength. It is joined by the same methods as PVC.

PB. A lightweight, flexible material, polybutylene can be used up to 210°F. It is used for both hot and cold plumbing water piping. It is joined by heat fusion or mechanical means, can be bent to a 10 diameter radius, and is provided in coils.

PE. Low-density polyethylene (LDPE) is a flexible lightweight tubing with good low-temperature properties. It is used in the food and beverage industry and for instrument tubing. It is joined by mechanical means such as compression fittings or push-on connectors and clamps.

High-density polyethylene (HDPE) is a tough, weather-resistant material used for large pipelines in the gas industry. Fabricated fittings are available. It is joined by heat fusion for large sizes; flare, compression, or insert fittings can be used on small sizes.

PP. Polypropylene is a lightweight plastic used for pressure applications and also for chemical waste lines, because it is inert to a wide range of chemicals. A broad variety of drainage fittings are available. For pressure uses, regular fittings are made. It is joined by heat fusion.

ABS. Acrylonitrile butadiene styrene is a high-strength, impact- and weather-resistant material. Certain formulations can be used for compressed air. ABS is also used in the food and beverage industry. A wide range of fittings is available. It is joined by solvent cement, threading, or flanging.

PVDF. Polyvinylidene fluoride is widely used for ultrapure water systems and in the pharmaceutical industry and has a wide temperature range. This material is over 20 times more expensive than PVC. It is joined by heat fusion, and fittings are made for this purpose. For smaller sizes, mechanical joints can be used.

PIPE-SUPPORTING ELEMENTS

Pipe-supporting elements consist of (1) hangers, which support from above; (2) supports, which bear load from below; and (3) restraints, such as anchors and guides, which limit or direct movement, as well as support loads. Pipe-supporting elements withstand all static and dynamic conditions including the following:

- Weight of pipe, valves, fittings, insulation, and fluid contents, including test fluid if using a heavier-than-normal media
- Occasional loads such as ice, wind, and seismic forces
- Forces imposed by thermal expansion and contraction of pipe bends and loops
- Frictional, spring, and pressure thrust forces imposed by expansion joints in the system
- Frictional forces of guides and supports
- Other loads, such as water hammer, vibration, and reactive force of relief valves
- Test load and force

In addition, pipe-supporting elements must be evaluated in terms of stress at the point of connection to the pipe and the building structure. Stress at the point of connection to the pipe is especially important for base elbow and trunnion supports, because the limiting and controlling parameter is usually not the strength of the structural member, but the localized stress and the point of attachment to the pipe. Loads on anchors, cast-in-place inserts, and other attachments to concrete should not be more than one-fifth the ultimate strength of the attachment, as determined by manufacturers' tests. All loads on the structure should be communicated to and coordinated with the structural engineer.

The ASME B31 standards establish criteria for the design of pipe-supporting elements and the Manufacturers Standardization Society of the Valve and Fittings Industry (MSS) has established standards for the design, fabrication, selection, and installation of pipe hangers and supports based on these codes.

MSS *Standard* SP-69 and the catalogs of many manufacturers illustrate the various hangers and components and provide information on the types to use with different pipe systems. Table 6 shows suggested pipe support spacing and Table 9 provides a maximum safe load for threaded steel rods.

The loads on most pipe-supporting elements are moderate and can be selected safely in accordance with manufacturers' catalog data and the information presented in this section; however, some loads and forces can be very high, especially in multistory buildings and for large-diameter pipe, especially where expansion joints are used at a high operating pressure. Consequently, a qualified engineer should design, or review the design of, all anchors and pipe-supporting elements, especially for the following:

Table 9 Capacities of ASTM A36 Steel Threaded Rods

Rod Diameter, in.	Root Area of Coarse Thread, in^2	Maximum Load,[a] lb
1/4	0.027	240
3/8	0.068	610
1/2	0.126	1130
5/8	0.202	1810
3/4	0.302	2710
7/8	0.419	3770
1	0.552	4960
1-1/4	0.889	8000

[a] Based on an allowable stress of 12,000 psi reduced by 25% using the root area in accordance with ASME *Standard* B31.1 and MSS *Standard* SP-58.

- Steam systems operating abovc 15 psig
- Hydronic systems operating above 160 psig or 250°F
- Risers over 10 stories or 100 ft
- Systems with expansion joints, especially for pipe diameters 3 in. and greater
- Pipe sizes over 12 in. diameter
- Anchor loads greater than 10,000 lb (10 kips)
- Moments on pipe or structure in excess of 1000 ft·lb

PIPE EXPANSION AND FLEXIBILITY

Temperature changes cause dimensional changes in all materials. Table 10 shows the coefficients of expansion for piping materials commonly used in HVAC. For systems operating at high temperatures, such as steam and hot water, the rate of expansion is high, and significant movements can occur in short runs of piping. Even though rates of expansion may be low for systems operating in the range of 40 to 100°F, such as chilled and condenser water, they can cause large movements in long runs of piping, which are common in

Table 10 Thermal Expansion of Metal Pipe

			Linear Thermal Expansion, in/100 ft		
	Saturated Steam Pressure, psig	Temperature, °F	Carbon Steel	Type 304 Stainless Steel	Copper
		−30	−0.19	−0.30	−0.32
		−20	−0.12	−0.20	−0.21
		−10	−0.06	−0.10	−0.11
		0	0	0	0
		10	0.08	0.11	0.12
		20	0.15	0.22	0.24
Vacuum	−14.6	32	0.24	0.36	0.37
Vacuum	−14.6	40	0.30	0.45	0.45
Vacuum	−14.5	50	0.38	0.56	0.57
Vacuum	−14.4	60	0.46	0.67	0.68
Vacuum	−14.3	70	0.53	0.78	0.79
Vacuum	−14.2	80	0.61	0.90	0.90
Vacuum	−14.0	90	0.68	1.01	1.02
Vacuum	−13.7	100	0.76	1.12	1.13
Vacuum	−13.0	120	0.91	1.35	1.37
Vacuum	−11.8	140	1.06	1.57	1.59
Vacuum	−10.0	160	1.22	1.79	1.80
Vacuum	−7.2	180	1.37	2.02	2.05
Vacuum	−3.2	200	1.52	2.24	2.30
	0	212	1.62	2.38	2.43
	2.5	220	1.69	2.48	2.52
	10.3	240	1.85	2.71	2.76
	20.7	260	2.02	2.94	2.99
	34.6	280	2.18	3.17	3.22
	52.3	300	2.35	3.40	3.46
	75.0	320	2.53	3.64	3.70
	103.3	340	2.70	3.88	3.94
	138.3	360	2.88	4.11	4.18
	181.1	380	3.05	4.35	4.42
	232.6	400	3.23	4.59	4.87
	666.1	500	4.15	5.80	5.91
	1528	600	5.13	7.03	7.18
	3079	700	6.16	8.29	8.47
		800	7.23	9.59	9.79
		900	8.34	10.91	11.16
		1000	9.42	12.27	12.54

distribution systems and high-rise buildings. Therefore, in addition to design requirements for pressure, weight, and other loads, piping systems must accommodate thermal and other movements to prevent the following:

- Failure of pipe and supports from overstress and fatigue
- Leakage of joints
- Detrimental forces and stresses in connected equipment

An unrestrained pipe operates at the lowest overall stress level. Anchors and restraints are needed to support pipe weight and to protect equipment connections. The anchor forces and the bowing of pipe anchored at both ends are generally too large to be acceptable, so general practice is to *never anchor a straight run of steel pipe at both ends*. Piping must be allowed to expand or contract due to thermal changes. Ample flexibility can be attained by designing pipe bends and loops or by including supplemental devices, such as expansion joints.

End reactions transmitted to rotating equipment, such as pumps or turbines, may deform the equipment case and cause bearing misalignment, which may ultimately cause the component to fail. Consequently, manufacturers' recommendations on allowable forces and movements that may be placed on their equipment should be followed.

PIPE BENDS AND LOOPS

Detailed stress analysis requires involved mathematical analysis and is generally performed by computer programs. However, such involved analysis is not required for most HVAC systems because the piping arrangements and temperature ranges at which they operate are simple to analyze.

L Bends

The guided cantilever beam method of evaluating L bends can be used to design L bends, Z bends, pipe loops, branch take-off connections, and some more complicated piping configurations.

Equation (4) may be used to calculate the length of leg BC needed to accommodate thermal expansion or contraction of leg AB for a guided cantilever beam (Figure 1).

$$L = \sqrt{\frac{3\Delta DE}{(144\text{in}^2/\text{ft}^2)S_A}} \qquad (5)$$

where

L = length of leg BC required to accommodate thermal expansion of long leg AB, ft
Δ = thermal expansion or contraction of leg AB, in.
D = actual pipe outside diameter, in.
E = modulus of elasticity, psi
S_A = allowable stress range, psi

For the commonly used A53 Grade B seamless or ERW pipe, an allowable stress S_A of 22,500 psi (see Table 1) can be used without overstressing the pipe. However, this can result in very high end reactions and anchor forces, especially with large-diameter pipe. Designing to a stress range S_A of 15,000 psi and assuming $E = 27.9 \times 10^6$ psi, Equation (4) reduces to Equation (5), which provides reasonably low end reactions without requiring too much extra pipe. In addition, Equation (5) may be used with A53 butt-welded pipe and B88 drawn copper tubing.

Thus, for A53 continuous (butt-) welded, seamless, and ERW pipe, and B88 drawn copper tubing,

$$L = 6.225\sqrt{\Delta D} \qquad (6)$$

The guided cantilever method of designing L bends assumes no restraints; therefore, care must be taken in supporting the pipe. For horizontal L bends, it is usually necessary to place a support near point B (see Figure 1), and any supports between points A and C must provide minimal resistance to piping movement; this is done by using slide plates or hanger rods of ample length, with hanger components selected to allow for swing no greater than 4°.

For L bends containing both vertical and horizontal legs, any supports on the horizontal leg must be spring hangers designed to support the full weight of pipe at normal operating temperature with a maximum load variation of 25%.

The force developed in an L bend that must be sustained by anchors or connected equipment is determined by the following equation:

$$F = \frac{12E_c I\Delta}{(1728\text{ in}^3/\text{ft}^3)L^3} \qquad (7)$$

where

F = force, lb
E_c = modulus of elasticity, psi
I = moment of inertia, in^4
L = length of offset leg, ft
Δ = deflection of offset leg, in.

In lieu of using Equation (6), for L bends designed in accordance with Equation (5) for 1 in. or more of offset, a conservative estimation of force is 500 lb per diameter inch (e.g., a 3 in. pipe would develop 1500 lb of force).

Z Bends

Z bends, as shown in Figure 2, are very effective for accommodating pipe movements. A simple and conservative method of sizing Z bends is to design the offset leg to be 65% of the values used for an L bend in Equation (5), which results in

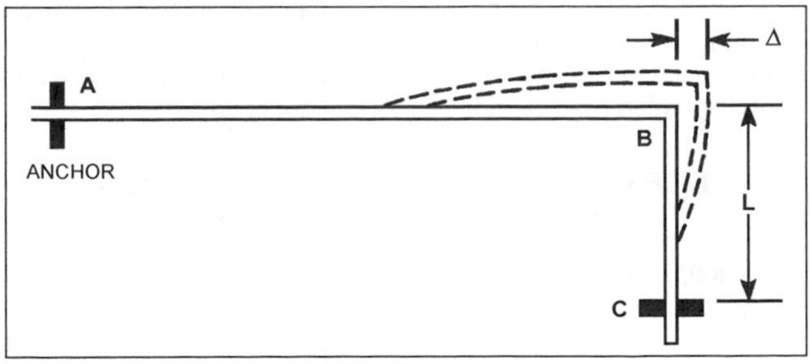

Fig. 1 Guided Cantilever Beam

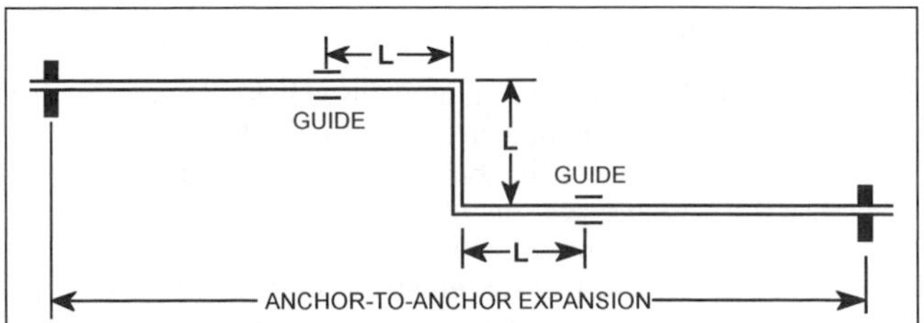

The distance from the guides, if used, to the offset should equal or exceed the length of the offset.
Offset piping must be supported with hangers, slide plates, and spring hangers similar to those for *L* bends.

Fig. 2 Z Bend in Pipe

Table 11 Pipe Loop Design for A53 Grade B Carbon Steel Pipe Through 400°F

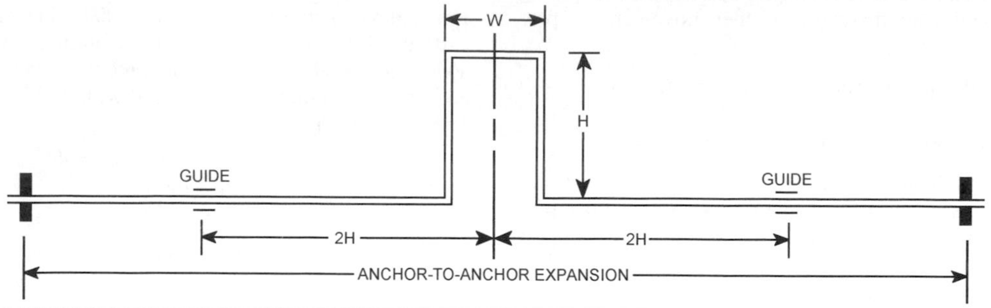

Pipe Size, in.	Anchor-to-Anchor Expansion, in. 2 W	H	4 W	H	6 W	H	8 W	H	10 W	H	12 W	H
1	2	4	3	6	3.5	7	4	8	4.5	9	5	10
2	3	6	4	8	5	10	5.5	11	6	12	7	14
3	3.5	7	5	10	6	12	6.5	13	7.5	15	8	16
4	4	8	5.5	11	6.5	13	7.5	15	8.5	17	9	18
6	5	10	6.5	13	8	16	9	18	10	20	11	22
8	5.5	11	7.5	15	9	18	10.5	21	12	24	13	26
10	6	12	8.5	17	10	20	11.5	23	13	26	14	28
12	6.5	13	9	18	11	22	12.5	25	14	28	15.5	31
14	7	14	9.5	19	11.5	23	13	26	15	30	16	32
16	7.5	15	10	20	12.5	25	14	28	16	32	17.5	35
18	8	16	11	22	13	26	15	30	17	34	18.5	37
20	8.5	17	11.5	23	14	28	16	32	18	36	19.5	39
24	9	18	12.5	25	14.5	29	17.5	35	19.5	39	21	42

Notes: *W* and *H* dimensions are feet.
L is determined from Equation (4). $W = L/5$ $H = 2W$ $2H + W = L$

Approximate force to deflect loop = 200 lb/diam. in.
For example, 8 in. pipe creates 1600 lb of force.

$$L = 4\sqrt{\Delta D} \tag{8}$$

where

L = length of offset leg, ft
Δ = anchor-to-anchor expansion, in.
D = pipe outside diameter, in.

The force developed in a Z bend can be calculated with acceptable accuracy as follows:

$$F = C_1\Delta(D/L)^2 \tag{9}$$

where

C_1 = 4000 lb/in.
F = force, lb
D = pipe outside diameter, in.
L = length of offset leg, ft
Δ = anchor-to-anchor expansion, in.

U Bends and Pipe Loops

Pipe loops or U bends are commonly used in long runs of piping. A simple method of designing pipe loops is to calculate the anchor-to-anchor expansion and, using Equation (5), determine the length *L* necessary to accommodate this movement. The pipe loop dimensions can then be determined using $W = L/5$ and $H = 2W$.

Note that guides must be spaced no closer than twice the height of the loop, and piping between guides must be supported, as described in the section on L Bends, when the length of pipe between guides exceeds the maximum allowable hanger spacing for the size pipe.

Table 11 lists pipe loop dimensions for pipe sizes 1 in. through 24 in. and anchor-to-anchor expansion (contraction) of 2 in. through 12 in.

No simple method has been developed to calculate pipe loop force; however, it is generally low. A conservative estimate is 200 lb per diameter inch (e.g., a 2 in. pipe will develop 400 lb of force and a 12 in. pipe will develop 2400 lb of force).

Cold Springing of Pipe

Cold springing or cold positioning of pipe consists of offsetting or springing the pipe in a direction opposite the expected movement. Cold springing is not recommended for most HVAC piping. Furthermore, *cold springing does not permit designing a pipe bend or loop for twice the calculated movement.* For example, if a particular L bend can accommodate 3 in. of movement from a neutral position, cold springing does not permit the L bend to accommodate 6 in. of movement.

Analyzing Existing Piping Configurations

Piping is best analyzed using a computer stress analysis program because these provide all pertinent data including stress, movements, and loads. Services can perform such analysis if programs are not available in-house. However, many situations do not require such detailed analysis. A simple, yet satisfactory method for single and multiplane systems is to divide the system with real or hypothetical anchors into a number of single-plane units, as shown in Figure 3, which can be evaluated as *L* and *Z* bends.

EXPANSION JOINTS AND EXPANSION COMPENSATING DEVICES

Although the inherent flexibility of the piping should be used to the maximum extent possible, expansion joints must be used where movements are too large to accommodate with pipe bends or loops or where insufficient room exists to construct a loop of adequate size. Typical situations are tunnel piping and risers in high-rise

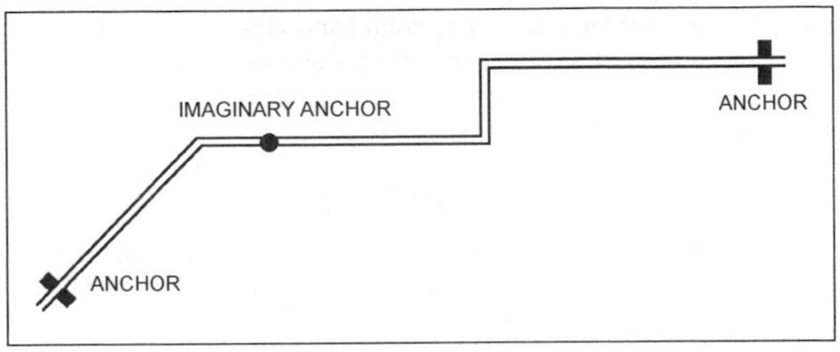

Fig. 3 Multiplane Pipe System

buildings, especially for steam and hot water pipes where large thermal movements are involved.

Packed and packless expansion joints and expansion compensating devices are used to accommodate movement, either axially or laterally.

In the **axial method** of accommodating movement, the expansion joint is installed between anchors in a straight line segment and accommodates axial motion only. This method has high anchor loads, primarily due to pressure thrust. It requires careful guiding, but expansion joints can be spaced conveniently to limit movement of branch connections. The axial method finds widest application for long runs without natural offsets, such as tunnel and underground piping and risers in tall buildings.

The **lateral** or **offset method** requires the device to be installed in a leg perpendicular to the expected movement and accommodates lateral movement only. This method generally has low anchor forces and minimal guide requirements. It finds widest application in lines with natural offsets, especially where there are few or no branch connections.

Packed expansion joints depend on slipping or sliding surfaces to accommodate the movement and require some type of seals or packing to seal the surfaces. Most such devices require some maintenance but are not subject to catastrophic failure. Further, with most packed expansion joint devices, any leaks that develop can be repacked under full line pressure without shutting down the system.

Packless expansion joints depend upon the flexing or distortion of the sealing element to accommodate movement. They generally do not require any maintenance, but maintenance or repair is not usually possible. If a leak occurs, the system must be shut off and drained, and the entire device must be replaced. Further, catastrophic failure of the sealing element can occur and, although likelihood of such failure is remote, it must be considered in certain design situations.

Packed expansion joints are preferred where long-term system reliability is of prime importance (using types that can be repacked under full line pressure) and where major leaks can be life threatening or extremely costly. Typical applications are risers, tunnels, underground pipe, and distribution piping systems. Packless expansion joints are generally used where even small leaks cannot be tolerated (such as for gas and toxic chemicals), where temperature limitations preclude the use of packed expansion joints, and for very large-diameter pipe where packed expansion joints cannot be constructed or the cost would be excessive.

In all cases, expansion joints should be installed, anchored, and guided in accordance with expansion joint manufacturers' recommendations.

Packed Expansion Joints

There are two types of packed expansion joints—packed slip expansion joints and flexible ball joints.

Packed Slip Expansion Joints. These are telescoping devices designed to accommodate axial movement only. Some sort of packing seals the sliding surfaces. The original packed slip expansion joint used multiple layers of braided compression packing, similar to the stuffing box commonly used with valves and pumps; this arrangement requires shutting and draining the system for maintenance and repair. Advances in design and packing technology have eliminated these problems, and most current packed slip joints use self-lubricating semiplastic packing, which can be injected under full line pressure without shutting off the system (see Figure 4). (Many manufacturers use asbestos-based packings, unless requested otherwise. Asbestos-free packings, such as flake graphite, are available and, although more expensive, should be specified in lieu of products containing asbestos.)

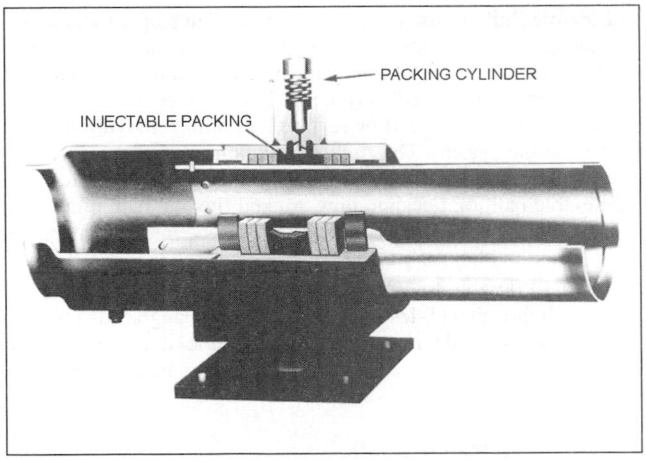

Fig. 4 Packed Slip Expansion Joint

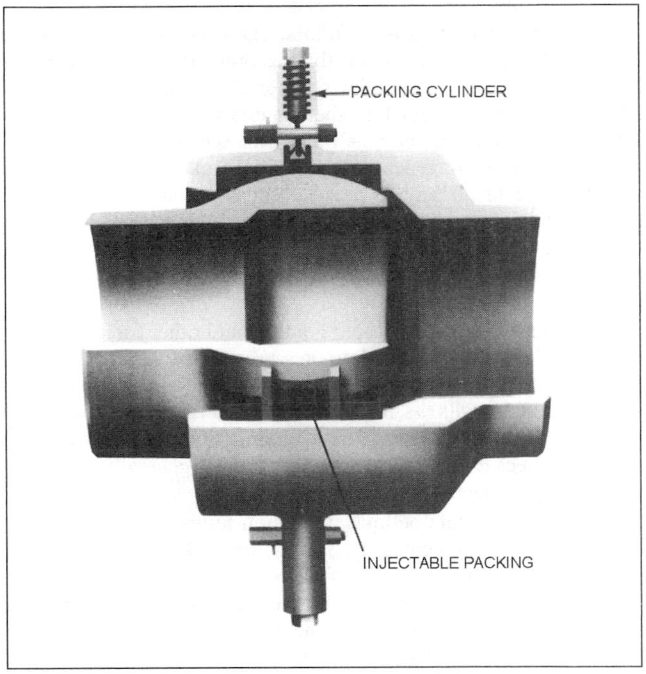

Fig. 5 Flexible Ball Joint

Standard packed slip expansion joints are constructed of carbon steel with weld or flange ends in sizes 1.5 to 36 in. for pressures up to 300 psig and temperatures up to 800°F. Larger sizes, higher temperature, and higher pressure designs are available. Standard single joints are generally designed for 4, 8, or 12 in. axial traverse; double joints with an intermediate anchor base can accommodate twice these movements. Special designs for greater movements are available.

Flexible Ball Joints. These joints are used in pairs to accommodate lateral or offset movement and must be installed in a leg perpendicular to the expected movement. The original flexible ball joint design incorporated only inner and outer containment seals that could not be serviced or replaced without removing the ball joint from the system. The packing technology of the packed slip expansion joint, explained previously, has been incorporated into the flexible ball joint design; now, packed flexible ball joints have self-lubricating semiplastic packing that can be injected under full line pressure without shutting off the system (see Figure 5).

Standard flexible ball joints are available in sizes 1-1/4 through 30 in. with threaded (1-1/4 to 2 in.), weld, and flange ends for pressures to 300 psig and temperatures to 750°F. Flexible ball joints are available in larger sizes and for higher temperature and pressure ranges.

Packless Expansion Joints

Metal bellows expansion joints, rubber expansion joints, and flexible hose or pipe connectors are some of the packless expansion joints available.

Metal Bellows Expansion Joints. These expansion joints have a thin-wall convoluted section that accommodates movement by bending or flexing. The bellows material is generally Type 304, 316, or 321 stainless steel, but other materials are commonly used to satisfy service conditions. Small-diameter expansion joints in sizes 3/4 through 3 in. are generally called *expansion compensators* and are available in all bronze or steel construction. Metal bellows expansion joints can generally be designed for the pressures and temperatures commonly encountered in HVAC systems and can also be furnished in rectangular configurations for ducts and chimney connectors.

Overpressurization, improper guiding, and other forces can distort the bellows element. For low-pressure applications, such distortion can be controlled by the geometry of the convolution or the thickness of the bellows material. For higher pressure, internally pressurized joints require reinforcing. Externally pressurized designs are not subject to such distortion and are not generally furnished without supplemental bellows reinforcing.

Single- and double-bellows expansion joints primarily accommodate axial movement only, similar to packed slip expansion joints. Although bellows expansion joints can accommodate some lateral movement, the **universal tied bellows expansion joint** accommodates large lateral movement. This device operates much like a pair of flexible ball joints, except that bellows elements are used instead of flexible ball elements. The tie rods on this joint contain the pressure thrust, so anchor loads are much lower than with axial-type expansion joints.

Rubber Expansion Joints. Similar to single-metal bellows expansion joints, rubber expansion joints incorporate a nonmetallic elastomeric bellows sealing element and generally have more stringent temperature and pressure limitations. Although rubber expansion joints can be used to accommodate expansion and contraction of the piping, they are primarily used as flexible connectors at equipment to isolate sound and vibration and eliminate stress at equipment nozzles.

Flexible Hose. This type of hose can be constructed of elastomeric material or corrugated metal with an outer braid for reinforcing and end restraint. Flexible hose is primarily used as a flexible connector at equipment to isolate sound and vibration and eliminate stress at equipment nozzles; however, flexible metal hose is well suited for use as an *offset-type expansion joint*, especially for copper tubing and branch connections off risers.

REFERENCES

ASHRAE. 1994. Safety code for mechanical refrigeration. *Standard* 15-1994.

ASME. 1983. Pipe threads, general purpose (inch). *Standard* B1.20.1. American Society of Mechanical Engineers, New York.

ASME. 1992. Refrigeration piping. *Standard* B31.5.

ASME. 1996. Building services piping. *Standard* B31.9.

ASME. 1996. Welded and seamless wrought steel pipe. *Standard* B36.10M.

ASME. 1998. Power piping. *Standard* B31.1.

ASME. 1998. Rules for construction of power boilers. *Boiler and Pressure Vessel Code*. Section I.

ASME. 1998. Qualification standard for welding and brazing procedures, welders, brazers, and welding and brazing operators. *Boiler and Pressure Vessel Code,* Section IX.

ASTM. 1996. Standard specification for seamless copper water tube. *Standard* B 88. American Society for Testing and Materials, West Conshohocken, PA.

ASTM. 1996. Standard practice for obtaining hydrostatic or pressure design basis for "fiberglass" (glass-fiber-reinforced thermosetting-resin) pipe and fittings. *Standard* D 2992.

ASTM. 1996. Standard specification for polybutylene (PB) plastic hot- and cold-water distribution systems. *Standard* D 3309 REV A.

ASTM. 1996. Standard specification for plastic inserts fittings for polybutylene (PB) tubing. *Standard* F 845.

ASTM. 1997. Standard specification for seamless carbon steel pipe for high-temperature service. *Standard* A 106.

ASTM. 1997. Standard specification for seamless copper tube for air conditioning and refrigeration field service. *Standard* B 280.

ASTM. 1998. Standard specification for pipe, steel, black and hot-dipped, zinc-coated welded and seamless. *Standard* A 53.

ASTM. 1998. Standard test method for obtaining hydrostatic design basis for thermoplastic pipe materials. *Standard* D 2837.

AWWA. 1996. Thickness design of ductile-iron pipe. *Standard* C150/A21.50. American Water Works Association, Denver, CO.

MSS. 1996. Pipe hangers and supports—Selection and application. *Standard* SP-69. Manufacturers Standardization Society of the Valve and Fittings Industry, Vienna, VA.

MSS. 1993. Pipe hangers and supports—Materials, design and manufacture. *Standard* SP-58.

NFPA. 1996. Installation of sprinkler systems. *Standard* 13. National Fire Protection Association, Quincy, MA.

NFPA/AGA. 1999. *National fuel gas code*. ANSI/NFPA *Standard* 70-99. National Fire Protection Association, Quincy, MA. ANSI/AGA *Standard* Z223.1-99. American Gas Association, Arlington, VA.

BIBLIOGRAPHY

ASTM. 1995. Standard specification for polybutylene (PB) plastic pipe (SDR-PR) based on outside diameter. *Standard* D 3000.

ASTM. 1996. Standard specification for poly(vinyl chloride) (PVC) plastic pipe, Schedules 40, 80, and 120. *Standard* D 1785. American Society for Testing and Materials, West Conshohocken, PA.

ASTM. 1996. Standard specification for polybutylene (PB) plastic pipe (SDR-PR) based on controlled inside diameter. *Standard* D 2662.

ASTM. 1996. Standard specification for polybutylene (PB) plastic tubing. *Standard* D 2666.

ASTM. 1997. Standard specification for chlorinated poly(vinyl chloride) (CPVC) plastic pipe, Schedules 40 and 80. *Standard* F 441.

Crane Co. 1988. Flow of fluids through valves, fittings, and pipe. *Technical Paper* 410. Joliet, IL.

CHAPTER 42

VALVES

FUNDAMENTALS

VALVES are the manual or automatic fluid-controlling elements in a piping system. They are constructed to withstand a specific range of temperature, pressure, corrosion, and mechanical stress. The designer selects and specifies the proper valve for the application to give the best service for the economic requirements.

Valves have some of the following primary functions:

- Starting, stopping, and directing flow
- Regulating, controlling, or throttling flow
- Preventing backflow
- Relieving or regulating pressure

The following service conditions should be considered before specifying or selecting a valve:

1. Type of liquid, vapor, or gas
   - Is it a true fluid or does it contain solids?
   - Does it remain a liquid throughout its flow or does it vaporize?
   - Is it corrosive or erosive?
2. Pressure and temperature
   - Will these vary in the system?
   - Should worst case (maximum or minimum values) be considered in selecting correct valve materials?
3. Flow considerations
   - Is pressure drop critical?
   - Should valve design be chosen for maximum wear?
   - Is the valve to be used for simple shutoff or for throttling flow?
   - Is the valve needed to prevent backflow?
   - Is the valve to be used for directing (mixing or diverting) flow?
4. Frequency of operation
   - Will the valve be operated frequently?
   - Will valve normally be open with infrequent operation?
   - Will operation be manual or automatic?

Nomenclature for basic valve components may vary from manufacturer to manufacturer and according to the application. Figure 1 shows representative names for various valve parts.

Body Ratings

The rating of valves defines the pressure-temperature relationship within which the valve may be operated. The valve manufacturer is responsible for determining the valve rating. ASME *Standard* B16.34, Valves—Flanged, Threaded, and Welding End, should be consulted, and a valve pressure class should be identified. Inlet pressure ratings are generally expressed in terms of the ANSI/ASME class ratings and range from ANSI Class 150 through 2500, depending on the style, size, and materials of construction, including seat materials. Automatic control valves are usually either Class 125 or Class 250. Tables in the standard and in various books show pressure ratings at various operating temperatures (ASME *Standard* B16.34; Lyons 1982; Ulanski 1991).

The preparation of this chapter is assigned to TC 6.1, Hydronic and Steam Equipment and Systems.

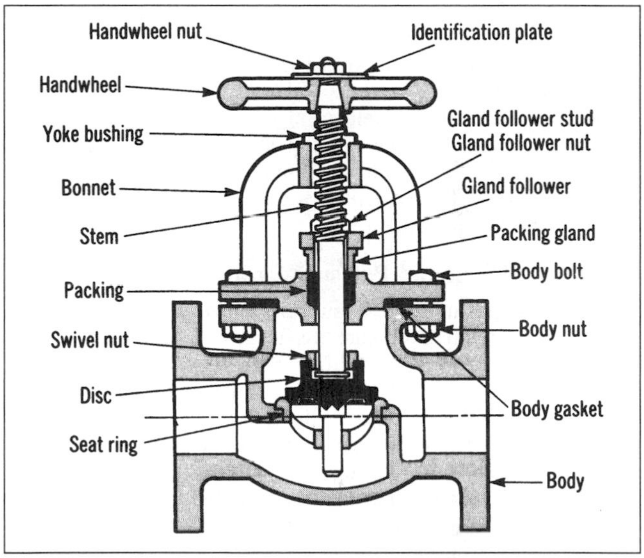

Fig. 1 Valve Components
(Courtesy Grinnell Corp.)

Materials

ASME *Standard* B16.34 addresses requirements for valves made from forgings, castings, plate, bar stock and shapes, and tubular products. This standard identifies acceptable materials from which valves can be constructed. In selecting proper valve materials, the valve body-bonnet material should be selected first and then the valve plug and seat trim.

Other factors that govern the basic materials selection include

- Pressure-temperature ratings
- Corrosion-resistance requirements
- Thermal shock
- Piping stress
- Fire hazard

Types of materials typically available include

- Carbon steel
- Ductile iron
- Cast iron
- Stainless steels
- Brass
- Bronze
- Polyvinyl chloride (PVC) plastic

Bodies. Body materials for small valves are usually brass, bronze, or forged steel and for larger valves, cast iron, cast ductile iron, or cast steel as required for the pressure and service. Listings of typical materials are given in Lyons (1982) and Ulanski (1991).

Seats. Valve seats can be machined integrally of the body material, press-fitted, or threaded (removable). Seats of different materials can be selected to suit difficult application requirements. The valve seat and the valve plug or disk are sometimes referred to as the valve trim and are usually constructed of the same material selected to meet the service requirements. The trim, however, is usually of a different material than the valve body. Replaceable composition disks are used in conjunction with the plug in some designs in order to provide adequate close-off.

Maximum permissible leakage ratings for control valve seats are defined in Fluid Controls Institute (FCI) *Standard* 70-2.

Stems. Valve stem material should be selected to meet service conditions. Stainless steel is commonly used for most HVAC applications, and bronze is commonly used in ball valve construction.

Stem Packings and Gaskets. Valve stem packings undergo constant wear due to the movement of the valve stem; and both the packings and body gaskets are exposed to pressure and pressure variations of the control fluid. Manufacturers can supply recommendations regarding materials and lubricants for specific fluid temperatures and pressures.

Flow Coefficient and Pressure Drop

Flow through any device results in some loss of pressure. Some of the factors affecting pressure loss in valves include changes in the cross section and shape of the flow path, obstructions in the flow path, and changes in direction of the flow path. For most applications, the pressure drop varies as the square of the flow when operating in the turbulent flow range. For check valves, this relationship is true only if the flow holds the valve in the full-open position.

For convenience in selecting valves, particularly control valves, manufacturers express valve capacity as a function of a flow coefficient C_v. By definition in the United States, C_v is the flow of water in gallons per minute (at 60°F) that causes a pressure drop of 1 psi across a fully open valve. Manufacturers may also furnish valve coefficients at other pressure drops. Flow coefficients apply only to water. When selecting a valve to control other fluids, be sure to account for differences in viscosity.

Figure 2 shows a typical test arrangement to determine the C_v rating with the test valve wide open. Globe valve HV-1 permits adjusting the supply gage reading (e.g., to 10 psi); HV-2 is then adjusted (e.g., to 9 psi return gage) to permit a test run at a pressure drop of 1 psi. A gravity storage tank may be used to minimize supply pressure fluctuations. The bypass valve permits fine adjustment of the supply pressure. A series of test runs is made with the weighing tank and a stopwatch to determine the flow rate. Further capacity test detail may be found in International Society for Measurement and Control (ISA) *Standard* S75.02.

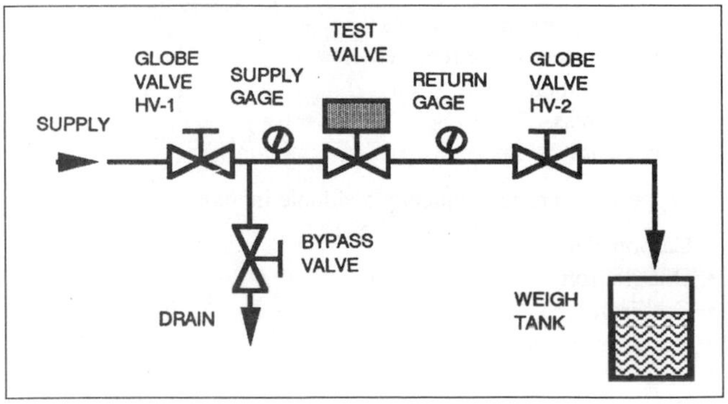

Fig. 2 Flow Coefficient Test Arrangement

Cavitation

Cavitation occurs when the pressure of a flowing fluid drops below the vapor pressure of that fluid (Figure 3). In this two-step process, the pressure first drops to the critical point, causing cavities of vapor to form. These are carried with the flow stream until they reach an area of higher pressure. The bubbles of vapor then suddenly collapse or implode. This reduction in pressure occurs when the velocity increases as the fluid passes through a valve. After the fluid passes through the valve, the velocity decreases and the pressure increases. In many cases, cavitation manifests itself as noise. However, if the vapor bubbles are in contact with a solid surface when they collapse, the liquid rushing into the voids causes high localized pressure that can erode the surface. Premature failure of the valve and adjacent piping may occur. The noise and vibration caused by cavitation have been described as similar to those of gravel flowing through the system.

Water Hammer

Water hammer is a series of pressure pulsations of varying magnitude above and below the normal pressure of water in the pipe. The amplitude and period of the pulsation depend on the velocity of the water as well as the size, length, and material of the pipe.

Shock loading from these pulsations occurs when any moving liquid is stopped in a short time. In general, it is important to avoid quickly closing valves in an HVAC system to minimize the occurrence of water hammer.

When flow stops, the pressure increase is independent of the working pressure of the system. For example, if water is flowing at 5 fps and a valve is instantly closed, the pressure increase is the same whether the normal pressure is 100 psig or 1000 psig.

Water hammer is often accompanied by a sound resembling a pipe being struck by a hammer—hence the name. The intensity of the sound is no measure of the magnitude of the pressure. Tests indicate that even if 15% of the shock pressure is removed by absorbers or arresters, adequate relief is not necessarily obtained.

Velocity of pressure wave and maximum water hammer pressure formulas may be found in the *Hydraulic Handbook* (Fairbanks Morse 1965).

Noise

Chapter 33 of the 1997 *ASHRAE Handbook—Fundamentals* points out that limitations are imposed on pipe size to control the level of pipe and valve noise, erosion, and water hammer pressure. One recommendation places a velocity limit of 4 fps for pipe 2 in. and smaller, and a pressure drop of 4 ft water/100 ft length for piping over 2 in. in diameter. Velocity-dependent noise in piping and piping systems results from any or all of four sources: turbulence, cavitation, release of entrained air, and water hammer (see Chapter 46 of the 1999 *ASHRAE Handbook—Applications*).

Some data are available for predicting hydrodynamic noise generated by control valves. ISA *Standard* 75.01 compiled prediction

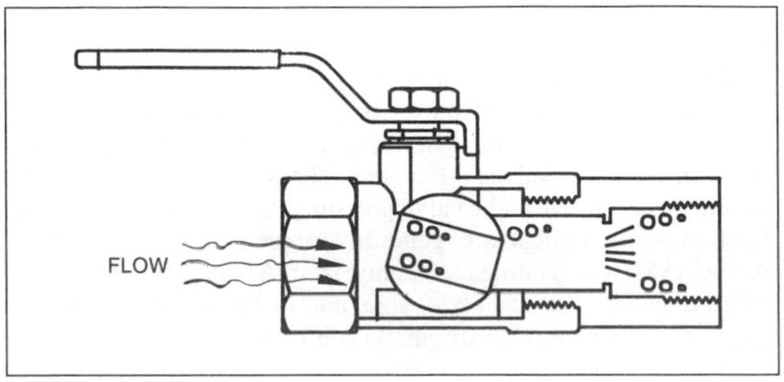

Fig. 3 Valve Cavitation Progress at Sharp Curves

correlations in an effort to develop control valves for reduced noise levels.

Body Styles

Valve bodies are available in many configurations depending on the desired service. Usual functions include stopping flow, allowing full flow, modulating flow between extremes, and directing flow. The operation of a valve can be automatic or manual.

The shape of bodies for automatic and manual valves is dictated by the intended application. For example, angle valves are commonly provided for radiator control. The principle of flow is the same for angle and straight-through valve configurations—the manufacturer provides a choice in some cases as a convenience to the installer.

The type or design of body connections is dictated primarily by the proposed conduit or piping material. Depending on material type, valves can be attached to piping in one of the following ways:

- Bolted to the pipe with companion flange.
- Screwed to the pipe, where the pipe itself has matching threads (male) and the body of the valve has threads machined into it (female).
- Welded, soldered, or sweated.
- Flared, compression, and/or various mechanical connections to the pipe where there are no threads on the pipe or the body.
- Valves of various plastic materials are fastened to the pipe if the valve body and the pipe are of compatible plastics.

MANUAL VALVES

Selection

Each valve style has advantages and disadvantages for the application. In some cases, the design documents provide inadequate information, so that selection is based on economics and local stock availability by the installer and not on what is really required. Good submittal practice and approval by the designer are required to prevent substitutions. The questions listed in the section on Fundamentals must be evaluated carefully.

Globe Valves

In a globe valve, flow is controlled by a circular disk forced against or withdrawn from an annular ring, or **seat**, that surrounds an opening through which flow occurs (Figure 4). The direction of movement of the disk is parallel to the direction of the flow through the valve opening (or seat) and normal to the axis of the pipe in which the valve is installed.

Globe valves are most frequently used in smaller diameter pipes but are available in sizes up to 12 in. They are used for throttling duty where positive shutoff is required. Globe valves for controlling service should be selected by class, and whether they are of the straight-through or angle type, composition disk, union or gasketed bonnet, threaded, and solder or grooved ends. Manually operated flow control valves are also available with fully guided V-port throttling plugs or needle point stems for precise adjustment.

Gate Valves

A gate valve controls flow by means of a wedge disk fitting against machined seating faces (Figure 5). The straight-through opening of the valve is as large as the full bore of the pipe, and the gate movement is perpendicular to the flow path.

Gate valves are intended to be fully open or completely closed. They are designed to permit flow or stop flow and should not be used to regulate or control flow. Various wedges for gate valves are available for specific applications. Valves in inaccessible locations may be provided with a chain wheel or with a hammer-blow operator. More detailed information is available from valve manufacturers.

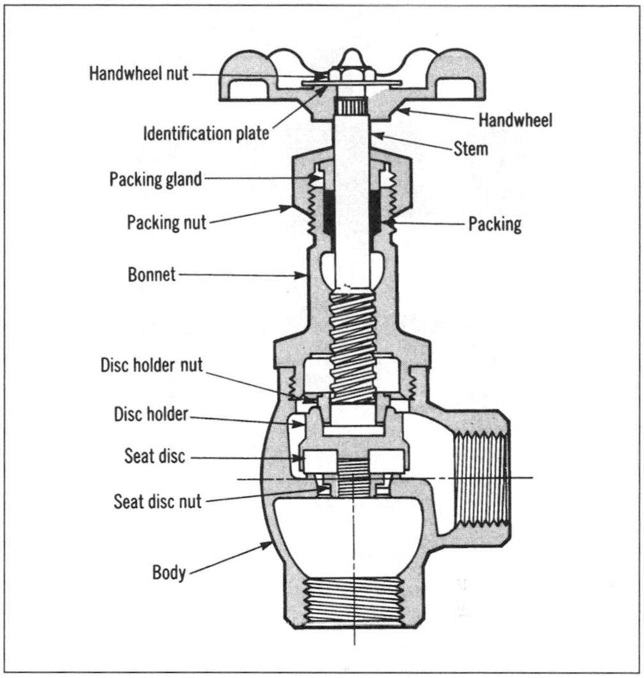

Fig. 4 Globe Valve
(Courtesy Grinnell Corp.)

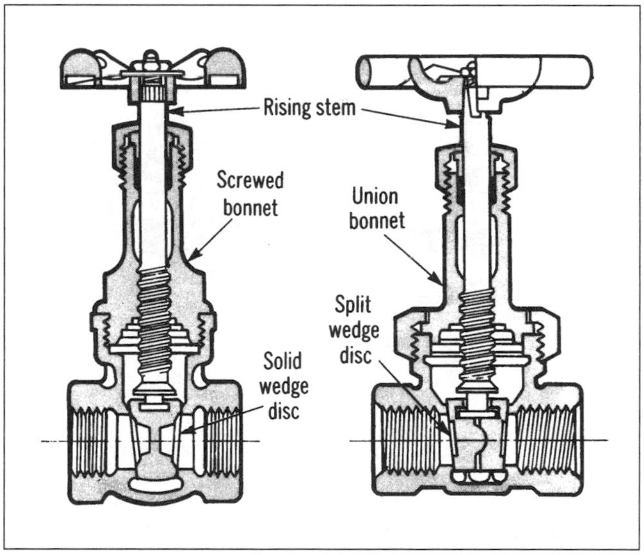

Fig. 5 Two Variations of Gate Valve

Plug Valves

A plug valve is a manual fluid flow control device (Figure 6). It operates from fully open to completely shut off within a 90° turn. The capacity of the valve depends on the ratio of the area of the orifice to the area of the pipe in which the valve is installed.

The cutaway view of a plug valve shows a valve with an orifice that is considerably smaller than the full size of the pipe. Lubricated plug valves are usually furnished in gas applications. A plug valve is selected as an on/off control device because (1) it is relatively inexpensive; (2) when adjusted, it holds its position; and (3) its position is clearly visible to the operator. The effectiveness of this valve as a flow control device is reduced if the orifice of the valve is fully ported (i.e., the same area as the pipe size).

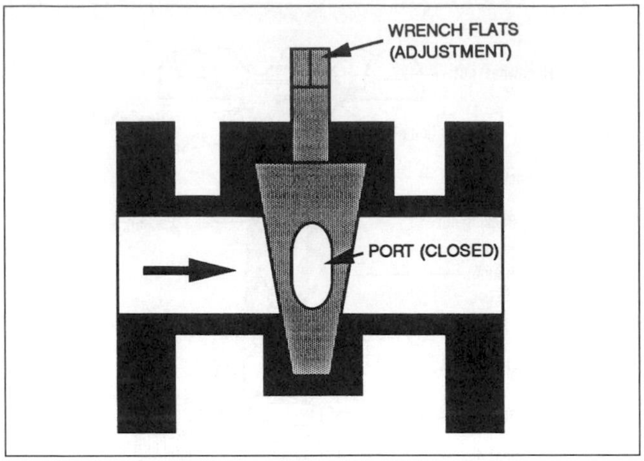

Fig. 6 Plug Valve

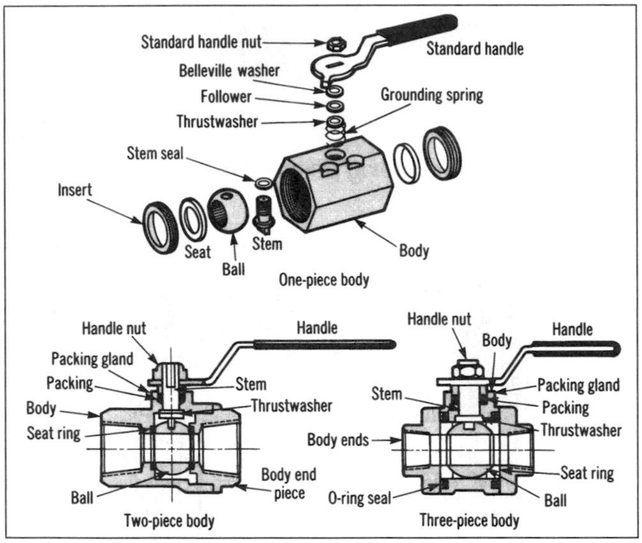

Fig. 7 Ball Valve

Ball Valves

A ball valve contains a precision ball held between two circular seals or seats. Ball valves have various port sizes. A 90° turn of the handle changes operation from fully open to fully closed. Ball valves for shutoff service may be fully ported. Ball valves for throttling or controlling and/or balancing service should have a reduced port with a plated ball and valve handle memory stop. Ball valves may be of one-, two-, or three-piece body design (Figure 7).

Butterfly Valves

A butterfly valve typically consists of a cylindrical, flanged-end body with an internal, rotatable disk serving as the fluid flow-regulating device (Figure 8). Butterfly valve bodies may be **wafer style**, which is clamped between two companion flanges whose bolts carry the pipeline tensile stress and place the wafer body in compression, or **lugged style**, with tapped holes in the wafer body, which may serve as a future point of disconnection. The disk's axis of rotation is the valve stem; it is perpendicular to the flow path at the center of the valve body. Only a 90° turn of the valve disk is required to change from the full-open to the closed position. Butterfly valves may be manually operated with **hand quadrants** (levers) or provided with an extended shaft for automatic operation by an actuator. Special attention should be paid to manufacturers' recommendations for sizing an actuator to handle the torque requirements.

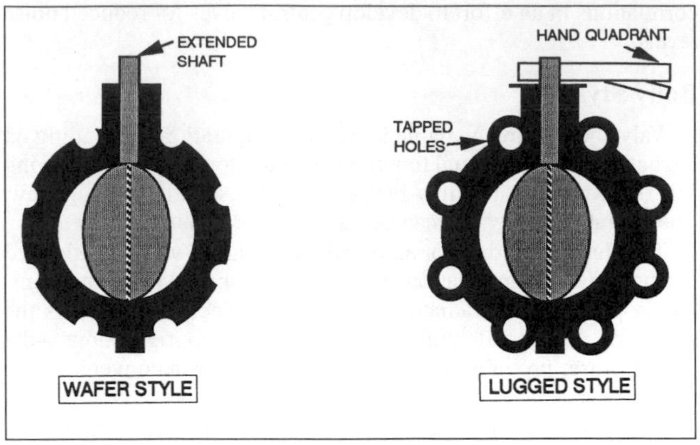

Fig. 8 Butterfly Valve

Simple and compact design, a low corresponding pressure drop, and fast operation characterize all butterfly valves. Quick operation makes them suitable for automated control, whereas the low pressure drop is suitable for high flow. Butterfly valve sizing for on/off applications should be limited to pipe sizing velocities given in Chapter 33 of the 1997 *ASHRAE Handbook—Fundamentals*; on the other hand, for throttling control applications, the valve coefficient sizing presented in the section on Automatic Valves must be followed.

Pinch Valves

Two styles of pinch valve bodies are normally used—the jacket pinch and the Saunders-type bodies used for slurry control in many industries. Pressure squeezing the flexible tube jacket of a pinch valve reduces its port opening to control flow. The Saunders type employs an actuator to manually or automatically squeeze the diaphragm against a weir-type port. These valves have limited HVAC application.

AUTOMATIC VALVES

Automatic valves are commonly considered as control valves that operate in conjunction with an automatic controller or device to control the fluid flow. The "control valve" as used here actually consists of a valve body and an actuator. The valve body and actuator may be designed so that the actuator is removable and/or replaceable, or the actuator may be an integral part of the valve body. This section covers the most common types of valve actuators and control valves with the following classifications:

- Two-way bodies (single- and double-seated)
- Three-way bodies (mixing and diverting)
- Ball valves
- Butterfly arrangements (two- and three-way)

Actuators

The valve actuator converts the controller's output, such as an electric or pneumatic signal, into the rotary or linear action required by the valve (stem), which changes the control variable (flow). Actuators cover a wide range of sizes, types, output capabilities, and control modes.

Sizes. Actuators range in physical size from small solenoid or clock motor self-operated types, to large pneumatic actuators with 100 to 200 in^2 of effective area.

Types. The most common types of actuators used on automatic valve applications are solenoid, thermostatic radiator, pneumatic, electric motor, electronic, and electrohydraulic.

Output (Force) Capabilities. Although the smallest actuators, designed for unitary commercial HVAC and residential control applications, are capable of only a very small output, larger pneumatic or electrohydraulic actuators are capable of great force. The overall force ranges from a few ounces to over 0.5 ton of force.

Pneumatic Actuators

Pneumatic or diaphragm valve actuators are available with diaphragm sizes ranging from 3 to 200 in^2. The design consists of a flexible diaphragm clamped between an upper and a lower housing. On direct-acting actuators, the upper housing and diaphragm create a sealed chamber (Figure 9). A spring opposing the diaphragm force is positioned between the diaphragm and the lower housing. Increasing air pressure on the diaphragm pushes the valve stem down and overcomes the force of the load spring to close a direct-acting valve. Springs are designated by the air pressure change required to open or close the valve. A 5 lb spring requires a 5 psi control pressure change at the actuator to operate the valve. Some valves have an adjustable spring feature; others are fixed. Springs for commercial control valves usually have ±10% tolerance, so the 5 lb spring setting is 5 psi ± 0.5 psi. Two valves in a control may be sequenced simply with adjustable actuator springs.

Reverse-acting valves may use a direct-acting actuator if they have reverse-acting valve bodies; otherwise, the actuator must be reverse-acting and constructed with a sealed chamber between the lower housing and the diaphragm.

The valve close-off point shifts as the supply and/or the differential pressure increases across a single-seated valve due to the fixed areas of the actuator and the valve seat. The manufacturer's close-off rating tables need to be consulted to determine if the actuator is of an adequate size or if a larger actuator or a pneumatic positioner relay is required.

A **pneumatic positioner relay** may be added to the actuator to provide additional force to close or open an automatic control valve (Figure 9). Sometimes called positive positioners or pilot positioners, pneumatic positioners are basically high-capacity relays that add air pressure to or exhaust air pressure from the actuator in relation to the stroke position of the actuator. Their application is limited by the supply air pressure available and by the actuator's spring.

Electric Actuators

Electric actuators usually consist of a double-wound electric motor coupled to a gear train and an output shaft connected to the valve stem with a cam or rack-and-pinion gear linkage (Figure 10). For valve actuation, the motor shaft typically drives through 160° of rotation. The use of damper actuators with 90° full stroke rotation is rapidly increasing in valve control applications. Gear trains are coupled internally to the electric actuators to provide a timed movement of valve stroke to increase operating torque and to reduce overshooting of valve movement. Gear trains can be fitted with limit switches, auxiliary potentiometers, etc., to provide position indication and feedback for additional system control functions.

In many instances, a linkage is required to convert rotary motion to the linear motion required to operate a control valve (except ball and butterfly valves). Electric valve actuators operate with two-position, floating, proportional electric, and electronic control systems. Actuators usually operate with a 24 V (ac) low-voltage control circuit. Actuator time to rotate (or drive full stroke) ranges from 30 s to 4 min, with 60 s being most common.

Electric valve actuators may have a spring return, which returns the valve to a normal position in case of power failure, or it may be powered with an electric relay and auxiliary power source. Since the motor must constantly drive in one direction against the return spring, spring return electric valve actuators generally have only approximately one-third of the torque output of non-spring return actuators.

Electrohydraulic Actuators

Hydraulic actuators combine characteristics of electric and pneumatic actuators. In essence, hydraulic actuators consist of a sealed housing containing the hydraulic fluid, pump, and some type of metering or control apparatus to provide pressure control across a piston or piston/diaphragm. A coil controlled by a low- to medium-level dc voltage usually activates the pressure control apparatus.

Solenoids

A solenoid valve is an electromechanical control element that opens or closes a valve on the energization of a solenoid coil. Solenoid valves are used to control the flow of hot or chilled water and steam and range in size from 1/8 to 2 in. pipe size. Solenoid actuators themselves are two-position control devices and are available for operation in a wide range of alternating current voltages (both 50 and 60 Hz) as well as direct current. Operation of a simple two-way, direct-acting solenoid valve in a deenergized state is illustrated in Figure 11.

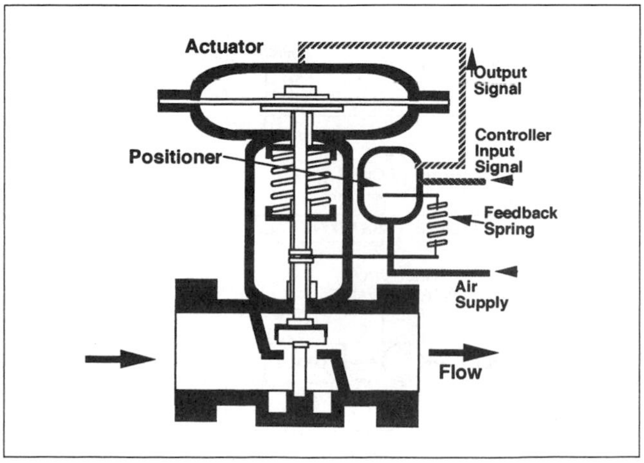

Fig. 9 Two-Way, Direct-Acting Control Valve with Pneumatic Actuator and Positioner

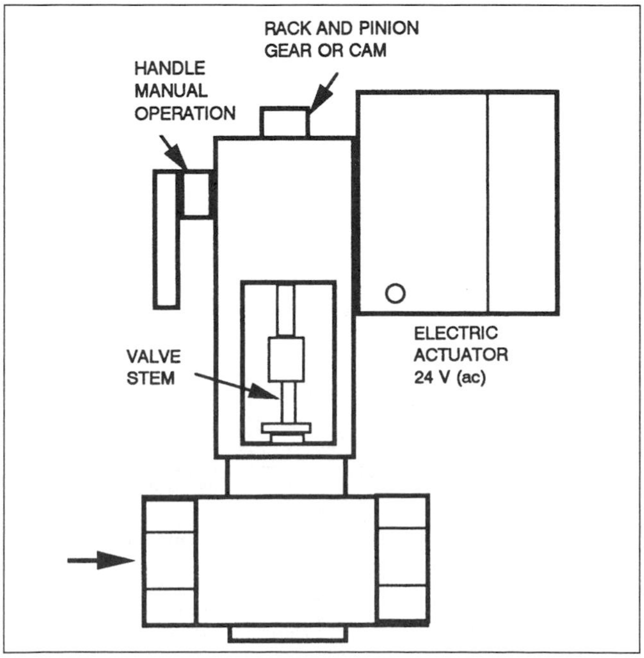

Fig. 10 Two-Way Control Valve with Electric Actuator

Thermostatic Radiator Valves

Thermostatic radiator valves are self-operated and do not require external energy. They control room or space temperature by modulating the flow of hot water or steam through free-standing radiators, convectors, or baseboard heating units. Thermostatic radiator valves are available for a variety of installation requirements with remote-mounted sensors or integral-mounted sensor and remote or integral set point adjustment (Figure 12).

Control of Automatic Valves

Computer-based control of automatic control valves is replacing older technologies and provides many benefits, including speed, accuracy, and data communication. However, care must be exercised in selecting the value of control loop parameters such as loop speed and dead band (allowable set-point deviation). High loop speed coupled with zero dead band can cause the valve-actuator to seek a new control position with each control loop cycle unless the actuator itself has some type of built-in protection against this. For example, a 1 s control loop with zero dead band could result in 30,000,000 repositions (corrections) in 1 year of service.

Computer-based control systems should be tuned to provide the minimum acceptable level of response and accuracy required for the application in order to achieve maximum valve and actuator service life.

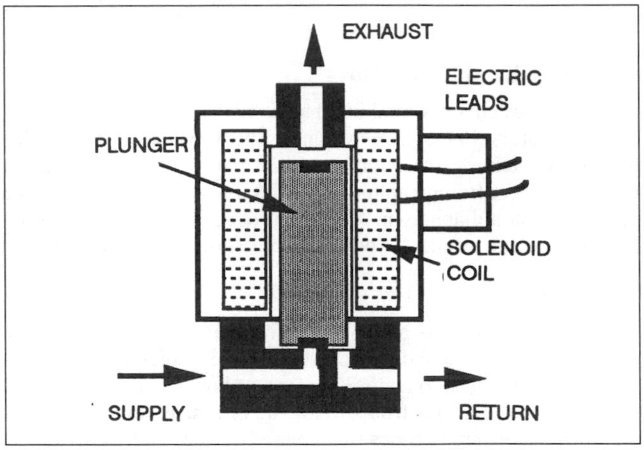

Fig. 11 Electric Solenoid Valve

Two-Way Valves (Single- and Double-Seated)

In a two-way automatic valve, the fluid enters the inlet port and exits the outlet port either at full or reduced volume, depending on the position of the stem and the disk in the valve. Two-way valves may be single- or double-seated.

In the single-seated valve, one seat and one plug-disk close against the stream. The style of the plug-disk varies depending on the requirements of the designer and the application. For body comparison, refer to Figure 8 in Chapter 37 of the 1997 *ASHRAE Handbook—Fundamentals*.

The double-seated valve is a special application of the two-way valve with two seats, plugs, and disks. It is generally applied to cases where the close-off pressure is too high for the single-seated valve.

Three-Way Valves

Three-way valves either mix or divert streams of fluid. Figure 13 shows some common applications for three-way valves. Figure 9 in Chapter 37 of the 1997 *ASHRAE Handbook—Fundamentals* shows typical cross sections of three-way mixing and diverting valves.

The **three-way mixing valve** blends two streams into one common stream based on the position of the valve plug in relation to the upper and lower seats of the valve. A common use is to mix chilled or hot water. The valve controls the temperature of the single stream leaving the valve.

The **three-way diverting or bypass valve** takes one stream of fluid and splits it into two streams for temperature control. In some limited applications, such as a cooling tower control, a diverting or

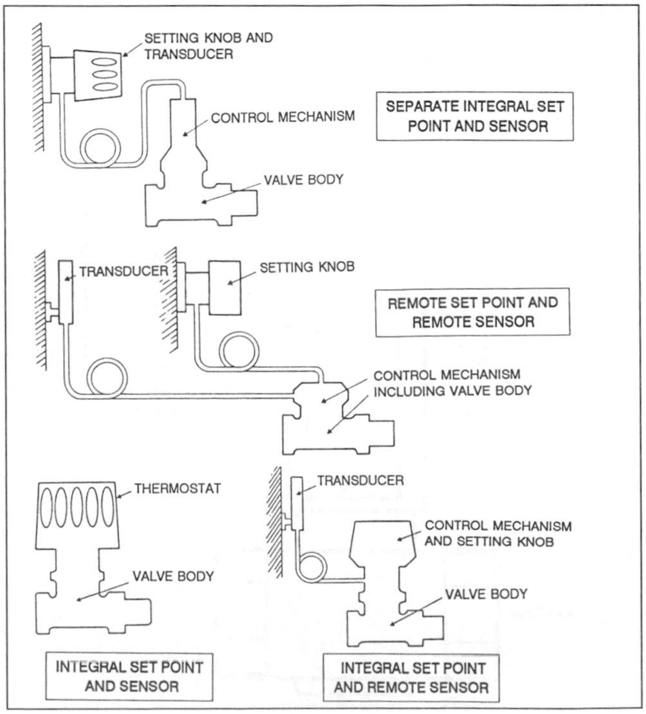

Fig. 12 Thermostatic Valves

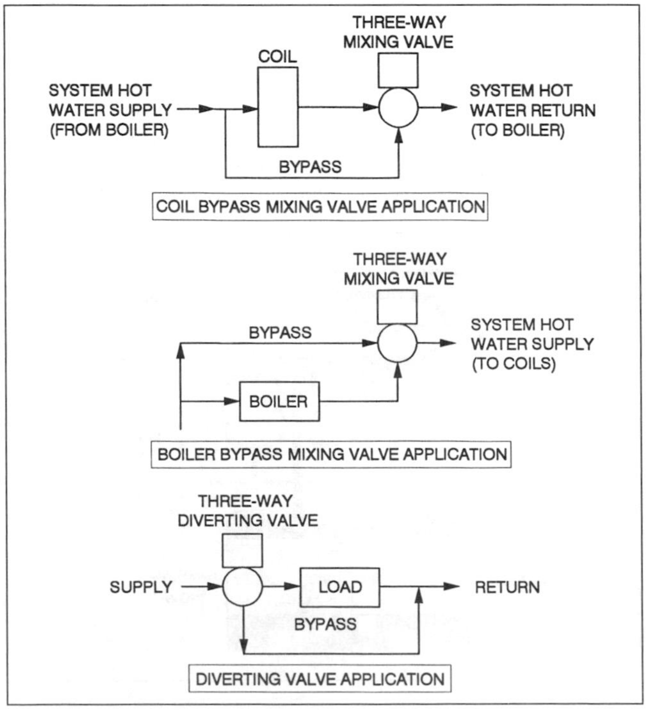

Fig. 13 Typical Three-Way Control Applications

bypass valve must be used in place of a mixing valve. In most cases, a mixing valve can perform the same function as a diverting or bypass valve if the companion actuator has a very high spring rate. Otherwise, water hammer or noise may occur when operating near the seat.

Special-Purpose Valves

Special-purpose valve bodies may be used on occasion. One type of four-way valve is used to allow separate circulation in the boiler loop and a heated zone. Another type of four-way valve body is used as a changeover refrigeration valve in heat pump systems to reverse the evaporator to a condenser function.

Float valves are used to supply water to a tank or reservoir or serve as a boiler feed valve to maintain an operating water level at the float level location (Figure 14).

Ball Valves

Ball valves coupled with rotary actuators have seen increasing use in HVAC control applications. A reduced port should be used on the ball valve, or the valve should be sized smaller than the piping system in order to achieve adequate control (pressure drop). In some cases, the packing system of ball valves has been redesigned to accommodate the modulating control action inherent in HVAC. The control characteristic for full-ported ball valves is equal percentage, but modified seats, ball ports, or inserts are available to provide other characteristics (e.g., linear, modified linear, etc.)

Butterfly Valves

In some applications, it is not possible to use standard three-way mixing or bypass valves because of size limitations or space constraints. In these cases, two butterfly valves are mounted on a piping tee and cross-linked to operate as either three-way mixing or three-way bypass valves (Figure 15). Butterfly valves have different flow characteristics from standard seat and disk-type valves, so they may be used only where their flow characteristics suffice.

Control Valve Flow Characteristics

Generally, valves control the flow of fluids by an actuator, which moves a stem with an attached plug. The plug seats within the valve port and against the valve seat with a composition disk or metal-to-metal seating. Based on the geometry of the plug, three distinct flow conditions can be developed (Figure 16):

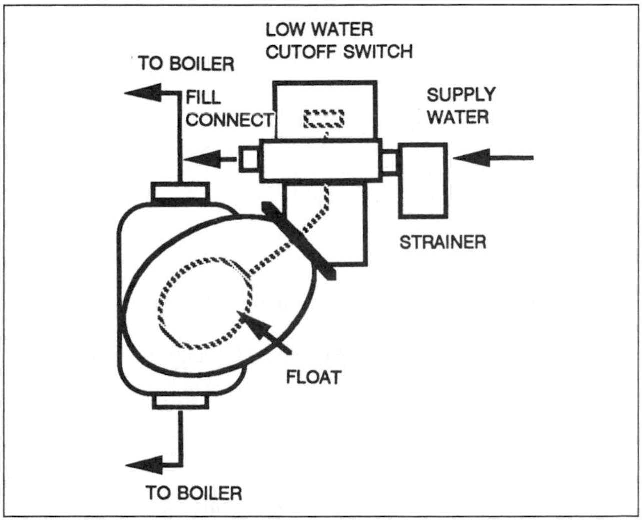

Fig. 14 Float Valve and Cutoff Steam Boiler Application

1. **Quick Opening.** When started from the closed position, a quick-opening valve allows a considerable amount of flow to pass for small stem travel. As the stem moves toward the open position, the rate at which the flow is increased per movement of the stem is reduced in a nonlinear fashion. This characteristic is used in two-position or on/off applications.
2. **Linear.** Linear valves produce equal flow increments per equal stem travel throughout the travel range of the stem. This characteristic is used on steam coil terminals and in the bypass port of three-way valves.
3. **Equal Percentage.** This type of valve produces an exponential flow increase as the stem moves from the closed position to the open. The term equal percentage means that for equal increments of stem travel, the flow increases by an equal percentage. For example, in Figure 16, if the valve is moved from 50 to 70% of full stroke, the percentage of full flow changes from 10 to 25%, an increase of 150%. Then, if the valve is moved from 80 to 100% of full stroke, the percentage of full flow changes from

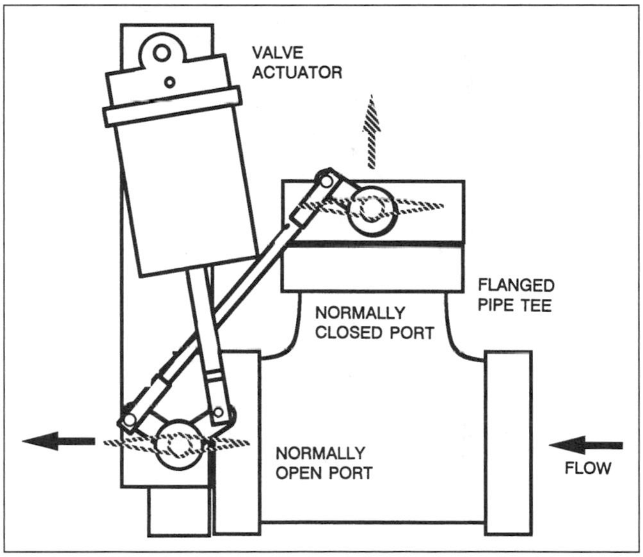

Fig. 15 Butterfly Valves—Diverting Tee Application

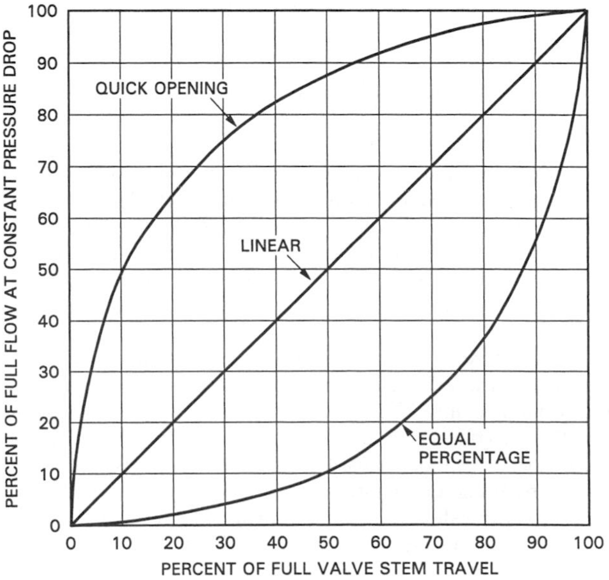

Fig. 16 Control Valve Flow Characteristics

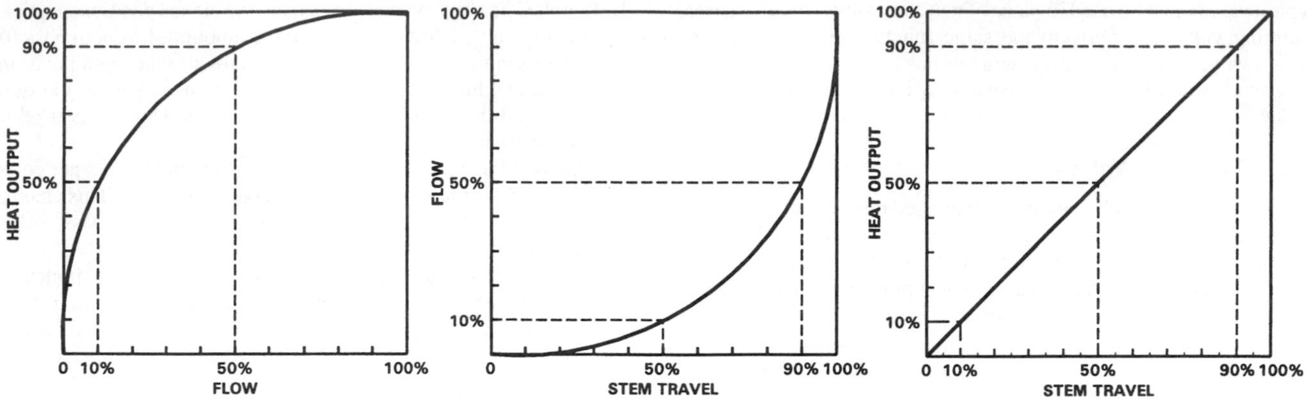

Fig. 17 Heat Output, Flow, and Stem Travel Characteristics of Equal Percentage Valve

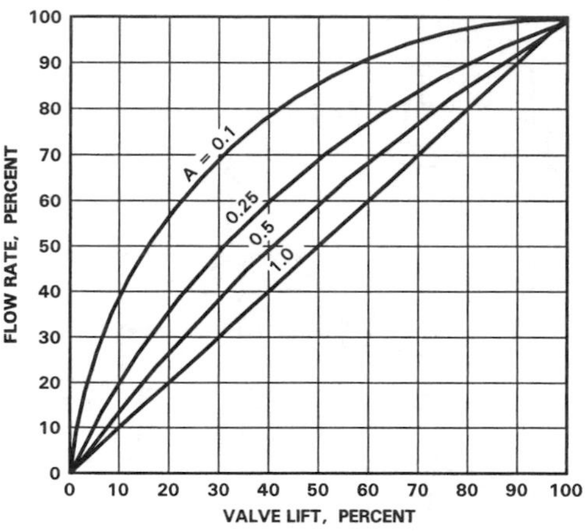

Fig. 18 Authority Distortion of Linear Flow Characteristics

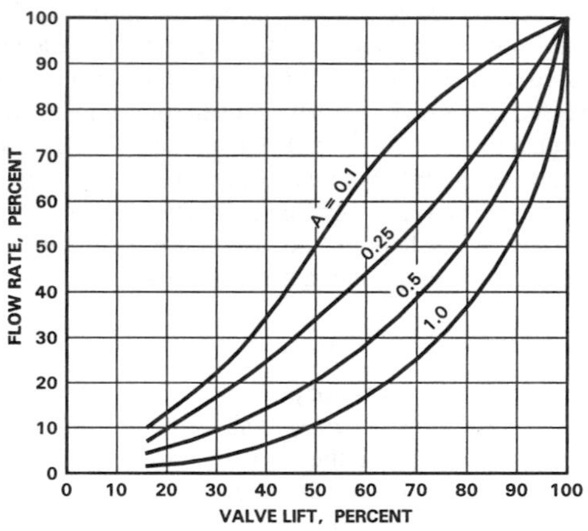

Fig. 19 Authority Distortion of Equal Percentage Flow Characteristic

40 to 100%, again, an increase of 150%. This characteristic is recommended for control on hot and chilled water terminals.

Control valves are commonly used in combination with a coil and another valve within a circuit to be controlled. The designer should combine the valve flow characteristics with coil performance curves (heating or cooling) because the resulting energy output profile of the circuit versus the stem travel improves (Figure 17). For a typical hydronic heating or cooling coil, the equal percentage results in the closest to a linear change and provides the most efficient control (Figure 17).

The three flow patterns are obtained by imposing a constant pressure drop across the modulating valve, but in actual conditions, the pressure drop across the valve varies between a maximum (when it is controlling) and a minimum (when the valve is near full open). The ratio of these two pressure drops is known as **authority**. Figures 18 and 19 show how linear and equal percentage valve flow characteristic are distorted as the control valve authority is reduced due to a reduction in valve pressure drop. The quick-opening characteristic, not shown, is distorted to the point that it approaches two-position or on/off control. The selection of the control valve pressure drop directly affects the valve authority and should be at least 25 to 50% of the system loop pressure drop (i.e., the pressure drop from the pump discharge flange, supply main, supply riser, supply branch, heat transfer coil, return branch, fittings, balancing valve, and return main to the pump suction flange). The location of the control valve in the system results in unique pressure drop selections for each control valve. A higher valve pressure drop allows a smaller valve pipe size and better control.

Control Valve Sizing

Liquids. A valve creates fluid resistance in a circuit to limit the flow of the fluid at a calculated pressure drop. Each passive element in a circuit creates a pressure drop according to the following general equation:

$$\Delta p = RQ^n\left(\frac{\rho}{\rho_w}\right) \tag{1}$$

where

Δp = pressure drop, psi
R = resistance
ρ = fluid density, lb/ft^3
ρ_w = density of water at 60°F, lb/ft^3
Q = volumetric flow, gpm
n = system coefficient

For turbulent flows, n is assumed to be 2, although for steel pipes $n = 1.85$.

For a valve, assuming $n = 2$, Equation (1) can be solved for flow:

$$Q = \sqrt{\left(\frac{\Delta p}{R}\right)\left(\frac{\rho_w}{\rho}\right)} \tag{2}$$

The term $\sqrt{1/R}$ can be replaced by the flow coefficient C_v, the ratio ρ/ρ_w is approximately 1 for water at temperatures below 250°F, and Equation (2) becomes

$$Q = C_v\sqrt{\Delta p} \tag{3}$$

or

$$Q = 0.67C_v\sqrt{\Delta h} \tag{4}$$

where Δh = pressure drop, ft of water.

The control valve size should be selected by calculating the required C_v to provide the design flow at an assumed pressure drop Δp. A pressure drop of 25 to 50% of the available pressure between the supply and return riser (pump head) should be selected for the control valve. This pressure drop gives the best flow characteristic as described in the section on Control Valve Flow Characteristics.

For liquids with a viscosity correction factor V_f,

$$Q = \frac{C_v}{V_f}\sqrt{\Delta p\left(\frac{\rho_w}{\rho}\right)} \tag{5}$$

Steam. For steam flow,

$$w_s = 2.1\frac{C_v}{K}\sqrt{\Delta p(P_1 + P_2)} \tag{6}$$

where

w_s = steam flow, lb/h
K = 1 + 0.0007 × (°F of superheat)
C_v = flow coefficient, gpm at Δp = 1 psi
P_1 = entering steam absolute pressure
P_2 = leaving steam absolute pressure
Δp = steam pressure drop across valve, $P_1 - P_2$

Note: Some manufacturers list the constant in Equation (6) as high as 3.2, but most agree on 2.1. As part of good practice, always confirm valve sizing with the manufacturer.

Steam reaches critical or sonic velocity when the downstream pressure is 58% or less of the absolute inlet pressure. If the downstream pressure is below the critical pressure, increasing the pressure drop produces no further increase in flow. As a result, when $P_2 \le 0.58P_1$, the following critical pressure drop formula is used:

$$C_v = \frac{w_s}{1.61P_1} \tag{7}$$

Applications

Automatically controlled valves are applied to control many different variables, the most common being temperature, humidity, flow, and pressure. However, a valve can be used directly to control only flow or pressure. When flow is controlled, a pressure drop is implied, and when pressure is controlled, some maximum flow rate is implied. These two factors must be considered in selecting control valves. For some typical valve applications, refer to Chapter 45 of the 1999 *ASHRAE Handbook—Applications*.

Although the discussion in this chapter applies to hot water, chilled water, and steam, control valves can be used with virtually any fluid. The fluid characteristics must be considered in selecting materials for the valve. The requirements are particularly strict for use with high-temperature water and high-pressure steam.

Steam is controlled in two ways:

1. When steam pressure is too high for use in a specific application, the pressure must be reduced by a **pressure-reducing valve** (PRV). This is normally a globe-type valve, because modulating control is required. The valve may be externally or internally piloted and is usually self-contained, using the steam pressure to drive the actuator. The load may vary, so it is sometimes desirable to use two or more valves in parallel, adjusted to open in sequence, for more accurate control.
2. Steam flow to a heat exchanger may be controlled in response to temperature or humidity requirements. In this case, an external control system is used with the steam valve as the controlled device. In selecting a steam valve, the maximum flow rate for the specific valve and entering steam pressure must be considered. These factors are determined from the critical pressure drop, which limits the flow.

Hot and chilled water are usually controlled in response to temperature or humidity requirements. When selecting a valve for controlling water flow, a pressure drop sufficiently large to allow the valve to control properly should be specified. The response of the heat exchanger coil to a change in flow is not linear; therefore, an equal percentage plug should be used, and the temperature of the water supply should be as high (hot water) or as low (chilled water) as required by the load conditions.

BALANCING VALVES

Two approaches are available for balancing hydronic systems: (1) a manual valve with integral pressure taps and a calibrated port, which permits field proportional balancing to the design flow conditions; (2) or an automatic flow-limiting valve selected to limit the circuit's maximum flow to the design flow.

Manual Balancing Valves

Manual balancing valves can be provided with the following features:

- Manually adjustable stems for valve port opening or a combination of a venturi or orifice and an adjustable valve
- Stem indicator and/or scale to indicate the relative amount of valve opening
- Pressure taps to provide a readout of the pressure difference across the valve port or the venturi/orifice
- Capability to be used as a shutoff for future service of the heat transfer terminal
- Locking device for field setting the maximum opening of a valve
- Body tapped for attaching drain hose

Manual balancing valves may have rotary, rising, or nonrising stems for port adjustment (Figure 20).

Meters with various scale ranges, a field carrying case, attachment hoses, and fittings for connecting to the manual balancing valve should be used to determine its flow by reading the differential pressure. Some meters employ analog measuring elements with direct-reading mechanical dual-element Bourdon tubes. Other meters are electronic differential pressure transducers with a digital data display.

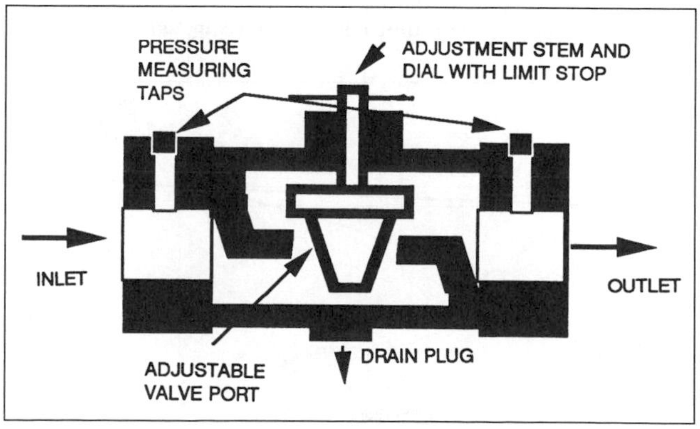

Fig. 20 Manual Balancing Valve

Many manufacturers of balancing valves produce circular slide rules to calculate circuit flow based on pressure difference readout across the balancing valve, its stem position, and/or the valve's flow coefficient. This calculator can also be used for selecting the size and setting of the valve when the terminal design flow conditions are known.

Automatic Flow-Limiting Valves

A **differential pressure-actuated flow control valve**, also called an automatic flow-limiting valve (Figure 21), regulates the flow of fluid to a preset value when the differential pressure across it is varied. This regulation (1) helps prevent an overflow condition in the circuit where it is installed and (2) aids the overall system balance when other components are changing (modulating valves, pump staging, etc.).

Typically, the valve body contains a moving element containing an orifice, which adjusts itself based on pressure forces so that the flow passage area varies.

The area of an orifice can be changed by either (1) a piston or cup moving across a shear plate or (2) increased pressure drop to squeeze the rubber orifice in rubber grommet valves.

A typical performance curve for the valve is shown in Figure 22. The flow rate for the valve is set. The flow curve is divided into three ranges of differential pressure: the start-up range, the control range, and the above-control range.

Balancing Valve Selection

The balancing valve is a flow control device that is selected for a lower pressure drop than an automatic control valve (5 to 10% of the available system pressure). Selection of any control valve is based on the pressure drop at maximum (design) flow to ensure that the valve provides control at all flow rates. A properly selected balancing valve can proportionally balance flow to its terminal with flow to the adjacent terminal in the same distribution zone. Refer to Chapter 36 of the 1999 *ASHRAE Handbook—Applications* for balancing details.

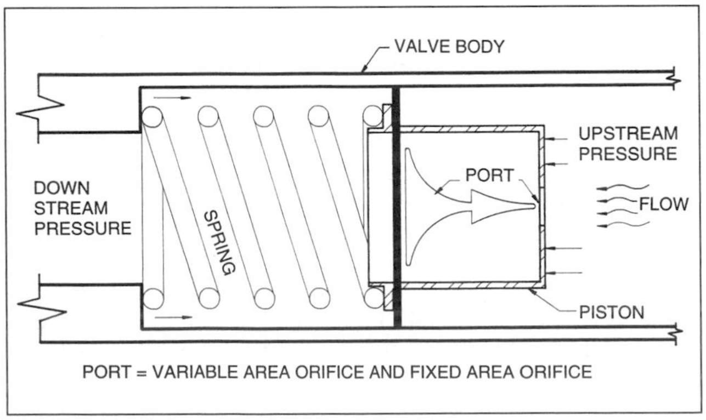

Fig. 21 Automatic Flow-Limiting Valve

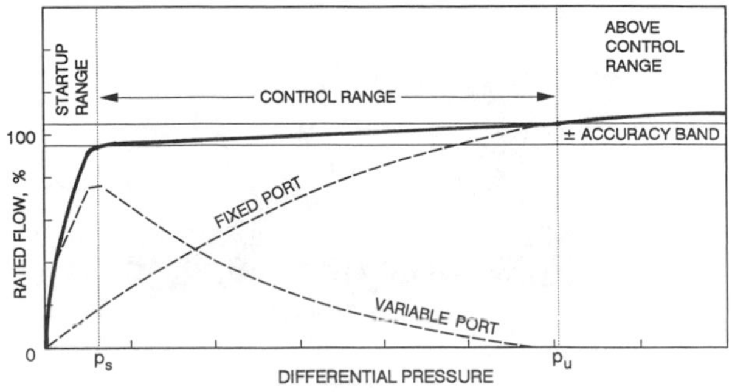

Fig. 22 Automatic Flow-Limiting Valve Curve

MULTIPLE-PURPOSE VALVES

Multiple-purpose valves are made in straight pattern or angle pattern. The valves can provide shutoff for servicing or can be partially closed for balancing. Pressure gage connections to read the pressure drop across the valve can be used with the manufacturer's calibration chart or meter to estimate the flow. Means are provided to return the valve to its as-balanced position after shutoff for servicing. The valve also acts as a check valve to prevent backflow when parallel pumps are used and one of the pumps is cycled off.

Figure 23 shows a straight pattern multiple-purpose valve designed to be installed 5 to 10 pipe diameters from the pump discharge of a hydronic system.

Figure 24 shows an angle pattern multiple-purpose valve installed 5 to 10 pipe diameters downstream of the pump discharge with a common gage and a push button trumpet valve manifold to measure the differential pressure across the strainer, pump, or multiple-purpose valve. From this, the flow can be estimated. The differential pressure across the pump suction strainer can also be estimated to determine whether the strainer needs servicing.

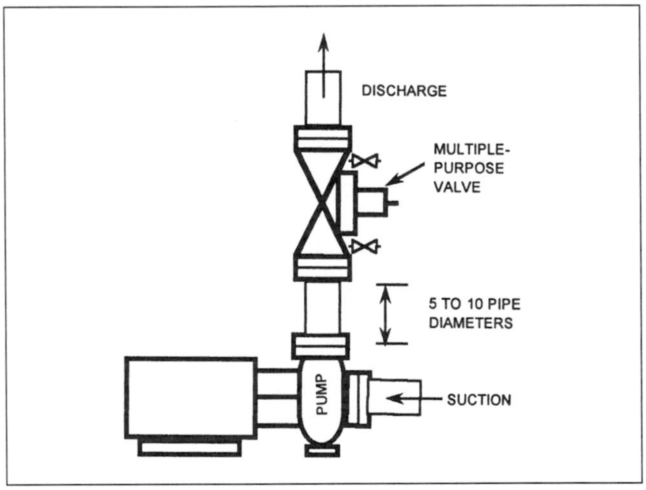

Fig. 23 Typical Multiple-Purpose Valve (Straight Pattern) on Discharge of Pump

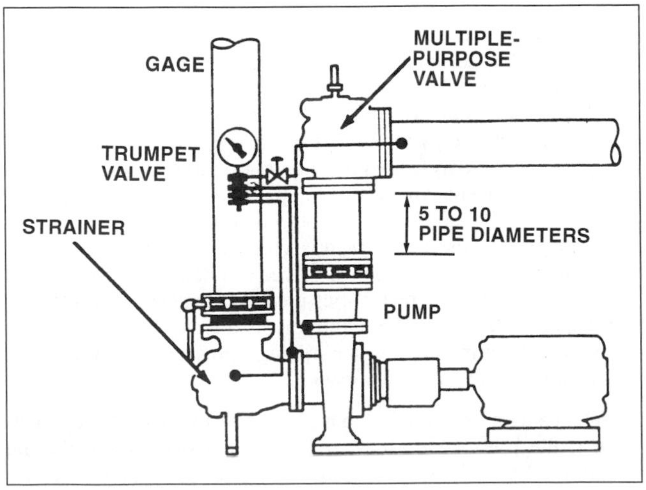

Fig. 24 Typical Multiple-Purpose Valve (Angle Pattern) on Discharge of Pump

SAFETY DEVICES

The terms safety valve, relief valve, and safety relief valve are sometimes used interchangeably, and although the devices generally provide a similar function (safety), they have important differences in their modes of operation and application in HVAC systems (Jordan 1998).

Safety valves open rapidly (pop-action). They are used for gases and vapors (e.g., compressed air and steam).

Relief valves open or close gradually in proportion to excessive pressure. They are used for liquids (e.g., unheated water).

Safety relief valves perform a dual function: they open rapidly (pop-action) for gases and vapors and gradually for liquids. Typical HVAC application is for heating water.

Temperature-actuated pressure relief valves (or temperature and pressure safety relief valves) are activated by excessive temperature or pressure. They are commonly used for potable hot water.

Application of these safety devices must comply with building codes and the ASME *Boiler and Pressure Vessel Code*. For the remainder of this discussion, the term "safety valve" is used generically to include any or all of the four types described.

Safety valve construction, capacities, limitations, operation, and repair are covered by the ASME *Boiler and Pressure Vessel Code*. For pressures above 15 psig, refer to Section I. Section IV covers steam boilers for pressures less than 15 psig. Unfired pressure vessels (such as heat exchange process equipment or pressure-reducing valves) are covered by Section VIII.

The capacity of a safety valve is affected by the equipment on which it is installed and the applicable code. Valves are chosen based on accumulation, which is the pressure increase above the maximum allowable working pressure of the vessel during valve discharge. Section I valves are based on 3% accumulation. Accumulation may be as high as 33.3% for Section IV valves and 10% for Section VIII. To properly size a safety valve, the required capacity and set pressure must be known. On a pressure-reducing valve station, the safety valve must have sufficient capacity to prevent an unsafe pressure rise if the reducing valve fails in the open position.

The safety valve set pressure should be high enough to allow the valve to remain closed during normal operation, yet allow it to open and reseat tightly when cycling. A minimum differential of 5 psi or 10% of inlet pressure (whichever is greater) is recommended.

When installing a safety valve, consider the following:

- Install the valve vertically with the drain holes open or piped to drain.
- The seat can be distorted if the valve is overtight or the weight of the discharge piping is carried by the valve body. A drip-pan elbow on the discharge of the safety valve will prevent the weight of the discharge piping from resting on the valve (Figure 25).
- Use a moderate amount of pipe thread lubricant (first 2 to 3 threads) on male threads only.
- Install clean flange connections with new gaskets, properly aligned and parallel, and bolted with even torque to prevent distortion.
- Wire cable or chain pulls attached to the test levers should allow for a vertical pull and their weight should not be carried by the valve.

Testing of safety valves varies between facilities depending on operating conditions. Under normal conditions, safety valves with a working pressure under 400 psig should be tested manually once per month and pressure-tested once each year. For higher pressures, the test frequency should be based on operating experience.

When steam safety valves require repair, adjustment, or set pressure change, the manufacturer or approved stations holding the ASME V, UV, and/or VR stamps must perform the work. Only the manufacturer is allowed to repair Section IV valves.

SELF-CONTAINED TEMPERATURE CONTROL VALVES

Self-contained or self-operated temperature control valves do not require an outside energy source such as compressed air or electricity (Figure 26). They depend on a temperature-sensing bulb and capillary tube filled with either an oil or a volatile liquid. In an **oil-filled** system, the oil expands as the sensing bulb is heated. This expansion is transmitted through the capillary tube to an actuator bellows in the valve top, which causes the valve to close. The valve opens as the sensing bulb cools and the oil contracts; a spring provides a return force on the valve stem.

A volatile liquid control system is known as a **vapor pressure** or **vapor tension** system. When the sensing bulb is warmed, some of

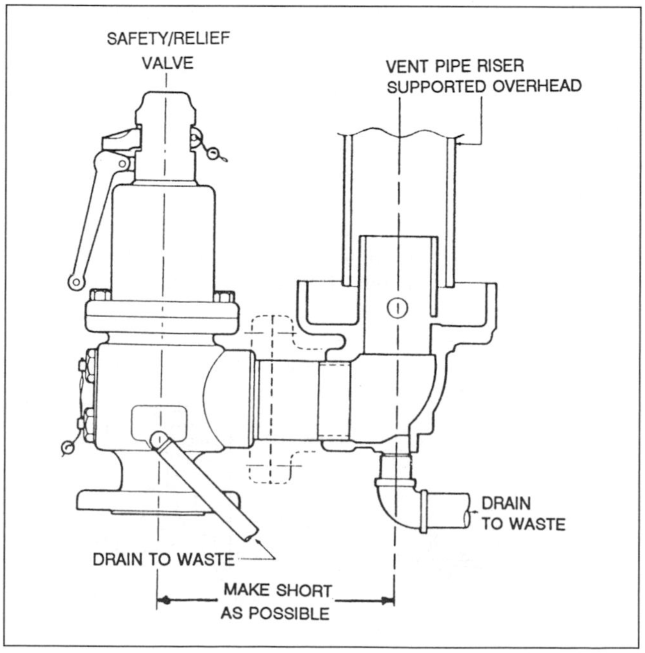

Fig. 25 Safety/Relief Valve with Drip-Pan Elbow

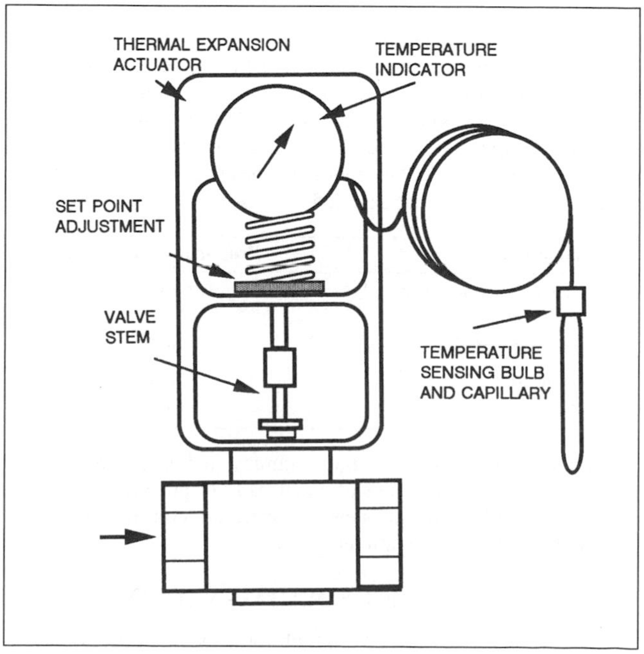

Fig. 26 Self-Operated Temperature Control Valve

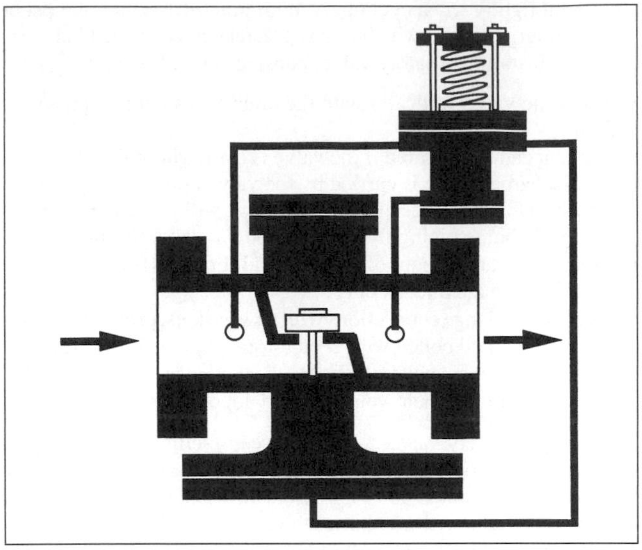

Fig. 27 Pilot-Operated Steam Valve

the volatile liquid vaporizes, causing an increase in the sealed system pressure. The pressure rise is transmitted through the capillary tube to expand the bellows, which then moves the valve stem and closes the valve. Thermal systems actuate the control valve either directly or through a pilot valve.

In a **direct-actuated** design, the control directly moves the valve stem and plug to close or open the valve. These valves must compensate for the steam pressure force acting on the valve seat by generating a greater force in the bellows to close the valve. An adjustable spring adjusts the temperature set point and provides the return force to move the valve stem upward as the temperature decreases.

A **pilot-operated valve** (Figure 27) uses a much smaller intermediate pilot valve that controls the flow of steam to a large diaphragm that then acts on the valve stem. This allows the control system to work against high steam pressures due to the smaller area of the pilot valve.

For self-contained temperature valves to operate as proportional controls, the bulb must sense a change in the temperature of the process fluid. The difference in temperature from no-load to maximum controllable load is known as the **proportional band**. Because the size of this proportional band can be varied depending on valve size, the accuracy is variable. Depending on the application, proportional bands of 2 to 18°F may be selected, as shown in the following table:

Application	Proportional Band, °F
Domestic hot water heat exchanger	6 to 14
Central hot water	4 to 7
Space heating	2 to 5
Bulk storage	4 to 18

Although their response time, accuracy, and ease of adjustment may not be as good as those of electrically or pneumatically actuated valves, self-contained steam temperature controls are widely accepted for many applications.

PRESSURE-REDUCING VALVES

Should steam pressure be too high for a specific process, a self-contained pressure-reducing valve (PRV) may be used to reduce this pressure, which will also increase the available latent heat. These valves may be direct-acting or pilot-operated (Figure 27), much like temperature control valves. To maintain set pressure, the downstream pressure must be sensed either through an internal port or an external line.

The amount of pressure drop below the set pressure that causes the valve to react to a load change is called **droop**. As a general rule, pilot-operated valves have less droop than direct-acting types. To properly size these valves, only the mass flow of steam, the inlet pressure, and the required outlet pressure must be known. Valve line size can be determined by consulting manufacturers' capacity charts.

Due to their construction, simplicity, accuracy, and ease of installation and maintenance, these valves have been specified for most steam-reducing stations.

Makeup Water Valves

A pressure-reducing valve is normally provided on a hydronic heating or cooling system to automatically fill the system with domestic or city water to maintain a minimum system pressure. This valve may be referred to as a fill valve, PRV fill valve, or automatic PRV makeup water valve and is usually located at or near the system expansion tank. Local plumbing codes may require a backflow prevention device where the city water connects to the building domestic water system (see the section on Backflow Prevention Devices).

CHECK VALVES

Check valves prevent reversal of flow, controlling the direction of flow rather than stopping or starting flow. Some basic types include swing check, ball check, wafer check, silent check, and stop-check valves. Most check valves are available in screwed and flanged body styles.

Swing check valves have hinge-mounted disks that open and close with flow (Figure 28). The seats are generally made of metal, while the disks may be of metallic or nonmetallic composition materials. Nonmetallic disks are recommended for fluids containing dirt particles or where tighter shutoff is required. The Y-pattern check valve has an access opening to allow cleaning and regrinding in place. Pressure drop through swing check valves is lower than that through lift check valves due to the straight-through design. Weight- or spring-loaded lever arm check valves are available to limit objectionable slamming or chattering when pulsating flows are encountered.

Lift check valves have a body similar in design to a globe or angle valve body with a similar disk seating. The guided valve disk is forced open by the flow and closes when flow reverses. Due to the body design, the pressure drop is higher than that of a swing check valve. Lift check valves are recommended for gas or compressed air or in fluid systems not having critical pressure drops.

Ball check valves are similar to lift checks, except that they use a ball rather than a disk to accomplish closure. Some ball checks are specifically designed for horizontal flow or vertical upflow installation.

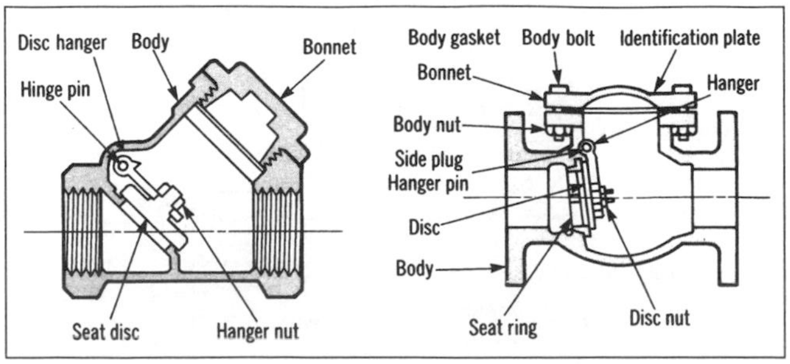

Fig. 28 Swing Check Valves
(Courtesy Grinnell Corp.)

Wafer check valves are designed to fit between pipe flanges similar to butterfly valves and are used in larger piping (4 in. diameter and larger). Wafer check valves have two basic designs: (1) dual spring-loaded flapper, which operates on a hinged center post, and (2) single flapper, which is similar to the swing check valve.

In **silent or spring-loaded check valves**, a spring positively and rapidly closes a guided, floating disk. This valve greatly reduces water hammer, which may occur with slow-closing check valves like the swing check. Silent check valves are recommended for use in pump discharge lines.

STOP-CHECK VALVES

Stop-check valves can operate as both a check valve and a stop valve. The valve stem does not connect to the guided seat plug, allowing the plug to operate as a conventional lift check valve when the stem is in the raised position. Screwing the stem down can limit the valve opening or close the valve. Stop-check valves are used for shutoff service on multiple steam boiler installations, in accordance with the ASME *Boiler and Pressure Vessel Code*, to prevent backflow of steam or condensate from an operating boiler to a shutdown boiler. They are mandatory in some jurisdictions. Local codes should be consulted.

BACKFLOW PREVENTION DEVICES

Backflow prevention devices prevent reverse flow of the supply in a water system. A **vacuum breaker** prevents back siphonage in a nonpressure system, while a **backflow preventer** prevents backflow in a pressurized system (Figure 29).

Selection

Vacuum breakers and backflow preventers should be selected on the basis of the local plumbing codes, the water supply impurities involved, and the type of cross-connection.

Impurities are classified as (1) contaminants (substances that could create a health hazard if introduced into potable water) and (2) pollutants (substances that could create objectionable conditions but not a health hazard).

Cross-connections are classified as nonpressure or pressure connections. In a nonpressure cross-connection, a potable water pipe connects or extends below the overflow or rim of a receptacle at atmospheric pressure. When this type of connection is not protected by a minimum air gap, it should be protected by an appropriate vacuum breaker or an appropriate backflow preventer.

In a pressure cross-connection, a potable water pipe is connected to a closed vessel or a piping system that is above atmospheric pressure and contains a nonpotable fluid. This connection should be protected by an appropriate backflow preventer only. Note that a pressure vacuum breaker should not be used alone with a pressure cross-connection.

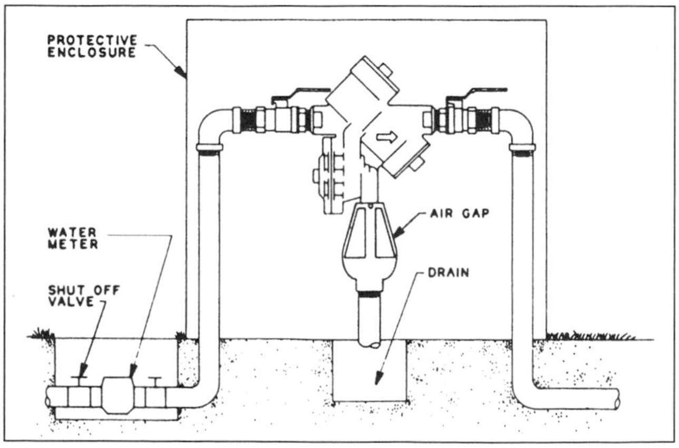

Fig. 29 Backflow Prevention Valve

Vacuum breakers should be corrosion-resistant. Backflow preventers, including accessories, components, and fittings that are 2 in. and smaller, should be made of bronze with threaded connections. Those larger than 2 in. should be made of bronze, galvanized iron, or fused epoxy-coated iron inside and out, with flanged connections. All backflow prevention devices should meet applicable standards of the American National Standards Institute, the Canadian Standards Association, or the required local authorities.

Installation

Vacuum breakers and backflow preventers equipped with atmospheric vents, or with relief openings, should be installed and located to prevent any vent or relief opening from being submerged. They should be installed in the position recommended by the manufacturer.

Backflow preventers may be double check valve (DCV) or reduced pressure zone (RPZ) types. Refer to manufacturers' information for specific application recommendations and code compliance.

STEAM TRAPS

For a description and diagram of these traps, refer to Chapter 10.

REFERENCES

ASME. 1996. Valves—Flanged, threaded, and welding end. ANSI/ASME *Standard* B16.34-1988. American Society of Mechanical Engineers, New York.

ASME. 1998. *Boiler and pressure vessel code.*

ASME. 1998. Rules for construction of power boilers. *Boiler and pressure vessel code*, Section I-98.

ASME. 1998. Rules for construction of heating boilers. *Boiler and pressure vessel code*, Section IV-98.

ASME. 1998. Rules for construction of pressure vessels. *Boiler and pressure vessel code*, Section VIII-98.

Fairbanks Morse Pump Company. 1965. *Hydraulic handbook.* Catalog C and #65-26313. Beloit, WI.

FCI. 1991. Control valve seat leakage. ANSI/FCI *Standard* 70-2-91. Fluid Controls Institute, Cleveland, OH.

ISA. 1985. Flow equations for sizing control valves. ANSI/ISA *Standard* S75.01-85 (R 1995). International Society for Measurement and Control, Research Triangle Park, NC.

ISA. 1996. Control valve capacity test procedures. ANSI/ISA *Standard* S75.02-96.

Jordan, C.H. 1998. Terminology of pressure relief devices. *Heating/Piping/Air Conditioning* 70(9):47-48.

Lyons, J.L. 1982. *Lyons' valve designer's handbook*, pp. 92-93, 209-10. ISBN 0-442-24963-2. Van Nostrand Reinhold.

Ulanski, W. 1991. Valve and actuator technology. ISBN 0-07-019477-7. McGraw-Hill, New York.

BIBLIOGRAPHY

ASHRAE. 1988. Practices for measurement, testing, adjusting and balancing of building heating, ventilation, air-conditioning and refrigeration systems. ANSI/ASHRAE *Standard* 111-1988.

ASHRAE. 1996. The HVAC commissioning process. *Guideline* 1-1996.

ASME. 1994. Pressure relief devices. ANSI/ASME *Performance Test Code* PTC 25-94. With Special Addenda (1998). American Society of Mechanical Engineers, New York.

Baumann, H.D. 1998. *Control valve primer*, 3rd ed. International Society of Measurement and Control (ISA), Research Triangle Park, NC.

Borden, G., Jr. and P.G. Friedmann, eds. *Control valves.* International Society of Measurement and Control (ISA), Research Triangle Park, NC.

CIBSE. 1985. *Automatic controls and their implications for system design. Applications Manual* AM1. The Chartered Institution of Building Services Engineers, London.

CIBSE. 1986. *Guide B*: Installation and equipment data.

Crane. Flow of fluids through valves, fittings and pipe. *Technical Paper* No. 410. Crane Valve Group, Long Beach, CA.

Gupton, G. 1987. *HVAC controls: Operation and maintenance*. ISBN 0-442-23732-4. Van Nostrand Reinhold, New York.

Haines, R.W. 1987. *Control systems for HVAC*. ISBN 0-442-23141-5. Van Nostrand Reinhold, New York.

IEC. 1987. Control valve terminology and general considerations. *Industrial-process control valves,* Part I. IEC *Publication* 60534-1. International Electrotechnical Commission, Geneva, Switzerland.

Matley, J. and Chemical Engineering. 1989. Valves for process control and safety. ISBN-0-07-607010-7. McGraw-Hill, New York.

McQuiston, F.C. and J.D. Parker. 1993. *Heating, ventilating, and air conditioning: Analysis and design*, 4th ed. ISBN 0-471-58107-0. John Wiley and Sons, New York.

Merrick, R.C. 1991. *Valve selection and specification guide*. ISBN 0-442-31870-7.

Miller, R.W. 1983. *Flow measurement engineering handbook.* ISBN 0-07-042045-9. McGraw-Hill, New York.

Zappe, R.W. 1987. *Valve selection handbook*, 2nd ed. ISBN 0-87201-918-7. Gulf Publishing, Houston, TX.

CHAPTER 43

HEAT EXCHANGERS

HEAT EXCHANGERS transfer heat from one fluid to another without the fluids coming in direct contact with each other. Heat transfer occurs in a heat exchanger when a fluid changes from a liquid to a vapor (evaporator), a vapor to a liquid (condenser), or when two fluids transfer heat without a phase change. The transfer of energy is caused by a temperature difference.

In most HVAC&R applications, heat exchangers are selected to transfer either sensible or latent heat. Sensible heat applications involve the transfer of heat from one liquid to another. Latent heat transfer results in a phase change of one of the liquids; transferring heat to a liquid by condensing steam is a common example.

This chapter describes some of the fundamentals, types, components, applications, selection criteria, and installation of heat exchangers. Chapter 3 of the 1997 *ASHRAE Handbook—Fundamentals* covers the subject of heat transfer. Specific applications of heat exchangers are detailed in other chapters of this and other volumes of the Handbook series.

FUNDAMENTALS

When heat is exchanged between two fluids flowing through a heat exchanger, the rate of heat transferred may be calculated using

$$Q = UA\Delta t_m \quad (1)$$

where

U = overall coefficient of heat transfer from fluid to fluid
A = heat transfer area of the heat exchanger associated with U
Δt_m = log mean temperature difference (LMTD)

For a heat exchanger with a constant U, the Δt_m is calculated as

$$\Delta t_m = C_f \frac{(T_1 - t_2) - (T_2 - t_1)}{\ln (T_1 - t_2)/(T_2 - t_1)} \quad (2)$$

where the temperature distribution is as shown in Figure 1 and C_f is a correction factor (less than 1.0) that is applied to heat exchanger configurations that do not follow a true counterflow design.

Figure 1 illustrates a **temperature cross**, where the outlet temperature of the heating fluid is less than the outlet temperature of the fluid being heated ($T_2 < t_2$). A temperature cross can only be obtained with a heat exchanger that has a 100% true counterflow arrangement.

The overall coefficient U is affected by the physical arrangement of the surface area A. For a given load, not all heat exchangers with equal surface areas perform equally. For this reason, load conditions must be defined when selecting a heat exchanger for a specific application.

The load for each fluid stream can be calculated as

$$Q = mc_p(t_{in} - t_{out}) \quad (3)$$

The value of Δt_m is an important factor in heat exchanger selection. If the value Δt_m is high, a relatively small heat exchange surface area is required for a given load. The economic impact is that the heat exchanger must be designed to accommodate the forces and movements associated with large temperature differences. When the **approach temperature** (the difference between T_2 and t_1) is small, Δt_m is also small and a relatively large A is required.

Chapter 3 of the 1997 *ASHRAE Handbook—Fundamentals* describes an alternative method of evaluating heat exchanger performance that involves the exchanger heat transfer effectiveness ε and number of exchanger transfer units (NTU). This method is based on the same assumptions as the logarithmic mean temperature difference method described previously.

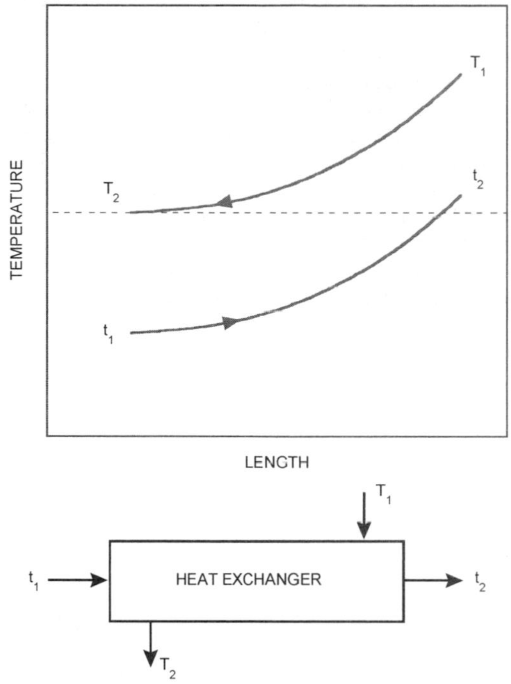

Fig. 1 Temperature Distribution in Counterflow Heat Exchanger

TYPES OF HEAT EXCHANGERS

Most heat exchangers for HVAC&R applications are counterflow shell-and-tube or plate units. While both types physically separate the fluids transferring heat, their construction is very different, and each has unique application and performance qualities.

The preparation of this chapter is assigned to TC 6.1, Hydronic and Steam Equipment and Systems.

Shell-and-Tube Heat Exchangers

Figure 2 illustrates the counterflow path of a shell-and-tube heat exchanger. The fluid at temperature T_1 enters one end of the shell, flows outside the tubes and inside the shell, and exits at the other end at temperature T_2. The other fluid flows inside the tubes, entering one end at temperature t_1 and exiting at the opposite end at temperature t_2.

In a shell-and-tube heat exchanger, a tube bundle assembly is welded or bolted inside a tubular shell. The bundle is constructed of metal tubes mechanically rolled or welded at one (U-tube) or both ends (straight-tube) into tubesheet(s) that function as headers. The shell is usually a length of pipe that has inlet and outlet connections located along one or more of its longitudinal centerlines.

The shell is flanged at one or both open ends to accommodate a head assembly. The tube bundle is positioned between the shell and head assemblies such that the tube wall of the bundle mechanically separates the two flow paths.

The tube bundle is assembled with tube supports, which are held together with tie rods and spacers. Units with liquid on the shell side have baffles for tube supports that direct the flow. Condensers must have baffles that have been notched on the bottom to allow the liquid condensate to flow freely to the exit nozzle.

The head assembly directs the other fluid across the tubesheet(s) into and out of the tube bundle. Head assemblies are designed with pass partitions to isolate sections of the tube bundle such that the fluid must traverse the length of the unit one, two, four or more times before exiting.

One of two types of head assemblies is mechanically attached to the shell. Units with multiple tube-side pass construction have a head with both an inlet and outlet connection bolted at one end with a welded cap (U-tube) or bolted reversing head (straight-tube) at the opposite shell end. Single-pass units have an inlet head attached at one shell end and an outlet head attached at the other end.

Many variations of the shell-and-tube design are available, some of which are described in the following paragraphs.

U-Tube. Figure 3 illustrates a U-tube removable-bundle shell-and-tube heat exchanger. These units are commonly called **converters**. Figures 4 and 5 illustrate modifications of the U-tube design.

Tank heaters are U-tube heat exchangers with the shell replaced by a mounting collar, which is welded to a tank. A hot fluid or steam flows inside the tubes heating the fluid in the tank by natural convection. The tank heater manufacturer should be consulted about optimizing the bundle length. While it is desirable for the bundle to significantly extend into the tank the designer must consider the need for additional bundle support.

Tank suction heaters differ from tank heaters because they have an additional opening that permits the fluid being heated to be pumped across the outside tube wall resulting in improved thermal performance.

Straight-Tube. Figures 6 and 7 illustrate two common designs of straight-tube, shell-and-tube exchangers, one with a fixed and the other with a removable tube bundle assembly.

Some straight-tube, shell-and-tube heat exchangers have a floating head bolted with a gasket to a floating tubesheet or a shell-side expansion joint. This configuration is expensive and is rarely specified in HVAC applications.

Shell-and-Coil. The tubes in this heat exchanger are coiled in a helical configuration around a small core. A spacer is placed

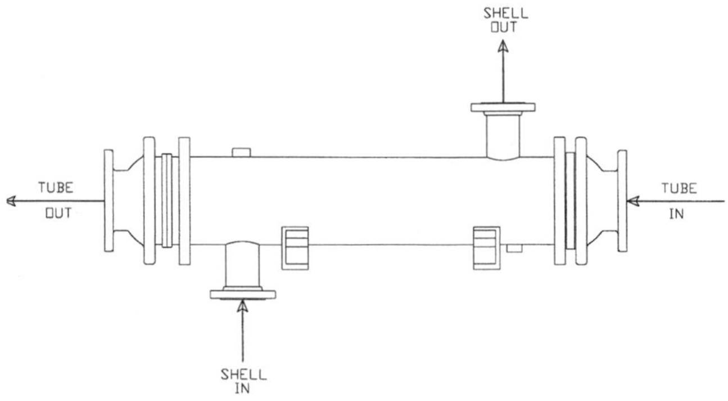

Fig. 2 Counterflow Path in Shell-and-Tube Heat Exchanger

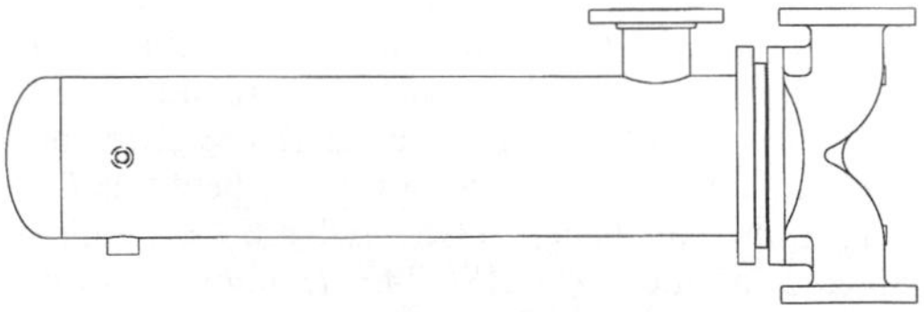

Fig. 3 U-Tube Shell-and-Tube Heat Exchanger with Removable Bundle Assembly and Cast "K" Pattern Flanged Head

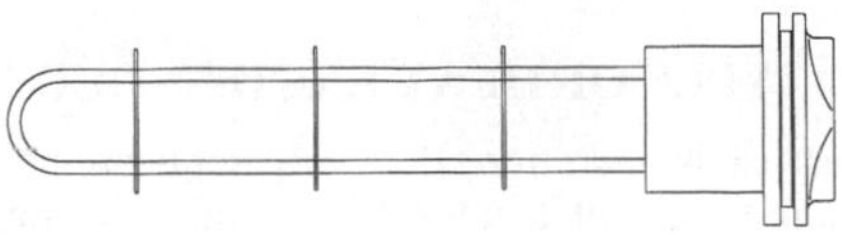

Fig. 4 U-Tube Tank Heater with Removable Bundle Assembly and Cast Bonnet Head

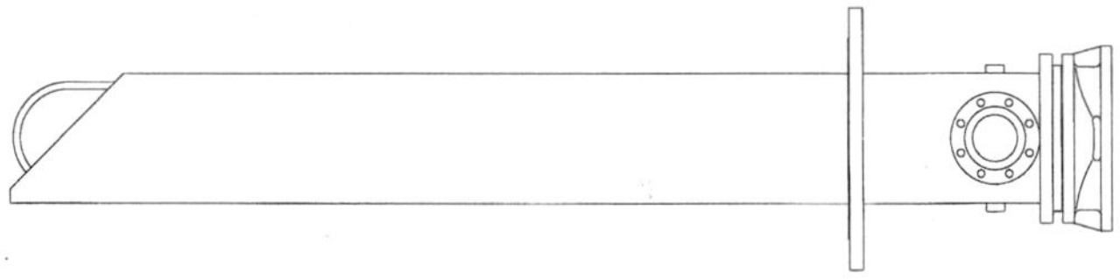

Fig. 5 U-Tube Tank Suction Heater with Removable Bundle Assembly and Cast Flanged Head

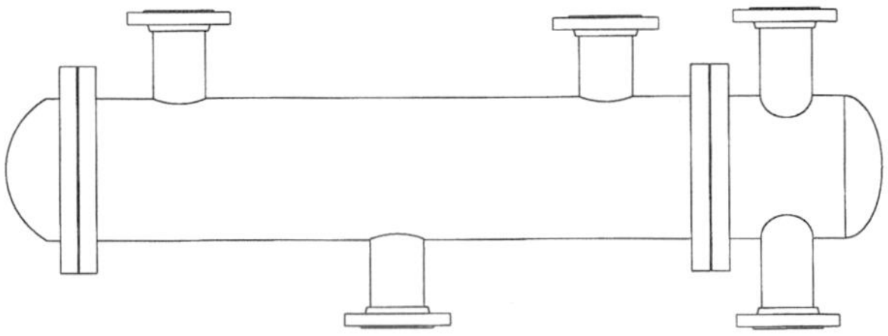

Fig. 6 Straight-Tube Fixed Tubesheet Shell-and-Tube Heat Exchanger with Fabricated Bonnet Heads and Split-Shell Flow Design

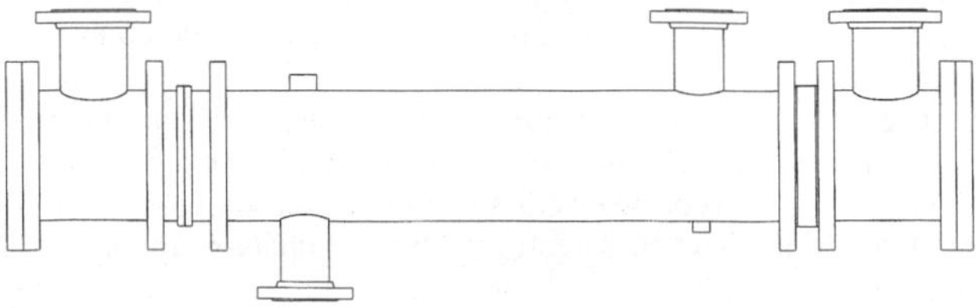

Fig. 7 Straight-Tube Floating Tubesheet Shell-and-Tube Heat Exchanger with Removable Bundle Assembly and Fabricated Channel Heads

between the tube layers. In some designs the tubes have an oval cross section. These heat exchangers are very compact and have a relatively large surface area for their size. Figure 4 in Chapter 37 illustrates a shell-and-coil heat exchanger.

Plate Heat Exchangers

Plate heat exchangers consist of metal plate pairs arranged to provide separate flow paths (channels) for two fluids. Heat transfer occurs across the plate walls. The exchangers have multiple channels in series that are mounted on a frame and clamped together. The rectangular plates have an opening or port at each corner. When assembled the plates are sealed such that the ports provide manifolds to distribute fluids through the separate flow paths. Figure 8 illustrates the flow paths.

The multiple plates, called a **plate pack**, are supported by a carrying bar and contained by pressure plates at each end. The design of the carrying bar and pressure plate permit the units to be opened for maintenance or the addition or removal of plate pairs. The adjoining plates are gasketed, welded, or brazed together.

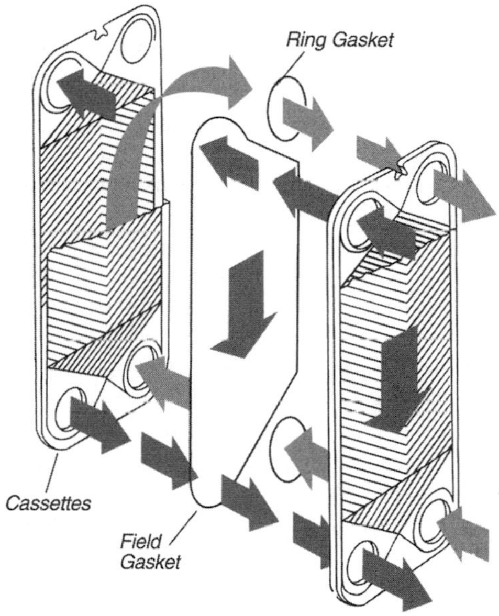

Fig. 8 Flow Path of Gasketed Plate Heat Exchanger

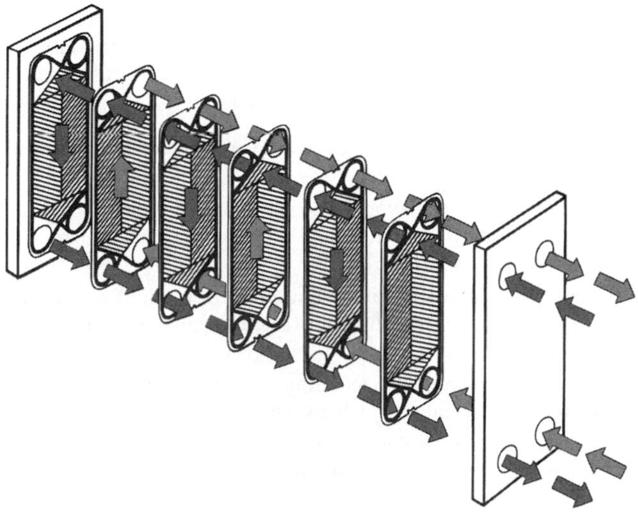

Fig. 9 Flow Path of Welded Plate Heat Exchanger

Gasketed plate heat exchangers are typically limited to design pressures of 300 psig. The type of gasket material used limits the operating temperature. Brazed plate units are designed for pressures up to 450 psig and temperatures up to 500°F.

Gasketed. The most common plate heat exchanger is the gasketed plate unit. Typically, nitrile butyl rubber (NBR) gaskets are used in applications up to 230°F. Ethylene-propylene terpolymer (EPDM) gaskets are available for temperatures up to 320°F. The gaskets are glued or clipped onto the plates. The gasket pattern on each plate creates the counterflow paths illustrated in Figure 8.

Welded. Two plates can be welded together at the edges into an assembly called a **cassette**. This flow channel contains fluids when appropriate gasket material is not available such as for handling corrosive fluids. The channels containing the non-aggressive fluids are sealed with standard gaskets. Welded units can also be used for refrigeration applications. Figure 9 shows the flow path of a welded plate heat exchanger.

Brazed. Brazed plate heat exchangers have neither gaskets nor frames. They consist of plates brazed together with a copper or nickel flux. This design can be a very cost-effective; however, the lack of maintenance access limits their application.

Double-Wall Heat Exchangers

Double-wall heat exchangers have a leakage path that warns of mechanical failure before fluids can be cross contaminated. Both shell-and-tube and plate heat exchangers are available in double-wall construction. The overall thermal performance of a double-wall unit is less than a comparable single-wall design. Double-wall units cost significantly more than single-wall units.

A double-wall U-tube unit (Figure 10) consists of a tube-in-tube design with double tubesheets. The outer tube is rolled into the inner tubesheet. The inner tube is finned or has grooves cut in it. It is rolled into the outer tubesheet to provide a vented leak path between tubesheets to provide a visible indication of a failure of either tube.

A double-wall plate heat exchanger (Figure 11) is constructed by welding two standard channel plates together at the four port openings to form a leak path between the plates should a plate fail.

COMPONENTS

Heat exchangers for HVAC applications should be constructed and labeled according to the applicable ASME *Boiler and Pressure Vessel Code* and rated for 150 psig at 375°F. Heat exchangers operating at elevated temperatures or pressures require special construction.

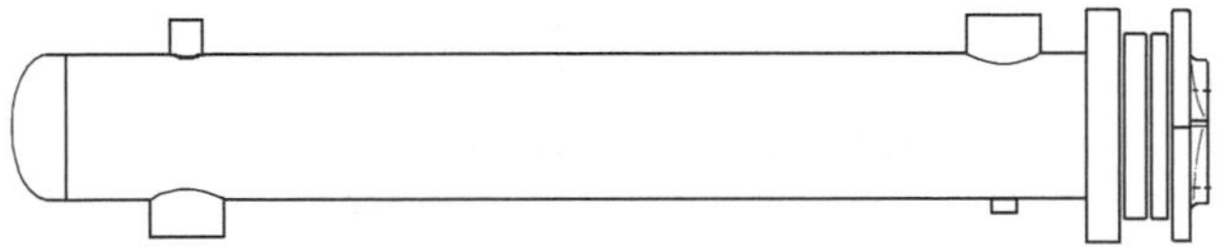

Fig. 10 Double-Wall U-Tube Heat Exchanger

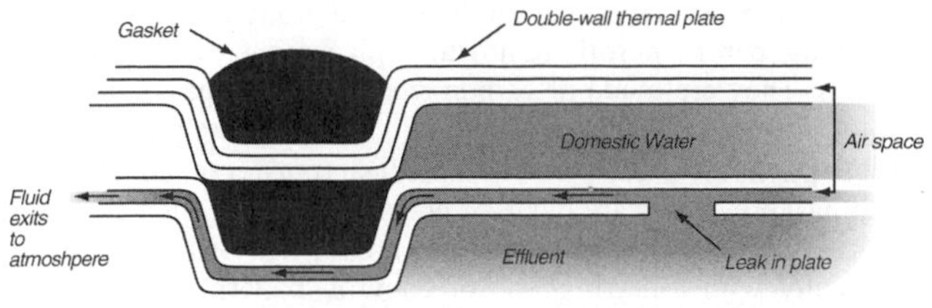

Fig. 11 Double-Wall Plate Heat Exchanger

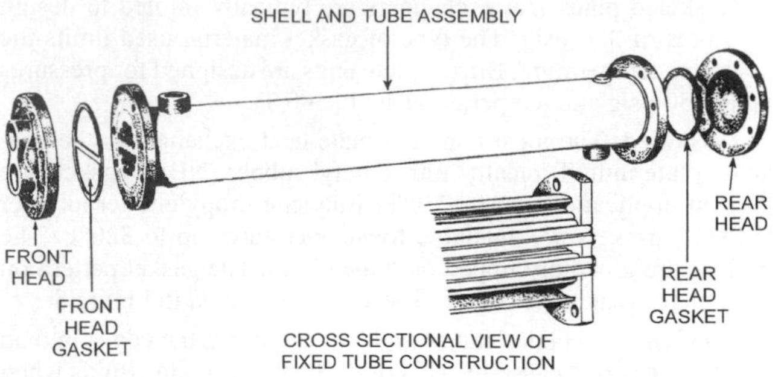

Fig. 12 Exploded View of Straight-Tube Heat Exchanger

Shell-and-Tube Components

Figure 12 illustrates the various components of a shell-and-tube exchanger, which include the following:

- **Shells** are usually made of steel pipe; brass and stainless steel are also used. The inlet and outlet nozzles can be made with standard flange openings in various orientations to suit piping needs. The nozzles are sized to avoid excessive fluid velocity and impingement on the tubes opposite a shell inlet connection.
- **Baffles, tube supports, tie rods, and spacers** are usually made of steel; brass and stainless steel are also available. The number and spacing of baffles controls the velocity and, therefore, a significant portion of the shell-side heat transfer coefficient and pressure drop.
- **Tubes** are usually made of copper; special grades of brass and stainless steel can be specified. The tube diameter, gage, and material affect the heat transfer coefficient and performance.
- **Tubesheets** are available in the same materials as baffles, although the materials do not have to be the same in a given heat exchanger. Tubesheets are drilled for a specific tube layout called **pitch**. The holes are sometimes serrated to improve the tube to tubesheet joint.
- **Heads** are usually cast iron or fabricated steel. Cast brass and cast stainless steel are available in limited sizes. Heads can be custom fabricated in most metals. The inlet and outlet nozzles can be made with standard flange openings. Figures 3, 4, and 5 illustrate three different head configurations that offer different levels of serviceability and ease of installation.

Plate Components

Figure 13 illustrates the various components of a gasketed plate and frame heat exchanger. The materials of construction and purpose of the components are as follows:

- **Fixed frame plates** are usually made of carbon steel. Single pass units have the inlet and outlet connections for both fluids located on the fixed frame plate. Connections are usually NPT or stud port design to accommodate ANSI flanges. NPT connections are carbon steel or stainless steel. Stud port connections can be lined with metallic or rubber-type materials to protect against corrosion.
- **Movable pressure plates** can be moved along the length of the carrying bar to permit removal, replacement, or addition, of plates. They are made of carbon steel. Multiple pass units have some connections located on the movable pressure plate.
- **Plate packs** are made up of multiple heat transfer (channel) plates and gaskets. Plates are made of pressable metals, such as 316 or 304 stainless steel or titanium. They are formed with corrugations typically with a herringbone or chevron pattern. The angle of these patterns affect the thermal performance and pressure drop of a given flow channel.

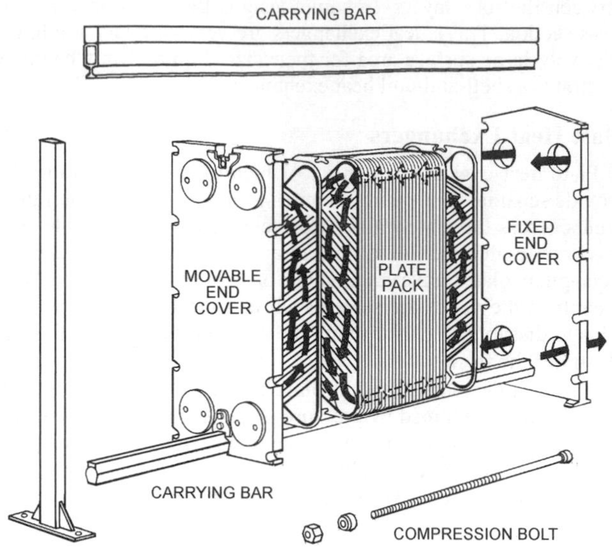

Fig. 13 Components of a Gasketed Plate Heat Exchanger

- **Compression bolts** compress the plate back between the moveable pressure and fixed frame plates. The dimension between the two is critical and is specified by the unit manufacturer for a given plate pack configuration.
- **Carrying and guide bars** support and align the channel plates. The upper bar is called a carrying bar, the lower a guide bar. They are made of stainless steel, aluminum, or carbon steel with zinc chromate finish.
- **Support columns** support the carrying and guide bars on larger plate heat exchangers.
- **Splashguards** are required in the United States by OSHA to enclose exterior channel plate and gasket surfaces. They are usually formed from aluminum.
- **Drip pans** made of stainless steel are often installed under plate heat exchangers to contain leakage on start-up or shut down, gasket failure, or condensation.

APPLICATION

Heat exchangers are used when the primary energy source is available for multiple purposes, uses a different medium, or its temperature or pressure is not in the design limits. Most of the following examples are discussed in other chapters and volumes of the *ASHRAE Handbook*. Heat exchangers are used:

- To condense the steam from a boiler to produce hot water for central water systems
- For service water for potable and non-potable applications, which is often heated by a converter and hot water or steam boilers, with or without a storage tank
- To meet special temperature requirement of portions of a system or to protect against freezing in isolated terminal units (coils) and cooling tower basins
- To isolate two systems operating at different pressures while transferring thermal energy between them
- In energy saving applications such as condensate cooling, vent condensing, boiler blow down, thermal storage, and chiller bypass (free cooling)
- In many refrigeration applications as evaporators, condensers, and liquid coolers

SELECTION CRITERIA

A heat exchanger is often selected by a computer program that optimizes the selection for the given design. A manufacturer should provide detailed selection guidance for both a shell-and-tube and plate exchanger for a given set of conditions.

Thermal/Mechanical Design

Shell-and-tube heat exchangers are designed first to be pressure vessels and second to transfer heat. Plate heat exchangers are designed to transfer heat efficiently within certain temperature and pressure limits.

Thermal Performance. The thermal performance of a heat exchanger is a function of the size and geometry of the heat transfer surface area. Different heat transfer surface materials also effect performance—copper has a higher coefficient of heat transfer than stainless steel.

Flow rates (velocity), viscosity, and thermal conductivity of the fluids are significant factors in determining the overall heat transfer coefficient U. In addition, the fluid to be heated should be on the tube side because the overall U of a shell-and-tube unit is often reduced if the fluid to be heated is on the shell side.

Properly selected shell-and-tube heat exchangers use tube pass options and shell-side baffle spacing to maximize velocity (turbulence) without causing tube erosion. The ability to maximize velocity on each side of a heat exchanger is particularly important when the flow rates of the two fluids are dissimilar. However, fluid velocity in shell-and-tube heat exchanger is limited to avoid tube erosion. U-tube exchangers have lower tube-side velocity limits than straight-tube units due to the thinner tube wall in the U bends.

Shell-and-tube heat exchangers can be constructed for split-shell flow design (Figure 6) to accommodate unusual conditions. Such units have one shell inlet connection and two outlet connections.

Plate heat exchangers typically have U-factors 3 to 5 times higher than shell-and-tube heat exchangers. The high turbulence created by the corrugated plate design increases convection and increases the U-factor. The plate design achieves a large temperature cross at a 2°F approach because of the counterflow fluid path and high U-factor.

Thermal Stress. Heat exchangers must accommodate the thermal stresses associated with large temperature differences. U-tube units offer superior economic performance over straight-tube units with removable tube bundles under extreme conditions. Units with fixed tubesheets do not handle large temperature differences well.

Gasketed plate units have a **differential pressure/temperature limitation** (DPTL), which is the maximum difference in operating pressure of the two fluids at a specific temperature. A unit rated for 300 psig at 260°F might have a DPTL of 220 psig at 200°F.

Pressure Drop. Fluid velocity and normal limitations on tube length tend to result in relatively low pressure drops in shell-and-tube heat exchangers. Plate units tend to have larger pressure drops unless the velocity is limited. Often a pressure drop limitation rather than a thermal performance requirement determines the surface area in a plate unit.

Fouling. Often excess surface area is specified to allow for scale accumulation on heat transfer surfaces without a significant reduction of performance. This fouling factor or allowance is applied when sizing the unit. Fouling allowance is better specified as a percentage of excess area rather than as a resistance to heat transfer.

Shell-and-tube exchangers with properly sized tubes can handle suspended solids better than plate units with narrow flow channels. The high fluid velocity and turbulence in plate exchangers make them less susceptible to fouling.

The addition of surface area (tube length) to a shell-and-tube exchanger does not affect fluid velocity, and, therefore, has little impact on thermal performance. This characteristic makes a fouling allowance practical. This is not the case in plate units. The number of parallel flow channels determines velocity in a plate heat exchanger. This means that as plate pairs are added to meet a load (heat transfer surface area) requirement, the number of channels increases and results in decreased fluid velocity. This lower velocity reduces performance and requires additional plate pairs, which further reduces performance.

Cost

On applications with temperature crosses and close approaches, plate heat exchangers usually have the lowest initial cost. Wide temperature approaches often favor shell-and-tube units. If the application requires stainless steel, the plate unit may be more economical.

Serviceability

Shell-and-tube heat exchangers have different degrees of serviceability. The type of header used facilitates access to the inside of the tubes. The heads illustrated in Figures 3, 6, and 7 can be easily removed without special pipe arrangements. The tube bundles in all of the shell-and-tube units illustrated, except the fixed-tube sheet unit (Figure 6), can be replaced after the head is removed if they are piped with proper clearance.

The diameter and configuration of the tubes are significant factors in whether the inside of the tubes of straight-tube units can be mechanically cleaned. Figure 7 shows a type of head that permits cleaning or inspection of the inside of tubes after the channel cover is removed.

Plate heat exchangers can be serviced by sliding the movable pressure plate back along the carrying bars. Individual plates can be removed for cleaning, regasketing, or replacement. Plate pairs can be added for additional capacity. Complete replacement plate packs can be installed.

Space Requirements

Cost-effective and efficient shell-and-tube heat exchangers have small diameter, long tubes. This configuration often challenges the designer when allocating the space required for service and maintenance. For this reason, many shell-and-tube selections have large diameters and short lengths. While this selection performs well, it often costs more than a smaller diameter unit with equal surface area. Caution should be taken to provide adequate maintenance clearance around heat exchangers. In the case of shell-and-tube units, space should be left clear so the tube bundle can be removed.

Plate heat exchangers tend to provide the most compact design in terms of surface area for a given space.

Steam

Most HVAC applications using steam are designed with shell-and-tube units. Plate heat exchangers are used in specialized industrial and food processes with steam.

INSTALLATION

Control. Heat exchangers are usually controlled by a valve with a temperature sensor. The sensor is placed in the flow stream of the fluid to be heated or cooled. The valve regulates the flow on the other side of the heat exchanger to achieve the sensor set-point temperature. Chapter 42 discusses control valves.

Piping. Heat exchangers should be piped such that air is easily vented. Pipes must be able to be drained and accessible for service.

Pressure Relief. Safety pressure relief valves should be installed on both sides between the heat exchanger and shut off valves to guard against damage from thermal expansion when the unit is not in service as well as to protect against overpressurization.

Flow Path. The intended flow path of each fluid on both sides of a heat exchanger design should be followed. Failure to connect to the correct inlet and outlet connections may reduce performance.

Condensate Removal. Heat exchangers that condense steam require special installation. Proper removal of condensate is particularly important. Inadequate drainage of condensate can result in a significant loss of capacity and mechanical failure.

The installation of a vacuum breaker aids in draining condensate particularly when modulating steam control valves are used. Properly sized and installed steam traps are critical. Chapter 10 discusses steam traps and condensate removal.

Insulation. Heat exchangers are often insulated. Chapter 22 of the 1997 *ASHRAE Handbook—Fundamentals* has further information on insulation.

CHAPTER 44

AIR-TO-AIR ENERGY RECOVERY

APPLICATIONS

AIR-TO-AIR energy recovery systems may be categorized according to their application as (1) process-to-process, (2) process-to-comfort, or (3) comfort-to-comfort. Typical air-to-air energy recovery applications are listed in Table 1.

Table 1 Applications for Air-to-Air Energy Recovery

Method	Typical Application
Process-to-process and Process-to-comfort	Dryers
	Ovens
	Flue stacks
	Burners
	Furnaces
	Incinerators
	Paint exhaust
	Welding
Comfort-to-comfort	Swimming pools
	Locker rooms
	Residential
	Smoking exhaust
	Operating rooms
	Nursing homes
	Animal ventilation
	Plant ventilation
	General exhaust

Process-to-Process

In process-to-process applications, heat is captured from the process exhaust stream and transferred to the process supply airstream. Equipment is available to handle process exhaust temperatures as high as 1600°F.

Process-to-process recovery devices generally recover only sensible heat and do not transfer latent heat (humidity), as moisture transfer is usually detrimental to the process. Process-to-process applications usually recover the maximum amount of energy. In cases involving condensable gases, less recovery may be desired in order to prevent condensation and possible corrosion.

Process-to-Comfort

In process-to-comfort applications, waste heat captured from a process exhaust heats the building makeup air during winter. Typical applications include foundries, strip coating plants, can plants, plating operations, pulp and paper plants, and other processing areas with heated process exhaust and large makeup air volume requirements.

Although full recovery is desired in process-to-process applications, recovery for process-to-comfort applications must be modulated during warm weather to prevent overheating of the makeup air. During summer, no recovery is required. Because energy is saved only in the winter and recovery is modulated during moderate weather, process-to-comfort applications save less energy over a year than do process-to-process applications.

Process-to-comfort recovery devices generally recover sensible heat only and do not transfer moisture between the airstreams.

Comfort-to-Comfort

In comfort-to-comfort applications, the heat recovery device lowers the enthalpy of the building supply air during warm weather and raises it during cold weather by transferring energy between the ventilation air supply and exhaust airstreams.

In addition to commercial and industrial energy recovery equipment, small-scale packaged ventilators with built-in heat recovery components known as heat recovery ventilators (HRVs) or energy recovery ventilators (ERVs) are available for residential and small-scale commercial applications.

Air-to-air energy recovery devices available for comfort-to-comfort applications may be sensible heat devices (i.e., transferring sensible energy only) or total heat devices (i.e., transferring both sensible energy and moisture). These devices are discussed further in the section on Technical Considerations.

ECONOMIC CONSIDERATIONS

Air-to-air energy recovery systems are used in new or retrofit applications. These systems should be designed for the maximum cost benefit or least life-cycle cost (LCC) expressed either over the service life or on an annual basis and with an acceptable payback period.

The annualized system owning, operating, and maintenance costs are discussed in the section on Uniform Annualized Costs Method in Chapter 35 of the 1999 *ASHRAE Handbook—Applications*. Although the capital cost and interest term in this method implies a simple value, it is in fact a complex function of the future value of money as well as all the design variables in the energy/heat exchanger. These variables include the mass of each material used, the cost of forming these materials into an energy/heat exchanger with a high effectiveness, the cost of auxiliary equipment and controls, and the cost of installation.

The **operating energy cost** for energy recovery systems involves functions integrated over time that include such variables as flow rate, pressure drop, fan efficiency, energy cost, and energy recovery rate. The calculations are quite complex because the air heating and/or cooling loads are, for a range of supply temperatures, time-dependent in most buildings. Time-of-use schedules for buildings often impose different ventilation rates for each hour of the day. The electrical utility charges often vary with the time of day, amount of energy used, and peak power load. For building ventilation air-heating applications, the peak heat recovery rate usually occurs at the outdoor supply temperature at which frosting control throttling

The preparation of this chapter is assigned to TC 5.5, Air-to-Air Energy Recovery.

must be imposed. Thus, unlike other HVAC designs, heat recovery systems should have a design temperature not at the ambient winter design temperature but rather at the temperature for maximum heat recovery rate.

The exchanger overall effectiveness ε should be high (see Table 2 for typical values); however, a high value of ε implies a high capital cost, even when the exchanger is designed to minimize the amount of materials used. Energy costs for fans and pumps are usually very important and accumulate operating cost even when the energy recovery system must be throttled back. For building ventilation, throttling may be required a large fraction of the time. Thus, the overall LCC minimization problem for optimal design may involve 10 or more independent design variables as well as a number of specified constraints and operating conditions [see, for example, Besant and Johnson (1995)].

In addition, comfort-to-comfort energy recovery systems often operate with much smaller temperature differences than do most auxiliary air-heating and -cooling heat exchangers. These small temperature differences imply the need for more accurate energy transfer models if the maximum cost benefit or lowest LCC is to be realized.

The **payback period** PP is best computed once the annualized costs have been evaluated. It is usually defined as

$$PP = \frac{\text{Capital cost and interest}}{\text{Annual operating energy cost saved}}$$

$$= \frac{(C_{s,\,init} - \text{ITC})}{C_e(1 - T_{inc})}\text{CRF}(i'',n) \qquad (1)$$

where

$C_{s,init}$ = initial system cost
ITC = investment tax credit for energy-efficient improvements
C_e = cost of energy to operate the system for one period
T_{inc} = net income tax rate where rates are based on the last dollar earned (i.e., the marginal rates) = (local + state + federal rate) − (federal rate) (local + state rate)
CRF = capital recovery factor
i'' = effective discount rate adjusted for energy inflation
n = total number of periods under analysis

The inverse of this term is usually called the return on investment (ROI). Well-designed energy recovery systems normally have a PP of less than 5 years; values less than 3 years are often realized.

Other economic factors include the following.

System Installed Cost. Initial installed HVAC system cost is often lower when using air-to-air energy recovery devices because mechanical refrigeration and fuel-fired heating equipment can be reduced in size. Thus, a more efficient HVAC system may also have a lower installed total HVAC cost. The installed cost of heat recovery systems becomes lower per unit of flow as the amount of outdoor air used for ventilation is increased.

Life-Cycle Cost. Air-to-air energy recovery cost benefits are best evaluated considering all capital, installation, operating, and energy-saving costs over the duration of the equipment life under its normal operating conditions in terms of a single cost relationship—the life-cycle cost. As a rule, neither the most efficient nor the least expensive energy recovery device will be most economical. The optimization of the life-cycle cost for maximum net savings may involve a large number of design variables, necessitating careful cost estimates and the use of an accurate model of the recovery system with all its design variables.

Energy Costs. The absolute cost of energy and the relative costs of various energy forms are major economic factors. High energy costs favor high levels of energy recovery. In regions where electrical costs are high relative to fuel prices, heat recovery devices with low pressure drops are preferable.

Other Conservation Options. Energy recovery should be evaluated against other cost-saving opportunities, including reducing or eliminating the primary source of waste energy through process modification.

Amount of Recoverable Energy. Economies of scale favor large installations. Equipment is commercially available for air-to-air energy recovery applications using 50 cfm and above. Although using equipment with higher effectiveness results in more recovered energy, equipment cost and space requirements also increase with effectiveness.

Grade of Exhaust Energy. High-grade (i.e., high-temperature) exhaust energy is generally more economical to recover than low-grade energy. Energy recovery is most economical for large temperature differences between the waste energy source and destination.

Coincidence and Duration of Waste Heat Supply and Demand. Energy recovery is most economical when the supply is coincident with the demand and both are relatively constant throughout the year. Thermal storage may be used to store energy if supply and demand are not coincident, but this adds cost and complexity to the system.

Proximity of Supply to Demand. Applications with a large central energy source and a nearby waste energy use are more favorable than applications with several scattered waste energy sources and uses.

Operating Environment. High operating temperatures or the presence of corrosives, condensable gases, and particulates in either airstream results in higher equipment and maintenance costs. Increased equipment costs result from the use of corrosion- or temperature-resistant materials, and maintenance costs are incurred by an increase in the frequency of equipment repair and washdown and additional air filtration requirements.

Effect on Pollution Control Systems. Removing process heat may reduce the cost of pollution control systems by (1) allowing less expensive filter bags to be used, (2) improving the efficiency of electronic precipitators, or (3) condensing out contaminant vapors, thus reducing the load on downstream pollution control systems. In some applications, recovered condensable gases may be returned to the process for reuse.

Effect on Heating and Cooling Equipment. Heat recovery equipment may reduce the size requirements for primary utility equipment such as boilers, chillers, and burners, as well as the size of piping and electrical services to them. Larger fans and fan motors (and hence fan energy) are generally required to overcome increased static pressure loss caused by the energy recovery devices. Auxiliary heaters may be required for frost control.

Effect on Humidifying or Dehumidifying Equipment. Selecting total energy recovery equipment results in the transfer of moisture from the airstream with the greater humidity ratio to the airstream with the lesser humidity ratio. This is desirable in many situations because humidification costs are reduced in cold weather and dehumidification loads are reduced in warm weather.

TECHNICAL CONSIDERATIONS

Ideal Air-to-Air Energy Exchange

An ideal air-to-air energy exchanger performs the following functions:

- Allows temperature-driven heat transfer between the participating airstreams
- Allows partial-pressure-driven moisture transfer between the two streams
- Totally blocks cross-stream transfer of air, other gases (in particular, pollutants), biological contaminants, and particulates

Heat transfer is widely recognized as an important vehicle for energy recovery from airstreams that carry waste heat. The role of moisture transfer as an energy recovery process is less well known and merits explanation.

Consider an air-to-air energy exchanger operating in a hot, humid environment; in view of the uncomfortable climate, the indoor air is conditioned. Many local ordinances require a specified number of outdoor air changes per hour. If the energy exchanger is a heat exchanger but not a moisture exchanger, it facilitates the cooling of outdoor ventilation air as it passes through the exchanger en route to the indoor space. Heat flows from the incoming outdoor air to the outgoing (and cooler) exhaust air drawn from the indoor conditioned space. This heat transfer process does very little to mitigate the high humidity that is carried into the indoor space by the outdoor ventilation air. A substantial amount of power will be required to dehumidify that air to reduce its moisture content to a level acceptable for comfort.

On the other hand, if the energy exchanger can transfer both heat and moisture, the highly humid outdoor air will transfer moisture to the less humid indoor air as the two streams pass through the exchanger. The lowered humidity of the entering ventilation air will allow a substantial savings of energy.

Forms of Energy Transfer in an Air-to-Air Energy Exchanger

Two generic types of air-to-air energy exchangers may be considered for service as heat and moisture exchangers. One of these is the **rotating energy wheel regenerator**, frequently called a **heat wheel** or an **energy wheel**. The other is the **permeable-walled flat-plate recuperator**. Although the end results are similar, the transport mechanisms in these two categories of devices are very different.

In a rotating energy wheel regenerator, a desiccant film coating the surfaces of the wheel absorbs moisture as the wheel passes through the more humid airstream. Once absorbed, the moisture rides the moving wheel until it reaches the less humid airstream, where it is desorbed from the film into the airstream.

In a moisture-transferring fixed-wall recuperator, the walls are made of a material that is permeable to water vapor. The moisture passes through the walls when there is a difference in the magnitude of the vapor pressures between the two airstreams.

Rotating energy wheel regenerators and permeable-walled flat-plate recuperators are prone to cross-stream leakage of air and other gases. Advances in microporous film technology have demonstrated that membranes can be synthesized which will nearly totally block the transfer of air and other gases, while providing relatively free passage for the water vapor. In light of this, the long-range commercial potential for the permeable-walled recuperator appears favorable.

Rate of Energy Transfer

The rate of energy transfer depends on the intrinsic characteristics of the energy exchanger and on the operating conditions. Intrinsic properties include the geometry of the exchanger (parallel flow/counterflow/cross-flow, number of passes, fins), the thermal conductivity of the walls separating the streams, and the permeability of the walls to the passage of various gases. As in a conventional heat exchanger, heat transfer between the airstreams is driven by cross-stream dry-bulb temperature differences. Energy is also transported piggyback-style between the streams by cross-stream mass transfer, which may include air, gases, and water vapor. In another mode of heat transfer, water vapor condenses into liquid in one of the two airstreams of the exchanger. The condensation process liberates the latent heat of condensation, which is transferred to the other stream as sensible heat. This two-step process encompassing release of latent energy followed by its subsequent transfer in the form of sensible heat is commonly called **latent heat transfer**.

Latent energy transfer between the airstreams occurs only when moisture is transferred from one stream to another without condensation, thereby maintaining the latent heat of condensation intact. Once the moisture has crossed from one airstream to the other, it may either remain in the vapor state or condense in the second stream, depending on the temperature of that stream.

Rotating energy wheel regenerators and permeable-walled flat-plate recuperators are used because of their moisture recovery function. The passage of air or other gases (e.g., pollutants) across the exchanger is a negative consequence. Cross-stream mass transfer may occur through leakage even when such transfer is unintended. The undetected presence of unintended leakage may alter the performance of the exchanger from its design value.

Conduction heat transfer differs in principle from the mass transfer. Certain phenomena relating to heat/mass exchanger performance must be recognized:

- The effectiveness for moisture transfer is likely not equal to the effectiveness for heat transfer.
- The total energy effectiveness is likely not equal to either the sensible effectiveness or the latent effectiveness.
- The total energy transfer may not equal the sum of the sensible and latent heat transfers.

Net total energy transfer and effectiveness need careful examination when the direction of sensible (temperature-driven) transfer is opposite to that of latent (moisture or water vapor) transfer. The following example illustrates the potential for misinterpretation of total energy transfer:

Example 1. Assume that the inlet supply air has a high temperature with low humidity ratio, the inlet exhaust air has a low temperature with high humidity ratio, and the supply and exhaust mass airflows are equal. Assume also that the energy exchanger has been tested to ASHRAE *Standard* 84, which rated the sensible heat transfer effectiveness at 50% and the latent (water-vapor) transfer effectiveness at 50%. Plotted on a psychrometric chart, the water vapor transfer is equal to the heat transfer but in the opposite direction. Mass airflow and enthalpy transfer balances are equal.

Simple examination of the plots shows that total energy transfer is zero because the change in enthalpy due to heat transfer is equal in magnitude but opposite in direction to that due to water vapor transfer. The net total energy effectiveness is shown to be zero. The direction and rate of heat and moisture energy transfer must be considered as separate, independent entities.

Cross-stream mass transfer can be driven by two independent types of pressure differences: (1) cross-stream total pressure differences and (2) cross-stream partial pressure differences. Air mass movement is driven primarily by total air pressure differences and is minimized by the bulk-flow resistance characteristics of the exchanger wall barrier. Moisture mass transfer, on the other hand, is driven by a combination of total pressure differences and vapor partial pressure differences. Cross-stream moisture transfer is maximized by the bulk-flow resistance and permeability of the exchanger wall barrier. High bulk-flow resistance retards viscous, bulk airflow and minimizes the effect of total pressure differences on air mass transfer. Total-pressure-driven mass transfer and partial-pressure-driven mass transfer may be mutually aiding or mutually opposing.

Heat, moisture, and air transfer rates are partially independent of and separate from one another. Heat is always driven cross-stream in a direction from higher to lower temperature. Air is predominantly driven cross-stream in a direction from higher to lower total pressure. Water vapor mass is driven cross-stream in an amount and direction influenced by several variables. Design and construction characteristics of the exchanger greatly influence whether the moisture mass is (1) transferred predominantly by riding on the cross-stream air mass or (2) separated from the air mass by a permeable desiccant, selective microporous membrane, or other moisture-separating device. The net effect of total and vapor pressure differences cross-stream will influence the net intensity and direction of moisture exchange.

Standard laboratory rating tests and predictive computer models give exchanger performance values individually for (1) heat transfer, (2) moisture transfer, (3) cross-stream air transfer, (4) average exhaust mass airflow, and (5) supply mass airflow leaving the exchanger. Effectiveness ratios for heat transfer and mass water vapor transfer must be separately determined by running rating tests for a given exchanger in a laboratory that is staffed and instrumented to accurately measure all performance parameters required in ASHRAE *Standard* 84 and ARI *Standard* 1060. It may be very difficult to adhere to any standard when field tests are made.

Performance Rating

ASHRAE *Standard* 84, Method of Testing Air-to-Air Heat Exchangers, (1) establishes a uniform method of testing for obtaining performance data; (2) specifies the data required, calculations to be used, and reporting procedures for testing the performance; and (3) specifies the types of test equipment for performing such tests. *Standard* 84 test methods minimize air leakage by specifying that all tests be conducted at zero pressure differential between supply outlet and exhaust inlet. Fan placement or effect on performance is not included in the standard. *Standard* 84 does not specify performance criteria for product certification or identify laboratories capable of performing these tests.

ARI *Standard* 1060, Rating Air-to-Air Energy Recovery Ventilation Equipment, is an industry-established standard for rating the performance of air-to-air heat/energy exchangers for use in energy recovery ventilation equipment. It establishes definitions, requirements for marking and nameplate data, and conformance conditions intended for the industry, including manufacturers, engineers, installers, contractors, and users. Air-to-air heat/energy exchangers for use in energy recovery ventilation equipment must be tested in accordance with ASHRAE *Standard* 84, except where modified by ARI *Standard* 1060. Standard temperature and humidity conditions at which equipment tests are to be conducted are specified. Published ratings must be reported for mass flow rate; pressure drop; and sensible, latent, and total effectiveness at the standard conditions specified. ARI *Certification Program* 1060 was established to verify ratings published by manufacturers.

In the field, air-to-air heat/energy exchanger performance may depart significantly from that measured under the idealized conditions in a test lab. System design and configuration will influence the amount and direction of (1) air leakage between airstreams or between the exchanger and its surroundings; (2) water vapor transfer between airstreams as a result of carryover, crossover, or leakage; (3) pressurization caused by fan placement; and (4) heat and vapor exchange to the surroundings.

Balanced mass airflows as required in the ASHRAE and ARI standard test methods are rarely achieved in field operation in air-handling systems. Fans are constant volume devices usually designed to run at a preset rpm. Significantly more mass airflow will be transported in cold (winter) conditions than in hot (summer) conditions. In packaged air handlers, it is impractical to expect unit fans to operate year-round at zero air leakage, zero pressure differential, and balanced mass airflows, as required by these standard test methods. The application designer should evaluate the probable actual (versus ideal laboratory standard) performance of the exchanger relative to placement of exhaust and supply fans in the equipment for all expected weather conditions.

The effectiveness of air-to-air heat exchangers is commonly measured in terms of

- Sensible energy (heat) transfer: dry-bulb temperature
- Latent energy (water vapor or moisture) transfer: humidity ratio
- Total energy (heat and moisture) transfer: enthalpy

ASHRAE *Standard* 84 defines effectiveness as

$$\varepsilon = \frac{\text{Actual transfer (of moisture or energy)}}{\text{Maximum possible transfer between airstreams}}$$

Referring to Figure 1,

$$\varepsilon = \frac{w_s(x_2 - x_1)}{w_{min}(x_3 - x_1)} = \frac{w_e(x_3 - x_4)}{w_{min}(x_3 - x_1)} \tag{2}$$

where

ε = moisture (or water vapor mass), sensible, or total effectiveness
x = humidity ratio W, dry-bulb temperature t, or enthalpy h at locations indicated in Figure 1
w_s = supply air mass flow
w_e = exhaust air mass flow
w_{min} = the smaller of w_s and w_e

The leaving supply air condition is

$$x_2 = x_1 + \varepsilon\left(\frac{w_{min}}{w_s}\right)(x_1 - x_3) \tag{3}$$

and the leaving exhaust air condition is:

$$x_4 = x_3 - \varepsilon\left(\frac{w_{min}}{w_e}\right)(x_1 - x_3) \tag{4}$$

Equations (2), (3), and (4) assume steady-state conditions; no heat transfer between the heat exchanger and its surroundings; and no gains from cross-leakage, fans, or frost control devices. Furthermore, condensation or frosting does not occur or is negligible. Those assumptions are generally true for larger commercial applications but not for heat recovery ventilators (HRVs). CAN/CSA-C439, Standard Methods of Test for Rating the Performance of Heat-Recovery Ventilators, is used to rate small (under 400 cfm) packaged ventilators with heat recovery. The rating terms used in CAN/CSA-C439 for HRVs are **energy recovery efficiency** (i.e., the actual energy transfer efficiency) and **apparent sensible effectiveness** (i.e., a measure of the temperature rise of the supply airstream, including that resulting from conversion of fan motor energy to heat, air leakage, and heat transfer across cases).

A number of variables can affect these performance factors, whether the device is designed to transfer total energy or just sensible heat. These variables include (1) water vapor partial pressure differences, (2) heat transfer area, (3) air velocity through the heat exchangers, (4) airflow arrangement or geometric configuration, (5) supply and exhaust air mass flow rates, and (6) method of frost

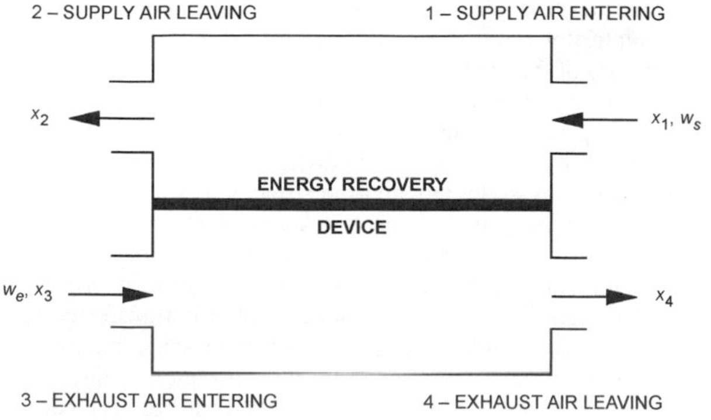

Fig. 1 Airstream Numbering Convention

control. The effect of frost control method on seasonal performance is discussed in Phillips et al. (1989a), and sensible versus latent heat recovery for residential comfort-to-comfort applications is addressed in Barringer and McGugan (1989b).

Sensible Versus Total Recovery

Sensible heat recovery devices do not transfer moisture. As water vapor changes state from vapor to liquid with no change in temperature, heat that is released into the same airstream is called the **latent heat of condensation**. Thus, air temperature is maintained in the airstream as a result of condensation. Because the airstream temperature is higher than it would have been without condensation, heat transfer is increased. Thus, latent heat is transferred only when the warmer airstream is cooled below its dew point and condensation occurs. In this case, latent heat is transferred, but moisture is not. Sensible heat exchangers should be used where latent heat transfer without moisture transfer is desired. Examples include indirect evaporative coolers, precool reheaters, and air dryers.

Total energy recovery devices (e.g., desiccant-coated rotary and permeable membrane energy exchangers) transfer both moisture and heat between the airstreams. The sensible (i.e., temperature) and latent (i.e., moisture) transfer effectivenesses for a given exchanger are different. Total heat exchange is desirable (1) in hot, humid climates where moisture transfer from supply air to the exhaust airstream reduces air-conditioning loads and (2) in cold, dry climates where moisture transfer from exhaust air to the supply airstream reduces humidification requirements. Barringer and McGugan (1989b) address sensible versus latent heat recovery for residential comfort-to-comfort applications.

Fouling

Fouling refers to an accumulation of dust or condensates on heat exchanger surfaces. By increasing the resistance to airflow and generally decreasing heat transfer coefficients, fouling reduces heat exchanger performance. The increased resistance increases fan power requirements and may reduce airflow.

Pressure drop across the heat exchanger core can be used as an indication of fouling and, with experience, may be used to establish cleaning schedules. Heat exchanger surfaces must be kept clean if system performance is to be maximized.

Corrosion

Process exhaust frequently contains substances requiring corrosion-resistant construction materials. If it is not known which materials are most corrosion-resistant for an application, the user and/or designer should examine on-site ductwork, review literature, and contact equipment suppliers prior to selecting materials. A corrosion study of heat exchanger construction materials in the proposed operating environment may be warranted if the installation costs are high and the environment is corrosive. Experimental procedures for such studies are described in an ASHRAE symposium (ASHRAE 1982). Often contaminants not directly related to the process are present in the exhaust airstream (e.g., welding fumes or paint carryover from adjacent processes).

Moderate corrosion generally occurs over time, roughening metal surfaces and increasing their heat transfer coefficients. Severe corrosion reduces overall heat transfer and can cause cross-leakage between airstreams due to perforation or mechanical failure.

Cross-Leakage

Cross-leakage, cross-contamination, or mixing between supply and exhaust airstreams may occur in air-to-air heat exchangers. It may be a significant problem if exhaust gases are toxic or odorous. Cross-leakage varies with heat exchanger type and design, airstream static pressure differences, and the physical condition of the heat exchanger (see Table 2).

Condensation and Freeze-Up

Condensation, ice formation, and/or frosting may occur on heat exchange surfaces. If entrance and exit effects are neglected, four distinct air/moisture regimes may occur as the warm airstream is cooled from its inlet condition to its outlet condition. First there is a dry region with no condensate. Once the warm airstream is cooled below its dew point, there is a condensing region, which results in wetting of the heat exchange surfaces. If the heat exchange surfaces fall below freezing, the condensation will freeze. Finally, if the warm airstream temperature is reduced below 32°F, sublimation causes frost to form. The locations of these regions and rates of condensation and frosting depend on the duration of frosting conditions; the airflow rates; the inlet air temperature and humidity; the heat exchanger core temperature; the heat exchanger effectiveness; the geometry, configuration, and orientation; and the heat transfer coefficients.

Sensible heat exchangers, which are ideally suited to applications in which heat transfer is desired but humidity transfer is not (swimming pools, kitchens, drying ovens), can benefit from the latent heat released by the exhaust gas when condensation occurs. One pound of moisture condensed transfers about 1050 Btu to the incoming air at room temperature.

Condensation increases the heat transfer rate and thus the sensible effectiveness; it can also increase pressure drops significantly in heat exchangers with narrow airflow passage spacings. Frosting fouls the heat exchanger surfaces, which initially improves energy transfer but subsequently restricts the exhaust airflow, which in turn reduces the energy transfer rate. In extreme cases, the exhaust airflow (and supply, in the case of heat wheels) can become blocked. Defrosting a fully blocked heat exchanger requires that the unit be shut down for an extended period. As water cools and forms ice, it expands, which may seriously damage the heat exchanger core.

For frosting or icing to occur, an airstream must be cooled below 32°F and below its dew point. Total heat exchangers transfer moisture from the airstream with the higher moisture content (usually the warmer airstream) to the airstream with the lower moisture content. As a result, frosting or icing occurs at lower supply air temperatures in enthalpy exchangers than in sensible heat exchangers. In enthalpy heat exchangers, which use chemical absorbents, condensation may cause the absorbents to deliquesce, permanently damaging the heat exchanger.

For these reasons, some form of freeze control must be incorporated into heat exchangers that are expected to operate under freezing conditions. Frosting and icing can be prevented by preheating the supply air; reducing the heat exchanger effectiveness (e.g., reducing heat wheel speed, tilting heat pipes, or bypassing part of the supply air around the heat exchanger). Alternatively, the heat exchanger may be periodically defrosted.

The performance of several freeze control strategies is discussed in ASHRAE 543-RP (Phillips et al. 1989a, 1989b). ASHRAE 544-RP (Barringer and McGugan 1989a, 1989b) discusses the performance of enthalpy heat exchangers. Many effective defrost strategies have been developed for residential air-to-air heat exchangers. These strategies may also be applied to commercial installations. Phillips et al. (1989c) describe frost control strategies and their impact on energy performance in various climates.

For sensible heat exchangers, system design should include drains to collect and dispose of condensation, which occurs in the warm airstream. In comfort-to-comfort applications, condensation may occur in the supply side in summer and in the exhaust side in winter.

Pressure Drop

The pressure drop for each airstream through a heat exchanger depends on many factors, including exchanger design, mass flow rate, temperature, moisture, and inlet and outlet air connections. The exchanger pressure drop must be overcome by fans or blowers.

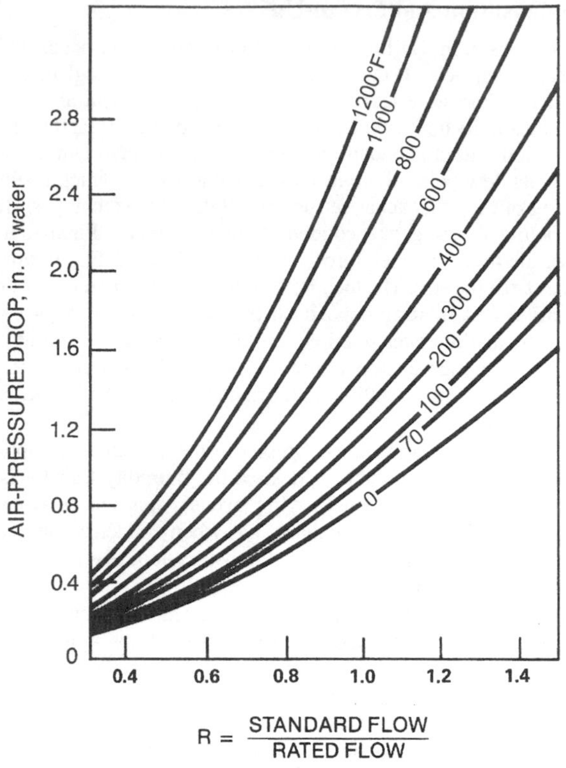

Fig. 2 Pressure Drop Versus Flow at Various Temperatures for Typical Plate Exchanger

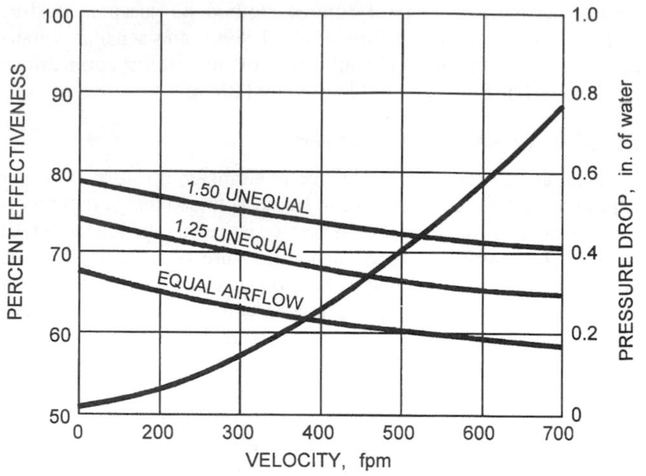

Fig. 3 Typical Flat-Plate Performance

When all other parameters are constant for a given exchanger, pressure drop increases

- With gas temperature (Figure 2)
- As plate contamination or fouling increases (e.g., due to condensation, frosting, or dust accumulation)
- If pressure differentials between airstreams deform airflow passages (plate-type exchangers)
- As barometric pressure increases
- With airflow velocity (Figure 3)

Face Velocity

Design face velocities for heat exchangers are based primarily on allowable pressure drop rather than on recovery performance. Recovery performance (effectiveness) decreases with increasing velocity, but the decrease in effectiveness is not as rapid as the increase in pressure drop (Figure 3). Low face velocities give lower pressure drop, higher effectiveness, and lower operating costs, but require larger units with higher capital costs and greater installation space.

Airflow Arrangements

Heat exchanger effectiveness depends to a great extent on the airflow direction and pattern of the supply and exhaust airstreams. Parallel-flow exchangers (Figure 4A), in which both airstreams move along heat exchange surfaces in the same direction, have a theoretical maximum effectiveness of 50%. Counterflow exchangers (Figure 4B), in which airstreams move along heat exchange surfaces in opposite directions, can have effectivenesses approaching 100%, but units designed for typical applications have a much lower effectiveness.

In practice, construction limitations favor designs in which the two airstreams interface in transverse (or cross-flow) configuration over much of the heat exchange surface (Figure 4C and Figure 4D). The normal range of effectiveness for cross-flow heat exchangers is 50 to 70% (Figure 4C) and 60 to 85% (Figure 4D).

Maintenance

The method used to clean a heat exchanger depends on the transfer medium or mechanism used in the heat exchanger and on the nature of the material to be removed. Grease buildup from kitchen exhaust, for example, is often removed with an automatic waterwash system. Other kinds of dirt may be removed by vacuuming, blowing compressed air through the passages, steam cleaning, manual spray cleaning, soaking the units in soapy water or solvents, or using soot blowers. The cleaning method should be determined at the design stage so that a compatible heat exchanger can be selected.

Cleaning frequency depends on the quality of the exhaust airstream. Residential and commercial HVAC systems generally require only infrequent cleaning; industrial systems, usually more. Equipment suppliers should be contacted regarding the specific cleaning and maintenance requirements of the systems being considered.

Filtration

Filters should be placed in both the supply and exhaust airstreams to reduce fouling and thus the frequency of cleaning. Exhaust filters are especially important if the contaminants are sticky or greasy or if particulates can plug airflow passages in the exchanger. Supply filters eliminate insects, leaves, and other foreign materials, thus protecting both the heat exchanger and the air-conditioning equipment. Snow or frost can block the air supply filter and cause severe problems. Supply air filters should be removed in very cold weather to avoid frosting blockage of the inlet. Steps to ensure a continuous flow of supply air should be incorporated into the design.

Controls

Heat exchanger controls may function to control frost formation or to regulate the amount of energy transferred between airstreams at specified operating conditions. For example, ventilation systems designed to maintain specific indoor conditions at extreme outdoor design conditions may require energy recovery modulation to prevent overheating ventilation supply air during cool to moderate weather or to prevent overhumidification. Modulation may be achieved by tilting heat pipes, changing rotational speeds of (or stopping) heat wheels, or bypassing part of one airstream around the

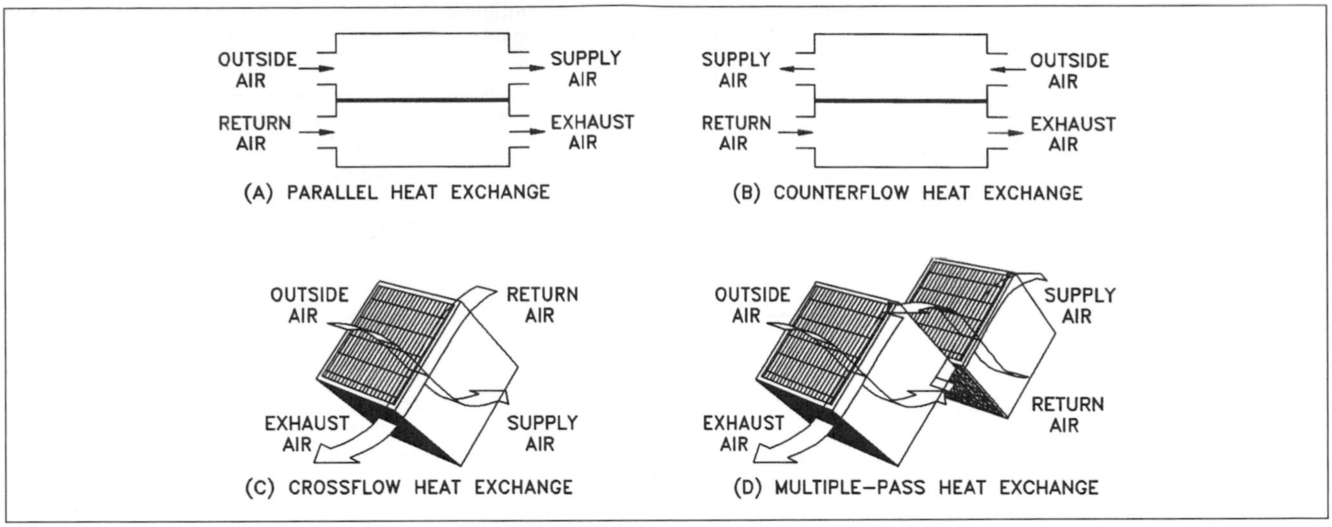

Fig. 4 Heat Exchanger Airflow Configurations

heat exchanger using face and bypass dampers (i.e., changing the supply-to-exhaust mass airflow ratio).

Indirect Evaporative Air Cooling

Exhaust air passing through a water spray absorbs water vapor until it becomes nearly saturated. As the water evaporates, it absorbs sensible energy from the air, lowering its temperature. This process follows a constant wet-bulb line on a psychrometric chart. Thus, the enthalpy of the airstream remains nearly constant, moisture content increases, and dry-bulb temperature decreases. The evaporatively cooled exhaust air can then be used to cool the supply air through an air-to-air heat exchanger. The heat exchanger may be applied either for year-round energy recovery or exclusively for its evaporative cooling benefits.

Indirect evaporative cooling has been applied with heat pipe heat exchangers, two-phase thermosiphon loops, and flat-plate heat exchangers for summer cooling (Scofield and Taylor 1986; Mathur 1990a, 1990b, 1992, 1993). Exhaust air or a scavenging airstream is cooled by passing it through a water spray, a wet filter, or other wetted media, resulting in a greater overall temperature difference between the supply and exhaust or scavenging airstreams and hence more heat transfer. Energy recovery is further enhanced by improved heat transfer coefficients due to wetted exhaust-side heat transfer surfaces. No moisture is added to the supply airstream, and there are no auxiliary energy inputs other than fan and water pumping power. The energy efficiency ratio (EER) of indirect evaporative cooling systems tends to be high, typically ranging from 30 to 70, depending on available wet-bulb depression.

Because less cooling is required with evaporative cooling, energy consumption and peak demand load are both reduced, yielding lower energy bills. Overall mechanical refrigeration system requirements are reduced, so that smaller mechanical refrigeration systems can be selected. In some cases, the mechanical system may be eliminated. Chapter 19 of this volume and Chapter 50 of the 1999 *ASHRAE Handbook—Applications* have further information on evaporative cooling.

Precooling Air Reheater

In some applications, such as ventilation in hot, humid climates, supply air is cooled below the desired delivery temperature to condense moisture and reduce the humidity. Using this overcooled supply air to precool outdoor air reduces the air-conditioning load, which allows refrigeration equipment to be downsized and eliminates the need to reheat the supply air with purchased energy.

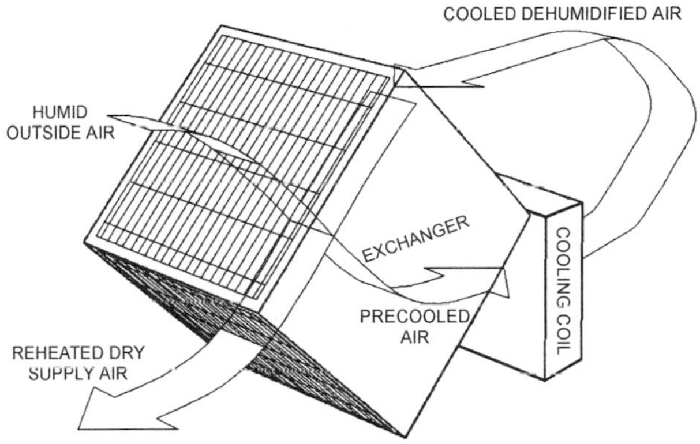

Fig. 5 Precooling Air Reheater

In this three-step process, illustrated in Figure 5, outdoor air passes through an air-to-air heat exchanger, where it is precooled by supply air leaving the cooling coil. It is then further cooled and dehumidified in the cooling coil. After leaving the cooling coil, it passes through the other side of the air-to-air heat exchanger, where it is reheated by the incoming supply air.

Fixed-plate, heat pipe, and rotary heat exchangers can be used to reheat precooled supply air. At part-load operating conditions, the amount of heat transferred from precooling to reheating may require modulation. This can be done using the heat rate control schemes noted in Table 2.

ENERGY RECOVERY CALCULATIONS

The rate of energy transfer to or from an airstream depends on the rate and direction of the heat transfer and on the rate and direction of the water vapor (moisture) transfer. Under customary design conditions, heat and water vapor transfer will be in the same direction, but the rate of heat transfer will not be the same as the rate of energy transfer by the cross-stream flow of water vapor. This is because the driving potentials for heat and mass transfer are different, as are the respective wall resistances for the two types of transport. Both transfer rates are dependent on exchanger construction characteristics. Equation (5) is used to determine the rate of energy transfer when sensible (temperature) and latent (moisture) energy transfer occurs, while Equation (6) is used for sensible-only energy transfer.

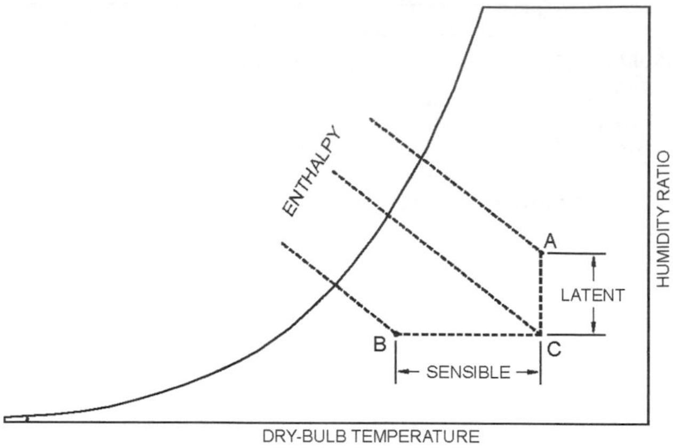

Fig. 6 Maximum Sensible and Latent Heat from Process A-B

$$q_{total} = 60Q\rho(h_{in} - h_{out}) \quad (5)$$

$$q_{sensible} = 60Q\rho c_p(t_{in} - t_{out}) \quad (6)$$

where

q_{total} = $q_{sensible}$ + q_{latent} = total energy transfer, Btu/h
$q_{sensible}$ = sensible heat transfer, Btu/h
Q = airflow rate, cfm
ρ = air density, lb/ft^3
c_p = specific heat of air = 0.24 Btu/lb·°F
t_{in} = dry-bulb temperature of air entering exchanger, °F
t_{out} = dry-bulb temperature of air leaving exchanger, °F
h_{in} = enthalpy of air entering heat exchanger, Btu/lb
h_{out} = enthalpy of air leaving heat exchanger, Btu/lb

The following general procedure may be used to determine energy recovered in air-to-air energy recovery applications.

Step 1. Calculate theoretical maximum moisture and energy transfer rates $w_{m,max}$ and q_{max}.

The airstream with the lower mass flow w_{min} limits heat and moisture transfer. Some designers specify and prefer working in scfm. In order to correctly calculate moisture or energy transfer rates, the designer must determine mass flow rates. For this reason, the designer must know whether airflow rates are quoted for the entry conditions specified or as scfm. If necessary, convert flow rates to mass flow rates (e.g., scfm or lb/min) to determine which airstream has the minimum mass.

If only sensible energy transfer occurs, the theoretical maximum rate of heat transfer q_{max}, using the airstream numbering convention from Figure 1, is $60\rho c_p Q_{min}(t_3 - t_1)$. If latent energy transfer occurs, the theoretical maximum energy transfer q_{max} is $60\rho Q_{min}(h_3 - h_1)$. The maximum moisture transfer rate $w_{m,max}$ is also implied by Equation (2) and is $w_{min}(W_3 - W_1)$, where W_3 and W_1 are the humidity ratios at state 3 and state 1.

The split between latent and sensible energy (enthalpy) potential flux can be determined by plotting the airstream conditions on a psychrometric chart as shown in Figure 6. Maximum sensible heat transfer is represented by a horizontal line drawn between the two dry-bulb temperatures, and maximum latent energy transfer is represented by the vertical line.

Step 2. Establish the moisture, sensible, and total effectivenesses ε_m, ε_s, and ε_t.

Each of these ratios is obtained from manufacturers' product data using input conditions and airflows for both airstreams. The effectiveness for equal airflows depends on (1) exchanger construction, including configuration, heat transfer material, moisture transfer properties, transfer surface area, airflow path, distance between heat transfer surfaces, and overall size; and (2) inlet conditions for both airstreams, including pressures, velocities, temperatures, and humidities. In applications with unequal airflow rates, the enthalpy change will be higher for the airstream with the lesser mass flow.

Each effectiveness should be verified by the manufacturer for the air inlet conditions. If the exchanger selected does not perform at the specified effectiveness, its impact on the project should be considered. The manufacturer should answer the following questions as well:

- Does the published sensible effectiveness result from tests with condensation in the exhaust airstream?
- Are the published effectivenesses for sensible and total energy transfer different or are they assumed to be equal?
- Are published airflow rates based on standard or actual temperature and barometric pressure at the fan?
- Has the exchanger performance been verified by an independent laboratory to meet ASHRAE *Standard* 84 criteria at the specified airflows and inlet conditions?

The pressure drop for each airstream should be determined from the manufacturer's data for the design conditions to calculate fan requirements.

Step 3. Calculate actual moisture and energy (sensible and total) transfer.

The amount of energy transferred is the product of the effectiveness ε for the airstream with the lesser mass flow rate and the theoretical maximum heat transfer determined in Step 1 using Equation (2):

$$w_m = \varepsilon_m w_{m,\,max} \quad (7a)$$

$$q_{actual} = \varepsilon q_{max} \quad (7b)$$

where ε and q may be for sensible or total energy transfer.

Step 4. Calculate leaving air conditions for each airstream.

If an enthalpy or moisture-permeable heat exchanger is used, moisture (and its inherent latent energy) is transferred between airstreams. If a sensible-only heat exchanger is used, and the warmer airstream is cooled below its dew point, the resulting condensed moisture transfers additional energy. When condensation occurs, latent heat is released, maintaining that airstream at a higher temperature than if condensation had not occurred. This higher air temperature (potential flux) increases the heat transfer to the other airstream. The assumption of no flows other than at states 1, 2, 3, and 4 in Equation (2) is not valid. In spite of this, the same definitions for sensible and total effectiveness are widely used because the energy flow in the condensate is relatively small in most applications. (Freezing and frosting are unsteady conditions that should be avoided unless a defrost cycle is included.) Equation (5) must be used to calculate the leaving air condition for airstreams in which inherent latent energy transfer occurs. Equation (6) may be used for an airstream if only sensible energy transfer is involved.

Step 5. Check the energy transfer balance between airstreams.

Total energy transferred from one airstream should equal total heat transferred to the other airstream. Calculate and compare the energy transferred to or from each airstream. Differences between these energy flows are usually due to measurement errors.

Step 6. Plot entering and leaving conditions on psychrometric chart.

Examine the plotted information for each airstream to verify that the performance is reasonable and accurate.

EXAMPLES

Example 2. Sensible Heat Recovery in Winter

In this example, 10,000 scfm of exhaust air at 75°F and 10% rh preheats an equal mass of outdoor air at 0°F and 60% rh (ρ = 0.087 lb/ft^3) using an air-to-air heat exchanger with moderate effectiveness (60%). Airflows are specified in scfm, so an air density of 0.075 lb/ft^3 for both airstreams is appropriate. Determine the leaving supply air conditions, calculate energy recovered, and check heat exchange balance.

1. Calculate the theoretical maximum heat transfer.

   An examination of the two inlet conditions plotted on a psychrometric chart (Figure 7) indicates that, because the exhaust air has low relative humidity, latent energy transfer will not occur. Using Equation (6),

$$q_{max} = 60 \times 0.075 \times 0.24 \times 10{,}000(75 - 0) = 810{,}000 \text{ Btu/h}$$

2. Establish the sensible effectiveness.

   From manufacturer's literature and certified performance test data, effectiveness is determined to be 60% at the design conditions.

3. Calculate actual heat transfer at given conditions.

$$q_{actual} = 0.6 \times 810{,}000 = 486{,}000 \text{ Btu/h recovered}$$

4. Calculate leaving air conditions.

   Because no moisture or latent energy transfer will occur, Equation (6) is used.

   a. Leaving supply air temperature t_2 is

$$t_2 = 0 + \left(\frac{486{,}000}{60 \times 0.24 \times 0.075 \times 10{,}000}\right) = 45°\text{F}$$

   b. Leaving exhaust air temperature t_4 is

$$t_4 = 75 + \left(\frac{-486{,}000}{60 \times 0.24 \times 0.075 \times 10{,}000}\right) = 30°\text{F}$$

5. Check performance.

$$q_s = 60 \times 0.24 \times 0.075 \times 10{,}000(45 - 0) = 486{,}000 \text{ Btu/h saved}$$

$$q_e = 60 \times 0.24 \times 0.075 \times 10{,}000(75 - 30) = 486{,}000 \text{ Btu/h recovered}$$

6. Plot conditions on psychrometric chart to confirm that no moisture exchange occurred (Figure 7).

Example 3. Sensible Heat Recovery in Winter with Water Vapor Condensation

In this application, 10,000 cfm of exhaust air at 75°F and 28% rh (ρ = 0.075 lb/ft^3) is used to preheat 9000 cfm of outdoor air at 14°F and 50% rh (ρ = 0.084 lb/ft^3) using a heat exchanger with an effectiveness of 70%. Determine the leaving supply air conditions, calculate energy recovered, and check energy exchange balance.

The supply airstream has a lower airflow rate than the exhaust airstream, and so it may appear that the supply airstream will limit heat transfer. However, determination of mass flow rates for the given entry conditions shows that the mass flow rate of the supply airstream (756 lb/min) is slightly greater than that of the exhaust airstream (750 lb/min), so exhaust is the limiting airstream. Nevertheless, because the mass difference is negligible, it is convenient to use supply air volume as the limiting airstream.

1. Calculate the theoretical maximum sensible heat transfer.

   The limiting airstream, the supply airstream, will be preheated in the heat exchanger, so it is not subject to condensation. Therefore, Equation (6) is used:

$$q_{max} = 60 \times 0.24 \times 0.084 \times 9000(75 - 14) = 664{,}000 \text{ Btu/h}$$

2. Select sensible effectiveness.

   From manufacturer's literature and performance test data, effectiveness is determined to be 70% at the design conditions.

3. Calculate actual heat transfer at design conditions using Equation (7b):

$$q_{actual} = 0.7 \times 664{,}000 = 465{,}000 \text{ Btu/h recovered}$$

4. Calculate leaving air conditions.

   a. Leaving supply air temperature is calculated using Equation (6).

$$t_2 = 14 + \left(\frac{465{,}000}{60 \times 0.24 \times 0.084 \times 9{,}000}\right) = 56.7°\text{F}$$

   b. Condensation will occur on the exhaust side, so the leaving exhaust air enthalpy is determined using Equation (5). The entering exhaust air enthalpy, determined for the wet-bulb temperature (55.4°F) using a psychrometric chart, is 23.5 Btu/lb.

$$h_4 = 23.5 + \left(\frac{-465{,}000}{60 \times 0.075 \times 10{,}000}\right) = 13.2 \text{ Btu/lb}$$

   The wet-bulb temperature corresponding to an enthalpy of 13.2 Btu/lb is 35°F.

5. Check performance.

$$q_s = 60 \times 0.24 \times 0.084 \times 9000(56.7 - 14) = 465{,}000 \text{ Btu/h saved}$$

$$q_e = 60 \times 0.075 \times 10{,}000(23.5 - 13.2) = 465{,}000 \text{ Btu/h recovered}$$

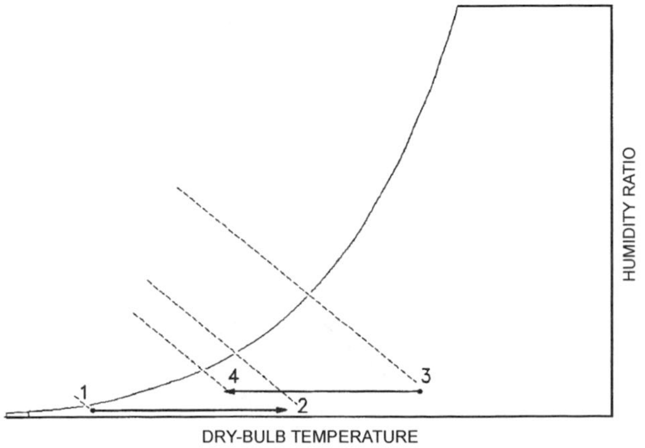

Fig. 7 Sensible Heat Recovery in Winter (Example 2)

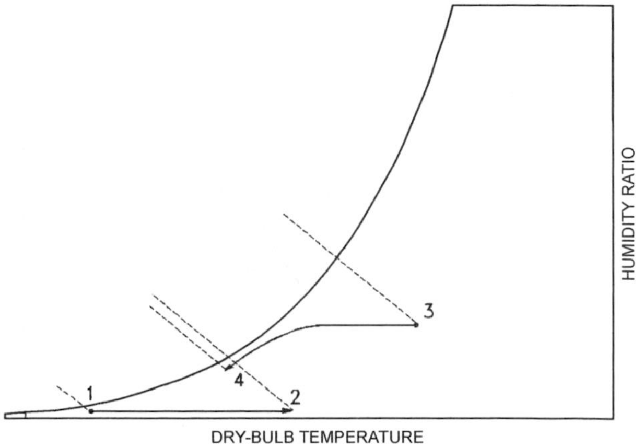

Fig. 8 Sensible Heat Recovery in Winter with Condensate (Example 3)

6. Plot conditions on psychrometric chart (Figure 8). Note that moisture will condense in the exhaust side of the heat exchanger.

Example 4. Total Heat Recovery in Summer

In this application, 10,000 cfm of exhaust air at 75°F and 63°F wet bulb ($\rho = 0.075$ lb/ft^3) is used to precool 8000 cfm of supply outdoor air at 95°F and 80°F wet bulb ($\rho = 0.071$ lb/ft^3) using a hygroscopic total energy exchanger. Determine the leaving supply air conditions, calculate energy recovered, and check energy exchange balance.

1. Calculate the theoretical maximum heat transfer.

2. The supply airstream is the lesser or limiting airstream for energy and moisture transfer. The hygroscopic energy exchanger will transfer moisture so the theoretical maximum heat transfer is calculated using Equation (5). Determine entering airstream enthalpies from psychrometric chart.

$$\text{Supply inlet } (95°\text{F db}, 80°\text{F wb}) \quad h = 43.7 \text{ Btu/lb}$$
$$\text{Exhaust inlet } (75°\text{F db}, 63°\text{F wb}) \quad h = 28.6 \text{ Btu/lb}$$

The intersection of sensible and latent heat transfer, shown as point A in Figure 9 (69.5°F wb, h = 33.8 Btu/lb).

$$q_{max} = 60 \times 0.071 \times 8000(43.7 - 28.6) = 515{,}000 \text{ Btu/h}$$
$$q_{latent} = 60 \times 0.071 \times 8000(43.7 - 33.8) = 337{,}000 \text{ Btu/h}$$
$$q_{sensible} = 60 \times 0.071 \times 8000(33.8 - 28.6) = 177{,}000 \text{ Btu/h}$$

3. Determine supply sensible and total effectiveness.

From manufacturer's selection data for the design conditions, the following effectiveness ratios are provided:

$$\varepsilon_{sensible} = 70\% \qquad \varepsilon_{total} = 56.7\%$$

4. Calculate energy transfer at design conditions.

$$q = 0.567 \times 515{,}000 = 292{,}000 \text{ Btu/h total recovered}$$
$$q_{sen} = -0.7 \times 177{,}000 = \underline{124{,}000 \text{ Btu/h sensible recovered}}$$
$$q_{lat} = 168{,}000 \text{ Btu/h latent recovered}$$

5. Calculate leaving air conditions.

Equation (6) is used to determine dry-bulb leaving conditions, and Equation (5) is used to determine leaving wet-bulb conditions.

a. Supply air conditions

$$t_2 = 95 + \left(\frac{-124{,}000}{60 \times 0.24 \times 0.071 \times 8000}\right) = 79.8°\text{F}$$
$$h_2 = 43.7 + \left(\frac{-292{,}000}{60 \times 0.071 \times 8000}\right) = 35.1 \text{ Btu/lb}$$

From the psychrometric chart, supply wet bulb = 71.1°F.

b. Exhaust air conditions

$$t_4 = 75 + \left(\frac{124{,}000}{60 \times 0.075 \times 0.24 \times 10{,}000}\right) = 86.5°\text{F}$$
$$h_4 = 28.6 + \left(\frac{292{,}000}{60 \times 0.075 \times 10{,}000}\right) = 35.1 \text{ Btu/lb}$$

From the psychrometric chart, exhaust wet bulb = 71.0°F.

6. Check total performance.

$$60 \times 0.071 \times 8000(43.7 - 35.1) = 292{,}000 \text{ Btu/h saved}$$
$$60 \times 0.075 \times 10{,}000(35.1 - 28.6) = 292{,}000 \text{ Btu/h recovered}$$

7. Plot conditions on psychrometric chart (Figure 9).

Example 5. Indirect Evaporative Cooling Recovery

In this application, 30,000 cfm of room air at 85°F and 63°F wb ($\rho = 0.075$ lb/ft^3) is used to precool 30,000 cfm of supply outdoor air at 101°F and 68°F wb ($\rho = 0.070$ lb/ft^3) using an aluminum fixed-plate heat exchanger and indirect evaporative cooling. The evaporative cooler increases the exhaust air to 90% rh before it enters the heat exchanger. Determine the leaving supply air conditions, calculate energy recovered, and check energy exchange balance.

First, determine the exhaust air condition entering the exchanger (i.e., after it is adiabatically cooled). Air at 85°F db, 63°F wb cools to 65°F db, 63°F wb as shown by the process line from point A to point 3 in Figure 10. In this problem the volumetric flows are equal, but the mass flows are not.

1. Calculate the theoretical maximum heat transfer.

Based on a preliminary assessment, the supply air is not expected to cool below its wet-bulb temperature of 68°F. Thus, use sensible heat Equation (6).

$$q_{max} = 60 \times 0.24 \times 0.070 \times 30{,}000(101 - 65) = 1{,}089{,}000 \text{ Btu/h}$$

2. Establish the sensible effectiveness.

From manufacturer's exchanger selection data for indirect evaporative coolers, an effectiveness of 78% is determined to be appropriate.

3. Calculate actual energy transfer at the design conditions.

$$q = 0.78 \times 1{,}089{,}000 = 849{,}000 \text{ Btu/h recovered}$$

4. Calculate leaving air conditions.

Because no moisture or latent heat is transferred, Equation (6) is used.

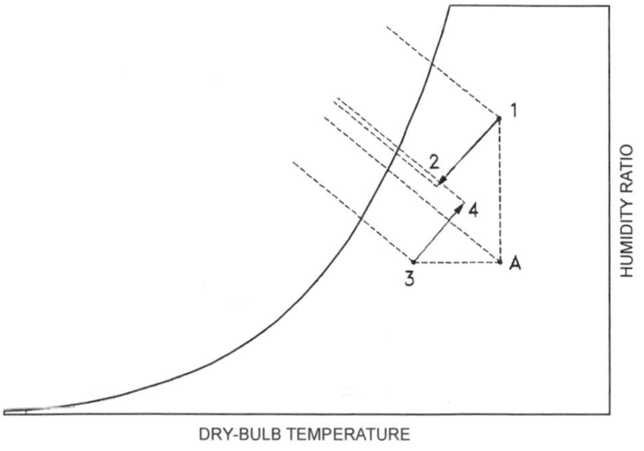

Fig. 9 Total Heat Recovery in Summer (Example 4)

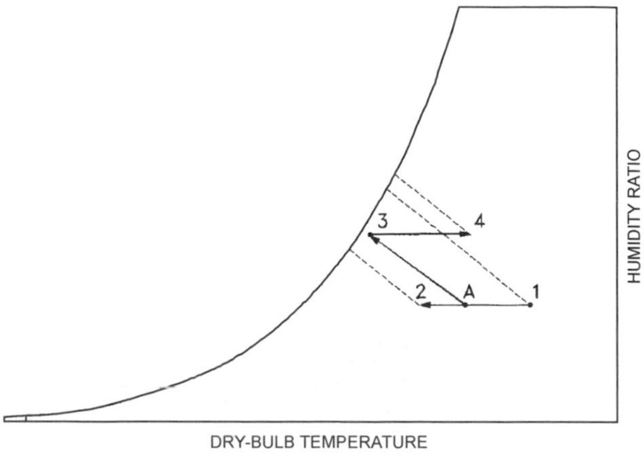

Fig. 10 Indirect Evaporative Cooling Recovery (Example 5)

a. Leaving supply air temperature is

$$t_2 = 101 + \left(\frac{-849{,}000}{60 \times 0.070 \times 0.24 \times 30{,}000}\right) = 72.9°F$$

b. Leaving exhaust air temperature is

$$t_4 = 65 + \left(\frac{849{,}000}{60 \times 0.24 \times 0.75 \times 30{,}000}\right) = 91.2°F$$

5. Check performance.

$$60 \times 0.070 \times 0.24 \times 30{,}000(101 - 72.9) = 849{,}000 \text{ Btu/h saved}$$

$$60 \times 0.075 \times 0.24 \times 30{,}000(91.2 - 65.0) = 849{,}000 \text{ Btu/h recovered}$$

6. Plot conditions on psychrometric chart (Figure 10), and confirm that no latent exchange occurred.

Example 6. Precooling Air Reheater Dehumidifier

In this application, 3300 cfm of outdoor supply air at 95°F and 80°F wb (ρ = 0.071 lb/ft^3) is used to reheat 3300 cfm of the same air leaving a cooling coil (exhaust) at 52.2°F and 52.1°F wb using a sensible heat exchanger as a precooling air reheater. The reheated air is to be between 75 and 78°F. In this application, the warm airstream is outdoor air and the cold airstream is the same air after it leaves the cooling coil. Determine the leaving precooled and reheated air conditions, calculate energy recovered, and check energy exchange balance.

1. Calculate the theoretical maximum energy transfer.

The air being reheated will have less mass than the outdoor air entering the precooler because moisture will condense from it as it passes through the precooler and cooling coil. Reheat is sensible heat only, so Equation (6) is used to determine the theoretical maximum energy transfer.

$$q_{max} = 60 \times 0.24 \times 0.071 \times 3300(95.0 - 52.2) = 144{,}000 \text{ Btu/h}$$

2. Establish the sensible effectiveness.

From the manufacturer's heat exchanger selection data, effectiveness is determined to be 58.4% at the designated operating conditions.

3. Calculate actual heat transfer at given conditions.

$$q = 0.584 \times 144{,}000 = 84{,}000 \text{ Btu/h}$$

4. Calculate leaving air conditions.

Because condensation is expected to occur as the outdoor airstream passes through the precooling side of the heat exchanger, Equation (5) is used to determine its leaving condition, which is the inlet condition for the cooling coil. The sensible heat transfer Equation (6) is used to determine the condition of air leaving the preheat side of the heat exchanger.

a. Precooler leaving air conditions

Entering enthalpy, determined from the psychrometric chart for 95°F db and 80°F wb, is 43.7 Btu/lb.

$$h_2 = 43.7 + \left(\frac{-84{,}000}{60 \times 0.071 \times 3300}\right) = 37.7 \text{ Btu/lb}$$

The wet-bulb temperature for saturated air with this enthalpy is 73.6°F. This is point 2 on the psychrometric chart (Figure 11), which is near saturation. Note that this precooled air will be further dehumidified by a cooling coil from point 2 to point 3 in Figure 11.

b. Reheater leaving air conditions

$$t_4 = 52.2 + \left(\frac{84{,}000}{60 \times 0.24 \times 0.071 \times 3300}\right) = 77.1°F$$

Entering enthalpy, determined from the psychrometric chart for 52.2°F db and 52.1°F wb, is 21.4 Btu/lb.

$$h_4 = 21.4 + \left(\frac{84{,}000}{60 \times 0.071 \times 3300}\right) = 27.4 \text{ Btu/lb}$$

The wet-bulb temperature for air with this temperature and enthalpy is 61.4°F.

5. Check performance.

$$60 \times 0.071 \times 3300(43.7 - 37.7) = 84{,}000 \text{ Btu/h precooling}$$

$$60 \times 0.071 \times 0.24 \times 3300(77.1 - 52.2) = 84{,}000 \text{ Btu/h reheat}$$

6. Plot conditions on psychrometric chart (Figure 11).

EQUIPMENT

Table 2 provides comparative data for common types of air-to-air energy recovery devices. The following sections describe the construction, operation, and unique features of the various devices.

FIXED-PLATE EXCHANGERS

Fixed surface plate exchangers have no moving parts. Alternate layers of plates, separated and sealed (referred to as the heat exchanger core), form the exhaust and supply airstream passages (Figure 12). Plate spacings range from 0.1 to 0.5 in., depending on the design and the application. Heat is transferred directly from the warm airstreams through the separating plates into the cool airstreams. Design and construction restrictions inevitably result in cross-flow heat transfer, but additional effective heat transfer surface arranged properly into counterflow patterns can increase heat transfer effectiveness.

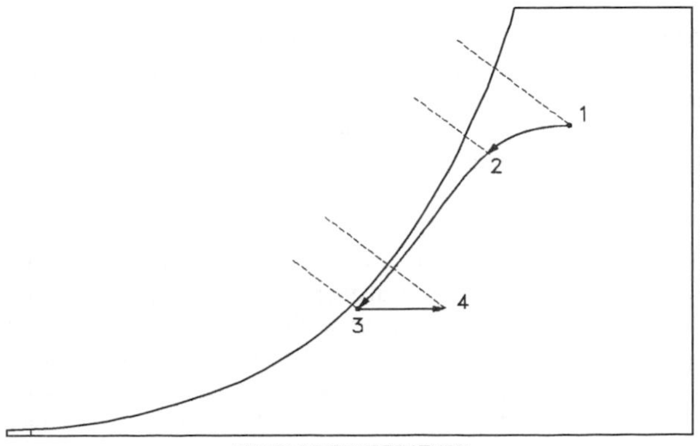

Fig. 11 Precooling Air Reheater Dehumidifier (Example 6)

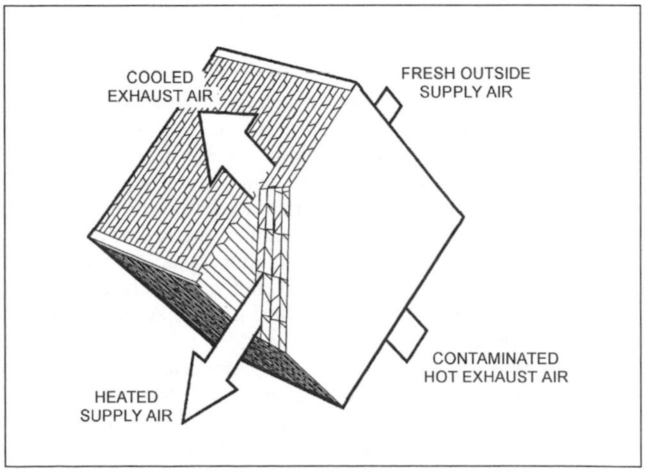

Fig. 12 Fixed-Plate Cross-Flow Heat Exchanger

Table 2 Comparison of Air-to-Air Energy Recovery Devices

	Fixed Plate	Rotary Wheel	Heat Pipe	Runaround Coil Loop	Thermosiphon	Twin Towers
Airflow arrangements	Counterflow Cross-flow Parallel flow	Counterflow Parallel flow	Counterflow Parallel flow	Counterflow Parallel flow	Counterflow Parallel flow	
Equipment size range, cfm	50 and up	50 to 70,000	100 and up	100 and up	100 and up	
Type of heat transfer (typical effectiveness)	Sensible (50 to 80%) Total (55 to 85%) Treated paper and poly membrane	Sensible (50 to 80%) Total (55 to 85%)	Sensible (45 to 65%)	Sensible (55 to 65%)	Sensible (40 to 60%)	Sensible (40 to 60%)
Face velocity, fpm (most common design velocity)	100 to 1000 (200 to 1000)	500 to 1000	400 to 800 (450 to 550)	300 to 600	400 to 800 (450 to 550)	300 to 450
Pressure drop, in. of water (most likely pressure)	0.02 to 1.8 (0.1 to 1.5)	(0.25 to 1.0)	(0.4 to 2.0)	(0.4 to 2.0)	(0.4 to 2.0)	0.7 to 1.2
Temperature range	−70 to 1500°F	−70 to 200°F	−40 to 95°F	−50 to 900°F	−40 to 104°F	−40 to 115°F
Typical mode of purchase	Exchanger only Exchanger in case Exchanger and blowers Complete system	Exchanger only Exchanger in case Exchanger and blowers Complete system	Exchanger only Exchanger in case Exchanger and blowers Complete system	Coil only Complete system	Exchanger only Exchanger in case	Complete system
Unique advantages	No moving parts Low pressure drop Easily cleaned	Latent (moisture mass) transfer Compact large sizes Low pressure drop	No moving parts except tilt Fan location not critical Allowable pressure differential up to 60 in. of water	Exhaust airstream can be separated from supply air Fan location not critical	No moving parts Exhaust airstream can be separated from supply air Fan location not critical	Latent transfer from remote airstreams Multiple units in a single system Efficient microbiological cleaning of both supply and exhaust airstreams
Limitations	Latent available in hygroscopic units only	Cold climates may increase service Cross-air contamination possible	Effectiveness limited by pressure drop and cost Few suppliers	High effectiveness requires accurate simulation model	Effectiveness may be limited by pressure drop and cost Few suppliers	Few suppliers
Cross-leakage	0 to 5%	1 to 10%	0%	0%	0%	0.025%
Heat rate control (HRC) schemes	Bypass dampers and ducting	Wheel speed control over full range	Tilt angle down to 10% of maximum heat rate	Bypass valve or pump speed control over full range	Control valve over full range	Control valve or pump speed control over full range

Normally, both latent heat of condensation [from moisture condensed as the temperature of the warm (exhaust) airstream drops below its dew point] and sensible heat are conducted through the separating plates into the cool (supply) airstream. Thus, energy is transferred but moisture is not. Recovering 80% or more of the available waste exhaust heat is not uncommon.

Design Considerations

Plate exchangers are available in many configurations, materials, sizes, and flow patterns. Many are modular, and modules can be arranged to handle almost any airflow, effectiveness, and pressure drop requirement. Plates are formed with integral separators (e.g., ribs, dimples, ovals) or with external separators (e.g., supports, braces, corrugations). Airstream separations are sealed by folding, multiple folding, gluing, cementing, welding, or any combination of these, depending on the application and manufacturer. Ease of access for examining and clean the heat transfer surfaces depends on the configuration.

Heat transfer resistance through the plates is small compared to the airstream boundary layer resistance on each side of the plates. Heat transfer efficiency is not substantially affected by the heat transfer coefficient of the plates. Aluminum is the most popular construction material for plates because of its nonflammability and durability. Polymer plate exchangers have properties that may improve heat transfer by breaking down the boundary layer and are popular because of their corrosion resistance and cost-effectiveness. Steel alloys are used for temperatures exceeding 400°F and for specialized applications where cost is not a key factor. Plate exchangers normally conduct sensible heat only; however, water-vapor-permeable materials, such as treated paper and new microporous polymeric membranes, are used to transfer moisture, thus providing total (enthalpy) energy exchange.

Most manufacturers offer plate exchangers in modular design. Modules range in capacity from 25 to 10,000 cfm and can be arranged into configurations exceeding 100,000 cfm. Multiple sizes and configurations permit selections to meet nearly all space and performance requirements.

Performance

Fixed-plate heat exchangers can economically achieve high sensible heat recovery and high total energy effectiveness because they have only a primary heat transfer surface area separating the airstreams and are therefore not inhibited by the additional secondary resistance (i.e., pumping liquid, condensing and vaporizing gases, or transporting a heat transfer medium) inherent in other exchanger types. Simplicity and lack of moving parts add to the reliability, longevity, low auxiliary energy consumption, and safety performance of these exchangers.

Differential Pressure/Cross-Leakage

One of the advantages of the plate exchanger is that it is a static device built so that little or no leakage occurs between airstreams.

As velocity increases, the pressure difference between the two airstreams increases exponentially. High differential pressures may deform the separating plates and, if excessive, can permanently damage the exchanger. This may reduce effectiveness, alter design fan airflow, and cause excessive air leakage. This is not normally a problem because the differential pressures in most applications are less than 4 in. of water. In applications requiring high air velocities, high static pressures, or both, select exchangers that are designed for these conditions.

Condensing Within Exhaust Airstreams

Most plate exchangers are equipped with condensate drains, which remove the condensate and also wastewater when a water-wash system is used. Heat recovered from a high-humidity exhaust is better returned to a building or process by a sensible heat exchanger rather than an enthalpy exchanger if humidity transfer is not desired.

Table 3 illustrates the effect of moisture content on the frost threshold for counterflow heat exchangers. Frosting can be controlled by preheating the incoming supply air, bypassing a portion of the incoming air, recirculating supply air through the exhaust side of the exchanger, or temporarily interrupting the supply air while maintaining exhaust. Bypassing reduces the ratio K in Table 3 and is generally more cost-effective than preheating.

For cross-flow plate heat exchangers, freezing and frost growth first occur at higher temperatures than those shown in Table 3. However, frost on cross-flow heat exchangers is less likely to block the exhaust airflow completely as with other types of exchangers. Generally, frost should be avoided unless a defrost cycle is included.

Microporous Permeable Membrane Fixed-Plate Exchangers

Fixed-plate heat exchangers can be fabricated of a permeable medium that transfers both moisture and heat from one airstream to the other. Media have been developed that minimize cross-leakage while maximizing moisture and energy transfer. Cross-stream mass transfer occurs without air transfer when the walls of an exchanger are made of such media. Suitable microporous permeable materials include cellulose, polymers, and other synthetic membranes.

ROTARY AIR-TO-AIR ENERGY EXCHANGERS

A rotary air-to-air energy exchanger, or **rotary enthalpy wheel**, has a revolving cylinder filled with an air-permeable medium having a large internal surface area. Adjacent supply and exhaust airstreams each flow through one-half the exchanger in a counterflow pattern (Figure 13). Heat transfer media may be selected to recover sensible heat only or total heat (sensible heat plus latent heat).

Sensible heat is transferred as the medium picks up and stores heat from the hot airstream and releases it to the cold one. Latent heat is transferred as the medium (1) condenses moisture from the airstream with the higher humidity ratio (either because the medium temperature is below its dew point or by means of absorption for liquid desiccants and adsorption for solid desiccants), with a simultaneous release of heat; and (2) releases the moisture through evaporation (and heat pickup) into the airstream with the lower humidity ratio. Thus, the moist air is dried while the drier air is humidified. In total heat transfer, both sensible and latent heat transfer occur simultaneously. Because rotary exchangers have a counterflow configuration and normally use small-diameter flow passages, they are quite compact and can achieve high transfer effectiveness.

Table 3 Frost Threshold Temperature for Various Counterflow Heat Exchanger Exhaust Air Conditions

Entering Exhaust Air		Frost Threshold Temperature t_1, °F			
		Ratio of Supply to Exhaust Airflow K			
t_3, °F	rh, %	0.5	0.7	1.0	2.0
60	30		15	23	32
60	40		15	23	32
60	50	−4		18	32
60	60	−9		13	32
70	30	−13		17	28
70	40	−21	−3	10	21
70	50	−27	−9		15
70	60	−32	−13	−1	10
75	30	−25	−4	10	23
75	40	−33	−12		15
75	50	−40	−20	−6	
75	60	−47	−26	−12	
80	30	−35	−11		19
80	40	−44	−20	−5	10
80	50	−53	−30	−14	
80	60	−62	−39	−23	−8
90	30	−58	−30	−11	
90	40			−24	−8
90	50				−20

Construction

Air contaminants, dew point, exhaust air temperature, and supply air properties influence the choice of materials for the casing, rotor structure, and medium of a rotary energy exchanger. Aluminum, steel, and polymers are the usual structural, casing, and rotor materials for normal comfort ventilating systems. Exchanger media are fabricated from metal, mineral, or man-made materials and provide either random flow or directionally oriented flow through their structures.

Random flow media are made by knitting wire into an open woven cloth or corrugated mesh, which is layered to the desired configuration. Aluminum mesh, commonly used for comfort ventilation systems, is packed in pie-shaped wheel segments. Stainless steel and monel mesh are used for high-temperature and corrosive applications. These media should only be used with clean, filtered airstreams because they plug easily. Random flow media also require a significantly larger face area than directionally oriented media for given values of airflow and pressure drop.

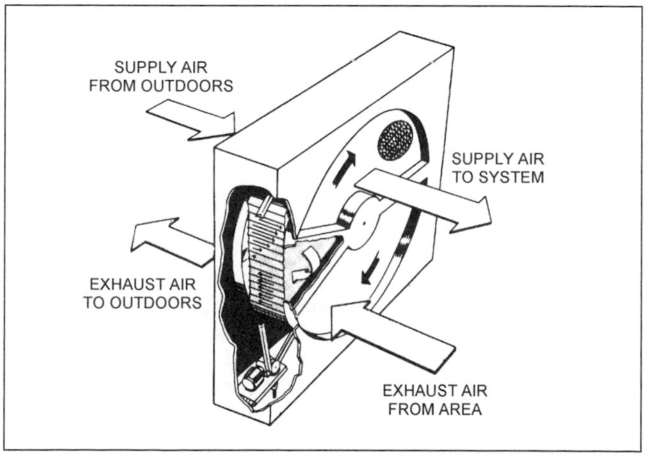

Fig. 13 Rotary Air-to-Air Energy Exchanger

Directionally oriented media are available in various geometric configurations. The most common consist of small (0.06 to 0.08 in.) air passages parallel to the direction of airflow. Air passages are very similar in performance regardless of their shape (triangular, hexagonal, or other). Aluminum foil, paper, plastic, and synthetic materials are used for low and medium temperatures. Stainless steel and ceramics are used for high temperatures and corrosive atmospheres.

Media surface areas exposed to airflow vary from 100 to over 1000 ft^2/ft^3, depending on the type of medium and physical configuration. Media may also be classified according to their ability to recover sensible heat only or total heat. Media for sensible heat recovery are made of aluminum, copper, stainless steel, and monel. Media for total heat recovery are fabricated from any of a number of materials and treated with a desiccant (typically zeolites, molecular sieves, silica gels, activated alumina, titanium silicate, synthetic polymers, lithium chloride, or aluminum oxide) to have specific moisture recovery characteristics.

Cross-Contamination

Cross-contamination, or mixing, of air between supply and exhaust airstreams occurs in all rotary energy exchangers by two mechanisms—carryover and leakage. **Carryover** occurs as air is entrained within the volume of the rotation medium and is carried into the other airstream. **Leakage** occurs because the differential static pressure across the two airstreams drives air from a higher to a lower static pressure region. Cross-contamination can be reduced by placing the blowers so that they promote leakage of outside air to the exhaust airstream. Carryover occurs each time a portion of the matrix passes the seals dividing the supply and exhaust airstreams. Because carryover from exhaust to supply may be undesirable, a **purge section** can be installed on the heat exchanger to reduce cross-contamination.

In many applications, recirculating some air is not a concern. However, critical applications such as hospital operating rooms, laboratories, and clean rooms require stringent control of carryover. Carryover can be reduced to less than 0.1% of the exhaust airflow with a purge section (ASHRAE 1974).

The theoretical carryover of a wheel without a purge section is directly proportional to the speed of the wheel and the void volume of the medium (75 to 95% void, depending on type and configuration). For example, a 10 ft diameter, 8 in. deep wheel with a 90% void volume operating at 14 rpm has a carryover volumetric flow of

$$\pi(10/2)^2(8/12)(0.9)(14) = 660 \text{ cfm}$$

If the wheel is handling a 20,000 cfm balanced flow, the percentage carryover is

$$\frac{660}{20{,}000} \times 100 = 3.3\%$$

The exhaust fan, which is usually located at the exit of the exchanger, should be sized to include leakage, purge, and carryover airflows.

Controls

Two control methods are commonly used to regulate wheel energy recovery. In the first, **supply air bypass control**, the amount of supply air allowed to pass through the wheel establishes the supply air temperature. An air bypass damper, controlled by a wheel supply air discharge temperature sensor, regulates the proportion of supply air permitted to bypass the exchanger.

The second method regulates the energy recovery rate by varying wheel rotational speed. The most frequently used **variable-speed drives** are (1) a silicon controlled rectifier (SCR) with variable-speed dc motor, (2) a constant speed ac motor with hysteresis coupling, and (3) an ac frequency inverter with an ac induction motor.

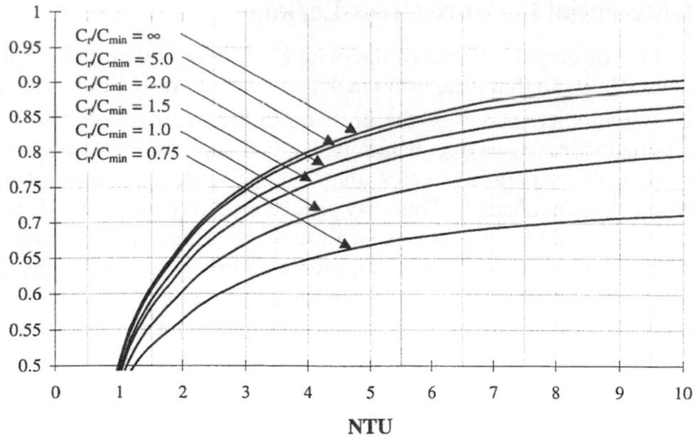

Fig. 14 Effectiveness of Counterflow Regenerator
(Shah 1981)

Figure 14 shows the effectiveness ε of a regenerative counterflow wheel versus number of transfer units (NTU). For sensible heat transfer only, with airflow balanced, convection-conduction ratio less than 4, and no leakage or cross-flow,

$$\text{NTU} = (UA)_{avg}/C_{min} \tag{8}$$

where

$(UA)_{avg}$ = product of modified overall heat transfer coefficient and heat exchange area, Btu/h·°F
C_{min} = minimum heat capacity rate of hot and cold airstreams, Btu/h·°F
C_r = heat capacity rate for air mass within rotary wheel, Btu/h·°F

Figure 14 also shows that regenerative counterflow rotary effectiveness increases with wheel speed (C_r is proportional to wheel speed); but there is no advantage in going beyond $C_r/C_{min} = 5$ because the carryover of contaminants increases with wheel speed. See Shah (1981) or Kays and Crawford (1993) for details.

Mathematical models to describe the sensible and total energy effectiveness of regenerator wheels with hygroscopic coatings are under development. Until these models become accepted, however, desiccant wheels should be tested under conditions defined by ASHRAE *Standard* 84.

A dead band control, which stops or limits the exchanger, may be necessary when no recovery is desired (e.g., when outside air temperature is higher than the required supply air temperature but below the exhaust air temperature). When the outside air temperature is above the exhaust air temperature, the equipment operates at full capacity to cool the incoming air. During very cold weather, it may be necessary to heat the supply air, stop the wheel, or, in the case of small systems, use a defrost cycle for frost control.

Maintenance

Rotary enthalpy wheels require little maintenance. The following maintenance procedures ensure best performance:

- Clean the medium when lint, dust, or other foreign materials build up, following the manufacturer's instructions for that medium. Media treated with a liquid desiccant for total heat recovery must not be wetted.
- Maintain drive motor and train according to the manufacturer's recommendations. Speed control motors that have commutators and brushes require more frequent inspection and maintenance

than do induction motors. Brushes should be replaced, and the commutator should be periodically turned and undercut.

- Inspect wheels regularly for proper belt or chain tension.
- Refer to the manufacturer's recommendations for spare and replacement parts.

COIL ENERGY RECOVERY (RUNAROUND) LOOPS

A typical coil energy recovery loop (Figure 15) places extended surface, finned tube water coils in the supply and exhaust airstreams of a building or process. The coils are connected in a closed loop via counterflow piping through which an intermediate heat transfer fluid (typically water or an antifreeze solution) is pumped.

This system operates for sensible heat recovery only. In comfort-to-comfort applications, energy transfer is seasonally reversible—the supply air is preheated when the outdoor air is cooler than the exhaust air and precooled when the outdoor air is warmer.

Freeze Protection

Moisture must not freeze in the exhaust coil air passage. A dual-purpose, three-way temperature control valve prevents the exhaust coil from freezing. The valve is controlled to maintain the temperature of the solution entering the exhaust coil at 30°F or above. This condition is maintained by bypassing some of the warmer solution around the supply air coil. The valve can also ensure that a prescribed air temperature from the supply air coil is not exceeded.

System Characteristics

Coil energy recovery loops are highly flexible and well suited to renovation and industrial applications. The loop accommodates remote supply and exhaust ducts and allows the simultaneous transfer of energy between multiple sources and uses. An expansion tank must be included to allow fluid expansion and contraction. A closed expansion tank minimizes oxidation when ethylene glycol is used.

Standard finned tube water coils may be used; however, these need to be selected using an accurate simulation model if high effectiveness values and low costs are to be realized (Johnson et al. 1995). Integrating runaround loops in buildings with variable loads to achieve maximum benefits may require combining the runaround simulation with building energy simulation (Dhital et al. 1995). Manufacturer's design curves and performance data should be used when selecting coils, face velocities, and pressure drops, but only when the design data are for the same temperature and operating conditions as in the runaround loop.

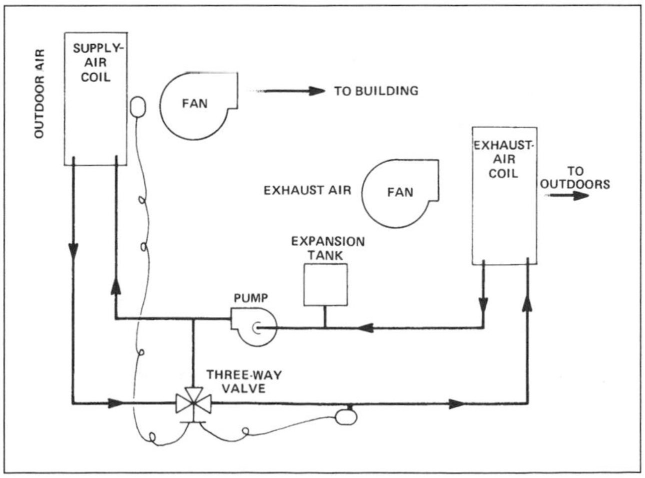

Fig. 15 Coil Energy Recovery Loop

Effectiveness

The coil energy recovery loop cannot transfer moisture from one airstream to another; however, indirect evaporative cooling can reduce the exhaust air temperature, which significantly reduces cooling loads. For the most cost-effective operation, with equal airflow rates and no condensation, typical effectiveness values range from 45 to 65%. Highest effectiveness does not necessarily give the greatest net cost saving.

The following example illustrates the capacity of a typical system:

Example 7. A waste heat recovery system heats 10,000 cfm of air from a 0°F design outdoor temperature to an exhaust dry-bulb temperature of 75°F and wet-bulb temperature of 60°F. (From Example 2, q_{max} = 810,000 Btu/h.) Air flows through identical eight-row coils at a 400 fpm face velocity. A 30% ethylene glycol solution flows through the coils at 26 gpm.

Figure 16 shows the effect of the outside air temperature on capacity, including the effects of the three-way temperature control valve. For this example, the capacity is constant for outside air temperatures below 18.5°F. This constant output occurs because the valve has to control the temperature of the fluid entering the exhaust coil to prevent frosting. As the exhaust coil is the source of heat and has a constant airflow rate, entering air temperature, liquid flow rate, entering fluid temperature (as set by the valve), and fixed coil parameters, energy recovered must be controlled to prevent frosting in the exhaust coil. Equation (2) may be used to calculate the sensible heat effectiveness.

When the three-way control valve operates at outside air temperatures of 18.5°F or lower, 414,000 Btu/h is recovered. At 18.5°F, the sensible heat effectiveness is 67.2%. At the 0°F design temperature, sensible effectiveness is 51% (ε = 414,000/810,000), and the leaving air dry-bulb temperature of the supply coil is 38.3°F. Above 60°F outside air temperature, the supply air is cooled with an evaporative cooler located upstream from the exhaust coil.

Typically, the sensible heat effectiveness of a coil energy recovery loop is independent of the outside air temperature. However, when the capacity is controlled (as in Example 7), the sensible heat effectiveness decreases.

Construction Materials

Coil energy recovery loops incorporate coils constructed to suit the environment and operating conditions to which they are exposed. For typical comfort-to-comfort applications, standard coil construction usually suffices. In process-to-process and process-to-comfort applications, the effect of high temperature, condensable

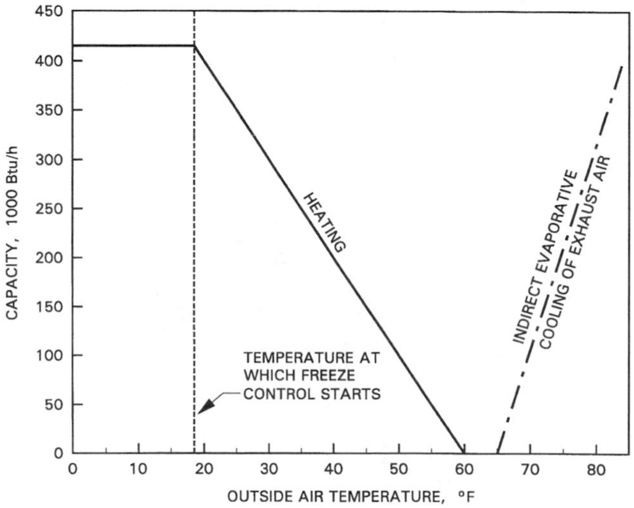

Fig. 16 Energy Recovery Capacity Versus Outside Air Temperature for Typical Loop

gases, corrosives, and contaminants on the coil(s) must be considered. At temperatures above 400°F, special construction may be required to ensure a permanent fin-to-tube bond. The effects of condensable gases and other adverse factors may require special coil construction and/or coatings. Chapter 21 and Chapter 23 discuss the construction and selection of coils in more detail.

Cross-Contamination

Complete separation of the airstreams eliminates cross-contamination between the supply and exhaust air.

Maintenance

Coil energy recovery loops require little maintenance. The only moving parts are the circulation pump and the three-way control valve. However, to ensure optimum operation, the air should be filtered, the coil surface cleaned regularly, the pump and valve maintained, and the transfer fluid refilled or replaced periodically. Fluid manufacturers or their representatives should be contacted for specific recommendations.

Thermal Transfer Fluids

The thermal transfer fluid selected for a closed-loop exchanger depends on the application and on the temperatures of the two airstreams. An inhibited ethylene glycol solution in water is commonly used when freeze protection is required. These solutions break down to an acidic sludge at temperatures above 275°F. If freeze protection is needed and exhaust air temperatures exceed 275°F, a nonaqueous synthetic heat transfer fluid should be used. Heat transfer fluid manufacturers and chemical suppliers should recommend appropriate fluids.

HEAT PIPE HEAT EXCHANGERS

A heat pipe heat exchanger is a passive energy recovery device. It has the outward appearance of an ordinary plate-finned water or steam coil, except that the tubes are not interconnected and the pipe heat exchanger is divided into evaporator and condenser sections by a partition plate (Figure 17). Hot air passes through the evaporator side of the exchanger, and cold air passes through the condenser side. Heat pipe heat exchangers are sensible heat transfer devices, but condensation on the fins does allow latent heat transfer, resulting in improved recovery performance.

Heat pipe tubes (Figure 18) are fabricated with an integral capillary wick structure, evacuated, filled with a suitable working fluid, and permanently sealed. The working fluid is normally a Class I refrigerant, but other fluorocarbons, water, and other compounds are used for applications with special temperature requirements.

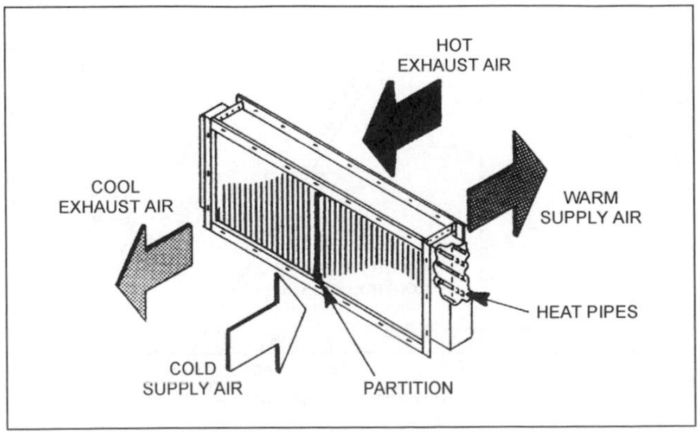

Fig. 17 Heat Pipe Heat Exchanger

Fin designs include continuous corrugated plate fin, continuous plain fin, and spiral fins. Modifying fin design and tube spacing changes pressure drop at a given face velocity.

Principle of Operation

Hot air flowing over the evaporator end of the heat pipe vaporizes the working fluid. A vapor pressure gradient drives the vapor to the condenser end of the heat pipe tube, where the vapor condenses, releasing the latent energy of vaporization. The condensed fluid is wicked or flows back to the evaporator, where it is revaporized, thus completing the cycle. Thus the heat pipe's working fluid operates in a closed-loop evaporation/condensation cycle that continues as long as there is a temperature difference to drive the process. Using this mechanism, heat transfer along a heat pipe is up to 1000 times faster than through copper (Ruch 1976).

Energy transfer in heat pipes is often considered isothermal. However, there is a small temperature drop through the tube wall, wick, and fluid medium. Heat pipes have a finite heat transfer capacity that is affected by such factors as wick design, tube diameter, working fluid, and tube (heat pipe) orientation relative to horizontal.

Construction Materials

HVAC systems use copper or aluminum heat pipe tubes with aluminum fins. For process-to-comfort applications with large temperature changes, tubes and fins are usually constructed of the same material to avoid problems with different thermal expansions of materials. Heat pipe heat exchangers for exhaust temperatures below 425°F are most often constructed with aluminum tubes and fins. Protective coatings designed for finned tube heat exchangers have permitted inexpensive aluminum to replace exotic metals in corrosive atmospheres; these coatings have a minimal effect on thermal performance.

Heat pipe heat exchangers for use above 425°F are generally constructed with steel tubes and fins. The fins are often aluminized to prevent rusting. Composite systems for special applications may be created by assembling units with different materials and/or different working fluids.

Operating Temperature Range

Selection of the proper working fluid for a heat pipe is critical to long-term operation. The working fluid should have high latent heat of vaporization, a high surface tension, and a low liquid viscosity over the operating range; it must be thermally stable at operating temperatures. Decomposition of the thermal fluids can cause the formation of noncondensable gases that deteriorate performance.

Cross-Contamination

Heat pipe heat exchangers typically have zero cross-contamination for pressure differentials between airstreams of up to 50 in. of water. Constructing a vented double-wall partition between the air-

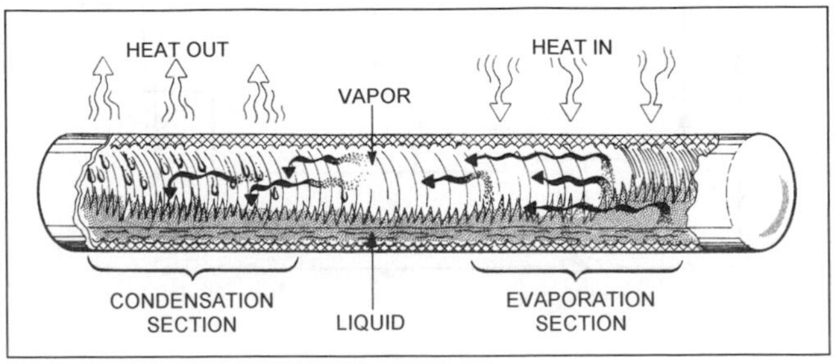

Fig. 18 Heat Pipe

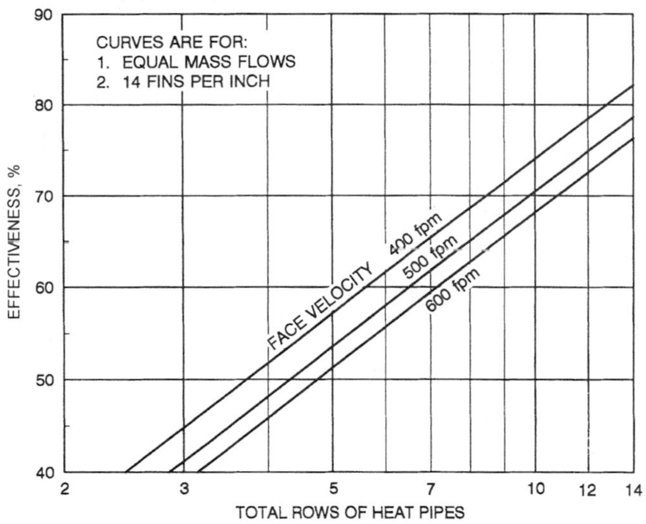

Fig. 19 Heat Pipe Exchanger Effectiveness

streams can provide additional protection against cross-contamination. If an exhaust duct is attached to the partition space, any leakage is withdrawn and exhausted from the space between the two ducts.

Performance

Heat pipe heat transfer capacity depends on design and orientation. Figure 19 presents a typical effectiveness curve for various face velocities and rows of tubes. As the number of rows increases, effectiveness increases at a decreasing rate. For example, doubling the number of rows of tubes in a 60% effective heat exchanger increases the effectiveness to 75%. The effectiveness of a counterflow heat pipe heat exchanger depends on the total number of rows such that two units in series yield the same effectiveness as a single unit of the same total number of rows. Series units are often used to facilitate handling, cleaning, and maintenance.

The heat transfer capacity of a heat pipe increases roughly with the square of the inside diameter of the pipe. For example, at a given tilt angle, a 1 in. inside diameter heat pipe will transfer roughly 2.5 times as much energy as a 5/8 in. inside diameter pipe. Consequently, heat pipes with large diameters are used for larger airflow applications and where level installation is required to accommodate both summer and winter operation.

Heat transfer capacity limit is virtually independent of heat pipe length, except for very short heat pipes. For example, a 4 ft long heat pipe has approximately the same capacity as an 8 ft pipe. Because the 8 ft heat pipe has twice the external heat transfer surface area of the 4 ft pipe, it will reach its capacity limit sooner. Thus, in a given application, it is more difficult to meet the capacity requirements as the heat pipes become longer. A system can be reconfigured to a taller face height and more numerous but shorter heat pipes to yield the same airflow face area while improving system performance.

The selection of fin design and spacing should be based on the dirtiness of the two airstreams and the resulting cleaning and maintenance required. For HVAC applications, 11 to 14 fins/in. is common. Wider fin spacings (8 to 10 fins/in.) are usually used for industrial applications. Heat pipe heat exchangers of the plate fin type can easily be constructed with different fin spacings for the exhaust and supply airstreams, allowing wider fin spacing on the dirty exhaust side. This increases design flexibility where pressure drop constraints exist and also prevents deterioration of performance due to dirt buildup on the exhaust side surface.

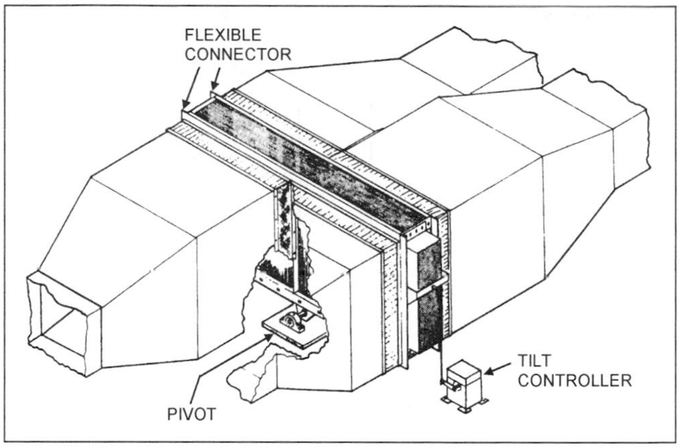

Fig. 20 Heat Pipe Heat Exchanger with Tilt Control

Controls

Changing the slope, or tilt, of a heat pipe controls the amount of heat it transfers. Operating the heat pipe on a slope with the hot end below (or above) the horizontal improves (or retards) the condensate flow back to the evaporator end of the heat pipe. This feature can be used to regulate the effectiveness of the heat pipe heat exchanger.

In practice, tilt control is effected by pivoting the exchanger about the center of its base and attaching a temperature-controlled actuator to one end of the exchanger (Figure 20). Pleated flexible connectors attached to the ductwork allow freedom for the small tilting movement (6° maximum).

The following are three functions for which tilt control may be desired:

- To change the operation from supply air heating to supply air cooling (i.e., to reverse the direction of heat flow) when seasonal changeover occurs.
- To modulate effectiveness to maintain desired supply air temperature. This kind of regulation is often required for large buildings to avoid overheating the air supplied to the interior zone.
- To decrease effectiveness to prevent frost formation at low outdoor air temperatures. With reduced effectiveness, the exhaust air leaves the unit at a warmer temperature and stays above frost-forming conditions.

Other devices, such as face and bypass dampers and preheaters, can be used for individual functions.

TWIN TOWER ENTHALPY RECOVERY LOOPS

In this air-to-liquid, liquid-to-air enthalpy recovery system, a sorbent liquid circulated between supply and exhaust contactor towers directly contacts both airstreams, transporting water vapor and energy between the airstreams (Figure 21). Supply air temperatures can be as high as 115°F or as low as –40°F. Any number of vertical and horizontal airflow contactor towers can be combined into a common system of any airflow capacity.

Leaving air passes through demister pads to remove entrained sorbent solution. Airstreams containing lint, animal hair, or other solids should be filtered upstream of the contactor towers. Wetted particles should be filtered from the sorbent solution, which minimizes particulate cross-contamination. Sorbent solutions (typically a halogen salt solution such as lithium chloride and water) are usually bactericidal and viricidal. Testing has shown that contactor towers can effectively remove up to 94% of atmospheric bacteria, a desirable feature in health care applications. Limited gaseous cross-contamination may occur. If either airstream contains gaseous contaminants, their effects on the sorbent solution should be investigated.

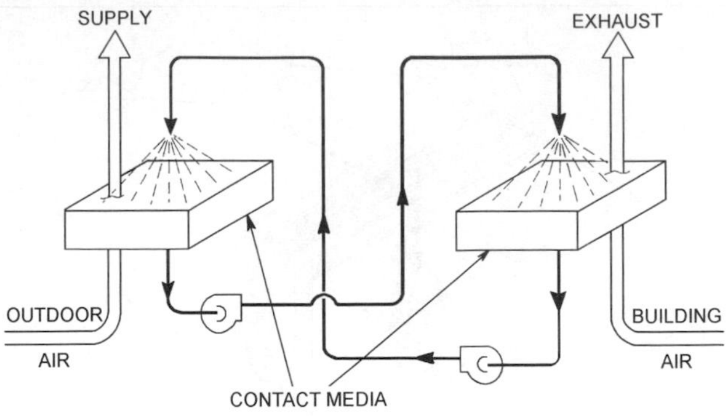

Fig. 21 Twin Tower Enthalpy Recovery Loop

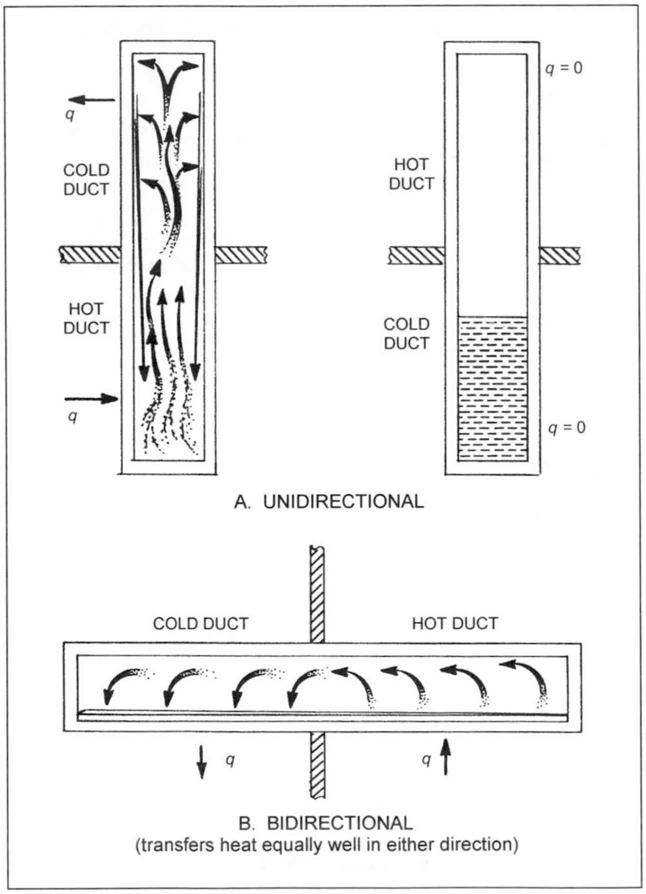

Fig. 22 Sealed Tube Thermosiphons

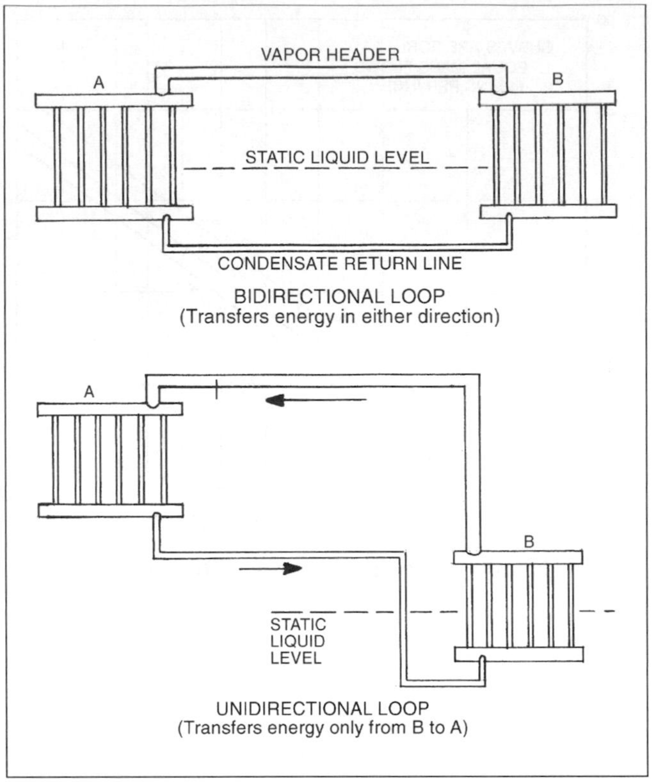

Fig. 23 Coil-Type Thermosiphon Loops

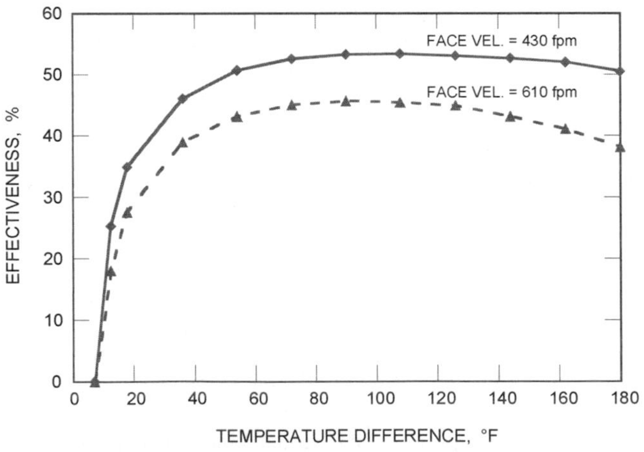

Fig. 24 Typical Performance of Two-Phase Thermosiphon Loop

In colder climates, moisture losses from the exhaust airstream may overdilute the sorbent solution. Heating the sorbent liquid entering the supply air contactor tower raises the discharge temperature and humidity of the leaving supply air, preventing overdilution. This, coupled with automatic makeup water addition, can maintain sorbent solution concentrations during cold weather, enabling the system to deliver air at a fixed humidity and temperature.

THERMOSIPHON HEAT EXCHANGERS

Two-Phase Thermosiphon Heat Exchangers

Two-phase thermosiphon heat exchangers are sealed systems that consist of an evaporator, a condenser, interconnecting piping, and an intermediate working fluid that is present in both liquid and vapor phases. Two types of thermosiphon are used—sealed tube (Figure 22) and coil type (Figure 23). In the sealed tube thermosiphon, the evaporator and the condenser are usually at opposite ends of a bundle of straight, individual thermosiphon tubes, and the exhaust and supply ducts are adjacent to each other (this arrangement is similar to that in a heat pipe system). In coil-type thermosiphons, evaporator and condenser coils are installed independently in the ducts and are interconnected by the working fluid piping (this configuration is similar to that of a coil energy recovery loop).

In thermosiphon systems, a temperature difference and gravity force a working fluid to circulate between the evaporator and condenser. As a result, thermosiphons may be designed to transfer heat equally in either direction (bidirectional), in one direction only (unidirectional), or in both directions unequally.

Although similar in form and operation to heat pipes, thermosiphon tubes are different in two ways: (1) they have no wicks and hence rely only on gravity to return the condensate to the evaporator, whereas heat pipes use capillary forces; and (2) they depend, at least initially, on nucleate boiling, whereas heat pipes vaporize the fluid from a large, ever-present liquid-vapor interface. As a result, thermosiphon heat exchangers may require a significant temperature difference to initiate boiling (McDonald and Shivprasad 1989). Thermosiphon tubes require no pump to circulate the working fluid. However, the geometric configuration must be such that liquid working fluid is always present in the evaporator section of the heat exchanger.

Thermosiphon loops differ from other coil energy recovery loop systems in that they require no pumps and hence no external power supply, and the coils must be appropriate for evaporation and condensation. Two-phase thermosiphon loops are used for solar water heating (Mathur 1990c). Figure 24 shows the performance of a thermosiphon loop (Mathur and McDonald 1986).

Principle of Operation

A thermosiphon is a sealed system containing a two-phase working fluid. Because part of the system contains vapor and part contains liquid, the pressure in a thermosiphon is governed by the liquid temperature at the liquid-vapor interface. If the surroundings cause a temperature difference between the two regions in a thermosiphon where liquid and vapor interfaces are present, the resulting vapor pressure difference causes vapor to flow from the warmer to the colder region. The flow is sustained by condensation in the cooler region and by evaporation in the warmer region. The condenser and evaporator must be oriented so that the condensate can return to the evaporator by gravity (Figure 22 and Figure 23).

REFERENCES

ARI. 1997. Rating air-to-air energy recovery ventilation equipment. *Standard* 1060-97. Air-Conditioning and Refrigeration Institute, Arlington, VA.

ASHRAE. 1974. Symposium on air-to-air heat recovery. *ASHRAE Transactions* 80(2):302-32.

ASHRAE. 1982. Symposium on energy recovery from air pollution control. *ASHRAE Transactions* 88(1):1197-1225.

ASHRAE. 1991. Method of testing air-to-air heat exchangers. *Standard* 84-1991.

Barringer, C.G. and C.A. McGugan. 1989a. Development of a dynamic model for simulating indoor air temperature and humidity. *ASHRAE Transactions* 95(2):449-60.

Barringer, C.G. and C.A. McGugan. 1989b. Effect of residential air-to-air heat and moisture exchangers on indoor humidity. *ASHRAE Transactions* 95(2):461-74.

Besant, R.W. and A.B. Johnson. 1995. Reducing energy costs using run-around systems. *ASHRAE Journal* 37(2):41-47.

CSA. 1988. Standard methods of test for rating the performance of heat-recovery ventilators. CAN/CSA-C439-88. Canadian Standards Association, Rexdale, ON.

Dhital, P., R. Besant, and G.J. Schoenau. 1995. Integrating run-around heat exchanger systems into the design of large office buildings. *ASHRAE Transactions* 101(2):979-99.

Kays, W.M. and M.E. Crawford. 1993. *Convective heat and mass transfer*, 3rd ed. McGraw-Hill, New York.

Johnson, A.B., R.W. Besant, and G.J. Schoenau. 1995. Design of multi-coil run-around heat exchanger systems for ventilation air heating and cooling. *ASHRAE Transactions* 101(2):967-78.

Mathur, G.D. 1990a. Indirect evaporative cooling using heat pipe heat exchangers. ASME Symposium, Thermal Hydraulics of Advanced Heat Exchangers, ASME Winter Annual Meeting, Dallas, TX, 79-85.

Mathur, G.D. 1990b. Indirect evaporative cooling using two-phase thermosiphon loop heat exchangers. *ASHRAE Transactions* 96(1):1241-49.

Mathur, G.D. 1990c. Long-term performance prediction of refrigerant charged flat plate solar collector of a natural circulation closed loop. ASME HTD 157:19-27. American Society of Mechanical Engineers, New York.

Mathur, G.D. 1992. Indirect evaporative cooling. *Heating/Piping/Air Conditioning* 64(4):60-67.

Mathur, G.D. 1993. Retrofitting heat recovery systems with evaporative coolers. *Heating/Piping/Air Conditioning* 65(9):47-51.

Mathur, G.D. and T.W. McDonald. 1986. Simulation program for a two-phase thermosiphon-loop heat exchanger. *ASHRAE Transactions* 92(2A):473-85.

McDonald, T.W. and D. Shivprasad. 1989. Incipient nucleate boiling and quench study. Proceedings of CLIMA 2000 1:347-52. Sarajevo, Yugoslavia.

Phillips, E.G., R.E. Chant, B.C. Bradley, and D.R. Fisher. 1989a. A model to compare freezing control strategies for residential air-to-air heat recovery ventilators. *ASHRAE Transactions* 95(2):475-83.

Phillips, E.G., R.E. Chant, D.R. Fisher, and B.C. Bradley. 1989b. Comparison of freezing control strategies for residential air-to-air heat recovery ventilators. *ASHRAE Transactions* 95(2):484-90.

Ruch, M.A. 1976. Heat pipe exchangers as energy recovery devices. *ASHRAE Transactions* 82(1):1008-14.

Scofield, M. and J.R. Taylor. 1986. A heat pipe economy cycle. *ASHRAE Journal* 28(10)35-40.

Shah, R.K. 1981. Thermal design theory for regenerators. In *Heat exchangers: Thermal-hydraulic fundamentals and design*. S. Kakec, A.E. Bergles, and F. Maysinger, eds. Hemisphere Publishing Corporation.

BIBLIOGRAPHY

Andersson, B., K. Andersson, J. Sundell, and P.A. Zingmark. 1992. Mass transfer of contaminants in rotary enthalpy exchangers. *Indoor Air* 93(3):143-48.

Dehli, F., T. Kuma, and N. Shirahama. 1993. A new development for total heat recovery wheels. Energy Impact of Ventilation and Air Infiltration, 14th AIVC Conference, Copenhagen, Denmark, 261-68.

Mathur, G.D. and T.W. McDonald. 1987. Evaporator performance of finned air-to-air two-phase thermosiphon loop heat exchangers. *ASHRAE Transactions* 93(2):247-57.

Ninomura, P.T. and R. Bhargava. 1995. Heat recovery ventilators in multifamily residences in the arctic. *ASHRAE Transactions* 101(2):961-66.

Phillips, E.G., D.R. Fisher, R.E. Chant, and B.C. Bradley. 1992. Freeze-control strategy and air-to-air energy recovery performance. *ASHRAE Journal* 34(12):44-49.

SMACNA. 1978. *Energy recovery equipment and systems*. Report.

Stauder, F.A. and T.W. McDonald. 1986. Experimental study of a two-phase thermosiphon-loop heat exchanger. *ASHRAE Transactions* 92(2A): 486-97.

CHAPTER 45

UNITARY AIR CONDITIONERS AND UNITARY HEAT PUMPS

THE AIR-CONDITIONING and Refrigeration Institute (ARI) defines a **unitary air conditioner** as one or more factory-made assemblies that normally include an evaporator or cooling coil and a compressor and condenser combination. It may include a heating function as well. ARI defines an **air-source unitary heat pump** as consisting of one or more factory-made assemblies, which normally include an indoor conditioning coil, compressor(s), and an outdoor coil. It must provide a heating function and possibly a cooling function as well. A **water-source heat pump** is a factory-made assembly that rejects or extracts heat to and from a water loop instead of from ambient air. A unitary air conditioner or heat pump having more than one factory-made assembly (e.g., indoor and outdoor units) is commonly called a **split system**.

Unitary equipment is divided into three general categories: residential, light commercial, and commercial. Residential equipment is single-phase unitary equipment with a cooling capacity of 65,000 Btu/h or less and is designed specifically for residential application. Light commercial equipment is generally three phase, with cooling capacity up to 135,000 Btu/h, and is designed for small businesses and commercial properties. Commercial unitary equipment has cooling capacity higher than 135,000 Btu/h and is designed for large commercial buildings.

In the development of unitary equipment, the following design objectives are considered: (1) user requirements, (2) application requirements, (3) installation, and (4) service.

User Requirements

The user primarily needs either space conditioning for comfort or a controlled environment for products or manufacturing processes. Cooling, dehumidification, filtration, and air circulation often meet those needs, although heating, humidification, and ventilation are also required in many applications.

Application Requirements

Unitary equipment is available in many secondary system configurations, such as

- **Single zone, constant volume**, which consists of one controlled space with one thermostat that controls to maintain a set point.
- **Multizone, constant volume**, which has several controlled spaces served by one unit that supplies air of different temperatures to different zones as demanded (Figure 1).
- **Single zone, variable volume**, which consists of several controlled spaces served by one unit. Supply air from the unit is at a constant temperature, with air volume to each space varied to satisfy space demands (Figure 2).

Such factors as size, shape, and use of the building; availability and cost of energy; building aesthetics (equipment located outdoors); and space available for equipment are considered to determine the type of unitary equipment best suited to a given application. In general, roof-mounted single-package unitary equipment is limited to five or six stories because duct space and available blower power become excessive in taller buildings. Indoor, single-zone equipment is generally less expensive to maintain and service than multizone units located outdoors.

The building load and airflow requirements determine equipment capacity, whereas the availability and cost of fuels determine the energy source. Control system requirements must be established, and any unusual operating conditions must be considered early in the planning stage. In some cases, custom-designed equipment (discussed in Chapter 8) may be necessary.

Manufacturers' literature has detailed information about geometry, performance, electrical characteristics, application, and operating limits. The system designer then focuses on selecting suitable equipment with the capacity for the application.

Installation

Unitary equipment is designed to keep installation costs low. The equipment must be installed properly so that it functions in accordance with the manufacturer's specifications. Interconnecting diagrams for the low-voltage control system should be documented for proper servicing in the future. Adequate planning for the installation of large, roof-mounted equipment is important because special rigging equipment is frequently required.

The refrigerant circuit must be clean, dry, and leak-free. An advantage of packaged unitary equipment is that proper installation minimizes the risk of field contamination of the circuit. Care must be taken to properly install split-system interconnecting tubing (e.g., proper cleanliness, brazing, and evacuation to remove moisture). Some residential split systems are provided with precharged line sets and quick-connection couplings, which reduce the risk of field contamination of the refrigerant circuit. Split systems should be charged according to the manufacturer's instructions.

In the installation of split systems, lines must be properly routed and sized to ensure good oil return to the compressor. Chapters 5 and 6 of the 1998 *ASHRAE Handbook—Refrigeration* have more details on appropriate refrigerant piping practices.

Unitary equipment must be located to avoid noise and vibration problems. Single-package equipment of over 20 ton capacity should be mounted on concrete pads if vibration control is a concern.

The preparation of this chapter is assigned to TC 7.6, Unitary and Room Air Conditioners and Heat Pumps.

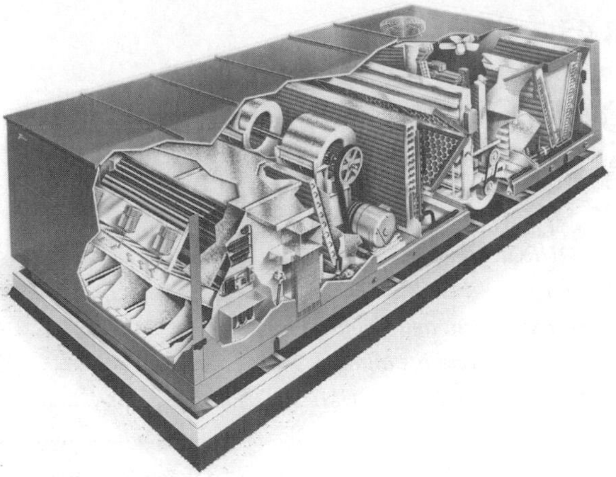

Fig. 1 Typical Rooftop Air-Cooled Single-Package Air Conditioner (Multizone)

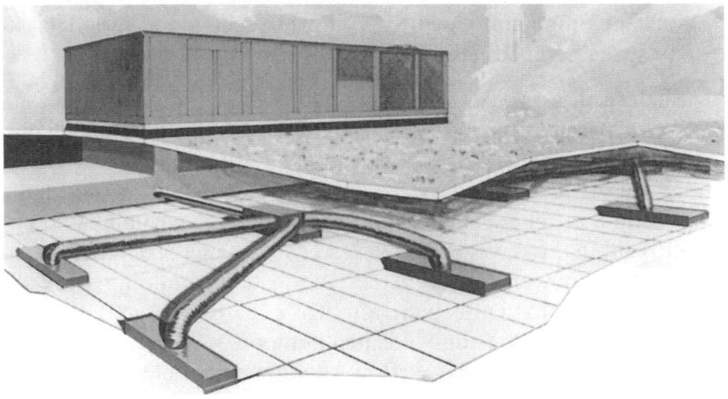

Fig. 2 Single-Package Air Equipment with Variable Air Volume

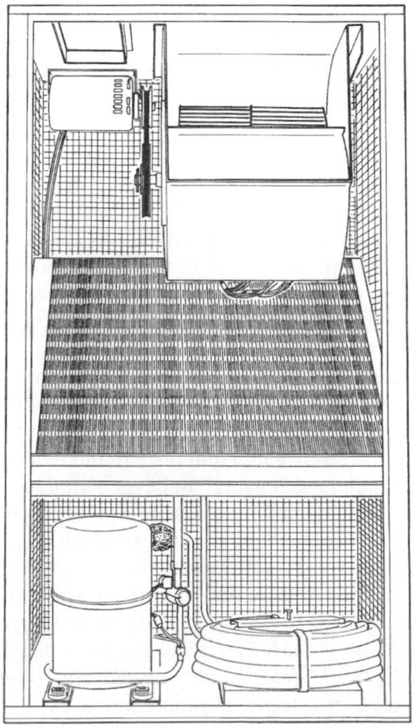

Fig. 3 Water-Cooled Single-Package Air Conditioner

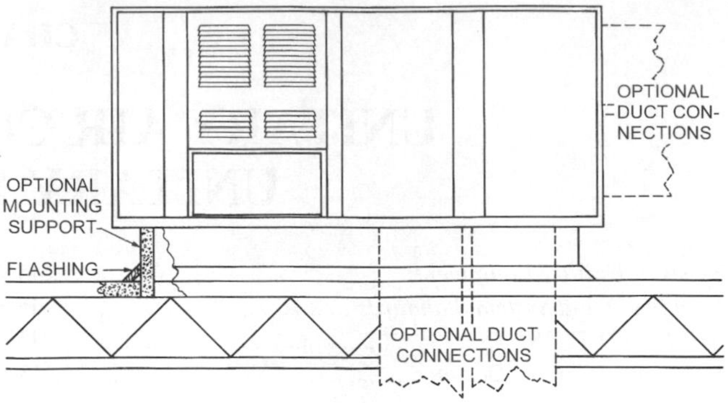

Fig. 4 Rooftop Installation of Air-Cooled Single-Package Unit

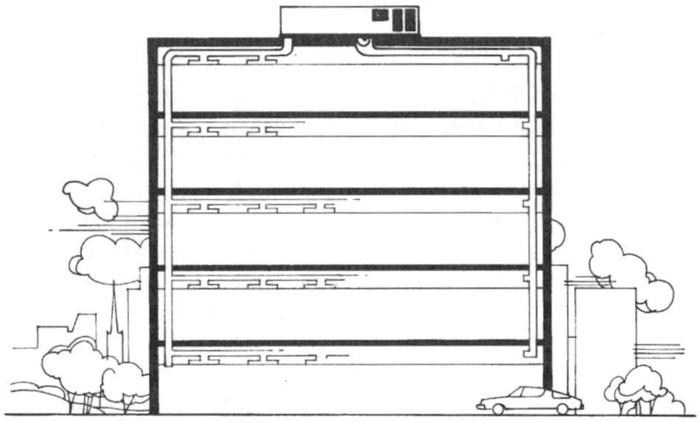

Fig. 5 Multistory Rooftop Installation of Single-Package Unit

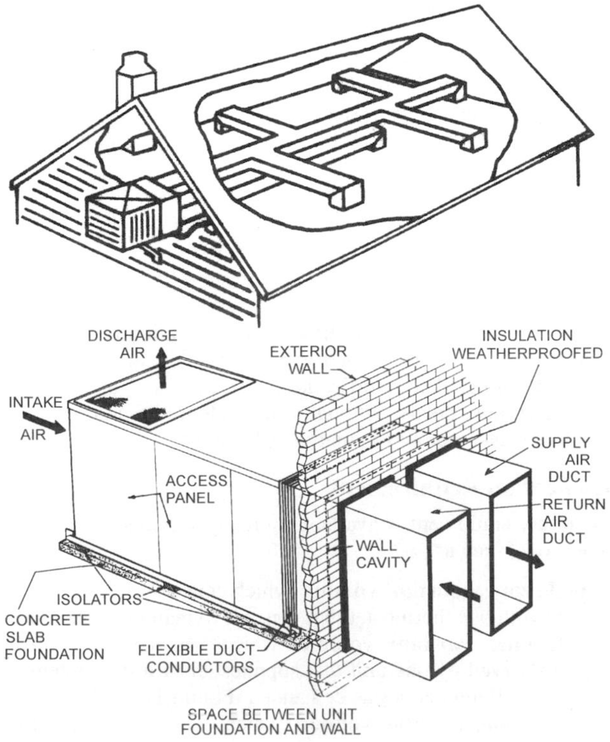

Fig. 6 Through-the-Wall Installation of Air-Cooled Single-Package Unit

Large-capacity equipment should be roof-mounted only after the structural adequacy of the roof has been evaluated. If they are located over occupied space, roof-mounted units with return fans that use ceiling space for the return plenum should have a lined return plenum according to the manufacturer's recommendations. Duct silencers should be used where low sound levels are desired. Weight and sound data are available from many manufacturers. Additional installation guidelines include the following:

- In general, install products containing compressors on solid, level surfaces.
- Avoid mounting products containing compressors (like remote units) on or touching the foundation of a house or building. A separate pad that does not touch the foundation is recommended to reduce any noise and vibration transmission through the slab.
- Do not box in outdoor air-cooled units with fences, walls, overhangs, or bushes. Doing so reduces the air-moving capability of the unit, thus reducing efficiency.
- For a split-system remote unit, choose an installation site that is close to the indoor portion of the system to minimize the pressure drop in the connecting tubing.
- Contact the unitary equipment manufacturer or consult the installation instructions for further information on installation procedures.

Unitary equipment should be listed or certified by nationally recognized testing laboratories to ensure safe operation and compliance with government and utility regulations. The equipment should also be installed to comply with the rating and application requirements of the agency standards to ensure that it performs according to industry criteria. Larger and more specialized equipment often does not carry agency labeling. However, power and control wiring practices should comply with the *National Electrical Code* (NFPA *Standard* 70). Local codes should be consulted before the installation is designed; local inspectors should be consulted before installation.

Service

A clear and accurate wiring diagram and a well-written service manual are essential to the installer and service personnel. Easy and safe service access must be provided in the equipment for periodic maintenance of filters and belts, cleaning, and lubrication. In addition, access for replacement of major components must be provided and preserved.

The availability of replacement parts aids proper service. Equitable warranty policies, covering 1 year of operation after installation, are offered by most manufacturers. Extended compressor warranties may be standard or optional.

Service personnel must be qualified to repair or replace mechanical and electrical components and to recover and properly recycle or dispose of any refrigerant removed from a system. They must also understand the importance of controlling moisture and other contaminants within the refrigerant circuit; they should know how to clean an hermetic system if it has been opened for service (see Chapter 6 of the 1998 *ASHRAE Handbook—Refrigeration*). Proper service procedures help ensure that the equipment will continue operating efficiently for its expected life.

TYPES OF UNITARY EQUIPMENT

Table 1 shows the types of unitary air conditioners available, and Table 2 shows the types of unitary heat pumps available. The following variations apply to some types and sizes of unitary equipment.

Arrangement. Major unit components for unitary air conditioners are arranged as shown in Table 1 and for unitary heat pumps as shown in Table 2.

Heat Rejection. Unitary air conditioner condensers may be air cooled, evaporatively cooled, or water cooled; the letters A, E, or W follow the ARI designation.

Table 1 ARI Classification of Unitary Air Conditioners

System Designation	ARI Type[a]	Heat Rejection	Arrangement	
Single package	SP-A	Air	Fan	Comp
	SP-E	Evap Cond	Evap	Cond
	SP-W	Water		
Refrigeration chassis	RCH-A	Air		Comp
	RCH E	Evap Cond	Evap	Cond
	RCH-W	Water		
Year-round single package	SPY-A	Air	Fan	
	SPY-E	Evap Cond	Heat	Comp
	SPY-W	Water	Evap	Cond
Remote condenser	RC-A	Air	Fan	
	RC-E	Evap Cond	Evap	Cond
	RC-W	Water	Comp	
Year-round remote condenser	RCY-A	Air	Fan	
	RCY-E	Evap Cond	Evap	Cond
	RCY-W	Water	Heat	
			Comp	
Condensing unit, coil alone	RCU-A-C	Air	Evap	Cond
	RCU-E-C	Evap Cond		Comp
	RCU-W-C	Water		
Condensing unit, coil and blower	RCU-A-CB	Air	Fan	Cond
	RCU-E-CB	Evap Cond	Evap	Comp
	RCU-W-CB	Air		
Year-round condensing unit, coil and blower	RCUY-A-CB	Air	Fan	
	RCUY-E-CB	Evap Cond	Evap	Cond
	RCUY-W-CB	Water	Heat	Comp

[a]Adding a suffix of "-O" following any of the above classifications indicates equipment not intended for use with field-installed duct systems.

Table 2 ARI Classification of Unitary Heat Pumps

System Designation	ARI Type[a]		Arrangement	
	Heating and Cooling	Heating Only		
Single package	HSP-A	HOSP-A	Fan	Comp
	HSP-W	HOSP-W	Indoor Coil	Outdoor Coil
Remote outdoor coil	HRC-A-CB	HORC-A-CB	Fan	
			Indoor Coil	Outdoor Coil
			Comp	
Remote outdoor coil with no indoor fan	HRC-A-C	HORC-A-C	Indoor Coil	Outdoor Coil
			Comp	
Split system	HRCU-A-CB	HORCU-A-CB	Fan	Comp
	HRCU-W-CB	HORCU-W-CB	Indoor Coil	Outdoor Coil
Split system, no indoor fan	HRCU-A-C	HORCU-A-C		Comp
			Indoor Coil	Outdoor Coil

[a] A suffix of "-O" following any of the above classifications indicates equipment not intended for use with field-installed duct systems.

Heat Source/Sink. Unitary heat pump outdoor coils are designated as air-source or water-source by an A or W, following ARI practice. The same coils that act as a heat sink in the cooling mode act as the heat source in the heating mode.

Unit Exterior. The unit exterior should be decorative for in-space application, functional for equipment room and ducts, and weatherproofed for outdoors.

Placement. Unitary equipment can be mounted on floors, walls, ceilings, roofs, or a pad on the ground.

Indoor Air. Equipment with fans may have airflow arranged for vertical upflow or downflow, horizontal flow, 90° or 180° turns, or multizone. Indoor coils without fans are intended for forced-air furnaces or blower packages. Variable volume blowers may be incorporated with some systems.

Location. Unitary equipment intended for indoor use may be placed in the conditioned space with plenums or furred-in ducts or concealed in closets, attics, crawl spaces, basements, garages, utility rooms, or equipment rooms. Wall-mounted equipment may be attached to or built into a wall or transom. Outdoor equipment may be mounted on roofs or concrete pads on the ground. Installations must conform with local codes.

Heat. Unitary systems may incorporate gas-fired, oil-fired, electric, hot water coil, or steam coil heating sections. In unitary heat pumps, these heating sections supplement the heating capability.

Ventilation Air. Outdoor air dampers may be built into the equipment to provide outdoor air for cooling or ventilation.

Desuperheaters. Desuperheaters may be applied to unitary air conditioners and heat pumps. These devices recover heat from the compressor discharge gas and use it to heat domestic hot water. The desuperheater usually consists of a pump, a heat exchanger, and controls, and it can produce about 5 to 6 gph of heated water per ton of air conditioning (heating water from 60 to 130°F). Because desuperheaters improve cooling performance and reduce the degrading effect of cycling during heating, they are best applied where cooling requirements are high and where a significant number of heating hours occur above the building's balance point (Counts 1985). While properly applied desuperheaters can improve cooling efficiency, they can also reduce space-heating capacity. This causes the unit to run longer, which reduces the cycling of the system above the balance point.

Ductwork. Unitary equipment is usually designed with fan capability for ductwork, although some units may be designed to discharge directly into the conditioned space.

Accessories. The manufacturer of any unitary equipment should be consulted before installing any accessories or equipment not specifically approved by the manufacturer. Such installations may not only void the warranty, but could cause the unitary equipment to function improperly or create fire or explosion hazards.

Combined Space-Conditioning/Water-Heating Systems

Unitary systems are available that provide both space conditioning and potable water heating. These systems are typically heat pumps, but some are available for cooling only. One type of combined system includes a full-condensing water-heating heat exchanger integrated into the refrigerant circuit of the space-conditioning system. Full-condensing system heat exchangers are larger than desuperheaters; they are generally sized to take the full condensing output of the compressor. Thus, they have much greater water-heating capacity. They also have controls that allow them to heat water year-round, either independently or coincidentally with space heating or space cooling. In spring and fall, the system is typically operated only to heat water.

Another type of combined system incorporates a separate, ancillary **heat pump water heater** (HPWH). The evaporator of this heater uses the return air (or liquid) stream of the space-conditioning system as a heat source. The HPWH thus cools the return stream during both space heating and cooling. In spring and fall, the space-conditioning blower (or pump) operates when water heating is needed.

As is the case with desuperheaters, simultaneous space and water heating reduces the output for space heating. This lower output is partially compensated for by the reduced cycling of the space-heating system above the balance point.

Combined systems can provide end users with significant energy savings and electric utilities with a significant reduction in demand. The overall performance of these systems is affected by the refrigerant charge and piping, the water piping, and the control logic and wiring. It is important, therefore, that the manufacturer's recommendations be closely followed. One special requirement is to locate the water-containing section(s) in areas not normally subjected to freezing temperatures.

Typical Unitary Equipment

Figures 1 through 6 show various types and installations of single-package equipment. Figure 7 shows a typical installation of a split-system, air-cooled condensing unit with indoor coil—the most widely used unitary cooling system. Figure 8 and Figure 9 also show split-system condensing units with coils and blower-coil units. Chapter 47 describes engine-driven heat pumps and air conditioners.

Many special light commercial and commercial unitary installations include a single-package air conditioner for use with variable air volume systems, as shown in Figure 2. These units are often equipped with a factory-installed system for controlling air volume in response to supply duct pressure (such as dampers or variable-speed drives).

Another example of a specialized unit is the **multizone unit** shown in Figure 1. The manufacturer usually provides all controls, including zone dampers. The air path in these units is designed so that supply air may flow through a hot deck containing a means of heating or through a cold deck, which usually contains a direct-expansion evaporator coil.

To make multizone units more efficient, a control is commonly provided that locks out cooling by refrigeration when the heating unit is in operation and vice versa. Another variation to improve efficiency is the three-deck multizone. This unit has a hot deck, a

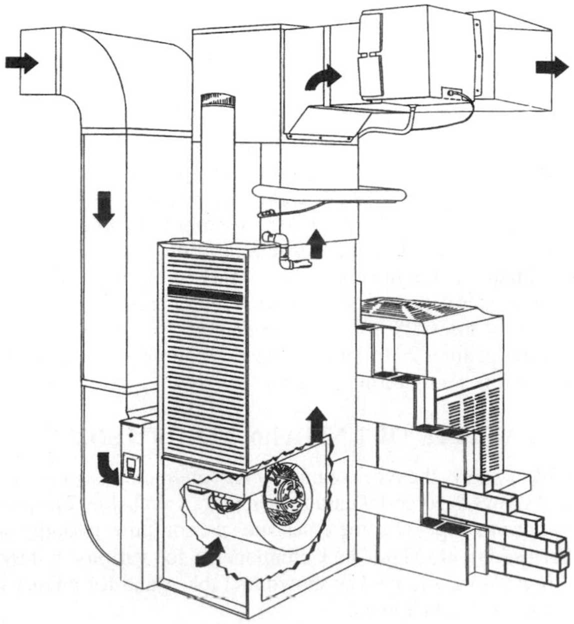

Fig. 7 Residential Installation of Split-System Air-Cooled Condensing Unit with Coil and Upflow Furnace

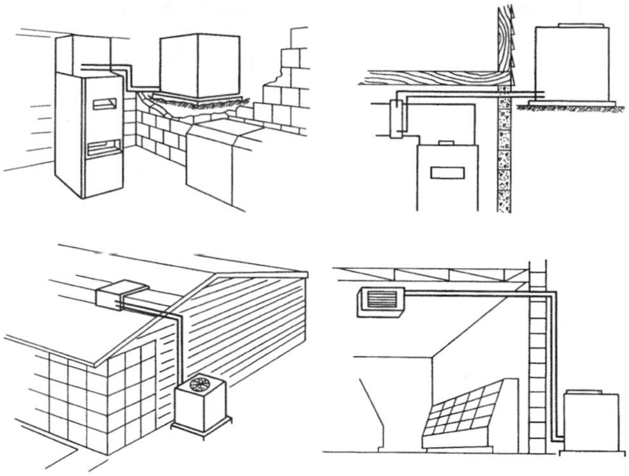

Fig. 8 Outdoor Installations of Split-System Air-Cooled Condensing Units with Coil and Upflow Furnace or with Indoor Blower-Coils

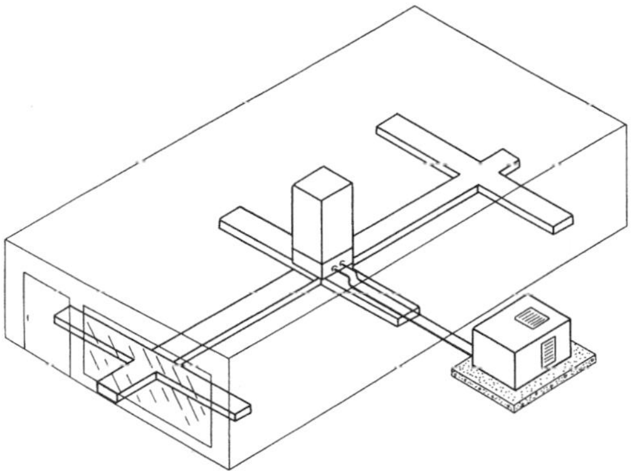

Fig. 9 Outdoor Installation of Split-System Air-Cooled Condensing Unit with Indoor Coil and Downflow Furnace

cold deck, and a neutral deck carrying return air. Hot and/or cold deck air mixes only with air in the neutral deck.

EQUIPMENT AND SYSTEM STANDARDS

Energy Conservation and Efficiency

In the United States, the Energy Policy and Conservation Act, (Public Law 95-163) requires the Federal Trade Commission (FTC) to prescribe an energy label for many major appliances, including unitary air conditioners and heat pumps. The National Appliance Energy Conservation Act (NAECA) (Public Law 100-12) provides minimum efficiency standards for major appliances, including unitary air conditioners and heat pumps.

The U.S. Department of Energy (DOE) testing and rating procedure is documented in Appendix M to Subpart 430 of Section 10 of the *Code of Federal Regulations*, Uniform Test Method for Measuring the Energy Consumption of Central Air Conditioners. This testing procedure provides a seasonal measure of operating efficiency for residential unitary equipment. The **seasonal energy efficiency ratio** (SEER) is the ratio of the total seasonal cooling output measured in Btu to the total seasonal watt-hours of input energy. This efficiency value is developed in the laboratory by conducting tests at various indoor and outdoor conditions, including a measure of performance under cyclic operation.

Seasonal heating mode efficiencies of heat pumps are similarly expressed as the ratio of the total heating output to the total seasonal input energy. This measure of efficiency is expressed as a **heating seasonal performance factor** (HSPF). In the laboratory, HSPF is determined from the test results at different conditions, including a measure of cyclic performance. The calculated HSPF depends not only on the measured equipment performance, but also on the climatic conditions and the heating load relative to the equipment capacity.

For HSPF rating purposes, the DOE has divided the United States into six climatic regions and has defined a range of maximum and minimum design loads. This division has the effect of producing about 30 different HSPF ratings for a given piece of equipment. DOE has established Region 4 (moderate northern climate) and the minimum design load as being the typical climatic region and building design load to be used for comparative certified performance ratings.

SEER, HSPF, and operating costs vary appreciably with equipment design and size and from manufacturer to manufacturer. SEER and HSPF values, size ranges, and unit operating costs for DOE-covered unitary air conditioners that are certified by ARI are published semiannually in the ARI *Directory of Certified Unitary Products*.

In the United States, the Energy Policy Act of 1992 requires unitary equipment with cooling capacities from 65,000 to 240,000 Btu/h to meet the minimum efficiency levels prescribed by ASHRAE *Standard* 90.1.

ARI Certification Programs

Equipment up to 135,000 Btu/h. ARI conducts three certification programs relating to unitary equipment up to 135,000 Btu/h, which are covered in ARI *Standard* 210/240 and ARI *Standard* 270. These standards include the performance requirements necessary for good equipment design. They also include the methods of testing established by ASHRAE *Standard* 37.

As part of its certification program, ARI publishes the *Directory of Certified Unitary Products*. Issued twice a year, this directory identifies certified products enrolled in one or more programs and lists the certified capacity and energy efficiency for each unit. Certification involves the annual audit testing of approximately 30% of the basic unitary equipment models of each participating manufacturer.

ARI *Standard* 210/240 established definitions and classifications, testing and rating methods, and performance requirements. Ratings are determined at ARI standard rating conditions, rated nameplate voltage, and prescribed discharge duct static pressures with rated evaporator airflow not exceeding 37.5 cfm per 1000 Btu/h. The standard requires that units dispose of condensate properly and have cabinets that do not sweat under cool, humid conditions. The ability to operate satisfactorily and restart at high ambient temperatures with low voltage is also tested.

For certification under ARI *Standard* 270, outdoor equipment is tested in accordance with Acoustical Society of America (ASA) *Standard* 92. Test results obtained on a one-third octave band basis are converted to a single number for application evaluation. Application principles are covered in ARI *Standard* 275.

Equipment over 135,000 Btu/h. Unitary air conditioners and heat pumps exceeding 135,000 Btu/h can be tested in accordance with ARI *Standard* 340/360. Unitary condensing units with capacities of 135,000 Btu/h or larger are covered by ARI *Standard* 365. ARI has a certification program for large unitary air-conditioning, heat pump, and condensing units with cooling capacities from 135,000 to 250,000 Btu/h. The ARI *Directory of Certified Applied Air-Conditioning Products*, published twice a year, contains the certified values for such equipment.

Safety Standards and Installation Codes

Approval agencies list unitary air conditioners complying with a standard like Underwriters Laboratories (UL) *Standard* 1995, Heating and Cooling Equipment (CSA *Standard* C22.2 No. 236). Other UL standards may also apply. An evaluation of the product determines that its design complies with the construction requirements specified in the standard and that the equipment can be installed in accordance with the applicable requirements of the *National Electrical Code*; ASHRAE *Standard* 15, Safety Code for Mechanical Refrigeration; NFPA *Standard* 90A, Installation of Air Conditioning and Ventilating Systems; and NFPA *Standard* 90B, Installation of Warm Air Heating and Air Conditioning Systems.

Tests determine that the equipment and all components will operate within their recognized ratings, including electrical, temperature, and pressure, when the equipment is energized at rated voltage and operated at specified environmental conditions. Stipulated abnormal conditions are also imposed under which the product must perform in a safe manner. The evaluation covers all operational features (such as electric space heating) that may be used in the product.

Products complying with the applicable requirements may bear the agency listing mark. An approval agency program includes the auditing of continued production at the manufacturer's factory.

AIR CONDITIONERS

Unitary air conditioners consist of factory-matched refrigerant circuit components that are applied in the field to fulfill the requirements of the user. The manufacturer often incorporates a heating function compatible with the cooling system and a control system that requires a minimum of field wiring.

A variety of products is available to meet the objectives of nearly any system. Many different heating sections (gas- or oil-fired, electric, or condenser reheat), air filters, and heat pumps, which are a specialized form of unitary product, are available. Such matched equipment, selected with compatible accessory items, requires little field design or field installation work.

REFRIGERANT CIRCUIT DESIGN

Chapters 21, 34, and 35 describe coil, compressor, and condenser designs. Chapters 2, 5, 6, and 7 of the 1998 *ASHRAE Handbook—Refrigeration* cover refrigerant circuit piping selection, chemistry, cleanliness, and lubrication. Proper coil circuiting is essential for adequate oil return to the compressor. Crankcase heaters are usually incorporated to prevent refrigerant migration to the compressor crankcase during shutdown. Oil pressure switches and pumpdown or pumpout controls are used when additional assurance of reliability is economical and/or required.

Safety Controls

High-pressure and high-temperature limiting devices, internal pressure bypasses, current limiting devices, and devices that limit compressor torque prevent excessive mechanical and electrical stresses. Low-pressure or temperature cutout controllers may be used to protect against loss of charge, coil freeze-up, or loss of evaporator airflow.

Flow Control Devices

Refrigerant flow is most commonly controlled either by a fixed metering device such as a short tube restrictor or capillary tube or by thermostatic expansion valves. Capillaries and short tube restrictors are simple, reliable, and economical, and they can be sized for peak performance at rating conditions. The evaporator may be overfed at high condensing temperatures and underfed at low condensing temperatures because of changing pressure differential across the fixed metering device. When such conditions exist, a less-than-optimum cooling capacity usually results. However, the degree of loss varies with the design of the condenser, the volume of the system, and the total refrigerant charge. The amount of unit charge is critical, and a capillary-controlled evaporator must be matched to the specific condensing unit.

Properly sized thermostatic expansion valves provide constant superheat and good control over a range of operating conditions. Superheat is adjusted to ensure that only superheated gas returns to the compressor, usually with 7 to 14°F superheat at the compressor inlet at normal rating conditions. This superheat setting may be higher at a lower outdoor ambient temperature (cooling tower water temperature for water-cooled products) or indoor wet-bulb temperature. Compressor loading can be limited with vapor-charged thermostatic expansion valves. Low discharge pressure (low ambient) operation causes a diminished pressure drop across valves and capillaries so that full flow is not maintained. Decreased capacity, low coil temperatures, and freeze-up can result unless low ambient condensing pressure control is provided.

Properly designed unitary equipment allows only a minimum amount of liquid refrigerant to return through the suction line to the compressor during abnormal operation. Normally, the heat absorbed in the evaporator vaporizes all the refrigerant and adds a few degrees of superheat. However, any conditions that increase refrigerant flow beyond the heat transfer capabilities of the evaporator can cause liquid carryover into the compressor return line. Such an increase may be caused by a poorly positioned thermal element of an expansion valve or by an increase in the condensing pressure of a capillary system, which may be caused by fouled condenser surfaces, excessive refrigerant charge, reduced flow of condenser air or water, or the higher temperature of the condenser cooling medium. Heat transfer at the evaporator may be reduced by dirty surfaces, low-temperature air entering the evaporator, or reduced airflow caused by a blockage in the air system.

Piping

Transient flow conditions are a special concern. During off periods, refrigerant migrates and condenses in the coldest part of the system. In an air conditioner within a cooled space, this area is typically the evaporator. When the compressor starts, the liquid tends to return to the compressor in slugs. The severity of **slugging** is affected by temperature differences, off time, component positions, and traps formed in suction lines. Various methods such as suction-line accumulators, specially designed compressors, the refrigerant pumpdown cycle, nonbleed port thermostatic expansion valves, liquid-line solenoid valves, or limited refrigerant charge are used to avoid equipment problems associated with excessive liquid return. Chapter 2 of the 1998 *ASHRAE Handbook—Refrigeration* has further information on refrigerant piping.

Strainers and filter dryers minimize the risk of foreign material restricting capillary tubes and expansion valves (e.g., small quantities of solder, flux, and varnish). Overheated and oxidized oil may dissolve in warm refrigerant and deposit at lower temperatures in capillary tubes, expansion valves, and evaporators. Filter dryers are highly desirable, particularly for split-system units, to remove any moisture that may have been introduced during installation or servicing. Moisture contamination can cause oil breakdown, motor insulation failure, and freezing or other restrictions at the expansion device.

Capacity Control

Buildings with high internal heat loads require cooling even at low outdoor temperatures. The capacity of air-cooled condensers can be controlled by changing airflow or flooding tubes with refrigerant. Airflow can be changed by using dampers, adjusting fan speed, or stopping some of the fan motors in a multifan system.

Suction pressures drop momentarily during start-up. Circuits with a low-pressure cutout controller may require a time-delay relay to bypass it momentarily to prevent nuisance tripping.

In cool weather, air conditioners operate for short periods only. If the weather is also damp, high levels and wide variances in humidity may occur. Properly designed capacity-controlled units operate for longer periods, which may improve humidity control and comfort. In any case, cooling equipment should not be oversized.

Units with two or more separate refrigerant circuits permit independent operation of the individual systems, which reduces capacity while better matching the changing load conditions. Larger, single-compressor systems may offer capacity reduction through the use of cylinder-unloading compressors, variable-speed or multispeed compressors, multiple compressors, or hot-gas bypass controls. At full-load operation, efficiency is unimpaired. However, reduced capacity operation may increase or decrease system efficiency, depending on the type of capacity-reduction method used.

Variable-speed or multispeed compressors and use of multiple compressors can improve efficiency at part-load operation. Cylinder unloading can increase or decrease system performance, depending on the particular method used. Multispeed and variable-speed compressors, multiple compressors, and cylinder unloading generally produce higher comfort levels through lower cycling and better matching of the capacity to the load. Hot-gas bypass does not reduce capacity efficiently, although it generally provides a wider range of capacity reduction. Units with capacity-reduction compressors usually have capacity-controlled evaporators; otherwise the evaporator coil temperatures may be too high to provide dehumidification. Capacity-controlled evaporators are usually split, with at least one of the expansion valves controlled by a solenoid valve. Evaporator capacity is reduced by closing the solenoid valve. The compressor capacity-reduction controls or the hot-gas bypass system then provides maximum dehumidification, while the evaporator coil temperature is maintained above freezing to avoid coil frosting. Chapter 2 of the 1998 *ASHRAE Handbook—Refrigeration* has details on hot-gas bypass.

AIR-HANDLING SYSTEMS

High airflow, low static pressure performance, simplicity, economics, and compact arrangement are characteristics that make propeller fans particularly suitable for nonducted air-cooled condensers. Small-diameter fans are direct-driven by four-, six-, or eight-pole motors. Low starting torque requirements allow the use of single-phase shaded pole and permanent split-capacitor (PSC) fan motors and simplify speed control for low outdoor temperature operation. Many larger units use multiple fans and three-phase motors. Larger diameter fans are belt driven at a lower rpm to maintain low tip speeds and quiet operation.

Centrifugal blowers meet the higher static pressure requirements of ducted air-cooled condensers, forced-air furnaces, and evaporators. Indoor airflow must be adjusted to suit duct systems and plenums while providing the required airflow to the coil. Some small blowers are direct-driven with multispeed motors. Large blowers are always belt-driven and may have variable-pitch motor pulleys for airflow adjustment. Vibration isolation reduces the amount of noise transmitted by bearings, motors, and blowers into cabinets. (See Chapter 18 for details of fan design and Chapters 16 and 17 for information on air distribution systems.)

Disposable fiberglass filters are popular because they are available in standard sizes at low cost. Cleanable filters offer economic advantages when cabinet dimensions are not compatible with common sizes. Electronic or other high-efficiency air cleaners are used when a high degree of cleaning is desired. Larger equipment frequently is provided with automatic roll filters or high-efficiency bag filters. (See Chapter 24 for additional details about filters.)

Provision for introducing outdoor air for economizer cooling and/or ventilation is made in many units; rooftop units are particularly adaptable for receiving outdoor air. Air-to-air heat exchangers can be used to reduce the energy losses from ventilation. Some units have automatically controlled dampers to permit cooling by outdoor air, which increases system efficiency.

ELECTRICAL DESIGN

Electrical controls for unitary equipment are selected and tested to perform their individual and interrelated functions properly and safely over the entire range of operating conditions. Internal line-break thermal protectors provide overcurrent protection for most single-phase motors, smaller sizes of three-phase motors, and hermetic compressor motors. These rapidly responding temperature sensors, embedded in motor windings, can provide precise locked rotor and running overload protection.

Branch-circuit, short-circuit, and ground-fault protection is commonly provided by fused disconnect switches. Time-delay fuses allow selection of fuse ratings closer to running currents and thus provide backup motor overload protection, as well as short-circuit and ground-fault protection. Circuit breakers may be used in lieu of fuses where their use conforms to appropriate code requirements.

Some larger compressor motors have dual windings and contactors for step starting. A brief delay when energizing contactors reduces the magnitude of inrush current.

The use of 24 V (NEC Class 2) control circuitry is common for room thermostats and interconnecting wiring between split systems. It offers advantages in temperature control, safety, and ease of installation. Electronic, communicating microprocessor thermostats and control systems are becoming common.

Motor speed controls are used to vary evaporator airflow of direct-drive fans, air-cooled condenser airflow for low outdoor temperature operation, and compressor speed to match load demand. Multitap motors and autotransformers provide one or more speed steps. Solid-state speed control circuits can provide a continuously variable speed range. However, motor bearings, windings, overload protection, and the motor suspension system must be suitable for operation over the full speed range.

In addition to speed control, solid-state circuits provide reliable temperature control, motor protection, and expansion valve refrigerant control. Complete temperature control systems are frequently included with the unit. Control system features such as automatic night setback, economizer control sequence, and zone demand control of multizone equipment contribute to improved comfort and energy savings. Chapter 45 of the 1999 *ASHRAE Handbook—Applications* has additional information on control systems.

MECHANICAL DESIGN

Cabinet height is important for rooftop and ceiling-suspended units. The size limitations of truck bodies, freight cars, doorways, elevators, and various rigging practices must be considered in large unit design. In addition, structural strength of both the unit and the crate must be adequate for handling, warehouse stacking, shipping, and rigging.

The following criteria are also important: (1) cabinet insulation must prevent excessive sweating in high-humidity ambient conditions; (2) insulated surfaces exposed to moving air should withstand air erosion; (3) air leakage around panels and at cabinet joints should be minimized; and (4) the cabinet insulation must be adequate to reduce energy transfer losses from the circulating airstream.

Also, cooling coil air velocities must be low enough to ensure that condensate is not blown off the coil. The drain pan must be sized to contain the condensate and must also be protected from high-velocity air. Service access must be provided for installation and repair. Versatility of application, such as multiple fan discharge directions and the ability to install piping from either side of the unit, is another consideration. Weatherproofing requires careful attention and testing.

ACCESSORIES

Using standard cataloged accessories, the designer can often incorporate unitary products in special applications. Typical examples (Figures 4, 6, 7, and 8) are plenum coil housings, return air filter grilles, and diffuser-return grilles for single-outlet units. Air duct kits offered for rooftop units (Figure 4) permit concentric or side-by-side ducting, as well as horizontal or vertical connections. Mounting curbs are available to facilitate unit support and roof flashing. Other accessories include high static pressure fan drives, controls for low outdoor temperature operation, and duct damper kits for control of outdoor air intakes and exhausts.

HEATING

It is important to install cooling coils downstream of furnaces so that condensation does not form inside the combustion and flue passages. Upstream cooling coil placement is permissible when the furnace has been approved for this type of application and designed to prevent corrosion. Burners, pilot flames, and controls must be protected from the condensate.

Chapter 23 describes hot water and steam coils used in unitary equipment, as well as the prevention of coil freezing from ventilation air in cold weather. Chapters 26, 28, and 31 discuss forced-air and oil- and gas-fired furnaces commonly used with, or included as part of, year-round equipment.

AIR-SOURCE HEAT PUMPS

Capacities of unitary air-source heat pumps range from about 1.5 to 30 tons, although there is no specific limitation. This equipment is used in residential, commercial, and industrial applications. Multi-unit installations are particularly advantageous because they permit zoning, which provides the opportunity for heating or cooling in each zone on demand. Application factors unique to unitary heat pumps include the following:

- The unitary heat pump normally fulfills a dual function—heating and cooling; therefore, only a single piece of equipment is required for year-round comfort. Some regions, especially central and northern Europe and parts of North America, have little need for cooling. Some manufacturers offer heating-only heat pumps for these areas and for special applications.
- A single source of energy can supply both heating and cooling requirements.
- Heat output can be as much as two to four times that of the purchased energy input.
- Vents and/or chimneys may be eliminated, thus reducing building costs.

In an air-source heat pump (Figure 10), the outdoor coil rejects heat to outdoor air when in the cooling mode and extracts heat from outdoor air when in the heating mode. Most residential applications consist of an indoor fan and coil unit, either vertical or horizontal, and an outdoor fan-coil unit. The compressor is usually located in the outdoor unit. Electric heaters are commonly included in the indoor unit to provide heat during defrost cycles and during periods of high heating demand that cannot be satisfied by the heat pump alone.

Add-On Heat Pumps

An air-source heat pump can be added to new or existing gas- or oil-fired furnaces. This unit, typically called an add-on, dual-fuel or hybrid heat pump, normally operates as a conventional heat pump. During extremely cold weather, the refrigerant circuit is turned off

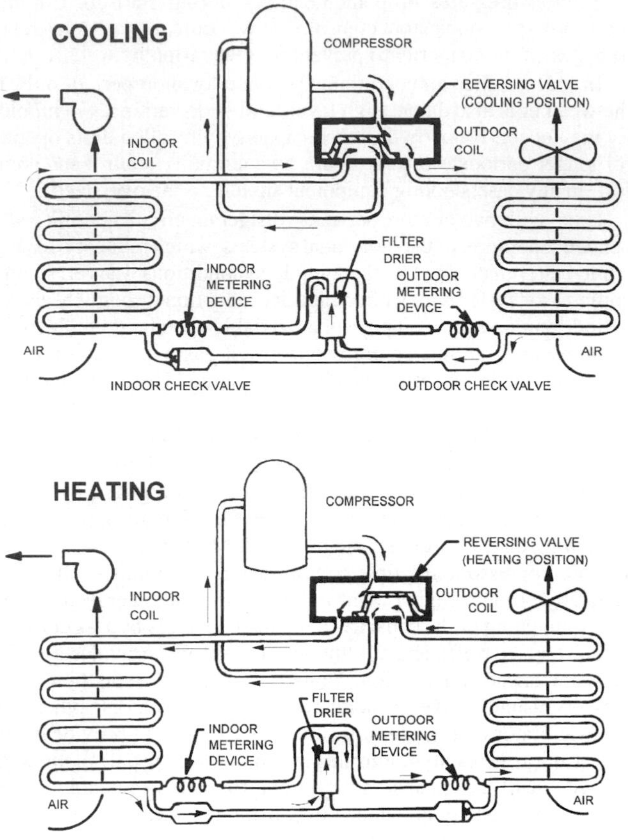

Fig. 10 Typical Schematic of Air-to-Air Heat Pump System

and the furnace provides the required space heating. These add-on heat pumps share the air distribution system with the warm air furnace. The indoor coil may be either parallel to or in series with the furnace. However, the furnace should never be upstream of the indoor coil if both systems are operated together.

Special controls are available that prevent simultaneous operation of the heat pump and the furnace in this configuration. Such operation raises the refrigerant condensing temperature, which could cause a compressor failure. In applications where the heat pump and furnace operate at the same time, the following conditions must be met: (1) the furnace and heat pump indoor coil must be arranged in parallel, or (2) the furnace combustion and flue passages must be designed to avoid condensation-induced corrosion during the cooling operation.

SELECTION

Figure 11 shows performance characteristics of a single-speed, air-source heat pump, along with the heating and cooling loads for a typical building. The heat pump heating capacity decreases as the ambient temperature decreases. This characteristic is opposite to the trend of the building load. The outdoor temperature at which the heat pump capacity equals the building load is called the **balance point**. When the outdoor temperature is below the balance point, supplemental heat (usually electric resistance) must be added to make up the difference, as shown by the shaded area. The COP shown in Figure 11 is for the refrigerant circuit only and does not include supplemental heat effects below the balance point.

In selecting the proper size heat pump, the cooling load for the building is calculated using standard practice. The heating balance

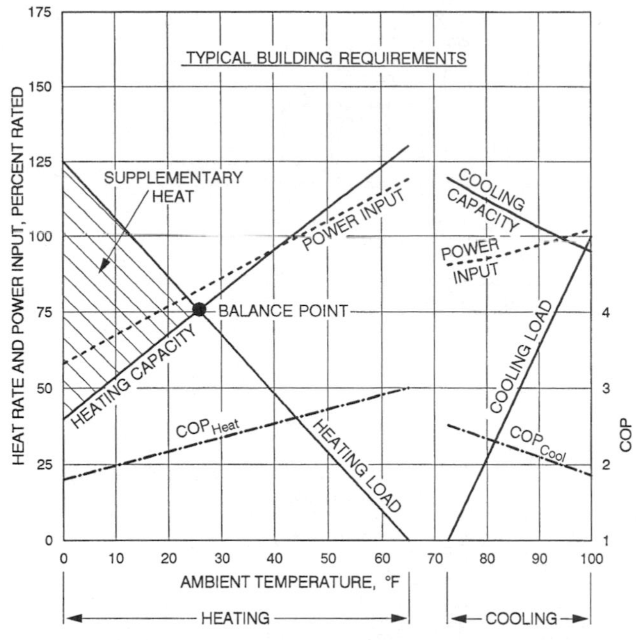

Fig. 11 Operating Characteristics of Single-Stage Unmodulated Heat Pump

point may be lowered by improving the thermal performance of the structure or by choosing a heat pump larger than the cooling load requires. Excessive oversizing of the cooling capacity causes excessive cycling, which results in uncomfortable temperature and humidity levels in cooling.

Use of variable-speed, multispeed, or multiple compressors and variable-speed fans can improve the matching of both heating and cooling loads over an extended range. This equipment can reduce cycling losses and provide improved comfort levels as well.

REFRIGERANT CIRCUIT AND COMPONENTS

Heat pump yearly operating hours are often up to five times those of a cooling-only unit. In addition, heating extends over a greater range of system operating conditions at higher stress conditions, so the design must be thoroughly analyzed to ensure maximum reliability. Improved components and protective devices contribute to better reliability, but the equipment designer must select components that are approved for the specific application.

For a reliable and efficient heat pump system, the following factors must be considered: (1) outdoor coil circuitry, (2) defrost and water drainage, (3) refrigerant flow controls, (4) refrigerant charge management, and (5) compressor selection.

Outdoor Coil Circuitry

When the heat pump is used for heating, the outdoor coil operates as an evaporator. The refrigerant in the coil is less dense than when the coil operates as a condenser. To avoid an excessive pressure drop during heating, the circuitry is usually a compromise between optimum performance as an evaporator and optimum performance as a condenser.

Defrost and Water Drainage

During colder outdoor temperatures, usually below 40 to 50°F, and high relative humidities (above 50%), the outdoor coil operates below the frost point of the outdoor air. The frost that builds up on the surface of the coil is usually removed by the **reverse-cycle defrost** method. In this method, the refrigerant flow in the system is reversed, and hot gas from the compressor flows through the outdoor coil, melting the frost. A typical defrost takes 4 to 10 min. The outdoor fan is normally off during defrost. Because the defrost is a transient process, capacity, power, and the pressures and temperatures of refrigerant in different parts of the system change throughout the defrost period (Miller 1989, O'Neal 1989a).

The performance of the heat pump during the defrost cycle can be enhanced in several ways. Improved defrost times and water removal can be achieved by ensuring that adequate refrigerant is routed to the lower refrigerant circuits in the outdoor coil. Properly sizing the defrost expansion device is critical for reducing defrost times and energy use (O'Neal 1989b). If the expansion device is too small, suction pressure can be below atmospheric, defrost times become long, and energy use is high. If the expansion device is too large, the compressor can be flooded with liquid refrigerant. During the conventional reverse-cycle defrost, there is a significant pressure spike at defrost termination. Starting the fan 30 to 45 s before defrost termination can minimize the spike (Anand et al. 1989). In cold climates, the cabinet should be installed above grade to provide good drainage during defrost and to minimize snow and ice buildup around the cabinet. During prolonged periods of severe weather, it may be necessary to clear ice and snow from around the unit.

Several methods are used to determine the need to defrost. One of the more common, simple, and reliable control methods is to initiate defrost at predetermined time intervals (usually 90 min). Demand-type systems detect a need for defrosting by measuring changes in air pressure drop across the outdoor coil or changes in temperature difference between the outdoor coil and the outdoor air. Microprocessors are applied to control this function, as well as numerous other functions (Mueller and Bonne 1980). Demand defrost control is preferred because it requires less energy than other defrost methods.

Refrigerant Flow Controls

Separate refrigerant flow controls are usually used for the indoor and outdoor coils. Because the refrigerant flow reverses direction between the heating and cooling mode of operation, a check valve bypasses in the appropriate direction around each expansion device. Capillaries, fixed orifices, thermostatic expansion valves, or electronically controlled expansion valves may be used; however, capillaries and fixed orifices require that greater care be taken to prevent excessive flooding of refrigerant into the compressor. A check valve is not needed when an orifice-type expansion device or a biflow expansion valve is used. The reversing valve is the critical additional component required to make a heat pump air-conditioning system.

Refrigerant Charge Management

Extra care is required to control compressor flooding and the storage of refrigerant in the system during both heating and cooling. The mass flow of refrigerant during cooling is greater than during heating. Consequently, the amount of refrigerant stored may be greater in the heating mode than in the cooling mode, depending on the relative internal volumes of the indoor and outdoor coils. Usually, the internal volume of the indoor coils ranges from 110 to 70% of the outdoor coil volume. The relative volumes can be adjusted so that the coils not only transfer heat but also manage the charge.

When capillaries or fixed orifices are used, the refrigerant may be stored in an accumulator in the suction line or in receivers that can remove the refrigerant charge from circulation when compressor floodback is imminent. Thermostatic expansion valves reduce the flooding problem, but storage may be required in the condenser. Use of accumulators and/or receivers is particularly important in split systems.

To maintain performance reliability, the amount of refrigerant in the system must be checked and adjusted in accordance with the manufacturer's recommendations, particularly when charging a heat pump. Manufacturer's recommendations for accumulator installation must be followed so that good oil return is assured.

Compressor Selection

Compressors are selected on the basis of performance, reliability, and probable applications of the unit. In good design practice, equipment manufacturers often consult with compressor manufacturers during both design and application phases of the unitary equipment to verify proper application of the compressor. Compressors in a heat pump operate over a wide range of suction and discharge pressures; thus, their design parameters, (e.g., refrigerant discharge temperatures, pressure ratios, clearance volume, and motor-overload protection) require special consideration. In all operating conditions, compressors should be protected against loss of lubrication, liquid floodback, and high discharge temperatures.

SYSTEM CONTROL AND INSTALLATION

The installation should follow the manufacturer's instructions. Because the supply air from a heat pump is usually at a lower temperature (typically 90 to 100°F) than that from most heating systems, ducts and supply registers should control air velocity and throw to minimize the perception of cool drafts.

Low-voltage heating/cooling thermostats control heat pump operation. Both models that switch automatically from heating to cooling operation and manual selection models are available. Usually, heating is controlled in two stages. The first stage controls heat pump operation, and the second stage controls supplementary heat. When the heat pump cannot satisfy the first stage's call for heat, supplementary heat is added by the second-stage control. The amount of supplementary heat is often controlled by an outdoor thermostat that allows additional stages of heat to be turned on only when required by the colder outdoor temperature.

Microprocessor technology has led to night setback modes and intelligent recovery schemes for morning warm-up on heat pump systems (see Chapter 45 of the 1999 *ASHRAE Handbook—Applications*).

WATER-SOURCE HEAT PUMPS

A water-source heat pump (WSHP) is a single-package reverse-cycle heat pump that uses water as the heat source when in the heating mode and as the heat sink when in the cooling mode. The water supply may be a recirculating closed loop, a well, a lake, or a stream. Water for closed-loop heat pumps is usually circulated at 2 to 3 gpm per ton of cooling capacity. A groundwater heat pump (GWHP) can operate with considerably less water flow. The main components of a WSHP refrigeration system are a compressor, a refrigerant-to-water heat exchanger, a refrigerant-to-air heat exchanger, refrigerant expansion devices, and a refrigerant-reversing valve. Figure 12 shows a schematic of a typical WSHP system.

Designs of packaged WSHPs range from horizontal units located primarily above the ceiling or on the roof, to vertical units usually located in basements or equipment rooms, to console units located in the conditioned space. Figure 13 and Figure 14 illustrate typical designs.

SYSTEMS

WSHPs are used in a variety of systems. These include the following:

1. Water-loop heat pump systems (Figure 15A)
2. Groundwater heat pump systems (Figure 15B)
3. Closed-loop surface water heat pump systems (Figure 15C)
4. Surface water heat pump systems (Figure 15D)
5. Ground-coupled heat pump systems (Figure 15E)

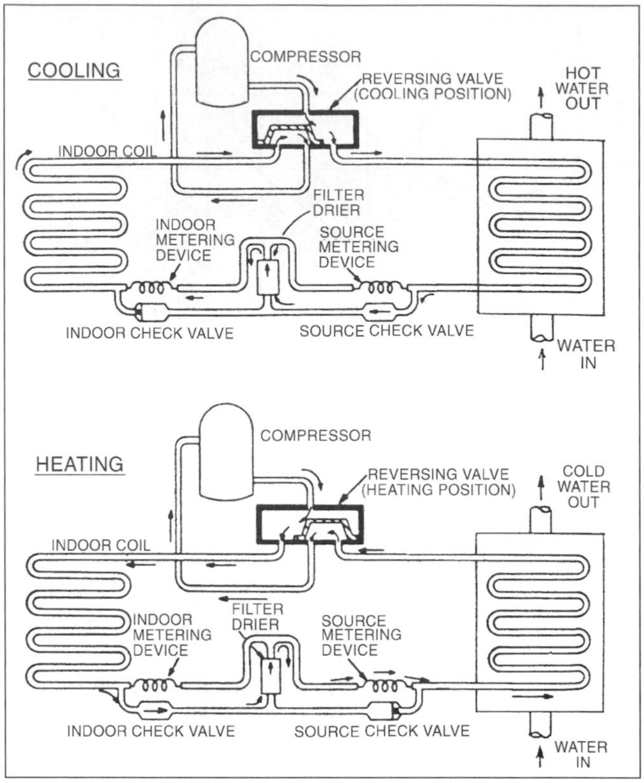

Fig. 12 Schematic of a Typical Water-Source Heat Pump System

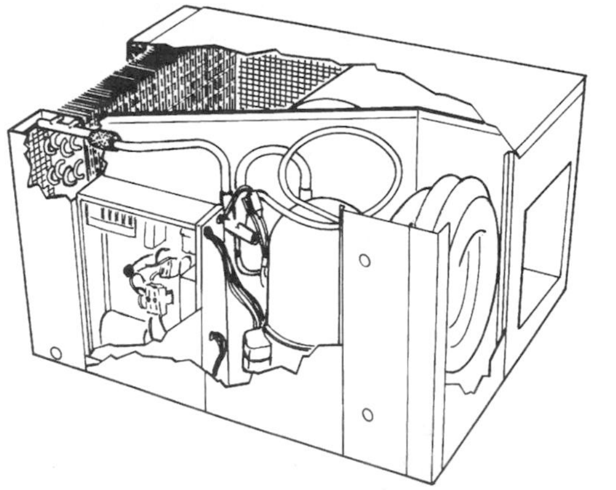

Fig. 13 Typical Horizontal Water-Source Heat Pump

A **water-loop heat pump** (WLHP) uses a circulating water loop as the heat source and the heat sink. When the loop water temperature exceeds a certain level during cooling operation, a cooling tower dissipates heat from the water loop into the atmosphere. When the loop water temperature drops below a prescribed level during heating operation, heat is added to the circulating loop water, usually with a boiler. In multiple-unit installations, some heat pumps may operate in the cooling mode while others operate in the heating mode, and controls are needed to keep the loop water temperature within the prescribed limits. Chapter 8 has more information on water-loop heat pumps.

A **groundwater heat pump** (GWHP) uses groundwater from a nearby well and passes it through the heat pump's water-to-refrigerant heat exchanger where it is warmed or cooled, depending on

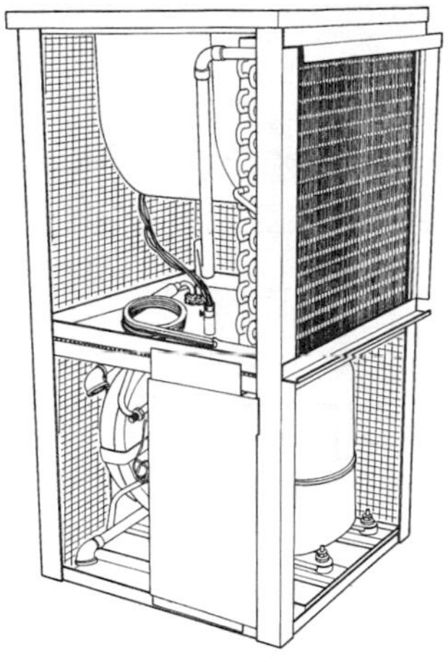

Fig. 14 Typical Vertical Water-Source Heat Pump

the operating mode. It is then discharged to a drain, a stream, or a lake, or it is returned to the ground through a reinjection well.

Many state and local jurisdictions have enacted ordinances relating to the use and discharge of groundwater. Because aquifers, the water table, and groundwater availability vary from region to region, these regulations cover a wide spectrum.

A **surface water heat pump** (SWHP) uses water from a nearby lake, stream, or canal. After passing through the heat pump heat exchanger, it is returned to the source or a drain several degrees warmer or cooler, depending on the operating mode of the heat pump. **Closed-loop** surface water heat pumps use a closed water or brine loop that includes pipes or tubing submerged in the surface water (river, lake, or large pond) that serves as the heat exchanger. The adequacy of the total thermal capacity of the body of water must be considered.

A **ground-coupled heat pump** (GCHP) system uses the earth as a heat source and sink. Usually, plastic piping is installed in either a shallow horizontal or deep vertical array to form the heat exchanger. The massive thermal capacity of the earth provides a temperature-stabilizing effect on the circulating loop water or brine. Installing this type of system requires detailed knowledge of the climate; the site; the soil temperature, moisture content, and thermal characteristics; and the performance, design, and installation of water-to-earth heat exchangers. The *Design/Data Manual for Closed-Loop Ground-Coupled Heat Pump Systems* (ASHRAE 1985) has detailed information on design of GCHP systems. Additional information on GCHP systems is presented in Chapter 31 of the 1999 *ASHRAE Handbook—Applications*.

Entering Water Temperatures

These various water sources provide a wide range of entering water temperatures to WSHPs. Not only do the entering water temperatures vary by water source, but they also vary by climate and time of year. Due to the wide range of entering water or brine temperatures encountered, it is not feasible to design a universal packaged product that can handle the full range of possibilities effectively. Therefore, WSHPs are rated for performance at a number of standard rating conditions.

PERFORMANCE CERTIFICATION PROGRAMS

ARI maintains the following certification programs for water-source heat pumps:

1. ARI *Standard* 320 for Water-Source Heat Pumps. A water-source heat pump is typically one of multiple units using fluid circulated in a common piping loop as a heat source/heat sink. The temperature of the loop fluid is usually mechanically controlled within a moderate temperature range of 60 to 90°F.

   Units tested in accordance with ARI *Standard* 320 have standard cooling ratings at 85°F entering water temperature and standard heating ratings at 70°F entering water temperature. Maximum operating conditions are checked at the upper temperature level at 90°F entering water temperature and at the lower temperature level at 60°F leaving water temperature. The range of test conditions covers the extremes of entering water temperatures typically encountered in WLHP systems.
2. ARI *Standard* 325 for Groundwater-Source Heat Pumps. A groundwater-source heat pump typically uses water pumped from a well, lake, or stream as a heat source/heat sink. The temperature of the water is related to climatic conditions and usually ranges from 45 to 75°F for deep wells.

   Units tested in accordance with ARI *Standard* 325 have standard cooling and heating ratings at both 50 and 70°F entering water temperature. Rated efficiencies include an allowance for water pumping power. Maximum operating conditions are checked at the upper temperature level at 75°F entering water temperature and at the lower temperature level at 45°F entering water temperature. These water temperatures bracket the range of groundwater temperatures found across the United States.
3. ARI *Standard* 330 for Ground-Source Closed-Loop Heat Pumps. A ground-source closed-loop heat pump typically uses fluid circulated through a subsurface piping loop as a heat source/heat sink. The heat exchange loop may be placed in horizontal trenches or vertical bores, or may be submerged in a body of surface water. The temperature of the fluid is related to climatic conditions and usually ranges from 25 to 100°F.

   Due to the low temperatures that may be encountered in closed-loop ground-coupled systems, units tested in accordance with ARI *Standard* 330 circulate an antifreeze solution instead of water. This rating procedure provides standard cooling ratings at 77°F entering fluid temperature and standard heating ratings at 32°F entering fluid temperature. Rated efficiencies include an allowance for power to circulate the fluid. Maximum operating conditions are checked at 100°F entering fluid temperature when operating in the cooling mode and 25°F entering fluid temperature when operating in the heating mode.

The ARI *Directory of Certified Applied Air-Conditioning Products* lists the cooling capacity, cooling efficiency, heating capacity, and heating COP of units rated in accordance with the three ARI standards.

Canadian Standards Association CAN/CSA *Standard* C446 for ground-source and water-source heat pump systems is essentially the same as ARI *Standards* 330 and 325.

There are no ARI certification programs for water-source heat pumps designed specifically for surface water applications.

EQUIPMENT DESIGN

Water-source heat pumps are designed to match differing levels of entering water temperatures by optimizing the relative sizing of the indoor refrigerant-to-air heat exchanger and the refrigerant-to-water heat exchangers and by matching the expansion devices to the refrigerant flow rates.

Compressors. WSHPs usually have single-speed compressors, although some high-efficiency models use multispeed compressors. Higher capacity equipment may use multiple compressors. The compressors may be of the reciprocating, rotary, or scroll type. Single-phase units are available at voltages of 115, 208, 230, and 265. All larger equipment is for three-phase power supplies with voltages of 208, 230, 460, or 575. The compressors are usually provided with electromechanical protective devices.

Indoor Air System. Console WSHP models are designed for free delivery of the conditioned air. Other models have ducting

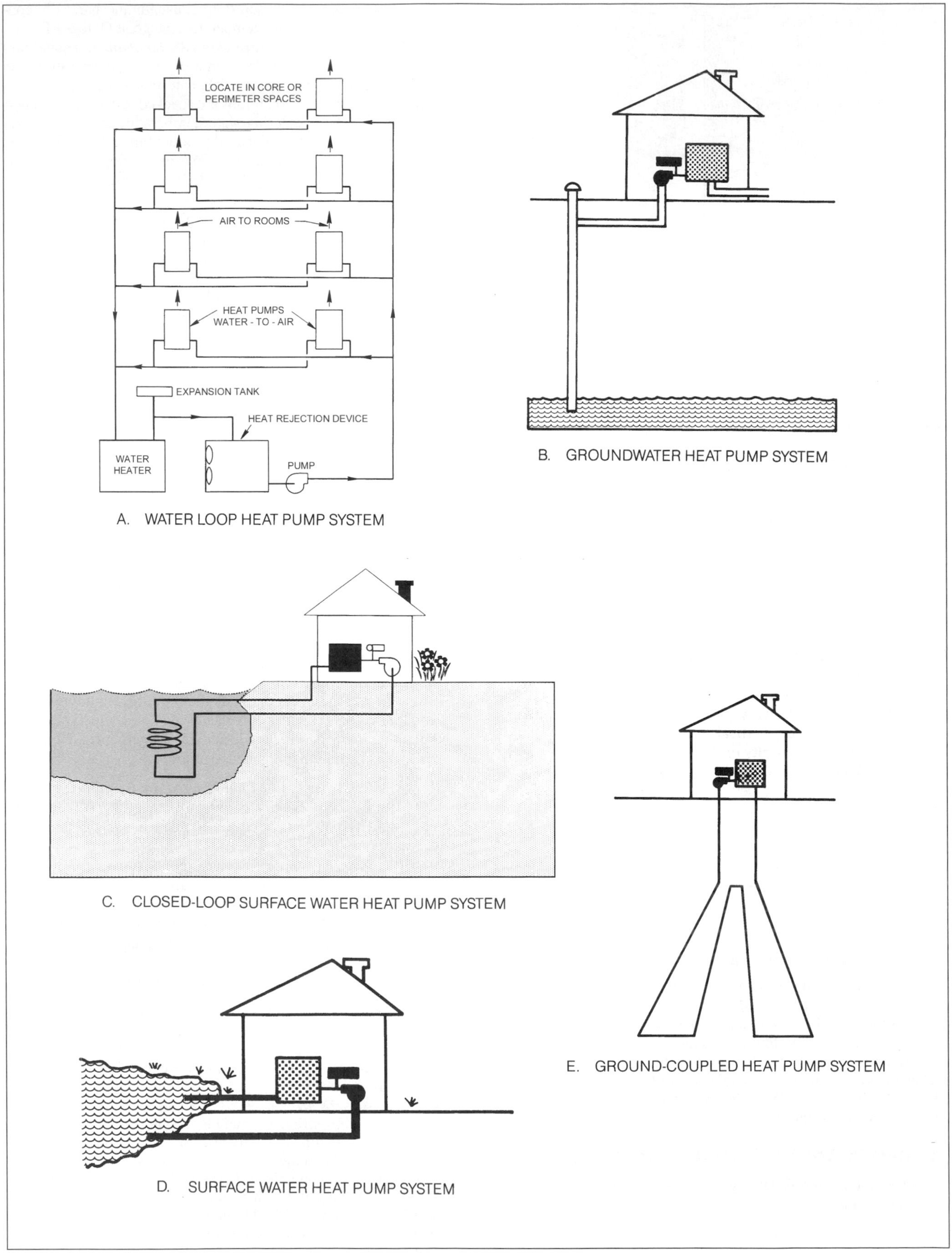

Fig. 15 Water-Source Heat Pump Systems

capability. Smaller WSHPs have multispeed, direct-drive centrifugal blower wheel fan systems. Large capacity equipment has belt-drive systems. All units have provisions for air filters of fiberglass, metal, or plastic foam.

Indoor Heat Exchanger. The indoor heat exchanger of WSHP units is a conventional plate-fin coil of copper tubes and aluminum fins. The tubing in the coil is circuited so that it can function effectively as an evaporator with the refrigerant flow in one direction and as a condenser when the refrigerant flow is reversed.

Refrigerant-to-Water Heat Exchanger. The heat exchanger, which couples the heat pump to source/sink water, is of the tube-in-tube, tube-in-shell, or brazed-plate type. It must function in either the condensing or evaporating mode, so special attention is given to refrigerant-side circuitry. Heat exchanger construction is usually of copper and steel, and the source/sink water is exposed only to the copper portions. Cupronickel options to replace the copper are usually available for use with brackish or corrosive water. Brazed-plate heat exchangers are usually constructed of stainless steel, which reduces the need for special materials.

Refrigerant Expansion Devices. WSHPs rated in accordance with ARI *Standard* 320 operate over a narrow range of entering water temperatures, so most use simple capillaries as expansion devices. Units rated according to ARI *Standard* 325 or 330 usually use thermostatic expansion valves for improved performance over a broader range of inlet fluid temperatures.

Refrigerant-Reversing Valve. The refrigerant-reversing valves in WSHPs are identical to those used in air-source heat pumps.

Condensate Disposal. Condensate, which forms on the indoor coil when cooling, is collected and conveyed to a drain system.

Controls. Console WSHP units have built-in operating mode selector and thermostatic controls. Ducted units use low-voltage remote heat/cool thermostats.

Size. Typical space requirements and weights of WSHPs are presented in Table 3.

Table 3 Space Requirements for Typical Packaged Water-Source Heat Pumps

Water-to-Air Heat Pump	Length × Width × Height, ft	Weight, lb
1.5 ton vertical unit	2.0 × 2.0 × 3.0	180
3 ton vertical unit	2.5 × 2.5 × 4.0	250
3 ton horizontal unit	3.5 × 2.0 × 2.0	250
5 ton vertical unit	3.0 × 2.5 × 4.0	330
11 ton vertical unit	3.5 × 3.0 × 6.0	720
26 ton vertical unit	3.5 × 5.0 × 6.0	1550

Note: See manufacturers' specification sheets for actual values.

Special Features. Certain models of WSHPs include

Desuperheater. Uses discharge gas in a special water/refrigerant heat exchanger to heat water for a building.

Capacity modulation. Multiple compressors, multispeed compressors, or hot-gas bypass may be used.

Variable air volume (VAV). Reduces fan energy usage and requires some form of capacity modulation.

Automatic water valve. Closes off water flow through the unit when the compressor is off and permits variable water volume in the loop, which reduces pumping energy.

Outdoor-air economizer. Cools directly with outdoor air to reduce or eliminate the need for mechanical refrigeration during mild or cold weather when outdoor humidity levels and outdoor air quality are appropriate.

Water-side economizer. Cools with loop water to reduce or eliminate the need for mechanical refrigeration during cold weather; requires a hydronic coil in the indoor air circuit that is valved into the circulating loop when loop temperatures are relatively low and there is a call for cooling.

Electric heaters. Used in WLHP systems that do not have a boiler as a source for loop heating.

REFERENCES

Anand, N.K., J.S. Schliesing, D.L. O'Neal, and K.T. Peterson. 1989. Effects of outdoor coil fan pre-start on pressure transients during the reverse cycle defrost of a heat pump. *ASHRAE Transactions* 95(2).

ARI. 1993. Commercial and industrial unitary air-conditioning and heat pump equipment. *Standard* 340/360-93. Air-Conditioning and Refrigeration Institute, Arlington, VA.

ARI. 1994. Commercial and industrial unitary air-conditioning condensing units. *Standard* 365-94.

ARI. 1994. Unitary air-conditioning and air-source heat pump equipment. ANSI/ARI *Standard* 210/240-94.

ARI. 1995. Sound rating of outdoor unitary equipment. *Standard* 270-95.

ARI. 1997. Application of sound rating levels of outdoor unitary equipment. *Standard* 275-97.

ARI. 1998. Ground source closed-loop heat pumps. *Standard* 330-98.

ARI. 1998. Ground water-source heat pumps. *Standard* 325-98.

ARI. 1998. Water-source heat pumps. *Standard* 320-98.

ARI. Semiannual. *Directory of certified applied air-conditioning products.*

ARI. Semiannual. *Directory of certified unitary products.*

ASA. 1990. Precision methods for the determination of sound power levels of discrete-frequency and narrow-band noise sources in reverberation rooms. ASA *Standard* 92-90 (ANSI *Standard* S12.32-90). Acoustical Society of America, New York.

ASHRAE. 1985. *Design/data manual for closed-loop ground-coupled heat pump systems.*

ASHRAE. 1988. Methods of testing for rating unitary air-conditioning and heat pump equipment. ANSI/ASHRAE *Standard* 37-1988.

ASHRAE. 1994. Safety code for mechanical refrigeration. ANSI/ASHRAE *Standard* 15-1994.

ASHRAE. 1999. Energy standard for buildings except low-rise residential buildings. ASHRAE *Standard* 90.1-1999.

Counts, D. 1985. Performance of heat pump/desuperheater water heating systems. *ASHRAE Transactions* 91(2B):1473-87.

CSA. 1994. Performance of ground source heat pumps. CAN/CSA *Standard* C446-94. CSA International, Etobicoke (Toronto).

CSA. 1995. Heating and cooling equipment. CSA *Standard* C22.2 No. 236-95.

DOE. Uniform test method for measuring the energy consumption of central air conditioners. 10 CFR 430, Appendix M. U.S. Department of Energy, Washington, D.C.

Miller, W.A. 1989. Laboratory study of the dynamic losses of a single speed, split system air-to-air heat pump having tube and plate fin heat exchangers. ORNL/CON-253. Oak Ridge National Laboratory, Oak Ridge, TN.

Mueller, D. and U. Bonne. 1980. Heat pump controls: Microelectronic technology. *ASHRAE Journal* 22(9).

NFPA. 1999. Installation of air conditioning and ventilating systems. ANSI/NFPA *Standard* 90A-99. National Fire Protection Association, Quincy, MA.

NFPA. 1999. Installation of warm air heating and air conditioning systems. ANSI/NFPA *Standard* 90B-99.

NFPA. 1999. *National electrical code.* ANSI/NFPA *Standard* 70-99.

O'Neal, D.L., N.K. Anand, K.T. Peterson, and S. Schliesing. 1989a. Determination of the transient response characteristics of the air-source heat pump during the reverse cycle defrost. *Final Report*, ASHRAE Research Project 479-TRP.

O'Neal, D.L., N.K. Anand, K.T. Peterson, and S. Schliesing. 1989b. Refrigeration system dynamics during the reverse cycle defrost. *ASHRAE Transactions* 95(2).

UL. 1995. Heating and cooling equipment. *Standard* 1995-95. Underwriters Laboratories, Northbrook, IL.

CHAPTER 46

ROOM AIR CONDITIONERS, PACKAGED TERMINAL AIR CONDITIONERS, AND DEHUMIDIFIERS

ROOM AIR CONDITIONERS

ROOM AIR CONDITIONERS are encased assemblies designed primarily for mounting in a window or through a wall. These units are designed for the delivery of cool or warm conditioned air to the room, either without ducts or with very short ducts (up to a maximum of about 48 in). Each unit includes a prime source of refrigeration and dehumidification and a means for circulating and filtering air; it may also include a means for ventilating and/or exhausting and heating.

The basic function of a room air conditioner is to provide comfort by cooling, dehumidifying, filtering or cleaning, and circulating the room air. It may also provide ventilation by introducing outdoor air into the room and/or exhausting room air to the outside. Room temperature may be controlled through a thermostat. The conditioner may provide heating by heat pump operation, electric resistance elements, or a combination of the two.

Figure 1 shows a typical room air conditioner in the cooling mode. Warm room air passes over the cooling coil and, in the process, gives up sensible and latent heat. The conditioned air is then recirculated in the room by a fan or blower.

The heat from the warm room air vaporizes the cold (low-pressure) liquid refrigerant flowing through the evaporator. The vapor then carries the heat to the compressor, which compresses the vapor and increases its temperature above that of the outdoor air. In the condenser, the hot (high-pressure) refrigerant vapor liquefies, giving up the heat from the room air to the outdoor air. Next, the high-pressure liquid refrigerant passes through a restrictor, which reduces its pressure and temperature. The cold (low-pressure) liquid refrigerant then enters the evaporator to repeat the refrigeration cycle.

SIZES AND CLASSIFICATIONS

Room air conditioners have line cords, which may be plugged into standard or special electric circuits. Most units in the United States are designed to operate at 115 V, 208 V, or 230 V; single-phase; 50 or 60 Hz power. Some units are rated at 265 V or 277 V, in which the chassis or chassis assembly must provide for permanent electrical connection. The maximum rating of 115 V units is generally 12 A, the maximum current permitted by the NEC for a single-outlet, 15 A circuit. Models designed for countries other than the United States are generally for 50 or 60 Hz systems, with typical design voltage ranges of 100 to 120 and 200 to 240 V, single-phase.

Popular 115 V models have capacities in the range of 5000 to 7000 Btu/h, and they are typically used in single-room applications. The larger capacity 115 V units are in the 12,000 to 15,000 Btu/h range. Capacities for 230, 208, or 230/208 V units range from 8000 to 36,000 Btu/h. Capacities for 265 or 277 V units range from 6000 to 17,000 Btu/h. These higher rated units are typically used in multiple-room installations.

Heat pump models are also available, usually for 208 or 230 V applications. These units are generally designed for reversed refrigerant cycle operation as the normal means of supplying heat, but they may incorporate electrical resistance heat to either supplement the heat pump capacity or to provide the total heating capacity when the outdoor temperature drops below a predetermined value.

Another type of heating model incorporates electrical heating elements in regular cooling units so that heating is provided entirely by electrical resistance heat.

DESIGN

The design of a room air conditioner is usually based on one or more of the following criteria, any one of which automatically limits the freedom of the designer in overall system design:

- Lowest initial cost
- Lowest operating cost (highest efficiency)
- Energy efficiency ratio (EER) or coefficient of performance (COP), as legislated by the federal and/or state governments

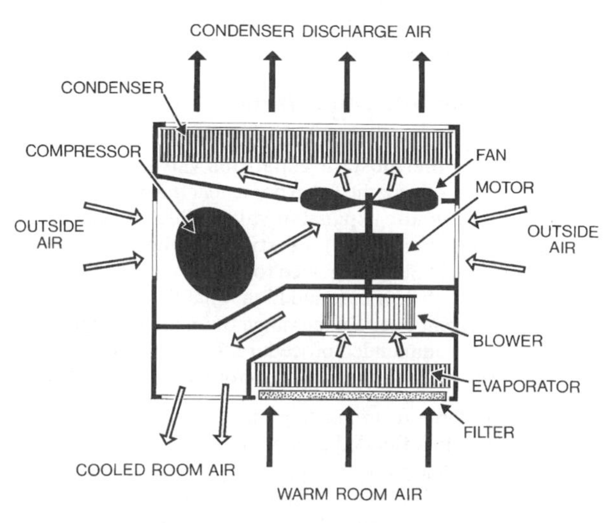

Fig. 1 Schematic View of Typical Room Air Conditioner

The preparation of this chapter is assigned to TC 7.6, Unitary and Room Air Conditioners and Heat Pumps.

- Low sound level
- Chassis size
- Unusual chassis shape (minimal depth, height, etc.)
- Amperage limitation (7.5 A, 12 A, etc.)
- Weight

The following combinations illustrate the effect of an initial design parameter on the various components:

Low Initial Cost. High airflow with minimum heat exchanger surface keeps the initial cost of a unit low. These units have a low-cost compressor, which is selected by analyzing various compressor and coil combinations and choosing the one that both achieves optimum performance and passes all tests required by Underwriters Laboratories (UL), the Association of Home Appliance Manufacturers (AHAM), and others. For example, a high-capacity compressor might be selected to meet the capacity requirement with a minimum heat transfer surface, but frost tests under maximum load may not be acceptable. These tests set the upper and lower limits of acceptability when low initial cost is the prime consideration.

Low Operating Cost. Large heat exchanger surfaces keep operating cost low. A compressor with a low compression ratio operates at low head and high suction pressure, which results in a high EER.

Compressors

Room air conditioner compressors range in capacity from about 4000 to 34,000 Btu/h. Design data are available from compressor manufacturers at the following standard rating conditions:

Evaporating temperature	45°F
Compressor suction temperature	95°F
Condensing temperature	130°F
Liquid temperature	115°F
Ambient temperature	95°F

Compressor manufacturers offer complete performance curves at various evaporating and condensing temperatures to aid in selection for a given design specification.

Evaporator and Condenser Coils

These coils are generally of the tube-and-plate-fin, tube-and-louvered-fin, or tube-and-spine-fin variety. Information on the performance of such coils is available from suppliers, and original equipment manufacturers usually develop data for their own coils. Design parameters to be considered when selecting coils are (1) cooling rate per unit area of coil surface ($Btu/h \cdot ft^2$); (2) dry-bulb temperature and moisture content of the entering air; (3) air-side friction loss; (4) internal refrigerant pressure drop; (5) coil surface temperature; (6) airflow; and (7) air velocity.

Restrictor Application and Sizing

There are essentially three types of restrictor devices available to the designer: (1) a **thermostatic expansion valve**, which maintains a constant amount of superheat at a point near the outlet of the evaporator; (2) an **automatic expansion valve**, which maintains a constant suction pressure; and (3) a **restrictor tube (capillary)**. The capillary is the most popular device for room air conditioner applications because of its low cost and high reliability, even though the control of refrigerant over a wide range of ambient temperatures is not optimal. A recommended procedure for optimizing charge balance, condenser subcooling, and restrictor sizing is as follows:

1. Use an adjustable restrictor (e.g., a needle valve), so that tests may be run with a flooded evaporator coil and various refrigerant charges to determine the optimum point of system operation.
2. Reset the adjustable restrictor to the optimum setting, remove it from the unit, and measure flow pressure with a flow comparator similar to that described in ASHRAE *Standard* 28.
3. Install a restrictor tube with the same flow rate as the adjustable restrictor. Usually, restrictor tubes are selected on the basis of cost, with shorter tubes generally being less expensive.

Fan Motor and Air Impeller Selection

The two types of motors generally used on room air conditioners are (1) the low-efficiency, shaded-pole type; and (2) the more efficient, permanent split-capacitor type, which requires the use of a run capacitor. Air impellers are usually of two types: (1) the forward-curved blower wheel and (2) the axial or radial flow fan blade. In general, blower wheels are used to move small to moderate amounts of air in a high-resistance system, and fan blades move moderate to high air volumes in low-resistance applications. Blower wheels and cross-flow fans also generate lower noise levels than do fan blades.

The combination of the fan motor and the air impeller is such an important part of the overall design that the designer should work closely with the manufacturers of both components. Performance curves are available for motors, blower wheels, and fans, but the data are for ideal systems not usually found in practice due to physical size, motor speed, and component placement limitations.

Electronics

Microprocessors monitor and control numerous functions for room air conditioners. These microelectronic controls offer digital displays and touch panels for programming desired temperature, on-off timing, modulated fan speeds, bypass capabilities, and sensing for humidity, temperature, and airflow control.

PERFORMANCE DATA

In the United States, an industry certification program under the sponsorship of the Association of Home Appliance Manufacturers (AHAM) covers the majority of room air conditioners and certifies the cooling and heating capacities and electrical input (in amperes) of each for adherence to nameplate rating. The following tests are specified by AHAM *Standard* RAC-1:

- Cooling capacity
- Heating capacity
- Maximum operating conditions (heating and cooling)
- Enclosure sweat
- Freeze-up
- Recirculated air quantity
- Moisture removal
- Ventilating air quantity and exhaust air quantity
- Electrical input (heating and cooling)
- Power factor
- Condensate disposal
- Application heating capacity
- Outside coil deicing

Efficiency

Efficiency for room air conditioners may be shown in either of two forms:

1. Energy efficiency ratio (EER—generally for cooling) (Capacity in Btu/h)/(Input in watts)
2. Coefficient of performance (COP—generally for heating) (Capacity in Btu/h)/(Input in watts × 3.413)

Sensible Heat Ratio

The ratio of sensible heat to total heat removal is a performance characteristic that is useful in evaluating units for specific conditions. A low ratio provides more dehumidification, and areas such as New Orleans and Phoenix might best be served with units having lower and higher ratios, respectively.

Energy Conservation and Efficiency

In the United States, two federal energy programs have increased the demand for higher efficiency room air conditioners. First, the Energy Policy and Conservation Act passed in December, 1975 (Public Law 94-163) requires the Federal Trade Commission (FTC) to prescribe an energy usage label for many major appliances, including room air conditioners. The program provides operating cost data at the point of sale. Second, the National Energy Conservation Policy Act (NECPA) passed in November, 1978 (Public Law 95-619) directs the Department of Energy (DOE) to establish minimum efficiency standards for major appliances, including room air conditioners.

The National Appliance Energy Conservation Act of 1987 (NAECA) provides a single set of minimum efficiency standards for major appliances, including room air conditioners. The room air conditioner portion of NAECA originally specified minimum efficiencies for 12 classes, based on physical conformation, with minimums ranging from 8 to 9 EER and applying to all units manufactured on or after January 1, 1990.

The DOE issued minimum efficiency standards that become effective on October 1, 2000 (Federal Register, September 24, 1997). Four additional classes have been created, two of which cover casement-type units. The new minimum standards range from 8 to 9.8 EER. For the most popular classes (cooling-only units with louvered sides ranging in capacity from less than 6000 to 20,000 Btu/h) the new minimum standards are either 9.7 or 9.8 EER.

All state and local minimum efficiency standards are automatically superseded in the U.S. by federal standards. Many other countries have or are considering minimum efficiency standards, so such standards should be sought as part of the design process.

Whether estimating potential energy savings associated with appliance standards or estimating consumer operating costs, the annual hours H of operation of a room air conditioner are important. These figures have been compiled from various studies commissioned by DOE and AHAM for every major city and region in the United States. The national average is estimated at 750 h per year.

The cost of operation is as follows:

$$C = RHW/1000$$

where

C = annual cost of operation, \$/year
R = average cost, \$/kWh
H = annual hours of operation
W = input, W

High-Efficiency Design

The EER can be affected by three design parameters. The first is **electrical efficiency**. Fan motor efficiency ranges from 25 to 65%; compressor motors range from 60 to 85%. The second parameter, **refrigerant cycle efficiency**, is increased by enlarging the heat transfer surface to minimize the difference between the refrigerant saturation temperature and the air temperature. This allows the use of a compressor with a smaller displacement and a high-efficiency motor. The third parameter is **air circuit efficiency**, which can be increased by minimizing the pressure drop across the heat transfer surface, which reduces the load on the fan motor.

Table 1 shows how room air conditioner cost and size are increased by raising the EER from 6.5 to Table 1 values. The increases, which are approximate, range from 40 to 60% in weight and 50 to 80% in volume; small-capacity units become less portable, and units larger than 20,000 Btu/h are less adaptable to window mounting.

Higher EERs are not the complete answer to reducing energy costs. More efficient use of energy can be accomplished by properly sizing the unit, keeping infiltration and leakage losses to a minimum, increasing building insulation, reducing unnecessary internal loading, providing effective maintenance, and balancing the load by use of a thermostat and thermostat setback.

Table 1 Effect of Increased EER over Base Value of 6.5

Capacity Range, Btu/h	EER, (Btu/h)/W	Weight Increment, Approx. %	Volume Increment, Approx. %	Price Increment, Approx. %
4000 to 10,000	10.0	39	48	50
12,000 to 20,000	10.5	52	60	37
24,000 to 27,000	11.0	58	80	40

SPECIAL FEATURES

Some room air conditioners are designed to minimize their extension beyond the building when mounted flush with the inside wall. Low-capacity models are usually smaller and less obtrusive than higher capacity models. Units are often installed through the wall, where they do not interfere with windows. Exterior cabinet grilles may be designed to harmonize with the architecture of various buildings.

Most units have adjustable louvers or deflectors to distribute the air into the room with satisfactory throw and without drafts. The louver design should eliminate recirculation of discharge air into the air inlet. Some units employ motorized deflectors for continuously changing the air direction. Discharge air velocities range from 300 to 1200 fpm, with low velocities preferred in rooms where people are at rest.

Most room air conditioners are designed to bring in outdoor air, exhaust room air, or both. Controls usually permit these features to function independently.

Temperature is controlled by an adjustable built-in thermostat. The thermostat and unit controls may operate in one of the following modes:

1. The unit is set to the cool position, and the thermostat setting is adjusted as needed. The circulation blower runs without interruption while the thermostat cycles the compressor on and off.
2. The unit uses a two- or three-stage thermostat, which reduces the blower speed as the room temperature approaches the set temperature, cycles the compressor off as the temperature drops further, and, if the temperature drops still further, finally cycles the blower off. As the room temperature rises, the sequence reverses.
3. The unit has, in addition to the control sequence in Item 2, an optional automatic fan mode in which both the blower and the compressor are cycled simultaneously by the thermostat; this mode of operation requires proper thermostat sensitivity. One of the advantages of this arrangement is improved humidity control because the moisture from the evaporator coil is not reevaporated into the room during the off cycle. Another advantage is lower operating cost because the blower motor does not operate during the off cycle. The effective EER may be increased an average of 10% by using the automatic fan mode (ORNL 1985).

Disadvantages of cycling fans with the compressor may be (1) varying noise level due to fan cycling and (2) deterioration in room temperature control.

Room air conditioners are simple to operate. Usually, one control selects the operating mode, while a second controls the temperature. Additional knobs or levers operate louvers, deflectors, the ventilation system, exhaust dampers, and other special features. The controls are usually arranged on the front of the unit or concealed behind a readily accessible door; however, they may also be arranged on the top or sides of the unit.

Filters on room air conditioners remove airborne dirt to provide clean air to the room and keep dirt off the cooling surfaces. Filters are made of expanded metal (with or without a viscous oil coating), glass fiber, or synthetic materials; they may be either disposable or reusable. A dirty filter reduces cooling and air circulation and frequently allows frost to accumulate on the cooling coil; therefore, filter location should allow easy monitoring, cleaning, and replacement.

Some units have louvers or grilles at the rear to enhance appearance and protect the condenser fins. Sometimes, these louvers separate the airstreams to and from the condenser and reduce recirculation. When provided, side louvers on the outside portion of the unit are an essential part of the condenser air system because they improve air movement to the condenser. Care should be taken not to obstruct the air passages through these louvers.

Room air conditioners, especially those parts exposed to the weather, require a durable finish. Some manufacturers use a special grade of plastic for weather-exposed parts. If the parts are metal, good practice calls for the use of phosphatized or zinc-coated steels with baked finishes and/or corrosion-resistant materials such as aluminum or stainless steel.

The sound level of a room air conditioner is an important factor, particularly when the unit is installed in a bedroom. A certain amount of sound can be expected due to the movement of air through the unit and the operation of the compressor. However, a well-designed room air conditioner is relatively quiet, and the sound emitted is relatively free of high-pitched and metallic noise. Usually, fan motors with two or more speeds are used in order to provide a slower fan speed for quieter operation. To avoid rattles and vibration in the building structure, units must be installed correctly (see the section on Installation and Service).

Excessive outdoor sound (condenser fan and compressor) can be irritating to neighbors. In the United States, many local and state outdoor noise ordinances are being considered, and some have been passed into law.

SAFETY CODES AND STANDARDS

United States. The *National Electrical Code* (NFPA *Standard* 70), ASHRAE *Standard* 15, and UL *Standard* 484 pertain to room air conditioners. Local regulations may differ with these standards, but the basic requirements are generally accepted throughout the United States.

Canada. CSA International developed the standard for Room Air Conditioners (CSA *Standard* C22.2 No. 117), which forms part of the *Canadian Electrical Code*.

International. Two useful documents that might assist the designer are (1) International Electrotechnical Commission (IEC) *Standard* 60335-2-40 and (2) International Organization for Standardization (ISO) *Standard* 5151.

Product Standards

CSA *Standard* C22.2 No. 117 and UL *Standard* 484 are similar in content. In *Standard* 484, the construction section involves such items as the unit enclosure (including materials), the unit's ability to protect against contact with moving and uninsulated live parts, and the means for unit installation or attachment. Attention is also given to the ability of the refrigeration system to withstand operating pressures, system pressure relief in the event of fire, and the toxicity of the refrigerant. Electrical considerations include supply connections, grounding, internal wiring and wiring methods, electrical spacings, motors and motor protection, uninsulated live parts, motor controllers and switching devices, air-heating components, and electrical insulating materials.

The performance section of the standard includes a rain test for determining the unit's ability to stand a beating rain without creating a shock hazard due to current leakage or insulation breakdown. Other tests include (1) leakage current limitations based on UL *Standard* C101; (2) measurement of input currents for the purpose of establishing nameplate ratings and for sizing the supply circuit for the unit; (3) temperature tests to determine whether components exceed their recognized temperature limits and/or electrical ratings (AHAM *Standard* RAC-1); and (4) pressure tests to ensure that excessive pressure does not develop in the refrigeration system.

Abnormal conditions are also considered, such as (1) failure of the condenser fan motor, which may result in the development of excessive pressure in the system; and (2) possible ignition of combustibles within or adjacent to the unit on air heater burnout. A static load test is also conducted on window-type room air conditioners to determine whether the mounting hardware can adequately support the unit. As part of normal production control, tests are conducted for refrigerant leakage, dielectric strength, and grounding continuity.

Plastic materials are receiving increased consideration in the design and fabrication of room air conditioners because of their ease in forming, inherent resistance to corrosion, and decorative qualities. When considering the use of plastic, the engineer should consider the tensile, flexural, and impact strength of the material; its flammability characteristics; and—from the standpoint of degradation—its resistance to water absorption and exposure to ultraviolet light, ability to operate at elevated temperatures, and thermal aging characteristics. From the standpoint of product safety, some of these factors are of lesser importance because failure of the part will not cause a hazard. However, for some parts (e.g., the bulkhead, base pan, and unit enclosure) that either support components or provide structural integrity, all of the preceding factors must be considered, and a complete analysis of the material must be made to determine whether it is suitable for the application.

INSTALLATION AND SERVICE

Installation procedures vary because units can be mounted in various ways. It is important to select the mounting for each installation that best satisfies the user and complies with applicable building codes. Common mounting methods include the following:

- **Inside flush mounting.** Interior face of conditioner is approximately flush with inside wall.
- **Balance mounting.** Unit is approximately half inside and half outside window.
- **Outside flush mounting.** Outer face of unit is flush with or slightly beyond outside wall.
- **Special mounting.** Examples include casement windows, horizontal sliding windows, office windows with swinging units (or swinging windows) to permit window washing, and transoms over doorways.
- **Through-the-wall mounts or sleeves.** This type of mounting is used for installing window-type chassis, complete units, or consoles in walls of apartment buildings, hotels, motels, and residences. Although very similar to window-mounted units, through-the wall models do not have side louvers for the condenser air; air comes from the rear of the unit.

Room air conditioners have become more compact because consumers prefer minimum loss of window light and minimum projection both inside and outside the structure. Several types of expandable mounts are now available for fast, dependable installation in single- and double-hung windows, as well as in horizontal sliding windows. Installation kits include all parts needed for structural mounting, such as gaskets, panels, and seals for weathertight assembly.

Adequate wiring and proper fuses must be provided for the service outlet. The necessary information is usually given on instruction sheets or stamped on the air conditioner near the service cord or on the serial plate. It is important to follow the manufacturer's recommendation for size and type of fuse. All units are equipped by the manufacturer with grounding-type plug caps on the service cord. Receptacles with grounding contacts correctly designed to fit these plug caps should be used when units are installed.

Units rated 265 V or 277 V must provide for permanent electrical connection with armored cable or conduit to the chassis or chassis assembly. Manufacturers usually provide an adequate cord and plug cap in the chassis assembly to facilitate installation and service.

One type of room air conditioner is the **integral chassis** design, with the outer cabinet fastened permanently to the chassis. Most of the electrical components can be serviced by partially dismantling the control area without removing the unit from the installation.

Another type of room air conditioner is the **slide-out chassis** design, which allows the outer cabinet to remain in place while the chassis is removed for service.

PACKAGED TERMINAL AIR CONDITIONERS

The Air-Conditioning and Refrigeration Institute (ARI) defines a packaged terminal air conditioner (PTAC) as a wall sleeve and a separate unencased combination of heating and cooling assemblies intended for mounting through the wall. A PTAC includes refrigeration components, separable outdoor louvers, forced ventilation, and heating by hot water, steam, or electric resistance. PTAC units with direct-fired gas heaters are also available from some manufacturers. PTACs designated cooling only need not include heating elements. A packaged terminal heat pump (PTHP) is a heat pump version of a PTAC that provides heat with a reverse-cycle operating mode. A PTHP should provide a supplementary heat source which could either be hot water, steam, electric resistance, or other source.

PTACs are designed primarily for commercial installations to provide the total heating and cooling functions for a room or zone and are specifically for through-the-wall installation. The units are mostly used in relatively small zones on the perimeter of buildings such as hotels and motels, apartments, hospitals, nursing homes, and office buildings. In larger buildings, they may be combined with nearly any system selected for environmental control of the building core.

PTACs and PTHPs are similar in design and construction. The most apparent difference is the addition of a refrigerant-reversing valve in the PTHP. Optional components that control the heating functions of the heat pump include an outdoor thermostat to signal the need for changes in heating operating modes, and, in the more complex designs, frost sensors, defrost termination devices, and base pan heaters.

SIZES AND CLASSIFICATIONS

Packaged terminal air conditioners are available in a wide range of rated cooling capacities, typically 6,000 to 18,000 Btu/h, with comparable levels of heating output.

The units are available as sectional types or integrated types. Both types include the following:

- Heating elements available in hot water, steam, electric, or gas heat
- Integral or remote temperature and operating controls
- Wall sleeve or box
- Removable (or separable) outdoor louvers
- Room cabinet
- Means for controlled forced ventilation
- Means for filtering air delivered to the room
- Ductwork

The assembly is intended for free conditioned-air distribution, but a particular application may require minimal ductwork with a total external static resistance up to 0.1 in. of water.

A sectional-type unit (Figure 2) has a separate cooling chassis; an integrated-type unit (Figure 3) has an electric or a gas heating option added to the chassis. Hot water or steam heating options are usually part of the cabinet or wall box.

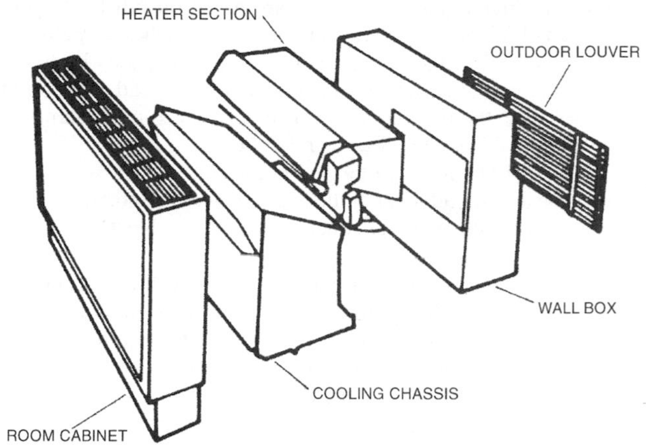

Fig. 2 Packaged Terminal Air Conditioner—Sectional Type

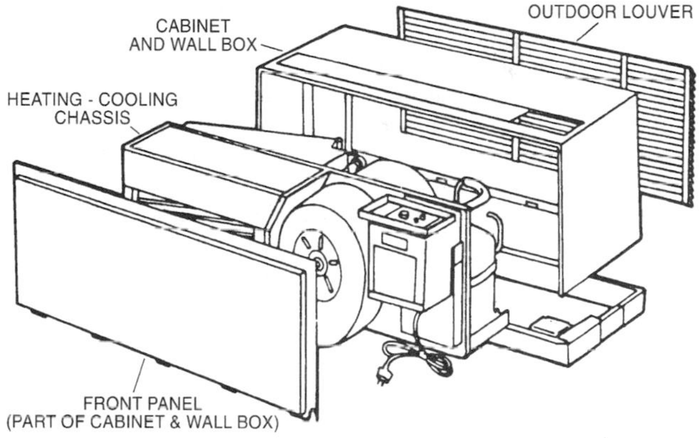

Fig. 3 Packaged Terminal Air Conditioner—Integrated Type

GENERAL DESIGN CONSIDERATIONS

Packaged terminal air conditioners and packaged terminal heat pumps allow the HVAC designer to integrate the exposed outdoor louver or grille with the building design. Various grilles are available to blend with or accent most construction materials. Because the product becomes part of the building's facade, the architect must consider the product during the conception of the building. The installation of wall sleeves is more of an ironworker's, mason's, or carpenter's craft. All-electric units dominate the market. Recent U.S. market statistics indicate that 45% are PTHPs, 49% are PTACs with electric resistance heat, and 6% involve other forms of heating.

All the energy of the all-electric versions is dispersed through the building through electrical wiring, so the electric designer and the electrical contractor play a major role. The final installation is reduced to sliding in the chassis and plugging the unit into an adjacent receptacle. For these all-electric units, the traditional HVAC contractor's contribution involving ducting, piping, and refrigeration systems is bypassed. This results in a low-cost installation and allows the actual installation of the PTAC/PTHP chassis to be deferred until just before occupancy.

When comparing a gas-fired PTAC to a PTAC with electric resistance heat or a PTHP, both operating and installation costs should be evaluated. Generally, a gas-fired PTAC is more expensive to install but less expensive to operate in the heating mode. A life-cycle cost comparison is recommended.

One main advantage of the PTAC/PTHP concept is that it provides excellent zoning capability. Units can be shut down or operated in a holding condition during unoccupied periods. Present equipment efficiency-rating criteria are based on full-load operation, so an efficiency comparison to other approaches may suffer.

The designer must also consider that the total capacity is the sum of the peak loads of each zone rather than the peak load of the building. Therefore, the total cooling capacity of the zonal system will exceed the total capacity of a central system.

Because PTAC units are located in the conditioned space, both appearance and sound level of the equipment are important considerations. Sound attenuation in ducting is not available with the free-discharge PTAC units.

The designer must also consider the added infiltration and thermal leakage load resulting from perimeter wall penetrations. These losses are accounted for during the on-cycle in the equipment cooling ratings and PTHP heating ratings, but during the off-cycle or with other forms of heating, they could be significant.

Widely dispersed PTAC units also present challenges relative to effective condensate disposal.

Most packaged terminal equipment is designed to fit into a wall aperture approximately 42 in. wide and 16 in. high. Although unitary products can increase in size with increasing cooling capacity, PTAC/PTHP units, regardless of cooling capacity, are usually constrained to a few cabinet sizes. The exterior of the equipment must be essentially flush with the exterior wall to meet most building codes. In addition, cabinet structural requirements and the slide-in chassis reduce the available area for outdoor air inlet and relief to less than a total of 3.5 ft^2. Manufacturers' specification sheets should be consulted for more accurate and detailed information.

Outdoor air recirculation must be minimized, and attention must be given to architectural appearance. These may result in an increase in the air-side pressure drop. With a cooling capacity range that usually spans about a 3 to 1 ratio, this causes difficulty in maintaining the efficiency of units at the higher levels of output.

DESIGN OF PTAC/PTHP COMPONENTS

Compressor. PTAC units are designed with single-speed compressors. Both reciprocating and rotary types are used. They normally operate on a single-phase power supply and are available in 208 V, 230 V, and 265 V versions. Smaller units are available in 115 V versions. The compressors are usually provided with electromechanical protective devices, with some of the more advanced models employing electronic protective systems.

Fan Motor(s). In some PTAC units, a single, double-shafted direct-drive fan motor drives both the indoor and outdoor air-moving devices. This motor usually has two speeds, which affect the equipment sound level, throw of the conditioned air, cooling capacity, efficiency, sensible-to-total heat capacity ratio, and condensate disposal.

Full-featured models have two fan motors: one moves the indoor air and the other brings in outdoor air. Two motors provide greater flexibility in locating components, because the indoor and outdoor fans are no longer constrained to the same rotating axis. They also allow different fan speeds for the indoor and outdoor systems. In this case, the outdoor fan motor is usually single speed, and the indoor fan motor has two or more speeds. Also, the designer has a broader selection of air-moving devices and can provide the user with a wider range of sound level and conditioned air throw options. Efficiency can be maintained at lower indoor fan speeds. When heating (other than with a heat pump), the outdoor fan motor can be switched off automatically to reduce electrical energy consumption, decrease infiltration and heat transmission losses through the PTAC unit, and prevent ice from entrapping the fan blade.

Indoor Air Mover. The airflow quantity, air-side pressure rise, available fan motor speed, and sound level requirements of the indoor air system of a PTAC indicate that a centrifugal blower wheel provides reasonable indoor air performance. In some cases, proprietary mixed-flow blowers are used. Dual-fan motor units permit the use of multiple centrifugal blower wheels or a cross-flow blower to provide more even discharge of the conditioned air.

Indoor Air Circuit. Packaged terminal air conditioners have an air filter of fiberglass, metal, or plastic foam, which removes large particles from the circulating airstream. In addition to improving the quality of the indoor air, this filter also reduces fouling of the indoor heat exchanger. The PTAC also provides mechanical means of introducing outdoor air into the indoor airstream. This air, which may or may not be filtered, controls infiltration and pressurization of the conditioned space.

Outdoor Air Mover. Outdoor air movers may be either centrifugal blower wheels, mixed-flow blowers, or axial flow fans.

Heat Exchangers. PTAC units may use conventional plate-fin heat exchangers, which have either copper or aluminum tubes. The fins are usually aluminum, which may be coated to retard corrosion. Because PTACs are generally restricted in physical size, performance improvements based on increasing heat exchanger size are limited. Therefore, to improve performance or reduce costs, some manufacturers install heat exchangers with performance enhancements on the air side (lanced fins, spine fin, etc.) and/or the refrigerant side (internal finning or rifling).

Refrigerant Expansion Device. Most PTACs use a simple capillary as an expansion device. Off-rating-point performance is improved if expansion valves are used.

Condensate Disposal. Condensate forms on the indoor coil when cooling. Some PTACs require that a drain be installed to convey the condensate to a disposal point. Other units spray the condensate on the outdoor coil, where it is evaporated and dispersed to the outdoor ambient air. This evaporative cooling of the condenser enhances performance, but the potentially negative effects of fouling and corrosion of the outdoor heat exchanger must be considered. This problem is especially severe in a coastal installation where salt spray could mix with the condensate and, after repeated evaporation cycles, build a corrosive, saltwater solution in the condensate sump.

A PTHP also produces condensate in the heating mode. If the outdoor coil operates below freezing, the condensate forms frost, which is melted during defrost. This water must be disposed of in some manner. If drains are used and the heat pump operates in below-freezing weather, the drain lines must be protected from freezing. Outside drains cannot be used in this case, unless they have drain heaters. Some PTHPs introduce the condensate formed during heating onto the indoor coil, which humidifies the indoor air at the expense of heating capacity. Inadequate condensate disposal can lead to overflow at the unit and potential staining of the building facade.

Controls. PTAC units usually have a built-in manual mode selector (cool, heat, fan only, and off) and a manual fan-speed selector. A thermostat adjustment is provided with set points usually identified in subjective terms such as *high*, *normal*, and *low*. Some units offer the option of automatic changeover from heat to cool functions. Most manufacturers offer low-voltage remote heat/cool thermostats. Some units incorporate electronic controls, which provide room temperature limiting, evaporator freeze-up protection, compressor lockout in the case of actual or impending compressor malfunction, and service diagnostic aids. Advanced master controls at a central location allow an operator to override the control settings registered by the occupant. These master controls may limit operation when certain room temperature limits are exceeded, adjust thermostat set points during unoccupied periods, and turn off certain units to limit peak electrical demand.

Wall Sleeves. A wall sleeve is a required part of a PTAC unit. It becomes an integral part of the building structure and must be designed with sufficient strength to maintain its dimensional integrity after installation. It must withstand the potential corrosive effects of other building materials, such as mortar, and must endure long-term exposure to the outdoor elements.

Outdoor Grille. The outdoor grille or louvers must be compatible with the architecture of the structure. Most manufacturers provide options in this area. A properly designed grille prevents birds, vermin, and outdoor debris from entering; impedes the entry of rain and snow; and, at the same time, provides adequate free area for the outdoor airstream to enter and exit with a minimum of recirculation.

Slide-In Components. The interface between the wall sleeve and the slide-in component chassis allows the components to be easily inserted and later removed for service and/or replacement. In the event of serious malfunction, the slide-in component can be quickly replaced with a spare, and the repair can be made off the premises. An adequate seal at the interface is essential to exclude wind, rain, snow, and insects without jeopardizing the slide-in/slide-out feature.

Indoor Appearance. Because the PTAC units are located within the conditioned space, their indoor appearance must blend with the decor of the room. Manufacturers provide a variety of indoor treatments, which include variations in shape, style, and materials.

HEAT PUMP OPERATION

Basic PTHP units operate in the heat pump mode down to an outdoor temperature just above the point that frosting of the outdoor heat exchanger would occur. When that outdoor temperature is reached, the heat pump mode is locked out, and other forms of heating are required. Some units include two-stage indoor thermostats and automatically switch from the heat pump mode to an alternate heat source if the space temperature drops too far below the first-stage set point. Some PTHPs use control schemes that extend the operation of the heat pump to lower temperatures. One approach permits heat pump operation down to outdoor temperatures just above freezing. If the outdoor coil frosts, it is defrosted by shutting down the compressor and allowing the outdoor fan to continue circulating outdoor air over the coil. Another approach permits heat pump operation to even lower outdoor temperatures by using a reverse-cycle defrost sequence. In those cases, the heat pump mode is usually locked out for outdoor temperatures below 10°F.

PERFORMANCE AND SAFETY TESTING

PTACs and PTHPs may be rated in accordance with ARI *Standard* 310/380, which is equivalent to CSA Intl. *Standard* C744.

ARI issues a *Directory of Certified Applied Air-Conditioning Products* semi-annually that lists the cooling capacity, efficiency, and heating capacity for each participating manufacturer's PTAC models. The listings of PTHP models also include the heating COP.

Cooling and heating capacities, as listed in the ARI directory, must be established in accordance with ASHRAE *Standard* 37, Methods of Testing for Rating Unitary Air-Conditioning and Heat Pump Equipment, or ASHRAE *Standard* 16, Method of Testing for Rating Room Air Conditioners and Packaged Terminal Air Conditioners. All standard heating ratings should be established in accordance with ASHRAE *Standard* 58, Method of Testing for Rating Room Air Conditioner and Packaged Terminal Air Conditioner Heating Capacity.

Additionally, PTAC units should be constructed in accordance with ASHRAE *Standard* 15 and should comply with the safety requirements of UL *Standard* 484.

DEHUMIDIFIERS

A domestic dehumidifier consists of a motor-compressor, a refrigerant condenser, an air-circulating fan, a refrigerated surface (refrigerant evaporator), a means for collecting and/or disposing of the condensed moisture, and a cabinet to house these components. A typical dehumidifier unit is shown in Figure 4.

The fan draws the moist room air through the cold coil and cools it below its dew point, removing moisture that either drains into the water receptacle or passes through a drain into the sewer. The cooled air next passes through the condenser, which reheats it. Then, with the addition of other unit-radiated heat, the air is discharged into the room at a higher temperature but a lower relative humidity. Continuous circulation of room air through the dehumidifier gradually reduces the relative humidity of the room air.

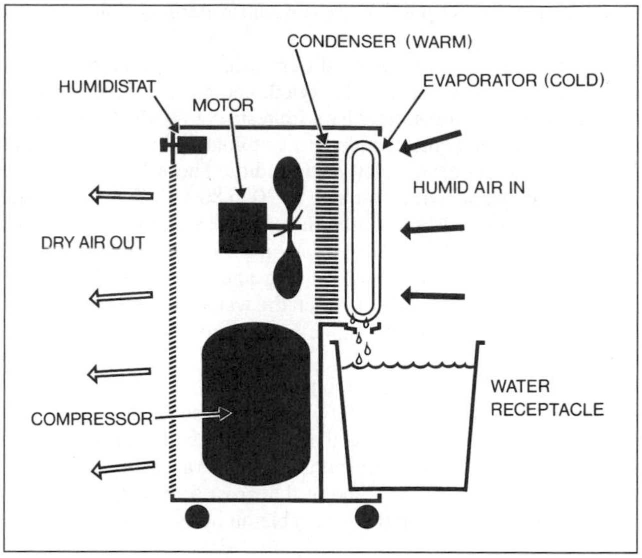

Fig. 4 Typical Dehumidifier Unit

DESIGN AND CONSTRUCTION

Dehumidifiers use hermetic motor-compressors of various ratings, depending on the design output of the unit. The refrigerant condenser is usually a conventional finned-tube coil. Refrigerant flow is usually controlled by a capillary tube, although some high-capacity dehumidifiers use an expansion valve.

A propeller fan, driven directly by a shaded-pole motor, moves air through the dehumidifier. The airflow through a typical dehumidifier ranges from 125 to 250 cfm, depending on the unit's moisture output capacity. Domestic dehumidifiers ordinarily maintain satisfactory humidity levels in an enclosed space when the airflow rate and unit placement move the entire air volume of the space through the dehumidifier once an hour. For example, an airflow rate of 200 cfm would provide one air change per hour in a room having a volume of 12,000 ft^3.

The refrigerated surface (evaporator) is usually a bare-tube coil, although finned-tube coils can be used if they are spaced to permit rapid runoff of water droplets. Vertically disposed bare-tube coils tend to collect smaller drops of water, promote quicker runoff, and cause less water reevaporation than do finned-tube cooling coils or horizontally arranged bare-tube coils. Continuous bare-tube coils, wound in the form of a flat circular spiral (sometimes consisting of two coil layers) and mounted with the flat dimension of the coil in the vertical plane, are a good design compromise because they have most of the advantages of the vertical bare-tube coil.

Evaporators are protected against corrosion by such finishes as waxing, painting, or anodizing (on aluminum). Waxing reduces the wetting effect that promotes condensate formation; however, tests on waxed evaporator surfaces versus nonwaxed surfaces show a negligible loss of capacity. Thin paint films do not have an appreciable effect on capacity.

Removable water receptacles, provided with most dehumidifiers, hold from 16 to 24 pints and are usually made of plastic to withstand corrosion. Ease of removal and handling without spillage are important. Most dehumidifiers also provide either a means of attaching a flexible hose to the water receptacle or a fitting provided specially for that purpose. A flexible hose permits direct gravity drainage to a sewer.

The more expensive dehumidifiers usually have higher output capacities, are more attractively styled, and have various auxiliary features. Cabinet enclosures vary with respect to aesthetic design. A humidity-sensing control added to automatically cycle the unit maintains a preselected relative humidity. These humidistats are normally adjustable within a range of 30 to 80% rh. The humidistat may also provide a detent setting that causes continuous running of the unit. Most humidistats provide an on-off line switch; some models also include an additional sensing and switching device that automatically turns the unit off when the water receptacle is full and requires emptying. This second device may or may not include an indicating light.

Dehumidifiers are designed to provide optimum performance at standard rating conditions of 80°F dry-bulb room temperature and 60% rh. When the room is less than 65°F dry bulb and 60% rh, the refrigerant pressure and corresponding evaporator surface temperature usually decrease to the point that frost forms on the cooling coil. This effect is especially noticeable on units having a capillary tube.

Some dehumidifiers are equipped with special defrost controls that cycle the compressor off under frosting conditions. This control is generally a bimetal thermostat that is strategically attached to the evaporator tubing, allowing the dehumidifying process to continue at a reduced rate when frosting conditions exist. The humidistat can sometimes be adjusted to a higher relative humidity setting, which reduces the number and duration of running cycles and permits reasonably satisfactory operation at low-load conditions. In many instances, especially in the late fall and early spring, supplemental heat must be provided from other sources to maintain temperature in the space above that at which frosting can occur.

Dehumidifiers are usually equipped with rollers or casters. Typical sizes of dehumidifiers vary as follows:

Width: 12 to 20 in.	Height: 12 to 24 in.
Depth: 10 to 20 in.	Weight: 35 to 75 lb

CAPACITY AND PERFORMANCE RATING

Dehumidifiers are available with moisture removal capacities of 11 to 60 pints per 24 h day. The input to domestic dehumidifiers varies from 200 to 800 W, depending on the output capacity rating. Dehumidifiers are operable from ordinary household electrical outlets (115 V, single-phase, 60 Hz).

AHAM *Standard* DH-1 establishes a uniform procedure for determining the rated capacity of dehumidifiers under certain specified test conditions and establishes other recommended performance characteristics as well. An industry certification program sponsored by AHAM covers the great majority of dehumidifiers and certifies the dehumidification capacity. The following tests are specified by *Standard* DH-1.

- **Dehumidifier Capacity.** The capacity is stated in volume per 24 hours. Units must be tested in a room maintained at the following test conditions:

Dry-bulb temperature	80°F
Wet-bulb temperature	69.6°F
Relative humidity	60%

- **Maximum Operation Conditions.** The test is conducted at 90°F dry bulb, 74.8°F wet bulb, and 50% rh. The unit runs continuously at this condition at 90 and 110% of rated voltage for 2 h, after which the power is cut off for 2 min. The unit should restart within 5 min and run continuously for 1 h.
- **Low-Temperature Test.** The unit is operated at 65°F dry bulb, 56.6°F wet bulb, and 60% rh for a period of 8 consecutive hours. If the unit has no defrost control, there should not be ice or frost remaining on the portion of the evaporator coil that is exposed to the entering airstream. If the unit is equipped with a defrost control, it should operate for at least 50% of the test period with no solid ice or frost remaining on the portion of the evaporator coil that is exposed to the entering airstream at the end of each cycle.

CODES

Dehumidifiers are designed to meet the safety requirements of UL *Standard* 474, *Canadian Electrical Code*, and ASHRAE *Standard* 15. UL-listed and CSA-International approved equipment customary have a label or data plate indicating approval. UL also publishes the Electric Appliance and Utilization Equipment Directory, which covers this type of appliance.

REFERENCES

AHAM. 1992. Dehumidifiers. *Standard* DH-1-1992. Association of Home Appliance Manufacturers, Chicago.

AHAM. 1992. Room air conditioners. *Standard* RAC-1-1992.

AHAM. Semi-annually. Directory of certified dehumidifiers.

AHAM. Semi-annually. Directory of certified room air conditioners.

ARI. Semi-annually. *Directory of certified applied air-conditioning products.*

ARI/CSA-Intl. 1993. Standard for packaged terminal air-conditioners and heat pumps. ARI *Standard* 310/380-93. CSA-Intl. *Standard* C744-93.

ASHRAE. 1988. Methods of testing for rating unitary air-conditioning and heat pump equipment. *Standard* 37-1988.

ASHRAE. 1994. Safety code for mechanical refrigeration. *Standard* 15-1994.

ASHRAE. 1996. Method of testing flow capacity of refrigerant capillary tubes. *Standard* 28-1996.

ASHRAE. 1999. Method of testing for rating room air conditioners and packaged terminal air conditioners. *Standard* 16-1999.

ASHRAE. 1999. Method of testing for rating room air conditioner and packaged terminal air conditioner heating capacity. *Standard* 58-1999.

CSA-Intl. 1998. Room air conditioners. *Canadian Electrical Code* C22.2 No. 117. Canadian Standards Association, Etobicoke, ON.

IEC. 1995. Safety of household and similar electrical appliances—Part 2: Particular requirements for electrical heat pumps, air conditioners, and dehumidifiers. IEC *Standard* 60335-2-40 (1995-05). International Electrotechnical Commission.

ISO. 1994. Non-ducted air conditioners and heat pumps—Testing and rating for performance. *Standard* 5151: 1994. International Organization for Standardization.

NFPA. 1998. *National electrical code, 1999 edition*. NFPA *Standard* 70-1998. National Fire Protection Association, Quincy, MA.

ORNL. 1985. *Report* ORNL-NSF-EP-85. Oak Ridge National Laboratory, Oak Ridge, TN.

UL. 1992. Safety leakage current for appliances. *Standard* C101-92. Underwriters Laboratories, Northbrook, IL.

UL. 1993. Dehumidifiers. *Standard* 474-93.

UL. 1993. Room air conditioners. *Standard* 484-93.

UL. 1998. Electrical appliance and utilization equipment directory

CHAPTER 47

ENGINE-DRIVEN HEATING AND COOLING EQUIPMENT

COMBUSTION ENGINES are used to drive space conditioning, liquid chilling, and refrigeration equipment. In many areas of the United States, the operational cost of an engine-driven system is less than the cost for an electric-driven system because the average cost of energy from fuel or natural gas is less per equivalent unit of energy than electricity. In addition, operation of engine drives, especially during the peak time of the day when electric demand charges are highest, result in additional savings. However, initial and maintenance costs are greater.

Technical advances in materials and engine design have resulted in the production of relatively small, long-lived, reliable engines with sufficient output to drive residential and small commercial/industrial systems. These engines provide lower fuel consumption, noise, and space requirements.

The overall efficiency of an engine-driven system is significantly improved when heat from the engine coolant and exhaust gas is recovered to meet auxiliary heat requirements such as space heating, hot water heating, process heat, domestic hot water, desiccant dehumidification, etc. State-of-the-art engine-driven equipment is designed for minimal service and maintenance. State-of-art controls permit efficient variable-speed operation and proper sequencing of the running characteristics.

This chapter presents the technical and operating data to assist an engineer in making an appropriate choice of equipment and system design based on cost and application. The section on Design and Installation in Chapter 7 includes additional information on engine-driven chillers and heat pumps.

The preparation of this chapter is assigned to TC 7.4, Combustion Engine Driven Heating and Cooling Equipment.

CHILLERS

An engine-driven chiller is driven in a conventional vapor compression mode by a natural gas, gasoline-derivative, or diesel-derivative engine rather than an electric motor. The components and thermodynamic cycle on the refrigeration side of the chiller are those common to all vapor compression chillers: a compressor, an evaporator, a condenser, and an expansion device (Figure 1). The prime movers are external to the compressor. Natural gas powered engines have the benefits of:

- Variable speed and capacity modulation capability
- High part-load efficiency
- Ability to efficiently recover high temperature waste heat for domestic water heating, steam generation or process use
- Reduced total operating costs

CHILLER COMPONENTS

Engines

Engines for chiller can be categorized into two types: diesel-derivative and gasoline-derivative. With diesel-derivative engines, the engine block and crankshaft remain in place for the life of the chiller. Most other components are replaced over time based on a schedule prescribed by the manufacturer. Gasoline-derivative engines are generally low power engines (less than 150 horsepower) that are fully replaced after a top end overhaul. Figure 2 provides a reference to estimate the engine size required for a given chiller capacity. Typically, about 1 horsepower per ton of cooling capacity is required.

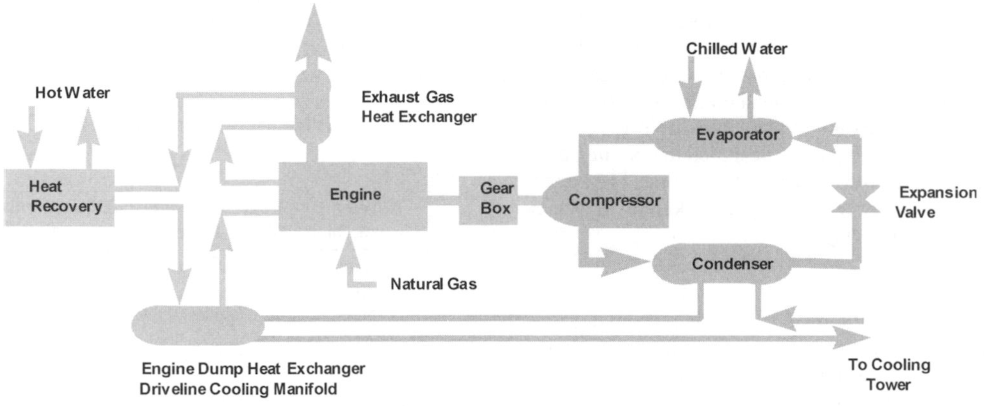

Fig. 1 Engine Driven Chiller with Heat Recovery

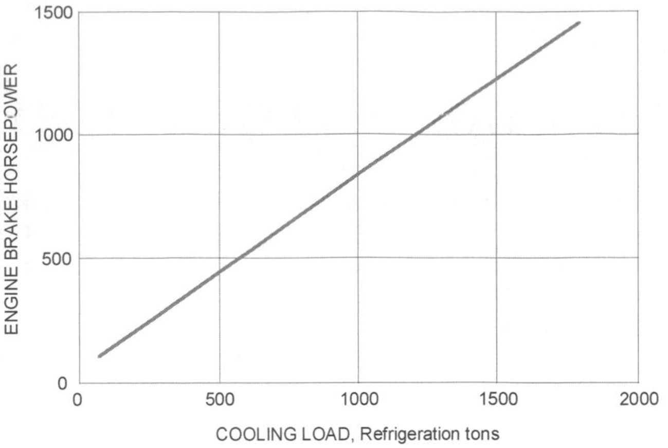

Fig. 2 Relationship Between Cooling Load and Engine Size

Table 1 Characteristics of Compressors for Engine-Driven Chillers

	Centrifugal	Screw	Reciprocating
Capacity range, tons	400 to 6000	100 to 4000	<200
Shaft speed range, rpm	3000 to 14000	1000 to 4000	1200 to 1800
Type of gear box	Speed increasing	Speed increasing	Direct drive or speed decreasing
Capacity control	Guide vanes	Slide valve	Cylinder unloading

Compressors

Three types of engine driven compressors for chiller application are available, differing usually by the cooling capacity:

Reciprocating compressors are generally used in applications of 200 tons of cooling or less. Part-load capacity is controlled by modulating engine speed down to about 30 to 50% of rated speed, which improves fuel economy. Some reciprocating engine chillers also use cylinder unloading to reduce capacity further.

Screw compressors are available in capacities from 100 to 1250 tons. Capacity control is achieved by varying the engine speed and adjusting the slide valve on the compressor.

Centrifugal compressors are most appropriate for large capacity systems and are available in sizes from 400 to over 2000 tons. Speed increasers operate the compressors at significantly higher speeds than the engine. Capacity is controlled by varying the speed of the compressor and/or by compressor inlet guide vane control. Table 1 summarizes the basic characteristics of the three types of chiller compressors. Chapter 34 has more information on compressors.

Engine Compressor Gear Boxes

Engines are typically mated to the compressor through a gear box, which may be internal or external to the compressor. Depending on the shaft speed of the compressor, a speed increasing or a speed decreasing gear box is used. Typical compressor shaft speeds as well as the type of gear box are indicated in Table 1.

Reciprocating engines are typically designed to operate between 1200 to 2400 rpm depending on the size of the engine. The larger the engine the slower the rated speed. The speed is kept as low as possible to extend the life of the engine.

Clutch

The clutch allows the engine to warm up before engaging and loading the compressor. The clutch also allows the engine to complete a cool down sequence as recommended by the engine manufacturer without the compressor being engaged. The clutch can be air-actuated or electric-actuated.

Table 2 COP of Engine-Driven Chillers

Heat Recovery Option	COP at Full Load
No heat recovery	1.2 to 2.0
Jacket water heat recovery	1.5 to 2.25
Jacket water and exhaust heat recovery	1.7 to 2.4

Dump Heat Exchangers

A dump heat exchanger transfers heat from the engine water jacket, engine oil, and compressor oil loops to the cooling tower. Both shell-and-tube or plate heat exchangers are used; but because plate heat exchangers have narrow passages, shell and tube heat exchangers are more commonly used to avoid plugging with debris from open cooling systems such as cooling towers.

RATING AND PERFORMANCE

Both engine-driven and electrically-driven chillers are rated according to ARI *Standard* 550/590 conditions as follows:

Chilled Water Conditions
44°F chilled water supply temperature
54°F chilled water return temperature
2.4 gpm/ton chilled water flow
Water-Cooled Condensers
85°F condenser water supply temperature
95°F condenser water return temperature
3.0 gpm/ton chilled water flow
Air-Cooled Condensers
95°F air supply temperature
20°F temperature differential between air supply and the condensing refrigerant
2°F refrigeration system heat loss to the condenser

Manufacturers offer performance curves for other than ARI standard conditions. As with any chiller, performance is largely a function of design conditions for the condenser and chilled water supply temperatures.

Table 2 provides a typical range of engine-driven chiller COPs with and without heat recovery. The COP is the cooling energy output divided by the fuel energy input. Engine fuel input is based on the higher heating value (HHV) of the fuel. For natural gas, the average HHV is 1020 Btu/ft^3.

The heat recovered from the jacket coolant and exhaust gas is added to the cooling load produced by the chiller, thereby increasing useful thermal output and the COP. Because no standards have been approved on how to calculate the COP of an engine-driven chiller when considering heat recovery, most manufacturers present COPs with and without heat recovery.

Typical COPs at part-load operation for engine-driven chillers are shown in Figure 3 and Figure 4. The variable speed provides a significant part-load performance advantage over single-speed chillers. The high part-load performance results in a higher integrated part-load value (IPLV) as defined by ARI *Standard* 550/590 as follows:

$$\text{IPLV} = 0.01A + 0.42B + 0.45C + 0.12D$$

where

Chiller Load, %	Entering Cond. Water Temp., °F	Manufacturer Rated COP	Part-Load Hours, %
100	85	*A*	1
75	75	*B*	42
50	65	*C*	45
25	65	*D*	12

The IPLV for engine chillers is significantly higher than its rated load performance. As an example, the full-load COP for a typical

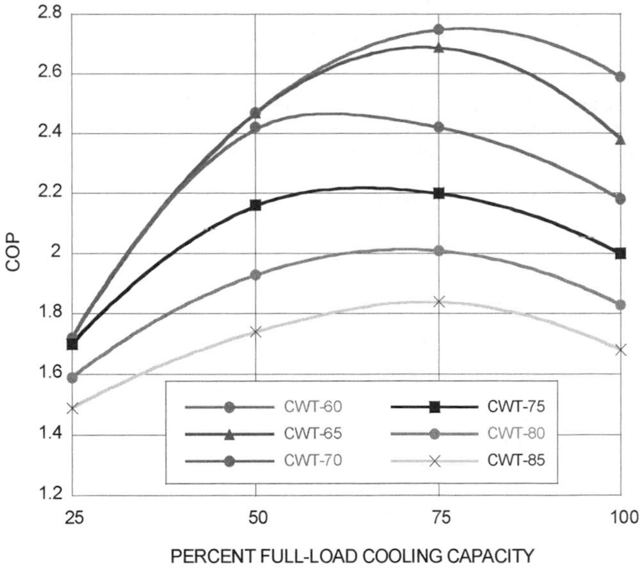

Fig. 3 Typical Engine-Driven Screw Chiller (500 Ton) COP Versus Part-Load and Entering Condenser Water Temperature (CWT)

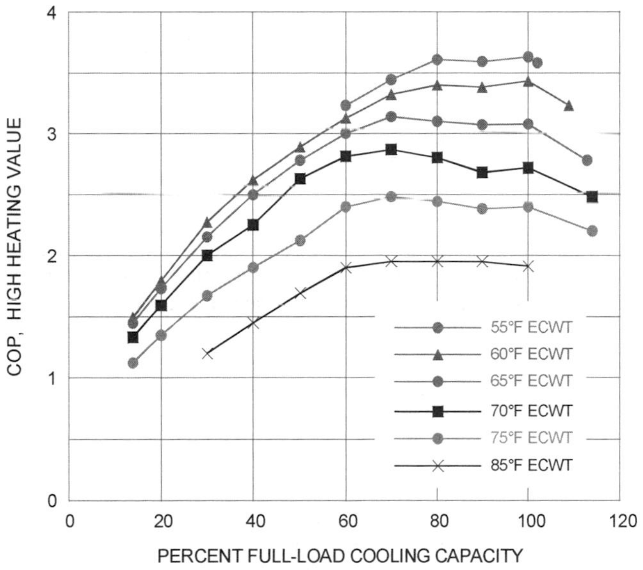

Fig. 4 Typical Engine-Driven Centrifugal Chiller COP Versus Part-Load and Entering Condenser Water Temperature (ECWT)

150 ton engine-driven screw compressor chiller would be 1.3 while the IPLV would be 1.8.

Engine Heat Rejection and Recovery

Energy in the fuel is released during combustion and is converted to shaft work and heat. The shaft work drives the compressor, while the heat is liberated from the engine through the coolant, exhaust gas, and surface radiation (see Figure 5). Approximately 75% of the total energy input is converted to heat. Much of this heat can be recovered from the engine exhaust and jacket coolant. The heat can be used to generate steam or hot water, or it can be used directly in certain industrial applications. To be cost effective, a heating load must be coincident with the cooling load. Examples of applications where this is possible include service water heating in hospitals, hot water for dish washing in restaurants, and a heat source for regenerating a desiccant cooling system. Other uses for recovered heat include space heating, reheating return air, preheating boiler water, and absorption chilling.

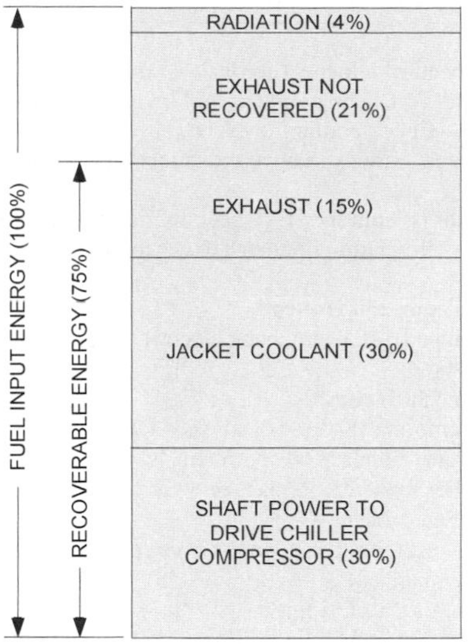

Fig. 5 Engine Energy Balance

Heat recovered from the engine jacket accounts for up to 30% of the energy input and is capable of producing 180°F hot water. To avoid thermal stress, the coolant temperature difference between inlet and outlet generally should not exceed 15°F. Almost all of the heat transferred to the engine coolant can be recovered, limited only by the efficiency of the heat exchanger, and assuming that there is a demand for the heat.

The other major source of heat is the engine exhaust. Exhaust temperatures of 850 to 1200°F are typical. Only a portion of the exhaust heat can be recovered because exhaust gas temperatures are generally kept above condensation thresholds. Most heat recovery units are designed for a 300 to 350°F exhaust outlet temperature to avoid the corrosive effects of condensation in the exhaust piping. Exhaust gas can generate hot water to about 230°F or low pressure steam of 15 psig.

By recovering heat in the jacket water and exhaust, approximately 75% of the fuel's energy can be effectively used. Figure 5 shows typical percentages of energy that can be recovered from the engine based on the fuel's heat content.

MAINTENANCE AND SERVICE

Engine-driven chillers, like any other type of cooler, require routine and preventive maintenance. Maintenance of the compressor is similar to the maintenance of electric-drive units, with the exception of the periodic changing of the gear box oil and the operation and care of the external lubricating pumps.

Engines require periodic servicing and parts replacement, depending on the severity of use and the type of engine. Maintenance and service tasks for engines are usually scheduled as a function of operating hours and equivalent full load hours (EFLH). Operating hours (run hours) are defined as the accumulated time of engine operation. EFLH indicates the severity of chiller operation and is defined as

$$\text{EFLH} = \frac{\text{Total ton-hours/year}}{\text{Max. continuous design capacity (tons)}}$$

Table 3 Typical Service and Overhaul for Diesel Engine

Daily
Walkaround Inspection – Inspect for leaks and loose connections
Lubrication System – Check crankcase oil level and oil filter diff. pressure
Cooling System – Check coolant level
Engine Air Cleaner – Check service indicator, Clean dust collector (if equipped)
Air Starting Motor (if equipped) – Check lubricator level/Empty collector jar
Aftercooling System – Drain condensate; Check intake manifold air temp.

Every Two Weeks or 250 Hours*
Batteries (if equipped) – Clean/Check electrolyte level

Every Month or 750 Hours*
Scheduled Oil Sampling (S-O-S) Analysis – Obtain
Lubrication System – Replace oil and filters
Auxiliary Oil Filter System – Replace oil
Crankcase Breather – Clean
Cooling System – Test for coolant additive concentration; Inspect/Clean radiator fins (if equipped)
Engine Valve Lash – Check/Adjust
Exhaust Valve Takeup and Cylinder Pressure Blowby (Crankcase) – Measure/Record
Carburetor/Governor Control Linkage – Check/Adjust/Lubricate
Spark Plugs – Clean/Check/Adjust gap
Ignition System – Inspect/Check/Adjust timing, inlet manifold temperature and air/fuel ratio
Gas Pressure Regulator – Drain water from drip leg; Check diff. pressure
Air Inlet and Exhaust Piping – Inspect
Belts and Hoses – Inspect/Replace
Fan Drive – Lubricate bearing
Engine Mounts – Inspect
Crankshaft Vibration Damper – Inspect

Every Month or 750 Hours (continued)*
Engine Protection Devices – Inspect for proper operation; Inspect wiring harness
Magnetic Pickups/Sensor (at first oil change only) – Check/Clean
Turbocharger – Inspect for proper operation
Aftercooler – Check for proper operation; Drain condensate; Clean fins

Every Two Months or 1500 Hours*
Auxiliary Oil System Filters – Replace elements

Every Six Months or 4000 Hours
Generator (if equipped) – Lubricate bearing
Magnetic Pickups/Sensor – Check/Clean
Driven Equipment – Inspect/Check/Lubricate as recommended by manufacturer
Jacket Water Pump – Inspect for proper operation
Starting Motor – Inspect for proper operation
Alternator – Inspect for proper operation

Top End and Overhaul – 12,000 to 15,000 Hours
Cylinder Head Assembly – Rebuild or Exchange
Cooling System – Clean/Flush coolant; Replace thermostats and lines
Water Pumps – Rebuild or Exchange
Oil Cooler and Aftercooler Cores – Clean/Test
Ignition Transformers – Test resistance
Carburetor – Inspect/Replace
Gas Pressure Regulator – Inspect/Replace
Prelube Pump – Inspect for proper operation
Turbochargers – Rebuild or Exchange
Exhaust Bypass Valve – Inspect/Clean

Overhaul – 24,000 to 30,000 Hours
Cylinder Head Assembly and Cylinder Packs – Rebuild or Exchange
Crankshaft, Camshaft, Camshaft Followers and Bearings, Damper, Gear Train Gears and Bushings, and Driven Unit Alignment – Inspect

*First perform previous service hour items.

For a typical chiller duty cycle, the EFLH is roughly one-half that of the total operating hours. Suggested maintenance schedules for two different manufacturers of natural gas-powered engine-driven chillers are presented in Table 3 and Table 4.

A preventive maintenance program should inspect for

- Leaks (a visual inspection facilitated by a clean engine)
- Abnormal sounds and odors
- Unaccountable speed changes
- Condition of fuel and lubricating filters

An analysis of lubricating oil is a low-cost method of determining the physical condition of the engine and a guide to maintenance procedures. Commercial laboratories routinely provide this service which includes measuring the concentration of various elements in the lubricating oil such as bearing metals, silicates, calcium, suspended and nonsuspended solids, water, and oil viscosity. The laboratory can assist in interpreting the readings and alert the user to impending problems.

CONTROLS AND OPERATION

Engine-driven chillers can be controlled locally or from a remote location equipped with a computer or a local area network. The control system performs a series of pre-start checks to verify that pressures, temperatures, and flows are within normal limits. After the readiness of these parameters is confirmed, the engine is started and allowed to reach normal operating temperature. During warm-up, engine speed is low and compressor loading is minimal. When the engine reaches normal operating temperature, loading on the compressor is gradually increased, maintaining the desired chilled water set point.

The microprocessor control maintains the set point by controlling both engine speed and compressor loading, a combination that gives many engine chillers a high turndown ratio (5:1 or greater). As the cooling demand decreases, engine speed is reduced until its low limit is reached. Further decreases in chilled water demand cause the compressor to be unloaded (in a reciprocating compressor, the cylinders are unloaded; in a screw compressor, the slide valve is closed; in a centrifugal compressor, the guide vanes are closed).

The control system also monitors key parameters to ensure safe operation. Safety shutdowns are generally initiated if low or high temperatures or pressures occur in the refrigerant, compressor oil, condenser water, or engine oil or coolant. Engine chiller shutdown follows an orderly, predetermined sequence of events including a reduction in engine speed and load in order to reduce thermal and stress shocks on the equipment.

Engine chiller controls may have a graphical interface to allow on-board diagnostics, monitoring of operating and performance parameters, and set-point control and scheduling. Most controls also provide an operating history of alarms and shutdowns to simplify service and troubleshooting. The ability to remotely monitor the chiller's operation and performance is another feature offered by some manufacturers.

DESIGN AND APPLICATION ISSUES

Noise and Vibration Control

Engine-driven machines installed indoors, even where the background noise is high, usually require noise attenuation and isolation from adjoining areas. Noise can be (1) airborne noise created by the engine, compressor, exhaust, and intake; and (2) structure-borne noise created by engine vibration transmitted through the floor and through pipes supported by the wall or ceiling.

Table 4 Typical Service and Overhaul for Gasoline Engine

Service Category	Interval, the First of ...	Item	Action	Est. Hours to Perform
A	750 EFLH or 1500 operating hours	Air filter	Replace	2
		Battery	Inspect	
		Timing	Check and adjust if necessary	
		Carburetor	Check and adjust if necessary	
		Exhaust drains	Check	
		Spark plugs	Replace	
		Ignition wires	Replace	
		Coupling	Inspect	
		Engine mounts	Inspect	
		Engine oil filter	Replace	
		Compressor shaft seal	Evaluate leak rate	
		Compressor oil level	Check	
		Dump heat exchanger strainer	Clean	
		Filter dryer cores	Replace (first season only, then as required)	
		General	Check for leaks and tighten electrical connections	
B	1500 EFLH or 3000 operating hours	Engine lube oil	Replace oil drum and drain pan	2 to 4
		PCV valve	Replace	
		Distributor cap	Replace	
		Rotor	Replace	
		Engine evaluation	Blowby and compression test (may be omitted first B service)	
		Dump heat exchanger	Check, clean if necessary	
		Condenser	Check, clean if necessary	
		Compressor lube oil	Take sample and log	
		Filter dryer	Check upstream moisture indicator	
C	6000 EFLH or 12,000 operating hours	Cylinder heads	Replace	8
D	7500 to 10,000 EFLH or 15,000 to 20,000 operating hours	Engine	Replace	16
E	Seasonal	Start-up	Follow procedure in manual	2 to 4
		Shutdown	Follow procedure in manual	2 to 4
F	As required (no set interval)	Engine valves	Adjust	1
		Compressor shaft seal	Replace	4
		Compressor oil	Replace	4
		Compressor oil filter	Replace	1
		Thermo mixing valve(s)	Replace elements	1

Table 5 Sound Reduction

		Approx. Sound Reduction, dBA
	Original machine (90 to 98 dB)	0
	Vibration isolators	2
	Baffle	5
	Absorption material only	5
	Rigid sealed enclosure	15 to 20
	Enclosure and isolators	25 to 30
	Enclosure, absorption, and isolators	35 to 40
	Double wall enclosure, absorption, and isolators	60 to 80

Sound levels measured in accordance with ARI *Standard* 575.

Exhaust noise is commonly attenuated with a silencer, which typically reduces the noise by 15 dB when measured 10 ft from the exhaust outlet. The muffler manufacturer can evaluate such factors as the number of cylinders and engine speeds to predict the specific effects of a muffler.

Vibration created noise can be minimized with engine vibration isolator mounts and flexible connections for piping. Sound absorbing materials or double wall construction can be applied to room enclosures to reduce noise. The approximate noise reduction level for various approaches is shown in Table 5.

Emissions and Permitting

Depending on local regulations, exhaust emissions from internal combustion engines is a concern of the equipment manufacturer. The emission characteristics of an engine are primarily affected by the ratio of air to fuel during combustion. In stoichiometric combustion, fuel reacts with the exact amount of air required to oxidize all carbon and hydrogen in the fuel. The combustion by-products include carbon dioxide, water, nitrogen dioxides, and sulfur dioxide. The exhaust theoretically contains no unburned fuel or unreacted oxygen. Because the air and fuel can never be completely mixed, combustion is never complete. Therefore, other contaminants such as unburned hydrocarbons and carbon monoxide become part of the exhaust emissions. Based on theoretical calculations, approximately 16 lb of air are required to burn 1 lb of natural gas. This ratio represents stoichiometric combustion. Chapter 17 of the 1997 *ASHRAE Handbook—Fundamentals* has more information on combustion.

A common method to represent the air-to-fuel ratio is to define the **relative air-to-fuel ratio** as

$$\lambda = \frac{\text{Operating air-fuel ratio}}{\text{Stoichiometric air-fuel ratio}}$$

A λ value of 1 indicates that the engine is operating at stoichiometric conditions. For values of λ less than 1.0, the engine is said to run rich because more fuel is in the cylinder than the air can theoretically oxidize. For values of λ greater than 1.0, the engine is said to run lean because more air is present than is required to theoretically oxidize the fuel.

Atmospheric Contaminants. Contaminants that can be produced in natural gas engines are classified in the following categories: NO_x, CO, HC, SO_x, and PM_{10}.

NO_x: Oxides of nitrogen consist of NO and NO_2 which are formed when N_2 and O_2 from the air react. The level of the reaction primarily depends on combustion temperature. A low combustion temperature will result in low NO_x levels, and vice versa, a high combustion temperature will result in high NO_x levels.

CO: Carbon monoxide is formed by incomplete combustion of the fuel, which occurs when air is insufficient or poorly mixed with the fuel to develop complete combustion.

HC: Hydrocarbons are unburned fuel (hydrogen and carbon). Hydrocarbons are the result of incomplete combustion because of inadequate air, insufficient mixing of the air and fuel, and inadequate combustion temperature. Natural gas is made up of several hydrocarbon gasses including methane, ethane, propane, and butane. Methane is the primary hydrocarbon representing about 95% of natural gas.

HC emissions are typically broken down into two categories: total hydrocarbons (THC), which include all HC gases in the exhaust stream; and non-methane hydrocarbons (NMHC), are the portion of the total hydrocarbons that does not include methane. Most regulating agencies only regulate non-methane hydrocarbons because they react with NO_x in the lower atmosphere, acting as the precursor in the formation of photochemical smog. NMHC gases are considered volatile organic compounds (VOCs) or reactive organic compounds (ROCs).

SO_x: Oxides of sulfur are formed when sulfur containing compounds in the fuel and air are oxidized in the combustion chamber. Sulfur levels in natural gas are negligible, therefore SO_x emissions from natural gas engines are extremely low. Oxides of sulfur enter the atmosphere and combine with water in the air forming H_2SO_3 (sulfurous acid) and H_2SO_4 (sulfuric acid).

PM_{10}: Particulate matter is formed during combustion of liquid fuels and engine lubricating oil. Particulates are often seen as black smoke coming from diesel truck engines. Particulate emissions from gaseous fueled engines are extremely low.

Figure 6 shows how the relative amount of each contaminant qualitatively changes with air/fuel ratio.

The two most common strategies used to meet emission regulations are (1) lean burn combustion and (2) stoichiometric combustion with three-way catalyst after-treatment.

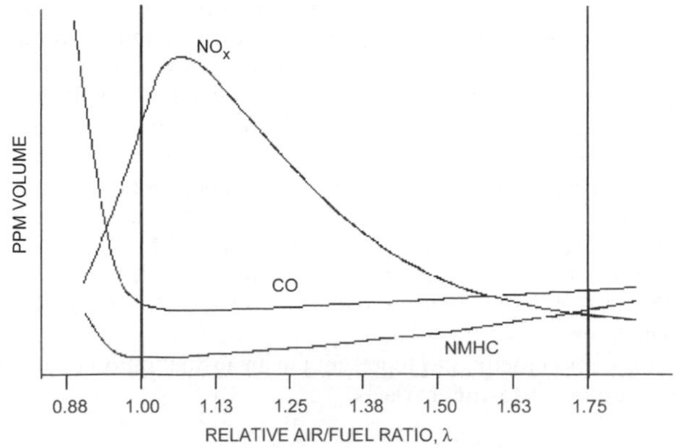

Fig. 6 Natural Gas Engine Emission Characteristics

Lean Burn Combustion. To the rich side ($\lambda < 1.0$) of stoichiometry, NO_x decreases significantly due to the lack of oxygen in the combustion chamber and the lower combustion temperature. On the lean side ($\lambda > 1.0$), the NO_x reaches a peak because the combustion temperature remains high and oxygen is abundant. At increasingly lean air/fuel ratios, the combustion temperature continues to fall, thus reducing NO_x levels.

Carbon monoxide levels are also lower in a lean combustion engine when compared to a stoichiometric engine because the fuel has plenty of oxygen to react with. Operation to the rich side causes a significant increase in CO because of the lack of sufficient oxygen to complete combustion. At a slightly lean point, CO output reaches a minimum because oxygen is sufficient and combustion temperatures are high. At leaner combustion air/fuel ratios, CO gradually increases due to poor combustion from low combustion temperatures and the lower flammability of the fuel.

Like CO, emissions of NMHC are higher at points rich of stoichiometry because of a lack of sufficient oxygen for combustion. NMHC emissions reach a minimum at a point slightly lean of stoichiometry, but gradually increase at higher air/fuel ratios due to poor combustion as a result of low combustion temperatures.

Overall, a lean combustion engine operating with λ between 1.0 and 1.5 allows the most effective combustion of the fuel and results in lowest overall levels of contaminants.

Stoichiometric Combustion with a Three-Way Catalytic Exhaust After-Treatment. Emissions from an engine can be reduced by chemically converting these contaminants into harmless, naturally occurring compounds. The most common method for achieving this is through the use of a catalytic converter. In a catalytic converter, the catalyst will either act as an oxidizing agent or a reducing agent.

A three-way catalyst contains both reduction catalyst materials and oxidation catalyst materials and converts NO_x, CO, and NMHCs to N_2, CO_2, and H_2O. Typical emission conversions for a three-way catalyst operating on a stoichiometric engine are

- 90%+ decrease in NO_x
- 80%+ decrease in CO
- 50%+ decrease in NMHC

These emission levels are satisfactory to meet the most stringent regulations, including the California South Coast Air Quality Management District. However, a very narrow air/fuel ratio operating range is necessary to maintain these percentages. Electronic air/fuel ratio controls are often required. In addition, stoichiometric combustion is necessary to provide exhaust gas temperatures high enough for effective catalyst operation.

Engine Room Ventilation and Combustion Air Requirements

Three to six percent of the fuel consumed by a gas engine is lost as heat radiated to the surrounding air. Additionally, heat lost from exhaust piping can easily equal engine radiated heat. Engine room ventilation must be sufficient to meet three basic requirements:

1. Limit the engine room temperature to less than 110°F in order to maintain rated engine power and reasonable temperatures for the engine operator and service personnel. The ventilation requirement can be calculated following the procedures in Chapter 28 of the 1997 *ASHRAE Handbook—Fundamentals* or estimated from Table 9 in Chapter 7.
2. Meet the requirements of ASHRAE *Standard* 15, Safety Code for Mechanical Refrigeration
3. Provide enough air for proper combustion. Ventilation air must be adequate to provide the engine with sufficient combustion air. About 4 cfm per brake horsepower is satisfactory.

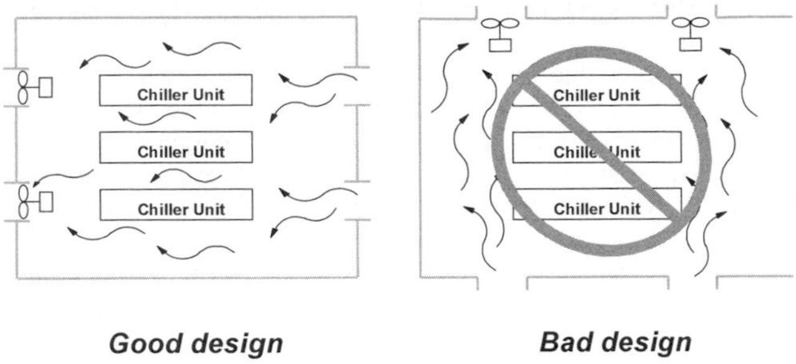

Fig. 7 Plan View of Engine Room Ventilation

As shown in Figure 7, proper ventilation provides an airflow path around the units. Also, cool air should enter low around the engines and remove the heated air from above the engines to encourage as much circulation as possible.

Technical Specifications

Engine-driven chiller manufacturers offer equipment with the same footprint as an electric chiller. The typical engine chiller can weigh up to 50% more than a conventional electric chiller depending on the type of prime mover. Diesel-derivative engines built for high combustion pressures add the greatest weight. Sound levels from engine chillers typically range from 90 to 100 dBA at full load without any enclosure. A removable sound enclosure reduces the full-load sound to that of a conventional electric chiller. Depending on the chiller operating load profile, a sound enclosure may not be needed in many cases because most of the operating time is spent at reduced engine operating speeds with lower noise levels.

Economics

An economic analysis by GRI (1997a) found that installing a mix of gas and electric chillers rather than all-gas or all-electric chillers provided the most attractive economics in most of the cities analyzed. The natural gas engine-driven chillers are operated at maximum capacity during periods with high peak electric demand charges, while the electric units operate during off-peak hours when electric rates are traditionally low.

Although engine chillers are premium priced products, they can be attractive investments yielding a return on investment of 20 to 50% in areas of the country with high electric rates and demand charges (GRI 1997b). Hospitals, colleges/universities, and manufacturing operations that require cooling are prime candidates for these products.

HEAT PUMPS AND AIR CONDITIONERS

EQUIPMENT DESCRIPTION AND DESIGN

Engine-driven heat pumps use an internal (reciprocating, Wankel, etc.) or external (Stirling) combustion engine to drive the compressor in a standard vapor-compression refrigeration cycle. Heat recovered from the engine jacket water and exhaust gas can be used to provide additional heating, dehumidification, or defrost capabilities. The inherent variable-speed capability of engines coupled with variable-speed or multi-speed indoor fans allow engine-driven heat pumps to modulate system heating and cooling capacities to closely match building loads and maintain reasonable indoor humidity levels.

Engine-driven heat pumps are commercially available in residential, light commercial, and commercial sizes and for use with natural gas and propane fuel. They are available with single or multiple indoor packages or units coupled to one outdoor unit.

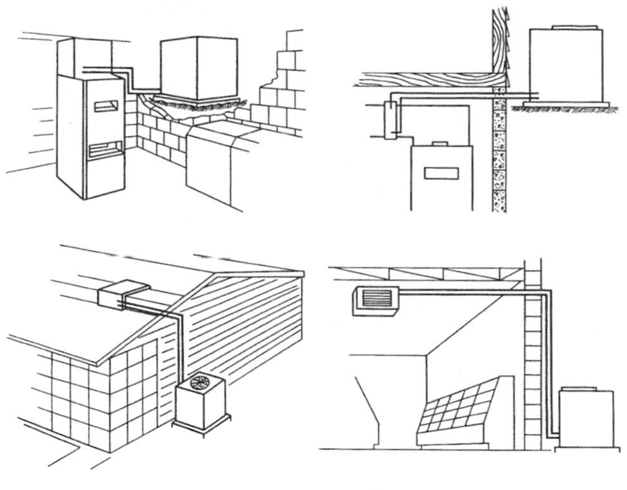

Fig. 8 Residential Engine-Driven Heat Pumps

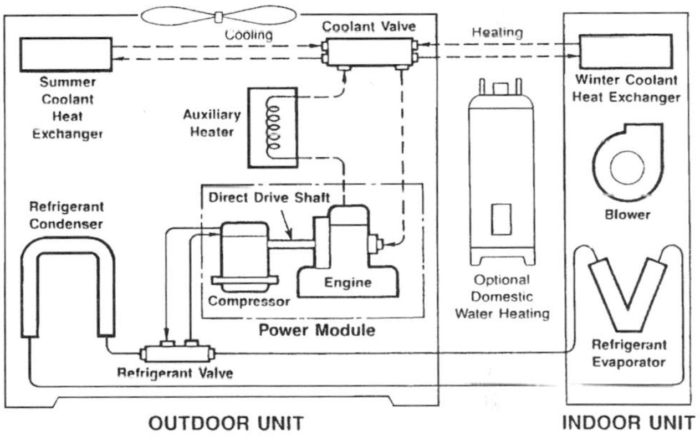

Fig. 9 Four-Pipe Heat Pump

A cross section of a typical engine-driven heat pump is shown in Figure 8. The outdoor unit houses the engine, compressor, and outside heat exchangers. The engine and compressor, combined with associated hardware (e.g., the exhaust gas recuperator, oil reservoir, intake air silencer, exhaust muffler, and auxiliary heater) are located in a common enclosure to reduce noise. This enclosure is ventilated to maintain reasonable internal air temperatures.

Engine-driven heat pumps use either a two-pipe or a four-pipe strategy, referring to the number of pipes running between the indoor and outdoor units. In a two-pipe system, only refrigerant circulates between the indoor and outdoor units. In a four-pipe system, a coolant also circulates between the indoor and outdoor units to carry heat recovered from the engine jacket water and exhaust gas (see Figure 9).

The indoor unit(s) houses the air handler and inside heat exchangers. A coolant-to-air heat exchanger is located downstream of the refrigerant coil for four-pipe systems. A multispeed or electronically commutated blower motor is controlled so that indoor airflow varies in direct proportion to engine speed. In the cooling mode, variable airflow is needed to provide a comfortable split between latent and sensible control of the indoor air. In the heating mode, variable indoor airflow allows the heat pump to deliver a constant air temperature throughout the operating range.

Except for the fuel piping and the coolant loop piping required for a four-pipe system, the installation of an engine-driven heat pump is similar to that of other split-system air-source heat pumps.

The compressor in the outdoor unit is either directly coupled to or belt driven by the engine. Engines vary in design, some having been designed specifically for use in engine-driven heat pumps and others having been modified from automotive engine designs. The engine/compressor suspension system is tuned to avoid resonances and to isolate vibrations throughout the normal operating speed range, which is generally between 1000 and 3000 rpm.

Reciprocating, automotive scroll-type, and rotary compressors are used in engine-driven heat pumps based on the design of specific units. The engine has inherent load-matching capability. Because engine speed can be varied over a broad range, the engine-driven heat pump can modulate output to meet the load. Modulation substantially reduces the cycling required compared to a fixed-capacity system. In addition, because cycling occurs at low-capacity operation, cycling rates are considerably lower (1.5 cycles per hour at 50% on-time) than those exhibited by single-speed systems (3.0 cycles per hour at 50% on-time).

RATINGS AND PERFORMANCE

Test procedures specific to engine-driven heat pumps are included in ANSI *Standard* Z21.40.4/CGA 2.94 (CSA International 1996). A diagram of test points for variable-speed engine-driven heat pumps is shown in Figure 10.

The test procedures for engine-driven heat pumps (CSA International 1996) were derived primarily from those previously developed for electric-motor-driven heat pumps. A study by Talbert et al. (1998), which was funded by ASHRAE, evaluated the relevance of engineering assumptions inherent in the electric heat pump procedures to the engine-driven heat pump test procedures. Several changes in the ANSI test procedures were recommended to better represent the performance characteristics of engine-driven heat pumps. The study also found that gas engine heat pumps are very tolerant of oversizing or undersizing because of their variable-speed operation.

Figure 11 shows the performance characteristics of a typical split-system engine-driven heat pump with one indoor unit. Figure 12 shows performance characteristics of a system with multiple

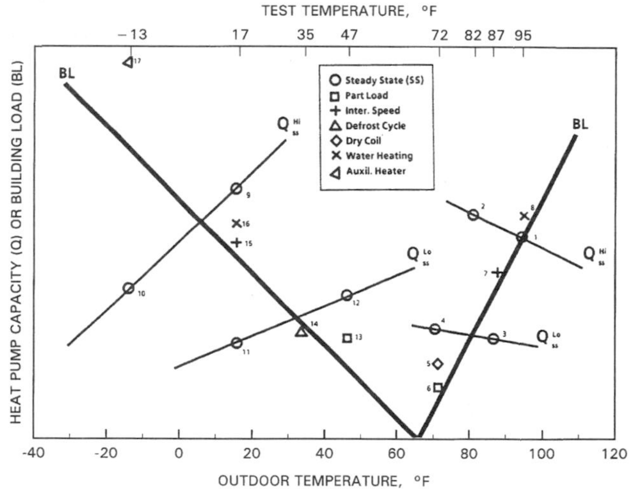

Fig. 10 Test Points for Variable-Speed Engine-Driven Heat Pumps (GRI 1986 and Talbert et al. 1987)

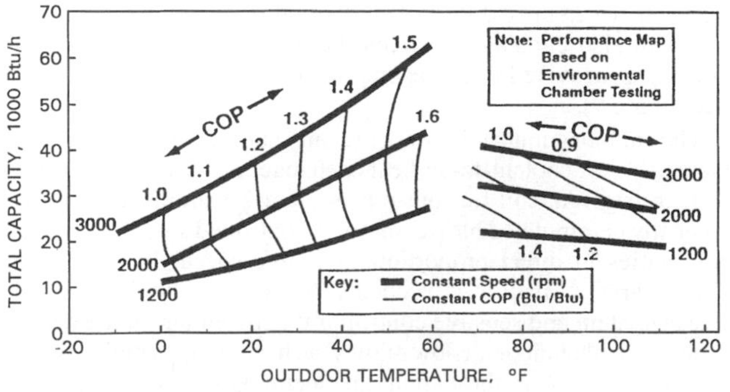

Fig. 11 Performance Characteristics of Split-System Engine-Driven Heat Pumps (GRI 1986)

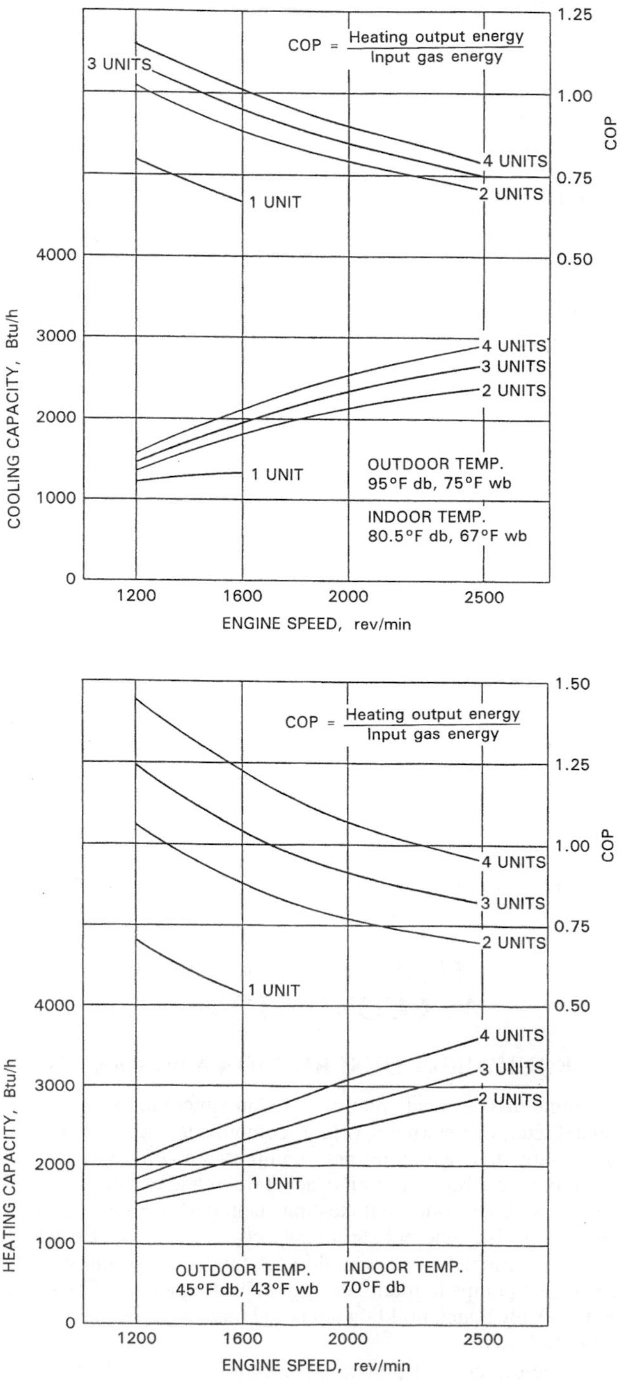

Fig. 12 Performance Characteristics of Multisplit Engine-Driven Heat Pump

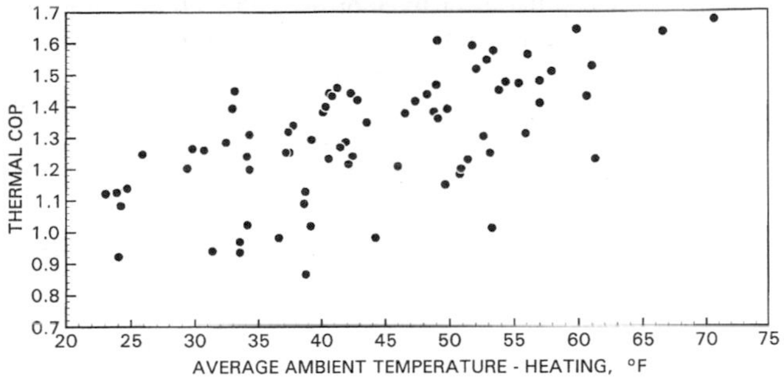

Fig. 13 Field Test Heating Performance (One Site)

indoor units. Heating and cooling capacities increase with increased engine speed and indoor fan speed. Efficiency, measured in terms of COP, is a strong function of outdoor temperature (COP is defined as the heating or cooling energy output divided by the energy input). As illustrated by the heating mode field test data in Figure 13, part-load operation at moderate ambient temperature yields the highest COP (GRI 1993). The same trend is true in the cooling mode; COP decreases as ambient temperature increases. Supplemental heating may be required in areas with severe cold climates, but it is not necessary in most climates if jacket and exhaust heat are used.

ENGINE HEAT REJECTION AND RECOVERY

An important feature of the engine-driven heat pump is the ability to recover heat energy from the engine jacket water and exhaust gas. In a two-pipe system in the heating mode, heat recovered from the engine is transferred to the refrigerant circuit through a double-walled heat exchanger, thus increasing the temperature and heat energy of the refrigerant. In a four-pipe system, a separate coolant circuit transports engine heat recovered through an exhaust gas-to-coolant recuperator in the outdoor unit to a coolant-to-air heat exchanger located in the indoor unit(s), thus increasing the heating capacity of the system.

Recovering engine heat increases the system heating output by as much as 25%. Due to the increased heating capacity and load matching with variable-speed capabilities, the balance point at which supplemental heating is required is 10 to 15°F lower than that of conventional, single-speed, air-source heat pumps.

In the cooling mode in a two-pipe system, engine heat is transferred to the refrigerant through the same double-walled heat exchanger used in the heating mode. The heat is then rejected through the outside heat exchanger along with heat carried by the refrigerant from the indoor air. In a four-pipe system, engine heat is transferred to a coolant loop connected to a second heat exchanger in the outside unit, where it is then rejected to the outside air.

MAINTENANCE AND SERVICE

Special maintenance tasks include checking oil and coolant levels, belt tension, and valve clearance; and changing spark plugs, oil, and air filters. A typical engine maintenance schedule is shown in Table 6. Maintenance intervals and tasks should follow those specified by the equipment manufacturer. Maintenance tasks should be performed by a trained service technician.

CONTROLS AND INSTALLATION

The installation should follow the manufacturer's instructions. Because the supply air from a heat pump is usually at a lower temperature (typically 90 to 100°F) than most heating systems, ducts and supply registers should control air velocity and throw to minimize the perception of cool drafts.

Table 6 Typical Maintenance Schedule for Engine-Driven Gas Heat Pump

Items	Recommendations: Check	Recommendations: Change
Oil level	—	1 year
Oil filter	—	1 year
Coolant level	1 year	5 years
Air filter	—	1 year
Spark plug	—	1 year
Valve clearance[a]	1 year	—
Indoor unit air filter	1 month	6 months

[a]Not necessary if engine is equipped with hydraulic valve lifters.

Low-voltage heating/cooling thermostats control heat pump operation. Models that switch automatically from heating to cooling operation and manual selection models are available. Usually, heating is controlled in two stages. The first stage controls heat pump operation, and the second stage controls supplementary heat. When the heat pump cannot satisfy the first stage's call for heat, supplementary heat is added by the second-stage control. The amount of supplementary heat is often controlled by an outdoor thermostat that allows additional stages of heat to be turned on only when required by the colder outdoor temperature.

Microprocessor technology has led to night setback modes and intelligent recovery schemes for morning warm-up on heat pump systems (Chapter 45 of the 1999 *ASHRAE Handbook—Applications* has further information).

A standard method of defrost is to briefly reverse the refrigeration cycle to send hot refrigerant gas through the outside heat exchanger to defrost the coil. For a two-pipe system with engine heat recovery, the defrost frequency is reduced because the average temperature in the outdoor unit's heat exchanger is higher. A hot-coolant defrost strategy uses a second heat exchanger located adjacent to the outdoor unit's heat exchanger. Coolant that has picked up heat from the engine is circulated through the secondary heat exchanger to defrost the main refrigerant heat exchanger.

In a four-pipe system, simultaneous heating and cooling can be used to counteract problems with cold air delivery during defrost operation. In defrost mode, the refrigerant cycle is reversed and operated in the cooling mode, providing hot refrigerant gas to the outdoor coil. The coolant loop between the indoor and outdoor units continues to operate in the heating mode, offsetting any cooling effect experienced due to the reversed refrigerant cycle.

DESIGN AND APPLICATION ISSUES

Heat pump yearly operating hours are often up to five times those of a cooling-only unit. In addition, heating extends over a greater range of operating conditions at higher stress conditions, so the design must be thoroughly analyzed to ensure maximum reliability. Improved components and protective devices contribute to better reliability, but the equipment designer must select components that are approved for the specific application.

For a reliable and efficient heat pump system, the following factors must be considered: (1) outdoor coil circuitry, (2) defrost and water drainage, (3) refrigerant flow controls, (4) refrigerant charge management, and (5) compressor selection.

Outdoor Coil Circuitry

The outdoor coil operates as an evaporator when the heat pump is used for heating. The refrigerant in the coil is less dense than when the coil operates as a condenser. To avoid an excessive pressure drop during heating, the circuitry is usually a compromise between optimum performance as an evaporator and optimum performance as a condenser.

Defrost and Water Drainage

During colder outdoor temperatures, usually below 40 to 50°F, and high relative humidities (above 50%), the outdoor coil operates below the frost point of the ambient air. The frost that forms on the surface of the coil is usually removed by the reverse-cycle defrost method. In this method, the refrigerant flow in the system is briefly reversed, and hot gas from the compressor flows through the outdoor coil, melting the frost. A typical defrost takes 4 to 10 min. The outdoor fan is normally off during defrost. Because the defrost is a transient process, capacity, power, and the pressures and temperatures of refrigerant in different parts of the system change throughout the defrost period (Miller 1989, O'Neal 1989a).

The performance of the heat pump during the defrost cycle can be enhanced in several ways. Improved defrost times and water removal can be achieved by ensuring that adequate refrigerant is routed to the lower refrigerant circuits in the outdoor coil. Properly sizing the defrost expansion device is critical for shorter defrost times and reducing energy use (O'Neal 1989b). If the expansion device is too small, suction pressure can be below atmospheric, defrost times become long, and energy use is high. If the expansion device is too large, the compressor can be flooded with liquid refrigerant. During the conventional reverse-cycle defrost, there is a significant pressure spike at defrost termination. Prestarting the fan 30 to 45 s before defrost termination can minimize the spike (Anand et al. 1989). In cold climates, the cabinet should be installed above grade to provide good drainage during defrost and to minimize snow and ice buildup around the cabinet. During prolonged periods of severe weather, it may be necessary to clear ice and snow from around the unit.

Several methods are used to determine the need to defrost. One of the more common, simple, and reliable control methods is to initiate defrost at predetermined time intervals (usually 90 min). Demand-type systems detect a need for defrosting by measuring changes in air pressure drop across the outdoor coil or changes in temperature difference between the outdoor coil and the outdoor air. Microprocessors are applied to control this function, as well as numerous other functions (Mueller and Bonne 1980). Demand defrost control is preferred because it requires less energy than other defrost methods.

Refrigerant Flow Controls

Separate refrigerant flow controls are usually used for the indoor and outdoor coils. Because the refrigerant flow reverses its direction between the heating and cooling mode of operation, a check valve bypasses in the appropriate direction around each expansion device. Capillaries, fixed orifices, thermostatic expansion valves, or electronically controlled expansion valves may be used; however, capillaries and fixed orifices require that greater care be taken to prevent excessive flooding of liquid refrigerant into the compressor. A check valve is not needed when an orifice-type expansion device or a biflow expansion valve is used. The reversing valve is the critical additional component required to make a heat pump air-conditioning system.

Refrigerant Charge Management

Refrigerant charge management requires extra care to control compressor flooding and the storage of refrigerant in the system during both heating and cooling. The mass flow of refrigerant during cooling is greater than during heating. Consequently, the amount of refrigerant stored may be greater in the heating mode than in the cooling mode, depending on the relative internal volumes of the indoor and outdoor coils. Usually, the internal volume of the indoor coils ranges from 70 to 110% of the outdoor coil volume. The relative volumes can be adjusted so that the coils not only transfer heat but also manage the charge.

When using capillaries or fixed orifices, the refrigerant may be stored in an accumulator in the suction line or in receivers that can remove the refrigerant charge from circulation when compressor floodback is imminent. Thermostatic expansion valves reduce the flooding problem, but storage may be required in the condenser. Use of accumulators and/or receivers is particularly important in split-system products.

To maintain performance reliability, the amount of refrigerant must be checked and adjusted in accordance with the manufacturer's recommendations, particularly when charging a heat pump. Manufacturer's recommendations for accumulator installation must also be followed so that good oil return is assured.

Compressor Selection

Compressors are selected on the basis of performance, reliability, and probable applications of the unit. In good design practice, equipment manufacturers often consult with compressor manufacturers during both design and application phases of the unitary equipment to verify proper application of the compressor. Compressors in a heat pump operate over a wide range of suction and discharge pressures; thus, their design parameters, such as refrigerant discharge temperatures, pressure ratios, clearance volume, and motor-overload protection require special consideration. In all operating conditions, compressors should be protected against loss of lubrication, liquid floodback, and high discharge temperatures.

REFRIGERATION EQUIPMENT

Engine-driven refrigeration equipment use the same mechanical vapor compression cycle as the engine-driven chillers discussed previously. This equipment is available from various manufacturers as custom and/or standard packaged products. Although engine-driven chillers are typically not operated to deliver fluid temperatures below 40°F, some manufacturers allow chillers with screw compressors to be operated to temperatures as low as 10°F. Figure 14 shows a typical effect of low temperature operation on the chilling capacity and energy input to the machine.

Because refrigeration equipment operates at low evaporator temperatures (+20 to –70°F), refrigerants such as ammonia and other cycles that provide improved efficiency over single-stage cycles are used. Besides the standard, single-stage vapor compression cycle, a multistage refrigeration cycle or cascade refrigeration cycle may be chosen. The multistage cycle is the most common cycle used to efficiently provide refrigeration from –10 to –70°F. The section on

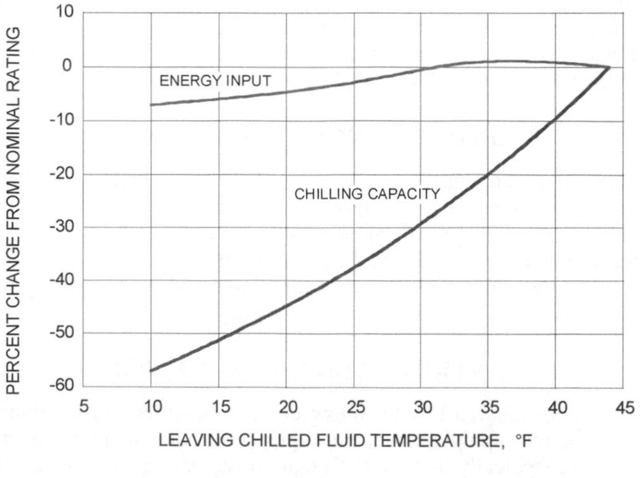

Fig. 14 Effect of Low Temperature Chilling on Nominal Rating of Engine-Driven Chiller with Screw Compressor

Table 7 Typical Efficiency of Engine-Driven Refrigeration Equipment (Ammonia Screw Compressor) (AGCC 1999)

	Sat. Suction Temp./Sat. Discharge Temp.		
	−40/95°F	−12/95°F	+20/95°F
Electric COP	1.32	2.66	4.62
Engine-driven COP without heat recovery	0.44	0.78	1.32
Engine-driven COP with jacket water heat recovery	0.74	1.08	1.62
Engine-driven COP with jacket water and exhaust heat recovery	0.89	1.23	1.77

Compression Refrigeration Cycles in Chapter 1 of the 1997 *ASHRAE Handbook—Fundamentals* describes this cycle.

A cascade system is used when two (or more) refrigerants in the same system provide an operating and/or capital cost advantages. Chapter 38 and Chapter 39 of the 1998 *ASHRAE Handbook—Refrigeration* describe cascade cycles in more detail.

As was the case with engine-driven chillers, heat recovery from the jacket coolant and exhaust gas boosts overall energy use and efficiency. Table 7 lists the coefficient of performance and impact of heat recovery for a range of conditions found in the industry.

REFERENCES

AGCC. 1999. Application engineering manual for engine-driven commercial and industrial refrigeration systems. American Gas Cooling Center, Washington, DC.

Anand, N.K., J.S. Schliesing, D.L. O'Neal, and K.T. Peterson. 1989. Effects of outdoor coil fan pre-start on pressure transients during the reverse cycle defrost of a heat pump. *ASHRAE Transactions* 95(2).

ARI. 1994. Method of measuring machinery sound within an equipment space. *Standard* 575-94. Air-Conditioning and Refrigeration Institute, Arlington, VA.

ARI. 1998. Standard for water chilling packages using the vapor compression cycle. *Standard* 550/590-98.

ASHRAE. 1994. Safety code for mechanical refrigeration. *Standard* 15-1994.

ASHRAE. 1988. Methods of testing for rating unitary air-conditioning and heat pump equipment. *Standard* 37-1988.

CSA International. 1996. Performance testing and rating of gas-fired, air-conditioning and heat pump appliances. ANSI *Standard* Z21.40.4-96/CGA 2.94-M96. CSA International, Cleveland, OH.

GRI. 1986. Gas-engine heat pump test procedures. *Topical Report* GRI-86/0083. Gas Research Institute, Chicago.

GRI. 1993. Home systems research house, gas heat pump heating characterization test results. *Topical Report* GRI-93/0027.

GRI. 1997a. Market opportunities for applied natural gas engine-driven hybrid chiller plants. *Topical Report* GRI-97/0026.

GRI. 1997b. Commercial gas cooling: An investment opportunity. *Topical Report* GRI-97/0140.

Miller, W.A. 1989. Laboratory study of the dynamic losses of a single speed, split system air-to-air heat pump having tube and plate fin heat exchangers. ORNL/CON-253. Oak Ridge National Laboratory, Oak Ridge, TN.

Mueller, D. and U. Bonne. 1980. Heat pump controls: microelectronic technology. *ASHRAE Journal* 22(9).

O'Neal, D.L., N.K. Anand, K.T. Peterson, and S. Schliesing. 1989a. Determination of the transient response characteristics of the air-source heat pump during the reverse cycle defrost. Final Report, ASHRAE Research Project 479-TRP.

O'Neal, D.L., N.K. Anand, K.T. Peterson, and S. Schliesing. 1989b. Refrigeration system dynamics during the reverse cycle defrost. *ASHRAE Transactions* 95(2).

Talbert, S.G., A.L. Rutz, and C.E. French. 1987. Recommended test procedures for rating the performance of gas engine-driven heat pumps. *ASHRAE Transactions* 93(2).

Talbert, S.G., W.G. Atterburg, T.A. Klausing, and F.E. Jakob. 1998. Relevance of existing heat pump testing and rating method assumptions to residential gas engine heat pumps. *ASHRAE Transactions* 104(1).

BIBLIOGRAPHY

Cornell, T.L., R.L. Hedrick, and W.W. Bassett. 1993. Performance characterization of an engine-driven gas heat pump in a single-family residence. *ASHRAE Transactions* 99(1):1430-37.

GRI. 1993. Field test of the York gas heat pump. *Final Report* GRI-92/0509. Gas Research Institute, Chicago.

Harnish, J.T., D.W. Procknow, F.E. Jakob, T.A. Klausing, C.E. French, and G. Nowakowski. 1991. Residential gas heat pump design and development. Cogeneration and Energy Conservation for the 90's Conference.

Kaneko, T., M. Obitani, and T. Imura. 1992. The performance of a four-ton gas-engine-driven heat pump. *ASHRAE Transactions* 98(1):989-93.

Kazuta, H. 1989. Development of small gas engine heat pumps. *ASHRAE Transactions* 95(1):982-90.

Nowakowski, G.A., M. Inada, and M.P. Dearing. 1992. Development and field testing of a high-efficiency engine-driven gas heat pump for light commercial applications. *ASHRAE Transactions* 98(1):994-1000.

Nowakowski, G.A. and R.W. Rasmussen. 1990. The development of controls for gas-engine-driven heat pumps. Proceedings of the 3rd International Energy Agency Heat Pump Conference, Tokyo.

Taira, K. 1992. Development of a 2.5-RT multiple-indoor unit gas engine heat pump. *ASHRAE Transactions* 98(1):982-88.

Yokoyama, T. 1992. Design considerations for gas-engine heat pumps. *ASHRAE Transactions* 98(1):975-81.

CHAPTER 48

CODES AND STANDARDS

THE Codes and Standards listed here represent practices, methods, or standards published by the organizations indicated. They are useful guides for the practicing engineer in determining test methods, ratings, performance requirements, and limits applying to the equipment used in heating, refrigerating, ventilating, and air conditioning. Copies of the standards can be obtained from most of the organizations listed in the Publisher column, from Global Engineering Documents at **global.ihs.com**, or from CSSINFO at **cssinfo.com**. Addresses of the organizations are given at the end of the chapter. A comprehensive database with over 250,000 industry, government, and international standards is at **www.nssn.org**.

Table 1 Codes and Standards Published by Various Societies and Associations

Subject	Title	Publisher	Reference
Air Conditioners	Commercial Applications Systems and Equipment (1993)	ACCA	ACCA Manual CS
	Residential Equipment and Selection (1995)	ACCA	ACCA Manual S
	Methods of Testing for Rating Ducted Air Terminal Units	ASHRAE	ANSI/ASHRAE 130-1996
	Ducted Air-Conditioners and Air-to-Air Heat Pumps—Testing and Rating for Performance	ISO	ISO 13253:1995
	Non-Ducted Air Conditioners and Heat Pumps—Testing and Rating for Performance	ISO	ISO 5151:1994
	Guidelines for Roof Mounted Outdoor Air-Conditioner Installations	SMACNA	SMACNA 1998
	Heating and Cooling Equipment (1995)	UL/CSA	UL 1995/C22.2 No. 236-M95
Central	Performance Standard for Split-System Central Air-Conditioners and Heat Pumps	CSA	CAN/CSA-C273.3-M91
	Performance Standard for Single Package Central Air-Conditioners and Heat Pumps	CSA	CAN/CSA-C656-M92
	Performance Standard for Rating Large Air Conditioners and Heat Pumps	CSA	CAN/CSA-C746-93
	Heating and Cooling Equipment (1995)	UL/CSA	UL 1995/C22.2 No. 236-M95
Gas Fired	Gas-Fired, Heat Activated Air Conditioning and Heat Pump Appliances	CSA	ANSI Z21.40.1 1996/CGA 2.91-M96
	Gas-Fired Work Activated Air Conditioning and Heat Pump Appliances (Internal Combustion)	CSA	ANSI Z21.40.2-1996/CGA 2.92-M96
	Performance Testing and Rating of Gas-Fired Air Conditioning and Heat Pump Appliances	CSA	ANSI Z21.40.4-1996/CGA 2.94-M96
Packaged Terminal	Standard for Packaged Terminal Air-Conditioners and Heat Pumps	ARI/CSA	ANSI/ARI 310/380-93/ CSA C744-93
Room	Room Air Conditioners	AHAM	ANSI/AHAM RA C-1-1992
	Method of Testing for Rating Room Air Conditioners and Packaged Terminal Air Conditioners	ASHRAE	ANSI/ASHRAE 16-1983 (RA 99)
	Method of Testing for Rating Room Air Conditioner and Packaged Terminal Air Conditioner Heating Capacity	ASHRAE	ANSI/ASHRAE 58-1986 (RA 99)
	Methods of Testing for Rating Room Fan-Coil Air Conditioners	ASHRAE	ANSI/ASHRAE 79-1984 (R 91)
	Performance Standard for Room Air Conditioners	CSA	CAN/CSA-C368.1-M90
	Room Air Conditioners	CSA	C22.2 No. 117-1970 (R1998)
	Room Air Conditioners (1993)	UL	ANSI/UL 484
Unitary	Application of Sound Rating Levels - Outdoor Unitary Equipment	ARI	ARI 275-97
	Commercial and Industrial Unitary Air-Conditioning and Heat Pump Equipment	ARI	ANSI/ARI 340/360-93
	Sound Rating of Outdoor Unitary Equipment	ARI	ARI 270-95
	Unitary Air-Conditioning and Air-Source Heat Pump Equipment	ARI	ANSI/ARI 210/240-94
	Method ofTesting for Rating Computer and Data Processing Room Unitary Air Conditioners	ASHRAE	ANSI/ASHRAE 127-1988
	Method of Rating Unitary Spot Air Conditioners	ASHRAE	ANSI/ASHRAE 128-1989
	Methods of Testing for Rating Heat-Operated Unitary Air-Conditioning Equipment for Cooling	ASHRAE	ANSI/ASHRAE 40-1980 (R 92)
	Methods of Testing for Rating Unitary Air-Conditioning and Heat Pump Equipment	ASHRAE	ANSI/ASHRAE 37-1988
	Methods of Testing for Rating Seasonal Efficiency of Unitary Air Conditioners and Heat Pumps	ASHRAE	ANSI/ASHRAE 116-1995
Air Conditioning	Comfort, Air Quality and Efficiency by Design (1997)	ACCA	ACCA Manual RS
	Commercial Applications Systems and Equipment (1993)	ACCA	ACCA Manual CS
	Load Calculation for Commercial Summer and Winter Air Conditioning (1988)	ACCA	ACCA Manual N
	Residential Load Calculation	ACCA	ACCA Manual J, 1986
	Environmental System Technology (1984)	NEBB	NEBB
	Installation of Air Conditioning and Ventilating Systems	NFPA	ANSI/NFPA 90A-1996
	Standard of Purity for Use in Mobile Air-Conditioning Systems	SAE	SAE J 1991-1989
	HVAC Systems Applications, 1st ed.	SMACNA	SMACNA 1987

Table 1 Codes and Standards Published by Various Societies and Associations (*Continued*)

Subject	Title	Publisher	Reference
	HVAC Systems—Duct Design, 3rd ed.	SMACNA	SMACNA 1990
	Heating and Cooling Equipment (1995)	UL/CSA	UL 1995/C22.2 No. 236-M95
Aircraft	Air Conditioning of Aircraft Cargo	SAE	SAE AIR 806B-1997
	Air Conditioning Systems for Subsonic Airplanes	SAE	ANSI/SAE ARP 85E-1991 (R 96)
	Aircraft Fuel Weight Penalty Due to Air Conditioning	SAE	SAE AIR 1168/8-1989
	Aircraft Ground Air Conditioning Service Connection	SAE	SAE AS 4262A-1997
	Air Cycle Air Conditioning Systems for Military Air Vehicles	SAE	ANSI/SAE ARP 4073-1993
	Control of Excess Humidity in Avionics Cooling	SAE	SAE ARP 1987A-1997
	Engine Bleed Air Systems for Aircraft	SAE	ANSI/SAE ARP 1796-1987
	Guide for Qualification Testing of Aircraft Air Valves	SAE	ANSI/SAE ARP 986C-1997
	Nomenclature, Aircraft Air-Conditioning Equipment	SAE	SAE ARP 147C-1978 (R 1992)
	Testing of Commercial Airplane Environmental Control Systems	SAE	SAE ARP 217C-1997
Automotive	Automotive Air-Conditioning Hose	SAE	ANSI/SAE J 51-1989
	Design Guidelines for Air Conditioning Systems for Off-Road Operator Enclosures	SAE	SAE J 169-1985
	Extraction and Recycle Equipment for Mobile Automotive Air-Conditioning Systems	SAE	SAE J 1990-1992
	Guide to the Application and Use of Passenger Car Air-Conditioning Compressor Face Seals	SAE	SAE J 1954-1995
	Information Relating to Duty Cycles and Average Power Requirements of Truck and Bus Engine Accessories	SAE	SAE J 1343-1981
	Rating Air Conditioner Evaporator Air Delivery and Cooling Capacities	SAE	ANSI/SAE J 1487-1985
	Service Hose for Automotive Air Conditioning	SAE	SAE J 2196-1992 (R1997)
	Test Method for Measuring Power Consumption of Air Conditioning and Brake Compressors for Trucks and Buses	SAE	ANSI/SAE J 1340-1981 (R1996)
Ships	Mechanical Refrigeration and Air-Conditioning Installations Aboard Ship	ASHRAE	ANSI/ASHRAE 26-1996
Air Curtains	Air Curtains for Entranceways in Food and Food Service Establishments	NSF	ANSI/NSF 37-1992
	Test Methods for Air Curtain Units	AMCA	AMCA 220-91
	Air Terminals	ARI	ARI 880-94
	Method of Testing for Rating the Performance of Air Outlets and Inlets	ASHRAE	ANSI/ASHRAE 70-1991
	Standard Methods for Laboratory Airflow Measurement	ASHRAE	ANSI/ASHRAE 41.2-1987 (R 92)
	Rating the Performance of Residential Mechanical Ventilating Equipment	CSA	CAN/CSA-C260-M90
	Direct Gas-Fired Door Heaters	CSA	ANSI Z83.17-1990 (R1998)
Air Diffusion	Air Distribution Basics for Residential and Small Commercial Buildings	ACCA	ACCA Manual T, 1989
	Test Code for Grilles, Registers and Diffusers	ADC	ADC 1062:GRD-84
	Method of Testing for Rating the Performance of Air Outlets and Inlets	ASHRAE	ANSI/ASHRAE 70-1991
	Method of Testing for Room Air Diffusion	ASHRAE	ANSI/ASHRAE 113-1990
Air Filters	Industrial Ventilation: A Manual of Recommended Practice, 23rd ed. (1998), Selection of Air Filtration Equipment, p. 4.9	AGCIH	AGCIH
	Method for Measuring Performance of Portable Household Electrical Cord Connected Room Air Cleaners	AHAM	ANSI/AHAM AC-1-1988
	Commercial and Industrial Air Filter Equipment	ARI	ARI 850-93
	Residential Air Filter Equipment	ARI	ARI 680-93
	Agricultural Cabs-Environmental Air Quality. Part 1: Definitions, Test Methods, and Safety Practices	ASAE	ASAE S525-1-1-1997
	Part 2: Pesticide Vapor Filters-Procedure and Performance Criteria	ASAE	ASAE S525-2-1997
	Gravimetric and Dust-Spot Procedures for Testing Air-Cleaning Devices Used in General Ventilation for Removing Particulate Matter	ASHRAE	ANSI/ASHRAE 52.1-1992
	Method of Testing General Ventilation Air-Cleaning Devices for Removal Efficiency by Particle Size	ASHRAE	ANSI/ASHRAE 52.2-1999
	Method for Sodium Flame Test for Air Filters	BSI	BS 3928
	Particulate Air Filters for General Ventilation—Requirements, Testing Marking	BSI	BS EN 779:1993
	Electrostatic Air Cleaners (1995)	UL	ANSI/UL 867-1997
	High-Efficiency, Particulate, Air Filter Units (1996)	UL	UL 586
	Test Performance of Air Filter Units (1994)	UL	ANSI/UL 900-1995
Air-Handling Units	Commercial Applications Systems and Equipment (1993)	ACCA	ACCA Manual CS
	Central Station Air-Handling Units	ARI	ANSI/ARI 430-89
	Direct Gas-Fired Make-Up Air Heaters	CSA	ANSI Z83.4-1991 (R1998)
Air Leakage	Air Leakage Performance for Detached Single-Family Residential Buildings	ASHRAE	ANSI/ASHRAE 119-1988 (R 94)
	Method of Determining Air Change Rates in Detached Dwellings	ASHRAE	ANSI/ASHRAE 136-1993
	Standard Practices for Air Leakage Site Detection in Building Envelopes and Air Retarder Systems	ASTM	ASTM E 1186-87 (R1998)
	Test Method for Determining Air Change in a Single Zone by Means of a Tracer Gas Dilution	ASTM	ASTM E 741-95

Table 1 Codes and Standards Published by Various Societies and Associations (*Continued*)

Subject	Title	Publisher	Reference
	Test Method for Determining Air Leakage Rate by Fan Pressurization	ASTM	ASTM E 779-87 (R1992)
	Test Method for Determining the Rate of Air Leakage Through Exterior Windows, Curtain Walls, and Doors Under Specified Pressure and Temperature Differences Across the Specimen	ASTM	ASTM E 1424-91
	Test Method for Determining the Rate of Air Leakage Through Exterior Windows, Curtain Walls, and Doors Under Specified Pressure Differences Across the Specimen	ASTM	ASTM E 283-91
	Test Method for Field Measurement of Air Leakage Through Installed Exterior Window and Doors	ASTM	ASTM E 783-93
Boilers	A Guide to Clean and Efficient Operation of Coal Stoker-Fired Boilers	ABMA	ABMA
	Boiler Water Limits and Steam Purity Recommendations for Watertube Boilers	ABMA	ABMA
	Boiler Water Requirements and Associated Steam Purity—Commercial Boilers	ABMA	ABMA
	Fluidized Bed Combustion Guidelines	ABMA	ABMA
	Guidelines for Industrial Boiler Performance Improvement	ABMA	ABMA
	Matrix of Recommended Quality Control Requirements	ABMA	ABMA
	Operation and Maintenance Safety Manual	ABMA	ABMA
	Recommended Design Guidelines for Stoker Firing of Bituminous Coals	ABMA	ABMA
	(Selected) Summary of Codes and Standards of the Boiler Industry	ABMA	ABMA
	Thermal Shock Damage to Hot Water Boilers as a Result of Energy Conservation Measures	ABMA	ABMA
	Commercial Applications Systems and Equipment (1993)	ACCA	ACCA Manual CS
	Methods of Testing for Annual Fuel Utilization Efficiency of Residential Central Furnaces and Boilers	ASHRAE	ANSI/ASHRAE 103-1993
	Boiler and Pressure Vessel Code (11 sections) (1998)	ASME	ASME
	Boiler, Pressure Vessel, and Pressure Piping Code	CSA	CSA B51-97
	Testing and Rating Standard for Heating Boilers (1989)	HYDI	IBR
	Prevention of Furnace Explosions/Implosions in Multiple Burner Boilers	NFPA	ANSI /NFPA 8502-1995
	Heating, Water Supply, and Power Boilers—Electric (1995)	UL	ANSI/UL 834-1998
Gas or Oil	Gas-Fired Low-Pressure Steam and Hot Water Boilers	CSA	ANSI Z21.13/CSA 4.9-1999
	Gas Utilization Equipment in Large Boilers	CSA	ANSI Z83.3-1971 (R 1995)
	Control and Safety Devices for Automatically Fired Boilers	ASME	ANSI/ASME CSD-1-1998
	Industrial and Commercial Gas-Fired Package Boilers	CSA	CAN1-3.1-77 (R 1996)
	Oil-Fired Steam and Hot-Water Boilers for Residential Use	CSA	B140.7.1-1976 (R 1991)
	Oil-Fired Steam and Hot-Water Boilers for Commercial and Industrial Use	CSA	B140.7.2-1967 (R 1991)
	Prevention of Furnace Explosions/Implosions in Multiple Burner Boilers	NFPA	ANSI/NFPA 8502-1995
	Single Burner Boiler Operations	NFPA	ANSI/NFPA 8501-1997
	Commercial-Industrial Gas Heating Equipment (1994)	UL	UL 795
	Oil-Fired Boiler Assemblies (1995)	UL	UL 726
	Standards and Typical Specifications for Deaerators, 6th ed. (1998)	HEI	HEI
Building Codes	ASTM Standards Used in Building Codes	ASTM	ASTM
	BOCA National Building Code, 14th ed. (1999)	BOCA	BOCA
	ICC International Property Maintenance Code, 1st ed. (1998)	ICC	ICC
	National Building Code of Canada (1995)	NRCC	NRCC
	International One- and Two-Family Dwelling Code (1998)	ICC	ICC
	International Energy Conservation Code (1998)	ICC	BOCA/ICBO/SBCCI
	Uniform Building Code, three volumes (1997)	ICBO	ICBO
	Directory of Building Codes and Regulations, State and City Volumes (annual)	NCSBCS	NCSBCS
	Standard Building Code (1997 with 1998 revisions)	SBCCI	SBCCI
Mechanical	Safety Code for Elevators and Escalators (plus two yearly supplements)	ASME	ANSI/ASME A 17.1-1996
	Natural Gas Installation Code	CSA	CAN/CGA-B149.1-M95
	Propane Installation Code	CSA	CAN/CGA-B149.2-M91
	Safety Code for Elevators	CSA	CAN/CSA-B44-94
	Uniform Mechanical Code (1997) (with Uniform Mechanical Code Standards)	ICBO/ IAPMO	ICBO/ IAPMO
	Standard Gas Code (1997)	SBCCI	SBCCI
	International Mechanical Code (1998)	ICC	ICC
	International Fuel Gas Code (1997)	ICC	ICC
	Standard Mechanical Code (1997)	SBCCI	SBCCI
Burners	Guidelines for Burner Adjustments of Commercial Oil-Fired Boilers	ABMA	ABMA
	Domestic Gas Conversion Burners	CSA	ANSI Z21.17-1998/CSA 2.7-M98
	Installation of Domestic Gas Conversion Burners	CSA	ANSI Z21.8-1994
	General Requirements for Oil Burning Equipment	CSA	CAN/CSA-B140.0-M87 (R 1998)
	Installation Code for Oil Burning Equipment	CSA	CAN/CSA-B139-M91
	Oil Burners: Atomizing-Type	CSA	CAN/CSA-B140.2.1-M90
	Pressure Atomizing Oil Burner Nozzles	CSA	B140.2.2-1971 (R 1996)

Table 1 Codes and Standards Published by Various Societies and Associations (*Continued*)

Subject	Title	Publisher	Reference
	Replacement Burners and Replacement Combustion Heads for Residential Oil Burners	CSA	B140.2.3-M1981 (R 1998)
	Vapourizing-Type Oil Burners	CSA	B140.1-1966 (R 1998)
	Commercial/Industrial Gas and/or Oil-Burning Assemblies with Emission Reduction Equipment (1993)	UL	ANSI/UL 2096-1995
	Commercial-Industrial Gas Heating Equipment (1994)	UL	UL 795-1994
	Oil Burners (1994)	UL	ANSI/UL 296-1995
	Waste Oil-Burning Air-Heating Appliances (1995)	UL	ANSI/UL 296A-1997
Chillers	Methods of Testing Liquid-Chilling Packages	ASHRAE	ASHRAE 30-1995
	Commercial Applications Systems and Equipment (1993)	ACCA	ACCA Manual CS
	Absorption Water-Chilling and Water Heating Packages	ARI	ARI 560-92
	Centrifugal and Rotary Screw Water-Chilling Packages	ARI	ANSI/ARI 550-92
	Positive Displacement Compressor Water-Chilling Packages	ARI	ANSI/ARI 590-92
	Performance Standard for Rating Packaged Water Chillers	CSA	C743-93
Chimneys	Design and Construction of Masonry Chimneys and Fireplaces	CSA	CAN/CSA-A405-M87
	Chimneys, Fireplaces, Vents, and Solid Fuel-Burning Appliances	NFPA	ANSI/NFPA 211-1996
	Medium Heat Appliance Factory-Built Chimneys (1995)	UL	ANSI/UL 959-1992
	Factory-Built Chimneys for Residential Type and Building Heating Appliance (1994)	UL	ANSI/UL 103-1995
Clean Rooms	Procedural Standards for Certified Testing of Cleanrooms, 2nd ed. (1996)	NEBB	NEBB
	Standard Practice for Continuous Sizing and Counting of Airborne Particles in Dust-Controlled Areas and Clean Rooms Using Instruments Capable of Detecting Single Sub-Micrometre and Larger Particles	ASTM	ASTM F 50-92 (R 1996)
Coils	Forced-Circulation Air-Cooling and Air-Heating Coils	ARI	ANSI/ARI 410-91
	Methods of Testing Forced Circulation Air Cooling and Air Heating Coils	ASHRAE	ASHRAE 33-1978
Comfort Conditions	Threshold Limit Values for Physical Agents (updated annually)	AGCIH	ACGIH
	Thermal Environmental Conditions for Human Occupancy	ASHRAE	ANSI/ASHRAE 55-1992
	Ergonomics—Determination of Metabolic Heat Production	ISO	ISO 8996:1990
	Comfort, Air Quality and Efficiency by Design (1997)	ACCA	ACCA Manual RS
	Ergonomics of the Thermal Environment—Estimation of the Thermal Insulation and Evaporative Resistance of a Clothing Ensemble	ISO	ISO 9920:1995
	Hot Environments—Estimation of the Heat Stress on Working Man, Based on the WBGT Index (Wet Bulb Globe Temperature)	ISO	ISO 7243:1989
	Moderate Thermal Environments—Determination of the PMV and PPD Indices and Specification of the Conditions for Thermal Comfort	ISO	ISO 7730:1994
Compressors	Compressors and Exhausters (reaffirmed 1986)	ASME	ANSI/ASME PTC 10-1997
	Displacement Compressors, Vacuum Pumps and Blowers	ASME	ANSI/ASME PTC 9-1970
	Safety Standard for Air Compressor Systems	ASME	ANSI/ASME B19.1-1995
	Safety Standard for Compressors for Process Industries	ASME	ANSI/ASME B19.3-1991
	Compressed Air and Gas Handbook, 5th ed. (1988)	CAGI	CAGI
Refrigerant	Ammonia Compressor Units	ARI	ANSI/ARI 510-93
	Method for Presentation of Compressor Performance Data	ARI	ARI 540-91
	Positive Displacement Condensing Units	ARI	ANSI/ARI 520-97
	Methods of Testing for Rating Positive Displacement Refrigerant Compressors and Condensing Units	ASHRAE	ANSI/ASHRAE 23-1993
	Safety Code for Mechanical Refrigeration	ASHRAE	ANSI/ASHRAE 15-1994
	Testing of Refrigerant Compressors	ISO	ISO 917:1989
	Refrigerant Compressors—Presentation of Performance Data	ISO	ISO 9309:1989
	Hermetic Refrigerant Motor-Compressors (1996)	UL/CSA	UL 984/CAN/ CSA-C22.2 No.140.2-M91
Computers	Method of Rating Computer and Data Processing Room Unitary Air Conditioners	ASHRAE	ANSI/ASHRAE 127-1988
	Protection of Electronic Computer/Data Processing Equipment	NFPA	ANSI/NFPA 75-1995
Condensers	Commercial Applications Systems and Equipment (1993)	ACCA	ACCA Manual CS
	Remote Mechanical-Draft Air-Cooled Refrigerant Condensers	ARI	ARI 460-94
	Remote Mechanical Draft Evaporative Refrigerant Condensers	ARI	ANSI/ARI 490-89
	Water-Cooled Refrigerant Condensers, Remote Type	ARI	ARI 450-93
	Methods of Testing for Rating Remote Mechanical-Draft Air-Cooled Refrigerant Condensers	ASHRAE	ASHRAE 20-1970
	Methods of Testing Remote Mechanical-Draft Evaporative Refrigerant Condensers	ASHRAE	ANSI/ASHRAE 64-1995
	Methods of Testing for Rating Water-Cooled Refrigerant Condensers	ASHRAE	ANSI/ASHRAE 22-1992
	Safety Code for Mechanical Refrigeration	ASHRAE	ANSI/ASHRAE 15-1994
	Steam Condensing Apparatus	ASME	ANSI/ASME PTC 12.2-1998
	Standards for Steam Surface Condensers, 9th ed.	HEI	HEI 1995

Table 1 Codes and Standards Published by Various Societies and Associations (*Continued*)

Subject	Title	Publisher	Reference
	Standards for Direct Contact Barometric and Low Level Condensers, 6th ed.	HEI	HEI 1995
Condensing Units	Commercial Applications Systems and Equipment (1993)	ACCA	ACCA Manual CS
	Commercial and Industrial Unitary Air-Conditioning Condensing Units	ARI	ARI 365-94
	Methods of Testing for Rating Positive Displacement Refrigerant Compressors and Condensing Units	ASHRAE	ANSI/ASHRAE 23-1993
	Heating and Cooling Equipment (1995)	UL/CSA	UL 1995/C22.2 No. 236-M95
Containers	Series I Freight Containers–Classifications, Dimensions, and Ratings	ISO	ISO 668:1995
	Series I Freight Containers–Specifications and Testing–Part 2: Thermal Containers	ISO	ISO 1496-2:1996
	Animal Environment in Cargo Compartments	SAE	SAE AIR 1600A-1997
Controls	Control Systems (1989)	AABC	National Standards, Ch 25
	Energy Management Control Systems Instrumentation	ASHRAE	ANSI/ASHRAE 114-1986
	BACnet—A Data Communication Protocol for Building Automation and Control Networks	ASHRAE	ANSI/ASHRAE 135-1995
	Performance Requirements for Electric Heating Line-Voltage Wall Thermostats	CSA	C273.4-M1978 (R1992)
	Temperature-Indicating and Regulating Equipment	CSA	C22.2 No. 24-93
	Control Centers for Changing Message Type Electric Signals (1996)	UL	UL 1433
	Limit Controls (1994)	UL	ANSI/UL 353-1995
	Primary Safety Controls for Gas- and Oil-Fired Appliances (1994)	UL	ANSI/UL 372-1994
	Solid State Controls for Appliances (1994)	UL	ANSI/UL 244A-1995
	Temperature-Indicating and -Regulating Equipment (1994)	UL	UL 873
	Tests for Safety-Related Controls Employing Solid-State Devices (1995	UL	UL 991
	Process Control Equipment (1998)	UL	UL 3121-1
	Electrical Controls for Household and Similar Use, Part 1: General Requirements (1993)	UL	UL 8730-1
	Part 2: Particular Requirements for Thermal Motor Protectors for Motor-Compressors of Hermetic and Semi-Hermetic Type (1995)	UL	UL 8730-2-4
Commercial and Industrial	Industrial Control and Systems General Requirements	NEMA	ANSI/NEMA ICS 1-1993
	Industrial Control and Systems, Controllers, Contactors, and Overload Relays Rated Not More than 2000 Volts AC or 750 Volts DC	NEMA	ANSI/NEMA ICS 2-1993
	Instructions for the Handling, Installation, Operation and Maintenance of Motor Control Centers	NEMA	NEMA ICS 2.3-1995
	Preventive Maintenance of Industrial Control and Systems Equipment	NEMA	NEMA ICS 1.3-1986 (R1991)
	Industrial Control Equipment (1993)	UL	UL 508
Residential	Automatic Gas Ignition Systems and Components	CSA	ANSI Z21.20-1997
	Gas Appliance Pressure Regulators	CSA	ANSI Z21.18-1995/CGA 6.3-M95
	Gas Appliance Thermostats	CSA	ANSI Z21.23-1993 (R1998)
	Manually Operated Gas Valves for Appliances, Appliance Connector Valves and Hose End Valves	CSA	ANSI Z21.15-1997/CGA 9.1-M97
	Manually-Operated Piezo Electric Spark Gas Ignition Systems and Components	CSA	ANSI Z21.77-1995/CGA 6.23-M95
	Hot-Water Immersion Controls	NEMA	NEMA DC-12-1985 (R 1991)
	Line-Voltage Integrally Mounted Thermostats for Electric Heaters	NEMA	NEMA DC 13-1991 (1997)
	Residential Controls—Electrical Wall-Mounted Room Thermostats	NEMA	NEMA DC 3-1989
	Residential Controls—Surface Type Controls for Electric Storage Water Heaters	NEMA	NEMA DC 5-1989
	Residential Controls—Temperature Limit Controls for Electric Baseboard Heaters	NEMA	NEMA DC 10-1983 (R 1989)
	Residential Controls—Class 2 Transformers	NEMA	NEMA DC 20-1992
	Safety Guidelines for the Application, Installation, and Maintenance of Solid State Controls	NEMA	NEMA ICS 1.1-1984 (R 1988)
	Electrical Quick-Connect Terminals (1995)	UL	ANSI/UL 310-1996
Coolers	Refrigeration Equipment	CSA	CAN/CSA-C22.2 No. 120-M91
	Unit Coolers for Refrigeration	ARI	ARI 420-94
Air	Methods of Testing Forced Convection and Natural Convection Air Coolers for Refrigeration	ASHRAE	ANSI/ASHRAE 25-1990
	Commercial Bulk Milk Dispensing Equipment	NSF	ANSI/NSF 20-1998
Drinking Water	Self-Contained, Mechanically-Refrigerated Drinking-Water Coolers	ARI	ARI 1010-94
	Methods of Testing for Rating Drinking-Water Coolers with Self-Contained Mechanical Refrigeration Systems	ASHRAE	ANSI/ASHRAE 18-1987 (R 97)
	Drinking-Water Coolers (1993)	UL	ANSI/UL 399-1992
Food and Beverage	Methods of Testing for Rating Bottled and Canned Beverage Vending Machines	ASHRAE	ANSI/ASHRAE 32.1-1997
	Methods of Testing for Rating Pre-Mix and Post-Mix Soft Drink Vending and Dispensing Equipment	ASHRAE	ANSI/ASHRAE 32.2-1997
	Refrigerated Vending Machines (1995)	UL	UL 541
	Manual Food and Beverage Dispensing Equipment	NSF	ANSI/NSF 18-1996
Liquid	Refrigerant-Cooled Liquid Coolers, Remote Type	ARI	ARI 480-95
	Methods of Testing for Rating Liquid Coolers	ASHRAE	ANSI/ASHRAE 24-1989

Table 1 Codes and Standards Published by Various Societies and Associations (*Continued*)

Subject	Title	Publisher	Reference
	Liquid Cooling Systems	SAE	SAE AIR 1811A-1997
Cooling Towers	Special Systems: Cooling Tower Performance Tests (1989)	AABC	National Standards, Ch 22
	Commercial Applications Systems and Equipment (1993)	ACCA	ACCA Manual CS
	Bioaerosols: Assessment and Control (1999)	AGCIH	AGCIH
	Atmospheric Water Cooling Equipment	ASME	ANSI/ASME PTC 23-1986 (R 97)
	Water-Cooling Towers	NFPA	ANSI/NFPA 214-1996
	Acceptance Test Code for Spray Cooling Systems (1985)	CTI	CTI ATC-133
	Acceptance Test Code for Water Cooling Towers: Mechanical Draft, Natural Draft Fan Assisted Types, Evaluation of Results, and Thermal Testing of Wet/Dry Cooling Towers (1990)	CTI	CTI ATC-105
	Certification Standard for Commercial Water Cooling Towers (1991)	CTI	CTI STD-201
	Code for Measurement of Sound from Water Cooling Towers (1981)	CTI	CTI ATC-128
	Fiberglass-Reinforced Plastic Panels for Application on Industrial Water-Cooling Towers (1986)	CTI	CTI STD-131
	Nomenclature for Industrial Water-Cooling Towers (1997)	CTI	CTI NCL-109
	Recommended Practice for Airflow Testing of Cooling Towers (1994)	CTI	CTI PFM-143
Crop Drying	Density, Specific Gravity, and Mass-Moisture Relationships of Grain for Storage	ASAE	ANSI/ASAE D241.4-1993
	Moisture Measurement—Forages	ASAE	ASAE S358.2-1993
	Moisture Measurement—Unground Grain and Seeds	ASAE	ASAE S352.2-1997
	Moisture Relationships of Plant-Based Agricultural Products	ASAE	ASAE D245.5-1995
	Resistance to Airflow of Grains, Seeds, Other Agricultural Products, and Perforated Metal Sheets	ASAE	ASAE D272.3-1996
Dehumidifiers	Commercial Applications Systems and Equipment (1993)	ACCA	ACCA Manual CS
	Bioaerosols: Assessment and Control (1999)	AGCIH	AGCIH
	Dehumidifiers	AHAM	ANSI/AHAM DH-1-1992
	Method of Testing for Rating Desiccant Dehumidifiers Utilizing Heat for the Regeneration Process	ASHRAE	ASHRAE 139-1998
	Dehumidifiers	CSA	C22.2 No. 92-1971 (R1992)
	Dehumidifiers (1993)	UL	ANSI/UL 474-1992
Desiccants	Method of Testing Desiccants for Refrigerant Drying	ASHRAE	ANSI/ASHRAE 35-1992
Documentation	Preparation of Operating and Maintenance Documentation for Building Systems	ASHRAE	ASHRAE Guideline 4-1993
Dryers	Method of Testing Liquid Line Refrigerant Driers	ASHRAE	ANSI/ASHRAE 63.1-1995
	Liquid-Line Driers	ARI	ANSI/ARI 710-86
Ducts and Fittings	Fibrous Glass Duct Liner Standards (1994)	NAIMA	NAIMA AH 124
	Hose, Air Duct, Flexible Nonmetallic, Aircraft	SAE	SAE AS 1501C-1994
	Ducted Electric Heat Guide for Air Handling Systems, 1st ed.	SMACNA	SMACNA 1971
	Factory-Made Air Ducts and Air Connectors (1996)	UL	UL 181
	Marine Rigid and Flexible Air Ducting (1986)	UL	ANSI/UL 1136-1986
Construction	Industrial Ventilation: A Manual of Recommended Practice, 23rd ed. (1998) Construction Guidelines for Local Exhaust Systems, p. 5.19	ACGIH	ACGIH
	Preferred Metric Sizes for Flat Metal Products	ASME	ANSI/ASME B32.3M-1984 (R 94)
	Sheet Metal Welding Code	AWS	ANSI/AWS D9.1-90
	Fibrous Glass Duct Construction Standards	NAIMA	NAIMA AH 116
	Fibrous Glass Duct Construction with 1-1/2" Duct Boards (1997)	NAIMA	NAIMA AH 120
	Fibrous Glass Residential Duct Construction Standards (1993)	NAIMA	NAIMA AH 119
	Fibrous Glass Duct Construction Standards, 6th ed.	SMACNA	SMACNA 1992
	HVAC Duct Construction Standards Metal and Flexible, 2nd ed.	SMACNA	SMACNA 1995
	Rectangular Industrial Duct Construction Standards, 1st ed.	SMACNA	SMACNA 1980
	Round Industrial Duct Construction Standards	SMACNA	SMACNA 1999
	Thermoplastic Duct (PVC) Construction Manual, 1st ed.	SMACNA	SMACNA 1974
Industrial	Rectangular Industrial Duct Construction Standards, 1st ed.	SMACNA	SMACNA 1980
	Round Industrial Duct Construction Standards	SMACNA	SMACNA 1999
Installation	Flexible Duct Performance and Installation Standards, 3rd ed. (1996)	ADC	ADC-91
	Installation of Air Conditioning and Ventilating Systems	NFPA	ANSI/NFPA 90A-1996
	Installation of Warm Air Heating and Air-Conditioning Systems	NFPA	ANSI/NFPA 90B-1996
Material Specifications	Specification for General Requirements for Flat-Rolled Stainless and Heat-Resisting Steel Plate, Sheet and Strip	ASTM	ASTM A 480/A 480M-98A
	Specification for General Requirements for Steel, Sheet, Carbon, and High-Strength, Low-Alloy, Hot-Rolled and Cold-Rolled	ASTM	ASTM A 568/A 568M-98
	Specification for General Requirements for Steel Sheet, Metallic-Coated by the Hot-Dip Process	ASTM	ASTM A 924/A 924M-97A
	Specification for Steel, Carbon (0.15 Maximum, Percent), Hot-Rolled Sheet and Strip Commercial	ASTM	ASTM A 569/A 569M-98

Table 1 Codes and Standards Published by Various Societies and Associations (*Continued*)

Subject	Title	Publisher	Reference
	Specification for Commercial Steel (CS) Sheet, Carbon (0.15 Maximum, Percent), Cold-Rolled	ASTM	ASTM A 366/A 366-97
	Specification for Steel Sheet, Zinc-Coated (Galvanized) or Zinc-Iron Alloy-Coated (Galvannealed) by the Hot-Dipped Process	ASTM	ASTM A 653/A 653-98
System Design	Commercial Low Pressure, Low Velocity Duct System Design (1990)	ACCA	ACCA Manual Q
	Installation Techniques for Perimeter Heating and Cooling, 11th ed.	ACCA	ACCA Manual 4
	Residential Duct Systems (1995)	ACCA	ACCA Manual D
	Closure Systems for Use with Rigid Air Ducts and Air Connectors (1994)	UL	UL 181A
	Closure Systems for Use with Flexible Air Ducts and Air Connectors (1995)	UL	UL 181B
Testing	Special Systems: Duct Testing (1989)	AABC	National Standards, Ch 23
	Flexible Air Duct Test Code	ADC	ADC FD-72 (R 1979)
	Method of Testing to Determine Flow Resistance in HVAC Ducts and Fittings	ASHRAE	ANSI/ASHRAE 120-1999
	Test Method for Measuring Acoustical and Airflow Performance of Duct Liner Materials and Prefabricated Silencers	ASTM	ASTM E 477-96
	HVAC Air Duct Leakage Test Manual, 1st ed.	SMACNA	SMACNA 1985
	HVAC Duct Systems Inspection Guide, 1st ed.	SMACNA	SMACNA 1989
Electrical	Voltage Ratings for Electrical Power Systems and Equipment	ANSI	ANSI C84.1-1989
	Test Method for Bond Strength of Electrical Insulating Varnishes by the Helical Coil Test	ASTM	ASTM D 2519-96
	Definite Purpose and Limited Duty Definite Purpose Magnetic Contactors	ARI	ANSI/ARI 780/790-97
	Canadian Electrical Code, Part I (18th ed.)	CSA	C22.1-1998
	Canadian Electrical Code, Part II—General Requirements	CSA	CAN/CSA-C22.2 No.0-M91 (R1997)
	Application Guide for Ground Fault Circuit Interrupters	NEMA	NEMA 280-1990
	Application Guide for Ground Fault Protective Devices for Equipment	NEMA	ANSI/NEMA PB 2.2-1988 (R 94)
	Enclosures for Electrical Equipment (1000 Volts Maximum)	NEMA	ANSI/NEMA 250-1997
	Industrial Control and Systems Enclosures	NEMA	ANSI/NEMA ICS 6-1993
	General Requirements for Wiring Devices	NEMA	NEMA WD 1-1983 (R 1989)
	Low Voltage Cartridge Fuses	NEMA	ANSI/NEMA FU 1-1986
	Molded Case Circuit Breakers and Molded Case Switches	NEMA	NEMA AB 1-1993
	Industrial Control and Systems Terminal Blocks	NEMA	NEMA ICS 4-1993
	National Electrical Code	NFPA	ANSI/NFPA 70-1999
	National Fire Alarm Code	NFPA	ANSI/NFPA 72-1996
	Compatibility of Electrical Connectors and Wiring	SAE	ANSI/SAE AIR 1329A-1988
Energy	Air Conditioning and Refrigerating Equipment Nameplate Voltages	ARI	ARI 110-97
	Comfort, Air Quality and Efficiency by Design (1997)	ACCA	ACCA Manual RS
	Energy Efficient Design of New Buildings Except Low-Rise Residential Buildings	ASHRAE	ASHRAE/IESNA 90.1-1999
	Energy-Efficient Design of New Low-Rise Residential Buildings	ASHRAE	ASHRAE 90.2-1993
	Energy Conservation in Existing Buildings	ASHRAE	ASHRAE 100-1995
	Standard Methods of Measuring and Expressing Building Energy Performance	ASHRAE	ANSI/ASHRAE 105-1984 (RA 99)
	International Energy Conservation Code (1998)	ICC	ICC
	Uniform Solar Energy Code (1997)	IAPMO	IAPMO
	Model Energy Code, Thermal Envelope Compliance Guide	NAIMA	NAIMA BI407
	Energy Management Guide for Selection and Use of Polyphase Motors	NEMA	NEMA MG 10-1994
	Energy Management Guide for Selection and Use of Single-Phase Motors	NEMA	NEMA MG 11-1977 (R 1997)
	Energy Systems Analysis and Management, 1st ed.	SMACNA	SMACNA 1997
	Energy Recovery Equipment and Systems, 2nd ed.	SMACNA	SMACNA 1991
	HVAC Systems Commissioning Manual, 1st ed.	SMACNA	SMACNA 1994
	Retrofit of Building Energy Systems and Processes	SMACNA	SMACNA 1982
	Energy Management Equipment (1994)	UL	ANSI/UL 916-1993
Exhaust Systems	Return and Exhaust Air Systems (1989)	AABC	National Standards, Ch 20
	Commercial Applications Systems and Equipment (1993)	ACCA	ACCA Manual CS
	Industrial Ventilation: A Manual of Recommended Practice, 23rd ed. (1998)	ACGIH	ACGIH
	Recirculation of Air from Industrial Process Exhaust Systems	AIHA	ANSI/AIHA Z9.7-1998
	Fundamentals Governing the Design and Operation of Local Exhaust Systems	ANSI	ANSI/AIHA Z9.2-1979 (R 1991)
	Laboratory Ventilation	AIHA	ANSI/AIHA Z9.5-1992
	Open-Surface Tanks—Ventilation and Operation	ANSI	ANSI/AIHA Z9.1-1991
	Safety Code for Design, Construction, and Ventilation of Spray Finishing Operations	AIHA	ANSI/AIHA Z9.3-1994
	Abrasive Blasting Operations—Ventilation and Safe Practices	AIHA	ANSI/AIHA Z9.4-1997
	Method of Testing Performance of Laboratory Fume Hoods	ASHRAE	ANSI/ASHRAE 110-1995
	Compressors and Exhausters	ASME	ANSI/ASME PTC 10-1997
	Mechanical Flue-Gas Exhausters	CSA	CAN 3-B255-M81 (R1996)
	Exhaust Systems for Air Conveying of Vapors, Gases, Mists, and Noncombustible Particulate Solids	NFPA	ANSI/NFPA 91-1999
	Draft Equipment (1993)	UL	UL 378

Table 1 Codes and Standards Published by Various Societies and Associations (*Continued*)

Subject	Title	Publisher	Reference
Expansion Valves	Thermostatic Refrigerant Expansion Valves	ARI	ARI 750-94
	Method of Testing Capacity of Thermostatic Refrigerant Expansion Valves	ASHRAE	ANSI/ASHRAE 17-1998
Fan-Coil Units	Industrial Ventilation: A Manual of Recommended Practice, 23rd ed. (1998) Installing Fan Coil Units, P. 8.4.11	ACGIH	ACGIH
	Room Fan-Coils and Unit Ventilators	ARI	ARI 440-97
	Methods of Testing for Rating Room Fan-Coil Air Conditioners	ASHRAE	ANSI/ASHRAE 79-1984 (R 91)
	Heating and Cooling Equipment (1995)	UL/CSA	UL 1995/C22.2 No. 236-M95
Fans	Commercial Low Pressure, Low Velocity Duct System Design (1990)	ACCA	ACCA Manual Q
	Residential Duct Systems (1995)	ACCA	ACCA Manual D
	Industrial Ventilation: A Manual of Recommended Practice, 23rd ed. (1998)	ACGIH	ACGIH
	Designation of Rotation and Discharge of Centrifugal Fans	AMCA	AMCA 99-2406-98
	Drive Arrangements for Centrifugal Fans	AMCA	AMCA 99-2404-98
	Drive Arrangements for Tubular Centrifugal Fans	AMCA	AMCA 99-2410-98
	Recommended Safety Practices for Users and Installers of Industrial and Commercial Fans	AMCA	AMCA 410-90
	Inlet Box Positions for Centrifugal Fans	AMCA	AMCA 99-2405-98
	Motor Positions for Belt or Chain Drive Centrifugal Fans	AMCA	AMCA 99-2407-66-98
	Industrial Process/Power Generation Fans: Site Performance Test Standard	AMCA	AMCA 803-96
	Standards Handbook	AMCA	AMCA 99-86
	Fans and Blowers	ARI	ARI 670-96
	Methods for the Measurement of Noise Emitted by Small Air-Moving Devices	ASA	ANSI S12.11-1987 (R 1997)
	Laboratory Methods of Testing Fans for Rating	ASHRAE/ AMCA	ANSI/ASHRAE 51-1998 ANSI/AMCA 210-85
	Laboratory Method of Testing In-Duct Sound Power Measurement Procedure for Fans	ASHRAE/ AMCA	ANSI/ASHRAE 68-1997 ANSI/AMCA 330-86
	Methods of Testing Fan Vibration—Blade Vibrations and Critical Speeds	ASHRAE	ANSI/ASHRAE 87.1-1992
	Fans	ASME	ANSI/ASME PTC 11-1984 (R 95)
	Fans and Ventilators	CSA	C22.2 No. 113-M1984 (R 93)
	Rating the Performance of Residential Mechanical Ventilating Equipment	CSA	CAN/CSA-C260-M90 (R1997)
	Acoustics—Method for the Measurement of Airborne Noise Emitted by Small Air-Moving Devices	ISO	ISO 10302:1996
	Electric Fans (1994)	UL	UL 507
Fenestration	Specification for Classification of the Durability of Sealed Insulating Glass Units	ASTM	ASTM E 774-97
	Standard Practice for Calculation of Photometric Transmittance and Reflectance of Materials to Solar Radiation	ASTM	ASTM 971-88 (R 1996)
	Practice for Determining the Load Resistance of Glazing	ASTM	ASTM E 1300-97
	Standard Tables for Terrestrial Direct Normal Solar Spectral Irradiance for Air Mass 1.5	ASTM	ASTM E 891-87 (R 1992)
	Standard Tables for Terrestrial Solar Spectral Irradiance at Air Mass 1.5 for a 37° Tilted Surface	ASTM	ASTM E 892-87 (R 1992)
	Test Method for Accelerated Weathering of Sealed Insulating Glass Units	ASTM	ASTM E 773-97
	Test Method for Solar Absorptance, Reflectance and Transmittance of Materials Using Integrating Spheres	ASTM	ASTM E 903-96A
	Test Method for Solar Photometric Transmittance of Sheet Materials Using Sunlight	ASTM	ASTM E 972-96
	Test Method for Solar Transmittance (Terrestrial) of Sheet Materials Using Sunlight	ASTM	ASTM E 1084-86 (R 1996)
	Energy Performance Evaluation of Swinging Doors	CSA	A453-95
	Windows	CSA	CAN/CSA-A440-M98
	Energy Performance Evaluation of Windows and Other Fenestration Systems	CSA	A440.2-98
	Window and Door Installation	CSA	A440.4
Filters	Comfort, Air Quality and Efficiency by Design (1997)	ACCA	ACCA Manual RS
	Industrial Ventilation: A Manual of Recommended Practice, 23rd ed. (1998)	ACGIH	ACGIH
	Flow-Capacity Rating and Application of Suction-Line Filters and Filter Driers	ARI	ANSI/ARI 730-86
	Specification for Octave-Band and Fractional-Octave-Band Analog and Digital Filters	ASA	ANSI S1.11-1986 (R 1998)
	Method of Testing Flow Capacity of Suction Line Filters and Filter Driers	ASHRAE	ANSI/ASHRAE 78-1985 (R 97)
	Method of Testing Liquid Line Filter-Drier Filtration Capability	ASHRAE	ANSI/ASHRAE 63.2-1996
	Exhaust Hoods for Commercial Cooking Equipment (1995)	UL	UL 710
	Grease Filters for Exhaust Ducts (1979)	UL	UL 1046
Fireplaces	Factory-Built Fireplaces (1996)	UL	UL 127
	Fireplace Stoves (1996)	UL	UL 737
Fire Protection	Standard Test Methods for Fire Teste of Building Construction and Materials	ASTM	ASTM E 119-98
	Standard Test Method for Surface Burning Characteristics of Building Materials	ASTM/NFPA	ASTM E 84-98/NFPA 255-1996
	BOCA National Fire Prevention Code, 11th ed. (1999)	BOCA	BOCA
	Uniform Fire Code	IFCI	IFCI 1997

Table 1 Codes and Standards Published by Various Societies and Associations (*Continued*)

Subject	Title	Publisher	Reference
	Fire-Resistance Tests—Elements of Building Construction Part 1: General Requirements	ISO	ISO 834:1975
	Fire-Resistance Tests—Methods of Test of Fire Doors and Shutters	ISO	ISO 3008:1976
	Reaction to Fire Tests—Ignitability of Building Products Using a Radiant Heat Source	ISO	ISO 5657:1997
	Fire-Resistance Tests—Service Installations In Buildings—Fire-Resisting Ducts	ISO	ISO 6944:1985
	Interconnection Circuitry of Noncoded Remote-Station Protective Signalling Systems	NEMA	NEMA SB 3-1969 (R 1989)
	Fire Doors and Fire Windows	NFPA	ANSI/NFPA 80-1995
	Fire Hazard Properties of Flammable Liquids, Gases, and Volatile Solids	NFPA	ANSI/NFPA 325-1994
	Fire Prevention Code	NFPA	ANSI/NFPA 1-1997
	Fire Protection for Laboratories Using Chemicals	NFPA	ANSI/NFPA 45-1996
	Fire Protection Handbook, 18th ed. (1996)	NFPA	NFPA
	Flammable and Combustible Liquids Code	NFPA	ANSI/NFPA 30-1996
	Health Care Facilities	NFPA	ANSI/NFPA 99-1996
	Installation of Sprinkler Systems	NFPA	NFPA 13-1996
	Life Safety Code	NFPA	ANSI/NFPA 101-1997
	National Fire Alarm Code	NFPA	ANSI/NFPA 72-1996
	National Fire Codes (issued annually)	NFPA	NFPA
	Methods of Fire Tests of Door Assemblies	NFPA	ANSI/NFPA 252-1995
	Standard Fire Prevention Code (1997)	SBCCI	SBCCI
	Fire, Smoke and Radiation Damper Installation Guide for HVAC Systems, 4th ed.	SMACNA	SMACNA 1992
	Fire Dampers (1999)	UL	UL 555
	Fire Tests of Building Construction and Materials (1997)	UL	ANSI/UL 263-1996
	Fire Tests of Door Assemblies (1997)	UL	UL 10B
	Fire Tests of Through-Penetration Firestops (1994)	UL	ANSI/UL 1479-1995
	Heat Responsive Links for Fire-Protection Service (1993)	UL	ANSI/UL 33-1995
Smoke Management	Commissioning Smoke Management Systems	ASHRAE	ASHRAE Guideline 5-1994
	Recommended Practice for Smoke Control Systems	NFPA	ANSI/NFPA 92A-1996
	Guide for Smoke Management Systems in Malls, Atria, and Large Areas	NFPA	ANSI/NFPA 92B-1995
	Ceiling Dampers (1996)	UL	UL 555C
	Smoke Dampers (1999)	UL	UL 555S
Freezers	Capacity Measurement and Energy Consumption Test Methods for Refrigerators, Combination Refrigerator-Freezers, and Freezers	CSA	CAN/CSA-C300-M91
	Energy Performance Standard for Food Service Refrigerators and Freezers	CSA	C827-98
	Refrigeration Equipment	CSA	CAN/CSA-C22.2 No. 120-M91
Commercial	Dispensing Freezers	NSF	ANSI/NSF 6-1996
	Ice Cream Makers (1993)	UL	ANSI/UL 621-1992
	Commercial Refrigerators and Freezers (1995)	UL	ANSI/UL 471-1996
	Ice Makers (1995)	UL	ANSI/UL 563-1995
Household	Household Refrigerators, Combination Refrigerator-Freezers and Household Freezers	AHAM	ANSI/AHAM HRF-1-1988
	Household Refrigerators and Freezers (1993)	UL/CSA	ANSI/UL 250-1997/ C22.2 No. 63-93
Fuels	Threshold Limit Values for Chemical Substances (updated annually)	AGCIH	AGCIH
	Standard Classification of Coals by Rank	ASTM	ANSI/ASTM D388-98
	Specification for Diesel Fuel Oils	ASTM	ANSI/ASTM D 975-97
	Specification for Fuel Oils	ASTM	ANSI/ASTM D 396-97
	Specification for Gas Turbine Fuel Oils	ASTM	ANSI/ASTM D 2880-98
	Reporting of Fuel Properties when Testing Diesel Engines with Alternative Fuels Derived from Biological Materials	ASAE	ASAE EP552-1996
Furnaces	Commercial Applications Systems and Equipment (1993)	ACCA	ACCA Manual CS
	Residential Equipment Selection (1995)	ACCA	ACCA Manual S
	Methods of Testing for Annual Fuel Utilization Efficiency of Residential Central Furnaces and Boilers	ASHRAE	ANSI/ASHRAE 103-1993
	Prevention of Furnace Explosions/Implosions in Multiple Burner Boilers	NFPA	ANSI /NFPA 8502-1995
	Standard Mechanical Code (1997)	SBCCI	SBCCI
	Heating and Cooling Equipment (1995)	UL/CSA	UL 1995/C22.2 No. 236-M95
	Residential Gas Detectors (1994)	UL	ANSI/UL 1484-1994
	Single and Multiple Station Carbon Monoxide Alarms (1996)	UL	UL 2034
Gas	Direct Vent Central Furnaces	CSA	ANSI Z21.64-1990
	Gas-Fired Central Furnaces	CSA	ANSI Z21.47-1998/CSA 2.3-M98
	Gas-Fired Duct Furnaces	CSA	ANSI Z83.9-1990, Z83.9A-1992
	Gas-Fired Gravity and Fan Type Direct Vent Wall Furnaces	CSA	ANSI Z21.44-1995
	Gas-Fired Gravity and Fan Type Floor Furnaces	CSA	Z21.48-1992

Table 1 Codes and Standards Published by Various Societies and Associations (*Continued*)

Subject	Title	Publisher	Reference
	Gas-Fired Gravity and Fan Type Vented Wall Furnaces	CSA	Z21.49-1992
	Gas Unit Heaters and Gas-Fired Duct Furnaces	CSA	ANSI Z83.8-1996/CGA-2.6-M96
	Gas-Fired Gravity and Fan Type Direct Vent Wall Furnaces	CSA	CAN1-2.19-M81 (R1996)
	Gas-Fired Gravity and Fan Type Vented Wall Furnaces	CSA	CAN/CGA-2.5-M86
	Industrial and Commercial Gas-Fired Package Furnaces	CSA	CGA 3.2-1976
	International Fuel Gas Code (1997)	ICC	ICC
	Standard Gas Code(1997)	SBCCI	SBCCI
	Commercial-Industrial Gas Heating Equipment (1994)	UL	UL 795
Oil	Standard Specification for Fuel Oils	ASTM	ANSI/ASTM D 396-97
	Test Method for Smoke Density in Flue Gases from Burning Distillate Fuels	ASTM	ANSI/ASTM D 2156-94
	Oil Burning Stoves and Water Heaters	CSA	B140.3-1962 (R1998)
	Oil-Fired Warm Air Furnaces	CSA	B140.4-1974 (R1998)
	Installation of Oil-Burning Equipment	NFPA	ANSI/NFPA 31-1997
	Oil-Fired Central Furnaces (1994)	UL	UL 727
	Oil-Fired Floor Furnaces (1994)	UL	ANSI/UL 729-1995
	Oil-Fired Wall Furnaces (1994)	UL	ANSI/UL 730-1995
Solid Fuel	Standard Classification of Coals by Rank	ASTM	ANSI/ASTM D975-98
	Installation Code for Solid-Fuel-Burning Appliances and Equipment	CSA	CAN/CSA-B365-M91
	Solid-Fuel-Fired Central Heating Appliances	CSA	CAN/CSA-B366.1-M91 (R1998)
	Solid-Fuel and Combination-Fuel Central and Supplementary Furnaces (1995)	UL	ANSI/UL 391-1997
Heaters	Gas-Fired Infrared Heaters	CSA	ANSI Z83.6-1990 (R1998)
	Threshold Limit Values for Chemical Substances (updated annually)	AGCIH	AGCIH
	Industrial Ventilation: A Manual of Recommended Practice, 23rd ed. (1998) Replacement Air Heating Equipment, p. 7.11	AGCIH	AGCIH
	Thermal Performance Testing of Solar Ambient Air Heaters	ASAE	ASAE S423-1993
	Air Heaters	ASME	ANSI/ASME PTC 4.3-1968 (R 91)
	Direct Gas-Fired Non-Recirculating Make-up Air Heaters	CSA	CAN1-3.7-77 (R1996)
	Gas-Fired Infra-Red Heaters	CSA	CAN1-2.16-M81 (R1996)
	Electric Air Heaters	CSA	C22.2 No.46-M1988 (R1996)
	Electric Duct Heaters	CSA	C22.2 No. 155-M1986 (R 1992)
	Portable Kerosine-Fired Heaters	CSA	CAN 3-B140.9.3 M86 (R1998)
	Standards for Closed Feedwater Heaters, 6th ed. (1998)	HEI	HEI
	Electric Dry Bath Heaters (1994)	UL	UL 875
	Electric Heating Appliances (1997)	UL	UL 499
	Electric Oil Heaters (1996)	UL	UL 574
	Oil-Burning Stoves (1993)	UL	ANSI/UL 896-1997
	Oil-Fired Air Heaters and Direct-Fired Heaters (1993)	UL	UL 733
Engine	Electric Engine Preheaters and Battery Warmers for Diesel Engines	SAE	SAE J 1310-1993
	Fuel Warmer—Diesel Engines	SAE	ANSI/SAE J 1422-1996
	Selection and Application Guidelines for Diesel, Gasoline, and Propane Fired Liquid Cooled Engine Pre-Heaters	SAE	SAE J 1350-1981
Nonresidential	Direct Gas-Fired Industrial Air Heaters	CSA	ANSI Z83.18-1990 (R1998)
	Gas-Fired Construction Heaters	CSA	ANSI Z83.7-1990 (R1999)
	Gas-Fired Unvented Commercial and Industrial Heaters	CSA	ANSI Z83.16-1982, Z83.16A-1984, Z83.16B-1989
	Portable Industrial Oil-Fired Heaters	CSA	B140.8-1967 (R 1998)
	Fuel-Fired Heaters—Air Heating—for Construction and Industrial Machinery	SAE	ANSI/SAE J 1024-1989
	Commercial-Industrial Gas Heating Equipment (1994)	UL	UL 795
	Electric Heaters for Use in Hazardous (Classified) Locations (1995)	UL	ANSI/UL 823-1996
Pool	Gas-Fired Pool Heaters	CSA	ANSI Z21.56-1998/CSA 4.7-M98
	Method of Testing and Rating Pool Heaters	ASHRAE	ASHRAE 146-1998
	Oil-Fired Service Water Heaters and Swimming Pool Heaters	CSA	B140.12-1976 (R1998)
Room	Vented Gas-Fired Space Heating Appliances	CSA	ANSI Z21.86-1998/CSA 2.32-M98
	Gas-Fired Room Heaters, Vol. II, Unvented Room Heaters	CSA	ANSI Z21.11.2-1996
	Gas-Fired Unvented Catalytic Room Heaters for Use with Liquefied Petroleum (LP) Gases	CSA	ANSI Z21.76-1994
	Fixed and Location-Dedicated Electric Room Heaters (1997)	UL	UL 2021
	Movable and Wall- or Ceiling-Hung Electric Room Heaters (1994)	UL	UL 1278
	Room Heaters, Solid Fuel-Type (1996)	UL	ANSI/UL 1482-1998
	Unvented Kerosene-Fired Room Heaters and Portable Heaters (1993)	UL	UL 647
Transport	Heater, Aircraft Internal Combustion Heat Exchanger Type	SAE	SAE AS 8040A-1996
	Heater, Airplane, Engine Exhaust Gas to Air Heat Exchanger Type	SAE	SAE ARP 86A-1952 (R 1992)
	Installation, Heaters, Airplane, Internal Combustion Heater Exchange Type	SAE	SAE ARP 266-1952 (R 1992)
	Motor Vehicle Heater Test Procedure	SAE	SAE J 638-1982 (R1993)
Unit	Gas Unit Heaters and Gas-Fired Duct Furnaces	CSA	ANSI Z83.8-1996/CGA 2.6-M96

Table 1 Codes and Standards Published by Various Societies and Associations (*Continued*)

Subject	Title	Publisher	Reference
	Oil-Fired Unit Heaters (1995)	UL	ANSI/UL 731-1995
Heat Exchangers	Remote Mechanical-Draft Evaporative Refrigerant Condensers	ARI	ANSI/ARI 490-89
	Method of Testing Air-to-Air Heat Exchangers	ASHRAE	ANSI/ASHRAE 84-1991
	Standard Methods of Test for Rating the Performance of Heat-Recovery Ventilators	CSA	CAN/CSA-C439-88
	Standards for Power Plant Heat Exchangers, 3rd ed. (1998)	HEI	HEI
	Standards of Tubular Exchanger Manufacturers Association, 7th ed. (1988)	TEMA	TEMA
Heating	Commercial Applications Systems and Equipment (1993)	ACCA	ACCA Manual CS
	Comfort, Air Quality and Efficiency by Design (1997)	ACCA	ACCA Manual RS
	Residential Equipment Selection (1995)	ACCA	ACCA Manual S
	Determining the Required Capacity of Residential Space Heating and Cooling Appliances	CSA	CAN/CSA-F280-M90
	Automatic Flue-Pipe Dampers for Use with Oil-Fired Appliances	CSA	B140.14-M1979 (R1991)
	Heater Elements	CSA	C22.2 No.72-M1984 (R1992)
	Advanced Installation Guide for Hydronic Heating Systems (1991)	HYDI	IBR 250
	Heat Loss Calculation Guide	HYDI	IBR H-21 (1984), IBR H-22 (1998)
	Installation Guide for Residential Hydronic Heating Systems, 6th ed. (1988)	HYDI	IBR 200
	Radiant Floor Heating (1993)	HYDI	IBR 400
	Environmental System Technology (1984)	NEBB	NEBB
	Pulverized Fuel Systems	NFPA	ANSI/NFPA 8503-1993
	Aircraft Electrical Heating Systems	SAE	ANSI/SAE AIR 860-1965 (R 1992)
	Performance Test for Air-Conditioned, Heated, and Ventilated Off-Road Self-Propelled Work Machines	SAE	SAE J 1503-1995
	Heating Value of Fuels	SAE	SAE J 1498-1990
	HVAC Systems Applications, 1st ed.	SMACNA	SMACNA 1987
	Electric Baseboard Heating Equipment (1994)	UL	ANSI/UL 1042-1995
	Electric Duct Heaters (1996)	UL	UL 1996
	Heating and Cooling Equipment (1995)	UL/CSA	UL 1995/C22.2 No. 236-M95
Heat Pumps	Commercial Applications Systems and Equipment (1993)	ACCA	ACCA Manual CS
	Geothermal Heat Pump Training Certification Program (1996)	ACCA	ACCA Training Manual
	Heat Pumps Systems, Principles and Applications, 2nd ed. (1984)	ACCA	ACCA Manual H
	Residential Equipment Selection (1995)	ACCA	ACCA Manual S
	Industrial Ventilation: A Manual of Recommended Practice, 23rd ed. (1998) HVAC Components and System Types, p. 8.3	ACGIH	ACGIH
	Commercial and Industrial Unitary Air-Conditioning and Heat Pump Equipment	ARI	ARI 340/360-93
	Ground Source Closed-Loop Heat Pumps	ARI	ANSI/ARI 330-93
	Ground Water-Source Heat Pumps	ARI	ANSI/ARI 325-93
	Water-Source Heat Pumps	ARI	ANSI/ARI 320-93
	Methods of Testing for Rating Unitary Air-Conditioning and Heat Pump Equipment	ASHRAE	ANSI/ASHRAE 37-1988
	Methods of Testing for Rating Seasonal Efficiency of Unitary Air-Conditioners and Heat Pumps	ASHRAE	ANSI/ASHRAE 116-1995
	Installation Requirements for Air-to-Air Heat Pumps	CSA	C273.5-1980 (R 1991)
	Performance Standard for Split-System Central Air-Conditioners and Heat Pumps	CSA	CAN/CSA-C273.3-M91
	Heating and Cooling Equipment (1995)	UL/CSA	UL 1995/C22.2 No. 236-M95
Gas Fired	Gas-Fired, Heat Activated Air Conditioning and Heat Pump Appliances	CSA	ANSI Z21.40.1-1996/CGA 2.91-M96, Z21.40.1a-1997/2.91a-M97
	Gas-Fired Work Activated Air Conditioning and Heat Pump Appliances (Internal Combustion)	CSA	ANSI Z21.40.2-1996/CGA 2.92-M96, Z21.40.2a-1997/2.92a-M97
	Performance Testing and Rating of Gas-Fired Air Conditioning and Heat Pump Appliances	CSA	ANSI Z21.40.4-1996/CGA 2.94-M96, Z21.40.4a-1998/2.94a-M98
Heat Recovery	Gas Turbine Heat Recovery Steam Generators	ASME	ANSI/ASME PTC 4.4-1981 (R1992)
	Energy Recovery Equipment and Systems, 2nd ed.	SMACNA	SMACNA 1991
Humidifiers	Method for Measuring Performance of Appliance Humidifiers	AHAM	ANSI/AHAM HU-1-1987
	Comfort, Air Quality and Efficiency by Design (1997)	ACCA	ACCA Manual RS
	Commercial Applications Systems and Equipment (1993)	ACCA	ACCA Manual CS
	Bioaerosols: Assessment and Control (1999)	AGCIH	AGCIH
	Central System Humidifiers for Residential Applications	ARI	ANSI/ARI 610-96
	Commercial and Industrial Humidifiers	ARI	ANSI/ARI 640-96
	Self-Contained Humidifiers for Residential Applications	ARI	ANSI/ARI 620-96
	Humidifiers (1993)	UL/CSA	UL 998/C22.2 No. 104-93
Ice Makers	Automatic Commercial Ice Makers	ARI	ARI 810-95
	Ice Storage Bins	ARI	ARI 820-95
	Methods of Testing Automatic Ice Makers	ASHRAE	ANSI/ASHRAE 29-1988 (RA 99)
	Refrigeration Equipment	CSA	CAN/CSA-C22.2 No. 120-M91
	Performance of Automatic Ice-Makers and Ice Storage Bins	CSA	C742-98

Table 1 Codes and Standards Published by Various Societies and Associations (*Continued*)

Subject	Title	Publisher	Reference
	Automatic Ice Making Equipment	NSF	ANSI/NSF 12-1992
	Ice Makers (1995)	UL	ANSI/UL 563-1997
Incinerators	Large Incinerators	ASME	ANSI/ASME PTC 33-1978 (R 1991)
	Incinerators and Waste and Linen Handling Systems and Equipment	NFPA	ANSI/NFPA 82-1994
	Residential Incinerators (1993)	UL	UL 791
Indoor Air Quality	Ventilation for Acceptable Indoor Air Quality	ASHRAE	ASHRAE 62-1999
	Bioaerosols: Assessment and Control (1999)	AGCIH	AGCIH
	Standard Practice for Continuous Sizing and Counting of Airborne Particles in Dust-Controlled Areas and Clean Rooms Using Instruments Capable of Detecting Single Sub-Micrometre and Larger Particles	ASTM	ASTM F 50-92 (R 1996)
	Practice for Referencing Suprathreshold Odor Intensity	ASTM	ASTM E 544-75 (R1997)
	Ambient Air—Determination of Mass Concentration of Nitrogen Dioxide—Modified Griess-Saltzman Method	ISO	ISO 6768:1985
	Air Quality- Presentation of Ambient Air Quality Data in Alphanumerical Form	ISO	ISO 7168:1985
	Comfort, Air Quality and Efficiency by Design (1997)	ACCA	ACCA Manual RS
	Indoor Air Quality—A Systems Approach	SMACNA	SMACNA 1998
Induction Units	Room Air-Induction Units	ARI	ARI 445-87 (RA 93)
	Frame Assignments for Alternating Current Integral-Horsepower Induction Motors	NEMA	NEMA MG1-1993
Insulation	Classification for Rating Sound Insulation	ASTM	ASTM E 413-87 (R 1994)
	Classification of Potential Health and Safety Concerns Associated with Thermal Insulation Materials and Accessories	ASTM	ASTM C 930-92
	Practice for Prefabrication and Field Fabrication of Thermal Insulating Fitting Covers for NPS Piping, Vessel Lagging, and Dished Head Segments	ASTM	ASTM C450-94
	Specification for Adhesives for Duct Thermal Insulation	ASTM	ASTM C 916-85 (R 1996)
	Specification for Fibrous Glass Duct Lining Insulation (Thermal and Sound Absorbing Material)	ASTM	ASTM C 1071-98
	Specification for Preformed Flexible Elastometric Cellular Thermal Insulation in Sheet and Tabular Form	ASTM	ASTM C534-94
	Specification for Cellular Glass Thermal Insulation	ASTM	ASTM C552-91
	Specification for Rigid, Cellular Polystyrene Thermal Insulation	ASTM	ASTM C578-95
	Specification for Unfaced Preformed Rigid Cellular Polyisocyanurate Thermal Insulation	ASTM	ASTM C591-94
	Specification for Faced or Unfaced Rigid Cellular Phenolic Thermal Insulation	ASTM	ASTM C1126-98
	Standard Practice for Determination of Heat Gain or Loss and the Surface Temperature of Insulated Pipe and Equipment Systems by the Use of a Computer Program	ASTM	ASTM C 680-89 (R 1995)
	Practice for Inner and Outer Diameters of Rigid Thermal Insulation for Nominal Sizes of Pipe and Tubing (NPS System))	ASTM	ASTM C 585-90 (R1998)
	Practice for Thermographic Inspection of Insulation Installations in Envelope Cavities of Frame Buildings	ASTM	ASTM C 1060-90 (R1997)
	Terminology Relating to Thermal Insulating Materials	ASTM	ASTM C 168-97
	Test Method for Steady-State Heat Flux Measurements and Thermal Transmission Properties by Means of the Guarded-Hot-Plate Apparatus	ASTM	ASTM C 177-97
	Test Method for Steady-State Heat Flux Measurements and Thermal Transmission Properties by Means of the Heat Flow Meter Apparatus	ASTM	ASTM C 518-98
	Test Method for Steady-State Heat Transfer Properties of Horizontal Pipe Insulations	ASTM	ASTM C 335-95
	Test Method for Steady-State and Thermal Performance of Building Assemblies by Means of a Guarded Hot Box	ASTM	ASTM C 236-89 (R 1993)
	Guidelines for Use of Thermal Insulation in Agricultural Buildings	ASAE	ASAE S401.2-1993
	Thermal Insulation—Definition of Terms	ISO	ISO 9229:1991
	National Commercial and Industrial Insulation Standards, 5th ed.	MICA	MICA 1999
Louvers	Test Methods for Louvers, Dampers and Shutters	AMCA	AMCA 500-89
	Laboratory Methods for Testing Dampers for Rating	AMCA	AMCA 500D-97
Lubricants	Methods of Testing the Floc Point of Refrigeration Grade Oils	ASHRAE	ANSI/ASHRAE 86-1994
	Classification of Industrial Fluid Lubricants by Viscosity System	ASTM	ANSI/ASTM D 2422-97
	Test Method for Mean Molecular Weight of Mineral Insulating Oils by the Cryoscopic Method	ASTM	ASTM D 2224-78 (R 1983)
	Test Method for Molecular Weight (Relative Molecular Mass) of Hydrocarbons by Thermoelectric Measurement of Vapor Pressure	ASTM	ASTM D 2503-92 (R1997)
	Test Method for Pour Point of Petroleum Products	ASTM	ASTM D 97-96A
	Petroleum Industry—Corrosiveness to Copper—Copper Strip Test	ISO	ISO 2160:1985
	Semiconductor Graphite	NEMA	NEMA CB 4-1989 (R1995)
Measurements	Industrial Ventilation: A Manual of Recommended Practice, 23rd ed. (1998)	AGCIH	AGCIH

Table 1 Codes and Standards Published by Various Societies and Associations (*Continued*)

Subject	Title	Publisher	Reference
	Engineering Analysis of Experimental Data	ASHRAE	ASHRAE Guideline 2-1986(R 96)
	Method of Measuring Solar-Optical Properties of Materials	ASHRAE	ANSI/ASHRAE 74-1988
	Standard Method for Measurement of Proportion of Lubricant in Liquid Refrigerant	ASHRAE	ASHRAE 41.4-1996
	Standard Method for Measurement of Moist Air Properties	ASHRAE	ANSI/ASHRAE 41.6-1994
	Standard Methods of Measuring and Expressing Building Energy Performance	ASHRAE	ANSI/ASHRAE 105-1984 (RA 99)
	Measurement of Industrial Sound	ASME	ANSI/ASME PTC 36-1985
	Measurement Uncertainty	ASME	ANSI/ASME PTC 19.1-1998
	Method for Establishing Installation Effects on Flowmeters	ASME	ANSI/ASME MFC-10M-1994
	Procedure for Bench Calibration of Tank Level Gaging Tapes and Sounding Rules	ASME	ANSI/ASME MC88.2-1974 (R 95)
	Specification for Temperature-Electromotive Force (EMF) Tables for Standardized Thermocouples	ASTM	ASTM E 230-96
	Standard Practice for Continuous Sizing and Counting of Airborne Particles in Dust-Controlled Areas and Clean Rooms Using Instruments Capable of Detecting Single Sub-Micrometre and Larger Particles	ASTM	ASTM F 50-92 (R 1996)
	Standard for Use of the International System of Units (SI): The Modern Metric System	IEEE/ ASTM	ANSI/IEEE/ASTM SI 10-97
	Test Methods for Water Vapor Transmission of Materials	ASTM	ASTM E 96-95
	Ergonomics—Determination of Metabolic Heat Production	ISO	ISO 8996:1990
	Ergonomics of the Thermal Environment—Estimation of the Thermal Insulation and Evaporative Resistance of a Clothing Ensemble	ISO	ISO 9920:1995
	Thermal Environments—Instruments and Methods for Measuring Physical Quantities	ISO	ISO 7726:1985
Fluid Flow	Standard Methods of Measurement of Flow of Liquids in Pipes Using Orifice Flowmeters	ASHRAE	ANSI/ASHRAE 41.8-1989
	Standard Calorimeter Test Method for Flow Measurement of a Volatile Refrigerant	ASHRAE	ANSI/ASHRAE 41.9-1988
	Application of Fluid Meters	ASME	ASME 19.5-1972
	Fluid Flow in Closed Conduits—Connections for Pressure Signal Transmissions Between Primary and Secondary Devices	ASME	ANSI/ASME MFC 8M-1988
	Glossary of Terms Used in the Measurement of Fluid Flow in Pipes	ASME	ANSI/ASME MFC-1M-1991
	Measurement of Fluid Flow by Means of Coriolis Mass Flowmeters	ASME	ANSI/ASME MFC-11M-1989 (R 94)
	Measurement of Fluid Flow in Pipes Using Orifice, Nozzle, and Venturi	ASME	ASME MFC 3M-1989 (R1995)
	Measurement of Fluid Flow in Pipes Using Vortex Flow Meters	ASME	ASME/ANSI MFC-6M-1998
	Measurement of Fluid Flow Using Small Bore Precision Orifice Meters	ASME	ANSI/ASME MFC-14M-1995
	Measurement of Liquid Flow in Closed Conduits by Weighting Method	ASME	ANSI/ASME MFC-9M-1988
	Measurement of Liquid Flow in Closed Conduits Using Transit-Time Ultrasonic Flowmeters	ASME	ANSI/ASME MFC-5M-1985 (R 1994)
	Measurement Uncertainty for Fluid Flow in Closed Conduits	ASME	ANSI/ASME MFC-2M-1983 (R 88)
	Measurement of Fluid Flow in Closed Conduits—Velocity Area Method Using Pitot Static Tubes	ISO	ISO 3966:1977
Gas Flow	Standard Methods for Laboratory Airflow Measurement	ASHRAE	ANSI/ASHRAE 41.2-1987 (R 92)
	Standard Method for Measurement of Flow of Gas	ASHRAE	ANSI/ASHRAE 41.7-1984 (R 91)
	Measurement of Gas Flow by Means of Critical Flow Venturi Nozzles	ASME	ASME/ANSI MFC-7M-1987 (R 92)
	Measurement of Gas Flow by Turbine Meters	ASME	ANSI/ASME MFC-4M-1986 (R 97)
Pressure	Standard Method for Pressure Measurement	ASHRAE	ANSI/ASHRAE 41.3-1989
	Gauges—Pressure Indicating Dial Type—Elastic Element	ASME	ANSI/ASME B40.1-1991
	Guide for Dynamic Calibration of Pressure Transducers	ASME	ANSI MC88.1-1972 (R 1995)
	Pressure Measurement	ASME	ANSI/ASME PTC 19.2-1987 (R 98)
Temperature	Standard Method for Temperature Measurement	ASHRAE	ANSI/ASHRAE 41.1-1986 (R 91)
	Temperature Measurement	ASME	ANSI/ASME PTC 19.3-1974 (R 98)
	Total Temperature Measuring Instruments (Turbine Powered Subsonic Aircraft)	SAE	SAE AS 793-1966 (R 1991)
Thermal	Method of Testing Thermal Energy Meters for Liquid Streams in HVAC Systems	ASHRAE	ANSI/ASHRAE 125-1992
	Standard Practice for Determining Thermal Resistance of Building Envelope Components from In-Situ Data	ASTM	ASTM C 1155-95
	Standard Practice for In-Situ Measurement of Heat Flux and Temperature on Building Envelope Components	ASTM	ASTM C 1046-95
	Test Method for Steady-State Heat Flux Measurements and Thermal Transmission Properties by Means of the Guarded-Hot-Plate Apparatus	ASTM	ASTM C 177-97
	Test Method for Steady-State Heat Flux Measurements and Thermal Transmission Properties by Means of the Heat Flow Meter Apparatus	ASTM	ASTM C 518-98
	Test Method for Thermal Performance of Building Assemblies by Means of a Calibrated Hot Box	ASTM	ASTM C 976-90 (R1996)
Mobile Homes and Recreational Vehicles	Residential Load Calculation (1986)	ACCA	ACCA Manual J
	Recreational Vehicle Cooking Gas Appliances	CSA	ANSI Z21.57-1993 (R1998)
	Mobile Homes	CSA	CAN/CSA-Z240 MH Series-92

Table 1 Codes and Standards Published by Various Societies and Associations (*Continued*)

Subject	Title	Publisher	Reference
	Oil-Fired Warm Air Heating Appliances for Mobile Housing and Recreational Vehicles	CSA	B140.10-1974 (R 1998)
	Recreational Vehicles	CSA	CAN/CSA-Z240 RV Series-M86
	Gas Supply Connectors for Manufactured Homes	IAPMO	IAPMO TSC 9-1997
	Manufactured Home Installations	NCSBCS	ANSI/NCSBCS A225.1-1994
	Recreational Vehicles	NFPA	NFPA 501C-1996 (ANSI A119.2)
	Plumbing System Components for Manufactured Homes and Recreational Vehicles	NSF	ANSI/NSF 24-1988 (R1996)
	Gas Burning Heating Appliances for Manufactured Homes and Recreational Vehicles (1995)	UL	UL 307B
	Gas-Fired Cooking Appliances for Recreational Vehicles (1993)	UL	UL 1075
	Liquid Fuel-Burning Heating Appliances for Manufactured Homes and Recreational Vehicles (1995)	UL	ANSI/UL 307A-1997
	Low Voltage Lighting Fixtures for Use in Recreational Vehicles (1994)	UL	ANSI/UL 234-1995
Motors and Generators	Steam Generating Units	ASME	ANSI/ASME PTC 4.1-1964 (R1991)
	Testing of Nuclear Air-Treatment Systems	ASME	ANSI/ASME N510-1989 (R1995)
	Nuclear Power Plant Air Cleaning Units and Components	ASME	ANSI/ASME N509-1989 (R1996)
	Test Methods for Film-Insulated Magnet Wire	ASTM	ANSI/ASTM D 1676-95
	Energy Efficiency Test Methods for Three-Phase Induction Motors	CSA	C390-98
	Motors and Generators	CSA	C22.2 No. 100-95
	Energy Management Guide for Selection and Use of Polyphase Motors	NEMA	NEMA MG 10-1994
	Energy Management Guide for Selection and Use of Single-Phase Motors	NEMA	NEMA MG 11-1977 (R1997)
	Motion/Position Control Motors, Controls, and Feedback Devices	NEMA	NEMA MG 7-1993
	Motors and Generators	NEMA	NEMA MG 1-1993
	Electric Motors (1994)	UL	UL 1004
	Electric Motors and Generators for Use in Division 1 Hazardous (Classified) Locations (1994)	UL	ANSI/UL 674-1993
	Overheating Protection for Motors (1987)	UL	UL 2111
	Standard Test Procedure for Polyphase Induction Motors and Generators	IEEE	IEEE 112-1996
Pipe, Tubing, and Fittings	Process Piping	ANSI	ANSI/ASME B31.3-1996
	Building Services Piping	ASME	ANSI/ASME B31.9-1996
	Pipe Threads, General Purpose (Inch)	ASME	ANSI/ASME B1.20.1-1983 (R 92)
	Power Piping	ASME	ANSI/ASME B31.1-1995
	Refrigeration Piping	ASME	ASME/ANSI B31.5-1992
	Scheme for the Identification of Piping Systems	ASME	ANSI/ASME A13.1-1996
	Standard Practice for Obtaining Hydrostatic or Pressure Design Basis for "Fiberglass" (Glass-Fiber-Reinforced Thermosetting-Resin) Pipe and Fittings	ASTM	ANSI/ASTM D 2992-96
	Standards of the Expansion Joint Manufacturers Association, Inc., 7th ed. (1998)	EJMA	EJMA
	Guideline for Quality Piping Installation	MCAA	MCAA
	Pipe Hangers and Supports—Materials, Design and Manufacture	MSS	MSS SP-58-93
	Pipe Hangers and Supports—Selection and Application	MSS	MSS SP-69-96
	Welding Procedure Specifications	NCPWB	NCPWB
	Electrical Nonmetallic Tubing (ENT)	NEMA	NEMA TC 13-1993
	Filament-Wound Reinforced Thermosetting Resin Conduit and Fittings	NEMA	NEMA TC 14-1984 (R 1986)
	National Fuel Gas Code	AGA/NFPA	ANSI Z223.1-1999/ANSI/NFPA 54
	Refrigeration Tube Fittings	SAE	SAE J 513-1997
	Seismic Restraint Manual—Guidelines for Mechanical Systems, 2nd ed.	SMACNA	SMACNA 1998
	Tube Fittings for Flammable and Combustible Fluids, Refrigeration Service, and Marine Use (1997)	UL	UL 109
Plastic	Specification for Acrylonitrile-Butadiene-Styrene (ABS) Plastic Pipe, Schedules 40 and 80	ASTM	ASTM D 1527-96A
	Specification for Chlorinated Polyvinyl Chloride (CPVC) Plastic Pipe, Schedules 40 and 80	ASTM	ASTM F 441/F 441M-97
	Specification for Plastic Insert Fittings for Polybutylene (PB) Tubing	ASTM	ASTM F 845-96
	Specification for Polybutylene (PB) Plastic Hot-Water Distribution Systems	ASTM	ASTM D 3309-96A
	Specification for Polybutylene (PB) Plastic Pipe (SDR-PR) Based on Controlled Inside Diameter	ASTM	ASTM D 26-96A
	Specification for Polybutylene (PB) Plastic Pipe (SDR-PR) Based on Outside Diameter	ASTM	ASTM D 3000-95A
	Specification for Polybutylene (PB) Plastic Tubing	ASTM	ASTM D 2666-96A
	Specification for Polyethylene (PE) Plastic Pipe, Schedule 40	ASTM	ASTM D 2104-96
	Specification for Polyvinyl Chloride (PVC) Plastic Pipe, Schedules 40, 80, and 120	ASTM	ASTM D 1785-96B
	Test Method for Obtaining Hydrostatic Design Basis for Thermoplastic Pipe Materials	ASTM	ASTM D 2837-98
	Electrical Plastic Tubing (EPT) and Conduit Schedule EPC-40 and EPC-80	NEMA	NEMA TC 2-1990

Table 1 Codes and Standards Published by Various Societies and Associations (*Continued*)

Subject	Title	Publisher	Reference
	Extra-Strength PVC Plastic Utilities Duct for Underground Installation	NEMA	NEMA TC 8-1990
	Fittings for ABS and PVC Plastic Utilities Duct for Underground Installation	NEMA	NEMA TC 9-1990
	PVC and ABS Plastic Utilities Duct for Underground Installation	NEMA	NEMA TC 6-1990
	Smooth-Wall Coilable Polyethylene Electrical Plastic Duct	NEMA	ANSI/NEMA TC 7-1990
	Plastics Piping Components and Related Materials	NSF	ANSI/NSF 14-1998
	Rubber Gasketed Fittings for Fire-Protection Service (1993)	UL	UL 213
Metal	Welded and Seamless Wrought Steel Pipe	ASME	ASME/ANSI B36.10M-1996
	Specification for Pipe, Steel, Black and Hot-Dipped, Zinc-Coated, Welded and Seamless	ASTM	ASTM A 53/53M-98A
	Specification for Seamless Carbon Steel Pipe for High-Temperature Service	ASTM	ASTMA 106-97A
	Specification for Seamless Copper Pipe, Standard Sizes	ASTM	ASTM B 42-98
	Specification for Seamless Copper Tube	ASTM	ASTM B 75-97
	Specification for Hand-Drawn Copper Capillary Tube for Restrictor Applications	ASTM	ASTM B 360-95
	Specification for Seamless Copper Tube for Air Conditioning and Refrigeration Field Service	ASTM	ASTM B 280-98
	Specification for Seamless Copper Water Tube	ASTM	ASTM B 88-96
	Specification for Welded Copper and Copper Alloy Tube for Air Conditioning and Refrigeration Service	ASTM	ASTM B 640-93
	Thickness Design of Ductile-Iron Pipe	AWWA	ANSI/AWWA C150/A21.50-96
	Fittings, Cast Metal Boxes, and Conduit Bodies for Conduit and Cable Assemblies	NEMA	NEMA FB 1-1993
	Polyvinyl-Chloride (PVC) Externally Coated Galvanized Rigid Steel Conduit and Intermediate Metal Conduit	NEMA	NEMA RN 1-1989
Plumbing	Uniform Plumbing Code (1997) (with IAPMO Installation Standards)	IAPMO	IAPMO
	International Plumbing Code (1997 with 1998 supplement)	ICC	ICC
	International Private Sewage Disposal Code (1997 with 1998 supplement)	ICC	ICC
	National Standard Plumbing Code (NSPC)	PHCC	NSPC 1996
	Standard Plumbing Code (1997)	SBCCI	SBCCI
Pumps	Centrifugal Pumps	ASME	ASME PTC 8.2-1990
	Displacement Compressors, Vacuum Pumps and Blowers	ASME	ANSI/ASME PTC 9-1970 (R1992)
	Liquid Pumps (R1997)	CSA	CAN/CSA-C.22.2 No. 108-M89
	Performance Standard for Liquid Ring Vacuum Pumps, 1st ed. (1987) (R1994)	HEI	HEI
	Centrifugal Pumps	HI	ANSI/HI 1.1-1.5 (1994)
	Centrifugal Pumps - Horizontal Baseplate Design	HI	ANSI/HI 1.3.4 (1997)
	Vertical Pumps	HI	ANSI/HI 2.1-2.5 (1994)
	Rotary Pumps	HI	ANSI/HI 3.1-3.5 (1994)
	Sealless Rotary Pumps	HI	ANSI/HI 4.1-4.6 (1994)
	Sealless Centrifugal Pumps	HI	ANSI/HI 5.1-5.6 (1994)
	Reciprocating Power Pumps	HI	ANSI/HI 6.1-6.5 (1994)
	Controlled Volume Pumps	HI	ANSI/HI 7.1-7.5 (1994)
	Direct Acting (Steam) Pumps	HI	ANSI/HI 8.1-8.5 (1994)
	Pumps—General Guideline	HI	ANSI/HI 9.1-9.6 (1994)
	Pumps—Polymer Material Selections	HI	ANSI/HI 9.3.3 (1997)
	Centrifugal and Vertical Pumps—NPSH Margin	HI	ANSI/HI 9.6.1 (1998)
	Centrifugal and Vertical Pumps—Allowable Operating Region	HI	ANSI/HI 9.6.3 (1997)
	Pump Intake Design	HI	ANSI/HI 9.8 (1998)
	Engineering Data Book, 2nd ed. (1990)	HI	HI
	Circulation System Components and Related Materials for Swimming Pools, Spas/Hot Tubs	NSF	ANSI/NSF 50-1996
	Swimming Pool Pumps, Filters and Chlorinators (1997)	UL	UL 1081
	Motor-Operated Water Pumps (1996)	UL	UL 778
	Pumps for Oil-Burning Appliances (1997)	UL	ANSI/UL 343-1998
Radiators	Testing and Rating Standard for Baseboard Radiation, 6th ed. (1990)	HYDI	IBR
	Testing and Rating Standard for Finned-Tube (Commercial) Radiation (1990)	HYDI	IBR
Receivers	Refrigerant Liquid Receivers	ARI	ARI 495-93
Refrigerants	Threshold Limit Values for Chemical Substances (updated annually)	AGCIH	AGCIH
	Refrigerant Recovery/Recycling Equipment	ARI	ARI 740-95
	Specifications for Fluorocarbon and Other Refrigerants	ARI	ARI 700-95
	Format for Information on Refrigerants	ASHRAE	ASHRAE Guideline 6-1996
	Method of Testing Flow Capacity of Refrigerant Capillary Tubes	ASHRAE	ANSI/ASHRAE 28-1996
	Number Designation and Safety Classification of Refrigerants	ASHRAE	ANSI/ASHRAE 34-1997
	Methods of Testing Discharge Line Refrigerant-Oil Separators	ASHRAE	ANSI/ASHRAE 69-1990
	Reducing Emission of Fully Halogenated Refrigerants in Refrigeration and Air-Conditioning Equipment and Systems	ASHRAE	ASHRAE Guideline 3-1996
	Refrigeration Oil Description	ASHRAE	ANSI/ASHRAE 99-1981 (R 87)

Table 1 Codes and Standards Published by Various Societies and Associations (*Continued*)

Subject	Title	Publisher	Reference
	Sealed Glass Tube Method to Test the Chemical Stability of Material for Use Within Refrigerant Systems	ASHRAE	ANSI/ASHRAE 97-1999
	Test Method for Acid Number of Petroleum Products by Potentiometric Titration	ASTM	ASTM D 664-95
	Test Method for Concentration Limits of Flammability of Chemical (Vapors and Gases)	ASTM	ASTM E 681-98
	Refrigerant-Containing Components for Use in Electrical Equipment	CSA	C22.2 No. 140.3-M1987 (R1993)
	Refrigerant-Containing Components and Accessories, Non-Electrical (1993)	UL	ANSI/UL 207-1994
	Refrigerants-Number Designation	ISO	ISO 817:1974
	Recommended Service Procedure for the Containment of HFC-134a	SAE	ANSI/SAE J 2211-1991
	HFC-134a Recycling Equipment for Mobile Air-Conditioning Systems	SAE	ANSI/SAE J 2210-1991
	CFC-12 (R-12) Extraction Equipment for Mobile Automotive Air-Conditioning Systems	SAE	ANSI/SAE J 2209-1992
	HFC-134a (R-134a) Service Hose Fittings for Automotive Air-Conditioning Service Equipment	SAE	SAE J 2197-1992 (R1997)
	Standard of Purity for Recycled HFC-134a for Use in Mobile Air-Conditioning Systems	SAE	ANSI/SAE J 2099-1991
	Recommended Service Procedure for the Containment of R-12, Highway Vehicle	SAE	SAE J 1989-1989
	Procedure for Retrofitting CFC-12 (R12) Mobile Air Conditioning Systems to HFC-134a (R134a)	SAE	ANSI/SAE J 1661-1993
	Field Conversion/Retrofit of Products to Change to an Alternate Refrigerant—Construction and Operation (1993)	UL	ANSI/UL 2170-1995
	Field Conversion/Retrofit of Products to Change to an Alternate Refrigerant—Insulating Material and Refrigerant Compatibility (1993)	UL	ANSI/UL 2171-1995
	Field Conversion/Retrofit of Products to Change to an Alternate Refrigerant—Procedures and Methods (1993)	UL	ANSI/UL 2172-1995
	Refrigerant Recovery/Recycling Equipment (1995)	UL	UL 1963
	Refrigerants (1994)	UL	UL 2182
Refrigeration	Safety Code for Mechanical Refrigeration	ASHRAE	ANSI/ASHRAE 15-1994
	Mechanical Refrigeration Code	CSA	B52-95
	Refrigeration Equipment	CSA	CAN/CSA-C22.2 No. 120-M91
	Equipment, Design and Installation of Ammonia Mechanical Refrigerating Systems	IIAR	ANSI/IIAR 2-1992
	Refrigerated Medical Equipment (1993)	UL	ANSI/UL 416-1996
Refrigeration Systems	Ejectors	ASME	ASME PTC 24-1976 (R 1982)
	Testing of Refrigerating Systems	ISO	ISO 916-1968
	Standards for Steam Jet Vacuum Systems, 4th ed. (1988) (R1995)	HEI	HEI
Transport	Mechanical Transport Refrigeration Units	ARI	ARI 1110-92
	Mechanical Refrigeration and Air-Conditioning Installations Aboard Ship	ASHRAE	ASHRAE 26-1996
	General Requirements for Application of Vapor Cycle Refrigeration Systems for Aircraft	SAE	SAE ARP 731B-1997
	Safety and Containment of Refrigerant for Mechanical Vapor Compression Systems Used for Mobile Air-Conditioning Systems	SAE	ANSI/SAE J 639-1994
Refrigerators	Method of Testing Open Refrigerators for Food Stores	ASHRAE	ANSI/ASHRAE 72-1998
	Methods of Testing Closed Refrigerators	ASHRAE	ANSI/ASHRAE 117-1992
Commercial	Energy Performance Standard for Commercial Refrigerated Display Cabinets and Merchandise	CSA	C657-95
	Energy Performance Standard for Food Service Refrigerators and Freezers	CSA	C827-98
	Gas Food Service Equipment	CSA	ANSI Z83.11-1996/CGA 1.8-M96
	Mobile Food Carts	NSF	ANSI/NSF 59-1997
	Food Equipment	NSF	ANSI/NSF 2-1996
	Commercial Refrigerators and Storage Freezers	NSF	ANSI/NSF 7-1997
	Commercial Refrigerators and Freezers (1995)	UL	ANSI/UL 471-1996
	Refrigerating Units (1994)	UL	ANSI/UL 427-1996
	Refrigeration Unit Coolers (1993)	UL	ANSI/UL 412-1992
Household	Refrigerators Using Gas Fuel	CSA	ANSI Z21.19-1990 (R1999)
	Household Refrigerators, Combination Refrigerator-Freezers and Household Freezers	AHAM	ANSI/AHAM HRF-1-1988
	Capacity Measurement and Energy Consumption Test Methods for Refrigerators, Combination Refrigerator-Freezers, and Freezers	CSA	CAN/CSA C300-M91
	Household Refrigerators and Freezers (1993)	UL/CSA	ANSI/UL 250-1997/C22.2 No. 63-93
Retrofitting			
Building	Retrofit of Building Energy Systems and Processes, 1st ed.	SMACNA	SMACNA 1982
Refrigerant	Procedure for Retrofitting CFC-12 (R12) Mobile Air Conditioning Systems to HFC-134a (R134a)	SAE	ANSI/SAE J 1661-1993
	Field Conversion/Retrofit of Products to Change to an Alternate Refrigerant—Construction and Operation (1993)	UL	ANSI/UL 2170-1995

Table 1 Codes and Standards Published by Various Societies and Associations (*Continued*)

Subject	Title	Publisher	Reference
	Field Conversion/Retrofit of Products to Change to an Alternate Refrigerant—Insulating Material and Refrigerant Compatibility (1993)	UL	ANSI/UL 2171-1995
	Field Conversion/Retrofit of Products to Change to an Alternate Refrigerant—Procedures and Methods (1993)	UL	ANSI/UL 2172-1995
Roof Ventilators	Commercial Low Pressure, Low Velocity Duct System Design (1990)	ACCA	ACCA Manual Q
	Power Ventilators (1994)	UL	ANSI/UL 705-1994
Safety Devices	Safety Devices for Protection Against Excessive Pressure Part 2: Bursting Disc Safety Devices	ISO	ISO 4126 2:1981
	Part 3: Safety Valves and Bursting Disc Safety Devices in Combination	ISO	ISO 4126-3: 1981
Solar Equipment	Method of Measuring Solar-Optical Properties of Materials	ASHRAE	ANSI/ASHRAE 74-1988
	Methods of Testing to Determine the Thermal Performance of Flat-Plate Solar Collectors Containing a Boiling Liquid	ASHRAE	ANSI/ASHRAE 109-1986 (R 96)
	Methods of Testing to Determine the Thermal Performance of Solar Collectors	ASHRAE	ANSI/ASHRAE 93-1986 (R 91)
	Methods of Testing to Determine the Thermal Performance of Solar Domestic Water Heating Systems	ASHRAE	ASHRAE 95-1981 (R 87)
	Methods of Testing to Determine the Thermal Performance of Unglazed Flat-Plate Liquid-Type Solar Collectors	ASHRAE	ANSI/ASHRAE 96-1980 (R 89)
	Reference Solar Spectral Irradiance at the Ground at Different Receiving Conditions—Part 1: Direct Normal and Hemispherical Solar Irradiance for Air Mass 1.5	ISO	ISO 9845-1:1992
	Solar Energy—Calibration of a Pyranometer Using a Pyrheliometer	ISO	ISO 9846:1993
	Solar Heating—Domestic Water Heating Systems—Part 2: Outdoor Test Methods for System Performance Characterization and Yearly Performance Prediction of Solar-Only Systems	ISO	ISO 9459-2:1995
	Solar Heating—Swimming Pool Heating Systems—Dimensions, Design and Installation Guidelines	ISO	ISO 12596:1995
	Solar Water Heaters—Elastomeric Materials for Absorbers, Connecting Pipes and Fittings—Method of Assessment	ISO	ISO 9808:1990
	Test Methods for Solar Collectors—Part 2: Qualification Test Procedures	ISO	ISO 9806-2:1995
	Test Methods for Solar Collectors—Part 3: Thermal Performance of Unglazed Liquid Heating Collectors (Sensible Heat Transfer Only) Including Pressure Drop	ISO	ISO 9806-3:1995
Solenoid Valves	Solenoid Valves for Use with Volatile Refrigerants	ARI	ARI 760-94
Sound Measurement	Threshold Limit Values for Physical Agents (updated annually)	AGCIH	AGCIH
	Method for the Calibration of Microphones (reaffirmed 1997)	ASA	ANSI S1.10-1966 (R 1997)
	Specification for Sound Level Meters	ASA	ANSI S1.4-1983, ANSI S1.4A-1985 (R1997)
	Test Method for Laboratory Measurement of Airborne Sound Transmission Loss of Building Partitions and Elements	ASTM	ASTM E 90-97
	Test Method for Measuring Acoustical and Airflow Performance of Duct Liner Materials and Prefabricated Silencers	ASTM	ASTM E 477-96
	Sound and Vibration Design and Analysis (1994)	NEBB	NEBB
Fans	Methods for Calculating Fan Sound Ratings from Laboratory Test Data	AMCA	AMCA 301-90
	Reverberant Room Method for Sound Testing of Fan	AMCA	AMCA 300-96
	Balance Quality and Vibration Levels for Fans	AMCA	AMCA 204-96
	Laboratory Methods of testing Air Circulator Fans for Rating	AMCA	AMCA 230-96
	Methods for the Measurement of Noise Emitted by Small Air-Moving Devices	ASA	ANSI S12.11-1987 (R 1997)
	Laboratory Method of Testing In-Duct Sound Power Measurement Procedure for Fans	ASHRAE/ AMCA	ANSI/ASHRAE 68-1986/ ANSI/AMCA 330-85
	Acoustics—Method for the Measurement of Airborne Noise Emitted by Small Air-Moving Devices	ISO	ISO 10302:1996
Other Equipment	Application of Sound Rating Levels of Outdoor Unitary Equipment	ARI	ARI 275-97
	Method of Measuring Machinery Sound Within Equipment Space	ARI	ARI 575-94
	Method of Measuring Sound and Vibration of Refrigerant Compressors	ARI	ARI 530-89
	Rating the Sound Levels and Sound Transmission Loss of Packaged Terminal Equipment	ARI	ANSI/ARI 300-88
	Sound Rating of Large Outdoor Refrigerating and Air-Conditioning Equipment	ARI	ARI 370-86
	Sound Rating of Non-Ducted Indoor Air-Conditioning Equipment	ARI	ARI 350-86
	Sound Rating of Outdoor Unitary Equipment	ARI	ARI 270-95
	Statistical Methods for Determining and Verifying Stated Noise Emission Values of Machinery and Equipment	ASA	ANSI S12.3-85 (R 1996)
	Sound Level Prediction for Installed Rotating Electrical Machines	NEMA	NEMA MG 3-1974 (R 1990)
Techniques	Methods for Measurement of Sound Emitted by Machinery and Equipment at Workstations and Other Specified Positions	ANSI	ANSI S12.43-1997
	Methods for Calculation of Sound Emitted by Machinery and Equipment at Workstations and Other Specified Positions from Sound Power Level	ANSI	ANSI S12.44-1997

Table 1 Codes and Standards Published by Various Societies and Associations (*Continued*)

Subject	Title	Publisher	Reference
	Criteria for Evaluating Room Noise	ASA	ANSI S12.2-1995
	Engineering Method for the Determination of Sound Power Levels of Noise Sources Using Sound Intensity	ASA	ANSI S12.12-1992 (R1997)
	Engineering Methods for the Determination of Sound Power Levels of Noise Sources for Essentially Free-Field Conditions over a Reflecting Plane	ASA	ANSI S12.34-1988 (R1997)
	Guidelines for the Use of Sound Power Standards and for the Preparation of Noise Test Codes	ASA	ANSI S12.30-1990 (R1997)
	Measurement of Sound Pressure Levels in Air	ASA	ANSI S1.13-1995
	Methods for Determination of Insertion Loss of Outdoor Noise Barriers	ASA	ANSI S12.8-1998
	Methods for the Determination of Sound Power Levels of Noise Sources in a Special Reverberation Test Room	ASA	ANSI S12.33-1990 (R1997)
	Precision Methods for the Determination of Sound Power Levels of Broad-Band Noise Sources in Reverberation Rooms	ASA	ANSI S12.31-1990 (R1996)
	Precision Methods for the Determination of Sound Power Levels of Discrete-Frequency and Narrow-Band Noise Sources in Reverberation Rooms	ASA	ANSI S12.32-1990 (R1996)
	Precision Methods for the Determination of Sound Power Levels of Noise Sources in Anechoic and Hemi-Anechoic Rooms	ASA	ANSI S12.35-1990 (R1996)
	Preferred Frequencies, Frequency Levels, and Band Numbers for Acoustical Measurements	ASA	ANSI S1.6-1984 (R1997)
	Procedure for the Computation of Loudness of Noise	ASA	ANSI S3.4-1980 (R1997)
	Procedures for Outdoor Measurement of Sound Pressure Level	ASA	ANSI S12.18-1994
	Reference Quantities for Acoustical Levels	ASA	ANSI S1.8-1989 (R1997)
	Survey Methods for the Determination of Sound Power Levels of Noise Sources	ASA	ANSI S12.36-1990 (R1997)
	Measurement of Industrial Sound	ASME	ASME/ANSI PTC 36-1985
	Test Method for Evaluating Masking Sound in Open Offices Using A-Weighted and One-Third Octave Band Sound Pressure Levels	ASTM	ASTM E 1573-93 (R1998)
	Test Method for Measurement of Sound in Residential Spaces	ASTM	ASTM E 1574-98
	Acoustics–Measurement of Sound Insulation in Buildings and of Building Elements—Part 1: Requirements for Laboratory Test Facilities with Suppressed Flanking Transmission	ISO	ISO 140-1: 1997
	Part 4: Field Measurements of Airborne Sound Insulation Between Rooms	ISO	ISO 140-4: 1978
	Part 5: Field Measurements of Airborne Sound Insulation of Facade Elements and Facade	ISO	ISO 140-5: 1990
	Part 6: Laboratory Measurements of Impact Sound Insulation of Floors	ISO	ISO 140-6: 1978
	Part 7: Field Measurements of Impact Sound Insulation of Floors	ISO	ISO 140-7: 1990
	Part 8: Laboratory Measurements of the Reduction of Transmitted Impact Noise by Floor Coverings on a Solid Standard Floor	ISO	ISO 140-8: 1978
	Acoustics—Determination of Sound Power Levels of Noise Sources Using Sound Intensity—Part 1: Measurement at Discrete Points	ISO	ISO 9614-1:1993
	Acoustics—Determination of Sound Power Levels of Noise Sources Using Sound Intensity—Part 2: Measurement by Scanning	ISO	ISO 9614-2:1996
	Acoustics—Method for Calculating Loudness Level	ISO	ISO 532:1975
	Procedural Standards for the Measurement and Assessment of Sound and Vibration (1994)	NEBB	NEBB
Terminology	Acoustical Terminology	ASA	ANSI S1.1-1994
	Terminology Relating to Environmental Acoustics	ASTM	ASTM C 634-98
Space Heaters	Methods of Testing for Rating Combination Space-Heating and Water-Heating Appliances	ASHRAE	ANSI/ASHRAE 124-1991
	Electric Air Heaters	CSA	C22.2 No. 46-M1988 (R1996)
	Fixed and Location-Dedicated Electric Room Heaters (1997)	UL	UL 2021
	Movable and Wall- or Ceiling-Hung Electric Room Heaters (1994)	UL	UL 1278
	Vented Gas-Fired Space Heating Appliances	CSA	ANSI Z21.86-1998/CSA 2.32-M98
	Gas-Fired Room Heaters, Vol. II, Unvented Room Heaters	CSA	ANSI Z21.11.2-1996, Z21.11.2A-1997, Z21.11.2B-1998
Symbols	Graphic Electrical/Electronic Symbols for Air-Conditioning and Refrigeration Equipment	ARI	ARI 130-88
	Graphic Symbols for Heating, Ventilating, and Air Conditioning	ASME	ANSI/ASME Y32.2.4-1949 (R 1998)
	Graphic Symbols for Pipe Fittings, Valves and Piping	ASME	ANSI/ASME Y32.2.3-1949 (R 1994)
	Graphic Symbols for Plumbing Fixtures for Diagrams used in Architecture and Building Construction	ASME	ANSI/ASME Y32.4-1977 (R 1994)
	Symbols for Mechanical and Acoustical Elements as used in Schematic Diagrams	ASME	ANSI/ASME Y32.18-1972 (R 1998)
	Graphic Symbols for Electrical and Electronic Diagrams	IEEE	ANSI/IEEE 315-1975 (R 1994)
	Recommended Practice for the Preparation and Use of Symbols	IEEE	IEEE 267-1966
	Standard Letter Symbols for Quantities Used in Electrical Science and Electrical Engineering	IEEE	IEEE 280-1985 (R1997)

Table 1 Codes and Standards Published by Various Societies and Associations (*Continued*)

Subject	Title	Publisher	Reference
	Standard for Logic Circuit Diagrams	IEEE	IEEE 991-1986 (R1994)
	Standard for Use of the International System of Units (SI): The Modern Metric System	IEEE/ ASTM	ANSI/IEEE/ASTM SI 10-1997
	Abbreviations for Use on Drawings and in Text	ASME	ANSI/ASME Y1.1-1989
	Safety Color Code	NEMA	ANSI/NEMA Z535.1-1997
Terminals, Wiring	Electrical Quick-Connect Terminals (1995)	UL	ANSI/UL 310-1996
	Equipment Wiring Terminals for Use with Aluminum and/or Copper Conductors (1994)	UL	ANSI/UL 486E-1994
	Splicing Wire Connectors (1997)	UL	ANSI/UL 486C-1998
	Wire Connectors and Soldering Lugs for Use with Copper Conductors (1997)	UL	ANSI/UL 486A-1998
	Wire Connectors for Use with Aluminum Conductors (1997)	UL	UL 486B
Testing and Balancing	Industrial Process/Power Generating Fans: Site Performance Test Standard	AMCA	AMCA 803-96
	Guideline for the HVAC Commissioning Process	ASHRAE	ASHRAE Guideline 1-1996
	Practices for Measurement, Testing, Adjusting, and Balancing of Building Heating, Ventilation, Air-Conditioning, and Refrigeration Systems	ASHRAE	ANSI/ASHRAE 111-1988
	Centrifugal Pump Test	HI	ANSI/HI 1.6-1994
	Vertical Pump Tests	HI	ANSI/HI 2.6-1994
	Rotary Pump Tests	HI	ANSI/HI 3.6-1994
	Reciprocating Pump Tests	HI	ANSI/HI 6.6-1994
	Pumps—General Guidelines (Including "Measurement of Airborne Sound")	HI	HI 9.1-9.6-1994
	Procedural Standards for Certified Testing of Cleanrooms, 2nd ed.	NEBB	NEBB 1996
	Procedural Standards for Testing, Adjusting, Balancing of Environmental Systems, 6th ed.	NEBB	NEBB 1998
	HVAC Systems Testing, Adjusting and Balancing, 2nd ed.	SMACNA	SMACNA 1993
Thermal Storage	Method of Testing Active Sensible Thermal Energy Storage Devices Based on Thermal Performance	ASHRAE	ANSI/ASHRAE 94.3-1986 (R 96)
	Method of Testing Active Latent-Heat Storage Devices Based on Thermal Performance	ASHRAE	ANSI/ASHRAE 94.1-1985 (R 91)
	Methods of Testing Thermal Storage Devices with Electrical Input and Thermal Output Based on Thermal Performance	ASHRAE	ANSI/ASHRAE 94.2-1981 (R 96)
	Practices for Measurement, Testing, Adjusting, and Balancing of Building Heating, Ventilation, Air-Conditioning, and Refrigeration Systems	ASHRAE	ANSI/ASHRAE 111-1988
Turbines	Steam Turbines	ASME	ANSI/ASME PTC 6-1996
	Wind Turbines	ASME	ANSI/ASME PTC 42-1988 (R 98)
	Specification for Gas Turbine Fuel Oils	ASTM	ANSI/ASTM D 2880-98
	Land Based Steam Turbine Generator Sets	NEMA	NEMA SM 24-1991
	Steam Turbines for Mechanical Drive Service	NEMA	ANSI/NEMA SM 23-1991
Valves	Methods of Testing Nonelectric, Nonpneumatic Thermostatic Radiator Valves	ASHRAE	ANSI/ASHRAE 102-1983 (R 89)
	Face-to-Face and End-to-End Dimensions of Valves	ASME	ANSI/ASME B16.10-1992
	Pressure Relief Devices	ASME	ANSI/ASME PTC 25-1994
	Valves—Flanged Threaded, and Welding End	ASME	ANSI/ASME B16.34-1996
	Relief Valves for Hot Water Supply Systems	CSA	ANSI Z21.22-1999/CSA 4.4-1998
	Control Valve Capacity Test Procedure	ISA	ANSI/ISA-S75.02-1996
	Flow Equations for Sizing Control Valves	ISA	ANSI/ISA-S75.01-1985 (R 1995)
	Industrial Valves—Part-Turn Valve Actuator Attachments—Part 1	ISO	ISO 5211-1:1977
	Industrial Valves—Part-Turn Valve Actuator Attachments—Part 2	ISO	ISO 5211-2:1979
	Industrial Valves—Part-Turn Valve Actuator Attachments—Part 3	ISO	ISO 5211-3:1982
	Metal Valves for Use in Flanged Pipe Systems—Face-to-Face Dimensions	ISO	ISO 5752:1982
	Safety Valves, Part 1: General Requirements	ISO	ISO 4126-1:1991
	Oxygen System Fill/Check Valve	SAE	SAE AS 1225A-1997
	Electrically Operated Valves (1994)	UL	UL 429
	Pressure Regulating Valves for LP-Gas (1994)	UL	UL 144
	Safety Relief Valves for Anhydrous Ammonia and LP-Gas (1997)	UL	UL 132
	Valves for Anhydrous Ammonia and LP-Gas (Other than Safety Relief) (1997)	UL	UL 125
	Valves for Flammable Fluids (1997)	UL	UL 842
Gas	Automatic Valves for Gas Appliances	CSA	ANSI Z21.21-1997/CGA 6.5-M95, Z21.21A-1998/CGA 6.5A-M98, Z21.21B-1999/CGA 6.5B-M99
	Combination Gas Controls for Gas Appliances (plus addenda 1998, 1999)	CSA	ANSI Z21.78-1997/CGA 6.20-M97
	Manually Operated Gas Valves for Appliances, Appliance Connection Valves, and Hose End Valves	CSA	ANSI Z21.15-1997/CGA 9.1-M97
	Relief Valves and Automatic Gas Shutoff Devices for Hot Water Supply Systems	CSA	ANSI Z21.22-1999/CSA 4.4-M99
	Large Metallic Valves for Gas Distribution (Manually Operated, NPS-2 1/2 to 12, 125 psig Maximum)	ASME	ANSI/ASME B16.38-1985 (R 1994)

Table 1 Codes and Standards Published by Various Societies and Associations (*Continued*)

Subject	Title	Publisher	Reference
	Manually Operated Metallic Gas Valves for Use in Gas Piping Systems up to 125 psig (Sizes 1/2 through 2)	ASME	ANSI/ASME B16.33-1990
	Manually Operated Thermoplastic Gas Shutoffs and Valves in Gas Distribution Systems	ASME	ANSI/ASME B16.40-1985 (R 1994)
Refrigerant	Refrigerant Access Valves and Hose Connectors	ARI	ARI 720-97
	Refrigerant Pressure Regulating Valves	ARI	ARI 770-94
	Solenoid Valves for Use with Volatile Refrigerants	ARI	ANSI/ARI 760-94
	Thermostatic Refrigerant Expansion Valves	ARI	ANSI/ARI 750-94
Vapor Retarders	Practice for Selection of Vapor Retarders for Thermal Insulation	ASTM	ASTM C 755-97
	Practice for Determining the Properties of Jacketing Materials for Thermal Insulation	ASTM	ASTM C 921-89 (R1996)
	Specification for Flexible, Low Permeance Vapor Retarders for Thermal Insulation	ASTM	ASTM C 1136-95
	Test Method for Water Vapor Transmission Rate of Flexible Barrier Materials Using an Infrared Detection Technique	ASTM	ASTM F 372-94
Vending Machines	Methods of Testing for Rating Bottled and Canned Beverage Vending Machines	ASHRAE	ANSI/ASHRAE 32.1-1997
	Methods of Testing for Rating Pre-Mix and Post-Mix Soft Drink Vending and Dispensing Equipment	ASHRAE	ANSI/ASHRAE 32.2-1997
	Vending Machines	CSA	C22.2 No.128-95
	Vending Machines for Food and Beverages	NSF	ANSI/NSF 25-1997
	Vending Machines (1995)	UL	ANSI/UL 751-1997
	Refrigerated Vending Machines (1995)	UL	UL 541
Vent Dampers	Automatic Vent Damper Devices for Use with Gas-Fired Appliances	CSA	ANSI Z21.66-1996/CSA 6.14-M96
	Vent or Chimney Connector Dampers for Oil-Fired Appliances (1994)	UL	UL 17-1995
Ventilation	Commercial Low Pressure, Low Velocity Duct System Design (1990)	ACCA	ACCA Manual Q
	Comfort, Air Quality and Efficiency by Design (1997)	ACCA	ACCA Manual RS
	Commercial Applications Systems and Equipment (1993)	ACCA	ACCA Manual CS
	Guide for Testing Ventilation Systems	ACGIH	ACGIH
	Industrial Ventilation: A Manual of Recommended Practice, 23rd ed. (1998)	ACGIH	ACGIH
	Ventilation for Acceptable Indoor Air Quality	ASHRAE	ASHRAE 62-1999
	Method of Testing for Room Air Diffusion	ASHRAE	ANSI/ASHRAE 113-1990
	Measuring Air Change Effectiveness	ASHRAE	ASHRAE 129-1997
	Method of Determining Air Change Rates in Detached Dwellings	ASHRAE	ANSI/ASHRAE 136-1993
	Design of Ventilation Systems for Poultry and Livestock Shelters	ASAE	ASAE D270.5-1994
	Design Values for Emergency Ventilation and Care of Livestock and Poultry	ASAE	ASAE EP282.2-1993
	Energy Ratings for Selection of Energy Efficient Agricultural Ventilation Fans	ASAE	ASAE S566-1997
	Heating, Ventilating and Cooling Greenhouses	ASAE	ASAE EP406.3-1997
	Residential Mechanical Ventilation Systems	CSA	CAN/CSA F326-M91
	Installation of Air Conditioning and Ventilating Systems	NFPA	ANSI/NFPA 90A-1996
	Parking Structures	NFPA	ANSI/NFPA 88A-1995
	Repair Garages	NFPA	ANSI/NFPA 88B-1997
	Ventilation Control and Fire Protection of Commercial Cooking Operations	NFPA	ANSI/NFPA 96-98
	Food Equipment	NSF	ANSI/NSF 2-1996
	Class II (Laminar Flow) Biohazard Cabinetry	NSF	NSF 49-1992
	(R) Test Procedure for Battery Flame Retardant Venting Systems	SAE	ANSI/SAE J 1495-1992
	Heater, Airplane, Engine Exhaust Gas to Air Heat Exchanger Type	SAE	SAE ARP 86A-1952 (R 1992)
	Aerothermodynamic Systems Engineering and Design	SAE	SAE AIR 1168/3-1990
Venting	Commercial Applications Systems and Equipment (1993)	ACCA	ACCA Manual CS
	Draft Hoods	CSA	ANSI Z21.12-1990 (R1998)
	National Fuel Gas Code	AGA/NFPA	ANSI Z223.1-1999/ANSI/NFPA 54
	Balance Quality and Vibration Levels for Fans	AMCA	AMCA 204-96
	Chimneys, Fireplaces, Vents and Solid Fuel-Burning Appliances	NFPA	ANSI/NFPA 211-1996
	Explosion Prevention Systems	NFPA	ANSI/NFPA 69-1997
	Smoke and Heat Venting	NFPA	ANSI/NFPA 204M-98
	Guide for Steel Stack Construction, 2nd ed.	SMACNA	SMACNA 1996
	Draft Equipment (1993)	UL	UL 378
	Gas Vents (1996)	UL	UL 441
	Low-Temperature Venting Systems, Type L (1995)	UL	ANSI/UL 641-1995
Vibration	Mechanical Vibration of Rotating and Reciprocating Machinery—Requirements for Instruments for Measuring Vibration Severity	ASA	ANSI S2.40-1984 (R 1997)
	Methods for Analysis and Presentation of Shock and Vibration Data	ASA	ANSI S2.10-1971 (R 1997)
	Selection of Calibrations and Tests for Electrical Transducers Used for Measuring Shock and Vibration	ASA	ANSI S2.11-1969 (R 1997)
	Techniques of Machinery Vibration Measurement	ASA	ANSI S2.17-1980 (R 1997)

Table 1 Codes and Standards Published by Various Societies and Associations (*Continued*)

Subject	Title	Publisher	Reference
	Vibrations of Buildings—Guidelines for the Measurement of Vibrations and Evaluation of Their Effects on Buildings	ASA	ANSI S2.47-1990 (R1997)
	Evaluation of Human Exposure to Whole-Body Vibration—Part 2: Continuous and Shock-Induced Vibrations in Buildings (1 to 80 Hz)	ISO	ISO 2631-2:1989
	Guidelines for the Evaluation of the Response of Occupants of Fixed Structures, Especially Buildings and Off-Shore Structures, to Low-Frequency Horizontal Motion (0.063 to 1 Hz)	ISO	ISO 6897:1984
	Procedural Standards for the Measurement and Assessment of Sound and Vibration (1994)	NEBB	NEBB
	Sound and Vibration Design and Analysis (1994)	NEBB	NEBB
Water Heaters	Gas Water Heaters, Vol. I, Storage Water Heaters with Input Ratings of 75,000 Btu per Hour or Less	CSA	ANSI Z21.10.1-1998/CSA 4.1-M98
	Gas Water Heaters, Vol. III, Storage, with Input Ratings Above 75,000 Btu per Hour, Circulating and Instantaneous Water Heaters	CSA	ANSI Z21.10.3-1998/CSA 4.3-M98
	Desuperheater/Water Heaters	ARI	ARI 470-95
	Method of Testing for Rating Commercial Gas, Electric, and Oil Water Heaters	ASHRAE	ANSI/ASHRAE 118.1-1993
	Method of Testing for Rating Residential Water Heaters	ASHRAE	ANSI/ASHRAE 118.2-1993
	Methods of Testing for Efficiency of Space-Conditioning/Water-Heating Appliances That Include a Desuperheater Water Heater	ASHRAE	ANSI/ASHRAE 137-1995
	Methods of Testing to Determine the Thermal Performance of Solar Domestic Water Heating Systems	ASHRAE	ASHRAE 95-1981 (R 87)
	Methods of Testing for Rating Combination Space-Heating and Water-Heating Appliances	ASHRAE	ANSI/ASHRAE 124-1991
	Construction and Test of Electric Storage-Tank Water Heaters	CSA	CAN/CSA-C22.2 No. 110-94
	Performance of Electric Storage Tank Water Heaters (R1997)	CSA	CAN/CSA-C191 Series-M90
	Oil Burning Stoves and Water Heaters	CSA	B140.3-1962 (R1998)
	Oil-Fired Service Water Heaters and Swimming Pool Heaters	CSA	B140.12-1976 (R1998)
	Water Heaters, Hot Water Supply Boilers, and Heat Recovery Equipment	NSF	NSF 5-1992
	Commercial-Industrial Gas Heating Equipment (1994)	UL	UL 795
	Electric Booster and Commercial Storage Tank Water Heaters (1995)	UL	UL 1453
	Household Electric Storage Tank Water Heaters (1996)	UL	ANSI/UL 174-1996
	Oil-Fired Storage Tank Water Heaters (1995)	UL	ANSI/UL 732-1997
Wood-Burning Appliances	Threshold Limit Values for Chemical Substances (updated annually)	AGCIH	AGCIH
	Installation Code for Solid Fuel Burning Appliances and Equipment	CSA	CAN/CSA-B365-M91
	Solid-Fuel-Fired Central Heating Appliances	CSA	CAN/CSA-B366.1-M91 (R1998)
	Chimneys, Fireplaces, Vents, and Solid Fuel-Burning Appliances	NFPA	ANSI/NFPA 211-1996
	Commercial Cooking, Rethermalization and Powered Hot Food Holding and Transport Equipment	NSF	ANSI/NSF 4-1997

Table 2 ABBREVIATIONS AND ADDRESSES

AABC	Associated Air Balance Council, 1518 K Street NW, Washington, D.C. 20005
ABMA	American Boiler Manufacturers Association, 950 North Glebe Road, Suite 160, Arlington, VA 22203-1824
ACCA	Air Conditioning Contractors of America, 1712 New Hampshire Avenue, NW, Washington, D.C. 20009
ACGIH	American Conference of Governmental Industrial Hygienists, 1330 Kemper Meadow Drive, Cincinnati, OH 45240
ADC	Air Diffusion Council, 104 South Michigan Avenue, Suite 1500, Chicago, IL 60603
AHAM	Association of Home Appliance Manufacturers, 20 North Wacker Drive, Suite 1600, Chicago, IL 60606
AIHA	American Industrial Hygiene Association, 2700 Prosperity Avenue, Suite 250, Fairfax, VA 22031
AMCA	Air Movement and Control Association International, Inc., 30 West University Drive, Arlington Heights, IL 60004-1893
ANSI	American National Standards Institute, 11 West 42nd Street, New York, NY 10036-8002
ARI	Air-Conditioning and Refrigeration Institute, 4301 North Fairfax Drive, Suite 425, Arlington, VA 22203
ASA	Acoustical Society of America, 120 Wall Street, New York, NY 10005-3993
ASAE	American Society of Agricultural Engineers, 2950 Niles Road, St. Joseph, MI 49085-9659
ASHRAE	American Society of Heating, Refrigerating and Air-Conditioning Engineers, Inc., 1791 Tullie Circle, NE, Atlanta, GA 30329
ASME Intl.	The American Society of Mechanical Engineers, 3 Park Avenue, New York, NY 10016-5990
ASTM	American Society for Testing and Materials, 100 Barr Harbor Drive, West Conshohocken, PA 19428-2959
AWS	American Welding Society, Inc., 550 N.W. LeJeune Road, Miami, FL 33126
AWWA	American Water Works Association, 6666 W. Quincy Avenue, Denver, CO 80235
BOCA	Building Officials and Code Administrators International, Inc., 4051 West Flossmoor Road, Country Club Hills, IL 60478-5795
BSI	British Standards Institution, 389 Chiswick High Road, London W4 4AL, England
CAGI	Compressed Air and Gas Institute, 1300 Sumner Avenue, Cleveland, OH 44115-2851
CSA	CSA International, 178 Rexdale Boulevard, Etobicoke (Toronto), ON M9W 1R3 Also available from CSA International, 8501 East Pleasant Valley Road, Cleveland, OH 44131
CTI	Cooling Tower Institute, P.O. Box 73383, Houston, TX 77273
EJMA	Expansion Joint Manufacturers Association, Inc., 25 North Broadway, Tarrytown, NY 10591-3201
HEI	Heat Exchange Institute, 1300 Sumner Avenue, Cleveland, OH 44115-2851
HI	Hydraulic Institute, 9 Sylvan Way, Parsippany, NJ 07054-3802
HYDI	Hydronics Institute, 35 Russo Place, P.O. Box 218, Berkeley Heights, NJ 07922
IAPMO	International Association of Plumbing and Mechanical Officials, 20001 Walnut Drive South, Walnut, CA 91789-2825
ICBO	International Conference of Building Officials, 5360 Workman Mill Road, Whittier, CA 90601
ICC	International Code Council, 5203 Leesburg Pike, Suite 708, Falls Church, VA 22041
IEEE	Institute of Electrical and Electronics Engineers, 445 Hoes Lane, P.O. Box 1331 Piscataway, NJ 08855-1331
IESNA	Illuminating Engineering Society of North America, 120 Wall Street, 17th floor, New York, NY 10005-4001
IFCI	International Fire Code Institute, 5360 Workman Mill Road, Whittier, CA 90601-2298
IIAR	International Institute of Ammonia Refrigeration, 1200 19th Street, NW, Suite 300, Washington, DC 20036-2412
ISA	Instrument Society of America, 67 Alexander Drive, P.O. Box 12777, Research Triangle Park, NC 27709
ISO	International Organization for Standardization, 1, rue de Varembé, Case postale 56, -1211 Genève 20, Switzerland
MCAA	Mechanical Contractors Association of America, 1385 Piccard Drive, Rockville, MD 20850-4329
MICA	Midwest Insulation Contractors Association, 2017 South 139th Circle, Omaha, NE 68144
MSS	Manufacturers Standardization Society of the Valve and Fittings Industry, Inc., 127 Park Street, N.E., Vienna, VA 22180-4602
NAIMA	North American Insulation Manufacturers Association, 44 Canal Center Plaza, Suite 310, Alexandria, VA 22314
PHCC	Plumbing-Heating-Cooling Contractors National Association, P.O. Box 6808, 180 South Washington, Falls Church, VA 22046-1148
NCPWB	National Certified Pipe Welding Bureau, 1385 Piccard Drive, Rockville, MD 20850
NCSBCS	National Conference of States on Building Codes and Standards, 505 Huntmar Park Drive, Suite 210, Herndon, VA 22070
NEBB	National Environmental Balancing Bureau, 8575 Grovemont Circle, Gaithersburg, MD 20877-4121
NEMA	National Electrical Manufacturers Association, 1300 North 17th Street, Suite 1847, Rosslyn, VA 22209
NFPA	National Fire Protection Association, 1 Batterymarch Park, P.O. Box 9101, Quincy, MA 02269-9101
NRCC	National Research Council of Canada, Client Services, M-20, 1200 Montreal Road, Ottawa, ON K1A 0R6, Canada
NSF	NSF International, P.O. Box 130140, Ann Arbor, MI 48113-0140
SAE Intl.	Society of Automotive Engineers International, 400 Commonwealth Drive, Warrendale, PA 15096-0001
SBCCI	Southern Building Code Congress International, Inc., 900 Montclair Road, Birmingham, AL 35213-1206
SMACNA	Sheet Metal and Air Conditioning Contractors' National Association, 4201 Lafayette Center Drive, Chantilly, VA 22021-1209
TEMA	Tubular Exchanger Manufacturers Association, Inc., 25 North Broadway, Tarrytown, NY 10591-3201
UL	Underwriters Laboratories Inc., 333 Pfingsten Road, Northbrook, IL 60062-2096

ASHRAE HANDBOOK

ADDITIONS AND CORRECTIONS

This section includes additional information and notes technical errors found between June 15, 1997 and April 1, 2000 in the inch-pound (I-P) edition of the *ASHRAE Handbooks*. Occasional typographical errors and nonstandard symbol labels will be corrected in future volumes. The most current list of Handbook Additions and corrections is also on the ASHRAE website. These corrections have been included in the 2000 *ASHRAE HandbookCD*.

The authors and editor encourage you to notify them if you find other technical errors. Please send corrections to: Handbook Editor, ASHRAE, 1791 Tullie Circle NE, Atlanta, GA 30329, or e-mail ashrae@ashrae.org.

1997 FUNDAMENTALS

p. 1.8, Equation (39). The equation for h should read $h_3 = h_4$ instead of $h_1 = h_2$.

p. 6.7. Change saturated solid/liquid enthalpy at –43°F to read –178.29 instead of –178.79.

p. 6.8. Change saturated solid/liquid enthalpy at –6°F to read –161.75 instead of –162.75.

p. 6.8. In the 9°F row, delete the negative sign from the values for evaporation enthalpy, saturated vapor enthalpy, evaporation entropy, and saturated vapor entropy.

p. 6.9. Change comma in saturated vapor entropy value at 56°F to a decimal so that it reads 2.1064.

p. 6.9. Change saturated vapor entropy at 58°F to read 2.1002 instead of 2.0002.

p. 6.9. Change saturated vapor specific volume at 77°F to read 699.80 instead of 794.40.

p. 6.11. Change saturated vapor entropy at 256°F to read 1.6916 instead of 1.6691.

p. 6.14, Equation (38). The second constant in the equation should read 26.142, not 26.412.

p. 6.14, Equation (40). The equation shoud read

$$\text{Enthalpy} \qquad h = h_a + \mu h_{as} \tag{40}$$

p. 8.7, Table 5. Oxygen consumed for Light work is < 1 ft^3/h.

p. 10.8, 2nd column. In the section on Heat Production, 7th line, add a minus sign to 40°F.

p. 15.11, 1st column, 4th line up. The AIHA standard number should read Z9.5.

p. 15.13, Equation (33). The constant for the equation is 0.25, not 0.025.

p. 18.8, Table 8, last line. Change the suction temperature to read 40°F instead of 47°F.

p. 21.4, Figure 6. Delete Column 4, with the heading, Total Volume of Capillaries, from the table.

p. 24.12, 1st column. In Line 2 of the section on Windows and Doors, delete "in Chapter 29."

p. 24.14, Figure 7. Dimensions on curves should read 3-1/2, 4, and 6 in., from bottom to top.

p. 25.9, Figure 6A. The $P_{OUTSIDE}$ arrow points to the P_{INSIDE} pressure line. It should point to the other pressure line instead.

p. 25.12, Equation (35). Move the multiplier of 60 from the numerator to the denominator to convert minutes to hours. Also, the specific heat (c_p) of air should read 0.24 Btu/lb·°F.

p. 26.3, 1st column. The extreme value distribution is a **double** exponential distribution, so the equation for F at the bottom of the column should read

$$F = -\frac{\sqrt{6}}{\pi}\left[0.5772 + \ln\ln\left(\frac{n}{n-1}\right)\right]$$

pp. 26.6–52, Tables 1A, 2A, and 3A. In the MWS/MWD to DB columns, the columns headed PWD should read MWD.

pp. 26.7–53, Tables 1B, 2B, and 3B. The humidity ratio (HR) units are grains of water vapor per pound of dry air (7000 gr = 1 lb).

p. 26.9. DP/MDB and HR data for Hartford, Brainard Field, Connecticut, should be corrected to read as follows:

DP/MDB AND HR								
0.4%			1%			2%		
DP	HR	MDB	DP	HR	MDB	DP	HR	MDB
73	121	81	71	116	79	70	110	78

p. 26.26. The latitude and longitude for the Salta Airport weather station in Argentina are S and W, respectively.

p. 26.52. The latitude and longitude for the Treinta y Tres weather station in Uruguay are S and W, respectively.

p. 27.4, Figure 1. To reflect the revised weather data in Chapter 26, the caption on the x-axis of the graph should read, 1% DESIGN HUMIDITY RATIO (W).

p. 27.12, Table 16. The values in the first and third degree-day column headings have been switched. The first value should read 2950, and the value in the last column should read 7433.

p. 28.8, 1st column. In the second paragraph below Equation (9) change *special allowance* to read, *lighting use factor.*

p. 28.10, 1st column. In the 4th line of the section titled Overloading or Underloading, change FLM to read F_{LM}.

p. 28.19. The R-value for B4 is 10.0, not 1.19.

pp. 28.26–27, Tables 18 and 19. The heading in the first column of these tables should read Wall Group.

p. 28.36. Lines 4 and 8 list values for the North (not East) exposure. Lines 5 and 9 list values for the South exposure, and lines 6 and 10 for the East exposure.

p. 28.38, 1st column, 2nd equation. Add "Btu/h" to the right side of the equation.

p. 28.47. The title of the table should read:

Table 33B Wall Types, Mass Evenly Distributed, for Use with Table 32

p. 28.48. Table 33C. The table is incorrect. The correct table is on the ASHRAE website and on page 26.48 of the 1993 *ASHRAE Handbook—Fundamentals*.

p. 28.54, 1st column, 2nd line up. Equation should read

$$C_2 = \frac{(94+74)}{2} - 85 = \text{daily average temperature correction}$$

p. 28.57, 1st column. In Equation (49), the letter l should be a λ.

p. 28.57, 2nd column. Under "Appliances," changc cnergy input to power input in the definition of q_{input}.

p. 29.13, Equation (6). Insert a multiplier of 100 into the right side of the equation.

p. 29.13, Figure 8. The captions for the two graphs should be reversed. The top graph is for 0.75 in. low-e insulating glass and the bottom graph is for 0.75 in. insulating glass.

p. 29.16, 2nd column. On the fifth line below the heading, "Determine Solar Angle," change Table 6 to read Table 8.

p. 29.28, 2nd column. The first sentence in the third paragraph up should refer to Chapter 26, not Chapter 24.

p. 29.37, Equation (49) and Equation (50). Change equations to read as follows:

$$\alpha_D = \sum_{j=0}^{5} a_j \cos^j \theta \qquad (49)$$

$$\tau_D = \sum_{j=0}^{5} t_j \cos^j \theta \qquad (50)$$

p. 29.37, Table 23. Change the word "Absorbtance" in the title caption to "Absorptance."

p. 29.37, Table 23. In the first column head change the caption from *i* to *j*. Also, the value of t_j at $j = 3$ is negative, or –7.07329.

p. 29.37, 2nd column. The second term in the numerator of the equation for solar azimuth ϕ should read sin δ, so the equation is

$$\cos\phi = \frac{\sin\beta \sin L - \sin\delta}{\cos\beta \cos L}$$

p. 30.5, Figure 1. The dimensions for the office floor plan are as follows:

Outside length	= 147.6 ft (45 m)
Outside width	= 98 ft (30 m)
Interior space length	= 124 ft (37.8 m)
Exterior space width	= 11.8 ft (3.6 m)
Example zone width	= 11.8 ft (3.6 m)

p. 30.12, 1st column, equation nomenclature. Change definition of *c* to read, *c* = fluid capacity, lb/h.

p. 30.19, Equation (42). Change θ to ϕ in two exponents.

p. 30.19, Table 2. Change σ to ϕ in the heading of the fifth column.

p. 30.24, 1st column. The tenth line below the heading, Neural Network Models, should refer to Figure 18, not Figure 20.

p. 32.9, Figure 9. The chart should include a shaded area to show the suggested range of air velocity and friction loss for design. Below about 17,000 cfm, the suggested limits for friction loss are 0.08 and 0.6 in. water gage per 100 ft of duct. Above 17,000 cfm the limiting factor is air velocity, which ranges from 1800 and 4000 fpm.

p. 33.19, Table 22. Change note to read, in part, "... a pressure drop of 0.3 in. of water; ..."

p. 35.2, last line of Table 2. Conversion of Fahrenheit to °R should read *x* + 459.67.

p. 39.1. The page numbers in the table of contents should read 39.#, not 38.#.

1998 REFRIGERATION

p. 8.5, Table 3. The specific heat above freezing for butter should be 0.52 Btu/lb·°F. Delete the value shown for the specific heat of butter below freezing.

p. 9.2, 1st line after Equation (4). Change *t* to θ.

p. 9.3, Table 2. The variable on the left side of the fifth equation (below the word *where*) should read *v*, not *w*.

p. 9.5, Equation (15). ω is in radians.

p. 9.5, Equation (20). Delete 2 from both the numerator and denominator in the first term on the right side. Then multiply the first term by 1.5 so the equation reads as follows:

$$E_o = 1.5\frac{\beta_1 + \beta_2 + \beta_1^2(1+\beta_2) + \beta_2^2(1+\beta_1)}{\beta_1\beta_2(1+\beta_1+\beta_2)} - \frac{[(\beta_1-\beta_2)^2]^{0.4}}{15} \qquad (20)$$

p. 9.6, Example 1. In Step 5, change the value of E_o to 2.06 to reflect the corrected Equation (20). As a result of this change, the value of *E* in the last equation in Step 5 is 1.45. Then in Step 8, θ = 3.38 h because *E* = 1.45.

p. 16.11, 1st column. Change the third sentence in the second paragraph in the section on Chilling and Freezing Variety Meats to read:

Quick Chilling. A better and more widely used method consists of quick ...

p. 18.4, 1st column. Change first heading to read "Irradiation of Freah Seafood."

p. 20.2, 1st column, 2nd para. Change second sentence to read, "Most of the shell's organic matter is protein."

p. 23.3, 1st column. Second heading should read, "Packaging, Loading, and Handling."

p. 23.3, Table 2. Beans, green should be in the third column. (Store at 50°F and 80 to 85% rh). Green beans suffer chilling injury if stored at 32 to 41°F.

p. 23.11, 2nd column, 7th line up. Change 45 cfm/lb to read 10 cfm/ton.

p. 24.2, 1st column. Insert the following paragraph to the top of the column before the paragraph on Flavor Fortification.

Evaporation-Water Removal. Removal of water follows juice extraction and preparation. Refer to the section on Concentration Methods for details.

p. 32.1, 1st para. Change the paragraph to read:

This chapter is a guide to specifying insulation systems for refrigeration piping, fittings, and vessels. It does not deal with HVAC systems or applications such as chilled water systems. Refer to Chapters 22, 23, 24, and 39 in the 1997 *ASHRAE Handbook—Fundamentals* for information about insulation and vapor barriers for these systems.

p. 32.3, Table 2. Change footnote a to read:

[a] Tested in accordance with ASTM E 84 for 1 in. thick insulation.

p. 34.2, 1st column, 7th line up. Change sentence to read,

The amount of control over each load source is indicated as an approximate percentage of the maximum reduction possible through effective design and operation.

p. 34.4, 1st column. In the equation for q_r, the third value should be 0.1714, not 0.714.

p. 45.27, 2nd column. In line 2 of the second paragraph in the section on Short Tube Restrictors, the length-to-diameter ratio should read 3 < *L*/*D* <20.

p. 47.12, Figure 17. Relabel valve on upper left of heat reclaim coil to read, 3-WAY SOLENOID VALVE. Relabel valve on upper right of heat reclaim coil to read, CHECK VALVE.

1999 HVAC APPLICATIONS

p. 15.1, Figure 1. Increments in the label for the horizontal scale should read: 0.1, 0.2, 0.5, 1, 2, 5, 10 μm.

p. 15.2, 1st column. FS 209 Class 10,000 particle count should not exceed 65 particles per cubic foot of a size 5 μm or larger.

p. 29.12, 1st column, 1st para. The last sentence should refer to Equation (25), not Equation (26).

p. 33.11, 1st column, last para. Change sentence to read, "Designers typically select a diffuser dimension to create an inlet Froude number of 1.0 or less."

p. 46.19, 2nd column. Replace text starting half-way down the column, Examples 3 and 4, and Figure 23 with the following.

For applications where the duct is exposed, the sound pressure level in an occupied space as a result of duct sound breakout can be obtained from

$$L_p = L_{w(out)} - 10\log(\pi r L) \qquad (14)$$

where

L_p = sound pressure level at specified point in the space, dB
$L_{w(out)}$ = sound power level of sound radiated from outside surface of duct walls given by Equation (7), dB
r = distance between duct and position at which L_p is being calculated, ft
L = length of duct sound radiating surface, ft

Equation (14) does not apply when the duct is in a plenum, concealed behind a suspended ceiling. In such cases, simply subtract the attenuation for ceilings listed in Table 23 instead of $10\log(\pi r L)$ and assume the sound field in the room is uniform.

Example 3. A 12 in. by 48 in. by 15 ft long rectangular supply duct above a mineral fiber lay-in tile ceiling is constructed of 22 gage sheet metal. Given the sound power levels in the duct, what are the sound pressure levels at a listener 5 ft from the duct?

Solution:

	Octave Band Center Frequency, Hz						
	63	**125**	**250**	**500**	**1000**	**2000**	**4000**
$L_{w(in)}$	90	85	80	75	70	65	60
$-TL_{out}$ (Table 20)	−19	−22	−25	−28	−31	−37	−43
10 log (S/A)	16	16	16	16	16	16	16
$L_{w(out)}$	87	79	71	63	55	44	33
− Ceiling tile (Table 23)	−13	−16	−18	−20	−26	−31	−36
L_p, dB	74	63	53	43	29	13	−3

The RC level associated with the above L_p values is RC 28 (LFV*b*).

Example 4. Repeat Example 3 using a round duct of equivalent airflow area: 27 in. in diameter, 22 gage, 15 ft long.

Solution:

	Octave Band Center Frequency, Hz						
	63	**125**	**250**	**500**	**1000**	**2000**	**4000**
$L_{w(in)}$	90	85	80	75	70	65	60
$-TL_{out}$ (Table 21)	−49	−50	−32	−30	−29	−25	−31
10 log (S/A)	14	14	14	14	14	14	14
$L_{w(out)}$	55	49	62	59	55	54	43
− Ceiling tile (Table 23)	−13	−16	−18	−20	−26	−31	−36
L_p, dB	42	33	44	39	29	23	7

The RC level associated with the above L_p values is RC 31(MF). The use of the round duct results in a slight increase in the RC level and eliminates the low-frequency rumble present in Example 3, but it introduces some mid-frequency roar that can be dealt with by another means. The two spectra are shown in Figure 23.

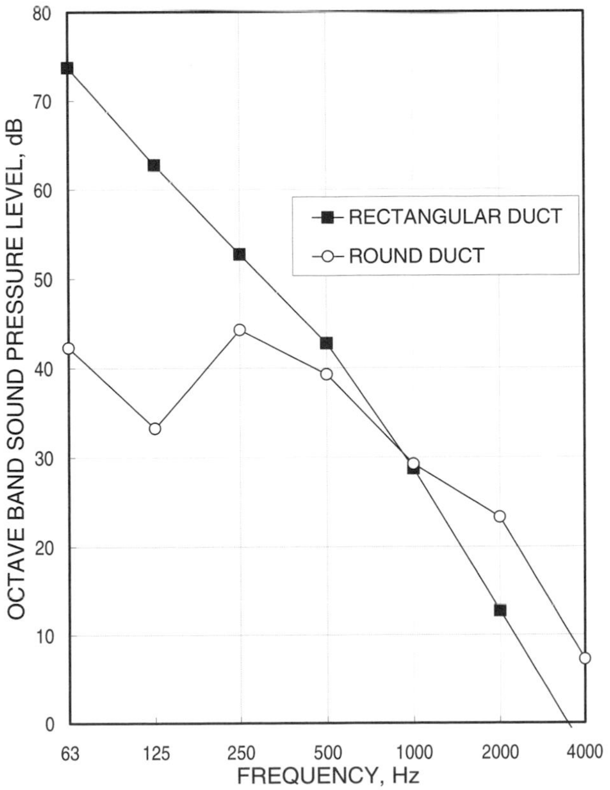

Fig. 23 Duct Breakout Sound Pressure Levels from Examples 3 and 4

p. 46.20, Table 23. Delete the footnote from the table.

p. 46.32, Equation (20). The last term in the equation should be deleted so that the equation reads as follows:

$$L_p = L_w + 10\log Q - 20\log d \qquad (20)$$

p. 50.9, 1st column. In the section on Apples change the last two sentences to read,

"Generally, a cooler designed to exchange air every 3 minutes (20 air changes per hour) is the largest that can be installed. This capacity provides a complete air change every 1 to 1.5 minutes (40 to 60 air changes per hour) when the storage is loaded."

p. 50.9, 2nd column, 11th line up. Delete "per hour."

p. 51.10, Equation (17). The value of B is calculated for sea level standard pressure.

COMPOSITE INDEX

ASHRAE HANDBOOK SERIES

This index covers the current Handbook series published by ASHRAE. The four volumes in the series are identified as follows:

F = 1997 Fundamentals

R = 1998 Refrigeration

A = 1999 Applications

S = 2000 Systems and Equipment

Alphabetization of the index is letter by letter; for example, **Heaters** precedes **Heat exchangers**, and **Floors** precedes **Floor slabs**.

The page reference for an index entry includes the book letter and the chapter number, which may be followed by a decimal point and the beginning page within the chapter. For example, the page number S31.4 means the information may be found in the 2000 Systems and Equipment volume, Chapter 31, beginning on page 4.

Each Handbook is revised and updated on a four-year cycle. Because technology and the interests of ASHRAE members change, some topics are not included in the current Handbook series but may be found in the earlier Handbook editions cited in the index.